Mader Concepts of Biology

ON TRACK . . . Various types of interactives and quizzing keep you motivated and on track in mastering the key concepts.

► **Practice Tests** include a pre- and post-test for each chapter that allow you to quickly gauge your understanding of the concepts. The tests include feedback for incorrect answers to guide your learning.

► **BioTutorials Animation Quizzes** present concepts through dynamic visuals. Watch biological processes and concepts come alive, and then take a quiz to determine your level of understanding.

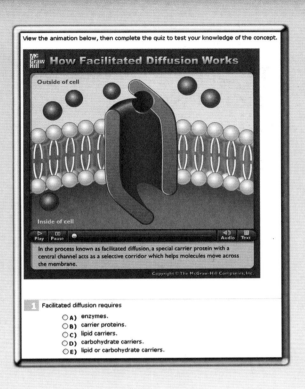

View the animation below, then complete the quiz to test your knowledge of the concept.

How Facilitated Diffusion Works

Outside of cell

Inside of cell

In the process known as facilitated diffusion, a special carrier protein with a central channel acts as a selective corridor which helps molecules move across the membrane.

Copyright © The McGraw-Hill Companies, Inc.

1 Facilitated diffusion requires
- A) enzymes.
- B) carrier proteins.
- C) lipid carriers.
- D) carbohydrate carriers.
- E) lipid or carbohydrate carriers.

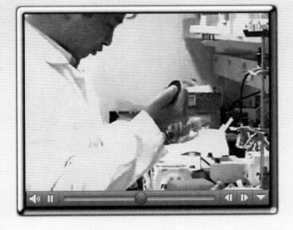

ScienCentral, Inc.

► **ScienCentral Video Quizzes** add relevancy to your study of biology. Learn how a photosynthesizing plant may one day keep your laptop and cell phone running longer!

► **Virtual Labs and Case Studies** provide unique online experiences. Step into the virtual lab to identify an unknown Eubacterium taken from a human mouth. Read and discuss case studies about medical, ecological, and social issues, including a case study about the effects of caffeine on the brain.

McGraw-Hill Higher Education

Concepts of
BIOLOGY

Sylvia S. Mader

 **McGraw-Hill**
Higher Education

Boston   Burr Ridge, IL   Dubuque, IA   New York   San Francisco   St. Louis
Bangkok   Bogotá   Caracas   Kuala Lumpur   Lisbon   London   Madrid   Mexico City
Milan   Montreal   New Delhi   Santiago   Seoul   Singapore   Sydney   Taipei   Toronto

McGraw-Hill
Higher Education

CONCEPTS OF BIOLOGY

Published by McGraw-Hill, a business unit of The McGraw-Hill Companies, Inc., 1221 Avenue of the Americas, New York, NY 10020.
Some ancillaries, including electronic and print components, may not be available to customers outside the United States.

 This book is printed on recycled, acid-free paper containing 10% postconsumer waste.

1 2 3 4 5 6 7 8 9 0 VNH/VNH 0 9 8

ISBN 978-0-07-340345-8
MHID 0-07-340345-8

Publisher: *Janice Roerig-Blong*
Executive Editor: *Michael S. Hackett*
Developmental Editor: *Rose M. Koos*
Marketing Manager: *Tamara Maury*
Senior Project Manager: *Jayne Klein*
Lead Production Supervisor: *Sandy Ludovissy*
Senior Media Project Manager: *Jodi K. Banowetz*
Designer: *Laurie B. Janssen*
Cover/Interior Designer: *Elise Lansdon*
Senior Photo Research Coordinator: *Lori Hancock*
Photo Research: *Evelyn Jo Hebert*
Supplement Producer: *Melissa M. Leick*
Art Studio and Compositor: *Electronic Publishing Services Inc., NYC*
Typeface: 9.5/12 Slimbach
Printer: *Von Hoffmann Press*
Cover images: *Clockwise from upper left: Gettyimages/3D4Medical.com, Gettyimages/Visuals Unlimited, Gettyimages/3D4Medical.com, Radius Images/Alamy, Gettyimages/Minden Pictures, Royalty-Free/CORBIS*

The credits section for this book begins on page C-1 and is considered an extension of the copyright page.

Library of Congress Cataloging-in-Publication Data

Mader, Sylvia S.
    Concepts of biology / Sylvia S. Mader. -- 1st ed.
        p. cm.
    Includes index.
    ISBN 978-0-07-340345-8 — ISBN 0-07-340345-8 (hard copy : alk. paper)   1. Biology.   I. Title.
QH308.2.M234 2009
570—dc22

                                                                2007039727

www.mhhe.com

Brief Contents

1   Biology, the Study of Life 2

PART I   Organisms Are Composed of Cells 18

2   Basic Chemistry and Cells 18
3   Organic Molecules and Cells 36
4   Structure and Function of Cells 54
5   Dynamic Activities of Cells 76
6   Pathways of Photosynthesis 94
7   Pathways of Cellular Respiration 112
Biological Viewpoints *Organisms Are Composed of Cells* 130

PART II   Genes Control the Traits of Organisms 132

8   Cell Division and Reproduction 132
9   Patterns of Genetic Inheritance 158
10   Molecular Biology of Inheritance 182
11   Regulation of Gene Activity 206
12   Biotechnology and Genomics 224
Biological Viewpoints *Genes Control the Traits of Organisms* 240

PART III   Organisms Are Related and Adapted to Their Environment 242

13   Darwin and Evolution 242
14   Speciation and Evolution 262
15   The History and Classification of Life on Earth 282
16   Evolution of Microbial Life 300
17   Evolution of Protists 322
18   Evolution of Plants and Fungi 340
19   Evolution of Animals 366

20   Evolution of Humans 398
Biological Viewpoints *Organisms Are Related and Adapted to Their Environment* 418

PART IV   Plants Are Homeostatic 420

21   Plant Organization and Homeostasis 420
22   Transport and Nutrition in Plants 440
23   Control of Growth and Responses in Plants 458
24   Reproduction in Plants 478
Biological Viewpoints *Plants Are Homeostatic* 492

PART V   Animals Are Homeostatic 494

25   Animal Organization and Homeostasis 494
26   Coordination by Neural Signaling 512
27   Sense Organs 534
28   Locomotion and Support Systems 552
29   Circulation and Cardiovascular Systems 572
30   Lymph Transport and Immunity 592
31   Digestive Systems and Nutrition 610
32   Gas Exchange and Transport in Animals 632
33   Osmoregulation and Excretion 648
34   Coordination by Hormone Signaling 664
35   Reproduction and Development 680
Biological Viewpoints *Animals Are Homeostatic* 706

PART VI   Organisms Live in Ecosystems 708

36   Population Ecology 708
37   Behavioral Ecology 724
38   Community and Ecosystem Ecology 740
39   Major Ecosystems of the Biosphere 762
40   Conservation of Biodiversity 778
Biological Viewpoints *Organisms Live in Ecosystems* 794

Preface

Biology—like no other discipline—uses concepts as a way to understand ourselves and the world we live in. An understanding of biological principles should be within the grasp of all those who decide to study biology. As a biology instructor, I am first and foremost motivated by the desire to help science-shy students gain a conceptual understanding of biology. This determination has inspired me through many years of authoring textbooks. In both my teaching and my writing, the non-major student has always been my primary focus.

Today we have many ways to reach out to students and increase their motivation so that they will understand the concepts of biology. I have watched the evolution of introductory biology textbooks as they moved from utilization of only a "second color" to beautifully illustrated figures in full color. This text uses the latest techniques to captivate students, enabling them to tap into their own resources and appreciate the world around them.

Concepts of Biology was written not only to present the major concepts of biology clearly and concisely but also to show the relationships between the concepts at various levels of complexity. Students also need to see how biological concepts pertain to their own lives in a meaningful way. This text fulfills these goals.

What Sets This Book Apart

Focus on Concepts

Concepts of Biology is organized around five major concepts of biology—all organisms are composed of cells, contain genes, evolve, are homeostatic, and live in ecosystems—and this organization facilitates the use of relationships to create a cohesive whole. Using the levels of biological organization as a model, this text begins with the cell and ends with ecosystems. Just as the levels of biological organization flow from one level to the next, relationships between biological themes and topics are emphasized throughout the book.

> *"The organization of the text around the major theories of Biology is a wise path to follow; it integrates the chapters into themes and points out the development of a theory.... If these chapters are representative of the text, you have a real winner on your hands in a very competitive marketplace."* —Paul E. Wanda, Southern Illinois University, Edwardsville

Emphasis on the concepts of biology begins with the part titles that are patterned after the major concepts of biology.

Part I: Organisms Are Composed of Cells
Part II: Genes Control the Traits of Organisms
Part III: Organisms Are Related and Adapted to Their Environment
Part IV: Plants are Homeostatic
Part V: Animals are Homeostatic
Part VI: Organisms Live in Ecosystems

Learning through Relevancy

Showing how concepts relate to each other and to student experiences are hallmarks of this text. An inviting and visually appealing story generates a desire to proceed with the chapter. These stories demonstrate that organisms and biological phenomena are of extreme interest in their own right. Each story ends with a lead into the rest of the chapter.

Each chapter is organized around concepts, and each concept is broken down into manageable topical sections. A brief introduction acclimates students to the concept, and the topical sections that follow provide evidence and lend support to the concept. This careful progression through the content, combined with transitions between topics, emphasizes how concepts are related and enables students to understand biology as more than isolated facts.

> *"The author has done an outstanding job of developing concepts in a logical, interesting, and student-friendly manner."* —Pamela L. Hanratty, Indiana University

The use of analogies helps students relate to and follow the discussion and, in like manner, numbers that refer to the correct part of an illustration promote the learning process. Each section concludes with a question (answered in Appendix A) that encourages students to apply what they have learned. When appropriate, the questions refer back to the opening story to actively engage students.

Applications are used throughout each chapter to show how biological concepts relate to students' lives, but the emphasis on applications is especially evident within the sections entitled, "How Biology Impacts Our Lives" and "How Science Progresses."

Each chapter ends with a "Connecting the Concepts" box that reviews the concepts of the chapter and tells how they relate to other concepts throughout the book. Each part ends with a "Biological Viewpoints" reading that provides students with take-home messages about the major concept emphasized in that part of the text.

Concise Writing, Instructional Art, and a Proven Pedagogical System

Writing Style

Concepts of Biology went through several stages of revision, but I maintained my clear and concise presentation of the major concepts as I worked to include as many relationships and applications as possible. Students will find the book easy to read and understand.

Instructional Art

Outstanding photographs and dimensional illustrations, vibrantly colored, are featured throughout *Concepts of Biology*. Accuracy and instructional value were primary considerations in the development of each figure. Students will learn from a variety of figure types, including process figures with numbered steps that relate to the text discussion, micro to macro representations, and the combination of photos and art.

Pedagogical System

Each chapter features numerous learning aids that were carefully developed to help students grasp challenging concepts. Examples include:

- **Learning Outcomes,** listed according to the chapter concepts, provide students with an overview of what they are to know.
- **Check Your Progress** questions at the end of each section help students assess and/or apply their understanding of a concept.

- A bulleted and illustrated **Summary** is organized according to the chapter concepts and helps students review the chapter.
- **Testing Yourself** offers another way to review the chapter concepts. Included are objective multiple-choice questions and Thinking Conceptually questions that ask students to apply their understanding of a concept.
- **Understanding the Terms** is a page-reference list of the boldface terms in the chapter. A matching exercise allows students to test their knowledge of the terms.
- **Thinking Scientifically** questions end the chapter. These questions apply directly to the chapter and ask students to design an experiment or explain some part of a hypothetical experiment.
- All questions are answered in Appendix A.

For a visual presentation of the key features of *Concepts of Biology,* see the Guided Tour on pages vi–xii.

I hope students will be motivated to develop a solid understanding of the biological concepts presented in *Concepts of Biology,* as well as an understanding of the relationships between the concepts. With this understanding of biology, students will continue to appreciate how biological concepts pertain to their decisions and actions long after they have completed their introductory biology course.

About the Author

Dr. Sylvia S. Mader has authored several nationally recognized biology texts published by McGraw-Hill. Educated at Bryn Mawr College, Harvard University, Tufts University, and Nova Southeastern University, she holds degrees in both Biology and Education. Over the years she has taught at University of Massachusetts, Lowell, Massachusetts Bay Community College, Suffolk University, and Nathan Mathew Seminars. Her ability to reach out to science-shy students led to the writing of her first text, *Inquiry into Life*, that is now in its twelfth edition. Highly acclaimed for her crisp and entertaining writing style, her books have become models for others who write in the field of biology.

Although her writing schedule is always quite demanding, Dr. Mader enjoys taking time to visit and explore the various ecosystems of the biosphere. Her several trips to the Florida Everglades and Caribbean coral reefs resulted in talks she has given to various groups around the country. She has visited the tundra in Alaska, the taiga in the Canadian Rockies, the Sonoran Desert in Arizona, and tropical rain forests in South America and Australia. A photo safari to the Serengeti in Kenya resulted in a number of photographs for her texts. She was thrilled to think of walking in Darwin's steps when she journeyed to the Galápagos Islands with a group of biology educators. Dr. Mader was also a member of a group of biology educators who traveled to China to meet with their Chinese counterparts and exchange ideas about the teaching of modern-day biology.

For My Children
—*Sylvia Mader*

Concepts of BIOLOGY *Guided Tour*

Getting the Most Out of This Textbook

The chapter organization, concise writing style, compelling art, and various pedagogical tools combine to help you build a cohesive understanding of the concepts—understanding how concepts are related to one another and to your daily experiences.

Learning Outcomes

Before you read a chapter, review the learning outcomes to focus your attention on the major concepts.

Chapter-Opening Story

Take time to read these stories that tell of interesting organisms or biological phenomena. When appropriate, the stories are referred to in the chapter and in the *Check Your Progress* questions, so you can understand the concepts of the chapter in context.

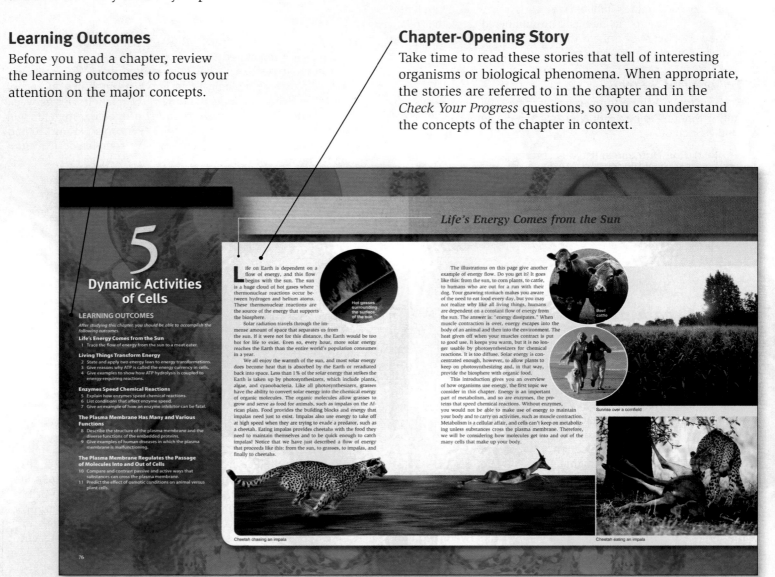

> "The making of any great textbook is not merely the compilation of facts and figures but, most importantly, the presentation of concepts, processes, and analogies in an understandable fashion. Mader's Concepts of Biology has achieved this yardstick."
> —Joseph D. Gar, West Kentucky Community and Technical College

Studying the Chapter

- Each main heading is written as a complete thought to clearly convey a major concept.

  - A short introduction highlights the main topics that will be discussed as part of the major concept.

    - The applicable learning outcomes are referenced so you can review them before and after you study the concept.

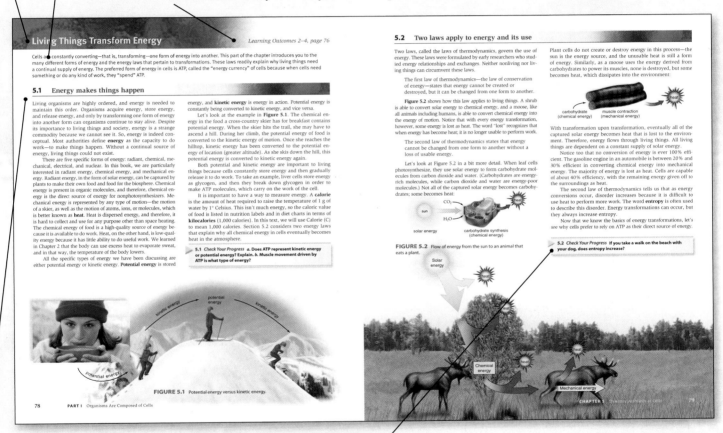

- The major concept is broken down into manageable topical and numbered sections. Each section begins with a heading that clearly states the topic and concludes with a segue into the next section to show the relationship between topics.

- Each topical section includes a *Check Your Progress* question (answered in Appendix A) so you can assess and apply your knowledge.

Applications and Overviews Emphasize How Concepts Are Related

Issues and happenings from the mainstream are discussed in the context of biological concepts to help you relate the concepts to something familiar.

How Biology Impacts Our Lives

Examines issues that affect our health and environment.

How Science Progresses

Discusses scientific research and advances that have helped us gain valuable biological knowledge.

HOW BIOLOGY IMPACTS OUR LIVES

5.12 Malfunctioning plasma membrane proteins can cause human diseases

Suppose you went to the doctor for a particular medical condition. Would you expect the doctor to say that your condition was due to your plasma membrane? That is what might happen if you wanted to know the exact cause of your illness. Let's take diabetes type 2, for example.

Diabetes Type 2 The typical diabetes type 2 patient is somewhat, or even grossly, overweight. The symptoms are unusual hunger and/or thirst; excessive fatigue; blurred vision; sores that do not heal; and frequent urination, especially at night. The doctor does a urinalysis and finds sugar in the urine, and yet the blood test shows insulin in the blood. Usually when we eat sugar, the pancreas, a gland that lies near the stomach, releases the hormone insulin into the bloodstream, and it travels to the cells, where it binds to its receptor protein. The binding of insulin signals a cell to send carriers to the plasma membrane that will transport glucose into the cells. In the case of diabetes type 2, the insulin binds to its receptor protein, but the number of carriers sent to the plasma membrane for glucose is not enough. The result is too much glucose in the blood, which spills over into the urine.

Patients can prevent, or at least control, diabetes type 2 by switching to a healthy diet and engaging in daily exercise. In addition to the symptoms mentioned, diabetics are at risk for ...cular disease.

...found yourself acci-...colors, you may have ...or blind. Most people ...ins in numerous folds ...in photoreceptor cells, ...shape. Cone cells are ...that responds to visible

...e three types of cones: ...different wavelengths ...requires activation of ...ne cells. When a cone ...which its photopigment ...um ion channels in the ...le, mostly males, have ...ck of functional red or ...ividuals have what is ...have difficulty distin-...ess common situation, ...ssing; such people may ...matic world.

...atient is a child, usually ...s experienced repeated ...orders a test that mea-...child's sweat, because ...weat than normal chil-...sily through a plasma

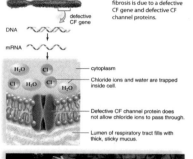

FIGURE 5.12 Cystic fibrosis is due to a defective CF gene and defective CF channel proteins.

chromosome 7

defective CF gene

DNA

mRNA

H₂O Cl⁻ — cytoplasm

Cl⁻ H₂O Cl⁻ H₂O — Chloride ions and water are trapped inside cell.

Defective CF channel protein does not allow chloride ions to pass through.

Lumen of respiratory tract fills with thick, sticky mucus.

membrane channel protein, but when their passage is not properly regulated, a thick mucus appears in the lungs and pancreas. The mucus clogs the lungs, causing breathing problems. It also provides fertile ground for bacterial growth. The result is frequent lung infections, which eventually damage the lungs and contribute to an early death. Also, thick digestive fluids may clog ducts leading from the pancreas to the small intestine. This prevents enzymes from reaching the small intestine, where they are needed to digest food. Digestive problems and slow growth result.

CF can be confirmed by doing a genetic test because we now have a test for the gene that causes CF. **Figure 5.12** reminds us that a genetic defect results in a defective protein—in this instance, an abnormal channel protein for chloride in the plasma membrane.

Sections 5.13 to 5.18 pertain to how molecules get into and out of the cell, including methods that utilize plasma proteins.

5.12 Check Your Progress Explain why plasma membrane protein malfunctions can manifest themselves differently.

CHAPTER 5 Dynamic Activities of Cells   **87**

HOW SCIENCE PROGRESSES

6.17 Destroying tropical rain forests contributes to global warming

Tropical rain forests occur near the equator. They can exist wherever temperatures are above 26°C and rainfall is regular and heavy (100–200 cm per year). Huge trees with buttressed trunks and broad, undivided, dark green leaves predominate. Nearly all of the plants in a tropical rain forest are woody, and woody vines are also abundant. There is no undergrowth except at clearings. Instead, orchids, ferns, and bromeliads live in the branches of the trees.

Despite the fact that tropical rain forests have dwindled from an original 14% to 6% of land surface today, they still make a substantial contribution to global CO₂ fixation. Taking into account all ecosystems, marine and terrestrial, photosynthesis produces organic matter that is 300 to 600 times the mass of the people currently living on Earth. Tropical rain forests contribute greatly to the uptake of CO₂ and the productivity of photosynthesis because they are the most efficient of all terrestrial ecosystems.

We have learned that organic matter produced by photosynthesizers feeds all living things and that photosynthesis releases oxygen (O₂), a gas that is needed to complete the process of cellular respiration. Does photosynthesis by tropical rain forests provide any other service that has significant worldwide importance? **Figure 6.17** projects a rise in the average global temperature during the 21st century due to the introduction of certain gases, chiefly carbon dioxide, into the atmosphere. The process of photosynthesis and also the oceans act as a sink for carbon dioxide. For at least a thousand years prior to 1850, atmospheric CO₂ levels remained fairly constant at 0.028%. Since the 1850s, when industrialization began, the amount of CO₂ in the atmosphere has increased to 0.036%.

Much like the panes of a greenhouse, CO₂ in our atmosphere traps radiant heat from the sun and warms the world. Therefore, CO₂ and other gases that act similarly are called greenhouse gases.

Without any greenhouse gases, Earth's temperature would be about 33°C cooler than it is now. Likewise, increasing the concentration of these gases causes an increase in global temperatures.

Burning fossil fuels adds CO₂ to the atmosphere. Have any other factors contributed to the increases in CO₂ in the atmosphere? Between 10 and 30 million hectares of rain forests are lost every year to ranching, logging, mining, and other means of developing forests for human needs. The clearing of forests often involves burning them, which is double trouble for global warming. Each year, deforestation in tropical rain forests accounts for 20–30% of all carbon dioxide in the atmosphere. At the same time, burning removes trees that would ordinarily absorb CO₂.

Some investigators have hypothesized that an increased amount of CO₂ in the atmosphere will cause photosynthesis to increase in the remaining portion of the forest. To study this possibility, they measured atmospheric CO₂ levels, daily temperature levels, and tree girth in La Selva, Costa Rica, for 16 years. The data collected demonstrated relatively lower forest productivity at higher temperatures. These findings suggest that, as temperatures rise, tropical rain forests may add to ongoing atmospheric CO₂ accumulation and accelerated global warming rather than the reverse. This is all the more reason to do what we can now to slow the process of global warming before it gets out of hand.

Some countries have programs to combat the problem of deforestation. In the mid-1970s, Costa Rica established a system of national parks and reserves to protect 12% of the country's land area from degradation. The current Costa Rican government wants to expand the goal by expanding protected areas to 25% in the near future. Similar efforts in other countries may help slow the ever-increasing threat of global warming.

6.17 Check Your Progress Are there advantages to preserving tropical rain forests outside the possibility of helping to prevent global warming?

FIGURE 6.17 Global warming: Past trends and future predictions. Burning forests make the maximum increase projected in the graph more likely.

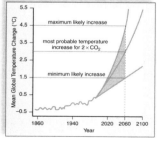

maximum likely increase

most probable temperature increase for 2 × CO₂

minimum likely increase

Mean Global Temperature Change (°C)

1860   1940   2020   2060   2100
Year

108

> "This (chapter 12) is current information that is interesting and cutting edge in the field of biology. Some of the topics are provocative and will excite students to further investigate these topics whether they are interested in genetics, the environment, politics, or economics."
> —Lori Bean, Monroe County Community College

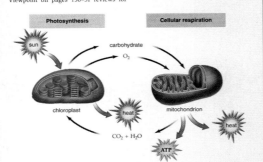
Connecting the Concepts

Each chapter ends with a *Connecting the Concepts* reading that reviews the concepts of the chapter and tells how they relate to other concepts in the book.

> *"The textbook has a clear and concise writing style that integrates relevance with basic concepts."*
> —Allan D. Nelson, Tarleton State University

> *"The text is easy to read and understand. The material ties various biological concepts together which helps students to see the big picture."*
> —Jerry W. Mimms, University of Central Arkansas

Biological Viewpoints

Each part ends with a *Biological Viewpoints* reading that briefly summarizes the take-home messages of the major concepts emphasized in that part.

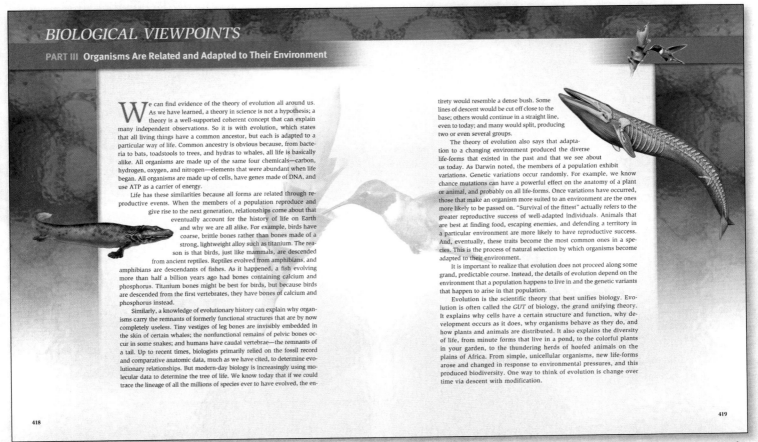

Dynamic and Accurate Illustrations Help You Visualize the Concepts

Multilevel Perspective

Illustrations depicting complex structures show macroscopic and microscopic views to help you see the relationships between increasingly detailed drawings.

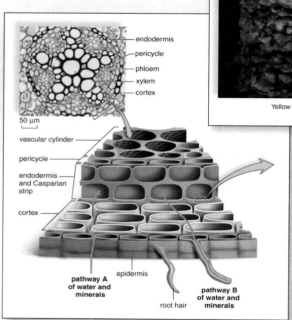

50 µm

- endodermis
- pericycle
- phloem
- xylem
- cortex

vascular cylinder

pericycle

endodermis and Casparian strip

cortex

epidermis

pathway A of water and minerals

root hair

pathway B of water and minerals

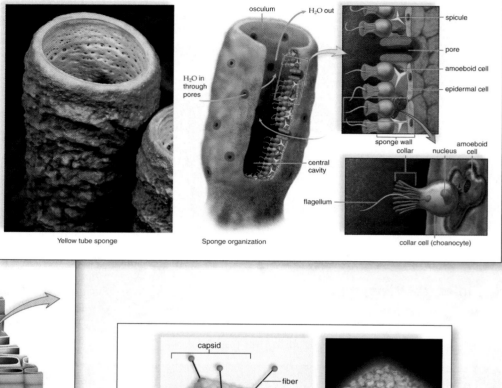

osculum — H₂O out

H₂O in through pores

Yellow tube sponge

Sponge organization

spicule
pore
amoeboid cell
epidermal cell

sponge wall
collar
nucleus
amoeboid cell

central cavity

flagellum

collar cell (choanocyte)

Combination Art

Drawings of structures are paired with micrographs to give you the best of both perspectives: the realism of photos and the explanatory clarity of line drawings.

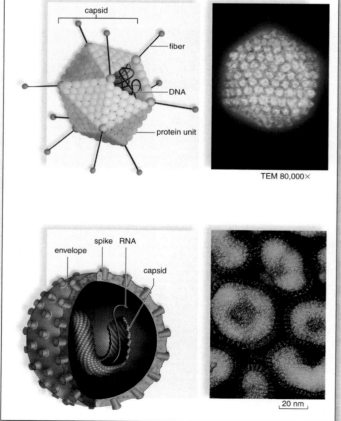

capsid

fiber

DNA

protein unit

TEM 80,000×

envelope spike RNA

capsid

20 nm

"The illustrations support the text strongly."
—Anju Sharma, Stevens Institute of Technology

Process Figures

Complex processes are broken down into a series of smaller steps that are easy to follow. Numbers guide you through the process.

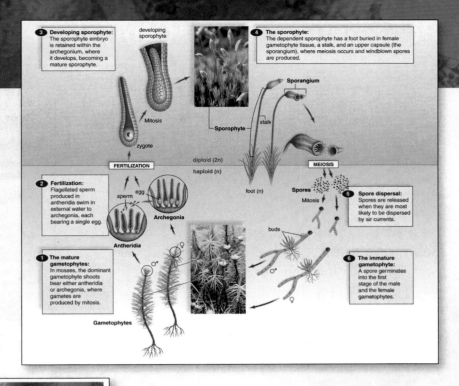

3 Developing sporophyte: The sporophyte embryo is retained within the archegonium, where it develops, becoming a mature sporophyte.

developing sporophyte

4 The sporophyte: The dependent sporophyte has a foot buried in female gametophyte tissue, a stalk, and an upper capsule (the sporangium), where meiosis occurs and windblown spores are produced.

Mitosis

Sporangium

Sporophyte

stalk

zygote

diploid (2n)

FERTILIZATION

haploid (n)

MEIOSIS

foot (n)

Spores

2 Fertilization: Flagellated sperm produced in antheridia swim in external water to archegonia, each bearing a single egg.

sperm

egg

Archegonia

Antheridia

Mitosis

buds

5 Spore dispersal: Spores are released when they are most likely to be dispersed by air currents.

1 The mature gametophyte: In mosses, the dominant gametophyte shoots bear either antheridia or archegonia, where gametes are produced by mitosis.

♂

♀

6 The immature gametophyte: A spore germinates into the first stage of the male and the female gametophytes.

Gametophytes

The numbered steps are coordinated with the narrative for an integrated approach to learning.

10.16 Elongation builds a polypeptide one amino acid at a time

During **elongation**, a polypeptide increases in length, one amino acid at a time. In addition to the participation of tRNAs, elongation requires elongation factors, which facilitate the binding of tRNA anticodons to mRNA codons at a ribosome.

Elongation is shown in **Figure 10.16**, where **1** a tRNA with an attached peptide is already at the P site, and a tRNA carrying its appropriate amino acid is just arriving at the A site. **2** Once the next tRNA is in place at the A site, the peptide will be transferred to this tRNA. This transfer requires energy and a ribozyme, which is a part of the larger ribosomal subunit. **3** Peptide bond formation occurs, and the peptide is one amino acid longer than it was before. **4** Finally, **translocation** occurs: The mRNA moves forward, and the peptide-bearing tRNA is now at the P site of the ribosome. The used tRNA exits from the E site. A new codon is at the A site, ready to receive another tRNA. Eventually, the ribosome reaches a stop codon, and termination occurs, during which the polypeptide is released.

The entire process of gene expression, consisting of transcription and translation, is reviewed in Section 10.17.

10.16 Check Your Progress With reference to Figure 10.10, what is the significance of finding these sequences in DNA: ATT, ATC, or ACT?

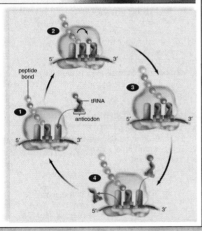

peptide bond

tRNA

anticodon

Color Consistency

Consistent use of color organizes information and clarifies concepts.

"The art, figures, and photos are excellent."
—Larry Szymczak, Chicago State University

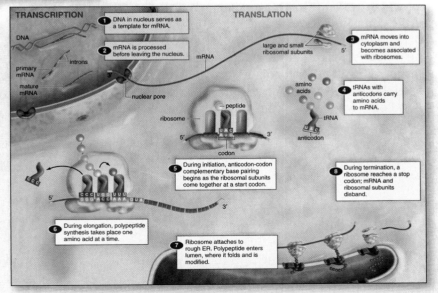

TRANSCRIPTION

TRANSLATION

DNA

1 DNA in nucleus serves as a template for mRNA.

2 mRNA is processed before leaving the nucleus.

primary mRNA

introns

mature mRNA

nuclear pore

mRNA

large and small ribosomal subunits

3 mRNA moves into cytoplasm and becomes associated with ribosomes.

amino acids

4 tRNAs with anticodons carry amino acids to mRNA.

tRNA

anticodon

peptide

ribosome

codon

5 During initiation, anticodon-codon complementary base pairing begins as the ribosomal subunits come together at a start codon.

8 During termination, a ribosome reaches a stop codon; mRNA and ribosomal subunits disband.

6 During elongation, polypeptide synthesis takes place one amino acid at a time.

7 Ribosome attaches to rough ER. Polypeptide enters lumen, where it folds and is modified.

Pedagogical Tools Help You Review and Assess Your Understanding

Chapter in Review

Concise, bulleted summaries, along with key illustrations, provide an excellent overview of the chapter concepts.

Testing Yourself

Section includes questions of varying degrees of difficulty—objective, multiple-choice questions to test knowledge and comprehension, and *Thinking Conceptually* questions to test your ability to apply the concepts.

Understanding the Key Terms

Each key term is page-referenced to help you master the scientific vocabulary.

Thinking Scientifically

Design an experiment or explain some part of a hypothetical experiment to take your understanding of the chapter concepts to a higher level.

ARIS Content

For that extra edge in your understanding of the concepts, visit the ARIS site for Mader, *Concepts of Biology*. Practice quizzes, BioTutorials, ScienCentral video quizzes, flashcards, and other media assets were all developed to help you succeed in your study of biology. www.mhhe.com/maderconcepts

"I think one of the strong points of this book is that it is written with students in mind, and not at a level that is too difficult or too easy."
—MaryLynne LaMantia, Golden West College

The Chapter in Review

Summary

Life's Energy Comes from the Sun
- Photosynthesis converts the energy of the sun into that of organic molecules.

Living Things Transform Energy

5.1 Energy makes things happen
- Energy is the capacity to do work.
- The five forms of energy are radiant, chemical, mechanical, electrical, and nuclear.

energy is

the

created or to another. be changed ble energy.

released.

Testing Yourself

Living Things Transform Energy

1. The _____ of energy would break the first law of thermodynamics.
   - a. creation
   - b. transformation
   - c. Both a and b are correct.
   - d. Neither a nor b is correct.
2. As a result of energy transformations,
   - a. entropy increases.
   - b. entropy decreases.
   - c. heat energy is gained.
   - d. energy is lost in the form of heat.
   - e. Both a and d are correct.
3. ATP is a modified
   - a. protein.
   - b. amino acid.
   - c. nucleotide.
   - d. fat.
4. **THINKING CONCEPTUALLY** How is the bulk of food you bring home from the store like glycogen, and how is ATP like the meal you prepare?

Enzymes Speed Chemical Reactions

5. The current mode
   - a. induced fit mod
   - b. activation mod
6. The active site of
   - a. is identical to t
   - b. is the part of th
   - c. can be used ov
   - d. is not affected
      and temperatu

Understanding the Terms

active site 82	hypertonic solution 89
active transport 90	hypotonic solution 89
aquaporin 89	induced fit model 82
bulk transport 90	isotonic solution 89
calorie 78	kilocalorie 78
carrier protein 86	kinetic energy 78
cholesterol 85	metabolic pathway 84
coenzyme 83	metabolism 82
cofactor 83	noncompetitive inhibition 84
competitive inhibition 84	osmosis 89
coupled reaction 81	phagocytosis 90
denatured 83	phospholipid bilayer 85
differentially permeable 86	pinocytosis 90
endergonic reaction 80	plasmolysis 89
endocytosis 90	potential energy 78
energy 78	receptor-mediated
energy of activation 82	endocytosis 90
entropy 79	receptor protein 86
enzyme 82	simple diffusion 88
exergonic reaction 80	sodium-potassium pump 90
exocytosis 90	solute 89

Thinking Scientifically

1. Use the first and second laws of thermodynamics to explain why each ecosystem needs a continuous supply of solar energy.
2. You have been assigned the task of designing an experiment to illustrate enzyme function. What factors, including environment, do you have to consider?

ARIS *Visit* www.mhhe.com/maderconcepts *for practice quizzes, animations, videos, and activities designed to help you master the material in this chapter.*

Acknowledgments

Many dedicated and talented individuals assisted in the development of *Concepts of Biology*. I am very grateful for the help of so many professionals at McGraw-Hill who were involved in bringing this book to fruition. In particular, let me thank Janice Roerig-Blong and Michael Hackett, the publisher and editor who steadfastly encouraged and supported this project. The developmental editor was Rose Koos, who was very devoted, and lent her talents and advice to all those who worked on this text. The project manager, Jayne Klein, faithfully and carefully steered the book through the publication process. Tamara Maury, the marketing manager, tirelessly promoted the text and educated the sales representatives on its message.

The design of the book is the result of the creative talents of Laurie Janssen and many others who assisted in deciding the appearance of each element in the text. Electronic Publishing Services followed my guidelines as they created and reworked each illustration, emphasizing pedagogy and beauty to arrive at the best presentation on the page. Evelyn Jo Hebert and Lori Hancock did a superb job of finding just the right photographs and micrographs.

My assistant, Beth Butler, worked faithfully as she helped proof the chapters and made sure all was well before the book went to press. As always, my family was extremely patient with me as I remained determined to meet every deadline on the road to publication. My husband, Arthur Cohen, is also a teacher of biology. The many discussions we have about the minutest detail to the gravest concept are invaluable to me.

I am very much indebted to the contributors and reviewers whose suggestions and expertise were so valuable as I developed *Concepts of Biology*.

360° Development

McGraw-Hill's 360° Development Process is an ongoing, never-ending, market-oriented approach to building accurate and innovative print and digital products. It is dedicated to continual large-scale and incremental improvement driven by multiple customer feedback loops and checkpoints. This is initiated during the early planning stages of our new products, and intensifies during the development and production stages, then begins again upon publication in anticipation of the next edition.

This process is designed to provide a broad, comprehensive spectrum of feedback for refinement and innovation of our learning tools, for both student and instructor. The 360° Development Process includes market research, content reviews, course- and product-specific symposia, accuracy checks, and art reviews. We appreciate the expertise of the many individuals involved in this process.

Contributors

Mark Bloom
Texas Christian University

Patrick L. Galliart
North Iowa Area Community College

Stephanie Harvey
Georgia Southwestern State University

Erica Kipp
Pace University

Tamara Lawson
Black Hills State University

Murray P. "Pat" Pendarvis
Southeastern Louisiana University

Jay Pitocchelli
Saint Anselm College

Gregory Pryor
Francis Marion University

Stephanie Songer
North Georgia College and State University

Michael Thompson
Middle Tennessee State University

Ancillary Authors

Instructor's Manual
Ryan L. Wagner
Millersville University of Pennsylvania

ARIS Practice Tests
Todd Kostman
University of Wisconsin–Oshkosh

ARIS Media Asset Correlations
Michael Windelspecht
Appalachian State University

Test Bank
Rob Dill
Bergen Community College

Lecture Outlines
Woody Moses
Highline Community College

Accuracy Checks
Patrick L. Galliart
North Iowa Area Community College

Jerry W. Mimms
University of Central Arkansas

Reviewers

Emily Allen
Gloucester County College

Kathy Pace Ames
Illinois Central College

Jason E. Arnold
Hopkinsville Community College

Dave Bachoon
Georgia College and State University

Andrei L. Barkovskii
Georgia College and State University

Lori Bean
Monroe County Community College

Mark G. Bolyard
Union University

Jason Brown
Young Harris College

Geralyn M. Caplan
Owensboro Community and Technical College

Carol E. Carr
John Tyler Community College

Misty Gregg Carriger
Northeast State Community College

Laurie-Ann Crawford
Hawkeye Community College

James Crowder
Brookdale Community College

Larry T. Crump
Joliet Junior College

John J. Dilustro
Chowan University

Toby Elberger
UConn–Stamford, Sacred Heart University

John A. Ewing, III
Itawamba Community College

Gregory S. Farley
Chesapeake College

Teresa G. Fischer
Indian River Community College

Patricia Flower
Miramar College

Joseph D. Gar
West Kentucky Community and Technical College

Nabarun Ghosh
West Texas A&M University
Jim R. Goetze
Laredo Community College
Andrew Goliszek
North Carolina A&T State University
Becky C. Graham
The University of West Alabama
Cary Guffey
Our Lady of the Lake University
James R. Hampton
Salt Lake Community College
Pamela L. Hanratty
Indiana University
Stephanie G. Harvey
Georgia Southwestern State University
Kendra Hill
South Dakota State University
B. K. Hull
Young Harris College
Troy W. Jesse
Broome Community College
H. Bruce Johnston
Fresno City College
Jacqueline A. Jordan
Clayton State University
Martin A. Kapper
Central Connecticut State University
Arnold J. Karpoff
University of Louisville
Dawn G. Keller
Hawkeye Community College
Diane M. Kelly
Broome Community College
Natasa Kesler
Seattle Central Community College
Dennis J. Kitz
Southern Illinois University, Edwardsville

Peter Kobella
Owensboro Community and Technical College
Anna Koshy
Houston Community College, Northwest
Todd A. Kostman
University of Wisconsin–Oshkosh
Jerome A. Krueger
South Dakota State University
James J. Krupa
University of Kentucky
Steven A. Kuhl
Lander University
Janice S. Lai
Seattle Central Community College
MaryLynne LaMantia
Golden West College
Thomas G. Lammers
University of Wisconsin–Oshkosh
Vic Landrum
Washburn University
Peggy Lepley
Cincinnati State Technical and Community College
Fordyce G. Lux III
Lander University
Janice B. Lynn
Auburn University, Montgomery
Michael P. Mahan
Armstrong Atlantic State University
Elizabeth A. Mays
Illinois Central College
TD Maze
Lander University
Jennifer Richter Maze
Lander University
Tiffany B. McFalls
Southeastern Louisiana University

Debra Meuler
Cardinal Stritch University
Thomas H. Milton
Richard Bland College
Jerry W. Mimms
University of Central Arkansas
Jeanne Mitchell
Truman State University
Brenda Moore
Truman State University
Allan D. Nelson
Tarleton State University
Jonas E. Okeagu
Fayetteville State University
Nathan Okia
Auburn University–Montgomery
Frank H. Osborne
Kean University
John C. Osterman
University of Nebraska–Lincoln
Surindar Paracer
Worcester State College
Ann V. Paterson
Williams Baptist College
Jay Pitocchelli
Saint Anselm College
Ramona Crain Popplewell
Grayson County College
Rongsun Pu
Kean University
Erin Rempala
San Diego Mesa College
John E. Rinehart
Eastern Oregon University
Dan Rogers
Somerset Community College
Jason F. Schreer
State University of New York at Potsdam
Gillian P. Schultz
Seattle Central Community College
Brian W. Schwartz
Columbus State University

Anju Sharma
Stevens Institute of Technology
Brian E. Smith
Black Hills State University
Maryann Smith
Brookdale Community College
Larry Szymczak
Chicago State University
Christopher Tabit
University of West Georgia
Season R. Thomson
Germanna Community College
Randall L. Tracy
Worcester State College
Anh-Hue Tu
Georgia Southwestern State University
George Veomett
University of Nebraska–Lincoln
Jyoti R. Wagle
Houston Community College System, Central
Ryan L. Wagner
Millersville University of Pennsylvania
Paul E. Wanda
Southern Illinois University, Edwardsville
Kelly J. Wessell
Tompkins Cortland Community College
Virginia White
Riverside Community College
Bob Wise
University of Wisconsin–Oshkosh
Michael L. Womack
Macon State College
Lan Xu
South Dakota State University
Alan Yauck
Middle Georgia College, Dublin Campus

General Biology Symposia

Every year McGraw-Hill conducts several General Biology Symposia, which are attended by instructors from across the country. These events are an opportunity for editors from McGraw-Hill to gather information about the needs and challenges of instructors teaching nonmajor-level biology courses. It also offers a forum for the attendees to exchange ideas and experiences with colleagues they might not have otherwise met. The feedback we have received has been invaluable, and has contributed to the development of *Concepts of Biology* and its ancillaries.

Norris Armstrong
University of Georgia
David Bachoon
Georgia College and State University
Sarah Bales
Moraine Valley Community College
Lisa Bellows
North Central Texas College

Joressia Beyer
John Tyler Community College
James Bidlack
University of Central Oklahoma
Mark Bloom
Texas Christian University
Paul Bologna
Montclair University

Bradford Boyer
Suffolk County Community College
Linda Brandt
Henry Ford Community College
Marguerite Brickman
University of Georgia
Art Buikema
Virginia Polytechnic Institute

Sharon Bullock
Virginia Commonwealth University
Raymond Burton
Germanna Community College
Nancy Butler
Kutztown University of Pennsylvania
Jane Caldwell
West Virginia University

Carol Carr
  John Tyler Community College
Kelly Cartwright
  College of Lake County
Rex Cates
  Brigham Young University
Sandra Caudle
  Calhoun Community College
Genevieve Chung
  Broward Community College
Jan Coles
  Kansas State University
Marian Wilson Comer
  Chicago State University
Renee Dawson
  University of Utah
Lewis Deaton
  *University of Louisiana
  at Lafayette*
Jody DeCamilo
  St. Louis Community College
Jean DeSaix
  *University of North Carolina
  at Chapel Hill*
JodyLee Estrada-Duek
  *Pima Community College,
  Desert Vista*
Laurie Faber-Foster
  *Grand Rapids
  Community College*
Susan Finazzo
  Broward Community College
Theresa Fischer
  *Indian River
  Community College*
Dennis Fulbright
  Michigan State University
Theresa Fulcher
  *Pellissippi State
  Technical College*
Steven Gabrey
  *Northwestern
  State University*
Cheryl Garett
  *Henry Ford
  Community College*
Farooka Gauhari
  *University
  of Nebraska–Omaha*
John Geiser
  Western Michigan University
Cindy Ghent
  Towson University
Julie Gibbs
  College of DuPage
William Glider
  *University
  of Nebraska–Lincoln*
Christopher Gregg
  Louisiana State University

Carla Guthridge
  Cameron University
Bob Harms
  *St. Louis Community
  College–Meramec*
Wendy Hartman
  *Palm Beach
  Community College*
Tina Hartney
  *California State
  Polytechnic University*
Kelly Hogan
  *University of North
  Carolina–Chapel Hill*
Eva Horne
  Kansas State University
David Huffman
  *Texas State University–
  San Marcos*
Shelley Jansky
  *University
  of Wisconsin–Stevens Point*
Sandra Johnson
  *Middle Tennessee
  State University*
Tina Jones
  *Shelton State Community
  College*
Arnold Karpoff
  University of Louisville
Jeff Kaufmann
  Irvine Valley College
Kyoungtae Kim
  Missouri State University
Michael Koban
  Morgan State University
Todd Kostman
  *University
  of Wisconsin–Oshkosh*
Steven Kudravi
  Georgia State University
Nicki Locascio
  Marshall University
Dave Loring
  *Johnson County
  Community College*
Janice Lynn
  Alabama State University
Phil Mathis
  *Middle Tennessee
  State University*
Mary Victoria McDonald
  University of Central Arkansas
Susan Meiers
  Western Illinois University
Daryl Miller
  *Broward Community
  College, South Campus*
Marjorie Miller
  Greenville Technical College

Meredith Norris
  *University of North Carolina
  at Charlotte*
Mured Odeh
  South Texas College
Nathan Olia
  *Auburn University–
  Montgomery*
Rodney Olsen
  Fresno City College
Alexander Olvido
  Virginia State University
Clark Ovrebo
  *University
  of Central Oklahoma*
Forrest Payne
  *University of Arkansas
  at Little Rock*
Nancy Pencoe
  University of West Georgia
Murray P. "Pat" Pendarvis
  *Southeastern
  Louisiana University*
Jennie Plunkett
  San Jacinto College
Scott Porteous
  Fresno City College
David Pylant
  *Wallace State
  Community College*
Fiona Qualls
  Jones County Junior College
Eric Rabitoy
  Citrus College
Karen Raines
  Colorado State University
Kirsten Raines
  San Jacinto College
Jill Reid
  *Virginia
  Commonwealth University*
Darryl Ritter
  Okaloosa-Walton College
Chris Robinson
  Bronx Community College
Robin Robison
  *Northwest Mississippi
  Community College*
Vickie Roettger
  *Missouri Southern
  State University*
Bill Rogers
  Ball State University
Vicki Rosen
  Utah State University
Kim Sadler
  *Middle Tennessee
  State University*
Cara Shillington
  Eastern Michigan University

Greg Sievert
  Emporia State University
Jimmie Sorrels
  *Itawamba
  Community College*
Judy Stewart
  *Community College
  of Southern Nevada*
Julie Sutherland
  College of DuPage
Bill Trayler
  *California
  State University–Fresno*
Linda Tyson
  Santa Fe Community College
Eileen Underwood
  *Bowling Green
  State University*
Heather Vance-Chalcraft
  East Carolina University
Marty Vaughan
  IUPUI–Indianapolis
Paul Verrell
  Washington State University
Thomas Vogel
  Western Illinois University
Brian Wainscott
  *Community College
  of Southern Nevada*
Jennifer Warner
  *University
  of North Carolina, Charlotte*
Scott Wells
  *Missouri Southern
  State University*
Robin Whitekiller
  *University
  of Central Arkansas*
Allison Wiedemeier
  *University
  of Illinois–Columbia*
Michael Windelspecht
  *Appalachian
  State University*
Mary Wisgirda
  *San Jacinto College,
  South Campus*
Tom Worcester
  *Mount Hood
  Community College*
Lan Xu
  *South Dakota
  State University*
Frank Zhang
  Kean University
Michelle Zjhra
  Georgia Southern University
Victoria Zusman
  Miami Dade College

Supplements

Dedicated to providing high-quality and effective supplements for instructors and students, the following supplements were developed for *Concepts of Biology*.

For Instructors

Laboratory Manual

ISBN (13) 978-0-07-329200-7
ISBN (10) 0-07-329200-1

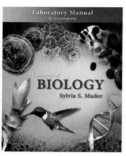

The *Concepts of Biology Laboratory Manual* is written by Dr. Sylvia S. Mader. With few exceptions, each chapter in the text has an accompanying laboratory exercise in the manual. Every laboratory has been written to help students learn the fundamental concepts of biology and the specific content of the chapter to which the lab relates, as well as gain a better understanding of the scientific method.

ARIS (Assessment, Review, and Instruction System)

 ARIS is a complete, online tutorial, electronic homework, and course management system designed for greater ease of use than any other system available. For students, ARIS contains self-study tools such as animations and interactive quizzes. This program enables students to complete their homework online, as assigned by their instructor. ARIS allows instructors to automatically grade and report easy-to-assign homework and quizzing, build their own assignments, track student progress, and share course materials with colleagues. ARIS also provides instructors with the ability to create or edit questions from the question bank, or import their own content. The fully integrated grade book can be downloaded to Excel, WebCT, or Blackboard.

Go to www.aris.mhhe.com to learn more.

Presentation Center

Build instructional materials wherever, whenever, and however you want!

Accessed from your textbook's ARIS website (www.mhhe.com/maderconcepts), an online digital library contains photos, artwork, animations, and other media types that can be used to create customized lectures, visually enhanced tests and quizzes, compelling course websites, or attractive printed support materials. All assets are copyrighted by McGraw-Hill Higher Education, but can be used by instructors for classroom purposes. The visual resources in this collection include:

- **Art** Full-color digital files of all illustrations in the book can be readily incorporated into lecture presentations, exams, or custom-made classroom materials. In addition, all files are pre-inserted into PowerPoint slides for ease of lecture preparation.
- **Photos** The photos collection contains digital files of photographs from the text, which can be reproduced for multiple classroom uses.
- **Tables** Every table that appears in the text has been saved in electronic form.
- **Animations** Numerous full-color animations illustrating important processes are also provided. Harness the visual impact of concepts in motion by importing these files into classroom presentations or online course materials.

Also residing on your textbook's ARIS website are:

- **PowerPoint Lecture Outlines** Ready-made presentations that combine art and lecture notes are provided for each chapter of the text.
- **PowerPoint Slides** For instructors who prefer to create their lectures from scratch, all illustrations, photos, and tables are pre-inserted by chapter into blank PowerPoint slides.

ScienCentral Videos

McGraw-Hill has teamed up with ScienCentral, Inc. to provide brief biology news videos for use in lecture or for student study and assessment purposes. A complete set of ScienCentral videos is located on this text's ARIS website. These active learning tools enhance a biology course by engaging students in real-life issues and applications such as developing new cancer treatments and understanding how methamphetamine damages the brain. ScienCentral, Inc., funded in part by grants from the National Science Foundation, produces science and technology content for television, video, and the Web.

McGraw-Hill: Biology Digitized Video Clips

ISBN (13) 978-0-312155-0
ISBN (10) 0-07-312155-X

McGraw-Hill is pleased to offer an outstanding presentation tool to text-adopting instructors—digitized biology video clips on DVD! Licensed from some of the highest-quality science video producers in the world, these brief segments range from about five seconds to just under three minutes in length and cover all areas of general biology from cells to ecosystems.

Engaging and informative, McGraw-Hill's digitized videos will help capture students' interest while illustrating key biological concepts and processes such as mitosis, how cilia and flagella work, and how some plants have evolved into carnivores.

Computerized Test Bank Online

A comprehensive bank of test questions is provided within a computerized test bank powered by McGraw-Hill's flexible electronic testing program EZ Test Online (www.eztestonline.com). EZ Test Online allows you to create paper and online tests or quizzes in this easy to use program!

Imagine being able to create and access your test or quiz anywhere, at any time, without installing the testing software. Now, with EZ Test Online, instructors can select questions from multiple McGraw-Hill test banks or author their own, and then either print the test for paper distribution or give it online.

Test Creation
- Author/edit questions online using the 14 different question-type templates
- Create question pools to offer multiple versions online—great for practice
- Export your tests for use in WebCT, Blackboard, PageOut, and Apple's iQuiz
- Sharing tests with colleagues, adjuncts, TAs is easy

Online Test Management
- Set availability dates and time limits for your quiz or test
- Assign points by question or question type with drop-down menu
- Provide immediate feedback to students or delay feedback until all finish the test
- Create practice tests online to enable student mastery
- Your roster can be uploaded to enable student self-registration

Online Scoring and Reporting
- Automated scoring for most of EZ Test's numerous question types
- Allows manual scoring for essay and other open-response questions
- Manual rescoring and feedback are also available
- EZ Test's grade book is designed to easily export to your grade book
- View basic statistical reports

Support and Help
- Flash tutorials for getting started on the support site
- Support Website: www.mhhe.com/eztest
- Product specialist available at 1-800-331-5094
- Online Training: http://auth.mhhe.com/mpss/workshops/

Student Response System

Wireless technology brings interactivity into the classroom or lecture hall. Instructors and students receive immediate feedback through wireless response pads that are easy to use and engage students. This system can be used by instructors to take attendance, administer quizzes and tests, create a lecture with intermittent questions, manage lectures and student comprehension through the use of the grade book, and integrate interactivity into their PowerPoint presentations.

For Students

ARIS (Assessment, Review, and Instruction System)

ARIS is an electronic study system that offers students a digital portal of knowledge. Students can readily access a variety of **digital learning objects** that include:

- Learning outcomes
- Chapter-level quizzing with pretest
- BioTutorial Animations with quizzing
- ScienCentral Videos with quizzing
- Narrated art for portable audio players

Electronic Books

If you or your students are ready for an alternative version of the traditional textbook, McGraw-Hill and VitalSource have partnered to bring you innovative and inexpensive electronic textbooks. By purchasing E-books from McGraw-Hill & VitalSource, students can save as much as 50% on selected titles delivered on the most advanced E-book platform available, VitalSource Bookshelf.

E-books from McGraw-Hill & VitalSource are smart, interactive, searchable, and portable. VitalSource Bookshelf comes with a powerful suite of built-in tools that allow detailed searching, highlighting, note taking, and student-to-student or instructor-to-student note sharing. In addition, the media-rich E-book for *Concepts of Biology* integrates relevant animations and videos into the textbook content for a true multimedia learning experience. E-books from McGraw-Hill & VitalSource will help students study smarter and quickly find the information they need. And they will save money. Contact your McGraw-Hill sales representative to discuss E-book packaging options.

How to Study Science

ISBN (13) 978-0-07-234693-0
ISBN (10) 0-07-234693-0

This workbook offers students helpful suggestions for meeting the considerable challenges of a science course. It gives practical advice on such topics as how to take notes, how to get the most out of laboratories, and how to overcome science anxiety.

Photo Atlas for General Biology

ISBN (13) 978-0-07-284610-2
ISBN (10) 0-07-284610-0

Atlas was developed to support our numerous general biology titles. It can be used as a supplement for a general biology lecture or laboratory course.

Contents

Preface iv
Guided Tour vi
Acknowledgments xiii
Supplements xvi

1 **Biology, the Study of Life 2**

Fire Ants Have a Good Defense 2

Organisms Are Characterized by Diversity and Unity 4
1.1 Life is diverse 4
1.2 Life has many levels of organization 4
1.3 Organisms share the same characteristics of life 6

Classification Helps Us Understand Diversity 8
1.4 Taxonomists group organisms according to evolutionary relationships 8

The Biosphere Is Organized 10
1.5 The biosphere is divided into ecosystems 10
1.6 Most of the biosphere's ecosystems are now threatened 10

Scientists Observe, Hypothesize, and Test 11
1.7 The natural world is studied by using scientific methods 11
1.8 Control groups allow for comparison of results 12
1.9 *How Biology Impacts Our Lives* DNA barcoding of life may become a reality 14

PART I   Organisms Are Composed of Cells 18

2 **Basic Chemistry and Cells 18**

Life Depends on Water 18

All Matter Is Composed of Chemical Elements 20
2.1 Six elements are basic to life 20
2.2 Atoms contain subatomic particles 21
2.3 *How Biology Impacts Our Lives* Radioactive isotopes have many medical uses 22

Atoms React with One Another to Form Molecules 23
2.4 After atoms react, they have a completed outer shell 23
2.5 An ionic bond occurs when electrons are transferred 24
2.6 A covalent bond occurs when electrons are shared 25
2.7 A covalent bond can be nonpolar or polar 26

2.8 A hydrogen bond can occur between polar molecules 26

The Properties of Water Benefit Life 27
2.9 Water molecules stick together: Cohesion 27
2.10 Water warms up and cools down slowly 28
2.11 Water dissolves other polar substances 28
2.12 Frozen water is less dense than liquid water 29

Living Things Require a Narrow pH Range 30
2.13 Acids and bases affect living things 30
2.14 The pH scale measures acidity and basicity 31
2.15 Buffers help keep the pH of body fluids relatively constant 31
2.16 *How Biology Impacts Our Lives* Acid deposition has many harmful effects 32

3 **Organic Molecules and Cells 36**

Plants and Animals Are the Same but Different 36

The Diversity of Organic Molecules Makes Life Diverse 38
3.1 The chemistry of carbon makes diverse molecules possible 38
3.2 Functional groups add to the diversity of organic molecules 39
3.3 Molecular subunits can be linked to form macromolecules 40

Carbohydrates Are Energy Sources and Structural Components 41
3.4 Simple carbohydrates provide quick energy 41
3.5 Complex carbohydrates store energy and provide structural support 42

Lipids Provide Storage, Insulation, and Other Functions 43
3.6 Fats and oils are rich energy-storage molecules 43
3.7 Other lipids have structural, hormonal, or protective functions 44

Proteins Have a Wide Variety of Vital Functions 45
3.8 Proteins are the most versatile of life's molecules 45
3.9 Each protein is a sequence of particular amino acids 45
3.10 The shape of a protein is necessary to its function 47

Nucleic Acids Are Information Molecules 48
3.11 The nucleic acids DNA and RNA carry coded information 48

3.12 *How Biology Impacts Our Lives* The Human Genome Project may lead to new disease treatments 49

3.13 The nucleotide ATP is the cell's energy carrier 50

4 Structure and Function of Cells 54

Cells: What Are They? 54

Cells Are the Basic Units of Life 56

4.1 All organisms are composed of cells 56

4.2 Metabolically active cells are small in size 57

4.3 *How Science Progresses* Microscopes allow us to see cells 58

4.4 Prokaryotic cells evolved first 59

4.5 Eukaryotic cells contain specialized organelles: An overview 60

Protein Synthesis Is a Major Function of Cells 62

4.6 The nucleus contains the cell's genetic information 62

4.7 The ribosomes carry out protein synthesis 63

4.8 The endoplasmic reticulum synthesizes and transports proteins and lipids 64

4.9 The Golgi apparatus modifies and repackages proteins for distribution 65

4.10 *How Science Progresses* Pulse-labeling allows observation of the secretory pathway 65

Vesicles and Vacuoles Have Varied Functions 66

4.11 Lysosomes digest macromolecules and cell parts 66

4.12 Peroxisomes break down long-chain fatty acids 66

4.13 Vacuoles have varied functions in protists and plants 66

4.14 The organelles of the endomembrane system work together 67

A Cell Carries Out Energy Transformations 68

4.15 Chloroplasts capture solar energy and produce carbohydrates 68

4.16 Mitochondria break down carbohydrates and produce ATP 68

4.17 *How Biology Impacts Our Lives* Malfunctioning mitochondria can cause human diseases 69

The Cytoskeleton Maintains Cell Shape and Assists Movement 70

4.18 The cytoskeleton consists of filaments and microtubules 70

4.19 Cilia and flagella contain microtubules 71

In Multicellular Organisms, Cells Join Together 72

4.20 Modifications of cell surfaces influence their behavior 72

5 Dynamic Activities of Cells 76

Life's Energy Comes from the Sun 76

Living Things Transform Energy 78

5.1 Energy makes things happen 78

5.2 Two laws apply to energy and its use 79

5.3 Cellular work is powered by ATP 80

5.4 ATP breakdown is coupled to energy-requiring reactions 81

Enzymes Speed Chemical Reactions 82

5.5 Enzymes speed reactions by lowering activation barriers 82

5.6 An enzyme's active site is where the reaction takes place 82

5.7 Enzyme speed is affected by local conditions 83

5.8 Enzymes can be inhibited noncompetitively and competitively 84

5.9 *How Biology Impacts Our Lives* Enzyme inhibitors can spell death 84

The Plasma Membrane Has Many and Various Functions 85

5.10 The plasma membrane is a phospholipid bilayer with embedded proteins 85

5.11 Proteins in the plasma membrane have numerous functions 86

5.12 *How Biology Impacts Our Lives* Malfunctioning plasma membrane proteins can cause human diseases 87

The Plasma Membrane Regulates the Passage of Molecules Into and Out of Cells 88

5.13 Simple diffusion across a membrane requires no energy 88

5.14 Facilitated diffusion requires a carrier protein but no energy 88

5.15 Osmosis can affect the size and shape of cells 89

5.16 Active transport requires a carrier protein and energy 90

5.17 Bulk transport involves the use of vesicles 90

6 Pathways of Photosynthesis 94

Color It Green 94

Photosynthesis Produces Food and Releases Oxygen 96

6.1 Photosynthesizers are autotrophs that produce their own food 96

6.2 In plants, chloroplasts carry out photosynthesis 97

6.3 Photosynthesis is a redox reaction that releases O_2 98

6.4 *How Science Progresses* Experiments showed that the O_2 released by photosynthesis comes from water 98

6.5 Photosynthesis involves two sets of reactions: The light reactions and the Calvin cycle reactions 99

First, Solar Energy Is Captured 100

6.6 Light reactions begin: Solar energy is absorbed by pigments 100

6.7 *How Science Progresses* Fall temperatures cause leaves to change color 100

6.8 Solar energy boosts electrons to a higher energy level 101

6.9 Electrons release their energy as ATP forms 101

6.10 During the light reactions, electrons follow a noncyclic pathway 102

6.11 The thylakoid membrane is organized to produce ATP and NADPH 103

Second, Carbohydrate Is Synthesized 104

6.12 The Calvin cycle uses ATP and NADPH from the light reactions to produce a carbohydrate 104

6.13 In plants, carbohydrate is the starting point for other molecules 105

C_3, C_4, and CAM Photosynthesis Thrive Under Different Conditions 106

6.14 C_3 photosynthesis evolved when oxygen was in limited supply 106

6.15 C_4 photosynthesis boosts CO_2 concentration for RuBP carboxylase 106

6.16 CAM photosynthesis is another alternative to C_3 photosynthesis 107

6.17 *How Science Progresses* Destroying tropical rain forests contributes to global warming 108

7 Pathways of Cellular Respiration 112

ATP Is Universal 112

Glucose Breakdown Releases Energy 114

7.1 Cellular respiration is a redox reaction that requires O_2 114

7.2 Cellular respiration has four phases—three phases occur in mitochondria 115

Carbon Dioxide and Water Are Produced During Glucose Breakdown 116

7.3 Glycolysis: Glucose breakdown begins 116

7.4 The preparatory reaction occurs before the citric acid cycle 118

7.5 The citric acid cycle: Final oxidation of glucose products 119

7.6 The electron transport chain captures much energy 120

7.7 The cristae create an H^+ gradient that drives ATP production 121

7.8 The ATP payoff can be calculated 122

Fermentation Is Inefficient 123

7.9 When oxygen is in short supply, the cell switches to fermentation 123

7.10 *How Biology Impacts Our Lives* Fermentation helps produce numerous food products 124

Metabolic Pathways Cross at Particular Substrates 125

7.11 Organic molecules can be broken down and synthesized as needed 125

7.12 *How Biology Impacts Our Lives* Exercise burns fat 126

Biological Viewpoints Organisms Are Composed of Cells 130

PART II Genes Control the Traits of Organisms 132

8 Cell Division and Reproduction 132

Cancer Is a Genetic Disorder 132

Cell Division Ensures the Passage of Genetic Information 134

8.1 Cell division is involved in both asexual and sexual reproduction 134

8.2 Prokaryotes reproduce asexually 135

Somatic Cells Have a Cell Cycle and Undergo Mitosis and Cytokinesis 136

8.3 The eukaryotic cell cycle is a set series of events 136

8.4 Eukaryotic chromosomes are visible during cell division 137

8.5 Mitosis maintains the chromosome number 138

8.6 Cytokinesis divides the cytoplasm 140

Cancer Is Uncontrolled Cell Division 141

8.7 Cell cycle control occurs at checkpoints 141

8.8 Signals affect the cell cycle control system 142

8.9 Cancer cells have abnormal characteristics 143

8.10 *How Biology Impacts Our Lives* Protective behaviors and diet help prevent cancer 144

Meiosis Produces Cells That Become the Gametes in Animals and Spores in Other Organisms 145

8.11 Homologous chromosomes separate during meiosis 145

8.12 Synapsis and crossing-over occur during meiosis I 146

8.13 Sexual reproduction increases genetic variation 146

8.14 Meiosis requires two division cycles 148

8.15 The life cycle of most multicellular organisms includes both mitosis and meiosis 150

8.16 Meiosis can be compared to mitosis 151

Chromosomal Abnormalities Can Be Inherited 152

8.17 An abnormal chromosome number is sometimes traceable to nondisjunction 152

8.18 Abnormal chromosome numbers cause syndromes 153

8.19 Abnormal chromosome structure also causes syndromes 154

9 Patterns of Genetic Inheritance 158

Inbreeding Leads to Disorders 158

Gregor Mendel Deduced Laws of Inheritance 160

9.1 A blending model of inheritance existed prior to Mendel 160

9.2 Mendel designed his experiments well 160

Single-Trait Crosses Reveal Units of Inheritance and the Law of Segregation 162

9.3 Mendel's law of segregation describes how gametes pass on traits 162

9.4 The units of inheritance are alleles of genes 163

Two-Trait Crosses Support the Law of Independent Assortment 164

9.5 Mendel's law of independent assortment describes inheritance of multiple traits 164
9.6 Mendel's results are consistent with the laws of probability 165
9.7 Testcrosses support Mendel's laws and indicate the genotype 166

Mendel's Laws Apply to Humans 167

9.8 Pedigrees can reveal the patterns of inheritance 167
9.9 Some human genetic disorders are autosomal recessive 168
9.10 Some human genetic disorders are autosomal dominant 169
9.11 *How Biology Impacts Our Lives* Genetic disorders may now be detected early on 170

Complex Inheritance Patterns Extend the Range of Mendelian Analysis 171

9.12 Incomplete dominance still follows the law of segregation 171
9.13 A gene may have more than two alleles 171
9.14 Several genes and the environment can influence a single multifactorial characteristic 172
9.15 One gene can influence several characteristics 173

Chromosomes Are the Carriers of Genes 174

9.16 Traits transmitted via the X chromosome have a unique pattern of inheritance 174
9.17 Humans have X-linked disorders 175
9.18 The genes on one chromosome form a linkage group 176
9.19 Frequency of recombinant gametes maps the chromosomes 176
9.20 *How Science Progresses* Thomas Hunt Morgan is commonly called "the fruit fly guy" 178

10 Molecular Biology of Inheritance 182

Arabidopsis Is a Model Organism 182

DNA Is the Genetic Material 184

10.1 DNA is a transforming substance 184
10.2 DNA, not protein, is the genetic material 184
10.3 DNA and RNA are polymers of nucleotides 186
10.4 DNA meets the criteria for the genetic material 187
10.5 DNA is a double helix 188

DNA Can Be Duplicated 190

10.6 DNA replication is semiconservative 190
10.7 Many different proteins help DNA replicate 191

Genes Specify the Makeup of Proteins 192

10.8 Genes are linked to proteins 192
10.9 The making of a protein requires transcription and translation 192
10.10 The genetic code for amino acids is a triplet code 193

10.11 During transcription, a gene passes its coded information to an mRNA 194
10.12 In eukaryotes, an mRNA is processed before leaving the nucleus 195
10.13 During translation, each transfer RNA carries a particular amino acid 196
10.14 Translation occurs at ribosomes in cytoplasm 197
10.15 Initiation begins the process of protein production 198
10.16 Elongation builds a polypeptide one amino acid at a time 198
10.17 Let's review gene expression 199

Mutations Are Changes in the Sequence of DNA Bases 200

10.18 Mutations affect genetic information and expression 200
10.19 *How Biology Impacts Our Lives* Many agents can cause mutations 201
10.20 *How Science Progresses* Transposons are "jumping genes" 202

11 Regulation of Gene Activity 206

Moth and Butterfly Wings Tell a Story 206

Gene Expression Is Controlled in Prokaryotic Cells 208

11.1 DNA-binding proteins turn genes on and off in prokaryotes 208

Control of Gene Expression in Eukaryotes Causes Specialized Cells 209

11.2 Eukaryotic cells are specialized 209
11.3 Plants are cloned from a single cell 209
11.4 Animals are cloned using a donor nucleus 210
11.5 *How Biology Impacts Our Lives* Animal cloning has benefits and drawbacks 211

Control of Gene Expression Is Varied in Eukaryotes 212

11.6 Chromatin is highly condensed in chromosomes 212
11.7 The genes in highly condensed chromatin are not expressed 212
11.8 DNA-binding proteins regulate transcription in eukaryotes 213
11.9 mRNA processing can affect gene expression 214
11.10 Control of gene expression also occurs in the cytoplasm 214
11.11 Synopsis of gene expression control in eukaryotes 215

Gene Expression Is Controlled During Development 216

11.12 Genes are turned on sequentially during development 216
11.13 Homeotic genes and apoptosis occur in a wide range of animals 217

Genetic Mutations Cause Cancer 218

11.14 When cancer develops, two types of genes are out of control 218

11.15 Faulty gene products interfere with signal transduction when cancer develops 219
11.16 Cancer develops and becomes malignant gradually 220

12 Biotechnology and Genomics 224

Witnessing Genetic Engineering 224
DNA Can Be Cloned 226
12.1 Genes can be isolated and cloned 226
12.2 Specific DNA sequences can be cloned 227
Organisms Can Be Genetically Modified 228
12.3 Bacteria are modified to make a product or perform a service 228
12.4 *How Science Progresses* Making cheese— Genetic engineering comes to the rescue 228
12.5 Plants are genetically modified to increase yield or to produce a product 229
12.6 *How Biology Impacts Our Lives* Are genetically engineered foods safe? 230
12.7 Animals are genetically modified to enhance traits or obtain useful products 231
12.8 A person's genome can be modified 232
12.9 *How Science Progresses* Gene therapy trials have varying degrees of success 233
The Human Genome Can Be Manipulated 234
12.10 The human genome has been sequenced 234
12.11 *How Biology Impacts Our Lives* New cures are on the horizon 235
12.12 Proteomics and bioinformatics are new endeavors 236
12.13 Functional and comparative genomics 236
Biological Viewpoints Genes Control the Traits of Organisms 240

PART III Organisms Are Related and Adapted to Their Environment 242

13 Darwin and Evolution 242

The "Vice Versa" of Animals and Plants 242
Darwin Developed a Natural Selection Hypothesis 244
13.1 Darwin made a trip around the world 244
13.2 Others had offered ideas about evolution before Darwin 245
13.3 Artificial selection mimics natural selection 246
13.4 Darwin formulated natural selection as a mechanism for evolution 246
13.5 *How Science Progresses* Wallace independently formulated a natural selection hypothesis 247
13.6 *How Science Progresses* Natural selection can be witnessed 248
The Evidence for Evolution Is Strong 249
13.7 Fossils provide a record of the past 249
13.8 Fossils are evidence for common descent 250
13.9 Anatomic evidence supports common descent 251

13.10 Biogeographic evidence supports common descent 252
13.11 Molecular evidence supports common descent 252
Population Genetics Tells Us When Microevolution Occurs 253
13.12 A Hardy-Weinberg equilibrium is not expected 253
13.13 Both mutations and sexual recombination produce variations 254
13.14 Nonrandom mating and gene flow can contribute to microevolution 254
13.15 The effects of genetic drift are unpredictable 255
13.16 Natural selection can be stabilizing, directional, or disruptive 256
13.17 *How Biology Impacts Our Lives* Stabilizing selection helps maintain harmful alleles 258

14 Speciation and Evolution 262

Hybrid Animals Do Exist 262
Evolution of Diversity Requires Speciation 264
14.1 Species have been defined in more than one way 264
14.2 Reproductive barriers maintain genetic differences between species 266
Origin of Species Usually Requires Geographic Separation 268
14.3 Allopatric speciation utilizes a geographic barrier 268
14.4 Adaptive radiation produces many related species 270
Origin of Species Can Occur in One Place 271
14.5 Speciation occasionally occurs without a geographic barrier 271
14.6 *How Science Progresses* Artificial selection produced corn 272
The Fossil Record Shows Both Gradual and Rapid Speciation 273
14.7 Speciation occurs at different tempos 273
14.8 *How Science Progresses* The Burgess Shale hosts a diversity of life 274
Developmental Genes Provide a Mechanism for Rapid Speciation 276
14.9 Gene expression can influence development 276
Speciation Is Not Goal-Oriented 278
14.10 Evolution is not directed toward any particular end 278

15 The History and Classification of Life on Earth 282

Motherhood Among Dinosaurs 282
The Fossil Record Reveals the History of Life on Earth 284
15.1 The geologic timescale is based on the fossil record 284

15.2 *How Science Progresses* The geologic clock can help put Earth's history in perspective 286

15.3 Continental drift has affected the history of life 286

15.4 Mass extinctions have affected the history of life 288

Systematics Traces Evolutionary Relationships 290

15.5 Organisms can be classified into categories 290

15.6 Linnaean classification reflects phylogeny 291

15.7 Certain types of data are used to trace phylogeny 292

15.8 Phylogenetic cladistics and evolutionary systematics use the same data differently 294

The Three-Domain Classification System Is Widely Accepted 296

15.9 This text uses the three-domain system of classifying organisms 296

16 Evolution of Microbial Life 300

At Your Service: Viruses and Bacteria 300

Viruses Reproduce in Living Cells 302

16.1 Viruses have a simple structure 302

16.2 Some viruses reproduce inside bacteria 302

16.3 *How Science Progresses* Viruses are responsible for a number of plant diseases 304

16.4 Viruses reproduce inside animal cells and cause diseases 305

16.5 The AIDS virus exemplifies RNA retroviruses 306

16.6 *How Biology Impacts Our Lives* Humans suffer from emerging viral diseases 307

The First Cells Originated on Early Earth 308

16.7 Experiments show how small organic molecules may have first formed 308

16.8 RNA may have been the first macromolecule 309

16.9 Protocells preceded the first true cells 310

Both Bacteria and Archaea Are Prokaryotes 311

16.10 Prokaryotes have particular structural features 311

16.11 Prokaryotes have a common reproductive strategy 312

16.12 How genes are transferred in bacteria 313

16.13 Prokaryotes have various means of nutrition 314

16.14 The cyanobacteria are ecologically important organisms 315

16.15 Some archaea live in extreme environments 316

16.16 *How Biology Impacts Our Lives* Prokaryotes have environmental and medical importance 317

16.17 *How Biology Impacts Our Lives* Disease-causing microbes can be biological weapons 318

17 Evolution of Protists 322

Protists Cause Disease Too 322

Protists May Represent the Oldest Eukaryotic Cells 324

17.1 Eukaryotic organelles probably arose by endosymbiosis 324

17.2 Protists are a diverse group 324

17.3 *How Science Progresses* How can the protists be classified? 326

Protozoans Are Heterotrophic Protists with Various Means of Locomotion 327

17.4 Protozoans called flagellates move by flagella 327

17.5 Protozoans called amoeboids move by pseudopods 328

17.6 Protozoans called ciliates move by cilia 329

17.7 Protozoans called sporozoans are not motile 330

Some Protists Have Moldlike Characteristics 331

17.8 The diversity of protists includes slime molds and water molds 331

Algae Are Photosynthetic Protists of Environmental Importance 332

17.9 The diatoms and dinoflagellates are significant algae in the oceans 332

17.10 Red algae and brown algae are multicellular 333

17.11 Green algae are ancestral to plants 334

17.12 *How Science Progresses* Life cycles among the algae have many variations 336

18 Evolution of Plants and Fungi 340

Some Plants Are Carnivorous 340

The Evolution of Plants Spans 500 Million Years 342

18.1 Evidence suggests that plants evolved from green algae 342

18.2 The evolution of plants is marked by four innovations 342

18.3 Plants have an alternation of generations life cycle 344

18.4 Sporophyte dominance was adaptive to a dry land environment 344

Plants Are Adapted to the Land Environment 346

18.5 Bryophytes are nonvascular plants in which the gametophyte is dominant 346

18.6 Ferns and their allies have a dominant vascular sporophyte 348

18.7 Most gymnosperms bear cones on which the seeds are "naked" 350

18.8 *How Biology Impacts Our Lives* Carboniferous forests became the coal we use today 352

18.9 Angiosperms are the flowering plants 353

18.10 The flowers of angiosperms produce "covered" seeds 354

18.11 *How Biology Impacts Our Lives* Flowering plants provide many services 356

Fungi Have Their Own Evolutionary History 358
18.12 Fungi differ from plants and animals 358
18.13 Fungi have mutualistic relationships with algae and plants 359
18.14 Fungi occur in three main groups 360
18.15 *How Biology Impacts Our Lives* Fungi have economic and medical importance 362

19 Evolution of Animals 366

The Secret Life of Bats 366
Key Innovations Distinguish Invertebrate Groups 368
19.1 Animals have distinctive characteristics 368
19.2 Animals most likely have a protistan ancestor 369
19.3 The traditional evolutionary tree of animals is based on seven key innovations 371
19.4 *How Science Progresses* Molecular data suggest a new evolutionary tree for animals 372
19.5 Some animal groups are invertebrates and some are vertebrates 373
19.6 Sponges are multicellular invertebrates 374
19.7 Cnidarians have true tissues 375
19.8 Free-living flatworms have bilateral symmetry 376
19.9 Some flatworms are parasitic 377
19.10 Roundworms have a pseudocoelom and a complete digestive tract 378
19.11 A coelom gives complex animal groups certain advantages 379
19.12 Molluscs have a three-part body plan 380
19.13 Annelids are the segmented worms 381
19.14 Arthropods have jointed appendages 382
19.15 Well known arthropods other than insects 383
19.16 Insects, the largest group of arthropods, are adapted to living on land 384
19.17 Echinoderms are radially symmetrical as adults 385
Further Innovations Allowed Vertebrates to Invade the Land Environment 386
19.18 Four features characterize chordates 386
19.19 Invertebrate chordates have a notochord as adults 386
19.20 The evolutionary tree of vertebrates is based on five key features 387
19.21 Jaws and lungs evolved among the fishes 388
19.22 Amphibians are tetrapods that can move on land 389
19.23 Reptiles have an amniotic egg and can reproduce on land 390
19.24 Birds have feathers and are endotherms 391
19.25 Mammals have hair and mammary glands 392
19.26 *How Biology Impacts Our Lives* Many vertebrates provide medical treatments for humans 394

20 Evolution of Humans 398

Lucy's Legacy 398
Humans Share Characteristics with All the Other Primates 400
20.1 Primates are adapted to live in trees 400
20.2 All primates evolved from a common ancestor 403
Humans Have an Upright Stance and Eventually a Large Brain 404
20.3 Early hominids could stand upright 404
20.4 Australopithecines had a small brain 406
20.5 *How Biology Impacts Our Lives* Origins of the genus *Homo* 407
20.6 Early *Homo* had a large brain 408
20.7 *How Science Progresses* Biocultural evolution began with *Homo* 409
***Homo sapiens* Is the Last Twig on the Primate Evolutionary Bush 410**
20.8 The Neandertal and Cro-Magnon people coexisted for 12,000 years 410
20.9 The particulars of *Homo sapiens* evolution are being studied 410
20.10 *How Science Progresses* Cro-Magnons made good use of tools 412
20.11 *How Science Progresses* Agriculture made modern civilizations possible 412
Today's Humans Belong to One Species 414
20.12 Humans have different ethnicities 414
Biological Viewpoints Organisms Are Related and Adapted to Their Environment 418

PART IV Plants Are Homeostatic 420

21 Plant Organization and Homeostasis 420

What Do Forests Have to Do with Global Warming? 420
Plants Have Three Vegetative Organs 422
21.1 Flowering plants typically have roots, stems, and leaves 422
21.2 Flowering plants are either monocots or eudicots 424
21.3 *How Biology Impacts Our Lives* Monocots serve humans well 425
The Same Plant Cells and Tissues Are Found in All Plant Organs 426
21.4 Plants have specialized cells and tissues 426
21.5 The three types of plant tissues are found in each organ 428
Plant Growth Is Either Primary or Secondary 430
21.6 Primary growth lengthens the root and shoot systems 430
21.7 Secondary growth widens roots and stems 432
21.8 *How Biology Impacts Our Lives* Wood has been a part of human history 433

Leaf Anatomy Facilitates Photosynthesis 434
21.9   Leaves are organized to carry on
       photosynthesis 434
Plants Maintain Internal Equilibrium 435
21.10 The organization of plants fosters
       homeostasis 435
21.11 Regulatory and other mechanisms help plants
       maintain homeostasis 435

22   Transport and Nutrition in Plants 440

Plants Can Adapt Too 440
**Plants Are Organized to Transport Water and
Solutes 442**
22.1   Transport begins in both the leaves
       and the roots of plants 442
22.2   *How Science Progresses* Competition for
       resources is one aspect of biodiversity 443
**Xylem Transport Depends on the Properties of
Water 444**
22.3   Water is pulled up in xylem by evaporation
       from leaves 444
22.4   Guard cells regulate water loss at leaves 446
22.5   *How Biology Impacts Our Lives* Plants can
       clean up toxic messes 447
**Phloem Function Depends on Membrane
Transport 448**
22.6   Phloem carries organic molecules 448
22.7   The pressure-flow model explains phloem
       transport 448
**Plants Require Good Nutrition and Therefore
Good Soil 450**
22.8   Certain nutrients are essential to plants 450
22.9   Roots are specialized for the uptake of water
       and minerals 452
22.10 Soil has distinct characteristics 453
22.11 Plants absorb minerals from the soil 453
22.12 Adaptations of plants help them acquire
       nutrients 454

23   Control of Growth and Responses in Plants 458

Recovering Slowly 458
**Plant Hormones Regulate Plant Growth
and Development 460**
23.1   Hormones act by utilizing signal transduction
       pathways 460
23.2   Auxins promote growth and cell
       elongation 460
23.3   Gibberellins control stem elongation 462
23.4   Cytokinins stimulate cell division and
       differentiation 463
23.5   Abscisic acid suppresses growth of buds
       and closes stomata 464
23.6   Ethylene stimulates the ripening of fruits 465
Plants Respond to Environmental Stimuli 466
23.7   Plants have many ways of responding
       to their external environment 466

23.8   Tropisms occur when plants respond to
       stimuli 466
23.9   Turgor and sleep movements are complex
       responses 468
23.10 Flowering is a response to the photoperiod
       in some plants 470
23.11 Response to the photoperiod requires
       phytochrome 471
23.12 Plants respond to the biotic environment 472
23.13 *How Biology Impacts Our Lives* Eloy Rodriguez
       has discovered many medicinal plants 474

24   Reproduction in Plants 478

With a Little Help 478
**Sexual Reproduction in Flowering Plants Is Suitable
to the Land Environment 480**
24.1   Plants have a sexual life cycle called
       alternation of generations 480
24.2   Pollination and fertilization bring gametes
       together during sexual reproduction 482
Seeds Contain a New Diploid Generation 484
24.3   A sporophyte embryo and its cotyledons
       develop inside a seed 484
24.4   The ovary becomes a fruit, which assists
       in sporophyte dispersal 485
24.5   With seed germination, the life cycle is
       complete 486
Plants Can Also Reproduce Asexually 487
24.6   Plants have various ways of reproducing
       asexually 487
24.7   Cloning of plants in tissue culture assists
       agriculture 488
Biological Viewpoints Plants Are Homeostatic 492

PART V   Animals Are Homeostatic 494

25   Animal Organization and Homeostasis 494

Staying Warm, Staying Cool 494
The Structure of Tissues Suits Their Function 496
25.1   Levels of biological organization are evident
       in animals 496
**Four Types of Tissues Are Common in the Animal
Body 496**
25.2   Epithelial tissue covers organs and lines
       body cavities 496
25.3   Connective tissue connects and supports
       other tissues 498
25.4   Muscular tissue is contractile and moves
       body parts 500
25.5   Nervous tissue communicates with and
       regulates the functions of the body's
       organs 501
25.6   *How Biology Impacts Our Lives* Will nerve
       regeneration reverse a spinal cord injury? 502

Organs, Composed of Tissues, Work Together in Organ Systems 503

25.7　Each organ has a specific structure and function 503

All Organ Systems Contribute to Homeostasis in Animals 504

25.8　Several organs work together to carry out the functions of an organ system 504

25.9　*How Science Progresses* Organs for transplant may come from various sources 506

25.10　Homeostasis is the constancy of the internal environment 507

25.11　Homeostasis is achieved through negative feedback mechanisms 508

26　Coordination by Neural Signaling 512

Getting a Head 512

Most Animals Have a Nervous System That Allows Responses to Stimuli 514

26.1　Invertebrates reflect an evolutionary trend toward bilateral symmetry and cephalization 514

26.2　Humans have well-developed central and peripheral nervous systems 516

Neurons Process and Transmit Information 517

26.3　Neurons are the functional units of a nervous system 517

26.4　Neurons have a resting potential across their membranes when they are not active 518

26.5　Neurons have an action potential across axon membranes when they are active 518

26.6　Propagation of an action potential is speedy 519

26.7　Communication between neurons occurs at synapses 520

26.8　Neurotransmitters can be stimulatory or inhibitory 520

26.9　Integration is a summing up of stimulatory and inhibitory signals 521

26.10　*How Biology Impacts Our Lives* Drugs that interfere with neurotransmitter release or uptake may be abused 522

The Vertebrate Central Nervous System (CNS) Consists of the Spinal Cord and Brain 524

26.11　The human spinal cord and brain function together 524

26.12　The cerebrum performs integrative activities 525

26.13　The other parts of the brain have specialized functions 526

26.14　The limbic system is involved in memory and learning as well as in emotions 527

The Vertebrate Peripheral Nervous System (PNS) Consists of Nerves 528

26.15　The peripheral nervous system contains cranial and spinal nerves 528

26.16　In the somatic system, reflexes allow us to respond quickly to stimuli 529

26.17　In the autonomic system, the parasympathetic and sympathetic divisions control the actions of internal organs 530

27　Sense Organs 534

The Eyes Have It 534

Sensory Receptors Respond to Stimuli 536

27.1　Sensory receptors can be divided into five categories 536

27.2　Sensory receptors communicate with the CNS 537

Chemoreceptors Are Sensitive to Chemicals 538

27.3　Chemoreceptors are widespread in the animal kingdom 538

27.4　Mammalian taste receptors are located in the mouth 538

27.5　Mammalian olfactory receptors are located in the nose 539

Photoreceptors Are Sensitive to Light 540

27.6　The vertebrate eye is a camera-type eye 540

27.7　*How Biology Impacts Our Lives* Protect your eyes from the sun 541

27.8　The lens helps bring an object into focus 541

27.9　*How Biology Impacts Our Lives* The inability to form a clear image can be corrected 542

27.10　The retina sends information to the visual cortex 542

Mechanoreceptors Are Involved in Hearing and Balance 544

27.11　The mammalian ear has three main regions 544

27.12　Hair cells in the inner ear detect sound vibrations 545

27.13　*How Biology Impacts Our Lives* Protect your ears from loud noises 546

27.14　The sense of balance occurs in the inner ear 546

27.15　*How Biology Impacts Our Lives* Motion sickness can be disturbing 548

27.16　Other animals respond to motion 548

28　Locomotion and Support Systems 552

Skeletal Remains Reveal All 552

Animal Skeletons Support, Move, and Protect the Body 554

28.1　Animal skeletons can be hydrostatic, external, or internal 554

28.2　Mammals have an endoskeleton that serves many functions 555

The Mammalian Skeleton Is a Series of Bones Connected at Joints 556

28.3　The bones of the axial skeleton lie in the midline of the body 556

28.4　The appendicular skeleton consists of bones in the girdles and limbs 558

28.5　*How Biology Impacts Our Lives* Avoidance of osteoporosis requires good nutrition and exercise 559

28.6 Bones are composed of living tissues 560

28.7 Joints occur where bones meet 561

28.8 *How Biology Impacts Our Lives* Joint disorders can be repaired 562

Animal Movement Is Dependent on Muscle Cell Contraction 563

28.9 Vertebrate skeletal muscles have various functions 563

28.10 Skeletal muscles contract in units 564

28.11 *How Biology Impacts Our Lives* Exercise has many benefits 565

28.12 A muscle cell contains many myofibrils 566

28.13 Sarcomeres shorten when muscle cells contract 566

28.14 Axon terminals bring about muscle contraction 567

28.15 Muscles have three sources of ATP for contraction 568

28.16 Some muscle cells are fast-twitch and some are slow-twitch 568

29 Circulation and Cardiovascular Systems 572

Not All Animals Have Red Blood 572

A Circulatory System Helps Maintain Homeostasis 574

29.1 A circulatory system serves the needs of cells 574

29.2 Some invertebrates do not have a circulatory system 574

29.3 Other invertebrates have an open or a closed circulatory system 575

29.4 All vertebrates have a closed circulatory system 576

The Mammalian Cardiovascular System Consists of the Heart and Blood Vessels 577

29.5 The mammalian heart has four chambers 577

29.6 The heartbeat is rhythmic 578

29.7 Blood vessel structure is suited to its function 579

29.8 Blood vessels form two circuits in mammals 580

29.9 Blood pressure is essential to the flow of blood in each circuit 581

29.10 *How Biology Impacts Our Lives* Blood vessel deterioration results in cardiovascular disease 582

29.11 *How Biology Impacts Our Lives* Cardiovascular disease can often be prevented 582

Blood Has Vital Functions 584

29.12 Blood is a liquid tissue 584

29.13 Blood clotting involves platelets 585

29.14 *How Science Progresses* Adult stem cells include blood stem cells 586

29.15 Capillary exchange is vital to cells 587

29.16 Blood types must be matched for transfusions 588

30 Lymph Transport and Immunity 592

AIDS Destroys the Immune System 592

The Lymphatic System Functions in Transport and Immunity 594

30.1 Lymphatic vessels transport lymph 594

30.2 Lymphatic organs defend the body 595

The Body's First Line of Defense Against Disease Is Nonspecific and Innate 596

30.3 Barriers to entry, complement proteins, and certain blood cells are first responders 596

30.4 *How Biology Impacts Our Lives* A fever can be beneficial 597

30.5 The inflammatory response is a localized response to invasion 598

The Body's Second Line of Defense Against Disease Is Specific to the Pathogen 599

30.6 The second line of defense targets specific antigens 599

30.7 Specific immunity can be active or passive 599

30.8 Lymphocytes are directly responsible for specific defenses 600

30.9 Antibody-mediated immunity involves B cells 601

30.10 Cell-mediated immunity involves several types of T cells 602

30.11 *How Biology Impacts Our Lives* Monoclonal antibodies have many uses 604

Abnormal Immune Responses Can Have Health Consequences 605

30.12 Tissue rejection makes transplanting organs difficult 605

30.13 Autoimmune disorders are long-term illnesses 605

30.14 *How Biology Impacts Our Lives* Allergic reactions can be debilitating and even fatal 606

31 Digestive Systems and Nutrition 610

How to Tell a Carnivore from an Herbivore 610

Animals Must Obtain and Process Their Food 612

31.1 A digestive system carries out ingestion, digestion, absorption, and elimination 612

31.2 Animals exhibit a variety of feeding strategies 613

31.3 A complete digestive tract has specialized compartments 614

31.4 Both mechanical and chemical digestion occur in the mouth 616

31.5 The esophagus conducts food to the stomach 617

31.6 Food storage and chemical digestion take place in the stomach 618

31.7 *How Biology Impacts Our Lives* Bacteria contribute to the cause of ulcers 618

31.8 In the small intestine, chemical digestion concludes, and absorption of nutrients occurs 619

31.9 The pancreas and the liver contribute to chemical digestion 620

31.10 The stomach and duodenum are endocrine glands 621

31.11 The large intestine absorbs water and prepares wastes for elimination 621

Good Nutrition and Diet Lead to Better Health 622

31.12 Carbohydrates are nutrients that provide immediate energy as well as fiber 622

31.13 Lipids are nutrients that supply long-term energy 623

31.14 Proteins are nutrients that supply building blocks for cells 623

31.15 Minerals have various roles in the body 624

31.16 Vitamins help regulate metabolism 625

31.17 *How Biology Impacts Our Lives* Nutritional labels allow evaluation of a food's content 626

31.18 *How Biology Impacts Our Lives* Certain disorders are associated with obesity 626

31.19 *How Biology Impacts Our Lives* Eating disorders appear to have a psychological component 628

32 Gas Exchange and Transport in Animals 632

Free-Diving Is Dangerous 632

Animals Have Gas-Exchange Surfaces 634

32.1 Respiration involves several steps 634

32.2 External respiration surfaces must be moist 634

32.3 Gills are an efficient gas-exchange surface in water 636

32.4 The tracheal system in insects permits direct gas exchange 637

32.5 The human respiratory system utilizes lungs as a gas-exchange surface 638

32.6 *How Biology Impacts Our Lives* Questions about tobacco, smoking, and health 639

Ventilation Precedes Transport 640

32.7 Breathing brings air into and out of the lungs 640

32.8 Our breathing rate can be modified 641

32.9 External and internal respiration require no energy 642

32.10 Hemoglobin is involved in transport of gases 643

32.11 *How Biology Impacts Our Lives* Respiratory disorders have resulted from breathing 9/11 dust 644

33 Osmoregulation and Excretion 648

Do Coral Reef Animals Regulate? 648

Metabolic Waste Products Have Different Advantages 650

33.1 The nitrogenous waste product of animals varies according to the environment 650

33.2 Many invertebrates have organs of excretion 651

Osmoregulation Varies According to the Environment 652

33.3 Aquatic vertebrates have adaptations to maintain the water-salt balance of their bodies 652

33.4 Terrestrial vertebrates have adaptations to maintain the water-salt balance of their bodies 653

The Kidney Is an Organ of Homeostasis 654

33.5 The kidneys are a part of the urinary system 654

33.6 The mammalian kidney contains many tubules 654

33.7 Urine formation requires three steps 656

33.8 *How Biology Impacts Our Lives* Urinalysis can detect drug use 657

33.9 The kidneys concentrate urine to maintain water-salt balance 658

33.10 Lungs and kidneys maintain acid-base balance 659

33.11 *How Science Progresses* The artificial kidney machine makes up for faulty kidneys 660

33.12 *How Biology Impacts Our Lives* Dehydration and water intoxication occur in humans 660

34 Coordination by Hormone Signaling 664

Pheromones Among Us 664

The Endocrine System Utilizes Chemical Signals 666

34.1 The endocrine and nervous systems work together 666

34.2 Hormones affect cellular metabolism 667

34.3 The vertebrate endocrine system includes diverse hormones 668

The Hypothalamus and Pituitary Are Central to the Endocrine System 670

34.4 The hypothalamus is a part of the nervous and endocrine systems 670

34.5 The anterior pituitary produces nontropic and tropic hormones 671

Hormones Regulate Metabolism and Homeostasis 672

34.6 The adrenal glands respond to stress 672

34.7 *How Biology Impacts Our Lives* Glucocorticoid therapy can lead to Cushing syndrome 673

34.8 The pancreas regulates the blood sugar level 674

34.9 *How Biology Impacts Our Lives* Diabetes is becoming a very common ailment 674

34.10 The pineal gland is involved in biorhythms 675

34.11 The thyroid regulates development and increases the metabolic rate 676

34.12 The thyroid and the parathyroids regulate the blood calcium level 676

35 Reproduction and Development 680

How to Do It on Land 680

Reproduction in Animals Is Varied 682

35.1 Both asexual and sexual reproduction occur among animals 682

35.2 Development in water and on land occurs among animals 683

Humans Are Adapted to Reproducing on Land 684

35.3 Testes are male gonads 684

35.4 Production of sperm and male sex hormones occurs in the testes 685

35.5 Ovaries are female gonads 686

35.6 Production of oocytes and female sex hormones occurs in the ovaries 687

35.7 The ovarian cycle drives the uterine cycle 688

35.8 *How Biology Impacts Our Lives* Sexual activity can transmit disease 689

35.9 *How Biology Impacts Our Lives* Numerous birth control methods are available 690

35.10 *How Biology Impacts Our Lives* Reproductive technologies are available to help the infertile 691

Vertebrates Have Similar Early Developmental Stages and Processes 692

35.11 Cellular stages of development precede tissue stages 692

35.12 Tissue stages of development precede organ stages 693

35.13 Organ stages of development occur after tissue stages 694

35.14 Cellular differentiation begins with cytoplasmic segregation 695

35.15 Morphogenesis involves induction also 696

Human Development Is Divided into Embryonic Development and Fetal Development 697

35.16 Extraembryonic membranes are critical to human development 697

35.17 Embryonic development involves tissue and organ formation 698

35.18 Fetal development involves refinement and weight gain 700

35.19 Pregnancy ends with the birth of the newborn 702

Biological Viewpoints Animals Are Homeostatic 706

PART VI Organisms Live in Ecosystems 708

36 Population Ecology 708

When a Population Grows Too Large 708

Ecology Studies Where and How Organisms Live in the Biosphere 710

36.1 Ecology is studied at various levels 710

Populations Are Not Static—They Change Over Time 711

36.2 Density and distribution are aspects of population structure 711

36.3 The growth rate results in population size changes 712

36.4 Survivorship curves illustrate age-related changes 712

36.5 Age structure diagrams divide a population into age groupings 713

36.6 Patterns of population growth can be described graphically 714

Environmental Interactions Influence Population Size 715

36.7 Density-independent factors affect population size 715

36.8 Density-dependent factors affect large populations more 716

The Life History Pattern Can Predict Extinction 717

36.9 Life history patterns consider several population characteristics 717

36.10 Certain species are more apt to become extinct than others 718

Human Populations Vary Between Overpopulation and Overconsumption 719

36.11 World population growth is exponential 719

36.12 Age distributions in MDCs and LDCs are different 720

37 Behavioral Ecology 724

For the Benefit of All 724

Both Innate and Learned Behavior Can Be Adaptive 726

37.1 Inheritance influences behavior 726

37.2 Learning can also influence behavior 728

37.3 Associative learning links behavior to stimuli 729

Reproductive Behavior Can Also Be Adaptive 730

37.4 Sexual selection can influence mating and other behaviors 730

Social Behavior Can Increase Fitness 732

37.5 Sociobiology studies the adaptive value of societies 732

Modes of Communication Vary with the Environment 734

37.6 Communication with others involves the senses 734

37.7 *How Science Progresses* Do animals have emotions? 736

38 Community and Ecosystem Ecology 740

Ridding the Land of Waste 740

A Community Contains Several Interacting Populations in the Same Locale 742

38.1 Competition can lead to resource partitioning 742

38.2 Predator-prey interactions affect both populations 744

38.3 Parasitism benefits one population at another's expense 746

38.4 Commensalism benefits only one population 746

38.5 *How Science Progresses* Coevolution requires interaction between two species 747

38.6 Mutualism benefits both populations 748

A Community Develops and Changes Over Time 749

38.7 *How Science Progresses* The study of island biogeography pertains to biodiversity 749

38.8 During ecological succession, community composition and diversity change 750

An Ecosystem Is a Community Interacting with the Physical Environment 752

38.9 Ecosystems have biotic and abiotic components 752

38.10 Energy flow and chemical cycling characterize ecosystems 753

38.11 Energy flow involves food webs 754

38.12 Ecological pyramids are based on trophic levels 755

38.13 Chemical cycling includes reservoirs, exchange pools, and the biotic community 756

38.14 The phosphorus cycle is sedimentary 756

38.15 The nitrogen cycle is gaseous 757

38.16 The carbon cycle is gaseous 758

39 Major Ecosystems of the Biosphere 762

Life Under Glass 762

On Land, the Biosphere Is Organized into Terrestrial Ecosystems 764

39.1 Major terrestrial ecosystems are characterized by particular climates 764

39.2 The tundra is cold and dark much of the year 765

39.3 Coniferous forests are dominated by gymnosperms 765

39.4 Temperate deciduous forests have abundant life 766

39.5 Temperate grasslands have extreme seasons 766

39.6 Savannas have wet-dry seasons 767

39.7 Deserts have very low annual rainfall 767

39.8 Tropical rain forests are warm with abundant rainfall 768

39.9 Solar radiation and winds influence climate 769

39.10 Topography and other effects also influence climate 770

Fresh Water and Salt Water Are Organized into Aquatic Ecosystems 771

39.11 Fresh water flows into salt water 771

39.12 Marine ecosystems include those of the coast and the ocean 772

39.13 Ocean currents affect climates 774

39.14 El Niño–Southern Oscillation alters weather patterns 774

40 Conservation of Biodiversity 778

Trouble in Paradise 778

Conservation Biology Focuses on Understanding and Protecting Biodiversity 780

40.1 Conservation biology is a practical science 780

40.2 Biodiversity is more than counting the total number of species 780

Biodiversity Has Direct Value and Indirect Value for Human Beings 782

40.3 The direct value of biodiversity is becoming better recognized 782

40.4 The indirect value of biodiversity is immense 784

The Causes of Today's Extinctions Are Known 785

40.5 Habitat loss is a major cause of wildlife extinctions 785

40.6 Introduction of alien species contributes to extinctions 786

40.7 Pollution contributes to extinctions 787

40.8 Overexploitation contributes to extinctions 787

40.9 Disease contributes to extinctions 788

Conservation Techniques Require Much Effort and Expertise 789

40.10 Habitat preservation is of primary importance 789

40.11 Habitat restoration is sometimes necessary 790

Biological Viewpoints Organisms Live in Ecosystems 794

Appendix A

Answers to Check Your Progress and Testing Yourself   A-1

Appendix B

Metric System   A-12

Appendix C

Periodic Table of the Elements   A-13

Glossary   G-1

Credits   C-1

Index   I-1

Concepts of
BIOLOGY

1

Biology, the Study of Life

LEARNING OUTCOMES

After studying this chapter, you should be able to accomplish the following outcomes.

Fire Ants Have a Good Defense

1 State why worker fire ants selflessly defend the colony and tirelessly work to keep the colony functioning.

Organisms Are Characterized by Diversity and Unity

2 Give examples that life is diverse.
3 List eleven levels of biological organization, stressing three levels in particular.
4 Discuss seven characteristics of life.
5 Show that evolution can account for the unity and diversity of life.

Classification Helps Us Understand Diversity

6 List the three taxonomic domains and the kingdoms in the domain Eukarya.

The Biosphere Is Organized

7 State, in order, the levels of organization that pertain to the biosphere.
8 Discuss the two ways that the populations of any community interact.
9 Discuss the manner in which humans threaten the organization of the biosphere.

Scientists Observe, Hypothesize, and Test

10 Divide the scientific method into four steps, and discuss each step.
11 Describe an experimental design that contains a control group.
12 Explain that DNA barcoding will be helpful for identifying species in ecosystems.

F ire ants have a red to reddish-brown color, but even so, they most likely take their name from the ability to STING. Their stinger protrudes from the rear, but in a split second, they can grab the skin of a person with their mandibles and position the stinger between their legs to sting from the front. The stinger injects a toxin into the tiny wound, and the result is a burning sensation. The next day, the person will end up with a white pustule at the site of the sting. The success of this defense mechanism is clear because most animals, including humans, try to stay away from bees, wasps, and ants—and any other animal that can sting.

Living usually in an open, grassy area, fire ants sting in order to defend their home, which is a mound of soil that they have removed from subterranean tunnels. They use the tunnels to travel far afield safely when searching for food, which they bring back to their nest mates. The queen and many worker ants live in chambers within the mound or slightly below it. The queen is much larger than the other members of the colony, and she has only one purpose: to produce many thousands of small, white eggs. The

Fire ant mound

Fire Ants Have a Good Defense

eggs develop into cream-colored, grublike larvae, which are lavishly tended by worker ants to keep them clean and well fed. When the larvae become encased by a hard covering, they are pupae. Inside a pupa, an amazing transformation takes place, and eventually, an adult ant breaks out. Most of these adults are worker ants, but in the spring, a few are winged "sexuals," which are male and female ants with the ability to reproduce. The sexuals remain inside the colony with nothing to do until the weather is cooperative enough for them to fly skyward to mate. A few of the fertilized females manage to survive the perils of an outside existence long enough to start another colony.

All of the ants in a colony have the same mother, namely the queen ant who produces the eggs. The workers are sterile, closely related sister ants. Because of their genetic relationship, we can view the members of a colony as a superorganism. The queen serves as the reproductive system, while the workers serve as the digestive, urinary, and, indeed, all the systems that keep the superorganism functioning. What fosters cooperation between the members of the superorganism? The answer is **pheromones**, chemicals secreted externally that influence the behavior and even the development of other species members. Fire ants, like other ants, produce several different pheromones that send messages when released into the air. The message could be "food is available" or "be alert for possible danger." The queen even releases pheromones that cause workers to attend her.

Why does it work, in a biological sense, for these sisters to spend their lives working away slavishly, mostly raising more sterile sisters and defending the colony with little regard for their own safety? It works because the few sexual females that survive their temporary existence on the outside pass the colony's joint genes on to future generations in new and different places. Any social system that allows an organism to pass on its genes is a successful one from an evolutionary point of view.

In this chapter, we will first study that organisms are characterized by diversity and unity. Their unity occurs because all organisms (including fire ants) share the same characteristics of life. Diversity occurs because organisms are adapted to different ways of life. (We have been looking at the fire ant's way of life.) Then, we will see that classification helps make sense of so much diversity, which is due to the different roles organisms play in ecosystems. Finally, we will examine how scientific understanding of life progresses by making observations and doing experiments.

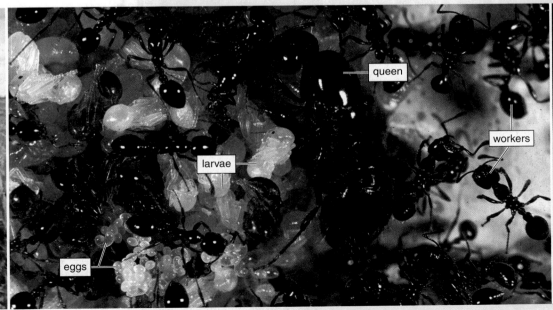

A fire ant colony (*Solenopsis invicta*).

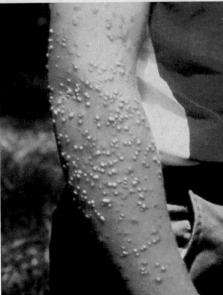

Pustules caused by fire ants.

In this part of the chapter, we will see that despite its diversity, life has unity because all living things display the same characteristics of life. Organisms have the same characteristics because each one can trace its ancestry to the first cell or cells. Diversity occurs because each type of organism is adapted to its particular way of life.

1.1 Life is diverse

The great variety of life on Earth often functions and behaves in ways strange to humans. For example, gastric-brooding frogs swallow their embryos and give birth to them later by throwing them up! Some species of puffballs, a type of fungus, are capable of producing trillions of spores when they reproduce. Fetal sand sharks kill and eat their siblings while still inside their mother. Some *Ophrys* orchids look so much like female bees that male bees try to mate with them. Octopuses and squids have small brains but remarkable problem-solving abilities. Some bacteria live out their entire lives in just 15 minutes, while bristlecone pine trees outlive ten generations of humans. Simply put, from the deepest oceanic trenches to the reaches of the atmosphere, life is abundant and varied.

Despite its diversity, biologists have managed to group living things, also called organisms, into the groups illus-

trated in **Figure 1.1**. Starting on the left, bacteria (and also archaea) are widely distributed, microscopic organisms with a very simple structure. A *Paramecium* is an example of a microscopic protist. Protists are larger in size and more complex than bacteria. The other organisms in Figure 1.1 are quite complex and easily seen with the naked eye. They can be distinguished by how they get their food. A morel is a fungus that digests its food externally. A sunflower is a photosynthetic plant that makes its own food, and a snow goose is an animal that ingests its food.

> **1.1** *Check Your Progress* Fire ants are most closely related to which organism in Figure 1.1?

FIGURE 1.1 Many diverse forms of life are found on planet Earth.

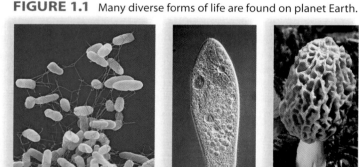

Bacteria   *Paramecium*   Morel   Sunflower   Snow goose

1.2 Life has many levels of organization

The unity of life is observable in that all forms of life are organized similarly. Notice that **Figure 1.2** lists eleven levels of biological organization, but even so, three levels of organization are particularly relevant: the cell, the multicellular organism, and the biosphere. A **cell** is the basic unit of structure and function of all living things. In a cell, **atoms**, which constitute basic building blocks of matter known as elements, combine with themselves or other atoms to form **molecules**. Some cells, such as unicellular paramecia, live independently. Other cells, such as those of the alga *Volvox,* cluster together in microscopic colonies.

An elephant is a multicellular organism in which similar cells combine to form a **tissue**; nerve tissue is a common tissue in animals. Tissues make up **organs**, as when various tissues combine to form the brain. Organs work together in **organ systems**; for example, the brain works with the spinal cord and a network

of nerves to form the nervous system. Organ systems are joined together to form a complete living thing, or **organism**.

The biosphere is the most complex level of biological organization beyond the individual organism. All the members of one species in a particular area belong to a population. For example, a nearby forest may have a population of gray squirrels and a population of white oaks. The populations of various animals and plants in the forest make up a community. The community of populations interacts with the physical environment and forms an ecosystem. Finally, all the Earth's ecosystems make up the biosphere.

The next section discusses all of the characteristics of life.

> **1.2** *Check Your Progress* *a.* What level of organization is a fire ant? *b.* What are the levels of organization beyond this one?

FIGURE 1.2 Levels of biological organization.

Biosphere
Regions of the Earth's crust, waters, and atmosphere inhabited by living things

Ecosystem
A community plus the physical environment

Community
Interacting populations in a particular area

Population
Organisms of the same species in a particular area

Organism
An individual; complex individuals contain organ systems

Organ System
Composed of several organs working together

Organ
Composed of tissues functioning together for a specific task

Tissue
A group of cells with a common structure and function

Cell
The structural and functional unit of all living things

Molecule
Union of two or more atoms of the same or different elements.

Atom
Smallest unit of an element composed of electrons, protons, and neutrons

Life cannot be given a simple definition. It is more feasible to discuss the characteristics that all organisms share. Here, we have elected to discuss seven characteristics displayed by all living things.

Order   As with all organisms, a fire ant's body is made up of highly ordered cells, tissues, organs, and organ systems that function together as a complete organism. The eye of a housefly (an insect, as is a fire ant) illustrates the orderliness of animal structures (**Fig. 1.3A**). Also, fire ant colonies form a population within a diverse community that includes the other plants and animals living in the area. The community is, in turn, part of an ecosystem of the biosphere.

Response to Stimuli   Living things interact with the environment as well as with other living things. Even unicellular organisms can respond to their environment. In some, the beating of microscopic hairs, and in others, the snapping of whip-like tails move them toward or away from light or chemicals. Multicellular organisms can manage more complex responses. A vulture can detect a carcass a mile away and soar toward dinner. A monarch butterfly can sense the approach of fall and begin its flight south, where resources are still abundant.

The ability to respond is reflected by the turning of a plant's leaves and stem toward the sun (**Fig. 1.3B**) and by an animal that darts safely away from danger. Appropriate responses help ensure the survival of the organism and allow it to carry on its daily activities. All together, these activities are termed the behavior of the organism. Organisms display a variety of behaviors as they search and compete for energy, nutrients, shelter, and mates. Many organisms display complex communication, hunting, and defensive behaviors as well.

Regulation of Internal Environment   To survive, it is imperative that an organism maintain a state of biological balance, or **homeostasis**. For example, temperature, moisture level,

FIGURE 1.3B   Plants respond to light.

acidity, and other physiological factors must remain within the tolerance range of the organism. Homeostasis is maintained by systems that monitor internal conditions and make routine and necessary adjustments.

Organisms have intricate feedback and control mechanisms that do not require any conscious activity. When a student is so engrossed in her textbook that she forgets to eat lunch, her liver releases stored sugar to keep her blood sugar level within normal limits. In this case, hormones regulate sugar storage and release, but in other instances, the nervous system is involved in maintaining homeostasis. Many organisms depend on behavior to regulate their internal environment. The same student may realize that she is hungry and decide to visit the local diner. Iguanas may raise their internal temperature by basking in the sun (**Fig. 1.3C**) or cool down by moving into the shade. Simi-

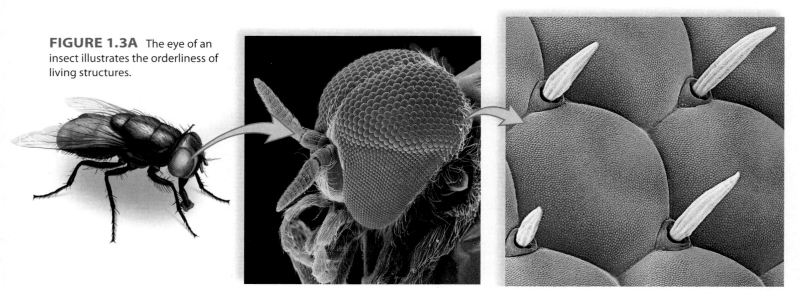

FIGURE 1.3A   The eye of an insect illustrates the orderliness of living structures.

FIGURE 1.3C Iguanas basking in the sun.

larly, fire ants move upward into the mound when the warmth of the sun is needed and move into their cooler subterranean passageways when the sun is too hot.

Acquisition of Materials and Energy
Living things cannot maintain their organization or carry on life's activities without an outside source of nutrients and energy (**Fig. 1.3D**). Food provides nutrients, which are used as building blocks or for energy. **Energy** is the capacity to do work, and it takes work to maintain the organization of the cell and the organism. When cells use nutrient molecules to make their parts and products, they carry out a sequence of chemical reactions. The term **metabolism** encompasses all the chemical reactions that occur in a cell.

The ultimate source of energy for nearly all life on Earth is the sun. Plants and certain other organisms are able to capture solar energy and carry on **photosynthesis**, a process that transforms solar energy into the chemical energy of organic nutrient molecules. All life on Earth acquires energy by metabolizing nutrient molecules made by photosynthesizers. This applies even to plants.

Reproduction and Development
Life comes only from life. Every type of living thing can **reproduce**, or make another organism like itself (Fig. 1.3D). Bacteria, protists, and other unicellular organisms simply split in two. In most multicellular organisms, the reproductive process begins with the pairing of a sperm from one partner and an egg from the other partner. The

FIGURE 1.3D Living things acquire materials and energy and they reproduce.

union of sperm and egg, followed by many cell divisions, results in an immature stage that grows and develops through various stages to become an adult.

Genetic Inheritance An embryo develops into a humpback whale or a purple iris because of a blueprint inherited from its parents. The instructions, or blueprint, for an organism's metabolism and organization are encoded in genes. The **genes**, which contain specific information for how the organism is to be ordered, are made of long molecules of DNA (deoxyribonucleic acid). DNA has a shape resembling a spiral staircase with millions of steps. Housed in this spiral staircase is the genetic code that is shared by all living things.

Evolutionary Adaptations **Adaptations** are modifications that make organisms suited to their way of life. Consider, for example, a hawk (Fig. 1.3D), which catches and eats rabbits. A hawk can fly, in part, because it has hollow bones to reduce its weight and flight muscles to depress and elevate its wings. When a hawk dives, its strong feet take the first shock of the landing, and its long, sharp claws reach out and hold onto the prey.

Adaptations come about through evolution. **Evolution** is the process by which a **species** (a group of similarly constructed organisms that successfully interbreed) changes through time. Just as you and your siblings can trace your ancestry from your parents to your grandparents and beyond, so species can trace their ancestry through common ancestors even to the very first cell. One of the basic tenets of the theory of evolution is the recognition that all present and past forms of life are descended from former forms of life. Further, descent from common ancestors explains why all organisms have the same basic characteristics.

As descent occurs from generation to generation, modifications arise. In other words, life-forms change over time. While the concept of evolution was not a new one in the 1800s, Charles Darwin laid the foundation for the modern theory of evolution by describing a mechanism, called **natural selection**, by which evolution occurs. Natural selection comes about in this manner: When a new variation arises that allows certain members of a species to capture more resources, these members tend to survive and to have more offspring than the other, unchanged members. Therefore, each successive generation includes more members with the new variation. In the end, most members of a species have the same adaptation to their environment. The society of insects, such as fire ants, arose because it was a successful way to live in their particular environment.

The theory of evolution can explain the diversity of life. Several species that have the same common ancestor become adapted to different environments. In this way, the diversity of life evolved over eons. Evolutionists try to discover the evolutionary history of organisms so that they know who is related to whom. This knowledge is used to classify organisms, the topic of the next part of the chapter.

> **1.3 Check Your Progress** In what ways do fire ants display the seven characteristics of life?

Classification, the topic of this part of the chapter, helps us understand diversity in at least two different ways. First, classification tells us what types of organisms have been found on Earth, both today and in the past. It also suggests how these organisms may be related because organisms are classified according to evolutionary relationships.

1.4 Taxonomists group organisms according to evolutionary relationships

Because life is so diverse, it is helpful to have a way to group organisms into categories. **Taxonomy** is the discipline of grouping organisms according to their evolutionary history and relationships. In this way, taxonomy helps make sense out of the bewildering variety of life on Earth and provides valuable insight into evolution. As more is learned about living things, including the evolutionary relationships between species, taxonomy changes.

The traditional classification categories, or taxa, going from least inclusive to most inclusive, are **species**, **genus**, **family**, **order**, **class**, **phylum**, **kingdom**, and **domain** (**Table 1.4**). Each successive classification category above species contains more types of organisms than the preceding one. Species placed within one genus share many specific characteristics and are the most closely related, while species placed in the same kingdom share only general characteristics with one another. For example, all species in the genus *Pisum* look pretty much the same—that is, like pea plants—but species in the plant kingdom can be quite varied, as is evident when we compare grasses to trees. By the same token, only humans are in the genus *Homo,* but many types of species, from tiny hydras to huge whales, are members of the animal kingdom. Species placed in different domains are the most distantly related.

Domains The domain is the largest classification category. Biochemical evidence tells us that there are three domains: **Bacteria**, **Archaea**, and **Eukarya**. Both bacteria and archaea are microscopic unicellular **prokaryotes**, which lack the membrane-bounded nucleus found in the **eukaryotes** of domain Eukarya.

Prokaryotes are structurally simple but metabolically complex (**Figs. 1.4A** and **1.4B**). Archaea are well known for living in aquatic environments that lack oxygen or are too salty, too hot, or

TABLE 1.4	Levels of Classification	
Category	**Human**	**Corn**
Domain	Eukarya	Eukarya
Kingdom	Animalia	Plantae
Phylum	Chordata	Anthophyta
Class	Mammalia	Monocotyledones
Order	Primates	Commelinales
Family	Hominidae	Poaceae
Genus	*Homo*	*Zea*
Species*	*H. sapiens*	*Z. mays*

*To specify an organism, you must use the full binomial name, such as *Homo sapiens*.

too acidic for most other organisms. Perhaps these environments are similar to those of the primitive Earth, and archaea may be representative of the first cells that evolved. Bacteria are variously adapted to living almost anywhere—in the water, soil, and atmosphere, as well as on our skin and in our mouths and large intestines. Although some bacteria cause diseases, others perform many valuable services, both environmentally and commercially. They are used to conduct genetic research in our laboratories, to produce innumerable products in our factories, and to purify water in our sewage treatment plants, for example.

Kingdoms Taxonomists are in the process of deciding how to categorize archaea and bacteria into kingdoms. Domain Eukarya, on the other hand, contains four kingdoms (**Fig. 1.4C**). **Protists** (kingdom Protista) range from unicellular forms to a few multicellular ones. Some are photosynthesizers, and some

Methanosarcina mazei, an archaeon   1.6 μm

FIGURE 1.4A Domain Archaea.

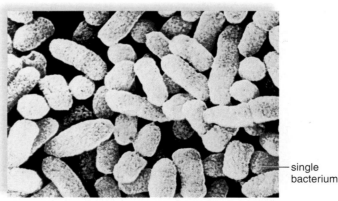

Escherichia coli, a bacterium   1.5 μm

FIGURE 1.4B Domain Bacteria.

must acquire their food. Common protists include the algae, protozoans, and water molds. Among the **fungi** (kingdom Fungi) are the familiar molds and mushrooms that, along with bacteria, help decompose dead organisms. **Plants** (kingdom Plantae) are multicellular photosynthetic organisms. Example plants include azaleas, zinnias, and pines. **Animals** (kingdom Animalia) are multicellular organisms that must ingest and process their food. Aardvarks, jellyfish, and zebras are representative animals.

Scientific Names Biologists use **binomial nomenclature** to assign each living thing a two-part name, called a scientific name. For example, the scientific name for mistletoe is *Phoradendron tomentosum.* The first word is the genus name, and the second word is the **specific epithet** of a species within a genus. The genus may be abbreviated (e.g., *P. tomentosum*), and the genus name alone may be given if the species is unknown (e.g., *Phoradendron* sp.).

Biologists universally use scientific names to avoid confusion. Common names tend to overlap and often are in the language of a particular country. Scientific names, which are in Latin, do not vary from one country to another. Just as organisms have a particular name, they play a particular role in the biosphere, as is discussed in the next part of the chapter.

> **1.4** *Check Your Progress* **Fire ants belong to what domain and what kingdom?**

DOMAIN EUKARYA

KINGDOM PROTISTA (protists)

Paramecium, a unicellular organism
- Algae, protozoans, slime molds, and water molds
- Complex single cell (sometimes filaments, colonies, or even multicellular)
- Absorb, photosynthesize, or ingest food

KINGDOM PLANTAE (plants)

Passiflora, passion flower, a flowering plant
- Mosses, ferns, conifers, and flowering plants (both woody and nonwoody)
- Multicellular with specialized tissues containing complex cells
- Photosynthesize food

KINGDOM FUNGI

Coprinus, a shaggy mane mushroom
- Molds, mushrooms, and yeasts
- Mostly multicellular fillaments with specialized, complex cells
- Absorb food

KINGDOM ANIMALIA (animals)

Vulpes, a red fox
- Sponges, worms, insects, fishes, frogs, turtles, birds, and mammals
- Multicellular with specialized tissues containing complex cells
- Ingest food

FIGURE 1.4C The four kingdoms in domain Eukarya.

The biosphere is the largest level of biological organization. In this part of the chapter, we will explore how the biosphere is organized and will note that chemicals cycle but energy flows through the biosphere. Therefore, the continued existence of organisms is dependent on the energy of the sun. It is of extreme concern that the biosphere is threatened by the activities of human beings.

1.5 The biosphere is divided into ecosystems

The organization of life extends to the biosphere, the zone of air, land, and water at the surface of the Earth where living organisms are found (see Fig. 1.2). Individual organisms belong to a **population**, which is all the members of a species within a particular area. The populations of a **community** interact among themselves and with the physical environment (e.g., soil, atmosphere, and chemicals), thereby forming an **ecosystem**. All the ecosystems on Earth comprise the **biosphere**.

Figure 1.5 depicts a grassland ecosystem inhabited by populations of rabbits, mice, snakes, hawks, and various types of plants. The interactions among the populations of the community lead to chemical cycling (gray arrows) and energy flow (yellow to red arrows), both of which begin when ❶ photosynthetic plants, algae, and some bacteria take in inorganic nutrients and solar energy to produce food in the form of organic nutrients. Chemical cycling and energy flow continue as ❷ rabbits and ❸ mice feed on plants and seeds; ❹ snakes feed on mice; and ❺ hawks feed on rabbits and snakes.

Chemicals cycle because, with ❻ death and decomposition, inorganic chemicals are made available to plants once more. Not so in the case of energy. With each transfer, some energy is lost as heat. Eventually, all the energy taken in by photosynthesizers has dissipated into the atmosphere.

Because energy flows and does not cycle, ecosystems could not stay in existence without a constant input of solar energy and the ability of photosynthesizers to absorb it. This explains why nearly all living things depend on plants for their existence.

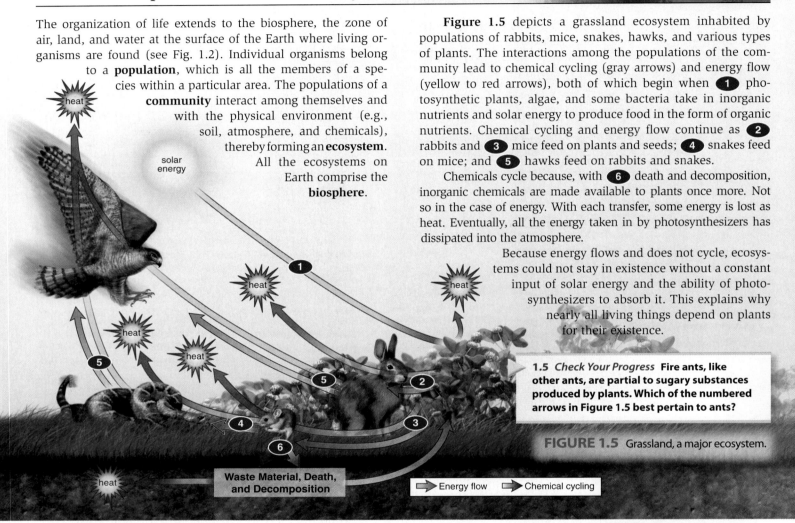

1.5 Check Your Progress Fire ants, like other ants, are partial to sugary substances produced by plants. Which of the numbered arrows in Figure 1.5 best pertain to ants?

FIGURE 1.5 Grassland, a major ecosystem.

1.6 Most of the biosphere's ecosystems are now threatened

Humans tend to modify existing ecosystems for their own purposes. We clear forests or grasslands to grow crops; later, we build houses on what was once farmland, and finally we convert small towns into cities. As coasts are developed, humans send sediment, sewage, and other pollutants into the sea. Human activities destroy valuable coastal wetlands, which serve as protection against storms and as nurseries for a myriad of invertebrates and vertebrates.

The two most biologically diverse ecosystems—tropical rain forests and coral reefs—are home to many organisms. The canopy of the tropical rain forest alone supports a variety of organisms, including orchids, insects, and monkeys. Coral reefs, which are found just offshore of the continents and islands of the Southern Hemisphere, are built up from the calcium carbonate skeletons of sea animals called corals. Reefs provide a habitat for many animals,

including jellyfish, sponges, snails, crabs, lobsters, sea turtles, moray eels, and some of the world's most colorful fishes. Like tropical rain forests, coral reefs are severely threatened as the human population increases in size. Aside from pollutants, overfishing and collection of coral for sale to tourists destroy the reefs.

In order to know how to preserve ecosystems and, indeed, to understand the natural world in general, scientists make observations and carry on experiments, as explained in the next part of the chapter.

1.6 Check Your Progress Fire ants from the tropics invade the temperate zone, and they prefer disturbed areas. Why could humans be called a fire ant's best friend?

This part of the chapter explains how scientists observe, hypothesize, and test to gather information and come to conclusions about the natural world. The natural world consists of all the matter and energy on planet Earth. We will outline the scientific method, give an example of a controlled study, and show how scientists use genetics to aid classification.

1.7 The natural world is studied by using scientific methods

Biology is the scientific study of life. Science differs from other ways of knowing and learning by its process, which can be adjusted to where and how a study is being conducted. Still, the **scientific process** often involves the use of the scientific method, which includes these four steps: observation, hypothesis, testing, and conclusion (**Fig. 1.7**).

Observation and Hypothesis Scientists use all of their senses to make **observations**. Scientists also extend the ability of their senses by using instruments; for example, the microscope enables us to see objects that could never be seen by the naked eye. Scientists also look up past studies at the library or on the Internet, or they may write or speak to others who are researching similar topics.

After making observations and gathering knowledge about a phenomenon, scientists develop a **hypothesis**, a possible explanation for a natural event. Scientists consider only explanations that can be tested.

Testing and Conclusion **Testing** a hypothesis involves either conducting an experiment or making further observations. To determine how to test a hypothesis, a scientist uses deductive reasoning. Deductive reasoning involves "if, then" logic. For example, a scientist might reason that if organisms are composed of cells, then microscopic examination of any part of an organism should reveal cells. We can also say that the scientist has made a prediction that

the hypothesis can be supported by doing microscopic studies. Making a prediction helps a scientist know what to do next.

The results of an experiment and/or further observations are referred to as the **data**. Mathematical data may be displayed in the form of a graph or a table. The data allow scientists to come to a conclusion. The **conclusion** indicates whether the results support or do not support the hypothesis. If appropriate, a conclusion should be accompanied by a probability of error, and in any case, the data never prove a hypothesis "true" because a conclusion is always subject to revision. On the other hand, it is possible to prove a hypothesis false. Science progresses, and even a conclusion that proves a hypothesis false can lead to a new hypothesis for another experiment, as represented by the double-headed arrow in Figure 1.7.

Scientists report their findings in scientific journals because experiments and observations must be repeatable—that is, the reporting scientist and any scientist who repeats the experiment must get the same results, or else the data are suspect.

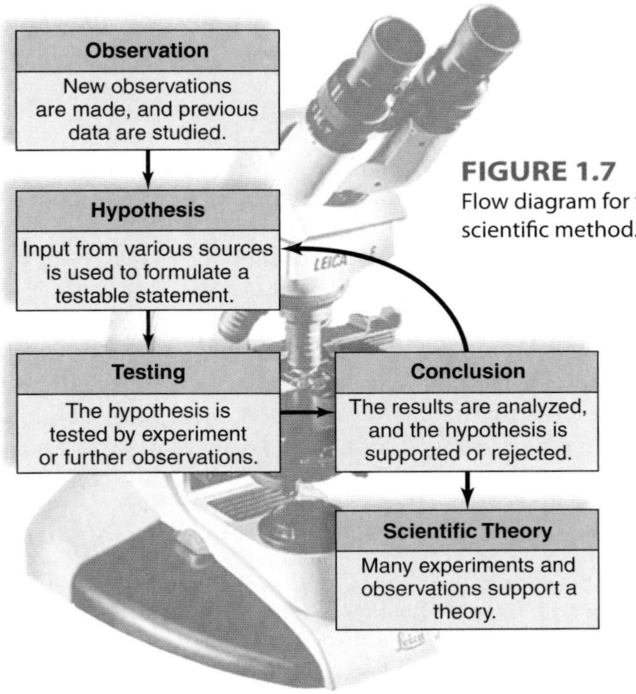

Scientific Theories The ultimate goal of science is to understand the natural world in terms of scientific theories. In ordinary speech, the word *theory* refers to a speculative idea. In contrast, a **scientific theory** is supported by a broad range of observations, experiments, and data, often from a variety of disciplines. The basic theories of biology to be studied in this text are:

Theory	Concept
Cell	All organisms are composed of cells, and new cells only come from preexisting cells.
Gene	Organisms contain coded information that dictates their form, function, and behavior.
Evolution	All living things have a common ancestor, but each is adapted to a particular way of life.
Homeostasis	The internal environment of an organism stays relatively constant—within a range protective of life.
Ecosystem	Organisms are members of populations that interact with each other and with the physical environment within a particular locale.

The experiment described in Section 1.8 would fall under "ecosystem" because it concerns ways to improve the yield of plants. This section also shows why experiments should contain a control group.

Observation
New observations are made, and previous data are studied.

Hypothesis
Input from various sources is used to formulate a testable statement.

Testing
The hypothesis is tested by experiment or further observations.

Conclusion
The results are analyzed, and the hypothesis is supported or rejected.

Scientific Theory
Many experiments and observations support a theory.

FIGURE 1.7
Flow diagram for the scientific method.

1.7 Check Your Progress You hypothesize that only the queen fire ant produces eggs. What type of data would support this hypothesis? Prove it false?

A group of investigators were concerned about the excessive use of nitrogen fertilizer to grow crops. About half the nitrogen fertilizer ever used has been applied since 1985. Rain washes fertilizer into local bodies of water before plants have a chance to absorb it. Much of the fertilizer moves slowly underground and can contaminate freshwater supplies and even the oceans for years to come. Too much nitrogen causes a massive loss of sea grasses, which are critical incubators of sea life. Algal blooms fed by nitrogen lead to massive fish kills in freshwater lakes. Also, nitrogen makes well water toxic to drink, especially for infants. Various illnesses in adults may result from years of drinking water laced with nitrogen.

Farmers need alternative ways to provide crops with nitrogen. The investigators doing this study knew that pea plants are legumes and that legumes deposit organic nitrogen in the soil. Organic nitrogen stays put and doesn't end up in water supplies as inorganic nitrogen does.

An Experiment When scientists perform an experiment, the environmental conditions must be kept constant, except for the **experimental variable**, which is deliberately changed. Therefore, the conditions for all pots of plants used in this experiment were kept exactly the same, and the wheat plants were systematically dried and weighed to determine the yield (biomass) in each of the pots. A controlled experiment involves test groups, which are exposed to the experimental variable, and a **control group**, which is not exposed to the experimental variable. The test groups should be as large as possible to eliminate the influence of undetected differences in the test subjects. The use of a control group ensures that the data from the test groups are due to the experimental variable, not to some unknown outside influence.

The investigators doing this study formulated this hypothesis:

Hypothesis A pigeon pea/winter wheat rotation will cause winter wheat production to increase as well as or better than the use of nitrogen fertilizer.

The investigators decided on the following experimental design (**Fig. 1.8A**):

Control Pots
- Winter wheat was planted in clay pots of soil that received no fertilization treatment—that is, no nitrogen fertilizer and no preplanting of pigeon peas.

Three Types of Test Pots
- Winter wheat was grown in clay pots in soil treated with nitrogen fertilizer equivalent to 45 kilograms per hectare (kg/ha).
- Winter wheat was grown in clay pots in soil treated with nitrogen fertilizer equivalent to 90 kg/ha.
- Pigeon pea plants were grown in clay pots in the summer. The pigeon pea plants were then tilled into the soil, and winter wheat was planted in the same pots.

Analysis of Results Figure 1.8A includes a color-coded bar graph that allows you to see at a glance the comparative amount of wheat obtained from each group of pots. After the first year, winter wheat yield was higher in test pots treated with nitrogen fertilizer than in the control pots. To the surprise of investigators, wheat production following summer planting of pigeon peas did not yield as high a yield as the control pots.

> **Conclusion** The hypothesis is not supported. Wheat yield following the growth of pigeon peas is not as great as that obtained with nitrogen fertilizer treatments.

The Follow-Up Experiment and Results The researchers decided to continue the experiment using the same design and the same pots as before, to see whether the buildup of residual soil nitrogen from pigeon peas would eventually increase wheat yield to a greater extent than the use of nitrogen fertilizer. This was their new hypothesis:

> **Hypothesis** A sustained pigeon pea/winter wheat rotation will eventually cause an increase in winter wheat production.

They predicted that wheat yield following three years of pigeon pea/winter wheat rotation will surpass wheat yield following nitrogen fertilizer treatment.

Analysis of Results After two years, the yield from pots treated with nitrogen fertilizer was not as much as it was the first year. Indeed, wheat yield in pots following a summer planting of pigeon peas was the highest of all treatments. After three years, wheat yield in pots treated with nitrogen fertilizer was more than in the control pots but not nearly as great as the yield in pots following summer planting of pigeon peas. Compared to the first year, wheat yield increased almost fourfold in pots having a pigeon pea/winter wheat rotation.

> **Conclusion** The hypothesis is supported. At the end of three years, the yield of winter wheat following a pigeon pea/winter wheat rotation was much better than for the other types of test pots.

To explain their results, the researchers suggested that the soil was improved by the buildup of the organic matter in the pots as well as by the addition of nitrogen from the pigeon peas. They published their results in a scientific journal[1], where their experimental method and results would be available to the scientific community.

Ecological Importance of This Study These experiments showed that the use of a legume, namely pigeon peas, to improve the soil produced a far better yield than the use of nitrogen fertilizer over the long haul. Legumes provide a home for bacteria that convert atmospheric nitrogen to a form usable

---

[1]Bidlack, J. E., Rao, S. C., and Demezas, D. H. 2001. Nodulation, nitrogenase activity, and dry weight of chickpea and pigeon pea cultivars using different *Bradyrhizobium* strains. *Journal of Plant Nutrition* 24:549–60.

Control pots and test pots of three types

Control pots
no fertilization treatment

Test pots
90 kg of nitrogen/ha

Test pots
Pigeon pea/winter wheat rotation

Test pots
45 kg of nitrogen/ha

Control Pots
■ = no fertilization treatment
Test Pots
■ = 45 kg of nitrogen/ha
■ = 90 kg of nitrogen/ha
■ = Pigeon pea/winter wheat rotation

Wheat Yield (grams/pot)

year 1   year 2   year 3

Results

FIGURE 1.8A Pigeon pea/winter wheat rotation study.

by the plant. The bacteria live in nodules on the roots (**Fig. 1.8B**). The products of photosynthesis move from the leaves to the root nodules; in turn, the nodules supply the plant with nitrogen compounds it can use to make proteins. The startling

increase in yield after three years of pigeon pea/winter wheat rotation can be explained in this way. When the pigeon pea plants were turned over into the soil, the decaying pigeon pea plants enriched the soil, and decomposition (especially of the roots) eventually made extra nutrients, including a goodly supply of nitrogen, available to winter wheat plants.

Rotation of crops, as was done in this study, is an important part of organic farming. In the long run, there are no adverse economic results when farmers switch from chemical-intensive to organic farming practices. But, because the cost benefit of making the transition may not be realized for several years, some researchers advocate that the process be gradual. For example, one-third of the farm could be changed to organic at a time. This study suggests, however, that organic farming will eventually be beneficial for the farmer. It will also benefit the environment!

In Section 1.9, we will examine a way that modern technology can possibly benefit society.

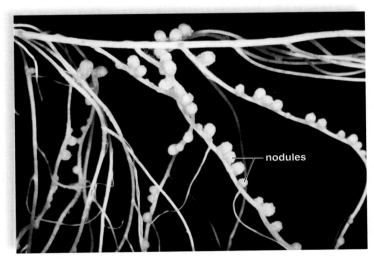

nodules

FIGURE 1.8B Root nodules.

1.8 Check Your Progress What would your test groups and your control groups be if you were testing whether a parasite could reduce the size of a fire ant colony?

1.9 DNA barcoding of life may become a reality

Imagine a mother who wants to know right away if the ants in her backyard are fire ants (**Fig. 1.9**). Right now, most biologists wouldn't be able to give her an immediate answer because only a few are trained to microscopically tell one ant from another. But this situation may soon change. The Consortium for the Barcode of Life (CBOL) is an international initiative devoted to developing DNA barcoding as the global standard in taxonomy. The CBOL proposes that any scientist will be able to identify a species with the use of a handheld barcoder. The CBOL believes that the DNA of an organism could be the basis for identifying its species, just as an 11-digit product code is used to identify the products sold in a supermarket. The idea is that a barcoder would tap into a database containing the barcodes for all species so far identified on planet Earth. A remnant of an ant's body, for example, could be put into a barcoder, and instantly you would know what the species is. Speedy identification using genetic barcodes would not only be a boon to ordinary citizens and taxonomists, but it would also benefit farmers who need to identify a pest attacking their crops, doctors who need to know the correct antivenom for snakebite victims, and college students who are expected to identify the plants, animals, and protists on an ecological field trip.

The idea of using barcodes to identify species is not new, but Paul Hebert and his colleagues at the University of Guelph in Canada are the first to believe it would be possible to use DNA differences to develop a barcode for each living thing. He has developed a method to barcode animals, and another researcher, John Kress, a plant taxonomist at the Smithsonian Institution in Washington, D.C., has developed a potential method for barcoding plant species. The CBOL is growing by leaps and bounds and now includes various biotech companies, museums, and universities, as well as the U.S. Food and Drug Administration and the U.S. Department of Homeland Security. Hebert has received a $3 million grant from the Gordon and Betty Moore Foundation to start the Biodiversity Institute of Ontario, which will be housed on the University of Guelph campus where he teaches.

Once the database has reached a certain size, The CBOL wants to enlist the help of a wide range of researchers, including biology students, to help with expanding the database. So far, scientists have identified only about 1.5 million species out of a potential 30 million. Clearly it would be a good idea to involve many more researchers in expanding the database so that easy identification of any species would become a reality.

FIGURE 1.9 This ant is the jumper ant, *Myrmecia pilosula*, and is not a fire ant.

> **1.9** *Check Your Progress* **Will two species of fire ants have different DNA barcodes?**

CONNECTING THE CONCEPTS

This chapter previews the rest of the text because it touches on all the major concepts of biology that we will be discussing in more depth in the chapters that follow. This chapter also explains the scientific process, the methodology that scientists use to gather data. What we know about biology and what we'll learn in the future result from objective observation and testing of the natural world. The ultimate goal of science is to understand the natural world in terms of theories—conceptual schemes supported by abundant research. This text is organized around the major theories of biology listed on page 11. The theory of evolution is called the unifying concept of biology because it pertains to all the different aspects of living things. For example, the theory of evolution enables scientists to understand the history of life, the variety of living things, and the anatomy, physiology, and development of organisms.

Science does not make ethical or moral decisions. The general public wants scientists to label certain research as "good" or "bad" and to predict whether any resulting technology, such as barcoding life, will primarily benefit or harm our society. Scientists should provide the public with as much information as possible when an issue such as global warming is being debated. Then they, along with other citizens, can help make intelligent decisions about what is most likely best for society. All men and women have a responsibility to decide how to use scientific knowledge so that it benefits living things, including the human species.

This textbook was written to help you understand the scientific process and learn the basic concepts of general biology so that you will be better informed. This chapter has introduced you to the levels of biological organization, from the cell to the biosphere. Chapter 2 begins the first part of the book whose chapters pertain to the cell theory. The cell theory states that all forms of life are cellular in nature. Just as there are levels of biological organization (atoms and molecules) that precede the cell, so the next chapter discusses the basic chemistry of cells. In other words, we cannot understand the cell without a knowledge of the atoms and molecules that make up the cell.

The Chapter in Review

Summary

Fire Ants Have a Good Defense

- For survival, fire ants acquire food, produce energy, and defend the colony and the queen in order to reproduce and pass on their genes.

Organisms Are Characterized by Diversity and Unity

1.1 Life is diverse

- There is a great variety of life on Earth.
- Some organisms are microscopic and simple in appearance.
- More complex organisms are distinguishable by how they acquire food. Fungi absorb their food, plants photosynthesize, and animals ingest their food.

1.2 Life has many levels of organization

- From the atomic level to organisms, each level is more complex than the preceding one.

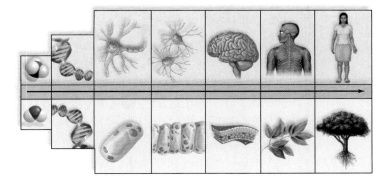

- Organisms also interact among themselves and with their environment in ecosystems and the biosphere.

1.3 Organisms share the same characteristics of life

- Living things display these seven characteristcs:
  - Order
  - Response to stimuli
  - Regulation of internal environment
  - Acquisition of materials and energy
  - Reproduction and development
  - Genetic inheritance
  - Evolutionary adaptations
- Evolution is the process by which species change over time.
- Charles Darwin told us that evolution has two aspects: descent from a common ancestor and adaptation to the environment by natural selection.

Classification Helps Us Understand Diversity

1.4 Taxonomists group organisms according to evolutionary relationships

- The classification categories are species (least inclusive), genus, family, order, class, phylum, kingdom, and domain (most inclusive).
- There are three domains: Bacteria, Archaea, and Eukarya.

- Domain Archaea and domain Bacteria contain prokaryotes (organisms without a membrane-bounded nucleus).

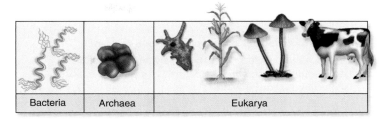

| Bacteria | Archaea | Eukarya |

- Domain Eukarya contains eukaryotes (organisms with a membrane-bounded nucleus).
- Kingdoms in domain Eukarya:
  - Protista—unicellular to multicellular organisms with various modes of nutrition
  - Fungi—molds and mushrooms
  - Plantae—multicellular photosynthesizers
  - Animalia—multicellular organisms that ingest food
- To classify an organism, two-part scientific names are used, consisting of the genus name and the specific epithet.

The Biosphere Is Organized

1.5 The biosphere is divided into ecosystems

- Individual organisms belong to a population.
- The populations of a community interact among themselves and with their physical environment to form an ecosystem.
- Ecosystems are either aquatic or terrestrial.
- Within an ecosystem, chemicals cycle, while energy flows but does not cycle.

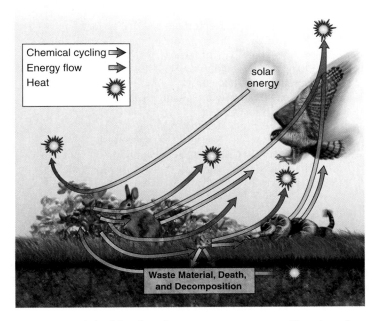

Chemical cycling
Energy flow
Heat
solar energy
Waste Material, Death, and Decomposition

1.6 Most of the biosphere's ecosystems are now threatened

- Diverse ecosystems, including tropical rain forests and coral reefs, are being destroyed by human activities.

Scientists Observe, Hypothesize, and Test

1.7 The natural world is studied by using scientific methods

- Biology is the scientific study of life.
- The scientific process involves use of the scientific method.
- The scientific method consists of four steps: observation, hypothesis, testing, and conclusion.
- A scientific theory is supported by many observations, experiments, and data.

1.8 Control groups allow for comparison of results

- In a scientific experiment, the experimental variable is deliberately chosen but the control group is not exposed to the experimental variable.

1.9 DNA barcoding of life may become a reality

- Researchers suggest that DNA differences between species can be used to develop a barcode for each living organism, and a handheld scanner could then identify an organism.

Testing Yourself

Organisms Are Characterized by Diversity and Unity

1. The level of organization that includes cells of similar structure and function would be
   a. an organ.   c. an organ system.
   b. a tissue.   d. an organism.
2. Which sequence represents the correct order of increasing complexity in living systems?
   a. cell, molecule, organ, tissue
   b. organ, tissue, cell, molecule
   c. molecule, cell, tissue, organ
   d. cell, organ, tissue, molecule
3. All of the chemical reactions that occur in a cell are called
   a. homeostasis.   c. heterostasis.
   b. metabolism.   d. cytoplasm.
4. The process of turning solar energy into chemical energy is called
   a. work.   c. photosynthesis.
   b. metabolism.   d. respiration.
5. What is the unifying theory in biology that explains the relationships of all living things?
   a. ecology   c. biodiversity
   b. evolution   d. taxonomy
6. Modifications that make an organism suited to its way of life are called
   a. ecosystems.   c. adaptations.
   b. populations.   d. None of these are correct.
7. **THINKING CONCEPTUALLY** Apply the concept that "the whole is more than the sum of its parts" to the cell, which is alive, and the parts of a cell, which are not alive.

Classification Helps Us Understand Diversity

8. Classification of organisms reflects
   a. similarities.
   b. evolutionary history.
   c. Neither a nor b is correct.
   d. Both a and b are correct.
9. Which of these exhibits an increasingly more inclusive scheme of classification?
   a. kingdom, phylum, class, order
   b. phylum, class, order, family
   c. class, order, family, genus
   d. genus, family, order, class

10. Humans belong to the domain
    a. Archaea.   c. Eukarya.
    b. Bacteria.   d. None of these are correct.
11. In which kingdom are you most likely to find unicellular organisms?
    a. kingdom Protista   c. kingdom Plantae
    b. kingdom Fungi   d. kingdom Animalia
12. The second word of a scientific name, such as *Homo sapiens*, is the
    a. genus.   d. species.
    b. phylum.   e. family.
    c. specific epithet.
13. Explain why the use of a common name, but not a scientific name, leads to confusion.

The Biosphere Is Organized

14. Which sequence represents the correct order of increasing complexity?
    a. biosphere, community, ecosystem, population
    b. population, ecosystem, biosphere, community
    c. community, biosphere, population, ecosystem
    d. population, community, ecosystem, biosphere
15. A population is defined as
    a. the number of species in a given geographic area.
    b. all of the individuals of a particular species in a given area.
    c. a group of communities.
    d. all of the organisms in an ecosystem.
16. In an ecosystem,
    a. energy flows and nutrients cycle.
    b. energy cycles and nutrients flow.
    c. energy and nutrients flow.
    d. energy and nutrients cycle.
17. An example of chemical cycling occurs when
    a. plants absorb solar energy and make their own food.
    b. energy flows through an ecosystem and becomes heat.
    c. hawks soar and nest in trees.
    d. death and decay make inorganic nutrients available to plants.
    e. we eat food and use the nutrients to grow or repair tissues.
18. Energy is brought into ecosystems by which of the following?
    a. fungi and other decomposers
    b. cows and other organisms that graze on grass
    c. meat-eating animals
    d. organisms that photosynthesize, such as plants
    e. All of these are correct.
19. **THINKING CONCEPTUALLY** How is a college campus, which is composed of buildings, students, faculty, and administrators, like an ecosystem?

Scientists Observe, Hypothesize, and Test

20. Which of these cannot be part of a conclusion regarding a hypothesis in scientific inquiry?
    a. proof   c. rejection
    b. support   d. All can be part of a conclusion.
21. Which term and definition are mismatched?
    a. data—factual information
    b. hypothesis—the idea to be tested
    c. conclusion—what the data tell us
    d. All of these are properly matched.
22. Which of these describes the control group in the pigeon pea/winter wheat experiment? The control group was
    a. planted with pigeon peas.
    b. treated with nitrogen fertilizer.
    c. not treated.
    d. not watered.
    e. Both c and d are correct.

adaptation   7
animal   9
Archaea   8
atom   4
Bacteria 8
binomial nomenclature   9
biology   11
biosphere   10
cell   4
class   8
community   10
conclusion   11
control group   12
data   11
domain   8
ecosystem   10
energy   7
Eukarya   8
eukaryote   8
evolution   7
experimental variable   12
family   8
fungi   9
gene   7
genus   8
homeostasis   6

hypothesis   11
kingdom   8
metabolism   7
molecule   4
natural selection   7
observation   11
order   8
organ   4
organism   4
organ system   4
pheromone   3
photosynthesis   7
phylum   8
plant   9
population   10
prokaryote   8
protist   8
reproduce   7
scientific process   11
scientific theory   11
species   7, 8
specific epithet   9
taxonomy   8
testing   11
tissue   4

Match the terms to these definitions:

a. _____ All of the chemical reactions that occur in a cell during growth and repair.

b. _____ Changes that occur among members of a species with the passage of time, often resulting in increased adaptation to the prevailing environment.

c. _____ Component in an experiment that is manipulated as a means of testing it.

d. _____ Process by which plants use solar energy to make their own organic food.

e. _____ Sample that goes through all the steps of an experiment but lacks the factor being tested.

1. An investigator spills dye on a culture plate and notices that the bacteria live despite exposure to sunlight. He decides to test if the dye is protective against ultraviolet (UV) light. He exposes one group of culture plates containing bacteria and dye and another group containing only bacteria to UV light. The bacteria on all plates die. Complete the following diagram:

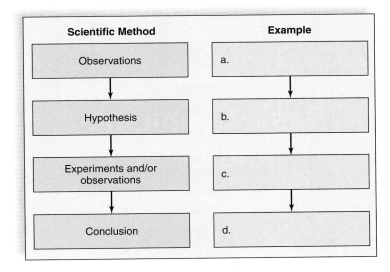

2. You want to grow large tomatoes and notice that a name-brand fertilizer claims to produce larger produce than a generic brand. How would you set test this claim?

ARIS *Visit* **www.mhhe.com/maderconcepts** *for practice quizzes, animations, videos, and activities designed to help you master the material in this chapter.*

2

Basic Chemistry and Cells

LEARNING OUTCOMES

After studying this chapter, you should be able to accomplish the following outcomes.

Life Depends on Water

1 Identify ways in which life depends on water.

All Matter Is Composed of Chemical Elements

2 Distinguish between matter, elements, and atoms.
3 Name the six elements that are basic to life.
4 Describe the structure of an atom.
5 Tell why an atom can have isotopes.
6 Give examples of how low levels and high levels of radiation can each be helpful in medicine.

Atoms React with One Another to Form Molecules

7 Explain the periodic table of the elements.
8 State, explain, and give examples of the octet rule.
9 Distinguish between an ionic bond, a covalent bond, and a hydrogen bond.
10 Be able to recognize and construct molecules that contain these bonds.

The Properties of Water Benefit Life

11 List and describe four properties of water that benefit organisms.

Living Things Require a Narrow pH Range

12 Distinguish between acids and bases.
13 Explain and use the pH scale.
14 Describe a buffer and tell how buffers assist organisms.
15 Discuss the harmful effects of acid deposition on lakes, forests, humans, buildings, and statues.

Scientists and laymen alike have often pondered the question, "Why did life arise on Earth?" It's long been thought that the answer, in part, must involve the presence of water. Three-fourths of the surface of our planet is covered by water. Water is so abundant that if the surface of the Earth were absolutely smooth, it would be covered by water. Only land that projects above the seas provides a terrestrial environment.

Cells most likely arose in the oceans. Any living system is 70–90% water, a medium in which chemical reactions can easily

Planet Earth has an abundance of water.

Life Depends on Water

occur. A watery environment supports and protects cells while providing an external transport system for chemicals. Homeostasis is also assisted by the ability of water to absorb and give off heat in a way that prevents rapid temperature changes. The abundance of terrestrial life correlates with the abundance of water; therefore, a limited variety of living things is found in the deserts, but much variety exists in the tropics, which receives, by far, the most rain. The tropics are also warm, and water helps maintain a constant year-round temperature day and night.

Do any of the other planets have life? To answer this question, astronomers first look for signs of water, because life as we know it does not exist without water. For several years, NASA has been looking for evidence of water on Mars, using unmanned space vehicles called rovers. In 2004, a rover named Opportunity discovered that Mars was at one time a wet planet. Now, scientists believe that evidence of life on Mars will also be found one day. This principle holds for the other planets as well. First, find evidence of water—then look for life!

The strength of the association between water and living things is observable in that all animals, whether aquatic or terrestrial, make use of water to reproduce. Animals that live in the sea or in fresh water can simply deposit their eggs and sperm in the water, where they join to form an embryo that develops in the water. The sperm of human beings, like those of many other terrestrial animals, are deposited inside the female, where they are protected from drying out. Then, as with most other mammals, the offspring develops within a fluid, called the amniotic fluid, while contained within the uterus. Amniotic fluid cushions the embryo and protects it against possible traumas, while maintaining a constant temperature. Later, it prevents limbs from sticking to the body and allows the fetus to move about.

Life as we know it cannot exist without a constant supply of water. This chapter discusses the properties of water that assist living things in maintaining homeostasis. It also covers basic chemistry necessary to understanding how the cell, and therefore the organism, functions. Some chemicals alter the properties of water and, in that way, threaten the ability of organisms to maintain homeostasis.

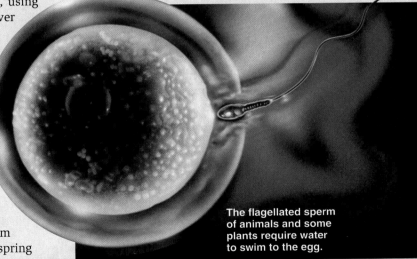

The flagellated sperm of animals and some plants require water to swim to the egg.

Of all the planets, only Earth has abundant water and abundant life.

Humans, like other animals, develop in a water environment.

Chemistry can be defined as the science of the composition, properties, and reactions of matter. In this part of the chapter, we consider the composition of matter—especially, the atoms found in matter. Atoms contain subatomic particles, one of which is associated with the occurrence of radioactive isotopes. We will see that radioactive isotopes have medical applications.

2.1 Six elements are basic to life

Turn the page, throw a ball, pat your dog, rake leaves; everything, from the water we drink to the air we breathe, is composed of matter. **Matter** refers to anything that takes up space and has mass. Although matter has many diverse forms—anything from molten lava to kidney stones—it only exists in three distinct states: solid, liquid, and gas.

Elements All matter, both nonliving and living, is composed of certain basic substances called elements. An **element** is a substance that cannot be broken down by chemical means into simpler substances with different properties. (A property is a physical or chemical characteristic, such as density, solubility, melting point, and reactivity.) It is quite remarkable that only 92 naturally occurring elements serve as the building blocks of matter. Other elements have been "human-made" and are not biologically important.

Both the Earth's crust and all organisms are composed of elements, but they differ as to which ones are predominant. Only six elements—carbon, hydrogen, nitrogen, oxygen, phosphorus, and sulfur—are basic to life. These elements make up about 95% of the body weight of organisms, such as the macaws in **Figure 2.1**. The macaws have gathered on a salt lick in South America. Salt contains the elements sodium and chlorine and is a common mineral sought after by many forms of life. A **mineral** is a solid substance formed by natural processes.

Every element has a name and also a symbol. The **atomic symbol** for sodium is Na because *natrium* means sodium in Latin. Chlorine, on the other hand, has the symbol Cl, which is consistent with its English name. Other elements, however, also take their symbol from Latin. For example, the symbol for iron is Fe because *ferrum* means iron in Latin. The symbols for the six elements basic to life are C, H, N, O, P, and S. Therefore, we can use the acronym CHNOPS to help us remember these six elements. As we shall discover in Chapter 3, the properties of the elements CHNOPS are essential to the uniqueness of cells and organisms. But other elements are also important to living things, including sodium, potassium, calcium, iron, and magnesium.

Atoms An **atom** is the smallest unit of an element that still retains the chemical and physical properties of the element. Therefore, it seems logical that an element and its atom have the same name. Notice the progression that we have now completed: Matter is composed of elements, and elements are composed of atoms. In the rest of the chapter, we will be speaking only of atoms because chemists consider atoms when discussing chemical reactions.

> **2.1 *Check Your Progress*** Water contains the elements hydrogen and oxygen. Which of the six elements basic to life are missing from water?

FIGURE 2.1 Elements that make up the Earth's crust and its organisms.

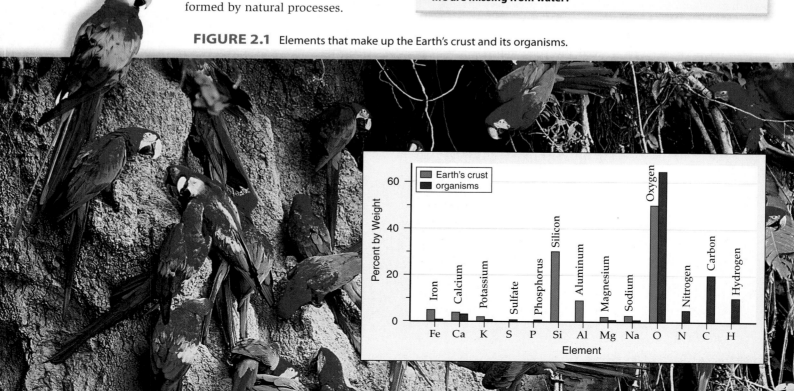

2.2 Atoms contain subatomic particles

Physicists have identified a number of subatomic particles—particles that are less complex than an atom but are components of an atom. The three best-known subatomic particles are positively charged **protons**, uncharged **neutrons**, and negatively charged **electrons** (**Table 2.2**). Protons and neutrons are located within the nucleus of an atom, and electrons move about the nucleus. The stippling in **Figure 2.2A** shows the probable location of the electrons in an atom that has ten electrons. In **Figure 2.2B**, the circles represent **electron shells**, the approximate orbital paths of electrons. The inner shell has the lowest energy level and can hold two electrons. The outer shell has a higher energy level and can hold eight electrons. An atom is most stable when the outer shell has eight electrons.

In science, a model is a useful simulation of a structure or process rather than the actual structure or process. Biologists find that the model of an atom shown in Figure 2.2B is sufficient for their purposes, and you will be asked to create such atomic models. Actually, today we know that most of an atom is empty space. If an atom could be drawn the size of a football field, the nucleus would be like a gumball in the center of the field, and the electrons would be tiny specks whirling about in the upper stands. Electrons don't have to always stay within certain shells. In our analogy to a football field, the electrons might very well stray outside the stadium at times.

Atomic Number All atoms of an element have the same number of protons. This is called the **atomic number**. For example, the atomic number for a carbon atom is 6. The atomic number not only tells you the number of protons, but it also tells you the number of electrons when the atom is electrically neutral. The model of a carbon atom in **Figure 2.2C** shows that an electrically neutral carbon atom has 6 protons and 6 electrons.

Atomic Mass Each atom has its own specific mass. The **atomic mass**, or mass number, of an atom depends on the presence of protons and neutrons, both of which are assigned one atomic mass unit (Table 2.2). Electrons are so small that their

TABLE 2.2 Subatomic Particles			
Particle	Electric Charge	Atomic Mass	Location
Proton	+1	1	Nucleus
Neutron	0	1	Nucleus
Electron	−1	0	Electron shell

mass is considered zero in most calculations. Therefore, the atomic mass for an atom is the number of protons plus the number of neutrons.

The term *atomic mass* is used rather than *atomic weight*, because mass is constant, whereas weight changes according to the gravitational force of a body. The gravitational force of the Earth is greater than that of the moon; therefore, substances weigh less on the moon, even though their mass has not changed. The atomic mass for most carbon atoms is 12. The atomic mass is written as a superscript to the left of the atomic symbol, and the atomic number is written as a subscript. This notation is illustrated for carbon in Figure 2.2C.

Isotopes **Isotopes** are atoms of a single element that differ in their number of neutrons. Isotopes have the same number of protons, but they have different atomic masses. For example, the element carbon has three common isotopes:

$$^{12}_{6}C \qquad ^{13}_{6}C \qquad ^{14}_{6}C*$$
$$*radioactive$$

Carbon 12 has six neutrons, carbon 13 has seven neutrons, and carbon 14 has eight neutrons. Unlike the other two isotopes, carbon 14 is unstable; it breaks down into atoms with lower atomic numbers. When it decays, it emits radiation in the form of radioactive particles or radiant energy. Therefore, carbon 14 is called a radioactive isotope. Biologists and other scientists have found many beneficial uses for radiation. For example, Melvin Calvin and his co-workers used carbon 14 to discover the sequence of reactions that occur during the process of photosynthesis. In Section 2.3, we will learn about various medical uses for radioactive isotopes.

> **2.2 Check Your Progress** Water contains the atoms hydrogen ($_1$H) and oxygen ($_8$O). Draw a model of each of these atoms.

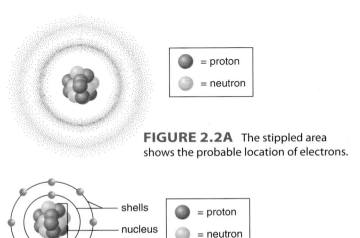

FIGURE 2.2A The stippled area shows the probable location of electrons.

FIGURE 2.2B The shells in this atomic model represent the average location of electrons.

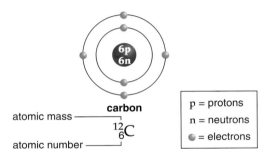

FIGURE 2.2C Atomic model of a carbon atom.

The importance of chemistry to biology and medicine is nowhere more evident than in the many medical uses of radioactive isotopes. Some of these applications require low levels of radiation, and others require high levels of radiation.

Low Levels of Radiation In a chemical reaction, the radioactive isotope of an element behaves the same as the stable isotopes of an element. This means that if you put a small amount of radioactive isotope in a sample, it becomes a **tracer** that can be used to detect molecular changes. The radiation given off by radioactive isotopes can be detected in various ways.

Specific tracers are used in imaging the body's organs and tissues. For example, after a patient drinks a solution containing a minute amount of iodine-131, it becomes concentrated in the thyroid gland. (The thyroid is the only organ to take up iodine-131, which it uses to make thyroid gland hormone.) The missing area (upper right) in the X-ray shown in **Figure 2.3A** indicates the presence of a tumor that does not take up radioactive iodine.

A procedure called positron-emission tomography (PET) is a way to determine the comparative activity of tissues. Radioactively labeled glucose, which emits a subatomic particle known as a positron, can be injected into the body. The radiation given off is detected by sensors and analyzed by a computer. The result is a color image that shows which tissues took up glucose and are metabolically active. The red areas surrounded by green in **Figure 2.3B**, indicate which areas of the brain are most active. PET scans of the brain are used to evaluate patients who have memory disorders of an undetermined cause and suspected brain tumors or seizure disorders that could possibly benefit from surgery. PET scans of the heart can detect signs of coronary artery disease and low blood flow to the heart muscle. For this procedure, the patient is injected with a radioisotope of the metallic element thallium (thallium-201). The more thallium taken up by the heart muscle, the better the blood supply to the heart. The thallium test is often done along with a stress test, in which the patient exercises on a treadmill or stationary bicycle. The heart is imaged at the end of the exercise period, when it should be working its hardest and receiving the greatest blood supply.

High Levels of Radiation Radioactive substances in the environment can harm cells, damage DNA, and cause cancer. For example, the release of radioactive particles following a nuclear power plant accident can have far-reaching and long-lasting effects on human health. However, high levels of radiation can also be put to good use. Radiation from radioactive isotopes has been used for many years to sterilize medical and dental products, and in the future it may be used to sterilize the U.S. mail in order to free it of possible pathogens, such as anthrax spores.

Rapidly dividing cells are particularly sensitive to damage by radiation. For this reason, some cancerous growths can be controlled or eliminated by irradiating the area containing the growth. Radiotherapy can be administered externally, as depicted in **Figure 2.3C**, or it can be given internally. Today, internal radiotherapy allows radiation to destroy only cancer cells, with little risk to the rest of the body. For example, iodine-131 is commonly implanted to treat thyroid cancer, probably the most successful kind of cancer treatment.

We have learned that isotopes vary by the number of neutrons, a type of subatomic particle. In the next part of the chapter, we will study how electrons, another subatomic particle, are involved in causing atoms to react with one another.

2.3 Check Your Progress If a radioactive isotope is incorporated into a molecule, would that molecule also be a tracer?

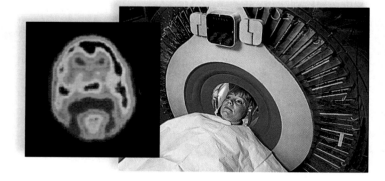

FIGURE 2.3A Detection of thyroid cancer.

larynx
thyroid gland
trachea

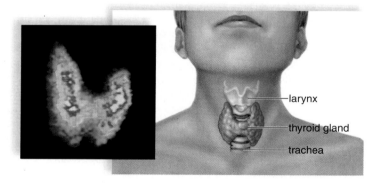

FIGURE 2.3B Detection of brain activity.

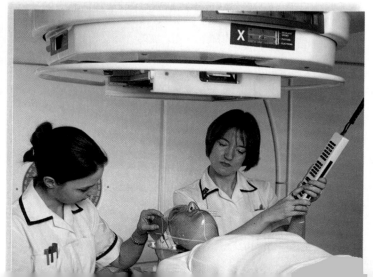

FIGURE 2.3C Radiation helps cure cancer.

Atoms React with One Another to Form Molecules

Learning Outcomes 7–10, page 18

The topic for this part of the chapter is how atoms react with one another. We will learn that the number of electrons in the outer shell determines whether atoms form an ionic bond or a covalent bond with one another. Another type of bond, called a hydrogen bond, will also be considered.

2.4 After atoms react, they have a completed outer shell

Once chemists discovered a number of the elements, they began to realize that the chemical and physical characteristics of atoms recur in a predictable manner. The periodic table was developed as a way to display the elements, and therefore the atoms, according to these characteristics. **Figure 2.4A** shows a portion of the periodic table of the elements. The period (horizontal row) tells you how many shells an atom has, and the group (vertical column marked by roman numeral) tells you how many electrons an atom has in its outer shell. For example, carbon in the second period (pink) has two shells, and being in group IV, it has four electrons in the outer shell.

A model can be drawn for each of the atoms in the periodic table. **Figure 2.4B** illustrates models for the six elements common to organisms, namely CHNOPS. For these atoms and all the others up through number 20 (calcium), each lower level is filled with electrons before the next higher level contains any electrons. Recall that the first shell (closest to the nucleus) can contain two electrons; thereafter, each additional shell can contain eight electrons.

If an atom has only one shell, as does H and He, the outer shell is complete when it has two electrons. Otherwise, the outer shell is most stable when it has eight electrons, a rule termed the **octet rule**. Atoms in group VIII of the periodic table are called the noble gases because they are normally nonreactive. Atoms with fewer than eight electrons in the outer shell react with other atoms in such a way that after the reaction, each has a stable outer shell.

Atoms can give up, accept, or share electrons in order to have eight electrons in the outer shell. In other words, the number of electrons in an atom's outer shell, called the **valence shell**, determines its chemical reactivity. The size of an atom can also affect reactivity. Both carbon and silicon have four outer electrons, but only the smaller carbon atom often bonds to other carbon atoms and forms long-chained molecules.

I	II	III	IV	V	VI	VII	VIII
1 **H** 1.008							2 **He** 4.003
3 **Li** 6.941	4 **Be** 9.012	5 **B** 10.81	6 **C** 12.01	7 **N** 14.01	8 **O** 16.00	9 **F** 19.00	10 **Ne** 20.18
11 **Na** 22.99	12 **Mg** 24.31	13 **Al** 26.98	14 **Si** 28.09	15 **P** 30.97	16 **S** 32.07	17 **Cl** 35.45	18 **Ar** 39.95
19 **K** 39.10	20 **Ca** 40.08	31 **Ga** 69.72	32 **Ge** 72.59	33 **As** 74.92	34 **Se** 78.96	35 **Br** 79.90	36 **Kr** 83.60

Groups

Periods

FIGURE 2.4A A portion of the periodic table of the elements. For a complete table, see Appendix C.

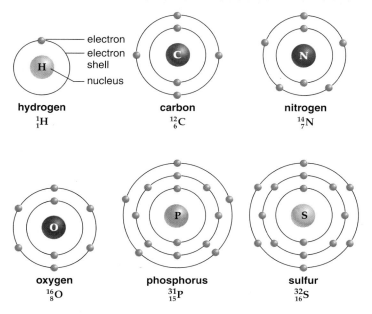

FIGURE 2.4B Models of the six elements that are predominant in living things.

Except for the noble gases, atoms routinely bond with one another. For example, oxygen does not exist in nature as a single atom, O; instead, two oxygen atoms are joined to form a molecule (O_2). Other naturally occurring molecules include hydrogen (H_2) and nitrogen (N_2). When atoms of two or more different elements bond together, the product is called a **compound**. Water (H_2O) is a compound that contains atoms of hydrogen and oxygen. A **molecule** is the smallest part of a compound that still has the properties of that compound.

One of the two fundamental ways that atoms bond with one another is studied in Section 2.5.

> **2.4 *Check Your Progress* *a.* When hydrogen bonds with oxygen to form water, how many electrons does each hydrogen require to achieve a completed outer shell? Explain. *b.* How many electrons does oxygen require to achieve a completed outer shell? Explain.**

2.5 An ionic bond occurs when electrons are transferred

Ions form when electrons are transferred from one atom to another. For example, sodium (Na), with only one electron in its third shell, tends to be an electron donor. Once it gives up this electron, the second shell, with eight electrons, becomes its outer shell. Chlorine (Cl), in contrast, tends to be an electron acceptor. Its outer shell has seven electrons, so if it acquires only one more electron, it has a completed outer shell. When a sodium atom and a chlorine atom come together, an electron is transferred from the sodium atom to the chlorine atom. Now, both atoms have eight electrons in their outer shells (**Fig. 2.5**).

This electron transfer, however, causes a charge imbalance in each atom. The sodium atom has one more proton than it has electrons; therefore, it has a net charge of +1 (symbolized by Na^+). The chlorine atom has one more electron than it has protons; therefore, it has a new charge of −1 (symbolized by Cl^-). Such charged particles are called **ions**.

Sodium (Na^+) and chloride (Cl^-) are not the only biologically important ions. Some, such as potassium (K^+), are formed by the transfer of a single electron to another atom; others, such as calcium (Ca^{2+}) and magnesium (Mg^{2+}), are formed by the transfer of two electrons.

Ionic compounds are held together by an attraction between negatively and positively charged ions, called an **ionic bond**. When sodium reacts with chlorine, an ionic compound called sodium chloride (NaCl) results. Sodium chloride is an example of a salt; it is commonly known as table salt because it is used to season food (Fig. 2.5). **Salts** can exist as dry solids, but when salts are placed in water, they release ions as they dissolve. NaCl separates into Na^+ and Cl^-. Ionic compounds are most commonly found in this separated (dissociated) form in living things because biological systems are 70–90% water.

Biologically important ions in the human body are listed in **Table 2.5**. The balance of these ions in the body is important to our health. Too much sodium in the blood can cause high blood pressure; too much or too little potassium results in heartbeat irregularities; and not enough calcium leads to rickets (bowed legs) in children. Bicarbonate ions are involved in maintaining the acid-base balance of the body.

Ions are less likely to occur when atoms form covalent bonds with one another, as discussed in the next section.

TABLE 2.5	Significant Ions in the Human Body	
Name	**Symbol**	**Special Significance**
Sodium	Na^+	Found in body fluids; important in muscle contraction and nerve conduction
Chloride	Cl^-	Found in body fluids
Potassium	K^+	Found primarily inside cells; important in muscle contraction and nerve conduction
Phosphate	PO_4^{3-}	Found in bones, teeth, and the high-energy molecule ATP
Calcium	Ca^{2+}	Found in bones and teeth; important in muscle contraction and nerve conduction
Bicarbonate	HCO_3^-	Important in acid-base balance

> **2.5 Check Your Progress** Knowing that oxygen is able to attract an electron to a greater degree than hydrogen, supply the correct charges for the ions that result when water breaks down like this: $H_2O \longrightarrow H + OH$

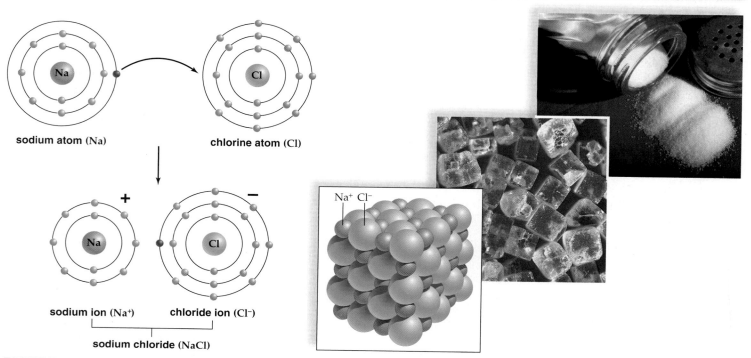

sodium atom (Na) chlorine atom (Cl)

sodium ion (Na^+) chloride ion (Cl^-)

sodium chloride (NaCl)

Na^+ Cl^-

FIGURE 2.5 Formation of sodium chloride (table salt).

2.6   A covalent bond occurs when electrons are shared

In a **covalent bond**, two atoms share electrons in such a way that each atom has an octet of electrons in the outer shell, or two electrons in the case of hydrogen. In a hydrogen atom, the outer shell is complete when it contains two electrons. If hydrogen is in the presence of a strong electron acceptor, such as oxygen, it gives up its electron to become a hydrogen ion (H^+). But if this is not possible, hydrogen can share an electron with another atom, and thereby have a completed outer shell. For example, one hydrogen atom will share with another hydrogen atom. Their two orbitals overlap, and the electrons are shared between them. Sharing is illustrated by drawing molecular models called electron models (**Fig. 2.6A**). When atoms share electron pairs, each one has a completed outer shell.

Bond Notations A common way to symbolize that atoms are sharing electrons is to draw a line between the two atoms, as in the structural formula H—H. In a molecular formula, the line is omitted, and the molecule is simply written as H_2 (Fig. 2.6A). Sometimes, atoms share more than one pair of electrons to complete their octets. A double covalent bond occurs when two atoms share two pairs of electrons. To show that oxygen gas (O_2) contains a double bond, the molecule can be written as O=O. It is also possible for atoms to form triple covalent bonds, as in nitrogen gas (N_2), which can be written as N≡N. Single covalent bonds between atoms are quite strong, but double and triple bonds are even stronger.

The molecule methane results when carbon binds to four hydrogen atoms (CH_4). In methane, each bond actually points to one corner of a tetrahedron. A ball-and-stick model is the best way to show this arrangement, while a space-filling model comes closest to showing the actual shape of the molecule (**Fig. 2.6B**). The shapes of molecules help dictate the roles they play in organisms.

Chemical Reactions Chemical reactions, such as those in photosynthesis, are very important to organisms. An overall equation for the photosynthetic reaction indicates that some bonds are broken and others are formed:

$$6\ CO_2\ +\ 6\ H_2O\ \longrightarrow\ C_6H_{12}O_6\ +\ 6\ O_2$$

carbon dioxide   water   glucose   oxygen

This equation says that six molecules of carbon dioxide react with six molecules of water to form one glucose molecule and six molecules of oxygen. The reactants (molecules that participate in the reaction) are shown on the left of the arrow, and the products (molecules formed by the reaction) are shown on the right. Notice that the equation is "balanced"—that is, the same number of each type of atom occurs on both sides of the arrow.

Note the glucose molecule in the equation above. It has six atoms of carbon, 12 atoms of hydrogen, and six atoms of oxygen bonded together to form one molecule. The structural formula for glucose is shown in Figure 3.4A. Some of the bonds in glucose are nonpolar, and others are polar. In Section 2.7, we will see what makes a covalent bond polar or nonpolar.

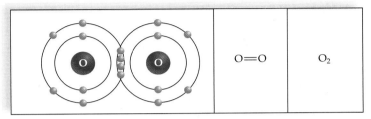

Hydrogen gas

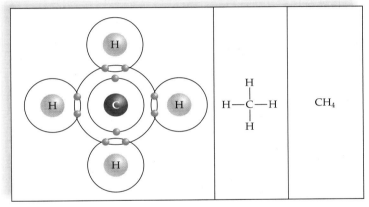

Oxygen gas

Electron Model	Structural Formula	Molecular Formula
Hydrogen gas	H—H	H_2
Oxygen gas	O=O	O_2
Methane	H—C—H (with H above and below)	CH_4

Methane

FIGURE 2.6A   Electron models and formulas representing covalently bonded molecules.

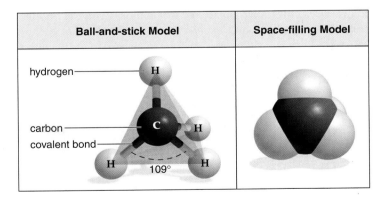

Ball-and-stick Model	Space-filling Model

hydrogen
carbon
covalent bond
109°

FIGURE 2.6B   Other types of molecular models—in this case, for methane (CH_4).

> **2.6 Check Your Progress a. Covalent bonds occur in water. Use overlapping atomic models to show the structure of water.**
> **b. Explain why water has the formula H_2O.**

2.7 A covalent bond can be nonpolar or polar

When the sharing of electrons between two atoms is fairly equal, the covalent bond is said to be a **nonpolar covalent bond**. All the molecules in Figure 2.6A, including methane (CH_4), are nonpolar. In a water molecule (H_2O), the sharing of electrons between oxygen and each hydrogen is not completely equal. The attraction of an atom for the electrons in a covalent bond is called its **electronegativity**. The larger oxygen atom, with the greater number of protons, is more electronegative than the hydrogen atom. The oxygen atom can attract the electron pair to a greater extent than each hydrogen atom can. The shape of a water molecule allows the oxygen atom to maintain a slightly

negative charge (δ^-, or "delta minus"), and the hydrogen atoms to maintain slightly positive charges (δ^+, or "delta plus"). The unequal sharing of electrons in a covalent bond creates a **polar covalent bond**, and in the case of water, the molecule itself is a polar molecule. **Figure 2.7** shows the electron model, the ball-and-stick model, and the space-filling model of a water molecule. The polarity of water molecules leads to the formation of hydrogen bonds, as discussed in the next section.

> **2.7 Check Your Progress** Why is water a polar molecule?

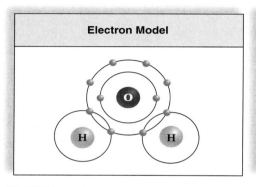
Electron Model

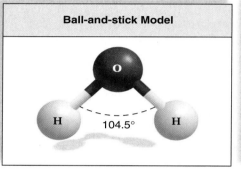
Ball-and-stick Model

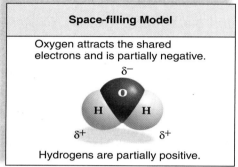
Space-filling Model

Oxygen attracts the shared electrons and is partially negative.

Hydrogens are partially positive.

FIGURE 2.7 Three models of water.

2.8 A hydrogen bond can occur between polar molecules

The polarity of water molecules causes the hydrogen atoms in one molecule to be attracted to the oxygen atoms in other water molecules. This attraction, although weaker than an ionic or covalent bond, is called a **hydrogen bond**. Because a hydrogen bond is easily broken, it is often represented by a dotted line (**Fig. 2.8**). Hydrogen bonding is not unique to water. Many biological molecules have polar covalent bonds involving an electropositive hydrogen and usually an electronegative oxygen or nitrogen. In these instances, a hydrogen bond can occur within the same molecule or between different molecules.

Hydrogen bonds are a bit like Velcro: Each tiny hook and loop is weak, but when hundreds of hooks and loops come together, they are collectively strong. Continuing the analogy, a Velcro fastener is easy to pull apart when needed, and in the same way, hydrogen bonds can be disrupted.

Hydrogen bonds between cellular molecules help maintain their proper structure and function. For example, hydrogen bonds hold the two strands of DNA together. When DNA makes a copy of itself, each hydrogen bond breaks easily, allowing the DNA to unzip. On the other hand, the hydrogen bonds, acting together, add stability to the DNA molecule. As we shall see, many of the important properties of water are the result of hydrogen bonding.

This completes our discussion of how atoms react to form molecules. In the next part of the chapter, we will be studying the properties of water.

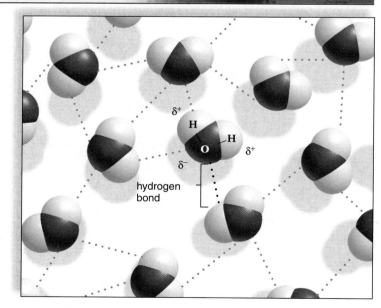

FIGURE 2.8 Hydrogen bonding between water molecules.

> **2.8 Check Your Progress** Like water (H_2O), ammonia (NH_3) is a polar molecule. *a.* Would you expect hydrogen bonding between ammonia molecules? *b.* Between ammonia and water molecules? *c.* Explain.

The introduction for this chapter stressed the association between water and living things. In this part of the chapter, we study four properties of water and show how these properties benefit organisms. We will see that water exhibits (1) cohesiveness, (2) a tendency to change temperature slowly, (3) an ability to dissolve other polar substances, and (4) an ability to expand as it freezes.

2.9 Water molecules stick together: Cohesion

Hydrogen bonding accounts for most of the properties of water that make life possible. For example, without hydrogen bonding, frozen water would melt at $-100°C$, and liquid water would boil at $-91°C$, making most of the water on Earth steam, and life unlikely. But because of hydrogen bonding, water is a liquid at temperatures typically found on the Earth's surface. It melts at $0°C$ and boils at $100°C$.

Hydrogen bonding is responsible for water's **cohesion**—the tendency of water molecules to cling together. Cohesion is apparent because water flows freely, and yet water molecules do not separate from each other. As a result of cohesion, water is an excellent transport medium, both outside of and within living organisms.

Unicellular organisms rely on external water to transport nutrient and waste molecules, but multicellular organisms often contain internal vessels in which water serves to transport nutrients and wastes. For example, the liquid portion of our blood, which transports dissolved and suspended substances throughout the body, is 90% water.

Cohesion contributes to the transport of water in plants. Plant roots absorb water from the soil, but the leaves are uplifted and exposed to solar energy. How is it possible for water to rise to the top of even very tall trees? A plant contains a system of vessels that reaches from the roots to the leaves (**Fig. 2.9**). Water transport in plants is somewhat like sucking water through a straw—or rather, a bundle of straws. The suction is supplied by the evaporation of water from leaves. Water evaporating from the leaves is immediately replaced with water molecules from the vessels. **Adhesion** of water to the walls of the vessels also helps prevent the water column from breaking apart.

The stronger the force between molecules in a liquid, the greater the surface tension. As with cohesion, hydrogen bonding causes water to have a high surface tension. This property makes it possible for humans to skip rocks on water. Water striders, a common insect, can even walk on the surface of a pond without breaking the surface.

Hydrogen bonding is responsible for the ability of water to warm up and cool down slowly, as discussed in the next section.

> **2.9 Check Your Progress** How is hydrogen bonding related to the cohesion of water?

FIGURE 2.9 Water as a transport medium in trees.

Water evaporates, pulling the water column from the roots to the leaves.

Water molecules cling together and adhere to sides of vessels in stems.

H_2O

Water enters a plant at root cells.

2.10 Water warms up and cools down slowly

One **calorie** is the amount of heat energy needed to raise the temperature of 1 gram (g) of water 1°C. In comparison, other covalently bonded liquids require input of about one-half calorie to rise in temperature 1°C. The many hydrogen bonds that link water molecules help water absorb heat, without a great change in temperature.

Converting 1 g of the coldest liquid water to ice requires the loss of 80 calories of heat energy. Water holds onto its heat, and its temperature falls more slowly than that of other liquids. This property of water is important, not only for aquatic organisms, but for all living things. Because the temperature of water rises and falls slowly, organisms are better able to maintain their normal internal temperatures and are protected from rapid temperature changes.

Converting 1 g of the hottest water to a gas requires an input of 540 calories of heat energy. Water has a high heat of vaporization because hydrogen bonds must be broken before water boils and water molecules vaporize (evaporate into the environment). Water's high heat of vaporization gives animals in a hot environment an efficient way to release excess body heat. When an animal sweats or gets splashed, body heat is used to vaporize the water, thus cooling the animal (**Fig. 2.10**).

Because of water's high heat capacity and high heat of vaporization, the temperatures along coasts are moderate. During the summer, the ocean absorbs and stores solar heat, and during the winter, the ocean slowly releases it. In contrast, the interior regions of continents experience abrupt changes in temperature. Water influences the lives of organisms, as in this example, and also affects their metabolism, as explained in the next section.

FIGURE 2.10 The bodies of organisms cool when their heat is used to evaporate water.

> **2.10** *Check Your Progress* In dry climates, evaporative coolers use a fan to draw air through a water-soaked fiber pad. Why does this work?

2.11 Water dissolves other polar substances

Because of its polarity, water facilitates chemical reactions, both outside and within living systems. It dissolves a great number of substances. A **solution** contains dissolved substances, which are then called **solutes**. When ionic salts—for example, sodium chloride (NaCl)—are put into water, the negative ends of the water molecules are attracted to the sodium ions, and the positive ends of the water molecules are attracted to the chloride ions. This causes the sodium ions and the chloride ions to dissociate in water (**Fig. 2.11**). Water is also a solvent for larger molecules that contain ionized atoms or are polar.

Those molecules that can attract water are said to be **hydrophilic**. When ions and molecules disperse in water, they move about and collide, allowing reactions to occur. Nonionized and nonpolar molecules that cannot attract water are said to be **hydrophobic**. Gasoline contains nonpolar molecules, and therefore it does not mix with water and is hydrophobic. Another property of water is discussed in Section 2.12: Ice floats on liquid water.

> **2.11** *Check Your Progress* Fats are nonpolar, but they can be physically dispersed in water by combining with molecules called emulsifiers. What property do emulsifiers have that fats lack?

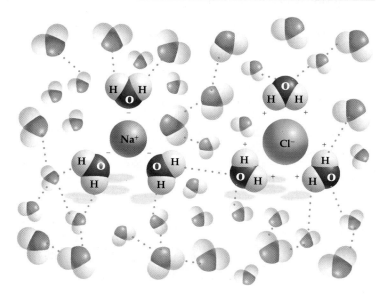

FIGURE 2.11 An ionic salt dissolves in water.

2.12 Frozen water is less dense than liquid water

Remarkably, water is more dense at 4°C than it is at 0°C. Most substances contract when they solidify, but water expands when it freezes. In ice, water molecules form a lattice, in which the hydrogen bonds are farther apart than they are in liquid water. This is why cans of soda burst when placed in a freezer, and why frost heaves make northern roads bumpy in the winter. It also means that ice is less dense than liquid water, and therefore ice floats (**Fig. 2.12A**).

If ice did not float on water, it would sink, and ponds, lakes, and perhaps even the ocean, would freeze solid, making life impossible in the water and also on land. Instead, bodies of water always freeze from the top down. The ice acts as an insulator to prevent the water below it from freezing and also to prevent the loss of heat to the external environment.

In a pond, the ice protects the protists, plants, and animals so that they can survive the winter (**Fig. 2.12B**). These animals, except for the otter, are ectothermic, which means that they take on the temperature of the outside environment. This might seem disadvantageous; however, water remains relatively warm because of its high heat capacity. During the winter, frogs and turtles hibernate and, in this way, lower their oxygen needs. Insects survive in air pockets. Fish, as you will learn later in this text, have an efficient means of extracting oxygen from the water, and they need less oxygen than the endothermic otter, which depends on muscle activity to warm its body.

This completes our study of how the properties of water affect living things. In the next part of the chapter, we discuss additions to water that challenge the ability of organisms to maintain homeostasis.

> **2.12** *Check Your Progress* **Eskimos use igloos built from blocks of ice to keep warm in winter. Why does this work?**

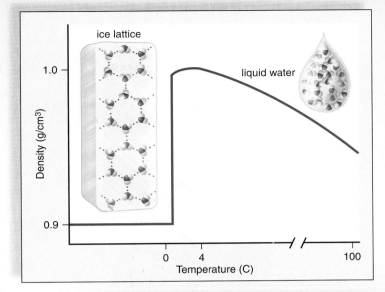

FIGURE 2.12A Ice is less dense than liquid water.

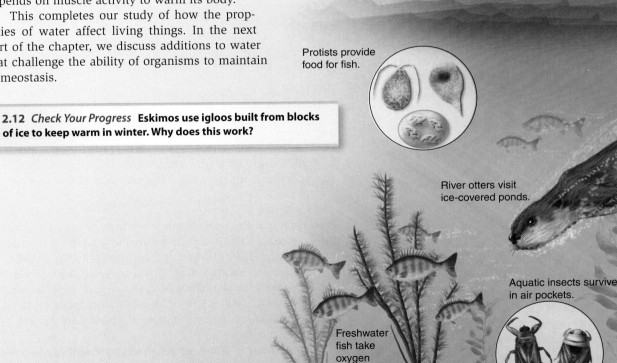

ice layer

Protists provide food for fish.

River otters visit ice-covered ponds.

Aquatic insects survive in air pockets.

Freshwater fish take oxygen from water.

FIGURE 2.12B A pond in winter.

Common frogs and pond turtles hibernate.

In this part of the chapter, we will learn how acids and bases differ from water and from each other. We will learn that living things are particularly sensitive to changes in the hydrogen ion concentration [H⁺] of water caused by the addition of acids and bases, which is measured in terms of pH. The body fluids of living things are buffered to keep the [H⁺] concentration relatively constant, but acid deposition has overcome the buffering ability of certain forests and lakes, which are dying as a consequence.

2.13 Acids and bases affect living things

FIGURE 2.13A Dissociation of water molecules.

When water dissociates, it releases an equal number of **hydrogen ions (H⁺)** and **hydroxide ions (OH⁻)**:

$$H{-}O{-}H \rightleftarrows H^+ + OH^-$$
$$\text{water} \qquad \text{hydrogen} \quad \text{hydroxide}$$
$$\text{ion} \qquad \text{ion}$$

Only a few water molecules at a time dissociate. The actual number of H⁺ is $(1 \times 10^{-7}$ moles/liter$)$[1], and an equal concentration of OH⁻ at $(1 \times 10^{-7}$ moles/liter$)$[1] is also present. Note that the amount of the two ions is the same—each has 10^{-7} moles/liter (**Fig. 2.13A**).

Acids: Excess Hydrogen Ions

When we eat acidic foods, the blood becomes more acidic. Lemon juice, vinegar, tomatoes, and coffee are all acidic foods. What do they have in common? **Acids** are substances that dissociate in water, releasing hydrogen ions (H⁺).[2] For example, hydrochloric acid (HCl) is an important inorganic acid that dissociates in this manner:

$$HCl \longrightarrow H^+ + Cl^-$$

Dissociation is almost complete; therefore, HCl is called a strong acid. If hydrochloric acid is added to a beaker of water, the number of hydrogen ions (H⁺) increases greatly (**Fig. 2.13B**).

Hydrochloric acid is produced in the stomach where protein is digested. Hydrochloric acid is capable of eating through most metals, and is highly toxic, burning on contact. However, a layer of mucus protects the stomach wall.

Bases: Excess Hydroxide Ions

When we take in basic substances, the blood becomes more basic. Milk of magnesia

FIGURE 2.13B Acids cause H⁺ to increase.

FIGURE 2.13C Bases cause OH⁻ to increase.

and ammonia are basic solutions familiar to most people. **Bases** are substances that either take up hydrogen ions (H⁺) or release hydroxide ions (OH⁻). For example, sodium hydroxide (NaOH) is an important inorganic base that dissociates in this manner:

$$NaOH \longrightarrow Na^+ + OH^-$$

Dissociation is almost complete; therefore, sodium hydroxide is called a strong base. If sodium hydroxide is added to a beaker of water, the number of hydroxide ions increases (**Fig. 2.13C**).

Sodium hydroxide is also known as lye or caustic soda. It is just as dangerous as a strong acid and can be used to etch aluminum. Contact with strong acids and bases should be avoided. For this reason, containers of these chemicals are marked with warning symbols.

Because living things are sensitive to the acidity and basicity of solutions, it is important for us to understand the pH scale, which measures [H⁺], as discussed next.

---

[1] In chemistry, a mole is defined as the amount of matter that contains as many objects (atoms, molecules, ions) as the number of atoms in exactly 12 g of ¹²C.

[2] A hydrogen atom contains one electron and one proton. A hydrogen ion has only one proton, so it is often simply called a proton.

> **2.13 Check Your Progress** Pure water contains an equal number of hydrogen ions (H⁺) and hydroxide ions (OH⁻).
> *a.* **Which one, H⁺ or OH⁻, increases with acids?** *b.* **With bases?**

2.14 The pH scale measures acidity and basicity

The **pH scale** is used to indicate the acidity or basicity (also called alkalinity) of a solution. The pH scale ranges from 0 to 14 (**Fig. 2.14**). A pH of 7 represents a neutral state, in which the hydrogen ion concentration $[H^+]$ equals the hydroxide ion concentration $[OH^-]$. A pH below 7 is an acidic solution when $[H^+]$ is greater than $[OH^-]$. A pH above 7 is basic when $[OH^-]$ is greater than $[H^+]$.

Moving down the pH scale from pH 14 to pH 0, each unit has ten times the $[H^+]$ of the previous unit. Moving up the scale from 0 to 14, each unit has ten times the $[OH^-]$ of the previous unit. The pH scale eliminates the use of cumbersome numbers. For example, in the following list, hydrogen ion concentrations are on the left, and the pH is on the right:

[H⁺] (moles per liter)		pH
0.000001	$= 1 \times 10^{-6}$	6
0.0000001	$= 1 \times 10^{-7}$	7
0.00000001	$= 1 \times 10^{-8}$	8

To further illustrate the relationship between $[H^+]$ and pH, consider the following question: Which of the pH values listed above indicates a higher hydrogen ion concentration $[H^+]$ than pH 7, and therefore would be an acidic solution? A number with a smaller negative exponent indicates a greater quantity of hydrogen ions; therefore, pH 6 is an acidic solution.

In most organisms, pH needs to be maintained within a narrow range. The pH of human blood is between 7.35 and 7.45; this is the pH at which our proteins, such as cellular enzymes, function properly. To maintain normal pH, blood is buffered, as we discuss in Section 2.15.

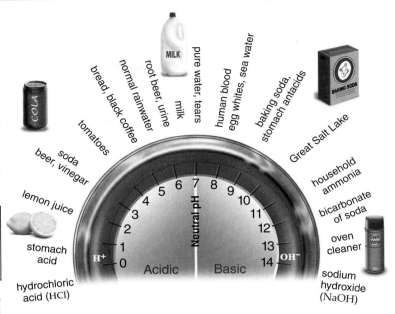

FIGURE 2.14 The pH scale.

2.14 Check Your Progress Pure water has a pH of 7. Rainwater normally has a pH of about 5.6. *a.* Is rainwater usually acidic or usually basic? *b.* Does normal rainwater have more or less H⁺ than pure water?

2.15 Buffers help keep the pH of body fluids relatively constant

A **buffer** resists changes in pH. Many commercial products, such as aspirin, shampoos, or deodorants, are buffered as an added incentive for us to buy them. Like blood, many other body fluids are buffered so the pH stays within a certain range. If blood pH rises much above pH 7.45, alkalosis is present, and if the pH lowers much below 7.35, acidosis is present. Weakness, cramping, and irritability are symptoms of alkalosis. Seizures, coma, and even death can result from acidosis. Normally, buffers take up excess hydrogen ions (H⁺) or hydroxide ions (OH⁻), thus preventing these occurrences.

Usually a buffer consists of a combination of chemicals. For example, carbonic acid (H_2CO_3) and bicarbonate(HCO_3) are two chemicals present in blood that keep pH within normal limits. Carbonic acid is a weak acid that releases bicarbonate and H⁺ when it dissociates:

$$H_2CO_3 \rightleftharpoons HCO_3^- + H^+$$
$$\text{carbonic acid} \qquad \text{bicarbonate}$$

When bases add hydroxide ions (OH⁻) to blood, they combine with the H⁺ from this reaction, and water forms. When acids add H⁺ to blood, carbonic acid simply re-forms:

$$H^+ + HCO_3^- \longrightarrow H_2CO_3$$

In addition to buffers, breathing helps maintain pH by ridding the body of CO_2, because the more CO_2 in the body, the more carbonic acid there is in blood. As powerful as the buffer and the respiration mechanisms are in maintaining pH, only the kidneys rid the body of a wide range of acidic and basic substances and otherwise adjust the pH. The kidneys are slower acting than the other two mechanisms, but they have a more powerful effect on pH. If the kidneys malfunction, alkalosis or acidosis are possible outcomes.

Ecosystems are buffered, but acid deposition can overcome their buffering ability, and the result is dead forests and trees, as described in the next section.

2.15 Check Your Progress When CO_2 enters the blood, it combines with water, and carbonic acid (H_2CO_3) results. *a.* When CO_2 enters the blood, does the pH go up or down? *b.* Why?

Normally, rainwater has a pH of about 5.6 because carbon dioxide in the air combines with water to produce a weak solution of carbonic acid. Acid deposition includes rain or snow that has a pH of less than 5, as well as dry acidic particles that fall to Earth from the atmosphere. When fossil fuels such as coal, oil, and gasoline are burned, sulfur dioxide (SO_2) and nitrogen oxides (NO_x) combine with water to produce sulfuric and nitric acids. These pollutants are generally found eastward of where they originated because of wind patterns. The use of very tall smokestacks causes them to be carried even hundreds of miles away (**Fig. 2.16**). For example, acid rain in southeastern Canada results from the burning of fossil fuels in factories and power plants in the midwestern United States.

Impact on Lakes Acid rain adversely affects lakes, particularly in areas where the soil is thin and lacks limestone (calcium carbonate, or $CaCO_3$), a buffer to acid deposition. Acid deposition leaches toxic aluminum from the soil and converts mercury deposits in lake bottom sediments to toxic methyl mercury, which accumulates in fish. People are now advised against eating fish from the Great Lakes because of high mercury levels. Hundreds of lakes are devoid of fish in Canada and New England, and thousands have suffered the same fate in the Scandinavian countries. Some of these lakes have no signs of life at all.

Impact on Forests The leaves of plants damaged by acid rain can no longer carry on photosynthesis as before. When plants are under stress, they become susceptible to diseases and pests of all types. Forests on mountaintops receive more rain than those at lower levels; therefore, they are more affected by acid rain (Fig. 2.16). Forests are also damaged when toxic chemicals such as aluminum are leached from the soil. These kill soil fungi that assist roots in acquiring the nutrients trees need. In New England, 1.3 million acres of high-elevation forests have been devastated.

Impact on Humans and Structures Humans may be affected by acid rain. Inhaling dry sulfate and nitrate particles appears to increase the occurrence of respiratory illnesses, such as asthma. Buildings and monuments made of limestone and marble break down when exposed to acid rain. The paint on homes and automobiles is likewise degraded.

> **2.16 Check Your Progress** The human digestive tract is protected by a layer of mucus. Still, consumers should be aware that the pH of cola drinks can be as low as pH 2. Explain the concern.

FIGURE 2.16 Environmental effects of acid rain.

Emissions from power plants and industrial facilities add acids to the atmosphere.

Lakes become sterile due to acid deposition.

Trees die due to acid deposition.

Statues corrode due to acid deposition.

CONNECTING THE CONCEPTS

Cells are made up of chemicals, and therefore it is necessary for us to understand the basic concepts of chemistry. An atom has a certain number of protons, neutrons, and electrons. An atom will react with other atoms to achieve a complete outer shell with eight electrons (or two electrons in the case of hydrogen). Some atoms transfer electrons in order to achieve a completed outer shell. In these molecules, such as sodium chloride (NaCl)

ions, charged atoms are attracted to one another by unlike charges.

In large part, cells are composed of the atoms carbon, hydrogen, nitrogen, oxygen, phosphorus, and sulfur. These atoms form covalent bonds with one another in which they share electrons in order to have a completed outer shell. Carbon, unlike the others, often bonds with itself, forming long chains of carbon atoms.

Water, which consists of polar molecules, can dissolve other polar substances. Other properties of water—cohesiveness, tendency to change temperature slowly, ability to expand as it freezes—are due to hydrogen bonding. Living things prefer a fairly neutral pH, which is the same pH as water.

Chapter 3 largely reviews the chemistry of carbon, which accounts for the formation of molecules that are unique to living things.

The Chapter in Review

Summary

Life Depends on Water

- Living systems are mostly composed of water, in which chemical reactions can easily occur.
- A watery environment supports and protects cells.
- Water assists homeostasis by helping the body maintain a constant temperature.
- All animals make use of water in reproduction.

All Matter Is Composed of Chemical Elements

2.1 Six elements are basic to life

- Both living and nonliving things are composed of matter.
- Matter takes up space; it can be a solid, a liquid, or a gas.
- All matter is composed of elements.
- Six elements—carbon, hydrogen, nitrogen, oxygen, phosphorus, and sulfur—play significant roles in all living organisms.
- An element has the same symbol as the atoms it contains.

2.2 Atoms contain subatomic particles

- The best-known subatomic particles are protons (positive charge), neutrons (uncharged), and electrons (negative charge).
- Electrons are located in shells that circle the nucleus.
- Atomic number is the number of protons in the nucleus of an atom.
- Atomic mass is the number of protons plus the number of neutrons in the nucleus.
- The isotopes of an element have the same number of protons, but they differ in atomic mass due to different numbers of neutrons.

2.3 Radioactive isotopes have many medical uses

- Low levels of radiation can be used as a tracer to detect molecular changes in the body.
- High levels of radiation can treat cancer by damaging rapidly dividing cancer cells.

Atoms React with One Another to Form Molecules

2.4 After atoms react, they have a completed outer shell

- The number of electrons in an atom's outer shell determines its reactivity with other atoms.
- The octet rule states that an atom's outer shell is most stable (least reactive) when it has eight electrons (or two electrons in the case of hydrogen).

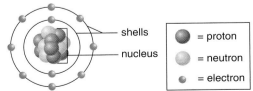

- shells
- nucleus

- = proton
- = neutron
- = electron

- A compound results when atoms of two or more different elements are bonded together.
- A molecule is the smallest part of a compound that has the properties of that compound.

2.5 An ionic bond occurs when electrons are transferred

- An ionic bond occurs when ionic compounds are held together by an attraction between negatively and positively charged ions.

2.6 A covalent bond occurs when electrons are shared

- A covalent bond occurs when two atoms share electrons in such a way that each atom has eight electrons in the outer shell (or two in the case of hydrogen).

2.7 A covalent bond can be nonpolar or polar

- A nonpolar covalent bond occurs when the sharing of electrons between two atoms is fairly equal (e.g., methane).
- A polar covalent bond occurs when the sharing of electrons is not equal, and so a molecule has a slightly negative charge and a slightly positive charge (e.g., water).

2.8 A hydrogen bond can occur between polar molecules

- A hydrogen bond is a weak attraction between a slightly positive hydrogen atom and a slightly negative atom of another molecule, or between atoms of the same molecule.
- Hydrogen bonds are individually weak and easily broken, but are collectively strong.

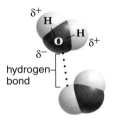

The Properties of Water Benefit Life

2.9 Water molecules stick together: Cohesion

- Hydrogen bonding is responsible for cohesion of water molecules.
- In living organisms, water serves as a transport medium for nutrients and wastes.
- Hydrogen bonding causes water to have a high surface tension.

2.10 Water warms up and cools down slowly

- Hydrogen bonding causes high heat capacity and high heat of vaporization in water.
- Water's high heat capacity protects living things from rapid changes in temperature.
- Water's high heat of vaporization helps organisms resist overheating.

2.11 Water dissolves other polar substances

- Because of polarity, water is a solvent; this property facilitates chemical reactions.
- Hydrophilic molecules (ionized and/or polar, such as salts) attract water.
- Hydrophobic molecules (nonionized and nonpolar, such as gasoline) do not attract water.

2.12 Frozen water is less dense than liquid water

- Frozen water expands and floats because hydrogen bonding becomes more rigid and open.
- Ice occurs at the top of ponds and lakes, protecting the water and organisms below it from freezing.

Living Things Require a Narrow pH Range

2.13 Acids and bases affect living things

- When water ionizes, it releases an equal number of hydrogen ions and hydroxide ions. The resulting concentration for each is 10^{-7} moles/liter.
- Acidic solutions contain excess hydrogen ions, H^+.
- Basic solutions contain excess hydroxide ions, OH^-.

2.14 The pH scale measures acidity and basicity

- Pure water has a neutral pH of 7.
- An acid has a pH below 7: $[H^+]$ is greater than $[OH^-]$.
- A base has a pH above 7: $[OH^-]$ is greater than $[H^+]$.
- Most organisms need to maintain pH within a narrow range (e.g., human blood pH is about 7.35–7.45).

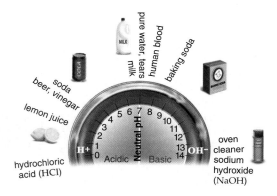

2.15 Buffers help keep the pH of body fluids relatively constant

- Illness results if the pH rises much above 7.45 (alkalosis) or dips much below 7.35 (acidosis).
- A buffer is a chemical or combination of chemicals that resists changes in pH and helps keep pH within normal limits.

2.16 Acid deposition has many harmful effects

- Acid deposition (i.e., acid rain) makes lakes acidic, killing fish and other aquatic organisms.
- In forests, acid rain damages leaves and makes plants unable to carry on photosynthesis; weakened plants become susceptible to pests and diseases.
- Acid rain leads to increased respiratory illnesses in humans and degradation of buildings, monuments, and painted structures.

Testing Yourself

All Matter Is Composed of Chemical Elements

1. CHNOPS are
   a. the only elements found in nonliving and living things.
   b. the only elements found in living things.
   c. elements basic to life.
   d. elements found in rocks.
   e. Both b and c are correct.
2. Which of the following is not a component of an atom?
   a. proton   c. neutron
   b. positron   d. electron

3. The atomic number tells you the
   a. number of neutrons in the nucleus.
   b. number of protons in the atom.
   c. atomic mass of the atom.
   d. number of its electrons if the atom has a neutral charge.
   e. Both b and d are correct.
4. Which of the subatomic particles contributes almost no weight to an atom?
   a. protons in the electron shells
   b. electrons in the nucleus
   c. neutrons in the nucleus
   d. electrons at various energy levels
5. Isotopes of the same element differ from each other only by the number of neutrons.
   a. True   b. False
6. Explain why iodine-131 and thallium-201 are considered radioactive isotopes, and describe how both can be used to detect health problems.

Atoms React with One Another to Form Molecules

7. This rule states that the outer electron shell is most stable when it contains eight electrons.
   a. stability rule   c. octet rule
   b. atomic rule   d. shell rule
8. How many electrons does nitrogen require to fill its outer shell?
   a. 0   c. 2
   b. 1   d. 3
9. When an atom gains electrons, it
   a. forms a negatively charged ion.
   b. forms a positively charged ion.
   c. forms covalent bonds.
   d. gains atomic mass.

nitrogen
$^{14}_{7}N$

10. An atom that has two electrons in the outer shell, such as calcium, would most likely
    a. share to acquire a completed outer shell.
    b. lose these two electrons and become a negatively charged ion.
    c. lose these two electrons and become a positively charged ion.
    d. bind with carbon by way of hydrogen bonds.
    e. bind with another calcium atom to satisfy its energy needs.
11. Molecules held together by _____ bonds tend to dissociate in biological systems due to the water content in those systems.
    a. covalent   c. hydrogen
    b. ionic   d. nitrogen
12. Which type of bond results from the sharing of electrons between atoms?
    a. covalent   c. hydrogen
    b. ionic   d. neutral
13. In the molecule CH_4,
    a. all atoms have eight electrons in the outer shell.
    b. all atoms are sharing electrons.
    c. carbon could accept more hydrogen atoms.
    d. All of these are correct.
14. In which of these are the electrons always shared unequally?
    a. double covalent bond   d. polar covalent bond
    b. triple covalent bond   e. ionic and covalent bonds
    c. hydrogen bond

15. An example of a hydrogen bond would be the
    a. bond between a carbon atom and a hydrogen atom.
    b. bond between two carbon atoms.
    c. bond between sodium and chlorine.
    d. bond between two water molecules.
16. Explain why the correct formula for ammonia is NH_3, not NH_4.

The Properties of Water Benefit Life

17. Water flows freely, but does not separate into individual molecules because water is
    a. cohesive.   c. hydrophobic.
    b. hydrophilic. d. adhesive.
18. Water can absorb a large amount of heat without much change in temperature, and therefore it has
    a. a high surface tension.
    b. a high heat capacity.
    c. ten times as many hydrogen ions.
    d. ten times as many hydroxide ions.
19. Which of these properties of water cannot be attributed to hydrogen bonding between water molecules?
    a. Water stabilizes temperature inside and outside the cell.
    b. Water molecules are cohesive.
    c. Water is a solvent for many molecules.
    d. Ice floats on liquid water.
    e. Both b and c are correct.
20. **THINKING CONCEPTUALLY** Explain why you would expect the blood of animals to be mostly water.

Living Things Require a Narrow pH Range

21. Acids
    a. release hydrogen ions in solution.
    b. cause the pH of a solution to rise above 7.
    c. take up hydroxide ions and become neutral.
    d. increase the number of water molecules.
    e. Both a and b are correct.
22. Which of these best describes the changes that occur when a solution goes from pH 5 to pH 8?
    a. The hydrogen ion concentration decreases as the solution goes from acidic to basic.
    b. The hydrogen ion concentration increases as the solution goes from basic to acidic.
    c. The hydrogen ion concentration decreases as the solution goes from basic to acidic.
23. When water dissociates, it releases
    a. equal amounts of H^+ and OH^-.
    b. more H^+ than OH^-.
    c. more OH^- than H^+.
    d. only H^+.
24. Rainwater has a pH of about 5.6; therefore, rainwater is
    a. a neutral solution.
    b. an acidic solution.
    c. a basic solution.
    d. It depends on whether the rainwater is buffered.
25. If a chemical accepted H^+ from the surrounding solution, the chemical could be
    a. a base.   d. None of these are correct.
    b. an acid.   e. Both a and c are correct.
    c. a buffer.

26. Compare a chemical buffer such as bicarbonate in the blood, the process of respiration, and kidney function with regard to how they regulate pH and how rapidly they respond to pH change.

Understanding the Terms

acid 30	hydrophobic 28
adhesion 27	hydroxide ion (OH^-) 30
atom 20	ion 24
atomic mass 21	ionic bond 24
atomic number 21	isotope 21
atomic symbol 20	matter 20
base 30	mineral 20
buffer 31	molecule 23
calorie 28	neutron 21
cohesion 27	nonpolar covalent bond 26
compound 23	octet rule 23
covalent bond 25	pH scale 31
electron 21	polar covalent bond 26
electronegativity 26	proton 21
electron shell 21	salt 24
element 20	solute 28
hydrogen bond 26	solution 28
hydrogen ion (H^+) 30	tracer 22
hydrophilic 28	valence shell 23

Match the terms to these definitions:
a. _____ Bond in which the sharing of electrons between atoms is unequal.
b. _____ Charged particle that carries a negative or positive charge(s).
c. _____ Molecules tending to raise the hydrogen ion concentration in a solution and to lower its pH numerically.
d. _____ The smallest part of a compound that still has the properties of that compound.
e. _____ A chemical or a combination of chemicals that maintains a constant pH upon the addition of small amounts of acid or base.

Thinking Scientifically

1. Natural phenomenon often require an explanation. Based on Figure 2.11 and Figure 2.12 (*top, left*), explain why the oceans don't freeze.
2. Melvin Calvin used radioactive carbon (as a tracer) to discover a series of molecules that form during photosynthesis. Explain why carbon behaves chemically the same, even when radioactive. (See Section 2.4.)

ARIS™ *Visit* **www.mhhe.com/maderconcepts** *for practice quizzes, animations, videos, and activities designed to help you master the material in this chapter.*

3

Organic Molecules and Cells

LEARNING OUTCOMES

After studying this chapter, you should be able to accomplish the following outcomes.

Plants and Animals Are the Same but Different

1 State at what levels of biological organization plants are the same as animals.

The Diversity of Organic Molecules Makes Life Diverse

2 List the features of carbon that result in the diversity of organic molecules.
3 Tell how macromolecules are assembled and disassembled.

Carbohydrates Are Energy Sources and Structural Components

4 Compare the structures of simple and complex carbohydrates.
5 Give the primary function of simple carbohydrates in organisms.

Lipids Provide Storage, Insulation, and Other Functions

6 Compare the structures of fats, phospholipids, and steroids.
7 State the primary function of fats and oils in cells.
8 Compare the functions of phospholipids and steroids in cells.

Proteins Have a Wide Variety of Vital Functions

9 Tell how amino acids are the same and how they can be different from one another.
10 List and discuss the four levels of a protein's structure.

Nucleic Acids Are Information Molecules

11 Tell how the functions of DNA and RNA are the same but different in cells.
12 Discuss the benefits of the Human Genome Project.
13 Relate the structure of ATP to its function in cells.

In what way is a plant like an animal? You're probably hard-pressed to think of ways plants and animals are alike, until you consider the lower rungs of biological organization, and then it becomes obvious that plants are indeed like animals. Vegetarians have no trouble sustaining themselves on plants, as long as they include a variety of plants in their diet. That's because plants and animals generally have the same molecules in their cells—namely, carbohydrates, lipids, proteins, and nucleic acids. When we feed on plants, we digest their macromolecules to smaller molecules, and then we use these smaller molecules to build our own types of carbohydrates, lipids, proteins, and nucleic acids.

"Same but different" will be a common theme in this chapter about the molecules of cells. For example, the genetic material for both plants and animals is the nucleic acid DNA. But each type of plant and animal has its own particular genes, even though the way genes function in cells is the same in all types of organisms.

Sameness is especially evident when animals acquire vitamins from plants and use them exactly as plants do in their own metabolism. You could go so far as to suggest that the inability of an animal to make vitamins is not disadvantageous, as long as it can get the vitamins it needs from plants. That way, an animal is not using up its own energy to make

a molecule it can get otherwise. Now it has more energy to use for growth, defense, and reproduction.

Vitamins assist enzymes, the molecules in cells that speed chemical reactions. Plants and animals have to build their own enzymes, but these enzymes function similarly. The enzymes needed to extract energy from nutrient molecules and form ATP, the energy currency of all cells, are the same in plants and animals. Plant cells have many more types of enzymes than do animals because they carry on photosynthesis to form their own food. Plant cells also produce molecules that allow them to protect themselves from predators, maintain an erect posture, and in general, be more colorful than most animals. The beautiful vegetables, fruits, and flowers shown here illustrate that plants can be very pleasing to the eye indeed.

In this chapter, we continue our look at basic chemistry by considering the types of molecules unique to living things. These are the molecules that account for the structure and function of all cells in any type of organism.

The Diversity of Organic Molecules Makes Life Diverse

Learning Outcomes 2–3, page 36

We begin our study of organic molecules by examining the chemistry of carbon because this atom makes the diverse organic molecules found in cells possible. Functional groups added to organic molecules increase their diversity and allow organic molecules to play particular roles in cells. Large organic molecules are modular, and the final size is dependent on how many subunits are joined end to end.

3.1 The chemistry of carbon makes diverse molecules possible

There are only four classes of molecules in any living thing: carbohydrates, lipids, proteins, and nucleic acids. Despite the limited number of classes, the so-called **organic molecules** in cells are quite diverse. A bacterial cell contains some 5,000 different organic molecules, and a plant or animal cell has twice that number. This diversity of organic molecules makes the diversity of life possible. Each of the organisms in **Figure 3.1** uses a carbohydrate as a structural molecule: A cactus uses cellulose to strengthen its cell walls, while a bacterium uses peptidoglycan for that purpose; a crab uses chitin to strengthen its shell.

Carbon is the essential ingredient in all organic molecules. Much as a salad chef first puts lettuce in a bowl, so organic molecules begin with carbon. A chef may then add other ingredients such as cucumbers or radishes to the lettuce to make different types of salads. So, the diversity of organic molecules comes about when different groups of atoms are added to carbon. Carbon is so versatile that an entire branch of chemistry, called **organic chemistry**, is devoted to it. Organic chemistry can be contrasted with inorganic chemistry as shown in **Table 3.1**.

TABLE 3.1 Organic Versus Inorganic Molecules

Organic Molecules	Inorganic Molecules
Always contain carbon bonded to other atoms	Usually contain positive and negative ions
Always covalent bonding	Usually ionic bonding
Often quite large, with many atoms	Always contain a small number of atoms
Usually associated with living organisms	Often associated with nonliving matter

Features of Carbon What is there about carbon that makes organic molecules the same and also different? Carbon is quite small, with a total of only six electrons: two electrons in the first shell and four electrons in the outer shell. To acquire four electrons to complete its outer shell, a carbon atom almost always shares electrons with—you guessed it—CHNOPS, the elements basic to living things (see Section 2.1).

Because carbon needs four electrons to complete its outer shell, it can share with as many as four other elements, and this spells diversity. But even more significant to the shape, and therefore the function, of organic molecules, is the fact that carbon often shares electrons with another carbon atom. The C—C bond is quite stable, and the result is carbon chains that can be quite long. Hydrocarbons are chains of carbon atoms bonded exclusively to hydrogen atoms:

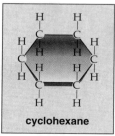

octane, a molecule in gasoline

Branching at any carbon atom is possible, and a hydrocarbon can also turn back on itself to form a ring compound when placed in water. One example is cyclohexane, used as an industrial solvent and in the manufacture of nylon.

Carbon can form double bonds with itself and other atoms. Double bonds restrict the movement of attached atoms, and in that way contribute to the shape of the molecule. Carbon is also capable of forming a triple bond with itself, as in acetylene, H—C≡C—H.

The diversity of organic molecules is further enhanced by the presence of particular functional groups, as discussed next.

cyclohexane

FIGURE 3.1 Each of these organisms uses a different type of structural carbohydrate.

Shell contains chitin.

Cell walls contain cellulose.

Cell walls contain peptidoglycan.

> **3.1 Check Your Progress** How do you know that plant cells contain organic molecules?

3.2 Functional groups add to the diversity of organic molecules

The carbon chain of an organic molecule is called its skeleton or backbone. The terminology is appropriate because just as a skeleton accounts for your shape, so does the carbon skeleton of an organic molecule account for its underlying shape. The reactivity of an organic molecule is largely dependent on the attached functional groups. A **functional group** is a specific combination of bonded atoms that always reacts in the same way, regardless of the particular carbon skeleton. As shown in **Figure 3.2A**, an *R* can be used to stand for the "rest" of the molecule because only the functional group is involved in reactions.

Notice that functional groups with a particular name and structure are found in certain types of compounds. For example, the addition of an —OH (hydroxyl group) to a carbon skeleton turns that molecule into an alcohol. When an —OH replaces one of the hydrogens in ethane, a 2-carbon hydrocarbon, it becomes ethanol, a type of alcohol that is consumable by humans. Whereas ethane, like other hydrocarbons, is **hydrophobic** (not soluble in water), ethanol is **hydrophilic** (soluble in water) because the —OH functional group is polar. Because cells are 70–90% water, the ability to interact with and be soluble in water profoundly affects the function of organic molecules in cells. Organic molecules containing carboxyl (acid) groups (—COOH) are polar, and when they ionize, they release hydrogen ions, making a solution more acidic:

$$-\text{COOH} \longrightarrow -\text{COO}^- + \text{H}^+$$

Functional groups determine the activity of a molecule in the body. You will see that alcohols react with carboxyl groups when a fat forms, and that carboxyl groups react with amino groups during protein formation. Notice in **Figure 3.2B** that the male sex hormone testosterone differs from the female sex hormone estrogen only by its attached groups. Yet, these molecules bring about the characteristics that determine whether an individual is male or female.

Isomers **Isomers** are organic molecules that have identical molecular formulas but a different arrangement of atoms. In essence, isomers are variations in the architecture of a molecule. Isomers are another example of how the chemistry of carbon leads to variations in organic molecules.

The two molecules below are isomers of one another; they have the same molecular formula but different functional groups. Therefore, we would expect them to react differently in chemical reactions.

glyceraldehyde	dihydroxyacetone
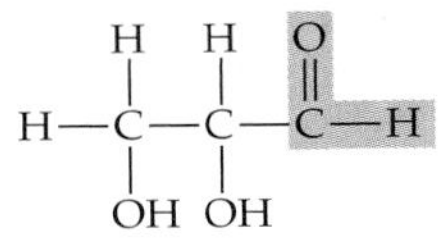 H H O | | || H—C—C—C—H | | OH OH	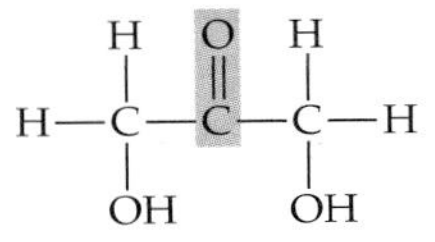 H O H | || | H—C—C—C—H | | OH OH

How small organic molecules join together to form much larger macromolecules found in cells is discussed in Section 3.3.

> **3.2 Check Your Progress** Oil in salad dressing separates from the watery vinegar portion. What do the oil molecules lack?

Functional Groups		
Group	**Structure**	**Compound**
Hydroxyl	$R-\text{OH}$	Alcohol Present in sugars, some amino acids
Carbonyl	$R-\overset{\displaystyle O}{\underset{\displaystyle H}{C}}$	Aldehyde Present in sugars
	$R-\overset{\displaystyle O}{C}-R$	Ketone Polar; present in sugar
Carboxyl (acidic)	$R-\overset{\displaystyle O}{\underset{\displaystyle OH}{C}}$	Carboxylic acid Present in fatty acids, amino acids
Amino	$R-\overset{\displaystyle H}{\underset{\displaystyle H}{N}}$	Amine Present in amino acids
Sulfhydryl	$R-\text{SH}$	Thiol Forms disulfide bonds Present in some amino acids
Phosphate	$R-O-\overset{\displaystyle O}{\underset{\displaystyle OH}{P}}-OH$	Organic phosphate Present in nucleotides, phospholipids

R = rest of molecule

FIGURE 3.2A Functional groups of organic molecules.

FIGURE 3.2B Functional groups in male and female sex hormones are highlighted.

3.3 Molecular subunits can be linked to form macromolecules

Carbohydrates, lipids, proteins, and nucleic acids are called macromolecules because of their large size. You are very familiar with these molecules because certain foods are known to be rich in them, as illustrated in **Figure 3.3A**. Even a late-night pizza, a quick hamburger, or an afternoon snack contains many of these essential molecules. When you digest these foods, they are broken down into the subunit molecules listed in **Table 3.3**. Your body then uses these subunits to build the macromolecules that make up your cells.

The largest of the macromolecules are called **polymers** because they are constructed by linking together a large number of the same type of subunits, called **monomers**. A protein can contain hundreds of amino acids, and a nucleic acid can contain hundreds of nucleotides. How can polymers get so large? Cells use the modular approach when constructing polymers. Just as a train increases in length when boxcars are hitched together one by one, so a polymer gets longer as monomers bond to one another.

In the top part of **Figure 3.3B,** notice how synthesis (the construction) of a macromolecule occurs. A cell uses a **dehydration reaction** to synthesize any type of macromolecule. In this reaction, the equivalent of a water molecule consisting of an —OH (hydroxyl group) and an —H (hydrogen atom) is removed as the reaction occurs. After water is removed, a bond now exists between the two monomers.

In the lower part of Figure 3.3B, notice how degradation (breaking down) of a macromolecule occurs. To degrade a macromolecule, our digestive tract, or any cell, uses an opposite type of reaction. During a **hydrolysis reaction**, an —OH group from water attaches to one subunit, and an —H from water attaches to the other subunit. (*Hydro* means "water," and *lysis* means "breaking apart.") In other words,

TABLE 3.3	Macromolecules	
Category	**Example**	**Subunit(s)**
Carbohydrates*	Polysaccharide	Monosaccharide
Lipids	Fat	Glycerol and fatty acids
Proteins*	Polypeptide	Amino Acid
Nucleic acids*	DNA, RNA	Nucleotide

*Polymers

water is used to break the bond holding subunits together. Biologists frequently refer to hydrolysis, sometimes called a hydrolytic reaction, so it is a term that you will want to be familiar with.

In order for these reactions, or almost any other type of reaction, to occur in a cell, an enzyme must be present. An **enzyme** is a molecule that speeds a reaction by bringing reactants together. The enzyme may even participate in the reaction, but it is unchanged by it. Frequently, monomers must be energized before they will bind together because synthesis of a macromolecule requires energy. On the other hand, hydrolysis of a macromolecule can release energy.

This completes our look at the chemistry of carbon. Now, in the next part of this chapter, we will study the structure and function of carbohydrates.

> **3.3 Check Your Progress** Cells, plant cells included, are always carrying on both dehydration and hydrolysis reactions. Why doesn't a cell become waterlogged from dehydration reactions?

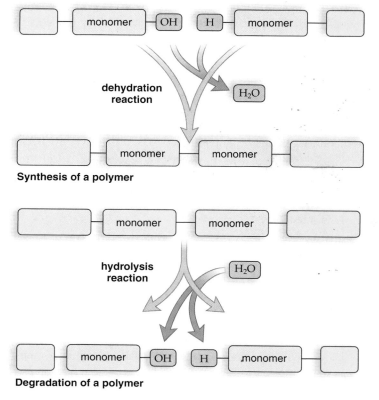

FIGURE 3.3B Synthesis and degradation of polymers.

FIGURE 3.3A All foods contain carbohydrates, lipids, proteins, and nucleic acids.

Carbohydrates Are Energy Sources and Structural Components

Learning Outcomes 4–5, page 36

The term *carbohydrate* refers to simple carbohydrates with one to three monomers, as well as to complex carbohydrates having many monomers. We will examine the structure and function of simple carbohydrates and then complex carbohydrates.

3.4 Simple carbohydrates provide quick energy

Carbohydrates are almost universally used as an immediate energy source in living things, but they also play structural roles in a variety of organisms (see Fig. 3.1). The majority of carbohydrates have a carbon to hydrogen to oxygen ratio of 1:2:1. Chain length varies from a few monosaccharides (sugars) to hundreds of sugars in complex carbohydrates. The long chains are therefore polymers.

Monosaccharides consist of only a single sugar molecule. A sugar can have a carbon backbone of three to seven carbons. It may also have many hydroxyl groups, and this polar functional group makes it soluble in water. **Glucose**, a common sugar with six carbon atoms, has a molecular formula of $C_6H_{12}O_6$. Notice in **Figure 3.4A** that there are several ways to represent glucose. When the carbon atoms are included, the molecule looks crowded, so a common practice is to omit the carbon atoms, or even to simply show the hexagon shape. You are supposed to imagine the molecule as flat, with the darkened region facing you. Certain atoms bonded to carbon are above the ring, and others are below it, as indicated.

Despite the fact that glucose has several isomers, such as fructose and galactose, we usually think of $C_6H_{12}O_6$ as glucose. This sugar is the major source of cellular fuel for all living things. Glucose is transported in the blood of animals, and it is the molecule that is broken down in nearly all types of cells to release its energy.

Ribose and **deoxyribose** are monosaccharides with five carbon atoms. They are of significance because they are found respectively in the nucleic acids RNA and DNA. RNA and DNA are discussed later in this chapter.

A **disaccharide** contains two monosaccharides that have joined during a dehydration reaction. **Figure 3.4B** shows how the disaccharide maltose (an ingredient used in brewing beer) arises when two glucose molecules bond together. When our hydrolytic digestive juices break this bond, the result is two glucose molecules.

Sucrose, or table sugar, is another disaccharide of special interest because it is the form in which sugar is transported in plants.

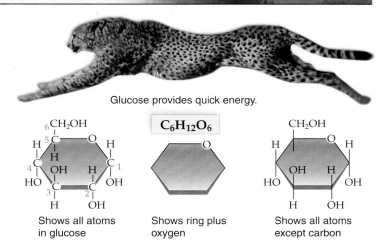

Glucose provides quick energy.

$C_6H_{12}O_6$

Shows all atoms in glucose — Shows ring plus oxygen — Shows all atoms except carbon

FIGURE 3.4A Three ways to represent glucose, a source of quick energy for this cheetah and all organisms.

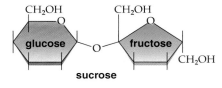

glucose   fructose CH_2OH

sucrose

Plants transport sucrose from cells carrying on photosynthesis to other parts of their bodies. Sucrose is also the sugar we use to sweeten our food. We acquire the sugar from plants such as sugarcane and sugar beets. You may also have heard of lactose, a disaccharide found in milk. Lactose is glucose combined with galactose. Individuals who are lactose intolerant cannot break this disaccharide down and therefore experience unpleasant digestive tract symptoms.

The sugar glucose is the subunit for complex carbohydrates, which are studied in Section 3.5.

> **3.4 Check Your Progress** Plants transport sucrose, but prefer glucose for metabolism. Why would you expect a plant to be able to convert the fructose in sucrose to glucose?

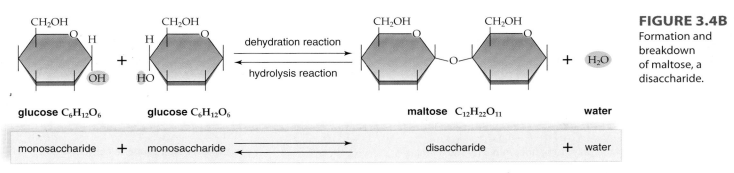

glucose $C_6H_{12}O_6$ +   **glucose** $C_6H_{12}O_6$ dehydration reaction ⇌ hydrolysis reaction   **maltose** $C_{12}H_{22}O_{11}$ + H_2O **water**

monosaccharide   +   monosaccharide   ⇌   disaccharide   +   water

FIGURE 3.4B Formation and breakdown of maltose, a disaccharide.

3.5 Complex carbohydrates store energy and provide structural support

Polysaccharides are polymers of monosaccharides. Some types of polysaccharides, such as glycogen in animals and starch in plants, function as short-term energy-storage molecules. They serve as storage molecules because they are not as soluble in water and are much larger than a simple sugar. Their large size prevents them from passing through the plasma membrane that forms a cell's boundary.

When an organism requires energy, polysaccharides are degraded to release sugar molecules. Their shape exposes the sugar linkages to the hydrolytic enzymes that can break them down.

Animals store glucose as **glycogen**. Notice in **Figure 3.5** that glycogen is highly branched. When a polysaccharide is branched, there is no main carbon chain because new chains occur at regular intervals. In our bodies and those of other vertebrates, liver cells contain granules where glycogen is stored until needed. The storage and release of glucose from liver cells is under the control of hormones. After we eat, the release of the hormone insulin from the pancreas promotes the storage of glucose as glycogen.

Plants store glucose as **starch**. Starch exists in two forms—one is nonbranched and the other is branched. The branched form is shown in Figure 3.5. Both the nonbranched and branched forms of starch serve as glucose reservoirs in plants. The cells of a potato contain granules in which starch resides during winter until energy is needed for growth in the spring. Plants, as well as animals, can hydrolyze starch and therefore can tap into these reservoirs for energy.

Some types of polysaccharides are structural polysaccharides, such as **cellulose** in plants, **chitin** in animals and fungi, and **peptidoglycan** in bacteria (see Fig. 3.1). The cellulose monomer is simply glucose, but in chitin, the monomer has an attached amino group. The structure of peptidoglycan is more complex because each monomer also has an amino acid chain.

Cellulose is the most abundant carbohydrate and, indeed, the most abundant organic molecule on Earth—plants produce over 100 billion tons of cellulose each year. Wood and cotton are cellulose plant products. Wood is used for construction, and cotton is used for cloth.

Microorganisms are able to digest the bond between glucose monomers in cellulose. The protozoans in the gut of termites allow them to digest wood. In cows and other ruminants, microorganisms break down cellulose in a special stomach pouch before the "cud" is returned to the mouth for more chewing and reswallowing. However, other animals have no means of hydrolyzing the bonds in cellulose; for them, cellulose serves as dietary fiber, which maintains regularity of elimination.

In the next part of this chapter, we will study the structure and function of lipids.

> **3.5 Check Your Progress** "Same but different" is illustrated by comparing the storage forms of glucose in plants and animals. What's the same and what's different?

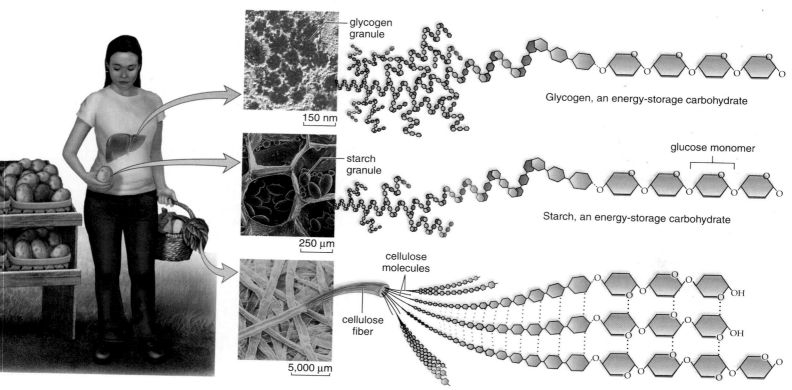

glycogen granule

150 nm

starch granule

250 μm

cellulose molecules

cellulose fiber

5,000 μm

Glycogen, an energy-storage carbohydrate

glucose monomer

Starch, an energy-storage carbohydrate

Cellulose, a structural carbohydrate

FIGURE 3.5 Some of the polysaccharides in plants and animals.

Lipids Provide Storage, Insulation, and Other Functions

Learning Outcomes 6–8, page 36

Most lipids are insoluble in water. Even so, their structure varies widely. Fats and oils, which function as long-term energy-storage molecules, are the best-known lipids. Phospholipids are constructed like fats but have a polar group that makes them soluble in water. Steroids and waxes have a structure and function that is different from fats and each other.

3.6 Fats and oils are rich energy-storage molecules

A variety of organic compounds are classified as **lipids**. Most lipids are insoluble in water due to their nonpolar, hydrocarbon chains that do not interact with water molecules. Fat, a well-known lipid, is used for both insulation and long-term energy storage by animals. Fat below the skin of marine mammals is called blubber; in humans, it is given slang expressions such as "spare tire" and "love handles." Plants use oil, instead of fat, for long-term energy storage. The difference in structure between fats and oils will be explored shortly.

Fats and **oils** contain two types of subunit molecules: glycerol and fatty acids. **Glycerol** is a compound with three —OH groups. A **fatty acid** consists of a long hydrocarbon chain with a —COOH (acid) group at one end. When a fat or oil forms, the acid portions of three fatty acids react with the —OH groups of glycerol during a dehydration reaction (**Fig. 3.6**). Because there are three fatty acids attached to each glycerol molecule, fats and oils are sometimes called **triglycerides**. Notice that the exposed polar groups in glycerol and fatty acids no longer appear in a triglyceride. Instead, the resulting molecule has exposed C—H groupings that do not mix with water. Despite the liquid nature of both cooking oils and water, cooking oils separate out of water, even after shaking. In Figure 3.6, note that synthesis also results in three molecules of water and that hydrolysis will degrade a fat to its original components.

Most of the fatty acids in cells contain 16 or 18 carbon atoms per molecule, although smaller ones are also found. Fatty acids are either saturated or unsaturated. **Saturated fatty acids** have no double bonds between the carbon atoms. The carbon chain is saturated, so to speak, with all the hydrogens that can be held. **Unsaturated fatty acids** have double bonds (see yellow highlight on Figure 3.6) in the carbon chain wherever the number of hydrogens is less than two per carbon atom.

In general, fats, which are solid at room temperature, are of animal origin and contain saturated fatty acids. Oils, which are liquids even in the refrigerator, are of plant origin and contain unsaturated fatty acids. A double bond creates a kink that prevents close packing of hydrocarbon chains and keeps oils soluble in the cold. Of interest, the feet of reindeer and penguins contain unsaturated oils, and this helps protect these exposed parts from freezing. Of interest also is the suggestion by nutritionists that replacing saturated fat in the diet with unsaturated oils, such as canola oil and olive oil, can benefit our cardiovascular health. In cardiovascular disease, deposits called plaque build up on arterial walls. Saturated fats participate in plaque formation.

Nearly all animals use fat for long-term energy storage. Gram per gram, fat stores more energy than glycogen. The C—H bonds of fatty acids make them a richer source of chemical energy than glycogen, because glycogen has many C—OH bonds. Also, fat droplets, being nonpolar, do not contain water. Small birds, such as the broad-tailed hummingbird, store a great deal of fat (about 0.15 g of fat per gram of body weight per day) before they start their long spring and fall migratory flights. If the same amount of energy were stored as glycogen, a bird would be too heavy to fly.

The next section considers the structure and function of phospholipids, steroids, and waxes.

> **3.6 Check Your Progress** Plants primarily store oils in seeds for use by the next generation of plants. Why don't adult plants store long-term energy, as animals do?

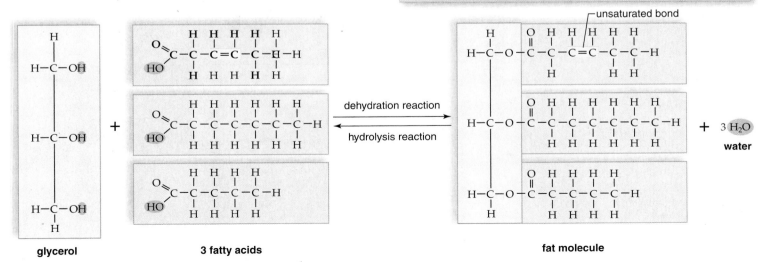

glycerol   3 fatty acids   fat molecule

FIGURE 3.6 Formation and breakdown of a fat.

3.7 Other lipids have structural, hormonal, or protective functions

The phospholipids, steroids, and waxes are also important lipids found in living things (**Fig. 3.7**). Like fats, **phospholipids** contain glycerol and three groups bonded to glycerol. In phospholipids, only two of these groups are fatty acids. After bonding to glycerol, the fatty acids form the hydrophobic tails of the molecule. The third group contains a polar phosphate group that becomes the polar head of a phospholipid. In a watery environment, phospholipids naturally form a bilayer in which the hydrophilic heads project outward and the hydrophobic tails project inward. The cell's plasma membrane consists of a phospholipid bilayer, as will be discussed in more detail in Chapter 4. A plasma membrane is absolutely essential to the structure and function of a cell.

Steroids are lipids that have an entirely different structure from that of fat. A steroid molecule has a skeleton of four fused carbon rings. Cholesterol is the steroid that stabilizes an animal's plasma membrane. It is also the precursor of several other steroids, such as the sex hormones testosterone and estrogen (see Fig. 3.2B). Among its many effects, testosterone is responsible for the generally greater muscle development of human males. For this reason, athletes of both sexes sometimes take anabolic steroids—testosterone or steroids that resemble testosterone—in an attempt to improve their athletic performance. This use of steroids is now banned by most athletic organizations. Anabolic steroid abuse can create serious health problems involving the kidneys and cardiovascular system, as

well as changes in sexual characteristics. Females take on male characteristics, and males become feminized.

Like saturated fats, cholesterol also participates in the formation of plaque along cardiovascular walls. Plaque can restrict blood flow and result in heart attacks or strokes.

In **waxes**, long-chain fatty acids bond with long-chain alcohols. Waxes are solid at normal temperatures. Being hydrophobic, they are also waterproof and resistant to degradation. In many plants, waxes, along with other molecules, form a protective waxy coating that retards the loss of water from all exposed parts. In many animals, waxes are involved in skin and fur maintenance. In humans, wax is produced by glands in the outer ear canal. Earwax contains cerumin, an organic compound that at the very least repels insects, and in some cases even kills them. It also traps dust and dirt, preventing them from reaching the eardrum.

Honeybees produce beeswax in glands on the underside of their abdomen. They then use this beeswax to make the six-sided cells of the comb where their honey is stored. Honey contains the sugars fructose and glucose, breakdown products of sucrose.

In the next part of this chapter, we consider the structure and function of proteins.

> **3.7 Check Your Progress** Why would you expect to find a waxy cuticle on exposed plant leaf cells, but not on internal leaf cells?

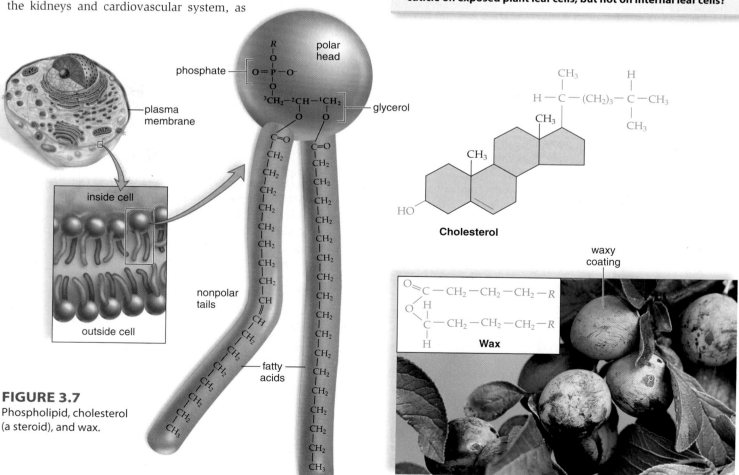

FIGURE 3.7
Phospholipid, cholesterol (a steroid), and wax.

Proteins are very significant macromolecules with amino acid subunits. The intimate connection between genes and proteins (discussed in the next part of the chapter) gives added importance to proteins. Each protein has a set sequence of amino acids that causes it to have a shape suitable to its specific role in cells.

3.8 Proteins are the most versatile of life's molecules

Proteins are of primary importance to the structure and function of cells. As much as 50% of the dry weight of most cells consists of proteins. Presently, over 100,000 proteins have been identified. Here are some of their many functions in animals:

Support Some proteins are structural proteins. Examples include the protein in spider webs; keratin, the protein that makes up hair and fingernails; and collagen, the protein that lends support to skin, ligaments, and tendons.

Metabolism Some proteins are enzymes. They bring reactants together and thereby speed chemical reactions in cells. They are specific for one particular type of reaction and can function at body temperature.

Transport Channel and carrier proteins in the plasma membrane allow substances to enter and exit cells. Other proteins transport molecules in the blood of animals—for example, **hemoglobin** is a complex protein that transports oxygen.

Defense Proteins called antibodies combine with disease-causing agents to prevent them from destroying cells and upsetting homeostasis, the relative constancy of the internal environment.

Regulation Hormones are regulatory proteins. They serve as intercellular messengers that influence the metabolism of cells. For example, the hormone insulin regulates the content of glucose in the blood and in cells, while growth hormone determines the height of an individual.

Motion The contractile proteins actin and myosin allow parts of cells to move and cause muscles to contract. Muscle contraction enables animals to move from place to place.

Proteins are such a major part of living organisms that tissues and cells of the body can sometimes be characterized by the proteins they contain or produce. For example, muscle cells contain large amounts of actin and myosin for contraction; red blood cells are filled with hemoglobin for oxygen transport; and support tissues, such as ligaments and tendons, contain the protein collagen, which is composed of tough fibers.

Despite their variety, all proteins are composed of amino acids, as discussed in Section 3.9.

> **3.8 Check Your Progress** In general, why would you expect a plant cell to have more varied enzymes than an animal cell?

3.9 Each protein is a sequence of particular amino acids

Proteins are macromolecules with **amino acid** monomers. **Figure 3.9A** shows how two amino acids join by a dehydration reaction between the carboxyl group of one and the amino group of another. The resulting covalent bond between two amino acids is called a **peptide bond**. The atoms associated with the peptide bond share the electrons unevenly because oxygen attracts electrons more than nitrogen. Therefore, the hydrogen attached to the nitrogen has a slightly positive charge (δ^+), while the oxygen has a slightly negative charge (δ^-).

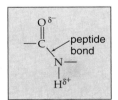

The polarity of the peptide bond means that hydrogen bonding is possible between the —CO of one amino acid and the —NH of another amino acid in a polypeptide.

A **peptide** is two or more amino acids bonded together, and a **polypeptide** is a chain of many amino acids joined by peptide bonds. A protein may contain more than one polypeptide chain; therefore, a single protein may have a very large number of amino acids. In 1953, Frederick Sanger developed a method to determine the amino acid sequence of a polypeptide. We now know the sequences of thousands of polypeptides, and it is clear that each type of polypeptide has its own particular sequence.

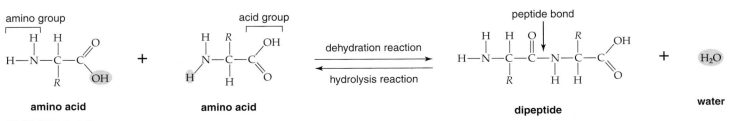

FIGURE 3.9A Formation and breakdown of a peptide.

Amino Acids The central carbon atom in an amino acid bonds to a hydrogen atom and also to three other groups of atoms. The name amino acid is appropriate because one of these groups is an —NH$_2$ (amino group), and another is a —COOH (an acid group). The third group is the R group for an amino acid:

amino   acid
group   H   group

$$H_2N—C—COOH$$

R

R = rest of molecule

Amino acids differ from one another according to their particular R group, shaded blue in **Figure 3.9B**. The R groups range in complexity from a single hydrogen atom to a complicated ring compound. Some R groups are polar and some are not. Also, the amino acid cysteine has an R group that ends with a —SH group, which often serves to connect one chain of amino acids to another by a **disulfide bond**, —S—S—. Several other amino acids commonly found in cells are shown in Figure 3.9B. Each protein has a definite sequence of amino acids, and this leads to the levels of organization described in Section 3.10.

3.9 Check Your Progress Why would you expect to find the same common amino acids in both plant and animal cells?

FIGURE 3.9B Amino acid diversity. The amino acids are shown in ionized form.

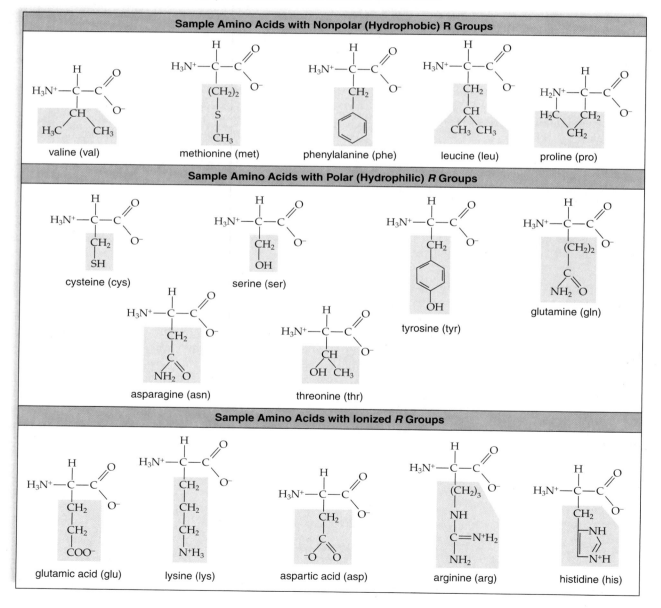

3.10 The shape of a protein is necessary to its function

When proteins are exposed to extremes in heat and pH, they undergo an irreversible change in shape called **denaturation**. For example, we are all aware that the addition of acid to milk causes curdling and that heating causes egg white, which contains a protein called albumin, to coagulate. Denaturation occurs because the normal bonding between the R groups has been disturbed. Once a protein loses its normal shape, it is no longer able to perform its usual function. Researchers hypothesize that an alteration in protein organization has occurred when Alzheimer disease and Creutzfeldt-Jakob disease (the human form of mad cow disease) develop.

Levels of Protein Organization The structure of a protein has at least three levels of organization and can have four levels (**Fig. 3.10**). The first level, called the primary structure, is the linear sequence of the amino acids joined by peptide bonds. Each particular polypeptide has its own sequence of amino acids. Just as an alphabet of 26 letters can form the sequence of many different words, so too can 20 amino acids form the sequence of many different proteins.

The secondary structure of a polypeptide comes about when it takes on a certain orientation in space. As mentioned, the peptide bond is polar, and hydrogen bonding is possible between the —CO of one amino acid and the —NH of another amino acid in a polypeptide. Due to hydrogen bonding, two possible shapes can occur: a right-handed spiral, called an alpha helix, and a folding of

the chain, called a pleated sheet. Fibrous proteins, such as those in hair and nails, exist as helices or pleated sheets.

Globular proteins have a tertiary structure as their final three-dimensional shape. In muscles, myosin molecules have a rod shape ending in globular (globe-shaped) heads. In enzymes, the polypeptide bends and twists in different ways. Invariably, the hydrophobic portions are packed mostly on the inside, and the hydrophilic portions are on the outside, where they can make contact with fluids. The tertiary shape of a polypeptide is maintained by various types of bonding between the R groups; covalent, ionic, and hydrogen bonding all occur.

Some proteins have only one polypeptide, and others have more than one polypeptide, each with its own primary, secondary, and tertiary structures. These separate polypeptides are arranged to give some proteins a fourth level of organization, termed the quaternary structure. Hemoglobin is a complex protein having a quaternary structure; most enzymes also have a quaternary structure.

In the next part of this chapter, we learn that the nucleic acid DNA makes up our genes and that each gene specifies the sequence of amino acids in a particular protein.

> **3.10 Check Your Progress** Why would you expect to find globular proteins, but not fibrous proteins, in plant cells? Explain.

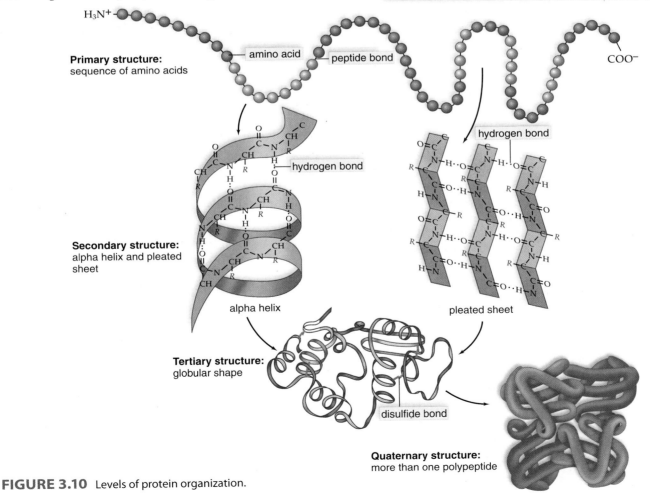

FIGURE 3.10 Levels of protein organization.

In this part of the chapter, we study the chemistry of DNA, RNA, and ATP. DNA and RNA are nucleic acids composed of many nucleotides, but ATP contains a single modified nucleotide. The functions of these molecules differ: DNA makes up genes, and RNA carries out the instructions of DNA during protein synthesis. ATP is the molecule that serves as a carrier of energy in cells.

3.11 The nucleic acids DNA and RNA carry coded information

DNA (deoxyribonucleic acid) and **RNA (ribonucleic acid)** are the **nucleic acids** in cells. Early investigators called them nucleic acids because they were first detected in the nucleus of cells. DNA is the genetic material, and each DNA molecule contains many genes. Genes code for the order in which amino acids are to be joined to form a protein. RNA is an intermediary, conveying coded information from DNA to direct protein synthesis.

A nucleic acid is a polymer of nucleotides. A **nucleotide** is a molecular complex of three types of molecules: a phosphate (phosphoric acid), a pentose (5-carbon) sugar, and a nitrogen-containing base (**Fig. 3.11A**).

Structure of RNA Figure 3.11B shows how the nucleotides are arranged in RNA, a single-stranded molecule. Notice how the sugar and phosphate molecules form the backbone of the molecule, while the bases project to the side. The 5-carbon sugar molecule is ribose, and this accounts for its name—ribonucleic acid. The nucleotides present in RNA and in DNA differ by their bases. The bases in RNA are guanine (G), adenine (A), cytosine (C), and uracil (U). These molecules are called bases because their presence raises the pH of a solution.

Structure of DNA The nucleotides in DNA contain the sugar deoxyribose, accounting for its name—deoxyribonucleic acid. DNA is double-stranded, as shown in **Figure 3.11C**. The ladder structure of DNA is so called because it shows that the sugar and phosphate molecules make up the sides of the ladder, and complementary paired bases make up the rungs of the ladder. The bases are held together by hydrogen bonds represented by dotted lines. The bases can be in any order, but between strands, thymine (T) is always paired with adenine (A), and guanine (G) is always paired with cytosine (C). **Complementary base pairing** is very important when DNA makes a copy of itself, a process called replication. Complementary base pairing also allows DNA to pass genetic information to RNA when RNA is formed using one DNA strand as a type of template. (When RNA forms, the base uracil, instead of thymine, pairs with adenine.) (**Table 3.11**) After RNA has the correct sequence of bases, it moves to where the protein is made in the cell. In the end, the sequence of bases in DNA determines the sequence of amino acids in a protein.

TABLE 3.11	DNA Structure Compared to RNA Structure	
	DNA	**RNA**
Sugar	Deoxyribose	Ribose
Bases	Adenine, guanine, thymine, cytosine	Adenine, guanine, uracil, cytosine
Strands	Double-stranded with base pairing	Single-stranded
Helix	Yes	No

The complete sequence of base pairs in human DNA has been determined. This is expected to result in new and novel treatments of human diseases, as discussed in Section 3.12.

> **3.11 Check Your Progress** If a base sequence in part of one strand of DNA is GATCCA, what is the complementary sequence of bases in the other strand?

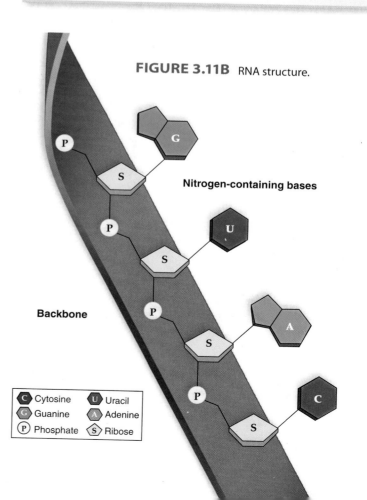

FIGURE 3.11B RNA structure.

Nitrogen-containing bases

Backbone

C Cytosine	U Uracil
G Guanine	A Adenine
P Phosphate	S Ribose

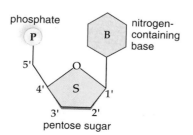

phosphate

nitrogen-containing base

pentose sugar

FIGURE 3.11A
One nucleotide.

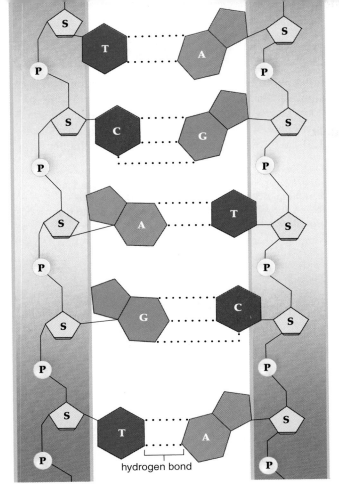

DNA: ladder configuration

DNA: double helix

DNA: space-filling model

FIGURE 3.11C DNA structure at three levels of complexity.

HOW BIOLOGY IMPACTS OUR LIVES

3.12 The Human Genome Project may lead to new disease treatments

The **Human Genome Project (HGP)** was a technological break-through with fortuitous consequences of the first magnitude. This ambitious project, which was sponsored by the United States Department of Energy and the National Institutes of Health, succeeded in sequencing the genome of our species—in other words, in determining the precise order of all three billion of the base pairs in human DNA. The project determined that the human genome includes between 20,000 and 25,000 genes, or DNA segments that provide instructions for building proteins. The next step is to locate these genes and determine what protein each one specifies.

Researchers hope that the genetic profile—a person's particular base sequence variations—can be used to predict potential illnesses before they occur. Our genes are a major factor in whether or not we will develop practically any illness. For example, sickle-cell disease arises from a change in the gene that specifies part of the hemoglobin molecule. Genes, plus the environment, cause many diseases. Diabetes type 2, for instance, often runs in families, but may be influenced by diet and exercise habits. Therefore, a person with a propensity for diabetes

type 2 can take the necessary steps to prevent its occurrence. Similarly, other conditions may be preventable. Even mental illnesses, such as depression and schizophrenia, may have a basis in a person's genetic makeup.

Improved treatment is expected to be another consequence of the HGP. We now realize that response to treatment varies between individuals because genetic profiles differ. Knowing a patient's genetic profile should allow physicians to select medications to treat illness, without causing unpleasant side effects and potentially dangerous adverse reactions.

The consequences of the HGP have gone beyond the human species. The genomes of other organisms commonly used in medical research have also been sequenced. These include the bacterium *Escherichia coli* and the lab mouse *Mus musculus*. We switch gears in Section 3.13 to consider the structure of ATP, the nucleotide that carries metabolic energy in cells.

▶ **3.12 Check Your Progress** What types of proteins do genes specify?

3.13 The nucleotide ATP is the cell's energy carrier

Adenosine triphosphate (ATP) is composed of the base adenine and the sugar ribose—a compound termed **adenosine**—plus three linked phosphate groups (**Fig. 3.13A**). ATP is a high-energy molecule because the last two phosphate bonds are unstable and easily broken. In cells, the last phosphate bond is usually hydrolyzed to form **adenosine diphosphate (ADP)** and a phosphate molecule Ⓟ.

The breakdown of ATP releases energy because the products of hydrolysis, ADP and Ⓟ, are more stable than the original reactant ATP. Cells couple the energy released by ATP breakdown to energy-requiring processes such as the synthesis of macromolecules. In muscle cells, the energy is used for muscle contraction, and in nerve cells, it is used for the conduction of nerve impulses. Just as you spend money when you pay for a product or a service, cells "spend" ATP when they need something done. Therefore, ATP is called the energy currency of cells (**Fig. 3.13B**). We will discuss more about ATP in Chapter 7.

> **3.13** *Check Your Progress* Glucose breakdown in cells leads to ATP buildup. ATP breakdown in muscle cells leads to movement. Show that your ability to move begins with the ability of plant cells to absorb solar energy.

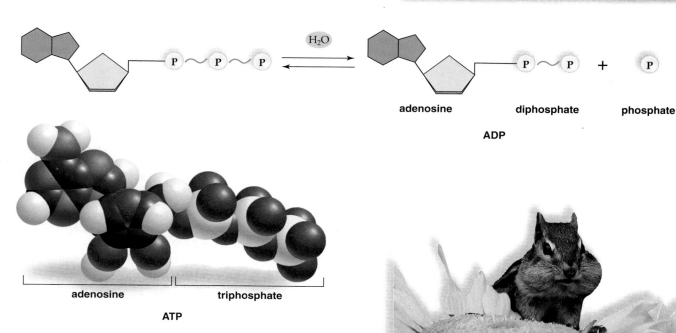

FIGURE 3.13A ATP hydrolysis releases energy.

FIGURE 3.13B Animals convert food energy to that of ATP.

CONNECTING THE CONCEPTS

What does the term *organic* mean? For some, organic means that food products have been grown without the use of chemicals or have been minimally processed. Biochemically speaking, organic refers to molecules containing carbon bonded to other atoms. The chemistry of carbon accounts for the molecules we associate with living things: carbohydrates, lipids, proteins, and nucleic acids.

Many of the macromolecules in cells are simply polymers of small organic molecules. Simple sugars are the monomers of complex carbohydrates; amino acids are the monomers of proteins; and nucleotides are the monomers of nucleic acids.

Diversity is still possible. Monomers exist in modified forms or can combine in slightly different ways; therefore, a variety of macromolecules can come about. Glucose monomers are linked differently in cellulose than in glycogen. One protein differs from another by the number and/or sequence of the same 20 amino acids.

The organic molecules discussed in this chapter contain the atoms CHNOPS sharing electrons. These molecules, as well as other small inorganic molecules and ions, are assembled into the components that make up cells. As discussed in Chapter 4, each component has a specific function necessary to the life of a cell.

The Chapter in Review

Summary

Plants and Animals Are the Same but Different

- Plants and animals have the same molecules in their cells: carbohydrates, lipids, proteins, and nucleic acids.

Category	Example	Subunit(s)
Carbohydrates	Polysaccharide	Monosaccharide
Lipids	Fat	Glycerol and fatty acids
Proteins	Polypeptide	Amino Acid
Nucleic acids	DNA, RNA	Nucleotide

The Diversity of Organic Molecules Makes Life Diverse

3.1 The chemistry of carbon makes diverse molecules possible

- Carbon needs four electrons to complete its outer shell; it can share with as many as four other elements.
- The carbon–carbon bond is very stable, and so carbon chains can be very long.
- Organic molecules are usually associated with living organisms.

3.2 Functional groups add to the diversity of organic molecules

- Functional groups are a specific combination of bonded atoms that always reacts in the same way.
- Isomers have identical molecular formulas but a different arrangement of atoms (or functional groups).
- Isomers react differently from one another in chemical reactions.

3.3 Molecular subunits can be linked to form macromolecules

- Carbohydrates, lipids, proteins, and nucleic acids are macromolecules.
  - The carbohydrate monomers are monosaccharides.
  - Lipid subunits are glycerol and fatty acids.
  - The protein monomers are amino acids.
  - The nucleic acid monomers are nucleotides.

Carbohydrates Are Energy Sources and Structural Components

3.4 Simple carbohydrates provide quick energy

- Monosaccharides (each composed of a single sugar molecule) are simple sugars.
- Glucose is a simple sugar and a major source of energy.
- Ribose and deoxyribose are 5-carbon sugars found in RNA and DNA, respectively.
- Disaccharides are formed from two monosaccharides joined during dehydration.
- Sucrose (table sugar) is a disaccharide.

3.5 Complex carbohydrates store energy and provide structural support

- Polysaccharides (polymers of monosaccharides that can be broken down to sugar molecules for energy) are:
  - Starch (stored glucose in plants)
  - Glycogen (stored glucose in animals)
- Polysaccharides used for structural support are:
  - Cellulose (in plants)
  - Chitin (in animals and fungi)
  - Peptidoglycan (in bacteria)

Lipids Provide Storage, Insulation, and Other Functions

3.6 Fats and oils are rich energy-storage molecules

- Fats and oils (triglycerides) contain three fatty acids attached to a glycerol molecule.
- Saturated fatty acids (no double bonds) are characteristic of solid fats found in animals.
- Unsaturated fatty acids (double bonds) are characteristic of liquid oils found primarily in plant seeds.

3.7 Other lipids have structural, hormonal, or protective functions

- Phospholipids are a plasma membrane component.
- Steroids serve as a plasma membrane component (cholesterol) or have a hormonal function (estrogen and testosterone).
- Waxes prevent water loss in plants and assist in skin and fur maintenance in animals (earwax, beeswax).

Proteins Have a Wide Variety of Vital Functions

3.8 Proteins are the most versatile of life's molecules

- Protein functions in animals include:
  - Support (structural proteins)
  - Metabolism (speed chemical reactions)
  - Transport (substances can move between cells; hemoglobin transports oxygen)
  - Defense (antibodies combine with antigens to remove them)
  - Regulation (hormones)
  - Motion (muscle contraction)

3.9 Each protein is a sequence of particular amino acids

- Amino acids have a central carbon attached to an amino group ($-NH_2$), an acid group ($-COOH$), and an R group.
- Amino acids differ according to their R group.
- A peptide consists of two amino acids bonded together.
- A peptide bond is the covalent bond between two amino acids.
- A polypeptide is a chain of amino acids joined by peptide bonds.

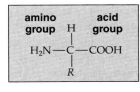

3.10 The shape of a protein is necessary to its function

- The primary structure is the linear sequence of amino acids in a protein.

- The secondary structure is the particular way a polypeptide folds—alpha helix or pleated sheet.
- The tertiary structure is a globular protein's final three-dimensional shape.
- Proteins that consist of more than one polypeptide may have a quaternary structure.

Nucleic Acids Are Information Molecules

3.11 The nucleic acids DNA and RNA carry coded information

- DNA is the genetic material of the cell.
- RNA functions in protein synthesis, conveying coded information from DNA to ribosomes.
- A nucleic acid is a polymer of nucleotides.
- A nucleotide is composed of a phosphate, a pentose sugar, and a nitrogen-containing base.
- RNA contains ribose; the bases guanine, adenine, cytosine, and uracil; and is single-stranded not helical.
- DNA contains deoxyribose; the bases guanine, adenine, cytosine, and thymine; and is double-stranded and helical.

- Complementary base pairing occurs as follows:
  - In DNA, T pairs with A; G pairs with C.
  - When RNA forms, A pairs with U (not T); G pairs with C.

3.12 The Human Genome Project may lead to new disease treatments

- The human genome has been sequenced.
- A genetic profile could be used to predict, prevent, or treat illnesses.
- The genomes of other organisms used in research have also been sequenced.

3.13 The nucleotide ATP is the cell's energy carrier (see art below)

- ATP is composed of adenine and ribose (adenosine) plus three phosphate groups (triphosphate)
- The last phosphate bond is hydrolyzed to form ADP + Ⓟ, and releases energy.
- Energy from ATP breakdown is used for the synthesis of macromolecules, muscle contraction, and nerve conduction.

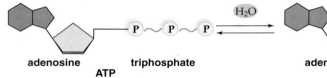

adenosine   triphosphate
ATP

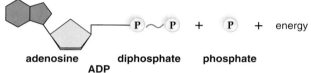

adenosine   diphosphate   phosphate
ADP

Testing Yourself

The Diversity of Organic Molecules Makes Life Diverse

1. Which of the following is an organic molecule?
   a. CH_4
   b. H_2O
   c. $C_6H_{12}O_6$
   d. O_2
   e. More than one of these is correct.

2. Which of these is not a characteristic of carbon?
   a. forms four covalent bonds
   b. bonds with other carbon atoms
   c. is sometimes ionic
   d. can form long chains
   e. sometimes shares two pairs of electrons with another atom

3. Organic molecules containing carboxyl groups are
   a. nonpolar.
   b. acidic.
   c. basic.
   d. More than one of these is correct.

4. Monomers are attached together to create polymers when a hydroxyl group and a hydrogen atom are _____ in a _____ reaction.
   a. added, dehydration
   b. removed, dehydration
   c. added, hydrolysis
   d. removed, hydrolysis

5. **THINKING CONCEPTUALLY** Based on the position of silicon in the periodic table of the elements, how is silicon like and how is it different from carbon? Explain why silicon does not form the same variety of molecules as carbon.

Carbohydrates Are Energy Sources and Structural Components

6. Which of the following is a disaccharide?
   a. glucose
   b. ribose
   c. fructose
   d. sucrose

7. Plants store glucose as
   a. maltose.
   b. glycogen.
   c. starch.
   d. None of these are correct.

8. Which of these makes cellulose nondigestible in humans?
   a. a polymer of glucose subunits
   b. a fibrous protein
   c. the linkage between the glucose molecules
   d. the peptide linkage between the amino acid molecules
   e. The carboxyl groups ionize.

9. **THINKING CONCEPTUALLY** After examining Figure 3.5, give reasons why cellulose and not starch is used as a structural component of plant cell walls.

Lipids Provide Storage, Insulation, and Other Functions

10. A triglyceride contains
    a. glycerol and three fatty acids.
    b. glycerol and three sugars.
    c. protein and three fatty acids.
    d. protein and three sugars.

11. A fatty acid is unsaturated if it
    a. contains hydrogen.
    b. contains carbon—carbon double bonds.
    c. contains a carboxyl (acidic) group.
    d. bonds to glycogen.
    e. bonds to a nucleotide.

12. Saturated fatty acids and unsaturated fatty acids differ in
    a. the number of double bonds present.
    b. their consistency at room temperature.
    c. the number of hydrogen atoms present.
    d. All of these are correct.

13. _____ is the precursor of _____.
    a. Estrogen, cholesterol
    b. Cholesterol, glucose
    c. Testosterone, cholesterol
    d. Cholesterol, testosterone and estrogen

14. Which of these is not a lipid?
    a. steroid
    b. fat
    c. polysaccharide
    d. wax
    e. phospholipids

15. Explain why phospholipids lend themselves to forming a bilayer membrane.

Proteins Have a Wide Variety of Vital Functions

16. Nearly all _____ are _____.
    a. proteins, enzymes
    c. enzymes, proteins
    b. sugars, monosaccharides
    d. sugars, polysaccharides
17. The difference between one amino acid and another is found in the
    a. amino group.
    d. peptide bond.
    b. carboxyl group.
    e. carbon atoms.
    c. *R* group.
18. The joining of two adjacent amino acids is called a
    a. peptide bond.
    c. covalent bond.
    b. dehydration reaction.
    d. All of these are correct.
19. Covalent bonding between *R* groups in proteins is associated with the _____ structure.
    a. primary
    c. tertiary
    b. secondary
    d. None of these are correct.
20. The three-dimensional structure of a protein that contains two or more polypeptides is the
    a. primary structure.
    c. tertiary structure.
    b. secondary structure.
    d. quaternary structure.

Nucleic Acids Are Information Molecules

21. Nucleotides
    a. contain a sugar, a nitrogen-containing base, and a phosphate group.
    b. are the monomers of fats and polysaccharides.
    c. join together by covalent bonding between the bases.
    d. are present in both DNA and RNA.
    e. Both a and d are correct.
22. Which of the following pertains to an RNA nucleotide, not a DNA nucleotide?
    a. contains the sugar ribose
    b. contains a nitrogen-containing base
    c. contains a phosphate molecule
    d. becomes bonded to other nucleotides by condensation
23. ATP
    a. is an amino acid.
    b. has a helical structure.
    c. is a high-energy molecule that can break down to ADP and phosphate.
    d. provides enzymes for metabolism.
    e. is most energetic when in the ADP state.

Understanding the Terms

adenosine   50
adenosine diphosphate
   (ADP)   50
adenosine triphosphate
   (ATP)   50
amino acid   45
carbohydrate   41
cellulose   42
chitin   42
complementary base
   pairing   48
dehydration reaction   40

denaturation   47
deoxyribose   41
disaccharide   41
disulfide bond   46
DNA (deoxyribonucleic
   acid)   48
enzyme   40
fat   43
fatty acid   43
functional group   39
glucose   41
glycerol   43
glycogen   42
hemoglobin   45
Human Genome Project
   (HGP)   49
hydrolysis reaction   40
hydrophilic   39
hydrophobic   39
isomer   39
lipid   43
monomer   40

monosaccharide   41
nucleic acid   48
nucleotide   48
oil   43
organic chemistry   38
organic molecule   38
peptide   45
peptide bond   45
peptidoglycan   42
phospholipid   44
polymer   40
polypeptide   45
polysaccharide   42
protein   45
ribose   41
RNA (ribonucleic acid)   48
saturated fatty acid   43
starch   42
steroid   44
triglyceride   43
unsaturated fatty acid   43
wax   44

Match the terms to these definitions:
a. _____ Class of organic compounds that includes monosaccharides, disaccharides, and polysaccharides.
b. _____ Class of organic compounds that tend to be insoluble in water.
c. _____ Macromolecules consisting of covalently bonded monomers.
d. _____ Molecules that have the same molecular formula but a different functional group.
e. _____ Two or more amino acids joined together by covalent bonding.

Thinking Scientifically

1. You hypothesize that the unsaturated oil content of temperate plant seeds will help them survive freezing temperatures compared to the saturated oil content of tropical plant seeds. a. How would you test your hypothesis? b. Assuming your hypothesis is supported, give an explanation.
2. Chemical analysis reveals that an abnormal form of an enzyme contains a polar amino acid where the normal form has a nonpolar amino acid. Formulate a hypothesis to test regarding the abnormal enzyme.

ARIS *Visit www.mhhe.com/maderconcepts for practice quizzes, animations, videos, and activities designed to help you master the material in this chapter.*

4

Structure and Function of Cells

LEARNING OUTCOMES

After studying this chapter, you should be able to accomplish the following outcomes.

Cells: What Are They?

1 Be aware of the importance of microscopes to our knowledge of cells.

Cells Are the Basic Units of Life

2 Cite the three tenets of the cell theory.
3 Compare and contrast the three commonly used types of microscopes.
4 Compare and contrast various types of cells.

Protein Synthesis Is a Major Function of Cells

5 Describe the structure and function of the nucleus, ribosomes, endoplasmic reticulum, and Golgi apparatus.

Vesicles and Vacuoles Have Varied Functions

6 Describe the structure and function of lysosomes and peroxisomes.
7 Describe the varied functions of vacuoles and/or vesicles in protists, plants, and animals.

A Cell Carries Out Energy Transformations

8 Compare and contrast the structure and function of chloroplasts and mitochondria.
9 Explain the endosymbiotic theory and the evidence that supports it.
10 Discuss in general the role of mitochondria in human diseases.

The Cytoskeleton Maintains Cell Shape and Assists Movement

11 Compare and contrast the structure and function of actin filaments, intermediate filaments, and microtubules.

In Multicellular Organisms, Cells Join Together

12 Describe the modifications of a cell's surface and how they function.

Imagine that you have never taken a biology course, and you are alone in a laboratory with a bunch of slides of plant and animal tissues and a microscope. The microscope is easy to use, and soon you are able to focus it and begin looking at the slides.

Your assignment is to define a cell. In order not to panic, you idly look at one slide after another, letting your mind wander. Was this the way Robert Hooke felt back in the 17th century, when he coined the word "cell"? What did he see? Actually, Hooke was using a light microscope, as you are, when he happened to look at a piece of cork. He drew what he saw like this:

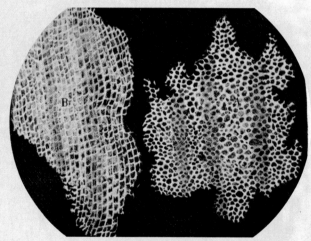

Hooke saw almost nothing except for outlines, which we know today are the cell walls of plant cells. Similarly, you can make out the demarcations between onion root cells in the micrograph on page 55. So, like Hooke, you might come to the conclusion that a cell is an entity, a unit of a larger whole.

Once you had such a definition for a cell, you might be able to conclude that cells are present in all the slides at your disposal—as in all the micrographs on these pages. But it certainly would take a gigantic leap to hypothesize that all organisms are composed of cells, and this didn't occur until almost 200 years after Hooke used the term cell. You can appreciate that science progresses slowly, little by little, and that a theory, such as the cell theory, becomes established only when an encompassing hypothesis is never found to be lacking. Indeed, it was only when Matthias Schleiden always saw cells in plant tissues, and

Cells: What Are They?

Theodor Schwann always saw cells in animal tissues, that they concluded, respectively, in the 1830s that plants and animals are composed of cells.

This chapter begins with an explanation of the cell theory and then progresses to considering the many attributes of cells. The cell theory was formulated before the electron microscope was invented and before the biochemical techniques used to study cells were developed. This improvement in technology tells us that cells have an intricate internal structure that allows them to carry on all of life's functions. In this chapter, we will be studying the structure and function of cell parts based on modern-day data.

Onion root cells

Rod-shaped bacteria

Euglena, a protist

Nerve cells

Microscopy was essential to formulating the cell theory because you cannot see a cell without a microscope. The small size of cells provides the surface area each cell needs to exchange molecules with the environment. Whereas the light microscope was essential to discovering that all organisms are composed of cells, our current knowledge of cells is dependent on the electron microscope, which reveals much more detail than the light microscope. The prokaryotic cell of archaea and bacteria has a simple structure, compared to the eukaryotic cell of protists, fungi, plants, and animals.

4.1   All organisms are composed of cells

The **cell theory** states that:

1. *A Cell Is the Basic Unit of Life*   This means that nothing smaller than a cell is alive. A unicellular organism exhibits the characteristics of life we discussed in Chapter 1. No smaller unit exists that is able to reproduce, respond to stimuli, remain homeostatic, grow and develop, take in and use materials from the environment, and adapt to the environment. In short, life has a cellular nature. On this basis, we can make two other deductions.

2. *All Living Things Are Made Up of Cells*   While it may be apparent that a unicellular organism is necessarily a cell, what about more complex organisms? Lilacs and rabbits, as well as most other visible organisms, are multicellular. **Figure 4.1A** illustrates that a lilac leaf is composed of cells, and **Figure 4.1B** illustrates that the intestinal lining of a rabbit is composed of cells. Is there any tissue in these organisms that is not composed of cells? For example, you might be inclined to say that bone does not contain cells. But if you were to examine bone tissue under a microscope, you would see that it, too, is composed of cells. Cells have distinct forms—a bone cell looks quite different from

a nerve cell, and they both look quite different from the cell of a lilac leaf. The cells that make up a multicellular organism are specialized in structure and function, but they all have certain parts in common. This chapter discusses those common components.

3. *New Cells Arise Only from Preexisting Cells*   This statement also requires some understanding; it certainly wasn't readily apparent to early investigators, who believed that organisms could arise from dirty rags, for example. Today, we know you cannot get a new lilac bush or a new rabbit without preexisting lilacs and rabbits. When lilacs, rabbits, or humans reproduce, a sperm cell joins with an egg cell to form a zygote, which is the first cell of a new multicellular organism.

The difficulty of formulating the cell theory was due to the microscopic size of cells. Section 4.2 shows that the microscopic size permits enough exchanges with the environment to sustain the life of a cell.

> **4.1** *Check Your Progress*   **A cell is alive, but its parts are not alive. Explain.**

Rabbit, an animal

Micrograph of intestine reveals cells.   140 μm

FIGURE 4.1B   Rabbit, with a photomicrograph of its intestinal lining below.

Lilac, a plant

Micrograph of leaf reveals cells.   50 μm

FIGURE 4.1A   Lilac leaf, with a photomicrograph below.

4.2 Metabolically active cells are small in size

Cells tend to be quite small. A frog's egg, at about 1 millimeter (mm) in diameter, is large enough to be seen by the human eye. But most cells are far smaller than 1 mm; some are even as small as 1 micrometer (µm)—one thousandth of a millimeter. Cell structures and macromolecules that are smaller than a micrometer are measured in terms of nanometers (nm). **Figure 4.2A** outlines the visual range of the eye, the light microscope, and the electron microscope. The discussion of microscopy on page 58 explains why the electron microscope allows us to see so much more detail than the light microscope does.

Why are cells so small? To answer this question, consider that a cell needs a surface area large enough to allow sufficient nutrients to enter and to rid itself of wastes. Small cells, not large cells, are more likely to have this adequate surface area. For example, cutting a large cube into smaller cubes provides a lot more surface area per volume. The calculations show that a 4-cm cube has a **surface-area-to-volume ratio** of only 1.5:1, whereas a 1-cm cube has a surface-area-to-volume ratio of 6:1 (**Fig. 4.2B**).

We would expect, then, that actively metabolizing cells would have to remain small. A chicken's egg is several centimeters in diameter, but the egg is not actively metabolizing. Once the egg is incubated and metabolic activity begins, the egg divides repeatedly without growth. Cell division restores the amount of surface area needed for adequate exchange of materials.

Further, cells that specialize in absorption have modifications that greatly increase the surface-area-to-volume ratio of the cell. The columnar epithelial cells along the surface of the intestinal wall have surface foldings called microvilli (sing., microvillus) that increase their surface area. See Figure 4.20B for a drawing of a cell with microvilli. Nerve cells and some large plant cells are long and thin, and this increases the ratio

of plasma membrane to cytoplasm. Nerve cells are shown on page 55.

Comparing the structure of cells is facilitated by the use of three basic types of microscopes, which are described in Section 4.3.

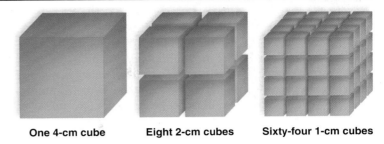

One 4-cm cube   Eight 2-cm cubes   Sixty-four 1-cm cubes

Total surface area (height × width × number of sides × number of cubes)		
96 cm^2	192 cm^2	384 cm^2
Total volume (height × width × length × number of cubes)		
64 cm^3	64 cm^3	64 cm^3
Surface-area-to-volume ratio per cube (surface area ÷ volume)		
1.5:1	3:1	6:1

FIGURE 4.2B Surface-area-to-volume relationships.

> **4.2 Check Your Progress** Why is your body made up of multitudes of small cells, instead of a single large cell?

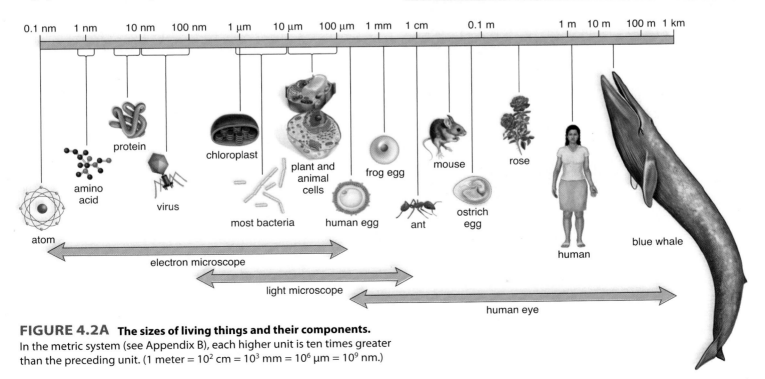

FIGURE 4.2A The sizes of living things and their components.
In the metric system (see Appendix B), each higher unit is ten times greater than the preceding unit. (1 meter = 10^2 cm = 10^3 mm = 10^6 µm = 10^9 nm.)

Because cells are so small, it is best to study them microscopically. A magnifying glass containing a single lens is the simplest version of a light microscope. However, such a simple device is not powerful enough to be of much use in examining cells. The **compound light microscope** is much more suitable. It has superior magnifying power because it uses a system of multiple lenses. As you can see in **Figure 4.3**, a condenser lens focuses the light into a tight beam that passes through a thin specimen (such as a drop of blood or a thin slice of an organ). An objective lens magnifies an image of the specimen, and another lens, called the ocular lens, magnifies it yet again. It is the image from the ocular lens that is viewed with the eye. The most commonly used compound light microscope is called

a bright-field microscope, because the specimen appears dark against a light background.

The compound light microscope is widely used in research, clinical, and teaching laboratories. However, the use of light to produce an image means that the ability to view two objects as separate—the resolution—is not as good as with an electron microscope. The resolution limit of a compound light microscope is 0.2 μm, which means that objects less than 0.2 μm apart will appear as a single object. Although there is no limit to the magnification that could be achieved with a compound light microscope, there is a definite limit to the resolution.

The wavelength of light is an important consideration in obtaining the best possible resolution with a compound light microscope. The shorter the wavelength of light, the better the resolution. This is why many compound light microscopes are equipped with blue filters. The shorter wavelength of blue light compared to white light improves resolution.

An electron microscope can produce an image of finer resolution than a light microscope because, instead of using light, it fires a beam of electrons at the specimen. Electrons have a shorter wavelength than does light. The essential design of an electron microscope is similar to that of a compound light microscope, but its lenses are made of electromagnets, instead of glass. Because the human eye cannot see at the wavelengths of electrons, the images produced by electron microscopes are projected onto a screen or viewed on a television monitor.

There are two types of electron microscopes: the transmission electron microscope and the scanning electron microscope. These microscopes didn't become widely used until about 1970. A **transmission electron microscope** passes a beam of electrons through a specimen (Fig. 4.3). Because electrons do not have much penetrating ability, the section must be very thin, indeed—usually between 50 and 150 nm. The transmission electron microscope can discern fine details, with a limit of resolution around 1.0 nm and a magnifying power up to 200,000×. A **scanning electron microscope** does not pass a beam through a specimen; rather, it collects and focuses electrons that are scattered from the specimen's surface and generates an image with a distinctive three-dimensional appearance (Fig. 4.3).

Scientists often preserve microscopic images; these are referred to as micrographs. A captured image from a light microscope is termed a light micrograph (LM), or a photomicrograph; there are also transmission electron micrographs (TEM) and scanning electron micrographs (SEM). The latter two are black-and-white in their original form, but are often colorized using a computer.

Prokaryotic cells, which are discussed next, are best viewed with electron microscopes because it is difficult to see prokaryotic cells with the light microscope.

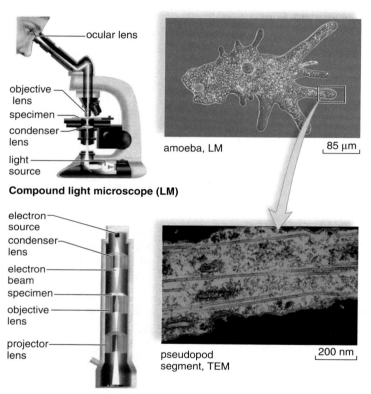

amoeba, LM   85 μm

Compound light microscope (LM)

pseudopod
segment, TEM   200 nm

Transmission electron microscope (TEM)

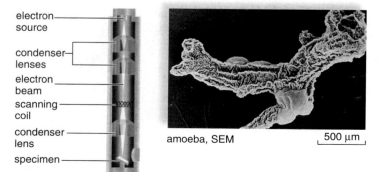

amoeba, SEM   500 μm

Scanning electron microscope (SEM)

FIGURE 4.3   Comparison of three microscopes.

4.3 *Check Your Progress   a.* To make cells larger, microscopically, increase the _____. *b.* To see more detail, increase the _____.

4.4 Prokaryotic cells evolved first

Fundamentally, two different types of cells exist. **Prokaryotic cells** (*pro*, before, and *karyon*, nucleus) are so named because they lack a membrane-bounded nucleus. The other type of cell, called a **eukaryotic cell**, has a nucleus. Prokaryotic cells are miniscule in size compared to eukaryotic cells (**Fig. 4.4A**). Prokaryotes are present in great numbers in the air, in bodies of water, in the soil, and also in and on other organisms.

Figure 4.4B shows the generalized structure of a bacterium. A **plasma membrane** is bounded by a **cell wall**, which in turn may be surrounded by a **capsule** and/or a gelatinous sheath called a **slime layer**. Motile bacteria usually have long, very thin flagella (sing., **flagellum**) that are composed of subunits of a protein called flagellin. Some bacteria also have **fimbriae**, which are short appendages that help them attach to an appropriate surface. The flagella, which rotate like propellers, rapidly move the bacterium in a fluid medium. Bacteria reproduce asexually by binary fission, but they can exchange DNA by way of the sex pili (sing., **sex pilus**), rigid tubular structures. Prokaryotes can also take up DNA from the external medium or by way of viruses.

Prokaryotes have a single chromosome (loop of DNA) located within a region called the **nucleoid**, which is not bounded by a plasma membrane. The semifluid **cytoplasm** has thousands of **ribosomes** for the synthesis of proteins. In addition, the photosynthetic cyanobacteria have light-sensitive pigments, usually within the membranes of flattened disks called thylakoids.

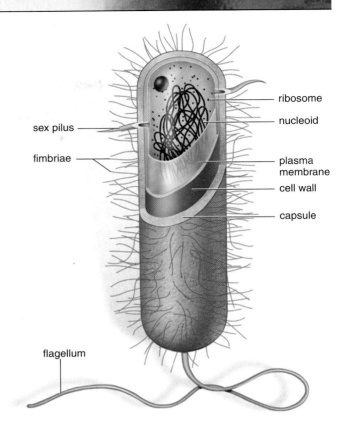

FIGURE 4.4B Prokaryotic cell structure.

FIGURE 4.4A Size comparison of a eukaryotic cell and a prokaryotic cell.

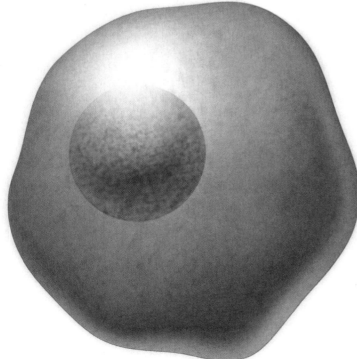

Eukaryotic cell

The two groups of prokaryotes—domain **Archaea** and domain **Bacteria**—are largely distinguishable by different nucleic acid base sequences. Bacteria are well known for causing serious diseases, such as tuberculosis, anthrax, tetanus, throat infections, and gonorrhea. However, they are important to the environment because they decompose the remains of dead organisms and contribute to the cycling of chemicals in ecosystems. Also, their great ability to synthesize molecules can be put to use for the manufacture of all sorts of products, from industrial chemicals to foodstuffs and drugs.

An overview of eukaryotic cell structure is given in Section 4.5, before we consider the structure and function of individual eukaryotic cell components.

4.4 *Check Your Progress* **Can you define or recognize a cell by the presence of a nucleus?**

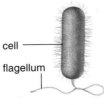

Prokaryotic cell

4.5 Eukaryotic cells contain specialized organelles: An overview

Organisms with eukaryotic cells, namely protists, fungi, plants, and animals, are members of domain **Eukarya,** the third domain of living things. Unlike prokaryotic cells, eukaryotic cells have a membrane-bounded nucleus, which houses their DNA.

Eukaryotic cells are much larger than prokaryotic cells, and therefore they have less surface area per volume than prokaryotic cells (see Fig. 4.4A). This difference in surface-area-to-volume ratio is not detrimental to the cells' existence because, unlike prokaryotic cells, eukaryotic cells are compartmentalized. They contain small structures called **organelles** that are specialized to perform specific functions. The organelles are located within the *cytoplasm,* a semifluid interior that is bounded by a plasma membrane. Originally, the term organelle referred only to membranous structures, but we will use it to include any well-defined subcellular structure. By that definition, ribosomes, which are the site of protein synthesis, are also organelles.

Figure 4.5A and **Figure 4.5B** illustrate cellular anatomy and the types of structures and organelles found in animal and plant cells. Not all eukaryotic cells contain all of the same organelles in the same abundance.

Organelles are in constant communication with one another. As an example, consider that the nucleus communicates with ribosomes in the cytoplasm, and that the major organelles of the **endomembrane system**—the *endoplasmic reticulum, Golgi apparatus,* and *lysosomes*—also communicate with one another. Each organelle has its own particular set of enzymes and synthesizes its own products, and the products move from one organelle to the other. The products are carried between organelles by little transport vesicles, membranous sacs that enclose the molecules and keep them separate from the cytoplasm.

Communication between the **energy-related organelles**—the *mitochondria* in plant and animal cells, and the *chloroplasts*

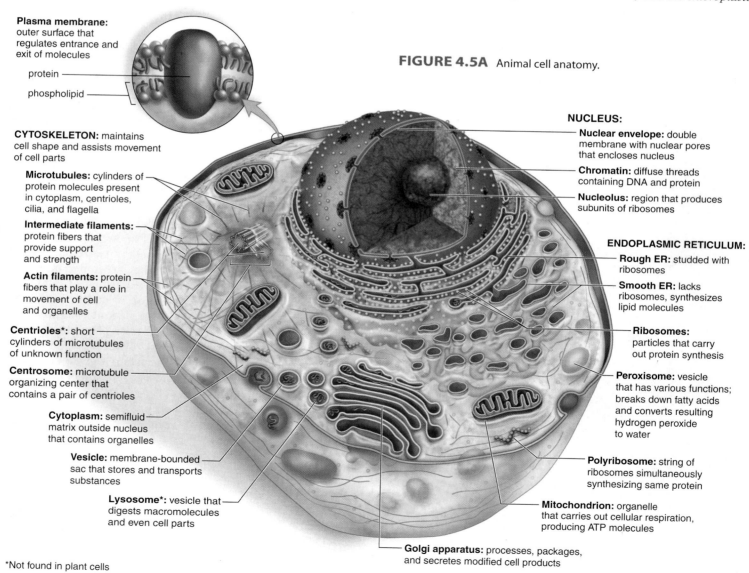

FIGURE 4.5A Animal cell anatomy.

Plasma membrane: outer surface that regulates entrance and exit of molecules

protein

phospholipid

CYTOSKELETON: maintains cell shape and assists movement of cell parts

Microtubules: cylinders of protein molecules present in cytoplasm, centrioles, cilia, and flagella

Intermediate filaments: protein fibers that provide support and strength

Actin filaments: protein fibers that play a role in movement of cell and organelles

Centrioles*: short cylinders of microtubules of unknown function

Centrosome: microtubule organizing center that contains a pair of centrioles

Cytoplasm: semifluid matrix outside nucleus that contains organelles

Vesicle: membrane-bounded sac that stores and transports substances

Lysosome*: vesicle that digests macromolecules and even cell parts

NUCLEUS:

Nuclear envelope: double membrane with nuclear pores that encloses nucleus

Chromatin: diffuse threads containing DNA and protein

Nucleolus: region that produces subunits of ribosomes

ENDOPLASMIC RETICULUM:

Rough ER: studded with ribosomes

Smooth ER: lacks ribosomes, synthesizes lipid molecules

Ribosomes: particles that carry out protein synthesis

Peroxisome: vesicle that has various functions; breaks down fatty acids and converts resulting hydrogen peroxide to water

Polyribosome: string of ribosomes simultaneously synthesizing same protein

Mitochondrion: organelle that carries out cellular respiration, producing ATP molecules

Golgi apparatus: processes, packages, and secretes modified cell products

*Not found in plant cells

in plant cells—is less obvious, but it does occur by other means than transport vesicles. Still, except for importing certain proteins, these organelles are self-sufficient. They even have their own genetic material, and their ribosomes resemble those of prokaryotic cells. This and other evidence suggests that the mitochondria and chloroplasts are derived from prokaryotes that took up residence in an early eukaryotic cell. Notice that an animal cell has only mitochondria, while a plant cell has both mitochondria and chloroplasts.

The **cytoskeleton** is a lattice of protein fibers that maintains the shape of the cell and helps organelles move within the cell. The protein fibers serve as tracks for the transport vesicles that are taking molecules from one organelle to another. In other words, the tracks direct and speed them on their way. The manner in which vesicles and other types of organelles move along these tracks will be discussed in more detail later in this chapter. Without a cytoskeleton, a eukaryotic cell would not have an efficient means of moving organelles and their products within the cell and, possibly, could not exist.

One type of protein fiber, called microtubules, project from the *centrosome*, the main microtubule organizer in eukaryotic cells. Animal cells, but not plant cells, have centrioles within the centrosome. Microtubules are present in centrioles and also in cilia and flagella, which are motile organelles more common in animal cells than in plant cells.

Other differences between animal and plant cells should be noted. Plant cells have a permeable but protective cell wall, in addition to a plasma membrane. All plant cells have a primary cell wall containing cellulose, but some plant cells also have a secondary cell wall that is reinforced with lignin, a stronger material. Animal cells do not have a cell wall.

In Figures 4.5A and 4.5B, each structure in an animal or plant cell has been given a particular color that will be used for this structure throughout the text.

4.5 *Check Your Progress* **Explain why early investigators were unable to make out the detail illustrated in Figures 4.5A and 4.5B.**

FIGURE 4.5B Plant cell anatomy.

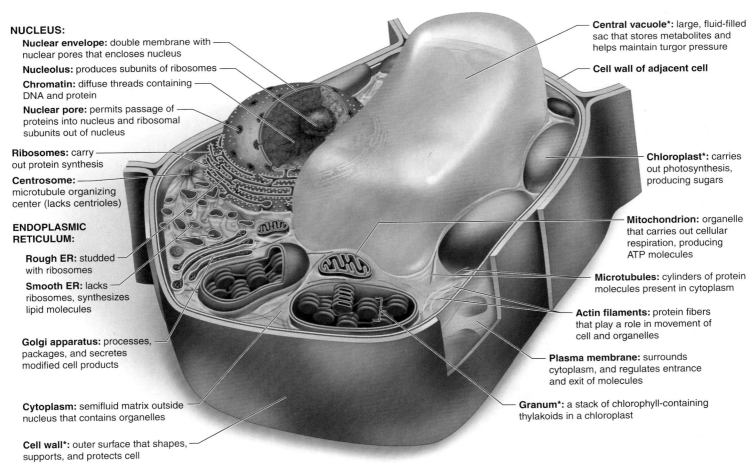

NUCLEUS:
Nuclear envelope: double membrane with nuclear pores that encloses nucleus
Nucleolus: produces subunits of ribosomes
Chromatin: diffuse threads containing DNA and protein
Nuclear pore: permits passage of proteins into nucleus and ribosomal subunits out of nucleus
Ribosomes: carry out protein synthesis
Centrosome: microtubule organizing center (lacks centrioles)

ENDOPLASMIC RETICULUM:
Rough ER: studded with ribosomes
Smooth ER: lacks ribosomes, synthesizes lipid molecules
Golgi apparatus: processes, packages, and secretes modified cell products
Cytoplasm: semifluid matrix outside nucleus that contains organelles
Cell wall*: outer surface that shapes, supports, and protects cell

*Not found in animal cells

Central vacuole*: large, fluid-filled sac that stores metabolites and helps maintain turgor pressure
Cell wall of adjacent cell
Chloroplast*: carries out photosynthesis, producing sugars
Mitochondrion: organelle that carries out cellular respiration, producing ATP molecules
Microtubules: cylinders of protein molecules present in cytoplasm
Actin filaments: protein fibers that play a role in movement of cell and organelles
Plasma membrane: surrounds cytoplasm, and regulates entrance and exit of molecules
Granum*: a stack of chlorophyll-containing thylakoids in a chloroplast

This part of the chapter discusses certain organelles of eukaryotic cells, namely the nucleus, the ribosome, the rough endoplasmic reticulum, and the Golgi apparatus, which are all involved in producing proteins that serve necessary functions in the cell. A technique called pulse-labeling can trace the location of labeled amino acids (in proteins) from the ribosomes, through the organelles mentioned, to the plasma membrane, where they are secreted.

4.6   The nucleus contains the cell's genetic information

The **nucleus** contains the genetic information that is passed on from cell to cell and from generation to generation. The ribosomes use this information to carry out protein synthesis.

The nucleus is a prominent structure, having a diameter of about 5 μm (**Fig. 4.6**). It generally has an oval shape and is located near the center of a cell. The nucleus contains **chromatin** in a semifluid matrix called the nucleoplasm. Chromatin looks grainy, but actually it is a network of strands that condenses and undergoes coiling into rodlike structures called **chromosomes** just before the cell divides. Chromatin, and therefore chromosomes, contains DNA, protein, and some RNA. Genes, composed of DNA, are units of heredity located on the chromosomes.

RNA, of which there are several forms, is produced in the nucleus. A **nucleolus** is a dark region of chromatin where a type of RNA called *ribosomal RNA (rRNA)* is produced and where rRNA joins with proteins to form the subunits of ribosomes, discussed in Section 4.7. Another type of RNA, called *messenger*

RNA (mRNA), acts as an intermediary for DNA and specifies the sequence of amino acids during protein synthesis. *Transfer RNA (tRNA)* is used in the assembly of amino acids during protein synthesis. The proteins of a cell determine its structure and functions; therefore, the nucleus is the command center for a cell.

The nucleus is separated from the cytoplasm by a double membrane known as the **nuclear envelope**. Even so, the nucleus communicates with the cytoplasm. The nuclear envelope has **nuclear pores** of sufficient size (100 nm) to permit the passage of ribosomal subunits and RNAs out of the nucleus into the cytoplasm, and the passage of proteins from the cytoplasm into the nucleus. High-power electron micrographs show that nonmembranous components associated with the pores form a nuclear pore complex.

> **4.6 Check Your Progress** Which of the photomicrographs on pages 54–55 are cells with nuclei? Explain.

FIGURE 4.6 Anatomy of the nucleus.

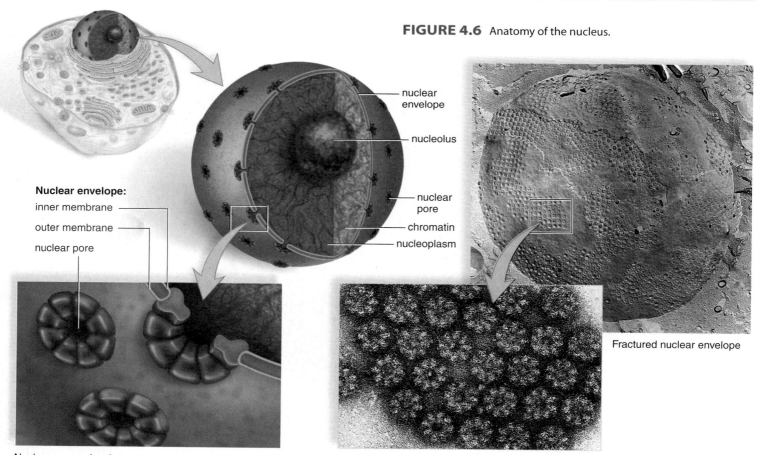

Nuclear envelope:
inner membrane
outer membrane
nuclear pore

nuclear envelope
nucleolus
nuclear pore
chromatin
nucleoplasm

Fractured nuclear envelope

Nuclear pores, drawing

Nuclear pores, TEM

4.7   The ribosomes carry out protein synthesis

Ribosomes are non-membrane-bounded particles where protein synthesis occurs. In eukaryotes, ribosomes measure about 20 nm by 30 nm, and in prokaryotes they are slightly smaller. In both types of cells, ribosomes are composed of two subunits, one large and one small. Each subunit has its own mix of proteins and rRNA.

The number of ribosomes in a cell varies, depending on the cell's functions. For example, pancreatic cells and those of other glands have many ribosomes because they produce secretions that contain proteins.

In eukaryotic cells, some ribosomes occur freely within the cytoplasm, either singly or in groups called polyribosomes. Other ribosomes are attached to the **endoplasmic reticulum (ER)**, a membranous system of flattened saccules (small sacs) and tubules, which is discussed more fully in Section 4.8. Ribosomes receive mRNA from the nucleus, and this nucleic acid carries a coded message from DNA indicating the correct sequence of amino acids in a protein. Proteins synthesized by cytoplasmic ribosomes often enter other organelles. For example, those synthesized by ribosomes attached to the ER end up in the lumen of the ER.

Attachment of Ribosomes to the ER   As shown in **Figure 4.7**, **1** after mRNA is produced and leaves the nucleus, **2** it joins with ribosomal subunits, and protein synthesis begins. **3** When a ribosome combines with a receptor at the ER, the protein enters the lumen of the ER through a channel in the receptor. **4** Exterior to the ER, the ribosome splits, releasing the mRNA, while the protein takes shape in the ER lumen.

As explained in Section 4.8, the endoplasmic reticulum is called rough ER if it has attached ribosomes and smooth ER if it does not have attached ribosomes.

4.7 *Check Your Progress*   **What role do the nuclear pores play in protein synthesis?**

FIGURE 4.7   Function of ribosomes.

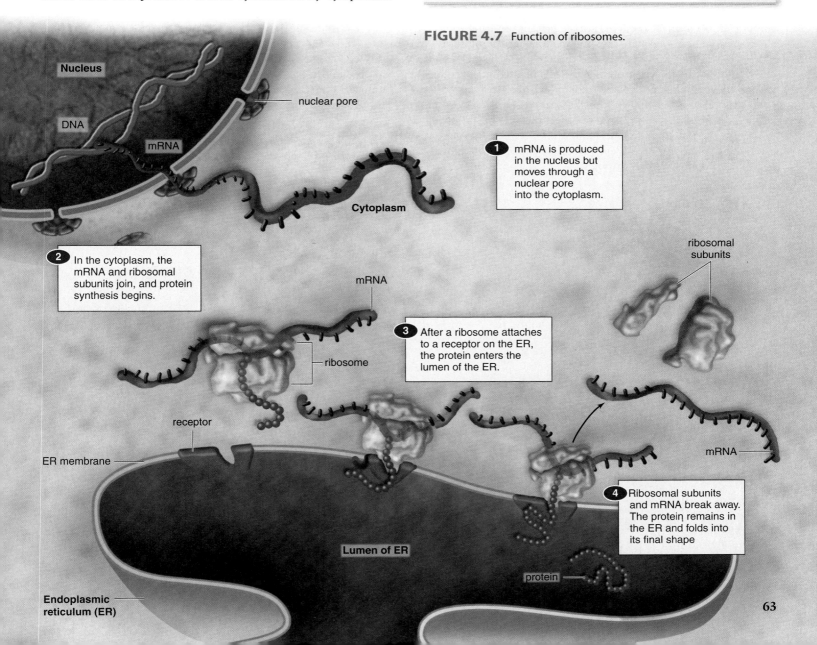

1 mRNA is produced in the nucleus but moves through a nuclear pore into the cytoplasm.

2 In the cytoplasm, the mRNA and ribosomal subunits join, and protein synthesis begins.

3 After a ribosome attaches to a receptor on the ER, the protein enters the lumen of the ER.

4 Ribosomal subunits and mRNA break away. The protein remains in the ER and folds into its final shape

Nucleus

nuclear pore

DNA

mRNA

Cytoplasm

ribosomal subunits

mRNA

ribosome

receptor

ER membrane

mRNA

Lumen of ER

protein

Endoplasmic reticulum (ER)

4.8 The endoplasmic reticulum synthesizes and transports proteins and lipids

The endoplasmic reticulum (ER) is physically continuous with the outer membrane of the nuclear envelope. It consists of many membranous channels and saccules, flattened vesicles that typically account for more than half of the total membrane within an average animal cell. The membrane of the ER is continuous and encloses a single internal space called the ER lumen (**Fig. 4.8**). The ER twists and turns as it courses through the cytoplasm, as if it were a long snake. This structure results in the ER having much more membrane than if it were simply one large sac. If you compare Figure 4.7 to Figure 4.8, you can see that Figure 4.7 shows only a small portion of the ER found in a cell.

The ER produces all the proteins and lipids for the membranes of the cell as well as most of the proteins that are secreted from the cell. Insulin is an example of a secreted protein in humans. Insulin is secreted by the pancreas into the blood and then circulates about the body.

Types of ER The ER is divided into the rough ER and the smooth ER. **Rough ER (RER)** is studded with ribosomes on the side of the membrane that faces the cytoplasm; therefore, it is correct to say that RER synthesizes proteins. As was discussed in Section 4.7, however, protein synthesis begins in ribosomes within the cytoplasm. Certain ribosomes migrate to the ER, even as a protein is being synthesized, and then the ribosome binds to a receptor on the ER. Once the ribosome has attached to the ER, the polypeptide enters the lumen of the ER, where it is often modified. Certain ER enzymes add carbohydrate (sugar) chains to proteins, which are then called glycoproteins. Other proteins assist the folding process that results in the final shape of the protein.

Smooth ER (SER), which is continuous with rough ER, does not have attached ribosomes. Smooth ER is abundant in gland cells, where it synthesizes lipids of various types. For example, cells that synthesize steroid hormones from cholesterol have much SER. The SER houses the enzymes needed to make cholesterol and modify it to produce the hormones. In the liver, SER, among other functions, adds lipid to proteins, forming the lipoproteins that carry cholesterol in the blood. Also, the SER of the liver increases in quantity when a person consumes alcohol or takes barbiturates on a regular basis, because SER contains the enzymes that detoxify these molecules.

The RER and SER, working together, produce membrane, which is composed of phospholipids and various types of proteins. Because the ER produces membrane, it can form the transport vesicles by which it communicates with the Golgi apparatus. Proteins to be secreted from the cell are kept in the lumen of the ER, but the ones destined to become membrane constituents become embedded in the membrane of the ER. **Transport vesicles** pinch off from the ER and carry membranes, proteins, and lipids, notably to the Golgi apparatus, where they undergo further modification. The products of the Golgi apparatus are utilized by the cell or secreted. The Golgi apparatus is discussed further in Section 4.9.

> **4.8 Check Your Progress** Is it correct to say that *all* ribosomes reside in the cytoplasm? Explain.

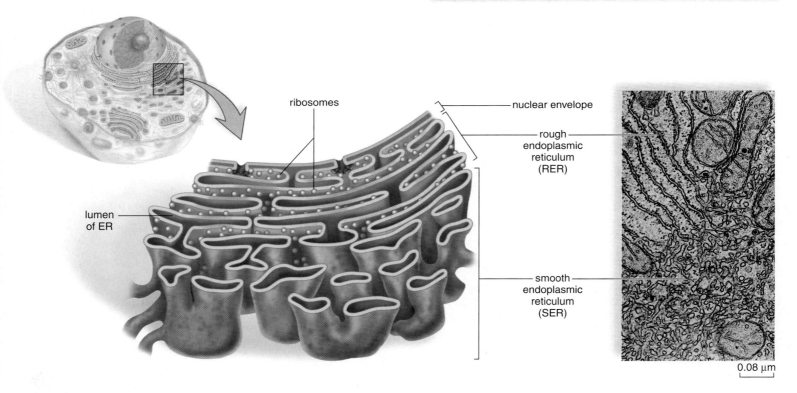

FIGURE 4.8 Rough ER (RER) and smooth ER (SER).

4.9 The Golgi apparatus modifies and repackages proteins for distribution

The **Golgi apparatus** is named for Camillo Golgi, who discovered its presence in cells in 1898. The Golgi apparatus typically consists of a stack of three to twenty slightly curved, flattened saccules whose appearance can be compared to a stack of pancakes (**Fig. 4.9**). One side of the stack (the inner or cis face) is directed toward the ER, and the other side of the stack (the outer or trans face) is directed toward the plasma membrane. Vesicles can frequently be seen at the edges of the saccules.

Protein-filled vesicles that bud from the rough ER and lipid-filled vesicles that bud from the smooth ER are received by the Golgi apparatus at its inner face. Thereafter, the apparatus alters these substances as they move through its saccules. For example, the Golgi apparatus contains enzymes that modify the carbohydrate chains first attached to proteins in the rough ER. It can exchange one sugar for another sugar. In some cases, the modified carbohydrate chain serves as a signal molecule that determines the protein's final destination in the cell.

The Golgi apparatus sorts and packages proteins and lipids in vesicles that depart from the outer face. In animal cells, some of these vesicles are lysosomes, which are discussed next. Other vesicles proceed to the plasma membrane, where they stay until a signal molecule triggers the cell to release them. Then they become part of the membrane as they discharge their contents during **secretion**. Secretion is also called *exocytosis* because the substance exits the cytoplasm. Pulse-labeling, discussed next, traces the path of a protein from synthesis to secretion.

> **4.9 Check Your Progress** How do proteins made by RER ribosomes become incorporated into a plasma membrane or secreted?

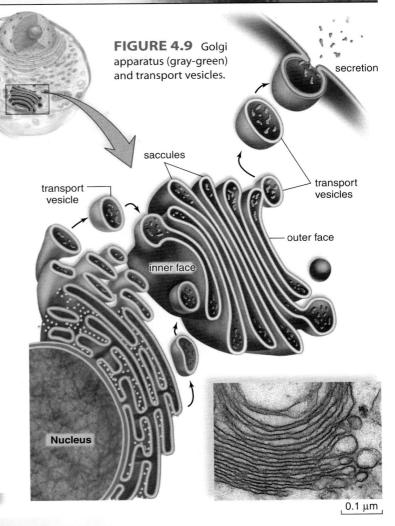

FIGURE 4.9 Golgi apparatus (gray-green) and transport vesicles.

secretion

saccules

transport vesicle

transport vesicles

outer face

inner face

Nucleus

0.1 μm

4.10 Pulse-labeling allows observation of the secretory pathway

The pathway of protein secretion was observed by George Palade and his associates using a pulse-chase technique. The rough ER was *pulse-labeled* by letting cells metabolize for a very short time with radioactive amino acids. Then the cells were given an excess of nonradioactive amino acids. This *chased* the labeled amino acids out of the ER into transport vesicles.

Electron microscopy techniques allowed these researchers to trace the fate of the labeled amino acids, as shown in **Figure 4.10**: ❶ The labeled amino acids were found in the ER, then in ❷ transport vesicles, and then in ❸ the Golgi apparatus, before appearing in ❹ vesicles at the plasma membrane and finally being released.

> **4.10 Check Your Progress** In the pulse-labeling procedure, radioactive atoms were used as _____.

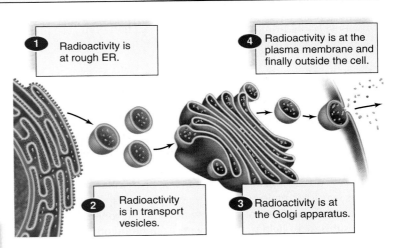

❶ Radioactivity is at rough ER.

❹ Radioactivity is at the plasma membrane and finally outside the cell.

❷ Radioactivity is in transport vesicles.

❸ Radioactivity is at the Golgi apparatus.

FIGURE 4.10 Observing the secretory pathway.

A cell has various types of vacuoles; many of them look the same in electron micrographs but actually have different functions. We have already mentioned the work of transport vesicles. Lysosomes contain powerful hydrolytic enzymes that digest macromolecules, even if they form cell parts. Peroxisomes are more specialized and assist mitochondria by breaking down lipids. Some of the vacuoles in protists and plants are unique to them and not found in other eukaryotes. This part of the chapter also summarizes the work of the endomembrane system.

4.11 Lysosomes digest macromolecules and cell parts

Lysosomes are membrane-bounded vesicles produced by the Golgi apparatus. They have a very low internal pH and contain powerful hydrolytic digestive enzymes. Lysosomes are important in recycling cellular material and digesting worn-out organelles, such as old peroxisomes (**Fig. 4.11**).

Sometimes macromolecules are engulfed (brought into a cell by vesicle formation) at the plasma membrane. When a lysosome fuses with such a vesicle, its contents are digested by lysosomal enzymes into simpler subunits that then enter the cytoplasm. Some white blood cells defend the body by engulfing bacteria, which are then enclosed within vesicles. When lysosomes fuse with these vesicles, the bacteria are digested.

Lysosomal storage diseases occur when a particular lysosomal enzyme is nonfunctional. Tay Sachs disease is one such condition, in which a newborn appears healthy but then gradually becomes nonresponsive, deaf, and blind before dying within a few months. The brain cells are filled with particles containing a type of lipid that could not be digested by lysosomes.

4.11 *Check Your Progress* **Some white blood cells have granules, now known to be lysosomes. Why would it be beneficial for these white blood cells to fight infection by engulfing viruses and bacteria?**

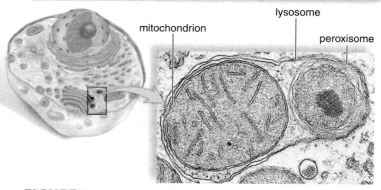

FIGURE 4.11 Lysosome fusing with and destroying spent organelles.

4.12 Peroxisomes break down long-chain fatty acids

Peroxisomes are small, membrane-bounded organelles that look very much like empty lysosomes. However, peroxisomes contain their own set of enzymes and carry out entirely different functions. Chiefly, peroxisomes bear the burden of breaking down excess quantities of long-chain fatty acids to products that can be metabolized by mitochondria for the production of ATP. In the process, they produce hydrogen peroxide (H_2O_2), a toxic molecule that is then broken down by the enzyme catalase to oxygen and water. Peroxisomes contain much catalase.

Peroxisomes also produce cholesterol and important phospholipids found primarily in brain and heart tissue. In germinating seeds, peroxisomes are called glyoxysomes because they convert fatty acids and lipids to sugars. The sugars are used as a source of energy by a germinating plant. It is fair to say that peroxisomes contribute to the energy metabolism of cells.

Normally, peroxisome size and number increase or decrease according to the needs of the cell. On rare occasions, long-chain fatty acids accumulate in cells because they are unable to enter peroxisomes for breakdown due to an inherited disorder. This leads to dramatic deterioration of the nervous system, as was depicted in the movie *Lorenzo's Oil*.

4.12 *Check Your Progress* **Why would you expect to find peroxisomes in the vicinity of mitochondria?**

4.13 Vacuoles have varied functions in protists and plants

Like vesicles, **vacuoles** are membranous sacs, but vacuoles are larger than vesicles. The vacuoles of some protists are quite specialized; they include contractile vacuoles for ridding the cell of excess water and digestive vacuoles for breaking down nutrients.

Vacuoles usually store substances. Plant vacuoles contain not only water, sugars, and salts, but also water-soluble pigments and toxic molecules. The pigments are responsible for the red, blue, or purple colors of many flowers and some leaves. The toxic substances help protect a plant from herbivorous animals.

Typically, plant cells have a large **central vacuole** that may occupy up to 90% of the volume of the cell (**Fig. 4.13**). This

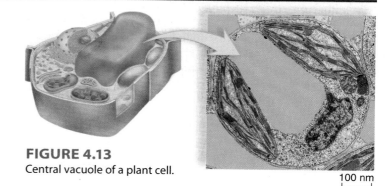

FIGURE 4.13
Central vacuole of a plant cell.

100 nm

vacuole is filled with a watery fluid called cell sap that gives added support to the cell. The central vacuole maintains turgor pressure (or hydrostatic pressure) in plant cells.

The central vacuole stores the same substances as other plant vacuoles and also waste products. A system to excrete wastes never evolved in plants, most likely because their metabolism is incredibly efficient and they produce little metabolic waste. What wastes they do produce are pumped across the membrane and stored permanently in the central vacuole. As organelles age and become nonfunctional, they fuse with the vacuole, where digestive enzymes break them down. This is a function carried out by lysosomes in animal cells.

> **4.13** *Check Your Progress* How is the central vacuole of plant cells similar to but different from the lysosomes of animals?

4.14   The organelles of the endomembrane system work together

The endomembrane system is a series of membranous organelles that work together and communicate by means of transport vesicles. It includes the endoplasmic reticulum (ER), the Golgi apparatus, lysosomes, and the transport vesicles.

Figure 4.14 shows how the components of the endomembrane system work together: **1** Proteins, produced in the rough ER, **2** are carried in transport vesicles to **3** the Golgi apparatus, which sorts the proteins and packages them into vesicles that transport them to various cellular destinations.

4 Secretory vesicles take the proteins to the plasma membrane, where they exit the cell when the vesicles fuse with the membrane. This is called secretion by exocytosis. For example, secretion into ducts occurs when the mammary glands produce milk or when the pancreas produces digestive enzymes. Similarly, lipids move from the smooth ER to the Golgi apparatus and can eventually be secreted.

5 In animal cells, lysosomes produced by the Golgi apparatus **6** fuse with incoming vesicles from the plasma membrane and digest macromolecules and debris. White blood cells are well-known for engulfing pathogens (e.g., disease-causing viruses and bacteria) that are then broken down in lysosomes.

This completes our study of the endomembrane system. The next part of the chapter considers the organelles involved in converting energy into forms useful to the cell.

> **4.14** *Check Your Progress* What parts of the cell are responsible for producing the proteins found in the endomembrane system?

FIGURE 4.14
The organelles of the endomembrane system.

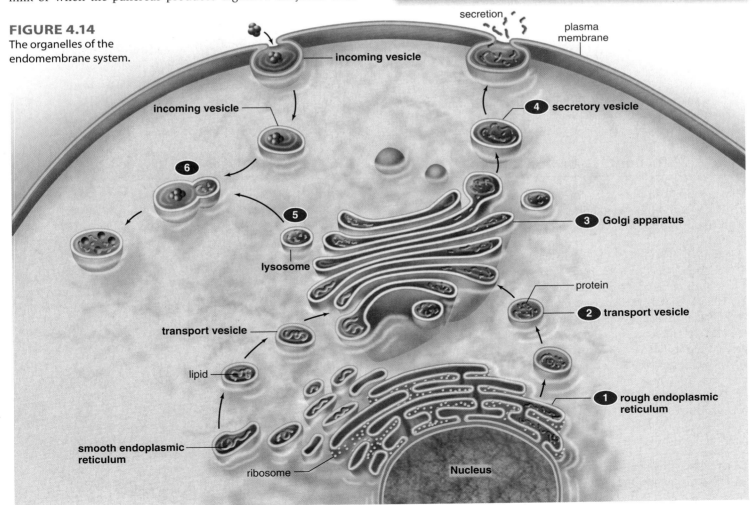

In this part of the chapter, we consider the structure and function of chloroplasts and mitochondria, the energy-related organelles. We also learn that malfunctioning mitochondria can cause human diseases.

4.15 Chloroplasts capture solar energy and produce carbohydrates

Chloroplasts are a type of plastid. **Plastids** are plant and algal organelles that are bounded by a double membrane and have a series of internal membranes separated by a ground substance. Plastids have DNA and are produced by division of existing plastids. Chloroplasts contain chlorophyll and carry on photosynthesis; the other types of plastids have a storage function.

Some algal cells have only one chloroplast, but some plant cells have as many as a hundred. Chloroplasts can be quite large, being twice as wide and as much as five times the length of a mitochondrion. Their structure is shown in **Figure 4.15**. The double membrane encloses a large space called the **stroma**, which contains **thylakoids**, disklike sacs formed from a third chloroplast membrane. A stack of thylakoids is a **granum**. The lumens of the thylakoids form a large internal compartment called the **thylakoid space**. Chlorophyll, as well as other pigments that capture solar energy, are located in the thylakoid membrane, and the enzymes that synthesize carbohydrates are located outside the thylakoid in the ground substance of the stroma.

The structure of chloroplasts, and the discovery that chloroplasts also have their own DNA and ribosomes, support the **endosymbiotic** (symbiosis means living together) **theory**, which states that chloroplasts are derived from photosynthetic bacteria that entered a eukaryotic cell approximately 1.6 billion years ago.

4.15 Check Your Progress How does the thylakoid membrane contribute to photosynthesis? Explain.

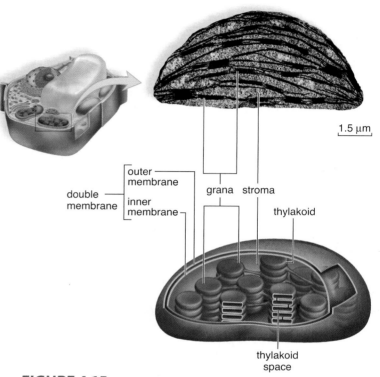

FIGURE 4.15 Chloroplast structure.

4.16 Mitochondria break down carbohydrates and produce ATP

Mitochondria can appear as in **Figure 4.16**, but they can also be longer and thinner, or shorter and broader, than this shape. Mitochondria are often found in cells where energy is most needed. For example, they are packed between the contractile elements of heart muscle cells and wrapped around the interior of a sperm's flagellum.

Mitochondria are also derived from bacteria. They arose some 1.8 billion years ago when a larger cell engulfed a bacterium that was good at using oxygen. Therefore, mitochondria have a double membrane. The inner membrane is highly folded into **cristae** that project into a **matrix**, which contains DNA, called mitochondrial DNA, and ribosomes. The cristae provide a very large surface area on which reactions can take place in an assembly-line fashion.

Mitochondria are often called the powerhouses of the cell because they produce most of the ATP used by the cell. That process, which also involves the cytoplasm, is called cellular respiration because oxygen is used and carbon dioxide is given off.

4.16 Check Your Progress Mitochondria and chloroplasts increase in number by splitting in two. Why is this to be expected?

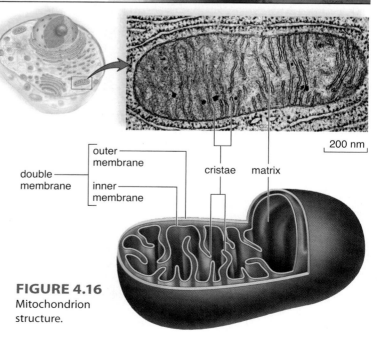

FIGURE 4.16
Mitochondrion structure.

4.17 Malfunctioning mitochondria can cause human diseases

Mitochondrial DNA (mtDNA) in humans has been sequenced, and we now know that it has 37 genes that code for either proteins needed for energy production or RNA involved in protein synthesis. Although we have known for some time that mitochondria have DNA, mtDNA mutations have only recently been linked to disease. The diseases are highly variable in terms of their onset and range of symptoms. Sometimes they occur in infancy, but they are more likely to develop later in life. If patients inherit a large percentage of bad mitochondria, they tend to get an mtDNA disease earlier and more severely. By contrast, those who inherit a large percentage of normal mitochondria either do not suffer to the same degree or get the disease later, in which case their symptoms may also be less severe.

mtDNA Diseases A general cause of mtDNA mutations has been proposed. Evidence suggests that during the production of ATP, mitochondria produce highly reactive forms of oxygen that combine with and destroy mtDNA. A vicious cycle can arise: A mutated mitochondrion reduces its energy production, which in turn leads to more highly reactive forms of oxygen and more mtDNA damage. mtDNA is subject to damage more than nuclear DNA because it is present in mitochondria, where highly reactive forms of oxygen are generated. mtDNA is thought to mutate ten times faster than nuclear DNA in an ordinary cell.

Consistent with this hypothesis, cells that need the most energy, and therefore have the most mitochondria, are also most likely to be affected by mitochondrial disease. The functioning of muscles (**Fig. 4.17**) and the central nervous system, kidneys, endocrine glands, and special senses, such as the eye and ears, is most affected by mitochondrial mutations and resultant diseases. As an example, let's consider MELAS (mitochondrial my-opathy, encephalopathy, lactic acidosis, stroke) syndrome. At first, the patient may simply feel tired; later, difficulty in walking and talking become apparent. Memory failure and dementia follow. A myriad of other conditions, such as diabetes, deafness, seizures, and pneumonia, can develop before death occurs.

People suffering from Parkinson disease or Alzheimer disease have been shown to have a higher mitochondrial mutation rate than do healthy people, and so the functioning of mitochondria may be implicated in these diseases. It is also possible that mutations caused by highly reactive forms of oxygen contribute to the aging process. Many more mtDNA mutations take place in people over 65 years of age than in younger people. Even so, many more factors are probably involved in the aging process.

Inheritance Pattern Once mtDNA mutations have arisen in a forebear, they can be inherited. However, mtDNA mutations are always inherited from the mother. The inheritance pattern is one-sided because when a sperm fertilizes an egg, the mitochondria come from the egg, not the sperm. Thereafter, cells do not create new mitochondria, and instead new mitochondria arise by splitting of old ones. So, your mtDNA was passed down to you from your mother, who received it from her mother, and so on. This means that all the mitochondria in your body are copies of the original ones in your mother's egg. Therefore, it is possible to use mtDNA to trace ancestry—to determine who is related to whom. mtDNA in teeth and bone bits was used to identify the remains of victims who died in the World Trade Center disaster of 9/11/2001.

It should be mentioned that nuclear DNA does code for some mitochondrial proteins, and this means that, on occasion, a mitochondrial disease can be due to mutations in nuclear DNA. These diseases, unlike mtDNA diseases, can be inherited from the father as well as the mother.

So far, there are no treatments for mtDNA diseases; however, an infertility treatment has been developed that seeks to correct disorders possibly associated with mitochondria. Cytoplasm, including mitochondria, from the cells of a younger woman is introduced into the eggs of an older woman. The eggs then undergo in vitro fertilization (IVF). During IVF, eggs are fertilized in laboratory glassware. Later, the eggs are placed in the uterus, where they develop normally. Because the mtDNA of the younger woman is in the egg and will be passed to all the cells of the new individual, the procedure is the first example of correcting DNA before life begins. Although controversial, the procedure has helped some 30 women worldwide carry a fetus to term and give birth to a healthy child.

This completes our study of the energy-related organelles. In the next part of the chapter, we consider the manner in which cells maintain their shape and move both their organelles and themselves.

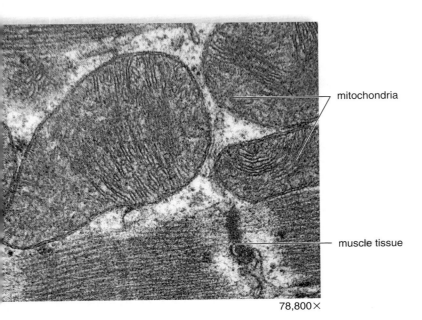

mitochondria

muscle tissue

78,800×

FIGURE 4.17 Mitochondria within a muscle cell.

> **4.17** *Check Your Progress* New nerve cells are rarely found in the body. Therefore, when you reach an advanced age, the nerve cells in your brain tend to be the same ones that you had when you were younger. Why are mtDNA brain diseases more likely to occur as we age?

As you know, the bones and muscles give an animal structure and produce movement. Similarly, we will see that the elements of the cytoskeleton maintain cell shape and cause the cell and its organelles to move. Cilia and flagella are also instrumental in producing movement; therefore, they are included in this part of the chapter as well.

4.18 The cytoskeleton consists of filaments and microtubules

The cytoskeleton contains actin filaments, intermediate filaments, and microtubules (**Fig. 4.18**). **Actin filaments** are long, extremely thin, flexible fibers (about 7 nm in diameter) that occur in bundles or mesh-like networks. Each actin filament contains two chains of globular actin monomers twisted about one another in a helical manner.

Actin filaments play a structural role when they form a dense, complex web just under the plasma membrane, to which they are anchored by special proteins. They are also seen in the microvilli that project from intestinal cells, and their presence most likely accounts for the ability of microvilli to alternately shorten and extend into the intestine. In plant cells, actin filaments apparently form the tracks along which chloroplasts circulate in a particular direction—a motion called cytoplasmic streaming. Also, the network of actin filaments lying beneath the plasma membrane accounts for the formation of **pseudopods** (false feet), extensions that allow certain cells to move in an amoeboid fashion. In our bodies, the presence of actin also permits certain white blood cells to crawl and move out of a blood vessel into the tissues, where they help defend against disease-causing agents. Actin filaments are also necessary to the contraction of muscle cells that allow all of us the freedom of locomotion.

Actin filaments interact with **motor molecules**, which are proteins that can attach, detach, and reattach farther along a filament. In the presence of ATP, myosin is a motor molecule that pulls actin filaments along in this way. Myosin has both a head and a tail. In muscle cells, the tails of several myosin molecules are joined to form a thick filament. In nonmuscle cells, cytoplasmic myosin tails are bound to membranes, but the heads still interact with actin:

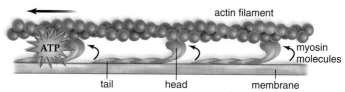

During animal cell division, the two new cells form when actin, in conjunction with myosin, pinches off the cells from one another (see Fig. 8.6A).

Intermediate filaments (8–11 nm in diameter) are intermediate in size between actin filaments and microtubules. They are a rope-like assembly of various fibrous polypeptides, each with a specific function according to the tissue. Some intermediate filaments support the nuclear envelope, whereas others support the plasma membrane and take part in the formation of cell-to-cell junctions. In the skin, intermediate filaments made of the protein keratin give great mechanical strength to skin cells.

Microtubules are made of a globular protein called tubulin. When assembly occurs, tubulin molecules come together as pairs, and these dimers arrange themselves in rows. Microtubules have 13 rows of tubulin dimers, surrounding what appears in electron micrographs to be an empty central core.

The regulation of microtubule assembly is under the control of a microtubule organizing center (MTOC). In most eukaryotic cells, the main MTOC is in the **centrosome**, which lies near the nucleus. Microtubules radiate from the centrosome, helping to maintain the shape of the cell and acting as tracks along which organelles can move. Whereas the motor molecule myosin is associated with actin filaments, the motor molecules kinesin and also dynein (not shown) are associated with microtubules.

Before a cell divides, microtubules disassemble and then reassemble into a structure called a spindle, which distributes chromosomes in an orderly manner. At the end of cell division, the spindle

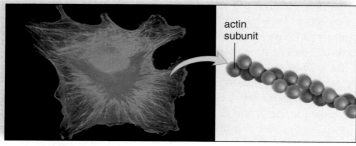

Actin filaments

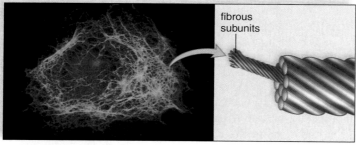

Intermediate filaments

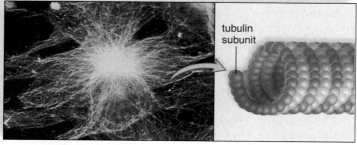

Microtubules

FIGURE 4.18 The three types of protein components of the cytoskeleton.

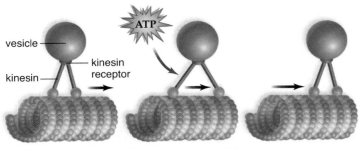

vesicle

kinesin

kinesin receptor

ATP

vesicle moves, not microtubule

disassembles, and microtubules reassemble once again into their former array. Plants have evolved various types of poisons that help prevent them from being eaten by herbivores. One of these, called colchicine, is a chemical that binds to tubulin and blocks the assembly of microtubules so that cell division is impossible.

The organelles that allow cells to move, namely cilia and flagella, contain microtubules, as discussed in Section 4.19.

4.18 Check Your Progress A cell is dynamic. What accounts for the ability of cell contents to move?

4.19 Cilia and flagella contain microtubules

Cilia and flagella (sing., **cilium**, flagellum) are whiplike projections of cells. Cilia move stiffly, like an oar, and flagella move in an undulating, snakelike fashion. Cilia are short (2–10 μm), and flagella are longer (usually no more than 200 μm). Unicellular protists utilize cilia or flagella to move about. In our bodies, ciliated cells are critical to respiratory health and our ability to reproduce. The ciliated cells that line our respiratory tract sweep debris trapped within mucus back up into the throat, which helps keep the lungs clean. Similarly, ciliated cells move an egg along the oviduct, where it can be fertilized by a flagellated sperm cell (**Fig. 4.19**).

A cilium and a flagellum have the same organization of microtubules within a plasma membrane covering. Attached motor molecules, powered by ATP, allow the microtubules in cilia and flagella to interact and bend, and thereby to move.

A particular genetic disorder illustrates the importance of normal cilia and flagella. Some individuals have an inherited defect that leads to malformed microtubules in cilia and flagella. Not surprisingly, they suffer from recurrent and severe respiratory infections, because the ciliated cells lining their respiratory passages fail to keep their lungs clean. They also are infertile due to the lack of ciliary action to move the egg in a female, or the lack of flagellar action by sperm in a male.

Centrioles Located in the centrosome, **centrioles** are short barrel-shaped organelles composed of microtubules. It's possible that centrioles give rise to **basal bodies** which lie at the base of and are believed to organize the microtubules in cilia and flagella. It's also possible that centrioles help organize the spindle, mentioned earlier, which is so necessary to cell division.

We have completed our study of eukaryotic organelles. In the next part of the chapter, we discuss the extracellular structures of cells and the matrix (packing material) that occurs between cells.

4.19 Check Your Progress How do cilia and flagella differ in structure and movement?

FIGURE 4.19 Cilia and flagella.

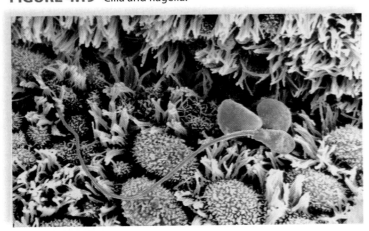

Flagellated sperm in oviduct lined by ciliated cells

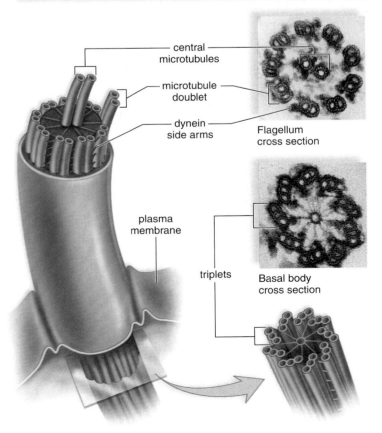

central microtubules

microtubule doublet

dynein side arms

Flagellum cross section

plasma membrane

triplets

Basal body cross section

Flagellum

Basal body

Cells have extracellular structures and a matrix between adjacent cells that take shape from material the cell produces and transports across its plasma membrane. Plants, and also prokaryotes, fungi, and most algae, have a fairly rigid cell wall. We will examine the plant cell wall in this part of the chapter. Then, we take a look at the cell surfaces of animal cells. Animal cells don't have a cell wall, but they do have junctions and/or a complex extracellular matrix.

4.20 Modifications of cell surfaces influence their behavior

Plant Cells All plants have a primary cell wall in which cellulose microfibrils are held together by noncellulose substances. The middle lamella containing pectin, a substance that usually attracts water, lies between plant cell walls. The middle lamella is a matrix (packing material) between adjacent cell walls.

Living plant cells are connected by plasmodesmata (sing., **plasmodesma**), numerous narrow, membrane-lined channels that pass through the cell wall (**Fig. 4.20A**). Cytoplasmic strands within these channels allow direct exchange of some materials between adjacent plant cells and, ultimately, all the cells of a plant. The plasmodesmata are large enough to allow only water and small solutes to pass freely from cell to cell.

Animal Cells Certain organs of vertebrate animals have junctions between their cells that allow them to behave in a coordinated manner (**Fig. 4.20B**). In an **anchoring junction,** two cells are joined together by intercellular filaments attached to internal cytoplasmic deposits, held in place by cytoskeletal filaments. The result is a sturdy but flexible sheet of cells. In some organs—such as the heart, stomach, and bladder, where tissues get stretched—anchoring junctions hold the cells together.

Adjacent cells are even more closely joined by **tight junctions,** by which plasma membrane proteins actually attach to each other, producing a zipperlike fastening. Tight junctions in the cells of the intestine prevent digestive juices from leaking into the abdominal cavity, and in the kidneys, urine stays within the kidney tubules, because the cells are joined by tight junctions.

Gap junctions, on the other hand, allow cells to communicate. A gap junction is formed when two identical plasma membrane channels join. The channel of each cell is lined by six plasma membrane proteins. A gap junction lends strength to the cells, but it also allows small molecules and ions to pass between them. Gap junctions allow the cells in heart muscle and smooth muscle to contract in a coordinated manner because they permit ions to flow between the cells.

Most animal cells have a matrix (packing material) between cells, which varies from quite flexible, as in cartilage, to rock solid, as in bone. Collagen and elastin fibers are two well-known structural proteins in the matrix. Collagen gives the matrix strength, and elastin gives it resilience.

An elaborate mixture of glycoproteins (proteins with short chains of sugars attached to them) is also present in the matrix. One glycoprotein attaches to a plasma membrane protein called integrin. Integrin spans the membrane and internally attaches to actin filaments of the cytoskeleton. Because of these connections, the matrix can influence cell migration patterns and help coordinate the behavior of all the cells in a particular tissue.

> **4.20** *Check Your Progress* **Strictly speaking, multicellular plants and animals are not composed only of cells. Why?**

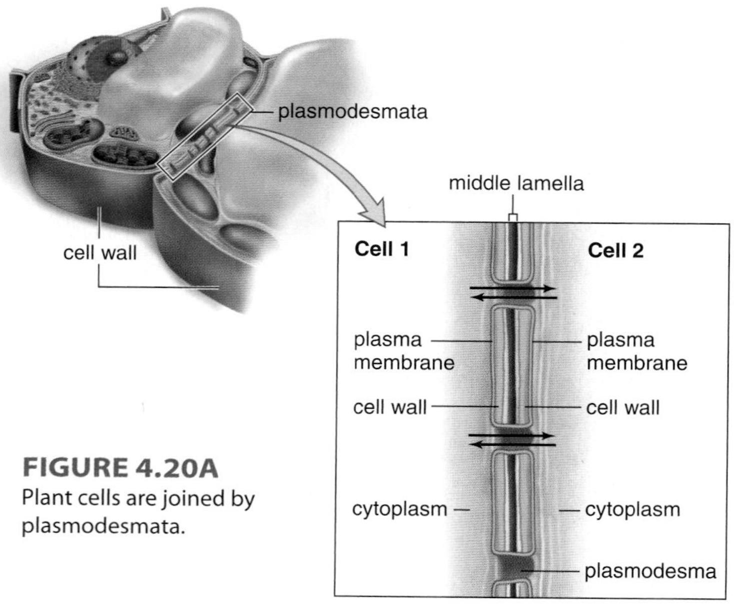

FIGURE 4.20B Animal cells are joined by three different types of junctions.

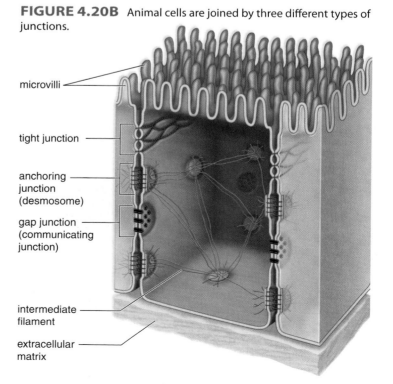

FIGURE 4.20A
Plant cells are joined by plasmodesmata.

Our knowledge of cell anatomy has been gathered by studying micrographs of cells. This has allowed cytologists (biologists who study cells) to arrive at a generalized picture of cells, such as those depicted for an animal and plant cell in Section 4.5. Eukaryotic cells, taken as a whole, contain several types of organelles, and the learning outcomes for the chapter suggest that you should know the structure and function of each one. A concept to keep in mind is that "structure suits function." For example, ribosome subunits move from the nucleus to the cytoplasm; therefore, it seems reasonable that the nuclear envelope has pores. Finding relationships between structure and function will give you a deeper understanding of the cell that will boost your memory capabilities.

Not all eukaryotic cells contain every type of organelle depicted. Cells actually have many specializations of structure for their particular functions. Because red blood cells lack a nucleus, more room is made available for molecules of hemoglobin, the molecule that transports oxygen in the blood. Muscle cells are tubular and specialized in contraction, while nerve cells have very long extensions that facilitate the transmission of impulses.

In Chapter 5, we continue our general study of the cell by considering some of the functions common to all cells. For example, all cells exchange substances across the plasma membrane and maintain a salt-water balance within certain limits. They also carry out enzymatic metabolic reactions, which either release energy or require energy. In that next chapter, we will also study the general principles of energy transformation.

The Chapter in Review

Summary

Cells: What Are They?

- Robert Hooke first defined the cell in the 17th century and since then we have learned about cells.

Cells Are the Basic Units of Life

4.1 All organisms are composed of cells

- The cell theory states the following:
  - A cell is the basic unit of life.
  - All living things are made up of cells.
  - New cells arise only from preexisting cells.

4.2 Metabolically active cells are small in size

- Cells must remain small in order to have an adequate amount of surface area per cell volume.

4.3 Microscopes allow us to see cells

- Compound light microscopes use lenses and focus light through a thin specimen.
- Transmission electron microscopes pass a beam of electrons through a thin specimen.
- Scanning electron microscopes collect and focus electrons on a specimen's surface and generate a three-dimensional image.

4.4 Prokaryotic cells evolved first

- Prokaryotic cells have the following characteristics:
  - Lack a membrane-bounded nucleus
  - Are simpler in structure than eukaryotic cells
  - Are members of domains Archaea and Bacteria

4.5 Eukaryotic cells contain specialized organelles: An overview

- Eukaryotic cells have the following characteristics:
  - Have a membrane-bounded nucleus
  - Contain organelles, structures specialized to perform specific functions

Protein Synthesis Is a Major Function of Cells

4.6 The nucleus contains the cell's genetic information

- Genes, composed of DNA, are located on chromosomes.
- RNA is produced in the nucleus.
  - rRNA is produced in the nucleolus; becomes ribosomes.
  - mRNA specifies the sequence of amino acids during protein synthesis.
  - tRNA is used in the assembly of amino acids during protein synthesis.
- Nuclear pores in the nuclear envelope permit communication between the nucleus and the cytoplasm.

4.7 The ribosomes carry out protein synthesis

- Ribosomes in the cytoplasm and the endoplasmic reticulum synthesize proteins.

4.8 The endoplasmic reticulum synthesizes and transports proteins and lipids

- The ER produces proteins (rough ER) and lipids (smooth ER) for membranes as well as proteins that are secreted from the cell.
- Transport vesicles from the ER carry proteins and lipids to the Golgi apparatus.

4.9 The Golgi apparatus modifies and repackages proteins for distribution

- Enzymes modify carbohydrate chains attached to proteins.
- Vesicles leave the Golgi apparatus and travel to the plasma membrane, where secretion occurs.

4.10 Pulse-labeling allows observation of the secretory pathway

- Electron microscopy confirms that labeled amino acids in the ER are transported in vesicles to the Golgi, and then appear in vesicles at the plasma membrane.

Vesicles and Vacuoles Have Varied Functions

4.11 Lysosomes digest macromolecules and cell parts
- Vesicles and vacuoles are membranous sacs.
- Lysosomes, which are produced by the Golgi apparatus, contain hydrolytic digestive enzymes.

4.12 Peroxisomes break down long-chain fatty acids
- Peroxisomes, which are membrane-bounded organelles resembling lysosomes, break down long-chain fatty acids.

4.13 Vacuoles have varied functions in protists and plants
- Vacuoles are larger than vesicles and usually store substances.
- Plant cells have a large central vacuole that stores water and sap and maintains turgor pressure.

4.14 The organelles of the endomembrane system work together
- The ER, Golgi apparatus, lysosomes, and other vesicles make up the endomembrane system.

A Cell Carries Out Energy Transformations

4.15 Chloroplasts capture solar energy and produce carbohydrates
- Chloroplasts carry on photosynthesis.
- Thylakoids (containing chlorophyll) capture solar energy and the stroma synthesize carbohydrates.

4.16 Mitochondria break down carbohydrates and produce ATP
- Mitochondria carry on cellular respiration.
- Matrix breaks down glucose products and the cristae produce ATP.

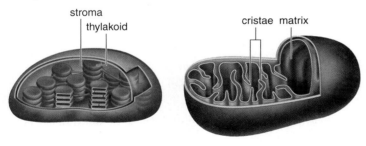

stroma
thylakoid
cristae matrix

4.17 Malfunctioning mitochondria can cause human diseases
- Mutations in mtDNA have been linked to various diseases.
- mtDNA can be used to trace ancestry.

The Cytoskeleton Maintains Cell Shape and Assists Movement

4.18 The cytoskeleton consists of filaments and microtubules
- Actin filaments are organized in bundles or networks.
- Intermediate filaments are ropelike assemblies of polypeptides.
- Microtubules are made of the globular protein tubulin. They act as tracks for organelle movement.
- The MTOC regulates microtubule assembly and is located in the centrosome.

4.19 Cilia and flagella contain microtubules
- Cilia (short) and flagella (long) are whiplike projections from the cell.
- Cilia and flagella grow from basal bodies, perhaps derived from centrioles.

In Multicellular Organisms, Cells Join Together

4.20 Modifications of cell surfaces influence their behavior
- Plants have cell walls; plant cells are joined by plasmodesmata.
- Animal cells are joined by anchoring junctions, tight junctions, and gap junctions.

Testing Yourself

Cells Are the Basic Units of Life

1. The cell theory states:
   a. Cells form as organelles and molecules become grouped together in an organized manner.
   b. The normal functioning of an organism depends on its individual cells.
   c. The cell is the basic unit of life.
   d. Only eukaryotic organisms are made of cells.
2. The small size of cells is best correlated with
   a. the fact that they are self-reproducing.
   b. their prokaryotic versus eukaryotic nature.
   c. an adequate surface area for exchange of materials.
   d. their vast versatility.
   e. All of these are correct.
3. Which of the following can only be viewed with an electron microscope?
   a. virus   c. bacteria
   b. chloroplast   d. human egg
4. Which of the following structures would be found in both plant and animal cells?
   a. centrioles   d. mitochondria
   b. chloroplasts   e. All of these are found in both types
   c. cell wall   of cells.
5. **THINKING CONCEPTUALLY** How does their comparative structure suggest that prokaryotic cells evolved before eukaryotic cells?
6. Eukaryotic cells compensate for a low surface-to-volume ratio by
   a. taking up materials from the environment more efficiently.
   b. lowering their rate of metabolism.
   c. compartmentalizing their activities into organelles.
   d. reducing the number of activities in each cell.
7. Secondary cell walls are found in _____ and contain
   _____.
   a. animals, ligand   c. plants, ligand
   b. animals, cellulose   d. plants, cellulose

Protein Synthesis Is a Major Function of Cells

8. What is synthesized by the nucleolus?
   a. mitochondria   c. transfer RNA
   b. ribosomal subunits   d. DNA
9. The organelle that can modify the sugars on a protein, determining the protein's destination in the cell is the
   a. ribosome.   c. Golgi apparatus.
   b. vacuole.   d. lysosome.
10. **THINKING CONCEPTUALLY** Explain how the structure of the rough endoplasmic reticulum suits its function.

Vesicles and Vacuoles Have Varied Functions

11. Plant vacuoles may contain
    a. flower color pigments.
    b. toxins that protect plants against herbivorous animals.
    c. sugars.
    d. All of these are correct.

12. _____ are produced by the Golgi apparatus and contain _____.
    a. Lysosomes, DNA   c. Lysosomes, enzymes
    b. Mitochondria, DNA   d. Nuclei, DNA

13. Vesicles from the ER most likely are on their way to
    a. the rough ER.   d. the plant cell vacuole only.
    b. the lysosomes.   e. the location suitable to
    c. the Golgi apparatus.   their size.

A Cell Carries Out Energy Transformations

14. Mitochondria
    a. are involved in cellular respiration.
    b. break down ATP to release energy for cells.
    c. contain grana and cristae.
    d. are present in animal cells but not in plant cells.
    e. All of these are correct.

15. The nonmembrane component of a mitochondrion is called the
    a. cristae.   c. matrix.
    b. thylakoid.   d. granum.

The Cytoskeleton Maintains Cell Shape and Assists Movement

16. Which of these are involved in the movement of the cell or parts of cells?
    a. actin   c. centrioles
    b. microtubules   d. All of these are correct.

17. Which of these statements is not true?
    a. Actin filaments are found in muscle cells.
    b. Microtubules radiate from the ER.
    c. Intermediate filaments sometimes contain keratin.
    d. Motor molecules that are moving organelles use microtubules as tracks.

In Multicellular Organisms, Cells Join Together

18. Which type of junction holds neighboring cells together so tightly that fluids cannot pass between them?
    a. anchoring   c. plasmodesmata
    b. gap   d. tight

Understanding the Terms

actin filament 70	centrosome 70
anchoring junction 72	chromatin 62
Archaea 59	chromosome 62
Bacteria 59	cilium 71
basal body 71	compound light
capsule 59	microscope 58
cell theory 56	cristae 68
cell wall 59	cytoplasm 59
central vacuole 66	cytoskeleton 61
centriole 71	endomembrane system 60

endoplasmic reticulum (ER) 63	plasma membrane 59
endosymbiotic theory 68	plasmodesma 72
energy-related organelle 60	plastid 68
Eukarya 60	prokaryotic cell 59
eukaryotic cell 59	pseudopod 70
fimbriae 59	ribosome 59
flagellum 59	rough ER (RER) 64
gap junction 72	scanning electron
Golgi apparatus 65	microscope 58
granum 68	secretion 65
intermediate filament 70	sex pilus 59
lysosome 66	smooth ER (SER) 64
matrix 68	stroma 68
microtubule 70	surface-area-to-volume
motor molecule 70	ratio 57
nuclear envelope 62	thylakoid 68
nuclear pore 62	thylakoid space 68
nucleoid 59	tight junction 72
nucleolus 62	transmission electron
nucleus 62	microscope 58
organelle 60	transport vesicle 64
peroxisome 66	vacuole 66

Match the terms to these definitions:
a. _____ Organelle, consisting of saccules and vesicles, that processes, packages, and distributes molecules about or from the cell.
b. _____ Especially active in lipid metabolism; always produces H_2O_2.
c. _____ Dark-staining, spherical body in the cell nucleus that produces ribosomal subunits.
d. _____ Internal framework of the cell, consisting of microtubules, actin filaments, and intermediate filaments.
e. _____ Allows prokaryotic cells to attach to other cells.

Thinking Scientifically

1. Utilizing Palade's procedure, described in Section 4.10, you decide to label and trace the base uracil (Fig. 3.11B). What type of molecule are you labeling, and where do you expect to find it in Figure 4.10?

2. After publishing your study from question 1, you are criticized for failing to trace uracil from mitochondria. Why might you have looked for uracil in mitochondria and what comparative difference between the nuclear envelope and the mitochondrial double membrane might justify your study as is?

ARIS™ Visit www.mhhe.com/maderconcepts for practice quizzes, animations, videos, and activities designed to help you master the material in this chapter.

5

Dynamic Activities of Cells

LEARNING OUTCOMES

After studying this chapter, you should be able to accomplish the following outcomes.

Life's Energy Comes from the Sun

1 Trace the flow of energy from the sun to a meat eater.

Living Things Transform Energy

2 State and apply two energy laws to energy transformations.
3 Give reasons why ATP is called the energy currency in cells.
4 Give examples to show how ATP hydrolysis is coupled to energy-requiring reactions.

Enzymes Speed Chemical Reactions

5 Explain how enzymes speed chemical reactions.
6 List conditions that affect enzyme speed.
7 Give an example of how an enzyme inhibitor can be fatal.

The Plasma Membrane Has Many and Various Functions

8 Describe the structure of the plasma membrane and the diverse functions of the embedded proteins.
9 Give examples of human diseases in which the plasma membrane is malfunctioning.

The Plasma Membrane Regulates the Passage of Molecules Into and Out of Cells

10 Compare and contrast passive and active ways that substances can cross the plasma membrane.
11 Predict the effect of osmotic conditions on animal versus plant cells.

Life on Earth is dependent on a flow of energy, and this flow begins with the sun. The sun is a huge cloud of hot gases where thermonuclear reactions occur between hydrogen and helium atoms. These thermonuclear reactions are the source of the energy that supports the biosphere.

Hot gasses surrounding the surface of the sun

Solar radiation travels through the immense amount of space that separates us from the sun. If it were not for this distance, the Earth would be too hot for life to exist. Even so, every hour, more solar energy reaches the Earth than the entire world's population consumes in a year.

We all enjoy the warmth of the sun, and most solar energy does become heat that is absorbed by the Earth or reradiated back into space. Less than 1% of the solar energy that strikes the Earth is taken up by photosynthesizers, which include plants, algae, and cyanobacteria. Like all photosynthesizers, grasses have the ability to convert solar energy into the chemical energy of organic molecules. The organic molecules allow grasses to grow and serve as food for animals, such as impalas on the African plain. Food provides the building blocks and energy that impalas need just to exist. Impalas also use energy to take off at high speed when they are trying to evade a predator, such as a cheetah. Eating impalas provides cheetahs with the food they need to maintain themselves and to be quick enough to catch impalas! Notice that we have just described a flow of energy that proceeds like this: from the sun, to grasses, to impalas, and finally to cheetahs.

Cheetah chasing an impala

Life's Energy Comes from the Sun

The illustrations on this page give another example of energy flow. Do you get it? It goes like this: from the sun, to corn plants, to cattle, to humans who are out for a run with their dog. Your gnawing stomach makes you aware of the need to eat food every day, but you may not realize why like all living things, humans are dependent on a constant flow of energy from the sun. The answer is: "energy dissipates." When muscle contraction is over, energy escapes into the body of an animal and then into the environment. The heat given off when your muscles contract is put to good use. It keeps you warm, but it is no longer usable by photosynthesizers for chemical reactions. It is too diffuse. Solar energy is concentrated enough, however, to allow plants to keep on photosynthesizing and, in that way, provide the biosphere with organic food.

This introduction gives you an overview of how organisms use energy, the first topic we consider in this chapter. Energy is an important part of metabolism, and so are enzymes, the proteins that speed chemical reactions. Without enzymes, you would not be able to make use of energy to maintain your body and to carry on activities, such as muscle contraction. Metabolism is a cellular affair, and cells can't keep on metabolizing unless substances cross the plasma membrane. Therefore, we will be considering how molecules get into and out of the many cells that make up your body.

Beef cattle

Sunrise over a cornfield

Cheetah eating an impala

Cells are constantly converting—that is, transforming—one form of energy into another. This part of the chapter introduces you to the many different forms of energy and the energy laws that pertain to transformations. These laws readily explain why living things need a continual supply of energy. The preferred form of energy in cells is ATP, called the "energy currency" of cells because when cells need something or do any kind of work, they "spend" ATP.

5.1 Energy makes things happen

Living organisms are highly ordered, and energy is needed to maintain this order. Organisms acquire energy, store energy, and release energy, and only by transforming one form of energy into another form can organisms continue to stay alive. Despite its importance to living things and society, energy is a strange commodity because we cannot see it. So, energy is indeed conceptual. Most authorities define **energy** as the capacity to do work—to make things happen. Without a continual source of energy, living things could not exist.

There are five specific forms of energy: radiant, chemical, mechanical, electrical, and nuclear. In this book, we are particularly interested in radiant energy, chemical energy, and mechanical energy. Radiant energy, in the form of solar energy, can be captured by plants to make their own food and food for the biosphere. Chemical energy is present in organic molecules, and therefore, chemical energy is the direct source of energy for nonphotosynthesizers. Mechanical energy is represented by any type of motion—the motion of a skier, as well as the motion of atoms, ions, or molecules, which is better known as **heat**. Heat is dispersed energy, and therefore, it is hard to collect and use for any purpose other than space heating. The chemical energy of food is a high-quality source of energy because it is available to do work. Heat, on the other hand, is low-quality energy because it has little ability to do useful work. We learned in Chapter 2 that the body can use excess heat to evaporate sweat, and in that way, the temperature of the body lowers.

All the specific types of energy we have been discussing are either potential energy or kinetic energy. **Potential energy** is stored energy, and **kinetic energy** is energy in action. Potential energy is constantly being converted to kinetic energy, and vice versa.

Let's look at the example in **Figure 5.1**. The chemical energy in the food a cross-country skier has for breakfast contains potential energy. When the skier hits the trail, she may have to ascend a hill. During her climb, the potential energy of food is converted to the kinetic energy of motion. Once she reaches the hilltop, kinetic energy has been converted to the potential energy of location (greater altitude). As she skis down the hill, this potential energy is converted to kinetic energy again.

Both potential and kinetic energy are important to living things because cells constantly store energy and then gradually release it to do work. To take an example, liver cells store energy as glycogen, and then they break down glycogen in order to make ATP molecules, which carry on the work of the cell.

It is important to have a way to measure energy. A **calorie** is the amount of heat required to raise the temperature of 1 g of water by 1° Celsius. This isn't much energy, so the caloric value of food is listed in nutrition labels and in diet charts in terms of **kilocalories** (1,000 calories). In this text, we will use Calorie (C) to mean 1,000 calories. Section 5.2 considers two energy laws that explain why all chemical energy in cells eventually becomes heat in the atmosphere.

> **5.1 Check Your Progress a. Does ATP represent kinetic energy or potential energy? Explain. b. Muscle movement driven by ATP is what type of energy?**

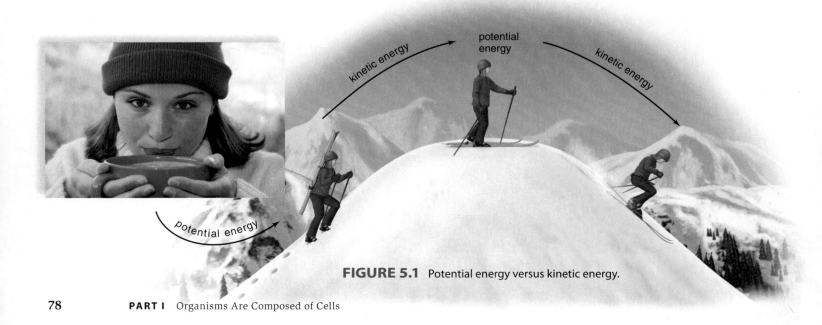

FIGURE 5.1 Potential energy versus kinetic energy.

5.2 Two laws apply to energy and its use

Two laws, called the laws of thermodynamics, govern the use of energy. These laws were formulated by early researchers who studied energy relationships and exchanges. Neither nonliving nor living things can circumvent these laws.

The first law of thermodynamics—the law of conservation of energy—states that energy cannot be created or destroyed, but it can be changed from one form to another.

Figure 5.2 shows how this law applies to living things. A shrub is able to convert solar energy to chemical energy, and a moose, like all animals including humans, is able to convert chemical energy into the energy of motion. Notice that with every energy transformation, however, some energy is lost as heat. The word "lost" recognizes that when energy has become heat; it is no longer usable to perform work.

The second law of thermodynamics states that energy cannot be changed from one form to another without a loss of usable energy.

Let's look at Figure 5.2 in a bit more detail. When leaf cells photosynthesize, they use solar energy to form carbohydrate molecules from carbon dioxide and water. (Carbohydrates are energy-rich molecules, while carbon dioxide and water are energy-poor molecules.) Not all of the captured solar energy becomes carbohydrates; some becomes heat:

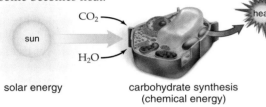

FIGURE 5.2 Flow of energy from the sun to an animal that eats a plant.

Plant cells do not create or destroy energy in this process—the sun is the energy source, and the unusable heat is still a form of energy. Similarly, as a moose uses the energy derived from carbohydrates to power its muscles, none is destroyed, but some becomes heat, which dissipates into the environment:

With transformation upon transformation, eventually all of the captured solar energy becomes heat that is lost to the environment. Therefore, energy flows through living things. All living things are dependent on a constant supply of solar energy.

Notice too that no conversion of energy is ever 100% efficient. The gasoline engine in an automobile is between 20% and 30% efficient in converting chemical energy into mechanical energy. The majority of energy is lost as heat. Cells are capable of about 40% efficiency, with the remaining energy given off to the surroundings as heat.

The second law of thermodynamics tells us that as energy conversions occur, disorder increases because it is difficult to use heat to perform more work. The word **entropy** is often used to describe this disorder. Energy transformations can occur, but they always increase entropy.

Now that we know the basics of energy transformations, let's see why cells prefer to rely on ATP as their direct source of energy.

> **5.2 Check Your Progress** If you take a walk on the beach with your dog, does entropy increase?

5.3 Cellular work is powered by ATP

Many of the appliances in your kitchen, such as the dishwasher, stove, and refrigerator, are powered by electricity. Cells, as mentioned earlier, use ATP (adenosine triphosphate) to power reactions. ATP is often called the energy currency of cells. Just as you use cash to purchase all sorts of products, a cell uses ATP to carry out nearly all of its activities, including synthesizing macromolecules, transporting ions across the plasma membranes, and causing organelles and cilia to move.

ATP is a nucleotide, the type of molecule that serves as a monomer for the construction of DNA and RNA. Its name, adenosine triphosphate, means that it contains the sugar ribose, the nitrogen-containing base adenine, and three phosphate groups (**Fig. 5.3A**, *top*). The three phosphate groups are negatively charged and repel one another. It takes energy to overcome their repulsion, and thus these phosphate groups make the molecule unstable.

ATP easily loses the last phosphate group because the breakdown products, ADP (adenosine diphosphate) and a separate phosphate group symbolized as ⓟ, are more stable than ATP. This reaction is written as: ATP ⟶ ADP + ⓟ. ADP can also lose a phosphate group to become AMP (adenosine monophosphate).

The continual breakdown and regeneration of ATP is known as the ATP cycle (Fig. 5.3A, *bottom*). ATP stores energy for only a short period of time before it is used in a reaction that requires energy. Then ATP is rebuilt from ADP + ⓟ. Each ATP molecule undergoes about 10,000 cycles of synthesis and breakdown every day. Our bodies use some 40 kg (about 88 lb) of ATP daily, and the amount on hand at any one moment is sufficiently high to meet only current metabolic needs.

ATP's instability, the very feature that makes it an effective energy donor, keeps it from being an energy storage molecule. Instead, the many H—C bonds of carbohydrates and fats make them the energy storage molecules of choice. Their energy is extracted during cellular respiration and used to rebuild ATP, mostly within mitochondria. Cellular respiration, during which glucose is broken down, is called an **exergonic reaction** because this process gives up energy. In other words, energy *exits* from cellular respiration. This energy is used to build up ATP.

The breakdown of one molecule of glucose permits the buildup of some 38 molecules of ATP. During cellular respiration, only 39% of the potential energy of glucose is converted to the potential energy of ATP; the rest is lost as heat.

The production of ATP is still worthwhile for the following reasons:

1. ATP is suitable for use in many different types of cellular reactions that only occur if energy is supplied. Such reactions are called **endergonic reactions**. In other words, energy must *enter* in order for the reaction to occur.

2. When ATP becomes ADP + ⓟ, the amount of energy released is more than the amount needed for a biological purpose, but not overly wasteful. Fireflies, for example, use ATP to produce light without producing excessive heat (**Fig. 5.3B**). If ATP were to break down on its own, its energy would be lost, but fortunately, ATP breakdown is coupled to reactions that require energy, as discussed next.

> **5.3** *Check Your Progress* Fireflies store little ATP, but they can produce light for an entire evening. Why?

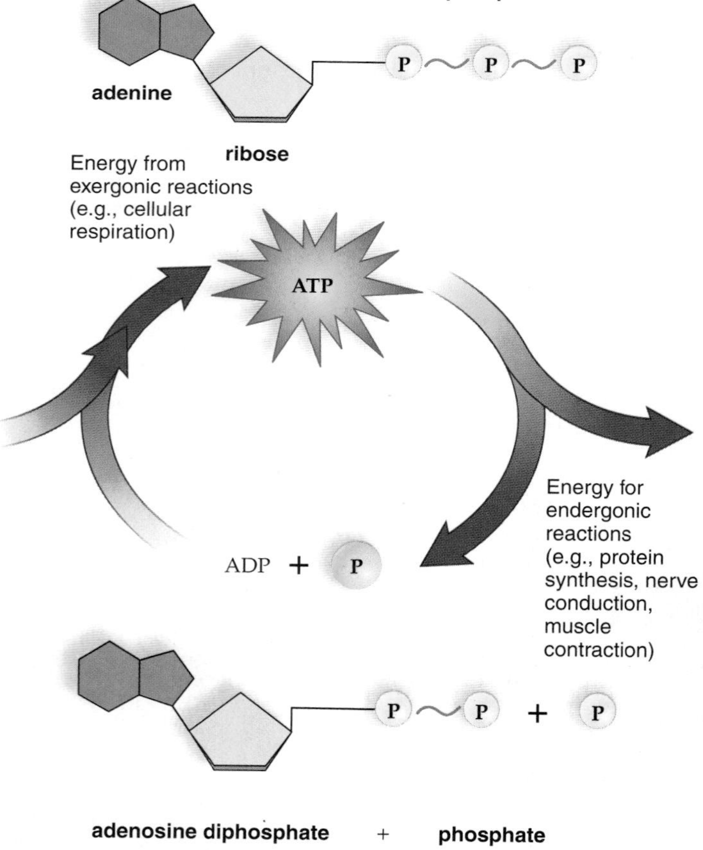

adenosine triphosphate

adenine

ribose

Energy from exergonic reactions (e.g., cellular respiration)

ATP

Energy for endergonic reactions (e.g., protein synthesis, nerve conduction, muscle contraction)

ADP + P

adenosine diphosphate + **phosphate**

FIGURE 5.3A The ATP cycle.

2.25×

FIGURE 5.3B Fireflies break down ATP to produce light.

How can the energy released by ATP hydrolysis be transferred to a reaction that requires energy, and therefore would not ordinarily occur? In other words, how does ATP act as a carrier of chemical energy? The answer is that ATP breakdown is coupled to the energy-requiring reaction. **Coupled reactions** are reactions that occur in the same place, at the same time, and in such a way that an energy-releasing (exergonic) reaction drives an energy-requiring (endergonic) reaction. Usually the energy-releasing reaction is the hydrolysis of ATP. Because the cleavage of ATP's phosphate group releases more energy than the amount consumed by the energy-requiring reaction, entropy will increase, and both reactions will proceed. The simplest way to represent a coupled reaction is like this:

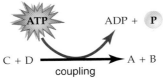

This reaction tells you that coupling occurs, but it does not show how coupling is achieved. A cell has two main ways to couple ATP hydrolysis to an energy-requiring reaction: ATP is used to energize a reactant, or ATP is used to change the shape of a reactant. Both can be achieved by transferring a phosphate group to the reactant.

For example, when an ion moves across the plasma membrane of a cell, ATP is hydrolyzed, and instead of the last phosphate group floating away, an enzyme attaches it to a protein. This causes the protein to undergo a change in shape that allows it to move the ion into or out of the cell. As a contrasting example, when a polypeptide is synthesized at a ribosome, an enzyme transfers a phosphate group from ATP to each amino acid in turn, and this transfer supplies the energy that allows an amino acid to bond with another amino acid.

Figure 5.4 shows how ATP hydrolysis provides the necessary energy for muscle contraction. During muscle contraction, myosin filaments pull actin filaments to the center of the cell, and the muscle shortens. **1** Myosin head combines with ATP (three connected green triangles) and takes on its resting shape. **2** ATP breaks down to ADP (two green triangles) plus Ⓟ. Now a change in shape allows myosin to attach to actin. **3** The release of ADP from myosin head, causes it to change its shape again and pull on the actin filament. The cycle begins again at **1**, when myosin head combines with ATP and takes on its resting shape. During this cycle, chemical energy has been transformed to mechanical energy, and entropy has increased.

Through coupled reactions, ATP drives forward energetically unfavorable processes that must occur to create the high degree of order essential for life. Macromolecules must be made and organized to form cells and tissues; the internal composition of the cell and the organism must be maintained; and movement of cellular organelles and the organism must occur if life is to continue.

This completes our discussion of energy transformations in cells. In the next part of the chapter, we will be studying metabolism in general.

5.4 Check Your Progress The ability of impalas to dash across the African plain obeys the second law of thermodynamics. How?

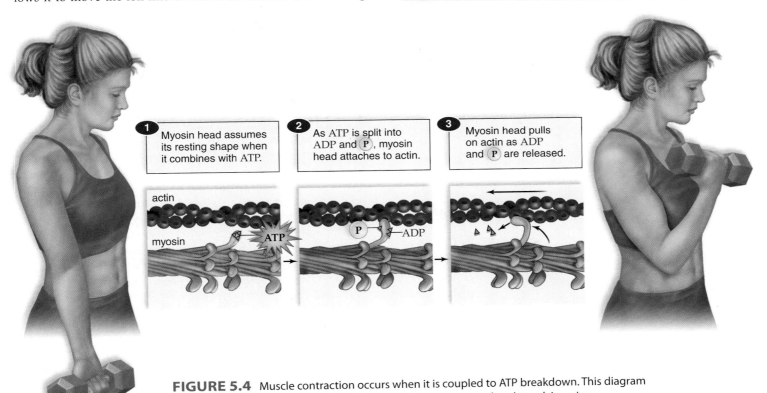

FIGURE 5.4 Muscle contraction occurs when it is coupled to ATP breakdown. This diagram shows the action of one myosin head but actually many myosin heads work in unison.

This part of the chapter involves a general study of **metabolism,** which includes all the chemical reactions that occur in a cell. Few reactions occur in a cell unless an enzyme is present to bring specific reactants together in a way that causes them to react. Therefore, a study of metabolism involves a study of enzymes and how they are affected by local conditions and regulated by inhibitors.

5.5 Enzymes speed reactions by lowering activation barriers

The food on your dinner plate doesn't break down into nutrient molecules until it enters your digestive tract, where it encounters enzymes. An **enzyme** is usually a protein molecule that functions as an organic catalyst to speed a chemical reaction without itself being affected by the reaction. Just like your digestive tract, cells contain many types of enzymes. Regardless of where they are, enzymes cause reactions to occur. However, enzymes can only speed reactions that would occur anyway, not energetically unfavorable reactions.

Imagine the graph in **Figure 5.5** as a roller coaster ride. To get the ride started, you have to push the car (the reactants) to the top of an incline. Then, just as the car will naturally fall, the reaction will occur. In the lab, heat is often used to increase the effective collisions between molecules so that the reaction can occur. When an enzyme is present, the **energy of activation** (E_a) is lower than it would be without the enzyme. Enzymes lower the energy of activation by bringing reactants together in an effective way *at body temperature*, as is discussed further in Section 5.6.

> **5.5 Check Your Progress** Each enzyme speeds only one type of reaction. Despite a different function, what do enzymes have in common?

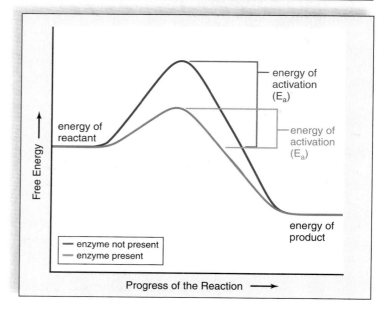

FIGURE 5.5 The energy of activation (E_a) is lower when an enzyme is involved.

5.6 An enzyme's active site is where the reaction takes place

Each enzyme is specific to the reaction that it speeds. That is why a cell needs so many different enzymes. The reactants in an enzymatic reaction are called the enzyme's **substrate(s).** Substrates are specific to a particular enzyme because they bind with an enzyme, forming an enzyme-substrate complex. Only one small part of the enzyme, called the **active site,** binds with the substrate(s) (**Fig. 5.6**).

At one time, it was thought that an enzyme and a substrate fit together like a key fits a lock, but now we know that the active site undergoes a slight change in shape to accommodate the substrate(s). This is called the **induced fit model** because the enzyme is induced to undergo a slight alteration to achieve optimum fit. The change in shape of the active site facilitates the reaction that now occurs. After the reaction has been completed, the product is released, and the active site returns to its original state, ready to bind to another substrate molecule. A cell needs only a small amount of enzyme because enzymes are not used up by the reaction; instead, they are used over and over again.

Some enzymes do more than simply bind with their substrate(s); they participate in the reaction. For example, trypsin digests protein by breaking peptide bonds. The active site of trypsin contains three amino acids with *R* groups that actually interact with members of the peptide bond—first to break the bond and then to introduce the components of water. This illustrates that the shape of an enzyme and the formation of the enzyme-substrate complex are critical to speeding the enzyme's reaction, as discussed further in Section 5.7.

FIGURE 5.6
Enzymatic action.

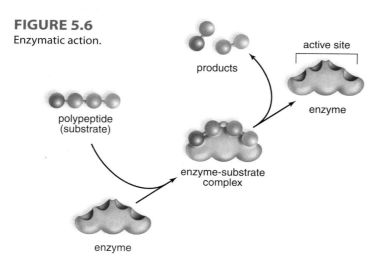

> **5.6 Check Your Progress** Every enzyme has an active site. How does the active site of one enzyme differ from another?

5.7 Enzyme speed is affected by local conditions

The rate of a reaction is the amount of product produced per unit time. Generally, enzymes work quickly, and in some instances they can increase the reaction rate more than 10 million times. To achieve the maximum rate, enough substrate should be available to fill the active sites of all enzyme molecules most of the time. Increasing the amount of substrate, and providing an adequate temperature and optimal pH, also increase the rate of an enzymatic reaction.

Substrate Concentration Molecules must come together in order to react. Generally, enzyme activity increases as substrate concentration increases because there are more chance encounters between substrate molecules and the enzyme. As more substrate molecules fill active sites, more product results per unit time. But when the enzyme's active sites are filled almost continuously with substrate, the enzyme's rate of activity cannot increase any more. Maximum rate has been reached.

Just as the amount of substrate can increase or limit the rate of an enzymatic reaction, so the amount of active enzyme can also increase or limit the rate of an enzymatic reaction.

Temperature Typically, as temperature rises, enzyme activity increases (**Fig. 5.7A**). This occurs because warmer temperatures cause more effective encounters between enzyme and substrate.

The body temperature of an animal seems to affect whether it is normally active or inactive. It has been suggested that the often cold temperature of a reptile's body (**Fig. 5.7B**) hinders metabolic reactions and may account for why mammals are more prevalent today. The generally warm temperature of a mammal's body (**Fig. 5.7C**) allows its enzymes to work at a rapid rate despite a cold outside temperature.

In the laboratory, if the temperature rises beyond a certain point, enzyme activity eventually levels out and then declines rapidly because the enzyme has been **denatured**. An enzyme's shape changes during denaturation, and then it can no longer bind its substrate(s) efficiently. Nevertheless, some prokaryotes can live in hot springs because their enzymes do not denature.

pH Each enzyme also has an optimal pH at which the rate of the reaction is highest. At this pH value, these enzymes have their nor-

FIGURE 5.7B If body temperature tends to be cold, as in reptiles, reaction rates are slow.

mal configurations. The globular structure of an enzyme is dependent on interactions, such as hydrogen bonding, between R groups. A change in pH can alter the ionization of these side chains and disrupt normal interactions; under extreme conditions of pH, denaturation eventually occurs. Again, the enzyme's shape has been altered so that it is unable to combine efficiently with its substrate.

Cofactors Many enzymes require the presence of an inorganic ion, or a nonprotein organic molecule, in order to be active; these necessary ions or molecules are called **cofactors**. The inorganic ions are metals such as copper, zinc, or iron. The nonprotein organic molecules are called **coenzymes**. These cofactors assist the enzyme and may even accept or contribute atoms to the reactions.

Vitamins are relatively small organic molecules that are required in trace amounts in the diets of humans and other animals for synthesis of coenzymes. The vitamin becomes part of a coenzyme's molecular structure. If a vitamin is not available, enzymatic activity decreases, and the result is a vitamin-deficiency disorder. For example, niacin deficiency results in a skin disease called pellagra, and riboflavin deficiency results in cracks at the corners of the mouth.

Inhibitors, which are discussed in Section 5.8, reduce the amount of product produced by an enzyme per unit time.

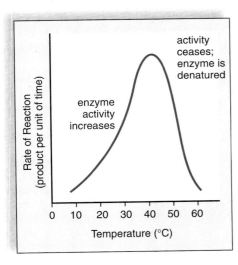

FIGURE 5.7A The effect of temperature on the rate of an enzymatic reaction.

Figure 5.7C If body temperature tends to be warm, as in mammals, reaction rates are increased.

> **5.7** *Check Your Progress* A pH of 1–2 is optimal for pepsin, a digestive enzyme. But like all human enzymes, pepsin works best at body temperature. Explain.

5.8 Enzymes can be inhibited noncompetitively and competitively

Figure 5.8 shows that reactions do not occur haphazardly in cells; they are usually part of ❶ a **metabolic pathway**, a series of linked reactions. Enzyme inhibition occurs when a molecule (the inhibitor) binds to an enzyme and decreases its activity. As shown, the inhibitor can be the end product of a metabolic pathway. This is beneficial because once sufficient end product of a metabolic pathway is present, it is best to inhibit further production to conserve raw materials and energy.

❷ Figure 5.8 also illustrates **noncompetitive inhibition** because the inhibitor (F, the end product) binds to the enzyme E_1 at a location other than the active site. The site is called an allosteric site. When an inhibitor is at the allosteric site, the active site of the enzyme changes shape.

❸ The enzyme E_1 is inhibited because it is unable to bind to A, its substrate. The inhibition of E_1 means that the metabolic pathway is inhibited and no more end product will be produced.

In contrast to noncompetitive inhibition, **competitive inhibition** occurs when an inhibitor and the substrate compete for the active site of an enzyme. Product will form only when the substrate, not the inhibitor, is at the active site. In this way, the amount of product is regulated.

Normally, enzyme inhibition is reversible, and the enzyme is not damaged by being inhibited. When enzyme inhibition is irreversible, the inhibitor permanently inactivates or destroys an enzyme. As discussed in Section 5.9, many metabolic poisons are irreversible enzyme inhibitors.

> **5.8 Check Your Progress** Enzyme inhibition can be dangerous, and yet it is used in the cell to regulate enzymes. Explain.

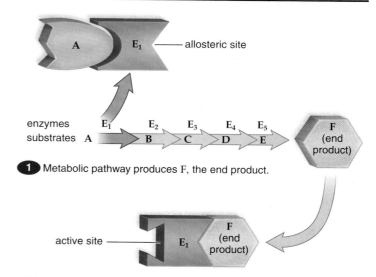

❶ Metabolic pathway produces F, the end product.

❷ F binds to allosteric site and the active site of E_1 changes shape.

❸ A cannot bind to E_1; the enzyme has been inhibited by F.

FIGURE 5.8 Metabolic pathways and noncompetitive inhibition. In the pathway, A–E are substrates, E_1–E_5 are enzymes, and F is the end product of the pathway.

HOW BIOLOGY IMPACTS OUR LIVES

5.9 Enzyme inhibitors can spell death

Cyanide gas was formerly used to execute people. How did it work? Cyanide can be fatal because it binds to a mitochondrial enzyme necessary for the production of ATP. MPTP (1-methyl-4-phenyl-1,2,3.6-tetrahydropyridine) is another enzyme inhibitor that stops mitochondria from producing ATP. The toxic nature of MPTP was discovered in the early 1980s, when a group of intravenous drug users in California suddenly developed symptoms of Parkinson disease, including uncontrollable tremors and rigidity. All of the drug users had injected a synthetic form of heroin that was contaminated with MPTP. Parkinson disease is characterized by the death of brain cells, the very ones that are also destroyed by MPTP.

Sarin is a chemical that inhibits an enzyme at neuromuscular junctions, where nerves stimulate muscles. When the enzyme is inhibited, the signal for muscle contraction cannot be turned off, so the muscles are unable to relax and become paralyzed. Sarin can be fatal if the muscles needed for breathing become paralyzed. In 1995, terrorists released sarin gas on a subway in Japan (**Fig. 5.9**). Although many people developed symptoms, only 17 died.

A fungus that contaminates and causes spoilage of sweet clover produces a chemical called warfarin. Cattle that eat the spoiled feed

die from internal bleeding because warfarin inhibits a crucial enzyme for blood clotting. Today, warfarin is widely used as a rat poison. Unfortunately, it is not uncommon for warfarin to be mistakenly eaten by pets and even very small children, with tragic results.

Many people are prescribed a medicine called Coumadin to prevent inappropriate blood clotting. For example, those who have received an artificial heart valve need such a medication. Coumadin contains a nonlethal dose of warfarin.

> **5.9 Check Your Progress** What would account for the survival of some people exposed to sarin?

FIGURE 5.9
The aftermath when sarin, a nerve gas that results in the inability to breathe, was released by terrorists in a Japanese subway in 1995.

In this part of chapter, we will study the structure of the plasma membrane and the functions of the many types of proteins found in the plasma membrane. The plasma membrane is not a passive boundary of the cell; it has many and various functions, some of which are dependent on the many proteins found in the membranes.

5.10 The plasma membrane is a phospholipid bilayer with embedded proteins

The plasma membrane marks the boundary between the outside and the inside of a cell. Its integrity and function are necessary to the life of the cell because it regulates the passage of molecules and ions into and out of the cell.

In both bacteria and eukaryotes, the plasma membrane is a fluid **phospholipid bilayer** that has the consistency of olive oil. Recall that the polar head of a phospholipid is hydrophilic, while the nonpolar tails are hydrophobic. The polar heads of the phospholipids face toward the outside of the cell and toward the inside of the cell, where there is a watery medium. The nonpolar tails face inward toward each other, where there is no water. Similarly, the embedded proteins have a hydrophobic region within the membrane and hydrophilic regions that extend beyond the surface of the membrane. The presence of these regions prevents embedded proteins from flipping, but they can move laterally. The **fluid-mosaic model** states that the protein molecules embedded in the membrane have a pattern (form a mosaic) within the phospholipid bilayer (**Fig. 5.10**). The pattern varies according to the particular membrane

and also within the same membrane at different times. **Cholesterol** molecules are steroids that lend support to the membrane; other steroids perform this function in the plasma membranes of plants.

Both phospholipids and proteins can have attached carbohydrate (sugar) chains. Molecules carrying such chains are called **glycolipids** and **glycoproteins**, respectively. Since the carbohydrate chains occur only on the outside surface, and since peripheral proteins occur only on the inner surface of the membrane, the two sides of the membrane are not identical.

In animal cells, the carbohydrate chains of proteins give the cell a "sugar coat," more properly called the glycocalyx. The glycocalyx protects the cell and has various other functions. For example, it facilitates adhesion between cells, reception of signal molecules, and cell-to-cell recognition. In Section 5.11, we explore the functions of all the types of embedded plasma membrane proteins.

> **5.10 Check Your Progress** A mosaic floor is made of small tiles of different colors, cemented in place. How is the structure of the plasma membrane similar to, but different from, a mosaic floor?

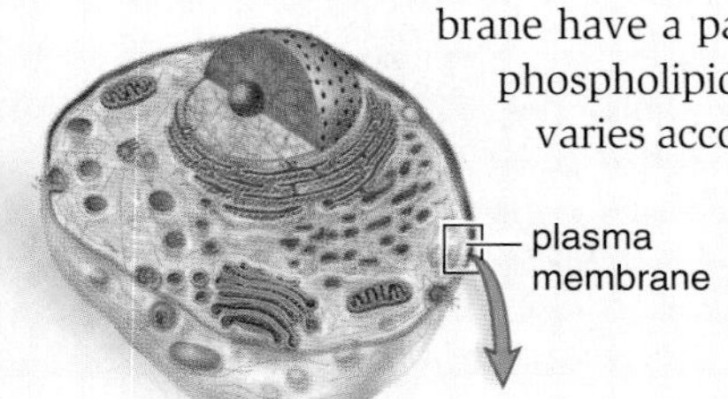

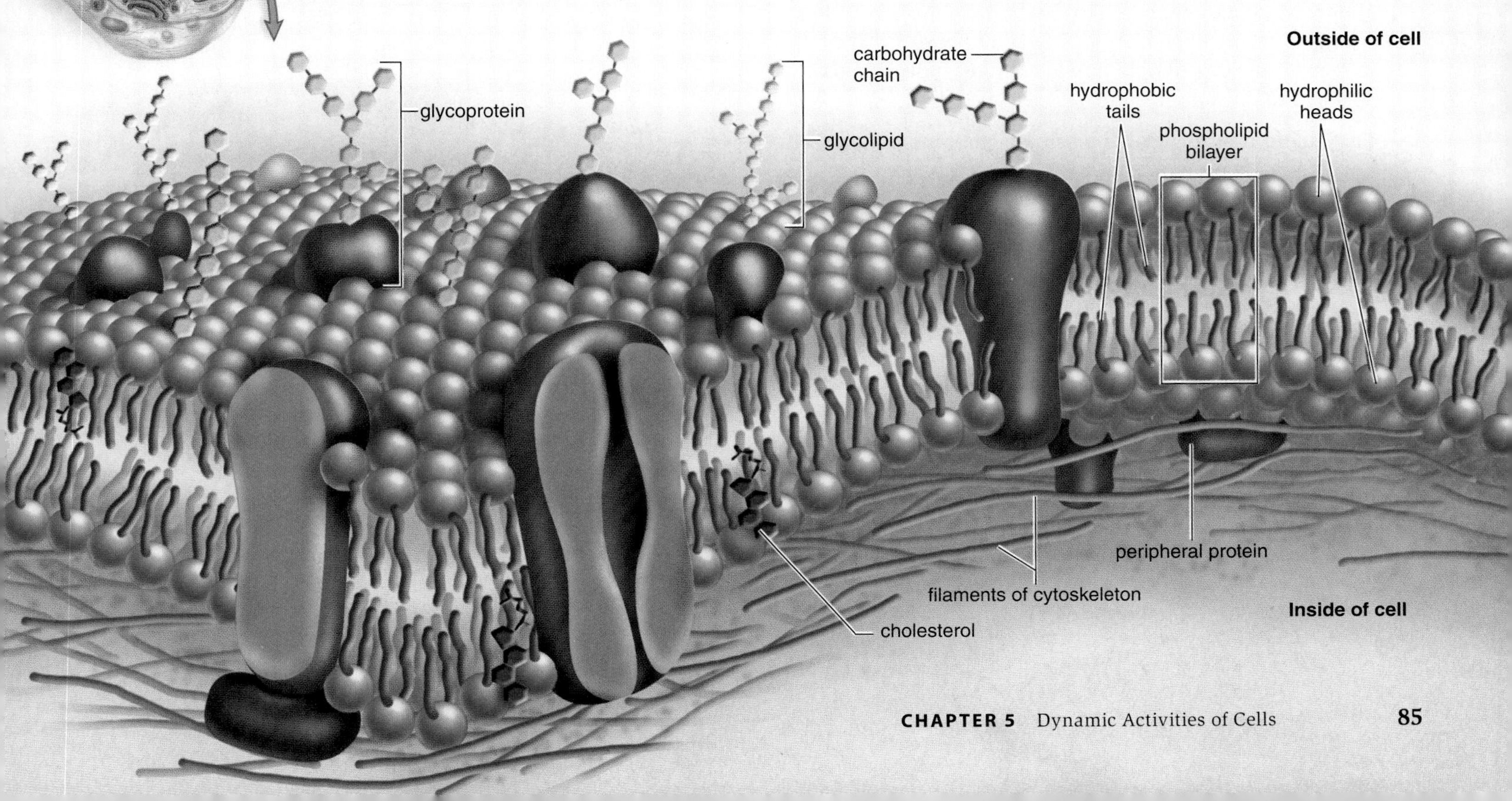

FIGURE 5.10 Fluid-mosaic model of plasma membrane structure.

plasma membrane

glycoprotein

carbohydrate chain

glycolipid

hydrophobic tails

hydrophilic heads

phospholipid bilayer

Outside of cell

peripheral protein

filaments of cytoskeleton

cholesterol

Inside of cell

The plasma membranes of different cells and the membranes of various organelles each have their own unique collections of proteins. The integral proteins largely determine a membrane's specific functions. As illustrated in **Figure 5.11**, the integral proteins can be of several types:

Channel proteins Channel proteins have a channel that allows molecules to simply move across the membrane. For example, a channel protein allows hydrogen ions to flow across the inner mitochondrial membrane. Without this movement of hydrogen ions, ATP would never be produced.

Carrier proteins Carrier proteins are also involved in the passage of molecules through the membrane. They combine with a substance and help it move across the membrane. For example, a carrier protein transports sodium and potassium ions across a nerve cell membrane. Without this carrier protein, nerve conduction would be impossible. The presence of carriers for some substances and not others means that the plasma membrane is **differentially permeable**—only certain substances can pass through.

Cell recognition proteins Cell recognition proteins are glycoproteins. Among other functions, these proteins help the body recognize foreign invaders so that an immune reaction can occur. Without this recognition, harmful organisms (pathogens) would be able to freely invade the body.

Receptor proteins Receptor proteins have a binding site for a specific molecule. The binding of this molecule causes the protein to change its shape and, thereby, bring about a cellular response. The coordination of the body's organs is totally dependent on signal molecules that bind to receptors. For example, the liver stores glucose after it is signaled to do so by insulin.

Enzymatic proteins Some plasma membrane proteins are enzymatic proteins that carry out metabolic reactions directly. Without the presence of enzymes, some of which are attached to the various membranes of the cell, a cell would never be able to perform the metabolic reactions necessary for its proper function.

Junction proteins As discussed in Section 4.20, proteins are also involved in forming various types of junctions between cells. The junctions assist cell-to-cell communication.

Section 5.12 discusses various illnesses that result when plasma membrane proteins fail to perform their usual functions.

> **5.11** *Check Your Progress* **Which types of plasma membrane proteins are directly involved in allowing substances to enter or exit the cell?**

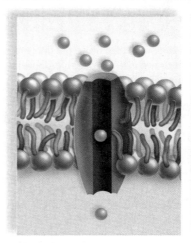
Channel protein

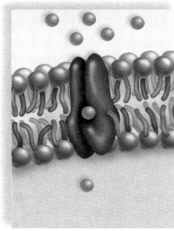
Carrier protein

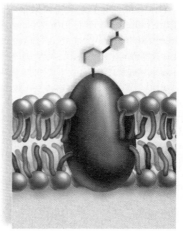
Cell recognition protein

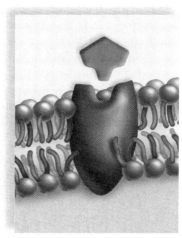
Receptor protein

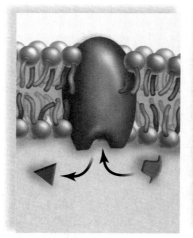
Enzymatic protein

Junction proteins

FIGURE 5.11 Functions of plasma membrane proteins.

5.12 Malfunctioning plasma membrane proteins can cause human diseases

Suppose you went to the doctor for a particular medical condition. Would you expect the doctor to say that your condition was due to your plasma membrane? That is what might happen if you wanted to know the exact cause of your illness. Let's take diabetes type 2, for example.

Diabetes Type 2 The typical diabetes type 2 patient is somewhat, or even grossly, overweight. The symptoms are unusual hunger and/or thirst; excessive fatigue; blurred vision; sores that do not heal; and frequent urination, especially at night. The doctor does a urinalysis and finds sugar in the urine, and yet the blood test shows insulin in the blood. Usually when we eat sugar, the pancreas, a gland that lies near the stomach, releases the hormone insulin into the bloodstream, and it travels to the cells, where it binds to its receptor protein. The binding of insulin signals a cell to send carriers to the plasma membrane that will transport glucose into the cells. In the case of diabetes type 2, the insulin binds to its receptor protein, but the number of carriers sent to the plasma membrane for glucose is not enough. The result is too much glucose in the blood, which spills over into the urine.

Patients can prevent, or at least control, diabetes type 2 by switching to a healthy diet and engaging in daily exercise. In addition to the symptoms mentioned, diabetics are at risk for blindness, kidney disease, and cardiovascular disease.

Color Blindness If you have ever found yourself accidentally wearing socks of two different colors, you may have endured a little teasing about being color blind. Most people have three types of photopigment proteins in numerous folds of plasma membrane located within certain photoreceptor cells, called cones because of their distinctive shape. Cone cells are located in the retina, the part of the eye that responds to visible light and allows us to see.

People with normal color vision have three types of cones: blue, green, and red, each activated by different wavelengths of visible light. The perception of color requires activation of a combination of these three types of cone cells. When a cone cell receives the wavelength of light to which its photopigment is sensitive, a signal is sent to close sodium ion channels in the plasma membrane. However, some people, mostly males, have inherited a mutation that results in a lack of functional red or green photopigment proteins. Such individuals have what is termed "red-green color blindness" and have difficulty distinguishing these two colors. In a much less common situation, both red and green photopigments are missing; such people may lack all color vision and see a monochromatic world.

Cystic Fibrosis (CF) The typical CF patient is a child, usually younger than three years of age, who has experienced repeated lung infections or poor growth. The doctor orders a test that measures the amount of salt (NaCl) in the child's sweat, because children with CF have more salt in their sweat than normal children. Usually, chloride ions (Cl⁻) pass easily through a plasma

FIGURE 5.12 Cystic fibrosis is due to a defective CF gene and defective CF channel proteins.

membrane channel protein, but when their passage is not properly regulated, a thick mucus appears in the lungs and pancreas. The mucus clogs the lungs, causing breathing problems. It also provides fertile ground for bacterial growth. The result is frequent lung infections, which eventually damage the lungs and contribute to an early death. Also, thick digestive fluids may clog ducts leading from the pancreas to the small intestine. This prevents enzymes from reaching the small intestine, where they are needed to digest food. Digestive problems and slow growth result.

CF can be confirmed by doing a genetic test because we now have a test for the gene that causes CF. **Figure 5.12** reminds us that a genetic defect results in a defective protein—in this instance, an abnormal channel protein for chloride in the plasma membrane.

Sections 5.13 to 5.18 pertain to how molecules get into and out of the cell, including methods that utilize plasma proteins.

5.12 Check Your Progress Explain why plasma membrane protein malfunctions can manifest themselves differently.

The plasma membrane is differentially permeable, meaning that certain substances can freely pass through the membrane while others are transported across. Basically, substances enter a cell in one of three ways: passive transport, active transport, and bulk transport and we discuss each of these ways in this part of the chapter. *Passive transport* moves substances from a higher to a lower concentration, and no energy is required. *Active transport* moves substances against a concentration gradient and requires energy. *Bulk transport* requires energy, but movement of the large substances involved is independent of concentration gradients.

5.13 Simple diffusion across a membrane requires no energy

Small, noncharged molecules, such as oxygen, carbon dioxide, glycerol, and alcohol are able to slip between the phospholipid molecules making up the plasma membrane. During **simple diffusion**, molecules move down their concentration gradient until equilibrium is achieved and they are distributed equally. Simple diffusion occurs because molecules are in motion, but it is a *passive* form of transport because a cell does not need to expend energy for it to happen.

Figure 5.13 demonstrates simple diffusion. Water is present on two sides of an artificial membrane, and red dye is added to one side. The dye particles move in both directions, but the net movement is toward the opposite side of the membrane (long arrow). Eventually, the dye is dispersed, with no net movement of dye in either direction.

Dissolved gases can diffuse readily through the phospholipid bilayer, and this is the mechanism by which oxygen enters cells and carbon dioxide exits them. Diffusion also allows oxygen to enter the blood from the air sacs of the lungs, and carbon dioxide to move in the opposite direction.

Facilitated diffusion, which is discussed next, is also a passive means of transport.

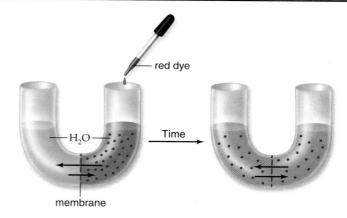

FIGURE 5.13 Some molecules can simply diffuse across a membrane.

> **5.13 Check Your Progress** Suppose the water in the U-shaped tube in Figure 5.13 is hot. How might this affect diffusion of the dye? Explain.

5.14 Facilitated diffusion requires a carrier protein but no energy

Certain molecules (e.g., glucose) cross membranes at a rate faster than expected based on their size and polarity. It is hypothesized that such molecules pass through the membrane by a passive form of transport called **facilitated diffusion**: They bind with a carrier protein, and then they diffuse rapidly across the membrane to the other side. The carrier proteins are thought to be specific. For example, among sugar molecules of the same size and polarity, glucose can cross the membrane hundreds of times faster than the other sugars because its passage is facilitated by a carrier.

A model for facilitated diffusion (**Fig. 5.14**) shows that after a carrier has assisted the movement of a molecule to the other side of the membrane, it is free to assist the passage of other similar molecules. Neither simple diffusion nor facilitated diffusion requires an expenditure of energy because the molecules are moving down their concentration gradient in the same direction they tend to move anyway.

In Section 5.15, we consider another special case of diffusion (i.e., the diffusion of water across the differentially permeable plasma membrane).

> **5.14 Check Your Progress** How is a carrier protein for facilitated diffusion like a turnstile?

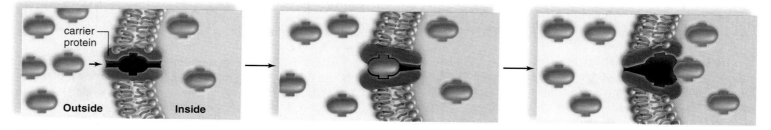

FIGURE 5.14 During facilitated diffusion, a carrier protein assists solute movement across the membrane.

5.15 Osmosis can affect the size and shape of cells

The diffusion of water across a differentially permeable membrane due to concentration differences is called **osmosis**. In certain cells, water diffuses into a cell more quickly than usual because it diffuses through a channel protein now called an **aquaporin**. First, let's consider that a **solution** contains both a solute and a solvent. In **Figure 5.15A**, salt is the **solute**, and the water is called the **solvent**. Solutes are solids, and solvents are liquids.

To illustrate osmosis, a thistle tube containing a 10% salt solution is covered at one end by a differentially permeable membrane and then placed in a beaker of 5% salt solution. The water can diffuse across the membrane, but the salt cannot. Water will move back and forth across the membrane, but the *net movement* is from the beaker to the tube. Theoretically, the solution inside the tube will rise until there is an equal concentration of water on both sides of the membrane.

As water enters and the solute does not exit, the level of the solution within the thistle tube rises. In the end, the concentration of solute in the thistle tube is less than 10%. Why? Because there is now less solute per unit volume. And the concentration of solute in the beaker is greater than 5%? Why? Because there is now more solute per unit volume.

How Osmosis Affects the Size and Shape of Cells

In the laboratory, cells are normally placed in **isotonic solutions** (*iso*, same as) in which the cell neither gains nor loses water—that is, the concentration of water is the same on both sides of the membrane (**Fig. 5.15B**). In medical settings, a 0.9% solution of sodium chloride (NaCl) is known to be isotonic to red blood cells; therefore, intravenous solutions usually have this concentration.

Cells placed in a **hypotonic solution** (*hypo*, less than) gain water. Outside the cell, the concentration of solute is less, and the concentration of water is greater, than inside the cell. Animal cells placed

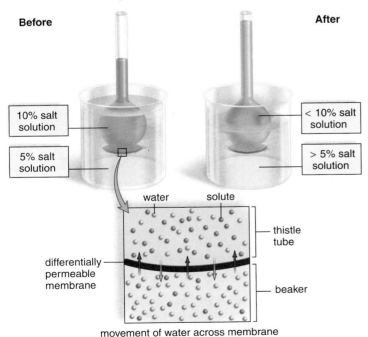

Before

10% salt solution

5% salt solution

After

< 10% salt solution

> 5% salt solution

water   solute

thistle tube

differentially permeable membrane

beaker

movement of water across membrane

FIGURE 5.15A During osmosis, net movement of water is toward greater solute concentration.

Solution	*Isotonic*	*Hypotonic*	*Hypertonic*
Animal cells	normal cell	cell swells, bursts	shriveled cell
Plant cells	normal cell	normal turgid cell	cytoplasm shrinks from cell wall

FIGURE 5.15B Osmosis in animal and plant cells.

in a hypotonic solution expand and sometimes burst. The term *lysis* refers to disrupted cells; *hemolysis*, then, is disruption of red blood cells (*hemo*, blood). Organisms that live in fresh water have to prevent their internal environment from gaining too much water. Many protozoans, such as paramecia, have contractile vacuoles that rid the body of excess water. Freshwater fishes excrete a large volume of dilute urine and take in salts at their gills, ensuring that their internal environment doesn't become hypotonic to their cells.

When a plant cell is placed in a hypotonic solution, the large central vacuole gains water, and the plasma membrane pushes against the rigid cell wall as the plant cell becomes *turgid*. The plant cell does not burst because the cell wall does not give way. Turgor pressure in plant cells is extremely important in maintaining the plant's erect position. If you forget to water your plants, they wilt due to a decrease in turgor pressure.

Cells placed in a **hypertonic solution** (*hyper*, more than) lose water. Outside the cell, the concentration of solute is more, and the concentration of water is less, than inside the cell. Animal cells placed in a hypertonic solution shrink. Marine fishes prevent their internal environment from becoming hypertonic to their cells by excreting salts across their gills. Hypertonicity can be put to good use. For example, meats are sometimes preserved by being salted. Bacteria are killed not by the salt, but by the lack of water in the meat.

When a plant cell is placed in a hypertonic solution, the plasma membrane pulls away from the cell wall as the large central vacuole loses water. This is an example of **plasmolysis**, shrinking of the cytoplasm due to osmosis.

In the next two sections, we will consider active transport, which, in contrast to the passive methods considered thus far, does require energy.

> **5.15 Check Your Progress** Salt spread on icy roads in winter has what effect on nearby vegetation in spring?

5.16 Active transport requires a carrier protein and energy

During **active transport**, molecules or ions move through the plasma membrane, accumulating on one side of the cell. For example, glucose is completely absorbed by the cells lining the digestive tract after you have eaten. Glucose is moved across the lining of the small intestine by a combination of facilitated diffusion and active transport. Facilitated diffusion works only as long as the concentration gradient is favorable, but active transport permits cells to remove the rest of the glucose into the body.

Most of the iodine that enters the body collects in the cells of the thyroid gland. In the kidneys, sodium can be almost completely withdrawn from urine by cells lining the kidney tubules. The movement of molecules against their concentration gradients requires both a carrier protein and ATP (**Fig. 5.16**). There-fore, cells involved in active transport, such as kidney cells, have a large number of mitochondria near their plasma membranes to generate ATP.

Proteins engaged in active transport are often called *pumps*. The **sodium-potassium pump**, vitally important to nerve conduction, undergoes a change in shape when it combines with ATP, and this allows it to combine alternately with sodium ions and potassium ions to move them across the membrane.

> **5.16** *Check Your Progress* How is a carrier protein for active transport like a turnstile?

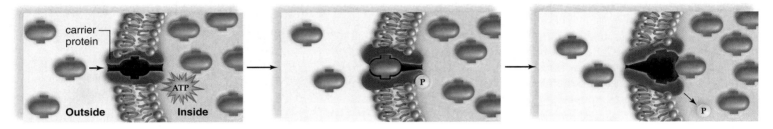

FIGURE 5.16 During active transport, a substance moves contrary to its concentration gradient. A protein carrier and energy are required.

5.17 Bulk transport involves the use of vesicles

Bulk transport occurs when fluid or particles are brought into a cell by vacuole formation, called **endocytosis** (**Fig. 5.17**), or out of a cell by evagination, called **exocytosis**. To imagine exocytosis, reverse the arrows in Figure 5.17.

Macromolecules, such as polypeptides, polysaccharides, or polynucleotides, are too large to be moved by carrier proteins. Instead, endocytosis takes them in and exocytosis takes them out of a cell. If the material taken in is large, such as a food particle or another cell, the process is called **phagocytosis**. Phagocytosis is common in unicellular organisms, such as amoebas. It also occurs in humans. Certain types of human white blood cells are amoeboid—that is, they are mobile like an amoeba, and are able to engulf debris such as worn-out red blood cells or bacteria. When an endocytic vesicle fuses with a lysosome in the cell, digestion occurs (see Fig. 4.11).

Pinocytosis occurs when vesicles form around a liquid or around very small particles. Cells that use pinocytosis to ingest substances include white blood cells, cells that line the kidney tubules and the intestinal wall, and plant root cells.

During **receptor-mediated endocytosis**, receptors for particular substances are found at one location in the plasma membrane. This location is called a coated pit because there is a layer of protein on its intracellular side. Receptor-mediated endocytosis is selective and much more efficient than ordinary pinocytosis. It is involved when some substances move from maternal blood into fetal blood at the placenta, for example.

In contrast to endocytosis, digestive enzymes and hormones are transported out of the cell by exocytosis. In cells that syn-

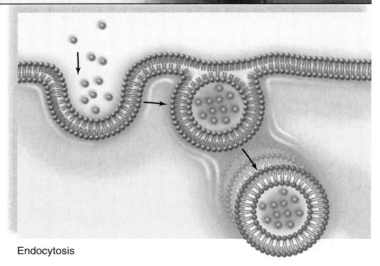

Endocytosis

FIGURE 5.17 Bulk transport into the cell is by endocytosis.

thesize these products, secretory vesicles accumulate near the plasma membrane. The vesicles release their contents only when the cell is stimulated by a signal received at the plasma membrane, a process called regulated secretion.

> **5.17** *Check Your Progress* Receptor-mediated endocytosis allows cholesterol to enter cells (along with the lipoprotein that transports cholesterol in the blood). If the receptor is faulty, cardiovascular disease results. Why?

Energy is the ability to do work, to bring about change, and to make things happen, whether it's a leaf growing or a human running. A cell is dynamic because it carries out enzymatic reactions, many of which release or require energy. Exchanges across the plasma membrane allow the cell to continue to perform its usual reactions. Few reactions occur in a cell without the presence of an enzyme because enzymes lower the energy of activation by bringing substrates together. Enzymes are proteins, and as such they are sensitive to environmental conditions, including pH, temperature, and any inhibitors present.

ATP, the universal energy "currency" of life, makes energy-requiring (endergonic) reactions go. Most often in cells, the exergonic breakdown of carbohydrates drives the buildup of ATP molecules. The metabolic pathways inside cells use the chemical energy of ATP to synthesize molecules, cause muscle contraction, and even allow you to read these words.

The plasma membrane is quite appropriately called the gatekeeper of the cell because its numerous proteins allow only certain substances to enter or exit. Also, its glycoproteins and glycolipids mark the cell as belonging to the organism. To know that the plasma membrane is malfunctioning in a person who has diabetes, cystic fibrosis, or high cholesterol is a first step toward curing these conditions.

In Chapter 6, we will see how photosynthesis inside chloroplasts transforms solar energy into the chemical energy of carbohydrates. And then in Chapter 7, we will discuss how carbohydrate products are broken down in mitochondria as ATP is built up. Chloroplasts and mitochondria are the cellular organelles that permit energy to flow from the sun through all living things.

The Chapter in Review

Summary

Life's Energy Comes from the Sun
- Photosynthesis converts the energy of the sun into that of organic molecules.

Living Things Transform Energy

5.1 Energy makes things happen
- Energy is the capacity to do work.
- The five forms of energy are radiant, chemical, mechanical, electrical, and nuclear.
- Potential energy is stored energy, while kinetic energy is energy in action.
- A calorie is the amount of heat needed to raise the temperature of 1 g of water by 1° Celsius.

5.2 Two laws apply to energy and its use
- First law of thermodynamics: Energy cannot be created or destroyed, but it can be changed from one form to another.
- Second law of thermodynamics: Energy cannot be changed from one form to another without a loss of usable energy.

5.3 Cellular work is powered by ATP
- When ATP breaks down to ADP + $\circled{P}$, energy is released.

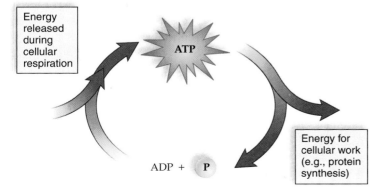

- Continual hydrolysis and regeneration of ATP is called the ATP cycle.
- An exergonic reaction releases energy.
- An endergonic reaction requires energy to occur.

5.4 ATP breakdown is coupled to energy-requiring reactions
- Reactions are coupled when an exergonic reaction drives an endergonic reaction.

Enzymes Speed Chemical Reactions

5.5 Enzymes speed reactions by lowering activation barriers
- Enzymes bring reactants together effectively at body temperature because they lower the energy of activation.

5.6 An enzyme's active site is where the reaction takes place
- The reactants in an enzymatic reaction are called the substrate.
- The active site is a small part of the enzyme that binds with the substrate(s) forming an enzyme-substrate complex.
- According to the induced fit model, the shape of the active site changes slightly to accommodate the substrate(s).
- Enzymes sometimes actively participate in reactions.

5.7 Enzyme speed is affected by local conditions
- Enzyme activity increases as substrate concentration increases until all active sites have been filled.
- As temperature rises, enzymatic activity increases until the temperature gets too high. Then the enzyme is denatured, and activity levels off and declines.
- Each enzyme has an optimal pH and may require cofactors or coenzymes for optimal reaction.
- Vitamins are required for synthesis of coenzymes.

5.8 Enzymes can be inhibited noncompetitively and competitively
- Enzyme inhibition occurs when a substance binds to an enzyme and decreases its activity.
- In noncompetitive inhibition, the inhibitor binds to an enzyme at the allosteric site.

- In competitive inhibition, an inhibitor and a substrate compete for the enzyme's active site.

5.9 Enzyme inhibitors can spell death
- Examples of enzyme inhibitors are cyanide, sarin gas, and warfarin (used as rat poison).

The Plasma Membrane Has Many and Various Functions

5.10 The plasma membrane is a phospholipid bilayer with embedded proteins
- Proteins form a mosaic pattern in the phospholipid bilayer of the plasma membrane.
- Cholesterol is a steroid found in animal plasma membranes.
- Phospholipids and proteins with attached carbohydrate chains are called glycolipids and glycoproteins, respectively.

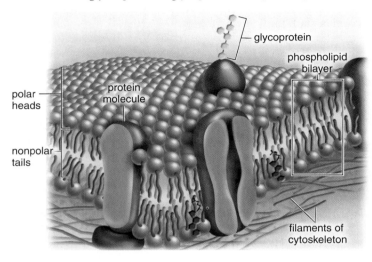

5.11 Proteins in the plasma membrane have numerous functions
- Channel proteins allow passage of molecules and carrier proteins assist the passage of molecules through the membrane.
- Cell recognition proteins are glycoproteins that help the body recognize foreign invaders.
- Receptor proteins bind specific signal molecules.
- Enzymatic proteins carry out metabolic reactions.
- Junction proteins assist cell-to-cell communication.

5.12 Malfunctioning plasma membrane proteins can cause human diseases
- Diabetes type 2, cystic fibrosis, and color blindness are caused by abnormal plasma membrane proteins.

The Plasma Membrane Regulates the Passage of Molecules Into and Out of Cells

5.13 Simple diffusion across a membrane requires no energy
- In simple diffusion, molecules in solution move down a concentration gradient until equally distributed.
- A solution contains a solute (usually a solid) and a solvent (usually a liquid).

5.14 Facilitated diffusion requires a carrier protein but no energy
- Carrier proteins facilitate the diffusion of nonlipid-soluble substances across a semipermeable membrane (e.g., plasma membrane).
- Molecules move down their concentration gradient.

5.15 Osmosis can affect the size and shape of cells
- Osmosis is diffusion of water across a semipermeable membrane (e.g., plasma membrane).
- Cells placed in an isotonic solution neither gain nor lose water.
- Cells placed in a hypotonic solution gain water.
- Cells placed in a hypertonic solution lose water.
- Expansion of a cell due to gain of water in a hypotonic solution is called turgor pressure. Shrinking of a cell due to loss of water in a hypertonic solution is called plasmolysis.

5.16 Active transport requires a carrier protein and energy
- During active transport, a substance moves against its concentration gradient.
- Protein carriers involved in active transport are called pumps (e.g., sodium-potassium pump).

5.17 Bulk transport involves the use of vesicles
- In endocytosis vesicles transport substances into the cell.
  - Phagocytosis: amoeboid cells engulf debris or bacteria.
  - Pinocytosis: vesicles form around a liquid or very small particles.
  - Receptor-mediated endocytosis is selective and more efficient than ordinary pinocytosis.
- In exocytosis, vesicles transport substances (e.g., digestive enzymes, hormones) out of the cell.

Testing Yourself

Living Things Transform Energy
1. The _____ of energy would break the first law of thermodynamics.
   a. creation   c. Both a and b are correct.
   b. transformation   d. Neither a nor b is correct.
2. As a result of energy transformations,
   a. entropy increases.
   b. entropy decreases.
   c. heat energy is gained.
   d. energy is lost in the form of heat.
   e. Both a and d are correct.
3. ATP is a modified
   a. protein.   c. nucleotide.
   b. amino acid.   d. fat.
4. **THINKING CONCEPTUALLY** How is the bulk of food you bring home from the store like glycogen, and how is ATP like the meal you prepare?

Enzymes Speed Chemical Reactions
5. The current model for enzyme action is called the
   a. induced fit model.   c. lock-and-key model.
   b. activation model.   d. active substrate model.
6. The active site of an enzyme
   a. is identical to that of any other enzyme.
   b. is the part of the enzyme where its substrate can fit.
   c. can be used over and over again.
   d. is not affected by environmental factors such as pH and temperature.
   e. Both b and c are correct.
7. Vitamins can be components of
   a. coreactants.   c. coenzymes.
   b. cosugars.   d. None of these are correct.

The Plasma Membrane Has Many and Various Functions

8. Cholesterol is found in the _____ of _____ cells.
   a. cytoplasm, plant
   b. plasma membrane, animal
   c. plasma membrane, plant
   d. None of these are correct.
9. Cystic fibrosis is caused by the malfunction of _____ channels.
   a. sodium ion   c. calcium ion
   b. chloride ion   d. potassium ion

The Plasma Membrane Regulates the Passage of Molecules Into and Out of Cells

10. Cells involved in active transport have a large number of _____ near their plasma membrane.
    a. vacuoles   c. actin filaments
    b. mitochondria   d. lysosomes
11. A coated pit is associated with
    a. simple diffusion.
    b. osmosis.
    c. receptor-mediated endocytosis.
    d. pinocytosis.
12. Phagocytosis is common in
    a. amoebas.   c. red blood cells.
    b. white blood cells. d. Both a and b are correct.
13. When a cell is placed in a hypotonic solution,
    a. solute exits the cell to equalize the concentration on both sides of the membrane.
    b. water exits the cell toward the area of lower solute concentration.
    c. water enters the cell toward the area of higher solute concentration.
    d. solute exits and water enters the cell.
    e. Both c and d are correct.
14. **THINKING CONCEPTUALLY** Our blood always has a greater concentration of solutes than does tissue fluid. Why is that important to the transport function of blood?

For questions 15–18, match the items to those in the key. Each answer may include more than one item.

KEY:
   a. simple diffusion   c. osmosis
   b. facilitated diffusion   d. active transport

15. Movement of molecules, including water, from high concentration to low concentration.
16. Requires a membrane.
17. Requires energy input.
18. Requires a protein.

Understanding the Terms

active site   82
active transport   90
aquaporin   89
bulk transport   90
calorie   78
carrier protein   86
cholesterol   85
coenzyme   83
cofactor   83
competitive inhibition   84
coupled reaction   81
denatured   83
differentially permeable   86
endergonic reaction   80
endocytosis   90
energy   78
energy of activation   82
entropy   79
enzyme   82
exergonic reaction   80
exocytosis   90
facilitated diffusion   88
fluid-mosaic model   85
glycolipid   85
glycoprotein   85
heat   78
hypertonic solution   89
hypotonic solution   89
induced fit model   82
isotonic solution   89
kilocalorie   78
kinetic energy   78
metabolic pathway   84
metabolism   82
noncompetitive inhibition   84
osmosis   89
phagocytosis   90
phospholipid bilayer   85
pinocytosis   90
plasmolysis   89
potential energy   78
receptor-mediated endocytosis   90
receptor protein   86
simple diffusion   88
sodium-potassium pump   90
solute   89
solution   89
solvent   89
substrate   82
vitamin   83

Match the terms to these definitions:
a. _____ Characteristic of the plasma membrane due to its ability to allow certain molecules but not others to pass through.
b. _____ Diffusion of water through the plasma membrane of cells.
c. _____ Higher solute concentration (less water) than the cytoplasm of a cell; causes cell to lose water by osmosis.
d. _____ Protein in plasma membrane that bears a carbohydrate chain.
e. _____ Process by which a cell engulfs a substance, forming an intracellular vacuole.

Thinking Scientifically

1. Use the first and second laws of thermodynamics to explain why each ecosystem needs a continuous supply of solar energy.
2. You have been assigned the task of designing an experiment to illustrate enzyme function. What factors, including environment, do you have to consider?

ARIS™ *Visit* www.mhhe.com/maderconcepts *for practice quizzes, animations, videos, and activities designed to help you master the material in this chapter.*

6

Pathways of Photosynthesis

LEARNING OUTCOMES

After studying this chapter, you should be able to accomplish the following outcomes.

Color It Green

1 Explain why leaves are not black.

Photosynthesis Produces Food and Releases Oxygen

2 List the types of organisms that carry on photosynthesis.
3 Identify the main parts of a chloroplast.
4 Show that photosynthesis is a redox reaction that produces a carbohydrate and releases O_2.
5 Describe an experiment showing that O_2 comes from water.
6 Divide photosynthesis into two sets of reactions, and associate each set with either capture of solar energy or reduction of carbon dioxide.

First, Solar Energy Is Captured

7 Explain why leaves are green, with reference to the electromagnetic spectrum.
8 Explain why leaves change color in the fall.
9 Trace the path of an excited electron from its absorption of solar energy to the production of ATP and NADPH.
10 Explain how the thylakoid membrane is organized to produce ATP and NADPH.

Second, Carbohydrate Is Synthesized

11 Describe the three phases of the Calvin cycle, and indicate when ATP and/or NADPH are involved.
12 Draw a diagram showing that G3P is a pivotal molecule in a plant's metabolic pathway.

C_3, C_4, and CAM Photosynthesis Thrive Under Different Conditions

13 Compare and contrast three modes of photosynthesis.
14 Discuss the possible effect of global warming on photosynthesis.

It's easy to show that plants do not use green light for photosynthesis. Simply put a sprig of the plant elodea in a glass jar, fill it with water (elodea lives in water), and shine a green light on it. NOTHING HAPPENS. But switch to a bright white light, and watch the bubbling. Bubbling is caused by oxygen gas escaping from the water. That's your evidence that photosynthesis is occurring. Plants always give off oxygen when they photosynthesize, for which we humans are mightily thankful.

White light, the visible light that shines down on us every day, contains different colors of light, from violet to blue, green, yellow, orange, and finally red. Plants use all the colors except green—and that's why we see them as green! Red algae are protists that live in the ocean, and like all the other types of algae,

Color It Green

they photosynthesize. Despite their name, some forms of red algae are dark colored, almost black, which means that they are able to use all the different colors in white light for photosynthesis. Does this mean that if plants weren't so wasteful and used green light for photosynthesis, in addition to all the other colors, they would appear black to us? Yes, it does. Look out the window and imagine that all the plants you see were black instead of green. Aren't we glad that photosynthesis on land is inefficient and wasteful of green light, making our world a sea of green!

How did it happen that plants do not use green light for photosynthesis? During the evolution of organisms, photosynthesizing bacteria floating in the oceans above the sediments possessed a pigment that could absorb and use green light. Natural selection, therefore, favored the evolution of a pigment that absorbed only blue and red light. The green pigment chlorophyll, which absorbs the blue and red ranges of light, evolved and became the photosynthetic pigment of plants. The inefficiency of chlorophyll doesn't matter on land, where light is readily available, so no plants on land ever evolved a more efficient pigment than chlorophyll. Plants are green and our world is beautiful because their primary photosynthetic pigment doesn't absorb green light!

Green leaves and variously colored eukaryotic algae carry

on photosynthesis in chloroplasts, as discussed in this chapter. Photosynthesis, which produces food in the form of carbohydrate and also oxygen for the biosphere, consists of two connected types of metabolic pathways: the light reactions and the Calvin cycle reactions. We will see how the absorption of solar energy during the light reactions drives the Calvin cycle reactions, which produces carbohydrate, the end product of photosynthesis. This chapter also shows that the process of photosynthesis is adapted to different environmental conditions.

In this part of the chapter, we will see that photosynthesizers not only produce food for the biosphere, but they are also the source of the fossil fuel our society burns to maintain our standard of living. The overall equation for photosynthesis shows the starting reactants and the end products of the process. But we will see that in actuality, photosynthesis requires two metabolic pathways: the light reactions that occur in thylakoid membranes and the Calvin cycle reactions that occur in the stoma.

6.1 Photosynthesizers are autotrophs that produce their own food

Photosynthesis converts solar energy into the chemical energy of a carbohydrate. Photosynthetic organisms, including plants, algae, and cyanobacteria, are called **autotrophs** because they produce their own organic food (**Fig. 6.1**). Photosynthesis produces an enormous amount of carbohydrate. So much that, if it could be instantly converted to coal and the coal loaded into standard railroad cars (each car holding about 50 tons), the photosynthesizers of the biosphere would fill more than 100 cars *per second* with coal.

No wonder photosynthetic organisms are able to sustain themselves and all other living things on Earth. With few exceptions, it is possible to trace any food chain back to plants and algae. In other words, producers, which have the ability to synthesize carbohydrates, feed not only themselves but also consumers, which must take in preformed organic molecules. Collectively, consumers are called **heterotrophs**. Both autotrophs and heterotrophs use organic molecules produced by photosynthesis as a source of building blocks for growth and repair and as a source of chemical energy for cellular work.

Our analogy about photosynthetic products becoming coal is apt because the bodies of many ancient plants did become the coal we burn today. This process began several hundred million years ago, and that is why coal is called a fossil fuel. We use coal in large part to produce electricity. The wood of trees also commonly serves as fuel. Then, too, the fermentation of plant materials produces alcohol, which can be used directly to fuel automobiles or as a gasoline additive.

Photosynthesis occurs in chloroplasts, the topic of Section 6.2.

> **6.1 Check Your Progress** It might seem as if plants are self-sustainable, as long as solar energy is available. Why?

FIGURE 6.1 Photosynthetic organisms.

Euglena, a protist

kelp, a protist

live oak, a tree

Gloeocapsa, a cyanobacterium

diatom, a protist

sunflower, a garden plant

moss, a plant

6.2 In plants, chloroplasts carry out photosynthesis

Photosynthesis takes place in the green portions of plants, particularly the leaves (**Fig. 6.2**). ❶ The leaves of a plant contain mesophyll tissue in which cells are specialized for photosynthesis. The raw materials for photosynthesis are water and carbon dioxide. The roots of a plant absorb water, which then moves in vascular tissue up the stem to a leaf by way of ❷ the leaf veins. ❸ Carbon dioxide in the air enters a leaf through small openings called **stomata** (sing., stoma).

After entering a leaf cell, carbon dioxide and water diffuse into **chloroplasts**, the organelles that carry on photosynthesis. ❹ A double membrane surrounds a chloroplast and its fluid-filled interior, called the ❺ **stroma**. A different membrane system within the stroma forms flattened sacs called **thylakoids**, which in some places are stacked to form ❻ **grana** (sing., granum), so called because they looked like piles of seeds to early microscopists. The space of each thylakoid is thought to be connected to the space of every other thylakoid within a chloroplast, thereby forming an inner compartment within chloroplasts called the thylakoid space.

The thylakoid membrane contains **chlorophyll** and other pigments that are capable of absorbing solar energy. This is the energy that drives photosynthesis. The stroma contains a metabolic pathway where carbon dioxide is first attached to an organic compound and then converted to a carbohydrate. Therefore, it is proper to associate the absorption of solar energy with the thylakoid membranes making up the grana and to associate the conversion of carbon dioxide to a carbohydrate with the stroma of a chloroplast.

Human beings, and indeed nearly all organisms, release carbon dioxide into the air. This is some of the same carbon dioxide that enters a leaf through the stomata and is converted to a carbohydrate. Carbohydrate, in the form of glucose, is the chief energy source for most organisms.

The overall equation examined in Section 6.3 shows only the beginning reactants and the end products of photosynthesis.

> **6.2** *Check Your Progress* **Which part of a chloroplast absorbs solar energy? Explain. Which part forms a carbohydrate? Explain.**

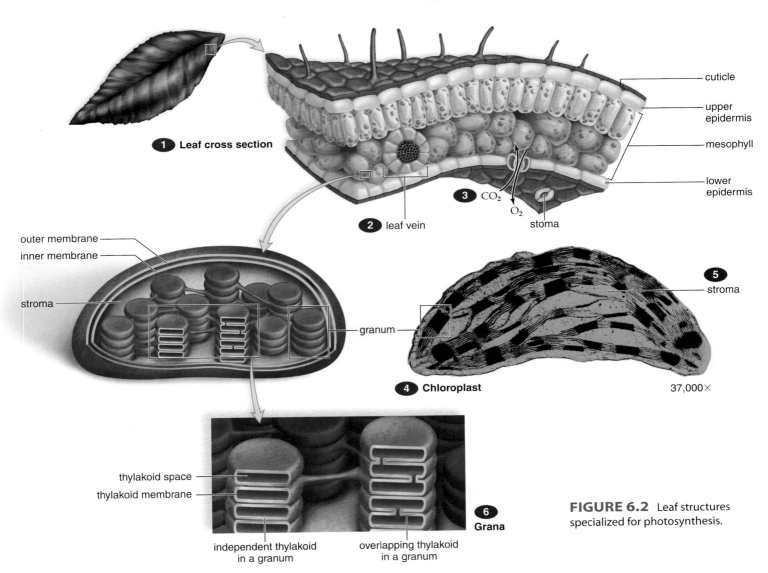

❶ **Leaf cross section**

cuticle

upper epidermis

mesophyll

lower epidermis

❸ CO₂

O₂

❷ leaf vein

stoma

outer membrane

inner membrane

stroma

granum

❺ stroma

❹ **Chloroplast**   37,000×

thylakoid space

thylakoid membrane

independent thylakoid in a granum

overlapping thylakoid in a granum

❻ **Grana**

FIGURE 6.2 Leaf structures specialized for photosynthesis.

6.3 Photosynthesis is a redox reaction that releases O_2

Understanding photosynthesis requires knowledge of oxidation and reduction. When oxygen (O) combines with a metal, such as iron or magnesium (Mg), oxygen receives electrons and forms ions that are negatively charged; the metal loses electrons and forms ions that are positively charged. When magnesium oxide ($Mg^{2+}O^{2-}$) forms, it is appropriate to say that magnesium has been *oxidized*, and that because of oxidation, it has lost electrons. On the other hand, oxygen has been *reduced* because it has gained negative charges (i.e., electrons).

Today, the terms oxidation and reduction are applied to many reactions, whether or not oxygen is involved. Very simply, **oxidation** is the loss of electrons, and **reduction** is the gain of electrons. Because oxidation and reduction go hand-in-hand, the entire reaction is called a **redox reaction**.

The terms oxidation and reduction also apply to covalent reactions in cells. In this case, however, oxidation is the loss of hydrogen atoms, and reduction is the gain of hydrogen atoms. A hydrogen atom contains one electron and one proton ($e^- + H^+$); therefore, when a molecule loses a hydrogen atom, it has lost an electron, and when a molecule gains a hydrogen atom, it has gained an electron.

Overall, photosynthesis is a redox reaction in which hydrogen atoms are transferred from water to carbon dioxide with the release of O_2 and the formation of glucose.

Instead of glucose, some prefer to show a generalized carbohydrate (CH_2O) as the end product of photosynthesis.

$$\text{Reduction}$$
$$CO_2 + H_2O \xrightarrow{\text{solar energy}} (CH_2O) + O_2$$
$$\text{Oxidation}$$

During photosynthesis, chloroplasts capture solar energy and convert it to the chemical energy of ATP molecules, which supply the energy necessary to reduce carbon dioxide. A coenzyme of oxidation-reduction called $NADP^+$ (nicotinamide adenine dinucleotide phosphate) is also active during photosynthesis. $NADP^+$ accepts electrons and a hydrogen ion derived from water. The reaction that reduces $NADP^+$ is:

$$NADP^+ + 2\ e^- + H^+ \longrightarrow NADPH$$

Later, NADPH helps reduce carbon dioxide to a carbohydrate (CH_2O). Section 6.4 reviews the experiments telling us that the oxygen released by photosynthesis comes from the oxidation of water.

6.3 Check Your Progress Oxidation and reduction go together. During photosynthesis, what is oxidized and what is reduced?

HOW SCIENCE PROGRESSES

6.4 Experiments showed that the O_2 released by photosynthesis comes from water

In 1930, C. B. van Niel of Stanford University found that oxygen given off by photosynthesis (**Fig. 6.4**) comes from water, not from carbon dioxide as had been originally thought. This was a first step in discovering the role of water in photosynthesis (see Section 6.10). That oxygen comes from water was proven by two separate experiments. When an isotope of oxygen, namely ^{18}O, was a part of carbon dioxide, the O_2 given off by a plant did not contain the isotope. On the other hand, when the isotope was a part of water, the isotope did appear in the O_2 given off by the plant.

$$CO_2 + 2\ H_2O \xrightarrow{\text{solar energy}} (CH_2O) + H_2O + O_2$$

The oxygen given off by photosynthesis goes into the atmosphere and serves as the source of oxygen for all organisms that carry on cellular respiration, even plants. Plants carry on cellular respiration both day and night, whether they are photosynthesizing or not.

In Section 6.5, we learn that photosynthesis actually requires two metabolic pathways: the light reactions and the Calvin cycle reactions.

6.4 Check Your Progress Why might you predict that the oxygen given off by photosynthesis comes from water?

FIGURE 6.4 Photosynthesis releases oxygen from water molecules.

6.5 Photosynthesis involves two sets of reactions: the light reactions and the Calvin cycle reactions

Researchers have known since the early 1900s that photosynthesis involves two sets of reactions: the **light reactions** and the **Calvin cycle reactions** (**Fig. 6.5A**). These two sets of reactions can be associated with the two parts of a chloroplast—namely, the stacks of thylakoids (grana) and the stroma. Chlorophyll molecules and other pigments that are able to absorb solar energy during daylight hours are located in the thylakoids. Enzymes that are able to speed the reduction of carbon dioxide, during both day and night, are located in the ground substance of the stroma.

Light Reactions The light reactions are so named because they only occur when solar energy is available (during daylight hours). During the light reactions, the chlorophyll molecules located within the thylakoid membranes absorb solar energy and use it to energize electrons. The energy of these electrons is captured and later used for ATP production.

Energized electrons are also taken up by NADP⁺, the coenzyme of oxidation-reduction mentioned earlier. After NADP⁺ accepts electrons, it becomes NADPH. The red arrows that run from the light reactions *to* the Calvin cycle in Figure 6.5A show that the ATP and NADPH produced in the light reactions are used by the Calvin cycle reactions.

Calvin Cycle Reactions The Calvin cycle reactions are named for Melvin Calvin, who received a Nobel Prize for discovering the enzymatic reactions that reduce carbon dioxide to a carbohydrate in the stroma of chloroplasts (**Fig. 6.5B**).

During the Calvin cycle reactions, CO_2 is taken up and then reduced to a carbohydrate that can be converted to glucose. The ATP and NADPH formed during the light reactions are needed to carry out this cyclical series of reactions.

FIGURE 6.5B Melvin Calvin in the laboratory.

The red arrows in Figure 6.5A from the Calvin cycle *to* the light reactions show that after the reactions are complete, ADP + ⓟ and NADP⁺ return to the light reactions, where they become ATP and NADPH once more.

In Section 6.6, we will consider what parts of visible light are most useful for the light reactions.

6.5 Check Your Progress What two molecules form a "bridge" linking the light reactions of photosynthesis with the Calvin cycle reactions?

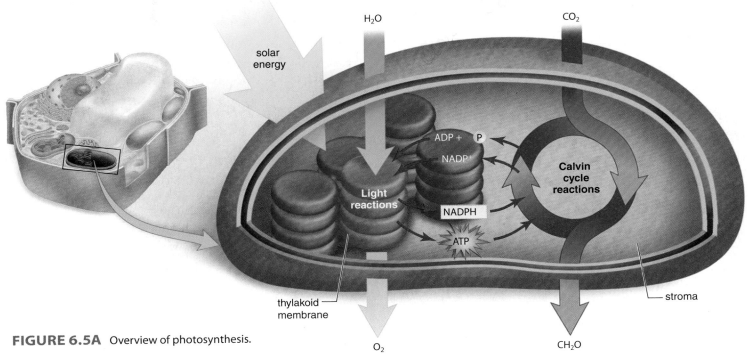

FIGURE 6.5A Overview of photosynthesis.

In this part of the chapter, we will study the properties of light and the light reactions of photosynthesis. We will see how the structure of the thylakoid membrane lends itself to absorbing solar energy and producing ATP and NADPH, needed by the Calvin cycle reactions to reduce carbon dioxide to a carbohydrate in the stroma.

6.6 Light reactions begin: Solar energy is absorbed by pigments

Solar energy (radiant energy from the sun) can be described in terms of its wavelength and its energy content. **Figure 6.6A** lists the different types of radiant energy, from the shortest wavelength, gamma rays, to the longest, radio waves. We are most interested in white, or *visible light*, because it is the type of radiation used for photosynthesis and for vision.

When visible light is passed through a prism, we can observe that it is made up of various colors. (Actually, of course, our brain interprets these wavelengths as colors.) The colors in visible light range from violet (the shortest wavelength) to blue, green, yellow, orange, and red (the longest wavelength). The energy content is highest for violet light and lowest for red light.

The pigments found within most types of photosynthesizing cells are **chlorophylls *a* and *b*** and **carotenoids**. These pigments are capable of absorbing various portions of visible light. The absorption spectrum for these pigments is shown in **Figure 6.6B**. Both chlorophyll *a* and chlorophyll *b* absorb violet, blue, and red light better than the light of other colors. The yellow or orange carotenoids are able to absorb light in the violet-blue-green range. Only in the fall, as discussed in Section 6.7, is it apparent that pigments other than chlorophyll assist in absorbing solar energy.

> **6.6 Check Your Progress** Why do leaves appear green to us?

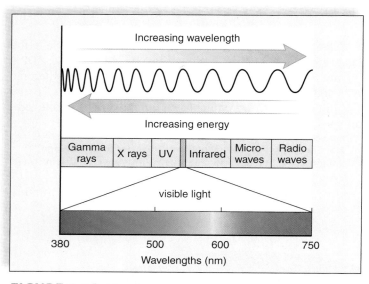

FIGURE 6.6A The electromagnetic spectrum includes visible light.

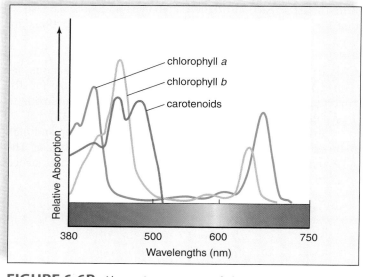

FIGURE 6.6B Absorption spectrum of photosynthetic pigments.

HOW SCIENCE PROGRESSES

6.7 Fall temperatures cause leaves to change color

The pigment chlorophyll is not very stable, and during the spring and summer, plant cells must keep using ATP molecules to rebuild it. As the hours of sunlight lessen in the fall, sufficient energy to rebuild chlorophyll is not available. Further, enzymes are working at a reduced speed because of the lower temperatures. Therefore, the amount of chlorophyll in leaves slowly disintegrates. When that happens, we begin to see yellow and orange pigments in the leaves.

In some trees, such as maples, certain pigments accumulate in acidic vacuoles, leading to

a brilliant red color. The brown color of certain oak leaves is due to wastes left in the leaves.

The role of light in photosynthesis is specific—it excites electrons (boosts them to a higher energy level) in photosystems during the light reactions, as described in the next section.

> **6.7 Check Your Progress** In tropical rain forests, where sunlight and temperatures stay fairly constant year-round, leaves do not change color. Explain.

6.8 Solar energy boosts electrons to a higher energy level

In the thylakoid membrane, chlorophyll molecules and other pigments that absorb solar energy form pigment complexes within photosystems which also include electron acceptor molecules. Two types of photosystems called **photosystem I (PS I)** and **photosystem II (PS II)** function in the same way.

Each pigment complex consists of antenna molecules and a reaction center (**Fig. 6.8**). Antenna molecules are light-absorbing accessory pigments, such as chlorophyll *b* and carotenoid pigments. They are called antenna molecules because they absorb light energy just as a radio antenna absorbs radio waves. The antenna molecules pass energy on to the reaction center. The reaction centers in PS I and II contain a special chlorophyll *a* molecule. The reaction center chlorophyll passes excited electrons on to the electron acceptor. Remember the old-fashioned "use a mallet to ring the bell, win a prize" carnival game? Similarly, solar energy has been used to launch electrons from the reaction center all the way to the electron receiver. Excited electrons then pass down a respiratory chain, releasing their energy, which is eventually converted to ATP molecules, as discussed in Section 6.9.

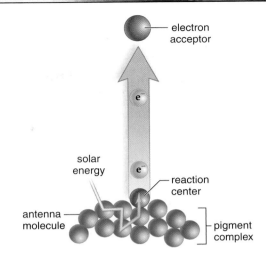

FIGURE 6.8 A general model of a photosystem.

> **6.8 Check Your Progress** What is the function of accessory pigments in a photosystem?

6.9 Electrons release their energy as ATP forms

Chloroplasts use electrons energized by solar energy to generate ATP. They do this by way of an **electron transport chain**, a series of membrane-bound carriers that pass electrons from one to another. High-energy electrons (e⁻) are delivered to the chain, and low-energy electrons leave it. Every time electrons are transferred to a new carrier, energy is released; this energy is ultimately used to produce ATP molecules (**Fig. 6.9**).

For many years, scientists did not know how ATP synthesis was coupled to the electron transport chain. Peter Mitchell, a British biochemist, received a Nobel Prize in 1978 for his model of how ATP is produced in both mitochondria and chloroplasts. The carriers of the electron transport chain are located within a membrane: thylakoid membranes in chloroplasts and cristae in mitochondria. Hydrogen ions (H^+) collect on one side of the membrane because they are pumped there by certain carriers of the electron transport chain. This establishes a *hydrogen ion (H^+) gradient* across the membrane.

ATP synthase complexes, which span the membrane, contain a channel that allows hydrogen ions to flow down their concentration gradient. The flow of hydrogen ions through the channel provides the energy needed for the ATP synthase enzyme to produce ATP from ADP + Ⓟ.

ATP production is comparable to a hydroelectric power plant. Water is trapped behind a dam, and its motion when released is used to generate electricity. Similarly, hydrogen ions are trapped behind the thylakoid membrane, and when they pass through an ATP synthase complex, their energy is used to generate ATP.

Section 6.10 pulls everything together. During the light reactions, water is split, releasing oxygen. Electrons are excited in PS II by solar energy before passing down an electron chain, and excited again in PS I before becoming a part of NADPH.

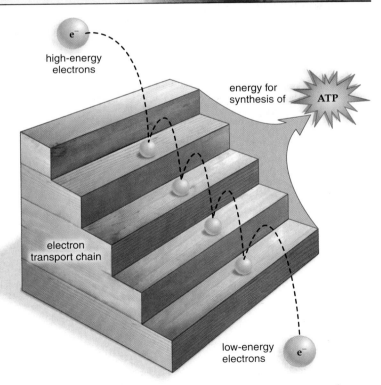

FIGURE 6.9 High-energy electrons (e⁻) release energy as they pass down an electron transport chain.

> **6.9 Check Your Progress** What kind of energy is the flow of hydrogen ions down their concentration gradient through an ATP synthase?

6.10　During the light reactions, electrons follow a noncyclic pathway

During the light reactions in the thylakoid membrane, electrons follow a *noncyclic pathway* that begins with PS II, so named because it was the second photosystem to be discovered (**Fig. 6.10**). **1** The absorbed solar energy in PS II is passed from one pigment to the other in the pigment complex, until it is concentrated in the reaction center, which contains a chlorophyll *a* molecule. **2** Electrons (e⁻) in the reaction center become so energized that they escape from the reaction center and **3** move to a nearby electron acceptor molecule.

PS II would disintegrate without replacement electrons; thus, **4** electrons are removed from water, which splits, releasing oxygen to the atmosphere. Notice that with the loss of electrons, water has been oxidized, and that indeed, the **5** oxygen released during photosynthesis does come from water. The oxygen escapes into the spongy mesophyll of a leaf and exits into the atmosphere by way of the stomata (see Fig. 6.2). **6** However, the hydrogen ions (H⁺) are trapped in the thylakoid space.

7 The electron acceptor sends the energized electrons down an electron transport chain. As the electrons pass from one carrier to the next, certain carriers pump hydrogen ions from the stroma into the thylakoid space. In this way, energy is captured and stored in the form of a hydrogen ion (H⁺) concentration gradient. **8** Later, as H⁺ flows down this gradient (from the thylakoid space into the stroma), ATP is produced.

When the PS I pigment complex absorbs solar energy, energized electrons leave its reaction center and are captured by a different **9** electron acceptor. (Low-energy electrons from the electron transport chain adjacent to PS II replace those lost by PS I.) This electron acceptor passes its electrons on to **10** NADP⁺ molecules. **11** Each NADP⁺ molecule accepts two electrons and an H⁺ to become a reduced form of the molecule—that is, NADPH.

Section 6.11 examines the thylakoid membrane. The steps of the light reactions occur in particles, a series of complexes that are present in the thylakoid membrane.

6.10 *Check Your Progress*　What molecule is the start of a noncyclic electron pathway, and what molecule is the end of the pathway?

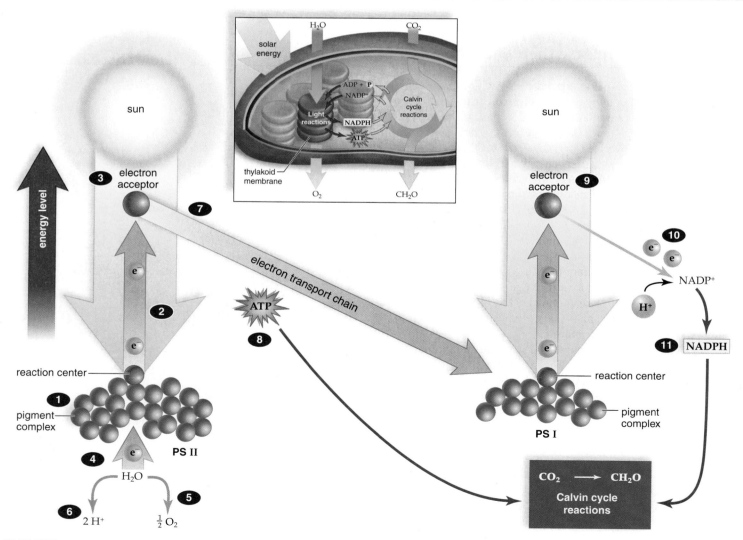

FIGURE 6.10　Noncyclic electron pathway in the thylakoid membrane; electrons move from water to NADP⁺.

6.11 The thylakoid membrane is organized to produce ATP and NADPH

Let's divide the molecular complexes in the thylakoid membrane (**Fig. 6.11**) into those that "get ready" and those that represent the "payoff."

Get Ready

1 PS II consists of a pigment complex that absorbs solar energy and passes electrons on to an electron-acceptor molecule. **2** PS II receives replacement electrons from water, which splits, releasing H^+ and oxygen (O_2).

3 The electron transport chain, consisting of a series of electron carriers such as cytochrome complexes, that pass electrons from PS II to PS I. (Notice, therefore, that PS I receives replacement electrons from the electron transport chain.) **4** Members of the electron transport chain also pump H^+ from the stroma into the thylakoid space. Eventually, an electron gradient is present: The thylakoid space contains much more H^+ than the stroma.

5 PS I consists of a pigment complex that absorbs solar energy and sends excited electrons on to an electron-acceptor molecule, which passes them to an enzyme called NADP reductase.

Payoff

6 NADP reductase, an enzyme, receives electrons and reduces $NADP^+$. $NADP^+$ combines with H^+ and becomes **7** NADPH.

8 H^+ flows down its concentration gradient through a channel in an ATP synthase complex. This complex contains an enzyme that then enzymatically binds ADP to (P), producing **9** ATP.

This method of producing ATP is called **chemiosmosis** because ATP production is tied to an H^+ gradient across a membrane.

We have now concluded our study of the light reactions, and in the next part of the chapter, we will study the Calvin cycle reactions.

> **6.11** *Check Your Progress* What is the end result of the "get ready" phase, and what is the end result of the "payoff" phase of the light reactions?

FIGURE 6.11 Organization of a thylakoid.

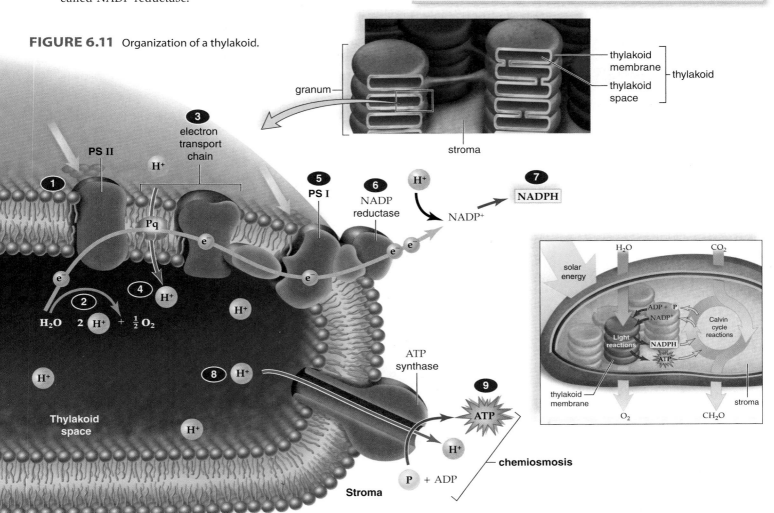

Second, Carbohydrate Is Synthesized

Learning Outcomes 11–12, page 94

In this part of the chapter, we study the Calvin cycle reactions, which occur in the stroma and use the ATP and the NADPH produced by the light reactions to reduce carbon dioxide to a carbohydrate. First, carbon dioxide is taken up (fixed) by a molecule of the cycle (RuBP) prior to its reduction to the molecule that is the carbohydrate product (G3P) of the cycle. It takes two molecules of G3P to form glucose, the molecule that is often thought of as the end product of photosynthesis.

6.12 The Calvin cycle uses ATP and NADPH from the light reactions to produce a carbohydrate

The Calvin cycle is a series of reactions that produces carbohydrate before returning to the starting point once more (**Fig. 6.12**). The cycle is named for Melvin Calvin, who, with colleagues, used the radioactive isotope ^{14}C as a tracer to discover its reactions.

The Calvin cycle reduces carbon dioxide (CO_2) from the atmosphere to produce carbohydrate. How does CO_2 get into the atmosphere? We, and most other organisms, take in O_2 from the atmosphere and release CO_2 to the atmosphere. The Calvin cycle includes these phases: (1) CO_2 fixation, (2) CO_2 reduction, and (3) regeneration of RuBP, the molecule that combines with and fixes CO_2 from the atmosphere (Fig. 6.12). The steps in the Calvin cycle are multiplied by three for reasons that will be explained.

CO_2 Fixation ❶ During the first phase of the Calvin cycle, CO_2 from the atmosphere combines with RuBP, a 5-carbon molecule, and a C_6 (6-carbon) molecule results. The enzyme that speeds this reaction, called **RuBP carboxylase**, is a protein that makes up 20–50% of the protein content in chloroplasts. The reason for its abundance may be that it is unusually slow (it processes only a few molecules of substrate per second compared to thousands per second for a typical enzyme), and so there has to be a lot of it to keep the Calvin cycle going.

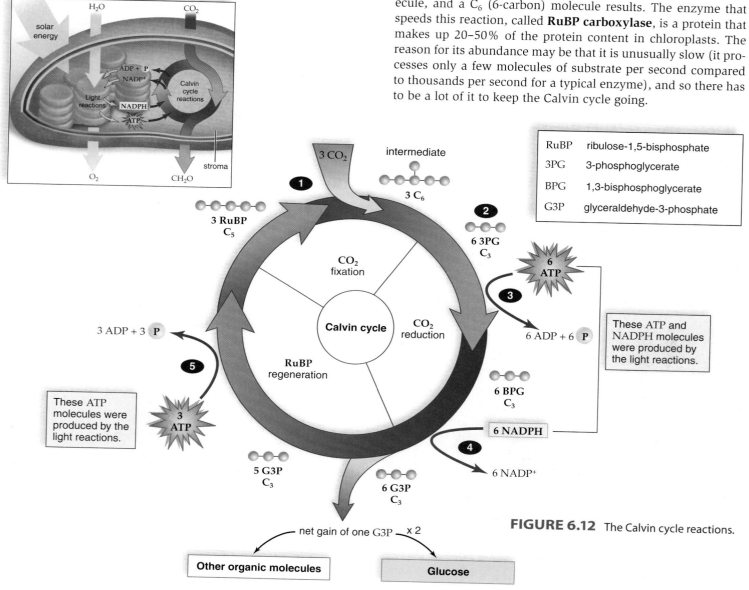

FIGURE 6.12 The Calvin cycle reactions.

RuBP	ribulose-1,5-bisphosphate
3PG	3-phosphoglycerate
BPG	1,3-bisphosphoglycerate
G3P	glyceraldehyde-3-phosphate

2 The C_6 molecule immediately splits into two C_3 molecules. (Remember that all steps are multiplied by three in Figure 6.12.) This C_3 molecule is called 3PG.

CO_2 Reduction **3** and **4** Each of the 3PG molecules undergoes reduction to G3P in two steps:

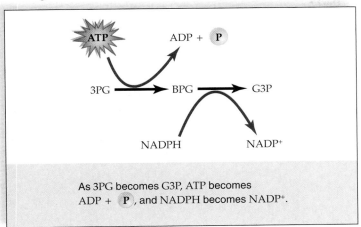

As 3PG becomes G3P, ATP becomes ADP + P, and NADPH becomes NADP⁺.

Notice that this is the sequence of reactions that uses ATP and NADPH from the light reactions. ATP becomes ADP + (P), and NADPH becomes NADP⁺.

This sequence signifies the reduction of CO_2 to a carbohydrate because $R—CO_2$ (3PG) has become $R—CH_2O$ (G3P). Energy and electrons are needed for this reduction reaction, and these are supplied by ATP and NADPH.

RuBP Regeneration The reactions in Figure 6.12 are multiplied by three because it takes three turns of the Calvin cycle to allow one G3P to exit. Why? Because, for every three turns of the Calvin cycle, five molecules of G3P are used to re-form three molecules of RuBP, and the cycle continues. Notice that 5×3 (carbons in G3P) = 3×5 (carbons in RuBP):

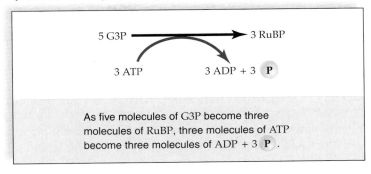

As five molecules of G3P become three molecules of RuBP, three molecules of ATP become three molecules of ADP + 3 P.

5 As this ATP produced by the light reaction breaks down, ADP + (P) results.

Section 6.13 shows that G3P is the starting point for many types of organic molecules produced by plant cells.

> **6.12 Check Your Progress** What are the fates of NADP⁺ and ADP + (P) from the Calvin cycle?

6.13 In plants, carbohydrate is the starting point for other molecules

G3P is the product of the Calvin cycle that can be converted to all sorts of organic molecules. Compared to animal cells, algae and plants have enormous biochemical capabilities. They use G3P for the purposes described in **Figure 6.13**.

1 Notice that glucose phosphate is among the organic molecules that result from G3P metabolism. This is of interest to us because glucose is the molecule that plants and other organisms most often metabolize to produce the ATP molecules they require. Glucose is blood sugar in human beings.

Glucose can be combined with **2** fructose (with the removal of phosphates) to form **3** sucrose, the transport form of sugar in plants.

4 Glucose phosphate is also the starting point for the synthesis of **5** starch and **6** cellulose. Starch is the storage form of glucose. Some starch is stored in chloroplasts, but most starch is stored in roots. Cellulose is a structural component of plant cell walls and becomes fiber in our diet because we are unable to digest it.

7 A plant can use the hydrocarbon skeleton of G3P to form fatty acids and glycerol, which are combined in plant oils such as the corn oil, sunflower oil, and olive oil used in cooking. **8** Also, when nitrogen is added to the hydrocarbon skeleton derived from G3P, amino acids are formed.

This completes our study of the Calvin cycle reactions, and in the next part of the chapter, we will consider how photosynthesis is adapted to occurring in various types of environments.

> **6.13 Check Your Progress** Glucose produced by a plant can be converted to what other molecules?

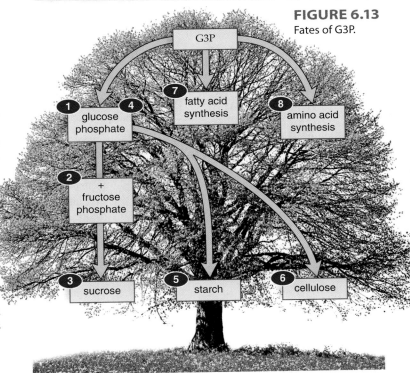

FIGURE 6.13
Fates of G3P.

C₃, C₄, and CAM Photosynthesis Thrive Under Different Conditions

Thus far, we have been observing C_3 photosynthesis, named for the number of carbons in the first observable molecule following the uptake of CO_2. C_4 photosynthesis is more advantageous when the weather is warm and CO_2 is in short supply inside leaves, due to closure of stomata. CAM, another type of photosynthesis, was first observed in desert plants, which need to conserve water. If limited CO_2 decreases photosynthesis, does an abundance of CO_2 increase photosynthesis? So far, studies seem contrary to such a hypothesis, and therefore we should keep as many tropical rain forests as possible because they help prevent global warming due to increasing CO_2 levels.

6.14 C_3 photosynthesis evolved when oxygen was in limited supply

Where temperature and rainfall tend to be moderate, plants carry on C_3 photosynthesis, and are therefore called C_3 plants. In a **C_3 plant**, the first detectable molecule after CO_2 fixation is a C_3 molecule, namely 3PG (**Fig. 6.14**). Look again at the Calvin cycle (see Fig. 6.12), and notice that the original C_6 molecule formed when RuBP carboxylase combines with carbon dioxide immediately breaks down to two 3PG, a C_3 molecule. It would be necessary to use a radioactive tracer just like Melvin Calvin did in order to determine that this molecule is the first detectable one following CO_2 uptake by RuBP carboxylase.

When stomata close due to lack of water, CO_2 decreases and O_2 increases in leaf spaces. In C_3 plants, this O_2 competes with CO_2 for the active site of RuBP carboxylase, the first enzyme of the Calvin cycle, and less C_3 is produced. Such decreases in yield are of concern to humans because many food crops are C_3 plants.

What can explain this apparent drawback in the efficiency of RuBP carboxylase? Photosynthesis, and therefore RuBP, would have evolved early in the history of life on Earth. At that time, oxygen was in limited supply, and the ability of RuBP carboxylase to combine with oxygen would not have been a problem. Photosynthesis itself caused O_2 to rise in the atmosphere, and now a plant has an advantage if it can prevent

FIGURE 6.14 Carbon dioxide fixation in C_3 plants as exemplified by these wildflowers.

RuBP from combining with O_2. As we shall see in Section 6.15, C_4 plants have such an advantage when the weather is warm.

6.14 Check Your Progress Explain the term "C_3 photosynthesis."

6.15 C_4 photosynthesis boosts CO_2 concentration for RuBP carboxylase

A modification of C_3 photosynthesis, called C_4 photosynthesis, is more efficient when CO_2 is scarce inside leaf spaces. In a **C_4 plant**, the first detectable molecule following CO_2 fixation is a C_4 molecule having four carbon atoms. C_4 plants are able to avoid the uptake of O_2 by RuBP carboxylase by increasing the amount of CO_2 available to the enzyme. Let's explore how C_4 plants do this.

The anatomy of a C_4 plant is different from that of a C_3 plant. In a C_3 leaf, mesophyll cells are arranged in parallel rows and contain well-formed chloroplasts (**Fig. 6.15A**, *left*). The Calvin cycle reactions occur in the chloroplasts of those mesophyll cells. In a C_4 leaf, chloroplasts are located in the mesophyll cells, but they are also located in bundle sheath cells surrounding the leaf vein. Further, the mesophyll cells are arranged concentri-

cally around the bundle sheath cells (Fig. 6.15A, *right*). This minimizes the amount of oxygen that can accumulate in the vicinity of the bundle sheath cells during the photosynthetic process.

In C_4 plants, the Calvin cycle reactions occur in the bundle sheath cells and not in the mesophyll cells. Therefore, CO_2 from the air is not fixed by the Calvin cycle. Instead, CO_2 is fixed by a C_3 molecule, and the C_4 that results is modified and then pumped into the bundle sheath cells (**Fig. 6.15B**). Now the C_4 molecule releases CO_2 to the Calvin cycle. This represents partitioning of pathways in space.

It takes energy to pump molecules, and you would expect the C_4 pathway outlined in Figure 6.15B to be disadvantageous. Yet, in warm climates when stomata close and CO_2 is in limited

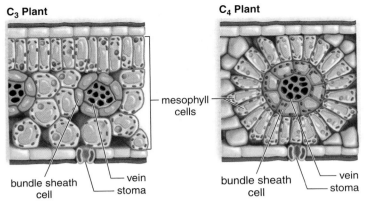

C₃ Plant **C₄ Plant**

mesophyll cells

bundle sheath cell — vein — stoma

bundle sheath cell — vein — stoma

FIGURE 6.15A Anatomy of a C₃ plant compared to a C₄ plant.

FIGURE 6.15B Carbon dioxide fixation in C₄ plants as exemplified by corn.

supply, the net photosynthetic rate of C₄ plants (e.g., sugarcane, corn, and Bermuda grass) is two to three times that of C₃ plants (e.g., wheat, rice, and oats). Why do C₄ plants enjoy such an advantage? The answer is that the availability of CO_2 is higher and the availability of O_2 is lower in bundle sheath cells than in mesophyll cells, and photorespiration is therefore negligible.

When the weather is moderate, C₃ plants ordinarily have the advantage, but when the weather becomes warm and CO_2 is less available, C₄ plants have their chance to take over, and we can expect them to predominate. In the early summer, C₃ plants such as Kentucky bluegrass and creeping bent grass are

predominant in lawns in the cooler parts of the United States, but by midsummer, crabgrass, a C₄ plant, begins to take over.

Section 6.16 discusses still another form of photosynthesis, called CAM photosynthesis, which is prevalent in desert plants.

> **6.15 Check Your Progress** Structure suits function. How do C₄ plants prevent exposure of RuBP carboxylase enzyme to oxygen?

6.16 CAM photosynthesis is another alternative to C₃ photosynthesis

CAM, which stands for crassulacean-acid metabolism, is an alternative to C₃ photosynthesis when the weather is hot and dry, as in deserts. **CAM photosynthesis** gets its name from the Crassulaceae, a family of flowering succulent (water-containing) plants that live in hot and arid regions of the world. CAM was first discovered in these plants, but now it is known to be prevalent among most succulent plants that grow in desert environments, including cactuses.

Whereas a C₄ plant represents partitioning in space—that is, carbon dioxide fixation occurs in spongy mesophyll cells, and the Calvin cycle reactions occur in bundle sheath cells—CAM is partitioning based on time. During the night, CAM plants use a C₃ molecule to fix some CO_2, forming C₄ molecules. These molecules are stored in large vacuoles in mesophyll cells. During the day, the C₄ molecules release CO_2 to the Calvin cycle when NADPH and ATP are available from the light reactions (**Fig. 6.16**).

Again, the primary advantage of this partitioning relates to the conservation of water. CAM plants open their stomata only at night, and so only then is atmospheric CO_2 available. During the day, the stomata are closed. This conserves water, but prevents more CO_2 from entering the plant.

Photosynthesis in a CAM plant is minimal because a limited amount of CO_2 is fixed at night. This amount of CO_2 fixation, however, does allow CAM plants to live under stressful conditions. Whenever water and carbon dioxide are plentiful, C₃ plants can compete well. But when temperatures are warmer and CO_2 is scarce inside leaves, C₄ plants become the better com-

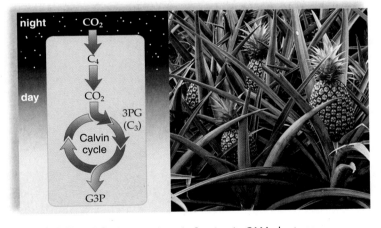

FIGURE 6.16 Carbon dioxide fixation in CAM plants as exemplified by pineapple.

petitor. Where the climate is hot and dry (deserts), stomata close up during the day and CAM plants become competitive also.

The occurrence of global warming is accompanied by an increased level of CO_2. Section 6.17 discusses how tropical rain forests can help prevent global warming.

> **6.16 Check Your Progress** How is CAM photosynthesis partitioned by the use of time?

6.17   Destroying tropical rain forests contributes to global warming

Tropical rain forests occur near the equator. They can exist wherever temperatures are above 26°C and rainfall is regular and heavy (100–200 cm per year). Huge trees with buttressed trunks and broad, undivided, dark green leaves predominate. Nearly all of the plants in a tropical rain forest are woody, and woody vines are also abundant. There is no undergrowth except at clearings. Instead, orchids, ferns, and bromeliads live in the branches of the trees.

Despite the fact that tropical rain forests have dwindled from an original 14% to 6% of land surface today, they still make a substantial contribution to global CO_2 fixation. Taking into account all ecosystems, marine and terrestrial, photosynthesis produces organic matter that is 300 to 600 times the mass of the people currently living on Earth. Tropical rain forests contribute greatly to the uptake of CO_2 and the productivity of photosynthesis because they are the most efficient of all terrestrial ecosystems.

We have learned that organic matter produced by photosynthesizers feeds all living things and that photosynthesis releases oxygen (O_2), a gas that is needed to complete the process of cellular respiration. Does photosynthesis by tropical rain forests provide any other service that has significant worldwide importance? **Figure 6.17** projects a rise in the average global temperature during the 21st century due to the introduction of certain gases, chiefly carbon dioxide, into the atmosphere. The process of photosynthesis and also the oceans act as a sink for carbon dioxide. For at least a thousand years prior to 1850, atmospheric CO_2 levels remained fairly constant at 0.028%. Since the 1850s, when industrialization began, the amount of CO_2 in the atmosphere has increased to 0.036%.

Much like the panes of a greenhouse, CO_2 in our atmosphere traps radiant heat from the sun and warms the world. Therefore, CO_2 and other gases that act similarly are called greenhouse gases.

Without any greenhouse gases, Earth's temperature would be about 33°C cooler than it is now. Likewise, increasing the concentration of these gases causes an increase in global temperatures.

Burning fossil fuels adds CO_2 to the atmosphere. Have any other factors contributed to the increases in CO_2 in the atmosphere? Between 10 and 30 million hectares of rain forests are lost every year to ranching, logging, mining, and other means of developing forests for human needs. The clearing of forests often involves burning them, which is double trouble for global warming. Each year, deforestation in tropical rain forests accounts for 20–30% of all carbon dioxide in the atmosphere. At the same time, burning removes trees that would ordinarily absorb CO_2.

Some investigators have hypothesized that an increased amount of CO_2 in the atmosphere will cause photosynthesis to increase in the remaining portion of the forest. To study this possibility, they measured atmospheric CO_2 levels, daily temperature levels, and tree girth in La Selva, Costa Rica, for 16 years. The data collected demonstrated relatively lower forest productivity at higher temperatures. These findings suggest that, as temperatures rise, tropical rain forests may add to ongoing atmospheric CO_2 accumulation and accelerated global warming rather than the reverse. This is all the more reason to do what we can now to slow the process of global warming before it gets out of hand.

Some countries have programs to combat the problem of deforestation. In the mid-1970s, Costa Rica established a system of national parks and reserves to protect 12% of the country's land area from degradation. The current Costa Rican government wants to expand the goal by expanding protected areas to 25% in the near future. Similar efforts in other countries may help slow the ever-increasing threat of global warming.

FIGURE 6.17 Global warming: Past trends and future predictions. Burning forests make the maximum increase projected in the graph more likely.

6.17 Check Your Progress Are there advantages to preserving tropical rain forests outside the possibility of helping to prevent global warming?

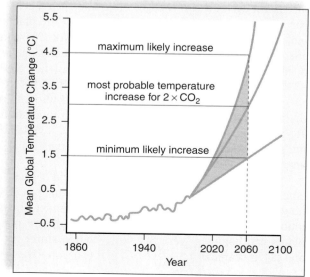

The overall equation for photosynthesis ($6\ CO_2 + 6\ H_2O \longrightarrow C_6H_{12}O_6 + 6\ O_2$) takes place in chloroplasts. This equation does not reflect that photosynthesis requires two separate sets of reactions: the light reactions (take place in thylakoid membrane) and the Calvin cycle reactions (take place in stroma).

The light reactions absorb solar energy and convert it into chemical forms of energy that drive the second set of reactions. NADPH carries electrons and ATP provides energy to reduce carbon dioxide to a carbohydrate dur-ing the Calvin cycle. C_3 photosynthesis was the first form of photosynthesis to evolve. But the first enzyme of the Calvin cycle, namely RuBP carboxylase, is inefficient in the presence of oxygen, and this led to the evolution of two other forms of photosynthesis—C_4 photo-synthesis (partitioning in space) and CAM photosynthesis (partitioning in time). Both of the alternative forms of photosynthesis are a means of supplying RuBP carboxylase with CO_2, while limiting its exposure to oxygen.

The details of photosynthesis should not cause us to lose sight of its great con-tribution to the biosphere. It keeps the biosphere functioning because it supplies energy, in the form of carbohydrates, to all organisms. Organisms have a way to tap into the energy provided by carbohydrates. It's called cellular respiration, and this is the subject of our next chapter. Cellular respiration is completed within mitochon-dria. Mitochondria are called the power-houses of the cell because they convert the energy of carbohydrates (and other or-ganic molecules) to that of ATP molecules, the energy currency of cells.

The Chapter in Review

Summary

Color It Green

- Leaves are green because the chlorophyll of plants does not absorb green light.

Photosynthesis Produces Food and Releases Oxygen

6.1 Photosynthesizers are autotrophs that produce their own food

- Photosynthesis converts solar energy to chemical energy.
- Producers (autotrophs) produce food for themselves and for consumers (heterotrophs).

6.2 In plants, chloroplasts carry out photosynthesis

- CO_2 enters a leaf through small openings called stomata.
- Chlorophyll and other pigments within the thylakoid membrane absorb solar energy.
- Conversion of CO_2 to carbohydrate occurs in the stroma, the enzyme-containing interior of chloroplasts.

6.3 Photosynthesis is a redox reaction that releases O_2

- Oxidation is the loss of electrons, and reduction is the gain of electrons.
- During photosynthesis, CO_2 is reduced and water is oxidized, resulting in carbohydrate and oxygen and an H_2O molecule:

$$CO_2 + 2\ H_2O \xrightarrow{\text{solar energy}} (CH_2O) + O_2$$

6.4 Experiments showed that the oxygen released during photosynthesis comes from water

- Two separate experiments using isotopes proved that oxygen comes from water, not from CO_2.

6.5 Photosynthesis involves two sets of reactions: the light reactions and the Calvin cycle reactions

- Light reactions only occur in thylakoids during the day when solar energy is available.
- Calvin cycle reactions are enzymatic reactions that reduce CO_2 to a carbohydrate in the stroma:

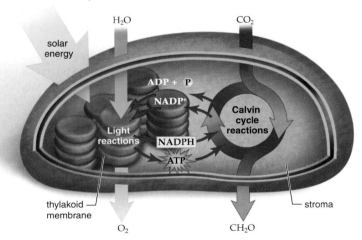

First, Solar Energy Is Captured

6.6 Light reactions begin: Solar energy is absorbed by pigments

- Chlorophylls a and b and carotenoids absorb violet, blue, and red light better than other portions of visible light.

6.7 Fall temperatures cause leaves to change color

- Less sunlight in fall means not as much solar energy to rebuild chlorophyll, which disintegrates, leaving yellow and orange pigments visible.

6.8 Solar energy boosts electrons to a higher energy level

- Within thylakoid membranes, pigment complexes in photosystems I and II absorb solar energy, which excites electrons in the complex.

- Energized electrons are passed by a reaction center chlorophyll *a* molecule to an electron acceptor.

6.9 Electrons release their energy as ATP forms
- Electron acceptors send energized electrons down an electron transport chain, a series of membrane-bounded electron carriers.
- An oxidation-reduction reaction occurs at each transfer, and energy is released.
- Carriers of the electron transport chain are found in the thylakoid membranes of chloroplasts.
- ATP is formed from the energy released during electron transfers.

6.10 During the light reactions, electrons follow a noncyclic pathway
- During the noncyclic electron pathway, electrons move from PS II down an electron transport chain to PS I, where they are re-energized and passed to NADP$^+$, which becomes NADPH.

6.11 The thylakoid membrane is organized to produce ATP and NADPH
- Get-ready phase:
  - PS II: Pigment complex plus electron acceptor. Water splits, releasing H$^+$ and O$_2$.
  - Members of the electron transport chain pump H$^+$ from the stroma to the thylakoid space; H$^+$ gradient results.
  - PS I absorbs solar energy; electrons eventually get passed to NADP reductase.
- Payoff phase (chemiosmosis):
  - NADP reductase passes electrons to NADP$^+$ and NADPH results. H$^+$ flows down concentration gradient through ATP synthase complex; ADP binds to P; ATP are produced.

Second, Carbohydrate Is Synthesized

6.12 The Calvin cycle uses ATP and NADPH from the light reactions to produce a carbohydrate
- CO$_2$ fixation: The enzyme RuBP carboxylase fixes CO$_2$ to RuBP, producing a C$_6$ molecule that immediately splits into two C$_3$ molecules (3PG).
- CO$_2$ reduction: Each 3PG is reduced to a G3P molecule.

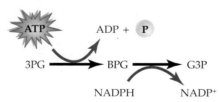

- RuBP regeneration: During three turns of the Calvin cycle, five molecules of G3P are used to re-form three molecules of RuBP. This step requires ATP energy.

6.13 In plants, carbohydrate is the starting point for other molecules
- G3P can be converted to all organic molecules needed by a plant.
- It takes two G3P molecules to make one glucose molecule.

C$_3$, C$_4$, and CAM Photosynthesis Thrive Under Different Conditions

6.14 C$_3$ photosynthesis evolved when oxygen was in limited supply
- C$_3$ photosynthesis occurs under conditions of moderate temperature and rainfall.

- RuBP carboxylase combines with O$_2$ when CO$_2$ supply is limited, and this reduces yield.

6.15 C$_4$ photosynthesis boosts CO$_2$ concentration for RuBP carboxylase
- C$_4$ plants grow where the climate is warm.
- Partitioning in space: CO$_2$ fixation occurs in spongy mesophyll; Calvin cycle occurs in bundle sheath cells, where RuBP is not exposed to O$_2$.

6.16 CAM photosynthesis is another alternative to C$_3$ photosynthesis
- CAM plants grow where the climate is hot and dry; stomata are closed during the day to conserve water.
- Partitioning in time: CO$_2$ is fixed at night; does not enter Calvin cycle until the next day.

Testing Yourself

Photosynthesis Produces Food and Releases Oxygen
1. The raw materials for photosynthesis are
   a. oxygen and water.
   b. oxygen and carbon dioxide.
   c. carbon dioxide and water.
   d. carbohydrates and water.
   e. carbohydrates and carbon dioxide.
2. _____ is reduced to _____ during photosynthesis.
   a. CO$_2$, oxygen   c. Water, oxygen
   b. Oxygen, CO$_2$ d. CO$_2$, carbohydrates
3. The light reactions
   a. take place in the stroma.   c. Both a and b are correct.
   b. consist of the Calvin cycle.   d. Neither a nor b is correct.
4. The function of light reactions is to
   a. obtain CO$_2$.
   b. make carbohydrate.
   c. convert light energy into usable forms of chemical energy.
   d. regenerate RuBP.
5. **THINKING CONCEPTUALLY** For the biosphere to have animal life, some animals have to eat plants. Explain.

First, Solar Energy Is Captured
6. When leaves change color in the fall, _____ light is absorbed for photosynthesis.
   a. orange range   c. violet-blue-green range
   b. red range   d. None of these are correct.
7. A photosystem contains
   a. pigments, a reaction center, and an electron receiver.
   b. ADP, (P), and hydrogen ions (H$^+$).
   c. protons, photons, and pigments.
   d. cytochromes only.
   e. Both b and c are correct.
8. PS I, PS II, and the electron transport chain are located in the
   a. thylakoid membrane.   c. outer chloroplast membrane.
   b. stroma.   d. cell's nucleus.
9. The final acceptor of electrons during the noncyclic electron pathway is
   a. PS I.   d. ATP.
   b. PS II.   e. NADP$^+$.
   c. water.

10. When electrons in the reaction center of PS II are passed to an acceptor molecule, they are replaced by electrons that come from
    a. oxygen.   c. carbon dioxide.
    b. glucose.   d. water.

11. During the light reactions of photosynthesis, ATP is produced when hydrogen ions move
    a. down a concentration gradient from the thylakoid space to the stroma.
    b. against a concentration gradient from the thylakoid space to the stroma.
    c. down a concentration gradient from the stroma to the thylakoid space.

Second, Carbohydrate Is Synthesized

12. The Calvin cycle reactions
    a. produce carbohydrate.
    b. convert one form of chemical energy into a different form of chemical energy.
    c. regenerate more RuBP.
    d. use the products of the light reactions.
    e. All of these are correct.

13. The Calvin cycle requires _____ from the light reactions.
    a. carbon dioxide and water   d. ATP and water
    b. ATP and NADPH   e. NADH and water
    c. carbon dioxide and ATP

14. Label the following diagram of a chloroplast.

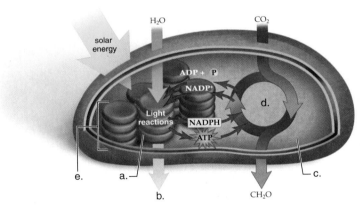

15. **THINKING CONCEPTUALLY** The overall equation for photosynthesis doesn't include ATP. Why is ATP needed?

C₃, C₄, and CAM Photosynthesis Thrive Under Different Conditions

16. C₄ photosynthesis
    a. occurs in plants whose bundle sheath cells contain chloroplasts.
    b. takes place in plants such as wheat, rice, and oats.
    c. is an advantage when the weather is warm.
    d. Both a and c are correct.

17. CAM photosynthesis
    a. is the same as C₄ photosynthesis.
    b. is an adaptation to cold environments in the Southern Hemisphere.
    c. is prevalent in desert plants that close their stomata during the day.
    d. stands for chloroplasts and mitochondria.

18. The different types of photosynthesis are dependent upon the timing and location of
    a. CO₂ fixation.   c. H₂O fixation.
    b. nitrogen fixation.   d. All of these are correct.

Understanding the Terms

ATP synthase complex   101   grana   97
autotroph   96   heterotroph   96
C₃ plant   106   light reactions   99
C₄ plant   106   oxidation   98
Calvin cycle reactions   99   photosynthesis   96
CAM photosynthesis   107   photosystem I (PS I)   101
carotenoid   100   photosystem II (PS II)   101
chemiosmosis   103   redox reaction   98
chlorophyll   97   reduction   98
chlorophyll a   100   RuBP carboxylase   104
chlorophyll b   100   stomata   97
chloroplast   97   stroma   97
electron transport chain   101 thylakoid   97

Match the terms to these definitions:
a. _____ Energy-capturing portion of photosynthesis that takes place in thylakoid membranes of chloroplasts and cannot proceed without solar energy; produces ATP and NADPH.
b. _____ Passage of electrons along a series of carrier molecules from a higher to a lower energy level; the energy released is used for the synthesis of ATP.
c. _____ Process usually occurring within chloroplasts whereby chlorophyll traps solar energy and carbon dioxide is reduced to a carbohydrate.
d. _____ Series of reactions in which carbon dioxide is fixed and reduced to G3P.
e. _____ Type of photosynthesis that fixes carbon dioxide at night to produce a C₄ molecule that releases carbon dioxide to the Calvin cycle during the day.

Thinking Scientifically

1. Elodea, a plant that lives in the water, is in a beaker of water. Bubbling occurs with white light, but not green. Tell what environmental conditions should be kept constant and suggest a control for this experiment.
2. The process of photosynthesis supports the cell theory. How?

ARIS *Visit* **www.mhhe.com/maderconcepts** *for practice quizzes, animations, videos, and activities designed to help you master the material in this chapter.*

7

Pathways of Cellular Respiration

LEARNING OUTCOMES

After studying this chapter, you should be able to accomplish the following outcomes.

ATP Is Universal

1 Relate the universality of ATP to its presence in the first cell.

Glucose Breakdown Releases Energy

2 Write the overall reaction for glucose breakdown, and show that it is a redox reaction.
3 Name and discuss the role of oxidation-reduction enzymes.
4 State the four phases of cellular respiration, and tell where each occurs in the cell.

Carbon Dioxide and Water Are Produced During Glucose Breakdown

5 Describe the pathways of cellular respiration (glycolysis, the preparatory reaction, the citric acid cycle, and the electron transport chain). Name the inputs and outputs of each pathway.
6 Explain the role of oxygen during cellular respiration.

Fermentation Is Inefficient

7 Compare and contrast fermentation with glycolysis.
8 Explain why fermentation is inefficient and results in oxygen debt.
9 Give examples of products made by fermenting yeasts and products made by fermenting bacteria.

Metabolic Pathways Cross at Particular Substrates

10 Show how catabolism of protein and fat utilizes the same pathways as glucose breakdown.
11 Explain how it is possible to become overweight by eating foods rich in sugar.
12 Use the graph from page 126 to show what type and duration of exercise are needed to burn fat.

ATP (adenosine triphosphate) is ancient—a molecular fossil, really—and it is universal. ATP was present 3.5 billion years ago when life began. Like other nucleotides, it probably formed in the early atmosphere and then rained down into the oceans to become incorporated into the first cells that evolved. Because all living cells are related to the first cell(s), ATP became universal. Whether you are a bacterium with undulating flagella, an ocelot hanging in a tree, a snail living in a garden, or a human striding past a giant cactus—your cells are making and using ATP, and so are the cells of the cactus and the other plants. When cells require energy to do work, they split ATP.

adenosine triphosphate

What is the secret of ATP? Why is ATP suited to be the energy currency of cells? ATP is a nucleotide, but unlike other nucleotides, it has three phosphate groups—and therein lies the secret of ATP's ability to supply energy. The phosphates repel one another because each has a negative charge. The breakdown of ATP to ADP relieves the repulsion and releases a significant amount of energy. Cells are able to couple the energy of ATP breakdown to reactions that require energy. Without coupling, all of the energy of breakdown would be lost as heat, and life would not exist.

We can imagine that ATP became incorporated into the first cells and that its breakdown releases energy. But then what happens? When ATP becomes ADP + Ⓟ, it has to become ATP again or else cells die. Enter the mitochondria. Mitochondria are aptly called the powerhouses of the cell because most ATP is made here from ADP + Ⓟ. Without mitochondria, a eukaryotic cell has a limited capacity to rebuild its

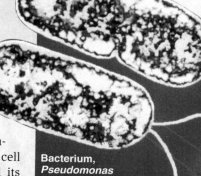

Bacterium, *Pseudomonas aeruginosa*

ATP Is Universal

ATP. The power plants of modern human society use all sorts of fuels—fossil fuels, nuclear energy, and renewable energy sources such as wind—to produce electricity. Similarly, mitochondria use glucose, fats, and even proteins as fuels to produce ATP.

ATP is unique among the cell's storehouse of chemicals; no other molecule performs its functions in such a solitary manner. Amino acids must join to make a protein, and nucleotides must join to make DNA or RNA, but ATP has the same structure and function in all cells. It provides the energy that makes all forms of life possible. Whether you go skiing, take an aerobics class, or just hang out, ATP molecules provide the energy needed for your muscles to contract. ATP is also needed for nerve conduction, protein synthesis, and any other reaction in a cell that requires energy. ATP molecules are produced during cellular respiration. **Cellular respiration**, the process by which cells harvest the energy stored in organic compounds, is the topic of this chapter.

Snails, *Achatina glutinosa*

Ocelot, *Felis pardalis*

Prickly pear cactus, *Opuntia echios gigantea*

This part begins with the overall equation for cellular respiration, which shows the reactants and the products of the process. In reality, however, cellular respiration requires four phases. One phase is anaerobic and takes place in the cytoplasm. The other three phases occur in the mitochondria when oxygen is available.

During cellular respiration, glucose and glucose breakdown products are oxidized by the removal of hydrogen atoms ($e^- + H^+$). Eventually, these high-energy electrons pass down an electron transport chain, leading to the production of ATP molecules.

7.1 Cellular respiration is a redox reaction that requires O_2

Cellular respiration is aptly named because just as you take in oxygen (O_2) and give off carbon dioxide (CO_2) during breathing, so does cellular respiration. In fact, cellular respiration, which occurs in all cells of the body, is the reason you breathe.

Oxidation of substrates, such as glucose, is a fundamental part of cellular respiration. In living things, as you know, oxidation occurs by the removal of hydrogen atoms ($e^- + H^+$). As cellular respiration occurs, hydrogen atoms are removed from glucose and the result is carbon dioxide (CO_2). Hydrogen atoms eventually reduce oxygen, forming water. The end result of cellular respiration is carbon dioxide (CO_2) and water (H_2O):

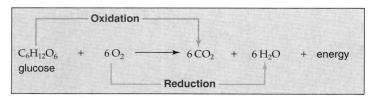

$$C_6H_{12}O_6 \ + \ 6\,O_2 \ \longrightarrow \ 6\,CO_2 \ + \ 6\,H_2O \ + \ \text{energy}$$

The breakdown of glucose during cellular respiration releases a lot of energy. If you mistakenly burn sugar in a pan, the energy escapes into the atmosphere as heat. A cell is more sophisticated than that. In a cell, glucose is broken down slowly—not all at once—and the energy given off isn't all lost as heat. Hydrogen atoms ($e^- + H^+$) are removed a few at a time, and this allows energy to be captured and used to make ATP molecules, largely in mitochondria (**Fig. 7.1**).

NAD$^+$ and FAD   The enzymes that carry out oxidation during cellular respiration are assisted by the coenzymes of oxidation-reduction, called **NAD$^+$** (nicotinamide adenine dinucleotide) and **FAD** (flavin adenine dinucleotide). When a substrate is oxidized, NAD$^+$ accepts two electrons plus a hydrogen ion (H$^+$), and NADH results:

$$NAD^+ + 2e^- + H^+ \longrightarrow NADH$$

FAD accepts two electrons and two hydrogen ions (H$^+$) to become FADH$_2$:

$$FAD + 2e^- + 2H^+ \longrightarrow FADH_2$$

The electrons received by NAD$^+$ and FAD are *high-energy electrons* that are usually carried to an electron transport chain (see Fig. 6.9). The energy captured as electrons move down the chain will be used for ATP production inside mitochondria.

An overview of the four phases that occur during cellular respiration is given in Section 7.2.

> **7.1 Check Your Progress** **Why could it be said that we breathe in order to produce ATP?**

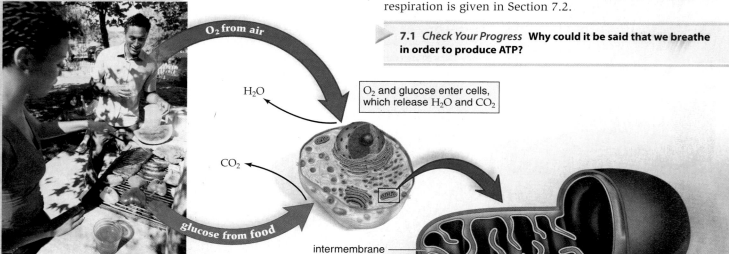

FIGURE 7.1   Cellular respiration produces ATP.

7.2 Cellular respiration has four phases—three phases occur in mitochondria

Cellular respiration involves four phases: glycolysis, the preparatory phase, the citric acid cycle, and the electron transport chain (**Fig. 7.2**). Glycolysis takes place outside the mitochondria and does not require the presence of oxygen, so glycolysis is called an **anaerobic** process. The other phases of cellular respiration take place inside the mitochondria, where oxygen is the final acceptor of electrons. Because they require oxygen, these phases are called **aerobic**.

During these phases, notice where CO_2 and H_2O, the end products of cellular respiration, are produced.

- **Glycolysis** is the breakdown of glucose to two molecules of pyruvate. Oxidation results in NADH, and there is enough energy left over for a net gain of 2 ATP molecules.
- The **preparatory (prep) reaction** takes place in the matrix of the mitochondria. Pyruvate is oxidized to a 2-carbon acetyl group, and CO_2 is released. Since glycolysis ends with two molecules of pyruvate, the prep reaction occurs twice per glucose molecule. Oxidation of pyruvate yields NADH.
- The **citric acid cycle** also takes place in the matrix of the mitochondria. As oxidation occurs, NADH and $FADH_2$ result, and more CO_2 is released. Because two acetyl groups enter the cycle per glucose molecule, the cycle turns twice. The citric acid cycle is able to produce one ATP per turn.

- The **electron transport chain (ETC)** is a series of electron carriers in the cristae of the mitochondria. NADH and $FADH_2$ give up electrons to the chain. Energy is released and captured as the electrons move from a higher-energy to a lower-energy state. Later, this energy will be used for the production of ATP by chemiosmosis. Oxygen (O_2) finally shows up here as the last acceptor of electrons from the chain. After oxygen receives electrons, it combines with hydrogen ions (H^+) and becomes water (H_2O).

Pyruvate, the end product of glycolysis, is a pivotal molecule; its further treatment is dependent on whether oxygen is available. If oxygen is available, pyruvate enters a mitochondrion and is broken down completely to carbon dioxide and water. If oxygen is not available, pyruvate is further metabolized in the cytoplasm by an anaerobic process called **fermentation**. Fermentation results in a net gain of only two ATP per glucose molecule. This is far fewer than the number produced by mitochondria.

The enzymatic reactions of glycolysis (also called the glycolytic pathway) are examined in Section 7.3.

7.2 Check Your Progress What is the benefit of oxidizing glucose slowly by removing hydrogen atoms?

FIGURE 7.2 The four phases of complete glucose breakdown.

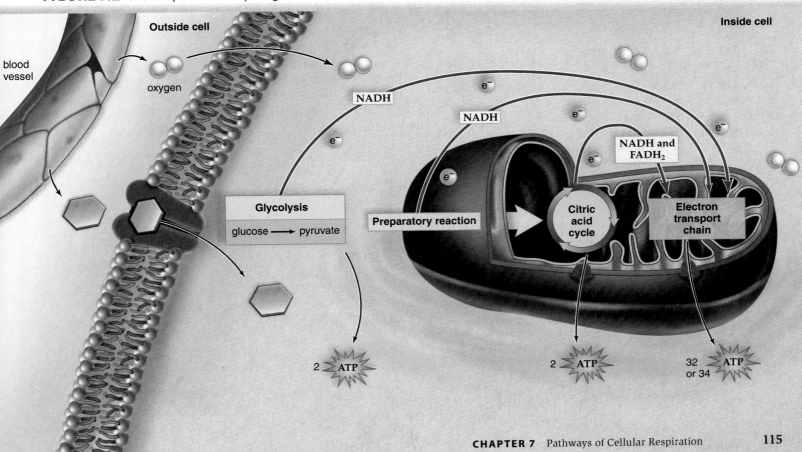

Carbon Dioxide and Water Are Produced During Glucose Breakdown

Learning Outcomes 5–6, page 112

Some of the enzymatic reactions that occur during cellular respiration lead to ATP production, and others lead to NADH and $FADH_2$. We will examine each of the four phases of cellular respiration to see when and how ATP, NADH, and $FADH_2$ are produced.

7.3 Glycolysis: Glucose breakdown begins

Glycolysis, which takes place within the cytoplasm outside the mitochondria, is the breakdown of glucose to two pyruvate molecules (**Fig. 7.3A,** on facing page). Since glycolysis occurs universally in organisms, it most likely evolved before the citric acid cycle and the electron transport chain. This may be why glycolysis occurs in the cytoplasm and does not require oxygen. There was no free oxygen in the early atmosphere of the Earth.

Glycolysis is a long series of reactions (Fig. 7.3A), and just as you would expect for a metabolic pathway, each step has its own enzyme. The pathway can be conveniently divided into the energy-investment steps and the energy-harvesting steps. During the energy-investment steps, ATP is used to "jump-start" glycolysis. Thereafter, more ATP than the input is made.

Energy-Investment Steps
1 As glycolysis begins, two ATP are used to activate glucose, a C_6 (6-carbon) molecule that **2** splits into two C_3 molecules, known as G3P. Each G3P has a phosphate group. From this point on, each C_3 molecule undergoes the same series of reactions.

Energy-Harvesting Steps
Oxidation of G3P now occurs by the removal of hydrogen atoms (e⁻ + H⁺). **3** In duplicate reactions, electrons are picked up by coenzyme NAD⁺, which becomes NADH:

$$2 \text{ NAD}^+ + 4 \text{ e}^- + 2 \text{ H}^+ \longrightarrow 2 \text{ NADH}$$

Now, each NADH molecule will carry two *high-energy electrons* to the electron transport chain, and become NAD⁺ again. Only a small amount of NAD⁺ need be present in a cell because, like other coenzymes, it is used over and over again.

4 The subsequent addition of inorganic phosphate results in two high-energy phosphate groups. **5** These phosphate groups are used to synthesize two ATP. This is called **substrate-level ATP synthesis** because an enzyme passes a high-energy phosphate to ADP, and ATP results (**Fig. 7.3B**). Notice that this is an example of coupling: An energy-releasing reaction is driving forward an energy-requiring reaction on the surface of the enzyme. **6** Oxidation occurs again but by removal of water (H_2O). **7** Substrate-level ATP synthesis occurs again, and **8** two molecules of pyruvate result. We will assume that oxygen is available and that these pyruvate molecules will enter a mitochondrion.

Net Gain of ATP
Two ATP were used to get glycolysis started, but two ATP were made in step **5**, and two more were made in step **7**. Therefore, there is a net gain of two ATP from glycolysis.

Inputs and Outputs of Glycolysis Altogether, the inputs and outputs of glycolysis are as follows:

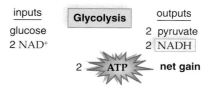

At this point, it is helpful to ask, "Where did the inputs come from, and what will happen to the outputs of glycolysis?" The food we eat contains glucose. It enters the bloodstream, is carried about the body, and then enters the body's cells. The coenzyme NAD⁺ and the molecules ADP and Ⓟ are always available in cells.

The output pyruvate enters mitochondria, where it is oxidized to CO_2 during the prep reaction and the citric acid cycle. NADH carries electrons to the electron transport chain, and the ATP can be used to jump-start glycolysis again.

So far, we have accounted for only two out of the 36 or 38 ATP formed per glucose molecule during cellular respiration. We have not seen the use of any oxygen or the production of any CO_2. It is clear we have a ways to go.

The second phase of cellular respiration is a single enzymatic reaction that occurs preparatory to the citric acid cycle, as discussed in Section 7.4

> **7.3 Check Your Progress** So far, what has happened to the energy that is in a glucose molecule?

FIGURE 7.3B
Substrate-level ATP synthesis.

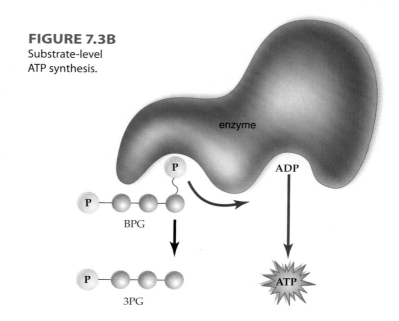

Glycolysis

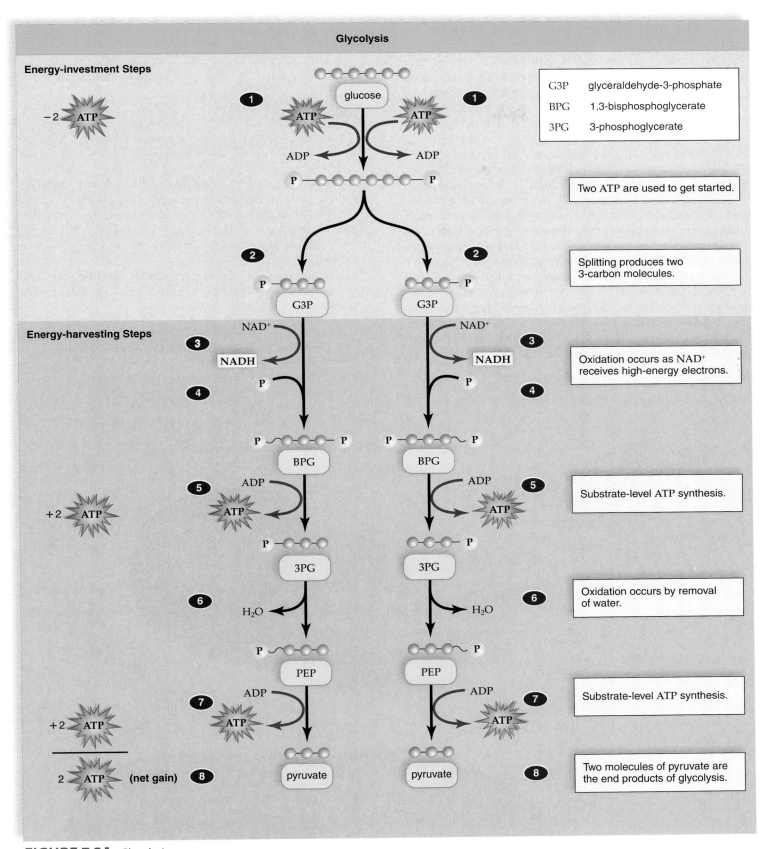

Energy-investment Steps

− 2 **ATP**

G3P   glyceraldehyde-3-phosphate
BPG   1,3-bisphosphoglycerate
3PG   3-phosphoglycerate

glucose

ATP   ATP

ADP   ADP

Two ATP are used to get started.

Splitting produces two 3-carbon molecules.

Energy-harvesting Steps

G3P   G3P

NAD⁺   NAD⁺

NADH   NADH

Oxidation occurs as NAD⁺ receives high-energy electrons.

BPG   BPG

ADP   ADP

+ 2 **ATP**

ATP   ATP

Substrate-level ATP synthesis.

3PG   3PG

H_2O H_2O

Oxidation occurs by removal of water.

PEP   PEP

ADP   ADP

+ 2 **ATP**

ATP   ATP

Substrate-level ATP synthesis.

2 **ATP** (net gain)

pyruvate   pyruvate

Two molecules of pyruvate are the end products of glycolysis.

FIGURE 7.3A Glycolysis.

The preparatory reaction, the citric acid cycle, and the electron transport chain, which are needed for the complete breakdown of glucose, take place within the mitochondria. A mitochondrion has a double membrane, with an intermembrane space between the outer and inner membranes. **Cristae** are folds of inner membrane that jut out into the **matrix**, the innermost compartment, which is filled with a gel-like fluid (**Fig. 7.4**). Just like a chloroplast, a mitochondrion is highly structured, and we would expect reactions to be located in particular parts of this organelle.

The enzymes that speed the prep reaction and the citric acid cycle are arranged in the matrix, and the electron transport chain is located in the cristae in a very organized manner. Most of the ATP from cellular respiration is produced in mitochondria; therefore, mitochondria are often called the powerhouses of the cell.

The **preparatory (prep) reaction** is so called because it occurs before the citric acid cycle. In this reaction, the 3-carbon pyruvate is converted to a 2-carbon (C_2) *acetyl group*, and CO_2 is given off. This is an oxidation reaction in which hydrogen atoms are removed from pyruvate and NADH is formed. One prep reaction occurs per pyruvate, so altogether, the prep reaction occurs twice per glucose molecule.

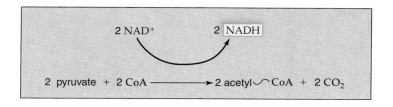

The 2-carbon acetyl group is combined with a molecule known as CoA. CoA will carry the acetyl group to the citric acid cycle. The two NADH carry electrons to the electron transport chain. What about the CO_2? In humans, CO_2 freely diffuses out of cells into the blood, which transports it to the lungs where it is exhaled.

The third phase of cellular respiration is the citric acid cycle, which occurs in the matrix of the mitochondria, as discussed in Section 7.5.

> **7.4 *Check Your Progress*** Account for the term **preparatory reaction. What is it prepping for?**

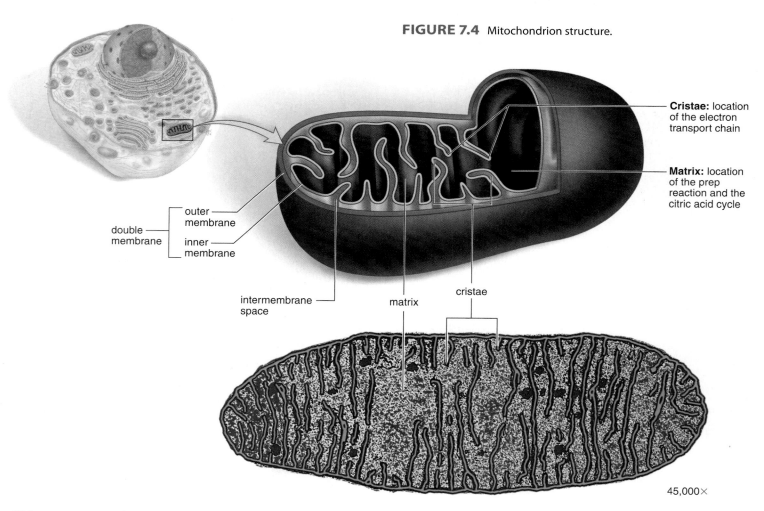

FIGURE 7.4 Mitochondrion structure.

Cristae: location of the electron transport chain

Matrix: location of the prep reaction and the citric acid cycle

double membrane

outer membrane

inner membrane

intermembrane space

matrix

cristae

45,000×

7.5 The citric acid cycle: Final oxidation of glucose products

The **citric acid cycle** is a metabolic pathway located in the matrix of mitochondria. The cycle consists of a series of reactions that return to their starting point once again (**Fig. 7.5**). The citric acid cycle is noteworthy because it produces a significant portion of the NADH and all of the $FADH_2$ that carry electrons to the electron transport chain. These are *high-energy electrons* that have been removed from glucose breakdown products.

Steps of the Cycle ❶ At the start of the citric acid cycle, an enzyme speeds the removal of an (C_2) acetyl group from CoA. ❷ The acetyl group joins with a 4-carbon (C_4) molecule, forming a 6-carbon (C_6) citrate molecule. ❸ , ❹ Following formation of citrate twice over an oxidation reaction occurs; both times NADH forms and CO_2 is released. ❺ A high-energy phosphate is transferred to an ADP, and one ATP molecule is produced by substrate-level ATP synthesis per turn. In substrate-level ATP synthesis, you may recall, an enzyme passes a high-energy phosphate to ADP.

❻ Additional oxidation reactions produce an $FADH_2$ and another NADH. ❼ The cycle has now returned to its starting point and is ready to receive another acetyl group.

Because the citric acid cycle turns twice for each original glucose molecule, the inputs and outputs of the citric acid cycle per glucose molecule are as follows:

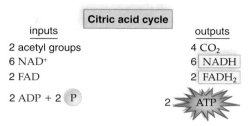

	Citric acid cycle	
inputs		outputs
2 acetyl groups		4 CO_2
6 NAD⁺		6 NADH
2 FAD		2 $FADH_2$
2 ADP + 2 Ⓟ		2 ⟡ATP

Production of CO_2 The six carbon atoms originally located in a glucose molecule have now become six molecules of CO_2. The prep reaction produces two CO_2, and the citric acid cycle produces four CO_2 per glucose molecule. We have already mentioned that this is the CO_2 we breathe out. In other words, the CO_2 we exhale comes from the mitochondria within our cells.

Thus far, we have broken down glucose to CO_2 and hydrogen atoms. NADH and $FADH_2$ are now carrying high-energy electrons. What will happen to the electrons? You already know that they will be passed to oxygen, but how? On to the story of the electron transport chain in the mitochondria, as discussed in Section 7.6.

> **7.5 Check Your Progress** What happens to the CO_2 produced by the prep reaction and the citric acid cycle in our bodies?

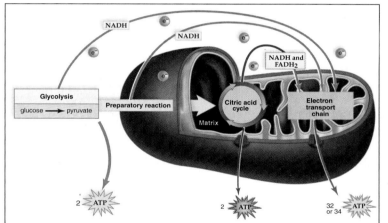

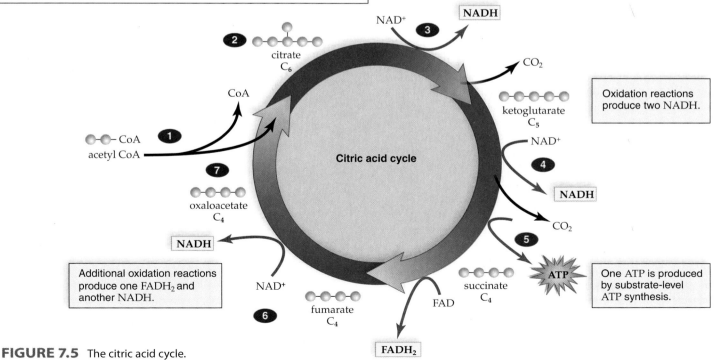

FIGURE 7.5 The citric acid cycle.

7.6 The electron transport chain captures much energy

The **electron transport chain (ETC)**, located in the cristae of the mitochondria (and the plasma membrane of aerobic prokaryotes), is a series of carriers that pass electrons from one to the other. The electrons that enter the electron transport chain are carried by NADH and FADH$_2$. **Figure 7.6** is arranged to show that high-energy electrons enter the chain, and low-energy electrons leave the chain.

Members of the Chain **1a** When NADH gives up its electrons, it becomes NAD$^+$; and when **1b** FADH$_2$ gives up its electrons, it becomes FAD. The next carrier gains the electrons and is reduced. This oxidation-reduction process continues, and each of the carriers, in turn, becomes reduced and then oxidized as the electrons move down the chain.

2 Many of the carriers are cytochrome molecules. A **cytochrome** is a protein that has a tightly bound heme group with a central atom of iron, the same as hemoglobin does. When the iron accepts electrons, it becomes reduced, and when it gives them up, it becomes oxidized. A number of poisons, such as cyanide, cause death by binding to and blocking the function of cytochromes. **3** As the pair of electrons is passed from carrier to carrier, energy is captured and eventually used to form ATP molecules.

What is the role of oxygen in cellular respiration and the reason we take in oxygen by breathing? Oxygen is the final acceptor of electrons from the electron transport chain. Oxygen receives the energy-spent electrons from the last of the carriers (i.e., cytochrome oxidase). **4** After receiving electrons, oxygen combines with hydrogen ions, and water forms:

$$\tfrac{1}{2}O_2 + 2\,e^- + 2\,H^+ \longrightarrow H_2O$$

The critical role of oxygen as the final acceptor of electrons during cellular respiration is exemplified by the fact that if oxygen is not present, the chain does not function, and the mitochondria produce no ATP. The limited capacity of the body to form ATP in a way that does not involve the electron transport chain means that death eventually results if oxygen is not available.

Cycling of Carriers Once NADH and FADH$_2$ have delivered electrons to the electron transport chain, they are "free" to return and pick up more hydrogen atoms. In the same manner, ADP is recycled in cells. The reuse of carriers increases cellular efficiency, since it does away with the need to synthesize them anew. Exactly how the delivery of high-energy electrons to the electron transport chain leads to ATP synthesis is explored in Section 7.7.

> **7.6 Check Your Progress** Electrons delivered to the electron transport chain by NADH eventually account for the production of three ATP. When FADH$_2$ delivers electrons to the chain, only two ATP are produced. How many ATP are produced as a result of 10 NADH plus 2 FADH$_2$ delivering electrons to the chain?

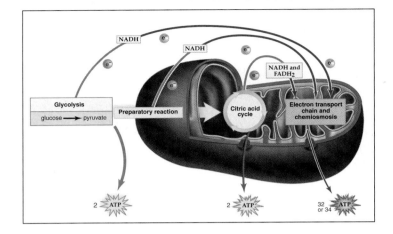

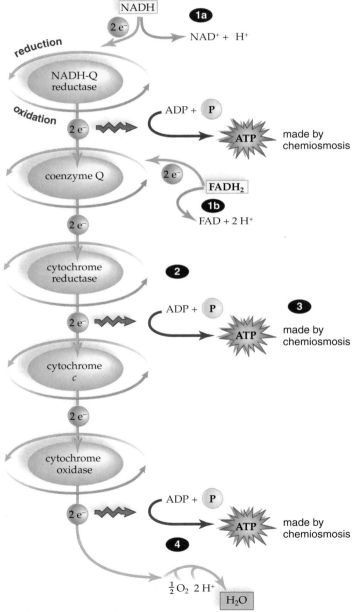

FIGURE 7.6 The electron transport chain.

7.7 The cristae create an H⁺ gradient that drives ATP production

Following the example of Section 6.11, we will divide the molecular complexes in the cristae into those that "get ready" and those that represent the "payoff."

Get Ready The carriers of the ETC are sequentially arranged in the cristae of mitochondria. The three protein complexes include the NADH-Q reductase complex, the cytochrome reductase complex, and the cytochrome oxidase complex. The two other carriers that transport electrons between the complexes are coenzyme Q and cytochrome c (**Fig. 7.7**).

The members of the electron transport chain accept electrons, which they pass from one to the other. What happens to the hydrogen ions (H^+) carried by NADH and $FADH_2$? Certain complexes of the ETC use the released energy to pump these hydrogen ions from the matrix into the intermembrane space of a mitochondrion. Vertical blue arrows in Figure 7.7 show which protein complexes of the electron transport chain pump H^+ into the intermembrane space. This establishes a strong electrochemical gradient; there are about ten times as many hydrogen ions in the intermembrane space as in the matrix.

Payoff The cristae contain an ATP synthase complex, through which hydrogen ions flow down a gradient from the intermembrane space into the matrix. As hydrogen ions flow from high to low concentration, the enzyme ATP synthase synthesizes ATP from ADP + Ⓟ. This process is called **chemiosmosis** because ATP production is tied to the establishment of an H^+ gradient.

The action of certain poisons that inhibit cellular respiration supports the chemiosmotic model. When one poison inhibits ATP synthesis, the H^+ gradient subsequently becomes larger than usual. When another poison makes the membrane leaky so that an H^+ gradient does not form, no ATP is made.

Once formed, ATP moves out of mitochondria and is used to perform cellular work, during which it breaks down to ADP and Ⓟ. Then these molecules are returned to the mitochondria for recycling. At any given time, the amount of ATP in a human would sustain life for only about a minute; therefore, ATP synthase must constantly produce ATP. It is estimated that the mitochondria produce our body weight in ATP every day.

Active tissues require greater amounts of ATP and have more mitochondria than less active cells. For example, in chickens the dark meat of the legs contains more mitochondria than the white meat of the breast. This suggests that chickens mainly walk or run, rather than flying about the barnyard. The color of dark meat is also due to the presence of blood vessels and a respiratory pigment called myoglobin that is found in muscles.

How many ATP are produced per glucose molecule by means of chemiosmosis? See Section 7.8, which calculates the number of ATP per each phase of cellular respiration and then totals the number of ATP per glucose molecule.

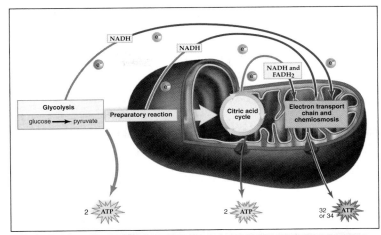

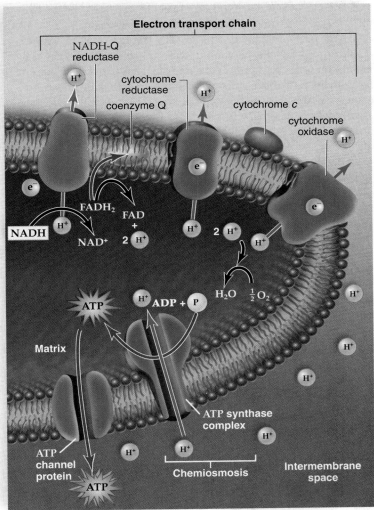

FIGURE 7.7 Organization and function of cristae.

> **7.7 Check Your Progress** What is the end result of the "get ready" phase, and what is the end result of the "payoff" phase in the mitochondria?

7.8　The ATP payoff can be calculated

Figure 7.8 calculates the ATP yield for the complete breakdown of glucose to CO_2 and H_2O during cellular respiration.

In the Cytoplasm ❶ Per glucose molecule, there is a net gain of two ATP from glycolysis, which takes place in the cytoplasm. These two ATP are produced by substrate-level ATP synthesis (see Fig. 7.3B).

In the Mitochondrion ❷ The prep reaction does not generate any ATP molecules directly. ❸ The citric acid cycle, which occurs in the matrix of mitochondria, directly accounts for two ATP per glucose molecule. These two ATP are formed by substrate-level ATP synthesis. ❹ This adds up to a subtotal of four ATP produced by substrate-level ATP synthesis.

❺ Most ATP is produced by the electron transport chain and chemiosmosis. Per glucose molecule, ten NADH and two $FADH_2$ take electrons to the electron transport chain.

❻ In some animal cells, NADH formed outside a mitochondrion by glycolysis cannot cross the inner mitochondrial membrane, but a "shuttle" mechanism allows its electrons to be delivered to the electron transport chain inside the mitochondrion. The cost to the cell is one ATP for each NADH that is shuttled to the ETC. This reduces the overall count of ATP produced as a result of glycolysis, in some animal cells, to four, instead of six, ATP.

For each NADH formed *inside* the mitochondrion by ❼ the prep reaction and ❽ the citric acid cycle, three ATP result, but ❾ for each $FADH_2$ from the citric acid cycle as expected, only two ATP are produced. Figure 7.6 explains the reason for this difference: $FADH_2$ delivers its electrons to the ETC after NADH, and therefore these electrons cannot account for as much ATP production.

❿ Altogether, 32 or 34 ATP are made as a result of electrons passing down the electron transport chain. ⓫ Therefore, the total number of ATP produced as a result of complete glucose breakdown is 36 or 38.

Efficiency of Cellular Respiration It is interesting to calculate how much of the energy in a glucose molecule eventually becomes available to the cell. The difference in energy content between the reactants (glucose and O_2) and the products (CO_2 and H_2O) is 686 kcal. An ATP phosphate bond has an energy content of 7.3 kcal, and if 36 ATP are produced during glucose breakdown, 36 phosphates are equivalent to a total of 263 kcal. Therefore, 263/686, or 39%, of the available energy is usually transferred from glucose to ATP. The rest of the energy is lost in the form of heat.

This concludes our discussion of the phases of cellular respiration. Next, in Section 7.9, we consider what the cell does when oxygen is lacking and mitochondria are not operational.

> **7.8 Check Your Progress** What part of Figure 7.8 was available to the first eukaryotic cells, assuming that these cells did not have mitochondria?

FIGURE 7.8 Energy yield per glucose molecule.

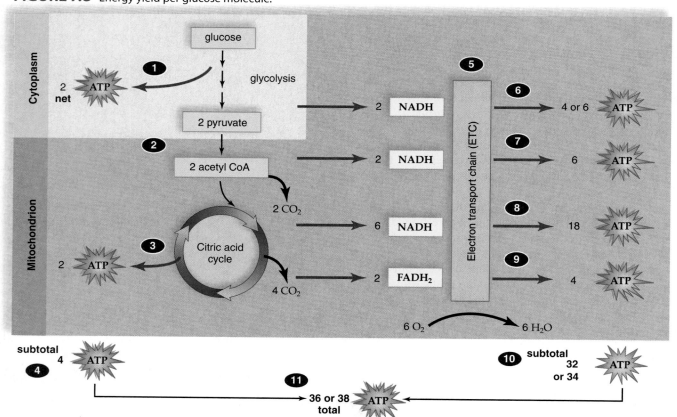

Fermentation Is Inefficient

Complete glucose breakdown demands an input of oxygen to accept electrons from the electron transport chain. If oxygen is not available, the ETC stops working, and some cells turn to fermentation, an anaerobic process. Fermentation produces only two ATP per glucose molecule, instead of 36 or 38, but even so, it is useful to a cell when a burst of energy is required, as when we run to escape a danger or to catch a bus. Humans also make use of fermentation by microorganisms to produce several different daily dietary products.

7.9 When oxygen is in short supply, the cell switches to fermentation

Fermentation continues to produce a limited amount of ATP by using organic molecules instead of oxygen as the final electron acceptor (**Fig. 7.9A**). In animal cells, including those of humans, the pyruvate formed by glycolysis accepts two hydrogen atoms and is reduced to lactate. Other types of organisms instead produce alcohol with the release of CO_2. Bacteria vary as to whether they produce an organic acid, such as lactate, or an alcohol and CO_2. Yeasts are good examples of organisms that generate ethyl alcohol and CO_2 as a result of fermentation.

Why is it beneficial for pyruvate to be reduced when oxygen is not available? This reaction regenerates NAD^+, which is required for the first step in the energy-harvesting phase of glycolysis. This NAD^+, therefore, allows glycolysis and substrate-level ATP synthesis to continue in the absence of oxygen.

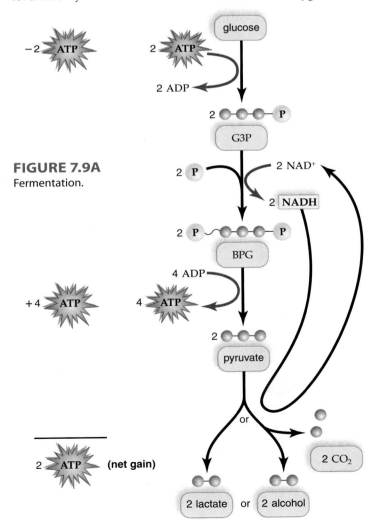

FIGURE 7.9A Fermentation.

Benefits Versus Drawbacks of Fermentation Despite its low yield of only two ATP, fermentation is essential to organisms because it is anaerobic. It allows yeasts to metabolize the sugar of grapes when oxygen is in short supply (**Fig. 7.9B**). In humans, it can provide a rapid burst of ATP—and is therefore more likely to occur in muscle cells than in other types of cells. When our muscles are working vigorously over a short period of time, as when we run, fermentation is a way to produce ATP even though oxygen is temporarily in limited supply.

Fermentation products are toxic to cells. Yeasts die from the alcohol they produce. In humans, blood carries away the lactate formed in muscles. Eventually, however, lactate begins to build up, changing the pH and causing the muscles to "burn," and eventually to fatigue so that they no longer contract. When we stop running, our bodies are in **oxygen debt**, as signified by the fact that we continue to breathe very heavily for a time. Recovery is complete when the liver reconverts lactate to pyruvate.

FIGURE 7.9B These grapes have a coating of yeasts.

Comparison of Yields Fermentation produces only two ATP by substrate-level ATP synthesis. These two ATP represent only a small fraction of the potential energy stored in a glucose molecule. As noted earlier, complete glucose breakdown during cellular respiration results in at least 36 ATP. Therefore, following fermentation, most of the potential energy a cell can capture from the oxidation of a glucose molecule is still waiting to be released:

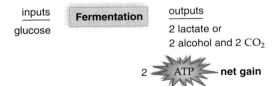

inputs	**Fermentation**	outputs
glucose		2 lactate or
		2 alcohol and 2 CO_2

2 ATP — **net gain**

Fermentation by microorganisms is put to use by humans to produce various products, as discussed in Section 7.10.

> **7.9 Check Your Progress** Birds feasting on grapes (Fig. 7.9B) sometimes act erratically, as if they were drunk. Is it possible that the grapes contain alcohol? Explain.

At the grocery store, you will find such items as bread, yogurt, soy sauce, pickles, and maybe even wine (**Fig. 7.10**). These are just a few of the many foods that are produced when microorganisms ferment (break down sugar in the absence of oxygen). Foods produced by fermentation last longer because the fermenting organisms have removed many of the nutrients that would attract other organisms. As mentioned, the products of fermentation can even be dangerous to the very organisms that produced them, as when yeasts are killed by the alcohol they produce.

Fermenting Yeasts Leaven Bread and Produce Alcohol

Baker's yeast, *Saccharomyces cerevisiae*, is added to bread for the purpose of leavening—the dough rises when the yeasts give off CO_2. The ethyl alcohol produced by the fermenting yeast evaporates during baking. The many different varieties of sourdough breads obtain their leavening from a starter composed of fermenting yeasts along with bacteria from the environment. Depending on the community of microorganisms in the starter, the flavor of the bread may range from sour and tangy, as in San Francisco–style sourdough, to a milder taste, such as that produced by most Amish friendship bread recipes.

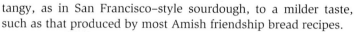

Ethyl alcohol is desired when yeasts are used to produce wine and beer. When yeasts ferment the carbohydrates of fruits, the end result is wine. If they ferment grain, beer results. A few specialized varieties of beer, such as traditional wheat beers, have a distinctive sour taste because they are produced with the assistance of lactic acid–producing bacteria, such as those of the genus *Lactobacillus*. Stronger alcoholic drinks (e.g., whiskey and vodka) require distillation to concentrate the alcohol content.

The acetic acid bacteria, including *Acetobacter aceti*, spoil wine. These bacteria convert the alcohol in wine or cider to acetic acid (vinegar). Until the renowned 19th-century scientist Louis Pasteur invented the process of pasteurization, acetic acid bacteria commonly caused wine to spoil. Although today we generally associate the process of pasteurization with making milk safe to drink, it was originally developed to reduce bacterial contamination in wine so that limited acetic acid would be produced.

Fermenting Bacteria That Produce Acid

Yogurt, sour cream, and cheese are produced through the action of various lactic acid bacteria that cause milk to sour. Milk contains lactose, which these bacteria use as a substrate for fermentation. Yogurt, for example, is made by adding lactic acid bacteria, such as *Streptococcus*

thermophilus and *Lactobacillus bulgaricus*, to milk and then incubating it to encourage the bacteria to act on lactose. During the production of cheese, an enzyme called rennin must also be added to the milk to cause it to coagulate and become solid.

Old-fashioned brine cucumber pickles, sauerkraut, and kimchi are pickled vegetables produced by the action of acid-producing, fermenting bacteria from the genera *Lactobacillus* and *Leuconostoc*, which can survive in high-salt environments. Salt is used to draw liquid out of the vegetables and aid in their preservation. The bacteria need not be added to the vegetables, because they are already present on the surfaces of the plants.

Soy Sauce Is a Combination Product

Soy sauce is traditionally made by adding a mold, *Aspergillus*, and a combination of yeasts and fermenting bacteria to soybeans and wheat. The mold breaks down starch, supplying the fermenting microorganisms with sugar they can use to produce alcohol and organic acids.

This completes our discussion of fermentation. Now let's move on to examining other metabolic pathways.

> **7.10 Check Your Progress** When lactate fermentation occurs, pyruvate is reduced to lactate. Which molecule, pyruvate or lactate, contains more hydrogen atoms?

FIGURE 7.10 Fermentation helps make the products shown on this page.

Carbohydrate, protein, and fat can be metabolized by entering degradative (catabolic) pathways at different locations. In this part of the chapter, we also examine why you are more likely to burn fat with prolonged exercise.

Catabolic pathways also provide metabolites needed for synthesis (anabolism) of various important substances. Therefore, catabolism and anabolism both use the same pools of metabolites.

7.11 Organic molecules can be broken down and synthesized as needed

Certain substrates reoccur in various key metabolic pathways, and therefore they form a **metabolic pool**. In the metabolic pool, these substrates serve as entry points for the degradation or synthesis of larger molecules (**Fig. 7.11**). Degradative reactions break down molecules and collectively participate in **catabolism**. Synthetic reactions build up molecules and collectively participate in **anabolism**.

Catabolism We already know that glucose is broken down during cellular respiration. When a fat is used as an energy source, it first breaks down to glycerol and three fatty acids. Then, as Figure 7.11 indicates, glycerol can enter glycolysis. The fatty acids are converted to acetyl CoA, which can enter the citric acid cycle. An 18-carbon fatty acid results in nine acetyl CoA molecules. Calculation shows that respiration of these can produce a total of 108 ATP molecules. For this reason, fats are an efficient form of stored energy—there are three long-chain fatty acids per fat molecule.

The carbon skeleton of amino acids can enter glycolysis, be converted to acetyl CoA, or enter the citric acid cycle directly. The carbon skeleton is produced in the liver when an amino acid undergoes **deamination**, removal of the amino group. The amino group becomes ammonia (NH_3), which enters the urea cycle and becomes part of urea, the primary excretory product of humans. Just where the carbon skeleton begins degradation depends on the length of the R group, since this determines the number of carbons left after deamination.

Anabolism The degradative reactions of catabolism drive the synthetic reactions of anabolism because they release energy that is required for anabolism to occur. For example, we have already mentioned that ATP breakdown drives synthetic reactions. But catabolism is also related to anabolism in another way. The substrates making up the pathways in Figure 7.11 can be used as starting materials for synthetic reactions. In other words, the macromolecules can be oxidized to substrates that can be used to synthesize other macromolecules. In this way, carbohydrate intake can result in the formation of fat. G3P of glycolysis can be converted to glycerol, and acetyl groups can be joined to form fatty acids. Fat synthesis follows. This explains why you gain weight from eating too much candy, ice cream, or cake.

Some substrates of the citric acid cycle can be converted to amino acids through transamination, the transfer of an amino group to an organic acid, forming a different amino acid. Plants are able to synthesize all of the amino acids they need. Animals, however, lack some of the enzymes necessary for synthesis of all amino acids. Adult humans, for example, can synthesize

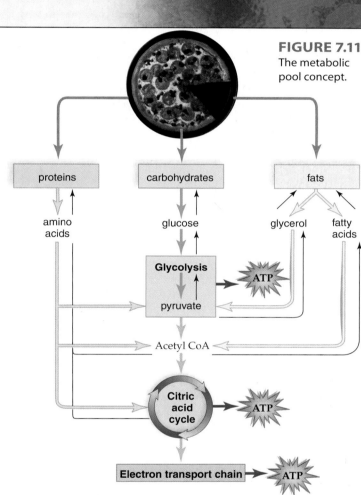

FIGURE 7.11 The metabolic pool concept.

eleven of the common amino acids, but they cannot synthesize the other nine. All amino acids are needed or else some proteins cannot be synthesized. It is quite possible for animals to suffer from protein deficiency if their diets do not contain adequate quantities of all the essential amino acids.

In the next section, we continue our study of the metabolic pool as we consider the different molecules that can be metabolized to build ATP. Weight loss is achieved if the fatty acids from fat molecules are burned, and only aerobic respiration burns the fatty acids from fat molecules.

> **7.11 Check Your Progress** In Chapter 3, you learned the terms "dehydration reaction" and "hydrolytic reaction." *a.* Which type of reaction is catabolic? Anabolic? *b.* Which term could be associated with ATP breakdown?

7.12   Exercise burns fat

The key to losing weight is to use up more calories than you take in. Simply reducing caloric intake is one approach to weight loss. However, it is quite difficult to sustain a very low-calorie diet for a long time, and doing so may result in nutritional deficiencies. Combining exercise with a sensible diet that satisfies the body's nutritional requirements appears to be the best long-term approach to weight management. Any form of exercise is preferable to inactivity because all types increase one's energy use—in other words, they all "burn calories" at a higher rate than the body does at rest. However, some forms of exercise may be more effective for weight loss than others.

Exercise Offers the Opportunity to Burn Fat Recall that a fat molecule is also called a triglyceride because it contains three fatty acids. **Figure 7.12A** suggests that muscle cells store both **1** fat and **4** glycogen, but the longer we exercise, the more depleted these stores become. Now muscle cells begin to burn either **2** fatty acids or **3** glucose from the blood. The fatty acids were deposited in the blood by adipose tissue, which makes us look fat because it does indeed store fat in the body. Figure 7.12A also suggests that the longer we exercise at about 70% effort, the more fatty acids are burned instead of glucose.

As mentioned in Section 7.11, fats are very energy-rich molecules. In fact, the amount of energy released from breaking down fat is more than double that derived from breaking down carbohydrate or protein. Still, simply taking a quick walk for a minimum of 30 minutes on most days can help burn fat. But keep in mind that long-term stored energy is the major function of fat in adipose tissue. Since fatty acids store so much energy, prolonged exercise is the best way to use up some of this stored energy.

Prolonged Aerobic Exercise Burns Fat The burning of fats requires oxygen; it cannot be done anaerobically. When a fat is broken down in an enzyme-catalyzed reaction, the glycerol can be used in glycolysis. But the fatty acids are broken down to form acetyl CoA, which feeds into the citric acid cycle. The citric

FIGURE 7.12B Aerobic exercise burns fat.

acid cycle continues only as long as oxygen is present to receive electrons from the electron transport chain. If oxygen runs out, as occurs when exercise is extremely vigorous, cells switch to fermentation, which is anaerobic, and fat burning decreases.

For this reason, people who want to lose weight usually have the most success with aerobic exercise (**Fig. 7.12B**). This includes activities requiring moderate exertion, such as brisk walking, bicycling, swimming, and jogging. Aerobic exercise is so named because the energy needed for it can be supplied mostly aerobically. Breathing and heart rate increase during exercise in order to supply the muscles with adequate oxygen. In fact, the heart rate is sometimes used as an indicator of the correct range of effort, which is why some people wear small heart rate monitors when they exercise. Likewise, breathing should increase, but not to the extent that one feels out of breath or unable to go on. Such extreme exertion pushes the body into fermentation.

Other Benefits of Aerobic Exercise Aerobic exercise has other benefits. The heart and lungs become better able to supply the muscles with oxygen. Skeletal muscles change—although they do not become very bulky, their blood supply and the number of mitochondria in the cells increase. The overall effect is that the muscles can use more energy, even when the body is at rest.

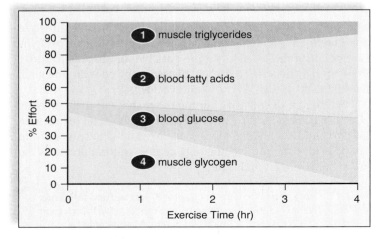

FIGURE 7.12A Sources of fuel for exercise.

> **7.12** *Check Your Progress* **Which activity would you recommend to someone who wants to lose weight: tennis, which requires bursts of energy, or swimming, which requires a steady pace? Explain.**

Energy from the sun flows through all living things with the participation of chloroplasts and mitochondria. Through the process of photosynthesis, chloroplasts in plants and algae capture solar energy and use it to produce carbohydrates, which are broken down to carbon dioxide and water in the mitochondria of all organisms. The energy released when carbohydrates (and other organic molecules) are oxidized is used to produce ATP molecules. When the cell uses ATP to do cellular work, all the captured energy dissipates as heat.

During cellular respiration, oxidation by removal of hydrogen atoms (e^- + H^+) from glucose or glucose products occurs during glycolysis, the prep reaction, and the citric acid cycle. The prep reaction and citric acid cycle release CO_2. The electrons are carried by NADH and $FADH_2$ to the electron transport chain (ETC) on the cristae of mitochondria. Oxygen serves as the final acceptor of electrons, and H_2O is produced. The pumping of hydrogen ions by the ETC into the intermembrane space leads to ATP production.

With this chapter, we have completed our study of the cell, and the Biological Viewpoint on pages 130–31 reviews for you some of the major concepts regarding the cell theory.

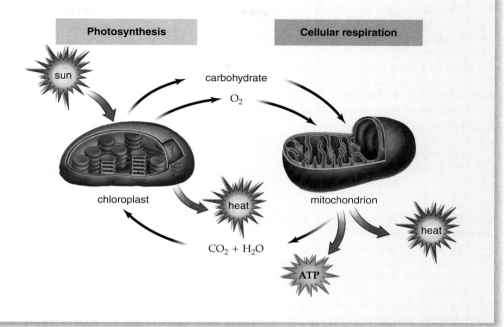

The Chapter in Review

Summary

ATP Is Universal

- ATP has the same structure and function in all cells.
- ATP breakdown provides energy for life.

Glucose Breakdown Releases Energy

7.1   Cellular respiration is a redox reaction that requires O_2

- During cellular respiration, glucose is oxidized to CO_2, which we breathe out.
- Oxygen, which we breathe in, is reduced to H_2O.

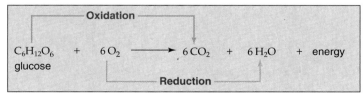

- As oxidation occurs, coenzymes NAD and FAD remove hydrogen atoms (e^- + H^+) from glucose.
- Slow release of energy allows it to be captured for ATP production.

7.2   Cellular respiration has four phases—three phases occur in mitochondria

- Glycolysis occurs in the cytoplasm and is anaerobic. The prep reaction, citric acid cycle, and electron transport chain (ETC) occur in the mitochondria and are aerobic.

- From the overall equation for cellular respiration, associate glucose with glycolysis, carbon dioxide (we breathe out) with the prep reaction and the citric acid cycle, and oxygen (we breathe in) and water with the ETC.
- Both glycolysis and the citric acid cycle produce minimal ATP by substrate-level ATP synthesis, and the ETC produces much ATP when oxygen is available.

Carbon Dioxide and Water Are Produced During Glucose Breakdown

7.3   Glycolysis: Glucose breakdown begins

- Energy investment: At the beginning of glycolysis, two ATP are used to activate glucose, which splits into two C_3 molecules.
- Energy harvesting: Removal of hydrogen atoms results in 2 NADH and four high-energy phosphate bonds. Four ATP result from substrate-level ATP synthesis.
- Result: net gain of two ATP—two used, four made—per glucose molecule.
- Glycolysis begins with C_6 glucose and ends with two C_3 pyruvate molecules.

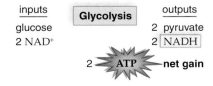

7.4 The preparatory reaction occurs before the citric acid cycle

- Pyruvate is converted to an acetyl group; NADH is made, and CO_2 is given off. Occurs twice per glucose.
- CoA carries the acetyl groups to the citric acid cycle.

7.5 The citric acid cycle: Final oxidation of glucose products

- Citrate begins and ends the cycle.
- NADH and $FADH_2$ are made as CO_2 is released.
- One ATP per cycle (two per glucose) results from substrate-level ATP synthesis.

7.6 The electron transport chain captures much energy

- NADH and $FADH_2$ bring electrons to the electron transport system.
- Electrons are passed from carrier to carrier, and energy is captured and used to form ATP. Three ATP are made per NADH, and two ATP are made per $FADH_2$.
- Oxygen (we breathe in) is the final acceptor of electrons from the electron transport chain. Oxygen is reduced to water.

7.7 The cristae create an H^+ gradient that drives ATP production

- Complexes in the cristae form the electron transport chain.
- As electrons move from one complex to another, H^+ are pumped into the intermembrane space, creating an H^+ gradient.
- Chemiosmosis: As H^+ flow down the H^+ gradient from the intermembrane space to the matrix, ATP is synthesized from ADP + Ⓟ.

7.8 The ATP payoff can be calculated

- Net gain of 2 ATP in the cytoplasm
- 2 ATP from citric acid cycle
- 32–34 ATP from the electron transport chain and chemiosmosis
- Total: Complete glucose breakdown produces a total of 36 or 38 ATP.

Fermentation Is Inefficient

7.9 When oxygen is in short supply, the cell switches to fermentation

- Fermentation is glycolysis followed by the reduction of pyruvate by NADH either to lactate or to alcohol and CO_2 depending on the organism.
- Fermentation results in only two ATP molecules, but provides a quick burst of ATP energy for short-term activity.
- Alcohol or lactate buildup is toxic to cells. In humans, lactate puts an individual into oxygen debt because oxygen is needed to completely metabolize it.

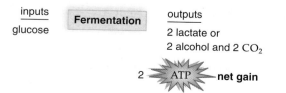

7.10 Fermentation helps produce numerous food products

- Yeasts are used to make bread and alcohol; bacteria are used to make yogurt, sour cream, cheese, pickles, and sauerkraut.

Metabolic Pathways Cross at Particular Substrates

7.11 Organic molecules can be broken down and synthesized as needed

- Carbohydrates, fats, and proteins can all be catabolized to produce ATP molecules.
- Amino acids from protein first have to undergo deamination, the removal of the amino group.
- During anabolism, molecules from the breakdown pathways can be used to build molecules. Example: Acetyl groups from carbohydrate breakdown can be used to build fat.

7.12 Exercise burns fat

- Burning more calories than are taken in results in weight loss.
- Prolonged aerobic exercise at moderate exertion is best for burning fats.

Testing Yourself

Glucose Breakdown Releases Energy

1. During cellular respiration, _____ is oxidized and _____ is reduced.
   a. glucose, oxygen
   b. glucose, water
   c. oxygen, water
   d. water, oxygen
   e. oxygen, carbon dioxide

2. The products of cellular respiration are energy and
   a. water.
   b. oxygen.
   c. water and carbon dioxide.
   d. oxygen and carbon dioxide.
   e. oxygen and water.

3. The correct order of events would be
   a. citric acid cycle, glycolysis, prep reaction.
   b. glycolysis, prep reaction, citric acid cycle.
   c. prep reaction, citric acid cycle, glycolysis.
   d. None of these are correct.

4. Relate the products and reactants of the overall equation for cellular respiration (Section 7.1) to the phases of cellular respiration (Section 7.2).

Carbon Dioxide and Water Are Produced During Glucose Breakdown

5. During the energy-harvesting steps of glycolysis, which are produced?
   a. ATP and NADH
   b. ADP and NADH
   c. ATP and NAD
   d. ADP and NAD

6. Which of the following is needed for glycolysis to occur?
   a. pyruvate
   b. glucose
   c. NAD^+
   d. ATP
   e. All of these are needed except a.

7. Acetyl CoA is the end product of
   a. glycolysis.
   b. the preparatory reaction.
   c. the citric acid cycle.
   d. the electron transport chain.

8. Which of these is not true of the prep reaction? The prep reaction
   a. begins with pyruvate and ends with acetyl CoA.
   b. produces more NADH than does the citric acid cycle.
   c. occurs in the mitochondria.
   d. occurs after glycolysis and before the citric acid cycle.

9. The strongest and final electron acceptor in the electron transport chain is
    a. NADH.   c. oxygen.
    b. FADH₂.   d. water.
10. The greatest contributor of electrons to the electron transport chain is
    a. oxygen.   d. the preparatory reaction.
    b. glycolysis.   e. fermentation.
    c. the citric acid cycle.
11. What is the name of the process that results in ATP production using the flow of hydrogen ions?
    a. substrate-level ATP synthesis   c. reduction
    b. fermentation   d. chemiosmosis
12. How many ATP molecules are produced when one NADH donates electrons to ETC?
    a. 1   c. 36
    b. 3   d. 10
13. Label this diagram of a mitochondrion, and state a function for each portion indicated.

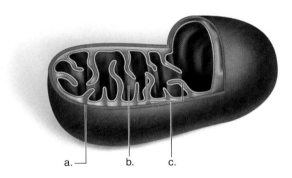

    a.___   b.   c.

14. **THINKING CONCEPTUALLY** If carbohydrate breakdown supplies the energy for ATP buildup, why does entropy increase as required by the second law of thermodynamics (see Section 5.2)?

Fermentation Is Inefficient

15. Fermentation occurs in the absence of
    a. CO₂.   c. oxygen.
    b. H₂O.   d. sodium.
16. Which are possible products of fermentation?
    a. lactate   c. CO₂
    b. alcohol   d. All of these are correct.
17. Which of the following is not true of fermentation? Fermentation
    a. has a net gain of only two ATP.
    b. occurs in the cytoplasm.
    c. donates electrons to the electron transport chain.
    d. begins with glucose.
    e. is carried on by yeast.
18. Fermentation does not yield as much ATP as cellular respiration does because fermentation
    a. generates mostly heat.
    b. makes use of only a small amount of the potential energy in glucose.
    c. creates by-products that require large amounts of ATP to break down.
    d. creates ATP molecules that leak into the cytoplasm and are broken down.

Metabolic Pathways Cross at Particular Substrates

19. Fatty acids are broken down to
    a. pyruvate molecules, which take electrons to the electron transport chain.
    b. acetyl groups, which enter the citric acid cycle.
    c. glycerol, which is found in fats.
    d. amino acids, which excrete ammonia.
    e. All of these are correct.
20. Deamination of an amino acid results in
    a. CO₂.   c. NH₂.
    b. H₂O.   d. O₂.
21. **THINKING CONCEPTUALLY** Why would you expect both photosynthesis and cellular respiration to use an electron transport chain to produce ATP? (See Sections 4.15 and 4.16.)

Understanding the Terms

aerobic 115
anabolism 125
anaerobic 115
catabolism 125
cellular respiration 113
chemiosmosis 121
citric acid cycle 115, 119
cristae 118
cytochrome 120
deamination 125
electron transport chain (ETC) 115, 120

FAD 114
fermentation 115, 123
glycolysis 115, 116
matrix 118
metabolic pool 125
NAD⁺ 114
oxygen debt 123
preparatory (prep) reaction 115, 118
pyruvate 115
substrate-level ATP synthesis 116

Match the terms to these definitions:
a. _____ A metabolic pathway that begins with glucose and ends with two molecules of pyruvate.
b. _____ Occurs due to the accumulation of lactate following vigorous exercise.
c. _____ Metabolic process that degrades molecules and tends to be exergonic.
d. _____ Flow of hydrogen ions down their concentration gradient through an ATP synthase, resulting in ATP production.
e. _____ Metabolic process by which cells harvest the energy stored in organic compounds.

Thinking Scientifically

1. You are able to extract mitochondria from the cell and remove the outer membrane. You want to show that the mitochondria can still produce ATP if placed in the right solution. The solution should be isotonic, but at what PH? Why?

ARIS *Visit* **www.mhhe.com/maderconcepts** *for practice quizzes, animations, videos, and activities designed to help you master the material in this chapter.*

I t is your first time ever in a laboratory, and your instructor hands you a slide containing one tiny drop of pond water to look at under the microscope. In the same manner as Leeuwenhoek, the inventor of the microscope, you peer through an eyepiece and see an amazing number of tiny creatures dancing across your field of vision. Inhaling sharply, you ask, "What are these?" Your lab partner answers, "I think they're cells!" To be specific, they are unicellular, microscopic organisms.

Next, your instructor tells you to use a toothpick to gently scrape the inside of your cheek and make a smear on a slide. The addition of a stain to the slide reveals a vast number of nonmotile cells under the microscope. We are, you conclude, multicellular animals.

Your laboratory experiences have allowed you to gather data to support the cell theory, which states that all organisms are composed of cells. No evidence runs contrary to this conclusion. A cell is the lowest level of biological organization to display the characteristics of life. What are they? The characteristics of life are to exhibit order; be responsive to environmental change; regulate the internal environment; acquire material and energy from the environment; reproduce and develop; store genetic information; and undergo developmental change.

A cell can do all of this, despite its size of only 10–100 μm in diameter, because it is alive. Why do cells remain so small? Because cells need a favorable surface area per volume to serve their need for acquiring energy and materials from the environment. In other words, cells obey the same laws of chemistry and physics that govern everything within the universe.

A cell is a chemical factory, and it is composed of chemicals. The little organelles of eukaryotic cells are constantly churning out products that keep them going—that is, keep them alive. Chloroplasts are the organelles in plants and some protists that capture solar energy and produce the carbohydrates

chloroplast

that serve as nutrients for all cells. The responsiveness of a plant or a protist is illustrated by its ability to position itself to soak up solar energy. Mitochondria are the organelles that convert carbohydrates to the energy currency of cells—namely, ATP molecules.

mitochondrion

Wait, you say, there is more to the cell theory than staying alive. What about reproduction and evolution? Already we have learned that cells come only from cells and that they are capable of self-reproduction. The implication is that the first living thing to evolve must have been a cell, or cells. Where did it come from? Scientists favor the simplest logical explanation and have amassed data to support the hypothesis that the first cell, or cells, must have come from preexisting organic chemicals. The chemical evolution that could have produced the first cell(s) is described in Chapter 16.

Around 1850, the German physician Rudolph Virchow was the first to witness cell division. Today, you might see clips of dividing cells on the Internet. We also know that dividing cells pass on their DNA to their offspring. Why do cells produce cells just like themselves, and organisms produce like organisms? DNA—the genetic material to be further studied in Part II—contains the instructions, or blueprint, that allow a cell to function as it does. At one time, even scientists believed a whole mouse could arise from dirty rags. But today, we know that cannot happen because cloth does not contain the DNA of a mouse.

Once the first cell(s) arose, chance mutations (genetic changes) led to all the other organisms we see about us today. Your cells are distantly related through evolution to the unicellular organisms you see under a microscope. The cell theory is a powerful synthesizer of biological knowledge, as are all the basic theories of biology we will be emphasizing in this text.

8

Cell Division and Reproduction

LEARNING OUTCOMES

After studying this chapter, you should be able to accomplish the following outcomes.

Cancer Is a Genetic Disorder

1 Explain which types of genes malfunction when cancer occurs.

Cell Division Ensures the Passage of Genetic Information

2 State when cell division occurs in multicellular eukaryotes, such as humans.
3 Describe how prokaryotes reproduce.

Somatic Cells Have a Cell Cycle and Undergo Mitosis and Cytokinesis

4 Describe the stages of the cell cycle.
5 Describe the phases of mitosis and the process of cytokinesis in animal and plant cells.

Cancer Is Uncontrolled Cell Division

6 State how the cell cycle is controlled and the importance of cell signaling molecules.
7 Describe the characteristics of cancer cells, and suggest ways to prevent the occurrence of cancer.

Meiosis Produces Cells That Become the Gametes in Animals and Spores in Other Organisms

8 Describe three ways genetic variation is ensured in the next generation.
9 Describe the phases of meiosis, and compare when meiosis occurs in the life cycle of animals to other organisms.
10 Compare the process and result of meiosis to those of mitosis.

Chromosomal Abnormalities Can Be Inherited

11 Draw diagrams to show the occurrence of nondisjunction during meiosis I and during meiosis II.
12 Describe Down syndrome, Turner syndrome, and Klinefelter syndrome, including their abnormalities in chromosome number.
13 Describe four abnormalities in chromosome structure, and relate them to human disorders.

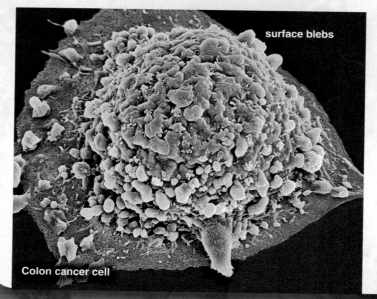

Cancer cell dividing

surface blebs

Colon cancer cell

We often think of diseases in terms of organs, and therefore it is customary to refer to lung cancer, or colon cancer, or pancreatic cancer. Cancer is present, however, when abnormal cells have formed a tumor. Exceptions are cancers of the blood, in which abnormal cells are coursing through the bloodstream. The cells of a tumor share a common ancestor—the first cell to become cancerous. Therefore, it is clear that cancer is a cellular disease.

Uncontrolled growth leading to a tumor is characteristic of multicellular organisms, not unicellular ones. The very mechanism that allows our bodies to grow and repair tissues is the one that turns on us and allows cancer to begin. Cancer is uncontrolled cell division.

Usually, cell division is confined to just certain cells of the body, called adult stem cells. For example, skin can replenish itself because the stem cells below the surface still have the ability to divide. In contrast, all embryonic cells can divide. How else would you get a newborn from a single fertilized egg? But something happens along the way: The cells undergo

Cancer Is a Genetic Disorder

specialization and become part of a particular organ. A mature multicellular organism contains many kinds of specialized cells in many different organs. Normally, these cells listen to their neighbors and participate in the operation of the organ. But when a cell becomes cancerous, it loses its specialization and becomes youthful again—it starts to divide and divide, until a tumor exists. The tumor interferes with the operation of the organ.

The cells in a multicellular organism contain complete copies of the genetic instructions inherited from the organism's parents, and they are passed to all the millions and millions of cells of the body. Some genes control whether cell division occurs. The control of cell division is useful to multicellular organisms because without it, we would be a bunch of embryonic cells with no particular purpose. When cell division genes mutate, cancer becomes possible. Therefore, cancer is a genetic disorder. Research

tumor

Lung cancer | 33.3 μm

tells us that cancer-causing mutations may be induced, for example, by chemicals or radiation that damage DNA; viruses that carry mutated genes into cells; or random errors that occur during DNA synthesis. A series of mutations is required before cells begin to grow abnormally and eventually become a tumor. In these cells, genetic alteration is obvious: Some chromosomes are present in three or four copies, rather than the usual two, and other chromosomes have been rearranged in various ways.

In this chapter, we will study cell division and how it is normally controlled, before examining the characteristics of cancer cells. We will also see how a special type of cell division contributes to the formation of the egg and sperm, which fuse during sexual reproduction. If abnormalities occur during the production of the egg and sperm, an offspring will be abnormal. A few such abnormalities are examined.

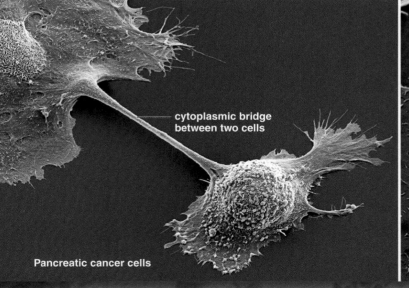

cytoplasmic bridge between two cells

Pancreatic cancer cells

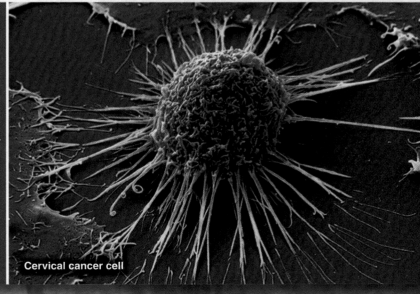

Cervical cancer cell

Cell Division Ensures the Passage of Genetic Information

Learning Outcomes 2–3, page 132

Cell division occurs when a parent cell divides and produces two new cells. Cell division occurs during both asexual reproduction and sexual reproduction. The importance of cell division is discussed in this part of the chapter, which also describes the asexual reproduction of bacteria.

8.1 Cell division is involved in both asexual and sexual reproduction

Cell division is essential to life, whether we are discussing multicellular organisms or unicellular organisms. We humans, like other multicellular organisms, begin life as a single cell. In nine short months, however, cell division has occurred over and over again so that we become trillions of body cells, called **somatic cells**. Even after we are born, somatic cell division does not stop—it continues as we grow (**Fig. 8.1A**), and when we are adults, it replaces worn-out or damaged tissues. Right now, your body is producing thousands of new red blood cells, skin cells, and the cells that line your respiratory and digestive tracts. If you suffer a cut, cell division helps repair the injury.

Cell division is also necessary for the reproduction of organisms (**Fig. 8.1B**). When an amoeba splits, two new amoebas are produced. An increase in the number of somatic cells or an increase in the number of unicellular organisms, such as an amoeba, are both forms of **asexual reproduction**. The resulting cells receive a copy of the parent cell's chromosomes and genes, and the new cells are identical to the parent cell and to each other.

Later in this chapter, we will discuss the production of egg and sperm, which are needed for **sexual reproduction**. The business of making sperm and eggs has been assigned to specialized cells called **germ cells** that are found exclusively in the testes of the male and the ovaries of the female. When a sperm and egg unite, the offspring receives a different combination of chromosomes than either parent and, therefore, a unique combination of genetic information. This means that the offspring of two parents are not identical to each

Amoebas reproduce asexually

Humans reproduce sexually

FIGURE 8.1B Organismal reproduction occurs either asexually or sexually.

other, or to their parents. The term sexual reproduction refers only to the production of new, independent organisms.

One way to emphasize the importance of cell division is to say that "all cells come from preexisting cells." You cannot have a new cell without a preexisting cell, and you cannot have a new organism without a preexisting organism. Cell division is necessary for the production of both new cells and new organisms. Why should that be? Because the nucleus of a cell is the repository for chromosomes. The passage of genetic information is absolutely necessary in order for each new cell to continue its existence.

Section 8.2 discusses the reproduction of prokaryotes, which reproduce by binary fission, a form of asexual reproduction.

Children grow

Tissues repair

FIGURE 8.1A Cell division occurs when multicellular organisms grow and when they repair their tissues.

> **8.1 Check Your Progress** Cancer occurs when somatic cells divide and produce a tumor. Is this a form of asexual reproduction or sexual reproduction?

8.2 Prokaryotes reproduce asexually

Cell division in prokaryotes (bacteria and archaea), which are unicellular, produces two new individuals. This is an example of asexual reproduction, in which the offspring are genetically identical to the parent. Reproduction consists of duplicating the single chromosome and distributing a copy to each of the daughter cells. Unless a mutation has occurred, the new cells are genetically identical to the parent cell.

Prokaryotes lack a nucleus and the other membranous organelles found in eukaryotic cells. Still, they do have a chromosome, which is composed of DNA and a limited number of associated proteins. The single chromosome of prokaryotes contains just a few proteins and is organized differently from eukaryotic chromosomes. A eukaryotic chromosome has many more proteins than a prokaryotic chromosome.

In electron micrographs, the bacterial chromosome appears as a dense, irregularly shaped region called the **nucleoid**, which is not enclosed by a membrane. When stretched out, the chromosome is a circular loop that may be up to a thousand times the length of the cell. No wonder it is folded when inside the cell.

Prokaryotes reproduce by binary fission. The process is termed **binary fission** because division (fission) produces two (binary) new cells. **Figure 8.2** shows the steps of binary fis-

sion. ❶ Before division takes place, the single chromosome is attached to a special plasma membrane site. ❷ The cell enlarges, and DNA replication occurs, so that there are two chromosomes. ❸ The cell wall and plasma membrane begin to indent. ❹ The chromosomes separate by an elongation of the cell that pulls them apart. ❺ During this period, new plasma membrane and cell wall develop and grow inward to divide the cell. When the cell is approximately twice its original length, the new cell wall and plasma membrane for each cell are complete.

Under favorable conditions, the bacterium *Escherichia coli*, which lives in our intestines, divides in about 20 minutes; this is called its generation time. In about seven hours, a single *E. coli* cell can produce over one million cells! The generation time of other bacteria varies, depending on the species and conditions.

This completes this part of the chapter. In the next part, we consider the life of a dividing cell in eukaryotes.

> **8.2** *Check Your Progress* **Contrast the function of asexual reproduction in unicellular and multicellular organisms.**

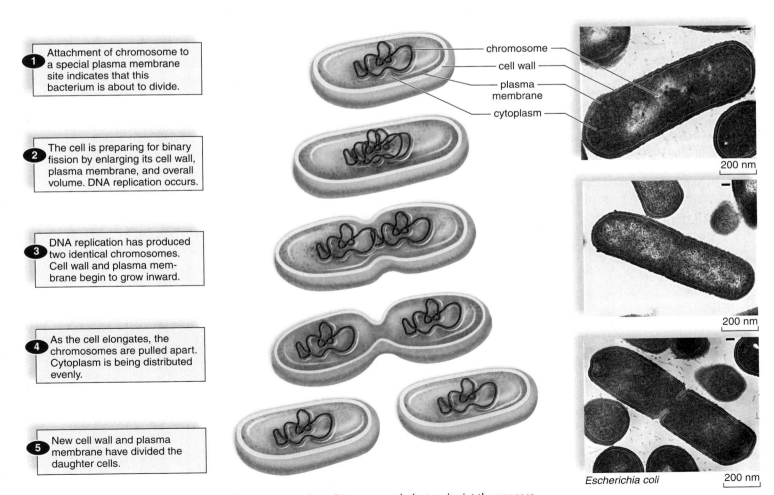

1 Attachment of chromosome to a special plasma membrane site indicates that this bacterium is about to divide.

2 The cell is preparing for binary fission by enlarging its cell wall, plasma membrane, and overall volume. DNA replication occurs.

3 DNA replication has produced two identical chromosomes. Cell wall and plasma membrane begin to grow inward.

4 As the cell elongates, the chromosomes are pulled apart. Cytoplasm is being distributed evenly.

5 New cell wall and plasma membrane have divided the daughter cells.

chromosome
cell wall
plasma membrane
cytoplasm

200 nm
200 nm
Escherichia coli   200 nm

FIGURE 8.2 Prokaryotes use binary fission to reproduce. Diagrams and photos depict the process.

In this part of the chapter, we describe the cell cycle—that is, the life of a cell that is metabolically active and undergoing normal cell division. We learn about the duplication of cellular contents before cell division begins and the behavior of chromosomes during nuclear division (mitosis). Finally, we examine the process of cytokinesis, when the cytoplasm divides.

8.3 The eukaryotic cell cycle is a set series of events

For cellular reproduction to be orderly, you would expect the cell's contents to be duplicated before cell division occurs. This is just what happens during the so-called cell cycle. The **cell cycle** is an orderly sequence of stages that takes place between the time a new cell has arisen from division of a parent cell and the point when it has given rise to two daughter cells. Duplication of cell contents occurs during the stage called interphase.

Interphase As **Figure 8.3A** shows, most of the cell cycle is spent in interphase. This is the time when a cell performs its usual functions, depending on its location in the body. The amount of time the cell takes for interphase varies widely. Differentiated cells, such as nerve and muscle cells, typically remain in interphase, and cell division is arrested. These cells are said to have entered a G_0 stage.

Adult stem cells are those cells in the mature body that have the ability to divide. For example, stem cells in red bone marrow continually produce all the types of blood cells in the body. In an adult stem cell, interphase may last for about 20 hours, which is 90% of the cell cycle. In contrast, embryonic cells complete the entire cell cycle in just a few hours.

DNA replication occurs in the middle of interphase and serves as a way to divide interphase into three stages: G_1, S, and G_2. G_1 is the stage before DNA is replicated, and G_2 is the stage following DNA replication. Originally, G stood for "gap," but now that we know how metabolically active the cell is, it is better to think of G as standing for "growth." Protein synthesis is very much a part of these growth stages.

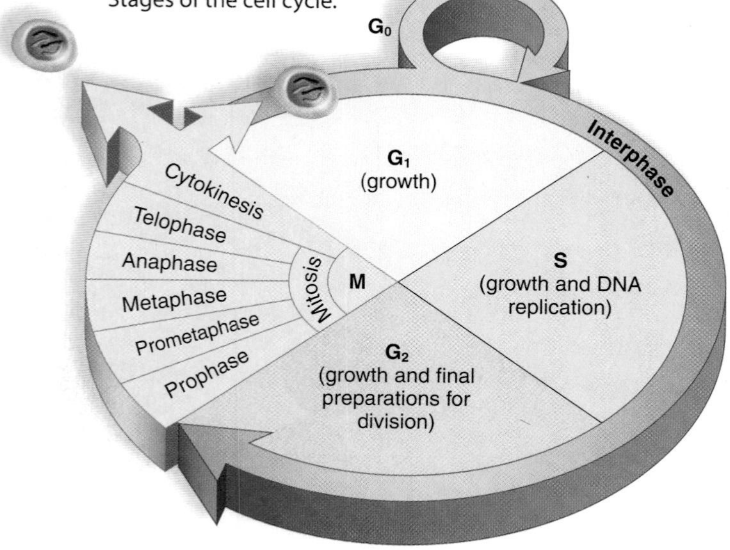

During G_1, a cell doubles its organelles (such as mitochondria and ribosomes) and accumulates materials that will be used for DNA replication. Following G_1, the cell enters the S stage. The S stands for synthesis, and certainly DNA synthesis is required for DNA replication. At the beginning of the S stage, each chromosome has one DNA double helix. At the end of this stage, each chromosome is composed of two **sister chromatids,** each having one double helix. Another way of expressing these events is to say that DNA replication results in duplicated chromosomes.

Following the S stage, G_2 extends from the completion of DNA replication to the onset of mitosis. During this stage, organelle replication continues, and the cell synthesizes proteins that will be needed for cell division, such as the protein found in microtubules. The role of microtubules in cell division is described in Section 8.5.

M (Mitotic) Stage Cell division occurs during the M stage, which encompasses both division of the nucleus and division of the cytoplasm. The type of nuclear division associated with the cell cycle is called **mitosis**, which explains why this stage is called the M stage. As a result of mitosis, the daughter nuclei are identical to the parent cell and to each other—they all have the same number and kinds of chromosomes. Division of the cytoplasm, which starts even before mitosis is finished, is called **cytokinesis** (**Fig. 8.3B**).

In Section 8.4, we look at the structure of chromosomes and how it changes preparatory to and during mitosis.

> **8.3** *Check Your Progress* Which part of the cell cycle—interphase or the M stage—would you expect to be curtailed in cancer cells? Why?

FIGURE 8.3A
Stages of the cell cycle.

G_0

Interphase

Cytokinesis
Telophase
Anaphase
Metaphase
Prometaphase
Prophase

Mitosis

M

G_1
(growth)

S
(growth and DNA replication)

G_2
(growth and final preparations for division)

FIGURE 8.3B Cytokinesis (shown here) is a noticeable part of the cell cycle.

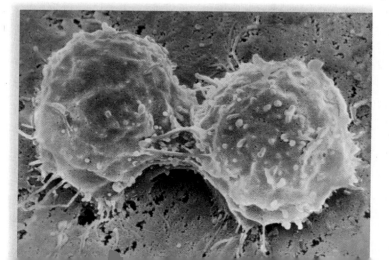

Cell division in eukaryotes involves division of the nucleus and division of the cytoplasm. The chromosomes in the nucleus of eukaryotes are associated with various proteins, including *histone proteins* that are especially involved in keeping them organized. When a eukaryotic cell is not undergoing division, the DNA (and associated proteins) within a chromosome is a tangled mass of thin threads called **chromatin**. Before nuclear division begins, chromatin becomes highly coiled and condensed, and it is easy to see the individual chromosomes.

Somatic Cells Are Diploid (2n) When the chromosomes are visible, it is possible to photograph and count them. Each species has a characteristic chromosome number (**Table 8.4**); for instance, human cells contain 46 chromosomes, corn has 20, and a goldfish has 94. This is the full or **diploid (2n) number** of chromosomes that is found in all somatic cells of the body. The diploid number includes two chromosomes of each kind.

During mitosis, a 2n nucleus divides to produce daughter nuclei that are also 2n. A dividing cell is called the **parent cell**, and the new cells are called the **daughter cells**. Before mitotic nuclear division takes place, DNA replicates, duplicating the chromosomes in the parent cell. As mentioned, each duplicated chromosome has two identical double helix molecules; each double helix is a chromatid, and the two identical chromatids are called sister chromatids (**Fig. 8.4A**). Sister chromatids are constricted and attached to each other at a region called the **centromere**.

TABLE 8.4	Diploid Chromosome Numbers of Some Eukaryotes	
Type of Organism	Name of Organism	Chromosome Number
Fungi	*Aspergillus nidulans* (mold)	8
	Neurospora crassa (mold)	14
	Saccharomyces cerevisiae (yeast)	32
Plants	*Vicia faba* (broad bean)	12
	Pisum sativum (garden pea)	14
	Zea mays (corn)	20
	Solanum tuberosum (potato)	48
	Nicotiana tabacum (tobacco)	48
	Ophioglossum vulgatum (Southern adder's tongue fern)	1,320
Animals	*Drosophila melanogaster* (fruit fly)	8
	Rana pipiens (leopard frog)	26
	Felis catus (cat)	38
	Homo sapiens (human)	46
	Pan troglodytes (chimp)	48
	Carassius auratus (goldfish)	94

During mitosis, the two sister chromatids separate at the centromere, and in this way, each duplicated chromosome gives rise to two daughter chromosomes. Each daughter chromosome has only one double helix molecule. The daughter chromosomes are distributed equally to the daughter cells. In this way, each daughter nucleus gets a copy of each chromosome that was in the parent cell (**Fig. 8.4B**).

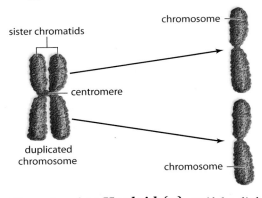

FIGURE 8.4B When sister chromatids separate, each daughter nucleus gets a chromosome.

Gametes Are Haploid (n) Half the diploid number, called the **haploid (n) number** of chromosomes, contains only one chromosome of each kind. Typically in the life cycle of animals, only sperm and eggs, collectively called **gametes,** have the haploid number of chromosomes.

In Section 8.5, we continue our discussion of mitosis, when duplicated chromosomes divide in such a way that each daughter cell does get a complete set of chromosomes and remains 2n.

8.4 Check Your Progress **a. Following mitosis, are diploid cells normally diploid (2n) or haploid (n)? b. Would the same hold true for cancer cells?**

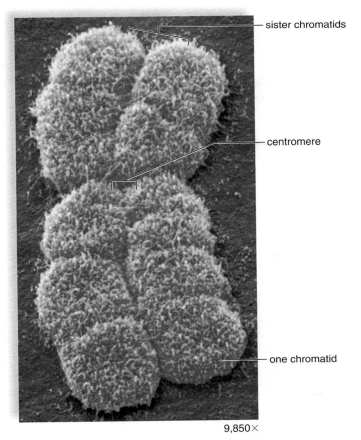

9,850×

FIGURE 8.4A A condensed duplicated chromosome.

Mitotic Spindle Before examining the phases of mitosis in **Figure 8.5,** take a look at the animal cell (*top*) and the plant cell (*bottom*) at interphase. Notice that both cells have a centrosome, but only the centrosome of the animal cell contains centrioles. Therefore, only animal cells will have asters during mitosis. An **aster** is an array of microtubules that radiate from the centrioles.

Before cell division begins, the centrosome divides. Then, the two resulting centrosomes organize the **mitotic spindle,** which consists of spindle fibers extending between the poles

(ends) of the spindle. The **spindle fibers,** each composed of microtubules, attach to the sister chromatids of each duplicated chromosome by way of protein molecules called **kinetochores.**

Mitotic Division Separation of chromaids during mitosis requires the phases described in Figure 8.5. Keep in mind that, although mitosis is divided into these phases, it is a continous process.

Note that some duplicated chromosomes are colored red and some are colored blue. The red chromosomes were inherited from

FIGURE 8.5 Phases of mitosis in animal cells and plant cells.

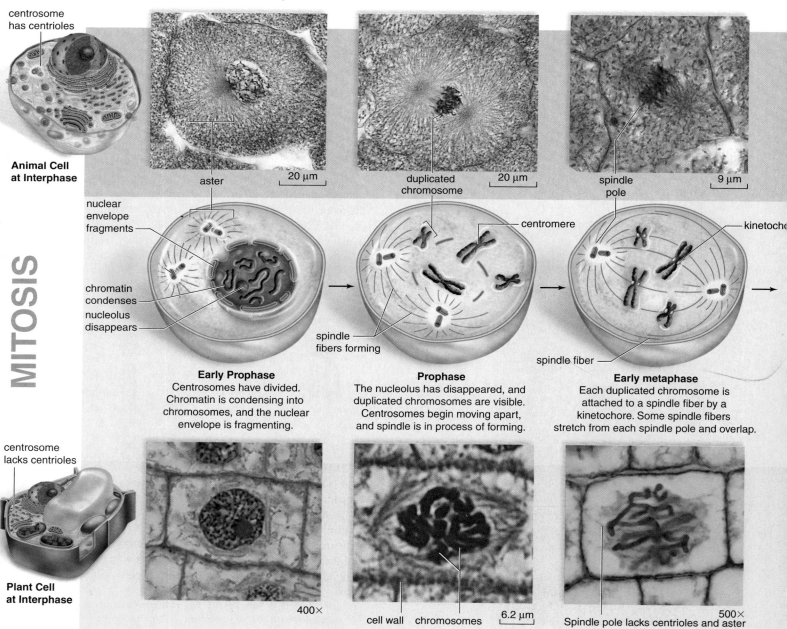

MITOSIS

Animal Cell at Interphase

centrosome has centrioles

aster — 20 µm

duplicated chromosome — 20 µm

spindle pole — 9 µm

nuclear envelope fragments

chromatin condenses

nucleolus disappears

spindle fibers forming

centromere

kinetochore

spindle fiber

Early Prophase
Centrosomes have divided. Chromatin is condensing into chromosomes, and the nuclear envelope is fragmenting.

Prophase
The nucleolus has disappeared, and duplicated chromosomes are visible. Centrosomes begin moving apart, and spindle is in process of forming.

Early metaphase
Each duplicated chromosome is attached to a spindle fiber by a kinetochore. Some spindle fibers stretch from each spindle pole and overlap.

Plant Cell at Interphase

centrosome lacks centrioles

400×

cell wall   chromosomes   6.2 µm

500×
Spindle pole lacks centrioles and aster

one parent, and the blue chromosomes from the other parent. Also, due to replication, each chromosome is composed of identical sister chromatids, held together at a centromere. The newly formed kinetochores attach each sister chromatid to a spindle fiber. They pull the sister chromatids apart toward opposite poles because they are motor molecules, which use a spindle fiber much as a train uses a track. Following separation, the sister chromatids are called daughter chromosomes.

Daughter Nuclei The resulting daughter nuclei receive one of each kind of chromosome. In Figure 8.5, the daughter nuclei have four chromosomes—they each have a red short,

a blue short, a red long, and a blue long—the same as the parent cell had. In other words, the daughter nuclei are identical to each other and to the parent nucleus. They contain a copy of the same chromosomes and, therefore, have the same genetic information.

Now that the nucleus has divided, we examine in Section 8.6 how the cytoplasm divides during cytokinesis.

8.5 Check Your Progress Imagine an adult stem cell that produces new skin cells. One daughter cell remains a stem cell. What happens to the other cell?

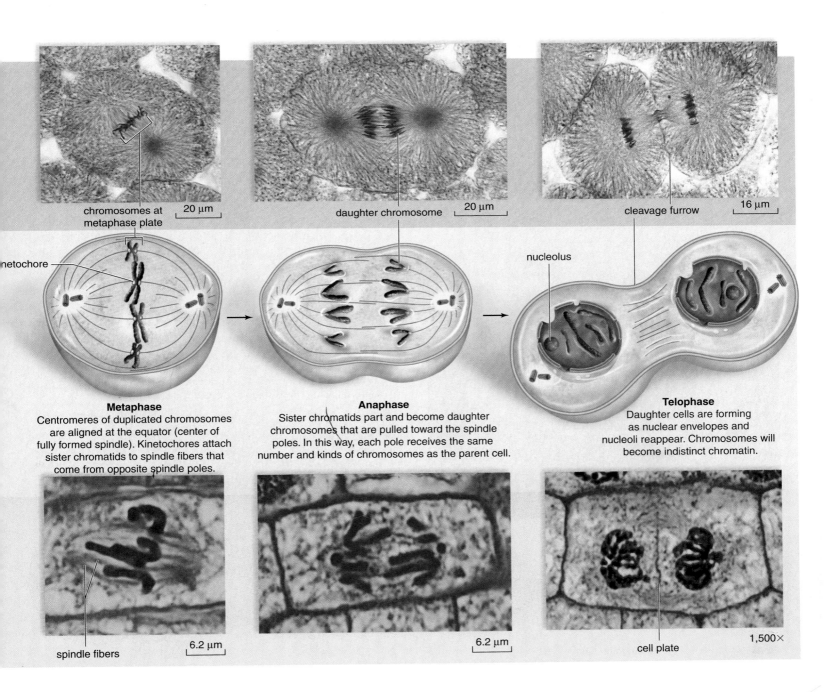

chromosomes at metaphase plate   20 μm

daughter chromosome   20 μm

cleavage furrow   16 μm

netochore

nucleolus

Metaphase
Centromeres of duplicated chromosomes are aligned at the equator (center of fully formed spindle). Kinetochores attach sister chromatids to spindle fibers that come from opposite spindle poles.

Anaphase
Sister chromatids part and become daughter chromosomes that are pulled toward the spindle poles. In this way, each pole receives the same number and kinds of chromosomes as the parent cell.

Telophase
Daughter cells are forming as nuclear envelopes and nucleoli reappear. Chromosomes will become indistinct chromatin.

spindle fibers   6.2 μm

6.2 μm

cell plate   1,500×

8.6 Cytokinesis divides the cytoplasm

Cytokinesis, meaning division of the cytoplasm, follows mitosis in most cells, but not all of them. When mitosis occurs but cytokinesis does not occur, the result is a multinucleated cell. For example, you will see later in this book that skeletal muscle cells in vertebrate animals and the embryo sac in a flowering plant are multinucleated.

Ordinarily, cytokinesis begins during telophase and continues after the nuclei have formed until there are two daughter cells. At that time, the M stage of the cell cycle is complete, and the daughter cells enter interphase, during which the cell grows and DNA replicates once again. In rapidly dividing mammalian stem cells, the cell cycle lasts for 24 hours. Mitosis and cytokinesis require only one hour of this time period.

Animal Cell Cytokinesis In animal cells, a cleavage furrow, which is an indentation of the membrane between the two daughter nuclei, begins at the start of telophase. The cleavage furrow deepens when a band of actin filaments, called the contractile ring, slowly forms a circular constriction between the two daughter cells. The action of the contractile ring can be likened to pulling a drawstring ever tighter about the middle of a balloon. A narrow bridge between the two cells is visible during telophase, and then the contractile ring continues to separate the cytoplasm until there are two independent daughter cells. This process is illustrated in **Figure 8.6A**. First, the cleavage furrow appears, and then the contractile ring tightens the constriction.

Plant Cell Cytokinesis In plant cells, cytokinesis occurs by a process different from that seen in animal cells. The rigid cell wall that surrounds plant cells does not permit cytokinesis by furrowing. Instead, cytokinesis in plant cells involves the building of new plasma membranes and cell walls between the daughter cells.

Cytokinesis is apparent when a small, flattened disk appears between the two daughter plant cells. Electron micrographs reveal that the disk is composed of vesicles (**Fig. 8.6B**). The Golgi apparatus produces these vesicles, which move along microtubules to the region of the disk. As more vesicles arrive and fuse, a cell plate can be seen. The **cell plate** is simply newly formed plasma membrane that expands outward until it reaches the old plasma membrane and fuses with it. The new membrane releases molecules that form the new plant cell walls. These cell walls are later strengthened by the addition of cellulose fibrils.

We have completed our study of mitosis and cytokinesis, and in the next part of the chapter, we study cell cycle control, because when it falters, cancer develops.

> **8.6 Check Your Progress** *a.* Cytokinesis in plant cells is more complex than in animal cells. Why? *b.* Compare the number and kinds of chromosomes in the nucleus of parent and daughter cells following the M stage of the cell cycle.

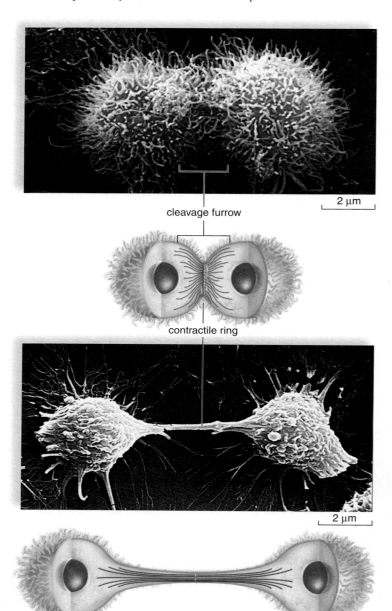

cleavage furrow

2 μm

contractile ring

2 μm

FIGURE 8.6A Cytokinesis in an animal cell.

© R. G. Kessel and C. Y. Shih, *Scanning Electron Microscopy in Biology. A Student's Atlas on Biological Organization*, 1974 Springer-Verlag, New York.

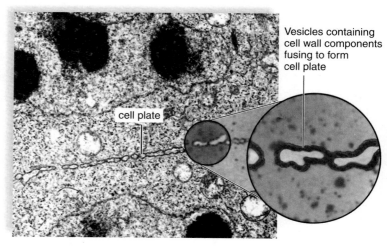

Vesicles containing cell wall components fusing to form cell plate

cell plate

FIGURE 8.6B Cytokinesis in plant cells.

In this part of the chapter, we study the cell cycle's control system, which ensures that the cell cycle occurs in an orderly manner. Cancer develops when the cell cycle control system is not functioning as it should. We also examine the characteristics of cancer cells and some ways you can help prevent cancer from developing.

8.7 Cell cycle control occurs at checkpoints

In order for a cell to reproduce successfully, the cell cycle must be controlled. The importance of cell cycle control can be appreciated by comparing the cell cycle to the events that occur in an automatic washing machine. The washer's control system starts to wash only when the tub is full of water, does not spin until the water has been emptied, delays the most vigorous spin until rinsing has occurred, and so forth. Similarly, the cell cycle's control system ensures that the G_1, S, G_2, and M stages occur in order and only when the previous stage has been successfully completed. The cell cycle has checkpoints that can delay the cell cycle until all is well. The cell cycle has many checkpoints, but we will consider only three: G_1, G_2, and the M (mitotic) checkpoints (**Fig. 8.7**).

1 The G_1 checkpoint is especially significant, because if the cell cycle passes this checkpoint, the cell is committed to divide. If the cell does not pass this checkpoint, it can enter G_0, during which it performs specialized functions but does not divide. If the DNA is damaged beyond repair, the internal signaling protein **p53** can stop the cycle at this checkpoint. First, p53 attempts to initiate DNA repair, but if that is not possible, it brings about the death of the cell by **apoptosis**, defined as programmed cell death.

2 The cell cycle hesitates at the G_2 checkpoint, ensuring that DNA has replicated. This prevents the initiation of the M stage unless the chromosomes are duplicated. Also, if DNA is damaged, as from exposure to solar radiation or X-rays, arresting the cell cycle at this checkpoint allows time for the damage to be repaired, so that it is not passed on to daughter cells. If repair is not possible, apoptosis occurs.

3 The M checkpoint occurs during the mitotic stage. The cycle hesitates at the M checkpoint to make sure the chromosomes are going to be distributed accurately to the daughter cells. The cell cycle does not continue until every duplicated chromosome is ready for the chromatids to separate.

Apoptosis During apoptosis, the cell progresses through a typical series of events that bring about its destruction. The cell rounds up and loses contact with its neighbors. The nucleus fragments, and the plasma membrane develops blisters. Finally, the cell breaks into frag-

ments, and its bits and pieces are engulfed by white blood cells and/or neighboring cells. A remarkable finding of the past few years is that cells routinely harbor the enzymes, now called caspases, that bring about apoptosis. These enzymes are ordinarily held in check by inhibitors, but are unleashed by either internal or external signals.

Cell division and apoptosis are two opposing processes that keep the number of cells in the body at an appropriate level. They are normal parts of growth and development. An organism begins as a single cell that repeatedly undergoes the cell cycle to produce many cells, but eventually some cells must die in order for the organism to take shape. For example, when a tadpole becomes a frog, the tail disappears as apoptosis occurs. In humans, the fingers and toes of an embryo are at first webbed, but later the webbing disappears as a result of apoptosis, and the fingers are freed from one another. Apoptosis is also helpful if an abnormal cell that could become cancerous appears. Death through apoptosis can prevent a tumor from developing.

As discussed in Section 8.8, internal and external signaling molecules help determine whether the cell cycle progresses through various checkpoints.

> **8.7 Check Your Progress** An altered p53 protein is found in most cancerous cells. Explain.

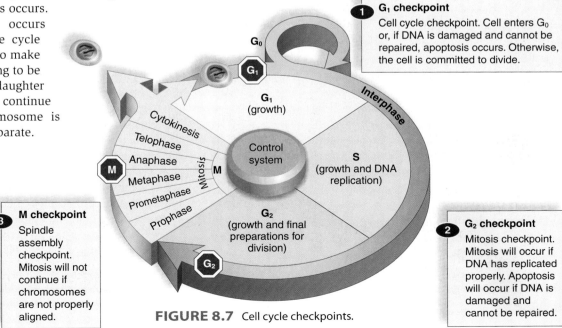

1 **G_1 checkpoint**
Cell cycle checkpoint. Cell enters G_0 or, if DNA is damaged and cannot be repaired, apoptosis occurs. Otherwise, the cell is committed to divide.

3 **M checkpoint**
Spindle assembly checkpoint. Mitosis will not continue if chromosomes are not properly aligned.

2 **G_2 checkpoint**
Mitosis checkpoint. Mitosis will occur if DNA has replicated properly. Apoptosis will occur if DNA is damaged and cannot be repaired.

G_0

G_1

G_1 (growth)

Interphase

Cytokinesis
Telophase
Anaphase
Metaphase
Prometaphase
Prophase

Mitosis

M

Control system

S (growth and DNA replication)

G_2 (growth and final preparations for division)

G_2

FIGURE 8.7 Cell cycle checkpoints.

The checkpoints of the cell cycle are controlled by internal and external signals. A **signal molecule** stimulates or inhibits an event. Internal signal molecules occur within a cell, while external signal molecules come from outside the cell. Inside the cell, enzymes called **kinases** remove phosphate from ATP and add it to another molecule. Just before the S stage, a protein called S-cyclin combines with a kinase called S-kinase, and synthesis of DNA takes place. Just before the M stage, M-cyclin combines with a kinase called M-kinase, and mitosis occurs. **Cyclins** are so named because their quantity is not constant. They increase in amount until they combine with a kinase, but this is a suicidal act. The kinase not only activates a protein that drives the cell cycle, but it also activates various enzymes, one of which destroys the cyclin (**Fig. 8.8A**).

Some external signal molecules, such as growth factors and hormones, stimulate cells to go through the cell cycle. An animal and its organs grow larger if the cell cycle occurs. Growth factors also stimulate repair of tissues. Even cells that are arrested in G_0 will finish the cell cycle if stimulated to do so by growth factors. For example, epidermal growth factor (EGF) stimulates the skin in the vicinity of an injury to finish the cell cycle, thereby repairing the damage. Hormones act on tissues at a distance, and some of them signal cells to divide. For example, at a certain time in the menstrual cycle of women, the hormone estrogen stimulates the cells lining the uterus to divide and prepare for implantation of a fertilized egg.

As shown in **Figure 8.8B**, ❶ during reception, an external signal molecule delivers a message to a specific receptor embedded in the plasma membrane of a receiving cell. ❷ The receptor relays the signal to proteins inside the cell's cytoplasm. The proteins form a pathway called the signal transduction pathway because they pass the signal from one to the other until it

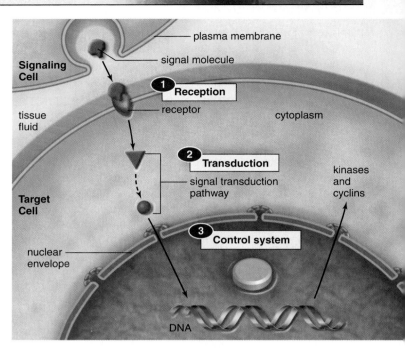

FIGURE 8.8B A cell-signaling pathway activates the control system to produce kinases and cyclins.

reaches the nucleus. ❸ The **control system** for cell division is located in the nucleus, and certain genes control whether kinases and cyclins are present in the cell.

The cell cycle can be inhibited by cells coming into contact with other cells, and by the shortening of chromosomes. In cell culture, cells will divide until they line a container in a one-cell-thick sheet. Then, they stop dividing, a phenomenon termed **contact inhibition**. Researchers are beginning to discover the external signal molecules that result in this inhibition. Some years ago, it was noted that mature mammalian cells in cell culture divide about 70 times, and then they die. Cells seem to "remember" the number of times they have divided, and they stop dividing when the usual number of cell divisions is reached. It is as if *senescence*, the aging of cells, is dependent on an internal clock that winds down and then stops. We now know that senescence is due to the shortening of telomeres. A **telomere** is a repeating DNA base sequence (TTAGGG) at the ends of the chromosomes that can be as long as 15,000 base pairs. Telomeres have been likened to the protective caps on the ends of shoelaces. However, other than keeping chromosomes from unraveling, telomeres stop chromosomes from fusing to each other. Each time a cell divides, some portion of a telomere is lost; when telomeres become too short, the chromosomes fuse and can no longer duplicate. Then the cell is "old" and dies by apoptosis.

If the cell cycle is not properly controlled, cancer can develop. The characteristics of cancer cells are described in Section 8.9.

> **8.8 Check Your Progress** Embryonic cells have an enzyme called telomerase, which rebuilds telomeres. Why would you predict that cancer cells have an active telomerase enzyme?

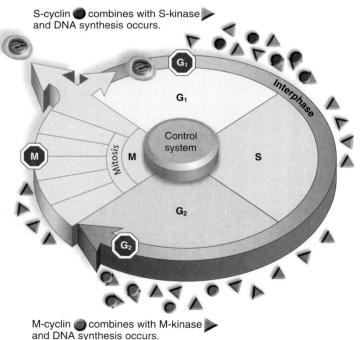

S-cyclin ● combines with S-kinase ▶ and DNA synthesis occurs.

M-cyclin ● combines with M-kinase ▶ and DNA synthesis occurs.

FIGURE 8.8A Internal signals of the cell cycle are kinases and cyclins.

As explained in the introduction to this chapter, **mutations** (DNA changes) due to different environmental assaults can result in cancer. Any tissue that already has a high rate of cell division is inherently more susceptible to mutations that can lead to cancer. In cancer cells, the cell cycle is out of control, and cellular reproduction occurs repeatedly without end. Cancers are classified according to their location. *Carcinomas* are cancers of the epithelial tissue that lines organs; *sarcomas* are cancers arising in muscle or connective tissue (especially bone or cartilage); and *leukemias* are cancers of the blood. In this section, we consider some general characteristics of cancer cells. Chapter 11 (Sections 11.14–11.16) explores specific changes that lead to cancerous growth.

Carcinogenesis, the development of cancer, can be gradual; it may be decades before a tumor is visible. Cancer cells have the following characteristics:

Cancer cells lack differentiation Cancer cells are nonspecialized and do not contribute to the functioning of a body part. A cancer cell does not look like a specialized epithelial, muscle, nervous, or connective tissue cell; instead, it looks distinctly abnormal. As mentioned, normal cells can enter the cell cycle about 70 times, and then they die. Cancer cells can enter the cell cycle repeatedly, and in this way they are immortal.

Cancer cells have abnormal nuclei The nuclei of cancer cells are enlarged and may contain an abnormal number of chromosomes. The chromosomes are also abnormal; some parts may be duplicated, or some may be deleted. In addition, gene amplification (extra copies of specific genes) is seen much more frequently than in normal cells. Ordinarily, cells with damaged DNA undergo apoptosis, but cancer cells fail to undergo apoptosis, even though they are abnormal.

Cancer cells form tumors Normal cells anchor themselves to a substratum and/or adhere to their neighbors. Then they exhibit contact inhibition and stop dividing. Cancer cells, on the other hand, have lost all restraint; they pile on top of one another and grow in multiple layers, forming a **tumor**. They have a reduced need for growth factors, and they no longer respond to inhibitory signals. As cancer develops, the most aggressive cell becomes the dominant cell of the tumor.

Cancer cells undergo metastasis and promote angiogenesis A *benign tumor* is usually encapsulated and, therefore, will never invade adjacent tissue. A cancerous tumor may progress from one to the other of the types shown in **Figure 8.9**. *Cancer in situ* is a tumor in its place of origin, but it is not encapsulated and will eventually invade surrounding tissues. Cancer cells produce enzymes that allow tumors to invade underlying tissues. *Invasive tumors* produce cancer cells that travel through the blood and lymph to start tumors elsewhere in the body. Malignancy is present once **metastasis** has established new *metastatic tumors* distant from the primary tumor.

Angiogenesis, the formation of new blood vessels, is required to bring nutrients and oxygen to a cancerous tumor.

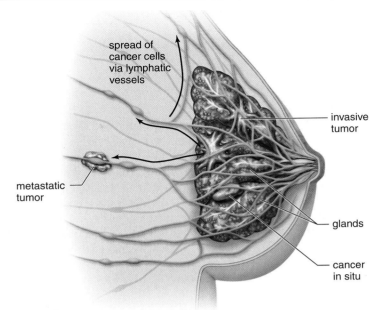

Types of cancerous tumors

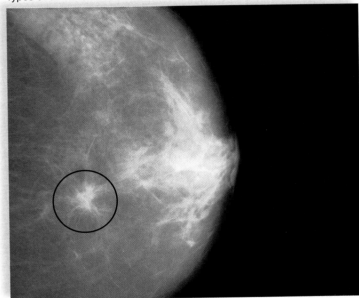

Mammogram showing tumor

FIGURE 8.9 Development of breast cancer.

Some modes of cancer treatment are aimed at preventing angiogenesis from occurring.

The patient's prognosis (probable outcome) is dependent on (1) whether the tumor has invaded surrounding tissues, and (2) whether there are metastatic tumors in distant parts of the body.

Section 8.10 suggests ways to protect yourself from developing cancer.

8.9 *Check Your Progress* Which characteristics of cancer cells can be associated with a loss of cell cycle control? Which are mutations that reach beyond loss of cell cycle control?

Evidence suggests that the risk of certain types of cancer can be reduced by adopting protective behaviors and the right diet.

Protective Behaviors The following behaviors help prevent cancer:

Don't smoke Cigarette smoking accounts for about 30% of all cancer deaths. Smoking is responsible for 90% of lung cancer cases among men and 79% among women—about 87% altogether. People who smoke two or more packs of cigarettes a day have lung cancer mortality rates 15–25 times greater than those of nonsmokers. Smokeless tobacco (chewing tobacco or snuff) increases the risk of cancers of the mouth, larynx, throat, and esophagus. Chances of cancer increase when smoking is accompanied by heavy alcohol use.

Use sunscreen Almost all cases of skin cancer are considered sun-related. Use a sunscreen of at least SPF 15 (**Fig. 8.10**), and wear protective clothing if you are going to be out during the brightest part of the day. Don't sunbathe on the beach or in a tanning salon.

Avoid radiation Excessive exposure to ionizing radiation can increase cancer risk. Even though most medical and dental X-rays are adjusted to deliver the lowest dose possible, unnecessary X-rays should be avoided. Radon gas from the radioactive decay of uranium in the Earth's crust can accumulate in houses and increase the risk of lung cancer, especially in cigarette smokers. It is best to test your home and take the proper remedial actions.

Be tested for cancer Do the shower check for breast cancer or testicular cancer. Have other exams done regularly by a physician.

Be aware of occupational hazards Exposure to several different industrial agents (nickel, chromate, asbestos, vinyl chloride, etc.) and/or radiation increases the risk of various cancers. Risk from asbestos is greatly increased when combined with cigarette smoking.

Carefully consider hormone therapy A new study conducted by the Women's Health Initiative found that combined estrogen-progestin therapy prescribed to ease the symptoms of menopause increased the incidence of breast cancer. And the risk outweighed the possible decrease in the number of colorectal cancer cases sometimes attributed to hormone therapy.

FIGURE 8.10
Sunscreen with SPF 15 minimizes skin cancer.

The Right Diet Statistical studies have suggested that people who follow certain dietary guidelines are less likely to have cancer. The following dietary precautions greatly reduce your risk of developing cancer:

Increase consumption of foods rich in vitamins A and C Beta-carotene, a precursor of vitamin A, is found in carrots, fruits, and dark-green, leafy vegetables. Vitamin C is present in citrus fruits. These vitamins are called antioxidants because in cells they prevent the formation of free radicals (organic ions having an unpaired electron) that can possibly damage DNA. Vitamin C also prevents the conversion of nitrates and nitrites into carcinogenic nitrosamines in the digestive tract.

Limit consumption of salt-cured, smoked, or nitrite-cured foods Consuming salt-cured or pickled foods may increase the risk of stomach and esophageal cancers. Smoked foods, such as ham and sausage, contain chemical carcinogens similar to those in tobacco smoke. Nitrites are sometimes added to processed meats (e.g., hot dogs and cold cuts) and other foods to protect them from spoilage. As mentioned previously, nitrites are converted to nitrosamines in the digestive tract.

Include vegetables from the cabbage family in the diet The cabbage family includes cabbage, broccoli, brussels sprouts, kohlrabi, and cauliflower. These vegetables may reduce the risk of gastrointestinal and respiratory tract cancers.

Be moderate in the consumption of alcohol The risks of cancer development rise as the level of alcohol intake increases. The strongest associations are with oral, pharyngeal, esophageal, and laryngeal cancer, but cancer of the breast and liver are also implicated. People who both drink and smoke greatly increase their risk for developing cancer.

Maintain a healthy weight The risk of cancer (especially colon, breast, and uterine cancers) is 55% greater among obese women, and the risk of colon cancer is 33% greater among obese men, compared to people of normal weight.

This completes our study of cell cycle control and the development of cancer.

8.10 *Check Your Progress* **Why do you think tobacco use increases the risk of other types of cancer besides lung cancer?**

Meiosis Produces Cells That Become the Gametes in Animals and Spores in Other Organisms

Learning Outcomes 8–10, page 132

This part of the chapter discusses meiosis, the type of nuclear division that participates in sexual reproduction. We discuss the ways in which meiosis causes variations among offspring, before considering the stages of meiosis. Exactly where meiosis occurs in the life cycle of organisms determines the adult chromosome number. Finally, we compare the stages of meiosis to those of mitosis.

8.11   Homologous chromosomes separate during meiosis

Human beings, like other vertebrates, engage in sexual reproduction in which two parents pass chromosomes to their offspring. Therefore, children are not exactly like either parent. In the next several sections, we will examine how sexual reproduction brings about the distribution of chromosomes to gametes (e.g., sperm and egg) in a way that ensures offspring will have the correct number of chromosomes and a unique combination of genetic information.

Let's begin by examining the chromosomes of one of the parents—for instance, the father. To view the chromosomes, a cell can be photographed just prior to division. The resulting picture of the chromosomes can be entered into a computer, and the chromosomes electronically arranged by numbered pairs, producing a **karyotype (Fig. 8.11)**. The members of a pair are called **homologous chromosomes**, or **homologues**, because they have the same size, shape, and constriction (location of the centromere). One homologue is inherited from the male parent, and the other is inherited from the female parent. Homologous chromosomes have the same characteristic banding pattern upon staining because they contain the same type of genetic information; for example, perhaps, finger length. However, one homologue including its sister chromatids may call for short fingers, while the other homologue and its sister chromatids may call for long fingers.

Both males and females normally have 23 *pairs* of chromosomes, but one of these pairs is unequal in males. The larger chromosome of this pair is the X chromosome, and the smaller is the Y chromosome. In contrast, females have two X chromosomes. The X and Y chromosomes are called the **sex chromosomes** because they contain the genes that determine gender. The other chromosomes, known as **autosomes**, include all the pairs of chromosomes except the X and Y chromosomes.

Unlike the individual, the sperm and the egg have only 23 chromosomes, which is the haploid (n) number of chromosomes. Why? Because a type of nuclear division called **meiosis** occurs during the production of the sperm and egg. Meiosis requires two divisions. During **meiosis I**, the chromosomes of each homologous pair separate, and the first set of daughter cells receives one member of each homologous pair. The chromosomes are still duplicated. During **meiosis II**, the sister chromatids of each duplicated chromosome separate, and each daughter cell receives a chromosome:

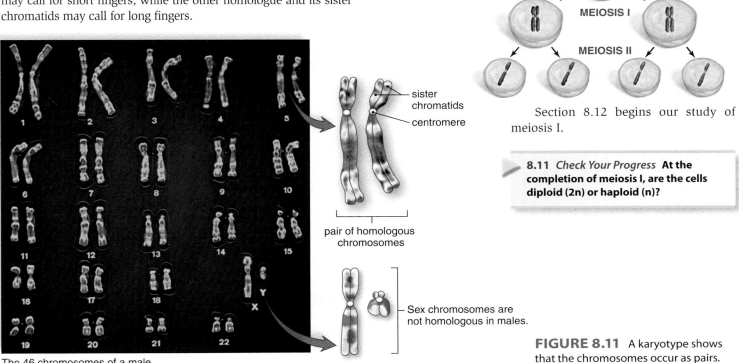

Section 8.12 begins our study of meiosis I.

> **8.11** *Check Your Progress* **At the completion of meiosis I, are the cells diploid (2n) or haploid (n)?**

The 46 chromosomes of a male

sister chromatids

centromere

pair of homologous chromosomes

Sex chromosomes are not homologous in males.

pair of homologous chromosomes

FIGURE 8.11 A karyotype shows that the chromosomes occur as pairs.

8.12 Synapsis and crossing-over occur during meiosis I

Prior to meiosis I, the chromosomes have duplicated, and each consists of two sister chromatids held together at the centromere. During meiosis I, two events occur that are not seen in mitosis: synapsis and crossing-over.

Synapsis The homologous chromosomes come together and line up side by side, much like two dancing partners that will stay together until the dance ends. The homologues are said to be in **synapsis**, which means "joined together." The homologues are held in place by a protein lattice that develops between them (**Fig. 8.12A**). Because each homologue has two sister chromatids, four chromosomes are in close association. Each set of four chromatids is called a **tetrad**.

Crossing-over During synapsis, the nonsister chromatids sometimes exchange genetic material, an event called **crossing-over. Figure 8.12B** shows what is meant by nonsister chromatids and how crossing-over occurs. The homologues carry genetic information for certain traits, such as finger length, eye color, and any number of other traits. The genetic information of nonsister chromatids can differ because they belong to the other homologue. For example, one set of nonsister chromatid could call for blue eyes, and the other set could call for brown eyes.

After the nonsister chromatids exchange genetic material during crossing-over, the sister chromatids may now carry different

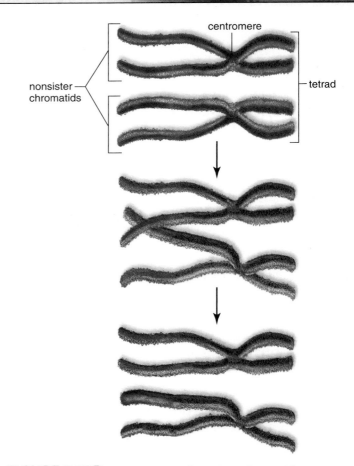

FIGURE 8.12B Crossing-over of nonsister chromatids.

genetic information represented by a change in color: One of the blue sister chromatids now has a red tip and one of the red sister chromatids now has a blue tip (Fig. 8.12B). Therefore, the gametes that form when these sister chromatids separate will have a better chance of being genetically different. Crossing-over increases the genetic variability of the gametes, and therefore, of the offspring.

In Section 8.13, we consider how meiosis I can be counted on to increase the genetic variability of the gametes.

> **8.12 Check Your Progress** Why does crossing-over between nonsister chromatids increase genetic variation, whereas crossing-over between sister chromatids would not increase genetic variation?

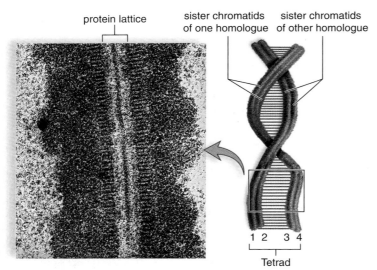

FIGURE 8.12A Synapsis of homologues.

8.13 Sexual reproduction increases genetic variation

In **Figure 8.13**, ❶ the parent cell has two pairs of homologues, which undergo synapsis soon after meiosis I begins. While the chromatids were in close association during synapsis, crossing-over occur between nonsister chromatids. ❷ Notice that two orientations are possible at the equator because either homologue can face either pole of the spindle. In the simplest of terms, with reference to Figure 8.13, the red chromosomes don't have to be on the left, and

the blue chromosomes don't have to be on the right. Therefore, it is said that the homologue pairs *align independently* at the equator.

❸ The homologues separate so that one chromosome from each pair goes to each daughter nucleus, and the daughter cells are haploid. ❹ All possible combinations of chromosomes can occur among the gametes. Therefore, **independent assortment** of homologues occurs during meiosis. In the simplest of terms, any short chromosome (blue or red) can be with any long chromosome (blue or red). The genetic variation brought about by independent assortment of chromosomes is increased by crossing-over.

Fertilization The union of male and female gametes during **fertilization** produces a **zygote**, the first cell of the new individual. As we have seen, the gametes produced by individuals, such as humans, have the same number of chromosomes, but the chromosomes may carry different genetic information due to independent assortment and crossing-over. In humans, each gamete has 23 chromosomes. Considering the fusion of unlike gametes due to independent assortment, it means that $(2^{23})^2$, or 70,368,744,000,000, chromosomally different zygotes are possible, even assuming no crossing-over. If crossing-over occurs once, then $(4^{23})^2$, or 4,951,760,200,000,000,000,000,000,000, genetically different zygotes are possible for every couple. Keep in mind that crossing-over can occur several times between homologues.

Significance of Genetic Variation Asexual reproduction passes on exactly the same combination of chromosomes and genes that the parent possesses. The process of sexual reproduction brings about genetic recombinations among members of a population. If a parent is already successful in a particular environment, is asexual reproduction advantageous? It would seem so, as long as the environment remains unchanged. However, if the environment changes, genetic variability among offspring, introduced by sexual reproduction, may be advantageous. Under these conditions, some offspring may have a better chance of survival and reproductive success than others in a population. For example, suppose the ambient temperature were to rise due to global warming. Perhaps a dog with genes for the least amount of fur may have an advantage over other dogs of its generation.

In a changing environment, asexual reproduction might saddle an offspring with a parent's disadvantageous gene combination. In contrast, sexual reproduction, with its reshuffling of genetic information due to meiosis and fertilization, might give a few offspring a better chance of survival when environmental conditions change.

This completes our study of how genetic variations come about among offspring due to the process of sexual reproduction. In Section 8.14, we study the stages of meiosis I and meiosis II.

8.13 *Check Your Progress* **The mating of relatives reduces possible variations. Why?**

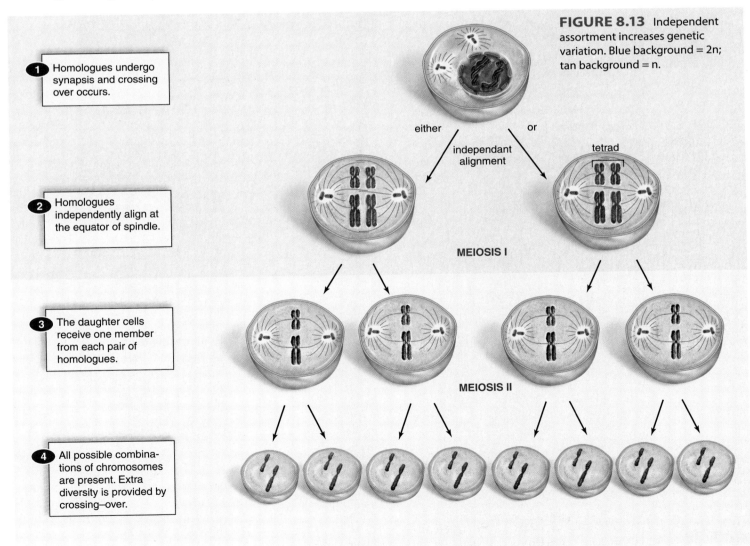

FIGURE 8.13 Independent assortment increases genetic variation. Blue background = 2n; tan background = n.

1. Homologues undergo synapsis and crossing over occurs.

either   or

independant alignment

tetrad

MEIOSIS I

2. Homologues independently align at the equator of spindle.

MEIOSIS II

3. The daughter cells receive one member from each pair of homologues.

4. All possible combinations of chromosomes are present. Extra diversity is provided by crossing-over.

8.14 Meiosis requires two division cycles

The same four phases of mitosis—prophase, metaphase, anaphase, and telophase—occur during both meiosis I (**Fig. 8.14A**) and meiosis II (**Fig. 8.14B**). During prophase I, the nuclear envelope fragments, the nucleolus disappears as the spindle appears, and the condensing homologues undergo synapsis. The formation of tetrads helps prepare the homologous chromosomes for separation; it also allows crossing-over to occur between nonsister chromatids. During metaphase I, tetrads are present and homologues align independently at the spindle equator. Following separation of the homologues during anaphase I and reformation of the nuclear envelopes during telophase, the daughter nuclei are haploid: Each daughter cell contains only one chromosome from each pair of homologues. The chromosomes are duplicated, and each still has two sister chromatids. No replication of DNA occurs during a period of time called **interkinesis**.

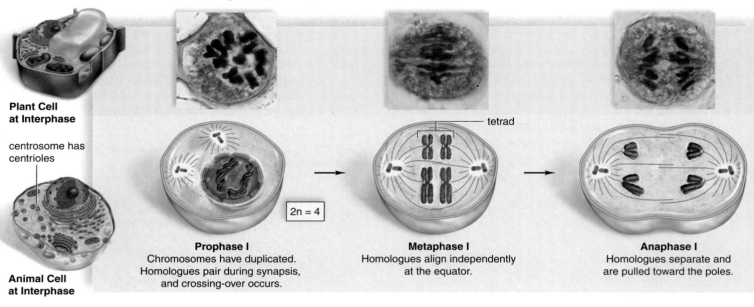

Plant Cell at Interphase

centrosome has centrioles

Animal Cell at Interphase

Prophase I
Chromosomes have duplicated. Homologues pair during synapsis, and crossing-over occurs.

2n = 4

tetrad

Metaphase I
Homologues align independently at the equator.

Anaphase I
Homologues separate and are pulled toward the poles.

FIGURE 8.14A Phases of meiosis I.

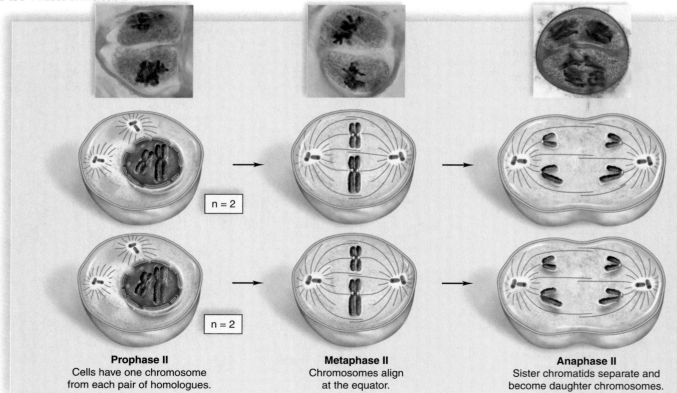

Prophase II
Cells have one chromosome from each pair of homologues.

n = 2

Metaphase II
Chromosomes align at the equator.

Anaphase II
Sister chromatids separate and become daughter chromosomes.

FIGURE 8.14B Phases of meiosis II. Blue background = 2n; tan background = n.

When you think about it, the events of meiosis II are the same as those for mitosis, except the cells are haploid. At the beginning of prophase II, a spindle appears, while the nuclear envelope fragments and the nucleolus disappears. Duplicated chromosomes (one from each pair of homologous chromosomes) are present, and each attaches to the spindle. During metaphase II, the duplicated chromosomes are lined up at the spindle equator. During anaphase II, sister chromatids separate and move toward the poles. Each pole receives the same number and kinds of chromosomes. In telophase II, the spindle disappears as nuclear envelopes form.

Now that we have a good working knowledge of meiosis, Section 8.15 discusses when it occurs in the life cycle of various types of organisms.

8.14 *Check Your Progress* **How would you recognize a diagram of plant cell meiosis?**

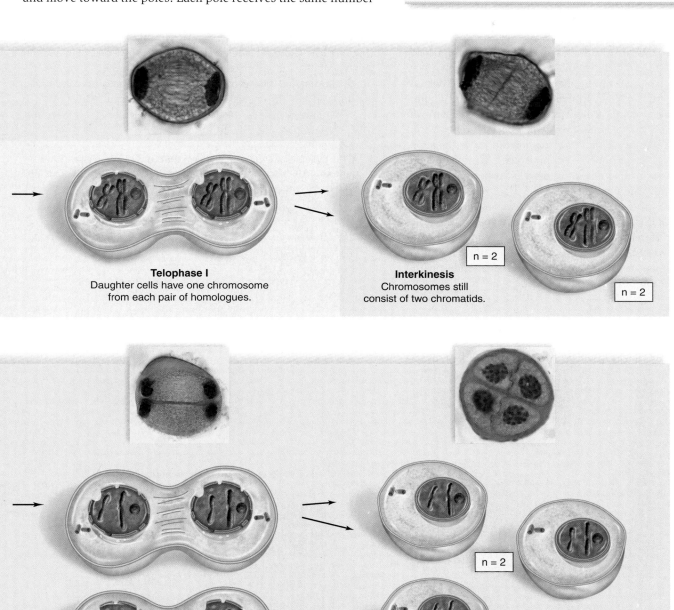

Telophase I
Daughter cells have one chromosome
from each pair of homologues.

n = 2

n = 2

Interkinesis
Chromosomes still
consist of two chromatids.

Telophase II
Spindle disappears, nuclei form,
and cytokinesis takes place.

n = 2

n = 2

Daughter cells
Meiosis results in four
haploid daughter cells.

8.15 The life cycle of most multicellular organisms includes both mitosis and meiosis

The term **life cycle** in sexually reproducing organisms refers to all the reproductive events that occur from one generation to the next. The human life cycle involves both mitosis and meiosis (**Fig. 8.15A**).

During development and after birth, mitosis is involved in the continued growth of the child and the repair of tissues at any time. As a result of mitosis, each somatic (body) cell has the diploid number of chromosomes (2n).

During gamete formation, meiosis reduces the chromosome number from the diploid to the haploid number (n) in such a way that the gametes (sperm and egg) have one chromosome derived from each pair of homologues. In males, meiosis is a part of **spermatogenesis**, which occurs in the testes and produces sperm. In females, meiosis is a part of **oogenesis**, which occurs in the ovaries and produces eggs. After the sperm and egg join during fertilization, the zygote has homologous pairs of chromosomes. The zygote then undergoes mitosis with differentiation of cells to become a fetus, and eventually a new human being.

Meiosis keeps the number of chromosomes constant between the generations, and it also, as we have seen, causes the gametes to be different from one another. Therefore, due to sexual reproduction, there are more variations among individuals. As we shall see in Chapter 13, evolution depends on the existence of variation between individuals.

Other Organisms In contrast to adult animals, which are always diploid, plants have an adult haploid phase that alternates with an adult diploid phase. The haploid generation, known as the gametophyte, may be larger or smaller than the diploid generation, called the sporophyte (**Fig. 8.15B**). The majority of plants, including pines, corn, and pea plants, are diploid most of the time, and the haploid generation is short-lived. In most fungi and algae, the zygote is the only diploid portion of the life cycle, and it undergoes meiosis (**Fig. 8.15C**). Therefore, the black mold that grows on bread and the green scum that floats on a pond are haploid. In plants, algae, and fungi, the haploid phase of the life cycle produces gamete nuclei without the need for meiosis because meiosis occurred earlier.

This completes our basic study of meiosis, and in Section 8.16, we compare meiosis to mitosis.

> **8.15** *Check Your Progress* **A sperm that carries a cancer-causing gene could lead to cancer in various organs of an offspring. Explain.**

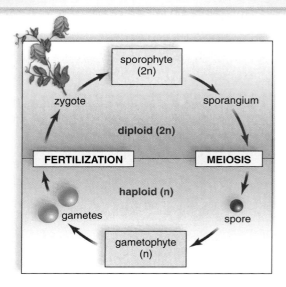

FIGURE 8.15B Life cycle of plants.

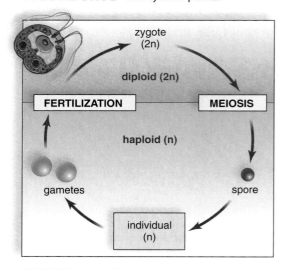

FIGURE 8.15C Life cycle of algae.

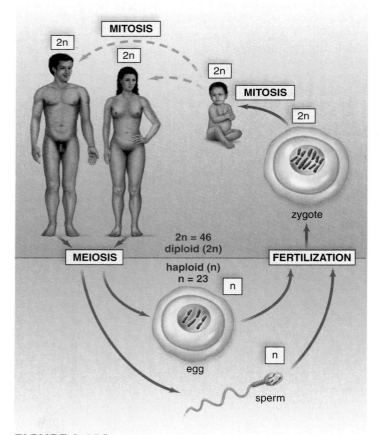

FIGURE 8.15A Life cycle of humans.

8.16 Meiosis can be compared to mitosis

Figure 8.16 compares meiosis to mitosis. Notice that:

- Meiosis requires two nuclear divisions, but mitosis requires only one nuclear division.
- Meiosis produces four daughter nuclei, and four daughter cells result following cytokinesis. Mitosis followed by cytokinesis results in two daughter cells.
- Following meiosis, the four daughter cells are haploid—they have half the chromosome number of the parent cell. Following mitosis, the daughter cells are diploid and have the same chromosome number as the parent cell.
- Following meiosis, the daughter cells are genetically dissimilar to each other and to the parent cell. Following mitosis, the daughter cells are genetically identical to each other and to the parent cell.

These differences are due to certain events:

- During meiosis I, tetrads form, and crossing-over occurs during prophase I. These events do not occur during mitosis.

- During metaphase I of meiosis, tetrads are at the equator. The homologues align at the spindle equator independently. During metaphase in mitosis, duplicated chromosomes align at the spindle equator.
- During anaphase I of meiosis, homologues separate, and duplicated chromosomes (with centromeres intact) move to opposite poles. During anaphase of mitosis, sister chromatids separate, becoming daughter chromosomes that move to opposite poles.

The events of meiosis II are just like those of mitosis except that in meiosis II, the daughter cells have the haploid number of chromosomes. Could abnormal meiosis cause the inheritance of an abnormal chromosome number? Section 8.17 shows how this is possible.

> **8.16 Check Your Progress** How are meiosis I and II like but different from mitosis?

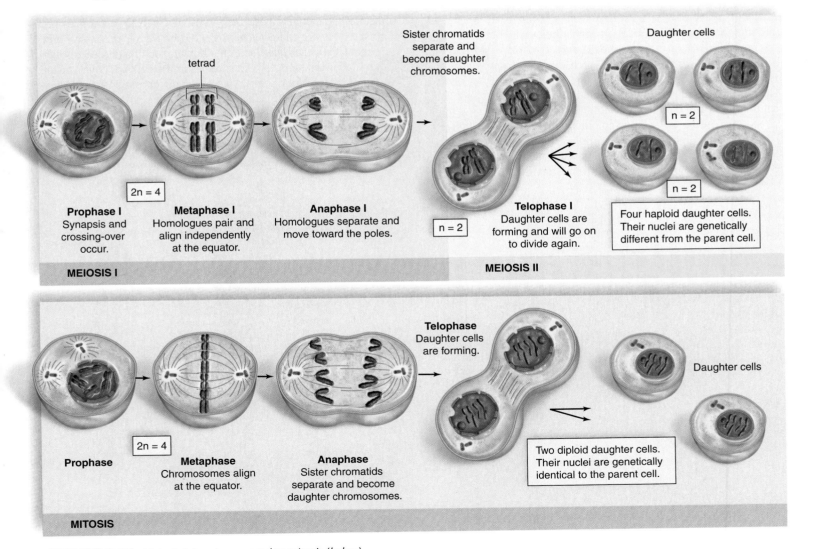

FIGURE 8.16 Meiosis (*above*) compared to mitosis (*below*).

Chromosomal mutations fall into two categories: change in chromosome number and change in chromosome structure. The presence of more than two sets of chromosomes (polyploidy) is common in plants. The gain or loss of a single chromosome (aneuploidy) occurs because of nondisjunction during meiosis and is exemplified by various syndromes in humans. Changes in chromosome structure include deletion, duplication, inversion, and translocation.

8.17   An abnormal chromosome number is sometimes traceable to nondisjunction

So far, we have mentioned three ways that genetic variation is increased in sexually reproducing organisms: independent assortment of homologous chromosomes, crossing-over, and gamete fusion during fertilization. Another way to increase the amount of genetic variation among individuals is through changes in chromosome number.

Changes in the chromosome number include polyploidy and aneuploidy. When a eukaryote has three or more complete sets of chromosomes, it is called a **polyploid**. More specifically, triploids (3n) have three of each kind of chromosome, tetraploids (4n) have four sets, pentaploids (5n) have five sets, and so on. Although polyploidy is not often seen in animals, it is a major evolutionary mechanism in plants, including many of our most important crops—wheat, corn, cotton, and sugarcane, as well as fruits such as watermelons, strawberries, bananas, and apples. Also, many attractive flowers, including chrysanthemums and daylilies, are polyploids. The strawberry on the left is an octaploid and much larger than the diploid one on the right:

An organism that does not have an exact multiple of the diploid number of chromosomes is an **aneuploid**.

When an individual has only one of a particular type of chromosome, **monosomy** (2n − 1) occurs. When an individual has three of a particular type of chromosome (2n + 1), **trisomy** occurs. The usual cause of monosomy and trisomy is nondisjunction during meiosis. **Nondisjunction** occurs during meiosis I when homologues fail to separate and both homologues go into the same daughter cell (**Fig. 8.17A**), or during meiosis II when the sister chromatids fail to separate and both daughter chromosomes go into the same gamete (**Fig. 8.17B**).

Monosomy and trisomy occur in both plants and animals. In animals, autosomal monosomies and trisomies are generally lethal, but a trisomic individual is more likely to survive than a monosomic one. The survivors are characterized by a distinctive set of physical and mental abnormalities, as in the human condition called trisomy 21 (see Section 8.18). Sex chromosome aneuploids have a better chance of producing survivors than do autosomal aneuploids.

Section 8.18 takes a look at some of the most well-known syndromes due to nondisjunction of chromosomes.

> **8.17 Check Your Progress** Why might problems arise if a person were to inherit three copies, instead of two copies, of a chromosome?

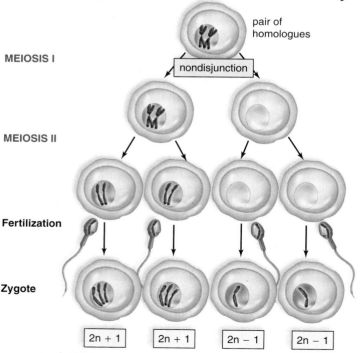

FIGURE 8.17A Nondisjunction of chromosomes during meiosis I of oogenesis, followed by fertilization with normal sperm.

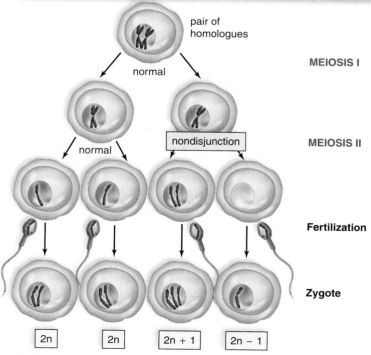

FIGURE 8.17B Nondisjunction of chromosomes during meiosis II of oogenesis, followed by fertilization with normal sperm.

A **syndrome** is a group of symptoms that occur together and comprise a particular disorder. Abnormal chromosome number is one cause of syndromes.

Trisomy 21 (Down Syndrome) The most common autosomal trisomy among humans is trisomy 21, also called Down syndrome. This syndrome is easily recognized by these characteristics: short stature, eyelid fold, flat face, stubby fingers, wide gap between the first and second toes, large, fissured tongue, round head, distinctive palm crease, heart problems, and mental retardation, which can sometimes be severe (**Fig. 8.18**). In addition, these individuals have an increased chance of developing Alzheimer disease later in life.

Over 90% of individuals with Down syndrome have three copies of chromosome 21. Usually, two copies are contributed by the egg; however, recent studies indicate that in 23% of the cases studied, the sperm contributed the extra chromosome. The chances of a woman having a child with Down syndrome increase rapidly with age. In women age 20–30, 1 in 1,400 births have Down syndrome, and in women 30–35, about 1 in 750 births have Down syndrome. It is thought that the longer the oocytes are dormant in the ovaries, the greater the chances of a nondisjunction event.

Although an older woman is more likely to have a Down syndrome child, most babies with Down syndrome are born to women younger than age 40 because this is the age group having the most babies. A karyotype of the individual's chromosomes can detect a Down syndrome child. However, young women are not routinely encouraged to undergo the procedures necessary to get a sample of fetal cells because the risk of complications is greater than the risk of having a Down syndrome child. Fortunately, a test based on substances in maternal blood can help identify fetuses who may need to be karyotyped.

FIGURE 8.18 Down syndrome is due to three copies of chromosome 21, shown in circle.

TABLE 8.18	Aneuploidy in Humans	
Chromosomes	**Syndrome**	**Frequency**
Autosomes		
Trisomy 21	Down	1/700
Trisomy 13[†]	Patau	1/5,000
Trisomy 18[†]	Edwards	1/10,000
Sex chromosomes, females		
XO, monosomy	Turner	1/5,000
XXX, trisomy[††]		1/700
Sex chromosomes, males		
XYY, trisomy	Normal	1/10,000
XXY, trisomy[††]	Klinefelter	1/500

[†]Structural abnormalities usually result in early death.

[††]A greater number of X chromosomes is possible.

Abnormal Sex Chromosome Inheritance Newborns with an abnormal X chromosome number are more likely to survive than those with an abnormal autosome number, because both males and females have only one functioning X chromosome. Any others become an inactive mass called a Barr body (after Murray Barr, the person who discovered it; see Section 11.7).

Turner syndrome females are born with only a single X chromosome. They tend to be short, with a broad chest and widely spaced nipples. These individuals also have a low posterior hairline and neck webbing. Their ovaries, oviducts, and uterus are very small and underdeveloped. Turner females do not undergo puberty or menstruate, and their breasts do not develop. However, some have given birth following in vitro fertilization using donor eggs. They usually are of normal intelligence and can lead fairly normal lives if they receive hormone supplements.

About 1 in every 700 females has an extra X chromosome, but they lack symptoms, showing the protective effect of X-inactivation—all but one of the X chromosomes is inactivated.

A male with **Klinefelter syndrome** has two or more X chromosomes in addition to a Y chromosome. The extra X chromosomes become inactivated. In Klinefelter males, the testes and prostate gland are underdeveloped, and facial hair is lacking. There may be some breast development. Affected individuals have large hands and feet and very long arms and legs. They are usually slow to learn but not mentally retarded, unless they inherit more than two X chromosomes. No matter how many X chromosomes are present, an individual with a Y chromosome is a male.

Table 8.18 summarizes human syndromes known to result from aneuploidy, some of which lead to early death because of the imbalance of genetic material. Chromosomes undergo changes in chromosome structure that also result in abnormalities among offspring. Some of these are examined in Section 8.19.

> **8.18** *Check Your Progress* **Table 8.18 lists XYY males. Would this chromosome inheritance be due to nondisjunction during meiosis I or meiosis II?**

Changes in chromosome structure occur in humans and lead to various syndromes, many of which are just now being discovered. Various agents in the environment, such as radiation, certain organic chemicals, or even viruses, can cause chromosomes to break. Ordinarily, when breaks occur in chromosomes, the two broken ends reunite and retain the same sequence of genes. Sometimes, however, the broken ends of one or more chromosomes do not rejoin in the same pattern as before, and the result is various types of chromosomal rearrangements.

Changes in chromosome structure include deletions, duplications, inversions, and translocations of chromosome segments (**Fig. 8.19**). A **deletion** occurs when an end of a chromosome breaks off or when two simultaneous breaks lead to the loss of an internal segment. Even when only one member of a pair of chromosomes is affected, a deletion often causes abnormalities.

A **duplication** is the presence of a particular chromosome segment more than once in the same chromosome. An **inversion** has occurred when a segment of a chromosome is turned 180 degrees. This reversed sequence of genes can lead to altered gene activities and to deletions and duplications.

A **translocation** is the movement of a chromosome segment from one chromosome to another, nonhomologous chromosome.

Syndrome Examples

Sometimes changes in chromosome structure can be detected in humans by doing a karyotype. They may also be discovered by studying the inheritance pattern of a disorder in a particular family.

Williams syndrome occurs when chromosome 7 loses a tiny end piece. Children who have this syndrome look like pixies, with turned-up noses, wide mouths, small chins, and large ears. Although their academic skills are poor, they exhibit excellent verbal and musical abilities. The gene that governs the production of the protein elastin is missing, and this affects the health of the cardiovascular system and causes their skin to age prematurely. Such individuals are very friendly but need an ordered life, perhaps because of the loss of a gene for a protein that is normally active in the brain.

Cri du chat (cat's cry) syndrome is seen when chromosome 5 is missing an end piece. The affected individual has a small head, is mentally retarded, and has facial abnormalities. Abnormal development of the glottis and larynx results in the most characteristic symptom—the infant's cry resembles that of a cat.

A person who has both of the chromosomes involved in a translocation has the normal amount of genetic material and is healthy, unless the chromosome exchange breaks an allele into two pieces. The person who inherits only one of the translocated chromosomes will no doubt have only one copy of certain alleles and three copies of certain other alleles. A genetic counselor begins to suspect a translocation has occurred when spontaneous abortions are commonplace, and family members suffer from various syndromes.

In 5% of cases, Down syndrome occurs because of a translocation between chromosomes 21 and 14 in an ancestor. Because of the inheritance of the unusual chromosome plus two copies of chromosome 21 the individual has three copies of certain genes and manifests Down syndrome. This cause of Down syndrome runs in families and is not related to the age of the mother.

Translocations can be responsible for certain types of cancer. In the 1970s, new staining techniques identified that a translocation from a portion of chromosome 22 to chromosome 9 was responsible for chronic myelogenous leukemia. In Burkitt lymphoma, a cancer common in children in equatorial Africa, a large tumor develops from lymph glands in the region of the jaw. This disorder involves a translocation from a portion of chromosome 8 to chromosome 14.

> **8.19 Check Your Progress** A woman with a normal karyotype reproduces with a man who has a translocation between chromosomes 21 and 14. Could their child have a normal karyotype?

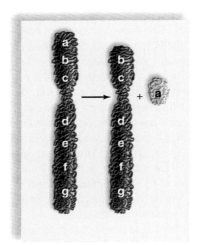

Inversion

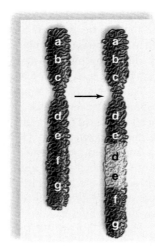

Translocation

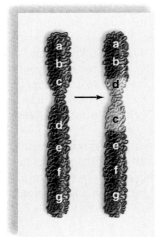

Deletion

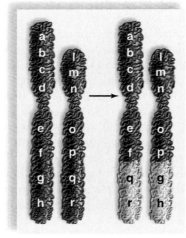

Duplication

FIGURE 8.19 Types of chromosomal mutations.

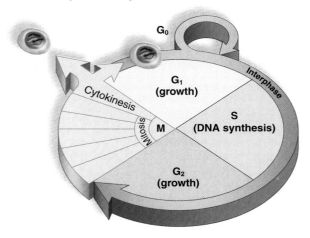

All cells receive DNA from pre-existing cells through the process of cell division. Cell division ensures that DNA is passed on to the next generation of cells and to the next generation of organisms.

The end product of ordinary cell division (i.e., mitosis) is two new cells, each with the same number and kinds of chromosomes as the parent cell. Mitosis is part of the cell cycle, and there are negative consequences if the cell cycle comes out of synchronization. Knowing how the cell cycle is regulated has contributed greatly to our knowledge of cancer and other disorders.

In contrast to mitosis, meiosis is part of the production of gametes, which have half the number of chromosomes as the parent cell. Through the mechanics of meiosis, sexually reproducing species have a greater likelihood of genetic variations among offspring than do asexually reproducing organisms. Some of these may be chromosomal abnormalities.

Genetic variations are essential to the process of evolution, which is discussed in Part III. In the meantime, Chapter 9 reviews the fundamental laws of genetics established by Gregor Mendel. Although Mendel had no knowledge of chromosome behavior, modern students have the advantage of being able to apply their knowledge of meiosis to their understanding of Mendel's laws. Mendel's laws are fundamental to understanding the inheritance of particular alleles on the chromosomes.

The Chapter in Review

Summary

Cancer Is a Genetic Disorder

- A cancer cell is genetically abnormal, loses specialization, and divides over and over again until a tumor forms.

Cell Division Ensures the Passage of Genetic Information

8.1 Cell division is involved in both asexual and sexual reproduction

- During asexual reproduction (for growth and repair of multicellular organisms and reproduction of unicellular ones), daughter cells receive a copy of the parental cell's chromosomes and genes.
- During sexual reproduction, an egg and sperm unite; the offspring receives a different combination of chromosomes than either parent.

8.2 Prokaryotes reproduce asexually

- Unicellular organisms reproduce by binary fission.

Somatic Cells Have a Cell Cycle and Undergo Mitosis and Cytokinesis

8.3 The eukaryotic cell cycle is a set series of events

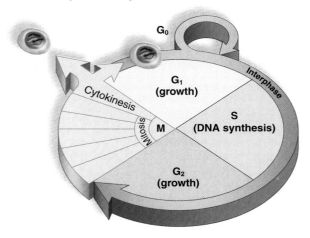

8.4 Eukaryotic chromosomes are visible during cell division

- Somatic cells are diploid (2n), meaning that two of each kind of chromosome are present.
- Gametes are haploid (n), meaning that the sperm and egg contain only one chromosome of each kind.

8.5 Mitosis maintains the chromosome number

- Chromosomes attach to spindle fibers (prophase) and align at metaphase plate (metaphase); sister chromatids separate and become chromosomes (anaphase) so that daughter cells (telophase) have the same number and kinds of chromosomes as the parental cell.

8.6 Cytokinesis divides the cytoplasm

- In animal cells, cytokinesis occurs by furrowing.
- In plant cells, cytokinesis involves the formation of a new plasma membrane and cell wall.

Cancer Is Uncontrolled Cell Division

8.7 Cell cycle control occurs at checkpoints

- If checkpoint G_1 is passed, the cell is ready to divide.
- Checkpoint G_2 ensures that DNA replicated properly.
- Checkpoint M ensures that chromosomes were distributed accurately to daughter cells.

8.8 Signals affect the cell cycle control system

- Internal signal molecules: cyclins combine with kinases that drive the cell cycle.
- External signaling molecules: growth factors and hormones come from outside the cell.

8.9 Cancer cells have abnormal characteristics

- Lack differentiation, have abnormal nuclei, form tumors, undergo metastasis (formation of tumors distant from primary tumor), and promote angiogenesis (formation of new blood vessels).

8.10 Protective behaviors and diet help prevent cancer

- Protective behaviors include avoiding certain substances and following a healthy diet.

Meiosis Produces Cells That Become the Gametes in Animals and Spores in Other Organisms

8.11 Homologous chromosomes separate during meiosis

- As a result of meiosis, one chromosome from each homologous pair is in the haploid egg and sperm (e.g., animals) and haploid spore (e.g., plants).

8.12 Synapsis and crossing-over occur during meiosis I

- During meiosis I, synapsis (homologues pair) and crossing-over (exchange of genetic material) between nonsister chromatids occurs.

8.13 Sexual reproduction increases genetic variation

- Crossing-over recombines genetic information and increases the variability of genetic inheritance on the chromosomes.
- Independent assortment of homologues increases the possible combination of chromosomes in the gametes.

8.14 Meiosis requires two division cycles

- Meiosis I: prophase I—homologues pair and crossing-over occurs; metaphase I—homologue pairs align at metaphase plate independently; anaphase I—homologues separate; telophase—daughter cells are haploid.
- Meiosis II: During stages designated by the roman numeral II, the chromatids of duplicated chromosomes from meiosis I separate, giving a total of four daughter cells for meiosis.

8.15 The life cycle of most multicellular organisms includes both mitosis and meiosis

- In animals, mitosis is involved in growth, while meiosis is part of spermatogenesis and oogenesis; gametes are the only haploid part of the life cycle.

8.16 Meiosis can be compared to mitosis

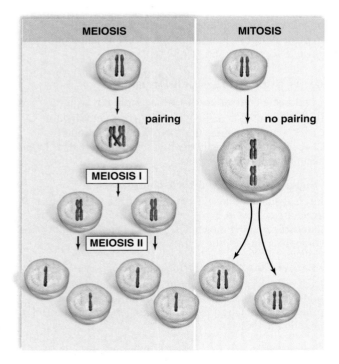

Chromosomal Abnormalities Can Be Inherited

8.17 An abnormal chromosome number is sometimes traceable to nondisjunction

- Nondisjunction occurs when homologues do not separate during meiosis I or when chromatids do not separate during meiosis II; nondisjunction causes monosomy or trisomy.

8.18 Abnormal chromosome numbers cause syndromes

- Down syndrome is an autosomal trisomy.
- Turner syndrome and Klinefelter syndrome result from abnormal sex chromosome inheritance.

8.19 Abnormal chromosome structure also causes syndromes

- Deletion: A segment of a chromosome is missing.
- Duplication: A segment occurs twice on same chromosome.
- Inversion: A segment has turned 180 degrees.
- Translocation: Segments have moved between nonhomologous chromosomes.

Testing Yourself

Cell Division Ensures the Passage of Genetic Information

1. What feature in prokaryotes substitutes for the lack of a spindle in eukaryotes?
   a. centrioles with asters
   b. fission instead of cytokinesis
   c. elongation of plasma membrane
   d. looped DNA
   e. presence of one chromosome
2. Which properly describes a prokaryotic chromosome? A prokaryotic chromosome
   a. is shorter and fatter.
   b. has a single loop of DNA.
   c. never replicates.
   d. contains many histones.
   e. All of these are correct.

Somatic Cells Have a Cell Cycle and Undergo Mitosis and Cytokinesis

3. The two identical halves of a duplicated chromosome are called
   a. chromosome arms.
   b. nucleosomes.
   c. chromatids.
   d. homologues.

For questions 4–7, match the descriptions to the terms in the key.

KEY:
   a. prophase
   b. metaphase
   c. anaphase
   d. telophase

4. The nucleolus disappears, and the nuclear envelope breaks down.
5. The spindle disappears, and the nuclear envelopes form.
6. Sister chromatids separate.
7. Chromosomes are aligned on the spindle equator.
8. Programmed cell death is called
   a. mitosis.
   b. meiosis.
   c. cytokinesis.
   d. apoptosis.

9. Label this diagram of a cell in early prophase of mitosis:

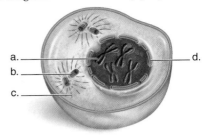

a. _____

b. _____

c. _____

d. _____

Cancer Is Uncontrolled Cell Division

10. Which of the following is not a feature of cancer cells?
    a. exhibit contact inhibition
    b. have enlarged nuclei
    c. stimulate the formation of new blood vessels
    d. are capable of traveling through blood and lymph
11. Which of these is not a behavior that could help prevent cancer?
    a. maintaining a healthy weight
    b. eating more dark-green, leafy vegetables, carrots, and various fruits
    c. not smoking
    d. maintaining estrogen levels through hormone replacement therapy
    e. consuming alcohol only in moderation

Meiosis Produces Cells That Become the Gametes in Animals and Spores in Other Organisms

12. Crossing-over occurs between
    a. sister chromatids of the same chromosomes.
    b. chromatids of nonhomologous chromosomes.
    c. nonsister chromatids of a homologous pair.
    d. two daughter nuclei.
    e. Both b and c are correct.
13. At the equator during metaphase I of meiosis, there are
    a. single chromosomes.
    b. unpaired duplicated chromosomes.
    c. homologous pairs.
    d. always 23 chromosomes.
14. During which phase of meiosis do homologous chromosomes separate?
    a. prophase II   d. anaphase I
    b. telophase I   e. anaphase II
    c. metaphase I
15. **THINKING CONCEPTUALLY** Use the events of meiosis to briefly explain why you and a sibling with the same parents have different characteristics.

Chromosomal Abnormalities Can Be Inherited

16. An individual can have too many or too few chromosomes as a result of
    a. nondisjunction.   d. amniocentesis.
    b. Barr bodies.   e. monosomy.
    c. trisomy.
17. Which of the following is the cause of Williams syndrome?
    a. inheritance of an extra chromosome 21
    b. deletion in chromosome 7
    c. chromosomal duplication
    d. translocation between chromosomes 2 and 20

Understanding the Terms

adult stem cell 136	kinase 142
aneuploid 152	kinetochore 138
angiogenesis 143	Klinefelter syndrome 153
apoptosis 141	life cycle 150
asexual reproduction 134	meiosis 145
aster 138	meiosis I 145
autosome 145	meiosis II 145
binary fission 135	metastasis 143
carcinogenesis 143	mitosis 136
cell cycle 136	mitotic spindle 138
cell division 134	monosomy 152
cell plate 140	mutation 143
centromere 137	nondisjunction 152
chromatin 137	nucleoid 135
contact inhibition 142	oogenesis 150
control system 142	p53 141
crossing-over 146	parent cell 137
cyclin 142	polyploid 152
cytokinesis 136	sex chromosome 145
daughter cell 137	sexual reproduction 134
deletion 154	signaling molecule 142
diploid (2n) number 137	sister chromatid 136
duplication 154	somatic cell 134
fertilization 147	spermatogenesis 150
gamete 137	spindle fiber 138
germ cell 134	synapsis 146
haploid (n) number 137	syndrome 153
homologous chromosome 145	telomere 142
homologue 145	tetrad 146
independent assortment 146	translocation 154
interkinesis 148	trisomy 152
inversion 154	tumor 143
karyotype 145	Turner syndrome 153
	zygote 147

Match the terms to these definitions:

a. _____ One of two genetically identical chromosome units that result from DNA replication.

b. _____ Central microtubule organizing center of cells, consisting of granular material. In animal cells, it contains two centrioles.

c. _____ Production of sperm in males by the process of meiosis and maturation.

d. _____ Programmed cell death that is carried out by enzymes routinely present in the cell.

Thinking Scientifically

1. Genetic testing shows that Mary has only 46 chromosomes, but the members of one homologous pair both came from her father. In which parent did nondisjunction occur? Explain.
2. Criticize the hypothesis that it would be possible to clone an individual by using an egg and sperm with the exact genetic makeup as those that produced the individual.

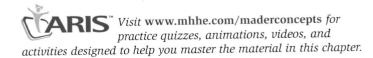

Visit **www.mhhe.com/maderconcepts** *for practice quizzes, animations, videos, and activities designed to help you master the material in this chapter.*

9

Patterns of Genetic Inheritance

LEARNING OUTCOMES

After studying this chapter, you should be able to accomplish the following outcomes.

Inbreeding Leads to Disorders

1 Use results of inbreeding to support the concept that traits are inherited.

Gregor Mendel Deduced Laws of Inheritance

2 Describe the state of genetics before Mendel and Mendel's experimental design.

Single-Trait Crosses Reveal Units of Inheritance and the Law of Segregation

3 Use the concept of alleles and the law of segregation to predict the results of single-trait crosses.

Two-Trait Crosses Support the Law of Independent Assortment

4 Use the law of independent assortment to predict the results of two-trait crosses.
5 Show how two laws of probability relate to the Punnett square.

Mendel's Laws Apply to Humans

6 Distinguish between autosomal and sex chromosomes.
7 Use a pedigree to determine whether a genetic disorder is autosomal dominant or autosomal recessive.
8 Explain the methods for testing fetal cells or an embryo for genetic disorders.

Complex Inheritance Patterns Extend the Range of Mendelian Analysis

9 Explain the inheritance pattern for and solve genetic problems pertaining to incomplete dominance, multiple alleles, polygenic inheritance, and pleiotropy.

Chromosomes Are the Carriers of Genes

10 Explain why more males than females have X-linked conditions, and solve X-linked genetics problems.
11 Explain why gene linkage is expected and how crossing-over can be used to map the chromosomes.

You buy a purebred puppy and all seems well, but over time it begins to develop debilitating physical ailments. Golden retrievers were bred to be beautiful, sturdy, friendly dogs with a cream- to golden-colored coat. They are also prone to hip dysplasia, a painful condition caused by malformed hip joints. In German shepherds, the body is longer than it is tall and has an outline of smooth curves rather than angles. German shepherds are also prone to hip dysplasia, as well as to gastric torsion and bloat. Swallowed air causes the stomach to swell and twist, leading to low blood pressure, shock, and damage to internal organs. English bulldogs and pugs look cute, but they may develop breathing and digestive problems due to a protruding lower jaw and a shortened upper jaw.

English bulldog

Inbreeding Leads to Disorders

What's the problem? The many different types of dogs available today are the result of inbreeding—the mating of related individuals for the purpose of achieving desired traits. The concept of "like begets like" was overdone, and along with the desirable traits came physical deformities. When dogs or humans reproduce, they pass on their genes—units of heredity that determine their characteristics, be they good or ill. Closely related individuals are more likely to pass along the same faulty genes, leading to deformities.

However, hip dysplasia is not a birth defect. Dogs are born with what appear to be normal hips and then develop the disease later. In addition, some dogs with the genetic tendency do not develop the condition, and the degree of hip dysplasia in others can vary. It appears that multiple genetic factors, plus environmental factors, are involved in determining the degree of hip dysplasia. This is also true of many human disorders—both genes and the environment seem to play a role.

Researchers have identified the possible contributing factors for hip dysplasia. A test group of Labrador retrievers, who were fed 25% less than normal, showed less hip dysplasia than a control group allowed to eat as much as they wanted. The researchers concluded that rapid weight gain can contribute to the development of hip dysplasia.

Collie

While the situation regarding genetic disorders in dogs is complicated, modern genetics can help. Breeders should keep careful records of matings and the results of these matings. From these records, they might be able to determine the pattern of a disorder's inheritance and which dogs should not be used for breeding. It is also possible today to screen for many genetic diseases, and dogs with the same but hidden (recessive) genetic faults need not be mated to each other. The breeding of dogs should be modernized to avoid producing dogs with physical deformities.

Modernization has led to the successful use of gene therapy in collies and Briard sheepdogs to cure blindness, another condition resulting from continual inbreeding. Researchers injected a harmless virus carrying the corrective gene beneath the retina and waited several months. When tested, the dogs could see through the eye that was treated, but not through the eye that was not treated. The researchers are hopeful that their work will be a step toward curing blindness in humans one day.

This chapter begins our study of inheritance by taking a look at the work of Gregor Mendel, who is often called the father of genetics.

German shepherd

German shepherd with hip dysplasia

The experiments performed by Gregor Mendel with garden peas refuted the blending concept of inheritance prevalent at the time. In contrast to a blending hypothesis, Mendel's work showed that inheritance is particulate and, therefore, that traits such as tall or short height always recur.

9.1 A blending model of inheritance existed prior to Mendel

Like begets like—zebras always produce zebras, never camels; pumpkins always produce seeds for pumpkins, never watermelons. It is apparent to anyone who observes such phenomena that parents pass hereditary information to their offspring. However, an offspring can be markedly different from either parent. For example, black-coated mice occasionally produce white-coated mice. The science of genetics provides explanations about not only the stability of inheritance, but also the variations that are observed between generations and organisms.

Virtually every culture in history has attempted to explain inheritance patterns. An understanding of these patterns has always been important to agriculture and animal husbandry, the science of breeding animals. But it was not until the 1860s that the Austrian monk Gregor Mendel developed the fundamental laws of heredity after performing a series of ingenious experiments with pea plants (**Fig. 9.1**). Previously, Mendel had studied science and mathematics at the University of Vienna, and at the time of his genetic research, he was a substitute natural science teacher at a local high school.

Various hypotheses about heredity had been proposed before Mendel began his experiments. In particular, investigators had been trying to support a blending concept of inheritance. Thus, most plant and animal breeders acknowledged that both sexes contribute equally to a new individual, and they felt that parents of contrasting appearance always produce offspring of intermediate appearance. According to this concept, a cross between plants with red flowers and plants with white flowers would yield only plants with pink flowers. When red and white flowers reappeared in future generations, the breeders mistakenly attrib-

uted this to an instability in the genetic material.

The blending model of inheritance had offered little help to Charles Darwin, the father of evolution. If populations contained only intermediate individuals and normally lacked variations, how could diverse forms evolve? However, the theory of inheritance eventually proposed by Mendel did account for the presence of variations among the members of a population, generation after generation.

Although Darwin was a contemporary of Mendel, Darwin never learned of Mendel's work because it went unrecognized until 1900. Therefore, Darwin was never able to make use of Mendel's research to support his theory of evolution, and his treatise on natural selection lacked a strong genetic basis.

FIGURE 9.1 Gregor Mendel examining a pea plant.

In Section 9.2, we begin to examine the data that allowed Mendel to arrive at a particulate, instead of blending, concept of inheritance.

> **9.1** *Check Your Progress* Inbreeding in dogs supports the genetic basis of inheritance and also the concept of evolution. Explain.

9.2 Mendel designed his experiments well

Mendel had a background suitable to his task. Aside from theoretical knowledge, Mendel knew how to cultivate plants. Most likely, his knowledge of mathematics prompted Mendel to use a statistical basis for his breeding experiments. He prepared for his experiments carefully and conducted preliminary studies with various animals and plants. He then chose to work with the garden pea, *Pisum sativum*.

The garden pea was a good choice. The plants are easy to cultivate, have a short generation time, and produce many offspring. A pea plant normally self-pollinates because the reproductive organs in the flower are completely enclosed by petals (**Fig. 9.2A**). ❶ As in all flowering plants, the reproductive organs in peas are the stamen and the carpel. A stamen produces sperm-bearing pollen in the anther, and the carpel produces egg-bearing ovules in the ovary. When Mendel wanted the plants to self-fertilize, he covered the flowers with a bag to ensure that only the pollen of that flower would reach the carpel of that flower. ❷ Even though pea plants

normally self-fertilize, they can be cross-pollinated by an experimenter who manually transfers pollen from an anther to the carpel. First, Mendel prevented self-fertilization by cutting away the anthers before they produced any pollen. ❸ Then he dusted that flower's carpel with pollen from another plant. ❹ Afterwards, the carpel developed into a pod containing peas. In the cross illustrated here, pollen from a plant that normally produces yellow peas was used to fertilize the eggs of a plant that normally produces green peas. These plants produced only yellow peas. To see the results of other crosses, it was necessary for Mendel to plant the peas and examine the next generation of plants.

Many varieties of pea plants were available, and Mendel chose 22 of them for his experiments. When these varieties *self-fertilized*, they were true-breeding—meaning that the offspring were like the parent plants and like each other. In contrast to his predecessors, Mendel studied the inheritance of relatively simple, clear-cut, and easily detected traits, such as seed shape, seed color, and flower color, and he observed no intermediate characteristics among the offspring (**Fig. 9.2B**).

As Mendel followed the inheritance of individual traits, he kept careful records, and he used his understanding of the mathematical laws of probability to interpret his results and to arrive at a theory that has been supported by innumerable experiments since. It is called a particulate theory of inheritance because the theory is based on the existence of minute particles, or he-reditary units, that we now call genes. Inheritance involves the reshuffling of the same genes from generation to generation.

Mendel clearly stated his conclusions in the form of certain laws. The first law is considered in Section 9.3.

> **9.2** *Check Your Progress* Why were true-breeding pea plants a better choice than true-breeding dogs for Mendel's studies?

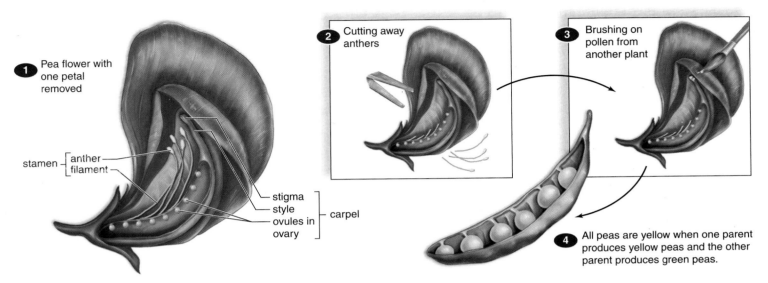

FIGURE 9.2A Garden pea anatomy and the cross-pollination procedure Mendel used.

Trait	Characteristics			F₂ Results*	
	Dominant		**Recessive**	**Dominant**	**Recessive**
Stem length	Tall		Short	 787	277
Pod shape	Inflated	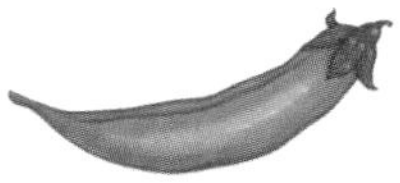	Constricted	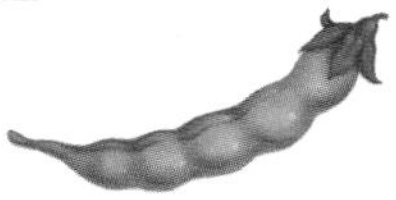 882	299
Seed shape	Round	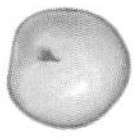	Wrinkled	 5,474	1,850
Seed color	Yellow		Green	 6,022	2,001
Flower color	Purple		White	 705	224
Pod color	Green	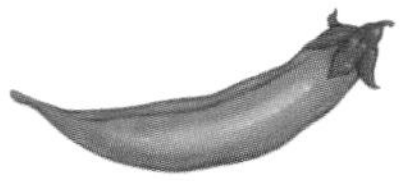	Yellow	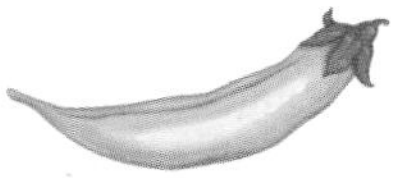 428	152

*All of these results give an approximate 3:1 ratio of dominant to recessive. For example, $\frac{787}{277} = \frac{3}{1}$.

FIGURE 9.2B Garden pea traits and crosses studied by Mendel. (See Section 9.3.)

After doing crosses that involved only one trait, such as height, Mendel proposed his law of segregation. The individual has two factors, called alleles today, for each trait, but the gametes have only one allele for each trait. This segregation process allows a recessive allele to be passed on to an offspring independent of the dominant allele.

9.3 Mendel's law of segregation describes how gametes pass on traits

After ensuring that his pea plants were true-breeding—for example, that his tall plants always had tall offspring and his short plants always had short offspring—Mendel was ready to perform a cross-fertilization experiment between two strains. For these initial experiments, Mendel chose varieties that differed in only one trait. If the blending theory of inheritance were correct, the cross should yield offspring with an intermediate appearance compared to the parents. For example, the offspring of a cross between a tall plant and a short plant should be intermediate in height.

Mendel called the original parents the *P generation* and the first generation the *F_1*, or first filial, *generation* purely for experimental purposes. He performed reciprocal crosses: First, he dusted the pollen of tall plants onto the stigmas of short plants, and then he dusted the pollen of short plants onto the stigmas of tall plants. In both cases, all F_1 offspring resembled the tall parent.

Certainly, these results were contrary to those predicted by the blending theory of inheritance. Rather than being intermediate, the F_1 plants were tall and resembled only one parent. Did these results mean that the other characteristic (i.e., shortness) had disappeared permanently? Apparently not, because when Mendel allowed the F_1 plants to self-pollinate, ¾ of the *F_2 generation* were tall, and ¼ were short, a 3:1 ratio (**Fig. 9.3**). Therefore, the F_1 plants were able to pass on a factor for shortness—it didn't just disappear. Perhaps the F_1 plants were tall because tallness was dominant to shortness?

Mendel counted many offspring. For this particular cross, he counted a total of 1,064 offspring, of which 787 were tall and 277 were short. In all the crosses that he performed, he found a 3:1 ratio in the F_2 generation. The characteristic that had disappeared in the F_1 generation reappeared in ¼ of the F_2 offspring (see Fig. 9.2B).

His mathematical approach led Mendel to interpret his results differently from previous breeders. He knew that the same ratio was obtained among the F_2 generation time and time again for the same type of cross involving the traits that he was studying. Eventually, Mendel arrived at this explanation: A 3:1 ratio among the F_2 offspring was possible if the F_1 parents contained two separate copies of each hereditary factor, one of these being dominant and the other recessive. The factors separated when the gametes were formed, and each gamete carried only one copy of each factor; random fusion of all possible gametes occurred upon fertilization. Only in this way would shortness reoccur in the F_2 generation.

After doing many F_1 crosses, called **monohybrid crosses** because the plants are hybrids in one way only, Mendel arrived at the first of his laws of inheritance—the law of segregation, which is a cornerstone of his particulate theory of inheritance. In Section 9.4, we restate the law of segregation using modern terminology.

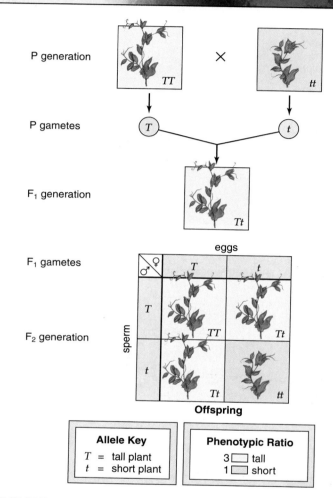

FIGURE 9.3 Monohybrid cross done by Mendel.

Allele Key
T = tall plant
t = short plant

Phenotypic Ratio
3 ☐ tall
1 ☐ short

The **law of segregation** states the following:

- Each individual has two factors for each trait.

- The factors segregate (separate) during the formation of the gametes.

- Each gamete contains only one factor from each pair of factors.

- Fertilization gives each new individual two factors for each trait.

▶ **9.3 Check Your Progress** Based on Mendel's study, explain why only some offspring of normal golden retrievers have hip dysplasia.

Mendel said that organisms received "factors" from their parents, but today we use the term "genes." Traits are controlled by **alleles**, alternative forms of a gene. The alleles occur on homologous chromosomes at a particular **gene locus** (**Fig. 9.4**). The **dominant allele** is so named because of its ability to mask the expression of the other allele, called the **recessive allele**. (Therefore, dominant does not mean the normal or most frequent condition.) The dominant allele is identified by a capital letter, and the recessive allele by the same but lowercase letter. Usually, the letter chosen has some connection to the trait itself. For example, when considering stem length in peas, the allele for tallness is *T*, and the allele for shortness is *t*.

As you learned in Chapter 8, meiosis is the type of cell division that reduces the chromosome number. During meiosis I, homologous chromosomes each having sister chromatids separate. During meiosis II, chromatids separate. Therefore, the process of meiosis explains Mendel's law of segregation and why there is only one allele for each trait in a gamete.

In Mendel's cross (see Fig. 9.3), the original parents (P generation) were true-breeding; therefore, the tall plants had two copies of the same allele for tallness (*TT*), and the short plants had two copies of the same allele for shortness (*tt*). When an organism has two identical alleles, as these had, we say it is **homozygous**. Because the parents were homozygous, all gametes produced by the tall plant contained the allele for tallness (*T*), and all gametes produced by the short plant contained the allele for shortness (*t*).

After cross-fertilization, all the individuals in the resulting F₁ generation had one allele for tallness and one for shortness (*Tt*). When an organism has two different alleles at a gene locus, we say that it is **heterozygous**. Although the plants of the F₁ generation had one of each type of allele, they were all tall. The allele that is expressed in a heterozygous individual is the domi-

nant allele. The allele that is not expressed in a heterozygote is the recessive allele.

You can see that two organisms with different allelic combinations for a trait can have the same outward appearance. (*TT* and *Tt* pea plants are both tall.) For this reason, it is necessary to distinguish between the alleles present in an organism and the appearance of that organism.

The word **genotype** refers to the alleles an individual receives at fertilization. Genotype may be indicated by letters or by short, descriptive phrases. Genotype *TT* is called homozygous dominant, genotype *tt* is called homozygous recessive, and genotype *Tt* is called heterozygous.

The word **phenotype** refers to the physical appearance of the individual. The homozygous dominant (*TT*) individual and the heterozygous (*Tt*) individual both show the dominant phenotype and are tall, while the homozygous recessive (*tt*) individual shows the recessive phenotype and is short. **Table 9.4** compares genotype with phenotype.

Continuing with the discussion of Mendel's cross (see Fig. 9.3), the F₁ plants produce gametes in which 50% have the dominant allele *T* and 50% have the recessive allele *t*. During the process of fertilization, we assume that all types of sperm (i.e., *T* or *t*) have an equal chance to fertilize all type of eggs (i.e., *T* or *t*). When this occurs, such a monohybrid cross will always produce a 3:1 (dominant to recessive) ratio among the offspring. Figure 9.2B gives Mendel's results for several monohybrid crosses, and you can see that the results were always close to 3:1.

The second of Mendel's laws will be examined in Section 9.5.

TABLE 9.4	Genotype Versus Phenotype	
Genotype	**Genotype**	**Phenotype**
TT	Homozygous dominant	Tall plant
Tt	Heterozygous	Tall plant
tt	Homozygous recessive	Short plant

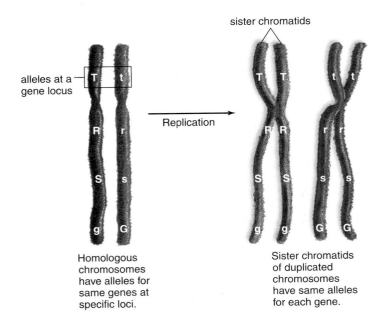

FIGURE 9.4 Occurrence of alleles on homologous chromosomes.

alleles at a gene locus

sister chromatids

Replication

Homologous chromosomes have alleles for same genes at specific loci.

Sister chromatids of duplicated chromosomes have same alleles for each gene.

9.4 Check Your Progress

1. For each of the following genotypes, list all possible gametes, noting the proportion of each for the individual.
   a. WW b. Ww c. Tt d. TT

2. In rabbits, if *B* = dominant black allele and *b* = recessive white allele, which of these genotypes (*Bb, BB, bb*) could a white rabbit have?

3. If a heterozygous rabbit reproduces with one of its own kind, what phenotypic ratio do you expect among the offspring? If there are 120 rabbits, how many are expected to be white?

After doing crosses that involved two traits, such as height and color of peas, Mendel proposed his law of independent assortment. Only one allele for each trait can occur in a gamete, but any allele for one trait can occur with any allele for another trait. Therefore, all possible phenotypes can occur among offspring.

9.5 Mendel's law of independent assortment describes inheritance of multiple traits

Mendel performed a second series of crosses in which true-breeding plants differed in two traits. For example, he crossed tall plants having green pods with short plants having yellow pods (**Fig. 9.5**). The F_1 plants showed both dominant characteristics. As before, Mendel then allowed the F_1 plants to self-pollinate. These F_1 crosses are called **dihybrid crosses** because the plants are hybrids in two ways. Mendel reasoned that two possible results could occur in the F_2 generation:

1. If the dominant factors (*TG*) always segregate into the F_1 gametes together, and the recessive factors (*tg*) always stay together, there would be two phenotypes among the F_2 plants—tall plants with green pods and short plants with yellow pods.

2. If the four factors segregate into the F_1 gametes independently, there would be four phenotypes among the F_2 plants—tall plants with green pods, tall plants with yellow pods, short plants with green pods, and short plants with yellow pods.

Figure 9.5 shows that Mendel observed four phenotypes among the F_2 plants, supporting the second hypothesis. Therefore, Mendel formulated his second law of heredity—the law of independent assortment.

The **law of independent assortment** states the following:

- Each pair of factors separates (assorts) independently (without regard to how the others separate).

- All possible combinations of factors can occur in the gametes.

Again, we know that the process of meiosis explains why the F_1 plants produced every possible type of gamete, and therefore four phenotypes appear among the F_2 generation of plants. As was explained in Figure 8.13, there are no rules regarding alignment of homologues at the equator—either homologue can face either spindle pole. Because of this, the daughter cells from meiosis I (and also meiosis II) have all possible combinations of alleles. The possible gametes are the two dominants (such as *TG*), the two recessives (such as *tg*), and the ones that have a dominant and a recessive (such as *Tg* and *tG*). When all possible sperm have an opportunity to fertilize all possible eggs, the phenotypic results of a dihybrid cross are always 9:3:3:1.

> **9.5 Check Your Progress** What are the four possible genotypes of a tall pea plant with green pods?

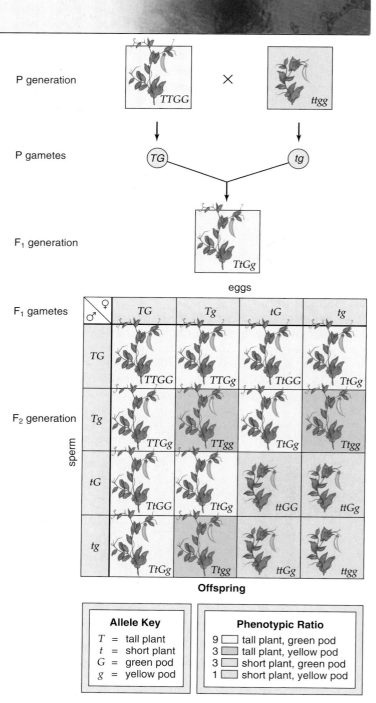

FIGURE 9.5 Dihybrid cross done by Mendel.

Allele Key

T	= tall plant
t	= short plant
G	= green pod
g	= yellow pod

Phenotypic Ratio

9	tall plant, green pod
3	tall plant, yellow pod
3	short plant, green pod
1	short plant, yellow pod

9.6 Mendel's results are consistent with the laws of probability

The diagram we have been using to calculate the results of a cross is called a **Punnett square.** The Punnett square allows us to easily calculate the chances, or the probability, of genotypes and phenotypes among the offspring. Like flipping a coin, an offspring of the cross illustrated in the Punnett square in **Figure 9.6** has a 50% (or ½) chance of receiving an E for unattached earlobe or an e for attached earlobe from each parent:

The chance of E = ½
The chance of e = ½

How likely is it that an offspring will inherit a specific set of two alleles, one from each parent? The product rule of probability tells us that we have to multiply the chances of independent events to get the answer:

1. The chance of *EE* $= ½ × ½ = ¼$
2. The chance of *Ee* $= ½ × ½ = ¼$
3. The chance of *eE* $= ½ × ½ = ¼$
4. The chance of *ee* $= ½ × ½ = ¼$

The Punnett square does this for us because we can easily see that each of these is ¼ of the total number of squares. How do we get the phenotypic results? The sum rule of probability tells us that when the same event can occur in more than one way, we can add the results. Because 1, 2, and 3 all result in unattached earlobes, we add them up to know that the chances of unattached earlobes is ¾, or 75%. The chances of attached earlobes is ¼, or 25%. The Punnett square doesn't do this for us—we have to add the results ourselves.

Another useful concept is the statement that "chance has no memory." This concept helps us know that each child has the same chances. So, if a couple has four children, each child has a 25% chance of having attached earlobes. This may not be significant if we are considering earlobes. It does become significant, however, if we are considering a recessive genetic disorder, such as cystic fibrosis, a debilitating respiratory illness. If a heterozygous couple has four children, each child has a 25% chance of inheriting two recessive alleles, and all four children could have cystic fibrosis.

We can use the product rule and the sum rule of probability to predict the results of a dihybrid cross, such as the one shown in Figure 9.5. The Punnett square carries out the multiplication for us, and we add the results to find that the phenotypic ratio is 9:3:3:1. We expect these same results for each and every dihybrid cross. Therefore, it is not necessary to do a Punnett square over and over again for either a monohybrid or a dihybrid cross. Instead, we can simply remember the probable results of 3:1 and 9:3:3:1. But we have to remember that the 9 represents the two dominant phenotypes together, the 3's are a dominant phenotype with a recessive, and the 1 stands for the two recessive phenotypes together. This tells you the probable phenotypic ratio among the offspring, but not the chances for each possible phenotype. Because the Punnett square has 16 squares, the chances are $^9/_{16}$ for the two dominants together,

Parents

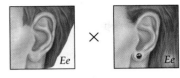

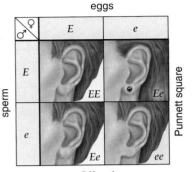
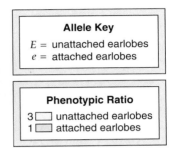

Allele Key
E = unattached earlobes
e = attached earlobes

Phenotypic Ratio
3 ▢ unattached earlobes
1 ▢ attached earlobes

FIGURE 9.6 Use of Punnett square to calculate probable results.

$^3/_{16}$ for the dominants with each recessive, and $^1/_{16}$ for the two recessives together.

Mendel counted the results of many similar crosses to get the probable results, and in the laboratory, we too have to count the results of many individual crosses to get the probable results for a monohybrid or a dihybrid cross. Why? Consider that each time you toss a coin, you have a 50% chance of getting heads or tails. If you tossed the coin only a couple of times, you might very well have heads or tails both times. However, if you toss the coin many times, you are more likely to finally achieve 50% heads and 50% tails.

Section 9.7 illustrates that the results of testcrosses are consistent with Mendel's laws and have the added advantage of determining the genotype of the heterozygote.

9.6 Check Your Progress

1. In humans, freckles is dominant over no freckles. A man with freckles reproduces with a woman with freckles, but the children have no freckles. What chance did each child have for freckles?

2. In fruit flies, long wings (*L*) is dominant over vestigial (short) wings (*l*), and gray body (*G*) is dominant over black body (*g*). Without doing a Punnett square, what phenotypic ratio is probable among the offspring of a dihybrid cross? What are the chances of an offspring with short wings and a black body?

3. In horses, *B* = black coat, *b* = brown coat, *T* = trotter, and *t* = pacer. A black pacer mated to a brown trotter produces a black trotter. Give all possible genotypes for this offspring.

One-trait Testcross

To confirm that the F₁ of his one-trait crosses were heterozygous, Mendel crossed his F₁ generation plants with true-breeding, short (homozygous recessive) plants. Mendel performed these so-called **testcrosses** because they allowed him to support the law of segregation. For the cross in **Figure 9.7A**, he reasoned that half the offspring should be tall and half should be short, producing a 1:1 phenotypic ratio. His results supported the hypothesis that alleles segregate when gametes are formed. In Figure 9.7A, the homozygous recessive parent can produce only one type of gamete—*t*—and so the Punnett square has only one column. The use of one column signifies that all the gametes carry a *t*. *When a heterozygous individual is crossed with one that is homozygous recessive, the probable results are always a 1:1 phenotypic ratio.*

Today, a one-trait testcross is used to determine if an individual with the dominant phenotype is homozygous dominant (e.g., *TT*) or heterozygous (e.g., *Tt*). Since both of these genotypes produce the dominant phenotype, it is not possible to determine the genotype by observation. **Figure 9.7B** shows that if the individual is homozygous dominant, all the offspring will be tall. Each parent has only one type of gamete, and therefore a Punnett square is not required to determine the results.

Two-trait Testcross

When doing a two-trait testcross, an individual with the dominant phenotype is crossed with one having the recessive phenotype. Suppose you are working with fruit flies in which:

L = long wings	*G* = gray bodies
l = vestigial (short) wings	*g* = black bodies

 You wouldn't know by examination whether the fly on the left was homozygous or heterozygous for wing and body color. In order to find out the genotype of the test fly, you cross it with the one on the right. You know by examination that this vestigial-winged and black-bodied fly is homozygous recessive for both traits.

If the test fly is homozygous dominant for both traits with the genotype *LLGG* it will form only one gamete: *LG*. Therefore, all the offspring from the proposed cross will have long wings and a gray body.

However, if the test fly is heterozygous for both traits with the genotype *LlGg*, it will form four different types of gametes:

Gametes: *LG*   *Lg*   *lG*   *lg*

and have four different offspring:

LlGg	*Llgg*	*llGg*	*llgg*

The presence of the offspring with vestigial wings and a black body shows that the test fly is heterozygous for both traits and has the genotype *LlGg*. Otherwise, it could not have this offspring. In general, you will want to remember that *when*

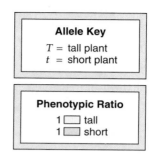

FIGURE 9.7A One-trait testcross, when the individual with the dominant phenotype is heterozygous.

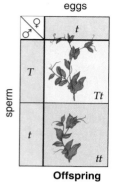

Allele Key

T = tall plant
t = short plant

Phenotypic Ratio

1 ☐ tall
1 ☐ short

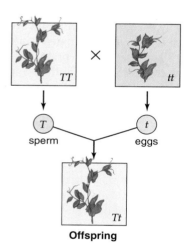
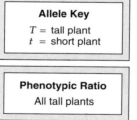

FIGURE 9.7B One-trait testcross, when the individual with the dominant phenotype is homozygous.

Allele Key

T = tall plant
t = short plant

Phenotypic Ratio

All tall plants

an individual heterozygous for two traits is crossed with one that is recessive for the traits, the offspring have a 1:1:1:1 phenotypic ratio.

We will observe in the next part of the chapter that Mendel's laws also apply to humans.

9.7 Check Your Progress

1. A heterozygous fruit fly (*LlGg*) is crossed with a homozygous recessive (*llgg*). What are the chances of offspring with long wings and a black body?

2. An individual with long fingers (*s*) has a father with short fingers (*S*). What is the genotype of the father?

3. In horses, trotter (*T*) is dominant over pacer (*t*). A trotter is mated to a pacer, and the offspring is a pacer. Give the genotype of all the horses.

9.8 Pedigrees can reveal the patterns of inheritance

Many human disorders are genetic in origin. Genetic disorders are medical conditions caused by alleles inherited from parents. Some of these conditions are due to the inheritance of abnormal recessive or dominant alleles on **autosomal chromosomes**, which includes all the chromosomes except the sex chromosomes.

When a genetic disorder is autosomal recessive, only individuals with the alleles *aa* have the disorder. When a genetic disorder is autosomal dominant, an individual with the alleles *AA* or *Aa* has the disorder. Genetic counselors often construct **pedigrees** to determine whether a condition is recessive or dominant.

In these patterns, males are designated by squares and females by circles. Shaded circles and squares are affected individuals. A line between a square and a circle represents a union. A vertical line going downward leads, in these patterns, to a single child. (If there are more children, they are placed off a horizontal line.) A pedigree shows the pattern of inheritance for a particular condition. Consider these two possible patterns of inheritance:

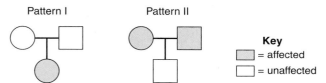

Pattern I   Pattern II

Key
▨ = affected
☐ = unaffected

Which pattern of inheritance (I or II) do you think pertains to an autosomal dominant characteristic, and which pertains to an autosomal recessive characteristic?

In pattern I, the child is affected, but neither parent is; this can happen if the condition is recessive and the parents are *Aa*. Notice that the parents are **carriers** because they appear normal but are capable of having a child with the genetic disorder. In pattern II, the child is unaffected, but the parents are affected. This can happen if the condition is dominant and the parents are *Aa*.

Figure 9.8A shows other ways to recognize an autosomal recessive pattern of inheritance, and **Figure 9.8B** shows other ways to recognize an autosomal dominant pattern of inheritance. In these pedigrees, generations are indicated by Roman numerals placed on the left side. Notice in the third generation of Figure 9.8A that two closely related individuals have produced three children, two of whom have the affected phenotype. This illustrates that reproduction between closely related persons increases the chances of children inheriting two copies of a potentially harmful recessive allele.

The inheritance pattern of alleles on the X chromosome follows different rules than those on the autosomal chromosomes, and you will learn to recognize this pattern also when we discuss it in Section 9.17. In the meantime, let's move on to Section 9.9, which discusses a few autosomal recessive disorders in humans.

> **9.8 Check Your Progress** How does Figure 9.8A demonstrate that dog breeders should keep careful pedigree records?

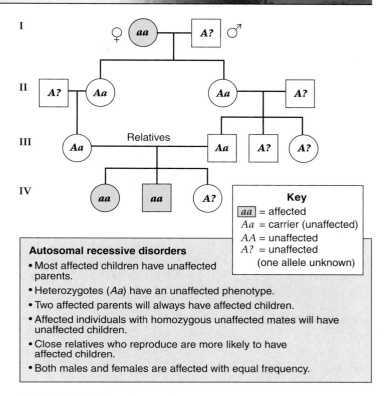

Autosomal recessive disorders
- Most affected children have unaffected parents.
- Heterozygotes (*Aa*) have an unaffected phenotype.
- Two affected parents will always have affected children.
- Affected individuals with homozygous unaffected mates will have unaffected children.
- Close relatives who reproduce are more likely to have affected children.
- Both males and females are affected with equal frequency.

Key
▨ *aa* = affected
Aa = carrier (unaffected)
AA = unaffected
A? = unaffected (one allele unknown)

FIGURE 9.8A Autosomal recessive pedigree.

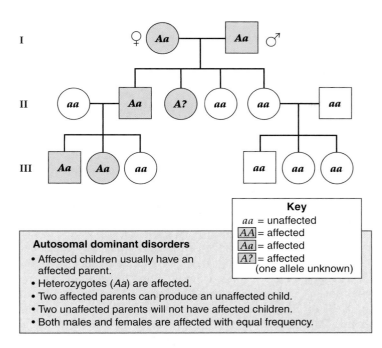

Autosomal dominant disorders
- Affected children usually have an affected parent.
- Heterozygotes (*Aa*) are affected.
- Two affected parents can produce an unaffected child.
- Two unaffected parents will not have affected children.
- Both males and females are affected with equal frequency.

Key
aa = unaffected
▨ *AA* = affected
▨ *Aa* = affected
▨ *A?* = affected (one allele unknown)

FIGURE 9.8B Autosomal dominant pedigree.

In humans, a number of genetic disorders are controlled by a single pair of alleles. Four of the best-known autosomal recessive disorders are Tay-Sachs disease, cystic fibrosis, phenylketonuria, and sickle-cell disease. Individuals can be carriers for these diseases.

Tay-Sachs Disease In a baby with Tay-Sachs disease, development begins to slow down between four and eight months of age, and neurological impairment and psychomotor difficulties then become apparent. The child gradually becomes blind and helpless, develops uncontrollable seizures, and eventually becomes paralyzed prior to dying. Tay-Sachs disease results from a lack of the enzyme hexosaminidase A (Hex A) and the subsequent storage of its substrate, a glycosphingolipid, in lysosomes. As the glycosphingolipid builds up in the lysosomes, it crowds the organelles and impairs their function, especially in the brain.

Carriers of Tay-Sachs disease have about half the level of Hex A activity found in homozygous dominant individuals. Prenatal diagnosis of the disease is possible following either amniocentesis or chorionic villi sampling. The gene for Tay-Sachs disease is located on chromosome 15.

Cystic Fibrosis Cystic fibrosis (CF) is the most common lethal genetic disease among Caucasians in the United States. Abnormal secretions related to the chloride ion channel characterize this disorder. One of the most obvious symptoms in CF patients is extremely salty sweat. In children with CF, the mucus in the bronchial tubes and pancreatic ducts is particularly thick and viscous, interfering with the function of the lungs and pancreas. To ease breathing, the thick mucus in the lungs has to be loosened periodically, but still the lungs frequently become infected. In the past few years, new treatments, including the administration of antibiotics by means of a nebulizer and a percussion vest to loosen mucus in the lungs (**Fig. 9.9**), have raised the average life expectancy for CF patients to as much as 35 years of age. Genetic testing for the recessive allele is possible if individuals want to know whether they are carriers.

Phenylketonuria Phenylketonuria (PKU) is the most commonly inherited metabolic disorder that affects nervous system development. Affected individuals lack an enzyme that is needed for the normal metabolism of the amino acid phenylalanine, so an abnormal breakdown product, a phenylketone, accumulates in the urine. Newborns are routinely tested in the hospital for elevated levels of phenylalanine in the blood. If elevated levels are detected, newborns are placed on a diet low in phenylalanine, which must be continued until the brain is fully developed (around the age of seven years), or else severe mental retardation occurs. Some doctors recommend that the diet continue for life, but in any case, a pregnant woman with phenylketonuria must be on the diet in order to protect her unborn child from harm. Many diet products, such as soft drinks, have warnings to phenylketonurics that the product contains the amino acid phenylalanine.

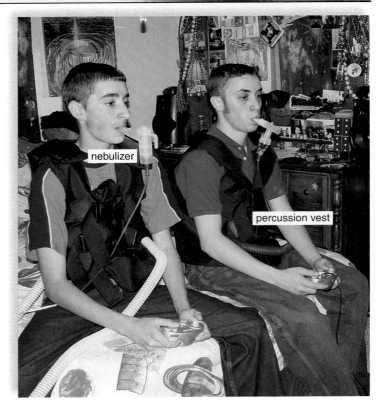

FIGURE 9.9 Cystic fibrosis therapy.

Sickle-cell Disease Sickle-cell disease occurs among people of African descent and is not usually seen among other racial groups. It is estimated that 1 in 12 African Americans are carriers for the disease. In individuals with sickle-cell disease, the red blood cells are shaped like sickles, or half-moons, instead of biconcave discs. An abnormal hemoglobin molecule (Hb^S) causes the defect. Normal hemoglobin (Hb^A) differs from Hb^S by one amino acid in the protein globin. The single change causes Hb^S to be less soluble than Hb^A.

A person with sickle-cell disease who has the genotype $Hb^S Hb^S$ exhibits a number of symptoms, ranging from severe anemia to heart failure. Individuals who are $Hb^A Hb^S$ have sickle-cell trait, in which sickling of the red blood cells occurs when the oxygen content of the blood is low. Presently, prenatal diagnosis for sickle-cell disease is possible. In the future, gene therapy may be available for these patients.

Two individuals with sickle-cell trait can produce children with three possible phenotypes. The chances of producing an individual with a normal genotype ($Hb^A Hb^A$) are 25%, sickle-cell trait ($Hb^A Hb^S$) 50%, and sickle-cell disease ($Hb^S Hb^S$) 25%. Because of the three possible phenotypes, some geneticists consider sickle-cell disease an example of incomplete dominance, an inheritance pattern to be discussed in Section 9.12. In the meantime, a few autosomal dominant disorders are discussed in Section 9.10.

> **9.9 Check Your Progress** What is the genotype of normal parents who have a child with cystic fibrosis?

A number of autosomal dominant disorders have been identified in humans. Three relatively common ones are neurofibromatosis, Huntington disease, and achondroplasia.

Neurofibromatosis Neurofibromatosis, sometimes called von Recklinghausen disease, is one of the most common genetic disorders and is seen equally in every racial and ethnic group throughout the world, many times in families with no history of the disorder. Once it appears in a family, it becomes a dominant trait.

At birth, or later, the affected individual may have six or more large, tan spots on the skin. Such spots may increase in size and number and get darker. Small, benign tumors (lumps) called neurofibromas, which arise from the fibrous coverings of nerves, may develop. In most cases, symptoms are mild, and patients live a normal life. In some cases, however, the effects are severe and include skeletal deformities, such as a large head, and eye and ear tumors that can lead to blindness and hearing loss. Many children with neurofibromatosis have learning disabilities and are hyperactive.

In 1990, researchers isolated the gene for neurofibromatosis and learned that it controls the production of a protein called neurofibromin, which normally blocks growth signals leading to cell division. Any number of mutations can lead to a neurofibromin that fails to block cell growth, and the result is the formation of tumors. Some mutations are caused by inserted DNA bases that do not belong in their present location.

Huntington Disease Huntington disease is a neurological disorder that leads to progressive degeneration of brain cells. **Figure 9.10A** shows that a portion of the brain involved in motor control atrophies, and this, in turn, causes severe muscle spasms that worsen with time (**Fig. 9.10B**). The disease is caused by a single mutated copy of the gene for a protein called huntingtin. Most patients appear normal until they are of middle age and have already had children, who may eventually also be stricken. Occasionally, the first sign of the disease in the next generation is seen in teenagers or even younger children. There is no effective treatment, and death comes 10–15 years after the onset of symptoms.

FIGURE 9.10B Patients with Huntington disease have neuromuscular spasms.

Several years ago, researchers found that the gene for Huntington disease was located on chromosome 4. A test was developed for the presence of the gene, but few people want to know if they have inherited the gene because there is no cure. At least now we know that the disease stems from a mutation that causes huntingtin to have too many copies of the amino acid glutamine. The normal version of the huntingtin protein has stretches of between 10 and 25 glutamines. If the huntingtin protein has more than 36 glutamines, it changes shape and forms large clumps inside neurons. Even worse, it attracts and causes other proteins to clump with it. One of these proteins, called CBP, ordinarily helps nerve cells survive. Researchers hope they may be able to combat the disease by boosting normal CBP levels.

Achondroplasia Achondroplasia is a common form of dwarfism associated with a defect in the growth of long bones. Individuals with achondroplasia have short arms and legs and a swayback, but a normal torso and head. About 1 in 25,000 people have achondroplasia. The condition arises when a gene on chromosome 4 undergoes a spontaneous mutation. Individuals who have achondroplasia are heterozygotes (*Aa*). The homozygous recessive (*aa*) genotype yields normal-length limbs. The homozygous dominant condition (*AA*) is lethal, and death generally occurs shortly after birth.

Potential parents often want to avoid having children with the disorders we have been discussing. Ways to detect genetic disorders before birth are reviewed in Section 9.11.

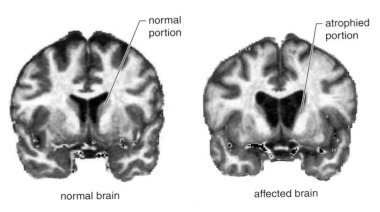

normal portion

atrophied portion

normal brain

affected brain

FIGURE 9.10A A normal brain compared to the brain of a patient affected by Huntington disease.

> **9.10** *Check Your Progress* **If a trait for blindness were dominant, could two blind collies have an offspring that was not blind?**

9.11 Genetic disorders may now be detected early on

A variety of procedures are available to test for genetic disorders such as the disorders discussed in this chapter.

Testing Fetal Cells
During **amniocentesis**, a long needle is passed through the abdominal and uterine walls to withdraw a small amount of the fluid that surrounds the fetus and contains a few fetal cells. Thereafter, genetic tests can be done on this fluid and on fetal chromosomes from the cells. During **chorionic villi sampling (CVS)**, a long, thin tube is inserted through the vagina into the uterus. Then a sampling of fetal cells is obtained by suction. The cells do not have to be cultured as they must be following amniocentesis, and testing can be done immediately.

Testing the Embryo
To test the embryo, it must have begun development in laboratory glassware through the process of in vitro fertilization (IVF). The physician obtains eggs from the prospective mother and sperm from the prospective father, and places them in the same receptacle, where IVF occurs. Then the zygote (fertilized egg) begins dividing. A single cell can be removed from the 8-celled embryo (**Fig. 9.11A**) and subjected to **preimplantation genetic diagnosis (PGD)**. Removing a single cell will not affect the developing embryo. Only healthy embryos that test negative for the genetic disorders of interest are placed in the mother's uterus, where they hopefully implant and continue developing.

Testing the Egg
Meiosis in females results in a single egg and at least two nonfunctional cells called **polar bodies**. Polar bodies, which later disintegrate, receive very little cytoplasm, but they do receive a haploid number of chromosomes. When a woman is heterozygous for a recessive genetic disorder, about half the polar bodies have received the mutated allele, and in these instances the egg received the normal allele. Therefore, if a polar body tests positive for a mutated allele, the egg received the normal allele. Only normal eggs are then used for IVF. Even if the sperm should happen to carry the mutation, the zygote will, at worst, be heterozygous. But the phenotype will appear normal (**Fig. 9.11B**).

If gene therapy becomes routine in the future, it's possible that eggs used for IVF could be given genes to treat genetic disorders, or even to control traits desired by the parents, such as musical or athletic ability, or intelligence. Such advanced genetic technologies will raise many moral and ethical issues and spur heated debates.

In the next part of the chapter, we examine various modes of inheritance, aside from autosomal recessive and autosomal dominant. We begin with incomplete dominance in Section 9.12.

> **9.11 Check Your Progress** The parents are carriers for cystic fibrosis but their baby is normal. State the (a) genotype of the cell in Figure 9.11A and (b) the allele in the polar body in Figure 9.11B.

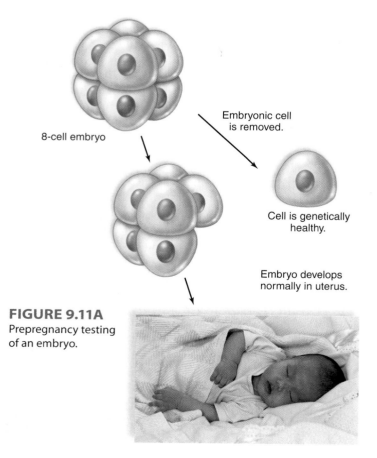

FIGURE 9.11A
Prepregnancy testing of an embryo.

8-cell embryo

Embryonic cell is removed.

Cell is genetically healthy.

Embryo develops normally in uterus.

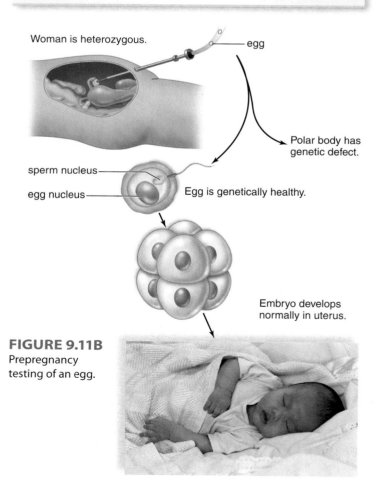

FIGURE 9.11B
Prepregnancy testing of an egg.

Woman is heterozygous.

egg

Polar body has genetic defect.

sperm nucleus

egg nucleus

Egg is genetically healthy.

Embryo develops normally in uterus.

In this part of the chapter, we see that Mendelian analysis can also be applied to complex patterns of inheritance, such as incomplete dominance, multiple alleles, polygenic inheritance, and pleiotropy. You will want to recognize each of these patterns of inheritance and to solve genetic problems concerning them.

9.12 Incomplete dominance still follows the law of segregation

When the heterozygote has an intermediate phenotype between that of either homozygote, **incomplete dominance** is exhibited. In a cross between a true-breeding, red-flowered four-o'clock strain and a true-breeding, white-flowered strain (**Fig. 9.12**), ❶ the offspring have pink flowers. But this is not an example of the blending theory of inheritance. When the plants with pink flowers self-pollinate, ❷ the offspring have a phenotypic ratio of 1 red-flower: 2 pink-flowers: 1 white-flower. The reappearance of all three phenotypes in this generation makes it clear that flower color is controlled by a single pair of alleles.

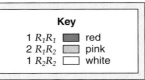

FIGURE 9.12 Incomplete dominance.

Key		
1 R_1R_1	■	red
2 R_1R_2	▨	pink
1 R_2R_2	□	white

It would appear that in R_1R_1 individuals, a double dose of pigment results in red flowers; in R_1R_2 individuals, a single dose of pigment results in pink flowers; and because the R_2R_2 individual produces no pigment, the flowers are white.

In humans, familial hypercholesterolemia (FH) is an example of incomplete dominance. An individual with two alleles for this disorder develops fatty deposits in the skin and tendons and may have a heart attack as a child. An individual with one normal allele and one FH allele may suffer a heart attack as a young adult, and an individual with two normal alleles does not have the disorder.

Perhaps the inheritance pattern of other human disorders should be considered one of incomplete dominance. For example, to detect the carriers of cystic fibrosis and Tay-Sachs disease, it is customary to determine the amount of enzyme activity of the gene in question. When the activity is one-half that of the dominant homozygote, the individual is a carrier. In other words, at the level of gene expression, the homozygotes and heterozygotes do differ in the same manner as four-o'clock plants.

An inheritance pattern called multiple alleles is discussed in Section 9.13.

> **9.12 Check Your Progress** If two carriers for FH conceive a baby, what are the potential phenotypes of the resulting offspring?

9.13 A gene may have more than two alleles

When a trait is controlled by **multiple alleles**, the gene exists in several allelic forms. But each person usually has only two of the possible alleles. For example, a person's ABO blood type is determined by multiple alleles. These alleles determine the presence or absence of antigens on red blood cells:

I^A = A antigen on red blood cells

I^B = B antigen on red blood cells

i = Neither A nor B antigen on red blood cells

The possible phenotypes and genotypes for blood type are as follows:

Phenotype	Genotype
A	I^AI^A, I^Ai
B	I^BI^B, I^Bi
AB	I^AI^B
O	ii

The inheritance of the ABO blood group in humans is also an example of **codominance** because both I^A and I^B are fully ex-

pressed in the presence of the other. Therefore, a person inheriting one of each of these alleles will have type AB blood.

This inheritance pattern differs greatly from Mendel's findings, since more than one allele is fully expressed. Both I^A and I^B are dominant over i. There are two possible genotypes for type A blood and two possible genotypes for type B blood. Use a Punnett square to confirm that reproduction between a heterozygote with type A blood and a heterozygote with type B blood can result in any one of the four blood types. Such a cross makes it clear that an offspring can have a different blood type from either parent, and for this reason, DNA fingerprinting is now used to identify the parents of an individual instead of blood type.

An inheritance pattern called polygenic inheritance is discussed in Section 9.14. When the environment is also involved, the inheritance pattern is called multifactorial.

> **9.13 Check Your Progress** In the past, blood type inheritance was used in paternity cases. Give an example that shows this method is imprecise.

9.14 Several genes and the environment can influence a single multifactorial characteristic

Polygenic inheritance occurs when a trait is governed by two or more genes—that is, sets of alleles. The individual has a copy of all allelic pairs, possibly located on many different pairs of chromosomes. Each dominant allele has a quantitative effect on the phenotype, and these effects are additive. The result is a continuous variation of phenotypes, resulting in a distribution that resembles a bell-shaped curve. The more genes involved, the more continuous are the variations and distribution of the phenotypes. In **Figure 9.14**, a cross between the genotypes *AABBCC* and *aabbcc* yields F$_1$ hybrids with the genotype *AaBbCc*. A range of genotypes and phenotypes results in the F$_2$ generation, and therefore a bell-shaped curve.

Multifactorial traits are controlled by polygenes subject to environmental influences. Recall that rapid weight gain possibly contributes to the occurrence of hip dysplasia in dogs, as discussed in the introduction to this chapter. In humans, skin color and disorders such as cleft lip and/or palate, clubfoot, congenital dislocations of the hip, hypertension, diabetes, schizophrenia, and even allergies and cancers are likely due to the combined action of many genes plus environmental influences. In recent years, reports have surfaced that all sorts of behaviors, including alcoholism, phobias, and even suicide, can be associated with particular genes. No doubt, behavioral traits are somewhat controlled by genes, but again, it is impossible at this time to determine to what degree. And very few scientists would support the idea that these behavioral traits are predetermined by our genes. The relative importance of genetic and environmental influences on the phenotype can vary, but in some instances the environment seems to have an extreme effect. In one interesting study, it was shown that cardiovascular disease is more prevalent among offspring whose biological or adoptive parents had cardiovascular disease. Can you suggest environmental reasons for the latter correlation, based on your study of Chapter 3?

These examples lend support to the belief that human traits controlled by polygenes are also subject to environmental influences. Therefore, many investigators are trying to determine what percentage of various traits is due to nature (inheritance) and what percentage is due to nurture (the environment). Some studies use twins separated since birth, because if identical twins in different environments share the same trait, that trait is most likely inherited. Identical twins are more similar in their intellectual talents, personality traits, and levels of lifelong happiness than are fraternal twins separated at birth. Biologists conclude that all behavioral traits are partly heritable, and that genes exert their effects by acting together in complex combinations susceptible to environmental influences.

Pleiotropy, discussed in Section 9.15, refers to a trait that affects many different tissues or organs of the body.

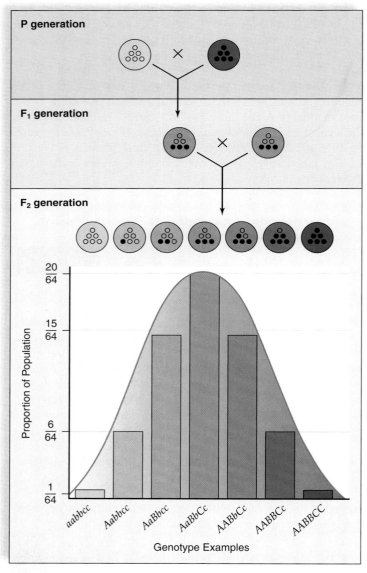

FIGURE 9.14 Polygenic inheritance: Dark dots stand for dominant alleles; the shading stands for environmental influences.

> **9.14 Check Your Progress**
>
> 1. What are the chances of an offspring inheriting full-blown FH when the parents are heterozygous?
>
> 2. In a paternity suit, the man has blood type AB. He could not be the father if the child has what blood type?
>
> 3. A child with type O blood is born to a mother with type A blood. What is the genotype of the child? The mother? What are the possible genotypes of the father?
>
> 4. A polygenic trait is controlled by three different loci. Give seven genotypes among the offspring that will result in seven different phenotypes when *AaBbCc* is crossed with *AaBbCc*.

9.15 One gene can influence several characteristics

Pleiotropy occurs when a single gene has more than one effect. For example, persons with Marfan syndrome have disproportionately long arms, legs, hands, and feet; a weakened aorta; poor eyesight; and other characteristics (**Fig. 9.15A**). All of these characteristics are due to the production of abnormal connective tissue. Marfan syndrome has been linked to a mutated gene (*FBN₁*) on chromosome 15 that ordinarily specifies a functional protein called fibrillin. Fibrillin is essential for the formation of elastic fibers in connective tissue. Without the structural support of normal connective tissue, the aorta can burst, particularly if the person is engaged in a strenuous sport, such as volleyball or basketball. Flo Hyman may have been the best American woman volleyball player ever, but she fell to the floor and died at the age of only 31 because her aorta gave way during a game. Now that coaches are aware of Marfan syndrome, they are on the lookout for it among very tall basketball players. Chris Weisheit, whose career was cut short after he was diagnosed with Marfan syndrome, said, "I don't want to die playing basketball."

Many other disorders, including porphyria and sickle-cell disease, are examples of pleiotropic traits. Porphyria is caused by a chemical insufficiency in the production of hemoglobin, the pigment that makes red blood cells red. The symptoms of porphyria are photosensitivity, strong abdominal pain, port-wine-colored urine, and paralysis in the arms and legs. Many members of the British royal family in the late 1700s and early 1800s suffered from this disorder, which can lead to epileptic convulsions, bizarre behavior, and coma.

In a person suffering from sickle-cell disease ($Hb^S Hb^S$), described in Section 9.9, the cells are sickle-shaped (**Fig. 9.15B**). The abnormally shaped sickle cells slow down blood flow and clog small blood vessels. In addi-

FIGURE 9.15B Sickle-shaped red blood cells.

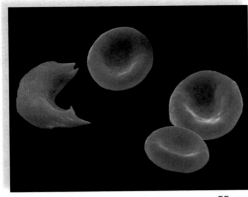

.55 μm

tion, sickled red blood cells have a shorter life span than normal red blood cells. Affected individuals may exhibit a number of symptoms, including severe anemia, physical weakness, poor circulation, impaired mental function, pain and high fever, rheumatism, paralysis, spleen damage, low resistance to disease, and kidney and heart failure.

Although sickle-cell disease is a devastating disorder, it provides heterozygous individuals with a survival advantage. People who have sickle-cell trait are resistant to the protozoan parasite that causes malaria. The parasite spends part of its life cycle in red blood cells feeding on hemoglobin, but it cannot complete its life cycle when sickle-shaped cells form and break down earlier than usual.

We have now finished our discussion of complex inheritance patterns, and in the next part of the chapter, we move on to consider that the genes are on the chromosomes.

> **9.15 Check Your Progress** Argue that cystic fibrosis (CF) should be considered a pleiotropic disorder.

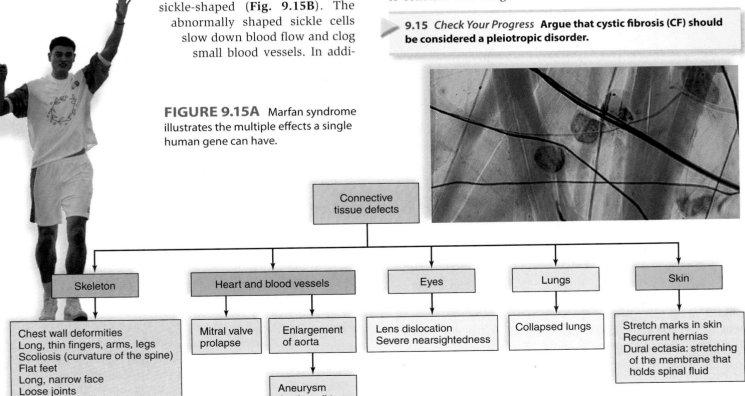

FIGURE 9.15A Marfan syndrome illustrates the multiple effects a single human gene can have.

Connective tissue defects

Skeleton
- Chest wall deformities
- Long, thin fingers, arms, legs
- Scoliosis (curvature of the spine)
- Flat feet
- Long, narrow face
- Loose joints

Heart and blood vessels
- Mitral valve prolapse
- Enlargement of aorta
  - Aneurysm
  - Aortic wall tear

Eyes
- Lens dislocation
- Severe nearsightedness

Lungs
- Collapsed lungs

Skin
- Stretch marks in skin
- Recurrent hernias
- Dural ectasia: stretching of the membrane that holds spinal fluid

White eye in *Drosophila*, the fruit fly, was the first gene to be definitively assigned to a chromosome. This gene is on the X chromosome, and therefore it has an unusual inheritance pattern, to be described in this part of the chapter. X-linked alleles also account for disorders in humans, particularly males.

The genes on a chromosome form a linkage group that tends to stay together during gamete formation. This reduces the possible variation among the gametes and offspring. A small number of recombinant gametes, due to crossing-over, can be used to map the chromosomes, as described.

9.16 Traits transmitted via the X chromosome have a unique pattern of inheritance

By the early 1900s, investigators had noted the parallel behavior of chromosomes and genes during meiosis (see Fig. 8.12), but they were looking for further data to support their belief that the genes were located on the chromosomes. The Columbia University group, headed by Thomas Hunt Morgan, performed the first experiments definitely linking a gene to a chromosome. This group worked with fruit flies (*Drosophila*). Fruit flies are even better subjects for genetic studies than garden peas: They can be easily and inexpensively raised in simple laboratory glassware; females mate and then lay hundreds of eggs during their lifetimes; and the generation time is short, taking only about ten days when conditions are favorable.

Drosophila flies have the same sex chromosome pattern as humans, and this facilitates our understanding of a cross performed by Morgan. Morgan took a newly discovered mutant male with white eyes and crossed it with a red-eyed female:

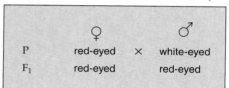

From these results, he knew that red eyes are the dominant characteristic and white eyes are the recessive characteristic. He then crossed the F_1 flies. In the F_2 generation, there was the expected 3 red-eyed: 1 white-eyed ratio, but it struck him as odd that all of the white-eyed flies were males:

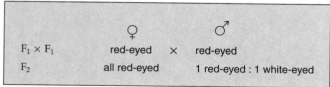

Obviously, a major difference between the male flies and the female flies was their sex chromosomes. Could it be possible that an allele for eye color was on the Y chromosome but not on the X? This idea could be quickly discarded because usually females have red eyes, and they have no Y chromosome. But perhaps an allele for eye color was on the X chromosome, and not on the Y chromosome. **Figure 9.16** indicates that this explanation matches the results obtained in the experiment. Therefore, the alleles must be on the chromosomes.

Notice that X-linked alleles have a different pattern of inheritance than alleles that are on the autosomes because the Y chromosome is lacking for these alleles, and the inheritance

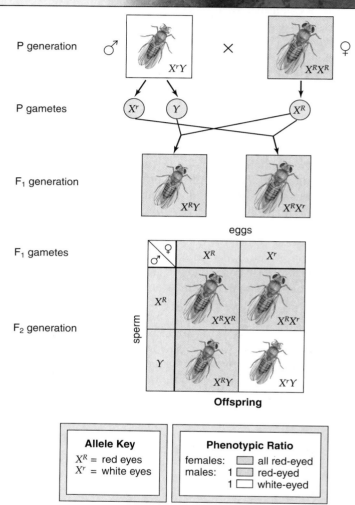

FIGURE 9.16 X-linked inheritance.

of a Y chromosome cannot offset the inheritance of an X-linked recessive allele. For the same reason, affected males always receive an X-linked recessive mutant allele from the female parent—they receive the Y chromosome from the male parent. X-linked inheritance disorders are seen in humans, as discussed in Section 9.17.

> **9.16 Check Your Progress** Examine the karyotype for a male in Figure 8.11 and give a reason why very few Y-linked alleles have been found.

Several X-linked recessive disorders occur in humans, including color blindness, muscular dystrophy, and hemophilia.

Color Blindness

In humans, the receptors for color vision in the retina of the eyes are three different classes of cone cells. Only one type of pigment protein is present in each class of cone cell; there are blue-sensitive, red-sensitive, and green-sensitive cone cells. The allele for the blue-sensitive protein is autosomal, but the alleles for the red- and green-sensitive proteins are on the X chromosome. About 8% of Caucasian men have red-green color blindness. Most of them see brighter greens as tans, olive greens as browns, and reds as reddish browns. A few cannot tell reds from greens at all. They see only yellows, blues, blacks, whites, and grays.

The pedigree in **Figure 9.17** shows the usual pattern of inheritance for color blindness and, indeed, any X-linked recessive disorder. More males than females exhibit the trait because recessive alleles on the X chromosome are expressed in males. The disorder often passes from grandfather to grandson through a carrier daughter.

Muscular Dystrophy

Muscular dystrophy, as the name implies, is characterized by a wasting away of the muscles. The most common form, Duchenne muscular dystrophy, is X-linked and occurs in about 1 out of every 3,600 male births.

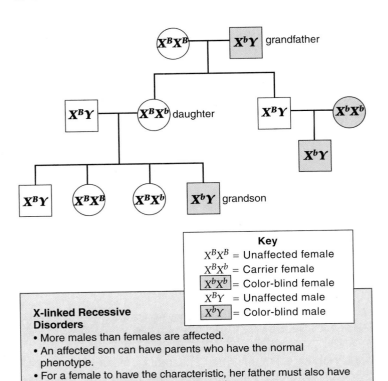

X-linked Recessive Disorders
- More males than females are affected.
- An affected son can have parents who have the normal phenotype.
- For a female to have the characteristic, her father must also have it. Her mother must have it or be a carrier.
- The characteristic often skips a generation from the grandfather to the grandson.
- If a woman has the characteristic, all of her sons will have it.

Key
$X^B X^B$ = Unaffected female
$X^B X^b$ = Carrier female
$X^b X^b$ = Color-blind female
$X^B Y$ = Unaffected male
$X^b Y$ = Color-blind male

FIGURE 9.17 X-linked recessive pedigree.

Symptoms, such as waddling gait, toe walking, frequent falls, and difficulty in rising, may appear as soon as the child starts to walk. Muscle weakness intensifies until the individual is confined to a wheelchair. Death usually occurs by age 20; therefore, affected males are rarely fathers. The recessive allele remains in the population through passage from carrier mother to carrier daughter.

The allele for Duchenne muscular dystrophy has been isolated, and it was discovered that the absence of a protein called dystrophin causes the disorder. Much investigative work determined that dystrophin is involved in the release of calcium from the sarcoplasmic reticulum in muscle fibers. The lack of dystrophin causes calcium to leak into the cell, which promotes the action of an enzyme that dissolves muscle fibers. When the body attempts to repair the tissue, fibrous tissue forms, and this cuts off the blood supply so that more and more cells die.

A test is now available to detect carriers of Duchenne muscular dystrophy. Also, various treatments been tried. Immature muscle cells can be injected into muscles, and for every 100,000 cells injected, dystrophin production occurs in 30–40% of muscle fibers. The allele for dystrophin has been inserted into thigh muscle cells, and about 1% of these cells then produced dystrophin.

Hemophilia

About 1 in 10,000 males is a hemophiliac. There are two common types of hemophilia: Hemophilia A is due to the absence or minimal presence of a clotting factor known as factor VIII, and hemophilia B (or Christmas disease) is due to the absence of clotting factor IX. Hemophilia is called the bleeder's disease because the affected person's blood either does not clot or clots very slowly. Although hemophiliacs bleed externally after an injury, they also bleed internally, particularly around joints. Hemorrhages can be stopped with transfusions of fresh blood (or plasma) or concentrates of the clotting protein. Also, clotting factors are now available as biotechnology products.

Knowing that organisms have many more genes than chromosomes allows us to conclude that more than one gene is on a chromosome. These genes form a linkage group, as discussed in Section 9.18.

9.17 Check Your Progress

1. Color blindness is an X-linked recessive trait. A female with normal vision and a color-blind male have a daughter who is color-blind. What are the genotypes of all the individuals involved?

2. In *Drosophila*, if a homozygous red-eyed female and a red-eyed male mated, what would be the possible genotypes of their offspring?

3. Which *Drosophila* cross would produce white-eyed males: (a) $X^R X^R \times X^r Y$ or (b) $X^R X^r \times X^R Y$? In what ratio?

9.18 The genes on one chromosome form a linkage group

After Thomas Morgan and his students had performed a great number of *Drosophila* crosses, they discovered many more mutants and were able to do various two-trait crosses. However, they did not always achieve the expected ratios among the offspring, due to **gene linkage**, the existence of several genes on the same chromosome. (Mendel, it turns out, was very lucky in that the pea traits he selected were always on different homologues.) The genes on a single chromosome form a **linkage group** because these genes tend to be inherited together.

Drosophila probably has thousands of different genes controlling all aspects of its structure, biochemistry, and behavior. Yet it has only four chromosomes. This paradox alone allows you to reason that each chromosome must carry a large number of genes. For example, it is now known that the genes controlling antennae type, wing length, leg length, body color, and eye color are all located on chromosome 2 (**Fig. 9.18**).

Figure 9.18 also illustrates a **linkage map**, because it tells you the relative distance between the gene loci on chromosome 2 of *Drosophila*. Geneticists are able to construct linkage maps by doing crosses and observing the number of recombinant gametes due to crossing-over, as is discussed in Section 9.19.

> **9.18 Check Your Progress** Why are linkage maps important to geneticists?

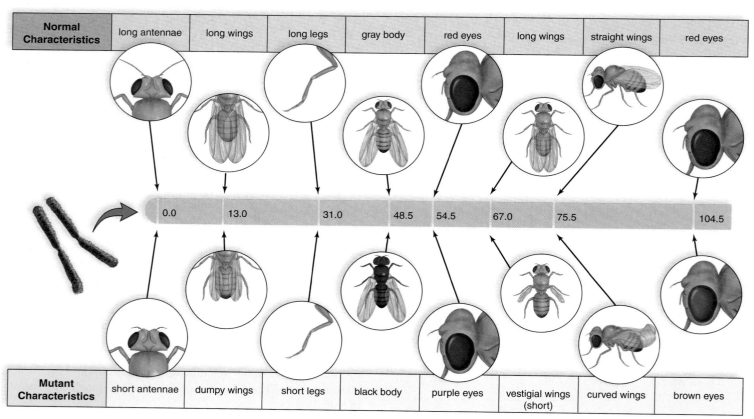

Normal Characteristics	long antennae	long wings	long legs	gray body	red eyes	long wings	straight wings	red eyes
	0.0	13.0	31.0	48.5	54.5	67.0	75.5	104.5
Mutant Characteristics	short antennae	dumpy wings	short legs	black body	purple eyes	vestigial wings (short)	curved wings	brown eyes

FIGURE 9.18 A simplified map of the genes on chromosome 2 of *Drosophila*.

9.19 Frequency of recombinant gametes maps the chromosomes

A linkage map can also be called a chromosome map because it tells the order of gene loci on chromosomes. To construct a chromosome map, investigators can sometimes rely on the frequency of crossing-over. Crossing-over, you recall, occurs during meiosis when pairs of homologues are in synapsis. During crossing-over, the nonsister chromatids of a tetrad exchange genetic material, and therefore alleles, and the result

is recombinant (recombined) gametes. Recombinant gametes occur in reduced number because crossing-over is infrequent. Still, recombinant gametes mean that when hybrids are crossed, all phenotypes will occur among the offspring, despite linkage.

To take an example, suppose you are doing a cross in which one parent is heterozygous for gray-body and red-eye

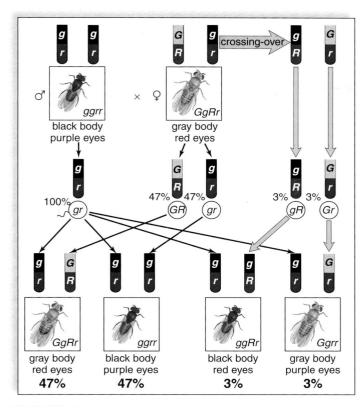

FIGURE 9.19 Example of incomplete linkage.

(*GgRr*) and the other is recessive for black-body and purple-eye (*ggrr*). Since the alleles governing these traits are both on chromosome 2 (see Fig. 9.18), you predict that the dominant alleles *GR* on one chromosome will stay together, and therefore half the offspring will be gray body with red eyes, and the other half will have the recessive traits, black body with purple eyes. Study **Figure 9.19**, *right* and satisfy yourself that these are the expected results.

When you perform the cross, you find that the ratio is almost 1:1, but not quite. A very small percentage of flies have recombinant phenotypes (Fig. 9.19, *left*). What you find is that some flies are *ggRr* and have black body and red eyes, and some are *Ggrr* and have gray body and purple eyes. In other words, the *G* and *R* split up instead of staying together. What happened? The answer is that *G* and *R* went into different gametes because crossing-over occurred. Crossing-over causes incomplete linkage and recombinant gametes. Recombinant gametes result in recombined phenotypes.

An examination of chromosome 2 shows that the two sets of alleles for this cross are very close together. Doesn't it stand to reason that the closer together two genes are, the less likely they are to cross over? This is exactly what various crosses have repeatedly shown. Therefore, the percentage of recombinant phenotypes can be used to map the chromosome. Map units are defined by the frequency of recombination (1% recombination equals one map unit). Therefore, the allele for black body and the allele for purple eyes are six map units apart.

Mapping the Chromosome

It is possible to use recombinant phenotype data to determine the distance between several alleles on a chromosome. For example, you can perform crosses that tell you the map distance between three pairs of alleles. If you know, for instance, that:

1. The distance between the black-body and purple-eye alleles = 6 map units.
2. The distance between the purple-eye and vestigial-wing alleles = 12.5 units.
3. The distance between the black-body and vestigial-wing alleles = 18.5 units.

Then the order of black-body, purple-eye, and vestigial-wing alleles must be as shown here:

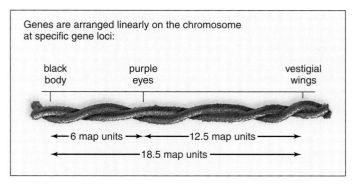

Because it is possible to map the chromosomes, the chromosome theory of inheritance includes the concept that alleles are arranged linearly along a chromosome at specific loci. While useful, molecular genetics (the topic of Chapter 10) tells us that this concept may be an oversimplification, and genes may actually overlap and share stretches of a chromosome.

Recombinant phenotype data have been used to map the chromosomes of *Drosophila*, but the possibility of using recombinant phenotype data to map human chromosomes is limited because it would only be possible to work with matings that have occurred by chance. This, coupled with the fact that humans tend not to have numerous offspring, means that additional methods must be used to sequence the genes on human chromosomes. Today, it is customary to mainly rely on molecular methods to map the human chromosomes, as is discussed in Section 13.11.

The achievements of T. H. Morgan, who worked with fruit flies, are discussed in Section 9.20.

9.19 Check Your Progress

1. When *AaBb* individuals are allowed to self-breed, the phenotypic ratio is just about 3:1. (a) What ratio was expected? (b) What may have caused the observed ratio?

2. Investigators performed crosses that indicated bar-eye and garnet-eye alleles are 13 map units apart, scallop-wing and bar-eye alleles are 6 units apart, and garnet-eye and scallop-wing alleles are 7 units apart. What is the order of these alleles on the chromosome?

9.20   Thomas Hunt Morgan is commonly called "the fruit fly guy"

As "The Star Spangled Banner" played to start the football game, the young genetics student silently smiled. In her research on the American Father of Genetics, Thomas Hunt Morgan (1866–1945), she had discovered an interesting piece of trivia from the history of biology. The Kentuckian T. H. Morgan came from a distinguished family lineage, including his father, Charles Hunt Morgan, who served as U.S. consul to Sicily; the famous financier J. P. Morgan; and the composer of "The Star Spangled Banner," Francis Scott Key.

In the footsteps of other great scientists, including Linnaeus, Darwin, and Mendel, the young Morgan was fascinated with nature and collecting bird eggs and fossils. At sixteen, he entered the State College of Kentucky and graduated with a degree in zoology in 1886. He completed his Ph.D in 1890 from Johns Hopkins University, specializing in sea spiders and morphology. After postdoctoral research and several years of teaching experience, Morgan accepted a position in experimental zoology at Columbia University in 1904 (**Fig. 9.20A**).

In 1908, Morgan began experimenting with the fruit fly (*Drosophila melanogaster*). The diminutive fly is still considered a classic model organism in genetics. It is believed to have arrived in the United States in the 1870s in banana shipments from Southeast Asia. It is a prolific breeder and easy to maintain in the laboratory. Morgan was awestruck with the genetics of the fruit fly. Visiting his wife shortly after the birth of their daughter, he regaled her with stories about fruit flies, finally asking, "How is the baby?"

As a result of questioning basic Darwinian evolution and Mendelian genetics, Morgan developed a more modern approach to genetics that included formulating the chromosomal theory of inheritance; mapping the genes of the fruit fly; revising the meanings of the terms mutation, recombination, assortment, and segregation; describing gene linkage, X-linked traits, and crossing-over; and creating the basic language of contemporary genetics. His fruit fly room at Columbia University stands today as a landmark in genetics.

T. H. Morgan possessed a complex character. Despite not easily sharing his innermost feelings, he was always willing to listen to criticism and welcomed his students' input. He had a wonderful relationship with his students, many of whom became geneticists. One of his students described Morgan's approach as "compounded with enthusiasm, combined with a strong critical sense, generosity, open-mindedness, and a remarkable sense of humor." Although a renowned researcher, Morgan was very uncomfortable with mathematics.

In 1928, Morgan retired from Columbia University and began working at The California Institute of Technology. He reorganized the biology department and promoted the interaction between biology, physics, and chemistry. As the result of his dedication and hard work, he received the Nobel Prize in Physiology or Medicine for his work in genetics. Morgan's influence on genetics and generations of scientists laid the foundations for modern genetics (**Fig. 9.20B**).

9.20 *Check Your Progress* Based on Mendel's law of segregation, cytologists studying cells, including mitosis and meiosis, hypothesized that the genes are on the chromosomes. Explain.

FIGURE 9.20B A modern investigator carries on the tradition of T. H. Morgan.

FIGURE 9.20A T. H. Morgan at work in the lab with one of his students.

Working with the garden pea, Mendel gave us two laws of genetics that apply to all organisms, including humans. The first of Mendel's laws tells us that an individual has two alleles, but the gametes have only one allele for every trait. The second law tells us that the gametes have all possible combinations of alleles. This increases variability among the offspring. Mendel was fortunate to be working with nonlinked genes because we

know today that the alleles on one chromosome do tend to stay together during the process of meiosis, except when crossing-over occurs. This observation has allowed researchers to map the chromosomes.

Certain patterns of inheritance, such as polygenic inheritance and X-linked inheritance, do not negate, but rather extend, the range of Mendelian analysis. Males are more apt than females to display an X-linked disorder because they receive

only one set of X-linked alleles from their mother. Their father gives them a Y chromosome, which is blank for these alleles.

Just as Mendelian genetics proposes, genes do have loci on the chromosomes, but today we know that genes are composed of DNA. In Chapter 10, we will learn that the sequence of the bases in each gene determines the sequence of amino acids in a protein. It is proteins that make us who we are.

The Chapter in Review

Summary

Inbreeding Leads to Disorders

- Closely related individuals are likely to pass on faulty genes, resulting in genetic disorders.

Gregor Mendel Deduced Laws of Inheritance

9.1   A blending model of inheritance existed prior to Mendel

- According to the blending concept, parents of contrasting appearance always produce offspring of intermediate appearance.

9.2   Mendel designed his experiments well

- The garden pea was an excellent experimental organism because it was easy to grow, had a short generation time, and produced many offspring.
- According to Mendel's particulate theory of inheritance, inheritance involves genes and a reshuffling of the same genes to offspring.

Single-Trait Crosses Reveal Units of Inheritance and the Law of Segregation

9.3   Mendel's law of segregation describes how gametes pass on traits

- An individual has two factors for each trait, which separate during gamete formation so that each gamete contains only one factor from each pair.
- Therefore, $Tt \times Tt$ gives these results:

Gametes	Phenotypic Ratio
T = tall plant	3 ☐ tall
t = short plant	1 ☐ short

9.4   The units of inheritance are alleles of genes

- Each trait has two alleles, and the dominant allele masks expression of the recessive allele.
- A homozygous organism has two copies of the same allele.
- A heterozygous organism has one of each type of allele at a gene locus.

- Genotype refers to genes of an individual.
- Phenotype refers to the physical appearance.

Two-Trait Crosses Support the Law of Independent Assortment

9.5   Mendel's law of independent assortment describes inheritance of multiple traits

- Each pair of factors assorts independently.
- All possible combinations of factors can occur in the gametes. For example, if $TtGg \times TtGg$, then:

Gametes	Phenotypic Ratio
TG	9 ☐ tall plant, green pod
Tg	3 ☐ tall plant, yellow pod
tG	3 ☐ short plant, green pod
tg	1 ☐ short plant, yellow pod

9.6   Mendel's results are consistent with the laws of probability

- The probability of genotypes and phenotypes can be calculated using a Punnett square.

9.7   Testcrosses support Mendel's laws and indicate the genotype

- A heterozygous individual crossed with a homozygous recessive individual produces a 1:1 phenotypic ratio in the offspring.
- A one-trait testcross determines whether a dominant phenotype is homozygous dominant or heterozygous.
- An individual heterozygous for two traits crossed with an individual recessive for those traits results in a 1:1:1:1 phenotypic ratio.

Mendel's Laws Apply to Humans

9.8   Pedigrees can reveal the patterns of inheritance

- Pedigrees show the patterns of inheritance for particular conditions. When a trait is recessive, the child may be affected but not the parents. When a trait is dominant, the child may be unaffected even though a parent is affected.
- Carriers appear normal but are capable of parenting a child with a genetic disorder.

9.9 Some human genetic disorders are autosomal recessive

- Tay-Sachs disease, cystic fibrosis, phenylketonuria, and sickle-cell disease are examples of autosomal recessive genetic disorders controlled by a single pair of alleles.

9.10 Some human genetic disorders are autosomal dominant

- Neurofibromatosis, Huntington disease, and achondroplasia are examples of autosomal dominant genetic disorders controlled by a single pair of alleles.

9.11 Genetic disorders may now be detected early on

- Amniocentesis and chorionic villi sampling test fetal cells.
- An embryonic cell can be tested following IVF prior to implantation.
- An egg can be tested for defects prior to IVF.

Complex Inheritance Patterns Extend the Range of Mendelian Analysis

9.12 Incomplete dominance still follows the law of segregation

- In incomplete dominance, a heterozygote has the intermediate phenotype between either homozygous parent (e.g., pink color in four o'clocks).
- In the F_2 generation, all three genotypes reappear.

9.13 A gene may have more than two alleles

- Multiple alleles control a trait when the gene exists in several allelic forms.
- The inheritance of ABO blood group in humans is an example of codominance.

9.14 Several genes and the environment can influence a single multifactorial characteristic

- In polygenic inheritance, a trait is governed by two or more sets of alleles, and continuous variation of phenotypes results in a bell-shaped curve.

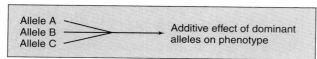

- Multifactorial traits are controlled by polygenes subject to environmental influences.

9.15 One gene can influence several characteristics

- Pleiotropy occurs when a single gene has more than one effect, and often leads to a syndrome.

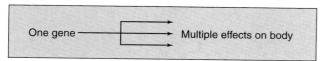

Chromosomes Are the Carriers of Genes

9.16 Traits transmitted via the X chromosome have a unique pattern of inheritance

- Fruit flies have been used to demonstrate X-linked inheritance, and the results are applicable to humans.

9.17 Humans have X-linked disorders

- X-linked recessive disorders in humans include color blindness, muscular dystrophy, and hemophilia.
- An X-linked recessive pedigree indicates that the trait can pass from grandfather through a carrier daughter to a grandson.

9.18 The genes on one chromosome form a linkage group

- Genes on the same chromosome that tend to be inherited together are known as a linkage group.

9.19 Frequency of recombinant gametes maps the chromosomes

- A direct relationship exists between the frequency of recombinant phenotypes and the distance between alleles (1% recombinants = 1 map unit).

Testing Yourself

Gregor Mendel Deduced Laws of Inheritance

1. Peas are good for genetics studies because they
   a. cannot self-pollinate.
   b. have a long generation time.
   c. are easy to grow.
   d. have fewer traits than most plants.
2. **THINKING CONCEPTUALLY** Explain how a "blending" model of inheritance would not support evolution, but Mendel's model does support evolution.

Single-Trait Crosses Reveal Units of Inheritance and the Law of Segregation

3. Which of the following is *not* a component of the law of segregation?
   a. Each gamete contains one factor from each pair of factors in the parent.
   b. Factors segregate during gamete formation.
   c. Following fertilization, the new individual carries two factors for each trait.
   d. Each individual has one factor for each trait.
4. If two parents with short fingers (dominant) have a child with long fingers, what is the chance their next child will have long fingers?
   a. no chance   c. ¼   e. ³/₁₆
   b. ½   d. ¹/₁₆
5. In humans, pointed eyebrows (*B*) are dominant over smooth eyebrows (*b*). Mary's father has pointed eyebrows, but she and her mother have smooth. What is the genotype of the father?
   a. *BB*   d. *BbBb*
   b. *Bb*   e. Any one of these is correct.
   c. *bb*

Two-Trait Crosses Support the Law of Independent Assortment

6. According to the law of independent assortment,
   a. all possible combinations of factors can occur in the gametes.
   b. only the parental combinations of gametes can occur in the gametes.
   c. only the nonparental combinations of gametes can occur in the gametes.
7. A testcross could be a cross between
   a. *aaBB* × *AABB*   c. *A?B?* × *aabb*
   b. *AABb* × *A?Bb*   d. *AaBb* × *AaBb*
8. Determine the probability that an *aabb* individual will be produced from an *AaBb* × *aabb* cross.
   a. 50%   c. 75%   e. 0%
   b. 25%   d. 100%

Mendel's Laws Apply to Humans

For questions 9–13, match the descriptions to the conditions in the key.

KEY:

    a. Tay-Sachs disease   d. sickle-cell disease
    b. cystic fibrosis   e. Huntington disease
    c. phenylketonuria

9. Autosomal dominant disorder.
10. The most common lethal genetic disorder among U.S. Caucasians.
11. Results from the lack of the enzyme hex A, resulting in the storage of its substrate in lysosomes.
12. Results from the inability to metabolize phenylalanine.
13. Late-onset neuromuscular genetic disorder.
14. Two affected parents have an unaffected child. The trait involved is
    a. autosomal recessive.
    b. incompletely dominant.
    c. controlled by multiple alleles.
    d. autosomal dominant.
15. **THINKING CONCEPTUALLY** A couple is concerned about preserving human life once fertilization has occurred. Which procedure would you recommend (fetal, embryonic, or egg testing) to detect a genetic disorder? Explain. (See Section 9.11.)

Complex Inheritance Patterns Extend the Range of Mendelian Analysis

16. If a man of blood group AB marries a woman of blood group A whose father was type O, what phenotypes could their children be?
    a. A only
    b. A, AB, B, and O
    c. AB only
    d. A, AB, and B
    e. O only
17. An anemic person has a number of problems, including lack of energy, fatigue, rapid pulse, pounding heart, and swollen ankles. This could be an example of
    a. pleiotropy.
    b. sex-linked inheritance.
    c. incomplete dominance.
    d. polygenic inheritance.
    e. codominance.

Chromosomes Are the Carriers of Genes

18. All of the genes on one chromosome are said to form a
    a. chromosomal group.
    b. recombination group.
    c. linkage group.
    d. crossing-over group.
19. Investigators found that a cross involving the mutant genes *a* and *b* produced 30% recombinants, a cross involving *a* and *c* produced 5% recombinants, and a cross involving *c* and *b* produced 25% recombinants. Which is the correct order of the genes?
    a. *a, b, c*
    b. *a, c, b*
    c. *b, a, c*
    d. Both *a* and *b* are correct.
    e. Both *b* and *c* are correct.
20. a. Determine the inheritance pattern of the trait possessed by the shaded squares (males). b. Then write in the genotype for the starred individual.

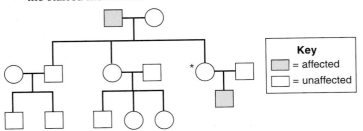

Key	
▦	= affected
☐	= unaffected

Understanding the Terms

allele 163
amniocentesis 170
autosomal chromosome 167
carrier 167
chorionic villi sampling (CVS) 170
codominance 171
dihybrid cross 164
dominant allele 163
gene linkage 176
gene locus 163
genotype 163
heterozygous 163
homozygous 163
incomplete dominance 171
law of independent assortment 164
law of segregation 162
linkage group 176
linkage map 176
monohybrid cross 162
multifactorial trait 172
multiple allele 171
pedigree 167
phenotype 163
pleiotropy 173
polar body 170
polygenic inheritance 172
preimplantation genetic diagnosis (PGD) 170
Punnett square 165
recessive allele 163
testcross 166

Match the terms to these definitions:

a. _____ Allele that exerts its phenotypic effect only in the homozygote; its expression is masked by a dominant allele.
b. _____ Alternative form of a gene that occurs at the same locus on homologous chromosomes.
c. _____ Allele that exerts its phenotypic effect in the heterozygote; it masks the expression of the recessive allele.
d. _____ Cross between an individual with the dominant phenotype and an individual with the recessive phenotype to see if the individual with the dominant phenotype is homozygous or heterozygous.
e. _____ Genes of an organism for a particular trait or traits; for example, *BB* or *Aa*.

Thinking Scientifically

1. You want to determine whether a newly found *Drosophilia* characteristic is dominant or recessive. Would you wait to cross this male fly with another of its own kind or cross it now with a fly that lacks the characteristic?
2. You want to test if the leaf pattern of a plant is influenced by the amount of fertilizer in the environment. What would you do?

⋎ARIS™ *Visit* www.mhhe.com/maderconcepts *for practice quizzes, animations, videos, and activities designed to help you master the material in this chapter.*

10

Molecular Biology of Inheritance

LEARNING OUTCOMES

After studying this chapter, you should be able to accomplish the following outcomes.

Arabidopsis Is a Model Organism

1 Describe a model organism.

DNA Is the Genetic Material

2 Describe Griffith's experiments and the Hershey and Chase experiments demonstrating that DNA is the genetic material.
3 Describe the structure of a DNA polymer and an RNA polymer.
4 Describe how the two strands of DNA are arranged in relation to one another.

DNA Can Be Duplicated

5 Summarize how DNA replicates and why the process is semiconservative.
6 Explain the complexities of DNA replication and why it is continuous in one strand and discontinuous in the other.

Genes Specify the Makeup of Proteins

7 Cite the early evidence indicating a link between DNA and the proteins of a cell.
8 Explain, in general, how a gene specifies the sequence of amino acids in a protein, and name the RNA molecules that participate in the process.
9 Describe the process of transcription.
10 Explain how tRNA and ribosomes are essential to the process of translation.
11 Diagram the processes of initiation and elongation in protein synthesis.

Mutations Are Changes in the Sequence of DNA Bases

12 Give examples of the different types of mutations and their possible effects.
13 Describe a transposon, and tell how a transposon can cause mutations.
14 Give several examples of environmental carcinogens that can cause cancer due to their effect on DNA.

Arabidopsis thaliana is a small flowering plant related to cabbage and mustard plants. *Arabidopsis* has no commercial value—in fact, it is a weed! However, it has become a model organism for the study of plant molecular genetics. Other model organisms in genetics are Mendel's peas and Morgan's fruit flies. Work with these models produces results that apply to many different organisms. For example, Mendel's two laws are generally applicable to all living things, and Morgan's discovery of X-linkage is applicable to nearly all sexually reproducing animals. *Arabidopsis* is a very useful model organism for these reasons:

- It is small, so many hundreds of plants can be grown in a small amount of space. *Arabidopsis* consists of a flat rosette of leaves from which grows a short flower stalk.

- Generation time is short. It only takes 5–6 weeks for plants to mature, and each one produces about 10,000 seeds!

- It normally self-pollinates, but it can easily be cross-pollinated. This feature facilitates gene mapping and the production of strains with multiple mutations.

Arabidopsis thaliana

Arabidopsis thaliana (enlarged drawing)

Arabidopsis Is a Model Organism

- The number of base pairs in its DNA is relatively small: 115,409,949 base pairs are distributed in 5 chromosomes (2n = 10) and 25,500 genes.

When Mendel worked with peas toward the end of the 19th century, he merely hypothesized that parents must pass genetic factors (now called alleles) to their offspring. A progression of genetic studies in the 20th century first established that DNA is the genetic material. Then scientists discovered the structure of DNA and how it functions in the cell. Today, the emphasis is on studying the specific function of individual genes, which are segments of DNA.

Every organism has its own sequence of bases in DNA, and we know the order of the bases for many organisms, including humans and *Arabidopsis*. Irradiating the seeds of *Arabidopsis* causes a mutation, a change in the normal sequence of bases. The creation of *Arabidopsis* mutants plays a significant role in discovering what each of its genes do. For example, if a mutant plant lacks stomata (openings in leaves), then we know that the affected gene influences the formation of stomata.

Think of Mendel working in an abbey garden, and then think of today's laboratory, where model organisms are studied with the aid of advanced, high-speed equipment. Amazing, too, is the recognition that, just like Mendel's peas, the work with *Arabidopsis* can assist our understanding of how human genes function. In this chapter, we begin our study of genetics at the molecular level by examining the nature of DNA and how it came to be discovered. Then we consider DNA replication, the activity of genes, and how they mutate, in that order.

Arabidopsis flower

Mutated flower

Mutated flower

A flat of *Arabidopsis*

Lab

During the first half of the 20th century, investigators established that DNA is the genetic material, and they also discovered the structure of DNA. The structure of DNA explains how it can store genetic information, replicate, and undergo mutations as required by the genetic material.

10.1   DNA is a transforming substance

During the late 1920s, the bacteriologist Frederick Griffith was attempting to develop a vaccine against *Streptococcus pneumoniae* (pneumococcus), a bacterium that causes pneumonia in mammals. In 1931, he noticed that when these bacteria are grown on culture plates, some, called S strain bacteria, produce shiny, smooth colonies, while others, called R strain bacteria, produce colonies that have a rough appearance. Under the microscope, S strain bacteria have a capsule (mucous coat), but R strain bacteria do not. **Figure 10.1** illustrates an experiment that Griffith performed. **1** He injected mice with the live S strain bacteria, and the mice died. **2** When he injected mice with the R strain, the mice did not die. In an effort to determine if the capsule alone was responsible for the virulence (ability to kill) of the S strain bacteria, **3** he injected mice with heat-killed S strain bacteria. The mice did not die. **4** Finally, Griffith injected the mice with a mixture of heat-killed S strain and live R strain bacteria. **5** Most unexpectedly, the mice died, and live S strain bacteria were recovered from the bodies! Griffith con-

cluded that a virulence-causing substance must have passed from the dead S strain bacteria to the live R strain bacteria causing the R strain bacteria to be *transformed.*

Griffith reasoned that the change in the phenotype of the R strain bacteria must be due to a change in their genotype. Indeed, couldn't the transforming substance that passed from S strain to R strain be genetic material? Questions such as this prompted investigators at the time to begin looking for the transforming substance in order to determine the chemical nature of the genetic material. Eventually, investigators were able to present evidence that DNA is the transforming material. Even so, many were not yet convinced, and further experiments were performed, such as those described in Section 10.2.

> **10.1 *Check Your Progress*** Assuming that S strain has the normal genotype and R strain is the mutant, what is DNA controlling?

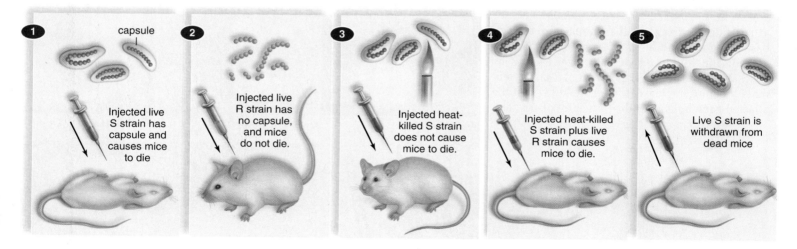

1 capsule — Injected live S strain has capsule and causes mice to die

2 Injected live R strain has no capsule, and mice do not die.

3 Injected heat-killed S strain does not cause mice to die.

4 Injected heat-killed S strain plus live R strain causes mice to die.

5 Live S strain is withdrawn from dead mice

FIGURE 10.1 Griffith's transformation experiment.

10.2   DNA, not protein, is the genetic material

During the 1950s, biologists were still performing experiments to determine the nature of the genetic material. Some believed it was the nucleic acid DNA, but others thought it was protein. In 1952, two experimenters, Alfred D. Hershey and Martha Chase, chose a virus, known as T2 bacteriophage, to determine which of the viral components—DNA or protein—entered the bacterium *Escherichia coli (E. coli)* and directed reproduction of more viruses.

Viruses such as T2 consist only of a protein coat, called a **capsid**, surrounding a DNA core (**Fig. 10.2A**). We now know that when T2 latches onto a bacterium, the tail contracts, allowing a tube to pass through the base plate and penetrate the bac-

terium. Only DNA enters the cell, and the capsid is left behind. Later, the bacterium releases many hundreds of new viruses. Why? Because phage DNA contains the genetic information necessary to cause the bacterium to produce new viruses.

Hershey and Chase Experiment In their experiment, Hershey and Chase relied on a chemical difference between DNA and protein to solve whether DNA or protein was the genetic material. In DNA, phosphorus is present but sulfur is not, and in protein, sulfur is present but phosphorus is not. They used radioactive ^{32}P to label the DNA core of the virus and ra-

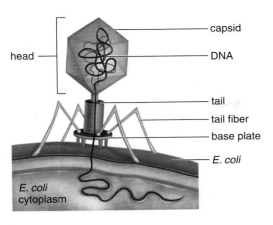

FIGURE 10.2A Structure of the T2 virus used by Hershey and Chase.

Labels: capsid, DNA, head, tail, tail fiber, base plate, *E. coli*, *E. coli* cytoplasm

dioactive ³⁵S to label the protein in the viral capsid. Recall that radioactive isotopes serve as labels (i.e., tracers) in biological experiments because it is possible to detect the presence of radioactivity by standard laboratory procedures.

Hershey and Chase did two separate experiments. In the first experiment (**Fig. 10.2B**), viral DNA was labeled with radioactive ³²P. **1** The viruses were allowed to attach to and inject their genetic material into bacterial cells. **2** Then the culture was agitated in a kitchen blender to remove whatever remained of the viruses on the outside of the bacterial cells. **3** Finally, the culture was centrifuged (spun at high speed) so that the bacterial cells collected as a pellet at the bottom of the centrifuge

tube. In this experiment, as you would predict, they found most of the ³²P-labeled DNA in the bacterial cells, not in the liquid medium. Why? Because the DNA had entered the cells.

In the second experiment (**Fig. 10.2C**), phage protein in capsids was labeled with radioactive ³⁵S. **1** The viruses were allowed to attach to and inject their genetic material into *E. coli* bacterial cells. **2** The culture was agitated in a kitchen blender to remove whatever remained of the viruses on the outside of the bacterial cells. **3** Finally, the culture was centrifuged so that the bacterial cells collected as a sediment at the bottom of the centrifuge tube. In this experiment, as you would predict, they found ³⁵S-labeled protein in the liquid medium but not in the bacterial cells. Why? Because the radioactive capsids remained on the outside of the cells and were removed by the blender.

These results indicated that the DNA (not the protein) of a virus enters the host, where viral reproduction takes place. *Therefore, DNA is the genetic material.* It transmits all the necessary genetic information needed to produce new viruses.

In Section 10.3, we begin a study of DNA and RNA structure.

> **10.2 Check Your Progress** It is possible to introduce a foreign gene into the cells of *Arabidopsis*. Suppose you wanted proof that the gene had entered the cells. What radioactive atom would you use to label the gene?

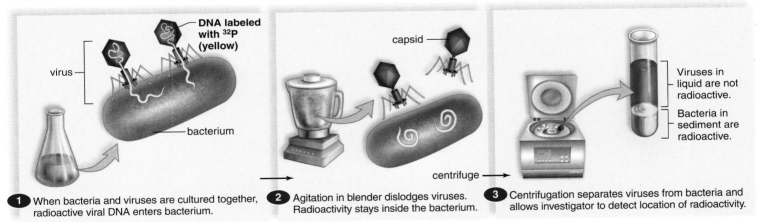

1 When bacteria and viruses are cultured together, radioactive viral DNA enters bacterium.

2 Agitation in blender dislodges viruses. Radioactivity stays inside the bacterium.

3 Centrifugation separates viruses from bacteria and allows investigator to detect location of radioactivity.

FIGURE 10.2B Hershey and Chase experiment I.

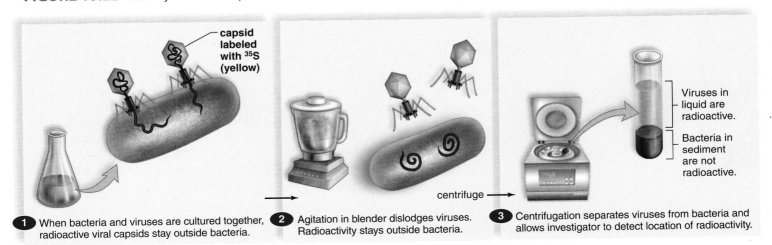

1 When bacteria and viruses are cultured together, radioactive viral capsids stay outside bacteria.

2 Agitation in blender dislodges viruses. Radioactivity stays outside bacteria.

3 Centrifugation separates viruses from bacteria and allows investigator to detect location of radioactivity.

FIGURE 10.2C Hershey and Chase experiment II.

During the same period that biologists were using viruses to show that DNA is the genetic material, biochemists were trying to determine the molecular configuration of nucleic acids. The term **nucleic acid** was coined in 1869 by the Swiss physician Johann Friedrich Miescher. Miescher removed nuclei from pus cells (these cells have little cytoplasm) and found that they contained a chemical he called *nuclein*. Nuclein, he said, was rich in phosphorus and had no sulfur, and these properties distinguished it from protein. Later, other chemists working with nuclein said that nuclein had acidic properties. Therefore, they decided to call the molecule nucleic acid.

Early in the 20th century, it was discovered that nucleic acids contain only **nucleotides**, molecules that are composed of a nitrogen-containing base, a phosphate, and a pentose (5-carbon sugar). **Figure 10.3A** shows how nucleotides are joined to form a polynucleotide: The sugar and phosphate portions of the nucleotides are covalently bonded to form the backbone of the molecule, and the bases project to the side.

DNA (deoxyribonucleic acid) is so named because its sugar content is a 5-carbon sugar called deoxyribose. DNA contains four different types of nucleotides. Two of the bases, **adenine (A)** and **guanine (G)**, have a double ring and are called purines. The other two bases, **thymine (T)** and **cytosine (C)**, have a single ring and are called pyrimidines (**Fig. 10.3B**).

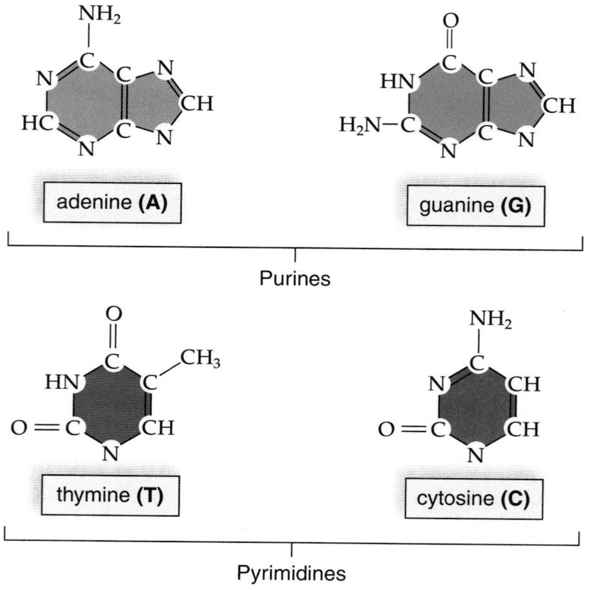

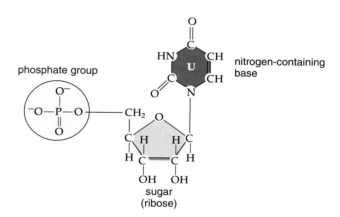

Purines

Pyrimidines

FIGURE 10.3B The four bases in DNA nucleotides.

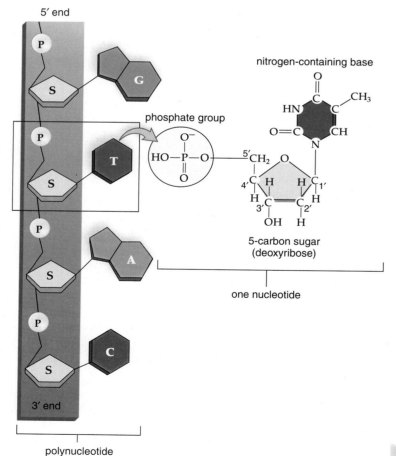

FIGURE 10.3A DNA is a polynucleotide—contains many nucleotides.

FIGURE 10.3C The uracil nucleotide in RNA replaces thymine in DNA.

Like DNA, **RNA (ribonucleic acid)** is also a polymer of nucleotides. RNA differs from DNA by its sugar content. The 5-carbon sugar in DNA is *deoxyribose*, while the 5-carbon sugar in RNA is *ribose*, which accounts for its name. Also, the base content of DNA and RNA differs slightly. In RNA, the base **uracil (U)** replaces thymine, so that the base content of RNA is cytosine, guanine, adenine, and uracil (**Fig. 10.3C**).

We will see that the function of RNA also differs from that of DNA. RNA serves as a helper to DNA to bring about protein synthesis. In Section 10.4, we continue our study of DNA structure, and also consider how the structure of DNA complements its function as the genetic material.

> **10.3** *Check Your Progress* **a. Do you predict that *Arabidopsis* contains both DNA and RNA? *b.* As a chemist, how would you distinguish the structure of DNA from RNA?**

At first, scientists were concerned that DNA might not fulfill the criteria for the genetic material. The genetic material must be:

1. Variable between species and *able to store information* that causes species to vary from one another. We know today that each species has its own sequence of DNA bases, and this sequence of bases stores information.

2. Constant within a species and *able to be replicated* with high fidelity during cell division, so that each and every member of a species contains the same information. We know today that every cell of an organism contains the same sequence of bases in DNA.

3. *Able to undergo rare changes*, called mutations, that provide the genetic variability that allows evolution to occur. We know today that a mutation is a change in the sequence of bases.

Chemists at the time thought that perhaps DNA had repeating units, each unit consisting of only four nucleotides—one for each of the four bases, like this: ATGC, ATGC, ATGC. . . . If so, the DNA of every species would always contain 25% of each kind of nucleotide. But with the development of new chemical techniques in the 1940s, it became possible for Erwin Chargaff, a chemist, to analyze in detail the base content of DNA's nucleotides. Here is a sample of Chargaff's data:

DNA Composition in Various Species (%)				
Species	**A**	**T**	**G**	**C**
Homo sapiens (human)	31.0	31.5	19.1	18.4
Drosophila melanogaster (fruit fly)	27.3	27.6	22.5	22.5
Zea mays (corn)	25.6	25.3	24.5	24.6
Neurospora crassa (fungus)	23.0	23.3	27.1	26.6
Escherichia coli (bacterium)	24.6	24.3	25.5	25.6

You can see that while some species—for example, *E. coli* and *Zea mays* (corn)—do have approximately 25% of each type of nucleotide, most do not. Further, the percentage of each type of nucleotide differs from species to species. Therefore, the nucleotide content of DNA is not fixed, and DNA does have the *variability* between species required of the genetic material. Within each species, however, DNA was also found to have the *constancy* required of the genetic material—that is, all members of a species have the same base composition.

Chargaff also discovered that regardless of the species, the percentage of A always equals the percentage of T, and the percentage of G equals the percentage of C. Chargaff's data suggest that A is always paired with T and G is always paired with C. Today, we call this **complementary base pairing**, and it occurs as illustrated in **Figure 10.4**. Note that hydrogen bonding between the bases is dependent on the nitrogen, oxygen, and hydrogen atoms attached to the purine and pyrimidine rings.

Does complementary base pairing suggest to you that DNA must have two backbones, each with attached bases?

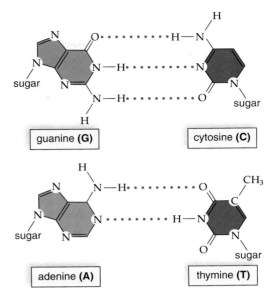

FIGURE 10.4 Complementary base pairing.

We will see that this was one of the major conclusions that allowed Watson and Crick to arrive at the double helix structure of DNA.

Chargaff's data also prompted researchers to discover that the paired bases can be in any order, accounting for why each species has a different percentage of paired bases. Here are some of the possible combinations:

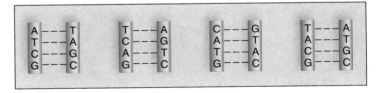

The variability that can be obtained is overwhelming. For example, it has been calculated that an average human chromosome contains about 140 million base pairs. Since any of the four possible nucleotides can be present at each nucleotide position, the total number of possible nucleotide sequences is to the 140-millionth power, or $4^{140,000,000}$. With so much variability possible, no wonder each species has its own base percentages!

Even though much was now known about DNA structure, it wasn't until the mid-1950s that Watson and Crick discovered that DNA is a double helix, as discussed in Section 10.5.

10.4 Check Your Progress *a.* In *Arabidopsis*, percentage of A = percentage of T, as it does in humans. Would you expect the same percentage of the other bases in both organisms? *b.* Why do A = T and C = G, regardless of the species? *c.* What happened to the normal sequence of bases in a eukaryotic ancestor to produce different eukaryotes?

10.5 DNA is a double helix

Researchers were racing against each other in the mid-1950s to discover the structure of DNA. Much had already been learned. They knew that DNA is a polymer of nucleotides in which the base A is probably paired with the base T and the base C is probably paired with the base G, but exactly how was the molecule put together? James D. Watson, an American, was on a postdoctoral fellowship at Cavendish Laboratories in Cambridge, England. While there, he began to work with the biophysicist Francis H. C. Crick, and together they constructed a model of DNA.

Watson and Crick's model was primarily based on the work of Rosalind Franklin and Maurice H. F. Wilkins at King's College of London. Franklin and Wilkins had studied the structure of DNA using X-ray crystallography. Franklin found that if a concentrated, viscous solution of DNA is made, it can be separated into fibers. Under the right conditions, the fibers are enough like a crystal (a solid substance whose atoms are arranged in a definite manner) that when X-rayed, an X-ray diffraction pattern results (**Fig. 10.5A**). The X-ray diffraction pattern of DNA suggested to Watson and Crick that DNA is a **double helix**. The helical shape is indicated by the crossed (X) pattern in the center of the photograph at right in Figure 10.5A. The dark portions at the top and bottom of the photograph indicate that some portion of the double helix is repeated.

Complementary base pairing suggests that DNA is double-stranded, with sugar-phosphate backbones on the outside and hydrogen paired bases on the inside. The helical arrangement determined by Watson and Crick conveniently made use of the mathematical measurements provided by the X-ray diffraction data for the spacing between the base pairs and for a complete turn of the double helix (**Fig. 10.5B**). Watson and Crick also noticed that the two strands of the molecule had to be antiparallel (run in opposite directions) to allow for complementary base pairing. Notice that the sugars in the right strand are upside down with respect to the sugars in the left strand.

The popular concept in biology, "structure suits function," can be applied to DNA. How does the structure of DNA suit its function as the genetic material? This was the next puzzle to be solved in the history of genetics. The double helix model suggests that the stability and variability of the molecule reside in the sequence of bases that is stable within a species but variable between species. The secret to how DNA functions as the genetic material, therefore, resides in the sequence of the bases. Watson and Crick immediately noted that the double helix model indicated a way that DNA could replicate, as needed, so that genetic information could pass from cell to cell and from generation to generation. We will see in Section 10.6 that complementary base pairing is the key to successful replication.

10.5 Check Your Progress DNA from *Arabidopsis* and from humans has the same X-ray diffraction pattern. Explain.

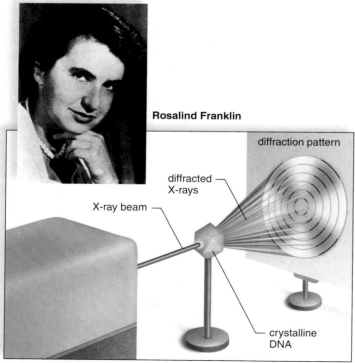

Rosalind Franklin

diffraction pattern

diffracted X-rays

X-ray beam

crystalline DNA

Procedure to obtain X-ray diffraction pattern of DNA

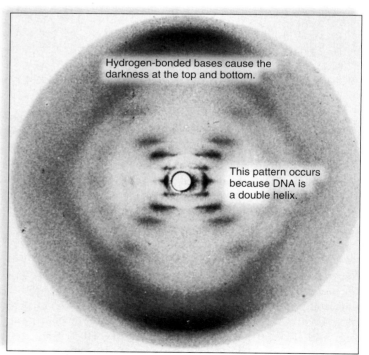

Hydrogen-bonded bases cause the darkness at the top and bottom.

This pattern occurs because DNA is a double helix.

Photograph of diffraction pattern

FIGURE 10.5A X-ray diffraction of DNA.

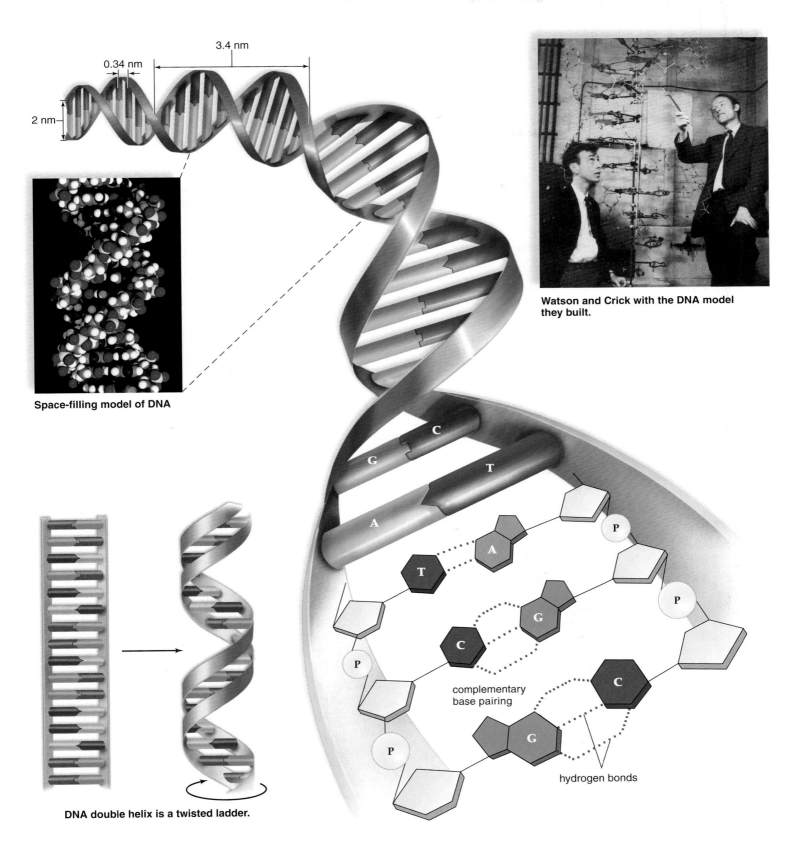

0.34 nm

3.4 nm

2 nm

Space-filling model of DNA

Watson and Crick with the DNA model they built.

complementary base pairing

hydrogen bonds

DNA double helix is a twisted ladder.

FIGURE 10.5B The Watson and Crick model of DNA.

In this part of the chapter, we consider that DNA replication results in two double helix molecules, and each new double helix molecule is a duplication of the preceding one. Your professor can choose to cover only the overview in Section 10.6, or also cover the more detailed presentation of DNA replication in Section 10.7.

10.6   DNA replication is semiconservative

The term **DNA replication** refers to the process of copying a DNA molecule. Following replication, there is usually an exact copy of the DNA double helix. As soon as Watson and Crick developed their double helix model, they commented, "It has not escaped our notice that the specific pairing we have postulated immediately suggests a possible copying mechanism for the genetic material."

During DNA replication, each original DNA strand of the parental molecule (original double helix) serves as a template for a new strand in a daughter molecule (**Fig. 10.6**). A **template** is a pattern used to produce a shape complementary to itself. DNA

replication is termed **semiconservative replication** because the template, or old strand, is conserved, or present, in each daughter DNA molecule (new double helix).

Replication requires the following steps:

1. *Unwinding.* The old strands that make up the parental DNA molecule are unwound and "unzipped" (i.e., the weak hydrogen bonds between the paired bases are broken). A special enzyme called helicase unwinds the molecule.

2. *Complementary base pairing.* New complementary nucleotides, always present in the nucleus, are positioned by the process of complementary base pairing.

3. *Joining.* The complementary nucleotides join to form new strands. Each daughter DNA molecule contains a template strand, or old strand, and a new strand.

Steps 2 and 3 are carried out by an enzyme complex called **DNA polymerase**. DNA polymerase works in a test tube as well as in cells.

In Figure 10.6, the backbones of the parental DNA molecule are bluish, and each base is given a particular color. Following replication, the daughter molecules each have a greenish backbone (new strand) and a bluish backbone (old strand). A daughter DNA double helix has the same sequence of bases that the parental DNA double helix had originally. Although DNA replication can be explained easily in this manner, it is actually a complicated process. Some of the more precise molecular events are discussed in Section 10.7.

DNA replication must occur before a cell can divide. Cancer, which is characterized by rapidly dividing cells, is sometimes treated with chemotherapeutic drugs that are analogs (have a similar, but not identical, structure) to one of the four nucleotides in DNA. When these are mistakenly used by the cancer cells to synthesize DNA, replication stops and the cells die off. Section 10.7 continues the study of DNA replication.

> **10.6 Check Your Progress** Investigators sequencing the DNA of *Arabidopsis* decide to use DNA only from the leaves. Is this okay, or do the roots have a different sequence?

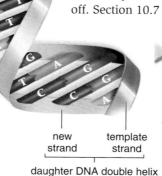

region of parental DNA double helix

region of replication: new nucleotides are pairing with those of template strands

region of completed replication

template strand   new strand

daughter DNA double helix

new strand   template strand

daughter DNA double helix

FIGURE 10.6 Semiconservative replication (simplified).

10.7   Many different proteins help DNA replicate

Watson and Crick realized that the strands in DNA had to be antiparallel to allow for complementary base pairing. This opposite polarity of the strands introduces complications for DNA replication, as we will now see. In **Figure 10.7**, ❶ take a look at a deoxyribose molecule, in which the carbon atoms are numbered. Use the structure to see that ❷ the DNA strand in the blue box runs opposite from the DNA strand in the green box. In other words, the strand in the blue box has a 5′ end at the top, and the strand in the green box has a 3′ end at the top. During replication, DNA polymerase has to join the nucleotides of the new strand so that the 3′ end is uppermost. Why? Because DNA polymerase can only join a nucleotide to the free 3′ end of the previous nucleotide, as shown. This also means that *DNA polymerase cannot start the synthesis of a new DNA chain.* Therefore, an RNA polymerase lays down a short amount of RNA, called an RNA primer, that is complementary to the template strand being replicated. After that, DNA polymerase can join DNA nucleotides to the 3′ end of the growing new strand.

❸ As a helicase enzyme unwinds DNA, one template strand can be copied in the direction of the replication fork. (Binding proteins serve to stabilize the newly formed, single-stranded regions.) ❹ This strand is called the leading new strand. The other template strand has to be copied in the direction away from the fork. Therefore, replication must begin over and over again as the DNA molecule unwinds. ❺ Replication of this so-called lagging new strand is, therefore, discontinuous, and it results in segments called ❻ Okazaki fragments, after the Japanese scientist Reiji Okazaki, who discovered them.

Replication is only complete when the RNA primers are removed. This works out well for the lagging new strand. While proofreading, DNA polymerase removes the RNA primers and replaces them with complementary DNA nucleotides. ❼ Another enzyme, called DNA ligase, joins the fragments. However, there is no way for DNA polymerase to replicate the 5′ ends of both new strands after RNA primers are removed. This means that DNA molecules get shorter as one replication follows another. The ends of eukaryotic DNA molecules have a special nucleotide sequence called a telomere. **Telomeres** do not code for proteins and, instead, are repeats of a short nucleotide sequence, such as TTAGGG.

Mammalian cells grown in a culture divide about 50 times and then stop. After this number of divisions, the loss of telomeres apparently signals the cell to stop dividing. Ordinarily, telomeres are only added to chromosomes during gamete formation by an enzyme called telomerase. This enzyme, unfortunately, is often mistakenly turned on in cancer cells, an event that contributes to the ability of cancer cells to keep on dividing without limit. Section 10.8 begins our discussion of what genes do.

FIGURE 10.7
DNA replication
(in depth).

10.7 *Check Your Progress* What causes DNA polymerase to add bases in the correct order during replication?

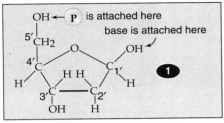

2 DNA polymerase attaches a new nucleotide to the 3′ carbon of the previous nucleotide.

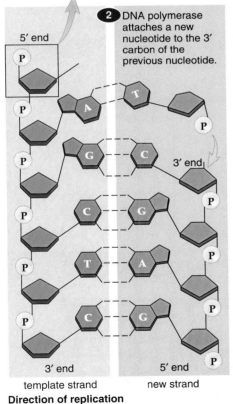

Deoxyribose molecule

OH ← P   is attached here
base is attached here
5′ CH₂
4′ C — O — C 1′ — OH
3′ C — C 2′

Direction of replication

template strand   new strand

Replication fork introduces complications

template strand
leading new strand
DNA polymerase
3 helicase at replication fork
5 lagging strand
template strand
6 Okazaki fragment
RNA primer
7 DNA ligase
DNA polymerase
parental DNA helix

Inborn errors of metabolism first suggested (and continue to illustrate) the connection between genes and proteins. In this part of the chapter, we see that gene expression requires two steps. During transcription, DNA is a template for RNA formation, and during translation, the sequence of bases in messenger RNA (mRNA) codes for the sequence of amino acids in a polypeptide. Transfer RNA (tRNA) and ribosomal RNA (rRNA) are also active during translation.

10.8 Genes are linked to proteins

Evidence began to mount in the 1900s that metabolic disorders can be inherited! An English physician, Sir Archibald Garrod, called them "inborn errors of metabolism." Investigators George Beadle and Edward Tatum, working with red bread mold, discovered what they called the "one gene, one enzyme hypothesis," based on the observation that a defective gene caused a defective enzyme. In order to test this idea further, other investigators decided to see if the hemoglobin in the red blood cells of persons with sickle-cell disease has a structure different from the hemoglobin in normal individuals (**Fig. 10.8**). They did find a structural difference. In one location, normal hemoglobin (Hb^A) contains the negatively charged amino acid glutamate, and in sickle-cell hemoglobin (Hb^S), the glutamate is replaced by the nonpolar amino acid valine. This causes Hb^S to be less soluble and to precipitate out of solution, especially when environmental oxygen is low. At these times, the Hb^S molecules stack up into long, semirigid rods that push against the plasma membrane and distort the red blood cell into the sickle shape.

By now, geneticists have confirmed many times over that proteins are the link between genotype and phenotype. Two other examples are the protein huntingtin, which is altered in Huntington disease, and a malfunctioning Cl^- channel protein in the plasma membranes of people with cystic fibrosis. Section 10.9 tells us that the linkage between genes and proteins requires two steps: transcription and translation.

> **10.8 Check Your Progress** Why does knowing the normal sequence of bases in a gene help those working with *Arabidopsis* identify mutant genes?

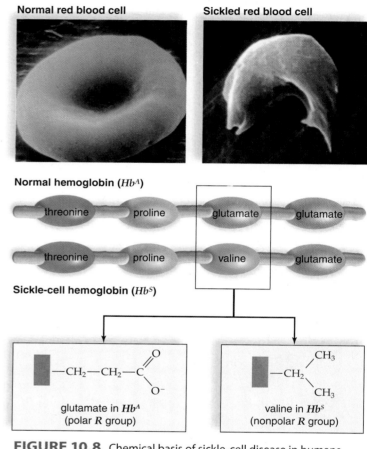

FIGURE 10.8 Chemical basis of sickle-cell disease in humans.

10.9 The making of a protein requires transcription and translation

A gene is said to be *expressed* when its product, a protein, is made and is functioning in a cell. A **gene** is a segment of DNA that specifies the amino acid sequence of a protein. Gene expression (production of a protein) requires two steps (**Fig. 10.9**). ❶ During **transcription**, DNA serves as a template for RNA formation. DNA is transcribed, monomer by monomer, into another type of polynucleotide (RNA). ❷ During **translation**, an RNA transcript directs the sequence of amino acids in a polypeptide. Like a translator who understands two languages, the cell changes a nucleotide sequence into an amino acid sequence.

DNA passes coded information to mRNA during transcription, as explained in Section 10.10.

> **10.9 Check Your Progress** *Arabidopsis* investigators want to know the function of each of the 25,000 *Arabidopsis* genes. Restate this in terms of proteins.

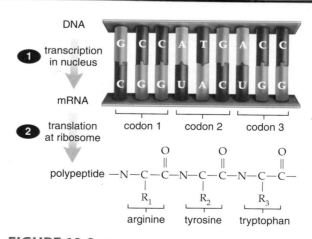

FIGURE 10.9 Overview of gene expression.

10.10 The genetic code for amino acids is a triplet code

As depicted in Figure 10.9, molecular biology tells us that the sequence of nucleotides in DNA specifies the order of amino acids in a polypeptide. It would seem, then, that there must be a **genetic code** for each of the 20 amino acids found in proteins. But can four nucleotides provide enough combinations to code for 20 amino acids? If each code word, called a **codon**, were made up of two bases, such as AG, there could be only 16 codons. But if each codon were made up of three bases, such as AGC, there would be 4^3, or 64, codons—more than enough to code for 20 amino acids:

number of bases in genetic code	1 2 3	4 16 64	number of different amino acids that can be specified

And indeed, this is the case—the genetic code is a **triplet code**. Each codon consists of three nucleotide bases and is expressed by three letters, such as AUC.

In 1961, Marshall Nirenberg and J. Heinrich Matthei performed an experiment that laid the groundwork for cracking the genetic code. First, they found that a cellular enzyme could be used to construct a synthetic RNA (one that does not occur in cells), and then they found that the synthetic RNA polymer could be translated in a test tube that contains the cytoplasmic contents of a cell. Their first synthetic RNA was composed only of uracil, and the protein that resulted was composed only of the amino acid phenylalanine. Therefore, the codon for phenylalanine was determined to be UUU. By translating just three nucleotides at a time, it was possible to assign an amino acid to each of the codons. **Figure 10.10** is a chart that lists all of the codons. To practice using this chart, find the rectangle where C is the first base and A is the second base. U, C, A, or G can be the third base. CAU and CAC are codons for histidine; CAA and CAG are codons for glutamine.

A number of important properties of the genetic code can be seen by careful inspection of Figure 10.10:

1. The genetic code is degenerate. This term means that most amino acids have more than one codon; for example, leucine, serine, and arginine have six different codons. The degeneracy of the code protects against the potentially harmful effects of mutations.

2. The genetic code is unambiguous. Each triplet codon has only one meaning.

3. The code has start and stop signals. There is only one start signal, but there are three stop signals.

Figure 10.9 illustrates that it is possible to use the principle of complementary base pairing to determine the original base sequence in DNA from a sequence of codons. Therefore, if the sequence of codons is CGG'UAC'UGG', the sequence of bases in DNA is GCC'ATG'ACC'.

The Genetic Code Is Universal
The genetic code (Fig. 10.10) is universal to all living things, with a few exceptions. For example, in 1979 researchers discovered that the genetic code used by chloroplasts and mammalian mitochondria differs slightly from the more familiar genetic code.

The universal nature of the genetic code provides strong evidence that all living things share a common evolutionary heritage. It also makes it possible to transfer genes from one organism to another. Many commercial and medicinal products, such as insulin, can be produced by inserting the correct gene into bacteria, which then express the gene. Genetic engineering has also produced some unusual organisms, such as mice that literally glow in the dark. In this case, the gene responsible for bioluminescence in jellyfish was placed into mouse embryos.

Now that we know what kind of information is passed to mRNA, we can examine the process of transcription in Section 10.11.

10.10 Check Your Progress Why don't all changes in DNA base sequence (mutations) result in an altered sequence of amino acids?

First Base	Second Base				Third Base
	U	**C**	**A**	**G**	
U	UUU phenylalanine	UCU serine	UAU tyrosine	UGU cysteine	U
	UUC phenylalanine	UCC serine	UAC tyrosine	UGC cysteine	C
	UUA leucine	UCA serine	UAA *stop*	UGA *stop*	A
	UUG leucine	UCG serine	UAG *stop*	UGG tryptophan	G
C	CUU leucine	CCU proline	CAU histidine	CGU arginine	U
	CUC leucine	CCC proline	CAC histidine	CGC arginine	C
	CUA leucine	CCA proline	CAA glutamine	CGA arginine	A
	CUG leucine	CCG proline	CAG glutamine	CGG arginine	G
A	AUU isoleucine	ACU threonine	AAU asparagine	AGU serine	U
	AUC isoleucine	ACC threonine	AAC asparagine	AGC serine	C
	AUA isoleucine	ACA threonine	AAA lysine	AGA arginine	A
	AUG *(start)* methionine	ACG threonine	AAG lysine	AGG arginine	G
G	GUU valine	GCU alanine	GAU aspartate	GGU glycine	U
	GUC valine	GCC alanine	GAC aspartate	GGC glycine	C
	GUA valine	GCA alanine	GAA glutamate	GGA glycine	A
	GUG valine	GCG alanine	GAG glutamate	GGG glycine	G

FIGURE 10.10 RNA codons.

10.11 During transcription, a gene passes its coded information to an mRNA

Suppose you have a woodworking encyclopedia in your bookcase, and you want to make a step stool. Rather than taking the entire set of encyclopedias out to the workshop, you might copy the instructions to make a step stool onto a sheet of paper and take just that to your workshop. We can liken DNA to the encyclopedia that contains instructions for all sorts of wood products. The sheet of paper becomes the mRNA molecule that has instructions for the step stool, which represents the protein to be made.

Transcription is the scientific term for copying a segment of DNA so that an RNA molecule results. Although all three classes of RNA are formed by transcription, we will focus on transcription to form **messenger RNA (mRNA)**, which takes instructions from DNA in the nucleus (the bookcase) to the ribosomes in the cytoplasm (the workshop).

An mRNA molecule is a copy of DNA because it has a sequence of bases complementary to a portion of one DNA strand; wherever A, T, G, or C is present in the DNA template, U, A, C, or G, respectively, is incorporated into the mRNA molecule

(**Fig. 10.11A**). A segment of the DNA double helix unwinds and unzips, and complementary RNA nucleotides pair with DNA nucleotides of the strand that is being transcribed. The nucleotides are joined together one at a time by **RNA polymerase**. The strand of DNA that is copied is called the the template strand, and the strand that is not transcribed is called the noncoding strand.

Transcription is initiated when RNA polymerase attaches to a region of DNA called a promoter. A **promoter** defines the start of a gene, the direction of transcription, and the strand to be transcribed. The RNA-DNA association is not as stable as the DNA double helix. Therefore, only the newest portion of an RNA molecule that is associated with RNA polymerase is bound to the DNA, and the rest dangles off to the side. Elongation of the mRNA molecule continues until RNA polymerase comes to a DNA stop sequence. The stop sequence causes RNA polymerase to stop transcribing the DNA and to release the mRNA molecule, now called an **mRNA transcript**.

Many RNA polymerase molecules can be working to produce mRNA transcripts at the same time. In **Figure 10.11B**, the transcripts get longer from left to right because transcription begins on the left. The many copies of the same mRNA molecule eventually result in many copies of the same protein, within a shorter period of time.

Transcription is complete, but mRNA has to be processed before we proceed to translation, as explained in Section 10.12.

> **10.11** *Check Your Progress* **What is the biological significance of transcription?**

FIGURE 10.11A
Transcription: synthesis of RNA.

noncoding strand

3′

RNA polymerase

DNA template strand

mRNA transcript

5′

to mRNA processing

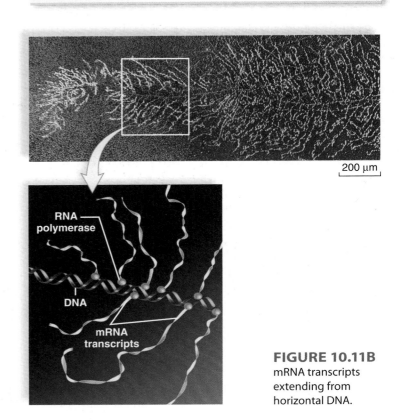

200 μm

RNA polymerase

DNA

mRNA transcripts

FIGURE 10.11B
mRNA transcripts extending from horizontal DNA.

10.12 In eukaryotes, an mRNA is processed before leaving the nucleus

Newly formed RNA molecules, called primary mRNA, are modified before leaving the eukaryotic nucleus, as shown in **Figure 10.12**. ❶ Note that the DNA at the top of Figure 10.12 is composed of exons and introns. ❷ Following transcription, primary mRNA, particularly in multicellular eukaryotes, is also composed of exons and introns. The exons of mRNA will be expressed, but the **introns**, which occur between the **exons**, will not and, therefore, introns were formerly labeled **junk DNA** by some investigators.

❸ Messenger RNA molecules receive a cap at the 5′ end and a tail at the 3′ end. The *cap* is a modified guanine (G) nucleotide that helps tell a ribosome where to attach when translation begins. The tail consists of a chain of 150–200 adenine (A) nucleotides. This so-called *poly-A tail* facilitates the transport of mRNA out of the nucleus and also inhibits degradation of mRNA by hydrolytic enzymes.

❹ The introns are removed by a process called RNA splicing. In prokaryotes, introns are removed by "self-splicing"—that is, the intron itself has the capability of enzymatically splicing itself out of a primary mRNA. In eukaryotes, the RNA splicing is done by spliceosomes, complexes that contain several kinds of ribonucleoproteins. A spliceosome cuts the primary mRNA and then rejoins the adjacent exons. ❺ An mRNA that has been processed and is ready to be translated is called mature mRNA. (RNAs with an enzymatic function are called **ribozymes**, and the presence of ribozymes in both prokaryotes and eukaryotes suggests that RNA could have preceded DNA in the evolutionary history of cells.)

Function of Introns Introns are far more common in eukaryotes than in prokaryotes, perhaps because prokaryotes do not have spliceosomes, and translation precedes directly after transcription. In humans, 95% or more of the average protein-coding gene is composed of introns. In general, as complexity increases, so does the proportion of non-protein-coding DNA sequences, such as introns. This phenomenon has piqued the interest of investigators who want to find a function for the introns that are removed. The sequencing of the human genome has shown that humans may have only about 25,000 coding genes, far less than the 100,000 formerly projected. How do humans make do with so few genes? Two hypotheses have been proposed:

1. It's possible that the presence of introns allows exons to be put together in different sequences so that various mRNAs and proteins can result from a single gene.

2. It's also possible that some introns regulate gene expression by feeding back to determine which coding genes are to be expressed and how they should be spliced. A heretofore overlooked RNA-signaling network may help achieve structural complexity far beyond anything seen in the unicellular world.

Now that mRNA has been processed, translation can begin. Another RNA, called transfer RNA (tRNA), also participates in translation, as explained in Section 10.13.

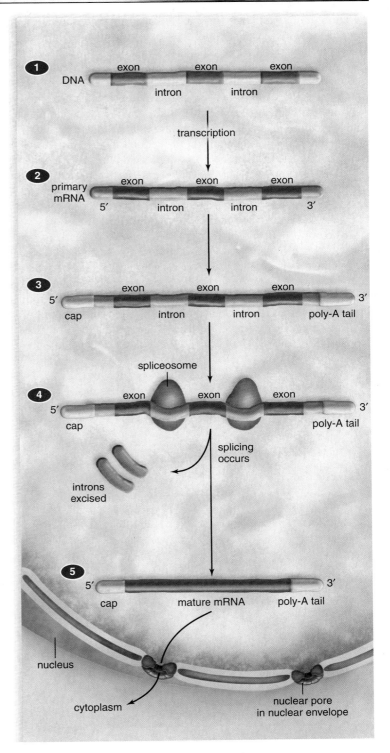

FIGURE 10.12 mRNA processing in eukaryotes.

10.12 Check Your Progress *Arabidopsis* and humans both have on the order of 25,000 genes, and both have introns. Do these data support or fail to support a hypothesis that the presence of introns leads to increased complexity?

To continue our analogy from Section 10.11, once the instructions from the encyclopedia are propped up on a table (ribosome) in the workshop (cytoplasm), the process of making the step stool (the polypeptide) will begin. The scientific term for making a protein according to the instructions of DNA is *translation*, because DNA and RNA are made of nucleotides, and polypeptides are made of amino acids. In other words, one language (nucleic acids) gets translated into another language (protein). During translation, the sequence of codons in the mRNA at a ribosome directs the sequence of amino acids in a polypeptide.

Transfer RNA (tRNA) molecules are like the tools you will use to make the step stool. The tRNA molecules transfer amino acids to the ribosomes. There is at least one tRNA molecule for each of the 20 amino acids found in proteins. **Figure 10.13A** shows the structure of a tRNA molecule. **1** The amino acid binds to one end of the molecule. **2** A tRNA molecule is a single-stranded nucleic acid that doubles back on itself to create regions where complementary bases are hydrogen-bonded to one another. **3** The opposite end of the molecule contains an **anticodon**, a group of three bases that is complementary to a specific **codon** of mRNA at a ribosome. For example, a tRNA that has the anticodon GAA binds to the codon CUU and carries the amino acid leucine.

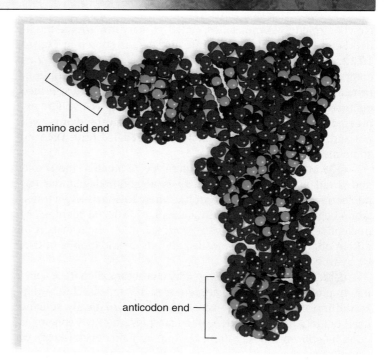

FIGURE 10.13B Space-filling model of tRNA molecule.

The structure of a tRNA molecule is generally drawn as a flat cloverleaf as in Figure 10.13A, but a space-filling model shows the molecule's three-dimensional shape (**Fig. 10.13B**).

How does the correct amino acid become attached to the correct tRNA molecule? This task is carried out by amino acid–activating enzymes, called aminoacyl-tRNA synthetases. Just as a key fits a lock, each enzyme has a recognition site for the amino acid to be joined to a particular tRNA. This is an energy-requiring process that uses ATP. Once the amino acid–tRNA complex is formed, it travels through the cytoplasm to a ribosome, where protein synthesis is occurring.

The Wobble Hypothesis In the genetic code, 61 codons encode for amino acids; the other three serve as stop sequences. Approximately 40 different tRNA molecules are found in most cells. There are fewer tRNAs than codons because some tRNAs can pair with more than one codon. In 1966, Francis Crick observed this phenomenon and called it the **wobble hypothesis**. He stated that the first two positions in a tRNA anticodon pair obey the A–U/G–C configuration. However, the third position can be variable. The wobble effect helps ensure that, despite changes in DNA base sequences, the correct sequence of amino acids will result in a protein.

Still another form of RNA, called ribosomal RNA (rRNA), participates in translation, as the overview of translation in Section 10.14 makes clear.

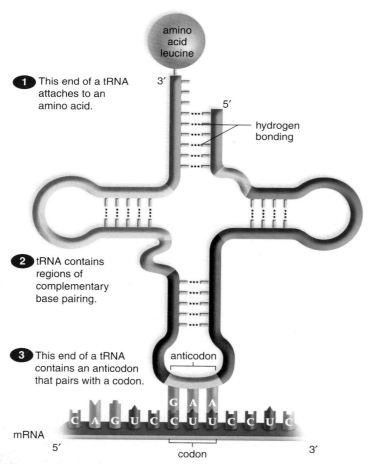

1 This end of a tRNA attaches to an amino acid.

amino acid leucine

3′

5′

hydrogen bonding

2 tRNA contains regions of complementary base pairing.

3 This end of a tRNA contains an anticodon that pairs with a codon.

anticodon

G A A

mRNA
5′
C A G U C C U U C C U C
3′

codon

FIGURE 10.13A Cloverleaf model of tRNA.

10.13 Check Your Progress *a.* Referring to Figure 10.10, what are the possible anticodons for tRNAs that pick up the amino acid arginine (arg)? *b.* Does this support the wobble hypothesis?

In eukaryotes, **ribosomal RNA (rRNA)** is produced from a DNA template in the nucleolus of a nucleus. The rRNA is packaged with a variety of proteins into two ribosomal subunits, one of which is larger than the other. Then the large and small subunits move separately through nuclear envelope pores into the cytoplasm, where they join as translation begins (**Fig. 10.14** *top, left*). Ribosomes can remain in the cytoplasm, or they can become attached to endoplasmic reticulum.

Both prokaryotic and eukaryotic cells contain thousands of ribosomes that serve as worktables where polypeptides are made. Ribosomes have a binding site for mRNA and three binding sites for transfer RNA (tRNA) molecules (Fig. 10.14 *top, right*). The tRNA binding sites facilitate complementary base pairing between tRNA anticodons and mRNA codons.

When a ribosome moves down an mRNA molecule, a polypeptide increases by one amino acid at a time (Fig. 10.14 *middle*). ➊ The first site on the right is for an incoming tRNA bearing a single amino acid. ➋ The middle site is for a tRNA that usually bears a polypeptide. ➌ The polypeptide is passed to the tRNA on the right, and ➍ the used tRNA leaves from the site on the left. ➎ Translation terminates at a stop codon. Once transcription is complete,

the polypeptide dissociates from the translation complex and adopts its normal shape.

In Chapter 3, we observed that a polypeptide twists and bends into a definite shape. This so-called folding process begins as soon as the polypeptide emerges from a ribosome and, often, so-called chaperone proteins are present in the cytoplasm and in the ER to make sure that all goes well. Some proteins contain only one polypeptide, and some contain more than one polypeptide. If so, the polypeptides join to produce the final three-dimensional structure of a functional protein.

Polyribosome As soon as the initial portion of mRNA has been translated by one ribosome, and the ribosome has begun to move down the mRNA, another ribosome attaches to this mRNA. Therefore, several ribosomes are often attached to and translating the same mRNA. The entire complex is called a **polyribosome** (Fig. 10.14 *bottom, right*).

We are now ready to begin a detailed look at the process of translation in Section 10.15.

> **10.14 Check Your Progress** Where in the cell are you apt to find all three types of RNA participating in protein synthesis?

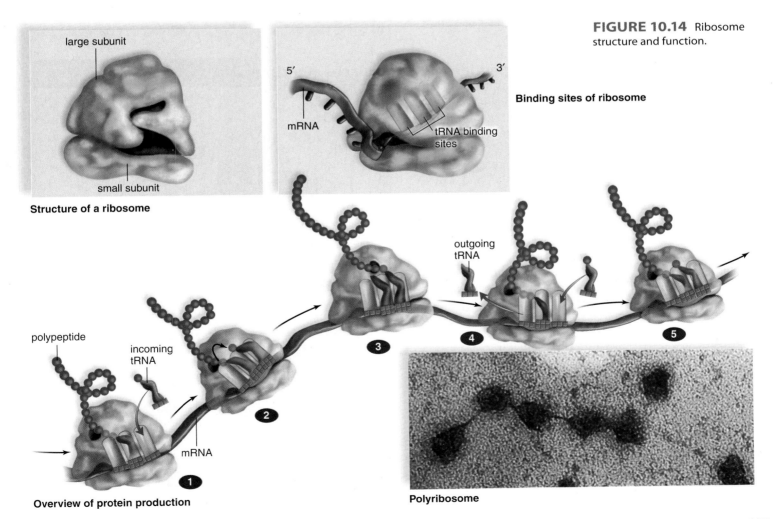

FIGURE 10.14 Ribosome structure and function.

large subunit

small subunit

Structure of a ribosome

5′ mRNA 3′

Binding sites of ribosome

tRNA binding sites

polypeptide

incoming tRNA

mRNA

outgoing tRNA

Overview of protein production

Polyribosome

10.15 Initiation begins the process of polypeptide production

Polypeptide synthesis occurs during the production of protein. Polypeptide synthesis involves three steps: initiation, elongation, and termination. Enzymes are required so that each of the three steps will function properly. The first two steps, initiation and elongation, require energy.

Initiation is the step that brings all the translation components together. Proteins called initiation factors are required to assemble the small ribosomal subunit, mRNA, initiator tRNA, and the large ribosomal subunit for the start of protein synthesis.

Initiation is shown in **Figure 10.15**. In prokaryotes, a small ribosomal subunit attaches to the mRNA in the vicinity of the *start codon* (AUG). The first, or initiator, tRNA pairs with this codon because its anticodon is UAC. Then, a large ribosomal subunit joins to the small subunit. Although similar in many ways, initiation in eukaryotes is much more complex.

As already discussed, a ribosome has three binding sites for tRNAs. One of these is called the *E (exit) site*, the second is the *P (peptide) site*, and the third is the *A (amino acid) site*. The initiator tRNA happens to be capable of binding to the P site, even though it carries only the amino acid methionine. The A site is for tRNA carrying the next amino acid, and the E site is for any tRNAs that are leaving a ribosome.

Following initiation, translation involves the process of elongation and ends with termination, as discussed in Section 10.16.

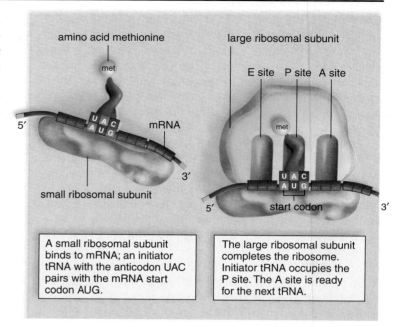

FIGURE 10.15 Initiation.

A small ribosomal subunit binds to mRNA; an initiator tRNA with the anticodon UAC pairs with the mRNA start codon AUG.

The large ribosomal subunit completes the ribosome. Initiator tRNA occupies the P site. The A site is ready for the next tRNA.

10.15 *Check Your Progress* **In the DNA of *Arabidopsis* (or any organism), what is the significance of finding the bases TAC in a row?**

10.16 Elongation builds a polypeptide one amino acid at a time

During **elongation**, a polypeptide increases in length, one amino acid at a time. In addition to the participation of tRNAs, elongation requires elongation factors, which facilitate the binding of tRNA anticodons to mRNA codons at a ribosome.

Elongation is shown in **Figure 10.16**, where ❶ a tRNA with an attached peptide is already at the P site, and a tRNA carrying its appropriate amino acid is just arriving at the A site. ❷ Once the next tRNA is in place at the A site, the peptide will be transferred to this tRNA. This transfer requires energy and a ribozyme, which is a part of the larger ribosomal subunit. ❸ Peptide bond formation occurs, and the peptide is one amino acid longer than it was before. ❹ Finally, **translocation** occurs: The mRNA moves forward, and the peptide-bearing tRNA is now at the P site of the ribosome. The used tRNA exits from the E site. A new codon is at the A site, ready to receive another tRNA. Eventually, the ribosome reaches a stop codon, and termination occurs, during which the polypeptide is released.

The entire process of gene expression, consisting of transcription and translation, is reviewed in Section 10.17.

10.16 *Check Your Progress* **With reference to Figure 10.10, what is the significance of finding these sequences in DNA: ATT, ATC, or ACT?**

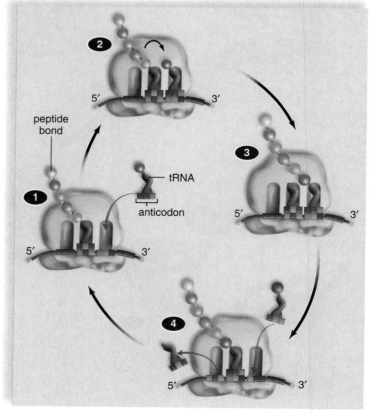

FIGURE 10.16 Elongation cycle.

10.17 Let's review gene expression

Gene expression requires two steps, called transcription and translation. **Figure 10.17** shows that in an eukaryotic cell, transcription occurs in the nucleus and translation occurs in the cytoplasm. ❶ and ❷ mRNA is produced and processed before leaving the nucleus. ❸–❻ After mRNA becomes associated with ribosomes, polypeptide synthesis occurs one base at a time. **Table 10.17** reviews the participants in gene expression.

Many ribosomes can be translating the same section of DNA at a time, and collectively these ribosomes are called a polysome. As discussed in Chapter 4, some ribosomes (polysomes) remain free in the cytoplasm, and some become attached to rough ER. Recall that the first few amino acids of a polypeptide act as a signal peptide that indicates where the polypeptide belongs in the cell, or if it is to be secreted from the cell. ❼ After the polypeptide enters the lumen of the ER by way of a channel, it is folded and further processed by the addition of sugars, phosphates, or lipids. ❽ In the meantime, the ribosome has reached a stop codon and termination occurs: Ribosomal units and mRNA are separated from one another and the polypeptide is released.

We have finished our examination of gene expression (the making of a protein), and in the next part of the chapter, we will study the biochemistry of mutations.

TABLE 10.17 Participants in Gene Expression

Name of Molecule	Special Significance	Definition
DNA	Genetic information	Sequence of DNA bases
mRNA	Has codons	Sequence of three RNA bases complementary to DNA
tRNA	Has an anticodon	Sequence of three RNA bases complementary to codon
rRNA	Located in ribosomes	Site of protein synthesis
Amino acid	Monomer of a polypeptide	Transported to ribosome by tRNA
Polypeptide	Enzyme, structural, or secretory product	Amino acids joined in a predetermined order

10.17 Check Your Progress What genetic information does DNA store?

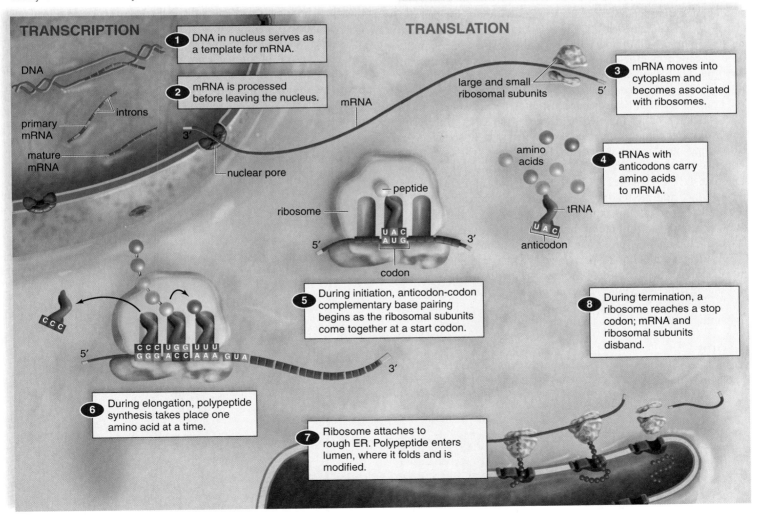

TRANSCRIPTION

DNA

1 DNA in nucleus serves as a template for mRNA.

introns

primary mRNA

mature mRNA

2 mRNA is processed before leaving the nucleus.

nuclear pore

TRANSLATION

large and small ribosomal subunits

mRNA

3 mRNA moves into cytoplasm and becomes associated with ribosomes.

amino acids

4 tRNAs with anticodons carry amino acids to mRNA.

tRNA

anticodon

peptide

ribosome

codon

5 During initiation, anticodon-codon complementary base pairing begins as the ribosomal subunits come together at a start codon.

8 During termination, a ribosome reaches a stop codon; mRNA and ribosomal subunits disband.

6 During elongation, polypeptide synthesis takes place one amino acid at a time.

7 Ribosome attaches to rough ER. Polypeptide enters lumen, where it folds and is modified.

FIGURE 10.17 Summary of gene expression in eukaryotes.

In molecular terms, a gene is a sequence of DNA bases, and a genetic mutation is a change in this sequence. Frameshift mutations can result in nonfunctional proteins and have powerful effects, and even point mutations, such as those in sickle-cell hemoglobin, can be serious (see Section 10.8). Agents that can cause mutations are surveyed in this part of the chapter also.

10.18 Mutations affect genetic information and expression

A **genetic mutation** is a permanent change in the sequence of bases in DNA. The effect of a DNA base sequence change on protein activity can range from no effect to complete inactivity. In general, there are two types of mutations: germ-line mutations and somatic mutations. Germ-line mutations are those that occur in sex cells and can be passed to subsequent generations. Somatic mutations occur in body cells, and therefore they may affect only a small number of cells in a tissue. Somatic mutations are not passed on to future generations, but they can lead to the development of cancer.

A mutation can affect the activity of a gene. **Point mutations** involve a change in a single DNA nucleotide and, therefore, a change in a specific codon. **Figure 10.18A** gives an example in which a single base change could have no effect or a drastic effect (in the form of a faulty protein), depending on the particular base change that occurs. You already know that sickle-cell disease is due to a single base change in DNA. Because a significant location in hemoglobin now contains valine instead of glutamate at one location, hemoglobin molecules form semirigid rods. The resulting sickle-shaped cells clog blood vessels and die off more quickly than normal-shaped cells.

Frameshift mutations occur most often because one or more nucleotides are either inserted or deleted from DNA. The result of a frameshift mutation can be a completely new sequence of codons and nonfunctional proteins. Here is how this occurs: The sequence of codons is read from a specific starting point, as in this sentence, THE CAT ATE THE RAT. If the letter C is deleted from this sentence and the reading frame is shifted, we read THE ATA TET HER AT—something that doesn't make sense. Cystic fibrosis involves a frameshift mutation that results from a faulty code for a chloride ion channel protein in the plasma membrane.

A single nonfunctioning protein can have a dramatic effect on the phenotype. Section 10.20 discusses **transposons**, which are movable genetic elements. For example, the human transposon *Alu* is responsible for hemophilia. When *Alu* inserts into the gene for clotting factor IX, it places a premature stop codon there.

One particular metabolic pathway in cells is as follows:

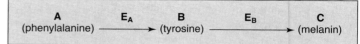

If a faulty code for enzyme E_A is inherited, a person is unable to convert molecule A to molecule B. Phenylalanine builds up in the system, and the excess causes the symptoms of the genetic disorder phenylketonuria (PKU). In the same pathway, if a person inherits a faulty code for enzyme E_B, then B cannot be converted to C, and the individual is an albino.

A rare condition called androgen insensitivity is due to a faulty receptor for androgens, which are male sex hormones such as testosterone. Although there is plenty of testosterone in the blood, the cells are unable to respond to it. Female instead of male external genitals form, and female secondary sex characteristics occur. The individual, who appears to be a normal female, may be prompted to seek medical advice when menstruation never starts. The person is XY rather than XX and does not have the internal sexual organs of a female (**Fig. 10.18B**).

Agents that cause mutations are considered in Section 10.19.

> **10.18 Check Your Progress** Why would a frameshift mutation in *Arabidopsis* affect a protein to a greater degree if it altered the base sequence early on?

FIGURE 10.18B
An XY person with androgen insensitivity.

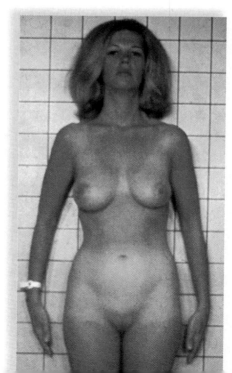

FIGURE 10.18A Types of point mutations.

Point mutations

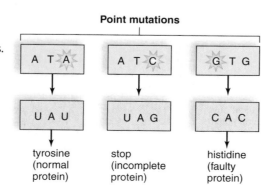

A T A	A T C	G T G
UAU	UAG	CAC
tyrosine (normal protein)	stop (incomplete protein)	histidine (faulty protein)

10.19 Many agents can cause mutations

Some mutations are spontaneous—they happen for no apparent reason—while others are due to environmental mutagens. An example of a spontaneous germ-line mutation in humans is achondroplasia, which results in a type of dwarfism and is a dominant trait (**Fig. 10.19A**). Spontaneous mutations due to DNA replication errors are rare. DNA polymerase, the enzyme that carries out replication, proofreads the new strand against the old strand and detects any mismatched nucleotides; usually, each

FIGURE 10.19A
Achondroplasia.

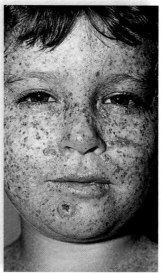

FIGURE 10.19B
Xeroderma pigmentosum.

is replaced with a correct nucleotide. In the end, only about one mistake occurs for every 1 billion nucleotide pairs replicated. Also, a cell has DNA repair enzymes that fix any mutations that occur independent of replication. The importance of repair enzymes is exemplified by individuals with the condition known as xeroderma pigmentosum (**Fig. 10.19B**). They lack some of the repair enzymes, and as a consequence, these individuals have a high incidence of skin cancer.

Environmental Mutagens A mutagen is an environmental agent that increases the chances of a mutation. Among the best-known mutagens are radiation and organic chemicals. Many mutagens are also **carcinogens**, meaning that they cause cancer. Scientists use the Ames test for mutagens to hypothesize that a chemical can be carcinogenic. In the Ames test, a histidine-requiring (His−) strain of bacteria is exposed to a chemical. If the chemical is mutagenic, the bacterium regains the ability to grow without histidine. A large number of chemicals used in agriculture and industry give a positive Ames test result. Examples are ethylene dibromide (EDB), which is added to leaded gasoline (to vaporize lead deposits in the engine and send them out the exhaust), and ziram, which is used to prevent fungal disease on crops. Some drugs, such as isoniazid (used to prevent tuberculosis), are mutagenic according to the Ames test. The mutagenic potency of AF-2, a food additive once widely used in Japan, and safrole, a natural flavoring agent that used to be added to root beers, caused them to be banned. Although most testing has been done on man-made chemicals, many naturally occurring substances (such as safrole) have been shown to be mutagenic. These include aflatoxin, produced in moldy grain and peanuts (and present in peanut butter at an average level of 2 parts per billion). Traces of nine different substances that give positive Ames test results have been found in fried hamburger.

Tobacco smoke contains a number of organic chemicals that are known carcinogens, and an estimated one-third of all cancer deaths are attributed to smoking. Lung cancer is the most frequent lethal cancer in the United States, and smoking is also implicated in the development of cancers of the mouth, larynx, bladder, kidneys, and pancreas. The greater the number of cigarettes smoked per day, the earlier the habit starts, and the higher the tar content, the greater chance a person has of developing cancer. When smoking is combined with drinking alcohol, the risk of these cancers increases even more.

Aside from chemicals, certain forms of radiation, such as X-rays and gamma rays, are called ionizing radiation because they create free radicals, which are ionized atoms with unpaired electrons. Free radicals react with and alter the structure of other molecules, including DNA. Ultraviolet (UV) radiation is easily absorbed by the pyrimidines in DNA. Wherever two thymine molecules exist next to one another, ultraviolet radiation may cause them to bond together, forming thymine dimers. A kink in the DNA results:

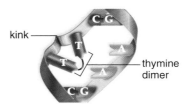

Usually, these dimers are removed from damaged DNA by repair enzymes, which constantly monitor DNA and fix any irregularities. One enzyme excises a portion of DNA that contains the dimer; another makes a new section by using the other strand as a template; and still another seals the new section in place. Because of the carcinogenic effect of X-rays, it is wise to avoid unnecessary exposure. When skin cancer develops because of sunbathing, repair enzymes have failed.

Transposons, which are discussed in Section 10.20, are also agents of mutations.

> **10.19** *Check Your Progress* **The more a person is exposed to environmental mutagens, such as those in cigarette smoke, the more likely it is that cancer will develop. Explain in biochemical terms.**

Barbara McClintock is shown in **Figure 10.20A** holding the Lasker Award for Basic Medical Research she won in 1981 for her study of transposons, which are sometimes called "jumping genes." When she began studying inheritance in corn (maize) plants, geneticists believed that each gene had a fixed locus on a chromosome. Thomas Morgan and his colleagues at Columbia University were busy mapping the chromosomes of *Drosophila,* but McClintock preferred to work with corn. In the course of her studies, she concluded that "controlling elements"—later called transposons—could undergo *transposition* and move from one location to another on the chromosome. If a transposon lands in the middle of a gene, it prevents the expression of that gene. Dr. McClintock said that because transposons are capable of suppressing gene expression, they could account for the pigment pattern of the corn strain popularly known as Indian corn (**Fig. 10.20B**).

Suppose, for example, that the expression of a normal gene results in a corn kernel that is purple:

What happens if transposition causes a transposon to land in the middle of this normal gene? The cells of the corn kernel are unable to produce the purple pigment, and the corn kernel is now white, instead of purple:

While mutations are usually stable, a transposition is very unstable. When the transposon jumps to another chromosomal location, some cells regain the ability to produce the purple pigment, and the result is a corn kernel with a speckled pattern, as shown in Figure 10.20B.

When McClintock first published her results in the 1950s, the scientific community paid little attention. Years later, when molecular genetics was well established, transposons were also discovered in bacteria, yeasts, plants, fruit flies, and humans. Geneticists now believe that transposons have the following effects:

1. Are involved in transcriptional control because they block transcription.

2. Can carry a copy of certain host genes with them when they jump. Therefore, they can be a source of chromosomal mutations such as translocations, deletions, and inversions.

3. Can leave copies of themselves and certain host genes before jumping. Therefore, they can be a source of a duplication, another type of chromosomal mutation.

4. Can contain one or more genes that make a bacterium resistant to antibiotics.

FIGURE 10.20A Barbara McClintock.

FIGURE 10.20B Transposons are responsible for the speckled or striped patterns in Indian corn.

Considering that transposition has a powerful effect on genotype and phenotype, it most likely has played an important role in evolution. For her discovery of transposons, McClintock was, in 1983, finally awarded the Nobel Prize in Physiology or Medicine. In her Nobel Prize acceptance speech, the 81-year-old scientist proclaimed that "it might seem unfair to reward a person for having so much pleasure over the years, asking the maize plant to solve specific problems, and then watching its responses."

> **10.20** *Check Your Progress* If a transposon results in a nonfunctioning enzyme in *Arabidopsis,* is it likely to have landed in an intron or an exon? Explain.

One of the most exciting periods of scientific activity in the history of biology occurred during the 30 short years between the 1930s and the 1960s. Due to several elegantly executed experiments, by the mid-1950s researchers realized that DNA, not protein, is the genetic material. Using all previously collected data concerning DNA structure, Watson and Crick were able to arrive at the legendary design of DNA—a double helix. Complementary base pairing explains the replication of DNA, how RNA molecules are made from a DNA template, and how protein synthesis comes about.

By studying the activity of genes in cells, geneticists have confirmed that proteins are the link between the genotype and the phenotype. In other words, you have blue, or brown, or hazel eye pigments because of the types of enzymes (proteins) contained within your cells. Always keep in mind this flow diagram:

DNA base sequence $\longrightarrow$ amino acid sequence $\longrightarrow$ enzyme $\longrightarrow$ organism structure

We now know the sequence of bases in human DNA, but it turns out that humans have far fewer genes than expected. A complicated organism such as a human being can make do with fewer genes if each gene has more than one function, according to how it is regulated. Regulation of gene activity, to be discussed in Chapter 11, has become a focal point of modern-day research.

The Chapter in Review

Summary

Arabidopsis Is a Model Organism

- *Arabidopsis thaliana* is a small flowering plant used in the study of plant molecular genetics.

DNA Is the Genetic Material

10.1 DNA is a transforming substance
- Frederick Griffith's experiments with bacteria suggested that a transforming substance was genetic material.

10.2 DNA, not protein, is the genetic material
- Hershey and Chase's experiments showed that DNA, not protein, entered bacterial cells and directed phage reproduction.

10.3 DNA and RNA are polymers of nucleotides
- DNA contains deoxyribose; adenine (A) pairs with thymine (T); cytosine (C) pairs with guanine (G).
- RNA contains ribose and the bases A, C, G, and uracil (U) instead of T.

10.4 DNA meets the criteria for the genetic material
- DNA varies between species, can store information, remains constant within a species, is replicated, and undergoes mutations.

10.5 DNA is a double helix
- Watson and Crick constructed the first double helix model of DNA, using Franklin and Wilkins's X-ray diffraction data.
- The double helix model suggests (1) a species has a stable sequence of bases; (2) the sequence can be variable between species; (3) how the replication of DNA occurs.

DNA Can Be Duplicated

10.6 DNA replication is semiconservative
- Semiconservative replication means that each new double helix contains an old strand and a new strand.
- The steps in replication are unwinding, complementary base pairing, and joining.
- DNA polymerase is used in pairing and joining.

10.7 Many different proteins help DNA replicate
- DNA strands must be antiparallel for complementary base pairing to occur.
- As DNA unwinds, replication is continuous for the leading strand but discontinuous for the lagging strand.

Genes Specify the Makeup of Proteins

10.8 Genes are linked to proteins
- The one gene, one enzyme hypothesis is based on the observation that a defective gene caused a defective enzyme.
- Proteins are the link between genotype and phenotype.

10.9 The making of a protein requires transcription and translation
- Gene expression has two steps:
  - In transcription, DNA is transcribed into mRNA. rRNA and tRNA are also made from a DNA template.
  - In translation, the mRNA transcript directs the amino acid sequence.

10.10 The genetic code for amino acids is a triplet code
- The genetic code is triplet degenerate, unambiguous, has start and stop signals, and is nearly universal.

10.11 During transcription, a gene passes its coded information to an mRNA
- During transcription, complementary base pairing occurs, and mRNA results when RNA polymerase joins bases.

10.12 In eukaryotes, an mRNA is processed before leaving the nucleus

- The exon is DNA that will be expressed; the intron is DNA that will not be expressed but may have a regulatory function.
- mRNA receives a cap and a tail.
- Introns are removed by RNA splicing.
- Mature mRNA is ready to be translated.

10.13 During translation, each transfer RNA carries a particular amino acid

- In cytoplasm, tRNA transfers amino acids to ribosomes.
- tRNA's anticodon base-pairs with mRNA's codon.

10.14 Translation occurs at ribosomes in cytoplasm

- Polypeptide synthesis occurs as a ribosome moves down mRNA.
- A ribosome has binding sites for mRNA and three tRNAs at the: E (exit) site, the P (peptide) site, and the A (amino acid) site.
- A polyribosome is composed of several ribosomes attached to and translating the same mRNA.

10.15 Initiation begins the process of polypeptide production

- In prokaryotes, ribosome subunits, mRNA, and initiator tRNA come together.
- Initiation is more complicated in eukaryotes.

10.16 Elongation builds a polypeptide one amino acid at a time

- During elongation, a tRNA at the P site passes a peptide to a tRNA-amino acid at the A site. Now translocation occurs: The ribosome moves forward and the peptide bearing tRNA is at the P site and the used tRNA exits from the E site. This process occurs over and over again.
- At termination, the ribosome reaches a stop codon, and the polypeptide is released.

10.17 Let's review gene expression

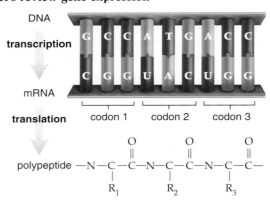

Mutations Are Changes in the Sequence of DNA Bases

10.18 Mutations affect genetic information and expression

- A genetic mutation is a permanent change in the sequence of DNA bases.
- Both point mutations and frameshift mutations can cause genetic disorders.

10.19 Many agents can cause mutations

- Spontaneous mutations are rare and happen for no apparent reason.
- Environmental mutagens include radiation (X-ray, gamma ray, and UV) and organic chemicals; many are carcinogens.

10.20 Transposons are "jumping genes"

- Transposons can block transcriptions and be a source of translocations, deletions, inversions, or duplications as well as having other effects.

Testing Yourself

DNA Is the Genetic Material

1. In the Hershey and Chase experiments, radioactive phosphorus was found
   a. outside the bacterial cells.
   b. inside the bacterial cells.
   c. both inside and outside the bacterial cells.
2. If 30% of an organism's DNA is thymine, then
   a. 70% is purine.　　　d. 70% is pyrimidine.
   b. 20% is guanine.　　e. Both c and d are correct.
   c. 30% is adenine.
3. In a DNA molecule, the
   a. backbone is sugar and phosphate molecules.
   b. bases are covalently bonded to the sugars.
   c. sugars are covalently bonded to the phosphates.
   d. bases are hydrogen-bonded to one another.
   e. All of these are correct.

DNA Can Be Duplicated

4. Because each daughter molecule contains one old strand of DNA, DNA replication is said to be
   a. conservative.　　c. semidiscontinuous.
   b. preservative.　　d. semiconservative.
5. During DNA replication, the parental strand ATTGGC would code for the daughter strand
   a. ATTGGC.　　c. TAACCG.
   b. CGGTTA.　　d. GCCAAT.

For questions 6–8, match the function in DNA replication with the enzyme in the key.

KEY:
   a. DNA helicase
   b. DNA polymerase
   c. DNA ligase

6. Addition of new complementary DNA nucleotides to the daughter strand.
7. Unwinds and unzips DNA.
8. Seals breaks in the sugar-phosphate backbone.
9. **THINKING CONCEPTUALLY** AZT, the well-known medicine for an HIV infection, is a DNA base analogue that hinders DNA replication. Explain why it works.

Genes Specify the Makeup of Proteins

10. Transcription produces _____, while translation produces
   _____.
   a. DNA, RNA　　　　　c. polypeptides, RNA
   b. RNA, polypeptides　　d. RNA, DNA
11. Which of the following statements does not characterize the process of transcription? Choose more than one answer if correct.
   a. RNA is made with one strand of the DNA serving as a template.
   b. In making RNA, the base uracil of RNA pairs with the base thymine of DNA.

c. The enzyme RNA polymerase synthesizes RNA.
d. RNA is made in the cytoplasm of eukaryotic cells.
12. Because there are more codons than amino acids,
a. some amino acids are specified by more than one codon.
b. some codons specify more than one amino acid.
c. some codons do not specify any amino acid.
d. some amino acids do not have codons.
13. If the sequence of bases in DNA is TAGC, then the sequence of bases in RNA will be
a. ATCG.   c. AUCG.
b. TAGC.   d. GCTA.
e. Both a and b are correct.
14. RNA processing
a. is the same as transcription.
b. is an event that occurs after RNA is transcribed.
c. is the rejection of old, worn-out RNA.
d. pertains to the function of transfer RNA during protein synthesis.
e. Both b and d are correct.
15. During protein synthesis, an anticodon on transfer RNA (tRNA) pairs with
a. DNA nucleotide bases.
b. ribosomal RNA (rRNA) nucleotide bases.
c. messenger RNA (mRNA) nucleotide bases.
d. other tRNA nucleotide bases.
e. Any one of these can occur.
16. Following is a segment of a DNA molecule. (Remember that the template strand only is transcribed.) What are (a) the RNA codons, (b) the tRNA anticodons, and (c) the sequence of amino acids in a protein?

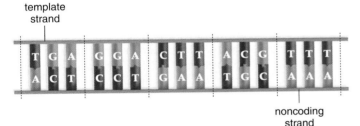

template strand

noncoding strand

Mutations Are Changes in the Sequence of DNA Bases

17. Transposable elements cause mutations by
a. altering DNA nucleotides.
b. removing segments of genes.
c. inserting themselves into genes.
d. breaking up RNA molecules.
18. A mutation involving the replacement of one DNA nucleotide base pair with another is called
a. a frameshift mutation.
b. a point mutation.
c. a transposon.
19. Give an example of a point mutation that would have no effect on the cell. Explain.
20. **THINKING CONCEPTUALLY** Mutations can cause cancer but, on the other hand, it is important for DNA to mutate. Explain.

Match the terms to these definitions:
a. _____ Enzyme that speeds the formation of mRNA from a DNA template.
b. _____ Noncoding segment of DNA that is transcribed, but the transcript is removed before mRNA leaves the nucleus.
c. _____ Process whereby the sequence of codons in mRNA determines (is translated into) the sequence of amino acids in a polypeptide.
d. _____ String of ribosomes simultaneously translating different regions of the same mRNA strand during protein synthesis.

Thinking Scientifically

1. How would you test your hypothesis that the genetic condition neurofibromatosis is due to a transposon?
2. Knowing that you can clone plants from a few cells in tissue culture, how would you determine if an isolated *Arabidopsis* gene causes a particular mutation?

ARIS *Visit* **www.mhhe.com/maderconcepts** *for practice quizzes, animations, videos, and activities designed to help you master the material in this chapter.*

11

Regulation of Gene Activity

LEARNING OUTCOMES

After studying this chapter, you should be able to accomplish the following outcomes.

Moth and Butterfly Wings Tell a Story

1 Relate the presence of eyespots to homeotic genes.

Gene Expression Is Controlled in Prokaryotic Cells

2 Describe an operon model as a means of regulating gene expression in prokaryotes.
3 Contrast and compare inducible operons with repressible operons.

Control of Gene Expression in Eukaryotes Causes Specialized Cells

4 Relate the specialization of cells to the activity of genes.
5 Relate totipotency to the ability to produce an entire organism.
6 Compare and contrast reproductive and therapeutic cloning.
7 Discuss the pros and cons of animal cloning.

Control of Gene Expression Is Varied in Eukaryotes

8 Describe chromosome structure and how it relates to control of gene expression in eukaryotes.
9 Explain the role of transcription activators and transcription factors in the eukaryotic nucleus.
10 Give examples of posttranscriptional control, translational control, and posttranslational control in eukaryotes.

Gene Expression Is Controlled During Development

11 Explain what keeps development so orderly.
12 Use the homeodomain to support the concept that living things are related through evolution.

Genetic Mutations Cause Cancer

13 Explain the antagonistic actions of proto-oncogenes and tumor suppressor genes.
14 Contrast the signal transduction pathways of proto-oncogenes and tumor suppressor genes.
15 Describe the development of cancer as a multistep process.

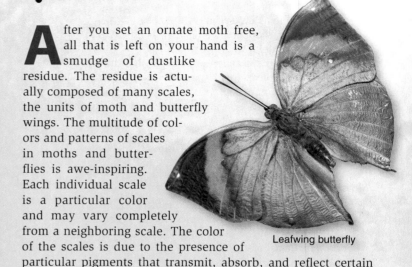
Leafwing butterfly

After you set an ornate moth free, all that is left on your hand is a smudge of dustlike residue. The residue is actually composed of many scales, the units of moth and butterfly wings. The multitude of colors and patterns of scales in moths and butterflies is awe-inspiring. Each individual scale is a particular color and may vary completely from a neighboring scale. The color of the scales is due to the presence of particular pigments that transmit, absorb, and reflect certain colors of light.

Most specialists who study insects agree that scales evolved from the bristles of an ancestor to moths and butterflies. Over time, the bristles became wide and flat and lost any sensory function. You might think that scales have an accessory and unnecessary function, but evidence suggests otherwise. For example, the easy detachment of scales may have made it easier for ancient moths and butterflies to escape from spiderwebs and other predators. The possible protective function of scales is strengthened by their role in forming eyespots, a rounded eye-like marking, on moth and butterfly wings.

eyespot
Bull's-eye moth

Other animals also have eyespots. For example, eyespots can be found on the tail of a redfish, the bodies of spiders and caterpillars, and occasionally on the back of a lynx's ear or the back of a cobra's hood. Eyespots confuse a potential predator and may divert attacks to body margins, thereby saving most of the animal from damage. Certainly an animal as delicate as a moth or butterfly needs all the help it can get to keep its body from being attacked.

Still more evidence suggests the importance of scales to the life of a moth or butterfly. Developmental biologists specializing in evolution have discovered that the same genes involved in building insect limbs are also involved in determining eyespot patterns. Eyespots are the result of basic organizer genes that lie at the center of development itself. Like that of any animal, the body organization of a moth or a butterfly is determined by ancient genes called homeotic genes. **Homeotic genes** are important switches for organizing differentiated cells into specific structures. This means that the amazing scales of moths and butterflies are a product of the genes that make them what they are. Without

these genes, there would be no scales and indeed no moths and butterflies. The two are intertwined. The next time a moth or butterfly flits by—think about it.

Homeotic genes are regulatory genes. Gene regulation plays an important role in the development of an organism before birth and in its health and welfare after it is born. This chapter surveys the basic rules of regulation in prokaryotes and eukaryotes before taking a look at abnormalities caused by the lack of proper regulation.

retractable antennae

eyespot

Caterpillar of citrus swallowtail butterfly

eyespots

eyespot

Indian spectacled cobra

Polyphemus moth hindwings

Gene regulation was first discovered in prokaryotes, and the study of operons in prokaryotes offers an opportunity to understand what is meant by gene regulation. This part of the chapter also introduces us to the concept that DNA-binding proteins are involved in gene regulation.

11.1 DNA-binding proteins turn genes on and off in prokaryotes

Because their environment is ever-changing, bacteria do not need the same enzymes (and possibly other proteins) all the time. For example, they need only the enzymes required to break down the nutrients available to them and the enzymes required to synthesize whatever metabolites are absent under the present circumstances.

In 1961, the French microbiologists François Jacob and Jacques Monod showed that *Escherichia coli* is capable of regulating the expression of its genes. They proposed the **operon** model to explain gene regulation in prokaryotes and later received a Nobel Prize for their investigations.

Note in **Figure 11.1A** that the operon model has these components:

A **regulator gene** is located outside the operon. The regulator gene codes for a repressor that, when active, binds to the operator and inactivates the operon.

A **promoter** is a short sequence of DNA where RNA polymerase first attaches when a gene is to be transcribed. Basically, the promoter signals the start of an operon.

An **operator** is a short portion of DNA where an active repressor binds. When an active repressor is bound to the operator, RNA polymerase cannot attach to the promoter, and transcription does not occur. In this way, the operator controls mRNA synthesis.

The **structural genes** may be one to several genes coding for the primary structure of the enzymes of a metabolic pathway that are transcribed as a unit.

Figure 11.1A and **Figure 11.1B** apply this information to the *lac* operon. The *lac* operon does not produce enzymes to digest the sugar lactose when it is not available to *E. coli*. Why not? Because the repressor binds to the operator, and this prevents RNA polymerase from binding to the promoter. When RNA polymerase cannot bind to the promoter, no mRNA transcripts are available for the production of lactose enzymes. On the other hand, when lactose is present, it binds to the repressor, inactivating it. Now RNA polymerase has room to attach to the promoter. The production of mRNA transcripts and lactose enzymes follows. In other words, a change in the shape of the repressor "induces" the operon to be active, and the *lac* operon is called an **inducible operon**.

In contrast to the *lac* operon, other bacterial operons, such as those that control amino acid synthesis, are usually functioning and producing enzymes most of the time. For example, in the *trp* operon, the regulator gene codes for a repressor that ordinarily is unable to attach to the operator. Therefore, RNA polymerase can bind to the promoter, and the genes needed to make the amino acid tryptophan are ordinarily expressed. When tryptophan is present, it binds to the repressor. A change in shape activates the repressor and allows it to bind to the operator. Now the operon is turned off. In other words, the *trp* operon must be repressed to turn it off, and therefore it is a **repressible operon**.

This completes our study of gene regulation in prokaryotes, and in the next part of the chapter, we will begin our study of gene regulation in eukaryotes.

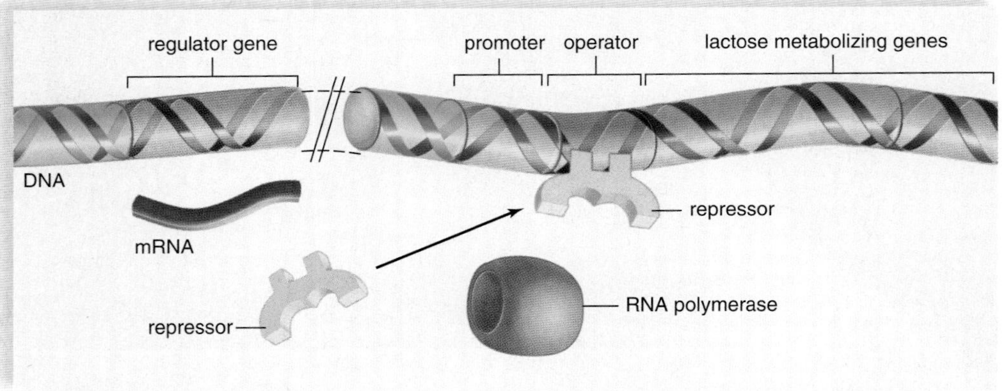

FIGURE 11.1A Inactive *lac* operon: Lactose enzymes are not produced.

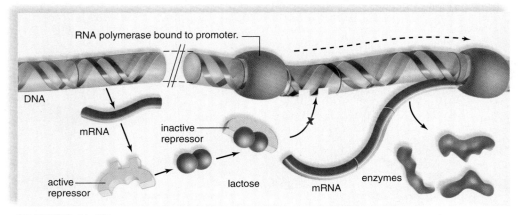

FIGURE 11.1B Active *lac* operon: Lactose enzymes are produced.

> **11.1 Check Your Progress** How does RNA polymerase know where to begin transcribing a prokaryotic gene?

Control of Gene Expression in Eukaryotes Causes Specialized Cells

The specialization of eukaryotic cells suggests that their genes are regulated—only some genes are functional in each cell. How might we show that all eukaryotic cells contain a full complement of potentially functional genes? Cloning of plants and animals provides this opportunity.

11.2 Eukaryotic cells are specialized

Unlike a prokaryote, which is a single cell, a multicellular organism, such as a human being, contains many types of cells that differ in structure and function. From our knowledge of cell division, we know that each and every cell in an organism contains a complete set of genes. But even so, each cell type contains its own mix of proteins that make it different from all other cell types. Therefore, only certain genes are turned on and active in human cells that perform specialized functions, such as nerve, muscle, gland, and blood cells.

Some of these active genes are called housekeeping genes because they govern functions that are common to many types of cells, such as glucose metabolism. But otherwise, the activity of selected genes accounts for the specialization of cells. In other words, gene expression is controlled in a cell, and this control accounts for its specialization (**Fig. 11.2**).

> **11.2** *Check Your Progress* **Homeotic genes do not normally function in adults. Explain.**

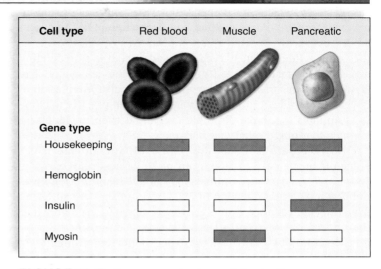

FIGURE 11.2 Gene expression in specialized cells.

11.3 Plants are cloned from a single cell

Theoretically, any cell, whether plant or animal, is **totipotent**, which means it has the ability to give rise to an entire organism. Totipotency in plants has been demonstrated for many years. For example, it is possible to grow a complete carrot plant from a tiny piece of phloem (**Fig. 11.3**). Today, plants can even be grown from a single cell. First, plant cell walls are removed by digestive enzyme action, resulting in naked cells, or protoplasts. The cell wall regenerates as cell division produces aggregates of cells called a callus. With proper stimulation, the callus differentiates into shoots and roots. Embryonic plants make their appearance and develop into plantlets. The new plants have the

same characteristics as the one that donated the 2n nuclei, and therefore they are clones of this plant. One advantage of propagating plants in the laboratory is the ability to grow a number of identical commercial plants in a small area.

Animal cloning occurs for two different purposes, as discussed in Section 11.4.

> **11.3** *Check Your Progress* **Theoretically, are animal cells totipotent? Do they contain all necessary genes, including homeotic genes, that can be activated if properly stimulated?**

FIGURE 11.3 Cloning carrots.

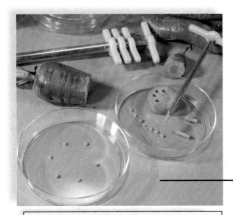
1 Tiny disks are obtained from carrot root.

2 Each disk produces an undifferentiated tissue mass called a callus.

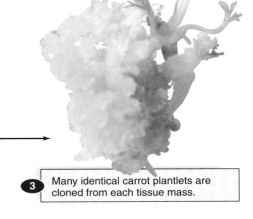
3 Many identical carrot plantlets are cloned from each tissue mass.

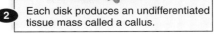

Just as with plant cells, the 2n nucleus from a donor animal can also be used for cloning. In **reproductive cloning**, the desired end is an individual that is exactly like the original individual. At one time, investigators found it difficult to have the nucleus of an adult cell "start over"—that is, overcome its specialization and direct the development of an animal. But in March 1997, Scottish investigators announced they had successfully cloned a Dorset sheep, which they named Dolly. How was their procedure different from all the others that had been attempted? Again, an adult nucleus was placed in an enucleated egg cell. However, the donor cell had been starved, which caused it to stop dividing and go into a resting stage (the G_0 stage of the cell cycle). The G_0 nucleus was amenable to cytoplasmic signals for initiation of development (**Fig. 11.4A**). Today it is common practice to clone farm animals that have desirable traits, and even to clone rare animals that might otherwise become extinct.

However, at this point, the cloning of farm animals is not efficient. In the case of Dolly, out of 29 clones, only one was successful. Also, cloned animals may not be healthy. Dolly was put down by lethal injection in 2003 because she was suffering from lung cancer and crippling arthritis. She had lived only half the normal life span for a Dorset sheep. In the United States, no federal funds can be used for experiments to reproductively clone human beings.

In human **therapeutic cloning**, the desired end is not an individual organism, but mature cells of various cell types (**Fig. 11.4B**). The purposes of therapeutic cloning are (1) to learn more about how specialization of cells occurs, and (2) to provide cells and tissues that could be used to treat human illnesses, such as diabetes, spinal cord injuries, and Parkinson disease.

Therapeutic cloning can be carried out in several ways. One procedure is the same as that used for reproductive cloning, except that embryonic cells, called *embryonic stem cells*, are isolated, and each is subjected to a treatment that causes it to develop into a particular type of cell, such as red blood cells, muscle cells, or nerve cells. Ethical concerns exist about this procedure because if the embryo had been allowed to continue development, it would have become a person. However, we now know that embryonic stem cells can also be obtained from the fluid surrounding an embryo. If you use these cells, you can simply treat them differently in order to produce specialized tissues.

Another way to carry out therapeutic cloning is to use *adult stem cells*, which are found in many organs of an adult's body. For example, the skin has stem cells that constantly divide and produce new skin cells, while the bone marrow has stem cells that produce new blood cells.

It is easiest to use adult stem cells for the tissue you want to produce, but this is not always possible. For example, to acquire nervous tissue stem cells, you have to take them from the brain. Recently, investigators circumvented the need to use adult stem cells by succeeding in obtaining specialized cells, namely skin cells, to return to an embryonic state. Obviously, the treatment of these cells to become tissues solves any ethical issues and any problems of availability. Knowledge of gene regulation is a must in order to achieve therapeutic and also reproductive cloning, the topic of the next section.

> **11.4 Check Your Progress** What is the difference between reproductive and therapeutic cloning?

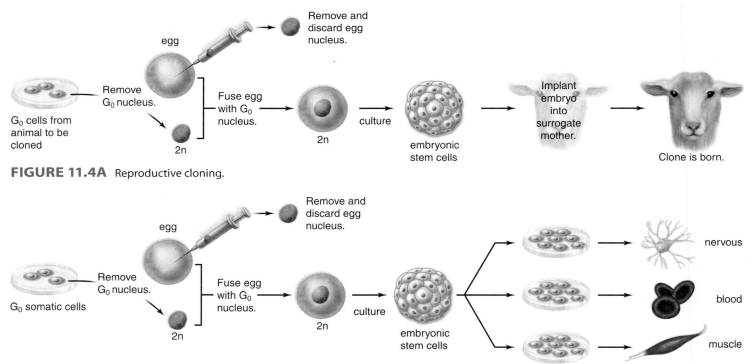

FIGURE 11.4A Reproductive cloning.

FIGURE 11.4B Therapeutic cloning.

11.5 Animal cloning has benefits and drawbacks

A number of Hollywood thrillers take advantage of the general public's lack of knowledge about cloning by portraying cloning as an evil process. In the movies, scientists clone dinosaurs, mammoths, and even saber-toothed tigers from prehistoric days. Sometimes, scientists clone evil characters from the past in order to take over the world. If considered pure entertainment, the movies are okay. However, many people take these movies seriously, and therefore consider cloning the work of dark hearts.

As mentioned in Section 11.4, there are two kinds of animal cloning. Reproductive cloning results in a replica of an individual (see Fig. 11.4A). Therapeutic cloning is used to produce stem cells that may be valuable in medical research (see Fig. 11.4B). Only with reproductive cloning is the zygote placed in a surrogate female for further development. If all goes well, an identical copy of the donor will be born at the end of the gestation period. Although this process sounds relatively simple, the failure rate is high.

The first mammal to be successfully cloned was Dolly, the sheep (see Section 11.4). Since Dolly, a number of mammals have been successfully cloned, including mice, rabbits, cats, dogs, pigs, deer, horses, cattle, mules, and rhesus monkeys (**Fig. 11.5A**).

The debate regarding animal cloning is intense. Both sides present valid arguments, and both sides are capable of powerful displays of emotion. Recent surveys by CNN and *Time* showed that 66% of Americans think animal cloning is immoral, 74% think cloning is against God's will, and although 49% do not mind eating cloned plants, only 33% will eat cloned animals.

Opponents of reproductive cloning present several valid scientific arguments. They contend that, because the donor's mitochondrial DNA is not passed to the clone, premature aging may occur; that the mutation rate is higher in clones and the regulation of gene expression is abnormal; that clones are prone to "large offspring syndrome"; and that the process of development can result in a clone that is different from the donor. For example, pigs that have been cloned are no more alike than siblings (**Fig. 11.5B**), and the cloned cat Carbon Copy is very different from its

FIGURE 11.5B
Cloned pigs.

donor, Rainbow (**Fig. 11.5C**). In addition to the tremendous failure rate at present, cloning is expensive; a cloned cat can cost as much as $50,000 to produce. Cloning also raises ethical issues concerning animal rights and the loss of individuality.

On the other hand, animal cloning has many advocates. The process of cloning increases our knowledge of gene interactions and embryological development. Cloning may be the only way, at present, to save endangered species. For example, cloning can save a species even if there are no females remaining, and can produce offspring when animals are infertile.

Therapeutic cloning can be used to develop and repair organs, combat cancer, and fight disease. However, therapeutic cloning often means that a human embryo is being used for its ability to create tissues. Recently, researchers have found embryonic cells in the liquid surrounding the embryo in the womb. If these cells can be successfully used to create tissues, therapeutic cloning utilizing an embryo will not be needed.

We have finished our study of cloning. In the next part of the chapter, we consider the methods by which genes are turned on or off in eukaryotic cells.

> **11.5 Check Your Progress** What's the difference between therapeutic cloning that utilizes an embryo and starting with embryonic cells taken from the fluid about an embryo in the womb?

FIGURE 11.5A
Cloned rhesus monkeys.

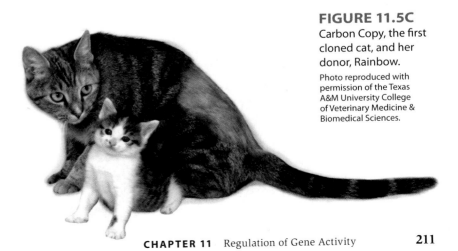

FIGURE 11.5C
Carbon Copy, the first cloned cat, and her donor, Rainbow.

Photo reproduced with permission of the Texas A&M University College of Veterinary Medicine & Biomedical Sciences.

In eukaryotes, regulation of gene expression occurs in the nucleus and in the cytoplasm. In the nucleus, we will see that genes occurring in highly condensed chromatin are not transcribed, and that DNA-binding proteins control the degree to which genes are transcribed. mRNA processing can alter which genes are sent to the cytoplasm and how quickly they are translated in the cytoplasm. Other methods of control also occur in the cytoplasm.

11.6 Chromatin is highly condensed in chromosomes

The DNA in eukaryotes is always associated with plentiful proteins, and together they make up the stringy material, called **chromatin**, that can be observed in the interphase nucleus. Previously, we discussed how chromatin is condensed to form chromosomes that are visible during cell division. DNA is periodically wound around a core of eight protein molecules so that it looks like beads on a string (**Fig. 11.6**). The protein molecules are **histones**, and each bead is called a **nucleosome**. The levels of chromatin organization seen in a nucleus can be related to the degree that the nucleosomes

coil as chromatin condenses. Just before cell division, chromatin coils tightly into a fiber that has six nucleosomes to a turn. Then the fiber loops back and forth, and condenses further to produce highly condensed chromosomes. Genes in highly condensed chromatin are not active, as discussed in Section 11.7.

> **11.6 Check Your Progress** What are the two major components of a nucleosome (see Fig. 11.6)?

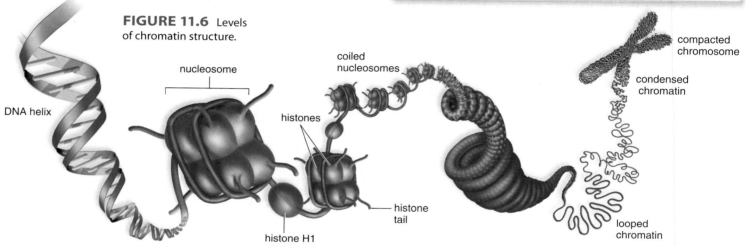

FIGURE 11.6 Levels of chromatin structure.

DNA helix · nucleosome · histones · histone H1 · coiled nucleosomes · histone tail · compacted chromosome · condensed chromatin · looped chromatin

11.7 The genes in highly condensed chromatin are not expressed

Some portions of chromatin are highly condensed, and these appear in electron micrographs as darkly stained portions called **heterochromatin** (**Fig. 11.7A**). A dramatic example of heterochromatin is seen in mammalian females. Females have a small, darkly staining mass of condensed chromatin adhering to the inner edge of the nuclear envelope. This structure, called

a **Barr body** after its discoverer Murray Barr, is an inactive X chromosome. On a random basis, one of the X chromosomes undergoes inactivation in the cells of female embryos. The inactive X chromosome does not produce gene products, and therefore female cells have a reduced amount of product from genes on the X chromosome.

How do we know that Barr bodies are inactive X chromosomes that are not producing gene product? To support this hypothesis, investigators have discovered that a female who is heterozygous for an X-linked recessive disorder has patches of normal cells and patches of abnormal cells. For example, women heterozygous for Duchenne muscular dystrophy have patches of normal muscle tissue and patches of degenerative muscle tissue. The female tortoiseshell cat also provides support for X-inactivation. In these cats, an allele for black coat color is on one X chromosome, and a corresponding allele for orange coat color is on the other X chromosome. The patches of black and orange in the coat can be related to which X chromosome is active and which is in the Barr bodies of the cells found in the patches (**Fig. 11.7B**).

FIGURE 11.7A Comparison of heterochromatin and euchromatin.

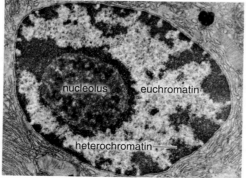

nucleolus · euchromatin · heterochromatin · 1 μm

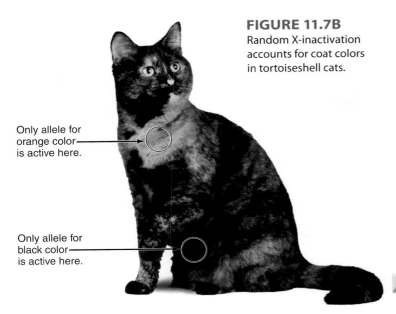

FIGURE 11.7B
Random X-inactivation accounts for coat colors in tortoiseshell cats.

Only allele for orange color is active here.

Only allele for black color is active here.

Euchromatin During interphase, most of the chromatin is not highly condensed. Instead, it is in a loosely condensed state called **euchromatin**. Whereas heterochromatin is inactive, euchromatin is potentially active in the sense that the genes located in euchromatin are subject to being expressed.

What regulates whether chromatin exists as heterochromatin or euchromatin? Histone molecules have tails, strings of amino acids that extend beyond the main portion of a nucleosome. In euchromatin, the histone tails tend to be acetylated and have attached acetyl groups ($-COCH_3$); in heterochromatin, the histone tails tend to bear methyl groups ($-CH_3$). Methylation of DNA can also occur, and when it does, genes are shut down from generation to generation.

Sections 11.8 to 11.11 consider how active genes present in euchromatin can be regulated.

11.7 Check Your Progress Would an individual with Klinefelter syndrome (XXY) have a Barr body?

11.8 DNA-binding proteins regulate transcription in eukaryotes

We have just observed the first step in **transcriptional control** in eukaryotes—the availability of DNA for transcription, as in euchromatin. The chromosomes within the developing egg cells of many vertebrates are called lampbrush chromosomes because they have many loops that appear to be bristles. (In olden days, lampbrushes were used to clean the inside glass of kerosene lamps.) Lampbrush chromosomes make genes available for transcription immediately following fertilization.

Transcription Factors and Transcription Activators

Although no operons like those of prokaryotic cells have been found in eukaryotic cells, transcription is controlled by DNA-binding proteins (**Fig. 11.8**). **1** Every cell contains many different types of **transcription factors**, proteins that help regulate transcription. A group of transcription factors binds to a promoter adjacent to a gene, but transcription may still not begin if transcription activators are also involved. **2** Often **transcription activators** are involved in promoting tran-

scription, and a different one is believed to regulate the activity of any particular gene. Transcription activators bind to regions of DNA called **enhancers.**

3 Enhancers can be quite a distance from the promoter, but bending of DNA can bring the transcription activator attached to the enhancer into the vicinity of the promoter. Mediator proteins may act as a bridge between transcription factors and the transcription activator at the promoter. Once a transcription activator makes contact with a transcription factor, RNA polymerase begins the transcription process.

mRNA processing also offers an opportunity to regulate gene expression—whether gene activity results in a product, as discussed in Section 11.9.

11.8 Check Your Progress Homeotic genes control development by turning on genes. What could they code for?

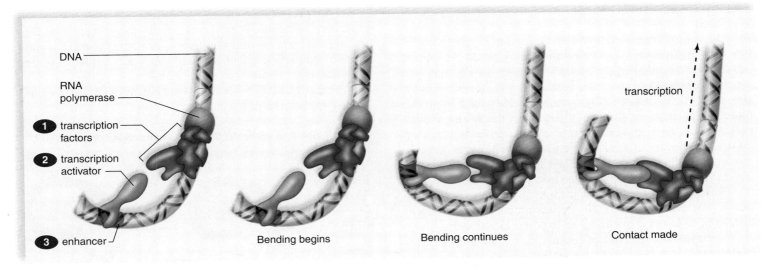

DNA

RNA polymerase

1 transcription factors

2 transcription activator

3 enhancer

Bending begins

Bending continues

transcription

Contact made

FIGURE 11.8 Activation of genes in eukaryotes.

11.9   mRNA processing can affect gene expression

Although not as significant as transcriptional control, there are other ways to control gene expression in eukaryotic cells. **Posttranscriptional control** occurs in the nucleus and involves (1) mRNA processing and (2) the speed with which mRNA leaves the nucleus.

During *mRNA processing,* introns (noncoding regions) are excised, and exons (expressed regions) are spliced together. When introns are removed from mRNA, differential splicing of exons can occur, and this affects gene expression. For example, both the hypothalamus and the thyroid gland produce a hormone called calcitonin, but the mRNA that leaves the nucleus is not the same in both types of cells. Radioactive labeling studies show that they vary because of a difference in mRNA splicing (**Fig. 11.9**). Evidence of different patterns of mRNA splicing is found in other cells, such as those that produce neurotransmitters, muscle regulatory proteins, and antibodies. It is not known how cells specify which form of mature mRNA they will make. Some researchers believe that RNAs regulate the process themselves because artificial RNAs can modify splicing patterns when they are added to cells being grown in petri dishes.

The length of time it takes mRNA transcripts to pass through a nuclear pore varies. Because the life span of an mRNA transcript is finite, the speed with which mRNA enters the cytoplasm can affect the amount of gene product formed in the cytoplasm. The final control over gene expression involves translation in the cytoplasm, as discussed in Section 11.10.

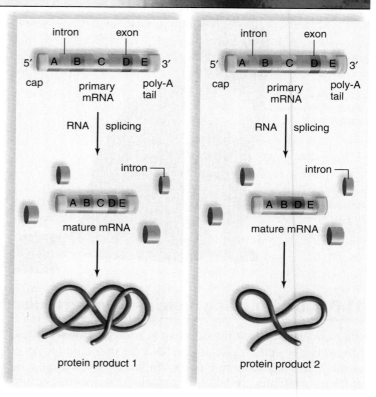

FIGURE 11.9 RNA splicing influences the particular protein product.

> **11.9 Check Your Progress** What happens to introns during mRNA processing?

11.10   Control of gene expression also occurs in the cytoplasm

Translational control begins when the processed mRNA molecule reaches the cytoplasm and before a protein product exists. For example, after being activated, an initiation factor, known as IF-2, inhibits the start of protein synthesis. Conditions inside a cell can also delay translation. For example, red blood cells do not produce hemoglobin unless heme, an iron-containing group, is available.

The longer an active mRNA remains in the cytoplasm before it is broken down, the more gene product is produced. During maturation, mammalian red blood cells eject their nuclei, and yet they continue to synthesize hemoglobin for several months after that. The necessary mRNAs must be able to persist all this time. The presence or absence of the cap at the 5′ end and the poly-A tail at the 3′ end of a mature mRNA transcript can determine whether translation takes place and how long the mRNA is active. Any influence that affects the length of the poly-A tail can lead to removal of the cap and destruction of mRNA.

Hormones cause the stabilization of certain mRNA transcripts. For example, an mRNA for vitellin persists for 3 weeks instead of 15 hours if the mRNA is exposed to estrogen.

Posttranslational control begins once a protein has been synthesized and become active. It represents the cell's last chance to influence gene expression.

Some proteins are not immediately active after synthesis. For example, at first bovine proinsulin is a single, long polypeptide that folds into a three-dimensional structure. Cleavage results in two smaller chains that are bonded together by disulfide (S—S) bonds (**Fig. 11.10**). Only then is active insulin present.

Some proteins are short-lived in cells because they are degraded or destroyed. For example, the cyclin proteins that control the cell cycle are only temporarily present. The cell has giant protein complexes, called proteasomes, that carry out the task of destroying proteins.

Section 11.11 summarizes all the means of control we have studied in this part of the chapter.

> **11.10 Check Your Progress** When does translational control begin?

FIGURE 11.10 Cleavage activates certain polypeptides.

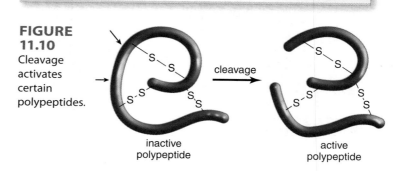

inactive polypeptide   active polypeptide

11.11 Synopsis of gene expression control in eukaryotes

We have observed various ways in which eukaryotes control gene expression, and these are summarized in **Figure 11.11**. Notice that control of gene activity in eukaryotes extends from transcription to protein activity:

1 *Chromatin structure.* Chromatin packing is used as a way to keep genes turned off. If genes are not accessible to RNA polymerase, they cannot be transcribed.

In the nucleus, highly condensed chromatin is not available for transcription, while more loosely condensed chromatin is available for transcription.

2 *Transcriptional control.* The degree to which a gene is transcribed into mRNA determines the amount of gene product.

Transposons (see Section 10.20) are DNA sequences that move between chromosomes and block transcription of particular genes. Various DNA-binding proteins are present in the nucleus and help determine when and if transcription begins. These DNA-binding proteins are: transcription factors, transcription activators, and enhancers.

3 *Posttranscriptional control.* Posttranscriptional control involves mRNA processing and how fast mRNA leaves the nucleus.

We now know that mRNA processing differences can affect gene expression by determining the type of protein product made by a cell. The speed with which a processed mRNA transcript enters the cytoplasm can determine the amount of gene product, particularly in the case of a short-lived mRNA transcript.

4 *Translational control.* Translational control occurs in the cytoplasm and affects when translation begins and how long it continues.

The presense of the 5′ cap and the length of the 3′ poly-A tail can influence the initiation of and the persistence of an mRNA transcript and, therefore, the amount of gene product. Also, hormones can stabilize mRNA transcripts so that translation lasts longer than otherwise.

5 *Posttranslational control.* Posttranslational control, which also takes place in the cytoplasm, occurs after protein synthesis. Only a functional protein is an active gene product.

The polypeptide product may have to undergo additional changes before it is biologically functional. A functional enzyme can be subject to feedback control, as discussed in Section 5.8, so that enzyme activity ceases for a time. Also, how long it takes for a protein to be broken down by a proteasome can affect how long a gene product is active.

We have not completely finished our study of gene control in eukaryotes. The next part of the chapter considers how genes are controlled during eukaryotic development.

> **11.11 Check Your Progress** When would you expect homeotic genes to be in loosely condensed chromatin?

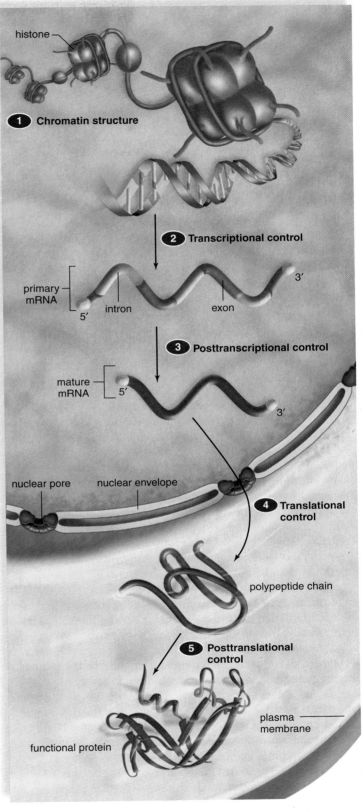

FIGURE 11.11 Levels at which control of gene expression occurs in eukaryotic cells.

Investigators studying the development of the fruit fly have discovered a series of genes that are sequentially active during development. These genes code for transcription factors, and therefore they are involved in gene control. Perhaps the most interesting of these developmental genes are the homeotic genes, because they contain a special sequence of bases, called a homeobox. The homeobox and its protein product, called a homeodomain, recur in a wide range of eukaryotes. Therefore, the homeotic genes are believed to be "ancient" genes.

11.12 Genes are turned on sequentially during development

For development to occur normally, genes have to be turned on or turned off in a particular sequence. First we will consider how patterns form during the development of vertebrates. To understand the concept of pattern formation, think about the pattern of your own body. Your vertebral column runs along the main axis, and your arms and legs are in certain locations. Experiments with *Drosophila melanogaster* (fruit fly), in particular, have contributed to our knowledge of pattern formation. Recall that fruit flies are convenient organisms to work with because a few pairs can produce hundreds of offspring in a couple of weeks, all within small bottles kept on a laboratory bench.

Investigators studying pattern formation in the fruit fly have discovered that some genes determine the animal's anterior/posterior (head/tail) axis and dorsal/ventral (back/belly) axis; others determine the fly's segmentation pattern; and still others, called homeotic genes (see Section 11.3), determine the fate of differentiated cells in the various segments.

Anterior/Posterior Axes One of the first events during development is the establishment of the body axes. In the *Drosophila* egg, there is a greater concentration of a protein called bicoid at one end. This end becomes the anterior region, where the head develops. (*Bicoid* means "two tailed," and a *bicoid* gene mutation can cause the embryo to have two posterior ends instead of an anterior and posterior end.) Researchers have cloned the *bicoid* gene and used it as a probe to establish that mRNA for the bicoid protein is present in a concentration gradient from the anterior end to the posterior end of the embryo.

Proteins that influence pattern formation, such as establishment of axes, are called **morphogens**. (Morphology is the study of structure.) Therefore, the bicoid gradient is a morphogen gradient. We know that the bicoid gradient switches on the expression of segmentation genes in *Drosophila*. A morphogen gradient is advantageous in that it can have a range of effects, depending on its concentration in a particular portion of the embryo.

Segmentation Pattern The next event in the development of *Drosophila* is the establishment of its segments. The bicoid gradient initiates a cascade in which a series of segmentation gene sets are turned on, one after the other.

Christiane Nusslein-Vollard and Eric Wieschaus received a Nobel Prize for discovering the segmentation genes in *Drosophila*. They exposed the flies to mutagenic chemicals and then performed innumerable crosses to map the mutated genes that caused segmental abnormalities. They went on to clone many of these genes. **Figure 11.12** illustrates their findings: ❶ The

FIGURE 11.12 Development in *Drosophila*, a fruit fly.

❶ Protein products of *gap* genes

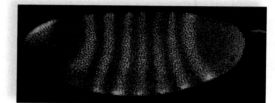

❷ Protein products of *pair-rule* genes

❸ Protein products of *segment-polarity* genes

first set of segmentation genes to be activated are *gap* genes, so called because if one of these genes mutates, there are *gaps*—that is, large blocks of segments are missing. ❷ Then the *pair-rule* genes become active, and the embryo has precisely 14 segments. If one of these mutates, the animal has half the number of segments. ❸ Next, the *segment-polarity* genes are expressed, and each segment has an anterior and posterior half.

Work with *Drosophila* has suggested how morphogenesis comes about. Sequential sets of master genes code for morphogen gradients that in turn activate the next set of master genes. How do morphogen gradients turn on genes? Morphogens are transcription factors that regulate which genes are active in which parts of the embryo, and in what order.

Now that we have discussed genes involved in establishing the anterior/posterior axis and the segmentation pattern, we move on to discuss homeotic genes in Section 11.13.

> **11.12** *Check Your Progress* Embryologists have long noted that development is orderly. How is orderly development achieved?

Homeotic genes act as main switches for the control of pattern formation, including the organization of differentiated cells into specific three-dimensional structures. During normal development of *Drosophila*, the homeotic genes are activated after the segmentation genes. As early as the 1940s, Edward B. Lewis was able to determine that certain homeotic genes controlled whether a particular segment would bear antennae, legs, or wings. A homeotic mutation caused these appendages to be misplaced—a mutant fly could have extra legs where antennae should be or a pair of wings instead of halteres, the organs that help stabilize the fly when in flight. A fly with the two pairs of wings resulted from inactivity of the first gene of the bithorax complex in a segment that normally would have produced halteres (**Fig. 11.13A**).

Homeotic genes have now been found in many other organisms, and surprisingly, they all contain the same particular sequence of nucleotides, called a **homeobox**. (Because homeotic genes contain a homeobox in mammals, they are called *Hox* genes.) The homeobox codes for a particular sequence of 60 amino acids called a **homeodomain**:

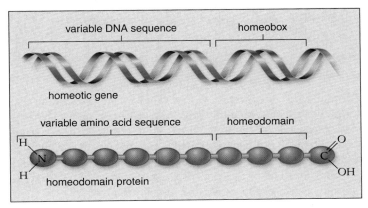

Homeotic genes, like many other developmental genes, code for transcription factors. The homeodomain protein is the part of a transcription factor that binds to DNA, but the other more variable sequences of a transcription factor determine which particular genes are turned on. Researchers envision that a homeodomain protein produced by one homeotic gene binds to

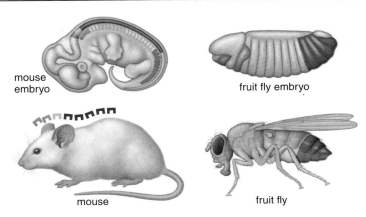

FIGURE 11.13B Homologous (color-coded) homeotic genes occur in the mouse and fruit fly.

and turns on the next homeotic gene, and so forth. This orderly process, in the end, determines the morphology of particular segments.

Mice and fruit flies have the same four clusters of homeotic genes located on four different chromosomes (**Fig. 11.13B**). In *Drosophila*, homeotic genes are located on a single chromosome. In both types of animals, homeotic genes are expressed from anterior to posterior in the same order. The first clusters determine the final development of anterior segments of the animal, while those later in the sequence determine the final development of posterior segments of the animal.

Since the homeotic genes of so many different organisms contain the same homeodomain, we know that this nucleotide sequence arose early in the history of life and has been largely conserved as evolution occurred. In general, researchers have been very surprised to learn how similar developmental genetics is in organisms ranging from yeasts to plants to a wide variety of animals. Certainly, the genetic mechanisms of development appear to be quite similar in all animals.

Apoptosis We have already discussed the importance of apoptosis (programmed cell death) in homeostasis and in preventing the occurrence of cancer. Apoptosis is also an important part of morphogenesis. During the development of a human, we know that apoptosis is necessary to the shaping of the hands and feet; if it does not occur, the child is born with webbing between its fingers and toes (syndactyly). When a cell-death signal is received, an inhibiting protein becomes inactive, allowing a cell-death cascade to proceed that ends in enzymes destroying the cell. Signaling between cells is a very important part of normal development.

This completes our discussion of genes that regulate development, and in the next part of the chapter, we discuss the development of cancer due to mutations that affect certain regulatory genes.

FIGURE 11.13A
The presence of two pairs of wings on a fruit fly represents a homeotic gene mutation.

11.13 *Check Your Progress* What is the significance of the fact that the same homeodomain exists in so many different organisms?

In this part of the chapter, we study genes, called proto-oncogenes and tumor suppressor genes, that control the cell cycle. These genes are a part of a regulatory pathway that stretches from the plasma membrane to the nucleus where they are located. Several proteins in these pathways must malfunction before cancer develops; therefore, it takes several years for cancer to develop.

11.14   When cancer develops, two types of genes are out of control

Recall that the cell cycle consists of interphase, followed by mitosis. *Cyclin* is a molecule that has to be present in order for a cell to proceed from interphase to mitosis.

p53 is a transcription factor normally instrumental in stopping the cell cycle when repair is needed. If repair is impossible, the p53 protein promotes apoptosis. Apoptosis is an important way carcinogenesis is prevented. When cancer develops, the cell cycle occurs repeatedly, in large part due to mutations in two types of genes:

1. **Proto-oncogenes** code for proteins that promote the cell cycle and inhibit apoptosis. They are often likened to the gas pedal of a car because they keep the cell cycle going (**Fig. 11.14A** *dull green*).

2. **Tumor suppressor genes** code for proteins that inhibit the cell cycle and promote apoptosis. They are often likened to the brakes of a car because they inhibit the cell cycle (Fig. 11.14A *dull yellow*).

Mutations in Proto-oncogenes

When proto-oncogenes mutate, they become cancer-causing genes, called **oncogenes**. These mutations can be called "gain-of-function" mutations because overexpression results in excess cyclin and inhibitors of *p53* (**Fig. 11.14B** *bright green*). Whatever a proto-oncogene does, an oncogene does it better. Here are some examples: (1) A proto-oncogene may code for a growth factor or for a receptor protein that receives a growth factor. When such a proto-oncogene becomes an oncogene, receptor proteins are easy to activate and may even be stimulated by a growth factor produced by the receiving cell itself. (2) Several proto-oncogenes code for Ras proteins, which are members of transduction pathways that stimulate transcription (see Section 11.15). *Ras* oncogenes are typically found in many

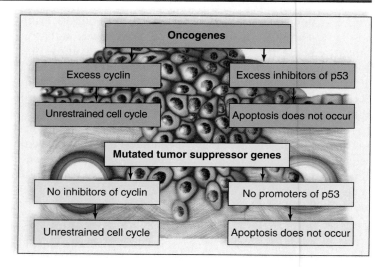

FIGURE 11.14B Abnormal actions of oncogenes (*bright green*) and mutated tumor suppressor genes (*bright yellow*) in cancer cells.

different types of cancers. (3) *Cyclin D* is a proto-oncogene that codes for a cyclin. When this gene becomes an oncogene, cyclin is readily available all the time. (4) A certain proto-oncogene codes for a protein that functions to make p53 unavailable. When this proto-oncogene becomes an oncogene, no matter how much p53 is made, none will be available. In general, when proto-oncogenes become oncogenes, apoptosis does not occur.

Mutations in Tumor Suppressor Genes

When tumor suppressor genes mutate, their products no longer inhibit the cell cycle or promote apoptosis (Fig. 11.14B *bright yellow*). Therefore, these mutations can be called "loss-of-function" mutations. Here are some examples: (1) The retinoblastoma protein (RB) inhibits the activity of a transcription activator for *cyclin D* and other genes whose products promote entry into the S phase of the cell cycle. When the tumor suppressor gene *p16* mutates, the RB protein is not functional, and the result is, again, too much active *cyclin D* in the cell, which continues to divide. The gene for retinoblastoma, a rare form of childhood eye cancer, is located on chromosome 13. (2) The protein Bax promotes apoptosis. When the tumor suppressor gene *Bax* mutates, the protein Bax is not present, and apoptosis is less likely to occur. The gene contains a run of eight consecutive guanine (G) bases, making it subject to mutation. When other tumor suppressor genes mutate, the result is also the loss of apoptosis.

In Section 11.15, we see that proto-oncogenes and tumor suppressor genes are a part of regulatory pathways.

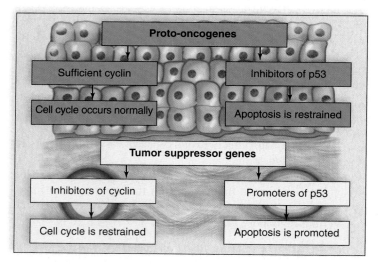

FIGURE 11.14A Normal actions of proto-oncogenes (*dull green*) and tumor suppressor genes (*dull yellow*) in normal cells.

11.14 Check Your Progress Why are proto-oncogenes and tumor suppressor genes considered antagonistic in action?

11.15 Faulty gene products interfere with signal transduction when cancer develops

All of the molecules discussed in Section 11.14 are a part of a signal transduction pathway that controls cell division. **Figure 11.15A** (*top*) illustrates a normal stimulatory pathway that contains a proto-oncogene. The pathway consists of a stimulating growth factor, the receptor, signal transducers, and a transcription factor. The transcription factor is necessary for turning on a proto-oncogene that codes for a protein that stimulates the cell cycle. For example, we mentioned that several proto-oncogenes code for Ras proteins that are members of stimulatory pathways.

Figure 11.15A (*bottom*) illustrates the same pathway when a mutation has caused a proto-oncogene to become an oncogene. For example, the result could be a Ras protein that overstimulates the cell cycle. A Ras protein has been implicated in about a one-third of all cancers in humans. The situation can get pretty complicated, however, because all the signal transducers are themselves proteins, coded for by genes. A mutation in any one of the genes that code for a member of a stimulatory pathway can lead to overstimulation of the cell cycle.

Figure 11.15B (*top*) illustrates a normal inhibitory pathway that contains exactly the same types of members as the stimulatory pathway, except that the external signal is an inhibiting growth factor, and the transcription factor turns on a tumor suppressor gene. The protein p53 is a transcription factor instrumental in stopping the cell cycle and activating chromosomal repair enzymes. If repair is impossible, the p53 protein goes on to promote apoptosis.

Figure 11.15B (*bottom*) illustrates the same pathway when a tumor suppressor gene has undergone a mutation. Perhaps, then, the mutated gene produces a protein that cannot inhibit the cell cycle, and the cell continues to divide, even if its chromosomes are abnormal. Lack of p53 is implicated in over half of the cancers in humans. Another example, as mentioned, is the retinoblastoma protein (RB). Mutation of tumor suppressor gene *p16* inactivates the RB protein, and the cell continues to divide.

In Section 11.16, we show that cancer develops slowly, and also briefly discuss the diagnosis and treatment of cancer.

> **11.15** *Check Your Progress* **What does the protein p53 do?**

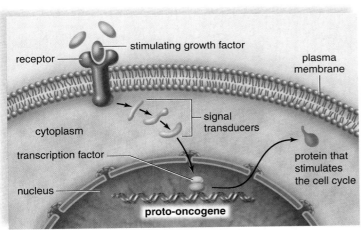

Normal stimulatory pathway: Cell cycle occurs.

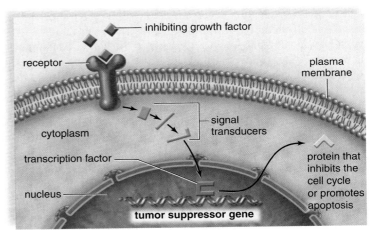

Normal inhibitory pathway: Cell cycle inhibited.

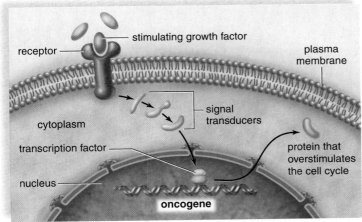

Abnormal stimulatory pathway: Cell cycle excellerates.

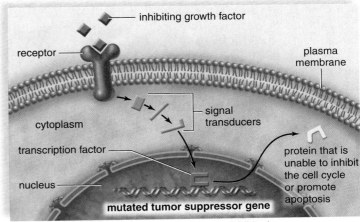

Abnormal inhibitory pathway: Cell cycle is not inhibited.

FIGURE 11.15A Signal transduction pathway, with a proto-oncogene (*top*) and an oncogene (*bottom*).

FIGURE 11.15B Signal transduction pathway, with a tumor suppressor gene (*top*) and a mutated tumor suppressor gene (*bottom*).

As we have seen, two types of signal transduction pathways are of fundamental importance to normal operation of the cell cycle and control of apoptosis. Therefore, it is not surprising that inherited or acquired defects in these pathways contribute to the development of cancer. However, because it takes several mutations to completely disrupt these pathways, **carcinogenesis**, the development of a malignant tumor, is a gradual process. **Figure 11.16** shows a possible sequence for carcinogenesis:

1 A single cell undergoes a mutation that causes it to begin to divide repeatedly. **2** Among the progeny of this cell, one cell mutates further and can now start a tumor, whose cells have further selective advantages. **3** A tumor is present, but it is called cancer in situ because it is contained within its place of origin. To grow larger than a pea, a tumor must have a well-developed capillary network to bring it nutrients and oxygen. The cells release growth factors that lead to **angiogenesis**, the formation of new blood vessels. A new investigative treatment for cancer uses drugs that break up the network of new capillaries in the vicinity of a tumor. **4** New mutations cause the tumor cells to have a disorganized internal cytoskeleton and to lack intact actin filament bundles. Now they are motile cells and can invade underlying tissues once they produce proteinase enzymes that degrade their basement membrane. **5** Cancer cells also develop the ability to invade lymphatic vessels and blood vessels, which take them to other parts of the body. Malignancy is present when cancer cells are found in nearby lymph nodes. **6** When cancer cells initiate new tumors far from the primary tumor, **metastasis** has occurred. Not many cancer cells achieve this feat (maybe 1 in 10,000), but those that successfully metastasize make the probability of complete recovery doubtful.

Diagnosis and Treatment of Cancer

The earlier cancer is detected, the more likely it is that treatment will be effective. The American Cancer Society publicizes seven warning signals that cancer is present, and a growing number of researchers are working on methods for testing body fluids to detect the early signs of cancer. Also available are routine tests, such as the Pap test for cervical cancer, mammography for breast cancer, and colonoscopy for colon cancer. A diagnosis of cancer can be confirmed by performing a biopsy or using various imaging procedures, such as an X-ray.

Surgery, radiation, and chemotherapy are the standard methods of cancer therapy today. Surgery alone is sufficient for cancer in situ. But because there is always the danger that some cancer cells were left behind, surgery is often preceded and/or followed by radiation therapy. Radiation and chemotherapy are based on the principle that dividing cancer cells are more susceptible to their effects than are other cells. Hair loss, digestive difficulties, and disruption of blood cell formation may occur because these treatments also affect areas of the body where normal cells are always dividing. Bone marrow transplants are sometimes done in conjunction with high doses of chemotherapy.

This completes our study of the genetics of cancer in this chapter.

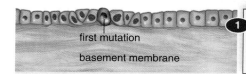

1. Cell (dark pink) acquires a mutation for repeated cell division.

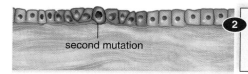

2. New mutations arise, and one cell (green) has the ability to start a tumor.

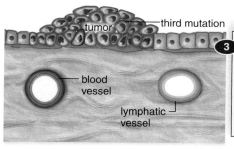

3. Cancer in situ. The tumor is at its place of origin. One cell (purple) mutates further. Angiogenesis occurs, and blood vessels arise and service the tumor.

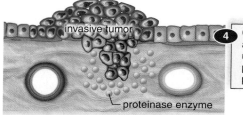

4. Cells have gained the ability to invade underlying tissues by producing a proteinase enzyme.

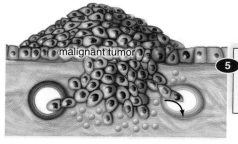

5. Cancer cells now have the ability to invade lymphatic and blood vessels.

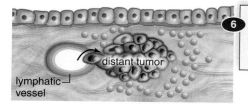

6. New metastatic tumors are found some distance from the original tumor.

FIGURE 11.16 Carcinogenesis.

11.16 Check Your Progress Why do radiation and chemotherapy result in hair loss, digestive difficulties, and disruption of blood cell formation?

Gene regulation determines the specialization of a cell because gene regulation determines which genes are expressed and/or the degree to which genes are expressed. Specialization of cells arises during the developmental process. In differentiated cells, such as nerve cells, muscle fibers, and reproductive cells, only certain genes are actively expressed.

Cancer cells have lost their specialization and divide continuously as if they were embryonic cells. In cancer cells, proto-oncogenes have become oncogenes, and tumor suppressor genes fail to be adequately expressed.

In Chapter 12, you will see how the basic knowledge of DNA structure, replication, and expression has contributed to a biotechnology revolution. We now know how to isolate and move genes between organisms of the same species and even different species. We have sequenced the DNA of humans and many other organisms. Soon we hope to know the sequence of all the human genes along the chromosomes as well. Knowledge of how genes are regulated can help explain how cells become specialized and how basic genetic differences may have arisen between species, such as between humans and chimpanzees.

The Chapter in Review

Summary

Moth and Butterfly Wings Tell a Story

- Patterns on the wings of moths and butterflies are due to homeotic genes—genes that are involved in pattern formation and organization of body parts.

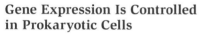

Gene Expression Is Controlled in Prokaryotic Cells

11.1 DNA-binding proteins turn genes on and off in prokaryotes

- The operon model explains gene regulation in prokaryotes.
- An operon includes a promoter, an operator, and structural genes.
- The *lac* operon is an inducible operon. When lactose is absent, the operon is turned off; when lactose is present, the operon is turned on.
- The *trp* operon is a repressible operon. When tryptophan is absent, enzymes needed to synthesize tryptophan are produced; when tryptophan is present, the operon is turned off.

Control of Gene Expression in Eukaryotes Causes Specialized Cells

11.2 Eukaryotic cells are specialized

- Specialization is due to the regulation of genes.

11.3 Plants are cloned from a single cell

- A plant cell illustrates totipotency—the nucleus contains a copy of all the genes needed for development of a new plant.

11.4 Animals are cloned using a donor nucleus

- The desired result of reproductive cloning is a new individual exactly like the original individual.
- The desired result of therapeutic cloning is mature cells of various types for medical purposes.
- Either embryonic stem cells or adult stem cells may be used in therapeutic cloning.

11.5 Animal cloning has benefits and drawbacks

- Reproductive cloning is useful in animal husbandry and can help save endangered species.
- Therapeutic cloning can provide valuable medical products and fight cancer and other diseases.

Control of Gene Expression Is Varied in Eukaryotes

11.6 Chromatin is highly condensed in chromosomes

- Nucleosomes (histones and DNA) coil as chromatin condenses.

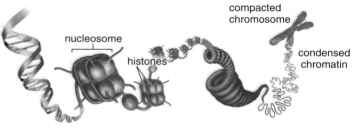

11.7 The genes in highly condensed chromatin are not expressed

- Highly condensed chromatin is called heterochromatin.
- Barr bodies are inactive X chromosomes not producing gene products.
- Euchromatin consists of loosely condensed chromatin; genes are expressed.

11.8 DNA-binding proteins regulate transcription in eukaryotes

- Transcription factors are proteins that help regulate transcription.
- Transcription activators and enhancers bind to regions of DNA and promote transcription.

11.9 mRNA processing can affect gene expression

- Posttranscriptional control of gene expression occurs in the nucleus and involves mRNA processing and the speed at which mRNA leaves the nucleus.
- Splicing of mRNA due to the removal of introns influences the particular protein product.

11.10 Control of gene expression also occurs in the cytoplasm

- Translational control begins when processed mRNA reaches the cytoplasm before there is a protein product.
- Posttranslational control begins once a protein has been synthesized and becomes active.

11.11 Synopsis of gene expression control in eukaryotes

- The five primary levels of gene expression control are chromatin structure, transcriptional control, posttranscriptional control, translational control, and posttranslational control.

Gene Expression Is Controlled During Development

11.12 Genes are turned on sequentially during development

- Development of the anterior/posterior body axes is one of the first events in *Drosophila* development.
- Establishment of segments in *Drosophila* is the second developmental event.
- Sequential activation of genes makes development orderly.

11.13 Homeotic genes and apoptosis occur in a wide range of animals

- Homeotic genes are developmental genes that do the following:
  - Control pattern formation
  - Code for transcription factors
  - Contain a homeobox that codes for a homeodomain
- Apoptosis is important in morphogenesis.

Genetic Mutations Cause Cancer

11.14 When cancer develops, two types of genes are out of control

- Proto-oncogenes promote the cell cycle and inhibit apoptosis.
- Tumor suppressor genes inhibit the cell cycle and promote apoptosis.
- Oncogenes and mutated tumor suppressor genes cause excess cyclin, which stimulates the cell cycle and makes p53 unavailable so that apoptosis does not occur.

11.15 Faulty gene products interfere with signal transduction when cancer develops

- A stimulatory signal transduction pathway turns on an oncogene whose product stimulates the cell cycle.
  - When cancer occurs the product of an oncogene leads to overstimulation of the pathway and the cell cycle.
- An inhibitory signal transduction pathway turns on a tumor suppressor gene whose product inhibits the cell cycle.
  - When cancer occurs, the product of a mutated tumor suppressor gene fails to turn on the pathway and fails to stop the cell cycle.

11.16 Cancer develops and becomes malignant gradually

- Carcinogenesis refers to tumor formation due to repeated mutations.
- Angiogenesis (formation of new blood vessels) provides nutrients to a growing tumor.
- Motile cells invade lymphatic and blood vessels.
- Metastasis has occurred when a new tumor forms far from the first tumor.

Testing Yourself

Gene Expression Is Controlled in Prokaryotic Cells

1. Label this diagram of an operon:

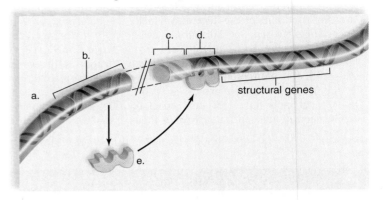

2. In operon models, the function of the promoter is to
   a. code for the repressor protein.
   b. bind with RNA polymerase.
   c. bind to the repressor.
   d. code for the regulator gene.
3. Which of these correctly describes the function of a regulator gene for the *lac* operon?
   a. prevents transcription from occurring
   b. a sequence of DNA that codes for the repressor
   c. prevents the repressor from binding to the operator
   d. keeps the operon off until lactose is present
   e. Both b and d are correct.

Control of Gene Expression in Eukaryotes Causes Specialized Cells

4. Specialized cells
   a. develop similarly.
   b. contain different genes.
   c. transcribe different genes.
   d. must undergo mutations.
   e. Both b and c are correct.
5. During reproductive cloning, a(n) _____ is placed into a(n) _____.
   a. enucleated egg cell, adult cell nucleus
   b. adult cell nucleus, enucleated egg cell
   c. egg cell nucleus, enucleated adult cell
   d. enucleated adult cell, egg cell nucleus
6. The major challenge to therapeutic cloning using adult stem cells is
   a. finding appropriate cell types.
   b. obtaining enough tissue.
   c. controlling gene expression.
   d. keeping cells alive in culture.
7. The product of therapeutic cloning differs from the product of reproductive cloning in that it is
   a. various mature cells, not an individual.
   b. genetically identical to the somatic cells from the donor.
   c. various mature cells with the ability to become an individual.
   d. an individual that is genetically identical to the donor of the nucleus.

Control of Gene Expression Is Varied in Eukaryotes

8. Which of these associations is mismatched?
   a. loosely packed chromatin—gene can be active
   b. transcription factors—gene is inactivated
   c. mRNA—translation can begin
   d. proteasomes—protein is inactive

For questions 9–13, match the examples to the gene expression control mechanisms in the key. Each answer can be used more than once.

KEY:
   a. chromatin structure
   b. transcriptional control
   c. posttranscriptional control
   d. translational control
   e. posttranslational control

9. Insulin does not become active until 30 amino acids are cleaved from the middle of the molecule.
10. The mRNA for vitellin is longer-lived if it is exposed to estrogen.
11. Genes in Barr bodies are inactivated.
12. Calcitonin is produced in both the hypothalamus and the thyroid gland, but in different forms due to exon splicing.
13. DNA-binding proteins are active.
14. **THINKING CONCEPTUALLY** A variety of mechanisms regulate gene expression in eukaryotic cells. What is the benefit and drawback of this arrangement?

Gene Expression Is Controlled During Development

15. The genes that determine which body parts form on each body segment of a fruit fly are called _____ genes.
   a. promoter   c. intron
   b. exon   d. homeotic
16. What developmental milestone is determined by a concentration gradient of Bicoid protein in fruit flies?
   a. anterior/posterior axis   c. limb and wing location
   b. dorsal/ventral axis   d. body segmentation
17. Which of the following is part of a transcription factor that binds to DNA?
   a. homeobox   c. homeotic gene
   b. homeodomain   d. Bicoid protein

Genetic Mutations Cause Cancer

For questions 18–21, choose two answers for each type gene.

KEY:
   a. cell cycle is promoted and apoptosis is inhibited
   b. cell cycle is inhibited and apoptosis is promoted
   c. signal transduction pathway produces a normal protein
   d. signal transduction pathway produces an abnormal protein
18. Tumor suppressor gene
19. Proto-oncogene
20. Mutated tumor suppressor gene
21. Oncogene
22. Sequence these events that lead to the development of cancer.
   a. Cells gain the ability to invade underlying tissues.
   b. Metastatic tumors occur.
   c. Cell division leads to a tumor.
   d. Blood vessels arise and service tumor.

22. Which association is incorrect?
   a. Mutations of both proto-oncogenes and tumor-suppressor gene lead to inactivity of p53.
   b. oncogenes–"gain of function" genes
   c. mutated tumor–suppressor genes–code for cyclin and proteins that inhibit the activity of p53
   d. mutated tumor–suppressor genes–"loss of function" mutations
   e. Both a and c are incorrect.

Understanding the Terms

angiogenesis 220	operon 208
Barr body 212	posttranscriptional
carcinogenesis 220	control 214
chromatin 212	posttranslational control 214
enhancer 213	promoter 208
euchromatin 213	proto-oncogene 218
heterochromatin 212	regulator gene 208
histone 212	repressible operon 208
homeobox 217	reproductive cloning 210
homeodomain 217	structural gene 208
homeotic gene 207, 217	therapeutic cloning 210
inducible operon 208	totipotent 209
metastasis 220	transcription activator 213
morphogen 216	transcriptional control 213
nucleosome 212	transcription factor 213
oncogene 218	translational control 214
operator 208	tumor suppressor gene 218

Match the terms to these definitions:
   a. _____ Regulation of gene expression that begins once there is an mRNA transcript.
   b. _____ Dark-staining body in the nuclei of female mammals that contains a condensed, inactive X chromosome.
   c. _____ Diffuse chromatin that is being transcribed.
   d. _____ Full genetic potential of a cell to become an organism.
   e. _____ Small basic protein associated with eukaryotic DNA in chromatin.

Thinking Scientifically

1. You receive much criticism for your conclusion that development in the mouse and fruit fly is similar because you have found several homeoboxes in the genes of both organisms. Why? See Section 11.13.
2. You are a skilled cytologist and want to show that a particular protein causes a cell to divide uncontrollably. What would you do? See Section 11.15.

ARIS *Visit* www.mhhe.com/maderconcepts *for practice quizzes, animations, videos, and activities designed to help you master the material in this chapter.*

12

Biotechnology and Genomics

LEARNING OUTCOMES

After studying this chapter, you should be able to accomplish the following outcomes.

Witnessing Genetic Engineering

1 Refer to the process of genetic engineering to show that the cells of all organisms function similarly.

DNA Can Be Cloned

2 Describe a procedure by which genes can be cloned.
3 Describe how sequences of DNA can be cloned.
4 List several applications for the polymerase chain reaction.

Organisms Can Be Genetically Modified

5 Give examples of genetically modified bacteria, plants, and animals.
6 Tell how genetic engineering came to the rescue of the cheese industry.
7 Discuss potential drawbacks for the planting and the consumption of genetically engineered corn plants.
8 Describe two gene therapy methods in humans, and give an example of each.
9 Discuss the status of gene therapy today.

The Human Genome Can Be Manipulated

10 Describe the method for sequencing the human genome, and state how many genes have been found.
11 Discuss the possible medical benefits derived from the Human Genome Project.
12 Explain the goals of functional and comparative genomics.
13 Define and discuss proteomics and bioinformatics.

Bioluminescent dinoflagellate

Genetic engineering has been around since 1973, so by now many genetically modified organisms (GMOs) have been produced. Fish and cows are now expressing foreign genes that make them grow larger. Pigs have been engineered to make their organs acceptable for transplant into humans. Strawberry and potato plants don't freeze, and soybeans are resistant to viral, bacterial, and fungal pathogens—all because they have been genetically engineered. Bacteria produce human insulin as well as other important medicines. And gene therapy in humans, which is the insertion of normal human genes to make up for ones that do not function properly, is already undergoing clinical trials.

With so many examples of GMOs, you might think it would be easy to "prove" to a friend that it is possible to transfer a gene from one organism to another—but how would you go about it? Well, first you need a gene that makes its appearance known visibly. How about a gene for bioluminescence? Some organisms, including fireflies, jellyfish, glowworms, beetles, and various fishes, can create their own light because they are bioluminescent. The advantages of bioluminescence are varied. Glowworms use their light to attract their prey, and fireflies use their ability to glow to attract mates. The gene for bioluminescence in jellyfish codes for a protein called green fluorescent protein (GFP), and when this gene is transferred to another organism, it glows!

Bioluminescent fish

Witnessing Genetic Engineering

The basic technique you would use to genetically engineer an organism is relatively simple. For example, to transfer the jellyfish gene for bioluminescence to a pig, first locate the gene among all the others in a jellyfish genome. Then fragment the DNA, and introduce the fragment that contains the bioluminescent gene into the embryo of a pig, mouse, or rabbit, for example. The result is a "glow-in-the-dark" organism.

Genes have no difficulty crossing the species barrier. Mammalian genes work just as well in bacteria, and an invertebrate gene, such as the bioluminescent gene, has no trouble functioning in mammals. The genes of any organism are composed of DNA, and the manufacture of a protein (and indeed, the function of that protein) is similar, regardless of the DNA source. Glowing pigs, mice, and rabbits are certainly living proof that genes can be transferred and also that all cells use basically the same machinery.

Bioluminescent mouse

Is it ethical to give a mouse a bioluminescent gene that makes it glow? Advocacy groups have even graver concerns about creating genetically modified organisms. Some worry that modified bacteria and plants might harm the environment. Others fear that products produced by GMOs might not be healthy for humans. Perhaps terrorists could use biotechnology to produce weapons of mass destruction. Finally, to what extent is it proper to improve the human genome? All citizens should be knowledgeable about genetics and biotechnology so that they can participate in deciding these issues.

In this chapter, we discuss gene cloning before studying how bacteria, plants, and animals have been genetically modified for purposes that benefit humans. Gene therapy occurs when humans are genetically modified in order to cure a genetic disorder. The sequencing of the human genome is finished and is expected to increase the possibility of treating and/or curing human genetic disorders. Comparative genomics is expected to shed light on our relationship to other animals. Proteomics and bioinformatics are new fields very much dependent on computer technologies.

Bioluminescent jellyfish   **Bioluminescent pigs**

Gene cloning can be done in one of two ways: through recombinant DNA technology or through the polymerase chain reaction (PCR). Recombinant DNA technology utilizing plasmids allows bacteria to be genetically modified. Plants, animals, and humans are modified by other means. So far, gene therapy trials have not met with marked success.

12.1 Genes can be isolated and cloned

In biology, **cloning** is the production of genetically identical copies of DNA, cells, or organisms through asexual means. **Gene cloning** is done to produce many identical copies of the same gene. Gene cloning requires **recombinant DNA (rDNA)**, which contains DNA from two or more different sources. To create rDNA, a technician needs a **vector**, by which the gene of interest will be introduced into a host cell, which is often a bacterium. One common vector is a **plasmid**, a small accessory ring of DNA found in bacteria. The ring is not part of the bacterial chromosome and replicates on its own.

Figure 12.1 traces the steps in cloning a gene. ❶ A **restriction enzyme** is used to cleave the plasmid. Hundreds of restriction enzymes occur naturally in bacteria, where they cut up any viral DNA that enters the cell. They are called restriction enzymes because they *restrict* the growth of viruses, but they also act as molecular scissors to cleave any piece of DNA at a specific site. For example, the restriction enzyme called *Eco*RI always cuts double-stranded DNA at this sequence of bases and in this manner:

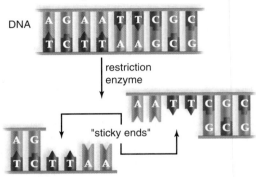

Notice that there is now a gap into which a piece of foreign DNA can be placed if it begins and ends in bases complementary to those exposed by the restriction enzyme. To ensure this, it is only necessary to cleave the foreign DNA; for example, a human chromosome that contains the gene for insulin (or a jellyfish chromosome that contains the gene for green fluorescent protein [GFP]) is cleaved with the same type of restriction enzyme.

❷ The enzyme **DNA ligase** is used to seal foreign DNA into the opening created in the plasmid. The single-stranded, but complementary, ends of a cleaved DNA molecule are called "sticky ends" because they can bind a piece of DNA by complementary base pairing. Sticky ends facilitate the pasting of the plasmid DNA with the DNA of the inserted gene. The use of both restriction enzymes and ligase allows researchers to cut and paste DNA strands at will. Now the vector is complete, and an rDNA molecule has been prepared.

❸ Some of the bacteria take up a recombinant plasmid, especially if the bacteria have been treated to make them more permeable.

❹ₐ Gene cloning occurs as the plasmid replicates on its own. Scientists clone genes for a number of reasons. They might want to determine the difference in base sequence between a normal gene and a mutated gene. Or, they might use the genes to genetically modify other organisms.

❹ᵦ The bacterium has been **genetically engineered** and is a **genetically modified organism (GMO)** that can make a product (e.g., insulin or GFP) it could not make before. For a human gene to express itself in a bacterium, the gene has to be accompanied by regulatory regions unique to bacteria.

Another way to clone a gene is to use the polymerase chain reaction, which is discussed in Section 12.2, along with various applications.

12.1 Check Your Progress Suppose you wanted to produce a number of bioluminescent pigs. Would you use the procedure shown in Figure 12.1 to produce *GFP* genes or the protein GFP?

FIGURE 12.1 Cloning a gene.

If only small pieces of identical DNA are needed, the **polymerase chain reaction (PCR)**, developed by Kary Mullis in 1985, can create copies of a segment of DNA quickly in a test tube. PCR is very specific—it amplifies (makes copies of) a targeted DNA sequence. The sequence of interest can be less than one part in a million of the total DNA sample!

PCR requires the use of DNA polymerase, the enzyme that carries out DNA replication, and a supply of nucleotides for the new DNA strands. PCR is a chain reaction because the targeted DNA is repeatedly replicated as long as the process continues. The colors in **Figure 12.2A** distinguish old DNA from new DNA. Notice that the amount of DNA doubles with each replication cycle.

PCR has been in use for years, and now almost every laboratory has automated PCR machines to carry out the procedure. Automation became possible after a heat-tolerant (thermostable) DNA polymerase was extracted from the bacterium *Thermus aquaticus*, which lives in hot springs. The enzyme can withstand the high temperature used to separate double-stranded DNA; therefore, replication does not have to be interrupted by the need to add more enzyme.

Following PCR, DNA can be subjected to **DNA fingerprinting**. Today, DNA fingerprinting is often carried out by detecting how many times a short sequence (two to five bases) is repeated. People differ by how many base repeat units (such as AGAA) they have at particular genome locations, but how can this be detected? Recall that PCR amplifies only a specific sequence of DNA. Therefore, the greater the number of repeat

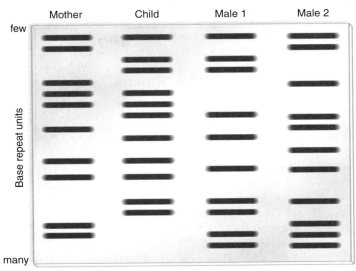

units at a location, the greater the amount of DNA that is amplified by PCR. During a process called *gel electrophoresis*, DNA fragments can be separated according to their size/charge ratios, and the result is a distinctive band pattern. If two DNA band patterns match, there is a high probability that the DNA came from the same person. **Figure 12.2B** shows how DNA fingerprinting can be used to decide paternity. It is customary to test for the number of specific repeat units at several locations to further define the individual.

Applications of PCR and DNA fingerprinting are limited only by our imagination. DNA amplified by PCR is often analyzed for various purposes. For example, mitochondrial DNA base sequences in modern living populations were used to decipher the evolutionary history of humans. Very little DNA is required for PCR to increase its quantity, and therefore it has even been possible to sequence DNA taken from mummified human brains. PCR analysis has been used to identify unknown soldiers and members of the Russian royal family. Paternity suits can be settled, and genetic disorders and even illegally poached ivory and whale meat can be identified, using this technology.

DNA fingerprinting has many other uses also. When the DNA matches that of a virus or mutated gene, a viral infection, genetic disorder, or cancer is present. Fingerprinting the DNA from a single sperm is enough to identify a suspected rapist. DNA fingerprinted from blood or tissues at a crime scene has been successfully used to convict criminals. DNA fingerprinting was extensively used in identifying the remains of victims of the September 11, 2001, terrorist attacks in the United States.

Genetic modification of bacteria, discussed in Section 12.3, usually utilizes recombinant plasmids.

FIGURE 12.2B　DNA fingerprinting of specific base repeat units at different genome locations establishes paternity: Male 1 is the father.

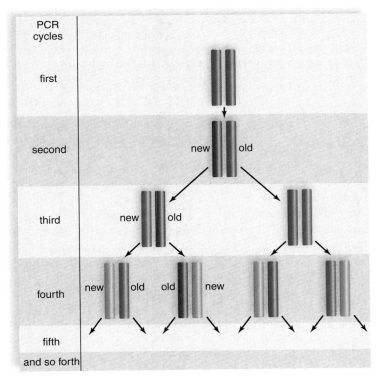

FIGURE 12.2A　Copies of DNA segments produced per PCR cycle per one original strand.

12.2　*Check Your Progress*　How could you use a single bioluminescent pig to get many copies of the *GFP* gene?

Bacteria, plants, and animals have all been genetically modified to produce commercial products useful to human beings. In addition, crops can be modified—for example, to be resistant to pests in order to increase yield. Gene therapy is the genetic modification of human beings in order to correct a mutation resulting in an illness. The status of gene therapy is also discussed in this part of the chapter.

12.3 Bacteria are genetically modified to make a product or perform a service

Genetically engineered bacteria have undergone genetic modifications to produce a product, called a **biotechnology product**. Genetic modification has occurred because the bacterium now contains a new and novel gene. Such an organism can also be referred to as a **transgenic organism**.

Genetically modified (GM) bacteria are grown in huge vats called bioreactors, and the gene product is collected from the medium. Biotechnology products on the market, produced by bacteria, include insulin, clotting factor VIII, human growth hormone, t-PA (tissue plasminogen activator), and hepatitis B vaccine. Transgenic bacteria have many other uses, as well. Some have been produced to promote the health of plants. For example, bacteria that normally live on plants and encourage the formation of ice crystals have been changed from frost-plus to frost-minus bacteria. Also, a bacterium that normally colonizes the roots of corn plants has now been endowed with genes (from another bacterium) that code for an insect toxin. The toxin protects the roots from insects.

GM bacteria can also perform various services. Bacteria can be selected for their ability to degrade a particular substance, and this ability can then be enhanced by genetic engineering (**Fig. 12.3**). For instance, naturally occurring bacteria that eat oil can be genetically engineered to do an even better job of cleaning up beaches after

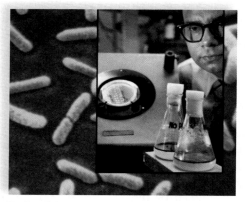

FIGURE 12.3 Producing GM bacteria in the laboratory.

oil spills. Bacteria can also remove sulfur from coal before it is burned and help clean up toxic waste dumps. One such strain was given genes that allowed it to clean up levels of toxins that would have killed other strains. Further, these bacteria were given "suicide" genes that caused them to self-destruct when the job was done.

Organic chemicals are often synthesized by having catalysts act on precursor molecules or by using bacteria to carry out the synthesis. Today, it is possible to go one step further and manipulate the genes that code for these enzymes. For instance, biochemists discovered a strain of bacteria that is especially good at producing phenylalanine, an organic chemical needed to make aspartame, the dipeptide sweetener better known as NutraSweet. They isolated, altered, and formed a vector for the appropriate genes so that various bacteria could be genetically engineered to produce phenylalanine.

Genetic engineering rescued the cheese-making industry, as discussed in Section 12.4.

12.3 Check Your Progress You have genetically modified bacteria to (1) express the *GFP* gene, (2) clean up an oil spill, and then (3) self-destruct. How could you be sure the bacteria did self-destruct?

HOW SCIENCE PROGRESSES

12.4 Making cheese—Genetic engineering comes to the rescue

In the past, the cheese-making industry was dependent upon a substance called rennet. Rennet consists of two enzymes: chymosin and bovine pepsin. These enzymes, especially chymosin, are essential for coagulating milk and converting it to cheese. Traditionally, rennet was collected from the stomach lining of calves, where chymosin and bovine pepsin ensure the proper digestion of milk. However, with a decline in the veal industry (veal is a meat derived from calves), a rennet shortage resulted. As the demand for rennet increased, scientists began testing various technologies to supply the substance, but they could find none that was satisfactory.

Finally, genetic engineering saved the day for the cheese industry. The gene responsible for the formation of chymosin was isolated from calf cells and cloned. The gene was then inserted into the genome of several organisms, including the bacterium *Escherichia coli*, the fungus *Aspergillus niger*, and the yeast *Kluyveromyces lactis*. The genetically engineered organisms produced copious

amounts of chymosin with great success. Researchers confirmed that the new chymosin product contained no toxins and no living recombinant organisms. After exhaustive testing, the U.S. Food and Drug Administration approved the use of genetically engineered chymosin in food. Cheese containing chymosin was the first food dependent on genetic engineering in United States history.

Cheese made with genetically engineered chymosin is indistinguishable from rennet-produced cheese. In addition, it meets the desires of strict vegetarians, since it contains no rennet. In North America today, more than 80% of cheese is made with genetically engineered chymosin.

Genetic modification of plants is utilized to increase productivity and to produce a product, as discussed in Section 12.5.

12.4 Check Your Progress GFP is a protein, and so is chymosin. Explain why both of these are proteins.

12.5 Plants are genetically modified to increase yield or to produce a product

Corn, potato, soybean, and cotton plants have been engineered to be resistant to either insect predation or widely used herbicides. Some corn and cotton plants are now both insect- and herbicide-resistant. In 2006, GMOs were planted on more than 252 million acres worldwide, an increase of over 13% from the previous year. If crops are resistant to a broad-spectrum herbicide and weeds are not, the herbicide can be used to kill the weeds. When herbicide-resistant plants were planted, weeds were easily controlled, less tillage was needed, and soil erosion was minimized.

Crops with other improved agricultural and food-quality traits are desirable (**Fig. 12.5A**). For example, crop production is currently limited by the effects of salinization on about 50% of irrigated lands. Salt-tolerant crops would increase yield on this land. Salt- and also drought- and cold-tolerant crops might help provide enough food for a world population that may nearly double by 2050. A salt-tolerant tomato has already been developed. First, scientists identified a gene coding for a channel protein that transports Na^+ across the vacuole membrane. Sequestering the Na^+ in a vacuole prevents it from interfering with plant metabolism. Then the scientists cloned the gene and used it to genetically engineer plants that overproduce the channel protein. The modified plants thrived when watered with a salty solution.

Potato blight is the most serious potato disease in the world. About 150 years ago, it was responsible for the Irish potato famine, which caused the deaths of millions of people. By placing a gene from a naturally blight-resistant wild potato into a farmed variety, researchers have now made potato plants that are invulnerable to a range of blight strains. In **Figure 12.5B**, the B.t.t. + potato plants produces an insecticide protein and are resistant to the Colorado potato beetle.

Some progress has also been made in increasing the food quality of crops. Soybeans have been developed that mainly produce the monounsaturated fatty acid oleic acid, a change that may improve human health.

Genetically modified plants requiring more than a single gene transfer are also expected to increase productivity. For ex-

FIGURE 12.5B Potato plant on left is nonresistant to the Colorado potato beetle, while plant on right is resistant.

ample, stomata might be altered to take in more carbon dioxide or lose less water. The efficiency of the enzyme RuBP carboxylase, which captures carbon dioxide in plants, could be improved. A team of Japanese scientists is working on introducing the C_4 photosynthetic cycle into rice. (As discussed in Chapter 6, C_4 plants do well in warm, dry weather.)

Genetic engineering of plants has also produced many products for human use, such as human hormones, clotting factors, and antibodies. One type of antibody made by corn can deliver radioisotopes to tumor cells, and another made by soybeans may be developed to treat genital herpes.

Section 12.6 makes it clear that people have two basic concerns about genetically modified foods: food safety and environmental impact.

> **12.5 Check Your Progress** Surprisingly, plants can be modified to produce any type of protein, even GFP. Explain why this is possible.

GM Crops of the Future	
Improved Agricultural Traits	
Disease-protected	Wheat, corn, potatoes
Herbicide-resistant	Wheat, rice, sugar beets, canola
Salt-tolerant	Cereals, rice, sugarcane
Drought-tolerant	Cereals, rice, sugarcane
Cold-tolerant	Cereals, rice, sugarcane
Improved yield	Cereals, rice, corn, cotton
Modified wood pulp	Trees
Improved Food Quality Traits	
Fatty acid/oil content	Corn, soybeans
Protein/starch content	Cereals, potatoes, soybeans, rice, corn
Amino acid content	Corn, soybeans

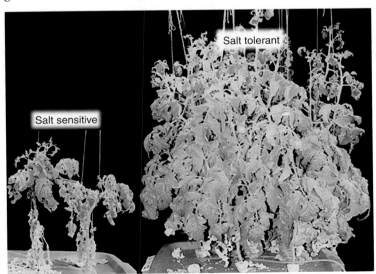

FIGURE 12.5A Genetically modified crops of the future.

A series of focus groups conducted by the Food and Drug Administration (FDA) in 2000 showed that although most participants believed genetically engineered foods might offer benefits, they also feared unknown long-term health consequences that might be associated with the technology. Conrad G. Brunk, a bioethicist at the University of Waterloo in Ontario, has said, "When it comes to human and environmental safety, there should be clear evidence of the absence of risks. The mere absence of evidence is not enough."

The discovery by activists that GM corn called StarLink had inadvertently made it into the food supply triggered the recall of taco shells, tortillas, and many other corn-based foodstuffs from U.S. supermarkets. Further, the makers of StarLink were forced to buy back StarLink from farmers and to compensate food producers at an estimated cost of several hundred million dollars. StarLink corn is a type of "BT" corn (**Fig. 12.6A**), so called because it contains a foreign gene taken from a common soil organism, *Bacillus thuringiensis*, whose insecticidal properties have been long known. About a dozen BT varieties, including corn, potato (see Fig. 12.5B), and even a tomato, have now been approved for human consumption. These strains contain a gene for an insecticide protein called CrylA. However, the makers of StarLink decided to use a gene for a related protein called Cry9C. They thought this molecule might slow down the development of pest resistance to BT corn. To get FDA approval for use in foods, the makers of StarLink performed the required tests. Like the other now-approved strains, StarLink wasn't poisonous to rodents, and its biochemical structure is not similar to those of most food allergens. But the Cry9C protein resisted digestion longer than the other BT proteins when it was put in simulated stomach acid and subjected to heat. Because this is a characteristic of most food allergens, StarLink was not approved for human consumption.

FIGURE 12.6B Herbicide-resistant soybean plants.

Scientists are now trying to devise more tests because they have not been able to determine conclusively whether Cry9C is an allergen. Also, at this point, it is unclear how resistant to digestion a protein must be in order to be an allergen, or what degree of sequence similarity to a known allergen is enough to raise concern.

Other scientists are concerned about the following potential drawbacks to the planting of BT corn: (1) resistance among populations of the target pest, (2) exchange of genetic material between the GM crop and related plant species, and (3) BT crops' impact on nontarget species. They feel many more studies are needed before it can be said for certain that BT corn has no ecological drawbacks.

Despite controversies, the planting of GM corn stayed steady in 2006. The USDA reports that U.S. farmers planted GM corn on 45% of all corn acres, about the same as in 2003. In all, U.S. farmers planted at least 150 million acres with mostly GM corn and soybeans (**Fig. 12.6B**). The public wants all genetically engineered foods to be labeled as such, but this may not be easy because, for example, most cornmeal is derived from both conventional and genetically engineered corn. So far, there has been no attempt to sort out one type of food product from the other.

In contrast to plants, special means are needed to genetically modify animal egg cells in order to achieve a genetically modified animal, as discussed next.

FIGURE 12.6A Genetically modified corn.

> **12.6** *Check Your Progress a.* **Which of the following is an engineered food: cheese made using chymosin produced by genetically engineered bacteria, and /or food made from BT corn containing an insecticide specified by a bacterial gene?** *b.* **Should both of these foods be tested for allergic properties?**

12.7 Animals are genetically modified to enhance traits or obtain useful products

Techniques have been developed to insert genes into the eggs of animals. It is possible to microinject foreign genes into eggs by hand, but another method uses vortex mixing. DNA and eggs are placed in an agitator with silicon-carbide needles. The needles make tiny holes, through which the DNA can enter. When these eggs are fertilized, the resulting offspring are transgenic animals. Using this technique, many types of animal eggs have taken up the gene for bovine growth hormone (bGH), and this has led to the production of larger fishes, cows, pigs, rabbits, and sheep.

Gene pharming, the use of transgenic animals to produce pharmaceuticals, is being pursued by a number of firms. Genes that code for therapeutic and diagnostic proteins are incorporated into an animal's DNA, and the proteins appear in the animal's milk (**Fig. 12.7A**). Plans are under way to produce medicines for the treatment of cystic fibrosis, cancer, blood diseases, and other disorders by this method. Figure 12.7A outlines the procedure for producing GM mammals. **1** The gene of interest (in this case, human growth hormone) is microinjected into donor eggs. **2** Following in vitro fertilization, the zygotes are placed in host females, where they develop. **3** After the transgenic female offspring mature, the product is secreted in their milk. Then, cloning can be used to produce many animals that produce the same product: **4** Donor enucleated eggs are fused with 2n transgenic nuclei. The eggs are coaxed to begin development in vitro. **5** Development continues in host females until the clones are born. **6** The female offspring are clones that have the same product in their milk.

Many researchers are using transgenic mice for various research projects. **Figure 12.7B** shows how this technology has demonstrated that a section of DNA called *SRY* (sex-determining region of the Y chromosome) produces a male animal. The *SRY* DNA was cloned, and then one copy was injected into one-celled mouse embryos with two X chromosomes. Injected embryos developed into males, but any that were not injected developed into females. Mouse models have also been created to study human diseases. An allele such as the one that causes cystic fibrosis can be cloned and inserted into mice embryonic stem cells, and occasionally a mouse embryo homozygous for cystic fibrosis will result. This embryo develops into a mutant mouse that has a phenotype similar to that of a human with cystic fibrosis. New drugs for the treatment of cystic fibrosis can then be tested in these mice.

Xenotransplantation is the use of animal organs, instead of human organs, in transplant patients. Scientists have chosen to work with pigs because they are prolific and have long been raised as a meat source. Pigs will be genetically modified to make their organs less likely to be rejected by the human body. The hope is that one day a pig organ will be as easily accepted by the human body as a blood transfusion from a person with the same blood type.

Gene therapy is discussed in the next two sections.

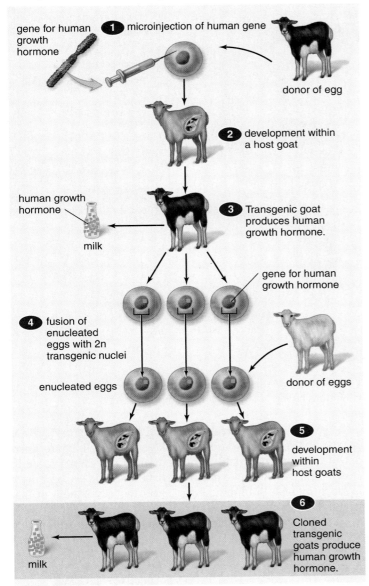

FIGURE 12.7A Procedure for producing many female clones that yield the same product.

> **12.7 Check Your Progress** In Figure 12.7A, only transgenic females produce milk. Why was cloning done instead of simply producing more GMO animals?

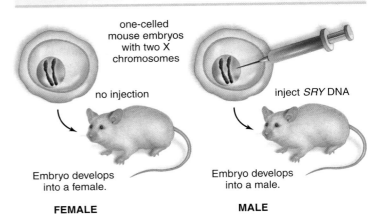

FIGURE 12.7B GM mice showed that maleness is due to *SRY* DNA.

12.8 A person's genome can be modified

The manipulation of an organism's genes can be extended to humans in a process called gene therapy. **Gene therapy** is the insertion of a foreign gene into human cells for the treatment of a disorder. Gene therapy has been used to cure inborn errors of metabolism as well as more generalized disorders, such as cardiovascular disease and cancer. **Figure 12.8** shows regions of the body that have received copies of normal genes by various methods of gene transfer. Viruses genetically modified to be safe can be used to ferry a normal gene into the body, and so can liposomes, which are microscopic globules of lipids specially prepared to enclose the normal gene. On the other hand, sometimes the gene is injected directly into a particular region of the body. In vivo gene therapy means the gene is delivered directly into the body, while ex vivo gene therapy means the gene is inserted into cells that have been removed and then returned to the body.

Ex Vivo Gene Therapy Children who have SCID (severe combined immunodeficiency) lack the enzyme ADA (adenosine deaminase), which is involved in the maturation of white blood cells. Therefore, these children are prone to constant infections and may die without treatment. To carry out gene therapy, bone marrow stem cells are removed from the bone marrow of the patient and infected with a virus that carries a normal gene for the enzyme. Then the cells are returned to the patient, where it is hoped they will divide to produce more blood cells with the same genes.

Familial hypercholesterolemia is a condition in which high levels of blood cholesterol make patients subject to fatal heart attacks at a young age. Through ex vivo gene therapy, a small portion of the liver is surgically excised and then infected with a virus containing a normal gene for the receptor before being returned to the patient. Patients are expected to experience lowered serum cholesterol levels following this procedure.

Investigators are also working on a cure for phenylketonuria (PKU), an inherited condition that can cause mental retardation. If detected early enough, the child can be placed on a special diet for the first few years of life, but this is very inconvenient. These investigators believe they will be able to inject the gene directly into the DNA of excised liver cells, which will then be returned to the patient.

In Vivo Gene Therapy Cystic fibrosis patients lack a gene that codes for the transmembrane carrier of the chloride ion. They often suffer from numerous and potentially deadly infections of the respiratory tract. In gene therapy trials, the gene needed to cure cystic fibrosis is sprayed into the nose or delivered to the lower respiratory tract by a virus or by liposomes. So far, this treatment has resulted in limited success.

Genes are also being used to treat medical conditions such as poor coronary circulation. Scientists have known for some time that VEGF (vascular endothelial growth factor) can cause the growth of new blood vessels. The gene that codes for this growth factor can be injected alone or within a virus into the heart to stimulate branching of coronary blood vessels. Patients report that they have less chest pain and can run longer on a treadmill.

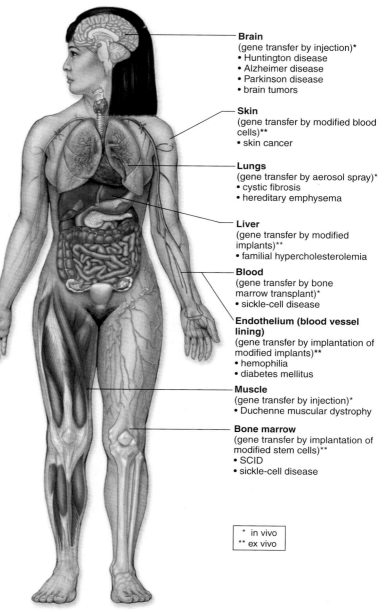

Brain
(gene transfer by injection)*
• Huntington disease
• Alzheimer disease
• Parkinson disease
• brain tumors

Skin
(gene transfer by modified blood cells)**
• skin cancer

Lungs
(gene transfer by aerosol spray)*
• cystic fibrosis
• hereditary emphysema

Liver
(gene transfer by modified implants)**
• familial hypercholesterolemia

Blood
(gene transfer by bone marrow transplant)*
• sickle-cell disease

Endothelium (blood vessel lining)
(gene transfer by implantation of modified implants)**
• hemophilia
• diabetes mellitus

Muscle
(gene transfer by injection)*
• Duchenne muscular dystrophy

Bone marrow
(gene transfer by implantation of modified stem cells)**
• SCID
• sickle-cell disease

* in vivo
** ex vivo

FIGURE 12.8 Sites of ex vivo and in vivo gene therapy to cure the conditions noted.

Gene therapy is increasingly being applied as a part of cancer therapy. Genes are used to make healthy cells more tolerant of chemotherapy and to make tumors more vulnerable to chemotherapy. The gene *p53* brings about apoptosis, and there is much interest in introducing it into cancer cells and, in that way, killing them off.

The likelihood of successful gene therapy is explored in Section 12.9.

> **12.8** *Check Your Progress* In DNA cloning (see Fig. 12.1), a plasmid is used as a vector (carrier) for the gene. What is the vector of choice for gene therapy?

12.9   Gene therapy trials have varying degrees of success

Dr. Theodore Friedmann, director of Molecular Genetics at the University of California–San Diego, is a well-known advocate of gene therapy. He says that medicine usually only treats the symptoms of a disorder, while gene therapy is capable of curing it. Still, he advises his colleagues to openly address gene therapy's difficulties, limitations, and failures. He says, "It's going to be difficult. Yet medicine has always had to work with imperfect knowledge and technology." If we take a look at a couple of gene therapy trials in detail, we can better appreciate Dr. Friedmann's words.

Severe Combined Deficiency Syndrome (SCID) As explained in Section 12.8, children with this disorder are constantly prone to life-threatening infections. One gene therapy trial aimed at SCID was conducted by Alain Fischer at the Hospital Necker, a children's hospital in Paris, and reported in the *New England Journal of Medicine* of April 18, 2002. First, Fischer and his colleagues perfected the virus they were going to use for a vector, and then they sought approval to start human trials. The researchers took bone marrow containing white blood cells from ten pediatric patients. In cell culture, they introduced the vector for the gene the patients needed. After several days, they injected the cells back into the patients. In 9 out of 10 cases, the treatment worked sufficiently to allow the children to leave the hospital and start to live normally with their parents.

After three years, 3 of the 10 children developed a severe complication, a type of leukemia, for which they were treated with massive doses of chemotherapy. One died, but the other two are doing well, as are the remaining seven.

The total cost for this gene therapy was between $30,000 and $50,000 per patient—about the same as for a heart transplant.

Skin Cancer The skin cancer melanoma accounts for only about 4% of skin cancer cases, but it is also the most serious and most aggressive type. In the United States, an estimated 62,190 new cases of melanoma were diagnosed in 2006, and approximately 7,910 people died of the disease.

To test the effectiveness of gene therapy against malignant melanoma, Dr. Steven A. Rosenberg of the National Cancer Institute and his colleagues conducted a clinical trial. **Figure 12.9** shows the procedure they followed. ❶ They drew a small sample of blood that contained white blood cells from the 17 patients included in the trial, and ❷ infected the cells with a virus in the laboratory. ❸ The virus carried a gene that gave the white blood cells special receptors for melanoma cancer cells. ❹ After injecting the white blood cells back into the patient, ❺ the white blood cells were expected to attach to the cancer cells and destroy them.

One month after receiving gene therapy, 15 patients still had 9% to 56% of their transgenic white blood cells. However, over one year later only two patients had sustained high levels of genetically altered white blood cells and remained disease-free. None of the patients experienced toxic side effects from the genetically modified cells.

The researchers plan to further perfect their technique. They want to improve viral delivery of the genes that code for

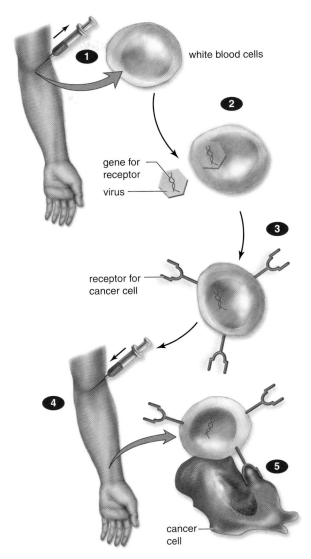

FIGURE 12.9 A procedure for modifying white blood cells to fight cancer.

the necessary receptors and develop white blood cells that can bind to tumor cells more tightly. In addition, the researchers believe it may be beneficial to modify white blood cells still more by inserting molecules that assist in directing them to cancerous tissues. Clinical trials are being conducted to enhance treatment by using total body radiation to deplete a patient's supply of nonaltered lymphocytes and then replace them with purely engineered cells.

This completes our study of gene transfers between organisms, and we will begin a study of the human genome in the next part of the chapter.

> **12.9 *Check Your Progress*** Suppose a virus ferrying genes into white blood cells also carried a gene for the GFP protein. How could scientists be sure the virus had entered the white blood cells?

We now know the sequence of the base pairs along the length of the human chromosomes. So far, researchers have found far fewer genes coding for proteins than expected, and they are busy studying how our genome differs from those of other organisms. Others expect to use the human genome information to develop better drugs; two new fields, proteomics and bioinformatics, will assist in this endeavor.

12.10   The human genome has been sequenced

In the previous century, researchers discovered the structure of DNA, how DNA replicates, and how protein synthesis occurs. Genetics in the 21st century concerns genomics, the study of **genomes**—our genes, and the genes of other organisms. As the result of the **Human Genome Project (HGP)**, a 13-year effort that involved both university and private laboratories around the globe, we now know the order of the 3 billion bases (A, T, C, and G) in our genome. This biological achievement has been likened to arriving at the periodic table of the elements in chemistry.

How did they do it? First, investigators developed a laboratory procedure that would allow them to decipher a short sequence of base pairs, and then instruments became available that could carry out this procedure automatically. Craig Venter is famous for his "shotgun" method of blasting the double helix into smaller fragments, using an automated sequencer to sequence them, and then using a supercomputer to arrange the sequenced DNA fragments into the original order. Whose DNA did they use? Sperm DNA was the material of choice because it has a much higher ratio of DNA to protein than other types of cells. (Recall that sperm do provide both X and Y chromosomes.) However, white cells from the blood of female donors was also used in order to include female-originated samples. The male and female donors were of European, African, American (both North and South), and Asian ancestry.

Many small regions of DNA that vary among individuals (e.g., polymorphisms) were identified during the HGP. Most of these are **single nucleotide polymorphisms (SNPs)** (individuals differ by only one nucleotide). Many SNPs have no physiological effect; others may contribute to the diversity of human beings or possibly increase an individual's susceptibility to disease and response to medical treatments.

Besides humans, a number of other organisms, called model organisms, have also had their genomes sequenced (**Table 12.10**). As discussed in Chapter 10, model organisms are used in genetic analysis because they have many genetic mechanisms and cellular pathways in common with each other and with humans. These organisms, such as mice, lend themselves to genetic experiments, including direct manipulation of their genomes. A surprising finding has been that genome size is not proportionate to the number of genes and does not correlate with the complexity of the organism. For example, the plant *Arabidopsis thaliana*, with a much smaller number of bases, has approximately the same number of genes as a human being. A gene was defined as a sequence of DNA bases that is transcribed into any type of RNA molecule. Most of these are mRNA molecules that direct protein synthesis. Determining that humans have 25,000 genes required a number of techniques, many of which relied on identifying RNAs in the cell and then working backward to find the DNA that can pair with that RNA. It is not yet known what each of the 25,000 human genes does specifically. Today, laboratory instruments called DNA sequencers can automatically analyze up to 2 million base pairs of DNA in a 24-hour period, and researchers are working on systems that can read as many as 1,700 bases a second! A DNA sequencer able to read an entire genome might be available for only $1,000 by 2014. The **Personal Genome Project** is under way, not only to produce faster sequences, but also to determine the possible benefits and drawbacks of sequencing every person's particular genome.

How our knowledge of the human genome will be used to help humans is the topic of Section 12.11.

> **12.10 Check Your Progress** How many bases are different from usual in an SNP?

TABLE 12.10   Genome Sizes of Humans and Some Model Organisms

Organism	*Homo sapiens* (human)	*Mus musculus* (mouse)	*Drosophila melanogaster* (fruit fly)	*Arabidopsis thaliana* (flowering plant)	*Caenorhabditis elegans* (roundworm)	*Saccharomyces cerevisiae* (yeast)
Estimated Size	2,900 million bases	2,500 million bases	180 million bases	125 million bases	97 million bases	12 million bases
Estimated Number of Genes	~25,000	~30,000	13,600	25,500	19,100	6,300
Average Gene Density	1 gene per 100,000 bases	1 gene per 100,000 bases	1 gene per 9,000 bases	1 gene per 4,000 bases	1 gene per 5,000 bases	1 gene per 2,000 bases
Chromosome Number	46	40	8	10	12	32

12.11 New cures are on the horizon

Now that we know the sequence of the bases in the DNA of all the human chromosomes, biologists all over the world believe this knowledge will result in rapid medical advances for ourselves and our children.

First prediction: Many new medicines will be available

Most drugs are proteins or small chemicals that are able to interact with proteins. Today's drugs were usually discovered in a hit-or-miss fashion, but now researchers will be able to take a more systematic approach to finding effective medicines. In a recent search for a medicine that makes wounds heal, researchers cultured skin cells with 14 proteins (found by chance) that can cause skin cells to grow. Only one of these proteins made skin cells grow and did nothing else. They expect this protein to become an effective drug for conditions such as venous ulcers, which are skin lesions that affect many thousands of people in the United States. Tests leading to effective medicines can be carried out with many more proteins that scientists will discover by examining the human genome.

Second prediction: Medicines will be safer due to genome scans

The use of a gene chip will quickly and efficiently provide knowledge of your genetic profile. A gene chip is an array of thousands of genes on one or several glass slides packaged together. After an individual's DNA is applied to the chip, a technician can note any mutant sequences in the individual's genes.

This knowledge is expected to make drugs safer to take. As you know, many drugs potentially have unwanted side effects. Why do some people and not others have one or more of the side effects? Most likely, this is because people have different genetic profiles. It is expected that physicians will be able to match patients to drugs that are safe for them on the basis of their genetic profiles.

One study found that various combinations of mutations can lead to the development of asthma. A particular drug, called albuterol, is effective and safe for patients with certain combinations of mutations and not others. This example and others show that many diseases are polygenic and that only a genetic profile can detect which mutations are causing a disease and how it should be treated.

Third prediction: A longer and healthier life will be yours

Preembryonic gene therapy may become routine once we discover the genes that contribute to a longer and healthier life. We know that the presence of free radicals causes cellular molecules to become unstable and cells to die. Certain genes are believed to code for antioxidant enzymes that detoxify free radicals. It could be that human beings with particular forms of these genes have more efficient antioxidant enzymes, and therefore live longer. If so, researchers should be able to locate these genes as well as others that promote a longer, healthier life.

FIGURE 12.11A Findings from the Human Genome Project could lead to a more carefree life.

Perhaps certain genetic profiles allow some people to live far beyond the normal life span. Researchers may find which genes allow individuals to live a long time and make them available to the general public. Then, many more people could live longer and healthier lives (**Fig. 12.11A**).

Fourth prediction: You will be able to design your children

Genome sequence data will be used to identify many more mutant genes that cause genetic disorders than are presently known. In the future, it may be possible to cure genetic disorders before a child is born by adding a normal gene to any egg that carries a mutant gene. Or an artificial chromosome, constructed to carry a large number of corrective genes, could automatically be placed in eggs. In vitro fertilization would have to be utilized in order to take advantage of such measures.

Genome sequence data can also be used to identify polygenic traits such as height, intelligence, or behavioral characteristics. A couple could decide on their own which genes they wish to use to enhance a child's phenotype. In other words, the sequencing of the human genome may bring about a genetically just society, in which all types of genes would be accessible to all parents (**Fig. 12.11B**).

Two new fields expected to assist in finding new drugs are discussed in Section 12.12.

FIGURE 12.11B The ability to design your children is predicted because of the Human Genome Project.

> **12.11** *Check Your Progress* A drug has been removed from the market because a small percentage of patients had a heart attack. Explain why only a few people were affected.

12.12　Proteomics and bioinformatics are new endeavors

Genes get a lot of attention, but it is proteins that perform most life functions. The translation of all coding genes results in a collection of proteins called the human **proteome**. **Proteomics** is the study of the structure, function, and interaction of cellular proteins. The analysis of proteomes is more challenging than the analysis of genomes for two reasons: Protein concentrations differ widely in cells, and researchers must be able to identify proteins, no matter whether one or thousands of copies are present in a cell. Any particular protein differs minute by minute in concentration, interactions, cellular location, and chemical modifications, among other features. Yet, to understand a protein, all these features must be analyzed. Computer modeling of the three-dimensional shape of cellular proteins is an important part of proteomics. The study of cellular proteins and how they function is essential to the discovery of better drugs. Most drugs are proteins or molecules that affect the function of proteins.

Genomics and proteomics produce raw data, and these fields depend on computer analysis to find significant patterns in the data. **Bioinformatics** is the application of computer technologies to the study of the genome. As a result of bioinformatics, scientists hope to find cause-and-effect relationships between various genetic profiles and genetic disorders caused by multifactorial genes. By correlating any sequence changes with resulting phenotypes, bioinformatics may find that desert regions of the genome do have functions. New computational tools will most likely be needed to accomplish these goals.

Other researchers are interested in comparing our genome to those of other organisms, and one such study is discussed in Section 12.13

> **12.12** *Check Your Progress* **If you were studying the protein GFP, what might you want to know about it?**

12.13　Functional and comparative genomics

The next step of the HGP is to find out how genes function to create different cells and different organisms. A surprising discovery has been that the genomes of all vertebrates are similar. Researchers were not surprised to learn that the genes of chimpanzees and humans are 98% alike, but they did not expect to find that our sequence is also 88% like that of a mouse. It's thought that the regulation of genes can explain why we have one set of traits and mice have another set, despite the similarity of our base sequences. One possibility is alternative gene splicing, which would cause us to differ from mice based on what types of proteins we manufacture and/or when and where certain proteins are present. It could be that gene deserts, 82 regions that comprise 3% of the genome where DNA has no identifiable genes, are involved in regulating the human genome to increase the possible number of proteins in human cells.

Comparing genomes is one way to determine how species have evolved and how genes and noncoding regions of the genome function. In one study, researchers compared the human genome to that of chromosome 22 in chimpanzees. They found three types of genes with different base sequences in chimpan-

zees and humans: a gene for proper speech development, several genes for hearing, and several for smell (**Fig. 12.13**). Genes necessary for speech development are thought to have played an important role in human evolution. Changes in hearing are also likely to have facilitated using language for communication between people. Changes in smell genes are a little more problematic. The investigators speculated that the olfaction genes may have affected dietary changes or sexual selection. Or, they may have been involved in other traits besides smell. The researchers who did this study were surprised to find that many of the other genes they located and studied are known to cause human diseases. They wondered if comparing genomes would be a way of finding other genes that are associated with human diseases. Investigators are taking all sorts of avenues to link human base sequence differences to illnesses.

> **12.13** *Check Your Progress* **What evidence presented in this chapter indicates that a protein can have the same function, whether in a jellyfish or a mammal?**

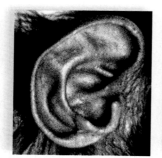
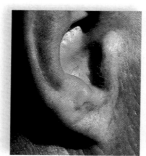
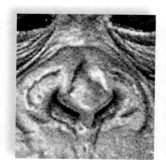
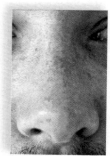

FIGURE 12.13 When comparing the genes of chimpanzees and humans, investigators found differences in the genes for speech, hearing, and smell. These photographs symbolize these differences.

Basic research into the nature and organization of genes in various organisms allowed geneticists to produce recombinant DNA molecules. A knowledge of transcription and translation also enabled scientists to manipulate the expression of foreign genes in organisms. These breakthroughs have spurred a biotechnology revolution. One result is that bacteria and eukaryotic cells are now used to produce vaccines, hormones, and growth factors for use in humans. Today, plants and animals are also engineered to make a product or to possess desired characteristics. Although some people are fearful about the consequences of biotechnology on ecological and human health, no serious problems have surfaced thus far. Biotechnology even offers the promise of curing human genetic disorders, such as muscular dystrophy, cystic fibrosis, hemophilia, and many others. It's possible that genomic research will discover the loci of many other genetic disorders. This information will be useful in order to cure a disorder using gene therapy. It will also allow us to determine people's genetic profiles for the purpose of prescribing medications and preventing future illness.

The alteration of species as a result of genetic changes is one of the definitions for evolution, the topic of Part III. Charles Darwin, who knew nothing about genes, was the first to present significant evidence that evolution does occur.

The Chapter in Review

Summary

Witnessing Genetic Engineering

- Transgenic organisms have received a foreign gene. Those who receive a gene for GFP glow.

DNA Can Be Cloned

12.1 Genes can be isolated and cloned

- A restriction enzyme cleaves DNA.
- DNA ligase seals the gene into the plasmid, which carries the foreign gene into the host cell.
- Gene cloning occurs and the bacteria make a new and different protein.

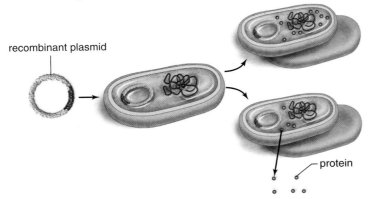

recombinant plasmid

protein

12.2 Specific DNA sequences can be cloned

- PCR makes copies of a specific DNA sequence.
- PCR has many uses, including DNA fingerprinting and evolutionary studies.

Organisms Can Be Genetically Modified

12.3 Bacteria are genetically modified to make a product or perform a service

- Recombinant DNA technology produces transgenic bacteria to manufacture medical and commercial products and perform services.

12.4 Making cheese—Genetic engineering comes to the rescue

- A decline in the veal industry led to a shortage of calves to supply rennet for making cheese.
- Chymosin produced by genetically engineered bacteria replaces rennet from calves in more than 80% of the cheese made today.

12.5 Plants are genetically modified to increase yield or to produce a product

- Certain crops have been engineered to resist disease, insects, or herbicides.
- Genetic engineering is being used to improve the agricultural and food qualities of certain crops.
- Some plants have been engineered to manufacture medical products.

12.6 Are genetically engineered foods safe?

- Some people fear long-term health or environmental problems could result from producing or consuming genetically engineered foods.

12.7 Animals are genetically modified to enhance traits or obtain useful products

- Genes can be inserted into the eggs of animals.
- Through gene pharming, transgenic animals produce pharmaceuticals.
- Transgenic mice are bred for research.
- Xenotransplantation is the use of animal organs in human transplant patients.

12.8 A person's genome can be modified

- Gene therapy can be done in two ways:
  - Using ex vivo therapy, cells or tissues are removed from the body, given a normal gene, and then reinserted into the body.
  - Using in vivo therapy, a gene is delivered directly into the body.

12.9 Gene therapy trials have varying degrees of success

- In a trial to cure SCID, seven out of ten children were immediately cured; two suffered a serious reaction but were cured; one suffered the reaction and died from it.

- In a trial to cure melanoma, one month later, 15 out of 17 patients still had some genetically modified white blood cells, but only two of these were apparently cured one year later.

The Human Genome Can Be Manipulated

12.10 The human genome has been sequenced

- The Human Genome Project (HGP) determined the order of bases in the human genome.
- Genomes of model organisms (i.e., mice, *Drosophila*, *Arabidopsis thaliana*) have been sequenced.
- The Personal Genome Project would enable individuals to have their own genome sequenced.

12.11 New cures are on the horizon

- Because of the HGP:
  - New and safer medicines will be available.
  - People may live longer and be healthier.
  - Genetic disorders may be corrected, even in gametes.

12.12 Proteomics and bioinformatics are new endeavors

- Proteomics is the study of the structure, function, and interaction of cellular proteins.
- The human proteome is the complete collection of proteins that humans produce.
- Bioinformatics is the application of computer technologies to the study of the genome.

12.13 Functional and comparative genomics

- One goal of functional genomics is to discover the function of regions where DNA has no identifiable genes.
- Comparative genomics focuses on determining how species are related and the function of genes and noncoding regions.

Testing Yourself

DNA Can Be Cloned

1. Which is not a clone?
   a. a colony of identical bacterial cells
   b. identical quintuplets
   c. a forest of identical trees
   d. eggs produced by oogenesis
   e. copies of a gene through PCR
2. These enzymes are needed to introduce foreign DNA into a vector.
   a. DNA gyrase and DNA ligase
   b. DNA ligase and DNA polymerase
   c. DNA gyrase and DNA polymerase
   d. restriction enzyme and DNA gyrase
   e. restriction enzyme and DNA ligase
3. Put the letters in the correct order to form a plasmid-carrying recombinant DNA.
   a. Use restriction enzymes
   b. Use DNA ligase
   c. Remove plasmid from parent bacterium
   d. Introduce plasmid into new host bacterium
4. Restriction enzymes found in bacterial cells are ordinarily used
   a. during DNA replication.
   b. to degrade the bacterial cell's DNA.
   c. to degrade viral DNA that enters the cell.
   d. to attach pieces of DNA together.

5. Recombinant DNA technology is used
   a. for gene therapy.
   b. to clone a gene.
   c. to combine DNA and protein.
   d. to clone a specific piece of DNA.
6. The restriction enzyme called *Eco*RI has cut double-stranded DNA in the following manner. The piece of foreign DNA to be inserted begins and ends by what base pairs?

7. Which of these would you not expect to be a biotechnology product?
   a. phospholipid   c. modified enzyme
   b. protein hormone   d. clotting enzyme
8. Which of the following is not required for the polymerase chain reaction?
   a. DNA polymerase   c. DNA sample
   b. RNA polymerase   d. nucleotides
9. Today, the polymerase chain reaction
   a. uses RNA polymerase.
   b. takes place in huge bioreactors.
   c. uses a heat-tolerant enzyme.
   d. makes lots of nonidentical copies of DNA.
   e. All of these are correct.
10. DNA fingerprinting can be used for which of these purposes?
    a. identifying human remains
    b. identifying infectious diseases
    c. finding evolutionary links between organisms
    d. solving crimes
    e. All of these are correct.
11. DNA amplified by PCR and then used for fingerprinting could come from
    a. any diploid or haploid cell.
    b. only white blood cells that have been karyotyped.
    c. only skin cells after they are dead.
    d. only purified animal cells.
    e. Both b and d are correct.
12. **THINKING CONCEPTUALLY** You have 30 dinosaur genes. Explain why it would be impossible to create a dinosaur, even if you use PCR to increase the number of genes.

Organisms Can Be Genetically Modified

13. Which of the following was the first type of food to be dependent on genetic engineering in the United States?
    a. BT corn   c. salt-tolerant tomato
    b. cheese containing   d. soybeans
       chymosin   e. None of these are correct.
14. Some fear that crops genetically modified to be resistant to herbicides and insects might
    a. cause wild plants to die also.
    b. kill off only beneficial insects.
    c. cause allergic reactions in people.
    d. cause supersize insects to evolve.
    e. All of these are correct.
15. Gene pharming uses
    a. genetically engineered farm animals to produce therapeutic drugs.

b. DNA polymerase to produce many copies of targeted genes.

c. restriction enzymes to alter bacterial genomes.

d. All of these are correct.

16. Gene therapy has been used to treat which of the following conditions?

   a. cystic fibrosis
   b. familial hypercholesterolemia
   c. severe combined immunodeficiency
   d. All of these are correct.

17. Gene therapy

   a. is still an investigative procedure.
   b. has met with no success.
   c. is only used to cure genetic disorders such as SCID and cystic fibrosis.
   d. makes use of viruses to carry foreign genes into human cells.
   e. Both a and d are correct.

18. **THINKING CONCEPTUALLY** Explain why gene therapy researchers prefer to genetically modify the stem cells of white blood cells, as opposed to the white blood cells themselves.

The Human Genome Can Be Manipulated

19. Because of the Human Genome Project, we know or will know

   a. the sequence of the base pairs of our DNA.
   b. the sequence of genes along the human chromosomes.
   c. the mutations that lead to genetic disorders.
   d. All of these are correct.
   e. Only a and c are correct.

20. Which of the following is not a likely outcome of the completed Human Genome Project?

   a. development of new medicines
   b. increase in human life span
   c. ability to design children
   d. All of these are likely outcomes of the completed Human Genome Project.

21. Which of these is mismatched?

   a. genome—all the genes of an individual
   b. proteome—all the proteins in an individual
   c. bioinformatics—all the genetic information present in the organism
   d. polymorphism—difference in DNA sequences

22. The field of comparative genomics is not concerned with

   a. function of genes products.
   b. number of genes in various organisms.
   c. who is related to whom.
   d. how to cure human genetic diseases.

23. Which of these is a true statement?

   a. The size of an organism does not correlate with the size of the genome.
   b. Genomes do not contain both coding and noncoding DNA.
   c. Bioinformations is not used as a tool for studying genomes.
   d. Alternative slicing of existing genes is not possible.

24. If you knew the sequence of genes on the chromosomes,

   a. it would reveal which genes are active in which cells.
   b. it would reveal how genes are regulated in a cell.
   c. more genes could be isolated and used for gene therapy.
   d. All of these are correct.

Understanding the Terms

bioinformatics 236
biotechnology product 228
cloning 226
DNA fingerprinting 227
DNA ligase 226
gene cloning 226
gene therapy 232
genome 234
genetically engineered 226
genetically modified organism (GMO) 226
Human Genome Project (HGP) 234
Personal Genome Project 234

plasmid 226
polymerase chain reaction (PCR) 227
proteome 236
proteomics 236
recombinant DNA (rDNA) 226
restriction enzyme 226
single nucleotide polymorphism (SNP) 234
transgenic organism 228
vector 226
xenotransplantation 231

Match the terms to these definitions:

a. _____ Bacterial enzyme that stops viral reproduction by cleaving viral DNA; used to cut DNA at specific points during production of recombinant DNA.

b. _____ All the genetic information of an individual or a species.

c. _____ Production of identical copies; in genetic engineering, the production of many identical copies of a gene.

d. _____ Self-duplicating ring of accessory DNA in the cytoplasm of bacteria.

Thinking Scientifically

1. Design an experiment based on Figure 12.7B that would allow you to determine where a dominant gene for "tailless" is located on mouse chromosome 10.

2. When doing a gene therapy study, what is the advantage of utilizing an ex vivo instead of an in vivo procedure? See Section 12.8.

ARIS *Visit* **www.mhhe.com/maderconcepts** *for practice quizzes, animations, videos, and activities designed to help you master the material in this chapter.*

BIOLOGICAL VIEWPOINTS

PART II Genes Control the Traits of Organisms

Like early breeders of plants and animals, we readily accept the principle that traits are inherited, and we have frequent discussions with friends and family about who in the past had our eye color or the shape of our nose or mouth. But many of us cannot readily grasp that we humans, like all organisms, contain coded information that dictates our form, function, and behavior. A remarkable series of discoveries, spurred by increasingly sophisticated technologies, has given us this modern-day statement of the gene theory: Organisms contain coded information that dictates their form, function, and behavior.

In 1860, Mendel merely deduced the existence of inheritable factors from his results of pea crosses, and he had no idea where these factors might be located or how they might function in the organism. Microscopy had improved by the 1900s, enabling researchers to observe the separation of chromosomes during meiosis. Only then did the idea begin to materialize that Mendel's factors are on the chromosomes. Both inheritable factors and chromosome pairs separate during the formation of gametes, so a parent passes only one of each kind of chromosome to an offspring. When Thomas Morgan began his breeding experiments with *Drosophila* around 1908, he observed that males, but not females, were apt to have white eyes. After cross upon cross, he concluded that the X chromosome carried hereditary units he called genes. Morgan even mapped *Drosophila* chromosomes and showed that genes are arranged linearly along a chromosome. Knowing that the genes are on the chromosomes tells the location of genes, but not what they are.

From 1930 to 1950, investigators used all sorts of creative ways to determine whether the protein or the DNA of a chromosome was the genetic material. At first, scientists were inclined to believe that protein was the genetic material until Hershey and Chase performed their famous experiments. They used radioactively tagged molecules to show that viruses insert their DNA into bacteria, and this molecule alone causes bacteria to produce more viruses. Therefore, DNA must be the genetic material. The chemist Chargaff did his bit by showing that the percentage of A = T and the percentage of G = C were always equal, but that each species has different percentages. DNA was variable as required for the genetic material of all organisms! Now that scientists knew DNA was the genetic material, they could begin to discover what DNA does.

Discovering the structure of DNA was absolutely critical to proposing how DNA works. Making use of available studies, Watson and Crick constructed a model of DNA in 1953, and by 1958 investigators could tell us how DNA replicates so that a copy can be passed to all the cells and offspring of an organism. Knowing that mutations lead to metabolic disorders helped geneticists realize that there must be a link between DNA and proteins. In 1961, investigators performed the experiments that laid the groundwork for cracking the genetic code. Every three bases in an mRNA transcript stand for an amino acid, and therefore an mRNA transcript tells the sequence of amino acids in a protein. In other words, DNA stores the information that allows cells to build their own proteins. Our proteins make us who we are! While the DNA always specifies proteins in the same way, diversity arises because the cellular proteins are different among organisms.

In recent years, the use of high-speed computers has allowed investigators to sequence the genome of many organisms, including humans. We are busy locating human genes and discovering the effects of many mutations. The fields of proteomics and bioinformatics have been initiated with the hope that many human conditions will be treatable or curable. Modern genetics contributes to most other fields of biology, and in Part III we will see how our newfound knowledge helps us understand the process of evolution and the history of life on Earth.

13

Darwin and Evolution

LEARNING OUTCOMES

After studying this chapter, you should be able to accomplish the following outcomes.

The "Vice Versa" of Animals and Plants

1 Give examples of adaptations, including how animals help plants and vice versa.

Darwin Developed a Natural Selection Hypothesis

2 Describe Darwin's trip aboard the HMS *Beagle* and some of the observations he made.
3 Name two early evolutionists who attempted to explain evolution but lacked a suitable mechanism.
4 Give examples of artificial selection carried out by human beings.
5 Explain Darwin's hypothesis for natural selection.
6 Give examples to show that natural selection results in adaptation to the environment.

The Evidence for Evolution Is Strong

7 Tell why fossils offer powerful evidence for common descent.
8 Discuss anatomic, biogeographic, and molecular evidence for common descent.

Population Genetics Tells Us When Microevolution Occurs

9 Use the Hardy-Weinberg principle to explain when microevolution occurs.
10 Explain how mutations, gene flow, nonrandom mating, genetic drift, and natural selection contribute to the process of microevolution.
11 Name three kinds of natural selection, and discuss the effect of each on a population.
12 Give an example to show that stabilizing selection can maintain harmful alleles in a population.

Adaptations provide powerful evidence for evolution. Bacteria that are able to survive and reproduce in the presence of an antibiotic have become adapted to their environment. Penguins are birds adapted to swimming in the ocean, and bats are mammals that can fly due to specific adaptations.

In certain instances, organisms are adapted to one another, and so it is with the animals that help plants reproduce. Plants must be pollinated in order to produce seeds and reproduce. Plants need help because they are immobile. Some depend on the wind to disperse pollen to other plants, but many depend on animals called pollinators. The pollinators carry pollen from one plant to the other—often to the same species of plant. A quick survey shows that plants and their pollinators are remarkably adapted to one another.

It might seem as if bees go to all flowers, but they don't. Bees visit only certain flowers—the ones that provide them with nectar, a surgery liquid that serves as their food. Bee-pollinated flowers advertise the presence of nectar. They are sweet-smelling and have ultraviolet shadings that lead bees to

packet of pollen

petal resembles
a female bumblebee

Bumblebee-pollinated flower,
Ophrys elegans

The "Vice Versa" of Animals and Plants

where nectar can be found. For their part, bees can see the shadings, and their feeding apparatus, called a proboscis, is the right size to reach down into a narrow floral tube where the nectar is located. Pollen clings to their hairy body, and as the bee moves from flower to flower of the species to which it is adapted, the pollen is distributed.

The orchid *Ophrys apifera* has a unique adaptation that causes a bumblebee to visit it. The center of the flower looks like a female bumblebee is resting there. Actually, the flower has a petal that resembles a bumblebee. Occasionally, a male bee tries to mate with the petal, and when it does, it gets dusted with pollen, which it takes to the next flower of only this species.

Moth-pollinated flowers are white, pale yellow, or pink—colors that are visible at night, when moths are active. The flowers also give off a strong, sweet perfume, which attracts moths. Moths don't land on flowers, but rather flap their wings rapidly—called hovering—in order to remain in one spot while they feed. The flowers have open margins that allow a hovering moth to reach the nectar with its long, specialized tongue, or proboscis. Hummingbirds also hover when they feed during

Butterfly-pollinated flower

the day from odorless, red flowers that curve backward. A hummingbird's long, thin beak can access the nectar through a slender floral tube.

Butterflies don't hover, so the flowers they feed from are colorful composites that provide a flat landing platform. Each individual flower of the composite has a floral tube that allows the long, thin butterfly proboscis to reach the nectar.

What would cause plants and their pollinators to be so suited to one another? Evolution, of course—genetic and phenotypic changes over many generations caused plants and their pollinators to **coevolve** until each was suited to the other! Why did it happen? Because plants use pollinators to reproduce and animals look for food to live; they cannot make their own.

We begin our study of evolution in this chapter by examining the work of Charles Darwin, who offered evidence that evolution consists of descent from a common ancestor and adaptation to the environment. Further, Darwin offered a mechanism for evolution he called natural selection. He called it natural selection because the environment, in a sense, chooses which members of a population reproduce, and in that way, adaptation to the environment is eventually achieved.

proboscis

moths hover

light-colored flower

Moth-pollinated flower

long thin beak

hummingbirds hover

floral tube with curved back margins

Hummingbird-pollinated flower

Darwin was persuaded that evolution occurs after taking a trip around the world as a naturalist aboard the HMS *Beagle*. Other scientists before Darwin had hypothesized that evolution occurs but had developed no mechanism. After studying artificial selection and the work of Thomas Malthus, Darwin—and later Alfred Wallace—suggested natural selection as a mechanism for evolution.

13.1 Darwin made a trip around the world

In December 1831, a new chapter in the history of biology began. A 22-year-old naturalist, Charles Darwin (1809–1882), set sail on the journey of a lifetime aboard the British naval vessel HMS *Beagle* (**Fig. 13.1**). Darwin's primary mission on this journey around the world was to expand the navy's knowledge of natural resources in foreign lands. The captain of the *Beagle*, Robert Fitzroy, also hoped that Darwin would find evidence to support the biblical account of creation. Contrary to Fitzroy's wishes, Darwin amassed observations that would eventually support another way of thinking and change the history of science and biology forever.

Rhea

Patagonian desert

Earth's strata contain fossils

Charles Darwin, age 31

Great Britain
Europe
North America
ATLANTIC OCEAN
PACIFIC OCEAN
Galápagos Islands
South America
Africa
INDIAN OCEAN
Australia
HMS *Beagle*

Tropical rain forest

Marine iguana

Woodpecker finch

FIGURE 13.1 *Middle*: Charles Darwin and the route of the HMS *Beagle*. Circles pinpoint highlights of Darwin's trip.

During the trip, Darwin made numerous observations. For example, he noted that the rhea of South America was suited to living on a plain and looked like the ostrich that lived in Africa. However, the rhea was not an ostrich. Why not? Because the rhea evolved in South America, while the ostrich evolved in Africa. Darwin also found that species varied according to whether they lived in the Patagonian desert or a lush tropical rain forest. Then, too, unique animals lived only on the Galápagos Islands, located off the coast of South America, not on the mainland. A marine iguana had large claws that allowed it to cling to rocks and a snout that enabled it to eat algae off rocks. One type of finch, lacking the long bill of a woodpecker, used a cactus spine to probe for insects. Why were these animals found only in the Galápagos Islands? Had they evolved there?

When Darwin explored the region that is now Argentina, he saw raised beaches for great distances along the coast. He thought it would have taken a long time for such massive movements of the Earth's crust to occur. While Darwin was making geologic observations, he also collected fossils that showed today's plants and animals resemble, but are not exactly like, their forebears. Darwin had

brought Charles Lyell's *Principles of Geology* on the *Beagle* voyage. This book said that weathering causes erosion and that, thereafter, dirt and rock debris are washed into the rivers and transported to oceans. When these loose sediments are deposited, layers of soil called **strata** (sing., stratum) result. The strata, which often contain fossils, are uplifted from below sea level to form land. Lyell's book went on to support a theory of **uniformitarianism**, which stated that geologic changes occur at a uniform rate. This idea of slow geologic change is still accepted today, although modern geologists realize that rates of change have not always been uniform. Darwin was convinced that the Earth's massive geologic changes are the result of slow processes and that, therefore, the Earth was old enough to have allowed *evolution* to occur. Before Darwin, other scientists had also believed in evolution, as explained in Section 13.2.

> **13.1 Check Your Progress** Look again at Figure 1.8, a diagram that illustrates the scientific method. Which part of the diagram applies to Darwin's approach so far?

13.2 Others had offered ideas about evolution before Darwin

Before Darwin, the worldview was forged by deep-seated beliefs that were not supported by the use of the scientific method. Prior to Darwin, laypeople believed that the Earth was only a few thousand years old and that species never changed in their attributes. Scientists were beginning to think that species change over time, but the mechanisms they suggested were not workable.

Darwin Was Aware of the Work of Other Scientists

A noted zoologist at the time, Georges Cuvier, founded the science of **paleontology**, the study of fossils (**Fig. 13.2A**). He knew that fossils showed a succession of different life-forms through time. To explain these observations, he hypothesized that whenever a new stratum showed a new mix of fossils, a local catastrophe had caused a mass extinction in that region. After each catastrophe, the region was repopulated by species from surrounding areas, and this accounted for the appearance of new fossils in the new stratum. The result of all these catastrophes was change appearing over time.

In contrast to Cuvier, Jean-Baptiste de Lamarck, an invertebrate biologist, used fossils to conclude that more complex organisms are descended from less complex organisms. To explain the process of adaptation to the environment, Lamarck offered the idea of *inheritance of acquired characteristics*, which proposes that use and disuse of a structure can bring about inherited change. One example Lamarck gave—and the one for which he is most famous—is that the long neck of a giraffe developed over time because animals stretched their necks to reach food high in trees and then passed on a longer neck to their offspring (**Fig. 13.2B**).

The inheritance of acquired characteristics has never been substantiated by experimentation. For example, if acquired characteristics were inherited, people who use tanning machines would have tan children, and people who have LASIK surgery to correct their vision would have children with perfect vision.

Observing artificial selection, described in Section 13.3, helped Darwin arrive at natural selection as a mechanism for evolution.

FIGURE 13.2A One of the animals that Cuvier reconstructed from fossils was the mastodon.

FIGURE 13.2B Lamarck thought the long neck of a giraffe was due to continued stretching in each generation.

> **13.2 Check Your Progress** Many scientists of Darwin's day formulated hypotheses, but these hypotheses have been rejected by science. When are hypotheses rejected by science?

13.3 Artificial selection mimics natural selection

Darwin made a study of **artificial selection**, a process by which humans choose, on the basis of certain traits, the animals and plants that will reproduce. For example, foxes are very shy and normally shun the company of people, but in forty years time, Russian scientists have produced silver foxes that now allow themselves to be petted and even seek attention (**Fig. 13.3A**). They did this by selecting the most docile animals to reproduce. The scientists noted that some physical characteristics changed as well. The legs and tails became shorter, the ears became floppier, and the coat color patterns changed. Artificial selection is only possible because the original population exhibits a range of characteristics, allowing humans to select which traits they prefer to perpetuate.

To take another example, several varieties of vegetables can be traced to a single ancestor that exhibits various characteristics. Chinese cabbage, brussels sprouts, and kohlrabi are all derived from one species of wild mustard (**Fig. 13.3B**). Cabbage was produced by selecting for reproduction only plants that had overlapping leaves; brussels sprouts came from crossing only

Chinese cabbage   Brussel sprouts   Kohlrabi

FIGURE 13.3B
These three plants came from the wild mustard plant through artificial selection.

Wild mustard

FIGURE 13.3A Artificial selection has produced domesticated foxes.

plants with certain types of buds; and kohlrabi was produced by crossing only the plants that had enlarged stems.

Darwin thought that a process of selection might occur in nature without human intervention. Using the process of artificial selection helped him arrive at the mechanism of natural selection, which allows evolution to occur.

In Section 13.4, we see that Darwin was influenced by Thomas Malthus when he formulated natural selection as a mechanism for evolution.

> **13.3** *Check Your Progress* **If you wanted to use artificial selection to achieve a particular type of flower, would you allow the flower to pollinate naturally?**

13.4 Darwin formulated natural selection as a mechanism for evolution

Darwin was very much impressed by an essay written by Thomas Malthus about the reproductive potential of human beings. Malthus had proposed that death and famine are inevitable because the human population tends to increase faster than the supply of food. Darwin applied this concept to all organisms and saw that available resources were insufficient for all members of a population to survive. For example, he calculated the reproductive potential of elephants. Assuming a life span of about 100 years and a breeding span of 30–90 years, a single female probably bears no fewer than six young. If all these young survive and continue to reproduce at the same rate, after only 750 years, the descendants of a single pair of elephants would number about 19 million! Each generation has the same reproductive potential as the previous generation. Therefore, Darwin hypothesized, there is a constant struggle for existence, and only certain members of a population survive and reproduce in each generation. What members might

those be? The members that have some advantage and are best able to compete successfully for limited resources.

Applying Darwin's thinking to giraffes, we can see that long-necked giraffes would be better able to feed off leaves in trees than short-necked giraffes. The longer neck gives giraffes an advantage that, in the end, would allow them to produce more offspring than short-necked giraffes. So, eventually, all the members of a giraffe population (individuals of a species in one locale) would have long necks. Or, what about bacteria living in an environment of antibiotics? The few bacteria that can survive in this environment have a tremendous advantage, and therefore their offspring will make up the next generation of bacteria, and this strain of bacteria will be resistant to the antibiotic.

Darwin called the process by which organisms with an advantage reproduce more than others of their kind **natural selection** because the environment (nature) selects which organisms reproduce. Darwin's hypothesis of natural selection consists of these components:

The members of a population have inheritable variations. For example, a wide range of differences exists among any group of human beings. Many of these variations are inheritable. Inheritance of variations is absolutely essential to Darwin's hypothesis, even though he did not know the means by which inheritance occurs.

A population is able to produce more offspring than the environment can support. The environment contains only so much food and water, places to live, potential mates, and so forth. The environment can't support all the offspring that a population can produce, and each generation is apt to be too large for the environment to support.

Only certain members of the population survive and reproduce. These members have an advantage suited to the environment that allows them to capture more resources than other members, as when long-necked giraffes are better able to browse on tree leaves. This advantage allows these members of the population to survive and produce more offspring. This is called differential reproduction because the members of a population differ as to how many surviving offspring they will have.

Natural selection results in a population adapted to the local environment. In each succeeding generation, an increasing proportion of individuals will have the adaptive characteristics—the characteristics suited to surviving and reproducing in that environment (**Fig. 13.4**).

Now it is possible to form a definition of evolution. **Evolution** consists of changes in a population over time due to the accu-

FIGURE 13.4 A flower and its pollinator are adapted to one another.

mulation of inherited differences. Evolution explains the unity and diversity of organisms. "Unity" means organisms share the same characteristics of life because they share a common ancestry, traceable even to the first cell or cells. "Diversity" comes about because each type of organism (each species) is adapted to one of the many different environments in the biosphere (e.g., oceans, deserts, mountains, etc.).

Independently, Alfred Wallace also arrived at natural selection as a mechanism for evolution, as explained next.

> **13.4 Check Your Progress** Based on Darwin's hypothesis for natural selection, explain how coevolution between a plant and its pollinator came about.

HOW SCIENCE PROGRESSES

13.5 Wallace independently formulated a natural selection hypothesis

Like Darwin, Alfred Russel Wallace (1823–1913) was a naturalist. While he was a schoolteacher at Leicester in 1844–1845, he met Henry Walter Bates, a biologist who interested him in insects. Together, they went on a collecting trip to the Amazon that lasted several years. Wallace's knowledge of the world's flora and fauna was further expanded by a tour he made of the Malay Archipelago from 1854 to 1862. Later, he divided the islands into a western group and an eastern group on the basis of their different plants and animals. The dividing line between these islands is a narrow but deep strait now known as the Wallace Line.

Just as Darwin had done, Wallace wrote articles and books that clearly showed his belief that species changed over time and that it was possible for new species to evolve. Later, he said that he had pondered for many years about a mechanism to explain the origin of a species. He, too, had read Malthus's essay on human population increases, and in 1858, while suffering an attack of malaria, the idea of "survival of the fittest" came upon him. He quickly completed an essay discussing a natural selection process, which he chose to send to Darwin for comment. Darwin was stunned upon its receipt. Here before him was the

hypothesis he had formulated as early as 1844, but never published. Darwin told his friend and colleague Charles Lyell that Wallace's ideas were so similar to his own that even Wallace's "terms now stand as heads of my chapters" in the book he had begun in 1856.

Darwin suggested that Wallace's paper be published immediately, even though he himself as yet had nothing in print. Lyell and others who knew of Darwin's detailed work substantiating the process of natural selection suggested that a joint paper be read to the Linnean Society. The title of Wallace's section was "On the Tendency of Varieties to Depart Indefinitely from the Original Type." Darwin allowed the abstract of a paper he had written in 1844 and an abstract of his book *On the Origin of Species* to be read. This book was published in 1859.

Modern investigators have shown that it is possible to observe the process of natural selection, as described in Section 13.6.

> **13.5 Check Your Progress** Did the work of Wallace lend support to the natural selection hypothesis?

large ground-dwelling finch

warbler-finch

cactus-finch

FIGURE 13.6A Finches on the Galápagos Islands.

Darwin had formed his natural selection hypothesis, in part, by observing the distribution of tortoises and finches on the Galápagos Islands. Tortoises with domed shells and short necks live on well-watered islands, where grass is available. Those with shells that flare up in front have long necks and are able to feed on tall trees. They live on arid islands, where treelike prickly-pear cactus is the main food source. Similarly, the islands are home to many different types of finches. The heavy beak of the large, ground-dwelling finch is suited to a diet of seeds. The beak of the warbler-finch is suited to feeding on insects found among ground vegetation or caught in the air. The longer, somewhat de-curved beak and split tongue of the cactus-finch are suited for probing cactus flowers for nectar (**Fig. 13.6A**).

Today, investigators, such as Peter and Rosemary Grant of Princeton University, are actually watching natural selection as it occurs. In 1973, the Grants began a study of the various finches on Daphne Major, near the center of the Galápagos Islands. The weather swung widely back and forth from wet years to dry years, and they found that the beak size of the ground finch, *Geospiza fortis*, adapted to each weather swing, generation after generation (**Fig. 13.6B**). These finches like to eat small, tender seeds that require a smaller beak, but when the weather turns dry, they have to eat larger, drier seeds, which are harder to crush. The birds that have a larger beak depth have an advantage and have more offspring. Therefore, among the next

generation of *G. fortis* birds, the beak size has more depth than the previous generation.

Among other examples, the shell of the marine snail *(Littorina obtusata)* has changed over time, probably due to being heavily hunted by crabs. Also, the beak length of the scarlet honeycreeper *(Vestiaria coccinea)* was reduced when the bird switched to a new source of nectar because its favorite flowering plants, the lobelloids, were disappearing.

A much-used example of natural selection is industrial melanism. Prior to the Industrial Revolution in Great Britain, light-colored peppered moths, *Biston betularia*, were more common than dark–colored peppered moths. It was estimated that only 10% of the moth population was dark at this time. With the advent of industry and an increase in pollution, the number of dark-colored moths exceeded 80% of the moth population. After legislation to reduce pollution, a dramatic reversal in the ratio of light-colored moths to dark-colored moths occurred. In 1994, one collecting site recorded a drop in the frequency of dark-colored moths to 19%, from a high of 94% in 1960.

The rise in bacterial resistance to antibiotics has occurred within the past 30 years or so. Resistance is an expected way of life now, not only in medicine, but also in agriculture. New chemotherapeutic and HIV drugs are required because of the resistance of cancer cells and HIV, respectively. Also, pesticides and herbicides have created resistant insects and weeds.

We have completed our study of natural selection as a mechanism for evolution. The next part of the chapter discusses the evidence for evolution.

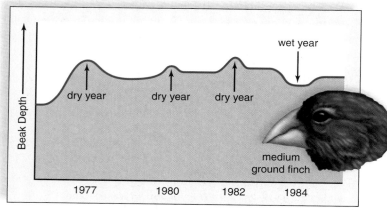

FIGURE 13.6B The beak depth of a ground finch varies from generation to generation, according to the weather.

13.6 *Check Your Progress* Bees rarely pollinate red flowers. Suppose you kept a population of bees locked up with a particular species of red flower for generation after generation, and then you released the bees into the wild. If natural selection occurred, what would you expect to happen?

The evidence for evolution is categorized according to its source. Evidence for common descent is based on fossils, comparative anatomy, biogeography, and molecular observations. Section 13.7 describes the evidence based on fossils.

13.7 Fossils provide a record of the past

The best evidence for evolution comes from fossils, the actual remains of organisms that lived on Earth between 10,000 and billions of years ago. Fossils are the traces of past life, such as trails, footprints, burrows, worm casts, or even preserved droppings. Fossils can also be such items as pieces of bone, impressions of plants pressed into shale, organisms preserved in ice, and even insects trapped in tree resin (which we know as amber).

Usually when an organism dies, the soft parts are either consumed by scavengers or decomposed by bacteria. This means that most fossils consist of hard parts, such as shells, bones, or teeth, because these are usually not consumed or destroyed. When a fossil is encased by rock, the remains are buried in sediment, then the hard parts are preserved by a process called mineralization, and finally, the surrounding sediment hardens to form rock. Most estimates suggest that less than 1% of past species have been preserved as fossils, and only a small fraction of these are found by humans.

More and more fossils have been found because researchers called **paleontologists** have been out in the field looking for them (**Fig. 13.7A**, *left*). Weathering and erosion of rocks produces an accumulation of particles that vary in size and nature and are called sediment. This process, called sedimentation, has been going on since the Earth was formed, and can take place on land or in bodies of water. Sediment becomes a stratum, a recognizable layer in a sequence of layers. Any given stratum is older than the one above it and younger than the one immediately below it (Fig. 13.7A, *right*). This allows investigators to know which fossils within the strata are older and which are younger.

Usually, paleontologists remove fossils from the strata to study them in the laboratory, and then they may decide to exhibit them (**Fig. 13.7B**). The **fossil record** is the history of life recorded by fossils and the most direct evidence we have that evolution has occurred. The species found in ancient sedimentary rock are not the species we see about us today.

FIGURE 13.7B Fossils are carefully cleaned, and organisms are reconstructed.

Darwin relied on fossils to formulate his theory of evolution, but today we have a far more complete record than was available to Darwin. The record tells us that, in general, life has progressed from the simple to the complex. Unicellular prokaryotes are the first signs of life in the fossil record, followed by unicellular eukaryotes and then multicellular eukaryotes. Among the latter, fishes evolved before terrestrial plants and animals. On land, nonflowering plants preceded the flowering plants, and amphibians preceded the reptiles, including the dinosaurs. Dinosaurs are directly linked to the evolution of birds, but only indirectly linked to the evolution of mammals, including humans.

Section 13.8 discusses the evidence for common descent based on the fossil record.

13.7 *Check Your Progress* Why would it be difficult to find fossils of flowers?

FIGURE 13.7A Paleontologists *(left)* carefully remove fossils from strata *(right)*.

Darwin used the phrase "descent with modification" to explain evolution. Because of descent, all living things can trace their ancestry to an original source. For example, you and your cousins have a **common ancestor** in your grandparents and also in your great grandparents, and so forth. In the end, it can be seen that one couple can give rise to a great number of descendants.

A **transitional fossil** is a common ancestor for two different groups of organisms, or it is closely related to the common ancestor for these groups. Transitional fossils allow us to trace the descent of organisms. Even in Darwin's day, scientists knew of the *Archaeopteryx lithographica* fossil, which was an intermediate between reptiles and birds. The dinosaurlike skeleton of these fossils had reptilian features, including jaws with teeth, and a long, jointed tail, but *Archaeopteryx* also had feathers and wings. **Figure 13.8A** shows a fossil of *Archaeopteryx* along with an artist's representation of the animal based on the fossil remains. Many

FIGURE 13.8B *Ambulocetus natans,* an ancestor of the modern toothed whale, and its fossil remains.

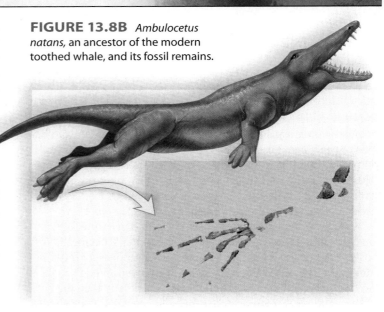

FIGURE 13.8A Fossil of *Archaeopteryx* and an artist's representation.

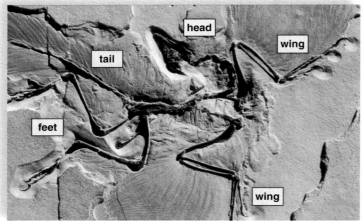

Archaeopteryx fossil

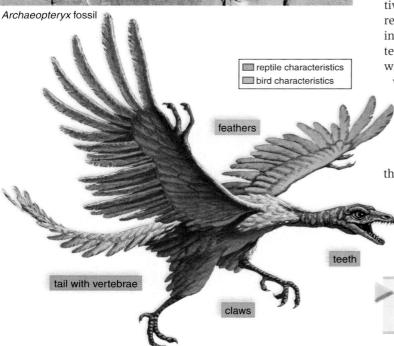

reptile characteristics
bird characteristics

feathers

tail with vertebrae

claws

teeth

more prebird fossils have been discovered recently in China. These fossils are progressively younger than *Archaeopteryx*: The skeletal remains of *Sinornis* suggest it had wings that could fold against its body like those of modern birds, and its grasping feet had an opposable toe, but it still had a tail. Another fossil, *Confuciusornis*, had the first toothless beak. A third fossil, called *Iberomesornis*, had a breastbone to which powerful flight muscles could attach. Such fossils show how the bird of today evolved.

It had always been thought that whales had terrestrial ancestors. Now, fossils have been discovered that support this hypothesis (see Fig. 14.1A). *Ambulocetus natans* (meaning the walking whale that swims) was the size of a large sea lion, with broad, webbed feet on its forelimbs and hindlimbs that enabled it to both walk and swim. It also had tiny hoofs on its toes and the primitive skull and teeth of early whales. **Figure 13.8B** is an artist's re-creation, based on fossil remains of *Ambulocetus*, which lived in freshwater streams. An older fossil, *Pakicetus*, was primarily terrestrial, and yet also had the dentition of an early toothed whale. A younger fossil, *Rodhocetus*, had reduced hindlimbs that would have been no help for either walking or swimming, but may have been used for stabilization during mating.

The origin of mammals is also well documented. The synapsids are mammal-like reptiles whose descendants diversified into different types of premammals. Slowly, mammal-like fossils acquired skeletal features that adapted them to live more efficiently on land. For example, the legs projected downward and not to the side as in reptiles. The earliest true mammals were shrew-sized creatures that have been unearthed in fossil beds about 200 million years old.

Section 13.9 discusses the evidence for common descent based on comparative anatomy.

13.8 Check Your Progress Suppose fossil hummingbirds had shorter, thicker beaks than at present. What would you expect to find about the flowers they pollinated?

Anatomic similarities exist between fossils and between living organisms. Darwin was able to show that a common descent hypothesis offers a plausible explanation for anatomic similarities among organisms. Structures that are anatomically similar because they are inherited from a recent common ancestor are called **homologous structures**. In contrast, **analogous structures** are structures that serve the same function, but they are not constructed similarly, nor do they share a *recent* common ancestry. The wings of birds and insects and the eyes of octopuses and humans are analogous structures. The presence of homology, not analogy, is evidence that organisms are closely related. Studies of comparative anatomy and embryologic development reveal homologous structures.

Comparative Anatomy   Vertebrate forelimbs are used for flight (birds and bats), orientation during swimming (whales and seals), running (horses), climbing (arboreal lizards), or swinging from tree branches (monkeys). Yet, all vertebrate forelimbs contain the same sets of bones organized in similar ways, despite their dissimilar functions (**Fig. 13.9A**). The most plausible explanation for this unity is that the basic forelimb plan belonged to a common ancestor for all vertebrates, and then the plan was modified as each type of vertebrate continued along its own evolutionary pathway.

Vestigial structures are fully developed in one group of organisms but reduced and possibly nonfunctional in similar groups. For example, modern whales have a vestigial pelvic girdle and legs because their ancestors walked on land. Most birds

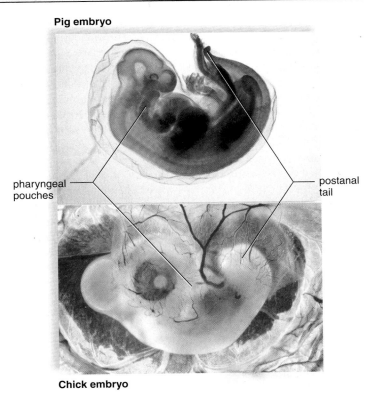

Pig embryo

pharyngeal pouches

postanal tail

Chick embryo

FIGURE 13.9B Vertebrate embryos have features in common, despite different appearances as adults.

have well-developed wings used for flight; however, some bird species (e.g., ostrich) have greatly reduced wings and do not fly. Similarly, snakes have no use for hindlimbs, and yet some have remnants of a pelvic girdle and legs. Humans have a tailbone but no tail. The presence of vestigial structures can be explained by the common descent hypothesis: Vestigial structures occur because organisms inherit their anatomy from their ancestors; they are traces of an organism's evolutionary history.

Embryological Evidence   The homology shared by vertebrates extends to their embryologic development. At some time during development, all vertebrates have a postanal tail and paired pharyngeal pouches (**Fig. 13.9B**). In fishes and amphibian larvae, these pouches develop into functioning gills. In humans, the first pair of pouches becomes the cavity of the middle ear and the auditory tube. The second pair becomes the tonsils, while the third and fourth pairs become the thymus and parathyroid glands. Why do terrestrial vertebrates develop and then modify structures such as pharyngeal pouches that have lost their original function? The most likely explanation is that fishes are ancestral to other vertebrate groups.

Section 13.10 discusses the evidence for common descent based on biogeography.

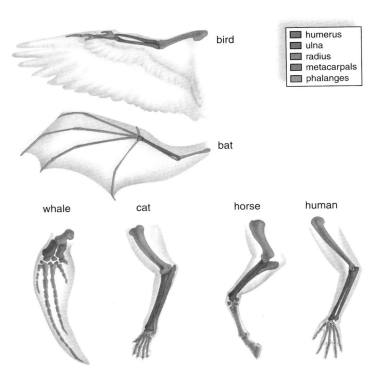

bird

| humerus |
| ulna |
| radius |
| metacarpals |
| phalanges |

bat

whale   cat   horse   human

FIGURE 13.9A   Despite differences in function, vertebrate forelimbs have the same bones.

> **13.9 Check Your Progress** What type ancestry would explain why two species of flowers have exactly the same type floral tube?

13.10   Biogeographic evidence supports common descent

Biogeography is the study of the distribution of plants and animals in different places throughout the world. Such distributions are consistent with the hypothesis that life-forms evolved in a particular locale. Therefore, you would expect a different mix of plants and animals whenever geography separates continents, islands, or seas. As mentioned, Darwin noted that South America lacked rabbits, even though the environment was quite suitable for them. He concluded that no rabbits lived in South America because rabbits evolved somewhere else and had no means of reaching South America. Instead, the Patagonian hare lives in South America. The Patagonian hare resembles a rabbit in anatomy and behavior but has the face of a guinea pig, from which it probably evolved in South America.

To take another example, both cactuses and euphorbia are succulent, spiny, flowering plants adapted to a hot, dry environment. Why do cactuses grow in North American deserts and euphorbia grow in African deserts, when each would do well on the other continent? It seems obvious that they just happened to evolve on their respective continents.

In the history of our planet, South America, Antarctica, and Australia were originally one continent. Marsupials (pouched mammals) arose at around the time Australia separated and drifted away on its own. Isolation allowed marsupials to diversify into many different forms suited to various environments of Australia. They were free to do so because there were few, if any, placental (modern) mammals in Australia. In South America, where there are placental mammals, marsupials are present but not as diverse. This supports the hypothesis that evolution is influenced by the mix of plants and animals in a particular continent—that is, by biogeography, not by design.

Section 13.11 discusses the evidence for common descent based on molecular evidence.

> **13.10** *Check Your Progress* **Explain the observation that the Galápagos Islands host many species of finches that are not found on the mainland.**

13.11   Molecular evidence supports common descent

Almost all organisms use the same basic biochemical molecules, including DNA (deoxyribonucleic acid), ATP (adenosine triphosphate), and many identical or nearly identical enzymes. Further, all organisms use the same DNA triplet code and the same 20 amino acids in their proteins. Since the sequences of DNA bases in the genomes of many organisms are now known, it has become clear that humans share a large number of genes with much simpler organisms. Also of interest, evolutionists who study development have found that many developmental genes are shared by animals ranging from worms to humans. It appears that life's vast diversity has come about by only a slight difference in the regulation of genes. The result has been widely divergent types of bodies. For example, a similar gene in arthropods and vertebrates determines the dorsal-ventral axis. Although the base sequences are similar, the genes have opposite effects. Therefore, in arthropods, such as fruit flies and crayfish, the nerve cord is ventral, whereas in vertebrates, such as chicks and humans, the nerve cord is dorsal. The nerve cord eventually gives rise to the spinal cord and brain.

When the degree of similarity in DNA base sequences or the degree of similarity in amino acid sequences of proteins is examined, the data are consistent with our knowledge of evolutionary descent through common ancestors. Cytochrome *c* is a molecule that is used in the electron transport chain of many organisms. Data show that the amino acid sequence of cytochrome *c* in a monkey differs from that in humans by only two amino acids, from that in a duck by 11 amino acids, and from that in a yeast by 51 amino acids (**Fig. 13.11**), as you might expect from anatomic data.

This completes our study of the evidence for the occurrence of evolution. The next part of the chapter discusses how it is possible to determine that evolution, on a small scale, has occurred.

> **13.11** *Check Your Progress* **Explain the observation that all organisms use DNA as their genetic material.**

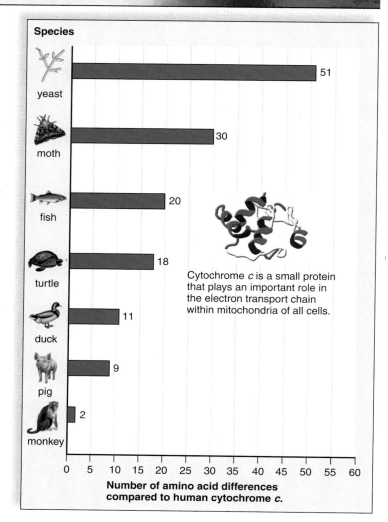

Cytochrome *c* is a small protein that plays an important role in the electron transport chain within mitochondria of all cells.

FIGURE 13.11   Biochemical differences indicate degree of relatedness.

The Hardy-Weinberg principle states that allele frequencies in a population, calculated by using the expression $p^2 + 2\,pq + q^2$, will stay constant generation after generation, unless evolution occurs. Usually, evolution, defined as an allele frequency change, does occur.

13.12   A Hardy-Weinberg equilibrium is not expected

It was not until the 1930s that population geneticists were able to apply the principles of genetics to populations and thereafter develop a way to recognize when evolution on a small scale, called **microevolution**, has occurred. The **gene pool** of a population is all the alleles in all the individuals making up the population. When the allele frequencies for a population change, microevolution has occurred. Microevolution does not necessarily result in a visible change.

Let's say that in a population of tortoises, 36% are homozygous dominant for long necks, 48% are heterozygous, and 16% are homozygous recessive for short necks. Therefore, in a population of 100 individuals, we have

36 LL, 48 Ll, and 16 ll

To determine the frequency of each allele, calculate its percentage from the total number of alleles in the population. For the dominant allele L, 120 L/200 total alleles = 0.6 L; for the recessive allele l, 80 l/200 total alleles = 0.4 l. The sperm and eggs produced by this population will contain these alleles in these frequencies. Assuming random mating (all possible gametes have an equal chance of combining with any other), we can calculate the ratio of genotypes in the next generation by using a Punnett square.

There is an important difference between a Punnett square used for a cross between individuals and the following one. Below, the sperm and eggs are those produced by the members of a population—not those produced by a single male and female. The results of the Punnett square indicate that the frequency for each allele in the next generation is the same as it was in the previous generation:

		eggs		
		0.6 *L*	0.4 *l*	**Genotype frequencies:**
sperm	0.6 *L*	0.36 *LL*	0.24 *Ll*	0.36 *LL* + 0.48 *Ll* + 0.16 *ll* = 1
	0.4 *l*	0.24 *Ll*	0.16 *ll*	

Therefore, sexual reproduction alone cannot bring about a change in allele frequencies. The potential constancy, or equilibrium state, of gene pool frequencies was independently recognized in 1908 by G. H. Hardy, an English mathematician, and W. Weinberg, a German physician. They used the binomial expression ($p^2 + 2\,pq + q^2$) to calculate the genotype and allele frequencies of a population (**Fig. 13.12**). From their findings they formulated the **Hardy-Weinberg principle**, which states that an equilibrium of allele frequencies in a gene pool will remain in effect in each succeeding generation of a sexually reproducing population as long as five conditions are met:

$$p^2 + 2\,pq + q^2$$

p^2 = frequency of homozygous dominant individuals *(AA)*

p = frequency of dominant allele *(A)*

q^2 = frequency of homozygous recessive individuals *(aa)*

q = frequency of recessive allele *(a)*

$2\,pq$ = frequency of heterozygous individuals *(Aa)*

Realize that $\quad p + q = 1$ (These are the only 2 alleles.)

$p^2 + 2\,pq + q^2 = 1$ (These are the only genotypes.)

Example: An investigator has determined by inspection that 16% of a human population has a recessive trait. What are the genotype and allele frequencies for this population?

Given: $\quad q^2 = 16\% = 0.16$ are homozygous recessive individuals

Therefore, $\quad q = \sqrt{0.16} = 0.4$ = frequency of recessive allele

$p = 1.0 - 0.4 = 0.6$ = frequency of dominant allele

$p^2 = (0.6)(0.6) = 0.36 = 36\%$ are homozygous dominant individuals

$2\,pq = 2(0.6)(0.4) = 0.48 = 48\%$ are heterozygous individuals

or

$2\,pq = 1.00 - 0.52 = 0.48 \qquad$ 84% have the dominant phenotype

FIGURE 13.12   Calculating gene pool frequencies.

1. No mutations: Allele changes do not occur, or changes in one direction are balanced by changes in the opposite direction.

2. No gene flow: Migration of alleles into or out of the population does not occur.

3. Random mating: Individuals pair by chance, not according to their genotypes or phenotypes.

4. No genetic drift: The population is very large, and changes in allele frequencies due to chance alone are insignificant.

5. No natural selection: No selective agent favors one genotype over another.

In real life, these conditions are rarely, if ever, met, and allele frequencies in the gene pool of a population do change from one generation to the next. Therefore, microevolution is expected to occur with each new generation. Mutations, and also sexual recombination, are possible causes of microevolution, as discussed in Section 13.13.

> **13.12 *Check Your Progress* How do you know when microevolution has occurred?**

13.13 Both mutations and sexual recombination produce variations

The Hardy-Weinberg principle recognizes mutation as a force that can cause allele frequencies to change in a gene pool and cause microevolution to occur. **Mutations**, which are permanent genetic changes, are the raw material for evolutionary change because without mutations, there could be no inheritable phenotypic variations among members of a population. The rate of mutations is generally very low—on the order of one per 100,000 cell divisions. Also, it is important to realize that evolution is not directed, meaning that no mutation arises because the organism "needs" one. For example, the mutation that causes bacteria to be resistant was already present before antibiotics appeared in the environment.

Mutations are the primary source of genetic differences among prokaryotes that reproduce asexually. Generation time is so short that many mutations can occur quickly, even though the rate is low, and since these organisms are haploid, any mutation that results in a phenotypic change is immediately tested by the environment. In diploid organisms, a recessive mutation can remain hidden and become significant only when a homozygous recessive genotype arises. The importance of recessive alleles increases if the environment is changing; it's possible that the homozygous recessive genotype could be helpful in a new environment, if not the present one. It's even possible that natural selection will maintain a recessive allele if the heterozygote has advantages (see Section 13.17).

In sexually reproducing organisms, sexual recombination is just as important as mutation in generating phenotypic differences, because sexual recombination can bring together a new and different combination of alleles. This new combination might produce a more successful phenotype. Success, of course, is judged by the environment and counted by the relative number of healthy offspring an organism produces.

Nonrandom mating and gene flow are possible causes of microevolution, as discussed in Section 13.14.

> **13.13** *Check Your Progress* Would you expect mutations to have helped flowers and their pollinators coevolve? Explain.

13.14 Nonrandom mating and gene flow can contribute to microevolution

Random mating occurs when individuals pair by chance. You make sure random mating occurs when you do a genetic cross on paper or in the lab, and cross all possible types of sperm with all possible types of eggs. **Nonrandom mating** occurs when only certain genotypes or phenotypes mate with one another. **Assortative mating** is a type of nonrandom mating that occurs when individuals mate with those having the *same* phenotype with respect to a certain characteristic. For example, flowers such as the garden pea usually self-pollinate—therefore, the same phenotype has mated with the same phenotype (**Fig. 13.14A**). Assortative mating can also be observed in human society. Men and women tend to marry individuals with characteristics such as intelligence and height that are similar to their own. Assortative mating causes homozygotes for certain gene loci to increase in frequency and heterozygotes for these loci to decrease in frequency.

Gene flow, also called gene migration, is the movement of alleles between populations. When animals move between populations or when pollen is distributed between species (**Fig. 13.14B**), gene flow has occurred. When gene flow brings a new or rare allele into the population, the allele frequency in the next generation changes. When gene flow between adjacent populations is constant, allele frequencies continue to change until an equilibrium is reached. Therefore, continued gene flow tends to make the gene pools similar and reduce the possibility of allele frequency differences between populations.

Genetic drift is a possible cause of microevolution, as discussed in Section 13.15.

> **13.14** *Check Your Progress* Create a scenario in which assortative mating causes flowers to become adapted to their pollinators.

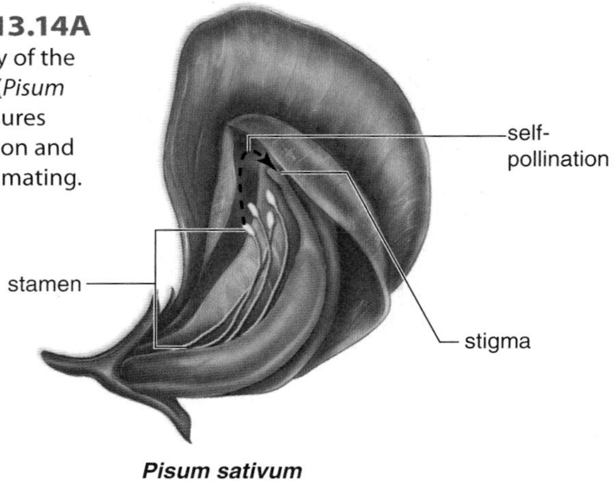

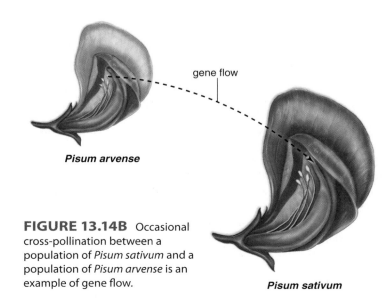

FIGURE 13.14A
The anatomy of the garden pea (*Pisum sativum*) ensures self-pollination and nonrandom mating.

self-pollination

stamen

stigma

Pisum sativum

gene flow

Pisum arvense

FIGURE 13.14B Occasional cross-pollination between a population of *Pisum sativum* and a population of *Pisum arvense* is an example of gene flow.

Pisum sativum

Genetic drift refers to changes in the allele frequencies of a gene pool due to chance rather than selection by the environment. Therefore, genetic drift does not necessarily result in adaptation to the environment, as does natural selection. For example, in California, there are a number of cypress groves, each a separate population. The phenotypes within each grove are more similar to one another than they are to the phenotypes in the other groves. Some groves have longitudinally shaped trees, and others have pyramidally shaped trees. The bark is rough in some colonies and smooth in others. The leaves are gray to bright green or bluish, and the cones are small or large. The environmental conditions are similar for all the groves, and no correlation has been found between phenotype and the environment across groves. Therefore, scientists hypothesize that these variations among the groves are due to genetic drift. We know of two mechanisms by which genetic drift could produce phenotypic similarities. They are called the bottleneck effect and the founder effect. Both of these require a very small population.

Small Versus Large Populations Although genetic drift occurs in populations of all sizes, a smaller population is more likely to show the effects of drift. Suppose the allele *B* (for brown) occurs in 10% of the members in a population of frogs. In a population of 50,000 frogs, 5,000 will have the allele *B*. If a hurricane kills off half the frogs, the frequency of allele *B* may very well remain the same among the survivors. On the other hand, 10% of a population with ten frogs means that only one frog has the allele *B*. Under these circumstances, a natural disaster could very well do away with that one frog, should half the population perish. Or, let's suppose that only five green frogs out of a ten-member population die. Now, the frequency of allele *B* will increase from 10% to 20% (**Fig. 13.15A**).

Bottleneck and Founder Effects When a species is subjected to near extinction because of a natural disaster (e.g., hurricane, earthquake, or fire) or because of overhunting, overharvesting, and habitat loss, it is as if most of the population has stayed behind and only a few survivors have passed through the neck of a bottle. This so-called **bottleneck effect** prevents the majority of genotypes from participating in the production of the next generation. The extreme genetic similarity found in cheetahs is believed to be due to a bottleneck effect. In a study of 47 different enzymes, each of which can come in several different forms, the sequence of amino acids in the enzymes was exactly the same in all the cheetahs. What caused the cheetah bottleneck is not known, but today they suffer from relative infertility because of the intense inbreeding that occurred after the bottleneck. Even if humans were to intervene and the population were to increase in size, without genetic variation, the cheetah could still become extinct. Other organisms pushed to the brink of extinction suffer a plight similar to that of the cheetah.

The **founder effect** is an example of genetic drift in which rare alleles, or combinations of alleles, occur at a higher frequency in a population isolated from the general population. Founding individuals could contain only a fraction of the total genetic diversity of the original gene pool. Which alleles the founders carry is dictated by chance alone. The Amish of Lancaster County, Pennsylvania, are an isolated group that was begun by German founders. Today, as many as 1 in 14 individuals carries a recessive allele that causes an unusual form of dwarfism (affecting only the lower arms and legs) and polydactylism (extra fingers) (**Fig. 13.15B**). In the general population, only one in 1,000 individuals has this allele.

Natural selection, either stabilizing, directional, or disruptive natural selection, is a possible cause of microevolution, as discussed in Section 13.16.

> **13.15** *Check Your Progress* Could genetic drift have set back the coevolution of flowers and their pollinators? Explain.

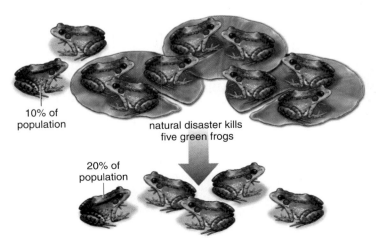

FIGURE 13.15A Chance events can cause allele frequency changes and genetic drift.

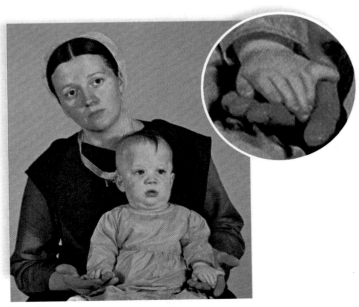

FIGURE 13.15B A rare form of dwarfism that is linked to polydactylism is seen among the Amish in Pennsylvania.

13.16 Natural selection can be stabilizing, directional, or disruptive

After outlining the process of natural selection in Section 13.4, we now wish to consider natural selection in a genetic context. Many traits are polygenic (controlled by many genes), and the continuous variation in phenotypes results in a bell-shaped curve. The most common phenotype is intermediate between two extremes. When this range of phenotypes is exposed to the environment, natural selection favors the one that is most adaptive under the present environmental circumstances. Natural selection acts much the same way as a governing board that decides which applying students will be admitted to a college. Some students will be favored and allowed to enter, while others will be rejected and not allowed to enter. Of course, in the case of natural selection, the chance to reproduce is the prize awarded. In this context, natural selection can be stabilizing, directional, or disruptive (**Fig. 13.16A**, *left*).

Stabilizing selection occurs when an intermediate phenotype is favored. It can improve adaptation of the population to those aspects of the environment that remain constant. With stabilizing selection, extreme phenotypes are selected against, and the intermediate phenotype is favored. As an example, consider that when Swiss starlings lay four to five eggs, more young survive than when the female lays more or less than this number. Genes determining physiological characteristics, such as the production of yolk, and behavioral characteristics, such as how long the female will mate, are involved in determining clutch size.

Human birth weight is another example of stabilizing selection. Through the years, hospital data have shown that human in-

fants born with an intermediate birth weight (3–4 kg) have a better chance of survival than those at either extreme (either much less or much greater than usual). When a baby is small, its systems may not be fully functional, and when a baby is large, it may have experienced a difficult delivery. Stabilizing selection reduces the variability in birth weight in human populations (**Fig. 13.16B**).

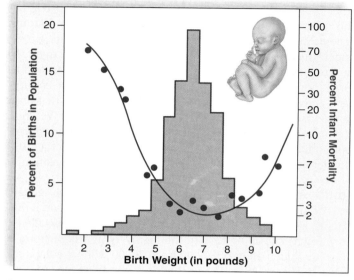

FIGURE 13.16B Stabilizing selection as exemplified by human birth weight.

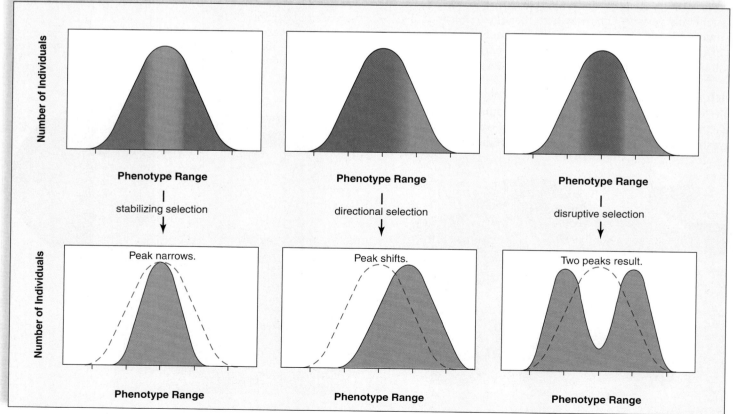

FIGURE 13.16A Phenotype range before and after three types of selection. Blue represents favored phenotype(s).

above waterfall

below waterfall

Experimental site

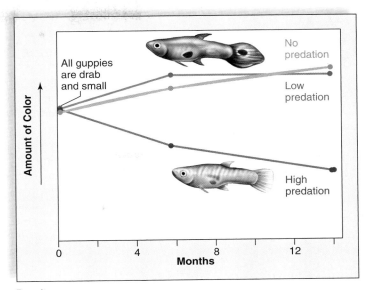
Result

FIGURE 13.16C Directional selection in guppies.

Directional selection occurs when an extreme phenotype is favored, and the distribution curve shifts in that direction. Such a shift can occur when a population is adapting to a changing environment (Fig. 13.16A, *middle*).

Two investigators, John Endler and David Reznick, both at the University of California, conducted a study of guppies, which are known for their bright colors and reproductive potential. These investigators noted that on the island of Trinidad, when male guppies are subjected to high predation by other fish, they tend to be drab in color and to mature early and at a smaller size. The drab color and small size are most likely protective against being found and eaten. On the other hand, when male guppies are exposed to minimal or no predation, they tend to be colorful, to mature later, and to attain a larger size.

Endler and Reznick performed many experiments, and one set is of particular interest. They took a supply of guppies from a high-predation area (below a waterfall) and placed them in a low-predation area (above a waterfall) (**Fig. 13.16C**). The waterfall prevented the predator fish (pike) from entering the low-predation area. They monitored the guppy population for 12 months, and during that year, the guppy population above the waterfall underwent directional selection (Fig. 13.16C). The male members of the population were now colorful and large in size. The members of the guppy population below the waterfall (the control population) were still drab and small.

In **disruptive selection**, two or more extreme phenotypes are favored over any intermediate phenotype (see Fig. 13.16A, *right*). For example, British land snails (*Cepaea nemoralis*) have a wide habitat range that includes low-vegetation areas (grass fields and hedgerows) and forests. In forested areas, thrushes feed mainly on light-banded snails, and the snails with dark shells become more prevalent. In low-vegetation areas, thrushes feed mainly on snails with dark shells, and light-banded snails become more prevalent. Therefore, these two distinctly different phenotypes are found in the population (**Fig. 13.16D**).

Stabilizing selection, discussed in Section 13.17, maintains the heterozygote, especially if the heterozygote has an advantage over the homozygote, as seen in sickle-cell disease.

> **13.16 *Check Your Progress*** If the flowers of a species are presently only one color and the pollinator prefers this color, is stabilizing selection occurring? Explain.

FIGURE 13.16D Disruptive selection in snails.

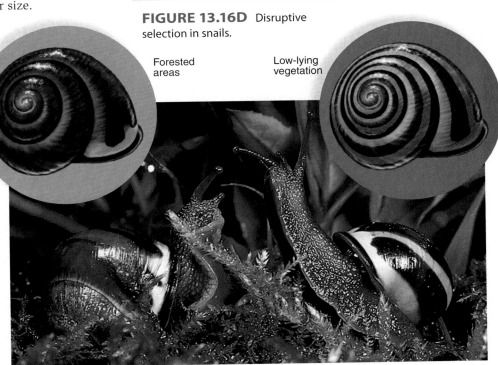
Forested areas

Low-lying vegetation

Variations are maintained in a population for any number of reasons. Mutation still creates new alleles, and recombination still recombines these alleles during gametogenesis and fertilization. Gene flow might still occur. If the receiving population is small and mostly homozygous, gene flow can be a significant source of new alleles. Genetic drift also occurs, particularly in small populations, and the end result may be contrary to adaptation to the environment. Even natural selection in the form of disruptive selection can promote polymorphism in a population. Here, we consider that heterozygote superiority can assist the maintenance of genetic, and therefore phenotypic, variations in future generations.

Sickle-Cell Disease Sickle-cell disease can be a devastating condition. Patients can have severe anemia, physical weakness, poor circulation, impaired mental function, pain and high fever, rheumatism, paralysis, spleen damage, low resistance to disease, and kidney and heart failure. In these individuals, the red blood cells are sickle-shaped and tend to pile up and block flow through tiny capillaries. The condition is due to an abnormal form of hemoglobin (Hb), the molecule that carries oxygen in red blood cells. People with sickle-cell disease (Hb^SHb^S) tend to die early and leave few offspring, due to hemorrhaging and organ destruction. Interestingly, however, geneticists studying the distribution of sickle-cell disease in Africa have found that the recessive allele (Hb^S) has a higher frequency in regions (purple color) where the disease malaria is also prevalent (**Fig. 13.17**). Malaria is caused by a protozoan parasite that lives in and destroys the red blood cells of the normal homozygote (Hb^AHb^A). Individuals with this genotype

also have fewer offspring, due to an early death or to debilitation caused by malaria.

People who are heterozygous (Hb^AHb^S) have an advantage over both homozygous genotypes because they don't die from sickle-cell disease and they don't die from malaria. The parasite causes any red blood cell it infects in these individuals to become sickle-shaped. Sickle-shaped red blood cells lose potassium, and this causes the parasite to die. **Heterozygote advantage** causes all three alleles to be maintained in the population. It's as if natural selection were a store owner balancing the advantages and disadvantages of maintaining the recessive allele Hb^S in the warehouse. As long as the protozoan that causes malaria is present in the environment, it is advantageous to maintain the recessive allele, as shown in the following table:

Genotype	Phenotype	Result
Hb^AHb^A	Normal	Dies due to malarial infection
Hb^AHb^S	Sickle-cell trait	Lives due to protection from both
Hb^SHb^S	Sickle-cell disease	Dies due to sickle-cell disease

Heterozygote advantage is also an example of stabilizing selection because the genotype Hb^AHb^S is favored over the two extreme genotypes, Hb^AHb^A and Hb^SHb^S. In the parts of Africa where malaria is common, one in five individuals is heterozygous (has sickle-cell trait) and survives malaria, while only 1 in 100 is homozygous, Hb^SHb^S, and dies of sickle-cell disease.

What happens in the United States where malaria is not prevalent? As you would expect, the frequency of the Hb^S allele is declining among African Americans because the heterozygote has no particular advantage in this country.

Cystic Fibrosis Stabilizing selection is also thought to have influenced the frequency of other alleles. Cystic fibrosis is a debilitating condition that leads to lung infections and digestive difficulties. In this instance, the recessive allele, common among individuals of northwestern European descent, causes the person to have a defective plasma membrane protein. The agent that causes typhoid fever can use the normal version of this protein, but not the defective one, to enter cells. Here again, heterozygote superiority caused the recessive allele to be maintained in the population.

This is the end of our discussion regarding microevolution of a population.

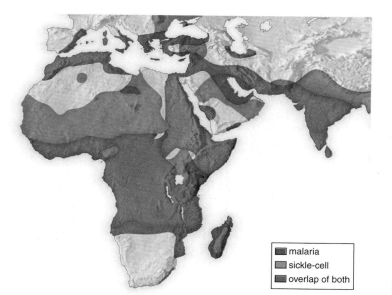

malaria
sickle-cell
overlap of both

FIGURE 13.17 Sickle-cell disease is more prevalent in areas of Africa where malaria is more common.

13.17 *Check Your Progress* **Could heterozygote advantage be used to show that natural selection does not always favor the dominant genotype?**

Darwin took a trip around the world as the naturalist aboard the HMS *Beagle*. During his trip, he collected fossils and made several observations that made him think evolution occurs. Darwin was aware of artificial selection, and he had read an essay by Malthus suggesting that the members of a population compete with one another for resources. Darwin began to see that a competitive edge would allow certain members of a population to survive and reproduce more than other members of the population. Assuming that advantageous traits are inheritable, future generations would eventually acquire adaptations to the local environment. Darwin called this process, by which a population adapts to its environment, natural selection because nature selects which members of a population will reproduce to a greater extent, just as a breeder selects which plants or animals will reproduce during artificial selection.

Evolution explains the unity and diversity of life. Life is unified because of common descent, and it is diverse because of adaptations to particular environments. Darwin used the expression "descent with modification" to explain evolution. Support for common descent includes transitional fossils, anatomic features (homologous structures, vestigial structures, and embryologic similarities), biogeographic data, and molecular evidence.

In the 1930s, biologists developed a way to apply the principles of genetics to evolution. Populations would be in a Hardy-Weinberg equilibrium (allele frequencies stay the same) if mutation, gene flow, nonrandom mating, genetic drift, and natural selection did not occur. However, these events do occur, and they are the agents of evolutionary change that lead to microevolution, recognizable by allele frequency changes. Mutations provide the raw material for evolution. Genetic drift results in allele frequency changes due to a chance event, as when only a few members of a population are able to reproduce because of a natural disaster or because they have founded a colony. Natural selection is the only agent of evolution that results in adaptation to the environment.

Chapter 14 concerns macroevolution, the manner in which new species arise. The origin of new species is essential to the history of life on Earth, which we consider in Chapter 15.

The Chapter in Review

Summary

The "Vice Versa" of Animals and Plants

- Plants and their pollinators are adapted to one another—the plant provides food, and the animal distributes pollen.

Darwin Developed a Natural Selection Hypothesis

13.1 Darwin made a trip around the world

- He observed that species change from place to place and through time.
- A book by Lyell convinced Darwin that the Earth had existed long enough for evolution to have occurred.

13.2 Others had offered ideas about evolution before Darwin

- Cuvier said catastrophes caused evolution to occur.
- Lamarck proposed the inheritance of acquired characteristics as a mechanism of evolution.

13.3 Artificial selection mimics natural selection

- Humans (not the environment) select certain characteristics to perpetuate.

13.4 Darwin formulated natural selection as a mechanism for evolution

- Natural selection has several components:
  - The members of a population have inheritable variations.
  - A population is able to produce more offspring than the environment can support.
  - Certain members of a population survive and reproduce because they have an advantage suited to the environment.
  - Natural selection results in a population adapted to its environment.
- Evolution can be defined as changes in a population over time due to an accumulation of inherited differences.

13.5 Wallace independently formulated a natural selection hypothesis

- Wallace was a naturalist who had also read Malthus and arrived at conclusions similar to those of Darwin.

13.6 Natural selection can be witnessed

- Natural selection has been observed in, for example, Galápagos tortoises and finches, peppered moths, and bacteria.

The Evidence for Evolution Is Strong

13.7 Fossils provide a record of the past

- Fossils are hard parts of organisms or other traces of life found in sedimentary rock.
- The fossil record indicates that life has progressed from simple to complex.

13.8 Fossils are evidence for common descent

- Transitional fossils have the characteristics of two different groups and thus provide clues as to the evolutionary relationships between organisms.

13.9 Anatomic evidence supports common descent

- Homologous structures are anatomically similar among organisms.
- Analogous structures have the same functions in different organisms but are not anatomically similar.
- Only homologous structures and not analogous structures indicate that organisms have a common ancestor.
- Organisms have vestigial structures despite their being reduced and nonfunctional because they were once functional in an ancestor.

- All vertebrates share the same embryonic features which are later modified for different purposes.

bat   bird

13.10 Biogeographic evidence supports common descent
- Plants and animals evolved in particular locations and therefore widely separated similar environments will contain different but similarly adapted organisms.

13.11 Molecular evidence supports common descent
- The degree of similarity of DNA base sequences or amino acid sequences shows a pattern of relatedness that is consistent with fossil record data.

Population Genetics Tells Us When Microevolution Occurs

13.12 A Hardy-Weinberg equilibrium is not expected
- Microevolution is evidenced by changes in gene pool allele frequencies.
- Hardy and Weinberg shows that it was possible to calculate the genotype and allele frequencies of a population by using the following:

$$p^2 + 2pq + q^2 = 1$$

- The Hardy-Weinberg principle states that microevolution does not occur as long as mutations, gene flow, nonrandom matings, genetic drift, and natural selection do not occur.
- Generally, allele frequencies do change between generations, and microevolution does occur.

13.13 Both mutations and sexual recombination produce variations
- Mutations are the primary source of genetic differences in prokaryotes.
- Sexual recombination and mutations are equally important in eukaryotes.

13.14 Nonrandom mating and gene flow can contribute to microevolution
- Assortative mating is a type of nonrandom mating in which individuals mate with those that have the same phenotype as they have for a particular characteristic.
- Gene flow results when alleles move between populations due to migration.

13.15 The effects of genetic drift are unpredictable
- Genetic drift refers to changes in allele frequency in a gene pool due to chance.
- The bottleneck effect prevents the majority of genotypes from participating in production of the next generation.
- The founder effect occurs when rare alleles contributed by the founders of a population occur at a higher frequency in isolated populations.

13.16 Natural selection can be stabilizing, directional, or disruptive
- In stabilizing selection, extreme phenotypes are selected against while intermediate phenotypes are favored.
- In directional selection, an extreme phenotype is favored.
- In disruptive selection, two or more extreme phenotypes are favored over the intermediate phenotype.

13.17 Stabilizing selection helps maintain harmful alleles
- Heterozygote advantage causes the sickle-cell allele to be maintained, even though the homozygous recessive is lethal.
- The allele for cystic fibrosis is believed to be maintained because a faulty membrane protein doesn't allow the typhoid bacterium to enter cells.

Testing Yourself

Darwin Developed a Natural Selection Hypothesis

1. Why was it helpful to Darwin to learn that Lyell had concluded the Earth was very old?
   a. An old Earth has more fossils than a new Earth.
   b. It meant there was enough time for evolution to have occurred slowly.
   c. It meant there was enough time for the same species to spread into all continents.
   d. Darwin said artificial selection occurs slowly.
   e. All of these are correct.
2. Which of these pairs is mismatched?
   a. Charles Darwin—natural selection
   b. Cuvier—series of catastrophes explains the fossil record
   c. Lamarck—uniformitarianism
   d. All of these are correct.
3. Which is most likely to be favored during natural selection, but not artificial selection?
   a. fast seed germination rate
   b. short generation time
   c. efficient seed dispersal
   d. lean pork meat production
4. Which of these is/are necessary to natural selection?
   a. variations
   b. differential reproduction
   c. inheritance of differences
   d. All of these are correct.
5. Natural selection is the only process that results in
   a. genetic variation.
   b. adaptation to the environment.
   c. phenotypic change.
   d. competition among individuals in a population.
6. **THINKING CONCEPTUALLY** The adaptive results of natural selection cannot be determined ahead of time. Explain.

The Evidence for Evolution Is Strong

7. The fossil record offers direct evidence for common descent because you can
   a. see that the types of fossils change over time.
   b. sometimes find common ancestors.
   c. trace the ancestry of a particular group.
   d. trace the biological history of living things.
   e. All of these are correct.

8. Which of the following is not an example of a vestigial structure?
   a. human tailbone   c. pelvic girdle in snakes
   b. ostrich wings   d. dog kidney
9. If evolution occurs, we would expect different biogeographic regions with similar environments to
   a. all contain the same mix of plants and animals.
   b. each have its own specific mix of plants and animals.
   c. have plants and animals with similar adaptations.
   d. have plants and animals with different adaptations.
   e. Both b and c are correct.
10. DNA nucleotide differences between organisms
    a. indicate how closely related organisms are.
    b. indicate that evolution occurs.
    c. explain why there are phenotypic differences.
    d. are to be expected.
    e. All of these are correct.

For questions 11–14, match the evolutionary evidence in the key to the description. Choose more than one answer if correct.

KEY:
   a. biogeographic evidence   c. molecular evidence
   b. fossil evidence   d. anatomic evidence
11. Islands have many unique species not found elsewhere.
12. All vertebrate embryos have pharyngeal pouches.
13. Distantly related species have more amino acid differences in cytochrome *c*.
14. Transitional links have been found between major groups of animals.
15. **THINKING CONCEPTUALLY** Why can researchers make decisions about who is related to whom using only DNA base sequence data? (See Section 13.11.)

Population Genetics Tells Us When Microevolution Occurs

For questions 16 and 17, consider that about 75% of white North Americans can taste the chemical phenylthiocarbamide. The ability to taste is due to the dominant allele *T*. Nontasters are *tt*. Assume this population is in Hardy-Weinberg equilibrium.
16. What is the frequency of *t*?
    a. 0.25   d. 0.09
    b. 0.70   e. 0.60
    c. 0.55
17. What is the frequency of heterozygous tasters?
    a. 0.50   c. 0.2475
    b. 0.21   d. 0.45
18. The offspring of better-adapted individuals are expected to make up a larger proportion of the next generation. The most likely explanation is
    a. mutations and nonrandom mating.
    b. gene flow and genetic drift.
    c. mutations and natural selection.
    d. mutations and genetic drift.
19. The Northern elephant seal went through a severe population decline as a result of hunting in the late 1800s. The population has rebounded but is now homozygous for nearly every gene studied. This is an example of
    a. negative assortative mating.
    b. migration.
    c. mutation.
    d. a bottleneck.
    e. disruptive selection.
20. When a population is small, there is a greater chance of
    a. gene flow.
    b. genetic drift.
    c. natural selection.
    d. mutations occurring.
    e. sexual selection.

Understanding the Terms

analogous structure   251
artificial selection   246
assortative mating   254
biogeography   252
bottleneck effect   255
coevolve   243
common ancestor   250
directional selection   257
disruptive selection   257
evolution   247
fossil record   249
founder effect   255
gene flow   254
gene pool   253
genetic drift   255

Hardy-Weinberg principle   253
heterozygote advantage   258
homologous structure   251
microevolution   253
mutation   254
natural selection   246
nonrandom mating   254
paleontologist   249
paleontology   245
stabilizing selection   256
stratum (pl., strata)   245
transitional fossil   250
uniformitarianism   245
vestigial structure   251

Match the terms to these definitions:
   a. _____ Outcome of natural selection in which extreme phenotypes are eliminated and the average phenotype is more common.
   b. _____ Change in the genetic makeup of a population due to chance (random) events.
   c. _____ Study of the geographic distribution of organisms.
   d. _____ Study of fossils that results in knowledge about the history of life.
   e. _____ Sharing of genes between two populations through interbreeding.

Thinking Scientifically

1. You decided to repeat the guppy experiment described in Section 13.16 because you want to determine what genotype changes account for the results. What might you do to detect such changes?
2. A cotton farmer applies a new pesticide against the boll weevil to his crop for several years. At first, the treatment was successful, but then the insecticide became ineffective and the boll weevil rebounded. Did evolution occur? Explain.

ARIS *Visit www.mhhe.com/maderconcepts for practice quizzes, animations, videos, and activities designed to help you master the material in this chapter.*

14

Speciation and Evolution

LEARNING OUTCOMES

After studying this chapter, you should be able to accomplish the following outcomes.

Hybrid Animals Do Exist

1   Define a hybrid, and tell why they do not usually occur in the wild.

Evolution of Diversity Requires Speciation

2   Compare and contrast the evolutionary species concept with the biological species concept.
3   List and give examples of five prezygotic isolating mechanisms and three postzygotic isolating mechanisms.

Origin of Species Usually Requires Geographic Separation

4   Describe and give examples of allopatric speciation.
5   Describe and give examples of adaptive radiation.

Origin of Species Can Occur in One Place

6   Relate sympatric speciation in plants to polyploidy.
7   Distinguish between autoploidy and alloploidy.
8   Explain the term allotetraploid with reference to *Zea mays*.

The Fossil Record Shows Both Gradual and Rapid Speciation

9   Compare and contrast the gradualistic model of speciation with the punctuated equilibrium model.
10 Use the fossil diversity of the Burgess Shale to support the punctuated equilibrium model.

Developmental Genes Provide a Mechanism for Rapid Speciation

11 Explain how differential gene expression relates to speciation.

Speciation Is Not Goal-Oriented

12 Use the evolution of the horse to show that evolution is not goal-oriented.

Liger

The immense liger, an offspring of a lion father and a tiger mother, really impressed Brian. Upon returning from the show, he immediately began researching for more information. To his surprise, he found that ligers are one of many hybridized species that have been recorded. His search led him to common hybrid websites that discussed mules, zorses, zonkys, and beefalos. He also discovered several strange hybrids, such as the wolphin, a cross between a false killer whale and a dolphin; a grolar, a cross between a grizzly bear and a polar bear; and a cama, a cross between a camel and a llama. Usually, in naming hybrids, the name of the male parent is used first. Thus, a zorse has a zebra father and a horse mother.

A hybrid results from breeding two closely related, but distinct, species. Lions and tigers meet this criterion, but hybrids between a cat and a rabbit would not exist because these animals are not closely re-

Hybrid Animals Do Exist

lated. Hybrids are usually the result of human activities, either direct intervention or the creation of unnatural conditions. For example, humans have mated female donkeys and male horses to develop mules for centuries. The vast majority of hybrids have resulted from placing animals in close proximity and unnatural conditions, such as a zoo. Most ligers are born in captivity, and reports of ligers in zoos can be traced back to the early 1800s.

Brian found that ligers are much larger than their parental stock. In fact, they are the largest felines in the world, measuring up to 12 feet tall when standing on their hind legs and weighing as much as 1,000 pounds. Their coat color is usually tan with tiger stripes on the back and hindquarters and lion cub spots on the abdomen. A liger can produce both the "chuff" sound of a tiger and the roar of a lion. Male ligers may have a modest lion mane or no mane at all. Most ligers have an affinity for water and love to swim. Generally, ligers have a gentle disposition; however, considering their size and heritage, handlers should be extremely careful.

In his search, Brian also discovered tigons, rare animals that have a tiger father and a lion mother. Generally, tigons are smaller than their parental stock. Most hybrids are sterile, but this is a rule, not a strict law. In recent years, Li-ligers (both parents are ligers), Li-tigons (father is a liger, mother is a tigon), Ti-ligers (father is a tiger, mother is a liger), and Ti-tigons (father is a tiger, mother is a tigon) have been produced. These unusual hybrids display a variety of lion and tiger traits.

Hybrids are usually sterile because their parents are two different species. This chapter is about speciation, the origin of species. It discusses how speciation occurs and how it may be observed during present times and in the fossil record.

Mules

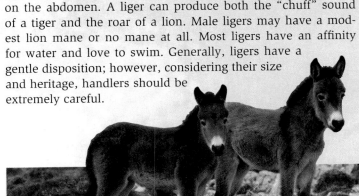

Zorses

Tigon

Macroevolution is the evolution of species, which can be defined using the evolutionary species or the biological species concept. In order for species to be biologically distinct, reproductive barriers are needed to maintain their genetic differences from other species. These barriers consist of prezygotic and postzygotic isolating mechanisms.

14.1 Species have been defined in more than one way

In Chapter 13, we defined microevolution as any allele frequency change within the gene pool of a population. **Macroevolution**, which is observed best within the fossil record, requires the origin of species, also called speciation. **Speciation** is the splitting of one species into two or more species or the transformation of one species into a new species over time. Speciation is the final result of changes in gene pool allele and genotypic frequencies. The diversity of life we see about us is absolutely dependent on speciation, so it is important to be able to define a species and to know when speciation has occurred. Before we defined a species as a type of living thing, but now we want to characterize a species in more depth.

The **evolutionary species concept** recognizes that every species has its own evolutionary history, at least part of which is in the fossil record. As an example, consider that the species depicted in **Figure 14.1A** are a part of the evolutionary history of toothed whales. Binomial nomenclature, discussed in Section 1.5, was used to name these ancestors of killer whales as well as the other species of toothed whales today. The two-part scientific name when translated from the Latin often tells you something about the organism. For example, the scientific name of the dinosaur, *Tyrannosaurus rex*, means "tyrant-lizard king."

The evolutionary species concept relies on traits, called diagnostic traits, to distinguish one species from another. As long as these traits are the same, fossils are considered members of the same species. Abrupt changes in these traits indicate the evolution of a new species in the fossil record. In summary, the evolutionary species concept states that members of a species share the same distinct evolutionary pathway and that species can be recognized by diagnostic trait differences.

One advantage of the evolutionary species concept is that it applies to both sexually and asexually reproducing organisms. However, a major disadvantage can occur when anatomic traits are used to distinguish species. The presence of variations, such as size differences in male and female animals, might make you think you are dealing with two species instead of one, and the lack of distinct differences could cause you to conclude that two fossils are the same species when they are not.

The evolutionary species concept necessarily assumes that the members of a species are reproductively isolated. If members of different species were to reproduce with one another, their evolutionary history would be mingled, not separate. By contrast, the **biological species concept** relies primarily on reproductive isolation rather than trait differences to define

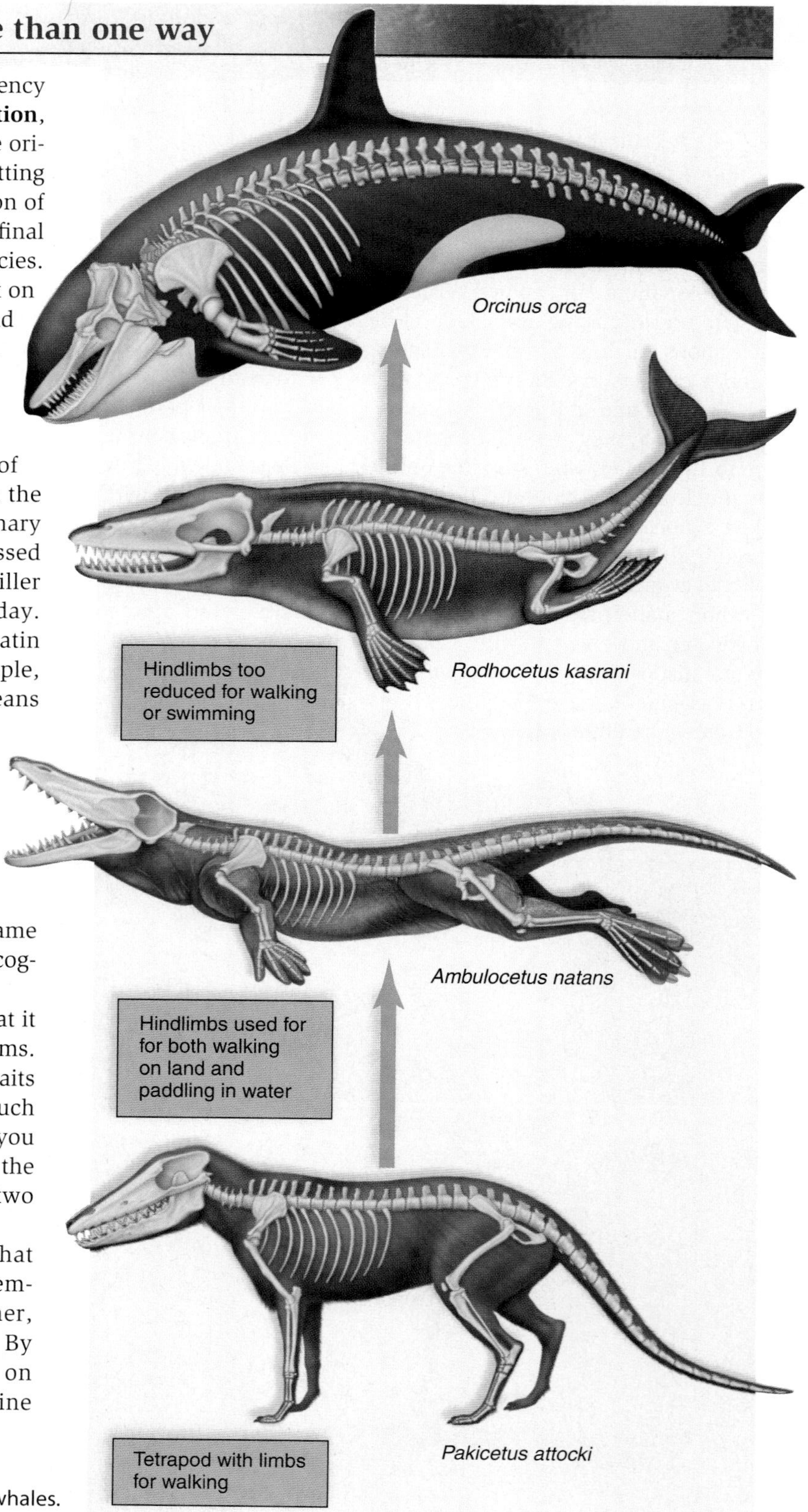

Orcinus orca

Rodhocetus kasrani

Hindlimbs too reduced for walking or swimming

Ambulocetus natans

Hindlimbs used for for both walking on land and paddling in water

Tetrapod with limbs for walking

Pakicetus attocki

FIGURE 14.1A Evolution of modern toothed whales.

pit-see

fitz-bew

che-bek or che-bek

Acadian flycatcher, *Empidonax virescens*   Willow flycatcher, *Empidonax trailli*   Least flycatcher, *Empidonax minimus*

FIGURE 14.1B Three species of flycatchers. The call of each bird is given on the photograph.

a species. In other words, although traits can help us distinguish species, the most important criterion, according to the biological species concept, is reproductive isolation—the members of a species have a single gene pool. While useful, the biological species concept cannot be applied to asexually reproducing organisms, organisms known only by the fossil record, or species that interbred when they lived near one another. The benefit of the concept is that it can designate species even when trait differences may be difficult to find. The flycatchers in **Figure 14.1B** are very similar, but they do not reproduce with one another; therefore, they are separate species. They live in different habitats. The Acadian flycatcher inhabits deciduous woods and wooded swamps, especially beeches; the willow flycatcher inhabits thickets, bushy pastures, old orchards, and willows; and the least flycatcher inhabits open woods, orchards, and farms. They also have different calls. Conversely, when anatomic differences are apparent, but reproduction is not deterred, only one species is present. Despite noticeable variations, humans from all over the world can reproduce with one another and belong to one species. The Massai of East Africa and the Eskimos of Alaska are kept apart by geography, but we know that, should they meet, reproduction between them would be possible (**Fig. 14.1C**).

The biological species concept gives us a way to know when speciation has occurred, without regard to anatomic differences. As soon as descendants of a group of organisms are able to reproduce only among themselves, speciation has occurred.

In recent years, the biological species concept has been supplemented by our knowledge of molecular genetics. DNA base sequence data and differences in proteins can indicate the relatedness of groups of organisms.

Section 14.2 introduces you to the various barriers that can keep species reproductively apart.

FIGURE 14.1C The Massai of East Africa (*left*) and the Eskimos of Alaska (*right*) belong to the same species.

14.1 Check Your Progress Should hybrid animals such as ligers be given their own scientific name and considered a separate species? Explain.

As mentioned in the previous section, for two species to be separate, they must be reproductively isolated—that is, gene flow must not occur between them. Isolating mechanisms that prevent successful reproduction from occurring are called reproductive barriers (**Fig. 14.2A**). In evolution, reproduction is successful only when it produces fertile offspring.

Prezygotic (before the formation of a zygote) **isolating mechanisms** are those that prevent reproductive attempts and make it unlikely that fertilization will be successful if mating is attempted. Scientists have identified several types of isolation that make it highly unlikely for particular genotypes to contribute to a population's gene pool:

Habitat isolation When two species occupy different habitats, even within the same geographic range, they are less likely to meet and attempt to reproduce. This is one of the reasons that the flycatchers in Figure 14.1B do not mate, and that red maple and sugar maple trees do not exchange pollen. In tropical rain forests, many animal species are restricted to a particular level of the forest canopy, and in this way they are isolated from similar species.

Temporal isolation Several related species can live in the same locale, but if each reproduces at a different time of year, they do not attempt to mate. Five species of frogs of the genus *Rana* are all found at Ithaca, New York (**Fig. 14.2B**). The species remain separate because the period of most active mating is different for each and because when-

ever there is an overlap, different breeding sites are used. For example, wood frogs are found in woodland ponds or shallow water, leopard frogs in lowland swamps, and pickerel frogs in streams and ponds on high ground.

Behavioral isolation Many animal species have courtship patterns that allow males and females to recognize one another. The male blue-footed boobie in **Figure 14.2C** does a dance unique to the species. Male fireflies are recognized by females of their species by the pattern of their flashings; similarly, female crickets recognize male crickets by their chirping. Many males recognize females of their species by sensing chemical signals called pheromones. For example, female gypsy moths have special abdominal glands from which they secrete pheromones (see Fig. 27.3) that are detected downwind by receptors on the antennae of males.

Mechanical isolation Inaccessibility of pollen to certain pollinators can prevent cross-fertilization in plants, and the sexes of many insect species have genitalia that do not match. When animal genitalia or plant floral structures are incompatible, reproduction cannot occur. Other characteristics can also make mating impossible. For example, male dragonflies have claspers that are suitable for holding only the females of their own species.

Gamete isolation Even if the gametes of two different species meet, they may not fuse to become a zygote. In animals, the sperm of one species may not be able to survive

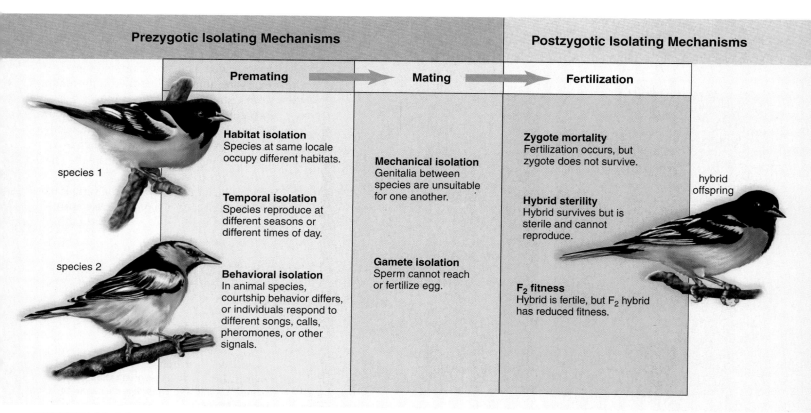

Prezygotic Isolating Mechanisms

Premating ➡ **Mating** ➡ **Fertilization**

species 1

Habitat isolation
Species at same locale occupy different habitats.

Temporal isolation
Species reproduce at different seasons or different times of day.

species 2

Behavioral isolation
In animal species, courtship behavior differs, or individuals respond to different songs, calls, pheromones, or other signals.

Mechanical isolation
Genitalia between species are unsuitable for one another.

Gamete isolation
Sperm cannot reach or fertilize egg.

Postzygotic Isolating Mechanisms

Zygote mortality
Fertilization occurs, but zygote does not survive.

Hybrid sterility
Hybrid survives but is sterile and cannot reproduce.

F_2 fitness
Hybrid is fertile, but F_2 hybrid has reduced fitness.

hybrid offspring

FIGURE 14.2A Reproductive barriers.

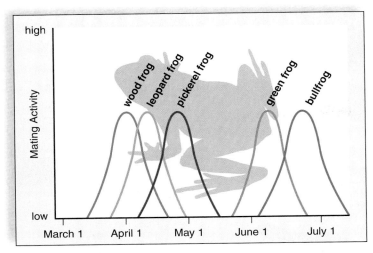

FIGURE 14.2B Mating activity peaks at different times of the year for these species of frogs.

FIGURE 14.2C Male blue-footed boobie doing a courtship dance for a female.

in the reproductive tract of another species, or the egg may have receptors only for sperm of its species. In plants, pollen grains are species specific and will not form a pollen tube for another species. Without a pollen tube, the sperm cannot successfully reach the egg.

Postzygotic (after the formation of a zygote) **isolating mechanisms** prevent hybrid offspring from developing or breeding, even if reproduction attempts have been successful.

Zygote mortality A hybrid zygote may not be viable, and so it dies. A zygote with two different chromosome sets may fail to go through mitosis properly, or the developing embryo may receive incompatible instructions from the maternal and paternal genes so that it cannot continue to exist.

Hybrid sterility The hybrid zygote may develop into a sterile adult. As is well known, a cross between a male horse and a female donkey produces a mule, which is usually sterile—it cannot reproduce (**Fig. 14.2D**). Sterility of hybrids generally results from complications in meiosis that lead to an inability to produce viable gametes. A cross between a cabbage and a radish produces offspring that cannot form gametes, most likely because the cabbage chromosomes and the radish chromosomes could not align during meiosis (see Section 14.5).

F_2 fitness Even if hybrids can reproduce, their offspring may be unable to reproduce. In some cases, mules are fertile, but their offspring (the F_2 generation) are not fertile.

Having discussed how to define a species and what keeps them apart, Section 14.3 discusses how species generally arise.

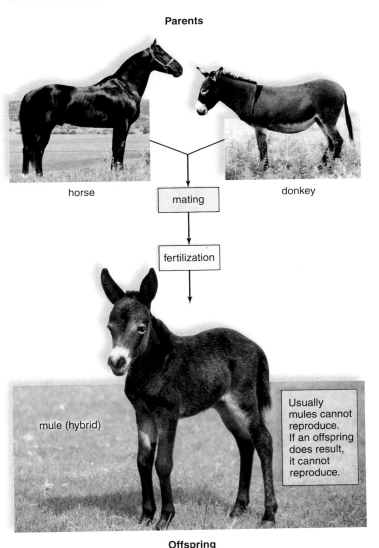

FIGURE 14.2D Mules cannot reproduce due to chromosome noncompatibility.

Parents

horse   mating   donkey

fertilization

mule (hybrid)

Usually mules cannot reproduce. If an offspring does result, it cannot reproduce.

Offspring

> **14.2 Check Your Progress** *a.* **Which of the prezygotic isolating mechanisms apparently keeps lions and tigers from mating in the wild? Explain.** *b.* **Which of the postzygotic isolating mechanisms is still working to a degree to keep lions and tigers separate species? Explain.**

Origin of Species Usually Requires Geographic Separation

Learning Outcomes 4–5, page 262

Geographic isolation fosters the genetic changes that result in reproductive isolation. Modern-day examples include the evolution of distinct forms of *Ensatina* salamanders in California. Adaptive radiation occurs when an ancestral species evolves into several new and different species, each adapted to a different environment. The evolution of a wide variety of honeycreepers on the Hawaiian Islands is an example of adaptive radiation.

14.3   Allopatric speciation utilizes a geographic barrier

In 1942, Ernst Mayr, an evolutionary biologist, published the book *Systematics and the Origin of Species*, in which he proposed the biological species concept and a process by which speciation could occur. He said that when members of a species become isolated, the subpopulations will start to differ because of genetic drift and natural selection over a period of time. Eventually, the two groups will be unable to mate with one another. At that time, they have evolved into new species. Mayr's hypothesis is termed **allopatric speciation** (allopatric means different country) because it requires that the subpopulations be separated by a geographic barrier.

Much data in support of allopatric speciation have since been discovered. **Figure 14.3A** features an example of allopatric speciation that has been extensively studied in California. An ancestral population of *Ensatina* salamanders lives in the Pacific Northwest. ❶ Members of this ancestral population migrated southward, establishing a series of subpopulations. Each subpopulation was exposed to its own selective pressures along the coastal mountains and the Sierra Nevada mountains. ❷ Due to the presence of the Central Valley of California, gene flow rarely occurs between the eastern populations and the western populations. ❸ Genetic differences increased from north to south, resulting in distinct forms of *Ensatina* salamanders in Southern California that differ dramatically in color and no longer interbreed.

Geographic isolation is even more obvious in other examples. The green iguana of South America is believed to be the common ancestor for both the marine iguana on the Galápagos Islands (to the west) and the rhinoceros iguana on Hispaniola, an island to the north. If so, how could it happen? Green iguanas are strong swimmers, so by chance, a few could have migrated to these islands, where they formed populations separate from each other and from the parent population back in South America. Each population continued on its own evolutionary path as new mutations, genetic drift, and different selection pressures occurred. Eventually, reproductive isolation developed, and the result was three species of iguanas that are reproductively isolated from each other.

A more detailed example of allopatric speciation involves sockeye salmon in Washington state. In the 1930s and 1940s, hundreds of thousands of sockeye salmon were introduced into Lake Washington. Some colonized an area of the lake near Pleasure Point Beach (**Fig. 14.3B**). Others migrated into the Cedar River (**Fig. 14.3C**). Andrew Hendry, a biologist at McGill University, is able to tell Pleasure Point Beach salmon from Cedar River salmon because they differ in shape and size due to the demands

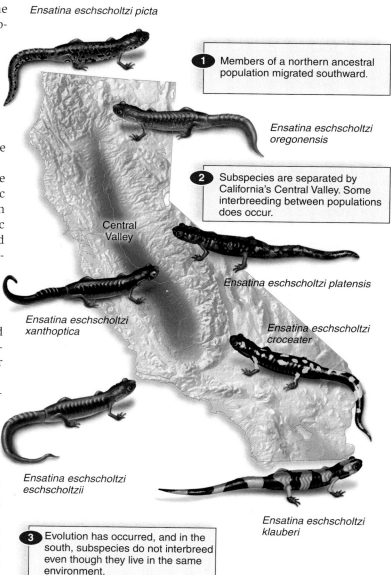

Ensatina eschscholtzi picta

❶ Members of a northern ancestral population migrated southward.

Ensatina eschscholtzi oregonensis

❷ Subspecies are separated by California's Central Valley. Some interbreeding between populations does occur.

Central Valley

Ensatina eschscholtzi platensis

Ensatina eschscholtzi xanthoptica

Ensatina eschscholtzi croceater

Ensatina eschscholtzi eschscholtzii

❸ Evolution has occurred, and in the south, subspecies do not interbreed even though they live in the same environment.

Ensatina eschscholtzi klauberi

FIGURE 14.3A Allopatric speciation among *Ensatina* salamanders.

of reproducing. In the river, where the waters are fast-moving, males tend to be more slender than those along the beach. A slender body is better able to turn sideways in a strong current, and the courtship ritual of a sockeye salmon requires this ma-

FIGURE 14.3B Sockeye salmon at Pleasure Point Beach, Lake Washington.

FIGURE 14.3C Sockeye salmon in Cedar River. The river connects with Lake Washington.

neuver. On the other hand, the females tend to be larger than those along the beach. This larger body helps them dig slightly deeper nests in the gravel beds on the river bottom. Deeper nests are not disturbed by river currents and remain warm enough for egg viability.

Hendry has an independent way of telling beach salmon from river salmon. Ear stones called otoliths reflect variations in water temperature while a fish embryo is developing. Water temperatures at Pleasure Point Beach are relatively constant compared to Cedar River temperatures. By checking otoliths in adults, Hendry found that a third of the sockeye males at Pleasure Point Beach had grown up in the river. Yet the distinction between male and female shape and size according to the two locations remains. Therefore, these males are not successful breeders along the beach. In other words, reproductive isolation has occurred.

As we have seen in sockeye salmon, a side effect to adaptive changes can be reproductive isolation. Another example is seen among *Anolis* lizards, which court females by extending a colorful flap of skin, called a "dewlap." The dewlap must be seen in order to attract mates. Therefore, populations of *Anolis* in a dim forest tend to evolve

light-colored dewlaps that reflect light, while populations in open habitats evolve dark-colored dewlaps. This change in dewlap color causes the populations to be reproductively isolated, because females distinguish males of their species by the color of the dewlap.

As populations become reproductively isolated, postzygotic isolating mechanisms may arise before prezygotic isolating mechanisms. As we have seen, when a horse and a mule reproduce, the hybrid or the offspring of a hybrid is not fertile. Therefore, natural selection would favor any variation in populations that prevents the occurrence of hybrids because most hybrids do not have offspring. Indeed, natural selection would favor the continual improvement of prezygotic isolating mechanisms until the two populations are completely reproductively isolated. The term reinforcement is given to the process of natural selection favoring variations that lead to reproductive isolation. An example of reinforcement has been seen in birds called the pied and collared flycatchers of the Czech Republic and Slovakia whenever both species occur in close proximity. Only here have pied flycatchers evolved a different coat color from the collared flycatchers. The difference in color helps the two species recognize and mate with their own species.

Adaptation to new environments can result in multiple species from a single ancestral species, as discussed in Section 14.4.

14.3 Check Your Progress Knowing that the coat colors of lions and tigers is adaptive to their habitats, construct a hypothetical scenario by which they evolved from an ancestral species.

Adaptive radiation occurs when a single ancestral species gives rise to a variety of species, each adapted to a specific environment. An *ecological niche* is where a species lives and how it interacts with other species. When an ancestral finch arrived on the Galápagos Islands, its descendants spread out to occupy various niches. Geographic isolation of the various finch populations caused their gene pools to become isolated. Because of natural selection, each population adapted to a particular habitat on its island. In time, the many populations became so genotypically different that now, when by chance they reside on the same island, they do not interbreed, and are therefore separate species. The finches use beak shape to recognize members of the same species during courtship. Rejection of suitors with the wrong type of beak is a behavioral type of prezygotic isolating mechanism.

Similarly, on the Hawaiian Islands, a wide variety of honeycreepers are descended from a common goldfinch-like ancestor that arrived from Asia or North America about 5 million years ago. Today, honeycreepers have a range of beak sizes and shapes for feeding on various food sources, including seeds, fruits, flowers, and insects (**Fig. 14.4**). Adaptive radiation also occurs among plants; a good example is the silversword alliance, which is discussed in the opening story for Chapter 22.

Adaptive radiation has occurred throughout the history of life on Earth when a group of organisms exploits a new environment. For example, with the demise of the di-

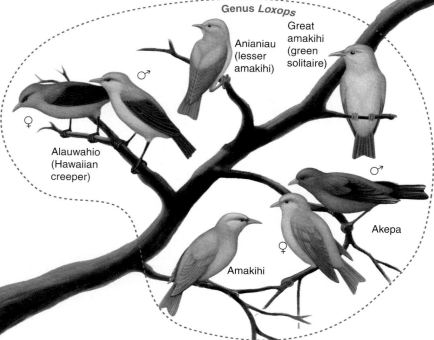

FIGURE 14.4 Adaptive radiation in Hawaiian honeycreepers.

* Extinct species or subspecies

nosaurs about 66 million years ago, mammals underwent adaptive radiation as they exploited niches previously occupied by the dinosaurs.

This completes our discussion of allopatric speciation. The next part of the chapter discusses speciation when there is no geographic barrier.

14.4 *Check Your Progress* **Five species of big cats are classified in a single genus: *Panthera leo* (lion), *P. tigris* (tiger), *P. pardus* (leopard), *P. onca* (jaguar), and *P. uncia* (snow leopard). What evidence would you need to show that this is a case of adaptive radiation?**

Speciation without the presence of a geographic barrier does occur, and the best examples are due to chromosome number changes in plants. Hybridization, followed by doubling of the chromosome number, can occur naturally or as a result of artificial selection. Such events must have occurred during the artificial selection of corn over the years.

14.5 Speciation occasionally occurs without a geographic barrier

Speciation without the presence of a geographic barrier is termed **sympatric speciation**. Sympatric speciation has been difficult to substantiate in animals. For example, two populations of the Meadow Brown butterfly, *Maniola jurtina*, have different distributions of wing spots. The two populations are both in Cornwall, England, and they maintain the difference in wing spots, even though there is no geographic boundary between them. But, as yet, no reproductive isolating mechanism has been found. In contrast, we know of instances in plants by which a postzygotic isolating mechanism has given rise to a new species within the range and habitat of the parent species. In other words, no geographic barrier was required. All instances in plants involve **polyploidy**, additional sets of chromosomes beyond the diploid (2n) number. Sympatric speciation is more common in flowering plants than in animals due to self-pollination. A polyploid plant can reproduce only with itself, and cannot reproduce with the parent (2n) population because not all the chromosomes would be able to pair during meiosis. Two types of polyploidy are known: autoploidy and alloploidy.

Autoploidy is seen in diploid plants when nondisjunction occurs during meiosis and the diploid species produces diploid gametes. If this diploid gamete fuses with a haploid gamete, a triploid plant results. A triploid (3n) plant is sterile and cannot produce offspring because the chromosomes cannot pair during meiosis. Humans have found a use for sterile plants because they produce fruits without seeds. **Figure 14.5A** contrasts a diploid banana with seeds to today's polyploid banana that produces no seeds. If two of the diploid gametes fuse, the plant is a tetraploid (4n) and the plant is fertile, so long as it reproduces with another of its own kind. The fruits of polyploid plants are much larger than those of diploid plants. The huge strawberries of today are produced by octaploid (8n) plants.

Alloploidy requires a more complicated process than autoploidy (**Fig. 14.5B**). The prefix "allo," which means different, is ap-

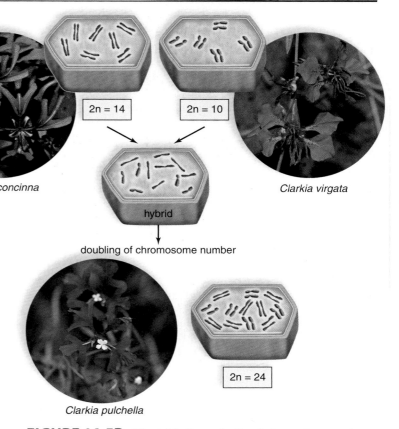

Clarkia concinna

2n = 14   2n = 10

hybrid

doubling of chromosome number

Clarkia virgata

Clarkia pulchella

2n = 24

FIGURE 14.5B Alloploidy: Reproduction between two species of *Clarkia* results in a sterile hybrid. Doubling of the chromosome number results in a fertile third *Clarkia* species.

propriate because the process begins when two different but related species of plants hybridize. Hybridization is followed by doubling of the chromosomes. For example, the Western wildflower, *Clarkia concinna*, is a diploid plant with fourteen chromosomes (seven pairs). The related species, *C. virgata*, is a diploid plant with ten chromosomes (five pairs). A hybrid of these two species is not fertile because seven chromosomes from one plant cannot pair evenly with five chromosomes from the other plant. However, meiosis occurs normally in the hybrid, *C. pulchella*, due to doubling of the chromosome number, which allows the chromosomes to pair during meiosis. Alloploidy also occurred during the evolution of the wheat plant, which is commonly used today to produce bread.

Hybridization by means of artificial selection, plus a doubling of the chromosome number, most likely occurred during the evolution of corn, as discussed in Section 14.6.

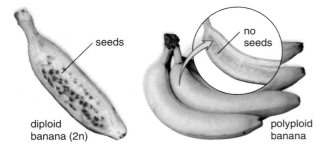

seeds

no seeds

diploid banana (2n)

polyploid banana

FIGURE 14.5A Autoploidy: The small, diploid-seeded banana is contrasted with the large, polyploid banana that produces no seeds.

14.5 *Check Your Progress* **What fossil evidence might support the hypothesis that the different species of cats arose sympatrically?**

When the world record for eating corn on the cob was set at 33½ ears in 12 minutes, the last thing on anyone's mind was the evolution of corn. Corn, also known as maize (*Zea mays*), represents one of the most remarkable plant-breeding achievements in the history of agriculture. Today, modern society literally reaps the benefits of corn as a domestic product.

Corn is America's number-one field crop, yielding approximately 9.5 billion bushels yearly. It is an important food source for both humans and livestock. Corn is a component of over 3,000 grocery products, including cereals, corn syrup, cornstarch, ice cream, soft drinks, chips, snack foods, and even peanut butter. It is also used in making glue, shoe polish, ink, soaps, and synthetic rubber. Recently, corn has been in the news as a source of ethanol to fuel our vehicles. The uses of corn seem to be limited only by our imaginations.

Modern corn bears little resemblance to its ancient ancestor, an inconspicuous wild grass called teosinte from southern Mexico. Teosinte is a drought-tolerant grass that produces reproductive spikes fairly close to the ground. Each spike is filled with two rows of small, triangular-shaped seeds enclosed in a tough husk. Each seed is encased and protected by a hard shell (**Fig. 14.6A**). Ancient peoples discovered that teosinte was a source of food and began selecting spikes to plant near their homes, close to irrigation systems. Thus, between 4000 and 3000 B.C., the hand of artificial selection began to shape the evolution of corn. The use of teosinte spread across Mesoamerica, opening the door for further development.

Archaeologists have uncovered corncobs distinctly different from teosinte at a 5,400-year-old site in the highlands of Oaxaca in southwestern Mexico. The corncob was only an inch long and possessed four rows of kernels, compared to an average corncob today that has 16 rows and 800 kernels. The attachment of kernels to the cob and the loss of the hard coat surrounding each kernel made corn even more dependent upon humans and susceptible to artificial selection. In order to be planted for the next season, healthy kernels had to be identified and removed from the cob. And since corn kernels lacked a tough outer coat, they had to be stored in a cool, dry place. Early farmers chose corn with the most desirable characteristics to plant each season.

Experimental hybridization soon followed, and many varieties of corn were developed. By A.D. 1070, corn had reached North America and was being grown by the Iroquois in New York. By the time Columbus visited the Americas, corn was being grown in a number of environments. Columbus even commented on the fields of corn and its great taste. We now know that corn is an allotetraploid, meaning it is 4n. Hybridization must have been followed by doubling of the chromosomes, accounting for why the ears of corn are now so large (**Fig. 14.6B**).

Today, there are hundreds of varieties of corn, representing the greatest diversity of any crop in the world. Corn can be found growing in high mountain regions or hot sunny areas, and on every continent except Antarctica. The height of a corn plant can vary from less than a meter to over 6 meters, and it can mature in 2 to 15 months. The ears of corn can vary in length from 12 to 115 cm. The six major types of corn primarily used today are parching corn, flint corn, dent corn, flour corn, sweet corn, and popcorn, each with its own set of unique characteristics.

Corn has come a long way from teosinte. Artificial selection has rendered a remarkable and diverse product that impacts all our lives. Although you may not be in a corn-on-the-cob eating contest, it is estimated that you consume over 11 ears of corn a year.

This completes our discussion of sympatric speciation. The next part of the chapter discusses speciation and the fossil record.

> **14.6** *Check Your Progress* Two plants are both hybrids, but only one of the plants is sterile. The other plant self-fertilizes to produce a larger fruit than either parent. Explain.

FIGURE 14.6B Corn (*Zea mays*) produces ears containing many edible kernels.

FIGURE 14.6A Teosinte (*Zea mexicana*) produces ears containing only a few tough kernels.

A gradualistic model of speciation can be contrasted with the punctuated equilibrium model. The gradualistic model predicts transitional links, while the punctuated equilibrium model predicts few, if any, transitional links in the fossil record.

14.7 Speciation occurs at different tempos

Many evolutionists believe, as Darwin did, that evolutionary changes occur gradually. Therefore, these evolutionists support a *gradualistic model*, which proposes that speciation occurs after populations become isolated, with each group continuing slowly on its own evolutionary pathway. These evolutionists often show the history of groups of organisms by drawing the type of diagram shown in **Figure 14.7A**. Note that in this diagram, an ancestral species has given rise to two separate species, represented by a slow change in plumage color. The gradualistic model suggests that it is difficult to indicate when speciation occurred because there would be so many transitional links. However, in some cases, it has been possible to trace the evolution of a group of organisms by finding transitional links.

After studying the fossil record, some paleontologists tell us that species can appear quite suddenly, and then they remain essentially unchanged phenotypically until they undergo extinction. Based on these findings, they developed a *punctuated equilibrium model* to explain the pace of evolution. This model says that periods of equilibrium (no change) are punctuated (interrupted) by speciation. **Figure 14.7B** shows this way of representing the history of evolution over time. This model suggests that transitional links are less likely to become fossils and less likely to be found. Moreover, speciation is apt to involve an isolated population at one locale, because a favorable genotype could spread more rapidly within such a population. Only when

this population expands and replaces other species is it apt to show up in the fossil record.

A strong argument can be made that it is not necessary to choose between these two models of evolution and that both could very well assist us in interpreting the fossil record. In a stable environment, a species may be kept in equilibrium by stabilizing selection for a long period. On the other hand, if the environment changes slowly, a species may be able to adapt gradually. If environmental change is rapid, a new species may arise suddenly before the parent species goes on to extinction. Because geologic time is measured in millions of years, the "sudden" appearance of a new species in the fossil record could actually represent many thousands of years. Using only a small rate of change (.0008/year), two investigators calculated that the brain size in the human lineage could have increased from 900 cm³ to 1,400 cm³ in only 135,000 years. This would appear to be a very rapid change in the fossil record. Actually, the record indicates that it took about 500,000 years, indicating that the real pace was slower than it could have been.

> **14.7 Check Your Progress** If a paleontologist were to find ligers in the fossil record, could she/he use that data to substantiate a gradualistic or a punctuated equilibrium model of evolution? Explain.

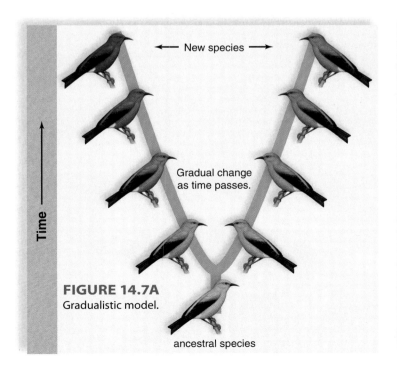

FIGURE 14.7A
Gradualistic model.

New species

Gradual change as time passes.

Time

ancestral species

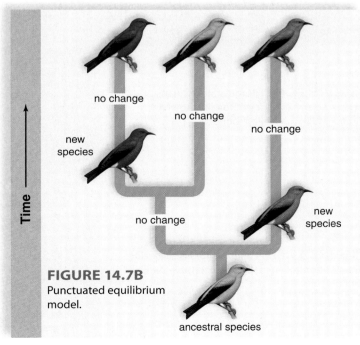

FIGURE 14.7B
Punctuated equilibrium model.

no change

no change

no change

new species

no change

new species

Time

ancestral species

14.8   The Burgess Shale hosts a diversity of life

Finding the Burgess Shale, a rock outcropping in Yoho National Park, British Columbia, was a chance happening. In 1909, Charles Doolittle Walcott of the Smithsonian Institute was out riding when his horse stopped in front of a rock made of shale. He cracked the rock open and saw the now-famous fossils of the animals depicted in **Figure 14.8A**. Walcott and his team began working the site and continued on their own for quite a few years. Around 1960, other paleontologists became interested in studying the Burgess Shale fossils.

As a result of uplifting and erosion, the intriguing fossils of the Burgess Shale are relatively common in that particular area. However, the highly delicate impressions and films found in the rocks are very difficult to remove from their matrix. Early attempts to remove the fossils involved splitting the rocks along their sedimentary plane and using rock saws. Unfortunately, these methods were literally "shots in the dark," and many valuable fossils were destroyed in the process. New methods, involving ultraviolet light to see the fossils and diluted acetic acid solutions to remove the matrix, have been more successful in freeing the fossils.

The fossils tell a remarkable story of marine life some 540 MYA (millions of years ago). In addition to fossils of organisms that had external skeletons, many of the fossils are remains of soft-bodied invertebrates; these are a great find because soft-bodied animals rarely fossilize. During this time, all organisms lived in the sea, and it is believed the barren land was subject to mudslides, which entered the ocean and buried the animals, killing them. Later, the mud turned into shale, and later still, an upheaval raised the shale. Before the shale formed, fine mud particles filled the spaces in and around the organisms so that the soft tissues were preserved and the fossils became somewhat three-dimensional.

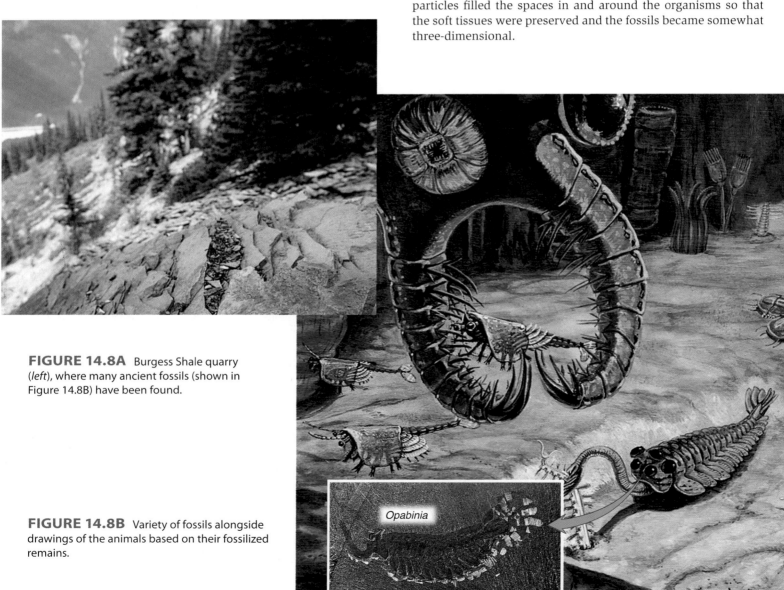

FIGURE 14.8A Burgess Shale quarry (*left*), where many ancient fossils (shown in Figure 14.8B) have been found.

FIGURE 14.8B Variety of fossils alongside drawings of the animals based on their fossilized remains.

Opabinia

The fossils tell us that the ancient seas were teaming with weird-looking, mostly invertebrate animals (**Fig. 14.8B**). All of today's groups of animals can trace their ancestry to one of these strange-looking forms, which include sponges, arthropods, worms, and tribolites, as well as spiked creatures and oversized predators. The animals featured in Figure 14.B have been assigned to these genera and are believed to be the type of animal mentioned:

Opabinia, a crustacean;
Thaumaptilon, a sea pen;
Vauxia, a sponge; and
Wiwaxia, a segmented worm

The vertebrates, like ourselves, are descended from *Pikaia*, the only one of the fossils that has a supporting rod called a notochord. (In vertebrates, the notochord is replaced by the vertebral column during development.)

Unicellular organisms have also been preserved at the Burgess Shale site. They appear to be bacteria, cyanobacteria, dinoflagellates, and other protists. Fragments of algae are preserved in thin, shiny carbon films. A technique has been perfected that allows the films to be peeled off the rocks.

Anyone can travel to Yoho National Park, look at the fossils, and get an idea of the types of animals that dominated the world's oceans for nearly 300 million years. Some of the animals had external skeletons, but many were soft-bodied. Interpretations of the fossils vary. Some authorities hypothesize that the great variety of animals in the Burgess Shale evolved within 20–50 million years, and therefore the site supports the hypothesis of punctuated equilibrium. Others believe that the animals started evolving much earlier and that we are looking at the end result of an adaptive radiation requiring many more millions of years to accomplish. Some investigators present evidence that all the animals are related to today's animals and should be classified as such. Others believe that several of them are unique creatures unrelated to the animals of today. Regardless of the controversies, the fossils tell us that speciation, diversification, and eventual extinction are part of the history of life.

This completes our discussion of the two models for speciation based on the fossil record. The next part of the chapter examines the genetic basis for possible rapid change in characteristics.

> **14.8 *Check Your Progress*** Could the Burgess Shale animals be used to substantiate that animals originated in the sea?

Thaumaptilon

Vauxia

Wiwaxia

Investigators have discovered genes that can bring about radical changes in body shapes and organs. For example, it is now known that the *Pax6* gene is involved in eye formation in all animals, and that homeotic (*Hox*) genes determine the location of repeated structures in all vertebrates.

14.9 Gene expression can influence development

Whether slow or fast, how could evolution have produced the myriad of animals in the Burgess Shale and, indeed, in the history of life? Or, to ask the question in a genetic context, how can genetic changes bring about such major differences in form? It has been suggested since the time of Darwin that the answer must involve development processes. In 1917, D'Arcy Thompson asked us to imagine an ancestor in which all parts are developing at a particular rate. A change in gene expression could stop a developmental process or continue it beyond its normal time. For instance, if the growth of limb bones were stopped early, the result would be shorter limbs, and if it were extended, the result would be longer limbs compared to those of an ancestor. Or, if the whole period of growth were extended, a larger animal would result, accounting for why some species of horses are so large today.

Using new kinds of microscopes and the modern techniques of cloning and manipulating genes, investigators have indeed discovered genes whose *differential expression* can bring about changes in body shapes and organs. This result suggests that these genes must date back to a common ancestor that lived more than 600 MYA (before the Burgess Shale animals), and that despite millions of years of divergent evolution, all animals share the same control switches for development.

Development of the Eye   The animal kingdom contains many different types of eyes, and it was long thought that each type would require its own set of genes. Flies, crabs, and other arthropods have compound eyes that have hundreds of individual visual units. Humans and all other vertebrates have a camera-type eye with a single lens. So do squids and octopuses. Humans are not closely related to either flies or squids, so wouldn't it seem as if all three types of animals evolved "eye" genes separately? Not so. In 1994, Walter Gehring and his colleagues at the University of Basel, Switzerland, discovered that a gene called *Pax6* is required for eye formation in all animals tested (**Fig. 14.9A**). Mutations in the *Pax6* gene lead to failure of eye development in both people and mice, and remarkably, the mouse *Pax6* gene can cause a compound eye to develop on the leg of a fruit fly (**Fig. 14.9B**).

Development of Limbs   Wings and arms are very different, but both humans and birds express the *Tbx5* gene in developing limb buds. *Tbx5* codes for a transcription factor that turns on the genes needed to make a limb. What seems to have changed as birds and humans evolved are the genes that *Tbx5* turns on. Perhaps in an ancestral tetrapod, the Tbx5 protein triggered the transcription of only one gene. In humans and birds, a few genes are expressed in response to Tbx5 protein, but the particular genes are different. There is also the question of timing. Changing the timing of gene expression, as well as which genes are expressed, can result in dramatic changes in shape.

Development of Overall Shape   Vertebrates have repeating segments, as exemplified by the vertebral column. Changes in the number of segments can lead to changes in overall shape. In general, *Hox* genes (homeotic genes) control the development of repeated structures along the main body axes of vertebrates. Shifts in where *Hox* genes are expressed in embryos are

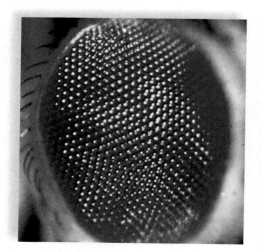 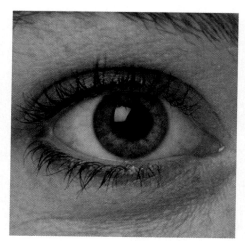 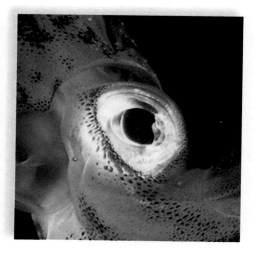

FIGURE 14.9A   *Pax6* is involved in eye development in a fly, a human, and a squid.

responsible for why the snake has hundreds of rib-bearing vertebrae and essentially no neck in contrast to other vertebrates, such as a chick (**Fig. 14.9C**). *Hox* genes have been found in all animals, and other shifts in the expression of these genes can explain why insects have just six legs and other arthropods, such as crayfish, have ten legs. In general, the study of *Hox* genes has shown how animal diversity is due to variations in the expression of ancient genes rather than to wholly new and different genes.

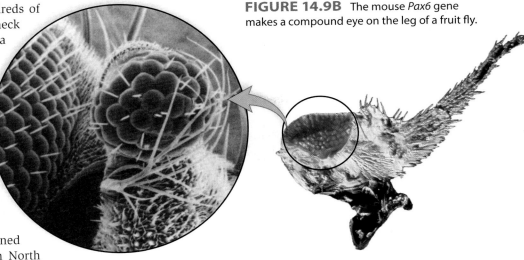

FIGURE 14.9B The mouse *Pax6* gene makes a compound eye on the leg of a fruit fly.

Pelvic Fin Genes The three-spined stickleback fish occurs in two forms in North American lakes. In the open waters of a lake, long pelvic spines help protect the stickleback from being eaten by large predators. But on the lake bottom, long pelvic spines are a disadvantage because dragonfly larvae seize and feed on young sticklebacks by grabbing them by their spines. The presence of short spines in bottom-dwelling stickleback fish can be traced to a reduction in the development of the pelvic-fin bud in the embryo, and this reduction is due to the altered expression of a particular gene.

Hindlimb reduction has occurred during the evolution of other vertebrates. The hindlimbs became greatly reduced in size as whales and manatees evolved from land-dwelling ancestors into fully aquatic forms. Similarly, legless lizards have evolved many times. The stickleback study has shown how natural selection can lead to major skeletal changes in a relatively short time.

Human Evolution The sequencing of genomes has shown us that our DNA base sequence is very similar to that of chimpanzees, mice, and, indeed, all vertebrates. Based on this knowledge and the work just described, investigators no longer expect to find new genes to account for the evolution of humans. Instead, they predict that differential gene expression and/or new functions for "old" genes will explain how humans evolved.

Mutations of developmental genes occur by chance, and in the next part of the chapter, we observe that evolution is not directed toward any particular end.

14.9 Check Your Progress Why does it seem that differential expression must occur during the development of ligers?

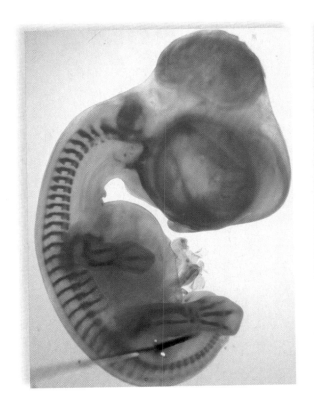

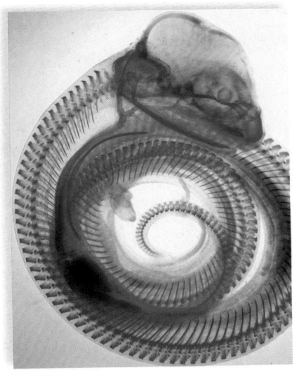

FIGURE 14.9C
Differential expression of *Hox6* genes causes a chick to have seven vertebrae (*purple*) and a snake to have many more vertebrae (*purple*).

Burke, A. C. 2000, *Hox* genes and the global patterning of the somitic mesoderm. In Somitogenesis. C. Ordahl (ed.) *Current Topics in Developmental Biology*, Vol. 47. Academic Press.

If the evolution of the horse (*Equus*) is examined carefully, we see that its ancestry was not directed to a particular end. The features of the modern horse represent adaptations to the environment in which it evolved.

14.10 Evolution is not directed toward any particular end

The evolution of the horse, *Equus*, has been studied since the 1870s, and at first the ancestry of this genus seemed to represent a model for gradual, straight-line evolution until its goal, the modern horse, had been achieved. Three trends were particularly evident during the evolution of the horse: increase in overall size, toe reduction, and change in tooth size and shape.

By now, however, many more fossils have been found, making it easier to tell that the lineage of a horse is complicated by the presence of many ancestors with varied traits. The tree in **Figure 14.10** is an oversimplification because each of the names is a genus that contains several species, and not all past genera in the horse family are included. It is apparent, then, that the ancestors of *Equus* form a thick bush of many equine species and that straight-line evolution did not occur. Because *Equus* alone remains and the other genera have died out, it might seem as if evolution was directed toward producing *Equus*, but this is not the case. Instead, each of these ancestral species was adapted to its environment. Adaptation occurs only because the members of a population with an advantage are able to have more offspring than other members. Natural selection is opportunistic, not goal-directed.

Fossils named *Hyracotherium* have been designated as the first probable members of the horse family, living about 57 MYA. These animals had a wooded habitat, ate leaves and fruit, and were about the size of a dog. Their short legs and broad feet with several toes would have allowed them to scamper from thicket to thicket to avoid predators. *Hyracotherium* was obviously well adapted to its environment because this genus survived for 20 million years. The first adaptive radiation of horses occurred about 35 MYA. The weather was becoming drier, and grasses were evolving. Eating grass requires tougher teeth, and an increase in size and longer legs would have permitted greater speed to escape enemies. The second adaptive radiation of horses occurred about 15 MYA, and by 10 MYA, the horse family was quite diversified, but only one species survives today. Modern horses evolved about 4 MYA from ancestors who had features that are adaptive for living on an open plain, such as large size, long legs, hoofed feet, and strong teeth.

> **14.10** *Check Your Progress* **There are only five species of cats in the genus *Panthera*. Does this represent a goal of evolution?**

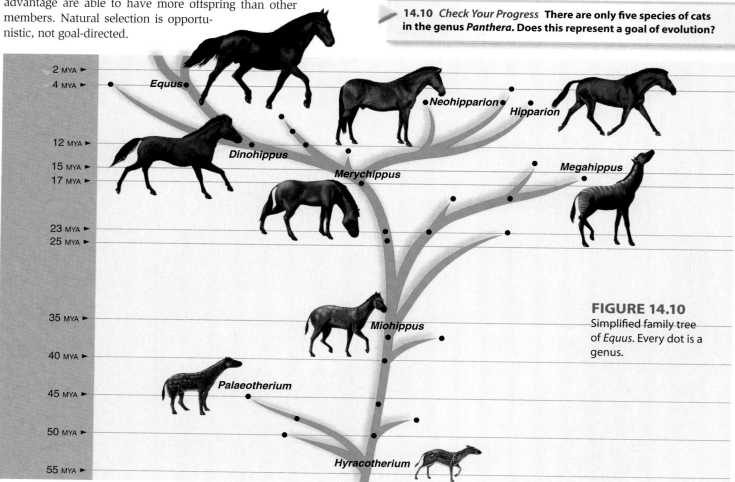

FIGURE 14.10
Simplified family tree of *Equus*. Every dot is a genus.

Macroevolution, the study of the origin and history of the species on Earth, is the subject of this chapter and the next. The biological species concept states that the members of a species have an isolated gene pool and can only reproduce with one another.

This chapter concerns speciation. Speciation usually occurs after two populations derived from a larger one are separated geographically. If a population of salamanders is suddenly divided by a barrier, each new population would become adapted to its particular environment over time. Eventually, the two populations might become so geneti-cally different that even if members of each population came into contact, they would not be able to produce fertile offspring. Because gene flow between the two populations would no longer be possible, the salamanders would be considered separate species. Aided by geographic separation, multiple species can repeatedly arise from an ancestral species, as when a common ancestor from the mainland led to 13 species of Galápagos finches, each adapted to its own particular environment.

Does speciation occur gradually, as Darwin supposed, or rapidly (in geologic time), as described by the punctu-ated equilibrium model? The fossils of the Burgess Shale support the punctuated equilibrium model. How can genetic changes bring about such major changes in form, whether fast or slow? Investigators have now discovered ancient genes whose differential expression can bring about changes in body shapes and organs.

Evolution is not directed toward any particular end, and the traits of the species alive today arose through common descent with adaptations to a local environment. The subject of Chapter 15 is the evolutionary history and classification of living organisms today.

The Chapter in Review

Summary

Hybrid Animals Do Exist

- A hybrid results from breeding two closely related, but distinct, species.
- Hybrids usually occur as a result of human intervention.

Evolution of Diversity Requires Speciation

14.1 Species have been defined in more than one way

- Macroevolution depends on speciation.
- Speciation occurs when one species splits into two or more species or when one species becomes a new species over time.
- According to the evolutionary species concept, every species has its own evolutionary history, and a species can be recognized by diagnostic traits.
- According to the biological species concept, members of a species are reproductively isolated from members of other species. They can only reproduce with members of their own species.

14.2 Reproductive barriers maintain genetic differences between species

- Prezygotic isolating mechanisms prevent reproductive attempts.
- Postzygotic isolating mechanisms prevent hybrid offspring from breeding.

Prezygotic Isolating Mechanisms		Postzygotic Isolating Mechanisms
Premating →	Mating →	Fertilization
Habitat isolation	Mechanical isolation	Zygote mortality
Temporal isolation	Gamete isolation	Hybrid sterility
Behavioral isolation		F_2 fitness

Origin of Species Usually Requires Geographic Separation

14.3 Allopatric speciation utilizes a geographic barrier

- When populations derived from a larger one are separated by a barrier, they will start to differ genetically and phenotypically.
- Following separation, postzygotic mechanisms followed by prezygotic mechanisms can develop over time.

14.4 Adaptive radiation produces many related species

- Several new species can evolve from an ancestral species when several populations adapt to fill different niches separated by geographic barriers.

Origin of Species Can Occur in One Place

14.5 Speciation occasionally occurs without a geographic barrier

- Sympatric speciation occurs without a geographic barrier.
- Polyploidy is present when plants have additional sets of chromosomes beyond the diploid (2n) number. The sudden occurrence of polyploidy is speciation because a polyploid cannot reproduce with parental 2n plants.
- Autoploidy occurs when a diploid gamete fuses with a haploid gamete, resulting in a triploid plant, which is sterile.
- Alloploidy occurs when two different but related species of plants hybridize, and then the chromosome number doubles.

14.6 Artificial selection produced corn

- Teosinte was the early ancestor of today's corn.
- Humans began practicing artificial selection to obtain desired traits from teosinte thousands of years ago.
- Today's corn is 4n (allotetraploid), which accounts for its large size.

The Fossil Record Shows Both Gradual and Rapid Speciation

14.7 Speciation occurs at different tempos

- According to the gradualistic model, speciation occurs gradually, perhaps due to a gradually changing environment.

- According to the punctuated equilibrium model, periods of equilibrium are interrupted by rapid speciation. Perhaps, if the environment changes rapidly, new species may suddenly arise.
- On occasion, fossil record data may fit one model of speciation and on another occasion, it may fit the other model.

14.8 The Burgess Shale hosts a diversity of life

- The Burgess Shale fossils represent Precambrian marine life from 600 MYA.

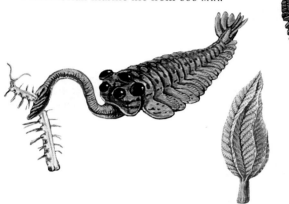

- Speciation, diversification, and eventual extinction are part of the history of life.

Developmental Genes Provide a Mechanism for Rapid Speciation

14.9 Gene expression can influence development

- Differential gene expression can bring about dramatic changes in body shapes and organs.
- Eye development, limb development, and shape determination are controlled by the same genes in different animals.
- It is hypothesized that differential gene expression and/or new functions for old genes can explain evolution, including human evolution.

Speciation Is Not Goal-Oriented

14.10 Evolution is not directed toward any particular end

- In horses, each ancestral species was adapted to its environment, but due to a changing environment, only *Equus* survived.
- Natural selection is opportunistic, not goal-oriented; adaptation occurs because members with an advantage can have more offspring.

Testing Yourself

Evolution of Diversity Requires Speciation

1. A biological species
   a. always looks different from other species.
   b. always has a different chromosome number from that of other species.
   c. is reproductively isolated from other species.
   d. never occupies the same niche in different environments.

For questions 2–7, indicate the type of isolating mechanism described in each scenario.

KEY:

   a. habitat isolation   e gamete isolation
   b. temporal isolation   f. zygote mortality
   c. behavioral isolation   g. hybrid sterility
   d. mechanical isolation   h. low F_2 fitness

2. Males of one species do not recognize the courtship behaviors of females of another species.
3. One species reproduces at a different time than another species.
4. A cross between two species produces a zygote that always dies.
5. Two species do not interbreed because they occupy different areas.
6. The sperm of one species cannot survive in the reproductive tract of another species.
7. The offspring of two hybrid individuals exhibit poor vigor.
8. Which of these is a prezygotic isolating mechanism?
   a. habitat isolation   d. zygote mortality
   b. temporal isolation   e. Both a and b are correct.
   c. hybrid sterility
9. Male moths recognize females of their species by sensing chemical signals called pheromones. This is an example of
   a. gamete isolation.   d. mechanical isolation.
   b. habitat isolation.   e. temporal isolation.
   c. behavioral isolation.
10. Which of these is mechanical isolation?
   a. Sperm cannot reach or fertilize an egg.
   b. Courtship pattern differs.
   c. The organisms live in different locales.
   d. The organisms reproduce at different times of the year.
   e. Genitalia are unsuitable to each other.
11. **THINKING CONCEPTUALLY** Regardless of how speciation occurs or how species are defined, what is required for separate species to be present?

Origin of Species Usually Requires Geographic Separation

12. Complete the following diagram illustrating allopatric speciation by using these phrases: genetic changes (used twice), geographic barrier, species 1, species 2, species 3.

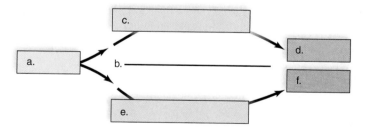

13. The creation of new species due to geographic barriers is called
   a. isolation speciation.   d. sympatric speciation.
   b. allopatric speciation.   e. symbiotic speciation.
   c. allelomorphic speciation.
14. The many species of Galápagos finches are each adapted to eating different foods. This is the result of
   a. gene flow.   d. genetic drift.
   b. adaptive radiation.   e. All of these are correct.
   c. sympatric speciation.

15. **THINKING CONCEPTUALLY** The Hawaiian Islands are some distance from any mainland, and the plants and animals on each island are unique. Only short distances separate the Florida Keys from each other and the mainland. The mainland and the Keys all contain the same species. Explain.

Origin of Species Can Occur in One Place

16. Allopatric, but not sympatric, speciation requires
    a. reproductive isolation.
    b. geographic isolation.
    c. spontaneous differences in males and females.
    d. prior hybridization.
    e. rapid rate of mutation.
17. Which of the following is not a characteristic of plant alloploidy?
    a. hybridization
    b. chromosome doubling
    c. self-fertilization
    d. All of these are characteristics of plant alloploidy.
18. Corn is an allotetraploid, which means that its.
    a. chromosome number is 4n.
    b. development resulted from hybridization.
    c. development required a geographic barrier.
    d. Both a and b are correct.

The Fossil Record Shows Both Gradual and Rapid Speciation

19. Transitional links are least likely to be found if evolution proceeds according to the
    a. gradualistic model.
    b. punctuated equilibrium model.
    c. Both a and b are correct.
    d. None of these are correct.
20. Adaptive raditation is only possible if evolution is punctuated.
    a. true
    b. false
21. Why are there no fish fossils in the Burgess Shale?
    a. The habitat was not aquatic.
    b. Fish do not fossilize easily because they do not have shells.
    c. The fossils of the Burgess Shale predate vertebrate animals.
    d. There are fish fossils in the Burgess Shale.

Developmental Genes Provide a Mechanism for Rapid Speciation

22. Which of the following can influence the rapid development of new types of animals?
    a. The influence of molecular clocks.
    b. A change in the expression of regulating genes.
    c. The sequential expression of genes.
    d. All of these are correct.
23. Which gene is incorrectly matched to its function?
    a. *Hox*—body shape
    b. *Pax6*—body segmentation
    c. *Tbx5*—limb development
    d. All of these choices are correctly matched.

Speciation Is Not Goal-Oriented

24. In the evolution of the modern horse, which was the goal of the evolutionary process?
    a. large size
    b. single toe
    c. Both a and b are correct.
    d. Neither a nor b is correct.
25. Which of the following was not a characteristic of *Hyracotherium*, an ancestral horse genus?
    a. small size
    b. single toe
    c. wooded habitat
    d. All of these are characteristics of *Hyracotherium*.

Understanding the Terms

adaptive radiation 270
allopatric speciation 268
alloploidy 271
autoploidy 271
biological species
    concept 264
evolutionary species
    concept 264
macroevolution 264
polyploidy 271
postzygotic isolating
    mechanism 267
prezygotic isolating
    mechanism 266
speciation 264
sympatric speciation 271

Match the terms to these definitions:
a. _____ Anatomic or physiologic difference between two species that prevents successful reproduction after mating has taken place.
b. _____ Evolution of many species from a common ancestor.
c. _____ Origin of new species due to the evolutionary process of descent with modification.
d. _____ Origin of new species between populations that are separated geographically.

Thinking Scientifically

1. You want to decide what definition of a species to use in your study. What are the advantages and disadvantages of the DNA bar code method (Section 1.9) and the evolutionary and biological species concept?
2. You decide to create a hybrid by crossing two species of plants. If the hybrid is a fertile plant that produces normal size fruit, what conclusion is possible?

ARIS *Visit www.mhhe.com/maderconcepts for practice quizzes, animations, videos, and activities designed to help you master the material in this chapter.*

15

The History and Classification of Life on Earth

LEARNING OUTCOMES

After studying this chapter, you should be able to accomplish the following outcomes.

Motherhood Among Dinosaurs

1 Describe the evidence that dinosaurs nested in the same manner as birds.

The Fossil Record Reveals the History of Life on Earth

2 Use the geologic timescale to trace macroevolution in broad outline.
3 Use a 24-hour day to show that most of the history of life on Earth pertains to unicellular organisms and life in the oceans.
4 Give two possible explanations for the mass extinctions noted in the geologic timescale.

Systematics Traces Evolutionary Relationships

5 Explain the binomial naming system, and name the eight main classification categories.
6 Explain why the Linnaean classification system forms a hierarchy.
7 Give an example that shows how Linnaean classification reflects phylogeny.
8 Use a phylogenetic tree to trace the ancestry of a group.
9 Explain how systematists use the fossil record, homology, and a molecular clock to trace phylogeny.
10 Contrast phylogenetic cladistics with evolutionary systematics.

The Three-Domain System Is Widely Accepted

11 Explain the rationale for the three-domain classification system.
12 Use the three-domain system to classify organisms.

Because dinosaurs are classified as reptiles, paleontologists at first assumed that dinosaurs behaved in the same manner as today's reptiles. For example, the female American alligator lays her eggs in a bowl-shaped nest made from vegetation and mud. She also covers the eggs with vegetation, which protects and keeps the eggs warm as it decays. The mother stays nearby, and when the young call out from inside the eggs, she opens the nest, allowing them to escape. The Nile crocodile is known for further helping her young by carrying the hatchlings down to the water in her mouth.

Within the past 25 years, similar bowl-shaped nests containing dinosaur eggs have been found in Mongolia, Argentina, and Montana. Paleontologists Jack Horner and Bob Makela of Montana State University discovered an entire colony of bowl-shaped nests in Montana. The nests contained fossilized eggs and bones along with eggshell fragments. The space between the nests was large enough for an adult parent to stand lengthwise. From this evidence, these researchers concluded that the baby dinosaurs stayed in the nest after hatching until they had grown large enough to walk around and fragment the eggshells. The spacing

Fossil nest of *Maiasaura* eggs

between the nests suggested that perhaps the mother fed the young. Such behavior would be more like that of a bird than an alligator or crocodile.

The nests belonged to a dinosaur the paleontologists called *Maiasaura*, meaning "good mother lizard." *Maiasaura* was a large dinosaur with a head that looked like that of a horse. The head had a skull crest, which could have served as a resonating chamber used when the dinosaurs communicated with one another. *Maiasaura* could stand and walk on either two or four legs and had a heavy, muscular tail. Perhaps the tail was used for

Fossil bones
of *Maiasaura*
hatchling

defense, or perhaps these dinosaurs were protected by their herd behavior. The remains of an enormous herd of *Maiasaura*, found by Horner and colleagues, are estimated to consist of nearly 30 million bones, representing 10,000 animals, in an area measuring about 1.6 mi^2. Herd behavior and a means of communication allowed Horner to conclude that these particular dinosaurs may have had some sort of social structure.

Since 1978, when Horner and Makela found the nests of *Maiasaura*, others have made similar discoveries. For example, in 1993, Mark Norrell of the American Museum of Natural History found nests in Mongolia that belonged to a dinosaur called *Oviraptor*, which was much smaller and more birdlike than *Maiasaura*. *Oviraptor* was less than 1.8 m long, lightly built, and fast-moving on long legs. Furthermore, the fossilized female parent was sitting on her eggs like a

chicken! It's believed this parent and her nest were buried by a fast-moving sandstorm.

The structural evidence that birds are dinosaurs is strong. Gerald Mayr of the Senckenberg Research Institute in Frankfurt, Germany, has stated that unique traits shared by *Archaeopteryx* and other early birdlike fossils are also present in dinoaurs, such as *Microraptor*, a gliding dinosaur. Beyond the structural evidence, we now have behavioral evidence that at least some dinosaurs nested in the same manner as birds, giving us additional indications that birds are the last surviving dinosaurs.

In this chapter, we will examine the history of life, as revealed by the fossil record, before considering how organisms are grouped according to their evolutionary relationships. Scientists use data regarding evolutionary relationships to construct diagrams depicting these relationships.

Microraptor, a winged gliding dinosaur

The geologic timescale gives a brief outline of the history of life on Earth, based on the fossil record. Meteorite bombardment and severe climate change most likely played a role in the mass extinctions that occurred during the history of life.

15.1 The geologic timescale is based on the fossil record

Because all life-forms evolved from the first cell or cells, life has a history, and this history is revealed by the fossil record. The geologic timescale, which was developed by both geologists and paleontologists, depicts the history of life based on the fossil record. We will be referring to the geologic timescale in future chapters as we study the evolution of various groups of organisms; therefore, it would be beneficial for you to become familiar with it now (**Table 15.1**).

Divisions of the Timescale The timescale divides the history of Earth into eras, then periods, and then epochs. The three eras (the Paleozoic, the Mesozoic, and the Cenozoic eras) span the greatest amounts of time, and the epochs have the shortest time frames. Notice that only the periods of the Cenozoic era are divided into epochs, meaning that more attention is given to the evolution of primates and flowering plants than to the earlier evolving organisms. Modern civilization is given its own epoch, despite the fact that humans have only been around about .04% of the history of life.

Dating Within the Timescale The timescale provides both relative dates and absolute dates. When you say, for example, "Flowering plants evolved during the Jurassic period," you are using relative time, because flowering plants evolved earlier or later than groups in other periods. If you use the dates that are given in millions of years (MYA), you are using absolute time. Absolute dates are usually obtained by measuring the amount of a radioactive isotope in the rocks surrounding the fossils.

Limitations of the Timescale Because the timescale tells when various groups evolved and flourished, it might seem that evolution has been a series of events leading only from the first cells to humans. This is not the case; for example, prokaryotes never declined and are still the most abundant and successful organisms on Earth. Even today, they constitute up to 90% of the total weight of living things.

Then, too, the timescale lists mass extinctions, but it doesn't tell when specific groups became extinct. **Extinction** is the total disappearance of a species or a higher group; a **mass extinction** occurs when a large number of species disappear in a few million years or less. For lack of space, the geologic timescale can't depict in detail what happened to the members of every group of organisms mentioned.

If we could trace the descent of all the millions of groups ever to have evolved, the entirety would resemble a dense bush. Some lines of descent would be cut off close to the base; some would continue in a straight line, even to today; and others would split, producing two or even several groups. The geologic timescale can't show the many facets, twists, and turns of the history of life.

How to Read the Timescale Using **Table 15.1**, you can trace the history of life by beginning with ❶ Precambrian time at the bottom of the timescale. The timescale indicates that the first cells (the prokaryotes) arose some 3,800 MYA. The prokaryotes evolved before any other group.

The Precambrian time was very long, lasting from the time the Earth first formed until 542 MYA. The fossil record during the Precambrian time is meager, but the fossil record from the Cambrian period onward is rich, as we know from our study of the Burgess Shale (see Section 14.8). This helps explain why the timescale usually does not show any periods until the Cambrian period of the Paleozoic era. (British geologists decided which strata formed an era or period and then named them. The names most often refer to places and ancient tribes in England. For example, the Ordovician period is named after an ancient tribe called the Ordovices.)

We can also use the timescale to check when certain groups evolved and/or flourished. For example, during ❷ the Ordovician period, the first simple plants appeared on land, and the first jawless and jawed fishes appeared in the sea. A mass extinction occurred at the end of the Ordovician period, as is often the case at the end of a period. The reasons for mass extinctions are diverse and will be examined in Section 15.4.

On the timescale, note ❸ the Carboniferous period. Rich coal deposits formed in England during this period. As we shall discuss, the climate conditions of the Carboniferous period were perfect for coal formation, not only in England but also in North America. This is the very coal that is burned today to fuel our modern way of life. The timescale tells us that reptiles appeared during the Carboniferous period, but of course it does not mention that reptiles are especially well adapted to living on land because they do not have to return to water to reproduce. This will be discussed in a later chapter.

❹ The evolution of dinosaurs was a significant event during the Mesozoic era. The dinosaurs (but not the mammals, which also appeared during the Mesozoic era) perished during the mass extinction at the end of the Cretaceous period. Why didn't mammals become extinct? Perhaps their small size and lack of specialization helped them survive. Once the dinosaurs departed, mammals underwent adaptive radiation to fill the niches left empty by the dinosaurs.

Table 15.1 introduces us to the evolution of life, but there is much more to be discussed in later chapters. In Section 15.2, the geologic timescale is converted to a 24-hour period for ease of judging the elapsed time between major events in the history of life.

> **15.1** *Check Your Progress* **When using the geologic timescale to trace the evolution of life, you read from the bottom up. Explain.**

TABLE 15.1 **The Geologic Timescale: Major Divisions of Geologic Time and Some of the Major Evolutionary Events of Each Time Period**

Era	Period	Epoch	Millions of Years Ago (MYA)	Plant Life	Animal Life
Cenozoic		Holocene	(0.01–0)	Human influence on plant life	Age of *Homo sapiens*
			Significant Mammalian Extinction		
	Quaternary	Pleistocene	(1.80–0.01)	Herbaceous plants spread and diversify.	Presence of Ice Age mammals. Humans appear.
		Pliocene	(5.33–1.80)	Herbaceous angiosperms flourish.	First hominids appear.
		Miocene	(23.03–5.33)	Grasslands spread as forests contract.	Apelike mammals and grazing mammals flourish; insects flourish.
		Oligocene	(33.9–23.03)	Many modern families of flowering plants evolve.	Browsing mammals and monkeylike primates appear.
	Tertiary	Eocene	(55.8–33.9)	Subtropical forests with heavy rainfall thrive.	All modern orders of mammals are represented.
		Paleocene	(65.5–55.8)	Flowering plants continue to diversify.	Primitive primates, herbivores, carnivores, and insectivores appear.
Mesozoic ④			**Mass Extinction: Dinosaurs and Most Reptiles**		
	Cretaceous		(145.5–65.5)	Flowering plants spread; conifers persist.	Placental mammals appear; modern insect groups appear.
	Jurassic		(199.6–145.5)	Flowering plants appear.	Dinosaurs flourish; birds appear.
			Mass Extinction		
	Triassic		(251–199.6)	Forests of conifers and cycads dominate.	First mammals appear; first dinosaurs appear; corals and molluscs dominate seas.
Paleozoic			**Mass Extinction**		
	Permian		(299–251)	Gymnosperms diversify.	Reptiles diversify; amphibians decline.
③	Carboniferous		(359.2–299)	Age of great coal-forming forests; ferns, club mosses, and horsetails flourish.	Amphibians diversify; first reptiles appear; first great radiation of insects.
			Mass Extinction		
	Devonian		(416–359.2)	First seed plants appear. Seedless vascular plants diversify.	First insects and first amphibians appear on land.
	Silurian		(443.7–416)	Seedless vascular plants appear.	Jawed fishes diversify and dominate the seas.
			Mass Extinction		
②	Ordovician		(488.3–443.7)	Nonvascular plants appear on land.	First jawless and then jawed fishes appear.
	Cambrian		(542–488.3)	Marine algae flourish.	All invertebrate phyla present; first chordates appear.
Precambrian Time ①			630	Soft-bodied invertebrates	
			1,000	Protists evolve and diversify.	
			2,200	First eukaryotic cells	
			2,700	O_2 accumulates in atmosphere.	
			3,800	First prokaryotic cells	
			4,570	Earth forms.	

15.2 The geologic clock can help put Earth's history in perspective

The time frames of the geologic timescale just discussed are so vast that they are hard for humans to relate to. An interesting perspective can be gained by comparing that timescale to a 24-hour period. The so-called 24-hour geologic clock makes it clear that the geologic timescale largely pertains to the most recent one-eighth of the Earth's existence (**Fig. 15.2**). The outer ring of the diagram shows the history of life as it would be measured on a 24-hour timescale, starting at midnight. The inner ring shows the actual years, starting at 4.6 BYA (billions of years ago).

The fossil record tells us that a very large portion of life's history was devoted to the evolution of unicellular organisms. Prokaryotes evolved around 5 A.M., and single-celled eukaryotes did not evolve until about 12 hours later at 5 P.M.! The first multicellular organisms finally appeared in the fossil record at just before 8 P.M. All this time, life remained in the oceans, and there was no terrestrial life until about 9:30 P.M. The demise of the dinosaurs occurred around 11 P.M., and the diversity of mammals started thereafter. Humans did not make the scene until less than a minute before midnight!

In Section 15.3, we learn that severe climate changes, as a result of continental drift, have contributed to extinctions and altered the distribution of life on Earth.

> **15.2 Check Your Progress** Account for the accumulation of free oxygen in the atmosphere before the evolution of plants.

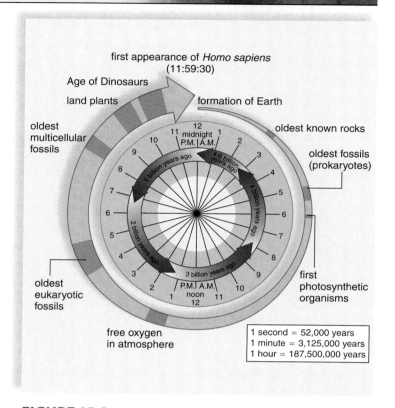

FIGURE 15.2 The 24-hour geologic clock.

15.3 Continental drift has affected the history of life

Prior to 1920, everyone thought the Earth's crust was immobile and the continents had always been in their present positions. But in that year, Alfred Wegener, a German meteorologist, presented data from a number of disciplines, including geology, to support a hypothesis that the continents move. He noted that the rocks in southeast Brazil and South Africa are very similar. This is explainable if the rocks formed in the same place at the same time. Wegener also noticed similarities in living things, which suggested that the continents had been joined at some time in the past. Or, consider that coal only forms under warm, wet conditions, and yet coal is found in Britain and also in the Antarctic, both of which cannot produce coal today. Perhaps the Antarctic and Britain were at one time nearer the equator, and since then, the continents have drifted apart.

Continental Drift We now know that continents are not fixed; rather, their positions and the positions of the oceans have changed over time (**Fig. 15.3A**). ❶ During the Permian period, the continents joined to form one supercontinent, called Pangaea. ❷ But during the Jurassic period, Pangaea separated into two large subcontinents, called Laurasia and Gondwana. ❸ Then, during the Cretaceous period, further separations occurred, forming the continents of today. ❹ Presently, the continents are still drifting in relation to one another. North America and Europe are drifting apart at a rate of about 2 cm per year, a difference not visible to the casual observer.

Continental drift explains why the coastlines of several continents are mirror images of each other—for example, the outline of the east coast of South America matches that of the west coast of Africa. The same geologic structures are found in places where the continents once touched. For example, a single mountain range runs through South America, Antarctica, and Australia.

Continental drift also explains the unique distribution patterns of some fossils. Seeds can't cross oceans, and yet fossils of the same species of seed fern (*Glossopteris*) have been found on all the southern continents. No suitable explanation could be found until scientists became aware of the formation of Pangaea. Then it seemed plausible that the plant could have

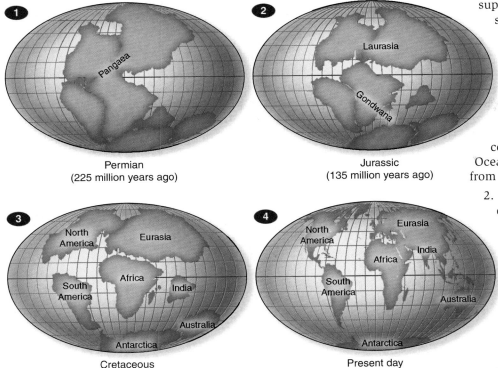

Permian
(225 million years ago)

Jurassic
(135 million years ago)

Cretaceous
(65 million years ago)

Present day

FIGURE 15.3A The continents have drifted through time.

spread to all the continents while they were joined as one. Continental drift also explains the distribution of placental mammals versus marsupials today. Marsupials evolved on Laurasia but spread to Gondwana just before those two land-masses split. A land bridge allowed marsupials to also spread to the Antarctica/Australia continent. In the meantime, pla-cental mammals arose on Laurasia and subsequently spread to South America, and then on to Africa and India. The pla-cental mammals couldn't get to Antarctica/Australia because the land bridge had disappeared by the time they arrived at the tip of South America. Wherever the placentals went, they outcompeted the marsupials, and the marsupials died out, except for the opossum. Antarctica moved to the South Pole, and the marsupials on that continent died from the cold. But Australia went north, near the equator, where the marsupials thrived. With no competition from placentals, Australian mar-supials underwent adaptive radiation to have the same type adaptations as placentals elsewhere: Kangaroos are grazers like cattle, koalas eat tree leaves as do giraffes, and marsupial dogs and Tasmanian devils are predators as are hyenas.

Plate Tectonics Why do the continents drift? The answer lies in the science of plate tectonics. The Earth's crust is frag-mented into slablike plates that float on a hot, liquefied metallic core that lies directly beneath the Earth's crust. The continents and the ocean basins are a part of these rigid plates. In other words, when a plate moves, so does an ocean basin and/or a continent, just as grocery items move on a conveyor belt at the

supermarket. The supermarket conveyor belt stays the same size because it rises at one end and drops out of sight at the other end. A tectonic plate stays the same size for the following reasons:

1. At deep oceanic ridges, seafloor spreading occurs as molten mantle rock rises and material is added to plates. Seafloor spreading causes the continents to move and causes the Atlantic Ocean to get wider, separating North America from Europe, for example.

2. Where the plates meet, the forward edge of one sinks into the mantle and is de-stroyed, creating a subduction zone. In the meantime, the continents collide, and the result is often a mountain range. For example, the Himalayas occur where India is colliding with Eurasia. The Himalayan Mountains are still being raised as India presses northward.

The place where two plates scrape past one another is called a transform boundary. The San Andreas fault in Southern California is at a transform boundary, and the movement of the two plates is re-sponsible for the many earthquakes in that region (**Fig. 15.3B**).

In Section 15.4, we observe that severe climate changes as a result of meteorite bombardment have contributed to the oc-curence of mass extinctions.

15.3 *Check Your Progress* **Climate permitting, would you expect to find dinosaur bones all over the globe? Explain.**

FIGURE 15.3B The San Andreas fault is a transform boundary.

15.4 Mass extinctions have affected the history of life

As **Figure 15.4** and Table 15.1 indicate, at least five mass extinctions—end of the Ordovician, Devonian, Permian, Triassic, and Cretaceous periods—have occurred throughout the history of life. Were these mass extinctions due to a gradual process brought on by day-to-day environmental pressures, or were they due to some dramatic, cataclysmic event? Two investigators, Nan Crystal Arens and Ian West, both of Hobart and William Smith Colleges, have proposed the press/pulse model of mass extinction. Presses are longer, steadier pressures that bring species to the brink of extinction, and pulses are catastrophic events that finally cause species to become extinct. A mass extinction occurs whenever a number of species have experienced "presses" and then been exposed to a sudden "pulse." Other investigators have been able to associate mass extinctions with certain cataclysmic events.

Meteorites A **meteorite** is a piece of rock from outer space that strikes the surface of the Earth and creates an impact crater. The result of a large meteorite striking Earth could be similar to that of a worldwide atomic bomb explosion: A cloud of dust might mushroom into the atmosphere, block out the sun, and cause plants to die. A decline in photosynthesis would lead to the death of animals.

In 1977, Walter and Luis Alvarez were the first to propose that a meteorite could cause a mass extinction. They found that Cretaceous clay contains an abnormally high level of iridium, an element that is rare in the Earth's crust but more common in meteorites. Therefore, they said that the mass extinction at the end of the Mesozoic era, called the Cretaceous-Tertiary (K-T) boundary, was due to a meteorite. Later, a large impact crater was found near the tip of the Yucatán Peninsula in Mexico. The date of this crater corresponded to the timing of the K-T extinction.

Similarly, a large meteorite impact crater in Canada is dated at the time of the Devonian mass extinction.

Climate Changes Severe climate change can cause an extinction. For example, we have mentioned that marsupials died

FIGURE 15.4 Mass extinctions.

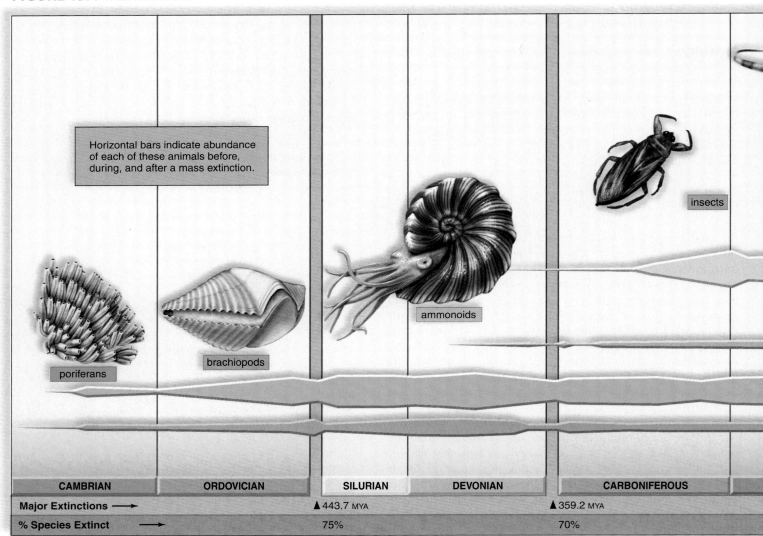

from the cold on the continent of Antarctica when it drifted to the South Pole.

The extinction at the end of the Permian period, sometimes called the Permian-Triassic (P-Tr) extinction, was quite severe; 90% of species in the ocean and 70% on land disappeared. We know that when Pangaea formed a single landmass, it stretched between and included both the North and South Poles. Perhaps glaciation at both poles caused severe climatic fluctuations around the globe, leading to species extinctions. However, Pangaea formed midway through the Permian period and not at the P-Tr boundary. Extreme volcanic eruptions are known to have occurred at the end of the Permian period. Volcanic ash could have blocked out the sun, and a worldwide decrease in temperature could have increased the formation of glaciers, leading to a drop in sea level. Greenhouse gases, such as carbon dioxide and methane, are also released by active volcanoes. These gases, in the long run, could have caused climate changes, such as global warming, that would have led to extinctions.

In 2006, a huge meteorite impact crater 482 kilometers across was discovered in a region of East Antarctica. The date of the crater suggests that the impact of a meteorite could have contributed to the P-Tr extinction. As yet, there is no generally accepted cause of the P-Tr mass extinction. Perhaps the formation of Pangaea could be considered a "press," while volcanism and meteorite impact are "pulses" that contributed to the P-Tr extinction.

Human Activities Some scientists believe we are currently in the midst of a mass extinction due to human activities, as discussed in Chapter 40. Ian West, mentioned earlier, suggests that the "press" of this modern-day extinction is due to our manipulation of the environment, such as modern agricultural methods, while the "pulse" is being provided by industrialization with its demand for energy and the resultant global warming.

This completes our discussion of the geologic timescale. The next part of the chapter discusses systematics, which attempts to discover how organisms are related through time.

15.4 *Check Your Progress* Humans did not become extinct during any of the five mass extinctions that have occurred on planet Earth. Explain.

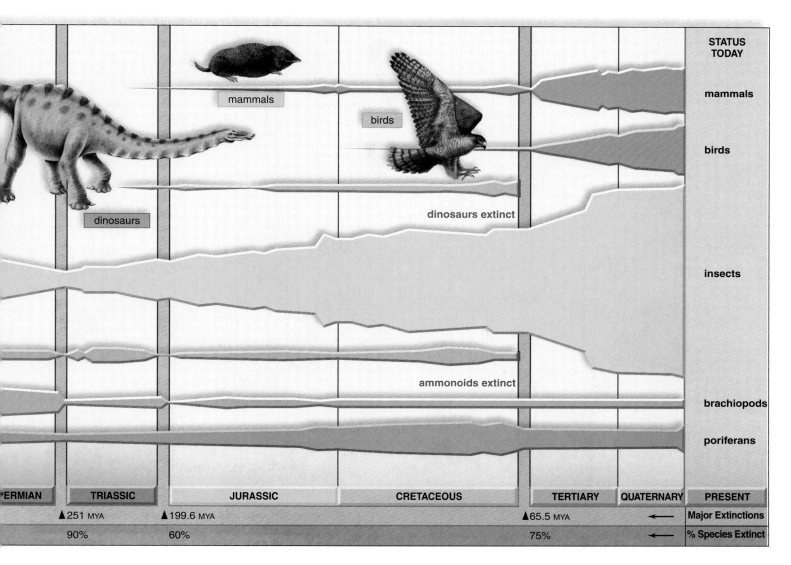

Systematics Traces Evolutionary Relationships

Classification is a part of systematics, which studies the diversity of organisms and their relationships as life evolved on Earth. Systematists rely on data from various sources to construct diagrams that reflect evolutionary relationships. In phylogenetic cladistics, a diagram called a cladogram is constructed, and in traditional evolutionary systematics, a phylogenetic tree is constructed

15.5 Organisms can be classified into categories

Linnaean classification is the grouping of extinct and living species into the following categories: **domain**, **kingdom**, **phylum**, **class**, **order**, **family**, **genus**, and **species**. A **taxon** (pl., taxa) is a group of organisms that fills a particular category of classification. There can be several species within a genus, several genera within a family, and so forth—the higher the category, the more inclusive it is. Thus, there is a hierarchy of categories. The organisms that fill a particular classification category are distinguishable from other organisms by a set of characteristics, or simply characters, that they share. A **character** is any trait, whether structural, molecular, reproductive, or behavioral, that distinguishes one group from another. (These types of data are discussed in more detail in Section 15.7.) Organisms in the same domain have general characters in common; those in the same species have quite specific characters in common.

In most cases, categories of classification can be divided into additional categories, as in superorder, order, suborder, and infraorder. Considering these, there are more than 30 categories of classification.

Taxonomy is the science of naming species. Taxonomy is a part of classification because a scientific name helps classify an organism. In the mid-18th century, Carolus Linnaeus, the father of taxonomy, gave us the binomial system of naming organisms. Each name is called binomial because it has two parts. The first word is the genus, and the second word is the specific epithet. For example, the scientific name *Parthenocissus quinquefolia* tells you the genus and specific epithet of the plant featured in **Figure 15.5**. From there, the boxes tell the other categories that are used to classify this plant in the Linnaean system.

Why is it preferable to use the scientific name, *Parthenocissus quinquefolia*, instead of the common name, Virginia creeper? Because the scientific name is always based on Latin, and common names often differ between countries and even within the same country. Classification categories hypothesize the evolutionary relationship between species, as shown in Section 15.6.

> **15.5 Check Your Progress** The different types of plants in genus *Parthenocissus* resemble each other closely. The four different types of organisms in domain Eukarya differ widely. Explain. (Hint: see Figures 1.4C and 15.9)

FIGURE 15.5 Hierarchy of taxa for *Parthenocissus quinquefolia*.

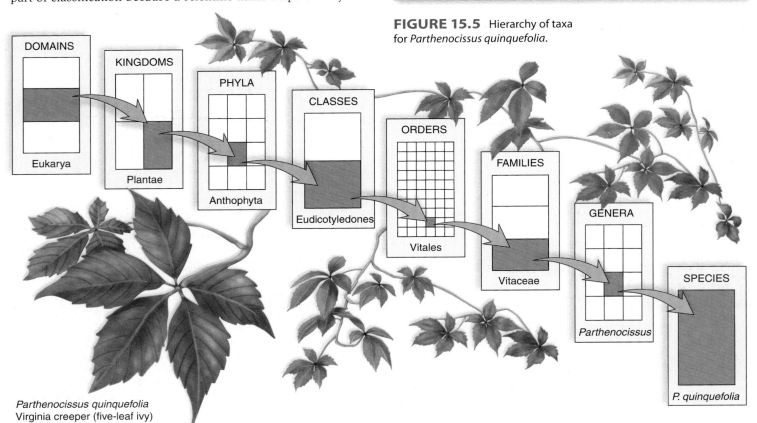

DOMAINS — Eukarya
KINGDOMS — Plantae
PHYLA — Anthophyta
CLASSES — Eudicotyledones
ORDERS — Vitales
FAMILIES — Vitaceae
GENERA — Parthenocissus
SPECIES — P. quinquefolia

Parthenocissus quinquefolia
Virginia creeper (five-leaf ivy)

Classification is a part of the broader field of **systematics**, which is the study of the diversity of organisms at all levels of biological organization. One goal of systematics is to determine **phylogeny**, or the evolutionary history of a group of organisms, which is often represented by a **phylogenetic (evolutionary) tree**, a diagram that indicates common ancestors and lines of descent (lineages). Each branch point in a phylogenic tree is a divergence from a **common ancestor**, a species that gives rise to two new groups. For example, this portion of an evolutionary tree shows that monkeys and apes share a common primate ancestor:

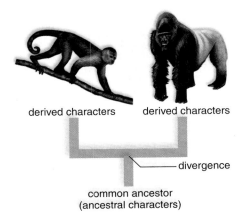

Divergence is presumed because monkeys and apes have their own individual characteristics (often called **derived characters**). For example, skeletal differences allow an ape to swing from limb to limb of a tree, while monkeys run along the tops of tree branches. The common primate ancestor to both monkeys and apes has **ancestral characters** that are shared by the ancestor as well as by monkeys and apes. For example, the common primate ancestor must have been able to climb trees, as can both monkeys and apes.

A phylogenetic tree has many branch points, and they can show that it is possible to trace the ancestry of a group of organisms farther and farther back in the past. For example, reindeer, monkeys, and apes all give birth to live young because they all have a common ancestor that was a placental mammal:

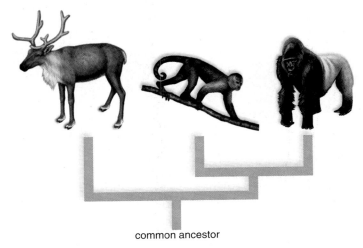

Classification is a part of systematics because classification categories list the unique characters of each taxon, which ideally reflect phylogeny. A species is most closely related to other species

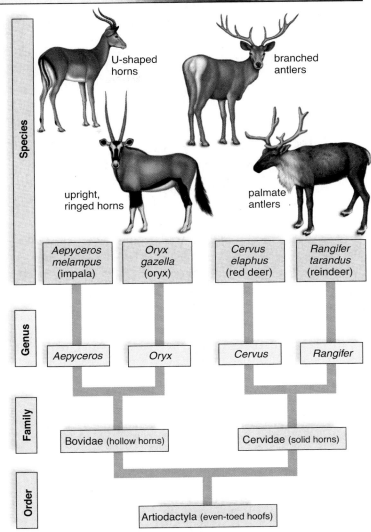

FIGURE 15.6 Linnaean classification and phylogeny.

in the same genus, then to genera in the same family, and so forth, from order to class to phylum to kingdom. When we say that two species (or genera, families, etc.) are closely related, we mean that they share a more recent common ancestor with each other than they do with members of other taxa. For example, all the animals in **Figure 15.6** are related because we can trace their ancestry back to the same order. The animals in the order Artiodactyla all have even-toed hoofs. Animals in the family Cervidae have solid horns, called antlers, but the horns in red deer (genus *Cervus*) are highly branched, while those in reindeer (genus *Rangifer*) are palmate (shaped like a hand). In contrast, animals in the family Bovidae have hollow horns, and unlike the Cervidae, both males and females have horns, although they are smaller in females.

In Section 15.7, we review the types of data that systematists collect in order to determine evolutionary relationships.

15.6 *Check Your Progress* **Knowing that fishes evolved from an ancestral vertebrate, amphibians from a fish, reptiles from an amphibian, and birds and mammals from a reptile, draw a simplified phylogenetic tree for vertebrates.**

Systematists gather all sorts of data in order to discover the evolutionary relationships between species. They rely heavily on the fossil record, homology, and molecular data to determine the correct sequence of common ancestors in any particular group of organisms. If you can propose common ancestors in phylogeny, you have a model by which evolution may have occurred and can classify organisms accordingly.

Fossil Record

It is possible to use the fossil record to trace the history of life in broad terms, and sometimes to trace the history of lineages. One of the advantages of fossils is that they can be dated. Unfortunately, it is not always possible to tell to which group, living or extinct, a fossil is related. For example, at present, paleontologists are discussing whether turtles are distantly or closely related to crocodiles. On the basis of his interpretation of fossil turtles, Olivier C. Rieppel of the Field Museum of Natural History in Chicago is challenging the conventional interpretation that turtles are ancestral (have traits seen in a common ancestor to all reptiles) and are not closely related to crocodiles, which evolved later. His interpretation is supported by molecular data that show turtles and crocodiles are closely related.

If the fossil record were more complete, there might be fewer controversies about the interpretation of fossils. One reason the fossil record is incomplete is that most fossils represent the harder body parts, such as bones and teeth. Soft parts are usually eaten or decayed before they have a chance to be fossilized. This may be one reason it has been difficult to discover when angiosperms (flowering plants) first evolved. A Jurassic fossil recently found, if accepted as an angiosperm by most botanists, may help pin down the date (**Fig. 15.7A**). As paleontologists continue to explore the world, the sometimes stingy fossil record will reveal some of its secrets.

Homology

As explained in Section 13.9, **homology** is character similarity that stems from having a common ancestor. Comparative anatomy, including embryological evidence, provides information regarding homology. **Homologous structures** are related to each other through common descent. The forelimbs of vertebrates are homologous because they contain the same bones organized in the same general way as in a common ancestor. For example, a horse has but a single digit and toe (the hoof), while a bat has four lengthened digits that support the membranous wings. Homologous structures are often linked to *divergent evolution*, which occurs when organisms having similar origins adapt to new environmental niches (see Section 14.4).

Deciphering homology is sometimes difficult because of convergent evolution. **Convergent evolution** is the acquisition of the same or similar characters in distantly related lines of descent. Similarity due to convergence is termed **analogy**; the wings of an insect and the wings of a bat are analogous. You may recall from Section 13.9 that **analogous structures** have the same function in different groups, but do not have a common ancestry. Both cacti and spurges are adapted similarly to a hot, dry environment, and both are succulent (thick, fleshy) with spiny leaves (**Fig. 15.7B**). However, the details of their flower structure indicate that these plants are not closely related. **Parallel evolution** is the acquisition of the same or a similar character in two or more related lineages without it being present in a common ancestor. Placental mammals and marsupials provide several examples. For example, the flying squirrel (a placental) and the flying phalanger (a marsupial), exemplify parallel evolution.

Molecular Data

Speciation occurs when mutations bring about changes in the base-pair sequences of DNA. Systematists, therefore, assume that when two species are closely related, a comparative study of their DNA will show few differences in base-pair sequences. Since DNA codes for amino acid sequences in proteins, it also follows that closely related species will have fewer differences in the amino acid sequences of their proteins.

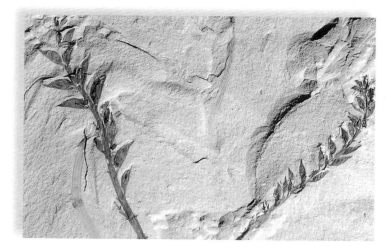

FIGURE 15.7A   A possible ancestral angiosperm.

Prickly pear cactus, *Opuntia*   Spurge, *Euphorbia*

FIGURE 15.7B   Convergent evolution.

FIGURE 15.7C Molecular data.

Because molecular data are numerical, they can sometimes sort out relationships obscured by inconsequential structural variations or convergent evolution. Software breakthroughs have made it possible to analyze nucleotide and amino acid sequences quickly and accurately using a computer. Also, these analyses are available through the Internet to anyone doing comparative studies. Therefore, each investigator doesn't have to start from scratch. Increased accuracy combined with the availability of past data has made molecular systematics the standard way to study the relatedness of groups of organisms today.

One study involving DNA differences produced the data shown in **Figure 15.7C**. Although the data suggest that chimpanzees are more closely related to humans than to other apes, in Linnaean classifications, humans and chimpanzees are placed in different families. Humans are in the family Hominidae, and chimpanzees are in the family Pongidae. In contrast, the rhesus monkey and the green monkey, which have more numerous DNA differences from each other, are placed in the same family (Cercopithecidae). To be consistent with the data, shouldn't humans and chimpanzees also be in the same family? Linnaean taxonomists, in particular, believe that humans are markedly different from chimpanzees because of adaptation to a different environment. Therefore, they judge it is justifiable to place humans in a separate family.

Mitochondrial DNA (mtDNA) mutates ten times faster than nuclear DNA. Thus, when determining the phylogeny of closely related species, investigators often choose to sequence mtDNA instead of nuclear DNA. One such study concerned North American songbirds. It had long been suggested that these birds diverged into eastern and western subspecies due to the presence of glaciers some 250,000–100,000 years ago. Sequencing of mtDNA allowed investigators to conclude that groups of North American songbirds diverged from one another about 2.5 MYA. Since the old hypothesis based on glaciation is apparently flawed, a new hypothesis is required to explain why eastern and western subspecies arose among songbirds.

Molecular Clocks When nucleic acid changes are neutral (not tied to adaptations) and accumulate at a fairly constant rate, these changes can be used as a **molecular clock** to determine when two species diverged from a common ancestor. For example, if neutral mutations occur every 100,000 years and two related species differ by five such mutations, it is possible that the two species diverged approximately 500,000 years ago.

Researchers used DNA sequence data to construct the tree in Figure 15.7C, and then they used the fossil record to assign the dates shown. They found that the comparative dates of evolution using a molecular clock and the actual dates from the fossil record were consistent with one another. Molecular clock data are always considered a hypothesis until checked by the fossil record.

Two schools of systematics are contrasted in Section 15.8.

15.7 *Check Your Progress* Would the bones of dinosaur limbs be homologous with those in the limbs of other vertebrates? Explain.

15.8 Phylogenetic cladistics and evolutionary systematics use the same data differently

Phylogenetic cladistics is a method of determining evolutionary relationships based on shared characters derived from a common ancestor. This school of systematics, based on the work of Willi Hennig, uses shared derived characters to classify organisms and arrange taxa in a diagram called a **cladogram**. A cladogram traces the evolutionary history of the group being studied. Let's see how it works.

The first step when constructing a cladogram is to draw a table that summarizes the characters of the taxa being compared (**Fig. 15.8A**). At least one species, but preferably several, are studied as an outgroup, a taxon that is distantly related to the study group. In this example, lancelets are the outgroup, and selected vertebrates are the study group. Any character found in both the outgroup and the study group is a shared ancestral character (e.g., notochord in embryo) presumed to have been present in a common ancestor to both the outgroup and the study group. Any character found in only one or in scattered taxa (e.g., long cylindrical body) is excluded from the cladogram because it probably doesn't pertain to evolutionary relationships. The other characters are shared derived characters—that is, they are homologies shared by certain taxa of the study group. In a cladogram, a **clade** is an evolutionary branch that includes a common ancestor, together with all its descendant species. A clade includes all taxa that have one or more unique shared derived characters not present in other groups of taxa.

The cladogram in **Figure 15.8B** has clades that differ in length because the first includes the other two, and so forth. ❶ Notice that the common ancestor at the root of the tree had one ancestral character: notochord in embryo. Then follow common ancestors that have ❷ vertebrae, ❸ lungs and a three-chambered heart, and finally ❹ amniotic egg and internal fertilization. Therefore, this is the sequence in which these characters evolved during the evolutionary history of vertebrates. These homologies also show which species are closely related to one another. All the taxa in the study group belong to the first clade because they all have vertebrae; newts, snakes, and lizards are in the clade that has lungs and a three-chambered heart; and only snakes and lizards have an amniotic egg and internal fertilization.

Cladists typically use many more characters than appear in our simplified cladogram. They also feel that a cladogram is a hypothesis that can be tested and either corroborated or refuted on the basis of additional data. These are the reasons that cladistics is now the accepted way to decipher evolutionary history.

Evolutionary Systematics Soon after Darwin published his book *On the Origin of Species*, **evolutionary systematics** began. It is the traditional method of using characters (but not always ancestry) to classify and determine evolutionary history. Evolutionary systematists mainly use structural data

	lancelet	eel	newt	snake	lizard
Notochord in embryo	X	X	X	X	X
Vertebrae		X	X	X	X
Lungs			X	X	X
Three-chambered heart			X	X	X
Internal fertilization				X	X
Amniotic egg				X	X
Four bony limbs			X		X
Long cylindrical body		X		X	

FIGURE 15.8A Data for constructing a cladogram.

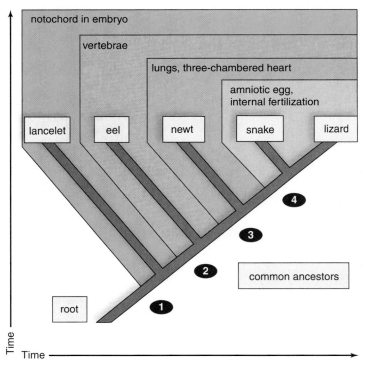

FIGURE 15.8B In a cladogram, a clade (colors) contains a common ancestor and all its descendents with shared derived characters.

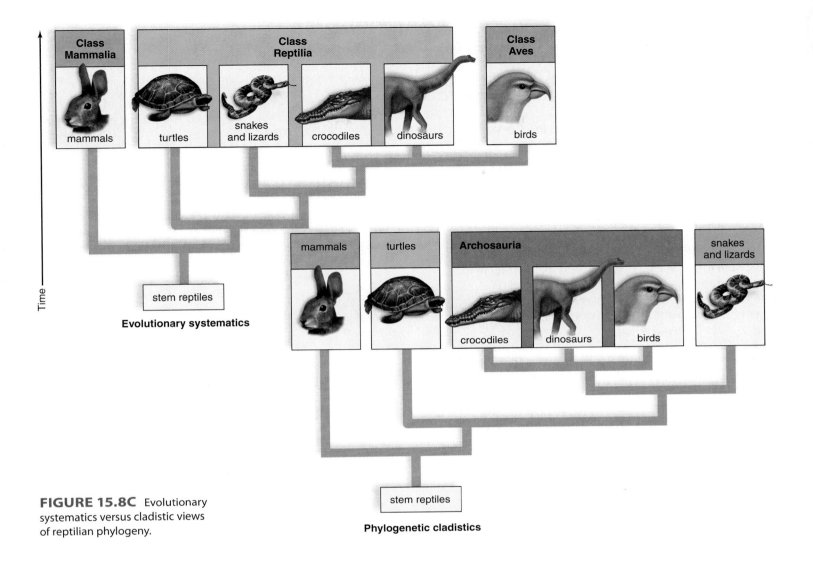

FIGURE 15.8C Evolutionary systematics versus cladistic views of reptilian phylogeny.

and the Linnaean system to classify organisms and construct phylogenetic trees. Evolutionary systematists differ from today's phylogenetic cladists largely by stressing both common ancestry *and* the degree of structural difference among divergent groups. Therefore, a group that has adapted to a new environment and shows a high degree of evolutionary change is not always classified with the common ancestor from which it evolved. In other words, evolutionary systematists do not necessarily include all the species that share a common ancestor in one taxon.

In the phylogenetic tree shown in **Figure 15.8C** (*top*), birds and mammals are placed in different classes because it is quite obvious to the most casual observer that mammals (having hair and mammary glands) and birds (having feathers) are quite different in appearance from one another and from reptiles (having scaly skin). The evolutionary systematist goes on to say that birds and mammals evolved from stem reptiles.

Cladists prefer the cladogram shown in Figure 15.8C (*bottom*). All the animals shown are in one clade because they all evolved from a common ancestor that laid eggs. Mammals (Mammalia) form a clade because they all have hair, mam-

mary glands, and three middle ear bones. Cladists believe that what an evolutionary systematist calls "reptiles" (which includes turtles, snakes, lizards, and crocodiles) is an incorrect classification because it does not include all the descendants from a common ancestor. That is, a proper clade would also include birds in the Reptilia because the fossil record indicates that early birds shared many characteristics with reptiles. It matters not that birds are now adapted to a different environmental niche. To indicate that crocodiles and birds, along with the dinosaurs, are closely related, a cladist would place them in a group of their own called Archosaurs. Figure 15.8 (*bottom*) indicates the common ancestors for all clades evolved in the order given.

This completes our study of how systematists study and depict evolutionary relationships. The next part of the chapter discusses the classification system used in this text.

15.8 *Check Your Progress* Cladists separate out crocodiles, dinosaurs, and birds and put them in their own group within the reptiles. Explain why.

This text uses the three-domain system of classification in which the domains are Archaea, Bacteria, and Eukarya. Protists, fungi, plants, and animals are all eukaryotes.

15.9   This text uses the three-domain system of classifying organisms

In the late 1970s, Dr. Carl Woese proposed that there are two groups of prokaryotes. Further, Woese said that the rRNA sequences of these two groups, called the bacteria and the archaea, are so fundamentally different from each other that they should be assigned to separate domains, a category of classification that is higher than the kingdom category. The phylogenetic tree shown in **Figure 15.9** is based on his rRNA sequencing data. The data suggest that the bacteria diverged first, followed by the archaea and then the eukaryotes. This means that the archaea and the eukaryotes are more closely related to each other than either is to the bacteria.

Domain   Bacteria ❶

Bacteria are a prokaryotic group that is so diversified and plentiful they are found in large numbers nearly everywhere on Earth. The cyanobacteria are photosynthetic, but most bacteria are heterotrophic. *Escherichia coli*, which lives in the human intestine, is heterotrophic, as are parasitic forms that cause human disease, such as *Clostridium tetani* (cause of tetanus) and *Bacillus anthracis* (cause of anthrax). Heterotrophic bacteria are beneficial in ecosystems because they are organisms of decay that, along with fungi, keep chemical cycling so that plants always have a source of inorganic nutrients.

Domain Archaea ❷ Like

bacteria, archaea are prokaryotic unicellular organisms that reproduce asexually. Archaea do not look that different from bacteria under the microscope, and the extreme conditions under which many species live has made it difficult to culture them. For example, the methanogens live in anaerobic environments, such as swamps and marshes and the guts of animals. The halophiles are salt-lovers, living in bodies of water such as the Great Salt Lake in Utah. The thermoacidophiles thrive in extremely hot, acidic environments, such as hot springs and geysers. The branched nature of diverse lipids in the archaeal plasma membrane could possibly help them live in extreme conditions.

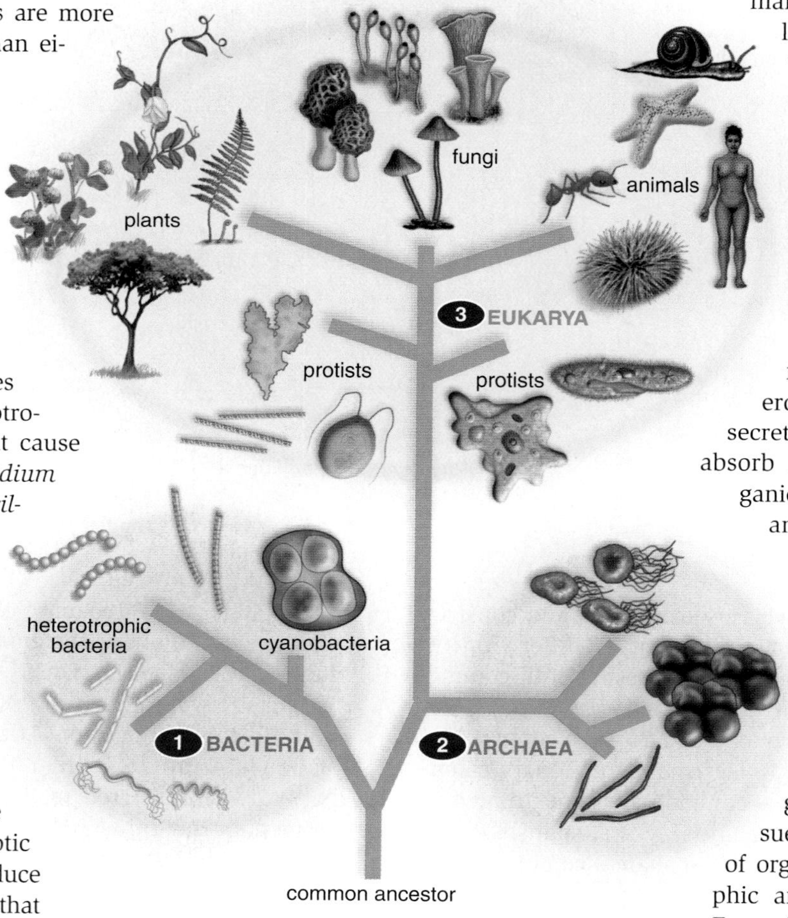

Domain Eukarya ❸ Eukaryotes are unicellular to multicellular organisms whose cells have a membrane-bounded nucleus. Sexual reproduction is common, and various types of life cycles are seen. Later in this text, we will study the four individual kingdoms that occur within the domain Eukarya: protists, fungi, plants, and animals. In the meantime, we can note that protists are a diverse group of organisms that are hard to classify and define. Although usually unicellular, some are filaments, colonies, or multicellular sheets. Their forms of nutrition are diverse, while others are heterotrophic by ingestion or absorption and some are photosynthetic. Algae, paramecia, and slime molds are representative protists.

Fungi are eukaryotes that form spores, lack flagella, and have cell walls containing chitin. Most are multicellular. Fungi are heterotrophic by absorption—they secrete digestive enzymes and then absorb nutrients from decaying organic matter. Mushrooms, molds, and yeasts are representative fungi. Despite appearances, molecular data suggest that fungi and animals are more closely related to each other than either group is to plants.

Plants are nonmotile eukaryotic multicellular organisms. They possess true tissues and the organ system level of organization. Plants are autotrophic and carry on photosynthesis. Examples include cacti, ferns, and cypress trees.

Animals are motile, eukaryotic, multicellular organisms. They also have true tissues and the organ system level of organization. Animals are heterotrophic by ingestion. Worms, whales, and insects are all examples of animals.

FIGURE 15.9 The three-domain system of classification.

> **15.9** *Check Your Progress* **If scientists were to discover a frozen dinosaur and could examine its tissues, would they expect to find the same tissues as in a human? Explain.**

The geologic timescale describes, in general, the history of life on Earth. Actually, if we could trace the descent of all the millions of groups ever to have evolved, the entirety would resemble a dense bush. The species that are alive today are the end product of all the changes that occurred on Earth as life evolved. What if other events had occurred? For example, what if the continents had not separated 65 MYA, what types of mammals, if any, would be alive today? Given a different sequence of environments, a different mix of plants and animals might very well have resulted.

Every known species that has evolved is given a two-part name consisting of a genus and a specific epithet. All sorts of data are used to classify organisms and develop tree diagrams that show evolutionary relationships among species. Cladistics is by now a widely accepted way to determine these evolutionary relationships.

Most biologists today have adopted the three-domain system of classifying species. The archaea are structurally similar to bacteria, but their rRNA differs from that of bacteria and is instead similar to that of eukaryotes. The domain Eukarya contains four kingdoms: protists, fungi, plants, and animals. Chapter 16 discusses the domains Archaea and Bacteria, and subsequent chapters review the four kingdoms within the domain Eukarya.

The Chapter in Review

Summary

Motherhood Among Dinosaurs

- Dinosaur nests have been found that resemble those built by modern crocodiles and birds. Evidence suggests that certain dinosaurs cared for their young much as birds do.

The Fossil Record Reveals the History of Life on Earth

15.1 The geologic timescale is based on the fossil record

- Eras, periods, and domains divide up the timescale.
- These divisions can be used to indicate the relative timing of events, but the MYA dates provide absolute timing.
- The timescale doesn't show the bush pattern of evolution.

15.2 The geologic clock can help put Earth's history in perspective

- The 24-hour geologic clock begins at 12 midnight and goes to the next midnight.
- Single-celled organisms were alone on the Earth for a large part of its history.
- There was no terrestrial life until 9:30 P.M., and humans did not appear until a minute before midnight.

15.3 Continental drift has affected the history of life

- The positions of continents and oceans have changed over time and are still changing.
- Plate tectonics explains the movements of Earth's crust.
  - Oceanic ridges form when molten mantle lava rises and material is added to plates.
  - Subduction zones occur where a plate sinks into the mantle and is destroyed.
  - The place where two plates meet and scrape past one another is called a transform boundary.

15.4 Mass extinctions have affected the history of life

- Mass extinctions occurred at the ends of the Ordovician, Devonian, Permian, Triassic, and Cretaceous periods.
- The press/pulse model suggests that species are pressured to the brink of extinction, and then a catastrophic event causes them to die out.

- Causes of catastrophic events include meteorites and climate changes due to continental drift or volcanism.

Systematics Traces Evolutionary Relationships

15.5 Organisms can be classified into categories

- The main Linnaean classification categories (taxons) are domain, kingdom, phylum, class, order, family, genus, and species.
- Members of a taxon share characters.
- Taxonomy begins with the naming of a species because the name tells the genus and specific epithet.

15.6 Linnaean classification reflects phylogeny

- Linnaean classification utilizes characters which reflect phylogeny.
- Systematics is the study of organism diversity at all levels of organization.
- Phylogeny is the evolutionary history of a group of organisms. A phylogenetic tree indicates common ancestors and lines of descent.
- Derived characters are the particular characteristics of a group, while ancestral characters are shared with a common ancestor.

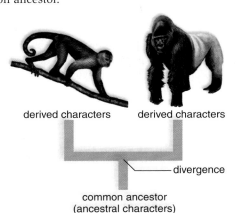

derived characters   derived characters

divergence

common ancestor
(ancestral characters)

15.7 Certain types of data are used to trace phylogeny

- The fossil record can trace the history of life and lineages.
- Homology refers to the similarity of characters inherited from a common ancestor.

- Few changes in DNA base-pair and amino acid sequences shows species are closely related.
- When nucleic acid changes are neutral and accumulate at a fairly constant rate, a molecular clock (number of base sequence changes per unit time) can indicate how long ago two species diverged from one another.

15.8 Phylogenetic cladistics and evolutionary systematics use the same data differently

- Phylogenetic cladistics uses shared derived characters and common ancestry to group organisms in clades.
- A clade is an evolutionary branch in a cladogram that includes the common ancestor and all descendant species.
- Evolutionary systematics uses characters (but not always ancestry) to classify organisms and construct phylogenetic trees.

The Three-Domain System Is Widely Accepted

15.9 This text uses the three-domain system of classifying organisms

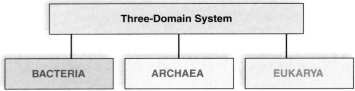

- Domain Bacteria is composed of a diverse and plentiful group of prokaryotes.
- Domain Archaea encompasses prokaryotes that are chemically different from bacteria and thrive in extreme environments.
- Domain Eukarya contains a wide variety of unicellular to multicellular organisms that all have a membrane-bounded nucleus but differ widely in their life cycles.
- Eukaryotes include protists, fungi, plants, and animals.

Testing Yourself

The Fossil Record Reveals the History of Life on Earth

For questions 1–4, match the phrases with a division of geologic time in the key.

KEY:

    a. Cenozoic era   c. Paleozoic era
    b. Mesozoic era   d. Precambrian

1. Dinosaur diversity; evolution of birds and mammals
2. Prokaryotes abound; eukaryotes evolve and become multicellular
3. Mammalian diversification
4. Invasion of land by plants
5. Continental drift helps explain
   a. mass extinctions.
   b. the distribution of fossils on the Earth.
   c. geologic upheavals such as earthquakes.
   d. climate changes.
   e. All of these are correct.
6. When we consider the history of Earth as if it occurred in 24 hours, terrestrial life first appeared around.
   a. 6:00 A.M.   d. 9:30 P.M.
   b. 9:00 A.M.   e. 11:55 P.M.
   c. 1:00 P.M.

Classification Reflects Evolutionary Relationships

7. Which is the scientific name of an organism?
   a. *Rosa rugosa*   d. *rugosa rugosa*
   b. *Rosa*   e. Both a and d are correct.
   c. *rugosa*
8. Classification of organisms reflects
   a. similarities.   c. Neither a nor b is correct.
   b. evolutionary history.   d. Both a and b are correct.
9. Which of these sequences exhibits an increasingly more-inclusive scheme of classification?
   a. kingdom, phylum, class, order
   b. phylum, class, order, family
   c. class, order, family, genus
   d. genus, family, order, class
10. Use the data from the following table to fill in the phylogenetic tree for vascular plants. (Plants with vascular tissue have transport tissue.)

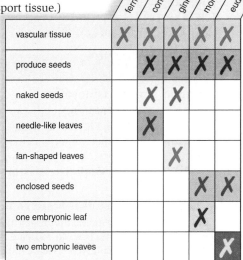

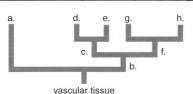

11. One benefit of the fossil record is
    a. that hard parts are more likely to fossilize.
    b. fossils can be dated.
    c. its completeness.
    d. that fossils congregate in one place.
    e. All of these are correct.
12. Which pair is mismatched?
    a. homology—character similarity due to a common ancestor
    b. molecular data—DNA strands match
    c. fossil record—bones and teeth
    d. homology—functions always differ
    e. molecular data—molecular clock
13. The discovery of common ancestors in the fossil record, the presence of homologies, and nucleic acid similarities help scientists decide
    a. how to classify organisms.
    b. the proper cladogram.
    c. how to construct phylogenetic trees.
    d. how evolution occurred.
    e. All of these are correct.

14. Molecular clock data are based on
    a. common adaptations among animals.
    b. DNA dissimilarities in living species.
    c. DNA fingerprinting of fossils.
    d. finding homologies among plants.
    e. All of these are correct.
15. In cladistics,
    a. a clade must contain the common ancestor plus all its descendants.
    b. derived characters help construct cladograms.
    c. data for the cladogram are presented.
    d. the species in a clade share homologous structures.
    e. All of these are correct.
16. In evolutionary systematics, birds are assigned to a different group from reptiles because
    a. they evolved from reptiles and couldn't be a monophyletic taxon.
    b. they are adapted to a different ecological niche compared to reptiles.
    c. feathers came from scales, and feet came before wings.
    d. all classes of vertebrates are only related by way of a common ancestor.
    e. All of these are correct.
17. **THINKING CONCEPTUALLY** DNA differences are expected to be consistent with evolutionary trees based on structure. Explain.

The Three-Domain System Is Widely Accepted

18. The three-domain classification system has recently been developed based on
    a. mitochondrial biochemistry and plasma membrane structure.
    b. cellular structure and rRNA sequence data.
    c. plasma membrane and cell wall structure.
    d. nuclear and mitochondrial biochemistry.
19. Which of these are domains? Choose more than one answer if correct.
    a. Bacteria   d. animals
    b. Archaea   e. plants
    c. Eukarya
20. Which of these are eukaryotes? Choose more than one answer if correct.
    a. Bacteria   d. animals
    b. Archaea   e. plants
    c. Eukarya
21. Which of these is a true statement?
    a. Eukaryotes are more closely related to bacteria than they are to archaea.
    b. Fungi, animals, and plants have different means of acquiring nutrients.
    c. The close relationship between bacteria and archaea places them in their own doman.
    d. Arachaea evolved during the Precambrian, but bacteria evolved during the Cambrian period.
    e. All of these are correct.
22. **THINKING CONCEPTUALLY** The adoption of the three-domain system emphasizes the increased use of genetic similarities and differences to classify organisms. Explain.

Understanding the Terms

analogous structure   292
analogy   292
ancestral character   291
character   290
clade   294
cladogram   294
class   290
classification   290
common ancestor   291
convergent evolution   292
derived character   291
domain   290
evolutionary systematics   294
extinction   284
family   290
genus   290
homologous structure   292

homology   292
kingdom   290
mass extinction   284
meteorite   288
molecular clock   293
order   290
parallel evolution   292
phylogenetic (evolutionary) tree   291
phylogenetic cladistics   294
phylogeny   291
phylum   290
species   290
systematics   291
taxon   290
taxonomy   290

Match the terms to these definitions:
    a. _____ Branch of biology concerned with identifying, describing, and naming organisms.
    b. _____ Diagram that indicates common ancestors and lines of descent.
    c. _____ Group of organisms that fills a particular classification category.
    d. _____ Concept that the rate at which mutational changes accumulate in certain types of genes is constant over time.
    e. _____ Group consisting of an ancestral species and all of its descendants, forming a distinct branch on a phylogenetic tree.

Thinking Scientifically

1. You were asked to supply an evolutionary tree of life and decided to use Figure 15.9. How is this tree consistent with evolutionary principles?
2. Explain the occurrence of living fossils, such as horseshoe crabs, that closely resemble their ancestors appearing in the fossil record.

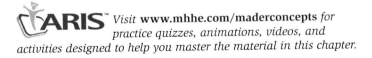 *Visit* **www.mhhe.com/maderconcepts** *for practice quizzes, animations, videos, and activities designed to help you master the material in this chapter.*

16

Evolution of Microbial Life

LEARNING OUTCOMES

After studying this chapter, you should be able to accomplish the following outcomes.

At Your Service: Viruses and Bacteria

1 Give examples to show that viruses and bacteria are useful to human beings.

Viruses Reproduce in Living Cells

2 Describe the structure of a virus and how a virus reproduces inside a bacterium.
3 Discuss the importance of viral infections in plants.
4 Describe how a virus, with a DNA genome, and HIV, with an RNA genome, invade and reproduce inside animal cells.
5 Define and give examples of emerging viral diseases, and account for their prevalence today.

The First Cells Originated on Early Earth

6 Outline two hypotheses for the origin of small organic molecules.
7 Discuss why RNA, instead of DNA or protein, may have been the first macromolecule.
8 Describe a manner by which the protocell may have evolved.

Both Bacteria and Archaea Are Prokaryotes

9 Explain the basis for the classification of prokaryotes.
10 Describe the structure of prokaryotes, including possible shapes and types of envelopes and appendages.
11 Discuss microbes as possible biological weapons.
12 Describe the reproductive strategy of prokaryotes, including the three mechanisms for genetic recombination.
13 Characterize the metabolic diversity of prokaryotes in terms of their need for oxygen and their means of acquiring food.
14 Describe how the structures of bacteria and archaea differ, and name the types of archaea.
15 Discuss the environmental and medical importance of prokaryotes.

Viruses are noncellular entities responsible for a number of diseases in plants, animals, and humans. In humans, for example, polio, smallpox, cervical cancer, and AIDS are all caused by viruses. Even so, viruses are useful tools in the biotechnology laboratory. Remember that Hershey and Chase used a T2 virus to support the hypothesis that DNA is the genetic material. The virus they used can infect bacteria and has the structure shown below. This virus is complex and contains both a so-called head and a tail. Today, recombinant viruses are used to store the genes of organisms. When a researcher wants to work with a particular gene, she or he selects a virus containing that gene, much as you would go to the library and choose a book containing an item of interest. Gene therapy also uses viruses to carry normal genes into the genomes of people with genetic disorders.

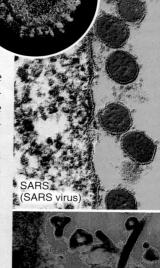

SARS
(SARS virus)

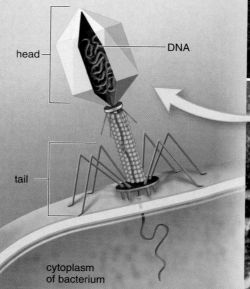

head

DNA

tail

cytoplasm
of bacterium

Ebola virus

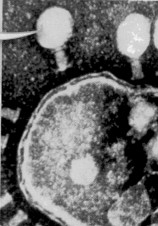

At Your Service: *Viruses and Bacteria*

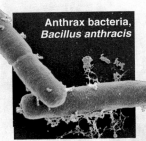

Anthrax bacteria, *Bacillus anthracis*

Bacteria and archaea are prokaryotes, organisms that are simple in structure but metabolically diverse. They can live under conditions that are too hot, too salty, too acidic, or too cold for eukaryotes. Through their ability to oxidize sulfides that spew forth from deep-sea vents and subsequently produce nutrients, they support communities of organisms in habitats where the sun never shines. Bacteria, but not archaea, also cause diseases, such as chlamydia, strep throat, food poisoning, and anthrax.

How are bacteria of service to humankind? Both on land and in the sea, bacteria are decomposers that digest dead plants and animals, generate oxygen, and recycle nutrients. Without the work of decomposers, life would soon come to a halt! Plants are unable to fix atmospheric nitrogen, but they need a source of nitrogen in order to produce proteins. Bacteria in the soil and those that live in nodules on plant roots reduce atmospheric nitrogen to a form that plants can use. Plants are also largely the source of amino acids that allow animals, including humans, to produce their own proteins.

Nodules where bacteria fix nitrogen

Humans use bacteria in the environment to mine minerals and degrade sewage. In addition, their vast ability to break down almost any substance has been applied to bioremediation, the biological cleanup of harmful chemicals called pollutants. Bacteria have been used to clean up oil spills,

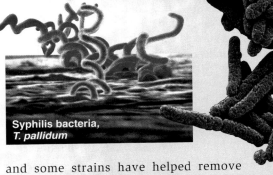

Syphilis bacteria, *T. pallidum*

and some strains have helped remove Agent Orange, a potent herbicide, from soil samples. Dual cultures of two types of bacteria have been shown to degrade PCBs, chemicals formerly used as coolants and lubricants.

E. coli
10,500×

While some bacteria cause deadly diseases, others produce antibiotics, such as streptomycin, that can help cure such illnesses. Due to the ease with which they can be grown and manipulated in the laboratory, bacteria are used to study basic life processes. The details of how DNA specifies the order of amino acids in a protein were first worked out by studying the process in *E. coli*. Through genetic engineering, bacteria now produce many commercial products, including insulin for diabetics. Such insulin is not allergenic because it is coded for by a human gene introduced into the bacterium. All in all, it is safe to say that we could not live without the services of bacteria.

This chapter focuses on the evolution of microbes, so-called because they are too small to be seen without the aid of a microscope. Archaea and bacteria are cellular microbes, while the viruses are noncellular. Still, viruses have a major impact on our lives, and we begin this chapter by describing their structure and mode of reproduction.

Sewage treatment plant

before   after

Bioremediation of an oil spill

Viruses are noncellular, and they reproduce only inside living cells. The life cycle of a bacteriophage describes, in general, the life cycle of a virus in any type of cell, including an animal cell. Of special concern are viruses that cause diseases in plants and animals, including humans.

16.1 Viruses have a simple structure

The size of a virus is comparable to that of a large protein macromolecule, ranging from 0.2 to 2 μm. Therefore, viruses are best studied through electron microscopy.

Many viruses can be purified and crystallized, and the crystals may be stored just as chemicals are stored. Still, viral crystals become infectious when the viral particles they contain are given the opportunity to invade a host cell. Viruses have a DNA or RNA genome, but they can reproduce only by using the metabolic machinery of a host cell. Viruses are a biological enigma. They are noncellular, and therefore they do not fit into current classification systems, which are devoted to categorizing the cellular organisms on Earth.

The following diagram summarizes viral structure:

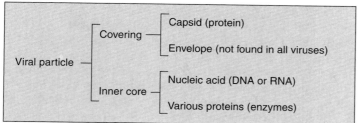

All viruses possess the same basic anatomy—an outer **capsid**, which is composed of protein, and an inner core of nucleic acid (DNA or RNA). A viral genome has as few as three and as many as 100 genes; a human cell contains 25,000 genes. The covering of a virus contains the capsid, which may be surrounded by an outer membranous envelope; if not, the virus is said to be naked. Naked viruses can be transmitted by contact with inanimate objects, such as desktops.

Figure 16.1A gives an example of a naked virus, while **Figure 16.1B** is an example of an enveloped virus. The envelope is actually a piece of the host's plasma membrane that also contains viral glycoprotein spikes. Enveloped viruses are usually transmitted by direct contact with an infected individual. Aside from its genome, a viral particle may also contain various proteins, especially enzymes such as the polymerases, which are needed to produce viral DNA and/or RNA.

Viruses are categorized by (1) their type of nucleic acid, which can be DNA or RNA, and whether it is single-stranded or double-stranded; (2) their size and shape (the capsid can have projecting fibers); and (3) the presence or absence of an outer envelope.

In Section 16.2, we examine the two possible life cycles of a bacteriophage.

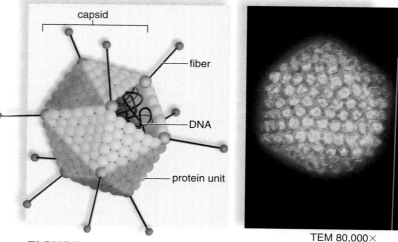

TEM 80,000×

FIGURE 16.1A Adenovirus, a naked virus, with a polyhedral capsid and a fiber at each corner.

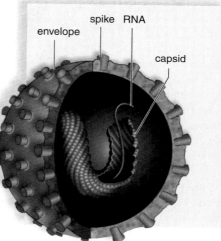

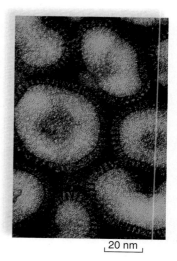

20 nm

FIGURE 16.1B Influenza virus, surrounded by an envelope with spikes.

16.1 Check Your Progress What part of a virus can be used to hold another organism's genes?

16.2 Some viruses reproduce inside bacteria

All sorts of cells, whether prokaryotic or eukaryotic, are susceptible to a viral infection. Viruses are specific. Specificity extends even to the type of cell infected by the virus. For example, tobacco mosaic virus infects only tobacco leaves, and adenoviruses attach to cells in our respiratory tract, causing colds. The proteins of the capsid determine the specificity of a virus because a spike must first combine with a particular protein in the plasma membrane of the host cell, and thereafter the virus enters the cell.

Bacteriophages, or simply phages, are viruses that parasitize bacteria. There are two types of bacteriophage life cycles, termed the lytic cycle and the lysogenic cycle (**Fig. 16.2**). Most types of bacteriophages have the lytic cycle, and only a few types are lysogenic. In the **lytic cycle**, viral reproduction occurs, and the host cell undergoes lysis, by which the cell breaks open to release the viral particles. In the **lysogenic cycle**, viral reproduction does not immediately occur, but reproduction may take place sometime in the future.

Lytic Cycle Figure 16.2 shows that the lytic cycle may be divided into five stages: attachment, penetration, biosynthesis, maturation, and release. ❶ During *attachment*, portions of the capsid combine with a receptor on the rigid bacterial cell wall in a lock-and-key manner. ❷ₐ During *penetration*, a viral enzyme digests away part of the cell wall, and viral DNA is injected into the bacterial cell. ❸ *Biosynthesis* of viral components begins after the virus brings about inactivation of host genes not necessary to viral replication. The virus takes over the machinery of the cell in order to carry out viral DNA replication and produce multiple copies of the capsid protein. ❹ During *maturation*, viral DNA and capsids assemble to produce several hundred

viral particles. Lysozyme, an enzyme coded for by a viral gene, is produced; this disrupts the cell wall, and ❺ the *release* of new viruses occurs. The bacterial cell dies as a result.

Lysogenic Cycle With the lysogenic cycle, the infected bacterium does not immediately produce phages but may do so sometime in the future. In the meantime, the phage is *latent*—not actively replicating. ❷ᵦ Following attachment and penetration, *integration* occurs. Viral DNA becomes incorporated into bacterial DNA with no destruction of host DNA. While latent, the viral DNA is called a *prophage*. The prophage is replicated along with the host DNA, and all subsequent cells, called lysogenic cells, carry a copy of the prophage. Certain environmental factors, such as ultraviolet radiation, can induce the prophage to enter the lytic stage of biosynthesis, followed by maturation and release.

In Section 16.3, we discuss viral diseases of plants, including those that reduce the yield of agricultural and horticultural crops.

16.2 *Check Your Progress* If you were a physician using a virus for the purpose of gene therapy, would you want the virus to undergo the lytic cycle or the lysogenic cycle?

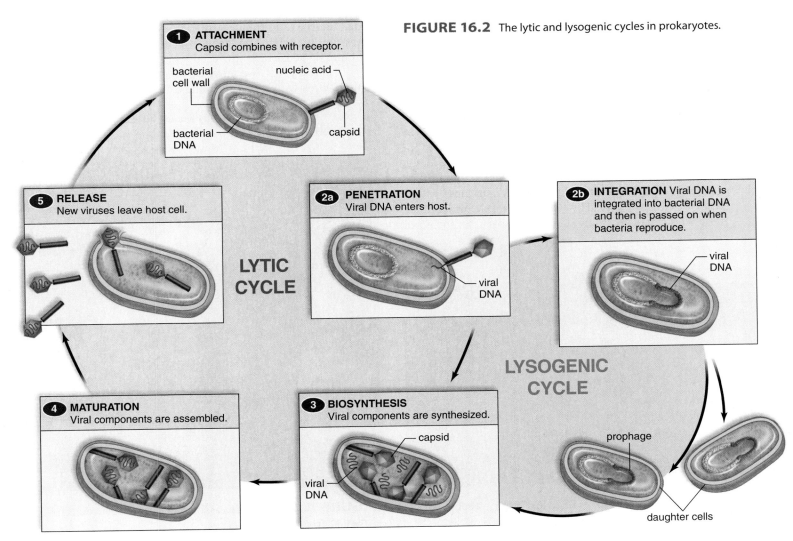

FIGURE 16.2 The lytic and lysogenic cycles in prokaryotes.

❶ ATTACHMENT
Capsid combines with receptor.

bacterial cell wall
nucleic acid
bacterial DNA
capsid

❺ RELEASE
New viruses leave host cell.

❷ₐ PENETRATION
Viral DNA enters host.

viral DNA

❷ᵦ INTEGRATION Viral DNA is integrated into bacterial DNA and then is passed on when bacteria reproduce.

viral DNA

LYTIC CYCLE

LYSOGENIC CYCLE

❹ MATURATION
Viral components are assembled.

❸ BIOSYNTHESIS
Viral components are synthesized.

capsid
viral DNA

prophage

daughter cells

Approximately 2,000 kinds of plant diseases have been attributed to viruses. The International Committee on Taxonomy of Viruses has identified and classified approximately 400 plant viruses. Three hundred more identified viruses are awaiting classification. Plant viruses are responsible for the loss of over 15 billion dollars annually by reducing the yield of important agricultural and horticultural crops. Tobacco mosaic virus (TMV) can infect a number of plants, including orchid, potato, tobacco, and tomato plants. The virus can remain viable for years on dried plant debris and is extremely tolerant of very high temperatures. TMV enters plants through wounds sustained in transplanting or pruning. It spreads rapidly once it is in the host. The virus interferes with chlorophyll production, and the infected plant develops unsightly light green, yellow, or white spots on its leaves and fruits (**Fig. 16.3A**). Tobacco products are the most common source of infection, and smokers can pass on the virus to plants by handling them. In fact, at one time, smokers were not allowed to work in ketchup factories because their touch could ruin tomatoes. Ironically, much of the early work with viruses involved TMV. In 1898, Martinus Beijerinck, working with what is now called TMV, determined that the disease was not caused by a bacterium but by a virus. The structure of viruses was discovered in the 1930s when Wendell Stanley studied TMV.

Plant viruses do not differ significantly in size and shape from bacteriophages and animal viruses. With the exception of three groups of DNA viruses, all of the other plant viruses are RNA viruses. The generalized symptoms of plants infected with a virus include: stunted growth; discoloration of leaves, flowers, and fruits; death of stems, leaves, and fruits; irregularities in fruit size; premature ripening of fruits; reduced sugar content of fruits; tumors; and leaf roll. Viruses seldom kill their plant hosts, but they weaken them, making them susceptible to opportunistic infections such as those caused by bacteria and fungi.

Plant viruses can be spread by a variety of mechanisms. Since the exterior surfaces of plants are protected by bark or cuticle and individual plant cells are protected by a cell wall, plant viruses have developed a number of means of infecting plants. Some plant viruses are transmitted by contaminated soil, pollen, seeds, and tubers. Plants that have fallen victim to wind damage and injury of any kind are also more susceptible to plant viruses. Insects spread many plant viruses, either by moving from plant to plant or by feeding. Sucking insects, such as leafhoppers and aphids, are responsible for transmitting the majority of plant viruses. Other organisms, including nematodes (roundworms) and parasitic plants, such as dodder, can also spread plant viruses.

Once a plant is infected with a virus, it can move from cell to cell through the plasmodesmata (cytoplasmic connections between adjacent cells). There is no cure for infected plants, but removing infected leaves and tree limbs may help. Scientists are presently developing varieties of plants that are resistant to viral diseases. In addition, controlling insect vectors can slow the spread of plant viruses.

In some instances, plants have been purposefully infected with a virus in order to produce traits considered desirable by gardeners. For example, some variegation in leaves and flowers can be brought about by viruses, as occurs in Rembrandt tulips (**Fig. 16.3B**). Unfortunately, the virus weakens the plant, and it does not live long. New genetic techniques using transposons, or jumping genes, are now used to attain desired streaking.

In Section 16.4, we move on to examine the life cycle of an animal virus with a DNA genome.

16.3 *Check Your Progress* **Why would you expect plant viruses to only infect plants?**

FIGURE 16.3B A virus is responsible for the variegation and streaking in Rembrandt tulips.

FIGURE 16.3A The tobacco mosaic virus (TMV) is responsible for discoloration in the leaves of tobacco plants.

16.4 Viruses reproduce inside animal cells and cause diseases

Viruses are always in the news: An oceanic cruise is preempted due to a mysterious virus; bioterrorists threaten to unleash a virus on an unsuspecting population; a hog cholera virus jeopardizes the pork industry; and viruses such as West Nile virus and avian flu virus are threatening the United States. Viral diseases have plagued humans and animals for thousands of years, causing much economic loss, suffering, and death. Although different in shape, host range, and genetic composition (DNA or RNA), viruses that invade human and animal cells use reproductive strategies similar to those of bacteriophages. As illustrated in **Figure 16.4**, replication of an animal virus with a DNA genome involves these steps:

1 *Attachment.* Glycoprotein spikes projecting through the envelope allow the virus to bind only to host cells having specific receptor surface proteins.

2 *Penetration.* After the viral particle is brought into the host cell, uncoating—the removal of the viral capsid—follows, and viral DNA is released into the host.

Biosynthesis. **3a** The capsid and other proteins are synthesized by host cell ribosomes according to viral DNA instructions. **3b** During viral replication, the virus instructs the host cell's enzymes to make many copies of the viral DNA.

4 *Maturation.* Viral proteins and DNA replicates are assembled to form new viral particles.

5 *Release.* In an enveloped virus, budding occurs, and the virus develops its envelope, which usually consists of the host's plasma membrane components and glycoprotein spikes that were coded for by the viral DNA.

Viruses are responsible for a number of diseases in animals and humans. Parvovirus causes severe gastrointestinal problems and perhaps death in dogs. Rabies affects many species of mammals, including humans. Viruses are responsible for several childhood maladies, including measles, mumps, chickenpox, warts, and viral pinkeye. More serious human viral diseases include polio, yellow fever, dengue fever, type 1 herpes (fever blisters), type 2 herpes (genital herpes), shingles, mononucleosis, hepatitis, HIV, smallpox, rubella, and various forms of flu and colds. Recently, several emerging viruses have captured the public's attention, including Ebola virus, hantavirus, and Lassa virus (see Section 16.6).

In recent years, viruses have been implicated in several forms of cancer in humans as well as other animals. These viruses, which are known as the tumor viruses, perhaps contribute to at least 15% of all human cancers worldwide. The virus that causes chronic hepatitis B is associated with a specific form of liver cancer. The Epstein-Barr virus is associated with Burkitt lymphoma, a malignant tumor of the jaw found in children in Central and West Africa. Human papillomavirus, which is responsible for genital warts, has been associated with cervical cancer. Two retroviruses have been associated with adult T-cell leukemia and hairy-cell leukemia. In the future, other viruses associated with cancer may be identified.

Viruses will continue to be in the headlines for many years. It is hoped that many of those headlines will address our better understanding of viruses and the great news that cures for dreaded viruses are at hand.

Next, in Section 16.5, we examine the life cycle of HIV, an animal retrovirus with an RNA genome.

> **16.4** *Check Your Progress* **Why is a researcher more apt to work with a bacteriophage than with a virus that infects human cells?**

FIGURE 16.4 Replication of an animal virus.

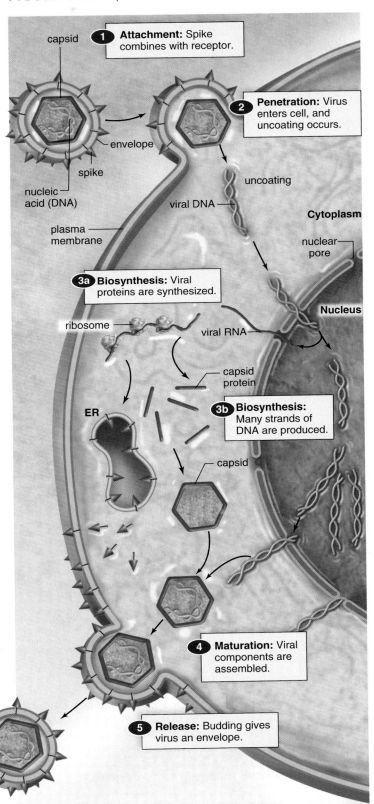

Like many other animal viruses, HIV (human immunodeficiency virus) has an envelope with spikes, a capsid, and a nucleic acid genome. The genome for an HIV virus consists of RNA, instead of DNA. In addition, HIV is a **retrovirus**, meaning that it uses reverse transcription from RNA into DNA in order to insert a complementary copy of its genome into the host's genome.

The events that occur in the reproductive cycle of an HIV virus are essentially the same as those for a DNA virus, but slightly different steps are needed because HIV is a retrovirus (**Fig. 16.5**):

1 *Attachment.* During attachment, the HIV virus binds to the plasma membrane. HIV has an envelope marker, and this marker allows the virus to bind to a receptor in the host-cell plasma membrane.

2 *Penetration.* After attachment, the HIV virus fuses with the plasma membrane, and the virus enters the cell. A process called uncoating removes the capsid, and RNA is released.

3 *Reverse transcription.* This event in the reproductive cycle is unique to retroviruses. The enzyme called reverse transcriptase makes a DNA copy of the retrovirus's RNA genetic material. Usually, in cells DNA is transcribed into RNA. Retroviruses can do the opposite only because they have this unique enzyme, from which they take their name. (*Retro* in Latin means reverse.)

4 *Integration.* The viral enzyme integrase now splices viral DNA into a host chromosome. The term **HIV provirus** refers to viral DNA integrated into host DNA. HIV is usually transmitted to another person by means of cells that contain proviruses. Also, proviruses serve as a latent reservoir for HIV during drug treatment. Even if drug therapy results in an undetectable viral load, investigators know that there are still proviruses inside infected lymphocytes.

5 *Biosynthesis.* When the provirus is activated, perhaps by a new and different infection, the normal cell machinery directs replication—the production of more viral RNA. The viral RNA brings about the synthesis of very long polypeptides. These polypeptides have to be cut into smaller pieces. This cutting process, called cleavage, depends on a third HIV enzyme, called protease.

6 *Maturation.* Capsid proteins, viral enzymes, and RNA can now be assembled to form new viral particles.

7 *Release.* During budding, the virus gets its envelope and envelope marker coded for by the viral genetic material.

HIV is an emerging viral disease in humans, as explained in Section 16.6.

> **16.5 Check Your Progress** At which of the steps of retroviral replication might a researcher try to find or create a medicine to prevent an HIV infection from succeeding?

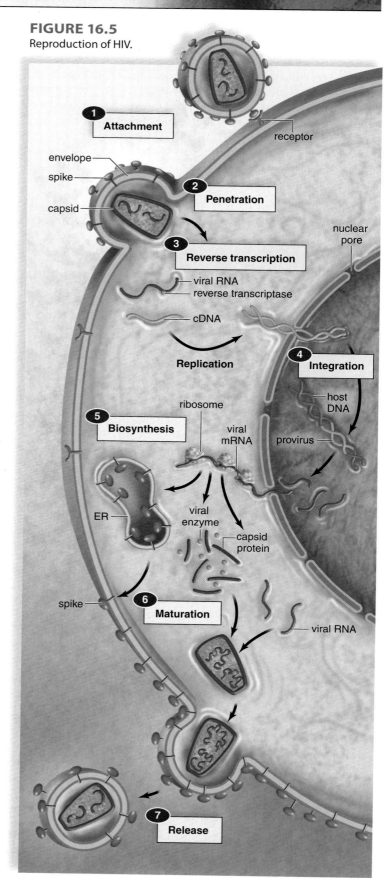

FIGURE 16.5
Reproduction of HIV.

16.6 Humans suffer from emerging viral diseases

Emergent diseases are ones that we newly recognize as being infectious. Sometimes the disease existed before but has begun to spread widely. The modern age of travel makes it convenient for a person to wake up in Bangkok in the morning and sleep in Los Angeles the same night. It also provides pathogens with unprecedented mobility. No longer are outbreaks limited to a small geographic region. The Severe Acute Respiratory Syndrome (SARS) virus causes high fever, body aches, and pneumonia. In 2003, its path of death was easily traced from Southeast Asia to Toronto, Canada. Its apparent mode of transmission is by droplet infection and direct contact, so some people protected themselves from SARS by wearing surgical masks (**Fig. 16.6A**).

In recent years, West Nile virus, which causes fever, headache, swollen lymph nodes, muscle weakness, disorientation, and possibly death, has emerged in Europe and the United States. It was originally described in Africa in 1937. The virus is easily transported by a mosquito vector to unsuspecting bird, horse, and human victims. Presently, there is much concern that avian influenza (or bird flu) might emerge. This disease arose in Southeast Asia, where markets are crowded with humans and animals, particularly domesticated chickens. There are several subtypes of the bird flu, some of which are devastating to birds and might be able to jump to humans. Human symptoms include cough, sore throat, muscle aches, eye infections, respiratory distress, and life-threatening complications. Scientists are cautioning that bird flu can reach pandemic proportions, but so far, the disease does not often spread from chickens to humans, nor is it efficiently transmitted among humans. Nevertheless, in some areas, chickens have been exterminated as a precaution (**Fig. 16.6B**).

Ebola is one of a number of viruses that can cause hemorrhagic fever. These extremely diabolical pathogens are highly contagious and can quickly cause intolerable fever and extensive tissue damage, leading to profuse internal bleeding, multiple organ failure, and certain death. The vector and reservoir for Ebola virus are unknown. There are three strains of Ebola virus, named for the regions where they were discovered (Zaire, Sudan, and Reston). The Zaire and Sudan strains were found in Africa, but the Reston strain was discovered in a 1989 shipment of 100 crab-eating monkeys, imported from the Philippines to Reston, Virginia. The movie thriller *Outbreak* was based on a book, *The Hot Zone*, whose story line concerned these monkeys.

FIGURE 16.6A Surgical masks provide protection against the transmission of SARS.

FIGURE 16.6B Exterminating possibly infected chickens may protect against bird flu.

Viruses are constantly in a state of evolutionary flux. A new pathogen can emerge through the acquisition of new surface antigens. The unsuspecting immune system, unable to recognize the new "signature" of the virus, does not mount a defense in time to defeat the pathogen. Thus, the virus successfully completes its life cycle, causes a disease, and spreads to other victims. When a virus jumps from animals to humans, as with bird flu and Ebola, it has changed its signature and can now infect humans. HIV infections began when the virus jumped from a primate to a human. HIV continually changes its signature, making it far more difficult to eradicate. By the time antibodies and vaccines are developed to combat the virus, it has again changed to an unrecognizable form.

Some viruses can easily move from animals to humans. Rabies can be spread by the bite of a rabid animal, such as a skunk, raccoon, bat, cat, dog, or even cow. In other instances, animal populations can serve as a reservoir for a disease that affects humans. For example, the deer mouse is the suspected reservoir for hantavirus, the cause of a severe respiratory disorder in humans. Viral diseases that have animal reservoirs are harder to control. Smallpox was able to be eradicated because this virus infects only humans.

Many viral diseases are transmitted by vectors, usually insects that carry pathogens from an infected individual or reservoir to a healthy individual. Mosquitoes serve as a common vector for several viral diseases, including St. Louis encephalitis, equine encephalitis, West Nile virus, and yellow fever. Many times, extensive efforts are required to control and eradicate the vector. If a virus originally transmitted by a rare species of mosquito becomes able to be transmitted by a more common species, the disease can emerge.

This completes our discussion of viruses, and in the next part of the chapter, we consider how the first cell, or cells, may have evolved.

16.6 Check Your Progress Explain the expression "emerging diseases." Give examples.

The first organic molecules could have originated in the atmosphere, or perhaps at hydrothermal vents. The discovery of RNA enzymes (ribozymes) suggests that RNA could have formed the first genes. A protocell (before the first true cell) surrounded by a plasma membrane could store genetic information and metabolize.

16.7 Experiments show how small organic molecules may have first formed

Organic molecules are necessary to the structure and function of cells. Two different hypotheses have been developed to explain how organic molecules could have formed from inorganic molecules. These hypotheses, called the prebiotic soup hypothesis and the iron-sulfur world hypothesis, have both been supported by experimental evidence.

Prebiotic Soup Hypothesis Early Earth had an atmosphere, but it was not the same as today's atmosphere. When the Earth formed, intense heat, produced by gravitational energy and radioactivity, resulted in several stratified layers. Heavier atoms of iron and nickel became the molten liquid core, and dense silicate minerals became the semiliquid mantle. Massive volcanic eruptions produced the first crust and the first atmosphere.

The early atmosphere most likely consisted mainly of these inorganic chemicals: water vapor (H_2O), nitrogen (N_2), and carbon dioxide (CO_2), with only small amounts of hydrogen (H_2), methane (CH_4), hydrogen sulfide (H_2S), and carbon monoxide (CO). Notice that this atmosphere contains no oxygen and, therefore, is a reducing atmosphere. This would have been fortuitous because oxygen (O_2) attaches to organic molecules, preventing them from joining to form larger molecules. In support of the hypothesis that small inorganic molecules such as these could have produced the first organic molecules, Stanley Miller and Harold Urey performed the ingenious experiment diagrammed in **Figure 16.7A**. ❶ They placed a mixture resembling a strongly reducing atmosphere—methane (CH_4), ammonia (NH_3), hydrogen (H_2), and water (H_2O)—in a closed system, and ❷ circu-

lated it past an electric spark (simulating lightning). ❸ After condensing the gases to a liquid, ❹ they heated it. ❺ After a week's run, Miller and Urey discovered that a variety of amino acids and organic acids had been produced. Since that time, other investigators have achieved similar results by using other, less-reducing combinations of gases dissolved in water.

These experiments lend support to the hypothesis that the Earth's first atmospheric gases could have reacted with one another to produce small organic molecules. Energy would have been required, but early Earth had abundant sources of energy in the form of lightning, volcanic activity, and intense radiation from a sun that had just formed. We know that there would have been plenty of time for synthesis to occur because the Earth is some 4.6 billion years old and the first cells are about 3.6 billion years old. Neither oxidation (there was no free oxygen) nor decay (there were no bacteria) would have destroyed the first molecules, and rainfall would have washed them into the ocean, where they accumulated for hundreds of millions of years. Therefore, the oceans would have been a thick, warm, prebiotic soup.

Iron-Sulfur World Hypothesis Other investigators are concerned that Miller and Urey used ammonia as one of the atmospheric gases. They point out that, whereas inert nitrogen gas (N_2) would have been abundant in the primitive atmosphere, ammonia (NH_3) would have been scarce. Where might NH_3 have been abundant? A team of researchers at the Carnegie Institution in Washington, D.C., believe they have found the answer: in hydrothermal vents on the ocean floor. These vents line the huge oceanic ridges, where molten magma wells up and adds material to the ocean floor (**Fig. 16.7B**). Cool water seeping through the vents is heated to a temperature as high as 350°C, and when it spews back out, it contains various mixed iron and nickel sulfides that can act as catalysts to change N_2 to NH_3. Even today, these conditions produce nutrient molecules for

FIGURE 16.7B Chemical evolution at hydrothermal vents.

plume of hot water rich in iron sulfides

hydrothermal vent

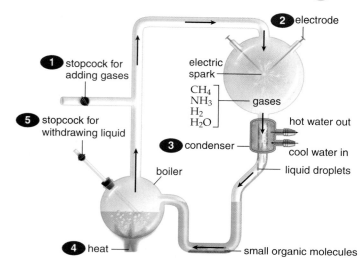

❷ electrode

❶ stopcock for adding gases

electric spark

CH_4
NH_3
H_2
H_2O — gases

❺ stopcock for withdrawing liquid

hot water out

❸ condenser

cool water in

liquid droplets

boiler

❹ heat

small organic molecules

FIGURE 16.7A Laboratory re-creation of chemical evolution in the atmosphere.

microorganisms that support a diverse community of other organisms, including huge clams and tube worms living in the vicinity of the oceanic ridges (see Fig. 16.13B).

A laboratory test of the iron-sulfur hypothesis worked perfectly. Under ventlike conditions, 70% of various N_2 sources were converted to NH_3 within 15 minutes. Two German organic chemists, Gunter Wachtershaüser and Claudia Huber, have gone one more step. They have shown that organic molecules will react and amino acids will form peptides in the presence of iron and nickel sulfides under ventlike conditions.

Aside from amino acids necessary for the production of proteins, the origin of cells is also dependent upon the presence of a genetic material, as discussed in Section 16.8.

16.7 Check Your Progress Many reactions occur in bacteria without the need for sources of energy such as heat, lightning, or radiation. Why did the first synthesis of organic molecules require intense energy?

16.8 RNA may have been the first macromolecule

Most scenarios for the origin of life recognize two stages: chemical evolution and biological evolution (**Fig. 16.8**). During chemical evolution, organic monomers arise from inorganic compounds, and polymers arise when monomers join together. Biological evolution begins when a plasma membrane surrounds the polymers, producing a **protocell**. A true cell has arisen when the cell reproduces in the same manner as today's cells.

RNA-First Hypothesis For several decades, scientists have been studying which of the three macromolecules—RNA, DNA, or protein—led to the origin of the first protocell. Certainly, amino acids were available to form proteins, and the necessary molecules to form nucleotides were also present in the prebiotic soup. Nucleotides could have then polymerized to form nucleic acids.

The possibility that proteins alone led to the first protocell is rejected by some because, as you know, proteins do not store genetic information. The genetic material must be able to (1) store genetic information and (2) replicate in order to transmit the genetic information to daughter cells. Scientists eventually discovered that RNA, not DNA, could have performed both functions by itself. RNA stores genetic information within the sequence of its bases when it participates in the formation of proteins. During the past two decades, scientists have been able to show that RNA can act as an enzyme, and therefore, could have replicated on its own.

Thomas Cech and Sidney Altman shared a Nobel Prize in 1989 because they discovered that RNA can be both a substrate and an enzyme. They had observed RNA acting as a ribozyme (*ribo* from ribonucleic acid and *zyme* from enzyme) during RNA processing. Ribozymes also join amino acids during protein synthesis.

It would also be absolutely critical that the first macromolecule have various enzymatic functions. The researchers at the Whitehead Institute for Biomedical Research have discovered that ribozymes are extremely versatile and can perform almost any metabolic function, including replication of RNA. They determined this by placing about 1,000 ribozymes in a test tube and selecting the ones that perform an enzymatic function the best. After selecting only the best for a particular function multiple times, the final group of ribozymes is much more efficient than the original group.

These findings have led these researchers to say that it was an "RNA world" some 4 BYA and that RNA chains were the first form of life! In this world, RNA molecules competed with each other for free nucleotides and were subject to natural selection. Only the most efficient RNA chains survived and became the genetic material for the protocell.

Now, as discussed in Section 16.9, we are ready for the origin of protocells, cells that would have preceded the first true cells.

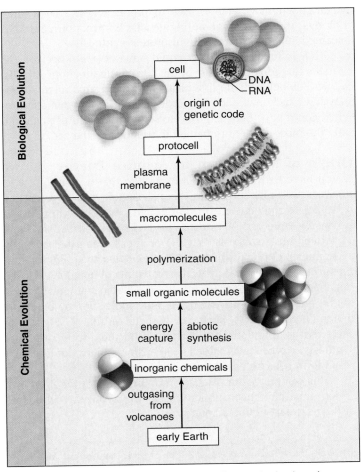

FIGURE 16.8 The origin of the first cell(s) can be broken down into these steps.

16.8 Check Your Progress *a.* What function can RNA perform that a protein cannot perform in cells? *b.* What function can RNA perform that DNA cannot perform in cells?

The next step in the origin of the first cell would be the formation of a protocell. Sidney Fox of the University of Miami suggests that once amino acids were present in the oceans, they collected in shallow puddles along the rocky shore. There, the heat of the sun could have caused them to congregate and become proteinoids. When Fox simulates the formation of proteinoids in the lab and returns them to water, they form microspheres, structures with many interesting properties: Microspheres resemble bacteria, they divide, and perhaps they are subject to selection. Although Fox believes they are a type of cell, others disagree.

Origin of Plasma Membrane First and foremost, the protocell would have had an outer membrane. The plasma membrane of today's cells separates the living interior from the nonliving exterior. There are two hypotheses about the origin of the first plasma membrane. The first hypothesis was suggested by Sidney Fox, who showed that if lipids are made available to microspheres, which after all are protein, they acquire a lipidprotein outer membrane (**Fig. 16.9A**).

The second hypothesis was formed in the early 1960s by the biophysicist Alec Bangham of the Animal Physiology Institute in Cambridge, England. He discovered that when he extracted lipids from egg yolks and placed them in water, the lipids naturally organized themselves into double-layered bubbles, roughly the size of a cell. Bangham's bubbles soon became known as **liposomes** (**Fig. 16.9B**). Later, Bangham, along with biophysicist David Deamer of the University of California, realized that liposomes might have provided life's first boundary. Perhaps liposomes with a phospholipid membrane engulfed early RNA molecules that had enzymatic abilities. The liposomes would have protected the molecules from their surroundings and concentrated them so they could react (and evolve) quickly and efficiently.

Origin of DNA Information System A protocell became a cell when it contained a DNA information system:

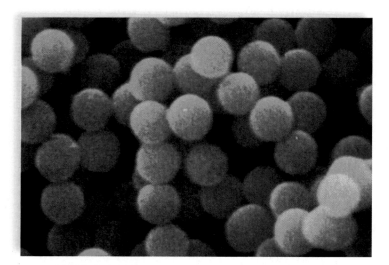

FIGURE 16.9A Microspheres, which are made of protein, could have acquired an outer lipid-protein membrane during the origin of the first cell.

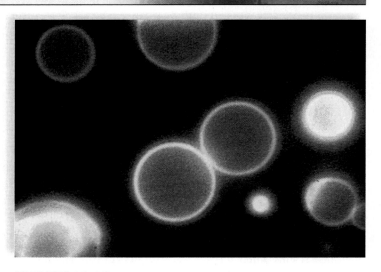

FIGURE 16.9B Liposomes, which are composed of lipids, have a double-layered outer membrane.

DNA ⟶ RNA ⟶ protein. How did it arise? Those who support an RNA world point out that ribozymes could have formed the first DNA molecules. To make DNA, a ribozyme could have acted in the same manner as the enzyme reverse transcriptase, which functions in retroviruses to produce DNA. Then DNA took over the function of storing genetic information, and RNA became its helper to bring about protein synthesis.

RNA is unique in that it could have also synthesized the proteins that took over most of the enzymatic functions in cells. Sidney Fox believes the first proteins making up a microsphere may have had enzymatic properties. These first enzymatic proteins would have been exposed to selective pressures such that only the most efficient remained to function in the cell.

Origin of Metabolism to Acquire Energy The cell would have had to carry on nutrition so that it could grow. If organic molecules formed in the atmosphere and were carried by rain into the ocean, nutrition would have been no problem because simple organic molecules could have served as food. This hypothesis suggests that the protocell was a heterotroph, an organism that takes in preformed food. On the other hand, if the protocell evolved at hydrothermal vents, it may have carried out chemosynthesis (see Section 16.7). Chemosynthetic bacteria obtain energy for synthesizing organic molecules by oxidizing inorganic compounds, such as hydrogen sulfide (H_2S), a molecule that is abundant at the vents. With the advent of photosynthesis, oxygen was added to the atmosphere, and cellular respiration became possible.

This completes our discussion of the origin of the first cells, which must have been prokaryotes. Prokaryotes are discussed in the next part of the chapter.

16.9 Check Your Progress *a.* Which organic components of a protocell would have allowed it to grow? *b.* Which components would have allowed it to reproduce?

Prokaryotes lack a true nucleus and other membranous organelles. They all reproduce by binary fission but have varied means of nutrition, including photosynthesis. The cyanobacteria release oxygen into the atmosphere, and the archaea are well-known for living in extreme environments. Prokaryotes are extremely plentiful and play an important role in the environment, in addition to causing human diseases.

16.10 Prokaryotes have particular structural features

Prokaryotes are unicellular organisms that generally range in size from 1 to 10 μm in length and from 0.7 to 1.5 μm in width. The term prokaryote means "before a nucleus," and reflects the observation that these organisms lack a eukaryotic nucleus. They also do not have membranous organelles. Both bacteria (domain Bacteria) and archaea (domain Archaea) are prokaryotes, but each is placed in its own domain because of molecular and cellular differences.

Figure 16.10A reviews the anatomy of a bacterium (see also Section 4.4). Although bacteria do not have a nucleus, they do have a dense area called a nucleoid, where a single chromosome consisting largely of a circular strand of DNA is found. Many bacteria also have accessory rings of DNA called *plasmids*. Plasmids can be extracted and used as vectors to carry foreign DNA into host bacteria during genetic engineering processes. Protein synthesis in a prokaryotic cell is carried out by thousands of ribosomes, which are smaller than eukaryotic ribosomes.

The outer envelope of a bacterium consists of a plasma membrane and a cell wall that is strengthened by *peptidoglycan*, a complex molecule containing a unique amino disaccharide. The cell wall may be surrounded by a layer of poly-saccharides. A well-organized layer is called a capsule, while a loosely organized one is called a slime layer. A capsule and/or slime layer can protect the bacterium from host defenses.

The appendages of bacteria include fimbriae, sex pili, and flagella. Fimbriae are short, bristlelike fibers that allow bacteria to adhere to surfaces. The fimbriae of *Neisseria gonorrhoeae* enable it to attach to host cells and cause the sexually transmitted disease gonorrhea. Sex pili are rigid tubular structures used by bacteria to pass DNA from cell to cell, as discussed in Section 16.12. Bacteria that possess flagella, which are composed of the protein flagellin wound in a helix, are capable of movement.

Common Shapes of Prokaryotes Three basic shapes occur among prokaryotes (**Fig. 16.10B**): Cocci (sing., coccus) are round or spherical; bacilli (sing., bacillus) are rod-shaped; and spirilla (sing., spirillum) are spiral- or helical-shaped. These three basic shapes may be augmented by particular arrangements or shapes of cells.

Now that we have reviewed the structural features of prokaryotes, we will discuss their means of reproduction in Section 16.11.

> **16.10 Check Your Progress** Name two features that would help bacteria be infectious.

flagellum

sex pilus

fimbriae

1 μm

ribosome

nucleoid

plasma membrane

cell wall

capsule

FIGURE 16.10A Anatomy of bacteria.

cocci

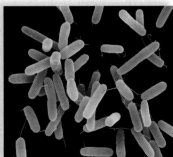

bacilli

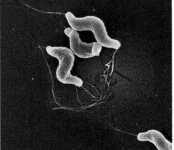

spirilla

FIGURE 16.10B The three shapes of prokaryotes. cocci = spheres; bacilli = rods; spirilla = curved

16.11 Prokaryotes have a common reproductive strategy

Prokaryotes reproduce asexually by means of **binary fission**, a process that results in two prokaryotes of nearly equal size. The process requires three steps: DNA replication, chromosome segregation, and cytokinesis. **Figure 16.11A** shows the process in bacteria. ❶ DNA replication begins at one location as the DNA double strand unzips. Each strand acts as a template for synthesis of a daughter strand by semiconservative DNA replication, the same as in eukaryotes. Each circular DNA strand is attached to the plasma membrane. ❷ The cell elongates, and the chromosomes segregate (separate). ❸ Cytokinesis requires that the plasma membrane invaginate to divide the cytoplasm. New cell wall formation also occurs. Mitosis, which requires the formation of a spindle apparatus, does not occur in prokaryotes.

Binary fission is asexual, and the offspring are at first genetically identical to the parent cell. But bacterial DNA has

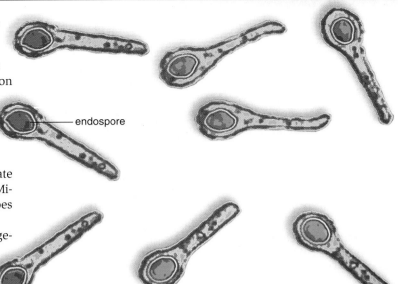

FIGURE 16.11B Endospores within *Clostridium tetani,* a bacterium.

a relatively high mutation rate, and bacteria have a generation time as short as 12 minutes under favorable conditions. Therefore, mutations are generated and passed on to offspring more quickly than in eukaryotes. Also, prokaryotes are haploid, and so mutations are immediately subjected to natural selection, which determines any possible adaptive benefit.

Formation of Endospores in Bacteria When faced with unfavorable environmental conditions, some bacteria form **endospores** (**Fig. 16.11B**). A portion of the cytoplasm and a copy of the chromosome dehydrate and are then encased by a heavy, protective spore coat. In some bacteria, the rest of the cell deteriorates, and the endospore is released. Spores survive in the harshest of environments—desert heat and dehydration, boiling temperatures, polar ice, and extreme ultraviolet radiation. They also survive for very long periods. When anthrax spores 1,300 years old germinate, they can still cause a severe infection (usually seen in cattle and sheep). Anthrax spores can be used as a bioterrorism weapon (see Section 16.17). In 2001, 22 cases of anthrax, including five deaths, occurred after spores were purposely sent through the mail. Humans also fear a deadly, but uncommon, type of food poisoning called botulism that is caused when endospores germinate inside cans of food (see Section 16.16). To germinate, the endospore absorbs water and grows out of the spore coat. Within a few hours, it becomes a typical bacterial cell, capable of reproducing once again by binary fission. Spore formation is not a means of reproduction, but it does allow bacteria to survive and to disperse to new places.

Although prokaryotes reproduce asexually, they can exchange genes. Bacteria use the three means discussed in Section 16.12.

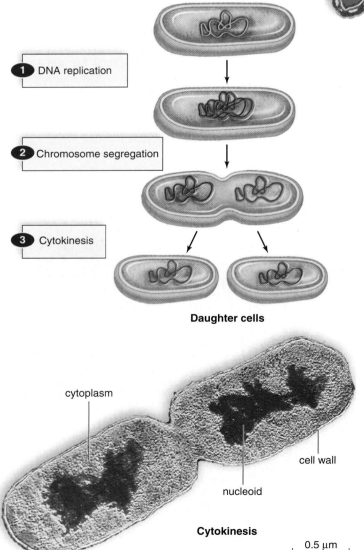

❶ DNA replication

❷ Chromosome segregation

❸ Cytokinesis

Daughter cells

cytoplasm

cell wall

nucleoid

Cytokinesis

0.5 µm

FIGURE 16.11A Binary fission results in two bacteria.

> **16.11** *Check Your Progress* **It is advantageous that bacteria reproduce rapidly and asexually when they are used to clean up oil spills. Explain.**

16.12   How genes are transferred in bacteria

Much of what we know about molecular genetics has come from studying the processes of DNA replication and gene expression in bacteria such as *Escherichia coli*, which lives in the human intestine. Section 16.10 discussed the structure of bacteria. We learned that bacteria, unlike viruses, are cellular and that they have a single chromosome composed of double-stranded DNA. The chromosome is actually quite large and condensed into a region called the nucleoid. Section 16.11 reviewed how bacteria reproduce by binary fission. At that time, DNA replicates and is pulled apart as the cell elongates. Formation of a new plasma membrane and cell wall divide the cell. Now we consider how a bacterium can acquire new genes from others of its own kind.

In eukaryotes, genetic recombination occurs as a result of sexual reproduction. But even though sexual reproduction does not occur among prokaryotes, three means of gene transfer take place in bacteria: transformation, conjugation, and transduction. In all three mechanisms, the donor is the cell that provides the genetic material for transfer, and the recipient is the cell that receives the material. The genes that allow bacteria to be resistant to antibiotics can be transferred by any one of these methods.

Transformation occurs when a recipient bacterium picks up (from its surroundings) free pieces of DNA secreted by live prokaryotes or released by dead prokaryotes (**Fig. 16.12A**). In Griffith's transformation experiment, illustrated in Figure 10.1, the one strain of *S. pneumoniae*, called the R strain, was not deadly until it picked up DNA released by the S strain. The DNA from the S strain transferred the ability to resist the host's immune system and, thereby, cause pneumonia. Transformation usually incorporates only a small amount of donor DNA into the chromosome of the recipient. Cells that do not naturally undergo transformation can be induced to do so by treatments that disrupt the cell wall.

Conjugation occurs between bacteria when the donor cell passes DNA to the recipient by way of a sex pilus, which temporarily joins the two bacteria. **Figure 16.12B** illustrates that a donor cell with a plasmid can initiate conjugation. A **plasmid** is a small circle of DNA that can replicate independently of the bacterial chromosome and is known for carrying antibiotic-resistant genes that confer resistance to bacteria. About 35% of the so-called F plasmid consists of genes that control the transfer of the plasmid to a recipient. Once the recipient receives a copy of the plasmid, it has the ability to transfer DNA to other bacteria.

During **transduction**, bacteriophages carry portions of bacterial DNA from a donor cell to a recipient (**Fig. 16.12C**). When a bacteriophage injects its DNA into the donor cell, the phage DNA takes over the machinery of the cell and causes it to produce more phage particles. During the lysogenic cycle in particular, a phage may incorporate a piece of the donor DNA and introduce it into recipients during subsequent rounds of infection.

Prokaryotes have a similar structure but have quite varied means of nutrition, as discussed in Section 16.13.

> **16.12** *Check Your Progress* Contrast the transfer of genes in bacteria to the process of achieving genetic variation via sexual reproduction.

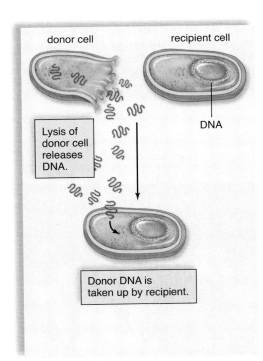

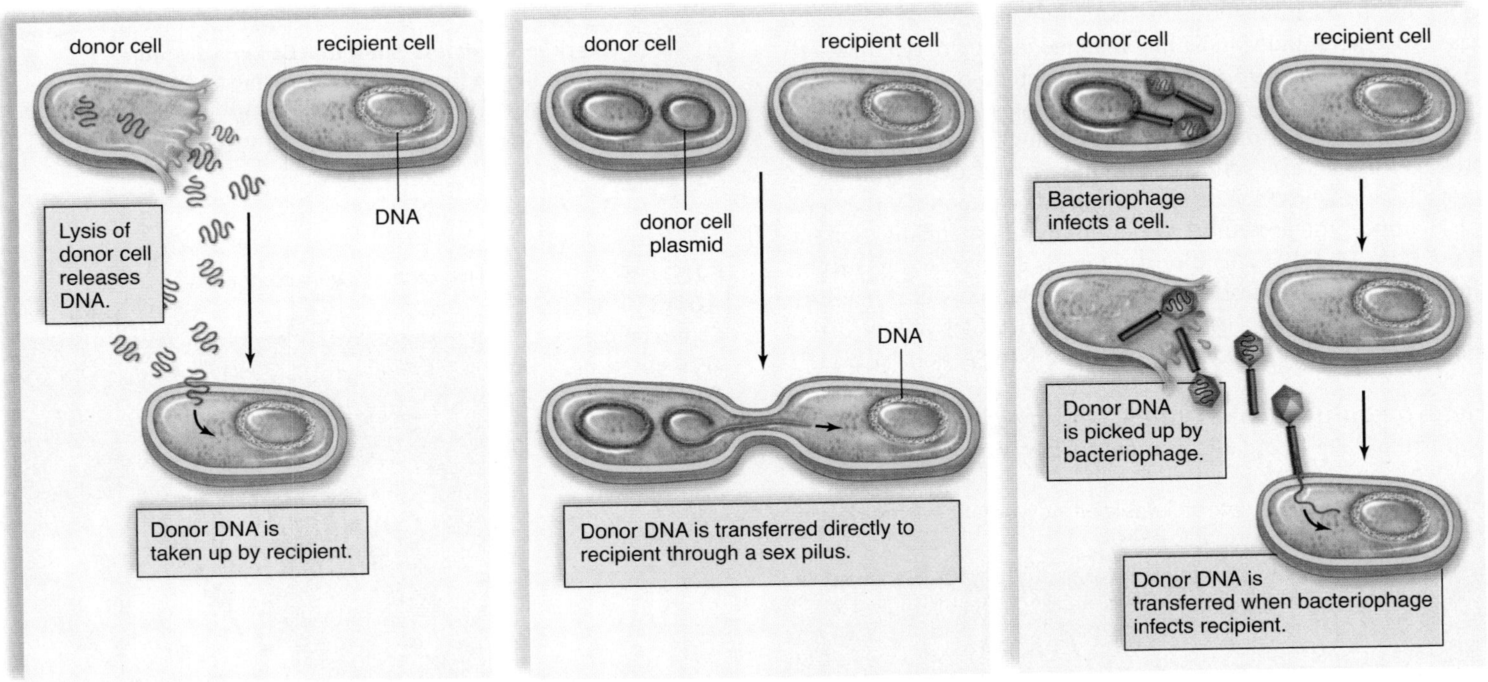

FIGURE 16.12A Gene transfer by transformation.

FIGURE 16.12B Gene transfer by conjugation.

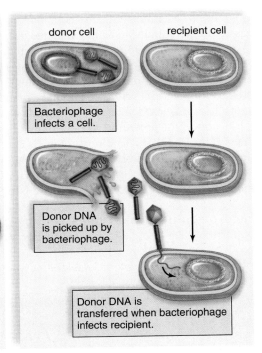

FIGURE 16.12C Gene transfer by transduction.

With respect to nutrient requirements, prokaryotes are not much different from other organisms. One difference, however, concerns their need for oxygen. Some prokaryotes are **obligate anaerobes**, meaning that they are unable to grow in the presence of free oxygen. A few serious illnesses—such as botulism, gas gangrene, and tetanus—are caused by anaerobic bacteria. Other prokaryotes, called **facultative anaerobes**, are able to grow in either the presence or the absence of gaseous oxygen. Most prokaryotes, however, are **aerobic** and, like animals, require a constant supply of oxygen to carry out cellular respiration.

Autotrophic Prokaryotes
Some prokaryotes produce their own organic nutrients. Some are photosynthetic and use solar energy to reduce carbon dioxide to organic compounds. There are two types of photosynthetic bacteria: those that evolved first and do not give off oxygen (O_2), and those that evolved later and do give off O_2. The green and purple sulfur bacteria carry on the first type of photosynthesis. These bacteria contain a different type of chlorophyll than plants and do not give off O_2, because they do not use water as an electron donor; instead, they can, for example, use hydrogen sulfide (H_2S):

$$CO_2 + 2H_2S \longrightarrow (CH_2O) + 2S + H_2O$$

These anaerobic bacteria live in the muddy bottom of bogs and marshes, where there is no O_2 (**Fig. 16.13A**). A type of bacteria called cyanobacteria (see Fig. 16.14) contains chlorophyll a, as do plants, and they carry on photosynthesis in the second way, just as green algae and plants do.

$$CO_2 + H_2O \longrightarrow (CH_2O) + O_2$$

Other autotrophic prokaryotes are **chemosynthetic**. They remove electrons from inorganic compounds, such as hydrogen gas, hydrogen sulfide, and ammonia, and use them to reduce CO_2 to an organic molecule. The nitrifying bacteria oxidize ammonia (NH_3) to nitrites (NO_2^-) and nitrites to nitrates (NO_3). Their metabolic abilities keep nitrogen cycling through ecosystems. Other prokaryotes oxidize sulfur compounds found at hydrothermal vents 2.5 km below sea level. The organic compounds they produce support the growth of a community of organisms found at vents (**Fig. 16.13B**). This discovery lends support to the suggestion that the first cells originated at hydrothermal vents (see Section 16.7). The archaea called methanogens, which reduce CO_2 to methane (CH_4), are also chemosynthetic (see Section 16.15).

Heterotrophic Prokaryotes
Most prokaryotes are heterotrophs that take in organic nutrients. They are also aerobic **sap-** rotrophs, which means that they secrete digestive enzymes into the environment for the breakdown of large organic molecules to smaller ones that can be absorbed. There is probably no natural organic molecule that cannot be digested by at least one prokaryotic species. In ecosystems, saprotrophic bacteria are called *decomposers*. They play a critical role in recycling matter and making inorganic molecules available to photosynthesizers.

The metabolic capabilities of heterotrophic prokaryotes have long been exploited by human beings. Bacteria are used commercially to produce chemicals, such as ethyl alcohol, acetic acid, butyl alcohol, and acetones. Prokaryotic action is also involved in the production of butter, cheese, sauerkraut, rubber, cotton, silk, coffee, and cocoa. Even antibiotics are produced by some bacteria.

Heterotrophs may be either free-living or symbiotic, meaning that they form mutualistic, commensalistic, or parasitic relationships. *Mutualism* exists when both partners benefit. Mutualistic bacteria live in the root nodules of soybean, clover, and alfalfa plants, where they receive organic nutrients and assist their hosts by reducing atmospheric nitrogen (N_2) for incorporation into organic compounds. Other mutualistic bacteria that live in human intestines release vitamins K and B_{12}, which we can use to help produce blood components. In the stomachs of cows and goats, special mutualistic prokaryotes digest cellulose, enabling these animals to feed on grass.

Commensalism often occurs when one population modifies the environment in such a way that a second population benefits. Obligate anaerobes can live in our intestines only because the bacterium *Escherichia coli* uses up the available oxygen. The *parasitic* bacteria cause disease, including human diseases.

The cyanobacteria, which carry on photosynthesis in the same manner as plants, are discussed in Section 16.14.

16.13 *Check Your Progress* Show that bacteria are more metabolically diverse than humans by stating (a) humans' need for oxygen and (b) humans' type of nutrition.

FIGURE 16.13B Some chemosynthetic prokaryotes live at hydrothermal vents.

Hydrothermal vent community

tubeworm ——

clam

FIGURE 16.13A Some anaerobic photosynthetic bacteria live in the muddy bottom of eutrophic lakes.

Formerly, the **cyanobacteria** were called blue-green algae and were classified with eukaryotic algae, but now they are known to be bacteria. Cyanobacteria are named for the blue-green pigment called phycocyanin that they contain. But they can also have other pigments that make them appear red, yellow, brown, or black, rather than only blue-green. Cyanobacteria photosynthesize in the same manner as plants and are believed to be responsible for first introducing oxygen into the early atmosphere. Without this event, animal evolution, as we know it, would not have occurred. Ancient seas contained stromatolites, which look like strange boulders but contain cyanobacteria. Some stromatolites are living today, and some contain fossils dated 2 BYA.

Cyanobacterial cells are rather large, ranging from 1 to 50 mm in width. They can be unicellular, colonial, or filamentous (**Fig. 16.14**). Cyanobacteria lack any visible means of locomotion, although some glide when in contact with a solid surface, and others oscillate (sway back and forth). Some cyanobacteria have a special advantage because they possess **heterocysts**, which are thick-walled cells without nuclei, where nitrogen fixation occurs. The ability to photosynthesize and also to fix atmospheric nitrogen (N_2) means that their nutritional requirements are minimal. They can serve as food for heterotrophs in ecosystems that are otherwise nutrient poor. Being bacteria, cyanobacteria reproduce by binary fission and can produce endospores that resist freezing and drying out. Spore formation means that cyanobacteria can come back when a dry lake receives water once again.

Cyanobacteria are common in fresh and marine waters, in soil, and on moist surfaces. But they are also found in harsh habitats, such as deserts, frozen lakes of Antarctica, extremely acidic, basic, or salty water, and even hot springs, where water temperatures approach 75°C. Cyanobacteria are the first photosynthetic organisms to appear on cooled lava after a volcanic eruption. It is hypothesized that they were the first colonizers of land during the course of evolution.

Cyanobacteria are symbiotic with a number of organisms, including some protists, plants, and animals. When living in these organisms, cyanobacteria often lose their cell walls and essentially function as chloroplasts inside the cells of their host. In association with fungi, they form **lichens** that can grow on rocks, buildings, and trees. A lichen is a symbiotic relationship in which the cyanobacterium provides organic nutrients to the fungus, while the fungus possibly protects and furnishes inorganic nutrients to the cyanobacterium. It is also possible that the fungus is parasitic on the cyanobacterium. Lichens help transform rocks into soil; other forms of life then may follow. Some lichens serve as bioindicators of air pollution.

Cyanobacteria are ecologically important in still another way. If care is not taken in disposing of industrial, agricultural, and human wastes, phosphates drain into lakes and ponds, resulting in a "bloom" of these organisms. The surface of the water becomes turbid, and light cannot penetrate to lower levels. When a portion of the cyanobacteria die off, the decomposing prokaryotes use up the available oxygen, causing fish to die from lack of oxygen.

Many archaea live in extreme environments, which are described in Section 16.15.

16.14 *Check Your Progress* **What service did cyanobacteria perform for animals in the ancient past that they still provide today?**

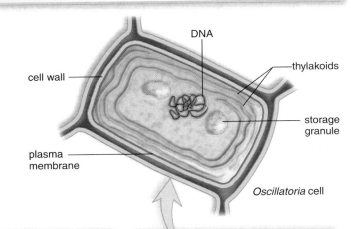

Oscillatoria cell

FIGURE 16.14 Diversity among the cyanobacteria.

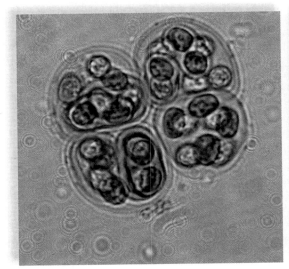

Gloeocapsa

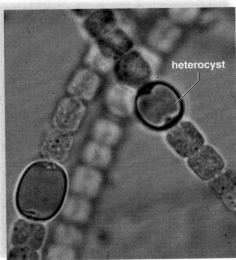

Anabaena

Oscillatoria

Scientists currently propose that the tree of life contains three domains: Archaea, Bacteria, and Eukarya. Because **archaea** and some bacteria are found in extreme environments (hot springs, thermal vents, and salt basins), they may have diverged from a common ancestor relatively soon after life began. Later, the eukarya are believed to have split off from the archaeal line of descent. Archaea and eukaryotes share some of the same ribosomal proteins (not found in bacteria), initiate transcription in the same manner, and have similar types of tRNA.

Structure and Function The plasma membranes of archaea contain unusual lipids that allow them to function at high temperatures. The archaea have also evolved diverse cell wall types, which facilitate their survival under extreme conditions. The cell walls of archaea do not contain peptidoglycan, as do the cell walls of bacteria. In some archaea, the cell wall is largely composed of polysaccharides, and in others, the wall is pure protein. A few have no cell wall.

Archaea have retained primitive and unique forms of metabolism. For example, some archaea are called **methanogens** because they have the unique ability to form methane. Most archaea are chemosynthetic, and a few are photosynthetic. This suggests that chemosynthesis predated photosynthesis during the evolution of prokaryotes. Archaea are sometimes mutualistic or even commensalistic, but none are parasitic—that is, archaea are not known to cause infectious diseases.

Types of Archaea Archaea are often discussed in terms of their unique habitats. The methanogens are found in anaerobic environments, such as swamps, marshes, and the intestinal tracts of animals (**Fig. 16.15A**). They couple the production of methane (CH_4) from hydrogen gas (H_2) and carbon dioxide (CO_2) to the formation of ATP. This methane, which is also called biogas, is released into the atmosphere, where it contributes to the greenhouse effect and global warming. About 65% of the methane in our atmosphere is produced by methanogenic archaea.

The **halophiles** are adapted to living in high salt concentrations (usually 12–15%; by contrast, the ocean is about 3.5% salt). Halophiles have been isolated from highly saline environments, such as the Great Salt Lake in Utah, the Dead Sea in the Mideast, solar salt ponds, and hypersaline soils (**Fig. 16.15B**). These archaea have evolved a number of mechanisms to thrive in high-salt environments. They depend on a pigment related to the rhodopsin in our eyes to absorb light energy to pump out chloride, and they use another similar type of pigment to synthesize ATP.

A third major type of archaea are the **thermoacidophiles** (**Fig. 16.15C**). These archaea are isolated from extremely hot and acidic environments, such as hot springs, geysers, hydrothermal vents, and around volcanoes. They reduce sulfur to sulfides, producing acidic sulfates, and these archaea grow best at pH 1 to 2. Thermoacidophiles survive best at temperatures above 80°C; some can even grow at 105°C (remember that water boils at 100°C)!

The prokaryotes have environmental and medical importance, as discussed in Section 16.16.

Methanosarcina mazei

FIGURE 16.15A Methanogen habitat and structure.

Great Salt Lake, Utah

Halobacterium salinarium

FIGURE 16.15B Halophile habitat and structure.

Boiling springs and geysers in Yellowstone National Park

Sulfolobus acidocaldarius

FIGURE 16.15C Thermoacidophile habitat and structure.

16.15 *Check Your Progress* Describe the evolutionary tree for the three domains.

16.16   Prokaryotes have environmental and medical importance

Prokaryotes Are Everywhere The prokaryotes are the most cosmopolitan of all life-forms on Earth. They can be found virtually everywhere, including in the oceans, in our intestines, in hot springs, and in the soil. Typically, 1 g of soil may contain over 100 million individual bacteria. NASA balloons have collected viable endospores of bacteria more than 30 km above the surface of the Earth, and living bacteria have been found in the deepest oceanic trenches. Although the prokaryotes are too small to be seen with the naked eye, they outweigh all of the eukaryotes on Earth as much as tenfold. Being found everywhere, the prokaryotes influence our lives in a myriad of ways.

Prokaryotes Were and Are Environmentally Important Since the early history of life on Earth, the prokaryotes have been a major influence on the environment. Ancient photosynthetic cyanobacteria altered Earth's primitive atmosphere by releasing copious amounts of their waste gas, oxygen. The presence of oxygen in the atmosphere, in turn, led to the evolution of cellular respiration and the rise of the diversity of life on Earth. Today, the descendants of these cyanobacteria continue to add valuable oxygen to the atmosphere. On the surface of the Earth, many species of bacteria serve as the principal creators of soil fertility. They help recycle the nutrients tied up in leaf clutter and animal corpses. Although nitrogen comprises 78% of the Earth's atmosphere, it cannot be used by organisms unless it is fixed to a usable form by nitrogen-fixing soil bacteria. These bacteria live on the roots of certain plants, such as legumes (peanuts, clover), acacias, and the tiny aquatic fern *Azolla*. Prokaryotes play an essential role in the carbon, nitrogen, sulfur, and phosphorus environmental cycles.

Prokaryotes Are Medically Important The vast majority of prokaryotic species are not pathogenic to humans. However, several prokaryotes have had a tremendous impact on human health since antiquity and continue to plague us today. Each year, millions of people suffer or die from bacterial infections. Several of the diseases caused by bacteria are listed in **Table 16.16**. Many species of pathogenic bacteria invade and destroy the tissue of their host; others produce powerful toxins (poisons).

Exotoxins are poisons secreted by bacteria. *E. coli* is capable of producing exotoxins that cause food poisoning and some forms of traveler's diarrhea. *Clostridium botulinum* produces a neurotoxin that is perhaps the most toxic substance on Earth. When canning, people may fail to heat food above the boiling point of water. Under these conditions, *Clostridium* can produce endospores that survive the canning process. These spores then germinate, and the cells that grow in the airless environment of the can produce the exotoxin that causes botulism. Endotoxins are components of bacteria. The endotoxins of *Salmonella* can cause food poisoning, and another species is responsible for typhoid fever.

Pathogenic bacteria seem to be winning the evolutionary arms race against antibiotics. Whereas penicillin was once 100% effective against hospital strains of *Staphylococcus au-*

Table 16.16 **Bacterial Diseases in Humans**	
Category	**Disease**
Sexually transmitted diseases	Syphilis, gonorrhea, chlamydia
Respiratory diseases	Strep throat, scarlet fever, tuberculosis, pneumonia, Legionnaires disease, whooping cough, inhalation anthrax
Skin diseases	Erysipelas, boils, carbuncles, impetigo, acne, infections of surgical or accidental wounds and burns, leprosy (Hansen disease)
Digestive tract diseases	Gastroenteritis, food poisoning, dysentery, cholera, peptic ulcers, dental caries
Nervous system diseases	Botulism, tetanus, leprosy, spinal meningitis
Systemic diseases	Plague, typhoid fever, diphtheria
Other diseases	Tularemia, Lyme disease

reus, today it is far less effective. New strains, such as MRSA (methicillin-resistant *Staphylococcus aureus*) are becoming serious health threats. Penicillin and tetracycline, long used to cure gonorrhea, now have a failure rate of more than 20% against certain strains of *Neisseria*. Pulmonary tuberculosis is on the rise, particularly among AIDS patients, the homeless, and the rural poor, and the strains are resistant to the usual combined antibiotic therapy. To keep antibiotics effective, the following steps are recommended:

Steps to Prevent Resistant Diseases
1. Never take an antibiotic for a viral infection, such as a cold or the flu.
2. Take antibiotics exactly and only as a doctor prescribes.
3. Wash your hands frequently and thoroughly.
4. Always handle food safely: • Keep your hands, utensils, and countertops clean. • Keep raw meat, poultry, and fish—or their juices—from contacting other foods. • Cook foods thoroughly and refrigerate foods promptly.
5. Get vaccinated when vaccines are available.
6. Exercise, eat right, drink lots of water, and get plenty of sleep.

The importance of microbes extends to their use as biological weapons, discussed in Section 16.17.

16.16 Check Your Progress What aspect of bacterial physiology enables bacteria to cycle nutrients in the environment?

Biological warfare is the use of viruses and bacteria, or their toxins, as weapons of war. In recent years, several nations have used genetic engineering to produce biological warfare agents that have an enhanced resistance to antimicrobial drugs and an altered pathogenic effect and incubation period. Bioterrorists prefer pathogens that are highly contagious, consistently produce a desired detrimental effect upon a population, have a short incubation period, and are easy to disseminate and deliver to a population. When dealing with these dangerous agents, protective clothing must be worn (**Fig. 16.17**).

In addition to humans, valuable animals and crops can serve as the targets of biological attacks. An attack upon a nation's cattle, pigs, or other domesticated animals could have serious consequences on the nation's food, animal products (wool, hides), and medicinal supplies (insulin, adrenaline, cortisone, vaccines). Agents that could be used against animals include anthrax, glanders, swine fever, hog cholera, foot-and-mouth disease, and fowl pest disease. Valuable crops, as well as commercially and medically important plants, can also be the targets of biological warfare. Several species of fungi and insects could be devastating to targeted plants. Several herbicides are considered biological weapons because they serve as bioregulators in plants.

The likely microbia agents to be used by bioterrorists are these:

> **Anthrax**, caused by the bacterium *Bacillus anthracis*, is a biological agent that is easy to acquire, grow, and disseminate. Anthrax occurs in three forms: inhalation anthrax, cutaneous anthrax, and gastrointestinal anthrax. Inhalation anthrax is the deadliest form to humans. It begins with flulike symptoms and, if not treated promptly, can lead to death in 24–36 hours after the onset of respiratory distress.

Smallpox is caused by the variola virus, a very dangerous and highly contagious airborne virus. Through diligent vaccination, smallpox has not been recorded since 1980. However, many young people have not received the vaccination, and the vaccine may have lost its effectiveness in others. After a 7–17 day incubation period, the disease begins as a fever, headache, and malaise. Patients are the most contagious 3–6 days after the onset of fever. Since many physicians have not seen smallpox, it can be easily misdiagnosed until the telltale lesions develop. Eventually, pus-filled lesions and blisters form on the patient's body. The pustules crust and form deep scars in survivors. In an unvaccinated population, smallpox has a 30% mortality rate.

Botulism, caused by the toxin of the anaerobic bacterium *Clostridium botulinum*, can be a lethal foodborne agent. Recent aerosol forms of the toxin have been developed. Botulinum toxins are some of the most lethal toxins known. They are easy to manufacture and weaponize, and represent a major threat to human populations. Initial symptoms are blurred vision, difficulty swallowing, and muscle weakness. The mortality rate from botulism is high, and death usually results from respiratory failure.

Plague, caused by the bacterium *Yersinia pestis*, has been called the Black Death and bubonic plague in the past, and has been responsible for millions of deaths. In a biological warfare scenario, the plague bacterium can be delivered by infected fleas, causing traditional bubonic plague, or by airborne droplets, causing the more deadly pneumonic plague.

Tularemia is caused by the bacterium *Francisella tularensis*. The disease has several forms, with the inhaled form being the most likely biological warfare candidate. Just 10–50 organisms inhaled by a human can cause an infection. A variety of symptoms may accompany tularemia, including fever, chills, headache, weakness, abdominal pain, vomiting, chest pain, and cutaneous ulcers. Pneumonia may develop in many victims, and death can result.

Hemorrhagic fevers, caused by several types of viruses, are characterized by high fever and severe, uncontrollable bleeding from several organs. Four deadly hemorrhagic fevers caused by virulent viruses are Crimean-Congo fever, Rift Valley fever, Marburg fever, and Ebola fever. Ebola is the best-known of these viruses.

The bacterial diseases anthrax, plague, botulism, and tularemia usually respond to specific antibiotics. Hemorrhagic fevers, if diagnosed soon enough, may respond to specific antiviral drugs, but these may be in short supply. Vaccines and preventives may be the best way to counter biological agents.

FIGURE 16.17 Bioterrorism represents a threat to our health.

> **16.17** *Check Your Progress*   **Smallpox is contagious. Why might a terrorist prefer to employ a contagious agent?**

Viruses are noncellular, disease-causing agents. As such, the medical significance of viruses cannot be underestimated. Nevertheless, humans use viruses for gene research and even for gene therapy by disarming their capability to cause disease.

Prokaryotes are cellular, but their structure is simpler than that of eukaryotes—they lack a nucleus and membranous organelles. Although there are significant structural differences between prokaryotes and eukaryotes, many biochemical similarities also exist between the two. Thus, the details of protein synthesis, first worked out in bacteria, are applicable to all cells, in-cluding those of humans. Today, transgenic bacteria routinely make products and otherwise serve the needs of human beings.

Many prokaryotes can live in environments that may resemble the habitats available when the Earth first formed. We find prokaryotes in such hostile habitats as swamps, the Dead Sea, and hot sulfur springs. The fossil record suggests that the prokaryotes evolved before the eukaryotes. Many investigators have performed experiments that suggest how a protocell may have preceded the evolution of the prokaryotic cell. Cyanobacteria are believed to have introduced oxygen into the Earth's ancestral atmosphere, and they may have been the first colonizers of the terrestrial environment. Most bacteria are decomposers that recycle nutrients in both aquatic and terrestrial environments. Clearly, humans are dependent on the past and present activities of prokaryotes.

All living things trace their ancestry to the prokaryotes, which contributed to the evolution of the eukaryotic cell. The mitochondria and chloroplasts of the eukaryotic cell are derived from bacteria that took up residence inside a nucleated cell, as we will discuss in Chapter 17. The rest of Part III pertains to the evolution of protists, plants, fungi, and animals, which are all eukaryotic organisms.

The Chapter in Review

Summary

At Your Service: Viruses and Bacteria

- Viruses are useful in gene research and in gene therapy.
- Bacteria generate oxygen, act as decomposers, function in bioremediation, and can produce antibiotics or other medicines.

Viruses Reproduce in Living Cells

16.1 Viruses have a simple structure

- A viral particle is composed of an outer protein capsid and an inner nucleic acid core.
- Viruses reproduce by using the metabolic machinery of the host cell.
- Viruses are categorized by whether their nucleic acid is DNA or RNA and single- or double-stranded; by their size and shape; and by the presence or absence of an outer envelope.

16.2 Some viruses reproduce inside bacteria

- Viruses are specific: Each infects a certain organism or tissue.
- Bacteriophages are viruses that parasitize bacteria.
- The lytic cycle has five stages: attachment, penetration, biosynthesis, maturation, and release.
- In the lysogenic cycle, integration occurs.

16.3 Viruses are responsible for a number of plant diseases

- Plant viruses are spread, particularly among injured plants, via contaminated tools, soil, pollen, seeds, or tubers, or by insects, nematodes, or parasitic plants.

16.4 Viruses reproduce inside animal cells and cause diseases

- The reproductive stages of an animal virus are similar to those of a bacteriophage: attachment, penetration, biosynthesis, maturation, and release.
- Diseases caused by viruses include parvovirus in dogs; rabies in mammals; measles, for example, in children; and some cancers.

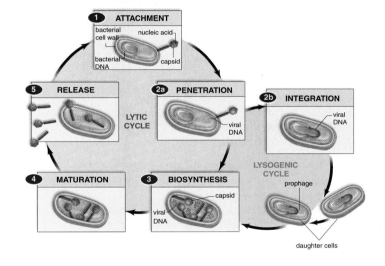

16.5 The AIDS virus exemplifies RNA retroviruses

- A retrovirus uses reverse transcription (from RNA to DNA) to insert a copy of its genome into the host genome.

16.6 Humans suffer from emerging viral diseases

- Primarily, viruses emerge when they are new to an area or jump from animals to humans.

The First Cells Originated on Early Earth

16.7 Experiments show how small organic molecules may have first formed

- The Miller and Urey experiment supports the prebiotic soup hypothesis: Gases from the early Earth's reducing atmosphere could have reacted to produce small organic molecules.
- The Wachtershaüser and Uber experiments support the iron-sulfur world hypothesis: Iron-sulfides at hydrothermal vents could have catalyzed reactions necessary to the formation of small organic molecules.

16.8 RNA may have been the first macromolecule
- Cech and Altman discovered that RNA can be both substrate and enzyme, supporting the hypothesis that only RNA was the first macromolecule.

16.9 Protocells preceded the first true cells
- Protocells arose when micromolecules were surrounded by a plasma membrane.
- Both microspheres and liposomes could possibly have acquired a plasma membrane.
- Protocell became a true cell when it contained a DNA information system and when it could carry on metabolism to acquire energy.

Both Bacteria and Archaea Are Prokaryotes

16.10 Prokaryotes have particular structural features
- Bacteria have a nucleoid that contains a single chromosome consisting of a circular strand of DNA.
- Ribosomes carry out protein synthesis.
- The outer envelope consists of a plasma membrane and a cell wall strengthened by peptidoglycan.
- Fimbriae, sex pili, or flagella may be present.
- Prokaryotes may be round, rod-shaped, or spiral.

16.11 Prokaryotes have a common reproductive strategy
- Prokaryotes reproduce asexually by binary fission.
- Endospore formation in bacteria occurs during unfavorable conditions.

16.12 How genes are transferred in bacteria
- Genes are transferred by transformation (pick up DNA from medium), conjugation (receive DNA via sex pilus), and transduction (receive DNA via virus).

16.13 Prokaryotes have various means of nutrition
- Obligate anaerobes cannot tolerate, while facultative anaerobes can tolerate, the presence of oxygen.
- Photosynthetics use solar energy, while chemosynthetic prokaryotes use inorganic compounds to make organic compounds.
- Bacteria are heterotrophs, which act as decomposers because they are saprotrophs.

16.14 The cyanobacteria are ecologically important organisms
- Cyanobacteria produce oxygen, fix atmospheric nitrogen, and form lichens.
- They are common in many aquatic and terrestrial habitats, including harsh environments.

16.15 Some archaea live in extreme environments
- Methanogens produce methane in anaerobic environments.
- Halophiles live in high-salt environments.
- Thermoacidophiles inhabit extremely hot, acidic environments.

16.16 Prokarotes have environmental and medical importance
- Prokaryotes produce oxygen and play roles in the carbon, nitrogen, sulfur, and phosphorus environmental cycles, which are essential to life.

- Prokaryotes cause many diseases for which antibiotics can become resistant.

16.17 Disease-causing microbes can be biological weapons
- Biological warfare is the use of viruses or bacteria, or their toxins, as weapons of war.

Testing Yourself

Viruses Reproduce in Living Cells

1. A virus contains
   a. a cell wall.
   b. a plasma membrane.
   c. nucleic acid.
   d. cytoplasm.
   e. More than one of these are correct.

2. Some scientists consider viruses nonliving because
   a. they do not locomote.
   b. they cannot reproduce independently.
   c. their nucleic acid does not code for protein.
   d. they are acellular.
   e. Both b and d are correct.

3. Which of these are found in all viruses?
   a. envelope, nucleic acid, capsid
   b. DNA, RNA, and proteins
   c. proteins and a nucleic acid
   d. proteins, nucleic acids, carbohydrates, and lipids
   e. tail fibers, spikes, and a rod shape

4. The five stages of the lytic cycle occur in this order:
   a. penetration, attachment, release, maturation, biosynthesis
   b. attachment, penetration, release, biosynthesis, maturation
   c. biosynthesis, attachment, penetration, maturation, release
   d. attachment, penetration, biosynthesis, maturation, release
   e. penetration, biosynthesis, attachment, maturation, release

5. Capsid proteins are synthesized during which phase of viral replication?
   a. replication
   b. biosynthesis
   c. assembly
   d. proteination
   e. All of these are correct.

6. **THINKING CONCEPTUALLY** What are the advantages of each type of life cycle?

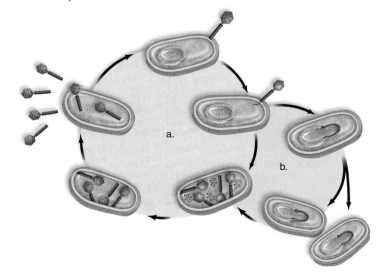

7. RNA retroviruses have a special enzyme that
   a. disintegrates host DNA.
   b. polymerizes host DNA.
   c. transcribes viral RNA to DNA.
   d. translates host DNA.
   e. repairs viral DNA.
8. Retroviruses
   a. only parasitize plant cells.
   b. have a reverse life cycle.
   c. include HIV.
   d. carry on anaerobic cellular metabolism.
   e. All of these are correct.

The First Cells Originated on Early Earth

9. The atmosphere in which life arose lacked
   a. carbon.   c. oxygen.
   b. nitrogen.   d. hydrogen.
10. The RNA-first hypothesis for the origin of cells is supported by the discovery of
    a. ribozymes.   c. polypeptides.
    b. proteinoids.   d. nucleic acid polymerization.
11. DNA genes may have arisen from RNA genes via
    a. DNA polymerase.   c. reverse transcriptase.
    b. RNA polymerase.   d. DNA ligase.
12. Liposomes (phospholipid droplets) are significant because they show that
    a. the first plasma membrane contained protein.
    b. a plasma membrane could have easily evolved.
    c. a biological evolution produced the first cell.
    d. there was water on the early Earth.
    e. the protocell had organelles.
13. Protocells probably obtained energy as
    a. photosynthetic autotrophs.   c. heterotrophs.
    b. chemosynthetics.   d. None of these are correct.
14. Which of these is an incorrect statement?
    a. The chemicals that Miller-Urey used to show that chemical evolution occurred in the atmosphere included ammonia (NH_3).
    b. Other experiments showed that nickel sulfides can act as a catalyst to change N_2 of the atmosphere to NH_3.
    c. Nickel sulfides were abundant in the early atmosphere.
    d. Both a and b are incorrect.

Both Bacteria and Archaea Are Prokaryotes

15. Bacterial cells contain
    a. ribosomes.   d. vacuoles.
    b. nuclei.   e. More than one of these are correct.
    c. mitochondria.
16. Which is not true of prokaryotes? They
    a. are living cells.
    b. lack a nucleus.
    c. all are parasitic.
    d. include both archaea and bacteria.
    e. evolved early in the history of life.
17. Bacterial endospores function in
    a. reproduction.   c. protein synthesis.
    b. survival.   d. storage.
18. Archaea differ from bacteria in that they
    a. have a nucleus.
    b. have membrane-bounded organelles.
    c. have peptidoglycan in their cell walls.
    d. are often photosynthetic.
    e. None of these are correct.

For questions 19–23, determine which type of organism is being described. Each answer in the key may be used more than once.

KEY:
a. bacteria   c. both bacteria and archaea
b. archaea   d. neither bacteria nor archaea

19. Peptidoglycan in cell wall.
20. Methanogens.
21. Sometimes parasitic.
22. Contain a nucleus.
23. Plasma membrane contains lipids.

Understanding the Terms

aerobic   314
anthrax   318
archaea   316
bacteriophage   303
binary fission   312
botulism   318
capsid   302
chemosynthetic   314
conjugation   313
cyanobacteria   315
emergent disease   307
endospore   312
facultative anaerobe   314
halophile   316
hemorrhagic fever   318
heterocyst   315
HIV provirus   306

lichen   315
liposome   310
lysogenic cycle   303
lytic cycle   303
methanogen   316
obligate anaerobe   314
plague   318
plasmid   313
protocell   309
retrovirus   306
saprotroph 314
smallpox   318
thermoacidophile   316
transduction   313
transformation   313
tularemia   318

Match the terms to these definitions:
a. _____ Bacteriophage life cycle in which the virus incorporates its DNA into that of the bacterium.
b. _____ Organism that contains chlorophyll and uses solar energy to produce its own organic nutrients.
c. _____ Organism that secretes digestive enzymes and absorbs the resulting nutrients back across the plasma membrane.
d. _____ Type of prokaryote that is most closely related to the Eukarya.
e. _____ Transfer of genetic material from one bacterium to another by way of a sex pilus.

Thinking Scientifically

1. While there are a few drugs that are effective against some viruses, they often produce a number of side effects by impairing the function of body cells. Most antibiotics (antibacterial drugs) do not cause side effects. Why would antiviral medications be more likely to produce side effects?
2. The bacterium *E. coli* is a model organism. What characteristics make *E. coli* particularly useful in genetic experiments?

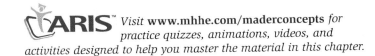 Visit **www.mhhe.com/maderconcepts** *for practice quizzes, animations, videos, and activities designed to help you master the material in this chapter.*

17

Evolution of Protists

LEARNING OUTCOMES

After studying this chapter, you should be able to accomplish the following outcomes.

Protists Cause Disease Too

1 Associate some common diseases with their protist vector.

Protists May Represent the Oldest Eukaryotic Cells

2 Draw and explain a diagram showing how mitochondria and chloroplasts evolved.
3 Show that protists are diverse by comparing size, mode of nutrition, reproduction, and symbiotic relationships.
4 Use an evolutionary tree of the protists to see how protists may be related to each other.
5 Discuss problems associated with classifying the protists.

Protozoans Are Heterotrophic Protists with Various Means of Locomotion

6 Use euglenoids to explain why protists are difficult to classify.
7 List the diseases caused by zooflagellates, and tell which of these diseases are common to temperate and tropical zones.
8 Distinguish amoebas from foraminiferans and from radiolarians.
9 Give examples of various types of ciliates, and tell how ciliates move, feed, and reproduce.
10 Describe the life cycle of *Plasmodium vivax*.

Some Protists Have Moldlike Characteristics

11 Compare and contrast slime molds with water molds.

Algae Are Photosynthetic Protists of Environmental Importance

12 Contrast the anatomy of diatoms and dinoflagellates, and tell why they are significant algae in the oceans.
13 Compare and contrast the anatomy and uses of red algae and brown algae.
14 Contrast the anatomy of five types of green algae, and describe how they reproduce.
15 Explain why green algae are not classified as plants.
16 Compare and contrast the three types of sexual life cycles among algae.

Many people relate disease to viruses, bacteria, and an occasional fungus. Little do they realize that members of the kingdom Protista cause disease too.

Malaria is caused by a protist that may have infected humans ever since they evolved. The infection causes recurring cycles of chills, fever, and sweating every few days. These symptoms are due to bursting of the red blood cells where the parasite's spores exist for a part of its life cycle. The protozoan that causes malaria was identified and named *Plasmodium* in 1880, and researchers learned in 1898 that the *Anopheles* mosquito transmits the protozoan from person to person. Even so, the administrators charged with building the Panama Canal in the early 1900s had other explanations. The name malaria means "bad air," and they still believed that breezes coming off swamps caused malaria. The best protection against malaria, they said, was a morally correct lifestyle. Malaria was finally brought under control in Panama when a young physician, Dr. William C. Gorgas, was given the resources to prevent *Anopheles* from breeding. Unfortunately,

The bite of the *Anopheles* mosquito transmits malaria.

malaria still affects millions around the world, particularly in South America and Africa.

Amoebic dysentery, which is characterized by bloody diarrhea, is caused by *Entamoeba histolytica*, an amoeboid protozoan. This infection is more likely to occur in the tropics, where the parasite is prevalent. It is associated with poor hygiene and filthy conditions because it is spread by food or water contaminated with feces.

Giardiasis is caused by a multiflagellated protozoan that adheres to the human intestinal lining by means of a sucking disk. The primary symptom of infection is extreme diarrhea. *Giardia* does not have a vector; instead, the protozoan is taken into the body by drinking contaminated water. Persons who drink shallow well water or water from a stream while camping or hiking, or those who accidentally ingest pool water while swimming, are subject to possible infection. Since 1971, *Giardia* has been the most commonly identified waterborne pathogen in the United States. You can protect yourself by only drinking water that has been properly filtered.

Flagellated protozoans in the genus *Trypanosoma* are well known for causing tropical diseases. Each disease is transmitted by a specific insect vector. Chagas disease occurs in Central and South America after an insect commonly called the "kissing bug" deposits feces containing the parasite *Trypanosoma cruzi* in its bite. Symptoms of Chagas disease include localized swelling, loss of strength, bone pain, anemia, and possible heart failure.

Misshapen red blood cells harbor the spores of malaria.

African sleeping sickness, caused by *Trypanosoma brucei*, has re-emerged as a serious health and economic problem in sub-Saharan Africa despite eradication efforts. The vector is the large, brown, and stealthy tsetse fly, named for the sound it makes while flying. Fever, lymph node swelling, and general malaise occur before the parasite makes its way to the brain. Neurological complications result in a stupor that accounts for the name sleeping sickness, Few recover from this disease, which occurs only in Africa.

Still other trypanosomes cause leishmaniasis, a disease transmitted by the bite of an infected female sand fly. This vector, about one-third the size of a mosquito, is a noiseless flyer that usually bites at night. The disease is common in tropical and subtropical countries, and therefore rare in the United States. Usually, leishmaniasis manifests itself as skin sores that heal within a few months, leaving noticeable scars. A more serious form spreads to the internal organs and is potentially fatal if untreated. Cutaneous strains from North and South America may destroy nasal and cheek mucosa and cause extreme facial disfigurement. Twenty cases of the cutaneous form and 12 cases of visceral infection were reported in soldiers during Operation Desert Storm from 1990 to 1991.

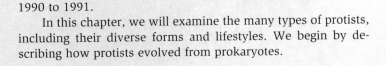

Biting female sand fly transmits leishmaniasis

leishmaniasis skin sores

In this chapter, we will examine the many types of protists, including their diverse forms and lifestyles. We begin by describing how protists evolved from prokaryotes.

Protists May Represent the Oldest Eukaryotic Cells

After reviewing the endosymbiotic theory, this part of the chapter discusses the complexity of protists, despite many existing as single cells. Because the classification of protists has not been finalized, this chapter categorizes them according to nutrition.

17.1 Eukaryotic organelles arose by endosymbiosis

Protists (kingdom Protista) are eukaryotes. The eukaryotic cell contains a nucleus and various membranous organelles. The **endosymbiotic theory**, introduced in Chapter 4, states that at least mitochondria and chloroplasts are derived from independent prokaryotic cells (**Fig. 17.1**). Observational data support this theory. For example, mitochondria resemble aerobic bacteria, and chloroplasts resemble cyanobacteria in size and structure. A protist, such as an amoeba, could have engulfed these prokaryotes. This would account for why mitochondria and chloroplasts have a double membrane; the outer membrane represents the vesicle that brought them into the cell, and the inner membrane is the original plasma membrane of the prokaryote.

Still, endosymbionts retained their ability to reproduce by binary fission and to make their own proteins, as do chloroplasts and mitochondria today. Evolutionary change brought about the mutualistic relationship that exists today.

All protists have mitochondria, but not all protists have plastids. This is explainable if you assume that cyanobacteria were taken up by some but not all members of a group. Some protists even have plastids with four membranes, suggesting that these plastids were originally part of independent protists! The point is that endosymbiosis was probably a common occurrence during the evolution of organisms.

Section 17.2 discusses the diversity of protists and the difficulties in finding relationships.

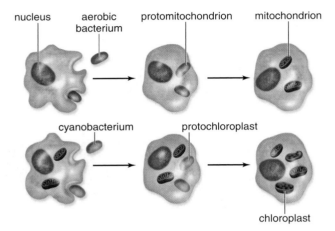

FIGURE 17.1 Origin of mitochondria (above) and chloroplasts (below).

Labels: nucleus, aerobic bacterium, protomitochondrion, mitochondrion, cyanobacterium, protochloroplast, chloroplast

> **17.1 Check Your Progress** Are the protistan parasites that cause amoebic dysentery, malaria, giardiasis, and leishmaniasis eukaryotes?

17.2 Protists are a diverse group

Protists vary in size from microscopic to macroscopic exceeding 200 m in length. Most protists are unicellular, but despite their small size, they have attained a high level of complexity. The amoeboids and ciliates possess unique organelles, such as a contractile vacuole that assists in water regulation.

Asexual reproduction by mitosis is the norm in protists. Sexual reproduction involving meiosis and spore formation generally occurs only in a hostile environment. **Spores** are haploid resting cells resistant to adverse conditions, and they can survive until favorable conditions return once more. Some protozoans form cysts, another type of resting stage. In parasites, a cyst often serves as a means of transfer to a new host.

While the protists have great medical importance because several of them cause diseases in humans, they are also of enormous ecological importance. Being aquatic, the photosynthesizers give off oxygen and function as producers in both freshwater and saltwater ecosystems. They are a major component of **plankton**, organisms that are suspended in the water and serve as food for heterotrophic protists and animals.

Protists enter symbiotic relationships ranging from parasitism to mutualism. For example, several parasitic protists cause disease in animals and humans. Coral reef formation is greatly aided by the presence of a symbiotic photosynthetic protist that lives in the tissues of coral animals.

The complexity and diversity of protists make finding relationships difficult, and biologists are in the process of developing an evolutionary tree that all can agree upon, as discussed in Section 17.3. In the meantime, this chapter groups the protists according to modes of nutrition, as shown in **Figure 17.2**. Primarily, protists include the **protozoans** (and **slime molds**), which are heterotrophic by ingestion, as are animals; the **water molds**, which are heterotrophic by absorption, as are fungi; and the **algae**, which are autotrophic, as are plants. Some protozoans and water molds are parasitic.

One proposed evolutionary tree of protists, out of many, is examined in Section 17.3.

> **17.2 Check Your Progress** Why do biologists find it difficult to reorganize kingdom Protista?

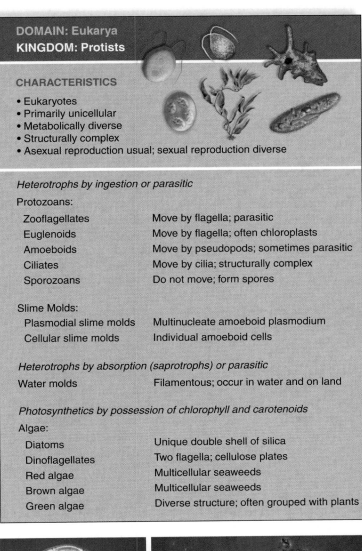

DOMAIN: Eukarya
KINGDOM: Protists

CHARACTERISTICS

- Eukaryotes
- Primarily unicellular
- Metabolically diverse
- Structurally complex
- Asexual reproduction usual; sexual reproduction diverse

Heterotrophs by ingestion or parasitic

Protozoans:

Zooflagellates	Move by flagella; parasitic
Euglenoids	Move by flagella; often chloroplasts
Amoeboids	Move by pseudopods; sometimes parasitic
Ciliates	Move by cilia; structurally complex
Sporozoans	Do not move; form spores

Slime Molds:

Plasmodial slime molds	Multinucleate amoeboid plasmodium
Cellular slime molds	Individual amoeboid cells

Heterotrophs by absorption (saprotrophs) or parasitic

Water molds	Filamentous; occur in water and on land

Photosynthetics by possession of chlorophyll and carotenoids

Algae:

Diatoms	Unique double shell of silica
Dinoflagellates	Two flagella; cellulose plates
Red algae	Multicellular seaweeds
Brown algae	Multicellular seaweeds
Green algae	Diverse structure; often grouped with plants

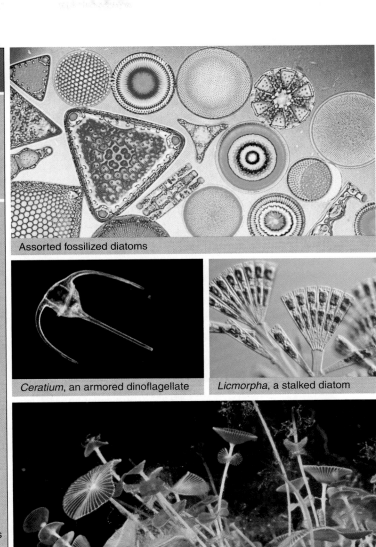

Assorted fossilized diatoms

Ceratium, an armored dinoflagellate

Licmorpha, a stalked diatom

Acetabularia, a single-celled green alga

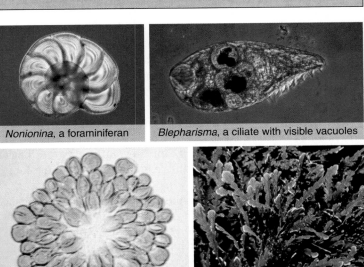

Nonionina, a foraminiferan

Blepharisma, a ciliate with visible vacuoles

Synura, a colony-forming golden alga

Bossiella, a coralline red alga

Amoeba proteus, a protozoan

FIGURE 17.2 Protist diversity.

17.3 How can the protists be classified?

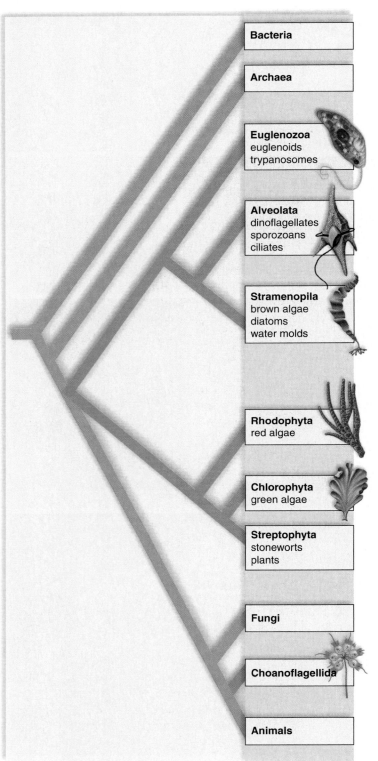

FIGURE 17.3 Proposed evolutionary tree of protists (blue branches) based on DNA and RNA sequencing.

As Section 17.2 observed, protists are traditionally grouped into those that are heterotrophic (protozoan and slime molds) and those that are photosynthetic (algae). In this chapter, we are also discussing the protists according to their mode of nutrition. This means that no attempt has been made to keep the most closely related protists together in a single group.

Today, the classification of protists is in a state of flux as biologists attempt to discover evolutionary relationships. Various evolutionary trees for the protists have been proposed based on different ways of interpreting the available data. Molecular studies, including the sequencing of DNA and RNA, have produced a number of hypothetical evolutionary trees, one of which is shown in **Figure 17.3**. When molecular data are consistent with electron microscopy studies, confidence increases that certain protists should be grouped together.

This much we know for sure: Lumping all the single-celled eukaryotes (protists) into a single kingdom is artificial and does not represent how evolution actually occurred. But how should the protists be grouped? Should they be placed in several kingdoms, thereby increasing the number of kingdoms in the domain Eukarya? Should some of them be placed in the other kingdoms—plants, animals, or fungi—or should new kingdoms be created that would include, say, the plants and some of the protists, or include the animals and some of the protists? For example, in Figure 17.3, the green algae called stoneworts are grouped with the plants and together called Streptophyta. The investigator who proposed this tree believed that stoneworts and plants should be in the same group because the sequencing of nucleic acids indicated that they shared a recent common ancestor. Notice also the last branch of Figure 17.3. It shows that the fungi and animals are believed to be closely related. However, the choanoflagellates, which are single-celled or colonial flagellates, each with an anterior end surrounded by a thin protoplasmic collar, are placed even closer to the animals. The reason is that choanoflagellates may have given rise to the animals.

The amoebas and slime molds discussed in this chapter are not in the tree at all. Why? Because there is little certainty where these protists should be placed. Some evolutionary trees place them in their own group, but on the branch leading to the fungi and animals.

Recent attempts to group the protists according to nucleic acid sequencing indicate that it is a powerful tool for sorting out how evolution occurred. Once we understand how the protists evolved, we will gain insight into the origins of plants, fungi, and animals, the other kingdoms in the domain Eukarya.

Section 17.4 begins our survey of protists with the protozoans.

> **17.3 Check Your Progress** Traditionally, protists are divided into the protozoans, slime molds, and algae. Give an example to show that Figure 17.3 does not support this method of classification.

Protozoans are heterotrophic protists that ingest their food, as do animals. Protozoans differ by their means of locomotion. Flagellates move by flagella, amoeboids move by pseudopods, ciliates move by cilia, and sporozoans are not motile.

17.4   Protozoans called flagellates move by flagella

The **zooflagellates** include thousands of species of mostly unicellular, heterotrophic protozoans that move by means of flagella.

Parasitic Zooflagellates A number of trypanosomes cause disease in humans. We will consider three examples. (1) *Trypanosoma brucei*, transmitted by the bite of the tsetse fly, is the cause of African sleeping sickness in humans. The area of the bite becomes an open sore, from which the trypanosomes move toward the lymphatic glands or remain in the bloodstream, where they divide every 5–7 hours. Weight loss and recurrent attacks of fever occur during this phase of the disease. The trypanosomes invade the central nervous system, and this leads to the typical symptoms of sleeping sickness—disturbed sleep cycle, change in personality, and coma. Many thousands of cases of human sleeping sickness are diagnosed each year. Fatalities or permanent brain damage are common. (2) Another trypanosome, *Trypanosoma cruzi*, causes Chagas disease in humans in Central and South America. Approximately 45,000 people die yearly from the severe cardiac and digestive problems caused by this parasite. (3) Leishmaniasis, characterized by skin sores and in some cases damage to the internal organs, is caused by a trypanosome transmitted by sand flies. These diseases are particularly troublesome in Africa and South America, and so far have been difficult to control.

Giardia lamblia is a zooflagellate whose cysts are transmitted by way of contaminated water. It attaches to the human intestinal wall by means of a sucking disk and causes severe diarrhea. *Giardia* is the most common flagellate of the human digestive tract and also lives in a variety of other mammals. Beavers seem to be an important reservoir of infection in the mountains of the western United States, and many cases of infection have been acquired by hikers who fill their canteens at beaver ponds.

Trichomonas vaginalis, a sexually transmitted zooflagellate, infects the vagina and urethra of women and the prostate, seminal vesicles, and urethra of men. Therefore, it is a common culprit of vaginitis in the United States.

Euglenoids The **euglenoids** include about 1,000 species of small (10–500 µm) freshwater unicellular organisms that typify the problem of classifying protists. One-third of all genera have chloroplasts; the rest do not. This may not be surprising when we consider that their chloroplasts are like those of green algae and are probably derived from them through endosymbiosis. A pyrenoid is a special region of the chloroplast where polysaccharides form. Euglenoids produce an unusual type of polysaccharide called paramylon. Those that lack chloroplasts ingest or absorb their food.

Euglenoids have two flagella, one of which typically is much longer than the other and projects out of an anterior, vase-shaped invagination. It is called a tinsel flagellum because it has hairs

on it. Near the base of this flagellum is an eyespot, which shades a photoreceptor for detecting light. Because euglenoids are bounded by a flexible pellicle composed of protein bands lying side by side, they can assume different shapes as the underlying cytoplasm undulates and contracts. A common euglenoid is *Euglena deces*, an inhabitant of freshwater ditches and ponds. A contractile vacuole allows this protist to rid its body of excess water (**Fig. 17.4**).

Having surveyed some of the flagellated protozoans, Section 17.5 moves on to examine the amoeboid protozoans.

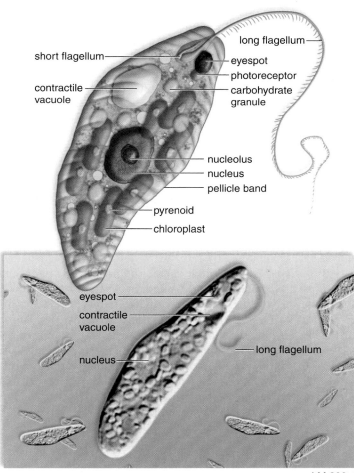

FIGURE 17.4 *Euglena*, a flagellate.

17.4 *Check Your Progress* **What is a common source of *Giardia* infections?**

17.5 Protozoans called amoeboids move by pseudopods

Pseudopods are extensions that form when cytoplasm streams in a particular direction. Protists that move by pseudopods usually live in aquatic environments. In oceans and freshwater lakes and ponds, they may be a part of the **zooplankton**, microscopic suspended organisms that feed on other organisms.

The **amoeboids** are protists that use pseudopods to move and also to ingest their food. Hundreds of species of amoeboids have been classified. *Amoeba proteus* is a commonly studied freshwater member of this group (**Fig. 17.5A**). When amoeboids feed, the pseudopods surround and **phagocytize** their prey, which may be algae, bacteria, or other protists. Digestion then occurs within a *food vacuole*. Freshwater amoeboids have *contractile vacuoles*, where excess water from the cytoplasm collects before the vacuole appears to "contract," releasing the water through a temporary opening in the plasma membrane.

Entamoeba histolytica is a parasitic amoeboid that lives in the human large intestine and causes amoebic dysentery. The ability of the organism to form cysts makes amoebic dysentery infectious. Complications arise when this parasite invades the intestinal lining and reproduces there. If the parasites enter the body proper, liver and brain involvement can be fatal.

The **foraminiferans** and the **radiolarians** have shells called tests, which are intriguing and beautiful. In the foraminiferans, the calcium carbonate test is often multichambered. The pseudopods extend through openings in the test, which covers the plasma membrane. Deposits of foraminiferans for millions of years, followed by geologic upheaval, formed the White Cliffs of Dover along the southern coast of England (**Fig. 17.5B**). In the radiolarians, the glassy silicon test is internal and usually has a radial arrangement of spines (**Fig. 17.5C**). The pseudopods are external to the test.

The tests of dead foraminiferans and radiolarians form a deep layer (700–4,000 m) of sediment on the ocean floor. The radiolarians lie deeper than the foraminiferans because their glassy test is insoluble at greater pressures. The presence of either or both is used as an indicator of oil deposits on land and sea. Their fossils date as far back as Precambrian times and are evidence of the antiquity of the protists. Because each geologic period has a distinctive form of foraminiferan, they can be used as index fossils to date sedimentary rock. The great Egyptian pyramids are built of foraminiferan limestone. One foraminiferan test found in the pyramids is about the size of a silver dollar. This species, known as *Nummulites*, has been found in deposits worldwide, including in central eastern Mississippi.

The next section, namely 17.6, will examine the protozoans known as ciliates.

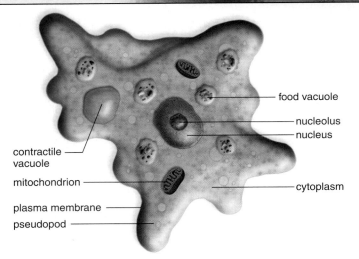

FIGURE 17.5A *Amoeba proteus*, an amoeboid.

food vacuole
nucleolus
nucleus
contractile vacuole
mitochondrion
cytoplasm
plasma membrane
pseudopod

250 µm

FIGURE 17.5B Foraminiferans, such as *Globigerina*, built the White Cliffs of Dover, England.

SEM 200×

FIGURE 17.5C Radiolarian tests.

> **17.5 Check Your Progress** Silica, or silicon dioxide (SiO_2), is a very hard solid. What role would you expect it to play in living things?

17.6 Protozoans called ciliates move by cilia

The **ciliates** consist of approximately 8,000 species of unicellular protists that range from 10 to 3,000 µm in size. Members of this phylum are called ciliates because they move by means of cilia. They are the most structurally complex and specialized of all protozoans. The majority of ciliates are free-living; however, several parasitic, sessile, and colonial forms exist.

The classic example of a ciliate is *Paramecium*. These unicellular ciliates are commonly found in ponds and ditches. Hundreds of cilia, which beat in a coordinated, rhythmic manner, project through tiny holes in a semirigid outer covering, or pellicle (**Fig. 17.6A**). Numerous oval capsules lying in the cytoplasm just beneath the pellicle contain **trichocysts**. Upon mechanical or chemical stimulation, trichocysts discharge long, barbed threads that are useful for defense and for capturing prey. Toxicysts are similar, but they release a poison that paralyzes prey.

When a paramecium feeds, food particles are swept down a gullet, below which food vacuoles form. Following digestion, the soluble nutrients are absorbed by the cytoplasm, and the nondigestible residue is eliminated at the anal pore.

During asexual reproduction, ciliates divide by transverse binary fission. Ciliates have two types of nuclei: a large macronucleus and one or more small micronuclei. The macronucleus controls the normal metabolism of the cell, while the micronuclei are concerned with reproduction. Sexual reproduction involves conjugation (**Fig. 17.6B**). The macronucleus disintegrates, and after the micronuclei undergo meiosis, two ciliates exchange a haploid micronucleus. Then the micronuclei give rise to a new macronucleus, which contains a copy of all the genes.

The ciliates are a diverse group of protozoans. Barrel-shaped didiniums expand to consume paramecia much larger than themselves. *Suctoria* have an even more dramatic way of getting food. They rest quietly on a stalk until a hapless victim comes along. Then they promptly paralyze it and use their tentacles like straws to suck it dry. *Stentor* may be the most elaborate ciliate, resembling a giant blue vase decorated with stripes (**Fig. 17.6C**). *Ichthyophthirius*, a ciliate, is responsible for a common disease in fishes called "ich." If left untreated, it can be fatal.

Section 17.7 gives examples of protozoans that are known as sporozoans because they form spores.

17.6 Check Your Progress What are some advantages of cilia?

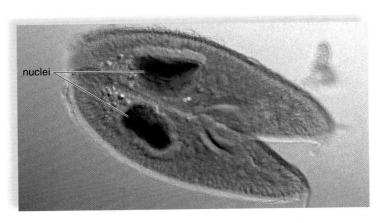

FIGURE 17.6B During conjugation, two paramecia first unite at oral areas.

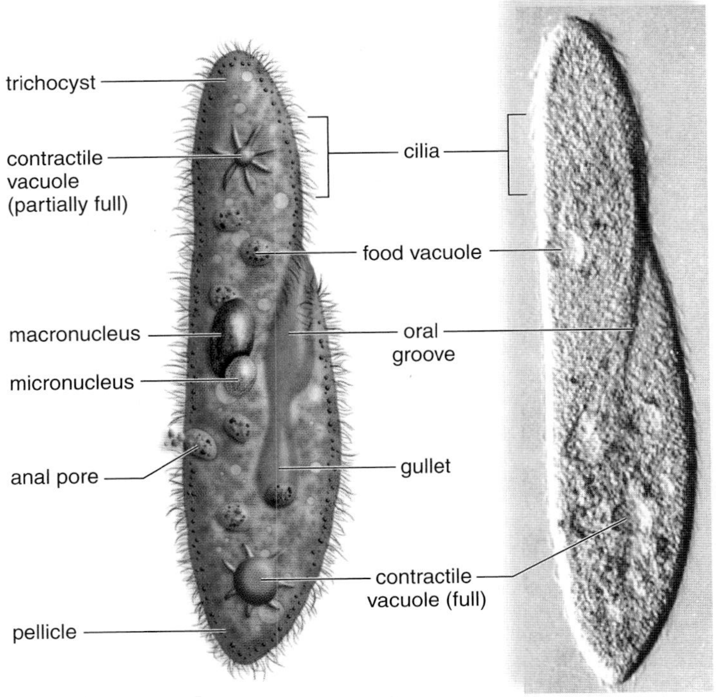

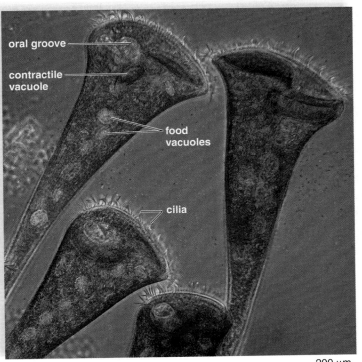

FIGURE 17.6C *Stentor*, a ciliate.

FIGURE 17.6A *Paramecium*, a ciliate.

17.7 Protozoans called sporozoans are not motile

The **sporozoans** consist of nearly 3,900 species of nonmotile, parasitic, spore-forming protozoans. Many sporozoans have multiple hosts.

Pneumocystis carinii causes the type of pneumonia seen primarily in AIDS patients. During its sexual reproduction, thick-walled cysts form in the lining of pulmonary air sacs. The cysts contain spores that successively divide until the cyst bursts and the spores are released. Each spore becomes a new mature organism that can reproduce asexually but may also enter the sexual stage and form cysts.

Today, approximately one million people die each year from **malaria,** a widespread disease caused by four types of sporozoan parasites in the genus *Plasmodium*. The disease is spread by a mosquito vector that passes the sporozoan to humans. International travel, coupled with new resistant forms of the vector and parasite, is presenting health professionals with formidable problems.

The life cycle of the sporozoan *Plasmodium vivax*, a common cause of malaria, is shown in **Figure 17.7.** The female *Anopheles* mosquito bites humans and other animals to acquire the protein she needs to produce eggs. ❶ If the mosquito picks up the parasite, the sexual phase of the life cycle occurs in her body. ❷ The next human the mosquito bites will now acquire the parasite, ❸ which begins its asexual phase in the liver. ❹ The chills and fever of malaria appear after red blood cells are infected and burst, ❺ releasing parasites and toxic substances into the blood. ❻ Some of these parasites become gametocytes that will be taken up by a mosquito.

Toxoplasma gondii, another sporozoan, causes toxoplasmosis, particularly in cats, but also in people. In pregnant women, the parasite can infect the fetus and cause birth defects and mental retardation; in AIDS patients, it can infect the brain and cause neurological symptoms.

This completes our discussion of protozoans, and the next part of the chapter discusses the protistan "molds."

17.7 Check Your Progress Why is malaria making a comeback?

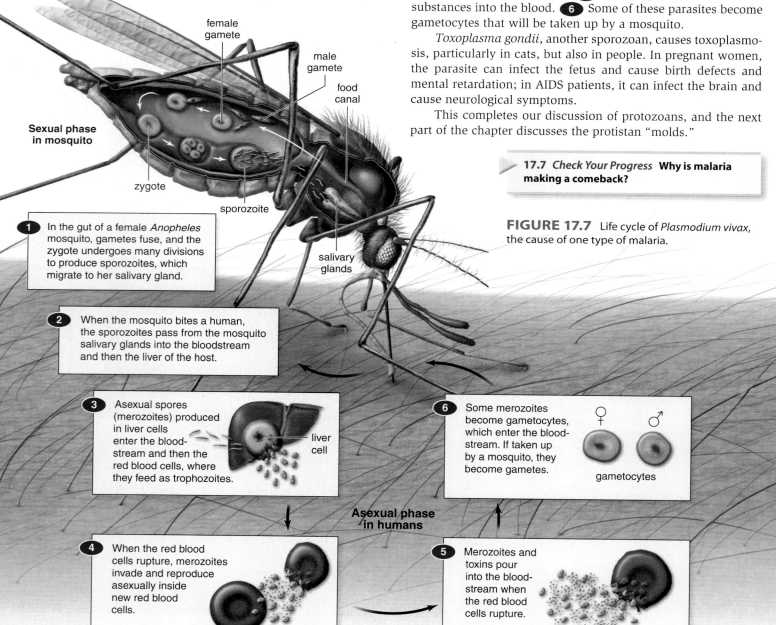

Sexual phase in mosquito

female gamete

male gamete

food canal

zygote

sporozoite

salivary glands

1 In the gut of a female *Anopheles* mosquito, gametes fuse, and the zygote undergoes many divisions to produce sporozoites, which migrate to her salivary gland.

2 When the mosquito bites a human, the sporozoites pass from the mosquito salivary glands into the bloodstream and then the liver of the host.

3 Asexual spores (merozoites) produced in liver cells enter the blood-stream and then the red blood cells, where they feed as trophozoites.

liver cell

4 When the red blood cells rupture, merozoites invade and reproduce asexually inside new red blood cells.

Asexual phase in humans

5 Merozoites and toxins pour into the bloodstream when the red blood cells rupture.

6 Some merozoites become gametocytes, which enter the bloodstream. If taken up by a mosquito, they become gametes.

gametocytes

FIGURE 17.7 Life cycle of *Plasmodium vivax,* the cause of one type of malaria.

In forests and woodlands, slime molds phagocytize, and therefore help dispose of bacteria and dead plant material. The two types of slime molds and the water molds are different in their structure, behavior, and nutrition from each other and from fungi, which we discuss in Chapter 18.

17.8 The diversity of protists includes slime molds and water molds

Plasmodial Slime Molds Usually, **plasmodial slime molds** exist as a plasmodium, a diploid, multinucleated, cytoplasmic mass enveloped by a slimy sheath that creeps along, phagocytizing decaying plant material in a forest or agricultural field (**Fig. 17.8**). Approximately 500 species of plasmodial slime molds have been described. Many species are brightly colored. At times that are unfavorable to growth, such as during a drought, the plasmodium develops many **sporangia**, reproductive structures that produce spores.

The spores produced by a sporangium can survive until moisture is sufficient for them to germinate. In plasmodial slime molds, spores release a haploid flagellated cell or an amoeboid

cell. Eventually, two of them fuse to form a zygote that feeds and grows, producing a multinucleated plasmodium once again.

Cellular Slime Molds In keeping with their name, **cellular slime molds** are so called because they exist as individual amoeboid cells. They are common in soil, where they feed on bacteria and yeasts. Their small size prevents them from being seen. Nearly 70 species of cellular slime molds have been described.

As the food supply runs out or unfavorable environmental conditions develop, the cells release a chemical that causes them to aggregate into a pseudoplasmodium. The pseudoplasmodium stage is temporary and eventually gives rise to a fruiting body, in which sporangia produce spores. When favorable conditions return, the spores germinate, releasing haploid amoeboid cells, and the asexual cycle begins again.

Water Molds The water molds usually live in the water, where they form furry growths when they parasitize fishes or insects and decompose remains. In spite of their common name, some water molds live on land and parasitize insects and plants. Nearly 500 species of water molds have been described. A water mold, *Phytophthora infestans*, was responsible for the 1840s potato famine in Ireland. However, most water molds are saprotrophic and live off dead organic matter. Another well-known water mold is *Saprolegnia*, which is often seen as a white, cottonlike mass on dead organisms.

Plasmodium, *Physarum*

Sporangia, *Hemitrichia* |— 1 mm —|

dead insect

filaments of water mold

Water molds have a filamentous body as do fungi, but their cell walls are largely composed of cellulose, whereas fungi have cell walls of chitin. The life cycle of water molds also differs from that of fungi. During asexual reproduction, water molds produce motile spores (2n zoospores), which are flagellated. The organism is diploid (not haploid as in the fungi), and meiosis produces gametes. The phylum name Oomycota refers to the enlarged tips (called oogonia) where eggs are produced.

Our survey of the protists shifts gears as we begin our discussion of algae, beginning with the diatoms and dinoflagellates in Section 17.9.

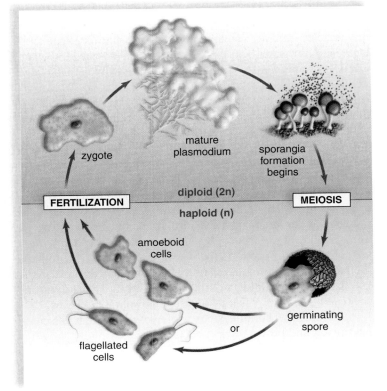

zygote

mature plasmodium

sporangia formation begins

diploid (2n)

FERTILIZATION

MEIOSIS

haploid (n)

amoeboid cells

germinating spore

or

flagellated cells

> **17.8 Check Your Progress** How did *Phytophthora infestans* change U.S. history?

FIGURE 17.8 Life cycle of plasmodial slime molds.

Algae Are Photosynthetic Protists of Environmental Importance

Learning Outcomes 12–16, page 322

Algae are the photosynthetic protists. Our survey of algae includes the golden brown diatoms and the variously colored dinoflagellates, which are major producers in the oceans; the red algae and the brown algae, which are multicellular; and the green algae, which are ancestral to plants. Some authorities classify the green algae with the plants.

17.9 The diatoms and dinoflagellates are significant algae in the oceans

Diatoms Most **diatoms** (approximately 11,000 species) are free-living photosynthetic cells that inhabit aquatic and marine environments. Diatoms are the most numerous unicellular algae in the oceans and freshwater environments. Diatoms are a significant part of the **phytoplankton**, photosynthetic organisms that are suspended in the water in both freshwater and marine ecosystems, where they serve as an important source of food and oxygen for heterotrophs.

The structure of a diatom is often compared to a hat box because the cell wall has two halves, or valves, with the larger valve acting as a "lid" that fits over the smaller valve (**Fig. 17.9A**). When diatoms reproduce asexually, each receives one old valve. The new valve fits inside the old one; therefore, new diatoms are smaller than the original ones. When they reproduce sexually, the size returns to normal.

The cell wall of a diatom has an outer layer of silica, a common ingredient in glass. The valves are covered with a great variety of striations and markings that form beautiful patterns when observed under the microscope. These are actually depressions or pores through which the organism makes contact with the outside environment. The remains of diatoms, called diatomaceous earth, accumulate on the ocean floor and are mined for use as filtering agents, soundproofing materials, components of reflective paints, and gentle polishing abrasives such as those found in silver polish and toothpaste.

Dinoflagellates The **dinoflagellates** (about 4,000 species) are usually bounded by protective cellulose plates impregnated with silicates (**Fig. 17.9B**). Typically, the organism has two flagella; one lies in a longitudinal groove with its distal end free, and the other lies in a transverse groove that encircles the organism. The longitudinal flagellum acts as a rudder, and the beating of the transverse flagellum causes the cell to spin as it moves forward.

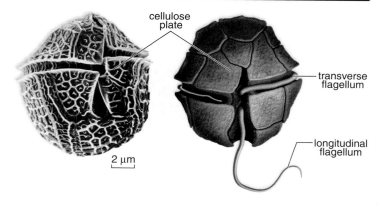

FIGURE 17.9B *Gonyaulax,* a dinoflagellate. This dinoflagellate is responsible for the poisonous "red tide" that sometimes occurs along the coasts.

The chloroplasts of a dinoflagellate vary in color from yellow-green to brown and some species, such as *Noctiluca,* are capable of bioluminescence (producing light). Being a part of the phytoplankton, the dinoflagellates are an important source of food for small animals in the ocean. They also live within the bodies of some invertebrates as symbionts. Symbiotic dinoflagellates lack cellulose plates and flagella and are called zooxanthellae. Corals, members of the animal kingdom, usually contain large numbers of zooxanthellae, which provide their hosts with organic nutrients while the corals in turn provide wastes that fertilize the algae. Some dinoflagellates lack chloroplasts and are heterotrophic; some of these are parasitic.

Like the diatoms, dinoflagellates are one of the most important groups of producers in marine environments. Occasionally, however, particularly in polluted waters in late summer, they undergo a population explosion and become more numerous than usual. At these times, their density can equal 30,000 in a single milliliter. When dinoflagellates, such as *Gonyaulax,* increase in number, they may cause a phenomenon called "**red tide.**" Massive fish kills can occur as the result of a powerful neurotoxin produced by these dinoflagellates. Humans who consume shellfish that have fed during a *Gonyaulax* outbreak may suffer from shellfish poisoning, which paralyzes the respiratory organs.

Section 17.10 continues our survey of the algae by discussing the red and brown algae.

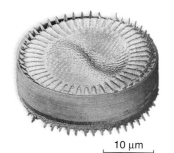

FIGURE 17.9A *Cyclotella,* a diatom. Diatoms live in "glass houses" because the outer visible valve, which fits over the smaller inner valve, contains silica.

> **17.9 Check Your Progress** What is a common characteristic in dinoflagellates?

Red Algae

The **red algae** include more than 5,000 species of multicellular organisms. These algae live primarily in warm seawater, both shallow and deep. Some grow attached to rocks in the intertidal zone, where they are exposed at low tide. Others can grow at depths exceeding 200 m, where light barely penetrates. Red algae are usually fairly small and delicate, although some species can exceed a meter in length.

Some forms of red algae are simple filaments, but most have complex branches with a feathery, flat, or expanded, ribbonlike appearance. Coralline algae are red algae whose cell walls are impregnated with calcium carbonate. In some instances, they contribute as much to the growth of coral reefs as do coral animals.

Red algae are economically important. Agar is a gelatin-like product made primarily from the algae *Gelidium* and *Gracilaria*. Agar is used commercially to make capsules for vitamins and drugs, as a material for making dental impressions, and as a base for cosmetics. In the laboratory, agar is a solidifying agent for a bacterial culture medium. When purified, it becomes the gel for electrophoresis, a procedure that separates proteins or nucleotides. Agar is also used in food preparation—as an anti-drying agent for baked goods and to make jellies and desserts set rapidly. Carrageenan, extracted from *Chondrus crispus* (**Fig. 17.10A**), is an emulsifying agent for the production of chocolate and cosmetics. The reddish-black wrappings around sushi rolls consist of processed blades from *Porphyra*, another red alga.

Brown Algae

The **brown algae** consist of over 1,500 species of seaweeds. The brown algae range from small forms with simple filaments to large, multicellular forms that may reach 100 m in length. Like the vast majority of brown algae, rockweed, *Fucus*, lives in cold ocean waters (**Fig. 17.10B**). The brown algae have chlorophylls *a* and *c* in their chloroplasts and a type of carotenoid pigment (fucoxanthin) that gives them their characteristic color. Reserve food is stored as a carbohydrate called *laminarin*.

The multicellular forms of green, red, and brown algae are called **seaweeds**, a common term for any large, complex alga. Brown algae are often observed along the rocky coasts in the north temperate zone, where they are pounded by waves as the tide comes in and are exposed to dry air as the tide goes out. They dry out slowly, however, because their cell walls contain a mucilaginous, water-retaining material.

Both *Laminaria*, commonly called kelp, and *Fucus*, known as rockweed, are brown algae that grow along the shoreline. In deeper waters, the giant kelps (*Macrocystis* and *Nereocystis*) often grow extensively in vast beds. Individuals of the genus *Sargassum* sometimes break off from their holdfasts and form floating masses. Brown algae not only provide food and habitat for marine organisms, but are harvested for human food and for fertilizer in several parts of the world. *Macrocystis* is the source of alginate (algin), a pectinlike material that is added to ice cream, sherbet, cream cheese, and other products to give them a smooth, stable consistency.

Laminaria is unique among the protists because members of this genus show tissue differentiation—that is, they transport organic nutrients by way of a tissue that resembles the phloem in land plants. Most brown algae have the alternation of generations life cycle, but some species of *Fucus* are unique in that meiosis produces gametes, and the adult is always diploid, as in animals.

The very versatile green algae are discussed in Section 17.11.

17.10 Check Your Progress What is agar?

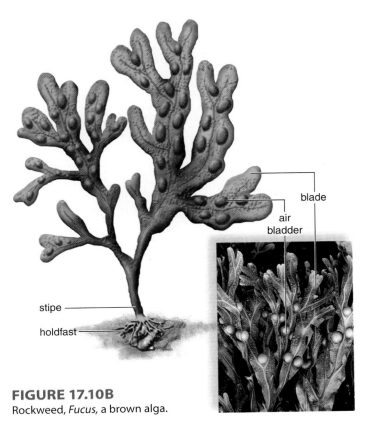

blade

air
bladder

stipe

holdfast

FIGURE 17.10B
Rockweed, *Fucus*, a brown alga.

FIGURE 17.10A *Chondrus crispus*, a red alga.

Some biologists classify green algae as plants because they have chlorophyll *a* and *b*, store excess carbohydrates as starch, and have cellulose in their cell walls. However, the green algae do not develop from an embryo protected by the organism, as do plants adapted to living on land.

The **green algae** include approximately 7,500 species. Although green algae contain chlorophyll, they are not always green; some possess pigments that give them an orange, red, or rust color. They inhabit a variety of environments, including oceans, freshwater environments, snowbanks, the bark of trees, and the backs of turtles. The green algae also form symbiotic relationships with fungi, plants, and animals. As discussed in Section 18.16, they associate with fungi in lichens. Members of phylum Chlorophyta occur in an abundant variety of forms. The majority of green algae are unicellular; however, filamentous and colonial forms exist. Some multicellular green algae are seaweeds that resemble lettuce leaves.

Chlamydomonas   An actively moving unicellular green alga called *Chlamydomonas* inhabits still, freshwater pools. Its fossil ancestors date back over a billion years. It has a definite cell wall and a single, large, cup-shaped chloroplast that contains a *pyrenoid*, a dense body where starch is synthesized. In many species, a bright red eyespot, or stigma, exists on the chloroplast, which is sensitive to light and helps bring the organism into the light, where photosynthesis can occur. Two long, whiplike flagella project from the anterior end of this alga and operate with a breaststroke motion.

Chlamydomonas most often reproduces asexually (**Fig. 17.11A**). During asexual reproduction, mitosis produces as many as 16 daughter cells still within the parent cell wall. Each daughter cell then secretes a cell wall and acquires flagella. The daughter cells escape by secreting an enzyme that digests the parent cell wall.

Chlamydomonas occasionally reproduces sexually when growth conditions are unfavorable. Gametes of two different mating types come into contact and join to form a zygote. A heavy wall forms around the zygote, and it becomes a resistant zygospore that undergoes a period of dormancy. When a zygospore germinates, it produces four zoospores by meiosis. **Zoospores** are flagellated spores typical of aquatic species.

Spirogyra   **Filaments** are end-to-end chains of cells that form after cell division occurs in only one plane. In some algae, the filaments are branched, and in others the filaments are unbranched. *Spirogyra* is an unbranched, filamentous green alga. Filamentous green algae often grow epiphytically (not taking in nutrients) on aquatic flowering plants; they also attach to rocks or other objects under water. Some filaments are suspended in the water.

Spirogyra is found in green masses on the surfaces of ponds and streams. It has ribbonlike, spiralled chloroplasts (**Fig. 17.11B**). During sexual reproduction, *Spirogyra* undergoes **conjugation**, a temporary union, during which the cells exchange genetic material. The two filaments line up parallel to each other, and the cell contents of one filament move into the cells of the other filament, forming diploid zygotes. Resistant zygospores survive the winter, and in the spring, they undergo meiosis to produce new haploid filaments.

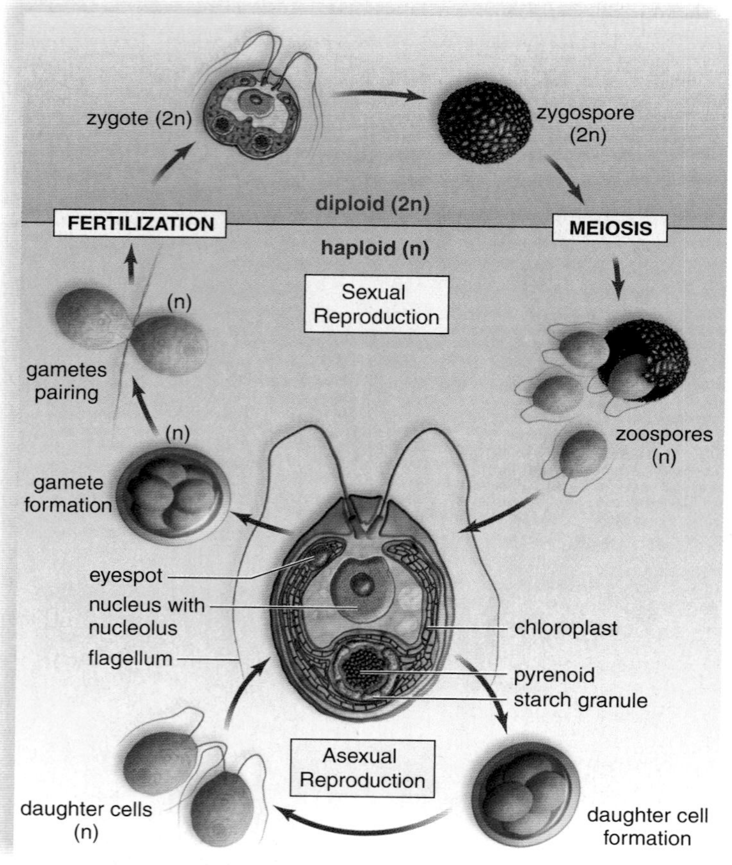

zygote (2n)

zygospore (2n)

diploid (2n)

FERTILIZATION

MEIOSIS

haploid (n)

Sexual Reproduction

(n)

gametes pairing

(n)

gamete formation

zoospores (n)

eyespot

nucleus with nucleolus

flagellum

chloroplast

pyrenoid

starch granule

Asexual Reproduction

daughter cells (n)

daughter cell formation

FIGURE 17.11A   Reproduction in *Chlamydomonas,* a motile green alga.

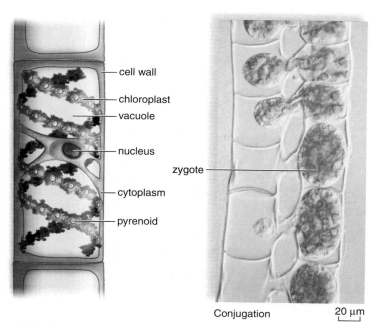

cell wall

chloroplast

vacuole

nucleus

zygote

cytoplasm

pyrenoid

Conjugation

20 μm

FIGURE 17.11B   Cell anatomy and conjugation in *Spirogyra,* a filamentous green alga.

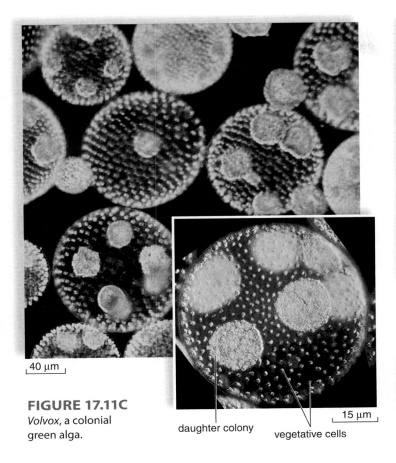

40 μm

FIGURE 17.11C
Volvox, a colonial green alga.

daughter colony   vegetative cells   15 μm

Ulva, several individuals   One individual

FIGURE 17.11D *Ulva*, a multicellular alga.

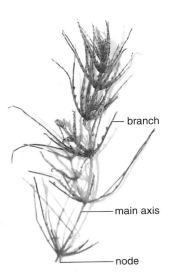

branch

main axis

node

Chara, several individuals   One individual

FIGURE 17.11E *Chara*, a stonewort.

Volvox Among the flagellated green algae, a number of forms are colonial, meaning that they exist in a **colony**, a loose association of independent cells. *Volvox* is a well-known colonial green alga. A *Volvox* colony is a hollow sphere with thousands of flagellated cells arranged in a single layer surrounding a watery interior. Each cell of a *Volvox* colony resembles a *Chlamydomonas* cell—perhaps it is derived from daughter cells that fail to separate following zoospore formation. In *Volvox*, the cells cooperate in that the flagella beat in a coordinated fashion. Some cells are specialized for reproduction, and each of these can divide asexually to form a new daughter colony (**Fig. 17.11C**). This daughter colony resides for a time within the parent colony, but then it leaves by releasing an enzyme that dissolves away a portion of the parent colony, allowing it to escape.

Ulva A multicellular green alga, *Ulva*, is commonly called sea lettuce because it lives in the sea and has a leafy appearance (**Fig. 17.11D**). The thallus (body) is two cells thick and can be as much as a meter long. *Ulva* has an alternation of generations life cycle like that of plants, except that both generations look exactly alike and the gametes both look the same.

Stoneworts The stoneworts are green algae that live in freshwater lakes and ponds. They are called stoneworts because some species, such as *Chara*, are encrusted with calcium carbonate deposits (**Fig. 17.11E**). The main axis of the alga, which can be over a meter long, is a single file of very long cells. Whorls of branches occur at multicellular nodes, regions between the giant cells of the main axis. Each of the branches is also a single file of cells.

Stoneworts basically have the same life cycle as *Chlamydomonas* (Fig. 17.11A). However, during sexual reproduction, they do produce male and female multicellular reproductive structures at the nodes. The male structure produces flagellated sperm, and the female structure produces a single egg. The diploid zygote is retained until it is enclosed by tough walls. DNA sequencing data suggest that among green algae, the stoneworts are most closely related to plants (see Fig. 17.3).

The three major types of life cycles are found among the algae, as discussed in Section 17.12.

17.11 *Check Your Progress* What evidence would a molecular biologist use to show that plants are most closely related to stoneworts?

Both asexual and sexual reproduction occur in algae, depending on the species and the environmental conditions. The types of life cycles seen in algae occur in other protists as well as in plants and animals.

Asexual Reproduction

When the environment is favorable to growth, asexual reproduction is a frequent mode of reproduction among protists. Asexual reproduction requires only one parent. The offspring are identical to this parent because they receive a copy of only this parent's genes. The new individuals are likely to survive and flourish if the environment is steady. Various modes of asexual reproduction occur, but growth alone produces a new individual. For example, in *Spiryogyra* and stoneworts, fragmentation of filaments produces a new individual.

Sexual Reproduction

With its genetic recombination due in part to fertilization and independent assortment of chromosomes, sexual reproduction is more likely to occur among protists when the environment is changing and is unfavorable to growth. Recombination of genes might produce individuals that are more likely to survive extremes in the environment—such as high or low temperatures, acidic or basic pH, or the lack of a particular nutrient.

Sexual reproduction requires two parents, each of which contributes chromosomes (genes) to the offspring by way of gametes. The gametes fuse to produce a diploid zygote. A reproductive cycle is *isogamous* when the gametes look alike (called isogametes) and *oogamous* when the gametes are dissimilar (called heterogametes). Usually, a small, flagellated sperm fertilizes a large egg with plentiful cytoplasm.

Meiosis occurs during sexual reproduction. Just *when* it occurs makes the sexual life cycles diagrammed in **Figure 17.12A–C** differ from one another. In these diagrams, the diploid phase is shown in blue, and the haploid phase is shown in tan. The haploid life cycle (**Fig. 17.12A**) most likely evolved first. In the **haploid life cycle**, the zygote divides by meiosis to form haploid spores that develop into haploid individuals. In algae, the spores are typically zoospores. The zygote is the only diploid stage in this life cycle, and the haploid individual gives rise to gametes. This form of sexual reproduction is seen in *Chlamydomonas* and a number of other algae, including stoneworts.

In **alternation of generations**, the sporophyte (2n) produces haploid spores by meiosis (**Fig. 17.12B**). A spore develops into a haploid gametophyte that produces gametes. The gametes fuse to form a diploid zygote, and the zygote develops into the sporophyte. This life cycle is characteristic of some algae (e.g., *Ulva* and *Laminaria*) and all plants. In *Ulva*, the haploid and diploid generations have the same appearance. In plants, they are noticeably different from each other. Also in plants, the zygote becomes an embryo protected by the female gametophyte. None of the algae protect the embryo as plants do.

In the **diploid life cycle**, which is also typical of animals, a diploid individual produces gametes by meiosis (**Fig. 17.12C**). Gametes are the only haploid stage in this cycle. They fuse to form a zygote that develops into the diploid individual. This life cycle is rare in algae but does occur in a few species of the brown alga *Fucus*.

> **17.12** *Check Your Progress* Why are algae not classified as plants?

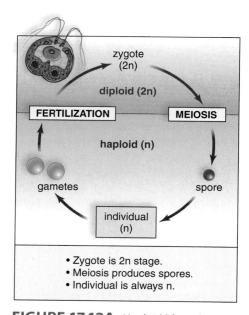

FIGURE 17.12A Haploid life cycle.

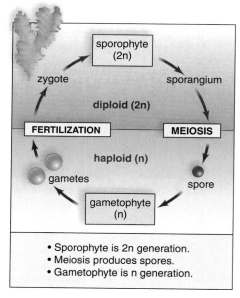

FIGURE 17.12B Alternation of generations.

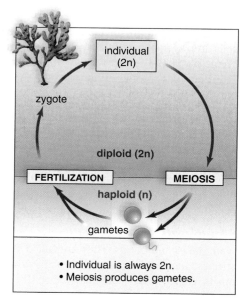

FIGURE 17.12C Diploid life cycle.

The protists we study today are not expected to include the direct ancestors to fungi, plants, and animals. Instead, they may be related to the other eukaryotic groups by way of common ancestors that have not been discovered in the fossil record. Today, nucleic acid sequencing alone tells us which group of protists are most closely related to the fungi, plants, and animals.

Perhaps today's protists represent an adaptive radiation experienced by the first eukaryotic cell to evolve. While certain structures present in eukaryotic cells may have been unique to them alone, others appear to be endosymbionts, present only because they were engulfed by a much larger cell. Mutualism is a powerful force that shaped the eukaryotic cell and also shapes all sorts of relationships in the living world. For example, we have already mentioned that mutualism between flowers and their pollinators has contributed to the success of flowering plants.

All possible forms of reproduction and nutrition are present among the protists, but each of the other eukaryotic groups specializes in a particular type of reproduction and a particular method of acquiring needed nutrients. Fungi, as we shall see, reproduce by means of windblown spores during both an asexual and sexual life cycle, and they are saprotrophic. Plants have the alternation of generations life cycle and are photosynthetic. Animals have the diploid life cycle and are heterotrophic. Chapter 18 pertains to the evolution of plants and fungi, while Chapter 19 discusses the evolution of animals.

The Chapter in Review

Summary

Protists Cause Disease Too

- Protist diseases are transmitted by contaminated food or water or by a vector.
- Some diseases caused by protists are malaria, giardiasis, amoebic dysentery, and African sleeping sickness.

Protists May Represent the Oldest Eukaryotic Cells

17.1 Eukaryotic organelles arose by endosymbiosis

- Mitochondria may be derived from aerobic bacteria; chloroplasts may be derived from cyanobacteria engulfed by a prokaryotic cell.

17.2 Protists are a diverse group

- Protists are diverse in cellular organization, means of nutrition, reproduction, and locomotion.
- Algae are photosynthetic; protozoans are heterotrophic—some ingest by endocytosis and some are parasitic; water molds are heterotrophic by absorption.

17.3 How can the protists be classified?

- DNA and RNA sequencing is now being used to classify protists.

Protozoans Are Heterotrophic Protists with Various Means of Locomotion

17.4 Protozoans called flagellates move by flagella

- Zooflagellates include the trypanosomes (causing such diseases as African sleeping sickness and leishmaniasis), *Giardia*, and *Trichomonas*.
- Euglenoids are flexible but often contain chloroplasts.

17.5 Protozoans called amoeboids move by pseudopods

- Pseudopods are extensions that form when cytoplasm streams forward.
- Examples of amoeboids include *Amoeba* and *Entamoeba*.

- Foraminiferans and radiolarians have tests that build up on the ocean floor and become available on land due to a geologic upheaval.

17.6 Protozoans called ciliates move by cilia

- *Paramecium*, a well-known example, is found in ponds.
- Paramecia reproduce asexually by binary fission or sexually by conjugation.

17.7 Protozoans called sporozoans are not motile

- Sporozoans are parasitic and spore-forming.
- *Plasmodium* causes malaria, a widespread disease in tropical countries.
- *Toxoplasma gondii* causes toxoplasmosis in AIDS patients and pregnant women.

Some Protists Have Moldlike Characteristics

17.8 The diversity of protists includes slime molds and water molds

- Plasmodial slime molds phagocytize decaying plant material.
- Cellular slime molds exist as individual amoeboid cells.
- Water molds form furry growths on insects or fishes.

Algae Are Photosynthetic Protists of Environmental Importance

17.9 The diatoms and dinoflagellates are significant algae in the oceans

- Diatoms have a cell wall, and dinoflagellates have protective cellulose plates impregnated with silica.
- Diatoms and dinoflagellates are marine producers and, as part of the phytoplankton, an important source of food for heterotrophs. Dinoflagellates are responsible for a toxic bloom called the red tide.

17.10 Red algae and brown algae are multicellular

- Red algae live in warm seawater, coralline red algae are a significant part of coral reefs; source of agar and used to wrap sushi rolls.

- Brown algae live along northern rocky coasts; provide food and habitat for marine organisms and are used for food and as fertilizers by humans.

17.11 Green algae are ancestral to plants

- Green algae photosynthesize in the same manner as green plants but are not classified as plants because they are not adapted to reproducing on land.
- Green algae diversity is exemplified by
  - *Chlamydomonas:* flagellated, unicellular, haploid life cycle
  - *Spirogyra:* filamentous, spiral chloroplast, conjugation
  - *Volvox:* colony of flagellated cells, daughter colonies develop inside adult
  - *Ulva:* multicellular, called sea lettuce, alternation of generation life cycle
  - Stoneworts: whorls of branches encrusted with calcium carbonate, DNA sequencing places these algae closest to plants

17.12 Life cycles among the algae have many variations

- Asexual reproduction occurs when the environment is favorable to growth; sexual reproduction occurs when the environment is changing and unfavorable to growth.
- Haploid life cycle (e.g., *Chlamydomonas*): meiosis produces spores and adult is haploid.
- Alternation of generation (e.g., *Ulva*): meiosis produces spores that become a haploid generation, egg and sperm unite to produce a zygote that become a diploid generation.
- Diploid life cycle (e.g., *Fucus*): meiosis produces egg and sperm and adult is diploid.

Testing Yourself

Protists May Represent the Oldest Eukaryotic Cells

1. Which of these sequences depicts a hypothesized evolutionary scenario?
   a. cyanobacteria—mitochondria
   b. Golgi—mitochondria
   c. mitochondria—cyanobacteria
   d. cyanobacteria—chloroplast
2. Which of the following pairs is matched correctly?
   a. slime mold—animal-like   c. algae—plantlike
   b. water mold—funguslike   d. protozoan—funguslike
3. Which of the following is not photosynthetic?
   a. algae   d. protozoans
   b. slime molds   e. More than one answer is correct.
   c. water molds
4. Determining how protists evolved will allow us to better understand the origin of
   a. plants.   d. bacteria.
   b. animals.   e. a, b, and c are all correct choices.
   c. fungi.
5. **THINKING CONCEPTUALLY** Why would you predict that mitochondria contain DNA that codes for mitochondrial proteins?

Protozoans Are Heterotrophic Protists with Various Means of Locomotion

6. What structure is related to the difficulty in classifying euglenoids?
   a. flagella   d. mitochondrion
   b. nucleus   e. c and d are both correct.
   c. chloroplast
7. Which of the following moves by flagella?
   a. *Paramecium*   d. Both a and b are correct.
   b. *Euglena*   e. None of the choices are correct.
   c. amoeba
8. Contractile vacuoles are found in _____ and function in _____.
   a. amoeboids, feeding
   b. amoeboids, water regulation
   c. ciliates, feeding
   d. ciliates, reproduction
   e. apicomplexans, attachment to host cells
9. Ciliates
   a. can move by pseudopods.
   b. are not as varied as other protists.
   c. have a gullet for food gathering.
   d. are closely related to the radiolarians.
10. List the four means of protozoan locomotion. Then, for each type, name the group(s) using this means of locomotion and give a unique characteristic of each group.
11. Considering your answer to question 10, what is surprising about the group called Alveolata in Figure 17.3?

Some Protists Have Moldlike Characteristics

12. Which is (are) found in slime molds but not in fungi?
   a. nonmotile spores   d. photosynthesis
   b. amoeboid vegetative cells   e. All of these are correct.
   c. zygote formation
13. Which is saprotrophic, as are fungi?
   a. cellular slime mold
   b. plasmodial slime mold
   c. water mold

Algae Are Photosynthetic Protists of Environmental Importance

14. Which of these pairs is mismatched?
   a. amoeboids—pseudopods
   b. sporozoans—disease agents
   c. algae—variously colored
   d. slime molds—trypanosomes
15. Which pair is properly matched?
   a. water mold—flagellate   c. *Plasmodium vivax*—mold
   b. trypanosome—protozoan   d. amoeboid—algae
16. Which of the following statements is incorrect?
   a. Unicellular protists can be quite complex.
   b. Euglenoids are motile but have chloroplasts.
   c. Plasmodial slime molds are amoeboid but have sporangia.
   d. *Volvox* is colonial but box-shaped.
   e. Both b and d are incorrect.
17. Dinoflagellates
   a. usually reproduce sexually.
   b. have protective cellulose plates.
   c. are insignificant producers of food and oxygen.
   d. have cilia instead of flagella.
   e. tend to be larger than brown algae.

For questions 18–22, match each organism to a characteristic. Answers can be used more than once and each organism can have more than one answer.

KEY:

  a. photosynthetic   d. closely related to plants
  b. protozoan   e. closely related to animals
  c. cause disease

18. Red algae
19. Ciliates
20. Brown algae
21. Amoeboids
22. Green algae
23. Which of these is not a green alga?
    a. *Volvox*   d. *Chlamydomonas*
    b. *Fucus*   e. *Ulva*
    c. *Spirogyra*
24. Which is not a characteristic of brown algae?
    a. multicellular   d. harvested for commercial
    b. chlorophylls *a* and *b*   reasons
    c. live along rocky coasts   e. contain a brown pigment
25. Which of these protists are not flagellated?
    a. *Volvox*   d. *Chlamydomonas*
    b. *Spirogyra*   e. trypanosomes
    c. dinoflagellates
26. Which is a false statement?
    a. Only protists that are heterotrophic and not photosynthetic are flagellated.
    b. Among protozoans, sporozoans are parasitic.
    c. Among protists, the haploid cycle is common.
    d. Ciliates exchange genetic material during conjugation.
    e. Slime molds have an amoeboid stage.
27. Which pair is properly matched?
    a. water mold—flagellate   c. *Plasmodium vivax*—mold
    b. trypanosome—protozoan d. amoeboid—algae
28. All of the following descriptions are true of brown algae except that they
    a. range in size from small   c. live on land.
       to large.   d. are photosynthetic.
    b. are a type of seaweed.   e. are usually multicellular.
29. Which of the following statements is false?
    a. Slime molds and water molds are protists.
    b. Some algae have flagella.
    c. Amoeboids have pseudopods.
    d. Among protists, only green algae ever have a sexual life cycle.
    e. Conjugation occurs among some green algae.
30. Which of the following is used to distinguish between algae and plants in this text?
    a. photosynthesis   d. aquatic habitat
    b. cell walls   e. chlorophyll
    c. embryonic development
31. In the haploid life cycle (e.g., *Chlamydomonas*),
    a. meiosis occurs following zygote formation.
    b. the adult is diploid.
    c. fertilization is delayed beyond the diploid stage.
    d. the zygote produces sperm and eggs.
32. In which life cycle does meiosis produce spores?
    a. haploid   d. Both a and b are correct.
    b. diploid   e. Both a and c are correct.
    c. alternation of generations

33. Give a reason why diatoms, dinoflagellates, red algae, and brown algae are useful, or otherwise significant, to human beings.
34. **THINKING CONCEPTUALLY** Considering your study of this chapter, what is surprising about the group called Stramenopila in Figure 17.3?

Understanding the Terms

algae 324	phytoplankton 332
alternation of generations 336	plankton 324
amoeboid 328	plasmodial slime mold 331
brown algae 333	protist 324
cellular slime mold 331	protozoan 324
ciliate 329	pseudopod 328
colony 335	radiolarian 328
conjugation 334	red algae 333
diatom 332	red tide 333
dinoflagellate 332	seaweed 333
diploid life cycle 336	slime mold 324
endosymbiotic theory 324	sporangium 331
euglenoid 327	spore 324
filament 334	sporozoan 330
foraminiferan 328	trichocyst 329
green algae 334	water mold 324
haploid life cycle 336	zooflagellate 327
malaria 330	zooplankton 328
phagocytize 328	zoospore 334

Match the terms to these definitions:

  a. _____ Cytoplasmic extension of amoeboid protists; used for locomotion and engulfing food.
  b. _____ Freshwater or marine unicellular protist with a cell wall consisting of two silica-impregnated valves; extremely numerous in phytoplankton.
  c. _____ Freshwater and marine organisms suspended on or near the surface of the water.
  d. _____ Part of plankton containing protozoans and other types of microscopic animals.
  e. _____ Complex unicellular protist that moves by means of cilia.

Thinking Scientifically

1. While studying a unicellular alga, you discover a mutant in which the daughter cells do not separate after mitosis. This gives you an idea about how filamentous algae may have evolved. Explain.
2. You are an investigator trying to discover a cure for malaria. Why might you decide to target human red blood cells? What might you want to learn about the merozoite stage of infection that is not known now (see #4 in Fig.17.7)?

ARIS™ *Visit* www.mhhe.com/maderconcepts *for practice quizzes, animations, videos, and activities designed to help you master the material in this chapter.*

18

Evolution of Plants and Fungi

LEARNING OUTCOMES

After studying this chapter, you should be able to accomplish the following outcomes.

Some Plants Are Carnivorous

1 Describe how the environment may have selected the lifestyle of carnivorous plants.

The Evolution of Plants Spans 500 Million Years

2 Draw an evolutionary tree for plants showing four significant innovations during their evolution.
3 Distinguish between the sporophyte and the gametophyte in the plant life cycle.
4 Associate the increased dominance of the sporophyte with plant adaptations to a dry land environment.

Plants Are Adapted to the Land Environment

5 Compare and contrast the adaptations of bryophytes, seedless vascular plants, and seed plants to the land environment.
6 Compare and contrast the life cycles of the moss, fern, pine, and the flowering plant, emphasizing reproductive adaptations to the land environment.
7 Discuss the significance of the carboniferous forest to today's world.
8 Discuss the benefits of plants, especially seed plants, to humans.

Fungi Have Their Own Evolutionary History

9 Compare and contrast the structure and terrestrial adaptations of fungi to those of plants.
10 Describe the structure of lichens and mycorrhizal fungi, and explain why they are considered mutualistic.
11 Name and describe the three groups of fungi and how they differ from one another.
12 Discuss the economic and medical aspects of fungi.

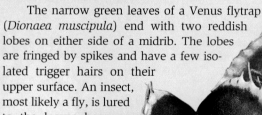
Venus flytrap with fly

We think of plants as largely minding their own business as they quietly photosynthesize their food. So it may come as a surprise that some plants are carnivorous—they feed on insects, or even on amphibians, birds, and mammals. Carnivorous plants are adapted to living in bogs, swamps, and marshes, where water collects, oxygen is limited, and decomposers are inhibited from recycling nutrients. These plants can survive where others cannot because they feed on animals, usually insects, as a source of nitrogen. We can think of carnivorous plants as a part of the great adaptive radiation of flowering plants into all sorts of environments on planet Earth. Let's look at three plant species among the 600 or so that are carnivorous.

The narrow green leaves of a Venus flytrap (*Dionaea muscipula*) end with two reddish lobes on either side of a midrib. The lobes are fringed by spikes and have a few isolated trigger hairs on their upper surface. An insect, most likely a fly, is lured to the leaves because the spikes are lined by a band of sweet-smelling nectar glands. When the fly touches one trigger hair twice or two hairs in rapid succession, a trap is sprung, and the spikes of the lobes become inter-

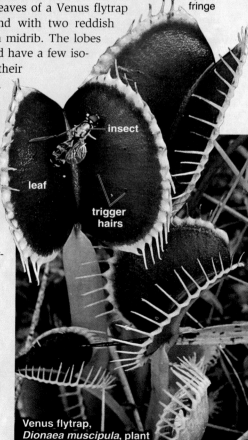
spiked fringe

insect

leaf

trigger hairs

Venus flytrap, *Dionaea muscipula*, plant

Venus flytrap flower

Some Plants Are Carnivorous

locked, enclosing the insect like the bars of a jail cell. This action, which takes only a half-second, involves a rapid loss of turgor pressure within the cells of the leaf. Digestive enzymes pour forth from glands on the leaf surface, breaking down the helpless victim. As a part of this remarkable adaptation, a small insect—not worth the energy to digest—walks free by just exiting between the spikes.

Sundew plants (e.g., *Drosera capensis*) are rather low-growing, so they are able to capture crawling insects as well as flying ones. The leaves are visually attractive, covered with hairs tipped with knobs that sparkle like dew in the sun. The insect gets stuck on the sticky hairs, and the knobs secrete mucuslike juices, which break down the insect. Rolling from the tip, the leaves enclose the prey, preventing it from escaping and hastening the digestive process.

Among the pitcher plants, the yellow trumpet pitcher (*Sarracenia flava*) stands over three feet tall, and its leaves form a pitcher. Just like the pitcher in your kitchen, this one is also filled with water—containing digestive juices, of course. The pitcher has a hood covered with glands that secrete nectar to attract insects, such as ants. Any inquisitive insect that leaves the hood to investigate the pitcher is greeted by downward-pointing hairs. And because the sides of the pitcher are slippery, the in-

sect loses its grip, tumbling into the lethal waters.

The carnivorous plants, like all plants, are adapted to living on land. Of all things, their flowers are pollinated by insects! The flowers produce seeds within fruits. In the three species we discussed, the fruit is a dry capsule that contains rather small seeds. Carnivorous plants are just a small part of the 280,000 known species that make up the kingdom Plantae. This chapter emphasizes the evolution of a reproductive strategy that allows flowering plants to live and be prevalent in all regions of the biosphere. Fungi, also discussed in this chapter, have ecological, economic, and medical importance.

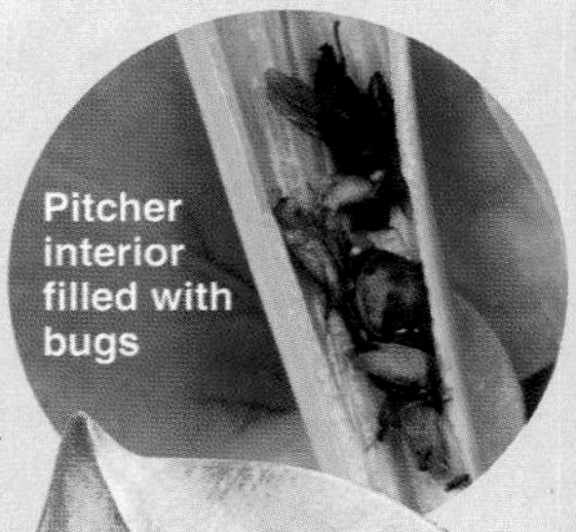

Pitcher interior filled with bugs

hood with nectar-producing glands

pitcher with pitfall trap

bulbs release digestive enzymes

sticky hairs

narrow leaf form

Sundew leaf enfolds prey

Cape sundew, *Drosera capensis*, plant

Yellow trumpet pitcher flower

Yellow trumpet pitcher, *Sarracenia flava*, plant

Plants evolved from green algae, which are adapted to living and reproducing in water. Four evolutionary innovations, corresponding with the four major groups of plants (bryophytes, ferns, pines, and flowering plants), represent adaptations useful to a plant's mode of living and reproducing on land.

18.1 Evidence suggests that plants evolved from green algae

Plants are multicellular, photosynthetic eukaryotes that range in size from the diminutive duckweed to the giant coastal redwoods of California. Plants are important ecologically, industrially, and medically. Members of the plant kingdom have an ancient and intriguing evolution.

Plants are believed to have evolved from freshwater green algal species over 500 million years ago. Scientists base this hypothesis on the following evidence: Both green algae and plants (1) contain chlorophylls *a* and *b* and various accessory pigments, (2) store excess carbohydrates as starch, and (3) have cellulose in their cell walls. In recent years, molecular systematists have compared the sequences of DNA bases coding for ribosomal RNA between organisms. The results suggest that plants are most closely related to a group of green algae known as stoneworts, and perhaps should be classified with them (see Fig. 17.3). **Figure 18.1** shows two representatives of the stoneworts, *Chara* and *Coleochaete*. Whereas algae live in an aquatic environment, plants live on the land, a dry environment. As we learn in the next section, over time plants have become increasingly adapted to a dry, terrestrial environment.

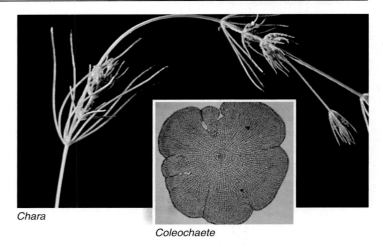

Chara

Coleochaete

FIGURE 18.1 Close algal relatives of plants.

> **18.1** *Check Your Progress* **What data would you collect to determine whether all carnivorous plants are closely related?**

18.2 The evolution of plants is marked by four innovations

Plant evolution is marked by four evolutionary innovations that can be conveniently associated with the four major groups of plants living on land today (**Fig. 18.2A**).

1 The nonvascular plants such as mosses (and the other three plant groups) nourish and protect a multicellular embryo that completes the life cycle. The embryo is protected by spe-

1 In mosses, the embryo is protected by a special structure, *right.*

3 In a pine tree, seeds disperse offspring, *right.*

2 A fern has vascular tissue, *right.*

4 In flowering plants, the seeds are enclosed in fruits, *right.*

FIGURE 18.2A Representatives of the four major groups of plants.

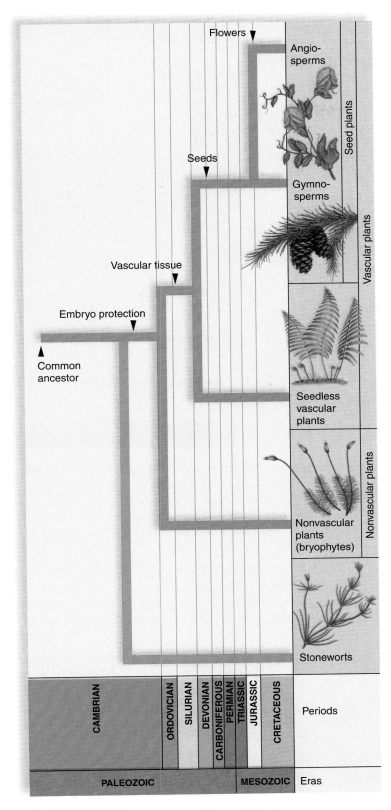

FIGURE 18.2B Evolutionary history of plants.

cialized tissues in the plant's body. This feature, which distinguishes plants from green algae, is an important adaptation to land because the embryo is thereby protected from drying out.

② The seedless vascular plants such as ferns (as well as the gymnosperms and angiosperms) have **vascular tissue**. Evolution

of vascular tissue solved the problem of transporting water and solute to cells when the plant body is surrounded by air, a dry environment. Vascular tissue has another advantage for plants. Vascular tissue has strong cell walls that allow plants with vascular tissue to attain a greater height and better access to sunlight.

③ The gymnosperms, which are primarily cone-bearing plants such as pine trees, and the angiosperms, flowering plants such as cherry trees, produce seeds. A seed contains an embryo and stored organic nutrients within a protective coat. When a seed is planted, it germinates (begins to grow), and a plant of the next generation emerges. Seeds are highly resistant structures, well suited for protecting a plant embryo from drying out until conditions are favorable for germination.

④ The fourth evolutionary innovation was the advent of the flower, a reproductive structure. Flowers attract pollinators, such as insects that are adapted to flying about in a dry environment, and they give rise to fruits, food for animals that can also help disperse the seeds. The special adaptations of flowering plants have allowed them to become far more diversified than any other group of plants.

All the innovations mentioned here are adaptations to a land existence. **Figure 18.2B** traces the evolutionary history of these plant adaptations, and the box below lists the characteristics of these groups. Section 18.3 discusses the life cycle utilized by all plants.

> **18.2** *Check Your Progress* Would you hypothesize that carnivorous plants evolved before or after most angiosperms? Explain.

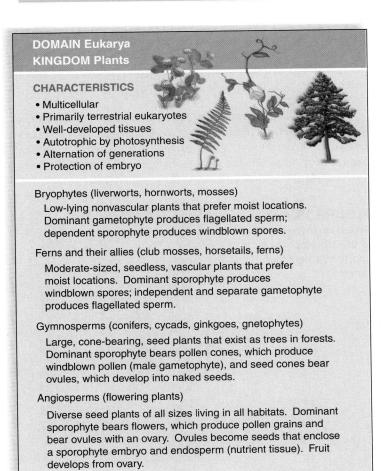

DOMAIN Eukarya
KINGDOM Plants

CHARACTERISTICS
- Multicellular
- Primarily terrestrial eukaryotes
- Well-developed tissues
- Autotrophic by photosynthesis
- Alternation of generations
- Protection of embryo

Bryophytes (liverworts, hornworts, mosses)
  Low-lying nonvascular plants that prefer moist locations. Dominant gametophyte produces flagellated sperm; dependent sporophyte produces windblown spores.

Ferns and their allies (club mosses, horsetails, ferns)
  Moderate-sized, seedless, vascular plants that prefer moist locations. Dominant sporophyte produces windblown spores; independent and separate gametophyte produces flagellated sperm.

Gymnosperms (conifers, cycads, ginkgoes, gnetophytes)
  Large, cone-bearing, seed plants that exist as trees in forests. Dominant sporophyte bears pollen cones, which produce windblown pollen (male gametophyte), and seed cones bear ovules, which develop into naked seeds.

Angiosperms (flowering plants)
  Diverse seed plants of all sizes living in all habitats. Dominant sporophyte bears flowers, which produce pollen grains and bear ovules with an ovary. Ovules become seeds that enclose a sporophyte embryo and endosperm (nutrient tissue). Fruit develops from ovary.

18.3   Plants have an alternation of generations life cycle

All plants have a life cycle that includes an **alternation of generations**. In this life cycle, two multicellular individuals alternate, each producing the other (**Fig. 18.3**). The two individuals are (1) a sporophyte, which represents the diploid generation, and (2) a gametophyte, which represents the haploid generation.

The **sporophyte** (2n) is so named for its production of spores by meiosis. A **spore** is a haploid reproductive cell that develops into a new organism without needing to fuse with another reproductive cell. In the plant life cycle, a spore undergoes mitosis and becomes a gametophyte.

The **gametophyte** (n) is so named for its production of gametes. In plants, eggs and sperm are produced by mitotic cell division. A sperm and egg fuse, forming a diploid zygote that undergoes mitosis and becomes the sporophyte.

Two observations are in order. First, meiosis produces haploid spores. This is consistent with the realization that the sporophyte is the diploid generation and spores are haploid reproductive cells. Second, mitosis occurs both as a spore becomes a gametophyte and again as a zygote becomes a sporophyte. Indeed, it is the occurrence of mitosis at these times that results in two generations.

A dominant sporophyte eventually allowed flowering plants to invade and be prevalent in dry, terrestrial environments, as discussed in Section 18.4.

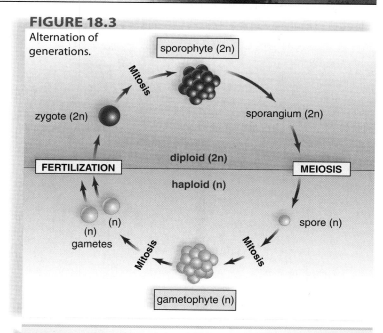

FIGURE 18.3
Alternation of generations.

18.3 Check Your Progress What does the gametophyte do in carnivorous plants?

18.4   Sporophyte dominance was adaptive to a dry land environment

Plants differ as to which generation is dominant—that is, more conspicuous. The appearances of the gametophyte and sporophyte in each group of plants are shown in **Figure 18.4A**. Only the sporophyte ever has vascular tissue for the transport of water and nutrients, and only plants with a dominant sporophyte attain significant height.

Notice that as the sporophyte gains in dominance, the gametophyte becomes microscopic. Microscopic size allows the gametophyte to be dependent on and protected by the generation that has vascular tissue. As the gametophyte becomes smaller among vascular plants, its dependence on the sporophyte increases.

Reproductive Adaptation to the Land Environment
Sporophyte dominance can be associated with an increasing adaptation for reproduction in a dry, terrestrial environment. To emphasize this concept, we will contrast features of fern adaptation to those of flowering plant adaptation. Ferns are seedless vascular plants with a dominant sporophyte. In ferns:

1. The sporophyte produces spores that disperse (scatter) separate gametophytes.

2. The gametophyte is a small, heart-shaped structure that has no vascular tissue and can dry out if the environment is not moist.

FIGURE 18.4A
Reduction in the size of the gametophyte as sporophyte becomes dominant.

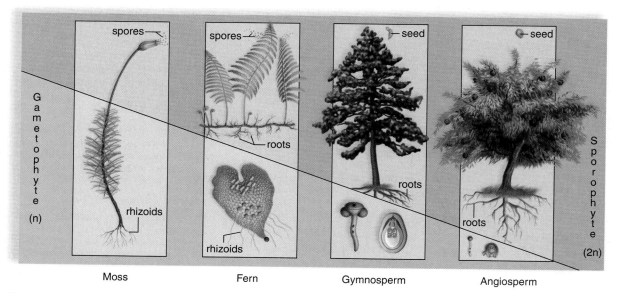

Moss   Fern   Gymnosperm   Angiosperm

3. Each **archegonium** (pl., archegonia) on the surface of a gametophyte produces an egg that is fertilized by a *flagellated sperm*, which must swim to the archegonium in a film of external water (**Fig. 18.4B**, *left*).

The water-dependent gametophyte makes it more difficult for ferns and related plants to spread to and live in dry environments.

Flowering plants are seed plants with a dominant sporophyte. In flowering plants:

1. The sporophyte produces seeds that disperse separate sporophytes protected by seed coats.

2. The female gametophyte is microscopic and retained and protected within an **ovule**, a sporophyte structure located within the sporophyte tissue of a flower (**Fig. 18.4B**, *right*).

3. The male gametophytes are *pollen grains* that are transported by wind, insects, or birds; therefore, they do not need external water to reach the egg. Following fertilization, the ovule becomes a seed.

In seed plants, all reproductive structures are protected from drying out in the terrestrial environment.

Other Adaptations to the Land Environment

Sporophyte dominance is accompanied by adaptations for water and nutrient transport and also for preventing water loss. The sporophyte is protected against drying out in ways other than the ability to transport water. The leaves and other exposed parts of the sporophyte plant are covered by a waxy cuticle (**Fig. 18.4C**). The **cuticle** is relatively impermeable and provides an effective barrier to water

cuticle

Stained photomicrograph of a leaf cross section

Vascular plant leaves have a cuticle and stomata.

stomata

400×
Falsely colored scanning electron micrograph of leaf surface

FIGURE 18.4C Features of the leaves of vascular plants.

loss, but it also limits gas exchange. Leaves and other photosynthesizing organs have little openings called stomata (sing., **stoma**) that let carbon dioxide enter while allowing oxygen and water to exit (Fig. 18.4C). A stoma is bordered by guard cells that regulate whether it is open or closed. A stoma closes when the weather is hot and dry, and this keeps water loss to a minimum.

Having given a broad overview of this chapter, we will now discuss each group of plants in turn. Section 18.5 discusses the bryophytes.

18.4 Check Your Progress Name two ways that increasing dominance of the sporophyte is an adaptation to the land environment.

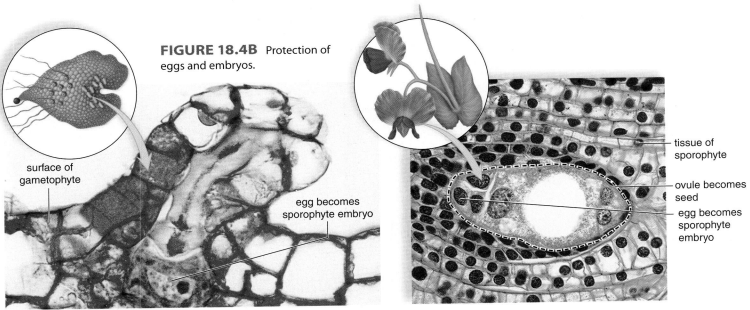

FIGURE 18.4B Protection of eggs and embryos.

surface of gametophyte

egg becomes sporophyte embryo

Archegonium in seedless plants

tissue of sporophyte

ovule becomes seed

egg becomes sporophyte embryo

Ovule in seed plants

In this part of the chapter, we survey the plant kingdom and discuss the diversity, adaptations, and economic value of the bryophytes, ferns, gymnosperms (cone-bearing plants), and angiosperms (flowering plants). This is the order of evolution among plants.

18.5 Bryophytes are nonvascular plants in which the gametophyte is dominant

The nonvascular plants lack a specialized means of transporting water and organic nutrients. Although they often have a "leafy" appearance, these plants do not have true roots, stems, and leaves—which, by definition, must contain true vascular tissue. Therefore, the nonvascular plants are said to have rootlike, stemlike, and leaflike structures. The term **bryophyte** (lowercase *b*) is a general term for nonvascular plants.

Approximately 24,000 species of nonvascular plants have been described. They are classified into three living groups: hornworts, liverworts, and mosses (**Fig. 18.5A**). Genetic and comparative evidence indicates that bryophytes diverged independently before the origin of vascular plants.

The Generations of Bryophytes In bryophytes, the gametophyte is the dominant generation, meaning that it is the generation we recognize as the plant. The female gametophyte produces eggs in archegonia, and the male gametophyte produces flagellated sperm in antheridia. The sperm swim to the vicinity of the egg in a continuous film of water. Following fertilization, the zygote becomes a sporophyte embryo that is protected from drying out within the archegonium. The embryo develops into a sporophyte that is attached to, and derives its nourishment from, the photosynthetic gametophyte. The sporophyte produces windblown spores that are resistant to drying out.

The lack of vascular tissue and the need for sperm to swim to archegonia in a film of water largely account for the limited height of bryophytes (usually no taller than a few centimeters). Nevertheless, some bryophytes compete well in harsh environments because the gametophyte can reproduce asexually, allowing them to spread into stressful and even dry habitats.

Diversity of Mosses Mosses (phylum Bryophyta) comprise the largest phylum of nonvascular plants, with over 15,000 species. The gametophytes of most mosses appear as small, leaflike structures arranged around a stemlike axis that sprouts rhizoids. Mosses can be found from the Antarctic through the tropics to parts of the Arctic. Although most prefer damp, shaded locations in the temperate zone, some survive in deserts, and others inhabit bogs and streams. In forests, they frequently form a mat that covers the ground and rotting logs. In dry environments, they may become shriveled, turn brown, and look completely dead. As soon as it rains, however, the plant becomes green and resumes metabolic activity.

The so-called copper mosses live only in the vicinity of copper and thus can serve as an indicator plant for copper deposits. Luminous mosses, which glow with a golden-green light, are found in caves, under the roots of trees, and in other dimly lit places. Although mosses do not grow well in polluted areas such as cities, many times they can be seen growing on bricks near moist ground.

Sphagnum, also called **peat** or peat moss, has commercial importance. Over 350 species of *Sphagnum* have been identified. The cells of this moss have a tremendous ability to absorb water, which is why gardeners often use peat moss to improve the water-holding capacity of the soil. Peat moss can also be used as fuel, and it was successfully used as a substitute for bandages during World War II.

The term *moss* is a misnomer for some plants. Many of the common "mosses" are not even nonvascular plants. Irish moss is an edible red alga that grows in leathery tufts along northern seacoasts. Reindeer moss is a lichen that serves as the dietary mainstay of reindeer and caribou in northern lands. Club mosses, discussed in section 18.6, are vascular plants, and Spanish moss, which hangs in grayish clusters from trees in the southeastern United States, is a flowering plant of the pineapple family.

FIGURE 18.5A Representative bryophytes.

Hornwort

Liverwort gametophyte

Moss gametophyte

Life Cycle of Mosses Figure 18.5B describes the life cycle of a typical temperate-zone moss. ❶ The mature gametophyte consists of shoots that bear antheridia and archegonia. ❷ An antheridium has an outer layer of sterile cells and an inner mass of cells that become flagellated sperm. An archegonium, which looks like a vase with a long neck, has an outer layer of sterile cells with a single egg located at the base. ❸ After fertilization, the sporophyte embryo is protected from drying out because it is located within the archegonium. ❹ The mature sporophyte lacks vascular tissue and is dependent on the gametophyte. It consists of a foot enclosed in female gametophyte tissue, a stalk, and an upper capsule (the **sporangium**), where windblown spores are produced by

meiosis. In some species, the sporangium can produce as many as 50 million spores. ❺ The spores disperse the gametophyte generation. ❻ A spore germinates into an algalike, branching filament of cells that precedes and produces the upright leafy shoots.

Having discussed the first plants on land, namely the bryophytes, we will now take a look at the ferns and their relatives in Section 18.6.

> **18.5** *Check Your Progress* **Name an advantage and disadvantage to the manner in which bryophytes reproduce on land.**

FIGURE 18.5B Moss life cycle, *Polytrichum* sp.

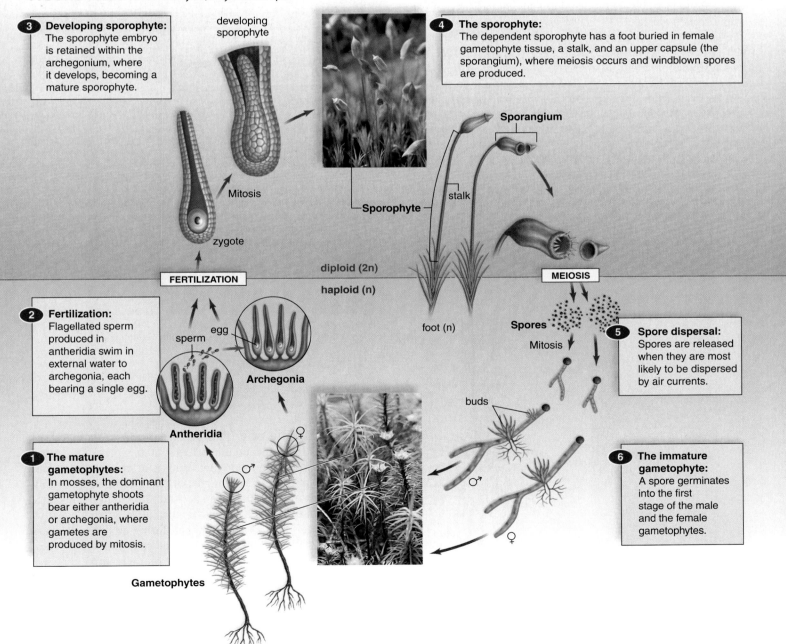

3 Developing sporophyte: The sporophyte embryo is retained within the archegonium, where it develops, becoming a mature sporophyte.

4 The sporophyte: The dependent sporophyte has a foot buried in female gametophyte tissue, a stalk, and an upper capsule (the sporangium), where meiosis occurs and windblown spores are produced.

2 Fertilization: Flagellated sperm produced in antheridia swim in external water to archegonia, each bearing a single egg.

1 The mature gametophytes: In mosses, the dominant gametophyte shoots bear either antheridia or archegonia, where gametes are produced by mitosis.

5 Spore dispersal: Spores are released when they are most likely to be dispersed by air currents.

6 The immature gametophyte: A spore germinates into the first stage of the male and the female gametophytes.

developing sporophyte

Mitosis

zygote

FERTILIZATION

sperm   egg

Archegonia

Antheridia

Gametophytes

Sporophyte   stalk

Sporangium

diploid (2n)

haploid (n)

foot (n)

MEIOSIS

Spores

Mitosis

buds

The evolution of vascular tissue was a significant innovation in plants as they became increasingly adapted to living on land. Vascular tissue allows water and solutes to be transported from the roots anchored in the soil to the leaves. Vascular tissue also enables plants to attain a height that allows the leaves to efficiently capture solar energy. The lack of vascular tissue accounts for the limited height of bryophytes. The seedless vascular plants, which have vascular tissue but do not produce seeds, were dominant from the late Devonian period through the Carboniferous period.

Diversity of Seedless Vascular Plants Today's seedless vascular plants, such as ferns, **horsetails,** and **club mosses** (**Fig. 18.6A**), have limited height. During the coniferous period, however, these plants were much taller and were a part of great swamp forests (see Fig. 18.8).

Ferns (phylum Pterophyta) include approximately 11,000 species. They range in size from minute aquatic species less than 1 cm in diameter to giant tropical tree ferns that exceed 20 m in height. Ferns are most abundant in warm, moist, tropical regions, but they can also be found in temperate regions and as far north as the Arctic Circle. Several species live in dry, rocky places, and others have adapted to an aquatic life.

The large and conspicuous leaves of ferns, called **fronds**, are commonly divided into leaflets. The cinnamon fern is named for spore-bearing fronds that become cinnamon-colored as the season progresses; those of the hart's tongue fern are straplike and leathery; and those of the maidenhair fern are broad, with subdivided leaflets (**Fig. 18.6B**). In nearly all ferns, the fronds first appear in a curled-up form called a fiddlehead, which unrolls as it grows.

Cinnamon fern, *Osmunda cinnamomea*

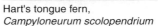

Hart's tongue fern, *Campyloneurum scolopendrium*

Maidenhair fern, *Adiantum pedatum*

FIGURE 18.6B Diversity of fern fronds.

Ground pine, *Lycopodium*

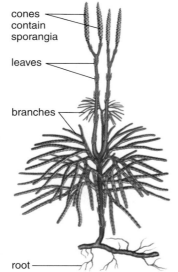

cones contain sporangia

leaves

branches

root

FIGURE 18.6A Sporophyte of a club moss.

Economic Value of Ferns At first, it may seem that ferns do not have much economic value, but they are often used by florists in decorative bouquets and as ornamental plants in the home and garden. Although not true wood, trunks from tropical tree ferns are often used as a building material because they resist decay. Ferns also have medicinal value; many Native Americans use ferns as an astringent during childbirth to stop bleeding, and the maidenhair fern is the source of an expectorant. Fern extracts were also used to expel intestinal parasites such as tapeworms.

The Environmental Protection Agency (EPA) has pointed out that the Boston fern can substantially remove formaldehyde from the air in closed rooms.

Life Cycle of Ferns The life cycle of a typical temperate-zone fern is shown in **Figure 18.6C**. ❶ The dominant sporophyte produces windblown spores by meiosis within ❷ sporangia that may be located within **sori** on the underside of the leaflets. ❸ The windblown spores disperse ❹ the gametophyte, the generation that lacks vascular tissue.

❺ The separate heart-shaped gametophyte produces flagellated sperm that swim in a film of water from the antheridium to the egg within the archegonium, where fertilization occurs.

❻ The sporophyte embryo is protected within the archegonium, where it gradually develops into a mature sporophyte.

In the history of the Earth, ferns preceded the gymnosperms, which are discussed in Section 18.7.

> **18.6** *Check Your Progress* **In what two ways is the fern life cycle dependent on external water?**

FIGURE 18.6C Fern life cycle.

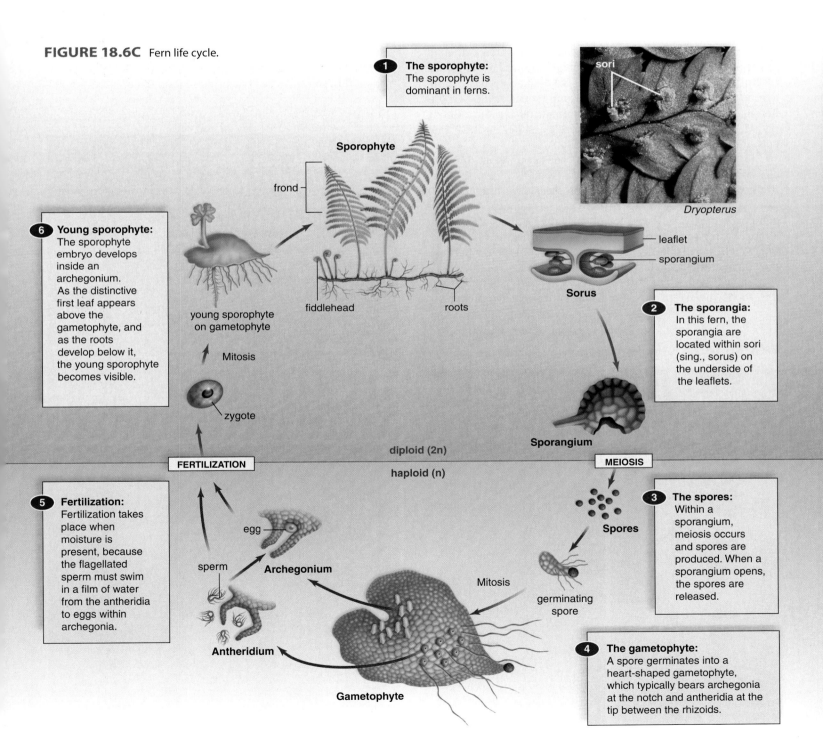

1 **The sporophyte:** The sporophyte is dominant in ferns.

Dryopterus

6 **Young sporophyte:** The sporophyte embryo develops inside an archegonium. As the distinctive first leaf appears above the gametophyte, and as the roots develop below it, the young sporophyte becomes visible.

2 **The sporangia:** In this fern, the sporangia are located within sori (sing., sorus) on the underside of the leaflets.

5 **Fertilization:** Fertilization takes place when moisture is present, because the flagellated sperm must swim in a film of water from the antheridia to eggs within archegonia.

3 **The spores:** Within a sporangium, meiosis occurs and spores are produced. When a sporangium opens, the spores are released.

4 **The gametophyte:** A spore germinates into a heart-shaped gametophyte, which typically bears archegonia at the notch and antheridia at the tip between the rhizoids.

Sporophyte — frond

fiddlehead — young sporophyte on gametophyte — roots

Mitosis

zygote

FERTILIZATION

diploid (2n)

haploid (n)

leaflet — sporangium — Sorus — Sporangium

MEIOSIS

Spores

germinating spore

Mitosis

sperm — egg — Archegonium — Antheridium — Gametophyte

18.7 Most gymnosperms bear cones on which the seeds are "naked"

The evolution of the seed was the next significant innovation in the vascular plants. The seed contains a sporophyte generation, along with stored food, within a protective seed coat. The ability of seeds to survive harsh conditions until the environment is again favorable for growth largely accounts for the dominance of seed plants today.

Diversity of Gymnosperms The four groups of living **gymnosperms** are **cycads, ginkgoes, gnetophytes,** and **conifers** (**Fig. 18.7A**). All of these plants have ovules and subsequently develop seeds that are exposed on the surface of cone scales or analogous structures. (Since the seeds are not enclosed by fruit, gymnosperms are said to have "naked seeds.") Early gymnosperms were present in the swamp forests of the Carboniferous period, and they became dominant during the Triassic period. Today, living gymnosperms are classified into 780 species, the most plentiful being the conifers.

Conifers (phylum Coniferophyta) consist of about 575 species of trees, many of them evergreens such as pines, spruces, firs, cedars, hemlocks, redwoods, cypresses, yews, and junipers. The name *conifer* signifies plants that bear cones, but other gymnosperm phyla are also cone-bearing. Vast areas of northern temperate regions are covered in evergreen coniferous forests. The tough, needlelike leaves of pines conserve water because they have a thick cuticle and recessed stomata.

The coastal redwood (*Sequoia sempervirens*), a conifer native to northwestern California and southwestern Oregon, is the tallest living vascular plant; it may attain nearly 100 m in height. Another conifer, the bristlecone pine (*Pinus longaeva*) of the White Mountains of California, is the oldest living tree; one is 4,900 years of age.

Economic Value of Conifers The wood of pines and other conifers is used extensively in construction. The wood consists primarily of transport tissue that lacks some of the more rigid cell types found in flowering trees. Therefore, it is considered a "soft" rather than a "hard" wood. Although called softwoods, some conifers, such as yellow pine, have wood that is actually harder than so-called hardwoods. The foundations of the 100-year-old Brooklyn Bridge are made of southern yel-

FIGURE 18.7A Gymnosperm diversity.

Cycad, *Encephalartos humlis*
Female plant with large seed cones

Ginkgo, *Ginkgo biloba*
Female maidenhair tree with seeds

Gnetophyte, *Ephedra*
Branched shrub with scalelike leaves

Conifer, *Picea*
Spruce tree with pollen cones and seed cones

low pine. Resin, produced naturally by pines to prevent insect and fungal invasion, is harvested commercially for a derived product called turpentine.

Life Cycle of Pines

Figure 18.7B shows the life cycle of a typical conifer, such as a pine. **①** Pine trees have two types of cones and produce two types of spores, an innovation by seed plants. This innovation leads to the production of pollen grains and seeds. **②** A megaspore mother cell within an ovule produces four megaspores by meiosis. Only one of these becomes a microscopic and dependent **female gametophyte**. Microspore mother cells produce microspores by meiosis, and they become the **male gametophytes**, which are windblown **pollen grains**. **③** During **pollination**, pollen grains are transported by wind to female gametophytes, and **④** the sperm they contain fertilize the eggs of the female gametophytes. *Note that no external water is needed to accomplish fertilization in a seed plant.*

⑤ The sporophyte embryo is enclosed within the ovule, which becomes a "naked" seed on the scale of the seed cone. The winged seeds are windblown and disperse the sporophyte, the generation that has vascular tissue.

The early gymnosperms were dominant and enjoyed great height during the Carboniferous period, as discussed in Section 18.8.

> **18.7** *Check Your Progress* Cite life cycle changes that represent seed plant adaptations, as exemplified by a pine tree, for reproducing on land.

FIGURE 18.7B
Pine life cycle.

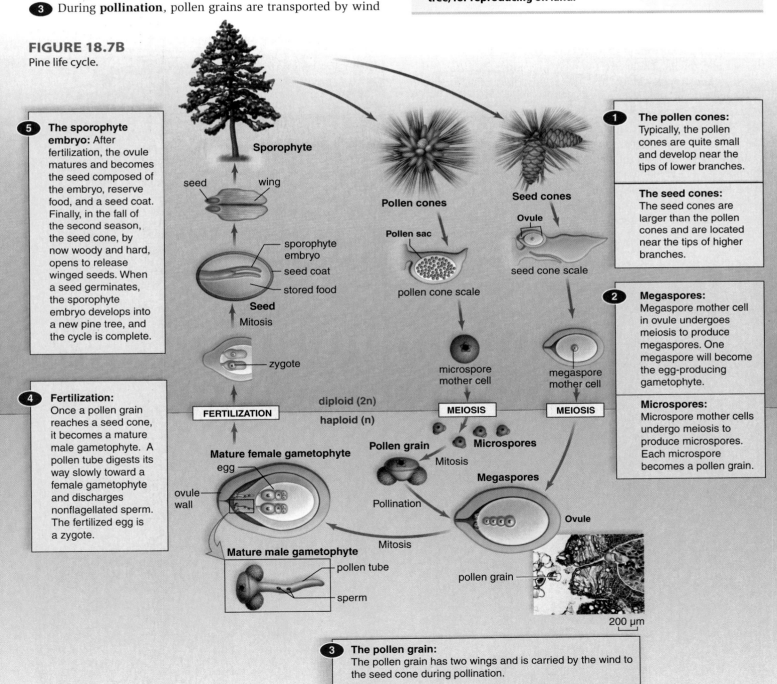

⑤ The sporophyte embryo: After fertilization, the ovule matures and becomes the seed composed of the embryo, reserve food, and a seed coat. Finally, in the fall of the second season, the seed cone, by now woody and hard, opens to release winged seeds. When a seed germinates, the sporophyte embryo develops into a new pine tree, and the cycle is complete.

④ Fertilization: Once a pollen grain reaches a seed cone, it becomes a mature male gametophyte. A pollen tube digests its way slowly toward a female gametophyte and discharges nonflagellated sperm. The fertilized egg is a zygote.

① The pollen cones: Typically, the pollen cones are quite small and develop near the tips of lower branches.

The seed cones: The seed cones are larger than the pollen cones and are located near the tips of higher branches.

② Megaspores: Megaspore mother cell in ovule undergoes meiosis to produce megaspores. One megaspore will become the egg-producing gametophyte.

Microspores: Microspore mother cells undergo meiosis to produce microspores. Each microspore becomes a pollen grain.

Sporophyte
seed — wing
sporophyte embryo
seed coat
stored food
Seed
Mitosis
zygote
diploid (2n)
haploid (n)
FERTILIZATION
Mature female gametophyte
egg
ovule wall
Mature male gametophyte
pollen tube
sperm

Pollen cones
Pollen sac
pollen cone scale
microspore mother cell
MEIOSIS
Microspores
Pollen grain
Mitosis
Pollination

Seed cones
Ovule
seed cone scale
megaspore mother cell
MEIOSIS
Megaspores
Mitosis
Ovule
pollen grain
200 µm

③ The pollen grain: The pollen grain has two wings and is carried by the wind to the seed cone during pollination.

18.8   Carboniferous forests became the coal we use today

Our industrial society runs on fossil fuels, such as **coal**. The term fossil fuel might seem odd at first, until one realizes that it refers to the remains of organic material from ancient times. During the Carboniferous period more than 300 MYA, a great swamp forest (**Fig. 18.8**) encompassed what is now northern Europe, the Ukraine, and the Appalachian Mountains in the United States. The weather was warm and humid, and the trees grew very tall. These were not the trees we know today; instead, they were related to today's seedless vascular plants: the club mosses, horsetails, and ferns! Club mosses today may stand as high as 30 cm, but their ancient relatives were 35 m tall and 1 m wide. The spore-bearing cones were up to 30 cm long, and some had leaves more than 1 m long. Horsetails too—at 18 m tall—were giants compared to today's specimens. The tree ferns were also taller than tree ferns found in the tropics today, and there were two other types of trees: seed ferns and early gymnosperms. "Seed fern" is a misnomer because it has been shown that these plants, which only resemble ferns, were actually a type of gymnosperm.

The amount of biomass was enormous, and occasionally the swampy water rose and the trees fell. Submerged trees do not decompose well, and their partially decayed remains became covered by sediment that sometimes changed to sedimentary rock. Sedimentary rock applied pressure, and the organic material then became coal, a fossil fuel. This process continued for millions of years, resulting in immense deposits of coal. Subsequent geologic upheavals raised the deposits to the level where they can be mined today.

With a change of climate, the trees of the Carboniferous period became extinct, and only their much smaller relatives survived to our time. Without these ancient forests, our life today would be far different because coal helped bring about our industrialized society.

Having discussed the gymnosperms, we will begin our discussion of flowering plants in Section 18.9.

> **18.8** *Check Your Progress* **How do we know what the Carboniferous forest was like?**

FIGURE 18.8 Swamp forest of the Carboniferous period.

Fossil seed fern

club mosses

horsetail

seed fern

early gymnosperm

fern

18.9 Angiosperms are the flowering plants

Angiosperms (phylum Anthophyta) are the flowering plants. The flowering plants evolved at the beginning of the Cenozoic era, some 65 MYA when the first flying insects appeared. Flowers and their pollinators evolved together, forging an alliance that continues today. The flower is an innovation of angiosperms and so, too, is the fruit. Whereas the flower is involved in securing the success of pollination so necessary to seed formation, the fruit serves as a means by which animals help with seed dispersal, as we shall discuss in Chapter 24. Angiosperms are an exceptionally large and successful group of plants, with 240,000 known species—six times the number of all other plant groups combined. Angiosperms live in all sorts of habitats, from fresh water to desert, and from the frigid north to the torrid tropics. It would be impossible to exaggerate the importance of angiosperms in our everyday lives. Angiosperms include all the hardwood trees of temperate deciduous forests and all the broadleaved evergreen trees of tropical forests. Also, all herbaceous (nonwoody) plants, such as grasses and most garden plants, are flowering plants. This means that all fruits, vegetables, nuts, herbs, and grains that are the staples of the human diet are angiosperms. As discussed in Section 18.11, they provide us with clothing, food, medicines, and many other commercially valuable products.

The flowering plants are called angiosperms because their ovules, unlike those of gymnosperms, are always enclosed within sporophyte tissues. In the Greek derivation of their name, *angio* ("vessel") refers to the ovary, which develops into a fruit, a unique angiosperm product that contains the seeds.

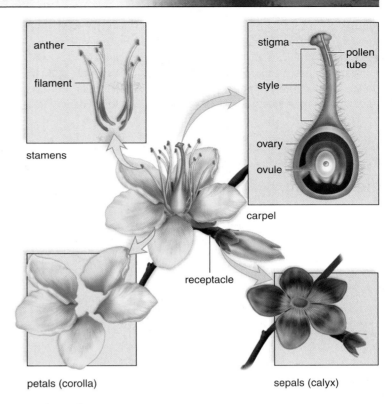

FIGURE 18.9 Generalized flower.

Angiosperm Diversity Most flowering plants belong to one of two groups: **Monocotyledones**, often shortened to simply the **monocots** (about 65,000 species), and **Eudicotyledones**, shortened to **eudicots** (about 175,000 species). It was discovered that some of the plants formerly classified as eudicots diverged before the evolutionary split that gave rise to the two major classes of angiosperms. These earlier evolving plants are not included in the designation eudicots.

Monocots and eudicots are named for their number of **cotyledons**. Cotyledons are seed leaves that contain nutrients and nourish the plant embryo. Monocots have only one cotyledon in their seeds. Common monocots include corn, tulips, pineapple, bamboo, and sugarcane. Monocot flower parts, such as petals, occur in threes or multiples of three. Eudicots possess two cotyledons in their seeds. Common eudicots include cactuses, strawberries, dandelions, poplars, and beans. Eudicot flower parts occur in fours or fives, or multiples thereof. The flower shown in **Figure 18.9** is a eudicot.

The Flower Although flowers vary widely in appearance, most have certain structures in common. The flower stalk expands slightly at the tip into a **receptacle**, which bears the other flower parts. These parts—sepals, petals, stamens, and the carpel (Fig. 18.9)—are attached to the receptacle in whorls (circles).

1. The **sepals**, collectively called the calyx, protect the flower bud before it opens. The sepals may drop off or may be colored like the petals. Usually, however, sepals are green and remain attached to the receptacle.

2. The **petals**, collectively called the corolla, are quite diverse in size, shape, and color. The petals often attract a particular pollinator.

3. Next are the **stamens**. Each stamen consists of two parts: the anther, a saclike container, and the filament, a slender stalk. Pollen grains develop from microspores produced in the anther.

4. At the very center of a flower is the **carpel**, a vaselike structure with three major regions: the **stigma**, an enlarged sticky knob; the **style**, a slender stalk; and the **ovary**, an enlarged base that encloses one or more ovules. The ovule becomes the seed, and the ovary becomes the fruit.

It can be noted that not all flowers have all these parts. A flower is said to be complete if it has all four parts; otherwise, it is called incomplete.

The life cycle of flowering plants is studied in Section 18.10.

> **18.9 Check Your Progress** *a.* **Based on the species shown on pages 340–41, are carnivorous plants monocots or eudicots?** *b.* **How would flower structure allow you to confirm which of the carnivorous plants are closely related?**

Figure 18.10 depicts the life cycle of a typical flowering plant. Like the gymnosperms, flowering plants produce two types of spores. **1** Microspores are produced in the pollen sacs of anthers, and megaspores are produced within the ovary of a carpel. **2** In pollen sacs within the anther, microspore mother cells produce microspores by meiosis. Megaspore mother cells located in ovules

within the ovary produce megaspores by meiosis. Each microspore becomes a pollen grain, but only one **megaspore** develops into an egg-bearing female gametophyte called the **embryo sac**. **3** In most angiosperms, the embryo sac has seven cells; one of these is an egg, and another contains two polar nuclei, so called because they came from opposite ends of the embryo sac.

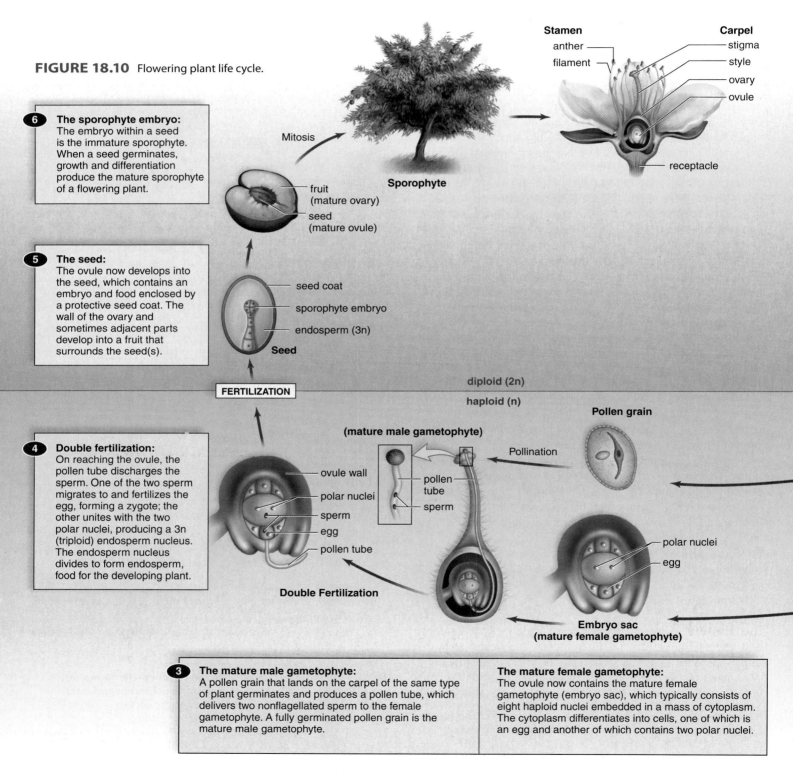

FIGURE 18.10 Flowering plant life cycle.

6 **The sporophyte embryo:**
The embryo within a seed is the immature sporophyte. When a seed germinates, growth and differentiation produce the mature sporophyte of a flowering plant.

5 **The seed:**
The ovule now develops into the seed, which contains an embryo and food enclosed by a protective seed coat. The wall of the ovary and sometimes adjacent parts develop into a fruit that surrounds the seed(s).

4 **Double fertilization:**
On reaching the ovule, the pollen tube discharges the sperm. One of the two sperm migrates to and fertilizes the egg, forming a zygote; the other unites with the two polar nuclei, producing a 3n (triploid) endosperm nucleus. The endosperm nucleus divides to form endosperm, food for the developing plant.

3 **The mature male gametophyte:**
A pollen grain that lands on the carpel of the same type of plant germinates and produces a pollen tube, which delivers two nonflagellated sperm to the female gametophyte. A fully germinated pollen grain is the mature male gametophyte.

The mature female gametophyte:
The ovule now contains the mature female gametophyte (embryo sac), which typically consists of eight haploid nuclei embedded in a mass of cytoplasm. The cytoplasm differentiates into cells, one of which is an egg and another of which contains two polar nuclei.

During pollination, a pollen grain is transported by various means from the anther to the stigma of a carpel, where it germinates. The **pollen tube** carries two sperm to the female gametophyte in the ovule. ④ During **double fertilization**, one sperm unites with an egg, forming a diploid zygote, and the other unites with the polar nuclei, forming a **triploid endosperm** that will be food for the embryo. Ultimately, the ovule becomes a seed that contains the sporophyte embryo and stored food enclosed within a seed coat.

⑤ In angiosperms, seeds are covered by a **fruit**, which is derived from an ovary and possibly accessory parts of the flower. Some fruits, such as apples and tomatoes, provide a fleshy covering for seeds, and other fruits, such as pea pods and acorns, provide a dry covering. ⑥ When a seed germinates, it becomes the mature sporophyte, a flowering plant.

Fruits The fruits of flowers protect and aid in the dispersal of seeds. Dispersal occurs when seeds are transported by wind, gravity, water, or animals to another location. Fleshy fruits may be eaten by animals, which transport the seeds to a new location and then deposit them when they defecate. Because animals live in particular habitats or have particular migration patterns, they are apt to deliver the fruit-enclosed seeds to a suitable location for seed germination and development of the plant.

Flowers and Diversification As discussed at the beginning of Chapter 13 on pages 242–243, plants and their specific pollinators, such as bees, wasps, flies, butterflies, moths, and even bats, are adapted to one another. Glands located in the region of the ovary produce nectar, a nutrient that pollinators gather as they go from flower to flower. The pollinator has mouthparts that are able to obtain the nectar from the base of the flower.

The fact that today there are some 240,000 species of flowering plants and over 700,000 species of insects suggests that the success of angiosperms has contributed to the success of insects, and vice versa. In recent years, the populations of bees and other pollinators have been declining worldwide. Consequently, some plants are endangered because they have lost their normal pollinator. The decline in pollinator populations has been caused by a variety of factors, including pollution, habitat loss, and emerging diseases. Although insecticides should not be applied to crops that are blooming, they frequently are. While beekeepers can quickly move their beehives, wild bees have no protection whatsoever. Then, too, widespread aerial applications to control mosquitoes, medflies, grasshoppers, gypsy moths, and other insects leave no region where wild insect pollinators can reproduce and repopulate. Our present "chemlawn" philosophy says that dandelions and clover, favored by bees, are weeds, and furthermore that lawns should be treated with pesticides. Few farms, suburbs, and cities provide a habitat where bees and butterflies like to live! Migratory pollinators, such as monarch butterflies and some hummingbirds, are also threatened because their nectar corridors no longer exist.

The success of flowering plants parallels their great usefulness to human beings, as discussed in Section 18.11.

①
The stamen:
An anther at the top of a stamen has four pollen sacs. Pollen grains are produced in pollen sacs.

The carpel:
The ovary at the base of a carpel contains one or more ovules. The contents of an ovule change during the flowering plant life cycle.

stigma

style

Anther

Carpel — **ovule**

ovary

pollen sac

microspore mother cell

megaspore mother cell

MEIOSIS

MEIOSIS

Mitosis

Microspores

Megaspores

Mitosis

degenerating megaspores

Ovule

② **Microspores:**
Microspore mother cells undergo meiosis to produce microspores. Each microspore becomes a pollen grain.

Megaspores:
Megaspore mother cell inside ovule undergoes meiosis to produce megaspores. One megaspore will become the egg-producing female gametophyte.

> **18.10** *Check Your Progress* **In angiosperms, the flower attracts insects that aid in pollination, and produces seeds enclosed by fruit. Do you expect this to be the case for carnivorous plants? Explain.**

Plants define the features of and are the producers in most eco-systems. Humans derive most of their sustenance from three flowering plants: wheat, corn, and rice (**Fig. 18.11A**). All three of these plants are in the grass family and are collectively, along with other species, called grains. Most of the Earth's 6.4 billion people have a simple way of life, growing their food on family plots. A virus or other disease could hit any one of these three plants and cause massive loss of life from starvation.

Wheat, corn, and rice originated and were first cultivated in different parts of the globe. Wheat is commonly used in the United States to produce flour and bread. It was first cultivated in the Middle East (Iran, Iraq, and neighboring countries) about 8000 B.C.; hence, it is thought to be one of the earliest cultivated plants. Wheat was brought to North America in 1520 by early settlers; now the United States is one of the world's largest producers of wheat.

Corn, or what is properly called maize, was first cultivated in Central America about 7,000 years ago. Maize developed from a plant called teosinte, which grows in the highlands of central Mexico. By the time Europeans were exploring Central America, over 300 varieties were already in existence—growing from Canada to Chile. We now commonly grow six major varieties of corn: sweet, pop, flour, dent, pod, and flint.

Rice originated several thousand years ago in southeastern Asia, where it grew in swamps. Today we are familiar with brown and white rice. Brown rice results when the seeds are threshed to remove the hulls, but the seed coat and complete embryo remain. If the seed coat and embryo are removed, leaving only the starchy endosperm, white rice results. Because the seed coat and embryo are a good source of vitamin B and fat-soluble vitamins, brown rice is the healthier choice. Today, rice is grown throughout the tropics and subtropics where water is abundant. It is also grown in some parts of the western United States by flooding diked fields with irrigation water.

FIGURE 18.11A Some species of grains.

Do you have an "addiction" to sugar? This simple carbohydrate comes almost exclusively from two plants—sugarcane (grown in South America, Africa, Asia, the southeastern United States, and the Caribbean) and sugar beets (grown mostly in Europe and North America). Each provides about 50% of the world's sugar.

Many foods are bland or tasteless without spices. In the Middle Ages, wealthy Europeans spared no cost to obtain spices from the Middle and Far East. In the 15th and 16th centuries, major expeditions were launched in an attempt to find better and cheaper routes for spice importation. The explorer Christopher Columbus convinced the queen of Spain that he would find a shorter route to the Far East by traveling west by ocean rather than east by land. Columbus's idea was sound, but he encountered a little barrier, the New World. Nevertheless, this discovery later provided Europe with a wealth of new crops, including corn, potatoes, peppers, and tobacco.

Our most popular drinks—coffee, tea, and cola—also come from flowering plants. Coffee originated in Ethiopia, where it was first used (along with animal fat) during long trips for sustenance and to relieve fatigue. Coffee as a drink was not developed until the 13th century in Arabia and Turkey, and it did not catch on in Europe until the 17th century. Tea is thought to have been developed somewhere in central Asia. Its earlier uses were almost exclusively medicinal, especially among the Chinese, who still drink tea for medical reasons. The drink as we now know it was not developed until the 4th century. By the mid-17th century, it had become popular in Europe. Cola is a common ingredient in tropical drinks and was used around the turn of the century, along with the drug coca (used to make cocaine), in the "original" Coca-Cola.

Plants have been used for centuries for a number of important household items, including the house itself. Lumber, the major structural portion in buildings, comes mostly from a variety of conifers: pine, fir, and spruce, among others. In the tropics, trees and even herbs provide important components for houses. In rural parts of Central and South America, palm leaves are preferable to tin for roofs, since they

grain head

Wheat plants, *Triticum*

Corn plants, *Zea*

ear

grain head

Rice plants, *Oryza*

last as long as 10 years and are quieter during a rainstorm (**Fig. 18.11B**, *bottom right*). In the Middle East, numerous houses along rivers are made entirely of reeds.

Rubber is another plant that has many uses today. The product was first made in Brazil from the thick, white sap (latex) of the rubber tree (Fig. 18.11B, *left*). Once collected, the sap is placed in a large vat, where acid is added to coagulate the latex. When the water is pressed out, the product is formed into sheets or crumbled and placed into bales. Much stronger rubber, such as that in tires, was made by adding sulfur and heating, a process called vulcanization; this produces a flexible material less sensitive to temperature changes. Today, though, much rubber is synthetically produced.

The 5,200-year-old remains of Ice Man found in the Alps was wearing a cape made of grass. Before the invention of synthetic fabrics, cotton and other natural fibers were our usual source of clothing (Fig. 18.11B, *bottom middle*). China is now the largest producer of cotton. The cotton fiber itself comes from filaments that grow on the seed. In 16th-century Europe, cotton was a little-understood fiber known only from stories brought back from Asia. Columbus and other explorers were amazed to see the elaborately woven cotton fabrics in the New World. But by 1800, Liverpool, England, was the world's center of cotton trade. (Interestingly, when Levi Strauss wanted to make a tough pair of jeans, he needed a stronger fiber than cotton, so he used hemp. Hemp is now known primarily as the source of a hallucinogenic drug—marijuana—though there has been a resurgence of its use in clothing.) Over 30 species of native cotton now grow around the world, including the United States.

An actively researched area today involves medicinal plants. Currently, about 50% of all pharmaceutical drugs have their origins from plants. The treatment of some cancers appears to rest in the discovery of new plants. Indeed, the National Cancer Institute (NCI) and most pharmaceutical companies have spent millions of dollars to send botanists out to collect and test plant samples around the world. Tribal medicine men, or shamans, of South America and Africa have already been of great importance in developing numerous drugs.

Over the centuries, malaria has caused far more human deaths than any other disease. After European scientists became aware that malaria can be treated with quinine, which comes from the bark of the cinchona tree, a synthetic form of the drug, chloroquine, was developed. But by the late 1960s, some of the malaria parasites, which live in red blood cells, had become resistant to the synthetically produced drug. Resistant parasites were first seen in Africa but are now showing up in Asia and the Amazon. Today, the only 100%-effective drug for malaria treatment must come directly from the cinchona tree, which is common to northeastern South America.

Numerous plant extracts continue to be misused for their hallucinogenic or other effects on the human body: coca for cocaine and crack, opium poppy for morphine, and wild yam for steroids.

In addition to all these uses of plants, we should not forget or neglect their aesthetic value (Fig. 18.11B, *top right*). Flowers brighten any yard, ornamental plants accent landscaping, and trees provide cooling shade during the summer and protect us from the winter wind. Plants also produce oxygen, which is so necessary for all animals and also plants themselves.

Now that we have finished our discussion of flowering plants, the next part of the chapter will survey the kingdom Fungi.

> **18.11 Check Your Progress** Plants provide us with a wider diversity of products than do animals. What does this say about the metabolic capability of plants?

Tulips, *Tulipa,* for beauty

Rubber, *Hevea*
for auto tires

Cotton, *Gossypium*
for cloth

Palm leaves, *Arecaceae*
for thatched roofs

FIGURE 18.11B Uses of plants.

Fungi feed on the substratum where they live and their structure is adaptive to their saprotrophic mode of nutrition. Their lifestyle allows fungi to form mutualistic relationships with algae in lichens and with plant roots in mycorrhizas. All fungi reproduce by means of windblown spores, but the appearance of the spore-bearing structure for each major group of fungi differs.

18.12   Fungi differ from plants and animals

Fungi (domain **Eukarya**, kingdom **Fungi**) are a structurally diverse group of eukaryotes that are strict heterotrophs. Unlike animals, fungi release digestive enzymes into the external environment and digest their food outside the body, while animals ingest their food and digest it internally. Most fungi are saprotrophs; they decompose the corpses of plants, animals, and microbes. Along with bacteria of decay, fungi play an important role in ecosystems by returning inorganic nutrients to the food producers—that is, photosynthesizers. Fungi can degrade even cellulose and lignin in the woody parts of trees. It is common to see fungi (brown rot or white rot) on the trunks of fallen trees. The body of a fungus can become large enough to cover acres of land. In fact, an 8,500-year-old fungus covers nearly 10 square kilometers of forest floor in northeast Oregon (and has been called "the humongous fungus among us").

The body of a fungus is a mass of filaments called a **mycelium** (**Fig. 18.12A**). Each of the filaments is a **hypha**. Some fungi have cross walls (called septa) that divide a hypha into a chain of cells. These hyphae are called septate. Septa have pores that allow cytoplasm and even organelles to pass from one cell to the other along the length of the hypha. Nonseptate fungi have no cross walls, and their hyphae are multinucleated. Hyphae give the mycelium quite a large surface area per volume of cytoplasm, and this facilitates absorption of nutrients into the body of a fungus. Hyphae grow at their tips, and the mycelium absorbs and then passes nutrients on to the growing tips.

Fungal cells are quite different from plant cells, not only because they lack chloroplasts, but also because their cell wall contains chitin rather than cellulose. Chitin, like cellulose, is a polymer of glucose, but in chitin, a nitrogen-containing amino group is attached to each glucose molecule. Chitin is the major structural component of the exoskeleton of arthropods, such as insects, lobsters, and crabs. The energy reserve of fungi is not starch, but glycogen, as in animals. Fungi are nonmotile and do not have flagella at any stage in their life cycle. They move toward a food source by growing toward it. Hyphae can grow as much as a kilometer a day!

Fungi are adapted to life on land by producing windblown spores during both asexual and sexual reproduction (**Fig. 18.12B**). In fungi, spores germinate into new mycelia. Sexual reproduction in fungi involves conjugation of hyphae from two different mating types (usually designated + and −). Often, the haploid nuclei from the two hyphae do not immediately fuse to form a zygote. The hyphae contain + and − nuclei for long periods of time. Eventually, the nuclei fuse to form a zygote that undergoes meiosis, followed by spore formation.

Fungi are differentiated on the basis of their mode of sexual reproduction. The fungal saprotrophic mode of nutrition promotes a mutualistic relationship with algae and plant roots, as discussed in Section 18.13.

18.12 Check Your Progress Describe how fungi differ from plants.

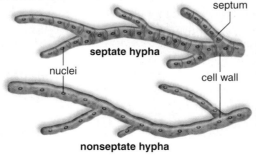

FIGURE 18.12A Fungal mycelia and hyphae.

Fungal mycelia on a corn tortilla

FIGURE 18.12B The fungus earthstar, releasing hordes of spores.
© R. G. Kessel and C. Y. Shih, "Living Images," 1982, Science Books International.

18.13 Fungi have mutualistic relationships with algae and plants

In a mutualistic relationship, two different species live together and help each other out. A **lichen** is a mutualistic association between a particular fungus and either cyanobacteria or green algae (**Fig. 18.13A**). The fungal partner is efficient at acquiring nutrients and moisture, and therefore lichens can survive in poor soils, as well as on rocks with no soil. The organic acids given off by fungi release minerals from rocks that can be used by the photosynthetic partner. Lichens are ecologically important because they produce organic matter and create new soil, allowing plants to invade the area.

Lichens occur in three varieties: compact crustose lichens, often seen on bare rocks or tree bark; shrublike fruticose lichens; and leaflike foliose lichens (Fig. 18.13A). Regardless of type, the body of a lichen has three layers: a thin, tough upper layer and a loosely packed lower layer, both formed by fungal hyphae, which shield a middle layer of photosynthetic cells. Specialized fungal hyphae that penetrate or envelop the photosynthetic cells transfer organic nutrients to the rest of the mycelium. The fungus not only provides minerals and water to the photosynthe-

no mycorrhizae   one type of mycorrhizae   several types of mycorrhizae

FIGURE 18.13B An experiment demonstrated the ability of mycorrhizal fungi to aid plant growth. Keeping plants with their mycorrihizae ensures growth.

sizer, but also offers protection from predation and desiccation. Lichens can reproduce asexually by releasing reproductive units that contain hyphae and an algal cell. At first, the relationship between the fungi and algae was likely a parasite-and-host interaction. Over evolutionary time, the relationship apparently became more mutually beneficial, although how to test this hypothesis is a matter of debate at the present time.

Mycorrhizal fungi form mutualistic relationships, called mycorrhizae, with the roots of most plants, helping them grow more successfully in dry or poor soils, particularly those deficient in inorganic nutrients (**Fig. 18.13B**). The fungal hyphae greatly increase the surface area from which the plant can absorb water and nutrients. It has been found beneficial to encourage the growth of mycorrhizal fungi when restoring lands damaged by strip mining or chemical pollution.

Mycorrhizal fungi may live on the outside of roots, enter between root cells, or penetrate root cells. The fungus and plant cells exchange nutrients; the fungus brings water and minerals to the plant, and the plant provides organic carbon to the fungus. Early plant fossils indicate that the relationship between fungi and plant roots is an ancient one, and therefore it may have helped plants adapt to life on dry land. The general public is not familiar with mycorrhizal fungi, but a few people consider truffles, a mycorrhizal fungus that grows in oak and beech forests, a gourmet delicacy. Pigs and dogs are trained to sniff them out in the woods, but truffles are also cultivated on the roots of seedlings.

The classification of fungi is primarily based on the shape of the structure that produces spores during sexual reproduction, as we shall see in Section 18.14.

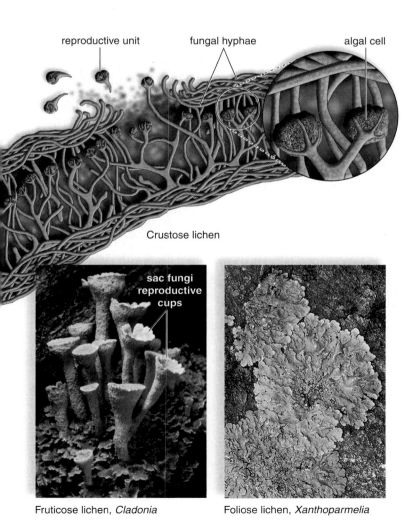

reproductive unit   fungal hyphae   algal cell

Crustose lichen

sac fungi reproductive cups

Fruticose lichen, *Cladonia*   Foliose lichen, *Xanthoparmelia*

FIGURE 18.13A Lichen arrangement and examples.

▶ **18.13 Check Your Progress** What are mycorrhizae?

18.14 Fungi occur in three main groups

The three main phyla of fungi are the zygospore fungi, the sac fungi, and the club fungi.

Zygospore Fungi The zygospore fungi (phylum Zygomycota) are mainly saprotrophs, but some are parasites of small soil protists or worms, and even of insects, such as the housefly. The black bread mold, *Rhizopus stolonifer* (**Fig. 18.14A**), is well known to many of us. In *Rhizopus*, the hyphae are specialized: Some are horizontal and exist on the surface of the bread; others grow into the bread to anchor the mycelium and carry out digestion; and **1** still others are stalks that bear sporangia. As in plants, a sporangium in fungi is a capsule that produces spores.

When *Rhizopus* reproduces sexually, **2** the ends of + strain and − strain hyphae join, **3** haploid nuclei fuse, and **4** a thick-walled zygospore results. The zygospore undergoes a period of dormancy before meiosis takes place. Following germination, **5** aerial hyphae, with sporangia at their tips, produce many spores. The spores, dispersed by air currents, give rise to new mycelia.

Sac Fungi The sac fungi (phylum Ascomycota) take their phylum name from the shape of the sexual reproductive structure, called an

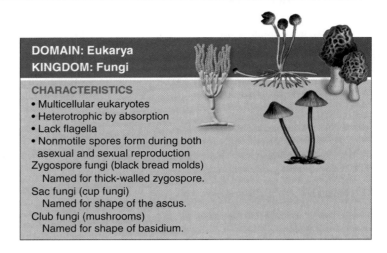

DOMAIN: Eukarya
KINGDOM: Fungi

CHARACTERISTICS
• Multicellular eukaryotes
• Heterotrophic by absorption
• Lack flagella
• Nonmotile spores form during both asexual and sexual reproduction
Zygospore fungi (black bread molds)
   Named for thick-walled zygospore.
Sac fungi (cup fungi)
   Named for shape of the ascus.
Club fungi (mushrooms)
   Named for shape of basidium.

ascus, where spores are produced by meiosis. Sac fungi account for nearly 75% of all described fungal species. Cup fungi, morels, and truffles have conspicuous ascocarps (**Fig. 18.14B**).

Many sac fungi are commonly called red bread molds (e.g., *Neurospora*). Powdery mildews grow on leaves, as do leaf curl

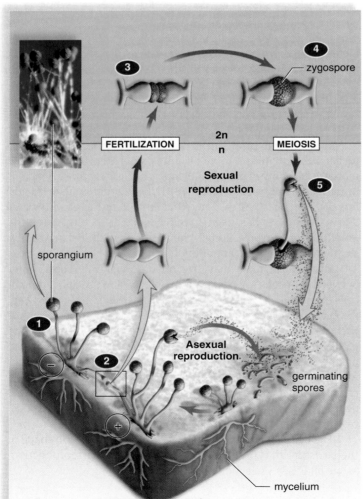

FIGURE 18.14A Black bread mold, *Rhizopus stolonifer.*

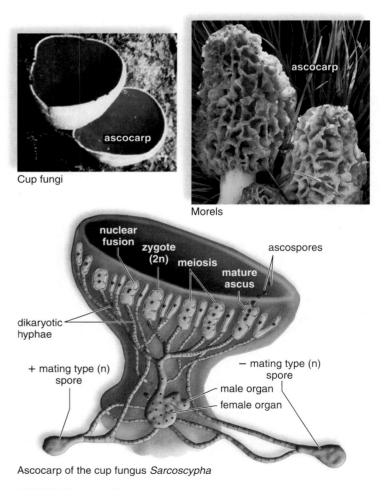

FIGURE 18.14B Sexual reproduction in sac fungi.

fungi; chestnut blight and Dutch elm disease destroy trees. Ergot, a parasitic sac fungus that infects rye, produces hallucinogenic compounds. These and other sac fungi usually reproduce by forming chains of asexual spores called conidia (sing., **conidium**) (**Fig. 18.14C**, *left*).

The term **yeasts** is generally applied to unicellular fungi, and many of these organisms are classified in the phylum Ascomycota. *Saccharomyces cerevisiae*, brewer's yeast, is representative of budding yeasts (Fig. 18.14C, *right*). Unequal binary fission occurs, and a small cell gets pinched off and then grows to full size. Sexual reproduction, which occurs when the food supply runs out, results in the formation of asci and ascospores.

When some yeasts ferment, they produce ethanol and carbon dioxide. In the wild, yeasts grow on fruits, and historically, the yeasts already present on grapes were used to produce wine. Today, selected yeasts are added to relatively sterile grape juice in order to make wine. Also, yeasts are added to prepare grains to make beer. Both the ethanol and carbon dioxide are retained in beers and sparkling wines; carbon dioxide is released from other wines. In breadmaking, the carbon dioxide produced by yeasts causes the dough to rise; the gas pockets are preserved as the bread bakes.

Club Fungi The phylum name of the club fungi (phylum Basidiomycota) comes from the shape of the sexual reproductive structure, called a basidium (pl., basidia), where spores are produced by meiosis. The basidia are located within a basidiocarp (**Fig. 18.14D**). When you eat a mushroom, you are consuming a basidiocarp. Shelf or bracket fungi found on dead trees are also basidiocarps. Less well-known basidiocarps are puffballs, bird's nest fungi, and stinkhorns. In puffballs, spores are produced inside parchmentlike membranes and then released through a pore or when the membrane breaks down. Stinkhorns resemble a mushroom with a spongy stalk and a compact, slimy cap. Stinkhorns emit an incredibly disagreeable odor. Flies are attracted by the odor, and when they linger to feed on the sweet jelly, they pick up spores that they later distribute.

Although club fungi occasionally produce conidia asexually, they usually reproduce sexually by forming basidia.

Section 18.15 emphasizes the economic and medical importance of fungi. Their ecological importance has already been discussed.

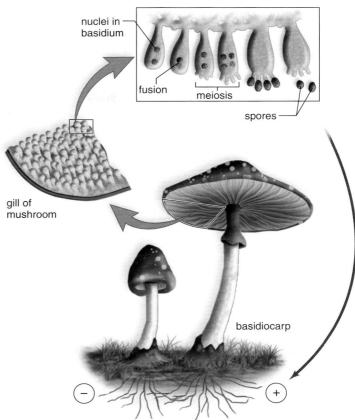

nuclei in basidium

fusion

meiosis

spores

gill of mushroom

basidiocarp

Sexual reproduction

Mushroom

Shelf fungi

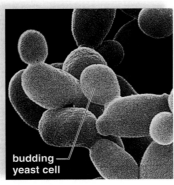

Giant puffball

FIGURE 18.14D Sexual reproduction in club fungi involves a basidiocarp of which three types are shown.

conidia

budding yeast cell

FIGURE 18.14C Asexual reproductive structures in sac fungi.

18.14 *Check Your Progress* **Provide a common example for each of the three phyla of fungi.**

18.15 Fungi have economic and medical importance

Economic Importance Fungi have great economic importance because they help us produce medicines and many types of foods. The mold *Penicillium* was the original source of penicillin, a breakthrough antibiotic that led to an important class of cillin antibiotics, which have saved millions of lives. As described in Section 18.14, yeast fermentation is utilized to make bread, beer, wine, and distilled spirits. Other types of fungal fermentation contribute to the manufacture of various cheeses as well as soy sauce from soybeans. Another commercial application is the use of fungi to soften the centers of certain candies.

In the United States, mushroom consumption has been steadily increasing. In 2006, total consumption of all mushrooms totalled 1.6 billion pounds—40% greater than in 2001. In addition to adding taste and texture to soups, salads, and omelets, and being used in stir-fry and on salads, mushrooms are an excellent low-calorie meat substitute containing lots of vitamins. Although there are thousands of mushroom varieties in the world, the white button mushroom, *Agaricus bisporus*, dominates the U.S. market. However, in recent years, sales of brown-colored variants have surged in popularity and have been one of the fastest-growing segments of the mushroom industry.

Fungal pathogens, which usually gain access to plants by way of the stomata or a wound, are a major concern for farmers. Serious crop losses occur each year due to fungal disease (**Fig. 18.15A**). As much as one-third of the world's rice crop is destroyed each year by rice blast disease. Corn smut is a major problem in the midwestern United States. Various rusts attack grains, and leaf curl is a disease of fruit trees.

Medical Importance Certain mushrooms are poisonous, and so wild mushrooms should be carefully chosen as a food source. *Amantia* spp. is a deadly wild mushroom known as the death angel. Ergot, a fungus that grows on grain, can cause ergotism in a person who eats contaminated bread. Ergotism is characterized by hysteria, convulsions, and sometimes death.

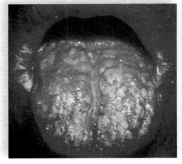
Thrush

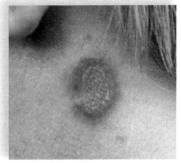
Ringworm

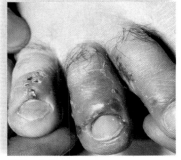
Athlete's foot

FIGURE 18.15B
Human fungal diseases.

Mycoses are diseases caused by fungi. Mycoses have three possible levels of invasion: Cutaneous mycoses only affect the epidermis, subcutaneous mycoses affect deeper skin layers, and systemic mycoses spread throughout the body by traveling in the bloodstream. Fungal diseases that can be contracted from the environment include rose gardener's disease from thorns, Chicago disease from old buildings, and basketweaver's disease from grass cuttings. Several opportunistic fungal infections now seen in AIDS patients stem from fungi that are always present in the body but take the opportunity to cause disease when the immune system becomes weakened.

Tineas are infections of the skin caused for the most part by fungi. Ringworm is a cutaneous infection contracted from soil. The fungal colony does not penetrate the skin but grows outward, forming a ring of inflammation. The center of the lesion begins to heal, producing a characteristic red ring surrounding an area of healed skin. Athlete's foot is a form of tinea that affects the foot, causing itching and peeling of the skin between the toes (**Fig. 18.15B**).

Candida albicans causes the widest variety of fungal infections. Disease occurs when antibacterial treatments kill off the microflora community, allowing *Candida* to proliferate. Vaginal *Candida* infections are commonly called "yeast infections" in women. Oral thrush is a *Candida* infection of the mouth common in newborns and AIDS patients (Fig. 18.15B). In individuals with inadequate immune systems, *Candida* can move throughout the body, causing a systemic infection that can damage the heart, brain, and other organs.

18.15 Check Your Progress Why would you expect heterotrophs, rather than photosynthesizers, to cause disease?

FIGURE 18.15A Plant fungal disease.

Both plants and fungi are multicellular organisms adapted to living on land. Four noteworthy innovations tell the story of plant adaptation to a terrestrial environment. Bryophytes, like all plants, protect the sporophyte embryo from drying out by retaining it within specialized tissues. Vascular tissue, which transports water and solutes within the plant body, is first seen among ferns and related seedless plants. The advent of seeds in gymnosperms allowed the sporophyte to be dispersed rather than the gametophyte, which does not have vascular tissue. Seed production only became possible with modifications of the life cycle that included the evolution of microscopic female and male gametophytes that are dependent on the sporophyte. No external water is needed for male gametophytes (pollen grains) to be transported to the female gametophyte. The evolution of the flower enlisted the help of animals to achieve pollination and to disperse seeds. Angiosperms are the most widely dispersed of the plants and can live in a wide variety of habitats on land.

Fungi are adapted to the land environment because they produce windblown spores within both asexual and sexual life cycles. Whereas plants are photosynthetic, fungi are saprotrophic. They release enzymes into the environment to digest organic remains and absorb the resultant nutrient molecules. Without photosynthesis carried out by algae and plants and without decomposition carried out by bacteria and fungi, animals could not exist. Animals are not essential to the biosphere, but plants and fungi are!

Animal evolution is the topic of Chapter 19. Even today, more groups of animals live in water than on land. Certain molluscs (e.g., snails), certain arthropods (e.g., insects), and many vertebrates (examples can be found among the amphibians, reptiles, birds, and mammals) live on land. Birds and mammals are the most successful of these groups today. Most people agree that birds are a continuation of dinosaur evolution, while mammals filled the ecological niches of the dinosaur groups that died out at the K-T boundary, as discussed in Chapter 15.

The Chapter in Review

Summary

Some Plants Are Carnivorous

- Carnivorous plants feed on insects, amphibians, birds, or mammals.
- Digestion of these animals provides these plants with nitrogen in an environment that lacks available nitrogen.

The Evolution of Plants Spans 500 Million Years

18.1 Evidence suggests that plants evolved from green algae

- Plants and green algae contain chlorophylls *a* and *b* and other pigments.
- Both plants and green algae store carbohydrates as starch.
- Both plants and green algae have cellulose in cell walls.

18.2 The evolution of plants is marked by four innovations

- The embryo is protected within the plant body.
- Vascular tissue developed.
- Seeds were produced.
- The flower evolved.

18.3 Plants have an alternation of generations life cycle

- The sporophyte is the diploid generation; it produces haploid spores by meiosis.
- The gametophyte is the haploid generation; it produces gametes (eggs and sperm) by mitosis.

18.4 Sporophyte dominance was adaptive to a dry land environment

- During the evolution of plants, the sporophyte became increasingly dominant as plants became increasingly adapted to life on land.

- An independent gametophyte and flagellated sperm make it difficult for ferns to live in dry environments.
- A dependent gametophyte and pollen carried by insects allow flowering plants to live in dry environments.
- The sporophyte of flowering plants is protected from drying out by a waxy cuticle interrupted by stomata.

Plants Are Adapted to the Land Environment

18.5 Bryophytes are nonvascular plants in which the gametophyte is dominant

- Hornworts, liverworts, and mosses are bryophytes.
- In the moss life cycle, the dependent sporophyte produces windblown spores; the dominant gametophyte produces flagellated sperm.

18.6 Ferns and their allies have a dominant vascular sporophyte

- Seedless vascular plants include ferns, club mosses, and horsetails.
- In the fern life cycle, the dominant sporophyte produces windblown spores; the separate gametophyte produces flagellated sperm.

18.7 Most gymnosperms bear cones on which the seeds are "naked"

- Gymnosperms include cycads, ginkgoes, gnetophytes, and conifers (evergreen trees).
- In the pine life cycle, the sporophyte is dominant.
- Male gametophytes (pollen) are windblown from the pollen cone to the seed cone.
- On the surface of seed cones, ovules, which contain female gametophytes, become seeds not enclosed by fruit.

18.8 Carboniferous forests became the coal we use today

- Coal is a fossil fuel produced in swamp forests in the Carboniferous period.

18.9 Angiosperms are the flowering plants

- Monocots have one cotyledon in the seed and flower parts in threes or multiples of three.
- Eudicots have two cotyledons in the seed and flower parts in fours, fives, or multiples thereof.
- The main parts of a flower are: the receptacle (bears other flower parts) and the sepals, petals, stamens, carpel, stigma, style, and ovary.

18.10 The flowers of angiosperms produce "covered" seeds

- In an ovule, the megaspore develops into the embryo sac (the female gametophyte). In the anther, microspores become pollen grains (the male gametophyte).
- Pollen is transported by wind or animals to the female gametophyte in the ovule.
- The ovule becomes a seed, which is covered by a fruit.

18.11 Flowering plants provide many services

- The three main food plants for humans are wheat, corn, and rice.
- Other plant products are lumber, rubber, cotton for fabric, pharmaceuticals, landscaping, and oxygen.

Fungi Have Their Own Evolutionary History

18.12 Fungi differ from plants and animals

- Fungi are saprotrophs that carry on external digestion.
- A mycelium, which is a mass of filaments called hyphae, makes up the fungal body.
- Fungi produce windblown spores during both sexual and asexual reproduction.

18.13 Fungi have mutualistic relationships with algae and plants

- A lichen is a mutualistic association between a fungus and cyanobacteria or green algae.
- Types of lichens are crustose, fruticose, and foliose.
- Mycorrhizal fungi form a mutualistic relationship with plant roots, thereby increasing root surface area as well as the water and nutrient absorption of the plant.

18.14 Fungi occur in three main groups

- During sexual reproduction, zygospore fungi (black bread mold) have a zygospore; sac fungi (cup fungi) have an ascus; and club fungi (mushrooms) have a basidium.

18.15 Fungi have economic and medical importance

- Fungi provide cillin antibiotics as well as mushrooms, a popular food.
- Some fungi cause diseases in crops (smuts and rusts) and in humans (tineas and *Candida* infections).

Testing Yourself

The Evolution of Plants Spans 500 Million Years

1. Which of the following is not a plant adaptation to land?
   a. recirculation of water
   b. protection of embryo in maternal tissue
   c. development of flowers
   d. creation of vascular tissue
   e. seed production

2. Plant spores are
   a. haploid and genetically different from each other.
   b. haploid and genetically identical to each other.
   c. diploid and genetically different from each other.
   d. diploid and genetically identical to each other.

3. Sporophyte dominance should be associated with
   a. windblown spores
   b. independent gametophytes
   c. pollen grains
   d. female gametophyte protected by sporophyte
   e. Both c and d.

4. **THINKING CONCEPTUALLY** The evolution of the vascular system allowed plants to grow tall. Why do tall plants have an advantage?

5. The gametophyte is the dominant generation in
   a. ferns.
   b. mosses.
   c. gymnosperms.
   d. angiosperms.
   e. More than one of these are correct.

Plants Are Adapted to the Land Environment

6. In bryophytes, sperm usually move from the antheridium to the archegonium by
   a. swimming.
   b. flying.
   c. insect pollination.
   d. worm pollination.
   e. bird pollination.

7. A fern sporophyte will develop on which region of the gametophyte?
   a. near the notch, where the archegonia are located
   b. near the tip, where the antheridia are located
   c. anywhere on the gametophyte
   d. All of these are correct.

8. How are ferns different from mosses?
   a. Only ferns produce spores for reproduction.
   b. Ferns have vascular tissue.
   c. In the fern life cycle, the gametophyte and sporophyte are both independent.
   d. Ferns do not have flagellated sperm.
   e. Both b and c are correct.

9. Which of the following is a seedless vascular plant?
   a. gymnosperm
   b. angiosperm
   c. fern
   d. monocot
   e. eudicot

10. In the life cycle of the pine tree, the ovules are found on
    a. needlelike leaves.
    b. seed cones.
    c. pollen cones.
    d. root hairs.
    e. All of these are correct.

11. Label the parts of the flower in the following illustration.

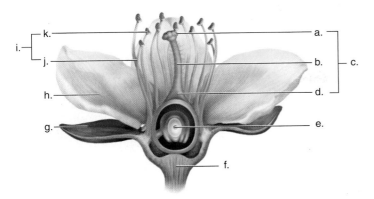

12. A seed contains a mature
    a. embryo. c. ovary.
    b. ovule. d. pollen grain.
13. Endosperm is produced by the union of the
    a. egg and polar nuclei. c. polar and sperm nuclei.
    b. egg and sperm nuclei. d. polar and egg nuclei.
14. Which of these plants contributed the most to our present-day supply of coal?
    a. bryophytes d. angiosperms
    b. seedless vascular plants e. Both b and c are correct.
    c. gymnosperms
15. **THINKING CONCEPTUALLY** Pollen cones are located on a tree's lower branches, they do not occur next to seed cones located on the upper branches of the same tree. What does this type of arrangement possibly prevent?

Fungi Have Their Own Evolutionary History

16. A mushroom is like a plant because it
    a. is a multicellular eukaryote.
    b. produces spores.
    c. is adapted to a land environment.
    d. is photosynthetic.
    e. All but d are correct.
17. Which feature is best associated with hyphae?
    a. strong, impermeable walls c. large surface area
    b. rapid growth d. pi gmented cells
    e. Both b and c are correct.
18. Symbiotic relationships of fungi include
    a. athlete's foot. d. Only b and c are correct.
    b. lichens. e. All three examples are correct.
    c. mycorrhizae.
19. A fungal spore
    a. contains an embryonic organism.
    b. germinates into an organism.
    c. is always windblown.
    d. is most often diploid.
    e. Both b and c are correct.
20. Conidia are formed
    a. asexually at the tips of special hyphae.
    b. during sexual reproduction.
    c. by all types of fungi except water molds.
    d. only when it is windy and dry.
    e. as a way to survive a harsh environment.
21. Which of the following diseases is (are) caused by *Candida*?
    a. oral thrush d. ringworm
    b. athlete's foot e. Both a and c are correct.
    c. vaginal yeast infection

21. **THINKING CONCEPTUALLY** Pine seedlings grow more vigorously if they are transplanted with some of their native soil. Explain.

Understanding the Terms

alternation of generations 344
angiosperm 353
archegonium 345
bryophyte 346
carpel 353
club moss 348
coal 352
conidium 361
conifer 350
cotyledon 353
cuticle 345
cycad 350
double fertilization 355
embryo sac 354
eudicot 353
Eudicotyledone 353
female gametophyte 351
fern 348
frond 348
fruit 355
fungus 358
gametophyte 344
ginkgo 350
gnetophyte 350
gymnosperm 350
horsetail 348
hypha 358
lichen 359
male gametophyte 351
megaspore 354
monocot 353
Monocotyledone 353
moss 346
mycelium 358
mycorrhizal fungi 359
ovary 353
ovule 345
peat 346
peduncle 353
petal 353
pollen grain 351
pollen tube 355
pollination 351
receptacle 353
sepal 353
sori 349
sporangium 347
spore 344
sporophyte 344
stamen 353
stigma 353
stoma 345
style 353
triploid endosperm 355
vascular tissue 343
yeast 361

Match the terms to these definitions:
a. _____ Tangled mass of hyphal filaments composing the vegetative body of a fungus.
b. _____ Diploid generation of the alternation of generations life cycle of a plant.
c. _____ Have one embryonic leaf and flower parts in threes.
d. _____ Mutualistic fungi that grow near the roots of vascular plants.
e. _____ Cone-bearing plants, including ginkgoes, gnetophytes, cycads, and conifers.

Thinking Scientifically

1. Which experimental group(s) would you expect to complete alternation of generations and why? (a) Mosses that are provided with water only to their rhizoids, or (b) mosses that are provided water to their rhizoids and, in addition, water is sprayed into the air.
2. An orchid produces flowers that attract particular male moths because the flowers resemble females of the same moth species. Why would you expect this to be an effective pollen dispersal strategy?

ARIS™ *Visit* **www.mhhe.com/maderconcepts** *for practice quizzes, animations, videos, and activities designed to help you master the material in this chapter.*

19

Evolution of Animals

LEARNING OUTCOMES

After studying this chapter, you should be able to accomplish the following outcomes.

The Secret Life of Bats

1 Argue that bats should be preserved.

Key Innovations Distinguish Invertebrate Groups

2 State the characteristics and trace the evolution of animals.
3 Relate seven significant innovations to an evolutionary tree of animals.
4 Compare and contrast the traditional evolutionary tree to the one based on molecular data.
5 Assign animals to the correct invertebrate or vertebrate group.
6 Compare and contrast the characteristics of sponges, cnidarians, flatworms, and roundworms.
7 Discuss the physical characteristics of flatworms that make them successful parasites.
8 Discuss the advantages of a coelom and how coelom development can be used to group animals.
9 Compare and contrast the characteristics of molluscs, annelids, arthropods, and echinoderms.
10 Discuss the advantages of segmentation and jointed appendages in arthropods.

Further Innovations Allowed Vertebrates to Invade the Land Environment

11 Name four features that distinguish chordates from other groups of animals.
12 Relate five significant innovations to an evolutionary tree of vertebrates.
13 Compare and contrast the characteristics of fishes, amphibians, reptiles, and mammals, particularly as they relate to living in water versus on land.
14 Discuss vertebrate contributions to human medicine.

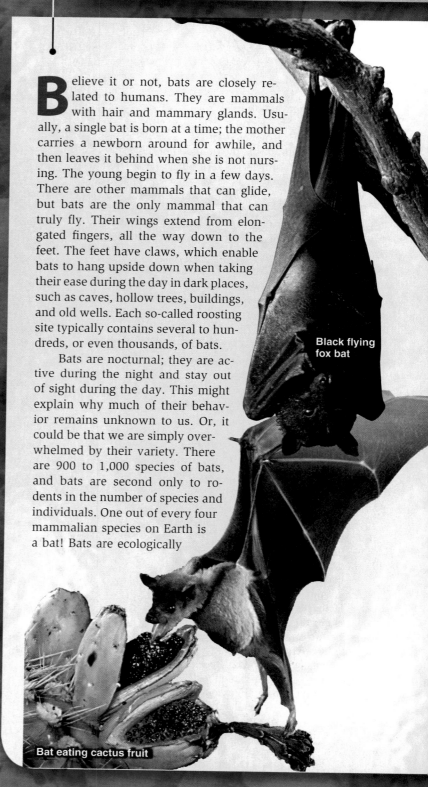

Believe it or not, bats are closely related to humans. They are mammals with hair and mammary glands. Usually, a single bat is born at a time; the mother carries a newborn around for awhile, and then leaves it behind when she is not nursing. The young begin to fly in a few days. There are other mammals that can glide, but bats are the only mammal that can truly fly. Their wings extend from elongated fingers, all the way down to the feet. The feet have claws, which enable bats to hang upside down when taking their ease during the day in dark places, such as caves, hollow trees, buildings, and old wells. Each so-called roosting site typically contains several to hundreds, or even thousands, of bats.

Bats are nocturnal; they are active during the night and stay out of sight during the day. This might explain why much of their behavior remains unknown to us. Or, it could be that we are simply overwhelmed by their variety. There are 900 to 1,000 species of bats, and bats are second only to rodents in the number of species and individuals. One out of every four mammalian species on Earth is a bat! Bats are ecologically

Black flying fox bat

Bat eating cactus fruit

The Secret Life of Bats

important, and that is one reason an effort to conserve bats is under way. Some bats feed on fruit, nectar, and pollen, and in so doing, they disperse pollen and also seeds, which pass through their digestive tract. Other bats feed on insects, greatly reducing the numbers of those that flit about during the night. These bats offer a way to biologically control insects at no trouble to ourselves and without using pesticides.

The face of a bat varies greatly; many species have odd-looking appendages on the snout and very large, elaborately convoluted ears. These appendages help them emit and receive sounds at a higher frequency than is audible to the human ear. After sending out these "ultrasounds," the returning echoes tell them the location of any nearby object. In other words, many bats use echolocation to find their prey.

Some bats feed on blood, and perhaps because of our interest in vampires, we know more about them. The common vampire bat, *Desmodus rotundus*, has a wingspan of nearly 20.5 cm, is about the size of an adult human's thumb, and weighs less

Vampire bat feeding off a sow

than 1.5 ounces. The bat finds its prey using echolocation, smell, and heat, and then uses its limbs as crutches to catapult toward a sleeping animal. Once the vampire bat finds a victim, it uses special sensors in its nose to locate a superficial vein. Contrary to popular myth, vampire bats do not suck the blood from their victims. Rather, using razor-sharp incisors, they painlessly open a small wound in their prey. A numbing chemical in the bat's saliva keeps the victim from waking up. The bat then uses its tongue to lap up the blood as it oozes from the wound or, if need be, repeatedly darts its tongue in and out of the wound. Typically, a vampire bat needs 2 tablespoons of blood per day to survive, but can consume up to 60% of its body weight during a 20-minute feeding. After feeding, the bat returns to its roost. The highly specialized stomach of the bat shunts the blood plasma to the kidneys for elimination, and only the red blood cells are used for nourishment. The saliva of vampire bats contains the most powerful anticoagulant known. In recent years, desmoteplase, a genetically engineered drug derived from the saliva of a vampire bat, has been successfully used in heart attack and stroke victims. Humans are rarely the victims of these infamous blood-eating winged parasites. Vampire bats usually feed upon the blood of cattle, pigs, horses, and large birds.

Bats are just one tiny part of the animal kingdom, the focus of this chapter. We begin by examining the characteristics that distinguish animals from other types of eukaryotes.

Vampire bat, *Desmodus rotundus*

Long-eared bat

Our survey of the animal kingdom begins with an examination of the characteristics of animals, their probable evolution from protists, and seven innovations that mark the evolution of animals. These innovations will be noted again as we study various groups of animals.

19.1 Animals have distinctive characteristics

Animals have distinctive characteristics that make them different from plants and fungi. Like both other groups, animals are multicellular eukaryotes, but unlike plants, which make their own food through photosynthesis, animals are heterotrophs and must acquire nutrients from an external source. Fungi are also heterotrophs, but fungi digest their food externally and absorb the breakdown products. Free-living animals ingest (eat) their food and digest it internally. Some parasitic animals absorb nutrient molecules from their host.

Animals usually carry on sexual reproduction and begin life only as a fertilized diploid egg. From this starting point, they undergo a series of developmental stages to produce an organism that has specialized tissues within organs that have specific functions. Two types of tissues in particular—muscles and nerves—characterize animals. The presence of these tissues allows an animal to perform a variety of flexible movements that help it search actively for food and prey on other organisms. Coordinated movements also allow animals to seek mates, shelter, and a suitable climate—behaviors that have resulted in the vast diversity of animals. The more than 30 animal phyla we recognize today are believed to have evolved from a single ancestor.

Figure 19.1 illustrates that a frog, like other animals, goes through a number of embryonic stages to become a larval form (the tadpole) with specialized organs, including muscular and nervous systems that enable it to swim. A larva is an immature stage that typically lives in a different habitat and feeds on different foods than the adult. By means of a change in body form called metamorphosis, the larva, which only swims, turns into a sexually mature adult frog that swims and hops. The aquatic tadpole lives on plankton, and the terrestrial adult typically feeds on insects and worms. A large African bullfrog will try to eat just about anything, including other frogs, as well as small fish, reptiles, and mammals.

Having looked at the distinctive characteristics of animals, the next section considers how they have evolved from a protistan ancestor.

> **19.1 Check Your Progress** List three features that show bats are animals.

Adult frog

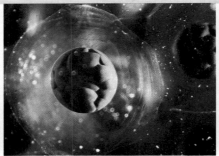

Stages in development, from zygote to embryo

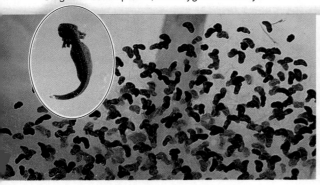

Stages in metamorphosis, from hatching to tadpole with legs

FIGURE 19.1 Developmental stages of a frog.

19.2 Animals most likely have a protistan ancestor

In Chapter 18, we mentioned evidence that plants evolved from green algae and most likely shared a common ancestor with stoneworts. What about animals? Did they evolve from a particular protozoan?

Two hypotheses have been put forth about the origins of animals. The **multinucleate hypothesis** suggests that animals arose from a ciliated protist in stages. First, the ciliate would have acquired multiple nuclei, and then it would have become multicellular. Like some ciliates today, this ancestor to the animals would have been bilaterally symmetrical. In other words, this hypothesis suggests that bilateral symmetry preceded radial symmetry in the history of animals. In a bilaterally symmetrical animal, only one longitudinal cut yields two roughly identical halves; in a radially symmetrical animal, any longitudinal cut produces two identical halves:

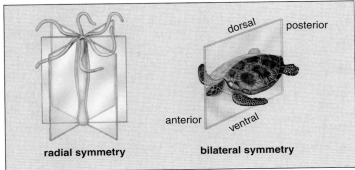

radial symmetry   **bilateral symmetry**

As you study the animal phyla in the next few pages, you will see that, contrary to this hypothesis, radially symmetrical animals most likely preceded bilaterally symmetrical animals in the history of life on Earth. The multinucleate hypothesis has also been rejected because molecular data based on ribosomal RNA (rRNA) sequences support a second hypothesis, called the **colonial flagellate hypothesis**. This hypothesis states that animals are descended from an ancestor that resembled a hollow, spherical colony of flagellated cells, perhaps related to today's choanoflagellates (see Fig. 17.3). As shown in **Figure 19.2**, ❶ the process would have begun with an aggregate of a few flagellated cells. ❷ A larger number of cells could have formed a hollow sphere. ❸ Individual cells within the colony would have become specialized for particular functions, such as reproduction. (A *Volvox* colony has this appearance and has cells with only a reproductive function.) ❹ Two tissue layers could have arisen by an infolding of certain cells into a hollow sphere. Certainly tissue layers do arise in this manner during the development of animals today. The colonial flagellate hypothesis is also attractive because it suggests that radial symmetry preceded bilateral symmetry in the history of animals, and this is probably the case. Finally, molecular data support the colonial flagellate hypothesis and go one step further by showing that all animals share the same flagellate ancestor.

Evolution of Body Plans As we discussed in Section 15.1, all of the various animal body plans were present by the Cambrian period. How could such diversity have arisen within a relatively short period of geologic time? As an animal develops, there are many possibilities regarding the number, position, size, and patterns of its body parts. Different combinations could have led to the great variety of animal forms in the past and present. We now know that slight shifts in genes called *Hox* (homeotic) genes are responsible for the major differences between animals that arise during development (see Fig. 14.9C). Perhaps changes in the expression of *Hox* developmental genes explains why all the animal phyla of today had representatives in the Cambrian seas.

As soon as animals evolved, they embarked on an evolutionary pathway that resulted in great diversity, which can be analyzed in terms of seven innovations, discussed in Section 19.3.

> **19.2 Check Your Progress** Does the colonial flagellate hypothesis pertain to bats?

FIGURE 19.2 The colonial flagellate hypothesis.

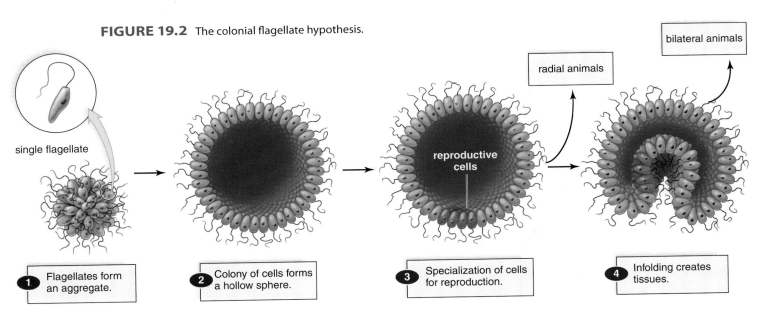

single flagellate

❶ Flagellates form an aggregate.

❷ Colony of cells forms a hollow sphere.

reproductive cells

❸ Specialization of cells for reproduction.

radial animals

bilateral animals

❹ Infolding creates tissues.

FIGURE 19.3A The traditional evolutionary tree of animals.

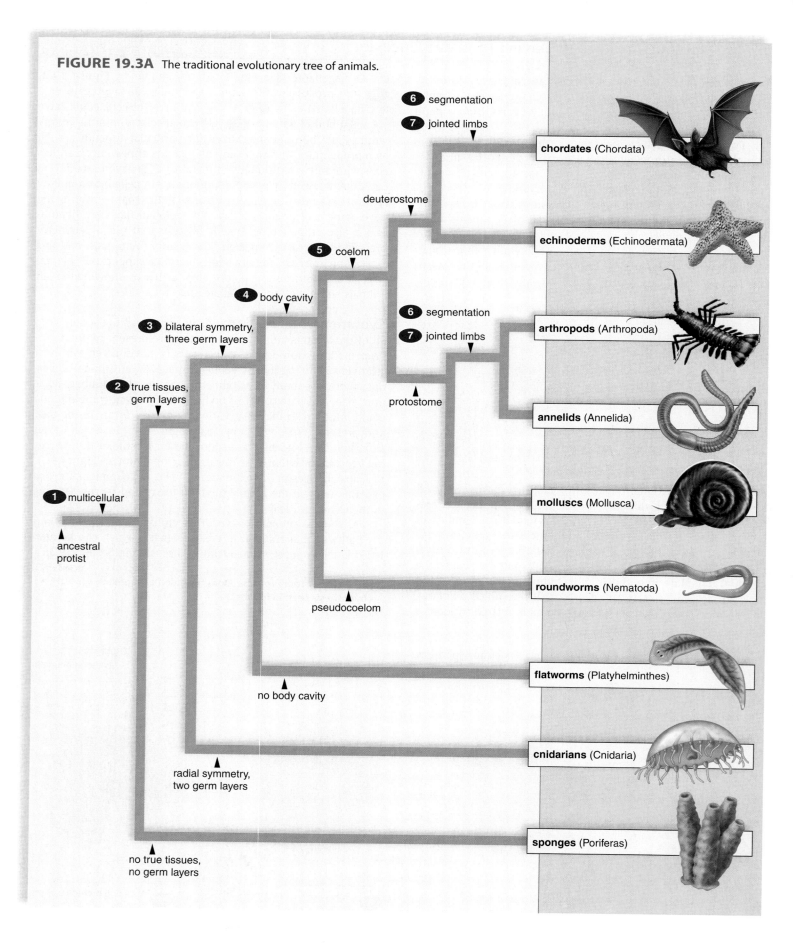

6 segmentation
7 jointed limbs

chordates (Chordata)

deuterostome

echinoderms (Echinodermata)

5 coelom

4 body cavity

6 segmentation
7 jointed limbs

arthropods (Arthropoda)

3 bilateral symmetry, three germ layers

protostome

annelids (Annelida)

2 true tissues, germ layers

molluscs (Mollusca)

1 multicellular

roundworms (Nematoda)

ancestral protist

pseudocoelom

flatworms (Platyhelminthes)

no body cavity

cnidarians (Cnidaria)

radial symmetry, two germ layers

sponges (Poriferas)

no true tissues, no germ layers

19.3   The traditional evolutionary tree of animals is based on seven key innovations

The first animals to evolve did not have a skeleton; therefore, the early history of animal evolution is not as complete as is desirable. Without an adequate fossil record, it has been impossible to trace the evolutionary history of animals with certainty. However, the developmental stages of today's living animals point to seven evolutionary innovations that could have allowed animals to become more complex (**Fig. 19.3A**).

Animals differ in biological organization. Sponges, like all animals, are ❶ *multicellular,* but they have no true tissues and, therefore, have the cellular level of organization. Cnidarians, such as *Hydra,* with two tissue layers in their body wall, have ❷ *true tissues,* which can be associated with two germ layers when they are embryos. The animals in all the other phyla have three embryonic germ layers, called ectoderm, endoderm, and mesoderm, and these shape their organs as they develop.

As mentioned in Section 19.2, animals differ in symmetry. Sponges are examples of animals that have no particular symmetry and are therefore called asymmetrical. **Radial symmetry,** as seen in *Hydra,* means that the animal is organized circularly, similar to a wheel. No matter where the animal is sliced longitudinally, two mirror images are obtained. ❸ **Bilateral symmetry,** as seen in flatworms, means that the animal has definite right and left halves; only a longitudinal cut down the center of the animal will produce mirror images. During the evolution of animals, the trend toward bilateral symmetry is accompanied by **cephalization,** localization of a brain and specialized sensory organs at the anterior end of an animal (the "head"). Active animals are benefited by a nervous and muscular system that allows them to go out and seek their food.

Animals differ with regard to ❹ a *body cavity.* Some animals have no body cavity and are acoelomate, as are flatworms (**Fig. 19.3B**). **Acoelomates** are packed solid with mesoderm. In contrast, a body cavity provides a space for the various internal organs. Roundworms are **pseudocoelomate,** and their body cavity is incompletely lined by mesoderm—that is, a layer of mesoderm exists beneath the body wall but not around the gut. Most animals have ❺ a true **coelom,** in which the body cavity is completely lined with mesoderm. In animals with such a coelom, mesentery, which is composed of strings of mesoderm, supports the internal organs. In **coelomates,** such as earthworms, lobsters, and humans, the mesoderm can interact not only with the ectoderm but also with the endoderm. Therefore, body movements are freer because the outer wall can move independently of the organs, and the organs have the space to become more complex. In animals without a skeleton, a coelom even acts as a so-called hydrostatic skeleton.

Animals can be nonsegmented or have ❻ **segmentation**—the repetition of body parts along the length of the body. Annelids (e.g., earthworm), arthropods (e.g., lobster), and chordates (e.g., humans) are segmented. To illustrate your segmentation, run your hand along your backbone, which is composed of a series of vertebrae. Segmentation leads to specialization of parts because the various segments can become differentiated for specific purposes. Independent movement of body parts leads to a greater diversity of body movements.

Two groups of animals, the arthropods and the chordates, have ❼ *jointed appendages,* which are particularly useful for movement on land. In vertebrates, we will use the term jointed limbs. A lobster has a jointed exoskeleton, while a human has a jointed endoskeleton. An endoskeleton allows an animal to grow larger than does an exoskeleton.

Notice that it would be correct to call each of the seven innovations a **homology** because each innovation is a similarity present in a common ancestor and all its descendants. A common ancestor is present at each junction of the tree.

The traditional evolutionary tree examined in this section is compared to one based on molecular data in Section 19.4.

> **19.3 Check Your Progress** Which of the seven innovations noted in the evolutionary tree of animals are visible in bats?

FIGURE 19.3B Types of body cavity.

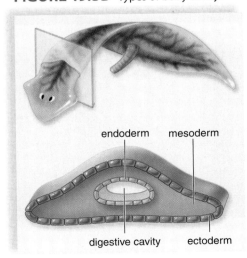

Acoelomate (flatworms)

endoderm   mesoderm

digestive cavity   ectoderm

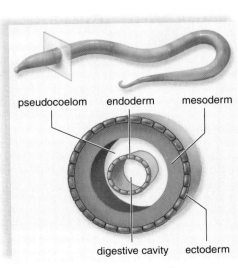

pseudocoelom   endoderm   mesoderm

digestive cavity   ectoderm

Pseudocoelomate (roundworms)

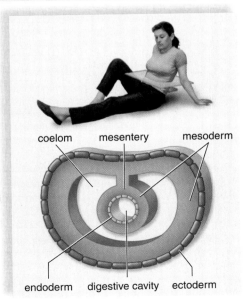

coelom   mesentery   mesoderm

endoderm   digestive cavity   ectoderm

Coelomate (molluscs, annelids, arthropods, echinoderms, chordates)

19.4 Molecular data suggest a new evolutionary tree for animals

Modern phylogenetic investigations take into account molecular data, primarily nucleotide sequences, when classifying animals. It is assumed that the more closely related two organisms are, the more nucleotide sequences they will have in common. An evolutionary tree based on molecular data is somewhat different from the one based only on anatomic characteristics that we have been discussing.

In the traditional tree, the protostomes are restricted to three phyla, which have a coelom: arthropods, annelids, and molluscs (see Fig. 19.3A). For example, flatworms, which are acoelomate, are not protostomes, whereas annelids are protostomes because they have a coelom that develops in a particular way.

Figure 19.4A shows an evolutionary tree based on molecular data. These data suggest that many more animal phyla should be designated protostomes because their rRNA sequences are so similar. However, the protostomes are divided into two groups. One group, which contains flatworms, molluscs, and annelids, has a particular type of immature stage, the trochophore larva and therefore is called trochozoans. The other group contains the roundworms and the arthropods. Both of these types of animals shed their outer covering as they grow; therefore, they are the molting animals (**Fig. 19.4B**).

Notice, too, that segmentation doesn't play a defining role in the evolutionary tree based on molecular data. In the traditional tree, the segmented worms (e.g., earthworm) are placed close to the arthropods, which are also segmented. In the new tree, the segmented worms are trochozoans, and the segmented arthropods are molting animals.

In Section 19.5, we present the groups of animals discussed in this text.

> **19.4 Check Your Progress** Does the proposed revision shown in Figure 19.4A affect the relationship between bats and other animal groups?

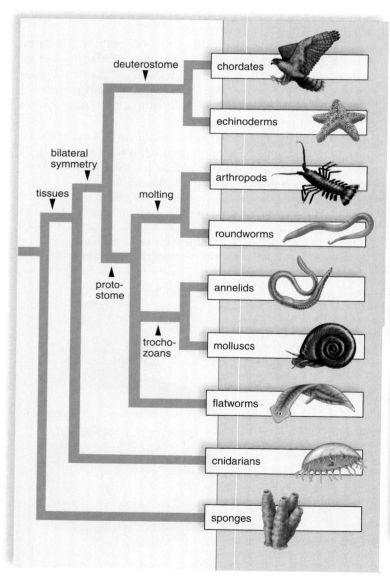

FIGURE 19.4A Proposed new evolutionary tree.

Ascaris, a roundworm

A millipede, an arthropod

FIGURE 19.4B Roundworms and arthropods are molting animals.

19.5 Some animal groups are invertebrates and some are vertebrates

The top of the table at the bottom of this page lists the characteristics of animals. All should be familiar to you, except possibly the diploid life cycle (see Section 17.12). When an organism has a diploid life cycle, the adult is always diploid, and only the gametes are haploid.

For convenience, the animal groups discussed in this text have been divided into **invertebrates** (those that do not have an endoskeleton of cartilage and bone) and **vertebrates** (those that do have such an endoskeleton). As you know, animals evolved in the sea, and surprising as it may seem, most animals still live in the water. Among the invertebrates, only the molluscs, anne-lids, and arthropods have terrestrial representatives. Among the vertebrates, the amphibians, reptiles, birds, and mammals have terrestrial representatives.

Now we will examine each group of animals listed in the table. Section 19.6 takes a look at sponges, representative of the first animals to have evolved.

> **19.5 Check Your Progress** Bats belong to which of the groups in the table below?

DOMAIN Eukarya
KINGDOM Animals

CHARACTERISTICS
- Multicellular
- Well-developed tissues (except sponges)
- Usually motile
- Heterotrophic by ingestion or absorption, generally a digestive cavity
- Diploid life cycle

Invertebrates

Sponges: (bony, glass, spongin) Multicellular*, asymmetrical, saclike body perforated by pores; internal cavity lined by food-filtering cells called choanocytes; spicules serve as internal skeleton. 5,150+

Cnidarians (Hydra, jellyfish, corals, sea anemones): Radially symmetrical with two tissue layers; sac body plan; tentacles with nematocysts. 10,000+

Flatworms (planarians, tapeworms, flukes): Bilateral symmetry with cephalization; three tissue layers and organ systems; acoelomate with incomplete digestive tract that can be lost in parasites; hermaphroditic. 20,000+

Roundworms (Ascaris, pinworms, hookworms, filarial worms): Pseudocoelom and hydroskeleton; complete digestive tract; plentiful, free-living forms in soil and water; parasites common. 25,000+

Molluscs (clams, snails, squids): Coelom, all have a foot, mantle, and visceral mass; foot is variously modified; in many, the mantle secretes a calcium carbonate shell as an exoskeleton; true coelom and all organ systems. 110,000+

Annelids (polychaetes, earthworms, leeches): Segmented with body rings and setae; cephalization in some polychaetes; hydroskeleton; closed circulatory system. 16,000+

Arthropods (crustaceans, spiders, scorpions, centipedes, millipedes, insects): Chitinous exoskeleton with jointed appendages undergoes molting; insects—most have wings—are most numerous of all animals. 1,000,000+

Echinoderms (sea stars, sea urchins, sand dollars, sea cucumbers): Radial symmetry as adults; unique water-vascular system and tube feet; endoskeleton of calcium plates. 7,000+

Chordates (tunicates, lancelets, vertebrates): All have notochord, dorsal tubular nerve cord, pharyngeal pouches, and postanal tail at some time; contains mostly vertebrates in which notochord is replaced by vertebral column. 56,000+

Vertebrates

Fishes (jawless, cartilaginous, bony): Endoskeleton, jaws, and paired appendages in most; internal gills; single-loop circulation; scales.

Amphibians (frogs, toads, salamanders): Jointed limbs; lungs; three-chambered heart with double-loop circulation; moist, thin skin.

Reptiles (snakes, turtles, crocodiles): Amniotic egg; rib cage in addition to lungs; three-chambered heart typical; scaly, dry skin; copulatory organ in males and internal fertilization.

Birds (songbirds, waterfowl, parrots, ostriches): Endothermy, feathers, and skeletal modifications for flying; lungs with air sacs; four-chambered heart.

Mammals (monotremes, marsupials, placental): Hair and mammary glands.

*After a character is listed, it is present in the rest, unless stated otherwise.
+Number of species.

While all animals are multicellular, **sponges** are the only animals to lack true tissues and to have a cellular level of organization. Actually, they have few cell types and no nerve or muscle cells to speak of. Most likely, sponges are out of the mainstream of animal evolution and represent an evolutionary dead end.

The Body of Sponges Sponges are placed in phylum Porifera because their saclike bodies are perforated by many pores (**Fig. 19.6**). Sponges are aquatic, largely marine animals that vary greatly in size, shape, and color. But, they all have a canal system of varying complexity that allows water to move through their bodies.

The interior of the canals is lined with flagellated cells called *collar cells*, or choanocytes. The beating of the flagella produces water currents that flow through the pores into the central cavity and out through the osculum, the upper opening of the body. Even a simple sponge only 10 cm tall is estimated to filter as much as 100 L of water each day. It takes this much water to supply the needs of the sponge. A sponge is a stationary **filter feeder**, also called a suspension feeder, because it filters suspended particles from the water by means of a straining device—in this case, the pores of the walls and the microvilli making up the collar of collar cells. Microscopic food particles that pass between the microvilli are engulfed by the collar cells and digested by them in food vacuoles.

Endoskeleton The skeleton of a sponge prevents the body from collapsing. All sponges have fibers of spongin, a modified form of collagen; a bath sponge is the dried spongin skeleton from which all living tissue has been removed. Today, however, commercial "sponges" are usually synthetic.

Typically, the endoskeleton of sponges also contains spicules—small, needle-shaped structures with one to six rays. Traditionally, the type of spicule has been used to classify sponges, in which case there are bony, glass, and spongin sponges. The success of sponges—they have existed longer than many other animal groups—can be attributed to their spicules. They have few predators because a mouth full of spicules is an unpleasant experience. Also, they produce a number of foul smelling and toxic substances that discourage predators.

Reproduction Sponges can reproduce both asexually and sexually. They reproduce asexually by fragmentation or by *budding*. During budding, a small protuberance appears and gradually increases in size until a complete organism forms. Budding produces colonies of sponges that can become quite large. During sexual reproduction, eggs and sperm are released into the central cavity, and the zygote develops into a flagellated larva that may swim to a new location. If the cells of a sponge are mechanically separated, they will reassemble into a complete and functioning organism! Like many less specialized organisms, sponges are also capable of regeneration, or growth of a whole from a small part.

In Section 19.7, we turn our attention to the cnidarians.

> **19.6** *Check Your Progress a.* Contrast the level of organization in sponges and bats. *b.* Are vampire bats filter feeders?

Yellow tube sponge

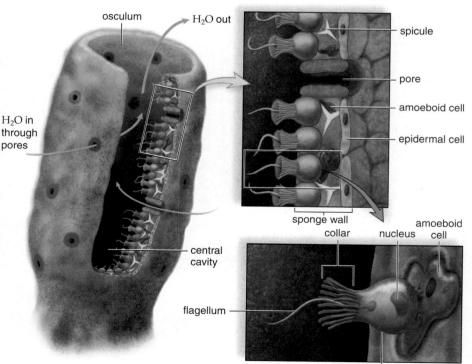

Sponge organization

FIGURE 19.6 Sponge anatomy.

19.7 Cnidarians have true tissues

The animals in the remaining phyla to be studied have true tissues. **Cnidarians** (phylum Cnidaria) are an ancient group of invertebrates with a rich fossil record. Most cnidarians live in the sea, but a few freshwater species exist. Although stationary, or at best slow moving, cnidarians have an effective means of capturing prey. They are radially symmetrical and capture their prey with a ring of tentacles that bear specialized stinging cells, called cnidocytes (**Fig. 19.7A**). Each cnidocyte has a capsule called a **nematocyst**, containing a long, spirally coiled, hollow thread. When the trigger of the cnidocyte is touched, the nematocyst is discharged. Some nematocysts merely trap a prey or predator; others have spines that penetrate the prey's body and inject paralyzing toxins before the prey is captured and drawn into the *gastrovascular cavity*. Cnidarians can digest a prey of fairly large size because extracellular digestion occurs in this cavity.

During development, cnidarians have only two germ layers (ectoderm and endoderm), and as adults, they have the *tissue level of organization*. Cnidarians are capable of coordinated movements because the ectodermal cells have contractile fibers that are stimulated by nerve cells that form a nerve net. Sensory cells, which receive external stimuli, also communicate with the nerve net.

Two basic body forms are seen among cnidarians—the polyp and the medusa. The mouth of the polyp is directed upward from the substrate, while the mouth of the medusa is directed downward. In any case, cnidarians have a *sac body plan* with only one opening. A medusa has much jellylike packing material, called *mesoglea*, and is commonly called a "jellyfish." Polyps are tubular and generally attach to a rock with some, but not as much, mesoglea (**Fig. 19.7B**).

Cnidarians, as well as other marine animals, have been the source of medicines, particularly drugs that counter inflammation. The other groups of animals to be studied also have true tissues, as do the cnidarians.

In Section 19.8, we look at the free-living flatworms.

> **19.7 Check Your Progress** *a.* **Suppose you wanted to show that cnidarians have two germ layers and bats have three germ layers. What type of study would you undertake?** *b.* **Compare the lifestyle of a cnidarian to that of a vampire bat.**

Sea anemone, a solitary polyp

Coral, a colonial polyp

Portuguese man-of-war, colony of modified polyps and medusae

Jellyfish, a medusa

FIGURE 19.7A Cnidarian diversity.

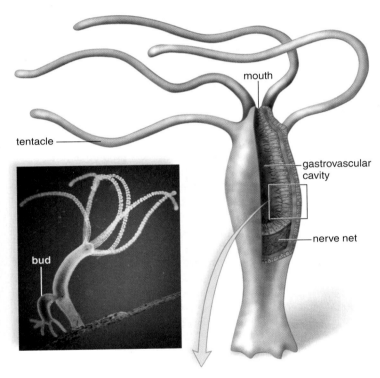

mouth

tentacle

gastrovascular cavity

nerve net

bud

tissue layers

gastrovascular cavity

flagella

mesoglea (packing material)

gland cell

sensory cell

cnidocyte

nematocyst

FIGURE 19.7B Anatomy of *Hydra*, a polyp.

The majority of animal phyla are bilaterally symmetrical, at least in some stage of their development. *Bilateral symmetry* is first seen among the **flatworms** (phylum Platyhelminthes). Like all the other animal phyla we will study, flatworms also have three germ layers: ectoderm, endoderm, and mesoderm. The body wall develops from ectoderm; the digestive cavity develops from endoderm; and mesoderm contributes to organ formation. Therefore, the presence of mesoderm, in addition to ectoderm and endoderm, gives bulk to the animal and leads to organ formation: Flatworms have the organ system level of organization. Nevertheless, flatworms have no body cavity. In the other animals to be studied, the organs lie in a body cavity that is lined by mesoderm and is called a coelum. Because the flatworms have no coelum, they are called *acoelomates*.

Free-living flatworms, called **planarians**, have several body systems, including the digestive system (**Fig. 19.8**). The animal captures food by wrapping itself around the prey, entangling it in slime, and pinning it down. Then the planarian extends a muscular pharynx and, by a sucking motion, tears up and swallows its food. The pharynx leads into a three-branched gastrovascular cavity where digestion begins. Digestion is finished inside the cells that line the gastrovascular cavity. The digestive tract is *incomplete* because it has only one opening, and undigested food passes out through the pharynx. Animals with only one opening have a sac body plan. Animals with two openings are said to have a tube-within-a-tube plan.

Living in fresh water, planarians have a well-developed excretory system composed of a series of interconnecting canals that run the length of the body on each side. Flame cells contain cilia that move back and forth, bringing water into the canals that empty at pores. The beating of the cilia reminded an early investigator of the flickering of a flame; therefore, he called them *flame cells*.

Planarians are **hermaphrodites**, meaning that they possess both male and female sex organs. The worms practice cross-fertilization: The penis of one is inserted into the genital pore of the other, and a reciprocal transfer of sperm takes place. The fertilized eggs hatch in 2–3 weeks as tiny worms. Development of the larva introduces bilateral symmetry; in other words, the larva was bilaterally symmetrical, and this type of symmetry is retained by the adult.

Planarians have a ladderlike nervous system. A small anterior brain and two lateral nerve cords are joined by cross-branches called transverse nerves. Planarians exhibit *cephalization*; aside from a brain, the "head" end has light-sensitive organs (the eyespots) and chemosensitive organs located on the auricles. The presence of mesoderm permits the development of three muscle layers—an outer circular layer, an inner longitudinal layer, and a diagonal layer—that allow for varied movement. A ciliated epidermis allows planarians to glide along a film of mucus.

Some parasitic flatworms are examined in Section 19.9.

> **19.8 Check Your Progress** *a.* **Do bats have the same symmetry as flatworms? Explain.** *b.* **Are bats hermaphroditic?**

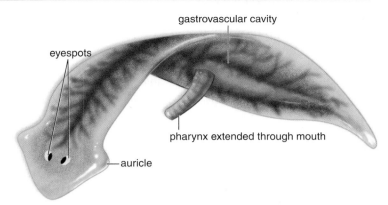

Digestive system

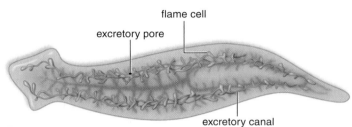

Excretory system

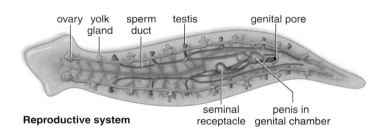

Reproductive system

Nervous system

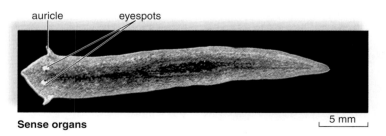

Sense organs

5 mm

FIGURE 19.8 Planarian anatomy.

Parasitic flatworms belong to two classes: the tapeworms and the flukes.

Tapeworms

As adults, tapeworms are endoparasites (internal parasites) of various vertebrates, including humans. They vary in length from a few millimeters to nearly 20 meters. Tapeworms have a tough body covering that is resistant to the host's digestive juices. The scolex is an anterior region, which bears hooks and suckers for attachment to the intestinal wall of the host. Behind the scolex, a series of reproductive units called proglottids contain a full set of female and male sex organs. After fertilization, the organs within a proglottid disintegrate, and it becomes filled with mature eggs. The eggs are eliminated in the feces of the host.

In the life cycle of *Taenia solium*, the pork tapeworm, a pig host alternates with a human host. The muscles of a pig become infected with bladder worms when pigs eat food contaminated with egg-containing feces. When humans eat infected pork that has not been thoroughly cooked, a bladder worm becomes a tapeworm attached to their intestinal wall (**Fig. 19.9A**). Most tapeworm carriers show no symptoms and usually become aware of the infection only after noticing tapeworm segments in their feces. Mild gastrointestinal symptoms, such as nausea or abdominal pain, can occur in infected individuals. In rare cases, where the tapeworm segments migrate into the appendix, pancreas, or bile duct, a person may experience sudden and severe abdominal discomfort.

Flukes

All flukes are endoparasites of various vertebrates. Their flattened and oval-to-elongated body is covered by a protective body wall. The anterior end of the animal has an oral sucker and at least one other sucker used for attachment to the host. Blood flukes (*Schistosoma* spp.) occur predominantly in the Middle East, Asia, and Africa. Adults are small (approximately 2.5 cm long) and may live for years in their human hosts. Nearly 800,000 persons die each year from an infection called schistosomiasis.

Adult humans become infected when they expose their skin to water that contains *Schistosoma* larvae released from a snail (**Fig. 19.9B**). Male and female flukes live in the veins of the human abdominal cavity. Here, they mate, and the females produce eggs. When the eggs penetrate the intestine or urinary bladder, they leave the body in feces or urine. The eggs hatch in water and become larvae that infect snails. Asexual reproduction occurs within the snails, and then the larval form that infects humans escapes the snails and enters the water.

Schistosomiasis is a debilitating disease because the eggs cause much tissue damage when they penetrate the walls of the veins of the small intestine or urinary bladder. The tissues hemorrhage, so that blood often appears in urine or feces. Even worse, many of the eggs produced by the female worms do not leave the veins, but are swept up in the circulatory system and deposited in the host's liver, where they are encapsulated.

This completes our study of flatworms; in Section 19.10, we consider the roundworms.

> **19.9 Check Your Progress** Compare the parasitism of vampire bats to that of tapeworms and flukes.

FIGURE 19.9A Tapeworm (*Taenia solium*) anatomy and life cycle.

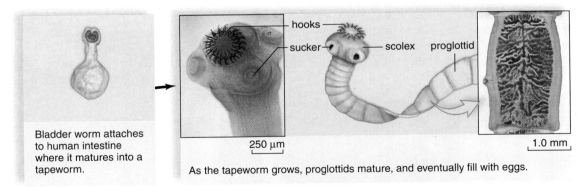

hooks — sucker — scolex — proglottid

250 μm · 1.0 mm

Bladder worm attaches to human intestine where it matures into a tapeworm.

As the tapeworm grows, proglottids mature, and eventually fill with eggs.

FIGURE 19.9B Sexual portion of blood fluke (*Schistosoma* spp.) life cycle.

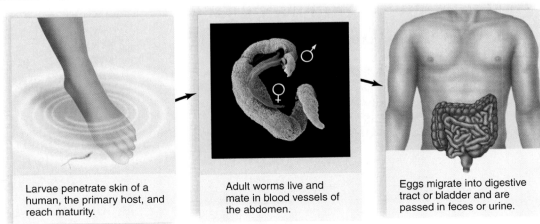

Larvae penetrate skin of a human, the primary host, and reach maturity.

Adult worms live and mate in blood vessels of the abdomen.

Eggs migrate into digestive tract or bladder and are passed in feces or urine.

Roundworms (phylum Nematoda) possess two anatomic features not seen in animals discussed previously: a *body cavity* and a *complete digestive tract*. The body cavity is a pseudocoelom and is incompletely lined with mesoderm (see Fig. 19.3B). The digestive tract is complete because it has both a mouth and an anus. Worms, in general, do not have a skeleton, but the fluid-filled pseudocoelom supports muscle contraction and enhances flexibility.

The roundworms are nonsegmented, meaning that they have a smooth outside body wall. Roundworms are generally colorless and less than 5 cm in length, and they occur almost everywhere—in the sea, in fresh water, and in the soil—in such numbers that thousands of them can be found in a small area. Many are free-living and feed on algae, fungi, microscopic animals, dead organisms, and plant juices, causing great agricultural damage. Parasitic roundworms live anaerobically in every type of animal and many plants. Several parasitic roundworms infect humans.

Ascaris Humans become infected with a roundworm called *Ascaris* (**Fig. 19.10A**) when eggs enter the body via uncooked vegetables, soiled fingers, or ingested fecal material and hatch in the intestines. The juveniles make their way into the cardiovascular system and are carried to the heart and lungs. From the lungs, the larvae travel up the trachea, where they are swallowed and, eventually, reach the intestines. There, the larvae mature and begin feeding on intestinal contents. A female *Ascaris* is very prolific, producing over 200,000 eggs daily. The eggs are eliminated with host feces.

Other Roundworm Parasites Trichinosis is a fairly serious human infection rarely seen in the United States. Humans acquire the disease when they eat meat that contains encysted larvae. Once in the digestive tract, the cysts release the larvae, which develop into adult worms, and the female burrows into the wall of the host's small intestine, where she deposits live larvae that are carried by the blood-stream to the skeletal muscles, where they encyst (**Fig. 19.10B**). The presence of adults in the small intestine causes digestive disorders, fatigue, and fever. After the larvae encyst in muscles, the symptoms include aching joints, muscle pain, and itchy skin.

Elephantiasis is caused by a roundworm called a filarial worm, which utilizes mosquitoes as a secondary host. The adult worms reside in human lymphatic vessels, which normally take up excess tissue fluid but are prevented from doing so by the presence of the worms. The limbs of an infected human can swell to an enormous size, even resembling those of an elephant (**Fig 19.10C**), hence the name of the disease. More common still is a disabling swelling of the scrotum in men. When a mosquito bites an infected person, it can transport larvae to a new host.

Other roundworm infections are more common in the United States. Children frequently acquire pinworm infections, and hookworm is seen in the southern states, as well as worldwide. A hookworm infection can be very debilitating because the worms attach to the intestinal wall and feed on blood. Good hygiene, proper disposal of sewage, thorough cooking of meat, and regular deworming of pets usually protect people from parasitic roundworms. A common fatal roundworm infection in dogs is due to the heartworm. The mosquito serves as the vector.

Section 19.11 interrupts our survey of animals to discuss the coelom, because animals differ according to the type of coelom.

> **19.10** *Check Your Progress* What does the presence of a complete digestive tract in both roundworms and bats tell us about their relatedness?

FIGURE 19.10A *Ascaris.*

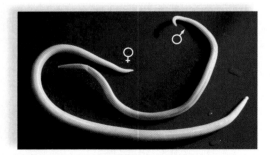

FIGURE 19.10B Encysted *Trichinella* larva.

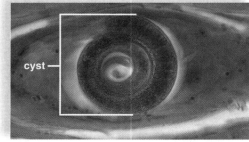

cyst

SEM 400×

FIGURE 19.10C Elephantiasis.

19.11   A coelom gives complex animal groups certain advantages

The remaining phyla to be studied are coelomates, which have a coelom, a body cavity completely lined by mesoderm, as described in Section 19.3. Developmental differences among the coelomates allow them to be divided into two groups: **protostomes** or **deuterostomes**. Molluscs, annelids, and arthropods are protostomes that have a coelom. As discussed previously, molecular data suggest that flatworms and roundworms are also protostomes. The flatworms do not have any type of body cavity for their internal organs, and roundworms have a pseudocoelom rather than a true coelom. The designation of echinoderms and chordates as deuterostomes is supported by both traditional and molecular data. Two major events during development (**Fig. 19.11**) can be used to distinguish protostomes from deuterostomes:

1. *Fate of blastopore.* As development proceeds, a hollow sphere of cells, called a blastula, forms, and the indentation that follows produces an opening called the blastopore. In protostomes (*proto*, before; *stome*, mouth), the mouth appears at or near the blastopore; in deuterostomes (*deutero*, second), the anus appears at or near the blastopore, and only later does a new opening form the mouth.

2. *Coelom formation.* Traditionally, all protostomes and deuterostomes have a coelom. However, the coelom develops differently in the two groups. In protostomes, the mesoderm arises from cells located near the embryonic blastopore, and a splitting occurs that produces the coelom. In deuterostomes, the coelom arises as a pair of mesodermal pouches from the wall of the primitive gut. The pouches enlarge until they meet and fuse, forming the coelom.

Advantages of a Coelom   A coelom offers many advantages. Body movements are freer because the outer wall can move independently of the enclosed organs. Also, the ample space of a coelom allows complex organs and organ systems to develop. For example, the digestive tract can coil and provide a greater surface area for absorption of nutrients. Coelomic fluid protects internal organs against damage and marked temperature changes. It can even assist in storage and transport of substances. Finally, fluid within the coelom can provide a hydrostatic skeleton—that is, muscular contraction pushes against the fluid and allows the animal to move.

As complex animals, coelomates have the organ level of organization. Like other animals, they evolved in the sea, but they have successful terrestrial representatives. Terrestrial existence requires breathing air, preventing desiccation, and having means of locomotion and reproduction that are not dependent on external water. The excretory system may be modified for excretion of a solid nitrogenous waste to help conserve water.

In Section 19.12, we begin our look at the coelomates with the molluscs.

19.11 Check Your Progress   **Are bats protostomes or deuterostomes?**

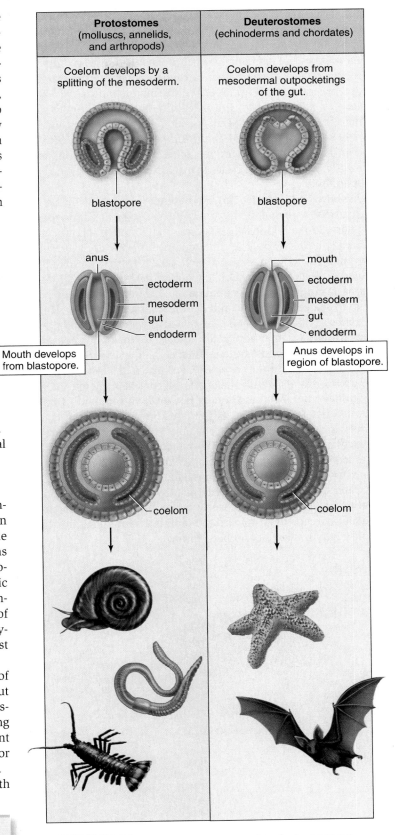

FIGURE 19.11   Protostomes compared to deuterostomes.

All molluscs (phylum Mollusca) have a body composed of at least three distinct parts: (1) The *foot* is the strong, muscular portion used for locomotion; (2) the *visceral mass* is the soft-bodied portion that contains internal organs; and (3) the *mantle* is a membranous, or sometimes muscular, covering that envelops the visceral mass (**Fig. 19.12A**). The mantle may secrete an exoskeleton called a *shell*. If a foreign body is placed between the mantle and the shell of an oyster (a mollusc), concentric layers of shell are deposited about the particle to form a pearl. Another feature often present in molluscs is a rasping, tongue-like *radula*, an organ that bears many rows of teeth and is used to obtain food.

As shown in **Figure 19.12B**, three common groups of molluscs are gastropods, cephalopods, and bivalves. **1** In **gastropods** (meaning stomach-footed), including snails and nudibranchs, the animal moves by muscle contractions that pass along its ventrally flattened foot. In snails, which are terrestrial, the mantle produces a shell, is richly supplied with blood vessels, and functions as a lung. **2** In **cephalopods** (meaning head-footed), including octopuses, squids, and nautiluses, the foot has evolved into tentacles about the head. The tentacles seize prey, and then a powerful beak and a radula tear it apart. Cephalopods possess well-developed nervous systems and complex sensory organs. Rapid movement and the secretion of a brown or black pigment from an ink gland help cephalopods escape their enemies. Octopuses have no shell, and squids have only a remnant of one concealed beneath the skin. **3** Clams, oysters, scallops, and mussels are called **bivalves** because their shells have two parts. A muscular foot projects ventrally from the shell. In a clam, such as the freshwater clam, the calcium carbonate shell has an inner layer of mother-of-pearl. The clam is a filter feeder. Food particles and water enter the mantle cavity by way of a siphon; mucous secretions cause smaller particles to adhere to the gills; and ciliary action sweeps them toward the mouth. Spent fluid exits the mantle cavity by way of another siphon.

Land snail

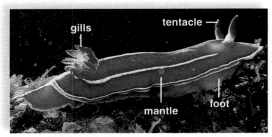
Three-stripe doris nudibranch

1 Gastropods

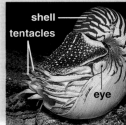
Two-spotted octopus   Chambered nautilus

2 Cephalopods

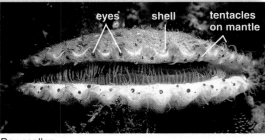
Bay scallop   Blue mussel

3 Bivalves

FIGURE 19.12B Three groups of molluscs.

Having studied the molluscs, the annelids are our topic in Section 19.13.

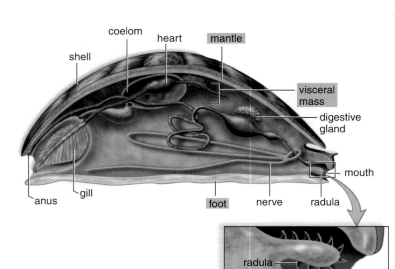

FIGURE 19.12A
Body plan of a typical mollusc.

19.12 *Check Your Progress* How many feature do bats share with molluscs?

Annelids (phylum Annelida) are *segmented*, as can be seen externally by the rings that encircle the body of an earthworm (**Fig. 19.13A**). Segmentation is also apparent throughout the body. Partitions called septa divide the well-developed, fluid-filled coelom, which is used as a hydrostatic skeleton to facilitate movement. The nervous system consists of a brain connected to a ventral nerve cord, with ganglia in each segment. The excretory system consists of **nephridia**, which are tubules in most segments that collect waste material and excrete it through an opening in the body wall. The complete digestive tract has led to many specialized organs from the mouth to the anus.

Phylum Annelida contains: oligochaetes, polychaetes, and leeches.

Oligochaetes　The earthworm is an oligochaete because it has few setae per segment. *Setae* are bristles that anchor the worm or help it move. Earthworms do not have a well-developed head, and they reside in soil, where there is adequate moisture to keep the body wall moist for gas exchange. They are scavengers that feed on leaves or any other organic matter, living or dead, that can conveniently be taken into the mouth along with dirt.

Polychaetes　Most annelids are polychaetes (having many setae per segment) that live in marine environments. **Figure 19.13B** shows a stationary polychaete, with tentacles that form a funnel-shaped fan. This animal is also known as a tube worm because it secretes and lives in a tube, from which it emerges to filter-feed. Water currents created by the action of cilia trap food particles that are directed toward the mouth. The clam worm *Nereis* is a polychaete with a pair of strong, chitinous jaws that extend with a part of the pharynx. In support of its predatory way of life, *Nereis* has a well-defined head region, with eyes and other sense organs.

Leeches　Leeches, in class Hirudinea, have no setae, but have the same body plan as other annelids. They are blood suckers that are able to keep blood flowing and prevent clotting by means of a powerful anticoagulant in their saliva known as *hirudin*. The medicinal leech is used to remove blood from tissues following surgery.

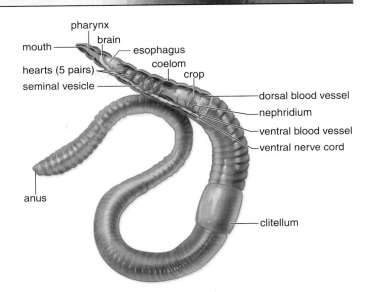

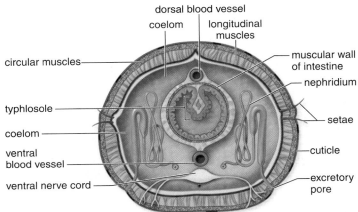

FIGURE 19.13A　Earthworm anatomy.

The diversity of annelids pales next to that of the arthropods, studied in Sections 19.14 to 19.16.

19.13 *Check Your Progress*　What anatomic feature shows that bats are segmented?

Christmas tree worm

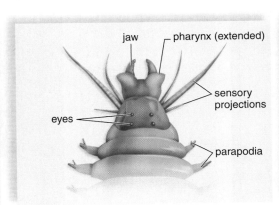

Clam worm

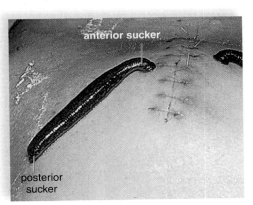

Medicinal leech

FIGURE 19.13B　Other annelids.

Arthropods (phylum Arthropoda) are extremely diverse. Over one million species have been discovered and described, but some experts suggest that as many as 30 million arthropods may exist—most of them insects. The success of arthropods can be attributed to the following six characteristics:

1. *Jointed appendages.* Basically hollow tubes moved by muscles, jointed appendages have become adapted to different means of locomotion, food gathering, and reproduction. Examples are the walking legs of a crayfish shown in **Figure 19.14A**. Modifications of appendages account for much of the diversity of arthropods.

2. *Exoskeleton.* A rigid but jointed exoskeleton is composed primarily of **chitin**, a strong, flexible, nitrogenous polysaccharide. The exoskeleton serves many functions, including protecting the body, preventing desiccation, serving as an attachment site for muscles, and aiding locomotion.

Because an exoskeleton is hard and nonexpandable, arthropods must undergo **molting**, or shedding of the exoskeleton, as they grow larger. During molting, arthropods are vulnerable and are attacked by many predators.

3. *Segmentation.* In many species, the repeating units of the body are called segments. In some arthropods, each segment has a pair of jointed appendages; in others, the segments are fused into a head, thorax, and abdomen.

4. *Well-developed nervous system.* Arthropods have a brain and a ventral nerve cord. The head bears various types of sense organs, including compound and simple eyes. Many arthropods also have well-developed touch, smell, taste, balance, and hearing capabilities. Arthropods display many complex behaviors and communication skills.

5. *Variety of respiratory organs.* Marine forms utilize gills; terrestrial forms have book lungs (e.g., spiders) or air tubes called tracheae. Tracheae serve as a rapid way to transport oxygen directly to the cells. The circulatory system is open, with the dorsal heart pumping blood into various sinuses throughout the body.

6. *Reduced competition through metamorphosis.* Many arthropods undergo a change in form and physiology as a larva becomes an adult. Metamorphosis allows the larva to have a different lifestyle than the adult (**Fig. 19.14B**). For example, larval crabs live among and feed on plankton, while adult crabs are bottom dwellers that catch live prey or scavenge dead organic matter. Among insects such as butterflies, the caterpillar feeds on leafy vegetation, while the adult feeds on nectar.

Now that we have introduced the arthropods, Section 19.15 takes a look at some major groups of arthropods other than insects.

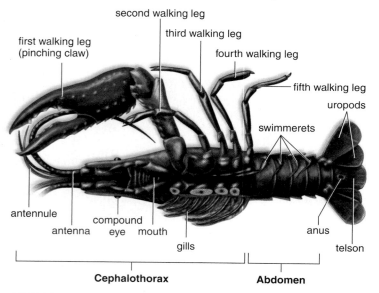

FIGURE 19.14A Exoskeleton and jointed appendages of a crayfish, an arthropod.

second walking leg
third walking leg
fourth walking leg
first walking leg (pinching claw)
fifth walking leg
uropods
swimmerets
antennule
compound eye
antenna
mouth
gills
anus
telson
Cephalothorax
Abdomen

19.14 *Check Your Progress* ***a.*** **Which of these six characteristics of arthropods would apply to bats?** ***b.*** **Explain the features that do not apply.**

Caterpillar, eating stage
Pupa, cocoon stage
Metamorphosis occurs
Emergence of adult
Butterfly, adult stage

FIGURE 19.14B Monarch butterfly metamorphosis.

Crustaceans, a name derived from their hard, crusty exoskeleton, are a group of largely marine arthropods that include crabs, barnacles, shrimps, and crayfish (**Fig. 19.15A**). Although crustacean anatomy is extremely diverse, the head usually bears a pair of compound eyes and five pairs of appendages. The first two pairs of appendages, called antennae and antennules, respectively, lie in front of the mouth and have sensory functions. The other three pairs are mouthparts used in feeding. In a crayfish (see Fig. 19.14A), the thorax bears five pairs of walking legs. The first walking leg is a pinching claw. The *gills* are situated above the walking legs. The abdominal segments are equipped with swimmerets, small, paddlelike structures. The last two segments bear the uropods and the telson, which make up a fan-shaped tail.

Crustaceans play a vital role in the food chain. Tiny crustaceans known as krill and also copepods are a major source of food for baleen whales, seabirds, and seals. Many species of lobsters, crabs, and shrimp are important in the seafood industry.

Centipedes, with a pair of appendages on every segment, are carnivorous, while millipedes, with two pairs of legs on most segments, are herbivorous (**Fig. 19.15B**). The head appendages of these animals are similar to those of insects, which are the largest group of arthropods, or indeed animals, as discussed in Section 19.16.

The **arachnids** include spiders, scorpions, ticks, mites, and harvestmen ("daddy longlegs") (**Fig. 19.15C**). Spiders have a narrow waist that separates the cephalothorax, which has four pairs of legs, from the abdomen. Spiders use silk threads for all sorts of purposes, from lining their nests to catching prey. The internal organs of spiders also show how they are adapted to a terrestrial way of life. Invaginations of the inner body wall form lamellae ("pages") of spiders' so-called book lungs. Scorpions are the oldest terrestrial arthropods (Fig. 19.15C). Ticks and mites are parasites. Ticks suck the blood of vertebrates and sometimes transmit diseases, such as Rocky Mountain spotted fever or Lyme disease. Like other arachnids, the first pair of appendages in a horseshoe crab are pinching structures used for feeding and defense (Fig. 19.15C).

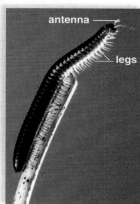

FIGURE 19.15B
Centipede and millipede.

Centipede

Millipede

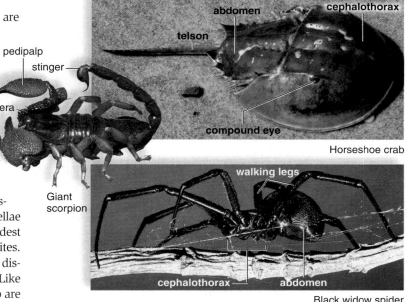
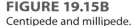

Horseshoe crab

Black widow spider

Giant scorpion

FIGURE 19.15C Spider and relatives.

> **19.15 Check Your Progress** Bats are terrestrial, as are what arthropods shown on this page?

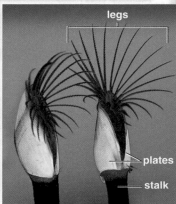

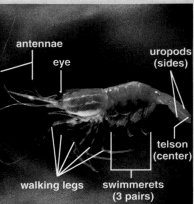

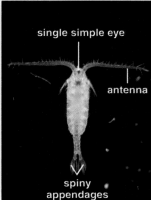

Sally lightfoot crab

Gooseneck barnacles

Red-backed cleaning shrimp

Copepod

FIGURE 19.15A Crustacean diversity.

19.16 Insects, the largest group of arthropods, are adapted to living on land

Insects are so numerous (probably over one million species) and so diverse that the study of this one group is a major specialty in biology called entomology (**Fig. 19.16**). Some insects show remarkable behavior adaptations, as exemplified by the social systems of bees, ants, termites, and other colonial insects.

Insects are adapted to an active life on land, although some have secondarily invaded aquatic habitats. The body is divided into a head, a thorax, and an abdomen. The head usually bears a pair of sensory antennae, a pair of compound eyes, and several simple eyes. The mouthparts are adapted to each species' particular way of life: A grasshopper has mouthparts that chew, and a butterfly has a long tubular proboscis for siphoning the nectar from flowers. The abdomen contains most of the internal organs; legs and/or wings are often attached to the thorax.

Wings enhance an insect's ability to survive by providing a way of escaping enemies, finding food, facilitating mating, and dispersing the offspring. The exoskeleton of an insect is lighter and contains less chitin than that of many other arthropods. The male has a penis, which passes sperm to the female. The female, as in the grasshopper, may have an ovipositor for laying the fertilized eggs. Some insects, such as butterflies, undergo complete metamorphosis, involving a drastic change in form.

Section 19.17 discusses the echinoderms, which are deuterostomes, as are the vertebrates.

> **19.16 Check Your Progress** Are the wings of insects and the wings of a bat analogous or homologous? Explain.

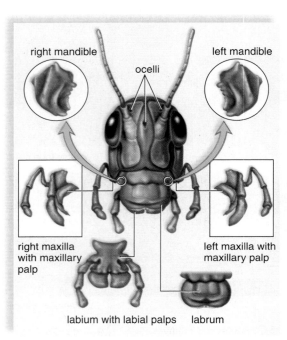

Mouthparts of a grasshopper

Mealybug

Grasshopper

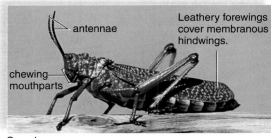

Dragonfly

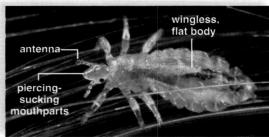

Head louse

Wasp

Butterfly

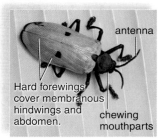

Beetle

Housefly

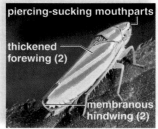

Leafhopper

FIGURE 19.16 Insect diversity.

Echinoderms (phylum Echinodermata) lack features associated with vertebrates, and yet we know they are related to chordates because they are both deuterostomes. The echinoderms are radially, not bilaterally, symmetrical as adults (**Fig. 19.17**). However, their larva is a free-swimming filter feeder with bilateral symmetry—it metamorphoses into the radially symmetrical adult. Also, adult echinoderms do not have a head, brain, or segmentation. Their nervous system consists of nerves in a ring around the mouth extending outward radially.

Echinoderm locomotion depends on a water vascular system. In the sea star, water enters this system through a sieve plate (Fig. 19.17). Eventually, it is pumped into many tube feet, expanding them. When the foot touches a surface, the center withdraws, producing suction that causes the foot to adhere to the surface. By alternating the expansion and contraction of its many tube feet, a sea star moves slowly along.

Echinoderms don't have a complex respiratory, excretory, or circulatory system. Fluids within the coelomic cavity and the water vascular system carry out many of these functions. For example, gas exchange occurs across the skin gills and the tube feet. Nitrogenous wastes diffuse through the coelomic fluid and the body wall.

In ecosystems, most echinoderms feed variously on organic matter in the sea or substratum, but sea stars prey upon crustaceans, molluscs, and other invertebrates. From the human perspective, sea stars cause extensive economic loss because they consume oysters and clams before they can be harvested. Fishes and sea otters eat echinoderms, and scientists favor echinoderms for embryological research.

This completes our study of the invertebrates, and the next part of the chapter studies the vertebrates.

> **19.17** *Check Your Progress*   *a.* **What feature would make you think that bats are not closely related to echinoderms?** *b.* **What feature would make you think they are?**

FIGURE 19.17 Echinoderm structure and diversity.

Sea star (starfish) anatomy

- stomach
- anus
- arm
- sieve plate
- endoskeletal plates
- eyespot
- gonad
- coelom
- ampulla
- tube feet
- digestive gland
- skin gill

spines

Sea urchin

Brittle star   Feather star

feeding tentacles

Sea cucumber

Sea lily

Sand dollar

Four characteristics distinguish the chordates from the other animal phyla. One of these is the presence of the notochord. The invertebrate chordates have a notochord as adults, but in the vertebrates, the notochord is replaced by vertebrae. After examining the key features of vertebrates, we examine each group of vertebrates in the order they evolved.

19.18 Four features characterize chordates

At some time during its life cycle, a **chordate** (phylum Chordata) has the four characteristics depicted in **Figure 19.18** and listed here:

1. A *dorsal supporting rod,* called a **notochord,** extends the length of the body. Vertebrates have an endoskeleton of cartilage or bone, including a vertebral column, that has replaced the notochord during development.

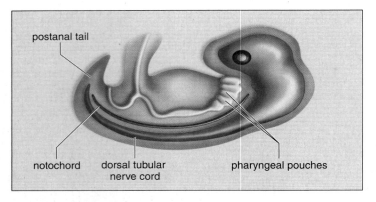

postanal tail

notochord · dorsal tubular nerve cord · pharyngeal pouches

FIGURE 19.18 The four chordate characteristics.

2. A *dorsal tubular nerve cord* contains a canal filled with fluid. In vertebrates, the nerve cord is protected by the vertebrae. Therefore, it is called the spinal cord because the vertebrae form the spine.

3. *Pharyngeal pouches* are seen only during embryonic development in most vertebrates. In the invertebrate chordates, the fishes, and some amphibian larvae, the pharyngeal pouches become functioning gills. Water passing into the mouth and the pharynx goes through the gill slits, which are supported by gill arches. In terrestrial vertebrates that breathe with lungs, the pouches are modified for various purposes. In humans, the first pair of pouches become the auditory tubes. The second pair become the tonsils, while the third and fourth pairs become the thymus gland and the parathyroids.

4. A *postanal tail* extends beyond the anus.

In Section 19.19, we contrast two groups of invertebrate chordates.

> **19.18 Check Your Progress** When would you expect bats to have all four chordate characteristics?

19.19 Invertebrate chordates have a notochord as adults

In a few of the invertebrate chordates, the notochord is never replaced by the vertebral column (**Fig. 19.19**). **Tunicates** (subphylum Urochordata) live on the ocean floor and take their name from a tunic that makes the adults look like thick-walled, squat sacs. They are also called sea squirts because they squirt water from one of their siphons when disturbed. The tunicate larva is bilaterally symmetrical and has the four chordate characteristics. Metamorphosis produces the sessile adult in which cilia move water into the pharynx and out numerous gill slits, the only chordate characteristic that remains in the adult.

Lancelets (subphylum Cephalochordata) are marine chordates only a few centimeters long. They look like a lancet, a small, two-edged surgical knife. Lancelets are found in the shallow water along most coasts, where they usually lie partly buried in sandy or muddy substrates with only their anterior mouth and gill apparatus exposed. They feed on microscopic particles filtered out of the constant stream of water that enters the mouth and exits through the gill slits. Lancelets retain the four chordate characteristics as adults. In addition, *segmentation* is present, as witnessed by the fact that the muscles are segmentally arranged and the dorsal tubular nerve cord has periodic branches.

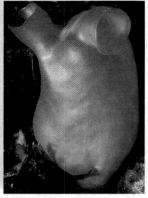

Tunicate   Lancelet

FIGURE 19.19 The invertebrate chordates.

Section 19.20 lays out five key features of the vertebrates.

> **19.19 Check Your Progress** Most chordates are vertebrates. Why shouldn't we simply change phylum Chordata to phylum Vertebrata?

19.20 The evolutionary tree of vertebrates is based on five key features

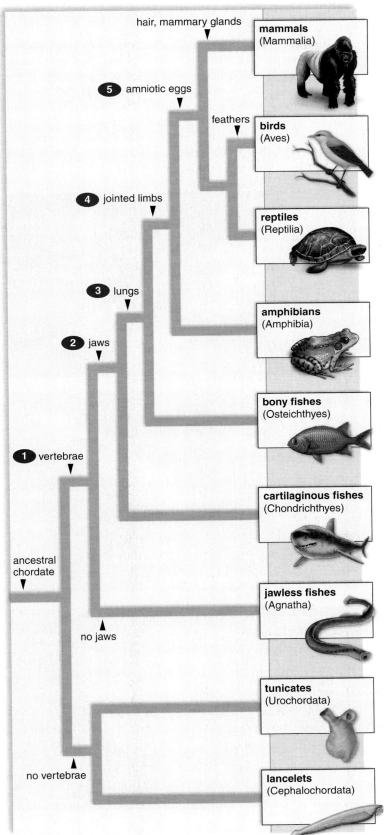

Figure 19.20 depicts the evolutionary tree of the chordates and previews the animal groups we will be discussing in the remainder of this chapter. The tunicates and lancelets are shown as invertebrate chordates because they don't have vertebrae. Figure 19.20 lists five derived characters that distinguish the rest of the vertebrates from the preceding ones. ❶ The first derived character, vertebrae, gives us an opportunity to discuss vertebrates in general.

Vertebrates The vertebrates are the fishes, amphibians, reptiles, birds, and mammals. The **vertebrae** of the vertebral column are their most obvious feature, signifying that vertebrates are segmented animals. The vertebral column is flexible because the vertebrae are separated by intervertebral disks, which cushion the vertebrae; the soft center of a disk presses on the spinal cord.

Vertebrates have an internal skeleton, a living *jointed endoskeleton*, with *paired appendages*. The skull of vertebrates, which protects the brain, is a part of the endoskeleton. Vertebrates show *cephalization;* their distinct head contains the brain and exhibits special sense organs, such as camera-type eyes. During the evolution of vertebrates, a complex brain increased in size and has attained its largest size in humans.

Vertebrates have a large coelom and *well-developed viscera:* The complex digestive system is complete, having both a mouth and an anus. The blood is contained entirely within blood vessels; therefore, the circulatory system is said to be closed. Vertebrates have efficient means of respiration and excretion. The respiratory system consists of gills or lungs, which are used to obtain oxygen from the environment. The kidneys are important excretory and water-regulating organs that conserve or rid the body of water as necessary.

Derived Characters Among Vertebrates The following other derived characters distinguish groups of vertebrates. ❷ The evolution of **jaws** separates the jawless fishes from all the other vertebrates. Jaws equip vertebrates for a predaceous way of life. ❸ Early bony fishes had **lungs**, a derived character that permitted life on land. ❹ Amphibians were the first chordate group to clearly have **jointed limbs**, in the same way that arthropods have jointed appendages, and to invade the land. We will see that fleshy fins with jointed bones evolved into these limbs. Amphibians, reptiles, birds, and mammals are **tetrapods** because they have four limbs: two anterior and two posterior limbs. ❺ The amnion, a membrane found in the **amniotic egg**, evolved in reptiles and is also present in birds and mammals. The special extraembryonic membranes of the shelled amniotic egg provided a means of reproduction suitable to land. Such membranes carry out all the functions needed to support the embryo as it develops into a young offspring capable of feeding on its own.

Each group of vertebrates will be examined in light of these five key features. Section 19.21 begins our survey with the fishes.

> **19.20 Check Your Progress** Which of the five vertebrate features would both bats and gorillas share, based on Figure 19.20?

FIGURE 19.20 Evolutionary tree of the chordates.

The first vertebrates were jawless fishes, which wiggled through the water and sucked up food from the ocean floor. Today, there are three living classes of fishes: jawless fishes, cartilaginous fishes, and bony fishes. The two latter groups have *jaws*, tooth-bearing bones of the head. Jaws are believed to have evolved from the first pair of gill arches, structures that ordinarily support gills (**Fig. 19.21A**). The presence of jaws permits a predatory way of life.

Jawless Fishes (Class Agnatha) Living representatives of the **jawless fishes** are cylindrical and up to a meter long. They have smooth, scaleless skin and no jaws or paired fins. The two groups of living jawless fishes are *hagfishes* and *lampreys*. The hagfishes are scavengers, feeding mainly on dead fishes, while some lampreys are parasitic. When parasitic, the oral disk of the lamprey (**Fig. 19.21B**) serves as a sucker. The lamprey attaches itself to another fish and taps into its circulatory system.

Cartilaginous Fishes (Class Chondrichthyes) These fishes, including the sharks (Fig. 19.21B), the rays, and the skates, have skeletons of cartilage, instead of bone. The small dogfish shark is often dissected in biology laboratories. The hammerhead shark is an aggressive predator that usually feeds on other fishes and invertebrates but has been known to attack people also. The largest sharks, the whale sharks, feed on small fishes and marine invertebrates and do not attack humans. Skates and rays are rather flat fishes that live partly buried in the sand and feed on mussels and clams.

Three well-developed senses enable sharks to detect their prey: (1) They are able to sense electric currents in water—even those generated by the muscle movements of animals; (2) they, and all other types of fishes, have a lateral line system, a series of cells that lie within canals along both sides of the body and can sense pressure waves caused by a fish or another animal swimming nearby; and (3) they have a keen sense of smell. Sharks can detect about one drop of blood in 115 L (25 gal) of water.

Bony Fishes (Class Osteichthyes) Bony fishes are by far the most numerous and diverse of all the vertebrates (Fig. 19.21B). Most of the bony fishes we eat, such as perch, trout, salmon, and haddock, are **ray-finned fishes**. Their fins, which are used to balance and propel the body, are thin and supported by bony spikes. Ray-finned fishes have various ways of life. Some, such

as herring, are filter feeders; others, such as trout, are opportunists; and still others, such as piranhas and barracudas, are predaceous carnivores.

Ray-finned fishes have a swim bladder, which usually serves as a buoyancy organ. The streamlined shape, fins, and muscle action of ray-finned fishes are all suited to locomotion in the water. Their skin is covered by bony scales that protect the body but do not prevent water loss. When fishes respire, the gills are kept continuously moist by the passage of water through the mouth and out the gill slits. As the water passes over the gills, oxygen is absorbed by the blood, and carbon dioxide is given

Lamprey, a jawless fish

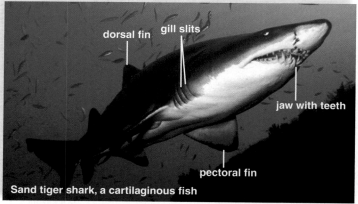

Sand tiger shark, a cartilaginous fish

Soldierfish, a bony fish

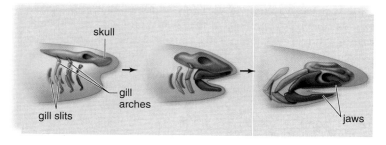

FIGURE 19.21A Evolution of jaws.

FIGURE 19.21B Diversity of fishes.

off. Ray-finned fishes have a single-circuit circulatory system. The heart is a simple pump, and the blood flows through the chambers, including a nondivided atrium and ventricle, to the gills. O_2-rich blood leaves the gills and goes to the body proper, eventually returning to the heart for recirculation.

Another type of bony fish is called the **lobe-finned fishes**. These fishes not only had fleshy appendages that could be adapted to land locomotion, but most also had a lung, which was used for respiration. Paleontologists have recently found a well-preserved transitional fossil from the Late Devonian period in Arctic Canada that represents an intermediate between lobe-finned fishes and tetrapods with limbs. The name of the fossil is *Tiktaalik roseae* (**Fig. 19.21C**, *left*). This fossil provides unique insights into how the legs of tetrapods arose (Fig. 19.21C, *right*).

Section 19.22 examines the amphibians.

> **19.21** *Check Your Progress* Evolution of jaws allowed animals to take up what way of life?

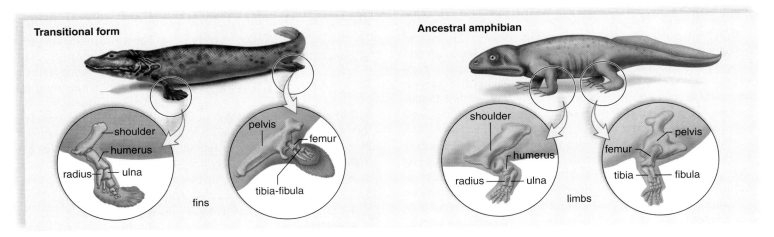

FIGURE 19.21C This transitional form links the lobes of lobe-finned fishes to the limbs of ancestral amphibians.

19.22   Amphibians are tetrapods that can move on land

Amphibians (class Amphibia), whose class name means living on both land and in the water, are represented today by frogs, toads, newts, and salamanders. Aside from jointed limbs, amphibians have other features not seen in bony fishes: eyelids for keeping their eyes moist, ears adapted to picking up sound waves, and a voice-producing larynx. The brain is larger than that of a fish. Adult amphibians usually have small lungs. Air enters the mouth by way of nostrils, and when the floor of the mouth is raised, air is forced into the relatively small lungs. Respiration is supplemented by gas exchange through the smooth, moist, and glandular skin. The amphibian heart has only three chambers, compared to the four of mammals. Mixed blood is sent to all parts of the body; some is sent to the skin, where it is further oxygenated.

Most members of this group lead an amphibious life—that is, the larval stage lives in the water, and the adult stage lives on the land. Figure 19.1 illustrates how the frog tadpole undergoes metamorphosis into an adult before taking up life on land. However, the adult usually returns to the water to reproduce. **Figure 19.22** compares the appearance of a frog to that of another amphibian, a salamander. In a frog, the head and trunk are fused, and the long hindlimbs are specialized for jumping. Frogs have smooth skin, and they live in or near fresh water; toads have stout bodies and warty skin, and they live in dark, damp places away from the water. Most salamanders have limbs that are set at right angles to the body and resemble the earliest fossil amphibians. They move like a fish, with a side-to-side, S-shaped motion.

Reptiles are the topic of Section 19.23.

> **19.22** *Check Your Progress* Are bats tetrapods? Explain.

FIGURE 19.22 Frogs and salamanders are well-known amphibians.

Tree frog   hindlimb   Barred tiger salamander   fleshy toes

19.23 Reptiles have an amniotic egg and can reproduce on land

Reptiles (class Reptilia) diversified and were most abundant between 245 and 66 MYA. These animals included the mammal-like reptiles, the ancestors of today's living mammals, and the dinosaurs, which became extinct, except for those that evolved into birds. Some dinosaurs are remembered for their great size. *Brachiosaurus*, a herbivore, was about 23 m (75 ft) long and about 17 m (56 ft) tall. *Tyrannosaurus rex*, a carnivore, was 5 m (16 ft) tall when standing on its hind legs. The bipedal stance of some dinosaurs was preadaptive for the evolution of wings in birds.

The reptiles living today are mainly alligators, crocodiles, turtles, snakes, lizards, and tuataras (**Fig. 19.23**). The body of a reptile is covered with hard, keratinized scales, which protect the animal from desiccation and from predators. Reptiles have well-developed lungs enclosed by a protective and functional rib cage. The heart has four chambers, but the septum that divides the two halves is incomplete in certain species; therefore, some exchange of O_2-rich and O_2-poor blood occurs.

Perhaps the most outstanding adaptation of the reptiles is their means of reproduction, which is suitable to a land existence. The penis of the male passes sperm directly to the female. Fertilization is internal, and the female lays leathery, flexible, shelled eggs. The *amniotic egg* made development on land possible and eliminated the need for a swimming larval stage during development. The amniotic egg has extraembryonic membranes that provide the developing embryo

with atmospheric oxygen (chorion), food (yolk sac), and water (amnion); it also removes nitrogenous wastes (allantois) and protects the embryo from drying out and from mechanical injury.

Fishes, amphibians, and reptiles are **ectotherms**, meaning that their body temperature matches the temperature of the external environment. If it is cold externally, they are cold internally; if it is hot externally, they are hot internally. Reptiles regulate their body temperatures by exposing themselves to the sun if they need warmth or by hiding in the shadows if they need cooling off.

Section 19.24 discusses the birds.

egg shell · yolk sac · albumin · amnion · embryo · chorion · allantois · air space

Amniotic egg

> **19.23 Check Your Progress** What is the advantage of an amniotic egg?

shell · beak · flipper

Green sea turtle

clawed foot

Gila monster, a venomous lizard

venom gland · fang · rattle

Diamondback rattlesnake

third eye (not visible) · scaly skin · tail

Tuatara, a living fossil

thick, scaly skin · tail · tongue · nostril

American alligator

FIGURE 19.23 Reptilian diversity.

Birds (class Aves) are characterized by the presence of *feathers*, which are modified reptilian scales. (Perhaps you have noticed the scales on the legs of a chicken.) However, birds lay a hard-shelled amniotic egg, rather than the leathery egg of reptiles. Ample data today indicate that birds are closely related to bipedal dinosaurs and that they should be classified as such.

Nearly every anatomic feature of a bird can be related to its ability to fly. The forelimbs are modified as wings. Bird flight requires an airstream and a powerful wing downstroke for lift, a force at right angles to the airstream (**Fig. 19.24A**). The hollow, very light bones are laced with air cavities. A horny beak has replaced jaws equipped with teeth, and a slender neck connects the head to a rounded, compact torso. Respiration is efficient, since the lobular lungs form anterior and posterior air sacs. The presence of these sacs means that the air moves one way through the lungs, and gases are continuously exchanged across respiratory tissues. Another benefit of air sacs is that they lighten the body and aid flying.

Birds have a four-chambered heart that completely separates O_2-rich blood from O_2-poor blood. Birds are **endotherms** and generate internal heat. Many endotherms can use metabolic heat to maintain a constant internal temperature. This may be associated with their efficient nervous, respiratory, and circulatory systems. Also, their feathers provide insulation. Birds have no bladder and excrete uric acid in a semidry state.

Birds have particularly acute vision and well-developed brains. Their muscle reflexes are excellent. These adaptations are suited to flight. An enlarged portion of the brain seems to be the area responsible for instinctive behavior. A ritualized courtship often precedes mating. Many newly hatched birds require parental care before they are able to fly away and seek food for themselves. A remarkable aspect of bird behavior is the seasonal migration of many species over very long distances. Birds navigate by day and night, whether it's sunny or cloudy, by using the sun and stars and even the Earth's magnetic field to guide them.

Traditionally, the classification of birds was particularly based on type of beak (**Fig. 19.24B**) and foot, and to some extent on habitat and behavior. A bald eagle's beak tears prey apart; a woodpecker's beak can drill in wood; a flamingo's beak strains food from water; a vulture's beak can grasp flesh; and a cardinal's beak can crack tough seeds.

Finally, in Section 19.25, we will discuss the mammals.

FIGURE 19.24A Bird flight.

> **19.24 *Check Your Progress*** Are birds more closely related to reptiles or to mammals? Explain.

Bald eagle

Pileated woodpecker

Flamingo

Turkey vulture

Cardinal

FIGURE 19.24B Types of bird beaks.

Mammals (class Mammalia) evolved during the Mesozoic era from mammal-like reptiles called therapsids. True mammals appeared during the Jurassic period, about the same time as the first dinosaurs. The first mammals were small, about the size of mice. During all the time the dinosaurs flourished (165 MYA), mammals were a minor group that changed little. Some of the earliest mammalian groups are still represented today by the monotremes and marsupials, but they are not abundant. The placental mammals that evolved later went on to live in many habitats.

The two chief characteristics of mammals are hair and milk-producing mammary glands. Mammals are endotherms, and many of their adaptations are related to temperature control. Hair, for example, provides insulation against heat loss and allows mammals to be active, even in cold weather.

Mammary glands enable females to feed (nurse) their young without leaving them to find food. Nursing also creates a bond between mother and offspring that helps ensure parental care while the young are helpless. In most mammals, the young are born alive after a period of development in the uterus, a part of the female reproductive system. Internal development shelters the young and allows the female to move actively about while the young are maturing.

Monotremes (Fig. 19.25A) are mammals that, like birds, have a *cloaca*, a terminal region of the digestive tract serving as a common chamber for feces, excretory wastes, and sex cells. They also lay hard-shelled amniotic eggs. They are represented by the spiny anteater and the duckbill platypus, both of which live in Australia. The female duckbill platypus lays her eggs in a burrow in the ground. She incubates the eggs, and after hatching, the young lick up milk that seeps from mammary glands on her abdomen. The spiny anteater has a pouch on the belly side formed by swollen mammary glands and longitudinal muscle. Hatching takes place in this pouch, and the young remain there for about 53 days. Then they stay in a burrow, where the mother spiny anteater periodically visits and nurses them.

The young of **marsupials** (Fig. 19.25A) begin their development inside the female's body, but they are born in a very immature condition. Newborns crawl up into a pouch on their mother's abdomen. Inside the pouch, they attach to the nipples of mammary glands and continue to develop. Frequently, more are born than can be accommodated by the number of nipples, and it's "first come, first served."

The Virginia opossum is the only marsupial that occurs north of Mexico. In Australia, however, marsupials underwent adaptive radiation for several million years without competition. Thus, marsupial mammals are now found mainly in Australia, with some in Central and South America as well. Among the herbivorous marsupials, koalas are tree-climbing browsers, and kangaroos are grazers. The Tasmanian wolf or tiger, thought to be extinct, was a carnivorous marsupial about the size of a collie dog.

The vast majority of living mammals are **placental mammals** (**Fig. 19.25B**). In these mammals, the extraembryonic membranes of the reptilian egg have been modified for internal development within the uterus of the female. The chorion contributes to the fetal portion of the placenta, while a part of the uterine wall contributes to the maternal portion. Here, nutrients, oxygen, and wastes are exchanged between fetal and maternal blood.

Mammals are adapted to life on land and have limbs that allow them to move rapidly. In fact, an evaluation of mammalian features leads us to the obvious conclusion that they lead active lives. The brain is well developed; the lungs are expanded not only by the action of the rib cage but also by the contraction of the diaphragm, a horizontal muscle that divides the thoracic cavity from the abdominal cavity; and the heart has four chambers. The internal temperature is constant, and hair, when abundant, helps insulate the body.

The mammalian brain is enlarged due to the expansion of the cerebral hemispheres that control the rest of the brain. The brain is not fully developed until after birth, and young learn to take care of themselves during a period of dependency on their parents.

FIGURE 19.25A Monotremes and marsupials.

Duckbill platypus, a monotreme of Australian streams

Virginia opossum, the only American marsupial

Koala, a tree-dwelling Australian marsupial

White-tailed deer, a forest-dwelling herbivore

African lioness, a grassland-dwelling carnivore

FIGURE 19.25B
Placental mammals.

Squirrel monkey, a tree-dwelling herbivore

Killer whale, a sea-dwelling carnivore

Placental mammals can be distinguished by their mode of locomotion and their way of obtaining food. For example, bats have membranous wings supported by digits; horses have long, hoofed legs; and whales have paddlelike forelimbs. The specific shape and size of the teeth may be associated with whether the mammal is an herbivore (eats vegetation), a carnivore (eats meat), or an omnivore (eats both meat and vegetation). For example, mice have continuously growing incisors; horses have large, grinding molars; and dogs have long canine teeth. The following are some of the major orders of placental mammals:

- The hoofed mammals are in the orders Perissodactyla (e.g., horses, zebras, tapirs, rhinoceroses; 17 species) and the Artiodactyla (e.g., pigs, cattle, deer, hippopotamuses, buffaloes, giraffes; 185 species) whose elongated limbs are adapted for running, often across open grasslands. Both groups of animals are herbivorous and have large, grinding teeth.

- Order Carnivora (270 species) includes dogs, cats, bears, raccoons, and skunks. The canines of these meat eaters are large and conical. Some carnivores are aquatic—namely, seals, sea lions, and walruses—and must return to land to reproduce.

- Order Primates (180 species) includes lemurs, monkeys, gibbons, chimpanzees, gorillas, and humans. Typically, primates are tree-dwelling fruit eaters, although some, like humans, are ground dwellers.

- Order Cetacea (80 species) includes the whales and dolphins, which are mammals despite their lack of hair or fur. Blue whales are the largest animal ever to live.

- Order Rodentia (1,760 species), the largest order, includes mice, rats, squirrels, beavers, and porcupines. Rodents have incisors that grow continuously.

- Order Chiroptera (900–1,000 species), the second largest order, include bats that feed on fruits, insects, or blood. Bats that feed on insects use echolocation to find their prey.

- Order Proboscidea (2 species) includes the elephants, the largest living land mammals, whose upper lip and nose have become elongated and muscularized to form a trunk.

- Order Lagomorpha (65 species) includes the herbivorous rabbits, hares, and pikas—animals that superficially resemble rodents. They also have two pairs of continually growing incisors, and their hind legs are longer than their front legs.

- Order Insectivora (419 species) includes the shrews and moles, which are mammals with short snouts that live primarily underground.

Aside from the many roles that animals play in ecosystems, they are also useful to humans as a source of medical treatments, as discussed in Section 19.26.

19.25 Check Your Progress Name two ways bats are unique among mammals.

Hundreds of pharmaceutical products come from vertebrates, and even animals that produce poisons and toxins give us medicines that benefit us. The Thailand cobra paralyzes its victim's nerves and muscles with a potent venom that eventually leads to respiratory arrest. However, that venom is also the source of the drug Immunokine, which has been used for 10 years in multiple sclerosis patients. Immunokine, which is almost without side effects, actually protects the patient's nerve cells from destruction by their immune system. A compound known as ABT-594, derived from the skin of the poison-dart frog (**Fig. 19.26**), is approximately 50 times more powerful than morphine in relieving chronic and acute pain, without the addictive properties. The southern copperhead snake, the cone snail, and the fer-de-lance pit viper are some of the unlikely vertebrates that either serve as the source of pharmaceuticals or provide a chemical model for the synthesis of effective drugs in the laboratory. These drugs include anticoagulants ("clot busters"), painkillers, antibiotics, and anticancer drugs.

A variety of friendlier vertebrates produce proteins that are similar enough to human proteins to be used for medical treatment. Until 1978, when recombinant DNA human insulin was produced, diabetics injected insulin purified from pigs. Currently, the flu vaccine is produced in fertilized chicken eggs. The production of these drugs, however, is often time-consuming, labor intensive, and expensive. In 2003, pharmaceutical companies used 90 million chicken eggs and took 9 months to produce the flu vaccine.

Some of the most powerful applications of genetic engineering can be found in the development of drugs and therapies for human diseases. In fact, this new biotechnology has actually led to a new industry: animal pharming. Animal pharming uses genetically altered vertebrates, such as mice, sheep, goats, cows, pigs, and chickens, to produce medically useful pharmaceutical products. The procedure is carried out as follows: The human gene for some useful product is inserted into the embryo of the vertebrate. That embryo is implanted into a foster mother, which gives birth to the transgenic animal, so called because it contains genes from two sources. An adult transgenic vertebrate produces large quantities of the pharmed product in its blood, eggs, or milk, from which the product can be easily harvested

and purified. The first such product, alpha 1 antitrypsin for the treatment of emphysema and cystic fibrosis, is now undergoing clinical trials. It is not yet being marketed because some trial patients experienced wheezing while taking the medication.

Xenotransplantation, the transplantation of vertebrate tissues and organs into human beings, is another benefit of genetically altered animals. There is an alarming shortage of human donor organs to fill the need for hearts, kidneys, and livers. The first animal-human transplant occurred in 1984, when a team of surgeons implanted a baboon heart into an infant, who unfortunately lived only a short while before dying of circulatory complications. In the late 1990s, two patients were kept alive using pig livers outside their bodies to filter their blood until a human organ was available for transplantation. Although baboons are genetically closer to humans than pigs, pigs are generally healthier, produce more offspring in a shorter time, and are already raised for food. Despite the fears of some, scientists think that viruses unique to pigs are unlikely to cross the species barrier and infect the human recipient. Currently, pig heart valves and skin are routinely used to treat humans. Miniature pigs, whose heart size is appropriate for humans, are being genetically engineered to make them less foreign to the human body in order to avoid rejection (Fig. 19.26).

The use of transgenic vertebrates for medical purposes does raise health and ethical concerns. Could a viral AIDS-like epidemic be unleashed by cross-species transplantation? What other unseen health consequences might there be? Is it ethical to change the genetic makeup of vertebrates in order to use them as drug or organ factories? Are we redefining the relationship between humans and other vertebrates to the detriment of both? These questions will continue to be debated as the research goes forward. Meanwhile, several U.S. regulatory bodies, including the Food and Drug Administration, have adopted voluntary guidelines for this new technology.

19.26 *Check Your Progress* A drug called Draculin was developed from the vampire bat's saliva to treat heart attack and stroke patients. Explain why this was possible.

Poison-dart frogs, source of a medicine

Pigs, source of organs

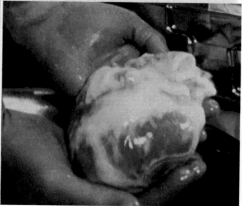

Pig heart for transplantation

FIGURE 19.26 Vertebrates used for medical purposes.

How do you measure success? As human beings, we may assume that vertebrate chordates, such as ourselves, are the most successful organisms. But depending on the criteria used, organisms that are in some ways less complex may come out on top!

For example, vertebrates are eukaryotes, which have been assigned to one domain, while the prokaryotes are now divided into two domains. In fact, the total number of prokaryotes is greater than the number of eukaryotes, and there are possibly more types of prokaryotes than any other living form. Thus, the unseen world is much larger than the seen world. Furthermore, prokaryotes are adapted to use most energy sources and to live in almost any type of environment.

As terrestrial mammals, humans might assume that terrestrial species are more successful than aquatic ones. However, if not for the myriad types of terrestrial insects, there would be more aquatic species than terrestrial ones on Earth. The adaptive radiation of mammals has taken place on land, and this might seem impressive to some. But actually, the number of mammalian species (4,800) is small compared to, say, the molluscs (110,000 species), which radiated in the sea.

The size and complexity of the brain is also sometimes cited as a criterion by which vertebrates are more successful than other living things. However, this very characteristic has been linked to others that make an animal prone to extinction. Studies have indicated that large animals have a long life span, are slow to mature, have few offspring, expend much energy caring for their offspring, and tend to become extinct if their normal way of life is destroyed. And finally, vertebrates, in general, are more threatened than other types of organisms by our present biodiversity crisis—a crisis brought on by the activities of the vertebrate with the most complex brain of all, *Homo sapiens*.

Chapter 20 traces the increase in complexity of the human brain by exploring the evolution of the primates, the order in which humans are classified.

The Chapter in Review

Summary

The Secret Life of Bats

- Bats disperse pollen and seeds, help control insects, and produce a powerful anticoagulant used in stroke victims.

Key Innovations Distinguish Invertebrate Groups

19.1 Animals have distinctive characteristics

- Animals are heterotrophs that must acquire nutrients from an external source; usually reproduce sexually; undergo developmental stages that produce specialized tissues and organs; and have both muscle and nerves.

19.2 Animals most likely have a protistan ancestor

- Animals may have arisen from a ciliated protist, or more likely, descended from an ancestor that resembled a colony of flagellated cells.
- Shifts in *Hox* gene expression in embryos are responsible for major differences between animals.

19.3 The traditional evolutionary tree of animals is based on seven key innovations

- The seven innovations are multicellularity, true tissues, bilateral symmetry, body cavity, coelom, segmentation, and jointed appendages.

19.4 Molecular data suggest a new evolutionary tree for animals

- Closely related organisms have more nucleotide sequences in common.

19.5 Some animal groups are invertebrates and some are vertebrates

- The familiar animal groups are cited in this section.

19.6 Sponges are multicellular invertebrates

- Sponges are multicellular but lack organized tissues and are filter feeders.

19.7 Cnidarians have true tissues

- Cnidarians have radial symmetry and a sac body plan; they have the tissue level of organization and two body forms: polyp and medusa.

19.8 Free-living flatworms have bilateral symmetry

- Flatworms are eucoelomates with three germ layers but no body cavity and a sac body plan.
- Their body systems include a ladderlike nervous system and an incomplete digestive tract.

19.9 Some flatworms are parasitic

- Both tapeworms and flukes are endoparasites of humans and other animals.

19.10 Roundworms have a pseudocoelom and a complete digestive tract

- Roundworms are nonsegmented, have a complete digestive tract, and have a pseudocoelom.

blastopore becomes mouth

19.11 A coelom gives complex animal groups certain advantages

- Molluscs, annelids, and arthropods are protostomes; the mouth appears at the blastopore; mesoderm splitting produces a coelom.
- Echinoderms and chordates are deuterostomes; the anus appears at the blastopore; mesodermal pouches form a coelom.

blastopore becomes anus

19.12 Molluscs have a three-part body plan
- All molluscs have a foot, a mantle, and a visceral mass.
- In gastropods (e.g., snails), the foot is flattened; in cephalopods (e.g., squid), the foot evolved into tentacles; and in bivalves (e.g., clam), the foot projects from the shell.

19.13 Annelids are the segmented worms
- Rings encircle the body, and septa divide the coelom.
- Oligochaetes have few setae per segment, reside in soil, and do not have a well-developed head.
- Polychaetes have many setae per segment.

19.14 Arthropods have jointed appendages
- Arthropods are very diverse and include about 30 million species, mostly insects.
- Six characteristics of arthropods are jointed appendages, exoskeleton, segmentation, well-developed nervous system, variety of respiratory organs, and metamorphosis.

19.15 Well known arthropods other than insects
- Crustaceans have a head and five pairs of walking legs.
- Arachnids have four pairs of walking legs attached to a cephalothorax.

19.16 Insects, the largest group of arthropods, are adapted to living on land
- Insects have wings for flying, but some are aquatic.
- Their body is divided into a head, thorax, and abdomen, with three pairs of legs attached to the thorax.

19.17 Echinoderms are radially symmetrical as adults
- Echinoderms are marine animals with no head, brain, or segmentation
- A water vascular system provides locomotion and helps carry out respiratory, excretory, and circulatory functions.

Further Innovations Allowed Vertebrates to Invade the Land Environment

19.18 Four features characterize chordates
- During the life cycle chordates have a notochord, a dorsal tubular nerve cord, pharyngeal pouches, and a postanal tail.

19.19 Invertebrate chordates have a notochord as adults
- The four chordate features are present in tunicate larvae and in adult lancelets.

19.20 The evolutionary tree of vertebrates is based on five key features
- Vertebrates have a jointed endoskeleton with paired appendages, cephalization, a well-developed coelom and viscera, a closed circulatory system, and efficient respiratory and excretory organs.
- Jaws allow all but jawless fishes to be predaceous.
- Lungs permit amphibians, reptiles, birds, and mammals to breathe air.
- Jointed limbs allowed many vertebrates to invade the land.
- The amniotic egg frees reptiles, birds, and mammals from needing external water to reproduce.

19.21 Jaws and lungs evolved among the fishes
- Jawless fishes were the first vertebrates.
- Cartilaginous fishes have jaws and a skeleton made of cartilage.

- Bony fishes have jaws and fins supported by bony spikes.

19.22 Amphibians are tetrapods that can move on land
- Amphibians have jointed limbs, eyelids, ears, a voice-producing larynx, and lungs to help the adult stage live on land.

19.23 Reptiles have an amniotic egg and can reproduce on land
- Reptiles lay a leathery-shelled amniotic egg, which contains extraembryonic membranes.

19.24 Birds have feathers and are endotherms
- Birds are adapted for flight.
- They have well-developed sense organs and lay hard-shelled amniotic eggs.

embryo

19.25 Mammals have hair and mammary glands
- Monotremes lay a hard-shelled amniotic egg.
- Marsupials have a pouch in which the newborn matures.
- Placental mammals retain their offspring inside a uterus until birth.

Testing Yourself

Key Innovations Distinguish Invertebrate Groups

1. Animals with a cellular level of organization have
   a. cells only.
   b. cells and tissues.
   c. cells and organs.
   d. cells, tissues, and organs.

2. Which of these sponge characteristics is not typical of animals?
   a. They practice sexual reproduction.
   b. They have the cellular level of organization.
   c. They have various symmetries.
   d. They have flagellated cells.
   e. Both b and c are not typical.

3. Cnidarians are considered to be organized at the tissue level because they contain
   a. ectoderm and endoderm.
   b. ectoderm.
   c. ectoderm and mesoderm.
   d. endoderm and mesoderm.
   e. mesoderm.

4. Unlike flatworms, roundworms have
   a. an internal skeleton and an incomplete digestive tract.
   b. an external skeleton and a tube-within-a-tube body plan.
   c. an internal skeleton and a body cavity.
   d. an external skeleton and a body cavity.
   e. a complete digestive tract and a body cavity.

5. Compared to an animal species that lacks a coelom, one that has a coelom
   a. is more flexible.
   b. has more complex organs.
   c. is more likely to tolerate temperature variations.
   d. Both a and b are correct.

6. A mollusc's shell is secreted by the
   a. foot.
   b. head.
   c. visceral mass.
   d. mantle.

7. Which of the following is not a feature of an insect?
   a. compound eyes
   b. eight legs
   c. antennae
   d. an exoskeleton
   e. jointed legs

8. Sea stars move by
   a. expansion and contraction of their tube feet.
   b. producing jets of water.
   c. using their feet as legs and hopping.
   d. taking advantage of water currents to carry them.
9. **THINKING CONCEPTUALLY** What traits of free-living flatworms would be advantageous to internal parasites?

Further Innovations Allowed Vertebrates to Invade the Land Environment

10. Which of the following is not a chordate characteristic?
    a. dorsal supporting rod, the notochord
    b. dorsal tubular nerve cord
    c. pharyngeal pouches
    d. postanal tail
    e. vertebral column
11. Which of the following is not a characteristic of vertebrates? Choose more than one answer if correct.
    a. All vertebrates have a complete digestive system.
    b. Vertebrates have a closed circulatory system.
    c. The sexes are usually separate in vertebrates.
    d. Vertebrates have a jointed endoskeleton.
    e. Most vertebrates never have a notochord.
12. Bony fishes are divided into which two groups?
    a. hagfishes and lampreys
    b. sharks and ray-finned fishes
    c. ray-finned fishes and lobe-finned fishes
    d. jawless fishes and cartilaginous fishes
13. Amphibians arose from
    a. tunicates and lancelets.
    b. cartilaginous fishes.
    c. jawless fishes.
    d. ray-finned fishes.
    e. bony fishes with lungs.
14. What indicates that birds are related to reptiles?
    a. Birds have scales, as well as feathers that are modified scales.
    b. Birds are ectothermic, as are reptiles.
    c. Birds lay leathery, shelled eggs, as do reptiles.
    d. Birds have an open circulatory system, as do all reptiles.
15. Which of the following is not an adaptation for flight in birds?
    a. air sacs
    b. modified forelimbs
    c. bones with air cavities
    d. acute vision
    e. well-developed bladder
16. Which of the following is a true statement? Choose more than one answer if correct.
    a. In all mammals, offspring develop completely within the female.
    b. All mammals have hair and mammary glands.
    c. All mammals have one birth at a time.
    d. All mammals are land-dwelling forms.
    e. All of these are true.
17. Which of the following animals does not produce an amniotic egg? Choose more than one answer if correct.
    a. bat
    b. duckbill platypus
    c. snake
    d. robin
    e. frog
18. **THINKING CONCEPTUALLY** Of what special significance are transitional fossils such as *Tiktaalik*, which preceded the amphibians?

Understanding the Terms

acoelomate 371	jaw 387
amniotic egg 387	jawless fish 388
amphibian 389	jointed limb 387
annelid 381	lancelet 386
arachnid 383	lobe-finned fish 389
arthropod 382	lungs 387
bilateral symmetry 371	mammal 392
bird 391	marsupial 392
bivalve 380	molting 382
cephalization 371	monotreme 392
cephalopod 380	multinucleate
chitin 382	hypothesis 369
chordate 386	nematocyst 375
cnidarian 375	nephridia 381
coelom 371	notochord 386
coelomate 371	placental mammal 392
colonial flagellate	planarian 376
hypothesis 369	protostome 379
crustacean 383	pseudocoelomate 371
deuterostome 379	radial symmetry 371
echinoderm 385	ray-finned fish 388
ectotherm 390	reptile 390
endotherm 391	roundworm 378
filter feeder 374	segmentation 371
flatworm 376	sponge 374
gastropod 380	tetrapod 387
hermaphrodite 376	tunicate 386
homology 371	vertebrae 387
insect 384	vertebrate 373
invertebrate 373	

Match the terms to these definitions:
   a. _____ Paired excretory tubules found in the earthworm and other invertebrates.
   b. _____ Strong but flexible nitrogenous polysaccharide found in the exoskeleton of arthropods.
   c. _____ Egg-laying mammal—for example, duckbill platypus and spiny anteater.
   d. _____ Dorsal supporting rod replaced by the vertebral column in vertebrates.
   e. _____ Animal possessing a body cavity completely lined by mesoderm.

Thinking Scientifically

1. For your senior project, you have decided to present evidence that sponges are animals. Describe the procedure you will use.
2. a. Most investigators today use what type data to determine relationships among animals? b. They might go on to substantiate it with what other type data?

ARIS *Visit* **www.mhhe.com/maderconcepts** *for practice quizzes, animations, videos, and activities designed to help you master the material in this chapter.*

20

Evolution of Humans

LEARNING OUTCOMES

After studying this chapter, you should be able to accomplish the following outcomes.

Lucy's Legacy

1 Describe the importance of finding "Lucy's Baby" to human evolution.

Humans Share Characteristics with All the Other Primates

2 List all the various types of primates.
3 Discuss four traits common to primates.
4 Arrange the groups of primates in an evolutionary tree that shows their relationships.

Humans Have an Upright Stance and Eventually a Large Brain

5 Name differences between humans and chimpanzees.
6 Describe the general characteristics of australopithecines.
7 Distinguish between australopithecines found in southern Africa and those found in eastern Africa.
8 Relate the origin of the genus *Homo* to a prolonged state of infancy.
9 Distinguish between the different types of early *Homo*, and tell which one migrated to Europe and Asia.
10 Discuss the rise of culture as an advantage that probably influenced the evolution of humans.

Homo sapiens Is the Last Twig on the Primate Evolutionary Bush

11 Give two possible reasons for the demise of the Neandertals.
12 Contrast three models explaining the evolution of *Homo sapiens*.
13 Relate brain development to the advancement of culture among Cro-Magnons.
14 Discuss the drawbacks and advantages to the rise of agriculture.

Today's Humans Belong to One Species

15 List three possible observations that might explain the different ethnic groups of humans.
16 Present evidence that all modern humans are members of the same species.

Barry was listening to *Sgt. Pepper's Lonely Hearts Club Band* when he read about a startling new fossil discovery known as Lucy. Like the Beatles, who shocked the world of music, Donald Johannsen shocked the world of human biology in 1974 when he discovered the remains of an ancient female that he nicknamed Lucy after the Beatles' song "Lucy in the Sky with Diamonds." Lucy is classified as *Australopithecus afarensis*, one of several species of australopithecines that lived in East Africa before humans evolved some 4.2 to 2.7 million years ago (MYA).

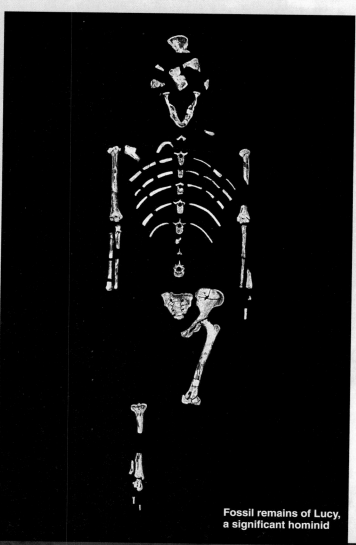

Fossil remains of Lucy, a significant hominid

Lucy stood nearly 1.1 m tall, weighed approximately 29 kg, and perhaps walked upright. More than 300 fossil specimens of *A. afarensis* have now been recorded, including incredible fossil footprints. The fossils indicate that *A. afarensis* was sexually dimorphic, meaning that the males were larger than the females.

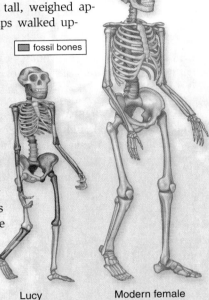

fossil bones

If Lucy did walk upright, her legs were not as straight as those of a modern-day human; her hips and knees were probably bent more like those of a chimp. So, there is considerable debate regarding whether Lucy was completely bipedal or whether she lived partially in trees. For sure, Lucy had a small brain, not much larger than that of a chimp, and had chimplike facial features as well.

Lucy   Modern female

Lucy was probably an omnivore. Her canine teeth were reduced, being more similar to the teeth of humans than those of gorillas. It is assumed that *A. afarensis* lived in small social groups that foraged the forest floor and grasslands.

Surprises always seem to be forthcoming in the field of human biology. In 2000, a team of scientists from the Max Planck Institute unearthed the fossilized remains of a 3.3-million-year-old juvenile *Australopithecus afarensis* locked in sandstone just 4 km from where Lucy had been discovered. The fossil is often called "Lucy's Baby," even though it is dated as tens of thousands of years older than Lucy. The preferred name by Dr. Alemseged, who discovered the fossil, is Selam.

It took five years of painstaking work for the sandstone to yield a complete skull, jaws with milk teeth, the hyoid bone, and most of the torso, spinal column, right arm, fingers, and leg bones, as well as a patella and complete left foot. This skeleton of Selam represents the most complete *Australopithecus* fossil ever found. Researchers estimate that these are the remains of a three-year-old female. They feel very fortunate because the fragile bones of infants and juveniles rarely survive the ravages of geologic time, and yet can provide great insight into the development of a species—in this case, ancient hominids.

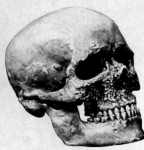

Fossil remains of Selam, child of the same species as Lucy

The structure of Selam's shoulder blade, clavicle, and fingers indicate that *A. afarensis* may have spent more time in trees than originally thought. The semicircular canals of the inner ear are similar to those of African apes, perhaps indicating that the members of this species were not as agile on two legs as humans. The braincase and jaw are apelike, and so is the delicate hyoid bone. Yet, the apelike hyoid bone indicates that *A. afarensis* had a limited capacity for vocalization.

Paleontologists and comparative anatomists have just begun to study the remains of Selam, and most likely the future will yield even more information, bringing the history of our ancient ancestors to life. This chapter traces the evolution of humans from the earliest primates to the first modern humans. It explains the occurrence of the large human population that now stresses the biosphere.

Sehelanthropus tchadensis
7–6 million years ago

Austrolopithecus boisei
3.6–2.9 million years ago

Homo habilis
2.3–1.4 million years ago

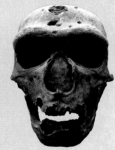

Homo neanderthalensis
200,000–30,000 years ago

Cro-Magnon
Homo sapiens
100,000–? years ago

After examining the characteristics of primates that distinguish them from other mammals, we will trace the evolution of various primate groups, starting with the first mammalian ancestor that entered trees. First the prosimians, then the monkeys, and finally the apes diverged from a hominid line of descent.

20.1 Primates are adapted to live in trees

The order **primates** includes prosimians, monkeys, apes, and humans (**Fig. 20.1A**). In contrast to other types of mammals, primates are adapted for an **arboreal** life—that is, a life spent in trees. The evolution of primates is characterized by trends toward mobile limbs, grasping hands, a flattened face with binocular vision, a large, complex brain, and a reduced reproductive rate. These traits are particularly useful for living in trees.

Mobile Forelimbs and Hindlimbs Primates have developed prehensile hands and feet, often with opposable thumbs and toes (**Fig. 20.1B**). In most primates, flat nails have replaced the claws of ancestral primates, and sensitive pads on the undersides of fingers and toes assist the grasping of objects. All primates have a thumb, but it is only truly opposable in Old World monkeys, great apes, and humans. Because an opposable thumb can touch each of the other fingers, the grip is both powerful and precise. In all but humans, primates with an opposable thumb also have an opposable toe.

The evolution of the primate limb was a very important adaptation for their life in trees. Mobile limbs with clawless opposable digits allow primates to freely grasp and release tree limbs. They also enable primates to easily reach out and bring food, such as fruit, to the mouth.

Stereoscopic Vision A foreshortened snout and a relatively flat face are also evolutionary trends in primates. These may be associated with a general decline in the importance of smell and an increased reliance on vision. In most primates, the eyes are located in the front, where they can focus on the same object from

FIGURE 20.1A Primate diversity.

PROSIMIANS

Ring-tailed lemur, *Lemus catta*

Tarsier, *Tarsius bancanus*

NEW WORLD MONKEY

White-faced monkey, *Cebus capucinus*

OLD WORLD MONKEY

Anubis baboon, *Papio anubis*

ASIAN APES

Orangutan, *Pongo pygmaeus*

White-handed gibbon, *Hylobates lar*

slightly different angles (**Fig. 20.1C**). The result is stereoscopic (three-dimensional) vision with good depth perception that permits primates to make accurate judgments about the distance and position of adjoining tree limbs.

Some primates, humans in particular, have color vision and greater visual acuity because the retina contains cone cells in addition to rod cells. Rod cells are activated in dim light, but the blurry image is in shades of gray. Cone cells require bright light, but the image is sharp and in color. The lens of the eye focuses light directly on the fovea, a region of the retina where cone cells are concentrated.

Large, Complex Brain
Sense organs are only as beneficial as the brain that processes their input. The evolutionary trend among primates is toward a larger and more complex brain. This is evident when comparing the brains of prosimians, such as lemurs and tarsiers, with those of apes and humans. The portion of the brain devoted to smell is smaller, and the portions devoted to sight have increased in size and complexity. Also, more of the brain is devoted to controlling and processing information received from the hands and the thumb. The result is good hand-eye coordination. A larger portion of the brain is devoted to communication skills, which support primates' tendency to live in social groups.

Reduced Reproductive Rate
One other trend in primate evolution is a general reduction in the rate of reproduction, as-

sociated with increased age at sexual maturity and extended life spans. Gestation is lengthy, allowing time for forebrain development. One birth at a time is the norm in primates; it is difficult to care for several offspring while moving from limb to limb in trees. The juvenile period of dependency is extended, and learned behavior and complex social interactions are emphasized.

Humans are a type of hominid, and the next section gives an overview of the evolution of hominids from primate ancestors.

20.1 *Check Your Progress* **Would Lucy have all the primate characteristics listed here? Explain.**

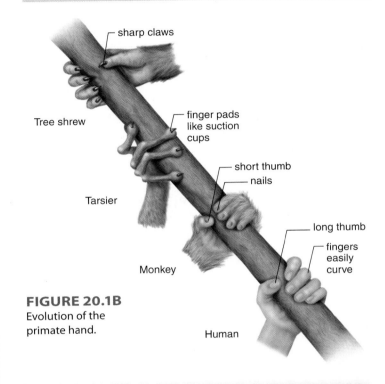

FIGURE 20.1B
Evolution of the primate hand.

AFRICAN APES

Chimpanzee, *Pan troglodytes*   Western lowland gorilla, *Gorilla gorilla*

Humans, *Homo sapiens*

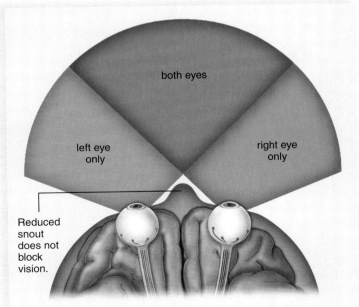

FIGURE 20.1C Stereoscopic vision.

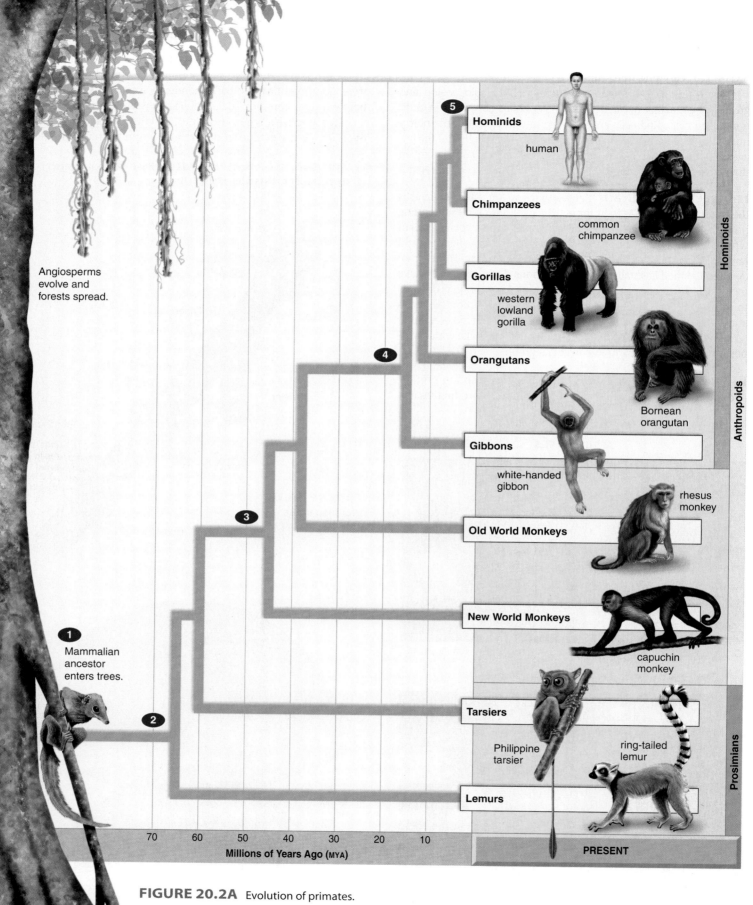

Angiosperms
evolve and
forests spread.

1
Mammalian
ancestor
enters trees.

2

3

4

5 Hominids

human

Chimpanzees

common
chimpanzee

Gorillas

western
lowland
gorilla

Orangutans

Bornean
orangutan

Gibbons

white-handed
gibbon

Old World Monkeys

rhesus
monkey

New World Monkeys

capuchin
monkey

Tarsiers

Philippine
tarsier

ring-tailed
lemur

Lemurs

Hominoids

Anthropoids

Prosimians

70   60   50   40   30   20   10

Millions of Years Ago (MYA)

PRESENT

FIGURE 20.2A Evolution of primates.

Figure 20.2A illustrates the sequence of primate evolution during the Cenozoic era. **❶** This evolutionary tree shows that all primates share one common mammalian ancestor and that the other types of primates diverged from the main line of descent (called a lineage) over time. **❷** Notice that **prosimians**, represented by lemurs and tarsiers, were the first types of primates to diverge.

❸ Today's **anthropoids** are classified into three superfamilies: New World monkeys, Old World monkeys, and **❹** the **hominoids** (apes and hominids). The New World monkeys often have long, prehensile (grasping) tails and flat noses, and Old World monkeys, which lack such tails, have protruding noses. Two of the well-known New World monkeys are the spider monkey and the capuchin, the "organ grinder's monkey." Some of the better-known Old World monkeys are the baboon, a ground dweller, and the rhesus monkey, which has been used in medical research.

Primate fossils similar to monkeys are first found in Africa, dated about 45 MYA. At that time, the Atlantic Ocean would have been too expansive for some of them to have easily made their way to South America, where the New World monkeys live today. It is hypothesized that a common ancestor to both the New World and Old World monkeys arose much earlier when a narrower Atlantic made crossing much more reasonable. The New World monkeys evolved in South America, and the Old World monkeys evolved in Africa.

Dated about 15 MYA, a fossil referred to as Proconsul (a nickname that means before Consul, a famous performing chimpanzee) is a probable transitional link between the monkeys and the apes. Proconsul was about the size of a baboon, and the size of its brain (165 cc) was also comparable. This fossil didn't have the tail of a monkey (**Fig. 20.2B**), but it walked as a quadruped on top of tree limbs as monkeys do. Primarily a tree dweller, Proconsul may have also spent time exploring nearby grasslands for food.

Proconsul was probably ancestral to the **dryopithecines**, from which all the apes arose. About 10 MYA, Africarabia (Africa plus the Arabian Peninsula) joined with Asia, and the apes migrated into Europe and Asia. In 1966, Spanish paleontologists announced the discovery of a specimen they named *Dryopithecus*, dated at 9.5 MYA, near Barcelona. The anatomy of these bones clearly indicates that *Dryopithecus* was a tree dweller and locomoted by swinging from branch to branch as the apes do today. The dryopithecines, now represented by a number of genera, are regarded as ancestral to the apes. **❺** The **hominids** (humans and species very closely related to humans), discussed in the next part of the chapter, can also trace their lineage to the dryopithecines.

> **20.2 Check Your Progress** With the help of Figure 20.2A, trace the path of evolution from a mammalian ancestor to Lucy.

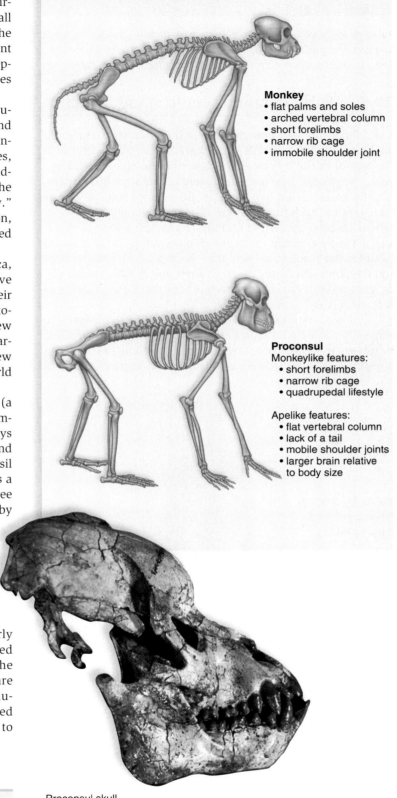

Monkey
- flat palms and soles
- arched vertebral column
- short forelimbs
- narrow rib cage
- immobile shoulder joint

Proconsul
Monkeylike features:
- short forelimbs
- narrow rib cage
- quadrupedal lifestyle

Apelike features:
- flat vertebral column
- lack of a tail
- mobile shoulder joints
- larger brain relative to body size

Proconsul skull

FIGURE 20.2B Monkey skeleton compared to Proconsul skeleton.

Paleontologists currently use evidence of standing erect as a way to distinguish hominids from apes. Which particular group of well-known hominids called australopithecines gave rise to early *Homo* is still being debated. A brain size of at least 600 cc is needed for a fossil to be considered an early *Homo*, best represented by *Homo habilis* and *Homo erectus*.

20.3 Early hominids could stand upright

The relationship of hominids to the other primates is shown in the box at the far right. DNA data have been used to determine the date of the split between the ape and the hominid lineage. When two lines of descent first diverge from a common ancestor, the genes of the two lineages are nearly identical. But as time goes by, each lineage accumulates genetic changes. Many genetic changes are neutral (not tied to adaptation) and accumulate at a fairly constant rate; such changes can be used as a kind of **molecular clock** to indicate the relatedness of the two groups and the point when they diverged from one another. Molecular data suggest that the split between the ape and hominid lineages occurred about 7 MYA (see Fig. 20.2A).

Genome studies show that humans and chimpanzees have almost identical DNA sequences as well as many common traits. However, there are also several distinct differences between humans and chimpanzees, as illustrated in **Figure 20.3A**. In

---

ORDER: Primates

• Adapted to an arboreal life
• Prosimians, monkeys, apes, hominids

Hominids (bipedal)
Early Hominids ⟶ *Sahelanthropus*, ardipithecines,
Later Hominids   Australopithecines

GENUS: *Homo* (humans)
Early *Homos* ⟶ *Homo habilis, Homo rudolfensis,*
Brain size greater   *Homo ergaster, Homo erectus*
than 600 cc; tool use
and culture

Later *Homos* ⟶ *Homo neandertalensis* (archaic human),
Brain size greater   *Homo sapiens*
than 1,000 cc; tool   (Cro-Magnon, modern human)
use and culture

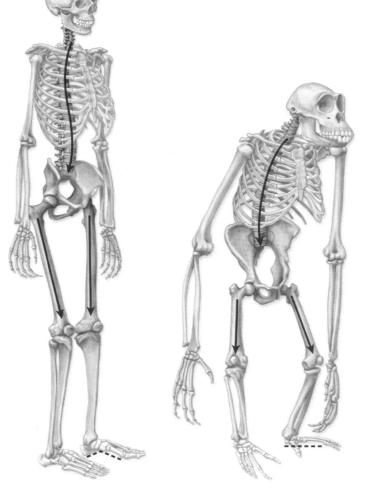

FIGURE 20.3A
Adaptations for standing erect.

Human spine exits from the center; ape spine exits from rear of skull.

Human spine is S-shaped; ape spine has a slight curve.

Human pelvis is bowl-shaped; ape pelvis is longer and more narrow.

Human femurs angle inward to the knees; ape femurs angle out a bit.

Human knee can support more weight than ape knee.

Human foot has an arch; ape foot has no arch.

humans, the spine exits inferior to the center of the skull, and this places the skull in the midline of the body. The longer, S-shaped spine of humans causes the trunk's center of gravity to be squarely over the feet. The broader pelvis and hip joint of humans keep them from swaying when they walk. The longer neck of the femur in humans causes the femur to angle inward at the knees. The human knee joint is modified to support the body's weight—that is, the femur is larger at the bottom, and the tibia is larger at the top. The human toe is not opposable; instead, the foot has an arch, which enables humans to walk long distances and run with less chance of injury.

Evolution of Bipedalism

The anatomy of humans is suitable for standing erect and walking on two feet, a characteristic called **bipedalism**. Humans are bipedal, while apes are quadrupedal (walk on all fours). Although bipedalism can lead to spinal strain and backaches, this disadvantage is most likely compensated for by the fact that an upright posture frees the hands for tool use.

Until recently, many scientists thought that hominids evolved in response to a dramatic change in climate that caused forests to be replaced by grassland. Now, some biologists suggest that the first hominid evolved even while it lived in trees, because they see no evidence of a dramatic shift in vegetation about 7 MYA. The first hominid's environment is now thought to have included some forest, some woodland, and some grassland. While still living in trees, the first hominids may have walked upright on large branches as they collected fruit from overhead. Then, when they began to forage on the ground, an upright stance would have made it easier for them to travel from woodland to woodland and/or to forage among bushes. Bipedalism may have had the added advantage of making it easier for males to carry food back to females. Or bipedalism may be associated with the need to carry a helpless infant from place to place.

Examples of the Earliest Hominids

In **Figure 20.3B**, early hominids are represented by *orange-colored bars*. The bars extend from the date of a species' appearance in the fossil record to the date it became extinct. Paleontologists have now found several fossils dated around the time the ape lineage and the human lineage are believed to have split, and one of these is *Sahelanthropus tchadensis*. Only the braincase has been found and dated at 7 MYA. Although the braincase is very apelike, a point at the back of the skull where the neck muscles would have attached suggests bipedalism. Also, the canines are smaller and the tooth enamel is thicker than those of an ape.

Another early hominid, *Ardipithecus ramidus*, is representative of the ardipithecines of 4.5 MYA. So far, only skull fragments of *A. ramidus* have been described. Indirect evidence suggests that the species was possibly bipedal, and that some individuals may have been 122 cm tall. The teeth seem intermediate between those of earlier apes and later hominids, which are discussed next.

Until recently, it was not possible to determine if these early hominids, and others not mentioned, are related to the later hominids and, indeed, whether they should be included in the human lineage. Recently, however, fossils dated 4 MYA do show a direct link between *A. ramidus* and the australopithecines, discussed next.

> **20.3** *Check Your Progress* Formerly, the criterion for classification as a hominid was a large brain; now the requirement is bipedalism. According to the new criterion, is Lucy a hominid? Why or why not?

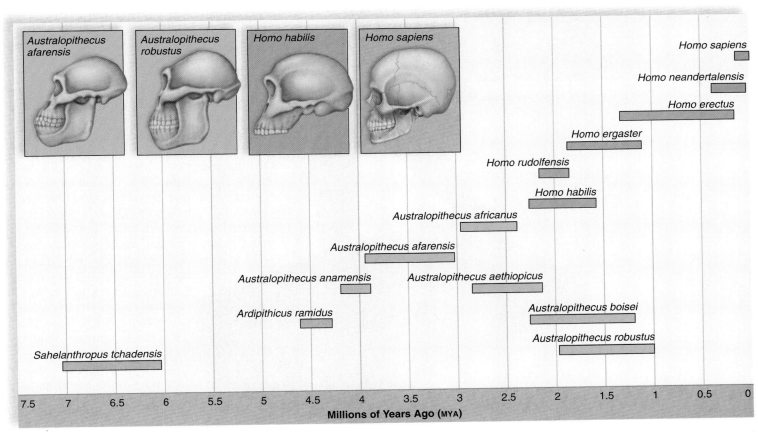

FIGURE 20.3B Human evolution.

The **australopithecines** are a group of hominids that evolved and diversified in Africa from 4 MYA until about 1 MYA. In Figure 20.3B (see page 405), the australopithecines are represented by *green-colored bars*.

The australopithecines had a small brain (an apelike characteristic) and walked erect (a human characteristic). Therefore, it seems that human characteristics did not evolve all together at the same time. Australopithecines give evidence of **mosaic evolution**, meaning that different body parts change at different rates and, therefore, at different times.

Australopithecines stood about 100–115 cm in height, and their brain averaged 370–515 cc—slightly larger than that of a chimpanzee. The forehead was low, and the face projected forward (**Fig. 20.4**). Tool use is not in evidence.

Some australopithecines were slight of frame and termed *gracile* (slender). Others were *robust* (powerful) and tended to have massive jaws because of their large grinding teeth. Their well-developed chewing muscles were anchored to a prominent bony crest along the top of the skull. The gracile types most likely fed on soft fruits and leaves, while the robust types had a more fibrous diet that may have included hard nuts. Therefore, the australopithecines show an adaptation to different ways of life.

Fossil remains of australopithecines have been found in both southern and eastern Africa. The exact relationship between these two groups is not known, and it is uncertain how the australopithecines are related to the next group of fossils we will discuss, namely, the early *Homo* species.

FIGURE 20.4 *Australopithecus afarensis.*

Adult fossilized footprints with those of a child to the side.

Reconstruction of Lucy at St. Louis Zoo

Fossils from South Africa The first australopithecine to be discovered was unearthed in southern Africa by Raymond Dart in the 1920s. This hominid, named ***Australopithecus africanus,*** is a gracile type. A second southern African specimen, *A. robustus,* is a robust type. Both *A. africanus* and *A. robustus* had a brain size of about 500 cc; variations in their skull anatomy are essentially due to their different diets.

These hominids walked upright. Nevertheless, the proportions of their limbs are apelike—that is, the arms are longer than the legs. Therefore, most paleontologists do not believe that *A. africanus* is ancestral to early *Homo,* discussed in Section 20.5.

Fossils from East Africa The earliest remains of an australopithecine (*Australopithecus anamensis*) have been found in Kenya and more recently in Ethiopia. Dated at 4 MYA, this species has characteristics that are anatomically intermediate between *Ardipithecus ramidus,* an early hominid, and *A. afarensis.* Lucy and Selam, described in the chapter introduction, are examples of the species *A. afarensis.* Although the brain was quite small (400 cc), the shapes and relative proportions of Lucy's limbs indicate that she stood upright and probably walked bipedally (Fig. 20.4). Even better evidence of bipedal locomotion comes from a trail of fossilized footprints in Laetoli dated about 3.7 MYA. The larger prints are double, as though a smaller-sized being was stepping in the footprints of another—and there are additional small prints off to the side, within hand-holding distance (Fig. 20.4).

A. *afarensis*, a gracile type, is believed to be ancestral to the robust types found in eastern Africa, *A. aethiopicus* and *A. boisei. A. boisei* had a powerful upper body and the largest molars of any hominid. *A. afarensis* is usually considered more directly related to early *Homo* than are the South African species. Section 20.5 offers an explanation for the brain enlargement seen in early *Homo*.

> **20.4** *Check Your Progress* Compare the australopithecines from southern Africa to those from eastern Africa to indicate why Lucy is the more likely ancestor for early *Homo*.

20.5 Origins of the genus *Homo*

Remains of australopithecines indicate that they spent part of their time climbing trees and that they retained many apelike traits. In some, the arms, like those of an ape, were long compared to the length of the legs. Then, too, *A. afarensis* had strong wrists and long, curved fingers and toes. These traits would have served well for climbing, and the australopithecines probably climbed trees for the same reason that chimpanzees do today: to gather fruits and nuts in trees and to sleep aboveground at night in order to avoid predatory animals, such as lions and hyenas.

Whereas our brain is about the size of a grapefruit, that of the australopithecines was about the size of an orange—and only slightly larger than that of a chimpanzee. There is no evidence that the australopithecines manufactured stone tools; presumably, they were not smart enough to do so.

We know that the genus *Homo* evolved from the genus *Australopithecus*, but several years ago Stephen Stanley of Johns Hopkins University concluded that this could not have happened as long as the australopithecines climbed trees every day. The obstacle relates to the way we, members of *Homo*, develop our large brain. Unlike other primates, we retain the high rate of fetal brain growth through the first year after birth. (That is why a one-year-old child has a very large head in proportion to the rest of its body.) The brain of other primates, including monkeys and apes, grows rapidly before birth, but immediately after birth the brain grows more slowly. As a result, an adult human brain is more than three times as large as that of an adult chimpanzee.

A continuation of the high rate of fetal brain growth eventually allowed the genus *Homo* to evolve from the genus *Australopithecus*. But the continued brain growth is linked to underdevelopment of the entire body. Although the human brain eventually becomes more complex, human babies are remarkably weak and uncoordinated. Such helpless infants must be carried about and tended. Human babies are unable to cling to their mothers the way chimpanzee babies can (**Fig. 20.5**).

The origin of the *Homo* genus entailed a great evolutionary compromise. Humans gained a large brain, but they were saddled with the largest interval of infantile helplessness in the entire class Mammalia. The positive value of a large brain must have outweighed the negative aspects of infantile helplessness, such as the inability of adults to climb trees while holding a helpless infant, or else genus *Homo* wouldn't have evolved.

FIGURE 20.5 Helplessness of a human infant.

Having a larger brain meant that humans were able to outsmart or ward off predators with weapons they were clever enough to manufacture.

Probably very few genetic changes were required to delay the maturation of *Australopithecus* and produce the large brain of *Homo*. The mutation of a regulatory gene, such as the *Hox* gene, that controls one or more other genes most likely could have delayed early maturation. As we learn more about the human genome, we will eventually uncover the particular gene or gene combinations that cause early *Homo* to have a large brain, and this will be a very exciting discovery. Brain enlargement and other characteristics of early *Homo* are discussed in Section 20.6.

> **20.5** *Check Your Progress* Use this section to associate the evolution of bipedalism with the evolution of a larger brain in humans.

Fossils designated as early *Homo* species are represented by *lavender-colored bars* in Figure 20.3B. These fossils appear in the fossil record somewhat earlier or later than 2 MYA. They all have a brain size of 600 cc or greater, their jaws and teeth resemble those of humans, and tool use is in evidence.

Homo habilis and *Homo rudolfensis*

Homo habilis and *Homo rudolfensis* are closely related and will be considered together. In general, *H. habilis* and *H. rudolfensis* have a more primitive anatomy than the other two fossils in this group.

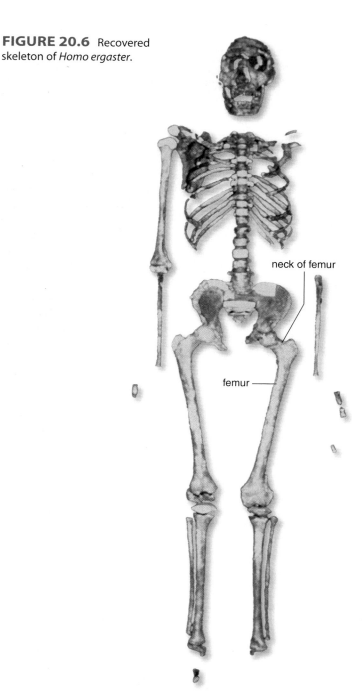

FIGURE 20.6 Recovered skeleton of *Homo ergaster*.

neck of femur

femur

H. rudolfensis was larger than *H. habilis*. Although the height of *H. rudolfensis* did not exceed that of the australopithecines, some of this species' fossils have a brain size as large as 800 cc, which is considerably larger than that of *A. afarensis*. The cheek teeth of these hominids tend to be smaller than even those of the gracile australopithecines. Therefore, it is likely that they were omnivorous and ate meat in addition to plant material. Certainly, these *Homo* species were the first to use tools.

Homo ergaster and *Homo erectus*

Homo ergaster evolved in Africa, perhaps from *H. rudolfensis*. Similar fossils found in Asia are different enough to be classified as **Homo erectus**. These fossils span the dates between 1.9 and 0.3 MYA. A Dutch anatomist named Eugene Dubois was the first to unearth *H. erectus* bones in Java in 1891, and since that time many other fossils belonging to both species have been found in Africa and Asia.

Compared to *H. rudolfensis*, *H. ergaster* had a larger brain (about 1,000 cc) and a flatter face with a nose that projected. This type of nose is adaptive for a hot, dry climate because it permits water to be removed before air leaves the body. The recovery of an almost complete skeleton of a 10-year-old boy indicates that *H. ergaster* was much taller than the hominids discussed thus far (**Fig. 20.6**). Males were 1.8 m tall, and females were 1.55 m. Indeed, these hominids stood erect and most likely had a *striding gait* like that of modern humans. The robust and, most likely, heavily muscled skeleton still retained some australopithecine features. Even so, the size of the birth canal indicates that infants were born in an immature state that required an extended period of care.

H. ergaster first appeared in Africa but then migrated into Europe and Asia sometime between 2 MYA and 1 MYA. Most likely *H. erectus* evolved from *H. ergaster* after *H. ergaster* arrived in Asia. In any case, such an extensive population movement is a first in the history of humankind and a tribute to the intellectual and physical skills of the species. They also had a knowledge of fire and may have been the first to cook meat.

Homo floresiensis

In 2004, scientists announced the discovery of the fossil remains of *Homo floresiensis*, another early *Homo* species. The 18,000-year-old fossil of a 1 m tall, 25 kg adult female was discovered on the island of Flores in the South Pacific. The specimen was the size of a three-year-old *Homo sapiens* but possessed a braincase only one-third the size of that of a modern human. Researchers suspect that this diminutive hominid and her peers evolved from normal-sized, island-hopping *H. erectus* populations that reached Flores about 840,000 years ago. Apparently, *H. floresiensis* used tools and fire.

A biocultural evolution that began with early *Homo* is discussed in Section 20.7.

> **20.6 Check Your Progress** What is significant about the migration of *H. ergaster* out of Africa?

Culture encompasses human activities and products that are passed on from one generation to another outside of direct biological inheritance. *Homo habilis* (and *H. rudolfensis*) could make the simplest of stone tools, called Oldowan tools after a location in Africa where the tools were first found. The main (core) tool could have been used for hammering, chopping, and digging. A flake tool was a type of knife sharp enough to scrape away hide and remove meat from bones. The diet of *H. habilis* most likely consisted of collected plants. But they probably had the opportunity to eat meat scavenged from kills abandoned by lions, leopards, and other large predators in Africa.

Homo erectus, who lived in Eurasia, also made stone tools, but the flakes were sharper and had straighter edges. They are called Acheulian tools for a location in France where they were first found. Their so-called multipurpose handaxes were large flakes with an elongated oval shape, a pointed end, and sharp edges on the sides. Supposedly they were hand-held, but no one knows for sure. *H. erectus* also made the same core and flake tools as *H. habilis*. In addition, *H. erectus* could have also made many other implements out of wood or bone and even grass, which can be twisted together to make string and rope. Excavation of *H. erectus* campsites dated 400,000 years ago have uncovered literally tens of thousands of tools.

H. erectus, like *H. habilis*, also gathered plants as food. However, *H. erectus* may have harvested large fields of wild plants that were growing naturally. The members of this species were not master hunters, but aside from scavenging meat, they could have hunted a bit. The bones of all sorts of animals litter the areas where they lived. Apparently, they ate pigs, sheep, rhinoceroses, buffalo, deer and many other smaller animals. *H. erectus* lived during the last Ice Age, but even so, moved northward. No wonder *H. erectus* is believed to have used fire. A campfire would have protected them from wild beasts and kept them warm at night. And the ability to cook would have made meat easier to eat. Plants can't provide much food in the dead of winter in northern climates, and so meat must have become a substantial part of the diet. It's even possible that the campsites of *H. erectus* were "home bases" where the women stayed behind with the children while the men went out to hunt. If so, these people may have been the first **hunter-gatherers** (**Fig. 20.7**)—that is, they hunted animals and gathered plants. This was a successful way of life that caused the hominid populations to increase from a few thousand australopithecines in Africa 2 MYA to hundreds of thousands of *H. erectus* by .3 MYA. The hunting and gathering way of life doesn't permit a population explosion, however. Children have to be carried long distances, and the men were frequently not around to father children.

Hunting does most likely encourage the development and spread of culture between individuals and generations. Those who could speak a language would have been able to cooperate better as they hunted and even as they sought places to gather food. Among animals, only humans have a complex language that allows them to communicate their experiences symbolically. Words stand for objects and events that can be pictured in the mind. The cultural achievements of *H. erectus* essentially began a new phase of human evolution, called **biocultural evolution**, in which natural selection is influenced by cultural achievements rather than by anatomic phenotype. *H. erectus* succeeded in new, colder environments because these individuals occupied caves, used fire, and became more capable of obtaining and eating meat as a substantial part of their diet.

We have now completed our discussion of early *Homo*. The next part of this chapter will consider later *Homo* evolution.

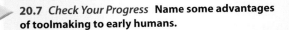

20.7 *Check Your Progress*   **Name some advantages of toolmaking to early humans.**

FIGURE 20.7   The *Homo erectus* people may have been hunter-gatherers.

Three different possible mechanisms are presently being debated regarding how Cro-Magnon, the first modern humans, replaced archaic humans, represented by the Neandertals, for example. The highly developed brain of Cro-Magnon no doubt helped these people achieve an advanced culture and form a society.

20.8 The Neandertal and Cro-Magnon people coexisted for 12,000 years

Later *Homos* are represented by *blue-colored bars* in Figure 20.3B. The later *Homos* include so-called archaic humans (e.g., *Homo neandertalensis*) and the first modern humans (e.g., Cro-Magnon, *Homo sapiens*). The Neandertals are an intriguing species of humans that lived between 200,000 and 28,000 years ago. Neandertal fossils are known from the Middle East and throughout Europe. Neandertals take their name from Germany's Neander Valley, where one of the first Neandertal skeletons, dated some 200,000 years ago, was discovered.

Surprisingly, the Neandertal brain was, on the average, slightly larger than that of modern humans (1,400 cc, compared with 1,360 cc in most humans today). The Neandertals had massive brow ridges and wide, flat noses. They also had a forward-sloping forehead and a receding lower jaw. Their nose, jaws, and teeth protruded far forward. Physically, the Neandertals were powerful and heavily muscled, especially in the shoulders and neck. The bones of Neandertals were shorter and thicker than those of Cro-Magnons. New fossils show that the pubic bone was long compared to that of Cro-Magnons. The Neandertals lived in Europe and the Near East during the last Ice Age, and their sturdy build could have helped conserve heat.

Archaeological evidence suggests that Neandertals were culturally advanced. Some Neandertals lived in caves; however, others probably constructed shelters. They manufactured a variety of stone tools, including spear points, which could have been used for hunting, and scrapers and knives, which would have helped in food preparation. They most likely successfully hunted bears, woolly mammoths, rhinoceroses, reindeer, and other contemporary animals. They used and could control fire, which probably helped them cook frozen meat and keep warm. They even buried their dead with flowers and tools and may have had a religion.

Cro-Magnons are named after a location in France where their remains were first found. Possibly, the Cro-Magnons entered Asia and Europe from Africa 100,000—60,000 years BP (**Fig. 20.8**). They probably migrated to western Europe about 40,000 years ago. Cro-Magnons had a thoroughly modern appearance, including lighter bones, flat high foreheads, domed skulls housing brains of 1,590 cc, small teeth, and a distinct chin. They were hunter-gatherers, as was *H. erectus*, but they hunted more efficiently. Section 20.9 discusses the possible evolutionary relationship between archaic humans and modern humans.

> **20.8 Check Your Progress** If the Neandertals and Cro-Magnons interbred, what type of fossils would you expect to find?

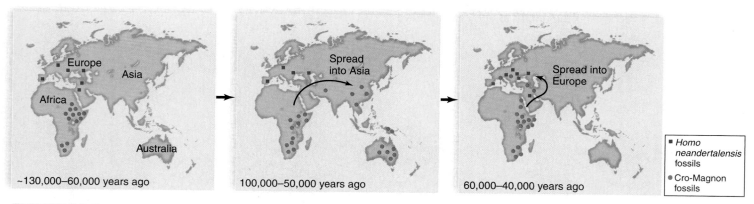

FIGURE 20.8 Possible migration patterns of Cro-Magnons from Africa.

20.9 The particulars of *Homo sapiens* evolution are being studied

Investigators are testing three hypotheses regarding the evolution of modern humans from **archaic humans**, who preceded the appearance of modern humans in Europe, Asia, and Africa.

Multiregional Continuity Model The multiregional continuity hypothesis (**Fig. 20.9A**) proposes that the first modern humans evolved more or less simultaneously in all major regions from archaic humans, who had evolved from *Homo erectus*

(Asia and Europe) or from *Homo ergaster* (Africa). It is further suggested that gene flow occurred between archaic humans in all locations, and therefore modern humans in all parts of the world should be phenotypically similar, but not completely the same. Also, the humans of each region should show a continuity of unique anatomic characteristics from about the time of the archaic species. This model is supported by fossil evidence. For example, some modern Chinese facial characteristics are seen in Asian archaic humans dating more than a 100,000 years BP (before the present). Further, like *H. erectus*, East Asians today commonly have shovel-shaped incisors, while Africans and Europeans rarely do. Many Europeans have relatively heavy brow ridges and a steep nose angle reminiscent of Neandertals.

Replacement Model

The replacement model is also called the **out-of-Africa hypothesis** because it proposes that modern humans evolved from archaic humans only in Africa, and then modern humans migrated to Europe and Asia, where they replaced the archaic species beginning about 100,000 years BP (**Fig. 20.9B**).

The replacement model is supported by the fossil record. The earliest remains of modern humans (Cro-Magnon), dated at least 130,000 years BP, have been found only in Africa. Modern humans are not found in Asia until 100,000 years BP and not in Europe until 60,000 years BP. Until earlier modern human fossils are found in Asia and Europe, the replacement model is supported.

The replacement model is also supported by DNA data. Several years ago, a study showed that the mitochondrial DNA of Africans is more diverse than the DNA of the people in Europe (and the world).

This is significant because if mitochondrial DNA has a constant rate of mutation, Africans should show the greatest diversity, since modern humans have existed the longest in Africa. Called the "Mitochondrial Eve" hypothesis by the press (note that this is a misnomer because no single ancestor in Africa is proposed), the statistics that calculated the date of the African migration were found to be flawed. Still, the raw data—which indicate a close genetic relationship among all Europeans—support the replacement model.

Assimilation Model

The assimilation model is based on both of the older models. It proposes that modern humans (Cro-Magnon) did evolve only in Africa and did migrate into Asia and Europe. Once there, they interbred with archaic humans, resulting in hybrid populations. The assimilation model goes on to say that the fossils dated 40,000 BP in Europe represent these hybrid populations, not pure Cro-Magnon.

The assimilation model is supported by the partial skeleton of a young male, dated 35,000 years BP, in a Romanian cave. The skeleton has features of both modern and archaic humans that could be explained by assuming that hybridization between modern humans and Neandertals, an archaic human, occurred. Further, a computer-based analysis of ten different human DNA sequences indicates that interbreeding has occurred for at least 600,000 years between people living in Asia, Europe, and Africa.

20.9 Check Your Progress Which model(s) is (are) dependent on a second migration (at a later date) from Africa?

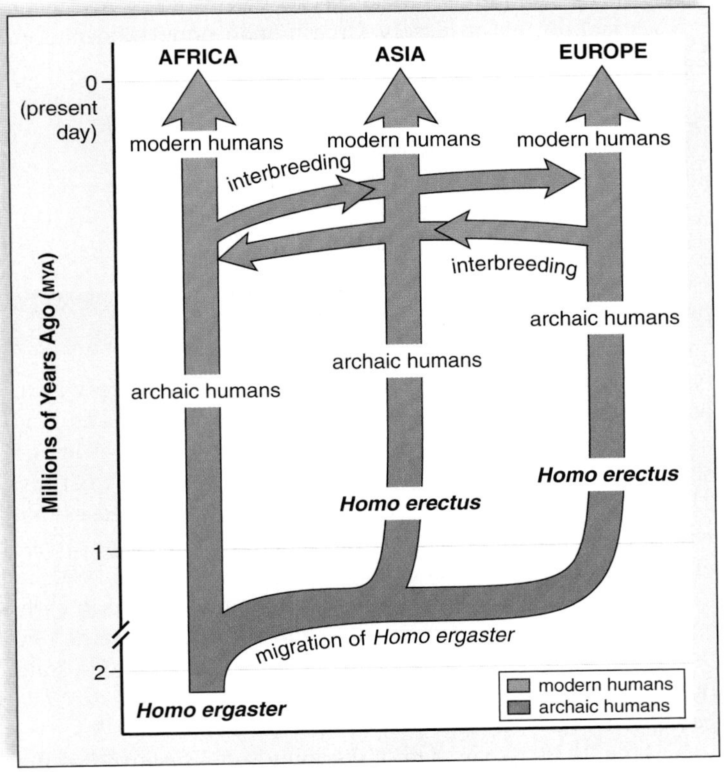

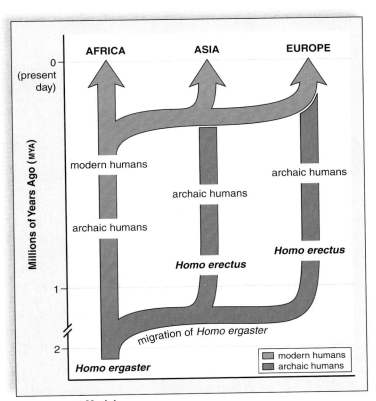

Multiregional Continuity Model

FIGURE 20.9A Multiregional continuity model: Modern humans evolved in Africa, Asia, and Europe.

Replacement Model

FIGURE 20.9B Replacement model: Modern humans evolved in Africa and then replaced archaic humans in Asia and Europe.

20.10 Cro-Magnons made good use of tools

During the last Ice Age, *Homo sapiens* had colonized all of the continents except Antarctica. Glaciation had caused a significant drop in sea level, and as a result, land bridges to the New World and Australia were available. No doubt, colonization was fostered by the combination of a larger brain and free hands with opposable thumbs that made it possible for Cro-Magnons to draft and manipulate tools and weapons of increasing sophistication. Cro-Magnons made advanced stone tools, including compound tools, as when stone flakes were fitted to a wooden handle. They may have been the first to make knifelike blades and to throw spears, enabling them to kill animals from a distance. They were such accomplished hunters that some researchers believe they may have been responsible for the extinction of many larger mammals, such as the giant sloth, the mammoth, the saber-toothed tiger, and the giant ox, during the late Pleistocene epoch. This event is known as the Pleistocene overkill.

A more highly developed brain may have also allowed Cro-Magnons to perfect a language composed of patterned sounds. Language greatly enhanced the possibilities for cooperation and a sense of cohesion within the small bands that were the predominant form of human social organization even for the Cro-Magnons. They combined hunting and fishing with the gathering of fruits, berries, grains, and root crops that grew in the wild. They also harvested wild plants.

The Cro-Magnons were extremely creative. They sculpted small figurines and jewelry out of reindeer bones and antlers. These sculptures could have had religious significance or been seen as a way to increase fertility. The most impressive artistic achievements of the Cro-Magnons were cave paintings, realistic and colorful depictions of a variety of animals, from woolly mammoths to horses, that have been discovered deep in caverns in southern France and Spain (**Fig. 20.10**). Maybe their place-

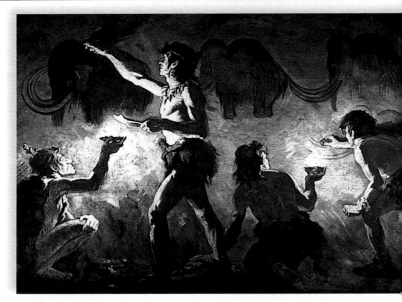

FIGURE 20.10 Cro-Magnons made cave paintings.

ment served a ritual purpose. Were they trying to capture the prowess of these animals as a way to assist their hunting abilities? Or were they celebrating and commemorating particularly successful hunting expeditions? Regardless, these paintings suggest that Cro-Magnons had the ability to think symbolically, as would be needed in order to speak.

Agriculture led to a very large human population and an advanced culture, as discussed in Section 20.11.

20.10 Check Your Progress What is the significance of the development of art by Cro-Magnons?

20.11 Agriculture made modern civilizations possible

Agriculture came into existence in at least three places: the Near East, the Far East, and Central and South America. At just about the same time in these locations, people gave up wandering in search of food and settled down to raise domesticated crops and animals. Although a date of about 10,000 years BP is usually quoted for the rise of agriculture, most likely full dependency on domestic crops and animals did not occur until the time people started making tools of bronze, instead of stone, about 4,500 years BP.

Anthropologists formerly thought that people turned to agriculture because the hunting and gathering way of life had its drawbacks. But this can't be the case, because hunting and gathering lasted for 90,000 years. In fact, evidence suggests that humans were far better off as foragers before they took up agriculture. Hunter-gatherers enjoyed a varied diet of thousands of

types of nutritious plants, seeds, fruits, and nuts. Today, wheat, corn, and rice provide most of the calories for humans, and each crop is deficient in certain essential proteins and amino acids. Also, agriculture caused people to live in closer quarters. This invited the spread of parasites and infectious diseases that foragers avoided by living in smaller numbers in larger areas. Studies of various skeletal evidence indicate that an increase in infectious diseases, malnutrition, and anemia occurred in early agricultural societies, compared to those of hunter-gatherers. Why, then, did people turn to agriculture as a way to sustain themselves? The answer is not known, but several reasons have been put forth.

About 12,000 years ago, a warming trend occurred as the Ice Age came to a close. A variety of big-game animals became extinct, including the saber-toothed cats, mammoths, and mastodons; this may have made hunting less productive. However, as the weather warmed, the glaciers retreated and left fertile valleys where rivers and streams were full of fish and the soil was good. The Fertile Crescent in Mesopotamia is one such region (**Fig. 20.11**). Here, fishing villages may have sprung up and caused people to settle down.

The people were already knowledgeable about what crops to plant. Hunter-gatherers in the Fertile Crescent knew that wheat and barley were good sources of grain for food. Over the past several thousand years, they had slowly begun to increase cultivation of these crops. First, they had slightly encouraged their growth with some tilling of the soil. They may even have selected seeds with desirable characteristics for propagation as they traveled about. Then, a chance mutation may have made these plants particularly suitable as a source of food. So now people began to till the good soil where they had settled and to systematically plant certain crops.

As people became more sedentary, they may have had more children, especially since the men were home more often. A population increase may have tipped the scales and caused them to adopt agriculture full-time, especially if agriculture could be counted on to provide food for hungry mouths. The availability of agricultural tools must have contributed to making agriculture worthwhile. The digging stick, the hoe, the sickle, and the plow were improved, and the introduction of iron some years later further promoted agriculture. Irrigation began as a way to control water supply, especially in semiarid areas and regions of periodic rainfall.

If evolutionary success is judged by population size, agriculture was extremely beneficial because it caused a rapid increase in human numbers all over the Earth. Also, agriculture ushered in civilization as we know it. When crops became bountiful, some people were freed from raising their own food, and they began to specialize in other ways of life in towns and then cities. These people became traders, shopkeepers, bakers, and teachers to name a few occupations. Others became the nobility, priests, and soldiers. Today, farming is highly mechanized, and cities are extremely large. However, we are on a treadmill. As the human population increases, we need new innovations in order to produce greater amounts of food. As soon as food production increases, populations grow once again, and the demand for food becomes still greater. Will there be a point when the population is greater than the food capacity? Perhaps that time is already upon us.

The human civilization that has arisen due to the advent of agriculture is now altering the global environment in a way that affects the evolution of other species. Other species are becoming extinct unless they are able to adapt to the presence of humans. It could be that biocultural evolution will be so harmful to the biosphere that the human species will eventually be driven to extinction also.

This completes our discussion of later *Homo*. The next part of the chapter considers the ethnicities of the human species today.

20.11 *Check Your Progress* **Describe agriculture as a treadmill from which we cannot escape.**

FIGURE 20.11 The Fertile Crescent, where agriculture began.

fertile crescent

TURKEY

IRAN

Euphrates River

Tigris River

SYRIA

IRAQ

JORDAN

SAUDI ARABIA

LIBYA

EGYPT

The diversity of modern day *Homo sapiens* is actually minimal and explained by adaptations to different climates.

20.12 Humans have different ethnicities

Human beings have been widely distributed about the globe ever since they evolved. As with any other species that has a wide geographic distribution, phenotypic and genotypic variations are noticeable between populations. Today, we say that people have different ethnicities (**Fig. 20.12**).

It has been hypothesized that human variations evolved as adaptations to local environmental conditions. One obvious difference among people is skin color. A darker skin is protective against the high UV intensity of bright sunlight. On the other hand, a whiter skin ensures vitamin D production when the UV intensity is low. Harvard University geneticist Richard Lewontin points out, however, that this hypothesis concerning the survival value of dark and light skin has never been tested.

Two correlations between body shape and environmental conditions have been noted since the 19th century. The first, known as Bergmann's rule, states that animals in colder regions of their range have a bulkier body build. The second, known as Allen's rule, states that animals in colder regions of their range have shorter limbs, digits, and ears. Both of these effects help regulate body temperature by increasing the surface-area-to-volume ratio in hot climates and decreasing the ratio in cold climates. For example, the Massai of East Africa tend to be slightly built with elongated limbs, while the Eskimos, who live in northern regions of the world, are bulky and have short limbs (see Fig. 14.1C).

Other anatomic differences among ethnic groups, such as hair texture, a fold on the upper eyelid (common in Asian peoples), or the shape of lips, cannot be explained as adaptations to the environment. Perhaps these features became fixed in different populations due simply to genetic drift. As far as intelligence is concerned, no significant disparities have been found among different ethnic groups.

Origin of Ethnic Groups The three hypotheses regarding the evolution of humans, discussed in Section 20.9, can be applied to the origin of ethnic groups. The multiregional continuity hypothesis suggests that different human populations evolved into modern humans, and therefore humans have different ethnicities despite gene flow. The replacement hypothesis, on the other hand, proposes that all modern humans have a relatively recent common ancestor—that is, the Cro-Magnons, who evolved in Africa and then spread into other regions. Paleontologists tell us that the variation among modern populations is considerably less than among archaic human populations some 250,000 years ago. If so, all ethnic groups evolved from the same single, ancestral population.

A comparative study of mitochondrial DNA shows that the differences among human populations are consistent with their having a common ancestor no more than a million years ago. Lewontin has also found that the genotypes of different modern populations are extremely similar. He examined variations in 17 genes, including blood groups and various enzymes, among major groups such as African, Asian, and European populations. (Fig. 20.12). He found that the great majority of genetic variation—85%—occurs within ethnic groups, not among them. In other words, the amount of genetic variation between individuals of the same ethnic group is greater than the variation between ethnic groups.

20.12 Check Your Progress What biological proof is there that we humans are all one species?

FIGURE 20.12
Ethnic groups.

One of the most unfortunate misconceptions concerning human evolution is the belief that paleontologists suggest that humans evolved from apes, specifically chimpanzees. On the contrary, we can trace our ancestry back to the first hominid and then back to the first hominoid, an ancestor to both humans and apes. Today's apes are our cousins, and we couldn't have evolved from our cousins because we are all contemporaries—living on Earth at the same time. Our relationship to apes is analogous to you and your first cousins being descended from your grandparents.

Aside from various anatomic differences related to human bipedalism and intelligence, a cultural evolution separates us from the apes. A hunter-gatherer society evolved when humans became able to make and use tools. That society then gave way to an agricultural economy about 12,000 to 15,000 years ago. The agricultural period extended from that time to about 200 years ago, when the Industrial Revolution began. Now, most people live in urban areas. Perhaps as a result, modern humans are for the most part divorced from nature and often endowed with the philosophy of exploiting and controlling nature.

Our cultural evolution has had far-reaching effects on the biosphere, especially since the human population has expanded to the point that it is crowding out many other species. Our degradation and disruption of the environment threaten the continued existence of many species, including our own. As discussed in Part VI of this text, however, we have recently begun to realize that we must work with, rather than against, nature if biodiversity is to be maintained and our own species is to continue to exist.

Before we examine the environment and the role of humans in ecosystems, we will study plant biology and the various organ systems of the human body. Humans need to keep themselves and the environment fit so that they and their species can endure.

The Chapter in Review

Summary

Lucy's Legacy

- The hominid *Australopithecus afarensis*, dubbed "Lucy," lived 3.9–3.2 MYA.
- *A. afarensis* walked upright, exhibited sexual dimorphism, and was probably an omnivore.
- An earlier fossil called Selam, found in 2000, is expected to yield more information.

Humans Share Characteristics with All the Other Primates

20.1 Primates are adapted to live in trees

- The order Primates encompasses prosimians, monkeys, apes, and humans.
- Primates are characterized by prehensile hands and feet, binocular vision, a large, complex brain, and a reduced reproductive rate.

20.2 All primates evolved from a common ancestor

- Other primates diverged from the human lineage over time.
- Anthropoids include monkeys, apes, and hominids.
- Hominoids are apes and hominids. The dryopithecines gave rise to the apes and eventually to the hominids.

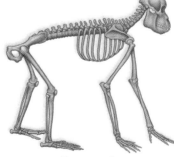

Proconsul

Humans Have an Upright Stance and Eventually a Large Brain

20.3 Early hominids could stand upright

- Hominids include humans and several extinct species.
- The split between apes and hominids occurred about 7 MYA.

- Humans and chimpanzees are genetically similar and have common traits, but distinct differences in spine position and shape, pelvis and hip size, femur length, and the design of the knee and toe.
- Human bipedalism may have evolved while the first hominids still lived in trees.
- Early hominids include *Sahelanthropus tchadensis* (7 MYA) and *Ardipithecus ramidus* (4.5 MYA)

20.4 Australopithecines had a small brain

- An australopithecine that lived in Africa from 4 MYA to 1 MYA could be a direct ancestor to humans.
- Change in body parts at different rates and times is called mosaic evolution.
- Robust and gracile types of australopithecines are adapted to different ways of life.
- *Australopithecus africanus*, discovered first in East Africa, stood upright and was bipedal and gracile.

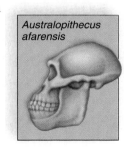

Australopithecus afarensis

20.5 Origins of the genus *Homo*

- A high rate of fetal brain growth allowed *Homo* to evolve from *Australopithecus*.
- Brain growth is linked to infantile helplessness and the inability of the parents to climb trees.

20.6 Early *Homo* had a large brain

- Genus *Homo* had a 600-cc brain size and jaws and teeth resembling humans; tool use is evident.
- *Homo habilis* and *H. rudolfensis* were omnivores with a brain size of 800 cc.
- *H. ergaster* had a 1,000-cc brain, larger than that of *H. erectus*.
- *H. floresiensis*, discovered in 2004, used tools and fire.

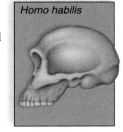

Homo habilis

20.7 Biocultural evolution began with *Homo*

- Biocultural evolution occurs when natural selection is based, in part, on cultural achievements.
  - *H. habilis* used Oldowan tools and gathered plants.
  - *H. erectus* used Acheulian tools, were hunter-gatherers, and used fire.
  - Hunting encourages the development and spread of culture.

Homo sapiens Is the Last Twig on the Primate Evolutionary Bush

20.8 The Neandertal and Cro-Magnon people coexisted for 12,000 years

- The Neandertals lived in Europe and the Near East; they had a brain size of 1,400 cc, built shelters, used stone tools, successfully hunted, used fire, and had burial ceremonies.
- The Cro-Magnons entered Asia and Europe from Africa; they had a brain size of 1,590 cc, were modern in appearance, and subsisted as hunter-gatherers.

20.9 The particulars of *Homo sapiens* evolution are being studied

- Three hypotheses have developed concerning the evolution of *Homo sapiens* from archaic humans:
  - The multiregional continuity model proposes that modern humans evolved in Asia, Africa, and Europe independently.
  - The replacement model hypothesizes that modern humans evolved in Africa and replaced archaic humans in Asia and Europe.
  - The assimilation model proposes that modern humans evolved only in Africa, migrated into Asia and Europe, and then interbred with archaic humans in those locations, resulting in hybrid populations.

20.10 Cro-Magnons made good use of tools

- The Cro-Magnons made advanced stone tools and were accomplished hunters.
- Their highly developed brain facilitated language, social organization, and artistic accomplishments.

20.11 Agriculture made modern civilizations possible

- Agriculture originated in the Near East, Far East, and Central and South America about the same time.
- Conditions favoring agriculture included a warming trend that caused glaciers to retreat and leave behind fertile valleys; prior knowledge about crops; increased population; the development of agricultural tools and irrigation; and the ability to store seeds and tubers for food and future crops.
- Bountiful crops freed some people to do other work.

Today's Humans Belong to One Species

20.12 Humans have different ethnicities

- Humans have a wide geographic distribution, and phenotypic and genotypic variations among human populations are evident.
- The replacement hypothesis proposes that all modern humans have a recent common ancestor.
- Genetic variations among individuals of the same ethnic group are greater than variations between ethnic groups.

Humans Share Characteristics with All the Other Primates

1. Which of the following lists the correct order of divergence from the main primate line of descent?
   a. prosimians, monkeys, gibbons, orangutans, African apes, humans
   b. gibbons, orangutans, prosimians, monkeys, African apes, humans
   c. monkeys, gibbons, prosimians, African apes, orangutans, humans
   d. African apes, gibbons, monkeys, orangutans, prosimians, humans
   e. *H. habilis, H. ergaster, H. neandertalensis*, Cro-Magnon
2. Stereoscopic vision is possible in primates due to
   a. the presence of cone cells.
   b. an enlarged brain.
   c. a shortened snout.
   d. None of these are correct.
3. Compare the features of New World monkeys and Old World monkeys.

Humans Have an Upright Stance and Eventually a Large Brain

4. The first humanlike feature to evolve in the hominids was
   a. a large brain.
   b. massive jaws.
   c. a slender body.
   d. bipedal locomotion.
5. The last common ancestor for African apes and hominids
   a. has been found, and it resembles a gibbon.
   b. was probably alive around 7 MYA.
   c. has been found, and it has been dated at 30 MYA.
   d. is not expected to be found because there was no such common ancestor.
   e. is now believed to have lived in Asia, not Africa.
6. Which of the following is NOT a characteristic of robust australopithecines?
   a. massive chewing muscles attached to a bony skull crest
   b. large brain size
   c. walked upright
   d. lived in southern Africa
   e. Both a and c are not characteristics of robust types.
7. Lucy is a(n)
   a. early *Homo*.
   b. australopithecine.
   c. ardipithecine.
   d. modern human.
8. Which of these characteristics is not consistent with the genus *Homo*?
   a. large brain size
   b. prolonged infancy
   c. life in the trees
   d. increased intelligence
   e. All of these are characteristics of genus *Homo*.
9. Which hominids could have inhabited the Earth at the same time?
   a. australopithecines and Cro-Magnons
   b. *Australopithecus robustus* and *Homo habilis*
   c. *Homo habilis* and *Homo sapiens*
   d. apes and humans
10. Compared to *H. habilis, H. erectus*
    a. had a smaller brain.
    b. was shorter.
    c. had a flatter nose.
    d. had a flatter face.

11. Which of these pairs is matched correctly?
    a. *Australopithecus afarensis*—bipedal but small brain
    b. *Homo habilis*—small brain, large teeth
    c. *Homo erectus*—larger brain, flatter face
    d. *Homo sapiens*—bipedal with projecting face
    e. Both a and c are correct.
12. This species was probably the first to use fire.
    a. *Homo habilis*   d. *Australopithecus robustus*
    b. *Homo erectus*   e. *Australopithecus afarensis*
    c. *Homo sapiens*
13. **THINKING CONCEPTUALLY** How do biocultural evolution and Darwinian evolution by natural selection differ?

Homo sapiens Is the Last Twig on the Primate Evolutionary Bush

14. If the multiregional continuity hypothesis is correct,
    a. hominid fossils in China after 100,000 BP would not be expected to resemble earlier fossils.
    b. hominid fossils in China after 100,000 BP would be expected to resemble earlier fossils.
    c. modern humans did not migrate out of Africa.
    d. Both b and c are correct.
    e. Both a and c are correct.
15. Mitochondrial DNA data support which hypothesis for the evolution of humans?
    a. multiregional continuity
    b. replacement
    c. assimilation
16. Complete this diagram of the replacement model by filling in the blanks.

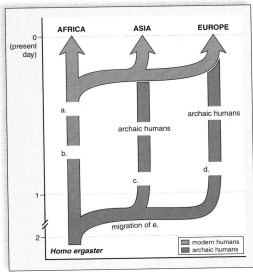

Replacement Model

17. Which of these pairs is NOT correctly matched?
    a. *H. erectus*—made tools   d. Cro-Magnon—good artist
    b. Neandertal—good hunter   e. *A. robustus*—fibrous diet
    c. *H. habilis*—controlled fire
18. The first *Homo* species to use art appears to be
    a. *H. neandertalensis*.   c. *H. habilis*.
    b. *H. erectus*.   d. *H. sapiens*.
19. The increased reliance on agriculture in some early societies led to increases in which of the following?
    a. infectious disease   c. anemia
    b. malnutrition   d. All of these are correct.

Today's Humans Belong to One Species

20. Which human characteristic is not thought to be an adaptation to the environment?
    a. bulky bodies of Eskimos
    b. long limbs of Africans
    c. light skin of northern Europeans
    d. hair texture of Asians
    e. Both a and b are correct.
21. **THINKING CONCEPTUALLY** How does genetic variation between ethnic groups compare to genetic variation within ethnic groups? What does this tell us about humans in general?

Understanding the Terms

anthropoid   403
arboreal   400
archaic human   410
australopithecine   406
Australopithecus
   africanus   406
biocultural evolution   409
bipedalism   405
Cro-Magnon   410
culture   409
dryopithecine   403

hominid   403
hominoid   403
Homo erectus   408
Homo ergaster   408
hunter-gatherer   409
molecular clock   404
mosaic evolution   406
out-of-Africa
   hypothesis   411
primate   400
prosimian   403

Match the terms to these definitions:
a. _____ Group of primates that includes monkeys, apes, and humans.
b. _____ The common name for the first fossils generally accepted as being modern humans.
c. _____ Type of early humans to first have a striding gait similar to that of modern humans.
d. _____ Member of a group containing humans and apes.

Thinking Scientifically

1. Bipedalism has many selective advantages, including the increased ability to spot predators and prey. However, bipedalism has one particular disadvantage—upright posture leads to a smaller pelvic opening, which makes giving birth to an offspring with a large head very difficult. This situation results in a higher percentage of deaths (of both mother and child) during birth in humans compared to other primates. How can you explain the selection for a trait, such as bipedalism, that has both positive and negative consequences for fitness?
2. How might you use biotechnology to show that humans today have Neandertal genes and, thereby, support the assimilation model discussed in Section 20.9?

ARIS *Visit* **www.mhhe.com/maderconcepts** *for practice quizzes, animations, videos, and* activities designed to help you master the material in this chapter.

BIOLOGICAL VIEWPOINTS

We can find evidence of the theory of evolution all around us. As we have learned, a theory in science is not a hypothesis; a theory is a well-supported coherent concept that can explain many independent observations. So it is with evolution, which states that all living things can trace their ancestry to a common source, but each is adapted to a particular way of life. Common ancestry is obvious because, from bacteria to bats, toadstools to trees, and hydras to whales, all life shares the same characteristics. For example, all organisms are made up of the same four chemicals—carbon, hydrogen, oxygen, and nitrogen—elements that were abundant when life began. All organisms are made up of cells, have genes made of DNA, and use ATP as a carrier of energy.

Life has these similarities because all forms are related through reproductive events. When the members of a population reproduce and give rise to the next generation over and over again, relationships come about that eventually account for the history of life on Earth and why we are all alike. For example, why do birds have coarse, brittle bones rather than bones made of a strong, lightweight alloy such as titanium? The reason is that birds, just like mammals, are descended from ancient reptiles. Reptiles evolved from amphibians, and amphibians are descendants of fishes. As it happened, a fish evolving more than half a billion years ago had bones containing calcium and phosphorus. Titanium bones might be best for birds, but because birds are descended from the first vertebrates, they have bones of calcium and phosphorus instead.

Similarly, a knowledge of evolutionary history can explain why organisms carry the remnants of formerly functional structures that are by now completely useless. Tiny vestiges of leg bones are invisibly embedded in the skin of certain whales; the nonfunctional remains of pelvic bones occur in some snakes; and humans have caudal vertebrae—the remnants of a tail. Up to recent times, biologists primarily relied on the fossil record and comparative anatomic data, much as we have cited, to determine evolutionary relationships. But modern-day biology is increasingly using molecular data to determine the tree of life. We know today that if we could trace the lineage of all the millions of species ever to have evolved, the entirety would resemble a dense bush. Some lines of descent would be cut off close to the base; others would continue in a straight line, even to today; and many would split, producing two or even several groups.

The theory of evolution also says that adaptation to a changing environment produced the diverse life-forms that existed in the past and that we see about us today. As Darwin noted, the members of a population exhibit variations. Genetic variations occur randomly. For example, we know chance mutations can have a powerful effect on the anatomy of a plant or animal, and probably on all life-forms. Once variations have occurred, those that make an organism more suited to an environment are the ones more likely to be passed on. "Survival of the fittest" actually refers to the greater reproductive success of well-adapted individuals. Animals that are best at finding food, escaping enemies, and defending a territory in a particular environment are more likely to have reproductive success. And, eventually, these traits become the most common ones in a species. This is the process of natural selection by which organisms become adapted to their environment.

It is important to realize that evolution does not proceed along some grand, predictable course. Instead, the details of evolution depend on the environment that a population happens to live in and the genetic variants that happen to arise in that population.

Evolution is the scientific theory that best unifies biology. Evolution is often called the *GUT* of biology, the grand *unifying theory*. It explains why cells have a certain structure and function, why development occurs as it does, why organisms behave as they do, and how plants and animals are distributed. It also explains the diversity of life, from minute forms that live in a pond, to the colorful plants in your garden, to the thundering herds of hoofed animals on the plains of Africa. From simple, unicellular organisms, new life-forms arose and changed in response to environmental pressures, and this produced biodiversity. One way to think of evolution is change over time via descent with modification.

21

Plant Organization and Homeostasis

LEARNING OUTCOMES

After studying this chapter, you should be able to accomplish the following outcomes.

What Do Forests Have to Do with Global Warming?

1 Relate the structure of leaves, specifically those of tropical rain forest trees, to the ability to soak up CO_2.

Plants Have Three Vegetative Organs

2 Contrast the general structure and function of roots, stems, and leaves.
3 List and describe five differences between monocots and eudicots.
4 Give several examples to show that monocots provide humans with varied products.

The Same Plant Cells and Tissues Are Found in All Plant Organs

5 Describe the location, structure, and function of epidermal tissue, ground tissue, and vascular tissue in angiosperms.
6 Describe the arrangement of tissues in a root, a stem, and a leaf.

Plant Growth Is Either Primary or Secondary

7 Compare and contrast primary growth in a root tip and in a shoot tip.
8 Describe secondary growth with emphasis on the stem.
9 Give examples to show that wood has been used for various purposes throughout human history.

Leaf Anatomy Facilitates Photosynthesis

10 Describe the organization, structure, and function of leaf tissues.

Plants Maintain Internal Equilibrium

11 Tell how plant cells receive the materials they need for growth and maintenance.
12 Describe the adaptations of plants that enable them to be homeostatic.

Much like the panes of a greenhouse, CO_2 in our atmosphere traps radiant heat from the sun and warms the world. Therefore, CO_2 and other gases that act similarly are called greenhouse gases. For at least a thousand years prior to 1850, atmospheric CO_2 levels remained fairly constant at 0.028%. Without these greenhouse gases, the Earth's temperature would have been about 33°C cooler. When industrialization began in the 1850s, the amount of CO_2 in the atmosphere increased to 0.036%. This is sufficient to cause an increase in global temperatures called global warming. Scientists tell us that the burning of fossil fuels is causing CO_2 and other greenhouse gases to enter the atmosphere. Due to global warming, the oceans are rising and could swamp coastal cities, such as New Orleans, New York, and Los Angeles. As temperatures rise, regions of suitable climate for various species will shift toward the poles and higher elevations. It's unlikely that most plants

What Do Forests Have to Do with Global Warming?

will be able to migrate northward at the pace required, and thus extinctions are expected. Farming will have to shift too, but soil tends to be less suitable for agriculture at higher altitudes and elevations, so food shortages could occur.

It's possible that tropical rain forests, which carry on a lot of photosynthesis, could help deter global warming because photosynthesis uses up CO_2. Unfortunately, every year an amount of rain forest the size of Panama is lost to ranching, logging, mining, and otherwise developing the forest for human needs. Worse yet, the clearing of forests often involves burning them. Each year, deforestation in tropical rain forests accounts for 20–30% of all CO_2 in the atmosphere. The consequence of burning forests is double trouble for global warming because burning a forest adds CO_2 to the atmosphere and, at the same time, removes trees that would ordinarily absorb CO_2. All countries, but especially those with tropical rain forests, should combat deforestation. In the mid-1970s, Costa Rica established a system of national parks and reserves to protect 12% of the country's land area from degradation, and the current Costa Rican government wants to increase protected areas to 25% in the near future. Similar efforts in other countries may help slow the ever-increasing threat of global warming.

People do not often think about plants' ability to deter global warming as another service they perform for us. This service is tied to the ability of plants to photosynthesize, which is dependent on the structure of plants, the topic of this chapter. We will see that the structure of plant roots, stems, and leaves allows plants to carry on photosynthesis. The huge trees in tropical rain forests are evergreen and ever capable of carrying on photosynthesis because abundant sunlight, water, and warmth are always present. This means that their leaves take up CO_2 throughout the entire year.

Plants Have Three Vegetative Organs

This part of the chapter introduces the organs of a plant by first discussing the shoot system, composed of stems and leaves, and then the root system, which lies beneath the surface. The anatomy of the two main groups of flowering plants, monocots and eudicots, can be contrasted in several different ways. Monocots are the smaller group, but they perform vital services for humans.

21.1 Flowering plants typically have roots, stems, and leaves

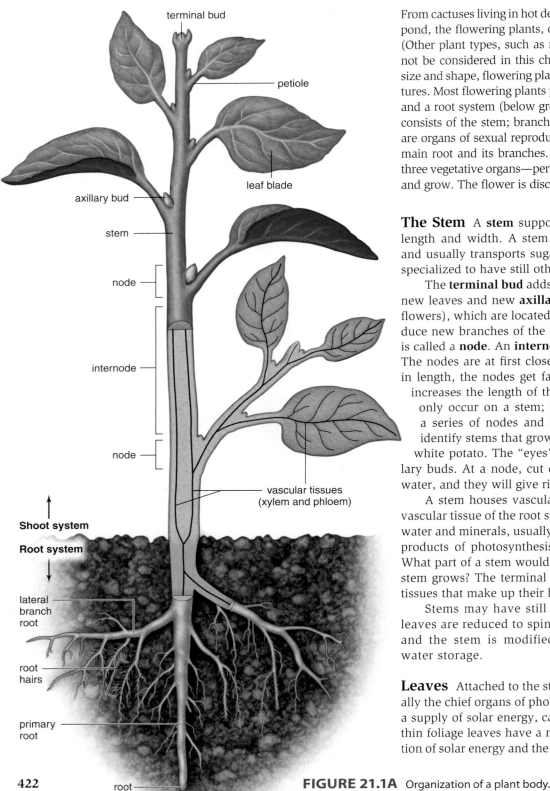

Shoot system

Root system

FIGURE 21.1A Organization of a plant body.

From cactuses living in hot deserts to water lilies growing in a nearby pond, the flowering plants, or angiosperms, are extremely diverse. (Other plant types, such as mosses, ferns, and gymnosperms, will not be considered in this chapter.) Despite their great diversity in size and shape, flowering plants share many common structural features. Most flowering plants possess a shoot system (above ground) and a root system (below ground) (**Fig. 21.1A**). The **shoot system** consists of the stem; branches; the leaves; and the flowers, which are organs of sexual reproduction. The **root system** consists of the main root and its branches. The stem, the leaf, and the root—the three vegetative organs—perform functions that allow a plant to live and grow. The flower is discussed in Section 24.1.

The Stem A **stem** supports a plant and allows it to grow in length and width. A stem also transports water to the leaves and usually transports sugars from the leaves. Some stems are specialized to have still other functions.

The **terminal bud** adds to the length of a stem and produces new leaves and new **axillary** (lateral) **buds**. **Axillary buds** (or flowers), which are located where leaves join the stem, can produce new branches of the stem (or flowers). This entire region is called a **node**. An **internode** is the region between the nodes. The nodes are at first close together, but as the stem increases in length, the nodes get farther apart. In other words, growth increases the length of the internodes. Nodes and internodes only occur on a stem; therefore, a stem can be defined as a series of nodes and internodes. This is a useful way to identify stems that grow underground, as does the stem of a white potato. The "eyes" of a white potato are actually axillary buds. At a node, cut out the axillary buds, place them in water, and they will give rise to a complete plant.

A stem houses vascular tissue that is continuous with the vascular tissue of the root system. The vascular tissue transports water and minerals, usually from the roots to the leaves, and the products of photosynthesis, usually in the opposite direction. What part of a stem would give rise to new vascular tissue as a stem grows? The terminal bud, of course. Plants produce new tissues that make up their bodies throughout their lives!

Stems may have still other functions. In the cactus, the leaves are reduced to spines in order to minimize water loss, and the stem is modified for photosynthesis and also for water storage.

Leaves Attached to the stem and its branches, **leaves** are usually the chief organs of photosynthesis, and as such they require a supply of solar energy, carbon dioxide, and water. Broad and thin foliage leaves have a maximum surface area for the collection of solar energy and the absorption of carbon dioxide. Leaves

stem

leaves

Spines of a cactus

tendril

Tendrils of a cucumber

Leaves of a Venus flytrap

FIGURE 21.1B Modified leaves adapt to a plant's environment.

receive water from the root system by way of vascular tissue that travels through the stem to the leaves. Photosynthesis by leaves allows a plant to grow, repair itself, and reproduce.

The wide portion of a foliage leaf is called the **blade**. The **petiole** is a stalk that attaches the blade to the stem. Some leaves do not have petioles and are instead attached directly to the stem. These leaves, such as those of the eucalyptus plant, are called sessile leaves.

Leaves are adapted to environmental conditions. Desert plants, such as a cactus, tend to minimize water loss by having reduced leaves. Climbing leaves, such as those of cucumbers, are modified into tendrils that can attach to nearby objects. The leaves of a few plants are specialized for catching insects. The Venus flytrap has hinged leaves that snap shut and interlock when an insect triggers sensitive hairs projecting from inside the leaves (**Fig. 21.1B**).

Some trees, called evergreens, retain their leaves, and others, called **deciduous**, lose their leaves during a particular season of the year.

Roots Roots anchor the plant in the soil, and they also absorb water and minerals from the soil for the entire plant. As a rule of thumb, the root system of a plant is at least equivalent in size and extent to its shoot system. Therefore, an apple tree has a much larger root system than a corn plant. Also, the extent of a root system depends on the environment. A single corn plant may have roots as deep as 2.5 m, while a mesquite tree that lives in the desert may have roots that penetrate to a depth of 20 m.

Just as cell division by the terminal bud increases the length of a stem, so too the root tip produces new cells and, in that way, increases the length of a root. The growth of a plant occurs at both ends, just as if you were to grow from your head and feet. The cells produced by a root tip become specialized tissues, just like those that are produced by the shoot tip, which is the terminal bud.

The cylindrical shape of a root tip and its slimy surface allow it to penetrate the soil as it grows and permit water to be absorbed from all sides. The slimy surface not only protects the root tip from abrasive soil particles, but also encourages the growth of beneficial soil bacteria. The absorptive capacity of a root is increased by its many **root hairs**, delicate extensions of

mature root cells in a special zone of the root tip. Root hairs are so numerous that they increase the absorptive surface of the root system tremendously. It has been estimated that a single rye plant has about 14 billion root-hair cells, and if placed end to end, the root hairs would stretch 10,626 km. Root hairs are constantly being replaced, and this same rye plant forms about 100 million new root-hair cells every day.

Figure 21.1C shows the diversity of roots. A carrot plant has one main taproot, which stores the products of photosynthesis, anchors the plant in the soil, and takes up water and minerals. Grasses have fibrous roots that cling to the soil in addition to taking up water and minerals. **Perennial plants** are able to regrow next season because their roots survive, even though the shoot system may have died back.

The specific anatomy of roots, stems, and leaves differs, whether a plant is a monocot or eudicot, as discussed next.

> **21.1 Check Your Progress** Trees living in tropical rain forests have big, broad leaves, while cactuses living in deserts have narrow leaves or only spines. Relate this observation to water availability.

Taproot

Fibrous root system

FIGURE 21.1C Taproot system (*left*) versus fibrous root system (*right*).

21.2 Flowering plants are either monocots or eudicots

Flowering plants are divided into two groups, depending on the number of **cotyledons**, or seed leaves, in the embryonic plant. The embryos of grasses and many other plants (e.g., tulips and daffodils) have one cotyledon. These plants are known as monocotyledons, or **monocots**. The embryos of kidney beans, peas, lima beans, and many other plants have two cotyledons. These plants are known as eudicotyledons, or **eudicots**. The cotyledons of eudicots supply nutrients when an embryo begins new growth, but the cotyledons of monocots act largely as a transfer tissue for nutrients derived from the endosperm, a storage tissue, before the true leaves begin photosynthesizing.

The term eudicot is a new one for botanists, who formerly compared monocots to a group called dicots. New findings about plant evolution have revealed that some of the plants formerly called dicots, such as water lilies, are so ancient that they arose before the angiosperms split into the monocots and eudicots. Therefore, the term dicotyledon is now obsolete.

Adult monocots and eudicots have other structural differences (**Fig. 21.2**). Some of these differences are observable with the unaided eye, while others require a microscope. For example, in the monocot root, vascular tissue occurs in a ring encircling a core of cells that comprise the pith. In the eudicot root, phloem, which transports organic nutrients, is located between the arms of xylem, which transports water and minerals and has a star shape.

In the monocot stem, the vascular bundles, which contain vascular tissue surrounded by a sheath, are scattered through-out the ground tissue. In a eudicot stem, the vascular bundles occur in a ring, which divides the ground tissue into cortex and the centrally located pith.

Visible to the naked eye, leaf veins are vascular bundles within a leaf. Monocots, such as grasses, exhibit parallel venation, and eudicots, such as maples, exhibit netted venation. Plants also differ by the number of flower parts and the number of apertures (thin areas in the wall) of pollen grains. Monocot flower parts are arranged in multiples of three, and eudicot flower parts occur in multiples of four or five. Eudicot pollen grains usually have three apertures, and monocot pollen grains usually have one aperture.

Although the division between monocots and eudicots may seem of limited importance, it does in fact affect many aspects of their structure. The eudicots are the larger group and include some of our most familiar flowering plants—from dandelions to oak trees. The monocots include grasses, lilies, orchids, and palm trees, as well as some of our most significant food sources —rice, wheat, and corn. Section 21.3 discusses these and other uses of monocots.

> **21.2 Check Your Progress** While walking in a forest, you notice a plant you have not seen before. You carefully collect a sample and examine it in the lab. The plant has a branching pattern of leaf venation. The vascular tissue is arranged in a ring in the stem, but not in the root. To which of the two groups of plants does this plant belong?

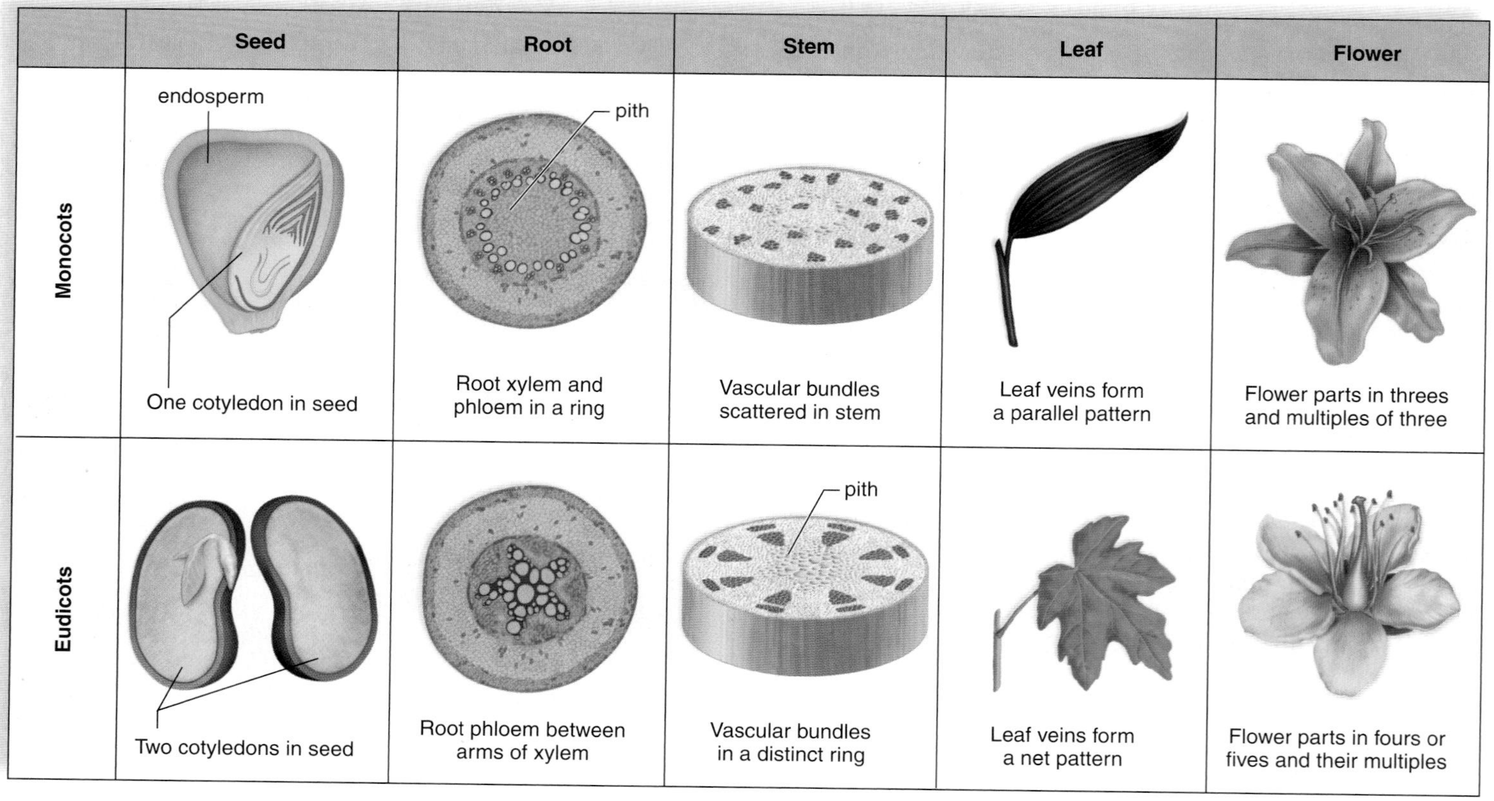

	Seed	Root	Stem	Leaf	Flower
Monocots	endosperm — One cotyledon in seed	pith — Root xylem and phloem in a ring	Vascular bundles scattered in stem	Leaf veins form a parallel pattern	Flower parts in threes and multiples of three
Eudicots	Two cotyledons in seed	Root phloem between arms of xylem	pith — Vascular bundles in a distinct ring	Leaf veins form a net pattern	Flower parts in fours or fives and their multiples

FIGURE 21.2 Monocots and eudicots differ structurally in several ways.

21.3   Monocots serve humans well

Although the monocots are a small group compared to eudicots, they have great importance. From cereal grains to alcoholic beverages, monocots play a significant role in all our lives, both past and present. After all, what would Italian cooking be without the monocot garlic?

Agricultural practices were in place over 10,000 years ago. The domestication of monocot plants included selective breeding in order to accumulate certain desirable traits in offspring. For example, you would probably never recognize "wild" corn because, through selective breeding, we have encouraged large, fleshy, starchy kernels that in no way resemble ancestral corn. Grains such as rice, wheat, corn, and barley (**Fig. 21.3**) are a chief source of calories for the majority of the world's people and their livestock. These grains are made into everything from flour to beverages. It is remarkable that wheat, corn, and rice are associated with different major cultures or civilizations—wheat with Europe and the Middle East, corn or maize with the Americas, and rice with the Far East.

Beer is also produced from cereal grain. The exact origin of beer is unclear, but it is believed to be over 10,000 years old. While most beer uses barley malt (dried young seedlings) to supply enzymes, the specific grain to be fermented varies geographically among all the beer-producing cultures. For example, wheat was used in Mesopotamia, rice in Asia, and sorghum in Africa. Sake, although called "rice wine," is actually beer produced from fermented rice (as opposed to a true wine, which is produced from grapes and other fruits). In the United States, various grains are used to make beer.

Three of the world's four most populous nations are rice-based societies—China, India, and Indonesia. Over 50% of the world's people depend on rice for about 80% of their calorie requirements. A diverse food, rice can be cooked and eated as is or can be used to produce breakfast cereals, desserts, rice cakes, and rice flour.

Rice, corn, and wheat are all used to make breakfast cereals. Do rice crispies, corn pops, and cream of wheat sound familiar? Corn is by far the most important crop plant in the United States, where about 80% of the corn produced goes to feed livestock. People in many developing countries, however, rely on corn for as much as 30% of the calories in their diets.

Bamboo, the common name for about 1,000 species of grass, ranges in height from 15 cm to 30–35 m. Depending on the species, bamboo can grow up to a foot a day. Eaten not only by pandas, young bamboo is also consumed as a vegetable (yes, those are actual bamboo shoots in Chinese food). Older bamboo is much tougher, and therefore harvested for making musical instruments, furniture, and acupuncture needles, as well as for roofing, flooring, and drainage pipes. In fact, about 73% of the population of Bangladesh lives in bamboo houses.

Finally, many of the flowers bought and sold are monocots, such as tulips, daffodils, and lilies. Floriculture, the cultivation and management of ornamental and flowering plants, is a multibillion-dollar industry in the United States alone.

This completes our preview of plant anatomy; the next section discusses the specialized cells and tissues of plants.

Rice plants, *Oryza*

Wheat plants, *Triticum*

Corn plants, *Zea*

Barley

FIGURE 21.3 Monocot variety.

> **21.3 Check Your Progress** Eudicots, not monocots, tend to be specialized for biotic pollination (e.g., by insects). Monocots evolved before eudicots. Do you think insects evolved around the time of the monocots or the eudicots? Explain.

The Same Plant Cells and Tissues Are Found in All Plant Organs

Learning Outcomes 5–6, page 420

The levels of biological organization apply to plants. For example, several cells form a tissue, and several tissues make up an organ. In this section, we examine plant cells and tissues, and how they are arranged in the organs of a plant.

21.4 Plants have specialized cells and tissues

Unlike humans, flowering plants grow in size their entire life because they have meristematic (embryonic) tissue composed of cells that divide. **Apical meristem** is located in the terminal bud of the shoot system and in the root tip. When apical meristem cells divide, one of the daughter cells remains a meristematic cell, and the other differentiates into one of the three types of primary tissues:

1. **Epidermal tissue.** Contains epidermal cells, which form the outer protective covering of a plant.

2. **Ground tissue.** Consists of parenchyma and other cell types that fill the interior of a plant.

3. **Vascular tissue.** The cells of xylem and phloem transport water and sugar in a plant and provide support.

Epidermal Tissue The entire body of a plant is covered by a layer of closely packed epidermal cells called the **epidermis.** The walls of epidermal cells that are exposed to air are covered with a waxy **cuticle** to minimize water loss. In leaves, the epidermis often contains **stomata** (sing., stoma) (**Fig. 21.4A**, *top left*). A stoma is a small opening surrounded by two modified epidermal cells called guard cells. When the stomata are open, CO_2 uptake and water loss occur. As mentioned in Section 21.1, roots have long, slender projections of epidermal cells called root hairs (Fig. 21.4A, *top right*). These hairs increase the surface area of the root for absorption of water and minerals.

In the trunk of a tree, the epidermis is replaced by cork, which is a part of bark. New cork cells are made by a meristem called cork cambium. As the new cork cells mature, they increase slightly in volume, and they become encrusted with *suberin*, a lipid material that both waterproofs them and causes them to die. These nonliving cells protect the plant and make it resistant to attack by fungi, bacteria, and animals. Lenticels, which are breaks in the periderm (cork plus cork cambium) function in gas exchange in some trees (Fig. 21.4A, *bottom*).

Ground Tissue Ground tissue forms the bulk of stems, leaves, and roots. Ground tissue contains three types of cells (**Fig. 21.4B**). **Parenchyma cells** are the least specialized of the cell types and are found in all the organs of a plant. When in a leaf, they may contain chloroplasts and carry on photosynthesis; in the cortex or pith region, they contain colorless plastids that store the products of photosynthesis. **Collenchyma cells** have thicker primary walls than parenchyma cells. The thickness is uneven, with the thicker areas usually found in the corners of the cell. Collenchyma cells particularly provide structural support in nonwoody plants. The familiar strands in celery stalks are composed mostly of collenchyma cells. **Sclerenchyma cells** have thick secondary cell walls

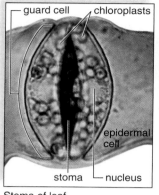

Stoma of leaf

guard cell — chloroplasts — epidermal cell — stoma — nucleus

corn seedling — root hairs — elongating root tip

Root hairs

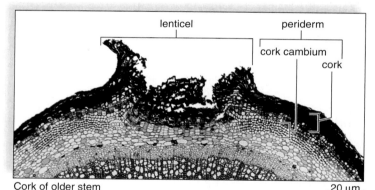

lenticel — periderm — cork cambium — cork

Cork of older stem   20 µm

FIGURE 21.4A Modifications of epidermal tissue.

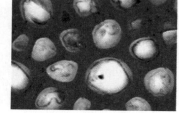

FIGURE 21.4B
Ground tissue cells.

Parenchyma cells with thin walls   50 µm

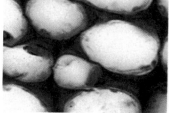

Collenchyma cells with thicker walls   50 µm

Sclerenchyma cells with very thick walls   50 µm

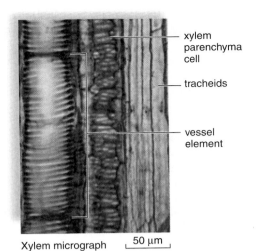

- xylem parenchyma cell
- tracheids
- vessel element

Xylem micrograph
50 μm

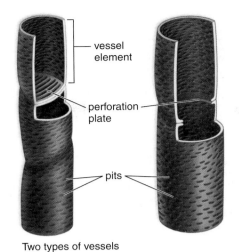

- vessel element
- perforation plate
- pits

Two types of vessels

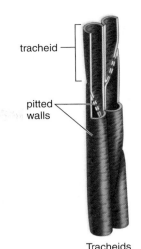

- tracheid
- pitted walls

Tracheids

FIGURE 21.4C
Xylem structure.

impregnated with *lignin*, a substance that makes plant cell walls tough and hard. If we compare a cell wall to reinforced concrete, cellulose fibrils would play the role of steel rods, and lignin would be analogous to the cement. Most sclerenchyma cells are nonliving at maturity; their primary function is to support the mature living regions of a plant. There are two main types of sclerenchyma cells: fibers and sclerids (also known as stone cells). The long fibers in plants make them useful for a number of purposes. For example, cotton and flax fibers can be woven into cloth, and hemp fibers can make strong rope. The gritty texture of a pear is due to the presence of stone cells throughout the fruit's flesh.

Vascular Tissue The xylem and phloem of vascular tissue have different functions. **Xylem** transports water and minerals from the roots to the leaves. Xylem contains conducting cells called vessel elements and tracheids (**Fig. 21.4C**). Both of these are hollow and nonliving, but tracheids are elongated and narrow, while vessel elements are wider and shorter. The perforated end walls of **vessel elements** are arranged to form a continuous pipeline for water and mineral transport. The end walls and side walls of **tracheids** have pits, depressions that allow water to move from one tracheid to another.

Phloem transports sugar, in the form of sucrose, and other organic compounds, such as hormones, usually from the leaves to the roots. The conducting cells of phloem are **sieve-tube members**, arranged to form a continuous sieve tube (**Fig. 21.4D**). Sieve-tube members contain cytoplasm but no nuclei. The term *sieve* refers to a cluster of pores in the end walls, collectively known as a sieve plate. Each sieve-tube member has a companion cell, which does have a nucleus. The two are connected by numerous **plasmodesmata** (strands of cytoplasm that extend through pores), and the nucleus of the companion cell may control and maintain the life of both cells. The companion cells are also believed to be involved in the transport function of phloem.

Vascular tissue, consisting of both xylem and phloem, extends from the root through the stem to the leaves and vice versa (see Fig. 21.1A). In the root, the vascular tissue is located in a central cylinder; in the stem, it is in vascular bundles; and in the leaves, the vascular bundles are referred to as veins.

In the next section, we will examine how plant tissues form the various plant organs.

> **21.4 Check Your Progress** Growing young plants are supported structurally but are also flexible. Which of the three types of ground tissue cells—parenchyma, collenchyma, or sclerenchyma—could best provide support without making the plant rigid?

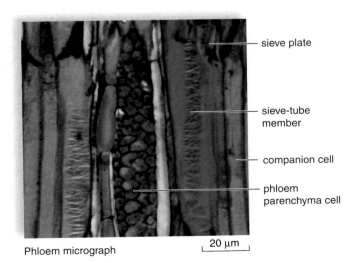

- sieve plate
- sieve-tube member
- companion cell
- phloem parenchyma cell

Phloem micrograph
20 μm

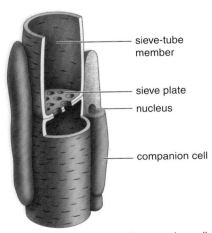

- sieve-tube member
- sieve plate
- nucleus
- companion cell

Sieve-tube member and companion cells

FIGURE 21.4D
Phloem structure.

Groups of cells that perform a similar function are called tissues. **Figure 21.5A** shows how epidermal, ground, and vascular tissues are arranged in a eudicot plant. Notice that in all three organs—namely, the leaf, the stem, and the root—the epidermal tissue forms the outer covering and the vascular tissue is embedded within ground tissue. In order to further demonstrate this characteristic organization of plant tissues, let's look at each organ in turn.

Leaf A typical eudicot leaf of a temperate-zone plant is shown in longitudinal section at the top of **Figure 21.5B**.

The upper epidermis often bears protective hairs and/or glands that secrete irritating substances to provide some protection against predation. These appendages and chemicals discourage insects from eating leaves. The upper and lower epidermis has an outer, waxy cuticle, which prevents water loss, and also prevents gas exchange because it is not gas permeable. Stomata (sing., stoma) are located in the epidermis. Carbon dioxide for photosynthesis enters a leaf at the stomata, and the by-product of photosynthesis, oxygen, exits a leaf at the stomata. Also, water evaporates and exits at the stomata as a gas.

The interior of a leaf is made of **mesophyll**, a ground tissue composed mostly of parenchyma cells that contain chloro-

plasts and carry on photosynthesis. Eudicot mesophyll has two distinct regions: palisade mesophyll, containing tightly packed elongated cells, and spongy mesophyll, containing irregular cells bounded by air spaces. The majority of chloroplasts are located within the palisade mesophyll region. The loosely packed arrangement of the cells in the spongy mesophyll increases the amount of surface area for gas exchange and water loss. The loss of water actually assists the transport of water, as we shall see in Chapter 22.

Notice how leaf veins terminate in the mesophyll. This termination is important because the leaf veins transport water and minerals through the xylem to a leaf and transport the product of photosynthesis, a sugar, away from the leaf through phloem.

Stem Plants that have nonwoody stems, such as zinnias and daisies, are termed **herbaceous** plants. As with leaves, the outermost tissue of a herbaceous stem is the epidermis, which is covered by a waxy cuticle to prevent water loss.

Ground tissue in a stem consists of the **cortex**, a narrow ring of parenchyma cells beneath the epidermis, and the central **pith**, which stores water and the products of photosynthesis. The cortex is sometimes green and carries on photosynthesis.

Herbaceous stems have distinctive vascular bundles containing xylem and phloem. In each bundle, xylem is typically found toward the inside of the stem, and phloem is found toward the outside. In the herbaceous eudicot stem, the vascular bundles are arranged in a distinct ring. In the herbaceous monocot stem, the vascular bundles are scattered throughout the ground tissue. Figure 21.5B (*middle*) contrasts herbaceous eudicot and monocot stems.

The vascular tissue of a stem supports the growth of the shoot system. The sclerenchyma cells of vascular tissue and the strong walls of vessel elements and tracheids help support the shoot system as it increases in length. Further, the vascular bundles bring water and minerals to the leaf veins, which distribute them to the mesophyll of leaves, the primary organs of photosynthesis in most plants. The vascular bundles also distribute the products of photosynthesis to the root system, where they may be stored as needed, and also to any immature leaves that are not yet photosynthesizing.

Root In a cross section of a mature root (Fig. 21.5B, *bottom*), the following specialized tissues are identifiable: The epidermis, which forms the outer layer of the root, usually consists of only a single layer of largely thin-walled, rectangular cells. When mature, many epidermal cells have root hairs.

Large, thin-walled parenchyma cells make up the cortex, the multiple layer of ground tissue cells located beneath the epidermis. The cells contain starch granules, and the cortex functions in food storage.

The **endodermis** is a single layer of rectangular cells that fit snugly together and form the outer tissue of the **vascular cylinder**. Further, a layer of impermeable suberin (the Casparian strip) on all but two sides forces water and minerals to pass through endodermal cells. In this way, the endodermis regu-

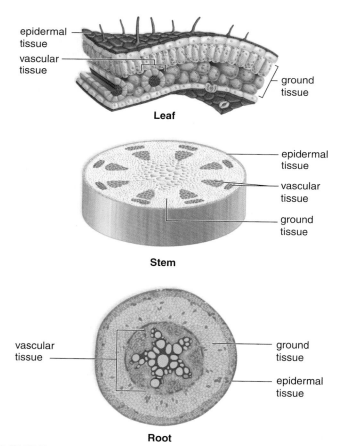

Leaf

epidermal tissue
vascular tissue
ground tissue

Stem

epidermal tissue
vascular tissue
ground tissue

Root

vascular tissue
ground tissue
epidermal tissue

FIGURE 21.5A Arrangement of plant tissues in the organs of eudicots. See also Figure 21.5B.

FIGURE 21.5B Internal structure of the leaf, stem, and root.

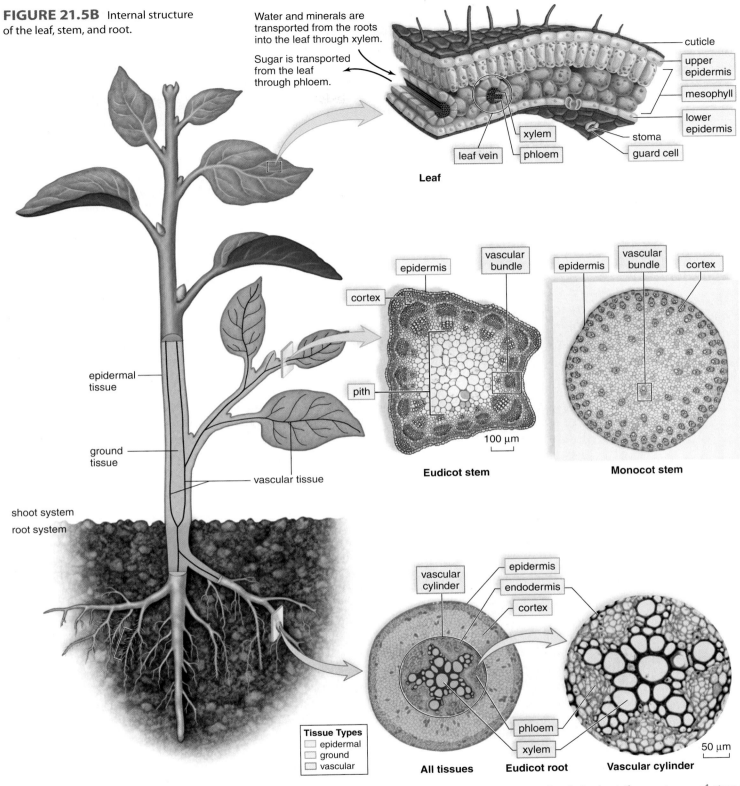

Water and minerals are transported from the roots into the leaf through xylem.

Sugar is transported from the leaf through phloem.

cuticle
upper epidermis
mesophyll
lower epidermis
stoma
guard cell
leaf vein
xylem
phloem

Leaf

epidermal tissue

ground tissue

vascular tissue

shoot system
root system

epidermis
cortex
pith
vascular bundle

100 μm

Eudicot stem

epidermis
vascular bundle
cortex

Monocot stem

vascular cylinder
epidermis
endodermis
cortex
phloem
xylem

All tissues   **Eudicot root**   **Vascular cylinder**

50 μm

Tissue Types
☐ epidermal
☐ ground
☐ vascular

lates the entrance of minerals into the vascular cylinder of the root. Just inside the endodermis layer is the **pericycle**, a layer of actively dividing cells from which lateral branch roots arise (see Fig. 21.6A). In the vascular cylinder of eudicot roots, xylem appears star-shaped because several arms of tissue radiate from a common center. The phloem is found in separate regions between the arms of the xylem.

This completes our in-depth look at the anatomy of stems, leaves, and roots. In the next part of the chapter, we will examine the growth of plants.

21.5 Check Your Progress Ground tissue is found in all three organs but has different names. Explain.

Primary growth refers to an increase in the length of stems and roots. Secondary growth refers to an increase in the girth of stems and roots. Only woody plants undergo secondary growth. Wood is a functional material for human beings.

21.6   Primary growth lengthens the root and shoot systems

Primary growth, which causes a plant to grow lengthwise, is centered in the apex (tip) of the shoot and of the root. As stated, these regions contain apical meristematic tissue. The term meristem is derived from the Greek word *merismos*, which means division. **Meristem** is a region of actively dividing cells. After new cells are produced by the process of mitosis, they go on to become the specialized tissues of a plant—epidermal tissue, ground tissue, and vascular tissue.

In contrast to animals, which grow to maturity and then stop growing in size, the existence of meristematic tissue allows a plant to keep on growing its entire life span. Also, growth in plants serves another function. Whereas animals often respond to external stimuli by moving a body part, plant organs grow toward or away from stimuli, as we shall see in Chapter 23.

Root System   The growth of many roots is continuous, pausing only when temperatures become too cold or when water becomes too scarce. **Figure 21.6A**, a longitudinal section of a eudicot root, reveals zones where cells are in various stages of differentiation as primary growth occurs. These zones are called the zone of cell division, the zone of elongation, and the zone of maturation.

The **zone of cell division** is protected by the root cap, which is composed of parenchyma cells and protected by a slimy sheath. As the root grows, root cap cells are constantly replaced as they are removed by rough soil particles. The zone of cell division contains the root apical meristem, where mitosis produces relatively small, many-sided cells having dense cytoplasm and large nuclei. These meristematic cells give rise to the primary meristems, called protoderm, ground meristem, and procambium. Eventually, the primary meristems develop, respectively, into the three mature primary tissues discussed previously: epidermis; ground tissue (cortex and pith); and vascular tissue.

The **zone of elongation** is the region where the root increases in length due to elongation of cells. In the zone of elongation, the cells lengthen, but they are not yet fully specialized. This region also gets longer due to the addition of cells produced by the zone of cell division.

The **zone of maturation** is the region that does contain fully differentiated cells. This region is recognizable because many of the epidermal cells bear root hairs. Here, also, the tissues of a

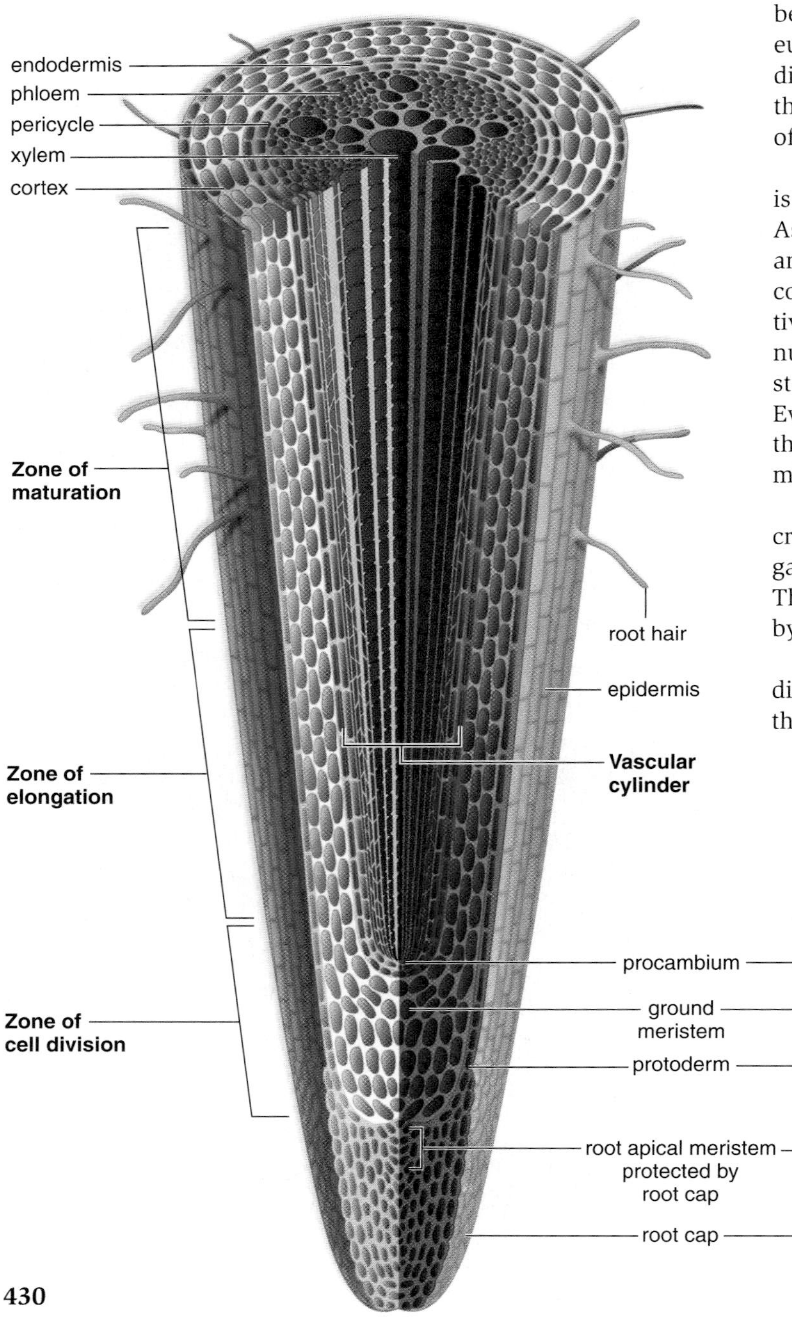

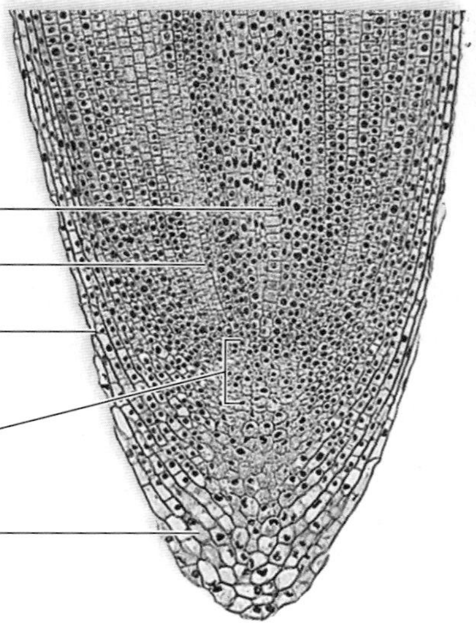

FIGURE 21.6A
Cells within a eudicot root tip.

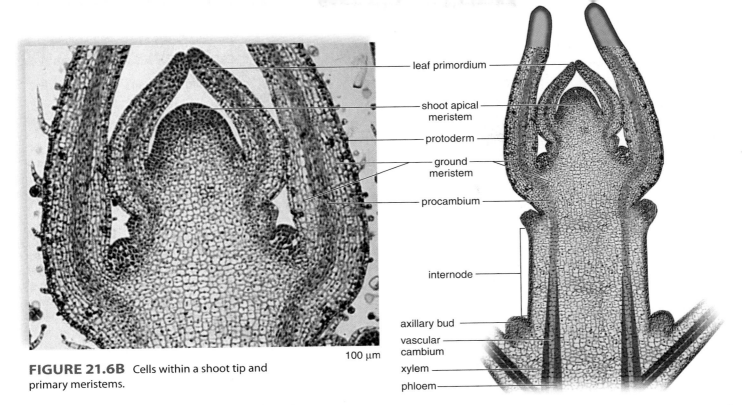

leaf primordium

shoot apical meristem

protoderm

ground meristem

procambium

internode

axillary bud

vascular cambium

xylem

phloem

FIGURE 21.6B Cells within a shoot tip and primary meristems.

100 µm

root—the epidermis, the cortex, and the vascular cylinder—can be readily distinguished from one another.

Shoot System The terminal bud includes the shoot apical meristem and also leaf primordia (young leaves) that differentiate from cells produced by the shoot apical meristem (**Fig. 21.6B**). The terminal bud doesn't have a protective covering comparable to the root cap. Instead, the leaf primordia fold over the apical meristem, providing protection. At the start of the season, the leaf primordia are, in turn, covered by terminal bud scales (scalelike leaves), but these drop off, or abscise, as growth continues.

It is not possible to make out in a growing shoot the same zones of growth seen in a root tip. The shoot apical meristem sometimes produces leaf primordia with such rapidity that even nodes and internodes cannot at first be distinguished. The stem increases in length as the internodes increase in length, and several internodes can increase at once.

The shoot apical meristem does give rise to the same primary meristems as in the root: protoderm, ground meristem, and procambium. These primary meristems, in turn, develop into the mature primary tissues of the plant body. In the stem, the protoderm becomes the epidermis of the stem and leaves. Ground meristem produces parenchyma cells that become the cortex and pith in the stem and mesophyll in the leaves. Procambium becomes vascular bundles in the stem and leaf veins in leaves.

The procambium differentiates into the first xylem cells, called primary xylem, and the first phloem cells, called primary phloem. Differentiation continues as certain cells become the first tracheids or vessel elements of the xylem within a vascular bundle. The first sieve-tube members of a vascular bundle do not have companion cells and are short-lived (some live only a day before being replaced). Mature vascular bundles contain fully differentiated xylem, phloem, and a lateral meristem called

vascular cambium, which is responsible for secondary growth. Vascular cambium is discussed more fully in Section 21.7.

Woody Twig The anatomy of a woody twig reviews for us the organization of a stem (**Fig. 21.6C**). The terminal bud contains the apical meristem and leaf primordia of the shoot tip protected by terminal bud scales, which are modified leaves. When leaves and flowers abscise, *leaf scars* and vascular *bundle scars* mark the spot of **abscission** (dropoff). Dormant axillary buds in this region can give rise to branches or flowers. Each spring, when growth resumes, terminal bud scales fall off and leave a scar that resembles a compact series of concentric circles. You can tell the age of a stem by counting these **terminal bud scale scars** because there is one for each year's growth.

The next section discusses the secondary growth of plants.

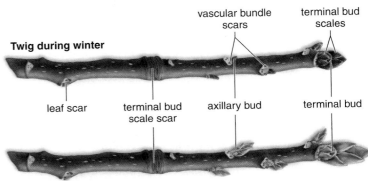

vascular bundle scars

terminal bud scales

Twig during winter

leaf scar

terminal bud scale scar

axillary bud

terminal bud

Twig during spring

FIGURE 21.6C Woody twig showing stem organization.

> **21.6 Check Your Progress** When examining a woody twig, what would a longer distance between two terminal bud scale scars indicate about environmental conditions that growing season?

Primary growth occurs in all plants and increases the length of a plant. **Secondary growth** occurs only in woody plants where it increases the girth of trunks, stems, branches, and roots. A woody plant, such as an oak tree, has both primary and secondary tissues. In a stem, primary tissues continue to form each year from primary meristems produced by the shoot apical meristem. Secondary tissues occur due to the growth of lateral meristems: vascular cambium and cork cambium.

As **Figure 21.7A** shows, secondary growth begins because of a change in the activity of vascular cambium. ❶ In herbaceous plants, vascular cambium is present between the xylem and phloem of each vascular bundle. In woody plants, the vascular cambium develops to form a ring of meristem that divides parallel to the surface of the plant, and produces ❷ secondary (new) xylem and phloem each year. Eventually, a woody eudicot stem has an entirely different organization from that of a herbaceous eudicot stem. ❸ A woody stem has three distinct areas: the bark, the wood, and the pith. Vascular cambium occurs between

the bark and the wood. Wood is actually secondary xylem that builds up year after year.

Figure 21.7A also shows the xylem rays and phloem rays in the cross section of a woody stem. Rays consist of parenchyma cells that permit lateral conduction of substances from the pith to the cortex and some storage of food. Rays are one to several cells thick and can be hundreds of cells in height. A phloem ray is a continuation of a xylem ray. Therefore, some phloem rays are much broader than other phloem rays.

Bark The **bark** of a tree contains periderm and phloem. The **periderm** is a secondary growth tissue that contains cork and cork cambium. Secondary phloem does not build up as xylem does, and only that year's phloem is found in the bark. The bark of a tree should not be removed because, without phloem, sugar cannot be transported. Removal of a ring of bark from a tree can be lethal to the tree. Bark has historically been used to make paper and bark cloth in some cultures. Additionally, bark is a source of medicines, as seems reasonable because it must possess chemicals that keep it from being eaten by herbivores. But since a tree cannot live without sufficient bark, one branch of research focuses on how much bark can be harvested without harming a tree.

Cork As discussed previously, **cork cambium** lies beneath the epidermis, but later it is part of the periderm, which replaces epidermis. Cork cambium divides and produces the cork cells that disrupt and replace the epidermis. Cork cells are impregnated with suberin, a waxy layer that makes them waterproof but also causes them to die. This is protective because it makes the stem less edible. The impermeable cork may be interrupted by **lenticels**, which are pockets of loosely arranged cork cells not impregnated with suberin. Gas exchange for processes such as photosynthesis can be accomplished at the lenticels after stems undergo secondary growth, and the epidermis is replaced by periderm.

Wood Secondary xylem that builds up year after year and increases the girth of trees is called **wood**. In trees that have a growing season, vascular cambium is dormant during the winter. In the spring, when moisture is plentiful and leaves require much water for growth, the secondary xylem contains wide vessel elements with thin walls. In this so-called *spring wood*, wide vessels transport sufficient water to the growing leaves. Later in the season, moisture is scarce, and the wood at this time becomes *summer wood*. Summer wood contains a lower proportion of vessels and a higher number of tracheids. Tracheids, with their thicker walls and smaller diameters, are less susceptible to water scarcity than the larger, less rigid vessel elements. At the end of the growing season, just before the cambium becomes dormant again, only heavy fibers containing sclerenchyma cells with especially thick secondary walls may develop.

When the trunk of a tree has spring wood followed by summer wood, the two together make up one year's growth, or an **annual ring**. You can tell the age of a tree by counting the annual rings. The outer annual rings, where transport occurs, are called *sapwood*; in older trees, the inner annual rings are called *heartwood* (**Fig. 21.7B**). Heartwood no longer functions in water transport. The

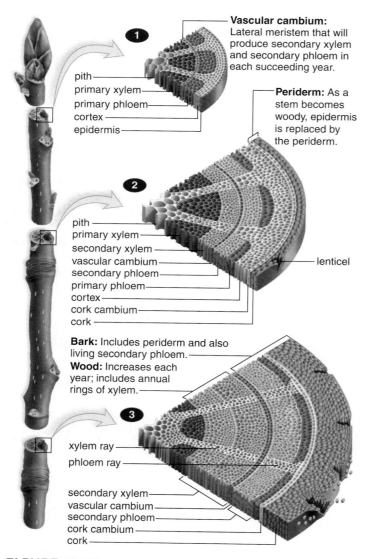

Vascular cambium: Lateral meristem that will produce secondary xylem and secondary phloem in each succeeding year.

pith
primary xylem
primary phloem
cortex
epidermis

Periderm: As a stem becomes woody, epidermis is replaced by the periderm.

pith
primary xylem
secondary xylem
vascular cambium
secondary phloem
primary phloem
cortex
cork cambium
cork

lenticel

Bark: Includes periderm and also living secondary phloem.
Wood: Increases each year; includes annual rings of xylem.

xylem ray
phloem ray

secondary xylem
vascular cambium
secondary phloem
cork cambium
cork

FIGURE 21.7A Diagrams of secondary growth of a stem.

cells become plugged with deposits, such as resins, gums, and other substances that inhibit the growth of bacteria and fungi. Heartwood may help support a tree, although some trees stand erect and live for many years after the heartwood has rotted away.

The annual rings are not only important in telling the age of a tree, but they can also serve as a historical record of environmental conditions at the time of growth. For example, if rainfall and other conditions were extremely favorable during a season, the annual ring may be wider than usual. If the tree was shaded on one side by another tree or building, the rings may be wider on the sunny side.

Human beings have found many uses for wood, from making houses to baseball bats, as discussed in Section 21.8.

> **21.7** *Check Your Progress* Will distinctive annual rings be visible if a tree experiences a fairly uniform amount of rainfall and temperature the entire year?

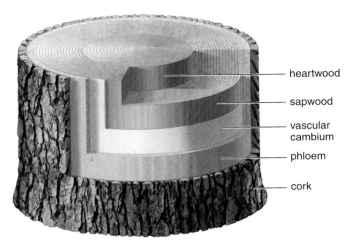

FIGURE 21.7B Layers of a woody stem, longitudinal view.

heartwood
sapwood
vascular cambium
phloem
cork

HOW BIOLOGY IMPACTS OUR LIVES

21.8 Wood has been a part of human history

Wood and people have had a long history and an intimate relationship. From newspapers, to building materials, to toys for children, wood has proven to be a durable, malleable, and indispensable part of our past and present (**Fig. 21.8**).

Wood pulp (a crushed mixture of cellulose, fiber cells, tracheids, and vessels) is the source of modern-day paper. Today, machines have made the papermaking process easier, faster, and less expensive than when the Chinese first processed paper by hand about A.D. 100.

Archaeological evidence suggests the first spears, used as far back as 400,000 years ago, were formed from one piece of wood, with the shaft fashioned into a point on one end. Over time, rocks, animal teeth, and eventually metal were used for the points of spears; however, the wood shaft remained. Used for

hunting various game animals and doing battle, these wooden spears were as diverse as the cultures that used them. In Hawaii, warriors used spears up to 15 feet long!

Since baseball began in the early 19th century, bats have been made of wood. Today, although many amateur baseball games are played with aluminum bats, the major leagues continue to use wooden bats. And while "the Babe" favored bats made from hickory trees, today's ball players use bats made from American ash. Children will play with any kind of wooden blocks.

This completes our discussion of plant growth and the wood of trees.

> **21.8** *Check Your Progress* What type of plant cells account for the strength of the wood in a baseball bat?

FIGURE 21.8 Humans use wood and wood products in daily life.

Leaves have an entirely different structure from stems and roots. Their structure is consistent with their function of carrying on photosynthesis.

21.9 Leaves are organized to carry on photosynthesis

Figure 21.9 shows a cross section of a generalized leaf of a temperate-zone eudicot plant. (Leaves from plants of arid or aquatic regions have modifications associated with their environment.) The top and bottom layers of epidermal tissue often bear protective hairs, sometimes modified as glands, that secrete irritating substances. These features may prevent the leaf from being eaten by insects. The epidermis characteristically secretes an outer, waxy cuticle that helps keep the leaf from drying out. The cuticle also prevents gas exchange because it is not gas permeable. However, the lower epidermis of eudicot leaves and both surfaces of monocot leaves contain stomata that allow gases to move into and out of the leaf. Photosynthesizing leaves take in CO_2 and give off O_2. Water loss also occurs at stomata, and each stoma has two guard cells that regulate its opening and closing. Stomata close when the weather is hot and dry.

The body of a leaf is composed of parenchyma cells that make up mesophyll tissue. Most eudicot leaves have two distinct regions: **palisade mesophyll,** containing elongated cells, and **spongy mesophyll,** containing irregular cells bounded by air spaces. The parenchyma cells of these layers have many chloroplasts and carry on most of the photosynthesis for the plant. The loosely packed arrangement of the cells in spongy mesophyll increases the amount of surface area for gas exchange and water loss.

Leaf veins bring water and minerals to the leaves from the stem and take the products of photosynthesis to the stem for distribution to other parts of the plant, where they fuel growth and repair or are stored until needed later. **Bundle sheaths** are layers of cells surrounding vascular tissue. Most bundle sheaths are parenchyma cells, sclerenchyma cells, or a combination of the two. The parenchyma cells help regulate the entrance and exit of materials into and out of the leaf vein.

This completes our study of leaves. In the next part of the chapter, we examine the phenomenon of homeostasis in plants.

> **21.9 *Check Your Progress*** Relate the horizontal orientation of eudicot leaves to the location of most stomata in the lower epidermis.

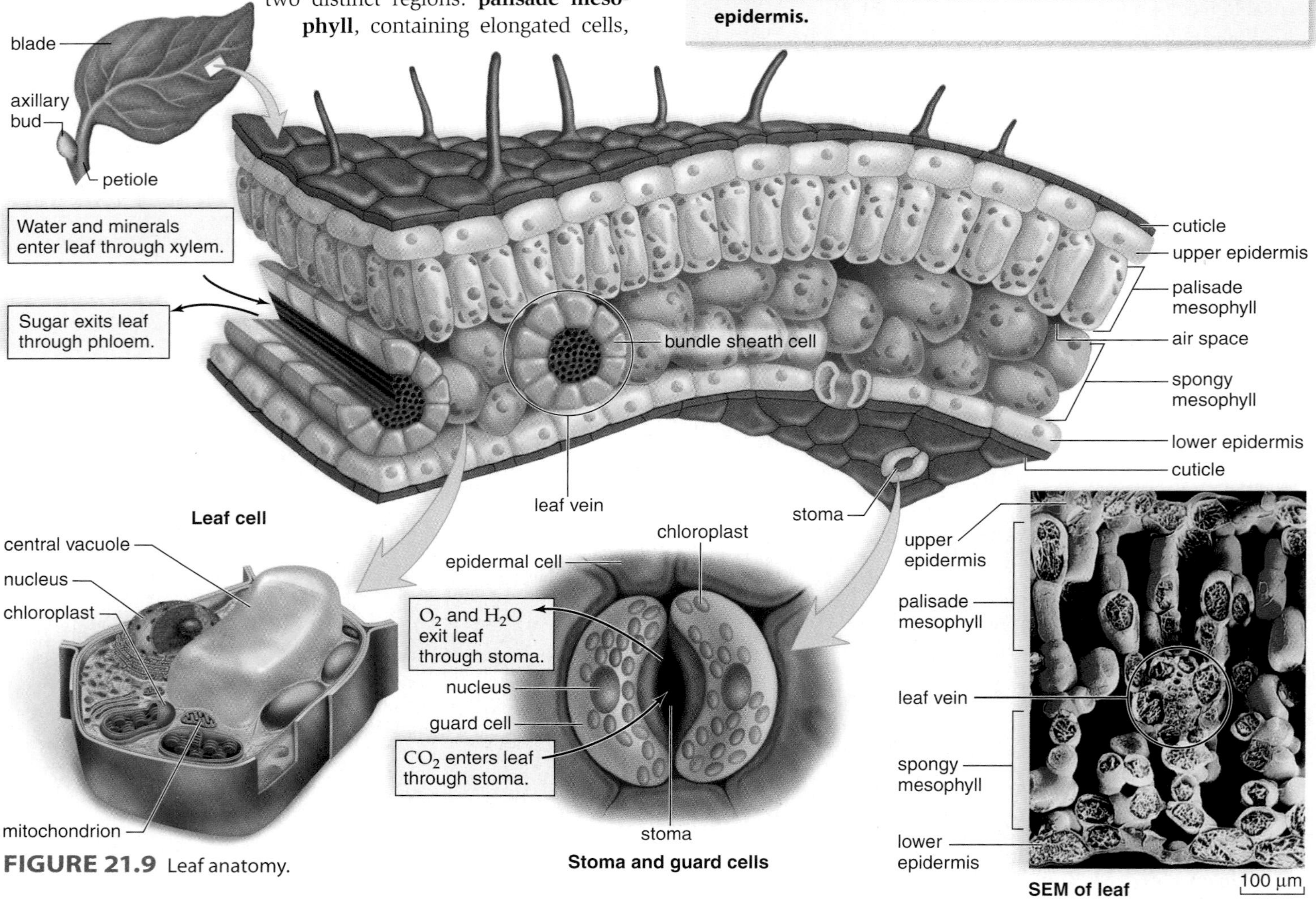

FIGURE 21.9 Leaf anatomy.

Water and minerals enter leaf through xylem.

Sugar exits leaf through phloem.

Leaf cell

central vacuole
nucleus
chloroplast
mitochondrion

blade
axillary bud
petiole

bundle sheath cell

leaf vein

epidermal cell
chloroplast
O_2 and H_2O exit leaf through stoma.
nucleus
guard cell
CO_2 enters leaf through stoma.
stoma

Stoma and guard cells

stoma

cuticle
upper epidermis
palisade mesophyll
air space
spongy mesophyll
lower epidermis
cuticle

upper epidermis
palisade mesophyll
leaf vein
spongy mesophyll
lower epidermis

SEM of leaf 100 µm

Plants, like other living things, have an organization that fosters **homeostasis**, the relative constancy of the internal environment. Regulation is usually required to keep internal conditions within tolerable limits, and in this section, we present evidence that plants do regulate their internal environment. As further evidence of a plant's ability to maintain homeostasis, we also preview some other homeostatic mechanisms plants use.

21.10   The organization of plants fosters homeostasis

The organization of plants allows photosynthesis to occur (**Fig. 21.10**). Only if leaf cells receive, by way of vascular tissue, the water and minerals absorbed by the roots and also CO_2, taken in at the stomata, will they be able to photosynthesize. What else do leaf cells need in order to carry on photosynthesis? Solar energy, of course. The structure of a stem also allows it to lift the leaves so they can absorb solar energy. Notice in Figure 21.10 how the palisade cells are arranged beneath the broad expanse of the epidermis, maximizing the number of cells exposed to solar energy.

Photosynthesis allows a plant to produce the building blocks and the ATP needed to maintain its structure and metabolism and, therefore, also ho-

meostasis. Nonphotosynthesizing cells receive sucrose by way of phloem, and thereafter, they too can produce the molecules needed to maintain homeostasis.

In our study of plant organization, we also observed that the epidermis in nonwoody plants and the cork in woody plants plays a role in homeostasis by protecting the interior of the plant from harm. Then, too, epidermal projections, including thorns and hairs, can discourage predators from feeding on plants.

Other plant homeostatic mechanisms are given in Section 21.11.

> **21.10** *Check Your Progress* **What type of leaf would be most advantageous for a plant exposed to limited light?**

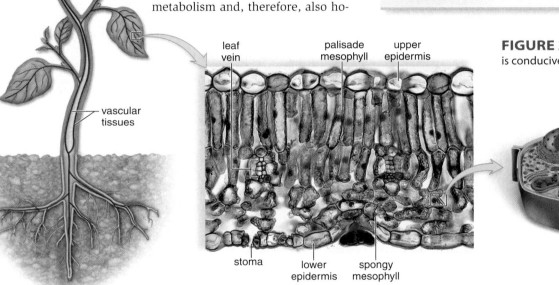
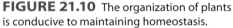

leaf vein · palisade mesophyll · upper epidermis · vascular tissues · stoma · lower epidermis · spongy mesophyll

FIGURE 21.10 The organization of plants is conducive to maintaining homeostasis.

21.11   Regulatory and other mechanisms help plants maintain homeostasis

This section gives various other examples of how plants maintain homeostasis.

Closing of Stomata A nonwoody plant remains upright only when water is available to maintain turgor pressure within its cells. Without adequate water, the plant first becomes limp and then dies. A plant has two ways of preventing water loss at the leaves: the cuticle, which has a waxy surface that keeps leaves from drying out, and the stomata, which have the ability to close. When a plant is water stressed, a hormone is released that causes the guard cells to change shape and close the stomata (**Fig. 21.11A**). This action conserves water, but how does closure of stomata affect photosynthesis? When

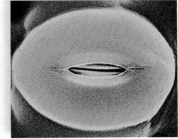

Stoma open · 25 μm · Stoma closed · 25 μm

FIGURE 21.11A Stomata open (*left*) and close (*right*) according to water availability.

stomata are closed, CO_2 cannot enter the leaf, and therefore photosynthesis may be put on hold. True, mitochondria give off CO_2, but without a source of oxygen from either chloroplasts or via the stomata, cellular respiration is also on hold. It would appear, then, that prevention of dehydration is of primary importance to plants, most likely because dehydration is equivalent to death.

Phloem Transport We have mentioned that phloem transports photosynthetic products about the body of a plant. Because the direction of flow can be from any source to any sink, distribution of sugar throughout a plant is possible (**Fig. 21.11B**, *upper left*). In the summer, the *source* is always mature leaves because they are producing sugar. Otherwise, the *sink* can be new leaves that have not started photosynthesizing yet or flowers and roots that lack chloroplasts and do not photosynthesize. Therefore, in the spring, if there are no mature photosynthesizing leaves, the stored carbohydrates of roots can be a source of sugars for the rest of the plant. In this way, mitochondria can carry on cellular respiration, giving plants the energy needed to remain alive and functioning.

Plant Hormones In animals, the nervous system and the endocrine system help coordinate responses to environmental stimuli. Plants rely exclusively on **hormones**, highly specific chemical signals between plant parts and cells, to respond to stimuli. Depending on the hormone, it may be transported through the xylem, the phloem, or from cell to cell. Auxins are a class of hormones, some of which are responsible for tropic responses. A **tropism** is a growth response toward or away from a particular stimulus. Among other responses, plants are able to bend toward the light and, in this way, better acquire solar energy for photosynthesis (Fig. 21.11B, *upper right*). When plants are trapped in a dark environment, as beneath a wooden porch, their shoot tips will emerge between the boards to reach the light. The capture of solar energy and photosynthesis, as we have established, is absolutely necessary to the life of most plants.

Defense Mechanisms We have already mentioned that leaf cells are covered by a thick and impervious cuticle and that epidermal projections, including thorns and hairs, can discourage hungry insects. If these defense mechanisms fail, many plants rely on chemical toxins—alkaloids such as morphine, quinine, taxol, and caffeine—as the next level of defense. If an insect injures a plant, a wound response causes the release of signaling molecules that travel throughout the plant. Subsequently, plant cells produce inhibitors of the pest's digestive enzymes. Or, consider the response of plants

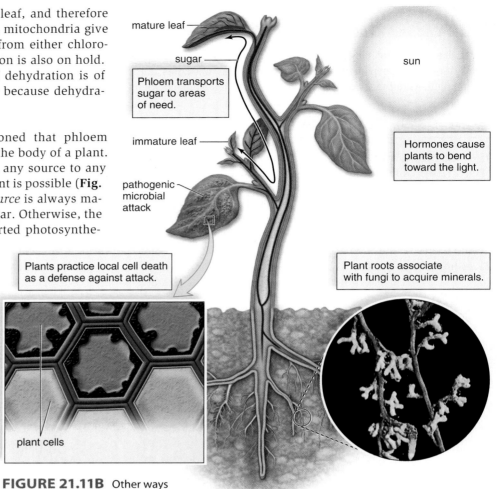

FIGURE 21.11B Other ways a plant can maintain homeostasis.

to pathogens, such as viruses. The tobacco mosaic virus, despite its name, attacks tomato, eggplant, pepper, and spinach plants, among others. A virus causes a hypersensitive response involving local cell death that seals off the pathogen (Fig. 21.11B, *lower left*). Now, homeostasis can continue in the rest of a plant's body.

Mutualistic Relationships Mutualistic symbiotic relationships permeate the kingdoms of life. In a mutualistic relationship, both organisms benefit. The roots of plants have a mutualistic relationship with fungi, which function as extensions of the root system (Fig. 21.11B, *lower right*). The fungi increase the surface area by which the roots absorb water and minerals from the soil. In turn, the plant supplies the fungus with food in the form of carbohydrates. Only if plant cells receive adequate amounts of minerals is homeostasis possible. For example, a plant needs phosphorus in order to produce adenosine triphosphate (ATP), the energy currency of cells.

> **21.11 Check Your Progress** Homeostasis is characteristic of all life, whether a single cell or an organism. Explain.

CONNECTING THE CONCEPTS

In Chapter 18, we saw how plants became adapted to reproducing on land. In an aquatic environment, drying out is less of a danger. Many other types of adaptations were also required. Because even humid air is drier than a living cell, the prevention of water loss is critical for plants. The epidermis and the cuticle it produces help prevent water loss and overheating in sunlight. Gas exchange in leaves depends on the presence of stomata, which close when a plant is water-stressed. The cork of woody plants is especially protective against water loss, but when cork is interrupted by lenticels, gas exchange is still possible.

In an aquatic environment, water buoys up organisms and keeps them afloat, but on land plants had to evolve a way to oppose the force of gravity. The stems of plants contain strong-walled sclerenchyma cells, tracheids, and vessel elements. The accumulation of secondary xylem allows a tree to grow in diameter and offers more support.

In an aquatic environment, water is available to all cells, but on land it is adaptive to have a means of water uptake and transport. In plants, the roots absorb water and have special extensions called root hairs that facilitate water uptake. Xylem transports water to all plant parts, including the leaves. In Chapter 22 we will see how the drying effect of air allows water to move from the roots to the leaves. Roots are buried in soil, where they can absorb water but cannot photosynthesize. In that chapter, we will also see how the properties of water allow phloem to transport sugars from the leaves to the roots and to any other plant part in need of sustenance. We will then discuss the inorganic nutrient needs of plants, which consist of substances they garner from their environment.

The Chapter in Review

Summary

What Do Forests Have to Do with Global Warming?

- Trees in tropical rain forests take up CO_2, but rain forests are being lost to clearing and developing.
- Clearing often involves burning, which adds large amounts of CO_2 to the atmosphere.

Plants Have Three Vegetative Organs

21.1 Flowering plants typically have roots, stems, and leaves

- A flowering plant has a root system and a shoot system; both grow at their tips.
- Stems support leaves, conduct materials to and from roots and leaves, and can help store water or plant products.
- Leaves carry on photosynthesis.
- Roots anchor a plant, absorb water and minerals, and can store the products of photosynthesis.

21.2 Flowering plants are either monocots or eudicots

- Monocots (e.g., grasses, lilies) have one cotyledon (seed leaf).
- Eudicots (e.g., dandelions, oak trees) have two cotyledons.
- Monocots and eudicots have structural differences in their roots, stems, leaf veins, and number of flowering parts.

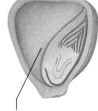

one cotyledon

two cotyledons

21.3 Monocots serve humans well

- Monocots are important as cereal grains and floriculture. Bamboo is used for housing, furniture, and flooring.

The Same Plant Cells and Tissues Are Found in All Plant Organs

21.4 Plants have specialized cells and tissues

- Apical meristem divides, producing epidermal, ground, and vascular tissue.
- Epidermal tissue contains epidermal cells (protected by a cuticle) and stomata.
- Ground tissue contains parenchyma, collenchyma, and sclerenchyma cells.
- Vascular tissue is composed of xylem (transports water and minerals from roots to leaves) and phloem (transports sugar from leaves to roots).
  - Xylem contains vessel elements and tracheids.
  - Phloem contains sieve-tube members and companion cells.

21.5 The three types of plant tissues are found in each organ

- A leaf has an upper and lower epidermis (with stomata), mesophyll composed mostly of parenchyma, and leaf veins.
- The stem of a nonwoody plant has epidermis, ground tissue in cortex and pith, and vascular bundles that are either scattered (monocot) or in a ring (eudicot).
- An eudicot root has an outer layer of epidermis, ground tissue in the cortex, and a vascular cylinder.

Plant Growth Is Either Primary or Secondary

21.6 Primary growth lengthens the root and shoot systems

- In a root, apical meristem produces a zone of cell division, a zone of elongation, and a zone of maturation.
- In a shoot system, apical meristem gives rise to new leaves, at nodes and internodes, and specialized tissues.

21.7 Secondary growth widens roots and stems

- Secondary growth occurs in woody plants.
- Secondary tissues develop from lateral meristems.
- Bark contains periderm and phloem.

- Cork arises from cork cambium and replaces epidermis; cork cells are waterproof.
- Wood is secondary xylem that builds up year after year; annual rings can tell the age of a tree and also the history of its environment.

21.8 Wood has been a part of human history

- Wood has been used for 400,000 years for making tools, weapons, building materials, and paper.

Leaf Anatomy Facilitates Photosynthesis

21.9 Leaves are organized to carry on photosynthesis

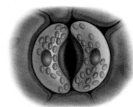
Stoma and guard cells

- Stomata allow water vapor and oxygen to escape and carbon dioxide to enter the leaf.
- Mesophyll (palisade and spongy) forms the body of a leaf and carries or photosynthesis.

Plants Maintain Internal Equilibrium

21.10 The organization of plants fosters homeostasis

- Homeostasis is the relative constancy of the internal environment of organisms.
- Plant organ systems assist homeostasis in plants by supplying leaf cells with the materials they need for photosynthesis.
- Photosynthesis allows a plant to produce the building blocks and the ATP needed to maintain its structure and metabolism, and therefore, also homeostasis.

21.11 Regulatory and other mechanisms help plants maintain homeostasis

- Stomata open and close according to water availability; they close to prevent water loss and, therefore, death of the plant.
- Phloem has the ability to transport sugar between source and sink.
- Plant hormones, such as auxin, cause tropic responses.
- Plant defense mechanisms include a waxy cuticle, thorns, toxins, and local death.
- Fungi on roots increase the root system of a plant, while the plant provides food for the fungus.

Testing Yourself

Plants Have Three Vegetative Organs

1. It is possible to distinguish between a stem and a root by looking for
   a. petioles on stems.   c. nodes on stems.
   b. petioles on roots.   d. nodes on roots.
2. Roots
   a. are the primary site of photosynthesis.
   b. give rise to new leaves and flowers.
   c. have a thick cuticle to protect the epidermis.
   d. absorb water and minerals.
   e. contain spores.
3. Which of these is an incorrect contrast between monocots (stated first) and eudicots (stated second)?
   a. one cotyledon—two cotyledons
   b. leaf veins parallel—net-veined leaves
   c. vascular bundles in a ring—vascular bundles scattered
   d. flower parts in threes—flower parts in fours or fives
   e. All of these are correct.

4. Monocots are the smaller group compared to eudicots but are of extreme importance because all the cereal grains are monocots.
   a. true
   b. false
5. Give the function of apical meristem and axillary buds.
6. What is a cereal grain?

The Same Plant Cells and Tissues Are Found in All Plant Organs

7. Sclerenchyma, parenchyma, and collenchyma are cells found in what type of plant tissue?
   a. epidermal   d. meristem
   b. ground   e. All but d are correct.
   c. vascular
8. Which of these cells in a plant is apt to be nonliving?
   a. parenchyma   d. epidermal
   b. collenchyma   e. guard cells
   c. sclerenchyma

In questions 9–13, match the function to the cell types in the key.

KEY:
   a. meristem   d. parenchyma
   b. sclerenchyma   e. epidermal
   c. vessel element

9. transport
10. support
11. cell division
12. photosynthesis or storage
13. protection

In questions 14–18, match the tissue to the organ. Answers can be used more than once.

KEY:
   a. root   c. leaf
   b. stem   d. All of these are correct.

14. mesophyll tissue
15. vascular tissue
16. endodermis
17. epidermis
18. pith
19. Which of these does not distinguish between xylem and phloem? A characteristic of xylem is mentioned first.
   a. vessel elements—sieve-tube members
   b. companion cell—sclerenchyma cell
   c. dead cells—cells contain cytoplasm
   d. transports water—transports sugar
20. Cortex is found in
   a. roots, stems, and leaves.   d. stems and leaves.
   b. roots and stems.   e. roots only.
   c. roots and leaves.
21. What is vascular tissue called in the root, stem, and leaf?

Plant Growth Is Either Primary or Secondary

22. New plant cells originate from
   a. the parenchyma.   d. the base of the shoot.
   b. the collenchyma.   e. the apical meristem.
   c. the sclerenchyma.
23. Root hairs are found in the zone of
   a. cell division.   d. apical meristem.
   b. elongation.   e. All of these are correct.
   c. maturation.

24. During secondary growth, a tree adds more xylem and phloem through the activity of the
    a. apical meristems.   c. intercalary meristems.
    b. vascular cambium.   d. cork cambium.
25. Between the bark and the phloem in a woody stem, there is a layer of meristem called
    a. cork cambium.   d. the zone of cell division.
    b. vascular cambium.   e. procambium preceding bark.
    c. apical meristem.
26. Bark is made of which of the following?
    a. phloem   d. cork cambium
    b. cork   e. All of these are correct.
    c. cortex
27. Annual rings are the number of
    a. internodes in a stem.
    b. rings of vascular bundles in a monocot stem.
    c. layers of xylem in a stem.
    d. bark layers in a woody stem.
    e. Both b and c are correct.
28. What part of a woody twig allows you to determine the amount of growth last year?
    a. leaf scar   c. vascular bundle scar
    b. terminal bud   d. axillary bud
29. Contrast primary growth with secondary growth by naming the tissue responsible for and the results of the growth.

Leaf Anatomy Facilitates Photosynthesis

30. Name a benefit to the plant for
    a. a flat thin blade.
    b. an epidermis that has closely packed cells covered by cuticle.
    c. palisade mesophyll containing chloroplasts next to upper epidermis.
    d. spongy mesophyll that is loosely packed.
    e. stomata that can open.

Plants Maintain Internal Equilibrium

31. **THINKING CONCEPTUALLY** Why is photosynthesis necessary to homeostasis in plants?

In questions 32–36, match the method of regulation to the descriptions in the key.

KEY:
    a. prevent pathogen invasion
    b. regulate growth
    c. aid metabolism and growth
    d. regulate water loss
    e. distribution of sugar according to need
32. phloem transport
33. stomata
34. plant hormones
35. defense mechanisms
36. mutualistic relationships
37. You would expect a plant living in a desert to maintain homeostasis by having
    a. deep roots.
    b. thick cuticles with sunken stomata.
    c. narrow leaves.
    d. spongy mesophyll lacking open spaces.

Understanding the Terms

abscission 431	perennial plant 423
annual ring 432	pericycle 429
apical meristem 426	periderm 432
axillary bud 422	petiole 423
bark 432	phloem 427
blade 423	pith 428
bundle sheath 434	plasmodesmata 427
collenchyma cell 426	root 423
cork cambium 432	root hair 423
cortex 428	root system 422
cotyledon 424	sclerenchyma cell 427
cuticle 426	secondary growth 432
deciduous 423	shoot system 422
endodermis 428	sieve-tube member 427
epidermal tissue 426	spongy mesophyll 434
epidermis 426	stem 422
eudicot 424	stomata 426
ground tissue 426	terminal bud 422
herbaceous 428	terminal bud scale scar 431
homeostasis 435	tracheid 427
hormone 436	tropism 436
internode 422	vascular cylinder 427
leaf 422	vascular tissue 426
lenticel 432	vessel element 427
meristem 430	wood 432
mesophyll 428	xylem 427
monocot 424	zone of cell division 430
node 422	zone of elongation 430
palisade mesophyll 434	zone of maturation 430
parenchyma cell 426	

Match the terms to these definitions:
   a. _____ Inner, thickest layer of a leaf; the site of most photosynthesis.
   b. _____ Seed leaf for embryonic plants; provides nutrient molecules before the leaves begin to photosynthesize.
   c. _____ Vascular tissue that contains vessel elements and tracheids.
   d. _____ Plant tissue forming a boundary between the cortex and the vascular cylinder in a root.
   e. _____ Member that joins with others in the phloem tissue of plants as a means of transport for nutrient sap.

Thinking Scientifically

1. Utilizing an electron microscope, how might you confirm structurally and biochemically that a companion cell communicates with its sieve-tube member? See Section 21.4 and Section 4.10.
2. For your senior project, you decide to make microscopic tissue slides confirming that root tips do have the three zones shown in Figure 21.6A. After growing many seedlings, what would you like your root tip slides to show?

ARIS™ *Visit www.mhhe.com/maderconcepts for practice quizzes, animations, videos, and activities designed to help you master the material in this chapter.*

22

Transport and Nutrition in Plants

LEARNING OUTCOMES

After studying this chapter, you should be able to accomplish the following outcomes.

Plants Can Adapt Too

1 Discuss the ways a plant would adapt to transporting water in a dry environment.

Plants Are Organized to Transport Water and Solutes

2 Describe the continuous pipelines for water transport and sugar transport in plants.

Xylem Transport Depends on the Properties of Water

3 Describe how roots absorb water and minerals.
4 Describe the cohesion-tension model of xylem transport and why most of the water that enters a plant typically evaporates at the leaves.
5 Show that water transport depends on the properties of water.
6 Describe how environmental factors influence the opening and closing of stomata.
7 Relate the environmental cleanup capability of plants to their ability to take up and concentrate minerals.

Phloem Function Depends on Membrane Transport

8 Describe the pressure-flow model of phloem transport.
9 Show that phloem transport depends on membrane transport.

Plants Require Good Nutrition and Therefore Good Soil

10 Name several macronutrients and micronutrients of plants.
11 Describe hydroponics and its use to determine essential nutrients of plants.
12 Describe how minerals cross the plasma membrane of cells.
13 Draw and explain a simplified soil profile.
14 Describe a simplified drawing showing how roots absorb positively charged mineral ions while negatively charged ions may be leached away.
15 Describe the mutualistic relationships that assist plants in acquiring nutrients from the soil.

Remember in Chapter 14 the many species of Darwin finches that evolved on the Galápagos Islands by adaptive radiation? Perhaps only a single finch species from the mainland made it to the islands, and then this species spread to the other islands of the archipelago. Today, there are many species of finches on the Galápagos Islands because the populations on the various islands were subjected to the founder effect and the process of natural selection. Due to natural selection, each isolated finch population became adapted in its own way to a particular island. Adaptation caused the size and shape of the beaks to vary, from strong and thick for cracking nuts and seeds to narrow and thin for feeding on insects.

Can adaptive radiation happen in plants too? Indeed it can, and one of the most spectacular examples occurs among 28 species of the silversword alliance (an alliance is an assemblage of closely related species). DNA analysis shows that all these species evolved from a single ancestor that migrated to the Hawaiian Islands from Baja California. After arriving, members of the ancestral species spread to the other islands, where they evolved into the many species of the silversword alliance. The lack of competitors and predators allowed the species of the alliance to evolve in any direction that was advantageous.

Dubautia latifolia competes for light.

Plants Can Adapt Too

Many species of the silversword alliance belong to the genus *Dubautia*, commonly called dubautia, which are featured on these pages. The species grow in a range of habitats that includes bogs, moist-to-wet forests, dry tree or shrub woodlands, and lava fields. The amount of annual rainfall varies from more than 1,230 cm (the wettest on Earth) to 40 cm. Like finches, the species have evolved structural differences—in this case, in their overall growth pattern and their leaf structure

Members of the alliance found in moist-to-wet forest habitats have modifications that show their need to compete for light. Either they have increased height (*D. knudsenii* and *D. waialealae*) or are vines (*D. latifolia*), while their leaves are thin with a comparatively large surface area.

Members of the alliance that inhibit open, more arid sites typically have features associated with conservation of water, such as reduced height and thick leaves with compact internal tissues. The surface area is reduced. The mat-forming clumps of the species *D. scabra* usually live on the lava fields of the largest island, Hawaii. Larger, shrubby species (*D. linearis, D. ciliolata,* and *D. menziesii*) are likely to be found in the dry areas of Hawaii and Maui.

Compare the microscopic cross sections of the two leaves below. In the leaf cross section on the left, note the loose organization of tissues, the thin cuticle, and the presence of thin-walled and open vessel elements in the xylem. Also, the stomata open into spacious cavities of spongy mesophyll. In the leaf cross section on the right, note the thick cuticle and the many thick-walled tracheids in the xylem. The spongy mesophyll lacks any spacious cavities. You know instantly which of these leaves is from a plant adapted to living in a moist environment and which is adapted to living in a dry environment.

In this chapter, you will learn how evaporation of water within the spongy mesophyll causes the movement of water in xylem from the leaves to the roots. At all costs, the water column within xylem must not break, or water transport will be compromised. In a dry environment, the presence of tracheids is an advantage because the thin water column in tracheids is less apt to break than the wide one in vessel elements. A thick cuticle prevents water loss. Fewer stomata and/or sunken stomata with hairs on the undersurface of leaves help regulate water evaporation so that the water column stays intact, ensuring that cells will continue to receive the water they need despite dry environmental conditions.

Dubautia scabra lives on lava fields.

Dubautia knudsenii   200 µm

Dubautia menziesii   200 µm

Plants Are Organized to Transport Water and Solutes

This part of the chapter reviews the structure and transport functions of xylem and phloem. Research has shown that the biodiversity of an ecosystem is in part related to the ability of plants to supply their cells with water and nutrients.

22.1   Transport begins in both the leaves and the roots of plants

Flowering plants are well adapted to living in a terrestrial environment and have a transport tissue, called **xylem**, that carries water and minerals from the roots to the leaves (**Fig. 22.1**). In addition to other types of cells, xylem contains two types of nonliving conducting cells: tracheids and vessel elements. **Tracheids** are tapered at both ends. These ends overlap with those of adjacent tracheids (see Fig. 21.4C). Pits located in adjacent tracheids allow water to pass from tracheid to tracheid. **Vessel elements** are long and tubular with perforation plates at each end. Vessel elements placed end to end form a completely hollow pipeline from the roots to the leaves. Xylem, with its strong-walled, nonliving sclerenchyma cells, gives trees much-needed internal support.

The process of photosynthesis results in sugars, which are used as a source of energy and building blocks for other organic molecules. **Phloem** is the type of vascular tissue that transports sugar to all parts of the plant. Roots buried in the soil cannot possibly carry on photosynthesis, but they still require a source of energy in order to carry on cellular activities. Vascular plants are able to transport the products of photosynthesis to regions that require them and/or that will store them for future use. Like xylem, phloem is also composed of several cell types. In flowering plants, the conducting cells of phloem are living **sieve-tube members**, each of which typically has a companion cell (see Fig. 21.4D). **Companion cells** can provide proteins to sieve-tube members, which contain cytoplasm but have no nucleus. The end walls of sieve-tube members are called sieve plates because they contain numerous pores. The sieve-tube members are aligned end to end, and with maturity the pores enlarge to allow for better flow.

Plant physiologists have performed numerous experiments to determine how water and minerals rise to the tops of very tall trees in xylem and how organic nutrients move from source to sink in phloem. Water is the main cellular component and constitutes a large part of **xylem sap** and **phloem sap**, as the water-based contents of xylem and phloem are called. We will see that transport in both xylem and phloem is dependent on the properties of water and the principles of osmosis.

As explained in Section 22.2, a study shows that the acquisition and transport of water and nutrients can affect biodiversity.

> **22.1 Check Your Progress** Vessel elements and tracheids are dead at functional maturity, but sieve-tube members are alive. Do you predict that transport of water or sugar requires a functional plasma membrane?

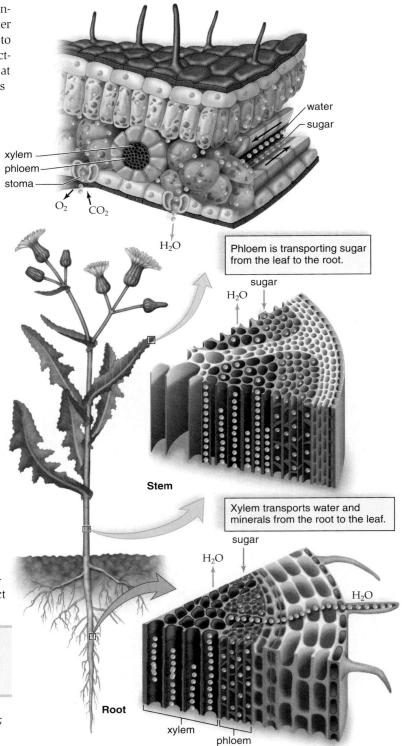

FIGURE 22.1 Plant transport system. (blue = phloem; pink = sugar; red = xylem; light blue = water)

Phloem is transporting sugar from the leaf to the root.

Xylem transports water and minerals from the root to the leaf.

From his earliest years in school, G. David Tilman of the University of Minnesota loved both mathematics and biology. In college, the biological issues that intrigued him most were concerned with the relationships of species to their environment, including the effects of interactions such as competition or predation. It seemed to him that biodiversity might relate to these interactions and, if so, that it might be possible to develop a mathematical theory to explain biodiversity.

G. David Tilman
University of Minnesota

The Minnesota grasslands in which Tilman works often harbor more than 100 plant species within an area of only a few hectares. Long-term experiments have shown that all these plant species are held in check (limited) by competition for the same resource, namely soil nitrogen. Mathematical models predict that the number of coexisting species can never be more than the number of resources that limit them. Tilman thought this must mean that the species are competing with one another on some other basis besides soil nitrogen.

To find out what this basis might be, Tilman and his colleagues performed a series of experiments. They found that the factor limiting these species was not the presence of insect or mammalian herbivores, light availability, or fire. Rather, the plant species differed in their ability to disperse to new sites. The researchers discovered this by planting over 50 different species in several plots and finding how well they could germinate, grow, and reproduce there. The best competitors for nitrogen were native bunchgrasses, which allocated 85% of their biomass to roots but only 0.5% of their biomass to seeds. Little bluestem (*Schizachyrium scoparium*) is an example of a bunchgrass (**Fig. 22.2A**). The best dispersers allocated 30% of their biomass to seeds, but only 40% of their biomass to roots. Bent grass (*Agrostis scabra*) is an example of a poor competitor for soil nitrogen but a good disperser (**Fig. 22.2B**).

A mathematical model showed that such allocation-based tradeoffs could explain the stable coexistence of a whole range of plant species that differ according to their abilities to compete for nitrogen and to disperse to new areas. Coexistence occurs because better competitors for soil nitrogen are poorer dispersers and therefore do not occupy all sites. Better dispersers are also better at finding and occupying all sites. Because their abilities to compete and disperse vary, a number of species can coexist.

Another issue of interest to Tilman is the relationship between biodiversity and the stability of an ecosystem. During a serious drought period in 1987–88, Tilman and his colleagues were annually sampling 207 permanent plots. They found that plots containing only one to four species had their productivity (total mass of living plants) fall to between $1/8$ and $1/16$ of the pre-drought level. But plots that contained 16 to 26 species were able to maintain their productivity at about $1/2$ the pre-drought level (**Fig. 22.2C**). This suggests that high biodiversity does buffer ecosystems against a disturbance and that it is wise to conserve the biodiversity of ecosystems in all areas—whether in Minnesota, New Jersey, Oregon, or the tropics.

In the next part of the chapter, we will study xylem transport of water and minerals.

22.2 Check Your Progress The researchers knew that biodiversity is restricted if species compete for the same limited resource, but they found that biodiversity increases if poor competitors can disperse to new locations. Explain why biodiversity increased under these circumstances.

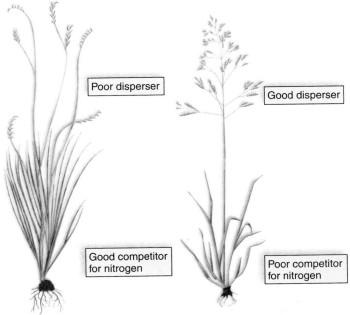

Poor disperser

Good disperser

Good competitor for nitrogen

Poor competitor for nitrogen

FIGURE 22.2A
Little bluestem.

FIGURE 22.2B
Bent grass.

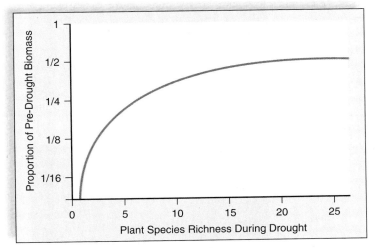

FIGURE 22.2C Biomass comparison.

Plants require a transport mechanism that can take water from the roots to even the tops of very tall trees. The cohesion-tension model relies on evaporation at the stomata and the chemical properties of water. Stomata close when plants become water stressed, but then carbon dioxide for photosynthesis cannot enter leaves. Absorption by roots offers the opportunity to use plants for ridding the soil of toxic mineral accumulations.

22.3 Water is pulled up in xylem by evaporation from leaves

As you know, the conducting cells in xylem are tracheids and vessel elements (**Fig. 22.3A**). Vessels constitute an open pipeline because the vessel elements have openings allowing flow from one to the other. The tracheids, which are elongated with tapered ends, form a less obvious means of transport, but water can move from one to the other because of pits, or depressions, where the secondary wall does not form.

How is it possible for water to rise to the top of a very tall plant? One contributing factor is **root pressure**. Water entering root cells creates a positive (internal) pressure compared to the water in the surrounding soil. Because root pressure primarily occurs at night, this water accumulation is more obvious during the early morning hours and may be confused with dew. Actually, however, it is **guttation**, during which drops of water are forced out of vein endings along the edges of leaves (**Fig. 22.3B**). This phenomenon is the result of root pressure. But, although root pressure may contribute to the upward movement of water in some instances, it is not believed to be the mechanism by which water can rise to the tops of very tall trees. For that explanation, we look to the cohesion-tension model.

Cohesion-Tension Model Once water enters xylem, it must be transported to all parts of the plant. Transporting water can be a daunting task, especially for some plants, such as redwood trees, which can exceed 90 m (almost 300 ft) in height.

The **cohesion-tension model** of xylem transport, outlined in **Figure 22.3C**, describes a mechanism for xylem transport that requires no expenditure of energy by the plant and is dependent on the properties of water. The term *cohesion* refers to the tendency of water molecules to cling together. Because of hydrogen bonding, water molecules interact with one another and form a continuous **water column** in xylem, from the leaves to the roots, that is not easily broken. In addition to cohesion, another property of water called *adhesion*, plays a role in xylem transport. Adhesion refers to the ability of water, a polar molecule, to interact with the molecules making up the walls of the vessels in xylem. Adhesion gives the water column extra strength and prevents it from slipping back.

FIGURE 22.3A Conducting cells of xylem.

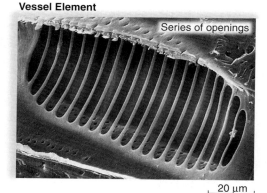

Vessel Element — Single, large opening — 20 μm

Vessel Element — Series of openings — 20 μm

Tracheids — pits — 50 μm

FIGURE 22.3B Guttation.

What Happens in the Leaf? Consider the structure of a leaf, as shown in Figure 22.3C. ➊ The stomata, while also found elsewhere, are most concentrated in the lower epidermis of a eudicot leaf, where they open directly into the spongy layer of the mesophyll. When the stomata are open, the cells of the spongy layer are exposed to the air, which can be quite dry. Water then evaporates as a gas or vapor from the spongy layer into the intercellular spaces. Evaporation of water through leaf stomata is called **transpiration**. At least 90% of the water taken up by the roots is eventually lost by transpiration. This means that the total amount of water lost by a plant over a long period of time is surprisingly large. A single *Zea mays* (corn) plant loses somewhere between 135 and 200 L of water through transpiration during a growing season. An average-sized birch tree with over 200,000 leaves will transpire up to 3,700 L of water per day during the growing season.

The water molecules that evaporate from cells into the intercellular spaces are replaced by other water molecules from the leaf veins. Because the water molecules are cohesive, transpiration exerts a *pulling force*, or *tension*, that draws the water column through the xylem to replace the water lost by leaf cells.

Note that the loss of water by transpiration is the mechanism by which minerals are transported throughout the plant body. Furthermore, the evaporation of water helps keep the temperature of leaf cells within normal limits through evaporative cooling.

There is an important consequence to the way water is transported in plants. When a plant is under water stress, the stomata close. Now the plant loses little water because the leaves are protected against water loss by the waxy **cuticle** of the upper and lower epidermis. When stomata are closed, however, carbon dioxide cannot enter the leaves, and many plants are unable to photosynthesize efficiently. Photosynthesis, therefore, requires an abundant supply of water so that stomata remain open, allowing carbon dioxide to enter.

What Happens in the Stem? Figure 22.3C shows that ➋ the tension in xylem created by evaporation of water at the leaves pulls the water column in the stem upward. Usually the water column in the stem is continuous because of the cohesive property of water molecules. The water molecules also adhere to the sides of the vessels. What happens if the water column within xylem breaks? The water column "snaps back" down the xylem vessel away from the site of breakage, making it more difficult for conduction to occur. Next time you use a straw to drink a soda, notice that pulling the liquid upward is fairly easy as long as there is liquid at the end of the straw. When the soda runs low and you begin to get air, it takes considerably more suction to pull up the remaining liquid. When preparing a vase of flowers, you should always cut the stems under water to preserve an unbroken water column and the life of the flowers.

What Happens in the Root? ➌ In the root (Fig. 22.3C), water enters xylem passively by osmosis (see Chapter 5) because xylem sap always has a greater concentration of solutes than do the root cells. The water column in xylem extends from the leaves down to the root. Water is pulled upward from the

roots due to the tension in xylem created by the evaporation of water at the leaves. The opening and closing of stomata, discussed in the next section, regulates the evaporation of water from the leaves.

> **22.3 *Check Your Progress*** Plants lose much water due to transpiration. What benefits does transpiration have?

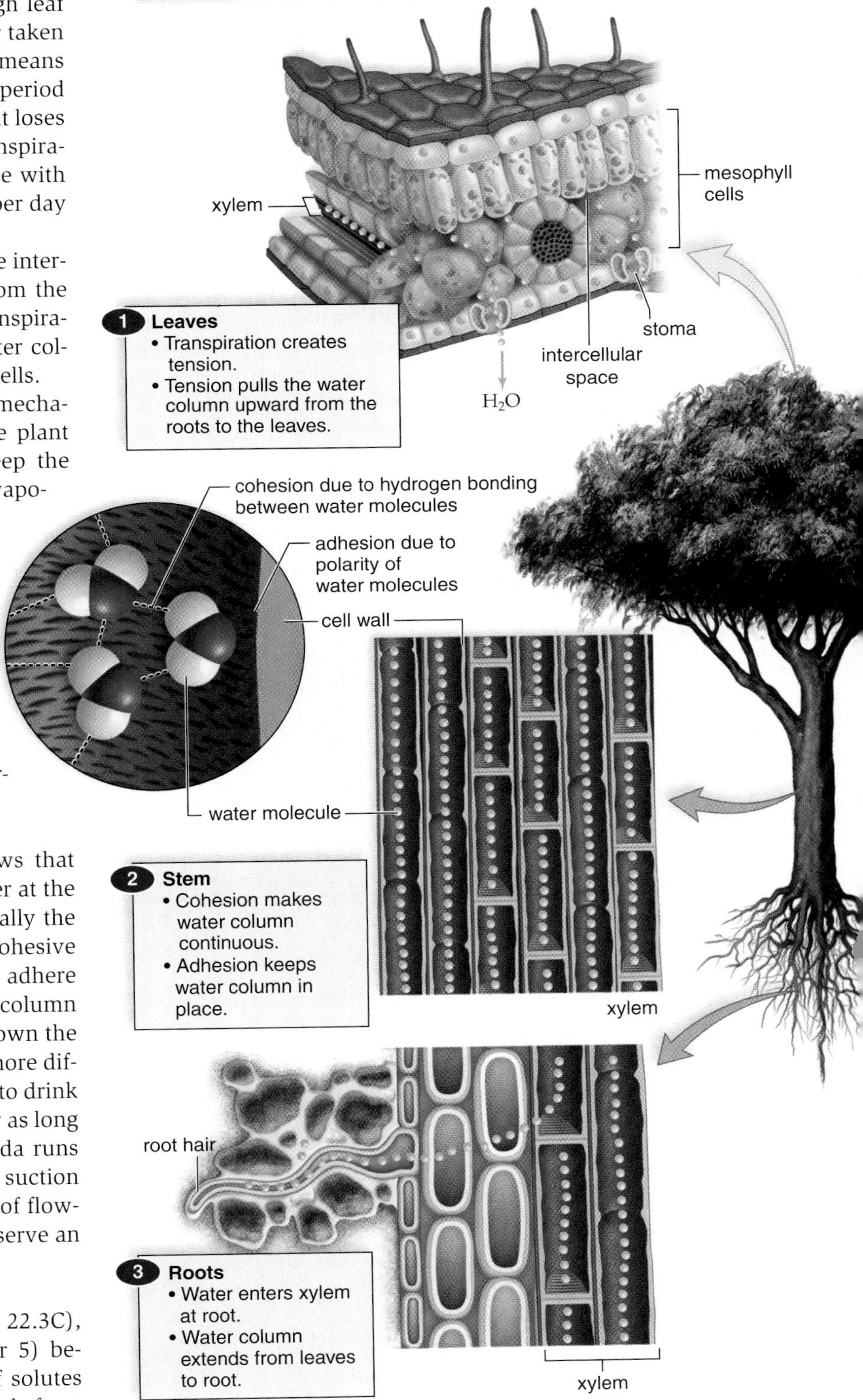

① Leaves
- Transpiration creates tension.
- Tension pulls the water column upward from the roots to the leaves.

mesophyll cells

xylem

stoma

intercellular space

H_2O

cohesion due to hydrogen bonding between water molecules

adhesion due to polarity of water molecules

cell wall

water molecule

② Stem
- Cohesion makes water column continuous.
- Adhesion keeps water column in place.

xylem

root hair

③ Roots
- Water enters xylem at root.
- Water column extends from leaves to root.

xylem

FIGURE 22.3C Cohesion-tension model of xylem transport.

22.4 Guard cells regulate water loss at leaves

Each **stoma** (pl., stomata) is a small pore in leaf epidermis bordered by modified epidermal cells called **guard cells**. When water enters the guard cells and turgor pressure increases, the stoma opens; when water exits the guard cells and turgor pressure decreases, the stoma closes. The guard cells are attached to each other at their ends, and the inner walls are thicker than the outer walls. When water enters, a guard cell's radial expansion is restricted because of cellulose microfibrils in the walls, but lengthwise expansion of the outer walls is possible. When the outer walls expand lengthwise, they buckle out from the region of their attachment, and the stoma opens.

Other factors aside from water are involved in controlling the turgor pressure of guard cells. To understand how these factors function, we need to look more closely at the opening of a stoma. Since about 1968, it has been clear that potassium ions (K^+) accumulate within guard cells when a stoma opens. In other words, the entrance of K^+ into guard cells creates an osmotic pressure that causes water to follow by osmosis. Then a stoma opens (**Fig. 22.4A**). Also interesting is the observation that hydrogen ions (H^+) accumulate outside guard cells, establishing not only a chemical gradient but also an electrical gradient because the inside of the cell is now negative. The electrical gradient allows K^+ to enter by way of a channel protein, and water follows by osmosis.

On the other hand, a stoma closes when turgor pressure decreases due to the exit of K^+ followed by the exit of water (**Fig. 22.4B**).

Three other factors, aside from water availability, regulate whether stomata open or close. (1) The presence of light causes stomata to open. Evidence suggests that the absorption of blue light by a flavin protein sets in motion the cytoplasmic response that leads to opening of stomata. (2) A high concentration of CO_2 causes stomata to close. Perhaps there is a receptor in the plasma membrane of guard cells that brings about inactivation of the H^+ pump when CO_2 concentration rises, as might happen when photosynthesis ceases. (3) Abscisic acid (ABA), produced by cells in wilting leaves, can cause stomata to close (see Section 23.5). Like CO_2, ABA most likely promotes the inhibition of the H^+ pump and, in that way, hinders the creation of conditions that cause water to enter guard cells by osmosis.

Interestingly enough, when plants are kept in the dark, stomata open and close just about every 24 hours. Circadian rhythms (a behavior that occurs nearly every 24 hours) and biological clocks that can keep time are areas of intense investigation.

Because xylem transports minerals, in addition to water, plants can be used to clean up toxic messes, as discussed next.

> **22.4 Check Your Progress** Stomata open even when a plant is kept in continuous darkness. What activity requiring gas exchange occurs in plant cells, even when it is dark? Explain.

FIGURE 22.4A
Opening of stoma.

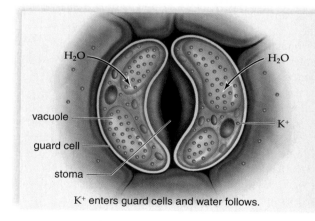

K+ enters guard cells and water follows.

25 μm

FIGURE 22.4B
Closing of stoma.

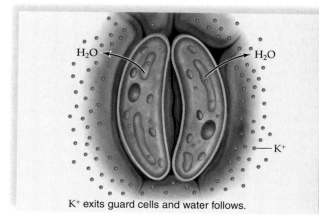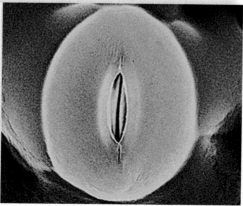

K+ exits guard cells and water follows.

25 μm

22.5   Plants can clean up toxic messes

Most trees planted along the edges of farms are intended to break the wind. But a mile-long stand of spindly poplars outside Amana, Iowa, serves a different purpose. It cleans pollution. The poplars act like vacuum cleaners, sucking up nitrate-laden runoff from a fertilized cornfield before this runoff reaches a nearby brook—and perhaps other waters. Nitrate runoff into the Mississippi River from midwestern farms is a major cause of the large "dead zone" of oxygen-depleted water that develops each summer in the Gulf of Mexico.

Before the trees were planted, the brook's nitrate levels were as much as ten times the amount considered safe. But then Louis Licht, a University of Iowa graduate student, had the idea that poplars, which absorb lots of water and tolerate pollutants, could help. In 1991, Licht tested his hunch by planting the trees along a field owned by a corporate farm. The brook's nitrate levels subsequently dropped more than 90%, and the trees have thrived, serving as a prime cleanup method known as **phytoremediation**.

The idea behind phytoremediation is not new; scientists have long recognized certain plants' abilities to absorb and tolerate toxic substances. But the idea of using these plants on contaminated sites has just gained support in the last decade. The plants clean up sites in two basic ways, depending on the substance involved. If the contaminant is organic, such as spilled oil, plants or the microbes living around their roots break down the substance. The remainder can either be absorbed by the plant or left in the soil or water. When the contaminant is inorganic, such as cadmium or zinc, the plants absorb the substance and trap it. The plants must then be harvested and disposed of or processed to reclaim the trapped contaminant.

Different plants work on different contaminants. The mulberry bush, for instance, is effective on industrial sludge; some grasses attack petroleum wastes; and sunflowers (together with soil additives) remove lead. Canola plants are grown in California's San Joaquin Valley to soak up excess selenium in the soil and help prevent an environmental catastrophe like the one that occurred there in the 1980s.

Back then, irrigated farming caused naturally occurring selenium to rise to the soil surface. When excess water was pumped onto the fields, some selenium would flow off into drainage ditches, eventually ending up in Kesterson National Wildlife Refuge. The selenium in ponds at the refuge accumulated in plants and fish and subsequently deformed and killed waterfowl, says Gary Bañuelos, a plant scientist with the U.S. Department of Agriculture who helped remedy the problem (**Fig. 22.5**). He recommended that farmers add selenium-accumulating canola plants to their crop rotations. As a result, selenium levels in runoff are being managed. Although the underlying problem of excessive selenium in soils has not been solved, says Bañuelos, "This is a tool to manage mobile selenium and prevent another unlikely selenium-induced disaster."

Phytoremediation has also helped clean up badly polluted sites, in some cases at a fraction of the usual cost. Edenspace Systems Corporation of Reston, Virginia, just concluded a phy-

FIGURE 22.5 Scientist Gary Bañuelos in a field of canola plants.

toremediation demonstration at a Superfund site on an Army firing range in Aberdeen, Maryland. The company successfully used mustard plants to remove uranium from the firing range, at as little as 10% of the cost of traditional cleanup methods. Depending on the contaminant involved, traditional cleanup costs can run as much as $1 million per acre, experts say.

Phytoremediation does have its limitations, however. One of them is its slow pace. Depending on the contaminant, it can take several growing seasons to clean a site—much longer than by conventional methods. "We normally look at phytoremediation as a target of one to three years to clean a site," notes Edenspace's Mike Blaylock. "People won't want to wait much longer than that."

Phytoremediation is also only effective at depths that plant roots can reach, making it useless against deep-lying contamination unless the contaminated soils are excavated. Phytoremediation will not work on lead and other metals unless chemicals are added to the soil. In addition, it is possible that animals may ingest pollutants by eating the leaves of the plants used in some projects.

Despite its shortcomings, experts see a bright future for this technology. David Glass, an independent analyst based in Needham, Massachusetts, predicts the business of phytoremediation will continue to grow considerably in future years. It is a promising solution to pollution problems but, says the EPA's Walter W. Kovalick, "It's not a panacea. It's another arrow in the quiver. It takes more than one arrow to solve most problems."

The next part of the chapter continues our study of transport in plants by considering the function of phloem.

> **22.5** *Check Your Progress* The authors of this article describe instances in which phytomediation will not work. When does it work?

Plants require a transport mechanism that can take sugar to all areas of a plant that are not photosynthesizing, be they immature leaves, flowers, or roots. The pressure-flow model relies on a positive pressure due to osmosis when sugar is actively transported into the sieve tubes of phloem.

22.6 Phloem carries organic molecules

Plants transport the organic molecules resulting from photosynthesis to the parts of plants that need them. This includes young leaves that have not yet reached their full photosynthetic potential, flowers that are in the process of making seeds and fruits, and roots, whose location in the soil prohibits them from carrying on photosynthesis.

As long ago as 1679, Marcello Malpighi studied the effects of **girdling**, the removal of a strip of bark from around a tree. If a tree is girdled below the level of the majority of its leaves, the bark swells just above the cut, and sugar accumulates in the swollen tissue. We know today that when a tree is girdled, the phloem is removed, but the xylem is left intact. Therefore, the results of this girdling experiment tell us that phloem is the tissue that transports sugars.

Radioactive tracer studies with carbon 14 (^{14}C) have confirmed that phloem transports organic molecules. When ^{14}C-labeled CO_2 is supplied to mature leaves, radioactively labeled sucrose is soon found moving down the stem into the roots. It's difficult to get samples from phloem sap without injuring the phloem, but this problem is solved by using aphids, small insects that are phloem feeders. The aphid drives its stylet, a sharp mouthpart that functions like a hypodermic needle, between the epidermal cells, and sap enters its body from a sieve-tube member (**Fig. 22.6**). If the aphid is anesthetized using ether, its body can be carefully removed, leaving the stylet. A researcher can then collect and analyze the phloem sap. The use of radioactive tracers and aphids has revealed that sap movement through phloem can be as fast as 60–100 cm per hour and possibly up to 300 cm per hour.

The mechanism of how organic molecules are transported in phloem is currently explained by the pressure-flow model, discussed in Section 22.7.

> **22.6 Check Your Progress** What organic molecules would you expect to find in phloem sap?

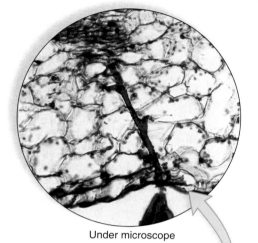

FIGURE 22.6
Aphid acquiring phloem sap.

Under microscope

waste due to feeding on phloem sap

An aphid feeding on a plant stem

22.7 The pressure-flow model explains phloem transport

The **pressure-flow model** is a current explanation for the movement of organic materials throughout the plant in phloem. Consider an experiment in which two bulbs are connected by a glass tube. The first bulb contains sucrose at a higher concentration than the second bulb. Each bulb is bounded by a differentially permeable membrane, and the entire apparatus is submerged in distilled water, which lacks ions.

As shown in **Figure 22.7A**, ❶ distilled water flows into the first bulb because it has the higher solute concentration. The entrance of water creates a positive *pressure*, and water *flows*

toward the second bulb ❷ . This flow not only drives water toward the second bulb, but it also provides enough force for water to move out through the membrane of the second bulb—even though the second bulb contains a higher concentration of solute than the distilled water ❸ .

In plants, the sieve tubes of phloem are analogous to the glass tube that connects the two bulbs. Sieve tubes are composed of sieve-tube members, each of which has a companion cell. It is possible that the companion cells assist the sieve-tube members in some way. The sieve-tube members align end to end, and at maturity

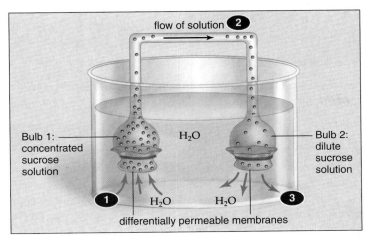

FIGURE 22.7A Pressure-flow experiment.

the wide pores in the sieve-tube plates allow movement of solutes from one cell to the other. Sieve tubes, therefore, form a continuous pathway for organic nutrient transport throughout a plant.

Flow Is from a Source to a Sink During the growing season, photosynthesizing leaves are producing sugar, which supplies the energy plants need to grow, repair tissues, and reproduce. Therefore, they are a **source** of sugar (e.g., sucrose). **Figure 22.7B** shows what happens in the source. ❶ This sugar is actively transported into the phloem. Transport is dependent on an *electrochemical gradient* established by a proton pump (H^+), a form of active transport. Sugar then travels across the membrane in conjunction with hydrogen ions (H^+), which are moving down their concentration gradient (see Fig. 22.9B). ❷ After sugar enters the sieve tubes, water follows passively by osmosis. ❸ The buildup of water within sieve tubes creates the positive pressure that starts a flow of the phloem contents. ❹ The roots (and other growth areas) are a **sink** for sugar, meaning that they are removing sugar and using it for cellular respiration. Also some plants, such as the sugar beet, the carrot, and the white potato, have modified organs for carbohydrate storage. ❺ After sugar is actively transported out of sieve tubes, water exits the phloem passively by osmosis and is taken up by the xylem. ❻ Xylem transports water to the mesophyll of the leaf where it is used for photosynthesis. ❼ Although up to 90% of water is transpired, some is used for photosynthesis, and some reenters the phloem by osmosis ❷ .

Transport of Sugar The pressure-flow model of phloem transport can account for any direction of flow in sieve tubes if we consider that the direction of flow is always from source to sink. In other words, phloem contents can move either up or down as appropriate for the plant at a particular time in its life cycle. For example, recently formed leaves can be a sink, and they will receive sucrose until they begin to photosynthesize maximally.

This completes our study of transport in plants, and the next part of the chapter discusses the nutrition of plants.

> **22.7 *Check Your Progress*** If carbohydrates stored by the roots are used as an energy source for flower production in the spring, what is the source and what is the sink?

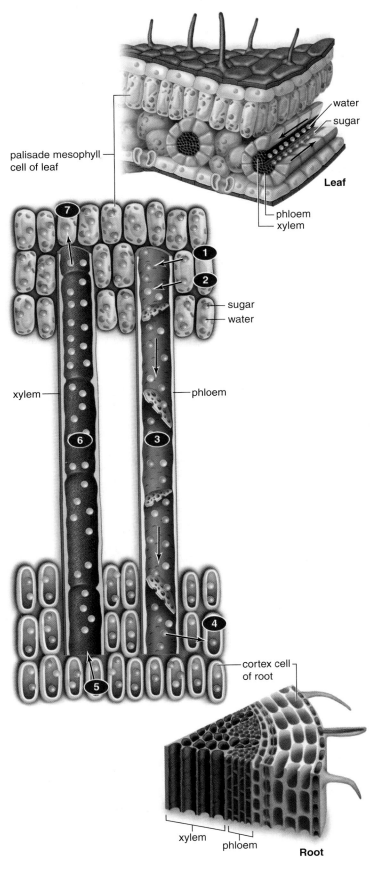

FIGURE 22.7B Pressure-flow model of phloem transport.

Nutrient requirements and the manner in which roots take up minerals from the soil are areas of study. The properties of soil affect the availability of minerals, and plants also have associations with other organisms that can enhance their success in acquiring nutrients.

22.8 Certain nutrients are essential to plants

The ancient Greeks believed plants were "soil-eaters" that somehow converted soil into plant material. Apparently to test this hypothesis, a 17th-century Dutchman named Jean-Baptiste Van Helmont planted a willow tree weighing 5 lb in a large pot containing 200 lb of soil. He watered the tree regularly for five years and then reweighed both the tree and the soil. The tree weighed 170 lb, and the soil weighed only a few ounces less than the original 200 lb. Van Helmont concluded that the tree's increase in weight was due primarily to the addition of water.

Although water is a vitally important nutrient for a plant, Van Helmont was unaware that most of the water entering a plant evaporates at the leaves. He was also unaware that CO_2 (taken in at the leaves, as shown in **Figure 22.8A**) combines with water

in the presence of sunlight to produce carbohydrates, the chief organic matter of plants.

Approximately 95% of a typical plant's dry weight (weight excluding free water) is carbon, hydrogen, and oxygen. Why? Because these are the elements that are found in most organic compounds, such as carbohydrates. Carbon dioxide supplies the carbon, and water (H_2O) supplies the hydrogen and oxygen for the organic compounds of a plant.

Minerals as Nutrients In addition to carbon, hydrogen, and oxygen, plants require certain other nutrients that are absorbed as minerals by the roots. A **mineral** is an inorganic substance usually containing two or more elements. Minerals are needed to help build molecules. For example, nitrogen is a major component of nucleic acids and proteins; magnesium is a component of chlorophyll, the main photosynthetic pigment; and iron is a building block of cytochrome molecules, which carry electrons during photosynthesis and cellular respiration. The major functions of various **essential nutrients** for plants are listed in **Table 22.8**. A nutrient is essential if (1) it has an identifiable role, (2) no other nutrient can substitute and fulfill the same role, and (3) a deficiency of this nutrient causes a plant to die or fail to complete its reproductive cycle. Essential nutrients are divided into **macronutrients** (needed in large quantity) and **micronutrients** (needed in trace amounts). The following diagram and slogan help us remember which are the macronutrients and which are the micronutrients for plants:

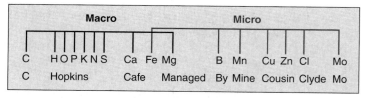

Macro		Micro	
C HOPKNS Ca Fe Mg		B Mn Cu Zn Cl Mo	
C Hopkins Cafe Managed		By Mine Cousin Clyde Mo	

Beneficial nutrients are another category of elements taken up by plants. Beneficial nutrients either are required for or enhance the growth of a particular plant. For example, horsetails require silicon as a mineral nutrient, and sugar beets show enhanced growth in the presence of sodium. Nickel is a beneficial nutrient in soybeans when root nodules are present. Aluminum is used by some ferns, and locoweeds, which are toxic to livestock, take up selenium.

When a plant is burned, its nitrogen component is given off as ammonia and other gases, but most other essential minerals remain in the ash. Still, the absence of nitrogen shows that the absence of a mineral from the ash cannot be relied on to determine whether a plant needs that mineral. The preferred method for determining the mineral requirements of a plant was developed at the end of the 19th century by the German plant physiologists Julius von Sachs and Wilhem Knop. The method,

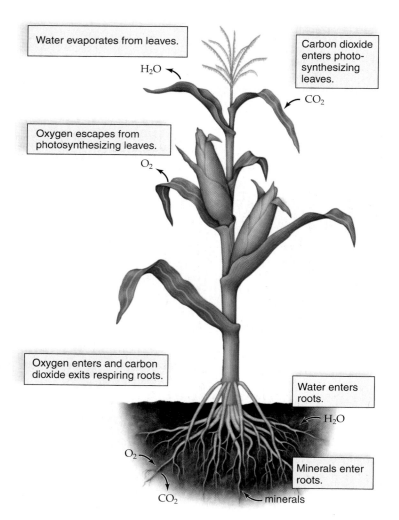

FIGURE 22.8A Overview of plant nutrition.

Labels in figure:
- Water evaporates from leaves. H_2O
- Carbon dioxide enters photosynthesizing leaves. CO_2
- Oxygen escapes from photosynthesizing leaves. O_2
- Oxygen enters and carbon dioxide exits respiring roots.
- Water enters roots. H_2O
- O_2
- CO_2 minerals
- Minerals enter roots.

TABLE 22.8		Some Essential Inorganic Nutrients in Plants	
Element	**Symbol**	**Form**	**Major Functions**
Macronutrients			
Carbon	C	CO_2	
Hydrogen	H	H_2O	Major component of organic molecules
Oxygen	O	O_2	
Phosphorus	P	$H_2PO_4^-$ HPO_4^{2-}	Part of nucleic acids, ATP, and phospholipids
Potassium	K	K^+	Cofactor for enzymes; functions in water balance and openings of stomata
Nitrogen	N	NO_3^-	Part of nucleic acids, proteins, chlorophyll, and coenzymes
Sulphur	S	SO_4^{2-}	Part of amino acids, some coenzymes
Calcium	Ca	Ca^{2+}	Regulates responses to stimuli and movement of substances through plasma membrane; involved in formation and stability of cell walls
Magnesium	Mg	Mg^{2+}	Part of chlorophyll; activates a number of enzymes
Micronutrients			
Iron	Fe	Fe^{2+} Fe^{3+}	Part of cytochrome needed for cellular respiration; activates some enzymes
Boron	B	BO_3^{3-} $B_4O_7^{2-}$	Role in nucleic acid synthesis, hormone responses, and membrane function
Manganese	Mn	Mn^{2+}	Required for photosynthesis; activates some enzymes such as those of the citric acid cycle
Copper	Cu	Cu^{2+}	Part of certain enzymes, such as redox enzymes
Zinc	Zn	Zn^{2+}	Role in chlorophyll formation; activates some enzymes
Chlorine	Cl	Cl^-	Role in water-splitting step of photosynthesis and water balance
Molybdenum	Mo	MoO_4^{2-}	Cofactor for enzyme used in nitrogen metabolism

Solution lacks nitrogen   Complete nutrient solution

Solution lacks phosphorus   Complete nutrient solution

Solution lacks calcium   Complete nutrient solution

FIGURE 22.8B Nutrient deficiencies.

called water culture or **hydroponics**, allows plants to grow well in water, instead of soil, if they are supplied with all the nutrients they need. The investigator omits a particular mineral and observes the effect on plant growth. If growth suffers, the omitted mineral is an essential nutrient (**Fig. 22.8B**). This method has been more successful for macronutrients than for micronutrients. For studies involving the latter, the water and the mineral salts used must be absolutely pure, but purity is difficult to attain, because even instruments and glassware can introduce micronutrients. It is also possible that the element in question may already be present in the seedling used in the experiment.

These factors complicate the process of determining essential plant micronutrients by means of hydroponics.

The next section discusses how roots are able to take up and concentrate minerals in their cells.

> **22.8 Check Your Progress** What element(s) in particular, aside from C, H, and O, is/are needed to form (*a*) proteins and (*b*) nucleic acids? How does a plant acquire these elements?

Absorption of water and minerals is one of the main functions of a root. In order for water to reach the xylem of a root, it must pass through the cortex in one of two ways, as **Figure 22.9A** shows. Via pathway A, water, along with minerals, can enter the root and travel through the cortex simply by passing between the porous cell walls. However, once the water reaches the endodermal layer, it can no longer pass between cells because bands of water-proofing suberin (called the **Casparian strip**) encircle the individual endodermal cells, and this forces water to pass through the endodermal cells before reaching the xylem. Alternately, via pathway B, water can enter epidermal cells at **root hairs** and then progress through cells across the cortex and endodermis of a root by means of cytoplasmic strands within plasmodesmata. Regardless of the pathway, water enters root cells when they have a higher solute concentration than does the soil solution.

Mineral Uptake In contrast to water, which passively crosses plasma membranes, minerals are actively taken up by plant cells. Plants possess an astonishing ability to concentrate minerals—that is, to take up minerals until they are many times more concentrated in the plant than in the surrounding medium. The concentration of certain minerals in roots is as much as 10,000 times greater than in the surrounding soil. Following their uptake by root cells, minerals move into the xylem and are transported into the leaves by the upward movement of water. Along the way, minerals can exit the xylem and enter those cells that require them. By what mechanism do minerals cross plasma membranes?

Recall that plant cells absorb minerals in the ionic form. For example, although the atmosphere is 79% nitrogen gas, plants require nitrogen in the form of nitrate (NO_3^-) or ammonium (NH_4^+). Phosphorus, another nutrient requirement for plants, is absorbed as phosphate (HPO_4^{2-}), potassium is absorbed as potassium ions (K^+), and so forth. While water can passively (no energy needed) cross a plasma membrane, ions need to be actively transported into plant cells because they are unable to cross the plasma membrane on their own.

It has long been known that plant cells expend energy to actively take up and concentrate mineral ions. **Figure 22.9B** describes how plants manage to do this. ❶ A proton pump hydrolyzes ATP and uses the energy released to transport hydrogen ions (H^+) out of the cell. This sets up an *electrochemical gradient*. ❷ The electrochemical gradient then drives positively charged ions such as K^+ through a channel protein into the cell. ❸ Negatively charged mineral ions are transported, along with H^+, by carrier proteins. Since H^+ is moving down its concentration gradient, no energy is required.

The properties of soil greatly affect the availability of mineral nutrients for plant use, as discussed in Section 22.10.

> **22.9** *Check Your Progress* **Review the structure of the plasma membrane in Section 5.10, page 85, and explain why the center of the plasma membrane is nonpolar, making it difficult for ions to cross the plasma membrane.**

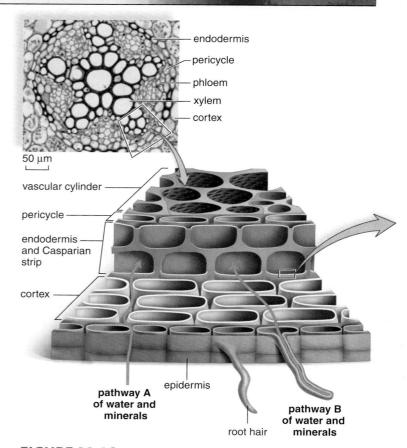

FIGURE 22.9A Pathways of water and minerals in a root.

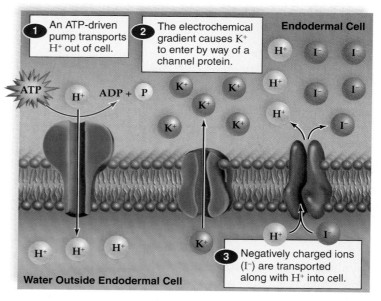

FIGURE 22.9B Transport of minerals across an endodermal plasma membrane in a root.

22.10 Soil has distinct characteristics

Soil is defined as a mixture of mineral particles (sand, silt, and clay), decaying organic material, living organisms, air, and water, which together support the growth of plants. The mineral particles vary in size: Sand particles are the largest; silt particles have an intermediate size; and clay particles are the smallest. Because sandy soils have many large particles, they have large spaces, and the water drains readily through the particles. In contrast to sandy soil, a soil composed mostly of clay particles has small spaces that fill completely with water. The type of soil called loam is composed of roughly one-third each sand, silt, and clay particles. This combination sufficiently retains water and nutrients, while still allowing the drainage necessary to provide air spaces. Loam is one of the most productive soils.

Humus Soil containing a high percentage of decomposing organic material is called **humus**. Humus mixes with the top layer of soil particles and augments beneficial soil characteristics. Plants do well in soils that contain 10–20% humus.

Humus causes soil to have a loose, crumbly texture that allows water to soak in without doing away with air spaces. After a rain, the presence of humus decreases the chances of runoff. Humus swells when it absorbs water and shrinks as it dries. This action helps aerate soil.

Soil that contains humus is nutritious for plants. When bacteria and fungi break down the organic matter in humus, inorganic nutrients are returned to plants.

Soil Profiles A **soil profile** is a vertical section of soil, from the ground surface to the unaltered rock below. Usually, a soil profile has parallel layers known as **horizons**. Mature soil generally has three horizons (**Fig. 22.10**). The A horizon is the uppermost (or

FIGURE 22.10
Simplified soil profile.

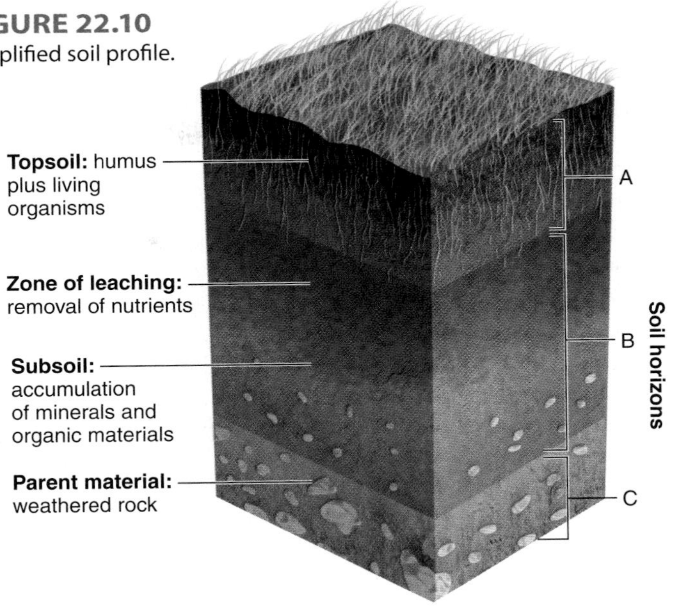

Topsoil: humus plus living organisms

Zone of leaching: removal of nutrients

Subsoil: accumulation of minerals and organic materials

Parent material: weathered rock

Soil horizons

A

B

C

topsoil) layer. It contains leaf litter, humus, and soil organisms, but minerals have drained into the B horizon. The B horizon has two parts. Minerals drain from the zone of leaching into subsoil, which accumulates both inorganic nutrients and organic materials. The C horizon is a layer of weathered and shattered rock.

Minerals interact with soil and this affects their availability for uptake by plants, as discussed next.

22.10 *Check Your Progress* **What are the benefits of humus in soil?**

22.11 Plants absorb minerals from the soil

In a good agricultural soil, the components come together in such a way that there are spaces of air and water. It is best if the soil contains particles of different sizes because only then will there be spaces for air. Roots take up oxygen from air spaces within the soil. Ideally, water clings to soil particles by capillary action and does not fill the spaces. That is why you should not overwater your houseplants!

Clay particles have a benefit that sand particles do not have. As Table 22.8 indicates, some minerals are negatively charged, and others are positively charged. Clay particles are negative, and they can retain positively charged minerals such as calcium (Ca^{2+}) and potassium (K^+), preventing these minerals from being washed away by leaching. Roots exchange H^+ for these minerals when they take them up (**Fig. 22.11**). Because clay particles are unable to retain negatively charged NO_3^-, the nitrogen content of soil is apt to be low.

Plants have mutualistic relationships that help them acquire minerals from the soil, as discussed in Section 22.12.

22.11 *Check Your Progress* **Some farmers do not remove the remains of last year's crops from agricultural lands. What are the benefits of this practice?**

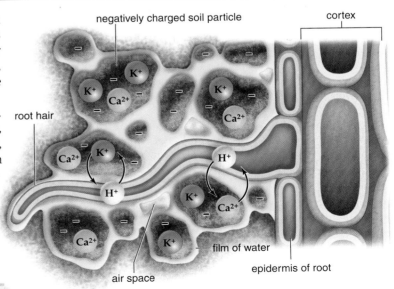

negatively charged soil particle

cortex

root hair

air space

film of water

epidermis of root

FIGURE 22.11 Mineral absorption.

Two mutualistic relationships assist roots in obtaining mineral nutrients. Root nodules involve a mutualistic relationship with bacteria, and mycorrhizae involve a mutualistic relationship with fungi.

Root Nodules Some plants, such as legumes, soybeans, and alfalfa, have roots colonized by *Rhizobium* bacteria, which can fix atmospheric nitrogen (N_2). They break the $N \equiv N$ bond and reduce nitrogen to NH_4^+ for incorporation into organic compounds. The bacteria live in **root nodules** and are supplied with carbohydrates by the host plant (**Fig. 22.12A**). The bacteria, in turn, furnish their host with nitrogen compounds.

Mycorrhizae The second type of mutualistic relationship, called mycorrhizae, involves fungi and almost any type of plant root (**Fig. 22.12B**). Only a small minority of plants do not have **mycorrhizae**, and these plants are most often limited as to the environment in which they can grow. The fungus increases the surface area available for mineral and water uptake and breaks down organic matter in soil, releasing nutrients that the plant can use. In return, the root furnishes the fungus with sugars and amino acids. Plants are extremely dependent on their mycorrhizal relationships. Orchid seeds, which are quite small and contain limited nutrients, do not germinate until a mycorrhizal fungus has invaded their cells. Nonphotosynthetic plants, such as Indian pipe, use nearby mycorrhizae to extract nutrients from the roots of a "host" tree.

Other Relationships Parasitic plants, such as dodders, broomrapes, and pinedrops, send out rootlike projections called haustoria that tap into the xylem and phloem of the host stem (**Fig. 22.12C**). Carnivorous plants, such as the Venus flytrap and the sundew, obtain some nitrogen and minerals when modified leaves capture and digest insects (**Fig. 22.12D**).

> **22.12 Check Your Progress** Under what soil conditions are plants apt to develop root nodules?

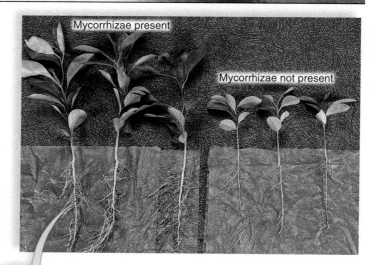

FIGURE 22.12B Mycorrhizae result in better growth.

mycorrhizae

dodder (brown)

FIGURE 22.12C
Dodder twists around host.

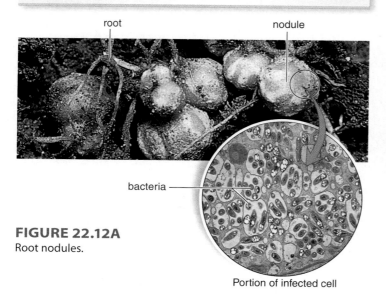

root   nodule

bacteria

Portion of infected cell

FIGURE 22.12A
Root nodules.

FIGURE 22.12D A Venus flytrap.

The land environment offers many advantages to plants, such as greater exposure to light and more availability of carbon dioxide for photosynthesis. (Water, even if clear, filters out light, and carbon dioxide concentration and rate of diffusion are lower in a water environment.) The evolution of a transport system was critical, however, in order for plants to make full use of these advantages. Only if a transport system is present can stems elevate plant leaves so that they are better exposed to solar energy and carbon dioxide in the air. A transport system brings water and minerals from the roots to the leaves and brings the products of photosynthesis down to the roots beneath the soil, where their cells depend on an input of organic molecules to remain alive. An efficient transport system also allows roots to penetrate deeply into the soil to absorb water and minerals.

In addition, the presence of a transport system allows materials to be distributed to those parts of the plant body that are growing most rapidly. For example, new leaves and flower buds would grow rather slowly if they had to depend on their own rate of photosynthesis. Height in vascular plants, due to the presence of a transport system, has other benefits aside from elevation of leaves. First, it is adaptive to have reproductive structures located where the wind can better distribute pollen and seeds. Second, once animal pollination came into existence, it was beneficial for flowers to be located where they would be more easily seen by animals.

Another benefit of a transport system is distribution of hormones that regulate plant responses to the environment, which involve growth, reproduction, and development. Chapter 23 explores the topic of plant hormones.

The Chapter in Review

Summary

Plants Can Adapt Too

- Adaptive radiation can happen in plants just as it occurred in Darwin's finches.
- The silversword alliance contains plants that are adapted to living in bogs, moist-to-wet forests, dry tree or shrub woodlands, and lava fields.

Plants Are Organized to Transport Water and Solutes

22.1 Transport begins in both the leaves and the roots of plants

- Xylem moves water and minerals from roots to leaves. Its conducting cells are tracheids and vessel elements.
- Phloem transports sugars from photosynthesizing leaves to all plant parts. Its conducting cells are sieve-tube members, which have companion cells.
- Xylem and phloem are in roots, stems, and leaves.

22.2 Competition for resources is one aspect of biodiversity

- Research concludes that competition for resources results in high biodiversity that buffers ecosystems against disturbances.

Xylem Transport Depends on the Properties of Water

22.3 Water is pulled up in xylem by evaporation from leaves

- The cohesion-tension model of xylem transport explains the movement of water upward.
  - In leaves, transpiration generates a tension that moves water upward from roots to leaves.
  - In the stem, cohesion due to hydrogen bonding makes the water column continuous, while adhesion keeps the water column in place.
  - Water enters the xylem at the root, and tension created at the leaves pulls water from the roots.

22.4 Guard cells regulate water loss at leaves

- A stoma opens when turgor pressure increases in the guard cells due to the entrance of K^+ and water.
- A stoma closes when turgor pressure decreases in the guard cells due to the exit of K^+ and water.

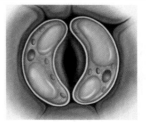 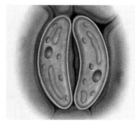

22.5 Plants can clean up toxic messes

- Phytoremediation is the absorption of toxic substances by certain plants.
- Plants or microbes living around the roots break down organic contaminants.
- Plants absorb inorganic contaminants, which can then be destroyed or reclaimed.

Phloem Function Depends on Membrane Transport

22.6 Phloem carries organic molecules

- Phloem sap contains sucrose, hormones, and some amino acids.

22.7 The pressure-flow model explains phloem transport

- Sugar is actively transported into phloem at a source, and water follows by osmosis.
- The resulting increase in pressure creates a flow, which moves water and sugar to a sink.
- Sugar is actively transported out of sieve tubes at a sink.

Plants Require Good Nutrition and Therefore Good Soil

22.8 Certain nutrients are essential to plants

- Plants need only inorganic nutrients to make all the organic compounds that comprise the plant body.
- Essential nutrients are either macronutrients or micronutrients.
- Beneficial nutrients are either required for or enhance plant growth.
- Hydroponics can help determine the essential nutrients in growing plants.

22.9 Roots are specialized for the uptake of water and minerals

- The root system absorbs water and minerals; nutrients cross the epidermis and cortex before entering the endodermis.
- In mineral uptake by roots, an ATP-driven proton pump transports H^+ out of the cell; an electrochemical gradient causes K^+ to enter through a channel protein; and negatively charged ions are transported along with H^+ into the cell.

22.10 Soil has distinct characteristics

- Soil is composed of minerals particles, organic matter, living organisms, air, and water.
- Humus is soil containing a high amount of decomposing organic matter.
- A soil profile shows horizons in a vertical section of soil, from the surface to the rock layer.

22.11 Plants absorb minerals from the soil

- Negatively charged soil retains positively charged ions. Roots exchange H^+ for these ions.

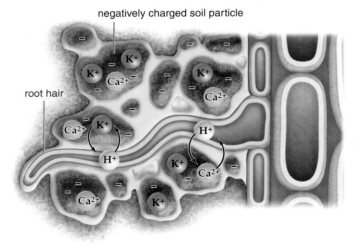

negatively charged soil particle

root hair

22.12 Adaptations of plants help them acquire nutrients

- Nitrogen-fixing bacteria live in root nodules.
- Mycorrhizal fungi increase the root surface area for water and mineral uptake.
- Some plants obtain nutrients by parasitizing other plants.
- Other plant leaves are modified to capture and digest insects, providing the plant with nutrients.

Plants Are Organized to Transport Water and Solutes

1. Xylem includes all of the following except
   a. companion cells.   c. tracheids.
   b. vessels.   d. dead cells.

2. **THINKING CONCEPTUALLY** Why would you expect leaves to be involved in the transport of both water and sugar?

Xylem Transport Depends on the Properties of Water

3. The process responsible for guttation is
   a. evaporation.   c. root pressure.
   b. cohesion.   d. transpiration.

4. What role do cohesion and adhesion play in xylem transport?
   a. Like transpiration, they create a tension.
   b. Like root pressure, they create a positive pressure.
   c. Like sugars, they cause water to enter xylem.
   d. They create a continuous water column in xylem.
   e. All of these are correct.

5. An opening in the leaf that allows gas and water exchange is called the
   a. lenticel.   d. guard cell.
   b. hole.   e. accessory cell.
   c. stoma.

6. What main force drives absorption of water, creates tension, and draws water through the plant?
   a. adhesion   d. transpiration
   b. cohesion   e. absorption
   c. tension

7. Stomata are usually open
   a. at night, when the plant requires a supply of oxygen.
   b. during the day, when the plant requires a supply of carbon dioxide.
   c. day or night if there is excess water in the soil.
   d. during the day, when transpiration occurs.
   e. Both b and d are correct.

8. By which process does phytoremediation work?
   a. Plants break down harmful chemicals.
   b. Microbes surrounding plants break down harmful chemicals.
   c. Plants take up and store harmful chemicals.
   d. Both a and b are correct.
   e. a, b and c are all correct.

9. **THINKING CONCEPTUALLY** Why does the xylem of summer wood (Section 21.7) and plants adapted to a dry environment (introduction to Chapter 22) contain more tracheids than vessel elements?

Phloem Function Depends on Membrane Transport

10. The sugar produced by mature leaves moves into sieve tubes by way of _____, while water follows by _____.
    a. osmosis, osmosis
    b. active transport, active transport
    c. osmosis, active transport
    d. active transport, osmosis

11. After sucrose enters sieve tubes,
    a. it is removed by the source.
    b. water follows passively by osmosis.
    c. it is driven by active transport to the source, which is usually the roots.
    d. stomata open so that water flows to the leaves.
    e. All of these are correct.

12. In contrast to transpiration, the transport of organic nutrients in the phloem
    a. requires energy input from the plant.
    b. always flows in one direction.
    c. results from tension.
    d. does not require living cells.
13. The pressure-flow model of phloem transport states that
    a. phloem contents always flow from the leaves to the root.
    b. phloem contents always flow from the root to the leaves.
    c. water flow brings sucrose from a source to a sink.
    d. water pressure creates a flow of water.
    e. Both c and d are correct.

Plants Require Good Nutrition and Therefore Good Soil

14. Which of these is not a mineral ion?
    a. NO_3^- d. Al^{3+}
    b. Mg^+ e. All of these are correct.
    c. CO_2
15. Which of these molecules is not a nutrient for plants?
    a. water   d. nitrogen gas
    b. carbon dioxide gas   e. None of these are nutrients.
    c. mineral ions
16. A nutrient element is considered essential if
    a. plant growth increases when the concentration of the element is reduced.
    b. plant growth suffers in the absence of the element.
    c. plants can substitute a similar element for the missing element with no ill effects.
    d. the element is a positive ion.
17. Which process is responsible for moving water from the soil into the xylem of a plant's roots?
    a. active transport   c. endocytosis
    b. diffusion   d. osmosis
18. The Casparian strip affects
    a. how water and minerals move into the vascular cylinder.
    b. vascular tissue composition.
    c. how soil particles function.
    d. how organic nutrients move into the vascular cylinder.
    e. Both a and d are correct.
19. Plants expend energy in order to take up
    a. carbon dioxide.   d. Both a and b are correct.
    b. minerals.   e. a, b, and c are all correct.
    c. water.
20. Which is a component of soil?
    a. mineral particles   d. air and water
    b. humus   e. All of these are correct.
    c. organisms
21. Soils rich in which type of soil particle have a high water-holding capacity?
    a. sand
    b. silt
    c. clay
    d. All soil particles hold water equally well.
22. Negatively charged clay particles attract
    a. K^+.   d. Both a and b are correct.
    b. NO_3^-.   e. Both a and c are correct.
    c. Ca^+.

23. Plants with mycorrhizae form a mutualistic relationship with
    a. algae.   d. protozoans.
    b. bacteria.   e. Both a and b are correct.
    c. fungi.

Understanding the Terms

beneficial nutrient 450	phytoremediation 447
Casparian strip 452	pressure-flow model 448
cohesion-tension model 444	root hair 452
companion cell 442	root nodule 454
cuticle 445	root pressure 444
essential nutrient 450	sieve-tube member 442
girdling 448	sink 449
guard cell 446	soil 453
guttation 444	soil profile 453
horizon 453	source 449
humus 453	stoma 446
hydroponics 451	tracheid 442
macronutrient 450	transpiration 445
micronutrient 450	vessel element 442
mineral 450	water column 444
mycorrhizae 454	xylem 442
phloem 442	xylem sap 442
phloem sap 442	

Match the terms to these definitions:
a. _____ Plants' loss of water to the atmosphere, mainly through evaporation at leaf stomata.
b. _____ Layer of impermeable lignin and suberin bordering four sides of root endodermal cells; causes water and minerals to enter endodermal cells before entering vascular tissue.
c. _____ Type of plant cell found in pairs, with one on each side of a leaf stoma.
d. _____ Liberation of water droplets from the edges and tips of leaves.
e. _____ Model explaining transport in sieve tubes of phloem.

Thinking Scientifically

1. Based on the data presented in Section 22.2, why would biologists suggest to farmers that polyculture (planting several species at a time) instead of monoculture (planting a single species) would better protect their fields from environmental assaults, such as insect attacks and drought?
2. Using hydroponics, design an experiment to determine if calcium is an essential plant nutrient. State the possible results.

ARIS™ Visit www.mhhe.com/maderconcepts for practice quizzes, animations, videos, and activities designed to help you master the material in this chapter.

23

Control of Growth and Responses in Plants

LEARNING OUTCOMES

After studying this chapter, you should be able to accomplish the following outcomes.

Recovering Slowly
1 Associate plant hormones with the recovery of the Mount St. Helens ecosystem.

Plant Hormones Regulate Plant Growth and Development
2 Define a hormone, and describe a signal transduction pathway.
3 Compare and contrast the effects of auxins, gibberellins, cytokinins, abscisic acid, and ethylene on plant growth and development.

Plants Respond to Environmental Stimuli
4 Show that gravitropism, phototropism, thigmotropism, turgor movements, and sleep movements are responses to particular stimuli.
5 Distinguish between a circadian rhythm and a biological clock.
6 Relate photoperiodism to flowering in short-day/long-night plants and in long-day/short-night plants.
7 Describe the role of phytochrome in flowering and other responses to varying amounts of light.
8 Describe ways in which plants respond to their biotic environment, including physical barriers, chemical toxins, systemic mechanisms, and relationships with animals.
9 Give examples to show that the chemical defenses of plants make good medicines for people.

On May 18, 1980, disaster struck in the state of Washington. Mount St. Helens, an active volcano, erupted with such force that a massive plume of debris shot high into the sky and then fell to Earth. The blast, which ranged from 220 to 670 miles per hour, knocked over trees or stripped away the forest canopy in an area of 350 square kilometers. Steamy, hot overflow from inside the volcano, coupled with avalanches of rocks and sediments brought on when the side of the mountain collapsed, produced a lifeless and barren scene. What had been a rich coniferous forest seemed completely destroyed. In some places, the ash was a meter thick.

Mount St. Helens before it erupted

Recovering Slowly

Now, almost 30 years later, the Mount St. Helens area is home to an increasing number and variety of plants. Some plants survived the eruption because they lived in areas protected by snow or deep gullies. Other plants managed to regrow from roots and push their way through the deep layer of ash and mud deposits. The surviving plants expanded and dominated the area during the first years following the eruption.

Next, plants that specialize in being the first to colonize disturbed areas joined the few survivors. In the worst-hit area, called the blast zone, great expanses of barren land are now dotted with small clumps of low-lying herbs. Fireweed (*Chamerion angustifolium*) takes its name from its ability to live at burn sites following forest fires. Prairie lupine (*Lupinus lepidus*) has root nodules where bacteria fix nitrogen. Therefore, lupines do not require nitrate in the soil. They actually enrich poor soil, allowing other species to subsequently establish themselves. Other early colonizers were grasses whose seeds were deposited after passing through the digestive tracts of elk and other animals moving through the area. Plants such as grasses spread to new areas by means of above- or belowground runners.

Trees also exist in the Mount St. Helens area, particularly along creeks or around water-filled craters where they can receive sufficient moisture. One species, the lodgepole pine (*Pinus contorta*), is common almost anywhere in western North America because it grows under adverse conditions such as cold, wet winters and warm, dry summers. It also grows in areas with very dry or nutrient-poor soils. Its winged seeds remain inside closed cones until intense heat triggers the covers to open and release the seeds, which are then dispersed by the wind. Another tree, the red alder, is a flowering hardwood tree. Both of these trees have roots covered by mutualistic fungal hyphae (see Fig. 22.12B), which assist in taking up water and minerals from the soil. Mycorrhizal roots (fungus roots) also produce auxin, a hormone common to plants that stimulates root growth.

It may take 200 to 500 years for the blast zone to become the rich forest it was before. This revelation should help all of us appreciate the beauty of a mature forest, and may make you wonder how plants do convert a barren landscape into a forest. In this chapter, we will see that plants produce hormones in response to stimuli, such as light and gravity. These hormones affect plant growth and movement by changing the rate and direction of cell expansion, differentiation, and division. They also influence seed germination and the growth and development of a plant, including flowers and fruit. We will explore the many roles of plant hormones and some plant responses to environmental stimuli that help them repopulate severely disturbed areas.

Prairie lupine, *Lupinus lepidus*

Fireweed, *Chamerion anustilfolium*

Mount St. Helens after it erupted

Plant Hormones Regulate Plant Growth and Development

This part of the chapter begins with an overview of hormones as signaling molecules. The overview is followed by a discussion of plant hormones (auxins, gibberellins, cytokinins, abscisic acid, and ethylene) and their effects on plant growth and development.

23.1 Hormones act by utilizing signal transduction pathways

Plant hormones are small organic molecules produced by the plant that regulate growth and development at very low concentrations. A hormone is produced and stored in one part of the plant, but it can travel within the vascular system or from cell to cell to another part of the plant. As we shall see, each hormone may have a variety of responses and may work with other hormones to bring about a specific response suitable to the particular environment.

Plant hormones are chemical signals, and as such, they are the first part of a signal transduction pathway. The signal transduction pathway in **Figure 23.1** is divided into three steps—reception, transduction, and response:

Reception The hormone is the signal that binds to a specific receptor. Each receptor has a particular shape that allows it to bind with only one kind of hormone molecule.

Transduction During transduction, a second messenger is formed or is released into the cytoplasm. The calcium ion, Ca^{2+}, has been identified as a common second messenger in plant cells. (Notice that the hormone is the first messenger, and therefore Ca^{2+} is the second messenger.)

Response The Ca^{2+} then combines with calcium-binding proteins, and the complex brings about the response. During this

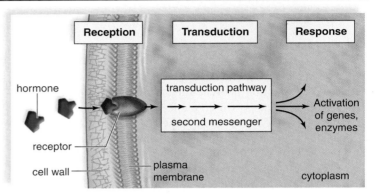

FIGURE 23.1 Signal transduction pathway.

step, the complex can activate an enzyme, and/or increase the permeability of the membrane, and/or activate a gene. The response consists of changes in the activity of the cell.

In Section 23.2, we discuss the effects of auxins on plant growth and development.

> **23.1 Check Your Progress** Based on this discussion, how is it that a hormone might cause a variety of responses?

23.2 Auxins promote growth and cell elongation

Auxins are a group of plant hormones that affect many aspects of plant growth and development. The most common naturally occurring auxin is indoleacetic acid (IAA), produced in shoot apical meristem and also found in young leaves, in flowers, and in fruits.

Effects of Auxin Apically produced or applied auxin prevents the growth of axillary buds, a phenomenon called **apical domi-**

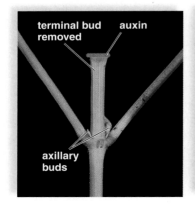

FIGURE 23.2A Auxin plays a role in apical dominance.

nance (**Fig. 23.2A**, *left*). When a terminal bud is removed deliberately or accidentally, the nearest axillary buds begin to grow, and the plant branches. Pruning the top (apical meristem) of a plant generally achieves a fuller look. This removes the source of auxin and, therefore, the apical dominance. More branching of the main body of the plant then occurs (Fig. 23.2A, *right*).

Horticulturists often apply IAA as a paste to plant cuttings to stimulate vigorous root formation. Auxin production by seeds also promotes the growth of fruit. As long as auxin is concentrated in leaves or fruits rather than in the stem, leaves and fruits do not fall off. Therefore, trees can be sprayed with auxin to keep mature fruit from falling to the ground.

The role of auxin in causing stems to bend toward a light source (called **phototropism**) has been studied for quite some time. A **coleoptile** is a protective sheath for the young leaves of the seedling. In 1881, Charles Darwin and his son found that phototropism will not occur if the coleoptile tip of a seedling is removed or covered by a black cap. They concluded that some influence that causes curvature is transmitted from the coleoptile tip to the rest of the shoot.

In a now-famous 1926 experiment, Frits Went cut off the coleoptile tips and placed them on agar (a gelatin-like material) (**Fig. 23.2B**). Auxin diffused from the tips into the agar, and then Went divided the agar. He put a small agar block to one side of a tipless coleoptile

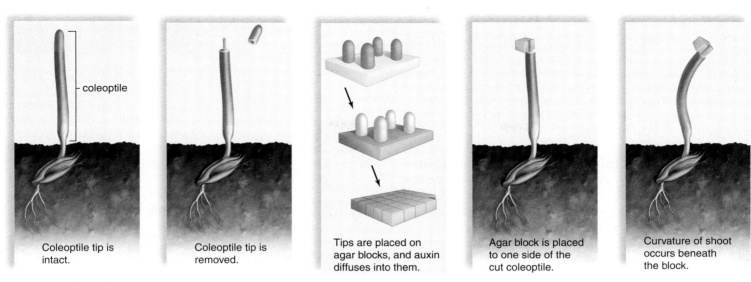

| Coleoptile tip is intact. | Coleoptile tip is removed. | Tips are placed on agar blocks, and auxin diffuses into them. | Agar block is placed to one side of the cut coleoptile. | Curvature of shoot occurs beneath the block. |

FIGURE 23.2B Demonstrating phototropism.

and found that the shoot curved away from that side. The bending occurred even though the seedlings were not exposed to light. Why? Because (1) the agar block released auxin to one side of the shoot, and (2) only the cells on that side experienced elongation, resulting in curvature of the shoot. The experiment showed that bending occurs because auxin is present on only one side of the stem.

How Auxin Brings About Phototropism

Figure 23.2C shows one proposed model for how auxin brings about phototropism. The model suggests that when auxin (the first messenger) moves to the shady side, it binds to receptors in the plasma membrane. Binding leads to the generation of at least three specific second messengers, still to be identified. ❶ One second messenger activates a proton (H⁺) pump, and the resulting acidic conditions

loosen the cell wall because hydrogen bonds are broken and cellulose fibrils are weakened. ❷ Another second messenger activates the Golgi apparatus. The Golgi apparatus then sends out vesicles laden with cell wall materials that will bolster the elongating cell wall. ❸ The third second messenger stimulates a DNA-binding protein that enters the nucleus and activates a particular gene. Activation of this gene leads to the production of growth factors. The result of these activities is elongation of the stem on the shady side so that it bends toward the light.

> **23.2 Check Your Progress** Generally, naturally occurring hormones break down soon after they have done their job. Is this beneficial? Explain.

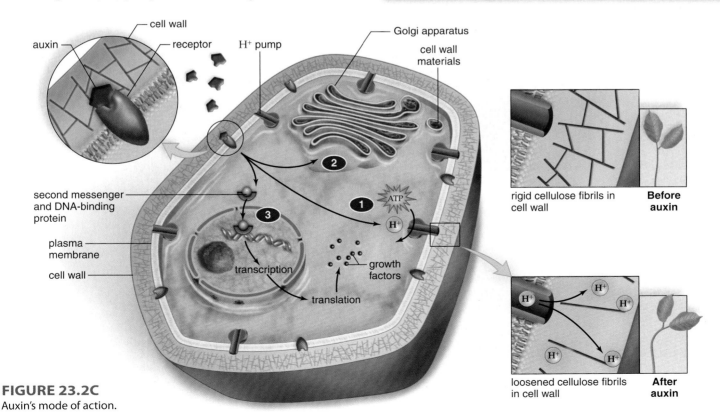

FIGURE 23.2C
Auxin's mode of action.

rigid cellulose fibrils in cell wall — **Before auxin**

loosened cellulose fibrils in cell wall — **After auxin**

Gibberellins are growth-promoting hormones that bring about internode elongation of stems. We know of about 136 gibberellins, and they differ chemically only slightly. The most common of these is gibberellic acid, GA$_3$ (the subscript designation distinguishes it from other gibberellins).

Effects of Gibberellins When gibberellins are applied externally to plants, the most obvious effect is stem elongation. In **Figure 23.3A** (*top*), the cyclamen plant on the left was not treated with gibberellins, while the plant on the right was. Gibberellins can cause dwarf plants to grow, cabbage plants to become 2 m tall, and bush beans to become pole beans. In Figure 23.3A (*bottom*), the plant that produced the grapes on the right was treated with gibberellins. This treatment causes an increase in the space between the grapes, allowing them to grow larger.

Gibberellins were discovered in 1926 when Ewiti Kurosawa, a Japanese scientist, was investigating a fungal disease of rice plants called "foolish seedling disease." The plants elongated too quickly, causing the stem to weaken and the plant to collapse. Kurosawa found that a fungus infecting the plants was producing an excess of a substance. Later, investigators isolated the substance and called it gibberellin, after the name of the fungus *Gibberella fujikuroi*. It wasn't until 1956 that gibberellic acid was isolated from a flowering plant, rather than from a fungus.

Sources of gibberellin in flowering plant parts are young leaves, roots, embryos, seeds, and fruits.

Commercially, gibberellins are used to break the **dormancy** (a time of low metabolic activity and arrested growth) of seeds and buds. After application, plants begin to grow, flowering occurs, or flowers grow larger. Gibberellins have been successfully used to produce larger seedless grapes and to improve rice production.

Action of Gibberellins Research with barley seeds has shown how GA$_3$ breaks the dormancy of seeds and buds. Barley seeds have a large, starchy endosperm, which must be hydrolyzed into sugars to provide energy for growth. After a tissue produces gibberellins, amylase, an enzyme that hydrolyzes starch, appears in cells. As shown in **Figure 23.3B**, it is hypothesized that ❶ GA$_3$ (the first messenger) attaches to a receptor in the plasma membrane. ❷ Then a second messenger, namely calcium ions (Ca^{2+}), binds to a protein. ❸ This complex activates the gene that codes for amylase. ❹ Amylase then acts on starch to release sugars, giving the embryo a source of energy to start growing.

Next, in Section 23.4, we discuss the effects of cytokinins on plant growth and development.

> **23.3** *Check Your Progress* **a.** In one experiment, GA$_3$ is applied to a dwarf plant, which then grows. **b.** In another experiment, GA$_3$ is applied to a different dwarf plant, which does not grow. Which plant might not produce gibberellin, and which plant might have a receptor that is unable to bind to gibberellin?

Stems elongate

Grape size increases

FIGURE 23.3A Effects of gibberellic acid (GA$_3$).

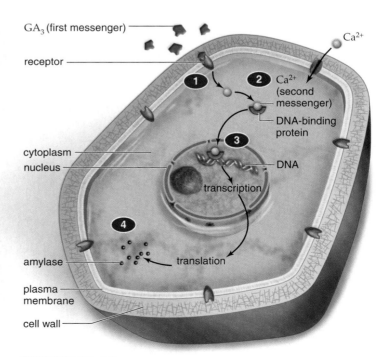

GA$_3$ (first messenger)

Ca^{2+}

receptor

❶ ❷ Ca^{2+} (second messenger)

DNA-binding protein

cytoplasm

nucleus

❸

DNA

transcription

❹

amylase

translation

plasma membrane

cell wall

FIGURE 23.3B GA$_3$'s mode of action.

Cytokinins are a class of plant hormones that, in combination with auxin, promote cell division, initiate growth, and bring about differentiation of cells. Cytokinins are compounds with a structure resembling adenine, one of the purine bases in DNA and RNA.

The cytokinins were discovered as a result of attempts to grow plant tissue and organs in culture vessels in the 1940s. Researchers found that cell division occurred when coconut milk (a liquid endosperm) and yeast extract were added to the culture medium. Although the effective agent or agents could not be isolated, they were collectively called cytokinins because cytokinesis means cell division. A naturally occurring cytokinin was not isolated until 1967. Because it came from the kernels of maize (*Zea*), it was called *zeatin*.

Cytokinins have been isolated from various plants, where they occur in the actively dividing tissues of roots and also in seeds and fruits. The cytokinins are produced in root apical meristem and then transported in xylem throughout the plant. A synthetic cytokinin called kinetin also promotes cell division. Cytokinins have been used to prolong the life of flower cuttings, as well as the freshness of vegetables in storage.

Tissue Culture   **Plant tissue culture** is the process of growing a plant from cells or tissues in laboratory glassware, rather than from the germination of seeds. The interactions of hormones are well exemplified by observing how the varying ratios of auxin and cytokinins affect the differentiation of plant tissues in culture. Researchers are well aware that the ratio of auxin to cytokinin and the acidity of the culture medium determine whether the plant tissue forms only an undifferentiated mass, called a callus (**Fig. 23.4A**), or whether the callus goes on to produce roots (**Fig. 23.4B**), vegetative shoots and leaves (**Fig. 23.4C**), or floral shoots (**Fig. 23.4D**).

These effects illustrate that each plant hormone rarely acts alone; it is the relative concentrations of hormones and their interactions that produce an effect. Modern researchers studying plant growth responses look for an interplay of hormones. They have reported that chemicals called *oligosaccharins* (fragments of short-chained sugars released from the cell wall) are effective in directing differentiation. They hypothesize that auxin and cytokinins are a part of a signal transduction pathway, which leads to the activation of enzymes that release these fragments from the cell wall.

Senescence   Cytokinins prevent senescence and initiate growth. When a plant organ, such as a leaf, loses its natural color, it is most likely undergoing an aging process called **senescence**. During senescence, large molecules within the leaf are broken down and transported to other parts of the plant. Senescence need not affect the entire plant at once; for example, as some plants grow taller, they naturally shed their lower leaves, and ripened fruits routinely separate from the parent plant.

Senescence of leaves can be prevented by applying cytokinins. Not only can cytokinins prevent the death of leaves, but they can also initiate leaf growth. Axillary buds begin to grow despite apical dominance when cytokinins are applied to them.

The next section discusses the effects of abscisic acid on plant growth and development.

> **23.4** *Check Your Progress*   **If you wanted to increase the size of a plant organ, you might apply both cytokinins and gibberellins. Explain.**

FIGURE 23.4A
A callus.

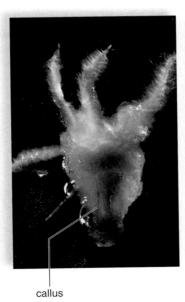

callus

FIGURE 23.4B
Callus produces roots.

callus

FIGURE 23.4C   Callus produces vegetative shoots and leaves.

callus

FIGURE 23.4D   Callus produces floral shoots.

Abscisic acid (ABA) is produced by any "green tissue" (that contains chloroplasts). ABA is also produced in monocot endosperm and roots, where it is derived from carotenoid pigments. Abscisic acid is sometimes called the stress hormone because it initiates and maintains seed and bud dormancy and brings about the closure of stomata.

It was once believed that ABA functioned in **abscission**, the dropping of leaves, fruits, and flowers from a plant. But although the external application of ABA promotes abscission, this hormone is no longer believed to function naturally in this process. Instead, the hormone ethylene seems to bring about abscission.

Dormancy   Recall that dormancy is a period of low metabolic activity and arrested growth. Dormancy occurs when a plant organ readies itself for adverse conditions by ceasing to grow (even though conditions at the time may be favorable for growth). For example, it is believed that ABA moves from leaves to vegetative buds in the fall, and thereafter these buds are converted to winter buds. A winter bud is covered by thick, hardened scales (**Fig. 23.5A**). A reduction in the level of ABA and an increase in the level of gibberellins are believed to break seed and bud dormancy. Then seeds germinate, and buds send forth leaves. In **Figure 23.5B**, corn kernels have begun to germinate on the developing cob because this maize mutant is deficient in ABA. Abscisic acid is needed to maintain the dormancy of seeds.

FIGURE 23.5A
Abscisic acid promotes the formation of winter buds.

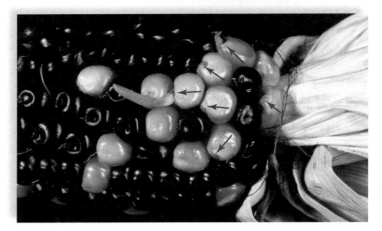

FIGURE 23.5B   Corn kernels start to germinate on the cob (see arrows) due to low abscisic acid.

Closing of Stomata   Abscisic acid brings about the closing of stomata when a plant is under water stress, as shown in **Figure 23.5C**:

Left: The stoma is open.

Middle: When ABA (the first messenger) binds to its receptor in the guard cell plasma membrane, the second messenger (Ca^{2+}) enters. Now, K^+ channels open, and K^+ exit the guard cells. After K^+ exit, so does water.

Right: The stoma closes.

Investigators have also found that ABA induces rapid depolymerization of actin filaments and formation of a new type of actin that is randomly oriented throughout the cell. This change in actin organization may be part of the signal transduction pathways involved in stomata closure.

In the next section, we discuss the effects of ethylene on plant growth and development.

> **23.5 *Check Your Progress***   *a.* **Why is abscisic acid sometimes referred to as an inhibitory hormone?** *b.* **What hormone has the opposite effect of ABA on seed and bud dormancy?**

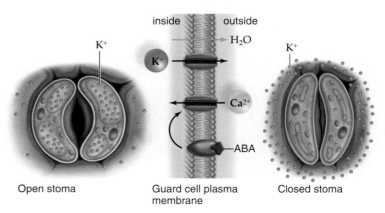

FIGURE 23.5C   Abscisic acid promotes closure of stomata.

Ethylene is a gas formed from the amino acid methionine. This hormone is involved in abscission and the ripening of fruits.

Ethylene Causes Abscission The presence of auxin, and perhaps gibberellin, probably initiates abscission. But once abscission has begun, ethylene stimulates certain enzymes, such as cellulase, which helps cause leaf, fruit, or flower drop. In **Figure 23.6A**, a ripe apple, which gives off ethylene, is under the bell jar on the right, but not under the bell jar on the left. As a result, only the holly plant on the right loses it leaves.

Ethylene Ripens Fruit In the early 1900s, it was common practice to prepare citrus fruits for market by placing them in a room with a kerosene stove. Only later did researchers realize that an incomplete combustion product of kerosene, namely **ethylene**, ripens fruit. It does so by increasing the activity of enzymes that soften fruits. For example, it stimulates the production of cellulase, which weakens plant cell walls. It also promotes the activity of enzymes that produce the flavor and smell of ripened fruits. And it breaks down chlorophyll, inducing the color changes associated with fruit ripening.

Ethylene moves freely through a plant by diffusion, and because it is a gas, ethylene also moves freely through the air. That is why a barrel of ripening apples can induce ripening of a bunch of bananas some distance away. Ethylene is released at the site of a plant wound due to physical damage or infection (which is why one rotten apple spoils the whole bushel).

The use of ethylene in agriculture is extensive. It is used to hasten the ripening of green fruits such as melons and honeydews, and is also applied to citrus fruits to attain pleasing colors before marketing. Normally, tomatoes ripen on the vine because the plants produce ethylene (**Fig. 23.6B**). Today, tomato plants can be genetically modified to not produce ethylene. This facilitates shipping because green tomatoes are

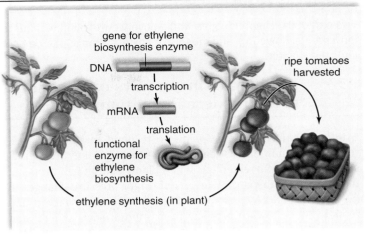

FIGURE 23.6B Wild-type tomatoes ripen on the vine after producing ethylene.

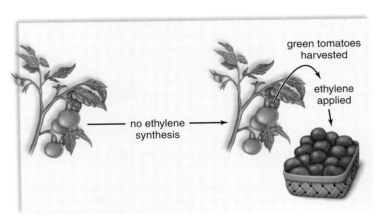

FIGURE 23.6C Tomatoes are genetically modified to produce no ethylene and stay green for shipping.

not subject to as much damage. Once the tomatoes have arrived at their destination, they can be exposed to ethylene so that they ripen (**Fig. 23.6C**).

Other Effects of Ethylene Ethylene is involved in axillary bud inhibition. Auxin, transported down from the apical meristem of the stem, stimulates the production of ethylene, and this hormone suppresses axillary bud development. Ethylene also suppresses stem and root elongation, even in the presence of other hormones.

This completes our discussion of plant hormones. The next part of the chapter explores plant responses to environmental stimuli.

> **23.6 Check Your Progress** Explain why ethylene is an effective hormone, even though it is a gas.

No abscission　　　　　Abscission

FIGURE 23.6A Ethylene promotes abscission.

This part of the chapter begins with an overview of plant responses to environmental stimuli. Then we discuss tropisms, turgor and sleep movements, flowering, and the role of phytochrome in plant responses. Finally, a discussion of plant responses to the biotic environment ends the chapter.

23.7 Plants have many ways of responding to their external environment

Many mechanisms enable plants to respond to external environmental conditions. Often, plant responses to environmental stimuli involve movement, as when plants exhibit heliotropism, the tracking of the sun by means of turgor pressure changes in the petiole (**Fig. 23.7**). At other times, growth allows shoots and roots to move toward or away from stimuli, such as light and gravity. Responses that recur regularly every 24 hours, as when stomata open and close without light cues, are called circadian rhythms. Events that recur every season, such as flowering, are often responses to the photoperiod (day or night length). Plants also respond to other living things, and we will discuss how they defend themselves against predators.

In the next section, we begin our study of plant responses to environmental stimuli by describing tropisms. Keep in mind that if a response requires growth and development, it also requires the participation of a hormone, one of those that we have just discussed.

> **23.7 Check Your Progress** Roots grow toward water. Explain why this is adaptive.

FIGURE 23.7 Heliotropism, sun tracking in the buttercup, *Ranunculus ficaria*.

23.8 Tropisms occur when plants respond to stimuli

Growth toward or away from a stimulus, such as gravity or light, is a **tropism**. Growth toward a stimulus is called a *positive tropism*, and growth away from it is called a *negative tropism*. For example, roots are positively gravitropic because they grow in the direction of gravity. Tropisms are due to differential growth—one side of an organ elongates faster than the other, and the result is a curving toward or away from the stimulus (see Section 23.2). A number of tropisms have been observed in plants, the three best-known being **gravitropism**, **phototropism**, and **thigmotropism**:

> Gravitropism: movement in response to gravity
>
> Phototropism: movement in response to a light stimulus
>
> Thigmotropism: movement in response to touch

Several other tropisms are chemotropism (chemicals), traumotropism (trauma), skototropism (dark), and aerotropism (oxygen).

Gravitropism Figure 23.8A shows that when an upright plant is placed on its side, the stem displays ❶ negative gravitropism because it grows upward, opposite the pull of gravity. Charles Darwin and his son were among the first to say that roots, in contrast to stems, show ❷ positive gravitropism. Further, they discovered that if the root cap is removed, roots no longer respond to gravity. ❸ Later, it was discovered that root cap cells contain sensors called **statoliths**, which are starch

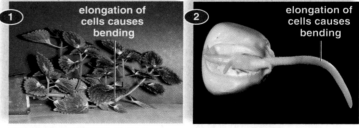

Negative gravitropism of stem   Positive gravitropism of root

FIGURE 23.8A Gravitropism.

Sedimentation of amyloplasts (arrows)   25 µm

grains located within amyloplasts, a type of plastid. Perhaps gravity causes the amyloplasts to settle to a lower part of the cell, where they come in contact with the endoplasmic reticulum

FIGURE 23.8B Positive phototropism in stems.

FIGURE 23.8D Coiling response of a morning glory plant, *Ipomoea*.

of a phosphate group from ATP (adenosine triphosphate) to a protein portion of the photoreceptor. ❸ The phosphorylated photoreceptor triggers a signal transduction pathway that, in some unknown way, leads to the binding of auxin. It is clearly adaptive for plants to have a way to increase their photosynthetic efficiency by bending and exposing their leaves to light.

Thigmotropism Unequal growth due to contact with solid objects is called thigmotropism. An example of this response is the coiling of the tendrils or the stems of plants, such as morning glory (**Fig. 23.8D**). These growth changes occur upon contact with a solid object. The cells in contact with the object grow less, while those on the opposite side elongate. Thigmotropism can be quite rapid; a tendril has been observed to encircle an object within 10 minutes. The response also endures; a couple of minutes of touching can bring about a response that lasts for several days. But sometimes the response can be delayed; tendrils touched in the dark will respond once they are illuminated. ATP, rather than light, can cause the response. Therefore, the need for light may simply be a need for ATP. Also, the hormones auxin and ethylene may be involved, since they can induce curvature of tendrils even in the absence of touch.

Turgor and sleep movements also allow plants to respond to external stimuli, as discussed in Section 23.9.

(ER). The ER then releases stored calcium ions (Ca^{2+}), and this leads to an influence of auxin on cell growth.

The hormone auxin is known to bring about the positive gravitropism of roots and the negative gravitropism of stems. The two types of tissues respond differently to auxin, which appears on the lower side of both stems and roots after gravity has been perceived. Auxin inhibits the growth of root cells; therefore, only the cells of the upper surface elongate so that the root curves downward. Auxin stimulates the growth of stem cells; therefore, the cells of the lower surface elongate, and the stem curves upward.

Phototropism As discussed in Section 23.2, positive phototropism of stems occurs because the cells on the shady side of the stem elongate due to the presence of auxin (**Fig. 23.8B**). Curving away from light is called negative phototropism. Roots, depending on the species examined, are either insensitive to light or exhibit negative phototropism.

Through the study of mutant plants, it is now known that phototropism occurs because plants respond to blue light (**Fig. 23.8C**). ❶ When blue light is absorbed, the pigment portion of a photoreceptor, called phototropin (phot), undergoes a conformation change. ❷ This change results in the transfer

23.8 Check Your Progress If a plant is in a horizontal position and rotated horizontally, would the stem or the root exhibit gravitropism? Explain.

FIGURE 23.8C
In the presence of blue light, a photoreceptor called phototropin (phot) initiates a signal transduction pathway.

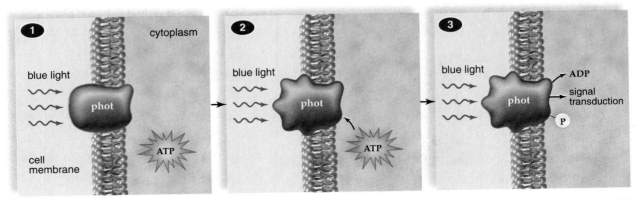

Recall that a plant cell exhibits turgor when it fills with water:

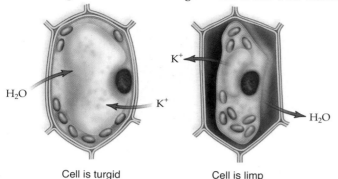

Cell is turgid   Cell is limp

In general, if water exits the many cells of a leaf, the leaf goes limp. Conversely, if water enters a limp leaf, and cells exhibit turgor, the leaf moves as it regains its former position. **Turgor movements** are dependent on turgor pressure changes in plant cells. In contrast to tropisms, turgor movements do not involve growth and are not related to the source of the stimulus.

Turgor movements can result from touch, shaking, or thermal stimulation. The sensitive plant, *Mimosa pudica*, has compound leaves, meaning that each leaf contains many leaflets. Touching one leaflet collapses the whole leaf (**Fig. 23.9A**). *Mimosa* is remarkable because the progressive response to the stimulus takes only a second or two.

The portion of a plant involved in controlling turgor movement is a thickening called a pulvinus at the base of each leaflet. A leaf folds when the cells in the lower half of the pulvinus, called the motor cells, lose potassium ions (K^+), and then water follows by osmosis. When the pulvinus cells lose turgor, the leaflets of the leaf collapse. An electrical mechanism may cause the response to move from one leaflet to another. The speed of an electrical charge has been measured, and the rate of transmission is about 1 cm/sec.

A Venus flytrap closes its trap in less than one second when three hairs at the base of the trap, called the trigger hairs, are touched by an insect. When the trigger hairs are stimulated by the insect, an electrical charge is propagated throughout the lobes of a leaf. Exactly what causes this electrical charge is being studied. Perhaps (1) the cells located near the outer region of the lobes rapidly secrete hydrogen ions into their cell walls, loosening them, and allowing the walls to swell rapidly by osmosis; or (2) perhaps the cells in the inner portion of the lobes and the midrib rapidly lose ions, leading to a loss of water by osmosis and collapse of these cells. In any case, it appears that turgor movements are involved.

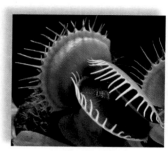

Venus flytrap, *Dionaea*

Sleep Movements and Circadian Rhythms Leaves that close at night are said to exhibit sleep movements. Activities such as sleep movements that occur regularly in a 24-hour cycle are called **circadian rhythms**. One of the most common

Before

pulvinus   vascular tissue

FIGURE 23.9A
Turgor movements in a mimosa plant.

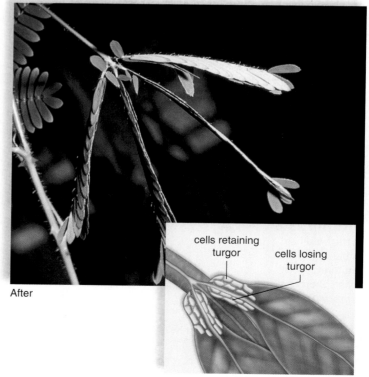

After

cells retaining turgor   cells losing turgor

Prayer plant (morning)   Prayer plant (night)

Morning glory (morning)   Morning glory (night)

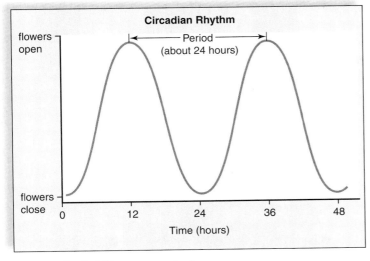

FIGURE 23.9B Circadian rhythms.

examples occurs in a houseplant called the prayer plant (*Maranta leuconeura*) because at night the leaves fold upward into a shape resembling hands at prayer (**Fig. 23.9B**). This movement is also due to changes in the turgor pressure of motor cells in a pulvinus located at the base of each leaf.

To take a few other examples, morning glory (*Ipomoea leptophylla*) is a plant that opens its flowers in the early part of the day and closes them at night. In most plants, stomata open in the morning and close at night, and some plants secrete nectar at the same time of the day or night. Figure 23.9B, *bottom*, shows how a circadian rhythm would appear if graphed for a morning glory plant.

To qualify as a circadian rhythm, the activity must (1) occur every 24 hours; (2) take place in the absence of external stimuli, such as in dim light; and (3) be able to be reset if external cues are provided. For example, if you take a transcontinental flight, you will likely suffer jet lag because your body will still be attuned to the day-night pattern of its previous environment. But after several days, you will most likely have adjusted and will be able to go to sleep and wake up according to your new time.

Biological Clock The internal mechanism by which a circadian rhythm is maintained in the absence of appropriate environmental stimuli is termed a **biological clock**. If organisms are sheltered from environmental stimuli, their biological clock keeps the circadian rhythms going, but the cycle extends. In prayer plants, for example, the sleep cycle changes to 26 hours when the plant is kept in constant dim light, as opposed to 24 hours when in traditional day/night conditions. Therefore, it is suggested that biological clocks are synchronized by external stimuli to 24-hour rhythms. The length of daylight compared to the length of darkness, called the photoperiod, sets the clock. Temperature has little or no effect. This is adaptive because the photoperiod indicates seasonal changes better than temperature changes. Spring and fall, in particular, can have both warm and cold days.

Work with *Arabidopsis* and other organisms suggests that the biological clock involves the transcription of a small number of "clock genes." One model proposes that the information-transfer system from DNA to RNA to enzyme to metabolite, with all its feedback controls, is intrinsically cyclical and could be the basis for biological clocks. In *Arabidopsis*, the biological clock involves about 5% of the genome. These genes control sleep movements, the opening and closing of stomata, the discharge of floral fragrances, and the metabolic activities associated with photosynthesis. The biological clock also influences seasonal cycles that depend on day-night lengths, including the regulation of flowering.

While circadian rhythms are outwardly very similar in all species, the clock genes that have been identified are not the same in all species. It would seem, then, that biological clocks have evolved several times to perform similar tasks.

Flowering, as a response to the photoperiod, is discussed next.

> **23.9 Check Your Progress** Many bat- and moth-pollinated plants open only at night and often produce scent during the evening only. Explain why this is adaptive.

23.10　Flowering is a response to the photoperiod in some plants

Many physiological changes in plants are related to a seasonal change in day length. Such changes include seed germination, the breaking of bud dormancy, and the onset of color changes associated with fall foliage. A physiological response prompted by changes in the length of day or night is called **photoperiodism**. In some plants, photoperiodism influences flowering; for example, violets and tulips flower in the spring, and asters and goldenrod flower in the fall.

In the 1920s, when U.S. Department of Agriculture scientists were trying to improve tobacco, they decided to grow plants in a greenhouse, where they could artificially alter the photoperiod. They came to the conclusion that plants can be divided into three groups:

1. **Short-day plants** flower when the day length is shorter than a critical length. (Examples are cocklebur, goldenrod, poinsettia, and chrysanthemum.)

2. **Long-day plants** flower when the day length is longer than a critical length. (Examples are wheat, barley, rose, iris, clover, and spinach.)

3. **Day-neutral plants** are not dependent on day length for flowering. (Examples are tomato and cucumber.)

The criterion for designating plants as short-day or long-day is not an absolute number of hours of light, but a critical number that either must be or cannot be exceeded. Spinach is a long-day plant that has a critical length of 14 hours; ragweed is a short-day plant with the same critical length. Spinach, however, flowers in the summer when the day length increases to 14 hours or more, and ragweed flowers in the fall, when the day length shortens to 14 hours or less. In addition, we now know that some plants require a specific sequence of day lengths in order to flower.

In 1938, K. C. Hammer and J. Bonner began to experiment with artificial lengths of light and dark that did not necessarily correspond to a normal 24-hour day. These investigators discovered that the cocklebur, a short-day plant, will *not* flower if a required long dark period is interrupted by a brief flash of white light. (Interrupting the light period with darkness has no effect.) On the other hand, a long-day plant will flower if an overly long dark period is interrupted by a brief flash of white light. They concluded that the length of the dark period, not the length of the light period, controls flowering. Of course, in nature, short days always go with long nights, and vice versa.

To recap, let's consider the figure on this page:

- Cocklebur is a short-day plant (**Fig. 23.10**, *left*). **1** When the night is longer than a critical length, cocklebur flowers. **2** The plant does *not* flower when the night is shorter than the critical length. **3** Cocklebur also does *not* flower if the longer-than-critical-length night is interrupted by a flash of light.

- Clover is a long-day plant (Fig. 23.10, *right*). **4** When the night is shorter than a critical length, clover flowers. **5** The plant does *not* flower when the night is longer than a critical length. **6** Clover does flower when a slightly longer-than-critical-length night is interrupted by a flash of light. These observations are explained in Section 23.11.

> **23.10 Check Your Progress** A plant is a long-day plant. Explain why the plant will still flower if the long day is interrupted by a period of darkness.

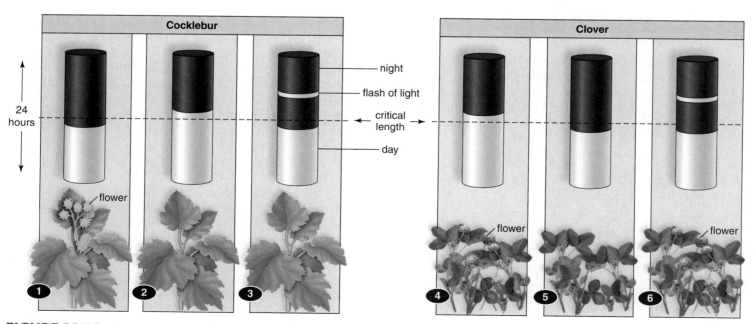

FIGURE 23.10 (*Left*) Flowering in a long-night (short-day) plant; (*right*) flowering in a short-night (long-day) plant.

If flowering is dependent on day and night length, plants must have some way to detect these periods. Many years of research by scientists at the U.S. Department of Agriculture led to the discovery of a plant pigment called phytochrome. (Although we will use phytochrome in the singular, several different phytochromes have been identified.) **Phytochrome** is a photoreceptor composed of two parts: The smaller part is a blue or blue-green pigment that absorbs red and far-red light, and the larger part is a protein active in transduction pathways because it is a kinase. As **Figure 23.11A** indicates:

P_r (phytochrome red) absorbs red light (of 660-nm wavelength) and is converted to P_{fr}.

P_{fr} (phytochrome far-red) absorbs far-red light (of 730-nm wavelength) and is converted to P_r.

Direct sunlight contains more red light than far-red light; therefore, P_{fr} is apt to be present in plant leaves during the day. In the shade and at sunset, there is more far-red light than red light; therefore, P_{fr} is converted to P_r as night approaches. There is a slow metabolic conversion of P_{fr} to P_r during the night.

Effects of Phytochrome Phytochrome conversion is the first step in a signal transduction pathway that results in flowering. In short-day (long-night) plants, the presence of P_{fr} inhibits flowering, and in long-day (short-night) plants, the presence of P_{fr} promotes flowering. It has now been shown that phytochrome in the P_{fr} form leads to the activation of a transcription factor in the cytoplasm. The complex migrates to the nucleus, where it binds to DNA and turns genes on or off. At one time, researchers hypothesized that a special flowering hormone, called florigen, might exist, but such a hormone has never been discovered.

The $P_r \longrightarrow P_{fr}$ conversion cycle is now known to control other growth functions in plants besides flowering, such as seed germination and stem elongation. The presence of P_{fr} indicates to some seeds that sunlight is present and conditions are favor-

Etiolation

Normal growth

FIGURE 23.11B Phytochrome control of growth pattern.

able for germination. This is why some seeds must be only partly covered with soil when planted. Germination of other seeds, such as those of *Arabidopsis*, is inhibited by light, so they must be planted deeper. Following germination, the presence of P_r indicates that stem elongation may be needed to reach sunlight. Seedlings grown in the dark etiolate—that is, the stem increases in length, and the leaves remain small. The plant tends to be light colored due to a decreased amount of chlorophyll (**Fig. 23.11B**). Once the seedling is exposed to sunlight and P_r is converted to P_{fr}, the seedling begins to grow normally—the leaves expand and become darker green, and the stem branches.

So far, we have discussed plant responses to the physical environment (e.g., gravity, light, and photoperiod). The next section takes a look at plant responses to the biotic environment.

> **23.11** *Check Your Progress* **Describe the signal transduction pathway as it applies to phytochrome.**

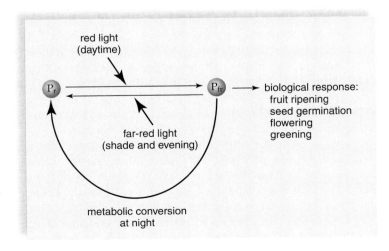

red light
(daytime)

P_r P_{fr} → biological response:
fruit ripening
seed germination
flowering
greening

far-red light
(shade and evening)

metabolic conversion
at night

FIGURE 23.11A Phytochrome conversion cycle.

Plants are always under attack by herbivores (animals that eat plants) and parasites. Fortunately, they have an arsenal of defense mechanisms to deal with insects and fungi, for example, (**Fig. 23.12A**).

First Line of Defense A plant's cuticle-covered epidermis and bark, if present, do a good job of discouraging attackers. The thorns of roses and the spines of cactuses are examples of other surface features that deter herbivores. Small hairs called trichomes, which project from the epidermis, may contain poisons. For example, under the slightest pressure, the stiff trichomes of the stinging nettle lose their tips, forming "hypodermic needles" that shoot a stinging chemical into an intruder.

Unfortunately, herbivores have ways around a plant's first line of defense. A fungus can invade a leaf by way of the stomata and set up shop inside a leaf, where it feeds on nutrients meant for the plant. Underground nematodes have sharp mouthparts to break through the epidermis of a root and establish a parasitic relationship, sometimes by way of a single cell, which enlarges and transfers carbohydrates to the animal. Similarly, the tiny insects called aphids have styletlike mouthparts that allow them to tap into the phloem of a nonwoody stem (see Fig. 22.6). These examples illustrate why plants need several other types of defenses not dependent on the outer surface.

Chemical Defenses The primary metabolites of plants, such as sugars and amino acids, are necessary to the normal workings of a cell, but plants also produce so-called **secondary metabolites** as a defense mechanism. Secondary metabolites were once thought to be waste products, but now we know that they are part of a plant's arsenal to prevent predation. Tannins, present in or on the epidermis of leaves, are defensive compounds that interfere with the outer proteins of bacteria and fungi. They also deter herbivores because of their astringent effect on the mouth and their interference with digestion. Some secondary metabolites, such as bitter nitrogenous substances called **alkaloids** (e.g., morphine, nicotine, and caffeine), are well-known to humans because we use them for our own purposes. The seedlings of coffee plants contain caffeine at a concentration high enough to kill insects and fungi by blocking DNA and RNA synthesis. Other secondary metabolites include the **cyanogenic glycosides**

Foxglove

(molecules containing a sugar group) that break down to cyanide and inhibit cellular respiration. Foxglove (*Digitalis purpurea*) produces deadly cardiac and steroid glycosides, which cause nausea, hallucinations, convulsions, and death in animals that ingest them. Section 23.13 tells about the search for secondary metabolites in tropical rain

Alfalfa plant bug

Fungus infection

Monarch caterpillar and butterfly

FIGURE 23.12A Plant predators.

forests, which have yielded many medicines for humans. Taxol, an unsaturated hydrocarbon, from the Pacific yew (*Taxus brevifolia*), is now a well-known cancer-fighting drug.

Even with regard to secondary metabolites, predators can be one step ahead of the plant. Monarch caterpillars are able to feed on milkweed plants, despite the presence of a poisonous glycoside, and they even store the chemical in their body. In this way, the caterpillar and the butterfly become poisonous to their own predators (Fig. 23.12A). Birds that become sick after eating a monarch butterfly know to leave them alone thereafter.

Wound Responses Wound responses illustrate that plants can make use of signal transduction pathways to produce chemical defenses only when they are needed. After a leaf is chewed or injured, a plant produces proteinase inhibitors, chemicals that destroy the digestive enzymes of a predator feeding on them. The proteinase inhibitors are produced throughout the plant, not just at the wound site. The growth regulator that brings about this effect is a small peptide called **systemin** (**Fig. 23.12B**). Systemin is produced in the wound area in response to the predator's saliva, but then it travels between cells to reach phloem, which distributes it about the plant. Signal transduction occurs in cells with systemin receptors, and the cells produce proteinase inhibitors. A chemical called jasmonic acid, and also possibly a chemical called salicyclic acid, are part of this signal transduction pathway. Salicyclic acid (a chemical also found in aspirin) has been known since the 1930s to bring about a phenomenon called systemic acquired resistance (SAR), the production of antiherbivore chemicals by defense genes. Recently, companies have begun marketing salicyclic acid and other similar compounds as a way to activate SAR in crops, including tomato, spinach, lettuce, and tobacco.

Hypersensitive Response (HR)

On occasion, plants produce a specific gene product that binds (like a key fits a lock) to a viral, bacterial, or fungal gene product made within the cell. This combination offers a way for the plant to "recognize" a particular pathogen. A signal transduction pathway now ensues, and the final result is a **hypersensitive response (HR)** that seals off the infected area and will also initiate the wound response just discussed.

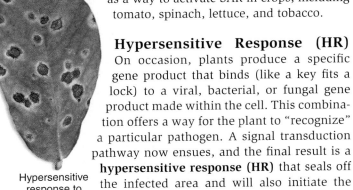

Hypersensitive response to fungus invasion

Indirect Defenses

Some defenses of plants are called indirect because they do not kill or discourage an herbivore outright. For example, female butterflies are less likely to lay their eggs on plants that already have butterfly eggs. So, because the leaves of some passion flowers (genus *Passiflora*) display physical structures resembling the yellow eggs of *Heliconius* butterflies, these butterflies do not lay eggs on this plant. Other plants produce hormones that prevent caterpillars from metamorphosing into adults and laying more eggs.

Certain plants attract the natural enemies of caterpillars feeding on them. They produce volatile molecules that diffuse into the air and advertise that food is available for a carnivore (an animal that eats other animals). For example, lima beans produce volatiles that attract carnivore mites only when they are being damaged by a spider mite. Corn and cotton plants release volatiles that attract wasps, which then inject their eggs into caterpillars munching on their leaves. The eggs develop into larvae that eat the caterpillars, not the leaves. The combined effect of a wide range of volatiles, some that attract predators of plant pests

and some that simply prevent egg laying, can result in as much as a 90% reduction in the number of viable eggs on leaves.

Relationships with Animals Mutualism is a relationship between two species in which both species benefit. As evidence that a mutualistic relationship can help protect a plant from predators, consider the bullhorn acacia tree, which provides a home for ants of the species *Pseudomyrmex ferruginea*. Unlike other acacias, this species has swollen thorns with a hollow interior where ant larvae can grow and develop. In addition to housing the ants, acacias provide them with food. The ants feed from nectaries at the base of leaves and eat fat- and protein-containing nodules called Beltian bodies, which are found at the tips of the

Mutualism between ants and a plant

leaves. In return, the ants constantly protect the plant by attacking and stinging any would-be herbivores because, unlike other ants, they are active 24 hours a day. Indeed, when the ants on experimental trees were removed, the acacia trees died.

Again, an herbivore can be one step ahead of the plant. Trees of the genus *Croton* have nectaries for ants, but unfortunately caterpillars of the butterfly *Thisbe irenea* also have nectaries for these ants. The caterpillars release chemicals that cause the ants to protect them while the caterpillars feast on the trees' leaves. Even worse, the caterpillars, besides eating the leaves, feed from the ant nectaries on the *Croton* trees.

Investigators, such as Eloy Rodriguez discussed next, have helped discover secondary metabolites that can be used as medicines in humans.

> **23.12 Check Your Progress** Are plants acting purposefully when they employ an antipredator defense?

FIGURE 23.12B
Wound response in tomato.

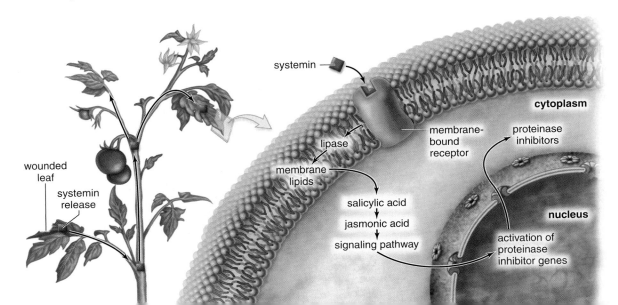

wounded leaf

systemin release

systemin

lipase

membrane lipids

membrane-bound receptor

cytoplasm

proteinase inhibitors

salicylic acid

jasmonic acid

signaling pathway

nucleus

activation of proteinase inhibitor genes

23.13   Eloy Rodriguez has discovered many medicinal plants

Eloy Rodriguez (**Fig. 23.13**), a Mexican-American biochemist formerly at the University of California–Irvine, now has an endowed chair, called the James A. Perkins Professor of Biology and Environmental Studies at Cornell University. A typical year for Rodriguez is seven months at Cornell, teaching biodiversity and tropical plant research, plus doing chemical research in his own lab. Then, for five months he does field research in the rain forests of South America and the Caribbean and the deserts of Africa. He involves students in all his activities. Rodriguez has spent 25 years traveling through the jungles and deserts to learn about medicinal plants used by native healers. A leader of 20th-century American ethnobotany (the study of plants used traditionally by indigenous people for food, medicine, shelter, and other purposes), Rodriguez is one of the first modern scientists to extract medicinal compounds from jojoba and candelilla in the laboratory. Yet, he reminds his students that these two plants were probably used medically by ancient desert dwellers as well.

Rodriguez points out that, without the participation of native peoples, Americans don't know where to begin to look for medicines in a tropical rain forest that contains 5,000 plant species. Indigenous people can point out which plants contain potential medicines because their ancestors, for many generations, have been using certain plants to heal diseases. He is concerned that only a small percentage of plants recognized as medicines have been studied by western pharmacologists. Of the entire 250,000 flowering plant species, only about 3–5% of them have been investigated for medicinal purposes, according to Rodriguez.

Rodriguez is working vigorously to save endangered plants, while filling the gap of knowledge between age-old native use of plants and their scientific investigation in modern labs. As he told a *Wildlife Conservation* journalist in 1991, "We've only just scratched the surface and we've discovered the first drug against malaria (quinine), and the first drug against the cough (codeine), and the first drug against cancer (vincristine)." Quinine was discovered in 1640 when Spanish colonists noticed Peruvians using the bark of the cinchona tree to treat malaria. Subsequently, quinine was isolated in 1820 and synthesized in 1944. Codeine is one of 10 alkaloids produced by the immature seed capsule of the opium poppy; morphine is another. Eli Lilly and Co. introduced the vinca alkaloids, a class of anticancer drugs, in the 1960s. These chemicals have all come from the Madagascar periwinkle.

By combining modern science with the age-old observations of indigenous people around the world, Rodriguez has learned that creosote (*Larrea tridentata*), a plant of the Sonoran desert that natives call "hedionda" or "bad little smeller," contains over 1,000 potential drugs. Native Americans have used creosote for generations to treat colds, chest infections, intestinal problems, menstrual pain, dandruff, toothaches, and other ills. Referring to its chemical properties, Rodriguez calls creosote a "botanical superstar."

FIGURE 23.13 Eloy Rodriguez believes that animals use plants to cure their illnesses and that we can learn to do the same.

Other potential superstars come from Africa. A colleague of Rodriguez, Harvard University anthropologist Richard Wrangham, discovered that sick chimps in Tanzania would chew on the pith of *Vernonia* plants. Subsequently, it was found that *Vernonia* pith contains chemicals with antiparasitic activity against microorganisms that infect both chimps and humans. The chemicals suppress the movement and egg-laying abilities of the parasitic worm *Schistosoma japonicum*. Rodriguez is delighted that one of the drugs discovered in studying the apes turned out to be an effective drug in humans. Rodriguez and Wrangham coined the word "zoopharmacognosy" to refer to animals' deliberate use of medicinal plants to treat their illnesses. As Rodriguez explained in *National Wildlife*,

> We think there is some learning and that knowledge is passed on—and that this has been going on for a long time. . . . Wild apes five or six million years ago were already using plants. . . . And as the human line evolved, we obviously learned from animals. We observed them. It gives us a peek into how we came about selecting medicinal plants. (1994. *National Wildlife* 32 [1]:46.)

Rodriguez obtained his Ph.D. at the University of Texas in 1975. He was named University of California–Irvine's first professor of phytochemistry in 1976. He came to Cornell in 1994 and has received many awards as an outstanding, hispanic educator.

23.13 *Check Your Progress* Why might a chemical produced by a plant to deter insect predators also be effective as a chemotherapeutic drug for cancer in humans?

Behavior in plants can be understood in terms of three different levels of organization. On the species level, plant responses that promote survival and reproductive success have evolved through natural selection. At the organismal level, hormones coordinate the growth and development of plant parts. And at the cellular level, hormones influence cellular metabolism.

We can illustrate these three perspectives by answering the question, Why do plants bend toward the light? At the species level, plants that bend toward the light will be able to produce more organic food and will have more offspring. At the organis-

mal level, light can cause the movement of auxin in certain plant parts, and when auxin moves from the lit side of a stem to the shady side, elongation occurs; thereafter, the plant bends toward the light. At the cellular level, after auxin is received by a plant cell, cellular activities cause its walls to expand.

The response of both the organism and the cell involve three steps: (1) reception of the stimulus, (2) transduction of the stimulus, and (3) response to the stimulus. In this chapter, we mainly discussed these steps at the cellular level. For example, reception of auxin by plasma membrane receptors is the first step, cellular activities

is the second step, and stretching of the cell wall is the third step.

In Chapter 24, we stress the organismal level by discussing how plants reproduce on land. Certainly we know that the manner in which flowering plants reproduce is an adaptation to the land environment. But we will consider the steps that permit plants to reproduce sexually and asexually. Sexual reproduction involves seed formation and embryo development. Asexual reproduction involving tissue culture of plants has become all the more important because it permits the introduction of improved traits by means of genetic engineering.

The Chapter In Review

Summary

Recovering Slowly

- Plant life on Mount St. Helens is slowly reappearing.
- Plant hormones play a role in the recovery.

Plant Hormones Regulate Plant Growth and Development

23.1 Hormones act by utilizing signal transduction pathways

- Hormones are small organic molecules that are produced in one part of a plant and travel to other parts, where they affect plant growth and development.
- A signal transduction pathway consists of reception, transduction, and response:

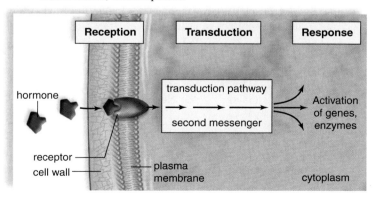

23.2 Auxins promote growth and cell elongation

- IAA (indoleacetic acid) is a natural auxin.
- IAA encourages apical dominance by preventing the growth of axillary buds.
- Went's experiment shows that auxin moves to the shady side of a plant where cells elongate, causing the plant to curve toward the light.

- One model proposes that auxin binds to a receptor, generating three second messengers and leading to elongation of the stem on the shady side.

23.3 Gibberellins control stem elongation

- Gibberellins are stimulatory growth-promoting hormones that cause stems to elongate and also break the dormancy of seeds and buds.
- In a signal transduction pathway in barley seeds, Ca^{2+} is the second messenger that binds to a protein; the complex then activates the gene for amylase.

23.4 Cytokinins stimulate cell division and differentiation

- Cytokinins promote cell division, prevent senescence, and initiate growth.
- Plant tissue culture shows that hormone concentrations and their interactions determine tissue differentiation.

23.5 Abscisic acid suppresses growth of buds and closes stomata

- ABA (abscisic acid) promotes formation of winter buds, promotes seed dormancy, and closes stomata when a plant is water stressed.

23.6 Ethylene stimulates the ripening of fruits

- Ethylene causes abscission (shedding of leaves, flowers, fruits), ripens fruit, and suppresses stem and root elongation.

Plants Respond to Environmental Stimuli

23.7 Plants have many ways of responding to their external environment

- Plant responses to the external environment can involve movement, a circadian rhythm, seasonal events, or a response to the biotic environment.
- The activity of hormones is involved in these responses.

23.8 Tropisms occur when plants respond to stimuli

- A tropism is plant growth toward or away from a unidirectional stimulus, such as light or gravity.
- Negative gravitropism is displayed by a stem; positive gravitropism is displayed by roots. Auxin is involved in both.
- In phototropism, plants curve toward or away from blue light received by a phototropin that initiates a signal transduction pathway. Auxin is involved.
- In thigmotropism, unequal growth results from contact with a solid object (i.e., coiling of tendrils). Auxin and ethylene may be involved.

> Gravitropism: movement in response to gravity
> Phototropism: movement in response to a light stimulus
> Thigmotropism: movement in response to touch

23.9 Turgor and sleep movements are complex responses

- Turgor movements (touch, shaking, thermal stimulation) depend on turgor changes.
- A circadian rhythm consists of periodic fluctuations corresponding to a 24-hour cycle.
- A biological clock is an internal mechanism that maintains the circadian rhythm in the absence of stimuli.

23.10 Flowering is a response to the photoperiod in some plants

- Photoperiodism refers to a physiological response to changes in the length of day or night.
  - Short-day plants flower when day length is shorter than a critical length.
  - Long-day plants flower when day length is longer than a critical length.
  - Day-neutral plants are not dependent on day length for flowering.

23.11 Response to the photoperiod requires phytochrome

- Phytochrome is a photoreceptor that responds to daylight. P_r absorbs red light; P_{fr} absorbs far-red light.
- P_{fr} initiates a signal transduction pathway that turns genes on or off and brings about these effects:

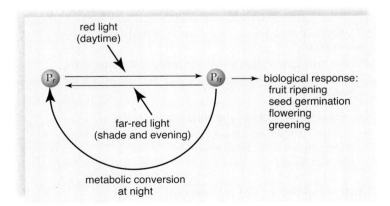

23.12 Plants respond to the biotic environment

- Plants' first line of defense includes a cuticle-covered epidermis, thorns, spines, and trichomes.
- Chemical toxins produced by plants help prevent predation (e.g., tannins, alkaloids, glycosides, terpenes).

- Wound responses involve a signal transduction pathway that leads to activation of genes for proteinase inhibitors.
- A hypersensitive response seals off the area infected by a virus, bacterium, or fungus.
- Indirect responses include the production of volatile molecules that attract carnivores to kill off herbivores or discourage egg laying on leaves.
- Mutualistic relationships are formed between certain plants and animals (e.g., the bullhorn acacia tree and ants).

23.13 Eloy Rodriguez has discovered many medicinal plants

- Ethnobotany is the study of plants used by indigenous people for food, medicine, shelter, and so on.
- Rodriguez is working to save endangered plants and to alert others to the medicinal values of plants.

Testing Yourself

Plant Hormones Regulate Plant Growth and Development

1. During which step of the signal transduction pathway is a second messenger released into the cytoplasm?
   a. reception
   b. response
   c. transduction
2. Which of the following plant hormones causes apical dominance?
   a. auxin   d. abscisic acid
   b. gibberellins   e. ethylene
   c. cytokinins
3. Internode elongation is stimulated by
   a. abscisic acid.   d. gibberellin.
   b. ethylene.   e. auxin.
   c. cytokinin.
4. Which of these is related to gibberellin activity?
   a. stem elongation
   b. initiation of bud dormancy
   c. repression of amylase production
   d. All of these are correct.
   e. Both b and c are correct.
5. _____ always promotes cell division.
   a. Auxin   c. Cytokinin
   b. Phytochrome   d. None of these are correct.
6. In the absence of abscisic acid, plants may have difficulty
   a. forming winter buds.   c. Both a and b are correct.
   b. closing the stomata.   d. Neither a nor b is correct.
7. Ethylene
   a. is a gas.
   b. causes fruit to ripen.
   c. is produced by the incomplete combustion of fuels such as kerosene.
   d. All of these are correct.
8. Which of the following plant hormones is responsible for a plant losing its leaves?
   a. auxin   d. abscisic acid
   b. gibberellins   e. ethylene
   c. cytokinins
9. Which is not a plant hormone?
   a. auxin   c. gibberellin
   b. cytokinin   d. All of these are plant hormones.

For questions 10–14, match each statement with a hormone in the key.

KEY:

    a. auxin   d. ethylene
    b. gibberellin   e. abscisic acid
    c. cytokinin

10. One rotten apple can spoil the barrel.
11. Cabbage plants bolt (grow tall).
12. Stomata close when a plant is water-stressed.
13. Sunflower plants all point toward the sun.
14. Coconut milk causes plant tissues to undergo cell division.
15. You bought green bananas at the grocery store this morning. However, you want a ripe banana for breakfast tomorrow morning. What could you do to accomplish this?

Plants Respond to Environmental Stimuli

16. Which of the following statements is correct?
    a. Both stems and roots show positive gravitropism.
    b. Both stems and roots show negative gravitropism.
    c. Only stems show positive gravitropism.
    d. Only roots show positive gravitropism.
17. The sensors in the cells of the root cap are called
    a. mitochondria.   d. chloroplasts.
    b. central vacuoles.   e. intermediate filaments.
    c. statoliths.
18. A student places 25 pea seeds in a large pot and allows the seeds to germinate in total darkness. Which of the following growth or movement activities would the seedlings exhibit?
    a. gravitropism, as the roots grow down and the shoots grow up
    b. phototropism, as the shoots search for light
    c. thigmotropism, as the tendrils coil around other seedlings
    d. Both a and c are correct.
19. Circadian rhythms
    a. require a biological clock.
    b. do not exist in plants.
    c. are involved in the tropisms.
    d. are involved in sleep movements.
    e. Both a and d are correct.
20. Plants that flower in response to long nights are
    a. day-neutral plants.   c. short-day plants.
    b. long-day plants.   d. impossible.
21. Short-day plants
    a. are the same as long-day plants.
    b. are apt to flower in the fall.
    c. do not have a critical photoperiod.
    d. will not flower if a short day is interrupted by bright light.
    e. All of these are correct.
22. A plant requiring a dark period of at least 14 hours will
    a. flower if a 14-hour night is interrupted by a flash of light.
    b. not flower if a 14-hour night is interrupted by a flash of light.
    c. not flower if the days are 14 hours long.
    d. not flower if the nights are longer than 14 hours.
    e. Both b and c are correct.
23. Phytochrome plays a role in
    a. flowering.   c. leaf growth.
    b. stem growth.   d. All of these are correct.
24. Phytochrome
    a. is a plant pigment.
    b. is present as P_{fr} during the day.
    c. activates DNA-binding proteins.
    d. is a photoreceptor.
    e. All of these are correct.

25. Primary metabolites are needed for _____ while secondary metabolites are produced for _____.
    a. growth, signal transduction
    b. normal cell functioning, defense
    c. defense, growth
    d. signal transduction, normal cell functioning
26. Which of the following is a plant secondary metabolite used by humans to treat disease?
    a. morphine   d. penicillin
    b. codeine   e. All but d are correct.
    c. quinine

Understanding the Terms

abscisic acid (ABA) 464	hypersensitive response (HR) 473
abscission 464	long-day plant 470
alkaloid 472	photoperiodism 470
apical dominance 460	phototropism 460, 466
auxin 460	phytochrome 471
biological clock 469	plant hormone 460
circadian rhythm 468	plant tissue culture 463
coleoptile 460	secondary metabolite 472
cyanogenic glycoside 472	senescence 463
cytokinin 463	short-day plant 470
day-neutral plant 470	statolith 466
dormancy 462	systemin 472
ethylene 465	thigmotropism 466
gibberellin 462	tropism 466
gravitropism 466	turgor movement 468

Match the terms to these definitions:
    a. _____ Biological rhythm with a 24-hour cycle.
    b. _____ Directional growth of plants in response to the Earth's gravity.
    c. _____ Dropping of leaves, fruits, or flowers from a plant.
    d. _____ Plant hormone producing increased stem growth between nodes; also involved in flowering and seed germination.
    e. _____ Relative lengths of daylight and darkness that affect the physiology and behavior of an organism.

Thinking Scientifically

1. Based on the data from Section 23.5, you hypothesize that abscisic acid (ABA) is responsible for the turgor pressure changes that permit a plant to track the sun (see Fig. 23.7). What observations could you make to support your hypothesis?
2. You formulate the hypothesis that the negative gravitropic response of stems is greater than the positive phototropism of stems. How would you test your hypothesis?

ARIS *Visit* www.mhhe.com/maderconcepts *for practice quizzes, animations, videos, and activities designed to help you master the material in this chapter.*

24

Reproduction in Plants

LEARNING OUTCOMES

After studying this chapter, you should be able to accomplish the following outcomes.

With a Little Help

1 Cite the two times in the life cycle of flowering plants when they might need a little help from animals, and describe that help.

Sexual Reproduction in Flowering Plants Is Suitable to the Land Environment

2 Explain an overall diagram of the flowering plant life cycle, with emphasis on adaptation to the land environment.
3 Label a diagram of a flower, and give a function for each part labeled.
4 Identify the female gametophyte and the male gametophyte of flowering plants.
5 Describe different means of pollination in flowering plants.
6 Give examples to show that flowers and their pollinators have coevolved.
7 Describe the outcome of double fertilization in flowering plants.

Seeds Contain a New Diploid Generation

8 Divide development of the embryo into six stages, and label the three main parts of a seed.
9 Give examples of fleshy and dry fruits. Distinguish between simple, compound, aggregate, and accessory fruits.
10 Compare and contrast germination of a bean plant and a corn plant. Compare a bean seed to a corn kernel.

Plants Can Also Reproduce Asexually

11 Give examples to show that plants can reproduce asexually.
12 Describe how tissue culture can be used to clone plants with desirable traits.

There are two times in the life cycle of a flowering plant when it might need a little help. The first is the time of pollination. How can a plant get its pollen from the male part of one flower to the female part of another flower? Some plants, such as the oak, rely on the wind. But others, such as roses, attract pollinators by the color of their petals and their sweet smell, both of which advertise the availability of nectar as food for the pollinator (see Section 24.2). The second time a flowering plant might need a little help is with the dispersal of its seeds. During dispersal, seeds are carried away from the parent plant to a site where they might have better growing conditions. In addition to achieving more room to grow, taking up residence in a new place may ensure that the species will survive should a disaster, such as a fire, devastate the plants in other locations.

In flowering plants, seeds are enclosed within a protective fruit, and some fruits release their seeds to be carried away by the wind. The seeds of an orchid are so small and light that they need no special adaptation for wind to carry them far away. The dandelion fruit, being slightly heavier, has a little parachute that allows wind currents to carry it away. The fruit of a maple tree contains two fairly large

seeds, but it too is able to be windblown because it is equipped with "wings" that can transport it up to 10 km from its parent.

Many flowering plants produce a fruit that is too heavy to be transported by wind. These plants need an alternative dispersal method, and again animals are willing to oblige. A fleshy fruit such as a berry encloses small seeds, while the fleshy part of, say, a peach encloses a single large seed, sometimes called a stone. Berries, peaches, apples, and cherries tend to be green and hidden from view by green leaves while they are developing. But when they become ripe, they take on an attractive color and scent. These changes entice an animal to eat the fruit, and later the seeds pass through the animal's digestive tract and are deposited some distance from the parent plant. Birds—and even animals as large as bears— enjoy eating berries such as blueberries, huckleberries, and rose hips. Raccoons eat fleshy fruits with large pits, and deer are known to eat crab apples.

Animals also eat nuts, in which a hard covering encloses the seed contents. When the seed is eaten, so is the embryo of the next generation. However, squirrels, as well as birds such as blue jays, store acorns and other nuts during the autumn to tide them over the long winter. Sometimes they forget where to go looking for the nuts, and in the meantime, the seeds enclosed by fruit have been dispersed.

The dry fruits of violets and trillium, among other plants, split open to release seeds that have a cap rich in oil and vitamins. The caps are prized by ants, which set about dragging the seeds back to the nest. Once there, the ants only eat the caps, and the rest of the seed stays intact until it germinates. Up to one-third of all the herbs in a deciduous forest of the United States are dispersed by ants!

Animals are also used to disperse seeds when the hooks and spines of clover, burdock, and cocklebur attach to their fur and are carried some distance away.

This chapter explores in some detail how the plant sexual life cycle is modified to permit the production of seeds enclosed by fruits, an evolutionary event that helps explain the success of flowering plants in a terrestrial environment. We will also discuss the development of the embryo within the seed and the structure of fruits. We will see that asexual reproduction permits humans to clone plants and their tissues for commercial purposes.

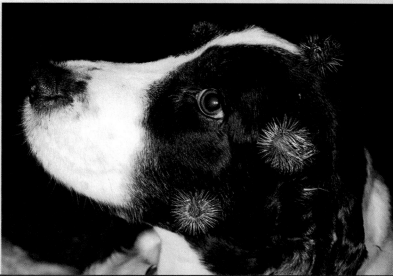

Sexual Reproduction in Flowering Plants Is Suitable to the Land Environment

Learning Outcomes 2–7, page 478

Sexual reproduction in flowering plants is centered in the flower, which produces seeds enclosed by a fruit. This part of the chapter reviews the structure of the flower, compares the life cycle of seedless and flowering plants, and stresses the events of pollination and double fertilization in the flowering plant life cycle.

24.1   Plants have a sexual life cycle called alternation of generations

The sexual life cycle of flowering plants, also called **angiosperms**, has contributed to their impressive ability to disperse and live in many different environments on land.

Flowers   The evolution of the flower accounts for the remarkable success of the angiosperms, because the flower produces seeds covered by a fruit. Let's begin with an examination of the flower. A typical flower has four whorls of modified leaves attached to a **receptacle** at the end of a flower stalk (**Fig. 24.1A**).

1. The **sepals**, which are the most leaflike of all the flower parts, are usually green, and they protect the bud as the flower develops. In the daylily, featured in **Figure 24.1B**, the sepals resemble petals.

2. An open flower has a whorl of **petals**, whose color and scent account for the attractiveness of many flowers. The size, shape, color, and scent of the petals play an important role in attracting pollinators, as we will discuss. Sometimes the petals fuse to form a floral tube that accommodates the mouthparts of a pollinator seeking **nectar**, a sweet liquid secreted by flowers. Wind-pollinated flowers may have no petals at all.

3. **Stamens** are the "male" portion of the flower. Each stamen has two parts: the **anther**, a saclike container, and the **filament**, a slender stalk. Pollen grains develop within the anther. For reproduction to occur, pollen grains must reach the female part of the flower because a mature pollen grain contains sperm cells.

FIGURE 24.1B   Anatomy of a daylily, *Hemerocallis* sp., a monocot flower (p = petal; s = sepal).

4. At the very center of a flower is the **carpel**, a vaselike structure that represents the "female" portion of the flower. A carpel usually has three parts: the **stigma**, an enlarged knob that is often sticky for capturing pollen; the **style**, a slender stalk; and the **ovary**, an enlarged base that encloses one or more **ovules**. When an egg within a mature ovule is fertilized, it develops into an embryonic plant. As the flower withers away, the mature ovule becomes the seed, and the ovary becomes the fruit. Fruit, as discussed in Section 24.4, helps disperse the seeds to a new location.

As we discussed in Section 21.2, flowering plants are divided into monocots and eudicots on the basis of several characteristics. One difference is that monocot flower parts occur in threes and multiples of three (Fig. 24.1B), while eudicot flower parts are in fours or fives and multiples of four or five (**Fig. 24.1C**).

Not all flowers have sepals, petals, stamens, and carpels. Those that do are said to be *complete*, and those that do not are said to be *incomplete*. Flowers that have both stamens and carpels are called *perfect* (bisexual) flowers; those with only stamens and those with only carpels are *imperfect* (unisexual) flowers. If staminate flowers and carpellate flowers are on one plant, the plant is *monoecious*. If staminate and carpellate flowers are on separate plants, the plant is *dioecious*. Holly trees are dioecious, and if red berries are a priority, it is necessary to acquire a plant with staminate flowers and another plant with carpellate flowers. Corn is an example of a monoecious plant.

Flowering Plant Life Cycle   Previously, we learned that plants have two multicellular stages in their life cycle, which is

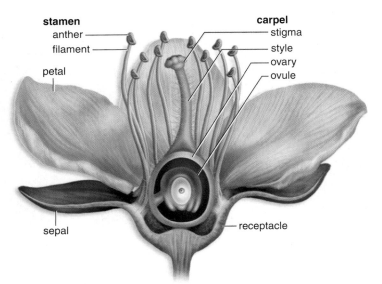

FIGURE 24.1A   Anatomy of a flower.

FIGURE 24.1C Anatomy of a festive azalea, *Rhododendron* sp., a eudicot flower (p = petal).

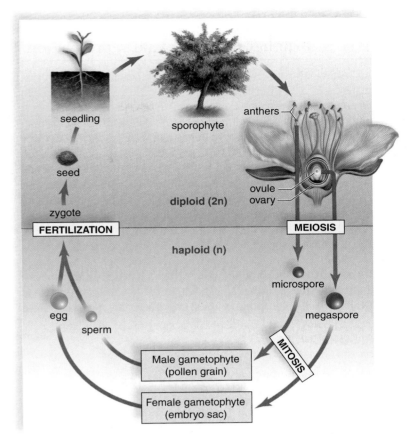

FIGURE 24.1D Alternation of generations in flowering plants.

therefore called an **alternation of generations**. In this life cycle, a diploid sporophyte alternates with a haploid gametophyte:

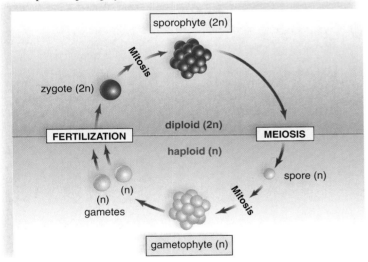

The **sporophyte** (2n) produces haploid spores by meiosis. The spores develop into gametophytes. The **gametophytes** (n) produce gametes. Upon fertilization, the cycle returns to the 2n sporophyte.

We also learned in Chapter 18 that as plant evolution occurred, the sporophyte gained in dominance, and the gametophyte became microscopic and dependent on the sporophyte. As you can see in Figure 18.4A, this statement is supported by the comparative size of the sporophyte and gametophytes in flowering plants, the last group of plants to evolve. However, the seed plants have an innovation that led to their remarkable adaptations for reproducing in the land environment. Seed plants, as opposed to nonseed plants, produce two types of spores (**Fig. 24.1D**). The microspore (little spore) develops into male gametophytes better known as **pollen grains**. The megaspore (big spore) develops into a female gametophyte. This innovation allows the female gametophyte to remain in the flower, where it is always protected from drying out. Pollen grains have strong walls that make them highly resistant to drying out.

First let's consider the pollen grains. As shown in Figure 24.2A, pollen grains are pro-

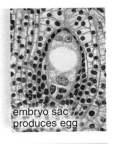

duced within the anther, where microspore mother cells undergo meiosis to produce microspores. Microspores become two-celled pollen grains; one of these (the generative cell) divides to produce two sperm cells. A pollen grain is either blown by the wind or carried by an animal to the stigma of the flower. Once a compatible pollen grain lands on the stigma, a pollen tube develops inside the style of a carpel. A germinated pollen grain is the mature male gametophyte. The two sperm cells move down the pollen tube to the egg-bearing female gametophyte. Notice that the style of a carpel protects the pollen tube from drying out as it delivers sperm cells to the female gametophyte.

How about the female gametophyte? As shown in Figure 24.2A, inside an ovule within an ovary, a megaspore mother cell undergoes meiosis to produce four megaspores, one of which divides to become the egg-bearing, seven-celled embryo sac. The **embryo sac** is the mature female gametophyte. Following fertilization, the ovule becomes a seed. The seed also contains stored food and is surrounded by a strong seed coat. The seeds are enclosed by a fruit, which develops from the ovary and aids in dispersing the seeds. When a seed germinates, a new sporophyte emerges and develops into the dominant sporophyte.

24.1 *Check Your Progress* Where would you look to find the gametophyte of flowering plants?

24.2 Pollination and fertilization bring gametes together during sexual reproduction

In Section 24.1, we reviewed sexual reproduction in flowering plants. We also discussed the parts of a flower; the production of pollen grains (male gametophytes) in the anthers of stamens; and the production of an embryo sac (female gametophyte) in an ovule located within the ovary of a carpel. The production of these two different gametophytes is paramount to the adaptation of seed plants to reproduction on land. It led to their ability to protect all stages of the life cycle from drying out in the land environment.

In this section, we discuss two other important events in the life cycle of flowering plants: pollination and double fertilization (**Fig. 24.2A**). These events help explain why flowering plants are able to disperse so well on land.

Pollination During **pollination**, pollen is transferred from the anther to the stigma so that an egg within the female gametophyte is fertilized. Self-pollination occurs if the pollen and stigma are from the same plant. Cross-pollination occurs when the pollen is from a member of the same species but not the same flower. Cross-pollination offers the best chance of the offspring having a different genotype from that of the parent. Plants have various means of achieving cross-pollination.

Some species of flowering plants—for example, the grasses and grains—rely on wind pollination, as do the gymnosperms, the other type of seed plant (**Fig. 24.2B**). Much of the plant's energy goes into making pollen to ensure that some pollen grains actually reach a stigma. Even the amount successfully transferred is staggering: A single corn plant may produce from 20 to 50 million grains a season. In corn, the flowers tend to be monoecious, and clusters of tiny male flowers move in the wind, freely releasing pollen into the air.

Most angiosperms rely on animals—be they insects (e.g., bumblebees, flies, butterflies, and moths), birds (e.g., hummingbirds), or mammals (e.g., bats)—to carry out pollination. The use

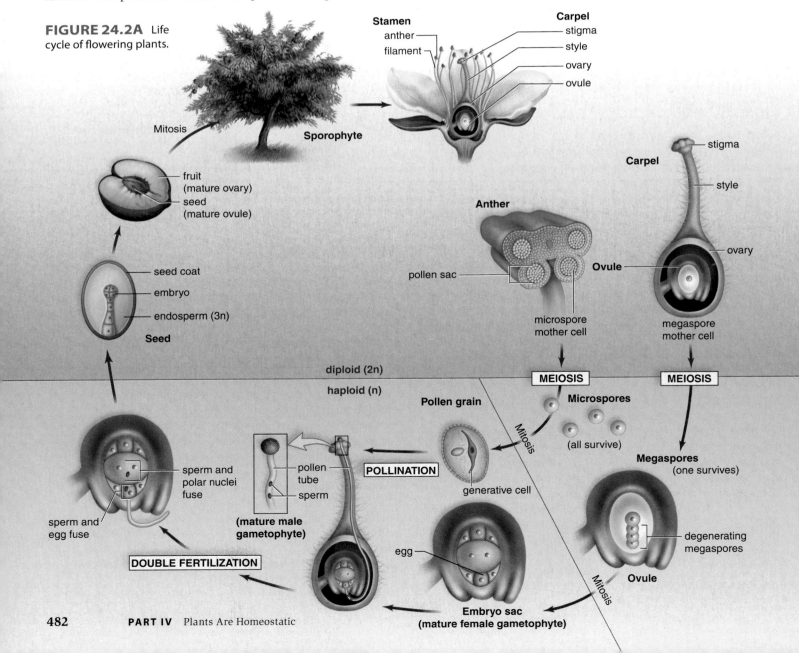

FIGURE 24.2A Life cycle of flowering plants.

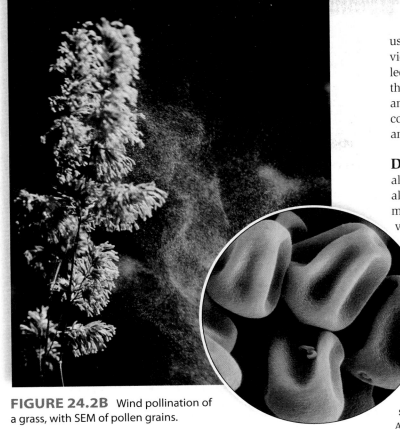

FIGURE 24.2B Wind pollination of a grass, with SEM of pollen grains.

use ultraviolet light in order to see, but bees are sensitive to ultraviolet light. A bee has a feeding proboscis of the right length to collect nectar from certain flowers and a pollen basket on its hind legs that allows it to carry pollen back to the hive. Because many fruits and vegetables are dependent on bee pollination, there is much concern today that the number of bees is declining due to disease and the use of pesticides.

Double Fertilization The process of **double fertilization** is also unique in angiosperms. It results in not only a zygote but also a food source for the developing zygote. Note the mature male gametophyte in Figure 24.2A. The generative cell has divided to produce two sperm cells, and a pollen tube is in the process of lengthening. This is in keeping with our expectation of the male gametophyte because, in plants, the gametophyte generation produces gametes. A pollen tube, which is an outgrowth of the inner wall of a pollen grain, digests its way through the tissue of the stigma and style of the carpel. The interval between pollen tube initiation and the time when the tube reaches the embryo sac in an ovule is quite variable, taking from a few hours to a few days.

When the pollen tube reaches the entry of the embryo sac, double fertilization occurs. Remember that the embryo sac is the female gametophyte, and as such, it produces an egg. Also as expected, one of the sperm unites with the egg, forming a 2n zygote. Unique to angiosperms, the other sperm unites with two polar nuclei centrally placed in the embryo sac, forming a 3n endosperm nucleus. This endosperm nucleus eventually develops into the **endosperm**, a nutritive tissue that the developing embryonic sporophyte will use as an energy source. Now the ovule begins to develop into a seed. One important aspect of seed development is formation of the seed coat from the ovule wall. A mature seed contains (1) the embryo, (2) stored food, and (3) the seed coat. **Figure 24.2C** shows a eudicot seed in which the embryo has already formed. The endosperm has been taken up by the **cotyledons**, or seed leaves.

This completes our discussion of seed production in flowering plants. Section 24.3 tells how the zygote develops into an embryonic sporophyte.

of animal pollinators is unique to flowering plants, and it helps account for why these plants are so successful on land. By the time flowering plants appear in the fossil record some 135 MYA, insects had long been present. For millions of years, then, plants and their animal pollinators have coevolved. **Coevolution** means that as one species changes, the other changes too, so that in the end, the two species are suited to one another. Plants with flowers that attracted a pollinator enjoyed an advantage because, in the end, they produced more seeds. Similarly, pollinators that were able to find and remove food from the flower were more successful. Today, we see that the reproductive parts of the flower are positioned so that the pollinator can't help but pick up pollen from one flower and deliver it to another. On the other hand, the mouthparts of the pollinator are suited to gathering the nectar from these particular plants.

Many examples of the coevolution between plants and their pollinators are given on pages 242–43. Here, we can note that bee-pollinated flowers are usually yellow, blue, or white because these are the colors bees can see. Bees respond to ultraviolet markings called nectar guides that help them locate nectar. Humans do not

24.2 Check Your Progress What is meant by double fertilization?

As we see it   As a bee sees it

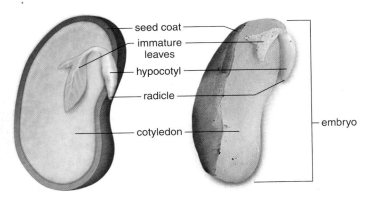

- seed coat
- immature leaves
- hypocotyl
- radicle
- cotyledon
- embryo

FIGURE 24.2C The parts of a bean seed, a eudicot.

This part of the chapter describes the stages of embryo development within a seed, the structure of fruits, and seed germination. With seed germination, the life cycle of flowering plants has come full circle.

24.3   A sporophyte embryo and its cotyledons develop inside a seed

Figure 24.3 shows the stages of development for a eudicot embryo. **1** The zygote stage is the beginning stage. **2** Then, the zygote divides repeatedly in different planes, forming several cells called a **proembryo**. Also formed is an elongated structure called a suspensor that has a basal cell. The suspensor, which anchors the embryo and transfers nutrients to it from the sporophyte plant, will disintegrate later.

3 During the **globular stage** the proembryo is largely a ball of cells. The root-shoot axis of the embryo is already established at this stage because the embryonic cells near the suspensor will become a root, while those at the other end will ultimately become a shoot.

The outermost cells of the plant embryo will become epidermal tissue. These cells divide with their cell plate perpendicular to the surface; therefore, they produce a single outer layer of cells. Recall that epidermal tissue protects the plant from desiccation and includes the stomata, which open and close to facilitate gas exchange and minimize water loss.

4 During the **heart stage**, the cotyledons appear as a result of rapid, local cell division, giving the embryo a heart shape. Monocots have one cotyledon, which in addition to storing certain nutrients, absorbs other nutrient molecules from the endosperm and passes them to the embryo. In eudicots, the cotyledons usually store the nutrient molecules the embryo uses.

5 As the embryo continues to enlarge and elongate, it takes on a torpedo shape, a period that is known as the **torpedo stage**. Now the root and shoot apical meristems are functional. The shoot apical meristem is responsible for aboveground growth, and the root apical meristem is responsible for belowground growth. Ground meristem, which gives rise to the bulk of the embryonic interior, is also present.

6 In the **mature embryo** of the final stage, the epicotyl is the portion between the cotyledons that contributes to shoot development. The hypocotyl is the portion below the cotyledons. It contributes to stem development and terminates in the radicle, or embryonic root. The cotyledons are quite noticeable in a eudicot embryo and may fold over. Procambium at the core of the embryo is destined to form the future vascular tissue. The embryo stops developing and becomes dormant within its seed coat (derived from the ovule wall). In flowering plants, mature seeds are enclosed by fruits, as discussed in the next section.

> **24.3 Check Your Progress** Why are both the seed coat and the embryo 2n?

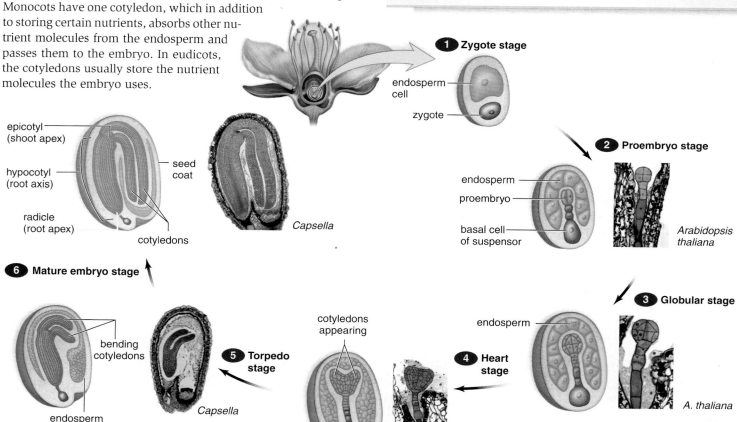

FIGURE 24.3 Development of embryo and its cotyledons in the seed of a eudicot.

24.4 The ovary becomes a fruit, which assists in sporophyte dispersal

A **fruit**—derived from an ovary and sometimes other flower parts—protects and helps disperse the next 2n sporophyte generation. How does a fruit help in seed dispersal? Often, fruits are an attractive and nutritious package that animals like to eat. Then they deposit the seeds some distance away.

As a fruit develops, the ovary wall thickens to become the **pericarp**, as labeled on the pea pod in **Figure 24.4**. The pericarp can have as many as three layers that encircle the seed: exocarp, mesocarp, and endocarp.

Fleshy Versus Dry Fruits

Figure 24.4 shows diverse types of fruit. **1** A pea pod is a dry fruit; at maturity, the pea pod breaks open on both sides to release the seeds. Peas and beans, you will recall, are legumes. The fruit of a legume is **dehiscent**—it splits along two sides when mature. **2** The fruit of a maple tree is dry and **indehiscent**—it does not split open.

Like legumes, cereal grains of wheat, rice, and corn are dry fruits. Sometimes such fruits are mistaken for seeds because a dry pericarp adheres to the seed within. These dry fruits are indehiscent—they don't split open. Humans gather grains before they are released from the plant and then process them to acquire their nutrients.

In some fruits, the mesocarp remains fleshy at maturity. Peaches and cherries are examples of fleshy fruits that have a hard, stony endocarp and are often, therefore, called stone fruits, although botanically they are known as drupes. This type of endocarp protects the seed so it can pass through the digestive system of an animal and remain unharmed. In a tomato, the entire pericarp is fleshy.

Simple Versus Aggregate and Multiple Fruits

Simple fruits are derived from the simple ovary of a single carpel or from the compound ovary of several fused carpels. A simple ovary has one chamber, and a compound ovary has a number of chambers; the exact number depends on the number of carpels that fused to make it up. If you cut open a tomato, you see several chambers of a compound ovary. Accessory fruits are fruits that form from other flower parts, in addition to the ovary. **3** A strawberry is an **accessory fruit** because the bulk of the fruit is not from the ovary, but from the receptacle. Similarly, only the core of an apple is derived from the ovary. If you cut an apple crosswise, it is obvious that an apple, like a tomato, came from a compound ovary with several chambers.

Both **aggregate** and **multiple fruits** are examples of compound fruits derived from several individual ovaries (Fig. 24.4). The strawberry is also an aggregate fruit, in which each ovary becomes a one-seeded fruit called an achene. **4** A raspberry is an example of an aggregate fruit because the flower had many separate carpels. **5** An example of a multiple fruit is a pineapple, which is derived from many individual flowers, each with its own carpel. During development, each separate developing fruit combines into a single larger fruit.

Following fruit and/or seed dispersal, seed germination, discussed in the next section, produces the next generation of flowering plants.

> **24.4 Check Your Progress** Why do plants expend resources (energy) to produce showy flowers and attractive (good food source) fruit?

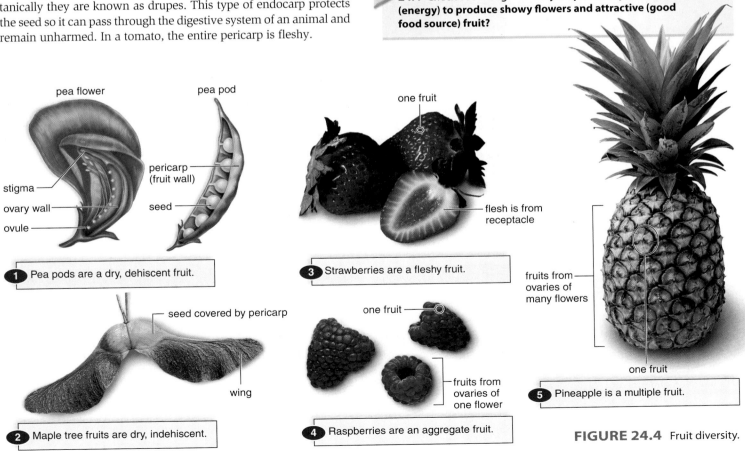

FIGURE 24.4 Fruit diversity.

pea flower · pea pod · stigma · ovary wall · ovule · pericarp (fruit wall) · seed

1 Pea pods are a dry, dehiscent fruit.

seed covered by pericarp · wing

2 Maple tree fruits are dry, indehiscent.

one fruit · flesh is from receptacle

3 Strawberries are a fleshy fruit.

one fruit · fruits from ovaries of one flower

4 Raspberries are an aggregate fruit.

fruits from ovaries of many flowers · one fruit

5 Pineapple is a multiple fruit.

Following dispersal, if conditions are right, seeds may **germinate** to form a seedling. Germination doesn't usually take place until there is sufficient water, warmth, and oxygen to sustain growth. These requirements help ensure that seeds do not germinate until the most favorable growing season has arrived. Some seeds do not germinate until they have been dormant for a period of time. For seeds, *dormancy* is the time during which no growth occurs, even though conditions may be favorable for growth. In the temperate zone, seeds often have to be exposed to a period of cold weather before dormancy is broken. Fleshy fruits (e.g., apples, pears, oranges, and tomatoes) contain inhibitors so that germination does not occur while the fruit is still on the plant. For seeds to take up water, bacterial action and even fire may be needed. Once water enters, the seed coat bursts and the seed germinates.

If the two cotyledons of a bean seed are parted, the rudimentary plant with immature leaves is exposed (**Fig. 24.5A**). As the eudicot seedling starts to grow, the shoot is hook-shaped to protect the immature leaves as they emerge from the soil. The cotyledons provide the new seedlings with enough energy to straighten and form true leaves. As the true leaves of the plant begin photosynthesizing, the cotyledons shrivel up.

A corn kernel is actually a fruit, and therefore its outer covering is the pericarp and seed coat combined (**Fig. 24.5B**). Inside is the single cotyledon. Also, the immature leaves and the root are covered, respectively, by a coleoptile and a coleorhiza. These sheaths are discarded when the seedling begins to grow.

This completes our discussion of sexual reproduction in flowering plants. The next part of the chapter discusses asexual reproduction in flowering plants.

24.5 Check Your Progress As a corn kernel (monocot) germinates, the immature shoot and root are covered by a sheath. What might be the function of the sheath?

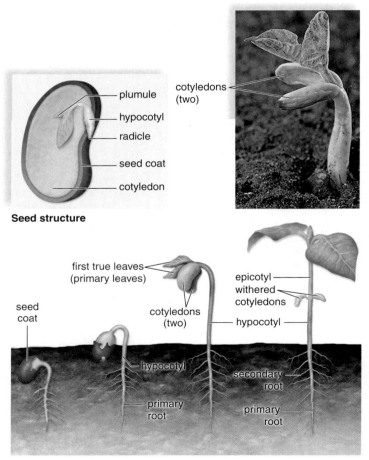

Seed structure

Germination and growth

FIGURE 24.5A Structure and germination of a common bean seed.

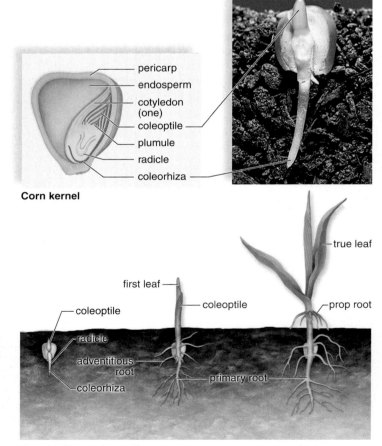

Corn kernel

Germination and growth

FIGURE 24.5B Structure and germination of a corn kernel.

We can observe the asexual reproduction of plants in the environment as well as in the laboratory, where the cloning of plants in tissue culture is commonplace. Cloning produces identical plants or plant tissues with highly desirable traits for agricultural and commercial purposes.

24.6   Plants have various ways of reproducing asexually

Unlike humans, who always reproduce sexually, plants can reproduce both sexually and asexually; the latter is also sometimes called **vegetative reproduction**. Asexual reproduction is more likely to produce an offspring that is exactly like the parent plant, and therefore is a type of cloning. Asexual reproduction is favored by agriculture when the parent plant already has desirable characteristics that should be maintained.

You may already be familiar with the examples of asexual reproduction in plants given in **Figure 24.6**. Plants can grow from the axillary buds of aboveground horizontal stems and various types of underground stems. Aboveground horizontal stems, called **stolons,** run along the ground. Complete strawberry plants can grow from axillary buds that appear at the nodes of a stolon.

Underground horizontal stems, called **rhizomes,** may be long and thin, as in sod-forming grasses, or thick and fleshy, as in irises. Rhizomes survive the winter and contribute to asexual reproduction because each node bears a bud. Irises grow from the buds of rhizomes, as do violets and many grasses.

Some rhizomes have enlarged portions called **tubers** that function in food storage. Potatoes are tubers, in which the eyes are axillary buds that mark the nodes. A bud has the potential to produce a new potato plant if it is planted with a portion of the swollen tuber.

Corms are bulbous underground stems that lie dormant during the winter, just as rhizomes do. They also produce new plants in the next growing season. Gladiolus corms are called bulbs by laypersons, but botanists reserve the term *bulb* for a structure composed of modified leaves attached to a short, vertical stem. Onions grow from bulbs, as do lilies, tulips, and daffodils.

Many different plants can be propagated from stem cuttings since the discovery that the plant hormone auxin can cause stems to produce roots.

We move from field to laboratory in the next section, as we discuss the cloning of plants in tissue culture.

> **24.6 *Check Your Progress* What are the possible benefits of asexual reproduction?**

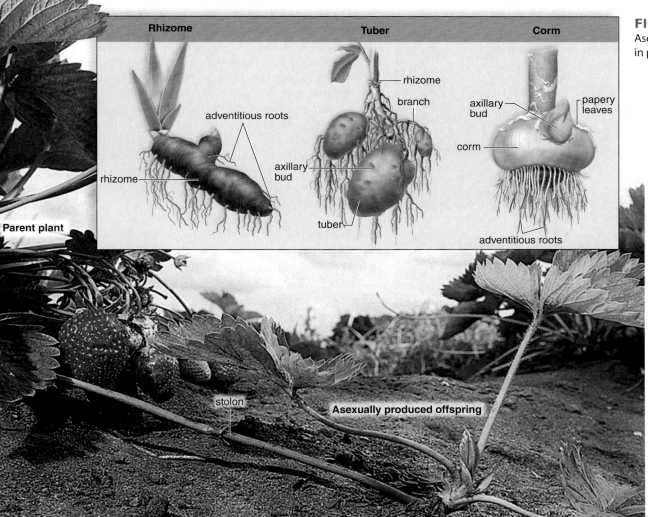

FIGURE 24.6
Asexual reproduction in plants.

Tissue culture is the growth of a tissue in an artificial liquid or solid culture medium. Somatic embryogenesis, meristem tissue culture, and anther tissue culture are methods of cloning plants due to the ability of plants to grow from single cells. Many plant cells are **totipotent,** which means that each one has the genetic capability of becoming an entire plant.

During *somatic embryogenesis,* hormones cause plant tissues to generate small masses of cells, from which many new genetically identical plants may grow. Thousands of little "plantlets" can be produced by using this method of plant tissue culture (**Fig. 24.7A**). Many important crop plants, such as tomato, rice, celery, and asparagus, as well as ornamental plants such as lilies, begonias, and African violets, have been produced using somatic embryogenesis. Plants generated from somatic embryos are not always genetically identical clones. They can vary

FIGURE 24.7B Producing whole plants from meristem tissue.

because of mutations that arise spontaneously during the production process. These mutations, called *somaclonal variations,* are another way to produce new plants with desirable traits. Somatic embryos can be encapsulated in hydrated gel, creating artificial "seeds" that can be shipped anywhere.

Meristem tissue can also be used as a source of plant cells. In this case, the resulting products are clonal plants that always have the same traits. In **Figure 24.7B**, culture flasks containing meristematic orchid tissue are rotated under lights. If the correct proportions of hormones are added to the liquid medium, many new shoots develop from a single shoot tip. When these are removed, more shoots form. Another advantage to producing identical plants from meristem tissue is that the plants are virus-free. (The presence of plant viruses weakens plants and makes them less productive.)

Anther tissue culture is a technique in which the haploid cells within pollen grains are cultured in order to produce haploid plantlets. Conversely, a diploid (2n) plantlet can be produced if chemical agents, to encourage chromosomal doubling, are added to the anther culture. Anther tissue culture is a direct way to produce plants that are certain to have the same characteristics.

Cell Suspension Culture
A technique called **cell suspension culture** allows scientists to extract chemicals (i.e., secondary metabolites) from plant cells in high concentration and without having to over-collect wild-type plants growing in their natural environments. These cells produce the same chemicals the entire plant produces. For example, cell suspension cultures of *Cinchona ledgeriana* produce quinine, which is used to treat leg cramping, a major symptom of malaria. And those of several *Digitalis* species produce digitalis, digitoxin, and digoxin, which are useful in the treatment of heart disease.

This completes our discussion of asexual reproduction of whole plants or their tissues.

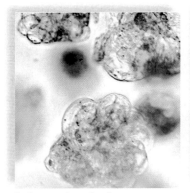

1. Protoplasts, naked cells

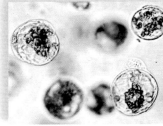

2. Cell wall regeneration

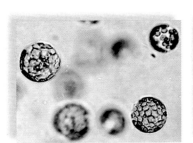

3. Aggregates of cells

4. Callus, undifferentiated mass

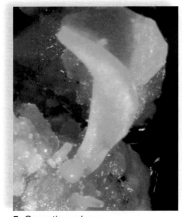

5. Somatic embryo

6. Plantlet

FIGURE 24.7A Somatic embryogenesis.

24.7 *Check Your Progress* **Which tissue culture technique would you use (a) to produce many diploid plants and (b) to collect secondary metabolites produced by plant cells?**

With this chapter, we bring to a close our study of plants. It is fitting that we end with a look at the reproduction of flowering plants. Life, as we know it, would not be possible without vascular plants—and specifically flowering plants, which now dominate the biosphere. *Homo sapiens* evolved in a world already dominated by flowering plants and, therefore, humans do not know a world without them. The earliest humans were mostly herbivores; they relied on foods they could gather for survival—fruits, nuts, seeds, tubers, roots, and so forth. Plants also provided protection from the environment, offering shelter from heavy rains and noonday sun. Later on, human civilizations could not have begun without the development of agriculture. Most of the world's population still relies primarily on three flowering plants—corn, wheat, and rice—for the majority of its sustenance. Sugar, coffee, spices of all kinds, cotton, rubber, and tea are plants that have even led to wars due to their importance to countries' economies.

Although we now live in an industrialized society, we are still dependent on plants and use them for many purposes. In fact, plants may be even more critical to our lives today than they were to our early ancestors on the African plains. For millions of urban dwellers, plants are their major contact with the natural world.

We grow plants not only for food and shelter, but also for their simple beauty. Also, plants now produce the substances needed to lubricate the engines of supersonic jets and to make cellulose acetate for films.

Currently, half of all pharmaceutical drugs have their origin in plants. The world's major drug companies are engaged in a frantic rush to collect and test plants from the rain forests for their drug-producing potential. Why the hurry? Because the rain forests may be gone before all the possible cures for cancer, AIDS, and other diseases have been found. Wild plants can not only help cure human ills, but can also serve as a source of genes for improving the quality of the plants that support our way of life.

In Part V, we study the animal systems, which may seem more familiar to you because humans are animals. However, we should not forget the dependence of animals on plants, a theme that returns in Part VI of this text.

The Chapter In Review

Summary

With a Little Help

- Wind and animals help plants accomplish pollination and disperse seeds.

Sexual Reproduction in Flowering Plants Is Suitable to the Land Environment

24.1 Plants have a sexual life cycle called alternation of generations

In flowering plants,
- The sexual life cycle occurs in the flower.
  - Male portion of flower consists of stamens.
  - Each stamen has an anther and a filament.
  - Female portion of flower consists of one or more carpels.
  - Each carpel consists of a stigma, style, and ovary.
  - Ovules are located in the ovary.
- The dominant sporophyte (2n) produces two types of spores by meiosis.
  - The microspore develops into a male gametophyte (pollen grain), which produces sperm.
  - The megaspore develops into a female gametophyte (embryo sac) contained within an ovule.

24.2 Pollination and fertilization bring gametes together during sexual reproduction

- Pollination transfers pollen from the anther to the stigma of a carpel.
- Wind pollination sometimes occurs, but pollination by animals is more common and helps account for the success of angiosperms.
- In double fertilization, two sperm reach the embryo sac. One sperm unites with the egg, and the other unites with two polar nuclei to form endosperm.
- The ovule wall becomes the seed coat that encloses the multicellular embryo and endosperm, a nutrient substance.

Seeds Contain a New Diploid Generation

24.3   A sporophyte embryo and its cotyledons develop inside a seed

- The stages of embryonic development are proembryo, globular, heart, torpedo, and mature embryo.
- Cotyledons are embryonic leaves that store food until the first leaves become functional. Monocots have one cotyledon; eudicots have two.

24.4 The ovary becomes a fruit, which assists in sporophyte dispersal

- Fruits may be grouped into various categories:
  - Dry fruits (beans, cereal grains)
  - Fleshy fruits (peach, cherry, tomato)
  - Simple fruits develop from a flower with a single ovary (grape, bean, wheat, maple) or a compound ovary (tomato, apple).
  - In aggregate fruits, many separate ovaries are from a single flower (strawberry, raspberry).
  - In multiple fruits, the ovaries are from separate flowers (pineapple).

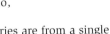
Fleshy fruit

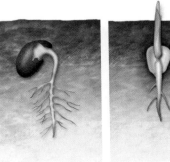

Eudicot seedling   Monocot seedling

24.5 With seed germination, the life cycle is complete

- Germination is regulated by water, warmth, and oxygen availability, among other factors.
- The embryo breaks out of the seed coat and becomes a seedling with leaves, stems, and roots.

Plants Can Also Reproduce Asexually

24.6 Plants have various ways of reproducing asexually

- Axillary buds on stems (either aboveground or underground) sometimes give rise to entire plants.
- Examples of asexual reproductive structures include stolons, rhizomes, tubers, corms, and suckers.
- Many plants can be propagated from stem cuttings.

24.7 Cloning of plants in tissue culture assists agriculture

- Tissue culture refers to the growth of tissue in an artificial liquid or solid culture medium.
  - A totipotent plant cell has the genetic capability of becoming an entire plant.
  - Cloning methods include somatic embryogenesis, use of meristem tissue, and anther culture.
  - Cell suspension culture is a way to obtain secondary metabolites directly from plant cells.

Testing Yourself

Sexual Reproduction in Flowering Plants Is Suitable to the Land Environment

1. Stigma is to carpel as anther is to
   a. sepal.
   b. stamen.
   c. ovary.
   d. style.
2. Which of the following is not a component of the carpel?
   a. stigma
   b. filament
   c. ovary
   d. ovule
   e. style
3. The flower part that contains ovules is the
   a. carpel.
   b. stamen.
   c. sepal.
   d. petal.
   e. seed.
4. Carpels
   a. are the female part of a flower.
   b. contain ovules.
   c. are the innermost part of a flower.
   d. are absent in some flowers.
   e. All of these are correct.
5. In plants,
   a. gametes become a gametophyte.
   b. spores become a sporophyte.
   c. both sporophyte and gametophyte produce spores.
   d. only a sporophyte produces spores.
   e. Both a and b are correct.
6. In plants, meiosis directly produces
   a. new xylem.
   b. phloem.
   c. spores.
   d. an egg.
   e. sperm.
7. In the life cycle of flowering plants, a microspore develops into
   a. a megaspore.
   b. a male gametophyte.
   c. a female gametophyte.
   d. an ovule.
   e. an embryo.
8. A pollen grain is
   a. a haploid structure.
   b. a diploid structure.
   c. first a diploid and then a haploid structure.
   d. first a haploid and then a diploid structure.

9. The megaspore is similar to the microspore in that both
   a. have the diploid number of chromosomes.
   b. become an embryo sac.
   c. become a gametophyte that produces a gamete.
   d. are necessary to seed production.
   e. Both c and d are correct.
10. The megaspore and the microspore
    a. both produce pollen grains.
    b. both divide meiotically.
    c. both divide mitotically.
    d. produce pollen grains and embryo sacs, respectively.
    e. All of these are correct.
11. The embryo of a flowering plant can be found in the
    a. pollen.
    b. anther.
    c. microspore.
    d. seed.
12. Which is the correct order of the following events: (1) megaspore becomes embryo sac, (2) embryo formed, (3) double fertilization, (4) meiosis?
    a. 1, 2, 3, 4
    b. 4, 1, 3, 2
    c. 4, 3, 2, 1
    d. 2, 3, 4, 1
13. Double fertilization refers to the formation of a _____ and a(n) _____.
    a. zygote, zygote
    b. zygote, pollen grain
    c. zygote, megaspore
    d. zygote, endosperm
14. Which of these pairs is incorrectly matched?
    a. polar nuclei—plumule
    b. egg and sperm—zygote
    c. ovule—seed
    d. ovary—fruit
    e. stigma—carpel
15. Label this diagram of a flower.

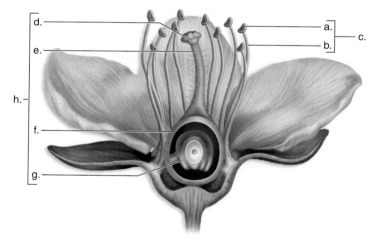

16. **THINKING CONCEPTUALLY** Would you expect a wind-pollinated plant or an animal-pollinated plant to produce more pollen? Explain.

Seeds Contain a New Diploid Generation

17. A seed is a mature
    a. embryo.
    b. ovule.
    c. ovary.
    d. pollen grain.
18. Globular, heart, and torpedo refer to
    a. embryo development.
    b. sperm development.
    c. female gametophyte development.
    d. seed development.
    e. Both b and d are correct.

19. A seed contains
    a. a seed coat.   d. cotyledon(s).
    b. an embryo.   e. All of these are correct.
    c. stored food.
20. The function of the flower is to _____, and the function of fruit is to _____.
    a. produce fruit; provide food for humans
    b. aid in seed dispersal; attract pollinators
    c. attract pollinators; assist in seed dispersal
    d. produce the ovule; produce the ovary
21. Fruits
    a. nourish embryo development.
    b. help with seed dispersal.
    c. signal gametophyte maturity.
    d. attract pollinators.
    e. signal when they are ripe.
22. In an apple, the bulk of the fruit is from the
    a. ovary.   c. pollen.
    b. style.   d. receptacle.
23. Which of these is not a fruit?
    a. walnut   d. peach
    b. pea   e. All of these are fruits.
    c. green bean
24. **THINKING CONCEPTUALLY** Seed germination sometimes requires exposure to cold temperatures. Explain the benefit to the plant.

Plants Can Also Reproduce Asexually

25. Asexual reproduction in flowering plants
    a. is unknown.   d. produces seeds also.
    b. is a rare event. e. is no fun.
    c. is common.
26. Plant tissue culture takes advantage of
    a. a difference in flower structure.   d. phototropism.
    b. sexual reproduction.   e. totipotency.
    c. gravitropism.
27. The term totipotent means
    a. that each plant cell can become an entire plant.
    b. hormones control all plant growth.
    c. all cells develop from the same tissue.
    d. None of these are correct.
28. **THINKING CONCEPTUALLY** Under what environmental conditions would it be advantageous for a plant to carry out asexual reproduction?

Understanding the Terms

accessory fruit   485
aggregate fruit   485
alternation of
    generations   481
angiosperm   480

anther   480
carpel   480
cell suspension
    culture   488
coevolution   483

corm   487
cotyledon   483
dehiscent   485
double fertilization   483
embryo sac   481
endosperm   483
filament   480
fruit   485
gametophyte   481
germinate   486
globular stage   484
heart stage   484
indehiscent   485
mature embryo   484
multiple fruit   485
nectar   480
ovary   480
ovule   480

petal   480
pollen grain   481
pollination   482
proembryo   484
receptacle   480
rhizome   487
sepal   480
sporophyte   481
stamen   480
stigma   480
stolon   487
style   480
tissue culture   488
torpedo stage   484
totipotent   488
tuber   487
vegetative
    reproduction   487

Match the terms to these definitions:
a. _____ Flower structure consisting of an ovary, a style, and a stigma.
b. _____ Mature male gametophyte in seed plants.
c. _____ In flowering plants, pollen-bearing portion of stamen.
d. _____ Haploid generation of the alternation of generations life cycle of a plant.
e. _____ Diploid generation of the alternation of generations life cycle of a plant.

Thinking Scientifically

1. You notice that a type of wasp has been visiting a flower type in your garden. What data about the wasp and flower would allow you to hypothesize that this wasp is a pollinator for this flower type?
2. You are a laboratory scientist who has discovered an unusual lettuce type and want to propagate it. What might you do if no seeds are available?

ARIS™ Visit **www.mhhe.com/maderconcepts** for practice quizzes, animations, videos, and activities designed to help you master the material in this chapter.

BIOLOGICAL VIEWPOINTS

PART IV Plants Are Homeostatic

Homeostasis is a concept that is most often applied to animals, but when you think about it, all organisms—be they single-celled or multicellular with a complex organization—have to be homeostatic. Homeostasis is present when the internal environment of the body remains within a range of normality. To take an example, think of a parenchyma cell within a leaf. Parenchyma cells, like all cells, require a moist environment so they don't dry out. What happens when the internal conditions within a leaf begin to dry out? The stomata close. All organisms have mechanisms to keep the internal environment livable for their cells. But homeostatic mechanisms have their limitations. If you forget to water a plant for weeks on end, the cells will eventually dry out, and the plant will most likely die.

Just as with animals, all of the organ systems of a plant participate in maintaining homeostasis because otherwise a plant cannot grow and reproduce. Reproduction is paramount to living things. The epidermal system, whether in a nonwoody or woody plant, protects the plant, but also allows exchanges with the external environment at both the roots and the leaves. Leaves have stomata that open to allow gas exchange, and roots have root hairs that allow water and minerals to be absorbed. Water is transported by xylem to all parts of a plant so that its organs do not dry out. Solar energy, as you know, is absorbed by the chlorophyll within a leaf. Following photosynthesis, phloem transports nutrients to all the cells that are not actively photosynthesizing at the moment. Plant cells use carbohydrates as a source of energy and building blocks to construct all the molecules they need to continue their existence.

No doubt about it, plant cells—and therefore plants themselves—actively exhibit the characteristics of life. Some people tend to forget that because plants don't have a nervous system; the job of coordinating the biological activities of a plant depends on the production of hormones. Many tissues in a plant produce hormones, but the apical meristems produce the hormone auxin, which has many effects on plants. For example, auxin helps plants respond to stimuli; when auxin accumulates on the shady side of a stem, the stem bends toward a light source. Hormones also regulate the growth and development of plants. Apply a little auxin to the cut end of a stem, and roots begin to sprout. At the cellular level, plant hormones control the activity of genes, much as the steroid hormones do in animal cells. This should not be surprising because plants and animals share a common ancestor that arose in the distant past. Plants and animals both exhibit increasing complexity as they develop, moving from the cellular to the tissue to the organismal levels of organization. Research is revealing more and more different types of organisms that make use of the same homeotic genes during development.

Woody plants in the temperate zone have a remarkable adaptation that allows them to survive unfavorable conditions. A lizard can move between the sun and the shade in order to maintain its temperature near normal, but a plant can't move. So when fall comes and winter threatens, trees and shrubs shut down. The decreasing photoperiod indicates the season, and hormones cause these plants to lose their leaves and otherwise suspend life's processes until the increasing photoperiod indicates that it is time to grow leaves once again. Stored nutrients in the roots give plants the energy for regrowth. Nonwoody perennials die back and start entirely anew from their underground roots; annuals depend solely on their seeds to survive as a species from season to season.

Plants form associations with other organisms in order to maintain homeostasis. Root nodules supply fixed nitrogen, and fungus roots (mycorrhizae) assist roots in gathering nutrients from the soil. Plants and insects interact both negatively and positively. On the one hand, plants are able to protect themselves from herbivorous insects in both physical and chemical ways in order to maintain homeostasis. On the other hand, many flowering plants depend on insects to help them complete their life cycle, or even to protect them from predators. Similarly, although fungi form favorable associations with plant roots, plants have mechanisms to prevent fungi from taking over their bodies. Without these mechanisms, homeostasis would be impossible.

Research into the homeostatic mechanisms of plants is ongoing. We need to know all we can about plants in order to increase their ability to survive. Why? Because all life, even our own, is dependent on plants. They are food producers for themselves and for most of the other organisms on planet Earth.

25

Animal Organization and Homeostasis

LEARNING OUTCOMES

After studying this chapter, you should be able to accomplish the following outcomes.

Staying Warm, Staying Cool

1 Compare and contrast thermoregulation in animals adapted to living in the tropics with those adapted to living in the Arctic Circle.

The Structure of Tissues Suits Their Function

2 State the levels of biological organization with reference to a particular organ system in complex animals.
3 Show that the structure of each organ in this system is appropriate to its function.

Four Types of Tissues Are Common in the Animal Body

4 Compare the main types of epithelial, connective, and muscular tissue, and relate their structure to their functions.
5 Describe the anatomy of a neuron, and relate its structure to its function.
6 Discuss the possibilities for curing spinal cord injuries.

Organs, Composed of Tissues, Work Together in Organ Systems

7 Show that the structure of human skin is appropriate to its functions.

All Organ Systems Contribute to Homeostasis in Animals

8 Show that all organ systems contribute to homeostasis.
9 Discuss three possible sources of organs for transplantation and the potential problems with each source.
10 Show that a negative, rather than positive, feedback mechanism maintains homeostasis.
11 Describe the feedback mechanism that allows humans to regulate their body temperature.

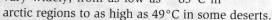

A nimals have ways of keeping their body temperature within normal limits. This is essential because the favorable temperature for metabolism is between 0° and 40°C. Yet, environmental temperature can vary widely, from as low as −65°C in arctic regions to as high as 49°C in some deserts.

Invertebrates, such as insects, and vertebrates, such as reptiles, usually live in warmer climates, where external temperatures can help them maintain their body temperature. Reptiles are especially known for modifying their behavior in order to stay warm or cool off. Basking in the sun while lying on a hot rock allows them to use radiant energy from the sun and heat from the rock to get warm. Animals that get their heat from the outside environment are said to be ectothermic. This doesn't mean that ectotherms don't also produce and make use of internal heat to stay warm. Some insects fan their wings in the early morning to warm the flight muscles so they can take off. The bodies of many nocturnal moths are covered with scales that help retain the heat of muscle contraction, so that they can be active at night when there is no sun. Still, ectotherms cannot rely completely on internal heat to stay warm.

Staying Warm, Staying Cool

Not so for the endotherms, as exemplified by birds and mammals. Endotherms rely on a high metabolic rate—heat from the inside—to stay warm. They devote as much as 80% of their basal metabolic rate to thermoregulation. Various structural modifications help keep them warm, including the feathers of birds and the fur and fat layers of mammals. Mammals, especially, are able to live where it is really cold, such as the Arctic Circle. Their stocky bodies, compared to close relatives living in warmer climates, gives them a reduced surface-area-to-volume ratio that helps keep the heat inside. For example, the snowshoe hare of Canada has a more compact body than the jackrabbit of the American Southwest. Birds and some mammals (e.g., caribou and foxes) have a vascular modification that keeps them from losing heat through exposed limbs. Arteries that carry blood to the limbs are surrounded by veins that bring blood back to the body core. Heat passes from outgoing to incoming vessels, rather than escaping from exposed surfaces. Then, too, lipids in plasma membranes are polyunsaturated and, thus, less apt to freeze.

Endotherms also use behavior modification as a way to stay warm. After all, when birds migrate south in the fall, they are relocating to warmer climates, and mammals such as groundhogs and black bears hibernate during the winter months. When groundhogs come out of their burrows, they are checking to see if it is time to stop hibernating. (It's a superstition that, if a groundhog does not see its shadow, winter is over and spring is under way.) During hibernation, animals go into a very deep, sleeplike state in which their heartbeat slows drastically.

Adaptations to stay warm help determine where various animals reside. They help account for why you do not find polar bears in the tropics or iguanas in the Arctic Circle! But animals have adaptations to stay cool, also. Reptiles and other invertebrates move to shady locations when the sun gets too hot. Endotherms, such as humans, become flushed as blood rushes to the skin, where cooling breezes and active sweat glands can dissipate heat. Evaporation of water at the tongue and mouth of dogs serves the same purpose. In this chapter, we focus on how animals maintain a stable internal environment, regardless of the external environment. But first we lay the groundwork for that discussion by describing the overall organization of animal bodies.

The Structure of Tissues Suits Their Function

Learning Outcomes 2–3, page 494

A review of the biological levels of organization serves as a backdrop to this chapter, which will consider the structure and function of various tissues, organs, and organ systems.

25.1 Levels of biological organization are evident in animals

An animal's body is composed of specialized cells that perform particular functions, as shown in **Figure 25.1**. **❶** Your body contains trillions of cells, of which there are approximately 100 different types. Cells of the same type occur within a tissue. **❷** A **tissue** is a group of similar cells performing a similar function. We will see that the four basic types of tissue in an animal's body are epithelial tissue, connective tissue, muscular tissue, and nervous tissue. **❸** An **organ** contains different types of tissues arranged in a certain fashion. In other words, the structure and function of an organ are dependent on the tissues it contains. That is why it is sometimes said that tissues, not organs, are the structural and functional units of the body. **❹** Several organs comprise an **organ system**, and the organs of the system work together to perform necessary functions for **❺** the **organism**.

The urinary system contains these organs: two kidneys, two ureters, a bladder, and a urethra (Fig. 25.1). The kidneys remove waste molecules from the blood and produce urine. The tubular shape of the ureters is suitable for passing urine to the bladder, which can store urine because it is expandable. Urine passes out of the body by way of the tubular urethra. The role of the urinary system is to produce, store, and rid the body of metabolic wastes. These vital functions are dependent on its organs, which in turn are dependent on the tissues making up the organs.

Other examples also show that an organ's function is dependent on its tissues. In the digestive system, the intestine absorbs nutrients. The cells of the tissue lining the lumen (cavity) of the small intestine have microvilli that increase the available surface area for absorption. Within muscular tissue, muscle cells shorten when they contract because they have intracellular components

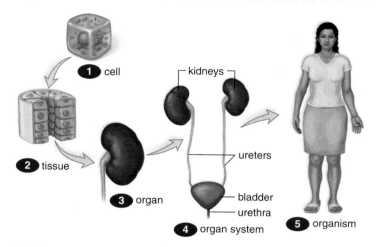

FIGURE 25.1 Levels of biological organization.

that move past one another. Within nervous tissue, nerve cells have long, slender projections that carry impulses to distant body parts. The biological axiom that "structure suits function," and vice versa, begins with the specialized cells within a tissue. And thereafter, this truism applies also to organs and organ systems.

In the next part of the chapter, we consider the four main types of animal tissues.

> **25.1** *Check Your Progress* **The outer surface of a dog is not adapted to losing heat in order to cool the body. Explain.**

Four Types of Tissues Are Common in the Animal Body

Learning Outcomes 4–6, page 494

This part of the chapter discusses the structure and function of the four tissue types in the body of a complex animal (one that has organ systems). These tissues are termed epithelial, connective, muscular, and nervous tissues.

25.2 Epithelial tissue covers organs and lines body cavities

Epithelial tissue, also called *epithelium*, forms the external and internal linings of many organs and covers the surface of the body. Therefore, in order for a substance to enter or exit the body at the digestive tract, the lungs, or the urinary tract, it must cross an epithelial tissue. Epithelial cells adhere to one another, but are generally only one cell thick. This characteristic enables them to fulfill a

protective function and yet allows substances to pass through to a tissue beneath them.

Epithelial cells are connected to one another by three types of junctions (see Fig. 4.20B): tight junctions, adhesion junctions, and gap junctions. Because epithelial cells are joined by tight junctions in the intestine, the gastric juices stay out of the body, and the same holds true in the kidneys, where tight junctions cause the urine to stay within the kidney tubules. Adhesion junctions allow epithelial cells in the skin to stretch and bend, while gap junctions permit the passage of molecules between adjacent cells. Epithelial cells are exposed on one side, but on the other side they have a basement membrane. The **basement membrane** is simply two thin layers of proteins that anchor the epithelium to underlying connective tissue.

The cells of epithelial tissue differ in shape (**Fig. 25.2**). ① Simple **squamous epithelium**, such as that lining the air spaces of the lungs and the **lumen** (open cavity) of blood vessels, is composed of a single layer of flattened cells attached to the basement membrane. ② Simple **cuboidal epithelium**, which lines the lumen of the kidney tubules, contains cube-shaped cells. ③ Simple **columnar epithelium** is a single layer of cells resembling rectangular pillars or columns, with nuclei usually located near the bottom of each cell. Simple columnar epithelium lines the lumen of the digestive tract.

Aside from cell type, epithelial tissue is classified according to the number of layers in the tissue. An epithelium is simple when it has one layer of cells and stratified when it has several layers of cells piled on top of one another. As we shall see, the outer layer of skin is **stratified** squamous epithelium, but these cells have been reinforced by keratin, a protein that provides strength. When an epithelium is pseudostratified, it appears to be layered, but true layers do not exist because each cell touches the baseline. An example is ④ **pseudostratified ciliated columnar epithelium**, which lines

the trachea (windpipe). Along the trachea, goblet cells produce mucus that traps foreign particles, and the upward motion of cilia carries the mucus to the back of the throat (pharynx), where it may be either swallowed or expelled. Smoking can cause a change in mucus secretion and inhibit ciliary action, resulting in an inflammatory condition called chronic bronchitis.

The lining of the urinary bladder is a transitional epithelium whose structure suits its function. When the walls of the bladder are relaxed, the transitional epithelium consists of several layers of cuboidal cells. When the bladder is distended with urine, the epithelium stretches, and the outer cells take on a squamous appearance. The cells are able to slide in relation to each other, while at the same time forming a barrier that prevents any part of urine from diffusing into other regions of the body.

When an epithelium secretes a product, it is said to be glandular. A **gland** can be a single epithelial cell, such as a mucus-secreting goblet cell, or a gland can contain many cells. Glands that secrete their product into ducts are called exocrine glands, and those that secrete their products into the bloodstream are called endocrine glands.

In the next section, we explore the structure and functions of the many types of connective tissues.

25.2 Check Your Progress In humans, sweat glands open onto the surface of the skin. What role do sweat glands play in temperature regulation?

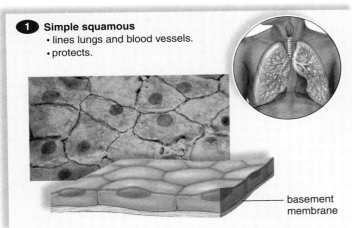

① **Simple squamous**
 • lines lungs and blood vessels.
 • protects.

② **Simple cuboidal**
 • lines kidney tubules, various glands.
 • absorbs molecules.

— basement membrane

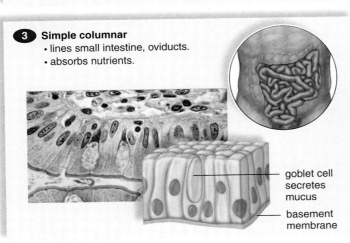

③ **Simple columnar**
 • lines small intestine, oviducts.
 • absorbs nutrients.

goblet cell secretes mucus

— basement membrane

④ **Pseudostratified ciliated columnar**
 • lines trachea.
 • sweeps impurities toward throat.

— cilia

goblet cell secretes mucus

— basement membrane

FIGURE 25.2 Types of epithelial tissue in vertebrates.

Connective tissue is the most abundant and widely distributed tissue in vertebrates. The many different types of **connective tissue** are all involved in binding organs together and providing support and protection. As a rule, connective tissue cells are widely separated by a **matrix**, a noncellular material that varies from solid to semifluid to fluid. The matrix usually has fibers, notably collagen fibers. Collagen is the most common protein in the human body, which gives you some idea of how prevalent connective tissue is. The fibers lend support and also make connective tissue resilient (able to adapt to changes).

Loose Fibrous and Related Connective Tissues Let's consider **loose fibrous connective tissue** first, and then compare the other types to it (**Fig. 25.3A**). ❶ This tissue occurs beneath an epithelium and connects it to other tissues within an organ. It also forms a protective covering for many internal organs, such as muscles, blood vessels, and nerves. Its cells are called **fibroblasts** because they produce a matrix that contains fibers, including collagen fibers and elastic fibers. The presence of loose fibrous connective tissue in the walls of the lungs and the arteries allows these organs to expand.

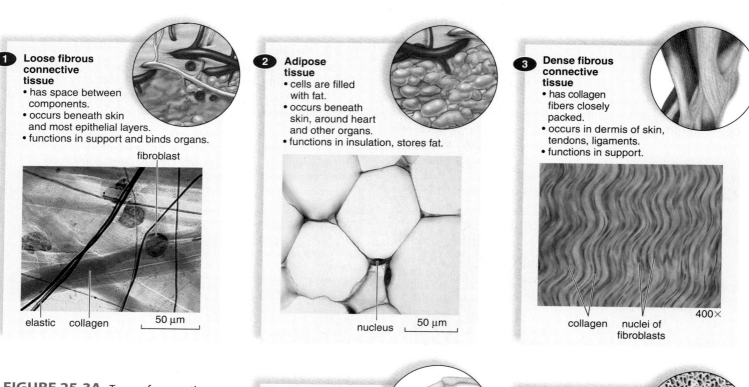

1 **Loose fibrous connective tissue**
• has space between components.
• occurs beneath skin and most epithelial layers.
• functions in support and binds organs.

fibroblast

elastic   collagen   50 μm

2 **Adipose tissue**
• cells are filled with fat.
• occurs beneath skin, around heart and other organs.
• functions in insulation, stores fat.

nucleus   50 μm

3 **Dense fibrous connective tissue**
• has collagen fibers closely packed.
• occurs in dermis of skin, tendons, ligaments.
• functions in support.

400×

collagen   nuclei of fibroblasts

FIGURE 25.3A Types of connective tissue in vertebrates.

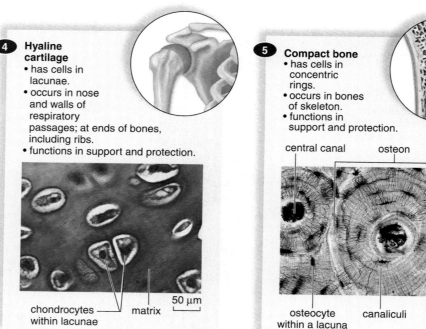

4 **Hyaline cartilage**
• has cells in lacunae.
• occurs in nose and walls of respiratory passages; at ends of bones, including ribs.
• functions in support and protection.

chondrocytes within lacunae   matrix   50 μm

5 **Compact bone**
• has cells in concentric rings.
• occurs in bones of skeleton.
• functions in support and protection.

central canal   osteon

osteocyte within a lacuna   canaliculi   320×

2 **Adipose tissue** is a type of loose connective tissue in which the fibroblasts enlarge and store fat, and there is limited matrix. Adipose tissue is located beneath the skin and around organs, such as the heart and kidneys. The body uses this stored fat for energy, insulation, and organ protection.

Compared to loose fibrous connective tissue, **3** **dense fibrous connective tissue** contains more collagen fibers, and they are packed closely together. This type of tissue has more specific functions than does loose fibrous connective tissue. For example, dense fibrous connective tissue is found in the dermis of the skin; in **tendons**, which connect muscles to bones; and in **ligaments**, which connect bones to other bones at joints.

4 In **cartilage**, the cells lie in small chambers called **lacunae**, separated by a matrix that is solid yet flexible. Unfortunately, because this tissue lacks a direct blood supply, it heals very slowly. **Hyaline cartilage**, the most common type of cartilage, contains only very fine collagen fibers. The matrix has a white, translucent appearance. Hyaline cartilage is found in the nose and at the ends of the long bones and the ribs, and it forms rings in the walls of respiratory passages. The human fetal skeleton is also made of this type of cartilage, which is later replaced by bone. Cartilaginous fishes, such as sharks, have a cartilaginous skeleton throughout their lives.

5 **Bone** is the most rigid connective tissue. It consists of an extremely hard matrix of inorganic salts, notably calcium salts, deposited around collagen fibers. The inorganic salts give bone rigidity, and the collagen fibers provide elasticity and strength, much as steel rods do in reinforced concrete. Compact bone, the most common type, consists of cylindrical structural units called osteons. Rings of hard matrix surround the central canal, which contains blood vessels. Bone cells (osteocytes) are located in empty spaces called lacunae between the rings of matrix. Blood vessels in the central canal carry nutrients that allow bone to renew itself. The nutrients can reach all of the cells because minute canals (canaliculi) containing thin extensions of the osteocytes connect osteocytes with one another and eventually with the central canal. These connections allow the osteocytes to have a constant supply of blood and nutrients.

Blood Blood is composed of several types of cells suspended in a liquid matrix called plasma. Blood is unlike other types of connective tissue in that the matrix (i.e., plasma) is not made by the cells (**Fig. 25.3B**). Some people do not classify blood as connective tissue; instead, they suggest a separate tissue category called vascular tissue.

Blood serves the body well. It transports nutrients and oxygen to cells and removes their wastes. It helps distribute heat and plays a role in fluid, ion, and pH balance. Also, various components of blood help protect us from disease, and blood's ability to clot prevents fluid loss.

Blood contains two types of cells. **Red blood cells** are small, biconcave, disk-shaped cells without nuclei. The presence of the red pigment hemoglobin makes the cells red and, in turn, makes the blood red. Hemoglobin combines with oxygen, and in this way, red blood cells transport oxygen. At a crime

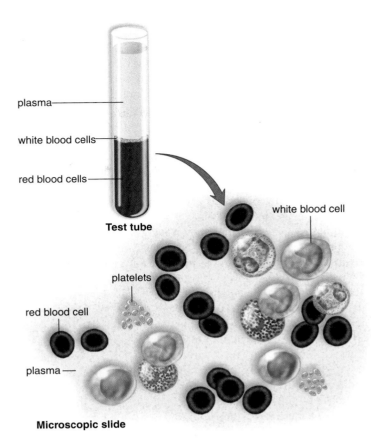

FIGURE 25.3B Composition of blood, a liquid tissue.

scene, the pigment portion of hemoglobin makes it difficult for the perpetrator to remove traces of blood. Forensic specialists can perform tests to confirm that a stain is due to hemoglobin. One test involves spraying the stain with luminal, a chemical that binds with blood and then glows in the dark. Other tests depend on the chemistry of red blood cells and other substances in blood that can identify the blood as belonging to a specific person.

White blood cells may be distinguished from red blood cells by the fact that they are usually larger, have a nucleus, and, without staining, would appear translucent. White blood cells fight infection in two primary ways: Some white blood cells are phagocytic and engulf infectious pathogens, while other white blood cells produce antibodies, molecules that combine with foreign substances to inactivate them.

Platelets are another component of blood, but they are not complete cells; rather, they are fragments of giant cells present only in bone marrow. When a blood vessel is damaged, platelets form a plug that seals the vessel, and injured tissues release molecules that help the clotting process.

Muscular tissue is the topic of Section 25.4.

25.3 Check Your Progress Which animal would you expect to have more adipose tissue—a polar bear living in the Arctic or a gila monster living in Arizona? Explain.

25.4 Muscular tissue is contractile and moves body parts

Muscular tissue and nervous tissue account for the ability of animals and their parts to move. **Muscular tissue** is also sometimes called *contractile tissue* because it contains contractile protein filaments, called actin and myosin filaments, that interact to produce movement. The three types of vertebrate muscles are skeletal, cardiac, and smooth (**Fig. 25.4**).

① **Skeletal muscle**, also called *voluntary muscle*, is attached by tendons to the bones of the skeleton, and when it contracts, the bones move. Contraction of skeletal muscle is under voluntary control and occurs faster than in the other muscle types. The cells of skeletal muscle, called fibers, are cylindrical and quite long—sometimes they run the length of the muscle. They arise during development when several cells fuse, resulting in one fiber with multiple nuclei. The nuclei are located at the periphery of the cell, just inside the plasma membrane. The fibers have alternating light and dark bands that give them a **striated** appearance. These bands are due to the placement of actin filaments and myosin filaments in the cell. The interaction of these filaments accounts for the ability of all three types of muscles to contract.

② **Cardiac muscle** is found only in the walls of the heart, and its contraction pumps blood and accounts for the heartbeat. Like skeletal muscle, cardiac muscle has striations, but the contraction of the heart is autorhythmic and involuntary. Cardiac muscle cells also differ from skeletal muscle cells in that they have a single, centrally placed nucleus. The cells are branched

and seemingly fused with one another, and the heart appears to be composed of one large interconnecting mass of muscle cells. Actually, cardiac muscle cells are separate and individual, but they are bound end to end at **intercalated disks** (gap junctions), areas where folded plasma membranes allow the contraction impulse to spread from one cell to the other.

③ **Smooth muscle** is so named because the cells lack striations. The spindle-shaped cells form layers in which the thick middle portion of one cell is opposite the thin ends of adjacent cells. Consequently, the nuclei form an irregular pattern in the tissue. Smooth muscle is not under voluntary control, and therefore is said to be *involuntary*. Smooth muscle is also sometimes called *visceral muscle* because it is found in the walls of the viscera (intestine, stomach, and other internal organs) and blood vessels. Smooth muscle contracts more slowly than skeletal muscle but can remain contracted for a longer time. When the smooth muscle of the intestine contracts, food moves along its lumen (central cavity). When the smooth muscle of the blood vessels contracts, blood vessels constrict, helping to raise blood pressure.

Nervous tissue, the last tissue to be studied, is described in Section 25.5.

> **25.4** *Check Your Progress* **Constriction of blood vessels by smooth muscle also occurs when mammals are cold. Explain.**

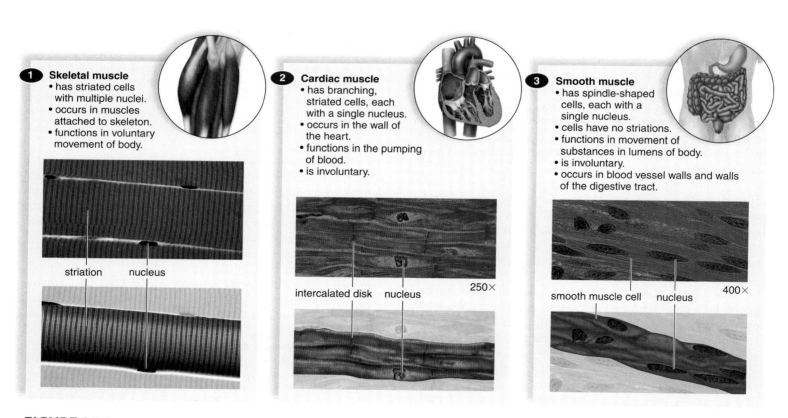

1 Skeletal muscle
- has striated cells with multiple nuclei.
- occurs in muscles attached to skeleton.
- functions in voluntary movement of body.

striation   nucleus

2 Cardiac muscle
- has branching, striated cells, each with a single nucleus.
- occurs in the wall of the heart.
- functions in the pumping of blood.
- is involuntary.

intercalated disk   nucleus   250×

3 Smooth muscle
- has spindle-shaped cells, each with a single nucleus.
- cells have no striations.
- functions in movement of substances in lumens of body.
- is involuntary.
- occurs in blood vessel walls and walls of the digestive tract.

smooth muscle cell   nucleus   400×

FIGURE 25.4 Types of muscular tissue.

25.5 Nervous tissue communicates with and regulates the functions of the body's organs

Nervous tissue coordinates body parts and allows an animal to respond to the environment. The nervous system depends on (1) sensory input, (2) integration of data, and (3) motor output to carry out its functions. Nerves conduct impulses from sensory receptors to the spinal cord and the brain, where integration occurs. The phenomenon called sensation occurs only in the brain, however. Nerves then conduct nerve impulses away from the spinal cord and brain to the muscles and glands, causing them to contract and secrete, respectively. In this way, a coordinated response to both internal and external stimuli is achieved.

A nerve cell is called a **neuron**. Every neuron has three parts: dendrites, a cell body, and an axon (**Fig. 25.5**). A *dendrite* is an extension that conducts signals toward the cell body. The *cell body* contains the major concentration of the cytoplasm and the nucleus of the neuron. An *axon* is an extension that conducts nerve impulses. The brain and spinal cord contain many neurons, whereas **nerves** contain only the axons of neurons. The dendrites and cell bodies of these neurons are located in the spinal cord or brain, depending on whether it is a spinal nerve or a cranial nerve. A nerve is like a land-based telephone trunk cable, because just like telephone wires, each axon is a communication channel independent of the others.

Neuroglia **Neuroglia** are cells that outnumber neurons as much as 50 to 1, and take up more than half the volume of the brain. Although the primary function of neuroglia is to support and nourish neurons, research is currently being conducted to determine how much they directly contribute to brain function. Various types of neuroglia are found in the brain. Microglia, astrocytes, and oligodendrocytes are shown in Figure 25.5. Microglia, in addition to supporting neurons, engulf bacterial and cellular debris. Astrocytes provide nutrients to neurons and produce a hormone known as glia-derived growth factor, which someday might be used as a cure for Parkinson disease and other conditions caused by neuron degeneration. (Parkinson is a movement disorder characterized by tremor, rigidity, slowness, and poor balance.) Oligodendrocytes form myelin, which acts much like the insulation of a telephone cable. Neuroglia do not have a long process, but even so, researchers are now beginning to gather evidence that they do communicate among themselves and with neurons! Neuroglia also form a plasma-like solution called **cerebrospinal fluid**, which supports and nourishes the brain and spinal cord.

Mature neurons have little capacity for cell division and seldom form tumors. The majority of brain tumors in adults involve actively dividing neuroglia. Most brain tumors have to be treated with surgery or radiation therapy because tight junctions between cells lining the blood vessels do not permit drugs to pass through to the brain.

In Section 25.6, we will explore whether nerve regeneration in the brain and spinal cord is possible.

> **25.5** *Check Your Progress* **How is the shape of a neuron appropriate to its function?**

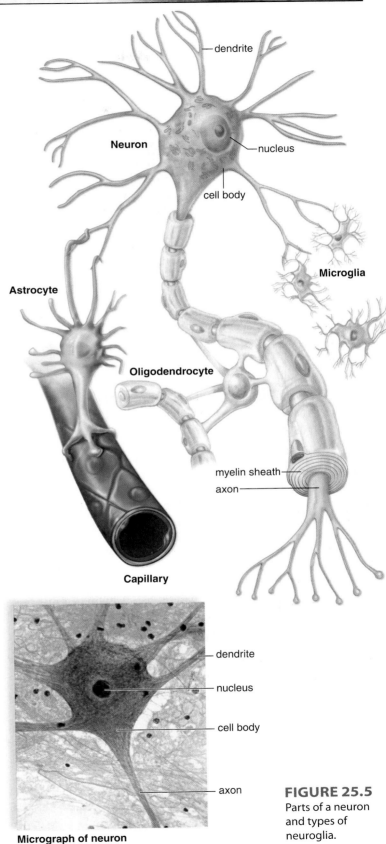

FIGURE 25.5
Parts of a neuron and types of neuroglia.

Micrograph of neuron

In 1995, Christopher Reeve, best known for his acting role as "Superman," was thrown headfirst from his horse, crushing his spinal cord just below the top two vertebrae (**Fig. 25.6A**). Immediately, his brain lost almost all communication with the portion of his body below the site of damage, and he could not move his arms and legs. Many years later, Reeve could move his left index finger slightly and could take tiny steps while being held upright in a pool. He had sensation throughout his body and could feel his wife's touch.

FIGURE 25.6A Reeve rode horses for enjoyment.

Reeve's improvement was not the result of cutting-edge drugs or gene therapy—it was due to exercise (**Fig. 25.6B**)! Reeve exercised as much as five hours a day, especially using a recumbent bike outfitted with electrodes that made his leg muscles contract and relax. The bike cost him $16,000. It could cost less if commonly used by spinal cord injury patients in their own homes. Reeve, who was an activist for the disabled, was pleased that insurance would pay for the bike about 50% of the time.

It is possible that Reeve's advances were the result of improved strength and bone density, which led to stronger nerve signals. Perhaps nerve stimulation and Reeve's intensive exercise brought back some of the normal communication between nerve cells. Reeve's physician, John McDonald, a neurologist at Washington University in St. Louis, is convinced that his axons were regenerating. The neuroscientist Fred Gage at the Salk Institute in La Jolla, California, has shown that exercise does enhance the growth of new cells in adult brains.

In humans, damaged axons within the central nervous system (CNS) do not regenerate and the result is permanent loss of nervous function. Not so in cold-water fishes and amphibians, where axon

FIGURE 25.6C Biochemists study the proteins needed for nerve regeneration.

regeneration in the CNS does occur. So far, investigators have identified several proteins that seem to be necessary in order for axon regeneration to occur in the CNS of these animals (**Fig. 25.6C**), but it will be a long time before biochemistry can offer a way to bring about axon regeneration in the human CNS. It is possible, though, that one day these proteins will be effective drugs when CNS injuries occur.

Reeve was convinced that stem cell therapy would one day allow him to be off his ventilator and functioning normally (**Fig. 25.6D**); unfortunately, he died in 2004 from an infection. So far, researchers have shown that both embryonic stem cells and bone marrow stem cells can differentiate into neurons in the laboratory. Bone marrow stem cells apparently can also become neurons when injected into the body.

In the next part of the chapter, we study skin as an example of an organ.

> **25.6** *Check Your Progress* **Explain why Reeve experienced sensations, but lacked motor control.**

FIGURE 25.6B Reeve received aqua therapy.

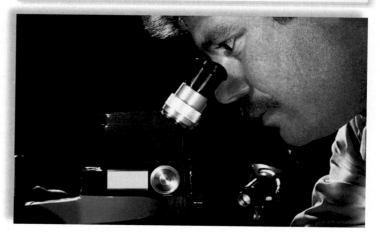

FIGURE 25.6D Cell biologists study the possibility of using stem cells for nerve regeneration.

The structure and function of the epidermis and dermis of the skin, along with the subcutaneous layer, are briefly studied in this part of the chapter.

25.7 Each organ has a specific structure and function

As stated in Section 25.1, an organ is a structural unit of an organism that performs particular functions. The structure and function of an organ is dependent on the tissues it contains. For example, the protective function of the **skin**, an organ in vertebrates, is enhanced by a thickened and keratinized epidermis covered by different derivatives according to the animal: The skin of fishes has numerous bony scales; amphibians have smooth skin covered with mucous glands; reptiles possess epidermal scales that vary in color and shape; and birds have scales on their legs, but feathers covering most of the rest of their body. The skin of mammals is characterized by the presence of hair and derivative structures, such as finger and toe nails. Scales, feathers, and hair are also involved in body temperature, as are sweat glands. Sweating in mammals not only helps regulate body temperature, it also allows the skin to excrete salts and even nitrogenous wastes. The sensory receptors in the skin provide animals with much knowledge about the outside world. In addition to these functions of the skin, mammalian skin cells manufacture precursor molecules that are converted to vitamin D after exposure to UV (ultraviolet) light.

Regions of the Skin In humans, skin has two regions, called the epidermis and the dermis. A subcutaneous layer, also called the hypodermis, lies between the skin and any underlying structures, such as muscle or bone. **Epidermis** is composed of stratified squamous epithelium. New cells derived from stem cells in the germinal layer become flattened and hardened as they are pushed to the surface. Hardening takes place because the cells produce keratin, a waterproof protein. A thick layer of

dead keratinized cells arranged in concentric spirals forms fingerprints and footprints. Specialized cells in the epidermis called melanocytes produce melanin, the pigment responsible for skin color. In a light-skinned person, tanning signifies that melanocytes are trying to protect the skin from the dangerous rays of the sun. Too much ultraviolet radiation can lead to skin cancer.

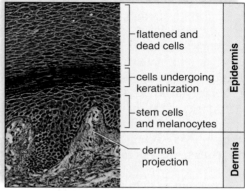

Photomicrograph of skin

The **dermis** is a region of dense fibrous connective tissue beneath the epidermis. With age and exposure to the sun, the number of collagen and elastic fibers in dermis decrease, resulting in wrinkles. A new treatment is Botox, a diluted form of a bacterial toxin that blocks innervation of muscles and reduces wrinkling. In addition to fibers, the dermis contains blood vessels, hair follicles, oil and sweat glands, and sensory receptors for touch, pressure, pain, hot, and cold (**Fig. 25.7**).

Technically speaking, the **subcutaneous layer** beneath the dermis is not a part of the skin. It is composed of loose connective tissue and adipose issue, which stores fat. A well-developed subcutaneous layer gives the body a rounded appearance, provides protective padding against external assaults, and reduces heat loss. Excessive development of the subcutaneous layer accompanies obesity.

The next part of the chapter will describe the organ systems of the human body.

FIGURE 25.7 Human skin anatomy.

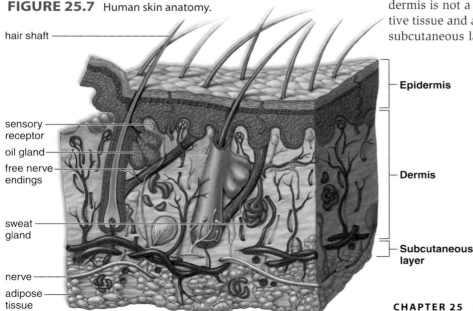

> **25.7** *Check Your Progress* Gila monsters, which live in the desert, have a thick epidermis. What trade-offs arise regarding temperature regulation versus retention of fluids because of the thick epidermis?

This part of the chapter reviews the organ systems in humans, emphasizing their contribution to homeostasis.

25.8 Several organs work together to carry out the functions of an organ system

An organ system is a collection of organs that work together to perform related roles in an organism. The functions of an organ system are dependent on its organs. For example, the function of the urinary system is to produce urine, store it, and then transport it out of the body. As described in Section 25.1, the kidneys produce urine, and then tubes called ureters transport it to the urinary bladder for storage until it is released from the body by way of a tube called the urethra.

The organs of vertebrates can be grouped in various ways. Here, we group them according to the following functions: control, sensory input and motor output, transport, maintenance, and reproduction. All of the body's systems are involved in maintaining **homeostasis**, the stability of the body's internal environment. Consider that the body has both an external environment and an internal environment. Vertebrates are able to regulate the internal environment so that it remains constant, despite fluctuations in the external environment.

Control The **nervous system** (**Fig. 25.8A,** *left*) consists of the brain, spinal cord, and associated nerves. The nerves conduct nerve impulses from receptors to the brain and spinal cord. They also conduct nerve impulses from the brain and spinal cord to the muscles and glands, allowing us to respond to both external and internal stimuli. (Sensory receptors and sense organs are sometimes considered part of the nervous system.)

The **endocrine system** (Fig. 25.8A, *right*) consists of the hormonal glands, which secrete chemicals that serve as messengers between body parts. Both the nervous and endocrine systems coordinate and regulate the functions of the body's other systems. The endocrine system also helps maintain the proper functioning of the male and female reproductive organs.

Both the nervous system and the endocrine system coordinate body parts. However, the nervous system is fast-acting, while the endocrine system is slower and has more lasting effects.

Sensory Input and Motor Output The **integumentary system** (**Fig. 25.8B,** *left*) consists of the skin and its accessory structures. Sensory receptors in the skin, as well as in organs, such as the eyes and ears, are sensitive to certain external stimuli. These receptors communicate with the brain and spinal cord by way of nerve fibers. Sensory receptors also provide us with information about our internal environment. These messages, sent by sensory receptors to the brain, cause the brain to bring about a response, called the motor output.

The **skeletal system** and the **muscular system** (Fig. 25.8B, *middle, right*) enable the body and its parts to move as a result of nerve stimulation. The skeleton, as a whole, serves as a place of attachment for the skeletal muscles. Contraction of muscles in the muscular system accounts for the actual movement of body parts.

The skeletal and muscular systems, along with the integumentary system, also protect and support the body. The bones of the skeleton protect body parts. For example, the skull forms a protective encasement for the brain, as does the rib cage for the heart and lungs. The skin and muscles assist in this endeavor because they are exterior to the bones.

Bones serve other functions as well. Red bone marrow produces blood cells, and bones serve as storage areas for calcium and phosphate salts.

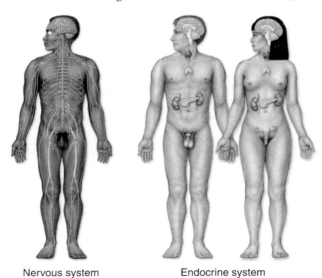

Nervous system   Endocrine system

FIGURE 25.8A The control systems.

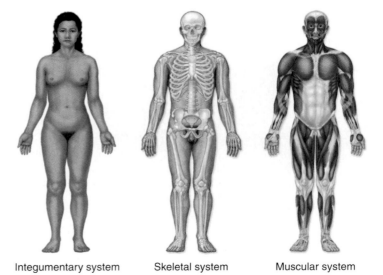

Integumentary system   Skeletal system   Muscular system

FIGURE 25.8B Sensory input and motor output.

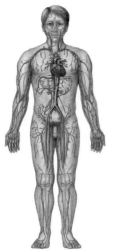
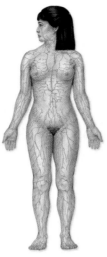

Cardiovascular system

Lymphatic and immune systems

FIGURE 25.8C
The body's transport systems.

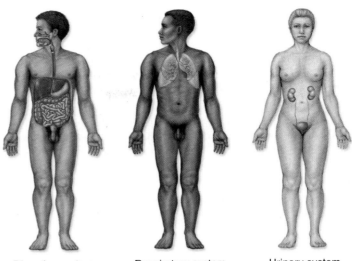

Digestive system

Respiratory system

Urinary system

FIGURE 25.8D The body's maintenance systems.

Transport The **cardiovascular system** (**Fig. 25.8C,** *left*) consists of the heart and the blood vessels that carry blood throughout the body. The pumping action of the heart propels blood into two circuits; one circuit takes blood to the lungs, and the other circuit takes blood to the body proper. The body's cells are surrounded by a liquid called **tissue fluid**; blood transports nutrients and oxygen to tissue fluid for the cells, and removes waste molecules excreted by cells from the tissue fluid. The internal environment of the body consists of the blood within the blood vessels and the tissue fluid that surrounds the cells.

The **lymphatic system** (Fig. 25.8C, *right*) consists of lymphatic vessels, which carry lymph, and lymphatic organs, including lymph nodes. Lymphatic vessels absorb fat from the digestive system and collect excess tissue fluid, which is returned to the blood in the cardiovascular system. In this way, the lymphatic system assists the cardiovascular system in maintaining blood pressure. The lymphatic organs have various other functions, but all are involved in defending the body against disease. Again, the cardiovascular system assists with this function. Certain blood cells in the lymph and blood are part of an **immune system**, which specifically protects the body from disease.

Maintenance Three systems (digestive, respiratory, and urinary) maintain the body by adding substances to and/or removing substances from the blood. If the composition of the blood remains constant, so does that of the tissue fluid.

The **digestive system** (**Fig. 25.8D,** *left*) consists of the various organs along the digestive tract together with associated organs, such as teeth, salivary glands, the liver, and the pancreas. The accessory organs produce digestive enzymes (salivary glands and pancreas) and bile (liver), which are sent to the digestive tract by way of ducts. The digestive system receives food and digests it into nutrient molecules that enter the blood. Nutrient molecules carried by the bloodstream sustain the body.

The **respiratory system** (Fig. 25.8D, *middle*) consists of the lungs and the tubes that take air to and from the lungs. The body has a way to store energy but has no way to store oxygen. Therefore, the respiratory system brings oxygen into the body and takes carbon dioxide out of the body through the lungs. It also exchanges gases with the blood.

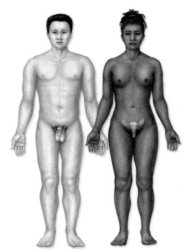

Reproductive system

FIGURE 25.8E
The reproductive system.

The **urinary system** (Fig. 25.8D, *right*) contains the kidneys and the urinary bladder along with tubes that transport urine. This system rids blood of wastes and also helps regulate the fluid level and chemical content of the blood.

Reproduction The **reproductive system** (**Fig. 25.8E**) involves different organs in the male and female. The male reproductive system consists of the testes, other glands, and various ducts that conduct semen to and through the penis. The testes produce sex cells called sperm. The female reproductive system consists of the ovaries, oviducts, uterus, vagina, and external genitals. The ovaries produce sex cells called eggs. When a sperm fertilizes an egg, an offspring begins to develop.

Organs for transplant are problematic, as discussed in Section 25.9.

> **25.8** *Check Your Progress* **Organ systems work together. Construct a scenario that shows that sensory input, the nervous system, and motor output are involved in behavioral modification to control body temperature.**

Human organs—or entire organ systems—sometimes cease to function due to injury or disease. Today, it is possible to successfully transplant various organs, and thus prolong a patient's life. Unfortunately, however, there are not enough human organ donors to go around. Thousands of patients die each year while waiting for an organ. It's no wonder, then, that scientists are suggesting we get organs from a source other than humans.

Xenotransplantation is the use of animal organs, instead of human organs, in transplant patients. You might think that apes, such as the chimpanzee or the baboon, would be a scientifically suitable species for this purpose. But apes are slow breeders and probably cannot be counted on to supply all the organs needed. Also, many people might object to using apes for this purpose. Regardless, a more suitable organ donor exists—the pig. In this country, animal husbandry has long included the raising of pigs as a meat source, and pigs are prolific. A female pig can become pregnant at six months of age and can have two litters a year, each averaging about ten offspring.

Ordinarily, the human body would violently reject transplanted pig organs. Genetic engineering, however, can change that, and pig valves have already been successfully used in human hearts. Someday, pig hearts may be used to keep a patient alive until a human organ becomes available.

As the possibility of xenotransplantation draws near, other concerns have been raised. Some experts fear that animals—and thus their organs—might be infected with viruses, akin to Ebola virus or the "mad cow" disease virus. After infecting a transplant patient, these viruses might spread into the general populace and begin an epidemic. Supporting this possibility is scientists' belief that HIV was originally spread to humans from monkeys when humans ate monkey meat. Advocates of using pigs for xenotransplantation point out that pigs have been around humans for centuries without infecting them with any serious diseases. Even so, a strain of inbred miniature pigs that do not produce porcine endogenous retrovirus (PERV) has been produced.

An alternative to xenotransplantation also exists. Just a few years ago, scientists believed that transplant organs had to come

FIGURE 25.9A Burn victim who received artificial skin as a transplant.

FIGURE 25.9B
Lab-grown bladder ready to serve as a transplant.

from humans or other animals. Now, however, tissue engineering is demonstrating that it is possible to make replacement organs in the lab. For example, two skin products have now been approved for use in humans. One is composed of dermal cells growing on a degradable polymer, which can be used to temporarily cover the wounds of burn patients while their own skin regenerates (**Fig. 25.9A**). The other utilizes only live human skin cells to treat diabetic leg and foot ulcers. Similarly, the damaged cartilage of a knee can be replaced with a tissue produced after chondrocytes are harvested from a patient. Soon to come are a host of other products, including replacement corneas, heart valves, bladder valves, and breast tissue.

Tissue engineers have also created cellular implants—cells that produce a useful product encapsulated within a narrow plastic tube or a capsule the size of a dime or quarter. The pores of the container are large enough to allow the product to diffuse out, but too small for immune cells to enter and destroy the cells. An implant whose cells secrete natural painkillers will survive for months in the spinal cord and can be easily withdrawn when desired. A "bridge to a liver transplant" is a bedside vascular apparatus. The patient's blood passes through porous tubes surrounded by pig liver cells. These cells convert toxins in the blood to nonpoisonous substances.

The goal of tissue engineering is to produce fully functioning organs for transplant, perhaps from embryonic or adult stem cells. After nine years, researchers have been able to produce a working urinary bladder in the laboratory (**Fig. 25.9B**). After being tested in laboratory animals, the bladder can now be implanted in humans whose own bladders have been damaged by accident or disease or will not function properly due to a congenital birth defect. Another group of scientists has been able to grow arterial blood vessels in the lab using a pig small intestine as the mold. The availability of engineered blood vessels would make heart bypass operations simpler to perform. Tissue engineers also hope to one day produce more complex internal organs, such as a liver or a kidney.

All of the body's organs contribute to homeostasis, as discussed in Section 25.10.

25.9 Check Your Progress In what ways are stem cell research, research into xenotransplantation, and tissue engineering in competition with one another?

You are probably most familiar with the use of the word *environment* in connection with the external environment. People today are very concerned about our external environment because of the need to keep pollution to a minimum in order to maintain the health of ecosystems and the organisms, including ourselves, that live in them. Your body, however, is composed of many cells, and its internal environment is the environment of the cells. Cells live in a liquid environment called tissue fluid, which is constantly renewed by exchanges with the blood (**Fig. 25.10**). Therefore, blood and tissue fluid are the internal environment of the body. Tissue fluid remains relatively constant only as long as blood composition remains near normal levels. Relatively constant means that the composition of both tissue fluid and blood usually falls within a certain range of normality.

Let us consider how the systems of the body contribute to maintaining homeostasis. The cardiovascular system conducts blood to and away from capillaries, where exchange occurs. The heart pumps the blood, and thereby keeps it moving toward the capillaries. Red blood cells transport oxygen and participate in the transport of carbon dioxide. White blood cells fight infection, and platelets participate in the clotting process. The lymphatic system is accessory to the cardiovascular system. Lymphatic capillaries collect excess tissue fluid and return it via lymphatic vessels to the cardiovascular system. Lymph nodes help purify lymph and keep it free of pathogens.

The digestive system takes in and digests food, providing nutrient molecules that enter the blood to replace those that are constantly being used by the body's cells. The respiratory system removes carbon dioxide from and adds oxygen to the blood. The chief regulators of blood composition are the kidneys and the liver. Urine formation by the kidneys is extremely critical to the body, not only because it rids the body of metabolic wastes, but also because the kidneys carefully regulate blood volume, salt balance, and pH. The liver, among other functions, regulates the glucose concentration of the blood. Immediately after glucose enters the blood, the liver removes the excess for storage as glycogen. Later, glycogen is broken down to replace the glucose that was used by body cells. In this way, the glucose composition of the blood remains constant. The liver also removes toxic chemicals, such as ingested alcohol and other drugs. The liver makes urea, a nitrogenous end product of protein metabolism.

The nervous system and the endocrine system regulate the other systems of the body. They work together to control body systems so that homeostasis is maintained. For example, they can cause the breathing rate to speed up or slow down. Likewise, they can speed the action of the heart or slow it down. In negative feedback mechanisms involving the nervous system, sensory receptors send nerve impulses to control centers in the brain, which then direct effectors to become active. Effectors can be muscles or glands. Muscles bring about an immediate change. Endocrine glands secrete hormones that bring about a slower, more lasting change that keeps the internal environment relatively stable.

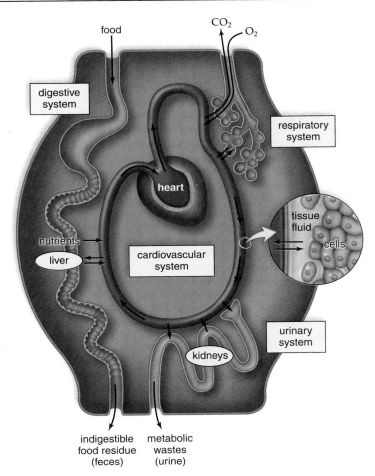

FIGURE 25.10   Process of achieving constancy in the internal environment (blood and tissue fluid).

Because of homeostasis, even though external conditions may change dramatically, internal conditions stay within a narrow range. One of the most obvious examples of homeostasis is body temperature. The temperature of the human body is maintained near 37°C (97° to 99°F), even if the surrounding temperature varies considerably from this temperature. Other examples of homeostasis include regulation of the body's water content, carbon dioxide concentration, pH, and glucose concentration. If you eat acidic foods, the pH of your blood still stays about 7.4, and even if you eat a candy bar, the amount of sugar in your blood remains at just about 0.1%. If the parameters of the blood fail to stay within a normal range, coma and death can result because cells can only continue to function when their needs are being met.

How negative feedback helps maintain homeostasis is described in Section 25.11.

25.10 *Check Your Progress* An important aspect of homeostasis is the constancy of a moderate body temperature (37°C). Explain, in terms of enzymatic reactions, why a moderate body temperature is important.

Negative feedback is the primary homeostatic mechanism that allows the body to keep the internal environment constant. The model for negative feedback shown in **Figure 25.11A** has two components: a sensor and a control center. The sensor detects a change in the internal environment (a stimulus); the control center initiates an effect that brings conditions back to normal again. Now the sensor is no longer activated. In other words, a negative feedback mechanism is present when the output of the system dampens the original stimulus.

Let's take a simple example. When the pancreas detects that the blood glucose level is too high, it secretes insulin, a hormone that causes cells to take up glucose. Now the blood sugar level returns to normal, and the pancreas is no longer stimulated to secrete insulin.

When conditions exceed their limits and negative feedback mechanisms cannot compensate, illness results. For example, if the pancreas is unable to produce insulin, the blood sugar level becomes dangerously high, and the individual can become seriously ill. The study of homeostatic mechanisms is, therefore, medically important.

Complex Examples

A home heating system is often used to illustrate how a more complex negative feedback mechanism works. You set the thermostat at, say, 20°C (68°F). This is the *set point*. The thermostat contains a thermometer, a sensor that detects when the room temperature is above or below the set point. The thermostat also contains a control center; it turns the furnace off when the room is warm and turns it on when the room is cool. When the furnace is off, the room cools a bit, and when the furnace is on, the room warms a bit. In other words, a negative feedback system results in controlled fluctuation above and below the set point.

In humans, the thermostat for body temperature is located in a part of the brain called the hypothalamus. When the core body temperature becomes higher than normal, the control center directs (via nerve impulses) the blood vessels of the skin to dilate (**Fig. 25.11B**, *left*). More blood is then able to flow near the surface of the body, where heat can be lost to the environment. In addition, the nervous system activates the sweat glands, and

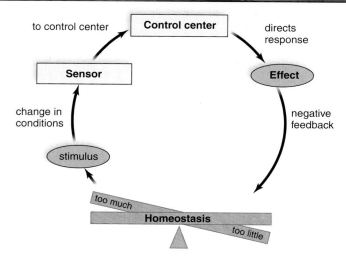

FIGURE 25.11A   Model of negative feedback.

the evaporation of sweat helps lower body temperature. Gradually, body temperature decreases to 37°C (98.6°F).

When the core body temperature falls below normal, the control center directs the blood vessels of the skin to constrict (Fig. 25.11B, *right*). This action conserves heat. If the core body temperature falls even lower, the control center sends nerve impulses to the skeletal muscles, and shivering occurs. Shivering generates heat, and, gradually, body temperature rises to 37°C (98.6°F). When the temperature rises to normal, the control center is inactivated.

Notice that a negative feedback mechanism prevents change in the same direction—in other words, body temperature does not get warmer and warmer, because warmth brings changes that decrease body temperature. Also, body temperature does not get colder and colder, because a body temperature below normal causes changes that bring the body temperature up.

> **25.11** *Check Your Progress*   **Create a feedback scenario that explains the behavior of a lizard trying to maintain its body temperature during the day in the tropics.**

FIGURE 25.11B
Regulation of body temperature by negative feedback.

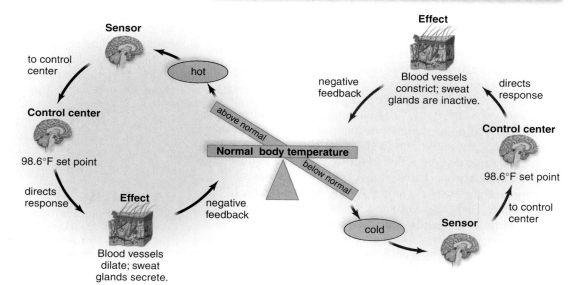

This chapter serves as an introduction to Part V because it not only reviews the types of tissues in the animal body, but it also reviews the organ systems we will be discussing. Two themes are introduced that will be referred to time and time again: (1) structure suits function, and (2) homeostasis. Structure suits function means that the function of an organ is reflected in its structure. For example, the function of the heart is to pump blood, and therefore it contains chambers for holding blood and has thick, muscular walls for pumping blood.

Homeostasis refers to the relative constancy of the internal environment. In complex animals, the internal environment is blood and tissue fluid. Let's take a familiar example of homeostasis in humans. After eating, the hormone insulin is released, and glucose is removed from the blood and stored in the liver as glycogen. In between eating, the hormone glucagon regulates glycogen breakdown so that the blood glucose level remains at just about 0.1%. In this way, glucose is constantly available to cells for the process of cellular respiration.

In complex animals, the nervous and endocrine systems also act together to regulate the actions of organs so that homeostasis is maintained. Sensory receptors gather information from the external and internal environments and convert it to a form that can be processed by the nervous system. Then the nervous system can direct the action of the other systems, sometimes even the endocrine system, so that homeostasis is maintained. Movement brought about by the skeletal and muscular systems may be required to achieve homeostasis. These and other examples of homeostasis will be discussed throughout the chapters of Part V. Chapters 26–28 are about the nervous, sensory, and musculoskeletal systems.

The Chapter in Review

Summary

Staying Warm, Staying Cool
- Ectothermic animals (e.g., reptiles) heat their bodies from the outside environment.
- Endothermic animals (e.g., birds, mammals) rely on heat produced inside the body.

The Structure of Tissues Suits Their Function

25.1 Levels of biological organization are evident in animals
In the scheme:

Cell ⟶ Tissues ⟶ Organs ⟶ Organ Systems ⟶ Organism

- A tissue is composed of similar cells performing a similar function.
- An organ is formed by different types of tissues arranged in a certain way.
- An organ system is made up of several organs that work together to perform necessary functions for the organism.

Four Types of Tissues Are Common in the Animal Body
The four animal tissues are:

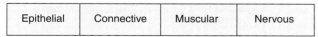

| Epithelial | Connective | Muscular | Nervous |

25.2 Epithelial tissue covers organs and lines body cavities
- By being one cell thick, epithelial tissue can be protective, and yet also allow substances to pass through.
- Epithelial cells connect to each other by tight junctions, adhesion junctions, and gap junctions.
- A basement membrane anchors epithelium to underlying connective tissue.
- Cells of the epithelium may be squamous, cuboidal, or columnar in shape; they also may be simple or stratified.
- One or more epithelial cells make up the glands.

25.3 Connective tissue connects and supports other tissues
- Connective tissue cells are separated by a matrix, which varies from solid (as in bone), to semifluid (as in cartilage), to fluid (as in blood).
- Types of connective tissue are loose fibrous, adipose, dense fibrous, cartilage, bone, and blood.

25.4 Muscular tissue is contractile and moves body parts
- Skeletal muscle, attached by tendons to bones, is voluntary.
- Cardiac muscle, found in the walls of the heart, is involuntary.
- Smooth muscle, found in the walls of viscera, is involuntary.

25.5 Nervous tissue communicates with and regulates the functions of the body's organs
- Nervous tissue coordinates body parts and allows animals to respond to their environment.
- The nervous system receives sensory input, integrates data, and brings about a response.
- A neuron has dendrites, a cell body, and an axon.
- Neuroglia support and nourish neurons; types include microglia, astrocytes, and oligodendrocytes.

25.6 Will nerve regeneration reverse a spinal cord injury?
- Nerve regeneration does not normally occur in the CNS.
- Research is proceeding on three main fronts:
  - Christopher Reeve advocated intense exercise.
  - Biochemists have identified proteins that may help.
  - Cell biologists are working with embryonic and adult stem cells.

Organs, Composed of Tissues, Work Together in Organ Systems

25.7 Each organ has a specific structure and function

- An organ performs functions that tissues alone cannot.
- The skin in vertebrates is an organ composed of an epidermis (stratified epithelium) and a dermis (dense fibrous connective tissue).
  - A subcutaneous layer (loose fibrous connective tissue) lies between the skin and underlying structures.

All Organ Systems Contribute to Homeostasis in Animals

25.8 Several organs work together to carry out the functions of an organ system

- Organ system functions include control, sensory input and motor output, transport, maintenance, and reproduction.
  - Control: nervous and endocrine systems
  - Sensory input/motor output: integumentary, skeletal, and muscular systems
  - Transport: cardiovascular and lymphatic systems
  - Maintenance: digestive, respiratory, and urinary systems
  - Reproductive: reproductive system.

25.9 Organs for transplant may come from various sources

- Human organs for transplant are in short supply.
- Xenotransplantation is the use of animal organs instead of human organs for transplant.
- Tissue engineering allows some replacement organs to be made in the lab.

25.10 Homeostasis is the constancy of the internal environment

- The internal environment of the body consists of blood and tissue fluid.
- Tissue fluid stays constant due to exchange of nutrients and wastes with the blood.

- All systems of the body contribute to homeostasis.
- Examples of homeostasis include body temperature, water content, CO_2 concentration, pH, and glucose concentration that stay within normal limits.

25.11 Homeostasis is achieved through negative feedback mechanisms

- A negative feedback mechanism has a sensor and a control center. The system's output dampens the original stimulus, preventing change in the same direction.

Testing Yourself

The Structure of Tissues Suits Their Function

1. A grouping of similar cells that perform a specific function is called a
   - a. sarcoma.
   - b. membrane.
   - c. tissue.
   - d. None of these are correct.
2. The microvilli on the cells lining the small intestine are adaptations to promote
   - a. absorption.
   - b. digestion.
   - c. movement.
   - d. secretion.

Four Types of Tissues Are Common in the Animal Body

3. Which tissue is more apt to line a lumen?
   - a. epithelial tissue
   - b. connective tissue
   - c. nervous tissue
   - d. muscular tissue
4. Tight junctions are most often associated with
   - a. connective tissue.
   - b. adipose tissue.
   - c. cartilage.
   - d. epithelium.
5. Which tissue has cells in lacunae?
   - a. epithelial tissue
   - b. cartilage
   - c. bone
   - d. smooth muscle
   - e. Both b and c are correct.
6. Blood is a(n) _____ tissue because it has a _____.
   - a. connective, gap junction
   - b. muscle, matrix
   - c. epithelial, gap junction
   - d. connective, matrix
7. A reduction in red blood cells would cause problems with
   - a. fighting infection.
   - b. carrying oxygen.
   - c. blood clotting.
   - d. None of these are correct.
8. White blood cells fight infection by
   - a. producing antibodies.
   - b. producing toxins.
   - c. engulfing pathogens.
   - d. More than one of these are correct.
   - e. All of these are correct.
9. **THINKING CONCEPTUALLY** Use the concept of "structure suits function" to discuss the structure of the neuron shown in Figure 25.5.

In questions 10–12, match each type of muscle tissue to as many terms in the key as possible.

KEY:
- a. voluntary
- b. involuntary
- c. striated
- d. nonstriated
- e. spindle-shaped cells
- f. branched cells
- g. long, cylindrical cells

10. Skeletal muscle
11. Smooth muscle
12. Cardiac muscle
13. Which choice is true of both cardiac and skeletal muscle?
    - a. striated
    - b. single nucleus per cell
    - c. multinucleated cells
    - d. involuntary control
14. The main part of a neuron that functions to conduct nerve impulses is the
    - a. astrocyte.
    - b. axon.
    - c. cell body.
    - d. dendrite.
15. Damaged axons in the _____ degenerate, and permanent nervous function loss occurs.
    - a. cranial nerves
    - b. spinal nerves
    - c. CNS
    - d. PNS
16. **THINKING CONCEPTUALLY** Many cancers develop from epithelial tissue. What are two attributes of this tissue type that make cancer more likely to develop?

Organs, Composed of Tissues, Work Together in Organ Systems

17. Which of the following is a function of skin?
    - a. temperature regulation
    - b. protection against water loss
    - c. collection of sensory input
    - d. protection from invading pathogens
    - e. All of these are correct.

18. Without melanocytes, skin would
    a. be too thin.   c. lack color.
    b. lack nerves.   d. None of these are correct.
19. What is the primary feature of human skin that makes it a good protection against most infectious agents, such as bacteria and viruses?

All Organ Systems Contribute to Homeostasis in Animals

20. The major function of tissue fluid is to
    a. provide nutrients and oxygen to cells and remove wastes.
    b. keep cells hydrated.
    c. prevent cells from touching each other.
    d. provide flexibility by allowing cells to slide over each other.
21. Which of these systems plays the biggest role in fluid balance?
    a. cardiovascular   c. digestive
    b. urinary   d. integumentary
22. Which of the following systems does not add or remove substances from the blood?
    a. digestive   d. respiratory
    b. cardiovascular   e. nervous system
    c. urinary
23. The skeletal system functions in
    a. blood cell production.   c. movement.
    b. mineral storage.   d. All of these are correct.
24. Which of these body systems contribute to homeostasis?
    a. digestive and urinary systems
    b. respiratory and nervous systems
    c. nervous and endocrine systems
    d. immune and cardiovascular systems
    e. All of these are correct.
25. Which of the following act as slow effectors in the negative feedback system?
    a. muscles   d. red blood cells
    b. epidermal cells   e. senses
    c. endocrine glands
26. The correct order for a negative feedback mechanism is:
    a. sensory detection, control center, effect brings about a change.
    b. control center, sensory detection, effect brings about a change.
    c. sensory detection, control center, effect causes no change.
    d. None of these are correct.
27. Which of the following is an example of negative feedback?
    a. Air conditioning goes off when room temperature lowers.
    b. Insulin decreases blood sugar levels after eating a meal.
    c. Heart rate increases when blood pressure drops.
    d. All of these are examples of negative feedback.
28. When a human being is cold, the blood vessels
    a. dilate, and the sweat glands are inactive.
    b. dilate, and the sweat glands are active.
    c. constrict, and the sweat glands are inactive.
    d. constrict, and the sweat glands are active.
    e. contract so that shivering occurs.
29. What are the benefits and possible drawbacks of using the pig for xenotransplantation?
30. **THINKING CONCEPTUALLY** Both the cardiovascular system and the nervous system pervade the body. Which one would you expect to have a pump and why?

Understanding the Terms

adipose tissue 499	muscular system 504
basement membrane 497	muscular tissue 500
bone 499	negative feedback 508
cardiac muscle 500	nerve 501
cardiovascular system 505	nervous system 504
cartilage 499	nervous tissue 501
cerebrospinal fluid 501	neuroglia 501
columnar epithelium 497	neuron 501
connective tissue 498	organ 496
cuboidal epithelium 497	organism 496
dense fibrous connective	organ system 496
tissue 499	platelet 499
dermis 503	pseudostratified ciliated
digestive system 505	columnar epithelium 497
endocrine system 504	red blood cell 499
epidermis 503	reproductive system 505
epithelial tissue 496	respiratory system 505
fibroblast 498	skeletal muscle 500
gland 497	skeletal system 504
homeostasis 504	skin 503
hyaline cartilage 499	smooth muscle 500
immune system 505	squamous epithelium 497
integumentary system 504	stratified 497
intercalated disk 500	striated 500
lacunae 499	subcutaneous layer 503
ligament 499	tendon 499
loose fibrous connective	tissue 496
tissue 498	tissue fluid 505
lumen 497	urinary system 505
lymphatic system 505	white blood cell 499
matrix 498	xenotransplantation 506

Match the terms to these definitions:
a. _____ Fibrous connective tissue that joins bone to bone at a joint.
b. _____ Outer region of the skin composed of stratified squamous epithelium.
c. _____ Having striations, as in cardiac and skeletal muscle.
d. _____ Self-regulatory state in which imbalances result in a fluctuation above and below a mean.

Thinking Scientifically

1. You are a histologist (person who specializes in tissues) working in the laboratory of a hospital. You are sent some tissues taken hastily and not identified from a person who died of AIDS. Your task is to identify the tissues. What would you do?
2. You hypothesize that more cases of skin cancer occur among people who frequent tanning salons than otherwise. Describe the study you would do to test the hypothesis.

ARIS *Visit www.mhhe.com/maderconcepts for practice quizzes, animations, videos, and activities designed to help you master the material in this chapter.*

26

Coordination by Neural Signaling

LEARNING OUTCOMES

After studying this chapter, you should be able to accomplish the following outcomes.

Getting a Head

1 Relate the presence of a head to the lifestyle and environment of an animal.

Most Animals Have a Nervous System That Allows Responses to Stimuli

2 Compare the nervous system in complex invertebrates and vertebrates to that of a planarian.
3 Divide the vertebrate brain into three parts, and relate the activities of each part to the complexity of the vertebrate brain.

Neurons Process and Transmit Information

4 Describe the anatomy of three types of neurons.
5 Describe the resting potential and the events of the action potential.
6 Describe the events and benefits of saltatory conduction.
7 Describe transmission of the nerve impulse from one neuron to the next.
8 Divide neurotransmitters into those that are excitatory and those that are inhibitory. Then describe the action of each group.
9 Describe integration as an activity that occurs at the level of the neuron.
10 Describe six drugs of abuse with emphasis on their deleterious effects.

The Vertebrate Central Nervous System (CNS) Consists of the Spinal Cord and Brain

11 Describe the structure and function of the human spinal cord.
12 Discuss the parts of the human brain with an emphasis on structure and function.
13 Describe the actions of the limbic system, including its involvement in memory.

The Vertebrate Peripheral Nervous System (PNS) Consists of Nerves

14 Define the somatic division, and describe the events of a reflex action.
15 Compare the parasympathetic and sympathetic divisions of the autonomic system.

How do you define a head? Whether you define a head as an anterior demarcation of the body or an anterior region that contains a brain and sense organs, it is clear that echinoderms, such as sea urchins, don't have one. Sea urchins depend on all those spines to protect them, rather than running away quickly on their many tube feet. What about the clam compared to the octopus? The inactive clam spends its life digging in the sand and does not have a head. But the octopus, in addition to all those creeping arms, does have a head. You might mistake the visceral hump for part of the head, but it is not. Its head is where you see the eyes, because that is where the brain is. An octopus needs its brain, eyes, and arms to go after its prey. A head is a definite advantage to a predator. Good eye-appendage coordination helps a lot too.

Sea urchin

Clam

Octopus

Getting a Head

Do whales have a head? Marine vertebrates, whether a fish, a whale, or a duckbill platypus, tend to have a head region that is not distinct from the rest of the body. If you were asked to cut off the head of a fish, where would you cut—the choice is yours! The more streamlined their bodies, the better aquatic animals can move in the water. If we compare the crab to a grasshopper, we can see that the land animal has a more definite head. The scavenger crab uses its claws to bring food that might float or swim away to its mouth, but the grasshopper has no claws. The herbivorous grasshopper has a head that can bob up and down to reach its stationary food on land. It uses all its legs for hopping. So, it appears that lifestyle within a particular environment plays a role in whether an animal has a head or not.

What makes animals have heads is an interesting study, but the real point is what type of nervous system does an animal have? Some animals have a few nerve cells here or there, as in a hydra, and they cannot do much of anything. But with a good brain, protected by a skull and connected to associated nerves and sense organs, an animal can lead a more complex life. It can receive stimuli from the environment, interpret those stimuli, and respond in an appropriate manner.

The vertebrate brain is divided into the hindbrain, the midbrain, and the forebrain. Fishes and amphibians rely on their midbrain to carry out complex behaviors, but in mammals especially, this function has been taken over by the forebrain. Humans have the best-developed forebrain of all the animals, and when you map it, you can see that an inordinate amount of space is allotted to the hands! When humans came down out of the trees and stood on two legs, they exposed their bellies, but freed their hands. With good eye-hand coordination and the ability to use tools, humans can flourish in almost any terrestrial environment. They can also decide on the best course of action in a wide variety of circumstances.

In this chapter, a comparison of animal nervous systems shows the manner in which the vertebrate nervous system may have evolved. The structure and function of neurons precedes consideration of the human nervous system.

Rainbow trout

Grasshopper

Dolphin

Crab

Student

An increase in complexity among animal nervous systems is observed as we consider the possible evolution of the vertebrate nervous system.

26.1 Invertebrates reflect an evolutionary trend toward bilateral symmetry and cephalization

In complex animals, the ability to survive is dependent on a nervous system that monitors internal and external conditions and makes appropriate changes to maintain homeostasis. By comparing the nervous system organization of simpler animals, we can discern evolutionary trends that may have led to the nervous system of vertebrates.

Invertebrate Nervous Organization Simple animals, such as sponges, which have the cellular level of organization, can respond to stimuli; the most common observable response is closure of the osculum (central opening). Hydras, which are cnidarians with the tissue level of organization, can contract and extend their bodies, move their tentacles to capture prey, and even turn somersaults. They have a **nerve net** that is composed of neurons in contact with one another and with contractile cells in the body wall (**Fig. 26.1A**, *left*). Sea anemones and jellyfishes, which are also cnidarians, seem to have two nerve nets. A fast-acting one allows major responses, particularly in times of danger, while the other one coordinates slower and more delicate movements.

Planarians, which are flatworms, have a nervous organization that reflects their bilateral symmetry. They possess two ventrally located lateral or longitudinal nerve cords (bundles of nerves) that extend from the cerebral ganglia to the posterior end of the body. Transverse nerves connect the nerve cords, as well as the cerebral ganglia, to the eyespots. The entire arrangement is a **ladderlike nervous system**. **Cephalization** has occurred, as evidenced by a concentration of ganglia and sensory receptors in a head region. A cluster of neurons is called a **ganglion** (pl., ganglia), and the anterior cerebral ganglia receive sensory information from photoreceptors in the eyespots and sensory cells in the auricles (Fig. 26.1A, *center*). The two lateral nerve cords allow rapid transfer of information from the cerebral ganglia to the posterior end, and the transverse nerves between the nerve cords keep the movement of the two sides coordinated. Bilateral symmetry plus cephalization are two significant trends in the development of a nervous organization that is adaptive for an active way of life. Also, the nervous organization in planarians foreshadows the organization of the nervous system in vertebrates.

In annelids (e.g., earthworm), arthropods (e.g., crab), and molluscs (e.g., squid), the nervous system shows further advances. The annelids and arthropods have the typical invertebrate nervous system. There is a brain and a ventral nerve cord that has a ganglion in each segment (Fig. 26.1A, *right*). The brain, which normally receives sensory information, controls

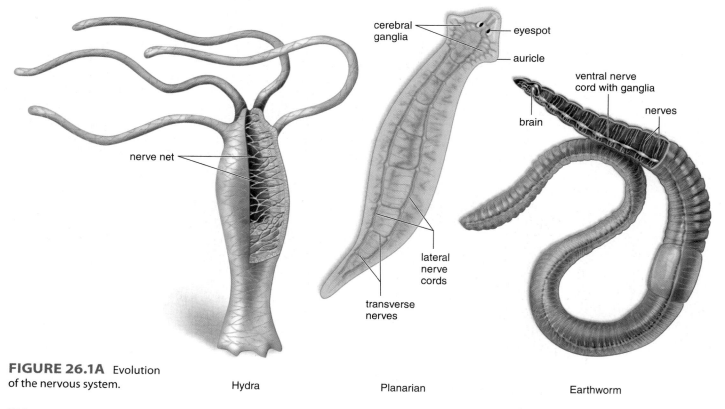

cerebral ganglia
eyespot
auricle
ventral nerve cord with ganglia
nerves
brain
nerve net
lateral nerve cords
transverse nerves

FIGURE 26.1A Evolution of the nervous system.

Hydra   Planarian   Earthworm

the activity of the ganglia and assorted nerves so that the muscle activity of the entire animal is coordinated. The crab and squid have a well-defined brain as well as well-developed sense organs, such as eyes. The presence of a brain and other ganglia in the body of all these animals indicates an increase in the number of neurons (nerve cells) among more complex invertebrates.

Vertebrate Nervous Organization

In vertebrates (e.g., cat), cephalization, coupled with bilateral symmetry, results in several types of paired sensory receptors, including the eyes, ears, and olfactory structures that allow the animal to gather information from the environment. Paired cranial and spinal nerves contain numerous nerve fibers. Vertebrates have many more neurons than do invertebrates. For example, an insect's entire nervous system may contain a total of about 1 million neurons, while a vertebrate's nervous system may contain many thousand to many billion times that number. A vertebrate's **central nervous system (CNS)**, consisting of a spinal cord and brain, develops from an embryonic dorsal neural tube. The spinal cord is continuous with the brain because the embryonic neural tube becomes the spinal cord posteriorly, while the vertebrate brain is derived from the enlarged anterior end of the neural tube. Ascending tracts carry sensory information to the brain, and descending tracts carry motor commands to the neurons in the spinal cord that control the muscles.

It is customary to divide the vertebrate brain into the hindbrain, midbrain, and forebrain (**Fig. 26.1B**). The hindbrain is the most ancient part of the brain. Nearly all vertebrates have a well-developed hindbrain that regulates motor activity below the level of consciousness. In humans, for example, the lungs and heart function even when we are sleeping. The medulla oblongata contains control centers for breathing and heart rate. Coordination of motor activity associated with limb movement, posture, and balance eventually became centered in the cerebellum.

The optic lobes are part of the midbrain, which was originally a center for coordinating reflexes involving the eyes and ears. Starting with the amphibians and

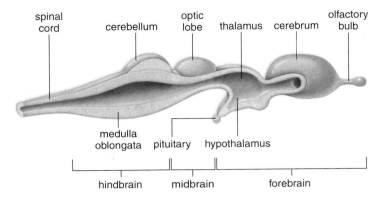

FIGURE 26.1B Organization of the vertebrate brain.

continuing in the other vertebrates, the forebrain processes sensory information. Originally, the forebrain was concerned mainly with the sense of smell. Later, the thalamus evolved to receive sensory input from the midbrain and the hindbrain and to pass it on to the cerebrum, the anterior part of the forebrain in vertebrates. In the forebrain, the hypothalamus is particularly concerned with homeostasis, and in this capacity, the hypothalamus communicates with the medulla oblongata and the pituitary gland.

The cerebrum, which is highly developed in mammals, integrates sensory and motor input and is particularly associated with higher mental capabilities. In humans, the outer layer of the cerebrum, called the cerebral cortex, is especially large and complex.

The next section gives an overview of the human nervous system.

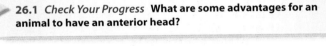

26.1 *Check Your Progress* What are some advantages for an animal to have an anterior head?

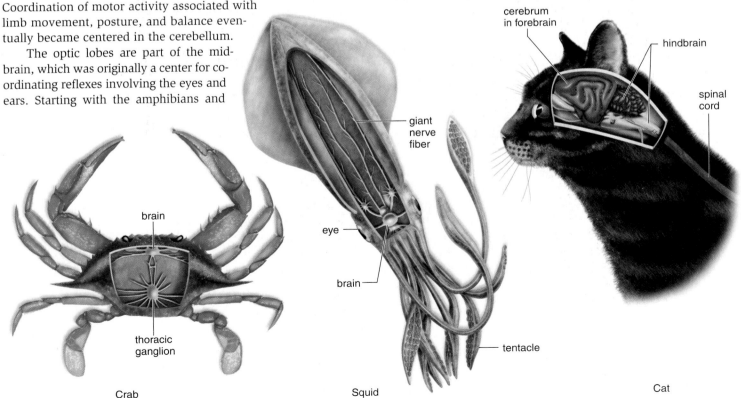

Crab

Squid

Cat

26.2 Humans have well-developed central and peripheral nervous systems

In humans, the central nervous system (CNS) consists of the brain and spinal cord (**Fig. 26.2**). The brain is enclosed in the skull, and the spinal cord is housed in the vertebral column. The **peripheral nervous system (PNS)** consists of all the nerves and ganglia that lie outside the CNS. All signals that enter and leave the CNS travel through paired nerves; those that connect to the spinal cord are called *spinal nerves*, whereas those attached to the brain are *cranial nerves*. Sensory pathways deliver information to the CNS. The somatic sensory axons (also called fibers) send signals from the skin and special sense organs (such as the eyes), and visceral sensory fibers convey information from the internal organs. Somatic motor fibers control skeletal muscles, and autonomic motor fibers control smooth and cardiac muscle, as well as the glands. The autonomic system is further divided into sympathetic and parasympathetic divisions, which will be discussed in Section 26.17.

The structural components of the human nervous system are complex, and the CNS and PNS must work in harmony to carry out three primary functions:

1. *Receive sensory input.* Sensory receptors in the skin and other organs respond to external and internal stimuli by generating nerve impulses that travel to the CNS.

2. *Perform integration.* The CNS sums up the input it receives from all over the body.

3. *Generate motor output.* Nerve impulses from the CNS go to the muscles and glands.

Muscle contractions and gland secretions are responses to stimuli received by sensory receptors. As an example, consider what happens in your body as you prepare to catch a ball thrown by a friend. Continual sensory input from the eyes and skeletal muscles informs the CNS of the position of the ball and the position of your hands and arms. The CNS sums up the incoming data and generates impulses to the muscles in your hands and arms so that they are properly positioned to catch the ball. Likewise, the CNS receives information about blood pressure from visceral sensory receptors and sends motor commands via the autonomic system to increase or decrease pressure as necessary.

The next part of the chapter describes the structure and function of nerve cells (neurons) in a nervous system.

> **26.2 Check Your Progress** In vertebrates, the central nervous system (CNS) consists of not only the brain in the head but also the spinal cord, which runs along the back. The spinal cord equates to what part of the planarian nervous system?

FIGURE 26.2 Organization of the nervous system in humans.

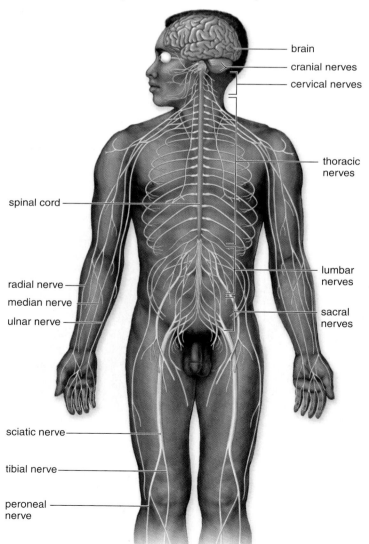

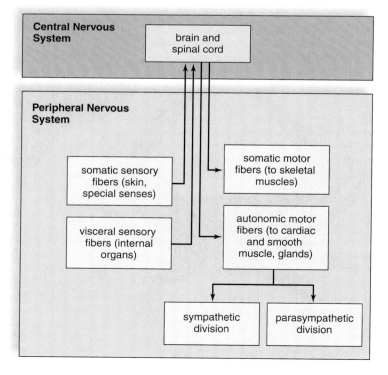

The structure of neurons precedes an examination of nerve impulse conduction along an axon and across a synapse. Drugs of abuse affect the action of neurotransmitter substances at a synapse.

26.3   Neurons are the functional units of a nervous system

Although complex, nervous tissue is composed of two principal types of cells. **Neurons**, also known as nerve cells, are the functional units of the nervous system. They receive sensory information, convey the information to an integration center such as the brain, and conduct signals from the integration center to effector structures such as the glands and muscles. **Neuroglia** are cells that provide support and nourishment to the neurons.

Neurons vary in appearance, depending on their function and location. They consist of three major parts: a cell body, dendrites, and an axon (**Fig. 26.3A**). The **cell body** contains a nucleus and a variety of organelles. The **dendrites** are short, highly branched processes that receive signals from the sensory receptors or other neurons and transmit them to the cell body. The **axon** is the portion of the neuron that conveys information to another neuron or to other cells. Axons can be bundled together to form nerves. For this reason, axons are often called **nerve fibers**. Many axons are covered by a white insulating layer called the **myelin sheath**.

Neuroglia, or glial cells, greatly outnumber neurons in the brain. There are several different types in the CNS, each with specific functions. Some (microglia) help remove bacteria and debris; others (astrocytes) provide metabolic and structural support directly to the neurons. The myelin sheath is formed from the membranes of tightly spiraled neuroglia. In the PNS, **Schwann cells** perform this function, leaving gaps called **nodes of Ranvier**, or neurofibril nodes. In the CNS, another type of neuroglia, called an oligodendrocyte, performs this function.

Types of Neurons Neurons can be classified according to their function and shape. **Motor (efferent) neurons** carry nerve impulses from the CNS to muscles or glands. The neuron shown in Figure 26.3A is a motor neuron. Like all motor neurons, it has many dendrites and a single axon. Motor neurons cause muscle fibers to contract or glands to secrete, and therefore they are said to innervate these structures.

Sensory (afferent) neurons take nerve impulses from sensory receptors to the CNS. The sensory receptor may be the end of a sensory neuron itself (a pain or touch receptor), or it may be a specialized cell that forms a synapse with a sensory neuron (e.g., the hair cells of the inner ear). In sensory neurons, the process that extends from the cell body divides into a branch that extends to the periphery and another that extends to the CNS (**Fig. 26.3B**). Since both of these extensions are long and myelinated and transmit nerve impulses, it is now generally accepted to refer to them as an axon.

Interneurons, also known as association neurons, occur entirely within the CNS. Interneurons parallel the structure of motor neurons (**Fig. 26.3C**) and convey nerve impulses between various parts of the CNS. Some lie between sensory neurons and motor neurons, and some take messages from one side of the spinal cord to the other, or from the brain to the spinal cord and vice

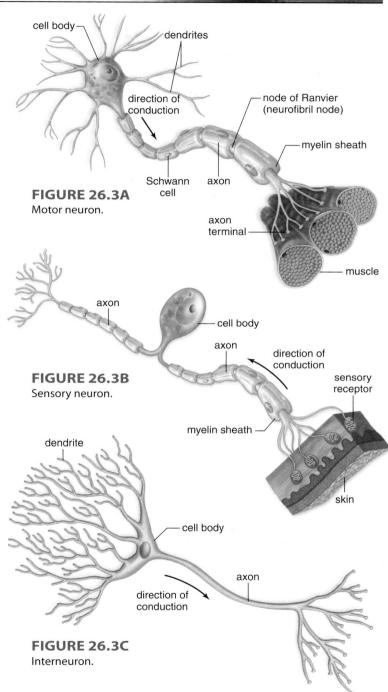

FIGURE 26.3A
Motor neuron.

FIGURE 26.3B
Sensory neuron.

FIGURE 26.3C
Interneuron.

versa. They also form complex pathways in the brain where the processes accounting for thinking, memory, and language occur.

> **26.3 *Check Your Progress*   A brain in the head consists of what type of cells?**

26.4 Neurons have a resting potential across their membranes when they are not active

Scientists have studied the activity of neurons by using excised axons and a voltmeter. Voltage, designated in millivolts (mV), is a measure of the electrical potential difference between two points. In the case of a neuron, the two points are the inside and the outside of the axon. When an electrical potential difference exists between the inside and outside of a cell (i.e., across the membrane), it is called a *membrane potential*, and we can say that there is *polarity* in the distribution of electrical charges. One side has more positive charges than the other side. When a neuron is not conducting an impulse, its **resting potential** is about −65 mV; the negative sign indicates that the inside of the cell is more negative than the outside (**Fig. 26.4**).

The existence of this membrane potential can be correlated with a difference in ion distribution on either side of the axon membrane. The unequal distribution of these ions is, in part, due to the activity of the sodium-potassium pump, which moves three sodium ions (Na^+) out of the neuron for every two potassium ions (K^+) it moves into the neuron. The membrane is more permeable to K^+ than to Na^+, and therefore more K^+ leaks out of the cell than Na^+ leaks in. There are also large, negatively charged proteins in the cytoplasm of the axon. All cells maintain a membrane potential, but neurons are unusual in that they alter their membrane potential when they transmit nerve impulses, as discussed in the next section.

26.4 Check Your Progress Cells use energy to pump ions. What does pumping accomplish?

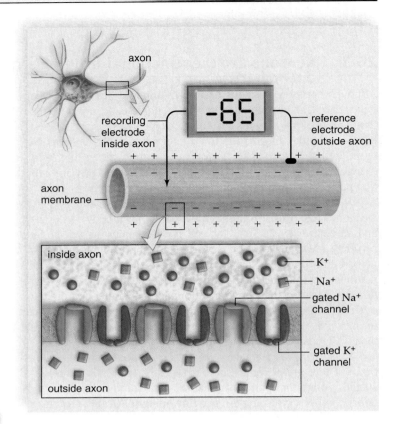

FIGURE 26.4 Resting potential: More Na^+ outside the axon and more K^+ inside the axon. Inside is −65 mV, relative to the outside.

26.5 Neurons have an action potential across axon membranes when they are active

An **action potential** is a rapid change in polarity across an axon membrane as the nerve impulse occurs. In order to visualize the rapid fluctuations in voltage during the action potential, researchers generally find it useful to graph the voltage changes over time (**Fig. 26.5A**). An action potential uses two types of gated ion channels in the axon membrane. During the first part of an action potential, a gated ion channel allows sodium (Na^+) to pass into the axon, and then another gated ion channel allows potassium (K^+) to pass out of the axon. In contrast to ungated ion channels, which constantly allow ions to cross the membrane, gated ion channels open and close in response to a stimulus.

If a stimulus causes the axon membrane to depolarize to a certain level, called **threshold**, an action potential occurs in an *all-or-none manner*. The strength of an action potential does not change; an intense stimulus can cause an axon to fire (start an axon potential) more often in a given time interval.

The Sodium Gates Open When an action potential begins, the gates of the sodium channels open, and Na^+ flows into the axon. As Na^+ moves to the inside of the axon, the membrane po-

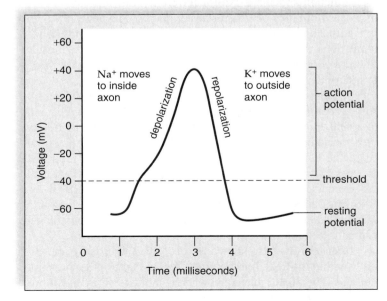

FIGURE 26.5A An action potential can be visualized as voltage changes over time.

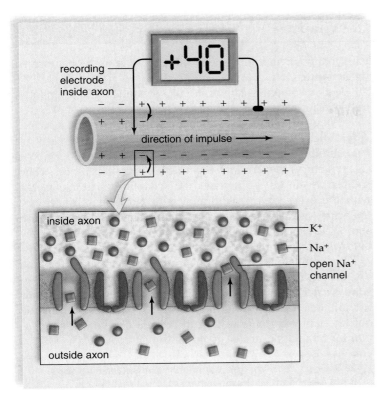

FIGURE 26.5B Action potential begins: Depolarization to +40 mV as Na⁺ gates open and Na⁺ moves to inside the axon.

tential changes from –65 mV to +40 mV. This is **depolarization** because the charge inside the axon changes from negative to positive (**Fig. 26.5B**). This reversal in polarity causes the sodium channels to close and the potassium channels to open.

The Potassium Gates Open Second, the gates of potassium channels open, and K⁺ flows out of the axon; the action

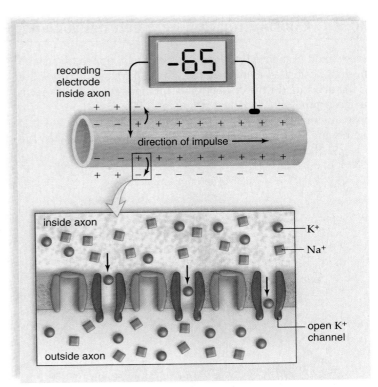

FIGURE 26.5C Action potential ends: Repolarization to –65 mV as K⁺ gates open and K⁺ moves to outside the axon.

potential changes from +40 mV back to –65 mV. This is **repolarization** because the inside of the axon becomes negative again as K⁺ exits the axon (**Fig. 26.5C**).

> **26.5** *Check Your Progress* A nerve impulse has two parts. *a.* During the first part, which ion moves where? *b.* During the second part, which ion moves where?

26.6 Propagation of an action potential is speedy

In nonmyelinated axons, the action potential travels down an axon one small section at a time, at a speed of about 1 m/second. As soon as an action potential has moved on, the previous section undergoes a **refractory period**, during which the Na⁺ gates are unable to open. Notice, therefore, that the action potential cannot move backward and instead always moves down an axon toward its terminals. When the refractory period is over, the sodium-potassium pump has restored the previous ion distribution by pumping Na⁺ to outside the axon and K⁺ to inside the axon.

In myelinated axons, the gated ion channels that produce an action potential are concentrated at the nodes of Ranvier. Just as taking giant steps during a game of "Simon Says" is more efficient, so ion exchange only at the nodes makes the action potential travel faster in nonmyelinated axons. *Saltar* in Spanish means "to jump," and so this mode of conduction is called **saltatory conduction**, meaning that the action potential "jumps" from node to node (**Fig. 26.6**). Speeds of 200 m/second (450 miles per hour) have been recorded.

The symptoms of multiple sclerosis, characterized by impaired motor skills, are due to demyelination. Passage of the

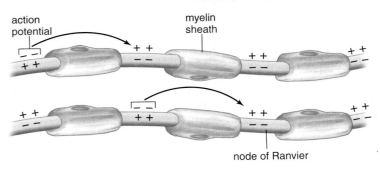

FIGURE 26.6 Saltatory conduction.

nerve impulse across a synapse is studied in Sections 26.7 through 26.9.

> **26.6** *Check Your Progress* The brain of a vertebrate communicates with the muscles much faster than the brain of an invertebrate. What structural difference probably exists between the two groups of animals?

26.7 Communication between neurons occurs at synapses

Every axon branches into many fine endings, tipped by a small swelling, called an axon terminal. Each terminal lies very close to the dendrite (or the cell body) of another neuron. This region of close proximity is called a **synapse** (**Fig. 26.7**). At a synapse, the membrane of the first neuron is called the *pre*synaptic membrane, and the membrane of the next neuron is called the *post*synaptic membrane. The small gap between the neurons is the **synaptic cleft**.

A nerve impulse cannot cross a synaptic cleft. Therefore, transmission across a synapse is carried out by molecules called **neurotransmitters**, which are stored in synaptic vesicles. ❶ When nerve impulses traveling along an axon reach an axon terminal, gated channels for calcium ions (Ca^{2+}) open, and calcium enters the terminal. Figure 26.7 traces this process. The sudden rise in Ca^{2+} stimulates the synaptic vesicles to merge with the presynaptic membrane. ❷ Now neurotransmitter molecules are released into the synaptic cleft and they diffuse across the cleft to the postsynaptic membrane. ❸ There they bind with specific receptor proteins.

Depending on the type of neurotransmitter and/or the type of receptor, the response of the postsynaptic neuron can be toward excitation or toward inhibition. Once a neurotransmitter has initiated a response in the postsynaptic neuron, it must be quickly removed from the cleft to prevent continuous stimulation (or inhibition), as discussed in Section 26.8.

1 After an action potential arrives at an axon terminal, Ca^{2+} enters, and synaptic vesicles fuse with the presynaptic membrane.

path of action potential

Ca^{2+}

axon terminal

synaptic vesicles enclose neuro-transmitter

synaptic cleft

2 Neuro-transmitter molecules are released and bind to receptors on the postsynaptic membrane.

neurotransmitter

presynaptic membrane

postsynaptic membrane

neuro-transmitter

receptor

Na^+

postsynaptic neuron

3 When an excitatory neuro-transmitter binds to a receptor, Na^+ diffuses into the postsynaptic neuron and, assuming threshold is reached, an action potential begins.

> **26.7 Check Your Progress** Communication between neurons is not dependent on the movement of ions. What does it depend on?

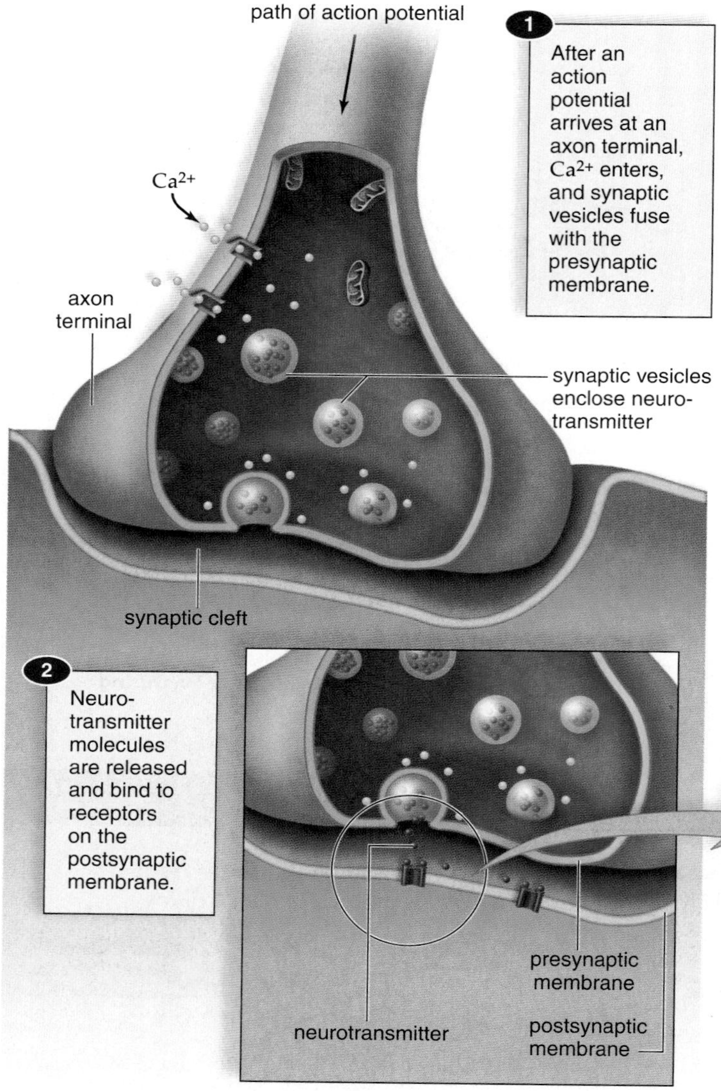

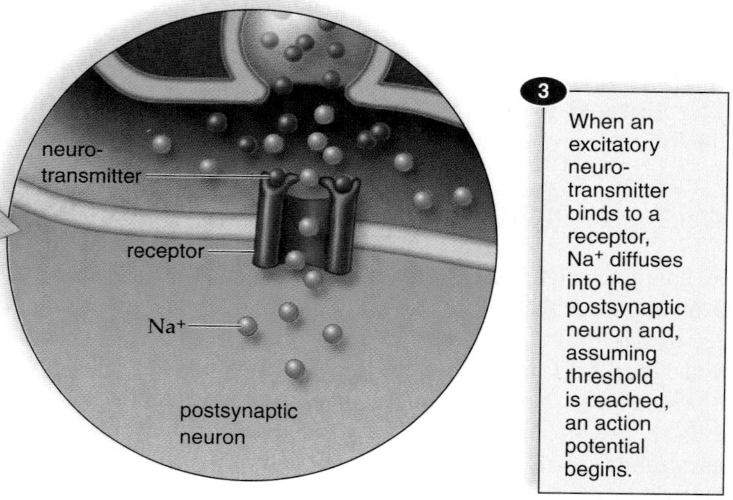

FIGURE 26.7 Synapse structure and function.

26.8 Neurotransmitters can be stimulatory or inhibitory

Among the more than 100 substances known, or suspected, to be neurotransmitters are acetylcholine, norepinephrine, dopamine, serotonin, and GABA (gamma aminobutyric acid). Various drugs, some of which alter mood, as discussed in Section 26.10, enhance or block the release of a neurotransmitter, mimic the action of a neurotransmitter or block the receptor, or interfere with the removal of a neurotransmitter from the synaptic cleft.

Acetylcholine (ACh) and **norepinephrine (NE)** are frequently mentioned and well-known neurotransmitters in both the CNS and the PNS. Alzheimer disease is associated with a deficiency of ACh in the CNS. In the PNS, ACh excites skeletal muscle but inhibits cardiac muscle. ACh has either an excitatory or inhibitory effect on smooth muscle or glands, depending on the particular organ. Botulism is a rare kind of food poisoning caused by the bacterium *Clostridium botulinum*, which produces botulin toxin. The toxin blocks the release of ACh at skeletal muscle synapses. Six hours to 8 days after eating contaminated food, usually canned improperly, the person may feel the effects of the toxin and may die if the respiratory muscles are affected. Botulin toxin is used as the drug Botox to paralyze facial skeletal muscles and thus reduce the appearance of wrinkles.

In the CNS, NE is important to dreaming, waking, and mood. Serotonin, another neurotransmitter, is involved in thermoregulation, sleeping, emotions, and perception. Reduced levels of NE and serotonin seem to be linked to depression, and the antidepressant drug Prozac blocks the removal of serotonin from a synapse. In the PNS, NE generally excites smooth muscle.

Dopamine and GABA are found primarily in the CNS. Dopamine, in particular, is involved in emotions, control of motor function, and attention. Parkinson disease is associated with a lack of dopamine in the brain. Many of the drugs that affect mood act by interfering with or increasing the effect of dopamine. GABA is an abundant inhibitory neurotransmitter in the CNS. The drug Valium binds to the receptors of GABA, thereby increasing its effects.

Neuromodulators are molecules that block the release of a neurotransmitter or modify a neuron's response to a neurotransmitter. The caffeine in coffee, chocolate, and tea keeps us awake by interfering with the effects of inhibitory neurotransmitters in the brain. Two well-known neuromodulators are substance P and endorphins. Substance P is released by sensory neurons when pain is present. Endorphins block the release of substance P and,

therefore, serve as natural painkillers. They are thought to be associated with the "runner's high" experienced by joggers and to be produced by the brain, not only in the presence of physical stress, but also emotional stress. The opiates—namely, codeine, heroin, and morphine—function similar to endorphins, and like them, they reduce pain and produce a feeling of well-being.

Clearing of Neurotransmitter from a Synapse Once a neurotransmitter has been released into a synaptic cleft and has initiated a response, it is removed from the cleft. The short existence of neurotransmitters at a synapse prevents continuous stimulation (or inhibition) of postsynaptic membranes.

In some synapses, the postsynaptic membrane contains enzymes that rapidly inactivate the neurotransmitter. For example, the enzyme **acetylcholinesterase (AChE)** breaks down acetylcholine. The enzyme GABA transaminase converts GABA into an inactive compound. In other synapses, the presynaptic membrane rapidly reabsorbs the neurotransmitter, possibly for repackaging in synaptic vesicles or for molecular breakdown.

> **26.8 Check Your Progress** People with Alzheimer disease produce less ACh than usual and may be treated with drugs that inhibit AChE activity. How would this help?

26.9 Integration is a summing up of stimulatory and inhibitory signals

Even though each individual synapse is excitatory or inhibitory, it is important to realize that a single neuron can have many synapses all over its dendrites and the cell body, as seen in the micrograph in **Figure 26.9**. Some neurons have as many as 10,000 synapses. Therefore, a neuron is on the receiving end of many excitatory and inhibitory signals. An excitatory neurotransmitter produces a signal that drives the neuron closer to threshold, and an inhibitory neurotransmitter produces a signal that drives the neuron further from threshold (see Fig. 26.5A). Excitatory signals have a depolarizing effect, and inhibitory signals have a hyperpolarizing effect.

Neurons integrate these incoming signals. **Integration** is the summing up of excitatory and inhibitory signals. If a neuron receives many excitatory signals (either from different synapses or from one synapse at a rapid rate), chances are the axon will transmit a nerve impulse. On the other hand, if a neuron receives both inhibitory and excitatory signals, the summing up of these signals may prohibit the axon from reaching threshold and firing.

In Figure 26.9, **1** the inhibitory signals received outweighed **2** the excitatory signals received by the neuron, and **3** threshold was never reached following integration. Threshold, as mentioned, must be reached, or else a nerve impulse does not start.

Drugs of abuse interfere with the passage of nervous impulses across a synapse, as discussed in Section 26.10.

> **26.9 Check Your Progress** The brain (most likely located in a head) is expected to carry on extensive integration. Explain.

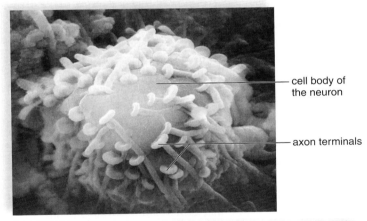

cell body of the neuron

axon terminals

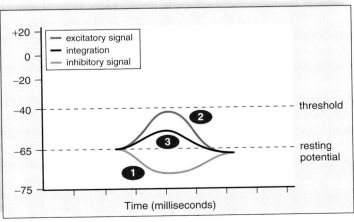

FIGURE 26.9 Synaptic integration.

26.10 Drugs that interfere with neurotransmitter release or uptake may be abused

Drug abuse is apparent when a person takes a drug at a dose level and under circumstances that increase the potential for a harmful effect. Addiction is present when more of the drug is needed to get the same effect, and withdrawal symptoms occur when the user stops taking the drug. This is true not only for teenagers and adults, but also for newborn babies of mothers who abuse and are addicted to drugs. Alcohol, drugs, and tobacco can all adversely affect the developing embryo, fetus, or newborn.

Alcohol

Alcohol consumption is the most socially accepted form of drug use worldwide. The approximate number of adults who consume alcohol in the United States on a regular basis is 65%. Of those, 5% say they are "heavy drinkers." Notably, 80% of college-age young adults drink. Unfortunately, so-called binge drinking has resulted in the deaths of many college students.

Alcohol (ethanol) acts as a *depressant* on many parts of the brain where it affects neurotransmitter release or uptake. For example, alcohol increases the action of GABA, which inhibits motor neurons, and it also increases the release of endorphins, which, as discussed, are natural painkillers. Depending on the amount consumed, the effects of alcohol on the brain can lead to a feeling of relaxation, lowered inhibitions, impaired concentration and coordination, slurred speech, and vomiting. If the blood level of alcohol becomes too high, coma or death can occur.

Chronic alcohol consumption can damage the frontal lobes, decrease overall brain size, and increase the size of the ventricles. Brain damage is manifested by permanent memory loss, amnesia, confusion, apathy, disorientation, or lack of motor coordination. Prolonged alcohol use can also permanently damage the liver, the major detoxification organ of the body, to the point that a liver transplant may be required.

Nicotine

When tobacco is smoked, nicotine is rapidly delivered to the CNS, especially the midbrain. There it binds to neurons, causing the release of dopamine, the neurotransmitter that promotes a sense of pleasure and is involved in motor control. In the PNS, nicotine also acts as a *stimulant* by mimicking acetylcholine and increasing heart rate, blood pressure, and muscle activity. Fingers and toes become cold because blood vessels have constricted. Increased digestive tract motility may account for the weight loss sometimes seen in smokers.

The physiologically and psychologically addictive nature of nicotine is well known. The addiction rate of smokers is about 70%. The failure rate in those who try to quit smoking is about 80–90% of smokers. Withdrawal symptoms include irritability, headache, insomnia, poor cognitive performance, the urge to smoke, and weight gain. Ways to quit smoking include applying nicotine skin patches, chewing nicotine gum, or taking oral drugs that block the actions of acetylcholine. The effectiveness of these therapies is variable. An experimental therapy involves "immunizing" the brain of smokers against nicotine. Injections cause the production of antibodies that bind to nicotine and prevent it from entering the brain. The effectiveness of this new therapy is not yet known.

Club and Date Rape Drugs

Methamphetamine and Ecstasy are considered club or party drugs. Methamphetamine (commonly called meth or speed) is a synthetic drug made by the addition of a methyl group to amphetamine. Because the addition of the methyl group is fairly simple, methamphetamine is often produced from amphetamine in clandestine, makeshift laboratories in homes, motel rooms, or campers. The number of toxic chemicals used to prepare the drug makes a former meth lab site hazardous to humans and to the environment. Over nine million people in the United States have used methamphetamine at least once in their lifetime. It is available as a powder or as crystals (crystal meth or ice).

The structure of methamphetamine is similar to that of dopamine, and its *stimulatory* effect mimics that of cocaine. It reverses the effects of fatigue, maintains wakefulness, and temporarily elevates the user's mood. The initial rush is typically followed by a state of high agitation that, in some individuals, leads to violent behavior. Chronic use can result in what is called an amphetamine psychosis, characterized by paranoia, auditory and visual hallucinations, self-absorption, irritability, and aggressive, erratic behavior. Excessive intake can lead to hyperthermia, convulsions, and death.

Ecstasy is the street name for MDMA (methylenedioxymethamphetamine), a drug with effects similar to those of methamphetamine. Also referred to as E, X, or the lung drug, it is taken as a pill that looks like an aspirin or candy. Many people using Ecstasy believe that it is totally safe if used with lots of water to counter its effect on body temperature. A British teen, Lorna Spinks, died after taking two high-strength Ecstasy pills, which caused her body temperature to rise to a fatal level. Also, Ecstasy has an overstimulatory effect on neurons that produce serotonin, which, like dopamine, elevates our mood. Most of the damage to these neurons can be repaired when the use of Ecstasy is discontinued, but some damage appears to be permanent.

Drugs with sedative effects, known as date rape or predatory drugs, include Rohypnol (roofies), Gamma-hydroxybutyric acid (GHB), and Ketamine (special K). These drugs can be given to an unsuspecting person, who then becomes vulnerable to sexual assault after the drug takes effect. Relaxation, amnesia, and disorientation occur after taking these drugs, which are popular at clubs because they enhance the effect of heroin and Ecstasy.

Cocaine

Cocaine is an alkaloid derived from the shrub *Erythroxylon coca*. Approximately 35 million Americans have used cocaine by sniffing/snorting, injecting, or smoking. Cocaine is a powerful *stimulant* in the CNS that interferes with the re-uptake of dopamine at synapses. The result is a rush of well-being that lasts from 5 to 30 minutes. People on cocaine sprees (or binges) take the drug repeatedly and at ever-higher doses. The result is sleeplessness, lack of appetite, increased sex drive, tremors, and "cocaine psychosis," a condition that resembles paranoid schizophrenia. During the crash period, fatigue, depression, and irritability are common, along with memory loss and confused thinking.

"Crack" is the street name given to cocaine that is processed to a free base for smoking. The term *crack* refers to the crackling sound heard when smoking. Smoking allows extremely high doses of the drug to reach the brain rapidly, providing an intense and immediate high, or "rush." Approximately eight million Americans use crack. Long-term use is expected to cause brain damage (**Fig. 26.10**).

Cocaine is highly addictive; related deaths are usually due to cardiac and/or respiratory arrest. The combination of cocaine and alcohol dramatically increases the risk of sudden death.

Heroin Heroin is derived from the resin or sap of the opium poppy plant, which is widely grown—from Turkey to Southeast Asia and in parts of Latin America. Heroin is a highly addictive drug that acts as a *depressant* in the nervous system. Drugs derived from opium are called opiates, a class that also includes morphine and codeine, both of which have painkilling effects.

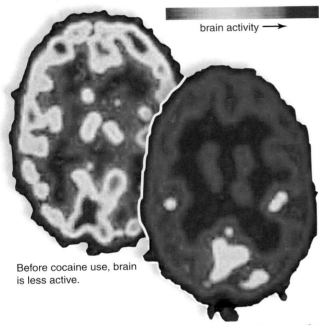

brain activity →

Before cocaine use, brain is less active.

After cocaine use, brain is more active.

Drug abuse

FIGURE 26.10 Drug use.

Heroin is the most abused opiate—it travels rapidly to the brain, where it is converted to morphine, and the result is a rush sensation and a feeling of euphoria. Opiates depress breathing, block pain pathways, cloud mental function, and sometimes cause nausea and vomiting. Long-term effects of heroin use are addiction, hepatitis, HIV/AIDS, and various bacterial infections due to the use of shared needles (Fig. 26.10). As with other drugs of abuse, addiction is common, and heavy users may experience convulsions and death by respiratory arrest.

Heroin can be injected, snorted, or smoked. Abusers typically inject heroin up to four times a day. It is estimated that four million Americans have used heroin some time in their lives, and over 300,000 people use heroin annually.

Marijuana The dried flowering tops, leaves, and stems of the Indian hemp plant, *Cannabis sativa,* contain and are covered by a resin that is rich in THC (tetrahydrocannabinol). The names *cannabis* and *marijuana* apply to either the plant or THC. Marijuana can be consumed, but usually it is smoked in a cigarette called a "joint." An estimated 22 million Americans use marijuana. Although the drug was banned in the United States in 1937, several states have legalized its use for medical purposes, such as lessening the effects of chemotherapy.

It seems that THC may mimic the actions of anandamide, a neurotransmitter that was recently discovered. Both THC and anandamide belong to a class of chemicals called cannabinoids. Receptors that bind cannabinoids are located in the hippocampus, cerebellum, basal ganglia, and cerebral cortex, brain areas that are important for memory, orientation, balance, motor coordination, and perception.

When THC reaches the CNS, the person experiences mild euphoria, along with alterations in vision and judgment. Distortions of space and time can also occur in occasional users. In heavy users, hallucinations, anxiety, depression, rapid flow of ideas, body image distortions, paranoia, and psychotic symptoms can result. The terms cannabis psychosis and cannabis delirium describe such reactions to marijuana's influence on the brain. Regular usage of marijuana can cause cravings that make it difficult to stop.

Treatment for Addictive Drugs Presently, treatment for addiction to drugs consists mainly of behavior modification. Heroin addiction can be treated with synthetic opiate compounds, such as methadone or suboxone, that decrease withdrawal symptoms and block heroin's effects. Unfortunately, inappropriate methadone use can be dangerous, as demonstrated by celebrity deaths associated with methadone overdose or taking methadone along with other drugs.

New treatment techniques include the administration of antibodies to block the effects of cocaine and methamphetamine. These antibodies would make relapses by former drug abusers impossible and could be used to treat overdoses. A vaccine for cocaine that would stimulate antibody production is being tested.

The next part of the chapter is about the central nervous system.

26.10 *Check Your Progress* How might a drug enhance a neurotransmitter, and how might another drug interfere with its action?

This part of the chapter studies the human central nervous system (CNS), consisting of the spinal cord and brain. The limbic system is also studied.

26.11   The human spinal cord and brain function together

The CNS consists of the spinal cord and the brain, where sensory information is received and motor control is initiated. The spinal cord and the brain are both protected by bone; the spinal cord is surrounded by vertebrae (see Fig. 26.15C), and the brain is enclosed by the skull. Both the spinal cord and the brain are wrapped in three protective membranes known as **meninges**. Meningitis (inflammation of the meninges) is a serious disorder caused by a number of bacteria or viruses that invade the meninges. The spaces between the meninges are filled with **cerebrospinal fluid**, which cushions and protects the CNS. Cerebrospinal fluid is contained in the central canal of the spinal cord and within the **ventricles** of the brain, which are interconnecting spaces that produce and serve as reservoirs for cerebrospinal fluid.

Spinal Cord   The **spinal cord** is a bundle of nervous tissue enclosed in the vertebral column; it extends from the base of the brain to the vertebrae just below the rib cage. The spinal cord has two main functions: (1) It is the center for many **reflex actions**, which are automatic responses to external stimuli, and (2) it provides a means of communication between the brain and the spinal nerves, which leave the spinal cord.

A cross section of the spinal cord reveals that it is composed of a central portion of **gray matter** and a peripheral region of white matter. The gray matter consists of cell bodies and unmyelinated fibers. It is shaped like a butterfly, or the letter H, with two dorsal (posterior) horns and two ventral (anterior) horns surrounding a central canal. The gray matter contains portions of sensory neurons and motor neurons, as well as short interneurons that connect sensory and motor neurons (see Fig. 26.16).

Myelinated long fibers of interneurons that run together in bundles called **tracts** give **white matter** its color. These tracts connect the spinal cord to the brain. These tracts are like a busy superhighway, by which information continuously passes between the brain and the rest of the body. Dorsally, the tracts are primarily ascending, taking information *to* the brain; ventrally, the tracts are primarily descending, carrying information *from* the brain. Because the tracts at one point cross over, the left side of the brain controls the right side of the body, and the right side of the brain controls the left side of the body.

If the spinal cord is severed as the result of an injury, paralysis results. If the injury occurs in the cervical (neck) region, all four limbs are usually paralyzed, a condition known as quadriplegia. If the injury occurs in the thoracic region, the lower body may be paralyzed, a condition called paraplegia.

Brain Ventricles   The brain contains four interconnected chambers called ventricles (**Fig. 26.11**). The two lateral ventricles are inside the cerebrum. The third ventricle is surrounded by the diencephalon, and the fourth ventricle lies between the cerebellum and the pons. Cerebrospinal fluid is continuously produced in the ventricles and circulates through them; it then flows out of the brain between the meninges.

Sections 26.12 through 26.14 describe various regions of the brain.

> **26.11 Check Your Progress** **The brain is very dependent on the spinal cord. Explain.**

FIGURE 26.11 The human brain.

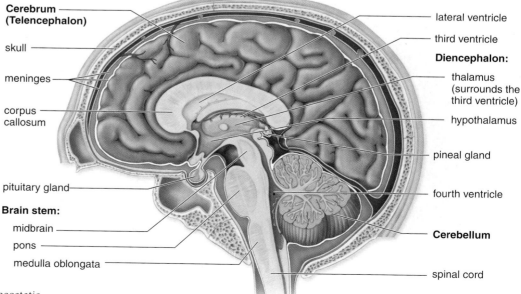

Labels:
Cerebrum (Telencephalon)
skull
meninges
corpus callosum
pituitary gland
Brain stem:
  midbrain
  pons
  medulla oblongata
lateral ventricle
third ventricle
Diencephalon:
  thalamus (surrounds the third ventricle)
  hypothalamus
pineal gland
fourth ventricle
Cerebellum
spinal cord

The **cerebrum** is the largest portion of the brain in humans. The cerebrum is the last center to receive sensory input and carry out integration before commanding voluntary motor responses. It communicates with and coordinates the activities of the other parts of the brain.

Cerebral Hemispheres The cerebrum is divided into two halves, called **cerebral hemispheres** (see Fig. 26.11). A deep groove called the longitudinal fissure divides the cerebrum into the right and left hemispheres. Each hemisphere receives information from and controls the opposite side of the body. Although the hemispheres appear the same, the right hemisphere is associated with artistic and musical ability, emotion, spatial relationships, and pattern recognition. The left hemisphere is more adept at mathematics, language, and analytical reasoning. The two cerebral hemispheres are connected by a bridge of tracts within the corpus callosum.

Shallow grooves called sulci (sing., sulcus) divide each hemisphere into lobes (**Fig. 26.12**). A *frontal lobe* is the anterior portion of a hemisphere and is associated with motor control, memory, reasoning, and judgment. For example, if a fire occurs, the frontal lobe enables you to decide whether to exit via the stairs or the window, or how to dress if the temperature plummets to subzero. The frontal lobe on the left side contains the *Broca area*, which organizes motor commands to produce speech.

The *parietal lobes* lie posterior to the frontal lobe and are concerned with sensory reception and integration, as well as taste. A *primary taste area* in the parietal lobe accounts for taste sensations.

The *temporal lobe* is located laterally. A primary auditory area in the temporal lobe receives information from our ears. The *occipital lobe* is the most posterior lobe. A *primary visual area* in the occipital lobe receives information from our eyes.

The Cerebral Cortex The **cerebral cortex** is a thin (less than 5 mm thick), but highly convoluted, outer layer of gray matter that covers the cerebral hemispheres. The convolutions increase the surface area of the cerebral cortex. The cerebral cortex contains tens of billions of neurons and is the region of the brain that accounts for sensation, voluntary movement, and all the thought processes required for learning and memory and for language and speech.

Two regions of the cerebral cortex are of particular interest. The **primary motor area** is in the frontal lobe just ventral to (before) the central sulcus. Voluntary commands to skeletal muscles begin in the primary motor area, and each part of the body is controlled by a certain section. The size of the section indicates the precision of motor control. For example, the face and hand take up a much larger portion of the primary motor area than does the entire trunk. The **primary somatosensory area** is just dorsal to the central sulcus in the parietal lobe. Sensory information from the skin and skeletal muscles arrives here, where each part of the body is sequentially represented in a manner similar to the primary motor area.

Basal Nuclei While the bulk of the cerebrum beneath the cerebral cortex is composed of white matter, masses of gray matter are located deep within the white matter. These so-called **basal nuclei** (formerly termed basal ganglia) integrate motor commands, ensuring that proper muscle groups are activated or inhibited. **Huntington disease** and **Parkinson disease**, which are both characterized by uncontrollable movements, are believed to be due to malfunctioning basal nuclei.

Having discussed the cerebrum, we move on to other parts of the brain in Section 26.13.

> **26.12 Check Your Progress** Would you expect the brain of a chimpanzee, humans' nearest relative, to have the same lobes as shown in Figure 26.12? How might the chimpanzees' brain differ?

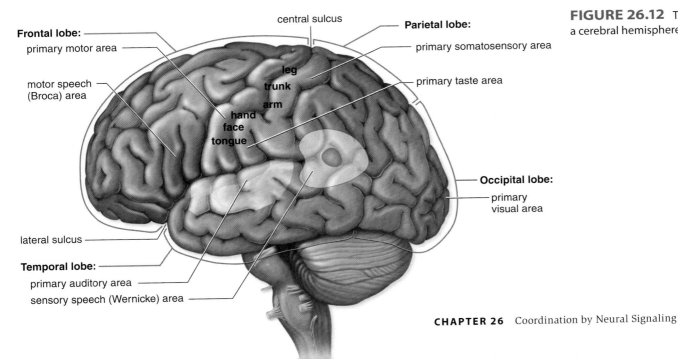

central sulcus

Parietal lobe:

Frontal lobe:

primary motor area

primary somatosensory area

FIGURE 26.12 The lobes of a cerebral hemisphere.

leg
trunk
arm
hand
face
tongue

motor speech (Broca) area

primary taste area

Occipital lobe:
primary visual area

lateral sulcus

Temporal lobe:
primary auditory area
sensory speech (Wernicke) area

The hypothalamus and the thalamus are in the **diencephalon**, a region that encircles the third ventricle. The **hypothalamus** forms the floor of the third ventricle. It is an integrating center that helps maintain homeostasis by regulating hunger, sleep, thirst, body temperature, and water balance. The hypothalamus controls the pituitary gland and, thereby, serves as a link between the nervous and endocrine systems.

The **thalamus** consists of two masses of gray matter located in the sides and roof of the third ventricle. It is on the receiving end for all sensory input except smell. Visual, auditory, and somatosensory information arrives at the thalamus via the cranial nerves and tracts from the spinal cord. The thalamus integrates this information and sends it on to the appropriate portions of the cerebrum. For this reason, the thalamus is often referred to as the "gatekeeper" for sensory information en route to the cerebral cortex. The thalamus is involved in arousal of the cerebrum, and it also participates in higher mental functions such as memory and emotions.

The **pineal gland**, which secretes the hormone melatonin, is located in the diencephalon. Presently, there is much interest in the role of melatonin in our daily rhythms; some researchers believe it may be involved in jet lag and insomnia. Scientists are also interested in the possibility that this hormone regulates the onset of puberty.

The **cerebellum** lies under the occipital lobe of the cerebrum and is separated from the brain stem by the fourth ventricle. It is the largest part of the hindbrain. The cerebellum has two portions that are joined by a narrow central portion. Each portion is primarily composed of white matter, which in longitudinal section has a treelike pattern. Overlying the white matter is a thin layer of gray matter that forms a series of complex folds.

The cerebellum receives sensory input from the eyes, ears, joints, and muscles about the present position of body parts, and it also receives motor output from the cerebral cortex about where these parts should be located. After integrating this information, the cerebellum sends motor impulses by way of the brain stem to the skeletal muscles. In this way, the cerebellum maintains posture and balance. It also ensures that all of the muscles work together to produce smooth, coordinated voluntary movements. The cerebellum assists the learning of new motor skills such as playing the piano or hitting a baseball. New evidence indicates that the cerebellum is important in judging the passage of time.

The **brain stem** contains the midbrain, the pons, and the medulla oblongata (see Fig. 26.11). The **midbrain** acts as a relay station for tracts passing between the cerebrum and the spinal cord or cerebellum. The tracts cross in the brain stem so that the right side of the body is controlled by the left portion of the brain and the left portion of the body is controlled by the right portion of the brain.

The brain stem also has reflex centers for visual, auditory, and tactile responses. The word **pons** means "bridge" in Latin, and true to its name, the pons contains bundles of axons traveling between the cerebellum and the rest of the CNS. In addition, the pons functions with the medulla oblongata to regulate breathing rate, and has reflex centers concerned with head movements in response to visual and auditory stimuli.

The **medulla oblongata** contains a number of reflex centers for regulating heartbeat, breathing, and blood pressure. It also contains the reflex centers for vomiting, coughing, sneezing, hiccuping, and swallowing. The medulla oblongata lies just superior to the spinal cord, and it contains tracts that ascend or descend between the spinal cord and higher brain centers.

The Reticular Activating System The reticular formation is a complex network of nuclei (masses of gray matter) and nerve fibers that extend the length of the brain stem (**Fig. 26.13**). The reticular formation is a major component of the reticular activating system (RAS), which receives sensory signals that it sends up to higher centers, and motor signals that it sends to the spinal cord.

The RAS arouses the cerebrum via the thalamus and causes a person to be alert. Apparently, the RAS can filter out unnecessary sensory stimuli, explaining why you can study with the TV on. If you want to awaken the RAS, surprise it with a sudden stimulus, like splashing your face with cold water; if you want to deactivate it, remove visual and auditory stimuli. General anesthetics function by artificially suppressing the RAS. A severe injury to the RAS can cause a person to be comatose, from which recovery may be impossible.

Several parts of the brain work together in the limbic system, discussed in the next section.

26.13 *Check Your Progress* The hypothalamus, which has sleep centers, communicates with the RAS. What might cause narcolepsy, the disorder characterized by brief periods of unexpected sleep?

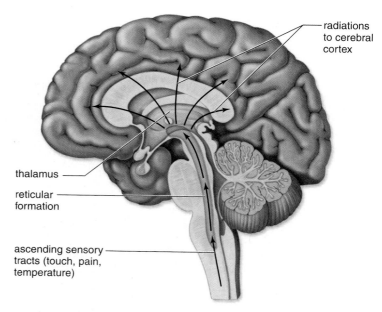

radiations to cerebral cortex

thalamus

reticular formation

ascending sensory tracts (touch, pain, temperature)

FIGURE 26.13 The reticular activating system.

26.14 The limbic system is involved in memory and learning as well as in emotions

The **limbic system** is a complex network of tracts and nuclei that incorporates portions of the cerebral lobes, the basal nuclei, and the diencephalon (**Fig. 26.14**). The limbic system blends higher mental functions and primitive emotions into a united whole. It accounts for why activities such as sexual behavior and eating seem pleasurable and also why, say, mental stress can cause high blood pressure.

Two significant structures within the limbic system are the hippocampus and the amygdala, which are essential for learning and memory. The **hippocampus**, a seahorse-shaped structure that lies deep in the temporal lobe, is well situated in the brain to make the frontal lobe aware of past experiences stored in various sensory areas. The **amygdala**, in particular, adds emotional overtones. The smell of smoke not only warns us that the hotel is on fire, it creates great anxiety. Because the frontal lobe is part of the limbic system, we may be able to calmly analyze the situation and walk to the nearest exit.

Learning and Memory Memory is the ability to hold a thought in mind or recall events from the past, ranging from a word we learned only yesterday to an early emotional experience that has shaped our lives. Learning takes place when we retain and use past memories.

The frontal lobe is active during short-term memory, as when we temporarily recall a telephone number. Some telephone numbers go into long-term memory. Think of a telephone number you know by heart, and see if you can bring it to mind without also thinking about the place or person associated with that number. Most likely, you cannot because, typically, long-term memory is a mixture of what is called semantic memory (numbers, words, etc.) and episodic memory (persons, events, etc.). Skill memory is a type of memory that can exist independent of episodic memory. Skill memory enables us to perform motor activities, such as riding a bike or playing ice hockey.

What parts of the brain are functioning when you remember something from long ago? As mentioned, our long-term memories are stored in bits and pieces throughout the sensory areas of the cerebral cortex. The hippocampus gathers this information together for use by the frontal lobe when we remember Uncle Frank or our summer holiday. Why are some memories so emotionally charged? Again, the amygdala is responsible for fear conditioning and associating danger with sensory information received from the thalamus and the cortical sensory areas.

Long-term potentiation (LTP) is an enhanced response at synapses seen particularly within the hippocampus. LTP is most likely essential to memory storage, but unfortunately, it sometimes causes a postsynaptic neuron to become so excited that it undergoes apoptosis, a form of cell death. This phenomenon, called *excitotoxicity*, is due to the action of glutamate, a neurotransmitter. When glutamate binds to the postsynaptic membrane, calcium may rush in too fast because of a receptor that is malformed due to a mutation. A gradual extinction of brain cells, particularly in the hippocampus, appears to be the underlying cause of **Alzheimer disease (AD)**, a condition characterized by gradual loss of memory as well as cognitive and behavioral changes.

In the brains of patients with AD, the neurons have neurofibrillary tangles (bundles of fibrous protein) surrounding the nucleus and protein-rich accumulations called amyloid plaques enveloping the axon branches. Although it is not yet known how excitotoxicity is related to the structural abnormalities of AD neurons, some researchers are trying to develop neuroprotective drugs that could guard brain cells against damage due to glutamate.

The next part of the chapter considers the peripheral nervous system.

> **26.14 Check Your Progress** A disconnect can occur between the amygdala and the portion of the cortex devoted to recognizing faces. People with this ailment recognize family members, but have no feelings for them. Explain.

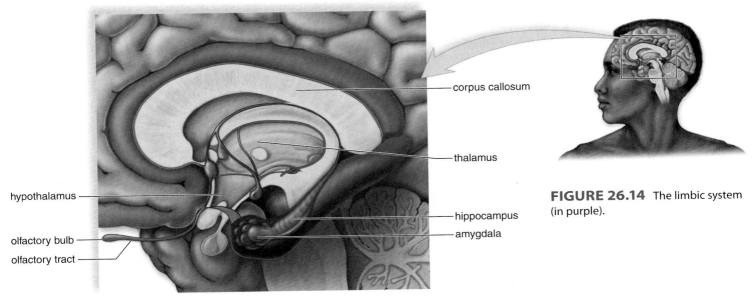

hypothalamus

olfactory bulb

olfactory tract

corpus callosum

thalamus

hippocampus

amygdala

FIGURE 26.14 The limbic system (in purple).

In this part of the chapter, a study of the structure of nerves precedes a description of reflex actions. We will also discuss the autonomic system, which is a part of the peripheral nervous system.

26.15 The peripheral nervous system contains cranial and spinal nerves

The peripheral nervous system (PNS) lies outside the central nervous system and contains **nerves**, which are bundles of axons. Axons within nerves are also called nerve fibers (**Fig. 26.15A**). Nerves are designated as cranial nerves when they arise from the brain and spinal nerves when they arise from the spinal cord. In any case, all nerves take impulses to and from the CNS. So, right now, your eyes are sending messages by way of a cranial nerve to the brain, allowing you to read this page, and your brain, by way of the spinal cord and a spinal nerve, will direct the muscles in your fingers to turn to the next page.

The **cranial nerves** are attached to the brain (**Fig. 26.15B**). Some of these are sensory nerves—that is, they contain only sensory nerve fibers. Some are motor nerves that contain only motor fibers, and others are mixed nerves that contain both sensory and motor fibers. Cranial nerves are largely concerned with the head, neck, and facial regions of the body. The vagus nerve, which arises from the brain stem—specifically, the medulla oblongata—has branches not only to the pharynx and larynx but also to most of the internal organs.

The **spinal nerves** are attached to the spinal cord. Each spinal nerve emerges from the spinal cord by two short branches, or roots (**Fig. 26.15C**). A spinal nerve separates the axons of sensory neurons from the axons of motor neurons. At the cord, the dorsal root contains the axons of sensory neurons, which conduct impulses to the spinal cord from sensory receptors. The cell body of a sensory neuron is in the **dorsal root ganglion**. The ventral root of a spinal nerve contains the axons of motor neurons, which conduct impulses away from the spinal cord to effectors that are muscle fibers or glands. These two roots join to form a spinal nerve. All spinal nerves are mixed nerves that contain many sensory and motor fibers. Each spinal nerve serves the particular region of the body in which it is located. For example, the intercostal muscles of the rib cage are innervated by the thoracic nerves.

Now that we have examined structure, let's move on to a description of nerve reflexes in Section 26.16.

FIGURE 26.15A
Anatomy of a nerve.

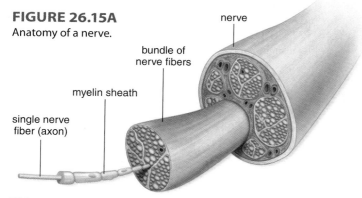

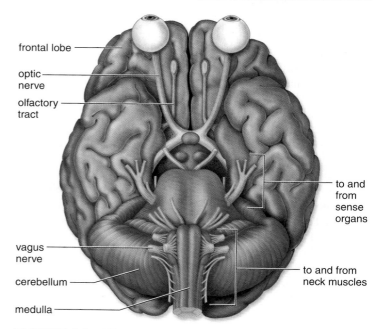

FIGURE 26.15B Ventral surface of brain showing the attachment of the cranial nerves (yellow).

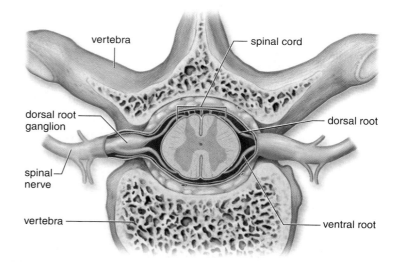

FIGURE 26.15C Cross section of the vertebral column and spinal cord, showing a spinal nerve.

> **26.15 Check Your Progress** Why is the peripheral nervous system (PNS) just as important as the central nervous system (CNS)?

26.16 In the somatic system, reflexes allow us to respond quickly to stimuli

The PNS has two divisions—somatic and autonomic—and we are going to consider the somatic system first. The nerves in the **somatic system** serve the skin, joints, and skeletal muscles. Therefore, the somatic system includes nerves that take (1) sensory information from external sensory receptors in the skin and joints to the CNS, and (2) motor commands away from the CNS to the skeletal muscles. The neurotransmitter acetylcholine (ACh) is active in the somatic system. In Chapter 28, we will see how axon terminals release ACh into neuromuscular junctions, after which ACh stimulates skeletal muscle fibers to contract (see Section 28.10).

Voluntary control of skeletal muscles always originates in the brain. Involuntary responses to stimuli, called **reflexes**, can involve either the brain or just the spinal cord. Reflexes enable the body to react swiftly to stimuli that could disrupt homeostasis. Flying objects cause our eyes to blink, and sharp pins cause our hands to jerk away, even without us having to think about it.

The Reflex Arc Figure 26.16 illustrates the path of a reflex that involves only the spinal cord. If your hand touches a sharp pin, sensory receptors in the skin generate nerve impulses that move along sensory axons through a dorsal root ganglion toward the spinal cord. Sensory neurons that enter the cord dorsally pass signals on to many interneurons in the gray matter of the spinal cord. Some of these interneurons synapse with motor neurons. The short dendrites and the cell bodies of motor neurons are also in the spinal cord, but their axons leave the cord ventrally. Nerve impulses travel along motor axons to an effector, which brings about a response to the stimulus. In this case, a muscle contracts so that you withdraw your hand from the pin. Various other reactions are possible—you will most likely look at the pin, wince, and cry out in pain. This whole series of responses is explained by the fact that some of the interneurons in the white matter of the cord carry nerve impulses in tracts to the brain. The brain makes you aware of the stimulus and directs subsequent reactions to the situation. You don't feel pain until the brain receives the information and interprets it! Visual information received directly by way of a cranial nerve may make you aware that your finger is bleeding. Then you might decide to look for a band-aid.

The autonomic system, discussed in Section 26.17, controls the internal organs.

> **26.16 Check Your Progress** *a.* **What part of the CNS is always active when a reflex action involving the limbs occurs?** *b.* **What part of the CNS is always active when we override a reflex action and do not react automatically?**

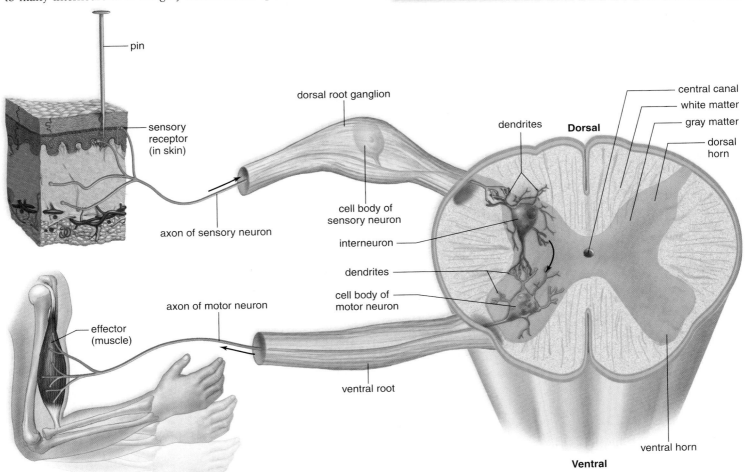

FIGURE 26.16 A reflex arc showing the path of a spinal reflex.

26.17 In the autonomic system, the parasympathetic and sympathetic divisions control the actions of internal organs

The **autonomic system** of the PNS automatically and involuntarily regulates the activity of glands and cardiac and smooth muscle. The system is divided into the parasympathetic and sympathetic divisions (**Fig. 26.17**). Activation of these systems generally causes opposite responses.

Although their functions are different, the two divisions share these same features: (1) They function automatically and usually in an involuntary manner; (2) they innervate all internal organs; and (3) they utilize two motor neurons and one ganglion for each impulse. The first neuron has a cell body within the CNS and a preganglionic fiber. The second neuron has a cell body within a ganglion and a postganglionic fiber.

Reflex actions, such as those that regulate blood pressure and breathing rate, are especially important to the maintenance of homeostasis. These reflexes begin when the sensory neurons in contact with internal organs send information to the CNS. They are completed by motor neurons within the autonomic system.

Parasympathetic Division The **parasympathetic division** includes a few cranial nerves (e.g., the vagus nerve) as well as axons that arise from the last portion of the spinal cord. The parasympathetic division, sometimes called the "housekeeping division," promotes all the internal responses we associate with a relaxed state. For example, it causes the pupil of the eye to constrict, promotes digestion of food, and retards the heartbeat. It's been suggested that the parasympathetic division be called the *rest-and-digest* system. The parasympathetic division utilizes the neurotransmitter ACh.

Sympathetic Division Axons of the **sympathetic division** arise from portions of the spinal cord. The sympathetic division is especially important during emergency situations and is associated with *fight or flight*. If you need to fend off a foe or flee from danger, active muscles require a ready supply of glucose and oxygen. On the one hand, the sympathetic division accelerates the heartbeat and dilates the bronchi, while at the same time it inhibits the digestive tract, since digestion is not an immediate necessity if you are under attack. The sympathetic division utilizes the neurotransmitter norepinephrine, which has a structure like that of epinephrine (adrenaline), an adrenal medulla hormone that usually increases heart rate and contractility.

> **26.17 Check Your Progress** In humans, the head runs everything. Develop a scenario to explain a response to a crisis situation.

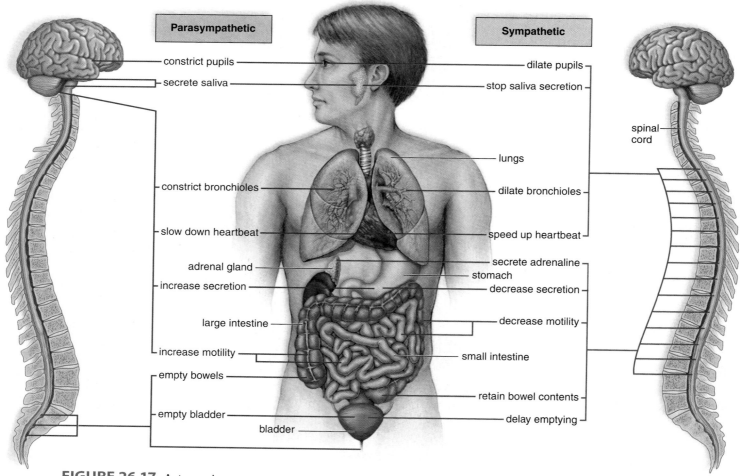

FIGURE 26.17 Autonomic system.

The human nervous system has just three functions: sensory input, integration, and motor output. Nerve impulses are the same in all neurons, so how is it that stimulation of the eyes causes us to see, and stimulation of the ears causes us to hear? Essentially, the central nervous system (CNS) carries out the function of integrating incoming data. The brain allows us to perceive our environment, to reason, and to remember. After sensory data have been processed by the CNS, motor output occurs. Muscles and glands are the effectors that allow us to respond to the original stimuli. Without the musculoskeletal system, discussed in Chapter 28, we would never be able to respond to a danger detected by our eyes and ears.

Similar to the wiring of a modern office building, the human peripheral nervous system (PNS) contains nerves that carry sensory input to the CNS and motor output to the muscles and glands. There is a division of labor among the nerves. The cranial nerves serve the face, teeth, and mouth; below the head, there is only one cranial nerve, the vagus nerve. All body movements are controlled by spinal nerves, and this is why paralysis may follow a spinal injury. Except for the vagus nerve, only spinal nerves make up the autonomic system, which controls the internal organs.

You might argue that sense organs, such as the eyes and ears discussed in Chapter 27, should be considered a part of the nervous system, since there would be no sensory nerve impulses without their ability to generate them. Our view of the world is dependent on the sense organs, which are sensitive to external and internal stimuli.

The Chapter in Review

Summary

Getting a Head

- Some animals do not have heads.
- Animals with a brain, a nervous system, and sense organs carry out complex behaviors.
- The vertebrate brain can be divided into a hindbrain, midbrain, and forebrain.

Most Animals Have a Nervous System That Allows Responses to Stimuli

26.1 Invertebrates reflect an evolutionary trend toward bilateral symmetry and cephalization

- Structures in invertebrate nervous systems include a nerve net (sponges), a ladderlike arrangement of nerve cords and ganglia (planarians), and a brain and ganglia (annelids, arthropods, molluscs).
- The vertebrate nervous organization is characterized by bilateral symmetry, cephalization, and increased number of neurons.

26.2 Humans have well-developed central and peripheral nervous systems

- The central nervous system (CNS) is composed of the spinal cord and brain.
- The peripheral nervous system (PNS) consists of the nerves and ganglia outside the CNS.

Neurons Process and Transmit Information

26.3 Neurons are the functional units of a nervous system

- Neurons receive and convey sensory information and conduct signals to glands and muscles.

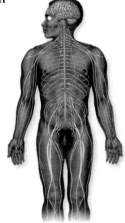

- Neuroglia support and nourish neurons. Schwann cells form myelin sheaths.
- A neuron has three parts: a cell body, dendrites, and an axon.
- The three types of neurons are motor, sensory, and interneurons.

26.4 Neurons have a resting potential across their membranes when they are not active

- In a resting potential, the axon is not conducting the impulse; there is more Na$^+$ outside the axon and more K$^+$ inside the axon.

26.5 Neurons have an action potential across axon membranes when they are active

- An action potential is a rapid change in polarity across the axon membrane as the nerve impulse occurs: Na$^+$ gates open, and Na$^+$ moves to inside the axon; K$^+$ gates open, and K$^+$ moves to outside the axon.

26.6 Propagation of an action potential is speedy

- In saltatory conduction, ion exchange occurs only at nodes, and the action potential jumps from node to node.

26.7 Communication between neurons occurs at synapses

- A synapse is a region of close proximity between an axon terminal and a dendrite.
- When neurotransmitter is released into a synaptic cleft, transmission of a nerve impulse occurs.

26.8 Neurotransmitters can be stimulatory or inhibitory

- Binding of neurotransmitter to receptors causes stimulation or inhibition.
- Acetylcholine (ACh) stimulates skeletal muscles; norepinephrine (NE) generally stimulates smooth muscle.

26.9 Integration is a summing up of stimulatory and inhibitory signals

- A neuron receives and integrates signals, which can be exitatory or inhibitory.
- An excitatory signal drives a neuron closer to threshold—a depolarizing effect.

- An inhibitory signal drives a neuron further from threshold—a hyperpolarizing effect.

26.10 Drugs that interfere with neurotransmitter release or uptake may be abused

- Drug addiction: More of drug is needed to get same effect.
- Methamphetamine and ecstasy are considered club or date rape drugs; cocaine is a powerful stimulant in the CNS; heroin is a depressant that converts to morphine in the brain; and marijuana can alter vision and judgment.

The Vertebrate Central Nervous System (CNS) Consists of the Spinal Cord and Brain

26.11 The human spinal cord and brain function together

- The spinal cord contains tracts that take messages to and from the brain.

26.12 The cerebrum performs integrative activities

- The cerebrum has two cerebral hemispheres connected by the corpus callosum; each cerebral hemisphere has four lobes: frontal, parietal, occipital, and temporal.
- The primary motor area sends out voluntary motor commands to skeletal muscles; the primary somatosensory area receives sensory information from the skin and skeletal muscles.
- Basal nuclei in the white matter integrate motor commands.

26.13 The other parts of the brain have specialized functions

- The hypothalamus controls homeostasis, while the thalamus sends sensory input to the cerebrum.
- The cerebellum coordinates skeletal muscle contractions.
- The medulla oblongata and pons contain centers for regulating breathing, heartbeat, and blood pressure.
- The reticular activating system arouses the cerebrum via the thalamus, causing alertness.

26.14 The limbic system is involved in memory and learning as well as in emotions

- The hippocampus is involved in storing and retrieving memories.
- The amygdala determines when a situation calls for "fear."

The Vertebrate Peripheral Nervous System (PNS) Consists of Nerves

26.15 The peripheral nervous system contains cranial and spinal nerves

- Cranial nerves take impulses to and from the brain.
- Spinal nerves take impulses to and from the spinal cord.

26.16 In the somatic system, reflexes allow us to respond quickly to stimuli

- The PNS is divided into the somatic system and the autonomic system.
- Nerves in the somatic system serve the skin, joints, and skeletal muscles.

- Some actions are due to reflexes, which are automatic and involuntary.

26.17 In the autonomic system, the parasympathetic and sympathetic divisions control the actions of internal organs

- The parasympathetic division governs responses that occur during times of relaxation.
- The sympathetic division is in charge of responses that occur during times of stress.

Most Animals Have a Nervous System That Allows Responses to Stimuli

1. Which of the following animals has a ladderlike nervous system?
   a. crab   c. jellyfish
   b. insect   d. planarian
2. Which of the following associations is not matched correctly?
   a. forebrain—processes sensory information
   b. hindbrain—controls breathing and heart rate
   c. midbrain—houses reflexes involving the eyes and ears
   d. All of these associations are matched correctly.

Neurons Process and Transmit Information

3. Which of the following neuron parts receive(s) signals from sensory receptors of other neurons?
   a. cell body   c. dendrites
   b. axon   d. Both a and c are correct.
4. Which of these would be covered by a myelin sheath?
   a. short dendrites   d. interneurons
   b. globular cell bodies   e. All of these are correct.
   c. long axons
5. What type of neuron lies completely in the CNS?
   a. motor neuron
   b. interneuron
   c. sensory neuron
6. When the action potential begins, sodium gates open, allowing Na^+ to cross the membrane. Now the polarity changes to
   a. negative outside and positive inside.
   b. positive outside and negative inside.
   c. neutral outside and positive inside.
   d. There is no difference in charge between outside and inside.
7. Repolarization of an axon during an action potential is produced by
   a. inward diffusion of NA^+.
   b. outward diffusion of K^+.
   c. inward active transport of Na^+.
   d. active extrusion of K^+.
8. A drug that inactivates acetylcholinesterase
   a. stops the release of ACh from presynaptic endings.
   b. prevents the attachment of ACh to its receptor.
   c. increases the ability of ACh to stimulate muscle contraction.
   d. All of these are correct.
9. The summing up of inhibitory and excitatory signals is called
   a. excitation.   c. depolarization.
   b. integration.   d. All of these are correct.
10. **THINKING CONCEPTUALLY** From an evolutionary perspective, it is not surprising that the nerve impulse makes use of a potential difference present in all plasma membranes. Explain.

The Vertebrate Central Nervous System (CNS) Consists of the Spinal Cord and Brain

11. Membranes surrounding the CNS are collectively called
    a. meninges.   c. vesicles.
    b. myelin.   d. None of these are correct.

12. Which of the following cerebral areas is not correctly matched with its function?
    a. occipital lobe—vision
    b. parietal lobe—somatosensory area
    c. temporal lobe—primary motor area
    d. frontal lobe—Broca motor speech area

13. The cerebellum
    a. coordinates skeletal muscle movements.
    b. receives sensory input from the joints and muscles.
    c. receives motor input from the cerebral cortex.
    d. All of these are correct.

14. The limbic system
    a. involves portions of the cerebral lobes and the diencephalon.
    b. is responsible for our deepest emotions, including pleasure, rage, and fear.
    c. is a system necessary to memory storage.
    d. is not directly involved in language and speech.
    e. All of these are correct.

15. **THINKING CONCEPTUALLY** Explain why you would expect learning to result in an increase in the number of synapses.

The Vertebrate Peripheral Nervous System (PNS) Consists of Nerves

16. The organs of the peripheral nervous system are the
    a. cranial and spinal nerves.
    b. brain and spinal cord.
    c. nerves and spinal cord.
    d. cranial nerves and brain.

17. Somatic is to skeletal muscle as autonomic is to
    a. cardiac muscle.   c. gland.
    b. smooth muscle.   d. All of these are correct.

18. Which of these statements about autonomic neurons is correct?
    a. They are motor neurons.
    b. Preganglionic neurons have cell bodies in the CNS.
    c. Postganglionic neurons innervate smooth muscles, cardiac muscle, and glands.
    d. All of these are correct.

19. The sympathetic division of the autonomic system will
    a. increase heart rate and digestive activity.
    b. decrease heart rate and digestive activity.
    c. cause pupils to constrict.
    d. None of these are correct.

Understanding the Terms

acetylcholine (ACh)   520
acetylcholinesterase (AChE)   521
action potential   518
Alzheimer disease (AD)   527
amygdala   527
autonomic system   530
axon   517
basal nuclei   525
brain stem   526
cell body   517
central nervous system (CNS)   515
cephalization   514
cerebellum   526
cerebral cortex   525
cerebral hemisphere   525
cerebrospinal fluid   524
cerebrum   525
cranial nerve   528
dendrite   517
depolarization   519
diencephalon   526
dorsal root ganglion   528
ganglion   514
gray matter   524
hippocampus   527
Huntington disease   525
hypothalamus   526
integration   521
interneuron   517
ladderlike nervous system   514
limbic system   527
medulla oblongata   526
memory   527
meninges   524
midbrain   526
motor (efferent) neuron   517
myelin sheath   517
nerve   528
nerve fiber   517
nerve net   514
neuroglia   517
neuron   517
neurotransmitter   520
nodes of Ranvier   517
norepinephrine (NE)   520
parasympathetic division   530
Parkinson disease   525
peripheral nervous system (PNS)   516
pineal gland   526
pons   526
primary motor area   525
primary somatosensory area   525
reflex   529
reflex action   524
refractory period   519
repolarization   519
resting potential   518
saltatory conduction   519
Schwann cell   517
sensory (afferent) neuron   517
somatic system   529
spinal cord   524
spinal nerve   528
sympathetic division   530
synapse   520
synaptic cleft   520
thalamus   526
threshold   518
tract   524
ventricle   524
white matter   524

Match the terms to these definitions:
  a. _____Automatic, involuntary response of an organism to a stimulus.
  b. _____Chemical stored at the ends of axons that is responsible for transmission across a synapse.
  c. _____System within the peripheral nervous system that regulates internal organs.
  d. _____Collection of neuron cell bodies usually outside the central nervous system.
  e. _____The largest part of the human brain consisting of two parts, each of which contains gray matter and white matter.

Thinking Scientifically

1. Knowing that the fight-or-flight response is initiated by the release of the neurotransmitter norepinephrine, how might it be possible to control the response in people who are stressed? What complications might ensue?
2. Hypothesize why a man with an amputated leg still feels pain as though it were coming from the missing limb.

ARIS Visit **www.mhhe.com/maderconcepts** for practice quizzes, animations, videos, and activities designed to help you master the material in this chapter.

27

Sense Organs

LEARNING OUTCOMES

After studying this chapter, you should be able to accomplish the following outcomes.

The Eyes Have It

1 Contrast the advantages of the compound eye with those of the camera-type eye.

Sensory Receptors Respond to Stimuli

2 List and describe five common types of sensory receptors.
3 Explain how the activity of sensory receptors results in sensation.

Chemoreceptors Are Sensitive to Chemicals

4 Compare and contrast the activities of vertebrate taste buds with those of olfactory cells.

Photoreceptors Are Sensitive to Light

5 Give a function for each part of the vertebrate eye.
6 Explain how the vertebrate eye functions with the brain to allow vision.
7 List disorders of the human eye that may be due to sun exposure, and suggest ways to reduce sun exposure.
8 Tell how it is possible to correct the inability to form a clear image.
9 Compare and contrast the action of rod and cone cells, and explain color vision.

Mechanoreceptors Are Involved in Hearing and Balance

10 Give a function for each part of the vertebrate ear.
11 Trace the path of sound waves in the ear, and explain how the organ of Corti functions.
12 Give examples of noise pollution, and explain the ear damage that loud noises can cause.
13 Compare and contrast rotational equilibrium with gravitational equilibrium.
14 Explain the causes of motion sickness.
15 Compare the lateral line system in fish with the ear in humans, and compare the activity of a statocyst to that of the utricle and saccule.

E yes pervade the animal kingdom, testifying to their usefulness in finding food, a mate, and a place to live—and assisting animals in general as they carry out their daily activities. The vertebrate eye contains the sensory receptors for light, called the rods and the cones. The rods function well in dim light but produce an image that is indistinct and lacks color. Turn off the lights tonight, wait a few minutes, and a shadowy-gray world will appear. Flip on the light, and your cones will take over, showing you a distinct, colorful world! Cones have terrific resolving power. The eyes of a hawk contain more than a million cones per cubic millimeter, allowing it to detect a tiny mouse scurrying among the underbrush of trees from a great height. At the other extreme, rods have replaced cones in the eyes of the tarsier, a rat-sized primate that is active only at night in the tropical rain forests of Southeast Asia. The attributes of an animal's eye correlate with its lifestyle.

A bumblebee gathering nectar at a flower has an entirely different type of eye from that of vertebrates. The eye of an insect is called compound because it has many visual units, each sending its own data to the brain, where a mosaic (compound) image is produced. The compound eye produces a crude image, but it is a taskmaster at detecting motion. The unusually rapid recovery of its light receptors makes this possible. The human eye can distinguish only about 24 images per second; after that, the images are fused into one. In contrast, the eye of an insect can distinguish different images at the rate of 330 per second. The fly sees your every move when you come after it with a flyswatter!

Visual Unit

photoreceptor cells

optic nerve fibers

Not only can the insect see color, but its eyes can also respond to ultraviolet rays. Many flowers have "nectar guides" that reflect ultraviolet light, and when a bee visits that flower, it does not see the color as much as the guides, which direct it to where the nectar is located. Some birds can respond to ultraviolet rays. In birds called budgies, investigators have found that fluorescent feathers, reflecting ultraviolet rays, are used to select a mate.

In contrast to the mosaic image of the compound eye shown in the circle, the vertebrate camera-type eye has one lens for its many photoreceptors, and the brain forms a single image after receiving data from the eyes. Resolving power is good, but the eye is relatively large and heavy, so only vertebrates and certain invertebrates have room for such an eye. The squid is one of these invertebrates. However, in the squid, the lens moves back and forth, while in human eyes, the lenses change shape to accommodate for the distance of an object. Among vertebrates, birds and humans are known for seeing color. Humans have cones of only three different colors (blue, green, and red), but see different shades, depending on which of these is stimulated. Birds, with cones of four to five different colors, have superior color vision to that of humans.

Previously it was thought that the compound eye and the camera-type eye evolved separately, and perhaps many times over in the animal kingdom. But now evo-devo geneticists (those who study development from an evolutionary perspective) tell us that the same genes are active whether an animal has a compound eye or a camera-type eye. Surprisingly, their conclusion is that all image-forming eyes can be traced to an original eye-bearing ancestor.

Mammals with two eyes facing forward have three-dimensional, or stereoscopic, vision. The visual fields overlap, and each eye is able to view an object from a different angle. Predators tend to have stereoscopic vision, and so do humans. Animals with eyes facing sideways, such as rabbits and zebras, don't have stereoscopic vision, but they do have panoramic vision, meaning that the visual field is very wide. Panoramic vision is useful to prey animals because it makes it easier for them to detect a predator sneaking up on them.

This chapter discusses the major types of animal sense organs from an evolutionary perspective. It stresses the chemoreceptors, the photoreceptors, and the mechanoreceptors.

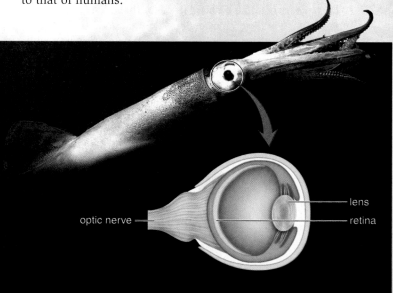

optic nerve — lens / retina

Sensory receptors detect certain types of stimuli, including chemical, pain, electromagnetic, temperature, and touch. Sensory receptors communicate by way of nerves with the central nervous system, which integrates nerve impulses and directs a response.

27.1 Sensory receptors can be divided into five categories

All animals have **sensory receptors** that allow them to respond to stimuli. Stimuli are environmental signals that tell us about the external or internal environment. Surprisingly, there are only five common categories of sensory receptors: chemoreceptors, pain receptors, electromagnetic receptors, thermoreceptors, and mechanoreceptors.

Chemoreceptors respond to chemical substances in the immediate vicinity. Taste and smell depend on this type of sensory receptor, but certain chemoreceptors in various other organs are sensitive to internal conditions. For example, chemoreceptors that monitor blood pH are located in the carotid arteries and aorta of humans. If the pH lowers, the breathing rate increases. As more carbon dioxide is expired, the blood pH rises.

Pain receptors (nociceptors) are sometimes classified as a type of chemoreceptor. However, pain receptors respond to excessive temperature and mechanical pressure in addition to a range of chemicals, some of which are released by damaged tissues. Pain receptors are protective because they alert us to possible danger. For example, without the pain of appendicitis, we might never seek the medical help needed to avoid a ruptured appendix.

Electromagnetic receptors are stimulated by changes in electromagnetic waves. Photoreceptors, present in the eyes of most animals, are sensitive to visible light energy. As mentioned in the introduction to this chapter, our eyes contain photoreceptors known as rod cells that result in black-and-white vision, while stimulation of photoreceptors known as cone cells results in color vision. The eyes of insects can detect ultraviolet radiation, and this helps them notice flowers, particularly the location of nectar. Several types of animals, including gray whales (**Fig. 27.1A**), migrate long distances from feeding to breeding areas and are believed to use the Earth's magnetic field as a type of compass to orient themselves.

Thermoreceptors are stimulated by changes in temperature. Humans have thermoreceptors located in the hypothalamus and the skin; those that respond when temperatures rise are

FIGURE 27.1B Thermoreceptors that are sensitive to infrared energy help pythons find their prey.

called heat receptors, and those that respond when temperatures lower are called cold receptors. Some snakes, such as pythons, have thermoreceptors located in pits near the mouth that detect the body heat of their prey up to 1–2 meters away (**Fig. 27.1B**). Some researchers classify thermoreceptors as electromagnetic receptors because they equate heat with infrared energy, which is part of the electromagnetic spectrum.

Mechanoreceptors are stimulated by mechanical forces, which most often result in pressure of some sort. When we hear, airborne sound waves are converted to fluid-borne pressure waves that can be detected by mechanoreceptors in the inner ear. The external ears of some bats are large for their size (**Fig. 27.1C**), and their inner ears are able to detect ultrasounds, sounds that are above the range humans can hear. These bats make ultrasonic clicking noises, and the echo of these sounds tells them where their prey is located in the dark.

FIGURE 27.1A Electromagnetic-receptors in gray whales help them migrate.

FIGURE 27.1C The ears of some bats use echolocation to find their prey.

The sense of balance and the sense of touch depend on mechanoreceptors. Also, in humans, pressoreceptors located in certain arteries detect changes in blood pressure, and stretch receptors in the lungs detect the degree of lung inflation. Proprioceptors are mechanoreceptors, which respond to the stretching of muscle fibers, tendons, joints, and ligaments, making us aware of the position of our limbs.

27.1 Check Your Progress Explain, with reference to Figure 6.6A, why eyes are classified as electromagnetic receptors.

27.2 Sensory receptors communicate with the CNS

Organisms need a way to become aware of the information collected by their sensory receptors. In complex animals, sensory receptors in the peripheral nervous system (PNS) send information to the brain and spinal cord of the central nervous system (CNS), which integrates sensory input before directing a motor response (**Fig. 27.2**). Sensory receptors transform the stimulus into nerve impulses that reach the cerebral cortex of the brain. When nerve impulses arrive at the cerebral cortex, **sensation,** which is the conscious perception of stimuli, occurs.

As we discussed in Chapter 26, sensory receptors are the first element in a reflex arc. However, we are only aware of a reflex action once input has reached the cerebral cortex. At that time, the brain integrates information received from various sensory receptors. For instance, if you burn yourself and quickly remove your hand from a hot stove, the brain receives information not only from your skin, but also from your eyes, nose, and all sorts of other sensory receptors.

Some sensory receptors are free nerve endings or encapsulated nerve endings, while others are specialized cells closely associated with neurons. If so, the plasma membrane of the sensory receptor can contain receptor proteins that react to the stimulus. For example, the receptor proteins in the plasma membrane of a chemoreceptor bind to certain molecules. When this happens, ion channels open, and ions flow across the plasma membrane. If the stimulus is sufficient, nerve impulses begin and are carried by a sensory nerve fiber within the PNS to the CNS (Fig. 27.2). The stronger the stimulus, the greater the frequency of nerve impulses. Nerve impulses that reach the spinal cord first are conveyed to the brain by ascending tracts. If nerve impulses finally reach the cerebral cortex, sensation and perception occur.

All sensory receptors initiate nerve impulses; the sensation that results depends on the part of the brain receiving the impulses. Nerve impulses that begin in the optic nerve eventually reach the visual areas of the cerebral cortex, and then we see objects. Nerve impulses that begin in the auditory nerve eventually reach the auditory areas of the cerebral cortex, and then we hear sounds. If it were possible to switch these nerves, stimulation of the eyes would result in hearing! On the other hand, when a blow to the eye stimulates photoreceptors, we "see stars" because nerve impulses from the eyes can only result in sight.

Before sensory receptors initiate nerve impulses, they carry out some **integration,** the summing up of signals. One type of integration is called **sensory adaptation,** a decrease in response to a stimulus. We have all had the experience of smelling an odor when we first enter a room and then later not being aware of it at all. Some authorities believe that when sensory adaptation occurs, sensory receptors have stopped sending impulses to the brain. Others believe that the reticular activating system (RAS) has filtered out the ongoing stimuli. You will recall that the RAS conveys sensory information from the brain stem, through the thalamus, to the cerebral cortex. The thalamus acts as a gatekeeper and only passes on information of immediate importance. Just as we can gradually become unaware of particular environmental stimuli, we can suddenly become aware of stimuli that may have been present for some time. This can be attributed to the workings of the RAS, which has synapses with many ascending sensory tracts.

The functioning of sensory receptors makes a significant contribution to homeostasis. Without sensory input, we would not receive information about our internal and external environments. This information, passed through nerves to the spinal cord to the brain, or directly to the brain, leads to appropriate reflex and voluntary actions that keep the internal environment constant.

We begin our discussion of specific types of receptors with chemoreceptors, in the next part of the chapter.

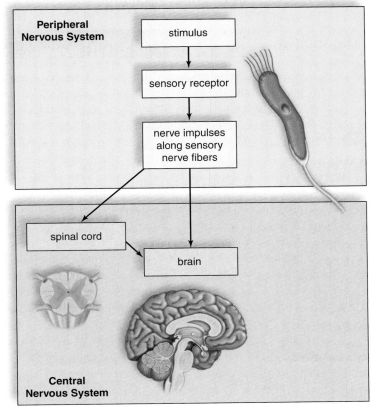

FIGURE 27.2 Nerve impulses from sensory receptors result in sensation and perception in the brain.

27.2 Check Your Progress Trace the path of nerve impulses from the human eyes to the brain, giving a function for each structure.

Chemoreceptors are prevalent throughout the animal kingdom. Pheromones are chemicals that provide a means of communicating with other members of a species. The taste and olfactory receptors of mammals provide chemical information about food sources and those that could adversely affect health.

27.3 Chemoreceptors are widespread in the animal kingdom

The sensory receptors responsible for taste and smell are termed chemoreceptors because they are sensitive to certain chemical substances in food, including liquids, and air. Chemoreception is found almost universally in animals and is, therefore, believed to be the most primitive sense. Chemoreceptors can be located in various places in animals. Studies suggest that the chemoreceptors of flatworms, such as planarians, are located in the auricles on the sides of the head. In the housefly, an insect, chemoreceptors are primarily on the feet. A fly literally tastes with its feet instead of its mouth. The antennae of insects detect airborne **pheromones**, which are chemical signals passed between members of the same species (**Fig. 27.3**).

In vertebrates, such as amphibians, chemoreceptors are located in the nose, mouth, and skin. Snakes and other vertebrates possess vomeronasal organs (VNO), a pair of pitlike organs located in the roof of the mouth. When a snake flicks its forked tongue, pheromones are carried to the VNO, and nerve impulses are sent to the brain for interpretation. In mammals, the receptors for taste are located in the mouth; the receptors for smell—and perhaps a VNO for detection of pheromones—are located in the nose. Section 27.4 discusses taste reception.

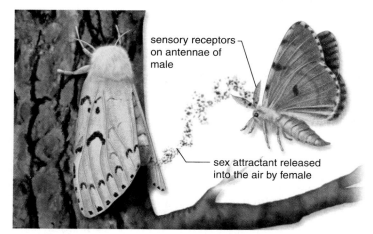

sensory receptors on antennae of male

sex attractant released into the air by female

FIGURE 27.3 A male moth responds to a species-specific sex attractant.

> **27.3 Check Your Progress** *a.* How are pheromones like any other chemical stimulus? *b.* How are they different?

27.4 Mammalian taste receptors are located in the mouth

In mammals, such as humans, taste receptors are a type of chemoreceptor located in **taste buds** (**Fig. 27.4**). In adult humans, approximately 3,000 taste buds are located primarily on the tongue **1**. Many taste buds lie along the walls of the papillae **2**, the small elevations on the tongue that are visible to the unaided eye. Isolated taste buds are also present on the hard palate, the pharynx, and the epiglottis.

Taste buds **3** open at a taste pore. A taste bud has supporting cells and a number of elongated taste cells that end in microvilli **4**. The microvilli, which project into the taste pore, bear recep-

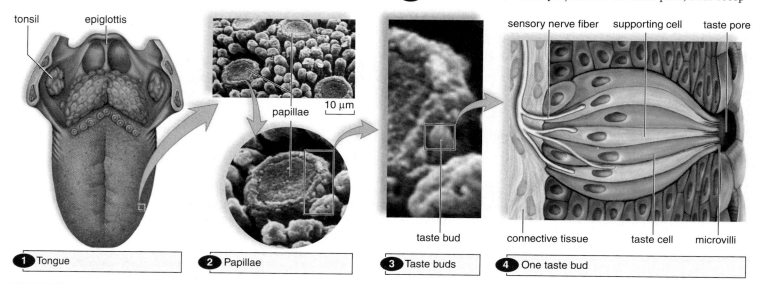

tonsil   epiglottis

papillae   10 μm

sensory nerve fiber   supporting cell   taste pore

taste bud   connective tissue   taste cell   microvilli

1 Tongue   **2** Papillae   **3** Taste buds   **4** One taste bud

FIGURE 27.4 Taste buds in humans.

tor proteins for certain molecules. When molecules dissolved in solution bind to receptor proteins, nerve impulses are generated in associated sensory nerve fibers. These nerve impulses go to the brain, including cortical areas that interpret them as tastes.

There are at least four primary types of taste (sweet, sour, salty, and bitter). A fifth taste, called unami, may exist for certain flavors of cheese, beef broth, and some seafood. Taste buds for each of these tastes are located throughout the tongue, although certain regions may be most sensitive to particular tastes: The tip of the tongue is most sensitive to sweet tastes; the margins to salty and sour tastes; and the rear of the tongue to bitter tastes. A

particular food can stimulate more than one of these types of taste buds. In this way, the response of taste buds can result in a range of sweet, sour, salty, and bitter tastes. The brain appears to survey the overall pattern of incoming sensory impulses and to take a "weighted average" of their taste messages as the perceived taste.

Section 27.5 discusses olfactory (smell) reception.

27.4 Check Your Progress Chemicals bind to chemoreceptors. Do you expect that light energy combines with photoreceptors in the same way?

27.5 Mammalian olfactory receptors are located in the nose

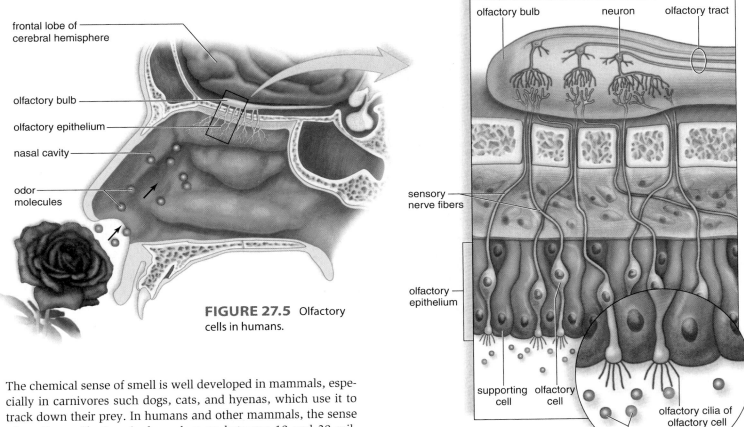

FIGURE 27.5 Olfactory cells in humans.

The chemical sense of smell is well developed in mammals, especially in carnivores such dogs, cats, and hyenas, which use it to track down their prey. In humans and other mammals, the sense of smell, or olfaction, is dependent on between 10 and 20 million **olfactory cells**. These structures are located within olfactory epithelium high in the roof of the nasal cavity (**Fig. 27.5**). Olfactory cells are modified neurons. Each cell ends in a tuft of about five olfactory cilia that bear receptor proteins for odor molecules. Each olfactory cell has only 1 out of 1,000 different types of receptor proteins. Nerve fibers from similar olfactory cells lead to the same neuron in the olfactory bulb, an extension of the brain. An odor contains many odor molecules that activate a characteristic combination of receptor proteins. A rose might stimulate certain olfactory cells, designated by blue and green in Figure 27.5, while a gardenia might stimulate a different combination. An odor's signature in the olfactory bulb is determined by which neurons are stimulated. When the neurons communicate this information via the olfactory tract to the olfactory areas of the cerebral cortex, we know we have smelled a rose or a gardenia.

Have you ever noticed that a certain aroma vividly brings to mind a certain person or place? A whiff of perfume may remind you of a specific person, or the smell of boxwood may remind you of your grandfather's farm. The olfactory bulbs have direct connections with the limbic system and its centers for emotions and memory. One investigator showed that when subjects smelled an orange while viewing a painting, they not only remembered the painting when asked about it later, but they also had many deep feelings about the painting.

The next part of the chapter concentrates on the vertebrate eye.

27.5 Check Your Progress Taste and smell are dependent on the combination of receptors stimulated. Correlate this observation with the ability to see shades of color.

The introduction to this chapter compared the compound eye of arthropods to the camera-type eye of a few invertebrates and all vertebrates. This part of the chapter discusses the structure and function of the vertebrate eye and how humans can protect their vision.

27.6   The vertebrate eye is a camera-type eye

The human eye is an elongated sphere about 2.5 cm in diameter and has three layers, or coats: the sclera, the choroid, and the retina (**Fig. 27.6**). The outer layer, the **sclera**, is an opaque, white, fibrous layer that covers most of the eye. A mucous membrane called the **conjunctiva** covers the exposed surface of the sclera and lines the inside of the eyelids. In front of the eye, the sclera becomes the **cornea**. The cornea is transparent, being composed of connective tissue with few cells and no blood vessels. Light rays pass through the cornea into the rest of the eye, and therefore, the cornea is called the window of the eye. However, the cornea plays an active role in vision by helping to focus light rays. Damage to the cornea is a frequent cause of blindness, but a damaged cornea is replaceable by a corneal transplant.

The middle, thin, dark-brown layer, the **choroid**, contains many blood vessels and a brown pigment that absorbs stray light rays. Toward the front of the eye, the choroid becomes the donut-shaped iris. The **iris** regulates the size of an opening called the pupil. The **pupil**, like the aperture on a camera lens, regulates the amount of light entering the eye. The color of the iris (color of the eyes) is dependent on its pigmentation. Heavily pigmented eyes are brown, while lightly pigmented eyes are green or blue. Behind the iris, the choroid thickens and forms the circular ciliary body. The **ciliary body**, consisting of many radiating folds, contains the ciliary muscles, which control the shape of the lens for near and far vision.

The **lens**, within a membranous capsule, lies directly behind the iris and the pupil. Attached to the ciliary body by suspensory ligaments, the lens divides the cavity of the eye into two compartments; the one in front of the lens is the anterior compartment, and the one behind the lens is the posterior compartment. A basic, watery solution called aqueous humor fills the anterior compartment. The aqueous humor provides a fluid cushion as well as nutrient and waste transport for the eye.

The third layer of the eye, the **retina**, is located in the posterior compartment, which is filled with a clear, gelatinous material called the vitreous humor. The retina contains the photoreceptors, called the rod cells and cone cells. The rods are very sensitive to light, but they do not see color; therefore, at night or in a darkened room, we see only shades of gray. The cones, which require bright light, are sensitive to different wavelengths of light, and therefore they give us the ability to distinguish colors. The retina has a very special region called the **fovea centralis**, where cone cells are densely packed. Light is normally focused on the fovea when we look directly at an object. This is helpful because vision is most acute in the fovea centralis. Sensory fibers form the optic nerve, which takes nerve impulses to the visual cortex of the brain. There are no rods and cones where the optic nerve exits the retina. Therefore, no vision is possible in this area, which is called a **blind spot**.

> **27.6 Check Your Progress** The compound eye of insects results in a mosaic image, while the camera-type eye results in one image. Explain.

FIGURE 27.6
Anatomy of the human eye.

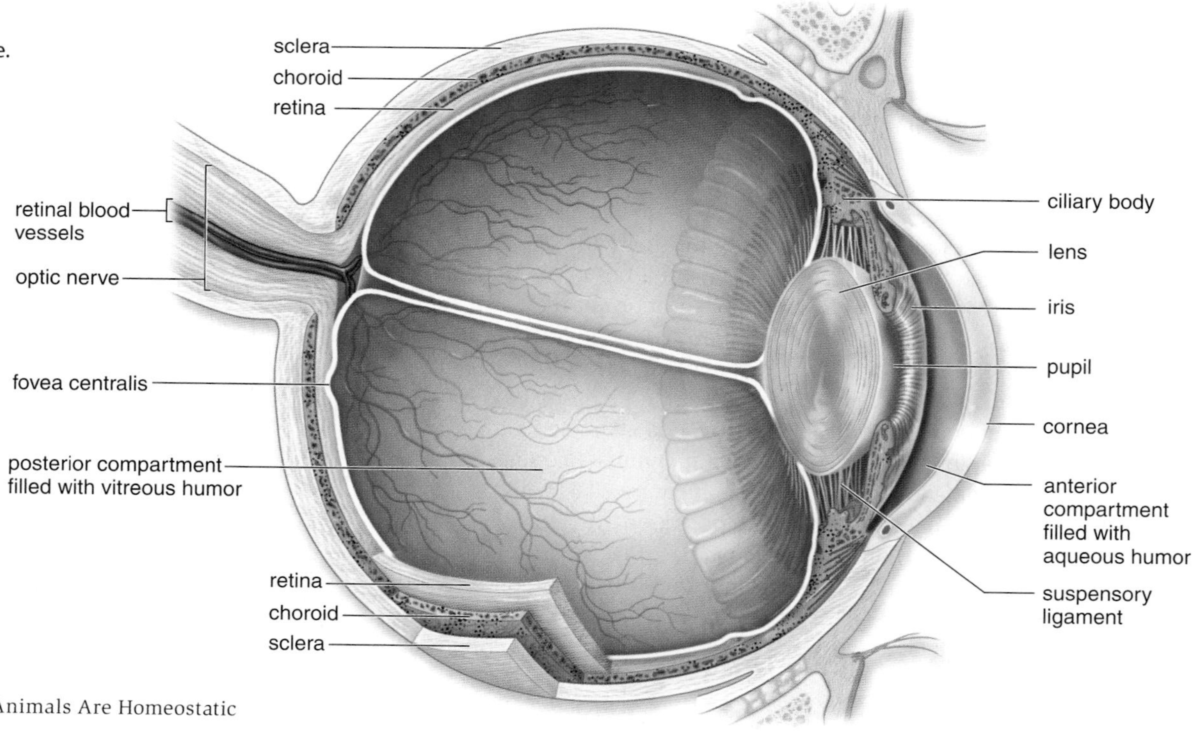

sclera — choroid — retina — retinal blood vessels — optic nerve — fovea centralis — posterior compartment filled with vitreous humor — retina — choroid — sclera — ciliary body — lens — iris — pupil — cornea — anterior compartment filled with aqueous humor — suspensory ligament

27.7   Protect your eyes from the sun

The most frequent causes of blindness are retinal disorders, glaucoma, and cataracts, in that order. Retinal disorders include diabetic retinopathy and macular degeneration. During retinopathy, capillaries to the retina burst, and blood spills into the vitreous humor. Careful regulation of blood glucose levels may protect against this condition. In macular degeneration, the cones are destroyed because thickened choroid vessels no longer function as they should. Glaucoma occurs when fluid builds up in the compartments and destroys the nerve fibers responsible for peripheral vision. People who have experienced acute glaucoma report that the eyeball feels as heavy as a stone. In cataracts, cloudy spots on the lens of the eye eventually pervade the whole lens. The milky, yellow-white lens scatters incoming light and blocks vision. Section 27.8 describes how cataracts are removed.

Accumulating evidence suggests that both macular degeneration and cataracts, which tend to occur in the elderly, are caused by long-term exposure to the ultraviolet rays of the sun. It is recommended, therefore, that everyone, especially those who live in sunny climates or work outdoors, wear sunglasses that absorb ultraviolet light. Large lenses worn close to the eyes offer further protection. The Sunglass

Association of America has devised the following system for categorizing sunglasses:

- Cosmetic lenses absorb 20% of UV-A (the type of radiation that reaches the Earth's surface) and 60% of visible light. Such lenses are worn for comfort, rather than protection.

- General-purpose lenses absorb at least 60% of UV-A, and 60–92% of visible light. They are good for outdoor activities in temperate regions.

- Special-purpose lenses block at least 60% of UV-A and 20–97% of visible light. They are good for bright sun combined with sand, snow, or water.

Health-care providers have found an increased incidence of cataracts in heavy cigarette smokers. The risk of cataracts doubles in men who smoke 20 cigarettes or more a day and in women who smoke 35 cigarettes or more a day. A possible reason is that smoking reduces the delivery of blood, and therefore nutrients, to the lens.

We continue our examination of the vertebrate eye in the next section by concentrating on the lens.

27.7 Check Your Progress Can any part of the eye be damaged and vision not be affected?

27.8   The lens helps bring an object into focus

When we look at an object, light rays pass through the pupil and focus on the retina. The image produced is much smaller than the object because light rays are bent (refracted) when they are brought into focus. Focusing mostly occurs at the cornea as light passes from an air medium to a fluid medium. The lens, however, provides additional focusing power as **visual accommodation** occurs for close vision. The shape of the lens is controlled by the **ciliary muscle** within the ciliary body. When we view a distant object, the ciliary muscle is relaxed, causing the suspensory ligaments attached to the ciliary body to be taut; therefore, the lens remains relatively flat (**Fig. 27.8A**). When we view a near object, the ciliary muscle contracts, releasing the tension on the suspensory ligaments, and the lens becomes more round due to its natural elasticity (**Fig. 27.8B**). Because close work requires contraction of the ciliary muscle, it very often causes muscle fatigue known as eyestrain. With normal aging, the lens loses its ability to accom-

modate for near objects; therefore, people frequently need reading glasses once they reach middle age.

Aging, or possibly exposure to the sun, also makes the lens subject to cataracts; the lens can become opaque, and therefore incapable of transmitting light rays. Currently, surgery is the only viable treatment for cataracts. First, a surgeon opens the eye near the rim of the cornea. Any one of several possible procedures are then used to remove the lens from its capsule. An artificial lens that can correct the patient's near and/or distant vision is inserted into the original lens capsule. Only minor vision corrections, such as one of those discussed in Section 27.9, are usually needed after cataract surgery today.

27.8 Check Your Progress The cornea, or even eyeglasses, cannot focus like a lens can. What can a lens do that eyeglasses cannot do?

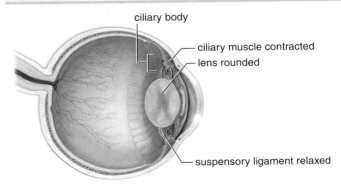

FIGURE 27.8A Focusing on a distant object.

FIGURE 27.8B Focusing on a near object.

27.9 The inability to form a clear image can be corrected

If you can see what is designated as size 20 letters from 20 feet away, you are said to have 20/20 vision. In **Figure 27.9**, ❶ people who can easily see a near object but have trouble seeing an optometrist's chart 20 feet away are said to be *nearsighted*, a condition called myopia. These individuals often have an elongated eyeball, and when they attempt to look at a distant object, the image is brought to focus in front of the retina. They can see close objects because the lens can compensate for the elongated eyeball. In order to see distant objects, nearsighted people can wear concave lenses, which diverge the light rays so that the image can be focused on the retina.

❷ People who can easily see the optometrist's chart 20 feet away but cannot easily see near objects are *farsighted*, a condition called hyperopia. They often have a shortened eyeball, and when they try to see near objects, the image is focused behind the retina. When the object is distant, the lens can compensate for the short eyeball. To see near objects, these individuals can wear a convex lens that increases the bending of light rays so that the image can be focused on the retina.

❸ When the cornea or lens is uneven, the image is fuzzy. This condition, called astigmatism, can be corrected by wearing an unevenly ground lens to compensate for the uneven cornea.

LASIK Surgery Rather than wearing glasses or contact lenses, many nearsighted people are now choosing to undergo LASIK eye surgery. LASIK stands for laser in-situ keratomileusis, which results in a reshaping of the cornea. Typically, adults affected by common vision problems (nearsightedness, farsightedness, or astigmatism) respond well to LASIK. An eye exam determines the present thickness and shape of the cornea and how much the cornea needs to be reshaped to achieve 20/20 vision. During the LASIK procedure, a small flap of conjunctiva is first lifted to expose the cornea. Then, the laser is used to remove tissue from the cornea. Each pulse of the laser removes a small amount of corneal tissue, allowing the surgeon to flatten or otherwise change the shape of the cornea. After the procedure, the flap of conjunctiva is put back in place and allowed to heal on its own. Most patients achieve vision that is close to 20/20, but the chances for improved vision are based, in part, on the condition of the eyes before surgery.

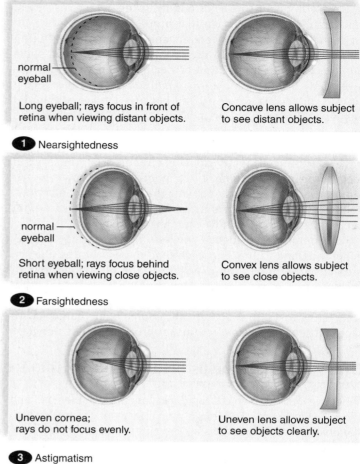

❶ Nearsightedness

Long eyeball; rays focus in front of retina when viewing distant objects.

Concave lens allows subject to see distant objects.

❷ Farsightedness

Short eyeball; rays focus behind retina when viewing close objects.

Convex lens allows subject to see close objects.

❸ Astigmatism

Uneven cornea; rays do not focus evenly.

Uneven lens allows subject to see objects clearly.

FIGURE 27.9 Common abnormalities of the eye with possible corrective lenses.

In the next section, let's see how the retina integrates stimuli before nerve impulses are sent to the brain.

> **27.9 Check Your Progress** In an octopus, the ciliary muscle moves the lens back and forth to bring about accommodation. What could cause accommodation to fail in an octopus?

27.10 The retina sends information to the visual cortex

So far, we have studied how the eye focuses an image on the retina. Now we will consider how vision is achieved. We will see that the visual system of humans does not merely record bits of light and dark like a camera. Instead, it constructs an image that helps us function in the environment.

As **Figure 27.10** shows, the retina has three layers of neurons: the rod cell and cone cell layer, the bipolar cell layer, and the ganglion cell layer. The rods and cones are in the layer closest to the

choroid. Examine the structure of a rod. The membrane disks of an outer segment contain numerous visual pigment molecules that absorb light. The plasma membrane contains ion channels. Synaptic vesicles are located at the synaptic endings of the inner segment.

The visual pigment in rods is a deep-purple pigment called rhodopsin. **Rhodopsin** is a complex molecule made up of the protein opsin and a light-absorbing molecule called *retinal*, which is a derivative of vitamin A. When a rod absorbs light,

To understand how we hear, first trace the path of sound waves through the outer and middle ear in Figure 27.11. Just as ripples travel across the surface of a pond, sound waves travel by the successive vibrations of molecules. Ordinarily, sound waves do not carry much energy, but when a large number of waves strike the tympanic membrane, it moves back and forth (vibrates) ever so slightly. The malleus takes the pressure from the tympanic membrane and passes it, by means of the incus, to the stapes that strikes the oval window. The stapes vibrates the membrane of the oval window with a force that has been multiplied about 20 times by the movement of the ossicles. This force allows sound waves to become fluid pressure waves in the inner ear. Eventually, the pressure waves disappear at the round window.

Figure 27.12 shows the mechanoreceptors for hearing at increasing levels of magnification. ❶ These receptors are located in the snail-shaped cochlea, a major part of the inner ear. ❷ A cross section of the cochlea reveals that they are located in the cochlear canal, one of three canals in the cochlea. When the stapes strikes the oval window, pressure waves move through the fluid of these canals. ❸ The **organ of Corti**, by which we hear, consists of little hair cells that occur along the length of the basilar membrane. The hair cells have extensions called stereocilia, which are embedded in the gelatinous tectorial membrane. ❹ When we hear, the fluid pressure waves in the inner ear cause the basilar membrane to vibrate and the stereocilia to bend because they are trapped in the tectorial membrane. The hair cells now generate nerve impulses that travel in the cochlear nerve to the brain stem. When these impulses reach the auditory areas of the cerebral cortex, they are interpreted as sound.

Sensory Coding   Each part of the organ of Corti is sensitive to different wave frequencies, or pitch. Think of the basilar membrane as a rope stretched between two posts. If you pluck the rope at one end, a wave of vibration travels down its length. Similarly, a sound causes a wave in the basilar membrane. If the wave reaches the tip of the organ of Corti, the brain interprets this as a low pitch, such as that of a tuba. If the wave remains near the base, the brain interprets this as a high pitch, such as a whistle. Thus, the pitch sensation we experience depends on which region of the basilar membrane vibrates and which area of the brain is stimulated. Researchers believe the brain interprets the tone of a sound based on the distribution of the hair cells stimulated.

Volume is a function of the size (amplitude) of sound waves. Loud noises cause the fluid within the vestibular canal to exert more pressure and the basilar membrane to vibrate to a greater extent. The brain interprets the resulting increased stimulation as volume.

All of us should protect our ears from loud noises, as explained in Section 27.13.

27.12 Check Your Progress Why are the sensory receptors for hearing classified as mechanoreceptors?

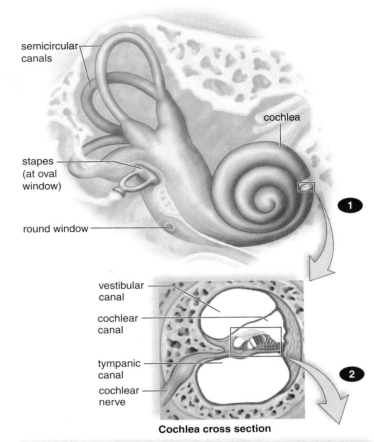

Cochlea cross section

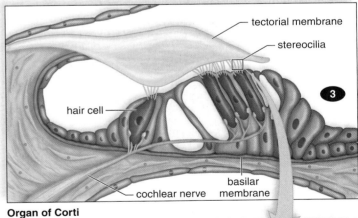

Organ of Corti

Stereocilia

2 µm

FIGURE 27.12 Mechanoreceptors for hearing.

Especially when we are children, the middle ear is subject to infections that can lead to hearing impairment if not treated promptly by a physician. With age, the mobility of the ossicles decreases, and in the condition called otosclerosis, new filamentous bone grows over the stirrup, impeding its movement and causing hearing loss. Surgical treatment is the only remedy for this type of deafness, which is called conduction deafness. Another type of hearing loss, called age-associated nerve deafness, results from stereocilia damage due to exposure to loud noises. This type of deafness is preventable (**Fig. 27.13A**), if care is taken.

In today's society, exposure to excessive noise is common. Noise is measured in decibels, and any noise above a level of 80 decibels could result in damage to the hair cells of the organ of Corti. Eventually, the stereocilia and then the hair cells disappear completely (**Fig. 27.13B**). Listening to city traffic for extended periods can damage hearing, and therefore it stands to reason that frequently attending rock concerts, constantly playing music loudly, or using earphones at high volume also damage hearing. The first hint of danger could be temporary hearing loss, a "full" feeling in the ears, muffled hearing, or tinnitus (e.g., ringing in the ears). If you have any of these symptoms, modify your listening habits immediately to prevent further damage. If exposure to noise is unavoidable, specially designed noise reduction earmuffs are available, and it is also possible to purchase earplugs made from

a compressible, spongelike material at the drugstore or a sporting-goods store. These earplugs are not the same as those worn for swimming, and they should not be used interchangeably.

Aside from loud music, noisy indoor or outdoor equipment, such as a rug-cleaning machine or a chain saw, is also damaging to hearing. Even motorcycles and recreational vehicles such as snowmobiles and motocross bikes can contribute to a gradual loss of hearing. Exposure to intense sounds of short duration, such as a burst of gunfire, can result in an immediate hearing loss. Hunters may experience a significant hearing reduction in the ear opposite the shoulder where they hold the rifle. The butt of the rifle offers some protection to the ear nearest the gun when it is shot.

Finally, people need to be aware that some medicines are ototoxic. Anticancer drugs, most notably cisplatin, and certain antibiotics (e.g., streptomycin, kanamycin, gentamicin) make the ears especially susceptible to hearing loss. Anyone taking such medications needs to be careful to protect his or her ears from any loud noises.

This completes our discussion of hearing; the next section goes on to consider our sense of balance.

> **27.13** *Check Your Progress* **It is best to realize that no device works as well as the normal ear. Why?**

FIGURE 27.13A
Normal hair cells in the organ of Corti.

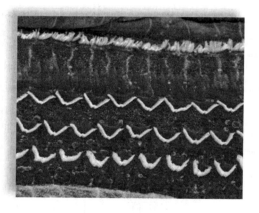

FIGURE 27.13B
Damaged hair cells in the organ of Corti.

27.14   The sense of balance occurs in the inner ear

The mechanoreceptors for hearing and for balance (equilibrium) are both located in the inner ear. The mechanoreceptors for balance detect rotational and/or angular movement of the head (**rotational balance**) and also straight-line movement of the head in any direction (**gravitational balance**).

Rotational balance involves the semicircular canals, which are arranged so that there is one in each dimension of space (**Fig. 27.14A**, *top*). The base of each of the three canals, called the ampulla, is slightly enlarged. Little hair cells, whose stereocilia are embedded within a gelatinous material called a cupula, are found within the ampullae. Because there are three semicircular canals, each ampulla responds to head movement in a different plane of space. As fluid (endolymph) within a semicir-

cular canal flows over and displaces a cupula, the stereocilia of the hair cells bend, and the pattern of impulses carried by the vestibular nerve to the brain changes (Fig. 27.14A, *bottom*). Continuous movement of fluid in the semicircular canals causes one form of motion sickness, discussed in Section 27.15.

Vertigo is dizziness and a sensation of rotation. It is possible to simulate a feeling of vertigo by spinning rapidly and stopping suddenly. When the eyes are rapidly jerked back to the midline position, the person feels like the room is spinning. This shows that the eyes are also involved in our sense of balance.

Gravitational balance depends on the **utricle** and **saccule**, two membranous sacs located in the vestibule (**Fig. 27.14B**, *top*). Both of these sacs contain little hair cells, whose stereocilia

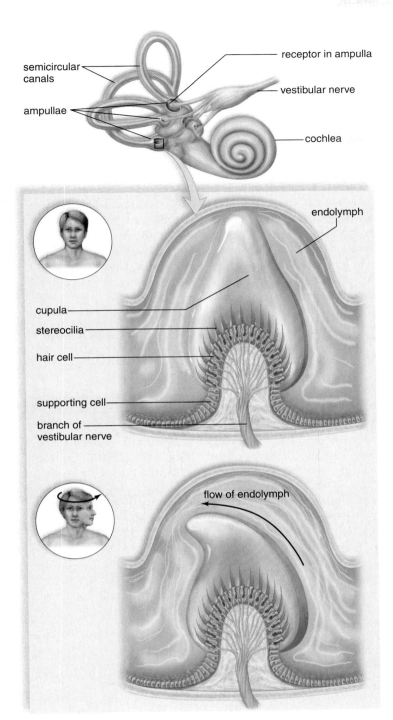

FIGURE 27.14A The receptors for rotational balance are in the ampullae of the semicircular canals.

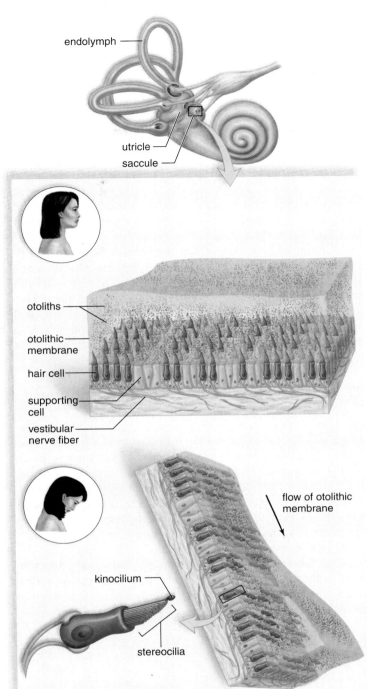

FIGURE 27.14B The receptors for gravitational balance are in the utricle and saccule of the vestibule.

are embedded within a gelatinous material called an otolithic membrane. Calcium carbonate ($CaCO_3$) granules, or *otoliths*, rest on this membrane. The utricle is especially sensitive to horizontal (back-forth) movements of the head, while the saccule responds best to vertical (up-down) movements.

When the head is still, the otoliths in the utricle and the saccule rest on the otolithic membrane above the hair cells (Fig. 27.14B, *bottom*). When the head moves in a straight line, the otoliths are displaced and the otolithic membrane sags, bending the stereocilia of the hair cells beneath. If the stereocilia move toward the largest stereocilium, called the kinocilium,

nerve impulses increase in the vestibular nerve. If the stereocilia move away from the kinocilium, nerve impulses decrease in the vestibular nerve. If you are upside down, nerve impulses in the vestibular nerve cease. These data tell the brain the direction of the movement of the head.

Section 27.15 explains the symptoms of motion sickness.

27.14 Check Your Progress The term gravitational balance can be correlated with the action of what structures in the utricle and saccule?

27.15 Motion sickness can be disturbing

Certain types of movement are likely to cause motion sickness. Seasickness may result from the rocking or swaying of a boat; if the water becomes rough enough, almost anyone can develop this type of motion sickness. Airsickness can occur due to abrupt changes in altitude during a plane trip or the jerky motions of turbulence. Motion sickness is also quite common for passengers in cars, buses, or trains. Motion sickness is the most common cause of vertigo; other symptoms include nausea, vomiting, and cold sweats. Fortunately, motion sickness subsides soon after the triggering motion ceases.

Motion sickness arises when the brain is bombarded with conflicting sensory input. For example, suppose you are trying to read a book while riding on a bus. As the vehicle goes up and down hills and around curves, starts and stops, and hits the occasional pothole, the mechanoreceptors of your inner ear send information about all these changes in position to your brain. So do the proprioceptors in your joints and muscles and the mecha-noreceptors in your skin. However, since your eyes are fixed on the pages of your book, the visual information that your brain receives says your position has not changed. The brain becomes overwhelmed, and motion sickness ensues.

One way to avoid motion sickness is to avoid reading while you are in motion; watch the scenery instead. (This is why many people who routinely become carsick as passengers experience no symptoms while driving.) If you are on a train, try to avoid facing backward. If you are on a plane, reserve a seat over the wings, where the plane is most stable. For the same reason, if you travel by ship, pick a cabin near the middle.

As discussed in Section 27.16, other animals have different types of receptors that respond to motion.

> **27.15** *Check Your Progress* **You would expect motion sickness to be mental. Explain.**

27.16 Other animals respond to motion

The **lateral line** system of fishes guides them in their movements and in locating other fish, including predators, prey, and mates. The system detects movements of nearby objects, much like the sensory receptors in the human inner ear detect motion. In bony fishes, the lateral line receptors are located within a canal that has openings to the outside (**Fig. 27.16A**). The receptor is a hair cell with cilia embedded in a gelatinous cupula. When the many cupulae bend due to pressure waves, the hair cells initiate nerve impulses.

Gravitational balance organs, called **statocysts**, are found in cnidarians, snails (molluscs), and lobsters and crabs (arthropods). These organs give information only about the position of the head; they are not involved in the sensation of movement.

When the head stops moving, a small particle called a statolith stimulates the cilia of the closest hair cells (**Fig. 27.16B**), and these cilia generate impulses that indicate the position of the head.

> **27.16** *Check Your Progress* **Whales do not have a lateral line system to help them locate moving objects. Explain.**

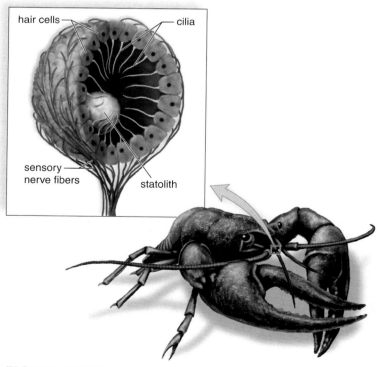

FIGURE 27.16A Lateral line system in fishes.

FIGURE 27.16B A statocyst of a lobster.

An animal's information exchange with the internal and external environments is dependent upon just a few types of sensory receptors. In this chapter, we have examined chemoreceptors, such as taste cells and olfactory cells; photoreceptors, such as eyes; and mechanoreceptors, such as the hair cells for hearing and balance.

The senses are not equally developed in all animals. For instance, male moths have chemoreceptors on the filaments of their antennae to detect minute amounts of an airborne sex attractant released by a female. This is certainly a more efficient method than searching for a mate by sight.

Birds that live in forested areas signal that a territory is occupied by singing, because it is difficult to see a bird in a tree, as most birders know. On the other hand, hawks have such a keen sense of sight that they are able to locate a small mouse in a field far below them. Insectivorous bats have an unusual adaptation for finding prey in the dark. They send out a series of sound pulses and listen for the echoes that come back. The time it takes for an echo to return indicates the location of an insect. A unique adaptation is found among the so-called electric fishes of Africa and Australia. They have electroreceptors that can detect disturbances in an electrical current they emit into the water. These disturbances indicate the location of obstacles and prey.

Through the evolutionary process, animals tend to rely on those stimuli and senses that are adaptive to their particular environment and way of life. In all cases, sensory receptors generate nerve impulses that travel to the brain, where sensation occurs. In mammals, and particularly human beings, integration of the data received from various sensory receptors results in perception of events occurring in the external environment.

The Chapter In Review

Summary

The Eyes Have It
- The compound eye of arthropods contains many visual units and is an excellent motion detector.
- The camera-type eye is found in all vertebrates and a few invertebrates.
- In stereoscopic vision, two eyes face forward, and the visual field overlaps.
- In panoramic vision, eyes face sideways, enabling animals to see predators more easily.

Sensory Receptors Respond to Stimuli
27.1 Sensory receptors can be divided into five categories
- Chemoreceptors, thermoreceptors, and mechanoreceptors are familiar types of sensory receptor.
- Pain receptors respond to, for example, excessive temperature or pressure and various chemicals.
- Electromagnetic receptors are stimulated by changes in electromagnetic waves (e.g., photoreceptors, UV radiation).

27.2 Sensory receptors communicate with the CNS
- Sensory receptors initiate nerve impulses that are transmitted to the spinal cord and/or brain.
- Sensation occurs when nerve impulses reach the cerebral cortex.
- Sensory adaptation is a type of integration by which response to a stimulus gradually decreases.

Chemoreceptors Are Sensitive to Chemicals
27.3 Chemoreceptors are widespread in the animal kingdom
- Chemoreceptors are responsible for taste and smell.
- Chemoreceptors are located in various places in various animals.

27.4 Mammalian taste receptors are located in the mouth
- Approximately 3,000 taste buds are present on the tongue.
- The four primary types of taste are sweet, sour, salty, and bitter; a fifth is unami.

27.5 Mammalian olfactory receptors are located in the nose
- Olfactory cells are modified neurons located high in the nasal cavity.
- Taste and smell depend on the combination of receptors stimulated.

Photoreceptors Are Sensitive to Light
27.6 The vertebrate eye is a camera-type eye
- The three layers of the eye are the sclera, the choroid, and the retina.
- The blind spot is the area containing no rods or cones, where the optic nerve exits the retina.

27.7 Protect your eyes from the sun
- The most frequent causes of blindness are retinal disorders, glaucoma, and cataracts.
- Macular degeneration and cataracts can be caused by long-term exposure to UV rays.
- Heavy cigarette smoking can also cause cataracts.

27.8 The lens helps bring an object into focus
- Light passes through the pupil and focuses on the retina.
- In visual accommodation, the lens rounds up to allow sight of near objects.
- With aging, the lens loses its ability to accommodate and becomes subject to cataracts.

27.9 The inability to form a clear image can be corrected
- For nearsighted people, a concave lens corrects trouble seeing far objects.
- For farsighted people, a convex lens corrects trouble seeing near objects.

- In astigmatism, the image is fuzzy and can be corrected by an unevenly ground lens.
- LASIK surgery can correct nearsightedness by reshaping the cornea with a laser.

27.10 The retina sends information to the visual cortex
- The first layer of the retina contains rods and cones:
  - Rod cells are the sensory receptors for dim light.
  - Cone cells are the sensory receptors for bright light and color.
- In color blindness, one type of cone is defective or deficient in number.
- The other two layers of the retina are composed of the bipolar cells and the ganglion cells.

Mechanoreceptors Are Involved in Hearing and Balance

27.11 The mammalian ear has three main regions
- The outer ear contains the pinna and auditory canal; the middle ear houses the tympanic membrane and ossicles; and the inner ear is the site of the semicircular canals, vestibule, and cochlea.
- The ear functions in both hearing and balance.
- Mechanoreceptors in the inner ear consist of hair cells with stereocilia.

27.12 Hair cells in the inner ear detect sound vibrations
- For hearing to occur, the tympanic membrane and ossicles amplify sound waves that strike the oval window membrane.
- In response to vibration, hair cells on the basilar membrane of the organ of Corti send impulses to the cerebral cortex, where they are interpreted as sound.

27.13 Protect your ears from loud noises
- Exposure to loud noise can damage hair cells and hearing.

27.14 The sense of balance occurs in the inner ear
- For rotational balance, mechanoreceptors in the semicircular canals detect rotational and/or angular movements of the head.
- For gravitational balance, mechanoreceptors in the utricle and saccule detect head movements in the vertical or horizontal plane.

27.15 Motion sickness can be disturbing
- Motion sickness occurs when the brain experiences conflicting sensory input.

27.16 Other animals respond to motion

- The lateral line system of fishes guides movement and locates other fish.
- Statocysts in certain arthropods indicate information about head position only.

statocyst

Testing Yourself

Sensory Receptors Respond to Stimuli

1. Chemoreceptors are involved in
   a. hearing.   d. vision.
   b. taste.   e. Both b and c are correct.
   c. smell.

2. Conscious perception of stimuli from the internal and external environment is called
   a. responsiveness.   c. sensation.
   b. interpretation.   d. accommodation.
3. **THINKING CONCEPTUALLY** Explain the expression "The sensory receptors are the window of the brain."

Chemoreceptors Are Sensitive to Chemicals

4. Tasting "sweet" versus "salty" is a result of
   a. activating different sensory receptors.
   b. activating many versus few sensory receptors.
   c. activating no sensory receptors.
   d. None of these are correct.

Photoreceptors Are Sensitive to Light

5. The thin, darkly pigmented layer that underlies most of the sclera is the
   a. conjunctiva.   c. retina.
   b. cornea.   d. choroid.
6. Which of these sequences is the correct path for light rays entering the human eye?
   a. sclera, retina, choroid, lens, cornea
   b. fovea centralis, pupil, aqueous humor, lens
   c. cornea, pupil, lens, vitreous humor, retina
   d. cornea, fovea centralis, lens, choroid, rods
   e. optic nerve, sclera, choroid, retina, humors
7. A blind spot occurs where the
   a. iris meets the pupil.
   b. retina meets the lens.
   c. optic nerve meets the retina.
   d. cornea meets the retina.
8. Which part of the eye is incorrectly matched with its function?
   a. pupil—admits light
   b. choroid—absorbs stray light rays
   c. fovea centralis—makes night vision possible
   d. optic nerve—transmits impulses to brain
   e. iris—regulates light entrance
9. During accommodation,
   a. the suspensory ligaments must be pulled tight.
   b. the lens needs to become more rounded.
   c. the ciliary muscle will be relaxed.
   d. All of these are correct.
10. Retinal is
    a. a derivative of vitamin A.
    b. sensitive to light energy.
    c. a part of rhodopsin.
    d. found in both rods and cones.
    e. All of these are correct.
11. A color-blind person has an abnormal type of
    a. rod.   d. cone.
    b. cochlea.   e. None of these are correct.
    c. cornea.
12. Which abnormality of the eye is incorrectly matched with its cause?
    a. astigmatism—either the lens or cornea is not even
    b. farsightedness—eyeball is shorter than usual
    c. nearsightedness—image focuses behind the retina
    d. color blindness—genetic disorder in which certain types of cones may be missing
13. **THINKING CONCEPTUALLY** What specific physiological defect can be corrected by both cataract surgery and LASIK surgery? Explain.

Mechanoreceptors Are Involved in Hearing and Balance

14. The middle ear communicates with the inner ear at the
    a. oval window.
    b. tympanic membrane.
    c. round window.
    d. Both a and c are correct.

15. The ossicle that articulates with the tympanic membrane is the
    a. malleus.   c. incus.
    b. stapes.   d. All of these are correct.

16. Which one of these wouldn't you mention if you were tracing the path of sound vibrations?
    a. auditory canal   d. cochlea
    b. tympanic membrane   e. ossicles
    c. semicircular canals

17. Which one of these correctly describes the location of the organ of Corti?
    a. between the tympanic membrane and the oval window in the inner ear
    b. in the utricle and saccule within the vestibule
    c. between the tectorial membrane and the basilar membrane in the cochlear canal
    d. between the outer and inner ear within the semicircular canals

18. Our perception of pitch is dependent upon the region of the _____ vibrated and the regions of the _____ stimulated.
    a. cochlea, spinal cord
    b. basilar membrane, spinal cord
    c. basilar membrane, auditory cortex
    d. cochlea, cerebellum

19. Loud noises generally lead to hearing loss due to damage to the
    a. outer ear.
    b. middle ear.
    c. inner ear.

20. Which part of the ear is incorrectly matched to its location?
    a. semicircular canals—inner ear
    b. utricle and saccule—outer ear
    c. auditory canal—outer ear
    d. cochlea—inner ear
    e. ossicles—middle ear

21. Which of these structures would assist you in knowing that you were upside down, even if you were in total darkness?
    a. utricle and saccule   c. semicircular canals
    b. cochlea   d. tectorial membrane

22. The lateral line system in fishes allows them to
    a. detect sound.
    b. locate other fish.
    c. maintain gravitational equilibrium.
    d. maintain rotational equilibrium.

Understanding the Terms

blind spot 540	outer ear 544
chemoreceptor 536	pain receptor 536
choroid 540	pheromone 538
ciliary body 540	pupil 540
ciliary muscle 541	retina 540
cochlea 544	rhodopsin 542
conjunctiva 540	rotational balance 546
cornea 540	saccule 546
electromagnetic receptor 536	sclera 540
fovea centralis 540	semicircular canal 544
gravitational balance 546	sensation 537
inner ear 544	sensory adaptation 537
integration 537	sensory receptor 536
iris 540	statocyst 548
lateral line 548	taste bud 538
lens 540	thermoreceptor 536
mechanoreceptor 536	tympanic membrane 544
middle ear 544	utricle 546
olfactory cell 539	vestibule 544
organ of Corti 545	visual accommodation 541
ossicle 544	

Match the terms to these definitions:
a. _____ Inner layer of the eyeball containing the photoreceptors—rod cells and cone cells.
b. _____ Outer, white, fibrous layer of the eye that surrounds the eye except for the transparent cornea.
c. _____ Receptor that is sensitive to chemical stimulation—for example, receptors for taste and smell.
d. _____ Structure in the vertebrate inner ear that contains auditory receptors.
e. _____ Canal system containing sensory receptors that allow fishes and amphibians to detect water currents and pressure waves from nearby objects.

Thinking Scientifically

1. Suggest a hypothesis that would explain why some people and not others have perfect pitch. How would you test your hypothesis?
2. How does LASIK surgery support the hypothesis that the cornea, and not the lens, provides most of the focusing when we see clearly.

ARIS *Visit* **www.mhhe.com/maderconcepts** *for practice quizzes, animations, videos, and activities designed to help you master the material in this chapter.*

28

Locomotion and Support Systems

LEARNING OUTCOMES

After studying this chapter, you should be able to accomplish the following outcomes.

Skeletal Remains Reveal All

1　Explain how skeletal remains can be used to determine the age, gender, and ethnicity of a deceased person.

Animal Skeletons Support, Move, and Protect the Body

2　Compare and contrast the three types of skeletons found in animals.
3　List and discuss the functions of the mammalian skeleton.

The Mammalian Skeleton Is a Series of Bones Connected at Joints

4　Describe the axial and the appendicular skeletons.
5　Explain how good nutrition when young can help avoid osteoporosis later on.
6　Describe the anatomy of a typical long bone.
7　Tell where you would find each type of joint, and describe the structure of a synovial joint.
8　Describe common joint disorders and how they can be repaired.

Animal Movement Is Dependent on Muscle Cell Contraction

9　List and discuss five functions of skeletal muscles.
10　Explain how skeletal muscles function in antagonistic pairs.
11　Explain degrees of skeletal muscle contraction in terms of motor units.
12　Describe the many benefits of exercise.
13　Describe the events that occur as a muscle contracts.
14　Compare and contrast the three sources of ATP for muscle contraction.
15　Compare and contrast fast- and slow-twitch muscle fibers.

D r. Sandra Bullock, a forensics expert, and her assistant Tom were standing in the tall grass of the empty lot at the corners of Marion and Washington Streets. The bones they were examining had obviously been bleached by the sun and scattered by passing dogs. "Let's get as many bones as we can find into the lab and try to identify them as one of the missing persons reported within the past year," said Sandra.

It is better for forensics if many bones, in good condition, are found. But even bones that are in poor condition, because of a fire or other catastrophe, can offer clues about the identity and history of a deceased person. Age can be approximated by examining the teeth, including whether any are missing. Infants aged 0–4 months have no teeth, of course; children about 6–10 years of age usually have missing "baby teeth"; and young adults acquire their last "wisdom teeth" around age 20. Older adults may have a number of missing or broken teeth.

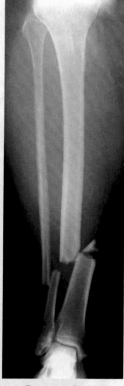

Fracture of both the tibia and fibula bones

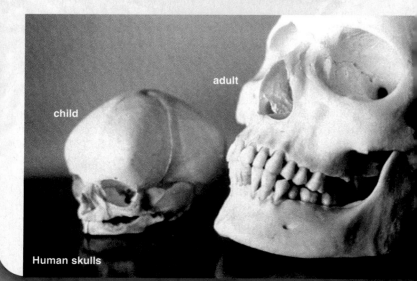

Human skulls

Skeletal Remains Reveal All

The condition of the long bones and the joints between the bones can also assist in telling how old a person was at the time of death. A thin, cartilaginous growth plate at the ends of the long bones is present during childhood. The growth plates begin to be replaced by bone during the teenage years. A smooth hipbone joint, rather than a rough one, indicates the individual is an adult of some years. Other joints also deteriorate with age. Over time, the hyaline cartilage capping the long bones becomes worn, yellowed, and brittle, and the amount of cartilage lessens. Also, as we age, the pads between the vertebrae are more apt to show damage.

If the skeletal remains include the individual's pelvic bones, these provide the best method for determining an adult's gender. To accommodate a fetus during pregnancy, the pelvis of a female is shallower and wider than that of a male, and a wider outlet allows a baby's head to pass when birth occurs. The long bones of the limbs give information about gender as well. In males, the long bones are thicker and more dense, and the points of attachment for the muscles are bigger and more noticeable. The skull of a male tends to have a square chin, and the eyebrow ridges above the eye sockets are more prominent. Also, males have a larger mastoid process (the lump behind the ear).

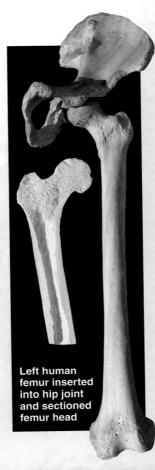

Left human femur inserted into hip joint and sectioned femur head

Determining the ethnic origin of skeletal remains can be difficult because many people today have a mixed racial heritage. But again, the bones, especially the skull, offer clues. Individuals of African descent have a greater distance between the eyes, the eye sockets are roughly rectangular, and the jaw is large and prominent. In Native Americans, the eye sockets are round, the cheek bones are prominent, and the palate is rounded. In Caucasians, the palate is U-shaped, and a suture line is apt to be visible.

After gathering all the data, Dr. Bullock hopes to use missing persons reports to assign a specific name to the bones, which have revealed all. In this chapter, we will survey the types of skeletons in the animal kingdom before concentrating on the bones and muscles of humans and how they move. In addition, the bones and muscles have many functions that contribute to homeostasis.

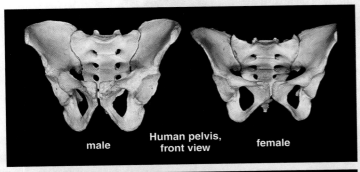

male **Human pelvis, front view** **female**

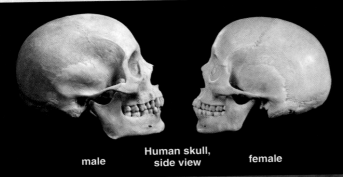

male **Human skull, side view** **female**

Forensic expert clears soil in mass grave in eastern Bosnia.

Animal Skeletons Support, Move, and Protect the Body

Three types of skeletons—hydrostatic skeletons, exoskeletons, and endoskeletons—are considered before we list the functions of the mammalian endoskeleton.

28.1   Animal skeletons can be hydrostatic, external, or internal

Skeletons serve as support systems for animals, providing rigidity, protection, and surfaces for muscle attachment. Three types of skeletons are found in animals: hydrostatic skeletons, exoskeletons, and endoskeletons.

Hydrostatic Skeleton   In animals that lack a hard skeleton, a fluid-filled gastrovascular cavity or a fluid-filled coelom can act as a hydrostatic skeleton. A **hydrostatic skeleton** offers support and resistance to the contraction of muscles so that mobility results. As an analogy, consider that a garden hose stiffens when filled with water, and that a water-filled balloon changes shape when squeezed at one end. Similarly, an animal with a hydrostatic skeleton can change shape and perform a variety of movements.

Hydras and flatworms (planarians) use their fluid-filled gastrovascular cavity as a hydrostatic skeleton. When muscle fibers at the base of epidermal cells in a hydra contract, the body or tentacles shorten rapidly. Planarians usually glide over a substrate with the help of muscular contractions that control the body wall and many cilia. Roundworms have a fluid-filled pseudocoelom and move in a whiplike manner when their longitudinal muscles contract. Earthworms are segmented and have septa that divide the coelom into compartments (**Fig. 28.1A**). Each segment has its own set of longitudinal and circular muscles and its own nerve supply, so each segment or group of segments may function independently. When circular muscles contract, the segments become thinner and elongate, just as a balloon would if you squeezed it. When longitudinal muscles contract, the segments become thicker and shorten, just as a balloon would if you pressed on it from both sides. By alternating circular muscle contraction and longitudinal muscle contraction and by using its setae to hold its position during contractions, the earthworm moves forward.

Even animals that have an exoskeleton or an endoskeleton move selected body parts by means of muscular hydrostats, meaning that fluid contained within certain muscle fibers assists movement of that part. Muscular hydrostats are used by clams to extend their muscular foot and by sea stars to extend their tube feet. Spiders depend on them to move their legs, and moths depend on them to extend their long tubular feeding apparatus. In vertebrates, movement of an elephant's trunk involves a muscular hydrostat that allows the trunk to reach as high as 23 feet, pick up a morsel of food, or pull down a tree.

Exoskeleton   Molluscs, arthropods, and vertebrates have rigid skeletons. The **exoskeleton** (external skeleton) of molluscs and arthropods protects and supports these animals and provides a location for muscle attachment. The strength of an exoskeleton can be improved by increasing its thickness and weight, but this might leave less room for internal organs. Molluscs, such as snails and clams, use a thick, nonmobile calcium carbonate shell primarily for protection against the environment and predators. A mollusc's shell can grow as the animal grows.

The exoskeleton of arthropods, such as insects and crustaceans, is composed of chitin, a strong, flexible nitrogenous polysaccharide. This exoskeleton protects against wear and tear, predators, and drying out—an important feature for arthropods that live on land. The exoskeleton of arthropods is particularly suitable for terrestrial life in another way. The jointed and movable appendages allow flexible movements. To grow, however, arthropods must molt to rid themselves of an exoskeleton that has become too small. While molting, arthropods are vulnerable to predators.

old exoskeleton

circular longitudinal septa fluid setae
muscles muscles

FIGURE 28.1A   The well-developed circular and longitudinal muscles of an earthworm push against a segmented, fluid-filled coelom.

circular muscles contracted   longitudinal muscles contracted

circular muscles contract, and anterior end moves forward

longitudinal muscles contract, and segments catch up

circular muscles contract, and anterior end moves forward

Endoskeleton Both echinoderms and vertebrates have an **endoskeleton** (internal skeleton). For example, the skeleton of a starfish (**Fig. 28.1B**) consists of plates of calcium carbonate embedded in the living tissue of the body wall. In contrast, the vertebrate endoskeleton is living tissue. Sharks and rays have skeletons composed only of cartilage. Other vertebrates, such as bony fishes, amphibians, reptiles, birds, and mammals, have endoskeletons composed of bone and cartilage. An endoskeleton grows with the animal, and molting is not required. It supports the weight of a large animal without limiting the space for internal organs. An endoskeleton also offers protection to vital internal organs, but is protected by the soft tissues around it. Injuries to soft tissue are easier to repair than injuries to a hard skeleton. The vertebrate endoskeleton is also jointed, allowing for complex movements such as swimming, jumping, flying, and running.

In the next section, we will examine the many functions of the mammalian endoskeleton.

FIGURE 28.1B
A starfish has an endoskeleton.

> **28.1** *Check Your Progress* Which of the three types of skeletons are likely to be in the fossil record and provide data for tracing the evolution of animals, including humans? Explain.

28.2 Mammals have an endoskeleton that serves many functions

The endoskeleton of a human makes up about 20% of the body weight. It is primarily composed of bone, but other types of connective tissue are present as well. Cartilage is found in various locations, such as between the vertebrae and in the nose, the outer ear, and the rib cage. Ligaments and tendons are composed of dense connective tissue. Ligaments join bone to bone, and tendons join muscle to bone at joints, the junctions between the bones. The skeleton performs many functions:

Bones protect the internal organs. The rib cage protects the heart and lungs; the skull protects the brain; and the vertebrae protect the spinal cord. The endocrine organs, such as the pituitary gland, pineal gland, thymus, and thyroid gland, are also protected by bone. The eyes and ears are protected by bone. The bony eye sockets ordinarily prevent blows to the head from reaching the eyes.

Bones provide a frame for the body. Our shape is dependent on the bones, which also support the body. The long bones of the legs and the bones of the pelvic girdle support the entire body when we are standing.

Bones assist all phases of respiration. The rib cage assists the breathing process, for example. Prior to inhalation, the rib cage lifts up and out (see Fig. 32.7A), and the diaphragm moves down, expanding the chest. Now air automatically flows into the lungs, enabling oxygen to enter the blood, where it is transported to the tissues. As we shall see in Section 29.12, some of the bones contain red bone marrow, which produces the red blood cells that transport oxygen. Without a supply of oxygen, the mitochondria could not efficiently produce ATP during cellular respiration. ATP, as you know, is needed for muscle contraction and nerve conduction, as well as for the many synthesis reactions that occur in cells.

FIGURE 28.2 Graceful movements are possible because muscles act on bones.

Bones store and release calcium. Some of the minerals found in blood are stored in bone. One of the these, calcium (Ca^{2+}), is important for its many functions. Calcium ions play a major role in muscle contraction and nerve conduction. Calcium ions also help regulate cellular metabolism and blood clotting. The storage of Ca^{2+} in the bones is under hormonal control. When you have plenty of Ca^{2+} in your blood, it is stored in bone, and when the level starts to fall, Ca^{2+} is removed from bone so that the blood level is always near normal.

Bones assist the lymphatic system and immunity. Red bone marrow produces not only the red blood cells but also the white blood cells. The white cells, which congregate in the lymphatic organs, such as the lymph nodes shown in Figure 25.9A, are involved in defending the body against pathogens and cancerous cells. Without the ability to withstand foreign invasion, the body may quickly succumb to disease and die.

Bones assist digestion. The jaws contain sockets for the teeth, which chew food, and a place of attachment for the muscles that move the jaws. Chewing breaks food into pieces small enough to be swallowed and chemically digested. Without digestion, nutrients would not enter the body to serve as building blocks for repair and a source of energy for the production of ATP.

The skeleton is necessary to locomotion. Locomotion is efficient in human beings because they have a jointed skeleton for the attachment of muscles that move the bones (**Fig. 28.2**). Our jointed skeleton allows us to seek out and move to a more suitable external environment in order to maintain the internal environment within reasonable limits.

The next section begins our study of the human skeleton.

> **28.2** *Check Your Progress* Which of the functions of bone still occur after death?

In this part of the chapter, we take a look at the bones of the axial and appendicular skeletons and the joints that occur between bones. Osteoporosis and joint disorders are also discussed.

28.3   The bones of the axial skeleton lie in the midline of the body

The **axial skeleton** consists of the bones in the midline of the body, and the **appendicular skeleton** consists of the limb bones and their girdles. The blue labels in **Figure 28.3A** point out the bones of the axial skeleton.

The Skull   The cranium and the facial bones form the **skull**, which protects the brain (**Fig. 28.3B**). In newborns, certain bones of the cranium are joined by membranous regions called **fontanels** (or "soft spots"), all of which usually close and become sutures by the age of two years. The bones of the cranium contain the sinuses, air spaces lined by mucous membrane that reduce the weight of the skull and give a resonant sound to the voice. Two sinuses, called the mastoid sinuses, drain into the middle ear. Mastoiditis, a condition that can lead to deafness, is an inflammation of these sinuses.

The major bones of the cranium have the same names as the lobes of the brain (see Fig. 26.12). On the top of the cranium, the frontal bone forms the forehead, and the parietal bones make up the sides of the skull. Below the much larger parietal bones, each temporal bone has an opening that leads to the middle ear. In the rear of the skull, the occipital bone curves to form the base of the skull. At the base of the skull, the spinal cord passes upward through a large opening called the **foramen magnum** and becomes the brain stem.

Certain cranial bones contribute to forming the face. The sphenoid bones account for the flattened areas on each side of the forehead, which we call the temples. The frontal bone not only forms the forehead, but it also has supraorbital ridges where the eyebrows are located. Glasses sit where the frontal bone joins the nasal bones.

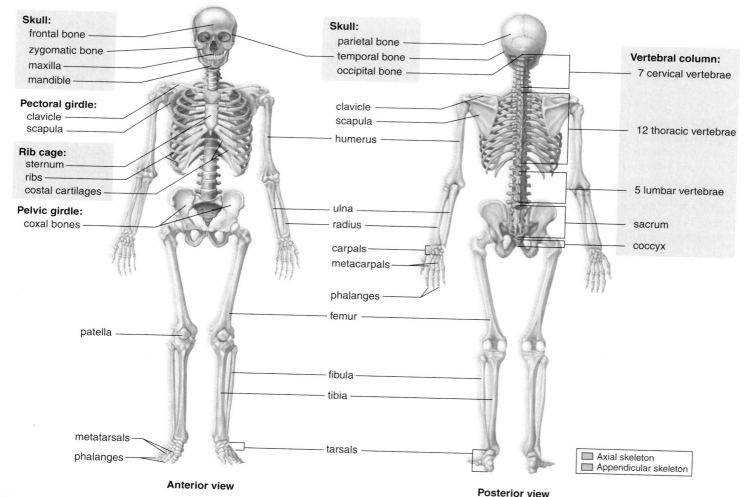

Skull:
  frontal bone
  zygomatic bone
  maxilla
  mandible

Pectoral girdle:
  clavicle
  scapula

Rib cage:
  sternum
  ribs
  costal cartilages

Pelvic girdle:
  coxal bones

patella

metatarsals
phalanges

Anterior view

Skull:
  parietal bone
  temporal bone
  occipital bone

clavicle
scapula
humerus

ulna
radius

carpals
metacarpals

phalanges

femur

fibula
tibia

tarsals

Vertebral column:
  7 cervical vertebrae

  12 thoracic vertebrae

  5 lumbar vertebrae

  sacrum
  coccyx

☐ Axial skeleton
☐ Appendicular skeleton

Posterior view

FIGURE 28.3A The human skeleton.

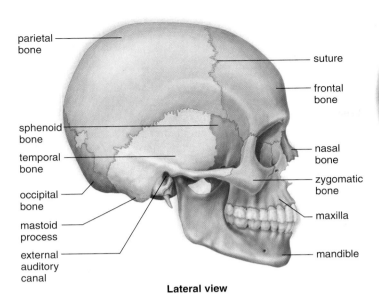

parietal bone
suture
frontal bone
sphenoid bone
nasal bone
temporal bone
zygomatic bone
occipital bone
maxilla
mastoid process
mandible
external auditory canal

Lateral view

frontal bone
parietal bone
temporal bone
nasal bone
zygomatic bone
maxilla
mandible

Frontal view

FIGURE 28.3B
Bones of the skull.

The most prominent of the facial bones are the mandible, the maxillae, the zygomatic bones, and the nasal bones. The mandible, or lower jaw, is the only freely movable portion of the skull, and its action permits us to chew our food. It also forms the "chin." Tooth sockets are located on the mandible and on the maxillae, which form the upper jaw and a portion of the hard palate. The zygomatic bones are the cheekbone prominences, and the nasal bones form the bridge of the nose. Other bones make up the nasal septum, which divides the nose cavity into two regions.

Whereas the outer ears are formed only by cartilage and not by bone, the nose is a mixture of bones, cartilage, and connective tissues. The lips and cheeks have a core of skeletal muscle.

The Vertebral Column
The head and trunk are supported by the **vertebral column**, which also protects the spinal cord and the roots of the spinal nerves. It is a longitudinal axis that serves either directly or indirectly as an anchor for all the other bones of the skeleton.

Twenty-four vertebrae make up the vertebral column (see Fig. 28.3A). The seven cervical vertebrae are located in the neck, and the twelve thoracic vertebrae are in the thorax. The five lumbar vertebrae are found in the small of the back. The five sacral vertebrae are fused to form a single sacrum. The coccyx, or tailbone, is composed of several fused vertebrae. Normally, the vertebral column has four curvatures that absorb shock and also provide more resilience and strength for an upright posture than could a straight column. *Scoliosis* is an abnormal lateral (sideways) curvature of the spine. Another well-known abnormal curvature results in a hunchback (kyphosis), and still another results in a swayback (lordosis), seen frequently in pregnant women.

Intervertebral disks, composed of fibrocartilage between the vertebrae, act as padding. They prevent the vertebrae from grinding against one another and absorb shock caused by movements such as running, jumping, and even walking. The presence of the disks allows the vertebrae to move as we bend forward, backward, and from side to side. Unfortunately, these disks become weakened with age and can herniate and rupture. Pain results if a disk presses against the spinal cord and/or spinal nerves. The body may heal itself, or the disk can be removed surgically. If removed, the vertebrae can be fused together, but this limits the flexibility of the body.

The Rib Cage
The thoracic vertebrae are a part of the rib cage. The rib cage also contains the ribs, the costal cartilages, and the sternum, or breastbone (**Fig. 28.3C**).

There are twelve pairs of ribs. The upper seven pairs are "true ribs" because they attach directly to the sternum. The lower five pairs do not connect directly to the sternum and are called the "false ribs." Three pairs of false ribs attach by means of a common cartilage, and two pairs are "floating ribs" because they do not attach to the sternum at all (Fig. 28.3C).

The rib cage demonstrates how the skeleton is protective but also flexible. The rib cage protects the heart and lungs; yet it swings outward and upward upon inspiration and then downward and inward upon expiration.

This completes our study of the axial skeleton; the next section considers the appendicular skeleton.

> **28.3 Check Your Progress** What could a forensics expert tell from the frontal bone, the mandible, the maxillae, the zygomatic bones, and the intervertebral disks?

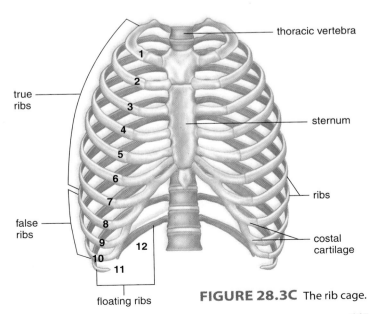

thoracic vertebra
true ribs
sternum
false ribs
ribs
costal cartilage
floating ribs

FIGURE 28.3C The rib cage.

28.4 The appendicular skeleton consists of bones in the girdles and limbs

The appendicular skeleton consists of the bones within the pectoral and pelvic girdles and the attached limbs (see Fig. 28.3A). The pectoral (shoulder) girdle and upper limbs are specialized for flexibility, but the pelvic girdle (hipbones) and lower limbs are specialized for strength. A total of 126 bones make up the appendicular skeleton.

The Pectoral Girdle and Upper Limbs The components of the **pectoral girdle** are only loosely linked together by ligaments (**Fig. 28.4A**). Each clavicle (collarbone) connects with the sternum in front and the scapula (shoulder blade) behind, but the scapula is largely held in place only by muscles. This allows it to freely follow the movements of the arm. The single long bone in the upper arm, the humerus, has a smoothly rounded head that fits into a socket of the scapula. The socket, however, is very shallow and much smaller than the head. Although this means that the arm can move in almost any direction, there is little stability. Therefore, this is the joint that is most apt to dislocate. The opposite end of the humerus meets the two bones of the forearm, the ulna and the radius, at the elbow. (The prominent bone in the elbow is the proximal part of the ulna.) When the upper limb is held so that the palm is turned frontward, the radius and ulna are about parallel to one

another. When the upper limb is turned so that the palm is next to the body, the radius crosses in front of the ulna, a feature that contributes to the easy twisting motion of the forearm.

The many bones of the hand increase its flexibility. The wrist has eight carpal bones, which look like small pebbles. From these, five metacarpal bones fan out to form a framework for the palm. The metacarpal bone that leads to the thumb is placed in such a way that the thumb can reach out and touch the other digits. (Digits is a term that refers to either fingers or toes.) Beyond the metacarpals are the phalanges, the bones of the fingers and the thumb. The phalanges of the hand are long, slender, and lightweight.

The Pelvic Girdle and Lower Limbs Two heavy, large coxal bones (hipbones) are joined at the pubic symphysis to form the **pelvic girdle** (**Fig. 28.4B**). The coxal bones are anchored to the sacrum, and together these bones form a hollow cavity called the pelvic cavity. The wider pelvic cavity in females accommodates childbearing. The weight of the body is transmitted through the pelvis to the lower limbs and then onto the ground. The largest bone in the body is the femur, or thighbone.

Distal to the thigh, the larger of the two bones, the tibia, has a ridge we call the shin. Both of the bones of the leg have a

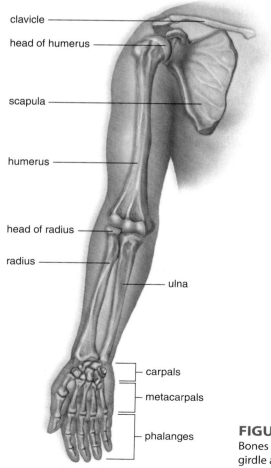

FIGURE 28.4A
Bones of the pectoral girdle and upper limb.

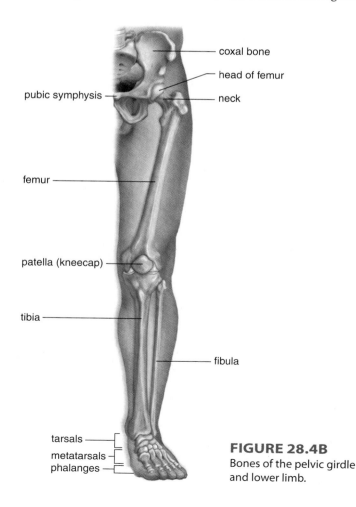

FIGURE 28.4B
Bones of the pelvic girdle and lower limb.

prominence that contributes to the ankle—the tibia on the inside of the ankle and the fibula on the outside of the ankle. Although there are seven tarsal bones in the ankle and heel, only one tarsal bone receives the body's weight and passes it on to the heel and the ball of the foot. If you wear high-heeled shoes, the weight is thrown toward the front of your foot.

The metatarsal bones participate in forming the arches of the foot. There is a longitudinal arch from the heel to the toes and a transverse arch across the foot. These provide a stable, springy base for the body. If the tissues that bind the metatarsals together become weakened, "flat feet" are apt to result. The bones of the toes are called phalanges, just as are those of the fingers, but in the foot the phalanges are stout and extremely sturdy.

Osteoporosis (brittle bones) is the topic of Section 28.5.

28.4 *Check Your Progress* **What could a forensics expert tell from the pelvic girdle?**

HOW BIOLOGY IMPACTS OUR LIVES

28.5 Avoidance of osteoporosis requires good nutrition and exercise

Throughout life, bones are continuously remodeled. While a child is growing, the rate of bone formation by bone cells called **osteoblasts** is greater than the rate of bone breakdown by bone cells called **osteoclasts**. The skeletal mass continues to increase until ages 20 to 30. After that, the rates of formation and breakdown of bone mass are equal until ages 40 to 50. Then, reabsorption begins to exceed formation, and the total bone mass slowly decreases. **Osteoporosis** is a condition in which the bones are weakened due to a decrease in the bone mass that makes up the skeleton. **Figure 28.5** shows the difference between normal bone and osteoporotic bone.

Osteoporosis is essentially a disease of aging. Over time, men are apt to lose 25% and women 35% of their bone mass. Sex hormones play an important role in maintaining bone strength. The level of testosterone (male sex hormone) in men begins to decrease steadily after age 18 so that it is significantly lower by age 65. The estrogen (female sex hormone) level in women begins to decline at about age 45. Fractures in women due to osteoporosis are more likely because men, in general, have more bone mass than women. The fractures in women usually involve the hip, vertebrae, long bones, and pelvis.

Everyone can take measures to avoid developing osteoporosis when they get older. Osteoblasts need calcium (Ca^{2+}) in order to form bone; therefore, adequate dietary calcium throughout life is an important protection against osteoporosis. The U.S. National Institutes of Health recommend a Ca^{2+} intake of 1,200–1,500 mg per day during puberty. Males and females require 1,000 mg per day until age 65 and 1,500 mg per day after age 65. A small daily amount of vitamin D is also necessary to absorb Ca^{2+} from the digestive tract. Exposure to sunlight is required to allow the skin to synthesize vitamin D.

Postmenopausal women should have an evaluation of their bone density. Presently, bone density is measured by a method called dual energy X-ray absorptiometry (DEXA). This test measures bone density based on the absorption of photons generated by an X-ray tube. If the bones are thin, it is worthwhile to try to improve bone density because even a slight increase can significantly reduce the risk of fractures. Exercise can build or maintain bone mass, but it must be weight-bearing exercise, such as dancing, walking, running, jogging, and tennis—activities that require you to be on your feet (Fig. 28.5). Lifting weights can also be beneficial. See Section 28.11 for a discussion of exercise.

normal bone

osteoporosis

FIGURE 28.5 Exercise can help prevent osteoporosis.

Medications for osteoporosis can slow or reverse the patient's bone loss. Bisphosphonates (i.e., Fosamax, Actonel, Boniva) are medications that inhibit the action of osteoclasts in bones. Calcitonin and parathyroid hormone are the body's two naturally occurring hormones for calcium homeostasis. Calcitonin, which causes calcium to be deposited in the bones, can be administered as a nasal spray or by injection. Like the bisphosphonates, it inhibits osteoclasts and slows the rate of bone thinning. The breast cancer drugs tamoxifen and raloxifene also stimulate the growth of new bone tissue.

Sex hormones are also used occasionally to treat osteoporosis. However, such therapy must be carefully monitored, because sex hormones may trigger the growth of certain reproductive tissue cancers.

In Section 28.6, we take a look at the anatomy of a long bone to illustrate principles of bone anatomy.

28.5 *Check Your Progress* **What is the rationale for taking calcium to prevent osteoporosis?**

28.6 Bones are composed of living tissues

As in **Figure 28.6**, ❶ when a long bone such as the humerus is split open, the longitudinal section shows that it is not solid but has a cavity, called the medullary cavity, bounded at the sides by compact bone and at the ends by spongy bone. The cavity of a long bone usually contains yellow bone marrow, which stores fat. Beyond the spongy bone is a thin shell of compact bone and finally a layer of ❷ hyaline cartilage, called **articular cartilage** when it occurs at articulations (joints). Articular cartilage is the "teflon coating" for the bones; it normally allows easy, frictionless movement between the bones of a joint.

Except for the articular cartilage on its ends, a long bone is completely covered by a layer of fibrous connective tissue called the periosteum. This covering contains blood vessels, lymphatic vessels, and nerves. Note in Figure 28.6 how a blood vessel penetrates the periosteum and gives off branches.

Compact bone makes up the shaft of a long bone. ❸ It contains many osteons (also called Haversian systems), where **osteocytes** derived from osteoblasts lie in tiny chambers called **lacunae**. The lacunae are arranged in concentric circles around central canals that contain branches of blood vessels and nerves. The lacunae are separated by a matrix of collagen fibers and mineral deposits, primarily calcium and phosphorous salts, as also discussed in Section 25.3.

❹ **Spongy bone** has numerous bony bars and plates separated by irregular spaces. Although lighter than compact bone, spongy bone is still designed for strength. Just as braces are used for support in buildings, the solid portions of spongy bone follow lines of stress. At the ends of long bones, the spaces in spongy bone are often filled with **red bone marrow**, a specialized tissue that produces blood cells. This is an additional way the skeletal system assists homeostasis. As you know, red blood cells transport oxygen, and white blood cells are a part of the immune system, which fights infection.

Also note the growth plate present near the end of the long bone in Figure 28.6. As long as a bone has a growth plate, it is capable of growing, because organized growth of bone in this region contributes to the length of the bone. The growth plate usually disappears when a person reaches maturity.

The next section explains the structure and function of joints, which occur where two bones meet.

> **28.6 *Check Your Progress*** Formulas are available to calculate a person's height based on the length of the bones. Which bones would a forensics expert use for this calculation?

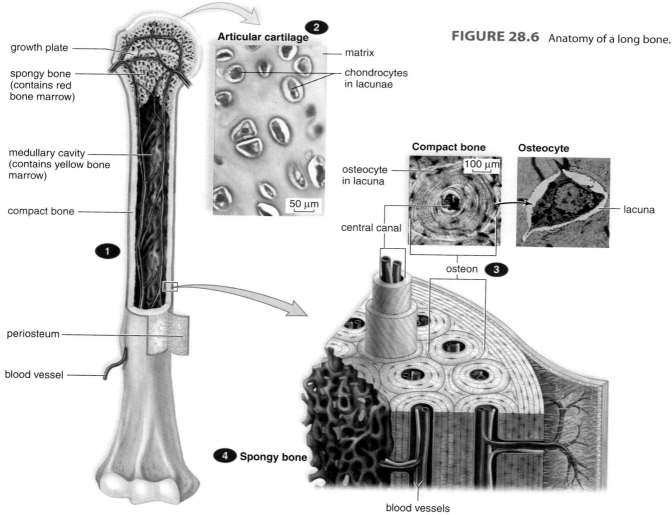

FIGURE 28.6 Anatomy of a long bone.

Bones articulate at the joints, which are classified as fibrous, cartilaginous, or synovial. Each of the three types of joints has a different appearance.

Fibrous joints, such as the sutures between the cranial bones, are immovable. Newborns have membranous regions, called "soft spots," where the cranial bones will come together and become fibrous joints.

Cartilaginous joints, which are connected by cartilage, tend to be slightly movable. The pubic symphysis, mentioned earlier, consists of fibrocartilage. The intervertebral disks are composed of fibrocartilage, and the ribs are joined to the rib cage by costal cartilages composed of hyaline cartilage.

Figure 28.7 illustrates examples of a freely movable **synovial joint**— ❶ a joint having a cavity lined with synovial membrane, which produces *synovial fluid*. Synovial fluid, which has an egg-white consistency, lubricates the joint. If the joint is stretched suddenly, the fluid does not immediately fill the joint and, in the meantime, the synovial membrane falls into the vacuum and a click is heard.

The absence of tissue between the articulating bones of a synovial joint allows them to be freely movable, but the joint has to be stabilized in some way. A synovial joint is stabilized by the joint capsule, a sleevelike extension of the periosteum of each articulating bone. Ligaments are fibrous bands that bind the two bones to one another and add even more stability. Tendons are fibrous tissue that connect muscle to bone and also help stabilize the joint.

The articulating surfaces of the bones are protected in several ways. First, the bones are covered by a layer of articular (hyaline) cartilage, described previously. Then, the **bursae,** which are fluid-filled sacs, ease friction between bone and overlapping muscles, or between skin and tendons. Inflammation of a bursa is called *bursitis.* **Menisci** are crescent-shaped pieces of cartilage in synovial joints that also ease friction between all parts of the joint. Injuries that involve the tearing of menisci are often called *torn cartilage.* Repair of these injuries is discussed in Section 28.8.

Two specific types of synovial joints are illustrated in Figure 28.7. ❷ **Ball-and-socket joints,** found at the hips and shoulders, allow movement in all planes, even rotational movement. Adduction occurs when limbs are moved toward the midline of the body. Abduction occurs when limbs are

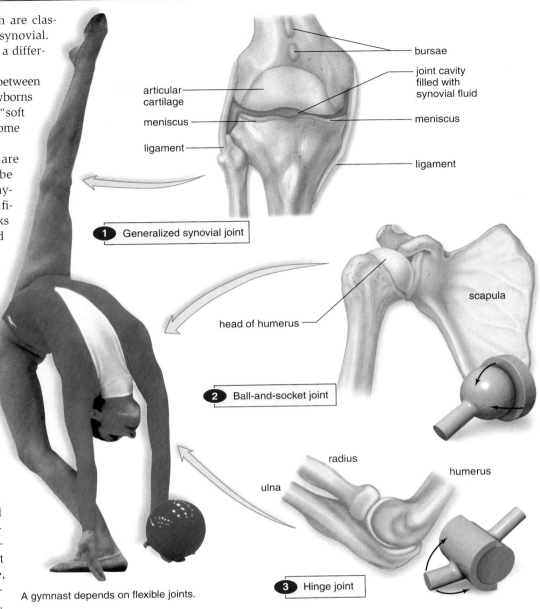

A gymnast depends on flexible joints.

1 Generalized synovial joint
— bursae
— joint cavity filled with synovial fluid
— meniscus
— ligament
articular cartilage
meniscus
ligament

2 Ball-and-socket joint
scapula
head of humerus

3 Hinge joint
radius
ulna
humerus

FIGURE 28.7 Synovial joints.

moved away from the midline of the body. ❸ **Hinge joints,** such as the elbow and knee joints, largely permit movement up and down in one plane only, like a hinged door. Flexion occurs when the angle decreases, as when the forearm moves upward. Extension occurs when the angle increases, as when the forearm moves downward.

In our hands are three other types of synovial joints: saddle (one bone fits inside another), gliding (the bones slide against one another), and condyloid (the convex surface of one bone fits in a depression of the other).

In Section 28.8, we will see how joints can be repaired.

28.7 *Check Your Progress* **What might a forensics expert tell from the condition of a body's synovial joints?**

28.8 Joint disorders can be repaired

To the young, otherwise healthy thirty-something athlete on the physician's exam table, the diagnosis must seem completely unfair. Perhaps he is a former football player, or she is a trained dancer. Whatever the sport or activity, the athlete is slender and fit, but knee pain and swelling are his or her constant companions. Examination of the knee shows the result of years of use and abuse while performing a sport. Bursitis or torn cartilage (menisci) and/or ligaments may have occurred. Also, the articular cartilage may have degenerated so that **arthritis** is present. Once repeated use has worn hyaline cartilage away, it does not grow back naturally. Exposed bone ends grind against one another, resulting in pain, swelling, and restricted movements that can cripple a person. In some cases, total joint replacement is called for, but some people have found glucosamine-chondroitin supplements beneficial as an alternative treatment. It is believed that glucosamine, an amino sugar, promotes the formation and repair of cartilage, while chondroitin, a carbohydrate and a cartilage component, promotes water retention and elasticity and inhibits enzymes that break down cartilage. Both compounds are naturally produced by the body.

If arthritis progresses, the exposed bone thickens and forms spurs that cause the bone ends to enlarge and restrict joint movement. Weight loss can ease arthritis. Taking off 3 lbs can reduce the load on a hip or knee joint by 9 to 15 lbs. Low-impact activities, such as biking and swimming, can help maintain muscle strength and stabilize joints. Many people also opt for arthroscopic surgery.

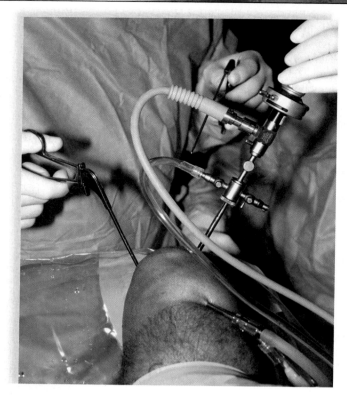

FIGURE 28.8 Arthroscopic surgery.

Arthroscopic Surgery Today, it is possible for surgeons to remove cartilage fragments, repair ligaments, or repair worn-away cartilage using a technique called arthroscopic surgery (**Fig. 28.8**). A small instrument bearing a tiny lens and light source is inserted into a joint, as are the surgical instruments. Fluid is then added to distend the joint and allow visualization of its structure. Usually, a monitor displays the surgery for the whole operating team to see. Arthroscopy is much less traumatic than surgically opening the knee with a long incision. The benefits of arthroscopy include the small incision, faster healing, a more rapid recovery, and less scarring. Because arthroscopic surgical procedures are often performed on an outpatient basis, the patient is able to return home the same day.

Replacing Cartilage Due to recent advances, the technique of tissue culture (growing cells outside the patient's body in a special medium) can be used so that a person's own hyaline cartilage can regenerate in the laboratory. Then, *autologous chondrocyte implantation (ACI)* takes place. As with all surgeries, there is a risk for postoperative complications, such as bleeding or infection. However, ACI surgery may offer young athletes the chance to restore essential hyaline cartilage and regain a healthy, functional knee joint.

In the ACI procedure, a piece of healthy hyaline cartilage from the patient's knee is first removed surgically. This piece of cartilage, about the size of a pencil eraser, is typically taken from an undamaged area at the top edge of the knee. Then the chondrocytes, living

cells of hyaline cartilage from this specimen, are grown outside the body in tissue culture medium. Millions of the patient's own cells can be grown to create a "patch" of living cartilage, a process that takes two to three weeks. Once the chondrocytes have grown, a pocket is created over the damaged area using the patient's own periosteum, the connective tissue that surrounds the bone. The periosteum pocket will hold the hyaline cartilage cells in place. Finally, the cells are injected into the pocket and left to grow.

As with all injuries to the knee, once the cartilage cells are firmly established, the patient still faces a lengthy rehabilitation period. The patient must use crutches or a cane for three to four months to protect the joint. Physical therapy will stimulate cartilage growth without overstressing the area being repaired. In six months, the athlete can return to light-impact training and jogging. Full workouts can be resumed about one year after surgery. However, most patients regain full mobility and a pain-free life after ACI surgery and do not have to undergo total knee replacement. This type of surgery is not recommended for the elderly or for overweight patients with arthritis.

This completes our discussion of bones. The next part of the chapter considers the skeletal muscles.

28.8 Check Your Progress Pain occurs when a piece of "torn cartilage" gets caught between the two bones of the knee joint. What two bones are these?

Animal Movement Is Dependent on Muscle Cell Contraction

In this part of the chapter, we turn our attention to the skeletal muscles. We consider the structure and function of whole muscle before concentrating on the structure and function of the muscle cell.

28.9 Vertebrate skeletal muscles have various functions

As noted in Figure 25.4, smooth muscle is involuntary muscle and is found in the walls of internal organs. Cardiac muscle is involuntary and makes up the wall of the heart. Skeletal muscle can be moved voluntarily and makes up the nearly 700 skeletal muscles, which account for approximately 40% of the weight of an average human. **Figure 28.9** illustrates several of the major muscles and their actions. The skeletal muscles perform many functions:

Skeletal muscles support the body. Muscle contraction opposes the force of gravity and allows us to remain upright.

Skeletal muscles make bones move. Muscle contraction accounts not only for movements of the arms and legs but also for movements of the eyes, facial expressions, and breathing.

Skeletal muscles help maintain a constant body temperature. Muscle contraction causes ATP to break down, releasing heat that is distributed about the body.

Skeletal muscle contraction assists movement in cardiovascular veins. The pressure of skeletal muscle contraction keeps blood moving in cardiovascular veins.

Skeletal muscles help protect internal organs and stabilize joints. Muscles pad the bones, and the muscular wall in the abdominal region helps protect the internal organs.

28.9 *Check Your Progress* **People tend to shiver when exposed to cold air. Explain this reaction.**

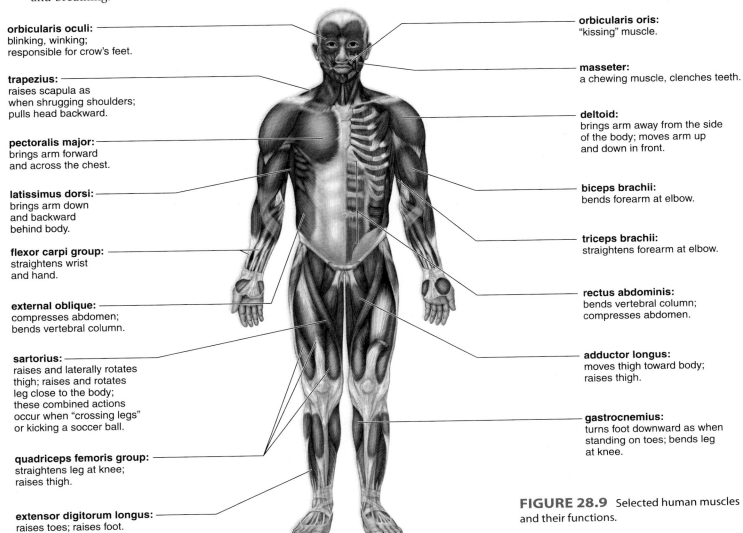

orbicularis oculi:
blinking, winking;
responsible for crow's feet.

trapezius:
raises scapula as
when shrugging shoulders;
pulls head backward.

pectoralis major:
brings arm forward
and across the chest.

latissimus dorsi:
brings arm down
and backward
behind body.

flexor carpi group:
straightens wrist
and hand.

external oblique:
compresses abdomen;
bends vertebral column.

sartorius:
raises and laterally rotates
thigh; raises and rotates
leg close to the body;
these combined actions
occur when "crossing legs"
or kicking a soccer ball.

quadriceps femoris group:
straightens leg at knee;
raises thigh.

extensor digitorum longus:
raises toes; raises foot.

orbicularis oris:
"kissing" muscle.

masseter:
a chewing muscle, clenches teeth.

deltoid:
brings arm away from the side
of the body; moves arm up
and down in front.

biceps brachii:
bends forearm at elbow.

triceps brachii:
straightens forearm at elbow.

rectus abdominis:
bends vertebral column;
compresses abdomen.

adductor longus:
moves thigh toward body;
raises thigh.

gastrocnemius:
turns foot downward as when
standing on toes; bends leg
at knee.

FIGURE 28.9 Selected human muscles and their functions.

Skeletal Muscles Work in Pairs

Skeletal muscles move the bones of the skeleton with the aid of bands of fibrous connective tissue called **tendons** that attach muscle to bone. In general, one muscle does most of the work of moving a bone, and that muscle is called a **prime mover**. When a muscle contracts, it shortens, and the tendon pulls on the bone. Therefore, muscles can only pull a bone; they cannot push it. Because of this, skeletal muscles must work in antagonistic pairs. If one muscle of an antagonistic pair flexes the joint and bends the limb, the other one extends the joint and straightens the limb. For example, the biceps brachii and the triceps brachii are antagonists; one bends the forearm, and the other straightens the forearm (**Fig. 28.10A**). If both of these muscles were to contract at once, the forearm would not move.

A Muscle Has Motor Units

A skeletal muscle has degrees of contraction because it is divided into motor units. A **motor unit** is composed of all the muscle fibers under the control of a single motor axon. The axon has branches that terminate at a number of muscle cells (fibers) of a muscle. Here, axon terminals release neurotransmitter molecules that cross a synapse, causing the muscle fiber to contract. We can liken a motor unit to a set of lights in a ceiling that is controlled by a single switch. A flip of the switch turns these lights on, much as a single axon causes its motor unit to contract. In other words, a motor unit obeys an "all-or-none law"—it either contracts or does not contract.

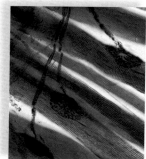

motor unit

The number of muscle fibers within a motor unit can vary. For example, in the ocular muscles that move the eyes, the innervation ratio is one motor axon per 23 muscle fibers, while in the gastrocnemius muscle of the leg, the ratio is about one motor axon per 1,000 muscle fibers. Thus, moving the eyes requires finer control than moving the legs.

When a motor unit is stimulated by a single stimulus, a contraction occurs that lasts only a fraction of a second. This response is called a **simple muscle twitch**. A muscle twitch is customarily divided into three stages: the latent period, or the period of time between stimulation and initiation of contraction; the contraction period, when the muscle shortens; and the relaxation period, when the muscle returns to its former length (**Fig. 28.10B**). If a motor unit is given a rapid series of stimuli, it can respond to the next stimulus without relaxing completely. Summation is increased muscle contraction until maximal sustained contraction, called *tetanus*, is achieved (**Fig. 28.10C**). Tetanus continues until the muscle fatigues due to depletion of energy reserves. Fatigue is apparent when a muscle relaxes even though stimulation continues. The tetanus of muscle cells is not the same as the infection called tetanus, which is caused by the bacterium *Clostridium tetani* and can cause death because the muscles, including the respiratory muscles, become fully contracted and do not relax.

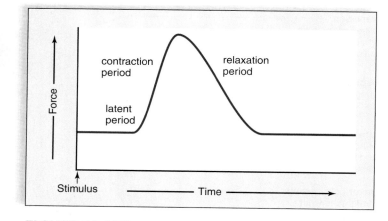

FIGURE 28.10B A single stimulus and a simple muscle twitch.

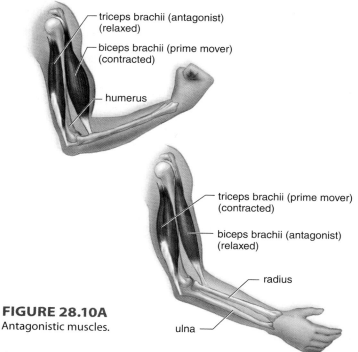

FIGURE 28.10A
Antagonistic muscles.

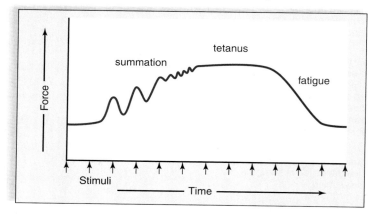

FIGURE 28.10C Multiple stimuli with summation and tetanus.

A whole muscle typically contains many motor units, much as a ceiling might contain several sets of lights. For maximum lighting, all the switches are turned on. So, as the intensity of nervous stimulation increases, more and more motor units in a muscle are activated. This phenomenon is known as *recruitment*. Maximum contraction of a muscle would require that all motor units be undergoing tetanic contraction. This rarely happens, or else they could all fatigue at the same time. Instead, some motor units are contracting maximally while others are resting, allowing sustained contractions to occur. One desirable effect of exercise is to achieve good "muscle tone," which is dependent on muscle contraction. When some motor units are always contracted but not enough to cause movement, the muscle is firm and solid.

The next section explains the benefits of exercising our muscles.

> **28.10** *Check Your Progress* You can lift this book with one arm. Using the same muscles, you can lift three such books. Explain why this is possible in terms of motor units.

HOW BIOLOGY IMPACTS OUR LIVES

28.11 Exercise has many benefits

Exercise programs improve muscular strength, muscular endurance, and flexibility. Exercise also improves cardiorespiratory endurance. The heart rate and capacity increase, and the air passages dilate so that the heart and lungs are able to support prolonged muscular activity. The blood level of high-density lipoprotein (HDL), the molecule that prevents the development of plaque in blood vessels, increases. Also, body composition—that is, the proportion of protein to fat—changes favorably when you exercise.

Exercise also seems to help prevent certain kinds of cancer. Cancer prevention involves eating properly, not smoking, avoiding cancer-causing chemicals and radiation, undergoing appropriate medical screening tests, and knowing the early warning signs of cancer. However, studies show that people who exercise are less likely to develop colon, breast, cervical, uterine, and ovarian cancers.

Physical training with weights can improve the density and strength of bones and the strength and endurance of muscles in all adults, regardless of age. Even men and women in their eighties and nineties can make substantial gains in bone and muscle strength that help them lead more independent lives. Exercise helps prevent osteoporosis (see Section 28.5) because it promotes the activity of osteoblasts in young as well as older people. The stronger the bones when a person is young, the less chance that person has of developing osteoporosis as he or she ages. Exercise helps prevent weight gain, not only because the level of activity increases but also because muscles metabolize faster than other tissues. As a person becomes more muscular, the body is less likely to accumulate fat.

Exercise relieves depression and enhances the mood. Some people report that exercise actually makes them feel more energetic, and that after exercising, particularly in the late afternoon, they sleep better that night. Self-esteem rises because of improved appearance, as well as other factors that are not well understood. For example, vigorous exercise releases endorphins, hormonelike chemicals that are known to alleviate pain and provide a feeling of tranquility.

A sensible exercise program is one that provides all of these benefits without the detriments of a too-strenuous program. Overexertion can actually be harmful to the body and might result in sports injuries, such as lower back strain or torn ligaments of the knees. The beneficial programs suggested in **Table 28.11** are tailored according to age.

Section 28.12 considers the anatomy of a muscle cell.

> **28.11** *Check Your Progress* We learned about oxygen debt in Chapter 7. Aerobic exercise avoids oxygen debt. Explain.

TABLE 28.11	A Checklist for Staying Fit		
Exercise	**Children, 7–12**	**Teenagers, 13–18**	**Adults, 19–55**
Amount	Vigorous activity 1–2 hours daily	Vigorous activity 1 hour, 3–5 days a week; otherwise, $1/2$ hour daily moderate activity	Vigorous activity 1 hour, 3 days a week; otherwise, $1/2$ hour daily moderate activity
Purpose	Free play	Build muscle with calisthenics	Exercise to prevent lower back pain: aerobics, stretching, yoga
Organized	Build motor skills through team sports, dance, swimming	Do aerobic exercise to control buildup of fat cells	Take active vacations: hike, bicycle, cross-country ski
Group	Encourage more exercise outside of physical education classes	Pursue tennis, swimming, horseback riding—sports that can be enjoyed for a lifetime	Find exercise partners: join a running club, bicycle club, or outing group
Family	Initiate family outings: bowling, boating, camping, hiking	Continue team sports: dancing, hiking, swimming	Initiate family outings: bowling boating, camping, hiking

28.12 A muscle cell contains many myofibrils

A whole muscle contains many long tubular muscle cells, whose anatomy is shown in **Figure 28.12**. Because a muscle cell has a slightly different structure from other cells, its parts are given special names. For example, the plasma membrane is called the **sarcolemma**. The sarcolemma of a muscle cell forms a T (for transverse) system. The T tubules penetrate, or dip down, into the cell so that they come in contact—but do not fuse—with expanded portions of modified endoplasmic reticulum, called the **sarcoplasmic reticulum**. These expanded portions serve as storage sites for calcium ions (Ca^{2+}), which are essential for muscle contraction. Also present in a muscle cell are many long, cylindrical organelles called **myofibrils**, which are the contractile portions of muscle cells. In cross section, a myofibril contains many contractile units called **sarcomeres**. Each sarcomere lies between two visible boundaries called Z lines.

Do you recall from Chapter 25 that skeletal muscle is striated—has bands of light and dark when seen with the light microscope? Notice that a skeletal muscle is striated because myofibrils and sarcomeres are striated. In the next section, we consider the manner in which a sarcomere contracts.

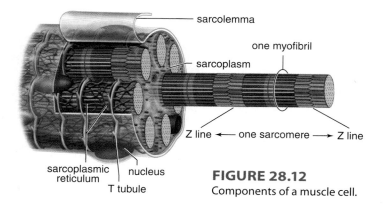

FIGURE 28.12
Components of a muscle cell.

> **28.12 Check Your Progress** A muscle cell contains many myofibrils, and a myofibril contains many sarcomeres. Explain.

28.13 Sarcomeres shorten when muscle cells contract

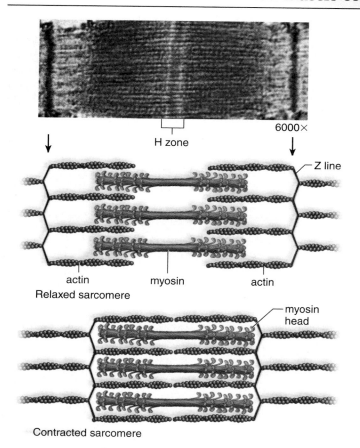

FIGURE 28.13A Contraction of a sarcomere.

The electron microscope reveals that skeletal muscle striations are due to the placement of protein filaments in sarcomeres. A sarcomere contains thick filaments made up of **myosin** and thin filaments made up of **actin**. A myosin filament has many globular heads. An area in the middle of a sarcomere, called the H zone, has only myosin filaments. The actin filaments are composed of long strands of globular actin molecules twisted about one another. The actin filaments are attached to the Z lines.

Figure 28.13A contrasts the appearance of a relaxed sarcomere with a contracted sarcomere. In a contracted sarcomere, the actin filaments are much closer to the center, and the H zone has all but disappeared. To achieve a contracted sarcomere, it is necessary for the actin filaments to slide past the myosin filaments. This occurs because the myosin heads pull the actin filaments toward the center of a sarcomere. When you play "tug of war," your hands grasp the rope, pull, let go, attach farther down the rope, and pull again. The myosin heads are like your hands—grasping, pulling, letting go, and then repeating the process. This model of muscle contraction is called the **sliding filament model**.

Sliding Filament Model **Figure 28.13B** pertains to only one myosin head, but actually many myosin heads act in unison to achieve the contraction of a sarcomere. The cycle of events shown occurs over and over again, and with each cycle, the actin filaments move nearer the center of the sarcomere, until the H zone all but disappears. ATP provides the energy for muscle contraction in a way that is not obvious. Each myosin head has a binding site for ATP, and the heads have an enzyme that splits ATP into ADP and Ⓟ. This activates the heads, making them ready to bind to actin. ADP and Ⓟ remain on the myosin heads while the heads attach to actin, forming cross-bridges. Release of ADP and Ⓟ causes the cross-bridges to bend sharply. This is the power stroke that pulls the actin filaments toward the middle of the sarco-

FIGURE 28.13B Role of ATP in muscle contraction.

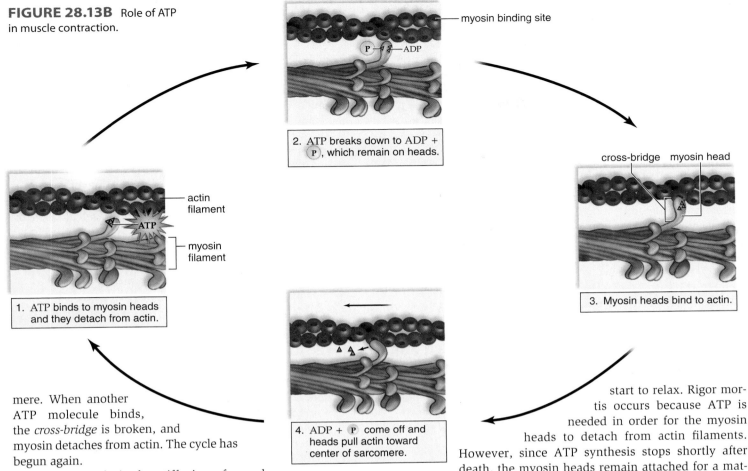

2. ATP breaks down to ADP + P, which remain on heads.

— myosin binding site

cross-bridge   myosin head

3. Myosin heads bind to actin.

actin filament

ATP

myosin filament

1. ATP binds to myosin heads and they detach from actin.

4. ADP + P come off and heads pull actin toward center of sarcomere.

mere. When another ATP molecule binds, the *cross-bridge* is broken, and myosin detaches from actin. The cycle has begun again.

Rigor mortis is the stiffening of muscles that occurs in a dead body. It is often used to estimate the time of death when a recently deceased body is discovered. At temperatures of 21–24°C (70–75°F), rigor mortis begins within one to three hours; maximum rigidity is reached 10–12 hours after death. Stiffness persists for 24 to 36 hours, and then the muscles start to relax. Rigor mortis occurs because ATP is needed in order for the myosin heads to detach from actin filaments. However, since ATP synthesis stops shortly after death, the myosin heads remain attached for a matter of hours, until deterioration sets in.

28.13 Check Your Progress We often think of rigor mortis as caused by a lack of ATP at death. But what events had to precede rigor mortis in order for it to be present?

28.14 Axon terminals bring about muscle contraction

Muscle cells (fibers) contract only because they are stimulated to do so by motor axons. A motor axon branches, and each branch terminates very close to a muscle cell. This region, called a *neuromuscular junction*, contains a *synaptic cleft* (**Fig. 28.14**). A nerve impulse traveling down an axon causes the axon terminals to release the neurotransmitter acetylcholine (ACh) (green). The sarcolemma of a muscle cell contains receptors for ACh molecules, and when these molecules bind to the receptors, a muscle action potential begins. The muscle action potential travels down the T tubules, and the close proximity of the T tubules to the sarcoplasmic reticulum causes it to release calcium (Ca^{2+}). The Ca^{2+} diffuses throughout the muscle cell and binds to actin filaments, exposing binding sites for myosin. Now the sarcomeres contract as long as ATP is present. You can actually watch this in the laboratory. Put a bit of skeletal muscle tissue on a slide, add Ca^{2+} and ATP, and suddenly the tissue shortens.

The next section discusses the sources of ATP for muscle contraction in the body.

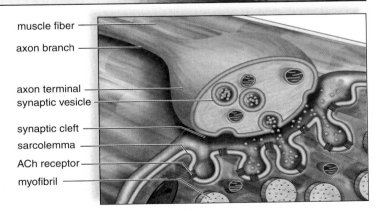

muscle fiber
axon branch

axon terminal
synaptic vesicle

synaptic cleft
sarcolemma
ACh receptor
myofibril

FIGURE 28.14 Neuromuscular junction (green = ACh).

28.14 Check Your Progress Compare a neuromuscular junction to a synapse (see Fig. 26.7).

28.15 Muscles have three sources of ATP for contraction

Muscle cells store limited amounts of ATP, but they have three ways of acquiring more ATP for contraction once this supply has been used up: the creatine phosphate pathway, fermentation, and cellular respiration.

Creatine Phosphate (CP) Pathway Creatine phosphate is a molecule that contains a high-energy phosphate. Creatine phosphate is only formed when a muscle cell is resting, and only a limited amount is stored. The simplest and most rapid way for muscle to produce ATP is to transfer the high-energy phosphate from CP to ADP.

This reaction occurs in the midst of sliding filaments, and therefore this method of supplying ATP is the speediest energy source available to muscles. The CP pathway is used at the beginning of exercise and during short-term, high-intensity exercise that lasts less than five seconds.

Fermentation Fermentation, as you know, produces two ATP from the anaerobic breakdown of glucose to lactate (see Fig. 7.9A). Fermentation, like the CP pathway, is fast-acting, but it results in the buildup of lactate, noticeable because it produces short-term muscle aches and fatigue upon exercising. Fermentation also results in **oxygen debt**, the oxygen required in part to complete the metabolism of lactate and restore cells to their original energy state.

Cellular Respiration Muscle cells have a rich supply of mitochondria where cellular respiration supplies ATP, usually from the breakdown of glucose whenever oxygen is available.

Section 28.16 explains why people can do well at one sport and not another.

> **28.15 Check Your Progress** The CP pathway doesn't use glucose as an energy source to produce ATP, as do fermentation and cellular respiration. What does it use?

28.16 Some muscle cells are fast-twitch and some are slow-twitch

Some athletes have more fast-twitch fibers, and some have more slow-twitch fibers. Fast-twitch fibers tend to rely on the creatine phosphate pathway and fermentation, while slow-twitch fibers tend to prefer cellular respiration, which is aerobic (**Fig. 28.16**).

Fast-Twitch Fibers ❶ Fast-twitch fibers are usually anaerobic and seem designed for strength because their motor units contain many fibers. They provide explosions of energy and are most helpful in sports activities such as sprinting, weight lifting, swinging a golf club, or throwing a shot. Fast-twitch fibers are light in color because they have fewer mitochondria, little or no myoglobin, and fewer blood vessels than slow-twitch fibers do. Fast-twitch fibers can develop maximum tension more rapidly than slow-twitch fibers can, and their maximum tension is greater. However, their dependence on anaerobic energy leaves them vulnerable to an accumulation of lactate, which causes them to fatigue quickly.

Slow-Twitch Fibers ❷ Slow-twitch fibers have a steadier tug and more endurance, despite having more units with a smaller number of fibers. These muscle fibers are most helpful in sports such as long-distance running, biking, jogging, and swimming. Because they produce most of their energy aerobically, they tire only when their fuel supply is gone. Slow-twitch fibers have many mitochondria and are dark in color because they contain myoglobin, the respiratory pigment found in muscles. They are also surrounded by dense capillary beds and draw more blood and oxygen than do fast-twitch fibers. Slow-twitch fibers have a low maximum tension, but the muscle fibers are highly resistant to fatigue. Because slow-twitch fibers have a substantial reserve of glycogen and fat, their abundant mitochondria can maintain steady, prolonged production of ATP when oxygen is available.

> **28.16 Check Your Progress** Could a forensics expert predict the probable sport of a deceased individual?

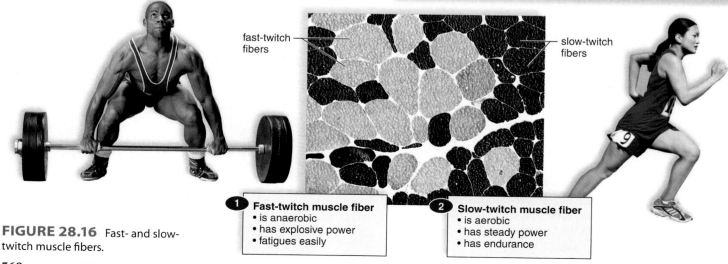

fast-twitch fibers

slow-twitch fibers

❶ **Fast-twitch muscle fiber**
• is anaerobic
• has explosive power
• fatigues easily

❷ **Slow-twitch muscle fiber**
• is aerobic
• has steady power
• has endurance

FIGURE 28.16 Fast- and slow-twitch muscle fibers.

This chapter gave us an opportunity to examine the body at both the macro and micro levels. The skeleton is easily observable at the macro level, and we learned the names of the bones making up the axial and appendicular portions of the skeleton. Then we considered the tissues of the bones and joints (compact bone, spongy bone, cartilage, fibrous connective tissue). At this level, we can understand the injuries that occur when we misuse our joints.

Similarly, we learned the names of various muscles and how they operate when we intentionally move our bones.

Although the body has three types of muscles, this chapter concentrates on the skeletal muscles. We can't understand how skeletal muscles contract and move the bones until we study skeletal muscles at the cellular level. A theme of "structure suits function" is observable at the macro level because whole muscles and bones have structures suitable to their functions. But this theme is even more observable at the micro level. A muscle cell is suited to its task because it contains contractile organelles called myofibrils. Myofibrils contain the filaments (actin and myosin) that account for muscle contraction. Imagine the satisfaction of the electron microscopists and biochemists who first solved the riddle of muscle contraction and were able to explain to forensics specialists why rigor mortis occurs. Without knowing how muscles contract at the cellular level, our understanding of muscles would be incomplete.

In Chapter 29, we continue our theme of homeostasis by studying the circulatory system, which is directly involved in homeostasis because blood and tissue fluid constitute the internal environment of the body. When blood and tissue fluid remain relatively constant, so do the cells making up the body tissues—including those of the bones and muscles.

The Chapter in Review

Summary

Skeletal Remains Reveal All

- Age, gender, and sometimes ethnicity can be determined by examining skeletal remains.

Animal Skeletons Support, Move, and Protect the Body

28.1 Animal skeletons can be hydrostatic, external, or internal

- A hydrostatic skeleton in animals that lack a hard skeleton is a fluid-filled gastrovascular cavity or coelom.
- An exoskeleton is a rigid external skeleton found in molluscs and arthropods.
- An endoskeleton is a rigid internal skeleton that protects the internal organs and is protected by soft tissues surrounding it; in vertebrates, the endoskeleton is jointed.

28.2 Mammals have an endoskeleton that serves many functions

- The skeleton is necessary to movement, protects internal organs, assists breathing, stores and releases calcium, and assists other systems.

The Mammalian Skeleton Is a Series of Bones Connected at Joints

28.3 The bones of the axial skeleton lie in the midline of the body

- The axial skeleton consists of the skull, vertebral column, rib cage, sacrum, and coccyx.
  - The cranium and facial bones of the skull protect the brain.
  - The vertebral column, composed of vertebrae separated by shock-absorbing disks, protects the spinal cord and nerves and anchors all other bones.
- The rib cage, composed of the ribs, the costal cartilages, and the sternum, protects the heart and lungs.

28.4 The appendicular skeleton consists of bones in the girdles and limbs

- The bones of the pectoral girdle (shoulder) and the upper limbs are adapted for flexibility.
- The pelvic girdle (hipbones) and lower limbs are adapted for strength and support.

28.5 Avoidance of osteoporosis requires good nutrition and exercise

- Protection against osteoporosis (weakened bones due to decreased bone mass) includes a lifetime diet of adequate calcium and weight-bearing exercise. Medications can slow or reverse bone loss.

28.6 Bones are composed of living tissues

- The long bone (e.g., humerus) has the following structures:
  - The medullary cavity contains yellow bone marrow.
  - Compact bone at the sides contains osteons separated by a hard matrix.
  - Spongy bone at the ends contains red bone marrow.
  - Articular cartilage covers the ends of a long bone.

28.7 Joints occur where bones meet

- Fibrous joints are immovable, cartilaginous joints are slightly movable, and synovial joints are freely movable.
- Synovial joints (e.g., ball-and-socket, hinge) are filled with synovial fluid, provide stability, and absorb shock.
  - Ball-and-socket joints allow movement in all planes, including rotation.
  - Hinge joints allow movement in one direction only.

- Other types of synovial joints include saddle, gliding, and condyloid joints.

28.8 Joint disorders can be repaired
- Arthroscopic surgery can remove cartilage fragments or repair ligaments or cartilage.
- Tissue culture (via ACI surgery) can help a patient regenerate hyaline cartilage.

Animal Movement Is Dependent on Muscle Cell Contraction

28.9 Vertebrate skeletal muscles have various functions
- Skeletal muscle supports the body, makes bones move, helps maintain a constant body temperature, assists blood movement in veins, helps protect internal organs, and stabilizes joints.

28.10 Skeletal muscles contract in units
- Muscles, attached to bones by tendons, work in antagonistic pairs.
- A whole muscle has many motor units:
  - Contraction involves twitch, summation, and tetanus.
  - Recruitment is activation of more units.
  - Tone requires some units always contracting.

28.11 Exercise has many benefits
- Improves muscular strength, endurance, and flexibility; improves cardiorespiratory endurance; helps prevent cancer; improves the density and strength of bones and the strength and endurance of muscles; relieves depression; enhances mood; and can fight disease, as in certain types of cancer.

28.12 A muscle cell contains many myofibrils
- Myofibrils, composed of units called sarcomeres, make up the contractile portion of a muscle cell.

28.13 Sarcomeres shorten when muscle cells contract
- Sarcomeres contain actin and myosin filaments.
- According to the sliding filament model, muscle contraction occurs when sarcomeres shorten and actin filaments slide past myosin filaments.

28.14 Axon terminals bring about muscle contraction
- Nerve impulses travel down motor neurons and stimulate muscle cells at neuromuscular junctions.

28.15 Muscles have three sources of ATP for contraction
- The creatine phosphate (CP) pathway is simple and rapid; fermentation (anaerobic) produces two ATP per glucose molecule; and cellular respiration (aerobic) produces ATP and uses glucose.

28.16 Some muscle cells are fast-twitch and some are slow-twitch
- Fast-twitch fibers are anaerobic and light in color; they supply short-term, explosive power; slow-twitch fibers are aerobic and dark in color; they are good for activities that require long-term endurance.

Testing Yourself

Animal Skeletons Support, Move, and Protect the Body

1. Unlike an exoskeleton, an endoskeleton
   - a. grows with the animal.
   - b. is composed of chitin.
   - c. is jointed.
   - d. protects internal organs.
2. _____ connect bone to bone, and _____ connect muscle to bone.
   - a. Ligaments, ligaments
   - b. Tendons, ligaments
   - c. Ligaments, tendons
   - d. None of these are correct.
3. Which of the following is not a function of the skeletal system?
   - a. production of blood cells
   - b. storage of minerals
   - c. involved in movement
   - d. storage of fat
   - e. production of body heat
4. All blood cells—red, white, and platelets—are produced by which of the following?
   - a. yellow bone marrow
   - b. red bone marrow
   - c. periosteum
   - d. medullary cavity

The Mammalian Skeleton Is a Series of Bones Connected at Joints

5. A component of the appendicular skeleton is the
   - a. rib cage.
   - b. skull.
   - c. femur.
   - d. vertebral column.
6. Which of the following is not a bone of the appendicular skeleton?
   - a. the scapula
   - b. a rib
   - c. a metatarsal bone
   - d. the patella
7. This bone is the only movable bone of the skull.
   - a. sphenoid
   - b. frontal
   - c. mandible
   - d. maxilla
   - e. temporal

For questions 8–14, match each bone to a location in the key. Answers can be used more than once.

KEY:
- a. arm (above forearm)
- b. forearm
- c. pectoral girdle
- d. pelvic girdle
- e. thigh
- f. leg (below thigh)

8. Ulna
9. Tibia
10. Clavicle
11. Femur
12. Scapula
13. Coxal bone
14. Humerus
15. _____ occupies the _____.
   - a. Cartilage, medullary cavity
   - b. Marrow, foramen magnum
   - c. Marrow, medullary cavity
   - d. None of these are correct.
16. The deterioration of a synovial joint over time can cause
   - a. arthritis.
   - b. a sprain.
   - c. a slipped disk.
   - d. tendonitis.
17. Spongy bone
   - a. contains osteons.
   - b. contains red bone marrow, where blood cells are formed.
   - c. lends no strength to bones.
   - d. contributes to homeostasis.
   - e. Both b and d are correct.

18. Which of these pairs is mismatched?
    a. slightly movable joint—vertebrae
    b. hinge joint—hip
    c. synovial joint—elbow
    d. immovable joint—sutures in cranium
    e. ball-and-socket joint—hip
19. After an examination, the doctor informs Isabella that she will have to have her baby by cesarean section. What skeletal abnormality is most likely?

Animal Movement Is Dependent on Muscle Cell Contraction

20. The biceps and triceps are considered
    a. synergists.   c. protagonists.
    b. antagonists.   d. None of these are correct.
21. Which of the following is not a function of the muscular system?
    a. hormone production
    b. heat production
    c. movement
    d. protection of internal organs
    e. All of these choices are functions of the muscular system.
22. To increase the force of muscle contraction,
    a. individual muscle cells have to contract with greater force.
    b. motor units have to contract with greater force.
    c. more motor units need to be recruited.
    d. All of these are correct.
    e. None of these are correct.
23. In a muscle fiber,
    a. the sarcolemma is connective tissue holding the myofibrils together.
    b. the sarcoplasmic reticulum stores calcium.
    c. both myosin and actin filaments have cross-bridges.
    d. there is a T system but no endoplasmic reticulum.
    e. All of these are correct.
24. The thick filaments of a muscle fiber are made up of
    a. actin.   c. fascia.
    b. troponin.   d. myosin.
25. A neuromuscular junction occurs between an axon terminal and
    a. a muscle fiber.   d. a sarcomere only.
    b. a myofibril.   e. Both a and d are correct.
    c. a myosin filament.
26. Nervous stimulation of muscles
    a. occurs at a neuromuscular junction.
    b. involves the release of ACh.
    c. results in impulses that travel down the T system.
    d. causes calcium to be released from the sarcoplasmic reticulum.
    e. All of these are correct.
27. **THINKING CONCEPTUALLY** Why do myosin heads have to be attached to actin during the power stroke of muscle contraction?
28. Which of the following statements about cross-bridges is false?
    a. They are composed of myosin.
    b. They bind to ATP after they attach to actin.
    c. They contain an ATPase.
    d. They split ATP before they attach to actin.
29. Which of these is the direct source of energy for muscle contraction?
    a. ATP   d. glycogen
    b. creatine phosphate   e. Both a and b are correct.
    c. lactic acid
30. Myoglobin content is higher in _____ -twitch fibers respiring _____.
    a. slow, aerobically   c. slow, anaerobically
    b. fast, aerobically   d. None of these are correct.

Understanding the Terms

actin   566
appendicular skeleton   556
arthritis   562
articular cartilage   560
axial skeleton   556
ball-and-socket joint   561
bursa   561
compact bone   560
endoskeleton   555
exoskeleton   554
fontanel   556
foramen magnum   556
hinge joint   561
hydrostatic skeleton   554
intervertebral disk   557
lacuna   560
meniscus   561
motor unit   564
myofibril   566
myosin   566
osteoblast   559
osteoclast   559
osteocyte   560
osteoporosis   559
oxygen debt   568
pectoral girdle   558
pelvic girdle   558
prime mover   564
red bone marrow   560
rigor mortis   567
sarcolemma   566
sarcomere   566
sarcoplasmic reticulum   566
simple muscle twitch   564
skull   556
sliding filament model   566
spongy bone   560
synovial joint   561
tendon   564
vertebral column   557

Match the terms to these definitions:
a. _____ Bone-forming cell.
b. _____ Muscle protein making up the thin filaments in a sarcomere; its movement shortens the sarcomere, yielding muscle contraction.
c. _____ Part of the skeleton that consists of the pectoral and pelvic girdles and the bones of the arms and legs.
d. _____ Movement of actin filaments in relation to myosin filaments, which accounts for muscle contraction.
e. _____ Portion of the skeleton that provides support and attachment for the arms.

Thinking Scientifically

1. You work in a morgue and do frequent autopsies. You know that it is possible to watch very thin muscle tissue contract under the microscope. How would you test the statement that rigor mortis is due to lack of ATP?
2. Exercise physiologists tell us that those who exercise use less oxygen per unit time and ferment less than those who do not exercise. Having the facilities of a medical laboratory, including oxygen tanks, how would you test this information? If your results agree, what explanation can you give?

ARIS™ *Visit* **www.mhhe.com/maderconcepts** *for practice quizzes, animations, videos, and activities designed to help you master the material in this chapter.*

29

Circulation and Cardiovascular Systems

LEARNING OUTCOMES

After studying this chapter, you should be able to accomplish the following outcomes.

Not All Animals Have Red Blood

1 Compare the respiratory pigments of various animals.

A Circulatory System Helps Maintain Homeostasis

2 State the overall function of a circulatory system.
3 Compare and contrast transport in animals with no circulatory system, an open system, and a closed system.
4 Relate differences in circulatory pathways to the way of life.

The Mammalian Cardiovascular System Consists of the Heart and Blood Vessels

5 Describe the anatomy of the heart, including its attached blood vessels.
6 Describe the heartbeat, and relate it to the cardiac cycle.
7 Compare the structure and function of arteries, capillaries, and veins.
8 Trace the path of blood in the pulmonary and systemic circuits.
9 Compare the velocity of blood and blood pressure in arteries, capillaries, and veins.
10 Explain the movement of blood in veins.
11 Relate the occurrence of hypertension to heart attack and stroke.

Blood Has Vital Functions

12 List and discuss four functions of blood.
13 Describe the composition of plasma and the structure and function of the formed elements.
14 Describe blood clotting as a series of three main steps.
15 Tell how adult stem cells might be used for the benefit of humankind.
16 Describe capillary exchange in the tissues.
17 Explain who can give blood to whom, utilizing the ABO system and the Rh system.

Our blood is red, as you no doubt have witnessed after suffering various cuts. We tend to think that most animals, whether vertebrates or invertebrates, are pretty much like ourselves. So, it comes as a surprise to learn that the blood of some invertebrates is green or blue, not red. The color of blood is dependent on the pigment that transports oxygen. The job of a respiratory pigment is to bind oxygen in areas of higher concentration (usually gas-exchange surfaces, such as lungs or gills) and to release it in areas of lower concentration, usually the tissues.

Vertebrates have red blood because their respiratory pigment, hemoglobin, is red when it is bound to oxygen. It is packaged inside blood cells, appropriately called red blood cells. Each subunit of hemoglobin consists of the protein globin plus an embedded heme group. The heme group contains an iron atom that binds to oxygen. When oxygen is attached to the iron, hemoglobin is red; when oxygen is not attached, hemoglobin is a sort of purplish color. The expression "blue blood" is used to refer to royalty because, in days gone by, their pale, untanned skin allowed the blue-tinged oxygen-poor blood in their veins to show through.

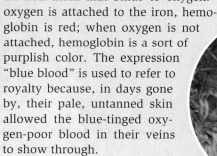

All vertebrates have red blood.

Not All Animals Have Red Blood

An earthworm (an invertebrate) has red blood, not because it contains hemoglobin, but because it contains giant, free-floating blood proteins bound to many dozens, even hundreds, of iron-containing heme groups. However, other annelids—tubeworms that live in the sea—have the respiratory pigment chlorocruorin. Chlorocruorin appears red when oxygenated, but green when deoxygenated!

Hemocyanin, the second most common oxygen-transporting pigment found among animals, uses copper-containing heme groups instead of iron-containing heme groups. Another big difference is that hemocyanin is dissolved in the blood rather than packaged in cells, as is the hemoglobin of vertebrates. Copper turns blue when oxygenated, so some invertebrates are truly blue-blooded. Hemocyanin is found in the blood of marine arthropods, such as lobsters and horseshoe crabs, and also in most molluscs, including squids. The heart of a giant squid pumps blue blood.

Which type of invertebrate is well known for having colorless blood with no respiratory pigment? The terrestrial insects, of course. They have no need of a respiratory pigment because little air tubes called tracheae take air directly to mitochondria just inside the muscle cells. The rapid delivery of oxygen-laden air to flight muscles is very adaptive because insects' mode of transportation on land is flying, which is energy-intensive.

Tube worms have green blood.

The occurrence of respiratory pigments does not appear strongly connected to evolutionary relationships, so it is hard to find a reason why some animals have red, some blue, and some green blood. One idea is that the pH of the environment affects the type of respiratory pigment. Hemocyanin is an excellent oxygen carrier, but it is very sensitive to pH changes. A very slight change toward acidity can cause hemocyanin to unload too early. Hemoglobin is not as sensitive to pH changes, so it becomes the better choice when the respiratory pigment is exposed to a different pH in the lungs compared to the tissues. In humans, the pH of the tissues is slightly lower than that of the lungs. Why? Because when carbon dioxide combines with the water in plasma, the liquid part of blood, it forms carbonic acid.

In this chapter, a comparison of animal circulatory systems precedes an in-depth examination of the mammalian cardiovascular system. The composition of blood will also be studied.

Marine lobsters have blue blood.

The blood of an insect is colorless.

The functions of a circulatory system are previewed before we compare how invertebrates and vertebrates provide their cells with nutrients and free them of wastes.

29.1   A circulatory system serves the needs of cells

Animals are multicellular, and most have a circulatory system that serves the needs of their cells. The circulatory system transports oxygen and nutrients, such as glucose and amino acids, to the cells. Then it picks up wastes, which are later excreted from the body by the lungs or kidneys.

Both gas exchange and nutrient-for-waste exchange occur across the walls of the smallest blood vessels, called capillaries (**Fig. 29.1**). No cell in the body of such an animal is far from a capillary. Of interest, cancer cells produce growth factors that cause angiogenesis, the growth of capillaries into a tumor. Without blood vessels, a tumor cannot continue to grow.

Some invertebrates depend on external fluids to service their cells, as described in Section 29.2.

> **29.1 Check Your Progress** How would you know the blood vessel featured in Figure 29.1 is that of a vertebrate?

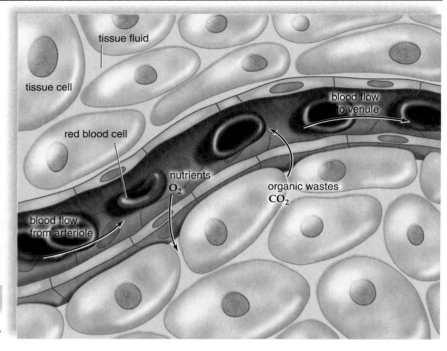

FIGURE 29.1 Exchanges of gases, nutrients, and wastes take place across capillary walls.

29.2   Some invertebrates do not have a circulatory system

Cnidarians, such as hydras, and flatworms, such as planarians, do not have a circulatory system (**Fig. 29.2**). Why not? The body of a hydra makes a circulatory system unnecessary. The cells are either part of an external layer, or they line the gastrovascular cavity. In either case, each cell is exposed to water and can independently exchange gases and get rid of wastes. The cells that line the gastrovascular cavity are specialized to carry out digestion. They pass nutrient molecules to other cells by diffusion. In a planarian, a trilobed gastrovascular cavity branches throughout the small, flattened body. No cell is very far from one of the three digestive branches, so nutrient molecules can diffuse from cell to cell. Similarly, diffusion meets the respiratory and excretory needs of the cells.

Some other invertebrates also lack a circulatory system. Pseudocoelomate invertebrates, such as roundworms, use the coelomic fluid of their body cavity for transport purposes as do the coelomate echinoderms, such as a sea star.

Other invertebrates have either an open or closed circulatory system, as we shall see in Section 29.3.

> **29.2 Check Your Progress** Do all animals have a respiratory pigment?

FIGURE 29.2
Invertebrates with a gastrovascular cavity.

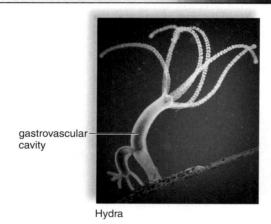

Hydra

Flatworm

All other invertebrates have a circulatory system in which a pumping heart sends a fluid into blood vessels. There are two types of circulatory fluid: **blood**, which is always contained within blood vessels, and **hemolymph**, which flows into a body cavity or cavities. Hemolymph is a mixture of blood and tissue fluid.

Hemolymph is found in animals that have an **open circulatory system**. For example, in most molluscs and arthropods, the heart pumps hemolymph containing the respiratory pigment hemocyanin via vessels into tissue spaces that are sometimes enlarged into saclike sinuses. Eventually, hemolymph drains back to the heart. In the grasshopper, an insect arthropod, the tubular heart pumps hemolymph into a dorsal aorta, which empties into the hemocoel (**Fig. 29.3A**). When the heart contracts, openings called ostia (sing., ostium) are closed; when the heart relaxes, the hemolymph is sucked back into the heart by way of the ostia. An open circulatory system, with its slow delivery of oxygen and nutrients to cells, is sufficient for a sluggish animal, such as a clam. But an insect such as a grasshopper is quite active. A grasshopper has colorless blood and doesn't depend on its open circulatory system to deliver oxygen to its muscles. Instead, it has numerous little air tubes, called tracheae, that open to the outside and take oxygen-laden air directly to its flight muscles. Flying is an adaptation to life on land, and so is the use of tracheae to deliver oxygen to muscles.

A **closed circulatory system** exists in annelids, such as earthworms. The heart pumps blood, which usually consists of cells and plasma, into a system of blood vessels (**Fig. 29.3B**). Valves prevent the backward flow of blood. In the segmented earthworm, five pairs of anterior hearts (aortic arches) pump blood into the ventral blood vessel (an artery), which has a branch called a lateral vessel in every segment of the worm's body. Blood moves through these branches into capillaries, where exchanges with tissue fluid take place. Blood then moves from small veins into the dorsal blood vessel (a vein). This dorsal blood vessel returns blood to the heart for repumping.

The earthworm has red blood that contains a respiratory pigment akin to hemoglobin. The pigment is dissolved in the blood and is not contained within cells. The earthworm has no specialized boundary, such as lungs, for gas exchange with the external environment. Gas exchange takes place across the body wall, which must always remain moist for this purpose.

While a slow mollusc such as a clam has an open circulatory system, rapid-moving molluscs such as squids and octopuses have a closed system. A closed system is more likely to deliver sufficient oxygen to muscles from respiratory gills than is an open system.

Section 29.4 examines the closed circulatory system of vertebrates.

> **29.3 *Check Your Progress*** In which type of circulatory system, open or closed, would blood move more quickly? Why?

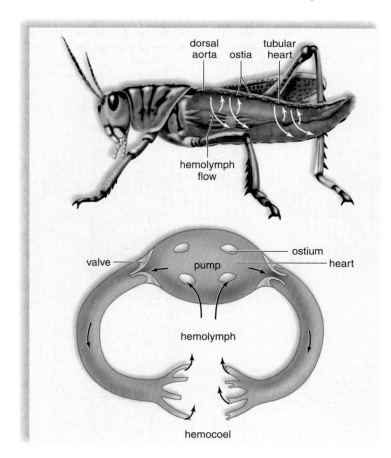

FIGURE 29.3A Open circulatory system in a grasshopper.

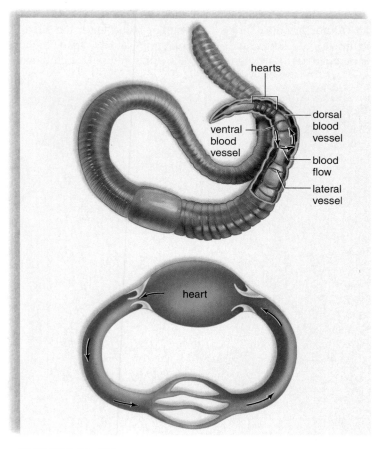

FIGURE 29.3B Closed circulatory system in an earthworm.

Two different types of circulatory pathways are seen among vertebrate animals. Fishes have a single-loop system in which the heart only pumps the blood to the gills. The other vertebrates have a two-circuit (double-loop) circulatory pathway. In the **systemic circuit**, the heart pumps blood to all parts of the body except for the lungs; the **pulmonary circuit** pumps blood to the lungs.

Fishes

In fishes, the heart has a single atrium and a single ventricle (**Fig. 29.4A**). The pumping action of the ventricle sends blood under pressure to the gills, where gas exchange with the external environment occurs. Fishes have an efficient means of respiration, and their blood is fully enriched with oxygen when it leaves the **gills**, the respiratory organ for aquatic organisms. But after passing through the gills, blood is no longer under pressure as it travels in the aorta to the rest of the body. This means the rate of oxygen delivery to the tissues is limited. Recall that ectotherms, such as fishes, are animals that get their heat from the environment. Because fishes have a reduced body temperature, the sluggish delivery of oxygen to the body proper is usually sufficient. The undulating movement of a fish's body helps move the blood back to the heart.

Amphibians and Reptiles

Amphibians were the first vertebrates to invade the land, and we see an evolutionary change that supports this change in environment (**Fig. 29.4B**). The single ventricle pumps blood in the pulmonary circuit to the lungs, the respiratory organ of land vertebrates. It also pumps blood in the systemic circuit to the rest of the body.

Although both O_2-rich and O_2-poor blood enter the single ventricle, it is kept somewhat separate because O_2-poor blood is pumped out of the ventricle to the lungs before O_2-rich blood enters and is pumped to the systemic circuit. Amphibians, such as frogs and salamanders, stay close to water, and their moist skin helps recharge their blood with oxygen.

In most reptiles, a septum partially divides the ventricle. In these animals, mixing of O_2-rich and O_2-poor blood is kept to a minimum. However, the opening between the two ventricles allows blood at times to bypass the lungs entirely. Both amphibians and reptiles are ectotherms, and reptiles in particular are quite idle compared to birds and mammals. Therefore, a slight mixing of O_2-rich and O_2-poor blood in the single ventricle is consistent with their way of life.

Birds and Mammals

In crocodilians (the reptiles most related to birds), the septum completely separates the ventricle. Note the two atria and two ventricles in the heart and the complete separation of the pulmonary and systemic circuits in birds and mammals (**Fig. 29.4C**). The right ventricle pumps blood under pressure to the lungs, and the larger left ventricle pumps blood under pressure to the rest of the body. This means that blood is under pressure when it goes to the lungs and while it is in the aorta, which distributes blood to the rest of the body. Birds and mammals are endotherms, and they locomote well on land. The pressure in the aorta means that the delivery of oxygen is adequate for their active way of life and for the maintenance of a warm internal temperature.

In the next part of the chapter, we consider the mammalian cardiovascular system.

> **29.4 Check Your Progress** Explain the use of blue and red in Figure 29.4A–C, with reference to hemoglobin.

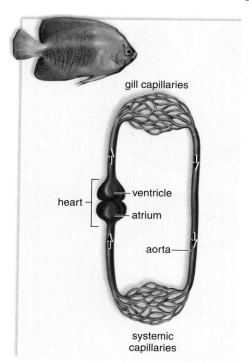

FIGURE 29.4A Single-loop circulatory pathway in fishes.

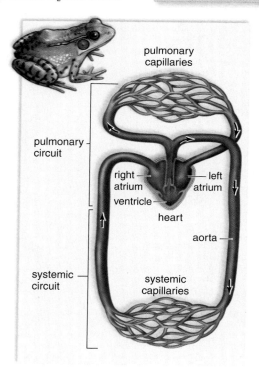

FIGURE 29.4B Two-circuit pathway in amphibians and most reptiles.

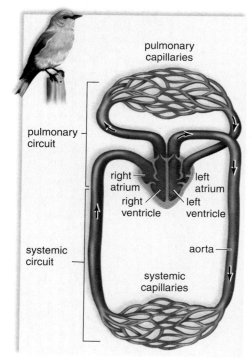

FIGURE 29.4C Complete separation of pulmonary and systemic circuits in birds, mammals, and some reptiles.

This part of the chapter describes the mammalian heart and the heartbeat. Then we will consider the pathways of blood and how blood flow is maintained, before concentrating on cardiovascular disease.

29.5   The mammalian heart has four chambers

All vertebrate animals have a closed circulatory system, which is called a **cardiovascular system** because it consists of a heart (*cardio*) and a system of blood vessels (*vascular*). The strong, muscular heart has four chambers.

A **septum** divides the heart into left and right sides. (Note that "right/left side of the heart" refers to how the heart is positioned in your body, not to the right side of a diagram.) The right side of the heart pumps O_2-poor blood to the lungs, and the left side of the heart pumps O_2-rich blood to the tissues. The septum is complete and prevents O_2-poor blood from mixing with O_2-rich blood.

Each side of the heart has two chambers. The upper, thin-walled chambers are called atria (sing., **atrium**)—thus, there is a right atrium and a left atrium (**Fig. 29.5**). The lower chambers are the thick-walled right and left **ventricles**. The atria receive blood; the ventricles pump blood away from the heart.

Valves occur between the atria and the ventricles, and between the ventricles and attached vessels. Because these valves close after the blood moves through, they keep the blood moving in the correct direction. The valves between the atria and ventricles are called the **atrioventricular valves**, and the valves between the ventricles and their attached vessels are called **semilunar valves**, because their cusps look like half moons.

The right atrium receives blood from attached veins (the superior and inferior venae cavae) that are returning O_2-poor blood to the heart from the tissues. After the blood passes through the atrioventricular valve (also called the *tricuspid valve*), the right ventricle pumps it through the *pulmonary semilunar valve* into the **pulmonary trunk** and **pulmonary arteries** that take it to the lungs. The **pulmonary veins** bring O_2-rich blood to the left atrium. After this blood passes through an atrioventricular valve (also called the *bicuspid valve*), the left ventricle pumps it through the aortic *semilunar valve* into the **aorta**, which takes it to the tissues.

Like mechanical valves, the heart valves are sometimes leaky; they may not close properly, permitting a backflow of blood. A **heart murmur** is often due to leaky atrioventricular valves, which allow blood to pass back into the atria after they

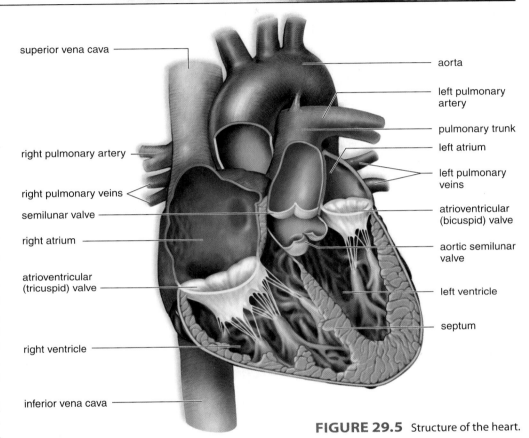

FIGURE 29.5 Structure of the heart.

have closed. The heart valves may also be affected by rheumatic fever, a bacterial infection that begins in the throat and spreads throughout the body. The bacteria attack various organs, including the heart valves. When damage is severe, the valve can be replaced with a synthetic valve or one taken from a pig's heart.

Another observation is in order. Some people associate O_2-poor blood with all veins and O_2-rich blood with all arteries, but this idea is incorrect: Pulmonary arteries and pulmonary veins are just the reverse. That is why the pulmonary arteries are colored blue and the pulmonary veins are colored red in Figure 29.5. Keep in mind that an **artery** is a vessel that takes blood away from the heart, and a **vein** is a vessel that takes blood to the heart, regardless of the blood's oxygen content.

Section 29.6 describes the events of a heartbeat.

> **29.5 Check Your Progress** Which side of the heart contains more O_2-poor hemoglobin, and which side contains more O_2-rich hemoglobin?

29.6 The heartbeat is rhythmic

The average human heart contracts, or beats, about 70 times a minute, or 2.5 billion times in a lifetime. Each heartbeat lasts about 0.85 seconds, called the **cardiac cycle**, and can be divided into three phases (**Fig. 29.6A**):

1 The atria contract (while the ventricles relax); 0.15 sec.

2 The ventricles contract (while the atria relax); 0.30 sec.

3 All chambers rest; 0.40 sec.

The term **systole** refers to contraction of the heart chambers, and the word **diastole** refers to relaxation of these chambers. Note that the heart is in diastole about 50% of the time. The short systole of the atria is appropriate, since the atria send blood only into the ventricles. It is the muscular ventricles that actually pump blood out into the cardiovascular system proper. The word *systole* used alone usually refers to the left ventricular systole. The volume of blood that the left ventricle pumps per minute into the systemic circuit is almost equivalent to the amount of blood in the body. During heavy exercise, the cardiac output can increase manyfold.

When the heart beats, the familiar "*lub-dub*" sound is heard as the valves of the heart close. The longer and lower-pitched *lub* is caused by vibrations of the heart when the atrioventricular valves close due to ventricular contraction. The shorter and sharper *dub* is heard when the semilunar valves close due to back pressure of blood in the arteries. The **pulse** is a wave effect that passes down the walls of the arterial blood vessels following ventricular systole and can be felt at various points externally.

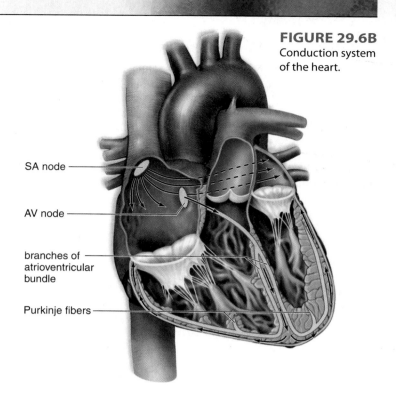

FIGURE 29.6B Conduction system of the heart.

The arterial pulse rate can be used to determine the heart rate, which is why taking your pulse is one of the first things a physician does during an examination.

The rhythmic contraction of the heart is due to the **cardiac conduction system** (**Fig. 29.6B**). Nodal tissue, which has both muscular and nervous characteristics, is a unique type of cardiac muscle. The SA (sinoatrial) node, located in the dorsal wall of the right atrium, initiates the heartbeat every 0.85 seconds. Therefore, the SA node is called the **cardiac pacemaker**. When the impulse reaches the AV (atrioventricular) node located in the base of the right atrium near the septum, it signals the ventricles to contract by way of large fibers terminating in the more numerous and smaller Purkinje fibers. Although the beat of the heart is intrinsic, it is regulated by the nervous system, which can increase or decrease the rate.

An **electrocardiogram (ECG)** is a recording of the electrical changes that occur in the heart during a cardiac cycle. When an ECG is being taken, electrodes placed on the skin are connected by wires to an instrument that detects the heart's electrical changes. Thereafter, a pattern appears that reflects the contractions of the heart. Various types of abnormalities can be detected by an ECG. Ventricular fibrillation (uncoordinated contractions) is of special interest because it can be caused by an injury or drug overdose. It is the most common cause of sudden cardiac death in a seemingly healthy person. Once the ventricles are fibrillating, they have to be defibrillated by applying a strong electric current for a short period of time.

This completes our study of the human heart. We begin our study of the blood vessels in Section 29.7.

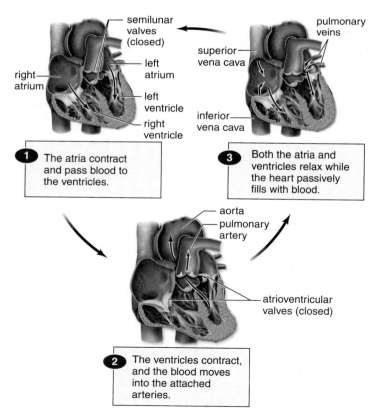

1 The atria contract and pass blood to the ventricles.

3 Both the atria and ventricles relax while the heart passively fills with blood.

2 The ventricles contract, and the blood moves into the attached arteries.

FIGURE 29.6A The phases of a heartbeat.

29.6 Check Your Progress With which node do you associate (*a*) atrial systole and (*b*) ventricular systole?

29.7 Blood vessel structure is suited to its function

The cardiovascular system has three types of blood vessels: arteries (and arterioles), which carry blood away from the heart to the capillaries; **capillaries**, which permit exchange of material with the tissues; and veins (and venules), which return blood from the capillaries to the heart (**Fig. 29.7A**).

Arteries have a much thicker wall than veins because of a well-developed middle layer consisting of smooth muscle and elastic tissue. The elastic tissue allows arteries to expand and accommodate the sudden increase in blood volume that results after each heartbeat. The well-developed smooth muscle prevents arteries from expanding too much.

Smaller arteries branch into a number of **arterioles**, which are just visible to the naked eye. The diameter of arterioles can

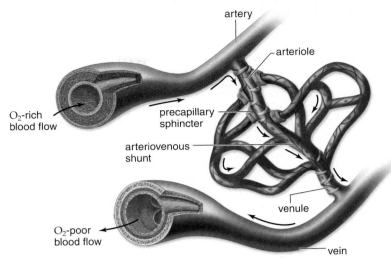

FIGURE 29.7B Anatomy of a capillary bed.

be regulated by the nervous system. When arterioles are dilated, more blood flows through them, and when they are constricted, less blood flows. The constriction of arterioles can also raise blood pressure.

Arterioles branch into capillaries. Capillaries are extremely narrow—about 8–10 mm wide—and have thin walls composed of a single layer of epithelium with a basement membrane. The thin walls of a capillary facilitate capillary exchange. Although each capillary is small, they form vast networks; their total surface area in humans is about 6,000 square meters. Since capillaries serve the cells, the heart and the other vessels of the cardiovascular system can be thought of as the means by which blood is conducted to and from the capillaries. Only certain capillary beds are open at any given time. For example, after eating, the capillary beds that serve the digestive system are open, and those that serve the muscles are closed. Each capillary bed has an arteriovenous shunt that allows blood to go directly from the arteriole to the venule, bypassing the bed (**Fig. 29.7B**). Contracted precapillary sphincter muscles prevent the blood from entering the capillary vessels.

Veins and venules take blood from the capillary beds to the heart. First, the **venules** (small veins) drain blood from the capillaries; then they join to form a vein. The middle layer of a vein (and venule) is thinner than that of an artery. This makes them subject to pressure exerted by skeletal muscles, and this pressure helps move blood in the veins. Also, veins often have **valves**, which allow blood to flow only toward the heart when open and prevent the backward flow of blood when closed.

The body has two blood pathways, which are described in Section 29.8.

> **29.7 Check Your Progress** What force helps move blood (*a*) in arteries and (*b*) in veins?

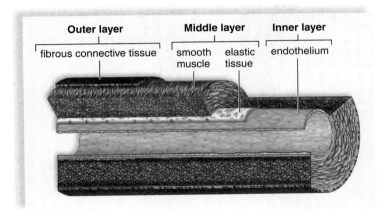

Artery

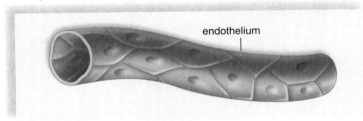

Capillary

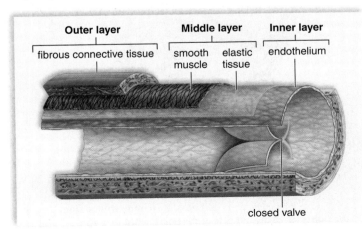

Vein

FIGURE 29.7A Types of blood vessels.

The human cardiovascular system includes two major circular pathways: the pulmonary circuit and the systemic circuit (**Fig. 29.8**).

The Pulmonary Circuit

In the pulmonary circuit, the path of blood can be traced as follows. O_2-poor blood from all regions of the body collects in the right atrium and then passes into the right ventricle, which pumps it into the pulmonary trunk. The pulmonary trunk divides into the right and left pulmonary arteries, which carry blood to the lungs. As blood passes through pulmonary capillaries, carbon dioxide is given off and oxygen is picked up. O_2-rich blood returns to the left atrium of the heart, through pulmonary venules that join to form pulmonary veins.

Notice in Figure 29.8 that in the pulmonary circuit, arteries contain O_2-poor blood and are colored blue. The pulmonary veins contain O_2-rich blood and are colored red.

The Systemic Circuit

In the systemic circuit, arteries contain O_2-rich blood and have a bright red color, but veins contain O_2-poor blood and appear dull red or, when viewed through the skin, blue. The aorta and the venae cavae (sing., **vena cava**) are the major blood vessels in the systemic circuit. To trace the path of blood to any organ in the body, you need only to start with the left ventricle and then mention the aorta, the proper branch of the aorta, the organ, and the vein returning blood to the vena cava, which enters the right atrium. For example, if you were tracing the path of the blood to and from the kidneys, you would mention the left ventricle, the aorta, the renal artery, the renal vein, the inferior vena cava, and the right atrium.

The coronary arteries (not shown in Figure 29.8) are extremely important because they serve the heart muscle itself. Failure of the coronary arteries to perform this function results in a heart attack because the heart is not nourished by the blood in its chambers. The coronary arteries arise from the aorta just above the aortic semilunar valve. They lie on the exterior surface of the heart, where they branch into arterioles and then capillaries. In the capillary beds, nutrients, wastes, and gases are exchanged between the blood and the tissues. The capillary beds enter venules, which join to form the cardiac veins, and these empty into the right atrium.

A **portal system** begins and ends in capillaries. The hepatic portal system takes blood from the intestines to the liver. The liver, an organ of homeostasis, modifies substances absorbed by the intestines, removes toxins and bacteria picked up from the intestines, and monitors the composition of the blood. Blood leaves the liver by way of the hepatic vein, which enters the inferior vena cava.

Although Figure 29.8 gives the impression that only arteries occur on the left side of the body and only veins occur on the right side of the body, this is not the case. In fact, all parts of the body contain all three types of blood vessels. For example, the iliac artery and vein run side by side into each leg. Similarly, each kidney receives both a renal artery and a renal vein. For both kidneys, the renal artery is taking blood into the arterioles/capillaries of the kidney, while the renal vein is draining the venules of a kidney.

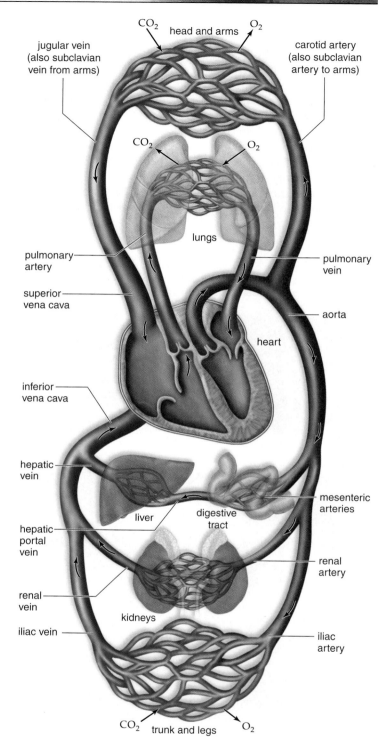

FIGURE 29.8 Path of blood in the body.

Blood flow in the arteries and veins is studied in Section 29.9.

> **29.8 *Check Your Progress*** Is it possible to move blood from the right side to the left side of the heart, without going through the lungs?

29.9 Blood pressure is essential to the flow of blood in each circuit

When the left ventricle contracts, blood is forced into the aorta and other systemic arteries under pressure. **Systolic pressure** results from blood being forced into the arteries during ventricular systole, and **diastolic pressure** is the pressure in the arteries during ventricular diastole. Human **blood pressure** can be measured with a traditional mercury manometer or with a digital manometer. Skill is required to accurately use a mercury manometer, but a digital manometer usually requires no training. With both types of manometers, a pressure cuff determines the amount of pressure required to stop the flow of blood through an artery. Blood pressure is normally measured on the brachial artery of the upper arm, but digital manometers often use other parts of the body, such as the wrist. A blood pressure reading consists of two numbers—for example, 120/80—that represent systolic and diastolic pressures, respectively.

Blood pressure accounts for the flow of blood from the heart to the capillaries. As blood flows from the aorta into the various arteries and arterioles, blood pressure falls. Also, the difference between systolic and diastolic pressure gradually diminishes. Notice in **Figure 29.9A** the fall of blood pressure and blood velocity in the capillaries. This may be related to the very high total cross-sectional area of the capillaries. It has been calculated that if all the blood vessels in a human were connected end to end, the total distance would reach around the Earth at the equator two times. A large portion of this distance would be due to the quantity of capillaries. The slow movement of blood in the capil-

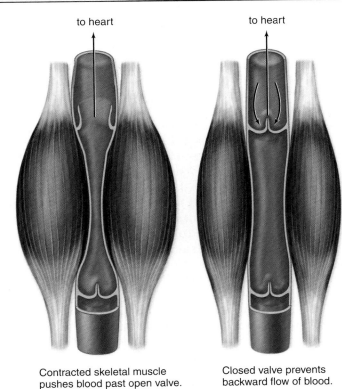

to heart   to heart

Contracted skeletal muscle pushes blood past open valve.

Closed valve prevents backward flow of blood.

FIGURE 29.9B How a valve affects the movement of blood in a vein.

laries provides time for the gas exchange and nutrient-for-waste exchange that occur across capillary walls.

Blood pressure in the veins is low and cannot move blood back to the heart, especially from the limbs. Instead, venous return depends upon three factors: skeletal muscle contraction, the presence of valves in veins, and respiratory movements. When the skeletal muscles near veins contract, they put pressure on the collapsible walls of the veins and on the blood contained in these vessels. Veins, however, have valves that prevent the backward flow of blood, and therefore pressure from muscle contraction is sufficient to move blood through the veins toward the heart (**Fig. 29.9B**). When a person inhales, the thoracic pressure falls and the abdominal pressure rises as the chest expands. This also aids the flow of venous blood back to the heart because blood flows in the direction of reduced pressure. Blood velocity increases slightly in the venous vessels due to a progressive reduction in the cross-sectional area as small venules join to form veins.

Varicose veins, abnormal dilations in superficial veins, develop when the valves of the veins become weak and ineffective due to backward pressure of the blood. Crossing the legs or sitting in a chair so that its edge presses against the back of the knees can contribute to the development of varicose veins in the legs. Varicose veins of the anal canal are known as hemorrhoids.

In Section 29.10, we begin our study of cardiovascular disease.

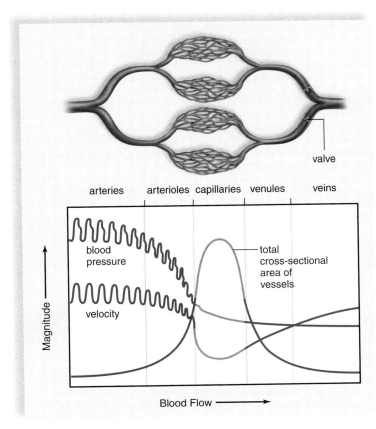

FIGURE 29.9A Velocity and blood pressure are related to the cross-sectional area of the blood vessels.

29.9 *Check Your Progress* Is blood pressure equal throughout the cardiovascular system? Explain.

29.10 Blood vessel deterioration results in cardiovascular disease

Cardiovascular disease (CD) is the leading cause of untimely death in Western countries. In the United States, it is estimated that about 20% of the population suffers from **hypertension**, which is high blood pressure. Hypertension is sometimes called a silent killer because it may not be detected until a stroke or heart attack occurs.

Heredity and lifestyle contribute to hypertension. For example, hypertension is often seen in individuals who have atherosclerosis, which occurs when **plaque** protrudes into the lumen of a vessel and interferes with the flow of blood (see Fig. 29.11). Atherosclerosis begins in early adulthood and develops progressively through middle age, but symptoms may not appear until an individual is 50 or older. To prevent its onset and development, the American Heart Association and other organizations recommend the health measures discussed in Section 29.11.

Plaque can cause a clot to form on the irregular arterial wall. As long as the clot remains stationary, it is called a *thrombus*, but when and if it dislodges and moves along with the blood, it is called an *embolus*. If *thromboembolism* is not treated, serious health problems can result. A cardiovascular accident, also called a **stroke**, often occurs when a small cranial arteriole bursts or is blocked by an embolus. Lack of oxygen causes a portion of the brain to die, and paralysis or death can result. A person is sometimes forewarned of a stroke by a feeling of numbness in the hands or the face, difficulty in speaking, or temporary blindness in one eye. If a coronary artery becomes completely blocked due to thromboembolism, a heart attack can occur, as described next.

The coronary arteries bring O_2-rich blood from the aorta to capillaries in the wall of the heart, and the cardiac veins return O_2-poor blood from the capillaries to the right ventricle. If the coronary arteries are narrow due to cardiovascular disease, the individual may first suffer from angina pectoris, chest pain that is often accompanied by a radiating pain in the left arm. When a coronary artery is completely blocked, a portion of the heart muscle dies due to lack of oxygen. This is known as a **heart attack**. Two surgical procedures are possible to correct a blockage or facilitate blood flow. In a coronary bypass operation, a portion of a blood vessel from another part of the body is sutured from the aorta to the coronary artery, past the point of obstruction (**Fig. 29.10**, *bottom left*). Now blood flows normally again from the aorta to the wall of the heart. In balloon angioplasty, a plastic tube is threaded through an artery to the

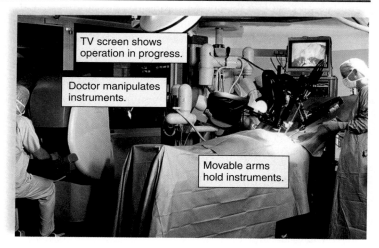

TV screen shows operation in progress.

Doctor manipulates instruments.

Movable arms hold instruments.

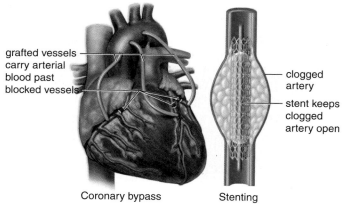

grafted vessels carry arterial blood past blocked vessels

clogged artery

stent keeps clogged artery open

Coronary bypass   Stenting

FIGURE 29.10 Treatment for clogged coronary artery.

blockage, and a balloon attached to the end of the tube is inflated to break through the blockage. A stent is often used to keep the vessel open (Fig. 29.10, *bottom right*).

Prevention of cardiovascular disease, as discussed in Section 29.11, is preferable to treatment.

> **29.10** *Check Your Progress* **What are the two major causes of a stroke, and how are they related?**

29.11 Cardiovascular disease can often be prevented

All of us can take steps to prevent cardiovascular disease. Certain genetic factors predispose an individual to cardiovascular disease, such as family history of heart attack under age 55, male gender, and ethnicity (African Americans are at greater risk). People with one or more of these risk factors need not despair, however. It only means that they should pay particular attention to the following guidelines for a heart-healthy lifestyle.

The Don'ts

Smoking Hypertension is well recognized as a major contributor to cardiovascular disease. When a person smokes, nicotine, the drug present in cigarette smoke, enters the bloodstream. Nicotine causes the arterioles to constrict and the blood pressure to rise. Restricted blood flow and cold hands are associated with smoking in most people. More serious is the need for the heart to pump harder to propel the blood through the lungs at a time when the oxygen-carrying capacity of the blood is reduced.

Drug Abuse Stimulants, such as cocaine and amphetamines, can cause an irregular heartbeat and lead to heart attacks and strokes in people who are using drugs, even for the first time. Intravenous drug use may result in a cerebral embolism.

Too much alcohol can destroy just about every organ in the body, the heart included. But investigators have discovered that people who take an occasional drink have a 20% lower risk of heart disease than do teetotalers. Two to four drinks a week is the recommended limit for men; one to three drinks for women.

Weight Gain Hypertension is prevalent in persons who are more than 20% above the recommended weight for their height. In these individuals, more tissues require servicing, and the heart sends the extra blood out under greater pressure. It may be harder to lose weight once it is gained, and therefore weight control should be a lifelong endeavor. Even a slight decrease in weight can bring about a reduction in hypertension, and a 4.5 kg weight loss doubles the chance that blood pressure can be normalized without drugs.

The Dos

Healthy Diet Diet influences the amount of cholesterol in the blood. Cholesterol is ferried by two types of plasma proteins, called LDL (low-density lipoprotein) and HDL (high-density lipoprotein). LDL (called "bad" lipoprotein) takes cholesterol from the liver to the tissues, and HDL (called "good" lipoprotein) transports cholesterol out of the tissues to the liver. When the LDL level in blood is high or the HDL level is abnormally low, plaque, which interferes with circulation, accumulates on arterial walls (**Fig. 29.11**; see Section 29.10).

Eating foods high in saturated fat (red meat, cream, and butter) and foods containing so-called trans fats (most margarines, commercially baked goods, and deep-fried foods) raises the LDL cholesterol level. Physicians advise people to replace these harmful fats with healthier ones, such as monounsaturated fats (olive and canola oils) and polyunsaturated fats (corn, safflower, and soybean oils).

Cold-water fish (e.g., halibut, sardines, tuna, and salmon) contain polyunsaturated fatty acids and especially omega-3 polyunsaturated fatty acids, which can reduce plaque.

Evidence is mounting to suggest a role for antioxidant vitamins (A, E, and C) in preventing cardiovascular disease. Antioxidants protect the body from free radicals that oxidize cholesterol and damage the lining of an artery, leading to a blood clot that can block blood vessels. Nutritionists believe that consuming at least five servings of fruits and vegetables a day may protect against cardiovascular disease.

Cholesterol Profile Starting at age 20, all adults are advised to have their cholesterol levels tested at least every five years. Even in healthy individuals, an LDL level above 160 mg/100 ml and an HDL level below 40 mg/100 ml are matters of concern. If a person has heart disease or is at risk for heart disease, an LDL level below 100 mg/100 ml is now recommended. Medications will most likely be prescribed for individuals who do not meet these minimum guidelines.

Exercise People who exercise are less apt to have cardiovascular disease. One study found that moderately active men who spent an average of 48 minutes a day on a leisure-time activity, such as gardening, bowling, or dancing, had one-third fewer heart attacks than their peers who spent an average of only 16 minutes each day being active. Exercise helps keep weight under control, may help minimize stress, and reduces hypertension.

The heart beats faster when exercising, but exercise slowly increases the heart's capacity. This means that the heart can beat more slowly when we are at rest and still do the same amount of work. One physician recommends that his cardiovascular patients walk for one hour, three times a week, and in addition, practice meditation and yogalike stretching and breathing exercises to reduce stress.

This completes our study of the cardiovascular system. We begin discussing blood in Section 29.12.

> **29.11** *Check Your Progress* *a.* What types of foods are protective against CD? *b.* What types of food should be avoided?

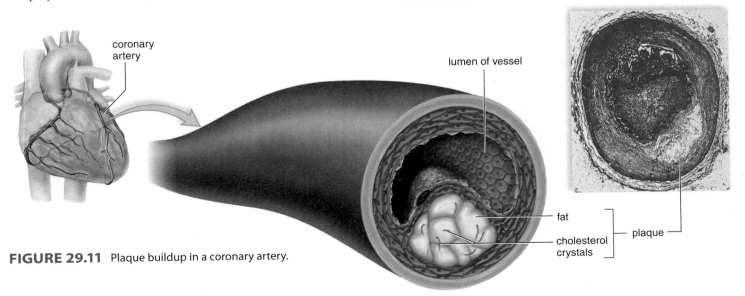

FIGURE 29.11 Plaque buildup in a coronary artery.

In this part of the chapter, we consider the composition and function of blood before taking up associated topics.

29.12 Blood is a liquid tissue

Blood's numerous functions include the following:

1. Transports substances to and from the capillaries, where exchanges with tissue fluid take place.
2. Helps defend the body against invasion by pathogens (e.g., disease-causing viruses and bacteria).
3. Helps regulate body temperature.
4. Forms clots, preventing a potentially life-threatening loss of blood.

In humans, blood has two main portions: the liquid portion, called plasma, and the formed elements, consisting of various cells and platelets (**Fig. 29.12**). The formed elements are manufactured continuously within the red bone marrow of certain bones, namely the skull, the ribs, the vertebrae, and the ends of the long bones.

Plasma is composed mostly of water (90–92%) and proteins (7–8%), but it also contains smaller quantities of many types of molecules, including nutrients, wastes, and salts. The salts and proteins are involved in buffering the blood, effectively keeping the pH near 7.4. They also maintain the blood's osmotic pressure so that water has an automatic tendency to enter blood capillaries. Several plasma proteins (e.g., prothrombin and fibrinogen) are involved in blood clotting, and others transport large organic molecules in the blood. Albumin, the most plentiful of the plasma proteins, transports bilirubin, a breakdown product of hemoglobin. Lipoproteins transport cholesterol.

The **formed elements** are red blood cells, white blood cells, and platelets. Among the formed elements, **red blood cells**, also called erythrocytes, transport oxygen. Red blood cells are small, biconcave disks that at maturity lack a nucleus and contain the respiratory pigment hemoglobin. There are 4–6 million red blood cells per mm^3 of whole blood, and each one of these cells contains about 250 million hemoglobin molecules. **Hemoglobin** contains iron, which combines loosely with oxygen; in this way, red blood cells transport oxygen. If the number of red blood cells is insufficient, or if the cells do not have enough hemoglobin, the individual suffers from **anemia** and has a tired, run-down feeling. The hormone *erythropoietin* stimulates the production of red blood cells. The kidneys produce erythropoietin when they act on a precursor made by the liver. Now available as a drug, erythropoietin is helpful to persons with anemia and has also been abused by athletes to enhance their performance.

Before they are released from the bone marrow into the blood, red blood cells lose their nuclei and begin to synthesize hemoglobin. After living about 120 days, they are destroyed, chiefly in the liver and the spleen, where they are engulfed by large phagocytic cells. When red blood cells are destroyed, hemoglobin is released. The iron is recovered and returned to the red bone marrow for reuse. Other portions of the molecules (i.e., heme) undergo chemical degradation and are excreted by the

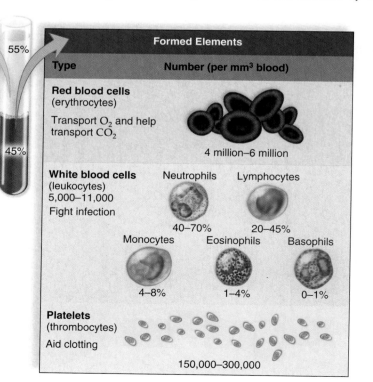

Plasma	
Type	**Function**
Water (90–92% of plasma)	Maintains blood volume; transports molecules
Plasma proteins (7–8% of plasma)	Maintain blood osmotic pressure and pH
Globulins	Transport; fight infection
Fibrinogen	Blood clotting
Salts (less than 1% of plasma)	Maintain blood osmotic pressure and pH; aid metabolism
Gases (O_2 and CO_2)	Cellular respiration
Nutrients (lipids, glucose, and amino acids)	Food for cells
Wastes (urea and uric acid)	End product of metabolism; excretion by kidneys
Hormones	Aid metabolism

55%

45%

Formed Elements	
Type	**Number (per mm^3 blood)**
Red blood cells (erythrocytes) Transport O_2 and help transport CO_2	4 million–6 million
White blood cells (leukocytes) 5,000–11,000 Fight infection	Neutrophils 40–70% Lymphocytes 20–45% Monocytes 4–8% Eosinophils 1–4% Basophils 0–1%
Platelets (thrombocytes) Aid clotting	150,000–300,000

FIGURE 29.12 Composition of blood.

liver as bile pigments in the bile. The bile pigments are primarily responsible for the color of feces.

White blood cells, also called leukocytes, help fight infections. White blood cells differ from red blood cells in that they are usually larger and have a nucleus, they lack hemoglobin, and without staining, they appear translucent. With staining, white blood cells appear light blue unless they have granules that bind with certain stains. The white blood cells that have granules also have a lobed nucleus. The agranular leukocytes have a spherical or indented nucleus and no granules. There are approximately 5,000–11,000 white blood cells per mm³ of blood. Growth factors are available to increase the production of all white blood cells, and these are helpful to people with low immunity, such as AIDS patients.

Red blood cells are confined to the blood, but white blood cells are able to squeeze between the cells of a capillary wall. Therefore, they are found in tissue fluid, lymph, and lymphatic organs. When an infection is present, white blood cells greatly increase in number. Many white blood cells live only a few days—they probably die while engaging pathogens. Others live months or even years.

When microorganisms enter the body due to an injury, an *inflammatory response*, characterized by swelling, reddening, heat, and pain, occurs at the injured site. Damaged tissue releases kinins, which dilate capillaries, and histamines, which increase capillary permeability. White blood cells called **neutrophils**, which are amoeboid, squeeze through the capillary wall and enter the tissue fluid, where they phagocytize foreign material. White blood cells called **monocytes** appear and are transformed into macrophages, large phagocytizing cells that release white blood cell growth factors. Soon, the number of white blood cells increases explosively. A thick, yellowish fluid called pus contains a large proportion of dead white blood cells that have fought the infection.

Lymphocytes, another type of white blood cell, also play an important role in fighting infection. Lymphocytes called **T cells** attack infected cells that contain viruses. Other lymphocytes, called **B cells**, produce antibodies. Each B cell produces just one type of antibody, which is specific for one type of antigen. An **antigen**, which is most often a protein but sometimes a polysaccharide, causes the body to produce an **antibody** to combine with the antigen. Antigens are present in the outer covering of parasites or in their toxins. When antibodies combine with antigens, the complex is often phagocytized by a macrophage. An individual is actively immune when a large number of B cells are all producing the antibody needed for a particular infection.

Blood clotting is our topic in the next section.

> **29.12 *Check Your Progress*** Why is it proper to call blood a liquid tissue?

29.13 Blood clotting involves platelets

Platelets result from fragmentation of large cells in the bone marrow called megakaryocytes. The blood contains 150,000–300,000 platelets per mm³. **Figure 29.13** shows the process of blood clotting. ❶ When a blood vessel in the body is damaged, ❷ platelets clump at the site of the puncture and partially seal the leak. Platelets and the injured tissues release a clotting factor called prothrombin activator that converts *prothrombin* to thrombin. This reaction requires calcium ions (Ca^{2+}). **Thrombin**, in turn, acts as an enzyme that severs two short amino acid chains from each *fibrinogen,* a plasma protein. These activated fragments then join end to end, forming long threads of **fibrin**. ❸ Fibrin threads wind around the platelet plug in the damaged area of the blood vessel and provide the framework for the clot. Red blood cells also are trapped within the fibrin threads; these cells make a clot appear red. A fibrin clot is present only temporarily. As soon as blood vessel repair is initiated, an enzyme called plasmin destroys the fibrin network and restores the fluidity of plasma.

If blood is allowed to clot in a test tube, a yellowish fluid develops above the clotted material. This fluid is called **serum**, and it contains all the components of plasma, except fibrinogen. Common blood tests often measure the amount of a substance in the serum, rather than in the blood.

Hemophilia is a well-known, inherited clotting disorder. Due to the absence of a particular clotting factor, the slightest bump can cause internal bleeding. Bleeding into the joints damages cartilage, and reabsorption of bone follows. Bleeding into the muscles causes muscle atrophy, and bleeding into the brain can lead to death.

Blood stem cells have the potential to help cure many human ills, as described in Section 29.14.

> **29.13 *Check Your Progress*** Why is it beneficial for clotting to require several steps?

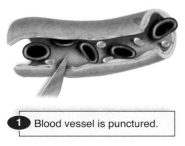
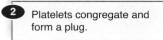
1. Blood vessel is punctured.

2. Platelets congregate and form a plug.

3. Fibrin threads form and trap red blood cells.

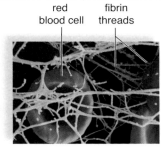
red blood cell / fibrin threads

FIGURE 29.13 Blood clotting.

Of the many scientific breakthroughs of the past few decades, none has been as controversial or divisive as stem cell technology. Few deny that this promising technique may potentially yield novel treatments for many devastating and debilitating disorders. However, ethical concerns and government-imposed restrictions have greatly inhibited research into new treatments and techniques and have delayed application of new technologies. Under current guidelines, embryonic stem cell research is limited to existing cell lines, and new embryonic stem cells may not be obtained. Because of these limitations, attention has focused on adult stem cells, a technology in use today that is both effective and much less controversial. In fact, you may be surprised to find that some adult stem cell technologies are already well-established and currently in use!

A stem cell is a cell that is capable of becoming different types of cells. While embryonic stem cells possess the ability to become virtually any cell type, adult stem cells are not quite as versatile because they can become only certain other types of cells. Adult stem cells may be found in many different tissues within the body, where they are mixed with normal cells. They generally remain inactive but may become active in the event of an injury, infection, or major tissue damage. Once activated, the stem cells divide rapidly and then develop into the needed cell type. The presence of adult stem cells is essential for normal growth, repair, and regeneration of adult tissues.

Adult stem cells have been identified in many tissues, including the liver, skin, muscle, and even within the brain, but the richest source is in the red bone marrow. More than 40 years ago, investigators discovered that hematopoietic stem cells in red bone marrow were capable of becoming a large number of different types of blood cells—including red blood cells, platelets, and many different types of white blood cells (**Fig. 29.14**). After they were discovered, adult stem cells were used to treat many white blood cell and immune system disorders, including leukemia, certain blood cancers, and anemia. To

this end, it is even possible to give someone a bone marrow transplant when their own has been destroyed by chemotherapy or radiation. The donated cells are injected into the recipient. If the transplant is successful, the cells find their way into the bone marrow and begin producing new white blood cells.

However, like any organ transplant, a bone marrow transplant poses the risk of rejection. This risk is generally much lower than for other types of organ transplants, and it can be further reduced by carefully matching recipients' tissue types to those of the donors. This obstacle is being further minimized through the use of umbilical cord blood, which is very rich in undifferentiated stem cells. Umbilical cord stem cells pose little risk of rejection, which can be reduced to almost zero when used to treat the individual from whom the cord blood stem cells were derived. Many companies now offer new parents the option of having their child's cord blood frozen and stored for potential future use.

Ethical concerns and government-imposed restrictions on embryonic stem cells have also spurred interest in the use of adult stem cells for tissue engineering. Currently, scientists are attempting to coax several types of adult stem cells into becoming other cell types, including neurons. Adult stem cells do not divide as vigorously in culture as do embryonic stem cells, but this difficulty is being addressed. If these efforts are successful, the ensuing technology has the potential to reduce the need for donated organs and tissues because the recipients of stem cells will repair their own organs.

In the next section, we will study capillary exchange, so necessary to the life of cells.

29.14 *Check Your Progress* **Explain why red bone marrow is a good source of adult stem cells.**

FIGURE 29.14 Hematopoietic cells (adult stem cells in red bone marrow) produce cells that become the various types of blood cells.

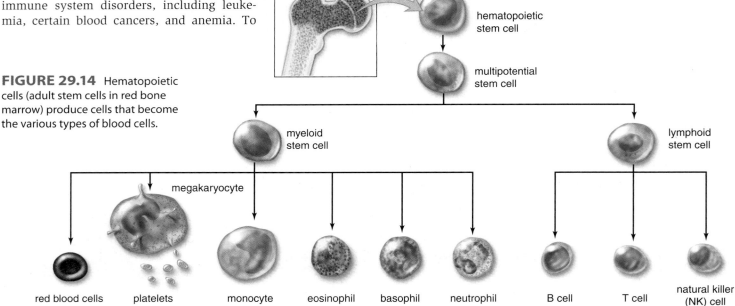

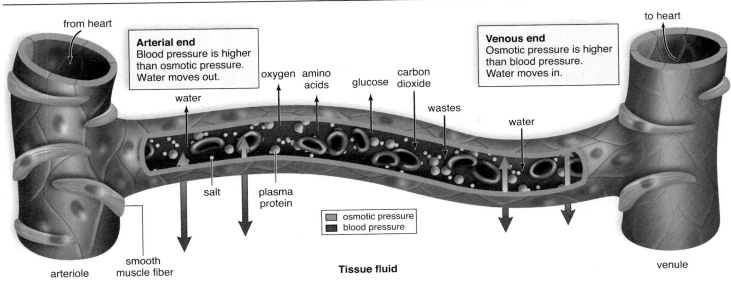

FIGURE 29.15A Capillary exchange.

Figure 29.15A illustrates capillary exchange between a systemic capillary and **tissue fluid**, the fluid between the body's cells. Blood that enters a capillary at the arterial end is rich in oxygen and nutrients, and it is under pressure created by the pumping of the heart. Two forces primarily control the movement of fluid through the capillary wall: blood pressure, which tends to cause water to move out of a capillary into the tissue fluid, and osmotic pressure, which tends to cause water to move from the tissue fluid into a capillary. At the arterial end of a capillary, blood pressure is higher than the osmotic pressure of blood (Fig. 29.15A). Osmotic pressure is created by the presence of salts and the plasma proteins. Because blood pressure is higher than osmotic pressure at the arterial end of a capillary, water exits a capillary at this end.

Midway along the capillary, where blood pressure is lower, blood pressure and osmotic pressure essentially cancel each other, and no net movement of water occurs. Solutes now diffuse according to their concentration gradient. Tissue fluid is always the area of lesser concentration of oxygen and nutrients because cells continually use them up. On the other hand, tissue fluid is always the area of greater concentration of carbon dioxide and wastes because cells generate wastes. Therefore, oxygen and nutrients (glucose and amino acids) diffuse out of the capillary, and carbon dioxide and other wastes diffuse into the capillary.

Red blood cells and almost all plasma proteins remain in the capillaries. The fluid and other substances that leave a capillary contribute to the tissue fluid. Because plasma proteins are too large to readily pass out of the capillary, tissue fluid tends to contain all the components of plasma, except much lesser amounts of protein.

At the venous end of a capillary, blood pressure has fallen to the point that osmotic pressure is greater than blood pressure, and water tends to move into the capillary. Almost the same amount of fluid that left the capillary returns to it, al-

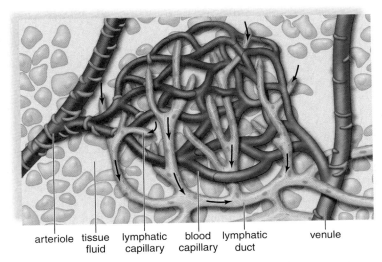

FIGURE 29.15B A lymphatic capillary bed lies near a blood capillary bed.

though some excess tissue fluid is always collected by the lymphatic capillaries (**Fig. 29.15B**). Tissue fluid contained within lymphatic vessels is called **lymph**. Lymph is returned to the systemic venous blood when the major lymphatic vessels enter the subclavian veins in the shoulder region. Lymphatic vessels begin at lymphatic capillaries in the tissues and end at the subclavian veins. The one-way lymphatic vessels play a vital role in maintaining blood pressure by returning fluid to the cardiovascular system.

Transfusions can help save lives, if blood types are carefully matched, as explained in Section 29.16.

29.15 *Check Your Progress* **What is the relationship between plasma, tissue fluid, and lymph?**

Many early blood transfusions resulted in illness and even death of some recipients. Eventually, it was discovered that only certain types of blood are compatible because red blood cell membranes carry specific proteins or carbohydrates that are antigens to blood recipients. An antigen is a foreign molecule, usually a protein, that the body reacts to. Several groups of red blood cell antigens exist, the most significant being the ABO system. Clinically, it is very important that the blood groups be properly cross-matched to avoid a potentially deadly transfusion reaction. In such a reaction, the recipient may die of kidney failure within a week.

ABO System

In the ABO system, the presence or absence of type A and type B antigens on red blood cells determines a person's blood type. For example, if a person has type A blood, the A antigen is on his or her red blood cells. This molecule is not an antigen to this individual, although it can be an antigen to a recipient who does not have type A blood.

In the ABO system, there are four types of blood: A, B, AB, and O. Within the plasma are antibodies to the antigens that are not present on the person's red blood cells. These antibodies are called anti-A and anti-B. This chart tells you what antibodies are present in the plasma of each blood type:

Blood Type	Antigen on Red Blood Cells	Antibody in Plasma
A	A	Anti-B
B	B	Anti-A
AB	A, B	None
O	None	Anti-A and anti-B

Because type A blood has anti-B and not anti-A antibodies in the plasma, a donor with type A blood can give blood to a recipient with type A blood (**Fig. 29.16A**). However, if type A blood is given to a type B recipient, agglutination occurs (**Fig. 29.16B**). Clumping of red blood cells, or agglutination, can cause blood to stop circulating in small blood vessels, and this leads to organ damage. It is also followed by hemolysis, or bursting of red blood cells, which if extensive, can cause the death of the individual.

Theoretically, which type blood would be accepted by all recipients? Type O blood has no antigens on the red blood cells and is sometimes called the universal donor. Which type blood could receive blood from any other blood type? Type AB blood has no anti-A or anti-B antibodies in the plasma and is sometimes called the universal recipient. In practice, however, it is not safe to rely solely on the ABO system when matching blood. Instead, samples of the two types of blood are physically mixed, and the result is microscopically examined before blood transfusions are done.

Rh System

Another important antigen in matching blood types is the Rh factor. Eighty-five percent of the U.S. population have this particular antigen on the red blood cells and are called Rh-positive. Fifteen percent do not have the antigen and are Rh-negative. Rh-negative individuals normally do not have antibodies to the Rh factor, but they may make them when exposed to the Rh factor. The designation of blood type usually also includes whether the person has or does not have the Rh factor on the red blood cells. This is done by attaching a minus (−) or Plus (+) sign to the blood type. For example, some people have A-negative blood, which is symbolized as A⁻.

Erythroblastosis Fetalis

During pregnancy, if the mother is Rh-negative and the father is Rh-positive, the child may be Rh-positive. The Rh-positive red blood cells may begin leaking across the placenta into the mother's cardiovascular system, since placental tissues normally break down before and at birth. Now, the mother produces anti-Rh antibodies. In this or a subsequent pregnancy with another Rh-positive baby, these antibodies may cross the placenta and destroy the child's red blood cells. Nowadays, this problem is prevented by giving Rh-negative women an Rh immunoglobulin injection midway through the first pregnancy and no later than 72 hours after giving birth to an Rh-positive child. This injection contains anti-Rh antibodies that attack any of the baby's red blood cells in the mother's blood before she starts making her own antibodies.

> **29.16 Check Your Progress** Why can't you give type A blood to a type B recipient?

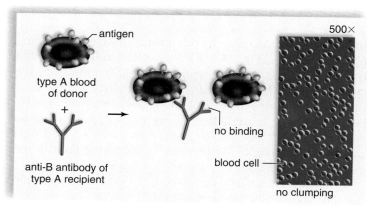

No agglutination

FIGURE 29.16A No agglutination occurs when the donor and recipient have the same type blood.

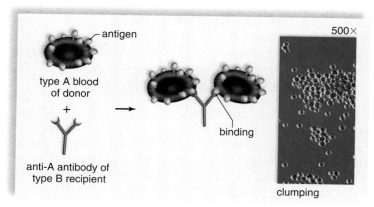

Agglutination

FIGURE 29.16B Agglutination occurs because blood type B has anti-A antibodies in the plasma.

CONNECTING THE CONCEPTS

It is possible to relate the type of cardiovascular system to the lifestyle of an animal. Some small, aquatic animals have no cardiovascular system—external water passing in and out of a gastrovascular cavity is sufficient to meet the needs of their cells. Invertebrates such as clams and lobsters have a cardiovascular system, but it utilizes hemolymph (blood plus tissue fluid) pumped into cavities of the body. Using the "structure suits function" theme, we can predict that these animals are relatively inactive because an open circulatory system is not an ef-

ficient method of transporting oxygen to cells, even if the circulatory fluid does contain a respiratory pigment. Grasshoppers are closely related to lobsters and have an active lifestyle—hopping and flying—and they have colorless blood. They utilize tracheae to deliver oxygen directly to their muscles. We traced the evolution of the two-circuit circulatory pathway in vertebrates and saw that a two-circuit pathway allows blood to pass to the lungs and to the tissues under pressure. This is particularly useful in birds and mammals, which maintain a warm body and an ac-

tive way of life on land. You will want to learn the particulars of the mammalian (human) cardiovascular system, keeping in mind the themes of structure suits function and homeostasis.

Body fluids make ideal culture media for the growth of infectious parasites, and these fluids often have ways to ward off an invasion. You already know that white blood cells are involved in these endeavors. Chapter 30 discusses how humans in particular are able to stay one step ahead of the microorganisms that want to take up residence in their bloodstream and cells.

The Chapter in Review

Summary

Not All Animals Have Red Blood
- Vertebrates have red blood because the respiratory pigment hemoglobin is red when it binds to oxygen.

A Circulatory System Helps Maintain Homeostasis

29.1 A circulatory system serves the needs of cells
- The circulatory system delivers oxygen and nutrients to cells and takes away carbon dioxide and other wastes.

29.2 Some invertebrates do not have a circulatory system
- Cnidarians, planarians, roundworms, and echinoderms lack a circulatory system.

29.3 Other invertebrates have an open or a closed circulatory system
- Open circulatory system: hemolymph (blood plus tissue fluid) is pumped into tissue spaces or a hemocoel.
- Closed circulatory system: blood is pumped into blood vessels.

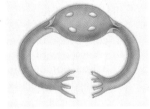
Open circulatory system

29.4 All vertebrates have a closed circulatory system
- Fishes have a single-loop circulatory pathway; heart has a single atrium and a single ventricle.
- Other vertebrates have a two-circuit circulatory pathway.
  - Amphibians and most reptiles: heart has two atria and a single ventricle.
  - Birds and mammals: heart has two atria and two ventricles.

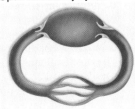
Closed circulatory system

The Mammalian Cardiovascular System Consists of the Heart and Blood Vessels

29.5 The mammalian heart has four chambers
- A septum separates the heart into right and left halves.
- Each side of the heart has an atrium and a ventricle.
- An artery takes blood away from the heart; a vein takes blood to the heart.
- Valves keep blood moving in the correct direction.

29.6 The heartbeat is rhythmic
- Systole is contraction; diastole is relaxation of heart chambers.
- During a single heartbeat, first the atria contract, and then the ventricles contract.
- Pulse: expansion of aorta following ventricular contraction.
- In the cardiac conduction system, the SA node initiates the heartbeat, and the AV node causes the ventricles to contract.

29.7 Blood vessel structure is suited to its function
- Blood pressure in arteries and arterioles carries blood away from the heart.
- Thin-walled capillaries permit exchange of material with the tissues.
- Skeletal muscle contraction returns blood in veins and venules to the heart.

29.8 Blood vessels form two circuits in mammals
- Pulmonary circuit: pulmonary arteries take O_2-poor blood to lungs; pulmonary veins return O_2-rich blood to heart.
- Systemic circuit: left ventricle sends O_2-rich blood to aorta; vena cava takes O_2-poor blood back to right atrium.
- Hepatic portal system begins at digestive tract and ends in liver.

29.9 Blood pressure is essential to the flow of blood in each circuit

- Systolic pressure is the pressure in arteries during ventricular systole.
- Diastolic pressure is the pressure in arteries during ventricular diastole.

29.10 Blood vessel deterioration results in cardiovascular disease

- Hypertension: high blood pressure, plaque obstructs coronary artery.
- Stroke: cranial arteriole bursts or is blocked by an embolus.
- Heart attack: blocked coronary artery.

29.11 Cardiovascular disease can often be prevented

- A heart-healthy lifestyle involves refraining from smoking or using drugs, controlling weight, following a healthy diet, monitoring cholesterol, and exercising regularly.

Blood Has Vital Functions

29.12 Blood is a liquid tissue

- Blood, which is composed of plasma and formed elements, transports, defends against pathogens, helps regulate body temperature, and forms clots.
- Plasma: mostly water and proteins along with some nutrients, wastes, and salts.
- Formed elements: red blood cells, white blood cells, and platelets.

29.13 Blood clotting involves platelets

- Platelets clump at the site of blood vessel damage, where they release clotting factor; long fibrin threads provide a framework for a blood clot.

29.14 Adult stem cells include blood stem cells

- Stem cells are capable of differentiating into different types of cells.
  - Hematopoietic stem cells are adult stem cells that are capable of producing all of the various types of blood cells.

29.15 Capillary exchange is vital to cells

- Capillary exchange in systemic tissues helps keep the internal environment constant.
- When blood reaches a capillary, water moves out at the arterial end due to blood pressure.
- Water moves in at the venous end of a capillary due to osmotic pressure.
- Between the arterial end and the venous end, nutrients diffuse out of and wastes diffuse into the capillary.
- Lymphatic capillaries collect excess tissue fluid (lymph) and return it to the cardiovascular system.

29.16 Blood types must be matched for transfusions

- ABO blood typing determines the presence or absence of A and B antigens on the surface of red blood cells.
- Clumping (agglutination): corresponding antigen and antibody are put together.
- The Rh factor is another important antigen in matching blood types.

Testing Yourself

A Circulatory System Helps Maintain Homeostasis

1. A circulatory system functions in
   - a. homeostasis.
   - b. transport of wastes.
   - c. transport of nutrients.
   - d. All of these are correct.

2. In insects with an open circulatory system, oxygen is taken to cells by
   - a. blood.
   - b. hemolymph.
   - c. tracheae.
   - d. capillaries.

3. Which one of these would you expect to be part of a closed, but not an open, circulatory system?
   - a. ostia
   - b. capillary beds
   - c. hemocoel
   - d. heart
   - e. All of these are correct.

4. **THINKING CONCEPTUALLY** Why is a closed circulatory system and four-chambered heart consistent with the lifestyle of birds and mammals and their ability to live in cold climates?

The Mammalian Cardiovascular System Consists of the Heart and Blood Vessels

5. Which of the following statements is true?
   - a. Arteries carry blood away from the heart, and veins carry blood to the heart.
   - b. Arteries carry blood to the heart, and veins carry blood away from the heart.
   - c. Arteries carry O_2-rich blood, and veins carry O_2-poor blood.
   - d. Arteries carry O_2-poor blood, and veins carry O_2-rich blood.

6. In humans, blood returning to the heart from the lungs returns to
   - a. the right ventricle.
   - b. the right atrium.
   - c. the left ventricle.
   - d. the left atrium.
   - e. both the right and left sides of the heart.

7. Systole refers to the contraction of the
   - a. major arteries.
   - b. SA node.
   - c. atria and ventricles.
   - d. major veins.
   - e. All of these are correct.

8. Which of the following lists the events of the cardiac cycle in the correct order?
   - a. contraction of atria, rest, contraction of ventricles
   - b. contraction of ventricles, rest, contraction of atria
   - c. contraction of atria, contraction of ventricles, rest
   - d. contraction of ventricles, contraction of atria, rest

9. Place the following blood vessels in order, from largest to smallest in diameter.
   - a. arterioles, capillaries, arteries
   - b. arteries, arterioles, capillaries
   - c. capillaries, arteries, arterioles
   - d. arterioles, arteries, capillaries
   - e. arteries, capillaries, arterioles

10. The best explanation for the slow movement of blood in capillaries is
    - a. skeletal muscles press on veins, not capillaries.
    - b. capillaries have much thinner walls than arteries.
    - c. there are many more capillaries than arterioles.
    - d. venules are not prepared to receive so much blood from the capillaries.
    - e. All of these are correct.

11. A stroke is caused by a(n)
   a. embolus in a cranial blood vessel.
   b. cranial blood vessel bursting.
   c. blockage of a coronary artery.
   d. Both a and b are correct choices.
   e. a, b, and c are all correct.
12. Which of the following is the best dietary choice for preventing cardiovascular disease?
   a. cholesterol   c. saturated fat
   b. polyunsaturated fat   d. trans fat

Blood Has Vital Functions

13. Which association is incorrect?
   a. white blood cells—infection fighting
   b. red blood cells—blood clotting
   c. plasma—water, nutrients, and wastes
   d. red blood cells—hemoglobin
   e. platelets—blood clotting
14. Which of the plasma proteins contributes most to osmotic pressure?
   a. albumin   c. erythrocytes
   b. globulins   d. fibrinogen
15. Red blood cells
   a. reproduce themselves by mitosis.
   b. live for several years.
   c. continually synthesize hemoglobin.
   d. are destroyed in the liver and spleen.
   e. More than one of these are correct.
16. When the oxygen capacity of the blood is reduced,
   a. the liver produces more bile.
   b. the kidneys release erythropoietin.
   c. the bone marrow produces more red blood cells.
   d. sickle-cell disease occurs.
   e. Both b and c are correct.
17. The last step in blood clotting
   a. is the only step that requires calcium ions.
   b. occurs outside the bloodstream.
   c. is the same as the first step.
   d. converts prothrombin to thrombin.
   e. converts fibrinogen to fibrin.
18. Stem cells are responsible for
   a. red blood cell production.
   b. white blood cell production.
   c. platelet production.
   d. the production of all formed elements.
19. In the tissues, nutrients and _____ are exchanged for _____ and other wastes.
   a. blood, oxygen   c. hemoglobin, tissue fluid
   b. oxygen, carbon dioxide d. None of these are correct.
20. **THINKING CONCEPTUALLY** Rita feels tired and run down, and the physician diagnoses iron-deficiency anemia. The physician doesn't prescribe erythropoietin. Why not?

Understanding the Terms

anemia 584
antibody 585
antigen 585
aorta 577
arteriole 579
artery 577
atrioventricular valve 577
atrium 577
B cell 585
blood 575
blood pressure 581
capillary 579
cardiac conduction
    system 578
cardiac cycle 578
cardiac pacemaker 578
cardiovascular system 577
closed circulatory
    system 575
diastole 578
diastolic pressure 581
electrocardiogram (ECG) 578
fibrin 585
formed element 584
gill 576
heart attack 582
heart murmur 577
hemoglobin 584
hemolymph 575
hemophilia 585
hypertension 582

lymph 587
lymphocyte 585
monocyte 585
neutrophil 585
open circulatory
    system 575
plaque 582
plasma 584
portal system 580
pulmonary artery 577
pulmonary circuit 576
pulmonary trunk 577
pulmonary vein 577
pulse 578
red blood cell 584
semilunar valve 577
septum 577
serum 585
stroke 582
systemic circuit 576
systole 578
systolic pressure 581
T cell 585
thrombin 585
tissue fluid 587
valve 579
vein 577
vena cava 580
ventricle 577
venule 579
white blood cell 585

Match the terms to these definitions:
   a. _____ Blood vessel that transports blood away from the heart.
   b. _____ The liquid portion of blood; contains nutrients, wastes, salts, and proteins.
   c. _____ The major systemic veins that take blood to the heart from the tissues.
   d. _____ Iron-containing respiratory pigment occurring in vertebrate red blood cells and in the blood plasma of many invertebrates.

Thinking Scientifically

1. What evidence from examining (a) dead vertebrates and (b) live vertebrates substantiates that their blood circulates?
2. You are a biochemist who decides to analyze the plasma composition of arterial versus venous blood. What data would be consistent with the function of capillaries? Explain.

ARIS *Visit* **www.mhhe.com/maderconcepts** *for practice quizzes, animations, videos, and activities designed to help you master the material in this chapter.*

30
Lymph Transport and Immunity

LEARNING OUTCOMES

After studying this chapter, you should be able to accomplish the following outcomes.

AIDS Destroys the Immune System

1 Relate the progression of opportunistic infections accompanying AIDS to the number of helper T cells in the body.

The Lymphatic System Functions in Transport and Immunity

2 Discuss four functions of the lymphatic system.
3 Describe the structure and function of the lymphatic vessels.
4 Give the chief functions of four lymphatic organs and three patches of lymphatic tissue.

The Body's First Line of Defense Against Disease Is Nonspecific and Innate

5 Group the first responders into four categories.
6 Discuss how a fever could be part of the body's first and second lines of defense.
7 Describe the inflammatory response in terms of four events.

The Body's Second Line of Defense Against Disease Is Specific to the Pathogen

8 Distinguish between a foreign antigen and a self-antigen.
9 Distinguish between active and passive immunity.
10 Name and describe the cells of the immune system.
11 Compare and contrast antibody-mediated immunity with cell-mediated immunity.
12 Discuss the research and medical uses of monoclonal antibodies.

Abnormal Immune Responses Can Have Health Consequences

13 Explain the role of MHC antigens in tissue rejection.
14 Name several autoimmune diseases, and tell the symptoms of each.
15 Distinguish between immediate and delayed allergic responses.

A person ravaged by **AIDS** (acquired immunodeficiency syndrome) can no longer fight off the onslaught of viruses, fungi, and bacteria that attack the body every day. AIDS results when **HIV** (immunodeficiency virus) damages and destroys the T cells of the immune system, which consists of all the types of cells that combat foreign substances. For years, the body is able to fight off the effects of HIV infection, but eventually the infection gains the upper hand. The number of T cells drops from the normal thousands (4,500 to 10,000) to less than a hundred as the immune system becomes helpless.

AIDS victim: Kaposi sarcoma is evident

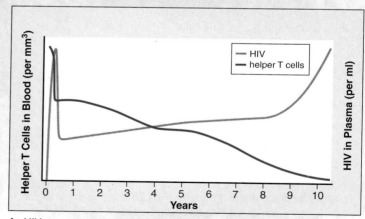

As HIV progresses, helper T cell numbers decline

The symptoms of an HIV infection begin with weight loss, chronic fever, cough, diarrhea, swollen glands, and shortness of breath, and progress to rare infections that would not cause disease in a person with a healthy immune system. Such infections are known as **opportunistic infections (OIs)** because they take advantage of a weakened immune system. The ap-

AIDS Destroys the Immune System

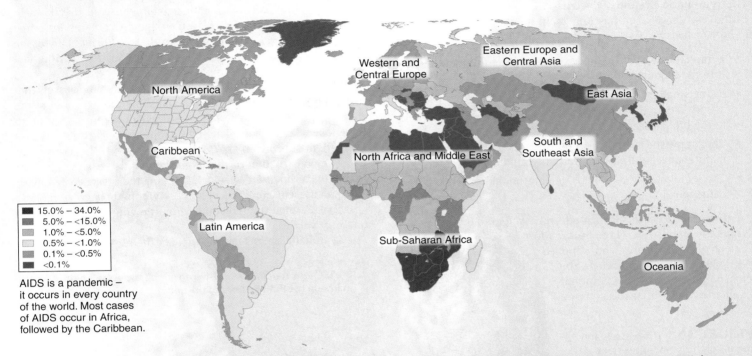

15.0% – 34.0%
5.0% – <15.0%
1.0% – <5.0%
0.5% – <1.0%
0.1% – <0.5%
<0.1%

North America

Caribbean

Latin America

Western and
Central Europe

Eastern Europe and
Central Asia

East Asia

North Africa and Middle East

South and
Southeast Asia

Sub-Saharan Africa

Oceania

AIDS is a pandemic –
it occurs in every country
of the world. Most cases
of AIDS occur in Africa,
followed by the Caribbean.

pearance of the following OIs can be associated with the helper T cell count:

- Shingles, a painful infection with chickenpox virus; helper T cell count of less than 500/mm^3.

- Candidiasis, a fungal infection of the mouth, throat, or vagina; helper T cell count of about 350/mm^3.

- Pneumocystis pneumonia, a fungal infection that causes the lungs to become useless as they fill with fluid and debris; helper T cell count of less than 200/mm^3.

- Kaposi sarcoma, a cancer of blood vessels due to human herpes virus 8; gives rise to reddish-purple, coin-sized spots and lesions on the skin; helper T cell count of less than 200/mm^3.

- Toxoplasmic encephalitis, a protozoan infection characterized by severe headaches, fever, seizures, and coma; helper T cell count of less than 100/mm^3.

Chicken pox virus

Candidiasis
(tongue)

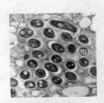

Pneumocystic
pneumonia

- *Mycobacterium avium* complex (MAC), a bacterial infection resulting in persistent fever, night sweats, fatigue, weight loss, and anemia; helper T cell count of less than 75/mm^3.

- Cytomegalovirus, a viral infection that leads to blindness, inflammation of the brain, and throat ulcerations; helper T cell count of less than 75/mm^3.

Worldwide, an estimated 40 million people are living with AIDS; more than 1 million are in the United States. Advances in treatment have reduced the serious complications of the disease and prolonged life. But now, new strains of the virus have emerged that are resistant to the drugs used for treatment.

Understanding why AIDS patients are so sick gives us a whole new level of appreciation for the workings of a healthy immune system, which protects us daily from serious illnesses. In this chapter, immunity will be broadly defined as involving the lymphatic system and any type of response to foreign substances entering the body. Homeostasis can only be maintained when the body is able to protect its own tissues and maintain its integrity.

The Lymphatic System Functions in Transport and Immunity

The lymphatic system consists of lymphatic vessels and lymphatic organs. Lymphatic vessels transport lymph, and lymphatic organs are involved in producing and storing cells that fight pathogens. Pathogens are disease-causing agents, such as viruses and bacteria.

30.1 Lymphatic vessels transport lymph

The **lymphatic system**, which is closely associated with the cardiovascular system, has four main functions that contribute to homeostasis:

- Lymphatic capillaries absorb excess tissue fluid and return it to the bloodstream.

- In the small intestines, lymphatic capillaries called **lacteals** absorb fats in the form of lipoproteins and transport them to the bloodstream.

- The lymphatic system is responsible for the production, maintenance, and distribution of lymphocytes.

- The lymphatic system helps defend the body against pathogens.

Lymphatic vessels form a one-way system that begins with lymphatic capillaries (**Fig. 30.1**). Most regions of the body are richly supplied with lymphatic capillaries—tiny, closed-ended vessels. Lymphatic capillaries take up excess tissue

fluid. The fluid inside lymphatic capillaries is called **lymph**. In addition to water and fat molecules, lymph contains the same ions, nutrients, gases, and proteins that are present in tissue fluid. It also contains defense molecules called **antibodies**, which are produced by lymphocytes.

The lymphatic capillaries join to form lymphatic vessels that merge before entering one of two ducts: the thoracic duct or the right lymphatic duct. The larger thoracic duct returns lymph to the left subclavian vein. The right lymphatic duct returns lymph to the right subclavian vein.

The construction of the larger lymphatic vessels is similar to that of cardiovascular veins, including the presence of valves. Skeletal muscle contraction forces lymph through lymphatic vessels, and then it is prevented from flowing backward by the one-way valves.

Lymphatic organs are the topic of the next section.

> **30.1 *Check Your Progress*** How do lymphatic vessels help maintain the fluid balance of the body?

FIGURE 30.1 The vessels and organs of the lymphatic system.

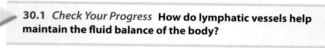

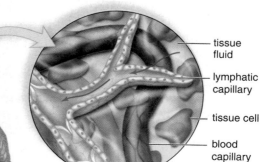

Right lymphatic duct: empties lymph into the right subclavian vein

Right subclavian vein: transports blood away from the right arm and the right ventral chest wall toward the heart

Axillary lymph nodes: located in the underarm region

Thoracic duct: empties lymph into the left subclavian vein

Inguinal lymph nodes: located in the groin region

Lymph nodes cleanse lymph and alert the immune system to pathogens.

Tonsil: patches of lymphatic tissue that help prevent the entrance of pathogens by way of the nose and mouth

Left subclavian vein: transports blood away from the left arm and the left ventral chest wall toward the heart

Red bone marrow: site for the origin of all types of blood cells

Thymus gland: lymphatic tissue where T cells mature and learn to tell "self" from "nonself"

Spleen: cleanses the blood of cellular debris and bacteria while resident T cells and B cells respond to the presence of antigens

tissue fluid

lymphatic capillary

tissue cell

blood capillary

valve

30.2 Lymphatic organs defend the body

The **lymphatic (lymphoid) organs** are shown in Figure 30.1 and **Figure 30.2**.

Red bone marrow is the site of stem cells that are ever capable of dividing and producing the various types of blood cells (see Fig. 29.14), including lymphocytes. Some of these become mature **B cells,** a major type of lymphocyte, in the bone marrow. In a child, most of the bones have red bone marrow, but in an adult, it is present only in the bones of the skull, the sternum (breastbone), the ribs, the clavicle (collarbone), the pelvic bones, the vertebral column, and the proximal heads of the femur and humerus.

The red bone marrow consists of a network of connective tissue fibers that supports the stem cells and their progeny. They are packed around thin-walled sinuses filled with venous blood. Differentiated blood cells enter the bloodstream at these sinuses.

The soft, bilobed **thymus gland** is located in the thoracic cavity between the trachea and the sternum ventral to the heart (Fig. 30.1). Immature **T cells,** the other major type of lymphocyte, migrate from the bone marrow through the bloodstream to the thymus, where they mature. The thymus also produces thymic hormones, such as thymosin, that are thought to aid in the maturation of T cells. The thymus varies in size, but it is largest in children and shrinks as we get older. In the elderly, it is barely detectable. When well developed, it contains many lobules.

Lymph nodes are small (about 1–25 mm in diameter), ovoid structures occurring along lymphatic vessels. Lymph nodes are named for their location. For example, inguinal lymph nodes are in the groin, and axillary lymph nodes are in the armpits. Physicians often feel for the presence of swollen, tender lymph nodes in the neck as evidence that the body is fighting an infection.

A lymph node has many open spaces called sinuses. As lymph courses through the sinuses, it is filtered by macrophages, which engulf debris and pathogens. Many B and T cells are also present in and around the sinuses of a lymph node.

Unfortunately, cancer cells sometimes enter lymphatic vessels and congregate in lymph nodes. Therefore, when a person undergoes surgery for cancer, it is a routine procedure to remove some lymph nodes and examine them to determine whether the cancer has spread to other regions of the body.

The **spleen** is located in the upper left side of the abdominal cavity posterior to the stomach. Most of the spleen is red pulp that filters the blood. Red pulp consists of blood vessels and sinuses, where macrophages remove old and defective blood cells. The spleen also has white pulp that is inside the red pulp and consists of little lumps of lymphatic tissue where B and T cells congregate.

The spleen's outer capsule is relatively thin, and an infection or a blow can cause the spleen to burst. Although the spleen's functions can be largely replaced by other organs, a person without a spleen is often slightly more susceptible to infections and may require antibiotic therapy indefinitely.

Patches of lymphatic tissue in the body include: the **tonsils,** located in the pharynx; **Peyer patches,** located in the intestinal wall; and the vermiform appendix, attached to the cecum. These structures encounter pathogens and antigens that enter the body by way of the mouth.

This completes our discussion of the lymphatic system. The next part of the chapter considers nonspecific defenses, also called nonspecific immunity.

> **30.2** *Check Your Progress* A person can have an HIV infection for many years before beginning to feel ill. What does this tell you about the ability of the red bone marrow to produce immature T cells, despite many mature ones being destroyed by an HIV infection?

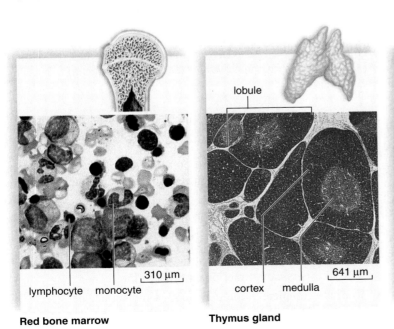
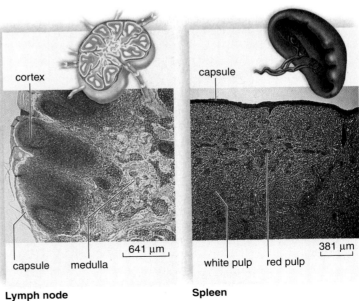

Red bone marrow

lymphocyte   monocyte   310 µm

Thymus gland

lobule

cortex   medulla   641 µm

Lymph node

cortex

capsule   medulla   641 µm

Spleen

capsule

white pulp   red pulp   381 µm

FIGURE 30.2 The lymphatic organs.

The Body's First Line of Defense Against Disease Is Nonspecific and Innate

Learning Outcomes 5–7, page 592

The first line of defense against disease consists of nonspecific defenses, also called nonspecific immunity. Included among nonspecific defenses are barriers to entry, complement proteins, certain blood cells, and the inflammatory response.

30.3 Barriers to entry, complement proteins, and certain blood cells are first responders

We are constantly exposed to pathogens in our food and drink and as we breathe air or touch objects, both living and inanimate, in our environment. The warm temperature and constant supply of nutrients in our bodies make them an ideal place for pathogens to flourish. Without a means of defense, we would be unable to prevent invasions by all sorts of pathogens and would be overrun. Fortunately, our body provides us with nonspecific defenses and specific defenses. Other organisms also have this ability to defend themselves, but none have been studied as much as humans because our defense mechanisms are very important to the field of medicine.

Immunity begins with the nonspecific defenses, which are summarized in **Figure 30.3A**. They include (1) barriers to entry such as the skin, (2) protective proteins such as complement and interferons, (3) phagocytes and natural killer cells, and (4) the inflammatory response, which is discussed in Section 30.5. Nonspecific defenses occur automatically because they are innate; however, no memory is involved—there is no recognition that an intruder has been attacked before. Later, we will discuss specific defenses, which are directed against particular pathogens and do exhibit memory.

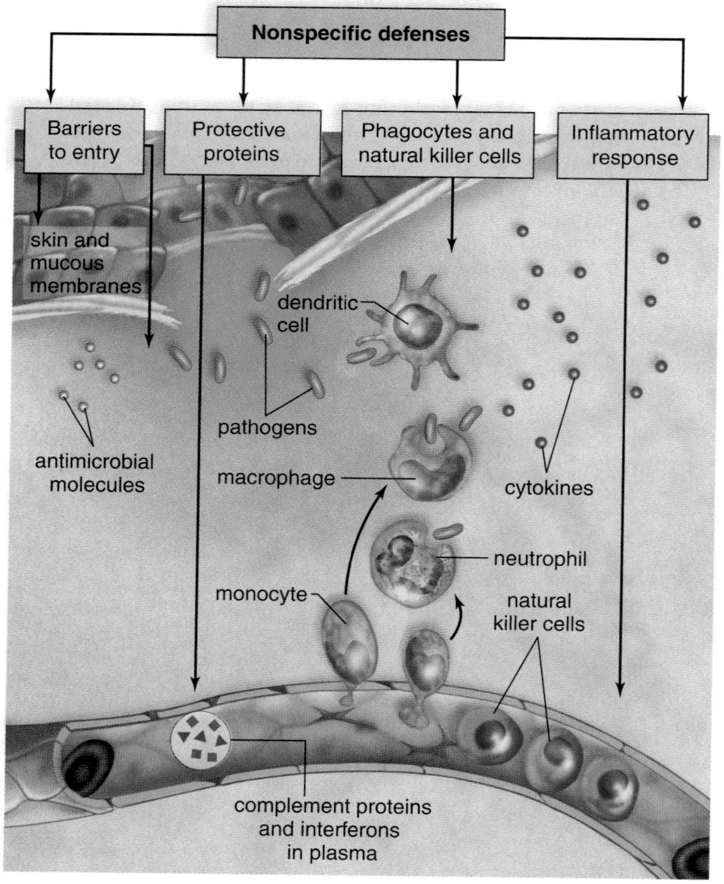

FIGURE 30.3A Overview of nonspecific defenses.

Barriers to Entry

Barriers to entry by pathogens include nonchemical, mechanical barriers, such as the skin and the mucous membranes lining the respiratory, digestive, and urinary tracts. For example, the upper layers of our skin are composed of dead, keratinized cells that form an impermeable barrier. But when the skin has been injured, one of the first concerns is the possibility of an infection. We are also familiar with the importance of sterility before an injection is given. The injection needle and the skin must be free of pathogens. This testifies to the importance of skin as a defense against invasion by a pathogen.

The mucus of mucous membranes physically ensnares microbes. The upper respiratory tract is lined by ciliated cells that sweep mucus and trapped particles up into the throat, where they can be swallowed or expectorated (coughed out). In addition, the various bacteria that normally reside in the intestine and other areas, such as the vagina, prevent pathogens from taking up residence.

Barriers to entry also include antimicrobial molecules. Oil gland secretions contain chemicals that weaken or kill certain bacteria on the skin; mucous membranes secrete lysozyme, an enzyme that can lyse bacteria; and the stomach has an acidic pH, which inhibits the growth of many types of bacteria, or may even kill them.

Protective Proteins

Complement is composed of a number of blood plasma proteins that "complement" certain immune responses, which accounts for their name. These proteins are continually present in the blood plasma but must be activated by pathogens to exert their effects. Complement helps destroy pathogens in three ways:

1. Enhanced inflammation. Complement proteins are involved in and amplify the inflammatory response because certain ones can bind to **mast cells** (type of white blood cell in tissues) and trigger histamine release, and others can attract phagocytes to the scene.
2. Some complement proteins bind to the surface of pathogens already coated with antibodies, which ensures that the pathogens will be phagocytized by a neutrophil or macrophage.

3. Certain other complement proteins join to form a **membrane attack complex** that produces holes in the surface of some bacteria and viruses. Fluids and salts then enter the bacterial cell or virus to the point that it bursts (**Fig. 30.3B**).

Interferons are **cytokines**, soluble proteins that affect the behavior of other cells. Interferons are made by virus-infected cells. They bind to the receptors of noninfected cells, causing them to produce substances that interfere with viral replication. Interferons, now available as a biotechnology product, are used to treat certain viral infections, such as a hepatitis C.

Phagocytes and Natural Killer Cells Several types of white blood cells are phagocytic. **Neutrophils** are cells that are able to leave the bloodstream and phagocytize (engulf) bacteria in connective tissues. They have various other ways of killing bacteria also. For example, their granules release antimicrobial peptides called defensins. **Eosinophils** are phagocytic, but they are better known for mounting an attack against animal parasites such as tapeworms that are too large to be phagocytized. The two most powerful of the phagocytic white blood cells are **macrophages** and macrophage-derived **dendritic cells**. They engulf pathogens, which are then destroyed by enzymes when their endocytic vesicles combine with lysosomes. Dendritic cells are found in the skin; once they devour pathogens, they travel to lymph nodes, where they stimulate natural killer cells or lymphocytes. Macrophages are found in all sorts of tissues, where they voraciously devour pathogens and then stimulate lymphocytes to carry on specific immunity.

Natural killer (NK) cells are large, granular lymphocytes that kill virus-infected cells and cancer cells by cell-to-cell contact. NK cells do their work while specific defenses are still mobilizing, and they produce cytokines that stimulate these cells.

What makes NK cells attack and kill a cell? First, they normally congregate in the tonsils, lymph nodes, and spleen, where they are stimulated by dendritic cells before they travel forth. Then, NK cells look for a self-protein on the body's cells. As may happen, if a

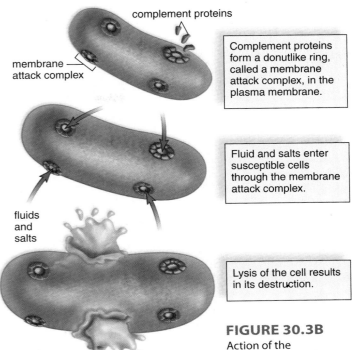

complement proteins

membrane attack complex

Complement proteins form a donutlike ring, called a membrane attack complex, in the plasma membrane.

Fluid and salts enter susceptible cells through the membrane attack complex.

fluids and salts

Lysis of the cell results in its destruction.

FIGURE 30.3B
Action of the complement system against a bacterium.

virus-infected cell or a cancer cell has lost its self-proteins, the NK cell kills it in the same manner used by cytotoxic T cells (see Fig. 30.10B). Unlike cytotoxic T cells, NK cells are not specific; they have no memory; and their numbers do not increase after stimulation.

A fever can be helpful to defense mechanisms, as discussed in Section 30.4.

> **30.3 Check Your Progress** Considering that HIV is a sexually transmitted disease, which of the nonspecific defense mechanisms would you expect to be helpful in combating an HIV infection?

HOW BIOLOGY IMPACTS OUR LIVES

30.4   A fever can be beneficial

A sudden illness, such as mononucleosis, may result in fever, an elevated body temperature. To treat or not to treat a fever is controversial within the medical field, but it is clear that in some instances, a fever may be beneficial. Several hypotheses have been proposed as to the benefits of a fever. Perhaps a fever is the body's way of informing us that something is wrong. Many bacterial and viral infections have few symptoms, and may be serious and difficult to treat by the time other symptoms are evident. The appearance of a fever may signal the brain that something is wrong and action is needed.

Alternatively, fever could be a part of our first line of defense. At times, a fever may directly participate in overcoming an illness. For example, a fever can contribute to the host's defense by providing an unfavorable environment for the invader. Some pathogens have very strict temperature requirements, and turning up the heat on them can slow their ability to multiply and thrive. In fact, supporting this hypothesis is the observation that increasing the body temperature in mice has been shown

to decrease death rates and shorten the recovery times linked to many infectious agents.

Other medical experts believe that the main function of a fever is to stimulate immunity. In support of this hypothesis, a fever has been shown to limit the growth of tumor cells more severely than that of normal body cells. This suggests either that tumor cells are directly sensitive to higher temperatures or that a fever stimulates immunity. Heat is a part of the inflammatory response, and perhaps its main function is to jump-start the response of the body.

While data concerning the benefits of fever are inconclusive, the general consensus is that an extreme fever should be treated, but milder cases may be best left alone.

The inflammatory response, discussed next, is also a nonspecific defense.

> **30.4 Check Your Progress** What part of the brain is possibly malfunctioning when we have a fever (see Section 25.11)?

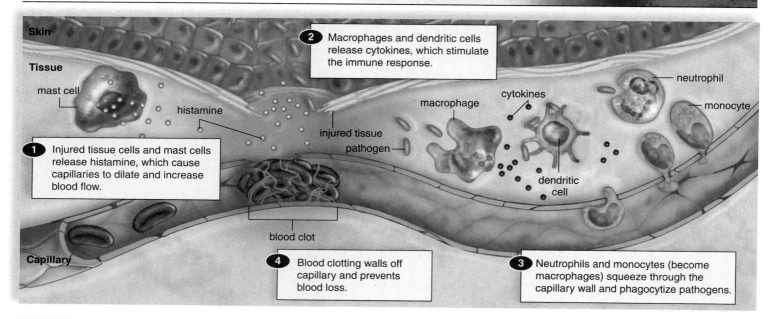

FIGURE 30.5A Inflammatory response.

Whenever tissue is damaged by physical or chemical agents or by pathogens, a series of events occurs that is known as the **inflammatory response**.

An inflamed area has four outward signs: redness, heat, swelling, and pain. All of these signs are due to capillary changes in the damaged area. **Figure 30.5A** illustrates the participants in the inflammatory response. ① Chemical mediators, such as **histamine**, released by damaged tissue cells, and mast cells cause the capillaries to dilate and become more permeable. Excess blood flow, due to enlarged capillaries, causes the skin to redden and become warm. Increased permeability of the capillaries allows proteins and fluids to escape into the tissues, resulting in swelling. The swollen area stimulates free nerve endings, causing the sensation of pain.

② Migration of phagocytes, namely neutrophils and monocytes, also occurs during the inflammatory response. Neutrophils and monocytes are amoeboid and can change shape to squeeze through capillary walls and enter tissue fluid. Also present are dendritic cells, notably in the skin and mucous membranes, and macrophages, both of which are able to devour many pathogens and still survive (**Fig. 30.5B**). ③ Macrophages also release colony-stimulating factors, cytokines that pass by way of the blood to the red bone marrow, where they stimulate the production and release of white blood cells, primarily neutrophils. As the infection is being overcome, some phagocytes die. These—along with dead tissue cells, dead bacteria, and living white blood cells—form pus, a whitish material. The presence of pus indicates that the body is trying to overcome an infection.

The inflammatory response can be accompanied by other responses to the injury. ④ A blood clot can form to seal a break in a blood vessel. Antigens, chemical mediators, dendritic cells, and macrophages move through the tissue fluid and lymph to the lymph nodes. There, B cells and T cells are activated to mount a specific defense to the infection.

Sometimes an inflammation persists, and the result is chronic inflammation that is often treated by administering anti-inflammatory agents such as aspirin, ibuprofen, or cortisone. These medications act against the chemical mediators released by the white blood cells in the damaged area.

We have now completed our discussion of nonspecific defenses. The next part of the chapter discusses specific defenses.

30.5 Check Your Progress Which of the responders active in the inflammatory response could stimulate other cells to respond to an HIV infection?

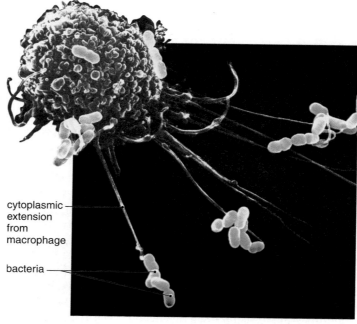

FIGURE 30.5B A macrophage engulfing bacteria. 6000×

In this part of the chapter, we continue our discussion of immunity by considering the body's second line of defense, which is composed of specific defenses against disease. Certain blood cells are particularly involved in specific defenses, also called specific immunity.

30.6 The second line of defense targets specific antigens

When nonspecific defenses have been inadequate to stem an infection, specific defenses come into play. Immunity is complete when a foreign substance is unable to cause an infection now or in the future, or when people are able to avoid cancer. *First*, a specific defense requires that the body be able to recognize a particular molecule, called an **antigen**. Some antigens are termed **foreign antigens** because the body does not produce them. Pathogens, such as bacteria, viruses, and transplanted tissues and organs, bear antigens the immune system usually recognizes as foreign. Other antigens are termed **self-antigens** because the body itself produces them. (The self-proteins mentioned earlier are self-antigens.) It is unfortunate when the immune system reacts to the body's own pancreatic cells (causing diabetes mellitus) or to nerve fiber sheaths (causing multiple sclerosis), but fortunate when the immune system can destroy the cancerous cells of a tumor. *Second*, after recognizing antigens, the immune system can respond to them. Unlike the nonspecific defenses, which occur immediately, it usually takes five to seven days to mount a specific defense. *Third*, the immune system can remember antigens it has met before. This is the reason, for example, that once we recover from the measles, we usually do not get the disease a second time. The reaction time for a nonspecific defense is always about the same, but if the body is immune, the reaction time for a specific defense is quite short the next time it encounters the same antigen.

Section 30.7 explains the difference between active and passive specific immunity.

> **30.6 Check Your Progress** The spike on an HIV virus (see Section 16.5) that allows it to bind to a T cell is designated Gp120. Would you expect Gp120 to be antigenic to humans? Explain.

30.7 Specific immunity can be active or passive

Active Immunity Active immunity develops naturally after a person is infected with a pathogen, such as a measles or chickenpox virus. However, active immunity is often induced when a person is well, so that possible future infection will not take place. Individuals can be immunized with a vaccine to keep from being infected by a specific organism (**Fig. 30.7A**). The **vaccine** contains antigens that cause the body to develop antibodies specific to these antigens. The United States is committed to immunizing all children against the common types of childhood disease; before children can enter school, they must show proof of immunization. Outbreaks of measles and mumps sometimes occur in college dormitories because students haven't been immunized or their immunity has worn off.

FIGURE 30.7A
Vaccines immunize children against diseases.

Passive Immunity Passive immunity occurs when an individual is given prepared antibodies (immunoglobulins) to combat a disease. Since these antibodies are not produced by the individual's plasma cells, passive immunity is short-lived. For example, newborn infants are passively immune to some diseases because antibodies have crossed the placenta from the mother's blood (**Fig. 30.7B**). These antibodies soon disappear, however, so that within a few months, infants become more susceptible to infections. Breast-feeding prolongs the natural passive immunity an infant receives from its mother because antibodies are present in the mother's milk.

Even though passive immunity does not last, it is sometimes used to prevent illness in a patient who has been unexpectedly exposed to an infectious disease. Artificial passive immunity is used in the emergency treatment of rabies, measles, tetanus, diphtheria, botulism, hepatitis A, and snakebites. Usually, the patient receives a gamma globulin injection (serum that contains antibodies), extracted from a large, diverse adult population.

The next section, Section 30.8, describes the two types of lymphocytes involved in specific defenses.

FIGURE 30.7B Breast-feeding provides infants with antibodies.

> **30.7 Check Your Progress** The test for HIV infection is based on detecting antibodies for HIV in the blood. These antibodies are ineffective in destroying HIV. Explain this on the basis of where HIV lives in the body.

Specific defenses primarily depend on the two types of lymphocytes, called B cells and T cells (**Fig. 30.8**). Both B cells and T cells are manufactured in the red bone marrow. B cells mature there, but T cells mature in the thymus. These cells are capable of recognizing antigens because they have specific **antigen receptors** that combine with antigens. B cells have **B cell receptors (BCR)**, and T cells have **T cell receptors (TCR)**. Each lymphocyte has receptors that will combine with only one type of antigen. If a particular B cell could respond to an antigen—a molecule projecting from the bacterium *Streptococcus pyogenes*, for example—it would not react to any other antigen. It is often said that the receptor and the antigen fit together like a lock and a key. Remarkably, diversification occurs to such an extent during maturation that there are specific B cells and/or T cells for any possible antigen we are likely to encounter during a lifetime.

B cells are responsible for **antibody-mediated immunity**. Once a B cell combines with an antigen, it gives rise to **plasma cells**, which produce specific antibodies. These antibodies can

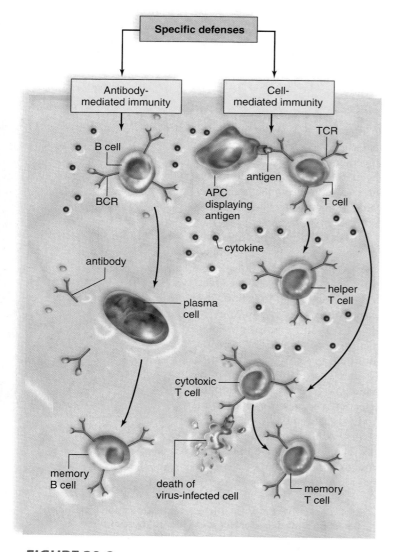

FIGURE 30.8 Overview of specific defenses.

TABLE 30.8	Cells Involved in Specific Defenses	
Cell		**Functions**
Macrophages		Phagocytize pathogens; inflammatory response and specific defense
Dendritic cells		Phagocytize pathogens in skin; play a role in specific defense
Lymphocytes		Responsible for specific defense
B cells		Produce plasma cells and memory cells
Plasma cells		Produce specific antibodies
Memory cells		Ready to produce antibodies in the future
T cells		Regulate immune response; produce cytotoxic T cells and helper T cells
Helper T cells		Regulate immunity
Cytotoxic T cells		Kill virus-infected and cancer cells

react to the same antigen as the original B cell. Therefore, an antibody has the same specificity as the BCR. Some progeny of the activated B cells become memory B cells, so called because these cells always "remember" a particular antigen and make us immune to a particular illness, but not to any other illness.

In contrast to B cells, T cells are responsible for **cell-mediated immunity**. T cells do not recognize an antigen until it is presented to them by an **antigen-presenting cell (APC)**. Macrophages and dendritic cells are APCs. T cells differentiate into either **helper T cells**, which release cytokines, or **cytotoxic T cells**, which attack and kill virus-infected cells and cancer cells. Some cytotoxic T cells become memory T cells, ever ready to defend against the same virus or kill the same type of cancer cell again. Significant immune cells are listed in **Table 30.8** for easy reference.

In Section 30.9, we concentrate on antibody-mediated immunity.

30.8 Check Your Progress Specific defense mechanisms are at work against HIV, but they are not effective. What is the chief way HIV desroys the immune system?

30.9 Antibody-mediated immunity involves B cells

The **clonal selection model** describes what happens when a *B cell receptor (BCR)* combines with an antigen (**Fig. 30.9A**). ❶ The antigen is said to "select" the B cell that will clone. Also, at this time, cytokines secreted by helper T cells stimulate B cells to clone. ❷ Defense by B cells is called *antibody-mediated immunity* because most members of a clone become plasma cells that produce specific antibodies. It is also called humoral immunity because these antibodies are present in blood and lymph. (A *humor* is any fluid normally occurring in the body.) ❸ Some progeny of activated B cells become **memory B cells**, which are the means by which long-term immunity is possible. Once the threat of an infection has passed, the development of new plasma cells ceases, and those present undergo apoptosis.

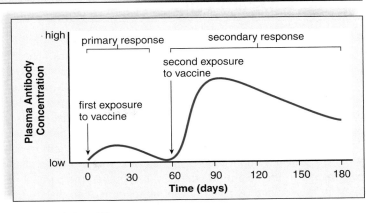

FIGURE 30.9B Antibody titers.

Immunization, described in Section 30.7, involves the use of vaccines to bring about clonal expansion, not only of B cells, but also of T cells. Traditionally, vaccines are the pathogens themselves, or their products, that have been treated so they are no longer virulent (able to cause disease). Vaccines against smallpox, polio, and tetanus have been successfully used worldwide. Today, it is possible to genetically engineer bacteria to mass-produce a protein from pathogens, and this protein can be used as a vaccine. This method was used to produce a vaccine against hepatitis B, a viral disease, and is being used to prepare a potential vaccine against malaria.

After a vaccine is given, it is possible to determine the antibody titer (the amount of antibody present in a sample of plasma). After the first exposure to a vaccine, a primary response occurs. For a period of several days, no antibodies are present; then the titer rises slowly, followed by first a plateau and then a gradual decline as the antibodies bind to the antigen or simply break down (**Fig. 30.9B**). After a second exposure, the titer rises rapidly to a plateau level much greater than before. The second exposure is called a "booster" because it boosts the antibody titer to a high level. The high antibody titer is expected to prevent disease symptoms when the individual is exposed to the antigen. Even years later, if the antigen enters the body, memory B cells quickly give rise to more plasma cells capable of producing the correct type of antibody.

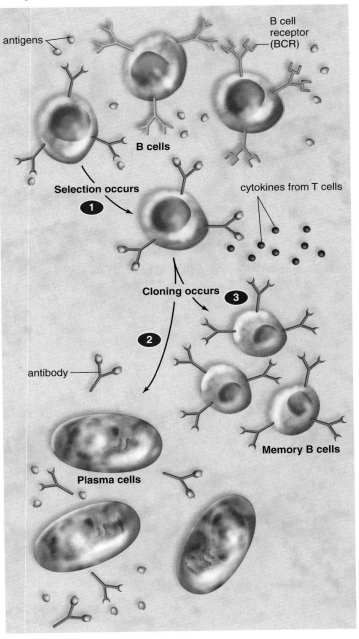

FIGURE 30.9A Clonal selection model as it applies to B cells.

Antibodies Another name for an antibody is **immunoglobulin (Ig)**. The most typical antibody, called IgG, is a Y-shaped molecule with two arms. IgG has constant regions, where the sequence of amino acids is set, and variable regions, where the sequence of amino acids varies between IgGs. The variable regions become hypervariable at their tips and form antigen-binding sites. The shape of the variable region is specific for a particular antigen.

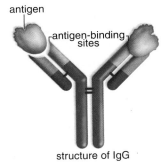

In Section 30.10, we discuss cell-mediated immunity.

> **30.9 Check Your Progress** Would antibodies produced by a plasma cell against the surface antigen Gp120 of an HIV infection be effective against the virus that causes measles? Explain.

T cells are formed in red bone marrow before they migrate to the thymus, a gland that secretes thymic hormones. These hormones stimulate T cells to develop *T cell receptors (TCRs)*. When a T cell leaves the thymus, it has a unique TCR, just as B cells have a BCR. Unlike B cells, however, T cells are unable to recognize an antigen without help. The antigen must be displayed to them by an antigen-presenting cell (APC), such as a dendritic cell or a macrophage. After phagocytizing a pathogen, APCs travel to a lymph node or the spleen, where T cells also congregate. In the meantime, the APC has broken the pathogen apart in a lysosome. A piece of the pathogen is then displayed in an **MHC (major histocompatibility complex) protein** on the cell's surface. MHC proteins are self-antigens because they mark cells as belonging to a particular individual and, therefore, make transplantation of organs difficult. MHC proteins differ by the sequence of their amino acids, and the immune system will attack as foreign any tissue that bears MHC antigens different from those of the individual.

In **Figure 30.10A**, the different types of T cells have specific TCRs represented by their different shapes and colors. ❶ A macrophage presents an antigen only to a T cell that has a TCR capable of combining with a particular antigen (colored green). ❷ Now, cloning occurs, and many copies of the selected T cell are produced. If an APC displays an antigen within the groove of an MHC I protein, the cloned T cells form cytotoxic T cells that will attack cells bearing the antigen (green) it now recognizes. (If an APC displays an antigen within the groove of an MHC II protein, the cloned T cells become helper T cells.) ❸ Some of the cloned cells become **memory T cells**.

❹ As the illness disappears, the immune response wanes, and active T cells become susceptible to apoptosis. As mentioned previously, apoptosis contributes to homeostasis by regulating the number of cells present in an organ, or in this case, in the immune system. When apoptosis does not occur as it should, T cell cancers (e.g., lymphomas and leukemias) can result. Also, in the thymus, any T cell that has the potential to destroy the body's own cells undergoes apoptosis.

Types of T Cells The two main types of T cells are cytotoxic T cells and helper T cells. Cytotoxic T cells have storage vacuoles containing perforins and storage vacuoles containing enzymes called granzymes (**Fig. 30.10B**). ❶ After a cytotoxic T cell binds to a virus-infected cell or cancer cell, it releases perforin molecules, which perforate the plasma membrane, forming a pore. ❷ Cytotoxic T cells then deliver granzymes into the pore, and these cause the cell to undergo osmotic destruction and die. Once cytotoxic T cells have released the perforins and granzymes, they move on to the next target cell. Cytotoxic T cells are responsible for so-called cell-mediated immunity.

Helper T cells play a critical role in coordinating nonspecific defenses and specific defenses, including both cell-mediated immunity and antibody-mediated immunity. When a helper T cell recognizes an antigen attached to an MHC protein, it secretes cytokines. The cytokines attract neutrophils, natural killer cells, and macrophages to

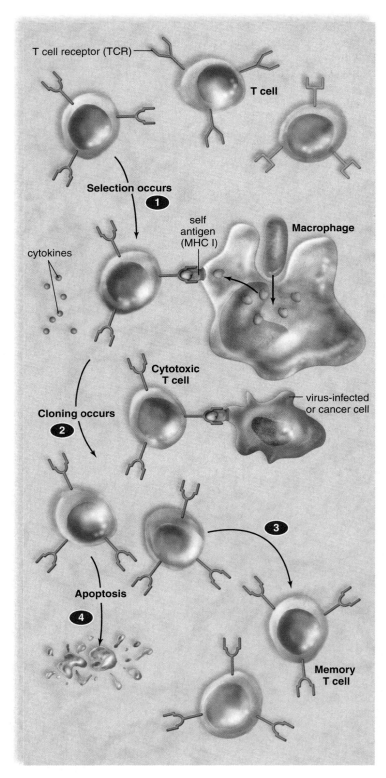

FIGURE 30.10A   Clonal selection model as it applies to T cells.

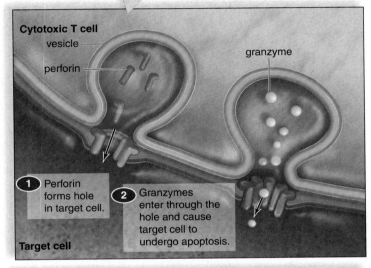

Cytotoxic T cell

vesicle

perforin

granzyme

1 Perforin forms hole in target cell.

2 Granzymes enter through the hole and cause target cell to undergo apoptosis.

Target cell

cytotoxic T cell

target cell (virus-infected or cancer cell)

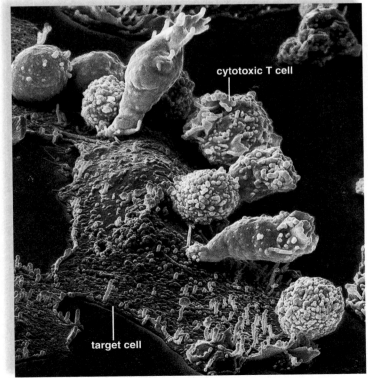

cytotoxic T cell

target cell

FIGURE 30.10B Cell-mediated immunity.

where they are needed. Cytokines stimulate phagocytosis of pathogens, as well as the clonal expansion of T and B cells.

Because more and more immune cells are recruited by helper T cells, the number of pathogens eventually begins to wane. But by now, memory T cells are available. Like memory B cells, memory T cells are long-lived, and their number is far greater than the original number of T cells that could recognize a specific antigen. Therefore, when the same antigen enters the body later on, the reaction is so rapid that no illness is detectable.

Cytokines and Cancer Therapy The term cytokine simply means a soluble protein that acts as a signaling molecule. Because cytokines stimulate white blood cells, they have been studied as a possible adjunct therapy for cancer. We have already mentioned that interferons are cytokines. Interferons are produced by virus-infected cells, and they signal other cells of the need to prevent infection. Interferon has been investigated as a possible cancer drug, but so far it has proven to be effective only in certain patients, and the exact reasons for this, as yet, cannot be discerned. Also, interferon has a number of side effects that limit its use.

Cytokines called interleukins are produced by white blood cells, and they act to stimulate other white blood cells. Scientists actively engaged in interleukin research believe that interleukins will soon be used in addition to vaccines, for the treatment of chronic infectious diseases, and perhaps for the treatment of cancer. Interleukin antagonists may also prove helpful in preventing skin and organ rejection, autoimmune diseases, and allergies.

When, and if, cancer cells carry an altered protein on their cell surface, they should be attacked and destroyed by cytotoxic T cells. Whenever cancer does develop, it is possible that the cytotoxic T cells have not been activated. In that case, interleukins might awaken the immune system and lead to the destruction of the cancer. In one technique being investigated, researchers first withdraw T cells from the patient and activate the cells by culturing them in the presence of an interleukin. The cells are then reinjected into the patient, who is given doses of interleukin to maintain the killer activity of the T cells.

Tumor necrosis factor (TNF) is a cytokine produced by macrophages that has the ability to promoted the inflammatory response and to cause the death of cancer cells. Like the interferons and interleukins, TNF stimulates the body's immune cells to fight cancer. TNF also directly affects tumor cells, damaging them and the blood vessels within the tumor. Without an adequate blood supply, a cancerous tumor cannot thrive. However, researchers are still uncertain about exactly how TNF destroys tumors. Researchers have found that TNF therapy is most effective and least toxic when directed at a specific tumor site, rather than administered throughout. Clinical trials are under way.

Monoclonal anitbodies, whose production is explained in Section 30.11, have added to the arsenal of specific defenses.

30.10 *Check Your Progress* What data could you offer (based on an HIV infection) that cytokines are absolutely essential to a healthy and effective immune system?

30.11   Monoclonal antibodies have many uses

Every plasma cell derived from a single B cell secretes antibodies against a specific antigen. These are called **monoclonal antibodies** because all of them are the same type. One method of producing monoclonal antibodies in vitro (outside the body in the laboratory) is depicted in **Figure 30.11**. B cells are removed from an animal (usually mice are used) and exposed to a particular antigen. ❶ The resulting plasma cells are fused with myeloma cells (malignant plasma cells that live and divide indefinitely). The fused cells are called hybridomas—*hybrid-* because they result from the fusion of two different cells, and *-oma* because one of the cells is a cancer cell. ❷ The hybridomas then secrete large amounts of the specified monoclonal antibody, which recognizes only a single antigen of interest.

Research Uses for Monoclonal Antibodies

The ability to quickly produce monoclonal antibodies in the laboratory has made them an important tool for academic research. Monoclonal antibodies are very useful because of their extreme specificity for only a particular molecule. A monoclonal antibody can be used to select out a specific molecule among many others, much like finding needles in a haystack. Now the molecule can be purified from all the others that are also present in a sample. In this way, monoclonal antibodies have simplified formerly tedious laboratory tasks, allowing investigators more time to focus on other priorities.

Medical Uses for Monoclonal Antibodies

Monoclonal antibodies also have many applications in medicine. For instance, they can now be used to make quick and certain diagnoses of various conditions. Today, a monoclonal antibody is used to signify pregnancy by detecting a particular hormone in the urine of a woman after she becomes pregnant. Thanks to this technology, pregnancy tests that once required a visit to a doctor's office and the use of expensive laboratory equipment can now be performed in the privacy and comfort of a woman's own home, at minimal expense.

Monoclonal antibodies can be used not only to diagnose infections and illnesses but also to fight them. Many bacteria and viruses possess unique proteins on their cell surfaces that make them easily recognized by an appropriate monoclonal antibody. When binding occurs with a certain monoclonal antibody but not another, a physician knows what type of infection is present. And because monoclonal antibodies can distinguish some cancer cells from normal tissue cells, they may also be used to identify cancers at very early stages when treatment can be most effective.

Finally, monoclonal antibodies have also shown promise as potential drugs to help fight disease. RSV, a common virus that causes serious respiratory tract infections in very young children, is now being successfully treated with a monoclonal antibody drug. The antibody recognizes a protein on the viral surface, and when it binds very tightly to the surface of the virus, the patient's own immune system can easily recognize the virus and destroy it before it has a chance to cause serious illness.

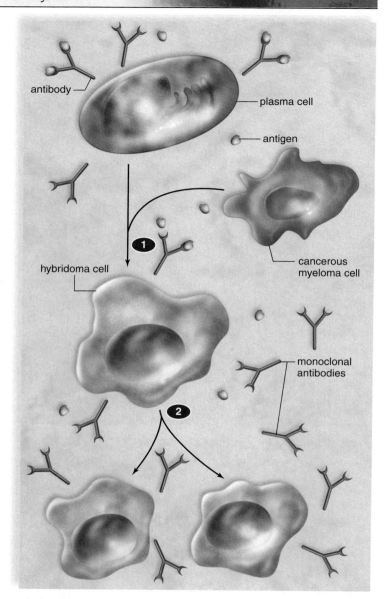

FIGURE 30.11 Production of monoclonal antibodies.

Other illnesses, such as cancer, are also being successfully treated with monoclonal antibodies. Since these antibodies are able to distinguish between cancerous and normal cells, they have been engineered to carry radioisotopes or toxic drugs to tumors so that cancer cells can be selectively destroyed without damaging other body cells. In short, monoclonal antibodies are helping scientists in their research and physicians in their attempts to diagnose and cure patients.

This completes our discussion of specific defenses. The next part of the chapter considers abnormal immune responses.

> **30.11** *Check Your Progress* Monoclonal antibodies are produced by hybridomas, but they do not cause cancer. Explain.

In certain instances, immunity works against the best interests of the body, as when it makes it more difficult to transplant organs from a donor to a recipient, causes autoimmune disorders, or leads to allergic reactions.

30.12 Tissue rejection makes transplanting organs difficult

Certain organs, such as skin, the heart, and the kidneys, could be transplanted easily from one person to another if the body did not attempt to reject them. Rejection occurs because antibodies and cytotoxic T cells bring about the destruction of foreign tissues in the body. When rejection occurs, the immune system is correctly distinguishing between self and nonself.

Organ rejection can be controlled by carefully selecting the organ to be transplanted and administering immunosuppressive drugs. It is best if the transplanted organ has the same type of MHC antigens as those of the recipient, because cytotoxic T cells recognize foreign MHC antigens. Two well-known immunosuppressive drugs, cyclosporine and tacrolimus, both act by inhibiting the production of certain cytokines that stimulate cytotoxic T cells.

Xenotransplantation, the transplantation of animal tissues and organs into human beings, is another way to solve the problem of rejection. Genetic engineering can make pig organs less antigenic by removing the MHC antigens. The ultimate goal is to make pig organs as widely accepted as blood type O. Other researchers hope that tissue engineering, including the production of human organs by using stem cells, will one day do away with the problem of rejection. Scientists have recently grown new heart valves in the laboratory using stem cells gathered from amniotic fluid following amniocentesis.

The cause of autoimmune disorders is not completely understood, as discussed in Section 30.13.

> **30.12 Check Your Progress** Would a patient with an HIV infection make a better or worse candidate for a transplant?

30.13 Autoimmune disorders are long-term illnesses

Autoimmune disorders are diseases that can be characterized by the failure of the immune system to distinguish between foreign antigens and the self-antigens that mark the body's own tissues. In an autoimmune disease, chronic inflammation occurs, and cytotoxic T cells or antibodies mistakenly attack the body's own cells as if they displayed foreign antigens.

Rheumatoid arthritis is a common autoimmune disorder that causes recurring inflammation in synovial joints (**Fig. 30.13A**). Complement proteins, T cells, and B cells all participate in deterioration of the joints, which eventually become immobile. This chronic inflammation gradually causes destruction of the delicate membrane and cartilage within the joint.

In myasthenia gravis, a well-understood autoimmune disease, antibodies attach to and interfere with the functioning of neuromuscular junctions, causing muscular weakness.

Systemic lupus erythematosus (lupus) is a chronic autoimmune disorder characterized by the presence of antibodies to the nuclei of the body's cells. Unlike rheumatoid arthritis or myasthenia gravis, lupus affects multiple tissues and organs, and is still very poorly understood. The symptoms vary somewhat, but most patients experience a characteristic skin rash (**Fig. 30.13B**), joint pain, and kidney damage. Lupus typically progresses to include many life-threatening complications.

The exact events that trigger an autoimmune disorder are not known. Some autoimmune disorders set in following a noticeable infection; for example, the heart damage following rheumatic fever is thought to be due to an autoimmune disorder triggered by the illness. Otherwise, most autoimmune disorders probably start after an undetected inflammatory response. However, the tendency to develop autoimmune disorders is known to be inherited in some cases, so there may be genetic causes as well.

Since little is known about the origin of autoimmune disorders, no cures are currently available. Regardless, most of these conditions can be managed over the long term with immunosuppressive drugs that control the various symptoms.

Allergic reactions, the topic of the next section, now affect a great number of people.

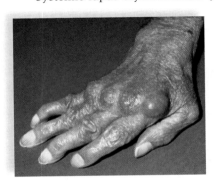

FIGURE 30.13A
Rheumatoid arthritis

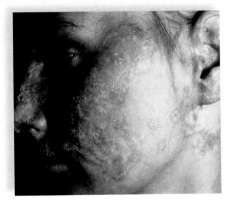

FIGURE 30.13B Systemic lupus; a butterfly-shaped rash appears on face.

> **30.13 Check Your Progress** Patients with an autoimmune disease are caught between a rock and a hard place. Explain.

30.14 Allergic reactions can be debilitating and even fatal

Allergies are hypersensitivities to substances, such as pollen, food, or animal hair, that ordinarily would do no harm to the body. The response to these antigens, called allergens, usually includes some degree of tissue damage.

An **immediate allergic response** can occur within seconds of contact with the antigen. The response is caused by antibodies known as IgE (**Table 30.14**). IgE antibody receptors are attached to the plasma membrane of mast cells in the tissues. When an allergen attaches to the IgE antibodies and the antibodies attach to their receptors on mast cells, the cells release histamine and other substances that bring about the allergic symptoms (**Fig. 30.14**). When pollen is an allergen, histamine combines with and stimulates the mucous membranes of the nose and eyes to release fluid, causing the runny nose and watery eyes typical of **hay fever**. If a person has **asthma**, the airways leading to the lungs constrict, resulting in difficult breathing accompanied by wheezing. When food contains an allergen, nausea, vomiting, and diarrhea result.

Drugs called antihistamines are used to treat allergies. These drugs compete for the same receptors on the nose, eyes, airways, and lining of the digestive tract cells that ordinarily combine with histamine. In this way, histamine is prevented from binding and causing its unpleasant symptoms. However, antihistamines are only partially effective because mast cells release other molecules, in addition to histamine, that cause allergic symptoms.

Anaphylactic shock is an immediate allergic response that occurs because the allergen has entered the bloodstream. Bee stings and penicillin shots are known to cause this reaction in some individuals because both inject the allergen into the blood. Anaphylactic shock is characterized by a sudden and life-threatening drop in blood pressure, due to increased permeability of the capillaries by histamine.

People with allergies produce ten times more IgE than people without allergies. A new treatment using injections of monoclonal IgG antibodies for IgEs is currently being tested in individuals with severe food allergies. More routinely, injections of the allergen are given so that the body will build up high quantities of IgG antibodies. The hope is that these will combine with allergens received from the environment before they have a chance to reach the IgE antibodies located in the membrane of mast cells and basophils.

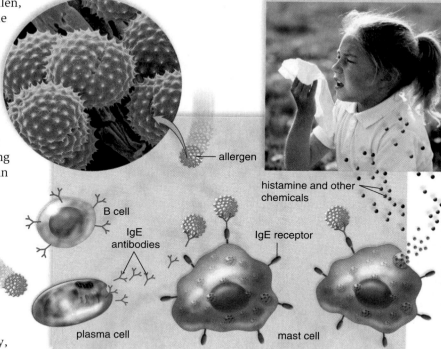

allergen

histamine and other chemicals

B cell

IgE antibodies

IgE receptor

plasma cell

mast cell

FIGURE 30.14 An allergic reaction.

A **delayed allergic response** is initiated by memory T cells at the site of allergen contact in the body. The allergic response is regulated by the cytokines secreted by both T cells and macrophages. A classic example of a delayed allergic response is the skin test for tuberculosis (TB). When the test result is positive, the tissue where the antigen was injected becomes red and hardened. This indicates prior exposure to tubercle bacilli, the cause of TB. Contact dermatitis, which occurs when a person is allergic to poison ivy, jewelry, cosmetics, and many other substances that touch the skin, is another example of a delayed allergic response.

30.14 Check Your Progress It is not correct to say that allergies are caused by IgE antibodies. Explain.

TABLE 30.14	Comparison of Immediate and Delayed Allergic Responses	
	Immediate Response	**Delayed Response**
Onset of Symptoms	Takes several minutes	Takes 1 to 3 days
Lymphocytes Involved	B cells	T cells
Immune Reaction	IgE antibodies	Cell-mediated immunity
Type of Symptoms	Hay fever, asthma, and many other allergic responses	Contact dermatitis (e.g., poison ivy)
Therapy	Antihistamine and adrenaline	Cortisone

The defense mechanisms of humans have been extensively studied, but little is known about these mechanisms in other animals. In humans, the levels of defense against invasion of the body by pathogens can be compared to how we protect our homes. Homes usually have external defenses, such as a fence, a dog, or locked doors. Similarly, the body has the skin and mucous membranes that act as barriers to prevent pathogens from entering the blood and lymph. Like a home alarm system, if an invasion does occur, a signal goes off. First, nonspecific defense mechanisms, such as the complement system and phagocytosis by white blood cells, come into play. Then, specific defense mechanisms, which are dependent on the activities of B cells and T cells, bring the infection to an end.

The cardiovascular and lymphatic systems are intimately connected in that tissue fluid, which becomes lymph in lymphatic vessels, is derived, in part, from the water and small molecules that leave the bloodstream. In the next few chapters, we discuss other systems that interact directly with the cardiovascular system. Blood is refreshed because the digestive, respiratory, and urinary systems make exchanges with the external environment. The food we eat is digested into nutrients that enter the blood; gas exchange in the lungs permits the removal of carbon dioxide from blood and the entrance of oxygen into blood; and the kidneys remove and excrete metabolic wastes taken from the blood. Homeostasis is critically dependent on these exchanges between blood and the external environment.

The Chapter in Review

Summary

AIDS Destroys the Immune System

- The HIV virus damages and destroys T cells, which promote the activity of other cells in the immune system.
- Opportunistic infections take hold when the immune system is damaged or destroyed.

The Lymphatic System Functions in Transport and Immunity

30.1 Lymphatic vessels transport lymph

- The lymphatic system is composed of lymphatic vessels and organs.
- Lymphatic capillaries absorb excess tissue fluid and fats and transport them to the bloodstream.
- The lymphatic system produces, maintains, and distributes lymphocytes, and helps defend the body against pathogens.

30.2 Lymphatic organs defend the body

- The red bone marrow is where all blood cells are made and where B cells mature.
- The thymus gland is where T cells mature.
- Lymph nodes cleanse the lymph of pathogens and debris.
- In the spleen, blood is cleansed of pathogens and debris.
- Tonsils, Peyer patches, and the appendix are patches of lymphatic tissue that encounter pathogens and antigens.

The Body's First Line of Defense Against Disease Is Nonspecific and Innate

30.3 Barriers to entry, complement proteins, and certain blood cells are first responders

- Barriers to entry include the skin, mucous membranes, resident bacteria, and antimicrobial molecules.

- Protective proteins are complement (blood plasma proteins) and interferons (cytokines).
- Phagocytes (neutrophils, eosinophils, macrophages, and dendritic cells) engulf pathogens, and natural killer cells kill virus-infected and cancer cells.

30.4 A fever can be beneficial

- Fever tells us that something is wrong.
- As a first line of defense, fever creates an unfavorable environment for an invader.
- As part of the second line of defense, fever may stimulate the immune system.

30.5 The inflammatory response is a localized response to invasion

- The inflammatory response involves mast cells, which release histamine to increase capillary permeability. Redness, warmth, swelling, and pain result.
- Neutrophils and macrophages enter tissue fluid and engulf pathogens.

The Body's Second Line of Defense Against Disease Is Specific to the Pathogen

30.6 The second line of defense targets specific antigens

- Specific immunity recognizes, responds to, and remembers an antigen.

30.7 Specific immunity can be active or passive

- Active immunity is long-lived; it develops naturally after infection or is induced by vaccination.
- Passive immunity is short-lived; it occurs when a person is given antibodies or when a mother's antibodies are passed to her baby by breast-feeding.

30.8 Lymphocytes are directly responsible for specific defenses

- Each B cell and T cell has an antigen receptor for a specific antigen.

- Altogether, B cells and T cells have antigen receptors for all possible antigens.
- B cells are responsible for antibody-mediated immunity.
- T cells are responsible for cell-mediated immunity.

30.9 Antibody-mediated immunity involves B cells

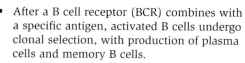
Plasma cell

- After a B cell receptor (BCR) combines with a specific antigen, activated B cells undergo clonal selection, with production of plasma cells and memory B cells.
- An antibody (immunoglobulin) is usually Y-shaped, with two binding sites for a specific antigen.

30.10 Cell-mediated immunity involves several types of T cells

- For a T cell receptor (TCR) to recognize an antigen, the antigen must be displayed to it by an antigen-presenting cell, along with an MHC protein. Thereafter, the activated T cell divides and, depending on the MHC, produces either cytotoxic T cells or helper T cells.

—TCR

Cytotoxic T cell

  - Cytotoxic T cells confer cell-mediated immunity.
  - Helper T cells coordinate cell-mediated immunity and antibody- mediated immunity by releasing cytokines.
- Memory T cells jump-start an immune reaction to an antigen they recognize.
- Cytokines, which stimulate other white blood cells, are being used for cancer therapy.

30.11 Monoclonal antibodies have many uses

- Besides being a valuable research tool in the lab, monoclonal antibodies are used for medical diagnosis and treatment, pregnancy testing, and cancer treatment.

Abnormal Immune Responses Can Have Health Consequences

30.12 Tissue rejection makes transplanting organs difficult

- Rejection of transplanted organs occurs because the immune system is correctly distinguishing between self and nonself.
- Organ rejection can be controlled by immunosuppressive drugs. Xenotransplantation and tissue engineering are alternatives.

30.13 Autoimmune disorders are long-term illnesses

- In autoimmune disorders, such as rheumatoid arthritis, myasthenia gravis, and lupus, the immune system mistakenly attacks the body's tissues.

30.14 Allergic reactions can be debilitating and even fatal

- Allergies are hypersensitivities to certain allergens.
- An immediate allergic response occurs within seconds and is caused by IgE antibodies.
- Anaphylactic shock is a dangerous immediate allergic response characterized by a sudden drop in blood pressure.
- A delayed allergic response occurs in 1 to 3 days after contact with the allergen.

Testing Yourself

The Lymphatic System Functions in Transport and Immunity

1. Which of the following is not a function of the lymphatic system?
   a. produces red blood cells
   b. returns excess fluid to the blood
   c. transports lipids absorbed from the digestive system
   d. defends the body against pathogens
2. Lymph nodes
   a. block the flow of lymph.
   b. contain B cells and T cells.
   c. decrease in size during an illness.
   d. filter blood.
3. B cells mature within
   a. the lymph nodes.　　c. the thymus.
   d. the spleen.　　　　　d. the bone marrow.
4. Which of the following is a function of the spleen?
   a. produces T cells
   b. removes worn-out red blood cells
   c. produces immunoglobulins
   d. produces macrophages
   e. regulates the immune system
5. **THINKING CONCEPTUALLY** Why would you expect the lymphatic system to be a one-way system?

The Body's First Line of Defense Against Disease Is Nonspecific and Innate

6. Defense mechanisms that function to protect the body against many infectious agents are called
   a. specific.　　　　c. barriers to entry.
   b. nonspecific.　　d. immunity.
7. Complement
   a. is a nonspecific defense mechanism.
   b. is involved in the inflammatory response.
   c. is a series of proteins present in the plasma.
   d. plays a role in destroying bacteria.
   e. All of these are correct.
8. Which cells will phagocytize pathogens?
   a. neutrophils　　c. mast cells
   b. macrophages　　d. Both a and b are correct..
9. _____ release histamines.
   a. Mast cells　　　c. Monocytes
   b. Neutrophils　　d. All of these are correct.
10. How would applying heat to an injured area benefit it?

The Body's Second Line of Defense Against Disease Is Specific to the Pathogen

11. People who have had chickenpox as children usually do not get the infection a second time because
    a. adults do not get chickenpox.
    b. the virus cannot enter the body twice.
    c. immune cells remember the virus and respond to it.
    d. the body has gained the ability to negate the effects of the virus.
12. Active immunity may be produced by
    a. having a disease.
    b. receiving a vaccine.
    c. receiving gamma globulin injections.
    d. Both a and b are correct.
    e. Both b and c are correct.

13. Which one of these does not pertain to B cells?
    a. have passed through the thymus
    b. have specific receptors
    c. are responsible for antibody-mediated immunity
    d. become plasma cells that synthesize and liberate antibodies
14. Unlike B cells, T cells
    a. require help to recognize an antigen.
    b. are components of a specific defense response.
    c. are types of lymphocytes.
    d. contribute to homeostasis.
15. Which one of these pairs is mismatched?
    a. helper T cells—increase during an HIV infection
    b. cytotoxic T cells—active in tissue rejection
    c. macrophages—activate T cells
    d. memory T cells—long-living line of T cells
    e. T cells—mature in thymus
16. The clonal selection model says that
    a. an antigen selects certain B cells and suppresses them.
    b. an antigen stimulates the multiplication of B cells that produce antibodies against it.
    c. T cells select those B cells that should produce antibodies, regardless of antigens present.
    d. T cells suppress all B cells except the ones that should multiply and divide.
    e. Both b and c are correct.
17. During a secondary immune response,
    a. antibodies are made quickly and in great amounts.
    b. antibody production lasts longer than in a primary response.
    c. antibodies of the IgG class are produced.
    d. lymphocyte cloning occurs.
    e. All of these are correct.
18. Which of the following would not be a participant in cell-mediated immune responses?
    a. helper T cells   d. cytotoxic T cells
    b. macrophages   e. plasma cells
    c. cytokines
19. **THINKING CONCEPTUALLY** How are the B and T cells like the U.S. Army Special Forces?

Abnormal Immune Responses Can Have Health Consequences

20. Which of the following is not an example of an autoimmune disease?
    a. multiple sclerosis   d. systemic lupus erythematosus
    b. myasthenia gravis   e. rheumatoid arthritis
    c. contact dermatitis
21. Anaphylactic shock is an example of
    a. an immediate allergic response.
    b. an autoimmune disease.
    c. a type of serum shock.
    d. a delayed allergic response.

Understanding the Terms

AIDS 592
allergy 606
anaphylactic
    shock 606
antibody 594
antibody-mediated
    immunity 600
antigen 599
antigen-presenting cell
    (APC) 600
antigen receptor 600
asthma 606
autoimmune disorder 605
B cell 595
B cell receptor (BCR) 600
cell-mediated
    immunity 600
clonal selection model 601
complement 596
cytokine 597
cytotoxic T cell 600
delayed allergic
    response 606
dendritic cell 597
eosinophil 597
foreign antigen 599
hay fever 606
helper T cell 600
histamine 598
HIV 592
immediate allergic
    response 606
immunity 596
immunization 601
immunoglobulin (Ig) 601
inflammatory response 598
interferon 597
lacteal 594
lymph 594
lymphatic (lymphoid)
    organ 595
lymphatic system 594
lymphatic vessel 594
lymph node 595
macrophage 597
mast cell 596
membrane attack
    complex 597
memory B cell 601
memory T cell 602
MHC (major
    histocompatibility
    complex) protein 602
monoclonal antibody 604
natural killer (NK) cell 597
neutrophil 597
opportunistic infection
    (OI) 592
Peyer patches 595
plasma cell 600
red bone marrow 595
self-antigen 599
spleen 595
T cell 595
T cell receptor (TCR) 600
thymus gland 595
tonsils 595
vaccine 599

Match the terms to these definitions:
a. _____ Antigens prepared in such a way that they can promote active immunity without causing disease.
b. _____ Fluid, derived from tissue fluid, that is carried in lymphatic vessels.
c. _____ Foreign substance, usually a protein or a polysaccharide, that stimulates immunity, such as by producing antibodies.
d. _____ Cell that matures in the thymus and exists in three varieties, one of which kills antigen-bearing cells outright.

Thinking Scientifically

1. Revisit Section 10.12 and formulate a hypothesis to account for the ability of lymphocytes to produce about two million different antibodies, despite having a limited number of genes for antibodies.
2. Design an experiment to test whether a drug known to suppress cell-mediated immunity has no affect on antibody-mediated immunity.

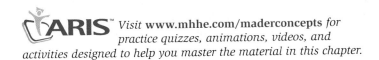 Visit www.mhhe.com/maderconcepts for practice quizzes, animations, videos, and activities designed to help you master the material in this chapter.

31

Digestive Systems and Nutrition

LEARNING OUTCOMES

After studying this chapter, you should be able to accomplish the following outcomes.

How to Tell a Carnivore from an Herbivore

1. Contrast the teeth and digestive tracts of carnivores and herbivores.

Animals Must Obtain and Process Their Food

2. List and discuss the four functions of a digestive system.
3. Describe, with examples, four different feeding strategies.
4. Compare the digestive tracts of animals.
5. Describe the anatomy of the human mouth, and contrast mechanical and chemical digestion in the mouth.
6. Describe the anatomy of the pharynx and changes that occur during swallowing.
7. Describe the anatomy of the esophagus and the process of peristalsis.
8. Describe the anatomy of the stomach and the function of gastric glands.
9. Substantiate that an infection is the cause of an ulcer.
10. Describe the chemical digestion that occurs as a result of pancreatic and gallbladder secretions.
11. Describe the anatomy of the small intestine in relation to its absorptive function.
12. Describe several functions of the pancreas and liver and three types of liver illnesses.
13. Describe how the secretion of hormones that influence digestion is controlled.
14. Describe the structure and function of the large intestine and common illnesses associated with the large intestine.

Good Nutrition and Diet Lead to Better Health

15. Compare the benefits and drawbacks of carbohydrates, fats, and proteins in the diet.
16. In general, discuss the need for minerals and vitamins in the diet.
17. Explain how to interpret a nutrition label.
18. Discuss how to control diabetes type 2 and cardiovascular disease through diet and exercise.
19. Define three types of eating disorders, and list the harmful consequences of each.

I magine that you are a zo-ologist who has just ar-rived to study the animals in a tropical rain forest, and someone brings you a skinless, decaying car-cass to identify. Where would you begin? First off, you might want to start with the teeth if they are still present. From the teeth you would be able to determine whether the animal was a carnivore (meat eater), an herbivore (plant eater), or an omnivore (meat and plant eater). The long, curved, and sharp canines of a carnivore, such as a tiger, are very good at stabbing and killing prey. The front incisors are short and pointed for scraping meat off bones. The posterior teeth (molars)

Horse

Tiger

How to Tell a Carnivore from an Herbivore

have jagged edges, similar to serrated knives, for slicing flesh.

You could move on from there to take a look at the digestive tract, which in a carnivore is short compared to the size of the animal and not very complicated. Digesting protein is a relatively easy matter, and that is why carnivores tend to bolt their food and swallow it quickly, with little prep work before it enters a large, simple (single-chambered) stomach. Carnivores do not eat all that often, so it is advantageous to have a large stomach they can fill up when they do eat. Digestion, which starts in the stomach, is finished in the short small intestine, where nutrients are absorbed. The large intestine (colon) of a carnivore is also simple and very short, as its only purpose is to absorb salt and water. The large intestine is cylindrical with a smooth appearance.

Herbivores, such as horses, have a more varied digestive tract than carnivores because their diets are so different. Some eat only veggies, some only fruit, and some only nuts, for example. If they eat grass or leaves, the face is muscular in order to do a lot of chewing, and a muscular tongue is good at moving food around. The front incisors are broad and have clipped edges like a spade. The canines may be short, or long for defense, or absent. The molars, herbivores' most important teeth, have flat surfaces for grinding. Grinding breaks up the strong walls of plant cells, but a lot of digestion is still needed, and it begins in the mouth while the animal is still chewing.

Cows, deer, and other herbivores that feed on grass are ruminants, with a large divided stomach where bacteria and protists are called upon to digest cellulose. Regard-

less of diet, the small intestine of a herbivore tends to be very long (10–12 times greater than body length), allowing space for digestion of plant material to continue in addition to absorption of nutrients. The large intestine is wider than the small intestine and also long enough to provide room for bacterial flora that live off the fiber that was not digested in the stomach or small intestine. In some herbivores (e.g., rabbits), the first section of the large intestine, called a cecum, is quite large, and serves as a primary or accessory place where microbes digest fiber. To make use of the resulting nutrients, some rabbits are known to re-ingest feces so that the nutrients can be absorbed by the small intestine the second time around.

Omnivores are generalists. They have anterior teeth for holding and cutting, but also canines for stabbing and tearing and flat molars with sharp blades. The digestive tract proportions tend to match those of a carnivore more than those of an herbivore. Of course, a zoologist would have no trouble at all recognizing the teeth and digestive tract of a human, which many say are more like those of an herbivore than an omnivore. Why? Because we have flat molars (along with massive chewing muscles), and although we do not have a multi-divided stomach, we do have long intestines. Still, it is debatable whether our teeth and digestive tract are more like that of a herbivore or a carnivore. After all, some of our ancestors did have longer canine teeth—maybe not as long as those of a tiger, but certainly longer than those of humans today.

This chapter stresses the digestive traits that animals have in common and those that distinguish one group of feeders from another. It also describes the structure and function of the human digestive system, from the mouth to the large intestine. Finally, it discusses good nutrition and offers suggestions for a healthy diet.

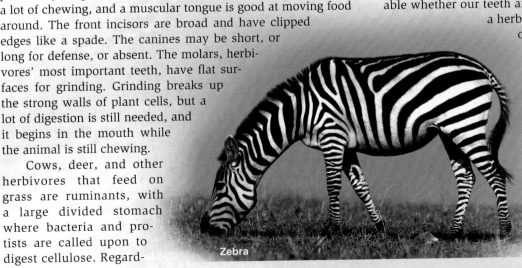

Wolf

Polar bear

Rabbit

Zebra

A comparison of animals' digestive systems and manner of securing food precedes a look at the human digestive system, from the mouth to the large intestine.

31.1 A digestive system carries out ingestion, digestion, absorption, and elimination

Digestion contributes to homeostasis by providing the body's cells with the nutrients they need to continue living. An animal's digestive tract accomplishes the following processes: ingestion, digestion, absorption, and elimination, as shown by the model of a salamander in **Figure 31.1A**. It is possible to associate ingestion with a mouth and elimination with an anus, but the locations of digestion and absorption can differ somewhat, depending on the animal.

Ingestion is the act of taking in food during the process of feeding. *Digestion* is often divided into mechanical digestion and chemical digestion. Mechanical digestion can be accomplished by chewing in animals with suitable mouthparts, such as teeth, but in animals that lack these features, mechanical digestion can occur elsewhere. For example, earthworms and birds have a gizzard, where the grinding action of pebbles and sand accomplishes mechanical digestion.

Chemical digestion requires enzymes, the protein molecules that speed a reaction at body temperature. Enzymes are secreted by the digestive tract itself or by accessory glands that send their juices into the tract by way of ducts. Enzymes are specific and usually work in small steps. For example, it takes two steps to digest polysaccharides (carbohydrate polymers), polypeptides of proteins, and nucleic acids to molecules small enough to be absorbed across the epithelial lining of the digestive tract, as illustrated in **Figure 31.1B**. First, polysaccharides are digested to disaccharides, and then disaccharides are digested to monosaccharides. Similarly, the polypeptides of proteins are first digested to peptides, and then peptides are broken down to amino acids. The first step in the breakdown of nucleic acids produces nucleotides, and then nucleotides become pentose sugars, phosphate, and bases during the second step. Fats are not polymers, and it takes only one step to digest them to monoglycerides (glycerol and one fatty acid) and fatty acids.

Absorption occurs when nutrients are taken into the body, often into the bloodstream. In humans, the small intestine, where absorption occurs, should really be called the "long intestine" because of its length. The wall of the small intestine is convoluted and contains

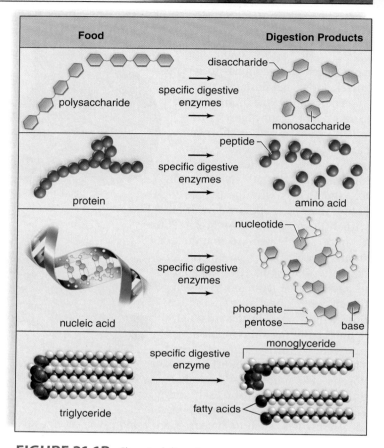

FIGURE 31.1B Chemical digestion.

even smaller projections to increase the available surface area for absorption to take place. Any molecules that are not absorbed by the digestive tract cannot be used by the animal. They are the waste products that pass through the anus during *elimination* (defecation).

Even though all digestive tracts carry out the same four functions, there are modifications according to the animal's way of life. As we saw in the introduction to this chapter, carnivores have adaptations for eating meat, which is easily digested, and herbivores have modifications for eating plant material, which needs a lot of chewing and other processing to break up the cellulose walls of plant cells. Humans, being omnivores, have modifications for eating both meat and plant material, but as we discussed, the human digestive system is more like that of an herbivore than a carnivore. The various ways animals acquire food are surveyed in Section 31.2.

> **31.1 Check Your Progress** Which of the four phases of digestion are more easily accomplished in a carnivore?

esophagus

intestine

mouth

stomach   anus

| Ingestion | Digestion | Absorption | Elimination |

FIGURE 31.1A The four phases of digestion.

31.2 Animals exhibit a variety of feeding strategies

Animals have at least four different feeding strategies to accompany the ingestion of food.

Bulk feeders bite off chunks of food, or even take in a whole organism, during one feeding. Humans are bulk feeders that usually use utensils to cut up food or bring bite-sized pieces of food to the mouth. By contrast, the expandable jaws of snakes allow them to swallow prey whole, even if the prey is much larger than the snake's head. A great blue heron stands on the water's edge and uses its swordlike beak to capture a fish prior to engulfing it whole (**Fig. 31.2A**). Bulk feeders do not have to be carnivores. An elephant is an herbivore that feeds on tree foliage, bark, leaves, twigs, loose fruits, and tall grasses. Bulk feeders tend to eat *discontinuously*, meaning that they have discrete periods of eating and not eating.

Filter feeders (also called suspension feeders) sift small food particles from the water. Like a clam (see Section 19.12), filter feeders often employ a ciliated surface to capture drifting particles from currents of water. Baleen whales are filter feeders that eat great quantities of small, shrimplike animals called krill. First the whale allows a large quantity of water to enter an open mouth. Then it lowers the baleen, a huge sieve made of keratin that hangs from the upper jaw (**Fig. 31.2B**). Finally, it uses its enormous tongue to force the water out, trapping the krill behind the baleen.

Fluid feeders include parasites that live in the body fluid of a host, as well as external parasites, which have a way to tap into the host's vascular system. Recall that aphids (see

FIGURE 31.2C Vampire bat, a fluid feeder.

FIGURE 31.2D
Caterpillar, a substrate feeder.

Section 22.6) use sharp mouthparts to reach the phloem sap of plants. Leeches have suckers and jaws that allow them to attach to and draw blood or hemolymph from a host (see Section 19.13). A vampire bat has fangs (actually pointed incisors) for piercing the skin of its host in order to get at its blood (**Fig. 31.2C**). A bat's saliva contains an anticoagulant called draculin, which runs into the bite; the bat then licks the flowing blood. Fluid feeders also include such pollinators as bees and hummingbirds, which feed on the nectar of flowers. Recall that pollinators have a mutualistic relationship with plants (see the introduction to Chapter 24).

Substrate feeders live on, or in, the material they eat. For example, caterpillars (**Fig. 31.2D**) live and feed on leaves, while maggots live in and feed on dead and decaying tissue. Unlike the heron and baleen whale, caterpillars and maggots are *continuous* feeders, as are earthworms. They eat a small amount all the time. Earthworms plow through dirt using a powerful muscular pharynx to suck in whatever is in their path. After extracting the nutrients, the earthworm's digestive system expels the undigested residue at the surface.

Section 31.3 explains the difference between an incomplete and a complete digestive system.

FIGURE 31.2A Great blue heron, a bulk feeder.

FIGURE 31.2B Baleen whale, a filter feeder.

> **31.2 *Check Your Progress*** Give exceptions to the generalizations that (*a*) only carnivores are bulk feeders and (*b*) only parasites are fluid feeders.

Most animals, including humans, have a *complete digestive tract* with both a mouth and an anus, but a few have an *incomplete digestive tract* with a single opening, usually called a mouth. The single opening is used as both an entrance for food and an exit for wastes.

Incomplete Digestive Tract Planarians, which are flatworms, have an incomplete digestive tract (**Fig. 31.3A**). It begins with a mouth and muscular pharynx, and then the tract, a gastrovascular cavity, branches throughout the body. Planarians are primarily carnivorous and feed largely on smaller aquatic animals as well as bits of organic debris. When a planarian is feeding, the pharynx actually extends beyond the mouth. The body is wrapped about the prey, and the pharynx sucks up minute quantities at a time. Digestive enzymes present in the tract allow some extracellular digestion to occur. Digestion is finished intracellularly by the cells that line the tract. No cell in the body is far from the digestive tract; therefore, diffusion alone is sufficient to distribute the nutrient molecules.

The digestive tract of a planarian is notable for its lack of specialized parts. It is saclike because the pharynx serves not only as an entrance for food but also as an exit for indigestible material. Specialization of parts does not occur under these circumstances.

Complete Digestive Tract in an Earthworm In contrast to planarians, earthworms, which are annelids, have a complete digestive tract (**Fig. 31.3B**). Earthworms are considered substrate feeders because they live underground and feed mainly on decayed organic matter in the soil. The muscular pharynx draws in food with a sucking action. Food then enters the **crop**, which is a storage area with thin, expansive walls. From there, food goes to the **gizzard**, which has thick, muscular walls that crush the food and pebbles that grind. Digestion is extracellular within an intestine. The surface area of digestive tracts is often increased for absorption of nutrient molecules, and in earthworms, this is accomplished by an intestinal fold called the **typhlosole**. Undigested remains pass out of the body at the anus.

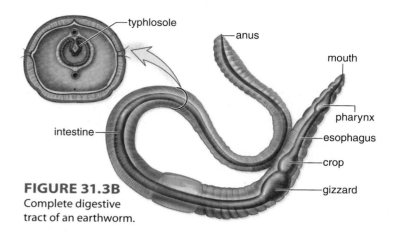

FIGURE 31.3B
Complete digestive tract of an earthworm.

Specialization of parts is obvious in the earthworm because the pharynx, the crop, the gizzard, and the intestine each have a particular function in the digestive process.

Complete Digestive Tract in Birds Some birds, such as the great blue heron, are discontinuous bulk feeders, and they have a thin-walled crop, which serves as a storage area. This is useful because food may not always be available, and the presence of a predator may make it essential to "eat and run." Therefore, it seems adaptive that birds do not chew their food but instead have a gizzard where food is broken apart mechanically (**Fig. 31.3C**). The stomach, which secretes digestive juices, precedes the gizzard in birds. Following the gizzard are the intestines. Attached to the large intestine are two long ceca; a **cecum** is a blind (has no opening) sac. The ceca of a herbivorous bird, such as one that feeds on seeds, are well-developed because they are fermentation organs where microorganisms help with the digestion of cellulose. In all birds, absorption occurs in the small intestine, and elimination occurs at the **cloaca**, which receives indigestible material and urine from the bladder and also gametes from the testes or ovaries.

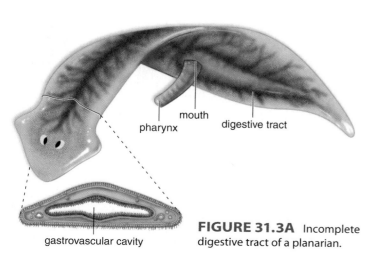

FIGURE 31.3A Incomplete digestive tract of a planarian.

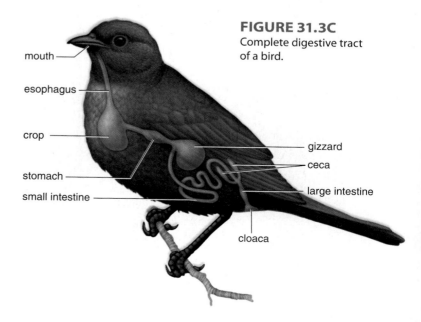

FIGURE 31.3C
Complete digestive tract of a bird.

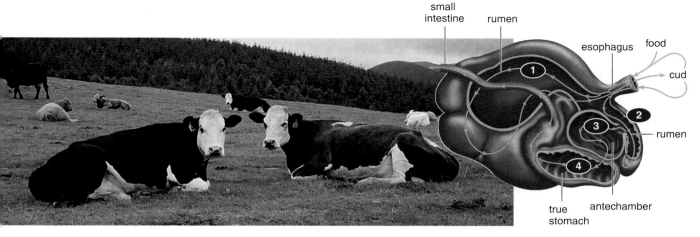

FIGURE 31.3D A ruminant's stomach.

Complete Digestive Tract in Mammalian Herbivores

The digestive tract of mammalian herbivores, such as cows and rabbits, is modified to provide a special compartment for microbes that can digest cellulose. No animal is able to produce an enzyme that can digest the cellulose of plant cell walls, but microbes do produce such enzymes. Mammalian herbivores can digest grass only by having a mutualistic relationship with these microorganisms. Mammalian herbivores called **ruminants,** such as cows, sheep, deer, and giraffes, have large divided stomachs with four chambers, one of which is called the rumen (**Fig. 31.3D**). ① The rumen is a large fermentation chamber that contains microbes and up to 50 gallons of undigested plant material. The location of a rumen at the entrance to the stomach allows partly digested food to be regurgitated and chewed again while the animal is at leisure. Because the animal "chews the cud," ② the rumen is able to do a better job when the food is re-swallowed, entering the rumen a second time. ③ Now, the partially digested food moves through another chamber before ④ it enters the true stomach and finally exits to the small intestine. The digestive tract of herbivores, in general, is long because the digestive action on cellulose also continues along the length of the tract.

Other mammalian herbivores, such as rabbits and horses, are not ruminants and instead make special use of a single, large cecum at the start of the large intestine (**Fig. 31.3E**, *top*). The cecum contains microorganisms that digest cellulose. Because the cecum is located some distance from the stomach, these animals do not regurgitate their food for more chemical digestion. However, rabbits and hares do re-ingest their food. They produce two kinds of fecal pellets—a firm, dark variety and a larger, soft, light-colored kind, which is not dropped but is instead eaten directly from the anus. Horses do not re-ingest, and their feces contain some undigested and indigestible remains. The appendix attached to the cecum in rabbits contains lymphatic tissue where lymphocytes respond to antigens, as discussed in Chapter 30.

Recall that carnivores, such as foxes, dogs, cats, lions, and tigers, have a much shorter digestive tract than do herbivores. Their food is much easier to digest, and if they have a cecum, it is small in size (Fig. 31.3E, *bottom*). Carnivores do not have to spend much time mechanically chewing their food and can go hunting instead.

We begin our study of the human digestive system by looking at the structure and function of the mouth in Section 31.4.

> **31.3 _Check Your Progress_** In what two ways is the small intestine the same in carnivores, herbivores, and omnivores?

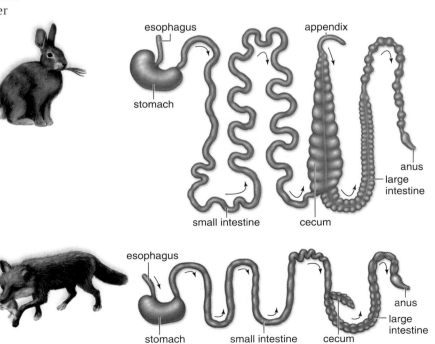

FIGURE 31.3E Digestive tract of a carnivore compared to a ruminant herbivore.

The mouth is the first part of the complete digestive system of humans (**Fig. 31.4A**). The mouth serves multiple functions, from allowing us to enjoy the taste and texture of food to fighting infection. The roof of the mouth separates the nasal cavities from the oral cavity. The roof has two parts: an anterior (toward the front) **hard palate** and a posterior (toward the back) **soft palate** (**Fig. 31.4B**). The hard palate contains several bones, but the interior of the soft palate is composed entirely of muscle. The soft palate ends in a finger-shaped projection called the uvula, which plays a role in snoring by obstructing the free flow of air through the respiratory passages. The tonsils are in the back of the mouth, on either side of the tongue, as well as in the nasopharynx, where they are called adenoids. The tonsils contain lymphatic tissue and help protect the body against infections. If the tonsils become inflamed, the person has **tonsillitis**. If tonsillitis recurs repeatedly, the tonsils may be surgically removed (called a tonsillectomy).

Salivary Glands Three pairs of **salivary glands** send juices (saliva), by way of ducts, to the mouth. Saliva has a neutral pH but is still one of the body's defenses against disease-causing pathogens because it contains antibacterial agents, including lysozyme. One pair of salivary glands lies at the sides of the face immediately below and in front of the ears. These glands swell when a person has the mumps, a disease caused by a viral infection. The glands have ducts that open on the inner surface of the cheek at the location of the second upper molar. Another pair of salivary glands lies beneath the tongue, and still another pair lies beneath the floor of the oral cavity. The ducts from these salivary glands open under the tongue. You can locate the openings if you use your tongue to feel for small flaps on the inside of your cheek and under your tongue. Saliva contains an enzyme called **salivary amylase** that begins the process of digesting starch.

The tongue, which is composed of skeletal muscle, mixes the chewed food with saliva. It then forms this mixture into a mass, called a bolus, in preparation for swallowing.

The Teeth We use our teeth to chew food into pieces convenient for swallowing. During the first two years of life, the 20 smaller deciduous, or baby, teeth appear. These are eventually replaced by 32 adult teeth of four types (Fig. 31.4B). Anteriorly, four chisel-shaped incisors function for biting. Flanking them on either side are two pointed canine teeth, or cuspids, which help tear food. Laterally, four flat premolar teeth, or bicuspids, are used for grinding food. The most posterior are the six molar teeth, designed for crushing and grinding food. The third pair of molars, called the wisdom teeth, sometimes fail to erupt. If they push on the other teeth or cause pain, they can be removed by a dentist or oral surgeon.

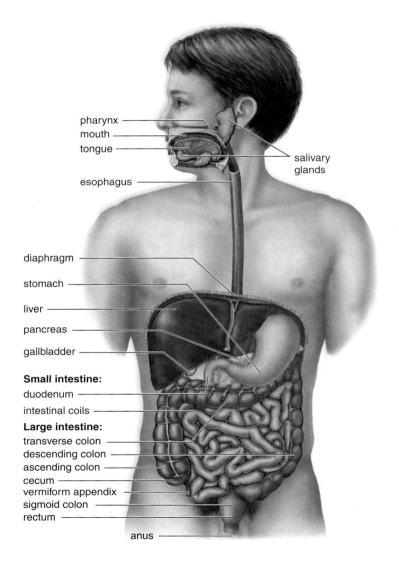

pharynx
mouth
tongue
salivary glands
esophagus
diaphragm
stomach
liver
pancreas
gallbladder
Small intestine:
duodenum
intestinal coils
Large intestine:
transverse colon
descending colon
ascending colon
cecum
vermiform appendix
sigmoid colon
rectum
anus

FIGURE 31.4A The human digestive tract.

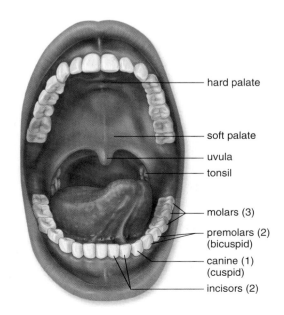

hard palate
soft palate
uvula
tonsil
molars (3)
premolars (2) (bicuspid)
canine (1) (cuspid)
incisors (2)

FIGURE 31.4B Adult teeth.

Tooth decay, called **dental caries**, or cavities, occurs when bacteria within the mouth metabolize sugar and give off acids, which erode teeth. Two measures can prevent tooth decay: eating a limited amount of sweets, and daily brushing and flossing the teeth. Fluoride treatments, particularly in children, can make the enamel stronger and more resistant to decay.

Gum Disease Brushing and flossing remove dental plaque, a sticky, colorless film that can cling to your teeth and line your gums. Bacteria accumulate in plaque and cause inflammation of the gums (gingivitis). Gingivitis can spread to the periodontal membrane, which lines the tooth sockets. A person then has periodontitis, characterized by a loss of bone and loosening of the teeth so that extensive dental work may be required. Stimulation of the gums, as recommended by a dentist, can be helpful in controlling this condition.

The consequences of gum infection may not end with loss of teeth. Researchers are discovering that inflammation, including gum disease, may increase the risk of serious health problems, such as heart attack, stroke, poorly controlled diabetes, and preterm labor.

Let's move on to the esophagus, the next part of the human digestive system, discussed in Section 31.5.

> **31.4** *Check Your Progress* No digestion of starch takes place in the mouth of a carnivore. Why is this consistent with their diet?

31.5 The esophagus conducts food to the stomach

The **esophagus** is a muscular tube that passes from the pharynx through the thoracic cavity and diaphragm into the abdominal cavity, where it joins the stomach (see Fig. 31.4A). The **pharynx** is a region that receives air from the nasal cavities and food from the mouth. The food passage and air passage cross in the pharynx because the trachea (windpipe) is anterior to (in front of) the esophagus. As shown in **Figure 31.5A**, swallowing is a process that occurs in the pharynx. **Swallowing** is a reflex action performed automatically, without conscious thought. During swallowing, ❶ the soft palate moves back to close off the **nasopharynx** (leads to nose), and ❷ the trachea moves up under the **epiglottis,** a flap that now covers the **glottis** (opening to air passages). A food bolus then enters the ❸ esophagus because the air passages are blocked—we do not breathe when we swallow.

The esophagus plays no role in the chemical digestion of food. Its sole purpose is to move the food bolus from the mouth to the stomach. A rhythmic contraction called **peristalsis** pushes the bolus along the digestive tract, much as squeezing a tube of toothpaste pushes toothpaste along (**Fig. 31.5B**). Peristalsis begins in the esophagus and continues in all the organs of the digestive tract.

The place where the esophagus enters the stomach is marked by a sphincter. **Sphincters** are muscles that encircle tubes and

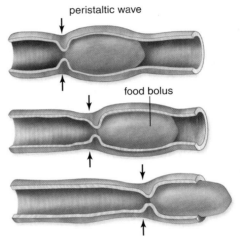

FIGURE 31.5B
Peristalsis, a rhythmic progressive wave of muscular contraction, pushes the food bolus along the digestive tract.

peristaltic wave

food bolus

act as valves; tubes close when sphincters contract, and they open when sphincters relax. Relaxation of the sphincter allows food to pass into the stomach, while contraction prevents the acidic contents of the stomach from backing up into the esophagus. The occasional backup, or reflux, of stomach contents into the esophagus usually causes no difficulty because it can be neutralized by saliva, which is slightly basic. Some people, however, develop gastroesophageal reflux disease.

Gastroesophageal Reflux Disease (GERD) GERD is apparent when refluxed stomach acid touches the lining of the esophagus and causes a burning sensation in the chest or throat called heartburn, or acid indigestion. Heartburn that occurs more than twice a week may be considered acid reflux disease.

Various medications—both prescription and nonprescription—are available to treat acid reflux. If these fail, surgery to strengthen the sphincter is available.

Section 31.6 explains the structure and digestive functions of the stomach.

> **31.5** *Check Your Progress* Do you predict that the wall of the esophagus contains muscular tissue? Which type—skeletal, cardiac, or smooth?

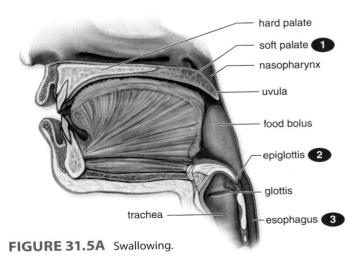

hard palate

soft palate ❶

nasopharynx

uvula

food bolus

epiglottis ❷

glottis

trachea

esophagus ❸

FIGURE 31.5A Swallowing.

31.6 Food storage and chemical digestion take place in the stomach

The *stomach* (**Fig. 31.6**) is a thick-walled, J-shaped organ that is continuous with the esophagus above and the duodenum of the small intestine below. The stomach (1) stores food, (2) initiates the digestion of protein, and (3) controls the movement of digested food into the small intestine. The stomach does not absorb nutrients; however, it does absorb alcohol because alcohol is fat-soluble and can pass through the plasma membrane of cells in the stomach wall.

The wall of the stomach contains three layers of smooth muscle—the oblique, circular, and longitudinal layers—that allow the stomach to stretch and to mechanically break down food into smaller fragments. The lining of the stomach has deep folds, called the rugae, which disappear as the stomach fills to an approximate capacity of 1 L. The stomach lining also has millions of gastric pits, which lead into gastric glands that produce gastric juice (Fig. 31.6). Gastric juice contains a precursor to the enzyme **pepsin,** which digests protein, plus hydrochloric acid (HCl) and mucus. This mucus forms a coating that protects the stomach from self-digestion. HCl causes the stomach to have high acidity (a pH of about 2), which is beneficial because it kills most of the bacteria present in food. Although HCl does not digest food, it does break down the connective tissue of meat and activates pepsin, which requires an acid pH.

Normally, the stomach empties in about two to six hours. When food leaves the stomach, it is a thick, soupy liquid called **chyme.** Chyme enters the small intestine in squirts by way of the pyloric sphincter, which acts like a valve, repeatedly opening and closing.

We diverge from our main topic in Section 31.7 to examine the cause of ulcers.

> **31.6 *Check Your Progress*** Would herbivores, carnivores, or omnivores have the most need for mechanical digestion and pepsin digestion in the stomach? Explain.

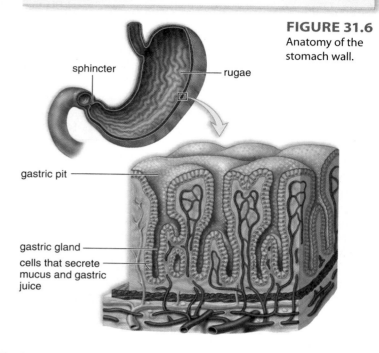

FIGURE 31.6 Anatomy of the stomach wall.

sphincter

rugae

gastric pit

gastric gland

cells that secrete mucus and gastric juice

HOW BIOLOGY IMPACTS OUR LIVES

31.7 Bacteria contribute to the cause of ulcers

In 1974, Dr. Barry Marshall, a brash, youthful physician trainee, made up his mind to pursue a research project with Dr. Robin Warren, a pathologist at the Royal Perth Hospital in Australia. Dr. Warren had observed the common presence of an unusual spiral-shaped bacterium in the biopsy tissue of many ulcer patients. Extensive research indicated that the bacteria belonged to a new, relatively unknown species known as *Helicobacter pylori.*

In 1983, Marshall proposed at an international conference that the cause of ulcers was not stress, diet, or excess stomach acid as had been previously thought, but a bacterial infection caused by *Helicobacter pylori.* Initially, the idea was met with scorn and ridicule. How could any bacterium survive the harsh, acidic environment of the stomach? The theory that ulcers were caused by stress and diet was well ingrained. Volumes of studies had been published establishing the prevailing dogma, and millions of dollars had gone into developing drugs to inhibit excess stomach acid production.

Marshall and Warren soon found their novel theory in danger, because they were unable to culture the strange spiral bacteria in laboratory animals. Finally, in desperation and defiance, Dr. Marshall decided to use himself as a human guinea pig and quaffed a tube of *Helicobacter pylori.* About one week later, he began vomiting and suffering the painful symptoms of gastritis, or stomach inflammation. Medical tests confirmed this condition, and revealed that his stomach was teeming with the bizarre bacteria. He treated himself by taking a combination of antibiotics to kill the bacteria and an acid-blocking drug to ease the symptoms. Within a few weeks, Dr. Marshall felt better. But more importantly, he had made his point. He challenged other researchers to prove his theory wrong, but before long, many of these very studies indeed supported his theory.

Today, ulcer sufferers are no longer resigned to following a bland diet and looking for ways to reduce stress in their lives. Instead of simply treating the symptoms of an ulcer, doctors can now treat the cause of the condition with a simple round of antibiotics, coupled with an acid-reducing drug to allow the stomach lining to heal. For their innovative thinking in making the connection between *Helicobacter pylori* and ulcers, Dr. Marshall and Dr. Warren were awarded the 2005 Nobel Prize in Physiology or Medicine.

Following the stomach, the small intestine is an important center for both chemical digestion and absorption, as discussed in Section 31.8.

> **31.7 *Check Your Progress*** Explain why Dr. Marshall ingested a tube of *Helicobacter pylori,* with reference to the statement, "A correlation does not necessarily mean causation."

31.8 In the small intestine, chemical digestion concludes, and absorption of nutrients occurs

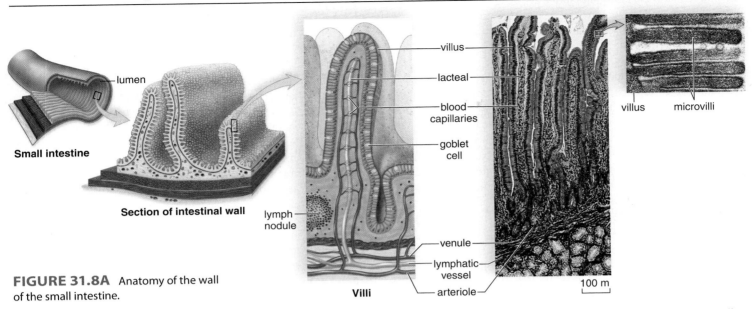

FIGURE 31.8A Anatomy of the wall of the small intestine.

The **small intestine** is quite long—about 6 meters in length (range: 4.6–9 m)—but it takes its name from its small diameter, which is about 2.5 cm. The first 25 cm of the small intestine is called the **duodenum**. A duct brings bile from the liver and gallbladder, and pancreatic juice from the pancreas, into the small intestine (see Fig. 31.9). **Bile** emulsifies fat—that is, it causes fat droplets to disperse in water. The intestine has a slightly basic pH because pancreatic juice contains sodium bicarbonate ($NaHCO_3$), which neutralizes chyme. The enzymes in pancreatic juice include **pancreatic amylase,** which digests starch to maltose; **trypsin,** which digests protein to peptides; **lipase,** which digests fat droplets to monoglycerides and fatty acids; and nucleases, which digest DNA and RNA to nucleotides.

Certain anatomic features increase the surface area of the small intestine. The wall of the small intestine contains finger-like projections called villi (sing., **villus**), which give the intestinal wall a soft, velvety appearance (**Fig. 31.8A**). The outer cells of a villus have thousands of microscopic extensions called **microvilli.** Just as the projecting threads of a terry-cloth towel absorb water, the microvilli greatly increase the surface area of the villus for the absorption of small nutrient molecules. The villi also bear the intestinal enzymes that finish the process of chemical digestion. As shown in **Figure 31.8B**, maltase digests maltose to glucose, and peptidases break down peptides to amino acids. Also present are nucleotidase enzymes that digest nucleotides to their component parts.

Nutrient molecules are absorbed into the vessels of a villus, which contains blood capillaries and a lymphatic capillary called a **lacteal.** Sugars (digested from carbohydrates) and amino acids (digested from proteins) enter the blood capillaries of a villus. Monoglycerides and fatty acids (digested from fats) enter the epithelial cells of the villi, and within these cells they are joined and packaged as lipoprotein droplets, which enter a lacteal (Fig. 31.8B, *right*). Lacteals are a part of the lymphatic system, a one-way system of lymphatic vessels that return lymph to the bloodstream (see Fig. 30.1).

The pancreas and the liver help the small intestine perform its functions, as explained in Section 31.9.

31.8 Check Your Progress Which types of macromolecules are broken down to nutrients that can be absorbed by the small intestine?

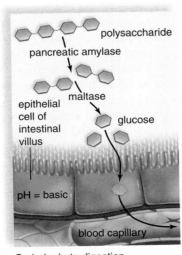

Carbohydrate digestion

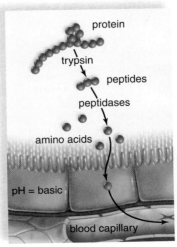

Protein digestion

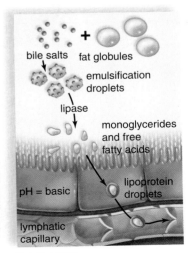

Fat digestion

FIGURE 31.8B Digestion and absorption of nutrients.

The pancreas, liver, and gallbladder are accessory digestive organs. **Figure 31.9** shows how the pancreatic duct from the pancreas and the common bile duct from the liver and gallbladder enter the duodenum.

The Pancreas   The **pancreas** lies deep in the abdominal cavity, behind the stomach. It is an elongated and somewhat flattened organ that has both an endocrine and an exocrine function. As an endocrine gland, it secretes insulin and glucagon, hormones that help keep the blood glucose level within normal limits. In this chapter, however, we are interested in its exocrine function. Most pancreatic cells produce pancreatic juice, which contains sodium bicarbonate ($NaHCO_3$) and digestive enzymes for all types of food. Sodium bicarbonate neutralizes acidic chyme from the stomach. Pancreatic amylase digests starch, trypsin digests protein, and lipase digests fat, as stated in Section 31.8.

The Liver   The **liver**, which is the largest gland in the body, lies mainly in the upper right section of the abdominal cavity, under the diaphragm (see Fig. 31.4A). The liver contains approximately 100,000 lobules that serve as its structural and functional units (Fig. 31.9). Triads, located between the lobules, consist of a bile duct, which takes bile away from the liver; a branch of the hepatic artery, which brings O_2-rich blood to the liver; and a branch of the hepatic portal vein, which transports nutrients from the intestines to the liver. The central veins of the lobules drain blood from the liver and form the hepatic veins, which enter the inferior vena cava.

In some ways, the liver acts as the gatekeeper to the blood. As blood in the hepatic portal vein from the intestines passes through the liver, the liver removes poisonous substances and detoxifies them. The liver also removes and stores iron and the vitamins A, B_{12}, D, E, and K. It manufactures the plasma proteins and helps regulate the quantity of cholesterol in the blood.

The liver maintains the blood glucose level at about 100 mg/100 mL (0.1%), even though a person eats intermittently. When insulin is present, any excess glucose present in the blood is removed and stored by the liver as glycogen. Between meals, glycogen is broken down to glucose, which enters the hepatic veins, and in this way, the blood glucose level remains constant.

If the supply of glycogen is depleted, the liver converts glycerol (from fats) and amino acids to glucose molecules. The conversion of amino acids to glucose requires the removal of amino groups. Via a complex metabolic pathway, the liver then combines ammonia from the amino groups with carbon dioxide to form urea. Urea is the usual nitrogenous waste product from amino acid breakdown in humans.

The liver produces bile, which is stored in the **gallbladder**. Bile has a yellowish-green color because it contains the bile pigment *bilirubin*, which is derived from the breakdown of hemoglobin, the red pigment of red blood cells. Bile also contains bile salts. Bile salts are derived from cholesterol, and they emulsify fat in the small intestine. When fat is emulsified, it breaks up into droplets, providing a much larger surface area that can be acted upon by a digestive enzyme from the pancreas.

Liver Disorders   When a person is jaundiced, the skin has a yellowish tint due to an abnormally large quantity of bile pigments in the blood. In hemolytic jaundice, red blood cells are broken down in abnormally large amounts; in obstructive jaundice, the bile duct is obstructed, or the liver cells are damaged. Obstructive jaundice often occurs when crystals of cholesterol precipitate out of the bile and form gallstones.

Jaundice can also result from a viral infection of the liver, called *hepatitis*. Hepatitis A is most often caused by eating contaminated food. Today, hepatitis B and C are mainly spread by the use of shared needles by drug addicts and unsterilized tools by tattoo artists. These three types of hepatitis can also be spread by sexual contact.

Cirrhosis is a chronic liver disease in which the organ first becomes fatty, and later liver tissue is replaced by inactive fibrous scar tissue. Alcoholics often get cirrhosis, most likely due, at least in part, to the excessive amounts of alcohol the liver is forced to break down.

The endocrine functions of the stomach and duodenum are reviewed in Section 31.10.

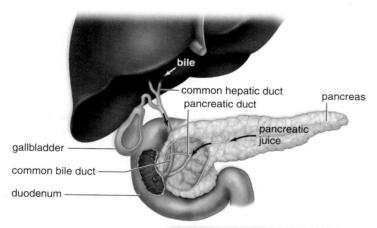

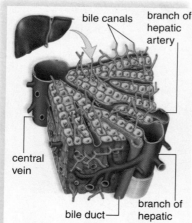

FIGURE 31.9 Liver, gallbladder, and pancreas.

> **31.9** *Check Your Progress* **When the doctor prescribes a medication, you are often told to take the medicine at regular intervals until it is gone. Why can't you take medicine once and be done with it?**

31.10 The stomach and duodenum are endocrine glands

Both the stomach and the duodenum function as endocrine glands, producing their own sets of hormones. The stomach produces **gastrin,** a hormone that increases the secretory activity of the gastric glands. The cells of the duodenal wall produce two other hormones that are of particular interest—**secretin** and CCK (**cholecystokinin**). Acid, especially HCl present in chyme, stimulates the release of secretin, while partially digested protein and fat stimulate the release of CCK. Soon after these hormones enter the bloodstream, the pancreas increases its output of pancreatic juice. At the same time, the liver increases its production of bile, and the gallbladder contracts to release stored bile (**Fig. 31.10**).

Section 31.11 describes the composition and function of the large intestine.

> **31.10** *Check Your Progress* How do the pancreas and the gallbladder "know" to send their secretions to the small intestine?

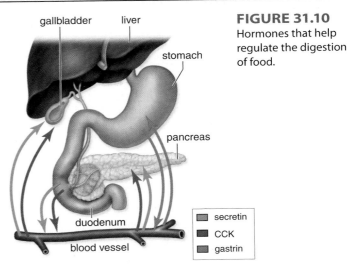

FIGURE 31.10 Hormones that help regulate the digestion of food.

- secretin
- CCK
- gastrin

31.11 The large intestine absorbs water and prepares wastes for elimination

The **large intestine**, which includes the cecum, the colon, the rectum, and the anus, is larger in diameter (6.5 cm) but shorter in length (1.5 m) than the small intestine. The large intestine absorbs water, salts, and some vitamins. It also stores indigestible material until it is eliminated. No digestion takes place in the large intestine.

The cecum in humans has a small projection called the **vermiform appendix** (**Fig. 31.11**). In humans and other animals, such as rabbits, the appendix plays a role in fighting infections. The appendix can become infected and filled with fluid, a condition called appendicitis. If an infected appendix bursts before it can be removed, a serious, generalized infection of the abdominal lining, called peritonitis, can result.

The **colon** is subdivided into the ascending, transverse, descending, and sigmoid colons (see Fig. 31.4A). The sigmoid colon enters the **rectum**, the last 20 cm of the large intestine. About 1.5 L of water enters the digestive tract daily as a result of eating and drinking. An additional 8.5 L enter the digestive tract each day carrying the various substances secreted by the digestive glands. About 95% of this water is absorbed by the small intestine, and much of the remaining portion is absorbed into the cells of the colon. If this water is not reabsorbed, diarrhea can lead to serious dehydration and ion loss, especially in children.

The large intestine has a large population of bacteria, notably *Escherichia coli*. These bacteria break down indigestible material, and they also produce some vitamins, such as vitamin K, which is necessary to blood clotting. Digestive wastes (feces) eventually leave the body through the anus, the opening of the anal canal. Feces are about 75% water and 25% solid matter.

The colon is subject to the development of **polyps**, small growths arising from the epithelial lining. Polyps, whether they are benign or cancerous, can be removed surgically. Some investigators believe that dietary fat increases the likelihood of colon cancer. Dietary fat causes an increase in bile secretion, and it could be that intestinal bacteria convert bile salts to substances that promote the development of colon cancer. By contrast, dietary fiber absorbs water and adds bulk, thereby preventing constipation and facilitating the movement of substances through the intestine, as we learn in Section 31.12.

This completes our study of animal digestive systems. The next part of the chapter discusses nutrition, beginning with carbohydrates.

> **31.11** *Check Your Progress* Our large intestine matches that of an herbivore, both in structure and function. Why might that be?

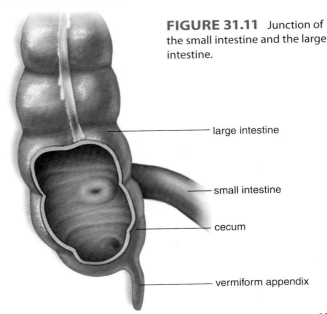

FIGURE 31.11 Junction of the small intestine and the large intestine.

- large intestine
- small intestine
- cecum
- vermiform appendix

This part of the chapter discusses the benefits of a balanced diet, which includes all necessary minerals and vitamins. Learning to read nutrition labels can assist in making sure food is nutritious as well as in limiting the number of calories. Eating disorders also need attention in order to ensure a healthy diet.

31.12 Carbohydrates are nutrients that provide immediate energy as well as fiber

Carbohydrates are present in food in the form of sugars, starch, and fiber. Fruits, vegetables, milk, and honey are natural sources of sugars. Glucose and fructose are monosaccharide sugars, and lactose (milk sugar) and sucrose (table sugar) are disaccharides. After being absorbed from the digestive tract into the bloodstream, all sugars are converted to glucose for transport in the blood and use by cells. Glucose is the preferred direct energy source in cells.

Plants store glucose as starch, and animals store glucose as glycogen. Good sources of starch are beans, peas, cereal grains, and potatoes. Starch is digested to glucose in the digestive tract, and any excess glucose is stored as glycogen. Although other animals likewise store glucose as glycogen in liver or muscle tissue (meat), little is left by the time an animal is eaten for food. Except for honey and milk, which contain sugars, animal foods do not contain carbohydrates.

Fiber includes various indigestible carbohydrates derived from plants. Food sources rich in fiber include beans, peas, nuts, fruits, and vegetables. Whole-grain products are also a good source of fiber, and are therefore more nutritious than food products made from refined grains. During *refinement*, fiber and also vitamins and minerals are removed from grains, so that primarily starch remains. For example, a slice of bread made from whole-wheat flour contains 3 g of fiber; a slice of bread made from refined wheat flour contains less than 1 g of fiber.

Technically, fiber is not a nutrient for humans because it cannot be digested to small molecules that enter the bloodstream. Insoluble fiber, however, adds bulk to fecal material, which stimulates movement in the large intestine, preventing constipation. Soluble fiber combines with bile acids and cholesterol in the small intestine and prevents them from being absorbed. In this way, high-fiber diets may protect against heart disease. The typical American consumes only about 15 g of fiber each day; the recommended daily intake of fiber is 25 g for women and 38 g for men. To increase your fiber intake, eat whole-grain foods, snack on fresh fruits and raw vegetables, and include nuts and beans in your diet (**Fig. 31.12**).

If you, or someone you know, has lost weight by following the Atkins or South Beach diet, you may think "carbs" are unhealthy and should be avoided. According to nutritionists, however, carbohydrates should supply a large portion of your energy needs. Evidence suggests that Americans are not eating the right kind of carbohydrates. In some countries, the traditional diet is 60–70% high-fiber carbohydrates, and these people have a low incidence of the diseases that plague Americans.

Some nutritionists hypothesize that the high intake of foods that are rich in refined carbohydrates and fructose sweeteners

TABLE 31.12   Reducing Dietary Sugars
To Reduce Dietary Sugar:
1. Eat fewer sweets, such as candy, soft drinks, ice cream, and pastry.
2. Eat fresh fruits or fruits canned without heavy syrup.
3. Use less sugar—white, brown, or raw—and less honey and syrups.
4. Avoid sweetened breakfast cereals.
5. Eat less jelly, jam, and preserves.
6. Drink pure fruit juices, not imitations.
7. When cooking, use spices such as cinnamon, instead of sugar, to flavor foods.
8. Do not put sugar in tea or coffee.
9. Avoid potatoes and processed foods made from refined carbohydrates, such as white bread, rice, and pasta. These foods are immediately broken down to sugar during digestion.

processed from cornstarch may be responsible for the prevalence of obesity in the United States. Because certain foods, such as donuts, cakes, pies, and cookies, are high in both refined carbohydrates and fat, it is difficult to determine which dietary component is responsible for the current epidemic of obesity among Americans. In any case, they are empty-calorie foods that provide sugars but no vitamins or minerals. **Table 31.12** tells how to reduce your sugar intake. Nutritionists also point out that consuming too much energy from any source contributes to body fat, which increases a person's risk of obesity and associated illnesses.

The inclusion of lipids in the diet requires careful consideration, as described in Section 31.13.

> **31.12 Check Your Progress** Which type of carbohydrate leads to poor health, and which type leads to good health?

FIGURE 31.12 Fiber-rich foods.

31.13 Lipids are nutrients that supply long-term energy

Like carbohydrates, **triglycerides** (fats and oils) supply energy for cells, but **fat** is stored for the long term in the body. Nutritionists generally recommend that people include unsaturated, rather than saturated, fats in their diets. Two unsaturated fatty acids (alpha-linolenic and linoleic acids) are *essential* in the diet. They can be supplied by eating fatty fish and by including plant oils, such as canola and soybean oils, in the diet. Delayed growth and skin problems can develop in people whose diets lack these essential unsaturated fatty acids.

Animal-derived foods, such as butter, meat, whole milk, and cheeses, contain saturated fatty acids. Plant oils contain unsaturated fatty acids; each type of oil has a particular percentage of monounsaturated and polyunsaturated fatty acids.

Cholesterol, a lipid, can be synthesized by the body. Cells use cholesterol to make various compounds, including bile, steroid hormones, and vitamin D. Plant foods do not contain cholesterol; only animal foods such as cheese, egg yolks, liver, and certain shellfish (shrimp and lobster) are rich in cholesterol. Elevated blood cholesterol levels are associated with an increased risk of cardiovascular disease, the number-one killer of Americans. A diet rich in cholesterol and saturated fats increases the risk of cardiovascular disease (see Section 31.18).

Statistical studies suggest that trans fatty acids (trans fats) are even more harmful than saturated fatty acids. Trans fatty acids arise when unsaturated oils are hydrogenated to produce a solid fat, as in shortening and some margarines. Trans fatty acids may reduce the function of the plasma membrane receptors that clear cholesterol from the bloodstream. Trans fatty acids are found in commercially packaged foods, such as cookies and crackers; in commercially fried foods, such as french fries; and in packaged snacks. **Table 31.13** tells you how to cut back on harmful lipids in your diet.

Dietary protein is our topic in Section 31.14.

> **31.13 Check Your Progress** Which types of lipids lead to good health, and which types lead to poor health?

TABLE 31.13 Reducing Harmful Lipids

To Reduce Dietary Saturated Fat:

1. Choose poultry, fish, or dry beans and peas as a protein source.

2. Remove skin from poultry and trim fat from red meats before cooking; place meat on a rack while cooking so that fat drains off.

3. Broil, boil, or bake rather than frying.

4. Limit your intake of butter, cream, trans fats, shortening, and tropical oils (coconut and palm oils).*

5. Use herbs and spices to season vegetables instead of butter, margarine, or sauces. Use lemon juice instead of salad dressing.

6. Drink skim milk instead of whole milk, and use skim milk in cooking and baking.

To Reduce Dietary Cholesterol:

1. Eat white fish and poultry in preference to cheese, egg yolks, liver, and certain shellfish (shrimp and lobster).

2. Substitute egg whites for egg yolks in both cooking and eating.

3. Include soluble fiber in the diet. Oat bran, oatmeal, beans, corn, and fruits, such as apples, citrus fruits, and cranberries, are high in soluble fiber.

*Although coconut and palm oils are from plant sources, they are mostly saturated fats.

31.14 Proteins are nutrients that supply building blocks for cells

Dietary **proteins** are digested to amino acids, which cells use to synthesize hundreds of cellular proteins. Of the 20 different amino acids, nine are *essential amino acids* that must be present in the diet. Children will not grow if their diets lack the essential amino acids. Eggs, milk products, meat, poultry, and most other foods derived from animals contain all nine essential amino acids and are considered "complete" or "high-quality" protein sources.

Foods derived from plants generally do not have as much protein per serving as those derived from animals, and each type of plant food generally lacks one or more of the essential amino acids. Therefore, most plant foods are "incomplete" or "low-quality" protein sources. Vegetarians, however, do not have to rely on animal sources of protein. To meet their protein needs, total vegetarians (vegans) can eat grains, beans, and nuts in various combinations. Also, tofu, soymilk, and other foods made from processed soybeans are complete protein sources. A balanced vegetarian diet is quite possible with a little planning.

According to nutritionists, protein should not supply the bulk of dietary calories. The average American eats about twice as much protein as he or she needs, and some people may be on a diet that encourages the intake of proteins, instead of carbohydrates, as an energy source. Also, bodybuilders should realize that excess amino acids are not always converted into muscle tissue. When amino acids are broken down, the liver removes the nitrogen portion (*deamination*) and uses it to form urea, which is excreted in urine. The water needed for excretion of urea can cause dehydration when a person is exercising and losing water by sweating. High-protein diets can also increase calcium loss in the urine and encourage the formation of kidney stones. Furthermore, high-protein foods often contain a high amount of fat.

In Section 31.15, we will study the mineral requirements of humans.

> **31.14 Check Your Progress** Why should we emphasize intake of vegetables, as opposed to protein?

The body needs about 20 elements called **minerals** for numerous physiological functions, including regulation of biochemical reactions, maintenance of fluid balance, and incorporation into certain structures and compounds. The body contains more than 5 g of each major mineral and less than 5 g of each trace mineral. **Table 31.15** lists both the major and trace minerals, and gives their functions and food sources. It also tells the health effects of too little or too much intake.

Occasionally, individuals (especially women) do not receive enough iron, calcium, magnesium, or zinc in their diets. Adult females need more iron in the diet than males (18 mg compared to 10 mg) if they are menstruating each month. *Anemia*, characterized by a run-down feeling due to insufficient red blood cells, results when the diet lacks sufficient iron.

Many people take calcium supplements, as directed by a physician, to counteract *osteoporosis*, a degenerative bone disease that affects an estimated one-quarter of older men and one-half of older women in the United States.

One mineral that people consume too much of is sodium. The recommended amount of sodium intake per day is 2,400 mg, while the average American takes in 4,000–4,700 mg each day. About one-third of the sodium we consume occurs naturally in foods, another one-third is added during commercial processing, and the last one-third is added either during home cooking or at the table in the form of table salt.

Vitamins are also a necessary component of any diet, as explained in Section 31.16.

> **31.15 Check Your Progress** Why should we use less salt in our food?

TABLE 31.15 Minerals

Mineral	Functions	Food Sources	Conditions Caused By: Too Little	Conditions Caused By: Too Much
Major Minerals				
Calcium (Ca^{2+})	Strong bones and teeth, nerve conduction, muscle contraction	Dairy products, leafy green vegetables	Stunted growth in children; low bone density in adults	Kidney stones; interferes with iron and zinc absorption
Phosphorus (PO_4^{3-})	Bone and soft tissue growth; part of phospholipids, ATP, and nucleic acids	Meat, dairy products, sunflower seeds, food additives	Weakness, confusion, pain in bones and joints	Low blood and bone calcium levels
Potassium (K^+)	Nerve conduction, muscle contraction	Many fruits and vegetables, bran	Paralysis, irregular heartbeat, eventual death	Vomiting, heart attack, death
Sodium (Na^+)	Nerve conduction, pH and water balance	Table salt	Lethargy, muscle cramps, loss of appetite	High blood pressure, calcium loss
Chloride (Cl^-)	Water balance	Table salt	Not likely	Vomiting, dehydration
Magnesium (Mg^{2+})	Part of various enzymes for nerve and muscle contraction, protein synthesis	Whole grains, leafy green vegetables	Muscle spasm, irregular heartbeat, convulsions, confusion, personality changes	Diarrhea
Trace Minerals				
Zinc (Zn^{2+})	Protein synthesis, wound healing, fetal development and growth, immune function	Meats, legumes, whole grains	Delayed wound healing, night blindness, diarrhea, mental lethargy	Anemia, diarrhea, vomiting, renal failure, abnormal cholesterol levels
Iron (Fe^{2+})	Hemoglobin synthesis	Whole grains, meats, prune juice	Anemia, physical and mental sluggishness	Iron toxicity disease, organ failure, eventual death
Copper (Cu^{2+})	Hemoglobin synthesis	Meat, nuts, legumes	Anemia, stunted growth in children	Damage to internal organs if not excreted
Iodine (I^-)	Thyroid hormone synthesis	Iodized table salt, seafood	Thyroid deficiency	Depressed thyroid function, anxiety
Selenium (SeO_4^{2-})	Part of antioxidant enzyme	Seafood, meats, eggs	Vascular collapse, possible cancer development	Hair and fingernail loss, discolored skin

31.16 Vitamins help regulate metabolism

Vitamins are organic compounds (other than carbohydrates, fats, and proteins, including amino acids) that regulate various metabolic activities and must be present in the diet. **Table 31.16** lists the major vitamins and some of their functions, food sources, and associated disorders.

Although many people think vitamins can enhance health dramatically, prevent aging, and cure diseases such as arthritis and cancer, there is no scientific evidence that vitamins are "wonder drugs." However, vitamins C, E, and A are believed to defend the body against free radicals, and therefore they are termed **antioxidants**. These vitamins are especially abundant in fruits and vegetables, and so it is suggested that we eat about 4½ cups of fruits and vegetables per day. Assuming an adequate diet, skin cells normally contain a precursor cholesterol mol-

ecule that is converted to vitamin D after UV exposure. Most milk today is fortified with vitamin D, which helps prevent the occurrence of rickets, characterized by defective mineralization of the skeleton.

Although many foods in the United States are now enriched, or fortified with vitamins, some individuals are still at risk for vitamin deficiencies, generally as a result of poor food choices. These include the elderly, young children, alcoholics, and low-income people. Section 31.17 helps you learn how to interpret nutrition panels to avoid various illnesses.

> **31.16 Check Your Progress** What food sources can be relied on to supply vitamins in the diet?

TABLE 31.16 Vitamins

Vitamin	Functions	Food Sources	Conditions Caused By: Too Little	Too Much
Water-Soluble Vitamins				
Vitamin C	Antioxidant; collagen synthesis for capillaries, bones, and teeth	Citrus fruits, green leafy vegetables, tomatoes	Scurvy, delayed wound healing, infections	Gout, kidney stones, diarrhea
Thiamine (vitamin B_1)	Coenzyme needed for cellular respiration	Whole grains, legumes, and nuts	Beriberi, anemia, muscle weakness	Absorption of other vitamins prevented
Riboflavin (vitamin B_2)	Aids cellular respiration, including oxidation of protein and fat	Nuts, dairy products, whole grains, poultry, green leafy vegetables	Dermatitis, blurred vision, growth failure	Unknown
Niacin (nicotinic acid)	Coenzyme needed for cellular respiration	Peanuts, poultry, whole grains, green leafy vegetables	Pellagra, diarrhea, mental disorders	High blood sugar and uric acid, vasodilation
Folacin (folic acid)	Coenzyme needed for production of hemoglobin and DNA	Dark-green leafy vegetables, nuts, beans, whole grains	Megaloblastic anemia, spina bifida	May mask B_{12} deficiency
Vitamin B_6	Coenzyme; synthesis of hormones and hemoglobin; CNS control	Whole grains, bananas, beans, poultry, nuts, green leafy vegetables	Rarely, convulsions, vomiting, seborrhea, muscle weakness	Insomnia, neuropathy
Pantothenic acid	Coenzyme A needed for oxidation of carbohydrates and fats	Nuts, beans, dark-green leafy vegetables, poultry, fruits, milk	Rarely, loss of appetite, mental depression, numbness	Unknown
Vitamin B_{12}	Coenzyme needed for synthesis of nucleic acids and myelin	Dairy products, fish, poultry, eggs, fortified cereals	Pernicious anemia	Unknown
Biotin	Coenzyme needed for metabolism of amino acids and fatty acids	Generally in foods; eggs	Skin rash, nausea, fatigue	Unknown
Fat-Soluble Vitamins				
Vitamin A	Antioxidant synthesized from beta-carotene; healthy eyes, skin, hair, and proper bone growth	Deep yellow/orange and green leafy vegetables, fruits, cheese, whole milk, butter, eggs	Night blindness, impaired growth of bones and teeth	Headache, nausea, hair loss, abnormal fetal development
Vitamin D	Steroid needed for development and maintenance of bones	Fortified milk, fish liver oil; exposure to sunlight	Rickets, bone decalcification and weakening	Calcification of soft tissues, diarrhea
Vitamin E	Antioxidant; prevents oxidation of vitamin A polyunsaturated fatty acids	Green leafy vegetables, fruits, nuts, and whole grains	Unknown	Diarrhea, headaches, fatigue, muscle weakness
Vitamin K	Synthesizes substances active in blood clotting	Green leafy vegetables, cauliflower	Easy bruising and bleeding	Interferes with anticoagulant drugs

31.17   Nutritional labels allow evaluation of a food's content

A Nutrition Facts panel, such as the one shown in **Figure 31.17**, provides specific dietary information about a product and general information about the nutrients the product contains.

The serving size is based on the typical serving size for the product, though not necessarily a proper serving size. If you are comparing **Calories (kcal)** and other data about products of the same type, you definitely want to be sure you are comparing the same serving sizes.

The total intake of calories depends on the amount consumed. Obviously, if you eat twice the serving size, you have taken in twice the number of Calories. The bottom of the panel lists the Calories per gram (g) of fat, carbohydrate, and protein. You can use this information to calculate the Calories per serving for each type of nutrient. The total for all nutrients should agree with the figure given for Calories at the top of the panel—in this case, 260.

The percent daily value (percent of the total amount needed in a 2,000 kcal diet) is calculated by comparing the specific information about this product with the information given at the bottom of the panel. For example, the product in Figure 31.17 has a fat content of 13 g, and the total daily recommended amount is less than 65 g, so 13/65 = 20%. *These percent daily values are not applicable for people who require more or less than 2,000 kcal per day.* A percent daily value for protein is generally not given because determining percent daily value would require expensive testing of the protein quality of the product by the manufacturer. Also, notice that there is a percent daily value for carbohydrates but not for sugars because there is no daily value for sugar.

Assuming the same serving size, you can use Nutrition Facts panels to compare two products of the same type. For example, if you wanted to reduce your caloric intake and increase your fiber and vitamin C intakes, comparing the panels from two different food products would allow you to see which one is lowest in Calories and highest in fiber and vitamin C.

A faulty diet can contribute to the occurrence of disease, as discussed in Section 31.18.

> **31.17 Check Your Progress** The percent daily values in a Nutrition Facts panel are based on a diet of how many Calories a day?

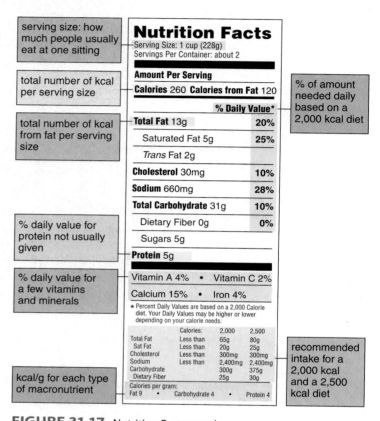

FIGURE 31.17 Nutrition Facts panel.

31.18   Certain disorders are associated with obesity

Nutritionists point out that consuming too many Calories from any source contributes to body fat, which increases a person's risk of obesity and associated illnesses. (Obesity is defined as weighing 30% more than the ideal body weight for your height and body build.) Still, foods such as donuts, cakes, pies, cookies, and white bread, which are high in refined carbohydrates (starches and sugars), and fried foods, which are high in fat, may very well be responsible for the current epidemic of obesity among Americans. Also implicated is the lack of exercise because of a sedentary lifestyle. Diabetes type 2 and cardiovascular disease are often seen in people who are obese.

Diabetes Type 2 Diabetes mellitus is indicated by the presence of glucose in the urine. Glucose has spilled over into the

urine because there is too high a level of glucose in the blood. Diabetes occurs in two forms. Diabetes type 1 is not associated with obesity. When a person has diabetes type 1, the pancreas does not produce insulin, and the patient must have daily insulin injections. In contrast, children, and more often adults, with diabetes type 2 are usually obese and display impaired insulin production and insulin resistance. In a person with insulin resistance, the body's cells fail to take up glucose, even when insulin is present. Therefore, the blood glucose level exceeds the normal level, and glucose appears in the urine.

Diabetes type 2 is increasing rapidly in most industrialized countries of the world. A healthy diet, increased physical activity, and weight loss have been seen to improve the ability of insulin to function properly in type 2 diabetics (**Fig. 31.18A**).

FIGURE 31.18A Exercising for good health.

FIGURE 31.18B Foods high in trans fats.

How might a poor diet contribute to the occurrence of diabetes type 2? Simple sugars in foods, such as candy and ice cream, immediately enter the bloodstream, as do sugars from the digestion of the starch within white bread and potatoes. When the blood glucose level rises rapidly, the pancreas produces an overload of insulin to bring the level under control. Chronically high insulin levels apparently lead to insulin resistance, increased fat deposition, and a high blood fatty acid level. High fatty acid levels can lead to increased risk of cardiovascular disease. It is well worth the effort to control diabetes type 2 because all diabetics, whether type 1 or type 2, are at risk for blindness, kidney disease, as well as cardiovascular disease.

Cardiovascular Disease In the United States, cardiovascular disease, which includes hypertension, heart attack, and stroke, is among the leading causes of death. Cardiovascular disease is often due to arteries blocked by plaque, which contains saturated fats and cholesterol. Cholesterol is carried in the blood by two types of lipoproteins: low-density lipoprotein (LDL) and high-density lipoprotein (HDL). LDL molecules are considered "bad" because they are like delivery trucks that carry cholesterol from the liver to the cells and to the arterial walls. HDL molecules are considered "good" because they are like garbage trucks that dispose of cholesterol. HDL transports cholesterol from the cells to the liver, which converts it to bile salts that enter the small intestine.

Consuming saturated fats, including trans fats, tends to raise LDL cholesterol levels, while eating unsaturated fats lowers LDL cholesterol levels. Beef, dairy foods, and coconut oil are rich sources of saturated fat. Foods containing partially hydrogenated oils (e.g., vegetable shortening and stick margarine) are sources of trans fats (**Fig. 31.18B**). Unsaturated fatty acids in olive and canola oils, most nuts, and coldwater fish tend to lower LDL cholesterol levels. Furthermore, coldwater fish (e.g., herring, sardines, tuna, and salmon) contain polyunsaturated fatty acids and especially *omega-3 unsaturated fatty acids*, which are believed to reduce the risk of cardiovascular disease. However, taking fish oil supplements to obtain omega-3s is not recommended without a physician's approval, because too much of these fatty acids can interfere with normal blood clotting.

The American Heart Association recommends limiting total cholesterol intake to 300 mg per day. This requires careful selection of the foods we include in our daily diets. For example, an egg yolk contains about 210 mg of cholesterol, which would be two-thirds of the recommended daily intake. Still, this doesn't mean eggs should be eliminated from a healthy diet, since the proteins in them are very nutritious; in fact, most healthy people can eat a couple of whole eggs each week without experiencing an increase in their blood cholesterol levels.

A physician can determine whether patients' blood lipid levels are normal. If a person's cholesterol and triglyceride levels are elevated, modifying the fat content of the diet, losing excess body fat, and exercising regularly can reduce them. If lifestyle changes do not lower blood lipid levels enough to reduce the risk of cardiovascular disease, a physician may prescribe special medications.

Eating disorders, as described in Section 31.19, can also jeopardize your health.

31.18 Check Your Progress How can we decrease our chances of acquiring diabetes type 2 and cardiovascular disease?

31.19 Eating disorders appear to have a psychological component

People with eating disorders are dissatisfied with their body image. Social, cultural, emotional, and biological factors all contribute to the development of an eating disorder. For example, it is possible that many women admire the extreme thinness of fashion models today. In a recent study at the University of Bath, two-thirds of participants favored thin models and were more likely to buy a product advertised by a thin model. In Madrid, steps were taken to safeguard the health of fashion models: They have to weigh a certain minimum according to their height, or they cannot participate in a fashion show.

Anorexia nervosa is a severe psychological disorder characterized by an irrational fear of getting fat that results in the refusal to eat enough food to maintain a healthy body weight. These people think they are fat when actually they are thin (**Fig. 31.19A**). A self-imposed starvation diet is often accompanied by occasional binge eating that is followed by purging and extreme physical activity to avoid weight gain. Binges usually include large amounts of high-calorie foods, and purging episodes involve self-induced vomiting and laxative abuse. About 90% of the people suffering from anorexia nervosa are young women; an estimated 1 in 200 teenage girls is affected.

A person with **bulimia nervosa** binge-eats, and then purges to avoid gaining weight (**Fig. 31.19B**). The binge-purge cyclic behavior can occur several times a day. People with bulimia nervosa can be difficult to identify because their body weights are often normal and they tend to conceal their binging and purging. Women are more likely than men to develop bulimia; an estimated 4% of young women suffer from this condition.

Other abnormal eating practices include binge-eating disorder and muscle dysmorphia. Many obese people suffer from **binge-eating disorder**, which is characterized by episodes of overeating that are not followed by purging. Stress, anxiety, anger, and de-

FIGURE 31.19B
Bulimia nervosa is characterized by occasional binge eating followed by purging.

pression can trigger food binges. Eating disorders, whether anorexia nervosa, bulimia nervosa, or simply binge eating, can lead to malnutrition, disability, and death.

Preoccupation with diet, bodybuilding activities, and body form characterize the disorder known as **muscle dysmorphia**. A person with this condition thinks his or her body is underdeveloped when actually it is extremely well developed (**Fig. 31.19C**). Each day, the person may spend hours in the gym working out on muscle-strengthening equipment. Unlike anorexia nervosa and bulimia, muscle dysmorphia affects more men than women.

Regardless of the eating disorder, early recognition and treatment are crucial. Treatment usually includes psychological counseling and antidepressant medications.

> **31.19 Check Your Progress** In what way might our society play a role in encouraging eating disorders?

FIGURE 31.19A
Anorexia nervosa patients think they are fat, even though they are thin.

FIGURE 31.19C
Muscle dysmorphia is also characterized by a distorted body image.

Assigning any organ to a particular body system seems arbitrary. Granted, a good argument can be made for declaring that the mouth, esophagus, stomach, and intestines are in the digestive system. But actually, even these organs assist other body systems. For example, the stomach and small intestine contribute to the endocrine system by producing hormones (see Section 31.10). And how could muscles and nerves function without a supply of calcium from the digestive tract? Or, for that matter, how could any other system in the body function without a supply of nutrients absorbed by the digestive tract and distributed by the cardiovascular system?

The liver has so many functions that it really belongs to many systems of the body. Doesn't it belong to the cardiovascular system because it produces plasma proteins and to the urinary system because it produces urea, as well as to the digestive system because it produces bile? The liver is a vital organ, meaning that we cannot live without it. Among its many functions, it detoxifies blood by removing and metabolizing poisonous substances. The liver has amazing regenerative powers, and in some instances can recover if the rate of regeneration exceeds the rate of damage. Otherwise, liver transplantation is usually the preferred treatment, but artificial livers have been developed and tried in a few

cases. As mentioned in Section 25.9, one type of artificial liver consists of a cartridge that contains liver cells. The patient's blood passes through the cellulose acetate tube of the cartridge and is serviced in the same manner as by a normal liver. In the meantime, the patient's liver has a chance to recover.

Three systems (digestive, respiratory, and urinary) refresh the blood directly because they communicate with the external environment. The digestive tract digests food to nutrient molecules that enter the blood. The lungs, as we will discuss in Chapter 32, exchange gases with the atmosphere and rid the blood of CO_2 before adding O_2 to blood.

The Chapter in Review

Summary

How to Tell a Carnivore from an Herbivore

- Carnivores are meat eaters with various teeth and a short, simple digestive system.
- Herbivores are plant eaters with grinding teeth and a complicated digestive system.

Animals Must Obtain and Process Their Food

31.1 A digestive system carries out ingestion, digestion, absorption, and elimination

- Ingestion is the intake of food.
- Digestion provides cells with nutrients.
- Absorption occurs as nutrient molecules are taken into the body, usually via the bloodstream.
- Elimination is the removal of unabsorbed molecules from the body.

Complete digestive tract

31.2 Animals exhibit a variety of feeding strategies

- A bulk feeder eats food in chunks or whole.
- A filter feeder collects and eats small particles from water.
- A substrate feeder lives on or in the food it eats.
- A fluid feeder consumes a fluid.

31.3 A complete digestive tract has specialized compartments

- An incomplete digestive tract has a single opening through which nutrients enter and exit.
- A complete digestive tract has a mouth and an anus.

- Specialized parts include a crop, gizzard, ceca, and cloaca in birds; a rumen in cows; and a cecum in rabbits.

31.4 Both mechanical and chemical digestion occur in the mouth

- Teeth chew food; saliva contains salivary amylase for digesting starch; and the tongue forms a food bolus for swallowing.

31.5 The esophagus conducts food to the stomach

- The esophagus is a muscular tube that passes from the pharynx to the stomach.
- During swallowing, the air passage is blocked off by the soft palate and epiglottis so that the food bolus enters the esophagus.
- After swallowing, peristalsis begins in the esophagus.

31.6 Food storage and chemical digestion take place in the stomach

- The stomach expands and stores food, which is churned and mixed with acidic gastric juices.
- Gastric juices contain pepsin, an enzyme that digests protein.

31.7 Bacteria contribute to the cause of ulcers

- Nobel prize winners Barry Marshall and Robin Warren discovered that infection by *Helicobacter pylori* causes ulcers.

31.8 In the small intestine, chemical digestion concludes, and absorption of nutrients occurs

- The duodenum receives bile from the gallbladder and pancreatic juice.
- Bile emulsifies fat and readies it for digestion by lipase.
- The pancreas produces pancreatic amylase (digests carbohydrates), trypsin (digests proteins), and lipase (digests fats).
- Intestinal enzymes complete digestion to small nutrient molecules, which are absorbed at the villi.

31.9 The pancreas and the liver contribute to chemical digestion

- The liver produces bile, destroys old blood cells, detoxifies blood, makes plasma proteins, stores glucose, and produces urea.
- Liver disorders include jaundice, hepatitis, and cirrhosis

31.10 The stomach and duodenum are endocrine glands

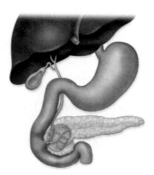

- The stomach produces gastrin to help regulate food digestion.
- The duodenum produces secretin and cholecystokinin (CCK).

31.11 The large intestine absorbs water and prepares wastes for elimination

- Besides water, the large intestine absorbs salts and some vitamins, and forms feces.

Good Nutrition and Diet Lead to Better Health

31.12 Carbohydrates are nutrients that provide immediate energy as well as fiber

- Sugars and starch provide energy for cells.
- Whole-grain carbohydrates are more nutritious; fiber is indigestible cellulose.

31.13 Lipids are nutrients that supply long-term energy

- Triglycerides supply energy; fat is stored.
- Alpha-linolenic and linoleic acids are essential fatty acids.
- Foods from animals, not plants, contain saturated fats and cholesterol.
- High intake of saturated fats, trans fats, and cholesterol is harmful to health.

31.14 Proteins are nutrients that supply building blocks for cells

- Proteins, which supply all essential amino acids, include meat, fish, poultry, eggs, nuts, soybeans, and cheese.

31.15 Minerals have various roles in the body

- Minerals regulate biochemical reactions, maintain fluid balance, and are incorporated into structures and compounds.

31.16 Vitamins help regulate metabolism

- Vitamins, obtained from most foods, regulate metabolism and physiological development; vitamins C, E, and A are antioxidants.

31.17 Nutritional labels allow evaluation of a food's content

- A Nutrition Facts panel provides information about nutrients as well as serving size, calories, and percent daily values.

31.18 Certain disorders are associated with obesity

- Obesity is defined as weighing 30% more than ideal body weight.
- Healthy diet, increased physical activity, and weight loss can improve insulin function in persons with type 2 diabetes.
- Saturated and trans fats are associated with high blood LDL-cholesterol levels that can lead to cardiovascular disease.

31.19 Eating disorders appear to have a psychological component

- Anorexia nervosa: distorted body image.
- Bulimia nervosa: binge-eating and then purging.
- Binge-eating disorder: overeating not followed by purging.
- Muscle dysmorphia: preoccupied with muscle development.

Testing Yourself

Animals Must Obtain and Process Their Food

1. The digestive system
   - a. breaks down food into usable nutrients.
   - b. absorbs nutrients.
   - c. eliminates waste.
   - d. All of these are correct.
2. Animals that feed discontinuously
   - a. have digestive tracts that permit storage.
   - b. are always filter feeders.
   - c. exhibit extremely rapid digestion.
   - d. have a nonspecialized digestive tract.
   - e. usually eat only meat.
3. The typhlosole within the gut of an earthworm compares best to which of these organs in humans?
   - a. teeth in the mouth
   - b. esophagus in the thoracic cavity
   - c. folds in the stomach
   - d. villi in the small intestine
   - e. the large intestine because it absorbs water
4. **THINKING CONCEPTUALLY** Argue that the digestive system is the most critical system to the life of an animal.
5. Tooth decay is caused by bacteria metabolizing _____ and giving off _____.
   - a. sugar, protein   c. acids, sugar
   - b. protein, sugar   d. sugar, acids
6. Food and air both travel through the
   - a. lungs.   c. larynx.
   - b. pharynx.   d. trachea.
7. Peristalsis occurs
   - a. from the mouth to the small intestine.
   - b. from the beginning of the esophagus to the anus.
   - c. only in the stomach.
   - d. only in the small and large intestines.
   - e. only in the esophagus and stomach.
8. The stomach
   - a. is lined with a thick layer of mucus.
   - b. contains sphincter glands.
   - c. has a pH of about 6.
   - d. digests pepsin.
   - e. More than one of these are correct.

9. Stomach ulcers result from
   a. a bacterial infection.
   b. a viral infection.
   c. excess stomach acid secretion.
   d. stressful conditions.
   e. recessive genes.
10. Which organ has both an exocrine and an endocrine function?
   a. liver   c. pancreas
   b. esophagus   d. cecum
11. Bile
   a. is an important enzyme for the digestion of fats.
   b. cannot be stored.
   c. is made by the gallbladder.
   d. emulsifies fat.
   e. All of these are correct.

In questions 12–14, match each statement to an answer in the key. Answers are used more than once; some statements may have more than one answer.

KEY:
   a. gastrin   d. All of these are correct.
   b. secretin   e. None of these are correct.
   c. CCK

12. Stimulates the gallbladder to release bile.
13. Stimulates the stomach to digest protein.
14. Secreted by the duodenum.
15. The lack of _____ activity would result in failure to maintain water balance.
   a. small intestinal   c. gallbladder
   b. large intestinal   d. stomach

Good Nutrition and Diet Lead to Better Health

For questions 16–20, choose the class of nutrient from the key that matches the description. Each answer may be used more than once.

KEY:
   a. carbohydrates   d. minerals
   b. lipids   e. vitamins
   c. proteins   f. water

16. Preferred source of direct energy for cells.
17. Include antioxidants.
18. Generally found in higher levels in animal sources than in plant sources.
19. An example is cholesterol.
20. Includes calcium, phosphorus, and potassium.
21. A percent daily value for sugar is not included in a nutrition label because
   a. sugar is not a nutrient.
   b. it is too difficult to determine the caloric value of sugar.
   c. a daily value is given for carbohydrates but not for sugars.
   d. sugar quality varies from product to product.
22. Anorexia nervosa is not characterized by
   a. a restrictive diet.
   b. binge eating followed by purging.
   c. an obsession with bodybuilding.
   d. a distorted body image so that the person feels fat even when emaciated.
23. **THINKING CONCEPTUALLY** Explain why vegetarians need not be concerned that their tissues will contain plant proteins.

Understanding the Terms

anorexia nervosa 628	lipase 619
antioxidant 625	liver 620
bile 619	microvillus 619
binge-eating disorder 628	mineral 624
bulimia nervosa 628	muscle dysmorphia 628
bulk feeder 613	nasopharynx 617
Calorie (kcal) 626	pancreas 620
carbohydrate 622	pancreatic amylase 619
cecum 614	pepsin 618
cholecystokinin 621	peristalsis 617
cholesterol 623	pharynx 617
chyme 618	polyp 621
cloaca 614	protein 623
colon 621	rectum 621
crop 614	ruminant 615
dental caries 617	salivary amylase 616
duodenum 619	salivary gland 616
epiglottis 617	secretin 621
esophagus 617	small intestine 619
fat 623	soft palate 616
fiber 622	sphincter 617
filter feeder 613	substrate feeder 613
fluid feeder 613	swallowing 617
gallbladder 620	tonsillitis 616
gastrin 621	triglyceride 623
gizzard 614	trypsin 619
glottis 617	typhlosole 614
hard palate 616	vermiform appendix 621
lacteal 619	villus 619
large intestine 621	vitamin 625

Match the terms to these definitions:
   a. _____ Essential requirement in the diet, needed in small amounts; often a part of coenzymes.
   b. _____ Lymphatic vessel in an intestinal villus; aids in the absorption of fats.
   c. _____ Muscular tube for moving swallowed food from the pharynx to the stomach.
   d. _____ Organ attached to the liver that stores and concentrates bile.

Thinking Scientifically

1. *a.* If you were testing the ability of pepsin to digest protein, what must your test tube contain? *b.* What control would you use? Explain. See Section 31.18.
2. Correlation studies, such as saturated fats in the diet increase the chances of cardiovascular disease, often lead to medical decisions. What criteria would you use to judge correlation studies? See Section 31.7.

ARIS™ *Visit* **www.mhhe.com/maderconcepts** *for practice quizzes, animations, videos, and activities designed to help you master the material in this chapter.*

32

Gas Exchange and Transport in Animals

LEARNING OUTCOMES

After studying this chapter, you should be able to accomplish the following outcomes.

Free-Diving Is Dangerous

1 Compare the ability to free-dive between aquatic mammals and humans.

Animals Have Gas Exchange Surfaces

2 Name the three events that occur during respiration.
3 Contrast the respiratory exchange surface of a hydra, a fish, an earthworm, an insect, and a vertebrate.
4 Show that countercurrent flow increases the efficiency of gills in extracting oxygen from water.
5 Explain how the tracheal system in insects accomplishes ventilation and exchange.
6 Trace the path of air in the human respiratory system.
7 List five or six reasons why smoking is dangerous to your health.

Ventilation Precedes Transport

8 Describe the mechanics of ventilation and the regulation of the breathing rate in humans.
9 Discuss the transport of O_2 and CO_2, and the regulation of pH in the blood, being sure to include the role of hemoglobin.
10 Describe the exchange of gases in the lungs and tissues, and relate these processes to the mechanisms of gas transport.
11 Based on the health consequences of breathing 9/11 dust, explain why the small particles in polluted urban air are dangerous to your health.

Many aquatic mammals can dive to great depths and stay submerged for some time. Yet they breathe air, just as we do, and therefore can't breathe when they dive—no oxygen tank is provided! The northern elephant seal (*Mirounga angustirostris*) has been observed to dive to a depth of 1,500 m. In comparison, a free-diving human can only dive to about 163 m. The Weddell seal (*Leptonychotes weddelli*) typically dives to only 300–400 m, but stays submerged for up to 15 minutes. The limit for humans is about 3 minutes.

Elephant seal

Weddell seal

Human free-diving

Free-Diving Is Dangerous

Let's look at how aquatic mammals, such as whales, seals, and dolphins, do it. First, they store oxygen before they dive. Blood doping (as when athletes have a blood transfusion before a competitive event) is natural to them because they have more blood cells and more blood per body weight than we do. Not only that, their muscles are chock full of myoglobin, a respiratory pigment that specializes in keeping oxygen where it is most needed—namely, in the muscles.

A special diving response occurs when aquatic mammals dive: (1) The heart rate slows down (called brachycardia) to about one-half to one-tenth the normal rate. (2) The peripheral blood vessels constrict, and the blood circulates to the heart and lungs only. (3) After the oxygen stored by myoglobin is used up, fermentation supplies ATP. The lactic acid produced is metabolized when the animal starts breathing again. (4) Finally, the spleen kicks in. In addition to cleansing the blood, the spleen acts as a storage area for red blood cells. As the water pressure increases during the dive, compression causes the spleen to release its supply of fully oxygenated red blood cells. This oxygen keeps the heart and brain going for a while longer.

Sperm whale

Bottlenose dolphin

Many humans want to free-dive—and they do. How far can we go in copying the aquatic mammals? Although we do not store oxygen to any great extent, it is helpful to warm up before diving into cold water. This makes sure the muscles are getting O_2-rich blood for as long as possible before the dive begins. Researchers also recommend that the mask not include the forehead—where cold receptors are located. They help bring on the diving response that will also occur in humans. You can practice bringing on the diving response by holding your breath (called the apneic time) with your face submerged in cold water. If the urge to breathe can be overcome long enough, even the spleen will kick in. To get the spleen to discharge its red blood cells, the only requirement is to not breathe for a certain length of time. Researchers have found that it doesn't matter whether a person is in the water or standing on solid ground.

Aquatic mammals apparently don't suffer from the "bends" as we do. In humans, N_2 from air inhaled just before the dive enters the blood. Surfacing rapidly after diving causes the N_2 to bubble from the blood, bursting capillaries, even in the brain. It has been shown in the laboratory that the lungs of elephant seals collapse on the way down and do not reinflate until after the seals have ascended to a shallow depth. The conclusion is that much of the nitrogen in their lungs doesn't have a chance to enter the bloodstream as they submerge. Presumably, other deep marine mammals experience similar lung collapse while diving.

Free-diving can be very dangerous to your life expectancy, even if you train with an expert. Therefore, free-diving is not recommended for humans—they lack the adaptations that aquatic mammals have and are instead highly adapted to living on land. In this chapter, we will learn how animals in the water and on land breathe and transport gases to and from their cells.

In this part of the chapter, a discussion of the events of respiration precedes a comparison of respiratory organs in animals. The damage to the human respiratory system due to smoking cigarettes is also included.

32.1 Respiration involves several steps

Respiration is the sequence of events that results in gas exchange between the body's cells and the environment. In terrestrial vertebrates, respiration includes these steps:

- **Ventilation** (i.e., breathing) includes inspiration (entrance of air into the lungs) and expiration (exit of air from the lungs).

- **External respiration** is gas exchange between the air and the blood within the lungs. Blood then transports oxygen from the lungs to the tissues (**Fig. 32.1**).

- **Internal respiration** is gas exchange between the blood and the tissue fluid. (The body's cells exchange gases with the tissue fluid.) The blood then transports carbon dioxide to the lungs (Fig. 32.1).

Gas exchange takes place by the physical process of diffusion. For external respiration to be effective, the gas-exchange region must be (1) moist, (2) thin, and (3) large in relation to the size of the body. Some animals, such as planarians, are small and shaped in a way that allows the surface of the animal to be the gas-exchange surface. Most complex animals have specialized external respiration surfaces, such as gills in aquatic animals and lungs in terrestrial animals. The effectiveness of diffusion is enhanced by vascularization (the presence of many capillaries), and delivery of oxygen to the cells is promoted when the blood contains a respiratory pigment, such as hemoglobin.

Regardless of the particular external respiration surface and the manner in which gases are delivered to the cells, in the end, oxygen enters mitochondria, where cellular respiration takes place. Without internal respiration, ATP production does not take place, and life ceases.

Section 32.2 gives a brief overview of animal external respiratory surfaces.

> **32.1 Check Your Progress** When animals dive, which of the three respiratory events does not occur?

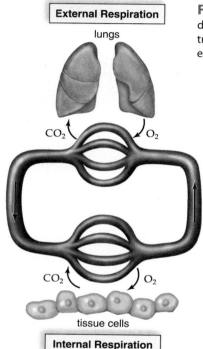

FIGURE 32.1 Blood delivers oxygen to cells and transports carbon dioxide to the external respiration surfaces.

32.2 External respiration surfaces must be moist

It is more difficult for animals to obtain oxygen from water than from air. Water fully saturated with air contains only a fraction of the amount of oxygen that would be present in the same volume of air. Also, water is more dense than air. Therefore, aquatic animals expend more energy carrying out gas exchange than do terrestrial animals. Fishes use as much as 25% of their energy output to respire, while terrestrial mammals use only 1–2% of their energy output for that purpose.

Hydras, which are cnidarians, and planarians, which are flatworms, have a large surface area in comparison to their size. This makes it possible for most of their cells to exchange gases directly with the environment. In hydras, the outer layer of cells is in contact with the external environment, and the inner layer can exchange gases with the water in the gastrovascular cavity (**Fig. 32.2A**). In planarians, the flattened body permits cells to exchange gases with the external environment.

The tubular shape of annelids (segmented worms) also provides a surface area adequate for external respiration. The earthworm, an annelid, is an example of a terrestrial invertebrate that is able to use its body surface for respiration because the capillaries come close to the surface (**Fig. 32.2B**). An earthworm keeps its body surface moist by secreting mucus and by releasing fluids from excretory pores. Further, the worm is behaviorally adapted to remain in damp soil during the day, when the air is driest. In addition to a tubular shape, aquatic polychaete worms have extensions of the body wall called parapodia, which are vascularized and used for gas exchange.

Aquatic invertebrates (e.g., clams and crayfish) and aquatic vertebrates (e.g., fish and tadpoles) have gills that extract oxygen from a watery environment. **Gills** are finely divided, vascularized outgrowths of the body surface or the pharynx (**Fig. 32.2C**). Various mechanisms are used to pump water across the gills, depending on the organism.

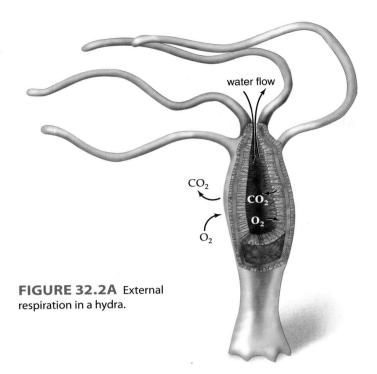

FIGURE 32.2A External respiration in a hydra.

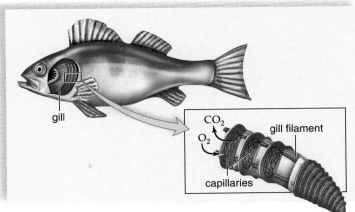

FIGURE 32.2C Fish have gills to assist external respiration.

Insects have a system of air tubes called tracheae through which oxygen is delivered directly to the cells without entering the blood (**Fig. 32.2D**). Tracheole fluid occurs at the end of the tracheae. Air sacs located near the wings, legs, and abdomen act as bellows to help move the air into the tubes through external openings.

Terrestrial vertebrates usually have **lungs**, which are vascularized outgrowths from the lower pharyngeal region. The tadpoles of frogs live in the water and have gills as external respiratory organs, but adult amphibians possess simple, saclike lungs. Most amphibians respire to some extent through the skin, and some salamanders depend entirely on the skin, which is kept moist by mucus produced by numerous glands on the surface of the body.

The lungs of birds and mammals are elaborately subdivided into small passageways and spaces, respectively (**Fig. 32.2E**). It has been estimated that human lungs have a total surface area that is at least 50 times the skin's surface area. Air is a rich source of oxygen compared to water; however, it does have a drying effect on external respiratory surfaces. A human loses about

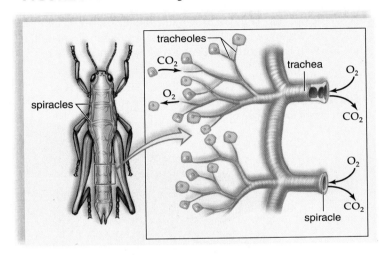

FIGURE 32.2D Insects have a tracheal system that delivers oxygen directly to their cells.

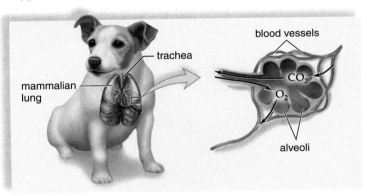

FIGURE 32.2E Vertebrates have lungs with a large total external respiration surface.

350 ml of water per day when the air has a relative humidity of only 50%. To keep the lungs from drying out, air is moistened as it moves through the passageways leading to the lungs.

In Section 32.3, we learn more about the structure and function of gills, such as those in fishes.

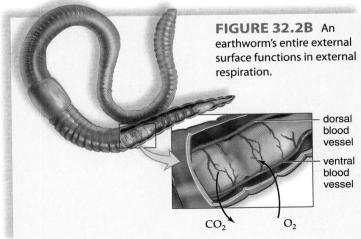

FIGURE 32.2B An earthworm's entire external surface functions in external respiration.

> **32.2 Check Your Progress** Per unit volume, air contains more oxygen than water, but breathing air causes one problem easily preventable in the water. What is it?

32.3 Gills are an efficient gas-exchange surface in water

Animals with gills use various means of ventilation. Among molluscs, such as a clam or a squid, water is drawn into the mantle cavity, where it passes through the gills. In crustaceans, such as crabs and shrimps, which are arthropods, the gills are located in thoracic chambers covered by the exoskeleton. The action of specialized appendages located near the mouth keeps the water moving. In fish, ventilation is brought about by the combined action of the mouth and gill covers, or opercula (sing., operculum). When the mouth is open, the opercula are closed and water is drawn in. Then the mouth closes, and the opercula open, drawing the water from the pharynx through the gill slits located between the gill arches.

As mentioned, the gills of bony fishes are outward extensions of the pharynx (**Fig. 32.3**). On the outside of the gill arches, the gills are composed of filaments that are folded into platelike lamellae. Fish use **countercurrent exchange** to transfer oxygen from the surrounding water into their blood. *Con*current flow would mean that O$_2$-rich water passing over the gills would flow in the same direction as O$_2$-poor blood in the blood vessels. This arrangement would result in an equilibrium point, at which only half the oxygen in the water would be captured. *Counter*current flow means that the two fluids flow in opposite directions. With countercurrent flow, as blood gains oxygen, it always encounters water having an even higher oxygen content. A countercurrent mechanism prevents an equilibrium point from being reached, and about 80–90% of the initial dissolved oxygen in water is extracted.

In Section 32.4, we study how the tracheal system in insects functions as a respiratory system.

> **32.3 Check Your Progress** The diving response is not seen in fishes. Explain.

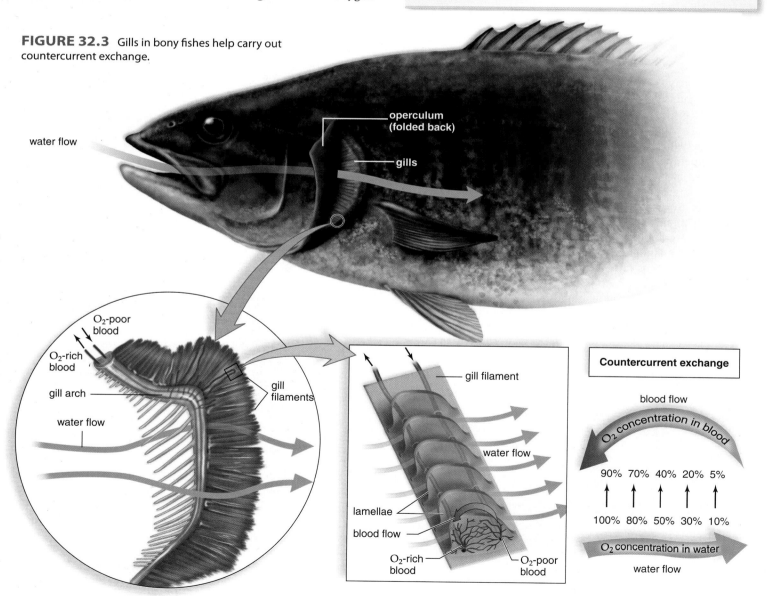

FIGURE 32.3 Gills in bony fishes help carry out countercurrent exchange.

32.4 The tracheal system in insects permits direct gas exchange

Arthropods are coelomate animals, but the coelom is reduced and the internal organs lie within a cavity called the hemocoel because it contains hemolymph, a mixture of blood and lymph. Hemolymph flows freely through the hemocoel, making circulation in arthropods inefficient. Many insects are adapted for flight, and their flight muscles require a steady supply of oxygen. Insects overcome the inefficiency of their blood flow by having a respiratory system that consists of **tracheae,** tiny air tubes that take oxygen directly to the cells (**Fig. 32.4**). Tracheae have a single layer of cells supported by spiral thickenings of a cuticle lining. The tracheae branch into even smaller tubules called tracheoles, which also branch and rebranch until finally the air tubes are only about 0.1 μm in diameter. There are so many fine tracheoles that almost every cell is near one. Also, the tracheoles indent the plasma membrane so that they terminate close to mitochondria. Therefore, O_2 can flow more directly from a tracheole to mitochondria, where cellular respiration occurs. The tracheae also dispose of CO_2.

The tracheoles are fluid-filled, but the larger tracheae contain air and open to the outside by way of spiracles (Fig. 32.4). Usually, the spiracle has some sort of closing device that reduces water loss, and this may be why insects have no trouble inhabiting drier climates. Recently, investigators presented evidence that tracheae actually expand and contract, thereby drawing air into and out of the system. This method is comparable to the way the human lungs expand to draw air into them. Otherwise, a tracheal system consisting of an expansive network of thin-walled tubes seems to be an entirely different mechanism of respiration from those used by other animals.

It's been suggested that a tracheal system that has no efficient method to improve flow is sufficient only in small insects. Larger insects have still another mechanism to ventilate—keep the air moving in and out—tracheae. Many larger insects have air sacs, which are thin-walled and flexible, located near major muscles. Contraction of these muscles causes the air sacs to empty, and relaxation causes the air sacs to expand and draw in air. Even so, insects still lack the efficient circulatory system of birds and mammals that is able to pump O_2-rich blood through arteries to all the cells of the body. This may be why insects remain small, despite the attempts of movies to make us think otherwise.

A tracheal system is an adaptation to breathing air, and yet, some insect larval stages and even some adult insects live in the water. In these instances, the tracheae do not receive air by way of spiracles. Instead, diffusion of oxygen across the body wall supplies the tracheae with oxygen. Mayfly and stonefly nymphs have thin extensions of the body wall called tracheal gills—the tracheae are particularly numerous in this area. This is an interesting adaptation because it dramatizes that tracheae function to deliver oxygen in the same manner as vertebrate blood vessels. The composition of the human respiratory system is explained in Section 32.5.

32.4 Check Your Progress In Section 32.1, we said that respiration has three events, but this is not so in insects. Explain.

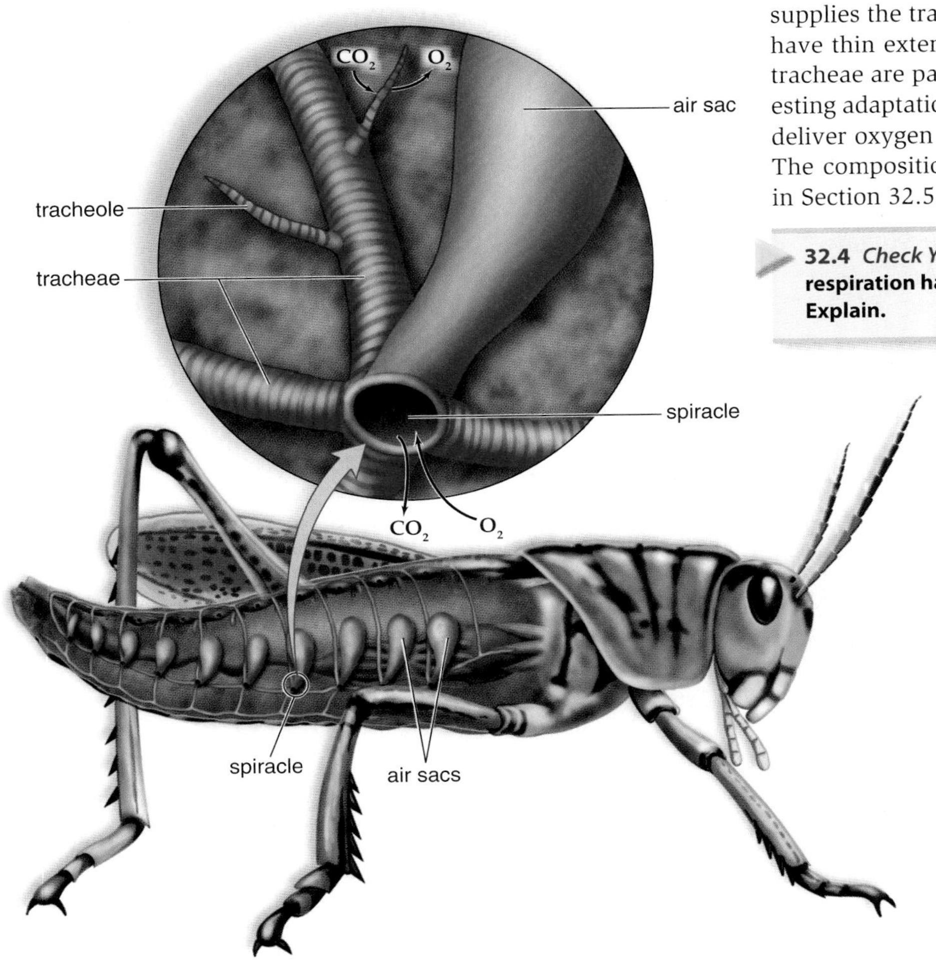

FIGURE 32.4 Tracheae in an insect.

32.5 The human respiratory system utilizes lungs as a gas-exchange surface

The human respiratory system includes all of the structures that conduct air in a continuous pathway to and from the lungs (Fig. 32.5, *left*). The lungs lie deep within the thoracic cavity, where they are protected from drying out. As air moves through the nose, the pharynx, the trachea, and the bronchi to the lungs, it is filtered so that it is free of debris, warmed, and humidified. By the time the air reaches the lungs, it is at body temperature and saturated with water. In the nose, hairs and cilia act as a screening device. In the trachea and the bronchi, cilia beat upward, carrying mucus, dust, and occasional small bits of food that "went down the wrong way" into the throat, where the accumulation may be swallowed or expectorated.

The hard and soft palates separate the nasal cavities from the mouth, but the air and food passages cross in the **pharynx**. This may seem inefficient, and there is danger of choking if food accidentally enters the trachea; however, this arrangement does have the advantage of letting you breathe through your mouth in case your nose is plugged up. In addition, it permits greater intake of air during heavy exercise, when greater gas exchange is required.

Air passes from the pharynx through the **glottis**, an opening into the **larynx**, or voice box. At the edges of the glottis, embedded in mucous membrane, are the **vocal cords**. These flexible and pliable bands of connective tissue vibrate and produce sound when air is expelled past them through the glottis from the larynx.

The larynx and the trachea remain open to receive air at all times. The larynx is held open by a complex of nine cartilages, among them the Adam's apple. Easily seen in many men, the Adam's apple resembles a small, rounded apple just under the skin in the front of the neck. The **trachea** is held open by a series of C-shaped, cartilaginous rings that do not completely meet in the rear. When food is being swallowed, the larynx rises, and the glottis is closed by a flap of tissue called the **epiglottis**. A backward movement of the soft palate covers the entrance of the nasal passages into the pharynx. The food then enters the esophagus, which lies behind the larynx.

The trachea divides into two primary **bronchi**, which enter the right and left lungs. Branching continues, eventually forming a great number of smaller passages called **bronchioles**. The two bronchi resemble the trachea in structure, but as the bronchial tubes divide and subdivide, their walls become thinner, and rings of cartilage are no longer present. Each bronchiole terminates in an elongated space enclosed by a multitude of air pockets, or sacs, called **alveoli**, which make up the lungs (Fig. 32.5, *right*). Internal gas exchange occurs between the air in the alveoli and the blood in the capillaries.

Lungs, and indeed many other organs, can be damaged by smoking, as described in Section 32.6.

32.5 Check Your Progress Does the path of air change when a person dives?

FIGURE 32.5 The human respiratory tract.

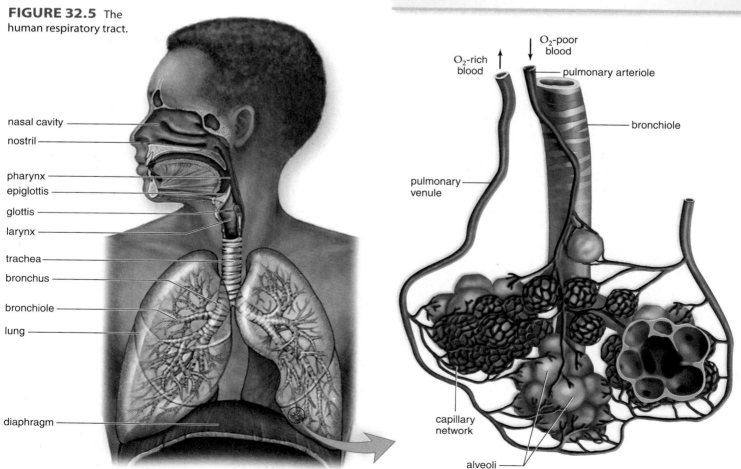

Is there a safe way to smoke? No. All cigarettes can damage the human body. Any amount of smoke is dangerous. Cigarettes are perhaps the only legal product whose advertised and intended use—smoking—is harmful to the body and causes cancer.

Is cigarette smoking really addictive? Yes. The nicotine in cigarette smoke causes addiction to smoking. Nicotine is an addictive drug (just like heroin and cocaine) for three main reasons: Small amounts make the smoker want to smoke more; smokers usually suffer withdrawal symptoms when they stop; and nicotine can affect the mood and nature of the smoker. The younger a person is when he or she begins to smoke, the more likely it is that an addiction to nicotine will develop.

Does smoking cause cancer? Yes. Tobacco use accounts for about one-third of all cancer deaths in the United States. Smoking causes almost 90% of lung cancers. Smoking also causes cancers of the larynx (voice box), oral cavity, pharynx (throat), and esophagus, and contributes to the development of cancers of the bladder, pancreas, cervix, kidney, and stomach. Smoking is also linked to the development of some leukemias.

How does cigarette smoke affect the lungs? All cigarette smokers have a lower level of lung function than nonsmokers. Cigarette smoking causes several lung diseases that can be just as dangerous as lung cancer: chronic bronchitis, in which the airways produce excess mucus, forcing the smoker to cough more; emphysema, a disease that slowly destroys a person's ability to breathe; and chronic obstructive pulmonary disease, a name that encompasses both chronic bronchitis and emphysema.

Why do smokers have "smoker's cough"? Cigarette smoke contains chemicals that irritate the air passages and lungs. When a smoker inhales these substances, the body tries to protect itself by producing mucus and stimulating coughing. Normally, cilia (tiny, hairlike formations that line the airways) beat outward and "sweep" harmful material out of the lungs. Smoke, however, decreases this sweeping action, so some of the poisons in the smoke remain in the lungs.

If you smoke but do not inhale, is there any danger? Yes. Wherever smoke touches living cells, it does harm. Even if smokers don't inhale, they are breathing the smoke secondhand and are still at risk for lung cancer. Pipe and cigar smokers, who often do not inhale, are at increased risk for lip, mouth, tongue, and several other cancers.

Does cigarette smoking affect the heart? Yes. Smoking increases the risk of heart disease, which is the number-one cause of death in the United States. Smoking, high blood pressure, high cholesterol, physical inactivity, obesity, and diabetes are all risk factors for heart disease, but cigarette smoking is the biggest risk factor for sudden heart death. Smokers who have a heart attack are more likely to die within an hour of the attack than nonsmokers. Cigarette smoke can cause harm to the heart at very low levels, much lower than the amount that causes lung disease.

How does smoking affect pregnant women and their babies? Smoking during pregnancy is linked to a greater chance of miscarriage, premature delivery, stillbirth, infant death, low birth weight, and sudden infant death syndrome (SIDS). Up to 10% of infant deaths would be prevented if pregnant women did not smoke. When a pregnant woman smokes, she is really smoking for two because the nicotine, carbon monoxide, and other dangerous chemicals in smoke enter her bloodstream and then pass into the baby's body, preventing the baby from getting essential nutrients and oxygen for growth.

What are some of the short-term and long-term effects of smoking cigarettes? Short-term effects include shortness of breath and a nagging cough, diminished ability to smell and taste, premature aging of the skin, and increased risk of sexual impotence in men. Smokers tend to tire easily during physical activity. Long-term effects include many types of cancer, heart disease, aneurysms, bronchitis, emphysema, and stroke. Smoking contributes to the severity of pneumonia and asthma.

What are the dangers of environmental tobacco smoke (ETS)? ETS causes about 3,000 lung cancer deaths and about 35,000 to 40,000 deaths from heart disease each year in healthy nonsmokers. Children whose parents smoke are more likely to suffer from asthma, pneumonia or bronchitis, ear infections, coughing, wheezing, and increased mucus production in the first two years of life than children who come from smoke-free households.

Are chewing tobacco and snuff safe alternatives to cigarette smoking? No. The juice from smokeless tobacco is absorbed directly through the lining of the mouth. This creates sores and white patches that often lead to cancer of the mouth. Smokeless tobacco users also greatly increase their risk of other cancers, including those of the pharynx (throat). Other effects of smokeless tobacco include harm to teeth and gums.

This completes our study of animal respiratory organs. In the next part of the chapter, we learn more about how the human respiratory system functions.

> **32.6 Check Your Progress** Name six illnesses associated with smoking cigarettes.

This part of the chapter is particularly concerned with how the human respiratory system functions. The processes of breathing and gas exchange in the lungs and tissues are stressed. We also discuss respiratory disorders, such as those seen in people exposed to 9/11 dust.

32.7 Breathing brings air into and out of the lungs

Terrestrial vertebrates ventilate their lungs by moving air into and out of the respiratory tract. Amphibians use positive pressure to force air into the respiratory tract. With the mouth and nostrils firmly shut, the floor of the mouth rises and pushes the air into the lungs. Reptiles, birds, and mammals use negative pressure to move air into the lungs and positive pressure to move it out. **Inspiration** (or inhalation) is the act of moving air into the lungs, and **expiration** (or exhalation) is the act of moving air out of the lungs.

Reptiles have jointed ribs that can be raised to expand the lungs, but mammals have both a rib cage and a diaphragm. The **diaphragm** is a horizontal muscle that divides the thoracic cavity (above) from the abdominal cavity (below). During inspiration in mammals, the rib cage moves up and out, and the diaphragm contracts and moves down (**Fig. 32.7A**). As the thoracic (chest) cavity expands and lung volume increases, air flows into the lungs due to decreased air pressure in the thoracic cavity and lungs. Also, inspiration is the active phase of breathing in reptiles and mammals.

During expiration in mammals, the rib cage moves down, and the diaphragm relaxes and moves up to its former position (**Fig. 32.7B**). No muscle contraction is required and expiration is the inactive phase of breathing in reptiles and mammals. During expiration, air flows out as a result of increased pressure in the thoracic cavity and lungs.

We can liken ventilation in reptiles and mammals to the way a bellows, used to fan a fire, functions (**Fig. 32.7C**). First, the handles of the bellows are pulled apart, decreasing the air pressure inside the

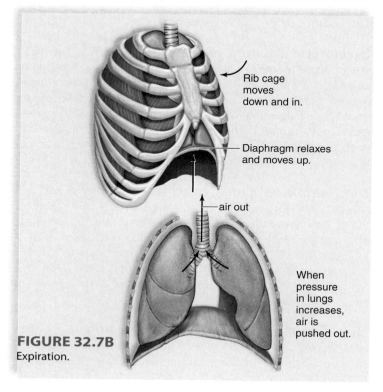

FIGURE 32.7B
Expiration.

Rib cage moves down and in.

Diaphragm relaxes and moves up.

air out

When pressure in lungs increases, air is pushed out.

bellows. This causes air to automatically flow into the bellows, just as air automatically enters the lungs because the rib cage moves up and out during inspiration. Then, when the handles of the bellows are pushed together, air automatically flows out because the air pressure increases inside the bellows. Similarly, air automatically exits the lungs when the rib cage moves down and in during expiration. The analogy is not exact, however, because no force is required for the rib cage to move down, and inspiration is the only active phase of breathing. Forced expiration can occur if we so desire, however.

All terrestrial vertebrates, except birds, use a *tidal ventilation mechanism*, so called because the air moves in and out by the same route. This means that the lungs of amphibians, reptiles,

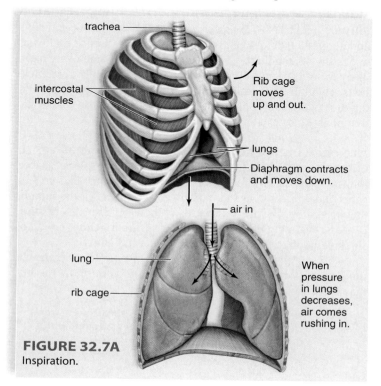

trachea

intercostal muscles

Rib cage moves up and out.

lungs

Diaphragm contracts and moves down.

air in

lung

rib cage

When pressure in lungs decreases, air comes rushing in.

FIGURE 32.7A
Inspiration.

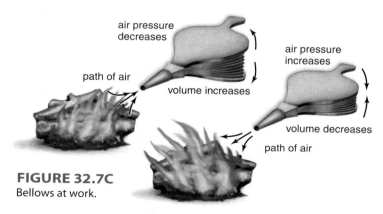

air pressure decreases

path of air

volume increases

air pressure increases

volume decreases

path of air

FIGURE 32.7C
Bellows at work.

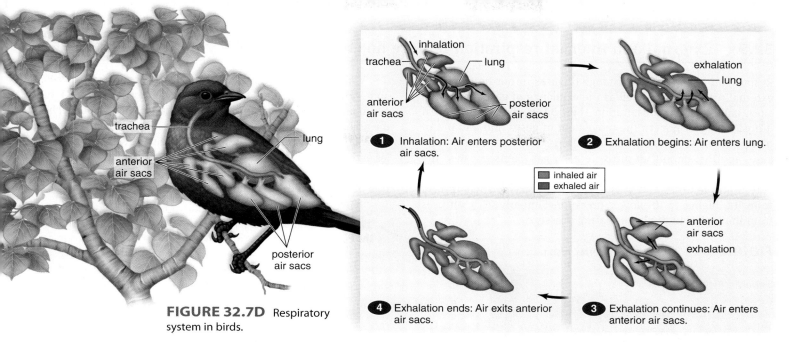

FIGURE 32.7D Respiratory system in birds.

① Inhalation: Air enters posterior air sacs.

② Exhalation begins: Air enters lung.

inhaled air
exhaled air

④ Exhalation ends: Air exits anterior air sacs.

③ Exhalation continues: Air enters anterior air sacs.

and mammals are not completely emptied and refilled during each breathing cycle. Because of this, the air entering mixes with used air remaining in the lungs. While this does help conserve water, it also decreases gas-exchange efficiency. In contrast, birds use a *one-way ventilation mechanism* (**Fig. 32.7D**). ① Incoming air is carried past the lungs by a trachea, which takes it to a set of posterior air sacs. ② The air then passes forward through the lungs into a set of ③ anterior air sacs. ④ From here, it is finally expelled.

Notice that fresh, O_2-rich air never mixes with used air in the lungs of birds, thereby greatly improving gas-exchange efficiency.

Section 32.8 explains how and when the usual breathing rate is altered.

32.7 Check Your Progress Penguins, being birds, have an advantage over mammals when diving because they have extra storage room for O_2-rich air. Where?

32.8 Our breathing rate can be modified

Normally, adults have a breathing rate of 12 to 20 ventilations per minute. The rhythm of ventilation is controlled by a **respiratory center** in the medulla oblongata of the brain. The respiratory center automatically sends out impulses by way of a spinal nerve to the diaphragm (phrenic nerve) and spinal nerves to the intercostal muscles of the rib cage (**Fig. 32.8**). Now inspiration occurs. When the respiratory center stops sending neuronal signals to the diaphragm and the rib cage, expiration occurs.

Although the respiratory center automatically controls the rate and depth of breathing, its activity can also be influenced by nervous input and chemical input. Following forced inhalation, stretch receptors in the alveolar walls initiate inhibitory nerve impulses that travel from the inflated lungs to the respiratory center. This stops the respiratory center from sending out nerve impulses.

The respiratory center is directly sensitive to the levels of hydrogen ions (H^+). However, when carbon dioxide (CO_2) enters the blood, it reacts with water and releases hydrogen ions. In this way, CO_2 participates in regulating the breathing rate. When hydrogen ions rise in the blood and the pH decreases, the respiratory center increases the rate and depth of breathing. The chemoreceptors in the **carotid bodies**, located in the carotid arteries, and in the **aortic bodies**, located in the aorta, will stimulate the respiratory center during intense exercise due to a reduction in pH and also if and when arterial oxygen decreases to 50% of normal.

External and internal respiration requires no energy, as explained in Section 32.9.

32.8 Check Your Progress Splenic contractions seem to help trained divers hold their breath. Explain.

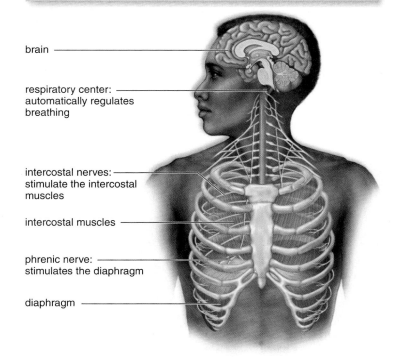

brain

respiratory center: automatically regulates breathing

intercostal nerves: stimulate the intercostal muscles

intercostal muscles

phrenic nerve: stimulates the diaphragm

diaphragm

FIGURE 32.8 Nervous control of breathing.

32.9   External and internal respiration require no energy

Respiration includes the exchange of gases in our lungs, called external respiration, as well as the exchange of gases in the tissues, called internal respiration (**Fig. 32.9**). The principles of diffusion largely govern the movement of gases into and out of blood vessels in the lungs and in the tissues. Gases exert pressure, and the amount of pressure each gas exerts is called the **partial pressure**, symbolized as P_{O_2} and P_{CO_2}. If the partial pressure of oxygen differs across a membrane, oxygen will diffuse from the higher to the lower pressure. Similarly, CO_2 diffuses from the higher to the lower partial pressure.

Ventilation causes the alveoli of the lungs to have a higher P_{O_2} and a lower P_{CO_2} than the blood in pulmonary capillaries, and this accounts for the exchange of gases in the lungs. When blood reaches the tissues, cellular respiration in cells causes the tissue fluid to have a lower P_{O_2} and a higher P_{CO_2} than the blood in the systemic capillaries and this accounts for the exchange of gases in the tissues.

In Section 32.10, we learn that hemoglobin plays a major role in the transport of gases in humans.

> **32.9** *Check Your Progress* Air contains nitrogen (N_2). On land, the P_{N_2} is always the same in the lungs and blood. Explain.

FIGURE 32.9 External and internal respiration.

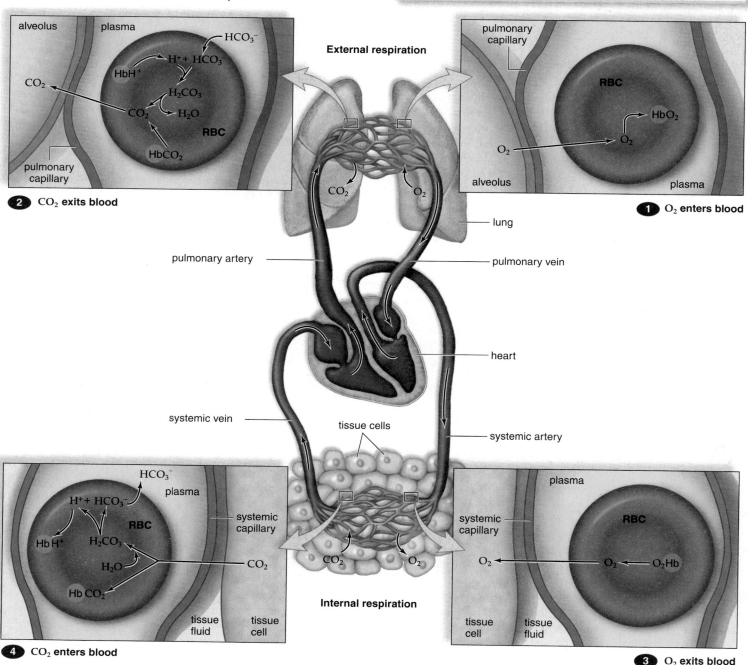

32.10 Hemoglobin is involved in transport of gases

External Respiration Most oxygen entering the pulmonary capillaries from the alveoli of the lungs combines with **hemoglobin (Hb)** in red blood cells (RBC) to form **oxyhemoglobin** (see Fig. 32.9, ❶):

$$Hb \quad + \quad O_2 \quad \longrightarrow \quad HbO_2$$

deoxyhemoglobin   oxygen   oxyhemoglobin

At the normal P_{O_2} in the lungs, hemoglobin is practically saturated with oxygen. Each hemoglobin molecule contains four polypeptide chains, and each chain is folded around an iron-containing group called **heme** (**Fig. 32.10**). It is actually the iron that forms a loose association with oxygen. Since there are about 250 million hemoglobin molecules in each red blood cell, each red blood cell is capable of carrying at least 1 billion molecules of oxygen. Unfortunately, carbon monoxide (CO) is an air pollutant that combines with hemoglobin more readily than does O_2, making Hb unavailable for O_2 transport. This is the reason that some homes are equipped with CO detectors.

As blood enters the lungs, a small amount of CO_2 is being carried by hemoglobin with the formula $HbCO_2$. Also, some hemoglobin is carrying hydrogen ions with the formula HbH^+. Most of the CO_2 in the pulmonary capillaries is carried as bicarbonate ions (HCO_3^-) in the plasma. As the free CO_2 from the following equation begins to diffuse out, this reaction is driven to the right:

$$H^+ \quad + \quad HCO_3^- \longrightarrow H_2CO_3 \longrightarrow H_2O \quad + \quad CO_2$$

hydrogen bicarbonate carbonic   water   carbon
ion   ion   acid   dioxide

The reaction occurs in red blood cells, where the enzyme **carbonic anhydrase** speeds the breakdown of carbonic acid (see Fig. 32.9, ❷).

What happens if you hyperventilate (breathe at a high rate) and, therefore, push this reaction far to the right? The blood will have fewer hydrogen ions, and alkalosis, a high blood pH, will result. In that case, breathing will be inhibited, but in the meantime, you may suffer various symptoms, from dizziness to muscle spasms. What happens if you hypoventilate (breathe at a low rate) and this reaction does not occur? Hydrogen ions build up in the blood, and acidosis will occur. Buffers may compensate for the low pH, and breathing will most likely increase. Otherwise, you may become comatose and die.

Internal Respiration Blood entering the systemic capillaries is a bright red color because RBCs contain oxyhemoglobin. Because the temperature in the tissues is higher and the pH is lower than in the lungs, oxyhemoglobin has a tendency to give up oxygen:

$$HbO_2 \longrightarrow Hb + O_2$$

Oxygen diffuses out of the blood into the tissues because the P_{O_2} of tissue fluid is lower than that of blood (see Fig. 32.9, ❸). The lower P_{O_2} is due to cells continuously using up oxygen in cellular respiration. After oxyhemoglobin gives

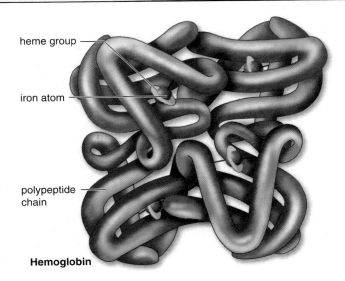

FIGURE 32.10 The iron atom of a heme group combines loosely with oxygen.

up O_2, it leaves the blood and enters tissue fluid, where it is taken up by cells.

Carbon dioxide, on the other hand, enters blood from the tissues because the P_{CO_2} of tissue fluid is higher than that of blood. Carbon dioxide, produced continuously by cells, collects in tissue fluid. After CO_2 diffuses into the blood, it enters the red blood cells, where a small amount combines with the protein portion of hemoglobin to form **carbaminohemoglobin** ($HbCO_2$). Most of the CO_2, however, is transported in the form of the **bicarbonate ion** (HCO_3^-). First, CO_2 combines with water, forming carbonic acid, and then this dissociates to a hydrogen ion (H^+) and HCO_3^-:

$$CO_2 \quad + \quad H_2O \longrightarrow H_2CO_3 \longrightarrow H^+ \quad + \quad HCO_3^-$$

carbon   water   carbonic   hydrogen bicarbonate
dioxide   acid   ion   ion

Carbonic anhydrase also speeds this reaction. The HCO_3^- diffuses out of the red blood cells to be carried in the plasma (see Fig. 32.9, ❹).

The release of H^+ from this reaction could drastically change the pH of the blood. However, the H^+ is absorbed by the globin portions of hemoglobin. Hemoglobin that has combined with H^+ is called reduced hemoglobin and has the formula HbH^+. HbH^+ plays a vital role in maintaining the normal pH of the blood. Blood that leaves the systemic capillaries is a dark maroon color because red blood cells contain reduced hemoglobin.

Respiratory disorders can result from breathing polluted air, as described in Section 32.11.

> **32.10 Check Your Progress** Transport of blood to only the heart and lungs during the diving response conserves oxygenated hemoglobin. Explain.

32.11 Respiratory disorders have resulted from breathing 9/11 dust

The self-cleaning mechanisms of the respiratory system can be overcome by breathing polluted urban air laded with particles. The dust particles at Ground Zero after the September 11, 2001, attack contained asbestos from fire-proofing materials, mercury from fluorescent lightbulbs, and lead from computers (**Fig. 32.11A**). The rescuers and bystanders breathed in this dust, unaware that their lungs could be permanently damaged. A number of possible disorders could occur in thousands of people who breathed the 9/11 dust.

Restrictive Pulmonary Disorders

In **restrictive pulmonary disorders**, lung capacity is reduced because the lungs have lost their elasticity. Inhaling particles such as asbestos, silica (sand), coal dust, and other pollutants can lead to pulmonary fibrosis, in which fibrous connective tissue builds up in the lungs. With a restrictive pulmonary disorder, the lungs cannot inflate sufficiently and are always tending toward deflation. It has been projected that one million deaths caused by asbestos exposure—mostly in the workplace—will occur in the United States between 1990 and 2020. Breathing asbestos and mercury is also associated with the development of cancer.

Obstructive Pulmonary Disorders

In **obstructive pulmonary disorders**, air does not flow freely in the airways, and the time it takes to inhale or exhale maximally is greatly increased. Several disorders, including chronic bronchitis, emphysema, and asthma, are collectively referred to as chronic obstructive pulmonary disease (COPD) because they tend to recur.

In **chronic bronchitis**, the airways are inflamed and filled with mucus. A cough that brings up mucus is common. The bronchi have undergone degenerative changes, including the loss of cilia and their normal cleansing action. Under these conditions, an infection is more likely to occur. Although smoking is the most frequent cause of chronic bronchitis, exposure to other pollutants, such as the dust at Ground Zero, can also cause this condition.

Emphysema is a chronic and incurable disorder in which the alveoli are distended and their walls damaged, so that the surface area available for gas exchange is reduced. Emphysema, which we know can be caused by smoking, is often preceded by chronic bronchitis. Air trapped in the lungs leads to alveolar damage and noticeable ballooning of the chest. The elastic recoil of the lungs is reduced, so not only are the airways narrowed, but the driving force behind expiration is also reduced. The victim is breathless and may have a cough. Lack of oxygen to the brain can make the person feel depressed, sluggish, and irritable.

Asthma is a disease of the bronchi and bronchioles that is marked by wheezing, breathlessness, and sometimes a cough and expectoration of mucus. The airways are unusually sensitive to specific irritants, which can include a wide range of allergens such as pollen, animal dander, dust, tobacco smoke, and industrial fumes. Even cold air can be an irritant. When exposed to the irritant, the smooth muscle in the bronchioles undergoes spasms. Most asthma patients have some degree of bronchial inflammation that further reduces the diameter. Special inhalers can control the symptoms of asthma.

Lung Cancer

Autopsies on smokers have revealed the progressive steps by which the most common form of lung cancer develops. The first event appears to be thickening and callusing of the cells lining the bronchi. (Callusing occurs whenever cells are exposed to irritants.) Then cilia are lost, making it impossible to prevent dust and dirt from settling in the lungs. Following this, cells with atypical nuclei appear in the callused lining. A tumor consisting of disordered cells with atypical nuclei is considered cancer in situ (at one location). A normal lung and a lung with cancerous tumors are shown in **Figure 32.11B**. A final step occurs when some of these cells break loose and penetrate other tissues, a process called metastasis. Now, the cancer has spread. The original tumor may grow until a bronchus is blocked, cutting off the supply of air to that lung. The entire lung then collapses, the secretions trapped in the lung spaces become infected, and pneumonia or a lung abscess results.

FIGURE 32.11A People are now ill from breathing dust at Ground Zero.

> **32.11** *Check Your Progress* In what ways is smoking cigarettes and cigars the same as breathing 9/11 dust?

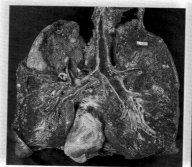
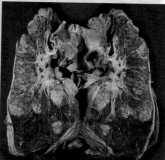

FIGURE 32.11B A normal lung (with the heart in place) compared to the lungs of a heavy smoker.

Assuming an adequate supply of O_2, the cells of most organisms continuously carry on cellular respiration and produce a comparable amount of CO_2. Without the ATP produced by cellular respiration, cells quickly die.

Respiration is the exchange of gases between an organism and its external environment. In small, thin, aquatic invertebrates, cells take care of their own respiratory needs. Oxygen diffuses from the water into the cells, and carbon dioxide diffuses out of the cells into the surrounding water. Complex invertebrates and vertebrates have respiratory organs, which usually consist of gills in aquatic forms and lungs in terrestrial forms.

Ventilation is the active movement of water across a respiratory surface (such as gills) or the movement of air into and out of lungs to increase the efficiency of gas exchange. The respiratory organs of complex organisms work in conjunction with a circulatory system to transport oxygen to cells and rid them of carbon dioxide.

We are well aware that breathing polluted air is dangerous to our health, and in this chapter we stressed that the human respiratory system is impaired by the foreign particles and molecules present in 9/11 dust and in cigarette smoke. We have much proof that cigarette smoke leads to blackened lungs and lung cancer—and that these effects are lessened by giving up the habit of smoking cigarettes and cigars.

In Chapter 33, we consider the urinary system, which also removes a metabolic waste from blood. Just as the lungs of the respiratory system remove carbon dioxide from blood, the kidneys remove urea, a nitrogenous end product of metabolism. The kidneys also adjust the salt-water balance and the pH of blood, as we shall see. These adjustments are critically important to homeostasis, the relative constancy of the internal environment.

The Chapter in Review

Summary

Free-Diving Is Dangerous

- Mammals that dive, such as whales and seals, have these advantages:
  - They have more blood cells and more blood per body weight than humans.
  - Their muscles have much myoglobin, a respiratory pigment that binds oxygen.
  - Their lungs collapse, and N_2 from the lungs doesn't enter the bloodstream.
- Humans experience the diving response just as aquatic mammals do:
  - The heart rate slows.
  - Blood circulates to the heart and lungs only.
  - The spleen releases stored O_2-rich red blood cells.
- Free-diving is dangerous and not recommended for humans!

Animals Have Gas-Exchange Surfaces

32.1 Respiration involves several steps

- In terrestrial vertebrates:
  - Ventilation is breathing.
  - External respiration is gas exchange between the air and the blood in the lungs.
  - Internal respiration is gas exchange between the blood and the tissue fluid.
  - The purpose of respiration is to deliver oxygen for cellular respiration.

External Respiration

lungs

CO_2 O_2

CO_2 O_2

tissue cells

Internal Respiration

32.2 External respiration surfaces must be moist

- Small aquatic animals can exchange gases directly with the external environment.
- Complex aquatic animals have gills.
- Earthworms use a moist body surface for gas exchange and, therefore, stay underground.
- Insects use tracheae to deliver oxygen directly to muscles.
- Terrestrial vertebrates usually have lungs, and air is moistened before it enters the lungs.

32.3 Gills are an efficient gas-exchange surface in water

- In clams, squids, and fishes, water moves across gills.
- Fish use countercurrent exchange to transfer oxygen efficiently from water into their blood.

32.4 The tracheal system in insects permits direct gas exchange

- The tracheae branch into smaller tracheoles, which also branch until the smallest fluid-filled tracheoles indent the plasma membrane to terminate close to mitochondria.
- In some insects, ventilation involves air sacs that expand to draw in air.

32.5 The human respiratory system utilizes lungs as a gas-exchange surface

- Air is warmed and humidified in the nose.
- The air and food passages cross in the pharynx, allowing us to breathe through our mouths.

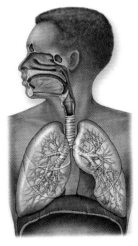

- The glottis opens into the larynx, and the trachea connects the larynx to the bronchi.
- Two bronchi lead to the right and left lungs and branch into bronchioles.
- Bronchioles end in alveoli, which make up the lungs.

32.6 Questions about tobacco, smoking, and health
- There is no safe way to smoke, and cigarette smoking is addictive.
- Short-term effects of smoking include a nagging cough, diminished sense of smell, and premature aging of the skin.
- Long-term effects on the lungs include cancer, heart disease, aneurysms, chronic obstructive pulmonary disease, and stroke.
- Smoking leads to a greater chance of premature birth, low birth weight, and stillbirth.

Ventilation Precedes Transport

32.7 Breathing brings air into and out of the lungs
- Inspiration consists of muscle contractions that lower the diaphragm and raise the ribs, followed by negative pressure that causes air to flow in.
- During expiration, the rib and diaphragm muscles relax, and air flows out due to increased pressure.

32.8 Our breathing rate can be modified
- Breathing is automatic, but decreased pH can cause the breathing center in the medulla oblongata to speed the rate.

32.9 External and internal respiration require no energy
- In the lungs, P_{CO_2} is higher in blood than in the lungs, and P_{O_2} is higher in the lungs; CO_2 diffuses out of blood into the lungs, and O_2 diffuses out of the lungs into the blood.
- In the tissues, P_{O_2} is higher in blood, and P_{CO_2} is higher in tissues; therefore, O_2 diffuses out of the blood, and CO_2 diffuses into the blood.

32.10 Hemoglobin is involved in transport of gases
- During external respiration:
  - Hb combines with O_2, forming HbO_2 (oxyhemoglobin).
  - HCO_3 forms H_2CO_3, which breaks down to water and CO_2, which exits blood.
- During internal respiration:
  - HbO_2 gives up its O_2, which enters tissue cells.
  - Hb combines with some CO_2 (carbaminohemoglobin).
  - CO_2 combines with water, forming H_2CO_3, which becomes H^+ and HCO_3^-.
  - Hb combines with some H^+ (reduced hemoglobin).

32.11 Respiratory disorders have resulted from breathing 9/11 dust
- In restrictive pulmonary disorders, lung capacity is reduced because the lungs have lost their elasticity.
- In obstructive pulmonary disorders, air does not flow freely in the airways; common types are chronic bronchitis, emphysema, and asthma.
- Lung cancer is characterized by thickening and callusing of cells lining the bronchi, loss of cilia, and formation of a tumor that eventually metastasizes.

Testing Yourself

Animals Have Gas Exchange Surfaces

1. Internal respiration refers to
   a. the exchange of gases between the air and the blood in the lungs.
   b. the movement of air into the lungs.
   c. the exchange of gases between the blood and tissue fluid.
   d. cellular respiration, resulting in the production of ATP.
2. Which is a requirement for using a body surface as a respiratory surface?
   a. having a high surface-area-to-volume ratio
   b. plentiful supply of water in the environment
   c. low metabolic activity
   d. thin, moist body wall
   e. Both a and d are correct.
3. One problem faced by terrestrial animals with lungs, but not by freshwater aquatic animals with gills, is that
   a. gas exchange involves water loss.
   b. breathing requires considerable energy.
   c. oxygen diffuses very slowly in air.
   d. the concentration of oxygen in water is greater than that in air.
   e. All of these are correct.
4. Countercurrent flow means that oxygen-_____ blood in the gills flows in the _____ direction as oxygen-rich water passing over the gills.
   a. rich, same   c. poor, same
   b. rich, opposite   d. poor, opposite
5. In which animal is the circulatory system not involved in gas transport?
   a. mouse   d. sparrow
   b. dragonfly   e. human
   c. trout
6. How is inhaled air modified before it reaches the lungs?
   a. It must be humidified.   c. It must be filtered.
   b. It must be warmed.   d. All of these are correct.
7. Food and air both travel through the
   a. lungs.   c. larynx.
   b. pharynx.   d. trachea.
8. What is the name of the structure that prevents food from entering the trachea?
   a. glottis   c. epiglottis
   b. septum   d. Adam's apple
9. If the digestive and respiratory tracts were completely separate in humans, there would be no need for
   a. swallowing.   d. a diaphragm.
   b. a nose.   e. All of these are correct.
   c. an epiglottis.
10. In tracing the path of air in humans, you would list the trachea
    a. directly after the nose.
    b. directly before the bronchi.
    c. after the pharynx.
    d. directly before the lungs.
    e. Both a and c are correct.
11. Which of these statements is anatomically incorrect?
    a. The nose has two nasal cavities.
    b. The pharynx connects the nasal cavity and mouth to the larynx.
    c. The larynx contains the vocal cords.
    d. The trachea enters the lungs.
    e. The lungs contain many alveoli.

In questions 12–16, match each description with a structure in the key.

KEY:

   a. pharynx   c. larynx   e. bronchi
   b. glottis   d. trachea   f. bronchioles

12. Branched tubes that lead from the bronchi to the alveoli.
13. Reinforced tube that connects the larynx with the bronchi.
14. Chamber behind the oral cavity and between the nasal cavity and larynx.
15. Opening into the larynx.
16. Divisions of the trachea that enter the lungs.

Ventilation Precedes Transport

17. Which animal breathes by positive pressure?
   a. fish   c. bird   e. planarian
   b. human   d. frog
18. Air enters the human lungs because
   a. atmospheric pressure is lower than the pressure inside the lungs.
   b. atmospheric pressure is greater than the pressure inside the lungs.
   c. although the pressures are the same inside and outside, the partial pressure of oxygen is lower within the lungs.
   d. the residual air in the lungs causes the partial pressure of oxygen to be lower than it is outside.
19. Which of these statements correctly goes with expiration rather than inspiration?
   a. The rib cage moves up and out.
   b. The diaphragm relaxes and moves up.
   c. Pressure in the lungs decreases, and air comes rushing out.
   d. The diaphragm contracts and lowers.
20. Birds have more efficient lungs than humans because the flow of air
   a. is the same during both inspiration and expiration.
   b. never backs up as it does in human lungs.
   c. is not hindered by a larynx.
   d. enters birds' bones.
21. In humans, the respiratory control center
   a. is located in the medulla oblongata.
   b. controls the rate of breathing.
   c. is stimulated by hydrogen ion concentration.
   d. All of these are correct.
22. Which equation occurs in pulmonary capillaries?
   a. $Hb + O_2 \longrightarrow HbO_2$
   b. $H^+ + Hb \longrightarrow HHb$
   c. $CO_2 + H_2O \longrightarrow H_2CO_3$
   d. $H_2CO_3 \longrightarrow H^+ + HCO_3^-$
   e. More than one of these is correct.
23. Hemoglobin assists the transport of gases by
   a. combining with oxygen.
   b. combining with CO_2.
   c. combining with H^+.
   d. being present in red blood cells.
   e. All of these are correct.
24. **THINKING CONCEPTUALLY** Knowing that proteins are sensitive to pH changes, hypothesize why hemoglobin binds to oxygen in the lungs but releases it in the tissues.
25. Most CO_2 is carried in the blood
   a. as HCO_3.   c. dissolved in plasma.
   b. on hemoglobin.   d. Both a and c are correct.

26. The chemical reaction that converts carbon dioxide to a bicarbonate ion takes place in
   a. the blood plasma.   c. the alveoli.
   b. red blood cells.   d. the hemoglobin molecule.
27. The enzyme carbonic anhydrase
   a. affects the heart rate.
   b. is found in red blood cells.
   c. is active once carbon dioxide enters the blood.
   d. attaches carbon dioxide to hemoglobin.
   e. Both b and c are correct.
28. **THINKING CONCEPTUALLY** The human body is able to go without breathing for only minutes but can go without eating for days. Explain.

Understanding the Terms

alveoli 638	heme 643
aortic body 641	hemoglobin (Hb) 643
asthma 644	inspiration 640
bicarbonate ion 643	internal respiration 634
bronchi 638	larynx 638
bronchiole 638	lungs 635
carbaminohemoglobin 643	obstructive pulmonary
carbonic anhydrase 643	disorder 644
carotid body 641	oxyhemoglobin 643
chronic bronchitis 644	partial pressure 642
countercurrent	pharynx 638
exchange 636	respiration 634
diaphragm 640	respiratory center 641
emphysema 644	restrictive pulmonary
epiglottis 638	disorder 644
expiration 640	trachea 638
external respiration 634	tracheae 637
gills 634	ventilation 634
glottis 638	vocal cord 638

Match the terms to these definitions:
   a. _____ In terrestrial vertebrates, the mechanical act of moving air in and out of the lungs; also called breathing.
   b. _____ Dome-shaped, muscularized sheet separating the thoracic cavity from the abdominal cavity in mammals.
   c. _____ Respiratory organ in most aquatic animals; in fish, an outward extension of the pharynx.
   d. _____ Stage during breathing when air is pushed out of the lungs.

Thinking Scientifically

1. Why might you hypothesize that the need to metabolize fats by a diabetic would lead to arterial blood with a higher pH, lower CO_2 content, and a higher O_2 content than normal? How would you test your hypothesis?
2. You are a physician who witnessed Christopher Reeve's riding accident described in Section 25.6. Why might you immediately use mouth to mouth resuscitation until mechanical ventilation becomes available?

 Visit **www.mhhe.com/maderconcepts** *for practice quizzes, animations, videos, and activities designed to help you master the material in this chapter.*

33

Osmoregulation and Excretion

LEARNING OUTCOMES

After studying this chapter, you should be able to accomplish the following outcomes.

Do Coral Reef Animals Regulate?

1 Discuss the osmoregulatory mechanisms of the animals depicted.

Metabolic Waste Products Have Different Advantages

2 Contrast the advantages of excreting ammonia, urea, and uric acid, and associate each with a particular environment.
3 Contrast the organs of excretion in planarians, earthworms, and arthropods.

Osmoregulation Varies According to the Environment

4 Contrast the ways in which aquatic vertebrates maintain the water-salt balance.
5 Contrast the manner in which terrestrial vertebrates in extreme environments maintain the water-salt balance.

The Kidney Is an Organ of Homeostasis

6 Compare the mammalian urinary system and the mammalian kidney to those of other vertebrates.
7 Describe the three primary steps in urine formation.
8 List all of the organs in the human body that are involved in excretion, and tell what each one excretes.
9 Describe how the mammalian kidney maintains the water-salt balance, and compare this to other vertebrates.
10 Describe how the mammalian kidney maintains the acid-base balance.
11 Describe how an artificial kidney performs the functions of the kidneys.
12 Describe the effects of not drinking enough water and drinking too much water.

Barracuda

Most of us don't ever think about whether the beautiful animals in a coral reef are osmoregulating—whether they are regulating the quantity of ions, such as Na^+, K^+, Ca^{2+}, Mg^{2+}, or Cl^-, in their body fluids. Some invertebrates (sea anemones and sea stars, for example) do not regulate. Their body fluids conform to the sea water around them, and they take what they get. Think of the energy they save. They're not pumping salt here and there to get their osmolarity just right. Remember osmosis? If a cell is in an isotonic solution and the concentration of salts is the same inside and out, there is no net movement of water. A cell in a hypotonic solution gains water, and a cell in a hypertonic solution loses water (see Fig. 5.15A). Invertebrate animals that live in the sea are isotonic to sea water. They have a lot of salts in their cells, but that is okay because their enzymes are adapted to this osmolarity.

Coral reef

Do Coral Reef Animals Regulate?

Representatives from only a handful of the approximately 34 animal phyla have managed to colonize land. What's the problem? We know that an appropriate respiratory organ is needed for living on land—gills won't do, but lungs are great. And we know that legs are much better for locomoting on land than are flippers. But an animal also needs a good osmoregulatory organ, such as the vertebrate kidney. In addition, an animal must have various accessory ways to reduce water loss to the dry air, or it will never make it on land.

The blood plasmas of vertebrates, such as marine fishes and sea turtles, are hypotonic to sea water, and when these animals live in a coral reef, they tend to gain salt. The vertebrate kidney is much better at conserving water on land than pumping out extra salts in the marine environment. But the fishes in the coral reef get a little help from their gills. In marine environments, gills pump salts out; in fresh water, they do the opposite.

Striped parrotfish

Sea turtles and seabirds also do just fine living in a marine environment. They have special salt glands that pump salts out. Ever see a green sea turtle cry? It's getting rid of salt by way of its lacrimal (tear) glands. With the exception of marine invertebrates, animals do spend a lot of energy regulating the osmolarity of their internal body fluids. Their enzymes demand it and won't function if the tonicity is not just right.

In this chapter, we learn how both aquatic and terrestrial animals excrete metabolic wastes, particularly nitrogenous wastes, and at the same time maintain their normal water-salt balance.

Christmas tree worms

Clownfish

Cup coral

Metabolic Waste Products Have Different Advantages

In this part of the chapter, we discuss **excretion,** the elimination of metabolic wastes, and discover that the nitrogenous waste excreted by an animal is an adaptation to the environment. Also, we take a look at the excretory organs of invertebrates.

33.1 The nitrogenous waste product of animals varies according to the environment

The breakdown of various molecules, including amino acids and nucleic acids, results in nitrogenous wastes. For simplicity's sake, we will limit our discussion to amino acid metabolism. When amino acids are broken down by the body to generate energy, or are converted to fats or carbohydrates, the amino groups ($-NH_2$) must be removed because they are not needed. Once the amino groups have been removed, they may be excreted from the body in the form of ammonia, urea, or uric acid, depending on the species. Removal of amino groups from amino acids requires a fairly set amount of energy. However, the amount of energy required to convert amino groups to ammonia, urea, or uric acid differs, as indicated in **Figure 33.1**.

Ammonia Amino groups removed from amino acids immediately form **ammonia** (NH_3) by the addition of a third hydrogen ion. Little or no energy is required to convert an amino group to ammonia by adding a hydrogen ion. Ammonia is quite toxic and can be a nitrogenous excretory product if a good deal of water is available to wash it from the body. Ammonia is excreted by most fishes and other aquatic animals whose gills and skin surfaces are in direct contact with the water of the environment.

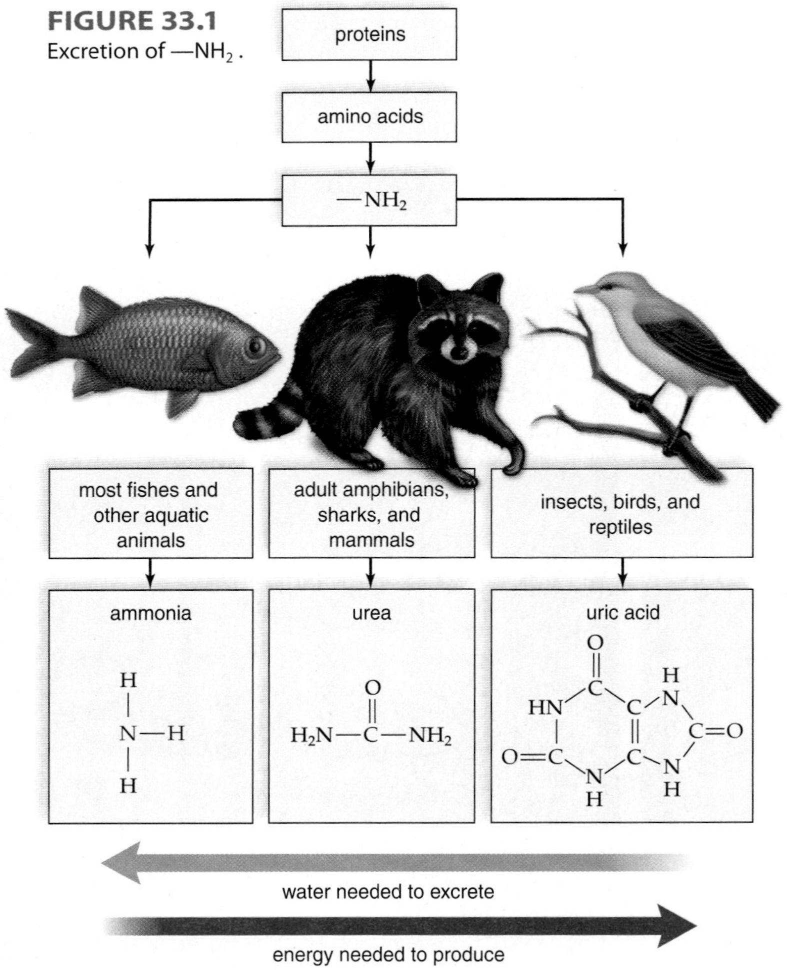

Urea Production of urea requires the expenditure of energy because it is produced in the liver by a set of energy-requiring enzymatic reactions, known as the urea cycle. In this cycle, carrier molecules take up carbon dioxide and two molecules of ammonia, finally releasing urea. **Urea** is much less toxic than ammonia and can be excreted in a moderately concentrated solution. This allows body water to be conserved, an important advantage for terrestrial animals with limited access to water. Sharks, adult amphibians, and mammals usually excrete urea as their main nitrogenous waste.

Uric Acid Uric acid is synthesized by a long, complex series of enzymatic reactions that requires expenditure of even more ATP than does urea synthesis. **Uric acid** is not very toxic, and it is poorly soluble in water. Poor solubility is an advantage if water conservation is needed, because uric acid can be concentrated even more readily than can urea. Uric acid is routinely excreted by insects, reptiles, and birds. In reptiles and birds, a dilute solution of uric acid passes from the kidneys to the *cloaca,* a common reservoir for the products of the digestive, urinary, and reproductive systems. The cloacal contents are refluxed into the large intestine, where water is reabsorbed. The white substance in bird feces is uric acid. Embryos of reptiles and birds develop inside completely enclosed shelled eggs. The production of insoluble, relatively nontoxic uric acid is advantageous for shelled embryos because all nitrogenous wastes are stored inside the shell until hatching takes place. Here again, the advantage of water conservation seems to counterbalance the disadvantage of energy expenditure for synthesis of an excretory molecule.

In general, it is possible to predict which metabolic waste product an animal excretes based on its anatomy and environment—but not always. For example, unlike most other birds, hummingbirds excrete more ammonia than uric acid. Excretion of ammonia, as stated, requires a lot of fluid to keep its toxicity under control. Recall that hummingbirds are fluid feeders, and as such, they have plenty of fluid available for excreting ammonia.

Section 33.2 surveys the excretory organs of invertebrates.

FIGURE 33.1
Excretion of $-NH_2$.

most fishes and other aquatic animals	adult amphibians, sharks, and mammals	insects, birds, and reptiles
ammonia	urea	uric acid

water needed to excrete →

energy needed to produce →

> **33.1 *Check Your Progress*** What is the advantage of excreting urea instead of (*a*) ammonia or (*b*) uric acid?

33.2 Many invertebrates have organs of excretion

Most animals have tubular excretory organs that regulate the water-salt balance of the body and excrete metabolic wastes into the environment. Here we give three examples among the invertebrates.

Planarians The planarians are flatworms that live in fresh water and have two strands of branching excretory tubules that open to the outside of the body through excretory pores (**Fig. 33.2**, *top*). Located along the tubules are bulblike **flame cells**, each of which contains a cluster of beating cilia that looks like a flickering flame under the microscope. The beating of flame-cell cilia propels fluid through the excretory tubules and out of the body. The system is believed to function in ridding the body of excess water and in excreting wastes.

Earthworms The body of an earthworm, an annelid, is divided into segments, and nearly every body segment has a pair of excretory structures called **nephridia**. Each nephridium is a tubule with a ciliated opening and an excretory pore (Fig. 33.2, *bottom*). As fluid from the coelom is propelled through the tubule by beating cilia, its composition is modified. For example, nutrient substances are reabsorbed and carried away by a network of capillaries surrounding the tubule. The urine of an earthworm contains metabolic wastes, salts, and water. Although the earthworm is considered a terrestrial animal, it excretes a very dilute urine. Each day, an earthworm may produce a volume of urine equal to 60% of its body weight. The excretion of ammonia is consistent with these data.

Arthropods Insects have a unique excretory system consisting of long, thin tubules called **Malpighian tubules** attached to the gut. Uric acid is actively transported from the surrounding hemolymph into these tubules, and water follows a salt gradient established by active transport of K^+. Water and other useful substances are reabsorbed at the rectum, but the uric acid leaves the body at the anus. Insects that live in water, or eat large quantities of moist food, reabsorb little water. But insects in dry environments reabsorb most of the water and excrete a dry, semisolid mass of uric acid.

The excretory organs of other arthropods are given different names, although they function similarly. In crustaceans (e.g., crabs, crayfish), nitrogenous wastes are generally removed by diffusion across the gills—for those species that have gills. Even so, crustaceans possess excretory organs called *green glands* located in the ventral portion of the head region. Fluid collects within the tubules from the surrounding blood of the hemocoel, but this fluid is modified by the time it leaves the tubules. The secretion of salts into the tubule regulates the amount of urine excreted. In shrimp and pillbugs, the excretory organs are located in the maxillary segments and are called *maxillary glands*. Spiders, scorpions, and other arachnids possess *coxal glands*, which are located near one or more appendages and used for excretion. Coxal glands are spherical sacs resembling annelid nephridia. Wastes are collected from the surrounding blood of the hemocoel and discharged through pores at one to several pairs of appendages.

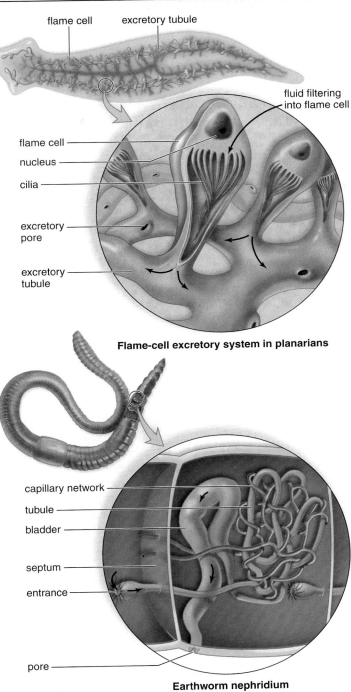

Flame-cell excretory system in planarians

Earthworm nephridium

FIGURE 33.2 Excretory organs in invertebrates.

In conclusion, invertebrates utilize tubules to rid the body of wastes and maintain a water-salt balance. On occasion, excretion also involves other organs, such as the rectum in the earthworm and the gills in crayfish.

The next part of the chapter considers osmoregulation in vertebrates.

33.2 *Check Your Progress* Earthworms have a thin skin for respiration. If they had thicker skin, would it affect the function of the nephridia?

Osmoregulation Varies According to the Environment

Learning Outcomes 4–5, page 648

We will study osmoregulation in vertebrates according to whether they live in the water or on land. Vertebrates utilize various mechanisms to osmoregulate their blood.

33.3 Aquatic vertebrates have adaptations to maintain the water-salt balance of their bodies

Marine vertebrates, unlike marine invertebrates, do **osmoregulate**—maintain particular ion concentrations in their blood. Therefore, few vertebrates have blood that is isotonic to sea water. Not so for the cartilaginous fishes (**Fig. 33.3A**), whose blood is isotonic to sea water, for reasons we will now discuss.

Cartilaginous Fishes The total concentration of the various ions in the blood of cartilaginous fishes is less than that in sea water. Their blood plasma is nearly isotonic to sea water because they pump it full of urea, and this molecule gives their blood the same tonicity as sea water. Cartilaginous fishes do regulate the concentration of other solutes in their blood and have rectal glands that rid the body of excess salt.

Marine Bony Fishes A marine environment, which is high in salts, is hypertonic to the blood plasma of bony fishes. Apparently, their common ancestor evolved in fresh water, and only later did some groups invade the sea. Therefore, marine bony fishes must avoid the tendency to become dehydrated (**Fig. 33.3B**). **1** As the sea washes over their gills, marine bony fishes lose water by osmosis. **2** To counteract this, they drink sea water almost constantly. On the average, marine bony fishes swallow an amount of water equal to 1% of their body weight every hour. This is equivalent to a human drinking about 700 ml of water every hour around the clock. But while they get water by drinking, this habit also causes these fishes to acquire salt. **3** To rid the body of excess salt, they actively transport it into the surrounding sea water at the gills. **4** The kidneys conserve water, and marine bony fishes produce a scant amount of isotonic urine.

Freshwater Bony Fishes The osmotic problems of freshwater bony fishes and the response to their environment are exactly opposite those of marine bony fishes (**Fig. 33.3C**).

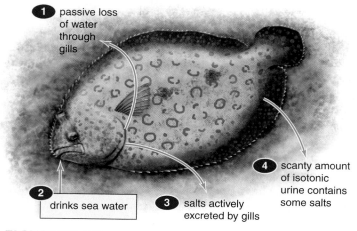

1 passive loss of water through gills

2 drinks sea water

3 salts actively excreted by gills

4 scanty amount of isotonic urine contains some salts

FIGURE 33.3B Osmoregulation in marine bony fishes.

1 Freshwater fishes tend to gain water by osmosis across the gills and the body surface. **2** As a consequence, these fishes never drink water. **3** They actively transport salts into the blood across the membranes of their gills. **4** They eliminate excess water by producing large quantities of dilute (hypotonic) urine. They discharge a quantity of urine equal to one-third their body weight each day.

Osmoregulation in terrestrial vertebrates is considered in Section 33.4.

> **33.3** *Check Your Progress* The blood plasma of a shark is isotonic to sea water. Would you expect a shark to lose water to the same degree as a freshwater bony fish or to actively excrete salt as a marine bony fish does?

FIGURE 33.3A The blood of sharks is isotonic to sea water.

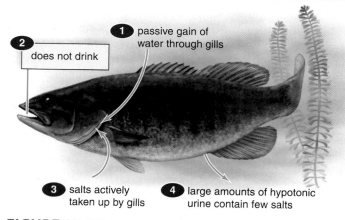

1 passive gain of water through gills

2 does not drink

3 salts actively taken up by gills

4 large amounts of hypotonic urine contain few salts

FIGURE 33.3C Osmoregulation in freshwater bony fishes.

33.4 Terrestrial vertebrates have adaptations to maintain the water-salt balance of their bodies

Desert mammals, such as the kangaroo rat, and seabirds, such as a seagull, illustrate different strategies for dealing with extreme terrestrial environments.

Kangaroo Rat Dehydration threatens all terrestrial animals, especially those that live in a desert, as does the kangaroo rat. Its fur prevents loss of water to the air, and during the day, it remains in a cool burrow. In addition, the kangaroo rat's nasal passage has a highly convoluted mucous membrane surface that captures condensed water from exhaled air. Exhaled air is usually full of moisture, which is why you can see it on cold winter mornings—the moisture in exhaled air is condensing.

As we shall see, humans mainly conserve water by producing urine that is hypertonic to blood plasma. The kangaroo rat forms a very concentrated urine—20 times more concentrated than its blood plasma. Also, its fecal material is almost completely dry.

Most terrestrial animals need to drink water occasionally to make up for the water lost from the skin and respiratory passages and through urination. However, the kangaroo rat is so adapted to conserving water that it can survive by using metabolic water derived from cellular respiration, and it never drinks water (**Fig. 33.4A**). The adaptations of the kangaroo rat allow it to remain in water-salt balance, even under desert conditions.

Seagulls, Reptiles, and Mammals Birds, reptiles, and mammals evolved on land, and their kidneys are especially good at conserving water. However, some animals have become secondarily adapted to living near or in the sea. They drink sea water and still manage to survive. If humans drink sea water, we lose more water than we take in just ridding the body of all that salt. Little is known about how whales manage to get rid of extra salt, but we know that their kidneys are enormous. Other animals have been studied, and we have learned that seabirds and

FIGURE 33.4B Marine birds and reptiles are apt to have salt glands to pump excess salt.

reptiles have salt glands that pump out salt (**Fig. 33.4B**). In the two types of animals we will mention, each has commandeered a gland meant for another purpose and used it to pump out the salt from blood plasma and leave behind the water, just as in a desalination plant.

In birds, salt-excreting glands are located near the eyes. The glands produce a salty solution that is excreted through the nostrils and moves down grooves on their beaks until it drips off. In marine turtles, the salt gland is a modified tear (lacrimal) gland, and in sea snakes, a salivary sublingual gland beneath the tongue gets rid of excess salt. The work of the gland is regulated by the nervous system. Osmoreceptors, perhaps located near the heart, are thought to stimulate the brain, which then orders the gland to excrete salt until the salt concentration in the blood decreases to a tolerable level.

This completes our study of osmoregulation in vertebrates. The next part of the chapter examines in detail the structure and function of the mammalian kidney.

> **33.4 Check Your Progress** Would the tonicity of the urine produced by a seagull be greater than that produced by a human? Explain.

Exhaled air is cooled and dried in long convoluted air passages.

Animal fur prevents evaporative loss of water at skin.

Urine is the most hypertonic known among animals.

Oxidation of food results in metabolic water.

Fecal pellets are dry.

FIGURE 33.4A Adaptations of a kangaroo rat to minimize water loss.

The Kidney Is an Organ of Homeostasis

Learning Outcomes 6–12, page 648

In this part of the chapter, we primarily study the organs of the mammalian urinary system, with an emphasis on the kidney. We learn how the kidney functions as an organ of homeostasis.

33.5   The kidneys are a part of the urinary system

In mammals, the **kidneys** are bean-shaped, reddish-brown organs, each about the size of a fist. They are located in the lower back on either side of the vertebral column just below the diaphragm, where they are partially protected by the lower rib cage. The right kidney is slightly lower than the left kidney. ❶ In **Figure 33.5, urine** made by the kidneys is conducted from the body by the other organs in the urinary system. ❷ Each kidney is connected to a **ureter**, a duct that takes urine from the kidney to ❸ the **urinary bladder**, where it is stored until it is voided from the body through ❹ the single **urethra**. In males, the urethra passes through the penis, and in females, the opening of the urethra is ventral to (in front of) the vagina. There is no connection between the genital (reproductive) and urinary systems in females, but there is a connection in males. In males, the urethra also carries sperm during ejaculation.

Other Vertebrates   Fishes and amphibians have an elongate kidney on each side of the abdominal cavity. The elongate kidney is only temporarily present during the development of reptiles and birds, and as adults, they have a kidney that resembles that of mammals. In all vertebrates, except for placental mammals, a duct from the kidney conducts urine to a *cloaca*, which you will recall is a common depository for indigestible remains, urine, and sex cells.

Section 33.6 is concerned with the macroscopic and microscopic anatomy of the mammalian kidney.

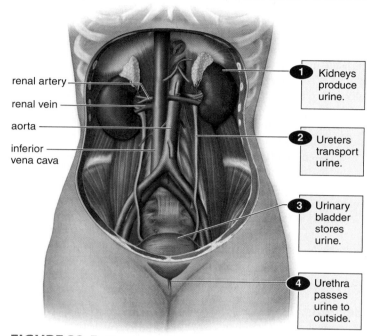

renal artery
renal vein
aorta
inferior vena cava

1 Kidneys produce urine.

2 Ureters transport urine.

3 Urinary bladder stores urine.

4 Urethra passes urine to outside.

FIGURE 33.5 The mammalian urinary system.

> **33.5** *Check Your Progress* Which of the organs shown in Figure 33.5 are organs of homeostasis, involved in osmoregulation and excretion?

33.6   The mammalian kidney contains many tubules

If a mammalian kidney is sectioned longitudinally, three major parts can be distinguished (**Fig. 33.6A,** *left*). The **renal cortex**, which is the outer region of a kidney, has a somewhat granular appearance. The **renal medulla** consists of six to ten cone-shaped renal pyramids that lie inside the renal cortex. The innermost part of the kidney is a hollow chamber called the **renal pelvis**. Urine collects in the renal pelvis before it is carried to the bladder by the ureter.

A kidney stone, or renal calculus, is a hard granule of phosphate, calcium, protein, or uric acid that forms in the renal pelvis. Many pass out of the body unnoticed. However, larger and jagged stones can block the renal pelvis or ureter, causing intense pain and damage.

Nephrons   All vertebrate kidneys contain tiny tubules called **nephrons** that produce urine. The mammalian kidney is composed of over 1 million nephrons. Some nephrons are located

primarily in the renal cortex, but others dip down into the renal medulla, as shown in Figure 33.6A, *right*. Each nephron is made of several parts (**Fig. 33.6B**). ❶ The blind end of a nephron is pushed in on itself to form a cuplike structure called the **glomerular capsule** (Bowman capsule). The outer layer of the glomerular capsule is composed of squamous epithelial cells; the inner layer is composed of specialized cells that allow easy passage of molecules. Leading from the glomerular capsule is a portion of the nephron known as ❷ the **proximal convoluted tubule**, which is lined by cells with many mitochondria and tightly packed microvilli. ❸ Next comes the **loop of the nephron** (loop of Henle), which has a descending limb and an ascending limb. ❹ The **distal convoluted tubule** follows the loop of the nephron. Several distal convoluted tubules enter one collecting duct. ❺ **Collecting ducts** transport urine down through the renal medulla and deliver it to the renal pelvis. The many loops of the nephron and collecting ducts give the

654   PART V Animals Are Homeostatic

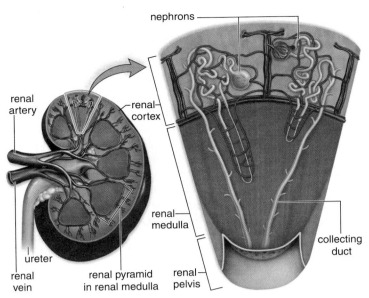

FIGURE 33.6A Macroscopic (*left*) and microscopic anatomy of the kidney.

pyramids of the renal medulla a striped appearance. In mammals, these structures are essential to producing urine that is hypertonic to blood plasma.

Each mammalian nephron has an extensive blood supply (Fig. 33.6B). From the dorsal aorta, the renal artery leads to many arterioles, one for each nephron. The arteriole, called an afferent arteriole, divides to form a capillary bed, the **glomerulus**, where liquid exits and enters the glomerular capsule. Reptiles and birds have a reduced glomerulus and, in this way, produce less urine for the excretion of uric acid. Blood from the glomerulus drains into an efferent arteriole, which subsequently branches into a second capillary bed around the tubular parts of the nephron. This capillary bed, called the **peritubular capillary network**, leads to venules that join to form the renal vein, a vessel that enters the inferior vena cava.

Sections 33.7 through 33.10 consider the functions of the kidney. Section 33.7 shows how urine is formed.

33.6 *Check Your Progress* Trace the path of a nephron from the glomerular capsule to the pelvis of the mammalian kidney.

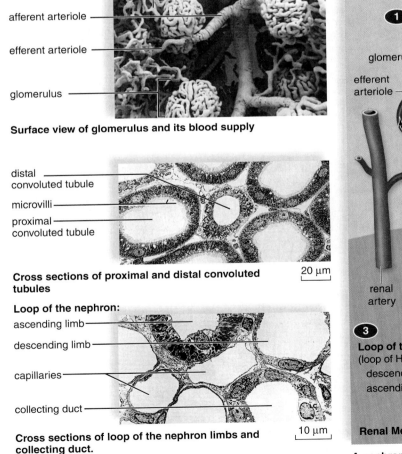

FIGURE 33.6B Nephron anatomy.

(top): © R. G. Kessel and R. H. Kardon, *Tissues and Organs: A Text-Atlas of Scanning Electron Microscopy,* 1979, W. H. Freeman, New York.

An average human produces 1–2 L of urine daily. Urine production requires three distinct steps (**Fig. 33.7**):

1. Glomerular filtration at the glomerular capsule
2. Tubular reabsorption at the convoluted tubules
3. Tubular secretion at the convoluted tubules

① **Glomerular filtration** is the movement of small molecules across the glomerular wall into the glomerular capsule as a result of blood pressure. When blood enters the glomerulus, blood pressure is sufficient to cause small molecules, such as water, nutrients, salts, and wastes, to move from the glomerulus to the inside of the glomerular capsule, especially since the glomerular walls are 100 times more permeable than the walls of most capillaries elsewhere in the body. The molecules that leave the blood and enter the glomerular capsule are called the **glomerular filtrate.** Plasma proteins and blood cells are too large to be part of this filtrate, so they remain in the blood as it flows into the efferent arteriole.

Glomerular filtrate is essentially protein-free, but otherwise it has the same composition as blood plasma. If this composition were not altered in other parts of the nephron, death from loss of nutrients (starvation) and loss of water (dehydration) would quickly follow. The total blood volume averages about 5 L, and this amount of fluid is filtered every 40 minutes. Thus, 180 L of filtrate is produced daily, some 60 times the amount of blood plasma in the body. Most of the filtered water is quickly returned to the blood; otherwise, a person would actually die from urination. Tubular reabsorption prevents this from happening.

② **Tubular reabsorption** takes place when substances move across the walls of the tubules into the associated peritubular capillary network. The osmolarity of the blood is essentially the same as that of the filtrate within the glomerular capsule, and therefore osmosis of water from the filtrate into the blood cannot yet occur. However, sodium ions (Na^+) are actively pumped into the peritubular capillary, and then chloride ions (Cl^-) follow passively. Now the osmolarity of the blood is such that water moves

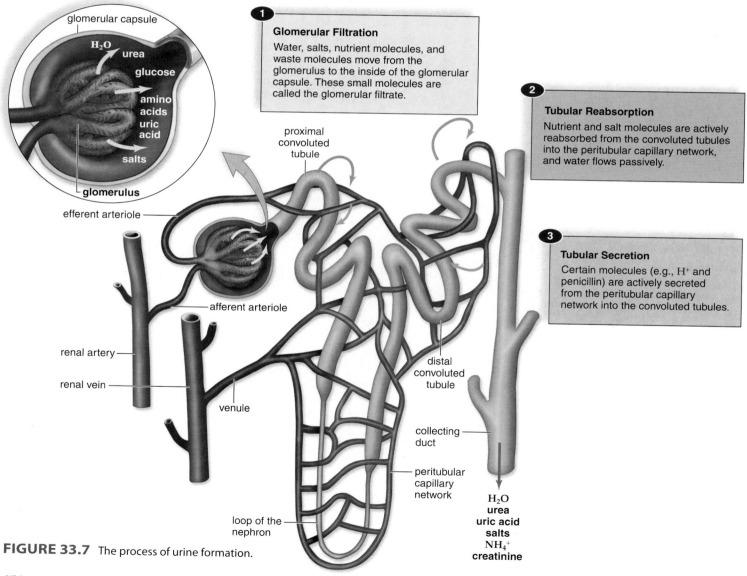

① Glomerular Filtration

Water, salts, nutrient molecules, and waste molecules move from the glomerulus to the inside of the glomerular capsule. These small molecules are called the glomerular filtrate.

② Tubular Reabsorption

Nutrient and salt molecules are actively reabsorbed from the convoluted tubules into the peritubular capillary network, and water flows passively.

③ Tubular Secretion

Certain molecules (e.g., H^+ and penicillin) are actively secreted from the peritubular capillary network into the convoluted tubules.

glomerular capsule
H_2O
urea
glucose
amino acids
uric acid
salts
glomerulus
efferent arteriole
afferent arteriole
renal artery
renal vein
venule
proximal convoluted tubule
distal convoluted tubule
collecting duct
peritubular capillary network
loop of the nephron
H_2O
urea
uric acid
salts
NH_4^+
creatinine

FIGURE 33.7 The process of urine formation.

passively from the tubule into the blood. About 60–70% of salt and water are reabsorbed at the proximal convoluted tubule.

Nutrients, such as glucose and amino acids, also return to the blood at the proximal convoluted tubule. This is a selective process, because only molecules recognized by carrier proteins in plasma membranes are actively reabsorbed. The cells of the proximal convoluted tubule have numerous microvilli, which increase the surface area, and numerous mitochondria, which supply the energy needed for active transport. Glucose is a molecule that ordinarily is reabsorbed completely because there is a plentiful supply of carrier molecules for it. However, if there is more glucose in the filtrate than there are carriers to handle it, glucose will exceed its renal threshold, or transport maximum. When this happens, the excess glucose in the filtrate appears in the urine. People with diabetes mellitus have an abnormally large amount of glucose in the blood and filtrate because the liver fails to store glucose as glycogen. The presence of glucose in the filtrate results in the absorption of less water; the increased thirst and frequent urination experienced by untreated diabetics are due to the reabsorption of less water into the peritubular capillary network.

Urea is a substance that is passively reabsorbed from the filtrate. At first, the concentration of urea within the filtrate is the same as that in blood plasma. But after water is reabsorbed, the urea concentration is greater than that in peritubular plasma. In the end, about 50% of the filtered urea is excreted.

3 **Tubular secretion** is the second way substances are removed from blood and added to tubular fluid. Substances such as uric acid, hydrogen ions, ammonia, creatinine, histamine, and penicillin are eliminated by tubular secretion. The process of tubular secretion helps rid the body of potentially harmful compounds that were not filtered into the glomerulus.

Section 33.8 considers how urinalysis can detect drug use.

> **33.7 Check Your Progress** The kidneys function on a take-back system. Explain.

HOW BIOLOGY IMPACTS OUR LIVES

33.8 Urinalysis can detect drug use

Since ancient times, urinalysis has been used to diagnose disease. As early as 600 B.C., Hindu physicians in India noted that the urine of a diabetic was sweet to the taste. In diabetes mellitus, blood glucose is abnormally high, either because insulin-secreting cells have been destroyed or because cell receptors do not respond to the insulin that is present. Thus, the filtrate level of glucose is extremely high, and the proximal convoluted tubule can't reabsorb it all. Tasting urine to diagnose diabetes persisted through the 1800s (thankfully, modern techniques have made it obsolete)! Similarly, the great Greek physician Hippocrates studied and wrote about urinalysis and noted that shaking the urine of renal failure patients produced frothy bubbles at the surface of the sample. We now know that the froth is a sign of proteinuria, or protein in the urine, an early indicator of chronic renal failure.

Today, the use of urinalysis has expanded beyond medical applications to include forensic diagnosis of drug use. Screening for illegal drug use is now mandated by federal and state agencies as a condition of employment, and most private employers now require it as well. A court can order that urinalysis be done if drug abuse is involved in the commission of a crime. The National Collegiate Athletic Association requires all student athletes to undergo drug testing, as do many high school athletic programs.

Urinalysis is not used to screen for the drugs themselves, but for drug metabolites, the breakdown products of drugs that have been consumed or injected. Once in the body, drugs are metabolized by the liver and filtered by the kidney. Thus, metabolites will be present in the urine of a drug abuser. Two types of techniques can detect metabolites. The first, a screening exam, involves placing a test strip into freshly voided urine. The test strip contains monoclonal antibodies (see page 604) specific for metabolites of street drugs. Strips can test for twelve or more different drugs at once, but most screen for five commonly abused drugs: marijuana, amphetamine, PCP, cocaine, and opiates such as heroin. Although urine-strip testing gives results within minutes, certain legal over-the-counter medications can give false positives. Should the sample test positive, it can be immediately sent for a second, more sophisticated chemical analysis, such as gas chromatography. Tracking long-term drug use may require using hair samples in addition to urine samples, since drug metabolites can be incorporated into the hair.

The urine specimen must be properly collected to avoid possible tampering by the individual being tested. Specimens can be deliberately diluted with sink or toilet water or contaminated with any number of additives. Bleach, drain cleaner, soft drinks, and so on, can be used, as well as products specifically sold for the purpose of "helping to beat a drug screen." Drug abusers may also attempt to substitute the urine of a "clean" individual for their own. Tampering may be prevented by requiring that a witness be present at all times while the sample is being collected and stored. In addition, proper documentation must accompany any urine sample. Chain-of-custody forms record each step of the handling of a specimen, from collection to disposal. This provides proof of everything that happens to the specimen, and prevents the specimen from being rejected as evidence in court proceedings.

The goal of any forensic urinalysis testing program should be to procure proper treatment and counseling for drug abusers so that they can overcome addiction and lead healthier and more productive lives.

Regulation of the water-salt balance is an important function of the kidney, as discussed in Section 33.9.

> **33.8 Check Your Progress** Drug testing for marijuana involves detecting the breakdown product tetrahydrocannabinol in the urine. At which step in urine formation did this chemical probably enter the urine?

The kidneys regulate the water-salt balance of the blood, and thereby also maintain the blood volume, the blood pressure, and the **osmolarity** (tonicity) of blood. **Figure 33.9A** traces the regulation of the water-salt balance in mammals. **1** Most of the water and salt (NaCl) present in the filtrate is reabsorbed across the wall of the proximal convoluted tubule. Reabsorption also occurs along the remainder of the nephron. The excretion of a hypertonic urine is dependent on the reabsorption of water and salt from the loop of the nephron and the collecting duct.

The long loop of the nephron, which typically penetrates deep into the renal medulla, is composed of a descending (going down) limb and an ascending (going up) limb. **2** Salt passively diffuses out of the lower portion of the ascending limb, but the upper, thick portion of the limb actively extrudes NaCl into the tissue of the outer renal medulla (Fig. 33.9A). Less and less NaCl is available for transport as fluid moves up the thick portion of the ascending limb. Because of these circumstances, there is an osmotic gradient within the tissues of the renal medulla: The concentration of NaCl is greater in the direction of the inner medulla. (Water cannot leave the ascending limb because the limb is impermeable to water, as indicated by the dark line in Fig. 33.9A.)

The wide arrow to the left in Figure 33.9A indicates that the innermost portion of the inner medulla has the highest concentration of solutes. This cannot be due to NaCl because active transport of NaCl does not start until fluid reaches the thick portion of the ascending limb. **3** Urea is believed to leak from the lower

portion of the collecting duct, and it is this molecule that contributes to the high solute concentration of the inner medulla.

4 Because of the osmotic gradient within the renal medulla, water leaves the descending limb along its entire length. This is a countercurrent mechanism: As water diffuses out of the descending limb, the remaining fluid within the limb encounters an even greater osmotic concentration of solute; therefore, water will continue to leave the descending limb from the top to the bottom.

Fluid enters the collecting duct from the distal convoluted tubule. This fluid is isotonic to the cells of the renal cortex. Thus, to this point, the net effect of reabsorption of water and salt is the production of a fluid that has the same osmolarity as blood plasma. However, the filtrate within the collecting duct also encounters the same osmotic gradient mentioned earlier (Fig. 33.9A). **5** Therefore, water diffuses out of the collecting duct into the renal medulla, and the urine within the collecting duct becomes hypertonic to blood plasma.

Antidiuretic hormone (ADH) released by the posterior lobe of the pituitary plays a role in water reabsorption at the collecting duct. To understand the action of this hormone, consider its name. *Diuresis* means increased amount of urine, and *antidiuresis* means decreased amount of urine. When ADH is present, more water is reabsorbed (blood volume and pressure rise), and a decreased amount of urine results. In practical terms, if an individual does not drink much water on a certain day, the posterior lobe of the pituitary releases ADH, causing more water to be reabsorbed and less urine to form. On the other hand, if an individual drinks a large amount of water and does not perspire much, ADH is not released. Now more water is excreted, and more urine forms. Diuretics, such as caffeine and alcohol, increase the flow of urine.

Hormones Control the Reabsorption of Salt
Usually, more than 99% of sodium (Na$^+$) filtered at the glomerulus is returned to the blood. Most sodium (67%) is reabsorbed at the proximal convoluted tubule, and a sizable amount (25%) is extruded by the ascending limb of the loop of the nephron. The rest is reabsorbed from the distal convoluted tubule and collecting duct.

Blood volume and pressure is, in part, regulated by salt reabsorption. When blood volume, and therefore blood pressure, is not sufficient to promote glomerular filtration, the kidneys secrete renin. **Renin** is an enzyme that changes angiotensinogen (a large plasma protein produced by the liver) into angiotensin I. Later, angiotensin I is converted to **angiotensin II**, a powerful vasoconstrictor that also stimulates the adrenal glands, which lie on top of the kidneys, to release aldosterone (**Fig. 33.9B**). **Aldosterone** is a hormone that promotes the excretion of potassium ions (K$^+$) and the reabsorption of sodium ions (Na$^+$) at the distal convoluted tubule. The reabsorption of sodium ions is followed by the reabsorption of water. Therefore, blood volume and blood pressure increase.

Atrial natriuretic hormone (ANH) is a hormone secreted by the atria of the heart when cardiac cells are stretched due to increased blood volume. ANH inhibits the secretion of renin by the juxtaglomerular apparatus and the secretion of aldosterone by the adrenal cortex. Its effect, therefore, is to promote the excretion of Na$^+$—that is, *natriuresis*. When Na$^+$ is excreted, so is water, and therefore blood volume and blood pressure decrease.

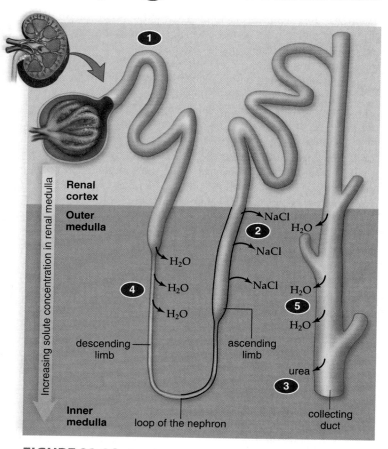

Increasing solute concentration in renal medulla

Renal cortex

Outer medulla

NaCl
2
H$_2$O
NaCl

H$_2$O
4 H$_2$O
NaCl H$_2$O
5
H$_2$O

descending limb

ascending limb

urea
3

Inner medulla

loop of the nephron

collecting duct

FIGURE 33.9A Regulation of water-salt balance in mammals.

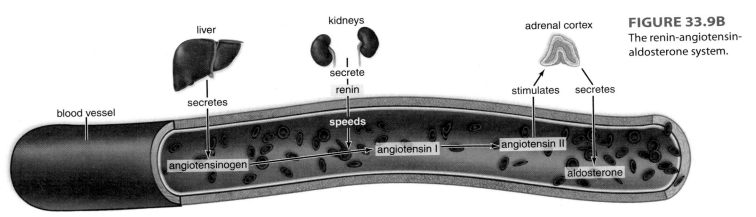

FIGURE 33.9B
The renin-angiotensin-aldosterone system.

These examples show that the body regulates the water balance in blood by controlling the excretion and the reabsorption of various ions. Sodium is an important ion in plasma that must be regulated, but the kidneys also excrete or reabsorb others, such as potassium ions (K^+), bicarbonate ions (HCO_3^-), and magnesium ions (Mg^{2+}), as needed.

The kidneys work with the lungs in order to maintain the pH balance, as explained in Section 33.10.

> **33.9 Check Your Progress** What does the renin-angiotensin-aldosterone system accomplish?

33.10 Lungs and kidneys maintain acid-base balance

The functions of proteins in the body—such as enzymes, hemoglobin, proteins of the electron transport chain, and others—are influenced by pH. Therefore, the regulation of pH is extremely important to good health.

The *bicarbonate (HCO_3^-) buffer system* works together with the breathing process to maintain the pH of the blood. Central to the mechanism is this reaction, which you have seen before:

$$H^+ + HCO_3^- \rightleftharpoons H_2CO_3 \rightleftharpoons H_2O + CO_2$$

The *excretion of carbon dioxide (CO_2) by the lungs* helps keep the pH within normal limits, because when CO_2 is exhaled, this reaction is pushed to the right, and hydrogen ions (H^+) are tied up in water. Indeed, when blood pH decreases, chemoreceptors in the carotid bodies (located in the carotid arteries) and in aortic bodies (located in the aorta) stimulate the respiratory control center, and the rate and depth of breathing increase. On the other hand, when blood pH begins to rise, the respiratory control center is depressed, and the amount of bicarbonate ion increases in the blood.

As powerful as these mechanisms are, only the kidneys can rid the body of a wide range of acidic and basic substances. The kidneys are slower acting than the buffer/breathing mechanism, but they have a more powerful effect on pH. For the sake of simplicity, we can think of the kidneys as reabsorbing bicarbonate ions and excreting hydrogen ions as needed to maintain the normal pH of the blood (**Fig. 33.10**).

If the blood is acidic, hydrogen ions move from the blood into the tubule and are excreted. Bicarbonate ions are reabsorbed. If the blood is basic, hydrogen ions are not excreted, and bicarbonate ions are not reabsorbed. The fact that urine is usually acidic (pH about 6) shows that usually an excess of hydrogen ions is excreted. Ammonia (NH_3), produced in tubule cells by the deamination of amino acids, provides a means of buffering these hydrogen ions in urine:

$$NH_3 + H^+ \longrightarrow NH_4^+$$

Phosphate provides another means of buffering hydrogen ions in urine. The presence of the ammonium ion (NH_4^+) is quite obvious in the diaper pail or kitty litter box by its odor.

Acidosis and Alkalosis The normal pH of arterial blood is around 7.4. Therefore, a person is said to have acidosis when the pH is below 7.34 and alkalosis when the pH is higher than 7.45. Abnormal breathing rates—too shallow or too rapid—can lead to acidosis and alkalosis, respectively. Alcohol consumption causes acidosis because alcohol is metabolized to acetic acid, which enters the blood. Similarly, diabetes can cause acidosis because of its effect on metabolism. Metabolic alkalosis is rare, but it can be caused by excessive intake of antacids or by an illness.

Section 33.11 describes how an artificial kidney can be of assistance when the kidneys fail.

> **33.10 Check Your Progress** Are there any lasting effects if the kidneys fail to regulate the pH?

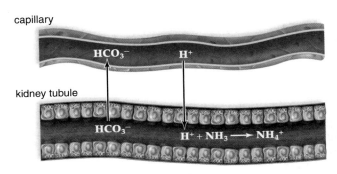

FIGURE 33.10 Excretion of these ions regulates pH.

33.11 The artificial kidney machine makes up for faulty kidneys

After a person suffers kidney damage, perhaps due to repeated infections, waste substances accumulate in the blood. This condition is called **uremia** because urea is one of the substances that accumulates. Although nitrogenous wastes can cause serious damage, the imbalance of ions is believed to lead to loss of consciousness and to heart failure. Patients in renal failure most often seek a kidney transplant, but in the meantime they undergo **hemodialysis**, utilizing an artificial kidney. Dialysis is the diffusion of dissolved molecules through a semipermeable membrane (an artificial membrane with pore sizes that allow only small molecules to pass through). Substances more concentrated in blood diffuse into the dialysis solution, which is called the dialysate, and substances more concentrated in the dialysate diffuse into

the blood (**Fig. 33.11**). Accordingly, the artificial kidney either extracts substances from blood, including waste products or toxic chemicals and drugs, or adds a substance to the blood—for example, bicarbonate ions (HCO_3^-) if the blood is acidic.

The first clinically useful artificial kidney was devised by William J. Kolff of the Netherlands in 1943. It consisted of 20 m of cellophane tubing wrapped around a drum. Rotation of the drum caused the blood to pass from one end of the tubing to the other. The lower half of the drum was immersed in a fluid. Kolff used an anticoagulant to prevent the patient's blood from clotting. The only patient died when all possible entry points for his blood had been utilized. Around 1960, Dr. Belding Scribner and Mr. Wayne Quinton in Seattle developed the arteriovenous shunt. Tubes permanently placed in an artery and vein of the arm are joined to a shunt leading to the artificial kidney, where fresh dialysate continuously circulates past many meters of tubing. Today's artificial kidney uses highly permeable dialysis tubing, making it possible to shorten dialysis time.

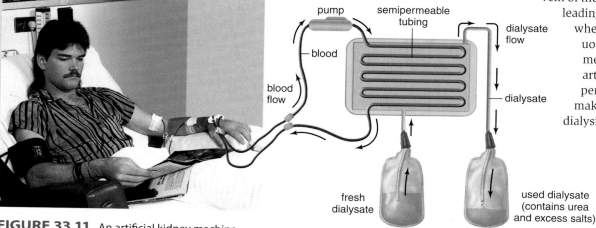

FIGURE 33.11 An artificial kidney machine.

> **33.11 Check Your Progress** Urea clearance is better in the artificial kidney machine than in the human kidney. Explain.

33.12 Dehydration and water intoxication occur in humans

Dehydration is a serious condition resulting from loss of water by cells. A common cause of dehydration is excessive sweating, perhaps during exercise, without replacing any of the water lost. Dehydration can also be a side effect of an illness that causes prolonged vomiting or diarrhea. The signs of moderate dehydration are a dry mouth, sunken eyes, and skin that will not bounce back after light pinching. If dehydration becomes severe, the pulse and breathing rate are rapid, the hands and feet are cold, and the lips are blue. To cure dehydration, physicians administer a low-sodium solution because water intake alone could lead to water intoxication.

Water intoxication is due to a gain of water by cells. The solute concentration in extracellular fluid decreases—that is, the tissue fluid becomes hypotonic to the cells, and water enters the cells. Water intoxication is not nearly as common in adults as is dehydration. One cause can be drinking too much water

during a marathon race. Marathoners who collapse and experience nausea and vomiting after a race are probably not suffering from a heart attack, but they may be suffering from water intoxication, which can lead to pulmonary edema and swelling in the brain. The cure, an intravenous solution containing high amounts of sodium, is the opposite of that for dehydration. Therefore, it is important that physicians be able to diagnose water intoxication in athletes who may have been drinking fluids for several hours.

> **33.12 Check Your Progress** In Chapter 4, we learned about the lysis (bursting) of cells and the crenation (shrinking) of cells due to osmosis. Associate these terms with dehydration and water intoxication.

Similar to its function in other organisms, we have seen that the human kidney plays three important roles in homeostasis: (1) excreting metabolic wastes, especially nitrogenous wastes; (2) maintaining the water-salt balance (osmoregulation); and (3) maintaining the acid-base balance, and thereby the osmolarity, of internal fluids.

The primary nitrogenous waste of humans is urea, produced by the liver and excreted by the kidneys. The kidneys are assisted to a limited degree by sweat glands in the skin, which excrete perspiration, a mixture of water, salt, and some urea.

If blood does not have the normal water-salt balance, blood volume and blood pressure are affected. Too low a concentration of Na^+ in the blood causes blood pressure to lower and activates the renin-angiotensin-aldosterone sequence. Then the kidneys increase Na^+ reabsorption, which is followed by the reabsorption of water. Subsequently, the blood osmolarity and blood pressure return to normal. Too high a concentration of Na^+ in the blood leads to the release of ADH from the posterior pituitary and the reabsorption of water. Osmolarity is restored, but blood pressure rises. Restricting salt in the diet is one way to prevent this cause of hypertension.

Two closely associated mechanisms offset short-term challenges to the acid-base balance: the blood bicarbonate (HCO_3^-) buffering system and the process of breathing. Usually, the excretion of carbon dioxide by the lungs helps keep the blood pH within normal limits. Only the kidneys, however, can rid the body of a wide range of acidic and basic substances—thus, the importance of the kidneys' ultimate control over blood pH cannot be overemphasized.

The Chapter in Review

Summary

Do Coral Reef Animals Regulate?

- Marine invertebrates don't need a regulatory mechanism because their body fluids are isotonic to sea water.
- The vertebrate kidney is better at conserving water on land than ridding the body of extra salt.
- In marine environments, gills pump out salts.
- Sea turtles and seabirds have special glands to pump out salt.

Metabolic Waste Products Have Different Advantages

33.1 The nitrogenous waste product of animals varies according to the environment

- Amino groups (—NH_2) removed from amino acids and nucleic acids are converted to ammonia, urea, or uric acid for excretion.
  - Aquatic animals excrete ammonia, which requires water but no energy to produce.
  - Sharks, adult amphibians, and mammals excrete urea, which requires energy to produce but less water to excrete.
  - Reptiles and birds excrete uric acid, which requires the most energy to produce.

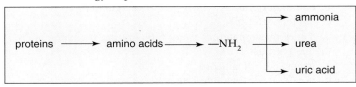

33.2 Many invertebrates have organs of excretion

- Planarians have flame cells, earthworms have nephridia, and insects have Malpighian tubules.

Osmoregulation Varies According to the Environment

33.3 Aquatic vertebrates have adaptations to maintain the water-salt balance of their bodies

- Cartilaginous fishes have blood that is isotonic to sea water because it contains plentiful urea.

- Marine bony fishes must offset osmotic loss of water by drinking water constantly and pumping out salt at the gills.
- Freshwater bony fishes must adjust for osmotic gain of water, so they don't drink water but pump in salt at the gills.

33.4 Terrestrial vertebrates have adaptations to maintain the water-salt balance of their bodies

- The kangaroo rat gets by on metabolic water due to adaptations that prevent loss of water
- Seagulls, reptiles, and mammals have salt-excreting glands.

The Kidney Is an Organ of Homeostasis

33.5 The kidneys are a part of the urinary system

- In mammals, the urinary system is composed of the kidneys, ureters, urinary bladder, and urethra.
- Fishes and amphibians have elongate kidneys; the adult kidneys in reptiles and birds resemble that of mammals. All vertebrates but mammals have a cloaca.

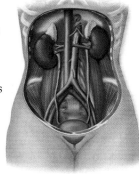

33.6 The mammalian kidney contains many tubules

- Divisions of a kidney are the renal cortex, renal medulla, and renal pelvis.
- A nephron (kidney tubule) has a glomerular capsule, proximal convoluted tubule, loop of the nephron, distal convoluted tubule, and collecting duct.

- The blood supply of a nephron consists of an afferent arteriole, glomerulus, efferent arteriole, peritubular capillary network, and venule.

33.7 Urine formation requires three steps
- During glomerular filtration, small molecules move into the glomerular capsule.
- Tubular reabsorption occurs at the proximal convoluted tubule and involves the reabsorption of most of the water, all nutrients, and other small molecules except urea.
- During tubular secretion, substances such as hydrogen ions, ammonia, creatinine, histamine, and penicillin are added to the filtrate.

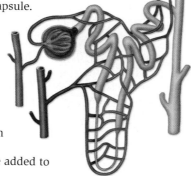

33.8 Urinalysis can detect drug use
- The urine of a diabetic is sweet and frothy due to the presence of glucose and protein.
- Today, urinalysis is used to detect drug use, but tampering with samples must be prevented.

33.9 The kidneys concentrate urine to maintain water-salt balance
- Most water and salts are reabsorbed at the proximal convoluted tubule.
- In mammals, the loop of the nephron concentrates urine.
  - In the ascending limb, the concentration of salt in the medulla increases, and urea leaks passively from the collecting duct.
  - Water increasingly leaves the descending limb and collecting duct.
  - Secretion of ADH regulates water reabsorption from the distal convoluted tubule and collecting duct.

33.10 Lungs and kidneys maintain acid-base balance
- Lungs excrete carbon dioxide, a substance that tends to make the blood acidic.
- The kidneys excrete H^+ via formation of ammonium ions (NH_4^+) and reabsorption of bicarbonate ions, as needed.
  - Acidosis is blood pH below 7.34.
  - Alkalosis is blood pH above 7.45.

33.11 The artificial kidney machine makes up for faulty kidneys
- During hemodialysis, molecules move across nonliving semipermeable tubing as needed to adjust the blood levels of salts and nitrogenous wastes and adjust pH.

33.12 Dehydration and water intoxication occur in humans
- Dehydration is due to loss of water from the cells and is treated by intake of a low-sodium solution.
- Water intoxication is due to gain in water by the cells and is treated by intravenous intake of a high-sodium solution.

Testing Yourself

Metabolic Waste Products Have Different Advantages
1. One advantage of urea excretion over uric acid excretion is that urea
   a. requires less energy than uric acid to form.
   b. can be concentrated to a greater extent.
   c. is not a toxic substance.
   d. requires no water to excrete.
   e. is a larger molecule.
2. Which of these pairs is mismatched?
   a. insects—excrete uric acid
   b. humans—excrete urea
   c. fishes—excrete ammonia
   d. birds—excrete ammonia
   e. All of these are correct.
3. Animals with which of these structures are most likely to excrete a semisolid nitrogenous waste?
   a. nephridia   d. flame cells
   b. Malpighian tubules   e. All of these are correct.
   c. human kidneys

Osmoregulation Varies According to the Environment
4. Freshwater bony fishes maintain water balance by
   a. excreting salt across their gills.
   b. periodically drinking small amounts of water.
   c. excreting a hypotonic urine.
   d. excreting wastes in the form of uric acid.
   e. Both a and c are correct.

The Kidney Is an Organ of Homeostasis
5. Urine transport follows which order?
   a. kidneys, urethra, urinary bladder, ureter
   b. kidneys, ureter, urinary bladder, urethra
   c. urinary bladder, kidneys, ureter, urethra
   d. urethra, kidneys, ureter, urinary bladder
6. When tracing the path of glomerular filtrate, the loop of the nephron follows which structure?
   a. collecting duct
   b. distal convoluted tubule
   c. proximal convoluted tubule
   d. glomerulus
   e. renal pelvis
7. When tracing the path of blood, the blood vessel that follows the renal artery is called the
   a. peritubular capillary network.   d. renal vein.
   b. efferent arteriole.   e. glomerulus.
   c. afferent arteriole.
8. Which of these materials is not filtered from the blood at the glomerulus?
   a. water   d. glucose
   b. urea   e. sodium ions
   c. protein
9. Absorption of the glomerular filtrate occurs at
   a. the convoluted tubules.
   b. only the distal convoluted tubule.
   c. the loop of the nephron.
   d. the collecting duct.

10. Which of the following materials would not be maximally reabsorbed from the glomerular filtrate?
    a. water
    b. glucose
    c. sodium ions
    d. urea
    e. amino acids

11. **THINKING CONCEPTUALLY** Structure suits function is exemplified by the cells lining the proximal convoluted tubule. These cells have microvilli and plentiful mitochondria. Explain.

12. In humans, water is
    a. found in the glomerular filtrate.
    b. reabsorbed from the nephron.
    c. in the urine.
    d. reabsorbed from the collecting duct.
    e. All of these are correct.

13. Sodium is actively extruded from which part of the nephron?
    a. descending portion of the proximal convoluted tubule
    b. ascending portion of the loop of the nephron
    c. ascending portion of the distal convoluted tubule
    d. descending portion of the collecting duct

14. The function of a long loop of the nephron in the process of urine formation is
    a. reabsorption of water.
    b. production of filtrate.
    c. reabsorption of solutes.
    d. secretion of solutes.

15. The presence of ADH (antidiuretic hormone) causes an individual to excrete
    a. less salt.
    b. less water.
    c. more water.
    d. more salt.
    e. Both a and c are correct.

16. Which of these is not likely to cause a rise in blood pressure?
    a. aldosterone
    b. antidiuretic hormone (ADH)
    c. renin
    d. atrial natriuretic hormone (ANH)

17. H_2CO_3 can be converted to
    a. $H^+ + HCO_3^-$.
    b. $H_2O + CO_2$.
    c. Both a and b are correct.
    d. Neither a nor b is correct.

18. To lower blood acidity,
    a. hydrogen ions are excreted, and bicarbonate ions are reabsorbed.
    b. hydrogen ions are reabsorbed, and bicarbonate ions are excreted.
    c. hydrogen ions and bicarbonate ions are reabsorbed.
    d. hydrogen ions and bicarbonate ions are excreted.
    e. urea, uric acid, and ammonia are excreted.

19. If a drug inhibits the kidneys' ability to reabsorb bicarbonate so that bicarbonate is excreted in the urine, the blood will become
    a. acidic.
    b. alkaline.
    c. first acidic and then alkaline.
    d. first alkaline and then acidic.

20. The use of an artificial kidney is known as
    a. filtration.
    b. excretion.
    c. hemodialysis.
    d. None of these are correct.

21. Water intoxication may result if
    a. an individual consumes too much water.
    b. tissue fluid becomes hypotonic to the cells.
    c. an individual suffers from prolonged vomiting or diarrhea.
    d. Both a and b are correct choices.
    e. a, b, and c are all correct choices.

22. **THINKING CONCEPTUALLY** Knowing that osmosis plays a role in water reabsorption by the nephron, why would you recommend a reduction in dietary salt for those with high blood pressure?

Understanding the Terms

aldosterone 658
ammonia 650
angiotensin II 658
antidiuretic hormone (ADH) 658
atrial natriuretic hormone (ANH) 658
collecting duct 654
dehydration 660
distal convoluted tubule 654
excretion 650
flame cell 651
glomerular capsule 654
glomerular filtrate 656
glomerular filtration 656
glomerulus 655
hemodialysis 660
kidneys 654
loop of the nephron 654
Malpighian tubule 651
nephridia 651

nephron 654
osmolarity 658
osmoregulate 652
peritubular capillary network 655
proximal convoluted tubule 654
renal cortex 654
renal medulla 654
renal pelvis 654
renin 658
tubular reabsorption 656
tubular secretion 657
urea 650
uremia 660
ureter 654
urethra 654
uric acid 650
urinary bladder 654
urine 654
water intoxication 660

Match the terms to these definitions:
a. _____ Blind, threadlike excretory tubule near the anterior end of an insect hindgut.
b. _____ Cuplike structure that is the initial portion of a nephron; site of glomerular filtration.
c. _____ Hormone secreted by the adrenal cortex that regulates the sodium and potassium ion balance of the blood.
d. _____ Enzyme released by the kidneys that leads to the secretion of aldosterone and a rise in blood pressure.
e. _____ Microscopic kidney unit that regulates blood composition by glomerular filtration, tubular reabsorption, and tubular secretion.

Thinking Scientifically

1. Suggest an anatomic study to substantiate that kangaroo rats are able to produce a very hypertonic urine, as stated in Figure 33.4A.
2. You have built an artificial kidney but want to improve its ability to clear the blood of urea. What could you do? See Figure 33.11.

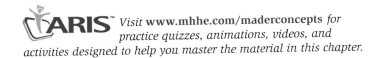 Visit **www.mhhe.com/maderconcepts** *for practice quizzes, animations, videos, and activities designed to help you master the material in this chapter.*

34

Coordination by Hormone Signaling

LEARNING OUTCOMES

After studying this chapter, you should be able to accomplish the following outcomes.

Pheromones Among Us

1 Discuss the available data showing that pheromones play a role in human behavior.

The Endocrine System Utilizes Chemical Signals

2 Compare the use of chemical signals by the nervous and endocrine systems.
3 Describe the control of hormonal secretion by negative feedback and the use of antagonistic hormones.
4 Contrast the manner in which peptide and steroid hormones affect cellular metabolism.
5 Name the major vertebrate endocrine glands, tell where they are located, and state their principal hormones.

The Hypothalamus and Pituitary Are Central to the Endocrine System

6 Contrast the anatomic and physiologic relationships of the hypothalamus to the posterior and anterior pituitary.
7 Describe the function of the hormones secreted by the posterior and anterior pituitary.
8 Describe the three-tier relationship between the anterior pituitary and the thyroid, adrenal cortex, and gonads.

Hormones Regulate Metabolism and Homeostasis

9 Contrast the adrenal medulla and adrenal cortex, including how they help us deal with stress.
10 Explain a possible relationship between glucocorticoid therapy and the development of Cushing syndrome.
11 Contrast the function of insulin with that of glucagon, and summarize the effects of diabetes.
12 Give reasons for the increasing prevalence of diabetes today.
13 Discuss the function of the thyroid gland and associated disorders.
14 Explain the antagonistic functions of calcitonin and parathyroid hormone.

Male and female peregrine falcons bow and "ee-chup," exchange food, and otherwise interact before they mate. Male elephant seals roar loudly, rear up on their foreflippers, and fight for territory and harems of females. Such courtship behaviors, no doubt, are strongly influenced by hormones.

But what about pheromones, each a mixture of small organic molecules wafting through the air? Pheromones are different from hormones because they are sent from one member of a species to another, and they modify the behavior and physiologic processes of the recipient. If the receptors on the antennae of a male silkworm moth detect a few molecules of a sex attractant, the moth takes off immediately toward the female that released it. The social insects, such as ants and bees, use pheromones to keep each member of the society performing its particular function. What keeps fire ant workers working and the queen producing all those eggs? Pheromones.

To what degree do pheromones, in addition to hormones, affect the behavior of vertebrates, even mammals? Dr. Milos Novotny runs the Institute for Pheromone Research at Indiana University, and

Parental care

Pheromones Among Us

he says there is no doubt about their importance in mammalian behavior. He can show that in mice, pheromones trigger inter-male aggression and dominance, readiness for mating, onset of puberty, and communication of stress to other members of a colony. It's possible that pheromones are involved in all sorts of mammalian behaviors, including the reproductive, maternal care, and migratory behaviors depicted here.

Like the mouse, humans apparently have an organ in the nose, called the vomeronasal organ (VNO), that can detect not only odors, but also pheromones. The neurons from this organ lead to the hypothalamus, the part of the brain that controls the release of so many hormones in the body, including the reproductive hormones. So, if we are receiving silent messages through the air from the other members of our species, could they be controlling our behavior, as occurs with the social insects? Well, maybe not ex-

Courtship behavior

actly. It's possible that our ability to reason and think could override unaccountable urges.

Previous data have suggested that human physiologic processes could possibly be controlled by pheromones. The dormitory effect, for example, refers to the ability of body odor to bring about synchronization of the menstrual cycle in women who live together. Recently, neurogeneticists at Rockefeller University and Yale University reported the isolation of a human gene, labeled *V1RL1*, which codes for a pheromone receptor. And other researchers, including David L. Berliner, have now identified a pheromone released by men that reduces premenstrual nervousness and tension in women. Perfumes containing the same chemical are now on sale! Berliner says, "Human beings communicate with each other with pheromones, just like any terrestrial animal. And they do it through the same organ (i.e., nose) that all terrestrial animals have." Once received, pheromones may initiate nerve impulses that go to the hypothalamus, and this chapter describes how the hypothalamus controls the secretion of hormones. We also inlcude a description of the endocrine system before studying the structure and function of the major glands that do the work of the endocrine system.

Aggressive behavior

Social behavior

Migratory behavior

The Endocrine System Utilizes Chemical Signals

This part of the chapter gives an overview of the endocrine system, including a listing of the mammalian hormones and how they bring about their effects.

34.1   The endocrine and nervous systems work together

The nervous system and the endocrine system work together to regulate the activities of other body systems (**Table 34.1**). The nervous system sends nerve impulses along a nerve fiber directly to a target organ. Nerve impulses cause the terminals of axons to release a neurotransmitter, which binds to plasma membrane receptors of muscles or glands. In **Figure 34.1**, ➊ The neurotransmitter is binding to the receptors of smooth muscle cells in the wall of an arteriole. Now the muscle cells will contract and arteriole constriction will occur. The endocrine system uses the blood vessels of the cardiovascular system to send **hormones,** chemical messengers, to target organs. In Figure 34.1, ➋ the target organ is the liver, and insulin from the pancreas is binding to receptors in the plasma membrane. Now the liver will store glucose as glycogen.

Axon conduction occurs rapidly, and so does diffusion of a neurotransmitter across a synapse. The rapid response of the nervous system is particularly useful if the stimulus is an external event that endangers our safety.

The **endocrine system,** which is composed of glands, functions more slowly. It takes time to deliver hormones, and it takes time for cells to respond, but the effect is longer lasting. It is important to remember that only certain cells, called target cells, can respond to a hormone. For a cell to respond, the hormone and receptor protein must bind together as a key fits a lock.

Negative Feedback Mechanisms Both the nervous system and the endocrine system make use of **negative feedback** mechanisms. If blood pressure falls, for example, sensory receptors signal a control center in the brain, which then sends nerve impulses to the arteriole walls so that they constrict, and blood pressure rises. Now the sensory receptors are no longer stimulated, and no nerve impulses are generated. Similarly, when the liver stores glucose as glycogen, the blood glucose level falls,

TABLE 34.1	Comparison of Nervous and Endocrine Systems	
	Nervous System	**Endocrine System**
Composed of	Neurons	Glands, usually
Delivery	Nerve impulse and neurotransmitter	Hormone
How Delivered	Axon and synapse	Bloodstream, usually
Target	Muscles and glands	Cells throughout body
Response	Rapid, short-lived	Slow, long-lasting
Controlled by	Negative feedback	Negative feedback

and the pancreas is no longer stimulated to release insulin. In negative feedback, the end result shuts down the stimulus.

Section 34.2 tells how hormones affect cellular metabolism.

> **34.1** *Check Your Progress* Neurotransmitters, hormones, and pheromones are all biological signals. Explain.

FIGURE 34.1 Modes of action of the nervous and endocrine systems.

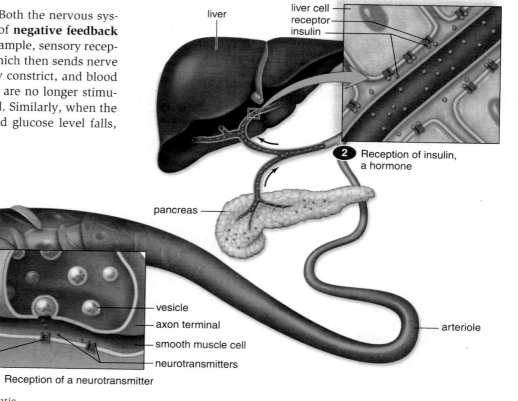

liver

liver cell
receptor
insulin

➋ Reception of insulin, a hormone

pancreas

arteriole

axon of nerve fiber

vesicle
axon terminal
smooth muscle cell
neurotransmitters

receptor

➊ Reception of a neurotransmitter

34.2 Hormones affect cellular metabolism

Hormones fall into two basic chemical classes: (1) The class **peptide hormone** will include not only peptides but also proteins, glycoproteins, or modified amino acids; (2) a **steroid hormone** always has the same complex of four carbon rings, but each one has different side groups.

Action of Peptide Hormones Figure 34.2A gives an example of peptide hormone activity in muscle cells. **1** The hormone epinephrine binds to a receptor in the plasma membrane. **2** The immediate effect is the formation of **cyclic adenosine monophosphate (cAMP)**, a molecule that contains one phosphate group attached to adenosine at two locations. **3** cAMP activates an enzyme cascade, a series of enzymes that amplify the effect of the molecule. **4** Finally, many molecules of glycogen are broken down to glucose, which enters the bloodstream.

Because a peptide hormone never enters the cell, the hormone is called the **first messenger**; cAMP, which sets the metabolic machinery in motion, is called the **second messenger**. To understand this terminology, let's imagine that a hormone-producing gland is like the home office that sends out a courier (the hormone) to a factory (the cell). The courier doesn't have a pass to enter the factory, so when he arrives there, he tells a supervisor through the screen door that the home office wants the factory to produce a particular product. The supervisor (cAMP, the second messenger) walks over and flips a switch

that starts the machinery (the enzymatic pathway), and a product is made.

Action of Steroid Hormones The only glands that produce steroid hormones are the adrenal cortex, the ovaries, and the testes. Thyroid hormones act similarly to steroid hormones, even though they have a different structure.

Steroid hormones do not bind to plasma membrane receptors (**Fig. 34.2B**). **1** Instead, they are able to enter the cell because they are lipids. **2** Once inside, steroid hormones bind to receptors, usually in the nucleus but sometimes in the cytoplasm. **3** Inside the nucleus, the hormone-receptor complex binds with DNA and activates transcription of certain portions of DNA. **4** Translation of messenger RNA (mRNA) transcripts at ribosomes results in enzymes and other proteins that can carry out a response to the hormonal signal.

Steroids act more slowly than peptides because it takes more time to synthesize new proteins than to activate enzymes already present in cells. Their action lasts longer, however.

Section 34.3 previews the hormones of the endocrine system.

> **34.2 Check Your Progress** When pheromones are received by a vomeronasal organ (VNO), nerve messages first go to the brain. Contrast this process with the activity of peptide hormones and steroid hormones.

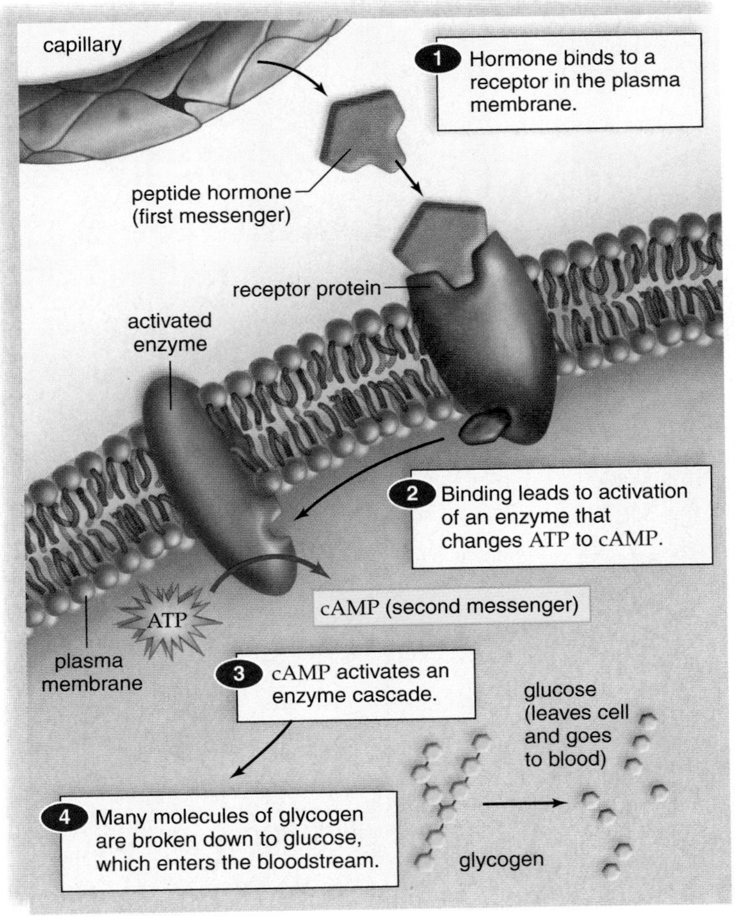

FIGURE 34.2A Action of peptide hormones.

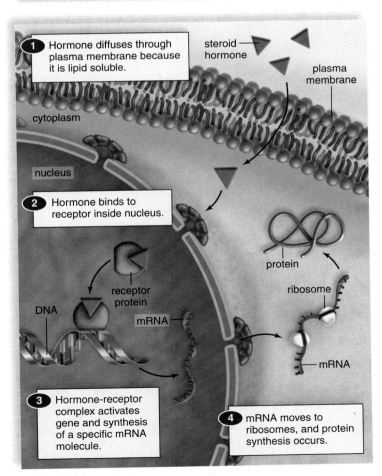

FIGURE 34.2B Action of steroid hormones.

Figure 34.3 depicts the locations of the major endocrine glands in the human body. The hypothalamus, a part of the brain, is in close proximity to the pituitary. The pineal gland is also located in the brain. The thyroid and parathyroids are in the neck, and the thymus gland lies just beneath the breastbone in the thoracic cavity. The adrenal glands and pancreas are in the abdominal cavity. The gonads include the ovaries, located in the pelvic cavity, and the testes, located outside this cavity in the scrotum.

Table 34.3 lists not only the glands, but also their hormones and the target tissues or organs they affect. Many hormones help mantain homeostasis, the relative constancy of the internal environment (the blood and tissue fluid). Several hormones (e.g., ADH and aldosterone) directly affect the osmolarity of the blood. Other hormones control the blood glucose and calcium levels. Similarly, the nervous system is intimately involved in homeostasis. An example of the close association between the endocrine and nervous systems appears in Table 34.3, where the hypothalamus, a part of the brain, is listed as a part of the endocrine system.

Other hormones influence the maturation and function of the reproductive organs. In fact, many people are most familiar with the effect of hormones on sexual function.

The secretion of hormones is controlled by negative feedback, discussed in Section 34.1, in conjunction with *antagonistic hormonal actions*. For example, negative feedback allows the parathyroid glands to secrete the hormone PTH when the blood Ca^{2+} level falls below normal and to cease secretion when the blood Ca^{2+} level rises. An antagonistic hormone called calcitonin, produced by the thyroid gland, now comes into play to lower the Ca^{2+} level once more. Then, too, while insulin lowers the blood glucose level, the hormone glucagon raises it. In subsequent sections of this chapter, we will point out other instances in which hormones are antagonistic to one another, and thereby help regulate the level of a substance in the blood.

The next part of the chapter discusses the actions of the hypothalamus and the anterior pituitary.

> **34.3** *Check Your Progress* **Not all species respond to all phero-mones, and not all body cells respond to all hormones. Explain.**

FIGURE 34.3 The human endocrine system.

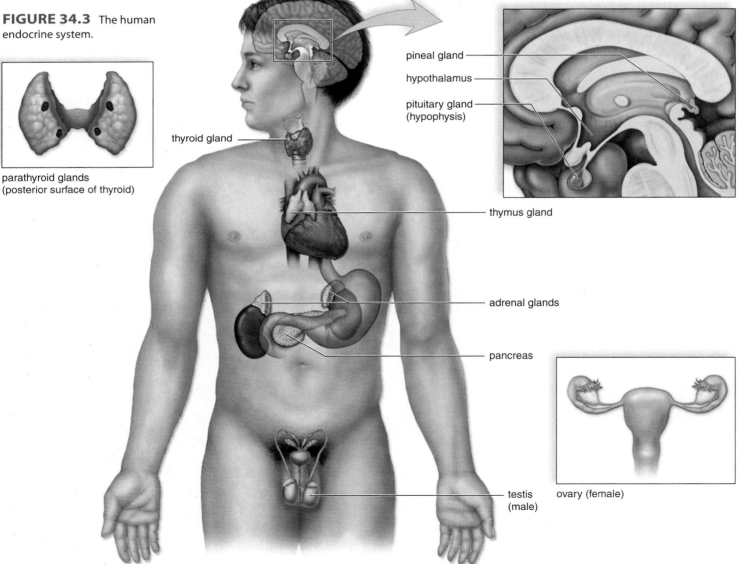

parathyroid glands
(posterior surface of thyroid)

thyroid gland

pineal gland

hypothalamus

pituitary gland
(hypophysis)

thymus gland

adrenal glands

pancreas

testis
(male)

ovary (female)

TABLE 34.3 Principal Endocrine Glands and Hormones

Endocrine Gland	Hormone Released	Chemical Class	Target Tissues/Organs	Chief Function(s) of Hormone
Hypothalamus	Hypothalamic-releasing and -inhibiting hormones	Peptide	Anterior pituitary	Regulates anterior pituitary hormones
Pituitary gland				
Posterior pituitary	Antidiuretic (ADH) (vasopressin)	Peptide	Kidneys	Stimulates water reabsorption by kidneys
	Oxytocin	Peptide	Uterus, mammary glands	Stimulates uterine muscle contraction and release of milk by mammary glands
Anterior pituitary	Thyroid-stimulating (TSH)	Glycoprotein	Thyroid	Stimulates thyroid
	Adrenocorticotropic (ACTH)	Peptide	Adrenal cortex	Stimulates adrenal cortex
	*Gonadotropic (FSH, LH)	Glycoprotein	Gonads	Stimulates egg and sperm production and sex hormone production
	Prolactin (PRL)	Protein	Mammary glands	Stimulates milk production
	Growth (GH)	Protein	Soft tissues, bones	Promotes cell division, protein synthesis, and bone growth
	Melanocyte-stimulating (MSH)	Peptide	Melanocytes in skin	Unknown function in humans; regulates skin color in lower vertebrates
Thyroid	Thyroxine (T_4) and triiodothyronine (T_3)	Iodinated amino acid	All tissues	Increases metabolic rate; regulates growth and development
	Calcitonin	Peptide	Bones, kidneys, intestine	Lowers blood calcium level
Parathyroids	Parathyroid (PTH)	Peptide	Bones, kidneys, intestine	Raises blood calcium level
Adrenal gland				
Adrenal cortex	Glucocorticoids (cortisol)	Steroid	All tissues	Raises blood glucose level; stimulates breakdown of protein
	Mineralocorticoids (aldosterone)	Steroid	Kidneys	Reabsorbs sodium and excretes potassium
	Sex hormones	Steroid	Gonads, skin, muscles, bones	Stimulates reproductive organs and brings about sex characteristics
Adrenal medulla	Epinephrine and norepinephrine	Modified amino acid	Cardiac and other muscles	Released in emergency situations; raises blood glucose level
Pancreas	Insulin	Protein	Liver, muscles, adipose tissue	Lowers blood glucose level; promotes formation of glycogen
	Glucagon	Protein	Liver, muscles, adipose tissue	Raises blood glucose level
Gonads				
Testes	*Androgens (testosterone)	Steroid	Gonads, skin, muscles, bones	Stimulates male sex characteristics
Ovaries	*Estrogens and progesterone	Steroid	Gonads, skin, muscles, bones	Stimulates female sex characteristics
Thymus gland	Thymosins	Peptide	T cells	Stimulates production and maturation of T cells
Pineal gland	Melatonin	Modified amino acid	Brain	Controls circadian and circannual rhythms; possibly involved in maturation of sexual organs

*Discussed in Chapter 35.

In this part of the chapter, we study the actions of the hypothalamus and the pituitary gland. The hypothalamus controls the anterior pituitary, which in turn stimulates various other glands.

34.4 The hypothalamus is a part of the nervous and endocrine systems

The hypothalamus controls the glandular secretions of the **posterior pituitary** (**Fig. 34.4,** #1–3, *left*) and the **anterior pituitary** (Fig. 34.4, #1–4, *right*). A stalklike structure connects the hypothalamus to the **pituitary gland,** which is about 1 cm in diameter.

Hypothalamus and Posterior Pituitary Neurons in the hypothalamus called **neurosecretory cells** produce the hormones antidiuretic hormone (ADH) and oxytocin. These hormones pass through axons into the posterior pituitary, where

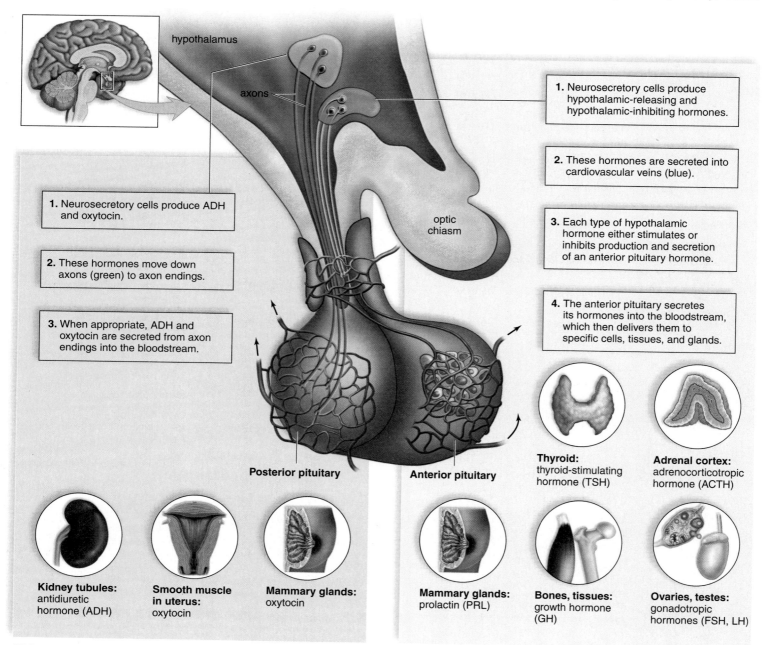

hypothalamus

axons

optic chiasm

1. Neurosecretory cells produce hypothalamic-releasing and hypothalamic-inhibiting hormones.

2. These hormones are secreted into cardiovascular veins (blue).

3. Each type of hypothalamic hormone either stimulates or inhibits production and secretion of an anterior pituitary hormone.

4. The anterior pituitary secretes its hormones into the bloodstream, which then delivers them to specific cells, tissues, and glands.

1. Neurosecretory cells produce ADH and oxytocin.

2. These hormones move down axons (green) to axon endings.

3. When appropriate, ADH and oxytocin are secreted from axon endings into the bloodstream.

Posterior pituitary

Anterior pituitary

Thyroid: thyroid-stimulating hormone (TSH)

Adrenal cortex: adrenocorticotropic hormone (ACTH)

Kidney tubules: antidiuretic hormone (ADH)

Smooth muscle in uterus: oxytocin

Mammary glands: oxytocin

Mammary glands: prolactin (PRL)

Bones, tissues: growth hormone (GH)

Ovaries, testes: gonadotropic hormones (FSH, LH)

FIGURE 34.4 The hypothalamus and pituitary gland.

they are stored in axon endings. Certain neurons in the hypothalamus are sensitive to the water-salt balance of the blood. When these cells determine that the blood is too concentrated, **antidiuretic hormone (ADH)** is released from the posterior pituitary. Upon reaching the kidneys, ADH causes water to be reabsorbed. As the blood becomes dilute, ADH is no longer released. This is another example of negative feedback.

The inability to produce ADH causes *diabetes insipidus* (watery urine), in which a person's body produces copious amounts of urine with a resultant loss of ions from the blood. The condition causes extreme thirst and can be corrected by the administration of ADH. ADH release is inhibited by the consumption of alcohol, which explains the frequent urination associated with drinking alcohol.

Oxytocin, the other hormone made in the hypothalamus and stored in the posterior pituitary, causes uterine contractions during childbirth and milk letdown when a baby is nursing. The more the uterus contracts during labor, the more nerve impulses reach the hypothalamus, causing oxytocin to be released until childbirth occurs. Similarly, the more a baby suckles, the more oxytocin is released until milk letdown occurs. In both instances, the release of oxytocin from the posterior pituitary is controlled by **positive feedback**—that is, the stimulus continues to bring about an effect that ever increases in intensity. Positive feedback is not a way to maintain stable conditions and homeostasis. Oxytocin may also play a role in propelling semen through the male reproductive tract and may affect feelings of sexual satisfaction and emotional bonding.

Hypothalamus and Anterior Pituitary The hypothalamus controls the anterior pituitary by producing **hypothalamic-releasing hormones** and in some instances hypothalamic-inhibiting hormones. For example, one of the hypothalamic-releasing hormones stimulates the anterior pituitary to secrete a thyroid-stimulating hormone, and a hypothalamic-inhibiting hormone prevents the anterior pituitary from secreting prolactin. These hormones reach the anterior pituitary by means of circulatory veins.

Section 34.5 centers in on the anterior pituitary.

> **34.4** *Check Your Progress* Some mothers report that hearing a baby cry can cause milk letdown. Create a scenario to explain milk letdown based on a pheromone.

34.5 The anterior pituitary produces nontropic and tropic hormones

Nontropic Hormones Some hormones produced by the anterior pituitary are controlled by the hypothalamus and are called **nontropic hormones** because they do not affect other endocrine glands (see Fig. 34.5, *right*).

Prolactin (PRL) causes the mammary glands in the breasts to develop and produce milk after childbirth.

Growth hormone (GH) promotes skeletal and muscular growth. It stimulates the rate at which amino acids enter cells and protein synthesis occurs. It also promotes fat metabolism.

Melanocyte-stimulating hormone (MSH) causes skin-color changes in vertebrates having melanophores, special skin cells that produce color variations. The concentration of this hormone in humans is very low.

Tropic Hormones Four hormones produced by the anterior pituitary (see Fig. 34.4) are called tropic hormones because they stimulate the activity of other endocrine glands:

Thyroid-stimulating hormone (TSH) stimulates the thyroid to produce thyroxine and triiodothyronine.

Adrenocorticotropic hormone (ACTH) stimulates the adrenal cortex to produce glucocorticoid.

Gonadotropic hormones (follicle-stimulating hormone, FSH, and luteinizing hormone, LH) stimulate the gonads—the testes in males and the ovaries in females—to produce gametes and sex hormones.

Figure 34.5 recognizes that the tropic hormones participate in a control system. For example, the hypothalamus secretes a

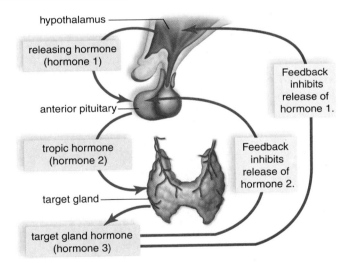

FIGURE 34.5 The hypothalamus-anterior pituitary-target gland control system.

releasing hormone that causes the anterior pituitary to release TSH, and TSH, in turn, stimulates the thyroid gland to secrete thyroxine. Thyroxine exerts feedback control over the activity of the first two structures and, in this way, controls its own blood level. The same type of control system pertains to ACTH and the gonadotropic hormones.

> **34.5** *Check Your Progress* Do the hypothalamus and the anterior pituitary have receptors for hormone 3 in Figure 34.5? Explain.

In this part of the chapter, we examine the function of certain other major endocrine glands. All of their hormones have specific functions that help regulate metabolism and homeostasis.

34.6 The adrenal glands respond to stress

Two **adrenal glands** sit atop the kidneys (see Fig. 34.3). Each gland is about 5 cm long and 3 cm wide and weighs about 5 g. Each adrenal gland consists of an inner portion called the **adrenal medulla** and an outer portion called the **adrenal cortex**. These portions, like the anterior pituitary and the posterior pituitary, have no physiological connection with one another. However, the hypothalamus exerts control over the activity of both portions of the adrenal glands. Stress of all types, including both emotional and physical trauma, prompts the hypothalamus to stimulate the adrenal glands.

Adrenal Medulla The hypothalamus initiates nerve impulses that travel by way of the brain stem, spinal cord, and sympathetic nerve fibers to the adrenal medulla, which then secretes its hormones. **Epinephrine** (adrenaline) and **norepinephrine** (noradrenaline) produced by the adrenal medulla rapidly bring about all the body changes that occur when an individual reacts to an emergency situation. The effects of these hormones are short-term (**Fig. 34.6**, *left*). Epinephrine and norepinephrine accelerate the breakdown of glucose to form ATP, trigger the mobilization of glycogen reserves in skeletal muscle, and increase the cardiac rate and force of contraction.

Adrenal Cortex The adrenal cortex secretes small amounts of male and female sex hormones in both sexes—that is, in the male, some male and female sex hormones are produced by the adrenal cortex, and in the female, some female and male sex hormones are produced by the adrenal cortex.

The adrenal cortex produces the **glucocorticoid** and the **mineralocorticoid** hormones. These hormones provide a long-term response to stress (Fig. 34.6, *right*). The hypothalamus, by means of corticotropin-releasing hormone, controls the anterior pituitary's secretion of ACTH, which in turn stimulates the adrenal cortex to secrete glucocorticoids.

Cortisol is a biologically significant glucocorticoid produced by the adrenal cortex. Cortisol raises the blood glucose level

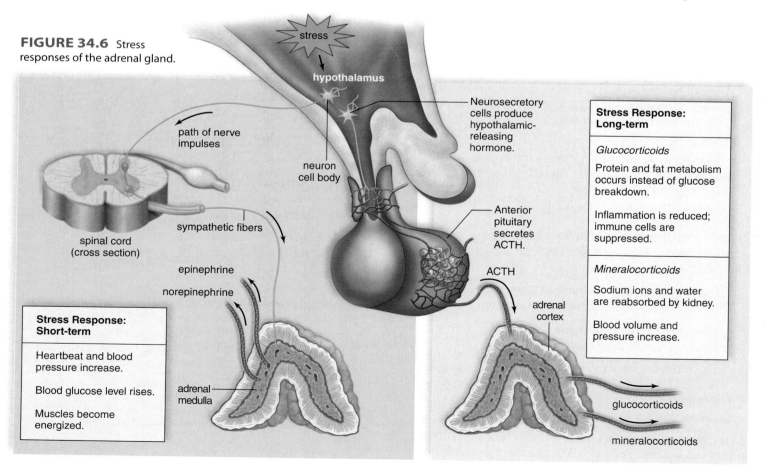

FIGURE 34.6 Stress responses of the adrenal gland.

stress

hypothalamus

path of nerve impulses

neuron cell body

Neurosecretory cells produce hypothalamic-releasing hormone.

spinal cord (cross section)

sympathetic fibers

epinephrine

norepinephrine

Anterior pituitary secretes ACTH.

ACTH

adrenal cortex

Stress Response: Short-term

Heartbeat and blood pressure increase.

Blood glucose level rises.

Muscles become energized.

adrenal medulla

Stress Response: Long-term

Glucocorticoids

Protein and fat metabolism occurs instead of glucose breakdown.

Inflammation is reduced; immune cells are suppressed.

Mineralocorticoids

Sodium ions and water are reabsorbed by kidney.

Blood volume and pressure increase.

glucocorticoids

mineralocorticoids

in at least two ways: (1) It promotes the breakdown of muscle proteins to amino acids, which are taken up by the liver from the bloodstream. The liver then breaks down these excess amino acids to glucose, which enters the blood. (2) Cortisol promotes the metabolism of fatty acids rather than carbohydrates, and this spares glucose.

Cortisol also counteracts the inflammatory response, which leads to the pain and swelling of joints in arthritis and bursitis. The administration of cortisol in the form of cortisone aids these conditions because it reduces inflammation. Very high levels of glucocorticoids in the blood can suppress the body's defense system, including the inflammatory response that occurs at infection sites. Cortisone and other glucocorticoids can relieve swelling and pain from inflammation, but by suppressing pain and immunity, they can also make a person highly susceptible to injury and infection.

Aldosterone is the most important of the mineralocorticoids. Aldosterone primarily targets the kidney, where it promotes renal absorption of sodium (Na^+) and renal excretion of potassium (K^+).

The secretion of mineralocorticoids is not controlled by the anterior pituitary. When the blood sodium level and, therefore, blood pressure are low, the kidneys secrete **renin**. Renin is an enzyme that converts the plasma protein angiotensinogen to angiotensin I, which later is changed to angiotensin II. Angiotensin II stimulates the adrenal cortex to release aldosterone. The effect of this process, called the renin-angiotensin-aldosterone system, is to raise blood pressure in two ways: (1) Angiotensin II constricts the arterioles, and (2) aldosterone causes the kidneys to reabsorb sodium. When the blood sodium level rises, water is reabsorbed, in part because the hypothalamus secretes ADH. Then blood pressure increases to normal.

Now that we know the function of the glucocorticoids, Section 34.7 emphasizes that glucocorticoid therapy can lead to the symptoms of a specific syndrome.

> **34.6 Check Your Progress** If an "alarm pheromone" stimulated the hypothalamus by way of the VNO, it would cause the release of which of the adrenal gland hormones first?

HOW BIOLOGY IMPACTS OUR LIVES

34.7 Glucocorticoid therapy can lead to Cushing syndrome

When the level of adrenal cortex hormones is high due to hypersecretion, a person develops **Cushing syndrome**. Excess cortisol results in a tendency toward diabetes mellitus as muscle protein is metabolized and subcutaneous fat is deposited in the midsection. The trunk is obese, while the arms and legs remain normal in size. Excess aldosterone and reabsorption of sodium and water by the kidneys lead to a basic blood pH and hypertension (high blood pressure). The face is moon-shaped due to edema (**Fig. 34.7**, *left*). Masculinization may occur in women because of excess adrenal male sex hormones. This condition, known as adrenogenital syndrome (AGS), is characterized by an increase in body hair, deepening of the voice, and beard growth in women.

Some people develop Cushing syndrome due to a tumor in the adrenal cortex or anterior pituitary. Normally, cortisol secretion fluctuates, with blood levels being highest in the early morning and falling throughout the day. In people with Cushing syndrome, cortisol levels remain much the same throughout the day. Additional tests may help doctors identify the location of the tumor that is causing Cushing syndrome. Radiation therapy or surgery is necessary to remove the tumor.

Other individuals develop Cushing syndrome because they are taking glucocorticoid drugs. Glucocorticoids are often prescribed for medical conditions such as asthma, rheumatoid arthritis, and lupus. Glucocorticoid therapy is also used to suppress the immune system in order to prevent rejection of a transplanted organ. If glucocorticoid therapy, in these instances, leads to Cushing syndrome, the physician will try to reduce the dosage, while still effectively continuing treatment.

Section 34.8 considers the functions of the pancreas.

> **34.7 Check Your Progress** Create a scenario by which a topical ointment containing glucocorticoid applied to the skin causes Cushing syndrome.

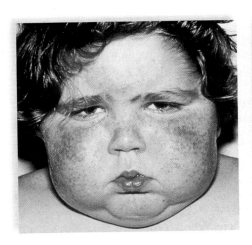
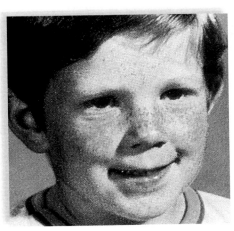

FIGURE 34.7 Patient with Cushing syndrome (*left*) before and (*right*) after treatment.

34.8 The pancreas regulates the blood sugar level

The **pancreas** is a slender, pale-colored organ that lies transversely in the abdomen between the kidneys and near the duodenum of the small intestine (see Fig. 34.3). The pancreas is rather lumpy in consistency and is composed of two types of tissue. Exocrine tissue produces and secretes digestive juices that travel by way of ducts to the small intestine. Endocrine tissue, called the **pancreatic islets** (islets of Langerhans), produces and secretes the antagonistic hormones insulin and glucagon directly into the blood. The majority of pancreatic tissues are exocrine in nature.

A negative feedback mechanism governs whether insulin or glucagon is secreted by the pancreas (**Fig. 34.8**). **Insulin** is secreted when the blood glucose level is high, which usually occurs just after eating. Insulin stimulates the uptake of glucose by all cells, but especially those of the liver, muscles, and adipose tissue. In liver and muscle cells, glucose is then stored as glycogen. Also in muscle cells, the breakdown of glucose supplies energy for protein metabolism, and in adipose tissue cells, the breakdown of glucose leads to fatty acids and glycerol for the formation of fat. In these ways, insulin lowers the blood glucose level.

Glucagon is secreted from the pancreas, usually between meals, when the blood glucose level is low. The major target tissues of glucagon are the liver and adipose tissue. Glucagon stimulates the liver to break down glycogen to glucose and to use fat and protein in preference to glucose as energy sources. Also, adipose tissue cells break down fat to glycerol and fatty acids. The liver takes these up and uses them as substrates for glucose formation. In these ways, glucagon raises the blood glucose level.

Section 34.9 explains that a malfunctioning pancreas is the cause of diabetes.

> **34.8** *Check Your Progress* **Explain the negative feedback mechanism that normally controls insulin secretion.**

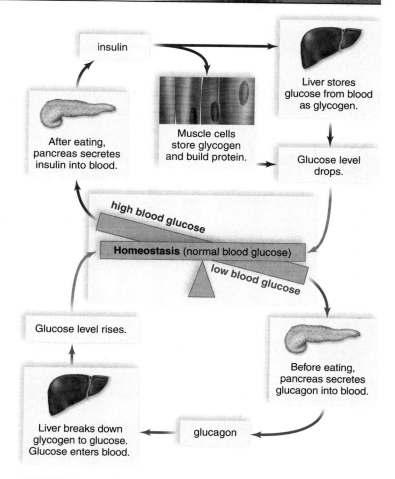

FIGURE 34.8 Regulation of blood glucose level.

34.9 Diabetes is becoming a very common ailment

Diabetes is a very serious ailment that is becoming more and more prevalent worldwide—21 million people have now been diagnosed with diabetes in the United States. Heredity is a risk factor, so you will want to investigate whether a close relative has diabetes. Also, the incidence of diabetes is higher in African Americans, Hispanics, and native Americans than in the general population. Obesity is an important risk factor as well. In particular, the following symptoms can help you decide if a physician should check you for diabetes:

- Frequent urination, especially at night
- Unusual hunger and/or thirst
- Unexplained change in weight
- Blurred vision
- Sores that do not heal
- Excessive fatigue

It is relatively easy to explain some of these symptoms. The blood glucose level in a diabetic is too high, and much of it spills over into the urine. Unusual thirst accompanies untreated diabetes because glucose in the urine causes the nephrons to lose a lot of water by osmosis. Hunger and excessive fatigue occur because the cells are starving for glucose in the midst of plenty. Without glucose inside mitochondria, cells cannot produce ATP, the energy currency of cells.

Diabetes Type 1 and Type 2 A physician must first determine whether a patient has diabetes type 1 or type 2. The number of people with diabetes type 2, in particular, is rising at an alarming rate. By 2025, the incidence of diabetes type 2 is expected to be double what it is now.

In the nondiabetic, the hormone insulin secreted by the pancreas causes tissue cells, particularly muscle and liver cells, to take up glucose so that the blood sugar level remains within the normal range. In diabetes type 1, the pancreas fails to secrete insulin because the cells that make insulin have died off. In diabetes type 2, the pancreas usually doesn't secrete enough insulin, but the real problem is that the cells are resistant to insulin. Receptors in the plasma membrane don't bind insulin properly, and this causes the plasma membrane to have a reduced number of transporters for glucose (i.e., carrier proteins for glucose). Without an adequate number of transporters, glucose enters the cell in reduced amount and stays in the blood instead.

Diabetes type 1 usually occurs after a viral infection. The immune system gears up to fight the infection by killing off cells that are harboring the virus. When the infection is over, the immune system keeps on killing cells, this time the cells of pancreatic islets that produce insulin. Diabetes type 1 is clearly a self-inflicted disease, in the sense that the body itself brings on the disease. In another way, diabetes type 2 is self-inflicted when it is caused by poor health habits. Diabetes type 2 was once considered an adult-onset disorder. Now, more and more children have it. Diabetes type 2 (not type 1) is associated with physical inactivity and excessive food intake leading to obesity, especially fat located in the abdominal region.

Treatment People with diabetes must regularly check their blood sugar level, usually before meals and at bedtime. Today, automatic lances prick the finger, and computerized devices can record readings automatically (**Fig. 34.9**). A blood sugar level between 70 and 110 mg/dL (milligrams per deciliter) is considered normal, but in diabetics, the blood sugar can rise to more than 126 mg/dL. Patients with diabetes type 1 must have daily insulin injections, which today don't present too much of a problem because nonallergic insulin and sterile needles are readily available. Why can't insulin be taken by mouth? Because the digestive juices would digest insulin to amino acids.

Medication is available for diabetes type 2, but a healthy diet and exercise can also help. Some intake of sugar with meals is fine, but these must be substituted for other carbohydrates,

FIGURE 34.9
Testing the blood sugar level.

not added on. Foods rich in complex carbohydrates, including dietary fiber, are preferred over easily digested foods such as many junk foods. Building up the muscles and using them regularly by, say, walking, riding a bike, or dancing uses up glucose and improves insulin efficiency. Walking, in particular, has been shown to increase the number of glucose transporters. All diabetics must work closely with a physician to tailor the correct regimen of medications, diet, and exercise for them.

It is well worth the effort to control diabetes because both types are associated with blindness, cardiovascular disease, and kidney disease. Nerve deterioration can lead to the inability to feel pain, particularly in the hands and feet. As a result, treatment of an infection may be delayed to the point that limb amputation is required.

Diabetes is definitely a disorder to avoid if at all possible. The very best course of action is to adopt a healthy lifestyle, the earlier in life the better. This is the way you can help your body maintain homeostasis, the relative constancy of the internal environment—in this case, a normal blood glucose level.

In Section 34.10, we discuss the pineal gland.

34.9 Check Your Progress Diabetes type 2 is the most common type of diabetes. Explain.

34.10 The pineal gland is involved in biorhythms

The **pineal gland**, which is located in the brain (see Fig. 34.3), produces the hormone **melatonin**, primarily at night. Melatonin is involved in our daily sleep-wake cycle; normally we grow sleepy at night when melatonin levels increase and awaken once daylight returns and melatonin levels are low. Also, because light suppresses melatonin secretion, the duration of secretion is longer in the winter than in the summer (**Fig. 34.10**). Daily 24-hour cycles such as this are called **circadian rhythms**, and circadian rhythms are controlled by an internal timing mechanism called a biological clock.

Animal research indicates that melatonin also regulates sexual development. In humans, researchers have noted that children whose pineal gland has been destroyed due to a brain tumor experience early puberty.

The important functions of the thyroid gland are examined in Section 34.11.

34.10 Check Your Progress Would you expect animals, including humans, to sleep more in the winter or in the summer? Explain.

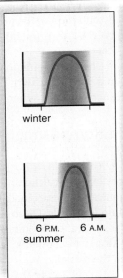

winter

6 P.M.   6 A.M.
summer

FIGURE 34.10 Melatonin production.

34.11 The thyroid regulates development and increases the metabolic rate

As with the adrenal cortex and the gonads, the thyroid gland is involved in a three-tier relationship with the hypothalamus and anterior pituitary (see Section 34.5). The **thyroid gland** is a large gland located in the neck (see Fig. 34.3). It consists of two distinct lobes connected by a slender isthmus. Each lobe has a large number of follicles filled with the thyroid hormones triiodothyronine (T_3), containing three iodine atoms, and **thyroxine** (T_4), containing four iodine atoms. These hormones increase the metabolic rate. They do not have a target organ; instead, they stimulate all the cells of the body to metabolize at a faster rate.

FIGURE 34.11C Exophthalmic goiter.

Hypo- and Hyperthyroidism

A low level of blood T_3 and T_4 results in hypothyroidism, indicated by the slowing of metabolic processes. If the thyroid fails to develop properly, a condition called congenital hypothyroidism (cretinism) results (**Fig. 34.11A**). Children with this condition are short and stocky and have had extreme hypothyroidism since infancy or childhood. Thyroid hormone therapy can initiate growth, but unless treatment is begun within the first two months of life, mental retardation results.

Hypothyroidism in an adult can result if the thyroid is unable to produce its hormones due to lack of iodine in the diet. Under these circumstances, the anterior pituitary increases its stimulation of the thyroid, which enlarges, resulting in a **simple goiter** (**Fig. 34.11B**). The introduction and use of iodized salt solved this problem in the United States.

A type of hypothyroidism called **myxedema** can occur without the appearance of goiter when

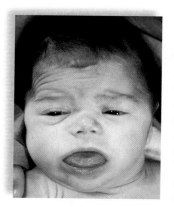

FIGURE 34.11A
Congenital hypothyroidism.

FIGURE 34.11B
Simple goiter.

either TRH from the hypothalamus or TSH from the anterior pituitary is lacking or when deterioration of the thyroid is present. Myxedema, which means thickness and puffiness of the skin, is accompanied by lethargy, weight gain, loss of hair, slower pulse rate, and lowered body temperature. Adequate doses of thyroid hormones restore normal function and appearance.

In the case of hyperthyroidism (oversecretion of thyroid hormone), also called Graves disease, the thyroid gland is overactive, and an **exophthalmic goiter** forms (**Fig. 34.11C**). The eyes protrude (exophthalmos) because of edema in the eye socket tissues and swelling of the muscles that move the eyes. The patient usually becomes hyperactive, nervous, and irritable, and suffers from insomnia. In addition, the individual may sweat and be sensitive to heat. Removal or destruction of a portion of the thyroid, by means of radioactive iodine, is sometimes effective in curing the condition. Hyperthyroidism can also be caused by a thyroid tumor, which is usually detected as a lump during physical examination. Treatment consists of surgery in combination with administration of radioactive iodine. The prognosis for most patients is excellent.

The thyroid, along with the parathyroids, regulates the blood calcium level, as discussed in Section 34.12.

> **34.11 Check Your Progress** Propose a hypothetical pathway that would explain how a pheromone could cause the release of thyroid hormones.

34.12 The thyroid and the parathyroids regulate the blood calcium level

The blood calcium level is regulated in part by **calcitonin**, a hormone secreted by the thyroid gland. The primary effect of calcitonin is to bring about the deposit of calcium in the bones when the blood calcium level rises. Thereafter, the blood calcium level falls. Osteoporosis is a bone disorder in which bones are weakened to the point that they fracture easily. Calcitonin as a medication to treat osteoporosis is not as effective as other medications on the market.

The **parathyroid glands** are embedded adjacent to the thyroid gland (see Fig. 34.3). Parathyroid hormone causes the blood phosphate (HPO_4^{2+}) level to decrease and the blood calcium

level to increase. PTH promotes the release of calcium from the bones when the blood calcium level lowers. PTH also promotes the reabsorption of calcium by the kidneys, where it activates vitamin D. Vitamin D, in turn, stimulates the absorption of calcium from the intestine. These effects cause the blood calcium level to rise so that the parathyroid glands no longer secrete PTH.

> **34.12 Check Your Progress** Calcitonin and PTH are antagonistic hormones. Explain.

The nervous system and the endocrine system are structurally and functionally related. The hypothalamus, a portion of the brain, controls the pituitary, an endocrine gland. The hypothalamus even produces the hormones that are released by the posterior pituitary. It also stimulates the release of hormones by the adrenal medulla. Neurosecretory cells in the hypothalamus also produce releasing hormones that control the activity of the anterior pituitary.

The nervous system is well known for bringing about an immediate response to environmental stimuli, as in the fight-or-flight reaction. Most often, the chemical signals released by the nervous system, called neurotransmitters, help maintain homeostasis. Heart rate, breathing rate, and blood pressure are all regulated to stay relatively constant. Hormones released by endocrine glands also help maintain homeostasis especially by keeping the levels of calcium, sodium, glucose, and other blood constituents within normal limits.

The endocrine system is slower acting than the nervous system and often regulates processes that occur over days or even months. Hormones secreted into the bloodstream control whole-body processes, such as growth and reproduction. In this chapter, we mentioned the effects of growth hormone, and in Chapter 35, we explore the activity of the hormones that influence reproduction.

The Chapter in Review

Summary

Pheromones Among Us

- Pheromones sent between species members modify behavior and physiological processes.
- Pheromones influence the behavior of social insects and possibly that of other animals, including humans.
- Humans have a vomeronasal (VNO) organ for receiving pheromones, as do mice. Neurons from this organ go to the hypothalamus.

The Endocrine System Utilizes Chemical Signals

34.1 The endocrine and nervous systems work together

- The nervous system acts quickly because it utilizes axons that release neurotransmitters for stimulation of a target organ (e.g., muscle).
- The endocrine system acts more slowly but is longer lasting; its glands release hormones that travel in the bloodstream to target organs.
- Both systems utilize negative feedback mechanisms.

34.2 Hormones affect cellular metabolism

- Peptide hormones are the first messengers received at a receptor in the plasma membrane. The second messenger changes the metabolism of the cell.
- Steroids enter the cell or nucleus, where they bind to a receptor. The complex stimulates a gene, resulting in a protein that carries out a response to the hormonal signal.

34.3 The vertebrate endocrine system includes diverse hormones

- Table 34.3 lists the principal glands and hormones.

- Hormones affect blood osmolarity, blood glucose and calcium levels, metabolism, and the maturation and function of reproductive organs.

The Hypothalamus and Pituitary Are Central to the Endocrine System

34.4 The hypothalamus is a part of the nervous and endocrine systems

- The hypothalamus produces the hormones released by the posterior pituitary.
  - Antidiuretic hormone (ADH) causes the reabsorption of water by the kidneys.
  - Oxytocin stimulates the uterus to contract during childbirth and promotes milk letdown while nursing.
- The hypothalamus produces hypothalamic-releasing and -inhibiting hormones that control the anterior pituitary.

34.5 The anterior pituitary produces nontropic and tropic hormones

- Nontropic hormones include prolactin (PRL), growth hormone (GH), and melanocyte-stimulating hormone (MSH).
- Tropic hormones are thyroid-stimulating hormone (TSH), adrenocorticotropic hormone (ACTH), and the gonadotropic hormones (FSH and LH).
- The hypothalamus stimulates the anterior pituitary to release hormones, which stimulate either the thyroid, adrenal cortex, or gonads to release hormones. These hormones feed back to control the hypothalamus and anterior pituitary.

Hormones Regulate Metabolism and Homeostasis

34.6 The adrenal glands respond to stress

- The adrenal medulla releases epinephrine and norepinephrine, which have short-term effects.

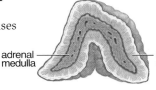

adrenal medulla — adrenal cortex

- The adrenal cortex secretes hormones that provide long-term response to stress.
  - Glucocorticoids regulate carbohydrate, protein, and fat metabolism, leading to increased blood glucose levels.
  - Mineralocorticoids help regulate sodium and potassium excretion.
- The adrenal cortex also releases a small amount of both male and female sex hormones in both sexes.

34.7 Glucocorticoid therapy can lead to Cushing syndrome
- Cushing syndrome is characterized by a tendency toward diabetes mellitus, subcutaneous fat accumulation in the trunk, and a moon-shaped face.

34.8 The pancreas regulates the blood sugar level
- Pancreatic islet cells release insulin after eating, normally causing uptake and storage of glucose, and thereby lowering blood glucose.
- Islet cells release glucagon between eating, causing the breakdown of glycogen to glucose, and thereby increasing blood glucose.

34.9 Diabetes is becoming a very common ailment
- In diabetes type 1, the pancreas fails to produce insulin, usually because pancreatic islet cells have died off following a viral infection.
- In diabetes type 2, the pancreas produces at least some insulin, but cells are unable to respond by taking up glucose.

34.10 The pineal gland is involved in biorhythms
- The hormone melatonin is involved in the sleep/wake cycle and may also regulate sexual development.

34.11 The thyroid regulates development and increases the metabolic rate
- The thyroid secretes triiodothyronine (T_3) and thyroxine (T_4), which increase the metabolic rate of all cells.
- Hypothyroidism is the cause of cretinism, simple goiter, and myxedema.
- Hyperthyroidism can lead to exophthalmic goiter, characterized by an enlarged thyroid and protruding eyes.

34.12 The thyroid and the parathyroids regulate the blood calcium level
- Calcitonin, secreted by the thyroid gland, causes blood calcium levels to decrease.
- Parathyroid glands cause the blood phosphate level to decrease and the blood calcium level to increase.

Testing Yourself

The Endocrine System Utilizes Chemical Signals

1. The nervous system and the endocrine system differ in
   a. how their signals are delivered in the body.
   b. the speed of the body's response to their signals.
   c. the use of negative feedback in controlling their signaling.
   d. Both a and b are correct.
   e. a, b, and c are all correct.
2. Which hormones typically cross the plasma membrane?
   a. peptide hormones   c. Both a and b are correct.
   b. steroid hormones   d. Neither a nor b is correct.

3. Peptide hormones
   a. are received by a receptor located in the plasma membrane.
   b. are received by a receptor located in the cytoplasm.
   c. bring about the transcription of DNA.
   d. Both b and c are correct.
4. Steroid hormones are secreted by
   a. the adrenal cortex.   d. Both a and b are correct.
   b. the gonads.   e. Both b and c are correct.
   c. the thyroid.
5. Steroid hormones
   a. bind to a receptor located in the plasma membrane.
   b. cause the production of cAMP.
   c. activate the protein kinase.
   d. stimulate the production of mRNA.

In questions 6–10, match the hormones to the correct gland in the key.

KEY:
   a. pancreas   d. thyroid
   b. anterior pituitary   e. adrenal medulla
   c. posterior pituitary   f. adrenal cortex

6. Cortisol
7. Growth hormone (GH)
8. Oxytocin storage
9. Insulin
10. Epinephrine
11. **THINKING CONCEPTUALLY** Caffeine inhibits the breakdown of cAMP in the cell. Referring to Figure 34.2A, how would this influence a stress response brought about by epinephrine?

The Hypothalamus and Pituitary Are Central to the Endocrine System

12. Which of the following statements about the pituitary gland is incorrect?
    a. The pituitary lies inferior to the hypothalamus.
    b. Growth hormone and prolactin are secreted by the anterior pituitary.
    c. The anterior pituitary and posterior pituitary communicate with each other.
    d. Axons run between the hypothalamus and the posterior pituitary.
13. _____ is released through positive feedback and causes _____.
    a. Insulin, stomach contractions
    b. Oxytocin, stomach contractions
    c. Oxytocin, uterine contractions
    d. None of these are correct.
14. Growth hormone is produced by the
    a. posterior adrenal gland.
    b. posterior pituitary.
    c. anterior pituitary.
    d. kidneys.
    e. None of these are correct.
15. The anterior pituitary controls the secretion(s) of
    a. both the adrenal medulla and the adrenal cortex.
    b. both the thyroid gland and the adrenal cortex.
    c. both the ovaries and the testes.
    d. Both b and c are correct.

16. Tropic hormones are hormones that affect other endocrine tissues. Which of the following would be considered a tropic hormone?
    a. calcitonin
    b. oxytocin
    c. glucagon
    d. melatonin
    e. None of these are correct.

Hormones Regulate Metabolism and Homeostasis

17. Both the adrenal medulla and the adrenal cortex are
    a. endocrine glands.
    b. in the same organ.
    c. involved in our response to stress.
    d. All of these are correct.
18. The blood cortisol level does not control the secretion of
    a. hypothalamic-releasing hormone from the hypothalamus.
    b. adrenocorticotropic hormone (ACTH) from the anterior pituitary.
    c. mineralocorticoids from the adrenal cortex.
    d. All of these are correct.
19. Lack of aldosterone will cause a blood imbalance of
    a. sodium.
    b. potassium.
    c. water.
    d. All of these are correct.
    e. None of these are correct.
20. Glucagon causes
    a. use of fat for energy.
    b. glycogen to be converted to glucose.
    c. use of amino acids to form fats.
    d. Both a and b are correct.
    e. None of these are correct.
21. Diabetes is associated with
    a. too much insulin in the blood.
    b. a blood glucose level that is too high.
    c. blood that is too dilute.
    e. All of these are correct.
22. The difference between diabetes type 1 and type 2 is that
    a. for diabetes type 2, insulin is produced but it is ineffective; type 1 results from lack of insulin production.
    b. treatment for type 2 involves insulin injections, while type 1 can be controlled, usually by diet.
    c. only type 1 can result in complications such as kidney disease, reduced circulation, or stroke.
    d. type 1 can be a result of lifestyle, and type 2 is thought to be caused by a virus or other agent.
23. Long-term complications of diabetes include
    a. blindness.
    b. kidney disease.
    c. circulatory disorders.
    d. All of these are correct.
    e. None of these are correct.
24. Which hormone and condition are mismatched?
    a. thyroxine—goiter
    b. parathyroid hormone—myxedema
    c. cortisol—Cushing syndrome
    d. insulin—diabetes
25. PTH causes the blood level of calcium to _____, and calcitonin causes it to _____.
    a. increase, not change
    b. increase, decrease
    c. decrease, also decrease
    d. decrease, increase
    e. not change, increase
26. **THINKING CONCEPTUALLY** Negative feedback controls the secretion of calcitonin by the thyroid gland. Explain how it would work. (Pattern your answer after Figure 34.8.)

Understanding the Terms

adrenal cortex   672
adrenal gland   672
adrenal medulla   672
adrenocorticotropic hormone (ACTH)   671
aldosterone   673
anterior pituitary   670
antidiuretic hormone (ADH)   671
calcitonin   676
circadian rhythm   675
cortisol   672
Cushing syndrome   673
cyclic adenosine monophosphate (cAMP)   667
endocrine system   666
epinephrine   672
exophthalmic goiter   676
first messenger   667
glucagon   674
glucocorticoid   672
gonadotropic hormone   671
growth hormone (GH)   671
hormone   666
hypothalamic-releasing hormone   671
insulin   674
melanocyte-stimulating hormone (MSH)   671
melatonin   675
mineralocorticoid   672
myxedema   676
negative feedback   666
neurosecretory cell   670
nontropic hormone   671
norepinephrine   672
oxytocin   671
pancreas   674
pancreatic islet   674
parathyroid gland   676
peptide hormone   667
pineal gland   675
pituitary gland   670
posterior pituitary 670
positive feedback   671
prolactin (PRL)   671
renin   673
second messenger   667
simple goiter   676
steroid hormone   667
thyroid gland   676
thyroid-stimulating hormone (TSH)   671
thyroxine (T_4)   676

Match the terms to these definitions:
a. _____ Organ in the neck; secretes several important hormones, including thyroxine and calcitonin.
b. _____ Organ that produces melatonin.
c. _____ Type of hormone that binds to a plasma membrane receptor; results in activation of enzyme cascade.
d. _____ Glucocorticoid secreted by the adrenal cortex that responds to stress on a long-term basis; reduces inflammation and promotes protein and fat metabolism.

Thinking Scientifically

1. Formulate a hypothesis to test the affect of calcitonin on osteoblasts (bone-forming cells) when the blood calcium level is high. (See Section 34.12.)

2. How would you test the hypothesis that melatonin levels affect when we fall asleep and when we wake up?

ARIS *Visit www.mhhe.com/maderconcepts for practice quizzes, animations, videos, and activities designed to help you master the material in this chapter.*

35

Reproduction and Development

LEARNING OUTCOMES

After studying this chapter, you should be able to accomplish the following outcomes.

How to Do It on Land

1 Describe the ways in which the human reproductive system is adapted to a land environment.

Reproduction in Animals Is Varied

2 Contrast asexual and sexual reproduction, and explain the advantages of each.
3 Compare the strategy of reproduction in water with that on land.

Humans Are Adapted to Reproducing on Land

4 Describe the organs of the male and female reproductive systems, emphasizing adaptations to the land environment
5 Describe significant sexually transmitted diseases.
6 Describe the ovarian cycle, including the manner in which it drives the three phases of the uterine cycle.
7 Describe methods of birth control and reproductive technologies to help the infertile.

Vertebrates Have Similar Early Developmental Stages and Processes

8 Describe the stages of fertilization and early development in a lancelet.
9 Relate the process of gastrulation to the establishment of the embryonic germ layers and later organ development.
10 Discuss the role of maternal determinants and induction in development.

Human Development Is Divided into Embryonic Development and Fetal Development

11 Contrast the function of extraembryonic membranes in chicks and in humans.
12 Describe the events of human development in terms of the embryonic and fetal periods of development, stressing the importance of the placenta.
13 Divide the process of birth into three stages, stating the significance of each stage.

Fishes and most amphibians, such as frogs, reproduce in the water, and they need no special reproductive adaptations to keep the gametes and the embryo from drying out. During so-called amplexus, a male frog clings to the body of a female; other than that, there is no contact between male and female. The male applies pressure to encourage the female to release her eggs and then covers the eggs with sperm as they emerge. The sticky eggs adhere to one another, and masses of eggs float freely away in the water or become fastened to a plant or an object under water, where they develop into swimming tadpoles.

How do reptiles do it? They are members of the first vertebrate class to reproduce on land. In American alligators, the male mounts the female and inserts a penis into her reproductive tract before releasing sperm. The job of a penis, essentially a tube that can stiffen, is to deliver sperm to the female in a way that pro-

Frogs mating

tects the sperm from drying out. The eggs of a female are protected deep inside her body. The female alligator lays her fertilized eggs, but they are protected by a shell, which keeps them from drying out. Inside the shell are a series of membranes that perform all sorts of services for the embryo, such as exchanging gases, collecting wastes, and providing a watery environment. That's right, alligators, and indeed all land vertebrates, still develop surrounded by water. In reptiles, the water is inside the shell. The evolution of the shelled egg was absolutely essential in order for vertebrates to reproduce on land. When baby alligators hatch, they are tiny beings only about 15–20 cm in length. But they are fully formed and able to survive on land, even though the mother may gently carry them inside her mouth to the water.

How do mammals do it? Most mammals, including humans,

Sea turtles mating

Sea turtle hatching

copulate just like the alligators do. The male mounts the female and inserts a penis into her reproductive tract to release sperm. Copulation, even in humans, is simply a way to protect the sperm from drying out. No need to worry about the eggs of the female—they never leave her body.

One big difference between mammals and reptiles is that the mammalian embryo develops inside the uterus of the female, and so mammals do not lay eggs. But the embryo is still surrounded by water and by the same membranes. The outer membrane develops into the placenta, the point of contact between mother and offspring through which nutrient-for-waste exchange occurs. When a human is born, the "water breaks," and for the first time, the baby has to breathe air. A slap on the behind can make sure breathing begins.

Whales are mammals, and all sorts of evidence points to the conclusion that whales evolved from land animals. Even though whales are marine, the males do have a penis; they do insert it into the reproductive tract of the female; they do reproduce internally; and they are born fully able to breathe air. As soon as a whale is born, it can swim, breathe, produce clicking sounds, and nurse (because all mammals have mammary glands that produce milk for the newborn).

In this chapter, we discuss principles of animal reproduction, and this is followed by a discussion of animal development.

Deer mating

Kangaroos mating

Whale penises

A brief survey of how animals reproduce is sufficient to realize that the process is quite varied. However, contrasting asexual with sexual reproduction and contrasting development in water and on land bring out general differences.

35.1 Both asexual and sexual reproduction occur among animals

Although the majority of animals reproduce sexually, a few groups of animals are also capable of asexual reproduction. In sexual reproduction, sex cells, or gametes, produced by the parents unite to form a genetically unique individual. In asexual reproduction, a single parent gives rise to offspring that are identical to the parent, unless mutations have occurred. The adaptive advantage of asexual reproduction is that organisms can reproduce rapidly and colonize a favorable environment quickly.

Asexual Reproduction Several phyla of invertebrates, including free-living sponges, cnidarians, flatworms, annelids, and echinoderms, can reproduce asexually. Sponges produce asexual gemmules that develop into new individuals. Cnidarians, such as hydras, can reproduce asexually by budding (**Fig. 35.1A**). A new individual arises as an outgrowth (bud) of the parent. *Obelia* is dimorphic and has an asexual colonial stage and a sexually reproducing medusa stage.

You can horizontally cut a planarian into as many as ten pieces in the laboratory and get ten new planarians. The parasitic flatworms reproduce asexually during certain stages of their complicated life cycles. Several annelids, such as earthworms and sandworms, also have the ability to regenerate from fragments. Fragmentation, followed by regeneration, is seen among sponges and echinoderms as well. If a sea star is chopped up, it has the potential to regenerate into several new individuals.

Several types of flatworms, roundworms, crustaceans, annelids, insects, fishes, lizards, and even some turkeys have the ability to reproduce parthenogenetically. **Parthenogenesis** is a modification of sexual reproduction in which an unfertilized egg develops into a complete individual. In honeybees, the queen bee stores sperm and can fertilize eggs or allow eggs to pass unfertilized as she lays them. The fertilized eggs become diploid females called workers, and the unfertilized eggs become haploid males called drones.

Sexual Reproduction Usually during sexual reproduction, the egg of one parent is fertilized by the sperm of another. The majority of animals are *dio*ecious, which means having separate sexes. *Mono*ecious, or hermaphroditic, organisms have both male and female sex organs in the same body. Some hermaphroditic organisms, such as tapeworms, practice self-fertilization, but the majority, such as earthworms, practice cross-fertilization. Sequential hermaphroditism, or sex reversal, also occurs. In coral reef fishes called wrasses, a male has a harem of several females. If the male dies, the largest female becomes a male.

Animals usually produce gametes in specialized organs called **gonads**. Sponges are an exception to this rule because the collar cells lining the central cavity of a sponge give rise to sperm and eggs. Hydras and other cnidarians produce only temporary gonads in the fall, when sexual reproduction occurs. Animals in other phyla have permanent reproductive organs. The gonads are **testes**, which produce sperm, and **ovaries**, which produce eggs. Eggs or sperm are derived from germ cells, which become specialized for this purpose during early development. Other cells in a gonad support and nourish the developing gametes or produce hormones necessary to the reproductive process. The reproductive system also consists of a number of accessory structures, such as ducts and storage areas, that aid in bringing the gametes together.

Many aquatic animals, such as the fish in **Figure 35.1B**, practice external fertilization. Palolo worms release their eggs

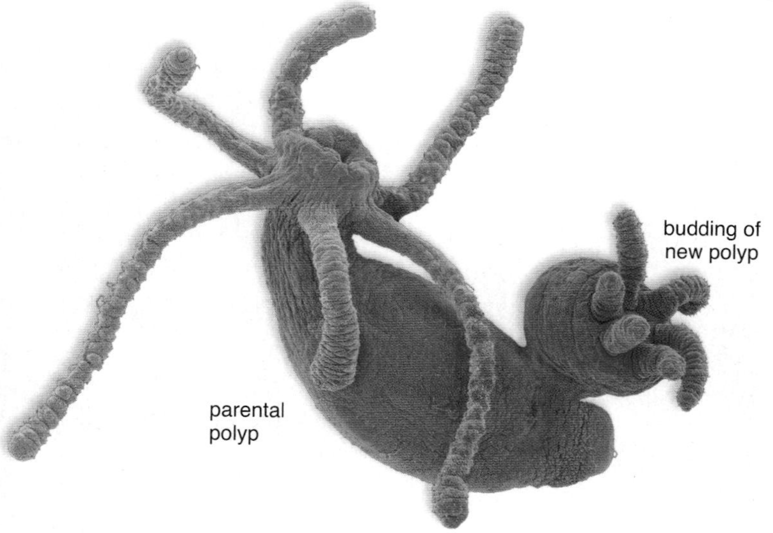

budding of
new polyp

parental
polyp

FIGURE 35.1A Hydras can reproduce asexually.

FIGURE 35.1B Orange-fin anemonefish, *Amphiprion chrysopterus,* tend an egg mass.

in the water only when the moon moves closer to the Earth, and the tides become somewhat higher than usual. Most likely, they have a biological clock that can sense the passage of time because their reproductive behavior is synchronized. Hundreds of thousands of palolo worms rise to the surface of the sea and release their eggs during a two- to four-hour period on two or three successive days of the year.

Copulation is sexual union to facilitate the reception of sperm by a female. In terrestrial vertebrates, males typically have a penis for depositing sperm into the vagina of females. But not so in birds, which lack a penis and vagina. A male transfers sperm to a female after placing his cloacal opening against hers. In damselflies, the male grasps the female at the back of her head with his tail, and she curls her tail forward to receive sperm previously deposited in a pouch by the male (**Fig. 35.1C**). Copulation doesn't occur.

In Section 35.2, we contrast development in water with that on land.

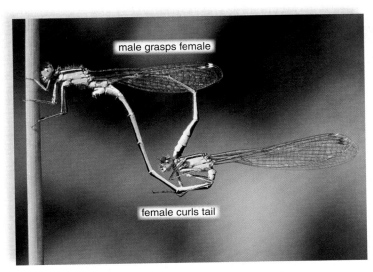

FIGURE 35.1C Azure damselflies mating on land.

35.1 *Check Your Progress* **Do all terrestrial males have a penis?**

35.2 Development in water and on land occurs among animals

Most aquatic animals deposit their eggs in the water, where they undergo development to become a **larva,** an immature stage. Since the larva has a different lifestyle, it is able to use a different food source than the adult. In seastars, the bilaterally symmetrical larva undergoes metamorphosis to become a radially symmetrical juvenile. Crayfish, on the other hand, do not have a larval stage; the egg hatches into a tiny juvenile with the same form as the adult. Some aquatic animals retain their eggs in various ways and release young able to fend for themselves. These animals are *ovoviviparous.* For example, oysters, which are molluscs, retain their eggs in the mantle cavity, and male sea horses, which are vertebrates, have a special brood pouch in which the eggs develop.

Garter snakes retain their eggs inside the abdomen until they hatch. Usually, reptiles (snakes, turtles, and crocodiles) lay a leathery-shelled egg containing **extraembryonic membranes** to serve the needs of the embryo and prevent drying out. One membrane surrounds an abundant supply of **yolk,** which is a nutrient-rich material. The shelled egg is a significant adaptation to the terrestrial environment. Birds lay and care for hard- shelled eggs with extraembryonic membranes. The newly hatched birds usually have to be fed before they are able to fly away and seek food for themselves (**Fig. 35.2**). Complex hormones and neural regulation are involved in the reproductive behavior of parental birds.

Among mammals, the duckbill platypus and spiny anteater lay shelled eggs. Marsupials and placental mammals do not lay eggs. In marsupials, the yolk sac membrane functions briefly to supply the unborn with nutrients acquired internally from the mother. Immature young finish their development within a pouch where they are nourished on milk. The placental mammals are termed *viviparous,* because they do not lay eggs and

FIGURE 35.2 Parenting in birds.

development occurs inside the female's body until offspring can live independently. Their **placenta** is a complex structure derived, in part, from the chorion, another of the reptilian extraembryonic membranes. The evolution of this type of placenta allows the developing young to internally exchange materials with the mother until the offspring can function on their own. Viviparity represents the ultimate in caring for the unborn, and in placental mammals, the mammary glands of the mother continues to supply the nutrient needs of her offspring after birth.

In the next part of the chapter, we study the reproductive system in human males and then females.

35.2 *Check Your Progress* **Do any other animals, aside from placental mammals, produce fully-developed young?**

In this part of the chapter, a description of the male reproductive system is followed by a description of the female reproductive system. In females, an ovarian cycle drives the uterine cycle so that menstruation occurs once a month. The control of sexually transmitted diseases and pregnancy is also discussed.

35.3 Testes are male gonads

In the human male, the paired **testes**, which produce sperm, are suspended within the scrotum (**Fig. 35.3**). The testes begin their development inside the abdominal cavity, but they descend into the scrotum as development proceeds. Normal sperm production is inhibited at body temperature; a slightly cooler temperature is required.

Sperm produced by the testes mature within the **epididymis**, a coiled tubule lying just outside each testis. Maturation seems to be required in order for the sperm to swim to the egg. Once the sperm have matured, they are propelled into the **vas deferens** by muscular contractions. Sperm are stored in both the epididymis and the vas deferens. When a male becomes sexually aroused, sperm enter first the ejaculatory duct and then the urethra, part of which is within the penis. (The urethra in males is part of both the urinary and the reproductive systems.)

The **penis** is a cylindrical organ that usually hangs in front of the scrotum. The penis has an enlarged tip normally covered by a layer of skin called the **foreskin**. Circumcision is the surgical removal of the foreskin. Three cylindrical columns of spongy, erectile tissue containing distensible blood spaces extend through the shaft of the penis. During sexual arousal, nervous reflexes cause an increase in arterial blood flow to the penis. This increased blood flow fills the blood spaces in the erectile tissue, and the penis, which is normally limp (flaccid), stiffens and increases in size. These changes are called an **erection**. Drugs have been developed to treat erectile dysfunction by increasing blood flow to the penis so that when a man is sexually excited, he can get and keep an erection.

Semen (seminal fluid) is a thick, whitish fluid that contains sperm and secretions from three types of glands. The **seminal vesicles** lie at the base of the bladder. As sperm pass from the vas deferens into the ejaculatory duct, these vesicles secrete a thick,

viscous fluid containing nutrients for possible use by the sperm. Just below the bladder is the **prostate gland**, which secretes a milky alkaline fluid believed to activate or increase the motility of the sperm. In older men, the prostate gland may become enlarged, thereby constricting the urethra and making urination difficult. Also, prostate cancer is the most common form of cancer in men. Slightly below the prostate gland, on either side of the urethra, is a pair of small glands called **bulbourethral glands**, which secrete mucus that has a lubricating effect.

Male Orgasm If sexual arousal reaches its peak, ejaculation follows an erection. The first phase of ejaculation is called emission. During *emission*, the spinal cord sends nerve impulses via appropriate nerve fibers to the epididymis and vas deferens. Their subsequent motility causes sperm to enter the ejaculatory duct, whereupon the seminal vesicles, prostate gland, and bulbourethral glands release their secretions.

During the second phase of ejaculation, called *expulsion*, rhythmic contractions of muscles at the base of the penis and within the urethral wall expel semen in spurts from the opening of the urethra. These contractions are a part of male **orgasm**, the physiological and psychological sensations that occur at the climax of sexual stimulation. Following ejaculation, a male typically experiences a time, called the refractory period, during which stimulation does not bring about an erection.

Section 35.4 describes the structure and function of the testes in more detail.

> **35.3 Check Your Progress** Sperm that are ejaculated onto, say, clothing die. Explain.

FIGURE 35.3
Male reproductive system.

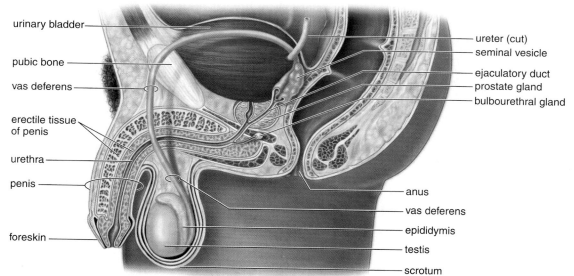

- urinary bladder
- pubic bone
- vas deferens
- erectile tissue of penis
- urethra
- penis
- foreskin
- ureter (cut)
- seminal vesicle
- ejaculatory duct
- prostate gland
- bulbourethral gland
- anus
- vas deferens
- epididymis
- testis
- scrotum

A testis is composed of compartments called lobules, each of which contains one to three tightly coiled **seminiferous tubules** (**Fig. 35.4**). A seminiferous tubule is packed with cells undergoing **spermatogenesis**, the production of sperm. Newly formed cells move away from the outer wall, increase in size, and undergo meiosis to become spermatids, which contain the haploid number of chromosomes. Spermatids then differentiate into sperm. Also present are **Sertoli cells**, which support, nourish, and regulate the production of sperm. A sperm has three distinct parts: a tail, a middle piece, and a head. The tail is a flagellum that allows a sperm to swim toward the egg, and the middle piece contains energy-producing mitochondria. The head contains a nucleus and is capped by a membrane-bounded acrosome. The acrosome contains enzymes that assist in allowing a sperm to enter an egg during fertilization. The ejaculated semen of a normal human male contains 40 million sperm per milliliter, ensuring an adequate number for fertilization to take place. Fewer than 100 sperm ever reach the vicinity of the egg, however, and only one sperm normally enters an egg.

Control of Testes Function   The hypothalamus has ultimate control of the testes' sexual function because it secretes a hormone called gonadotropin-releasing hormone, or GnRH, that stimulates the anterior pituitary to produce the gonadotropic hormones: In males, **follicle-stimulating hormone (FSH)** promotes spermatogenesis in the seminiferous tubules. **Lutein-**

izing hormone (**LH**) in males was formerly called interstitial cell–stimulating hormone (ICSH) because it controls the production of testosterone by the interstitial cells, which are scattered in the spaces between the seminiferous tubules.

Testosterone, the main sex hormone in males, is essential for the normal development and functioning of the sexual organs. Testosterone is also necessary for the maturation of sperm. In addition, testosterone brings about and maintains the male **secondary sex characteristics** that develop at the time of **puberty**, the time of life when sexual maturity is attained.

Testosterone causes males to develop noticeable hair on the face, chest, and occasionally other regions of the body, such as the back. Testosterone also leads to the receding hairline and pattern baldness that occur in males. Testosterone is responsible for the greater muscular development in males. Males are generally taller than females and have broader shoulders and longer legs relative to trunk length. The deeper voice of males compared to females is due to males having a larger larynx with longer vocal cords. Because the so-called Adam's apple is a part of the larynx, it is usually more prominent in males than in females.

We begin our study of the female reproductive system in Section 35.5.

35.4 *Check Your Progress*   **How do the sperm from the seminiferous tubules reach the penis?**

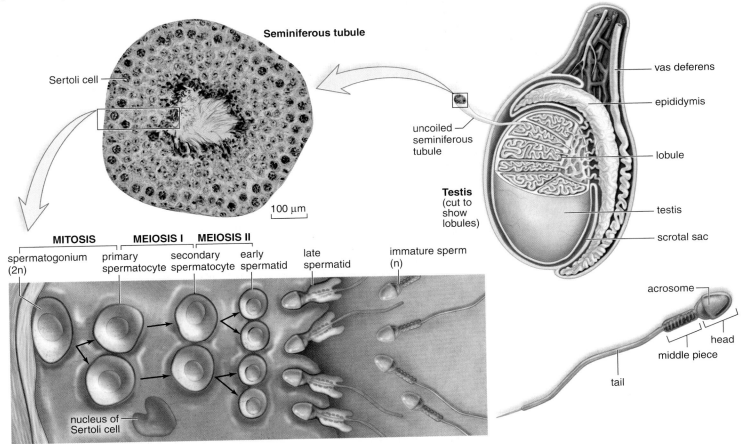

Seminiferous tubule

Sertoli cell

100 µm

vas deferens

epididymis

uncoiled seminiferous tubule

lobule

Testis (cut to show lobules)

testis

scrotal sac

MITOSIS	MEIOSIS I	MEIOSIS II			

spermatogonium (2n)   primary spermatocyte   secondary spermatocyte   early spermatid   late spermatid   immature sperm (n)

nucleus of Sertoli cell

Spermatogenesis

acrosome

head

middle piece

tail

FIGURE 35.4 Seminiferous tubules, where sperm are produced via the process of spermatogenesis.

In the human female, the **ovaries**, which produce one **oocyte** each month, lie in shallow depressions, one on each side of the upper pelvic cavity (**Fig. 35.5A**). The **oviducts**, also called uterine or fallopian tubes, extend from the ovaries to the uterus; however, the oviducts are not attached to the ovaries. Instead, they have fingerlike projections called **fimbriae** (sing., fimbria) that lie over the ovaries. When an oocyte bursts from an ovary during ovulation, it usually is swept into an oviduct by the combined action of the fimbriae and the beating of cilia that line the oviducts. Fertilization, if it occurs, normally takes place in an oviduct, and the developing embryo is propelled slowly by ciliary movement and tubular muscle contraction to the uterus. The **uterus** is a thick-walled muscular organ about the size and shape of an inverted pear. The narrow end of the uterus is called the **cervix**. The embryo completes its development after embedding itself in the uterine lining, called the **endometrium**. If, by chance, the embryo should embed itself in another location, such as an oviduct, a so-called ectopic pregnancy results.

A small opening in the cervix leads to the vaginal canal. The **vagina** is a tube at a 45-degree angle to the small of the back. The mucosal lining of the vagina lies in folds, and the vagina can distend. This is especially important when the vagina serves as the birth canal, and it can also facilitate sexual intercourse, when the vagina receives the penis. Several types of bacteria normally reside in the vagina and create an acidic environment. This environment protects against the possible growth of pathogenic bacteria, but sperm survive better in the basic environment provided by semen.

External Genital Organs and Orgasm In the female, the external genital organs are known collectively as the **vulva** (**Fig. 35.5B**). The mons pubis and two sets of skin folds called the

FIGURE 35.5B Vulva.

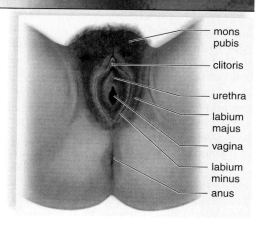

mons pubis
clitoris
urethra
labium majus
vagina
labium minus
anus

labia minora (sing., labium minus) and labia majora (sing., labium majus) are on either side of the urethral and vaginal openings. Beneath the labia majora, pea-sized greater vestibular glands (Bartholin glands) open on either side of the vagina. They keep the vulva moist and lubricated during intercourse.

At the juncture of the labia minora is the **clitoris**, which is homologous to the penis in males. The clitoris has a shaft of erectile tissue and is capped by a pea-shaped glans. The many sensory receptors in the clitoris allow it to function as a sexually sensitive organ. The clitoris has twice as many nerve endings as the penis. Orgasm in the female is a release of neuromuscular tension in the muscles of the genital area, vagina, and uterus.

Section 35.6 describes the structure and function of the ovaries in more detail.

35.5 Check Your Progress How does the design of the female reproductive tract facilitate reproduction on land?

FIGURE 35.5A
Female reproductive system.

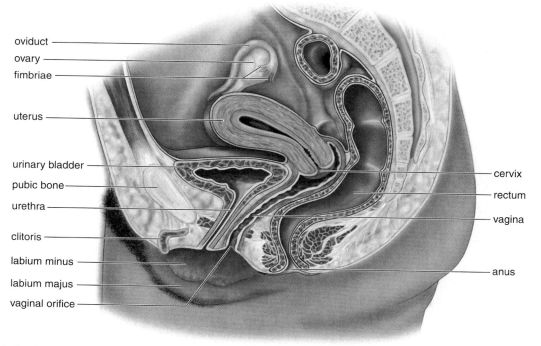

oviduct
ovary
fimbriae
uterus
urinary bladder
pubic bone
urethra
clitoris
labium minus
labium majus
vaginal orifice
cervix
rectum
vagina
anus

35.6 Production of oocytes and female sex hormones occurs in the ovaries

Oocyte Production An oocyte is produced as a **follicle** changes from a primary to a secondary to a vesicular (Graafian) follicle (**Fig. 35.6A**). ❶ Epithelial cells of a primary follicle surround a primary oocyte. ❷ Pools of follicular fluid surround the oocyte in a secondary follicle. ❸ In a vesicular follicle, a fluid-filled cavity increases to the point that the follicle wall balloons out on the surface of the ovary.

❹ The vesicular follicle bursts, releasing an oocyte surrounded by a clear membrane. This process is referred to as **ovulation**. For the sake of convenience, the released oocyte is often called an egg. ❺ Once a vesicular follicle has lost its oocyte, it develops into a **corpus luteum**, a glandlike structure.

Oocyte maturation requires **oogenesis**, a form of meiosis depicted in **Figure 35.6B**, is initiated and continues. The primary oocyte divides, producing two haploid cells. One cell is a secondary oocyte, and the other is called the first polar body. At ovulation, the secondary oocyte enters an oviduct, where it remains viable and capable of being fertilized for only 12–24 hours. The sperm survive for 24 and perhaps up to 72 hours. If **fertilization** occurs, a sperm enters the secondary oocyte, and then the oocyte completes meiosis. An egg with 23 chromosomes and a second polar body results. When the sperm nucleus unites with the egg nucleus, a zygote with 46 chromosomes is produced. ❻ If zygote formation and pregnancy do not occur, the corpus luteum begins to degenerate after about ten days.

Hormone Production The ovaries also produce the female sex hormones, **estrogens** (collectively called estrogen) and **progesterone**, during the ovarian cycle. Estrogen, in particular, is essential for the normal development and functioning of the female reproductive organs. Estrogen is also largely responsible for the secondary sex characteristics in females, including body hair and fat distribution. In general, females have a more rounded appearance than males because of a greater accumulation of fat beneath their skin. Also, the pelvic girdle enlarges so that females have wider hips than males, and the thighs converge at a greater angle toward the knees. Both estrogen and progesterone are required for breast development as well.

Section 35.7 explains the hormonal relationship between the ovarian and uterine cycles.

> **35.6 Check Your Progress** Which structures in Figure 35.6A produce estrogen and/or progesterone?

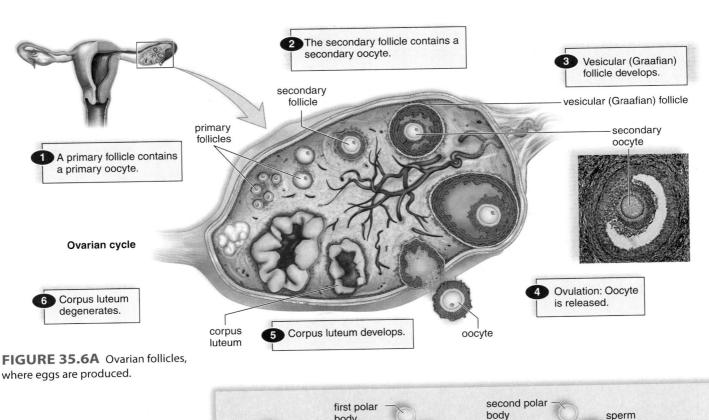

❷ The secondary follicle contains a secondary oocyte.

❸ Vesicular (Graafian) follicle develops.

vesicular (Graafian) follicle

secondary oocyte

secondary follicle

primary follicles

❶ A primary follicle contains a primary oocyte.

Ovarian cycle

❻ Corpus luteum degenerates.

corpus luteum

❺ Corpus luteum develops.

oocyte

❹ Ovulation: Oocyte is released.

FIGURE 35.6A Ovarian follicles, where eggs are produced.

first polar body

second polar body

sperm

MEIOSIS I → **MEIOSIS II** →

Sperm enters, and meiosis II goes to completion.

primary oocyte (46 chromosomes)

secondary oocyte (23 chromosomes)

egg

FIGURE 35.6B Oogenesis.

Ovarian Cycle The ovaries go through the same series of events called the **ovarian cycle** each month. In **Figure 35.7**, the ovarian cycle begins with a ① **follicular phase**, during which the anterior pituitary produces the follicle-stimulating hormone (FSH). FSH promotes the development of a follicle that secretes estrogen and some progesterone. As the estrogen level in the blood rises, it exerts feedback control over the anterior pituitary secretion of FSH so that the follicular phase comes to an end.

Presumably, the high level of estrogen in the blood causes the hypothalamus to suddenly secrete a large amount of gonadotrophic-releasing hormone (GnRH). This leads to a surge in LH production by the anterior pituitary and to ② ovulation at about the fourteenth day of a 28-day cycle (Fig. 35.7).

During the second half, or ③ **luteal phase**, of the ovarian cycle, luteinizing hormone (LH) promotes the development of the corpus luteum, which primarily secretes progesterone. As the blood level of progesterone rises, it exerts feedback control over anterior pituitary secretion of LH so that the corpus luteum begins to degenerate. As the luteal phase comes to an end, the low levels of progesterone and estrogen in the body cause menstruation to begin, as discussed next.

Uterine Cycle The female sex hormones produced during the ovarian cycle (estrogen and progesterone) affect the endometrium of the uterus, causing the cyclical series of events known as the **uterine cycle** (Fig. 35.7). Twenty-eight-day cycles are divided as follows:

During *days 1–5*, the level of female sex hormones in the body is low, causing the endometrium to disintegrate and its blood vessels to rupture. A flow of blood passes out of the vagina during ④ **menstruation**, also known as the menstrual period.

During *days 6–13*, increased production of estrogen by an ovarian follicle causes the endometrium to thicken and to become vascular and glandular. This is called the ⑤ **proliferative phase** of the uterine cycle.

Ovulation usually occurs on the fourteenth day of the 28-day cycle.

During *days 15–28*, increased production of progesterone by the corpus luteum causes the endometrium to double in thickness and the uterine glands to mature, producing a thick mucoid secretion. This is called the ⑥ **secretory phase** of the uterine cycle. The endometrium is now prepared to receive the developing embryo. If pregnancy does not occur, the corpus luteum degenerates, and the cycle begins again.

During menstruation, arteries that supply the lining constrict, and the capillaries weaken. Blood spilling from the damaged vessels detaches layers of the lining, not all at once but in random patches. Endometrium, mucus, and blood descend from the uterus, through the vagina, creating menstrual flow. Fibrinolysin, an enzyme released by dying cells, prevents the blood from clotting. Menstruation lasts from three to five days, as the uterus sloughs off the thick lining that was three weeks in the making.

We switch gears in Section 35.8 to take a look at sexually transmitted diseases.

> **35.7 Check Your Progress** During the uterine cycle, the endometrium builds up and then is sloughed off each month. Why does the uterus need to be prepared monthly?

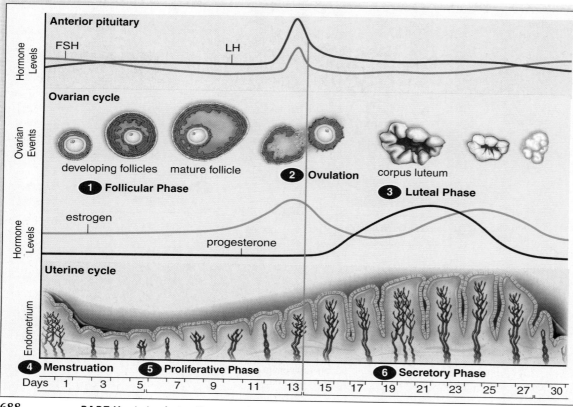

FIGURE 35.7 Female hormone levels during the ovarian and uterine cycles.

35.8 Sexual activity can transmit disease

A **sexually transmitted disease (STD)** is passed from one person to another during sexual relations. Prevention of STDs depends on the consistent use of safe sex practices as described in **Table 35.8**. We have already discussed AIDS (see pages 592–593), an STD caused by the human immunodeficiency virus (HIV), and there are many more STDs caused by viruses, bacteria, or protozoans.

Genital herpes is caused by the **herpes simplex virus (HSV)**. There are two types of HSV. HSV-1 usually affects the mouth, causing "fever blisters" or "cold sores." HSV-2 usually infects the genitals but sometimes the anus. In genital herpes, clusters of small blisters arise, rupture, and form painful sores. The sores heal in several days, but the virus persists in nearby nerve tissue and reemerges to cause periodic outbreaks of blisters and sores. Antibiotics are not effective against viral infections. Antiviral drugs such as Acyclovir cannot cure herpes at this time either, but such drugs may reduce the number and duration of outbreaks an infected individual experiences.

Human papillomavirus (HPV) causes a common STD. Like HSV, different versions of HPV exist. Many types cause warts in areas such as the genitals or rectum that can be removed by freezing, burning with a laser, or surgery. However, a few types of HPV do not cause visible warts, but cause cervical cancer instead. Women aged 18–65 are encouraged to have an annual Pap smear, in which a sample of cells is removed from the cervix and examined to see if any are transforming into cancer cells. In 2006, the FDA approved a vaccine against HPV, called Gardasil, for females between the ages of 9 and 26. The vaccine prevents infection by the two most common cancer-causing types of HPV and the two types that most frequently cause genital warts.

Chlamydial infection, the most common STD in the U.S., frequently occurs in the urethra or cervix, but may also exist in the rectum or throat. Usually, men with chlamydial infections experience painful urination and an abnormal discharge from the penis. Four out of five women infected with chlamydia do not experience any symptoms, although some report painful urination, vaginal bleeding, and/or abdominal pain. Chlamydial infection is a common cause of pelvic inflammatory disease (PID) in women. If a woman has a chlamydial infection, she can pass it to her baby during birth, resulting in eye infection or pneumonia. Antibiotics such as doxycycline are prescribed to treat the disease.

Syphilis is caused by the spirochete bacterium *Treponema pallidum*. An open sore called a chancre may appear at or near the site where the bacteria entered the body. After the bacteria increase in number, a rash may develop over the entire body or just on the hands and feet; mouth sores may occur; and the individual may feel generally ill and feverish. An apparent recovery may then take place, with no obvious symptoms for several years. However, if syphilis is not eventually treated, the bones, heart, and/or brain may be affected. If a woman with syphilis is pregnant, she may pass the bacterium to her child prior to birth, causing severe birth defects or even death. The antibiotic penicillin is still successfully used to treat syphilis.

TABLE 35.8 Some Guidelines for Preventing the Spread of STDs

1. Abstain from sexual intercourse or develop a long-term monogamous (always the same person) relationship with a person who is free of STDs.

2. Refrain from having multiple sex partners or having a relationship with a person who has multiple sex partners.

3. Be aware that having sexual relations with an intravenous drug user is risky because the behavior of this group puts them at risk for certain sexually transmitted diseases.

4. Avoid anal intercourse because HIV has easy access through the lining of the rectum.

5. Always use a latex condom if your partner has not been free of STDs for the past five years.

6. Avoid oral sex because this may be a means of transmitting AIDS and other STDs.

7. Stop, if possible, the habit of injecting drugs, and if you cannot stop, at least always use a sterile needle.

Gonorrhea is caused by the bacterium *Neisseria gonorrhoeae*, sometimes referred to as *gonococcus*. Especially in men, the bacteria may infect the urethra, causing painful urination and pus-filled urine. In women, cervical infection may go unnoticed for several months. Meanwhile, the bacteria spread throughout the reproductive organs, often leading to PID. If a woman has gonorrhea, she may transmit the bacterium to her baby's eyes during vaginal birth; if untreated, blindness will result. Although penicillin has long been used to treat gonorrhea, resistant strains of *N. gonorrhoeae* have appeared in recent years.

The protozoan *Trichomonas vaginalis* causes **trichomoniasis**. Usually the parasite infects the urethra in men and the vagina in women. Symptoms are more obvious in women. An infected woman (but not a man) typically notices a frothy, smelly, yellow-green discharge. Trichomoniasis in a pregnant woman may result in a premature and/or underweight baby. The infection is treated with the antiprotozoan drug metronidazole.

Section 35.9 surveys birth control measures for those who want to prevent pregnancy.

35.8 Check Your Progress Which of the sexually transmitted diseases described here cannot be cured by antibiotics because they are caused by viruses?

35.9 Numerous birth control methods are available

Aside from the methods and devices pictured in **Figure 35.9** and listed in **Table 35.9**, you should be aware that so-called "morning-after" pills can be taken after unprotected intercourse. A kit called Preven contains four synthetic progesterone pills; two are taken up to 72 hours after unprotected intercourse, and two more are taken 12 hours later. The medication upsets the normal uterine cycle, making it difficult for an embryo to implant itself in the endometrium. A recent study estimated that the medication was 85% effective in preventing unintended pregnancies.

In addition, mifepristone, better known as RU-486, is a pill that is presently used to cause the loss of an implanted embryo by blocking the progesterone receptors of endometrial cells. Without functioning receptors for progesterone, the endometrium sloughs off, carrying the embryo with it. When taken in conjunction with a prostaglandin to induce uterine contractions, RU-486 is 95% effective. It is possible that some day this medication will also be a "morning-after pill," taken when menstruation is late without evidence that pregnancy has occurred.

Section 35.10 surveys reproductive technologies for those who have lowered fertility.

> **35.9 Check Your Progress** What does "effectiveness" mean in Table 35.9?

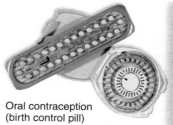

Oral contraception
(birth control pill)

Female condom

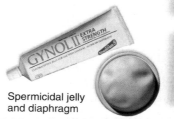

Spermicidal jelly
and diaphragm

Intrauterine device

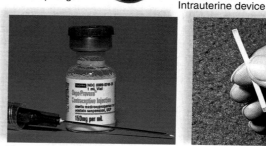

Depo-Provera injection

Contraceptive implant

FIGURE 35.9 Contraceptive devices.

TABLE 35.9	Common Birth Control Methods	
Name	**Procedure**	**Effectiveness**
Abstinence	Refrain from sexual intercourse	100%
Sterilization		
Vasectomy	Vasa deferentia cut and tied	Almost 100%
Tubal ligation	Oviducts cut and tied	Almost 100%
Combined estrogen/progesterone (available as a pill, injectable, or vaginal ring and patch)	Pill is taken daily; injectable and ring last a month; patch is replaced weekly	About 100%
Progesterone (only available as a tube implant and injectable)	Implant lasts three years; injectable lasts three weeks	About 95%
Intrauterine device (IUD)	Newest device contains progesterone and lasts up to five years	More than 90%
Vaginal sponge	Sponge permeated with spermicide is inserted into vagina	About 90%
Diaphragm	Latex cup inserted into vagina to cover cervix	With jelly, about 90%
Cervical cap	Latex cap held by suction over cervix	Almost 85%
Male condom	Latex sheath fitted over erect penis	About 85%
Female condom	Polyurethane liner fitted inside vagina	About 85%
Coitus interruptus	Penis withdrawn before ejaculation	About 75%
Jellies, creams, foams	Spermicidal products inserted into vagina	About 75%
Natural family planning	Day of ovulation determined by record keeping or various other methods of testing	About 70%
Douche	Vagina cleaned after intercourse	Less than 70%

35.10 Reproductive technologies are available to help the infertile

Infertility is the inability of a couple to achieve pregnancy after one year of regular, unprotected intercourse. The American Medical Association estimates that 15% of all couples are infertile. The cause of infertility can be attributed to the male (40%), the female (40%), or both (20%).

Sometimes the causes of infertility may be corrected by medical intervention so that couples can have children. If no obstruction is apparent and body weight is normal, it is possible for females to take fertility drugs, which are gonadotropic hormones that stimulate the ovaries and bring about ovulation. Such hormone treatments may cause multiple ovulations and multiple births.

When reproduction does not occur in the usual manner, many couples adopt a child. Others sometimes try one of the assisted reproductive technologies (ART) developed to increase the chances of pregnancy. In these cases, sperm and/or eggs are often retrieved from the testes and ovaries, and fertilization takes place in a clinical or laboratory setting.

Artificial Insemination by Donor (AID)
During artificial insemination, sperm are placed in the vagina by a physician. Sometimes a woman is artificially inseminated by her partner's sperm. This is especially helpful if the partner has a low sperm count, because the sperm can be collected over a period of time and concentrated so that the sperm count is sufficient to result in fertilization. Often, however, a woman is inseminated by sperm acquired from a donor. At times, a combination of partner and donor sperm is used.

A variation of AID is *intrauterine insemination (IUI)*. In IUI, fertility drugs are given to stimulate the ovaries, and then the donor's sperm is placed in the uterus, rather than in the vagina.

If the prospective parents wish, sperm can be sorted into those believed to be X-bearing or Y-bearing to increase the chances of having a child of the desired sex. First, the sperm are dosed with a DNA-staining chemical. Because the X chromosome has slightly more DNA than the Y chromosome, it takes up more dye. When a laser beam shines on the sperm, the X-bearing sperm shine a little more brightly than the Y-bearing sperm. A machine sorts the sperm into two groups on this basis. Parents can expect about a 65% success rate for males and about 85% for females.

In Vitro Fertilization (IVF)
During IVF, fertilization, the union of a sperm and an egg to form a zygote, occurs in laboratory glassware (**Fig. 35.10**). Ultrasound machines can now spot follicles in the ovaries that hold immature eggs; therefore, the latest method is to forgo the administration of fertility drugs and retrieve immature eggs by using a needle. The immature eggs are then brought to maturity in glassware before concentrated sperm are added. After about two to four days, the embryos are ready to be transferred to the uterus of the woman, who is now in the secretory phase of her uterine cycle. If desired, the embryos can be tested for a genetic disease, and only those found

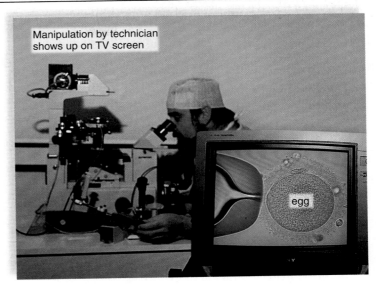

Manipulation by technician shows up on TV screen

egg

FIGURE 35.10 Fertilization can occur in the laboratory.

to be free of disease will be used. If implantation is successful, development is normal and continues to term.

Gamete Intrafallopian Transfer (GIFT)
Recall that the term **gamete** refers to a sex cell, either a sperm or an egg. Gamete intrafallopian transfer was devised to overcome the low success rate (15–20%) of in vitro fertilization. The method is exactly the same as for in vitro fertilization, except the eggs and the sperm are placed in the oviducts immediately after they have been brought together. GIFT has the advantage of being a one-step procedure for the woman—the eggs are removed and reintroduced all in the same time period. A variation on this procedure is to fertilize the eggs in the laboratory and then place the zygotes in the oviducts.

Surrogate Mothers
In some instances, women are contracted and paid to have babies. These women are called surrogate mothers. The sperm and even the egg can be contributed by the contracting parents.

Intracytoplasmic Sperm Injection (ICSI)
In this highly sophisticated procedure, a single sperm is injected into an egg. It is used effectively when a man has severe infertility problems.

If all the assisted reproductive technologies discussed were employed simultaneously, it would be possible for a baby to have five parents: (1) sperm donor, (2) egg donor, (3) surrogate mother, (4) contracting mother, and (5) contracting father.

The next part of the chapter moves on to discuss development before birth.

35.10 Check Your Progress Which of the potential parents listed in the last paragraph has the best claim to the child? Explain your reasoning.

In this part of the chapter, we consider certain principles of development. First, we observe that the first few stages of development are the same in all animals. Then, we consider that differentiation (specialization) of cells occurs during these stages.

35.11   Cellular stages of development precede tissue stages

The cellular stages of development are (1) cleavage resulting in a multicellular embryo and (2) formation of the blastula. **Cleavage** is cell division without growth. DNA replication and mitotic cell division occur repeatedly, and the cells get smaller with each division. In other words, cleavage increases only the number of cells; it does not change the original volume of the egg cytoplasm. As a model organism that demonstrates cleavage, we will use the lancelet, a chordate.

As shown in **Figure 35.11**, cleavage of a lancelet zygote is equal and results in uniform cells that form a **morula**, which is a ball of cells. The 16-cell morula resembles a mulberry and continues to divide, forming a blastula. A **blastula** is a hollow ball of cells having a fluid-filled cavity called a **blastocoel**. The blastocoel forms when the cells of the morula extrude Na^+ into extracellular spaces and water follows by osmosis. The water collects in the center, and the result is a hollow ball of cells.

The zygotes of other animals, such as a frog, chick, or human, which are vertebrates, also undergo cleavage and form a morula. In frogs, cleavage is not equal because of the presence of yolk. When yolk is present, the zygote and embryo exhibit polarity, and the embryo has an animal pole and a vegetal pole. The animal pole of a frog embryo is deep gray in color because the cells contain melanin granules, and the vegetal pole is yellow because the cells contain yolk.

Similarly, all vertebrates have a blastula stage, but the blastula can look different from that of a lancelet. Birds lay a hard-shelled egg containing plentiful yolk. Because yolk-filled eggs do not participate in cleavage, the blastula is a layer of cells that spreads out over the yolk. The blastocoel is a space that separates these cells from the yolk:

Chick blastula (cross section)   blastocoel

The blastula of humans resembles that of the chick embryo, and yet this resemblance is not related to the amount of yolk because the human egg contains little yolk. Rather, this resemblance can be explained by the evolutionary history of these two animals. Because both birds and mammals are related to reptiles, all three groups develop similarly, despite a difference in the amount of yolk in their eggs.

As seen in Section 35.12, the tissue stages of development follow the cellular stages of development.

35.11 *Check Your Progress* A lancelet develops in the water, and its zygote has minimal yolk; a bird embryo develops on land and has much yolk; a human develops in the uterus and has minimal yolk. Explain.

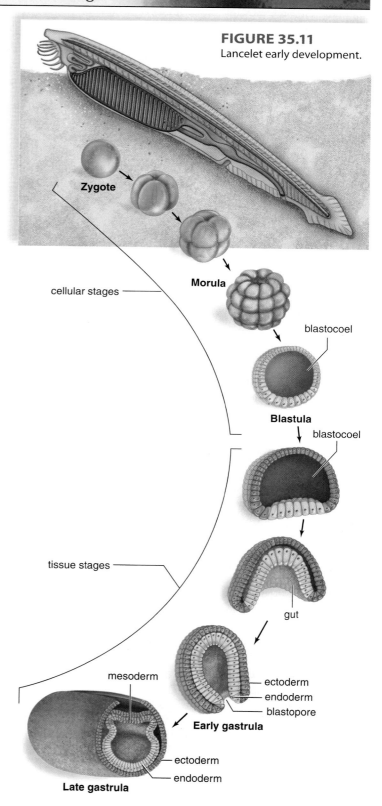

FIGURE 35.11
Lancelet early development.

Zygote

cellular stages

Morula

blastocoel

Blastula

blastocoel

tissue stages

gut

mesoderm

ectoderm
endoderm
blastopore
Early gastrula

ectoderm
endoderm
Late gastrula

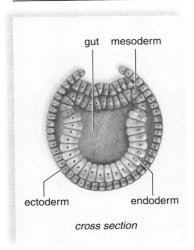

gut   mesoderm

ectoderm   endoderm

cross section

Lancelet late gastrula

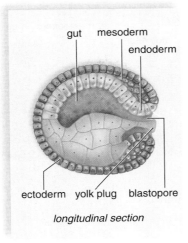

gut   mesoderm
endoderm

ectoderm   yolk plug   blastopore

longitudinal section

Frog late gastrula

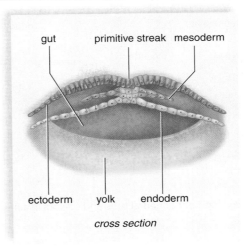

gut   primitive streak   mesoderm

ectoderm   yolk   endoderm

cross section

Chick late gastrula

FIGURE 35.12
Comparative development of mesoderm.

The tissue stages of development are (1) the early gastrula and (2) the late gastrula. During early gastrulation, cells migrate into the blastocoel, creating a double layer of cells (see Fig. 35.11). Cells continue to migrate during other stages of development also, sometimes traveling quite a distance before reaching a destination, where they continue developing. Extracellular proteins and cytoskeletal elements participate in allowing migration to occur. As cells migrate, they "feel their way" by changing their pattern of adhering to extracellular proteins.

An early gastrula has two layers of cells. The outer layer is called the **ectoderm**, and the inner layer is called the **endoderm**. The endoderm borders a primitive gut. The pore, or hole, created by invagination is the **blastopore**, and in a lancelet, the blastopore eventually becomes the anus. **Gastrulation** is not complete until three layers of cells that will develop into adult organs are produced. In addition to ectoderm and endoderm, the late gastrula has a middle layer of cells called the **mesoderm** (see Fig. 35.11).

Figure 35.12 compares the lancelet, frog, and chick late gastrula stages. In the lancelet, mesoderm formation begins as outpocketings from the primitive gut. These outpocketings will grow in size until they meet and fuse, forming two layers of mesoderm. The space between them is the coelom. The coelom is a body cavity lined by mesoderm that contains internal organs. (In humans, the coelom becomes the thoracic and abdominal cavities of the body.)

In the frog, the cells containing yolk do not participate in gastrulation, and therefore they do not invaginate. Instead, a slitlike blastopore is formed when the animal pole cells begin to invaginate from above, forming endoderm. Animal pole cells also move down over the yolk, to invaginate from below. Some yolk cells remain temporarily in the region of the blastopore, and are called the yolk plug. Mesoderm forms when cells migrate between the ectoderm and endoderm. Later, splitting of the mesoderm creates the coelom.

The chick egg contains so much yolk that endoderm formation does not occur by invagination. Instead, an upper layer of cells becomes ectoderm, and a lower layer becomes endoderm. Mesoderm arises as cells invaginate along the edges of a longitudinal furrow in the midline of the embryo. Because of its appearance, this furrow is called the *primitive streak*. Later, the newly formed mesoderm splits to produce a coelomic cavity.

Ectoderm, mesoderm, and endoderm are called the embryonic **germ layers**. No matter how gastrulation takes place, the result is the same—three germ layers are formed. It is possible to relate the development of future organs to these germ layers (**Table 35.12**).

Section 35.13 explains the organ stages of development, which follow the tissue stages of development.

35.12 *Check Your Progress* **What event, once complete, initiates the organ level of organization during development?**

TABLE 35.12	Embryonic Germ Layers
Embryonic Germ Layer	**Vertebrate Adult Structures**
Ectoderm (outer layer)	Nervous system; epidermis of skin and derivatives of the epidermis (hair, nails, glands); tooth enamel, dentin, and pulp; epithelial lining of oral cavity and rectum
Mesoderm (middle layer)	Muscular and skeletal systems; dermis of skin; cardiovascular system; urinary system; lymphatic system; reproductive system—including most epithelial linings; outer layers of respiratory and digestive systems
Endoderm (inner layer)	Epithelial lining of digestive tract and respiratory tract, associated glands of these systems; epithelial lining of urinary bladder; thyroid and parathyroid glands

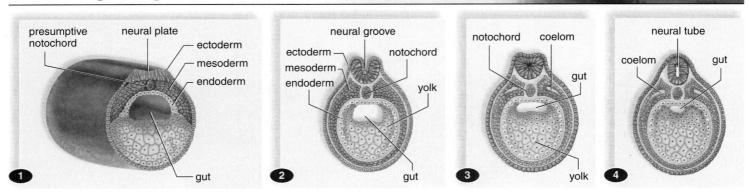

FIGURE 35.13A Development of neural tube and coelom in a frog embryo.

The organs of an animal's body develop from the three embryonic germ layers. In this section, we explain this process, beginning with how the nervous system develops from the ectoderm.

The newly formed mesoderm cells lie along the main longitudinal axis of the animal and coalesce to form a *presumptive notochord* because these cells mark the location of the notochord. The notochord persists in lancelets, but in frogs, chicks, and humans, it is later replaced by the vertebral column (which explains why these animals are called vertebrates).

Figure 35.13A shows cross sections of frog development to illustrate the formation of the neural tube. The nervous system starts to develop from midline ectoderm located just above the presumptive notochord. ❶ At first, a thickening of cells, called the **neural plate**, appears along the dorsal surface of the embryo. ❷ Then neural folds develop on either side of a neural groove. ❸ The coelom appears, and ❹ the **neural tube** is complete. At this point, the embryo is called a **neurula**. Later, the anterior end of the neural tube develops into the *brain*, and the rest becomes the *spinal cord*. In addition, the neural crest is a band of cells that develops where the neural tube pinches off from the ectoderm. Neural crest cells migrate to various locations, where they contribute to the formation of skin and muscles as well as the adrenal medulla and the ganglia of the peripheral nervous system.

Midline mesoderm cells that did not contribute to the formation of the notochord now form two longitudinal masses of tissue. These two masses become blocked off into somites, which are serially arranged along both sides along the length of the notochord. Somites give rise to muscles associated with the axial skeleton and to the vertebrae. The serial origins of axial muscles and the vertebrae testify that vertebrates are segmented animals. Lateral to the somites, the mesoderm splits, forming the mesodermal lining of the coelom.

Figure 35.13B will help you relate the formation of the vertebrate structures and organs discussed in this section to the three embryonic layers of cells: the ectoderm, the mesoderm, and the endoderm. A primitive gut tube is formed by endoderm as the body itself folds into a tube. The heart, too, begins as a simple tubular pump formed from mesoderm. Organ formation continues until the germ layers have given rise to the specific organs listed in Table 35.12.

While the first few stages of development occur, cellular differentiation begins, as explained in Section 35.14.

35.13 Check Your Progress How is a notochord different from a neural tube?

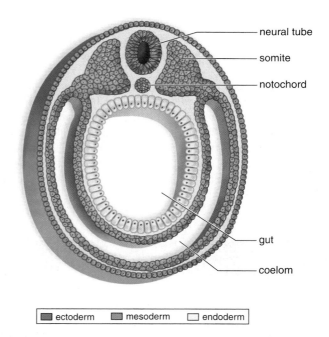

FIGURE 35.13B Vertebrate embryo, cross section.

35.14   Cellular differentiation begins with cytoplasmic segregation

Development requires growth, cellular differentiation, and morphogenesis. We have already seen that growth occurs when there is an increase in the number of cells, which then become larger. And we know that differentiation occurs when cells become specialized in structure and function—that is, a muscle cell looks and acts differently than a nerve cell. **Morphogenesis**, on the other hand, produces the shape and form of the body. One of the earliest indications of morphogenesis is cell movement. Later, morphogenesis includes **pattern formation**, the way that tissues and organs are arranged in the body. Apoptosis, or programmed cell death, first discussed in Section 8.7, is an important part of pattern formation. In this section, we consider what event might precede cellular differentiation and morphogenesis.

Cytoplasmic Segregation Cellular differentiation must begin long before we can recognize specialized types of cells. Ectodermal, endodermal, and mesodermal cells in the gastrula look quite similar, but they must be different because they develop into different organs. An examination of the frog's egg shows that it is not uniform (**Fig. 35.14**). ❶ The egg is polar and has both an anterior/posterior axis and a dorsal/ventral axis, which can be correlated with the **gray crescent**, a gray area that appears after the sperm fertilizes the egg. Hans Spemann, who received a Nobel Prize in 1935 for his extensive work in embryology, showed that ❷ if the gray crescent is divided equally by the first cleavage, each experimentally separated daughter

cell develops into a complete embryo. ❸ However, if the egg divides so that only one daughter cell receives the gray crescent, only that cell becomes a complete embryo. This experiment allows us to speculate that the gray crescent must contain particular chemical signals that are needed in order for development to proceed normally. We now know that an egg contains substances called maternal determinants that influence the course of development. As mitosis occurs, maternal determinants are parceled out, a process known as cytoplasmic segregation:

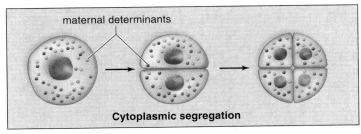

Cytoplasmic segregation

Cytoplasmic segregation helps determine how the various cells of the morula will develop.

How organ formation comes about is explained in Section 35.15.

> **35.14** *Check Your Progress* A cell that does not receive a portion of the gray crescent never undergoes gastrulation. Therefore, development is halted at the tissue level. Explain.

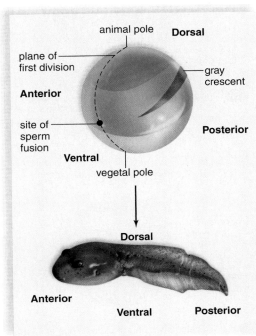

❶ Frog's egg is polar and has axes.

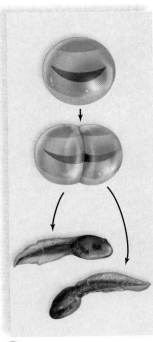

❷ Each cell receives a part of the gray crescent.

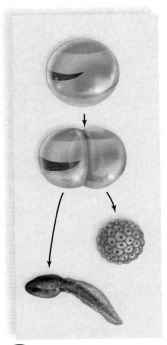

❸ Only the cell on the left receives the gray crescent.

FIGURE 35.14
Cytoplasmic influence on development.

As development proceeds, specialization of cells and formation of organs are influenced not only by maternal determinants, but also by signals given off by neighboring cells. **Induction** is the ability of one embryonic tissue to influence the development of another tissue.

Hans Spemann showed that a frog embryo's gray crescent becomes the dorsal lip of the blastopore, where gastrulation begins. Since this region is necessary for complete development, he called the dorsal lip of the blastopore the primary organizer. The cells closest to Spemann's primary organizer become endoderm; those farther away become mesoderm; and those farthest away become ectoderm. This suggests that a molecular concentration gradient may act as a chemical signal to induce germ layer differentiation.

The gray crescent of a frog's egg marks the dorsal side of the embryo, where the mesoderm becomes notochord and ectoderm becomes the nervous system. In a classic experiment, Spemann and his colleague Hilde Mangold showed that presumptive (potential) notochord tissue induces the formation of the nervous system. Two experiments are described in **Figure 35.15**. In the first experiment, ❶ presumptive nervous system tissue, located just above the presumptive notochord, is cut out and ❷ transplanted to the belly region of the embryo; ❸ it does not form a neural tube. In the second experiment, ❶ presumptive notochord tissue is cut out and ❷ transplanted beneath what would be belly ectoderm; ❸ this ectoderm differentiates into

neural tissue. Still other examples of induction are now known. In 1905, Warren Lewis studied the formation of the eye in frog embryos. He found that an optic vesicle, which is a lateral outgrowth of developing brain tissue, induces overlying ectoderm to thicken and become a lens. The developing lens, in turn, induces an optic vesicle to form an optic cup, where the retina develops. This suggests that induction, which involves signaling molecules, is the process that explains differentiation and morphogenesis.

Homeotic Genes Section 11.14 discusses the role of homeotic genes in development. Most likely, signaling molecules during the process of induction turn on homeotic genes that code for homeotic proteins, which are involved in morphogen gradients. A surprising finding has been that the homeotic proteins of so many different organisms, both invertebrate and vertebrate, contain the same homeodomain, a sequence of amino acids that binds to DNA and turns on other genes.

We have completed our discussion of the principles of animal development. In the next part of the chapter, we study the stages of human development.

> **35.15** *Check Your Progress* **In what ways is induction a part of morphogenesis?**

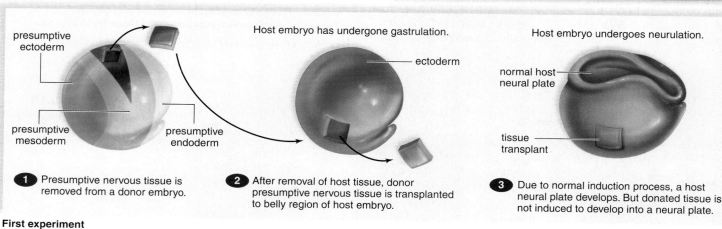

1 Presumptive nervous tissue is removed from a donor embryo.

2 After removal of host tissue, donor presumptive nervous tissue is transplanted to belly region of host embryo.

3 Due to normal induction process, a host neural plate develops. But donated tissue is not induced to develop into a neural plate.

Host embryo has undergone gastrulation.

Host embryo undergoes neurulation.

First experiment

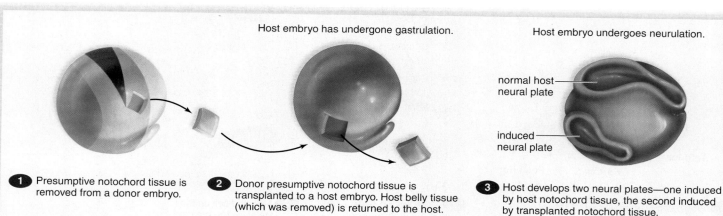

1 Presumptive notochord tissue is removed from a donor embryo.

2 Donor presumptive notochord tissue is transplanted to a host embryo. Host belly tissue (which was removed) is returned to the host.

3 Host develops two neural plates—one induced by host notochord tissue, the second induced by transplanted notochord tissue.

Host embryo has undergone gastrulation.

Host embryo undergoes neurulation.

Second experiment

FIGURE 35.15 Control of nervous system development.

The extraembryonic membranes are described first in this part of the chapter, because they play a significant role in human development. Human development is divided into embryonic and fetal development. Fetal development ends with the birth of the newborn.

35.16 Extraembryonic membranes are critical to human development

In humans, the length of time from conception (fertilization followed by **implantation**) to birth (parturition) is approximately nine months (266 days). It is customary to calculate the time of birth by adding 280 days to the start of the last menstrual period, because this date is usually known, whereas the day of fertilization is usually unknown.

In humans, pregnancy, or gestation, is the time during which the mother carries the developing embryo. Biologists divide human development into embryonic development (months 1 and 2) and fetal development (months 3–9). During **embryonic development**, the major organs are formed, and during fetal development, these structures are refined.

Development can also be divided into trimesters, with each one characterized by specific developmental accomplishments. During the first trimester, embryonic and early fetal development occur. The second trimester is characterized by the development of organs and organ systems. By the end of the second trimester, the fetus appears distinctly human. In the third trimester, the fetus grows rapidly, and the major organ systems become functional. An infant born one, or perhaps two, months prematurely has a reasonable chance of survival.

Before we consider human development chronologically, we must understand the placement of **extraembryonic membranes**. Extraembryonic membranes are best understood by considering their function in reptiles and birds. In reptiles, these membranes made development on land possible. If an embryo develops in the water, the water supplies oxygen for the embryo and takes away waste products. The surrounding water prevents desiccation, or drying out, and provides a protective cushion. For an embryo that develops on land, all these functions are performed by the extraembryonic membranes.

In the chick, the extraembryonic membranes develop from extensions of the germ layers, which spread out over the yolk. **Figure 35.16** (*top*) shows the chick surrounded by the membranes. The **chorion** lies next to the shell and carries on gas exchange. The **amnion** contains the protective amniotic fluid, which bathes the developing embryo. The **allantois** collects nitrogenous wastes, and the **yolk sac** surrounds the remaining yolk, which provides nourishment.

Humans (and other mammals) also have extraembryonic membranes (Fig. 35.16, *bottom*). The chorion develops into the fetal half of the placenta, where nutrient for waste exchange occurs between the fetus and mother (see Section 35.18). The yolk sac, which lacks yolk, is the first site of blood cell formation; the blood vessels of the allantois become the umbilical blood vessels; and the amnion contains fluid to cushion and protect the embryo, which develops into a fetus. Therefore, the function of the membranes in humans has been modified to suit internal development, but their very presence indicates our relationship to birds and to reptiles. It is interesting to note that all chordate ani-

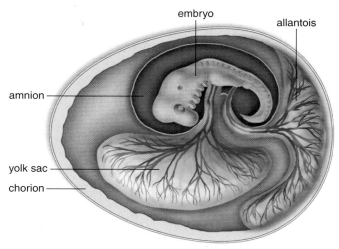

Chick

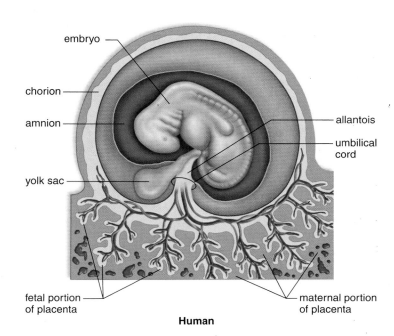

Human

FIGURE 35.16 Extraembryonic membranes.

mals develop in water—either in bodies of water or surrounded by amniotic fluid within a shell or uterus.

Embryonic human development (first two months) is described in Section 35.17.

> **35.16 Check Your Progress** The extraembryonic membranes, which first evolved in reptiles, have been modified in mammals for what purpose?

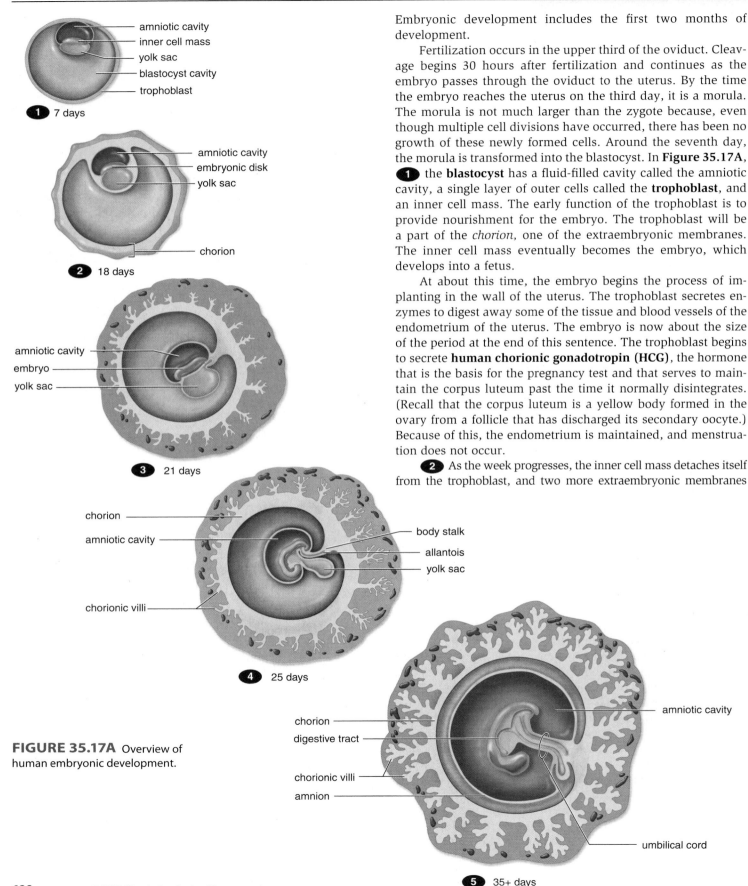

FIGURE 35.17A Overview of human embryonic development.

Embryonic development includes the first two months of development.

Fertilization occurs in the upper third of the oviduct. Cleavage begins 30 hours after fertilization and continues as the embryo passes through the oviduct to the uterus. By the time the embryo reaches the uterus on the third day, it is a morula. The morula is not much larger than the zygote because, even though multiple cell divisions have occurred, there has been no growth of these newly formed cells. Around the seventh day, the morula is transformed into the blastocyst. In **Figure 35.17A**, ❶ the **blastocyst** has a fluid-filled cavity called the amniotic cavity, a single layer of outer cells called the **trophoblast**, and an inner cell mass. The early function of the trophoblast is to provide nourishment for the embryo. The trophoblast will be a part of the *chorion,* one of the extraembryonic membranes. The inner cell mass eventually becomes the embryo, which develops into a fetus.

At about this time, the embryo begins the process of implanting in the wall of the uterus. The trophoblast secretes enzymes to digest away some of the tissue and blood vessels of the endometrium of the uterus. The embryo is now about the size of the period at the end of this sentence. The trophoblast begins to secrete **human chorionic gonadotropin (HCG)**, the hormone that is the basis for the pregnancy test and that serves to maintain the corpus luteum past the time it normally disintegrates. (Recall that the corpus luteum is a yellow body formed in the ovary from a follicle that has discharged its secondary oocyte.) Because of this, the endometrium is maintained, and menstruation does not occur.

❷ As the week progresses, the inner cell mass detaches itself from the trophoblast, and two more extraembryonic membranes form. In humans, the *yolk sac*, which forms below the inner cell mass, has no nutritive function as it does in chicks, but it is the first site of blood cell formation. However, the *amnion* and its cavity are where the embryo (and then the fetus) develops. In humans, amniotic fluid acts as an insulator against cold and heat and also absorbs shock, such as that caused by the mother exercising.

Gastrulation occurs during the second week. The inner cell mass now has flattened into the **embryonic disk**, composed of two layers of cells: ectoderm above and endoderm below. Once the embryonic disk elongates to form the primitive streak, the third germ layer, mesoderm, forms by invagination of cells along the streak. The trophoblast is reinforced by mesoderm and becomes the chorion. It is possible to relate the development of future organs to these germ layers (see Table 35.12).

3 Two important organ systems make their appearance during the third week as the embryonic disk becomes the embryo. The nervous system is the first organ system to be visually evident. At first, a thickening appears along the entire dorsal length of the embryo; then the neural folds appear. When the neural folds meet at the midline, the neural tube, which later develops into the brain and the nerve cord, is formed. After the notochord is replaced by the vertebral column, the nerve cord is called the spinal cord.

Although the embryo has undergone quite a bit of development, the woman, as yet, probably doesn't realize she is pregnant. Many authorities suggest that sexually active females always practice good health habits to protect a possible developing embryo from harm. These habits include avoiding alcohol, cigarettes, illegal drugs, and any medication not approved by a physician. Aside from a positive pregnancy test, the early symptoms of pregnancy are nausea, breast swelling and tenderness, and fatigue.

The second organ system to appear during the third week of development is the cardiovascular system as the heart begins to form. At first, there are right and left heart tubes; when these fuse, the heart begins pumping blood, even though the chambers of the heart are not fully formed. The veins enter posteriorly, and the arteries exit anteriorly from this largely tubular heart, but later the heart twists so that all major blood vessels are located anteriorly.

4 At four weeks, the embryo is barely larger than the height of this print. A bridge of mesoderm called the body stalk connects the caudal (tail) end of the embryo with the chorion, which has treelike projections called **chorionic villi**. The chorionic villi are a part of the placenta. The fourth extraembryonic membrane, the *allantois*, is contained within this stalk, and its blood vessels become the umbilical blood vessels. **5** The head and the tail then lift up; the body stalk and the yolk sac fuse to become the **umbilical cord**, which connects the developing embryo to the placenta.

Little flippers called limb buds appear at the fifth week (**Fig. 35.17B**); later, the arms and the legs develop from the limb buds, and even the hands and the feet become apparent. At the same time—during the fifth week—the head enlarges, and the sense organs become more prominent. It is possible to make out the developing eyes, ears, and even the nose.

During the sixth to eighth weeks of development, the embryo becomes easily recognizable as human. Concurrent with brain development, the head achieves its normal relationship with the body as a neck region develops. The nervous system is developed well enough to permit reflex actions, such as a startle response to touch. At the end of this period, the embryo is about 38 mm long and weighs no more than an aspirin tablet; even so, all organ systems have been established.

Section 35.18 is concerned with fetal development.

35.17 *Check Your Progress* The increasing complexity of the embryo mirrors the increasing complexity of the animal kingdom as evolution occurred. Explain.

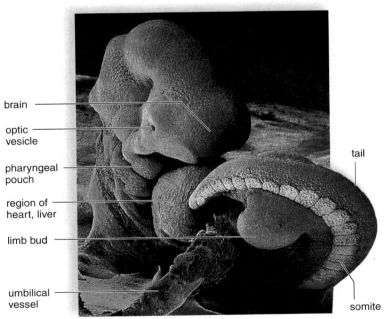

brain
optic vesicle
pharyngeal pouch
region of heart, liver
limb bud
umbilical vessel
tail
somite

FIGURE 35.17B Human embryo at fifth week.

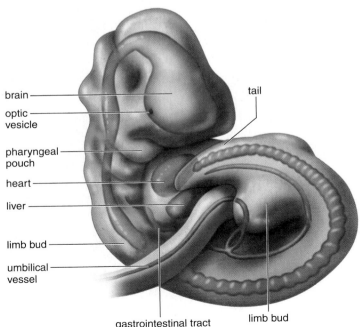

brain
optic vesicle
pharyngeal pouch
heart
liver
limb bud
umbilical vessel
tail
gastrointestinal tract
limb bud

Fetal development includes the third through the ninth months of development. The placenta is now fully developed (**Fig. 35.18A**). Aside from an increase in weight, many of the physiological changes in the mother are due to the placental hormones that support fetal development. For example, progesterone decreases uterine motility, but smooth muscle relaxation also brings about the heartburn and constipation so often experienced by pregnant women.

The **placenta** has a fetal side contributed by the chorion, the outermost extraembryonic membrane, and a maternal side consisting of uterine tissues. The treelike chorionic villi are surrounded by maternal blood; yet maternal and fetal blood never mix because exchange always takes place across the villi. Carbon dioxide and other wastes move from the fetal side to the maternal side of the placenta, and nutrients and oxygen move from the maternal side to the fetal side. Simple diffusion and active transport are at work. The umbilical cord stretches between the placenta and the fetus. The umbilical blood vessels are an extension of the fetal circulatory system and simply take fetal blood to and from the placenta. Harmful chemicals can also cross the placenta. This is of particular concern during the embryonic period, when various structures are first forming. Each organ or part seems to have a sensitive period during which a substance can alter its normal development. For example, if a woman takes the drug thalidomide, a tranquilizer, between days 27 and 40 of her pregnancy, the infant is likely to be born with deformed limbs.

At the beginning of the third month, the fetal head is still very large, the nose is flat, the eyes are far apart, and the ears are well formed (**Fig. 35.18B**). Head growth now begins to slow as the rest of the body increases in length. Epidermal refinements, such as fingernails, nipples, eyelashes, eyebrows, and hair on the head, appear.

Cartilage begins to be replaced by bone as ossification centers appear in most of the bones. Cartilage remains at the ends of the long bones, and ossification is not complete until age 18 to 20 years. The skull has six large membranous areas called **fontanels**, which permit a certain amount of flexibility as the head passes through the birth canal and allow rapid growth of the brain during infancy. Progressive fusion of the skull bones causes the fontanels to close, usually by two years of age.

Sometime during the third month, it is possible to distinguish males from females. Researchers have discovered a series of genes on the chromosomes that cause the differentiation of gonads into testes. Ovaries develop because these genes are lacking. Once these have differentiated, they produce the sex hormones that influence the differentiation of the genital tract.

At this time, either testes or ovaries are located within the abdominal cavity, but later, in the last trimester of fetal development, the testes descend into the scrotal sacs (scrotum). Sometimes the testes fail to descend, and in that case, an operation may be done later to place them in their proper location.

During the fourth month, the fetal heartbeat is loud enough to be heard when a physician applies a stethoscope to the mother's abdomen, and fetal movement can be felt by women who have previously been pregnant. By the end of this month, the fetus is about 6 in. long and weighs about 6 oz.

During the fifth through seventh months (**Figs. 35.18C** and **35.18D**), fetal movement increases. What is at first only a flut-

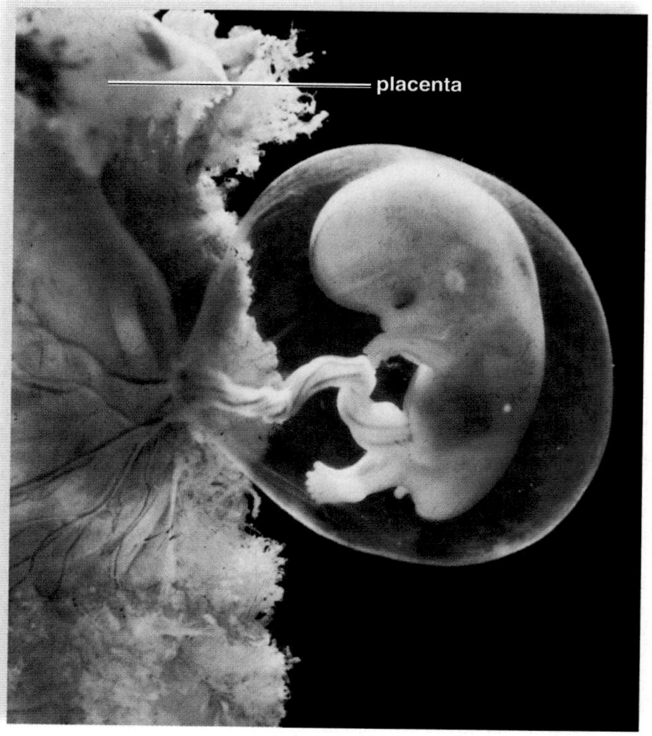

placenta

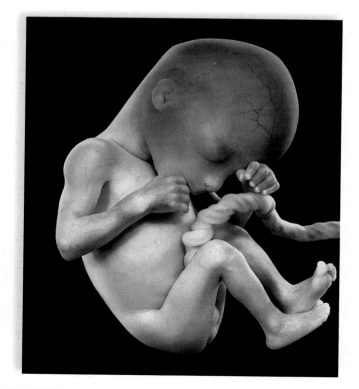

FIGURE 35.18A 9-week fetus.

FIGURE 35.18B 12- to 16-week fetus.

tering sensation turns into kicks and jabs as the fetal legs grow and develop. The wrinkled, translucent, pink-colored skin is covered by a fine down called *lanugo*. This, in turn, is coated with a white, greasy, cheeselike substance called *vernix caseosa*, which probably protects the delicate skin from the amniotic fluid. The eyelids will now fully open.

At the end of the seventh month, the fetus's length has increased to about 15 in., and the weight is about 3 lb. It is possible that, if born now, the baby will survive.

Of interest, the increasing size of the uterus during pregnancy contributes to an improvement in the mother's respiratory functions because breathing depth typically increases by 40%. The uterus comes to occupy most of the abdominal cavity, reaching nearly to the breastbone. This increase in size pushes the intestines out of the way and also widens the thoracic cavity. The fall in the mother's blood carbon dioxide level creates a favorable gradient for the flow of carbon dioxide from fetal blood to maternal blood at the placenta.

As the end of development approaches, the fetus usually assumes the fetal position, with the head bent down and in contact with the flexed knees. Eventually, the body rotates so that the head points toward the cervix. If the fetus does not turn, a breech birth (rump first) is likely. It is very difficult for the cervix to expand enough to accommodate the baby's body in this position, and asphyxiation of the baby is more likely to occur. Thus, a **cesarean section** (incision through the abdominal and uterine walls) may be prescribed for delivery of the fetus.

At the end of nine months, the fetus is about 20½ in. long and weighs about 7½ lb. This latest weight gain is largely due to

FIGURE 35.18D 24-week fetus.

an accumulation of fat beneath the skin. Full-term babies have the best chance of survival (**Fig. 35.18E**).

The stages of birth are described in Section 35.19.

> **35.18 *Check Your Progress*** Explain the prominence of weight gain during later development, rather than development of the systems.

FIGURE 35.18E Newborn (40 weeks).

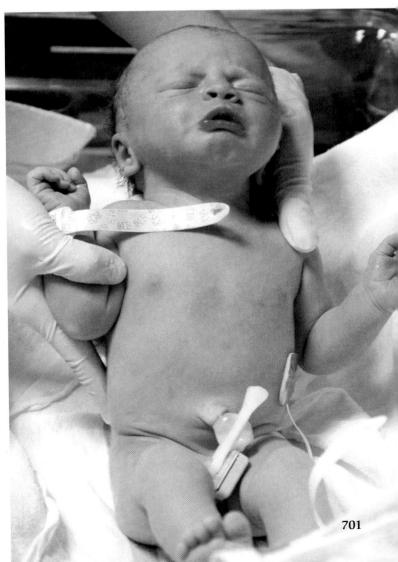

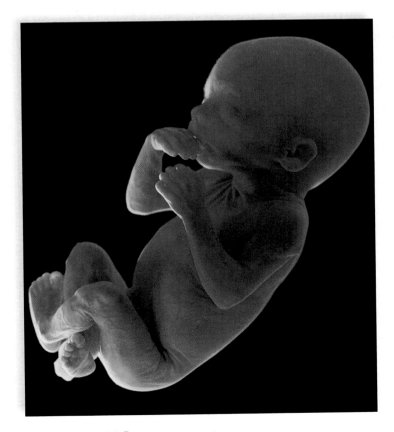

FIGURE 35.18C 20- to 28-week fetus.

35.19 Pregnancy ends with the birth of the newborn

The uterus undergoes contractions throughout pregnancy. At first, these are light, lasting about 20–30 seconds and occurring every 15–20 minutes. Near the end of pregnancy, the contractions may become stronger and more frequent so that a woman thinks she is in labor. However, the onset of true labor is marked by uterine contractions that occur regularly every 15–20 minutes and last for 40 seconds or longer.

Prior to or at the first stage of the birth process, there may be a "bloody show" caused by expulsion of a mucous plug from the cervical canal. This plug prevents bacteria and sperm from entering the uterus during pregnancy. Birth has these three stages:

Stage 1 During the first stage of labor, the uterine contractions occur in such a way that the cervical canal slowly disappears as the lower part of the uterus is pulled upward toward the baby's head. This process is called effacement, or "taking up the cervix." With further contractions, the baby's head acts as a wedge to assist cervical dilation (**Fig. 35.19**). If the amniotic membrane has not already ruptured, it is apt to do so during this stage, releasing the amniotic fluid, which leaks out the vagina (an event sometimes referred to as "breaking water"). The first stage of parturition ends once the cervix is dilated completely.

Stage 2 During the second stage of parturition, the uterine contractions occur every 1–2 minutes and last about one minute each. They are accompanied by a desire to push, or bear down.

As the baby's head gradually descends into the vagina, the desire to push becomes greater. When the baby's head reaches the exterior, it turns so that the back of the head is uppermost (Fig. 35.19). Because the vaginal orifice may not expand enough to allow passage of the head, an **episiotomy** is often performed. This incision, which enlarges the opening, is sewn together later. As soon as the head is delivered, the baby's shoulders rotate so that the baby faces either to the right or the left. At this time, the physician may hold the head and guide it downward, while one shoulder and then the other emerges. The rest of the baby follows easily.

Once the baby is breathing normally, the umbilical cord is cut and tied, severing the child from the placenta. The stump of the cord shrivels and leaves a scar, which is the navel.

Stage 3 The placenta, or **afterbirth**, is delivered during the third stage of parturition (Fig. 35.19). About 15 minutes after delivery of the baby, uterine muscular contractions shrink the uterus and dislodge the placenta. The placenta is then expelled into the vagina. As soon as the placenta and its membranes are delivered, the third stage of parturition is complete.

> **35.19 Check Your Progress** With birth, the digestive, respiratory, and urinary systems begin to function, whereas they were not functioning before birth. Explain.

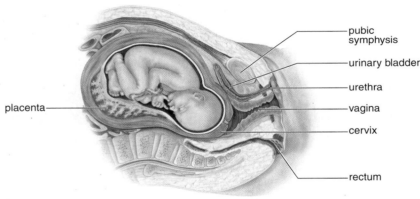

placenta

pubic symphysis
urinary bladder
urethra
vagina
cervix
rectum

9-month-old fetus

ruptured amniotic sac

First stage of birth: cervix dilates

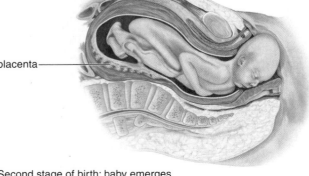

placenta

Second stage of birth: baby emerges

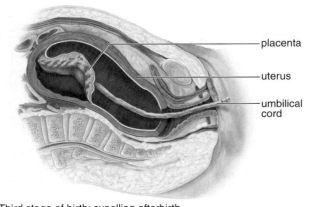

placenta
uterus
umbilical cord

Third stage of birth: expelling afterbirth

FIGURE 35.19 The three stages of birth.

CONNECTING THE CONCEPTS

Reproduction in animals is varied, depending on their complexity and the environment in which they live. Sexual reproduction always produces a zygote that undergoes development to become a functioning offspring. We have observed that animals go through the same early embryonic stages of morula, blastula, gastrula, and so forth. The set sequence of these stages is due to the expression of genes; therefore, once again we are called upon to study organisms at the cellular level of organization.

We have seen that hormones are signals that affect cellular metabolism. Like the hormones in plants, steroid hormones in animals turn on the expression of genes. Signal transduction pathways also occur during development. Transduction means the signal has been transformed into an event that has an effect on the cell.

A set sequence of signaling molecules is produced as development occurs. Each new signal in the sequence turns on a specific gene or, more likely, a sequence of genes. Developmental biology is now making a significant contribution to the field of evolution. Homeotic (pattern formation) genes have been discovered in many different types of organisms. As we discussed in Chapter 14, this suggests that homeotic genes arose early in the history of life, and mutations in these genes could possibly account for macroevolution—the appearance of new species, or even higher taxons.

The Chapter in Review

Summary

How to Do It on Land

- Unlike animals that reproduce in water, animals that reproduce on land need adaptations to protect the gametes and the embryo from drying out.

Reproduction in Animals Is Varied

35.1 Both asexual and sexual reproduction occur among animals

- In asexual reproduction, a single parent produces offspring that are genetically identical to the parent.
- In sexual reproduction, the egg of one parent is fertilized by the sperm of another, producing offspring that are genetically different from the parents.

35.2 Development in water and on land occurs among animals

- Many aquatic animals have a larval stage.
- Reptiles and birds develop in shelled eggs with a yolk and extraembryonic membranes.
- Ovoviviparous animals retain their eggs in the body until hatching fully developed offspring.
- Placental mammals are viviparous, meaning that complete development occurs in the body of the female.

Humans Are Adapted to Reproducing on Land

35.3 Testes are male gonads

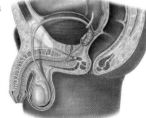

- The human male reproductive system includes the testes, epididymides, vasa deferentia, and urethra.
- The penis is the organ of sexual intercourse.
- Semen is composed of sperm and secretions from the seminal vesicles, prostate gland, and bulbourethral glands.
- Male orgasm results in ejaculation of semen from the penis.

35.4 Production of sperm and male sex hormones occurs in the testes

- The testes contain seminiferous tubules, which produce sperm, and interstitial cells, which produce testosterone.

- Testosterone is the hormone that influences sex organ function, sperm maturation, and the male secondary sex characteristics.

35.5 Ovaries are female gonads

- The human female reproductive system includes the ovaries, oviducts, uterus, and vagina.
- The female external genital area includes the vaginal opening, clitoris, labia minora, and labia majora.
- Female orgasm culminates in uterine and oviduct contractions.

35.6 Production of oocytes and female sex hormones occurs in the ovaries

- Production of an egg occurs as a follicle changes from primary to secondary to a vesicular follicle.
- The ovaries produce estrogen and progesterone, which maintain female sex characteristics.

35.7 The ovarian cycle drives the uterine cycle

- Ovarian cycle occurs.
  - During the follicular phase, FSH causes maturation of a follicle that secretes estrogen and some progesterone.
  - During the luteal phase, LH promotes development of the corpus luteum, which secretes progesterone.
- The uterine cycle occurs.
  - Estrogen causes uterine lining to thicken.
  - Progesterone causes the lining to become secretory.

35.8 Sexual activity can transmit disease

- Common STDs include AIDS, herpes simplex virus (HSV), human papillomavirus (HPV), chlamydial infections, syphilis, and gonorrhea.

35.9 Numerous birth control methods are available

- "Morning-after" pills can be taken after unprotected intercourse.
- Birth control methods vary in their effectiveness.

35.10 Reproductive technologies are available to help the infertile

- Assisted reproductive technologies include artificial insemination by donor (AID), in vitro fertilization (IVF), gamete intrafallopian transfer (GIFT), use of a surrogate mother, and intracytoplasmic sperm injection (ICSI).

Vertebrates Have Similar Early Developmental Stages and Processes

35.11 Cellular stages of development precede tissue stages

- Cleavage results in a multicellular embryo (morula), and then a blastula forms.

35.12 Tissue stages of development precede organ stages

- A mature gastrula has three germ layers: ectoderm, endoderm, and mesoderm.

35.13 Organ stages of development occur after tissue stages

- Each germ layer develops into specific organs.
- The nervous system develops from ectoderm just above the notochord; the neural plate appears, and then neural folds become the neural tube.
- At the neurula stage, cross sections of all chordate embryos are similar in appearance.

35.14 Cellular differentiation begins with cytoplasmic segregation

- Growth refers to an increase in cell number and size.
- Cellular differentiation means that cells specialize in structure and function.
- Cytoplasmic segregation and induction are two mechanisms that help explain cellular differentiation.

35.15 Morphogenesis involves induction also

- Induction is the ability of one embryonic tissue to influence the development of another tissue.
- Signaling molecules turn on homeotic genes that code for homeotic proteins, which may be involved in causing morphogen gradients.

Human Development Is Divided into Embryonic Development and Fetal Development

35.16 Extraembryonic membranes are critical to human development

- Extraembryonic membranes (chorion, amnion, allantois, and yolk sac) make internal development possible.

35.17 Embryonic development involves tissue and organ formation

- Embryonic development spans the first two months of development, from fertilization through the acquisition of organ systems.

35.18 Fetal development involves refinement and weight gain

- Fetal development includes the third through the ninth months of development.
- Exchanges at the placenta supply the fetus with oxygen and nutrients and take away carbon dioxide and wastes.
- During the third and fourth months, the skeleton becomes ossified, and the fetus's sex is distinguishable.

- During the fifth through seventh months, fetal movement begins, and the fetus continues to grow and gain weight.

35.19 Pregnancy ends with the birth of the newborn

- Stage 1: Uterine contractions begin; cervix dilates.
- Stage 2: Uterine contractions occur every 1–2 minutes; baby is born; umbilical cord is cut.
- Stage 3: Uterine muscle contractions shrink the uterus and dislodge the placenta, which is expelled.

Testing Yourself

Reproduction in Animals Is Varied

1. Internal fertilization
   a. can prevent the drying out of gametes and zygotes.
   b. must take place on land.
   c. is practiced by humans.
   d. requires that males have a penis.
   e. Both a and c are correct.
2. The evolutionary significance of the placenta is that it allowed
   a. animals to retain their eggs until they hatched.
   b. offspring to be born in an immature state and then develop outside the mother.
   c. offspring to exchange materials with the mother while developing inside the mother.
   d. mammals to develop the ability to produce milk.
3. **THINKING CONCEPTUALLY** Some animals, including hydras, only reproduce sexually when environmental conditions are less than optimal. Explain.

Humans Are Adapted to Reproducing on Land

4. In human males, sterility results when the testes are located inside the abdominal cavity instead of the scrotum because
   a. the sperm cannot pass through the vas deferens.
   b. sperm production is inhibited at body temperature.
   c. the sperm cannot travel the extra distance.
   d. digestive juices destroy the sperm.
5. In the testes, the _____ produce(s) the sperm.
   a. seminiferous tubules   c. seminal vesicles
   b. bulbourethral glands   d. prostate gland
6. The release of the oocyte from the follicle is caused by
   a. a decreasing level of estrogen.
   b. a surge in the level of follicle-stimulating hormone.
   c. a surge in the level of luteinizing hormone.
   d. progesterone released from the corpus luteum.
7. Which of the following is not an event of the ovarian cycle?
   a. FSH promotes the development of a follicle.
   b. The endometrium thickens.
   c. The corpus luteum secretes progesterone.
   d. Ovulation of an egg occurs.
8. Which of the following may cause an eye infection in newborns delivered to infected mothers?
   a. chlamydia   c. AIDS
   b. gonorrhea   d. Both a and b are correct.
9. Which contraceptive method is most effective?
   a. abstinence   c. IUD
   b. diaphragm   d. Depo-Provera injection
10. **THINKING CONCEPTUALLY** Giving men additional testosterone has been shown to decrease sperm production enough to be considered a successful form of birth control. Use negative feedback to find an explanation.

Vertebrates Have Similar Early Developmental Stages and Processes

11. Gastrulation is influenced by the
    a. nervous system.   c. amount of yolk.
    b. environment.   d. presence of hormones.
12. Which of the germ layers is best associated with development of the heart?
    a. ectoderm   d. neurula
    b. mesoderm   e. All of these are correct.
    c. endoderm
13. Which of these stages is mismatched?
    a. cleavage—cell division
    b. blastula—gut formation
    c. gastrula—three germ layers
    d. neurula—nervous system
    e. Both b and c are mismatched.
14. Differentiation is equivalent to which term?
    a. morphogenesis   c. specialization
    b. growth   d. gastrulation
15. Which process refers to the shaping of the embryo and involves cell migration?
    a. cleavage   c. growth
    b. differentiation   d. morphogenesis
16. Place the following in the correct order to describe development: blastula, cleavage, gastrula, morula, neurula, zygote.

Human Development Is Divided into Embryonic Development and Fetal Development

17. In humans, the structure that develops into part of the placenta is the
    a. amnion.   d. chorion.
    b. yolk sac.   e. Both a and c are correct
    c. allantois.
18. In human development, which part of the blastocyst develops into a fetus?
    a. morula   d. chorion
    b. trophoblast   e. yolk sac
    c. inner cell mass
19. In humans, the fetus
    a. has four extraembryonic membranes.
    b. has developed organs and is recognizably human.
    c. is dependent upon the placenta for excretion of wastes and acquisition of nutrients.
    d. All of these are correct.

Understanding the Terms

afterbirth   702
allantois   697
amnion   697
blastocoel   692
blastocyst   698
blastopore   693
blastula   692
bulbourethral gland   684
cervix   686
cesarean section   701
chlamydial infection   689
chorion   697
chorionic villi   699

cleavage   692
clitoris   686
copulation   683
corpus luteum   687
ectoderm   693
embryonic development   697
embryonic disk   699
endoderm   693
endometrium   686
epididymis   684
episiotomy   702
erection   684

estrogen   687
extraembryonic membrane   683, 697
fertilization   687
fetal development   700
fimbria   686
follicle   687
follicle-stimulating hormone (FSH)   685
follicular phase   688
fontanel   700
foreskin   684
gamete   691
gastrulation   693
germ layer   693
gonad   682
gonorrhea   689
gray crescent   695
herpes simplex virus (HSV)   689
human chorionic gonadotropin (HCG)   698
human papillomavirus (HPV)   689
implantation   697
induction   696
larva   683
luteal phase   688
luteinizing hormone (LH)   685
menstruation   688
mesoderm   693
morphogenesis   695
morula   692
neural plate   694
neural tube   694
neurula   694
oocyte   686

oogenesis   687
orgasm   684
ovarian cycle   688
ovary   682, 686
oviduct   686
ovulation   687
parthenogenesis   682
pattern formation   695
penis   684
placenta   683, 700
progesterone   687
proliferative phase   688
prostate gland   684
puberty   685
secondary sex characteristic   685
secretory phase   688
semen (seminal fluid)   684
seminal vesicle   684
seminiferous tubule   685
Sertoli cell   685
sexually transmitted disease (STD)   689
spermatogenesis   685
syphilis   689
testes   682, 684
testosterone   685
trichomoniasis   689
trophoblast   698
umbilical cord   699
uterine cycle   688
uterus   686
vagina   686
vas deferens   684
vulva   686
yolk   683
yolk sac   697

Match the terms to these definitions:
a. _____ Release of an oocyte from the ovary.
b. _____ Organ that produces gametes; the ovary, which produces eggs, and the testis, which produces sperm.
c. _____ Ability of a chemical or a tissue to influence the development of another tissue.
d. _____ Primary tissue layer of a vertebrate embryo.

Thinking Scientifically

1. State a hypothesis that would explain why prostate cancer is common in men. How would you test your hypothesis?
2. Mary has had several early miscarriages and her doctor prescribed progesterone therapy the next time she got pregnant. Explain.

ARIS™ Visit www.mhhe.com/maderconcepts for practice quizzes, animations, videos, and activities designed to help you master the material in this chapter.

BIOLOGICAL VIEWPOINTS

PART V Animals Are Homeostatic

You would need to take many precautions if you decided to hike to the South Pole. Antarctica is a very inhospitable place. No plants and animals live there to serve as a source of food. It's desolate, dangerous, and nothing but ice. Ice bridges, just large enough for a sled, stretch across deep crevasses, and if you fall into the water below, you freeze in minutes. You can get frostbitten if your skin is exposed to the subfreezing air. Therefore, you would be sure to purchase the latest equipment and plan to wear the best available protective clothing. Ultimately however, you would depend on the ability of your body to maintain homeostasis, the relative constancy of the internal environment despite changes in the external environment. Swim the English Channel, climb Mt. Washington, cross the Sahara desert by camel, or hike to the South Pole; if you are healthy, internal conditions stay just about normal because the body has mechanisms that maintain homeostasis.

A French physiologist, Claude Bernard, was the first to recognize, in 1859, that while an animal lives in an external environment, the cells of the body have their own internal environment that is always kept relatively constant. Bernard was working with the liver and noted its ability to release sugar as needed to keep the blood level constant. It wasn't until 1932 that Walter Cannon, an American, used the term homeostasis for physiological processes that keep internal conditions constant, despite an unfavorable change in external conditions. Cannon concluded that a dynamic interplay takes place between external events tending to change the internal environment and physiological activities that counteract the possibility. For example, a sudden drop in environmental temperature will prompt the body to limit blood flow to the skin so that its core temperature stays within a normal range.

All body systems of animals play a role in homeostasis: The digestive system absorbs nutrients into the bloodstream; the respiratory system exchanges gases with the blood; and the cardiovascular system transports these nutrients and gases to tissue fluid, the fluid that bathes the cells. We can conclude, then, that blood and tissue fluid make up the internal environment of animals. The concentration of substances in the blood and tissue fluid must remain relatively constant if animals are to survive. The lymphatic system returns excess tissue fluid to the cardiovascular system, and this helps maintain blood pressure. The lymphatic system also works with the immune system to protect us from disease. No internal organs are more involved in homeostasis than the kidneys, which have the ultimate responsibility of maintaining the normal pH and water-salt balance of the blood. Other systems assist the kidneys; for example, the excretion of carbon dioxide by the lungs helps keep the pH of the blood within normal limits, but only the kidneys can fine-tune the pH of blood.

Animals have two systems—the nervous system and the endocrine system—that regulate the other systems. The nervous system is fast acting because it utilizes nerve impulses that travel along nerve fibers to stimulate particular muscles or glands. The endocrine system acts more slowly because its glands release hormones into the blood, which transports them to target organs or tissues. Once there, hormones cause cells to alter their metabolism; although hormones act more slowly, their effects last longer.

The nervous and endocrine systems communicate with each other and also with the organs of the body to produce a well-ordered whole, in which homeostasis is usually maintained. The hypothalamus, a portion of the brain, communicates with the anterior pituitary, causing it to release or not release its hormones The hypothalamus can also communicate with the autonomic nervous system, whose parasympathetic division keeps organs functioning in a way that promotes healthy internal conditions. The sympathetic division is the "fight or flight" system that keeps animals out of harm's way.

We have observed that both the nervous and endocrine systems utilize negative feedback mechanisms. For example, if blood pressure lowers, the brain directs contraction of smooth muscle within blood vessel walls. The blood vessels constrict, causing blood pressure to rise to an acceptable level. Now the brain ceases to stimulate the smooth muscle of blood vessel walls. Similarly when the blood glucose level rises, the pancreas secretes insulin into the bloodstream. After insulin brings about a lower blood glucose level, the pancreas no longer secretes insulin.

The behavior of an animal that is dependent on the nervous and musculoskeletal systems can also help maintain homeostasis. If we feel thirsty, we drink water until we are no longer thirsty; if we are too cold, we put on clothing until we feel suitably warm. Thus, there is a negative feedback aspect to behavior also!

36

Population Ecology

LEARNING OUTCOMES

After studying this chapter, you should be able to accomplish the following outcomes.

When a Population Grows Too Large

1 Show that overpopulation is injurious in many ways.

Ecology Studies Where and How Organisms Live in the Biosphere

2 Name and compare the ecological levels of study.

Populations Are Not Static—They Change Over Time

3 Contrast population density with population distribution, and give three patterns of population distribution.
4 Calculate the growth rate when given the birthrate and the death rate per 1,000 individuals.
5 List four factors that affect biotic potential.
6 Describe three types of survivorship curves and what they indicate about the life history of a species.
7 Compare exponential growth to logistic growth with reference to the carrying capacity of the environment.

Environmental Interactions Influence Population Size

8 Give examples of density-independent factors and density-dependent factors, telling how they relate to population size.

The Life History Pattern Can Predict Extinction

9 Contrast the characteristics of an opportunistic population with those of an equilibrium population.
10 List factors that help determine whether a population will become extinct.

Human Populations Vary Between Overpopulation and Overconsumption

11 Explain why the world population is still undergoing exponential growth.
12 Contrast the population growth and age distributions of the more-developed and the less-developed countries.
13 Contrast the environmental impact of the more-developed and the less-developed countries.

White-tailed deer, which live from southern Canada to below the equator in South America, are prolific breeders. In one study, investigators found that two male and four female deer produced 160 offspring in six years. Theoretically, the number could have been 300 because a large proportion of does (female deer) breed their first year, and once they start breeding, produce about two young each year of life.

A century ago, the white-tailed deer population across eastern United States was less than half a million. Today, it is well over 200 million deer—even more than existed when Europeans first arrived to colonize America. This dramatic increase in population size can probably be attributed to a lack of predators. For one thing, hunting is tightly controlled by government agencies, and in some areas, it is banned altogether because of the danger it poses to the general public. Similarly, the natural predators of

Buck

deer, such as wolves and mountain lions, are now absent from most regions. This too can be traced to a large human population that is fearful of large predators because they could possibly attack humans and domestic animals.

We like to see a mother with her fawns by the side of the road or scampering off into woods with tails raised to show off the white underside. Or, we find it thrilling to see a large buck (male deer) with majestic antlers partially hidden in the woods. But the sad reality is that, in those areas where deer populations have become too large, the deer suffer from starvation as they deplete their own food supply. For example, after deer hunting was banned on Long Island, New York, the deer population quickly outgrew available food resources. The animals became sickly and weak and weighed so little that their ribs, vertebrae, and pelvic bones were visible through their skin.

Then, too, a very large deer population causes humans many problems. A homeowner is dismayed to see new plants decimated and

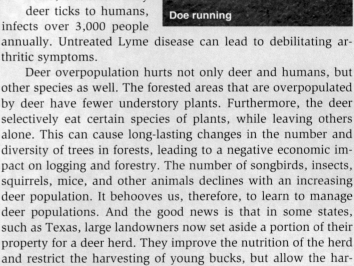

Doe running

evergreen trees damaged due to the munching of deer. The economic damage that large deer populations cause to agriculture, landscaping, and forestry exceeds a billion dollars per year. More alarming, a million deer-vehicle collisions take place in the U.S. each year, resulting in over a billion dollars in insurance claims, thousands of human injuries, and hundreds of human deaths. Lyme disease, transmitted by deer ticks to humans, infects over 3,000 people annually. Untreated Lyme disease can lead to debilitating arthritic symptoms.

Deer overpopulation hurts not only deer and humans, but other species as well. The forested areas that are overpopulated by deer have fewer understory plants. Furthermore, the deer selectively eat certain species of plants, while leaving others alone. This can cause long-lasting changes in the number and diversity of trees in forests, leading to a negative economic impact on logging and forestry. The number of songbirds, insects, squirrels, mice, and other animals declines with an increasing deer population. It behooves us, therefore, to learn to manage deer populations. And the good news is that in some states, such as Texas, large landowners now set aside a portion of their property for a deer herd. They improve the nutrition of the herd and restrict the harvesting of young bucks, but allow the harvesting of does. The result is a self-sustaining herd that brings them economic benefits—they charge others for the privilege of hunting on their land.

In this chapter, we examine the general characteristics of populations. You will learn how the size, distribution, and age structure of a population can change over time and what factors influence populations. You will see that, like the deer in eastern North America, human populations too may suffer the consequences of overpopulation.

Doe and fawn

Ecology Studies Where and How Organisms Live in the Biosphere

As was previously shown in Figure 1.2, the levels of biological organization extend beyond the individual and include the population, community, ecosystem, and biosphere. Ecology studies these higher levels of organization.

36.1 Ecology is studied at various levels

In 1866, the German zoologist Ernst Haeckel coined the word ecology from two Greek roots (*oikos*, home, and *-logy*, study of). He said that **ecology** is the study of the interactions of organisms with other organisms and with the physical environment. Haeckel also pointed out that ecology and evolution are intertwined because ecological interactions are selection pressures that result in evolutionary change, which in turn affects ecological interactions.

Ecology, like so many biological disciplines, is wide-ranging. At one of its lowest levels, ecologists study how the individual organism is adapted to its environment. For example, they study how a fish is adapted to and survives in its **habitat** (the place where the organism lives) (**Fig. 36.1**). Most organisms do not exist singly; rather, they are part of a **population**, defined as all the organisms within an area belonging to the same species and interacting with the environment. At this level of study, ecologists are interested in factors that affect the growth and regulation of population size.

A **community** consists of all the various populations interacting at a locale. In a coral reef, there are numerous populations of algae, corals, crustaceans, fishes, and so forth. At this level, ecologists want to know how interactions such as predation and competition affect the organization of the community. An **ecosystem** encompasses a community of populations as well as the abiotic environment (e.g., the availability of sunlight for plants). Ecosystems rarely have distinct boundaries and are not totally self-sustaining. Usually, a transition zone called an *ecotone*, composed of a mixture of organisms from adjacent ecosystems, exists between ecosystems. The **biosphere** encompasses the zones of the Earth's land, water, and air where living organisms are found. Taking the global view, the entire biosphere is an ecosystem, a place where organisms interact among themselves and with the physical and chemical environments. These interactions help maintain ecosystems and, in turn, the biosphere.

In the next part of the chapter, we begin our study of population dynamics.

> **36.1 Check Your Progress** What ecological levels are affected by deer overpopulation?

FIGURE 36.1
Ecological levels.

| Organism | → | Population | → | Community | → | Ecosystem |

Coral reef ecosystem

Populations Are Not Static—They Change Over Time

Population density and distribution are studied before we consider population growth and survivorship. Population growth affects the age distribution and also the type of growth curve.

36.2 Density and distribution are aspects of population structure

Density Once population size has been estimated, it is possible to calculate the **population density**, which is the number of individuals per unit area. For example, it has been estimated that there are 73 persons per square mile in the United States. Population density figures make it seem as if individuals are uniformly distributed, but this is often not the case. For example, we know full well that most people in the United States live in cities, where the number of people per unit area is dramatically higher than in the country. And even within a city, more people live in particular neighborhoods than others, Furthermore, such distributions can change over time. Therefore, basing ecological models solely on population density, as has often been done in the past, can lead to misleading results.

Distribution **Population distribution** is the pattern of dispersal of individuals across an area of interest. The availability of resources can affect where populations of a species are found. **Resources** are nonliving (abiotic) and living (biotic) components of an environment that support living organisms. Light, water, space, mates, and food are some important resources for populations. **Limiting factors** are those environmental aspects that particularly determine where an organism lives. For example, trout live only in cool mountain streams, where the oxygen content is high, but carp and catfish are found in rivers near the coast because they can tolerate warm waters, which have a low concentration of oxygen. The timberline is the limit of tree growth in mountainous regions or in high latitudes. Trees cannot grow above the high timberline because of low temperatures and the fact that water remains frozen most of the year. The distribution of organisms can also be due to biotic factors. In Australia, the red kangaroo does not live outside arid inland areas because it is adapted to feeding on the grasses that grow there.

Three descriptions—*clumped, random,* and *uniform*—are often used to characterize observed patterns of distribution. Suppose you considered the distribution of a species across its full **range**, that portion of the globe where the species can be found. For example, red kangaroos live in Australia. On that scale, you would expect to find a clumped distribution because organisms are located in areas suitable to their adaptations. That is why red kangaroos live in grasslands, and catfish live in warm river water near the coast.

Within a smaller area, such as a single body of water or a single forest, the availability of resources again influences which distribution pattern is common for a particular population. For example, a study of the distribution of hard clams in a bay on the south shore of Long Island, New York, showed that clam abundance is associated with sediment shell content. Indeed, it might be possible to use this information to transform areas that have few clams into high-abundance areas. Distribution

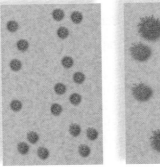
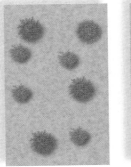
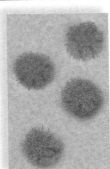

| Clumped | Random | Uniform |

Mature desert shrubs

FIGURE 36.2A Distribution patterns of the creosote bush.

patterns need not be constant. A study of the creosote bush, a desert shrub (**Fig. 36.2A**), revealed that the distribution changed from clumped to random to uniform as the plants matured. As time passed, competition for belowground resources caused the distribution pattern to become uniform. Over their range, Cape gannet populations are clumped, but in a nesting colony, the birds are uniformly distributed (**Fig. 36.2B**).

In Section 36.3, we learn how to calculate the growth rate of a population.

> **36.2 Check Your Progress** There are over 30 deer per square mile in Pennsylvania. This is an estimate of deer population _____.

FIGURE 36.2B Nesting colony of Cape gannets off the coast of New Zealand.

36.3 The growth rate results in population size changes

A population's annual growth rate is dependent upon the number of individuals born each year, the number of individuals that die each year, and annual immigration and emigration. Usually, it is possible to assume that immigration and emigration are equal and need not be considered in the calculation. Therefore, in a simple example, if the number of births is 30 per year, and the number of deaths is 10 per year per 1,000 individuals, the growth rate is 2.0%:

$$(30 - 10)/1,000 = 0.02 = 2.0\%$$

This population will grow because the number of births exceeds the number of deaths. On the other hand, if the number of deaths exceeds the number of births, the value of the growth rate is negative, and the population will shrink.

The **biotic potential** of a population is the highest possible growth rate and is achieved when resources are unlimited.

Whether the biotic potential is high or low depends primarily on the following factors:

1. Usual number of offspring per reproduction
2. Chances of survival until age of reproduction
3. How often each individual reproduces
4. Age at which reproduction begins

For example, a pig, which produces many offspring that quickly mature to produce more offspring, has a much higher biotic potential than a rhinoceros, which produces only one or two offspring per infrequent reproductive event (**Fig. 36.3**).

In Section 36.4, we see that populations have patterns of survivorship.

> **36.3 Check Your Progress** How does decreased hunting affect the growth rates of deer populations?

FIGURE 36.3 Biotic potential varies.

36.4 Survivorship curves illustrate age-related changes

The population growth rate does not take into account that the individuals of a population are in different stages of their life span. **Cohort** is the term used to describe population members that are the same age and have the same chances of surviving. Some investigators study population dynamics and construct **life tables** that show how many members of a cohort are still alive after certain intervals of time. For example, **Figure 36.4A** is a life table for Dall sheep. The cohort contains 1,000 individuals. The table tells us that after one year, 199 individuals have died. Another way to express this same statistic, however, is to consider that 801 individuals are still alive—have survived—after one year. **Survivorship** is the probability of cohort members surviving to particular ages.

If we plot the number surviving at each age, a survivorship curve is produced (**Fig. 36.4B**). The results of such investigations show that each species has a particular survivorship curve. Three typical survivorship curves, numbered I, II, and III, are seen (Fig. 36.4B). Mammals, represented here by the Dall sheep, usually have a type I survivorship curve; they survive well past

Age (years)	Number of survivors at beginning of year	Number of deaths during year
0–1	1,000 1,000 − 199 =	199
1–2	801	12
2–3	789	13
3–4	776	12
4–5	764	30
5–6	734	46
6–7	688	48
7–8	640	69
8–9	571	132
9–10	439	187
10–11	252	156
11–12	96	90
12–13	6	3
13–14	3	3
14–15	0	

FIGURE 36.4A A life table for Dall sheep.

the midpoint of the life span, and death does not come until near the end of the life span. On the other hand, the type III curve is typical of a population, such as oysters, in which most individuals will probably die very young. This type of survivorship curve occurs in many invertebrates, fishes, and humans in less-developed countries. In the type II curve, survivorship decreases at a constant rate throughout the life span; this pattern is typical of hydras, many songbirds, small mammals, and some invertebrates, for which death is usually unrelated to age.

Much can be learned about the life history of a species by studying its life table and the survivorship curve that can be constructed based on this table. For example, a type III survivorship curve indicates that since death probably comes early for most members, only a few live long enough to reproduce. How do you think the other two types of survivorship curves affect reproduction?

Other types of information are also available from studying life tables. In the life table for a plant called blue grass, per capita seed production increases as plants mature, and then seed production drops off. The survivorship curve for blue grass shows that most individuals survive until six to nine months and then the chances of survivorship diminish at an increasing rate.

Populations also have age structure diagrams, as discussed in Section 36.5.

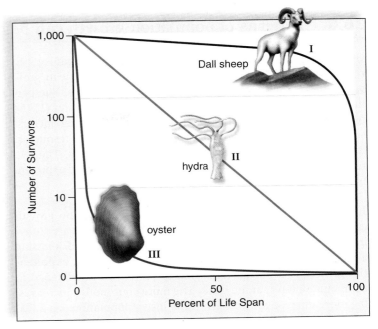

FIGURE 36.4B Three typical survivorship curves.

> **36.4 Check Your Progress** How would the survivorship curve change for the Long Island deer population that was protected from hunting for 30 years, if it became heavily hunted?

36.5 Age structure diagrams divide a population into age groupings

When the individuals in a population reproduce repeatedly, several generations may be alive at any given time. From the perspective of population growth, a population contains three major age groups: prereproductive, reproductive, and postreproductive. Populations differ according to what proportion of the population falls in each age group. At least three patterns are possible, as shown in the **age structure diagrams** in **Figure 36.5**.

When the prereproductive group is the largest of the three groups, the birthrate is higher than the death rate, and a *pyramid-shaped* diagram is expected. Under such conditions, even if the growth for that year were matched by the deaths for that year, the population would continue to grow in the following years. Why? Because there are more individuals entering than leaving the reproductive years. Eventually, as the size of the reproductive group equals the size of the prereproductive group, a *bell-shaped*

diagram results. The postreproductive group is still the smallest, however, because of mortality. If the birthrate falls below the death rate, the prereproductive group becomes smaller than the reproductive group. The age structure diagram is then *urn-shaped*, because the postreproductive group is the largest.

Age distribution reflects the past and future history of a population. Because a postwar baby boom occurred in the United States between 1946 and 1964, the postreproductive group will soon be the largest group in this country.

Typically, the pattern of population growth results in one of two growth curves described in Section 36.6.

> **36.5 Check Your Progress** If fewer prereproductive deer than normal were killed during the hunting season, what effect would this have on the shape of the deer age structure diagram?

FIGURE 36.5
Age structure diagrams.

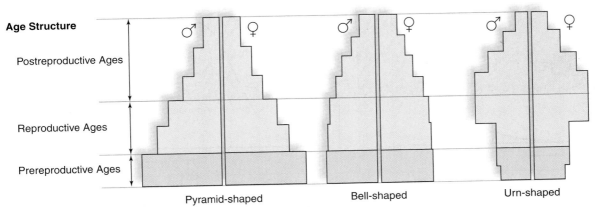

Age Structure

Postreproductive Ages

Reproductive Ages

Prereproductive Ages

Pyramid-shaped   Bell-shaped   Urn-shaped

One of the characteristics of life is the ability of organisms to reproduce. All populations have an enormous ability to grow when the rate of reproduction exceeds the death rate. The particular pattern of a population's growth is dependent on (1) the biotic potential of a population combined with other factors, such as their age structure, and (2) the availability of resources, including environmental factors such as nesting sites. The two fundamental patterns of population growth are termed exponential growth and logistic growth.

Exponential Growth The capacity for population growth can be quite dramatic, as witnessed by ecologists who are studying the growth of a population of insects capable of infesting and taking over an area. Under these circumstances, **exponential growth** is expected. This growth pattern can be likened to compound interest at the bank: The more your money increases, the more interest you will get. If the insect population has 2,000 individuals and the per capita rate of increase is 20% per month, there will be 2,400 insects after one month, 2,880 after two months, 3,456 after three months, and so forth. An exponential pattern of population growth results in a J-shaped curve (**Fig. 36.6A**). Notice that a J-shaped curve has these phases:

> **Lag phase** Growth is slow because the number of individuals in the population is small.

> **Exponential growth phase** Growth is accelerating due to biotic potential (see Section 36.3).

Usually, exponential growth can only continue as long as resources in the environment are unlimited. When the number of individuals in the population approaches the number that can be supported by available resources, competition among individuals for these resources increases.

Logistic Growth As resources decrease, population growth levels off, and a pattern of population growth called **logistic growth** is expected. Logistic growth results in an S-shaped growth curve (**Fig. 36.6B**). Notice that an S-shaped curve has these phases:

> **Lag phase** Growth is slow because the number of individuals in the population is small.

> **Exponential growth phase** Growth is accelerating due to biotic potential).

> **Deceleration phase** The rate of population growth slows because of increased competition among individuals for available resources.

> **Stable equilibrium phase** Although fluctuations can occur, little if any growth takes place because births and deaths are about equal.

The stable equilibrium phase is said to occur at the carrying capacity of the environment. The **carrying capacity** is the total number of individuals the resources of the environment can support for an extended period of time. This number is not a constant and varies with the circumstances and the environment. For example, a large island can support a larger population of penguins than a small island, because the smaller island has a limited amount of space for nesting sites. When the number of nesting sites is inadequate, some birds do not reproduce.

Applications Our knowledge of logistic growth has practical applications. The model predicts that exponential growth will occur only when population size is much lower than the carrying capacity. So, if humans are using a fish population as a continuous food source, it would be best to maintain that population size in the exponential phase of growth when biotic potential is having its full effect and the birthrate is the highest. If people overfish, the fish population will sink into the lag phase, and it may be years before exponential growth recurs. On the other hand, if people are trying to limit the growth of a pest, it is best to reduce the carrying capacity rather than to reduce the population size. Reducing the population size only encourages exponential growth to begin once again. For example, farmers can reduce the carrying capacity for a pest by alternating rows of different crops instead of growing one type of crop throughout the entire field.

Regulation of population size is discussed in the next part of the chapter.

36.6 Check Your Progress If there were abundant but limited food resources for the deer living in a region where hunting was not allowed any longer, what population growth pattern would likely occur?

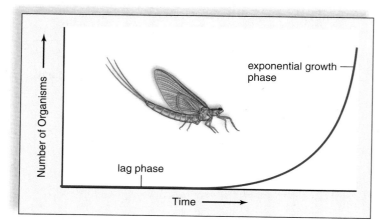

FIGURE 36.6A Exponential growth.

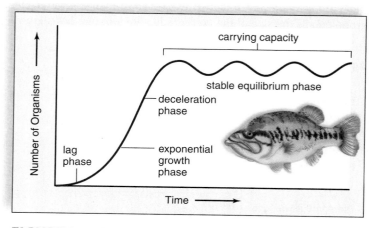

FIGURE 36.6B Logistic growth.

Regulation of population size can be achieved by density-dependent factors (e.g., predators) but not by density-independent factors (e.g., natural disasters). The latter only sporadically reduce population size.

36.7 Density-independent factors affect population size

So far, we have observed that a population has a particular density and a particular growth pattern resulting in a population size. In addition, ecologists have long recognized that environmental interactions play an important role in population size. Abiotic environmental factors include droughts (lack of rain), freezes, hurricanes, floods, and forest fires. Any one of these natural disasters can cause individuals to die and lead to a sudden and catastrophic reduction in population size. However, such an event does not necessarily kill a larger percentage of a dense population compared to a less dense population. Therefore, an abiotic factor is usually a **density-independent factor**, meaning that the percentage of individuals killed remains the same regardless of the population size. In other words, the intensity of the effect does not increase with increased population size. The red line in **Figure 36.7A** shows that mortality percentage (percentage killed) remains the same, regardless of the density of the population.

An example of a density-independent factor is a drought on the Galápagos Islands that caused the population size of one of Darwin's finches (*Geospiza fortis*) to decline from 1,400 to 200 individuals. (The drought caused a reduction in the availability of seeds this species ate.) This is a reduction of 86% of the original population size. Assuming no competition between members of the population, drought, in this instance, is acting as a density-independent factor. We can assume, then, that if the population began with 2,800 individuals, the population would reduce to 400 members, which is the same percentage reduction.

Natural disasters such as hurricanes and floods can have a drastic effect on a population and cause sudden and catastrophic reductions in population size. However, a flash flood does not necessarily kill a larger percentage of a dense population than a less dense population. Therefore, such a flood cannot be counted on to regulate population size, keeping it within the carrying capacity of the environment. Nevertheless, as with our first example, the larger the population, the greater the number of individuals probably affected. In **Figure 36.7B**, the impact of

Low density of mice: 3/5 killed = 60%

High density of mice: 12/20 killed = 64%

FIGURE 36.7B Density-independent effect of a flood.

a flash flood on a low-density population of mice living in a field (mortality rate of 3/5, or 60%) is similar to the impact on a high-density population (mortality rate of 12/20, or 64%).

In Section 36.8, we consider biotic (living) factors, whose effect does increase with the density of the population (see blue line in Fig. 36.7A). Such factors do regulate population size, keeping it within the carrying capacity of the environment.

FIGURE 36.7A
Percentage that die per density of population.

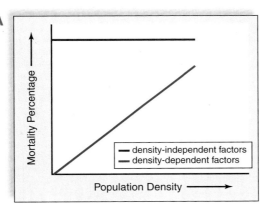

Graph axes: Mortality Percentage (vertical) vs. Population Density (horizontal)
— density-independent factors
— density-dependent factors

> **36.7 Check Your Progress** What are some examples of density-independent factors in a region overpopulated by deer?

Biotic factors tend to be **density-dependent factors** because the percentage of the population affected does increase as the density of the population increases, as represented by the blue line on the graph in Figure 36.7A. Competition, predation, and parasitism are all biotic factors that increase in intensity as the density increases. We will discuss these interactions between populations again in Chapter 38 because they influence community composition and diversity.

Competition can occur when members of the same species attempt to use needed resources (such as light, food, or space) that are in limited supply. As a result, not all members of the population can have access to the resource to the degree necessary to ensure survival or reproduction. As a theoretical example of a density-dependent effect, let's consider a woodpecker population in which members have to compete for nesting sites. Each pair of birds requires a tree hole to raise offspring. In **Figure 36.8A** (*left*), if there are more holes than breeding pairs, each pair can have a hole in which to lay eggs and rear young birds. But if there are fewer holes than breeding pairs, each pair must compete to acquire a nesting site (Fig. 36.8A, *right*). Pairs that fail to gain access to holes will be unable to contribute new members to the population.

As an actual example of competition, four male and 21 female reindeer (*Rangifer*) were released on St. Paul Island in the Bering Sea off Alaska in 1911. St. Paul Island had a completely undisturbed environment, with little hunting pressure and no predators. The herd grew exponentially to about 2,000 reindeer

in 1939, overgrazed the habitat, and then abruptly declined to about eight animals in 1950.

Predation occurs when one living organism, the predator, eats another, the prey. In the broadest sense, predation can include not only animals such as lions, which kill zebras, but also filter-feeding blue whales, which strain krill from the ocean waters; parasitic ticks, which suck blood from their hosts; and even herbivorous deer, which browse on trees and bushes. The effect of predation on a prey population generally increases as the population grows more dense, because prey are easier to find when hiding places are limited. Consider a field inhabited by a population of mice (**Fig. 36.8B**). Each mouse must have a hole in which to hide to avoid being eaten by a hawk. If there are 102 mice, but only 100 holes, two mice will be left out in the open. It might be hard for the hawk to find only two mice in the field. If neither mouse is caught, then the predation rate is 0/2 = 0%. However, if there are 200 mice and only 100 holes, there is a greater chance that hawks will be able to find some of these 100 mice without holes. If half of the *exposed* mice are caught, the predation rate is 50/100 = 50%. Therefore, increasing the density of the available prey has increased the proportion of the population preyed upon.

In the next part of the chapter, we consider how ecologists predict the possible extinction of a species.

> **36.8** *Check Your Progress*   What are some examples of density-dependent factors in a region overpopulated by deer?

FIGURE 36.8A Density-dependent effects: competition.

Every bird has a nesting site.   Every bird does not have a nesting site.

FIGURE 36.8B Density-dependent effects: predation.

2/100 mice cannot hide.   100/200 mice cannot hide.

The life history of a species is based on the population demographics we have just studied and also attributes that result in these demographics. The life history pattern can be used to predict the possibility of extinction.

36.9 Life history patterns consider several population characteristics

Populations vary in terms of the number of births per reproduction, the age at reproduction, the life span, and the probability of living the entire life span. Such particulars are part of a species' **life history**, and life histories often involve trade-offs. For example, each population is able to capture only so much of the available energy, and how this energy is distributed between its life span (short versus long), reproduction events (few versus many), care of offspring (little versus much), and so forth has evolved over the years. Natural selection shapes the final life history of individual species, and therefore it is not surprising that even related species, such as frogs and toads, may have different life history patterns if they occupy different types of environments.

Analysis of exponential and logistic population growth patterns suggests that some populations follow an opportunistic pattern and others follow an equilibrium pattern. An **opportunistic population** tends to live in a fluctuating and/or unpredictable environment. The population remains small until favorable conditions promote exponential growth. The members of the population are small in size, mature early, have a short life span, and provide limited parental care for a great number of offspring. Density-independent effects dramatically affect population size, which is large enough to survive an event that threatens to annihilate it. The population has a high dispersal capacity. Various types of insects and weeds are the best examples of opportunistic species (**Fig. 36.9A**), but there are others. A cod is a rather large fish, weighing up to 12 kg and measuring nearly 2 m in length—but the cod releases gametes in vast numbers, the zygotes form in the sea, and the parents make no further investment in developing offspring. Of the 6–7 million eggs released by a single female cod, only a few will become adult fish.

In contrast, some environments are relatively stable and predictable, allowing population size to remain fairly stable. **Equilibrium populations** exhibit logistic population growth, and the size of the population remains close to, or at, the carrying capacity. Resources are relatively scarce, and the individuals best able to compete—those with phenotypes best suited to the environment—have the largest number of offspring. Members allocate energy to their own growth and survival, and to the growth and survival of their few offspring. Therefore, they are fairly large, slow to mature, and have a relatively long life span (**Fig. 36.9B**). The size of equilibrium populations tends to be regulated by density-dependent effects. The best possible examples include long-lived plants (saguaro cacti, oaks, cypress, and pine), birds of prey (hawks and eagles), and large mammals (whales, elephants, bears, and gorillas).

The application of life history patterns to possible extinction is considered in Section 36.10.

> **36.9 *Check Your Progress*** Which type of species best describes a deer: opportunistic or equilibrium?

Opportunistic Pattern

- Small individuals
- Short life span
- Fast to mature
- Many offspring
- Little or no care of offspring

Equilibrium Pattern

- Large individuals
- Long life span
- Slow to mature
- Few and large offspring
- Much care of offspring

FIGURE 36.9A Dandelions are an opportunistic species.

FIGURE 36.9B Mountain gorillas are an equilibrium species.

36.10 Certain species are more apt to become extinct than others

Extinction is the total disappearance of a species or higher group. Which species shown in Figures 36.9A and 36.9B, the dandelion or the mountain gorilla, is apt to become extinct? Because the dandelion matures quickly, produces many offspring at one time, and has seeds that are dispersed widely by wind, it can more easily withstand a local decimation than can the mountain gorilla.

A study of equilibrium species shows that three other factors—namely, size of geographic range, degree of habitat tolerance, and size of local populations—can help determine whether an equilibrium species is in danger of extinction. **Figure 36.10** compares several equilibrium species on the basis of these three factors. The highlighted text in Figure 36.10 indicates the chances and causes of extinction. The mountain gorilla has a restricted geographic range, narrow habitat tolerance (few preferred places to live), and small local population. This combination of characteristics makes the mountain gorilla very vulnerable to extinction. The possibility of extinction increases depending on whether a species is similar to the gorilla in one, two, or three ways.

The Bengal tiger is endangered because, presently, there are only small local populations, despite the animal having an extensive geographic range and a broad habitat tolerance. What happened? Human beings have restricted the tigers to their present locations. Habitat loss is the greatest threat to wildlife globally.

The results of such population studies can assist conservationists and others who are trying to preserve biodiversity. **Metapopulations** are local populations connected to one another by the movement of individuals between them. One way to create metapopulations is to provide corridors, areas that connect the local populations. This idea is being explored by conservationists as a way to preserve species by providing enough space for a large population to exist.

In the next part of the chapter, we switch gears to consider the demographics of the human population.

> **36.10** *Check Your Progress* **What factors determine the risk of extinction for species that cannot emigrate from one region to another?**

FIGURE 36.10 Highlighted factors indicate vulnerability to extinction for an equilibrium species.

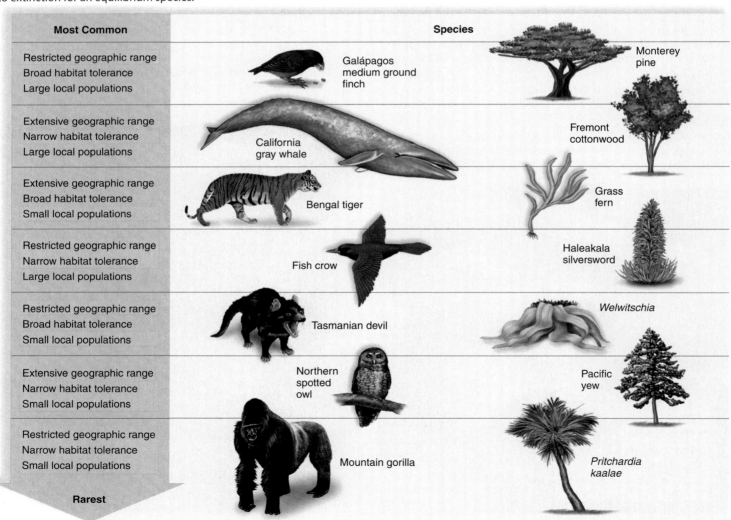

Most Common	Species
Restricted geographic range / Broad habitat tolerance / Large local populations	Galápagos medium ground finch — Monterey pine
Extensive geographic range / Narrow habitat tolerance / Large local populations	California gray whale — Fremont cottonwood
Extensive geographic range / Broad habitat tolerance / Small local populations	Bengal tiger — Grass fern
Restricted geographic range / Narrow habitat tolerance / Large local populations	Fish crow — Haleakala silversword
Restricted geographic range / Broad habitat tolerance / Small local populations	Tasmanian devil — Welwitschia
Extensive geographic range / Narrow habitat tolerance / Small local populations	Northern spotted owl — Pacific yew
Restricted geographic range / Narrow habitat tolerance / Small local populations	Mountain gorilla — Pritchardia kaalae

Rarest

The human population continues to increase, but demographics differ for the more-developed and the less-developed countries.

36.11 World population growth is exponential

The world's population has risen steadily to a present size of about 6.5 billion people (**Fig. 36.11**). Prior to 1750, the growth of the human population was relatively slow, but as more reproducing individuals were added, growth increased, until the curve began to slope steeply upward, indicating that the population was undergoing exponential growth. The number of people added annually to the world population peaked at about 87 million around 1990, and currently it is a little over 79 million per year. This is roughly equal to the current populations of Argentina, Ecuador, and Peru combined.

The potential for future population growth can be appreciated by considering the **doubling time**, the length of time it takes for the population size to double. Currently, the doubling time is estimated to be 53 years. Such an increase in population size would put extreme demands on our ability to produce and distribute resources. In 53 years, the world would need double the amount of food, jobs, water, energy, and so on just to maintain the present standard of living.

Many people are gravely concerned that the amount of time needed to add each additional billion persons to the world population has become shorter and shorter. The first billion didn't occur until 1800; the second billion was attained in 1930; the third billion in 1960; and today there are over 6.6 billion. Only when the number of young women entering the reproductive years is the same as those leaving those years behind can there be zero population growth, when the birthrate equals the death rate, and population size remains steady. The world's population may level off at 8, 10.5, or 14.2 billion, depending on the speed at which the growth rate declines.

More-Developed and Less-Developed Countries

The countries of the world can be divided into two groups. In the **more-developed countries (MDCs)**, typified by countries in North America, Europe, Japan, and Australia, population growth is low, and the people enjoy a good standard of living. In the **less-developed countries (LDCs)**, such as countries in Latin America, Africa, and Asia, population growth is expanding rapidly, and the majority of people live in poverty.

The MDCs doubled their populations between 1850 and 1950. This was largely due to a decline in the death rate, the development of modern medicine, and improved socioeconomic conditions. The decline in the death rate was followed shortly thereafter by a decline in the birthrate, so that populations in the MDCs experienced only modest growth between 1950 and 1975. This sequence of events (i.e., decreased death rate followed by decreased birthrate) is termed a **demographic transition**. Yearly growth of the MDCs as a whole has now stabilized at less than 1%.

Although the death rate began to decline steeply following World War II with the importation of modern medicine from the MDCs, the birthrate remained high. The yearly growth of the LDCs peaked at 2.5% between 1960 and 1965. Since that time, a demographic transition has occurred: The decline in the death rate slowed, and the birthrate fell. The yearly growth rate is now 1.5%. Still, because of past exponential growth, the population of the LDCs may explode from 6.7 billion today to 8 billion in 2050. Today, 88% of the world's population lives in Asia (India and China), Africa, and Latin America.

The age distribution of MDCs and LDCs can predict future growth, as explained in Section 36.12.

> **36.11 Check Your Progress** What effect would using injectable birth control on a wild deer population have on the doubling time of the population?

FIGURE 36.11 World population growth over time.

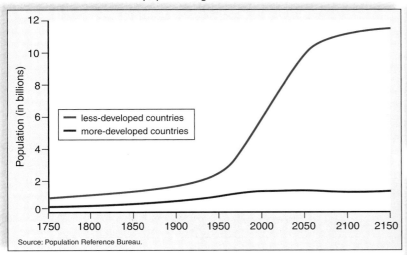

Source: Population Reference Bureau.

Living conditions in more-developed countries

Living conditions in less-developed countries

The populations of the MDCs and LDCs can be divided into three age groups: prereproductive, reproductive, and postreproductive (**Fig. 36.12**). Currently, the LDCs are experiencing a population momentum because they have more women entering the reproductive years than older women leaving them.

Laypeople are sometimes under the impression that if each couple has two children, **zero population growth** (no increase in population size) will take place immediately. However, most countries today will continue growing due to the age structure of the population. If there are more young women entering the reproductive years than there are older women leaving them, so-called **replacement reproduction** will still result in population growth.

Many MDCs have a stable age structure, but most LDCs have a youthful profile—a large proportion of the population is younger than 15. For example, in Nigeria, 49% of the population is under age 15. On average, Nigerian women have 5.9 children. As a result of these rates, the population of Nigeria is expected to increase from 135 million presently to 300 million in 2050. The population of the continent of Africa is projected to increase from 924 million to 2 billion between 2002 and 2050. This means that the LDC populations will still expand, even after replacement reproduction is attained. The more quickly replacement reproduction is achieved, however, the sooner zero population growth will result.

Population Growth and Environmental Impact

Population growth is putting extreme pressure on each country's social organization, the Earth's resources, and the biosphere. Since the population of the LDCs is still growing at a significant rate, it might seem that their population increase will be the greater cause of future environmental degradation. But this is not necessarily the case because the MDCs consume a much larger proportion of the Earth's resources than do the LDCs. For example, the MDCs account for only about one-fourth of the world population, but 90% of the hazardous waste production (pollution) because they consume much greater amounts of fossil fuel, metal, and paper than do LDCs. This consumption leads to environmental degradation, which is also of utmost concern. The environmental impact (E.I.) of a population is measured not only in terms of population size, but also in terms of resource consumption per capita and the pollution that results because of this consumption. In other words:

$$E.I. = \text{population size} \times \text{resource consumption per capita}$$
$$= \text{population per unit of resource used}$$

Therefore, there are two possible types of overpopulation: The first is simply due to population growth, and the second is due to increased resource consumption caused by population growth. The first type of overpopulation is more obvious in LDCs, and the second type is more obvious in MDCs where the per capita consumption is so much higher. For example, an average family in the United States, in terms of per capita resource consumption and waste production, is the equivalent of 30 people in India. We need to realize, therefore, that only a limited number of people can be sustained anywhere near the standard of living of the MDCs.

36.12 *Check Your Progress* **How could a deer population continue to grow if each deer "couple" in the population produced only two offspring?**

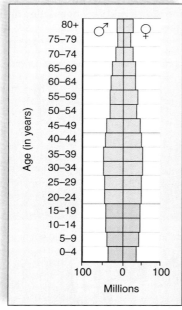
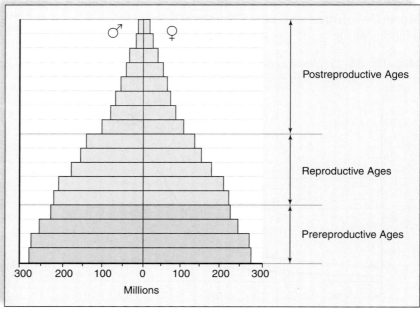

FIGURE 36.12 Age structures in more-developed and less-developed countries.

More-developed countries (MDCs)   Less-developed countries (LDCs)

Modern ecology began with descriptive studies by 19th-century naturalists. In fact, an early definition of the field was "scientific natural history." However, modern ecology has now grown from a simple descriptive field to an experimental, predictive science.

Much of the success in the development of ecology as a predictive science has come from studies of populations and the creation of models that examine how populations change over time. The simplest models are based on population growth when resources are unlimited. This results in exponential population growth, a type only rarely seen in nature. Pest species may exhibit exponential growth until they run out of resources. Because so few natural populations exhibit exponential growth, population ecologists realized they must incorporate resource limitation into their models. The simplest models that account for limited resources result in logistic growth. Populations that exhibit logistic growth cease growth when they reach the environmental carrying capacity.

Many modern ecological studies are concerned with identifying the factors that limit population growth and set the environmental carrying capacity. A combination of careful descriptive studies, experiments done in nature, and sophisticated models has allowed ecologists to accurately predict which factors have the greatest influence on population growth.

We continue this study in Chapter 37 as we see how the behavior of an organism is adapted to the environment. Adaptations to the environment, as you know, result in reproductive success leading possibly to an increase in population size, or at least maintenance of population size, in that environment.

The Chapter In Review

Summary

When a Population Grows Too Large

- Overpopulation can be detrimental to a deer herd by exhausting the local food resources.
- Hunting can help keep a deer herd from becoming overpopulated.

Ecology Studies Where and How Organisms Live in the Biosphere

36.1 Ecology is studied at various levels

- Ecology is the study of the interactions of organisms with other organisms and with the physical environment.
- The levels of ecology are organism, population, community, ecosystem, and biosphere.

Populations Are Not Static—They Change Over Time

36.2 Density and distribution are aspects of population structure

- Population density refers to the number of individuals per unit area.
- Population distribution is the pattern of dispersal of individuals across an area.
- Resources are the components of the environment that support living organisms.
- Limiting factors, such as nutrient availability, are aspects of the environment that determine where an organism lives.

36.3 The growth rate results in population size changes

- Yearly birth and death rates mainly determine a population's growth rate.
- A population's biotic potential is its highest possible growth rate when resources are unlimited.

36.4 Survivorship curves illustrate age-related changes

- A cohort is all the members of the population of the same age with the same chance of surviving.

- Survivorship is the probability of cohort members surviving to a particular age.
- Three general survivorship patterns are recognized.

36.5 Age structure diagrams divide a population into age groupings

- The three age groups are prereproductive, reproductive, and postreproductive.
- The shape of the age distribution diagram reflects the past and future history of a population.

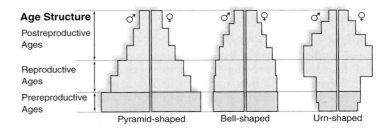

36.6 Patterns of population growth can be described graphically

- Exponential growth is population growth that accelerates over time and results in a J-shaped curve.
- Logistic growth is population size that stabilizes when the carrying capacity has been reached, resulting in an S-shaped curve.

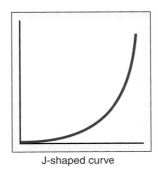

J-shaped curve

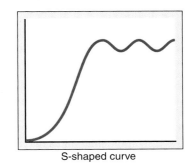

S-shaped curve

Environmental Interactions Influence Population Size

36.7 Density-independent factors affect population size

- Abiotic factors that affect population size include droughts, freezes, hurricanes, floods, and forest fires.
- Density-independent means that mortality (% killed) remains the same, regardless of density.

36.8 Density-dependent factors affect large populations more

- Biotic factors that affect population size include competition, predation, and parasitism.
- Density-dependent means that mortality increases as the density of the population increases.

The Life History Pattern Can Predict Extinction

36.9 Life history patterns consider several population characteristics

- Opportunistic species exhibit exponential population growth; small individuals have a short life span and may die before reproducing.
- Equilibrium species exhibit logistic population growth; large individuals have a long life span, and most of their offspring survive to reproductive age.
- Opportunistic species are adapted to a fluctuating environment, and equilibrium species are adapted to a relatively stable environment.

Opportunistic Pattern	Equilibrium Pattern
• Small individuals • Short life span • Fast to mature • Many offspring • Little or no care of offspring	• Large individuals • Long life span • Slow to mature • Few and large offspring • Much care of offspring
Example: dandelions	Example: mountain gorillas

36.10 Certain species are more apt to become extinct than others

- Extinction is the total disappearance of a species or higher group.
- In addition to life history pattern, size of geographic range, degree of habitat tolerance, and size of local populations determine the risk of extinction for species that cannot migrate.

Human Populations Vary Between Overpopulation and Overconsumption

36.11 World population growth is exponential

- The current world population is 6.6 billion people.
- Based on the predicted doubling time, the world population will require double the amount of current resources in 53 years.
- MDCs are characterized by low population growth and a good standard of living but high resource consumption and waste production.
- LDCs are known for rapidly expanding populations and poverty conditions; their population could increase by 3 billion by 2050.

36.12 Age distributions in MDCs and LDCs are different

- Replacement reproduction occurs when there are more young women entering their reproductive years than older women leaving them.
- Environmental impact is measured based on population size, resource consumption, and pollution.

Ecology Studies Where and How Organisms Live in the Biosphere

1. Which of the following levels of ecological study involves both abiotic and biotic components?
   - a. organisms
   - b. populations
   - c. communities
   - d. ecosystem
   - e. All of these are correct.

2. Place the following levels of organization in order, from lowest to highest.
   - a. community, ecosystem, population, organism
   - b. organism, community, population, ecosystem
   - c. population, ecosystem, organism, community
   - d. organism, population, community, ecosystem

Populations Are Not Static—They Change Over Time

3. The distribution of the human population is
   - a. variable.
   - b. clumped.
   - c. random.
   - d. uniform.

4. If the human birthrate were reduced to 15 per 1,000 per year and the death rate remained the same (9 per thousand), what would be the growth rate?
   - a. 9%
   - b. 6%
   - c. 10%
   - d. 0.6%
   - e. 15%

5. A population's maximum growth rate is also called its
   - a. carrying capacity.
   - b. biotic potential.
   - c. growth curve.
   - d. replacement rate.

6. Which of the following statements about a plant species is not relevant for determining its biotic potential?
   - a. It produces 10 kg of mass per year.
   - b. It produces its first flowers at five years of age.
   - c. 50% of seedlings grow into mature plants.
   - d. On average, 100 seedlings are produced by each plant every year.

7. If a population has a type I survivorship curve (most of its members live the entire life span), which of the following events would you also expect?
   - a. a single reproductive event per adult
   - b. most individuals reproduce
   - c. sporadic reproductive events
   - d. reproduction occurring near the end of the life span
   - e. None of these are correct.

8. Exponential growth is best described by
   - a. steep unrestricted growth.
   - b. an S-shaped growth curve.
   - c. a constant rate of growth.
   - d. growth that levels off after rapid growth.
   - e. Both b and d are correct.

9. When the carrying capacity of the environment is exceeded, the population typically
   - a. increases, but at a slower rate.
   - b. stabilizes at the highest level reached.
   - c. decreases.
   - d. dies off entirely.

10. **THINKING CONCEPTUALLY** The range of a plant that reproduces asexually by runners would be different from one that reproduces by windblown seeds. Explain.

Environmental Interactions Influence Population Size

11. Which of these is a density-independent factor?
    a. competition   c. weather
    b. predation   d. resource availability

For statements 12–14, indicate the type of factor in the key exemplified by the scenario.

KEY:
    a. density-independent   b. competition
       factor   c. predation

12. A severe drought destroys the entire food supply of a herd of gazelle.
13. Only the swiftest coyotes are able to catch the limited supply of rabbits available as a food source. The remaining animals are not strong enough to reproduce.
14. Deer in a forest damage a dense thicket of oak saplings more severely than a few young oak trees.

The Life History Pattern Can Predict Extinction

15. Which of the following is not a feature of an opportunistic life history pattern?
    a. many offspring   c. long life span
    b. little or no care   d. small individuals
       of offspring   e. fast to mature

16. An equilibrium life history pattern does not include
    a. large individuals.   c. few offspring.
    b. long life span.   d. little or no care of offspring.

17. **THINKING CONCEPTUALLY** Under what conditions is it advantageous for a population to produce a large number of small, uncared-for offspring and not large, well-cared-for offspring?

Human Populations Vary Between Overpopulation and Overconsumption

18. The human population
    a. is undergoing exponential growth.
    b. is not subject to environmental resistance.
    c. fluctuates from year to year.
    d. grows only if emigration occurs.
    e. All of these are correct.

19. The current doubling time for the world's population is 53 years. Therefore, if the population is 6 billion people today, what will it be in 159 years if the doubling rate does not change?
    a. 18 billion   c. 36 billion
    b. 48 billion   d. 24 billion

20. Decreased death rate followed by decreased birthrate has occurred in
    a. MDCs.   c. MDCs and LDCs.
    b. LDCs.   d. neither MDCs nor LDCs.

21. The overall reason human population in less-developed countries is expected to explode in the near future is
    a. annual growth rates are very high.
    b. death rates are near zero.
    c. families are getting larger.
    d. the population is in an exponential growth stage.

22. A pyramid-shaped age distribution means that
    a. the prereproductive group is the largest group.
    b. the population will grow for some time in the future.
    c. the country is more likely an LDC than an MDC.
    d. fewer women are leaving the reproductive years than entering them.
    e. All of these are correct.

Understanding the Terms

age structure diagram   713
biosphere   710
biotic potential   712
carrying capacity   714
cohort   712
community   710
demographic transition   719
density-dependent factor   716
density-independent
   factor   715
doubling time   719
ecology   710
ecosystem   710
equilibrium population   717
exponential growth   714
extinction   718
habitat   710
less-developed country
   (LDC)   719
life history   717
life table   712
limiting factor   711
logistic growth   714
metapopulation   718
more-developed country
   (MDC)   719
opportunistic
   population   717
population   710
population density   711
population distribution   711
range   711
replacement
   reproduction   720
resource   711
survivorship   712
zero population
   growth   720

Match the terms to these definitions:
a. _____ Due to industrialization, a decline in the birthrate following a reduction in the death rate so that the population growth rate is lowered.
b. _____ Group of organisms of the same species occupying a certain area and sharing a common gene pool.
c. _____ Growth, particularly of a population, in which the increase occurs in the same manner as compound interest.
d. _____ Maximum population growth rate under ideal conditions.

Thinking Scientifically

1. The right whale population remains dangerously small, despite many decades of complete protection. Formulate a hypothesis based on the four factors listed in Section 36.3 to explain this observation. How would you test your hypothesis?
2. You are a river manager charged with maintaining the flow through the use of dams so that trees, which have equilibrium life histories, can continue to grow along the river. What would you do?

ARIS *Visit* **www.mhhe.com/maderconcepts** *for practice quizzes, animations, videos, and activities designed to help you master the material in this chapter.*

37

Behavioral Ecology

LEARNING OUTCOMES

After studying this chapter, you should be able to accomplish the following outcomes.

For the Benefit of All

1 Relate the highly organized mole rat society to increasing tunnel excavation efficiency.

Both Innate and Learned Behavior Can Be Adaptive

2 Describe experiments showing that behavior is inherited.
3 Describe advantageous circumstances for innate behaviors.
4 Describe experiments showing that behavior can be learned.
5 Describe advantageous circumstances for learned behaviors.
6 Distinguish between classical conditioning, operant conditioning, insight learning, and habituation.

Reproductive Behavior Can Also Be Adaptive

7 Describe the possible effects of female choice on the phenotype of males.
8 Describe how a dominance hierarchy or territoriality could be the end result of male competition.

Social Behavior Can Increase Fitness

9 Compare the reproductive advantages and disadvantages of living in a society.
10 Compose a list of similarities and differences to contrast species with different social structures.
11 Show that altruistic behavior can be self-serving.

Modes of Communication Vary with the Environment

12 Give examples of the different types of communication, and state possible advantages and disadvantages of each type.
13 Present data to show that animals may have emotions.

The mole rat is a rodent, as are the more familiar mice, rats, hamsters, and gerbils. But mole rats are not as closely related to the rodents mentioned as they are to porcupines, guinea pigs, and chinchillas. Porcupines vary in size, but some are almost as large as beavers. They lead a solitary life in forests, deserts, and grasslands. They have a good defense for their nocturnal lifestyle—quills. If a porcupine engages in combat, the quills become embedded in the skin or even a vital organ of the assailant. Chinchillas and guinea pigs are about half the size of a rabbit. Chinchillas are known for their soft thick fur, which is much thicker than that of a rabbit. They live in the Andes mountains of South America where they are so agile that they can jump up to five feet above their heads. Chinchillas live in large colonies, but the organization is not rigid and there are no dominant males or females. Guinea pigs are social and live in small groups consisting of a male, several females, and their young. Their natural habitat is open, grassy areas, and they seek shelter in naturally protected areas or burrows deserted by other animals. Porcupines, chinchillas, and guinea pigs are usually nocturnal, preferring to be active in the dark, They feed on shrubs, grasses, tubers, or fruits. Their adaptations are also seen in mole rats.

Chinchillas

Porcupine

Mole rats (*Heterocephalus glaber*) have a different appearance from their relatives because they have no fur and, indeed, are quite naked except for a few hairs. Also, a mole rat is tiny—only about 7.5 cm long. Mole rats spend almost their entire lives underground and are social to the extreme. In fact, their social behavior is more reminiscent of social insects, such as ants and bees, than most other mammals. Unfortunately, no one knows how much their behavior depends on the secretion of pheromones passed between members of the colony. However, we do know that they recognize each other by smell. Colony members visit the designed "toilet chamber" within the tunnel system to roll around in excrement and ensure that they smell as though they belong.

An average-sized mole rat colony contains about 75 animals, and naked mole rats are so dependent on being part of a colony that an individual kept in isolation will die. The mole rat society has a rigid hierarchical structure. At the top of the hierarchy is the queen, the only female who reproduces. At the bottom are workers of both sexes, who do not reproduce, but instead work tirelessly to help the queen and excavate long and intricate tunnels. When digging tunnels, they line up nose to tail and operate in conveyor-belt style. A digger at the front uses its teeth to break through new soil. Behind the first one, sweepers use their feet to whisk the dirt backwards. The last one in line kicks the dirt up onto the surface of the ground. While the workers dig, soldiers protect them from threats, such as predators and interloping mole rats from other colonies. Soldiers can be male or female and are physically larger than the workers.

The tunnels enable the mole rats to obtain the roots and tubers that they rely on for food and moisture. Without efficient tunnel digging, a mole rat can't live, and it takes several mole rats working together to efficiently dig a tunnel. Some research-

ers believe this explains why mole rats and some other burrowing animals are social.

The queen helps the colony not just by producing young but also by inspecting the tunnels. If the queen senses that more work needs to be done in a particular area of the tunnel system, she forces other colony members into that place.

Naked mole rats digging a burrow

The queen keeps a harem of a few very close relatives with whom she mates. The result of this inbreeding is that all members of the colony share an unusually high degree of genetic similarity: About 80% of their genes are identical. This means that the members of the group see to the propagation of their genes when they help the rest of the colony survive. Interestingly, when the queen is pregnant, the mole rats—of both sexes—develop teats, even though only the queen produces milk to nurse the pups. Pheromones could be at work. The entire colony is devoted to the queen and her offspring. Once the pups are born, the queen delegates much of their care to workers. Such cohesion keeps the colony together.

As we can see in mole rats, behavior is adapted to how and where animals live. The term behavioral ecology recognizes this phenomenon and serves as the title for this chapter. One day, studies will most likely determine how much of animal behavior is innate and how much is learned, a topic often called the "nature versus nurture" question.

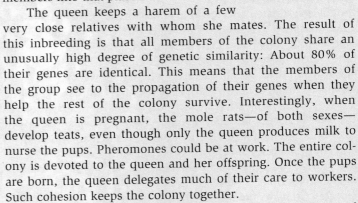

Guinea pigs

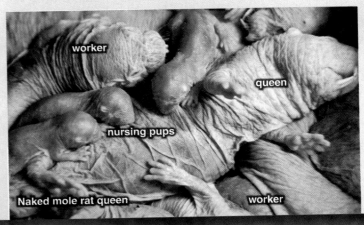

worker

queen

nursing pups

Naked mole rat queen

worker

Both Innate and Learned Behavior Can Be Adaptive

Studies involving various animals conclude that genetics (nature) plays a role in behavior. Other studies show that learning (nurture) is also involved in behavior. Various types of learning are discussed in this section.

37.1 Inheritance influences behavior

The "nature versus nurture" question asks to what extent our genes (nature) and environmental influences (nurture) affect behavior. **Behavior** encompasses any action that can be observed and described. The anatomy and physiology of an animal determine what types of behavior are possible for that animal. Therefore, genes, which control anatomy and physiology, must also control behavior. Experiments have been done to discover the degree to which genetics controls behavior, and many of the results support the hypothesis that some behaviors are innate—that is, they have a genetic basis. Apparently, genes influence the development of neural and hormonal mechanisms that control behavior.

Experiments with Lovebirds Lovebirds are small, green and pink African parrots that nest in tree hollows. There are several closely related species of lovebirds in the genus *Agapornis* that build their nests differently (**Fig. 37.1A**). Fischer lovebirds, *Agapornis fischeri*, use their bills to cut large leaves (or in the laboratory, pieces of paper) into long strips. Then they carry the strips in their bills to the nest, where they weave them with others to make a deep cup. Peach-faced lovebirds, *Agapornis roseicollis*, cut somewhat shorter strips, and they carry them to the nest in a very unusual manner. They pick up the strips in their bills and then insert them into their rump feathers. In this way, they can carry several of these short strips with each trip to the nest, while Fischer lovebirds can carry only one longer strip at a time.

Researchers hypothesized that if the behavior for obtaining and carrying nesting material is inherited, hybrid lovebirds might show intermediate behavior. When the two species of birds were mated, the hybrid birds had difficulty carrying nesting materials. They cut strips of intermediate length and then attempted to tuck the strips into their rump feathers. However, they did not push the strips far enough into the feathers, and so the strips always came out as they walked or flew. Hybrid birds eventually (about three years in this study) learned to carry the cut strips in their beak, but they still briefly turned their head toward their rump before flying off. These studies support the hypothesis that behavior has a genetic basis.

Experiments with Snakes and Snails Several experiments using the garter snake, *Thamnophis elegans*, have been conducted to determine whether food preference has a genetic basis. There are two different types of garter snake populations in California. Inland populations are aquatic and commonly feed underwater on frogs and fish. Coastal populations are terrestrial and feed mainly on slugs. In the laboratory, inland adult snakes

Fischer lovebird with nesting material in its beak.

FIGURE 37.1A Nest-building behavior in lovebirds.

Peach-faced lovebird with nesting material in its rump feathers.

refused to eat slugs, while coastal snakes readily did so. The experimental results of matings between snakes from the two populations (inland and coastal) show that their newborns have an overall intermediate incidence of slug acceptance.

Differences between slug acceptors and slug rejectors appear to be inherited, but what physiological difference is there between the two populations? A clever experiment answered this question. When snakes eat, their tongues carry chemicals to an odor receptor in the roof of the mouth. They use tongue flicks to recognize their prey. Even newborns will flick their tongues at cotton swabs dipped in fluids taken from prey. In this experiment, swabs were dipped in slug extract, and the number of tongue flicks were counted for newborn inland snakes and coastal snakes. Coastal snakes had a higher number of tongue flicks than inland snakes (**Fig. 37.1B**). Apparently, inland snakes do not eat slugs because they are not sensitive to their smell. A genetic difference between the two populations of snakes results in a physiological difference in their nervous systems. Although hybrids showed a great deal of variation in the number of tongue flicks, they were generally intermediate, as predicted by the genetic hypothesis.

The nervous and endocrine systems are both responsible for the coordination of body systems. The studies with garter snakes suggest that the nervous system is involved in determining behavior. Is the endocrine system also involved in behavior? Research studies answer this question in the affirmative. For example, the egg-laying behavior in the marine snail *Aplysia* involves a set sequence of movements. Following copulation, the animal extrudes long strings of more than a million egg cases. It takes the egg case string in its mouth, covers it with mucus, waves its head back and forth to wind the string into an irregular mass, and attaches the mass to a solid object, such as a rock. Several years ago, scientists isolated and analyzed an egg-laying hormone (ELH) that can cause a snail to lay eggs even if it has not mated. ELH was found to be a small protein of 36 amino acids that diffuses into the circulatory system and excites the smooth muscle cells of the reproductive duct, causing them to contract and expel the egg string. Using recombinant DNA techniques, the investigators isolated the ELH gene. The gene's product turned out to be a protein with 271 amino acids. The protein can be cleaved into as many as 11 possible products, and ELH is one of these. ELH alone, or in conjunction with these other products, is thought to control all the components of egg-laying behavior in *Aplysia*.

Experiments with Humans Human twins, on occasion, have been separated at birth and raised under different environmental conditions. Studies of such twins show that they have similar food preferences and activity patterns, and that they even select mates with similar characteristics. These twin studies lend support to the hypothesis that at least certain types of human behavior are primarily influenced by nature (i.e., the genes).

Section 37.2 cites studies that indicate learning modifies behavior.

37.1 *Check Your Progress* If tunnel excavating in mole rats is an inherited trait, what would happen to a mole rat colony if a mutation in the queen reduced the ability of mole rats to cooperate in excavating tunnels?

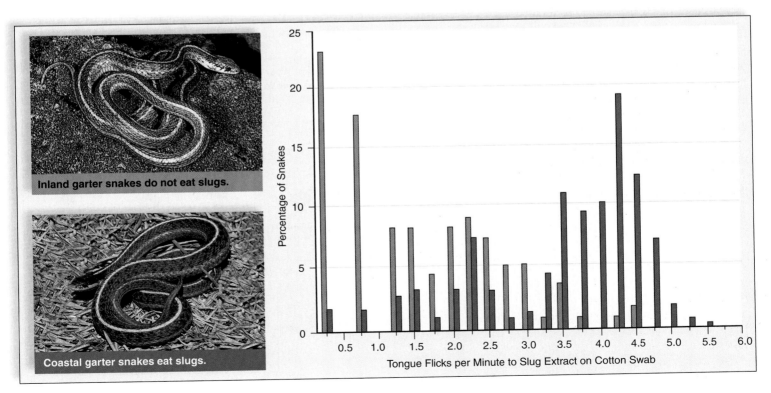

FIGURE 37.1B Feeding behavior in garter snakes.

37.2 Learning can also influence behavior

Even though genetic inheritance serves as a basis for behavior, it is possible that environmental influences (nurture) also affect behavior. For example, behaviorists originally believed that some behaviors were **fixed action patterns (FAP)** elicited by a sign stimulus. But then they found that many behaviors formerly thought to be FAPs improve with practice—for example, by learning. In this context, **learning** is defined as a durable change in behavior brought about by experience.

Learning in Birds Laughing gull chicks' begging behavior appears to be an FAP, because it is always performed the same way in response to the parent's red bill (the sign stimulus) (**Fig. 37.2**). **1** A chick directs a pecking motion toward the parent's bill, grasps it, and strokes it downward. A parent can bring about the begging behavior by swinging its bill gently from side to side. After the chick responds, the parent regurgitates food onto the floor of the nest. If need be, the parent then encourages the chick to eat. This interaction between the chicks and their parents suggests that the begging behavior involves learning. To test this hypothesis, eggs were collected in the field and hatched in a dark incubator to eliminate visual stimuli before the test. For testing purposes, diagrammatic pictures of gull heads were painted on small cards, and each chick was allowed to make about a dozen pecks at the model. The chicks were returned to the nest, and then each was retested daily. **2** The tests showed that on the average, only one-third of the pecks by a newly hatched chick strike the model. But one day after hatching, more than half of the pecks are accurate, and **3** two days after hatching, the accuracy reaches a level of more than 75%. Investigators concluded that improvement in motor skills, as well as visual experience, strongly affect the development of chick begging behavior.

Imprinting Imprinting is considered a form of learning. Imprinting was first observed in birds when chicks, ducklings, and goslings followed the first moving object they saw after hatching. This object is ordinarily their mother, but investigators found that birds can seemingly be imprinted on any object—a human or a red ball—if it is the first moving object they see during a *sensitive period* of two to three days after hatching. The term sensitive period (also called the critical period) means that the behavior develops only during this time.

A chick imprinted on a red ball follows it around and chirps whenever the ball is moved out of sight. Social interactions between parent and offspring during the sensitive period seem key to normal imprinting. For example, female mallards cluck during the entire time imprinting is occurring, and it could be that vocalization before and after hatching is necessary to normal imprinting.

Song Learning White-crowned sparrows sing a species-specific song, but the males of a particular region have their own dialect. Birds were caged in order to test the hypothesis that young white-crowned sparrows learn how to sing from older members of their species.

Three groups of birds were tested. Birds in the first group *heard no songs at all*. When grown, these birds sang a song, but it was not fully developed. Birds in the second group *heard tapes of white-crowns singing*. When grown, they sang in that dialect, as long as the tapes had been played during a sensitive period from about age 10–50 days. White-crowned sparrows' dialects (or other species' songs) played before or after this sensitive period had no effect on the birds. Birds in a third group did not hear tapes and instead were *given an adult tutor*. These birds sang a song of even a different species—no matter when the tutoring began—showing that social interactions apparently assist learning in birds.

Investigators have identified various types of learning, as explained in Section 37.3.

> **37.2 Check Your Progress** How would you test whether worker mole rats have to learn (at least in part) to excavate tunnels?

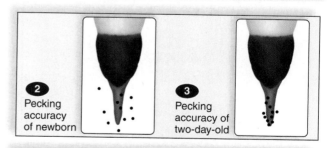

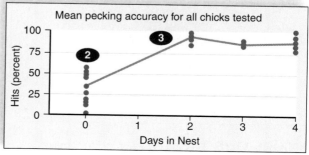

FIGURE 37.2 Begging behavior in laughing gulls improves with experience.

A change in behavior that involves an association between two events is termed **associative learning**. For example, birds that get sick after eating a monarch butterfly no longer prey on monarch butterflies, even though they may be readily available. Or the smell of fresh-baked bread may entice you, even though you have just eaten. If so, perhaps you associate the taste of bread with a pleasant memory, such as being at home. Examples of associative learning include classical conditioning and operant conditioning.

Classical Conditioning In **classical conditioning**, the presentation of two different types of stimuli (at the same time) causes an animal to form an association between them. The best-known laboratory example of classical conditioning was conducted by the Russian psychologist Ivan Pavlov (**Fig. 37.3**). Pavlov observed that dogs salivate when presented with food; in this case, **1** food is an unconditioned stimulus that brings about **2** an unconditioned response (saliva) (left-hand side of Figure 37.3). Then Pavlov rang a bell whenever the dogs were fed. In time, the dogs began to salivate at the sound of the bell, without any food being present. As shown on the right-hand side of Figure 37.3, **3** the sound of the bell is a conditioned stimulus that brings about **4** the conditioned response—that is, salivation even though no food is present.

Classical conditioning suggests that an organism can be trained—that is, conditioned—to associate any response with any stimulus. Unconditioned responses are those that occur naturally, as when salivation follows the presentation of food. Conditioned responses are those that are learned, as when a dog learns to salivate when it hears a bell. Advertisements attempt to use classical conditioning to sell products. Why do commercials pair attractive people with a product being advertised? Advertisers hope that consumers will associate attractiveness with the product and that this pleasant association will cause them to buy the product.

Some types of classical conditioning can be helpful. For example, it's been suggested that you hold children on your lap when reading to them. Why? Because they will associate a pleasant feeling with reading.

Operant Conditioning During **operant conditioning**, a stimulus-response connection is strengthened. Most people know that it is helpful to give an animal a reward, such as food or affection, when teaching it a trick. When we go to an animal show, it is quite obvious that trainers use operant conditioning. They present a stimulus, say, a hoop, and then give a reward (food) for the proper response (jumping through the hoop). The reward need not always be immediate. In latent operant conditioning, an animal makes an association without the immediate reward, as when squirrels make a mental map of where they have hidden nuts.

B. F. Skinner became well known for studying this type of learning in the laboratory. In the simplest type of experiment performed by Skinner, a caged rat happens to press a lever and is rewarded with sugar pellets, which it avidly consumes. There-

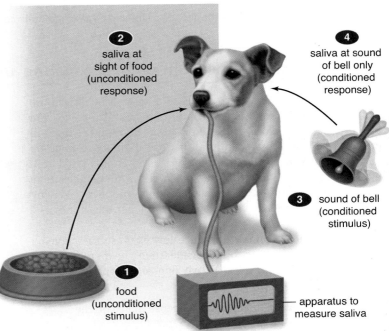

FIGURE 37.3 After classical conditioning, a dog salivates at the sound of a bell and no food is present.

after, the rat regularly presses the lever whenever it wants a sugar pellet. In more sophisticated experiments, Skinner even taught pigeons to play ping-pong by reinforcing desired responses to stimuli. This technique also can be applied to child rearing. It's been suggested that parents who give positive reinforcement for good behavior will be more successful than parents who punish undesirable behavior.

Other Means of Learning In addition to the modes of learning already discussed, animals may learn through insight, imitation, and habituation. **Insight learning** occurs when an animal suddenly solves a problem without any prior experience with the problem. The animal appears to adapt prior experience to solving the problem. For example, chimpanzees have been observed stacking boxes to reach bananas in laboratory settings.

Many organisms learn through observation and imitation. For example, Japanese macaques learn to wash sweet potatoes before eating them by imitating others.

Habituation occurs when an animal no longer responds to a repeated stimulus. Deer ignoring traffic as they graze on the side of a busy highway demonstrate habituation.

The next part of the chapter discusses sexual selection as a part of reproductive behavior.

37.3 *Check Your Progress* Suppose an experimenter could show that older mole rat workers share food with younger workers who are new to tunnel excavation. What type of learning would that be for the younger workers?

It is proposed that sexual selection leads to female choice and male competition, which are observed in, for example, the Raggiana Bird of Paradise.

37.4   Sexual selection can influence mating and other behaviors

We have observed that both innate and learned behaviors are adaptive. Investigations suggest that reproductive behavior can also be adaptive. **Sexual selection** refers to adaptive changes in males and females that lead to an increased ability to secure a mate. Sexual selection in males may result in an increased ability to compete with other males for a mate, while females may select a male with the best **fitness** (ability to produce surviving offspring). In that way, the female increases her own fitness.

Female Choice Females produce few eggs, so the choice of a mate becomes a serious consideration. In a study of satin bowerbirds (**Fig. 37.4A**), two opposing hypotheses regarding female choice were tested:

1. *Good genes hypothesis:* Females choose mates on the basis of traits that improve the chance of survival.
2. *Runaway hypothesis:* Females choose mates on the basis of traits that improve male appearance. The term "runaway" pertains to the possibility that the trait will be exaggerated in the male until its mating benefit is checked by the trait's unfavorable survival cost.

As investigators observed the behavior of satin bowerbirds, they discovered that females usually chose aggressive males as mates. Was this because inherited aggressiveness improves the chance of survival, or because aggressive males are good at stealing blue feathers from other males (since females prefer blue feathers as bower decorations)? The data did not clearly support either hypothesis.

Another bird species, the Raggiana Bird of Paradise, is remarkably *dimorphic*, meaning that males and females differ in size and other traits. The males are larger than the females and have beauti-

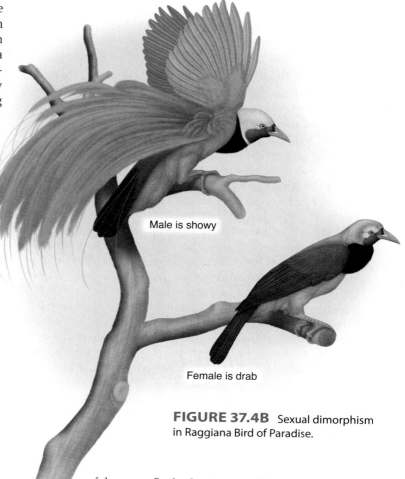

Male is showy

Female is drab

FIGURE 37.4B Sexual dimorphism in Raggiana Bird of Paradise.

ful orange flank plumes, most likely because for generations, females have always chosen this type of male. In contrast, the females are drab (**Fig. 37.4B**). Female choice can explain why male birds are more ornate than females. Consistent with the two hypotheses above, it is possible that the remarkable plumes of the male signify health and vigor to the female. Or, it's possible that females choose the flamboyant males on the basis that their sons will have an increased chance of being selected by females. Some investigators have hypothesized that extravagant features in males could indicate that they are relatively parasite-free. In studies of barn swallows, females also chose males with the longest tails, and investigators have shown that males that are relatively free of parasites have longer tails than otherwise.

Male Competition Males can father many offspring because they continuously produce sperm in great quantity. We expect males to compete in order to inseminate as many females as possible. **Cost-benefit analyses** have been done to determine if the **benefit** of access to mating is worth the **cost** of competition among males.

FIGURE 37.4A Male bowerbird uses blue objects to attract a female to its bower.

FIGURE 37.4C
Male baboon
displaying full threat.

Baboons, a type of Old World monkey, live together in a troop. Males and females have separate **dominance hierarchies** in which a higher-ranking animal has greater access to resources than a lower-ranking animal. Dominance is decided by confrontations, resulting in one animal giving way to the other.

Baboons are dimorphic; the males are larger than the females, and they can threaten other members of the troop with their long, sharp canines (**Fig. 37.4C**). One or more males become dominant by frightening the other males. However, the male baboon pays a cost for his dominant position. Being larger means that he needs more food, and being willing and able to fight predators means that he may get hurt, and so forth. Is there a reproductive benefit to his behavior? Yes, in that dominant males do indeed monopolize females when they are most fertile. Nevertheless, there may be other ways to father offspring. A male may act as a helper to a female and her offspring; then, the next time she is in estrus, she may mate preferentially with him instead of with a dominant male. Or subordinate males may form a friendship group that opposes a dominant male, making him give up a receptive female.

Scientists are able to track an animal in the wild in order to determine its home range, or **territory**. **Territoriality** includes the type of behavior needed to defend a territory against competitors. Baboons travel within a home range, foraging for food each day and sleeping in trees at night. Dominant males decide where and when the troop will move. If the troop is threatened, dominant males protect the troop as it retreats and attack intruders when necessary. Vocalization and displays, rather than outright fighting, may be sufficient to defend a territory. In songbirds, for example, males use singing to announce their willingness to defend a territory. Other males of the species then become reluctant to use the same area.

Red deer stags (males) on the Scottish island of Rhum compete to be the harem master of a group of hinds (females) that mate only with them. The reproductive group occupies a territory that the harem master defends against other stags. Harem masters first attempt to repel challengers by roaring. If the challenger remains, the two lock antlers and push against one another (**Fig. 37.4D**). If the challenger then withdraws, the master pursues him for a short distance, roaring the whole time. If the challenger wins, he becomes the harem master.

A harem master can father two dozen offspring at most, because he is at the peak of his fighting ability for only a short time. But there is a cost to being a harem master. Stags must be large and powerful in order to fight; therefore, they grow faster and have less body fat. During bad times, they are more likely to die of starvation, and in general, they have shorter lives. Harem master behavior will persist in the population only if its cost (reduction in the number of offspring because of a shorter life) is less than its benefit (increased number of offspring due to harem access).

The next part of the chapter discusses aspects of social behavior.

37.4 Check Your Progress A male peacock displays a fan of large, brightly colored tailfeathers to females. Explain the evolution of this trait as an example of sexual selection.

Male is roaring.

Males are fighting.

FIGURE 37.4D Competition between male red deer.

The principles of evolution apply to social behavior. When animals live in groups, the benefits must outweigh the costs, or the behavior would not exist.

37.5   Sociobiology studies the adaptive value of societies

Sociobiology applies the principles of evolutionary biology to the study of social behavior in animals. Sociobiologists hypothesize that living in a society is adaptive because the reproductive benefit is greater than the reproductive cost. The reproductive benefit contributes to the fitness of the individual. Fitness, in this context, means the lifetime reproductive success of the individual—that is, the number of fertile offspring the individual produces.

Group Living   Group living does have its benefits, including helping an animal avoid predators, rear offspring, and find food. A group of impalas is more likely than a solitary one to hear an approaching predator. Many fish moving rapidly in many directions might distract a would-be predator. Weaverbirds form giant colonies that help protect them from predators, and may also allow them to share information about food sources. Primates belonging to the same troop signal one another when they have found an especially bountiful fruit tree. Lions working together are able to capture large prey, such as zebras and buffalo. Pair bonding of trumpet manucodes helps the birds raise their young because, due to their particular food source, the female cannot rear as many offspring alone as she can with the male's help.

Group living also has its disadvantages. When animals are crowded together into a small area, disputes can arise over access to the best feeding places and sleeping sites. Dominance hierarchies are one way to apportion resources, but this puts subordinates at a disadvantage. Among red deer, sons are preferred because, as a harem master, sons will result in a greater number of grandchildren. However, sons, being larger than daughters, need to be nursed more frequently and for a longer period of time. Subordinate females do not have access to enough food resources to adequately nurse sons, and therefore they tend to rear daughters, not sons. Still, like the subordinate males in a baboon troop, subordinate females in a red deer harem may be better off in terms of fitness if they stay with a group, despite the cost involved.

Living in close quarters also exposes individuals to illness and parasites that can easily pass from one animal to another. However, social behavior helps offset some of the proximity disadvantages. For example, baboons and other social primates invest much time in grooming one another, an activity that most likely helps them remain healthy. Likewise, humans have developed extensive medical care to help offset the health problems that arise from living in the densely populated cities of the world.

Altruism   **Altruism** is a behavior that has the potential to decrease the fitness of an individual while benefiting the reproductive success of other members of the society. Altruism does not exist if the inclusive fitness of the individual is enhanced by the behavior.

Genes are passed from one generation to the next in two quite different ways. The first way is direct: A parent passes a gene directly to an offspring. The second way is indirect: A relative that reproduces passes the gene to the next generation. *Direct selection* is adaptation to the environment due to the reproductive success of an individual. *Indirect selection*, also called **kin selection**, is adaptation to the environment due to the reproductive success of the individual's relatives. Together, direct selection and indirect selection make up **inclusive fitness**. In other words, an individual's inclusive fitness includes its direct offspring and the offspring of its relatives. Let's look at examples of inclusive fitness among various societies.

In army ants, like many other insect societies, reproduction is often limited to only one pair, the queen and her mate, the only male in the colony. In addition, the society has three different sizes of sterile female workers. The smallest workers (3 mm), called the nurses, take care of the queen and larvae, feeding them and keeping them clean (**Fig. 37.5A**). The intermediate-sized workers, constituting most of the population, go out on raids to collect food. The soldiers (14 mm), with huge heads and powerful jaws, run along the sides and rear of raiding parties, protecting the column of ants from attack by intruders. Are the sterile workers being altruistic? Among social insects, the queen is diploid (2n), but her mate is haploid (n). If the queen has had only one mate, the sister workers are more closely related to each

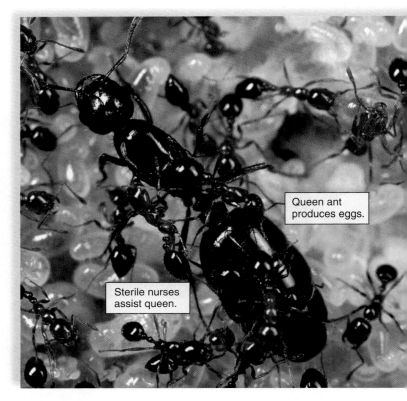

Queen ant produces eggs.

Sterile nurses assist queen.

FIGURE 37.5A   In an ant colony, only the queen reproduces.

other. Sisters share, on average, 75% of their genes because they inherit 100% of their father's alleles. But their potential offspring would share, on average, only 50% of their genes. Therefore, a worker can achieve greater inclusive fitness by helping her mother (the queen) produce additional sisters than by directly reproducing. Under these circumstances, behavior that appears altruistic is more likely to evolve.

Kin selection can also occur among animals whose offspring receive only a half set of genes from both parents. Consider that your brother or sister shares 50% of your genes, your niece or nephew shares 25%, and so on. Therefore, the survival of two nieces (or nephews) is worth the survival of one sibling, assuming they both go on to reproduce.

Among chimpanzees in Africa, a female in estrus frequently copulates with several members of the same group, and the males make no attempt to interfere with each other's matings. How can they be acting in their own self-interest? Genetic relatedness appears to underlie their apparent altruism. Members of a group share more than 50% of their genes in common because members never leave the territory in which they are born.

Reciprocal Altruism In some bird species, offspring from a previous clutch of eggs may stay at the nest to help their parents rear the next batch of offspring. In a study of Florida scrub jays, the number of fledglings produced by an adult pair doubled when they had helpers. Mammalian offspring are also observed to help their parents. In **Figure 37.5B**, a meerkat is acting as a babysitter for its young sisters and brothers while their mother is away. Among jackals in Africa, solitary pairs managed to rear an average of 1.4 pups, whereas pairs with helpers reared 3.6 pups per breeding season.

What are the benefits of staying behind to help? First, a helper is contributing to the survival of its own kin. Therefore, the helper actually gains a fitness benefit. Second, a helper is more likely than a nonhelper to inherit a parental territory—including other helpers. Helping, then, involves making a minimal, short-term reproductive sacrifice in order to maximize future reproductive potential. Therefore, helpers at the nest are also practicing a form of **reciprocal altruism**. Reciprocal altruism also occurs in animals that are not necessarily closely related. In this event, an animal helps or cooperates with another animal with no immediate benefit. However, the animal that was helped will repay the debt at some later time. Reciprocal altruism usually occurs in groups of animals that are mutually dependent. Cheaters in reciprocal altruism are recognized and not reciprocated in future events. For example, reciprocal altruism occurs in vampire bats that live in the tropics. Bats returning to the roost after feeding share their blood meal with other bats in the roost. If a bat fails to share blood with one that previously shared blood with it, the cheater bat will be excluded from future blood sharing.

Applications to Humans Sociobiologists interpret human behavior according to these same principles. Human infants are born helpless and have a much better chance of developing properly if both parents contribute to the effort. Perhaps this explains why the human female, unlike other primates, is continuously amenable to sexual intercourse. Under these circumstances, the male is more likely to remain and help care for the offspring. In any case, it has been suggested that parental love

FIGURE 37.5B Meerkat acting as a babysitter for younger siblings.

is selfish in that it promotes the likelihood that an individual's genes will be present in the next generation's gene pool.

Studies of other human cultures also lend themselves to sociobiological interpretations. Among primitive African tribes, one man often had several wives. This is advantageous not only to the male, but also to the females. In Africa, sources of protein are scarce, and this arrangement allows a female to nurse her baby longer. In contrast, in remote rural regions of Nepal, brothers sometimes even today have the same wife. There, the environment is hostile, and two men are better able to provide the everyday necessities for one family. Since the men are brothers, they help each other look after common genes.

Some people object to interpreting human behavior based on evolutionary fitness, and they stress that modern human behavior need not be determined by relatedness, pointing out that people adopt and care for children who are not related to them.

The next part of the chapter reviews means of communication between animals living in social groups.

> **37.5 Check Your Progress** Porcupines, guinea pigs, and chinchillas are all very vocal, so it's possible that mole rats are vocal too. In other types of animals, sentinels give an alarm call only when close relatives are nearby; otherwise, they remain silent. Would a mole rat soldier ever remain silent?

Animals that live in social groups communicate with one another by signaling. Various means of communication are discussed in this part of the chapter. Some researchers believe that the signaling of animals indicates they have emotions.

37.6 Communication with others involves the senses

Animals exhibit a wide diversity of social behaviors. Some animals are largely solitary and join with a member of the opposite sex only for the purpose of reproduction. Others pair, bond, and cooperate in raising offspring. Still others form a **society** in which the species members are organized in a cooperative manner, extending beyond sexual and parental behavior. We have already mentioned the social groups of baboons and red deer. Social behavior in these and other animals requires that they communicate with one another.

Communication is an action by a sender that may influence the behavior of a receiver. The communication can be purposeful, but it does not have to be. Bats send out a series of sound pulses and then listen for the corresponding echoes in order to find their way through dark caves and locate food at night. Some moths have the ability to hear these sound pulses, and they begin evasive tactics when they sense that a bat is near. Are the bats purposefully communicating with the moths? No; bat sounds are simply a cue to the moths that danger is near.

Types of communication include chemical, language, and tactile.

Chemical Communication

Chemical signals have the advantage of being effective both night and day and from long distances, allowing protection to the communicator. The term **pheromone** designates chemical signals in low concentration that are passed between members of the same species (see the introduction to Chapter 34). Some animals are capable of secreting different pheromones, each with a different meaning. Female moths have special abdominal glands that secrete chemicals, which are detected downwind by receptors on male antennae. The antennae are especially sensitive, and this ensures that only male moths of the correct species (not predators) will be able to detect the pheromones.

Ants and termites mark their trails with pheromones. Cheetahs and other cats mark their territories by depositing urine, feces, and anal gland secretions at the boundaries (**Fig. 37.6A**). Klipspringers (small antelope) use secretions from a gland below the eye to mark twigs and grasses of their territory.

Language Communication

Language uses sound, including the spoken word, and visual gestures or written symbols to communicate with another individual. Many investigators have used various methods to have chimpanzees learn a human language, with some success. Chimpanzees are unable to make many of the sounds humans make, but they can apparently read individual words and use sign language, as illustrated in **Figure 37.6B**.

Sound (auditory) **communication** has some advantages over other kinds of communication. It is faster than chemical communication, and it too is effective both night and day. Further, sound communication can be modified not only by loudness but also by

pattern, duration, and repetition. In an experiment with rats, a researcher discovered that an intruder can avoid attack by increasing the frequency with which it makes an appeasement sound.

Male crickets have calls, and male birds have songs for a number of different occasions. For example, birds may have one song for distress, another for courting, and still another for marking territories. Sailors have long heard the songs of humpback whales transmitted through the hull of a ship. But only recently has it been shown that the song has six basic themes, each with its own phrases, that can vary in length and be interspersed with sundry cries and chirps. The purpose of the song is probably sexual, serving to advertise the availability of the singer. Bottlenose dolphins have one of the most complex languages in the animal kingdom.

Language can be the ultimate auditory communication. Only humans have the biological ability to produce a large number of different sounds and to put them together in many different ways. Nonhuman primates have at most only 40 different vocalizations, each having a definite meaning, such as the one meaning "baby on the ground," which is uttered by a baboon when a baby baboon falls out of a tree. Although chimpanzees can be taught to use an artificial language, they never progress beyond the speech capability of a two-year-old child. It has also been difficult to prove that chimps understand the concept of grammar or can use their language to reason. It still seems that humans possess a communication ability unparalleled by other animals.

Visual communication is most often used by species that are active during the day. Contests between males make use of threat postures and possibly prevent outright fighting, a behavior that might result in reduced fitness. A male baboon displaying full threat is an awesome sight that establishes his dominance and

FIGURE 37.6A Chemical communication: cheetah spraying urine to mark its territory.

FIGURE 37.6B Communication by language can include written symbols and gestures. This chimpanzee points to himself when a shirt has his name on it.

keeps peace within the baboon troop (see Fig. 37.4C). Hippopotamuses perform territorial displays that include mouth opening.

Many animals use complex courtship behaviors and displays. The plumage of a male Raggiana Bird of Paradise allows him to put on a spectacular courtship dance to attract a female, giving her a basis on which to select a mate (see Fig. 37.4B). Territorial and courtship displays are exaggerated and always performed in the same way so that their meaning is clear.

Visual communication allows animals to signal others of their intentions without needing to provide any auditory or chemical messages. The body language of students during a lecture provides an example. Some students lean forward in their seats and make eye contact with the instructor. They want the instructor to know they are interested and find the material of value. Others lean back in their chairs and look at the floor or doodle. These students indicate they are not interested in the material. Teachers can use students' body language to determine whether they are effectively presenting the material and make changes accordingly.

FIGURE 37.6C Tactile communication: baboons engaged in grooming.

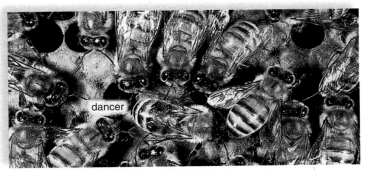

FIGURE 37.6D Tactile communication: waggle dance of bees.

Other human behaviors also send visual clues to others. The hairstyle and dress of a person or the way he or she walks and talks are ways to send messages to others. Some studies have suggested that women are more apt to dress in an appealing manner and be sexually inviting when they are ovulating. People who dress in black, move slowly, fail to make eye contact, and sit alone may be telling others that they do not want social contact. Psychologists have long tried to understand how visual clues can be used to better understand human emotions and behavior. Similarly, researchers are using body language in animals to suggest that they, as well as humans, have emotions (see Section 37.7).

Tactile Communication Tactile communication occurs when one animal touches another. For example, laughing gull chicks peck at the parent's bill to induce the parent to feed them (see Fig. 37.2). A male leopard nuzzles the female's neck to calm her and to stimulate her willingness to mate. In primates, grooming—one animal cleaning the coat and skin of another—helps cement social bonds within a group (**Fig. 37.6C**).

Honeybees use a combination of communication methods, but especially tactile ones, to impart information about the environment. When a foraging bee returns to the hive, it performs a **waggle dance** that indicates the distance and the direction of a food source (**Fig. 37.6D**). As the bee moves between the two loops of a figure 8, it buzzes noisily and shakes its entire body in so-called waggles. Outside the hive, the dance is done on a horizontal surface, and the straight run indicates the direction of the food. Inside, it is dark, and other bees have to follow and touch the dancer in order to get the message. Inside, the angle of the straight run to that of the direction of gravity is the same as the angle of the food source with the sun. In other words, a 40° angle to the left of vertical means that food is 40° to the left of the sun.

Bees can use the sun as a compass to locate food because they have a biological clock, which allows them to compensate for the movement of the sun in the sky. A biological clock is an internal means of telling time. Today, we know that the "ticking" of the clock, in both insects and mammals (including humans), requires alterations in the expression of a gene called *period*.

In the last section of this chapter, we consider whether animals have emotions.

37.6 Check Your Progress The singing of male birds marks their territory, preventing other birds from nesting in the same territory. How would you determine whether this behavior is adaptive?

In recent years, investigators have become interested in determining whether animals have emotions. The body language of animals can be interpreted to suggest that animals do have feelings. When wolves reunite, they wag their tails to and fro, whine, and jump up and down; elephants vocalize—emit their "greeting rumble"—flap their ears, and spin about. Many young animals play with one another or even with themselves, as when dogs chase their own tails. On the other hand, upon the death of a friend or parent, chimps are apt to sulk, stop eating, and even die. It seems reasonable to hypothesize that animals are "happy" when they reunite, "enjoy" themselves when they play, and are "depressed" over the loss of a close friend or relative. Even people who rarely observe animals usually agree about what an animal must be feeling when it exhibits certain behaviors. (**Fig. 37.7**).

In the past, scientists found it expedient to collect data only about observable behavior and to ignore the possible mental state of the animal. Why? Because emotions are personal, and no one can ever know exactly how another animal is feeling. B. F. Skinner, whose research method is described earlier in this chapter, regarded animals as robots that become conditioned to respond automatically to a particular stimulus. He and others never considered that animals might have feelings. But now, some scientists believe they have sufficient data to suggest that at least other vertebrates and/or mammals do have feelings, including fear, joy, embarrassment, jealousy, anger, love, sadness, and grief. And they believe that those who hypothesize otherwise should have to present the opposing data.

Perhaps it would be reasonable to consider the suggestion of Charles Darwin, the father of evolution, who said that animals are different in degree rather than in kind. This means that all animals can, say, feel love, but perhaps not to the degree that humans can. B. Würsig watched the courtship of two right whales. They touched, caressed, rolled side-by-side, and eventually swam off together. He wondered if their behavior indicated they felt love for one another. When you think about it, it is unlikely that emotions first appeared in humans with no evolutionary homologies in animals.

Iguanas, but not fish and frogs, tend to stay where it is warm. M. Cabanac has found that warmth makes iguanas experience a rise in body temperature and an increase in heart rate. These are biological responses associated with emotions in humans. Perhaps the ability of animals to feel pleasure and displeasure is a mental state that rises to the level of consciousness.

Neurobiological data support the hypothesis that other animals, aside from humans, are capable of enjoying themselves when they perform an activity such as play. Researchers have found a high level of dopamine in the brain when rats play, and the dopamine level increases even when rats anticipate the opportunity to play. Certainly even the staunchest critic is aware that many different species of animals have limbic systems and are capable of fight-or-flight responses to dangerous situations.

FIGURE 37.7
Emotions in animals.

Is the rabbit afraid?   Is the young chimp comforted?

Can we go further and suggest that animals feel fear even when no physiological response has yet occurred?

Laboratory animals may be too stressed to provide convincing data on emotions; we have to consider that emotions evolved under an animal's normal environmental conditions. This makes field research more useful. It is possible to fit animals with devices that transmit information on heart rate, body temperature, and eye movements as they go about their daily routine. Such information will help researchers learn how animal emotions might correlate with their behavior, just as emotions influence human behavior. One possible definition describes emotion as a psychological phenomenon that helps animals direct and manage their behavior.

M. Bekoff, who is prominent in the field of animal behavior, encourages us to be open to the possibility that animals have emotions. He states:

> By remaining open to the idea that many animals have rich emotional lives, even if we are wrong in some cases, little truly is lost. By closing the door on the possibility that many animals have rich emotional lives, even if they are very different from our own or from those of animals with whom we are most familiar, we will lose great opportunities to learn about the lives of animals with whom we share this wondrous planet.[1]

> **37.7 Check Your Progress** When a dog snarls, it might be feeling what emotion? When a dog frolics, it might be feeling what emotion? Is your answer anthropomorphic (pertaining to human characteristics)?

[1]Bekoff, M. Animal emotions: Exploring passionate natures. October 2000. *Bioscience* 50:10, page 869.

Birds build nests, dogs bury bones, cats chase quick-moving objects, and snakes bask in the sun. All animals, including humans, behave—they respond to stimuli, both from the physical environment and from other individuals of the same or different species. Behaviorists are scientists who seek answers to two types of questions: What is the neurophysiological mechanism of the behavior, and how is the behavior beneficial to the organism? Using an evolutionary approach, behaviorists generate hypotheses that can be tested to better understand how the behavior increases individual fitness (i.e., the capacity to produce surviving offspring).

Both types of behavioral studies—those concerned with neurophysiology and those concerned with biological fitness—recognize that behavior has a genetic basis. Inheritance determines, for example, that hawks hunt by using vision rather than smell, and that cats, not dogs, climb trees. These behaviors are adaptive to the ecological environment in which the animals live. Inheritance produces the behavioral variations that are subject to natural selection, resulting in the most adaptive responses to the environment.

The study of behavior does not settle the nature-versus-nurture question because we now know that behavior can be modified by experience. Songbirds are born with the ability to sing, but which song or dialect they sing is strongly dependent on the songs they hear from their parents and siblings. A new chimpanzee mother naturally cares for her young, but she is a better mother if she observes other females in her troop raising their young.

Certain behaviors come with a cost. Why should subordinate members of a baboon troop, or subordinate females in a red deer harem, remain in a situation that seemingly leads to reduced fitness? And why should older offspring help younger offspring? The evolutionary answer is that, in the end, the benefits outweigh the costs. Otherwise, the behavior would not continue. The evolutionary approach to studying behavior has proved fruitful in helping us understand why birds sing their melodious songs, dolphins frolic in groups, and wondrous Raggiana Birds of Paradise display to females in their particular ecological environment.

The Chapter in Review

Summary

For the Benefit of All

- The social behavior of naked mole rats is somewhat like that of ants and bees.
- Mole rat society has a rigid hierarchical structure for work and reproduction.

Both Innate and Learned Behavior Can Be Adaptive

37.1   Inheritance influences behavior

- Behavior is any action that can be observed and described.
- Lovebird experiments and studies of human twins support the hypothesis that behavior has a genetic basis (nature).

37.2   Learning can also influence behavior

- Learning (nurture) causes a durable change in behavior.
- In laughing gull chicks, visual experience and improvement in motor skills affect the development of begging behavior.
- Imprinting is a form of learning that allows baby birds to follow the first moving object they see during a sensitive period.
- Social interactions assist song learning in birds.

37.3   Associative learning links behavior to stimuli

- Classical conditioning involves two types of stimuli that occur at the same time (e.g., Pavlov's dogs).

- In operant conditioning, positive reinforcement is provided for desired behavior.
- Insight learning appears to have occurred when an animal solves a problem without prior experience.
- Habituation has taken place when an animal no longer responds to a repeated stimulus.

Reproductive Behavior Can Also Be Adaptive

37.4   Sexual selection can influence mating and other behaviors

- Sexual selection refers to adaptive changes in males and females that lead to increased ability to get a mate.
- The good genes hypothesis suggests that females choose mates based on traits that improve survival.
- The runaway hypothesis suggests that females choose mates based on traits that improve male appearance.
- Sexual dimorphism means that males and females differ in size and other traits.
- Territoriality includes the behaviors needed to defend a territory.

Social Behavior Can Increase Fitness

37.5 Sociobiology studies the adaptive value of societies

- Sociobiology applies the principles of evolutionary biology to the study of social behavior.
- Only if benefits (in terms of individual reproductive success) outweigh costs will societies come into existence.
- Fitness is the lifetime reproductive success of an individual.
- Group living has both advantages and disadvantages.
- Altruism is a behavior that benefits the reproductive success of another individual, but decreases the reproductive success of the altruist.
- When animals act altruistically, they may be increasing their own inclusive fitness.

Modes of Communication Vary with the Environment

37.6 Communication with others involves the senses

- Communication is a signal by a sender that affects the behavior of a receiver.
- Chemical communication involves pheromones.
- Language communication uses sound, including the spoken word, and visual communication, including posturing, body language, or eye contact.
- Tactile communication occurs via touch.

37.7 Do animals have emotions?

- The body language of certain animals suggests that they do have emotions, and some research backs this up.

Testing Yourself

Both Innate and Learned Behavior Can Be Adaptive

1. Behavior is
   a. any action that is learned.
   b. all responses to the environment.
   c. any action that can be observed and described.
   d. all activity that is controlled by hormones.
   e. unique to birds and humans.
2. Egg-laying hormone causes snails to lay eggs, implicating which system in this behavior?
   a. digestive   d. lymphatic
   b. endocrine   e. respiratory
   c. nervous
3. Which of the following is not an example of a genetically based behavior?
   a. Inland garter snakes do not eat slugs, while coastal populations do.
   b. One species of lovebird carries nesting strips one at a time, while another carries several.
   c. Human twins have similar activity patterns.
   d. Snails lay eggs in response to egg-laying hormone.
   e. Wild foxes raised in captivity are not capable of hunting for food.
4. How would the following graph differ if pecking behavior in laughing gulls were a fixed action pattern not influenced by learning?
   a. It would be a diagonal line with an upward incline.
   b. It would be a diagonal line with a downward incline.
   c. It would be a horizontal line.
   d. It would be a vertical line.
   e. None of these are correct.

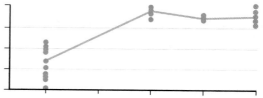

5. The benefits of imprinting (following and obeying the mother) generally outweigh the costs (following the wrong object instead of the mother) because
   a. an animal that has been imprinted on the wrong object can be reimprinted on the mother.
   b. imprinting behavior never lasts more than a few months.
   c. animals in the wild rarely imprint on anything other than their mother.
   d. animals that imprint on the wrong object generally die before they pass their genes on.
6. In white-crowned sparrows, social experience exhibits a very strong influence over the development of singing patterns. What observation led to this conclusion?
   a. Birds only learned to sing when they were trained by other birds.
   b. The window during which birds learn from other birds is wider than that during which they learn from tape recordings.
   c. Birds could learn different dialects only from other birds.
   d. Birds that learned to sing from a tape recorder could change their song when they listened to another bird.
7. Which of the following best describes classical conditioning?
   a. gradual strengthening of stimulus-response connections that are seemingly unrelated
   b. type of associative learning in which there is no contingency between response and reinforcer
   c. learning behavior in which an organism follows the first moving object it encounters
   d. learning behavior in which an organism exhibits a fixed action pattern from the time of birth
8. Using treats to train a dog to do a trick is an example of
   a. imprinting.   c. vocalization.
   b. tutoring.   d. operant conditioning.
9. Why does operant conditioning work? An animal
   a. is punished when it does not perform the behavior.
   b. associates the ringing of a bell with the behavior.
   c. is rewarded when it performs the behavior.
   d. is imprinted on the first moving object it sees.

Reproductive Behavior Can Also Be Adaptive

10. In the Raggiana Bird of Paradise, the female bird chooses a mate that is the most flamboyant. This supports
    a. the good genes hypothesis.
    b. the runaway hypothesis.
    c. Either hypothesis could be true.
    d. Neither hypothesis is true.
11. Which of the following are costs that a dominant male baboon must pay in order to gain a reproductive benefit?
    a. He requires more food and must travel longer distances.
    b. He requires more food and must care for his young.
    c. He is more prone to injury and requires more food.
    d. He is more prone to injury and must care for his young.
    e. He must care for his young and travel longer distances.

12. A red deer harem master may die earlier than other deer because he is
    a. likely to be expelled from the herd and cannot survive alone.
    b. more prone to disease because he interacts with so many animals.
    c. more likely to starve to death.
    d. apt to place himself between a predator and the herd to protect the herd.
13. **THINKING CONCEPTUALLY** Males typically compete for mates and females typically choose between males for mates. Explain why these behaviors evolved.

Social Behavior Can Increase Fitness

14. Which of the following is not a benefit of group living?
    a. increased availability of food
    b. protection from predators
    c. increased success in rearing offspring
    d. protection from illness and parasites
15. Which of the following does not contribute to inclusive fitness?
    a. direct selection   d. Both b and c are correct.
    b. indirect selection   e. All of these are correct.
    c. kin selection
16. On the average, how may genes do first cousins have in common?
    a. $\frac{1}{8}$ c. $\frac{1}{4}$
    b. $\frac{1}{2}$ d. $\frac{1}{16}$
17. **THINKING CONCEPTUALLY** Some birds do not produce offspring for the first couple of years. Why would it be beneficial for them to help their parent reproduce or practice reciprocal altruism?

Modes of Communication Vary with the Environment

For problems 18–21, match the type of communication in the key with its description. Answers may be used more than once.

KEY:
    a. chemical communication
    b. auditory communication
    c. visual communication
    d. tactile communication
18. Aphids (insects) release an alarm pheromone when they sense they are in danger.
19. Male peacocks exhibit an elaborate display of feathers to attract females.
20. Ground squirrels give an alarm call to warn others of the approach of a predator.
21. Laughing gull chicks peck at a parent's bill to stimulate feeding.
22. The hypothesis that rats enjoy playing is supported by the finding that
    a. rat neurotransmitter levels change in anticipation of play.
    b. rats, if allowed, will play for hours on end.
    c. rats continue to play throughout their lives.
    d. rats that play are less likely to be aggressive.
23. Which of these statements best explains the evolution of a behavior that attracts mates but could also attract a predator?
    a. Both are equally possible.
    b. Attracting a mate is more likely than attracting a predator.
    c. Attracting a predator is more likely than attracting a mate.
    d. You have good genes or else you couldn't attract a mate.
    e. Both b and d are correct.

Understanding the Terms

altruism 732	insight learning 729
associative learning 729	kin selection 732
behavior 726	learning 728
benefit 730	operant conditioning 729
classical conditioning 729	pheromone 734
communication 734	reciprocal altruism 733
cost 730	sexual selection 730
cost-benefit analysis 730	society 734
dominance hierachy 731	sociobiology 732
fitness 730	sound communication 734
fixed action pattern	territoriality 731
(FAP) 728	territory 731
habituation 729	visual communication 734
imprinting 728	waggle dance 735
inclusive fitness 732	

Match the terms to these definitions:
a. _____ Behavior related to defending a particular area; often used for the purpose of feeding, mating, and caring for young.
b. _____ Social interaction that benefits others but has the potential to decrease the lifetime reproductive success of the member exhibiting the behavior.
c. _____ Signal by a sender that may influence the behavior of a receiver.
d. _____ Chemical substance secreted into the environment by one organism that may influence the behavior of another of the same species.

Thinking Scientifically

1. You have observed that New York rats keep their distance from Limburger cheese, which has a disagreeable odor, and do not approach to eat it, while Florida rats readily approach and eat the cheese. Design an experiment that would test that this behavior is genetically controlled.
2. Among meerkats, sentries stand guard on rocks and serve as lookouts while others feed. Determine if the sentry is closely related to the animals who are eating. How would you decide whether sentry behavior is altruistic?

ARIS *Visit* www.mhhe.com/maderconcepts *for practice quizzes, animations, videos, and activities designed to help you master the material in this chapter.*

38

Community and Ecosystem Ecology

LEARNING OUTCOMES

After studying this chapter, you should be able to accomplish the following outcomes.

Ridding the Land of Waste

1 Distinguish between scavengers and decomposers, and state their roles in ridding the land of waste.

A Community Contains Several Interacting Populations in the Same Locale

2 Show that competition leads to resource partitioning.
3 Define predation, and discuss predator-prey population size dynamics.
4 Discuss various types of prey defenses, and give examples of two types of mimicry.
5 Contrast parasitism, commensalism, and mutualism, giving examples of each.

A Community Develops and Changes Over Time

6 Contrast community composition and diversity.
7 Compare and contrast the stages of primary and secondary succession.

An Ecosystem Is a Community Interacting with the Physical Environment

8 Define an organism's habitat and its niche.
9 Characterize the biotic components of a community according to their food source.
10 Compare and contrast energy flow through an ecosystem with chemical cycling within an ecosystem.
11 Compare and contrast a grazing food web with a detrital food web.
12 Explain the shape of an ecological pyramid, whether considering biomass or energy.
13 Describe the function of the reservoir, exchange pool, and biotic community in a biogeochemical cycle.
14 State the major steps of the phosphorus, nitrogen, and carbon cycles.

I f it weren't for scavengers and decomposers, dead organic matter would accumulate in deep piles so that you wouldn't even be able to step outside your door! But thanks to decomposition, that doesn't happen. As soon as an animal dies, all of the microorganisms in its gut set to work, and decomposition begins. The faint smell of decay very quickly attracts scavengers, which seek out dead animals and eat them.

Vultures

One well-known scavenger is the vulture, a bird equipped with an excellent sense of smell that helps it locate an animal carcass from miles away. Vultures attract a lot of attention because they are rather large, usually black birds with bald heads devoid of feathers. In some rural mountainous areas of the world, humans rely on vultures' assistance with "sky burials," in which a human corpse is cut into small pieces and left for the scavenging birds to eat. In Tibet, this practice is necessary because the frozen ground is too hard to bury the dead, and wood is too scarce for cremation. This ritual is also based on an appreciation for nutrient recycling; the deceased is providing nutrients that ultimately sustain all living beings.

Vultures assist in a sky burial.

740

Ridding the Land of Waste

Besides vultures, other scavengers include beetles, earthworms, and some insects. We can't include carrion beetles and flies because they are attracted to decomposing animals not as a source of food but as a nutrient-rich place to lay their eggs. Fly larvae are called maggots, and almost everyone has seen them crawling around on dead meat left to rot. Dung beetles are not scavengers either, but they are of interest because they eat and raise their young in animal feces. They have no need to eat other foods or drink water, because feces provide them with complete nutrition. By helping recycle fecal nutrients back into the land and by improving the hygiene of livestock pastures, dung beetles provide $380 million worth of services to farmers in the United States every year.

Decomposers are not scavengers either because they do not take into their bodies pieces of dead animals or plants, as scavengers do. Instead, they are saprotrophs that secrete their digestive juices into the environment and then absorb the nutrients. The most abundant and widespread saprotrophs in the world are microbes, predominantly bacteria and fungi. In order to compete with scavengers for the nutrients in decaying matter, microbes often secrete chemicals that are foul-smelling, distasteful, and/or toxic. According to biologist Dan Janzen, this is why fruits rot, seeds mold, and meat spoils. If you have ever avoided eating a moldy strawberry or drinking curdled milk, the microbes have won.

Decomposers perform a service for the entire biosphere when they recycle nutrients to plants, which produce food for themselves directly and for all living things indirectly. Of increasing interest is the use of microbial decomposers to clean up oil spills and obtain energy from leftover organic matter. For example, microbes can convert wood chips, animal (e.g., cow) feces, and all kinds of organic matter into high-energy biogas, which can be used as a source of methane for heating, cooking, and generating electricity. In Rwanda, there are "poop-powered" prisons in which prisoners' feces are digested into biogas, and in San Francisco, biogas may soon be made from the huge quantity of pet feces produced within the city.

Biogas plant

In this chapter, we will look at how the populations in a community interact, as when decomposers compete with scavengers for food. Then we will consider how populations interact not only among themselves, but also with the physical environment in ecosystems. Ecosystems are very much dependent on the work of scavengers and decomposers for nutrient recycling. We rely on the process of nutrient cycling every day for agriculture, and also to treat sewage and purify water. Our society can also tap into the leftover high-energy compounds produced by decomposers. Someday, your house may be heated, your food may be cooked, or your electronics may be powered using methane (biogas) obtained from decomposing matter!

Dung beetles rolling dung

Dung beetles assist in dung cleanup.

Before   After

Microbes assist in oil cleanup.

A Community Contains Several Interacting Populations in the Same Locale

Learning Outcomes 2–5, page 740

Step into a forest and look around. Most likely, you will notice populations of animals interacting with plants and other animals. Insects will be feeding on the leaves of trees, or visiting flowers or even grasses. If you are lucky, you might see a hawk dart from a tree on the edge of the forest and grab a rodent running through an adjoining meadow. All the populations interacting in the forest form a **community.** This part of the chapter considers various ways that populations interact within a community, including competition, predation, and various symbiotic relationships.

38.1 Competition can lead to resource partitioning

Competition is rivalry between populations for the same resources, such as light, space, nutrients, or mates. In the 1930s, G. F. Gause grew two species of *Paramecium* in one test tube containing a fixed amount of bacterial food. Although populations of each species survived when grown in separate test tubes, only one species, *Paramecium aurelia*, survived when the two species were grown together (**Fig. 38.1A**). *P. aurelia* acquired more of the food resource and had a higher population growth rate than did *P. caudatum*. Eventually, as the *P. aurelia* population grew and obtained an increasingly greater proportion of the food, the number of *P. caudatum* individuals decreased, and the population died out. This experiment and others helped ecologists formulate the **competitive exclusion principle,** which states that no two species can occupy the same niche at the same time. The **ecological niche** of an organism is the role it plays in its community, including its **habitat** (where the organism lives) and its interactions with other organisms and the environment. See Figure 38.9 for a more complete description of niche.

Competition for resources does not always lead to localized extinction of a species. Multiple species can coexist in communi-ties by partitioning resources. In another laboratory experiment using other species of *Paramecium*, Gause found that two species could survive in the same test tube if one species consumed bacteria at the bottom of the tube and the other ate bacteria suspended in solution. This **resource partitioning** decreased competition between the two species. What could have been one niche became two more-specialized niches due to species differences in feeding behavior.

When species of ground finches live on the same island of the Galápagos, their beak sizes differ, and each species feeds on a different-sized seed (**Fig. 38.1B**). When the ground finches live on different islands, the beak size does not differ and each species feeds on the same preferred range of seeds. Such so-called **character displacement** is often viewed as evidence that competition and resource partitioning do take place.

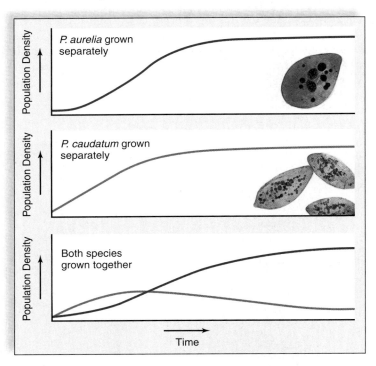

FIGURE 38.1A Competition only occurs between two species of *Paramecium* when they are grown together.

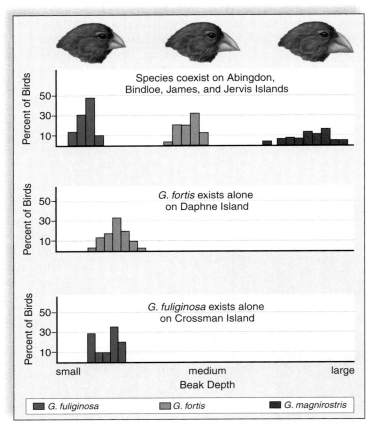

FIGURE 38.1B Character displacement occurs in finches when they coexist, compared to when they exist separately.

The niche specialization that permits coexistence of multiple species can be very subtle. Species of warblers that live in North American forests are all about the same size, and all feed on budworms, a type of caterpillar found on spruce trees. Robert MacArthur recorded the length of time each warbler species spent in different regions of spruce canopies to determine where each species did most of its feeding. He discovered that each species primarily used different parts of the tree canopy and, in that way, had a more specialized niche (**Fig. 38.1C**). As another example, consider that three types of birds—swallows, swifts, and martins—all eat flying insects and parachuting spiders. These birds even frequently fly in mixed flocks. But each type of bird has a different nesting site and migrates at a slightly different time of year.

In all these cases of niche specialization, we have merely assumed that what we observe today is due to competition in the past. Some ecologists are fond of saying that in doing so we have invoked the "ghosts of competition past." Are there any instances in which competition has actually been observed? Joseph Connell has studied the distribution of barnacles on the Scottish coast, where a small barnacle (*Chthamalus stellatus*) lives on the high part of the intertidal zone, and a large barnacle (*Balanus balanoides*) lives on the lower part (**Fig. 38.1D**). Why should that be, when their free-swimming larvae attach themselves to rocks at any point in the intertidal zone? The answer is that the faster-growing *Balanus* crowds out *Chthamalus* in the lower tidal zone, but cannot do so in the upper tidal zone because it is more susceptible to drying out than *Chthamalus*.

In Section 38.2, we consider predation, which is another common type of interaction between populations in a community.

38.1 Check Your Progress The activities of decomposers illustrate that competitors are not always passive toward another competitor. Explain.

FIGURE 38.1C Niche specialization occurs among five species of coexisting warblers.

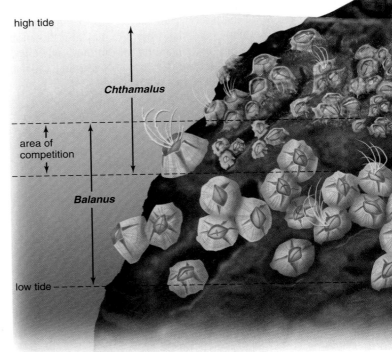

FIGURE 38.1D Competition occurs between two species of barnacles.

Predation occurs when one living organism, called the predator, feeds on another, called the prey. In its broadest sense, predaceous consumers include not only animals such as lions, which kill zebras, but also filter-feeding blue whales, which strain krill from ocean waters; herbivorous deer, which browse on trees and shrubs; parasitic ticks, which suck blood from their victims; and *parasitoids*, which are wasps that lay their eggs inside the body of a host. The resulting larvae feed on the host, sometimes causing death. Parasitism can be considered a type of predation because one individual obtains nutrients from another (see Section 38.3). Predation and parasitism are expected to increase the abundance of the predator and parasite at the expense of the abundance of the prey or host.

Predator-Prey Population Dynamics

Do predators reduce the population density of prey? In another classic experiment, G. F. Gause reared the ciliated protozoans *Paramecium caudatum* (prey) and *Didinium nasutum* (predator) together in a culture medium. He observed that *Didinium* ate all the *Paramecium* and then died of starvation. In nature, we can find a similar example. When a gardener brought prickly-pear cactus to Australia from South America, the cactus spread out of control until millions of acres were covered with nothing but cacti. The cacti were brought under control when a moth from South America, whose caterpillar feeds only on the cactus, was introduced. The caterpillar was a voracious predator on the cactus, efficiently reducing the cactus population. Now both cactus and moth are found at greatly reduced densities in Australia.

This raises an interesting point: The population density of the predator can be affected by the prevalence of the prey. In other words, the predator-prey relationship is actually a two-way street. In that context, consider that at first the biotic potential (maximum reproductive rate) of the prickly-pear cactus was maximized, but factors that oppose biotic potential came into play after the moth was introduced. And the biotic potential of the moth was maximized when it was first introduced, but the carrying capacity decreased after its food supply was diminished.

Sometimes, instead of remaining in a steady state, predator and prey populations first increase in size and then decrease. We can appreciate that an increase in predator population size is dependent on an increase in prey population size. But what causes a decrease in population size instead of the establishment of a steady population size? At least two possibilities account for the reduction: (1) Perhaps the biotic potential (reproductive rate) of the predator is so great that its increased numbers overconsume the prey, and then as the prey population declines, so does the predator population; or (2) perhaps the biotic potential of the predator is unable to keep pace with the prey, and the prey population overshoots the carrying capacity and suffers a crash. Now the predator population follows suit because of a lack of food. In either case, the result will be a series of peaks and valleys, with the predator population size lagging slightly behind that of the prey population.

A famous example of predator-prey cycles occurs between the snowshoe hare and the Canadian lynx, a type of small predatory cat (**Fig. 38.2A**). The snowshoe hare is a common herbivore in the coniferous forests of North America, where it feeds on terminal twigs of various shrubs and small trees. The Canadian lynx feeds on snowshoe hares but also on ruffed grouse and spruce grouse, two types of birds. Studies have revealed that the hare and lynx populations cycle regularly, as graphed in Figure 38.2A. Investigators at first assumed that the lynx brings about a decline in the hare population and that this accounts for the cycling. But others have noted that the decline in snowshoe hare abundance was accompanied by low growth and reproductive rates, which could be signs of a food shortage. Experiments were done to test whether factor (1), predation, or factor (2), lack of food, caused the decline in the hare population. The results suggest that both factors combined to produce a low hare population and the cycling effect.

Prey Defenses

Prey defenses are mechanisms that thwart the possibility of being eaten by a predator. Prey species have evolved a variety of mechanisms that enable them to avoid predators, including heightened senses, speed, protective armor, protective spines or thorns, tails and appendages that break off, and chemical defenses.

FIGURE 38.2A

Predator-prey interaction between a snowshoe hare and a lynx.

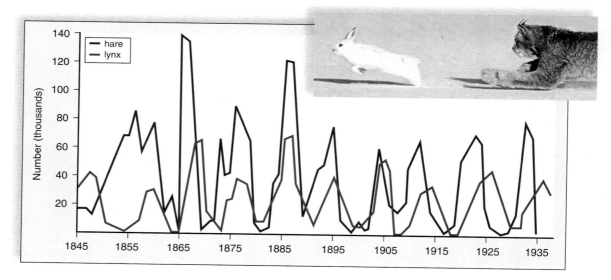

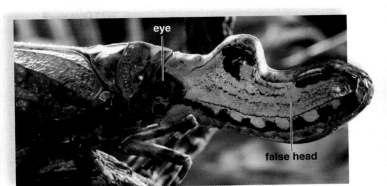

Camouflage

Warning coloration

Fright

FIGURE 38.2B Antipredator defenses.

Consider, for the moment, only animals. There are a number of ways that a predator deceives its prey and that prey avoids capture by predators. One common strategy is **camouflage**, or the ability to blend into the background. Some animals have cryptic coloration that blends them into their surroundings. For example, flounders can take on the same coloration as their background (**Fig. 38.2B**, *left*). Many examples of protective camouflage are known: Walking sticks look like twigs; katydids look like sprouting green leaves; some caterpillars resemble bird droppings; and some insects and moths blend into the bark of trees.

Another common antipredator defense among animals is *warning coloration*, which tells the predator that the prey is potentially dangerous. As a warning to possible predators, poison-dart frogs are brightly colored (Fig. 38.2B, *middle*). Also, many animals, including caterpillars, moths, and fishes, possess false eyespots that confuse or startle another animal. Other animals have elaborate anatomic structures that cause *fright*. The South American lantern fly has a large false head with false eyes, making it resemble the head of an alligator (Fig. 38.2B, *right*). However, warning coloration and a fright defense are not always false. A porcupine certainly looks formidable, and for good reason. Its arrowlike quills have barbs that dig into the predator's flesh and penetrate even deeper as the enemy struggles after being impaled. In the meantime, the porcupine runs away.

Association with other prey is another common strategy that may help avoid capture. Flocks of birds, schools of fish, and herds of mammals stick together as protection against predators. Baboons that detect predators visually, and antelopes that detect predators by smell, sometimes forage together, gaining double protection against stealthy predators. The gazellelike springboks of southern Africa jump stiff-legged 2–4 m into the air when alarmed. Such a jumble of shapes and motions might confuse an attacking lion, allowing the herd to escape.

Mimicry **Mimicry** occurs when one species resembles another that possesses an overt antipredator defense. A mimic that lacks the defense of the organism it resembles is called a Batesian mimic (named for Henry Bates, who discovered the phenomenon). Once an animal experiences the defense of the model, it remembers the coloration and avoids all animals that look similar. **Figure 38.2C** (*top*) shows two insects (flower fly and longhorn beetle) that resemble a yellow jacket wasp but lack the wasp's

ability to sting. Classic examples of Batesian mimicry include the scarlet kingsnake mimicking the venomous coral snake and the viceroy butterfly mimicking the foul-tasting monarch butterfly.

There are also examples of species that have the same defense and resemble each other. Many stinging insects—bees, wasps, hornets, and bumblebees—have the familiar black and yellow bands. Once a predator has been stung by a black and yellow insect, it is wary of that color pattern in the future. Mimics that share the same protective defense are called Müllerian mimics, after Fritz Müller, who suggested that this, too, is a form of mimicry. In Figure 38.2C (*bottom*), the bumblebee is a Müllerian mimic of the yellow jacket wasp because both of them can sting.

The last three relationships we will explore—parasitism, commensalism, and mutualism—are symbiotic relationships in which two species are intimately related. Let's begin with parasitism in Section 38.3.

38.2 Check Your Progress Could scavengers be considered predators when they eat decomposing meat? How so?

flower fly

longhorn beetle

bumblebee

yellow jacket

FIGURE 38.2C Mimicry: All of these insects have the same coloration.

38.3 Parasitism benefits one population at another's expense

Parasitism is similar to predation in that an organism, called the **parasite,** derives nourishment from another, called the **host.** Parasitism is an example of a **symbiosis,** an association in which at least one of the species is dependent on the other (**Table 38.3**).

Viruses, such as HIV, that reproduce inside human lymphocytes are always parasitic, and parasites occur in all of the kingdoms of life as well. Bacteria (e.g., strep infection), protists (e.g., malaria), fungi (e.g., rusts and smuts), plants (e.g., mistletoe), and animals (e.g., tapeworms and fleas) all have parasitic members. While small parasites can be endoparasites (pinworms), larger ones are more likely to be ectoparasites (leeches), which remain attached to the exterior of the body by means of special-ized organs and appendages. The effects of parasites on the health of the host can range from slightly weakening them to actually killing them over time. When host populations are at a high density, parasites readily spread from one host to the next, causing intense infestations and a subsequent decline in host density. Parasites that do not kill their host can still play a role in reducing the host's population density because an infected host is less fertile and becomes more susceptible to another cause of death.

In addition to nourishment, host organisms also provide their parasites with a place to live and reproduce, as well as a mechanism for dispersing offspring to new hosts. Many parasites have both a primary and a secondary host. The secondary host may be a vector that transmits the parasite to the next primary host. Usually both hosts are required in order to complete the life cycle. The association between parasite and host is so intimate that parasites are often specific and even require certain species as hosts.

We take a look at commensalism in Section 38.4 and mutualism in Section 38.6.

TABLE 38.3	Symbiotic Relationships
Interaction	**Expected Outcome**
Parasitism	Abundance of parasite increases, and abundance of host decreases.
Commensalism	Abundance of one species increases, and the other is not affected.
Mutualism	Abundance of both species increases.

> **38.3 Check Your Progress** Explain why decomposers are not parasites.

38.4 Commensalism benefits only one population

Commensalism is a symbiotic relationship between two species in which one species is benefited and the other is neither benefited nor harmed.

Instances are known in which one species provides a home and/or transportation for the other species. Barnacles that attach themselves to the backs of whales and the shells of horseshoe crabs are provided with both a home and transportation. It is possible, though, that the movement of the host is impeded by the presence of the attached animals, and therefore some scientists are reluctant to call these examples of commensalism.

Epiphytes, such as Spanish moss and some species of orchids and ferns, grow in the branches of trees, where they receive light, but they take no nourishment from the trees. Instead, their roots obtain nutrients and water from the air. Clownfishes live within the waving tentacles of sea anemones (**Fig. 38.4**). Because most fishes avoid the anemones' poisonous tentacles, clownfishes are protected from predators. Perhaps this relationship borders on mutualism, because the clownfishes may actually attract other fishes on which the anemone can feed.

Commensalism often turns out, on closer examination, to be an instance of either mutualism or parasitism. Cattle egrets are so named because these birds stay near cattle, which flush out their prey—insects and other animals—from vegetation. The relationship becomes mutualistic when egrets remove ectoparasites from the cattle. Remoras are fishes that attach themselves to the bellies of sharks by means of a modified dorsal fin acting as a suction cup. Remoras benefit by getting a free ride, and they also feed on a shark's leftovers. However, the shark benefits when remoras remove its ectoparasites. To some, it seems like wasted effort to

FIGURE 38.4 A clownfish living among a sea anemone's tentacles.

try to classify symbiotic relationships into the three categories of parasitism, commensalism, and mutualism. Often, the amount of harm or good two species seem to do to one another is dependent on what the investigator chooses to measure.

We will take a break and discuss coevolution, a hallmark of symbiotic relationships, in Section 38.5.

> **38.4 Check Your Progress** How would you determine that remoras are commensalistic and not parasitic on a shark?

Coevolution is present when two species adapt in response to selective pressure imposed by the other. In other words, an evolutionary change in one species results in an evolutionary change in the other. Organisms in symbiotic associations, which include parasitism, commensalism, and mutualism (see Table 38.3), are especially prone to the process of coevolution. As an example of coevolution, flowers pollinated by animals have features that attract them. Therefore, a butterfly-pollinated flower is often a composite, containing many individual flowers; the broad expanse provides room for the butterfly to land, and the butterfly has a proboscis that it inserts into each flower in turn. In this case, coevolution is highly beneficial to the flower because the insect will carry pollen to other flowers of the same type.

Coevolution also occurs between predators and prey. For example, a cheetah sprints forward to catch its prey, and this behavior might be selective for those gazelles that are able to run away or jump high in the air (such as a springbok) to avoid capture. The adaptation of the prey may very well put selective pressure on the predator to adapt to the prey's defense mechanism. Hence, an "arms race" can develop.

The process of coevolution has been studied in the cuckoo, a social parasite that reproduces at the expense of other birds. A cuckoo lays its eggs in another bird's nest, and when the cuckoo egg hatches, the other bird ends up caring for it. It is odd to see a small bird feeding a cuckoo nestling several times its size. How did this strange behavior develop? Investigators discovered that in order to "trick" a host bird, the adult cuckoo has to (1) lay an egg that mimics the host's egg, (2) lay its egg very rapidly (only 10 seconds are required) in the afternoon while the host is away from the nest, and (3) leave most of the eggs in the nest because hosts will desert a nest that has only one egg in it. (The cuckoo chick hatches first and is adapted to removing any other eggs in the nest [**Fig. 38.5**].) It seems that the host birds may very well next evolve a way to distinguish the cuckoo from their own young.

Coevolution can take many forms. In the case of *Plasmodium*, a cause of malaria, the sexual portion of the life cycle occurs within mosquitoes (the vector), and the asexual portion occurs in humans. The human immune system uses surface proteins to detect pathogens, and *Plasmodium* has numerous genes for surface proteins. But it is capable of changing its surface proteins repeatedly, and in this way it stays one step ahead of the host's immune system. HIV has a similar capability, which has added to the difficulty of producing an AIDS vaccine.

The relationship between parasite and host can include the ability of parasites to seemingly manipulate the behavior of their hosts in self-serving ways. Ants infected with the lance fluke (but not those uninfected) mysteriously cling to blades of grass with their mouthparts. There, the infested ants are eaten by grazing sheep, and the flukes are transmitted to the next host in their life cycle. Similarly, when snails of the genus *Succinea* are parasitized by worms of the genus *Leucochloridium*, they are eaten by birds. As the worms mature, they invade the snail's eyestalks, making them resemble edible caterpillars. Now the

birds eat the snails, and the parasites release their eggs, which complete development inside the urinary tracts of birds.

It used to be thought that as host and parasite coevolved, each would become more tolerant of the other since, if the opposite occurred, the parasite would soon run out of hosts. Parasites could first become commensal, or harmless to the host. Then, given enough time, the parasite and host might even become mutualists. In fact, the evolution of the eukaryotic cell by endosymbiosis is predicated on the supposition that bacteria took up residence inside a larger cell, and then the parasite and cell became mutualists.

However, this argument is too complex for some; after all, no organism is capable of "looking ahead" at its evolutionary fate. Rather, if an aggressive parasite could transmit more of itself in less time than a benign one, aggressiveness would be favored by natural selection. On the other hand, other factors, such as the life cycle of the host, can determine whether aggressiveness is beneficial or not. For example, a benign parasite of newts will do better than an aggressive one. Why? Because newts take up solitary residence outside ponds in the forest for six years, and parasites have to wait that long before they are likely to meet up with another possible host.

Section 38.6 considers mutualism.

> **38.5** *Check Your Progress* Suggest a study to determine whether the fungus of a lichen is mutualistic with the algae of a lichen and/or parasitic on the algae.

Cuckoo's egg looks like host's egg

FIGURE 38.5
Social parasitism in the cuckoo.

Cuckoo pushes out host's eggs after hatching first

Mutualism is a symbiotic relationship in which both members benefit. As with other symbiotic relationships, it is possible to find numerous examples among all organisms. Bacteria that reside in the human intestinal tract acquire food, but they also provide us with vitamins, molecules we are unable to synthesize for ourselves. Termites would not be able to digest wood if not for the protozoans that inhabit their intestinal tracts and digest cellulose. Mycorrhizae are mutualistic associations between the roots of plants and fungal hyphae. The hyphae improve the uptake of nutrients for the plant, protect the plant's roots against pathogens, and produce plant growth hormones. In return, the plant provides the fungus with carbohydrates. Some sea anemones make their home on the backs of crabs. The crab uses the stinging tentacles of the sea anemone to gather food and to protect itself; the sea anemone gets a free ride that allows it greater access to food than other anemones. Lichens can grow on rocks because their fungal member conserves water and leaches minerals that are provided to the algal partner, which photosynthesizes and provides organic food for both populations. However, it's been suggested that the fungus is parasitic, at least to a degree, on the algae.

In tropical America, the bullhorn acacia tree is adapted to provide a home for ants of the species *Pseudomyrmex ferruginea* (see page 473). Unlike other acacias, this species has swollen thorns with a hollow interior, where ant larvae can grow and develop. In addition to housing the ants, acacias provide them with food. The ants feed from nectaries at the base of the leaves and eat fat- and protein-containing nodules called Beltian bodies, found at the tips of the leaves. The ants constantly protect the plant from herbivores and other plants that might shade it because, unlike other ants, they are active 24 hours a day.

The relationship between plants and their pollinators, mentioned previously, is a good example of mutualism. Perhaps the relationship began when herbivores feasted on pollen. The provision of nectar by the plant may have spared the pollen and, at the same time, allowed the animal to become an agent of pollination. By now, pollinator mouthparts are adapted to gathering the nectar of a particular plant species, and this species is dependent on the pollinator for dispersing pollen (see pages 478–79). The mutualistic relationships between flowers and their pollinators are examples of coevolution (see Section 38.5). A striking example is the flower of the orchid, *Ophrys apitera*. This flower resembles the body of a bumblebee, and its odor mimics the pheromones of a female bee. Therefore, male bees are attracted to the flower, and when they attempt to mate with it, they become covered with pollen, which they transfer to the next orchid flower (see page 242). The orchid is dependent upon bees for pollination because neither wind nor other insects pollinate these flowers.

The outcome of mutualism is an intricate web of species interdependencies critical to the community. For example, in areas of the western United States, the branches and cones of whitebark pine are turned upward, meaning that the seeds do not fall to the ground when the cones open. Birds called Clark's nutcrackers eat the seeds of whitebark pine trees and store them in the ground (**Fig. 38.6A**). Therefore, Clark's nutcrackers are critical seed dispersers for the trees. Also, grizzly bears find the stored seeds and

FIGURE 38.6A Clark's nutcrackers store and disperse the seeds of whitebark pine trees.

consume them. Whitebark pine seeds do not germinate unless their seed coats are exposed to fire. When natural forest fires in the area are suppressed, whitebark pine trees decline in number, and so do Clark's nutcrackers and grizzly bears. When lightning-ignited fires are allowed to burn, or prescribed burning is used in the area, the whitebark pine populations increase, as do the populations of Clark's nutcrackers and grizzly bears.

Cleaning symbiosis is a symbiotic relationship in which crustaceans, fish, and birds act as cleaners for a variety of vertebrate clients. Large fish in coral reefs line up at cleaning stations and wait their turn to be cleaned by small fish that even enter the mouths of the large fish (**Fig. 38.6B**). Whether cleaning symbiosis is an example of mutualism has been questioned because of the lack of experimental data. If clients respond to tactile stimuli by remaining immobile while cleaners pick at them, then cleaners may be exploiting this response by feeding on host tissues, as well as on ectoparasites.

This completes our study of interactions between populations in a community, including symbiotic relationships. In the next part of the chapter, we consider community composition and diversity and how these attributes can change over time.

38.6 *Check Your Progress* **What is the benefit for protozoans living in the gut of a termite?**

FIGURE 38.6B
Cleaning symbiosis occurs when small fish clean large fish.

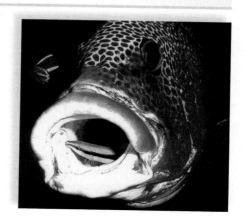

Communities differ in their composition and diversity, which is obvious if you compare a tropical rain forest with a coniferous forest. A hypothesis called island biogeography, discussed in this part, suggests that just as climate can affect composition and diversity, so can the amount of space available to the community. The composition and diversity of a community can also change over time, a fact that is dramatically illustrated by ecological succession, also discussed in this part of the chapter.

HOW SCIENCE PROGRESSES

38.7   The study of island biogeography pertains to biodiversity

Would you expect larger coral reefs to have a greater number of species, called **species richness**, than smaller coral reefs? The area (space) occupied by a community can have a profound effect on its biodiversity. American ecologists Robert MacArthur and E. O. Wilson developed a general **model of island biogeography** to explain and predict the effects of (1) distance from the mainland and (2) size of an island on community diversity.

Imagine two new islands that, as yet, contain no species at all. One of these islands is near the mainland, and one is far from the mainland (**Fig. 38.7A**). Which island will receive more immigrants from the mainland? Most likely, the near one because it's easier for immigrants to get there. Similarly, imagine two islands that differ in size (**Fig. 38.7B**). Which island will be able to support a greater number of species? The large one, because its greater amount of resources can support more populations, while species on the smaller island may eventually face extinction due to scarce resources. MacArthur and Wilson studied the biodiversity on many island chains, including the West Indies, and discovered that species richness does correlate positively with island distance from mainland and island size. They developed a model of island biogeography that takes into account both factors. An equilibrium is reached when the rate of species immigration matches the rate of species extinction due to limited space (**Fig. 38.7C**). Notice that the equilibrium point is highest for a large island that is near the mainland. The equilibrium could be dynamic (new species keep on arriving, and new extinctions keep on occurring), or the composition of the community could remain steady unless disturbed.

Biodiversity Conservationists note that the trends graphed in Figure 38.7C in particular apply to their work because humans often create preserved areas surrounded by farms, towns, and cities, or even water. For example, in Panama, Barro Colorado Island (BCI) was created in the 1910s when a river was dammed to form a lake. As predicted by the model of island biogeography, BCI lost species because it was a small island that had been cut off from the mainland. Among the species that became extinct were the top predators on the island, namely the jaguar, puma, and ocelot. Thereafter, medium-sized terrestrial mammals, such as the coatimundi, increased in number. Because the coatimundi is an avid predator of bird eggs and nestlings, soon there were fewer bird species on BCI, even though the island is large enough to support them.

The model of island biogeography suggests that the larger the conserved area, the better the chance of preserving more species. Is it possible to increase the amount of space without using more area? Two possibilities come to mind. If the environment has patches, it has a greater number of habitats—and thus greater diversity. As gardeners, we are urged to create patches in our yards if we wish to attract more butterflies and birds! One way to introduce patchiness is through stratification, the use of layers. Just as a high-rise apartment building allows more human families to live in an area, so can stratification within a community provide more and different types of living space for different species.

> **38.7 Check Your Progress** How is stratification observed in Figure 38.1C?

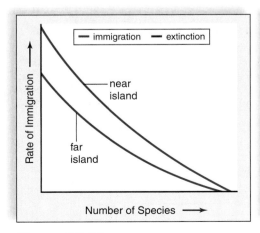

Figure 38.7A Island distance from mainland and rate of species migration.

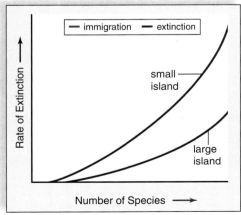

FIGURE 38.7B Island size and rate of species extinction.

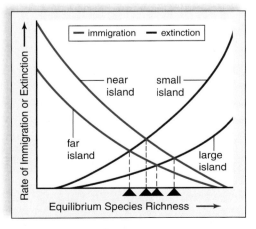

FIGURE 38.7C Equilibrium model of species richness.

38.8 During ecological succession, community composition and diversity change

When comparing various communities, two fundamental characteristics that are examined are composition and diversity. The *composition* of a community is a thorough listing of the various species in a particular community. For example, pictorially, it is evident that broadleaved trees are numerous in a temperate deciduous forest (see Fig. 39.4), while succulent cacti and nonsucculent shrubs are numerous in some deserts (see Fig. 39.7). The animal inhabitants are also different.

The *diversity* of a community goes beyond composition because it includes not only a listing of species but also the abundance of each species. To take an extreme example: A forest in West Virginia has, among other species, 76 yellow poplar trees but only one American elm. If we were simply walking through this forest, we might miss seeing the American elm. If, instead, the forest had 36 poplar trees and 41 American elms, the forest would seem more diverse to us, and indeed it would be more diverse. The greater the diversity, the greater the number—and the more even the distribution—of species.

The composition and diversity of a community can change over time due to various disturbances. Disturbances can range in severity from a storm blowing down a patch of trees, to a beaver damming a pond, to a volcanic eruption. We know from observation that, following these disturbances, changes will take place in the plant and animal communities; often, we'll wind up with the same kind of community with which we started due to ecological succession.

Ecological Succession A series of species replacements in a community following a disturbance is called ecological succession. *Primary succession* occurs in areas where no soil is present, such as following a volcanic eruption or a glacial re-treat. *Secondary succession* begins in areas where soil is present. **Figure 38.8A** shows the stages of secondary succession in a large conifer plantation in central New York State. The process of regrowth proceeds through these stages: grasses, low shrubs, tall shrubs, shrub-trees, low trees, and finally tall trees. Approximately the same stages are observed when a cultivated field, such as a cornfield in New Jersey, returns to its natural state (**Fig. 38.8B**):

1. During the first year, only the remains of corn plants are seen.

2. During the second year, wild grasses have invaded the area.

3. By the fifth year, the grasses look more mature, and sedges have joined them.

4. During the tenth year, there are goldenrod plants, shrubs (blackberry), and juniper trees.

5. After twenty years, the juniper trees are mature, and birch and maple trees are present, in addition to the blackberry shrubs.

The first species to begin secondary succession are called **pioneer species;** these are usually plants that tend to invade disturbed areas. Then the area progresses through the series of stages depicted in Figures 38.8A and B. In other words, we observe a series that begins with grasses and proceeds from shrub stages to a mixture of shrubs and trees, until finally there are only trees. Ecologists have tried to determine the processes and mechanisms by which these changes take place—and whether these processes always end at the same point of community composition and diversity.

FIGURE 38.8A Secondary succession in a conifer plantation.

| grass | → | low shrub | → | tall shrub | → | shrub-tree | → | low tree | → | tall tree |

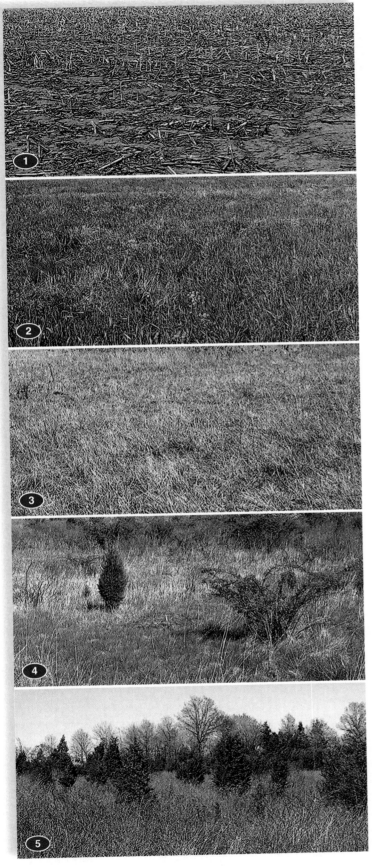

FIGURE 38.8B Secondary succession in a cultivated cornfield.

Models of Succession
In 1916, F. E. Clements proposed the *climax-pattern model* of succession, which suggests that succession in a particular area will always lead to the same type of community, called a **climax community.** Clements believed that climate, in particular, determines whether a desert, a type of grassland, or a particular type of forest results. This is the reason, he said, that coniferous forests occur in northern latitudes, deciduous forests in temperate zones, and tropical rain forests in the tropics. Secondarily, he believed that soil conditions might also affect the results. Shallow, dry soil might produce a grassland where a forest would otherwise be expected, or the rich soil of a riverbank might produce a woodland where a prairie would be expected.

Further, Clements believed that each stage facilitated the invasion and replacement by organisms of the next stage. As in the examples given, shrubs can't arrive until grasses have made the soil suitable for them. Each successive community prepares the way for the next, so that grass-shrub-forest development occurs sequentially. This is known as the *facilitation model* of succession.

Aside from the facilitation model, there is also an *inhibition model.* That model predicts that colonists hold onto their space and inhibit the growth of other plants until the colonists die or are damaged. Still another model, called the *tolerance model,* predicts that different types of plants can colonize an area at the same time. Sheer chance determines which seeds arrive first, and successional stages may simply reflect the length of time it takes species to mature. This alone could account for the grass-shrub-forest sequence that is often seen. The length of time it takes for trees to develop might give the impression that plant communities develop in a recognizable series, from the simple to the complex. But in reality, the models we have mentioned are not mutually exclusive, and succession is probably a multiple, complex process.

Although it may not have been apparent to early ecologists, we now recognize that the most outstanding characteristic of natural communities is their dynamic nature. Also, it seems obvious to us now that the most complex communities most likely consist of habitat patches that are at various stages of succession. Each successional stage has its own mix of plants and animals, and if a sample of all stages is present, community diversity is greatest. Further, we do not know if succession continues to a certain end point, because the process may not be complete anywhere on the face of the Earth.

This completes our study of community composition and diversity and how these attributes change. In the next part of the chapter, we broaden our horizons and consider the level of biological organization called an **ecosystem,** in which populations interact among themselves and with the physical environment.

> **38.8 Check Your Progress** If decomposers were in short supply, could succession occur? Why or why not?

An Ecosystem Is a Community Interacting with the Physical Environment

Learning Outcomes 8–14, page 740

When we study a community, we are considering only the populations of organisms that make up that community. But when we study an ecosystem, we are concerned with the living community along with its physical environment. **Ecology** is the study of the interactions of organisms with each other and with the physical environment. In this part of the chapter, we study the composition of ecosystems and how they are characterized by energy flow and chemical cycling.

38.9 Ecosystems have biotic and abiotic components

An ecosystem possesses both living (biotic) and nonliving (abiotic) components. The abiotic components include resources, such as sunlight and inorganic nutrients, and conditions, such as type of soil, water availability, prevailing temperature, and amount of wind. The biotic components of an ecosystem are influenced by the abiotic components, as when the force of the wind has affected the growth of a tree in **Figure 38.9** (*left*). Each biotic component has an *ecological niche* whose aspects are listed in Figure 38.9 for plants and animals.

Biotic Components of an Ecosystem Among the biotic components of an ecosystem, **autotrophs** require only inorganic nutrients and an outside energy source to produce organic nutrients for their own use and indirectly for all the other members of a community. They are called **producers** because they produce food. Photosynthetic organisms produce most of the organic nutrients for the biosphere. Algae of all types contain chlorophyll and carry on photosynthesis in freshwater and marine habitats. Algae make up the phytoplankton, which are photosynthesizing organisms suspended in water. Green plants, such as trees in forests and corn plants in fields, are the dominant photosynthesizers on land.

Some autotrophic bacteria are chemosynthetic. They obtain energy by oxidizing inorganic compounds such as ammonia, nitrites, and sulfides, and they use this energy to synthesize organic compounds. Chemosynthesizers have been found supporting communities in some caves and also at hydrothermal vents along deep-sea oceanic ridges.

Heterotrophs need a preformed source of organic nutrients. They are called **consumers** because they consume food.

Herbivores are animals that graze directly on plants or algae. In terrestrial habitats, insects are small herbivores, and antelopes and bison are large herbivores. In aquatic habitats, zooplankton, which is composed of protozoans and tiny invertebrates, are small herbivores while some fishes, as well as manatees, are large herbivores. **Carnivores** feed on other animals; for example, birds that feed on insects are carnivores, and so are hawks that feed on birds. **Omnivores** are animals that feed on both plants and animals. Chickens, raccoons, and humans are omnivores.

Scavengers, such as jackals and vultures (see chapter introduction), feed on the dead remains of animals and also on plants that have recently begun to decompose. **Detritus** is a term that refers to organic remains in the water and soil that are in the final stages of decomposition. Marine fan worms take detritus from the water, while clams take it from the substratum. Earthworms, some beetles, and termites feed on detritus in the soil. Bacteria and fungi, including mushrooms, are the **decomposers** that use their digestive secretions to chemically break down dead organic matter, including animal wastes in the external environment. Notice that decomposers produce detritus. Without decomposers, plants would be completely dependent only on physical processes, such as the release of minerals from rocks, to supply them with inorganic nutrients.

In Section 38.10, we establish that energy flows and chemicals cycle through the biotic components of ecosystems.

> **38.9 Check Your Progress** Why is it correct to say that autotrophs support all the biotic components (including decomposers) of an ecosystem?

FIGURE 38.9 Niche specifications of plants compared to animals.

Aspects of Niche for Plants

- Season of year for growth and reproduction
- Sunlight, water, and soil requirements
- Relationships with other organisms
- Effect on abiotic environment

Aspects of Niche for Animals

- Time of day for feeding and season of year for reproduction
- Habitat and food requirements
- Relationships with other organisms
- Effect on abiotic environment

The diagram in **Figure 38.10A** illustrates that every ecosystem is characterized by two fundamental phenomena: energy flow and chemical cycling. Energy is lost from the biosphere, but inorganic nutrients are not. They recycle within and between ecosystems. Decomposers return some proportion of inorganic nutrients to autotrophs, and other portions are imported or exported between ecosystems in global cycles.

Energy flow begins when producers absorb solar energy, and chemical cycling begins when producers take in inorganic nutrients from the physical environment. Thereafter, via photosynthesis, producers make organic nutrients (food) directly for themselves and indirectly for the other populations of the ecosystem. Energy flows through an ecosystem via photosynthesis because, as organic nutrients pass from one component of the ecosystem to another, as when an herbivore eats a plant or a carnivore eats an herbivore, a portion of those nutrients is used as an energy source. Eventually, the energy dissipates into the environment as heat. Therefore, the vast majority of ecosystems cannot exist without a continual supply of solar energy.

Only a portion of the organic nutrients made by producers is passed on to consumers because plants use organic molecules to fuel their own cellular respiration. Similarly, only a small percentage of nutrients consumed by lower-level consumers, such as herbivores, is available to higher-level consumers, or carnivores. As **Figure 38.10B** demonstrates, a certain amount of the

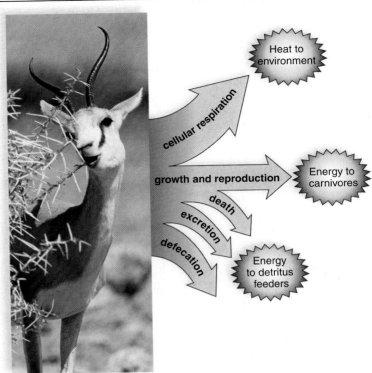

FIGURE 38.10B   Energy balances for an herbivore.

food eaten by an herbivore is never digested and is eliminated as feces. Metabolic wastes are excreted as urine. Of the assimilated energy, a large portion is used during cellular respiration for the production of ATP, and thereafter it becomes heat. Only the remaining energy, which is converted into increased body weight or additional offspring, becomes available to carnivores.

The elimination of feces and urine by a heterotroph, and indeed the death of all organisms, does not mean that organic nutrients are lost to the ecosystem; instead, they are made available to scavengers and decomposers. Decomposers convert the organic nutrients, such as glucose, back into inorganic chemicals, such as carbon dioxide and water, and release them to the soil or the atmosphere. Chemicals complete their cycle within an ecosystem when inorganic chemicals are absorbed by the producers from the atmosphere or from the soil.

The first law of thermodynamics states that energy cannot be created (or destroyed). This explains why ecosystems are dependent on a continual outside source of energy. The second law states that, with every transformation, some energy is degraded into a less available form, such as heat. For example, because plants carry on cellular respiration, only 55% of the original energy absorbed by plants is available to an ecosystem.

In Section 38.11, we apply the principles of energy flow and chemical cycling to an actual ecosystem.

38.10 *Check Your Progress*   What happens to the chemicals within glucose once decomposers have broken it down to inorganic chemicals, such as carbon dioxide and water?

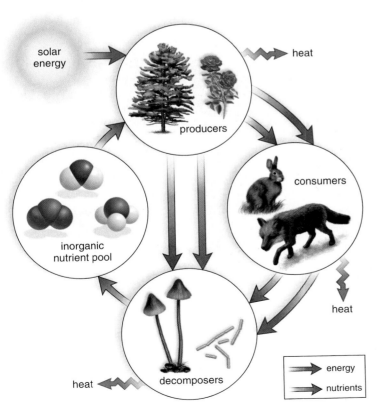

FIGURE 38.10A   Energy flow and chemical cycling in an ecosystem.

38.11　Energy flow involves food webs

The principles discussed in the previous section can now be applied to an actual ecosystem—a forest of 132,000 m² in New Hampshire. The various interconnecting paths of energy flow may be represented by a **food web,** a diagram that describes trophic (feeding) relationships. **Figure 38.11** (*top*) is a graz-

ing food web because it begins with a producer, specifically ❶ the oak tree. ❷ Insects, in the form of caterpillars, feed on leaves, while ❸ mice, rabbits, and deer feed on plant material at or near the ground. Mice also feed on nuts, while ❹ birds, collectively, are omnivorous, feeding on both nuts

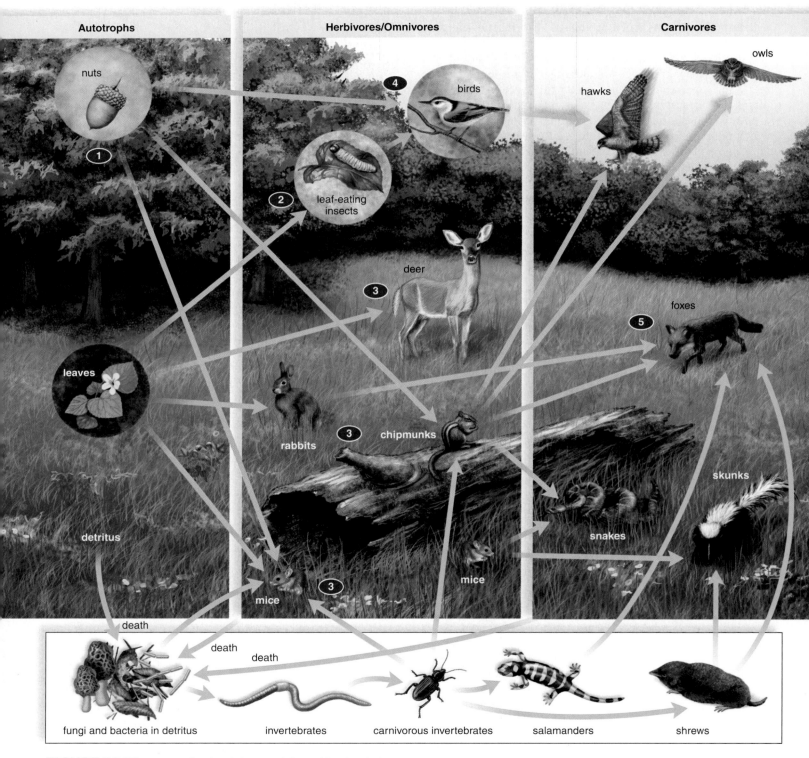

FIGURE 38.11　Grazing food web (*top*) and detrital food web (*bottom*).

and caterpillars. Herbivores and omnivores all provide food for
a number of different carnivores.

Figure 38.11 (*bottom*) is a **detrital food web,** which begins
with detritus. Detritus is food for soil organisms such as earthworms. Earthworms are, in turn, eaten by carnivorous invertebrates, and these may be consumed by salamanders or shrews.
Because the members of a detrital food web may become food
for aboveground carnivores, the detrital and grazing food webs
are joined.

We naturally tend to think that aboveground plants such as
trees are the largest storage form of organic matter and energy,
but this is not necessarily the case. In this particular forest, the organic matter lying on the forest floor and mixed into the soil contains over twice as much energy as the leaf matter of living trees.
Therefore, more energy in a forest may be funneling through the
detrital food web than through the grazing food web.

Trophic Levels The arrangement of the species in Figure
38.11 suggests that organisms are linked to one another in a
straight line, according to feeding relationships, or who eats
whom. Diagrams that show a single path of energy flow in an
ecosystem are called **food chains.** For example, in the grazing
food web, we could find this food chain:

$$\text{leaves} \longrightarrow \text{caterpillars} \longrightarrow \text{birds} \longrightarrow \text{hawks}$$

And in the detrital food web, we could find this food chain:

$$\text{detritus} \longrightarrow \text{earthworms} \longrightarrow \text{carnivore}$$

A **trophic level** is composed of organisms that occupy the
same position within a food web or chain. In the grazing food web
in Figure 38.11 (*top*), going from left to right, the trees are producers (first trophic level), the first series of animals are primary consumers (second trophic level), and the next group of animals are
secondary consumers (third trophic level). Energy moves from
trophic level to level, but some is lost between the various levels
because energy has been used to do biological work. As Section
38.12 shows, if we successively diagram the relative energy content of each trophic level of a food web, a pyramid results.

38.11 Check Your Progress What type of organisms are found
in detritus at the start of a detrital food chain?

38.12 Ecological pyramids are based on trophic levels

An **ecological pyramid** is a graphic representation of the number
of organisms, the biomass, or the relative energy content of the
various trophic levels in an ecosystem. For example, **Figure 38.12**
shows the biomass content of each trophic level of a bog in Silver
Springs, Florida. Data regarding the biomass or energy content of
trophic levels places the producer trophic level at the base of the
pyramid, and each succeeding trophic level, which has less biomass and energy, follows thereafter. In general, only about 10%
of the energy of one trophic level is available to the next trophic
level because of energy losses between trophic levels. Therefore, if
an herbivore population consumes 1,000 kg of plant material, only
about 100 kg is converted to herbivore tissue, 10 kg to first-level
carnivores, and 1 kg to second-level carnivores. The so-called 10%
rule of thumb explains why ecosystems have only a few trophic
levels and why a food web can support only a few carnivores.

Ecological pyramids are helpful for explaining energy loss in
an ecosystem, but they oversimplify energy flow. Most likely, a
pyramid based on the number of organisms in each trophic level
wouldn't work. For example, in Figure 38.11A, each tree would
contain numerous caterpillars, so there would be more herbivores
than autotrophs! Similarly in aquatic ecosystems, such as some
lakes and open seas where algae are the only producers, the
herbivores may have a greater biomass than the producers when
they are measured, because the algae are consumed at a high
rate. In any case, ecological pyramids based on grazing food webs
don't account for energy that passes to decomposers. The energy
content between the autotroph level and the herbivore level is dis-
proportionate in Figure 38.12 because much of the energy in a bog
flows through the detrital food web, not the grazing food web.

This completes our discussion of energy flow in an ecosystem. In Section 38.13, we examine in more detail how chemicals
cycle through ecosystems.

FIGURE 38.12 Ecological pyramid based on the biomass content
of bog populations could also be used to represent an energy pyramid.

38.12 Check Your Progress In an ecological pyramid, what
trophic level (if any) accounts for decomposers?

38.13 Chemical cycling includes reservoirs, exchange pools, and the biotic community

The pathways by which chemicals circulate through ecosystems involve both living (biotic) and nonliving (geologic) components; therefore, they are known as **biogeochemical cycles.** In the next sections of this chapter, we describe three of the biogeochemical cycles: the phosphorus, nitrogen, and carbon cycles. A biogeochemical cycle may be sedimentary or gaseous. The phosphorus cycle is a sedimentary cycle; the chemical is absorbed from the soil by plant roots, passed to heterotrophs, and eventually returned to the soil by decomposers. The carbon and nitrogen cycles are gaseous, meaning that the chemical returns to and is withdrawn from the atmosphere as a gas.

Chemical cycling involves the components of ecosystems shown in **Figure 38.13.** A *reservoir* is a source normally unavailable to producers, such as the carbon present in calcium carbonate shells on ocean bottoms. An *exchange pool* is a source from which organisms do generally take chemicals, such as the atmosphere or soil. Chemicals move along food chains in a *biotic community*, perhaps never entering an exchange pool.

Human activities (purple arrow) remove chemicals from reservoirs and exchange pools and make them available to the

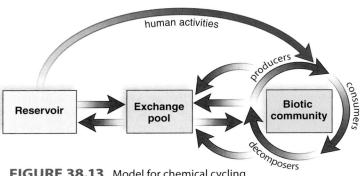

FIGURE 38.13 Model for chemical cycling.

biotic community. In this way, human activities result in pollution because they upset the normal balance of nutrients for producers in the environment.

We will study the phosphorus cycle in Section 38.14.

> **38.13 Check Your Progress** Which portion of Figure 38.13 represents the abiotic environment?

38.14 The phosphorus cycle is sedimentary

Figure 38.14 depicts the phosphorus cycle. ❶ Phosphorus, trapped in oceanic sediments, moves onto land after a geologic upheaval. ❷ On land, the very slow weathering of rocks places ❸ phosphate ions (PO_3^- and HPO_4^{2-}) in the soil. ❹ Some of these become available to plants, which use phosphate in a variety of molecules, including phospholipids, ATP, and the nucleotides that become a part of DNA and RNA. ❺ Animals eat producers and incorporate some of the phosphate into their teeth, bones, and shells, which take many years to decompose. ❻ However, eventually the death and decay of all organisms and also the decomposition of animal wastes make phosphate ions available to producers once again. Because the available amount of phosphate is already being used within food chains, phosphate is usually a limiting inorganic nutrient for plants—that is, the lack of it limits the size of populations in ecosystems.

❼ Some phosphate naturally runs off into aquatic ecosystems, where algae acquire phosphate from the water before it becomes trapped in sediments. Phosphate in marine sediments does not become available to producers on land again until a geologic upheaval exposes sedimentary rocks on land. Now, the cycle begins again.

Human Activities and the Phosphorus Cycle

❽ Human beings boost the supply of phosphate by mining phosphate ores for producing fertilizer and detergents. Runoff of phos-

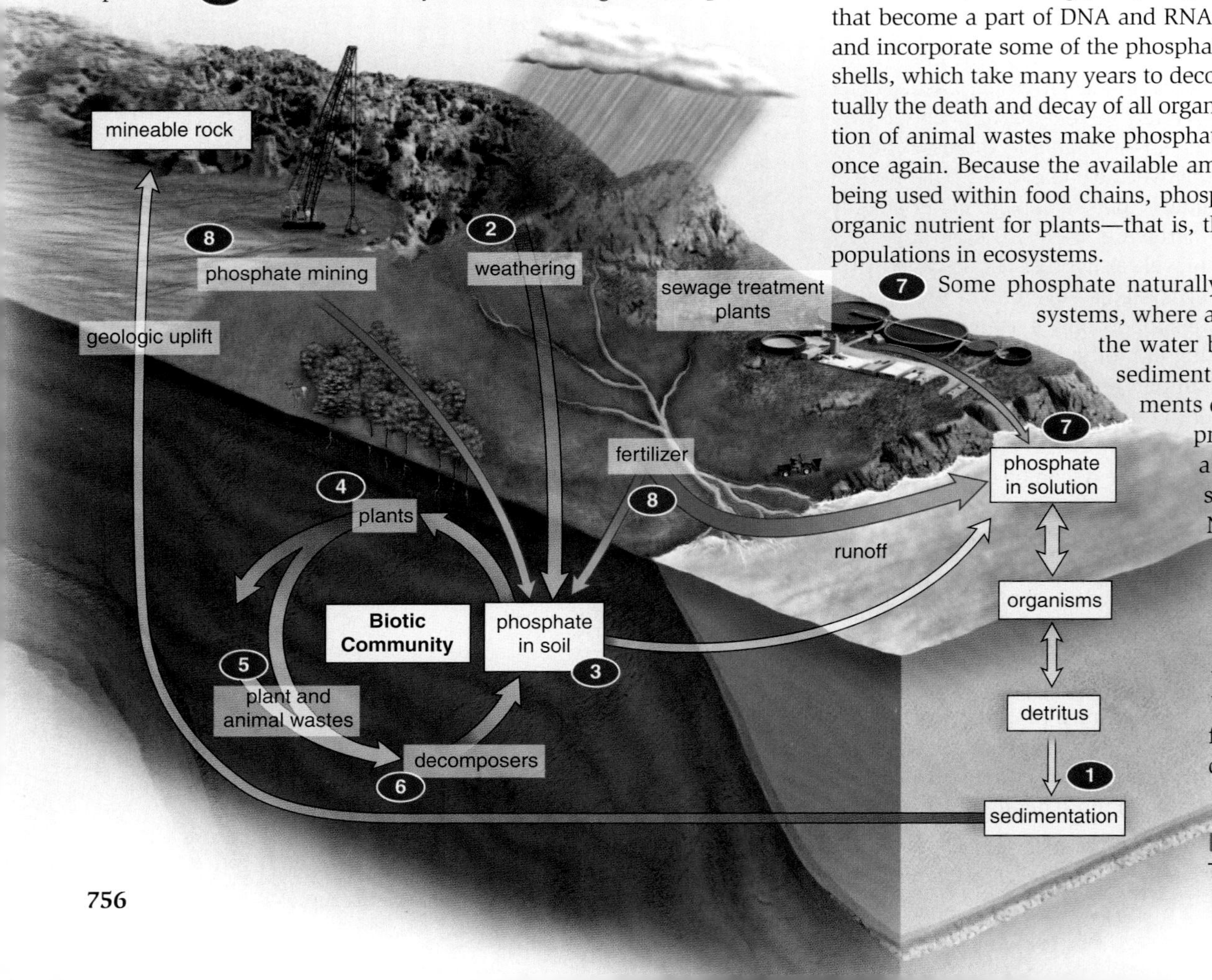

FIGURE 38.14
The phosphorus cycle.

mineable rock

❽ phosphate mining

geologic uplift

❷ weathering

sewage treatment plants

fertilizer

❹ plants

❽

❼

phosphate in solution

runoff

Biotic Community

phosphate in soil

❸

organisms

❺ plant and animal wastes

detritus

decomposers

❻

❶

sedimentation

phate and nitrogen due to fertilizer use, animal wastes from livestock feedlots, and discharge from sewage treatment plants results in **eutrophication** (overenrichment) of waterways.

Section 38.15 considers the nitrogen cycle.

38.14 Check Your Progress How do the phosphate ions in animal bones become available to producers?

38.15 The nitrogen cycle is gaseous

Nitrogen gas (N_2) makes up about 78% of the atmosphere, but plants cannot make use of nitrogen in its gaseous form. Therefore, nitrogen can be a nutrient that limits the amount of growth in an ecosystem.

Ammonium (NH_4^+) Formation and Use
In the nitrogen cycle, **1** N_2 **(nitrogen) fixation** occurs when nitrogen gas (N_2) is converted to ammonium (NH_4^+), a form plants can use (**Fig. 38.15**). Some cyanobacteria in aquatic ecosystems and some free-living bacteria in soil are able to fix atmospheric nitrogen in this way. Other nitrogen-fixing bacteria live in nodules on the roots of legumes, such as beans, peas, and clover. They make organic compounds containing nitrogen available to the host plants so that the plant can form proteins and nucleic acids.

Nitrate (NO_3^-) Formation and Use
2 Plants can also use nitrates (NO_3^-) as a source of nitrogen. The production of nitrates during the nitrogen cycle is called **nitrification.** Nitrification can occur in two ways: (1) Nitrogen gas (N_2) is converted to NO_3^- in the atmosphere when cosmic radiation, meteor trails, and lightning provide the high energy needed for nitrogen to react with oxygen. (2) Ammonium (NH_4^+) in the soil from various sources, including decomposition of organisms and animal wastes, is converted to NO_3^- by nitrifying bacteria in soil. Specifically, NH_4^+ (ammonium) is converted to NO_2^- (nitrite), and then NO_2^- is converted to NO_3^- (nitrate). **3** During the process of assimilation, plants take up NH_4^+ and NO_3^- from the soil and use these ions to produce proteins and nucleic acids.

Notice in Figure 38.15 that the subcycle involving the biotic community, which occurs on land and in the ocean, need not depend on the presence of nitrogen gas at all.

Formation of Nitrogen Gas
4 **Denitrification** is the conversion of nitrate back to nitrogen gas, which then enters the atmosphere. Denitrifying bacteria living in the anaerobic mud of lakes, bogs, and estuaries carry out this process as a part of their own metabolism. In the nitrogen cycle, denitrification would counterbalance nitrogen fixation if not for human activities.

Human Activities and the Nitrogen Cycle
5 Humans significantly alter the transfer rates in the nitrogen cycle by producing fertilizers from N_2—in fact, they nearly double the fixation rate. Fertilizer, which also contains phosphate, runs off into lakes and rivers and results in an overgrowth of algae and rooted aquatic plants. When the algae die off, enlarged populations of decomposers use up all the oxygen in the water, and the result is a massive fish kill.

Acid deposition occurs because nitrogen oxides (NO_x) and sulfur dioxide (SO_2) enter the atmosphere from the burning of fossil fuels. Both these gases combine with water vapor to form acids that eventually return to the Earth. Acid deposition has drastically affected forests and lakes in northern Europe, Canada, and the northeastern United States because their soils are naturally acidic and their surface waters are only mildly alkaline (basic). Acid deposition reduces agricultural yields and corrodes marble, metal, and stonework.

Section 38.16 is about the carbon cycle.

38.15 Check Your Progress Nitrate production by soil microbes is called _____.

FIGURE 38.15 The nitrogen cycle.

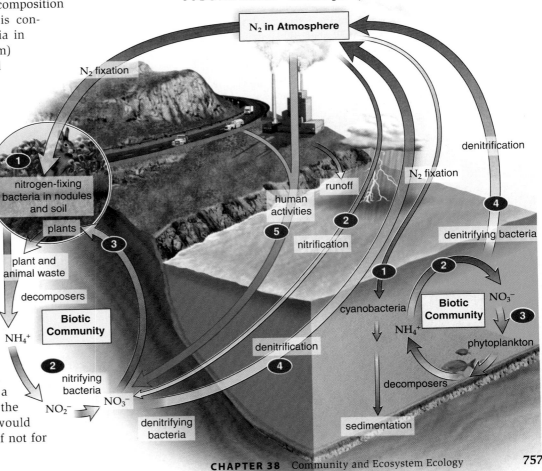

In the carbon cycle, organisms in both terrestrial and aquatic ecosystems exchange carbon dioxide (CO_2) with the atmosphere (**Fig. 38.16**). Therefore, the CO_2 in the atmosphere is the exchange pool for the carbon cycle. On land, plants take up CO_2 from the air, and through photosynthesis, they incorporate carbon into nutrients that are used by autotrophs and heterotrophs alike. ❶ When organisms, including plants, respire, carbon is returned to the atmosphere as CO_2. ❷ CO_2 then recycles to plants by way of the atmosphere.

In aquatic ecosystems, the exchange of CO_2 with the atmosphere is indirect. ❸ Carbon dioxide from the air combines with water to produce bicarbonate ion (HCO_3^-), a source of carbon for algae that produce food for themselves and for heterotrophs. ❹ Similarly, when aquatic organisms respire, the CO_2 they give off becomes HCO_3^-. The amount of bicarbonate in the water is in equilibrium with the amount of CO_2 in the air.

Reservoirs Hold Carbon Living and dead organisms contain organic carbon and serve as one of the reservoirs for the carbon cycle. The world's biotic components, particularly trees, contain 800 billion tons of organic carbon, and an additional 1,000–3,000 billion metric tons are estimated to be held in the remains of plants and animals in the soil. ❹ Ordinarily, decomposition of animals returns CO_2 to the atmosphere.

Some 300 MYA, plant and animal remains were transformed into coal, oil, and natural gas, the materials we call fossil fuels. Another reservoir for carbon is the inorganic carbonate that accumulates in limestone and in calcium carbonate shells. Many marine organisms have calcium carbonate shells that remain in bottom sediments long after the organisms have died. Geologic forces change these sediments into limestone.

Human Activities and the Carbon Cycle ❺ More

CO_2 is being deposited in the atmosphere than is being removed largely due to the burning of fossil fuels and the destruction of forests to make way for farmland and pasture. When we humans do away with forests, we reduce a reservoir and also the very organisms that take up excess carbon dioxide. Today, the amount of CO_2 released into the atmosphere is about twice the amount that remains in the atmosphere. It's believed that much of this has been dissolving into the ocean.

Carbon dioxide, and other gases as well, are being emitted due to human activities. The other gases include nitrous oxide (N_2O) from fertilizers and animal wastes and methane (CH_4) from bacterial decomposition that takes place particularly in the guts of animals, in sediments, and in flooded rice paddies. These gases are known as **greenhouse gases** because, just like the panes of a greenhouse, they allow solar radiation to pass through but hinder the escape of infrared rays (heat) back into space. This phenomenon has come to be known as the **greenhouse effect.** The greenhouse gases are contributing significantly to an overall rise in the Earth's ambient temperature, a trend called **global warming.** The global climate has already warmed about 0.6°C since the Industrial Revolution. Computer models are unable to consider all possible variables, but the Earth's temperature may rise 1.5–4.5°C by 2100 if greenhouse emissions continue at the current rates.

It is predicted that, as the oceans warm, temperatures in the polar regions will rise to a greater degree than in other regions. If so, glaciers will melt, and sea level will rise, not only due to this melting but also because water expands as it warms. Increased rainfall is likely along the coasts, while dryer conditions are expected inland. Coastal agricultural lands, such as the deltas of Bangladesh and China, will be inundated with sea water, and billions of dollars will have to be spent to keep coastal cities such as New Orleans, New York, Boston, Miami, and Galveston from disappearing into the sea.

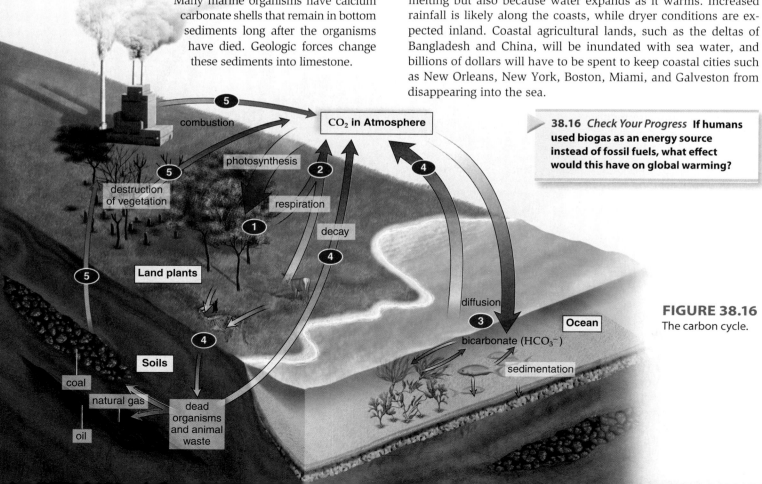

38.16 *Check Your Progress* **If humans used biogas as an energy source instead of fossil fuels, what effect would this have on global warming?**

FIGURE 38.16
The carbon cycle.

Community ecology is concerned with how populations of different species interact with each other. The number of individuals in each population is influenced by negative interactions such as competition, predation, and parasitism. But positive interactions such as mutualism are also fairly common in nature (especially for plants) and are presumed to increase or maintain population sizes. Perhaps one of the most important recent discoveries about communities is that they are highly dynamic, meaning that the number of species, kinds of species, and sizes of populations within most communities are constantly changing. This dynamic quality is well demonstrated by the process of ecological succession.

Instead of studying the composition of communities, some ecologists concentrate on the movement of energy and nutrients through communities. The physical environment has a large influence on energy flow and chemical cycling. Therefore, our study of communities must include the abiotic environment. Human activities also influence the operation of ecosystems. For example, burning fossil fuels and trees is adding carbon dioxide to the atmosphere.

Carbon dioxide and other greenhouse gases allow the sun's rays to pass through, but they absorb and reradiate heat back to the Earth, which may be leading to global warming. Transfer rates in both the phosphorus and nitrogen cycles are affected when we produce fertilizers and detergents. Nitrogen and phosphorus runoff causes eutrophication in aquatic ecosystems. The resulting pollution brought on by human activities affects the functioning of the biosphere, the largest ecosystem of all. In Chapter 39, we take a look at the specific types of ecosystems that make up the biosphere.

The Chapter in Review

Summary

Ridding the Land of Waste

- Scavengers consume dead organic matter, and decomposers recycle nutrients within ecosystems.

decomposers

A Community Contains Several Interacting Populations in the Same Locale

38.1 Competition can lead to resource partitioning

- The competitive exclusion principle states that no two species can occupy the same niche at the same time.
- An ecological niche is the role an organism plays in its community, including both habitat and interactions.
- Resource partitioning decreases competition for resources between two species.
- Characteristics tend to diverge when similar species belong to the same community, a phenomenon known as character displacement.

38.2 Predator-prey interactions affect both populations

- Predator-prey populations cycle: More predators/fewer prey result in fewer predators/more prey.
- Prey defenses include senses, speed, protective body parts, and chemicals.
- Ways to deceive predators include camouflage, cryptic coloration, and mimicry.

consumers

38.3 Parasitism benefits one population at another's expense

- Parasitism is a type of symbiosis; other types are commensalism and mutualism.

38.4 Commensalism benefits only one population

- Although commensalism benefits only one species, closer examination sometimes reveals more of a mutualistic or parasitic relationship.

38.5 Coevolution requires interaction between two species

- Coevolution occurs when two species adapt in response to selective pressure imposed by each other.
- Symbionts are especially prone to coevolution.

38.6 Mutualism benefits both populations

- Mutualism results in an intricate web of species interdependencies critical to the community.

A Community Develops and Changes Over Time

38.7 The study of island biogeography pertains to biodiversity

- The size of an island and its distance from a population source affect biodiversity.
- An environment with patches has more habitats and, therefore, more biodiversity.

38.8 During ecological succession, community composition and diversity change

- Ecological succession involves a series of species replacements in a community.
  - Primary succession occurs where there is no soil present.
  - Secondary succession occurs where soil is present and certain plant species can begin to grow.
  - A climax community forms when stages of succession lead to a particular type of community.

An Ecosystem Is a Community Interacting with the Physical Environment

38.9 Ecosystems have biotic and abiotic components

- Biotic components are living components.
  - Autotrophs are producers.
  - Heterotrophs are consumers: types of consumers include herbivores, carnivores, and omnivores.
  - Scavengers feed on dead remains of animals and plants.
  - Detritus is composed of the organic remains of decomposition found in water and soil.
  - Decomposers include bacteria and fungi.
- Abiotic components are resources (e.g., sunlight, inorganic nutrients) and conditions (e.g., soil, water, temperature, wind).

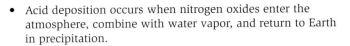

inorganic nutrient pool

38.10 Energy flow and chemical cycling characterize ecosystems

- Energy flows through ecosystems because as food passes from one population to the next, each population makes energy conversions that result in a loss of usable energy.
- Chemicals cycle because they pass from one population to the next until decomposers return them once more to the environment where producers can take them up again.

38.11 Energy flow involves food webs

- A food web is an interconnecting path of energy flow that describes trophic (feeding) relationships.
  - A grazing food web begins with a producer, such as an oak tree.
  - A detrital food web begins with detritus.
  - Grazing and detrital food webs are joined.
- A food chain is a single path of energy flow.
- A trophic level is composed of organisms that occupy the same position within a food web.

producers

38.12 Ecological pyramids are based on trophic levels

- An ecological pyramid is a graphic representation of the number of organisms, biomass, or energy content of trophic levels.

38.13 Chemical cycling includes reservoirs, exchange pools, and the biotic community

- Biogeochemical cycles may be sedimentary (phosphorus cycle) or gaseous (carbon and nitrogen cycles).
- Chemical cycling involves a reservoir, an exchange pool, and a biotic community.

38.14 The phosphorus cycle is sedimentary

- Geologic upheavals move phosphorus from the ocean to land.
- Slow weathering of rocks returns phosphorus to the soil.
- Most phosphorus is recycled within a community, and phosphorus is a limiting nutrient.

38.15 The nitrogen cycle is gaseous

- Plants cannot use nitrogen gas (N_2) from the atmosphere.
- During nitrogen fixation, N_2 converts to ammonium, making nitrogen available to plants.
- Nitrification is the production of nitrates, while denitrification is the conversion of nitrate back to N_2, which enters the atmosphere.
- Human activities increase transfer rates in the nitrogen cycle.

- Acid deposition occurs when nitrogen oxides enter the atmosphere, combine with water vapor, and return to Earth in precipitation.

38.16 The carbon cycle is gaseous

- CO_2 in the atmosphere is an exchange pool for the carbon cycle; terrestrial and aquatic plants and animals exchange CO_2 with the atmosphere.
- Living and dead organisms serve as reservoirs for the carbon cycle because they contain organic carbon.
- Human activities increase the levels of CO_2 and other greenhouse gases in the atmosphere, which contributes to global warming.

Testing Yourself

A Community Contains Several Interacting Populations in the Same Locale

1. According to the competitive exclusion principle,
   a. one species is always more competitive than another for a particular food source.
   b. competition excludes multiple species from using the same food source.
   c. no two species can occupy the same niche at the same time.
   d. competition limits the reproductive capacity of species.
2. Resource partitioning pertains to
   a. niche specialization.
   b. character displacement.
   c. increased species diversity.
   d. the development of mutualism.
   e. All but d are correct.

For statements 3–7, indicate the type of interaction in the key that is described in each scenario.

KEY:
   a. competition   d. commensalism
   b. predation   e. mutualism
   c. parasitism

3. An alfalfa plant gains fixed nitrogen from the bacterial species *Rhizobium* in its root system, while *Rhizobium* gains carbohydrates from the plant.
4. Both foxes and coyotes in an area feed primarily on a limited supply of rabbits.
5. Roundworms establish a colony inside a cat's digestive tract.
6. A fungus captures nematodes as a food source.
7. An orchid plant lives in the treetops, gaining access to sun and pollinators, but not harming the trees.
8. **THINKING CONCEPTUALLY** You want to reintroduce a predator into an area. What concern might you have?

A Community Develops and Changes Over Time

9. The model of island biogeography is pertinent to
   a. only islands surrounded by water.
   b. explaining community composition and diversity.
   c. explaining the intermediate disturbance hypothesis.
   d. why exotics are such a problem today.
   e. All of these are correct.
10. Mosses growing on bare rock will eventually help to create soil. These mosses are involved in _____ succession.
   a. primary
   b. secondary
   c. tertiary

11. Assume that a farm field is allowed to return to its natural state. By chance, the field is first colonized by native grasses, which begin the succession process. This is an example of which model of succession?
    a. climax pattern   c. facilitation
    b. tolerance   d. inhibition

An Ecosystem Is a Community Interacting with the Physical Environment

12. The ecological niche of an organism
    a. is the same as its habitat.
    b. includes how it competes and acquires food.
    c. is specific to the organism.
    d. is usually occupied by another species.
    e. Both b and c are correct.
13. In what way are decomposers like producers?
    a. Either may be the first member of a grazing or a detrital food chain.
    b. Both produce oxygen for other forms of life.
    c. Both require nutrient molecules and energy.
    d. Both are present only on land.
    e. Both produce organic nutrients for other members of ecosystems.
14. When a heterotroph takes in food, only a small percentage of the energy in that food is used for growth. The remainder is
    a. not digested and   c. given off as heat.
       eliminated as feces.   d. All of these are correct.
    b. excreted as urine.   e. None of these are correct.
15. During chemical cycling, inorganic nutrients are typically returned to the soil by
    a. autotrophs.   c. decomposers.
    b. detritivores.   d. tertiary consumers.
16. In a grazing food web, carnivores that eat herbivores are
    a. producers.   c. secondary consumers.
    b. primary consumers.   d. tertiary consumers.
17. Choose the statement that is true concerning this food chain:
    grass ⟶ rabbits ⟶ snakes ⟶ hawks
    a. Each predator population has a greater biomass than its prey population.
    b. Each prey population has a greater biomass than its predator population.
    c. Each population is omnivorous.
    d. Each population returns inorganic nutrients and energy to the producer.
    e. Both a and c are correct.
    f. Both a and b are correct.
18. Which of the following is a sedimentary biogeochemical cycle?
    a. carbon
    b. nitrogen
    c. phosphorus
19. Which of the following is not a component of the nitrogen cycle?
    a. proteins   d. photosynthesis
    b. ammonium   e. bacteria in root nodules
    c. decomposers
20. How do plants contribute to the carbon cycle?
    a. When plants respire, they release CO_2 into the atmosphere.
    b. When plants photosynthesize, they consume CO_2 from the atmosphere.
    c. When plants photosynthesize, they provide oxygen to heterotrophs.
    d. Both a and b are correct.

21. **THINKING CONCEPTUALLY** Could several different photosynthetic and nonphotosynthetic aquatic species living in the same pond have the same trophic level? Explain.

Understanding the Terms

acid deposition 757	grazing food web 754
autotroph 752	greenhouse effect 758
biogeochemical cycle 756	greenhouse gases 758
camouflage 745	habitat 742
carnivore 752	herbivore 752
character displacement 742	heterotroph 752
climax community 751	host 746
commensalism 746	mimicry 745
community 742	model of island
competition 742	biogeography 749
competitive exclusion	mutualism 748
principle 742	nitrification 757
consumer 752	N_2 (nitrogen) fixation 757
decomposer 752	omnivore 752
denitrification 757	parasite 746
detrital food web 755	parasitism 746
detritus 752	pioneer species 750
ecological niche 742	predation 744
ecological pyramid 755	producer 752
ecology 752	resource partitioning 742
ecosystem 751	scavenger 752
eutrophication 757	species richness 749
food chain 755	symbiosis 746
food web 754	trophic level 755
global warming 758	

Match the terms to these definitions:
a. _____ Assemblage of populations interacting with one another within the same environment.
b. _____ Place where an organism lives and is able to survive and reproduce.
c. _____ Role an organism plays in its community, including its habitat and its interactions with other organisms.
d. _____ Symbiotic relationship in which both species benefit in terms of growth and reproduction.

Thinking Scientifically

1. As per Figure 14.1B, you observe three species of *Empidonax* flycatchers in the same general area, and you hypothesize that they occupy different niches. How could you substantiate your hypothesis?
2. In order to improve species richness, you decide to add phosphate to a pond. How might you determine how much phosphate to add in order to avoid eutrophication?

ARIS *Visit* **www.mhhe.com/maderconcepts** *for practice quizzes, animations, videos, and activities designed to help you master the material in this chapter.*

39

Major Ecosystems of the Biosphere

LEARNING OUTCOMES

After studying this chapter, you should be able to accomplish the following outcomes.

Life Under Glass

1 Discuss the practicality of building a spaceship that contains a biosphere for travel to another planet.

On Land, the Biosphere Is Organized into Terrestrial Ecosystems

2 Describe the terrestrial ecosystems discussed with reference to the climate, the soil, and the types of plants and animals that live there.

3 Discuss the factors that determine and affect the climate of an ecosystem.

Fresh Water and Salt Water Are Organized into Aquatic Ecosystems

4 Explain how aquatic ecosystems are classified, including the different types of wetlands.

5 Contrast oligotrophic lakes with eutrophic lakes.

6 Contrast the shallow waters with the deep-sea waters of the oceans.

7 Describe coral reefs, and explain why they occur in shallow waters.

8 Divide the open ocean into three zones, and describe each one, including its marine life.

9 Describe ocean currents in general terms and the benefits of currents, such as the Gulf Stream.

10 Describe the events of an El Niño and why they cause concern.

J ust a 20-minute drive north of Tucson, in the desert of Arizona, a futuristic glass structure emerges from the surrounding landscape. Called Biosphere 2, the structure was built to help establish a new field called ecological engineering and to investigate the interaction and evolution of ecosystems enclosed within a heavily subsidized environment. Pumps kept the hydrologic cycle going, blowers provided atmospheric movement, and huge "lungs" relieved the pressure that builds up in a glass-enclosed system.

The name, Biosphere 2, means that the structure was modeled after the Earth's biosphere, or Biosphere 1. Biosphere 2 contained representative living organisms, as well as nonliving components of five major natural biomes (a tropical rain forest, a savanna, a coastal fog desert, a mangrove wetland, and an ocean) plus two anthrogenic biomes where eight biospherians lived for two years. Their contact with Biosphere 1 was extremely limited.

The cost of more than $200 million to build Biosphere 2 was funded by billionaire Edward P. Bass. The structure, when sealed, was not entirely airtight, but it was 30 times less leaky than the space shuttle, and thousands of times less leaky than a typical skyscraper. Ecosystems did not flourish as they had been

ecosystems ecosystems

human habitat

intensive agriculture

lung

Biosphere 2 © G. Kenny

planned, but this gave information and insight into the workings of their analogs in nature. CO_2 concentrations soared, and O_2 concentrations dipped to below normal during the first years, indicating the importance of the world's oceans in absorbing CO_2 and the world's forests in producing O_2. Soil microbes and the building's concrete components also played a role in creating the imbalance of these gases in Biosphere 2. Some of the crops did not bear fruit because of pests, diseases, and the lack of pollinators. Oxygen was piped in, but many vertebrates died off. When cockroaches proliferated, toads and geckoes were introduced from the outside to eat them. The biospherians were stressed by living within a confined structure.

Still, the success of the Biosphere 2 project can be measured by the quality of the observations and extensive data now available to the scientific community. Biosphere 2, it turns out, was immensely valuable to many scientific disciplines, and it inspired quite a few scientific publications. Some papers presented computer models to describe the hydrologic balance and the heat versus humidity within the system. Other papers told of the changes that occurred in the rain forest, mangrove, ocean, and agronomic ecosystems in this CO_2-rich environment.

The overall ecological message is clear—we should appreciate and try hard to understand the workings of the ecosystems within Biosphere 1 better. While Biosphere 2 was a start, more experimentation needs to be done before a self-sustaining habitat can be established on the Moon or Mars. Biosphere 2 was only the first step toward achieving such a system.

Biosphere harvest

© A. Alling

Biosphere agriculture

Biosphere feast

© A. Alling

This part of the chapter discusses selected terrestrial ecosystems. Because climate plays a significant role in determining the characteristics of an ecosystem, we also discuss factors that affect climate.

39.1 Major terrestrial ecosystems are characterized by particular climates

Each major terrestrial ecosystem (sometimes called a **biome**) has a particular mix of plants and animals that are adapted to living under certain environmental conditions, of which climate is an overriding influence. **Climate** consists of the average yearly temperature and precipitation of a region. When terrestrial ecosystems are plotted according to their climate, a particular distribution pattern results. The distribution of a number of terrestrial ecosystems is shown in **Figure 39.1**. However, we are going to concentrate on the selected ecosystems listed in **Table 39.1**. Be sure to note where the ecosystems listed in Table 39.1 are distributed. Even though Figure 39.1 shows definite demarcations, keep in mind that ecosystems gradually change from one type to the other. Also, although we will be discussing different ecosystems separately, we should remember that each one has inputs from and outputs to all the other terrestrial and aquatic ecosystems of the biosphere.

We begin our description of major biomes with the tundra in Section 39.2.

39.1 *Check Your Progress* **Why would you expect greater biodiversity in a rain forest than in a desert?**

TABLE 39.1	Selected Terrestrial Ecosystems
Name	**Characteristic**
Tundra	Around North Pole; very cold (−12°C to −6°C); little precipitation (less than 25 cm/year); permafrost (permanent ice) year-round within a meter of surface.
Taiga (coniferous forest)	Large northern biome just below the Arctic Circle; temperature is below freezing for half the year; moderate precipitation (30–85 cm/year); long nights in winter and long days in summer.
Temperate deciduous forest	Eastern half of United States, Canada, Europe, and parts of Russia; four seasons of the year with hot summers and cold winters; ample precipitation (75–150 cm/year).
Grasslands	Savanna in Africa and temperate grassland elsewhere; hot in summer and cold in winter (U.S.); moderate precipitation; good soil for agriculture.
Tropical rain forest	Located near the equator in Latin America, Southeast Asia, and West Africa; warm (20–25°C) and heavy precipitation (190 cm/year).
Desert	Northern and Southern Hemispheres at 30° latitude; hot (38°C) days and cold (7°C) nights; low precipitation (less than 25 cm/year).

FIGURE 39.1 Pattern of ecosystem distribution on land.

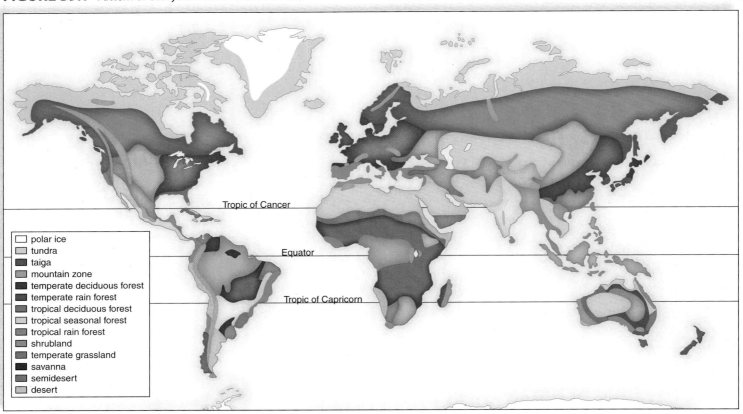

Tropic of Cancer

Equator

Tropic of Capricorn

☐ polar ice
☐ tundra
☐ taiga
☐ mountain zone
☐ temperate deciduous forest
☐ temperate rain forest
☐ tropical deciduous forest
☐ tropical seasonal forest
☐ tropical rain forest
☐ shrubland
☐ temperate grassland
☐ savanna
☐ semidesert
☐ desert

39.2 The tundra is cold and dark much of the year

The **Arctic tundra** ecosystem, which encircles the Earth just south of the ice-covered polar seas in the Northern Hemisphere, covers about 20% of the Earth's land surface (**Fig. 39.2**). (A similar ecosystem, called the alpine tundra, occurs above the timberline on mountain ranges.) The Arctic tundra is cold and dark much of the year. Its winters are extremely long, cold, and harsh, and its summers are short (6–8 weeks). Because rainfall amounts to only about 20 cm a year, the tundra could possibly be considered a desert, except that melting snow creates a landscape of pools and mires in the summer, especially because so little evaporates. Only the topmost layer of soil thaws; the **permafrost** beneath this layer is always frozen, and therefore drainage is minimal. The available soil in the tundra is nutrient-poor.

Trees are not found in the tundra because the growing season is too short, their roots cannot penetrate the permafrost, and they cannot become anchored in the boggy soil of summer. In the summer, the ground is covered with short grasses and sedges, as well as numerous patches of lichens and mosses. Dwarf woody shrubs, such as dwarf birch, and flowers seed quickly while there is plentiful sun for photosynthesis.

A few animals live in the tundra year-round. For example, the mouselike lemming stays beneath the snow; the ptarmigan, a grouse, burrows in the snow during storms; and the musk ox conserves heat because of its thick coat and short, squat body. Other animals that live in the tundra include snowy owls, lynx, voles, Arctic foxes, and snowshoe hares. In the summer, the tundra is alive with numerous insects and birds, particularly shorebirds and waterfowl that migrate

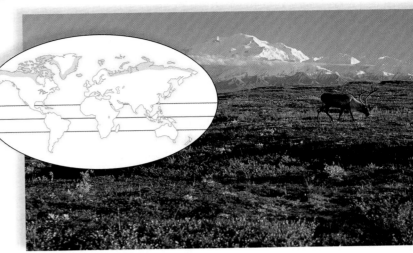

FIGURE 39.2 Tundra, the northern-most ecosystem.

inland. Caribou in North America and reindeer in Asia and Europe also migrate to and from the tundra, as do the wolves that prey upon them. Polar bears are common near the coastal regions.

The next section discusses the plants and animals of a coniferous forest.

> **39.2 Check Your Progress** Give reasons why Biosphere 2 didn't have an Arctic tundra.

39.3 Coniferous forests are dominated by gymnosperms

Coniferous forests are found in three locations: in the **taiga**, which extends around the world in the northern part of North America and Eurasia; near mountaintops (where it is called a montane coniferous forest); and along the Pacific coast of North America, as far south as northern California.

The taiga, also called boreal (northern) forest, exists south of the tundra and covers approximately 11% of the Earth's landmasses (**Fig. 39.3**). The needlelike leaves of its cone-bearing trees can withstand the weight of heavy snow. There is a limited understory of plants, but the floor is covered with low-lying mosses and lichens beneath a layer of needles. Birds harvest the seeds of the conifers, and bears, deer, moose, beavers, and muskrats live around the cool lakes and along the streams. Wolves prey on these larger mammals.

The coniferous forest that runs along the west coasts of Canada and the United States is sometimes called a **temperate rain forest.** Winds moving in off the Pacific Ocean lose their moisture when they meet the coastal mountain range. The plentiful rainfall and rich soil have produced some of the tallest conifer trees ever in existence, including the coastal redwoods. This forest is also called an old-growth forest because some trees are more than 1,000 years old. It truly is an evergreen forest because mosses, ferns, and other plants grow on all the tree trunks. Squirrels, lynx, and numerous species of amphibians, reptiles, and birds inhabit the temperate rain forest.

FIGURE 39.3 Taiga, a northern coniferous forest.

Let's move on in Section 39.4 to see what a temperate deciduous forest is like.

> **39.3 Check Your Progress** What type of trees would be needed to simulate the taiga in Biosphere 2?

39.4 Temperate deciduous forests have abundant life

Temperate deciduous forests are found south of the taiga in eastern North America, eastern Asia, and much of Europe (**Fig. 39.4**). The climate in these areas is moderate, with relatively high rainfall (75–150 cm per year). The seasons are well defined, and the growing season ranges between 140 and 300 days. The trees, which include oak, beech, sycamore, and maple, have broad leaves and are termed deciduous trees; they lose their leaves in the fall and grow them in the spring. In southern temperate deciduous forests, evergreen magnolia trees can be found.

The tallest trees form a canopy, an upper layer of leaves that are the first to receive sunlight. Even so, enough sunlight penetrates to provide energy for another layer of trees, called understory trees. Beneath these trees are shrubs that may flower in the spring before the trees have put forth their leaves. Still another layer of plant growth—mosses, lichens, and ferns—resides beneath the shrub layer. This *stratification* provides a variety of habitats for insects and birds. Ground life is also plentiful. Squirrels, cottontail rabbits, shrews, skunks, woodchucks, and chipmunks are small herbivores. These and ground birds, such as turkeys, pheasants, and grouse, are preyed on by red foxes. White-tailed deer and black bears have increased in number in recent years. In contrast to the taiga, amphibians and reptiles occur in this ecosystem because the winters are not as cold. Frogs and turtles prefer an aquatic existence, as do the beavers and muskrats, which are mammals.

Autumn fruits, nuts, and berries provide a supply of food for the winter. The leaves, after turning brilliant colors and falling to the ground, contribute to a rich layer of humus. The minerals within the rich soil are washed far into the ground by spring rains, but the deep tree roots capture these and bring them back up into the forest system again.

Section 39.5 allows us to contrast a temperate forest with a temperate grassland.

FIGURE 39.4 Temperate deciduous forest in the fall.

> **39.4 Check Your Progress** To simulate a temperate forest for Biosphere 2, would it be enough to simply bring in many types of deciduous trees?

39.5 Temperate grasslands have extreme seasons

The **temperate grasslands** include the Russian steppes, the South American pampas, and the North American prairies (**Fig. 39.5**). In these grasslands, winters are bitterly cold, and summers are hot and dry. When traveling across the United States from east to west, temperate deciduous forest transitions into *tall-grass prairie* roughly along the border between Illinois and Indiana. The tall-grass prairie requires more rainfall than does the *short-grass prairie*, which occurs near deserts. Large herds of bison—estimated at hundreds of thousands—once roamed the prairies, as did herds of pronghorn antelope. Now, small mammals, such as mice, prairie dogs, and rabbits, typically live belowground, but usually feed aboveground. Hawks, snakes, badgers, coyotes, and foxes feed on these mammals. However, virtually all of these grasslands have been converted to agricultural lands because of their fertile soils.

Section 39.6 describes savannas, another type of grassland.

> **39.5 Check Your Progress** If you put a simulated prairie in Biosphere 2, would you give it the same amount of rainfall as the deciduous forest? See Table 39.1.

FIGURE 39.5 Temperate grassland in the summer.

39.6 Savannas have wet-dry seasons

Savannas occur in regions where a relatively cool dry season is followed by a hot rainy season. The largest savannas are in central and southern Africa; other savannas exist in Australia, Southeast Asia, and South America (**Fig. 39.6**). The savanna is characterized by large expanses of grasses with sparse populations of trees. The plants of the savanna have extensive and deep root systems that enable them to survive drought and fire. One tree that can survive the severe dry season is the thorny flat-topped acacia, which sheds its leaves during a drought. The African savanna supports the greatest variety and number of large herbivores of all the biomes (Fig. 39.6). Elephants and giraffes are browsers that feed on tree vegetation. Antelopes, zebras, wildebeests, water buffalo, and some rhinoceroses are grazers that feed on grasses. Any plant litter that is not consumed by grazers is attacked by a variety of small organisms, among them termites. Termites build towering nests in which they tend fungal gardens, their source of food. The herbivores support a large population of carnivores. Lions and hyenas sometimes hunt in packs, cheetahs hunt singly by day, and leopards hunt singly by night.

Deserts are not always barren, as noted in Section 39.7.

> **39.6 Check Your Progress** If Biosphere 2 housed a simulated savanna ecosystem suitable for giraffes, elephants, and zebras, what type of plants would be needed?

FIGURE 39.6 The African savanna.

39.7 Deserts have very low annual rainfall

Deserts are usually found at latitudes of about 30° in both the Northern and Southern Hemispheres (**Fig. 39.7**). The winds that descend in these regions lack moisture, and the annual rainfall is less than 25 cm. Days are hot because a lack of cloud cover allows the sun's rays to penetrate easily, but nights are cold because heat escapes easily into the atmosphere.

The Sahara Desert, which stretches all the way from the Atlantic coast of Africa to the Arabian peninsula, and a few other deserts have little or no vegetation. But most have a variety of plants. Desert plants are highly adapted to survive long droughts, extreme heat, and extreme cold. Adaptations to these conditions include thick epidermal layers, water-storing stems and leaves, and the ability to set seeds quickly in the spring. The best-known desert perennials in North America are the succulent, spiny-leafed cactuses, which have stems that store water and carry on photosynthesis.

Some animals are adapted to the desert environment. To conserve water, many desert animals are nocturnal or burrowing and have a protective outer body covering. A desert has numerous insects, which pass through the stages of development when there is rain. Reptiles, especially lizards and snakes, are perhaps the most characteristic group of vertebrates found in deserts, but running birds (e.g., the roadrunner) and rodents (e.g., the kangaroo rat) are also well known. Larger mammals, such as the kit fox, prey on the rodents, as do hawks.

Life is plentiful in a tropical rain forest, as we shall see in Section 39.8.

FIGURE 39.7 Desert with some vegetation.

> **39.7 Check Your Progress** Are the nights in the desert outside the Biosphere 2 structure hot or cold?

39.8 Tropical rain forests are warm with abundant rainfall

In the **tropical rain forests** of South America, Africa, and the Indo-Malayan region near the equator, the temperature is always warm (between 20° and 25°C), and rainfall is plentiful (a minimum of 190 cm per year). This may be the richest ecosystem, in terms of both number of different kinds of species and their abundance. The diversity of species is enormous—a 10-km² area of tropical rain forest may contain 1,500 species of flowering plants, including the trees.

A tropical rain forest has a complex structure, with many levels of life, including the forest floor, the understory, and the canopy. Sunlight is filtered out by the canopy, and the plants of the forest floor, such as ferns, are tolerant of minimal light. The understory consists of shorter trees that receive some light and bear epiphytes. **Epiphytes** are plants that grow on other plants but usually have roots of their own that absorb moisture and minerals leached from their hosts; others catch rain and debris in hollows produced by overlapping leaf bases. The most common epiphytes are related to pineapples, orchids, and ferns. The canopy, topped by the crowns of tall trees, is the most productive level of the tropical rain forest (**Fig. 39.8A**). Some of the broadleaf evergreen trees grow to 15–50 m or more. These tall trees often have trunks buttressed at ground level to prevent them from toppling over. Lianas, or woody vines, which encircle the tree as it grows, also help strengthen the trunk.

Although some animals live on the forest floor (e.g., pacas, agoutis, peccaries, and armadillos), most live in the trees (**Fig. 39.8B**). Insect life is so abundant that the majority of species have not been identified yet. Termites play a vital role in the decomposition of woody plant material, and ants are found everywhere, particularly in the trees. The various birds, such as hummingbirds, parakeets, parrots, and toucans, are often beautifully colored. Amphibians and reptiles are well represented by many types of frogs, snakes, and lizards.

FIGURE 39.8A Levels of life in a tropical rain forest.

Lemurs, sloths, and monkeys are well-known primates that feed on the fruits of the trees. The largest carnivores are the big cats—the jaguars in South America and the leopards in Africa and Asia.

This completes our discussion of terrestrial ecosystems. The next section discusses factors that determine climate.

> **39.8 Check Your Progress** In the tropical rain forest ecosystem of Biosphere 2, the plants growing on other plants high up in the canopy are called _____.

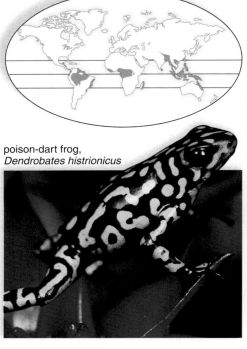

poison-dart frog,
Dendrobates histrionicus

brush-footed butterfly,
Anartia amalthea linnaeus

blue and gold macaw,
Ara ararauna

cone-headed katydid,
Panacanthus cuspidatus

lemur,
Propithecus verreauxi

ocelot,
Felis pardalis

arboreal lizard,
Calotes calotes

FIGURE 39.8B Representative animals of the tropical rain forests of the world.

Because the Earth is a sphere, the sun's rays are more direct at the equator and more spread out at the polar regions (**Fig. 39.9A,** *left*). Therefore, the tropics are warmer than the temperate regions.

The tilt of the Earth as it orbits the sun causes one pole or the other to be closer to the sun (except at the spring and fall equinoxes, when the sun aims directly at the equator), and this accounts for the seasons that occur in all parts of the Earth, but not at the equator (Fig. 39.9A, *right*). When the Northern Hemisphere is having winter, the Southern Hemisphere is having summer, and vice versa.

If the Earth were standing still and were a solid, uniform ball, all air movements—which we call winds—would be in two directions. Warm equatorial air would rise and move directly to the poles, creating a zone of lower pressure that would be filled by cold polar air moving equatorward.

However, because the Earth rotates on its axis daily and its surface consists of continents and oceans, the overall flows of warm and cold air are modified into three large circulation cells in each hemisphere (**Fig. 39.9B,** *large arrows*). At the equator, the sun heats the air and evaporates water. The warm, moist air rises, cools, and loses most of its moisture as rain. The greatest amounts of rainfall on Earth are near the equator. The rising air flows toward the poles, but at about 30° north and south latitude, it sinks toward the Earth's surface and reheats. As the air descends and warms, it is very dry, creating zones of low rainfall. The great deserts of Africa, Australia, and the Americas occur at these latitudes. At the Earth's surface, the air flows both poleward and equatorward. At about 60° north and south latitude, the air rises and cools, producing additional zones of high rainfall. This moisture supports the great forests of the temperate zone. Part of this rising air flows equatorward, and part continues poleward, descending near the poles, which are zones of low precipitation.

Besides affecting precipitation, the spinning of the Earth also affects the direction of the winds. Periods of calm, called the doldrums, occur at the equator (Fig. 39.9B, *small red arrows*). In the Northern Hemisphere, large-scale winds generally bend clockwise,

FIGURE 39.9B
Wind circulation as air moves from the equator to the poles and back again.

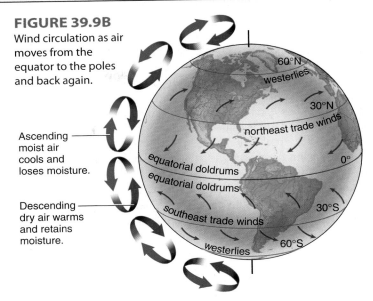

Ascending moist air cools and loses moisture.

Descending dry air warms and retains moisture.

and in the Southern Hemisphere, they bend counterclockwise. The curving pattern of the winds, ocean currents, and cyclones results because the Earth rotates in an eastward direction. At about 30° north latitude and 30° south latitude, the winds blow from the east-southeast in the Southern Hemisphere and from the east-northeast in the Northern Hemisphere (the east coasts of continents at these latitudes are wet). These are called trade winds because sailors depended on them to fill the sails of their trading ships. Between 30° and 60° north and south latitudes, strong winds, called the prevailing westerlies, blow from west to east. The west coasts of the continents at these latitudes are wet, as is the Pacific Northwest, where a massive evergreen forest is located.

Section 39.10 describes how topography affects climate.

> **39.9** *Check Your Progress* In contrast to the "real" biosphere, in Biosphere 2 the prevailing weather, or _____, is entirely controlled by humans.

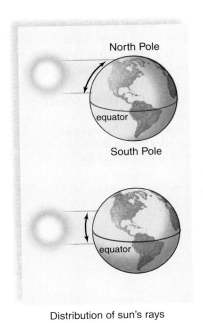

Distribution of sun's rays

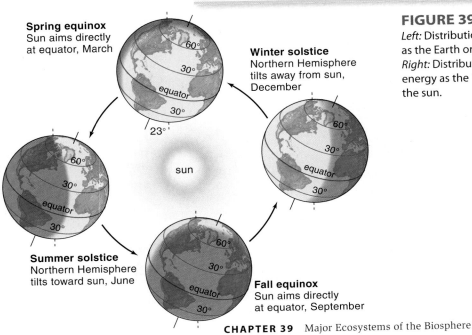

Spring equinox
Sun aims directly at equator, March

Winter solstice
Northern Hemisphere tilts away from sun, December

Summer solstice
Northern Hemisphere tilts toward sun, June

Fall equinox
Sun aims directly at equator, September

sun

FIGURE 39.9A
Left: Distribution of sun's rays as the Earth orbits the sun. *Right:* Distribution of solar energy as the Earth orbits the sun.

Topography refers to the surface features of the land. Mountains are topographic features that affect climate, and therefore the distribution of ecosystems. For example, traveling from the equator to the North Pole, you might see first a tropical rain forest, followed by a temperate deciduous forest, a coniferous forest, and tundra in that order because of the change in temperature. This same sequence is seen when ascending a mountain (**Fig. 39.10A**). The coniferous forest of a mountain is called a montane coniferous forest, and the tundra near the peak of a mountain is called an alpine tundra.

Mountains also affect precipitation. As air blows up and over a coastal mountain range, it rises and releases its moisture as it cools. One side of the mountain, called the windward side, receives more rainfall than the other side, called the leeward side. On the leeward side, the air descends, picks up moisture, and produces clear weather (**Fig. 39.10B**). The difference between the windward side and the leeward side can be quite dramatic. In the Hawaiian Islands, for example, the windward side of a mountain receives more than 750 cm of rain a year, while the leeward side, which is in a **rain shadow,** gets on the average only 50 cm of rain and is generally sunny. In the United States, the western side of the Sierra Nevada Mountains is lush, while the eastern side is a semidesert.

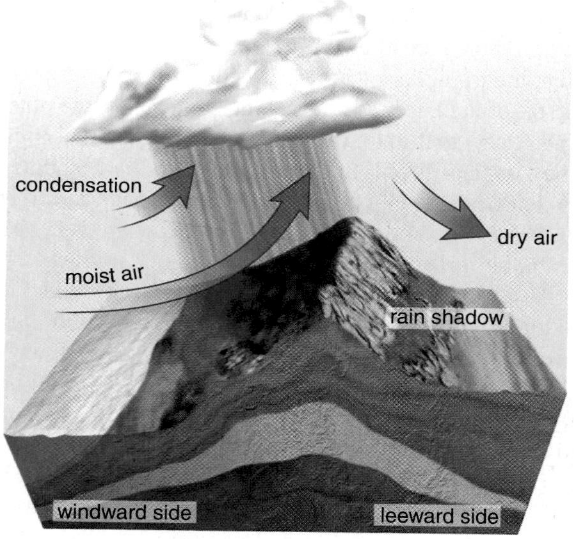

Nearby Bodies of Water The temperature of the oceans is more stable than that of the landmasses. Ocean water gains or loses heat more slowly than terrestrial environments do. This gives coasts a unique weather pattern that is not observed inland. During the day, the land warms more quickly than the ocean, and the air above the land rises. Then a cool sea breeze blows in from the ocean. At night, the reverse happens; the breeze blows from the land toward the sea.

FIGURE 39.10B Formation of a rain shadow.

India and some other countries in southern Asia have a **monsoon** climate, in which wet ocean winds blow onshore for almost half the year. During spring, the land heats more rapidly than the waters of the Indian Ocean, resulting in a temperature differential between the land and the ocean and a gigantic circulation of air: Warm air rises over the land, and cooler air comes in off the ocean to replace it. As the warm air rises, it loses its moisture, and the monsoon season begins. As just discussed, rainfall is particularly heavy on the windward side of hills. Cherrapunji in northern India receives an average of 1,090 cm of rain per year because of its high altitude. By November, the weather pattern has reversed. The land is now cooler than the ocean; therefore, dry winds blow from the Asian continent across the Indian Ocean. In the winter, the air over the land is dry, the skies cloudless, and temperatures pleasant. The chief crop of India is rice, which starts to grow when the monsoon rains begin.

In the United States, people often speak of the "lake effect," meaning that in the winter, arctic winds blowing over the Great Lakes become warm and moisture-laden. As these winds rise and lose their moisture, snow begins to fall. Places such as Buffalo, New York, get heavy snowfalls due to the lake effect, and snow is on the ground there for an average of 90–140 days every year.

The next part of the chapter discusses aquatic ecosystems.

> **39.10** *Check Your Progress* **When going from the equator to the South Pole, you would not reach a region that corresponds to the coniferous forest and tundra of the Northern Hemisphere. Explain.**

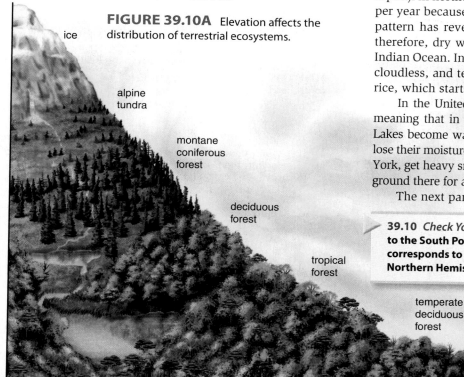

FIGURE 39.10A Elevation affects the distribution of terrestrial ecosystems.

ice

alpine tundra

montane coniferous forest

deciduous forest

tropical forest

temperate deciduous forest

coniferous forest

tundra

ice

Increasing Altitude

Increasing Latitude

Fresh Water and Salt Water Are Organized into Aquatic Ecosystems

An overview of aquatic ecosystems is followed by a more detailed description of marine ecosystems, including those of the coast and the open ocean.

39.11 Fresh water flows into salt water

Figure 39.11A shows how a freshwater ecosystem joins a saltwater ecosystem. Fresh water flows within **streams** and **rivers** and is contained, at least temporarily, in **lakes** and **ponds**. Mountain streams have cold, clear water that flows over waterfalls and rapids (Fig. 39.11A). **1** Here, the clawed feet of a long-legged stonefly larva help it hold onto the stones in the streambed. The streams join to form a river that flows gently enough for **2** trout to exist in occasional pools of oxygen-rich water. **3** Carp are fish adapted to water that contains little oxygen and much sediment, as might be found in a lake. The river meanders across broad, flat valleys, and finally empties into the ocean. At its mouth, the river divides into the many muddy channels of a delta. **4** Blue crabs frequent such areas. Nearby, salt marshes, which are characterized by the presence of rushes, reeds, and other grasses, are extremely productive ecosystems. They provide food and habitats for fish, waterfowl, and other wildlife.

In general, wetlands (lands that are wet some part of the year) directly absorb storm waters and also absorb overflows from lakes and rivers. In this way, they protect farms, cities, and towns from the devastating effects of floods. They also purify waters by filtering them and by diluting and breaking down toxic wastes and excess nutrients. Unfortunately, humans have the habit of filling in wetlands. Between the 1780s and the 1980s, approximately 53% of the original wetlands in the contiguous 48 states of the United States were lost. On average, this amounts to a loss of about 60 acres of wetlands per hour during that 200-year time span.

Lakes are often classified by their nutrient status (**Fig. 39.11B**). **Oligotrophic lakes** are nutrient-poor, having a small amount of organic matter and low productivity. **Eutrophic lakes** are nutrient-rich, having plentiful organic matter and high productiv-

FIGURE 39.11B
Types of lakes.

Oligotrophic lake

Eutrophic lake

ity. Eutrophic lakes are usually situated in naturally nutrient-rich regions, or are enriched by agricultural or urban and suburban runoff. Oligotrophic lakes can become eutrophic through large inputs of nutrients, a process called **eutrophication.**

Marine ecosystems are discussed in Section 39.12.

39.11 *Check Your Progress* **What type of ecosystem could help purify waste water inside Biosphere 2?**

FIGURE 39.11A A freshwater and saltwater ecosystem.

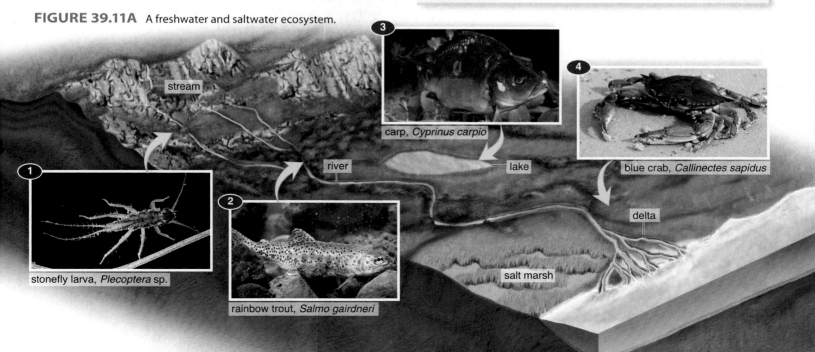

stream

carp, *Cyprinus carpio*

blue crab, *Callinectes sapidus*

river

lake

delta

stonefly larva, *Plecoptera* sp.

rainbow trout, *Salmo gairdneri*

salt marsh

Coastal Ecosystems Border the Oceans Salt marshes, discussed previously, and also mudflats and mangrove swamps featured in **Fig. 39.12A** are ecosystems that occur at a delta. Mangrove swamps develop in subtropical and tropical zones, while marshes and mudflats occur in temperate zones. These ecosystems are often designated as an estuary. So are coastal bays, fjords (an inlet of water between high cliffs), and some lagoons (a body of water separated from the sea by a narrow strip of land). Therefore, the term estuary has a very broad definition. An **estuary** is a partially enclosed body of water where fresh water and sea water meet and mix as a river enters the ocean.

Organisms living in an estuary must be able to withstand constant mixing of waters and rapid changes in salinity. But those organisms adapted to the estuarine environment find an abundance of nutrients. An estuary acts as a nutrient trap because the sea prevents the rapid escape of nutrients brought by a river. As the result of usually warm, calm waters and plentiful nutrients, estuaries are biologically diverse and highly productive.

Phytoplankton and shore plants thrive in nutrient-rich estuaries, providing an abundance of food and habitat for animals. It is estimated that nearly two-thirds of marine fishes and shellfish spawn and develop in the protective and rich environment of estuaries, making the estuarine environment the nursery of the sea. An abundance of larval, juvenile, and mature fish and shellfish attract a number of predators, such as reptiles, birds, and fishers of various types.

Rocky shores (Fig. 39.12A, *bottom*) and *sandy shores* are constantly bombarded by the sea as the tides roll in and out. The **intertidal zone** (**Fig. 39.12B**) lies between the high- and low-tide marks. In the upper portion of the intertidal zone, barnacles are glued so tightly to the stone by their own secretions that their calcareous outer plates remain in place, even after the enclosed shrimplike animal dies. In the midportion of the intertidal zone, brown algae, known as rockweed, may overlie the barnacles. Below the intertidal zone, macroscopic seaweeds, which are the main photosynthesizers, anchor themselves to the rocks by holdfasts.

Organisms cannot attach themselves to shifting, unstable sands on a sandy beach; therefore, nearly all the permanent residents dwell underground. They either burrow during the day and surface to feed at night, or they remain permanently within their burrows and tubes. Ghost crabs and sandhoppers (amphipods) burrow themselves above the high-tide mark and feed at night when the tide is out. Sandworms and sand (ghost) shrimp remain within their burrows in the littoral zone and feed on detritus whenever possible. Still lower in the sand, clams, cockles, and sand dollars are found. A variety of shorebirds visit the beaches and feed on various invertebrates and fishes.

Mudflat

Mangrove swamp

Rocky shore

FIGURE 39.12A Coastal ecosystems.

FIGURE 39.12B

Ocean ecosystems.

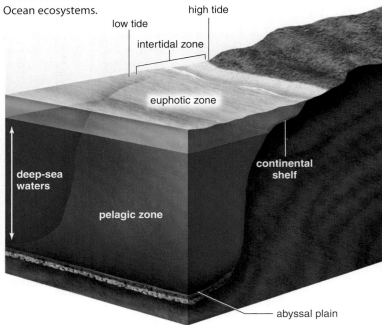

Oceans Shallow ocean waters (called the *euphotic zone*) contain a greater concentration of organisms than the rest of the sea (see Fig. 39.12B). Here, **phytoplankton,** (i.e., algae) is food not only for **zooplankton** (i.e., protozoans and microscopic animals) but also for small fishes. These attract a number of predatory and commercially valuable fishes. On the continental shelf, seaweed can be found growing, even on outcroppings as the water gets deeper. Clams, worms, and sea urchins are preyed upon by sea stars, lobsters, crabs, and brittle stars.

Coral reefs are areas of biological abundance just below the surface in shallow, warm, tropical waters (see Fig. 36.1). Their chief constituents are stony corals, animals that have a calcium carbonate (limestone) exoskeleton, and calcareous red and green algae. Corals provide a home for microscopic algae called *zooxanthellae*. The corals, which feed at night, and the algae, which photosynthesize during the day, are mutualistic and share materials and nutrients. The algae need sunlight, and this may be the reason coral reefs form only in shallow, sunlit waters.

A reef is densely populated with life. The large number of crevices and caves provide shelter for filter feeders (sponges, sea squirts, and fanworms) and for scavengers (crabs and sea urchins). The barracuda, moray eel, and shark are top predators in coral reefs. There are many types of small, beautifully colored fishes. These become food for larger fishes, including snappers that are caught for human consumption.

Most of the ocean lies within the **pelagic zone,** divided as noted in **Figure 39.12C.** The *epipelagic zone* lacks the inorganic nutrients of shallow waters, and therefore it does not have as high a concentration of phytoplankton, even though the surface is sunlit. Still, the photosynthesizers are food for a large assembly of zooplankton, which then become food for schools of various fishes. A number of porpoise and dolphin species visit and feed in the epipelagic zone. Whales too are mammals found in this zone. Baleen whales strain krill (small crustaceans) from the water, and toothed sperm whales feed primarily on the common squid.

Animals in the deeper waters of the *mesopelagic zone* are carnivores, which are adapted to the absence of light, and tend to be translucent, red-colored, or even luminescent. There are luminescent shrimps, squids, and fishes, including lantern and hatchet fishes. Various species of zooplankton, invertebrates, and fishes migrate from the mesopelagic zone to the surface to feed at night.

The deepest waters of the *bathypelagic zone* are in complete darkness except for an occasional flash of bioluminescent light. Carnivores and scavengers inhabit this zone. Strange-looking fishes with distensible mouths and abdomens and small, tubular eyes feed on infrequent prey.

It once was thought that few animals exist on the *abyssal plain* (see Fig. 39.12B) because of the intense pressure and the extreme cold. Yet, many invertebrates survive there by feeding on debris floating down from the mesopelagic zone. Sea lilies (crinoids) rise above the seafloor; sea cucumbers and sea urchins crawl around on the sea bottom; and tube worms burrow in the mud.

The flat abyssal plain is interrupted by enormous underwater mountain chains called oceanic ridges. Along the axes of the ridges, crustal plates spread apart, and molten magma rises to fill the gap. At **hydrothermal vents,** sea water percolates through cracks and is heated to about 350°C, causing sulfate to react

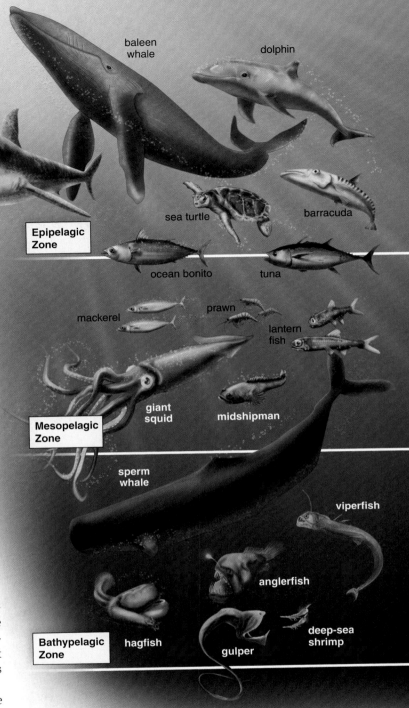

FIGURE 39.12C Ocean inhabitants in divisions of pelagic zone.

with water and form hydrogen sulfide (H_2S). Chemoautotrophic bacteria that obtain energy from oxidizing hydrogen sulfide exist freely or mutualistically within the tissues of organisms. They are the start of food chains for an ecosystem that includes huge tube worms, clams, crustaceans, echinoderms, and fishes. This ecosystem can exist where light never penetrates because, unlike photosynthesis, chemosynthesis does not require light energy.

> **39.12 Check Your Progress** Biosphere 2 contains a small-scale ocean that is sunlit throughout and relatively shallow. Thus, it replicates the _____ zone.

39.13 Ocean currents affect climates

Climate is driven by the sun, but the oceans play a major role in redistributing heat in the biosphere. Air takes on the temperature of the water below, and warm air moves from the equator to the poles. In other words, the oceans make the winds blow. When the wind blows strongly and steadily across a great expanse of ocean for a long time, friction from the moving air begins to drag the water along with it. Once the water has been set in motion, its momentum, aided by the wind, keeps it moving in a steady flow called a current. Because the ocean currents eventually strike land, they move in a circular path—clockwise in the Northern Hemisphere and counterclockwise in the Southern Hemisphere (**Fig. 39.13**). As the currents flow, they take warm water from the equator to the poles. One such current, called the Gulf Stream, brings tropical Caribbean water to the east coast of North America and the higher latitudes of western Europe. Without the Gulf Stream, Great Britain, which has a relatively warm temperature, would be as cold

as Greenland. In the Southern Hemisphere, another major ocean current warms the eastern coast of South America.

Also in the Southern Hemisphere, a current called the Humboldt Current flows toward the equator. The Humboldt Current carries phosphorus-rich cold water northward along the west coast of South America. During a process called **upwelling,** cold offshore winds cause cold, nutrient-rich waters to rise and take the place of warm, nutrient-poor waters. In South America, the enriched waters cause an abundance of marine life that supports the fisheries of Peru and northern Chile. Birds feeding on these organisms deposit their droppings on land, where they are mined as guano, a commercial source of phosphorus.

> **39.13** *Check Your Progress* **If the ocean in Biosphere 2 exhibited upwelling, what would rise to take the place of warm, nutrient-poor water?**

FIGURE 39.13 Ocean currents.

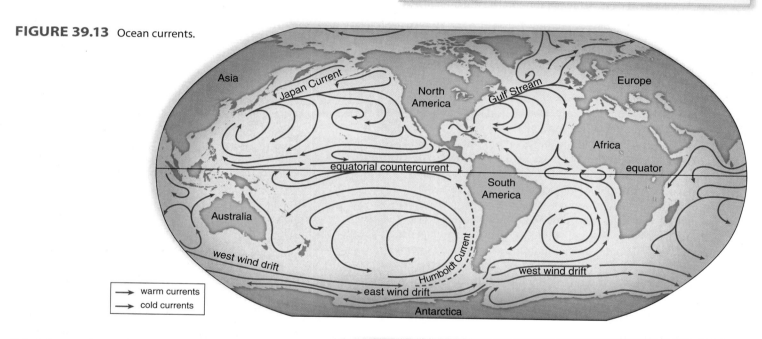

39.14 El Niño–Southern Oscillation alters weather patterns

When the Humboldt Current is not as cool as usual, upwelling of nutrients does not occur, stagnation results, the fisheries decline, and climate patterns change globally. This phenomenon, called an **El Niño–Southern Oscillation,** has a profound effect on weather; a severe El Niño affects the weather over three-quarters of the globe.

During an El Niño, southern California is hit by storms and even hurricanes, and the deserts of Peru and Chile receive so much rain that flooding occurs. A jet stream (strong wind currents) can carry moisture into Texas, Louisiana, and Florida, with flooding a near certainty. Or, the winds can turn northward and deposit snow in the mountains along the West Coast so that flooding occurs in the spring. Some parts of the United States, however, benefit from an El Niño. The Northeast is warmer than usual; few, if any, hurricanes hit

the East Coast; and there is a lull in tornadoes throughout the Midwest.

Eventually, an El Niño dies out, and normal conditions return. The normal cold-water state off the coast of Peru is known as La Niña (the girl). Since 1991, the sea surface has been almost continuously warm, and two record-breaking El Niños have occurred. What could be causing more of the El Niño state than the La Niña state? Some scientists are seeking data to relate this environmental change to global warming—a rise in environmental temperature due to greenhouse gases in the atmosphere (see Section 38.16).

> **39.14** *Check Your Progress* **Ordinarily, strong winds assist the movement of waters away from the coast so that cold, nutrient-rich water rises to the surface. Do you predict these winds are as strong as usual during an El Niño?**

The Earth's diverse ecosystems have resulted from interactions between the biotic communities and the abiotic environment. Organisms have helped create the chemical and physical conditions of streams, lakes, and oceans. The soils of terrestrial ecosystems and the sediments of aquatic ecosystems are structured largely by the activities of organisms.

Over geologic time, the biosphere has been changing constantly. Changes in the sun's radiation output and in the tilt of the Earth's axis have altered the pattern of solar energy reaching the Earth's surface. Geologic processes have also modified conditions for life. The drifting of continents has changed the arrangement of landmasses and oceans. Mountain ranges have been thrust up or eroded. Through these changing conditions, life has been evolving, and the structure of the Earth's ecosystems has evolved as well. In the last few million years, humans appeared and learned to exploit the Earth's ecosystems.

Modern humans have transformed vast areas of many of the terrestrial biomes into farmland, cities, highways, and other developments. Through our use of resources and release of pollutants, we have now become an agent of global importance. Yet, we still depend on the biodiversity that exists in the Earth's ecosystems, and on the interactions of other organisms within the biosphere. These interactions influence climate, patterns of nutrient cycling and waste processing, and basic biological productivity. The Earth's biotic diversity also provides enjoyment and inspiration to millions of people, who spend billions of dollars to visit coral reefs, deserts, rain forests, and even the Arctic tundra. Unfortunately, as we shall see in Chapter 40, many of the Earth's ecosystems may not survive in their present form for our descendants' benefit. Although most of us value Earth's biodiversity, human activities are threatening many species with extinction.

The Chapter in Review

Summary

Life Under Glass

- Biosphere 2 tested how ecosystems interact in a closed system.
- Biosphere 2 provided a lot of data about a CO_2-rich biosphere.

On Land, the Biosphere Is Organized into Terrestrial Ecosystems

39.1 Major terrestrial ecosystems are characterized by particular climates

- The major terrestrial ecosystems are tundra, taiga, temperate deciduous forest, grassland, desert, and tropical rain forest.

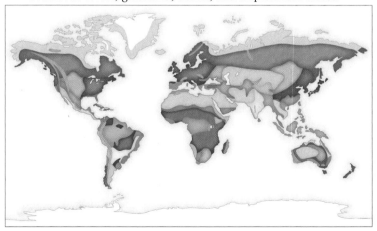

39.2 The tundra is cold and dark much of the year

- In the Arctic tundra (light blue), winters are long and summers are short.

39.3 Coniferous forests are dominated by gymnosperms

- Taiga (dark blue), found south of the tundra, has less rainfall than other forests.

- Temperate rain forests (brown), occurring along the west coasts of Canada and the U.S., have plentiful rain and rich soil.

39.4 Temperate deciduous forests have abundant life

- Temperate deciduous forests (red) occur south of the taiga in eastern North America, eastern Asia, and Europe; they are characterized by moderate climate, high rainfall, and seasons.

39.5 Temperate grasslands have extreme seasons

- The Russian steppes, South American pampas, and North American prairies are all grasslands (violet) known for very cold winters and hot, dry summers.

39.6 Savannas have wet-dry seasons

- The largest savannas (purple) exist in central and southern Africa; they are characterized by large expanses of grasses with scattered trees and many and varied large herbivores and carnivores.

39.7 Deserts have very low annual rainfall

- Deserts (tan) exist in the Northern and Southern Hemispheres at latitudes of about 30°; they are characterized by hot days, cold nights, and low annual rainfall.

39.8 Tropical rain forests are warm with abundant rainfall

- Tropical rain forests (olive green) exist in equatorial regions of South America, Africa, and the Indo-Malayan region; they are characterized by wide species diversity and abundance.
- The three levels of a rain forest are the forest floor (sparse vegetation), the understory (small plants that can live in shade), and the canopy (the most productive level)

39.9 Solar radiation and winds influence climate

- Climate is the average yearly temperature and precipitation of a region.

- The distribution of solar energy and global wind patterns caused by the spherical Earth and the rotation of Earth around the sun affect temperature (seasons), amounts of rainfall, and how winds blow.

39.10 Topography and other effects also influence climate

- Elevation affects temperature, and therefore mountains affect climate just as latitude does.
- Mountains in the path of winds affect rainfall.
- Atmospheric circulations between the ocean and the landmasses influence regional climate conditions.

Fresh Water and Salt Water Are Organized into Aquatic Ecosystems

39.11 Fresh water flows into salt water

- Freshwater ecosystems include lakes, ponds, streams, rivers, and wetlands (marshes, swamps, and bogs).
- Lakes are classified as oligotrophic if they are nutrient-poor with low productivity and eutrophic if they are nutrient-rich with high productivity.

39.12 Marine ecosystems include those of the coast and the ocean

- Coastal ecosystems include estuaries and intertidal zones.
  - Shallow ocean waters are sunlit and contain many inorganic nutrients for photosynthesizers, phytoplankton, and zooplankton.
  - Ocean waters are divided into the epipelagic zone, mesopelagic zone, and bathypelagic zone.

39.13 Ocean currents affect climates

- Winds cause a steady flow of water; currents move clockwise in the Northern Hemisphere and counterclockwise in the Southern Hemisphere.
- The Gulf Stream brings tropical water to the eastern coast of North America and to western Europe.
- The Humboldt Current brings cold water north to western South America.
- Upwelling occurs when cold, nutrient-rich waters rise, resulting in plentiful fisheries.

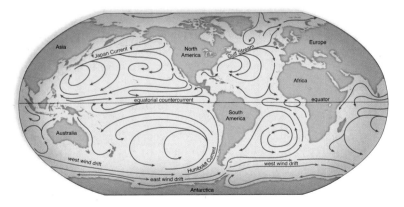

39.14 El Niño–Southern Oscillation alters weather patterns

- During an El Niño–Southern Oscillation, the Humboldt Current is not cool; upwelling does not occur; fisheries decline; and climate patterns change over the entire globe.

On Land, the Biosphere Is Organized into Terrestrial Ecosystems

1. Which of the following statements is not true of the Arctic tundra?
   a. Soil below the uppermost layer stays frozen year-round.
   b. Annual rainfall is low.
   c. Trees cannot grow because the growing season is too short.
   d. Large mammals are absent.
2. All of these phrases describe the tundra except
   a. low-lying vegetation.   c. short growing season.
   b. northernmost biome.   d. many different types of species.
3. Which biome is characterized by a coniferous forest with low average temperature and moderate rainfall?
   a. taiga   d. tropical rain forest
   b. savanna   e. temperate forest
   c. tundra
4. The major environmental factor that determines whether a prairie has tall grass or short grass is
   a. annual rainfall.
   b. mean annual high temperature.
   c. mean annual low temperature.
   d. soil nutrient levels.
   e. types of grazers present.
5. All of the following phrases describe a tropical rain forest except
   a. nutrient-rich soil.
   b. many arboreal plants and animals.
   c. canopy composed of many layers.
   d. broad-leaved evergreen trees.
6. Epiphytes are most likely to be found in
   a. deserts.   d. tropical rain forests.
   b. tundra.   e. shrublands.
   c. grasslands.
7. The seasons are best explained by
   a. the distribution of temperature and rainfall in biomes.
   b. the tilt of the Earth as it orbits about the sun.
   c. the daily rotation of the Earth on its axis.
   d. the fact that the equator is warm and the poles are cold.
8. Which of these influences the location of a particular biome?
   a. latitude
   b. average annual rainfall
   c. average annual temperature
   d. altitude
   e. All of these are correct.
9. Whether ascending a mountain or traveling from the equator to the North Pole, you would observe terrestrial biomes in which order?
   a. coniferous forest, tropical rain forest, temperate deciduous forest, tundra
   b. tropical rain forest, coniferous forest, temperate deciduous forest, tundra
   c. tropical rain forest, temperate deciduous forest, coniferous forest, tundra
   d. tropical rain forest, coniferous forest, tundra, temperate deciduous forest
   e. tundra, temperate deciduous forest, tropical rain forest, coniferous forest

10. Which side of a mountain range lies in a rain shadow?
    a. windward
    b. leeward
11. A monsoon climate is produced
    a. when a landmass heats more rapidly than the ocean.
    b. when the ocean heats more rapidly than a landmass.
    c. in regions of the Earth that are tilting toward the sun.
    d. in regions of the Earth that are tilting away from the sun.
12. Rising temperatures at the equator are expected to intensify the effect of global warming in the temperate zone and at the poles. See Section 39.9 and explain.

Fresh Water and Salt Water Are Organized into Aquatic Ecosystems

13. Which of these is a function of a wetland?
    a. purifies water
    b. is an area where toxic wastes can be broken down
    c. helps absorb overflow and prevents flooding
    d. is a home for organisms that are links in food chains
    e. All of these are correct.
14. An oligotrophic lake
    a. is nutrient-rich.
    b. is cold.
    c. is likely to be found in an agricultural or urban area.
    d. has poor productivity.
15. Runoff of fertilizer and animal wastes from a large farm that drains into a lake would be an example of which process?
    a. fall overturn   c. spring overturn
    b. eutrophication   d. upwelling
16. **THINKING CONCEPTUALLY** Why might a large popluation of decomposers in an oligotrophic lake cause it to have no fish?
17. In an ocean, the highest concentrations of phytoplankton would be found in the
    a. intertidal zone.
    b. euphotic zone.
    c. pelagic zone.
    d. All areas would have roughly equal concentrations.
18. Which zone of the oceanic province is completely dark?
    a. epipelagic zone   c. bathypelagic zone
    b. mesopelagic zone   d. euphotic zone
19. Energy for the food chain near hydrothermal vents comes from
    a. dead organisms that fall down from above.
    b. highly efficient photosynthetic phytoplankton.
    c. chemoautotrophic bacteria.
    d. heat given off by the vents.
20. The directions of ocean currents are influenced by
    a. wind.   d. lunar cycles.
    b. landmasses.   e. Both a and b are correct choices.
    c. rivers.
21. The mild climate of Great Britain is best explained by
    a. the winds called westerlies.
    b. the spinning of Earth on its axis.
    c. Great Britain being a mountainous country.
    d. the flow of ocean currents.
22. Which of the following describes a change in weather conditions in the United States that can result from an El Niño–Southern Oscillation?
    a. fewer tornadoes in the Midwest
    b. less snow in the Northeast
    c. fewer hurricanes on the East Coast
    d. flooding in the southcentral and southeastern regions
    e. All of these choices are correct.

23. **THINKING CONCEPTUALLY** Why might estuaries disappear if sea levels rise due to global warming?

Understanding the Terms

Arctic tundra 765	permafrost 765
biome 764	phytoplankton 773
climate 764	pond 771
coral reef 773	rain shadow 770
desert 767	river 771
El Niño–Southern	savanna 767
Oscillation 774	stream 771
epiphyte 768	taiga 765
estuary 772	temperate deciduous
eutrophication 771	forest 766
eutrophic lake 771	temperate grassland 766
hydrothermal vent 773	temperate rain forest 765
intertidal zone 772	topography 770
lake 771	tropical rain forest 768
monsoon 770	upwelling 774
oligotrophic lake 771	zooplankton 773
pelagic zone 773	

Match the terms to these definitions:
a. _____ Portion of ocean where fresh water and salt water mix as they meet.
b. _____ Terrestrial ecosystem that is a coniferous forest extending in a broad belt across northern Eurasia and North America.
c. _____ Terrestrial ecosystem that is grassland in Africa, characterized by a few trees and a severe dry season.
d. _____ Enrichment of water by inorganic nutrients used by phytoplankton.
e. _____ Plant that takes its nourishment from the air because of its placement in other plants.

Thinking Scientifically

1. Global warming might make it possible to test the hypothesis that climate determines the terrestrial ecosystem. How so?
2. Hypothesize why nearby logging, which causes dirt runoff in coastal areas, might interfere with the health of coral.

ARIS *Visit* **www.mhhe.com/maderconcepts** *for practice quizzes, animations, videos, and activities designed to help you master the material in this chapter.*

40

Conservation of Biodiversity

LEARNING OUTCOMES

After studying this chapter, you should be able to accomplish the following outcomes.

Trouble in Paradise

1 Give reasons why Hawaii is becoming a "lost" paradise.

Conservation Biology Focuses on Understanding and Protecting Biodiversity

2 Define and characterize the field of conservation biology.
3 Distinguish between an endangered and a threatened species.
4 Define biodiversity according to four levels of study.
5 State the significance of "hotspots" when conserving species.

Biodiversity Has Direct Value and Indirect Value for Human Beings

6 Discuss the direct value of biodiversity with reference to (1) medicinal, (2) agricultural, and (3) consumptive use values.
7 Discuss the indirect value of biodiversity with reference to (1) the disposal of wastes within biogeochemical cycles; (2) the provision of fresh water, prevention of soil erosion, and regulation of climate by natural areas; and (3) ecotourism.

The Causes of Today's Extinctions Are Known

8 Discuss the five major reasons for the loss of biodiversity.

Conservation Techniques Require Much Effort and Expertise

9 Explain the basis for deciding which species to preserve and what methodology to use.
10 Explain the purpose and the methodology for restoring an ecosystem that has been seriously damaged.

O ver 7 million tourists a year visit the Hawaiian Islands. Among the plentiful attractions are many species of native plants and animals found nowhere else on Earth. But sadly, about 63% of these species are at risk of extinction, and many others have already become extinct. What is causing such a tragedy?

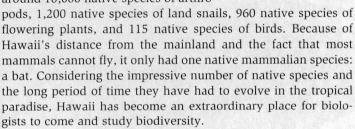

Gardenia

Originally, the archipelago had around 10,000 native species of arthropods, 1,200 native species of land snails, 960 native species of flowering plants, and 115 native species of birds. Because of Hawaii's distance from the mainland and the fact that most mammals cannot fly, it only had one native mammalian species: a bat. Considering the impressive number of native species and the long period of time they have had to evolve in the tropical paradise, Hawaii has become an extraordinary place for biologists to come and study biodiversity.

Waikiki, Hawaii

Trouble in Paradise

Biologists were preceded by a long list of other people who came to live in Hawaii. Polynesians arrived more than 1,000 years ago, and Europeans arrived in the late 1700s. These people brought with them—both accidentally and intentionally—over 5,000 species of nonnative animals and plants. Mammals such as pigs, rats, cats, goats, and mongooses, as well as plants such as myrtle trees, Koster's curse (a weedy shrub), and fountain grass, are alien organisms, which have caused much destruction in Hawaii.

The damage caused by feral pigs has gained a lot of attention in particular, and these animals are a good example of the problems created by alien species. The pigs voraciously eat all kinds of plants and animals, and create disturbed areas, allowing other aliens to come in. Kahili ginger, banana poke, and strawberry guava are examples of alien plants that thrive in pig-damaged areas. Furthermore, the water held in pig wallows allows mosquitoes to breed and spread

Feral pig

avian malaria. Both the mosquitoes and the parasite that causes avian malaria have wreaked havoc on Hawaii's native bird populations.

Nonnative species are just part of the reason for the biodiversity crisis in Hawaii. Humans have depleted natural resources, created pollution, and overpopulated fragile coastal areas. Hawaii has some of the fastest-growing cities in the United States; the demand for fresh water is so high that frequent water-use restrictions exist in some areas; and almost a third of the landfills are at, or will soon reach, capacity. Hawaii is fast becoming a lost paradise with degraded and damaged vistas.

As you will see in this chapter, conservation biologists are working diligently to preserve Earth's biodiversity and natural resources. Whether it is too late to protect Hawaii's remaining native species remains to be seen. One thing is for sure: The Hawaiian paradise urgently needs protection from the very people who want so much to visit and appreciate it.

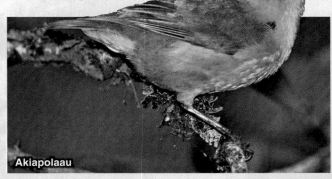

Akiapolaau

Akohekohe

Hibiscus

Conservation Biology Focuses on Understanding and Protecting Biodiversity

Conservation biology is concerned with saving biodiversity, which has various levels of complexity. We should pay attention to preserving species in areas where biodiversity is the most complex.

40.1 Conservation biology is a practical science

Conservation biology is a relatively new discipline that studies all aspects of biodiversity with the goal of conserving natural resources for this generation and all future generations. Conservation biology is unique in that it is concerned with both the development of scientific concepts and the application of these concepts to the everyday world. A primary goal is the management of biodiversity for sustainable use by humans. To achieve this goal, conservation biologists are interested in, and come from, many subfields of biology that only now have been brought together into a cohesive whole:

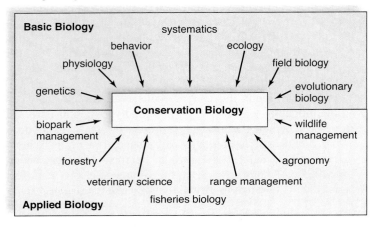

Like a physician, a conservation biologist must be aware of the latest findings, both theoretical and practical, and be able to use this knowledge to diagnose the source of trouble and suggest a suitable treatment. Often, it is necessary for conservation biologists to work with government officials at both the local and federal levels. Public education is another important duty of conservation biologists.

Conservation biology is a unique science in another way. It unabashedly supports the following ethical principles: (1) Biodiversity is desirable for the biosphere and, therefore, for humans; (2) extinctions, due to human actions, are therefore undesirable; (3) the complex interactions in ecosystems support biodiversity and are desirable; and (4) biodiversity brought about by evolutionary change has value in and of itself, regardless of any practical benefit.

Conservation biology has emerged in response to a crisis—never before in the history of the Earth are so many extinctions expected in such a short period of time. Estimates vary, but at least 10–20% of all species now living most likely will become extinct in the next 20–50 years, unless immediate action is taken. It is urgently important, then, that all citizens understand the concept of biodiversity, the value of biodiversity, the likely causes of present-day extinctions, and what could be done to prevent extinctions from occurring.

To protect biodiversity, **bioinformatics,** the science of collecting and analyzing biological information, is applied. Throughout the world, molecular, descriptive, and biogeographical information on organisms is being collected. Eventually, this information will be used to help us understand and protect biodiversity.

In Section 40.2, we learn about the various levels of understanding biodiversity, other than number of species.

> **40.1 Check Your Progress** What are the main goals of conservation biology?

40.2 Biodiversity is more than counting the total number of species

At its simplest level, **biodiversity** is the variety of life on Earth. **Figure 40.2A** accounts for the species that have so far been described, but it has been estimated that between 10 and 50 million species may exist in all. If this is the case, many species are still to be found and described. Of the described species, nearly 1,200 in the United States and 40,000 worldwide are in danger of **extinction,** the total disappearance of a species or higher group. An **endangered species** is in peril of immediate extinction throughout all or most of its range. Examples of endangered species include the black lace cactus, armored snail, hawksbill sea turtle, California condor, West Indian manatee, and snow leopard. **Threatened species** are organisms that are likely to become endangered in the foreseeable future. Examples of threatened species include the Navaho sedge, puritan tiger beetle, gopher tortoise, bald eagle, gray wolf, and Louisiana black bear.

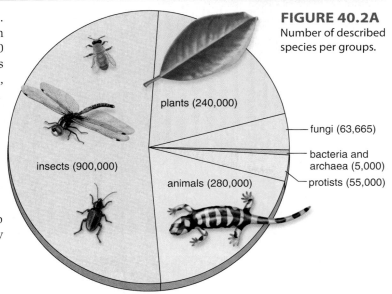

FIGURE 40.2A
Number of described species per groups.

plants (240,000)

fungi (63,665)

bacteria and archaea (5,000)

insects (900,000)

animals (280,000)

protists (55,000)

Levels of Understanding Biodiversity To develop a meaningful understanding of life on Earth, we need to know more about species than their total number. Ecologists also study biodiversity as an attribute of three other levels of biological organization: genetic diversity, ecosystem diversity, and landscape diversity.

Genetic diversity refers to variations among the members of a population. Populations with high genetic diversity are more likely to have some individuals that can survive a change in the structure of their ecosystem. For example, the 1846 potato blight in Ireland, the 1922 wheat failure in the Soviet Union, and the 1984 outbreak of citrus canker in Florida were all made worse by limited genetic variation among these crops. If a species' populations are quite small and isolated, it is more likely to eventually become extinct because of a loss of genetic diversity. As organisms become endangered and threatened, they lose their genetic diversity.

Ecosystem diversity is dependent on the interactions of species at a particular locale. Although past conservation efforts frequently concentrated on saving particular charismatic species, such as the California condor, the black-footed ferret, or the spotted owl, this is a shortsighted approach. A better approach is to conserve species that play a critical role in an ecosystem. Saving an entire ecosystem can save many species, and the contrary is also true—disrupting an ecosystem threatens the existence of more than one species. For example, opossum shrimp, *Mysis relicta*, were introduced into Flat-head Lake in Montana and its tributaries as food for salmon. But the shrimp ate so much zooplankton that there was, in the end, far less food for the salmon and ultimately for the grizzly bears and bald eagles as well (**Fig. 40.2B**).

Landscape diversity involves a group of interacting ecosystems; within one landscape, for example, there may be plains, mountains, and rivers. Any of these ecosystems can be so fragmented that they are connected by only patches (remnants) or strips of land that allow organisms to move from one ecosystem to the other. Fragmentation of the landscape reduces reproductive capacity and food availability and can disrupt seasonal behaviors.

Distribution of Biodiversity Biodiversity is not evenly distributed throughout the biosphere; therefore, protecting particular areas will help save more species. Biodiversity is highest at the tropics, and it declines toward the poles whether considering terrestrial, freshwater, or marine ecosystems. Also, for example, more species are found in the coral reefs of the Indonesian archipelago than in coral reefs west of this archipelago.

Some regions of the world are called **biodiversity hotspots** because they contain unusually large concentrations of species. Although hotspots harbor about 44% of all known higher plant species and 35% of all terrestrial vertebrate species, they are present in only about half of the Earth's landmass. The island of Madagascar, the Cape region of South Africa, Indonesia, the coast of California, and the Great Barrier Reef of Australia are all biodiversity hotspots.

One surprise of late has been the discovery that rain forest canopies and the deep-sea benthos have many more species than formerly thought. Some conservationists refer to these two areas as biodiversity frontiers.

The direct and indirect value of biodiversity is explained in the next part of the chapter.

40.2 *Check Your Progress* **What is the difference between an endangered species and a threatened species?**

FIGURE 40.2B Disrupted ecosystem in Flat-head Lake, Montana.

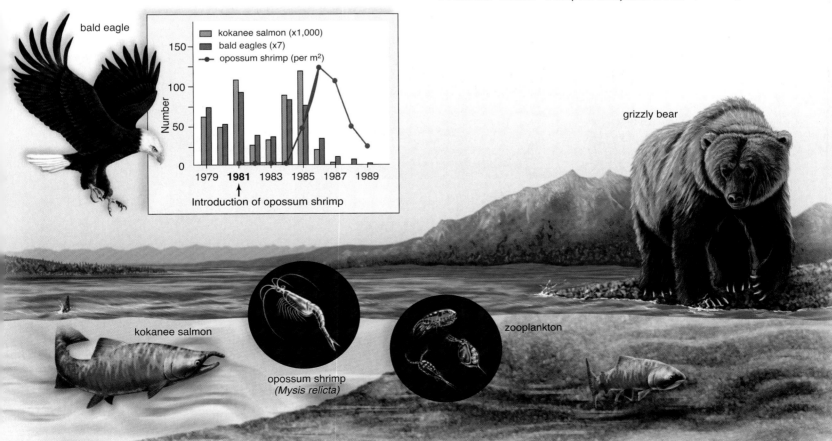

Biodiversity Has Direct Value and Indirect Value for Human Beings

Learning Outcomes 6–7, page 778

The direct value of biodiversity is seen in the observable services of individual wild species. The indirect value is evidenced by the many services provided by ecosystems.

40.3 The direct value of biodiversity is becoming better recognized

Conservation biology strives to reverse the trend toward the possible extinction of tens of thousands of living things. To bring this about, it is necessary to make all people aware of the various ways that biodiversity has direct value and indirect value. Direct value is discussed in this section and exemplified in **Figure 40.3**. Section 40.4 discusses the indirect value of wildlife. Following are some of the ways that wildlife has direct value:

Wildlife Has Medicinal Value Most of the prescription drugs used in the United States, valued at over $200 billion, were originally derived from living organisms. The rosy periwinkle from Madagascar is an excellent example of a tropical plant that has provided us with useful medicines (Fig. 40.3, *top left*). Potent chemicals from this plant are now used to treat two forms of cancer: leukemia and Hodgkin disease. Because of these drugs, the survival rate for childhood leukemia has gone from 10% to 85%, and Hodgkin disease is usually curable. Researchers tell us that, judging from the success rate in the past, an additional 328 types of drugs are yet to be found in tropical rain forests, and the value of this resource to society is probably $147 billion.

You may already know that the antibiotic penicillin is derived from a fungus and that certain species of bacteria produce the antibiotics tetracycline and streptomycin. These drugs have been indispensable in the treatment of diseases, including sexually transmitted diseases such as gonorrhea and syphilis.

Leprosy is among the diseases for which there is, as yet, no cure. The bacterium that causes leprosy will not grow in the laboratory, but scientists have discovered that it grows naturally in the nine-banded armadillo (Fig. 40.3, *bottom left*). Having a source for the bacterium may make it possible to find a cure for leprosy. The blood of horseshoe crabs contains a substance called limulus amoebocyte lysate, which is used to ensure that medical devices such as pacemakers, surgical implants, and prosthetic devices are free of bacteria. Blood is taken from 250,000 crabs a year, and then they are returned to the sea unharmed.

Wildlife Has Agricultural Value Crops such as wheat, corn, and rice are derived from wild plants that have been modified to be high producers. The same high-yield, genetically similar strains tend to be grown worldwide. When rice crops in

FIGURE 40.3 Direct value of diverse wildlife.

Wild species, such as the rosy periwinkle, *Catharanthus roseus,* are sources of many medicines.

Wild species, including many marine species, provide us with food.

Wild species, such as the nine-banded armadillo, *Dasypus novemcinctus,* play a role in medical research.

Africa were being devastated by a virus, researchers grew wild rice plants from thousands of seed samples until they found one that contained a gene for resistance to the virus. They then used these wild plants in a breeding program to transfer the gene into high-yield rice plants. If this variety of wild rice had become extinct before it was discovered, rice cultivation in Africa might have collapsed.

Biological pest controls—natural predators and parasites—are often preferable to chemical pesticides. When a rice pest called the brown planthopper became resistant to pesticides, farmers began to use natural brown planthopper enemies instead. The economic savings were calculated at well over $1 billion. Similarly, cotton growers in Cañete Valley, Peru, found that pesticides were no longer working against the cotton aphid because of resistance. Research identified natural predators, such as the ladybug, that cotton farmers are now using to an ever greater degree (Fig. 40.3, *bottom left*). Again, savings have been enormous.

Most flowering plants are pollinated by animals, such as bees, wasps, butterflies, beetles, birds, and bats (Fig. 40.3, *top left*). The honeybee, *Apis mellifera*, has been domesticated, and it pollinates almost $10 billion worth of food crops annually in the United States. However, the danger of this dependency on a single species is exemplified by mites, which have now wiped out more than 23% of the commercial honeybee population in the United States. Where can we get resistant bees? From the wild, of course. The value of wild pollinators to the U.S. agricultural economy has been calculated at $4.1–$6.7 billion a year.

Wildlife Has Consumptive Use Value

Humans have had much success cultivating crops, keeping domesticated animals, growing trees in plantations, and so forth. But so far, aquaculture, the growing of fish and shellfish for human consumption, has contributed only minimally to human welfare. Instead, most freshwater and marine harvests depend on the catching of wild animals, such as crustaceans (e.g., lobsters, shrimps, and crabs), mammals (e.g., whales), and fishes (e.g., trout, cod, tuna, and flounder). Obviously, these aquatic organisms are an invaluable biodiversity resource.

The environment provides a variety of other products that are sold in the marketplace worldwide, including wild fruits and vegetables, skins, fibers, beeswax, and seaweed. Also, by hunting and fishing, some people obtain their meat directly from the environment. In one study, researchers calculated that the economic value of the wild pig in the diet of native hunters in Sarawak, East Malaysia, was approximately $40 million per year.

Similarly, many trees are still felled in the natural environment for their wood. Researchers have calculated that a species-rich forest in the Peruvian Amazon is worth far more if the forest is used for fruit and rubber production than for timber production (Fig. 40.3, *right*). Fruit and the latex needed to produce rubber can be brought to market for an unlimited number of years, whereas once the trees are gone, no more products can be harvested.

Section 40.4 discusses the indirect value of biodiversity.

> **40.3 Check Your Progress** From a "consumptive" perspective, why is it important to preserve aquatic biodiversity?

Wild species, such as certain bats (e.g., *Leptonycteris curasoae)*, are pollinators of agricultural and other plants.

Wild species, including ladybugs, *Coccinella*, play a role in the biological control of agricultural pests.

Wild species, such as rubber trees, *Hevea*, can provide a product indefinitely if the forest is not destroyed.

Ecosystems perform many services for modern humans, who increasingly live in cities. The services discussed in this section are said to be indirect because they are pervasive and not easily discernible.

Biogeochemical Cycles Help Dispose of Waste

The biodiversity within ecosystems contributes to the workings of the water, carbon, phosphorus, and nitrogen cycles. We are dependent on these cycles for fresh water, removal of carbon dioxide from the atmosphere, uptake of excess soil nitrogen, and provision of phosphate. When human activities upset the usual workings of biogeochemical cycles, the dire environmental consequences include the release of excess pollutants that are harmful to us. Technology is unable to artificially contribute to or create any of the biogeochemical cycles.

As discussed in the introduction to Chapter 38, if not for decomposition, waste would soon cover the entire surface of our planet. We can build sewage treatment plants, but they are expensive, and few of them break down solid wastes completely to inorganic nutrients. It is less expensive and more efficient to water plants and trees with partially treated wastewater and let soil bacteria cleanse it completely. Biological communities are also capable of breaking down and immobilizing pollutants, such as heavy metals and pesticides, that humans release into the environment. A review of wetland functions in Canada assigned a value of $50,000 per hectare (25 acres, or 10,000 m²) per year to the ability of natural areas to purify water and take up pollutants.

Natural Areas Provide Fresh Water, Prevent Soil Erosion, and Regulate Climate

Few terrestrial organisms are adapted to living in a salty environment—they need fresh water. We can remove salt from sea water to obtain fresh water, but the cost of desalination is about four to eight times the average cost of fresh water acquired via the water cycle. Humans use fresh water in innumerable ways, including for drinking and for irrigating their crops. Freshwater ecosystems such as rivers and lakes also provide us with fish and other types of organisms for food.

Forests and other natural ecosystems exert a "sponge effect" (**Fig. 40.4**). The leaves of trees acquire water from the roots and then release it at a regular rate. The water-holding capacity of forests and wetlands reduces the possibility of flooding. The value of a marsh outside Boston, Massachusetts, has been estimated at $72,000 per hectare per year solely on its ability to reduce floods. Forests release water slowly for days or weeks after the rains have ceased. Rivers flowing through forests in West Africa release twice as much water halfway through the dry season, and between three and five times as much at the end of the dry season, as do rivers bordered by coffee plantations.

Intact ecosystems naturally retain soil and prevent erosion. In Pakistan, the world's largest dam, the Tarbela Dam, is losing its storage capacity of 13.5 billion cubic meters many years sooner than expected because silt is building up behind the dam due to deforestation. At one time, the Philippines were exporting $240.8 million worth of oysters, mussels, clams, and cockles each year. Now, silt carried down rivers following deforestation is smothering the mangrove ecosystem that serves as a nursery for the sea.

Globally, forests ameliorate the climate because they take up carbon dioxide. The leaves of trees use carbon dioxide when they photosynthesize, and the bodies of the trees store carbon. When trees are cut and burned, carbon dioxide is released into the atmosphere. Carbon dioxide makes a significant contribution to global warming, which is expected to be stressful for many plants and animals. If temperatures become warmer, only a small percentage of wildlife will be able to move northward to find weather suitable for them.

Ecotourism Is Enjoyed by Many

In the United States, nearly 100 million people enjoy vacationing in a natural setting. To do so, they spend $4 billion each year on fees, travel, lodging, and food. Many tourists want to go sport fishing, whale watching, boat riding, hiking, birdwatching, and the like (Fig. 40.4). Others merely want to immerse themselves in the beauty and serenity of a natural environment.

The next part of the chapter discusses the causes of extinctions.

> **40.4** *Check Your Progress* **What are ecotourists in Hawaii interested in seeing?**

FIGURE 40.4 Tourists (*inset*) love to visit natural ecosystems, such as this forest, which has indirect value because of its water-holding capacity and its ability to take up carbon dioxide.

Researchers have identified the major causes of extinction. They are, in order of significance, habitat loss, introduction of alien species, pollution, overexploitation, and disease. This part of the chapter discusses each cause in turn.

40.5 Habitat loss is a major cause of wildlife extinctions

To stem the tide of extinction due to human activities, it is first necessary to identify the causes. Based on the records of 1,880 threatened and endangered wild species in the United States, it has been found that habitat loss was involved in 85% of the cases (**Fig. 40.5A**). Other significant causes of extinction are introduction of alien species, pollution, overexploitation, and disease. In Figure 40.5A, the percentages add up to more than 100% because most species are imperiled for more than one reason. Macaws are a good example of a species in decline due to a combination of factors.

Habitat loss has occurred in all ecosystems, but concern has now centered on tropical rain forests and coral reefs because they are particularly rich in species. A sequence of events in Brazil offers a fairly typical example of how rain forest is converted to land uninhabitable for wildlife. The construction of a major highway first provided a way to reach the interior of the forest. Small towns and industries sprang up along the highway, and roads branching off the main highway gave rise to even more roads. The result was fragmentation of the once immense forest. The government offered subsidies to anyone willing to take up

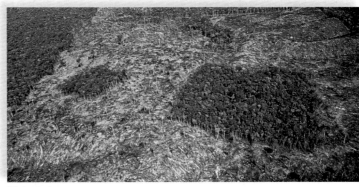

Distant view

Close-up view

FIGURE 40.5B Destruction of a rain forest in Brazil.

residence in the forest, and people began to cut and burn trees in patches (**Fig. 40.5B**). Tropical soils contain limited nutrients, but when the trees are burned, nutrients are released that support a lush growth so that cattle can be grazed for about three years. Once the land has been degraded, farmers move on to another portion of the forest and start over again.

Loss of habitat also affects freshwater and marine biodiversity. Coastal degradation is mainly due to the large concentration of people living on or near the coast. Already, 60% of coral reefs have been destroyed or are on the verge of destruction; it is possible that all coral reefs may disappear during the next 40 years. Mangrove forest destruction is also a problem; Indonesia, with the most mangrove acreage, has lost 45% of its mangroves, and the percentage is even higher for other tropical countries. Wetland areas, estuaries, and seagrass beds are also being rapidly destroyed by the actions of humans.

Alien species are also a significant cause of extinctions, as discussed in Section 40.6.

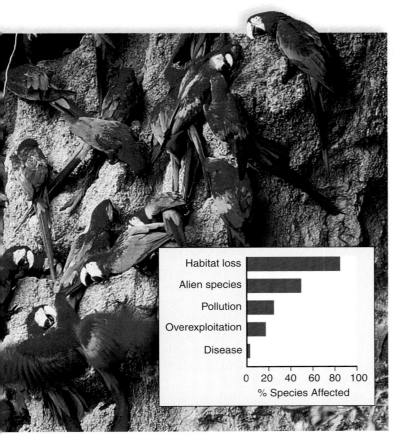

FIGURE 40.5A Macaws, *Ara macao*, and other species are endangered for the reasons graphed here.

> **40.5 Check Your Progress** What is the main cause of coastal degradation in places such as Hawaii?

Ecosystems around the globe are characterized by unique assemblages of organisms that have evolved together in one location. Migrating to a new location is not usually possible because of barriers such as oceans, deserts, mountains, and rivers. Humans, however, have introduced **alien species,** nonnative members, into new ecosystems through the following means:

Colonization Europeans, in particular, brought various familiar species with them when they colonized new places. For example, the pilgrims brought the dandelion to the United States as a familiar salad green. In addition, they introduced pigs that have become feral, reverting to their wild state. In some parts of the United States, feral pigs are very destructive, just as they are in Hawaii.

Horticulture and agriculture Some aliens now taking over vast tracts of land have escaped from cultivated areas. Kudzu is a vine from Japan that the U.S. Department of Agriculture thought would help prevent soil erosion. The plant now covers much of the landscape in the South, including even walnut, magnolia, and sweet gum trees (**Fig. 40.6A**). The water hyacinth was introduced to the United States from South America because of its beautiful flowers. Today, it clogs up waterways and diminishes natural diversity.

Accidental transport Global trade and travel accidentally bring many new species from one country to another. Researchers found that the ballast water released from ships into Coos Bay, Oregon, contained 367 marine species from Japan. The zebra mussel from the Caspian Sea was accidentally introduced into the Great Lakes in 1988. It now forms dense beds that squeeze out native mussels. Other

FIGURE 40.6B Mongooses, introduced into Hawaii, prey on the native birds.

organisms accidentally introduced into the United States include the Formosan termite, the Argentinian fire ant, and the nutria, a type of rodent.

Alien species can disrupt food webs. As mentioned earlier, opossum shrimp introduced into a lake in Montana added a trophic level that in the end meant less food for bald eagles and grizzly bears (see Fig. 40.2B). Introduction of alien species, sometimes called exotic species, plays a role in nearly 50% of extinctions (see Fig. 40.5A).

Aliens on Islands Islands are particularly susceptible to environmental discord caused by the introduction of alien species. Islands have unique assemblages of native species that are closely adapted to one another and cannot compete well against aliens. Myrtle trees, *Myrica faya*, introduced into the Hawaiian Islands from the Canary Islands, are symbiotic with a type of bacterium that is capable of nitrogen fixation. This feature allows the species to establish itself on nutrient-poor volcanic soil, a distinct advantage in Hawaii. Once established, myrtle trees halt the normal succession of native plants on volcanic soil.

The brown tree snake has been inadvertently introduced onto a number of islands in the Pacific Ocean. The snake eats eggs, nestlings, and adult birds. On Guam, it has reduced ten native bird species to the point of extinction. On the Galápagos Islands, black rats have reduced populations of giant tortoises, while goats and feral pigs have changed the vegetation from highland forest to pampaslike grasslands and destroyed stands of cactus. In Australia, mice and rabbits have stressed native marsupial populations. Mongooses introduced into the Hawaiian Islands to control rats also prey on native birds (**Fig. 40.6B**).

Section 40.7 shows that pollution also contributes to extinctions.

FIGURE 40.6A Kudzu, a vine from Japan, has displaced many native plants in the southern United States.

40.6 *Check Your Progress* **What effect do alien species have on biodiversity?**

40.7　Pollution contributes to extinctions

So far, we have discussed habitat loss and alien species as causes of extinction. Now, we will discuss pollution, which is a factor in 24% of extinctions (see Fig. 40.5A). In the present context, **pollution** can be defined as any environmental change that adversely affects the lives and health of living things. Biodiversity is particularly threatened by the following types of environmental pollution:

Acid deposition Both sulfur dioxide from power plants and nitrogen oxides in automobile exhaust are converted to acids when they combine with water vapor in the atmosphere. These acids return to Earth as either wet deposition (acid rain or snow) or dry deposition (sulfate and nitrate salts). Acid deposition causes trees to weaken and increases their susceptibility to disease and insects. Many lakes in the northern United States are now lifeless because of the effects of acid deposition.

Eutrophication Lakes are also under stress due to over-enrichment. When lakes receive excess nutrients due to runoff from agricultural fields and wastewater from sewage treatment, algae begin to grow in abundance. An algal bloom is apparent as a green scum or excessive mats of filamentous algae. Upon death, the decomposers break down the algae, but in so doing, they use up oxygen. A decreased amount of oxygen is available to fish, leading sometimes to a massive fish kill.

Ozone depletion The ozone shield is a layer of ozone (O_3) in the stratosphere, some 50 km above the Earth. The ozone shield absorbs most of the wavelengths of harmful ultraviolet (UV) radiation so that they do not strike the Earth. The cause of ozone depletion can be traced to chlorine atoms (Cl^-) that come from the breakdown of chlorofluoro-carbons (CFCs). Severe ozone shield depletion can impair crop and tree growth and also kill plankton (microscopic plant and animal life) that sustain oceanic life.

Organic chemicals Our modern society uses organic chemicals in all sorts of ways. Organic chemicals called nonylphenols are used in products ranging from pesticides to dishwashing detergents, cosmetics, plastics, and spermicides. These chemicals mimic the effects of hormones and, in that way, most likely harm wildlife. Salmon are born in fresh water but mature in salt water. After investigators exposed young fish to nonylphenol, they found that 20–30% were unable to make the transition between fresh and salt

FIGURE 40.7 Global warming leads to the "bleaching" of corals in reefs.

water. Nonylphenols cause the pituitary to produce prolactin, a hormone that may prevent saltwater adaptation.

Global warming Recall that certain gases, such as carbon dioxide and methane, are known as greenhouse gases because, just like the panes of a greenhouse, they allow solar radiation to pass through but hinder the escape of its heat back into space. Data collected around the world show a steady rise in the concentration of the various greenhouse gases due to the burning of fossil fules. The Earth may warm to temperatures never before experienced by living things.

As the Earth warms, sea level will rise because glaciers will melt and water expands as it warms. A 1-meter rise in sea level could inundate 25–50% of U.S. coastal wetlands. The growth of corals is very dependent on mutualistic algae living in their walls. When the temperature rises by 4°, corals expel their algae and are said to be "bleached" (**Fig. 40.7**). Almost no growth or reproduction occurs until the algae return. Global warming could very well cause many extinctions on land also. As temperatures rise, regions of suitable climate for various species will shift toward the poles and higher elevations. The present assemblages of species in ecosystems will be disrupted as some species migrate northward, leaving others behind.

Overexploitation, discussed in Section 40.8, also contributes to the occurrence of extinctions.

> **40.7** *Check Your Progress* **What is pollution?**

40.8　Overexploitation contributes to extinctions

Overexploitation occurs when the number of individuals taken from a wild population is so great that the population becomes severely reduced in number. Overexploitation accounts for 17% of extinctions (see Fig. 40.5A). A positive feedback cycle explains overexploitation: The smaller the population, the more valuable its members, and the greater the incentive to capture the few remaining organisms. Poachers and participants in organized crime are very active in collecting and selling endangered and threat-

ened species because it has become so lucrative. The overall international value of trading wildlife species is $20 billion, of which $8 billion is attributed to the illegal sale of rare species.

Markets for decorative plants and exotic pets support both legal and illegal trade in wild species. Rustlers dig up rare cactuses, such as the single-crested saguaro, and sell them to gardeners for as much as $15,000 each. Parakeets and macaws are among the birds taken from the wild for sale to pet owners. For every bird delivered alive, many more have died in the process. The same holds true for tropical fish, which often come from the coral reefs of Indonesia and the Philippines. Divers dynamite reefs or use plastic squeeze-bottles of cyanide to stun them; in the process, many fish and valuable corals die.

Declining species of mammals, such as the Siberian tiger, are still hunted for their hides, tusks, horns, or bones. Because of its rarity, a single Siberian tiger is now worth more than $500,000—its bones are pulverized and used as a medicinal powder. The horns of rhinoceroses become ornate carved daggers, and their bones are ground up to sell as a medicine. The ivory of an elephant's tusk is used to make art objects, jewelry, or piano keys. The fur of a Bengal tiger sells for as much as $100,000 in Tokyo.

The U.N. Food and Agricultural Organization tells us that humans have now overexploited 11 of 15 major oceanic fishing areas. Fish are a renewable resource if harvesting does not exceed the ability of the fish to reproduce. However, our society continues to use larger and more efficient fishing fleets to decimate fishing stocks. Pelagic species such as tuna are captured by purse-seine fishing, in which a very large net surrounds a school of fish, and then the net is closed in the same manner as a drawstring purse. Up to thousands of dolphins that swim above schools of tuna are often captured and then killed in this type of net. Other fishing boats drag huge trawling nets, large enough to accommodate 12 jumbo jets, along the seafloor to capture bottom-dwelling fish (**Fig. 40.8**). Only large fish are kept; undesirable small fish and sea turtles are discarded, dying, back into the ocean. Trawling has been called the marine equivalent of clear-cutting trees because, after the net goes by, the sea bottom is devastated. Today's fishing practices don't allow fisheries to recover. Cod and haddock, once the most abundant bottom-dwelling fish along the northeast coast of the United States, are now often outnumbered by dogfish and skate.

A marine ecosystem can be disrupted by overfishing, as exemplified on the U.S. West Coast. When sea otters began to decline in numbers, investigators found that they were being eaten by orcas (killer whales). Usually orcas prefer seals and

FIGURE 40.8 Large nets catch massive numbers of fish.

sea lions to sea otters, but they began eating sea otters when few seals and sea lions could be found. What caused a decline in seals and sea lions? Their preferred food sources—perch and herring—were no longer plentiful due to overfishing. Ordinarily, sea otters keep the population of sea urchins, which feed on kelp, under control. But with fewer sea otters around, the sea urchin population exploded and decimated the kelp beds. Thus, overfishing set in motion a chain of events that detrimentally altered the food web of an ecosystem.

Disease is another possible cause of extinctions, as explained in Section 40.9.

> **40.8** *Check Your Progress* Reducing the numbers of certain species of Hawaiian fish due to overfishing is an example of _____.

40.9 Disease contributes to extinctions

Scientists tell us the number of pathogens that cause diseases is on the rise, threatening human health and also that of wildlife. Due to the encroachment of humans on their habitat and other

FIGURE 40.9 The Harlequin toad is near extinction due to a fungal pathogen.

general interventions, wildlife have been exposed to emerging diseases. For example, canine distemper was spread from domesticated dogs to lions in the African Serengeti, causing population declines. Avian influenza likely emerged from domesticated fowl (e.g., chicken) populations and will probably lead to the deaths of millions of wild birds.

Pollution can weaken organisms so that they are more susceptible to disease. Almost half of sea otter deaths along the coast of California are now due to infectious diseases. Pollution, most likely, plays a role in the worldwide decline of amphibians due to disease (**Fig. 40.9**).

The next part of the chapter discusses habitat preservation and restoration.

> **40.9** *Check Your Progress* The rapid decline of amphibians worldwide is at least partially due to _____.

To preserve species, it is necessary to preserve their habitat, or possibly restore a habitat. Conservation aims for sustainable development, which allows multiple uses of the land.

40.10 Habitat preservation is of primary importance

Preservation of a species' habitat is of primary concern, but first we must decide which species to preserve. As mentioned previously, the biosphere contains biodiversity hotspots, relatively small areas having a concentration of endemic (native) species not found anyplace else. In the tropical rain forests of Madagascar, 93% of the primate species, 99% of the frog species, and over 80% of the plant species are endemic to Madagascar. Preserving these forests and other hotspots will save a wide variety of organisms.

Keystone species are species that influence the viability of a community, although their numbers may not be excessively high. The extinction of a keystone species can lead to other extinctions and a loss of biodiversity. For example, bats are designated a keystone species in tropical forests of the Old World. They are pollinators that also disperse the seeds of trees. When bats are killed off and their roosts destroyed, the trees fail to reproduce. The grizzly bear is a keystone species in the northwestern United States and Canada (**Fig. 40.10A**). Bears disperse the seeds of berries; as many as 7,000 seeds may be in one dung pile. Grizzlies kill the young of many hoofed animals and, thereby, keep their populations under control. Grizzlies are also a principal mover of soil when they dig up roots and prey upon hibernating ground squirrels and marmots. Other keystone species are beavers in wetlands, bison in grasslands, alligators in swamps, and elephants in grasslands and forests.

The grizzly bear population is actually a **metapopulation,** a population subdivided into several small, isolated populations. Habitat fragmentation by humans contributes to the creation of metapopulations. Originally, there were probably 50,000–100,000 grizzlies south of Canada, but this number has been reduced because communities have encroached on their home range and bears have been killed by frightened homeowners. Now there are six virtually isolated subpopulations, totaling about 1,000 individuals. The Yellowstone National Park population numbers 200, but the others are even smaller.

Saving metapopulations sometimes requires determining which of the populations is the source and which are sinks. A **source population** is one that lives in a favorable area, and its birthrate is most likely higher than its death rate. Individuals from source populations move into **sink populations,** where the environment is not as favorable and where the birthrate equals the death rate at best. When trying to save the northern spotted owl, conservationists determined that it was best to avoid having owls move into sink habitats. The northern spotted owl reproduces successfully in old-growth rain forests of the Pacific Northwest (**Fig. 40.10B**), but not in nearby immature forests that are in the process of recovering from logging. Distinct boundaries that hindered the movement of owls into these sink habitats proved beneficial in maintaining source populations.

Landscape Preservation May Be Necessary Grizzly bears inhabit a number of different types of ecosystems, including plains, mountains, and rivers. Saving any one of these types of ecosystems alone would not be sufficient to preserve grizzly bears. Instead, it is necessary to save diverse ecosystems that are at least connected by corridors. You will recall that a landscape encompasses different types of ecosystems. An area called the Greater Yellowstone Ecosystem, where bears are free to roam, has now been defined. It contains millions of acres in Yellowstone National Park; state lands in Montana, Idaho, and Wyoming; five different national forests; various wildlife refuges; and even private lands.

Landscape protection for one species is often beneficial for other wildlife that share the same space. The last of the contiguous 48 states' harlequin ducks, bull trout, westslope cutthroat trout, lynx, pine martens, wolverines, mountain caribou, and great gray owls are found in areas occupied by grizzlies. The recent return of gray wolves has occurred in this territory also. Then, too, the grizzly range overlaps with 40% of Montana's vascular plants of special conservation concern.

40.10 Check Your Progress A population of native plants subdivided by towns built by humans is a _____.

FIGURE 40.10B Old-growth forest, home of the northern spotted owl, *Strix occidentalis caurina.*

FIGURE 40.10A Landscape preservation will help grizzly bears, *Ursus arctos horribilis,* survive.

Restoration ecology is a new subdiscipline of conservation biology that seeks scientific ways to return ecosystems to their former state. Three principles have so far emerged: First, it is best to begin as soon as possible before remaining fragments of the original habitat are lost. These fragments are sources of wildlife and seeds from which to restock the restored habitat. Second, once the natural history of the habitat is understood, it is best to use biological techniques that mimic natural processes to bring about restoration. This might take the form of using controlled burns to bring back grassland habitats, biological pest controls to rid the area of alien species, or bioremediation techniques to clean up pollutants. Third, the goal is **sustainable development,** the ability of an ecosystem to maintain itself while providing services to human beings. We will use the Everglades ecosystem to illustrate these principles.

The Everglades The Everglades, located in southern Florida, is a vast sawgrass prairie, interrupted occasionally by a hardwood tree island. Within these islands, both temperate and tropical evergreen trees grow amongst dense and tangled vegetation. Mangroves are found along sloughs (creeks) and at the shoreline. The prop roots of red mangroves protect over 40 different types of juvenile fishes as they grow to maturity. During the wet season, from May to November, animals disperse throughout the region, but in the dry season, from December to April, they congregate wherever pools of water are found. Alligators are famous for making "gator holes," where water collects and fish, shrimp, crabs, birds, and a host of living things survive until the rains come again. The Everglades once supported millions of large and beautiful birds, including herons, egrets, the white ibis, and the roseate spoonbill. **Figure 40.11** shows various animals that live in the Everglades.

At the turn of the 20th century, settlers began to drain land in central Florida to grow crops in the newly established Everglades Agricultural Area (EAA). A large dike was used to keep water in a large lake called Lake Okeechobee. The dike prevents water from overflowing its banks and moving slowly southward. Water is contained not only in the lake but also in three so-called conservation areas established to the south of the lake. Water must be conserved to irrigate the farmland and to recharge the Biscayne aquifer (underground river), which supplies drinking water for the cities on the east coast of Florida. The Everglades National Park receives only water that is discharged artificially from a conservation area, and the discharge is according to the convenience of humans rather than according to the natural wet/dry season of southern Florida. Largely because of this, the Everglades are now dying, as witnessed by declining bird populations. The birds, which used to number in the millions, now number in the thousands.

A restoration plan has been developed that will sustain the Everglades ecosystem, while maintaining the services society requires. The Everglades is to receive a more natural flow of water from Lake Okeechobee. This will require flooding the EAA and growing only crops such as sugarcane and rice that can tolerate these conditions. This has the benefit of stopping the loss of topsoil and preventing possible residential development in the area. There will also be an extended buffer zone between an expanded Everglades and the urban areas on Florida's east coast. The buffer zone will contain a contiguous system of interconnected marsh areas, detention reservoirs, seepage barriers, and water treatment areas. This plan is expected to stop the decline of the Everglades, while still allowing agriculture to continue and providing water and flood control to the eastern coast. Sustainable development will maintain the ecosystem indefinitely and still meet human needs.

Florida panther, *Puma concolor coryi*

American alligator, *Alligator mississippiensis*

White ibis, *Eudocimus albus*

Roseate spoonbill, *Ajaia ajaja*

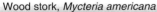

Wood stork, *Mycteria americana*

FIGURE 40.11 A variety of animals make their home in the Everglades.

40.11 *Check Your Progress* **What is the goal of restoration ecologists in preserved areas of the Hawaiian Islands that have been heavily damaged?**

Our industrial societies are over-using the environment to the point of exhaustion. Forests throughout tropical, temperate, and subarctic regions are being harvested and cut for timber at unsustainable rates. Urban sprawl is completely replacing natural ecosystems in highly populated regions. Fresh waters are being diverted for agricultural and urban uses to the extent that riverbeds and lake beds are becoming dry in places. Dams are being constructed for hydropower and irrigation with little consideration for their impact on aquatic life. Alien species of plants, animals, and microbes are being released into new environments with little or no restraint. Marine fisheries are being exploited by major fishing nations at unsustainable levels.

All these actions, and others, are reducing biodiversity, which we now realize is a resource of enormous economic value. If properly managed, sustainable yields of food and fiber can be obtained from many natural lands and waters. Modern genetic engineering technologies make the genes of millions of wild species available for use in breeding improved crops and domestic animals, and as biological control agents. Enjoyment of nature can also enrich human life enormously.

As natural forests, grasslands, streams, lakes, and seas are degraded, human society must expend greater amounts of nonrenewable energy and materials to substitute for benefits that biodiversity provides at no cost. Lost species, and ultimately lost ecosystems, cannot be replaced. Biodiversity is, therefore, a nonrenewable resource. The goal of conservation biology is to protect, restore, and use this resource wisely. To that end, the vision of conservation biology is a world where:

1. Leaders are committed to long-term environmental protection and to international leadership and cooperation in addressing the world's environmental problems.

2. An environmentally literate citizenry has the knowledge, skills, and ethical values needed to achieve sustainable development.

3. Market prices and economic indicators reflect the full environmental and social costs of human activities.

4. A new generation of technologies contributes to the conservation of resources and the protection of the environment.

5. The landscape sustains natural systems, maximizes biological diversity, and uplifts the human spirit.

6. Human numbers are stabilized, all people enjoy a decent standard of living through sustainable development, and the global environment is protected for future generations.[1]

[1]Modified from *The Report of the National Commission on the Environment*, 1993.

The Chapter in Review

Summary

Trouble in Paradise

- Conservation problems in Hawaii include extinction of some plant and animal species, damage by feral pigs and other nonnative species, depletion of natural resources, and pollution.

Conservation Biology Focuses on Understanding and Protecting Biodiversity

40.1 Conservation biology is a practical science

- Conservation biology studies biodiversity with the goal of conserving natural resources and preserving biodiversity.
- Bioinformatics is the science of collecting and analyzing biological information and using it to understand and protect biodiversity.

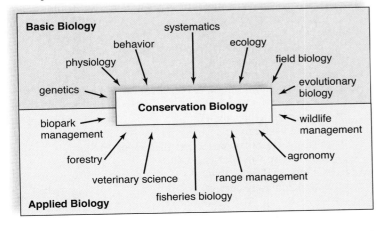

40.2 Biodiversity is more than counting the total number of species

- Biodiversity refers to the variety of life on Earth; additional levels are genetic diversity, ecosystem diversity, and landscape diversity.
- Extinction is the total disappearance of a species or higher group.
- Endangered species are at immediate risk of extinction.
- Threatened species are likely to soon become endangered.
- Diversity is highest at the tropics and declines toward the poles; biodiversity hotspots contain large concentrations of species.

Biodiversity Has Direct Value and Indirect Value for Human Beings

40.3 The direct value of biodiversity is becoming better recognized

- Many prescription drugs were originally derived from living organisms, and still more drugs may be derived from rain forest species.
- Biodiversity can help protect crops against disease and save billions of dollars for a nation's agricultural economy.
- The environment produces wild fruits, vegetables, meats, and fish, and also sustainable products such as rubber.

40.4 The indirect value of biodiversity is immense

- Biodiversity contributes to the successful workings of the water, carbon, phosphorus, and nitrogen cycles.
- Freshwater ecosystems provide fresh water and fish; forests soak up and release water at a regular rate and take up carbon dioxide; and intact ecosystems retain soil.
- Ecotourists enjoy vacations in natural settings.

The Causes of Today's Extinctions Are Known

40.5 Habitat loss is a major cause of wildlife extinctions

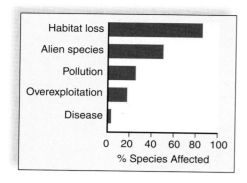

- Habitat loss is the cause of 85% of threatened/endangered cases.
- Coastal degradation destroys freshwater and marine biodiversity.

40.6 Introduction of alien species contributes to extinctions

- Alien species are nonnative species introduced through colonization, horticulture and agriculture, and accidental transport.
- Islands are particularly susceptible to disturbances by introduced aliens.

40.7 Pollution contributes to extinctions

- Pollution is any change in the environment that adversely impacts living things.
- Forms of environmental pollution include acid deposition, eutrophication, ozone depletion, organic chemicals, and global warming.

40.8 Overexploitation contributes to extinctions

- Overexploitation is the removal of a number of individuals so that the population is severely reduced.
- The market for exotic plants and pets supports the legal and illegal trade of wild species.
- Declining species are still being hunted, and fisheries are being overfished.

40.9 Disease contributes to extinctions

- Pathogens are on the rise, subjecting wildlife to emerging diseases.

Conservation Techniques Require Much Effort and Expertise

40.10 Habitat preservation is of primary importance

- Biological hotspots must be preserved.
- Keystone species influence the viability of a community; their extinction can lead to other extinctions and loss of biodiversity.
- A metapopulation is subdivided into small, isolated populations, sometimes due to habitat fragmentation.
- Preserving keystone species requires landscape preservation, which will lead to preservation of other species as well.

40.11 Habitat restoration is sometimes necessary

- Restoration ecology seeks scientific ways to return ecosystems to their former state through the use of natural processes and with the goal of sustainable development.

Conservation Biology Focuses on Understanding and Protecting Biodiversity

1. Which of these would not be within the realm of conservation biology?
   a. helping to manage a national park
   b. a government board charged with restoring an ecosystem
   c. writing textbooks and/or popular books on the value of biodiversity
   d. introducing endangered species back into the wild
   e. All of these are concerns of conservation biology.
2. Most likely, ecosystem performance improves
   a. the more diverse the ecosystem.
   b. as long as selected species are maintained.
   c. as long as species have both direct and indirect value.
   d. if extinctions are diverse.
   e. Both b and c are correct.
3. Biodiversity hotspots
   a. have few populations because the temperature is too hot.
   b. contain a large proportion of the Earth's species even though their area is small.
   c. are always found in tropical rain forests and coral reefs.
   d. are sources of species for the ecosystems of the world.
   e. All except a are correct.
4. Which of these pairs does not show a contrast in the number of species?
   a. temperate zone—tropical zone
   b. hotspots—cold spots
   c. rain forest canopy—rain forest floor
   d. pelagic zone—deep-sea benthos
5. **THINKING CONCEPTUALLY** Draw an energy pyramid (see Section 38.12) with and without the shrimp in Figure 40.2, and use the pyramids to show that the salmon received less energy after the shrimp were introduced.

Biodiversity Has Direct Value and Indirect Value for Human Beings

6. The value of wild pollinators to the U.S. agricultural economy has been calculated to be $4.1–$6.7 billion a year. What is the implication?
   a. Society could easily replace wild pollinators by domesticating various types of pollinators.
   b. Pollinators may be valuable, but that doesn't mean any other species won't provide us with valuable services also.
   c. If we did away with all natural ecosystems, we wouldn't be dependent on wild pollinators.
   d. Society doesn't always appreciate the services that wild species provide naturally and without any fanfare.
   e. All of these statements are correct.
7. Consumptive use value
   a. means we should think of conservation in terms of the long run.
   b. means we are placing too much emphasis on living things that are useful to us.
   c. means some organisms, other than crops and farm animals, are valuable as products.
   d. is a type of direct value.
   e. Both c and d are correct.

8. The services provided to us by ecosystems are unseen. This means
   a. they are not valuable.
   b. they are noticed particularly when the service is disrupted.
   c. biodiversity is not needed in order for ecosystems to keep functioning as before.
   d. we should be knowledgeable about ecosystems and protect them.
   e. Both b and d are correct.
9. Which of the following is not a function that ecosystems can perform for humans?
   a. purification of water
   b. immobilization of pollutants
   c. reduction of soil erosion
   d. removal of excess soil nitrogen
   e. breakdown of heavy metals
10. Which of these is not an indirect value of a species?
    a. participates in biogeochemical cycles
    b. participates in waste disposal
    c. helps provide fresh water
    d. prevents soil erosion
    e. All of these are indirect values.

The Causes of Today's Extinctions Are Known

11. The most significant cause of the loss of biodiversity is
    a. habitat loss.   d. disease.
    b. pollution.   e. overexploitation.
    c. alien species.
12. Eagles and bears feed on spawning salmon. If shrimp are introduced that compete with salmon for food,
    a. the salmon population will decline.
    b. the eagle and bear populations will decline.
    c. only the shrimp population will decline.
    d. all populations will increase in size.
    e. Both a and b are correct.
13. Global warming has nothing to do with
    a. habitat loss.
    b. introduction of alien species into new environments.
    c. pollution.
    d. overexploitation.
    e. Global warming pertains to all of these.
14. Which of these is not expected because of global warming?
    a. changes in the composition of ecosystems
    b. the bleaching and drowning of coral reefs
    c. rise in sea levels and loss of coastal wetlands
    d. preservation of pests because cold weather reduces their population size
    e. All of these are expected.
15. Which of the following associations is not correct?
    a. excess nutrients—eutrophication
    b. carbon dioxide—ozone depletion
    c. sulfur dioxide—acid deposition
    d. methane—global warming
16. **THINKING CONCEPTUALLY** Considering Figure 38.11, why would you expect a predator, such as a mongoose, to prey on more than one type of organism?

Conservation Techniques Require Much Effort and Expertise

17. Why is a grizzly bear a keystone species existing as a metapopulation?
    a. Grizzly bears require many thousands of miles of preserved land because they are large animals.

b. Grizzly bears have functions that increase biodiversity, but presently the population is subdivided into isolated subpopulations.
    c. When grizzly bears are preserved, so are many other types of species within a diverse landscape.
    d. Grizzly bears are a source population for many other types of organisms across several population types.
    e. All of these statements are correct.
18. A population in an unfavorable area with a high infant mortality rate would be a
    a. metapopulation.   c. sink population.
    b. source population.   d. new population.
19. The goal of restoration ecology is
    a. returning damaged ecosystems to their former states.
    b. maximizing direct values of ecosystems.
    c. sustainable development.
    d. Both a and c are correct choices.
    e. a, b, and c are all correct choices.

Understanding the Terms

alien species 786		landscape diversity 781
biodiversity 780		metapopulation 789
biodiversity hotspot 781		overexploitation 787
bioinformatics 780		pollution 787
conservation biology 780		restoration ecology 790
ecosystem diversity 781		sink population 789
endangered species 780		source population 789
extinction 780		sustainable development 790
genetic diversity 781		threatened species 780
keystone species 789		

Match the terms to these definitions:
a. _____ A rather small area with an unusually large concentration of species.
b. _____ A subdivided population in isolated patches of habitat.
c. _____ A population that has a positive growth rate and net emigration rate to other locations.
d. _____ A population that is found in an unfavorable area where at best the birthrate equals the death rate.

Thinking Scientifically

1. Hypothesize what effect overharvesting has on genetic diversity (see Section 13.15), even if the population is protected for a while and allowed to return to its normal level. How would you test your hypothesis?
2. A conservationist wants to rescue a species of songbirds from extinction by gathering a few eggs and raising a population that s/he can reintroduce into the wild. Criticize the plan.

ARIS™ *Visit* **www.mhhe.com/maderconcepts** *for practice quizzes, animations, videos, and activities designed to help you master the material in this chapter.*

BIOLOGICAL VIEWPOINTS

PART VI Organisms Live in Ecosystems

If you take a scenic vacation, you might decide to visit a lush, coniferous forest in the Canadian Rockies. Any outdoor place such as a forest, a desert, or a lake is an ecosystem where populations interact among themselves and with the physical environment. Interactions include bark beetle larvae feeding on the cambium layer of pine trees, and their exchange of gases with the atmosphere. Beetles give off carbon dioxide, eventually taken in by the tree, and the tree gives off oxygen needed by the beetles.

Suppose the bark beetle population is running rampant and is killing all the pine and spruce trees in the forest. To study the situation, an ecologist would want to analyze the bark beetle interactions further at this locale. If any predators or competitors of the beetle are missing, this might explain why population growth is out of control. On the other hand, mutualistic interactions can assist and increase the size of the two interacting populations. Modern ecologists are devoted to understanding population growth so they can predict which biotic, and also abiotic, factors might affect population sizes under different situations.

Visiting the same ecosystem for several years in a row might allow you to notice what changes occur. As you know from the introduction to Chapter 28, a volcanic eruption can destroy much of the surrounding forest. It may take a while, even up to 200 years, but the forest is expected to return to normal through the process of ecological succession. The new community may not be exactly as it was before, but it will contain a number of different populations. Diversity will increase as more niches become available for organisms to fill. Although no two species can occupy the same niche at the same time, resource partitioning decreases competition and allows more species to live in the same general area. Several species of insects can feed off one tree because each one prefers eating just certain parts of the tree, from the leaves to the roots.

Through studying the interactions of populations, two aspects of ecosystems are fundamental: Chemicals cycle within ecosystems, and energy flows through them. A population of bark beetle larvae provides food for woodpeckers, which in turn are eaten by hawks. When trees, bark beetles, and birds die off, decomposers return the chemicals within dead remains to photosynthesizers once again. In the meantime, the energy has dissipated into the environment and cannot be recycled. Ecosystems continue to function only because photosynthesizers, such as trees, continually absorb solar energy when they produce food for themselves directly and for all the other populations in an ecosystem indirectly.

Cycling of chemicals not only involves food webs, but also the abiotic components of ecosystems, such as the atmosphere above and the sediment below. Human activities affect the transfer rates of biogeochemical cycles and, therefore, the functioning of the biosphere, the largest ecosystem of all. For example, pumping water from the ground is not normally part of the water cycle, and burning fossil fuels and trees alters the carbon cycle, so that global warming is predicted. Transfer rates in the phosphorus and nitrogen cycles are affected when we produce fertilizers. Runoff of fertilizers into aquatic ecosystems causes overenrichment. Pollution can be defined as a change in transfer rates that leads to the degradation of human health or plant and animal life.

As forests, grasslands, streams, lakes, and seas are degraded, human society uses additional nonrenewable energy and materials to substitute for the benefits that biodiversity can provide at no cost. Now that we recognize the value of biodiversity, efforts are being made to prevent the forecasted mass extinction of unprecedented numbers of species. The new field of conservation biology involves many subfields of biology, all dedicated to the preservation of biodiversity for sustainable appreciation and use by human beings. We now understand that in preserving the biosphere for other species, we are preserving it for ourselves.

Appendix A | Answer Key

CHAPTER 1

Check Your Progress

1.1 Snow goose. **1.2** a. Organism; b. Population, community, ecosystem, and biosphere. **1.3** Fire ants (1) have levels of biological organization; (2) respond to stimuli, such as animal invaders; (3) regulate colony temperature by relocating when necessary; (4) acquire food for the colony; (5) help the queen so that their genes are passed to future generations; and (6) have evolved a social system that works for them. **1.4** Domain Eukarya and kingdom Animalia. **1.5** 2; **1.6** Humans disturb natural ecosystems. **1.7** Observation data; observing that a worker ant produces eggs. **1.8** Test group: colonies exposed to the parasite; control group: colonies not exposed to the parasite. **1.9** For DNA barcoding to be successful, all species, no matter how closely related, will have a different barcode.

Testing Yourself

1. b; 2. c; 3. b; 4. c; 5. b; 6. c; 7. Being alive is a property possessed by the cell that is not arrived at by summing up the parts of a cell. 8. d; 9. d; 10. c; 11. a; 12. c; 13. The common name of an organism can vary from place to place and, on occasion, has changed over time, but an organism's scientific name is universal and constant. 14. d; 15. b; 16. a; 17. d; 18. d; 19. A college campus has a location, as does an ecosystem. The populations of students, faculty, and administrators communicate with each other and the physical environment (the buildings). 20. a; 21. d; 22. c

Understanding the Terms

a. metabolism; b. evolution; c. experimental design; d. photosynthesis; e. control group

Thinking Scientifically

1. a. Bacteria don't die in sunlight when dye is present. b. Dye is protective against UV radiation. c. Experiment consists of exposing control and test groups to UV light. d. Hypothesis is not supported. 2. Plant the same species of tomato plants in three large plots. All plots receive the same treatment, except plot 1, your control, receives no fertilizer; plot 2 receives the name brand fertilizer in the same quantity as plot 3 which receives the generic brand. Weigh the tomatoes from each plot and sum up the weight to calculate which plot results in the most yield.

CHAPTER 2

Check Your Progress

2.1 Carbon, nitrogen, phosphorus, sulfur. **2.2** See Figure 2.4B, page 23, in text. **2.3** Yes. Tracer experiments usually use molecules that contain radioactive isotopes. **2.4** a. One. Hydrogen has one shell, which is complete with two electrons; b. Two. Oxygen has two shells with six valence electrons in the outer shell. Therefore, oxygen requires two more electrons for a completed outer shell. **2.5** H^+, OH^-. **2.6** a. See Figure 2.7, page 26, in text; b. This is the formula that gives each atom a completed outer shell. **2.7** Because oxy-

gen bonds more tightly to electrons than hydrogen. **2.8** a. Yes; b. Yes; c. Electropositive hydrogens are attracted to either electronegative oxygen or nitrogen. **2.9** The existence of hydrogen bonds between individual water molecules makes water a sort of "super" molecule, and the individual molecules stay together. **2.10** The air loses heat as it causes water in the pad to evaporate. **2.11** Polarity makes the emulsifiers hydrophilic. **2.12** The blocks of ice trap heat inside and prevent it from escaping to the environment. **2.13** a. H^+; b. OH^-. **2.14** a. Acidic; b. More H^+. **2.15** a. Down; b. Carbonic acid releases H^+, and this causes the pH to decrease. **2.16** The corrosive effect of a low-pH drink can be damaging to the digestive tract.

Testing Yourself

1. c; 2. b; 3. e; 4. d; 5. a; 6. Iodine-131 and thallium-201 are radioactive isotopes because they, unlike other isotopes of iodine and thallium, emit ionizing radiation. Iodine-131 can be used to diagnose a thyroid gland tumor because the thyroid gland takes up the element iodine. The element thallium is taken up by heart muscle tissue and can be used to evaluate the blood supply to the heart. 7. c; 8. d; 9. a; 10. c; 11. b; 12. a; 13. b; 14. d; 15. d; 16. Because nitrogen needs three more electrons in the outer shell to be stable; each hydrogen contributes one electron. 17. a; 18. b; 19. c; 20. Blood is transported in a tube (blood vessel) and water fills a tube due to its cohesive and adhesive properties. 21. a; 22. a; 23. a; 24. b; 25. c; 26. A bicarbonate buffer combines immediately with H^+ or OH^-, normalizing the blood pH. Respiration, by removing carbon dioxide from the blood, provides a slower response that decreases H_2CO_3 concentration. The kidneys excrete H^+ and provide the slowest of the three responses, but the kidenys also have the ability to cause the greatest overall change in the pH level.

Understanding the Terms

a. polar covalent bond; b. ion; c. acid; d. molecule; e. buffer

Thinking Scientifically

1. Na^+Cl^- interrupts hydrogen bonding enough to prevent the formation of the ice lattice that forms during freezing. 2. Chemical behavior is dependent on the number of electrons in the outer shell, not the number of neutrons in the nucleus.

CHAPTER 3

Check Your Progress

3.1 All living things contain organic molecules. **3.2** Hydrophilic functional groups, which would make them water soluble. **3.3** It uses the water for hydrolysis reactions. **3.4** Fructose is an isomer of glucose. **3.5** Both store glucose as a complex carbohydrate, but plants store it as starch, and animals store it as glycogen. **3.6** Plants produce glucose through photosynthesis, but animals do not photosynthesize. **3.7** Internal leaf cells normally live in a fluid environment. **3.8** Plant cells carry out more varied reactions than an animal cell does. **3.9** Animals can be sustained by

feeding on plants. **3.10** Globular proteins are enzymes present in all organisms, fibrous proteins are structural proteins, and plants do not use proteins as structural compounds. **3.11** The complementary DNA sequence would be CTAGGT. **3.12** In subsection 3.8, we learned that proteins function in support, metabolism, transport, defense, regulation, and motion. **3.13** Plant cells absorb solar energy and produce glucose. Glucose in food leads to ATP buildup in cells. In muscle cells, ATP breakdown leads to muscle contraction and movement.

Testing Yourself

1. e; 2. c; 3. b; 4. b; 5. Like carbon, silicon has four outer electrons and can form four covalent bonds, as in SiO_2, the main component of sand. Carbon has two, but silicon has three, shells of electrons. Because of its larger size, silicon rarely forms chains, nor does it bond to four different types of atoms. 6. d; 7. c; 8. c; 9. Cellulose chains can form fibers because they lie side-by-side and hydrogen bonds form between them. The branching observed in starch makes fiber formation more unlikely. 10. a; 11. b; 12. d; 13. d; 14. c; 15. The hydrophilic heads interact with fluids, allowing the hydrophobic tails to orient toward each other. 16. c; 17. c; 18. d; 19. a; 20. d; 21. e; 22. a; 23. c

Understanding the Terms

a. carbohydrate; b. lipid; c. polymer; d. isomer; e. peptide

Thinking Scientifically

1. a. Subject the seeds of temperate and tropical plants, for which you know the amount and kind of oil content, to a range of temperatures from above freezing to below freezing for an extended length of time. Plant the seeds and compare the percentage of survivals per type of plant. b. The presence of unsaturated oils in temperate plant seeds may be an adaptation to the environment. 2. Possible hypothesis: (1) The abnormal enzyme will not produce as much product per unit time as the normal enzyme. (2) The abnormal enzyme will have a different shape from the normal enzyme due to changes in organization.

CHAPTER 4

Check Your Progress

4.1 The parts individually cannot perform all the functions of a living cell. **4.2** Small cells have a larger surface-area-to-volume ratio and are better able to exchange materials. **4.3** a. magnification; b. resolution. **4.4** No, because prokaryotic cells do not have a nucleus. **4.5** The electron microscope, not the light microscope, allows us to "see" this amount of detail, and it was not available until the 20th century. **4.6** The eukaryotes: Euglena (a protist), onion root cells, kidney tubule cells, muscle cells, nerve cells, and liver cells. Only eukaryotes, not prokaryotes (bacteria and archaea), have a nucleus. **4.7** First the ribosomal subunits, and then mRNA, pass from the nucleus to the cytoplasm, by way of the nuclear pores. They combine when protein synthesis begins. **4.8** Yes.

Polyribosomes are in the cytoplasm. Other ribosomes are temporarily attached to the ER, and only the synthesized polypeptides enter the lumen of the ER. **4.9** Synthesized polypeptides enter the lumen of the RER, and the RER sends vesicles to the Golgi apparatus. The Golgi apparatus sends vesicles to the plasma membrane. **4.10** Tracers, see Section 3.3. **4.11** Lysosomes combine with and digest microorganisms brought into the cell by vesicle formation. **4.12** Peroxisomes pass fatty acid breakdown products to mitochondria for further metabolism. **4.13** Lysosomes only digest; the plant cell central vacuole has storage and structural functions, as well as digestive functions. **4.14** Genes located in the nucleus code for the polypeptides that are produced by rough ER. These polypeptides enter the ER and are found throughout the endomembrane system. **4.15** Thylakoid membrane captures solar energy because it contains chlorophyll. **4.16** They both were once independent organisms and have their own DNA. **4.17** MtDNA mutations, due to highly reactive forms of oxygen, likely accumulate as a nerve cell ages. **4.18** Actin filaments, intermediate filaments, and microtubules. **4.19** The only anatomical difference is length; cilia are short and flagella are long. Cilia move like an oar, and flagella move in an undulating, snakelike fashion. **4.20** Both plants and animals contain extracellular components and a matrix (packing material) between their cells.

Testing Yourself

1. c; 2. c; 3. a; 4. d; 5. The much smaller size of prokaryotic cells and their structural simplicity (they lack a nucleus and other membranous organelles) suggests they evolved first. 6. c; 7. c; 8. b; 9. c; 10. RER: studded with rbosomes—suits protein synthesis; long and tubular—enzymes to modify protein could be attached to its walls; forms vesicles—vesicles function in transport; 11. d; 12. c; 13. c; 14. a; 15. c; 16. b; 17. b; 18. d

Understanding the Terms

a. Golgi apparatus; b. proxisome; c. nucleolus; d. cytoskeleton; e. fimbriae

Thinking Scientifically

1. Labeling RNA. You expect to find RNA in the nucleus, passing through a nuclear pore, and on RER. 2. Mitochondria have their own DNA and make their own proteins; therefore, they have RNA. The mitochondrial double membrane does not have any pores.

CHAPTER 5

Check Your Progress

5.1 a. ATP represents potential energy because the energy is present in chemical bonds; b. Muscle movement is kinetic energy because motion is occurring. **5.2** Yes; any motion or cellular reaction that requires the breakdown of ATP increases entropy. **5.3** As long as cellular respiration is possible, ATP is constantly replenished by joining ADP with Ⓟ. **5.4** Because ATP breakdown is coupled to the energy-requiring reaction of muscle contraction, the overall reaction increases entropy. **5.5** Every enzyme is a protein that lowers the energy of activation for its reaction. **5.6** The active site of each enzyme accommodates only its particular substrate. **5.7** The optimum pH gives the active site the correct shape to bind the substrate; body temperature increases the number of encounters between the substrate and the enzyme. **5.8** Inhibition of enzyme requires an abundance of an enzyme product and is easily reversed if the amount of product decreases. **5.9** The dosage of sarin would have to be enough to inhibit all the pertinent enzymes. **5.10** Like the colored tiles, the proteins in a plasma membrane are different. But unlike the tiles, which are cemented in place, the proteins are free to move from side to side in the fluid phospholipid layer. **5.11** Channel proteins and carrier proteins. **5.12** Effect depends on function of protein.

5.13 Diffusion increases because diffusion is dependent on the motion of molecules, which increases as temperature rises. **5.14** A turnstile provides a way for people to pass through a barrier, and the carrier protein provides a way for a molecule to pass through the plasma membrane. **5.15** Hypertonicity of the soil beside the road causes plants to wilt and die. **5.16** Passage through a turnstile is one-way, and people (molecules) can enter but not exit. **5.17** Cholesterol builds up in the blood, where it can cause health problems.

Testing Yourself

1. a; 2. e; 3. c; 4. Food in bulk is stored food in your pantry; glycogen is stored energy in the liver and muscles; meals provide only enough energy till you eat again; ATP is just enough energy for a reaction. 5. a; 6. e; 7. c; 8. b; 9. b; 10. b; 11. c; 12. d; 13. c; 14. Solutes retain the liquid portion of blood in the vessels, allowing the blood to flow. 15. a, b, c; 16. b, c, d; 17. d; 18. b, d

Understanding the Terms

a. differentially permeable; b. osmosis; c. hypertonic solution; d. glycoprotein; e. phagocytosis

Thinking Scientifically

1. Ecosystems need a source of energy because energy cannot be created. They need a continous supply because with each energy transformation, heat is lost; eventually, all the energy taken in is lost. 2. You need to decide the proper enzyme versus substrate concentrations, type of glassware, amount of time needed for the reaction, how to vary the temperature and the pH, and how to test for the product.

CHAPTER 6

Check Your Progress

6.1 Plants can use the carbon dioxide and water from cellular respiration to photosynthesize food. **6.2** Thylakoids absorb solar energy because they contain chlorophyll; carbohydrates form in the stroma, because it contains enzymes. **6.3** Water is oxidized, and carbon dioxide is reduced. **6.4** The cell has a mechanism by which it moves hydrogens from one molecule to another. It has no mechanism to move carbons from one molecule to another. **6.5** NADPH and ATP are formed in the light reactions and used in the Calvin cycle reactions. **6.6** We see leaves as green because leaves do not absorb green light, but instead reflect it. **6.7** Leaf color changes are simply a response to seasonal light and temperature changes, and these do not occur in the tropics. **6.8** Accessory pigments permit absorption of a greater range of wavelengths than does chlorophyll a alone. **6.9** The energy of motion—i.e., kinetic energy. **6.10** At the start, chlorophyll a. At the end, NADPH. **6.11** "Get ready": NADP reductase has electrons, and H^+ gradient is present. "Payoff": NADPH and ATP. **6.12** They return to the light reactions. **6.13** Glucose can become sucrose, starch, or cellulose. **6.14** C_3 photosynthesis takes its name from first detectable molecule following carbon dioxide fixation in a plant cell. **6.15** Bundle sheath cells (the location of the Calvin cycle) are not exposed to the open spaces of spongy mesophyll. **6.16** CO_2 is fixed at night but does not enter the Calvin cycle until the next day. **6.17** The diversity of life in tropical rain forests alone argues for their preservation.

Testing Yourself

1. c; 2. d; 3. d; 4. c; 5. Only plants produce organic food, and if animals only ate animals, food would run out. 6. a; 7. a; 8. a; 9. e; 10. d; 11. a; 12. e; 13. b; 14. a. thylakoid membrane; b. O_2; c. stroma; d. Calvin cycle reactions; e. granum, see Figure 6.5A, page 99, in text. 15. Solar energy is unable to participate directly in the chemical reactions that reduce carbon dioxide to carbohydrate, but ATP can do so. 16. d; 17. c; 18. a.

Understanding the Terms

a. light reactions; b. electron transport chain; c. photosynthesis; d. Calvin cycle reactions; e. CAM photosynthesis

Thinking Scientifically

1. Using the same sample of elodea in the same beaker will control most factors, except temperature, which should remain constant. For a control, do the same experiment without elodea. 2. Photosynthesis takes place in leaf cells, showing that cells, and not organs, are the basic units of a plant.

CHAPTER 7

Check Your Progress

7.1 Cellular respiration produces ATP. The oxygen we breathe in is necessary to cellular respiration, and the carbon dioxide we breathe out is a by-product of cellular respiration. **7.2** Allows energy to be captured for production of ATP molecules. **7.3** Pyruvate still contains most of the energy. NADH contains some of the energy, and there is a net gain of two ATP. Some energy was lost as heat, and this heat will help warm our bodies. **7.4** The reaction is getting ready for the citric acid cycle, the final breakdown of glucose to carbon dioxide. **7.5** It is transported to the lungs, where it is exhaled. **7.6** As many as 34; see Section 7.8. **7.7** "Get ready": H^+ gradient is established. "Pay off": ATP is produced. **7.8** Only glycolysis, because the other metabolic pathways occur in mitochondria. **7.9** Grapes are sometimes coated with yeasts that can start fermenting sugars (and producing alcohol) if the weather is conducive. **7.10** Lactate. **7.11** a. Hydrolytic reactions are catabolic; b. A hydrolytic reaction occurs when ATP breaks down to ADP + Ⓟ (catabolic). **7.12** Engage in an exercise like swimming because submaximal aerobic exercise burns fat.

Testing Yourself

1. a; 2. c; 3. b; 4. Glucose breakdown begins with glycolysis; oxygen becomes water at the end of the ETC; and carbon dioxide is produced by the prep reaction and citric acid cycle. Some of the energy released is converted to ATP by glycolysis (2 ATP), citric acid cycle (2 ATP per two turns), and 32-34 ATP by the ETC. 5. a; 6. e; 7. b; 8. b; 9. c; 10. c; 11. d; 12. b; 13. a. Inner membrane partitions the mitochondrion into the intermembrane space and the matrix; b. Matrix is location of preparatory reaction and citric acid cycle; c. Cristae contain electron transport chain and ATP synthase complex. 14. Only 39% of available energy becomes ATP and the rest becomes heat. 15. c; 16. d; 17. c; 18. b; 19. b; 20. c; 21. Chloroplasts and mitochondria were originally independent prokaryotes. Prokaryotes evolved from a common ancestor, which must have used an ETC.

Understanding the Terms

a. glycolysis; b. oxygen debt; c. catabolism; d. chemiosmosis; e. cellular respiration

Thinking Scientifically

1. Acid, because for ATP to be produced, H^+ must flow through the ATP synthase complex.

CHAPTER 8

Check Your Progress

8.1 Asexual reproduction. **8.2** Asexual reproduction in unicellular organisms produces new individuals. It is used for growth and repair in multicellular ones. **8.3** Interphase, because the cells are not differentiated and do not help with the work of the organ. Instead, they simply keep dividing. **8.4** a. Cells are normally 2n; b. Cancer cells will not necessarily be 2n because mitosis may not be orderly. **8.5** It differentiates and functions like a skin cell does. **8.6** a. New cell walls are needed for each daughter cell; b. Parent and daughter cells have the same number and kinds of chromosomes following the M stage of the cell cycle. **8.7** Without a normal p53 protein, damaged cells do not undergo apoptosis and instead keep on dividing. **8.8** Cancer cells are like embryonic cells because they can divide indefinitely. **8.9** Loss of cell cycle control: lack of differentiation; abnormal nuclei; form tumors. Beyond loss of control: angiogenesis and metastasis. **8.10** Tobacco smoke contains chemicals that increase the chances of mutations and these are absorbed into the blood at the lungs (see page 133). **8.11** Haploid, because each daughter cell receives one member from each homologous pair of chromosomes. **8.12** Sister chromatids have the same genetic information. Non-sister chromatids have genetic information for the same traits, but the specifics are different. For example, one sister chromatid could call for freckles, and the other could call for no freckles. **8.13** The chromosomes of relatives are more likely to carry some of the same genetic information. Therefore, fertilization may bring together like homologues. **8.14** Plant cells do not have centrioles or asters. They do have cell walls. **8.15** Following fertilization, mitosis ensures that all cells of the body get a complete set of genes. **8.16** Meiosis I has the same phases as mitosis, but the resulting cells are haploid. Meiosis II has the same phases as mitosis, but the resulting cells are haploid. **8.17** An excess of genetic material would cause metabolic abnormalities. **8.18** Meiosis II because the only way for YY to occur is through the nonseparation of chromatids during meiosis II. **8.19** The child would inherit a chromosome that was either too short or too long, therefore the homologues for this chromosomes would look different in a karyotype.

Testing Yourself

1. c; 2. b; 3. c; 4. a; 5. d; 6. c; 7. b; 8. d; 9. a. chromatid of chromsome; b. centriole; c. spindle fiber or aster; d. nuclear envelope (fragment). 10. a; 11. d; 12. c; 13. c; 14. d; 15. Due to crossing-over and independent alignment, the chromosomes in the egg and sperm carry different genetic information. 16. a; 17. b

Understanding the Terms

a. sister chromatid; b. centrosome; c. spermatogenesis; d. apoptosis

Thinking Scientifically

1. Both parents because the egg had to be missing this chromosome and the sperm had to have two of this chromosome. 2. While theoretically possible, it would be practically impossible to find two gametes with the exact same genetic information that produced this indiviudal.

CHAPTER 9

Check Your Progress

9.1 The selection of certain traits (controlled by genes) brings about recognizable species changes. **9.2** Compared to dogs, pea plants are smaller, have a shorter maturation time, have traits that are easier to distinguish, can both self- and cross-pollinate, and produce many more offspring. **9.3** Hip dysplasia only occurs when offspring inherit two recessive alleles for hip dysplasia. **9.4** (1) a. all W; b. ½ W, ½ w; c. ½ T, ½ t; d. all T; (2) bb; (3) 3 black rabbits: 1 white rabbit; 40 rabbits/120 are white. **9.5** TtGg, TTGG, TtGG, TTGg. **9.6** (1) 75%; (2) 9:3:3:1, $\frac{1}{16}$; (3) BbTt. **9.7** (1) ¼; (2) Ss; (3) Tt x tt, tt. **9.8** Without careful records, the breeder may not know that all the individuals in row III are cousins who could carry the same recessive allele for, say, hip dysplasia. **9.9** Each parent is heterozygous. **9.10** Yes, if each were heterozygous. **9.11** a. a; b. a. **9.12** The phenotypes of the potential offspring are: ¼ are normal; ½ may suffer a heart attack as young adults; and ¼ may suffer a heart attack in childhood. **9.13** All four phenotypes are possible from the cross $I^A i$ x $I^B i$, making paternity determination impossible. **9.14** (1) 25%; (2) Type O; (3) Child: ii, mother: $I^A i$; father: $I^A i$, ii, $I^B i$; (4) see Figure 9.14, page 172, in text. **9.15** In CF, faulty chloride channels occur throughout the body and lead to problems in several organs, including the skin, lungs, and pancreas. **9.16** The Y chromosome is quite small. **9.17** (1) mother $X^B X^b$, father $X^b Y$, daughter $X^b X^b$; (2) $X^R X^R$, $X^R Y$; (3) (b); 1:1. **9.18** Linkage maps show how frequently certain traits may be inherited together. **9.19** (1) a. 9:3:3:1; b. linkage; (2) bar-eye, scallop-wing, garnet-eye. **9.20** Genes and chromosomes behave similarly during mitosis and meiosis. For example, with regard to mitosis: Two genes for each trait (two chromosomes of each type) ar in the adult; only one gene for each trait (only one chromosome from each pair) are in the gametes.

Testing Yourself

1. c; 2. The blending model of inheritance results in no variations, whereas Mendel's model does result in variations among the offspring. Evolution requires variations. 3. d; 4. c; 5. b; 6. a; 7. c; 8. b; 9. e; 10. b; 11. a; 12. c; 13. e; 14. d; 15. Test the egg because the egg is not fertilized unless it is normal. 16. d; 17. a; 18. c; 19. b; 20. a. X-linked recessive; b. XAXa

Understanding the Terms

a. recessive allele; b. allele; c. dominant allele; d. test cross; e. homozygous

Thinking Scientifically

1. Cross it now with a fly that lacks the characteristic. Most likely, the fly is heterozygous and only a single autosomal mutation has occurred. Therefore, the cross will be Aa x aa with 1:1 results. If the characteristic disappears in males, cross two F1 flies to see if it reappears; it could be X-linked (see Section 9.16). 2. Give plants with a particular leaf pattern different amounts of fertilizer from none (your control) to over-enriched, and observe the results. Keep other conditions, such as amount of water, the same for all.

CHAPTER 10

Check Your Progress

10.1 The presence of a capsule. **10.2** ^{32}P, because DNA contains phosphorus. **10.3** a. Yes, all organisms contain both DNA and RNA; b. DNA contains the sugar deoxyribose and the base thymine. RNA contains the sugar ribose and the base uracil in place of thymine. **10.4** a. No, every species has its own sequence of bases, leading to different percentages; b. Complementary base pairing can occur in only one way; c. As mutations occurred, the base sequence changed and new species arose. **10.5** DNA is a double helix in all organisms. **10.6** All the cells of Arabidopsis have the same sequence because of DNA replication during the growth process. **10.7** Complementary base pairing. **10.8** Mutations are due to a change in the base sequence of a gene's DNA. **10.9** Arabidopsis investigators want to know which gene goes with which protein and the function of each protein. **10.10** The genetic code is degenerate, and not all base changes specify different amino acids. **10.11** Transcription forms mRNA transcipts that carry genetic information from DNA in the nucleus to the ribosomes, where the message is translated. **10.12** They fail to support the hypothesis because Arabidopsis is a much simpler organism than a human being. A further hypothesis would be needed, such as "only in animals do introns function to increase complexity." **10.13** a. The anticodons are GCA, GCG, GCU, GCC, UCU, UCC; b. Yes, because it is the third position that varies. **10.14** At a ribosome attached to ER, or as a part of a polyribosome. **10.15** These bases are complementary to AUG, the start codon of a mRNA transcript. Therefore, they indicate the start of a gene. **10.16** A gene stops at any one of these. **10.17** The sequence of amino acids in the proteins of a cell. **10.18** A greater number of codons in mRNA and a greater number of amino acids would be incorrect. **10.19** The more one is exposed to environmental mutagens, the more likely a change in the base sequence of a gene will occur, leading to the development of cancer. **10.20** An exon because introns are removed during mRNA processing.

Testing Yourself

1. b; 2. c; 3. e; 4. d; 5. c; 6. b; 7. a; 8. c; 9. DNA analogues prevent the formation of new DNA molecules, and each new virus requires a DNA molecule. 10. b; 11. d; 12. a; 13. c; 14. b; 15. c; 16. a. ACU CCU GAA UGC AAA; b. UGA GGA CUU ACG UUU; c. threonine—proline—glutamate—cysteine—lysine; 17. c; 18. b; 19. A change from ATG to ATA has no effect because both of the resulting codons code for tyrosine. 20. Without mutations, new life forms cannot evolve.

Understanding the Terms

a. RNA polymerase; b. intron; c. translation; d. polyribosome

Thinking Scientifically

1. Determine if any transposon base sequence occurs in the sequence for the neurofibromatosis gene. 2. Using tissue culture and DNA technology, place the gene in Arabidopsis cells and look for the mutation in cloned adult plants.

CHAPTER 11

Check Your Progress

11.1 A promoter (short sequence of DNA) where RNA polymerase binds signals the start of a gene. **11.2** These genes control early development of the organism and are normally turned off in adult cells. **11.3** Yes, because an animal cell can develop into a new individual. **11.4** Reproductive cloning produces new individuals, therapeutic cloning produces mature cells for medical purposes. **11.5** Using an embryo stops short the development of a human being. Embryonic cells floating in the liquid of the womb are no longer part of an embryo, which continues its normal development to term. **11.6** A nucleosome includes DNA and histone proteins. **11.7** Yes, the individual would possess a Barr body containing an inactive X chromosome. **11.8** Transcription factors or transcription activators. **11.9** Introns are removed during processing. **11.10** As soon as mRNA reaches the cytoplasm. **11.11** During early development of an organism. **11.12** Genes get turned on sequentially, and this makes development orderly. **11.13** It shows that this nucleotide sequence developed early in the evolution of life, and that these ancient sequences are essential to development. **11.14** Proto-oncogenes promote the cell cycle and inhibit apoptosis; tumor suppressor genes inhibit the cell cycle and promote apoptosis. **11.15** It regulates the cell cycle and initiates chromosome repair. If repair is impossible, it initiates apoptosis. **11.16** These therapies kill dividing cells all over the body, not just cancer cells.

Testing Yourself

1. a. DNA; b. regulator gene; c. promoter; d. operator; e. repressor; 2. b; 3. b; 4. c; 5. b; 6. c; 7. a; 8. b; 9. e; 10. d; 11. a; 12. c; 13. b; 14. Benefit is refinement of control; drawback is more chances of losing control; 15. d; 16. a; 17. a; 18. b, c 19. a, c; 20. a, d; 21. a, d; 22. c, d, a, b; 23. In cancer, cell division is out of control; 24. c

Understanding the Terms

a. posttranscriptional control; b. Barr body; c. euchromatin; d. totipotency; e. histone

Thinking Scientifically

1. You should also show that the genes lead to similar developmental stages. 2. Culture normal cells in the presence of the protein and observe the results. Culture the same types of cells in the same manner but without the protein and observe the results.

CHAPTER 12

Check Your Progress

12.1 You would need to produce many copies of the GFP gene for insertion into pig embryos. **12.2** Use PCR to target the gene in pig DNA and make copies. **12.3** Examine the area in dim light to detect bioluminescence. Seeing none tells you the bacteria are not present. **12.4** Both are products of genes, and gene expression results in a protein product. **12.5** As long as the gene is turned on in plant cells, it will be expressed, regardless of its origin because plants and animals use the same cellular machinery. **12.6** a. Strictly speaking, both are genetically engineered foods because both contain a protein made by an engineered organism; b. It is probably not necessary to test the cheese, because chymosin produced by bacteria is exactly the same molecule as chymosin produced by calves. However, the corn should be tested because foods don't ordinarily contain a protein that has insecticidal properties. **12.7** Not all GMO animals will be females, but all cloned animals in Figure 12.7A will be females. **12.8** A virus. **12.9** The virus-infected cells would glow in the dark. **12.10** One—hence their name. **12.11** The people who had heart attacks have base pair differences (mutations) that make them vulnerable to heart attacks. **12.12** You would want to know its structure/shape at the four levels of protein organization; when it is normally present in a cell; how it interacts with other proteins; all of its functions; and whether there are any abnormal versions of the protein. **12.13** GFP can cause both a jellyfish and a mammal to glow.

Testing Yourself

1. d; 2. e; 3. c, a, b, d; 4. c; 5. b; 6. begins: AATT; ends: TTAA; 7. a; 8. b; 9. c; 10. e; 11. a; 12. Each gene performs only one function, so despite increased number of genes, only 30 gene functions would be available. 13. b; 14. c; 15. a; 16. d; 17. e; 18. Genetically modified stem cells pass on their modification to their offspring (more white blood cells) indefinitely. 19. d; 20. d; 21. c; 22. b; 23. a; 24. c

Understanding the Terms

a. restriction enzyme; b. genome; c. cloning; d. plasmid

Thinking Scientifically

1. Use restriction enzymes to fragment the DNA of chromosome 10. Inject these fragments, one at a time, into groups of one-celled mouse embryos. Following development, see which mice have no tails. (This question is hypothetical.) 2. An ex vivo study allows you to perfect your procedure in the lab and possibly avoid harm to the patient.

CHAPTER 13

Check Your Progress

13.1 Darwin made observations that would help him formulate a hypothesis. **13.2** When they are proven false. **13.3** No, you would pollinate it by hand with the pollen of your choice. **13.4** Any plant that provided food (nectar) to a pollinator had an advantage that resulted in more offspring than other members of the population because its pollen was being distributed. Similarly, any animal having a feeding apparatus that could access the nectar had an advantage that resulted in more offspring with the same trait. **13.5** Yes. Whenever independent sources come

to the same conclusion based on the data they have collected, the hypothesis is supported. **13.6** The few bees that can pollinate red flowers would reproduce more than the other members of the population. Eventually, the bees would become adapted to getting nectar from the red flower. **13.7** Flowers are delicate and decompose quickly, not allowing time for them to become fossils. **13.8** The flowers would have shorter and thicker floral tubes. **13.9** The flowers probably had a common ancestor. **13.10** These species of finches evolved on the Galápagos Islands. **13.11** All organisms are related through common descent from the first cell or cells. **13.12** Microevolution has occurred when allele frequency changes are observed due to mutations, gene flow, nonrandom mating, genetic drift, and natural selection. **13.13** Yes; any flower mutation that increases the chances of being pollinated and any mutation in the pollinator that increases the chances of getting food results in more offspring with some adaptation. **13.14** Imagine a pollinator that pollinates a particular species and can only see flowers of a certain color. Only flowers of that color would produce seeds, and soon all flowers of that species would be that color. **13.15** Yes, because genetic drift causes a change in allele frequencies, but does not necessarily contribute to adaptation. **13.16** Yes, because a flower of any other color has little chance of being pollinated. **13.17** Yes, because, in this instance, natural selection favors the heterozygote. Natural selection favors whichever genotype/phenotype is most advantageous in a particular environment

Testing Yourself

1. b; 2. d; 3. c; 4. d; 5. b; 6. Neither the occurrence of mutations nor the changing of environmental conditions are known ahead of time. 7. e; 8. d; 9. e; 10. e; 11. a; 12. d; 13. c; 14. b; 15. The same data sequences were inherited from a common ancestor. 16. a; 17. a; 18. c; 19. d; 20. b

Understanding the Terms

a. stabilizing selection; b. genetic drift; c. biogeography; d. palentology; e. gene flow

Thinking Scientifically

1. If you know the genotype of the various colors, you could use the Hardy-Weinberg equation to calculate changes in gene pool frequencies. Otherwise, you could base sequence before and after the experiment to determine a change in the genome. 2. Yes; due to natural selection of boll weevils resistant to the insecticide.

CHAPTER 14

Check Your Progress

14.1 No, because ligers share the ancestry of both lions and tigers, and they are not normally produced in the wild. **14.2** a. Habitat isolation; lions live on the plains, and tigers live in forests; b. F2 fitness; ligers are usually sterile (due to mispairing of chromosomes during meiosis). **14.3** Members of the ancestral species separated into those that lived on the plains and those that lived in the forest. Over time, each became adapted to its habitat, including a change in coat color. Now, lions and tigers do not mate because they rarely come into contact, and also because the coat color allows them to recognize members of their own species. **14.4** Show that each of the cats is adapted to fill a different niche. **14.5** The fossil record would have to show fossils of the different types of cats existing in the same location at the same time—before they begin to appear in various locations. **14.6** The chromosomes of the sterile plant are unable to pair during meiosis, accounting for its sterility. The chromosomes of the self-fertilizing plant have doubled, allowing it to self fertilize, but the plant is now a tetra-

ploid that produces a larger fruit. **14.7** A gradualistic model because the liger could be a transitional link. **14.8** Yes, because these animals are dated before any land forms are found in the fossil record. **14.9** Ligers are bigger than lions and tigers. **14.10** No, evolution is not goal oriented.

Testing Yourself

1. c; 2. c; 3. b; 4. f; 5. a; 6. e; 7. h; 8. e; 9. c; 10. e; 11. Genetic differences; 12. a. species 1; b. geographic barrier; c. genetic changes; d. species 2; e. genetic changes; f. species 3. 13. b; 14. b; 15. Allopatric speciation was possible on the Hawaiian Islands, but not on the Florida Keys. 16. b; 17. c; 18. d; 19. b; 20. b; 21. c; 22. b; 23. b; 24. d; 25. b.

Understanding the Terms

a. postzygotic isolating mechanism; b. adaptive radiation; c. speciation; d. allopatric speciation

Thinking Scientifically

1. Both the biological species concept and DNA barcodes provide a way to identify species without the need to examine them anatomically, but the DNA barcoding method is faster and unequivocal. The evolutionary species concept allows you to trace the history of an organism in the fossil record, and the biological species concept allows you to determine how species are kept separate. 2. Their chromosomes are compatible, and the two species are very closely related. It's doubtful they should be considered different species.

CHAPTER 15

Check Your Progress

15.1 When tracing the evolution of life, you need to start with the past and move toward the present. The dates are in mya; therefore, the larger numbers in the timescale are more distant from the present than are the smaller numbers. **15.2** The first photosynthesizers were single-celled eukaryotes called algae, which appeared before plants evolved. **15.3** Yes, because dinosaurs evolved at the start of the Mesozoic era, when the continents were all still joined. **15.4** Humans didn't evolve until the Holocene epoch, long after the last mass extinction. The possibility exists that humans could become extinct due to their own activities. **15.5** Only plants that resemble virgina creeper are in the same genus. Domain Eukarya contains the plants, animals, fungi, and protists. **15.6** See Figure 19.20. **15.7** Yes, because dinosaurs are reptiles, a type of vertebrate. **15.8** Derived characters indicate that these reptiles are more closely related to each other than to the other reptiles. **15.9** Yes, because we are closely related to dinosaurs.

Testing Yourself

1. b; 2. d; 3. a; 4. c; 5. e; 6. d; 7. a; 8. b; 9. d; 10. a. Ferns; b. seed plants; c. naked seeds; d. needle-like leaves, Conifers; e. fan-shaped leaves, Gingkos; f. enclosed seeds; g. one embryonic leaf, Monocots; h. two embryonic leaves, Eudicots. 11. b; 12; d; 13. e; 14. b; 15. e; 16. b; 17. New and different structures arise due to DNA differences. 18. b; 19. a, b, c; 20. c,d,e; 21. b; 22. The three-domain system is based on differences/similarities in the sequencing of rRNA. This is backed up by differences in structure.

Understanding the Terms

a. taxonomy; b. phylogenetic tree; c. taxon; d. molecular clock; e. clade

Thinking Scientifically

1. The tree shows that all life forms have a common source and how they are related, despite the occurrence of divergence, which gives rise to different groups of organisms. 2. The specialized environmental niche of these organisms is the same as it was when they first evolved.

CHAPTER 16

Check Your Progress

16.1 The genes become incorporated into the virus's nucleic acid core. **16.2** The virus should be lysogenic and should never undergo the lytic cycle. **16.3** Only plant cells have receptors for plant viruses. **16.4** To avoid unintentionally infecting animal, including human, cells. **16.5** Today's medicines prevent reverse transcription and biosynthesis. It would be good to prevent attachment in the first place and also maturation. **16.6** An emerging disease is one that is new to humans or one that spreads to new areas, perhaps by a new vector, or one that now infects a large number of people. **16.7** As yet, there were no enzymes to allow reactions at mild temperatures to occur. **16.8** a. RNA can store, replicate, and transmit genetic information; b. RNA can perform enzymatic functions, including those for replicating itself. **16.9** a. Proteins (enzymes) allow a cell to acquire energy and grow; b. Genetic information (either in the form of RNA or DNA) is needed for reproduction. **16.10** The capsule protects bacteria from the host defenses, and the fimbriae help bacteria adhere to parts of the human body. **16.11** Binary fission will quickly produce a large number of bacteria, which will have an ability to feed on oil. **16.12** During sexual reproduction, each parent passes on one copy of all its genes to an offspring. During transformation, conjugation, and transduction, a fully formed bacterium receives a few genes from another fully formed bacterium. **16.13** a. All humans are aerobic; b. All humans are heterotrophs. **16.14** Cyanobacteria are believed to be the first organisms to release oxygen into the atmosphere. Cyanobacteria still release oxygen, just as algae and plants do. **16.15** Bacteria and archaea arose from a common ancestor; eukaryotes arose from archaea. **16.16** Bacteria are saprotrophs whose digestive enzymes break down materials in the environment. **16.17** A contagious agent can spread through the populace after only a few people are infected.

Testing Yourself

1. c; 2. e; 3. c; 4. d; 5. b; 6. In life cycle (a), the virus is immediately produced and can go on to infect other cells. In life cycle (b), the virus is being replicated along with the host cells and is protected from exposure to host defenses. 7. c; 8. c; 9. c; 10. a; 11. c; 12. b; 13. c; 14. c; 15. a; 16. c; 17. b; 18. e; 19. a; 20. b; 21. a; 22. d; 23. b

Understanding the Terms

a. lysogenic; b. photosutotroph; c. saprotroph; d. archaea; e. conjugation

Thinking Scientifically

1. Viruses replicate inside human cells, and therefore, medications aimed at a virus can interfere with the workings of human cells. 2. The bacterium E. coli has the same genetic machinery as all other cells, including eukaryotic cells. They reproduce quickly and a large number can be kept in a small container. Since bacteria are haploid, mutations can be immediately observed.

CHAPTER 17

Check Your Progress

17.1 Yes, all protists are eukaryotes and have a nucleus. **17.2** The protists are a diverse group of organisms that vary tremendously in cellular organization, size, mode of nutrition, and means of locomotion. **17.3** The group called stramenopila contains both brown algae and water molds. **17.4** Contaminated water is a common source of Giardia infections. **17.5** A structural role. Silica is found in the tests of radiolarians. (It is also found in the cell walls of some plants and in the spicules of some sponges.) **17.6** Cilia give an organism the ability to move through the water and to direct water and food particles into a gullet for digestion. **17.7** The parasite that causes malaria is becoming resistant to some medicines. In addition, the vector for malaria, the Anopheles mosquito, is becoming resistant to pesticides. **17.8** As a result of the Irish potato famine, a large number of people emigrated from Ireland to the United States. **17.9** They are bound by protective cellulose plates that are impregnated by silicates. They also, typically, have two flagella. **17.10** A gelatinous product derived from some species of red algae. It is used as a solidifying agent for bacterial cultures. **17.11** Nucleic acid sequencing data. **17.12** None protect the embryo in the same manner as plants do.

Testing Yourself

1. d; 2. c; 3. a; 4. e; 5. Mitochondria were at one time free-living heterotrophic bacteria. 6. c; 7. b; 8. b; 9. c; 10. Flagella; trypanosomes cause disease, euglenoids contain chloroplasts. Pseudopods; amoeboids sometimes cause disease, foraminiferans and radiolarians form tests. Cilia; ciliates are complex and have trichocysts and undergo conjugation. None; sporozoans cause diseases. 11. Ciliates and sporozoans are traditionally separate groups based on structure, but they are grouped together in the Alveolata. This means their DNA shows that they are closely related. 12. b; 13. c; 14. d; 15. b; 16. d; 17. b; 18. a; 19. b; 20. a; 21. b, c; 22. a, d; 23. b; 24. b; 25. b; 26. a; 27. b; 28. c; 29. d; 30. c; 31. a; 32. e; 33. Diatoms have a two valve shell of silica and become diatomaceous earth. Dinoflagellates have two flagella and cellulose plates and are responsible for "red tides." Red algae are variously structured seaweeds and are sources of agar and carrageean. Brown algae are seaweeds harvested for food and are a source of alginate (algin). 34. Brown algae, diatoms, and water molds are traditionally separate groups based on mode of nutrition and/or structure, but they are grouped together in the Stramenopila. This means that their DNA shows they are closely related.

Understanding the Terms

a.pseudopod; b. diatom; c. plankton; d. zooplankton; e. ciliate

Thinking Scientifically

1. If single cells do not separate, and if each cell divides in a way that allows the cells to join end on end, the end result could be a filament. 2. Merozoites enter red blood cells, and if you knew by what process they enter red blood cells, you might be able to develop a way to stop them from doing so.

CHAPTER 18

Check Your Progress

18.1 Data on structural and molecular characters, including nucleotide base sequences. **18.2** After, because they have a specialization beyond the ordinary. **18.3** As in all plants, the gametophyte produces the gametes. **18.4** (1) The gametophyte doesn't depend on water to accomplish reproduction, and all stages of the life cycle are protected from drying out. (2) The sporophyte has vascular tissue and is protected from drying out by a cuticle interrupted by stomata. **18.5** Advantages: The sporophyte embryo is protected from drying out, and the sporophyte produces windblown spores that are resistant to drying out. Disadvantage: The sperm are flagellated and need an outside source of moisture in order to swim to the egg. **18.6** The independent gametophyte generation lacks vascular tissue, and it produces flagellated sperm. **18.7** (1) Water is not required for fertilization because pollen grains (male gametophytes) are windblown, and (2) ovules protect female gametophytes and become seeds that disperse the sporophyte, the generation that has vascular tissue. **18.8** From the fossil record. **18.9** a. With five petals (Venus flytrap and sundew) and four petals (pitcher plants), carnivorous plants are eudicots; b. Venus flytrap and sundew appear to be related. **18.10** Yes, insects do pollinate carnivorous plants, and seeds are produced and enclosed by fruit. The fruit is a dry capsule that contains seeds. **18.11** The metabolic capabilities of plants far surpass those of animals. **18.12** Fungi are heterotrophic, not photosynthetic, and the cell walls of fungi contain chitin, not cellulose. **18.13** Mycorrhizas result from the mutualistic relationship between a fungus and the roots of certain plants. The fungus helps bring water and minerals to the plant, and the plant provides organic carbon to the fungus. **18.14** Phylum Zygomycota, black bread mold; phylum Ascomycota, truffles; phylum Basidiomycota, mushrooms. **18.15** Photosynthesizers make their own organic food, while heterotrophs need an outside source of organic food.

Testing Yourself

1. a; 2. a; 3. e; 4. Tall plants have better access to sunlight. 5. b; 6. a; 7. a; 8. e; 9. c; 10. b; 11. a. stigma; b. style; c. carpel; d. ovary; e. ovule; f. receptacle; g. sepal; h. petal; i. stamen; j. filament; k. anther. 12. a; 13. c; 14. e; 15. Pollen is less likely to be moved from the pollen cone to the seed cone of the same tree. 16. e; 17. e; 18. e; 19. a; 20. a; 21. e; 22. Mycorrhiza fungi are more likely to go along with the seedling in the native soil.

Understanding the Terms

a. mycelium; b. sporophyte; c. monocotyledone; d. mycorrhizal fungi; e.gymnosperm

Thinking Scientifically

1. Only group (b) because mosses require a film of moisture in order for flagellated sperm to swim to an egg. 2. When male moths attempt to mate with these flowers, they carry pollen only between flowers of this type.

CHAPTER 19

Check Your Progress

19.1 Bats acquire nutrients from an outside source, ingest food and digest it internally, and have muscles and nerves. **19.2** Yes, because all animals can trace their ancestry to the same protistan ancestor. **19.3** Externally, we can observe that bats are bilaterally symmetrical and that they have jointed appendages. We can presume that they have the other homologies because they are chordates. **19.4** No, the chordates, which include bats, have not changed their position in the new tree. **19.5** Eukarya, Animalia, Chordata, Vertebrata, Mammals. **19.6** a. Sponges have the cellular level of organization, and bats have the organ-system level of organization; b. No, vampire bats simply have a liquid food, namely blood. **19.7** a. Microscopically, compare the development of cnidarians to that of bats; b. Cnidarians are predators that feed on live protists and small animals. Vampire bats are external parasites that feed on the blood of a host. **19.8** a. Yes; a longitudinal cut would divide the body in two equal halves; b. Bats are not hermaphrodites; the sexes are separate. **19.9** Vampire bats are external parasites; tapeworms and flukes are internal parasites. **19.10** Bats are expected to be more closely related to roundworms than to animals that have an incomplete digestive tract. **19.11** Deuterostomes, as are all vertebrates. **19.12** Both bats and molluscs are multicellular, have three germ layers, have bilateral symmetry, and are eucoelomates with internal organs. **19.13** The sequential vertebrae of the spinal column. **19.14** a. Bats have jointed appendages, are segmented, and have a well-developed nervous system; b. Bats have an endoskeleton, not an exoskeleton; they all breathe by lungs; and they have reduced competition by diversifying. Some are fruit eaters, some are insect eaters, and some, such as vampire bats, are parasites. **19.15** Spiders, scorpions, millipedes, and centipedes. **19.16** Analogous, because insects and bats do not have a recent common ancestor. **19.17** a. Echinoderms are radially symmetrical; b. Both types of animals are deuterostomes. **19.18** As embryos. **19.19** Two groups of chordates are invertebrates. **19.20** Vertebrae, jaws, lungs, limbs, amniotic egg. **19.21** Predaceous; actively ingesting chunks of food. **19.22** Yes; they have four limbs. **19.23** It provided a means for reptiles, birds, and mammals to reproduce on land in the absence of water. **19.24** Data support the hypothesis that birds are dinosaurs, making them reptiles. **19.25** They are the only mammal that can fly (as opposed to glide) and the only mammal to hang by their feet alone. **19.26** Vampire bat saliva contains a powerful anticoagulant that can be used to keep blood from clotting.

Testing Yourself

1. a; 2. b; 3. a; 4. e; 5. d; 6. d; 7. b; 8. a; 9. Flatworms are small, have a large absorptive surface, and are hermaphroditic. 10. e; 11. e; 12. c; 13. e; 14. a; 15. e; 16. b; 17. e; 18. Traditional fossils show how evolution occurred.

Understanding the Terms

a. nephridia; b. chitin; c. monotreme; d. notochord; e. coelomate

Thinking Scientifically

1. As per Section 19.1, study whole sponges and determine how they acquire food, whether they move, how they reproduce, and what developmental stages they have. Microscopically, determine the structure and function of their cells. 2. a. DNA/RNA sequencing; b. The data mentioned in Section 15.7: Fossil record and homologies in anatomy and development.

CHAPTER 20

Check Your Progress

20.1 Yes; her brain was small but still a bit larger than that of a chimpanzee, for example. **20.2** Mammalian ancestor to promisians; tarsier to first anthropoids; Old World monkeys to first apes; orangutans to gorillas to chimpanzees; and chimpanzees to hominids (Lucy). **20.3** Lucy's features, including bipedalism, means that she is a hominid, even though her brain is small. **20.4** Southern African specimens were bipedal, but the arms were longer than the legs. Lucy, an East African species, was bipedal, and the arms were shorter than the legs. **20.5** Bipedalism evolved because the human infant is helpless and needs to be carried. An extended infancy allows the brain to grow larger after birth. **20.6** This migration accounts for how the first Homo species arrived in Europe and Asia. **20.7** Toolmaking, in particular, allowed humans to hunt and prepare food and clothing. These cooperative activities may have required language. **20.8** Fossils that were intermediate between the two, or had features of both. **20.9** Replacement model and assimilation model. **20.10** It represents symbolic thinking. **20.11** The advent of agriculture allowed the human population to increase in size, and as the population continues to increase, agriculture is required to produce more and more food. **20.12** People from any two ethnic groups produce fertile offspring.

Testing Yourself

1.a; 2. c; 3. See page 403; 4. d; 5. b; 6. b; 7. b; 8. e; 9. b; 10. d; 11. e; 12. b; 13. Darwinian evolution is dependent on genetic differences, and biocultural evolution is dependent on advances that cannot be inherited. The ability to learn is inherited and learning is necessary to biocultural evolution. 14. d; 15. b; 16. a. modern humans; b. archaic humans; c. Homo erectus;

d. Homo erectus; e. Homo ergaster. 17. c; 18. d; 19. d; 20. d; 21. See page 414. It tells us that all humans are one species.

Understanding the Terms

a. anthropoid; b. Cro-Magnon; c. Homo ergaster; d. hominoid

Thinking Scientifically

1. The benefits of bipedalism must have outweighed the cost or else bipedalism wouldn't have eveled.
2. Sequence the Neandertal genome using DNA from Neandertal bones, and compare the sequence to the human genome of today. Look for sequences present in both genomes.

CHAPTER 21

Check Your Progress

21.1 When water is available, leaves can grow larger, maximizing solar absorption. When water is not available, narrow leaves help conserve water. **21.2** Eudicot. **21.3** The lack of insect pollination in monocots indicates that insects evolved at the time of eudicots, not monocots. **21.4** Collenchyma cells have areas of thicker primary cell walls, without having an entirely rigid secondary cell wall. **21.5** Ground tissue in a leaf is called mesophyll; in a stem, cortex and pith; and in the root, cortex. **21.6** Environmental conditions were favorable for primary growth that growing season. **21.7** Annual rings will probably not be visible because they occur when a tree experiences a change in rainfall during growing seasons that are interrupted by nongrowing seasons. **21.8** The presence of vessel elements, tracheids, and sclerenchyma cells make wood strong. **21.9** When leaves are horizontal, the lower epidermis is shaded and tends to lose less water when the stomata are open. **21.10** The leaf should be as large and thin as possible to increase the surface area for solar absorption. **21.11** Life cannot continue unless internal conditions remain within an optimum range.

Testing Yourself

1. c; 2. d; 3. c; 4. a; 5. Terminal bud is at the shoot tip and produces cells that add to the length of the stem and become new leaves and new axillary buds. Axillary bud activity produces new branches (and flowers). 6. The carpel, specifically the ovule, produces the seed, and the ovary produces the fruit. 7. b; 8. c; 9. c; 10. b; 11. a; 12. d; 13. e; 14. c; 15. d; 16. a; 17. d; 18. b; 19. b; 20. b; 21. Root = vascular cylinder, stem = vascular bundle, leaf = vein; 22. e; 23. c; 24. b; 25. b; 26. e; 27. c; 28. b; 29. Apical meristem (shoot tip and root tip meristem) is responsible for primary growth and the result is increase in the length of stem and root. Vascular cambium is responsible for secondary growth and the result is increase in girth. Girth increases because secondary xylem builds up as annual rings (wood). 30. a. Broader expanse to collect sunlight; b. Prevents loss of water; c. Collects sunlight; d. Allows gas exchange; e. Allows carbon dioxide to enter and water to exit. 31. Photosynthesis allows a plant to produce the building blocks and ATP it needs to maintain metabolism and its structure. 32. e; 33. d; 34. b; 35. a; 36. c; 37. a.

Understanding the Terms

a. mesophyll; b. cotyledon; c. xylem; d. endodermis; e. sieve-tube member

Thinking Scientifically

1. Confirm that plamodesmata do run between companion cells and sieve-tube members. Same as Palade, use labeled amino acids to show that proteins pass by way of plasmodesmata from companion cells to sieve-tube members. 2. Zone of cell division slides should show small cells dividing; zone of elongation slides should show cells that are longer

than the previous zone—no cell division is occurring; zone of maturation should show mature cells with root hairs.

CHAPTER 22

Check Your Progress

22.1 We will see that transport of sugar requires a functional plasma membrane. **22.2** The species that were good at dispersing had a way to compensate for being poor competitors. **22.3** Transpiration provides a means of transport for water and minerals and helps maintain internal leaf temperatures by providing evaporative cooling. **22.4** Cellular respiration, which requires oxygen, occurs in plant cells, even in the dark. **22.5** Phytomediation works when plants are capable of taking up and trapping the particular pollutant in their bodies. **22.6** Sucrose and hormones, for example, are expected. Surprisingly, a number of amino acids are also present. **22.7** The roots are the source, and the flowers are the sink. **22.8** a. Nitrogen and sulfur are needed to form protein, and all plant roots take up nitrate (NO_3^-) and sulfate (SO_4^{2-}) from the soil; b. Nitrogen and phosphate (HPO_4^{2-}) are needed to make nucleic acids, and plant roots also take up phosphate from the soil. **22.9** The nonpolar tails of phospholipid molecules make the center of the plasma membrane nonpolar. **22.10** Humus improves soil aeration, soil texture, increases water-holding capacity, decomposes to release nutrients for plant growth, and helps retain positively charged minerals and make them available for plant uptake. **22.11** Helps prevent soil erosion; helps retain moisture; and as the remains decompose, nutrients are returned to the soil. **22.12** Plants develop root nodules when the amount of nitrate in the soil is not sufficient to support growth.

Understanding the Terms

a. transpiration; b. Casparian strip; c. guard cell; d. guttation; e. pressure-flow model

Testing Yourself

1. a; 2. Leaves must receive water for photosynthesis, and they produce sugar as a result of photosynthesis. 3. c; 4. d; 5. c; 6. d; 7. e; 8. c; 9. Tracheids have a narrow water column that is less susceptible to breakage under the increased tension required to draw water from the roots when water is in short supply. 10. d; 11. b; 12. a; 13. c; 14. c; 15. d; 16. b; 17. b; 18. a; 19. b; 20. e; 21. c; 22. e; 23. c

Thinking Scientifically

1. Due to diversity, some species may not be susceptible to the assault. Also, the susceptible plants may be protected by the unsusceptible. For example, insects may not find all the susceptible plants, or most of the available water will go to those plants that need it. 2. Divide a large number of identical plants into control and experimental groups. Both groups are to receive the same treatment, including all necessary nutrients, but the experimental group will not be given any calcium. It is expected that only the experimental group will suffer any ill effects. If only the control or if both groups do poorly, some unknown variable is affecting the results.

CHAPTER 23

Check Your Progress

23.1 The second messenger can bring about various cellular activities in the same or different types of plant cells. **23.2** Yes, because if the hormone remained, it would continue to trigger a response long after the response was no longer needed. For example, it would bring about bending after the light source was no longer present. **23.3** The plant in (a) most likely doesn't produce gibberellin, and the receptor in (b) is more likely defective. **23.4** You could apply cytokinins to increase the number of cells and gibberellins to increase the size of the cells. **23.5** a. Abscisic acid maintains dormancy and closes stomata; b. Gibberellins have the opposite effect. **23.6** To be an effective hormone, a molecule needs only to combine with its receptor. **23.7** It is adaptive for roots to grow toward water because it enhances their ability to extract water and dissolve minerals from the soil for plant tissues. **23.8** Rotating horizontally will prevent the statoliths from settling and triggering differential growth. Therefore, neither the root nor the shoot is expected to curve up or down. **23.9** These animals are nocturnal, so it would be a waste of energy to produce scent during the day. **23.10** The plant is responding to a short night, not to the length of the day. **23.11** Red light converts P_r to P_{fr}; P_{fr} binds to a transcription factor; and the complex moves to the nucleus, where it binds to DNA so that genes are turned on or off. **23.12** No; adaptations come about by natural selection, which is not purposeful. **23.13** A chemical can interfere with metabolism, whether it is taken up by an insect or by a cancer cell, because all cells have certain common metabolic pathways.

Testing Yourself

1. c; 2. a; 3. d; 4. a; 5. c; 6. c; 7. d; 8. e; 9. d; 10. d; 11. b; 12. e; 13. a; 14. c; 15. Place the banana in a closed container with a ripened fruit. 16. d; 17. c; 18. d; 19. e; 20. c; 21. b; 22. e; 23. a; 24. e; 25. b; 26. c

Understanding the Terms

a. circadian rhythm; b. gravitropism; c. abscission; d. gibberellin; e. phototropism

Thinking Scientifically

1. Use a plant that tracks the sun as your experimental material. Make tissue slides to confirm the presence of a pulvinus, as in Figure 23.9. Apply ABA to live pulvinus tissue under the microscope to test for the results described in Figure 23.9. 2. Shine a light underneath a plant growing on its side (see Fig. 23.8A, upper left). If the stem now curves down, the phototropic response is greater than the gravitropic response and your hypothesis is not supported.

CHAPTER 24

Check Your Progress

24.1 Pollen grains (the male gametophytes) are visible when the anther releases them. To find the female gametophyte, you would have to microscopically examine the contents of an ovule just before fertilization takes place. **24.2** One sperm unites with the egg to form an embryo and the other combines with two polar nuclei forming a triploid (3n) endosperm cell. **24.3** The ovule is a sporophyte structure produced by the female parent. Therefore, the wall (becomes seed coat) is 2n. The embryo inside the ovule is the product of fertilization and is, therefore, 2n. **24.4** Showy, colorful flowers attract pollinators, ensuring seed formation from dispersal agents, which transport seeds away from the parent plant. **24.5** The sheaths protect the shoot and root apical meristems from damage as they push through the soil. **24.6** Advantages to asexual reproduction include: (1) the newly formed plant is often nutritionally supported by the parent plant until it is established; (2) if the parent is ideally suited for the environment of a given area, the offspring will be as well; and (3) if pollination is unlikely, asexual reproduction is a good alternative. **24.7** a. Either somatic embryogenesis or anther culture. If using anther culture, you would have to promote chromosome doubling to get 2n cells; b. Cell suspension culture would allow you to collect chemicals produced by a plant.

Testing Yourself

1. b; 2. b; 3. a; 4. e; 5. d; 6. c; 7. b; 8. a; 9. e; 10. c; 11. d; 12. b; 13. d; 14. a; 15. a. anther; b. filament; c. stamen; d. stigma; e. style; f. ovary; g. ovule; h. carpel. 16. A wind pollinated plant produces more pollen because the method of pollen transfer is less efficient. 17. b; 18. a; 19. e; 20. c; 21. b; 22. d; 23. e; 24. The need for a period of cold weather helps ensure that the seeds germinate when the weather will be favorable to continued growth. 25. c; 26. e; 27. a; 28. When the environment is not changing or when male and female plants are not close together.

Understanding the Terms

a. carpel; b. pollen grain; c. anther; d. gametophyte; e. sporophyte

Thinking Scientifically

1. Study (a) the anatomy of the wasp and flower, trying to determine if the mouth parts of the wasp are suitable for collecting nectar from this flower; (b) the appearance of the flower in sunlight/ultraviolet light to determine suitability to the vision of the wasp; and (c) the behavior of the wasp to see if it is compatible to that of the flower. 2. Protoplasts can be made from leaf cells and then cultured to grow entire plants. These plants are expected to produce seeds that you can use to propagate the plant outside.

CHAPTER 25

Check Your Progress

25.1 Fur traps heat and does not release it. **25.2** Evaporative cooling; water in sweat absorbs heat energy to become vapor. **25.3** Polar bear; adipose tissue helps keep the polar bear warm and is a source of nutrients when food is scarce. **25.4** Blood vessels taking blood to the surface constrict in order to reduce heat loss through the skin. **25.5** Dendrites are numerous and gather input from other neurons; the long axon is appropriate for conveying nervous impulses some distance. **25.6** Neither Reeve's brain nor the axons of neurons sending messages to his brain were affected by the accident. However, the cell bodies and axons of the neurons controlling his muscles were damaged by the accident. **25.7** The thick skin of a gila monster helps prevent the loss of water, rather than helping to regulate body temperature. However, a thick skin would help lessen the gain of heat in a warm environment. **25.8** Sensory receptors respond to a change in body temperature and communicate this information to the brain, which then commands motor output to perform the behavior that will warm or cool the body, as needed. **25.9** The ultimate aim of each is to provide parts for the human body. Most likely, stem cell research will at first aim to provide only tissues, but ultimately, by employing the same techniques as tissue engineering, stem cell research could lead to providing organs that will not be subject to rejection. This, of course, is also the aim of xenotransplantation. **25.10** Enzymatic reactions are necessary to the life of a cell, and enzymes function best at a moderate body temperature. **25.11** An early morning cool body temperature causes the lizard to move into the sun, where it stays until the body temperature is too hot, causing the lizard to seek shade. Thus, its behavior alternates back and forth.

Testing Yourself

1. c; 2. a; 3. a; 4. d; 5. e; 6. d; 7. b; 8. d; 9. The dendrites offer wide surface area for the reception of stimuli and the long dendrite is suitable for the transmission of stimuli. 10. a, c, g; 11. b, d, e; 12. b, c, f; 13. a; 14. b; 15. c; 16. Epithelial cells are exposed to mutagens (agents that cause mutations) in the environment. Also, high rate of cell division means that spontaneous mutations may occur that lead to cancer. 17. e; 18. c; 19. The epidermis is composed of stratified epithelium, which provides an impenetrable barrier to invasion by microorganisms. 20. a; 21. b; 22. e; 23. d; 24. e; 25. c; 26. a; 27. a; 28. e; 29. Organs from pigs are the right size and are available in great number; possible, however, that an animal virus could pass from pig to human. 30. You would expect the cardiovascular system to have a pump (the heart) to move blood through tubular vessels.

Understanding the Terms

a. ligament; b. epidermis; c. striated; d. homeostasis

Thinking Scientifically

1. Examine the tissue visually, trying to determine the particular organ before preparing the microscope slides in the same manner as known tissue slides from AIDS patients. Compare the appearance of the two sets of slides to properly match the unknown with the known tissues. 2. Test two groups: (1) People who visit tanning salons, say two or more times a week. (2) People who never visit tanning salons. Find out how many people in each group have been treated for skin cancer. Compare the percentages and determine if the difference is significant.

CHAPTER 26

Check Your Progress

26.1 A head with sense organs allows an animal to search for food and avoid predators. **26.2** The spinal cord is comparable to a planarian's lateral nerve cords. **26.3** Neurons and neuroglia. **26.4** Na^+ is pumped to the outside of an axon, and K^+ is pumped to the inside. **26.5** a. Na^+ moves to the inside of the axon; b. K^+ moves to the outside of an axon. **26.6** Saltatory conduction. **26.7** Neurotransmitters that cross the synaptic cleft. **26.8** Inhibition of AChE would cause ACh to remain in a synapse. **26.9** The neurons of the brain receive signals directly or indirectly via the spinal cord from all the rest of the nervous system. **26.10** A drug that interferes with neurotransmitter breakdown enhances its action. A drug that prevents a neurotransmitter from binding to its receptor interferes with its activity. **26.11** The spinal cord relays messages back and forth between the brain and the many nerves in the body. **26.12** A chimpanzee's brain would be very similar to ours but smaller. **26.13** When sleeping occurs, the sleep center is active and the RAS is inactive. **26.14** The amygdala adds emotional overtones to experiences. **26.15** Without the PNS, the CNS would neither receive stimuli nor be able to direct a response to the stimuli. **26.16** a. spinal cord; b. brain. **26.17** Sensing danger because of input from sensory receptors, the brain sends nerve impulses via the spinal cord to the nerve fibers of the sympathetic division, which brings about appropriate responses to the crisis situation.

Testing Yourself

1. d; 2. d; 3. d; 4. c; 5. b; 6. a; 7. b; 8. c; 9. b; 10. New characters often arise by modifications of previously evolved characters. 11. a; 12. c; 13. d; 14. e; 15. Learning requires the formation fo new associations and this is mirrored in the brain by the formation of new synapses. 16. a; 17. b; 18. d; 19. c

Understanding the Terms

a. reflex; b. neurotransmitter; c. autonomic system; d. ganglion; e. cerebrum

Thinking Scientifically

1. Administer a medication that interferes with the reception of norepinephrine at a synapse. The patient may not respond properly to a real danger. 2. Severed sensory neurons are still releasing neurotransmitters in the spinal cord, resulting in messages to the brain that are interpreted as pain in the limb.

CHAPTER 27

Check Your Progress

27.1 Eyes are sensitive to visible light rays and sometimes to ultraviolet rays, both of which are part of the electromagnetic spectrum. **27.2** Eyes receive stimuli and initiate nerve impulses; the optic nerve sends impulses to the brain; and the visual cortex of the brain interprets the stimuli, resulting in formation of an image. **27.3** a. Pheromones combine with a chemoreceptor that initiates a nerve impulse; b. Pheromones are signals sent by one member of a species to affect the behavior of another member. **27.4** No, photoreceptors contain pigments that absorb light. **27.5** The color we see is dependent upon which combination of cones is stimulated. **27.6** The compound eye has many independent visual units, but the retina of the camera-type eye is one large visual unit. **27.7** All parts of the eye are needed for our sense of vision. **27.8** A lens can accommodate—change shape as needed. **27.9** Malfunctioning ciliary muscles could cause accommodation problems in an octopus. **27.10** a. Ganglion cell layer, bipolar cell layer, rod and cone cell layer; b. Rd and cone cell layer, bipolar cell layer, ganglion cell layer (whose axons form the optic nerve). **27.11** They offer some protection to sensory receptors and prevent them from drying out. **27.12** Bending of hair cells is a mechanical event. **27.13** This knowledge can encourage people to preserve their hearing. **27.14** Otoliths rest on the otolithic membrane because of gravity. **27.15** All sensations are mental. Without the functioning of the brain, we have no sensations. **27.16** Whales are mammals, which evolved on land where a lateral line system cannot function because it utilizes pressure waves.

Testing Yourself

1. e; 2. c; 3. The brain requires input from sensory receptors in order to produce sensations about the world at large. 4. a; 5. d; 6. c; 7. c; 8. c; 9. b; 10. e; 11. d; 12. c; 13. Both procedures can correct an inability to focus properly; The first with an artificial lens and the second by changing the shape of the cornea. 14. a; 15. a; 16. c; 17. c; 18. c; 19. c; 20. b; 21. a; 22. b

Understanding the Terms

a. retina; b. sclera; c. chemoreceptor; d. organ of Corti; e. lateral line

Thinking Scientifically

1. One possible answer: The size of the auditory cortex is larger in those who have perfect pitch. Test the pitch ability of subjects, and then stimulate the brain directly to determine the size of the auditory cortex. 2. LASIK surgery only corrects the shape of the cornea in order to achieve 20/20 vision.

CHAPTER 28

Check Your Progress

28.1 Exoskeletons and endoskeletons are likely to be a part of the fossil record because they do not decompose, as do the soft hydrostatic skeletons of animals. **28.2** The bones provide a frame for the body after death. Even disconnected bones can help a forensics expert determine what the person looked like. **28.3** Age (condition of mandible, maxillae, teeth, and intervertebral disks), gender (mandible, frontal bone, and supraorbital ridges), and ethnicity (zygomatic bones). **28.4** Gender of the individual and the condition of the pubic symphysis, which gives an indication of age. **28.5** Compact bone contains a hard matrix and in order to lay down matrix, osteoblasts require calcium. **28.6** A long bone in the appendicular skeleton, such as the humerus or the femur, would be suitable for calculating height. **28.7** As we age, the joints deteriorate, so their condition can be used to roughly indicate age. **28.8** The femur and the tibia are the two bones of the knee joint. **28.9** During shivering, the skeletal muscles contract quite rapidly, generating heat, which can help maintain normal body temperature. **28.10** More motor units are recruited to lift three books. **28.11** The term aerobic implies that enough oxygen is entering the body to prevent oxygen debt from occurring. **28.12** A myofibril is a long, cylindrical structure in a muscle cell. A sarcomere is a section of a myofibril. **28.13** The events shown in Figure 28.13B #1–3. **28.14** A neuromuscular junction occurs between an axon terminal and a muscle cell; a synapse occurs between an axon terminal of one neuron and either the dendrite or cell body of the next neuron. **28.15** The CP pathway uses creatine phosphate. **28.16** To a degree—for example, a weight lifter would have larger bones with larger protuberances. If the musculature could be observed, it would be well developed.

Testing Yourself

1. a; 2. c; 3. e; 4. b; 5. c; 6. b; 7. c; 8. b; 9. f; 10. c; 11. e; 12. c; 13. d; 14. a; 15. c; 16. a; 17. e; 18. b; 19. Pelvic girdle is too small for a normal delivery. 20. b; 21. a; 22. c; 23. b; 24. d; 25. a; 26. e; 27. Unless they were attached, myosin couldn't cause the actin filaments to move shortening sarcomeres. 28. b; 29. a; 30. a

Understanding the Terms

a. osteoblast; b. actin; c. appendicular skeleton; d. sliding filament model; e. pectoral girdle

Thinking Scientifically

1. Remove muscle tissue from a corpse in rigor mortis, slice it thin. While watching under the microscope, flood your slide with ATP and necessary ions to see if muscle contraction occurs. 2. Acquire two test groups: aerobic instructors and confirmed couch potatoes. Oxygen tanks supply their only air, while they are running on a tread mill. Those who routinely exercise have more mitochondria than those who do not exercise, accounting for why the first group uses less oxygen and has less lactate (actic acid) in their blood.

CHAPTER 29

Check Your Progress

29.1 It contains red blood cells; they are red because they contain the respiratory pigment hemoglobin. **29.2** No; for example, in hydras, each cell makes exchanges with the environment. In roundworms, coelomic fluid transports substances. **29.3** Closed; because the blood flows freely into the tissues. **29.4** When the iron of hemoglobin combines with oxygen, at the gills or lungs, the pigment is red. After giving up oxygen in the tissues, hemoglobin has a bluish color when viewed through the skin. **29.5** The right side of the heart contains more O_2-poor hemoglobin and the left side of the heart contains more O_2-rich hemoglobin. **29.6** a. SA node activity is responsible for atrial systole; b. AV node activity is responsible for ventricular systole. **29.7** a. Blood pressure moves blood in arteries; b. Mechanical pressure exerted by skeletal muscle contraction helps move blood in veins. **29.8** No, try as you will in Figure 29.8. This ensures that all blood passes through the lungs. **29.9** No, it is highest close to the heart and falls off dramatically after moving through the capillaries. **29.10** A blood clot (thrombus) can block an arteriole in the brain, or hypertension can cause a cerebral blood vessel to burst. In both cases, a portion of the brain is robbed of oxygen and subsequently dies. **29.11** a. The diet should include monounsaturated fatty acids and omega-3 polyunsaturated fatty acids as well as fruits and vegetables that contain antioxidants; b. Foods high in cholesterol (e.g., eggs) and commercially prepared food high in saturated fats and trans fats should be avoided. **29.12** Blood is composed of blood cells (plus platelets) in a liquid matrix called plasma. **29.13** The steps prevent clotting from occurring unnecessarily. **29.14** Red bone marrow contains the stem cells that divide to produce all the types of cells in the blood (see Section 25.3). These types of stem cells are easier to retrieve than the stem cells of other adult tissues. **29.15** Plasma loses water in a capillary. Most of this water returns to plasma, but some becomes tissue fluid. Excess tissue fluid is absorbed by lymphatic vessels and becomes lymph. **29.16** A type B recipient has anti-A antibodies in the plasma, and they will react with the donor's red blood cells, causing agglutination.

Testing Yourself

1. d; 2. c; 3. b; 4. Efficient delivery of oxygen to the muscles allows birds and mammals to have an active lifestyle and generate heat to maintain a warm body temperature. 5. a; 6. d; 7. c; 8. c; 9. b; 10. c; 11. d; 12. b; 13. b; 14. a; 15. d; 16. b; 17. e; 18. d; 19. b; 20. Erthropoietin increases the number of red blood cells, and Rita's problem is lack of iron in her diet.

Understanding the Terms

a. atery; b. plasma; c. vena cava; d. hemoglobin

Thinking Scientifically

1. By dissecting animals, you will see three different types of blood vessels to the valves in the heart leading to arteries and in the veins leading to the heart. A deep cut to a vertebrate limb draws bright, red arterial blood under pressure; pressing on a vein causes it to expand on the far side. 2. The amount of amino acids, sugar, and oxygen is higher in arterial blood. and the amount of bicarbonate ion is higher in venous blood.

CHAPTER 30

Check Your Progress

30.1 They return excess tissue fluid to cardiovascular veins. Without the return of this fluid, the tissues would become water-logged, blood pressure would drop dramatically, and blood circulation would falter. **30.2** As long as the red bone marrow can produce more helper T cells than are being destroyed by an HIV infection, the person can fight off infections. **30.3** All three categories are helpful. The lining of the vagina is protective; interferons, macrophages, and natural killer cells should be helpful as well. **30.4** Hypothalamus. **30.5** Macrophages and dendritic cells, because they activate T cells. **30.6** Yes, because it is specific to the virus and foreign to humans. **30.7** HIV lives inside T cells and is rarely exposed to the antibodies in the blood. **30.8** HIV attacks and lives inside helper T cells. Destruction of helper T cells follows (see Section 16.5). Therefore, as the infection progresses, fewer T cells are available to perform their usual functions, and the immune system fails. **30.9** No; each antibody is effective only against one specific antigen. **30.10** HIV destroys helper T cells, and the number of cytokine-secreting helper T cells declines. Eventually, the immune system is ineffective, and the person, if untreated, dies of OIs. **30.11** Cancer is a genetic disorder, and monoclonal antibodies do not affect genes. **30.12** The compromised immune system would make rejection less of an issue, but the patient would be susceptible to all sorts of possible pathogen infections due to the surgery. **30.13** If the disease is untreated, the quality of life suffers. If the disease is treated with immunosuppressive drugs, patients may be more susceptible to pathogenic infections. **30.14** When the allergen combines with IgE, mast cells release histamine and other substances that bring on the allergic symptoms.

Testing Yourself

1. a; 2. b; 3. d; 4. b; 5. It collects excess tissue fluid at the blood capillaries and returns it to the subclavian veins of the cardiovascular system. 6. b; 7. e; 8. d; 9. a; 10. Fever creates an unfavorable environment for

an invader and may stimulate the immune system.
11. c; 12. d; 13. a; 14. a; 15. a; 16. b; 17. e; 18. e;
19. B and T cells are specifically selected to fight a
particular enemy. 20. c; 21. a

Understanding the Terms
a. vaccine; b. lymph; c. antigen; d. T cell

Thinking Scientifically
1. Hypothesis: each type of antibody is coded for by a
different sequence of exons from the same gene or
genes. 2. Control group receives vaccines and pathogens for same. Test groups receive drug plus vaccine
and pathogens for same. Observe whether the control
group of test groups become ill.

CHAPTER 31

Check Your Progress
31.1 In particular, digestion (both mechanical and
chemical) and absorption are more easily accomplished in a carnivore. **31.2** a. Elephants are herbivorous, bulk feeders; b. Bees are pollinators that feed on
the nectar of flowers. **31.3** The length of the small
intestine may vary but in all these animals, it carries
out chemical digestion and absorbs nutrients.
31.4 The diet of a carnivore consists of protein, and
therefore starch digestion is not required in the mouth
of a carnivore. **31.5** Since peristalsis is a rhythmic contraction and the esophagus is an internal organ, its
wall must contain smooth muscle. **31.6** A carnivore
could best use mechanical and pepsin digestion
because its food is meat and it uses its teeth for seiz-
ing and killing prey, not chewing it. **31.7** Finding the
bacterium in the stomach did not necessarily mean
that this bacterium causes ulcers. **31.8** Carbohydrates,
proteins, fats, and nucleic acids are digested to their
unit molecules and absorbed by the small intestine.
31.9 The liver breaks down the medicine for excretion,
so it is necessary to keep taking it in order to maintain
a certain level in the body. **31.10** The small intestine
"tells" them to do so by releasing secretin and CCK
31.11 Most likely, humans were largely herbivores
before they became omnivores. **31.12** Low-fiber,
refined carbohydrates lead to poor health; high-fiber,
whole-grain carbohydrates lead to good health.
31.13 Oils containing unsaturated fatty acids lead to
good health. Fats containing saturated fatty acids and/
or trans fatty acids, in particular, lead to poor health.
31.14 Vegetables supply nutrients and do not overtax
the body's metabolism the way protein does.
31.15 Salts increase the osmolarity of blood and cause
more water to be absorbed by the kidneys, leading to
hypertension. **31.16** Whole grains, fruits, and vegetables, in general, supply vitamins in the diet.
31.17 2,000 calories per day, an average amount.
31.18 Have a healthy diet and exercise. **31.19** People
with eating disorders might be striving to be thin, as
valued by our society.

Testing Yourself
1. d; 2. a; 3. d; 4. Life is sustained by a source of
energy, and the digestive system provides the nutrients that provide energy to animals. 5. d; 6. b; 7. b;
8. e; 9. a; 10. c; 11. a; 12. c; 13. a; 14. b; 15. b; 16. a;
17. e; 18. c; 19. b; 20. d; 21. c; 22. c; 23. The source of
amino acids is of no consequence because the DNA of
each cell specifies the types of proteins for that cell.

Understanding the Terms
a. vitamin; b. lacteal; c. esophagus; d. gallbladder

Thinking Scientifically
1. Pepsin, HCL, substrate (e.g., piece of cooked egg
white), water. Omit the pepsin. If digestion still
occurs, pepsin may not be the cause of digestion.
2. The use of a control group and a large number of
participants makes the correlation mre certain.

CHAPTER 32

Check Your Progress
32.1 Ventilation does not occur. **32.2** Air has a drying
effect, and respiratory surfaces have to be moist. The
body of a terrestrial animal provides this moisture.
32.3 Gills are meant to function in the water, and
therefore a diving response is needed. **32.4** Larger
insects have a means of ventilating the tracheae, but
external respiration (exchange of gases with incoming
air), and internal respiration (exchange of gases in the
tissues) comprise a single event in insects. **32.5** No,
the path of air always stays the same. **32.6** Chronic
bronchitis, emphysema, cancer, aneurysms, stroke,
miscarriage, and many more. **32.7** Penguins can store
air in the posterior air sacs. **32.8** When the spleen
contracts, more red blood cells (more hemoglobin)
enter the blood. This not only raises the amount of
available oxygen, but the hemoglobin also helps control pH (see Section 32.10). **32.9** The concentration of
N_2 remains the same because the body is not using it
up like O_2 and not producing it like CO_2. **32.10** Oxygenated hemoglobin will last longer because it is not
being transported to the tissues. When myoglobin has
released all its oxygen, the tissues then carry on fermentation, which does not require oxygen (see Section 7.9). **32.11** In both cases, pollutants, including
particles, overcome the cleansing capability of the
respiratory system, and the same types of lung disorders result, such as chronic bronchitis, emphysema,
and cancer.

Testing Yourself
1. c; 2. e; 3. a; 4. d; 5. b; 6. d; 7. b; 8. c; 9. d; 10. c; 11.
b; 12. f; 13. d; 14. a; 15. b; 16. e; 17. d; 18. b; 19. b;
20. b; 21. d; 22. a; 23. e; 24. The shape of hemoglobin
changes when the pH changes from near neutral in the
lungs to slightly acidic in the tissues, and this causes it
to unload its oxygen. 25. d; 26. b; 27. e; 28. The body
has a limited capacity to store oxygen and has a better
ability to store energy.

Understanding the Terms
a. ventilation; b. diaphragm; c. gills; d. expiration

Thinking Scientifically
1. Fat metabolism results in acids that enter the bloodstream; a lower pH stimulates the respiratory center
and causes increased breathing, which lowers the
CO_2, but raises the O_2 blood level. Test the blood of a
diabetic group and a normal group for these levels
and compare. 2. A severed spinal cord prevents the
medulla oblongata from communicating with the rib
cage and diaphragm via the phrenic nerve and intercostal nerves.

CHAPTER 33

Check Your Progress
33.1 a. Urea is not as toxic as ammonia, and it does
not require as much water to excrete; b. It takes less
energy to prepare urea than uric acid. **33.2** No, the
workings of the nephridia stay the same, regardless of
respiration. **33.3** Since a shark is isotonic to sea water,
it will not lose water to its environment. **33.4** Most
likely, the tonicity of a seagull's urine is about the
same as that of a human because they rid the body of
salt using a salt gland, not kidneys. **33.5** The kidneys.
33.6 Glomerular capsule, proximal convoluted tubule,
loop of nephron, distal convoluted tubule, collecting
duct, renal pelvis. **33.7** All small molecules enter the
filtrate, and the blood takes back what it needs.
33.8 Secretion. **33.9** It fine-tunes the reabsorption of
sodium ions. **33.10** Yes, death can follow because
enzymes need the correct pH in order to function
properly. **33.11** New dialysate continuously enters the
tubing, and therefore urea continuously moves from
the blood to the dialysate, with no backward flow into
the blood. **33.12** Lysis is possible with water intoxication, and crenation is possible with dehydration.

Testing Yourself
1. a; 2. d; 3. b; 4. c; 5. b; 6. c; 7. c; 8. c; 9. a; 10. d;
11. These cells reabsorb most of the contents of the
nephron and need increased surface area and energy
to better pump molecules back into the blood. 12. e;
13. b; 14. a; 15. b; 16. d; 17. c; 18. a; 19. a; 20. c;
21. d; 22. The presence of salt in the blood causes
more water to be passively reabsorbed, and the resulting increase in blood volume contributes to high
blood pressure.

Understanding the Terms
a. Malpighian tubule; b. glomerular capsule; c. aldosterone; d. renin; e. nephron

Thinking Scientifically
1. A microscopic study of their kidneys should reveal
that they have a reduced glomerulus and a very long
loop of the nephron. 2. Use the pump to increase the
pressure of the blood passing through the tubule,
increase the length of the tubule, and increase the
rapidity with which the dialysis fluid passes through
the apparatus.

CHAPTER 34

Check Your Progress
34.1 They all send a chemical message that is received
by a receptor. **34.2** Hormones act on body cells
directly; peptide hormones stimulate an enzyme cascade; and steroid hormones stimulate protein synthesis. **34.3** A pheromone can't affect an organism, and a
hormone can't affect a cell, unless a particular receptor is present. **34.4** When a baby is hungry, it might
release a pheromone that is received by the mother's
VNO. The VNO would stimulate the hypothalamus,
which would release oxytocin. **34.5** Yes; otherwise,
negative feedback by way of hormone 3 would not be
possible. **34.6** Epinephrine and norepinephrine by the
adrenal medulla. **34.7** The glucocorticoid is absorbed
into the skin and enters the blood, which transports it
about the body, resulting in Cushing syndrome.
34.8 Insulin is released when the level of glucose rises
in the blood. Once the level drops, the stimulus for
insulin secretion has been removed. **34.9** Diabetes
type 2 can be brought on by an unhealthy diet, and
many people today prefer fast foods to healthy foods.
34.10 Winter; melatonin secretion starts earlier in the
PM and lowers later in the AM because the nights are
longer. **34.11** Pheromone to VNO to hypothalamus to
anterior pituitary to thyroid gland and release of hormones. **34.12** Calcitonin causes the blood calcium to
fall, and PTH causes it to rise.

Testing Yourself
1. d; 2. b; 3. a; 4. d; 5. d; 6. f; 7. b; 8. c; 9. a; 10. e;
11. Caffeine would increase the stress effect of epinephrine because cAMP, which drives the effect, would be
slower to breakdown. 12. c; 13. c; 14. c; 15. d; 16. e;
17. d; 18. c; 19. d; 20. d; 21. b; 22. a; 23. d; 24. b;
25. b; 26. High blood calcium causes the thyroid to
secrete calcitonin, which leads to the uptake of calcium
by bone. Calcium level drops and thyroid is no longer
stimulated to release calcium.

Understanding the Terms
a. thyroid gland; b. pineal gland; c. peptide hormone;
d. cortisol

Thinking Scientifically
1. Calcitonin, being a peptide hormone, stimulates the
metabolism of osteoblasts to form bone utilizing calcium. 2. Use two groups of volunteers; one group consists of the "night owls" and the other group are
"early birds." Collect blood samples from all volunteers when they are typically sleepy and when they

typically wake up. If all goes well, the melatonin level rises earlier at night and lowers earlier in the day for the early birds.

CHAPTER 35

Check Your Progress

35.1 No, but successful terrestrial animals all have some means of passing sperm to the female without exposing the sperm to dry air. **35.2** Yes, many other animals have a way of retaining fertilized eggs until they develop into offspring prepared to live on their own. **35.3** Sperm must be kept moist to survive. **35.4** They travel through the epididymis, the vas deferens, the ejaculatory duct, and, finally, the urethra. **35.5** The vagina receives the penis, which delivers the sperm; therefore, the sperm do not dry out. The egg is fertilized in the oviduct, where body fluids provide moisture. The embryo develops in the uterus, where it cannot dry out. **35.6** Follicles and the corpus luteum. **35.7** Because the ovaries produce one egg a month. If that egg is fertilized, the uterus is ready to receive it. **35.8** AIDS caused by human immunodeficiency virus (HIV), genital herpes caused by herpes simplex virus (HSV), and genital warts caused by human papillomavirus (HPV). **35.9** Percentage of pregnancies that will be prevented by those using this method of birth control. **35.10** Perhaps the surrogate mother because the child was born to her. **35.11** Lancelets develop quickly into a free-living aquatic larva; birds develop inside a shelled egg and need a food source (the yolk); human embryos develop in the uterus and get nutrients from the mother. **35.12** The establishment of the germ layers during gastrulation. **35.13** The notochord is formed from mesoderm and is a supportive structure that lies underneath the neural tube, which is formed from ectoderm. In vertebrates, the notochord is replaced by the vertebral column, and the neural tube becomes the nervous system. **35.14** Without gastrulation, organ development never occurs. **35.15** Induction causes cells to move and form a particular structure. **35.16** To make internal development possible; the chorion is a part of the placenta, for example. **35.17** Sponges have the cellular level of organization; cnidarias (e.g., hydras) have the tissue level of organization; flatworms (e.g., planarian) have the organ level of organization; and all animals thereafter have the systems level of organization. **35.18** The systems are just about fully formed during embryonic development, even though the embryo is tiny. **35.19** Before birth, the placenta supplied the needs of the body.

Testing Yourself

1. e; 2. c; 3. Sexual reproduction produces variations among the offspring and one of these may be phenotypes that can deal with the new conditions. 4. b; 5. a; 6. c; 7. b; 8. d; 9. a; 10. A high blood testosterone level shuts down the anterior pituitary production of gonadotropic hormones, reducing sperm production. 11. c; 12. b; 13. b; 14. c; 15. d; 16. zygote, cleavage, morula, blastula, gastrula, neurula. 17. d; 18. c; 19. d

Understanding the Terms

a. ovulation; b. gonad; c. induction; d. germ layer

Thinking Scientifically

1. Testosterone causes increased cell division in the prostate. Microscopically, observe the effect of testosterone on cells taken from the prostate. 2. Progesterone maintains the uterine lining, and if the placenta doesn't begin producing it when it should, a woman could lose the embryo embedded in the lining.

CHAPTER 36

Check Your Progress

36.1 The population, habitat, community, and ecosystem levels are all affected. **36.2** Density. **36.3** It increases the growth rates of the deer populations because more deer survive to reproduce, and the deer that do reproduce breed more during their lifetimes. **36.4** Before hunting, the average life span of individuals in the population was long, resulting in a type I survivorship curve. After hunting was allowed, many individuals in the population would die before reproducing, resulting in a type III survivorship curve. **36.5** The survival of prereproductive individuals would cause the age structure diagram to become more pyramid-shaped. **36.6** Exponential growth would occur until competition for resources slowed the population growth to meet the carrying capacity. **36.7** Severe natural disasters (such as hurricanes) and weather-related food shortages are two examples. **36.8** Competition for food and habitat resources are two examples. **36.9** Equilibrium species, because of their large body size, relatively long life span, and the few offspring produced per deer each year. **36.10** The size of the geographic range, the degree of habitat tolerance, and the size of local populations. **36.11** It would lengthen the doubling time, because the deer would add fewer offspring to the population. **36.12** If there are many young, prereproductive deer already in the population, they will mature and produce more offspring than needed for zero population growth.

Testing Yourself

1. d; 2. d; 3. b; 4. d; 5. b; 6. a; 7. b; 8. a; 9. c; 10. The offspring of a plant that reproduces by runners would remain near the parent. The offspring of a plant that reproduces by windblown seeds could be taken far away from the parent. 11. c; 12. a; 13. b; 14. c; 15. c; 16. d; 17. When the environment is unstable due to density-independent and density-dependent factors, a few of the many small, uncared for offspring might have a better chance of survival and dispersal to a favorable environment. 18. a; 19. a; 20. a; 21. d; 22. e

Understanding the Terms

a. demographic transition; b. population; c. exponential growth; d. biotic potential

Thinking Scientifically

1. The right whale has only one offspring per reproduction and chances of death before maturity is good. The right whale begins reproducing well after maturity, and reproduces infrequently. To test your hypothesis, you would have to observe a captive population or tag individuals in the wild and observe them from a distance. 2. Determine the original normal flow of the river and maintain the flow as close to normal as possible.

CHAPTER 37

Check Your Progress

37.1 Because all members of the colony have the same mother, they would all inherit the gene, and the colony would die out as tunnel excavating ceased. **37.2** Observe a mole rat in its colony from the time of birth and see if efficiency in building tunnels improves over time. **37.3** Operant conditioning because the younger workers would be rewarded for tunnel work. **37.4** This trait evolved because females choose males with large, brightly colored tailfeathers, perhaps on the basis that such birds are healthier. **37.5** No, because all mole rats are close relatives. **37.6** Raise birds that are unable to perform this behavior and then compare the fitness of birds that sing to mark their territory with those that do not. **37.7** Fear or anger; happiness. Yes, because the answer is based on what a human being might feel.

Testing Yourself

1. c; 2. b; 3. e; 4. c; 5. c; 6. c; 7. a; 8. d; 9. c; 10. c; 11. c; 12. c; 13. Because females produce few gametes and have few offspring during their lifetime, they place an emphasis on quality of offspring. Because males produce many gametes all the time, they place an emphasis on quantity of offspring. 14. d; 15. e; 16. a; 17. They would be increasing their inclusive fitness when they help their parents at a time they cannot be reproducing themselves. Reciprocal altruism would ensure that they will successfully reproduce when the time comes. 18. a; 19. c; 20. b; 21. d; 22. a; 23. b

Understanding the Terms

a. territoriality; b. altruism; c. communication; d. pheromone

Thinking Scientifically

1. Mate the two types of rats and test how the offspring react to limburger cheese. If they show an intermediate response, such as being willing to approach the cheese but still not eating it, then the behavior may be genetically controlled. 2. Observe sentry behavior more cloesly. Recent observations have shown that sentries are the first ones to reach safety when a predator is spotted, and meerkats only serve as sentries after they have eaten.

CHAPTER 38

Check Your Progress

38.1 Decomposers release chemicals that make dead material smell and taste unpleasant to many other animals. Some animals, such as scavengers, are not repulsed, however. **38.2** When scavengers eat decomposing meat, they are also eating the decomposers that are devouring the food source before them. **38.3** Parasites use nutrients meant for the host organism. If the host dies, the parasite is also threatened with death. **38.4** Study the feeding behavior of a remora when not attached to a shark. If it can survive independently, it is more likely commensalistic with the shark. **38.5** Study the distribution of the lichen versus the algae alone. If the fungus is mutualistic, the lichen is more plentiful or has a wider distribution than the algae alone. If parasitic, the algae is more plentiful or has a wider distribution when independent of the lichen. **38.6** The protozoans are provided with food and habitat. **38.7** Each species of warblers is largely found in a particular portion of the spruce trees. **38.8** Succession would essentially stop. Most chemical nutrients would be tied up in dead organisms making new growth very limited. **38.9** Only autotrophs can use an outside energy source to produce organic food required by all the biotic components of an ecosystem. **38.10** The inorganic chemicals are taken up by plants and the recycling of chemicals in an ecosystem begins again. **38.11** Bacteria and fungi of decay. **38.12** None; ecological pyramids do not include decomposers. **38.13** The reservoir and the exchange pool are abiotic. **38.14** Decomposers gradually break down the bones making phosphate ions and other chemicals available. **38.15** Nitrification. **38.16** It would decrease greenhouse emissions and reduce the threat of global warming.

Testing Yourself

1. c; 2. e; 3. e; 4. a; 5. c; 6. b; 7. d; 8. Will the area be able to meet the niche requirements of this predator? (see Fig. 38.9); 9. b; 10. a; 11. b; 12. e; 13. c; 14. d; 15. c; 16. c; 17. b; 18. c; 19. d; 20. d; 21. No; Photosynthetic species are always at the first trophic level and nonphotosynthetic species are always at a higher level.

Understanding the Terms

a. community; b. habitat; c. ecological niche; d. mutualism

Thinking Scientifically

1. Observe the birds carefully to see if they differ in ways suggested by Figure 38.9. 2. Fill a large container with water from the pond. Add phosphate slowly over several days or months, and when you see growth,

calculate the amount of phosphate you need for the pond.

CHAPTER 39

Check Your Progress

39.1 Rain forests receive much rainfall and solar energy; therefore, they produce enough food for many organisms. The plants in a tropical rain forest provide many different types of niches for many different organisms. These two factors help account for why tropical rain forests exhibit species richness. **39.2** It would be difficult to simulate the winter of a tundra and to stock it with animals as large as caribou. **39.3** Coniferous trees are soft wood. **39.4** No, you need the rich soil, the understory plants, and the many animals. **39.5** No, prairies receive less rainfall than do forests. **39.6** Grasses and scattered trees. **39.7** Cold. **39.8** Epiphytes. **39.9** Climate. **39.10** Look at the distribution of landmasses in this chapters various maps. They are shifted toward the northern latitudes. **39.11** Wetlands. **39.12** Euphotic. **39.13** Cold, nutrient-rich water. **39.14** No, the winds are weaker than usual.

Testing Yourself

1. d; 2. d; 3. a; 4. a; 5. a; 6. d; 7. b; 8. e; 9. c; 10. b; 11. a; 12. Hot air from the equator moves toward the poles. 13. e; 14. d; 15. b; 16. Decomposers use up any available oxygen, leaving none for the fish. 17. b; 18. c; 19. c; 20. e; 21. d; 22. e; 23. Sea level would be too deep for revious estuaries to exist, and new ones may not develop further inland due to development of coastal regions.

Understanding the Terms

a. estruary; b. taiga; c. savanna; d. eutrophication; e. epiphyte

Thinking Scientifically

1. You will be able to see if a rising temperature affects the distribution of biomes in the biosphere. 2. Coral houses microscopic algae and if the dirty water blots out the sun, the algae will die.

CHAPTER 40

Check Your Progress

40.1 To preserve natural resources/biodiversity. **40.2** An endangered species is at immediate risk of extinction, whereas a threatened species is likely to become endangered soon. **40.3** Reduced aquatic biodiversity would limit the amount and types of seafood we consume. **40.4** Native animals and plants in their natural environment. **40.5** A large number of people living along the coast. **40.6** They reduce biodiversity by causing other species to become extinct. **40.7** Any change in the environment that negatively impacts organisms. **40.8** Overexploitation. **40.9** Disease. **40.10** Metapopulation. **40.11** To return the area to its former state (i.e., a thriving ecosystem).

Testing Yourself

1. e; 2. a; 3. b; 4. b; 5. The added shrimp pushes the salmon to a higher place in the pyramid, indicating that less energy is now available to them. 6. d; 7. e; 8. e; 9. e; 10. e; 11. a; 12. e; 13. d; 14. e; 15. b; 16. Food webs show many connections between different populations. 17. b; 18. c; 19. e

Understanding the Terms

a. biodiversity hotspot; b. metapopulation; c. source population; d. sink population

Thinking Scientifically

1. Overharvesting reduces genetic diversity due to the bottleneck effect. As with the cheetah, determine how many loci are now homozygous. 2. Besides having produced a population with limited genetic diversity, none of the problems that brought the species to near extinction (see Sections 40.5–40.9) have been solved.

Appendix B | Metric System

Unit and Abbreviation	Metric Equivalent	Approximate English-to-Metric Equivalents	Units of Temperature
Length			
nanometer (nm)	$= 10^{-9}$ m (10^{-3} µm)		
micrometer (µm)	$= 10^{-6}$ m (10^{-3} mm)		
millimeter (mm)	$= 0.001$ (10^{-3}) m		
centimeter (cm)	$= 0.01$ (10^{-2}) m	1 inch = 2.54 cm 1 foot = 30.5 cm	
meter (m)	$= 100$ (10^{2}) cm $= 1,000$ mm	1 foot = 0.30 m 1 yard = 0.91 m	
kilometer (km)	$= 1,000$ (10^{3}) m	1 mi = 1.6 km	
Weight (mass)			
nanogram (ng)	$= 10^{-9}$ g		
microgram (µg)	$= 10^{-6}$ g		
milligram (mg)	$= 10^{-3}$ g		
gram (g)	$= 1,000$ mg	1 ounce = 28.3 g 1 pound = 454 g	
kilogram (kg)	$= 1,000$ (10^{3}) g	= 0.45 kg	
metric ton (t)	$= 1,000$ kg	1 ton = 0.91 t	
Volume			
microliter (µl)	$= 10^{-6}$ l (10^{-3} ml)		
milliliter (ml)	$= 10^{-3}$ liter $= 1$ cm^3 (cc) $= 1,000$ mm^3	1 tsp = 5 ml 1 fl oz = 30 ml	
liter (l)	$= 1,000$ ml	1 pint = 0.47 liter 1 quart = 0.95 liter 1 gallon = 3.79 liter	
kiloliter (kl)	$= 1,000$ liter		

°C	°F	
100	212	Water boils at standard temperature and pressure.
71	160	Flash pasteurization of milk.
57	134	Highest recorded temperature in the United States, Death Valley, July 10, 1913.
41	105.8	Average body temperature of a marathon runner in hot weather.
37	98.6	Human body temperature.
13.7	56.66	Human survival is still possible at this temperature.
0	32.0	Water freezes at standard temperature and pressure.

To convert temperature scales:

$$°C = \frac{(°F - 32)}{1.8}$$

$$°F = 1.8\,(°C) + 32$$

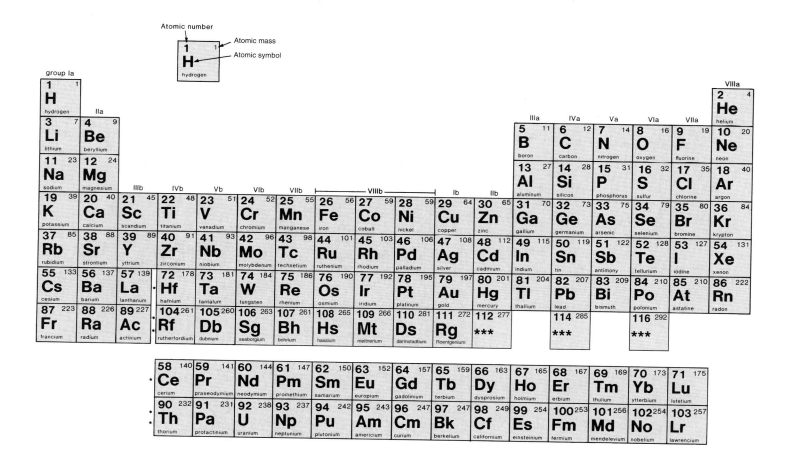

Glossary

A

abscisic acid (ABA) Plant hormone that causes stomata to close and that initiates and maintains dormancy. 464

abscission Dropping of leaves, fruits, or flowers from a plant. 431, 464

absorption Taking in of substances by cells or membranes. 612

accessory fruit Fruit, or an assemblage of fruits, whose fleshy parts are derived from tissues other than the ovary (e.g., strawberry). 485

acetylcholine (ACh) (uh-seet-ul-koh-leen) Neurotransmitter active in both the peripheral and central nervous systems. 520

acetylcholinesterase (AChE) (uh-seet-ul-koh-luh-nes-tuh-rays) Enzyme that breaks down acetylcholine bound to postsynaptic receptors within a synapse. 521

acid Molecules tending to raise the hydrogen ion concentration in a solution and to lower its pH numerically. 30

acid deposition The return to Earth in rain or snow of sulfate or nitrate salts of acids produced by commercial and industrial activities. 757

acoelomate Animal that has no body cavity (i.e., tapeworm). 371

actin (ak-tin) One of two major proteins of muscle; makes up thin filaments in myofibrils of muscle fibers. See myosin. 566

actin filament Cytoskeletal filaments of eukaryotic cells composed of the protein actin; also refers to the thin filaments of muscle cells. 70

action potential Electrochemical changes that take place across the axomembrane; the nerve impulse. 518

active site Region on the surface of an enzyme where the substrate binds and where the reaction occurs. 82

active transport Use of a plasma membrane carrier protein and energy to move a substance into or out of a cell from lower to higher concentration. 90

adaptation Organism's modification in structure, function, or behavior suitable to the environment. 7

adaptive radiation Evolution of several species from a common ancestor into new ecological or geographical zones. 270

adenine (A) (ad-uh-neen) One of four nitrogen bases in nucleotides composing the structure of DNA and RNA. 186

adenosine Portion of ATP and ADP that is composed of the base adenine and the sugar ribose. 50

adenosine diphosphate (ADP) (ah-den-ah-seen dy-fahs-fayt) Nucleotide with two phosphate groups that can accept another phosphate group and become ATP. 50

adenosine triphosphate (ATP) (ah-den-ah-seen try-fahs-fayt) Nucleotide with three phosphate groups. The breakdown of ATP into ADP + Ⓟ makes energy available for energy-requiring processes in cells. 50

adhesion Attachment of cells, as in water adhering to vessel walls of plants. 27

adipose tissue Connective tissue in which fat is stored. 499

adrenal cortex (uh-dree-nul kor-teks) Outer portion of the adrenal gland; secretes mineralocorticoids such as aldosterone and glucocorticoids such as cortisol. 672

adrenal gland (uh-dree-nul) An endocrine gland that lies atop a kidney, consisting of the inner adrenal medulla and the outer adrenal cortex. 672

adrenal medulla (uh-dree-nul muh-dul-uh) Inner portion of the adrenal gland; secretes the hormones epinephrine and norepinephrine. 672

adrenocorticotropic hormone (ACTH) (uh-dree-noh-kawrt-ih-koh-troh-pik) Hormone secreted by the anterior lobe of the pituitary gland that stimulates activity in the adrenal cortex. 671

adult stem cell Cells in a mature body that have the ability to divide; found in red bone marrow. 136

aerobic Phase of cellular respiration that requires oxygen. 115, 314

afterbirth Placenta and the extraembryonic membranes, which are delivered (expelled) during the third stage of parturition. 702

age structure diagram In demographics, a display of the age groups of a population; a growing population has a pyramid-shaped diagram. 713

aggregate fruit Fruit developed from several separate carpels of a single flower. 485

AIDS Disease caused by a retrovirus and transmitted via body fluids; characterized by failure of the immune system. 592

aldosterone (al-dahs-tuh-rohn) Hormone secreted by the adrenal cortex that decreases sodium and increases potassium excretion; raises blood volume and pressure. 658, 673

algae Type of protist that carries on photosynthesis; unicellar forms are a part of phytoplankton, and multicellular forms are called seaweed. 324

alien species Nonnative species that migrate or are introduced by humans into a new ecosystem; also called exotics. 786

alkaloid Bitter-tasting nitrogenous compounds that are basic (e.g., caffeine). 472

allantois (uh-lan-toh-is) Extraembryonic membrane that contributes to the formation of umbilical blood vessels in humans. 697

allele (uh-leel) Alternative form of a gene; alleles occur at the same locus on homologous chromosomes. 163

allergy Immune response to substances that usually are not recognized as foreign. 606

allopatric speciation Origin of new species between populations that are separated geographically. 268

alloploidy Polyploid organism that contains the genomes of two or more different species. 271

alternation of generations Life cycle, typical of plants, in which a diploid sporophyte alternates with a haploid gametophyte. 336, 344, 481

altruism Social interaction that has the potenetial to decrease the lifetime reproductive success of the member exhibiting the behavior. 732

alveolus (pl., alveoli) Air sac of a lung. 638

Alzheimer disease (AD) Brain disorder characterized by a general loss of mental abilities. 527

amino acid Organic molecule having an amino group and an acid group, which covalently bonds to produce peptide molecules. 45

ammonia Colorless gas, which has a penetrating odor and is soluble in water. 650

amniocentesis Procedure in which a sample of amniotic fluid is removed through the abdominal wall of a pregnant woman. Fetal cells in it are cultured before doing a karyotype of the chromosomes. 170

amnion (am-nee-ahn) Extraembryonic membrane that forms an enclosing, fluid-filled sac. 697

amniotic egg Egg that has an amnion, as seen during the development of reptiles, birds, and mammals. 387

amoeboid Cell that moves and engulfs debris with pseudopods. 328

amphibian Member of a class of vertebrates that includes frogs, toads, and salamanders;

they are still tied to a watery environment for reproduction. 389

amygdala (uh-mig-duh-luh) Portion of the limbic system that functions to add emotional overtones to memories. 527

anabolism Metabolic process by which larger molecules are synthesized from smaller ones; anabolic metabolism. 125

anaerobic Growing or metabolizing in the absence of oxygen. 115

analogous structure Structure that has a similar function in separate lineages but differs in anatomy and ancestry. 251, 292

analogy Similarity of function but not of origin. 292

anaphylactic shock Severe systemic form of allergic reaction involving bronchiolar contriction, impaired breathing, vasodilation, and a rapid drop in blood pressure with a threat of circulatory failure. 606

ancestral character Structural, physiological, or behavioral trait that is present in a common ancestor and all members of a group. 291

anemia (uh-nee-mee-uh) Inefficiency in the oxygen-carrying ability of blood due to a shortage of hemoglobin. 584

aneuploid Individual whose chromosome number is not an exact multiple of the haploid number for the species. 152

angiogenesis (an-jee-oh-jen-uh-sis) Formation of new blood vessels; one mechanism by which cancer spreads. 143, 220

angiosperm Flowering plant that produces seeds within an ovary that develops into a fruit; therefore, the seeds are covered. 353, 480

angiotensin II Hormone produced from angio-tensinogen (a plasma protein) by the kidneys and lungs; raises blood pressure. 658

animal Multicellular, heterotrophic organism belonging to the animal kingdom. 9

annelid Member of a phylum of invertebrates that contains segmented worms, such as the earthworm and the clam worm. 381

annual ring Layer of wood (secondary xylem) usually produced during one growing season. 432

anorexia nervosa (a-nuh-rek-see-uh nur-voh-suh) Eating disorder characterized by a morbid fear of gaining weight. 628

anterior pituitary (pih-too-ih-tair-ee) Portion of the pituitary gland that is controlled by the hypothalamus and produces six types of hormones, some of which control other endocrine glands. 670

anther In flowering plants, pollen-bearing portion of stamen. 480

anthrax Infectious disease caused by a spore-forming bacterium (*Bacillus anthracis*); transmissible from animals to man; characterized by lesions on the lungs. 318

anthropoid Group of primates that includes monkeys, apes, and humans. 403

antibody (an-tih-bahd-ee) Protein produced in response to the presence of an antigen; each antibody combines with a specific antigen. 585, 594

antibody-mediated immunity Specific mechanism of defense in which plasma cells

derived from B cells produce antibodies that combine with antigens. 600

anticodon (an-tih-koh-dahn) Three-base sequence in a transfer RNA molecule base that pairs with a complementary codon in mRNA. 196

antidiuretic hormone (ADH) (an-tih-dy-uh-ret-ik) Hormone secreted by the posterior pituitary that increases the permeability of the collecting ducts in a kidney. 658, 671

antigen (an-tih-jun) Foreign substance, usually a protein or a polysaccharide, that stimulates the immune system to produce antibodies. 585, 599

antigen-presenting cell (APC) Cell that displays the antigen to the cells of the immune system so they can defend the body against that particular antigen. 600

antigen receptor Receptor proteins in the plasma membrane of immune system cells whose shape allows them to combine with a specific antigen. 600

antioxidant Substances, such as vitamins C, E, and A, which defend the body against free radicals. 625

aorta (ay-or-tuh) Major systemic artery that receives blood from the left ventricle. 577

aortic body Sensory receptor in the aortic arch sensitive to the O_2, CO_2, and H^+ content of the blood. 641

apical dominance Influence of a terminal bud in suppressing the growth of lateral buds. 460

apical meristem In vascular plants, masses of cells in the root and shoot that reproduce and elongate as primary growth occurs. 426

apoptosis (ap-uh-toh-sis, -ahp-) Programmed cell death involving a cascade of specific cellular events leading to death and destruction of the cell. 141

appendicular skeleton (ap-un-dik-yuh-lur) Portion of the skeleton forming the pectoral girdles and upper extremities and the pelvic girdle and lower extremities. 556

aquaporin Protein membrane channel through which water can diffuse. 89

arachnid Group of arthropods that contains spiders and scorpions. 383

arboreal Living in trees. 400

Archaea One of the three domains of life; contains prokaryotic cells that often live in extreme habitats and have unique genetic, biochemical, and physiological characteristics; its members are sometimes referred to as archaea. 8, 59, 316

archaic human Regionally diverse descendants of *H. erectus* that lived in Africa, Asia, and Europe; considered by some to be a separate species. 410

archegonium Egg-producing structures, as in the moss life cycle. 345

Arctic tundra Biome that encircles the Earth just south of ice-covered polar seas in the Northern Hemisphere. 765

arteriole (ar-teer-ee-ohl) Vessel that takes blood from an artery to capillaries. 579

artery Vessel that takes blood away from the heart to arterioles; characteristically possessing thick, elastic, muscular walls. 577

arthritis Inflammation of one or more joints. 562

arthropod Member of a phylum of invertebrates that contains, among other groups, crustaceans and insects that have an exoskeleton and jointed appendages. 382

articular cartilage (ar-tik-yuh-lur) Hyaline cartilaginous covering over the articulating surface of the bones of synovial joints. 560

artificial selection Change in the genetic structure of populations due to selective breeding by humans. 246

asexual reproduction Reproduction that requires only one parent and does not involve gametes. 134

associative learning Acquired ability to associate two stimuli or between a stimulus and a response. 729

assortative mating Individuals tend to mate with those that have the same phenotype with respect to certain characteristics. 254

aster Short, radiating fibers about the centrioles at the poles of a spindle. 138

asthma (az-muh, as-) Condition in which bronchioles constrict and cause difficulty in breathing. 606, 644

atom Smallest particle of an element that displays the properties of the element. 4, 20

atomic mass Mass of an atom equal to the number of protons plus the number of neutrons within the nucleus. 21

atomic number Number of protrons within the nucleus of an atom. 21

atomic symbol One or two letters that represent the name of an element—e.g., H stands for a hydrogen atom. 20

ATP synthase complex Complex formed of enzymes and their carrier proteins; functions in the production of ATP in chloroplasts and mitochondria. 101

atrial natriuretic hormone (ANH) (ay-tree-ul nay-tree-yoo-ret-ik) Hormone secreted by the heart that increases sodium excretion and, therefore, lowers blood volume and pressure. 658

atrioventricular valve Valve located between the atrium and the ventricle. 577

atrium (ay-tree-um) One of the upper chambers of the heart, either the left atrium or the right atrium, that receives blood. 577

australopithecine One of several species of *Australopithecus*, a genus that contains the first generally recognized hominids. 406

Australopithecus africanus Hominid that lived between 3.6 and 3 MYA, e.g., Lucy, discovered at Hadar, Ethiopia, in 1974. 406

autoimmune disorder Disorder that results when the immune system mistakenly attacks the body's own tissues. 605

autonomic system (aw-tuh-nahm-ik) Branch of the peripheral nervous system that has control over the internal organs; consists of the sympathetic and parasympathetic systems. 530

autoploidy Polyploid organism that contains a duplicated genome of the same species. 271

autosomal chromosome Any chromosome of a type that is the same in male and females of a species. 167

autosome (aw-tuh-sohm) Any chromosome other than the sex chromosomes. 145

autotroph Organism that can capture energy and synthesize organic nutrients from inorganic nutrients. 96, 752

auxin Plant hormone regulating growth, particularly cell elongation; most often indoleacetic acid (IAA). 460

axial skeleton (ak-see-ul) Portion of the skeleton that supports and protects the organs of the head, the neck, and the trunk. 556

axillary bud Bud located in the axil of a leaf. 422

axon (ak-sahn) Elongated portion of a neuron that conducts nerve impulses typically from the cell body to the synapse. 517

B

Bacteria One of the three domains of life; contains prokaryotic cells that differ from archaea because they have their own unique genetic, biochemical, and physiological characteristics. 8, 59

bacteriophage Virus that infects bacteria. 303

ball-and-socket joint The most freely movable type of joint (e.g., the shoulder or hip joint). 561

bark External part of a tree, containing cork, cork cambium, and phloem. 432

Barr body Dark-staining body (discovered by M. Barr) in the nuclei of female mammals that contains a condensed, inactive X chromosome. 212

basal body Cytoplasmic structure that is located at the base of and may organize cilia or flagella. 71

basal nuclei (bay-sul) Subcortical nuclei deep within the white matter that serve as relay stations for motor impulses and produce dopamine to help control skeletal muscle activities. 525

base Molecules tending to lower the hydrogen ion concentration in a solution and raise the pH numerically. 30

basement membrane Layer of nonliving material that anchors epithelial tissue to underlying connective tissue. 497

B cell Lymphocyte that matures in the bone marrow and, when stimulated by the presence of a specific antigen, gives rise to antibody-producing plasma cells. 585, 595

B cell receptor (BCR) Molecule on the surface of a B cell that binds to a specific antigen. 600

beneficial nutrient In plants, element that is either required or enhances the growth and production of a plant. 450

benefit Positive outcome. 730

behavior Observable, coordinated responses to environmental stimuli. 726

bicarbonate ion Ion that participates in buffering the blood; the form in which carbon dioxide is transported in the bloodstream. 643

bilateral symmetry Body plan having two corresponding or complementary halves. 371

bile Secretion of the liver that is temporarily stored and concentrated in the gallbladder before being released into the small intestine, where it emulsifies fat. 619

binary fission Bacterial reproduction into two daughter cells without the utilization of a mitotic spindle. 135, 312

binge-eating disorder Condition characterized by overeating episodes that are not followed by purging. 628

binomial nomenclature Two-part scientific name of an organism. The first part designates the genus, the second part the specific epithet. 9

biocultural evolution Phase of human evolution in which cultural events affect natural selection. 409

biodiversity hotspot Regions of the world that contains unusually large concentrations of species. 781

biogeochemical cycle (by-oh-jee-oh-kem-ih-kul) Circulating pathway of elements such as carbon and nitrogen involving exchange pools, storage areas, and biotic communities. 756

biogeography Study of the geographical distribution of organisms. 252

bioinformatics Computer technologies used to study the genome. 236, 780

biological clock Internal mechanism that maintains a biological rhythm in the absence of environmental stimuli. 469

biological species concept The concept that defines species as groups of populations that have the potential to interbreed and that are reproductively isolated from other groups. 264

biology Scientific study of life. 11

biome One of the biosphere's major terrestrial communities, characterized in particular by certain climatic conditions and particular types of plants. 764

biosphere (by-oh-sfeer) Zone of air, land, and water at the surface of the Earth, in which living organisms are found. 10, 710

biotechnology product Product created by using biotechnology techniques. 228

biotic potential Maximum reproductive rate of an organism, given unlimited resources and ideal environmental conditions. Compare with environmental resistance. 712

bipedalism Walking erect on two feet. 405

bird Endothermic vertebrate that has feathers and wings, is often adapted for flight, and lays hard-shelled eggs. 391

bivalve Type of mollusc with a shell composed of two valves; includes clams, oysters, and scallops. 380

blade Broad expanded portion of a plant leaf that may be single or compound. 423

blastocoel Fluid-filled cavity of a blastula. 692

blastocyst (blas-tuh-sist) Early stage of human embryonic development that consists of a hollow, fluid-filled ball of cells. 698

blastopore Opening into the primitive gut formed at gastrulation. 693

blastula Hollow, fluid-filled ball of cells occurring during animal development prior to gastrula formation. 692

blind spot Region of the retina lacking rods or cones where the optic nerve leaves the eye. 540

blood Type of connective tissue in which cells are separated by a liquid called plasma. 575

blood pressure Force of blood pushing against the inside wall of an artery. 581

bone Connective tissue having protein fibers and a hard matrix of inorganic salts, notably calcium salts. 499

bottleneck effect Cause of genetic drift; occurs when a majority of genotypes are prevented from participating in the production of the next generation as a result of a natural disaster or human interference. 255

botulism Acute food poisoning caused by the ingestion of food containing the spore-forming bacterium *Clostridium botulinum*. Characterized by muscle weakness and paralysis, blurred vision, and a high mortality rate. 318

brain stem Portion of the brain consisting of the medulla oblongata, pons, and midbrain. 526

bronchi (sing., bronchus) One of two major divisions of the trachea leading to the lungs. 638

bronchiole (brahng-kee-ohl) Smaller air passages in the lungs that begin at the bronchi and terminate in alveoli. 638

brown alga Marine photosynthetic protists with a notable abundance of xanthophyll pigments, this group includes well-known seaweeds of northern rocky shores. 333

bryophyte Member of one of three phyla of nonvascular plants—the mosses, liverworts, and hornworts. 346

buffer Substance or group of substances that tend to resist pH changes of a solution, thus stabilizing its relative acidity and basicity. 31

bulbourethral gland (bul-boh-yoo-ree-thrul) Either of two small structures located below the prostate gland in males; each adds secretions to semen. 684

bulimia nervosa (byoo-lee-mee-uh, -lim-ee-, nur-voh-suh) Eating disorder characterized by binge eating followed by purging via self-induced vomiting or use of a laxative. 628

bulk feeder Animal that eats relatively large pieces of food. 613

bulk transport Movement of elements in an organism in large amounts. 90

bundle sheath Sheath formed from tightly packed cells that is located around the veins of a leaf. 434

bursa (bur-suh) Saclike, fluid-filled structure, lined with synovial membrane, that occurs near a joint. 561

C

C₃ plant Plant that fixes carbon dioxide via the Calvin cycle; the first stable product of C_3 photosynthesis is a 3-carbon compound. 106

C₄ plant Plant that fixes carbon dioxide to produce a C_4 molecule that releases carbon dioxide to the Calvin cycle. 106

calcitonin (kal-sih-toh-nin) Hormone secreted by the thyroid gland that increases the blood calcium level. 676

Calorie (kcal) Amount of heat energy required to raise the temperature of 1 g of water 1°C. 28, 78, 626

Calvin cycle reaction Portion of photosynthesis that takes place in the stroma of chloroplasts and can occur in the dark; it uses the products of the light reactions to reduce CO_2 to a carbohydrate. 99

camouflage Method of hiding from predators in which the organism's behavior, form, and pattern of coloration allow it to blend into the background and prevent detection. 745

CAM photosynthesis Crassulacean-acid metabolism; plant fixes carbon dioxide at night to produce a C_4 molecule that releases carbon dioxide to the Calvin cycle during the day. 107

capillary (kap-uh-lair-ee) Microscopic vessel connecting arterioles to venules; exchange of substances between blood and tissue fluid occurs across their thin walls. 579

capsid Protein coat or shell that surrounds a virion's nucleic acid. 184, 302

capsule Gelatinous layer surrounding the cells of blue-green algae and certain bacteria. 59

carbaminohemoglobin Hemoglobin carrying carbon dioxide. 643

carbohydrate Class of organic compounds that includes monosaccharides, disaccharides, and polysaccharides. 41, 622

carbonic anhydrase (kar-bahn-ik an-hy-drays, -drayz) Enzyme in red blood cells that speeds the formation of carbonic acid from the reactants water and carbon dioxide. 643

carcinogen (kar-sin-uh-jun) Environmental agent that causes mutations leading to the development of cancer. 201

carcinogenesis (kar-suh-nuh-jen-uh-sis) Development of cancer. 143, 220

cardiac conduction system System of specialized cardiac muscle fibers that conduct impulses from the SA node to the chambers of the heart, causing them to contract. 578

cardiac cycle One complete cycle of systole and diastole for all heart chambers. 578

cardiac muscle Striated, involuntary muscle found only in the heart. 500

cardiac pacemaker Mass of specialized cardiac muscle tissue that controls the rhythm of the heartbeat; the SA node. 578

cardiovascular system Organ system in which blood vessels distribute blood under the pumping action of the heart. 505, 577

carnivore (kar-nuh-vor) Consumer in a food chain that eats other animals. 752

carotenoid Yellow or orange pigment that serves as an accessory to chlorophyll in photosynthesis. 100

carotid body (kuh-raht-id) Structure located at the branching of the carotid arteries; contains chemoreceptors sensitive to the O_2, CO_2, and H^+ content in blood. 641

carpel Ovule-bearing unit that is a part of a pistil. 353, 480

carrier Heterozygous individual who has no apparent abnormality but can pass on an allele for a recessively inherited genetic disorder. 167

carrying capacity Maximum number of individuals of any species that can be supported by a particular ecosystem on a long-term basis. 714

cartilage Connective tissue in which the cells lie within lacunae separated by a flexible proteinaceous matrix. 499

Casparian strip Layer of impermeable lignin and suberin bordering four sides of root endodermal cells; prevents water and solute transport between adjacent cells. 452

catabolism Metabolic process that breaks down large molecules into smaller ones; catabolic metabolism. 125

cecum (see-kum) Small pouch that lies below the entrance of the small intestine and is the blind end of the large intestine. 614

cell Smallest unit that displays the properties of life; always contains cytoplasm surrounded by a plasma membrane. 4

cell body Portion of a neuron that contains a nucleus and from which dendrites and an axon extend. 517

cell cycle Repeating sequence of cellular events that consists of interphase, mitosis, and cytokinesis. 136

cell division Process by which cells multiply; meiosis. 134

cell-mediated immunity Specific mechanism of defense in which T cells destroy antigen-bearing cells. 600

cell plate Structure across a dividing plant cell that signals the location of new plasma membranes and cell walls. 140

cell suspension culture Plant tissue extraction of chemicals performed in a laboratory without the need of disturbing plants in their natural environments. 488

cell theory One of the major theories of biology; states that all organisms are made up of cells and cells come only from preexisting cells. 56

cellular respiration Metabolic reactions that use the energy primarily from carbohydrates but also from fatty acid or amino acid breakdown to produce ATP molecules. 113

cellular slime mold Free-living amoeboid cells that feed on bacteria and yeasts by phagocytosis and aggregate to form a plamodium that produces spores. 331

cellulose (sel-yuh-lohs, -lohz) Polysaccharide that is the major complex carbohydrate in plant cell walls. 42

cell wall Structure that surrounds a plant, protistan, fungal, or bacterial cell and maintains the cell's shape and rigidity. 59

central nervous system (CNS) Portion of the nervous system consisting of the brain and spinal cord. 515

central vacuole In a plant cell, a large, fluid-filled sac that stores metabolites. During growth, it enlarges, forcing the primary cell wall to expand and the cell surface-area-to-volume ratio to increase. 66

centriole (sen-tree-ohl) Cellular structure, existing in pairs, that possibly organizes the mitotic spindle for chromosomal movement during mitosis and meiosis. 71

centromere (sen-truh-meer) Constriction where sister chromatids of a chromosome are held together. 137

centrosome Central microtubule organizing center of cells. In animal cells, it contains two centrioles. 70

cephalization Having a well-recognized anterior head with a brain and sensory receptors. 371, 514

cephalopod Type of mollusc in which a modified foot develops into the head region; includes squid, cuttlefish, octopus, and nautilus. 380

cerebellum (ser-uh-bel-um) Part of the brain located posterior to the medulla oblongata and pons that coordinates skeletal muscles to produce smooth, graceful motions. 526

cerebral cortex (suh-ree-brul, ser-uh-brul korteks) Outer layer of cerebral hemispheres; receives sensory information and controls motor activities. 525

cerebral hemisphere One of the large, paired structures that together constitute the cerebrum of the brain. 525

cerebrospinal fluid Fluid found in the ventricles of the brain, in the central canal of the spinal cord, and in association with the meninges. 501, 524

cerebrum (sair-uh-brum, suh-ree-brum) Main part of the brain consisting of two large masses, or cerebral hemispheres; the largest part of the brain in mammals. 525

cervix (sur-viks) Narrow end of the uterus, which projects into the vagina. 686

cesarean section Birth by surgical incision of the abdomen and uterus. 701

character Any structural, chromosomal, or molecular feature that distinguishes one group from another. 290

character displacement Tendency for characteristics to be more divergent when similar species belong to the same community than when they are isolated from one another. 742

chemiosmosis Ability of certain membranes to use a hydrogen ion gradient to drive ATP formation. 103, 121

chemoreceptor (kee-moh-rih-sep-tur) Sensory receptor that is sensitive to chemical stimuli—for example, receptors for taste and smell. 536

chemosynthetic Organism able to synthesize organic molecules by using carbon dioxide as the carbon source and the oxidation of an inorganic substance (such as hydrogen sulfide) as the energy souurce. 314

chitin Strong but flexible nitrogenous polysaccharide found in the exoskeleton of arthropods. 42, 382

chlamydial infection (kluh-mid-ee-uh) Sexually transmitted disease, caused by the bacterium *Chlamydia trachomatis*; can lead to pelvic inflammatory disease. 689

chlorophyll, chlorophyll *a*, chlorophyll *b* Green pigment that absorbs solar energy and is important in algal and plant photosynthesis; occurs as chlorophyll *a* and chlorophyll *b*. 97, 100

chloroplast Membrane-bounded organelle in algae and plants with chlorophyll-containing membranous thylakoids; where photosynthesis takes place. 68, 97

cholecystokinin Polypeptide used as a hormone and neurotransmitter, secreted by some brain neurons and cells of the digestive tract. 621

cholesterol One of the major lipids found in animal plasma membranes; makes the membrane impermeable to many molecules. 85, 623

chordate Member of the phylum Chordata, which includes lancelets, tunicates, fishes, amphibians, reptiles, birds, and mammals; characterized by a notochord, dorsal tubular nerve cord, pharyngeal gill pouches, and post-anal tail at some point in the life cycle. 386

chorion (kor-ee-ahn) Extraembryonic membrane that contributes to placenta formation. 697

chorionic villi (kor-ee-ahn-ik vil-eye) Treelike extensions of the chorion that project into the maternal tissues at the placenta. 699, 700

chorionic villi sampling (CVS) Removal of cells from the chorionic villi portion of the placenta. Karyotyping is done to determine if the fetus has a chromosomal abnormality. 170

choroid (kor-oyd) Vascular, pigmented middle layer of the eyeball. 540

chromatin (kroh-muh-tin) Network of fine threads in the nucleus that are composed of DNA and proteins. 62, 137, 212

chromosome (kroh-muh-som) Chromatin condensed into a compact structure. 62

chronic bronchitis Obstructive pulmonary disorder that tends to recur; marked by inflamed airways filled with mucus and degenerative changes in the bronchi, including loss of cilia. 644

chyme Thick, semiliquid food material that passes from the stomach to the small intestine. 618

ciliary body (sil-ee-air-ee) Structure associated with the choroid layer that contains ciliary muscle and controls the shape of the lens of the eye. 540

ciliary muscle Within the ciliary body of the vertebrate eye, the ciliary muscle controls the shape of the lens. 541

ciliate Complex unicellular protist that moves by means of cilia and digests food in food vacuoles. 329

cilium (pl., cilia) (sil-ee-um) Short, hair-like projection from the plasma membrane, occurring usually in large numbers. 71

circadian rhythm Regular physiological or behavioral event that occurs on an approximately 24-hour cycle. 468, 675

citric acid cycle Cycle of reactions in mitochondria that begins with citric acid; it breaks down an acetyl group as CO_2, ATP, NADH, and $FADH_2$ are given off; also called the Krebs cycle. 115, 119

clade Taxon or other group consisting of an ancestral species and all of its descendants, forming a distinct branch on a phylogenetic tree. 294

cladogram In cladistics, a branching diagram that shows the relationship among species in regard to their shared derived characters. 294

class One of the categories, or taxa, used by taxonomists to group species; class is the taxon above the order level. 8, 290

classical conditioning Type of learning whereby an unconditioned stilumus that elicits a specific response is paired with a neutral stimulus so that the response becomes conditioned. 729

classification Way of organizing and retrieving information about species. 290

cleavage Cell division without cytoplasmic addition or enlargement; occurs during the first stage of animal development. 692

climate Weather condition of an area, including especially prevailing temperature and average/yearly rainfall. 764

climax community In ecology, community that results when succession has come to an end. 751

clitoris Small, erectile, female organ located in the vulva; homologous to the penis. 686

cloaca Posterior portion of the digestive tract in certain vertebrates that receives feces and urogenital products. 614

clonal selection model Concept that an antigen selects which lymphocyte will undergo clonal expansion and produce more lymphocytes bearing the same type of antigen receptor. 601

cloning Production of identical copies; can be either the production of identical individuals or, in genetic engineering, the production of identical copies of a gene. 226

closed circulatory system Blood is confined to vessels and is kept separate from the interstitial fluid. 575

club moss Type of seedless vascular plant that is also called ground pine because it has the appearance of a miniature pine tree. 348

cnidarian Invertebrate in the phylum Cnidaria existing as either a polyp or medusa with two tissue layers and radial symmetry. 375

coal Fossil fuel formed millions of years ago from plant material that did not decay. 352

cochlea (kohk-lee-uh, koh-klee-uh) Portion of the inner ear that resembles a snail's shell and contains the spiral organ, the sense organ for hearing. 544

codominance Inheritance pattern in which both alleles of a gene are equally expressed. 171

codon Three-base sequence in messenger RNA that causes the insertion of a particular amino acid into a protein or termination of translation. 193, 196

coelom (see-lum) Embryonic body cavity lying between the digestive tract and body wall that becomes the thoracic and abdominal cavities. 371

coelomate Animal with a eucoelom. 371

coenzyme (koh-en-zym) Nonprotein organic molecule that aids the action of the enzyme to which it is loosely bound. 83

coevolve (coeveolution) Interaction of two species such that each influences the evolution of the other species. 243, 483

cofactor Nonprotein adjunct required by an enzyme in order to function; many cofactors are metal ions, others are coenzymes. 83

cohesion Clinging together of water cells. 27

cohesion-tension model Explanation for upward transportation of water in xylem, based upon transpiration-created tension and the cohesive properties of water molecules. 444

cohort Group of individuals having a statistical factor in common, such as year of birth, in a population study. 712

coleoptile Protective sheath that covers the young leaves of a seedling. 460

collecting duct Duct within the kidney that receives fluid from several nephrons; the reabsorption of water occurs here. 654

collenchyma cell Plant tissue composed of cells with unevenly thickened walls; supports growth of stems and petioles. 426

colon (koh-lun) The major portion of the large intestine, consisting of the ascending colon, the transverse colon, and the descending colon. 621

colonial flagellate hypothesis The proposal first put forth by Haeckel that protozoans descended from colonial protists; supported by the similarity of sponges to flagellated protists. 369

colony Loose association of cells that remain independent for most funtions. 335

columnar epithelium Type of epithelial tissue with cylindrical cells. 497

commensalism Symbiotic relationship in which one species is benefited, and the other is neither harmed nor benefited. 746

common ancestor Ancestor held in common by at least two lines of descent. 250, 291

communication Signal by a sender that influences the behavior of a receiver. 734

community Assemblage of populations interacting with one another within the same environment. 10, 710, 742

compact bone Type of bone that contains osteons consisting of concentric layers of matrix and osteocytes in lacunae. 560

companion cell Cell associated with sieve-tube members in phloem of vascular plants. 442

competition Interaction between two organisms in which both require the same limited resource, which results in harm to both. 742

competitive exclusion principle Theory that no two species can occupy the same niche. 742

competitive inhibition Form of enzyme inhibition where the substrate and inhibitor are both able to bind to the enzyme's active site; each complexes with the enzyme. Only when the substrate is at the active site will product form. 84

complement Collective name for a series of enzymes and activators in the blood, some of which may bind to antibody and may lead to rupture of a foreign cell. 596

complementary base pairing Hydrogen bonding between particular bases. In DNA, thymine (T) pairs with adenine (A), and guanine (G) pairs with cytosine (C); in RNA, uracil (U) pairs with A, and G pairs with C. 48, 187

compound Substance having two or more different elements united chemically in fixed ratio. 23

compound light microscope Consists of a two-lens system, one above the other, to magnify an object. 58

conclusion Statement made following an experiment as to whether the results support the hypothesis. 11

conidium Structure that asexually produces fungal spores. 361

conifer Member of a group of cone-bearing gymnosperm plants that includes pine, cedar, and spruce trees. 350

conjugation Transfer of genetic materials from one cell to another. 313, 334

conjunctiva Delicate membrane that lines the eyelid protecting the sclera. 540

connective tissue Type of tissue characterized by cells separated by a matrix that often contains fibers. 498

conservation biology Scientific discipline that seeks to understand the effects of human activities on species, communities, and ecosystems and to develop practical approaches to preventing the extinction of species and the destruction of ecosystems. 780

consumer Organism that feeds on another organism in a food chain; primary consumers eat plants, and secondary consumers eat animals. 752

contact inhibition In cell culture, the point where cells stop dividing when they become a one-cell-thick sheet. 142

control group Sample that goes through all the steps of an experiment but lacks the factor or is not exposed to the factor being tested; a standard against which results of an experiment are checked. 12

control system Controls cell division in the nucleus. 142

convergent evolution Similarity in structure in distantly related groups due to adaptation to the environment. 292

copulation Sexual union between a male and a female. 683

coral reef Area of biological abundance found in shallow, warm, tropical waters on and around coral formations. 773

cork cambium Lateral meristem that produces cork. 432

corm Underground, upright plant stem, in which food is stored, usually in the form of starch. 487

cornea (kor-nee-uh) Transparent, anterior portion of the outer layer of the eyeball. 540

corpus luteum (kor-pus loot-ee-um) Yellow body that forms in the ovary from a follicle that has discharged its secondary oocyte; it secretes progesterone and some estrogen. 687

cortex In plants, ground tissue bounded by the epidermis and vascular tissue in stems and roots; in animals, outer layer of an organ such as the cortex of the kidney or adrenal gland. 428

cortisol (kor-tuh-sawl) Glucocorticoid secreted by the adrenal cortex that responds to stress on a long-term basis; reduces inflammation and promotes protein and fat metabolism. 672

cost Negative outcome. 730

cost-benefit analysis A weighing-out of the costs and benefits (in terms of contributions to reproductive success) of a particular strategy or behavior. 730

cotyledon Seed leaf for embryo of a flowering plant; provides nutrient molecules for the developing plant before photosynthesis begins. 424, 483

countercurrent exchange Fluids flow side-by-side in opposite directions, as in the exchange of fluids in the kidneys. 636

coupled reaction Reaction that occurs simultaneously; one is an exergonic reaction that releases energy, and the other is an endergonic reaction that requires an input of energy in order to occur. 81

covalent bond (coh-vay-lent) Chemical bond in which atoms share one pair of electrons. 25

cranial nerve Nerve that arises from the brain. 528

cristae Short, fingerlike projections formed by the folding of the inner membrane of mitochondria. 68, 118

Cro-Magnon Common name for the first fossils to be designated *Homo sapiens*. 410

crop Part of the digestive tract in birds and some invertebrates that stores or digests food. 614

crossing-over Exchange of segments between nonsister chromatids of a tetrad during meiosis. 146

crustacean Member of a group of marine arthropods that contains, among others, shrimps, crabs, crayfish, and lobsters. 383

cuboidal epithelium Type of epithelial tissue with cube-shaped cells. 497

culture Total pattern of human behavior; includes technology and the arts and is dependent upon the capacity to speak and transmit knowledge. 409

Cushing syndrome (koosh-ing) Condition resulting from hypersecretion of glucocorticoids; characterized by thin arms and legs and a "moon face," and accompanied by high blood glucose and sodium levels. 673

cuticle Waxy layer covering the epidermis of plants that protects the plant against water loss and disease-causing organisms. 345, 426, 445

cyanobacteria Photosynthetic bacterium that contains chlorophyll and releases oxygen; formerly called a blue-green algae. 315

cyanogenic glycoside Plant compound that contains sugar; produces cyanide. 472

cycad Type of gymnosperm with palmate leaves and massive cones; cycads are most often found in the tropics and subtropics. 350

cyclic adenosine monophosphate (cAMP) (sy-klik, sih-klik) ATP-related compound that acts as the second messenger in peptide hormone transduction; it initiates activity of the metabolic machinery. 667

cyclin Protein that regularly increases and decreases in concentration during the cell cycle. 142

cytochrome Any of several iron-containing protein molecules that serve as electron carriers in photosynthesis and cellular respiration. 120

cytokine (sy-tuh-kyn) Type of protein secreted by a T lymphocyte that stimulates cells of the immune system to perform their various functions. 597

cytokinesis (sy-tuh-kyn-ee-sus) Division of the cytoplasm following mitosis and meiosis. 136

cytokinin Plant hormone that promotes cell division; often works in combination with auxin during organ development in plant embryos. 463

cytoplasm (sy-tuh-plaz-um) Contents of a cell between the nucleus and the plasma membrane that contains the organelles. 59

cytosine (C) (sy-tuh-seen) One of four nitrogen bases in nucleotides composing the structure of DNA and RNA. 186

cytoskeleton Internal framework of the cell, consisting of microtubules, actin filaments, and intermediate filaments. 61

cytotoxic T cell (sy-tuh-tahk-sik) T lymphocyte that attacks and kills antigen-bearing cells. 600

D

data Facts or pieces of information collected through observation and/or experimentation. 11

daughter cell Cell that arises from a parental cell by mitosis or meiosis. 137

day-neutral plant Plant whose flowering is not dependent on day length, e.g., tomato and cucumber. 470

deamination Removal of an amino group (—NH_2) from an amino acid or other organic compound. 125

deciduous Plant that sheds leaves annually. 423

decomposer Organism, usually a bacterium or fungus, that breaks down organic matter into inorganic nutrients that can be recycled in the environment. 752

dehiscent Opening of an anther, fruit, or other structure, which permits the release of reproductive bodies inside. 485

dehydration Removal of water. 660

dehydration reaction Chemical reaction resulting in a covalent bond with the accompanying loss of a water molecule. 40

delayed allergic response Allergic response initiated at the site of the allergen by sensitized T cells, involving macrophages and regulated by cytokines. 606

deletion Change in chromosome structure in which the end of a chromosome breaks off or two simultaneous breaks lead to the loss of an internal segment; often causes abnormalities (e.g., cri du chat syndrome). 154

demographic transition Due to industrialization, a decline in the birthrate following a reduction in the death rate so that the population growth rate is lowered. 719

denatured (denaturation) Loss of an enzyme's normal shape so that it no longer functions; caused by a less than optimal pH or temperature. 47, 83

dendrite (den-dryt) Branched ending of a neuron that conducts signals toward the cell body. 517

dendritic cell Antigen-presenting cell of the epidermis and mucous membranes. 597

denitrification Conversion of nitrate or nitrite to nitrogen gas by bacteria in soil. 757

dense fibrous connective tissue Type of connective tissue containing many collagen fibers packed together, and found in tendons and ligaments, for example. 499

density-dependent factor Biotic factor, such as disease or competition, that affects population size according to the population's density. 715

density-independent factor Abiotic factor, as fire or flood, that affects population size independent of the population's density. 715

dental caries (kar-eez) Tooth decay that occurs when bacteria within the mouth metabolize sugar and give off acids that erode teeth; a cavity. 617

deoxyribose Pentose sugar found in DNA that has one less hydroxyl group than ribose. 41

depolarization Loss in polarization, as when a nerve impulse occurs. 519

derived character Structural, physiological, or behavioral trait that is present in a specific lineage and is not present in the common ancestor for several lineages. 291

dermis Region of skin that lies beneath the epidermis. 503

desert Ecological biome characterized by a limited amount of rainfall; deserts have hot days and cool nights. 767

detrital food web (dih-tryt-ul) Complex pattern of interlocking and crisscrossing food chains that begins with detritus. 755

detritus Patrially decomposed organic matter derived from tissues and animal wastes. 752

deuterostome Group of coelomate animals in which the second embryonic opening is associated with the mouth; the first embryonic opening, the blastopore, is associated with the anus. 379

diaphragm (dy-uh-fram) Dome-shaped horizontal sheet of muscle and connective tissue that divides the thoracic cavity from the abdominal cavity. 640

diastole (dy-as-tuh-lee) Relaxation period of a heart chamber during the cardiac cycle. 578

diastolic pressure (dy-uh-stahl-ik) Arterial blood pressure during the diastolic phase of the cardiac cycle. 581

diatom Golden-brown alga with a cell wall in two parts, or valves; significant part of phytoplankton. 332

diencephalon (dy-en-sef-uh-lahn) Portion of the brain in the region of the third ventricle that includes the thalamus and hypothalamus. 526

differentially permeable Ability of plasma membranes to regulate the passage of substances into and out of the cell, allowing some to pass through and preventing the passage of others. 86

digestive system Organ system that includes the mouth, esophagus, stomach, small intestine, and large intestine (colon), which receives food and digests it into nutrient molecules. Also has associated organs: teeth, tongue, salivary glands, liver, gallbladder, and pancreas. 505

dihybrid cross Single genetic cross involving two different traits, such as flower color and plant height. 164

dinoflagellate Photosynthetic unicellular protist with two flagella, one whiplash and the other located within a groove between protective cellulose plates; a significant part of phytoplankton. 332

diploid life cycle Presence of two of each type of chromosome at interphase of the cell cycle. 336

diploid (2n) number Cell condition in which two of each type of chromosome are present in the nucleus. 137

directional selection Outcome of natural selection in which an extreme phenotype is favored, usually in a changing environment. 257

disaccharide (dy-sak-uh-ryd) Sugar that contains two units of a monosaccharide (e.g., maltose). 41

disruptive selection Outcome of natural selection in which the two extreme phenotypes are favored over the average phenotype, leading to more than one distinct form. 257

distal convoluted tubule Final portion of a nephron that joins with a collecting duct; associated with tubular secretion. 654

DNA (deoxyribonucleic acid) Nucleic acid polymer produced from covalent bonding of nucleotide monomers that contain the sugar deoxyribose; the genetic material of nearly all organisms. 48, 186

DNA fingerprinting The use of DNA fragment lengths resulting from restriction enzyme cleavage to identify particular individuals. 227

DNA ligase (ly-gays) Enzyme that links DNA fragments; used during production of recombinant DNA to join foreign DNA to vector DNA. 226

DNA polymerase During replication, an enzyme that joins the nucleotides complementary to a DNA template. 190

DNA replication Synthesis of a new DNA double helix prior to mitosis and meiosis in eukaryotic cells and during prokaryotic fission in prokaryotic cells. 190

domain The primary taxonomic group above the kingdom level; all living organisms may be placed in one of three domains. 8, 290

dominance hierarchy Organization of animals in a group that determines the order in which the animals have access to resources. 731

dominant allele (uh-leel) Allele that exerts its phenotypic effect in the heterozygote; it masks the expression of the recessive allele. 163

dormancy In plants, a cessation of growth under conditions that seem appropriate for growth. 462

dorsal-root ganglion (gang-glee-un) Mass of sensory neuron cell bodies located in the dorsal root of a spinal nerve. 528

double fertilization In flowering plants, one sperm joins with polar nuclei within the embryo sac to produce a 3n endosperm nucleus, and another sperm joins with an egg to produce a zygote. 355, 483

double helix Double spiral; describes the three-dimensional shape of DNA. 188

doubling time Number of years it takes for a population to double in size. 719

dryopithecine Tree dweller; ancestral to apes. 403

duodenum (doo-uh-dee-num) First part of the small intestine where chyme enters from the stomach. 619

duplication Change in chromosome structure in which a particular segment is present more than once in the same chromosome. 154

E

echinoderm Phylum of marine animals that includes sea stars, sea urchins, and sand

dollars; characterized by radial symmetry and a water vascular system. 385

ecological niche Role an organism plays in its community, including its habitat and its interactions with other organisms. 742

ecological pyramid Pictorial graph based on the biomass, number of organisms, or energy content of various trophic levels in a food web—from the producer to the final consumer populations. 755

ecology Study of the interactions of organisms with other organisms and with the physical and chemical environment. 710, 752

ecosystem (ek-oh-sis-tum, ee-koh-) Biological community together with the associated abiotic environment; characterized by energy flow and chemical cycling. 10, 710, 751

ecosystem diversity Variety of species in a particular locale, dependent on the species interactions. 781

ectoderm Outermost primary tissue layer of an animal embryo; gives rise to the nervous system and the outer layer of the integument. 693

ectotherm Organism having a body temperature that varies according to the environmental temperature. 390

electrocardiogram (ECG) (ih-lek-troh-kar-dee-uh-gram) Recording of the electrical activity associated with the heartbeat. 578

electromagnetic receptor Sensory receptor that detects energy of different wavelengths, such as electricity, magnetism, and light. 536

electron Negative subatomic particle, moving about in an energy level around the nucleus of an atom. 21

electronegativity Ability of an atom to attract electrons toward itself in a chemcial bond. 26

electron shell Concentric energy levels in which electrons orbit. 21

electron transport chain (ETC) Passage of electrons along a series of membrane-bounded carrier molecules from a higher to lower energy level; the energy released is used for the synthesis of ATP. 101, 115, 120

element Substance that cannot be broken down into substances with different properties; composed of only one type of atom. 20

El Niño-Southern Oscillation Warming of water in the Eastern Pacific equatorial region such that the Humboldt Current is displaced, with possible negative results such as reduction in marine life. 774

elongation During DNA replication or the formation of mRNA during transcription, elongation is the step whereby the bases are added to the new or "daughter" strands of DNA or the mRNA transcript strand is lengthened, respectively. 198

embryonic development Period of development from the second through eighth weeks. 697

embryonic disk Stage of embryonic development following the blastocyst stage that has two layers; one layer will be endoderm, and the other will be ectoderm. 699

embryo sac Female gametophyte of flowering plants that produces an egg cell. 354, 481

emergent disease Disease that is newly mutated or a newly opportunistic strain of an existing disease. 307

emphysema (em-fih-see-muh) Degenerative lung disorder in which the bursting of alveolar walls reduces the total surface area for gas exchange. 644

endangered species A species that is in peril of immediate extinction throughout all or most of its range (e.g., California condor, snow leopard). 780

endergonic reaction Chemical reaction that requires an input of energy; opposite of exergonic reaction. 80

endocrine system Organ system involved in the coordination of body activities; uses hormones as chemical signals secreted into the bloodstream. 504, 666

endocytosis Process by which substances are moved into the cell from the environment by phagocytosis (cellular eating) or pinocytosis (cellular drinking); includes receptor-mediated endocytosis. 90

endoderm Innermost primary tissue layer of an animal embryo that gives rise to the linings of the digestive tract and associated structures. 693

endodermis Plant root tissue that forms a boundary between the cortex and the vascular cylinder. 428

endomembrane system A collection of membranous structures involved in transport within the cell. 60

endometrium Mucous membrane lining the interior surface of the uterus. 686

endoplasmic reticulum (ER) (en-duh-plaz-mik reh-tik-yuh-lum) System of membranous saccules and channels in the cytoplasm, often with attached ribosomes. 63

endoskeleton Protective internal skeleton, as in vertebrates. 555

endosperm In angiosperms, the 3n tissue that nourishes the embryo and seedling and is formed as a result of a sperm joining with two polar nuclei. 483

endospore Spore formed within a cell; certain bacteria form endospores. 312

endosymbiotic theory Possible explanation of the evolution of eukaryotic organelles by phagocytosis of prokaryotes. 68, 324

endotherm Animal that maintains a constant body temperature independent of the environmental temperature. 391

energy Capacity to do work and bring about change; occurs in a variety of forms. 7, 78

energy of acticvation Energy that must be added in order for molecules to react with one another. 82

energy-related organelle Mitochondria in plant and animal cells, chloroplasts in plant cells. 61

entropy Measure of disorder or randomness. 79

enzyme (en-zym) Organic catalyst, usually a protein, that speeds a reaction in cells due to its particular shape. 40, 82

enzyme inhibition Means by which cells regulate enzyme activity; may be competitive or noncompetitive inhibition. 84

eosinophil (ee-oh-sin-oh-fill) White blood cell containing cytoplasmic granules that stain with acidic dye. 597

epidermal tissue Exterior tissue, usually one cell thick, of leaves, young stems, roots, and other parts of plants. 426

epidermis In plants, tissue that covers roots and leaves and stems of nonwoody organism; in animals, the outer protective region of the skin. 426, 503

epididymis (ep-uh-did-uh-mus) Coiled tubule next to the testes where sperm mature and may be stored for a short time. 684

epiglottis (ep-uh-glaht-us) Structure that covers the glottis during the process of swallowing. 617, 638

epinephrine (ep-uh-nef-rin) Hormone secreted by the adrenal medulla in times of stress; adrenaline. 672

epiphyte Plant that takes its nourishment from the air because of its placement in other plants gives it an aerial position. 768

episiotomy (ih-pee-zee-aht-uh-mee) Surgical procedure performed during childbirth in which the opening of the vagina is enlarged to avoid tearing. 702

epithelial tissue Type of tissue that lines hollow organs and covers surfaces; epithelium. 596

equilibrium population Population demonstrating a life history pattern in which members exhibit logistic population growth and the population size remains at or near the carrying capacity. Its members are large in size, slow to mature, have a long life span, have few offspring, and provide much care to offspring (e.g., bears, lions). 717

erection Increase in blood flow to the penis during sexual arousal, causing the penis to stiffen and become erect. 684

esophagus (ih-sahf-uh-gus) Muscular tube for moving swallowed food from the pharynx to the stomach. 617

essential nutrient In plants, substance required for normal growth, development, or reproduction. 450

estrogen (es-truh-jun) Female sex hormone that helps maintain sex organs and secondary sex characteristics. 687

estuary Portion of the ocean located where a river enters and fresh water mixes with salt water. 772

ethylene Plant hormone that causes ripening of fruit and is also involved in abscission. 465

euchromatin Chromatin that is extended and accessible for transcription. 213

eudicot Abbreviation of eudicotyledon. Flowering plant group; members have two embryonic leaves (cotyledons), net-veined leaves, vascular bundles in a ring, flower parts in fours or fives and their multiples, and other characteristics. 353, 424

Eudicotyledone Class of flowering plants, characterized by two embryonic leaves (cotyledons), net-veined leaves, vascular bundles in a ring, flower parts in fours or fives and their multiples, and other characteristics. 353

euglenoid Flagellated and flexible freshwater unicellular protist that usually contains chloroplasts and has a semirigid cell wall. 327

Eukarya One of the three domains of life, consisting of organisms with eukaryotic cells and further classified into the kingdoms Protista, Fungi, Plantae, and Animalia. 8

eukaryotic cell (eukaryote) Type of cell that has a membrane-bounded nucleus and membranous organelles. 8, 59

eutrophication Enrichment of water by inorganic nutrients used by phytoplankton. Often, overenrichment caused by human activities leads to excessive bacterial growth and oxygen depletion. 757, 771

eutrophic lake Lake containing many nutrients and decaying organisms, often tinted green with algae. 771

evolution Descent of organisms from common ancestors with the development of genetic and phenotypic changes over time that make them more suited to the environment. 7, 247

evolutionary species concept Every species has its own evolutionary history, which is partly documented in the fossil record. 264

evolutionary systematics School of systematics that takes into consideration the degree of difference between derived characters to construct phylogenetic trees. 294

excretion Elimination of metabolic wastes by an organism at exchange boundaries such as the plasma membrane of unicellular organisms and excretory tubules of multicellular animals. 650

exergonic reaction Chemical reaction that releases energy; opposite of endergonic reaction. 80

exocytosis Process in which a intracellular vesicle fuses with the plasma membrane so that the vesicle's contents are released outside the cell. 90

exon A segment of a gene that codes for a protein. 195

exophthalmic goiter (ex opp THOWL mick GOI turr) Enlargement of the thyroid gland accompanied by an abnormal protrusion of the eyes. 676

exoskeleton Protective external skeleton, as in arthropods. 554

experimental variable A value that is expected to change as a result of an experiment; represents the factor that is being tested by the experiment. 12

expiration (ek-spuh-ray-shun) Act of expelling air from the lungs; also called exhalation. 640

exponential growth Growth at a constant rate of increase per unit of time; can be expressed as a constant fraction or exponent. 714

external respiration Exchange of oxygen and carbon dioxide between alveoli and blood. 634

extinction Total disappearance of a species or higher group. 284, 718, 780

extraembryonic membrane (ek-struh-em-bree-ahn-ik) Membrane that is not a part of the embryo but is necessary to the continued existence and health of the embryo. 683, 697

F

facilitated diffusion Passive transfer of a substance into or out of a cell along a concentration gradient by a process that requires a carrier. 88

facultative anaerobe Prokaryote that is able to grow in either the presence or the absence of gaseous oxygen. 314

FAD Flavin adenine dinucleotide; a coenzyme of oxidation-reduction that becomes $FADH_2$ as oxidation of substrates occurs, and then delivers electrons to the electron transport chain in mitochondria during cellular respiration. 114

family One of the categories, or taxa, used by taxonomists to group species; the taxon above the genus level. 8, 290

fat Organic molecule that contains glycerol and fatty acids and is found in adipose tissue. 43, 623

fatty acid Molecule that contains a hydrocarbon chain and ends with an acid group. 43

feedback inhibition Mechanism for regulating metabolic pathways in which the concentration of the product is kept within a certain range until binding shuts down the pathway, and no more product is produced. 84

female gametophyte In seed plants, the gametophyte that produces an egg; in flowering plants, an embryo sac. 351

fermentation Anaerobic breakdown of glucose that results in a gain of two ATP and end products such as alcohol and lactate. 115, 123

fern Member of a group of plants that have large fronds; in the sexual life cycle, the independent gametophyte produces flagellated sperm, and the vascular sporophyte produces windblown spores. 348

fertilization Union of a sperm nucleus and an egg nucleus, which creates a zygote. 147, 687

fetal development Period of development from the ninth week through birth. 700

fiber Structure resembling a thread; also, plant material that is nondigestible. 622

fibroblast Cell in connective tissues that produces fibers and other substances. 498

filament End-to-end chains of cells that form as cell division occurs in only one plane; in plants, the elongated stalk of a stamen. 334, 480

filter feeder Method of obtaining nourishment by certain animals that strain minute organic particles from the water in a way that deposits them in the digestive tract. 374, 613

fimbria (pl., fimbriae) (fim-bree-uh) Fingerlike extension from the oviduct near the ovary. 59, 686

first messenger Chemical signal such as a peptide hormone that binds to a plasma membrane receptor protein and alters the metabolism of a cell because a second messenger is activated. 667

fitness Ability of an organism to reproduce and pass its genes to the next fertile generation; measured against the ability of other organisms to reproduce in the same environment. 730

fixed action pattern (FAP) Innate behavior pattern that is stereotyped, spontaneous, independent of immediate control, genetically encoded, and independent of individual learning. 726

flagellum (pl., flagella) (fluh-jel-um) Slender, long extension that propels a cell through a fluid medium. 59

flame cell Found along excretory tubules of planarians; functions in propulsion of fluid through the excretory canals and out of the body. 651

flatworm Unsegmented worm lacking a body cavity; phylum Platyhelminthes. 376

fluid feeder Animal that gains needed nutrients by sucking nutrient-rich fluids from another living organism. 613

fluid-mosaic model Model for the plasma membrane based on the changing location and pattern of protein molecules in a fluid phospholipid bilayer. 85

follicle (fahl-ih-kul) Structure in the ovary that produces a secondary oocyte and the hormones estrogen and progesterone. 687

follicle-stimulating hormone (FSH) Hormone secreted by the anterior pituitary gland that stimulates the development of an ovarian follicle in a female or the production of sperm in a male. 685

follicular phase First half of the ovarian cycle, during which the follicle matures and much estrogen (and some progesterone) is produced. 688

fontanel (fahn-tun-el) Membranous region located between certain cranial bones in the skull of a fetus or infant. 556, 700

food chain The order in which one population feeds on another in an ecosystem, from detritus (detrital food chain) or producer (grazing food chain) to final consumer. 755

food web In ecosystems, complex pattern of interlocking and crisscrossing food chains. 754

foramen magnum (fuh-ray-mun mag-num) Opening in the occipital bone of the vertebrate skull through which the spinal cord passes. 556

foraminiferan Member of the phylum Foraminifera bearing a calcium carbonate test with many openings through which pseudopods extend. 328

foreign antigen Organism does not produce this type of antigen. 599

foreskin Skin covering the glans penis in uncircumcised males. 684

formed element Constituent of blood that is either cellular (red blood cells and white blood cells) or at least cellular in origin (platelets). 584

fossil record History of life recorded from remains from the past. 249

founder effect Cause of genetic drift due to colonization by a limited number of individuals who, by chance, have different gene frequencies than the parent population. 255

fovea centralis Region of the retina consisting of densely packed cones; responsible for the greatest visual acuity. 540

frameshift mutation Alternation in a gene due to deletion of a base, so that the reading "frame" is shifted; can result in a nonfunctional protein. 200

frond Leaf of a fern. 348

fruit Flowering plant structure consisting of one or more ripened ovaries that usually contain seeds. 355, 485

functional group Specific cluster of atoms attached to the carbon skeleton of organic molecules that enters into reactions and behaves in a predictable way. 39

fungus (pl., fungi) Saprotrophic decomposer; the body is made up of filaments called hyphae that form a mass called a mycelium. 9, 358

G

gallbladder Organ attached to the liver that serves to store and concentrate bile. 620

gamete (ga-meet, guh-meet) Haploid sex cell; the egg or a sperm, which join in fertilization to form a zygote. 37, 691

gametophyte Haploid generation of the alternation of generations life cycle of a plant; produces gametes that unite to form a diploid zygote. 344, 481

ganglion Collection or bundle of neuron cell bodies usually outside the central nervous system. 514

gastrin Peptide hormone secreted by stomach that stimulates gastric acid secretion. 621

gastropod Mollusc with a broad flat foot for crawling (e.g., snails and slugs). 380

gastrulation Stage of animal development during which germ layers form, at least in part, by invagination. 693

gene Unit of heredity existing as alleles on the chromosomes; in diploid organisms, typically two alleles are inherited—one from each parent. 7, 192

gene cloning Production of one or more copies of the same gene. 226

gene flow Sharing of genes between two populations through interbreeding. 254

gene linkage Existence of several alleles on the same chromosome. 176

gene locus Specific location of a particular gene on homologous chromosomes. 163

gene pool Total of all the genes of all the individuals in a population. 253

gene therapy Correction of a detrimental mutation by the addition of normal DNA and its insertion in a genome. 232

genetically engineered Alteration of genomes for medical or industrial purposes. 226

genetically modified organism (GMO) Organism that carries the genes of another organism as a result of DNA technology. 226

genetic code Universal code that has existed for eons, specifies protein synthesis in the cells of all living things. Each codon consists of three letters standing for the DNA nucleotides that make up one of the 20 amino acids found in proteins. 193

genetic diversity Variety among members of a population. 781

genetic drift Mechanism of evolution due to random changes in the allelic frequencies of a population; more likely to occur in small populations or when only a few individuals of a large population reproduce. 255

genetic mutation Alteration in chromosome structure or number and also an alteration in a gene due to a change in DNA composition 200

genome Full set of genetic information of a species or a virus. 234

genotype (jee-nuh-typ) Genes of an individual for a particular trait or traits; often designated by letters, for example, *BB* or *Aa*. 163

genus One of the categories, or taxa, used by taxonomists to group species; contains those species that are most closely related through evolution. 8, 290

germ cell During zygote development, cells that are set aside from the somatic cells and that will eventually undergo meiosis to produce gametes. 134

germinate Beginning of growth of a seed, spore, or zygote, especially after a period of dormancy. 486

germ layer Primary tissue layer of a vertebrate embryo—namely, ectoderm, mesoderm, or endoderm. 693

gibberellin Plant hormone promoting increased stem growth; also involved in flowering and seed germination. 462

gills Respiratory organ in most aquatic animals; in fish, an outward extension of the pharynx. 576, 634

ginkgo Member of phlyum Ginkgophyte; maidenhair tree. 350

girdling Removing a strip of bark from around a tree. 448

gizzard Muscular part of the digestive tract that grinds food in some animals. 614

gland Epithelial cell or group of epithelial cells that are specialized to secrete a substance. 497

global warming Predicted increase in the Earth's temperature, due to human activities that promote the greenhouse effect. 758

globular stage Stage of development of a sporophyte embryo; root-shoot axis is established and dermal tissue is formed. 484

glomerular capsule (gluh-mair-yuh-lur) Double-walled cup that surrounds the glomerulus at the beginning of the nephron. 654

glomerular filtrate Filtered portion of blood contained within the glomerular capsule. 656

glomerular filtration Movement of small molecules from the glomerulus into the glomerular capsule due to the action of blood pressure. 656

glomerulus (gluh-mair-uh-lus, gloh-mair-yuh-lus) Cluster; for example, the cluster of capillaries surrounded by the glomerular capsule in a nephron, where glomerular filtration takes place. 655

glottis (glaht-us) Opening for airflow in the larynx. 617, 638

glucagon (gloo-kuh-gahn) Hormone secreted by the pancreas that causes the liver to break down glycogen and raises the blood glucose level. 674

glucocorticoid (gloo-koh-kor-tih-koyd) Type of hormone secreted by the adrenal cortex that influences carbohydrate, fat, and protein metabolism; see cortisol. 672

glucose (gloo-kohs) Six-carbon sugar that organisms degrade as a source of energy during cellular respiration. 41

glycerol Three-carbon carbohydrate with three hydroxyl groups attached; a component of fats and oils. 43

glycogen (gly-koh-jun) Storage polysaccharide that is composed of glucose molecules joined in a linear fashion but having numerous branches. 42

glycolipid Lipid in plasma membranes that bears a carbohydrate chain attached to a hydrophobic tail. 85

glycolysis Anaerobic breakdown of glucose that results in a gain of two ATP. 115, 116

glycoprotein Protein in plasma membranes that bears a carbohydrate chain. 85

gnetophyte Member of one of the four phyla of gymnosperms; Gnetophyta has only three living genera, which differ greatly from one another—e.g., *Welwitschia* and *Ephedra*. 350

Golgi apparatus (gohl-jee) Organelle, consisting of saccules and vesicles, that processes, packages, and distributes molecules about or from the cell. 65

gonad (goh-nad) Organ that produces gametes; the ovary produces eggs, and the testis produces sperm. 682

gonadotropic hormone (goh-nad-uh-trahp-ic, -troh-pic) Chemical signal secreted by the anterior pituitary that regulates the activity of the ovaries and testes; principally, follicle-stimulating hormone (FSH) and luteinizing hormone (LH). 671

gonorrhea (gahn-nuh-ree-uh) Sexually transmitted disease caused by the bacterium *Neisseria gonorrhoeae* that can lead to pelvic inflammatory disease. 689

granum (pl., grana) Stack of chlorophyll-containing thylakoids in a chloroplast. 68, 97

gravitational balance Maintenance of balance when the head and body are motionless. 546

gravitropism Growth response of roots and stems of plants to the Earth's gravity; roots demonstrate positive gravitropism, and stems demonstrate negative gravitropism. 466

gray crescent Gray area that appears in an amphibian egg after being fertilized by the sperm; thought to contain chemical signals that turn on the genes that control development. 524

gray matter Nonmyelinated axons and cell bodies in the central nervous system. 524

grazing food web Complex pattern of interlocking and crisscrossing food chains that begins with populations of autotrophs serving as producers. 754

green algae Members of a diverse group of photosynthetic protists; contain chlorophylls *a* and *b* and have other biochemical characteristics like those of plants. 334

greenhouse effect Reradiation of solar heat toward the Earth because gases such as carbon dioxide, methane, nitrous oxide, and water vapor allow solar energy to pass through toward the Earth but block the escape of heat back into space. 758

greenhouse gases Gases that are involved in the greenhouse effect. 758

ground tissue Tissue that constitutes most of the body of a plant; consists of parenchyma, collenchyma, and sclerenchyma cells that

function in storage, basic metabolism, and support. 426

growth hormone (GH) Substance secreted by the anterior pituitary; controls size of individual by promoting cell division, protein synthesis, and bone growth. 671

guanine (G) (gwah-neen) One of four nitrogen-containing bases in nucleotides composing the structure of DNA and RNA; pairs with cytosine. 186

guard cell One of two cells that surround a leaf stoma; changes in the turgor pressure of these cells cause the stoma to open or close. 446

guttation Liberation of water droplets from the edges and tips of leaves. 444

gymnosperm Type of woody seed plant in which the seeds are not enclosed by fruit and are usually borne in cones, such as those of the conifers. 350

H

habitat Place where an organism lives and is able to survive and reproduce. 710, 742

habituation Simplest form of learning, in which an animal learns not to respond to irrelevant stimuli. 729

halophile Type of archaea that lives in extremely salty habitats. 316

haploid life cycle Presence of one of each type of chromosome during meiosis of the cell cycle. 336

haploid (n) number (hap-loyd) The n number of chromosomes—half the diploid number; the number characteristic of gametes, which contain only one set of chromosomes. 137

hard palate (pal-it) Bony, anterior portion of the roof of the mouth. 616

Hardy-Weinberg principle Law stating that the gene frequencies in a population remain stable if evolution does not occur due to nonrandom mating, selection, migration, and genetic drift. 253

hay fever Seasonal variety of allergic reaction to a specific allergen. Characterized by sudden attacks of sneezing, swelling of nasal mucosa, and often asthmatic symptoms. 606

heart attack Damage to the myocardium due to blocked circulation in the coronary arteries; also called a myocardial infarction (MI). 582

heart murmur Clicking or swishy sounds, often due to leaky valves. 577

heart stage Stage of development of a sporophyte embryo; cotyledons appear. 484

heat Type of kinetic energy; captured solar energy eventually dissipates as heat in the environment. 78

helper T cell T lymphocyte that secretes cytokines that stimulate all kinds of immune system cells. 600

hemodialysis (he-moh-dy-al-uh-sus) Cleansing of blood by using an artificial membrane that causes substances to diffuse from blood into a dialysis fluid. 660

hemoglobin (Hb) (hee-muh-gloh-bun) Iron-containing pigment in red blood cells that combines with and transports oxygen. 45, 584, 643

hemolymph Circulatory fluid that is a mixture of blood and tissue fluid; seen in animals that have an open circulatory system, such as molluscs and arthropods. 575

hemophilia Most common of the severe clotting disorders caused by the absence of a blood clotting factor. 585

hemorrhagic fever Acute fever caused by several types of viruses; characterized by high fever and uncontrollable bleeding from several organs. 318

herbaceous Nonwoody stem. 428

herbivore (hur-buh-vor) Primary consumer in a grazing food chain; a plant eater. 752

hermaphrodite Animal having both male and female sex organs. 376

herpes simplex virus (HSV) Virus that causes genital herpes, a sexually transmitted disease. 689

heterochromatin Highly compacted chromatin that is not accessible for transcription. 212

heterocyst Cyanobacterial cell that synthesizes a nitrogen-fixing enzyme when nitrogen supplies dwindle. 315

heterotroph Organism that cannot synthesize organic molecules from inorganic nutrients and therefore must take in organic nutrients (food). 96, 752

heterozygote advantage Situation in which individuals heterozygous for a trait have a selective advantage over those who are homozygous; an example is sickle-cell anemia. 258

heterozygous Possessing unlike alleles for a particular trait. 163

hinge joint Type of joint that allows movement as a hinge does, such as the movement of the knee. 561

hippocampus (hip-uh-kam-pus) Portion of the limbic system where memories are stored. 527

histamine (his-tuh-meen, -mun) Substance, produced by basophils in blood and mast cells in connective tissue, that causes capillaries to dilate. 598

histone Protein molecule responsible for packing chromatin. 212

HIV Virus responsible for AIDS. 592

HIV provirus Viral DNA that has been integrated into host cell DNA. 306

homeobox 180-nucleotide sequence located in all homeotic genes. 217

homeodomain Conserved DNA-binding region of transcription factors encoded by the homeobox of homeotic genes. 217

homeostasis (hoh-mee-oh-stay-sis) Maintenance of normal internal conditions in a cell or an organism by means of self-regulating mechanisms. 6, 435, 504

homeotic gene Gene that controls the overall body plan by controlling the fate of groups of cells during development. 207, 217

hominid Member of the family Hominidae, which contains australopithecines and humans. 403

hominoid Member of the superfamily Hominoidae, which includes apes, humans, and their recent ancestors. 403

Homo erectus Hominid who used fire and migrated out of Africa to Europe and Asia. 408

Homo ergaster Extinct hominid; some palenontologists separate it from *H. erectus*,

some do not and consider it a part of the African line of *H. erectus*. 408

homologous chromosome (hoh-mahl-uh-gus, huh-mahl-uh-gus) Member of a pair of chromosomes that are alike and come together in synapsis during prophase of the first meiotic division. 145

homologous structure Structure that is similar in two or more species because of common ancestry. 251, 292

homologue Member of a homologous pair of chromosomes. 145

homology Similarity of different species due to evolutionary differentiation from the same ancestors. 292, 371

homozygote advantage Situation in which individuals homozygous for a trait have a selective advantage over those who are heterozygous. 258

homozygous Possessing two identical alleles for a particular trait. 163

horizon Major layer of soil visible in vertical profile; for example, topsoil is the A horizon. 453

hormone Chemical signal produced in one part of an organism that controls the activity of other parts. 436, 666

horsetail Division of seedless vascular plants having only one genus (*Equisetum*) in existence today; characterized by rhizomes, scalelike leaves, strobili, and tough, rigid stems. 348

host Organism that provides nourishment and/or shelter for a parasite. 746

human chorionic gonadotropin (HCG) (kor-ee-ahn-ik goh-nad-uh-trahp-in, -troh-pin) Hormone produced by the chorion that functions to maintain the uterine lining. 698

Human Genome Project (HGP) Initiative to determine the complete sequence of the human genome and to analyze this information. 49, 234

human papillomavirus (HPV) Virus that causes genital warts, a common sexually transmitted disease. Also is linked to cervical cancer. 689

humus Decomposing organic matter in the soil. 453

hunter-gatherer Human that hunted animals and gathered plants, for food. 409

Huntington disease Genetic disease marked by progressive deterioration of the nervous system due to deficiency of a neurotransmitter. 525

hyaline cartilage Cartilage whose cells lie in lacunae separated by a white translucent matrix containing very fine collagen fibers. 499

hydrogen bond Weak bond that arises between a slightly positive hydrogen atom of one molecule and a slightly negative atom of another or between parts of the same molecule. 26

hydrogen ion (H⁺) Hydrogen atom that has lost its electron and therefore bears a positive charge. 30

hydrolysis reaction (hy-drahl-ih-sis re-ak-shun) Splitting of a compound by the addition of water, with the H^+ being incorporated in one fragment and the OH^- in the other. 40

hydrophilic (hy-druh-fil-ik) Type of molecule that interacts with water by dissolving in water and/or forming hydrogen bonds with water molecules. 28, 39

hydrophobic (hy-druh-foh-bik) Type of molecule that does not interact with water because it is nonpolar. 28, 39

hydroponics Technique for growing plants by suspending them with their roots in a nutrient solution. 451

hydrostatic skeleton Fluid-filled body compartment that provides support for muscle contraction resulting in movement; seen in cnidarians, flatworms, roundworms, and segmented worms. 554

hydrothermal vent Hot springs in the seafloor along ocean ridges where heated seawater and sulfate react to produce hydrogen sulfide; here, chromosynthetic bacteria support a community of varied organisms. 773

hydroxide ion (OH⁻) One of two ions that results when a water molecule dissociates; it has gained an electron, and therefore bears a negative charge. 30

hypersensitive response (HR) Plants respond to pathogens by selectively killing plant cells to block the spread of the pathogen. 473

hypertension Elevated blood pressure, particularly the diastolic pressure. 582

hypertonic solution Higher solute concentration (less water) than the cytoplasm of a cell; causes cell to lose water by osmosis. 89

hypha Filament of the vegetative body of a fungus. 358

hypothalamic-inhibiting hormone (hy-poh-thuh-lah-mik) One of many hormones produced by the hypothalamus that inhibits the secretion of an anterior pituitary hormone. 671

hypothalamus (hy-poh-thal-uh-mus) Part of the brain located below the thalamus that helps regulate the internal environment of the body and produces releasing factors that control the anterior pituitary. 526

hypothesis (hy-pahth-ih-sis) Supposition that is formulated after making an observation; it can be tested by obtaining more data, often by experimentation. 11

hypotonic solution Lower solute (more water) concentration than the cytoplasm of a cell; causes cell to gain water by osmosis. 89

I

immediate allergic response Allergic response that occurs within seconds of contact with an allergen, caused by the attachment of the allergen to IgE antibodies. 606

immune system All the cells in the body that protect the body against foreign organisms and substances and also cancerous cells. 505

immunity Ability of the body to protect itself from foreign substances and cells, including disease-causing agents. 596

immunization (im-yuh-nuh-zay-shun) Use of a vaccine to protect the body against specific disease-causing agents. 601

immunoglobulin (Ig) (im-yuh-noh-glahb-yuh-lin, -yoo-lin) Globular plasma protein that functions as an antibody. 601

implantation Attachment and penetration of the embryo into the lining of the uterus (endometrium). 697

imprinting Learning to make a particular response to only one type of animal or object. 726

inclusive fitness Fitness that results from personal reproduction and from helping nondescendant relatives reproduce. 732

incomplete dominance Inheritance pattern in which the offspring has an intermediate phenotype, as when a red-flowered plant and a white-flowered plant produce pink-flowered offspring. 171

indehiscent Remaining closed at maturity, as are many fruits. 485

independent assortment Alleles of unlinked genes segregate independently of each other during meiosis so that the gametes contain all possible combinations of alleles. 146

induced fit model Change in the shape of an enzyme's active site that enhances the fit between the active site and its substrate(s). 82

inducible operon In a catabolic pathway, an operon causes transcription of the genes controlling a group of enzymes. 208

induction Ability of a chemical or a tissue to influence the development of another tissue. 696

inflammatory reaction Tissue response to injury that is characterized by redness, swelling, pain, and heat. 598

initiation During DNA replication or transcription, initiation is the step whereby DNA replication begins or transcription begins; this step is catalyzed by specific enzymes such that the DNA "unravels" and forms a bubble. 198

inner ear Portion of the ear consisting of a vestibule, semicircular canals, and the cochlea where equilibrium is maintained and sound is transmitted. 544

insect Member of a group of arthropods in which the head has antennae, compound eyes, and simple eyes; a thorax has three pairs of legs and often wings; and an abdomen has internal organs. 384

insight learning Ability to apply prior learning to a new situation without trial-and-error activity. 729

inspiration (in-spuh-ray-shun) Act of taking air into the lungs; also called inhalation. 640

insulin (in-suh-lin) Hormone secreted by the pancreas that lowers the blood glucose level by promoting the uptake of glucose by cells and the conversion of glucose to glycogen by the liver and skeletal muscles. 674

integration Summing up of excitatory and inhibitory signals by a neuron or by some part of the brain. 521, 537

integumentary system Organ system consisting of skin and various organs, such as hair, which are found in skin. 504

intercalated disks Region that holds adjacent cardiac muscle cells together and that appears as dense bands at right angles to the muscle striations. 500

interferon (in-tur-feer-ahn) Antiviral agent produced by an infected cell that blocks the infection of another cell. 597

interkinesis Period of time between meiosis I and meiosis II during which no DNA replication takes place. 148

intermediate filament Ropelike assemblies of fibrous polypeptides in the cytoskeleton that provide support and strength to cells; so called because they are intermediate in size between actin filaments and microtubules. 70

internal respiration Exchange of oxygen and carbon dioxide between blood and tissue fluid. 634

interneuron Neuron located within the central nervous system that conveys messages between parts of the central nervous system. 517

internode In vascular plants, the region of a stem between two successive nodes. 422

intertidal zone Region along a coastline where the tide recedes and returns. 772

intervertebral disk (in-tur-vur-tuh-brul) Layer of cartilage located between adjacent vertebrae. 557

intron Segment of a gene that does not code for a protein. 195

inversion Change in chromosome structure in which a segment of a chromosome is turned around 180°; this reversed sequence of genes can lead to altered gene activity and abnormalities. 154

invertebrate Referring to an animal without a serial arrangement of vertebrae. 373

ion (eye-un, -ahn) Charged particle that carries a negative or positive charge. 24

ionic bond (eye-ahn-ik) Chemical bond in which ions are attracted to one another by opposite charges. 24

iris (eye-ris) Muscular ring that surrounds the pupil and regulates the passage of light through this opening. 540

isomer Moleculees with the same molecular formula but a different structure, and therefore a different shape. 39

isotonic solution Solution that is equal in solute concentration to that of the cytoplasm of a cell; causes cell to neither lose nor gain water by osmosis. 89

isotope (eye-suh-tohp) One of two or more atoms with the same atomic number but a different atomic mass due to the number of neutrons. 21

J

jaw Tooth-bearing bone of the head. 387

jawless fish Type of fish that has no jaws; includes today's hagfishes and lampreys. 388

jointed limb Vertebrate appendage with one or more movable segment. 387

junk DNA Portion of a DNA sequence in which no function has yet been identified. 195

K

karyotype (kar-ee-uh-typ) Duplicated chromosomes arranged by pairs according to their size, shape, and general appearance. 145

keystone species Species whose activities significantly affect community structure. 789

kidneys Paired organs of the vertebrate urinary system that regulate the chemical composi-

tion of the blood and produce a waste product called urine. 654

kilocalorie Caloric value of food; 1,000 calories. 78

kinase Enzyme that activates another enzyme by adding a phosphate group. 142

kinetic energy Energy associated with motion. 78

kinetochore Disk-shaped structure within the centromere of a chromosome to which spindle microtubules become attached during mitosis and meiosis. 138

kingdom One of the categories used to classify organisms; the category above phylum. 8, 290

kin selection Indirect selection; adaptation to the environment due to the reproductive success of an individual's relatives. 732

Klinefelter syndrome Condition caused by the inheritance of XXY chromosomes. 153

L

lacteal (lak-tee-ul) Lymphatic vessel in an intestinal villus; it aids in the absorption of lipids. 594, 619

lacunae Small pit or hollow cavity, as in bone or cartilage, where a cell or cells are located. 499, 560

ladderlike nervous system In planarians, two lateral nerve cords joined by transverse nerves. 514

lake Body of fresh water, often classified by nutrient status, such as oligotrophic (nutrient-poor) or eutrophic (nutrient-rich). 771

lancelet Invertebrate chordate with a body that resembles a lancet and has the four chordate characteristics as an adult. 386

landscape diversity Variety of habitat elements within an ecosystem (e.g., plains, mountains, and rivers). 781

large intestine Last major portion of the digestive tract, extending from the small intestine to the anus and consisting of the cecum, the colon, the rectum, and the anal canal. 621

larva Immature form in the life cycle of some animals; it sometimes undergoes metamorphosis to become the adult form. 683

larynx (lar-ingks) Cartilaginous organ located between the pharynx and the trachea that contains the vocal cords; also called the voice box. 638

lateral line Canal system containing sensory receptors that allow fishes and amphibians to detect water currents and pressure waves from nearby objects. 548

leaf Lateral appendage of a stem, highly variable in structure, often containing cells that carry out photosynthesis. 422

learning Relatively permanent change in behavior that results from practice and experience. 726

lens Clear, membranelike structure found in the eye behind the iris; brings objects into focus. 540

lenticel Frond of usually numerous, slightly raised, somewhat spongy, groups of cells in the bark of woody plants. Permits gas exchange between the interior of a plant and the external atmosphere. 432

less-developed country (LDC) Country that is becoming industrialized; typically, population growth is expanding rapidly, and the majority of people live in poverty. 719

lichen Symbiotic relationship between certain fungi and algae, in which the fungi possibly provide inorganic food or water and the algae provide organic food. 315, 359

life cycle Recurring pattern of genetically programmed events by which individuals grow, develop, maintain themselves, and reproduce. 150

life history Adaptations in characteristics that influence an organism's biology, such as how many offspring it produces, its survival, and factors such as age and size that determine its reproductive maturity. 717

life table Table that shows, at each age, the probability of an organism dying before the next year passes. Survivorship curves can then be established based on this data. 712

ligament Tough cord or band of dense fibrous connective tissue that joins bone to bone at a joint. 499

light reaction Portion of photosynthesis that captures solar energy and takes place in thylakoid membranes of chloroplasts; it produces ATP and NADPH. 99

limbic system Association of various brain centers, including the amygdala and hippocampus; governs learning and memory and various emotions, such as pleasure, fear, and happiness. 527

limiting factor Resource or environmental condition that restricts the abundance and distribution of an organism. 711

linkage group Alleles of different genes that are located on the same chromsome and tend to be inherited together. 176

linkage map Depicts the distances between loci as well as the order in which they occur on the organism. 176

lipase Fat-digesting enzyme secreted by the pancreas. 619

lipid (lip-id, ly-pid) Class of organic compounds that tends to be soluble only in nonpolar solvents such as alcohol; includes fats and oils. 43

liposome Droplet of phospholipid molecules formed in a liquid environment. 310

liver Large, dark red internal organ that produces urea and bile, detoxifies the blood, stores glycogen, and produces the plasma proteins, among other functions. 620

lobe-finned fish Type of fish with limblike fins. 389

logistic growth Population increase that results in an S-shaped curve; growth is slow at first, steepens, and then levels off due to environmental resistance. 714

long-day plant Plant that flowers when day length is longer than a critical length (e.g., wheat, barley, clover, spinach). 470

loop of the nephron (nef-rahn) Portion of the nephron lying between the proximal convoluted tubule and the distal convoluted tubule that functions in water reabsorption. 654

loose fibrous connective tissue Tissue composed mainly of fibroblasts widely separated by a matrix containing collagen and elastic fibers. 498

lumen Cavity inside any tubular structure, such as the lumen of the digestive tract. 497

lungs Paired, cone-shaped organs within the thoracic cavity; function in internal respiration and contain moist surfaces for gas exchange. 387, 635

luteal phase Second half of the ovarian cycle, during which the corpus luteum develops and much progesterone (and some estrogen) is produced. 688

luteinizing hormone (LH) Hormone produced by the anterior pituitary gland that stimulates the development of the corpus luteum in females and the production of testosterone in males. 685

lymph (limf) Fluid, derived from tissue fluid, that is carried in lymphatic vessels. 587, 594

lymphatic (lymphoid) organ Organ other than a lymphatic vessel that is part of the lymphatic system; includes lymph nodes, tonsils, spleen, thymus gland, and bone marrow. 595

lymphatic system Organ system consisting of lymphatic vessels and lymphatic organs that transports lymph and lipids and aids the immune system. 505, 594

lymphatic vessel Vessel that carries lymph. 594

lymph node Mass of lymphatic tissue located along the course of a lymphatic vessel. 595

lymphocyte (lim-fuh-syt) Specialized white blood cell that functions in specific defense; occurs in two forms—T lymphocyte and B lymphocyte. 585

lysogenic cycle Bacteriophage life cycle in which the virus incorporates its DNA into that of a bacterium; occurs preliminary to the lytic cycle. 303

lysosome (ly-suh-sohm) Membrane-bounded vesicle that contains hydrolytic enzymes for digesting macromolecules. 66

lytic cycle Bacteriophage life cycle in which the virus takes over the operation of the bacterium immediately upon entering it and subsequently destroys the bacterium. 303

M

macroevolution Large-scale evolutionary change, such as the formation of new species. 264

macronutrient Essential element needed in large amounts for plant growth, such as nitrogen, calcium, or sulfur. 450

macrophage (mak-ruh-fayj) Large phagocytic cell derived from a monocyte that ingests microbes and debris. 597

malaria Serious infectious illness caused by the parasitic protozoan Plasmodium. Malaria is characterized by bouts of high chills and fever that occur at regular intervals. 330

male gametophyte In seed plants, the gametophyte that produces sperm; a pollen grain. 351

Malpighian tubule Blind, threadlike excretory tubule near the anterior end of an insect's hindgut. 651

mammal Homeothermic vertebrate characterized especially by the presence of hair and mammary glands. 392

marsupial Member of a group of mammals bearing immature young nursed in a marsupium, or pouch (e.g., kangaroo and opossum). 392

mass extinction Episode of large-scale extinction in which large numbers of species disappear in a few million years or less. 284

mast cell Cell to which antibodies attach, causing it to release histamine, thus producing allergic symptoms. 596

matrix (may-triks) Unstructured semifluid substance that fills the space between cells in connective tissues or inside organelles. 68, 118, 498

matter Anything that takes up space and has mass. 20

mature embryo Sporophyte embryo after all of the stages of development. 484

mechanoreceptor (mek-uh-noh-rih-sep-tur) Sensory receptor that responds to mechanical stimuli, such as that from pressure, sound waves, and gravity. 536

medulla oblongata (muh-dul-uh ahb-lawng-gah-tuh) Part of the brainstem that is continuous with the spinal cord; controls heartbeat, blood pressure, breathing, and other vital functions. 526

megaspore One of the two types of spores produced by seed plants; develops into a female gametophyte (embryo sac). 354

meiosis, meiosis I, meiosis II (my-oh-sis) Type of nuclear division that occurs as part of sexual reproduction in which the daughter cells receive the haploid number of chromosomes in varied combinations. 145

melanocyte-stimulating hormone (MSH) Substance that causes melanocytes to secrete melanin in lower vertebrates. 671

melatonin (mel-uh-toh-nun) Hormone, secreted by the pineal gland, that is involved in biorhythms. 675

membrane attack complex Group of complement proteins that form channels in a microbe's surface, thereby destroying it. 597

memory Capacity of the brain to store and retrieve information about past sensations and perceptions; essential to learning. 527

memory B cell Forms during a primary immune response but enters a resting phase until a secondary immune response occurs. 601

memory T cell T cell that differentiates during an initial infection and responds rapidly during subsequent exposure to the same antigen. 602

meninges (sing., meninx) (muh-nin-jeez) Protective membranous coverings about the central nervous system. 524

meniscus (pl., menisci) (muh-nis-kus,-kee, -sy) Cartilaginous wedges that separate the surfaces of bones in synovial joints. 561

menstruation (men-stroo-ay-shun) Loss of blood and tissue from the uterus at the end of a uterine cycle. 688

meristem Undifferentiated, embryonic tissue in the active growth regions of plants. 430

mesoderm Middle primary tissue layer of an animal embryo that gives rise to muscle, several internal organs, and connective tissue layers. 693

mesophyll Inner, thickest layer of a leaf consisting of palisade and spongy mesophyll; the site of most of photosynthesis. 428

messenger RNA (mRNA) Type of RNA formed from a DNA template that bears coded information for the amino acid sequence of a polypeptide. 194

metabolic pathway Series of linked reactions, beginning with a particular reactant and terminating with an end product. 84

metabolic pool Metabolites that are the products of and/or substrates for key reactions in cells, allowing one type of molecule to be changed into another type, such as carbohydrates converted to fats. 125

metabolism All of the chemical reactions that occur in a cell. 7, 182

metapopulation Population subdivided into several small and isolated populations due to habitat fragmentation. 718, 789

metastasis (muh-tas-tuh-sis) Spread of cancer from the place of origin throughout the body; caused by the ability of cancer cells to migrate and invade tissues. 143

meteorite Particles that reach Earth from space, which survive entry through Earth's atmosphere. 288

methanogen Type of archaea that lives in oxygen-free habitats, such as swamps, and releases methane gas. 316

MHC (major histopcompatibility complex) protein Protein marker that is a part of cell-surface markers anchored in the plasma membrane, which the immune system uses to identify "self." 602

microevolution Change in gene frequencies between populations of a species over time. 253

micronutrient Essential element needed in small amounts for plant growth, such as boron, copper, and zinc. 450

microtubule (my-kro-too-byool) Small cylindrical structure that contains 13 rows of the protein tubulin about an empty central core; present in the cytoplasm, centrioles, cilia, and flagella. 70

microvillus Cylindrical process that extends from an epithelial cell of a villus of the intestinal wall and serves to increase the surface area of the cell. 619

midbrain Part of the brain located below the thalamus and above the pons; contains reflex centers and tracts. 526

middle ear Portion of the ear consisting of the tympanic membrane, the oval and round windows, and the ossicles; where sound is amplified. 544

mimicry Superficial resemblance of two or more species; a mechanism that avoids predation by appearing to be noxious. 745

mineral Naturally occurring inorganic substance containing two or more elements; certain minerals are needed in the diet. 20, 450, 624

mineralocorticoid (min-ur-uh-loh-kor-tih-koyd) Type of hormone secreted by the adrenal cortex that regulates salt and water balance, leading to increases in blood volume and blood pressure. 672

mitochondrion (mite-oh-KAHN-dree-uhn) Membrane-bounded organelle in which ATP molecules are produced during the process of cellular respiration. 68

mitosis (my-toh-sis) Type of cell division in which daughter cells receive the exact chromosomal and genetic makeup of the parent cell; occurs during growth and repair. 136

mitotic spindle Microtubule structure that brings about chromosomal movement during nuclear division. 138

model of island biogeography Model developed by ecologists Robert MacArthur and E. O. Wilson to explain the effects of distance from the mainland and size of an island on its diversity. 749

molecular clock Mutational changes that accumulate at a presumed constant rate in regions of DNA not involved in adaptation to the environment. 293, 404

molecule Union of two or more atoms of the same element; also, the smallest part of a compound that retains the properties of the compound. 4, 23

molting Periodic shedding of the exoskeleton in arthropods. 382

monoclonal antibody One of many antibodies produced by a clone of hybridoma cells that all bind to the same antigen. 604

monocot Abbreviation of monocotyledon. Flowering plant group; among other characteristics, members have one embryonic leaf, parallel-veined leaves, and scattered vascular bundles. 353, 424

Monocotyledon Class of flowering plants characterized by one embryonic leaf, parallel-veined leaves, and scattered vascular bundles. 353

monocyte (mahn-uh-syt) Type of agranular white blood cell that functions as a phagocyte and an antigen-presenting cell. 585

monohybrid cross Single genetic cross involving only one trait, such as flower color. 162

monomer Small molecule that is a subunit of a polymer—e.g., glucose is a monomer of starch. 40

monosaccharide (mahn-uh-sak-uh-ryd) Simple sugar; a carbohydrate that cannot be decomposed by hydrolysis (e.g., glucose). 41

monosomy One less chromosome than usual. 152

monotreme Egg-laying mammal (e.g., duckbill platypus and spiny anteater). 392

monsoon Climate in India and southern Asia caused by wet ocean winds that blow onshore for alomst half the year. 770

more-developed country (MDC) Country that is industrialized; typically, population growth is low, and the people enjoy a good standard of living. 719

morphogen Protein that is part of a gradient that influences morphogenesis. 216

morphogenesis Emergence of shape in tissues, organs, or entire embryo during development. 695

morula Spherical mass of cells resulting from cleavage during animal development prior to the blastula stage. 692

mosaic evolution Concept that human characteristics did not evolve at the same rate; e.g., some body parts are more humanlike than others in early hominids. 406

moss Type of bryophyte. 346

motor (efferent) neuron Nerve cell that conducts nerve impulses away from the central nervous system and innervates effectors (muscle and glands). 517

motor molecule Protein that moves along either actin filaments or microtubules and translocates organelles. 70

motor unit Motor neuron and all the muscle fibers it innervates. 564

mRNA transcript Faithful copy of the sequence of bases in DNA. 194

multifactorial trait Trait or illness determined by several genes and the environment. 172

multinucleate hypothesis Proposal that animals arose from ciliated protists in stages; ancestor would have been bilaterally symmetrical. 369

multiple allele (uh-leelz) Inheritance pattern in which there are more than two alleles for a particular trait; each individual has only two of all possible alleles. 171

multiple fruit Cluster of mature ovaries produced by a cluster of flowers, as in a pineapple. 485

muscle dysmorphia Mental state where a person thinks his or her body is underdeveloped, and becomes preoccupied with body-building and diet; affects more men than women. 628

muscular system System of muscles that produces movement, both within the body and of its limbs; principal components are skeletal muscle, smooth muscle, and cardiac muscle. 504

muscular tissue Type of tissue composed of fibers that can shorten and thicken. 500

mutation Alteration in chromosome structure or number and also an alteration in a gene due to a change in DNA composition. 143, 254

mutualism Symbiotic relationship in which both species benefit in terms of growth and reproduction. 748

mycelium Tangled mass of hyphal filaments composing the vegetative body of a fungus. 358

mycorrhizal fungi (mycorrhizae) Mutually beneficial symbiotic relationship between a fungus and the roots of vascular plants. 359, 454

myelin sheath (my-uh-lin) White, fatty material, derived from the membrane of Schwann cells, that forms a covering for nerve fibers. 517

myofibril (my-uh-fy-brul) Contractile portion of muscle cells that contains a linear arrangement of sarcomeres and shortens to produce muscle contraction. 566

myosin (my-uh-sin) One of two major proteins of muscle; makes up thick filaments in myofibrils of muscle fibers. See actin. 566

myxedema (mik-sih-dee-muh) Condition resulting from a deficiency of thyroid hormone in an adult. 576

N

N₂ (nitrogen) fixation Process whereby free atmospheric nitrogen is converted into compounds, such as ammonium and nitrates, usually by bacteria. 757

NAD+ Nicotinamide adenine dinucleotide; coenzyme of oxidation-reduction that accepts electrons and hydrogen ions to become NADH + H+ as oxidation of substrates occurs. During cellular respiration, NADH carries electrons to the electron transport chain in mitochondria. 114

nasopharynx (nay-zoh-far-ingks) Region of the pharynx associated with the nasal cavity. 617

natural killer (NK) cell Lymphocyte that causes an infected or cancerous cell to burst. 597

natural selection Mechanism resulting in adaptation to the environment. 7, 246

nectar Sucrose-rich fluid secreted from phloem in plants; attracts pollinators. 480

negative feedback Mechanism of homeostatic response by which the output of a system suppresses or inhibits activity of the system. 508, 666

nematocyst In cnidarians, a capsule that contains a threadlike fiber whose release aids in the capture of prey. 375

nephridia Segmentally arranged, paired excretory tubules of many invertebrates, as in the earthworm. 381, 651

nephron (nef-rahn) Microscopic kidney unit that regulates blood composition by glomerular filtration, tubular reabsorption, and tubular secretion. 654

nerve Bundle of nerve fibers outside the central nervous system. 501, 528

nerve fiber Axon; conducts nerve impulses away from the cell. They are classified as either myelinated or unmyelinated based on the presence or absence of a myelin sheath. 517

nerve net Diffuse, noncentralized arrangement of nerve cells in cnidarians. 514

nervous system Organ system consisting of the brain, spinal cord, and associated nerves that coordinates the other organ systems of the body. 504

nervous tissue Tissue that contains nerve cells (neurons), which conduct impulses, and neuroglial cells, which support, protect, and provide nutrients to neurons. 501

neural plate Region of the dorsal surface of the chordate embryo that marks the future location of the neural tube. 694

neural tube Tube formed by closure of the neural groove during development. In vertebrates, the neural tube develops into the spinal cord and brain. 694

neuroglia Nonconducting nerve cells that are intimately associated with neurons and function in a supportive capacity. 501, 517

neuron Nerve cell that characteristically has three parts: dendrites, cell body, and axon. 501, 517

neurosecretory hormone Peptide hormone (ADH and oxytocin) released by the hypo-thalamus in the brain; influences blood pH. 670

neurotransmitter Chemical stored at the ends of axons that is responsible for transmission across a synapse. 520

neurula The early embryo during the development of the neural tube from the neural plate, marking the first appearance of the nervous system; the next stage after the gastrula. 694

neutron (noo-trahn) Neutral subatomic particle, located in the nucleus and having a weight of approximately one atomic mass unit. 21

neutrophil (noo-truh-fil) Granular leukocyte that is the most abundant of the white blood cells; first to respond to infection. 585, 597

nitrification Process by which nitrogen in ammonia and organic molecules is oxidized to nitrites and nitrates by soil bacteria. 757

node In plants, the place where one or more leaves attach to a stem. 422

node of Ranvier (rahn-vee-ay) Gap in the myelin sheath around a nerve fiber. 517

noncompetitive inhibiton Form of enzyme inhibition where the inhibitor binds to an enzyme at a location other than the active site; while at this site, the enzyme shape changes, the inhibitor is unable to bind to its substrate, and no product forms. 84

nondisjunction Failure of homologous chromosomes or daughter chromosomes to separate during meiosis I and meiosis II, respectively. 152

nonpolar covalent bond Bond in which the sharing of electrons between atoms is fairly equal. 26

nonrandom mating Mating among individuals on the basis of their phenotypic similarities or differences, rather than mating on a random basis. 254

nontropic hormone Peptide hormone produced by the anterior pituitary and controlled by the hypothalamus (i.e., prolactin, growth hormone, and melanocyte-stimulating hor-mone). 671

norepinephrine (NE) (nor-ep-uh-nef-rin) Neurotransmitter of the postganglionic fibers in the sympathetic division of the autonomic system; also, a hormone produced by the adrenal medulla. 520, 672

notochord Cartilaginous-like supportive dorsal rod in all chordates sometime in their life cycle; replaced by vertebrae in vertebrates. 386

nuclear envelope Double membrane that surrounds the nucleus and is connected to the endoplasmic reticulum; has pores that allow substances to pass between the nucleus and the cytoplasm. 62

nuclear pore Opening in the nuclear envelope that permits the passage of proteins into the nucleus and ribosomal subunits out of the nucleus. 62

nucleic acid Polymer of nucleotides; both DNA amd RNA are nucleic acids. 48, 186

nucleoid An irregularly shaped region in the prokaryotic cell that contains its genetic material. 59, 135

nucleolus (noo-klee-uh-lus, nyoo-) Dark-staining, spherical body in the cell nucleus that produces ribosomal subunits. 62

nucleosome In the nucleus of a eukaryotic cell, a unit composed of DNA wound around a core of eight histone proteins, giving the appearance of a string of beads. 212

nucleotide Monomer of DNA and RNA consisting of a 5-carbon sugar bonded to a nitrogen-containing base and a phosphate group. 48, 186

nucleus (noo-klee-us, nyoo-) Membrane-bounded organelle that contains chromosomes and controls the structure and function of the cell. 62

O

obligate anaerobe Prokaryote unable to grow in the presence of free oxygen. 314

observation Step in the scientific method by which data are collected before a conclusion is drawn. 11

obstructive pulmonary disorder Characterized by airflow restriction in the airways; includes chronic bronchitis, emphysema, and asthma. 644

octet rule States that an atom other than hydrogen tends to form bonds until it has eight electrons in its outer shell; an atom that already has eight electrons in its outer shell does not react and is inert. 23

oil Substance, usually of plant origin and liquid at room temperature, formed when a glycerol molecule reacts with three fatty acid molecules. 43

olfactory cell (ahl-fak-tuh-ree, -tree, ohl-) Modified neuron that is a sensory receptor for the sense of smell. 539

oligotrophic lake Lake with few nutrients, usually very blue. 771

omnivore (ahm-nuh-vor) Organism in a food chain that feeds on both plants and animals. 752

oncogene (ahng-koh-jeen) Cancer-causing gene. 218

oocyte Immature egg that is undergoing meiosis; upon completion of meiosis, the oocyte becomes an egg. 686

oogenesis (oh-uh-jen-uh-sis) Production of an egg in females by the process of meiosis and maturation. 150, 687

open circulatory system Arrangement of internal transport in which blood bathes the organs directly, and there is no distinction between blood and interstitial fluid. 575

operant conditioning Learning that results from rewarding or reinforcing a particular behavior. 729

operator In an operon, the sequence of DNA that binds tightly to a repressor, and thereby regulates the expression of structural genes. 208

operon Group of structural and regulating genes that function as a single unit. 208

opportunistic infection Infection that has an opportunity to occur because the immune system has been weakened. 592

opportunistic population Population demonstrating a life history pattern in which members exhibit exponential population growth. Its members are small in size, mature early, have a short life span, produce many offspring, and provide little or no care to offspring (e.g., dandelions). 717

order One of the categories, or taxa, used by taxonomists to group species; the taxon above the family level. 8, 290

organ Combination of two or more different tissues performing a common function. 4, 496

organelle (or-guh-nel) Small membranous structure in the cytoplasm having a specific structure and function. 60

organic chemistry The study of carbon compounds; chemistry of the living world. 38

organic molecule Type of molecule that contains carbon and hydrogen—and often contains oxygen also. 38

organism Individual living thing. 4, 496

organ of Corti Structure in the vertebrate inner ear that contains auditory receptors (also called spiral organ). 545

organ system Group of related organs working together. 4, 496

orgasm Physiological and psychological sensations that occur at the climax of sexual stimulation. 684

osmolarity Total solute concentration of a solution; measure of water concentration in that the higher the solution osmolarity, the lower the water concentration. 658

osmoregulate Regulation of the water volume, or solute concentration, by sensory receptors. 652

osmosis (ahz-moh-sis, ahs-) Diffusion of water through a selectively permeable membrane. 89

ossicle (ahs-ih-kul) One of the small bones of the middle ear—malleus, incus, and stapes. 256

osteoblast (ahs-tee-uh-blast) Bone-forming cell. 559

osteoclast (ahs-tee-uh-klast) Cell that causes erosion of bone. 559

osteocyte (ahs-tee-uh-syt) Mature bone cell located within the lacunae of bone. 560

osteoporosis Condition in which bones break easily because calcium is removed from them faster than it is replaced. 559

outer ear Portion of ear consisting of the pinna and auditory canal. 544

out-of-Africa hypothesis Proposal that modern humans originated only in Africa; then they migrated out of Africa and supplanted populations of early *Homo* in Asia and Europe about 100,000 years ago. 411

ovarian cycle (oh-vair-ee-un) Monthly follicle changes occurring in the ovary that control the level of sex hormones in the blood and the uterine cycle. 688

ovary In animals, the female gonad, the organ that produces eggs, estrogen, and progesterone; in flowering plants, the base of the pistil that protects ovules and, along with associated tissues, becomes a fruit. 353, 480, 682, 686

overexploitation When the number of individuals taken from a wild population is so great that the population becomes severely reduced in numbers. 787

oviduct Tube that transports oocytes to the uterus; also called a uterine tube. 686

ovulation (ahv-yuh-lay-shun, ohv-) Release of a secondary oocyte from the ovary; if fertilization occurs, the secondary oocyte becomes an egg. 687

ovule In seed plants, a structure where the megaspore becomes an egg-producing female gametophyte and which develops into a seed following fertilization. 345, 480

oxidation Loss of one or more electrons from an atom or molecule; in biological systems, generally the loss of hydrogen atoms. 98

oxygen debt Amount of oxygen needed to metabolize lactate, a compound that accumulates during vigorous exercise. 123, 568

oxyhemoglobin (ahk-see-hee-muh-gloh-bin) Compound formed when oxygen combines with hemoglobin. 643

oxytocin (ahk-sih-toh-sin) Hormone released by the posterior pituitary that causes contraction of uterus and milk letdown. 671

P

p53 For control of cell division, the *p53* gene halts the cell cycle when DNA mutates and is in need of repair. 141

pain receptor Sensory receptor that is sensitive to chemicals released by damaged tissues or excess stimuli of heat or pressure. 536

paleontologist Individual who studies fossils and the history of life. 249

paleontology Study of fossils that results in knowledge about the history of life. 245

palisade mesophyll Layer of tissue in a plant leaf containing elongated cells with many chloroplasts. 434

pancreas (pang-kree-us, pan-) Internal organ that produces digestive enzymes and the hormones insulin and glucagon. 620, 674

pancreatic amylase Enzyme that digests starch to maltose. 619

pancreatic islets (islets of Langerhans) Masses of cells that constitute the endocrine portion of the pancreas. 674

parallel evolution Similarity in structure in related groups that cannot be traced to a common ancestor. 292

parasite Species that is dependent on a host species for survival, usually to the detriment of the host species. 746

parasitism Symbiotic relationship in which one species (the *parasite*) benefits in terms of growth and reproduction to the detriment of the other species (the *host*). 746

parasympathetic division That part of the autonomic system that is active under normal conditions; uses acetylcholine as a neurotransmitter. 530

parathyroid gland (par-uh-thy-royd) Gland embedded in the posterior surface of the thyroid gland; it produces parathyroid hormone. 676

parenchyma cell Plant tissue composed of the least-specialized of all plant cells; found in all organs of a plant. 426

parent cell Cell that divides so as to form daughter cells. 137

Parkinson disease Progressive deterioration of the central nervous system due to a deficiency in the neurotransmitter dopamine. 525

parthenogenesis Development of an egg cell into a whole organism without fertilization. 682

partial pressure Pressure exerted by each gas in a mixture of gases. 642

pattern formation Positioning of cells during development that determines the final shape of an organism. 695

peat Organic fuel consisting of the partially decomposed remains of peat mosses that accumulate in bogs. 346

pectoral girdle (pek-tur-ul) Portion of the skeleton that provides support and attachment for an arm; consists of a scapula and a clavicle. 558

pelagic zone Open portion of the sea. 773

pelvic girdle Portion of the skeleton to which the legs are attached; consists of the coxal bones. 558

penis External organ in males through which the urethra passes; also serves as the organ of sexual intercourse. 684

pepsin Enzyme secreted by gastric glands that digests proteins to peptides. 618

peptide Two or more amino acids joined together by covalent bonding. 45

peptide bond Type of covalent bond that joins two amino acids. 45

peptide hormone Type of hormone that is a protein, a peptide, or derived from an amino acid. 667

peptidoglycan Unique molecule found in bacterial cell walls. 42

perennial plant Flowering plant that lives more than one growing season because the underground parts regrow each season. 423

pericycle Layer of cells surrounding the vascular tissue of roots; produces branch roots. 429

periderm Protective tissue that replaces epidermis; includes cork, cork cambium. 432

peripheral nervous system (PNS) (puh-rif-ur-ul) Nerves and ganglia that lie outside the central nervous system. 516

peristalsis (pair-ih-stawl-sis) Wavelike contractions that propel substances along a tubular structure, such as the esophagus. 617

peritubular capillary network (pair-ih-too-byuh-lur) Capillary network that surrounds a nephron and functions in reabsorption during urine formation. 654

permafrost Permanently frozen ground, usually occurring in the tundra, a biome of Arctic regions. 765

peroxisome Enzyme-filled vesicle in which fatty acids and amino acids are metabolized to hydrogen peroxide that is broken down to harmless products. 66

Personal Genome Project Sequencing of every person's particular genome. 234

petal A flower part that occurs just inside the sepals; often conspicuously colored to attract pollinators. 353, 480

petiole Part of a plant leaf that connects the blade to the stem. 423

Peyer patches Lymphatic organs located in the small intestine. 595

phagocytize To ingest extracellular particles by engulfing them, as do amoeboid cells. 328

phagocytosis (fag-uh-sy-toh-sis) Process by which amoeboid-type cells engulf large substances, forming an intracellular vacuole. 90

pharynx (far-ingks) Portion of the digestive tract between the mouth and the esophagus that serves as a passageway for food and also for air on its way to the trachea. 617, 638

phenotype (fee-nuh-typ) Visible expression of a genotype—for example, brown eyes or attached earlobes. 163

pheromone Chemical signal released by an organism that affects the metabolism or influences the behavior of another individual of the same species. 3, 538, 734

phloem Vascular tissue that conducts organic solutes in plants; contains sieve-tube elements and companion cells. 427, 442

phloem sap Solution of sugars, nutrients, and hormones found in the phloem tissue of a plant. 442

phospholipid (fahs-foh-lip-id) Molecule that forms the bilayer of the cell's membranes; has a polar, hydrophilic head bonded to two nonpolar, hydrophobic tails. 44

phospholipid bilayer Comprises the plasma membrane; each polar, hydrophilic head is bonded to two nonpolar, hydrophobic tails; contains embedded proteins. 85

photoperiodism Relative lengths of daylight and darkness that affect the physiology and behavior of an organism. 470

photosynthesis Process occurring usually within chloroplasts whereby chlorophyll-containing organelles trap solar energy to reduce carbon dioxide to carbohydrate. 7, 96

photosystem I (PSI) and **photosystem II (PSII)** Photosynthetic unit where solar energy is absorbed and high-energy electrons are generated; contains a pigment complex and an electron acceptor. 101

phototropism Growth response of plant stems to light; stems demonstrate positive phototropism. 460, 466

pH scale Measurement scale for hydrogen ion concentration. 31

phylogenetic cladistics School of systematics that uses derived characters to determine monophyletic groups and construct cladograms. 294

phylogenetic (evolutionary) tree Diagram that indicates common ancestors and lines of descent among a group of organims. 291

phylogeny Evolutionary history of a group of organisms. 291

phylum One of the categories, or taxa, used by taxonomists to group species; the taxon above the class level. 8, 290

phytochrome Photoreversible plant pigment that is involved in photoperiodism and other responses of plants such as etiolation. 471

phytoplankton Part of plankton containing organisms that photosynthesize, releasing oxygen to the atmosphere and serving as food producers in aquatic ecosystems. 332, 773

phytoremediation The use of plants to restore a natural area to its original condition. 447

pineal gland (pin-ee-ul, py-nee-ul) Endocrine gland located in the third ventricle of the brain; produces melatonin. 526, 675

pinocytosis Process by which vesicle formation brings macromolecules into the cell. 90

pioneer species First species to colonize an area devoid of life. 750

pith Parenchyma tissue in the center of some stems and roots. 428

pituitary gland Endocrine gland that lies just inferior to the hypothalamus; consists of the anterior pituitary and posterior pituitary. 670

placenta Organ formed during the development of placental mammals from the chorion and the uterine wall; allows the embryo, and then the fetus, to acquire nutrients and rid itself of wastes; produces hormones that regulate pregnancy. 683, 700

placental mammal Member of mammalian subclass characterized by the presence of a placenta during the development of an offspring. 392

plague Acute, contagious disease caused by the bacterium *Pasteurella pestis*, usually transmitted by rodents to humans. 318

planarian Free-living flatworm with a ladderlike nervous system. 376

plankton Freshwater and marine organisms that are suspended on or near the surface of the water; includes phytoplankton and zooplankton. 324

plant Multicellular, usually photosynthetic, organism belonging to the plant kingdom. 9

plant hormone Chemical signal that is produced by various plant tissues and coordinates the activities of plant cells. 460

plant tissue culture Process of growing plant cells in the laboratory. 463

plaque (plak) Accumulation of soft masses of fatty material, particularly cholesterol, beneath the inner linings of the arteries. 582

plasma (plaz-muh) Liquid portion of blood; contains nutrients, wastes, salts, and proteins. 584

plasma cell Cell derived from a B cell that is specialized to mass-produce antibodies. 600

plasma membrane Membrane surrounding the cytoplasm that consists of a phospholipid bilayer with embedded proteins; functions to regulate the entrance and exit of molecules from the cell. 59

plasmid (plaz-mid) Self-replicating ring of accessory DNA in the cytoplasm of bacteria. 226, 313

plasmodesma In plants, a cytoplasmic strand that extends through pores in the cell wall and connects the cytoplasm of two adjacent cells. 11, 72, 427

plasmodial slime mold Free-living mass of cytoplasm that moves by pseudopods on a forest floor or in a field, feeding on decaying plant material by phagocytosis; reproduces by spore formation. 331

plasmolysis Contraction of the cell contents due to the loss of water. 89

plastid Organelles of plants and algae that are bounded by a double membrane and contain internal membranes and/or vesicles (i.e., chloroplasts, chromoplasts, leucoplasts). 68

platelet Cell fragment that is necessary to blood clotting; also called a thrombocyte. 499

pleiotropy Inheritance pattern in which one gene affects many phenotypic charactertistics of the individual. 173

point mutation Alteration in a gene due to a change in a single nucleotide; results of this mutation vary. 200

polar body In oogenesis, a nonfunctional product; two to three meiotic products are of this type. 170

polar covalent bond Bond in which the sharing of electrons between atoms is unequal. 26

pollen grain In seed plants, the sperm-producing male gametophyte. 351, 481

pollen tube In seed plants, a tube that forms when a pollen grain lands on the stigma and germinates. The tube grows, passing between the cells of the stigma and the style to reach the egg inside an ovule, where fertilization occurs. 355

pollination In seed plants, the delivery of pollen to the vicinity of the egg-producing female gametophyte. 351, 482

pollution Any environmental change that adversely affects the lives and health of living things. 787

polygenic inheritance Pattern of inheritance in which a trait is controlled by several allelic pairs; each dominant allele contributes to the phenotype in an additive and like manner. 172

polymer Macromolecule consisting of covalently bonded monomers; for example, a polypeptide is a polymer of monomers called amono acids. 40

polymerase chain reaction (PCR) (pahl-uh-muh-rays, -rayz) Technique that uses the enzyme DNA polymerase to produce millions of copies of a particular piece of DNA. 227

polyp (pahl-ip) Small, abnormal growth that arises from the epithelial lining. 621

polypeptide Polymer of many amino acids linked by peptide bonds. 45

polyploid Having a chromosome number that is a multiple greater than twice that of the monoploid number. 152, 271

polyribosome (pahl-ih-ry-buh-sohm) String of ribosomes simultaneously translating regions of the same mRNA strand during protein synthesis. 197

polysaccharide (pahl-ee-sak-uh-ryd) Polymer made from sugar monomers; the polysaccharides starch and glycogen are polymers of glucose monomers. 42

pond Freshwater basin, smaller than a lake. 771

pons (pahnz) Portion of the brain stem above the medulla oblongata and below the midbrain; assists the medulla oblongata in regulating the breathing rate. 526

population Organisms of the same species occupying a certain area. 10, 710

population density The number of individuals per unit area or volume living in a particular habitat. 711

population distribution The pattern of dispersal of individuals living within a certain area. 711

portal system Pathway of blood flow that begins and ends in capillaries, such as the portal system located between the small intestine and liver. 580

positive feedback Mechanism in which the stimulus initiates reactions that lead to an increase in the stimulus. 671

posterior pituitary Portion of the pituitary gland that stores and secretes oxytocin and antidiuretic hormone produced by the hypothalamus. 670

posttranscriptional control Gene expression following translation regulated by the way mRNA transcripts are processed. 214

posttranslational control Gene expression following translation regulated by the activity of the newly synthesized protein. 214

postzygotic isolating mechanism Anatomical or physiological difference between two species that prevents successful reproduction after mating has taken place. 267

potential energy Stored energy as a result of location or spatial arrangement. 78

predation Interaction in which one organism (the *predator*) uses another (the *prey*) as a food source. 744

predator Organism that practices predation. 744

preimplantation genetic diagnosis (PGD) Testing of a cell from an 8-celled embryo for genetic defects prior to implantation into the mother. 170

preparatory (prep) reaction Reaction that oxidizes pyruvate with the release of carbon dioxide; results in acetyl CoA and connects glycolysis to the citric acid cycle. 115, 118

pressure-flow model Explanation for phloem transport; osmotic pressure following active transport of sugar into phloem brings about a flow of sap from a source to a sink. 448

prezygotic isolating mechanism Anatomical or behavioral difference between two species that prevents the possibility of mating. 266

primary motor area Area in the frontal lobe where voluntary commands begin; each section controls a part of the body. 525

primary somatosensory area (soh-mat-uh-sens-ree, -suh-ree) Area dorsal to the central sulcus where sensory information arrives from skin and skeletal muscles. 525

primate Member of the order Primate; includes prosimians, monkeys, apes, and hominids, all of whom have adaptations for living in trees. 400

prime mover Muscle most directly responsible for a particular movement. 564

producer Photosynthetic organism at the start of a grazing food chain that makes its own food (e.g., green plants on land and algae in water). 752

proembryo Smaller portion of divided sporophyte embryo that, after dividing repeatedly, becomes the embryo of a plant. 484

progesterone (proh-jes-tuh-rohn) Female sex hormone that helps maintain sex organs and secondary sex characteristics. 687

prokaryotic cell (prokaryote) Organism that lacks the membrane-bounded nucleus and membranous organelles typical of eukaryotes. 8, 59

prolactin (PRL) (proh-lak-tin) Hormone secreted by the anterior pituitary that stimulates the production of milk from the mammary glands. 671

proliferative phase Phase of the uterine cycle in which there is an increased production of estrogen, causing the endometrium to thicken. 688

promoter In an operon, a sequence of DNA where RNA polymerase binds prior to transcription. 194, 208

prosimian Group of primates that includes lemurs and tarsiers, and may resemble the first primates to have evolved. 403

prostate gland (prahs-tayt) Gland located around the male urethra below the urinary bladder; adds secretions to semen. 684

protein Molecule consisting of one or more polypeptides. 45, 623

proteome Collection of proteins resulting from the translation of genes into proteins. 236

proteomics The study of all proteins in an organism. 236

protist Member of the kingdom Protista. 8, 324

protocell In biological evolution, a possible cell forerunner that became a cell once it could reproduce. 309

proton Positive subatomic particle, located in the nucleus and having a weight of approximately one atomic mass unit. 21

proto-oncogene (proh-toh-ahng-koh-jeen) Normal gene that can become an oncogene through mutation. 218

protostome Group of coelomate animals in which the first embryonic opening (the blastopore) is associated with the mouth. 379

protozoan Heterotrophic, unicellular protist that moves by flagella, cilia, or pseudopodia, or is immobile. 324

proximal convoluted tubule Highly coiled region of a nephron near the glomerular capsule, where tubular reabsorption takes place. 654

pseudocoelomate Animal with a body cavity lying between the digestive tract and body wall that is incompletely lined by mesoderm. 371

pseudopod Cytoplasmic extension of amoeboid protists; used for locomotion and engulfing food. 70, 328

pseudostratified ciliated columnar epithelium Appearance of layering in some epithelial cells when, actually, each cell touches a baseline and true layers do not exist. 497

puberty Period of life when secondary sex changes occur in humans; marked by the onset of menses in females and sperm production in males. 685

pulmonary artery (pool-muh-nair-ee, puul-) Blood vessel that takes blood away from the heart to the lungs. 577

pulmonary circuit Circulatory pathway that consists of the pulmonary trunk, the pulmonary arteries, and the pulmonary veins; takes O_2-poor blood from the heart to the lungs and O_2-rich blood from the lungs to the heart. 576

pulmonary trunk Large blood vessel that divides into the pulmonary arteries; takes blood away from the heart to the lungs. 577

pulmonary vein Blood vessel that takes blood from the lungs to the heart. 577

pulse Vibration felt in arterial walls due to expansion of the aorta following ventricle contraction. 578

pupil (pyoo-pul) Opening in the center of the iris of the eye. 540

pyruvate End product of glycolysis; its further fate, involving fermentation or entry into a mitochondrion, depends on oxygen availability. 115

R

radial symmetry Body plan in which similar parts are arranged around a central axis, like spokes of a wheel. 371

radiolarian Member of the phylum Actinopoda bearing a glassy silicon test, usually with a radial arrangement of spines; pseudopods are external to the test. 328

rain shadow Leeward side (side sheltered from the wind) of a mountainous barrier, which receives much less precipitation than the windward side. 770

range That portion of the globe where a certain species can be found. 711

ray-finned fish Group of bony fishes with fins supported by parallel bony rays connected by webs of thin tissue. 388

receptacle Area where a flower attaches to a floral stalk. 353, 480

receptor-mediated endocytosis Selective uptake of molecules into a cell by vacuole formation after they bind to specific receptor proteins in the plasma membrane. 90

recessive allele (uh-leel) Allele that exerts its phenotypic effect only in the homozygote; its expression is masked by a dominant allele. 163

reciprocal altruism The trading of helpful or cooperative acts, such as helping at the nest, by individuals—the animal that was helped will repay the debt at some later time. 733

recombinant DNA (rDNA) DNA that contains genes from more than one source. 226

rectum (rek-tum) Terminal end of the digestive tube between the sigmoid colon and the anus. 621

red algae Marine photosynthetic protists with a notable abundance of phycobilin pigments; include coralline algae of coral reefs. 333

red blood cell Formed element that contains hemoglobin and carries oxygen from the lungs to the tissues; erythrocyte. 499, 584

red bone marrow Blood-cell-forming tissue located in the spaces within spongy bone. 560, 595

red tide Occurs frequently in coastal areas and is often associated with population blooms of dinoflagellates. Dinoflagellate pigments are responsible for the red color of the water. Under these conditions, the dinoflagellates often produce saxitoxin, which can lead to paralytic shellfish poisoning. 332

redox reaction Oxidation-reduction reaction; one molecule loses electrons (oxidation) while another molecule simultaneously gains electrons (reduction). 98

reduction Chemical reaction that results in addition of one or more electrons to an atom, ion, or compound. Reduction of one substance occurs simultaneously with oxidation of another. 98

reflex Automatic, involuntary response of an organism to a stimulus. 529

reflex action An action performed automatically, without conscious thought (e.g., swallowing). 524

refractory period (rih-frak-tuh-ree) Time following an action potential when a neuron is unable to conduct another nerve impulse. 519

regulator gene In an operon, a gene that codes for a protein that regulates the expression of other genes. 208

renal cortex (ree-nul kor-teks) Outer portion of the kidney that appears granular. 654

renal medulla (ree-nul muh-dul-uh) Inner portion of the kidney that consists of renal pyramids. 654

renal pelvis Hollow chamber in the kidney that lies inside the renal medulla and receives freshly prepared urine from the collecting ducts. 654

renin (ren-in) Enzyme released by kidneys that leads to the secretion of aldosterone and a rise in blood pressure. 658

replacement reproduction Population in which each person is replaced by only one child. 720

repolarization Recovery of a neuron's polarity to the resting potential after the neuron ceases transmitting impulses. 519

repressible operon Operon that is normally active because the repressor is normally inactive. 208

reproduce To produce a new individual of the same kind. 7

reproductive cloning Genetically identical to the original individual. 210

reproductive system Organ system that contains male or female organs and specializes in the production of offspring. 505

reptile Member of a class of terrestrial vertebrates with internal fertilization, scaly skin, and an egg with a leathery shell; includes snakes, lizards, turtles, and crocodiles. 390

resource In economic terms, anything with potential use in creating wealth or giving satisfaction. 711

resource partitioning Mechanism that increases the number of niches by apportioning the supply of a resource such as food or living space between species. 742

respiration Sequence of events that results in gas exchange between the cells of the body and the environment. 634

respiratory center Group of nerve cells in the medulla oblongata that send out nerve impulses on a rhythmic basis, resulting in involuntary inspiration on an ongoing basis. 641

respiratory system Organ system consisting of the lungs and tubes that bring oxygen into the lungs and take carbon dioxide out. 505

resting potential Polarity across the plasma membrane of a resting neuron due to an unequal distribution of ions. 518

restoration ecology Subdiscipline of conservation biology that seeks ways to return ecosystems to their former state. 790

restriction enzyme Bacterial enzyme that stops viral reproduction by cleaving viral DNA; used to cut DNA at specific points during production of recombinant DNA. 226

restrictive pulmonary disorder Lung capacity is reduced because of loss of elasticity causing dificulty in airflow. 644

retina (ret-n-uh, ret-nuh) Innermost layer of the eyeball that contains the rod cells and the cone cells. 540

retrovirus RNA virus containing the enzyme reverse transcriptase that carries out RNA to DNA transcription. 306

rhizome Rootlike underground stem. 487

rhodopsin (roh-dahp-sun) Light-absorbing molecule in rod cells and cone cells that contains a pigment and the protein opsin. 542

ribose Pentose sugar found in RNA. 41

ribosomal RNA (rRNA) (ry-buh-soh-mul) Type of RNA found in ribosomes where protein synthesis occurs. 197

ribosome (ry-buh-sohm) RNA and protein in two subunits; site of protein synthesis in the cytoplasm. 59

ribozyme Enzyme that carries out mRNA processing. 195

rigor mortis Contraction of muscles at death due to lack of ATP. 567

river Freshwater channel that flows eventually to the oceans. 771

RNA (ribonucleic acid) (ry-boh-noo-klee-ik) Nucleic acid produced from covalent bonding of nucleotide monomers that contain the sugar ribose; occurs in three forms: messenger RNA, ribosomal RNA, and transfer RNA. 48, 186

RNA polymerase (pahl-uh-muh-rays) During transcription, an enzyme that joins nucleotides complementary to a DNA template. 194

root The usually descending axis of a plant, normally below ground, which anchors the plant and serves as the major point of entry for water and minerals. 423

root hair Extension of a root epidermal cell that increases the surface area for the absorption of water and minerals. 423, 452

root nodule Structure on plant root that contains nitrogen-fixing bacteria. 454

root pressure Osmotic pressure caused by active movement of minerals into root cells; serves to elevate water in xylem for a short distance. 444

root system Includes the main root and any and all of its lateral (side) branches. 422

rotational balance Maintenance of balance when the head and body are suddenly moved or rotated. 546

rough ER (RER) Membranous system of tubules, vesicles, and sacs in cells; has attached ribosomes. 64

roundworm Member of the phylum Nematoda with a cylindrical body that has a complete digestive tract and a pseudocoelom; some forms are free-living in water and soil; many are parasitic. 378

RuBP carboxylase Enzyme that is required fr carbon dioxide fixation (atmospheric CO_2 attaches to RuBP) in the Calvin cycle. 104

ruminant Cow and related mammals wherein digestion of cellulose occurs in an extra stomach, rumen, from which partially digested material can be ejected back into the mouth. 615

S

saccule (sak-yool) Saclike cavity in the vestibule of the inner ear; contains sensory receptors for gravitational equilibrium. 546

salivary amylase (sal-uh-vair-ee am-uh-lays, -layz) Secreted from the salivary glands; the first enzyme to act on starch. 616

salivary gland Gland associated with the mouth that secretes saliva. 616

salt Compound produced by a reaction between an acid and a base. 24

saltatory conduction Movement of nerve impulses from one neurolemmal node to another along a myelinated axon. 519

saprotroph Organism that secretes digestive enzymes and absorbs the resulting nutrients back across the plasma membrane. 314

sarcolemma (sar-kuh-lem-uh) Plasma membrane of a muscle fiber; also forms the tubules of the T system involved in muscular contraction. 566

sarcomere (sar-kuh-mir) One of many units, arranged linearly within a myofibril, whose contraction produces muscle contraction. 566

sarcoplasmic reticulum (sar-kuh-plaz-mik rih-tik-yuh-lum) Smooth endoplasmic reticulum of skeletal muscle cells; surrounds the myofibrils and stores calcium ions. 566

saturated fatty acid Fatty acid molecule that lacks double bonds between the atoms of its carbon chain. 43

savanna Terrestrial biome that is a grassland in Africa, characterized by few trees and a severe dry season. 767

scanning electron microscope Beam of electrons scans over a specimen point by point and builds up an image on a fluorescent screen. 58

scavenger Animal that specializes in the consumption of dead animals. 752

Schwann cell Cell that surrounds a fiber of a peripheral nerve and forms the myelin sheath. 517

scientific process Process by which scientists formulate a hypothesis, gather data by observation and experimentation, and come to a conclusion. 11

scientific theory Concept supported by a broad range of observations, experiments, and conclusions. 11

sclera (skleer-uh) White, fibrous, outer layer of the eyeball. 540

sclerenchyma cell Plant tissue composed of cells with heavily lignified cell walls; functions in support. 427

seaweed Multicellular forms of red, green, and brown algae found in marine habitats. 333

secondary growth In vascular plants, an increase in stem and root diameter made possible by cell division of the lateral meristems. 432

secondary metabolite Molecule not directly involved in growth, development, or reproduction of an organism; in plants, these molecules, which include nicotine, caffeine, tannins, and menthols, can discourage herbivores. 472

secondary sex characteristic Trait that is sometimes helpful but not absolutely necessary for reproduction and is maintained by the sex hormones in males and females. 685

second messenger Chemical signal such as cyclic AMP that causes the cell to respond to the first messenger—a hormone bound to a plasma membrane receptor. 667

secretin Peptide hormone secreted by upper small intestine; stimulates pancreas to secrete bicarbonate into small intestine. 621

secretion In the cell, release of a substance by exocytosis from a cell that may be a gland or part of a gland; in the urinary system, movement of certain molecules from blood into the distal tubule of a nephron so that they are added to urine. 65

secretory phase Phase of the uterine cycle in which there is an increased production of progesterone causing the endometrium to double in thickness, producing a thick mucoid secretion. 688

segmentation Repetition of body parts as segments along the length of the body; seen in annelids, arthropods, and chordates. 371

self-antigen Antigen that is produced by an organism. 599

semen (seminal fluid) (see-mun) Thick, whitish fluid consisting of sperm and secretions from several glands of the male reproductive tract. 684

semicircular canal (sem-ih-sur-kyuh-lur) One of three tubular structures within the inner ear that contain sensory receptors responsible for the sense of rotational equilibrium. 544

semiconservative replication Duplication of DNA resulting in two double helix molecules, each having one parental and one new strand. 190

semilunar valve (sem-ee-loo-nur) Valve resembling a half moon located between the ventricles and their attached vessels. 577

seminal vesicle (sem-uh-nul) Convoluted, saclike structure attached to the vas deferens near the base of the urinary bladder in males; adds secretions to semen. 684

seminiferous tubule (sem-uh-nif-ur-us) Long, coiled structure contained within chambers of the testis; where sperm are produced. 685

senescence Sum of the processes involving aging, decline, and eventual death of a plant or plant part. 463

sensation Conscious awareness of a stimulus due to nerve impulse sent to the brain from a sensory receptor by way of sensory neurons. 537

sensory adaptation Phenomenon of a sensation becoming less noticeable once it has been recognized by constant repeated stimulation. 537

sensory (afferent) neuron Nerve cell that transmits nerve impulses to the central nervous system after a sensory receptor has been stimulated. 517

sensory receptor Structure that receives either external or internal environmental stimuli and is a part of a sensory neuron or transmits signals to a sensory neuron. 536

sepal Outermost, sterile, leaflike covering of the flower; usually green in color. 353, 480

septum Wall between two cavities; in the human heart, a septum separates the right side from the left side. 577

Sertoli cell Type of cell in seminiferous tubules with FSH receptors; helps nourish and support developing sperm. 685

serum (seer-um) Light yellow liquid left after clotting of blood. 585

sex chromosome Chromosome that determines the sex of an individual; in humans, females have two X chromosomes, and males have an X and Y chromosome. 145

sex pilus in a bacterium, elongated, hollow appendage used to transfer DNA to other cells. 59

sexually transmitted disease (STD) Illness communicated primarily or exclusively through sexual contact. 689

sexual reproduction Reproduction involving meiosis, gamete formation, and fertilization; produces offspring with chromosomes inherited from each parent with a unique combination of genes. 134

sexual selection Changes in males and females, often due to male competition and female selectivity, leading to increased fitness. 730

shoot system Aboveground portion of a plant consisting of the stem, leaves, and flowers. 422

short-day plant Plant that flowers when day length is shorter than a critical length (e.g., cocklebur, poinsettia, and chrysanthemum). 470

sieve-tube member Member that joins with others in the phloem tissue of plants as a means of transport for nutrient sap. 427, 442

signaling molecule Molecule that stimulates or inhibits an event in the cell cycle. 142

simple diffusion Movement of molecules or ions from a region of higher to lower concentration; it requires no energy and tends to lead to an equal distribution. 88

simple goiter (goy-tur) Condition in which an enlarged thyroid produces low levels of thyroxine. 676

simple muscle twitch Contraction of a whole muscle in response to a single stimulus. 564

single nucleotide polymorphism (SNP) Site present in at least 1% of the population at which individuals differ by a single nucleotide. These can be used as genetic markers to map unknown genes or traits. 234

sink In the pressure-flow model of phloem transport, the location (roots) from which sugar is constantly being removed. Sugar will flow to the roots from the source. 449

sink population Population that is found in an unfavorable area where at best the birthrate equals the death rate; sink populations receive new members from source populations. 789

sister chromatid One of two genetically identical chromosomal units that are the result of DNA replication and are attached to each other at the centromere. 136

skeletal muscle Striated, voluntary muscle tissue that comprises skeletal muscles; also called striated muscle. 500

skeletal system System of bones, cartilage, and ligaments that works with the muscular system to protect the body and provide support for locomotion and movement. 504

skin Outer covering of the body; can be called the integumentary system because it contains organs such as sense organs. 503

skull Bony framework of the head, composed of cranial bones and the bones of the face. 556

sliding filament model An explanation for muscle contraction based on the movement of actin filaments in relation to myosin filaments. 556

slime mold Protists that decompose dead material and feed on bacteria by phagocytosis; vegetative state is mobile and amoeboid. 324

small intestine In vertebrates, the portion of the digestive tract that precedes the large intestine; in humans, consists of the duodenum, jejunum, and ileum. 619

smallpox Acute contagious viral disease characterized by skin pustules. 318

smooth ER (SER) Membranous system of tubules, vesicles, and sacs in eukaryotic cells; lacks attached ribosomes. 64

smooth muscle Nonstriated, involuntary muscle tissue found in the walls of internal organs. 500

society Group in which members of species are organized in a cooperative manner, extending beyond sexual and parental behavior. 734

sociobiology Application of evolutionary principles to the study of social behavior of animals, including humans. 732

sodium-potassium pump Carrier protein in the plasma membrane that moves sodium ions out of and potassium ions into cells; important in nerve and muscle cells. 90

soft palate (pal-it) Entirely muscular posterior portion of the roof of the mouth. 616

soil Accumulation of inorganic rock material and organic matter that is capable of supporting the growth of vegetation. 453

soil profile Vertical section of soil from the ground surface to the unaltered rock below. 453

solute Substance that is dissolved in a solvent, forming a solution. 28, 89

solution Fluid (the solvent) that contains a dissolved solid (the solute). 28, 89

solvent Liquid portion of a solution that serves to dissolve a solute. 89

somatic cell (soh-mat-ik) Body cell; excludes cells that undergo meiosis and become a sperm or egg. 134

somatic system That portion of the peripheral nervous system containing motor neurons that control skeletal muscles. 529

sorus (pl., sori) Dark spot on the underside of fern fronds that is a collection of spore-producing structures. 349

sound communication Form of communication between animals using auditory (sound) effects. 734

source In the pressure-flow model of phloem transport, the location (leaves) of sugar production. Sugar will flow from the leaves to the sink. 449

source population Population that can provide members to other populations of the species because it lives in a favorable area, and the birthrate is most likely higher than the death rate. 789

speciation Origin of new species due to the evolutionary process of descent with modification. 264

species Group of similarly constructed organisms capable of interbreeding and producing fertile offspring; organisms that share a common gene pool; the taxon at the lowest level of classification. 7, 8, 290

species richness Number of species in a community. 749

specific epithet In the binominal system of taxonomy, the second part of an organism's name; it may be descriptive. 9

spermatogenesis (spur-mat-uh-jen-ih-sis) Production of sperm in males by the process of meiosis and maturation. 150, 685

sphincter (sfingk-tur) Muscle that surrounds a tube and closes or opens the tube by contracting and relaxing. 617

spinal cord Part of the central nervous system; the nerve cord that is continuous with the base of the brain plus the vertebral column that protects the nerve cord. 524

spinal nerve Nerve that arises from the spinal cord. 528

spindle fiber Microtubule structure that brings about chromosomal movement during nuclear division. 138

spleen Large, glandular organ located in the upper left region of the abdomen; stores and purifies blood. 595

sponge Invertebrate animal of the phylum Porifera; pore-bearing filter feeder whose inner body wall is lined by collar cells. 374

spongy bone Porous bone found at the ends of long bones where red bone marrow is sometimes located. 560

spongy mesophyll Layer of tissue in a plant leaf containing loosely packed cells, increasing the amount of surface area for gas exchange. 428, 434

sporangium (pl., sporangia) Structure that produces spores. 331, 347

spore Asexual reproductive or resting cell capable of developing into a new organism without fusion with another cell, in contrast to a gamete. 324, 344

sporophyte Diploid generation of the alternation of generations life cycle of a plant; produces haploid spores that develop into the haploid generation. 344, 481

sporozoan Spore-forming protist that has no means of locomotion and is typically a parasite with a complex life cycle having both sexual and asexual phases. 330

squamous epithelium Type of epithelial tissue that contains flat cells. 497

stabilizing selection Outcome of natural selection in which extreme phenotypes are eliminated and the average phenotype is conserved. 256

stamen In flowering plants, the portion of the flower that consists of a filament and an anther containing pollen sacs where pollen is produced. 353, 480

starch Storage polysaccharide found in plants that is composed of glucose molecules joined in a linear fashion with few side chains. 42

statocyst Gravitational equilibrium organ found in some invertebrates; gives information about head position. 548

statolith Sensors found in root cap cells that cause a plant to demonstrate gravitropism. 466

stem Usually the upright, vertical portion of a plant, that transports substances to and from the leaves. 422

steroid (steer-oyd) Type of lipid molecule having a complex of four carbon rings; examples are cholesterol, progesterone, and testosterone. 44

steroid hormone Type of hormone that has a complex of four carbon rings but different side chains from other steroid hormones. 667

stigma In flowering plants, portion of the pistil where pollen grains adhere and germinate before fertilization can occur. 353, 480

stolon Stem that grows horizontally along the ground and may give rise to new plants where it contacts the soil—e.g., the runners of a strawberry plants. 487

stoma (pl., stomata) Small opening between two guard cells on the underside of leaf epidermis through which gases pass. 97, 345, 426, 446

stratified As in the outer layer of skin, having several layers. 497

stratum (pl., strata) Ancient layer of sedimentary rock; results from slow deposition of silt, volcanic ash, and other materials. 245

stream Freshwater channel, smaller than a river. 771

striated Having bands; in cardiac and skeletal muscle, alternating light and dark crossbands produced by the distribution of contractile proteins. 500

stroke Condition resulting when an arteriole in the brain bursts or becomes blocked by an embolism; also called cerebrovascular accident. 582

stroma Fluid within a chloroplast that contains enzymes involved in the synthesis of carbohydrates during photosynthesis. 68, 97

structural gene Gene that codes for an enzyme in a metabolic pathway. 208

style Elongated, central portion of the pistil between the ovary and stigma. 353, 480

subcutaneous layer A sheet that lies just beneath the skin and consists of loose connective and adipose tissue. 503

substrate Reactant in a reaction controlled by an enzyme. 82

substrate feeder Organism that lives in or on its food source. 613

substrate-level ATP synthesis Process in which ATP is formed by tranferring a phosphate from a metabolic substrate to ADP. 116

surface-area-to-volume ratio Ratio of a cell's outside area to its internal volume. 57

survivorship Probability of newborn individuals of a cohort surviving to particular ages. 712

sustainable development Management of an ecosystem so that it maintains itself while providing services to human beings. 790

swallowing Muscular movement of the pharynx and esophagus to take the food bolus from the mouth to the stomach. 617

symbiosis Relationship that occurs when two different species live together in a unique way; it may be beneficial, neutral, or detrimental to one and/or the other species. 746

sympathetic division The part of the autonomic system that usually promotes activities associated with emergency (fight-or-flight) situations; uses norepinephrine as a neurotransmitter. 530

sympatric speciation Origin of new species in populations that overlap geographically. 271

synapse (sin-aps, si-naps) Junction between neurons consisting of the presynaptic (axon) membrane, the synaptic cleft, and the postsynaptic (usually dendrite) membrane. 520

synapsis (sih-nap-sis) Pairing of homologous chromosomes during prophase I of meiosis I. 146

synaptic cleft (sih-nap-tik) Small gap between presynaptic and postsynaptic membranes of a synapse. 520

syndrome Group of symptoms that appear together and tend to indicate the presence of a particular disorder. 153

synovial joint (sih-noh-vee-ul) Freely movable joint in which two bones are separated by a cavity. 561

syphilis (sif-uh-lis) Sexually transmitted disease caused by the bacterium *Treponema pallidum* that, if untreated, can lead to cardiac and central nervous system disorders. 389

systematic Study of the diversity of organisms to classify them and determine their evolutionary relationships. 291

systemic circuit Blood vessels that transport blood from the left ventricle and back to the right atrium of the heart. 576

systemin In plants, an 18-amino-acid peptide that is produced by damaged or injured leaves that leads to the wound response. 472

systole (sis-tuh-lee) Contraction period of the heart during the cardiac cycle. 578

systolic pressure (sis-tahl-ik) Arterial blood pressure during the systolic phase of the cardiac cycle. 581

T

taiga Terrestrial biome that is a coniferous forest extending in a braod belt across northern Eurasia and North America. 765

taste bud Sense organ containing the receptors associated with the sense of taste. 538

taxon Group of organisms that fills a particular classification category. 290

taxonomy Branch of biology concerned with identifying, describing, and naming organisms. 8, 290

T cell Lymphocyte that matures in the thymus. Cytotoxic T cells kill antigen-bearing cells outright; helper T cells release cytokines that stimulate other immune system cells. 585, 595

T cell receptor (TCR) Molecule on the surface of a T cell that can bind to a specific antigen fragment in combination with an MHC molecule. 600

telomere (tel-uh-meer) Tip of the end of a chromosome. 142, 191

temperate deciduous forest Forest found south of the taiga; characterized by deciduous trees such as oak, beech, and maple, moderate climate, relatively high rainfall, stratified plant growth, and plentiful ground life. 766

temperate grassland Grazing, fire, and drought restrict tree growth in this terrestrial biome. 766

temperate rain forest Coniferous forest—e.g., that running along the west coast of Canada and the United States—characterized by plentiful rainfall and rich soil. 765

template (tem-plit) Pattern or guide used to make copies; parental strand of DNA serves as a guide for the production of daughter DNA strands, and DNA also serves as a guide for the production of messenger RNA. 190

tendon Strap of fibrous connective tissue that connects skeletal muscle to bone. 499, 564

terminal bud Bud that develops at the apex of a shoot. 422

terminal bud scale scar Marking from the shedding of terminal bud scales; age of a stem can be determined by number of scars. 431

territoriality Marking and/or defending a particular area against invasion by another species member; area often used for the purpose of feeding, mating, and caring for young. 731

territory Area occupied and defended exclusively by an animal or group of animals. 731

testcross Cross between an individual with the dominant phenotype and an individual with the recessive phenotype. The resulting phenotypic ratio indicates whether the dominant phenotype is homozygous or heterozygous. 166

testis (pl., testes) male gonad that produces sperm and the male sex hormones. 682, 684

testing Means in which to determine the accuracy of a prediction, as by conducting experiments. 11

testosterone (tes-tahs-tuh-rohn) Male sex hormone that helps maintain sexual organs and secondary sex characteristics. 685

tetrad Homologous chromosomes, each having sister chromatids that are joined during meiosis. 146

tetrapod Four-footed vertebrate; includes amphibians, reptiles, birds, and mammals. 387

thalamus (thal-uh-mus) Part of the brain located in the lateral walls of the third ventricle that serves as the integrating center for sensory input; it plays a role in arousing the cerebral cortex. 526

therapeutic cloning Used to create mature cells of various cell types. Also, used to learn about specialization of cells and provide cells and tissue to treat human illnesses. 210

thermoacidophile Type of archaea that lives in hot, acidic, aquatic habitats, such as hot springs or near hydrothermal vents. 316

thermoreceptor Sensory receptor that is sensitive to changes in temperature. 536

thigmotropism In plants, unequal growth due to contact with solid objects, as the coiling of tendrils around a pole. 466

threatened species Species that is likely to become an endangered species in the foreseeable future (e.g., bald eagle, gray wolf, Lousiana black bear). 780

threshold Electrical potential level (voltage) at which an action potential or nerve impulse is produced. 518

thrombin (thrahm-bin) Enzyme that converts fibrinogen to fibrin threads during blood clotting. 585

thylakoid Flattened sac within a granum whose membrane contains chlorophyll and where the light reactions of photosynthesis occur. 68, 97

thylakoid space Inner compartment of the thylakoid. 68

thymine (T) (thy-meen) One of four nitrogen-containing bases in nucleotides composing the structure of DNA; pairs with adenine. 186

thymus gland Lymphatic organ, located along the trachea behind the sternum, involved in the maturation of T lymphocytes in the thymus gland. Secretes hormones called thymosins, which aid the maturation of T cells and perhaps stimulate immune cells in general. 595

thyroid gland Endocrine gland in the neck that produces several important hormones, including thyroxine, triiodothyronine, and calcitonin. 676

thyroid-stimulating hormone (TSH) Substance produced by the anterior pituitary that causes the thyroid to secrete thyroxine and triiodothyronine. 671

thyroxine (T₄) (thy-rahk-sin) Hormone secreted from the thyroid gland that promotes growth and development; in general, it increases the metabolic rate in cells. 676

tissue Group of similar cells that perform a common function. 4, 496

tissue culture Process of growing tissue artificially in usually a liquid medium in laboratory glassware. 488

tissue fluid Fluid that surrounds the body's cells; consists of dissolved substances that leave the blood capillaries by filtration and diffusion. 505, 587

tonsillitis Infection of the tonsils that causes inflammation and can spread to the middle ears. 616

tonsils Partially encapsulated lymph nodules located in the pharynx. 595

topography Surface features of the Earth. 770

torpedo stage Stage of development of a sporophyte embryo; embryo has a torpedo shape and the root and shoot apical meristems are present. 484

totipotent Cell that has the full genetic potential of the organism, including the potential to develop into a complete organism. 209, 488

tracer Substance having an attached radioactive isotope that allows a researcher to track its whereabouts in a biological system. 22

trachea (tray-kee-uh) In birds and mammals, passageway that conveys air from the larynx to the bronchi; also called the windpipe. 638

tracheae In insects, air tubes located between the spiracles and the tracheoles. 637

tracheid In flowering plants, type of cell in xylem that has tapered ends and pits through which water and minerals flow. 427, 442

tract Bundle of myelinated axons in the central nervous system. 524

transcription Process whereby a DNA strand serves as a template for the formation of mRNA. 192

transcription activator Protein that initiates transcription by RNA polymerase and thereby starts the process that results in gene expression. 213

transcriptional control Control of gene expression during the transcriptional phase determined by mechanisms that control whether transcription occurs or the rate at which it occurs. 213

transcription factor In eukaryotes, protein required for the initiation of transcription by RNA polymerase. 213

transduction Exchange of DNA between bacteria by means of a bacteriophage. 313

transfer RNA (tRNA) Type of RNA that transfers a particular amino acid to a ribosome during protein synthesis; at one end, it binds to the amino acid, and at the other end it has an anticodon that binds to an mRNA codon. 196

transformation Taking up of extraneous genetic material from the environment by bacteria. 313

transgenic organism Free-living organism in the environment that has had a foreign gene inserted into it. 228

transitional fossil A fossil that bears a resemblance to two groups that in present day are classified separately. 250

translation Process whereby ribosomes use the sequence of codons in mRNA to produce a polypeptide with a particular sequence of amino acids. 192

translational control Gene expression regulated by the activity of mRNA transcripts. 214

translocation Movement of a chromosomal segment from one chromosome to another nonhomologous chromosome, leading to abnormalities (e.g, Down syndrome). 154, 198

transmisssion electron microscope Similar to the scanning electron microscope but image is colored by a computer. 58

transpiration Plant's loss of water to the atmosphere, mainly through evaporation at leaf stomata. 445

transport vesicle Vesicle formed in the ER that carries proteins and lipids to the Golgi apparatus. 64

transposon DNA sequence capable of randomly moving from one site to another in the genome. 200

trichocyst Found in ciliates; contains long, barbed threads useful for defense and capturing prey. 329

trichomoniasis Sexually transmitted disease caused by the parasitic protozoan *Trichomonas vaginalis*. 689

triglyceride (trih-glis-uh-ryd) Neutral fat composed of glycerol and three fatty acids. 43, 623

triplet code Each sequence of three nucleotide bases in the DNA of genes stands for a particular amino acid. 193

triploid endosperm In flowering plants, nutritive storage tissue that is derived from an egg uniting with a polar nuclei during double fertilization. 355

trisomy One more chromosome than usual. 152

trophic level Feeding level of one or more populations in a food web. 755

trophoblast Outer cells of a blastocyte that help form the placenta and other extraembryonic membranes. 698

tropical rain forest Biome near the equator in South America, Africa, and the Indo-Malay regions; characterized by warm weather, plentiful rainfall, a diversity of species, and mainly tree-living animal life. 768

tropism In plants, a growth response toward or away from a directional stimulus. 436, 466

trypsin Protein-digesting enzyme secreted by the pancreas. 619

tuber Enlarged, short, fleshy underground stem—e.g., potato. 487

tubular reabsorption Movement of primarily nutrient molecules and water from the contents of the nephron into blood at the proximal convoluted tubule. 656

tubular secretion Movement of certain molecules from blood into the distal convoluted tubule of a nephron so that they are added to urine. 657

tularemia Infectious disease caused by bacterium *Pasteurella tularensis;* characterized by headache, chills, and fever. 318

tumor (too-mur) Cells derived from a single mutated cell that has repeatedly undergone cell division; benign tumors remain at the site of origin, and malignant tumors metastasize. 143

tumor suppressor gene Gene that codes for a protein that ordinarily suppresses cell division; inactivity can lead to a tumor. 218

tunicate Type of primitive invertebrate chordate. 386

turgor movement In plant cells, pressure of the cell contents against the cell wall when the central vacuole is full. 468

Turner syndrome Condition caused by the inheritance of a single X chromosome. 153

tympanic membrane (tim-pan-ik) Located between the outer and middle ear where it receives sound waves; also called the eardrum. 544

typhlosole Expanded dorsal surface of long intestine of earthworms, allowing additional surface for absorption. 614

U

umbilical cord Cord connecting the fetus to the placenta through which blood vessels pass. 699

uniformitarianism Belief espoused by James Hutton that geological forces act at a continous, uniform rate. 245

unsaturated fatty acid Fatty acid molecule that has one or more double bonds between the atoms of its carbon chain. 43

upwelling Upward movement of deep, nutrient-rich water along coasts; it replaces surface waters that move away from shore when the direction of prevailing winds shifts. 774

uracil (U) (yoor-uh-sil) The base in RNA that replaces thymine found in DNA; pairs with adenine. 186

urea Main nitrogenous waste of terrestrial amphibians and most mammals. 650

uremia High level of urea nitrogen in the blood. 660

ureter (yoor-uh-tur) One of two tubes that take urine from the kidneys to the urinary bladder. 654

urethra (yoo-ree-thruh) Tubular structure that receives urine from the bladder and carries it to the outside of the body. 654

uric acid Main nitrogenous waste of insects, reptiles, and birds. 650

urinary bladder Organ where urine is stored before being discharged by way of the urethra. 654

urinary system Organ system consisting of the kidneys and urinary bladder; rids the body of nitrogenous wastes and helps regulate the water-salt balance of the blood. 505

urine Liquid waste product made by the nephrons of the vertebrate kidney through the processes of glomerular filtration, tubular reabsorption, and tubular secretion. 654

uterine cycle (yoo-tur-in, -tuh-ryn) Monthly-occurring changes in the characteristics of the uterine lining (endometrium). 688

uterus (yoo-tur-us) Organ located in the female pelvis where the fetus develops; also called the womb. 686

utricle (yoo-trih-kul) Saclike cavity in the vestibule of the inner ear that contains sensory receptors for gravitational equilibrium. 546

V

vaccine Antigens prepared in such a way that they can promote active immunity without causing disease. 599

vacuole Membrane-bounded sac, larger than a vesicle; usually functions in storage and can contain a variety of substances. In plants, the central vacuole fills much of the interior of the cell. 66

vagina Organ that leads from the uterus to the vestibule and serves as the birth canal and organ of sexual intercourse in females. 686

valence shell Outer shell of an atom. 23

valve Membranous extension of a vessel of the heart wall that opens and closes, ensuring one-way flow. 579

vascular cylinder In dicot roots, a core of tissues bounded by the endodermis, consisting of vascular tissues and pericycle. 427

vascular tissue Transport tissue in plants consisting of xylem and phloem. 343, 426

vas deferens (vas def-ur-unz, -uh-renz) Tube that leads from the epididymis to the urethra in males. 684

vegetative reproduction In seed plants, reproduction by means other than by seeds; in other organisms, reproduction by vegetative spores, fragmentation, or division of the somatic body. 487

vector (vek-tur) In genetic engineering, a means to transfer foreign genetic material into a cell (e.g., a plasmid). 226

vein Vessel that takes blood to the heart from venules; characteristically has nonelastic walls. 577

vena cava Large systemic vein that returns blood to the right atrium of the heart in tetrapods; either the superior or inferior vena cava. 580

ventilation Process of moving air into and out of the lungs; also called breathing. 634

ventricle (ven-trih-kul) Cavity in an organ, such as a lower chamber of the heart or the ventricles of the brain. 524, 577

venule (ven-yool, veen-) Vessel that takes blood from capillaries to a vein. 579

vermiform appendix Small, tubular appendage that extends outward from the cecum of the large intestine. 621

vertebrae Series of bones that enclose and protect the dorsal nerve cord. 387

vertebral column (vur-tuh-brul) Series of joined vertebrae that extends from the skull to the pelvis. 557

vertebrate Chordate in which the notochord is replaced by a vertebral column. 373

vessel element Cell that joins with others to form a major conducting tube found in xylem. 427, 442

vestibule (ves-tuh-byool) Space or cavity at the entrance of a canal, such as the cavity that lies between the semicircular canals and the cochlea. 544

vestigial structure Remains of a structure that was functional in some ancestor but is no longer functional in the organism in question. 251

villus (pl., villi) (vil-us) Small, fingerlike projection of the inner small intestinal wall. 619

visual accommodation Ability of the eye to focus at different distances by changing the curvature of the lens. 541

visual communication Form of communication between animals using their bodies, includes fighting. 734

vitamin Essential requirement in the diet, needed in small amounts. They are often part of coenzymes. 83, 625

vocal cord Fold of tissue within the larynx; creates vocal sounds when it vibrates. 638

vulva External genitals of the female that surround the opening of the vagina. 686

W

waggle dance Figure-eight dance performed by honeybees to indicate locations of nectar sources. 735

water column In plants, water molecules joined together in xylem from the leaves to the roots. 444

water intoxication Gain in water in cells in which the tissue fluid becomes hypotonic to the cells; caused by drinking too much water. 660

water mold Filamentous organisms having cell walls made of cellulose; typically decomposers of dead freshwater organisms, but some are parasites of aquatic or terrestrial organisms. 324

wax Sticky, solid, waterproof lipid consisting of many long-chain fatty acids usually linked to long-chain alcohols. 44

white blood cell Leukocyte, of which there are several types, each having a specific function in protecting the body from invasion by foreign substances and organisms. 499, 585

white matter Myelinated axons in the central nervous system. 524

wobble hypothesis Ability of the 5'-most nucleotide of an anticodon to interact with more than one nucleotide at the 3'-end of codons. 196

wood Secondary xylem that builds up year after year in woody plants and becomes the annual rings. 432

X

xenotransplantation Use of animal organs, instead of human organs, in human transplant patients. 231, 506

xylem Vascular tissue that transports water and mineral solutes upward through the plant body; it contains vessel elements and tracheids. 427, 442

xylem sap Solution of inorganic nutrients moved from a plant's roots to its shoots through xylem tissue. 442

Y

yeast Unicellular fungus that has a single nucleus and reproduces asexually by budding or fission, or sexually through spore formation. 361

yolk Dense nutrient material in the egg of a bird or reptile. 683

yolk sac Extraembryonic membrane that encloses the yolk of birds; in humans, it is the first site of blood cell formation. 697

Z

zero population growth No growth in population size. 720

zone of cell division In plants, the part of the young root that includes the root apical meristem and the cells just posterior to it; cells in this zone divide every 12–36 hours. 430

zone of elongation In plants, the part of the young root that lies just posterior to the zone of cell division; cells in this zone elongate, causing the root to lengthen. 430

zone of maturation In plants, the part of the root that lies posterior to the zone of elongation; cells in this zone differentiate into specific cell types. 430

zooflagellate Nonphotosynthetic protist that moves by flagella; zooflagellates typically enter into symbiotic relationships, and some are parasitic. 327

zooplankton Part of plankton containing protozoans and other types of microscopic animals. 328, 773

zoospore A motile spore. 334

zygote (zy-goht) Diploid cell formed by the union of sperm and egg; the product of fertilization. 147

Credits

Line Art

Chapter 22

Figure 22.2a: Courtesy of G. David Tilman of the University of Minnesota.

Chapter 27

Page 543, TA 27.10: Reproduced from Ishihara's Tests for Colour Deficiency published by KANEHARA TRADING INC., located at Tokyo in Japan. But tests for colour deficiency cannot be conducted with this material. For accurate testing, the original plates should be used.

Chapter 36

Figure 36.12: United Nations Population Division, 2002.

Chapter 37

Figure 37.1b: Data from S.J. Arnold, "The Microevolution of Feeding Behavior" in Foraging Behavior: Ecology, Ethological, and Psychology Approaches, edited by A. Kamil and T. Sargent, 1980, Garland Publishing Company, New York, NY.

Chapter 38

Figure 38.1a: Data from G.F. Gause, The Struggle for Existence, 1934, Williams & Wilkins Company, Baltimore, MD. **Figure 38.2a:** Data from D.A. MacLulich, Fluctuations in the Numbers of the Varying Hare (Lepus americanus), University of Toronto Press, Toronto, 1937, reprinted 1974.

Chapter 40

Figure 40.2b: Redrawn from "Shrimp Stocking, Salmon Collapse, and Eagle Displacement" by C.N. Spencer, B.R. McClelland and J.A. Stanford, Bioscience, 41(1): 14-21. Copyright © 1991 American Institute of Biological Sciences.

Photographs

Chapter 1

Openers(Ant colony, eggs): © Dr. Sanford D. Porter; (Mound): © S.B. Vinson–TAMU; (Arm with ant stings): © Texas Dept. of Agriculture; (Salmonella): © Dr. Dennis Kunkel/Phototake; (Paramecium): © M. Abbey/Visuals Unlimited; 1.1(Morel): © Royalty-Free Corbis; (Sunflower): © Dave Thompson/Life File/Getty Images; (Snow goose): © Charles Bush Photography; 1.3A(left): © Eye of Science/Photo Researchers, Inc.; 1.3A(right): © SPL/Photo Researchers, Inc.; 1.3B: © Runk/Schoenberger/ Grant Heilman Photography; 1.3C: © George D. Lepp/Corbis; 1.3D: © John Cancalosi/Peter Arnold, Inc.; 1.4A: © Ralph Robinson/Visuals Unlimited; 1.4B: © A.B. Dowsett/SPL/Photo Researchers, Inc.; 1.4C (Paramecium): © Michael Abby/Visuals Unlimited; (Flower): © Pat Pendarvis; (Mushroom): © Rob Planck/Tom Stack; (Fox): © Royalty-Free/Corbis; 1.7:Courtesy Leica Microsystems Inc.; 1.8A (all): Courtesy Jim Bidlack; 1.8B: © Dr. Jeremy Burgess/Photo Researchers, Inc.; 1.9: © Alex Wild 2005.

Chapter 2

Openers(Earth): © Royalty-Free/Corbis; (Waterfall): © Scenics of America/ PhotoLink/Getty Images; (Fertilization): © Nestle/Petit Format/Photo Researchers, Inc.; (Planets): © IAU/Martin Kommesser/Handout/epa/ Corbis; (Embryo): © Lennart Nilsson, "A Child is Born" 1990 Delacorte Press, p. 85; 2.1: © Gunter Ziesler/Peter Arnold, Inc.; 2.3A: 2007 SIU BIOMED COMM/Custom Medical Stock Photo, All Rights Reserved; 2.3B(right): © Hank Morgan/Rainbow; 2.3B(left): © Mazzlota et al./Photo Researchers, Inc.; 2.3C(Radiographers): © SPL/Photo Researchers, Inc.; 2.5(Crystals): © Charles M. Falco/Photo Researchers, Inc.; 2.5(Salt): © Tony Freeman/PhotoEdit; 2.9: © Claude Nuridsany & Marie Perennou/ Photo Researchers, Inc.; 2.10: © Grant Taylor/Stone/Getty Images; 2.16(Smokestacks): © Charles O'Rear/Corbis; (Statue): © Ray Pfortner/ Peter Arnold, Inc.; (Forest): © Frederica Georgia/Photo Researchers, Inc.; (Lake): © Roger Evans/Photo Researchers, Inc.

Chapter 3

Openers(Strawberries): © Royalty-Free/Corbis; (Raspberries,Pineapple): © C Squared Studios/Getty Images; (Fruit): © Maximilian Stock Ltd./Photo Researchers, Inc.; (Vegetables): © Photolink/Getty Images; (Flowers): © Adam Jones/Photo Researchers, Inc.; 3.1(Cactus): © Brand X Pictures/ PunchStock; (Crab): © Ingram Publishing/Alamy; (Bacterium): © H. Pol/ CNRI/SPL/Photo Researchers, Inc.; 3.2B(Man): © Royalty-Free/Corbis; (Woman): © Royalty-Free/Corbis; 3.3A: © The McGraw Hill Companies, Inc./John Thoeming, photographer; 3.4A: © Steve Bloom/Taxi/Getty; 3.5(Glycogen): © Don W. Fawcett/Photo Researchers, Inc.; (Starch): © Jeremy Burgess/SPL/Photo Researchers, Inc.; (Cellulose): © Science Source/J.D. Litvay/Visuals Unlimited; 3.7: © Das Fotoarchiv/Peter Arnold, Inc.; 3.8: © Duomo/Corbis; 3.11C: © PhotoDisk Red/Getty Images; 3.13B: © Jennifer Loomis/Animals Animals / Earth Scenes.

Chapter 4

Openers(Hooke's sketches): © Science VU/Zeiss/Visuals Unlimited; (Bacteria): © David Scharf/SPL/Photo Researchers, Inc.; (Euglena): © T. E. Adams/Visuals Unlimited; (Root cells): © Eye of Science/Photo Researchers, Inc.; (Nerve cells): © Dr. Dennis Kunkel/Visuals Unlimited; 4.1A(top): © Geoff Bryant/Photo Researchres, Inc.; (bottom): Courtesy Ray F. Evert/University of Wisconsin Madison; 4.1B (top): © Barbara J. Miller/ Biological Photo Service; (bottom): Courtesy O. Sabatakou and E. Xylouri-Frangiadaki; 4.3(left, Micrograph): © Robert Brons/Biological Photo Service; (center, TEM): © M. Schliwa/Visuals Unlimited; (right, SEM): © Kessel/Shih/Peter Arnold, Inc.; 4.6(bottom left, Nuclear pores): Courtesy Ron Milligan/Scripps Research Institute; (top right, Freeze-fracture): Courtesy E.G. Pollock; 4.8: © R. Bolender & D. Fawcett/Visuals Unlimited; 4.9: Courtesy Charles Flickinger, from Journal of Cell Biology 49:221-226, 1971, Fig. 1 page 224; 4.11: Courtesy Daniel S. Friend; 4.13: © Newcomb/ Wergin/Biological Photo Service; 4.15: © Dr. Jeremy Burgess/Photo Researchers, Inc.; 4.16: Courtesy Dr. Keith Porter; 4.17: © Dr. Fred Hossler/ Visuals Unlimited; 4.18(Actin): © M. Schliwa/Visuals Unlimited; (Intermediate): © K.G. Murti/Visuals Unlimited; (Microtubule): © K.G. Murti/Visuals Unlimited; 4.19:(Sperm): © Y. Nikas/Photo Researchers, Inc.; (Flagellum, Basal body): © William L. Dentler/Biological Photo Service.

Chapter 5

Openers(Hot gases): © John Chumack/Photo Researchers, Inc.; (Cheetah chasing impala): © Steve Bloom/Getty Images; (Cornfield): © Teri Dixon/ Getty Images; (Cattle): © Ed Lallo/ZUMA/Corbis; (Seniors): © Horizon International Images Limited/Alamy; (Cheetah with kill): © PhotoAlto/ PunchStock; 5.1: © GoodShoot / SuperStock RF; 5.3B: © Darwin Dale/Photo

Researchers, Inc.; 5.7B: © James Watt/Visuals Unlimited; 5.7C: © Creatas/PunchStock; 5.9: © Sygma/Corbis; 5.12: Courtesy Cystic Fibrosis Foundation.

Chapter 6

Openers(Both trees): © Kent Foster/Photo Researchers, Inc.; 6.1(Euglena): © T. E. Adams/Visuals Unlimited; (Tree): © Connie Coleman/Stone/Getty Images; (Sunflower): © Royalty-Free/Corbis; (Kelp): © Chuck Davis/Stone/Getty Images; (Cyanobacteria): © Sherman Thomas/Visuals Unlimited; (Diatom): © Ed Reschke/Peter Arnold; (Moss): © Bruce Iverson; 6.2: © Dr. George Chapman/Visuals Unlimited; 6.4: © B. Runk/S. Schoenberger/Grant Heilman Photography; 6.5B: Courtesy Lawrence Berkeley National Lab; 6.7: © Brand X Pictures/PunchStock; 6.13: © Herman Eisenbeiss/Photo Researchers, Inc.; 6.14: © Brand X Pictures/PunchStock; 6.15B: © Nigel Cattlin/Photo Researchers, Inc.; 6.16: © S. Alden/PhotoLink/Getty Images; 6.17: © Fridmar Damm/zefa/Corbis.

Chapter 7

Openers(Ocelot): © Martin Wendler/Peter Arnold, Inc.; (Bacteria): © Dr. Linda Stannard, UCt/Photo Researchers, Inc.; (Snails): © Bob Evans/Peter Arnold, Inc.; (Giant cactus): © Wolfgang Kaehler/Corbis; 7.1: © Royalty-Free/Corbis; 7.4: © Dr. Donald Fawcett and Dr. Porter/Visuals Unlimited; 7.9B: © Grant Heilman/Grant Heilman Photography; 7.10 (Wine, cheese, bread,): © The McGraw Hill Companies, Inc./John Thoeming, photographer; (Yogurt; vinegar still life): © The McGraw Hill Companies, Inc./Bruce M. Johnson, photographer; 7.11: © C Squared Studios/Getty Images; 7.12B: © PhotoLink/Getty Images; p.130: © Cre8tive Studios/Alamy.

Chapter 8

Openers(Dividing cancer cell): © SPL/Photo Researchers, Inc.; (Lung cancer tumor): Moredun Scientific/Photo Researchers, Inc.; (Colon, pancreatic, and cervical cancer cells): Steve Gschmeissner/Photo Researchers, Inc.; 8.1A(Mother, daughter): © Royalty-Free/Corbis; 8.1A (Gecko): © Kevin Hanley/Animals Animals Earth Scenes; 8.1A (Regeneration): © McDonald Wildlife Photography/Animals Animals Earth Scenes; 8.1B: (Family): © BananaStock/PunchStock; 8.1B(Amoeba): © Biophoto Associates/Photo Researchers, Inc.; 8.2(all): © Stanley C. Holt/Biological Photo Service; 8.3B: © Prof. P. Motta & D. Palermo/Photo Researchers, Inc.; 8.4A: © Andrew Syred/Photo Researchers, Inc.; 8.5 (top-Interphase, Prophase, Metaphase, Anaphase, Telophase): © Ed Reschke; (Prometaphase): © Michael Abbey/Photo Researchers, Inc.; 8.5 (bottom-(Interphase, Late Prophase): © Ed Reschke; (Prophase, Metaphase, Anaphase): © R. Calentine/Visuals Unlimited; (Telophase): © Jack M. Bostrack/Visuals Unlimited; 8.6A(both): © R.G. Kessel and C.Y. Shih, "Scanning Electron Microscopy in Biology: A Students' Atlas on Biological Organization," 1974 Springer-Verlag, New York; 8.6B: © B.A. Palevitz & E.H. Newcomb/Biological Photo Service; 8.9: © Breast Scanning Unit, Kings College Hospital, London/SPL/Photo Researchers, Inc. 8.10: © image100/Alamy; 8.11: © CNRI/SPL/Photo Researchers, Inc.; 8.12A: Courtesy of Dr. D. Von Wettstein; 8.14A (all): © Ed Reschke; 8.14B (all): © Ed Reschke; 8.17: © Leonard Lessin/Peter Arnold, Inc.; 8.18 (Girl): © Jose Carrilo/PhotoEdit; (Chromosome 21): © CNRI/SPL/Photo Researchers, Inc.

Chapter 9

Openers(German shepherd running): © J. M. Labat/F. Roquette/Peter Arnold, Inc.; (German shepherd in cart): © Cecile Podlaseck; (Collie) © Thorsten Milse/Robert Harding World Imagery/Corbis; (Bulldog): © Photodisc/Getty Images; 9.1: © National Geographic Image Sales; 9.9: © Pat Pendarvis; 9.10A(both): Courtesy Jane S. Paulsen/University of Iowa; 9.10B(both): © Steve Uzzell; 9.11A(both): © Brand X/SuperStock RF; 9.15A(Athlete): © AP/Wide World Photos; 9.15A(Loose fibrous tissue): © Ed Reschke; 9.15B: © Dr. Stanley Flegler/Getty Images; 9.20A: © Courtesy of the Archives, California Institute of Technology; 9.20B: © Phototake, Inc./Alamy; 9.20(Fruit flies): © Biology Media/Photo Researchers, Inc.

Chapter 10

Openers(Flat): © Inga Spence/Visuals Unlimited; (Lab): © Vo Trung Dung/Corbis Sygma; (All *Arabidopsis* plant/flower photos): Courtesy Elliot

Meyerowitz/California Institute of Technology; 10.5A(Franklin): © Photo Researchers, Inc.; (DNA model): © Science Source/Photo Researchers, Inc.; 10.5B(Double helix): © Kenneth Eward/Photo Researchers, Inc.; (Watson, Crick): © A. Barrington Brown/Photo Researchers, Inc.; 10.8(both): © Stan Flegler/Visuals Unlimited; 10.11B: © Oscar L. Miller/Photo Researchers, Inc.; 10.13B: Courtesy University of California Lawrence Livermore National Library and the U.S. Department of Energy; 10.14: Courtesy Alexander Rich; 10.18B: Courtesy Dr. Howard Jones, Eastern VA Medical School; 10.19A: © AP Photo/The Republic, Darron Cummings; 10.19B: © Ken Greer/Visuals Unlimited; 10.20(Barbara McClintock): Courtesy Cold Springs Harbor Archives; (Indian corn): © David Young-Wolf/PhotoEdit; (Close-up kernels): © John N.A. Lott/Biological Photo Service.

Chapter 11

Openers(Butterfly, moth): © Creatas/PunchStock;(Cobra): © Joe McDonald/Visuals Unlimited; (Caterpillar): © Getty Images; (Moth hindwings): © Digital Vision Ltd.; 11.3(left): © Barry Runk/STAN/Grant Heilman; (center): © Grant Heilman; (right): © E. Webber/Visuals Unlimited; 11.5A: © Jack Smith/AP Images; 11.5B: © AP/Wide World Photos; 11.5C: Photo courtesy of the College of Veterinary Medicine, Texas A & M; 11.7A: Courtesy of Dr. Stephen Wolfe; 11.7B: © Chanan Photo 2004; 11.12(All): Courtesy Steve Paddock, Howard Hughes Medical Research Institute; 11.13A: Courtesy E.B. Lewis; Summary: © Stockbyte.

Chapter 12

Openers(Dinoflagellate): © Joe Scott/Visuals Unlimited; (Viper fish): © OSF/Animals Animals/Earth Scenes; (Jellyfish): © R. Jackman/OSF/Animals Animals/Earth Scenes; (Pigs): Courtesy Norrie Russell, The Roslin Institute; (Mouse): © Eye of Science/Photo Researchers, Inc.; 12.3 (both): Courtesy General Electric Research & Development; 12.5A: © Dr. Eduardo Blumwald; 12.5B: Courtesy Monsanto: 12.6A: © Nick Gregory/Alamy; 12.6B: Courtesy Monsanto; Table 12.10 (Mouse): © David A. Northcott/Corbis; (Human): © Image Source/JupiterImages; (Drosophila): © Herman Eisenbeiss/Photo Researchers, Inc.; (*Arabidopsis*): © Brad Mogen/Visuals Unlimited; (Roundworm): © 2007 J.L. Carson//Custom Medical Stock Photo, All Rights Reserved; (Yeast): © David Scharf/Photo Researchers, Inc.; 12.11A: © Ron Chapple/FPG International/Getty Images; 12.11B: © Steven Jones/FPG International/Getty Images; 12.13(Chimp mouth): © Digital Vision/Getty Images; (Human mouth): © Royalty Free/Getty Images; (Chimp ear): © Digital Vision/Getty Images; (Human ear): © The McGraw-Hill Companies, Inc./Bob Coyle, photographer; (Chimp nose): © Digital Vision/Getty Images; (Human nose): © Royalty-free/Getty Images; Biological Viewpoint page 240–241 (DNA): © Radius Images/Alamy.

Chapter 13

Openers(Bat): © Merlin D. Tuttle/Bat Conservation International; (Bee, orchid): © Helmut Presser; (Butterfly): © Robert Maier/Animals/Animals/Earth Scenes; (Moth): © Andrew Darrington/Alamy; (Hummingbird): © Anthony Mercieca/Photo Researchers, Inc.;13.1(Strata): © Gary J. James/Biological Photo Service; (Desert): © C. Luiz Claudio Marigo/Peter Arnold; (Rain forest): © Charles Benes/Index Stock Imagery; (Rhea): © Wolfgang Kaehler/Corbis; (Iguana): © Galen Rowell/Corbis; (Finch): © D. Parer & E. Parer-Cook/Ardea; (Darwin): © Carolina Biological/Visuals Unlimited; 13.2A: © James H. Bailey/National Geographic Image Sales; 13.2B: © Daryl Balfour/Photo Researchers, Inc.; 13.3A(both): Courtesy of Lyudmila N. Trut, Institute of Cytology & Genetics, Siberian Dept. of the Russian Academy of Sciences; 13.3B (Chinese cabbage, Brussel sprouts, cabbage): Courtesy W. Atlee Burpee Company; (Mustard): © Jack Wilburn/Animals Animals/Earth Scenes; 13.4: © Comstock; 13.6A(Ground finch): © Adrienne T. Gibson/Animals Animals/Earth Scenes; (Warbler finch): © Joe McDonald/Animals Animals/Earth Scenes; (Cactus finch): © Leonard Lee Rue/Animals Animals/Earth Scenes; 13.6(moths): © Michael Wilmer Forbes Tweedie/Photo Researchers, Inc.; 13.7A: © Richard T. Nowitz/Corbis; 13.7B: © Louis Psihoyos/Corbis; 13.8A(Fossil): © Jean-Claude Carton/Bruce Coleman Inc.; (Art): © Joe Tucciarone; 13.8B: © J.G.M. Thewissen, http://darla.neoucom.edu/DEPTS/ANAT/Thewissen/; 13.9B (Chick, pig): © Carolina Biological Supply/Phototake; 13.15B: Courtesy Victor McKusick; 13.16C(left): © Helen Rodd; 13.16D: © Bob Evans/Peter Arnold, Inc.

Chapter 14

Openers(Liger): © Andy Carvin; (Mules): © BIOS Klein & Hubert/Peter Arnold, Inc.; (Zebroid): © Michele Burgess/Visuals Unlimited; (Tigon): © One World Images/Alamy; 14.1B(Acadian): © Karl Maslowski/Visuals Unlimited; (Least): © Stanley Maslowski/Visuals Unlimited; (Willow): © Ralph Reinhold/Animals Animals/Earth Scenes; 14.1C(Massai): © Sylvia Mader; (Eskimo): © B & C Alexander/Photo Researchers, Inc.; 14.2C: © Barbara Gerlach/Visuals Unlimited; 14.2D(Horse): © Superstock, Inc.; (Mule): © Jorg & Petra Wegner/Animals Animals Earth Scenes; (Donkey): © Robert J. Erwin/Photo Researchers, Inc.; 14.3: © Jonathan Losos; 14.5B(C. concinna): © Gerald & Buff Corsi/Visuals Unlimited; (C. virgata): ©: Dr. Dean Wm. Taylor/Jepson Herbarium, UC Berkeley; (C. pulchella): © J. L. Reveal; 14.5A(Bunch): © Stockdisc/PunchStock; (Diploid): © Randy C. Ploetz; (Sliced insert): © The McGraw Hill Companies, Inc./Evelyn Jo Hebert, photographer; 14.6A: © Courtesy of John Doebley; 14.6B: © Kingsley Stern; 14.8A: © Boehm Photography; 14.8B(Opabinia): © A. J. Copley/Visuals Unlimited; (Thaumaptilon): © Simon Conway Morris, University of Cambridge; (Wiwaxia): © Albert Copley/Visuals Unlimited; (Vauxia): © Alan Siruinikoff/Photo Researchers, Inc.; 14.9A(Fly eye): © Carolina Biological Supply/Photo Researchers, Inc.; (Human eye): © Vol. OS02/PhotoDisc/Getty Images; (Squid eye): © Aldo Brando /Peter Arnold, Inc.; 14.9B: Courtesy of Walter Gehring, reprinted with permission from *Induction of Ectopic Eyes by Target Expression of the Eyeless Gene in Drosophila*, G. Halder, P. Callaerts, Walter J. Gehring, *Science* Vol. 267, © 24 March 1995 American Association for the Advancement of Science; 14.9C(both): A. C. Burke, 2000.

Chapter 15

Openers(Fossilized nest, Maiosaura fossil): Courtesy Museum of the Rockies; (Painting): © Joe Tucciarone/Photo Researchers, Inc.; 15.7A: Courtesy Dr. David Dilcher, Florida Museum of Natural History, University of Florida; 15.7B: (Cactus): © John D. Cunningham/Visuals Unlimited; (Spurge): © John Shaw/Tom Stack & Associates.

Chapter 16

Openers(Ebola): © RYABCHIKOVA-VOISIN/Photo Researchers, Inc.; (Coronavirus): © Eye of Science/Photo Researchers, Inc.; (Coronavirus close up): © Science VU/Visuals Unlimited; (T. pallidum): © Science Source/Photo Researchers, Inc.; (E. coli): © David Scharf/SPL/Photo Researchers, Inc.; (Virus capsid): © Lee Simon/Photo Researchers, Inc.; (Root nodules): © Dr. Jeremy Burgess/Photo Researchers, Inc.; (Water treatment plant): © Vol. 29/PhotoDisc/Getty Images; (Anthrax): © Dr. Gary Gaugler/Photo Researchers, Inc.; (Bioremediation): Ken Graham/Accent Alaska; (Rocks): © Science VU/Visuals Unlimited; 16.1A: © Dr. Hans Gelderblom/Visuals Unlimited; 16.1B: © K.G. Murti/Visuals Unlimited; 16.3A: © Brad Mogen/Visuals Unlimited; 16.3B: © Holt Studios International Ltd/Alamy; 16.6A(Masks): © Wolfgang Schmidt/Peter Arnold, Inc.; 16.6B: © Vassil Donev/epa/Corbis; 16.7B: © Ralph White/Corbis; 16.9A: © Science VU/Visuals Unlimited; 16.9B: Courtesy Dr. David Deamer; 16.10A(Bacillus): © Ralph A. Slepecky/Visuals Unlimited; (E. coli): 16.10A: Harley W. Moon, U.S. Dept. of Agriculture; 16.10B(Cocci): © Dr. David M. Phillips/Visuals Unlimited; (Spirilla): © Dr. Gary D. Gaugler/Phototake; (Bacilli): © Dr. Dennis Kunkel/Visuals Unlimited; 16.11A: © CNRI/SPL/Photo Researchers, Inc.; 16.11B: © Alfred Pasieka/SPL/Photo Researchers, Inc.; 16.13A: © Michael P. Gadomski/Photo Researchers, Inc.; 16.13B: © Science VU/Visuals Unlimited; 16.14(Anabaena): © Philip Sze/Visuals Unlimited; (Gloeocapsa): © Runk/Schoenberger/Grant Heilman Photography; (Oscillatoria): © Tom Adams/Visuals Unlimited; 16.15A(Swamp): © Susan Rosenthal/Corbis; (Inset): © Ralph Robinson/Visuals Unlimited; 16.15B(Inset): From J.T. Staley, et al., *Bergey's Manual of Systematic Bacteriology*, Vol. 13, © 1989 Williams and Wilkins Col, Baltimore. Prepared by A.L. Usted Photography. Dept. of Biophysics, Norwegian Institute of Technology; 16.15B(Lake): © John Sohlden/Visuals Unlimited; 16.15C(Geysers): © Jeff Lepore/Photo Researchers, Inc.; (Inset): Courtesy Dennis W. Grogan, University of Cincinnati; 16.17: © AP Photo/Kenneth Lambert.

Chapter 17

Openers(Mosquito): © CDC/PHIL/Corbis; (Malaria): © Dr. Gopal Murti/Photo Researchers, Inc.; (Sandfly): © Sinclair Stammers/Photo Researchers, Inc.; (Leishmaniasis): © BSIP/Photo Researchers, Inc.; 17.2(Synura): © Dr. Ronald W. Hoham; (Nonionina): © Astrid & Hanns-Frieder Michler/Photo Researchers, Inc.; (Blepharisma): © Eric Grave/Photo Researchers, Inc.; (Bossiella): © Daniel V. Gotschall/Visuals Unlimited; (Diatoms): © M.I. Walker/Photo Researchers, Inc.; (Ceratium): © D.P. Wilson/Photo Researchers, Inc.; (Licmorpha): © Biophoto Associates/Photo Researchers, Inc.; (Acetabularia): © Linda L. Sims/Visuals Unlimited; (Amoeba): © Michael Abbey/Visuals Unlimited; 17.5B (Cliffs): © Rick Ergenbright/Corbis; (Inset): © Manfred Kage/Peter Arnold, Inc.; 17.4: © Michael Abbey/Visuals Unlimited; 17.5C(Tests): © Dr. Richard Kessel & Dr. Gene Shih/Visuals Unlimited; 17.6A: © CABISCO/Phototake; 17.6B: © Manfred Kage/Peter Arnold, Inc.; 17.6C: © Eric Grave/Photo Researchers, Inc.; 17.8A(Physarum): © CABISCO/Visuals Unlimited; (Hemitrichia): © V. Duran/Visuals Unlimited; 17.8B: © James Richardson/Visuals Unlimited; 17.9A: © Dr. Ann Smith/Photo Researchers, Inc.; 17.9B(left): © Biophoto Associates/Photo Researchers, Inc.; 17.10A: © Walter Hodge/Peter Arnold, Inc.; 17.10B: © D.P Wilson/Eric & David Hosking/Photo Researchers, Inc.; 17.11B: © M.I. Walker/Science Source/Photo Researchers, Inc.; 17.11C(top): © John D. Cunningham/Visuals Unlimited; 17.11C(lower right): © Cabisco/Visuals Unlimited; 17.11D(Ulva): © William E. Ferguson; 17.11E(Several *Chara*): © Dr. John D. Cunningham/Visuals Unlimited; 17.11E(Single *Chara*): © Kingsley Stern.

Chapter 18

Openers(Backlit trap): © Jerome Wexler/Visuals Unlimited; (Flower): © Barry Rice/Visuals Unlimted; (Open trap): © David Sieren/Visuals Unlimited; (Sundew plant): © Barry Rice/Visuals Unlimited; (Sundew leaf) © Dr. Jeremy Burgess/Photo Researchers, Inc.; (Pitcher plants): © Milton H. Tierney, Jr./Visuals Unlimited; (Pitcher plant open): © David Sieren/Visuals Unlimited; (Pitcher plant flowers): © Jeffrey Lepore/Photo Researchers, Inc.; 18.1(Inset): © T. Mellichamp/Visuals Unlimited; (Chara): © Heather Angel/Natural Visions; 18.2A(Moss micrograph): © Ed Reschke; (Moss): © John Shaw/Tom Stack & Associates; (Fern): © William E. Ferguson; (Fern micrograph): Courtesy Graham Kent; (Cone): © Kent Dannen/Photo Researchers, Inc.; (Seed): Courtesy Graham Kent; (Cherry blossom): © E. Webber/Visuals Unlimited; (Cherries): © Richard Shiell/Earth Scenes/Animals Animals; 18.4B(Archegonium): © Triarch/Visuals Unlimited; (Embryo sac): © Ed Reschke; 18.4C(Leaf cross section): © Kingsley Stern; (Stomata): © Andrew Syred/SPL/Photo Researchers, Inc.; 18.5A(Hornwort): © Barry Runk/Stan/Grant Heilman Photography; (Liverwort): © Hal Horwitz/Corbis; (Moss): © Steven P. Lynch; 18.5B(Sporophyte): © Heather Angel/Biofotos; (Gametophyte): © Bruce Iverson; 18.6A: © Steve Solum/Bruce Coleman, Inc.; 18.6B(Cinnamon): © James Strawser/Grant Heilman Photography; 18.6B(Harts tongue): © Walter H. Hodge/Peter Arnold, Inc.; (Maidenhair): © Larry Lefever/Jane Grushow/Grant Heilman Photography; 18.6C: © Matt Meadows/Peter Arnold, Inc.; 18.7A(Cycad): © D. Cavagnaro / Visuals Unlimited; (Ginkgo): © Kingsley Stern (Ephedra): © Virginia Weinland/Photo Researchers, Inc.; (Conifer): © D. Giannechini/Photo Researchers, Inc.; 18.7B: © Phototake; 18.8: © Sinclair Stammers/SPL/Photo Researchers, Inc.; 18.11A(Wheat): © Earl Roberge/Photo Researchers, Inc.; (Corn ear): © Doug Wilson/Corbis; (Corn plants): © Adam Hart-Davis/SPL/Photo Researchers, Inc.; (Rice plant): © Royalty-free/Corbis; (Rice grains): © Dex Image/Getty RF; 18.11B(Rubber): © Steven King/Peter Arnold, Inc.; (Cotton): © Dale Jackson/Visuals Unlimited; (Tulip): © Photodisc Blue/Getty RF; (Thatch): © Danny Lehman/Corbis; 18.12A: © Gary R. Robinson/Visuals Unlimited; 18.12B (Spores): From C.Y. Shih and R.G. Kessel,"Living Images" Science Books International, Boston, 1982; (Fungus): © Jeffrey Lepore/Photo Researchers, Inc.; 18.13A (Fruticose): © Stephen Sharnoff/Visuals Unlimited; (Foliose): © Kerry T. Givens/Tom Stack & Associates; 18.13B: © R. Roncadori/Visuals Unlimited; 18.14A: © Barry Runk/Stan/Grant Heilman Photography; 18.14B(Cup fungus): © Robert Calentine/Visuals Unlilmited; (Morel): © Michael Viard/Peter Arnold, Inc.; 18.14C(Yeast): © David Philips/Visuals Unlimited; (Conidia): © David Philips/Visuals Unlimited; 18.14D(Mushroom): © Biophoto Associates/Photo Researchers, Inc.; (Shelf fungus): © Inga Spence/Tom Stack & Associates; (Puffball): © L. West/Photo Researchers, Inc.; 18.15A: © Kingsley Stern; 18.15B(Thrush): © Everett S.

Beneke/Visuals Unlimited; (Ringworm): © John Hadfield/SPL/Photo Researchers, Inc.; (Athlete's foot): © P. Marazzi/SPL/Photo Researchers, Inc.

Chapter 19

Openers(Flying fox bat): © Austin J. Stevens/Animals Animals/Earth Scenes; (Bat eating cactus): © Dr. Merlin D. Tuttle/Photo Researchers, Inc.; (Vampire bat feeding): © Stephen Dalton/OSF/Animals Animals/Earth Scenes; (Vampire bat): © Michael Fogden/Animals Animals/Earth Scenes; (Long-eared bat): © Press-Tige/OSF/Animals Animals/Earth Scenes; 19.1(Adult frog): © Dwight Kuhn; (Top row, zygote to embryo): © Cabisco/Phototake; (Jelly egg mass, embryo inset, tadpoles [no legs, with legs]): © Dwight Kuhn; 19.4B(Roundworm): © Arthur Siegelman/Visuals Unlimited; (Millipede): © John MacGregor/Peter Arnold, Inc.; 19.6: © Andrew J. Martinez/Photo Researchers, Inc.; 19.7A(Sea anemone): © Azure Computer & Photo Services/Animals Animals/Earth Scenes; 19.7A(Coral): © Ron Taylor/Bruce Coleman, Inc.; 19.7A(Man-of-war): © Runk/Schoenberger/Grant Heilman Photography; 19.7A(Jellyfish): © Under Watercolours; 19.7B: © CABISCO/Visuals Unlimited; 19.8: © Tom E. Adams/Peter Arnold, Inc.; 19.9A(Scolex): © James Webb/Phototake; (Proglottid): © John D. Cunningham/Visuals Unlimited; 19.9B: © SPL/Photo Researchers, Inc.; 19.10A: © Lauritz Jensen/Visuals Unlimited; 19.10B: © James Solliday/Biological Photo Service; 19.10C: © Vanessa Vick/The New York Times/Redux; 19.12B(Snail): © Farley Bridges; (Nudibranch): © Kenneth W. Fink/Bruce Coleman, Inc.; (Octopus): © Ken Lucas/Visuals Unlimited; (Nautilus): © Douglas Faulkner/Photo Researchers, Inc.; (Scallop): Courtesy of Larry S. Roberts; (Mussel): © Fred Whitehead/Animals/Animals/Earth Scenes; 19.13B(Fan worm): © James H. Carmichael; (Leech): © St. Bartholomews Hospital/SPL/Photo Researchers, Inc.; 19.14B(All): © John Shaw/Tom Stack & Associates; 19.15A(Crab): © Michael Lustbader/Photo Researchers, Inc.; (Barnacles): © Kjell Sandved/Butterfly Alphabet; (Shrimp): © Bruce Robinson/Corbis; (Copepod): © Kim Taylor/Bruce Coleman, Inc.; 19.15B(Crab): © Jana R. Jirak/Visuals Unlimited; (Scorpion): © Tom McHugh/Photo Researchers, Inc.; (Spider): © Ken Lucas; 19.15C(Centipede): © David M. Dennis/Animals Animals/Earth Scenes; (Millipede): © Geof de Feu/Imagestate; 19.16(Mealybug, dragonfly, leafhopper): © Farley Bridges; (Grasshopper): © Chris Mattison/Frank Lane Picture Agency/Corbis; (Beetle): © Wolfgang Kaehler/Corbis; (Butterfly): © McDonald Wildlife Photography/Animals Animals/Earth Scenes; (Louse): © Darlyne A. Murawski/Peter Arnold, Inc.; (Housefly): © L. West/Bruce Coleman, Inc.; (Wasp): © Johnathan Smith; Cordaiy Photo Library/Corbis; 19.17(Sea urchins): © Randy Morse/Animals Animals/Earth Scenes; (Brittle star): © Jim Greenfield/Imagequestmarine.com; (Feather star): © Carl Roessler/Tom Stack & Associates; (Sea cucumber): © Alex Kerstitch/Visuals Unlimited; (Sea lily): © P. Danna/Peter Arnold, Inc.; (Sand dollar): © Roger Steene/Imagequestmarine.com; 19.19(Tunicate): © Rick Harbo; (Lancelet): © Heather Angel; 19.21B(Lamprey): © Heather Angel; (Shark): © James Watt/Animals Animals/Earth Scenes; (Soldierfish): © Ron & Valerie Taylor/Bruce Coleman, Inc.; 19.22B(Frog): © Joe McDonald/Visuals Unlimited; (Salamander): © Suzanne L. Collins & Joseph T. Collins/Photo Researchers, Inc.; 19.23(Sea turtles): © H. Hall/OSF/Animals Animals/Earth Scenes; (Gila monster): © Joe McDonald/Visuals Unlimited; (Rattlesnake): © Joel Sartorie/National Geographic/Getty Images; (Tuatara): © Nathan W. Cohen/Visuals Unlimited; (Alligator): © OS21/PhotoDisc/Getty Images; 19.24A(Eagles flying): © Daniel J. Cox; 19.24B(Eagle): © Thomas Kitchin/Tom Stack & Associates; (Woodpecker): © Joel McDonald/Corbis; (Flamingo): © Brian Parker/Tom Stack & Associates; (Cardinal): © Kirtley Perkins/Visuals Unlimited; (Vulture): © Robert Comport/Animals Animals/Earth Scenes; 19.25A(Platypus): © D. Parer & E. Parer-Cook/Ardea; (Opossum): © Leonard Lee Rue/Photo Researchers, Inc.; (Koala): © Fritz Prenzel/Animals Animals/Earth Scenes; 19.25B(Deer): © Stephen J. Krasemann/Photo Researchers, Inc.; (Lioness): © Stephen J. Krasemann/DRK Photo; (Monkey): © Gerald Lacz/Animals Animals/Earth Scenes; (Whale): © Mike Bacon/Tom Stack & Associates; 19.26(Frog): © Mark Smith/Photo Researchers, Inc.; (Pig): © Allan Friedlander/SuperStock; (Heart): © Account Phototake/Phototake.

Chapter 20

Openers: (Lucy): © Associated Press; (S. tchadensis): © Prof. Michel Brunet; (A. boisei): © Science VU/NMK/Visuals Unlimited; (H. habilis, H. neanderthalensis): © Herve Conge/ISM/Phototake; (Cro-Magnon): © 2007 Educational Images Ltd./Custom Medical Stock Photo, All Rights Reserved; (Selam): © Dr. Zeresenay Alemseged, 20.1A(Lemur): © Frans Lanting/Minden Pictures; (Tarsier): © Doug Wechsler; (Monkey): © C.C. Lockwood/DRK Photo; (Baboon): © St. Meyers/Okapia/Photo Researchers, Inc.; (Orangutan, humans): © Tim Davis/Photo Researchers, Inc.; (Gibbon): © Hans & Judy Beste/Animals Animals/Earth Scenes; (Chimpanzee, gorilla): © Martin Harvey/Peter Arnold, Inc.; 20.2B: © National Museums of Kenya; 20.4(Lucy): © Dan Dreyfus and Associates; (Footprints): © John Reader/Photo Researchers, Inc.; 20.5: © Ryan McVay/Getty Images; 20.6: © National Museums of Kenya; 20.10: Transp. #608 Courtesy Dept. of Library Services, American Museum of Natural History; 20.12 (top left, top middle, top right): © Vol. 105/PhotoDisc/Getty; (bottom #1): © Royalty-Free/Corbis; (bottom #2): © Mark Hamel/Photo Researchers, Inc.; (bottom #3): © Vol. 116/PhotoDisc/Getty; (bottom #4, #5): © Vol. 105/PhotoDisc/Getty Images; p.418: © Tom Vezo/Getty Images.

Chapter 21

Openers(Rainforest): © James J. Stachecki/Animals Animals/Earth Scenes; (Burning): © Digital Vision/PunchStock; 21.1B(Cactus): © Patti Murray/Animals Animals/Earth Scenes; (Cucumber): © Michael Gadomski/Photo Researchers, Inc; (Flytrap): © CABISCO/Phototake; 21.1C(Fibrous root): © Ed Degginger/Color Pic; (Taproot): © G.R. Roberts/Natural Sciences Image Library; 21.3(Rice plants): © Royalty-Free/Corbis; (Rice grain heads): © Dex Image/Getty Royalty Free; (Wheat): © Earl Roberge/Photo Researchers, Inc.; (Corn ear): © Doug Wilson/Corbis; (Corn plants): © Adam Hart-Davis/SPL/Photo Researchers, Inc.; (Barley plants): © Sundell Larsen/Getty Images; (Barley grains): © C. Sherburne/Photolink/Getty Images; 21.4A(Root hairs): © Runk Schoenberger/Grant Heilman Photography; (Stoma): © J.R. Waaland/Biological Photo Service; (Cork): © Kingsley Stern; 21.4B(Parenchyma): © Biophoto Associates/Photo Researchers, Inc.; (Collenchyma): © Biophoto Assoc./Photo Researchers, Inc.; (Sclerenchyma): © Biophoto Assoc./Photo Researchers, Inc.; 21.4C: © J. Robert Waaland/Biological Photo Service; 21.4D: © George Wilder/Visuals Unlimited; 21.5B(Eudicot): © Ed Reschke; (Monocot): © CABISCO/Phototake; (Vascular cylinder): © CABISCO/Phototake; 21.6A: Courtesy Ray F. Evert/University of Wisconsin Madison; 21.6B(both): © J. Robert Waaland/Biological Photo Service; 21.8(left): © Imagesource/JupiterImages; (center): © Brand X/Fotosearch; (right): © Thinkstock Imaes/JupiterImages; 21.9: © Jeremy Burgess/SPL/Photo Researchers, Inc.; 21.10: Courtesy Ray F. Evert/University of Wisconsin Madison; 21.11A(both): © Jeremy Burgess/SPL/Photo Researchers, Inc.; 21.11B: © D. H. Marx/Visuals Unlimited.

Chapter 22

Openers(all): © Dr. Gerald D. Carr; 22.2(Tilman): ©: Tim Rummelhoff/Tim Rummelhoff Photography; 22.3A(top, center): Courtesy Wilfred A. Cote, from H.A. Core, W.A. Cote, and A.C. Day, "Wood: Structure and Identification" 2/e; (bottom): Courtesy W.A. Cote, Jr., N.C. Brown Center for Ultrastructure Studies, SUNY-ESF; 22.3B: © Ed Reschke/Peter Arnold, Inc.; 22.4A,B: © Jeremy Burgess/SPL/Photo Researchers, Inc; 22.5: Courtesy Gary Banuelos/Agriculture Research Service/USDA; 22.6(top): © Bruce Iverson/SPL/Photo Researchers, Inc.; (bottom): From M.H. Zimmerman "Movement of Organic Substances in Trees" in SCIENCE 133 (13) January 1961 page 667, Fig. 3 page 73, © 1961 AAAS; 22.8(all): Courtesy Mary E. Doohan; 22.9A: © CABISCO/Phototake; 22.12A(Nodule): © Dwight Kuhn; (Circle): © E.H. Newcomb & S.R. Tardon/Biological Photo Service; 22.12B(top): © Runk Schoenberger/Grant Heilman; (circle): © Dana Richter/Visuals Unlimited; 22.12C(Dodder): © Kevin Schafer/Corbis; 22.12D(both): © B. Runk/S. Schoenberger/Grant Heilman Photography.

Chapter 23

Openers(Before): © Courtesy of the U.S. Geological Survey, Harry Glicken, photographer; (After): © Courtesy of the U.S. Geological Survey, Lyn Topinka, photographer; (Fireweed): © Courtesy of the U.S. Geological

Survey; (Lupine): © James Hughes/Visuals Unlimited; 23.2A(both): Courtesy Prof. Malcolm B. Wilkins; 23.3A(Grapes): © Sylvan Whittwer/ Visuals Unlimited; (Cyclamen): © Robert E. Lyons/Visuals Unlimited; 23.4A: Courtesy Prof. Dr. Hans-Ulrich Koop, from "Plant Cell Reports," 17:601-604; 23.4B (all): Courtesy Alan Darvill and Stefan Eberhard, Complex Carbohydrate Research Center, University of Georgia; 23.5A: © John Solden/Visuals Unlimited; 23.5B: Courtesy of Donald R. McCarthy, from "Molecular Analysis of Viviparous-1: An Abscisic Acid-Insensitive Mutant of Maize," *The Plant Cell*, v. 1, 523–532, C1989 American Society of Plant Physiologists; 23.6A(both): © Kingsley Stern; 23.7: © Kim Taylor/ Bruce Coleman, Inc.; 23.8A(top left): © Kingsley Stern; (top right): Courtesy Malcolm Wilkins, Botany Department, Glasgow University; 23.8A(bottom): © BioPhot; 23.8B: © Maryann Frazier/Photo Researchers, Inc.; 23.8D: © John D. Cunningham/Visuals Unlimited; 23.9A(Before, after): © John Kaprielian/Photo Researchers, Inc.; 23.9(Flytrap): © P. Goetgheluck/Peter Arnold, Inc.; 23.9B(Morning glory, morning): © BIOS A. Thais/Peter Arnold, Inc.; (Morning glory, night): © BIOS Pierre Huguet/ Peter Arnold, Inc.; (Prayer plant, both): © Tom McHugh/Photo Researchers, Inc.; 23.11B(both): © Grant Heilman Photography; 23.12A(Leaf curl): © Kingsley Stern; (Monarch caterpillar): © Frans Lanting/Minden Pictures; (Monarch adult): © Dwight Kuhn; Courtesy of U. S. Department of Agriculture/Agricultural Research Service, Photo by Scott Bauer; (Foxglove): © Steven P. Lynch; 23.12B(Leaf): © Nigel Cattlin/Photo Researchers, Inc.; (Acacia ant): © David Thompson/OSF/Animals Animals/ Earth Scenes; 23.13: © Goldberg Diego /Corbis Sygma.

Chapter 24

Openers(Chipmunk): © Tom McHugh/Photo Researchers, Inc.; (Bird): © Marie Read/Animals Animals/Earth Scenes; (Duck): © Jennifer Loomis/ Animals Animals/Earth Scenes; (Dog): © Scott Camazine/Photo Researchers, Inc.; 24.1B: © Farley Bridges; 24.1C: © Pat Pendarvis; p.481(Pollen grain): Courtesy Graham Kent; (Embryo sac): © Ed Reschke; 24.2B(Grass): © 2007 John Bolivar/Custom Medical Stock Photo, All Rights Reserved; (Pollen): © Microworks Color/Phototake; 24.2(both flowers): © Heather Angel; 24.2C: © Dwight Kuhn; 24.3(Proembryo, Globular, Heart): Courtesy Dr. Chun-Ming Liu; 24.3(Torpedo): © Biology Media/ Photo Researchers, Inc.; 24.3(Embryo): Jack Bostrack/Visuals Unlimited; 24.4(Winged seeds): © James Mauseth; (Strawberries): © Royalty-Free/ Corbis; (Raspberries): © Runk/Schoenberger/Grant Heilman Photography, Inc.; (Pineapple): © BJ Miller/Biological Photo Service; 24.5A: © Ed Reschke; 24.5B: © James Mauseth; 24.6: © G.I. Bernard/Animals Animals/ Earth Scenes; 24.7A(all): Courtesy Prof. Dr. Hans-Ulrich Koop, from "Plant Cell Reports", 17:601-604; 24.7B: © Kingsley Stern; Biological Viewpoint, p.492: © Brian Sullivan/Getty Images.

Chapter 25

Openers(Rainforest): © Steven P. Lynch; (Snake): © Marian Bacon/Animals Animals/Earth Scenes; (Dragonfly): © John Foxx/Stockbyte/Getty Images; (Iguana): © Digital Vision Ltd.; (Tortoise): © Digital Vision/PunchStock; (Caribou Herd): © Johnny Johnson/Animals Animals/Earth Scenes; (Terns): © Norbert Rosing/Animals Animals/Earth Scenes; (Caribou bull): © Johnny Johnson/Animals Animals/Earth Scenes; (Wolverine): © PhotoDisc Collection/Getty Images; (Hare): © PhotoDisc Collection/ Getty Images; 25.1(all), 25.3A(Loose fibrous, adipose, hyaline, compact bone): © Ed Reschke; (Dense fibrous): © The McGraw-Hill Companies, Inc./Dennis Strete, Photographer; 25.4(Skeletal, cardiac muscle): © Ed Reschke; (Smooth muscle): © The McGraw-Hill Companies, Inc./Dennis Strete, Photographer; 25.5: © Ed Reschke; 25.6A: © AP Photo/Elliot D. Novak; 25.6B: © Diana De Rosa–Press Link; 25.6C: © Vol. 154/Corbis; 25.6D: © Vol. 12/Corbis; 25.7: © John D. Cunningham/Visuals Unlimited; 25.9A: © Dennis Oda, Honolulu Star-Bulletin; 25.9B: © AP Images/Brian Walker.

Chapter 26

Openers(Clam): © Andrew J. Martinez/Photo Researchers, Inc.; (Octopus): © Bruce Watkins/Animals Animals/Earth Scenes; (Crab): © Michael Lustbader/Photo Researchers, Inc.; (Grasshopper): © Masterfile (Royalty-Free Div.); (Student): © BananaStock/JupiterImages; (Trout): © Tom & Pat Leeson/Photo Researchers, Inc.; (Dolphin): © Gerard Lacz/Animals

Animals/Earth Scenes; (Sea urchins): © Randy Morse/Animals Animals/ Earth Scenes; 26.9: Courtesy Dr. E.R. Lewis, University of California Berkeley; 26.10: © Reuters/Corbis; 26.10 (Two brains): © Science VU/ Visuals Unlimited; (Shooting up): © Science VU/Visuals Unlimited.

Chapter 27

Openers(Squid): © Jeff Rotman/Photo Researchers, Inc.; (Zebra): © Ingrid Van Den Berg/Animals Animals/Earth Scenes; (Bee): © Comstock; (Woman): © Roy Morsch/Corbis; (Insect eye view): © James Gould; 27.1A: © Jeff Foott; 27.1B: © David M. Dennis/Animals Animals/Earth Scenes; 27.1C: © Dr. Merlin D. Tuttle/Bat Conservation International/Photo Researchers, Inc.; 27.4(all): © Omikron/SPL/Photo Researchers, Inc.; 27.10: © Lennart Nilsson, from "The Incredible Machine"; 27.12: © P. Motta/SPL/ Photo Researchers, Inc.; 27.13A, B: Courtesy Dr. Yeohash Raphael, the University of Michigan, Ann Arbor.

Chapter 28

Openers(Female skull): © Ralph Hutchings/Visuals Unlimited; (Male skull): © Ralph Hutchings/Visuals Unlimited; (Child, adult skulls): © 2007 Educational Images Ltd./Custom Medical Stock Photo, All Rights Reserved; (Pelvic bones): © L. Bassett/Visuals Unlimited; (Femur): © Dr. Fred Hossler/Visuals Unlimited; (Forensic expert): © Fehim Demir/epa/Corbis; (Fracture): © Scott Camazine/Phototake; 28.1(Exoskeleton): © Michael Fogden/OSF/Animals Animals/Earth Scenes; 28.1B: © Dynamic Graphics Group/PunchStock; 28.2: © Julie Lemberger/Corbis; 28.5(Tennis player): © Susan Mullane/NewSport/Corbis; (Bone, both): © Michael Klein/Peter Arnold, Inc.; 28.6(Articular cartilage, compact bone): © Ed Reschke; (Osteocyte): © Biophoto Associates/Photo Researchers, Inc.; 28.7: © Gerard Vandystadt/Photo Researchers, Inc.; 28.8: © Peter Skinner/Photo Researchers, Inc.; 28.10(Motor unit): © Victor B. Eichler; 28.13A: © Biology Media/Photo Researchers, Inc.; 28.16(left): © Lawrence Manning/Corbis; (center): © G. W. Willis/Visuals Unlimited; (right): © Royalty-Free/Corbis.

Chapter 29

Openers(Lionfish): © Brand X Pictures/PunchStock; (Panda): © Digital Vision; (Feather duster): © Diane R. Nelson; (Lobster): Ken Lucas/Visuals Unlimited; (Wasps): © Carson Baldwin, Jr./Animals Animals/Earth Scenes; 29.2(Hydra): © CABISCO/Visuals Unlimited; (Flatworm): © Barry Runk/ Stan/Grant Heilman Photography; 29.10: © Pascal Goethgheluck/SPL/ Photo Researchers, Inc.; 29.11: © Biophoto Associates/Photo Researchers, Inc.; 29.13: © Eye of Science/Photo Researchers, Inc.; 29.16A, B: © J. C. Revy/Phototake.

Chapter 30

Openers(AIDS patient): © A. Ramey/PhotoEdit; (Chickenpox virus): © George Musil/Visuals Unlimited; (Pneumocystic pneumonia): © Dr. Dennis Kunkel/Visuals Unlimited; (Candidiasis): © Everett S. Beneke/Visuals Unlimited; 30.2(Bone marrow): © R. Calentine/Visuals Unlimited; (Thymus, spleen): © Ed Reschke/Peter Arnold, Inc.; (Lymph node): © Fred E. Hossler/ Visuals Unlimited; 30.5B: © Bohringer Ingelheim International, photo by Lennart Nilsson; 30.7A: © Michael Newman/PhotoEdit; 30.7B: © Digital Vision/Getty Images; 30.10B: © Bohringer Ingelheim International, photo by Lennart Nilsson; 30.13A: © Richard Anderson; 30.13B: © Dr. Ken Greer/ Visuals Unlimited; 30.14(Pollen): © David Scharf/SPL/Photo Researchers, Inc.; (Girl): © Damien Lovegrove/SPL/Photo Researchers, Inc.

Chapter 31

Openers (Tiger): © Ken Cole/Animals Animals/ Earth Scenes; (Rabbit): © John Gerlach/Animals Animals/Earth Scenes; (Horse): © Yva Momatiuk/John Eastcott/Getty Images; (Zebra): © Jean-Michel Labat/Peter Arnold, Inc.; (Polar bear): © Tom Walker/Visuals Unlimited; (Wolf): © Hans Reinhard/zefa/Corbis; 31.2A: © Arthur Morris/Visuals Unlimited; 31.2B: James D. Watt/Visuals Unlimited; 31.2C: © Michael & Patricia Fogden/Minden Pictures; 31.2D: © Brian Kenney/OSF/Animals Animals/Earth Scenes; 31.3D: © Ardea London Ltd.; 31.8A (left): © Manfred Kage/Peter Arnold, Inc.; (right): Photo by Susumu Ito, from Charles Flickinger *Medical Cell Biology* W.B. Saunders, 1979; 31.12: © Amiard/photocuisine/Corbis; 31.18A: Ryan McVay/Getty Images; 31.18B: © Benjamin F. Fink, Jr./Brand X/Corbis; 31.19B: © Donna Day/Stone/Getty Images; 31.19C: © Royalty-Free/Corbis.

Chapter 32

Openers(Sperm whale): © Brandon Cole/Visuals Unlimited; (Elephant seal): © Bruce Watkins/Animals Animals/Earth Scenes; (Weddell seal): © Mark Chappell/Animals Animals/Earth Scenes; (Diver): © Buzz Pictures/Alamy; (Dolphin): © Stephen Frink/Corbis; 32.11A: © Frank Schwere/Getty Images; 32.11B(Both): © Martin M. Rotker, 2007.

Chapter 33

Openers(Barracuda): © Tobias Bernhard/Jupiter Images; (Reef): © Gregory Ochocki/Photo Researchers, Inc.; (Coral): © Jeff Rotman/Getty Images; (Christmas tree worm): © Espen Rekdal/SeaPics.com; (Parrotfish): © Hal Beral/Grant Heilman Photography; (Clownfish): © Dave B. Fleetham/Visuals Unlimited; 33.3A: © Norbert Wu; 33.4A(Kangaroo rat): © Bob Calhoun/Bruce Coleman, Inc.; 33.4B: © Eric Hosking/Photo Researchers, Inc.; 33.6B(top): © R.G. Kessel and R H. Kardon, *Tissues and Organs: A Text-Atlas of Scanning Electron Microscopy*, 1979, W. H. Freeman, New York; (center, bottom): © 1966 Academic Press, from A.B. Maunsbach, Journal of Ultrastructure Research Vol. 15:242-282; 33.11: © SIU/Peter Arnold, Inc.

Chapter 34

Openers(Humans): © David Raymer/Corbis; (Lions): © K. Ammann/Bruce Coleman, Inc.; (Ants): © Dr. Mark Moffett/Minden Pictures; (Wildebeest): © Alamy; (Sea lions): © NGS/Getty Images; 34.7(both): "Atlas of Pediatric Physical Diagnosis," Second Edition by Zitelli & Davis, 1992. Mosby-Wolfe Europe Limited, London, UK; 34.9: © Vol. 83/PhotoDisc/Getty Images; 34.10: © James Darell/Stone/Getty Images; 34.11A: © Medical-on-Line/Alamy; 34.11B: © Bruce Coleman, Inc./Alamy; 34.11C: © Dr. P. Marazzi/SPL/Photo Researchers, Inc.

Chapter 35

Openers(Frogs): © Hans Pfletschinger/Peter Arnold, Inc.; (Kangaroos): © Martin Harvey/Corbis; (Turtle hatching): © Kevin Schafer/Corbis; (Whale penises): © Phillip Colla/SeaPics.com; (Sea turtles): © Michael Patrick O'Neill/Photo Researchers, Inc.; (Elks): © Leonard Lee Rue, Jr./Photo Researchers, Inc.; 35.1A: © Dr. Dennis Kunkel/Visuals Unlimited; 35.1B: © Kelvin Aitken/Peter Arnold, Inc.; 35.1C: © Herbert Kehrer/zefa/Corbis; 35.2: © Anthony Mercieca/Photo Researchers, Inc.; 35.4: © Ed Reschke; 35.6A: © Ed Reschke/Peter Arnold, Inc.; 35.8: © Vol. 62/Corbis; 35.9(Birth control pill, female condom, spermicidal jelly, diaphragm, Depo-Provera): © The McGraw-Hill Companies, Inc/Bob Coyle photographer; 35.9(IUD): © The McGraw-Hill Companies, Inc./Vincent Ho photographer; (Implant): © Population Council/Karen Tweedy-Holmes; 35.10: © CC Studio/SPL/Photo Researchers, Inc.; 35.11, 35.13: © PhotoDisc/Getty Images; 35.17B, 35.18A: © Lennart Nilsson, *A Child is Born*, Dell Publishing; 35.18B: © John Watney/Photo Researchers, Inc.; 35.18C: © James Stevenson/SPL/Photo Researchers, Inc.; 35.18D: © Lennart Nilsson, "A Child is Born" 1990 Delacorte Press, p.111(top frame), 35.18E: © Dennis MacDonald/PhotoEdit; p.706: © 3D4Medicalcom/Getty Images.

Chapter 36

Openers (Doe and fawn): © Stephen J. Krasemann/Photo Researchers, Inc.; (Deer running): © John Cancalosi/Peter Arnold, Inc.; (Buck): © Dominique Braud/Animals Animals/Earth Scenes; 36.1: © David Hall/Photo Researchers, Inc.; 36.2A: © Richard Weymouth Brooks/Photo Researchers, Inc.; 36.2B: © Karen Fuller/Alamy; 36.3(Pigs): © age fotostock/SuperStock; (Rhinos): © Royalty-Free/Corbis; 36.9A: © Ted Levin/Animals Animals/Earth Scenes; 36.9B: © Martin Harvey/Gallo Images/Corbis; 36.11 (top): © Royalty-Free/Corbis; (bottom): © Ben Osborne/OSF/Animals Animals/Earth Scenes.

Chapter 37

Openers(Porcupine): © Mark Boulton/Photo Researchers, Inc.; (Chinchillas): © J.M. Labat/P. Rocher/Peter Arnold, Inc.; (Guinea pigs): © Jorg & Petra Wegner/Animals Animals/Earth Scenes; (Naked mole rat queen): © Jennifer Jarvis/Visuals Unlimited; (Naked mole rats digging): © M. J. O'Riain & J. Jarvis/Visuals Unlimited; 37.1A(top): © Joe McDonald; (bottom): Courtesy Refuge for Saving the Wildlife, Inc.; 37.1B(top): © R. Andrew Odum/Peter Arnold, Inc.; (bottom): © John Sullivan/Monica Rua/Ribbitt Photography; 37.4A: © T & P Gardner/Bruce Coleman, Inc.; 37.4C: © Barbara Gerlach/Visuals Unlimited; 37.4D: (left): © Y. Arthus-Bertrand/Peter Arnold, Inc.; (right): © Neil McIntyre/Getty Images; 37.5A: © Mark Moffett/Minden Pictures; 37.5B: © J & B Photo/Animals Animals/Earth Scenes: 37.6A: © Gregory G. Dimijian/Photo Researchers, Inc.; 37.6B: © Susan Kuklin/Photo Researchers, Inc.; 37.6C: © Punchstock; 37.6D: © OSF/Animals Animals/Earth Scenes; 37.7(left): © Alan Carey/Photo Researchers, Inc.; (right): © Tom McHugh/Photo Researchers, Inc.

Chapter 38

Openers(Vultures, top): © Jorge Sierra/OSF/Animals Animals; (Dung beetles): © Anthony Bannister/Animals Animals/Earth Scenes; (Dung heap): © Lex Hes/ABPL /Animals Animals/Earth Scenes; (Vultures, bottom): © Owen Franken/Corbis; (Biogas plant): © Christiane Eisler/Peter Arnold, Inc.; (Bioremediation): © Science VU/Visuals Unlimited; (Before/After): © Science VU/Visuals Unlimited; 38.2A: © Alan Carey/Photo Researchers, Inc.; 38.2B(Warning): © Zig Leszczynski/Animals Animals/Earth Scenes; (Camouflage): © Gustav Verderber/Visuals Unlimited; (Fright): © National Audubon Society/A. Cosmos Blank/Photo Researchers, Inc.; 38.2C: (Flower fly, longhorn beetle, yellow jacket): © Edward S. Ross; (Bumblebee): © James H. Robinson/Photo Researchers, Inc.; 38.4: © Dave B. Fleetham/Visuals Unlimited; 38.5: Courtesy Dr. Ian Wyllie; 38.6A: © C. C. Lockwood/Animals Animals; 38.6B: © Bill Wood/Bruce Coleman, Inc.; 38.8B(all): © Breck P. Kent/Animals Animals/Earth Scenes; 38.9(left): © Kevin Schafer/kevinshafer.com; (right): © Herbert Spichtinger/zefa/Corbis; 38.10: © George D. Lepp/Photo Researchers, Inc.

Chapter 39

Openers(Biosphere 2): © G. Kenny; (Agriculture): Biosphere 2 staff photo; (Harvest, feast): © A. Alling; 39.2: © Michio Hoshino/Minden Pictures; 39.3: © Ken Cole/Animals Animals/Earth Scenes; 39.4: © Altrendo Nature/Getty Images; 39.5: © Gordon & Cathy Illg/Animals Animals/Earth Scenes; 39.6: © Darla G. Cox; 39.7: © John Shaw/Bruce Coleman, Inc.; 39.8B(Frog): © Art Wolfe/Photo Researchers, Inc.; (Macaw): © Tony Craddock/SPL/Photo Researchers, Inc.; (Katydid): © Dr. James L. Castner; (Ocelot): © Martin Wendler/Peter Arnold, Inc.; (Butterfly): © Kjell Sandved/Butterfly Alphabet; (Lemur): © Erwin & Peggy Bauer/Bruce Coleman, Inc.; (Lizard): © Kjell Sandved/Butterfly Alphabet; 39.11A(1): © Kim Taylor/Bruce Coleman, Inc.; (2): © William H. Mullins/Photo Researchers, Inc.; (3): © Robert Maier/Animals Animals/Earth Scenes; (4): © Gerlach Nature Photography/Animals Animlas/Earth Scenes; 39.11B(Eutrophic): © Michael Gadomski/Animals Animals/Earth Scenes; (Oligotrophic): © Roger Evans/Photo Researchers, Inc.; 39.12A(Mudflats): © John Eastcott/Yva Momatiuk/Animals Animals/Earth Scenes; (Mangrove swamp): © Dr. James L. Castner; (Rocky shore): © Brandon Cole/Visuals Unlimited.

Chapter 40

Openers(Waikiki): © Mark Karrass/Corbis; (Gardenia, Hibiscus): © Forest & Kim Starr; (Akiapolaau, Akohehohe, feral pig): © Jack Jeffrey Photography; 40.3: (Periwinkle): © Kevin Schaefer/Peter Arnold, Inc.; (Armadillo): © John Cancalosi/Peter Arnold, Inc.; (Fishermen): © Herve Donnezan/Photo Researchers, Inc.; (Bat): © Merlin D. Tuttle/Bat Conservation International; (Man): © Bryn Campbell/Stone/Getty; (Ladybug) © Anthony Mercieca/Photo Researchers, Inc.; 40.4(Forest): © Salvatore Vasapolli/Animals Animals/Earth Scenes; (Birdwatchers): © Steve Maslowski/Photo Researchers, Inc.; 40.5A: © Gunter Ziesler/Peter Arnold, Inc.; 40.5B(Distant): (c) R.O. Bierregaard; (Close-up): Courtesy Thomas G. Stone, Woods Hole Research Center; 40.6A: © Chuck Pratt/Bruce Coleman, Inc.; 40.6B: © Chris Johns/National Geographic Society Image Collection; 40.7: Courtesy Walter C. Jaap/Florida Fish & Wildlife Conservation Commission; 40.8: © Shane Moore/Animals Animals/Earth Scenes; 40.9: © Dr. Paul A. Zahl/Photo Researchers, Inc.; 40.10A: © Gerard Lacz/Peter Arnold, Inc.; 40.10B(Forest): © Art Wolfe; (Owl): © Pat & Tom Leeson/Photo Researchers, Inc.; 40.11(Panther): © Tom & Pat Leeson/Photo Researchers, Inc.; (Alligator): © Fritz Polking/Visuals Unlimited; (Ibis): © Stephen G. Maka; (Spoonbill): © Kim Heacox/Peter Arnold, Inc.; (Wood storks): © Millard H. Sharp/Photo Researchers, Inc.; Biological Viewpoint, pp. 794-795(forest): © Royalty-Free/Corbis; (pine cones): Siede Preis/Getty Images.

Index

Note: Page references followed by *f* and *t* refer to figures and tables respectively.

A

Abdomen
of crayfish, 382*f*
of insect, 384
Abduction, 561
Abiotic factors, in population control, 715, 715*f*
ABO blood type, 171, 588
Abortions, spontaneous, 154
Abscisic acid (ABA), 446, 464, 464*f*
Abscission, 431, 464, 465
Absorption, in digestive process, 612, 612*f*
Absorption spectrums, of photosynthetic pigments, 100, 100*f*
ABT-594, 394
Abyssal plain, 772*f*, 773
Acacia trees, 473, 473*f*, 767
Acetabularia, 325*f*
Acetic acid bacteria, 124
Acetylcholine (ACh), 520–521, 521*f*, 522, 529, 567
Acetylcholinesterase (AchE), 521
Acetyl CoA, 118, 119, 119*f*, 125, 125*f*, 126
Acetylene, 38
ACh. *See* Acetylcholine
Acheulian tools, 409
Achondroplasia, 169, 201, 201*f*
ACI. *See* Autologous chondrocyte implantation
Acid(s), 30, 30*f*
Acid-base balance, regulation of, 507, 584, 643, 659
Acid deposition, 32, 32*f*, 757, 787
Acidosis, 31, 643, 659
Acid rain. *See* Acid deposition
Acoelomates, 371, 371*f*, 376
Acon terminal, 520, 520*f*
ACTH. *See* Adrenocorticotropic hormone
Actin
and cell motion, 71
in cell structure, 60*f*, 61*f*, 70, 70*f*
in muscle contraction, 81, 81*f*, 500, 566–567, 566*f*, 567*f*
Action potential
mechanism of, 518–519, 518*f*, 519*f*
speed of, 519, 519*f*
Active immunity, 599
Active site, of enzyme, 82, 82*f*
Active transport, 90, 90*f*
Actonel, 559
Acyclovir, 689
ADA (adenosine deaminase), 232
Adam's apple, 638, 685
Adaptation
in angiosperms, 423, 423*f*
as characteristic of life, 7
early theories of, 245
and evolution, 419
innate behaviors and, 726–727

to land environment, in plants, 344–345
learned behaviors and, 728–729
in pollinators, 242–243, 242*f*, 243*f*
reproductive behaviors and, 730–731
social behaviors and, 732–733, 732*f*, 733*f*
Adaptive radiation, 270, 270*f*, 340, 440–441
Addiction
drug abuse and, 522, 523
smoking and, 639
treatment of, 523
Adduction, 561
Adductor longus, 563*f*
Adenine (A), 48, 48*f*, 48*t*, 186, 186*f*, 187, 187*f*, 193, 193*f*
Adenoids, 616
Adenosine diphosphate (ADP), 50, 50*f*, 80, 80*f*, 566–567, 567*f*
Adenosine monophosphate (AMP), 80
Adenosine triphosphate. *See* ATP
Adenoviruses, 302, 302*f*
ADH. *See* Antidiuretic hormone
Adhesion, of water in plant vessels, 27, 27*f*, 444
Adhesion junctions, in animal cells, 501–502
Adipose tissue, 498*f*, 499
ADP (adenosine diphosphate), 50, 50*f*, 80, 80*f*, 566–567, 567*f*
Adrenal cortex, 667, 669*t*, 672–673, 672*f*
Adrenal glands, 668, 668*f*, 669*t*, 672–673, 672*f*
Adrenaline. *See* Epinephrine
Adrenal medulla, 669*t*, 672, 672*f*
Adrenocorticotropic hormone (ACTH), 669*t*, 670*f*, 671, 672, 672*f*
Adrenogenital syndrome (AGS), 673
Adult stem cells, 132, 136, 586
Advertising, and classical conditioning, 729
Aerobic exercise, 126, 126*f*
Aerobic process, 115
Aerobic prokaryotes, 314
AF-2, 201
Afferent arteriole, 655, 655*f*, 656*f*
Afferent neurons. *See* Sensory (afferent) neurons
Aflatoxin, 201
Africa, population growth in, 720
African apes, 401*f*
African sleeping sickness, 323, 327
Afterbirth, 702, 702*f*
Agar, 333
Agaricus bisporus, 362
Age
age structure diagrams, 713, 713*f*
distributions, in more- *vs.* less-developed countries, 720, 720*f*
and risk of Down syndrome child, 153
Age-associated nerve deafness, 546

Agent Orange, 301
Agglutination, of blood, 588, 588*f*
Aggregate fruits, 485
Agnatha. *See* Jawless fishes
Agranular leukocytes, 585
Agriculture
development of, 412–413, 413*f*
fungal diseases, 362, 362*f*
and introduction of alien species, 786
pesticide resistance, 248
plant viruses and, 304
wildlife, value of, 782–783
AGS. *See* Adrenogenital syndrome
AID. *See* Artificial insemination by donor
AIDS. *See* HIV/AIDS
Air pollution, and acid rain, 32, 32*f*
Airsickness, 548
Albumin, 390*f*, 584
Albuterol, 235
Alcohol
and acidosis, 659
and cancer, 144, 201
and cirrhosis, 620
as diuretic, 658
effects of, 522, 583, 671
fermentation and, 123, 123*f*, 124
functional groups and, 39
Alcoholic drinks, production of, 124
Aldosterone, 658, 659*f*, 673
Alemseged, Dr., 399
Algae, 332–336
blooms, 315, 325, 757
blue-green, 315 (*See also* Cyanobacteria)
brown, 325*f*, 333, 333*f*
characteristics of, 325*f*
classification of, 324
diatoms, 325*f*, 332, 332*f*
dinoflagellates, 325*f*, 332, 332*f*
green, 334–335, 335*f*, 773
characteristics of, 325*f*
classification of, 326, 326*f*
evolution of, 369
mutualism in, 359
life cycle of, 150, 150*f*, 333, 334, 334*f*, 336, 336*f*
as photoautotroph, 752
red, 94–95, 325*f*, 333, 333*f*, 773
Algin (alginate), 333
Alien species, and extinctions, 779, 785*f*, 786, 786*f*
Alkaloids, as plant defense, 472
Alkalosis, 31, 643, 659
Allantois, 390*f*, 697, 697*f*, 698*f*, 699
Allele(s), 145, 163, 163*f*
dominant, 163
harmful, heterozygote advantage and, 258, 258*f*
multiple, 171
recessive, 163, 254
x-linked, 174, 174*f*
Allele frequency
calculation of, 253, 253*f*
changes in, causes, 253–257

Hardy-Weinberg principle, 253
Allen's rule, 414
Allergic reactions, 606, 606*f*, 606*t*
Alliance, definition of, 440
Alligators, 282, 390, 390*f*, 680–681
Allopatric speciation, 268–270
Alloploidy, 271, 271*f*
All-or-none manner, of action potential, 518, 518*f*
Allosteric site, 84, 84*f*
Alpha 1 antitrypsin, 394
Alpha helix, 47, 47*f*
Alpine tundra, 765, 770, 770*f*
Alternation of generations, 336, 344, 344*f*, 481
Altman, Sidney, 309
Altruism, 732–733
Alveolata, classification of, 326*f*
Alveoli, 638, 638*f*
Alzheimer disease, 47, 69, 153, 520, 527
Amantia, 362
Ambulocetus natans, 250, 250*f*, 264*f*
American Heart Association, 627
Ames test, 201
Amino acids
common, 46*f*
metabolism of, 650, 650*f*
peptide formation, 45, 45*f*
R groups of, 46, 46*f*
sources of, 301
structure of, 46, 46*f*
synthesis of, 125
Aminoacyl-tRNA synthetases, 196
Amino groups, 39*f*
Amish, genetic abnormalities in, 255, 255*f*
Ammonia
and origin of organic molecules, 308–309, 308*f*
as polar molecule, 26*f*
as waste product, 650, 650*f*
Ammonium, in nitrogen cycle, 757, 757*f*
Amniocentesis, 168, 170
Amnion, 387, 390*f*, 697, 697*f*, 698*f*, 699
Amniotic cavity, 698, 698*f*
Amniotic egg, 387, 390, 390*f*, 391, 392
Amniotic fluid, 19, 697, 699, 702, 702*f*
Amniotic membrane, 702, 702*f*
Amoeba proteus, 325*f*, 328
Amoebic dysentery, 323, 328
Amoeboids, 324, 325*f*, 328, 328*f*
AMP (adenosine monophosphate), 80
Amphetamines
amphetamine psychosis, 522
effects of, 583
urine tests for, 657
Amphibians
characteristics, 373, 389, 389*f*
circulatory system, 576, 576*f*
evolution of, 387*f*
excretory system, 650
kidneys, 654
nerve regeneration in, 502
reproduction, 680
respiration, 635, 640

Amplexus, 680, 680f
Ampulla, 546, 547f
Amygdala, 527, 527f
Amylase, 462, 462f
Amyloid plaques, 527
Amyloplasts, 466–467, 466f
Anabaena, 315f
Anabolic steroids, 44
Anabolism, 125
Anaerobes
    facultative, 314
    obligate, 314
Anaerobic process, 115
Anal canal, 621
Analogous structures, 251, 292
Analogy, 292
Anandamide, 523
Anaphase
    in meiosis
        anaphase I, 148, 148f–149f,
            150, 151f
        anaphase II, 148f–149f, 149
    in mitosis, 136f, 138f–139f, 150,
        151f
Anaphylactic shock, 606
Anatomy, comparative, as evidence for
    evolution, 251, 251f
Ancestor, common
    evidence for, 250–252
    phylogeny and, 291, 291f
Ancestral character, 291
Anchoring junctions, in animal cells,
    72, 72f
Androgen insensitivity, 200, 200f
Androgens, 669t
Anemia, 584, 624
Aneuploidy, 152
Angina pectoris, 582
Angiogenesis, 143, 220, 220f, 574
Angiosperms, 353–357
    adaptation in, 423, 423f
    alternation of generations in, 336,
        344, 344f, 481
    cells and tissues, 426–429,
        426f–429f
    characteristics of, 343
    diversity in, 353
    eudicots. See Eudicots
    evolution of, 292, 343, 343f, 353
    growth
        primary, 430–431, 430f, 431f
        secondary, 432–433, 432f, 433f
    homeostasis in, 435–436, 435f, 436f
    importance of, 353
    leaves of. See Leaf
    life cycle, 353–356, 354f–355f, 418f,
        480–481
    monocots. See Monocots
    ornamental, 425
    pollination, 351, 351f, 354f–355f,
        355, 478–479, 482–483, 483f
        (See also Pollinators)
    roots. See Root system
    seed dispersal strategies, 478–479
    sexual reproduction in, 345, 480–486
    stem of
        structure and function, 422, 422f
        tissues, 428, 428f, 429f
    structure of, 422–423, 422f, 423f
    success of, insects and, 355
    sugar transport in, 448–449, 448f,
        449f
    uses of, 356–357, 356f, 357f
    water and solute transport in, 442,
        442f, 444–445, 445f
    water regulation in, 446, 446f
Angiotensin I and II, 658, 659f, 673
Angiotensinogen, 658, 659f, 673
ANH. See Atrial natriuretic hormone
Animal(s), 366–394
    animal pharming, 394

aquatic
    development in, 683
    excretory system, 650, 650f
    mammals, diving in, 632–633
    reproduction, 682–683, 682f
    respiration in, 634
biological weapons and, 318
characteristics of, 368
classification of, 9, 9f, 296, 373
cloning of
    benefits and drawbacks, 211,
        211f
    ethics of, 211
    types, 210, 210f
communication, 734–735, 734f, 735f
of coniferous forest, 765
of desert, 767, 767f
development, 217, 252, 368, 368f
emotions in, 736, 736f
energy storage in, 42, 42f, 43, 43f
in Everglades, 790, 790f
evolution of, 369–372
    body plans, 369
    colonial flagellate hypothesis,
        369, 369f
    evolutionary tree, 370f, 372, 372f
    key innovations, 370f, 371
    molecular data, 372
    multinucleate hypothesis, 369
feeding strategies, 613, 613f
of fresh water ecosystems, 771, 771f
genetically-modified, 231, 231f
hibernation, 495
homeostasis in, 494, 507–508
hormones. See Hormones
major phyla, 373
of marine ecosystem, 772, 773, 773f
medicinal plant use, 474
ovoviviparous, 683
plant-animal mutualism, 473, 473f
as pollinators. See Pollinators
reproduction
    asexual, 682, 682f
    overview of, 680–681
    sexual, 682–683, 682f, 683f
of savanna, 767, 767f
seed dispersal and, 478–479, 485
similarity to plants, 36–37
skeleton types, 554–555, 554f, 555f
species, number of, 780f
of temperate deciduous forest, 766
of temperate grassland, 766, 766f
tissues, 496–501
    connective tissue, 498–499, 498f
    epithelial tissue, 496–497, 497f
    muscular tissue, 500, 500f
    nervous tissue, 501, 501f
of tropical rain forest, 768, 768f
of tundra, 765
as type of organism, 4, 4f
viral diseases, 305–307
viviparous, 683
Animal cells
    anatomy, 60f, 61
    cell cycle in, 136–140, 136f,
        138f–139f, 140f
    chromosome numbers, 137t
    cytokinesis in, 140, 140f
    junctions in, 72, 72f
    mitosis in, 138–139, 138f–139f
Animal pole, of embryo, 692
Annelids, 381, 381f, 514–515, 514f
    blood of, 573
    characteristics of, 373
    circulatory system, 575
    evolution of, 370f, 372f
    excretory system, 651
    reproduction, 682
    respiration in, 634
Annual rings, 432–433
Anolis, speciation in, 269, 269f

Anopheles mosquito, 322, 322f, 330, 330f
Anorexia nervosa, 628, 628f
Ant(s)
    altruism in, 732–733, 732f
    Argentinian fire ant, 786
    army, 732–733, 732f
    chemical communication, 734
    fire, 2–3, 2f, 3f, 6, 7
    jumper, 14
    lance fluke infection in, 747
    mutualism in, 473, 473f, 748
    plant-animal mutualism in, 473,
        473f
    seed dispersal and, 479
Antagonistic hormonal actions, 668
Anteater, spiny, 392
Antenna
    of crayfish, 382f, 383
    of insects, 384, 384f
Antenna molecules, of photosystem,
    101, 101f
Antennule, 382f, 383
Anterior compartment, of eye, 540, 540f
Anterior pituitary, 669t, 670f, 671, 685,
    707
Anther, 353, 353f, 354f–355f, 480, 480f,
    481, 481f, 482f
Anther culture, 488
Antheridium, 346, 347, 347f, 349, 349f
Anthrax, 318
Anthropoids, 403
Antibiotics
    bacterial production of, 301, 314
    and hearing damage, 546
    resistance to
        in bacteria, 246, 248, 313, 317
        in protists, 357
Antibodies (immunoglobulin), 585, 594,
    601, 601f, 606, 606f, 607t
    monoclonal, 604, 604f, 657
Antibody-mediated immunity, 600,
    600f, 601, 601f
Anticoagulants
    in bats, 367
    in leeches, 381
Anticodons, 196, 196f
Antidiuretic hormone (ADH), 658, 669t,
    670–671, 670f
Antigen(s), 585, 588, 599
Antigen-presenting cell (APC), 600,
    600f, 602, 602f
Antigen receptors, 600
Antihistamines, 606
Anti-inflammatory drugs, 598
Antioxidants, 144, 583, 625
Antiviral drugs, herpes and, 689
Anus, 616f, 621
Anvil (incus), 544, 544f, 545
Aorta, 577, 577f, 580, 580f
Aortic bodies, 641, 659
APC. See Antigen-presenting cell
Ape(s), 400, 400f, 401, 401f, 403. See
    also Chimpanzees; Gorillas
Aphids, 448, 448f, 472, 783
Apical dominance, 460, 460f
Apical meristem, 426, 430, 430f, 431,
    431f, 463, 484
Apoptosis
    and cancer, 218, 602
    cell cycle checkpoints and, 141
    cell senescence and, 142
    development and, 141, 217
    in pattern formation, 695
    in T cells, 602
Appendages. See also Limb(s)
    jointed
        in arthropods, 382, 382f
        evolution of, 370f, 371
        in vertebrates, 387
    paired, as vertebrate characteristic,
        387

Appendicitis, 621
Appendicular skeleton, 556,
    558–559, 558f
Appendix, vermiform, 595, 621, 621f
Apple, structure of, 485
Aquaculture, 783
Aquaporin, 89
Aquatic animals
    development in, 683
    excretory system, 650, 650f
    mammals, diving in, 632–633
    reproduction, 682–683, 682f
    respiration in, 634
Aquatic ecosystems, 771–774
Aquatic invertebrates, respiration in,
    634
Aquatic vertebrates
    osmoregulation in, 652, 652f
    respiration in, 634
Aqueous humor, 540
Arabidopsis thaliana
    embryo development in, 484, 484f
    genome of, 234, 234t
    as model organism, 182–183, 182f,
        183f
    photoperiodism and, 469, 471
Arachnids, 383, 383f, 651
Archaea
    as domain, 8, 8f
    as earliest creatures, 8
    evolution of, 316
    in extreme environments, 8, 316,
        316f
    kingdoms of, 8
    nucleic acid base sequences, 59
    as prokaryote, 311
    species, number of, 780f
    structure and function, 316
    types of, 316, 316f
Archaeopteryx, 283
Archaeopteryx lithographica, 250, 250f
Archaic humans, evolution of, 410–411
Archegonium, 345, 345f, 346, 347, 347f,
    349, 349f
Arctic tundra, 765, 765f
Ardipithecus ramidus, 405, 405f, 406
Area (space), and species richness, 749
Arens, Nan Crystal, 288
Argentinian fire ant, 786
Arginine, 46f
Armadillo, nine-banded (*Dasypus
    novemcinctus*), 782, 782f
Army ants, 732–733, 732f
ART. See Assisted reproductive
    technologies
Arterioles, 579, 579f
Artery(ies), 577, 579, 579f
Arthritis, 562
Arthropods, 372f, 382–384, 514–515,
    514f
    arachnids, 383, 383f, 651
    blood of, 573
    characteristics of, 373, 382, 382f
    circulatory system, 575
    crustaceans. See Crustaceans
    evolution of, 370f, 372f
    excretory system, 651
    exoskeleton, 554
    respiration in, 636
Arthroscopic surgery, 562, 562f
Articular cartilage, 560, 560f, 561
Artificial insemination by donor (AID),
    691
Artificial selection, 246, 246f, 272, 272f
Artiodactyla, 290, 290f, 393
Ascaris, 372f, 378, 378f
Ascending colon, 616f, 621
Ascomycota, 360–361, 360f
Ascus, 360, 360f
Asexual reproduction
    advantages and disadvantages of, 147

in algae, 334, 334*f*, 336, 336*f*
in animals, 682, 682*f*
binary fission, 135, 135*f*, 312, 312*f*
cell division in, 134, 135, 135*f*
in ciliates, 329, 329*f*
in diatoms, 332
in dinoflagellates, 325
in fungi, 361, 361*f*
in lichens, 359
in plants, 487–488, 487*f*, 488*f*
in protists, 324
in sponges, 374
Asian Apes, 400*f*
A (amino acid) site, 198, 198*f*
Asparagine, 46*f*
Aspartic acid, 46*f*
Aspergillus nidulans, chromosome number, 137*t*
Aspirin, 598
Assimilation model, of human evolution, 411
Assisted reproductive technologies (ART), 691, 691*f*
Association neurons, 517, 517*f*
Associative learning, 729, 729*f*
Assortative mating, and microevolution, 254
Aster, 138, 138*f*–139*f*
Asthma, 235, 606, 644
Astigmatism, 542, 542*f*
Astrocytes, 501, 501*f*, 517
Atherosclerosis, 582
Athlete's foot, 362, 362*f*
Atkins diet, 622
Atmosphere
    carbon dioxide levels, 420–421
    early, 308, 308*f*, 315, 317
    photosynthesis and, 310
Atom(s)
    characteristics of, 21, 21*f*
    definition of, 20
    as level of organization, 4, 5*f*
    molecule formation, 23
    periodic table, 23, 23*f*
    subatomic particles, 21, 21*f*, 21*t*
Atomic mass, 21, 21*f*, 21*t*
Atomic models, 23, 23*f*
Atomic number, 21, 21*f*
Atomic symbol, 20
ATP (adenosine triphosphate)
    ATP cycle, 80, 80*f*
    breakdown of
        coupling to energy-requiring reactions, 81, 81*f*, 112
        energy released in, 50, 50*f*, 80, 80*f*
    in Calvin cycle, 104–105, 104*f*, 105*f*
    as energy currency, 50, 50*f*, 80, 252
    function of, 50
    and muscle contraction, 81, 81*f*, 566–567, 567*f*, 568
    origin of, 112
    production of
        in cellular respiration. *See* Cellular respiration
        in fermentation, 123–124, 123*f*
        in photosynthesis, 98, 101–103, 101*f*–103*f*
    structure of, 50, 80, 112
    universality of, 113
ATP cycle, 80, 80*f*
ATP synthase complexes, 101, 121, 121*f*
Atrial natriuretic hormone (ANH), 658
Atrioventricular valves, 577, 577*f*, 578, 578*f*
Atrium, heart, 577, 577*f*
Attachment stage, of viral life cycle, 303, 305, 305*f*, 306, 306*f*
Auditory canal, 544, 544*f*
Auditory (sound) communication, 734
Auditory tube, 544, 544*f*

Australia, alien species in, 786
Australopithecines, 406–407, 406*f*
Australopithecus aethiopicus, 405*f*, 407
Australopithecus afarensis, 398–399, 398*f*, 399*f*, 405*f*, 406–407, 406*f*
Australopithecus africanus, 405*f*, 406
Australopithecus anamensis, 405*f*, 406
Australopithecus boisei, 405*f*, 407
Australopithecus robustus, 405*f*, 406
Autoimmune disorders, 605, 605*f*
Autologous chondrocyte implantation (ACI), 562
Autonomic system, 530, 530*f*
Autoploidy, 271, 271*f*
Autosomal chromosomes, 145, 167
Autosomal dominant genetic disorders, 167, 167*f*, 169, 169*f*
Autosomal recessive genetic disorders, 167–168, 167*f*, 168*f*
Autosomes, 145, 167
Autotrophs
    definition of, 96
    in ecosystem, 752, 755, 755*f*
    prokaryotes, 314
Auxins, 436, 459, 460–461, 460*f*, 461*f*, 463, 465, 492
Avian influenza (bird flu), 307, 307*f*
AV (atrioventricular) node, 578, 578*f*
Axial skeleton, 556–557, 556*f*, 557*f*
Axillary buds, 422, 422*f*
Axillary lymph nodes, 594*f*, 595
Axons, 501, 501*f*, 517, 517*f*, 521*f*, 528, 528*f*
    motor axons, 564, 567, 567*f*
    regeneration of, 502
Azalea (*Rhododendron*), 481
Azure damselflies, 683, 683*f*

B

Baboons, 403, 731, 731*f*, 734–735
Baby boom, 713
Bacillus (bacilli), 311, 311*f*
Bacillus anthracis, 296, 318
Bacteria
    antibiotic resistance, 246, 248, 313, 317
    bacteriophages and, 184, 185*f*
    as biological weapons, 318
    chemoautotrophic, 310, 773
    chromosomes, 311
    commercial uses, 314
    cyanobacteria. *See* Cyanobacteria
    denitrifying, 757
    and disease, 301, 317, 317*t*
    as domain, 8, 8*f*, 296
    endospore formation, 312, 312*f*
    functions performed by, 8
    generation time, 135, 312
    genetically engineered, 228, 228*f*
    gene transfer in, 313, 313*f*
    habitats of, 8
    in human intestines, 748
    human uses of, 301
    kingdoms of, 8
    mutation rate, 312
    nitrogen-fixing, 12–13, 301, 315, 317
    nucleic acid base sequences, 59
    parasitism in, 746
    as prokaryote, 311
    rod-shaped, 55*f*
    species, number of, 780*f*
    in stomach, 618
    structure of, 59, 59*f*, 311, 311*f*
    as type of organism, 4, 4*f*
Bacteriophages
    life cycle, 303, 303*f*
    T2, 184–185, 185*f*
Balance, sense of, 537, 544, 546–548, 547*f*
Balanus balanoides, 743, 743*f*
Bald eagle, 391*f*

Baleen whales, 613, 613*f*
Ball-and-socket joints, 561, 561*f*
Ball-and-stick model, 25, 25*f*, 26*f*
Balloon angioplasty, 582, 582*f*
Bamboo, 425
Bangham, Alec, 310
Bañuelos, Gary, 447, 447*f*
Bark, 432, 432*f*
Barley, 425, 425*f*
Barnacles, 383, 383*f*, 743, 743*f*
Barn swallows, 730
Barr, Murray, 212
Barr bodies, 212
Barriers, as nonspecific immune defense, 596, 596*f*
Barro Colorado Island, 749
Bartholin glands (greater vestibular glands), 686
Basal bodies, 71, 71*f*
Basal ganglia. *See* Basal nuclei
Basal nuclei (basal ganglia), 525, 525*f*, 527
Baseball bats, 433
Basement membrane, 497
Bases, 30, 30*f*
Basidiocarp, 361, 361*f*
Basidiomycota, 361
Basidium (basidia), 361, 361*f*
Basilar membrane, 545, 545*f*
Basketweaver's disease, 362
Basophils, 584*f*
Bass, Edward P., 762
Bat(s), 366–367, 366*f*, 367*f*
    echolocation, 536, 536*f*, 734
    as keystone species, 789
    as placental mammal, 393
    as pollinators, 783, 783*f*
    vampire, 613, 613*f*, 733
Bates, Henry Walter, 247, 745
Batesian mimic, 745
Bathypelagic zone, 773, 773*f*
Bax gene, 218
B cell receptors (BCR), 600, 600*f*, 601, 601*f*
B cells, 585, 595, 600, 600*f*, 600*t*, 601, 601*f*, 604
BCR. *See* B cell receptors
Beadle, George, 192
Beagle (ship), 244, 244*f*
Beaks, of birds, 391, 391*f*
Bean seed, structure of, 483*f*, 486*f*
Bears, grizzly (*Ursus arctos horribilis*), 789, 789*f*
Bee(s)
    beeswax, 44
    biological clock in, 735
    honeybees (*Apis mellifera*), 682, 735, 735*f*, 783
    mimicry in, 745, 745*f*
    as pollinators, 243, 243*f*
Beer, brewing of, 124, 361, 425
Beetles, 741, 741*f*, 745, 745*f*, 794
Behavior
    courtship, 735
    definition of, 6, 726–730
    genetically-based, 726–727, 726*f*, 727*f*
    genetic *vs.* environmental factors in, 172
    imprinting, 728
    learning, in animals, 728–729, 728*f*, 729*f*
    "nature *vs.* nurture" question, 726–730
Behavioral isolation, as reproductive barrier, 266, 267*f*
Beijerinck, Martinus, 304
Bekoff, M., 736
Bell-shaped age structure diagrams, 713, 713*f*
Beltian bodies, 473, 748

The Bends, 633
Beneficial nutrients, 450
Bengal tigers, 718, 718*f*, 788
Benign tumor, 143
Bent grass (*Agrostis scabra*), 443, 443*f*
Bergmann's rule, 414
Berliner, David L., 665
Bernard, Claude, 706
Beta-carotene, 144
Bicarbonate, 24*t*, 31, 643, 758, 758*f*
Biceps brachii, 563*f*, 564, 564*f*
Bicoid, 216
bicoid gene, 216
Bicuspids (premolar teeth), 616, 616*f*
Bicuspid valve, 577, 577*f*
Biepharisma, 325*f*
Bilateral symmetry, 369, 371, 376, 514, 515
Bile, 619, 620
Bile salts, 620
Bilirubin, 584, 620
Binary fission, 135, 135*f*, 312, 312*f*
Binge-eating disorder, 628
Binocular vision, in primates, 400–401, 401*f*
Binomial nomenclature, 9
Biocultural evolution, 409, 412–413, 412*f*
Biodiversity. *See also* Extinctions
    biodiversity hotspots, 781, 789
    changes in, 750, 750*f*
    competition for resources and, 443
    definition of, 780
    direct value of, 782–783, 782*f*, 783*f*
    distribution of, 781
    and ecosystem stability, 443
    habitat and, 749, 749*f*, 785
    indirect value of, 784, 784*f*
    levels of understanding, 781
    preservation of, 718, 789–790, 789*f*
Biodiversity hotspots, 781, 789
Biogas, 741
Biogeochemical cycles, 756–758, 756*f*, 784
Biogeography, evidence for evolution, 252
Bioinformatics, 236, 780
Biological clock, 469, 735
Biological evolution, 309, 309*f*
Biological species concept, 264–265
Biological weapons, 318, 318*f*
Biology
    basic theories of, 11, 11*t*
    definition of, 11
Bioluminescence
    in dinoflagellates, 325
    genetic engineering and, 224–225, 224*f*, 225*f*
Biomass, ecological pyramid and, 755, 755*f*
Bioreactors, 228
Bioremediation, 301, 301*f*, 741*f*
Biorhythms, 675, 675*f*
Biosphere
    aquatic ecosystems, 771–774
    definition of, 4, 10
    ecological study of, 710
    human impact on, 10
    as level of organization, 4, 5*f*
    organization of, 10
    terrestrial ecosystems, 764–768, 764*t*
Biosphere 2, 762–763
Biosynthesis stage, of viral life cycle, 303, 305, 305*f*, 306, 306*f*
Biotechnology products, 228. *See also* Genetically modified organisms (GMOs)
Biotic community, in chemical cycling, 756, 756*f*
Biotic factors, in population regulation, 715*f*, 716, 716*f*

Biotic potential, 712
Biotin, dietary requirements, 625t
Bipedalism, evolution of, 405, 406–407, 406f
Bipolar cell layer, of retina, 542–543, 543f
Birch trees, water loss in, 445
Bird(s)
  ancestors of, 250, 283
  beak types, 391, 391f
  behavior
    genetically-based, 726, 726f
    learned, 728, 728f
  characteristics of, 373, 391, 391f
  circulatory system, 576, 576f
  communication in, 734
  development, 692
  digestive system, 614, 614f
  energy storage, 43
  evolution of, 251, 288f–289f, 387f, 391
  excretory system, 650, 650f
  extraembryonic membranes, 697, 697f
  eyes, 535
  on Galápagos Islands, 245, 248, 248f, 270, 715, 718f, 742, 742f
  group living in, 732, 745
  in Hawaiian Islands, 779, 779f
  kidneys of, 654, 655
  mate selection in, 730, 730f
  parental behavior in, 683, 683f
  and pet trade, 787–788
  reciprocal altruism in, 733
  reproduction, 683
  resource partitioning in, 742–743, 742f, 743f
  respiration in, 635, 640–641, 641f
  songbird phylogeny, 293
  speciation in, 269, 270, 270f
  symbiosis in, 748
  territoriality in, 731
  of tropical rain forest, 768
Bird flu (avian influenza), 307, 307f
Birth, 702, 702f
  breech, 701
  cesarean section, 701
  hormones in, 671
  premature, 697
Birth control, methods of, 690, 690f, 690t
Birth weight, stabilizing selection in, 256, 256f
Bisphosphonates, 559
Bivalves, 380, 380f
Black bread mold (Rhizopus stolonifer), 360, 360f
Black Death. See Plague
Black widow spider, 383f
Bladder. See Urinary bladder
Blade, of leaf, 422f, 423
Blastocoel, 692, 692f
Blastocyst, 698, 698f
Blastopore, 379, 379f, 692f, 693, 693f
Blastula, 692, 692f
Blaylock, Mike, 447
Blending model of inheritance, 160
Blindness, causes of, 541
Blind spot, 540
Blood. See also Clotting; Red blood cells; White blood cells
  calcium level, regulation of, 676
  color of, 572–573
  composition of, 499, 499f, 584, 584f
  as connective tissue, 499
  functions of, 499, 584
  glucose levels, regulation of, 507, 620
  and homeostasis, 507, 508
  pH level, 31, 659
    regulation of, 507, 584, 643, 659

phosphate levels, regulation of, 676
  stem cells, 586, 586f
  water content of, 27
Blood clots, immune system and, 598
Blood doping, 633
Blood flukes, 377, 377f
Blood pressure, 581, 581f
  arterioles and, 579, 579f
  and blood flow, 581, 587, 587f
  hypertension, 582–583
  kidneys and, 656
  measurement of, 581, 581f
  regulation of, 537, 666, 673, 707
  water-salt balance and, 658
Blood type, 588, 588f
  alleles responsible for, 171, 588, 588f
  and erythroblastosis fetalis, 588
Blood vessels
  and cardiovascular disease, 582
  engineered, 506
  and homeostasis, 495
  pulmonary circuit, 576, 576f, 580, 580f
  systemic circuit, 576, 576f, 580, 580f
  types and functions, 579, 579f
"Blue bloods," 572
Blue crabs, 771, 771f
Blue-footed boobies, 266, 267f
Blue grass, 713
Blue-green algae, 315. See also Cyanobacteria
Body axes, establishment of, 216
Body cavity
  evolution of, 370f, 371
  in roundworms, 378
  types of, 371, 371f
Body plans
  evolution of, 369
  sac, 375, 376
  tube-within-a-tube, 376
Body shape, environmental conditions and, 414
Body stalk, 698f, 699
Body temperature, regulation of, 507, 508, 508f
Bolus, 616, 617
Bond(s)
  chemical reactivity, determinants of, 23
  covalent, 25, 25f
    nonpolar, 26
    notation for, 25
    polar, 26
  hydrogen, 26, 26f
    notation for, 26
    in water, 27–28
  ionic, 24, 24f
  peptide, 45, 45f
Bone(s). See also Skeleton
  compact, 560, 560f
  composition of, 560, 560f
  as connective tissue, 498f, 499
  exercise and, 565
  forensics, 552–553
  functions of, 504
  growth of, 560
  osteoporosis, 624, 676
    prevention of, 559, 559f, 565
    treatment of, 676
  spongy, 560, 560f
Bone marrow
  red, 136, 560, 560f
  transplantation of, 586
  yellow, 560, 560f
Boniva, 559
Bonner, J., 470
Bony fishes, 387f, 388–389, 388f, 652, 652f
Book lungs, 382, 383
Boreal (northern) forest, 765

Bossiella, 325f
Botox, 503, 521
Bottleneck effect, 255, 255f
Bottlenose dolphins, 734
Botulism, 314, 318, 520–521
Bovidae, 290, 290f
Bovine growth hormone (bGH), 231, 231f
Bowman capsule. See Glomerular capsule
BPG, 116f
Brachiopods, evolution of, 288f–289f
Brachiosaurus, 390
Brachycardia, 633
Brain. See also Central nervous system
  development of, 694
    hominids, 407
  human
    evolution of, 407, 513
    organization of, 525–527, 525f, 526f
    ventricles of, 524, 524f
  of mammals, 392
  PET scans of, 22, 22f
  of primates, 401
  protective membranes, 524
  vertebrate, organization of, 513, 513f, 515, 515f
Brain stem, 524f, 526
Brain tumors, 501
Braxton Hicks contractions, 702
Brazil, rain forest destruction in, 785, 785f
Breadmaking, yeast and, 361
Breaking water, 702
Breast cancer, 143f, 144
Breast feeding, and immunity, 599, 599f
Breathing. See Ventilation
Breech birth, 701
Bright-field microscopes, 58
Bristlecone pine (Pinus longaeva), 350
Brittle star, 385f
Broca area, 525, 525f
Bronchioles, 638, 638f
Bronchitis, chronic, 644
Bronchus (bronchi), 638, 638f
Brown algae, 325f, 333, 333f
Brown planthopper, 783
Brown tree snake, 786
Brunk, Conrad G., 230
Bryophytes, 346–347, 346f
  characteristics of, 342–343, 342f, 343, 346
  evolution of, 343, 343f
  life cycle, 346, 347, 347f
BT corn, 230, 230f
Bubonic plague. See Plague
Buds
  axillary, 422, 422f
  terminal, 422, 422f
Buffers, pH and, 31
Bulb, plant, 487
Bulbourethral gland, 684, 684f
Bulimia nervosa, 628, 628f
Bulk feeders, 613, 613f
Bulk transport, 90, 90f
Bullhorn acacia, 473, 473f, 748
Bullock, Sandra, 552–553
Bumblebees, mimicry in, 745, 745f
Bunchgrass, 443
Bundle scars, 431, 431f
Bundle sheath(s), 434, 434f
Bundle sheath cells, 106, 107f
Burgess shale, 274–275, 274f–275f, 284
Burkitt lymphoma, 154, 305
Bursae, 561, 561f
Bursitis, 561
Butterflies, 384, 384f
  coevolution in, 747
  metamorphosis in, 382, 382f
  mimicry in, 745

  monarch, 745
  as pollinators, 243, 243f
  speciation in, 271
  viceroy, 745
  wings, 206–207, 206f, 207f

C

Cabanac, M., 736
Cabbage family, and cancer prevention, 144
Cactuses
  evolution of, 252
  leaves of, 423, 423f
  photosynthesis in, 107
  phylogeny, 292, 292f
  rare, trade in, 787
Caenorhabditis elegans, 234t
Caffeine
  as drug, 521, 658
  as plant defense, 472
Calcitonin, 668, 669t
Calcium
  as biologically important ion, 24t
  blood levels, regulation of, 676
  dietary requirements, 559, 624, 624t
  storage, in bones, 555
California
  coast, as biodiversity hotspot, 781
  speciation in, 268, 268f
California gray whales, 718f
Callus, 209, 209f, 463, 463f
calorie, definition of, 78
Calorie (C), definition of, 78
Calvin, Melvin, 99, 99f
Calvin cycle reactions
  in C4 plants, 106, 107f
  in CAM plants, 107, 107f
  overview of, 99, 99f, 102f
  phases of, 104–105, 104f, 105f
Calyx, 353, 353f
Cambrian period, 284, 285t, 369
Camera-type eye, 534–535
Camouflage, as prey defense, 745, 745f
cAMP. See Cyclic adenosine monophosphate
CAM photosynthesis, 107, 107f
Canaliculi, 498f, 499
Cancer
  angiogenesis, 143, 220, 220f, 574
  anticancer drugs, 472, 474, 546, 559, 603
  apoptosis and, 218, 602
  behaviors and, 144
  breast, 143f, 144
  Burkitt lymphoma, 154, 305
  carcinogenesis, 143, 220, 220f
  carcinogens, 201
  carcinomas, 143
  causes of, 144, 154
  as cell division malfunction, 132–133, 132f, 143, 218–220
  cervical, 305, 689
  colon, 132f, 144, 621
  diagnosis of, 220, 604
  diet and, 144
  as genetic disorder, 133
  genetic mutation and, 218–220
  immortality of, 191
  leukemias. See Leukemias
  liver, 305
  lung, 133f, 144, 201, 639, 644, 644f
  lymph nodes and, 595
  malignancy, 143
  metastasis, 143, 143f, 220, 220f, 644
  mutations causing, sources of, 133
  pancreatic, 133f
  prevention of, 565
  prostate, 684
  proto-oncogene mutations and, 218, 218f, 219, 219f
  sarcomas, 143, 592f, 593

in situ, 143, 143f, 220, 220f
skin
    DNA repair and, 201
    treatment of, 233, 233f
    UV radiation and, 503
    xeroderma pigmentosum
      and, 201
smoking and, 144, 201, 639
treatment of, 143, 190, 220, 232, 233,
    233f, 357, 586, 603, 604, 782
tumor suppressor gene mutations
    and, 218, 218f, 219, 219f
Cancer cells, characteristics of, 143
Candida albicans, 362
Candidiasis, 593, 593f
Canine teeth (cuspids), 616, 616f
Cannabinoids, 523
Cannabis sativa, 523
Cannon, Walter, 706
Cape Gannet, 711, 711f
Capillaries, 574, 574f, 579, 579f
Capillary beds, 579, 579f, 580
Capillary exchange, 587, 587f
Capsid, 184, 185f, 302, 302f, 305, 305f
Capsule, in prokaryotic cells, 59, 59f,
    311, 311f
Capuchin monkeys, 403
Carbaminohemoglobin, 643
Carbohydrates
    as class of organic molecule, 38
    in diet, 622, 622t
    as energy source, 125, 126
    production of, in photosynthesis,
      97, 98, 104–105, 104f, 105f
    structure and function
      complex carbohydrates, 42, 42f
      simple carbohydrates, 41, 41f
    subunits of, 40f
Carbon
    atomic model of, 21, 21f
    characteristics of, 38
    and life, 20, 20f
    as organic molecule base, 38
Carbon chains, 38, 38f, 39
Carbon cycle, 758, 758f, 784
Carbon dioxide
    and acid-base balance, 659
    atmospheric levels, 420–421
    carbon cycle and, 758, 758f
    in cellular respiration, 114, 115,
      118, 119
    forests' uptake of, 784
    as greenhouse gas, 108, 420–421,
      784, 787
    and photosynthesis, 97, 98,
      104–107, 104f, 105f, 450
    in respiration, 642, 642f, 643
Carbonic acid, 31
Carbonic anhydrase, 643
Carboniferous period, 284, 285t, 348,
    350, 351, 352, 352f
Carbon monoxide (CO), and
    respiration, 643
Carbonyl groups, 39f
Carboxyl groups, 39, 39f
Carcinogenesis, 143, 220, 220f
Carcinogens, 201
Carcinomas, 143
Cardiac conduction system, 578, 578f
Cardiac cycle, 578, 578f
Cardiac muscle, 500, 500f
Cardiac pacemaker, 578
Cardiac veins, 582
Cardinal, 391f
Cardiovascular disease
    causes of, 43, 44, 582, 627
    prevention of, 582–583
    smoking and, 639
    treatment of, 232, 582, 582f
Cardiovascular system. *See also* Heart
    development of, 699

and exercise, 565
and homeostasis, 507
in humans, 505, 505f
pulmonary circuit, 576, 576f,
    580, 580f
systemic circuit, 576, 576f, 580, 580f
Carnivores
    digestive system, 611, 615, 615f
    in ecosystem, 752, 755, 755f
    as order, 393
    teeth of, 393, 610
Carnivorous plants, 340–341, 341f, 342f,
    454, 454f
Carotenoids
    absorption spectrum of, 100, 100f
    in photosystem, 101
Carotid artery, 580f
Carotid bodies, 641, 659
Carp, 771, 771f
Carpal bones, 556f, 558, 558f
Carpel, 353, 353f, 354f–355f, 480, 480f,
    482f, 483
Carrageenan, 333
Carrier(s), of genetic disorders, 167
Carrier proteins, 86, 86f, 88, 88f
Carrying capacity, 714, 714f
Cartilage, 498f, 499, 555
    articular, 560, 560f, 561
    hyaline, 498f, 499, 560, 560f,
      561, 562
    replacement of, 562
    torn, 561, 562
Cartilaginous fishes, 387f, 388, 388f,
    652, 652f
Cartilaginous joints, 561
Casparian strip, 452, 452f
Caspases, 141
Cat *(Felix catus),* 137t
Catabolism, 125, 125f
Cataracts, 541
Caterpillars
    defenses, 745
    plant defenses against, 472, 473
    as substrate feeders, 613, 613f
Cattle egrets, 746
Caudal vertebrae, 418
Caustic soda (sodium hydroxide), 30
Cave paintings, 412, 412f
CBOL. *See* Consortium for the
    Barcoding of Life
CBP, 169
CCK (cholecystokinin), 621, 621f
Cech, Thomas, 309
Cecum (ceca), 611, 614, 615, 615f, 616f,
    621, 621f
Cell(s). *See also* Animal cells;
    Eukaryotic cells; Prokaryotic cells
    as basic unit of life, 56
    differentiation. *See* Differentiation
    energy use by, 78
      ATP cycle and, 50, 50f, 80–81,
        80f, 81f
      efficiency of conversion, 79
    as level of organization, 4, 5f,
      496, 496f
    origin of, 131, 308–310
    osmosis and, 89, 89f
    preexisting cells as only source
      of, 56
    senescence, 142, 191
    size of, 57, 57f
    surface-area-to-volume ratio of,
      57, 57f
Cell body, of neuron, 501, 501f, 517,
    517f, 521, 521f
Cell cycle, 136–140, 136f, 138f–139f, 140f
    control checkpoints, 141, 141f
    control signals, 142, 142f
Cell division, 134–140
    and cancer, 132–133, 132f, 143,
      218–220

in eukaryotes, 135–140, 136f,
    138f–139f, 140f, 142, 142f, 143
    microtubules in, 71
    in prokaryotes, 134, 135, 135f
    in sexual reproduction, 134
Cell-mediated immunity, 600, 600f,
    602–603, 603f
Cell plate, 140, 140f
Cell sap, 67
Cell suspension culture, 488
Cell theory
    as basic theory of biology, 11t
    history of, 54–55
    overview, 130–131
    principles of, 56
Cellular implants, 506
Cellular respiration, 114–122
    ATP payoff of, 122, 122f
    citric acid cycle, 115, 115f, 119, 119f,
      122, 122f
    efficiency of, 122
    electron transport chain, 115, 115f,
      120–121, 121f, 122, 122f
    equation for, 114
    evolution of, 317
    as exergonic reaction, 80
    glycolysis, 115, 115f, 116, 117f,
      122, 122f
    and muscle contraction, 568
    overview, 68, 114–115, 114f, 115f
    phases of, 115, 115f
    preparatory reaction, 115, 118,
      122, 122f
    in respiration process, 634
Cellular slime molds, 325f, 331
Cellular stage of development, 692, 692f
Cellulase, 465
Cellulose, 38f, 42, 42f, 61
Cell wall
    in archaea, 316
    in bacteria, 311, 311f
    in fungi, 358
    in plants, 61, 61f, 72, 72f
    in prokaryotic cells, 59, 59f
Cenozoic era, 284, 285t, 353, 403
Centipedes, 383, 383f
Central nervous system (CNS), 515,
    515f, 516, 516f, 524
    neurotransmitters in, 521
    sensory receptor communication
      with, 537, 537f
Central vacuole, 61f, 66–67, 66f, 89, 89f
Centrioles, 60f, 61, 71, 138, 138f
Centromere, 137, 137f, 145
Centrosomes, 60f, 61, 61f
    in cell cycle, 138, 138f
    function of, 70
Cephalization, 371, 376, 387, 514, 515
Cephalopods, 380, 380f
Cephalothorax, 382f, 383, 383f
Ceratium, 325f
Cerebellum, 515f, 524f, 526, 528f
Cerebral cortex, 525, 527
Cerebral hemispheres, 525, 525f
Cerebral lobes, 527
Cerebrospinal fluid, 501, 524
Cerebrum, 515, 515f, 524f, 525, 525f
Cerumin, 44
Cervical cancer, 305, 689
Cervical cap, 690t
Cervical nerves, 516f
Cervical vertebrae, 556f, 557
Cervidae, 290, 290f
Cervix, 686, 686f, 702, 702f
Cesarean section, 701
Cetaceans, 393
CF. *See* Cystic fibrosis
CFCs (chlorofluoro-carbons), 787
CF gene, 87, 87f
C₄ plants, 106–107, 107f
Chagas disease, 323, 327

Channel proteins, 86, 86f, 89
Chaperone proteins, 197
Chara, 335, 335f, 341, 341f
Character
    ancestral, 291
    definition of, 290
    derived, 291, 387
Character displacement, 742–743, 742f,
    743f
Chargaff, Erwin, 187, 241
Chase, Martha, 184–185, 185f,
    240–241, 300
Checkpoints, in cell cycle, 141, 141f
Cheese
    fungi and, 362
    and genetic engineering, 228
Cheetahs, 255, 734, 734f
Chemical communication, 734, 734f
Chemical cycling, in ecosystems, 10,
    10f, 753, 753f, 756–758, 756f
Chemical digestion, 612, 612f, 616,
    619, 619f
Chemical energy, as energy type, 78
Chemical equations, for
    photosynthesis, 25
Chemical evolution, 309, 309f
Chemical reactivity
    determinants of, 23
    of water, 28
Chemiosmosis, 103, 103f, 121, 121f,
    122, 122f
Chemistry, definition of, 20
Chemoautotrophs
    bacteria, 310, 773
    prokaryotes, 314
Chemoheterotrophs, prokaryotic, 314
Chemoreceptors, 536, 537, 538–539,
    538f, 539f
Chemosynthetic autotrophs, 752
Chestnut blight, 360–361
Chewing tobacco, 639
Chicago disease, 362
Chicken(s)
    development, 692, 693, 693f, 694
    meat color in, 121
Chicken pox, 305, 593, 593f
Chidocytes, 375, 375f
Childbirth. *See* Birth
Children
    genetic designing of, 235
    operant conditioning and, 729
Chimpanzees, 401f
    altruism in, 733
    chromosome number, 137t
    communication, 734, 735f
    emotions, 736, 736f
    evolution of, 402f
    genome, vs. humans, 404
    genome of, 236, 236f
    human-chimpanzee differences,
      404–405, 404f
    learning in, 729
Chinchillas, 724, 724f
Chiroptera, as order, 393
Chitin, 38f, 42, 358, 382, 554
Chlamydial infection, 689
Chlamydomonas, 334, 334f, 336
Chloride
    as biologically important ion, 24t
    dietary requirements, 624
Chloride ion channel, 168
Chlorocruorin, 573
Chlorofluoro-carbons (CFCs), 787
Chlorophyll
    a, 100, 100f, 101, 102
    b, 100, 100f, 101
    breakdown of, 465
    and fall leaf color, 100
    in light reactions, 99
    location of, 68, 97
    in photosystem, 101

Chlorophyta, 326f, 334
Chloroplasts
    in angiosperms, 428
    ATP generation in, 101, 101f
    in C4 plants, 106, 107f
    communication by, 60–61
    function of, 68, 101
    genetic code, 193
    origin of, 61, 68, 324, 324f
    and photosynthesis, 68, 97, 97f
    in plant cell anatomy, 61f
    structure of, 68, 68f
Chloroquine, 357
CHNOPS, 20. See also Carbon;
        Hydrogen; Nitrogen; Oxygen;
        Phosphorus; Sulfur
Choanocytes, 374, 374f
Choanoflagellates, 369
Cholecystokinin (CCK), 621, 621f
Cholesterol
    blood levels, testing of, 583
    and cardiovascular disease, 583,
        623, 627
    in diet, 623, 623f
    in phospholipid bilayer, 85, 85f
    structure of, 44, 44f
Chondrichthyes. See Jawless fishes
Chondrocytes, 562
Chondroitin, 562
Chondrus crispus, 333, 333f
Chordates
    characteristics of, 373, 386, 386f
    evolution of, 370f, 372f
    invertebrate, 386, 386f
    vertebrate. See Vertebrate(s)
Chorion, 390f, 392, 683, 697, 697f,
        699, 700
Chorionic villi, 698f, 699, 700
Chorionic villi sampling (CVS),
        168, 170
Choroid, 540, 540f
Christmas tree worm, 381
Chromatids
    in mitosis, 138–139, 138f–139f
    sister, 137, 137f, 138f–139f, 139,
        145, 145f
Chromatin
    in cell anatomy, 60f, 61f, 62f
    definition of, 137
    and gene expression, 212–213,
        215, 215f
    structure of, 62, 212f
Chromosome(s)
    abnormalities
        of chromosome number,
            152–153, 152f, 153f, 153t
        of sex chromosomes, 153, 154t
        of structure, 154, 154t
    autosomal, 145, 167
    bacterial, 311
    daughter, 137, 138f–139f, 139
    eukaryotic, 62
        formation of, 212, 212f
    in reproduction, 137, 137f
    separation in meiosis, 145, 145f
    gene linkage on, 176
    as gene location, 174
    homologous, 145, 145f, 146, 146f
        alleles on, 163, 163f
    human, number of, 137, 137t,
        145, 145f
    independent alignment of, 146
    independent assortment of,
        146–147, 147f
    lampbrush, 213, 213f
    mapping of, 176–177, 176f
    prokaryotic, 59, 135, 135f
Chromosome maps, 176–177, 176f
Chromosome number
    abnormalities of, 152–153, 152f,
        153f, 153t

diploid, 137, 137t
    haploid (n), 137
Chronic bronchitis, 644
Chronic myelogenous leukemia, 154
Chronic obstructive pulmonary disorder
        (COPD), 644
Chthamalus stellatus, 743, 743f
Chyme, 618, 619
Chymosin, 228
Cilia, 61, 71, 71f
    in invertebrate excretory systems,
        651, 651f
    in oviducts, 686
    of Paramecium, 329, 329f
    in respiratory system, 638, 638f, 639
Ciliary body, 540, 540f
Ciliary muscle, 541, 541f
Ciliates, 324, 325f, 329, 329f
Cinchona ledgeriana, 488
Cinchona tree, 357
Cinnamon fern (osmunda
        cinnamomea), 348, 348f
Circadian rhythms, 468–469, 469f
    biological clock and, 469
    definition of, 466
    pineal gland and, 675, 675f
    and plant stoma, 446
Circulatory systems
    of amphibians, 389
    in birds, 391
    closed, 575, 575f
    double-loop, 576, 576f
    echinoderms, 385
    of fish, 389
    functions of, 574
    of invertebrates, 574–575, 574f, 575f
    open, 575, 575f
    single-loop, 576, 576f
    vertebrates, 387, 576, 576f (See also
        Cardiovascular system)
Circumcision, 684
Cirrhosis, 620
Cisplatin, 546
Citric acid cycle, 115, 115f, 119, 119f,
        122, 122f
Civilization, origin of, 413
Clade, 294
Cladograms, 294, 294f, 295f
Clams, 380, 512, 512f, 575, 613, 636, 711
Clam worm (Nereis), 381, 381f
Clarkia, alloploidy in, 271, 271f
Clark's nutcrackers, 748, 748f
Class, 8, 8t, 290, 290f
Classical conditioning, 729, 729f
Classification of organisms, 290, 290f,
        292–293, 292f, 293f. See also
        Phylogeny; Taxonomy
    Linnean, 290, 291–292, 291f, 293,
        294–295
    taxa, 8, 8t, 290, 290f
    three-domain system, 296, 296f
    viruses, 302
Clavicle, 556f, 558, 558f
Clay, in soil, 453
Cleaning symbiosis, 748, 748f
Cleavage, 692, 692f, 698
Cleavage furrow, 138f–139f, 140, 140f
Clements, F. E., 751
Climate
    change in, and mass extinction,
        288–289
    of coniferous forest, 765
    of desert, 767
    Earth's rotation and, 769, 769f
    ocean currents and, 774
    regulation of, by natural areas,
        784, 784f
    of savanna, 767
    solar energy and, 769, 769f
    of temperate deciduous forest, 766
    of temperate grassland, 766, 766f

and terrestrial ecosystems, 764
    topography and, 770, 770f
    of tropical rain forest, 768, 768f
    of tundra biosystem, 765
Climax community, 751
Climax-pattern model of
        succession, 751
Clitoris, 686, 686f
Cloaca, 392, 612f, 614, 614f, 650,
        654, 683
Clock genes, 469
Clonal selection model, 601, 601f
Cloning
    of animals
        benefits and drawbacks,
            211, 211f
        ethics of, 211
        types, 210, 210f
    definition of, 226
    of DNA sequences, 227, 227f
    of humans, 210, 226, 226f
    of plants, 209, 209f, 488, 488f
    therapeutic, 210, 210f, 211
Closed circulatory system, 575, 575f
Clostridium botulinum, 317, 318,
        520–521
Clostridium tetani, 296, 564
Clotting. See also Hemophilia
    inhibition of, 84
    process of, 585, 585f
Clotting factor(s), synthetic, 175
Clotting factor IX, 175, 200
Clotting factor VIII, 175
Clover, 470, 470f
Clownfishes, 746, 746f
Club drugs, 522
Club fungi, 361
Club mosses, 346, 348, 348f, 352, 352f
Clumped population distribution,
        711, 711f
Cnidarians, 375, 375f, 514, 548
    characteristics of, 373
    evolution of, 370f, 372f
    reproduction, 682
    respiration in, 634
CNS. See Central nervous system
Coal, origin of, 96, 284, 285f, 758
Coastal ecosystems, 772, 772f, 785
Coastal redwoods (Sequoia
        sempervirens), 350
Cocaine
    cocaine psychosis, 522
    effects of, 522–523, 523f, 583
    urine tests for, 657
Coccus (cocci), 311, 311f
Coccyx, 556f, 557
Cochlea, 544, 545, 545f
Cochlear canal, 545f
Cochlear nerve, 545f
Cocklebur, 470, 470f
Cod, 717, 788
Codeine, neurotransmitters and,
        521, 523
Codominance, 171
Codons, 193, 193f, 196, 196f
    start, 198
    stop, 197, 198
Coelom, 371, 371f, 379, 379f, 381, 381f,
        387, 693, 693f, 694, 694f
Coelomates, 379
Coenzyme(s), 83
Coenzyme Q, 120f, 121, 121f
Coevolution, 242–243, 483, 747, 747f
Cofactors, 83
Coffee, 356
Coffee plants, defenses, 472
Cohesion, of water molecules, 27, 444
Cohesion-tension model, 444–445, 445f
Cohorts, 712
Coitus interruptus, 690t
Cola, 356

Colchicine, 71
Cold, as viral disease, 305
Cold receptors, 536
Coleochaete, 341, 341f
Coleomates, 371, 371f
Coleoptile, 460
Collagen, 72, 498, 498f, 499
Collar cells, 374, 374f
Collecting ducts, 654–655, 655f, 656f
Collenchyma cells, 426–427, 426f
Colon, 616f, 621, 621f
    cancer of, 132f, 144, 621
    polyps in, 621
Colonial flagellate hypothesis, 369, 369f
Colonization, and alien species
        introduction, 786
Colony, green algae, 334–335, 335f
Coloration, warning, 745, 745f
Color blindness, 87, 175, 175f, 543
Color vision, 400–401, 401f, 543. See
        also Cone cells
Columbus, Christopher, 356, 357
Columnar epithelium, 497, 497f
Commensalism, 314, 746, 746f, 746t
Common ancestor
    evidence for, 250–252
    phylogeny and, 291, 291f
Communication
    by animals, 734–735, 734f, 735f
    chemical, 734, 734f
    definition of, 734
    of nucleus with cytoplasm, 62
    between organelles, 60–61
    tactile, 735, 735f
    visual, 734–735
Community
    change over time, 749–751
    characteristics of, 750
    climax, 751
    definition of, 4, 10, 710
    ecological study of, 710
    interaction of populations in,
        742–748
    as level of organization, 4, 5f
Compact bone, 498f, 499, 560, 560f
Companion cells, 427f
Comparative anatomy, as evidence for
        evolution, 251, 251f
Comparative genomics, 236, 236f
Competition for resources
    and biodiversity, 443
    and population control, 716, 716f
    and resource partitioning, 742–743,
        742f, 743f
Competitive exclusion principle, 742
Competitive inhibition, 84, 84f
Complement, 596–597, 596f, 597f
Complementary base pairing, 48, 187,
        188, 189f, 190
Complete digestive tracts, 614–615,
        614f, 615f
Composition, of community, 750
Compound, definition of, 23
Compound eye, 534–535
Compound light microscopes, 58, 58f
Conclusion, in scientific method, 11,
        11f, 12
Condenser lens, 58, 58f
Conditioning
    classical, 729, 729f
    operant, 729
Condoms, 690f, 690t
Conduction deafness, 546
Condyloid joints, 561
Cone cells, 87, 401, 534, 535, 536,
        542–543, 543f
Cone snail, 394
Confuciusornis, 250
Congenital hypothyroidism (cretinism),
        676, 676f
Conidium (conidia), 361

Conifer(s), 350, 350*f*, 432
Coniferous forest, 764*t*, 765, 765*f*, 770, 770*f*
Conjugation
   in bacteria, 313, 313*f*
   in ciliates, 329, 329*f*
Conjunctiva, 540
Connective tissue, 498–499, 498*f*
Connell, Joseph, 743
Conservation biology, 780, 780*f*, 782
Conservation of energy, law of, 79
Consortium for the Barcoding of Life (CBOL), 14
Consumers, 752, 753, 753*f*
Contact dermatitis, 606
Contact inhibition, 142, 143
Continental drift, 286–287, 287*f*
Continental shelf, 772*f*
Contractile ring, 140, 140*f*
Contractile tissue. *See* Muscle(s)
Contractile vacuole, 328, 328*f*
Contraction, of muscle, 81, 81*f*, 564–565, 566–568, 566*f*, 567*f*
Control groups, 12
Control signals, in cell cycle, 142, 142*f*
Control systems
   for cell division, 142, 142*f*, 143
   of human body, 504, 504*f*
Convergent evolution, 292
COPD (Chronic obstructive pulmonary disorder), 644
Copepods, 383, 383*f*
Copper, dietary requirements, 624
Copper mosses, 346
Copulation, 683
Coral
   body form, 375*f*
   global warming and, 787, 787*f*
   zooxanthellae and, 325
Coral reefs, 773, 785, 785*f*
   bleaching of, 787, 787*f*
   formation of, 324
   human impact on, 10
Coral snake, 745
Cork, 426, 426*f*, 432, 432*f*
Cork cambium, 426, 426*f*, 432, 432*f*
Corms, 487, 487*f*
Corn (*Zea mays*), 356*f*, 425*f*
   artificial selection and, 272, 272*f*
   chromosome number, 137*t*
   flowers, 482
   history of, 272, 272*f*, 356
   hormones and, 464, 464*f*
   kernel
      germination of, 486, 486*f*
      structure of, 486, 486*f*
   McClintock's research on, 202, 202*f*
   plant defenses, 473
   types, 356
   uses, 425
   water loss in, 445
Cornea, 540, 540*f*
Corn smut, 362
Corolla, 353, 353*f*
Coronary arteries, 580, 582
Coronary bypass operation, 582, 582*f*
Corpus callosum, 524*f*, 525, 527*f*
Corpus luteum, 687, 687*f*, 688, 688*f*, 698
Cortex, in angiosperms, 428, 429*f*, 430*f*
Corticotropin-releasing hormone, 672
Cortisol, 672–673
Cortisone, 598, 673
Costal cartilages, 556*f*
Costa Rica, forest preservation in, 108, 421
Cost-benefit analyses, in male competition for mate, 730–731
Cotton (*Gossypium*), 357, 357*f*, 473
Cotton aphids, 783
Cotyledons, 353, 424, 424*f*, 483, 483*f*, 486, 486*f*

Coumadin, 84
Countercurrent exchange, in gills, 636, 636*f*
Coupled reactions, 81, 81*f*
Courtship behavior, 735
Covalent bonds, 25, 25*f*
   nonpolar, 26
   notation for, 25
   polar, 26
Cows, digestion in, 42
Coxal bones, 556*f*, 558, 558*f*
Crabs, 383, 383*f*, 514–515, 514*f*, 548, 748
Crack cocaine, 523
Cranial nerves, 516, 516*f*, 528, 528*f*
Crayfish, 382, 382*f*, 383
   development, 683
Creatine phosphate (CP) pathway, and muscle contraction, 568
Creosote (*Larrea tridentata*), 474, 711, 711*f*
Cretaceous period, 285*t*, 286, 288, 289*f*
Cretinism (congenital hypothyroidism), 676, 676*f*
Creutzfeldt-Jakob disease, 47
Crick, Francis H. C., 188, 189*f*, 191, 196, 241
Crickets, 734
Cri du chat syndrome, 154
Crimean-Congo fever, 318
Criminal investigations, DNA fingerprinting, 227
Cristae, 68, 114*f*, 118, 118*f*, 120, 121, 121*f*
Critical period, in imprinting, 728
Crocodiles, 282, 292
Crocodilians, circulatory system of, 576
Cro-Magnons, 399*f*, 410, 410*f*, 411, 412, 412*f*, 414
Crop (agricultural product), rotation of, 13
Crop (digestive tract component), 614, 614*f*
Crossing-over, 146–147, 146*f*
   and chromosome mapping, 176–177, 176*f*
Cross-pollination, 482
Crostose lichens, 359, 359*f*
Croton, 473
Crustaceans, 383, 383*f*
   excretory system, 651
   reproduction, 682
   respiration in, 636
   symbiosis in, 748
Cry9C protein, 230, 230*f*
C₃ plants, 106, 106*f*
Cuboidal epithelium, 497, 497*f*
Cuckoo, 747, 747*f*
Cud, chewing of, 615
Culture
   definition of, 409
   spread of, 409
Cup fungi, 360, 360*f*
Cupula, 546, 547*f*, 548
Cushing syndrome, 673, 673*f*
Cuspids (canine teeth), 616, 616*f*
Cuticle
   in angiosperms, 426, 428, 434, 434*f*, 435, 445
   in sporophyte plants, 345, 345*f*
Cutin, 428
Cuvier, Georges, 245
CVS. *See* Chorionic villi sampling
Cyanide, 84, 120, 472
Cyanobacteria, 315, 315*f*
   mutualism in, 359
   photosynthesis in, 296, 314, 315
   structure of, 59
Cycads, 350, 350*f*
Cyclic adenosine monophosphate (cAMP), 667, 667*f*

Cyclin(s), 142, 142*f*, 218
   M-cyclin, 142, 142*f*
   S-cyclin, 142, 142*f*
Cyclin D gene, 218
Cyclohexane, 38, 38*f*
Cyclosporine, 605
Cysteine, 46*f*
Cystic fibrosis (CF)
   cause and symptoms, 87, 87*f*, 168, 192, 200
   heterozygote advantage and, 258
   inheritance pattern, 171
   research on, 231
   treatment of, 168*f*, 232
Cysts, protozoan, 324, 327, 328, 330
Cytochrome(s), 120
Cytochrome *c*, 120*f*, 121, 121*f*, 252
Cytokines, 596*f*, 597, 598, 598*f*, 600, 601, 601*f*, 602–603, 605
Cytokinesis, 136, 136*f*, 140, 140*f*
Cytokinins, 463
Cytomegalovirus, 593
Cytoplasm
   in eukaryotic cells, 60, 60*f*, 61*f*
   in prokaryotic cells, 59
   translational control in, 214, 214*f*, 215, 215*f*
Cytoplasmic segregation, 695, 695*f*
Cytoplasmic streaming, 70
Cytosine (C), 48, 48*f*, 48*t*, 186, 186*f*, 187, 187*f*, 193, 193*f*
Cytoskeleton, 60*f*, 61, 70, 70*f*
Cytotoxic T cells, 600, 600*f*, 600*t*, 602, 602*f*, 603, 603*f*, 605

D

Daddy longlegs (harvestmen), 383
Dall sheep, 712–713, 712*f*, 713*f*
Dandelions, 717, 717*f*, 786
Dart, Raymond, 406
Darwin, Charles
   on emotion in animals, 736
   on evolution, 250, 251
   life and work of, 244–245, 244*f*
   Mendel and, 160
   on natural selection, 7, 246–247, 248
   On the Origin of Species, 247
   plant research, 466
Darwin's finches (*Geospiza fortis*), 715, 715*f*
Data, in scientific method, 11
Date rape drugs, 522
Daughter cells, 137, 148, 148*f*–149*f*, 151, 151*f*
Daughter chromosomes, 137, 138*f*–139*f*, 139
Daylilies (*Hemerocallis*), 480*f*
Day-neutral plants, 470
Deafness, types of, 546
Deamer, David, 310
Deamination, 125
Death angel, 362
Deceleration phase, of S-shaped curve, 714, 714*f*
Decibel level, and hearing damage, 546
Deciduous trees, 423
Decomposers, 314, 740, 741, 741*f*, 752, 753, 753*f*
Deductive reasoning, in scientific method, 11
Deer, white-tailed, 393*f*, 708–709, 708*f*, 709*f*
Defecation (elimination), in digestion process, 612, 612*f*
Defense mechanisms
   plant epidermal projections, 436
   plant toxins, 71, 436, 472
   stinging, 2, 3*f*
Deforestation, of tropical rain forests, 421
Dehydration, 660

Dehydration reaction, 40, 40*f*, 43, 43*f*, 45, 45*f*
Delayed allergic response, 606, 606*t*
Deletions, 154, 154*f*
Deltoid, 563*f*
Demographic transition, 719
Denaturation
   of enzyme, 83
   of protein, 47
Dendrite, 501, 501*f*, 517, 517*f*
Dendritic cells, 597, 598, 600, 600*t*, 602
Dengue fever, 305
Denitrifying bacteria, 757
Dense fibrous connective tissue, 498*f*, 499
Density-dependent factors, in population regulation, 715*f*, 716, 716*f*, 717
Density-independent factors, in population regulation, 715, 715*f*, 717
Dental caries, 617
Deoxyribonucleic acid. *See* DNA
Deoxyribose, 41, 48, 48*t*, 186, 186*f*, 191, 191*f*
Depolarization, in action potential, 518–519, 519*f*
Depo-Provera, 690*f*
Derived characters, 291, 387
Dermatitis, contact, 606
Dermis, 503, 503*f*
Descending colon, 616*f*, 621
Descent with modification, 250
Desert, 764*f*, 764*t*, 767, 767*f*, 769
Detrital food webs, 754*f*, 755
Detritus, 752, 754*f*, 755
Deuterostomes, 372*f*, 379, 379*f*, 385
Development, 692–702
   in animals, 217, 252, 368, 368*f*
   apoptosis and, 141, 217
   cellular differentiation, 695, 695*f*
   as characteristic of life, 7
   in deuterostomes, 379, 379*f*
   and evolution, 275–276, 275*f*, 276*f*
   in frogs, 368, 368*f*
   gene expression
      control of, 216–217
      and evolution, 276–277, 276*f*, 277*f*
   in humans, 697–702
      cellular stage, 692
      embryonic development, 697, 697*f*, 698–699, 698*f*, 699*f*
      extraembryonic membranes, 697, 697*f*
      fetal development, 697, 700–701, 700*f*, 701*f*
      organ stages, 694
      tissue stage, 693
   morphogenesis, 695, 696, 696*f*
   in protostomes, 379, 379*f*
   sustainable, 790
   in vertebrates, 692–696
      cellular stage, 692, 692*f*
      organ stages, 694, 694*f*
      tissue stage, 692*f*, 693, 693*f*
Devonian period, 285*t*, 288, 288*f*
Dewlaps, 269, 269*f*
DEXA. *See* Dual energy X-ray absorptiometry
Diabetes
   diabetes insipidus, 671
   prevalence of, 674
   retinopathy in, 541
   symptoms of, 674
   treatment of, 675
   type 1, 626, 674–675
   type 2, 674–675
      and acidosis, 659
      causes of, 599
      diagnosis of, 657

Diabetes (continued)
    diet and, 626–627
    genetic basis of, 49
    plasma membrane proteins
      and, 87
    prevention of, 87
    symptoms of, 657
Diagnostic traits, 264
Dialysate, 660, 660f
Diamondback rattlesnake, 390f
Diaphragm (anatomical feature), 638f,
    640, 641f
Diaphragm (birth control device), 690t
Diastole, 578
Diastolic pressure, 581
Diatom(s), 325f, 332, 332f
Diatomaceous earth, 332
Dicots, 424
Didinium nasutum, 744
Didiniums, 329
Diencephalon, 524f, 526, 527
Diet. See also Food
    and cancer, 144
    carbohydrates, 622, 622t
    and cardiovascular disease, 583
    cholesterol in, 623, 623f
    dietary requirements, 624, 624t
    eating disorders, 628, 628f
    essential amino acids, 623
    fat, 583, 621, 623, 623t
    fiber, 622
    fish in, 583, 623, 627
    fruit in, 625
    lipids, 623, 623t
    minerals in, 624, 624t
    and osteoporosis, 559
    protein, 623
    vegetables in, 625
    vegetarian diets, 623
    vitamins in, 625, 625t
Differential permeability, of plasma
    membrane, 86
Differentiation, 143, 695, 695f, 696. See
    also Specialization
Diffusion
    facilitate, 88, 88f
    simple, 88, 88f
Digestion
    chemical, 612, 612f, 616, 619, 619f
    in digestive process, 612, 612f
    mechanical, 612, 616
    overview of, 612, 612f
Digestive system
    annelids, 381
    birds, 614, 614f
    carnivores, 611, 615, 615f
    complete digestive tracts, 614–615,
      614f, 615f
    disease of, bacterial, 317t
    earthworms, 614, 614f
    flatworms, 376, 376f
    herbivore mammals, 615, 615f
    and homeostasis, 507
    human, bacteria in, 748
    humans, 616–621, 616f
      characteristics of, 611, 612
      esophagus, 617, 617f
      large intestine, 621, 621f
      liver, 620
      overview, 505, 505f
      pancreas, 620, 620f
      small intestine, 619, 619f
      stomach, 618, 618f
    incomplete digestive tracts,
      614, 614f
    omnivores, 611
    phases of digestion, 612, 612f
    planarians, 614, 614f
    roundworms, 378
    ruminants, 611
    vertebrates, 387

Digitalis, 488
Digitoxin, 488
Digoxin, 488
Dihybrid crosses, 164–177, 164f, 166f
    definition of, 164
    phenotypic results, 164–165,
      164f, 165f
Dinoflagellates, 325f, 332, 332f
Dinosaurs
    behavior, 282–283
    birds as descendants of, 283
    evolution of, 284, 285t, 390
    extinction of, 285t, 286, 286f
    size of, 390
Dioecious animals, 682
Dioecious plants, 480
Dionaea muscipula (Venus flytrap),
    340–341, 340f
Diplid (2n) chromosome number,
    137, 137t
Diploid life cycle, in animals, 373
Diploid phase
    in algae, 334, 334f, 336, 336f
    in angiosperms, 354f–355f, 482f
    in ferns, 349, 349f
    in fungi, 360f
    in mosses, 347f
    in pine trees, 351, 351f
    in plant life cycle, 150, 150f,
      344, 344f
Directional selection, 256f, 257
Direct selection, 732
Disaccharides, 41, 41f
Disease. See also specific diseases
    amoeboids and, 328
    arachnid-borne, 383
    bacteria and, 301, 317, 317t
    emergent, 307
    and extinctions, 785f, 788
    fungi and, 362, 362f
    genes and. See Genetic disorders
    lysosomal storage diseases, 66
    mitochondrial DNA mutations
      and, 69
    mycoses, 362, 362f
    obesity and, 626–627
    parasitic flatworms, 377, 377f
    parasitic roundworms, 378, 378f
    plasma membrane proteins and,
      87, 87f
    protists and, 322–323, 324, 328,
      330, 330f
    protozoans and, 323, 327
    sporozoans and, 330, 330f
    treatment, genetic profiles and, 49
    viral, 300
      in animals, 305–307
      in plants, 304, 304f
Disruptive selection, 256f, 257, 257f
Distal convoluted tubule, 654,
    655f, 656f
Distemper, 788
Diuretics, 658
Divergent evolution, 292
Diversity. See also Biodiversity
    in humans, 414, 414f
    of life, 4, 4f, 38, 38f
DNA (deoxyribonucleic acid)
    bacterial, mutation rate, 312
    deletions, 154, 154f
    digestion of, 619
    duplications, 154, 154f
    evolution of, 195, 310
    function of, 48, 241
    and genetic inheritance, 7
    history of research on, 184–185,
      187, 188, 188f, 193
    hydrogen bonding in, 26
    inversions, 154, 154f
    mitochondrial. See
      Mitochondrial DNA

    nucleotide sequences, possible
      number of, 187
    and phylogenetic research,
      292–293, 293f
    proofreading of, 191, 201
    replication. See Replication
    similarity in all organisms, 252
    steroid hormones and, 667, 667f
    structure
      bases, 48, 49t, 186, 186f
      chemical structure, 48, 49f,
        186, 186f
      complementary base pairing, 48,
        187, 188, 189f
      double helix structure, 48, 48f,
        188, 189f, 241
      and function, 188
      genetic code, 193, 193f
      overview of, 48
      thymine dimers, 201, 201f
      translocations, 154, 154f
DNA barcoding, 14
DNA-binding proteins
    and gene regulation in eukaryotes,
      213, 213f
    and gene regulation in prokaryotes,
      208, 208f
DNA fingerprinting, 227, 227f
DNA ligase, 191, 226
DNA polymerase, 190, 191, 191f,
    201, 227
DNA repair
    cell cycle checkpoints and, 141
    genetic mutation and, 201
DNA replication, 190–191, 190f, 191f
    cell cycle controls and, 142
    complementary base pairing and,
      48, 190
    in eukaryotic cell cycle, 136,
      136f, 137
    in prokaryotic cell cycle, 312, 312f
    as semiconservative replication,
      190, 190f
DNA sequencers, 234
Dogs, genetic disorders in, 154–155,
    155f, 156f
Dolly the sheep, 210, 211
Doldrums, 769
Domain(s), 8, 8t, 290, 290f. See also
    Archaea; Bacteria; Eukaryotes
    three domain system, 296, 296f
Dominance hierarchies, 731, 732
Dominant alleles, 163
Dopamine, 520, 521, 522, 736
Dormancy
    in plants, 462, 464, 464f, 493
    in seeds, 486
Dormitory effect, 665
Dorsal root ganglion, 528, 528f
Dorsal tubular nerve cord, as chordate
    characteristic, 386, 386f
Double fertilization, in angiosperms,
    354f–355f, 355, 482f, 483, 483f
Double helix structure of DNA, 48, 49f,
    188, 189f, 241
Doubling time, 719
Douche, as birth control method, 690t
Down syndrome, 152, 153, 153f,
    153t, 154
Draculin, 613
Dragonfly, 384f
Drosophilia melanogaster (fruit flies)
    chromosome map of, 176, 176f, 177
    chromosome number, 137t
    development, 216–217, 216f
    gene linkage in, 176
    genetic studies with, 174, 174f
    genome of, 234t
    Morgan's research with, 178, 240
Drug(s). See Pharmaceuticals
Drug abuse. See also specific drugs

    and addiction, 522, 523
    and cardiovascular disease, 583
    hallucinogenic drugs, 357
    neurotransmitter-interfering drugs,
      522–523
    urinalysis testing, 657
Drupes, 485
Dryopithecines, 403
Dryopithecus, 403
Dual energy X-ray absorptiometry
    (DEXA), 559
Dubautia, 440, 440f–441f
Dubois, Eugene, 408
Duchenne muscular dystrophy, 175, 212
Duckbill platypus, 392, 392f, 683
Dung beetles, 741, 741f
Duodenum, 619, 621, 621f
Duplications, 154, 154f
Dwarfism
    achondroplasia, 169, 201, 201f
    among Amish, 255, 255f
Dynein, 70
Dysentery, amoebic, 323, 328
Dystrophin, 175

E

Eardrum, 544, 544f
Early metaphase, 138f–139f
Early prophase, 138f–139f
Ears. See also Hearing
    and balance, 544, 546–548, 547f
    damage to, 546, 546f
    mammal, anatomy, 544, 544f
    sound detection in, 545, 545f
Earth
    atmosphere. See Atmosphere
    rotation of, and climate, 387,
      769, 769f
Earthstar fungus, 358f
Earthworms
    anatomy, 381f
    blood, 573, 575
    characteristics of, 381
    circulatory system, 575, 575f
    classification of, 381
    digestive system, 614, 614f
    and energy flow in ecosystems, 755
    excretory system, 651, 651f
    hydrostatic skeleton, 544f, 554
    nervous system, 514–515, 514f
    reproduction, 682
    respiration in, 634, 635f
    as scavenger, 741
Earwax, 544
East wind drift, 774f
Eating disorders, 628, 628f
Ebola virus, 300f, 305, 307, 318
Ecdysozoa, 372, 372f
Echinoderms, 385, 385f
    characteristics of, 373
    endoskeleton, 555
    evolution of, 370f, 372f
    reproduction, 682
Echolocation, 536, 536f, 734
Ecological engineering, 762
Ecological niches
    and adaptive radiation, 270
    and resource partitioning, 742–743,
      742f, 743f
Ecological pyramid, 755, 755f
Ecological succession, 750–751,
    750f, 751f
Ecology
    definition of, 752
    subjects of study, 710, 710f
EcoRI, 226
Ecosystems, 752–753
    abiotic components, 752
    aquatic, 771–774
    biotic components, 752
    buffering mechanisms, 31–32

chemical cycling in, 756–758, 756f
definition of, 4, 10
diversity of, and biodiversity, 781
ecological pyramid, 755, 755f
ecological study of, 710
energy flow in, 753–755, 753f, 754f
human impact on, 10
as level of organization, 4, 5f
pond, in winter, 29, 29f
stability of, and biodiversity, 443, 794–795
study of, 794–795
terrestrial, 764–768, 764t
Ecosystem theory, as basic theory of biology, 11t
Ecotone, 710
Ecotourism, 784, 784f
Ecstasy, 522
Ectoderm, 371, 371f, 376, 692f, 693, 693f, 693t, 694, 694f, 695, 696, 696f, 699
Ectoparasites, 746
Ectopic pregnancy, 686
Ectotherms, 390, 494
EDB. See Ethylene dibromide
Edenspace Systems Corporation, 447
Effacement, 702
Efferent arteriole, 655, 655f, 656, 656f
Efferent neurons. See Motor (efferent) neurons
Egg(s). See also Oocyte
amniotic, 387, 390, 390f, 391, 392
in angiosperms, 480, 480f, 482f, 483
frog, structure of, 695, 695f
genetic testing of, 170, 170f
shelled, evolution of, 681, 683
Egg-laying hormone (ELH), 727
Ejaculation, 684, 685
Ejaculatory duct, 684, 684f
Elastin, 72, 154
Electrical energy, as energy type, 78
Electrocardiogram (ECG), 578
Electrochemical gradients, and root uptake of minerals, 452, 452f
Electromagnetic receptors, 536, 536f
Electromagnetic spectrum, 100
Electron(s), 21, 21f, 21t
Electron acceptor, in photosystem, 101–103, 101f–103f
Electronegativity, 26
Electron microscopes, 57, 57f, 58, 58f
Electron models, 25, 25f, 26f
Electron shells, 21, 21f
Electron transport chains (ETCs)
in cellular respiration, 115, 115f, 120–121, 121f, 122, 122f
in photosynthesis, 101–103, 101f–103f
Elements
definition of, 20
and life, 20, 20f
periodic table, 23, 23f
Elephantiasis, 378, 378f
Elephants, 736
Elephant seals, 632, 632f, 633, 664
ELH. See Egg-laying hormone
ELH gene, 727
Eli Lilly and Co., 474
Elimination (defecation), in digestion process, 612, 612f
El Niño–Southern Oscillation, 774
Elongation, in translation, 198, 198f, 199f
Elongation factors, 198
Embolus, 582
Embryo
of angiosperm, development of, 484, 484f
animal pole, 692
of nonvascular plants, 342–343, 342f
vegetal pole, 692

Embryology, evidence for evolution, 251, 251f
Embryonic development, in humans, 697, 697f, 698–699, 698f, 699f
Embryonic disk, 699
Embryonic stem cells, 136, 210, 586
Embryo sac, in angiosperms, 354, 354f–355f, 481, 481f, 482f, 483
Emergent disease, 307
Emission, in male orgasm, 684
Emotional bonding, hormones and, 671
Emotions, of animals, 736, 736f
Emphysema, 644
Encephalitis, 307
toxoplasmic, 593
Endangered species, 780, 787
Endergonic reactions, 80
Endler, John, 257
Endocarp, 485
Endocrine glands, 497, 507, 621, 668, 668f
Endocrine system
adrenal glands, 668, 668f, 669t, 672–673, 672f
and behavior, 727
gonads, 668, 668f
and homeostasis, 507, 706–707
hypothalamus and, 668, 668f, 669f, 670–671, 670f, 672–673, 672f, 676
nervous system and, 666, 666f, 666t, 668
ovaries, 668, 668f, 669t
overview of, 504, 504f, 668, 668f
pancreas, 668, 668f, 669t, 674, 674f (See also Pancreas)
parathyroid glands, 668, 668f, 669t, 676
pineal gland, 524f, 526, 668, 668f, 669t, 675, 675f
pituitary glands, 515, 515f, 524f, 669t, 670–671, 670f
testes in, 668, 668f, 669t
thymus gland, 594f, 595, 595f, 602, 668, 668f, 669t
thyroid gland, 668, 668f, 669t, 676, 676f
Endocrine tissue, 674
Endocytosis, 90, 90f
receptor-mediated, 90
Endoderm, 371, 371f, 376, 692f, 693, 693f, 693t, 694, 694f, 695, 696, 696f, 699
Endodermis, in angiosperms, 428–429, 429f, 430f
Endomembrane system, 60, 67, 67f
Endometrium, 686, 686f
Endoparasites, 746
Endoplasmic reticulum (ER). See also Endomembrane system
in cell anatomy, 60, 60f, 61f
in lipid synthesis, 64
in protein synthesis, 64
ribosome attachment to, 63, 63f, 64, 64f
rough, 60f, 61f, 64, 64f, 67, 67f
smooth, 60f, 61f, 64, 64f, 67, 67f
structure of, 64
Endorphins, 521, 522
Endoskeleton, 555, 555f
advantages of, 371
jointed, as vertebrate characteristic, 387
of sponge, 374
Endosperm, 483
Endospores, 312, 312f
Endosymbionts, 324
Endosymbiotic theory, 68, 324
Endotherms, 391, 495
Endotoxins, 317

Energy
acquisition of, as characteristic of life, 7, 7f
ATP as currency of, 50, 50f, 80, 252
conservation of, 79
conversion, efficiency of, 79
definition of, 7, 78
first law of thermodynamics, 79, 753
flow of, in ecosystems, 10, 10f, 753–755, 753f, 754f
forms of, 78
glucose as source of, 41, 41f, 42
kinetic, 78, 78f
organelles and, 60–61, 68–69
organisms' need for, 78
potential, 78, 78f
second law of thermodynamics, 79
storage, 80
in animals, 42, 42f, 43, 43f
in plants, 42, 42f, 43, 43f
Sun as ultimate source of, 7, 10, 10f, 76–77, 96
units of, 78
Energy of activation (E$_a$), 82, 82f
Energy-related organelles, 60–61
Enhancers, 213, 213f
Ensatina, speciation in, 268, 268f
Entamoeba histolytica, 323, 328, 328f
Entomology, 384
Entropy, 79
Envelope, viral, 302, 302f
Enveloped viruses, 302, 302f
Environment. See also Ecosystems
and body shape, 414
and human evolution, 414
and human population growth, 720
impact on genetic expression, 172
internal, regulation of. See Homeostasis
"nature vs. nurture" question, 726–730
plant response to, 466–473
in population control, 715, 715f
and prokaryotes, importance of, 317
protists, importance of, 324
tree rings as record of, 433
Environmental impact, in more- vs. less-developed countries, 720
Environmental Protection Agency (EPA), 348
Environmental tobacco smoke (ETS), 639
Enzymatic proteins, 86, 86f
Enzyme(s), 82–84
active site, 82, 82f
in clearing of neurotransmitters, 521
coenzymes, 83
cofactors, 83
definition of, 40, 82
denaturation of, 83
DNA repair, 201
function of, 45, 82
induced fit model of, 82
inhibitors, 84, 84f
in protein synthesis, 198
rate of reaction, factors in, 83
restriction, 226, 226f
similarity in all organisms, 252
Enzyme-substrate complex, 82, 82f
Eocene epoch, 285t
Eosinophils, 584f, 597
EPA. See Environmental Protection Agency
Epicotyl, 484, 484f
Epidermal growth factor (EGF), 142
Epidermis
of angiosperms, 426, 426f, 428, 428f, 429f, 434, 434f, 472
of humans, 503, 503f
Epididymis, 684, 684f
Epiglottis, 617, 617f, 638, 638f

Epinephrine (adrenaline), 667, 669t, 672, 672f
Epipelagic zone, 773, 773f
Epiphytes, 746, 768, 768f
Episiotomy, 702
Epithelial tissue, 496–497, 497f
Epochs, 284, 285t
Epstein-Barr virus, 305
Equatorial countercurrent, 774f
Equilibrium population, 717, 717f, 718, 749, 749f
Equine encephalitis, 307
Equus, evolution of, 278, 278f
ER. See Endoplasmic reticulum
Eras, 284, 285t
Erection, 684
Ergot, 361, 362
Ergotism, 362
Erosion, prevention of, 784
Erythroblastosis fetalis, 588
Erythrocytes. See Red blood cells
Erythropeietin, 584
Erythroxylon coca, 522
Escherichia coli, 8f
commensalism in, 314
exotoxins, 317
generation time, 135
genome sequencing of, 49
Hershey-Chase research with, 184–185, 185f
as heterotroph, 296
in large intestine, 621
E (exit) site, 198, 198f
Esophagus, 617, 617f
Essential amino acids, 623
Estrogen, 669t
and bone strength, 559
and cell cycle control, 142
functional groups and, 39, 39f
and ovarian cycle, 688, 688f
production of, 687
and uterine cycle, 688, 688f
Estrogen-progestin therapy, and cancer, 144
Estuaries, 772, 785
ETC. See Electron transport chains
Ethane, functional groups and, 39
Ethanol, functional groups and, 39
Ethics
cloning, 211
genetic engineering, 225
science and, 14
stem cell research, 210, 211, 586
transgenic organisms, 394
Ethnic groups, origin of, 414
Ethyl alcohol, fermentation and, 124
Ethylene, 464, 465, 465f, 467
Ethylene dibromide (EDB), 201
ETS. See Environmental tobacco smoke
Euchromatin, 212f, 213
Eudicots, 353, 424, 424f
embryo development in, 484, 484f
flowers, 424, 424f, 480, 481f
leaves, 424, 424f, 428, 428f, 429f, 434, 434f
roots, 424, 424f, 429, 430–431, 430f
seeds, 424f
stems, 424, 424f, 428, 429f
Eudicotyledones, 353
Euglena, 55f
Euglena deces, 327
Euglenoids, characteristics of, 325f, 327, 327f
Euglenoza, classification of, 326f
Eukarya domain, 8–9, 9f, 296
Eukaryotes
cell cycle, 136–140, 136f, 138f–139f, 140f
control checkpoints, 141, 141f
control signals, 142, 142f

Eukaryotes (continued)
  cell division, 135–140, 136f,
    138f–139f, 140f
    control system for, 142, 142f, 143
  chromosomes, 62
    formation of, 212, 212f
    in reproduction, 137, 137f
    separation in meiosis, 145, 145f
  definition of, 8
  as domain, 8–9, 9f, 296
  evolution of, 286, 286f, 316
  gene expression control in, 209–217
    and cell specialization, 209
    chromatin structure and,
      212–213, 215, 215f
    cloning. See Cloning
    in development, 216–217
    gene splicing and, 236
    posttranscriptional control, 214,
      214f, 215, 215f
    posttranslational controls, 214,
      215, 215f
    synopsis of, 215, 215f
    transcriptional control, 213,
      213f, 215, 215f
    translational controls, 214, 214f,
      215, 215f
  initiation in, 198
  posttranscriptional processing in,
    195, 195f
Eukaryotic cells
  origin of, 61, 747
  ribosomes in, 63
  size of, 60
  specialization of, 132–133, 209, 209f
  structure of, 59, 60–61, 60f, 61f
  totipotency of, 209
Euphorbia, evolution of, 252
Euphotic zone, 772f, 773
Eustachian tube, 544, 544f
Eutrophication, 757, 771, 787
Eutrophic lakes, 771, 771f
Everglades, restoration of, 790
Everglades Agricultural Area (EAA), 790
Evergreens, 350, 423
Evolution. See also Microevolution;
    Natural selection; Speciation
  and adaptation, 419
  of animals, 369–372
    body plans, 369
    colonial flagellate hypothesis,
      369, 369f
    evolutionary tree, 370f, 372, 372f
    key innovations, 370f, 371
    molecular data, 372
    multinucleate hypothesis, 369
  archaea, 316
  as basic theory of biology, 11t
  biocultural, 409, 412–413, 412f
  biological, 309, 309f
  of birds, 251, 288f–289f, 387f, 391
  of body plans, 369
  of brachiopods, 288f–289f
  of cells, 131
  of cellular respiration, 317
  as characteristic of life, 7
  chemical, 309, 309f
  coevolution, 242–243, 483, 747, 747f
  common ancestor, evidence for,
    250–252
  convergent, 292
  cyanobacteria, 315
  Darwin on, 250, 251
  definition of, 7
  developmental genes and, 275–276,
    275f, 276f
  of dinosaurs, 284, 285t
  directionlessness of, 254, 419
  divergent, 292
  of DNA, 195, 310
  early research on, 245–247

of Equus, 278, 278f
eukaryotes, 316
of eukaryotes, 286, 286f
evidence for, 249–252, 418–419
of fish, 284, 285t, 387f
  genetic code as evidence of, 193
  glycolysis and, 116
as grand unifying theory (GUT), 419
and homeotic genes, 217
of humans, 398–414
  assimilation model, 411
  biocultural, 409, 412–413, 412f
  bipedalism, 405, 406–407, 406f
  environment and, 414
  evidence of, 277
  evolutionary tree, 402f, 403
  in geological timescale, 286, 286f
  hominid predecessors, 398–399,
    398f, 405–407, 405f, 406f
  human-chimpanzee differences,
    404–405, 404f
  monkeys and, 415
  multiregional continuity model,
    410–411, 411f
  origin of Homo genus, 407
  replacement model, 411, 411f
  vestigial structures, 251
of insects, 288f–289f
of jaws, 387, 388, 388f
of life, 18–19, 308–310
macroevolution, 264
of mammals, 284, 285t, 288f–289f
of marsupials, 392
of monkeys, 415
of nervous system, 514
parallel, 292
of photosynthesis, 95
of placental mammals, 287, 291
of plants, 95, 284, 285t, 341–342,
  341f, 342f, 343, 343f, 348,
  350, 353
of poriferans, 288f–289f
of primates, 400
of prokaryotes, 284, 285t, 286, 286f
of protists, 324, 326, 326f
of protocells, 309, 309f, 310, 310f
of reptiles, 284, 285t, 387f, 390
speciation and, 264
transposition and, 202
undirected nature of, 278, 278f
as unifying concept of biology, 14
of vertebrates
  ancestral chordate, 275
  embryological evidence,
    251, 251f
  evolutionary tree, 387, 387f
Evolutionary species concept, 264
Evolutionary systematics, 294–295, 295f
Evolutionary (phylogenetic) trees, 291,
  291f, 295, 295f
Exchange pool, in chemical cycling,
  756, 756f
Excitotoxicity, 527
Excretory system. See also Urinary
    system
  in annelids, 381
  echinoderms, 385
  in flatworms, 376, 376f
  invertebrates, 651, 651f
  waste product types and
    characteristics, 650, 650f
Exercise
  aerobic, 126, 126f
  benefits of, 126, 565
  and cardiovascular disease, 583
  fat burning in, 126, 126f
  and nerve regeneration, 502, 502f
  and osteoporosis, 559, 559f
  recommendations on, 565t
Exergonic reactions, 80
Exocarp, 485

Exocrine glands, 497
Exocrine tissue, 674
Exocytosis, 65, 65f, 67, 67f, 90
Exons, 195, 195f, 214, 214f
Exophthalmic goiter, 676, 676f
Exoskeleton, 382, 384, 544f, 554
Exotoxins, 317
Experimental variables, 12
Expiration, 640, 640f
Exploitation of resources
  and extinctions, 785f, 787–788
  and sustainable development, 790
Exponential growth, in populations,
  714, 714f, 719, 719f
Exponential growth phase
  of J-shaped curve, 714, 714f
  of S-shaped curve, 714, 714f
Expulsion, in male orgasm, 684
Extensor digitorum longus, 563f
External fertilization, 682–683, 682f
External oblique, 563f
External respiration
  gas exchange in, 642, 642f,
    643, 643f
  overview of, 634, 634f
  types of, 634–635, 635f
Extinctions. See also Biodiversity
  causes of, 785–788
  current rate of, 780
  definition of, 284, 718
  mass
    causes of, 288–289
    in geological timescale, 284,
      285t, 288f–289f
    press/pulse theory of, 288, 289
    P-Tr (Permian-Triassic)
      extinction, 289
  Pleistocene overkill, 412
  vulnerability to, 718, 718f
Extraembryonic membranes, 683,
  697, 697f
Ex vivo gene therapy, 232
Eye(s). See also Vision
  bird, 535
  camera-type, 534–535
  compound, 534–535
  development of, 276, 276f, 277f
  evolution of, 535
  formation of, 696
  human, 534–535, 540–543
    abnormalities, common,
      542, 542f
    anatomy, 540, 540f
    blindness, causes of, 541
    communication with visual
      cortex, 542–543
    eyestrain, 541
    focussing of, 541, 541f
    20/20 vision, 542
    vision correction, 542, 542f
  insect, 534, 536
  vertebrate, 534–535, 540–543
  visual range of, 57, 57f
Eyespots
  in Chlamydomonas, 334, 334f
  false, 745, 745f
  in flatworms, 376
  on moths and butterflies, 206–207,
    206f, 207f

F

Facial bones, 557, 557f
Facilitate diffusion, 88, 88f
Facilitation model of succession, 751
Facultative anaerobes, 314
FAD (flavin adenine dinucleotide), 114,
  119, 119f, 120, 120f, 122
FADH₂, 115, 115f, 119, 119f, 120, 120f,
  121, 121f, 122, 122f
Fallopian tubes. See Oviducts
False labor, 702

Familial hypercholesterolemia, 171, 232
Family, as taxon, 8, 8t, 290, 290f
Family planning, natural, 690t
FAP. See Fixed action patterns
Farsightedness (hyperopia), 542, 542f
Fast-twitch muscle fibers, 568, 568f
Fats
  in body
    burning, with exercise, 126, 126f
    formation of, 125
  in diet, 583, 621, 623, 623t
  digestion of, 619, 619f
  as energy source, 125, 126
  structure of, 43, 43f
Fatty acids, 43, 43f
FBN₁ gene, 173
Feather star, 385f
Feces, composition of, 621
Feedback and control mechanisms, in
  homeostasis, 6
Feeding strategies, of animals, 613, 613f
Female(s), human
  external genitals, 686, 686f
  and iron in diet, 624
  mate selection, 730, 730f
  and osteoporosis, 559
  reproductive system, 686–688,
    686f, 687f
Female gametophyte, 351, 351f,
  354f–355f, 481, 481f, 482, 483
Femur, 556f, 558, 558f
Fer-de-lance pit viper, 394
Fermentation
  ATP yield from, 115, 123
  foods produced by, 124, 124f,
    361, 362
  in metabolism, 115, 123, 123f
  and muscle contraction, 568
Ferns, 348–349, 348f
  in Carboniferous period, 352, 352f
  characteristics of, 343
  diversity of, 348, 348f
  economic value, 348–349
  life cycle, 344–345, 349, 349f
Fertile Crescent, 413, 413f
Fertilization
  double, in angiosperms, 354f–355f,
    355, 482f, 483, 483f
  external, 682–683, 682f
  and genetic variation, 147
  in humans, 686, 687, 698
Fertilizers, and chemical cycling,
  756–757, 757
Fetal development, in humans, 697,
  700–701, 700f, 701f
Fever, 597
Fiber, in diet, 622
Fibers, plant, 427
Fibrillin, 173
Fibrin, 585, 585f
Fibrinolysin, 688
Fibroblasts, 498, 498f
Fibrous joints, 561
Fibula, 556f, 558f, 559
Fiddleheads, 348
Fight or flight response, 530
Filament(s)
  in algae, 334
  of fungus (hypha), 358, 358f
  in stamen, 353, 353f, 354f–355f,
    480, 480f, 482f
Filarial worm, 378
Filter feeders, 374, 613, 613f
Fimbriae, 59, 59f, 311, 311f, 686, 686f
Finches
  on Galápagos islands, 715, 715f,
    718f, 742, 742f
  natural selection in, 248, 248f
  speciation in, 270
Fingerprints, 503
Fire ants, 2–3, 2f, 3f, 6, 7

Fireweed *(Chamerion augustifolium)*, 459
First law of thermodynamics, 79, 753
First messenger, 667, 667*f*
First polar body, 687, 687*f*
Fischer, Alain, 233
Fischer lovebirds *(Agapornis fischeri)*, 726, 726*f*
Fish
  bony, 387*f*, 388–389, 388*f*, 652, 652*f*
  cartilaginous, 387*f*, 388, 388*f*, 652, 652*f*
  characteristics of, 373
  circulatory system, 576, 576*f*
  in diet, 583, 623, 627
  evolution of, 284, 285*t*, 387*f*
  excretory system, 650, 650*f*
  group living in, 732, 745
  jawless, 387*f*, 388, 388*f*
  kidneys of, 654
  lateral line system in, 388, 548, 548*f*
  lobe-finned, 389
  nerve regeneration in, 502
  in ocean ecosystem, 773, 773*f*
  overfishing, 788, 788*f*
  and pet trade, 788
  ray-finned, 388–389
  reproduction, 680, 682
  respiration in, 634, 635*f*, 636, 636*f*
  speciation in, 268–269, 269*f*
  symbiosis in, 748, 748*f*
Fish crow, 718*f*
Fish oil supplements, 627
Fitness, and sexual selection, 730
Fitzroy, Robert, 244, 244*f*
5′ end, 191, 191*f*, 195, 195*f*
Fixed action patterns (FAP), 728, 728*f*
Flagellin, 311
Flagellum (flagella), 59, 59*f*, 61, 71, 71*f*, 311, 311*f*
  in Chlamydomonas, 334, 334*f*
  in dinoflagellates, 325, 325*f*
  in sponge collar cells, 374, 374*f*
  tinsel, 327
Flame cells, 376, 376*f*, 651, 651*f*
Flamingo, 391*f*
Flat-head Lake, Montana, 781, 781*f*
Flatworms, 371, 371*f*
  characteristics of, 373
  chemoreceptors in, 538
  circulatory system, lack of, 574, 574*f*
  evolution of, 370*f*, 372*f*
  free-living, 376
  hydrostatic skeleton, 554
  parasitic, 377, 377*f*
  reproduction, 682
  respiration in, 634
Flexor carpi group, 563*f*
Flooding, natural regulation of, 784
Florida scrub jay, 733
Florigen, 471
Flounders, 745, 745*f*
Flower(s)
  anatomy of, 480, 480*f*
  complete and incomplete, 353, 353*f*, 480
  in eudicots, 424, 424*f*, 480, 481*f*
  evolution of, 343, 343*f*
  in monocots, 424, 424*f*, 480, 480*f*
  perfect and imperfect, 480
  petals, 353, 353*f*, 480, 480*f*
  photoperiod and, 470, 470*f*
  structure, generalized, 353, 353*f*
Flower fly, 745, 745*f*
Flowering plants. *See* Angiosperms
Flu, as viral disease, 305
Fluid feeders, 613, 613*f*
Fluid-mosaic model, 85
Flukes, 377, 377*f*
Fluoride treatments, for teeth, 617

Flycatchers, speciation in, 265, 265*f*
Focusing, of eye, 541, 541*f*
Folacin, dietary requirements, 625*t*
Folic acid, dietary requirements, 625*t*
Foliose lichens, 359, 359*f*
Follicle, in oocyte production, 687, 687*f*, 688, 688*f*
Follicle-stimulating hormone (FSH), 669*t*, 670*f*, 671, 685, 688, 688*f*
Follicular phase, of ovarian cycle, 688, 688*f*
Fontanels, 556, 700
Food. *See also* Diet
  allergic reaction to, 606
  angiosperms and, 356, 356*f*
  caloric value of, 78
  carcinogens in, 201
  corn, development of, 272, 272*f*, 356
  fermentation and, 124, 124*f*, 361, 362
  fungi, 361, 362
  genetically engineered, 228, 229, 230, 230*f*
  global warming and, 421
  macromolecules in, 40, 40*f*
  polyploid plants, 271, 271*f*
  and population, sustainable, 413
Food and Drug Administration, U.S., 228, 230, 394
Food chains, 755
Food vacuole, 328, 328*f*, 329, 329*f*
Food webs
  alien species and, 786
  detrital, 754*f*, 755
  and energy flow in ecosystems, 754–755, 754*f*
  grazing, 754–755, 754*f*
Foolish seedling disease, 462
Foot, of mollusc, 380, 380*f*
Foot-and-mouth disease, 318
Foramens magnum, 556
Foraminiferans, 328, 328*f*
Forebrain, 513, 515, 515*f*
Foreign antigens, 599
Forelimbs, of vertebrates, as homologous structures, 251, 251*f*, 292
Forem, 292
Forensics, 552–553
Foreskin, 684, 684*f*
Forest(s)
  acid rain and, 32, 32*f*
  boreal (northern), 765
  and carbon cycle, 758
  in Carboniferous period, 352, 352*f*
  climate amelioration by, 784
  coniferous, 764*t*, 765, 765*f*, 770, 770*f*
  economic exploitation of, 783
  energy flow in, 754–755, 754*f*
  montane coniferous, 765, 770, 770*f*
  temperate deciduous, 764*f*, 764*t*, 766, 766*f*, 770, 770*f*
  temperate rain, 764*f*, 765
  tropical deciduous, 764*f*
  tropical rain, 764*f*, 764*t*, 768, 768*f*, 770, 770*f*, 785, 785*f*, 789
    and global warming, 108, 108*f*, 421
    human impact on, 10
    medicines and, 472, 474, 474*f*
  tropical seasonal, 764*f*
  water-holding ability, 784, 784*f*
Formed elements, of blood, 584, 584*f*
Formosan termite, 786
Fosamax, 559
Fossil
  Burgess shale, 274–275, 274*f*–275*f*
  early research on, 245, 245*f*
  as evidence for evolution, 249–250, 249*f*

and evolutionary species concept, 264
  speciation and, 273–275
  transitional, 250, 250*f*
Fossil fuels
  and carbon cycle, 758
  and global warming, 108, 420
  origin of, 758
Fossil record
  continental drift and, 286–287
  evolution, evidence of, 249, 371
  geological timescale and, 284, 285*t*
  hominids, 398–399, 398*f*, 406–407, 407*f*
  phylogenetic research and, 292, 292*f*
  primates, 403, 403*f*
Founder effect, 255, 255*f*
Fovea centralis, 540, 540*f*, 543
Fowl pest disease, 318
Fox, Sidney, 310
Foxes
  digestion in, 615, 615*f*
  domesticated, 246, 246*f*
Foxglove *(Digitalis purpurea)*, 472
Frameshift mutations, 200
Francisella tularensis, 318
Franklin, Rosalind, 188, 188*f*
Free-diving, by humans, 632, 633
Free radicals
  detoxification of, 144, 235
  effect of, 201, 583
Fremont cottonwood, 718*f*
Freshwater bony fishes, 652, 652*f*
Fresh water ecosystems, 771, 771*f*, 784, 785
Friedmann, Theodore, 233
Frogs, 389, 389*f*
  development, 368, 368*f*, 692, 693, 693*f*, 694, 694*f*, 696, 696*f*
  egg, structure of, 695, 695*f*
  reproduction, 680, 680*f*
  reproductive barriers in, 266
Fronds, 348, 348*f*
Frontal bone, 556, 556*f*, 557*f*
Frontal lobe, 525, 525*f*, 527, 528*f*
Fruit, 354*f*–355*f*, 355
  anatomy of, 485, 485*f*
  development of, 481, 482*f*, 485
  in diet, 625
  evolution of, 353
  types of, 485, 485*f*
Fruit flies. *See Drosophila melanogaster*
Fruiting body, 331
Fruticose lichens, 359, 359*f*
FSH. *See* Follicle-stimulating hormone
Functional groups, 39, 39*f*
Fungus (fungi), 358–362
  characteristics of, 358, 360
  chromosome numbers, 137*t*
  classification of, 296, 326*f*
  club, 361
  and disease, 362, 362*f*
  economic importance, 362
  as kingdom, 9, 9*f*
  life cycle, 150, 358, 360, 360*f*
  main phyla, 360–361
  mutualism in, 358, 358*f*, 436, 436*f*
  mycorrhizal, 359, 359*f*, 454, 454*f*, 459, 748
  parasitism in, 746
  sac, 360–361, 360*f*
  species, number of, 780*f*
  as type of organism, 4, 4*f*
  zygospore, 360

G

G_0 checkpoint, 141, 141*f*
G_0 stage, 136, 136*f*, 142
G_1 checkpoint, 141, 141*f*
G_1 stage, 136, 136*f*

G_2 checkpoint, 141, 141*f*
G_2 stage, 136, 136*f*
GABA, 521, 522
GABA transaminase, 521
Gage, Fred, 502
Gain-of-function mutations, 218
Galápagos Islands
  alien species on, 786
  birds on, 245, 248, 248*f*, 270, 715, 718*f*, 742, 742*f*
  lizards on, 245, 268
  resource partitioning on, 742, 742*f*
Gallbladder, 619, 620, 620*f*, 621, 621*f*
Gamete(s)
  in algae life cycle, 336, 336*f*
  as haploid, 137
  heterogametes, 336
  isogametes, 336
  isolation, as reproductive barrier, 266–267
  production of, 150, 150*f*, 682
Gamete intrafallopian transfer (GIFT), 691
Gametocytes, 330*f*
Gametophytes, 150, 150*f*, 344, 344*f*
  in algae, 336, 336*f*
  in angiosperms, 345, 354*f*–355*f*, 481, 481*f*
  in bryophytes, 346, 347, 347*f*
  in ferns, 344–345, 349, 349*f*
  in pines, 351, 351*f*
Gamma-hydroxybutyric acid (GHB), 522
Ganglion (ganglia), 514
Ganglion cell layer, of retina, 542–543, 543*f*
gap genes, 216, 216*f*
Gap junctions, in animal cells, 72, 72*f*, 501–502
Gardasil, 689
Garden peas. *See Pisum sativum*
Garrod, Archibald, 192
Garter snakes *(Thamnophis elegans)*, 683, 726–727, 727*f*
Gas chromatography, 657
Gas denitrification, 757, 757*f*
Gas exchange. *See also* Respiration
  countercurrent exchange, in gills, 636, 636*f*
  in external respiration, 642, 642*f*, 643, 643*f*
  in internal respiration, 642, 642*f*, 643, 643*f*
  respiration surface, characteristics of, 634
Gas gangrene, 314
Gastric juice, 618, 618*f*
Gastric pits, 618, 618*f*
Gastrin, 621, 621*f*
Gastrocnemius, 563*f*
Gastroesophageal reflux disease (GERD), 617
Gastropods, 380, 380*f*
Gastrula
  early, 692*f*, 693
  late, 692*f*, 693
Gastrulation, 692*f*, 693, 699
Gated ion channels, and action potential, 518–519, 518*f*, 519*f*
Gause, G. F., 742, 744
Gazelles, 747
Gehring, Walter, 276
Gel electrophoresis, 227, 227*f*, 333
Gelidium, 333
Gene(s)
  chromosome as location of, 174
  clock genes, 469
  definition of, 7, 192, 234
  developmental, and evolution, 276–277, 276*f*, 277*f*
  in eukaryotic cells, 62

Gene(s), continued
    function of, 48
    homeotic, 207, 217, 217f, 696
    human
        number of, 49, 195, 234
        sequencing of, 49
        "nature vs. nurture" question,
            726–730
        structural, 208, 208f
Gene amplification, in cancer cells, 143
Gene chips, 235
Gene cloning
    polymerase chain reaction (PCR),
        227, 227f
    recombinant DNA technology,
        226, 226f
Gene deserts, 236
Gene expression. See also Transcription;
        Translation
    definition of, 192
    environmental factors and, 172
    overview of, 192, 192f, 199, 199f
Gene expression control
    in eukaryotes, 209–217
        and cell specialization, 209
        chromatin structure and,
            212–213, 215, 215f
        cloning. See Cloning
        in development, 216–217
        gene splicing and, 236
        posttranscriptional control, 214,
            214f, 215, 215f
        posttranslational controls, 214,
            215, 215f
        synopsis of, 215, 215f
        transcriptional control, 213,
            213f, 215, 215f
        translational controls, 214, 214f,
            215, 215f
    in prokaryotes, 208, 208f
Gene flow, and microevolution,
        254, 254f
Gene linkage, 176
Gene locus, 163, 163f
Gene migration. See Gene flow
Gene pharming, 231
Gene pool, 253
Generations, alternation of, in
        angiosperms, 336, 344, 344f, 481
Generation time, of bacteria, 135, 312
Gene regulation. See Gene expression
        control
Gene splicing, and gene expression
        regulation, 236
Gene theory
    as basic theory of biology, 11t
    history of, 184–185, 187, 240–241
    modern statement of, 240
Gene therapy
    applications of, 232, 232f
    ex vivo, 232
    future of, 235
    overview, 224, 225
    success of, 233
    viruses and, 300
    in vitro fertilization and, 170
    in vivo, 232
Genetically modified organisms (GMOs)
    animals, 231, 231f
    bacteria, 228, 228f
    food products, 228, 229, 229f,
        230, 230f
    overview, 224–225
    plants, 229, 229f, 230, 230f
Genetic code, 193, 193f
Genetic disorders
    autosomal dominant, 167, 167f,
        169, 169f
    autosomal recessive, 167–168,
        167f, 168f
    cancer as, 133

chromosome number abnormalities,
        152–153, 152f, 153f, 153t
    chromosome structure
        abnormalities, 154, 154f
    curing of, 235
    and disease, 49, 69, 87
    in dogs, 154–155, 155f, 156f
    incomplete dominant, 171
    pleiotropy, 173, 173f
    sex chromosome abnormalities,
        153, 153t
    X-linked, 175, 175f
Genetic diversity. See Genetic variation
Genetic drift, and microevolution,
        253, 255
Genetic engineering. See also Cloning;
        Gene therapy; Genetically
        modified organisms
    animal pharming, 394
    applications, 228–233
    and bacteria, 301
    genetic code and, 193
    overview, 224
    viruses and, 300
    and xenotransplantation, 506
Genetic inheritance
    blending model of, 160
    as characteristic of life, 7
    codominance, 171
    environmental factors and, 172
    incomplete dominance, 171, 171f
    Mendel's experiments, 160–162,
        160f, 161f, 162f, 164, 164f,
        166, 166f
    multifactorial traits, 172
    multiple alleles, 171
    particulate theory of, 160f, 161,
        162–163
    pleiotropy, 173, 173f
    polygenic, 172, 172f
    X-linked alleles, 174, 174f
Genetic material
    characteristics of, 187
    DNA established as, 184–185, 187
Genetic mutation, 200–203
    and cancer, 133, 218–220
    causes of, 133, 201–202
    definition of, 200
    degeneracy of genetic code and, 193
    frameshift, 200
    gain-of-function, 218
    germ-line, 200
    loss-of-function, 218
    and microevolution, 253, 254
    mutagens, 133, 201
    and natural selection, 419
    point, 200, 200f
    in proto-oncogenes, 218, 218f,
        219, 219f
    rate of, 254
    somaclonal variations, 488
    somatic, 200
    spontaneous, 201
    transposons and, 200, 202, 202f
    in tumor-suppressing genes, 218,
        218f, 219, 219f
Genetic testing, 168, 170, 175
    amniocentesis, 168, 170
    chorionic villi sampling (CVS),
        168, 170
    preimplantation diagnosis, 170, 170f
Genetic variation
    advantages and disadvantages
        of, 147
    and biodiversity, 781
    chromosome number changes,
        152–153, 152f, 153f, 153t
    fertilization and, 147
    sexual reproduction and, 146–147,
        147f, 150
Gene transfer, in bacteria, 313, 313f

Genital herpes, 689
Genital organs
    female, 686, 686f
    male, 684, 684f
Genital tract, differentiation of, 700
Genital warts, 305, 689
Genome(s). See also Human genome;
        Human genome project
    analysis of, 236
    comparative genomics, 236, 236f
    of model organisms, 234, 234t
    sequencing of, 234
    size of, 234, 234t
    of vertebrates, similarity of,
        236, 236f
    viral, 302
Genotype, definition of, 163
Gentamicin, 546
Genus, 8, 8t, 290, 290f
Geological timescale, 284, 285t,
        286, 286f
Geology, uniformitarianism in, 245
GERD. See Gastroesophageal reflux
        disease
Germ cells, 134
Germination, 486, 486f
Germ layers
    in animals, 371, 371f, 693, 693f,
        693t, 694, 694f, 696, 696f
    in cnidarians, 375
    in flatworms, 376
Germ-line mutations, 200
Gestation
    in humans, 697 (See also
        Development, in humans;
        Pregnancy)
    in primates, 401
GH. See Growth hormone
GHB. See Gamma-hydroxybutyric acid
Giant kelp, 333
Giardia lamblia, 327
Giardiasis, 323
Gibberellic acid (GA₃), 462, 462f
Gibberellins, 462, 462f, 465
Gibbons, evolution of, 400f, 402f
GIFT. See Gamete intrafallopian transfer
Gila monster, 390f
Gills
    in arthropods, 382
    in crayfish, 382f, 383
    in fish, 388–389, 576
    structure and function of, 634, 635f,
        636, 636f, 651
    tracheal, 637
Gingivitis, 617
Ginko trees, 350, 350f
Giraffes, adaptation in, 245, 245f, 246
Girdling, 448
Gizzard, 612, 614, 614f
Gland
    definition of, 497
    endocrine, 497
    exocrine, 497
    secretions, and immune system, 596
Glanders, 318
Glass, David, 447
Glaucoma, 541
Glia-derived growth factor, 501
Glial cells. See Neuroglia
Gliding joints, 561
Global warming
    and extinctions, 787
    forests and, 784
    greenhouse gases and, 420–421, 758
    and ocean currents, 774
    projected rise in temperature,
        108, 108f
    and P-Tr (Permian-Triassic)
        extinction, 289
    tropical rain forests and, 108
Globin, 572

Globular stage, of angiosperm embryo
        development, 484, 484f
Gloeocapsa, 315f
Glomerular capsule (Bowman capsule),
        654, 655f
Glomerular filtration, 656–659
Glomerulus, 655, 655f, 656
Glottis, 617, 617f
Glucagon, 620, 668, 669t, 674, 674f
Glucocorticoids, 669t, 672–673,
        672f, 673
Glucosamine-chondroitin therapy, 562
Glucose
    blood levels, regulation of, 507, 620
    breakdown of, ATP created in, 80
    in cellular respiration, 114, 114f,
        116, 117f
    production of, in photosynthesis,
        97, 98
    storage of, 42, 42f
    structure and function of, 41, 41f
Glutamate, 192, 527
Glutamic acid, 46f
Glutamine, 46f
Glycerol, 43
Glycocalyx, 85, 311
Glycogen, 42, 42f
Glycolipids, 85
Glycolysis, 115, 115f, 116, 117f,
        122, 122f
Glycoproteins, 64, 85
Glycosides, as plant defense, 472
Glyoxysomes, 66
GMOs. See Genetically modified
        organisms
Gnetophytes, 350, 350f
GnRH. See Gonadotropin-releasing
        hormone
Goiter, simple, 676, 676f
Goldfish, chromosome number, 137t
Golgi, Camillo, 65
Golgi apparatus. See also
        Endomembrane system
    in cell anatomy, 60, 60f, 61f
    cytokinesis and, 140
    in lipid synthesis, 65
    in phototropism, 461, 461f
    in protein synthesis, 65
    structure of, 65, 65f
Gonad(s)
    in endocrine system, 668, 668f, 669t
    functions of, 682
Gonadotropic hormones (FSH, LH),
        669t, 670f, 671, 685, 691
Gonadotropin-releasing hormone
        (GnRH), 685, 688
Gondwana, 286, 287f
Gonoccus, 689
Gonorrhea, 311, 317, 689, 782
Gonyaulax, 325
Gooseneck barnacles, 383f
Gorgas, William C., 322
Gorillas, 401f, 402f, 717, 717f, 718, 718f
Graafian (vesicular) follicle, 687, 687f
Gracilaria, 333
Gracile, defined, 406
Gradualistic model of speciation,
        273, 273f
Grains, 356, 356f
    fungal diseases of, 362
    refinement of, 622
Grand unifying theory (GUT), evolution
        as, 419
Grant, Peter and Rosemary, 248
Granum (grana), 61f, 68, 68f, 97,
        97f, 99
Granzymes, 602, 603f
Grass fern, 718f
Grasshoppers, 384, 384f, 575, 575f
Grassland ecosystem, 10, 10f, 764t
    temperate grassland, 764f, 766, 766f

Graves disease, 676
Gravitational balance, 546–547, 547*f*
Gravitropism, 466–467, 466*f*
Gravity, plant responses to, 466–467, 466*f*
Gray crescent, 695, 695*f*
Gray matter, 524
Gray whales, 536*f*, 718*f*
Grazing food web, 754–755, 754*f*
Great Barrier Reef, 781
Great blue heron, 614
Greater vestibular glands (Bartholin glands), 686
Greater Yellowstone Ecosystem, 789
Great Lakes, and lake effect snow, 770
Green algae, 334–335, 334*f*, 335*f*, 773
   characteristics of, 325*f*
   classification of, 326, 326*f*
   evolution of, 369
   mutualism in, 359
Green glands, 651
Greenhouse effect, 758
Greenhouse gases, 108, 289, 316, 420–421, 758, 787
Green sea turtle, 390*f*
Griffith, Frederick, 184, 184*f*, 313
Grizzly bears (*Ursus arctos horribilis*), 789, 789*f*
Grooming, 735, 735*f*
Ground finch, natural selection in, 248
Ground meristem, 430, 430*f*, 431, 431*f*, 484
Ground pine (*Lycopodium*), 348*f*
Ground tissues, 426–427, 426*f*, 428
Group living, adaptive benefits of, 732
Growth, angiosperms
   primary, 430–431, 430*f*, 431*f*
   secondary, 432–433, 432*f*, 433*f*
Growth factors, and cell cycle control, 142
Growth hormone (GH), 669*t*, 670*f*, 671
Growth plate, in bone, 560, 560*f*
G3P, 104*f*, 105, 105*f*, 116, 117*f*
Guam, alien species on, 786
Guanine (G), 48, 48*t*, 48*t*, 186, 186*f*, 187, 187*f*, 193, 193*f*
Guard cells, 426, 426*f*, 434, 434*f*, 446, 446*f*
Guinea pigs, 724
Gulf Stream, 774, 774*f*
Gullet, of *Paramecium*, 329, 329*f*
Gum disease, 617
Guppies, directional selection in, 257, 257*f*
GUT (grand unifying theory), evolution as, 419
Guttation, 444, 444*f*
Gymnosperms, 350–351, 350*f*
   in Carboniferous forest, 352, 352*f*
   characteristics of, 342*f*, 343
   coniferous forest, 765, 765*f*
   diversity of, 350, 350*f*
   evolution of, 343, 343*f*
   life cycle, 351, 351*f*

H

Habitat
   and biodiversity, 749, 749*f*, 785
   and competition for resources, 742, 742*f*
   ecological study of, 710
   fragmentation of, 789
   isolation of, as reproductive barrier, 266
   loss of, 718, 785, 785*f*
   preservation of, 789, 789*f*
   restoration of, 790
   size of, and biodiversity, 749, 749*f*
   stratification of, 749
Habituation, 729
Haddock, 788
Haeckel, Ernst, 710

Hagfish, 388
Hair, of mammals, 392
Hair cells, in ear, 545, 546–547, 546*f*, 547*f*
Hairy-cell leukemias, 305
Haleakala silversword, 718*f*
Hallucinogenic plants, 357
Halophiles, 296, 316, 316*f*
Hammer (malleus), 544, 544*f*, 545
Hammer, K. C., 470
Hand
   bones of, 558, 558*f*
   joints of, 561
Hantavirus, 305, 307
Haploid (*n*) chromosome number, 137
Haploid phase
   in algae, 334, 334*f*, 336, 336*f*
   in angiosperms, 354*f*–355*f*, 482*f*
   in ferns, 349, 349*f*
   in fungi, 360*f*
   in mosses, 347*f*
   in pine trees, 351, 351*f*
   in plant life cycle, 150, 150*f*, 344, 344*f*
Hard palate, 616, 616*f*
Hardwood, 350
Hardy, G. H., 253
Hardy-Weinberg principle, 253
Harlequin toad, 788
Hart's tongue fern (*Campyloneurum scolopendrium*), 348, 348*f*
Harvestmen (daddy longlegs), 383
Haustoria, 454
Haversian systems, 560
Hawaiian honeycreepers, speciation in, 270, 270*f*
Hawaiian Islands
   extinction of native species in, 778–779, 786
   rain patterns in, 770
Hay fever, 606
Hazardous waste production, in more-vs. less-developed countries, 720
HCG. *See* Human chorionic gonadotropin
HDL. *See* High-density lipoprotein
HDL (high-density lipoprotein), 565, 583, 627
Head, of animals, 512–513
Head louse, 384*f*
Health. *See also* Disease; Medicine
   acid rain and, 32
   radiation and, 22
   smoking and, 639, 644
Hearing. *See also* Ears
   damage to, 546, 546*f*
   process of, 545, 545*f*
   receptors for, 536, 536*f*, 544, 545
Heart
   anatomy of, 577, 577*f*
   cardiac cycle, 578, 578*f*
   defects in, 577, 578
   development of, 699, 699*f*
   PET scans of, 22
Heart attack, 580, 582
Heartbeat, 578, 578*f*
Heart murmur, 577
Heart stage, of angiosperm embryo development, 484, 484*f*
Heartwood, 432–433, 433*f*
Heat, as energy source, 78
Heat receptors, 536
Hebert, Paul, 14
Helicase, 190, 191, 191*f*
Helicobacter pylori, 618
Heliotropism, 466, 466*f*
Helper T cells, 600, 600*f*, 600*t*, 601, 601*f*, 602–603
Hematopoietic stem cells, 586, 586*f*
Heme groups, 572, 643, 643*f*
Hemocoel, 637
Hemocyanin, 573, 575

Hemodialysis, 660, 660*f*
Hemoglobin
   and gas exchange, 643, 643*f*
   HBA, 168, 192, 258
   HBS, 168, 192, 258
   as protein, 45
   reduced, 643
   in sickle cell disease, 192, 192*f*, 258
   structure and function of, 572, 584
Hemolymph, 575
Hemolysis, 89
Hemophilia, 175, 200, 585
   Hemophilia A, 175
   Hemophilia B, 175
Hemorrhagic fevers, 307, 318
Hemorrhoids, 581
Hemp, 357
Hendry, Andrew, 268–269
Hennig, Willi, 294
Hepatic artery, 620
Hepatic portal vein, 620
Hepatic veins, 620
Hepatitis, 305, 620
   Hepatitis A, 620
   Hepatitis B, 305, 601, 620
   Hepatitis C, 620
Herbaceous plants, 428, 432
Herbicides, resistance to, 248
Herbivores
   digestive system, 611, 615, 615*f*
   in ecosystem, 752, 753, 753*f*, 755, 755*f*
   teeth of, 393, 611
Hermaphrodites, 376, 682
Heroin, and neurotransmitters, 521, 523
Herpes
   type 1, 305
   type 2, 305
Herpes simplex virus (HSV), 689
Hershey, Alfred D., 184–185, 185*f*, 240–241, 300
Heterochromatin, 212–213, 212*f*
Heterocysts, 315
Heterogametes, 336
Heterotrophs, 96, 752
Heterozygous individuals, 163
   testcross for, 166, 166*f*
Hex A. *See* Hexosaminidase A
Hexosaminidase A (Hex A), 168
HGP. *See* Human Genome Project
Hibernation, 495
High-density lipoprotein (HDL), 565, 583, 627
Hindbrain, 513, 515, 515*f*
Hind limbs, evolution and, 277
Hinge joints, 561, 561*f*
Hip bones (coxal bones), 556*f*, 558, 558*f*
Hip dysplasia, in dogs, 154–155, 156*f*
Hippocampus, 527, 527*f*
Hippocrates, 657
Hippopotamuses, 735
Hirudin, 381
Histamines, 585, 598, 598*f*, 606, 606*f*, 607*t*
Histidine, 46*f*
Histones, 137, 212, 212*f*
Histone tail, 212*f*, 213
HIV/AIDS, 306, 306*f*
   immune system and, 592–593, 592*f*
   mutation of, 307
   opportunistic infections in, 362, 529*f*, 592–593, 593*f*
   prevalence, 593
   spread to humans, 506
   symptoms, 592
HIV provirus, 306
Hodgkin disease, 782
Hog cholera, 318
Holocene epoch, 285*t*
Homeobox, 217

Homeodomain, 217
Homeostasis. *See also* Osmoregulation
   in animals, 494, 507–508
   as basic theory of biology, 11*t*
   as characteristic of life, 6–7, 7*f*
   definition of, 6
   digestion and, 612
   endocrine system and, 668
   lymphatic system and, 594
   negative feedback mechanisms in, 508, 508*f*
   organ systems and, 504
   overview, 706–707
   in plants, 435–436, 435*f*, 436*f*, 492–493
   reflex actions and, 530
   water and, 19
Homeotic genes, 207, 217, 217*f*, 696
Hominids
   characteristics of, 404
   early, examples of, 405
   evolution of, 402*f*, 403, 405*f*
Hominoids, 403
Homo. See also Human(s)
   origin of, 407
   species, 405*f*, 408
Homo erectus, 405*f*, 408, 409, 411*f*
Homo ergaster, 405*f*, 408, 408*f*, 411, 411*f*
Homo floresiensis, 408
Homo habilis, 399*f*, 405*f*, 408, 409
Homologous chromosomes, 145, 145*f*, 146, 146*f*
   alleles on, 163, 163*f*
Homologous structures, 251, 251*f*, 292
Homologues. *See* Homologous chromosomes
Homology, in evolution of animals, 371
Homo neanderthalensis, 399*f*, 405*f*, 410
Homo rudolfensis, 405*f*, 408
Homo sapiens, 405*f*, 410–411, 412, 412*f*, 414, 414*f*
Homozygous dominant individuals, 163
   testcross for, 166, 166*f*
Homozygous individuals, 163
Homozygous recessive individuals, 163
Honeybees (*Apis mellifera*), 682, 735, 735*f*, 783
Hoofed mammals, 393
Hooke, Robert, 54
Hookworms, 378
Hormones. *See also* Plant hormones
   antagonistic actions, 668
   and cell cycle control, 142
   digestive system and, 621, 621*f*
   endocrine system and, 504, 504*f*
   functions of, 45, 666, 666*f*, 666*t*
   glia-derived growth factor, 501
   nontropic, 671
   peptide, action of, 667, 667*f*
   as protein, 45
   sex
      female, 687
      male, 685
   steroid, action of, 667, 667*f*
   target cells, 666
   thyroid, 667
   tropic, 671, 671*f*
   water-salt balance and, 658–659
Hormone therapy, and cancer, 144
Horner, Jack, 282–283
Hornets, mimicry in, 745, 745*f*
Hornworts, 346, 346*f*
Horse(s), 393
Horseshoe crabs, 383*f*, 573, 782
Horsetails, 348, 348*f*
Horticulture, and introduction of alien species, 786
Host, 746
Houseflies, 384*f*
   chemoreceptors in, 538
   eye of, 6, 6*f*

Housekeeping genes, 209, 209*f*
Hox g, 407
Hox genes, 217, 276–277, 277*f*, 369
HPV. *See* Human papillomavirus
HR. *See* Hypersensitive response
HSV. *See* Herpes simplex virus
Huber, Claudia, 309
Human(s). *See also* under specific topics
  acid rain, health effects of, 32
  adaptive social behavior, 733
  amino acid synthesis in, 125
  babies, weakness of, 407, 407*f*
  behavior, genetically-based, 727
  birth weight, stabilizing selection in, 256, 256*f*
  chromosomes, number of, 137, 137*t*, 145, 145*f*
  cloning of, 210, 226, 226*f*
  diversity in, 414, 414*f*
  early, characteristics of, 408
  ethnic groups, origin of, 414
  free-diving by, 632, 633
  genes
    number of, 49, 195, 234
    sequencing of, 49
  genome, vs. chimpanzee, 404
  ions, biologically important, 24, 24*t*
  language and, 734
  life cycle, 150, 150*f*
  organ systems of, 504–505, 504*f*, 505*f*
  pharyngeal pouches in, 386
  pheromones and, 665
  primate characteristics and, 400–401
  proteome, 236
  respiratory system. *See* Respiratory system
  as single species, 265, 265*f*
  sky burials, 740, 740*f*
  teeth, 611, 616–617, 616*f*
  viral diseases, 306–307, 306*f*, 307*f*
  visual communication, 735
  and wildlife. value of, 782–783, 782*f*, 783*f*
Human activity. *See also* Pollution
  and chemical cycling, 756–757, 756*f*, 757, 757*f*, 758, 758*f*
  and mass extinction, 289
Human chorionic gonadotropin (HCG), 698
Human genome, similarity to other vertebrates, 236, 236*f*
Human Genome Project (HGP), 48, 234, 235
Human papillomavirus (HPV), 305, 689
Humboldt Current, 774, 774*f*
Humerus, 556*f*, 558, 558*f*
Hummingbirds
  excretory system, 650
  as pollinators, 243
Humoral immunity, 601
Humpback whales, 734
Humus, 453
Hunchback (kyphosis), 557
Hunter-gatherers, 409
Hunting and gathering, 412–413
Huntingtin, 169, 192
Huntington's disease, 169, 169*f*, 192, 525
Hyaline cartilage, 498*f*, 499, 560, 560*f*, 561, 562
Hybridized species
  animals, 262–263, 262*f*, 263*f*, 267, 267*f*
  corn, development of, 272, 272*f*
  plants, alloploidy in, 271, 271*f*
  sterility of, 263, 267, 267*f*, 271
Hybridomas, 604, 604*f*
Hydras, 514, 514*f*, 554
  circulatory system, lack of, 574, 574*f*
  reproduction, 682

respiration in, 634, 635*f*
  survivorship curve, 713, 713*f*
Hydrochloric acid, 30
  in stomach, 618
Hydrogen
  carbon binding with, 38
  and life, 20, 20*f*
Hydrogen bonds, 26, 26*f*
  notation for, 26
  in water, 27–28
Hydrogen ion gradients, 101–103
Hydrogen ions, and pH, 30, 30*f*
Hydrolysis reaction, 40, 40*f*, 43*f*, 45*f*
Hydrolytic reaction. *See* Hydrolysis reaction
Hydrophilic molecules, 28, 39
Hydrophobic molecules, 28, 39
Hydroponics, 450–451
Hydrostatic skeleton, 379, 554, 554*f*
Hydrothermal vents, 773
  and origin of organic molecules, 308–309, 308*f*
  prokaryotes in, 314, 314*f*
Hydroxide ions, and pH, 30, 30*f*
Hydroxyl groups, 39, 39*f*
Hyman, Flo, 173
Hyperopia (farsightedness), 542, 542*f*
Hypersensitive response (HR), 473
Hypertension, 582–583
Hyperthyroidism, 676
Hypertonic solutions, 89, 89*f*
Hyperventilation, 643
Hypha, 358, 358*f*, 748
Hypodermis, 503
Hypothalamus, 515, 515*f*, 524*f*, 526, 527*f*, 536
  in endocrine system, 668, 668*f*, 669*t*, 670–671, 670*f*, 672–673, 672*f*, 676
  and homeostasis, 707
  homeostasis and, 508
  in nervous system, 670–671, 670*f*
  and ovarian cycle, 688
  pheromones and, 665
  testes, control of, 685
Hypothesis, in scientific method, 11, 11*f*, 12
Hypothyroidism, 676
Hypotonic solutions, 89, 89*f*
Hypoventilation, 643
Hyracotherium, 278, 278*f*
H zone, 566, 566*f*

I

IAA. *See* Indoleacetic acid
Iberomesornis, 250
Ibuprofin, 598
Ice Age, ending of, 412–413
Ich (fish disease), 329
Ichthyophthirius, 329
ICSH. *See* Interstitial cell-stimulating hormone
ICSI. *See* Intracytoplasmic sperm injection
IF-2, 214
Ig. *See* Immunoglobulin
IgE, 606, 606*f*, 607*t*
IgG, 601, 601*f*, 606
Iguanas, 6, 7*f*, 268, 736
Immediate allergic response, 606, 606*t*
Immune system
  abnormalities in, 605–606
  AIDS and, 592–593, 592*f*
  allergic reactions, 606, 606*f*, 606*t*
  antibody-mediated immunity, 600, 600*f*, 601, 601*f*
  autoimmune disorders, 605, 605*f*
  barriers to entry, 596, 596*f*
  cell-mediated immunity, 600, 600*f*, 602–603, 603*f*
  fever, 597

  in humans, 505
  inflammatory response, 585, 596*f*, 598, 598*f*
  lymphatic system and, 594–595, 594*f*, 595*f*
  monoclonal antibodies, 604, 604*f*, 657
  natural killer cells, 596*f*, 597
  nonspecific defenses, 596–598, 596*f*
  phagocytes, 596*f*, 597, 598
  protective proteins, 596–597, 596*f*, 597*f*
  specific immunity, 599–604
    active, 599
    cells involved in, 600*t*
    passive, 599
  tissue rejection, 605
  white blood cells and, 499
Immunization, 599, 599*f*, 601
Immunoglobulin (Ig; antibodies), 585, 594, 601, 601*f*, 606, 606*f*, 607*t*
Immunokine, 394
Immunosuppressive drugs, 605
Impalas, 732
Implantation, 697, 698
Imprinting, 728
Incisors, 616, 616*f*
Inclusive fitness, 732
Incomplete digestive tracts, 614, 614*f*
Incomplete dominance, 171, 171*f*
Incus (anvil), 544, 544*f*, 545
Independent alignment, of chromosomes, 146
Independent assortment
  of chromosomes, 146–147, 147*f*
  Mendel's law of, 164–166
India, climate, 770
Indian corn, 202, 202*f*
Indigenous peoples, and medicinal plants, 474
Indirect plant defenses, 473
Indirect selection, 732–733
Indoleacetic acid (IAA), 460
Indonesia
  as biodiversity hotspot, 781
  habitat loss in, 785
Induced fit model, 82
Inducible operons, 208
Induction, 696, 696*f*
Industrialization, and global warming, 108
Industrial melanism, 248, 248*f*
Industrial Revolution, and industrial melanism, 248, 248*f*
Infants, human, helplessness of, 407
Inferior vena cava, 580, 580*f*, 620
Inflammatory response, 585, 596*f*, 598, 598*f*
  counteracting of, 673
Influenza virus, 302*f*
Ingestion, 612, 612*f*
Inguinal lymph nodes, 594*f*, 595
Inheritance. *See* Genetic inheritance
Inhibition, of enzyme activity, 84, 84*f*
Inhibition of succession, 751
Initiation, of translation, 198, 198*f*, 199*f*
Initiation factors, 198
Initiator tRNA, 198, 198*f*
Inner cell mass, 698, 698*f*
Inner ear, 544, 544*f*, 545, 545*f*
Inorganic chemistry, 38, 38*t*
Inorganic nutrient pool, and energy flow in ecosystems, 753, 753*f*
Insect(s). *See also* specific insects
  anatomy, generalized, 384, 384*f*
  blood of, 573
  chemoreceptors in, 538, 538*f*
  circulatory system, 575, 575*f*
  as disease vectors, 307, 322, 322*f*, 330, 330*f*, 378
  evolution of, 288*f*–289*f*

  excretory system, 650, 650*f*, 651
  eye of, 534, 536
  mimicry in, 745, 745*f*
  nervous system, 515
  and pheromones, 664
  plant viruses and, 304
  as pollinators, 242–243, 242*f*, 243*f*, 353, 783
  reproduction, 384, 682
  respiration in, 635, 635*f*, 637, 637*f*
  as scavengers, 741
  species, number of, 780*f*
  success of, plants and, 355
  of tropical rain forest, 768
Insecticides, and pollinators, 355
Insectivora, 393
Insight learning, 729
Insomnia, 526
Inspiration, 640, 640*f*
Insulin
  and antagonistic hormonal action, 668
  and diabetes, 626–627
  function of, 42, 620, 669*t*, 674–675, 674*f*
  from pigs, 394
  synthetic, 193, 228
Integration
  of neural stimuli, 521, 521*f*, 537
  of visual stimuli, 543
Integration stage, in viral life cycle, 303, 306, 306*f*
Integrin, 72
Integumentary system, in humans, 504, 504*f*
Intercalated disks, in cardiac muscle, 500, 500*f*
Intercostal muscles, 640*f*, 641*f*
Intercostal nerves, 641*f*
Interferons, 596*f*, 597, 603
Interkinesis, 148, 148*f*–149*f*
Interleukins, 603
Intermediate filaments, 60*f*, 70, 70*f*
Intermembrane space, in mitochondria, 114*f*
Internal environment, regulation of. *See* Homeostasis
Internal respiration
  gas exchange in, 642, 642*f*, 643, 643*f*
  overview of, 634, 634*f*
International Committee on Taxonomy of Viruses, 304
Interneurons, 517, 517*f*, 524
Internode, of angiosperm, 422, 422*f*
Interphase, 136, 136*f*
Interstitial cell-stimulating hormone (ICSH), 685
Intertidal zone, 772, 772*f*
Intervertebral disks, 557
Intestinal tract. *See* Digestive system
Intracytoplasmic sperm injection (ICSI), 691
Intrauterine device (IUD), 690*f*, 690*t*
Introns
  function of, 195
  posttranscriptional processing, 195, 195*f*, 214, 214*f*
Invasive tumors, 143, 143*f*
Inversions, 154, 154*f*
Invertebrates
  aquatic, respiration in, 634
  chordates, 386, 386*f*
  circulatory systems, 574–575, 574*f*, 575*f*
  excretory system, 651, 651*f*
  homeostasis in, 494
  major phyla, 373
  nervous system, 514–515, 514*f*
  osmoregulation in, 648
  reproduction, 682, 682*f*

In vitro fertilization (IVF), 691
    gene therapy, 170
    preimplantation diagnosis, 170, 170f
In vivo gene therapy, 232
Involuntary muscle, 500, 500f
Iodine
    dietary requirements, 624
    and hypothyroidism, 676
Iodine-131, 22
Ion(s), biologically, 24, 24t
Ionic bonds, 24, 24f
Iris, 540, 540f
Irish moss, 346
Irish potato famine, 229, 331
Iron, dietary requirements, 624, 624t
Iron-sulfur world hypothesis,
    308–309, 308f
Island(s), alien species on, 779, 786
Island biogeography, model of, 749, 749f
Islets of Langerhans. See Pancreatic
    islets
Isogametes, 336
Isogamous, defined, 336
Isolating mechanisms
    postzygotic, 267, 267f, 269
    prezygotic, 266–267, 266f, 267f, 269
Isomers, 39, 39f
Isoniazid, 201
Isotonic solutions, 89, 89f
Isotopes
    definition of, 21
    radioactive, 21–22 (See also
        Radiation)
IUD. See Intrauterine device
IVF. See In vitro fertilization

J

Jackals, 733
Jacob, François, 208
Janzen, Dan, 741
Japan current, 774f
Jasmonic acid, 473, 473f
Jaundice, 620
Jaw(s), evolution of, 387, 388, 388f
Jawless fishes, 387f, 388, 388f
Jellyfish, 375, 375f, 514
Jet lag, 526
Johannsen, Donald, 398–399
Joint(s)
    in arthropods, 382, 382f
    evolution of, 370f, 371
    repair of, 562, 562f
    structure and function of, 561, 561f
    types of, 561
    in vertebrates, 387
Joint capsules, 561
Jointed endoskeleton, as vertebrate
    characteristic, 387
J-shaped curve, characteristics of,
    714, 714f
Jugular vein, 580f
Jumper ants, 14
Junction proteins, 86, 86f
Jurassic period, 285t, 286, 287f, 392

K

Kanamycin, 546
Kangaroo Rat, 653, 653f
Kangaroos, 392
Kaposi sarcoma, 592f, 593
Karyotypes, 145, 145f, 153, 154
Katydids, 745
Kelp, 333
Keratin, 503
Ketamine (special K), 522
Key, Francis Scott, 178
Keystone species, 789
Kidneys
    and acid-base balance, 659
    artificial, 660, 660f
    function of, 387

and homeostasis, 507, 706
human
    function of, 654, 654f
    structure of, 654–655, 655f
    in urinary system, 654, 654f
    urine concentration, 658–659,
        658f, 659f
    urine formation, 557f, 656–657
    structure of, 654–655, 655f
Kidney stones, 654
Killer whale (orca; Orcinus orca), 264f,
    393f, 788
Kilocalories, 78
Kinase(s), 142, 142f
M-kinase, 142, 142f
S-kinase, 142, 142f
Kinesin, 70, 70f
Kinetic energy, 78, 78f
Kinetin, 463
Kinetochores, 138–139
Kingdom, 8–9, 8t, 290, 290f
Kinins, 585
Kinocilium, 547, 547f
Kin selection, 732–733
Klinefelter syndrome, 153, 153t
Klipspringers, 734
Knop, William, 450–451
Koalas, 392, 392f
Kolff, William J., 660
Kovalick, Walter W., 447
Kress, John, 14
Krill, 383, 613
Kudzu, 786, 786f
Kurosawa, Ewiti, 462
Kyphosis (hunchback), 557

L

Labia majora, 686, 686f
Labia minora, 686, 686f
Labor, 702
Lac operon, 208, 208f
Lactate, 123, 123f
Lacteals, 594, 619, 619f
Lactose, 41
Lacunae, 498f, 499
Ladderlike nervous system, 514
Ladybugs (Coccinella), 783, 783f
Lagging new strand, 191, 191f
Lagomorphs, 393
Lag phase
    of J-shaped curve, 714, 714f
    of S-shaped curve, 714, 714f
Lake effect snow, 770
Lake Okeechobee, 790
Lakes, 771, 771f
    acid rain and, 32, 32f
    eutrophication of, 787
Lamarck, Jean-Baptiste de, 245
Laminaria, 333, 336
Laminarin, 333
Lampbrush chromosomes, 213, 213f
Lampreys, 388, 388f
Lance fluke, 747
Lancelets, 386, 386f, 387f, 692, 692f,
    693, 693f
Landscape diversity, and
    biodiversity, 781
Landscape preservation, 789
Language
    in animals, 734–735, 735f
    human development of, 409, 412
La Niña, 774
Lantern flies, 745, 745f
Lanugo, 701
Large intestine, human, 621, 621f
Larva
    fire ant, 3, 3f
    trochophore, 372
Larval stage, in aquatic animals, 683
Larynx (voice box), 617, 638, 638f
LASIK surgery, 542

Lassa virus, 305
Latent operant conditioning, 729
Lateral line system, 388, 548, 548f
Latissimus dorsi, 563f
Laughing gulls, 728, 728f, 735
Laurasia, 286, 287f
Law of independent assortment,
    164–166
Law of segregation (Mendel), 162–163,
    162f, 171
Laws of probability, Mendel's laws
    and, 165
Laws of thermodynamics, 79
LDCs. See Less-developed countries
LDL (low-density lipoprotein), 583, 627
Leading new strand, 191, 191f
Leaf (leaves)
    cells, 434f
    color of, 94–95, 100
    eudicots, 424, 424f, 428, 428f, 429f,
        434, 434f
    monocots, 424, 424f, 434
    sessile, 423
    structure and function, 422–423,
        422f, 423f, 434, 434f
    tissues, 428, 428f, 429f
    of vascular plants, structure of,
        345, 345f
    veins in, 97, 97f, 424, 424f, 428,
        429f, 434, 434f, 435f
    and water transport in plants,
        445, 445f
Leaf curl fungi, 360–361
Leafhopper, 384f
Leaf primordia, 431, 431f
Leaf scars, 431, 431f
Learning
    in animals, 728–729, 728f, 729f
        associative learning, 729, 729f
        insight learning, 729
    brain areas in, 527
Leeches, 381, 381f, 613
Left atrium, 577, 577f
Left hemisphere, 525
Left ventricle, 577, 577f
Legumes, nitrogen fixing in, 12–13
Leishmaniasis, 323, 323f, 327
Lemurs, 400f, 402f
Lens, of eye, 540, 540f, 541, 541f
Lenticels, 426, 426f, 432, 432f
Leopards, 735
Leprosy, 782
Less-developed countries (LDCs)
    age distributions in, 720, 720f
    population growth in, 719, 719f
Leucine, 46f
Leukemias, 133f, 143
    chronic myelogenous, 154
    hairy-cell, 305
    T-cell, 305
    treatment of, 586, 782
Leukocytes. See White blood cells
Lewis, Edward B., 217
Lewis, Warren, 696
Lewontin, Richard, 414
Lianas, 768, 768f
Lichens, 315, 334, 359, 359f, 748
Licht, Louis, 447
Liemorpha, 325f
Life
    cell as basic unit of, 56
    characteristics of, 6–7, 130
    diversity of, 4, 4f, 38, 38f
    elements basic to, 20, 20f
    history of, 284, 285t
    levels of organization, 4, 5f
    molecules of. See Organic molecules
    origin of, 18–19, 308–310
    on other planets, 19
    pH range required by, 31
    unity of, 4

water and, 18–19, 19f, 27–29
Life cycle
    algae, 150, 150f, 333, 334, 334f,
        336, 336f
    angiosperms, 353–356, 354f–355f,
        418f, 480–481
    bacteriophages, 303, 303f
    blood flukes, 377, 377f
    bryophytes, 346, 347, 347f
    ferns, 349, 349f
    fungi, 150, 358, 360, 360f
    gymnosperms, 351, 351f
    humans, 150, 150f
    mitosis and meiosis in, 150, 150f,
        344, 344f
    mosses, 347, 347f
    pine trees, 351, 351f
    plants, 150, 150f, 344, 344f
    tapeworms, 377, 377f
    viruses, 303, 305, 305f, 306, 306f
    water molds, 331
Life history, 717
Life tables, 712, 712f
Ligaments, 498f, 499, 555, 561, 561f
Ligers, 262–263, 262f
Light, and phototropism, 467, 467f
Light micrograph (LM), 58
Light microscopes, 57, 57f, 58, 58f
Light reactions, in photosynthesis,
    99–103
    absorption of solar energy, 100, 100f
    overview, 99, 99f
    production of ATP and NADPH,
        101–103, 101f–103f
Lignin, 61, 427
Lima beans, plant defenses, 473
Limb(s). See also Appendages; Joint(s)
    development of, 276
    forelimbs of vertebrates, as
        homologous structures, 251,
        251f, 292
    lower, bones of, 558–559, 558f
    of primates, 400, 401f
    upper, bones of, 558, 558f
Limb buds, 699, 699f
Limbic system, 527, 527f, 539
Limiting factors, on population, 711
Limulus amoebocyte lysate, 782
Linkage groups, 176
Linkage maps, 176, 176f
Linnaeus, Carolus, 290
Linnean classification of organisms,
    290, 291–292, 291f, 293, 294–295
Lions, 393f, 732, 788
Lipase, 619, 619f
Lipids
    as class of organic molecule, 38
    in diet, 623, 623t
    fats and oils, 43, 43f
    phospholipids, structure of, 44, 44f
    steroids, structure of, 44, 44f
    subunits of, 40f
    synthesis of
        endomembrane system in,
            67, 67f
        endoplasmic reticulum in, 64
        Golgi apparatus in, 65
        waxes, structure of, 44, 44f
Lipoproteins, 584
Liposomes, 310, 310f
Little bluestem (Schizachyrium
    scoparium), 443, 443f
Liver
    cancer of, 305
    disorders of, 620
    functions of, 580, 619, 620, 650,
        658, 659f
    glycogen storage in, 42
    and homeostasis, 507
    structure of, 620
Liverworts, 346, 346f

Lizards, 390, 390f
  on Galápagos Islands, 245, 268
  reproduction, 682
  speciation in, 268, 269, 269f
LM. See Light micrograph
Loam, 453
Lobe(s), of brain, 525, 525f
Lobe-finned fishes, 389
Lobsters, 548, 548f, 573, 573f
Lodgepole pine (Pinus contorta), 459
Logistic growth, 714, 714f, 717
Long-day plants, 470, 471
Longhorn beetle, 745, 745f
Long-term memory, 527
Long-term potentiation (LTP), 527
Loop of Henle. See Loop of the nephron
Loop of the nephron (loop of Henle),
  654, 655f, 656f, 658, 658f
Loose fibrous connective tissue,
  498, 498f
Lophotrochozoa, evolution of, 372, 372f
Lordosis (swayback), 557
Lorenzo's Oil (film), 66
Loss-of-function mutations, 218
Low-density lipoprotein (LDL), 583, 627
LTP. See Long-term potentiation
Lucy, 398–399, 398f, 406, 406f
Lucy's baby, 399, 406
Lumbar nerves, 516f
Lumbar vertebrae, 556f, 557
Lumber, 356–357
Lumen
  of blood vessels, 497
  of ER, 63, 63f, 64, 64f
Lungs
  cancer of, 133f, 144, 201, 639,
    644, 644f
  evolution of, 387, 389
  gas exchange in, 88
  human, 638, 638f
    diseases of, 644, 644f
    smoking and, 639
  structure of, 634, 635, 635f
Lupus, 605, 605f
Luteal phase, of ovarian cycle, 688, 688f
Luteinizing hormone (LH), 669t, 670f,
  671, 685, 688, 688f
Lycopodium (Ground pine), 348f
Lye (sodium hydroxide), 30
Lyell, Charles, 245, 247
Lyme disease, 383, 709
Lymph, 505, 587, 594
Lymphatic capillaries, 587, 587f,
  594, 594f
Lymphatic duct, 594
Lymphatic organs, 594f, 595, 595f
Lymphatic system, 505, 505f, 594–595
  bones and, 555
  functions of, 594
  and homeostasis, 507
  organs, 594f, 595, 595f
Lymphatic vessels, 594, 594f
Lymph nodes, 505, 594f, 595, 595f
Lymphocytes, 584f, 585, 594, 595f,
  600–603, 600t
Lynx, Canadian, 744, 744f
Lysine, 46f
Lysis, definition of, 89
Lysogenic cells, 303
Lysogenic cycle, 303, 303f
Lysosomal storage diseases, 66
Lysosomes, 60, 60f, 66, 66f. See also
  Endomembrane system
Lysozyme, 303, 596, 616
Lytic cycle, 303, 303f

M

MAC. See Mycobacterium avium
  complex
MacArthur, Robert, 743, 749
Macaws (Ara macao), 785, 785f

Macrocystis, 333
Macroevolution, 264
Macromolecules, 40, 40f
Macronucleus, of ciliate, 329, 329f
Macronutrients, 450, 451t
Macrophages, 585, 596, 596f, 597, 598,
  598f, 600, 600t, 602–603, 602f
Macular degeneration, 541
Madagascar, 781, 789
Madagascar periwinkle, 474
Maggots, 613, 741
Magnesium, dietary requirements, 624
Maiasaura, 283, 283f
Maidenhair fern (Adiantum pedantum),
  348, 348f
Maintenance systems, in humans,
  505, 505f
Maize, 272
Major histocompatibility complex
  (MHC) protein, 602
Makela, Bob, 282–283
Malaria
  avian, 779
  sickle-cell disease and, 173,
    258, 258f
  spread of, 322, 322f, 323f, 330, 330f
  treatment of, 357, 474, 488
  vaccine for, 601
Male
  competition for mates,
    730–731, 731f
  orgasm, 684
  reproductive system, human,
    684–685, 684f, 685f
Male gametophyte, 351, 351f, 354f–355f,
  481, 481f, 483
Malignancy, 143
Malleus (hammer), 544, 544f, 545
Malpighi, Marcello, 448
Malpighian tubules, 651
Maltase, 619, 619f
Malthus, Thomas, 246, 246f, 247
Maltose, formation and breakdown of,
  41, 41f
Mammal(s)
  aquatic, diving in, 632–633
  characteristics of, 373, 392–393
  chemoreceptors in, 538–539,
    538f, 539f
  circulatory system, 576, 576f
  ear of, anatomy, 544, 544f
  evolution of, 284, 285t, 288f–289f,
    387f, 392
  excretory system, 650
  homeostasis in, 495
  kidneys, structure of, 654–655, 654f
  marsupials, 287, 392, 392f
  monotremes, 392, 392f
  origin of, 250
  osmoregulation in, 653
  pheromones and, 664–665
  placental, 392–393, 393f
    evolution of, 287, 291
    reproduction, 683
  reproduction, overview of, 681
  respiration in, 635, 635f
  skeleton, functions of, 555
  smell receptors in, 539, 539f
  taste receptors in, 536,
    538–539, 538f
  terrestrial, respiration in, 634, 635f
Mammary glands, 392
Manatees, evolution of, 277
Mandible, 557, 557f
Mangold, Hilde, 696
Mangrove swamps, 772, 772f, 785
Manometers, 581
Mantle, of mollusc, 380, 380f
Marburg fever, 318
Marfan syndrome, 173, 173f
Marijuana

effects of, 523
  urine tests for, 657
Marine bony fishes, 652, 652f
Marine ecosystems, 772–773, 772f,
  773f, 785
Marine snail (Aplysia), 727
Mars, life on, 19
Marshall, Barry, 618
Marsupials, 287, 392, 392f
Martins, 743
Masseter, 563f
Mass extinctions. See Extinctions
Mass number. See Atomic mass
Mast cells, 596, 598, 598f
Mastoiditis, 556
Mastoid process, 557f
Mastoid sinuses, 556
Maternal determinants, 695
Mating
  assortative, 254
  in birds, 391
  courtship behavior, 735
  nonrandom, 254
  random, 254
Matrix
  in animal cell, 72, 72f
  in connective tissue, 498, 498f
  mitochondrial, 68, 118, 118f, 119,
    121, 121f
Matter, definition of, 20
Matthei, J. Heinrich, 193
Maturation stage, of viral life cycle, 303,
  305, 305f, 306, 306f
Mature embryo stage, in angiosperms,
  484, 484f
Mature mRNA, 195, 195f, 215f
Maxillae, 557, 557f
Maxillary glands, 651
Mayr, Ernst, 268
McClintock, Barbara, 202, 202f
McDonald, John, 502
M-cyclin, 142, 142f
MDCs. See More-developed countries
MDMA (methylenedioxy-
  methamphetamine), 522
Meadow Brown butterfly, 271
Mealybug, 384f
Measles, 305, 599
Mechanical digestion, 612, 616
Mechanical energy, as energy type, 78
Mechanical isolation, as reproductive
  barrier, 267
Mechanoreceptors, 536, 544, 545
Median nerve, 516f
Medical imaging, 22, 22f
Medicine. See also Disease;
    Pharmaceuticals
  medicinal plants, 348, 357
  monoclonal antibodies, 604
  prokaryotes, important, 317, 317t
  radiation in, 22, 22f
  transplants
    bone marrow, 586
    and tissue rejection, 605
    xenotransplantation, 231, 394,
      506, 605
  value of wildlife, 782, 782f
  vertebrate products in, 394, 394f
Medulla oblongata, 515f, 524f,
  526, 641
Medullary cavity, 560, 560f
Medusa body form, in cnidarians,
  375, 375f
Meerkats, 733, 733f
Megakaryocytes, 585
Megaspores, 351, 351f, 354, 354f–355f,
  481, 481f, 482f
Meiosis, 145–151
  chromosome separation in, 145, 145f
  in life cycle, 150, 150f

meiosis I, 145, 145f, 146, 146f, 147f,
    148, 148f–149f, 151, 151f
  nondisjunction in, 152, 152f
  in oogenesis, human, 687, 687f
  meiosis II, 145, 145f, 147f,
    148f–149f, 149, 151, 151f
  nondisjunction in, 152, 152f
  in oogenesis, human, 687, 687f
  Mendel's laws and, 163, 164
  nondisjunction in, 152, 152f
  in plant life cycle, 344, 344f, 481,
    481f, 482f
  vs. mitosis, 151, 151f
Melanin, 200, 503
Melanocytes, 503
Melanocyte-stimulating hormone
  (MSH), 669t, 671
Melanoma, treatment of, 233, 233f
MELAS (mitochondrial myopathy,
  encephalopathy, lactic acidosis,
  stroke) syndrome, 69
Melatonin, 526, 669t, 675, 675f
Membrane attack complex, 596
Membrane potential, 518
Memory
  brain areas, 527
  types of, 527
Memory B cells, 600, 600t, 601, 601f
Memory T cells, 600, 600t, 600t, 602,
  602f, 603
Mendel, Gregor, 160–162, 160f, 161f,
  162f, 164, 164f, 166, 166f, 240
Mendel's laws
  human genetic inheritance and,
    167–170
  law of independent assortment,
    164–166
  law of segregation, 162–163,
    162f, 171
  and laws of probability, 165
Meninges, 524, 524f
Meningitis, 524
Meniscus (menisci), 561, 561f
Menstrual cycle, 665, 688, 688f
Menstruation. See Menstrual cycle
Meristem, 430, 432, 432f
  apical, 426, 430, 430f, 431, 431f,
    463, 484
  and cloning of plants, 488, 488f
  ground, 430, 430f, 431, 431f, 484
Merozoites, 330f
Mesocarp, 485
Mesoderm, 371, 371f, 376, 379, 692f,
  693, 693f, 693t, 694, 694f, 695,
  696, 696f, 699
Mesoglea, 375, 375f
Mesopelagic zone, 773, 773f
Mesophyll
  in angiosperms, 428, 429f, 434
  in C3 plants, 97, 97f, 102,
    106, 107f
  in C4 plants, 106, 107f
  palisade, 428, 434, 434f, 435f
  spongy, 428, 434, 434f, 435f
Mesozoic era, 284, 285t
Messenger RNA (mRNA). See mRNA
Metabolic pathways. See also Cellular
  respiration; Fermentation
  regulation of, 84, 84f
Metabolic pool, 125, 125f
Metabolism
  as characteristic of life, 7
  definition of, 7
  origin of, 310
Metacarpal bones, 556f, 558, 558f
Metamorphosis
  in arthropods, 382, 382f
  in butterflies, 382, 382f
  in tunicates, 386
Metaphase
  in meiosis

metaphase I, 148, 148f–149f, 150, 151f
metaphase II, 148f–149f, 149
in mitosis, 136f, 138f–139f, 151f
Metaphase plate, 138f–139f
Metapopulations, 718, 789
Metastasis, 143, 143f, 220, 220f, 644
Metastatic tumors, 143, 143f
Metatarsal bones, 556f, 558f, 559
Meteorites, and mass extinction, 288, 289
Methadone, 523
Methamphetamine, 522
Methane, as greenhouse gas, 316, 758, 787
Methanogens, 296, 314, 316, 316f
Methanosarcina mazei, 8f
Methionine, 46f, 465
MHC (major histocompatibility complex) protein, 602, 605
Microbes
    classification of, 301
    as decomposers, 741
Microevolution
    causes of, 253–257
    detection of, 253
    Hardy-Weinberg principle, 253
Microglia, 501, 501f, 517
Micrographs, definition of, 58
Micronucleus, of ciliate, 329, 329f
Micronutrients, 450, 451t
Microscopes
    resolution of, 58, 58f
    types of, 58, 58f
    visual range of, 57, 57f
Microspheres, and origin of cellular life, 310, 310f
Microspores, 351, 351f, 354, 354f–355f, 481, 481f, 482f
Microtubule organizing center (MTOC), 70
Microtubules, 60f, 61, 61f, 70–71, 70f
Microvilli
    actin in, 71
    in intestines, 57, 70, 619, 619f
    in taste buds, 538–539, 538f
Midbrain, 513, 515, 515f, 524f, 526
Middle ear, 544, 544f, 545, 546
Middle lamella, 72, 72f
Miescher, Johann Friedrich, 186
Mifepristone, 690
Migration, of birds, 391
Miller, Stanley, 308, 308f
Millipedes, 372f, 383, 383f
Mimicry, as prey defense, 745, 745f
Mimosa pudica, 468, 468f
Mineralocorticoids, 669t, 672–673, 672f
Minerals
    definition of, 20
    dietary requirements, 624, 624t
    plant absorption of, 453, 453f
    as plant nutrients, 450–451, 451t
    root uptake of, 452, 452f
Miocene epoch, 285t
Mitchell, Peter, 101
Mites, 383
Mitochondria
    ATP generation in, 68, 101, 112–113, 114–115, 115, 115f, 118–122
    in cell anatomy, 60f, 61f
    communication by, 60–61
    function of, 68
    in muscle, 69f, 121
    origin of, 61, 63, 324, 324f
    structure of, 68, 68f, 114f, 118, 118f
Mitochondrial DNA, 68
    human evolution and, 411, 414
    mammalian, genetic code, 193
    mutation in
        and disease, 59
        inheritance patterns, 69
    and phylogenetic research, 293

Mitochondrial Eve hypothesis, 411
Mitosis, 136–137, 136f, 138–139, 138f–139f
    in life cycle, 150, 150f
    M-stage checkpoint, 141, 141f
    in plant life cycle, 344, 344f, 481f, 482f
    vs. meiosis, 151, 151f
Mitotic spindle, 138
M-kinase, 142, 142f
Model(s)
    atomic, 23, 23f
    molecular, 25, 25f, 26f
    in science, 21
Model of island biogeography, 749, 749f
Model organisms, in genome sequencing, 234, 234t
Molars, 616, 616f
Molds. *See* Slime molds; Water molds
Molecular biology, evidence for evolution, 252, 252f
Molecular clocks, and phylogenetic research, 293
Molecular data, and phylogenetic research, 292–293, 293f
Molecular models, 25, 25f, 26f
Molecule(s)
    definition of, 23
    formation of, 23
    hydrophilic, 28, 39
    hydrophobic, 28, 39
    as level of organization, 4, 5f
Mole rats, 724–725
Molluscs, 380, 380f
    blood of, 573
    characteristics of, 373
    circulatory system, 575
    evolution of, 370f, 372f
    exoskeleton, 554
    nervous system, 514–515, 514f
    respiration in, 636
Molting, in arthropods, 382
Monarch butterfly, 745
Monarch caterpillars, 472, 472f
Mongooses, as alien species, 786, 786f
Monkeys
    baboons, 403, 731, 731f, 734–735
    characteristics of, 400–401
    evolution of, 403, 403f, 415
    new world, 400f, 402f, 403
    old world, 400, 400f, 402f, 403
    skeletal system, 403
Monocots, 353, 424, 424f, 425, 425f
    embryo development in, 484
    flowers, 424, 424f, 480, 480f
    leaves, 424, 424f, 434
    roots, 424, 424f
    seeds, 424f
    stems, 424f, 428, 429f
Monocotyledones, 353
Monocytes, 584f, 585, 595f, 598
Monod, Jacques, 208
Monoecious animals, 682
Monoecious plants, 480
Monohybrid crosses, 162, 162f
Monomers, 40, 40f
Mononucleosis, 305
Monosaccharides, 41
Monosomy, 152
Monotremes, 392, 392f
Monsaccharides, 41
Monsoons, origin of, 770
Mons pubis, 686, 686f
Montane coniferous forest, 765, 770, 770f
Monterey Pine, 718f
More-developed countries (MDCs)
    age distributions in, 720, 720f
    population growth in, 719, 719f
Morels, 360
Morgan, Charles H., 178

Morgan, J. P., 178
Morgan, Thomas H., 174, 176, 178, 178f, 240
Morning-after pill, 690
Morning glory plants (*Ipomoea*), 467, 467f, 469, 469f
Morphine
    and neurotransmitters, 521, 523
    as plant defense, 472, 474
Morphogenesis, 695, 696, 696f
Morphogens, 216
Morula, 692, 692f, 698
Mosaic evolution, 406
Mosquitoes, as disease vectors, 307, 322, 322f, 330, 330f, 378
Moss(es), 346, 346f
    diversity of, 346
    life cycle, 347, 347f
Mothers, surrogate, 691
Moths
    chemical communication, 734
    chemoreceptors in, 538, 538f
    industrial melanism in, 248, 248f
    natural selection in, 248, 248f
    pheromones in, 266
    as pollinators, 243
    wings, 206–207, 206f, 207f
Motion sickness, 548
Motor axons, 564, 567, 567f
Motor molecules, 70
Motor (efferent) neurons, 517
Motor output system, in humans, 504, 504f
Motor speech area, 525f
Mountain(s), and climate, 770, 770f
Mountain gorillas, 717, 717f, 718, 718f
Mountain zone ecosystem, 764f
Mount St.. Helens, 458–459, 458f–459f
Mouse (*Mus musculus*)
    genome, sequencing of, 49, 234f
    transgenic, in research, 231, 231f
Mouth, and digestion, 616, 616f
MPTP, 84
mRNA (messenger RNA)
    function of, 62, 214, 241
    mature, 195, 195f, 215f
    posttranscriptional processing of, 195, 195f, 214, 214f, 215, 215f
    primary, 195, 195f, 215f
    in transcription, 192, 192f, 194, 194f, 199, 199f
    in translation, 62, 63, 63f, 196–198, 196f, 197f, 199, 199f
    and translational controls, 214
mRNA transcript, 194, 194f
MRSA (methicillin-resistant *Staphylococcus aureus*), 317
MSH. *See* Melanocyte-stimulating hormone
M stage (mitotic stage), 136–137, 136f, 138–139, 138f–139f
M stage (mitotic stage) checkpoint, 141, 141f
MTOC. *See* Microtubule organizing center
Mucous membranes, immune system and, 596, 596f
Mucus, in stomach, 618
Mudflats, 772, 772f
Mules, 263, 267, 267f
Müller, Fritz, 745
Müllerian mimics, 745
Multifactorial traits, 172
Multinucleate hypothesis, 369
Multiple alleles, 171
Multiple fruits, 485
Multiple sclerosis, 394, 519, 599
Multiregional continuity model
    of human ethnic diversity, 414
    of human evolution, 410–411, 411f
Mumps, 305, 599, 616

Muscle(s)
    as animal characteristic, 368
    cardiac, 500, 500f
    contraction, 81, 81f, 564–565, 566–568, 566f, 567f
    development of, 694
    skeletal, 500, 500f, 563–568
        antagonistic pairs, 564, 564f
        contraction of, 81, 81f, 564–565, 566–568, 566f, 567f
        fast- and slow-twitch fibers, 568, 568f
        functions, 563, 563f
        motor units, 564–565, 564f
        muscle cells, 566, 566f
        neuromuscular junction, 567, 567f
    smooth, 500, 500f
    striation in, 500, 500f, 566
Muscle cells, 566, 566f
Muscle dysmorphia, 628, 628f
Muscle fiber, 500, 500f
Muscle tissue, 500, 500f
Muscular dystrophy, 175
Muscular hydrostats, 554
Muscular system
    and homeostasis, 507
    in humans, 504, 504f
Mushrooms, 361, 361f, 362
Mussels, 380, 380f
Mutagens, 133, 201
Mutation. *See* Genetic mutation
Mutualism, 746t, 748, 748f
    in fungi, 358, 358f, 359, 436, 436f
    herbivore digestive system and, 615
    in lichens, 359
    and plant nutrition, 454, 454f
    in prokaryotes, 314
Myasthenia gravis, 605
Mycelium, 358, 358f
Mycobacterium avium complex (MAC), 593
Mycorrhizae, 359, 359f, 454, 454f, 459, 748
Mycorrhizal fungi, 359, 359f
Mycoses, 362, 362f
Myelin sheath, 501, 501f, 517, 517f
Myofibrils, 566, 566f
Myoglobin, 633
Myopia (nearsightedness), 542, 542f
Myosin, 70, 81, 81f, 500, 566–567, 566f, 567f
Myrmecia pilosula, 14
Myrtle trees (*Myrica faya*), 786
Myxedema, 676

N

NAD⁺, 114, 116, 117f, 118, 119f, 120, 120f, 123, 123f
NADH, 114, 115, 115f, 116, 117f, 118, 119, 119f, 120, 120f, 121, 121f, 122, 122f
NADP⁺, 98, 99, 102–103, 102f, 103f
NADPH
    in Calvin cycle, 104–105, 104f, 105f
    production in light reactions, 98, 99, 101–103, 102f, 103f
NADP reductase, 103, 103f
Naked mole rats (*Heterocephalus glaber*), 725, 725f
Naked seeds, 350, 351
Naked viruses, 302, 320f
Nasal bones, 557, 557f
Nasopharynx, 617, 617f
National Cancer Institute (NCI), 357
National Collegiate Athletic Association, 657
Natriuresis, 658
Natural killer cells, 596f, 597

Natural selection
    Darwin on, 7, 246–247, 248
    directional selection, 256f, 257
    disruptive selection, 256f, 257, 257f
    evidence for, 248, 248f
    and evolution of genetic
      material, 195
    life history and, 717
    and microevolution, 253, 256–258
    overview of, 419
    principles of, 7, 247
    reinforcement of speciation by, 269
    stabilizing selection, 256, 256f, 258
    undirected nature of, 278
    Wallace on, 247
"Nature vs. nurture" question, 726–730
Nautiluses, 380, 380f
Navel, 702
NCI. See National Cancer Institute
NE. See Norepinephrine
Nearsightedness (myopia), 542, 542f
Nectar, 480
Negative feedback mechanisms
    in endocrine system, 666, 674
    in homeostasis, 508, 508f, 707
    in nervous system, 666
Negative gravitropism, 466–467, 466f
Negative phototropism, 467
Negative tropism, 466, 466f
Neisseria gonorrhoeae, 311, 317, 689
Nematocysts, 375, 375f
Nematoda, 378, 378f, 472
Nephridium (nephridia), 381, 381f,
    651, 651f
Nephrons, 654–655, 655f, 658, 658f
Nereocystis, 333
Nerve(s). See also Neurons
    anatomy of, 501, 501f, 528, 528f
    as animal characteristic, 368
    cranial, 516, 516f
    nerve impulses, 518–521, 518f,
      519f, 520f
    regeneration of, 502, 502f
    spinal, 516, 516f
    transverse, 376, 376f
Nerve cells, 55*
Nerve cord, dorsal tubular, as chordate
    characteristic, 386, 386f
Nerve fibers, 517, 528
Nerve net, 514, 514f
Nervous system
    in arthropods, 382
    bacterial diseases, 317t
    and behavior, 727
    development of, 694, 694f, 699
    in echinoderms, 385
    endocrine system and, 666, 666f,
      666t, 668
    evolution of, 514
    in flatworms, 376, 376f
    formation of, 696
    and homeostasis, 507, 706–707
    in humans, 504, 504f, 516, 516f,
      641, 641f
    in invertebrates, 514–515, 514f
    ladderlike, 514
    and muscle contraction, 567
    in vertebrates, 515, 515f
    central nervous system (CNS),
      515, 515f, 516, 516f, 521, 524
    peripheral nervous system
      (PNS), 516, 516f, 528–530,
      528f, 529f, 530f
    autonomic system, 530, 530f
    somatic system, 529, 529f
Nervous tissue, 501, 501f
Neural crest, 694
Neural folds, 699
Neural plate, 694, 694f
Neural tube, 694, 694f, 699
Neurofibrillary tangles, 527

Neurofibril nodes. See Nodes of Ranvier
Neurofibromatosis (von Recklinghausen
    disease), 169
Neurofibromin, 169
Neuroglia, 501, 501f, 517
Neuromodulators, 521
Neuromuscular junction, 567, 567f
Neurons
    action potential
      mechanism of, 518–519,
        518f, 519f
      speed of, 519, 519f
    communication between
      (neurotransmitters),
      520–521, 520f
    excitotoxicity, 527
    functions of, 517
    integration of stimuli, 521, 521f
    refractory period, 519
    resting potential, 518, 518f
    structure of, 501, 501f, 517, 517f
    types of, 517
Neurospora crassa, 137t
Neurotransmitters, 520–521, 520f
    clearing of, 521
    drugs and, 522–523
Neursecretory cells, 670–671, 670f
Neurula, 694, 694f
Neutrons, 21, 21f, 21t
Neutrophils, 584f, 585, 596, 596f, 597,
    598, 602–603
Newts, 389
New world monkeys, 400f, 402f, 403
Niacin, 83, 625t
Niche specialization, 742–743,
    742f, 743f
Nicotiana tabacum (tobacco), 137t
Nicotine
    effects on body, 522
    as plant defense, 472
Nicotinic acid, dietary
    requirements, 625t
Nigeria, population growth in, 720
La Niña, 774
Nirenberg, Marshall, 193
Nitrate, in nitrogen cycle, 757, 757f
Nitrification, 757, 757f
Nitrites, and cancer, 144
Nitrogen
    blood levels, and The Bends, 633
    carbon binding with, 38
    and life, 20, 20f
    plant competition for, 443
Nitrogen cycle, 757, 757f, 784
Nitrogen fertilizers, 12–13, 13f
Nitrogen fixation, 757, 757f
Nitrogen-fixing bacteria, 12–13, 301,
    315, 317, 454
Nitrogenous waste products, types and
    characteristics, 650, 650f
Nitrogen oxides, and acid deposition,
    757, 787
Nitrous oxide, 758
Nobel Prize, 99, 101, 178, 202, 208, 216,
    309, 618, 695
Noble gases, 23
Nociceptors. See Pain receptors
Nodal tissue, 578, 578f
Node of angiosperm, 422, 422f
Nodes of Ranvier, 517, 517f, 519, 519f
Nomenclature, binomial, 9
Noncoding strand, 194, 194f
Noncompetitive inhibition, 84, 84f
Noncyclic pathway, in light
    reactions, 101
Nondisjunction, and chromosome
    abnormalities, 152–153, 152f
Nonionina, 325f
Nonpolar covalent bonds, 26
Nonrandom mating, and
    microevolution, 253, 254

Nonspecific immune defenses,
    596–598, 596f
Nontropic hormones, 671
Nonvascular plants. See Bryophytes
Nonylphenols, 787
Noradrenaline. See Norepinephrine
Norepinephrine (noradrenaline; NE),
    520–521, 530, 669t, 672, 672f
Norrell, Mark, 283
Northern elephant seal (Mirounga
    angustirostris), 632, 632f, 633
Northern spotted owl (Strix occidentalis
    caurina), 718f, 789, 789f
Nose, in respiration, 638, 638f
Notochords, 275, 386, 386f, 694, 694f
Novotny, Milos, 664–665
Nuclear energy, as energy type, 78
Nuclear envelope, 60f, 61f, 62, 62f
Nuclear pores, 61f, 62, 62f
Nucleases, 619
Nucleic acids. See also DNA; RNA
    as class of organic molecule, 38
    definition of, 48
    digestion of, 612, 612f
    origin of term, 48, 186
    subunits of, 40f
Nuclein, 186
Nucleoid, 59, 135, 311, 311f
Nucleolus, 60f, 61f, 62, 62f
Nucleoplasm, 62, 62f
Nucleosomes, 212, 212f
Nucleotidase enzymes, 619
Nucleotides
    digestion of, 619
    genetic code, 193
    structure of, 48, 48f, 186
Nucleus
    anatomy of, 62, 62f
    in animal cell, 60f
    of cancer cell, 143
    communication with cytoplasm, 62
    division of. See Mitosis
    of neuron, 501, 501f
    in plant cell, 61f
Nummulites, 328
Nursing, hormones in, 671
Nusslein-Vollard, Christiane, 216
Nutrients, plant
    beneficial nutrients, 450
    essential nutrients, 450–451, 451f
    and mutualistic relationships,
      454, 454f
Nutrition, 622–628. See also Diet
    carbohydrates, 622, 622t
    lipids, 623, 623t
    minerals, 624, 624t
    nutrition labels, 626, 626f
    in prokaryotes, 314, 314f
    protein, 623
    vitamins, 625, 625t

O

Obelia, 682
Obesity
    causes of, 622, 626
    and diabetes, 674, 675
    diseases associated with, 626–627
Objective lens, 58, 58f
Obligate anaerobes, 314
Observation, in scientific method,
    11, 11f
Obstructive pulmonary disorders, 644
Occipital bone, 556, 556f, 557f
Occipital lobe, 525, 525f
Ocean(s)
    benthos, as biodiversity frontier, 781
    as carbon dioxide sink, 108
    currents, 769, 774, 774f
    diatomaceous earth and, 332
    as ecosystem, 773, 773f
    sediment, 328

temperature conservation properties
    of, 28
Oceanic ridges, 773
Octaploid plants, 271
Octet rule, 23
Octopuses, 380, 380f, 512, 512f, 575
Ocular lens, 58, 58f
Oil(s), structure of, 43, 43f
Oil glands, 503, 503f
OIs. See Opportunistic infections
Okazaki, Reiji, 191
Okazaki fragments, 191, 191f
Oldowan tools, 409
Old world monkeys, 400, 400f, 402f, 403
Olfactory bulb, 515f, 527f, 528f, 539f
Olfactory epithelium, 539f
Olfactory receptors, 539, 539f
Olfactory tract, 527f, 539f
Oligocene epoch, 285t
Oligochaetes, 381
Oligodendrocytes, 501, 501f, 517
Oligosaccharins, 463
Oligotrophic lakes, 771, 771f
Omega-3 unsaturated fatty acids, 627
Omnivores, 393, 610, 611, 752
Oncogenes, 218, 218f, 219, 219f
One-trait testcrosses, 166, 166f
One-way ventilation mechanism,
    641, 641f
On the Origin of Species (Darwin), 247
Oocyte, human, 686, 687, 687f
Oogamous, defined, 336
Oogenesis, 150, 687, 687f
Oomycota, 331
Open circulatory system, 575, 575f
Operant conditioning, 729
Operators, 208, 208f
Operculum, 636, 636f
Operons, 208, 208f
Ophrys apifera, 243, 243f
Opiates, 523, 657
Opium, 523
Opossums, 287, 392, 392f
Opossum shrimp (Mysis relicta),
    781, 781f
Opportunistic infections (OIs), AIDS
    and, 529f, 592–593, 593f
Opportunistic populations, 717, 717f
Opportunity Mars rover, 19
Opposable thumbs, in primates, 400, 401f
Opsin, 542–543
Optic lobe, 515f
Optic nerve, 528f
Oral thrush, 362, 362f
Orangutans, 400f, 402f
Orbicularis oculi, 563f
Orbicularis oris, 563f
Orca (killer whale; Orcinus orca), 264f,
    393f, 788
Order
    as characteristic of life, 6
    as taxon, 8, 8t, 290, 290f
Ordovician period, 284, 285t, 288, 288f
Organ(s). See also Organ systems
    artificial, 506, 506f
    as level of organization, 4, 5f, 379,
      496, 496f
    structure and function of, 503
    transplantation of, 506–507
Organelles
    in animal cells, 60f
    communication between, 60–61
    definition of, 60
    energy-related, 60–61, 68–69
    origin of, 61
    in plant cells, 61f
    replication, in cell cycle, 136
    variation in, 73
Organic, definition of, 50
Organic chemicals, environmental
    impact of, 787

Organic chemistry, 38, 38t
Organic farming, and rotation of crops, 13
Organic molecules
    backbone of, 39
    carbon chain of, 38, 38f, 39
    classes of, 38
    diversity of, 38, 38f
    functional groups of, 39, 39f
    isomers, 39, 39f
    origin of, 308–309, 308f
Organisms
    characteristics of, 6–7, 6f, 7f
    diversity of, 4, 4f, 38, 38f
    elements basic to, 20, 20f
    energy needs of, 78
    as level of organization, 4, 5f, 496, 496f
    molecules of. See Organic molecules
    pH range required by, 31
Organization, levels of, in organisms, 4, 5f
Organ of Corti, 545, 546, 546f
Organ systems
    and homeostasis, 507
    as level of organization, 4, 5f, 376, 496, 496f
    overview of, 504–505, 504f, 505f
Orgasm
    female, 686
    male, 684
Orphrys apitera, 748
Oscillatoria, 315f
Osmoregulation
    hormones and, 668
    in mammals, 656–659
    overview of, 648
    in vertebrates
        aquatic, 652, 652f
        terrestrial, 653, 653f
Osmosis, 89, 89f
Osteichthyes. See Bony fishes
Osteoblasts, 559
Osteoclasts, 559
Osteocytes, 560, 560f
Osteons, 498f, 499, 560, 560f
Osteoporosis, 624, 676
    prevention of, 559, 559f, 565
    treatment of, 676
Ostium (ostia), 575, 575f
Otolith(s), 269, 547, 547f
Otolithic membrane, 547, 547f
Otosclerosis, 546
Outbreak (film), 307
Outer ear, 544, 544f, 545
Out-of-Africa hypothesis, 411
Oval window, 544, 544f, 545, 545f
Ovarian cycle, 688, 688f
Ovary(ies)
    angiosperm, 353, 353f, 354f–355f, 480, 480f, 481, 482f
    human, 686, 686f
        development of, 700
        in endocrine system, 668, 668f, 669t
        functions of, 682
        and hormones, 667, 687
Overexploitation, and extinctions, 785f, 787–788
Overpopulation, types of, 720
Oviducts, 686, 686f
Oviraptor, 283
Ovoviviparous animals, 683
Ovulation, 686, 687, 687f, 688, 688f
Ovules, 345, 345f, 353, 353f, 354, 354f–355f, 480, 480f, 481, 482f
Oxidation
    in cellular respiration, 114, 116, 117f, 119
    definition of, 98

Oxygen
    carbon binding with, 38
    in cellular respiration, 120, 120f
    in early atmosphere, 315, 317
    and life, 20, 20f
    as photosynthesis product, 98, 102
    in respiration, 642, 642f, 643
Oxygen debt, 123, 123f, 568
Oxyhemoglobin, 643
Oxytocin, 669t, 670–671, 670f
Oysters, 380, 683, 713, 713f
Ozone depletion, 787

P
p16 gene, 219
p53, 141, 218, 219
Pacific yew (Taxus brevifolia), 472, 718f
Pain receptors (nociceptors), 536
pair-rule genes, 216, 216f
Pakicetus, 250
Pakicetus attocki, 264f
Pakistan, erosion in, 784
Palade, George, 65
Paleocene epoch, 285t
Paleontology, 245, 249
Paleozoic era, 284, 285t
Palisade mesophyll, 428, 434, 434f, 435f
Palm trees (Arecaceae), 357, 357f
Palolo worms, reproduction, 682–683
Pancreas
    cancer of, 133f
    in endocrine system, 668, 668f, 674, 674f
    functions of, 619, 620, 620f, 669t, 674, 674f
Pancreatic amylase, 619, 619f, 620
Pancreatic islets (islets of Langerhans), 674
Pancreatic juice, 620
Pangaea, 286, 287f
Pantothenic acid, dietary requirements, 625t
Papillae, 538, 538f
Parallel evolution, 292
Paramecium, 4, 4f, 329, 329f, 742, 742f
Paramecium aurelia, 742, 742f
Paramecium caudatum, 742, 742f, 744
Paramylon, 327
Paraplegia, 524
Parapodia, 634
Parasite, 746
Parasitism, 746, 746t
    in arachnids, 383
    and coevolution, 747
    ectoparasites, 746
    endoparasites, 746
    in flatworms, 377, 377f
    fluid feeders, 613, 613f
    and population control, 716
    as predation, 744
    prey defenses, 744–745, 745f
    in prokaryotes, 314
    in roundworms, 378
Parasitoids, 744
Parasympathetic division, 530, 530f
Parathyroid glands, 668, 668f, 669t, 676
Parathyroid hormone (PTH), 668, 669t, 676
Parenchyma cells, 426, 426f, 428, 430, 431, 432, 434
Parent cells, 137
Parietal bone, 556, 556f, 557f
Parietal lobes, 525, 525f
Parkinson disease, 69, 501, 521, 525
Parthenocissus quinquefolia (Virginia creeper), 290, 290f
Parthenogenesis, 682
Partially hydrogenated oils, 627
Partial pressure, and respiration, 642
Particulate theory of inheritance, 161, 162–163

Parturition, 702, 702f
Parvovirus, 305
Passion flowers (Passiflora), 473
Passive immunity, 599
Passive transport, 88, 88f
Pasteur, Louis, 124
Patagonian hare, 252
Patella, 556f
Pattern formation, 695
Pavlov, Ivan, 729
Pax6 gene, 276, 276f, 277f
PCBs, 301
PCP, urine tests for, 657
PCR. See Polymerase chain reaction
Peach-faced lovebirds (Agapornis roseicollis), 726, 726f
Pearls, 380
Peas. See Pisum sativum (garden pea)
Peat, 346
Peat moss, 346
Pectin, 72
Pectoral girdle, 556f, 558, 558f
Pectoralis major, 563f
Pedigrees, 167, 167f
Pelagic zone, 772f, 773, 773f
Pellagra, 83
Pellicle, 329, 329f
Pelvic fin genes, 277
Pelvic girdle, 556f, 558, 558f
Pelvic inflammatory disease (PID), 689
Penetration stage, of viral life cycle, 303, 305, 305f, 306, 306f
Penicillin, 362, 782
Penicillium, 362
Penis
    anatomy of, 684, 684f
    function of, 680–681
Pentaploids, 152
Pepsin, 618
Peptidases, 619, 619f
Peptide(s), definition of, 45
Peptide bonds, 45, 45f
Peptide hormones, action of, 667, 667f
Peptidoglycan, 38f, 42, 311
Percussion vests, 168, 168f
Peregrine falcons, 664
Perennial plants, 423
Perforins, 602, 603f
Pericarp, 485, 485f
Pericycle, 429
Periderm, 426f, 432, 432f
Period gene, 735
Periodic table, 23, 23f
Periodontitis, 617
Periosteum, 560, 560f
Peripheral nervous system (PNS), 516, 516f. See also Sensory receptors
Perissodactyla, 393
Peristalsis, 617
Peritonitis, 621
Peritubular capillary network, 655, 655f, 656, 656f
Permafrost, 765
Permian period, 285t, 286, 287f, 288, 288f–289f, 289
Permian-Triassic (P-Tr) extinction, 289
Peroneal nerve, 516f
Peroxisomes, 60f, 66
Personal Genome Project, 234
Peruvian Amazon, economic exploitation of, 783
PERV. See Porcine endogenous retrovirus
Pest control, biological, 783
Pesticides, resistance to, 248
PET. See Positron-emission tomography
Petals, 353, 353f, 480, 480f
Petiole, 422f, 423
Pet trade, 787–788
Peyer patches, 595

pH
    of blood, 31, 659
        regulation of, 507, 584, 643, 659
    buffers, 31
    and enzymatic reactions, 83
    and life, required range for, 31
    pH scale, 31, 31f
    of small intestine, 619
    of stomach, 618
    of urine, 659
Phages. See Bacteriophages
Phagocytes, immune system and, 596f, 597, 598
Phagocytosis, 90, 328
Phalanges, 556f, 558, 558f, 559
Pharmaceuticals. See also Drug abuse
    anticancer drugs, 472, 474, 546, 559, 603
    anti-inflammatory drugs, 598
    antiviral drugs, and herpes, 689
    cell suspension culture and, 488
    cnidarians as source of, 375, 375f
    gene pharming, 231
    immunosuppressive drugs, 605
    medicinal plants, 357, 472, 474, 474f
    research
        Human Genome Project and, 235
        proteomics and, 236
    tropical rain forest as source of, 472, 474, 474f
    vertebrate products, 394, 394f
    wildlife, importance of, 782
Pharyngeal pouches, 386, 386f, 699f
Pharynx, 617, 617f, 638, 638f
Phenotype(s)
    definition of, 163
    frequency distribution of, 256, 256f
    genetic mutation and, 200
Phenylalanine, 46f, 168, 200
Phenylketonuria (PKU), 168, 200, 232
Pheromones
    chemoreceptors and, 538, 538f
    as communication, 734
    humans and, 665
    and insect behavior, 3, 664
    mammals and, 664–665, 725
    as reproductive barrier, 266
Philippines, erosion in, 784
Phloem, 424, 424f, 427, 427f, 429, 429f, 430f, 431, 431f, 432, 432f, 436, 436f, 442
Phloem rays, 432, 432f
Phloem sap, 442
Phosphate
    as biologically important ion, 24t
    blood levels, regulation of, 676
Phosphate groups, 39f
Phospholipid(s), structure of, 44, 44f
Phospholipid bilayer
    proteins embedded in, 85–87, 85f, 86f
    structure of, 85, 85f
Phosphorus
    carbon binding with, 38
    dietary requirements, 624
    and life, 20, 20f
Phosphorus cycle, 756–757, 756f, 784
Photoautotrophs, 314, 752
Photomicrograph, 58
Photoperiod
    and biological clock, 469
    definition of, 469
    plant response to, 466, 470, 470f, 493
Photoperiodism, 470–471, 470f, 471f
Photopigment proteins, 87
Photoreceptors, 536, 540. See also Cone cells; Rod cells

Photosynthesis
 absorption spectrums of
  photosynthetic pigments,
  100, 100*f*
 ATP production in, 101–103,
  101*f*–103*f*
 C3, 106, 106*f*
 C4, 106–107, 107*f*
 Calvin Cycle reactions
  in C4 plants, 106, 107*f*
  in CAM plants, 107, 107*f*
  overview of, 99, 99*f*, 102*f*
  phases of, 104–105, 104*f*, 105*f*
 CAM, 107, 107*f*
 chemical equation for, 25
 chloroplasts in, 68, 97, 97*f*
 in cyanobacteria, 296, 314, 315
 equation for, 98
 evolution of, 95
 leaf structure and, 434, 434*f*,
  435, 435*f*
 light reactions in, 99–103
  absorption of solar energy,
   100, 100*f*
  overview, 99, 99*f*
  production of ATP and NADPH,
   101–103, 101*f*–103*f*
 and oxygenation of atmosphere, 310
 oxygen released in, 98, 102
 photosystems in, 101–103, 102*f*, 103*f*
 phytoplankton and, 332
 as redox reaction, 98
 as source of energy on Earth, 7, 10,
  10*f*, 76–77, 96
 stomata and, 435–436
 tropisms and, 436, 436*f*
 and water, 98, 102, 445
Photosynthetic cyanobacteria, 59
Photosynthetic organisms, 752
Photosynthetic pigments, absorption
 spectrums of, 100, 100*f*
Photosynthetic plants, as type of
 organism, 4, 4*f*
Photosystem I (PS I), 101–103, 102*f*, 103*f*
Photosystem II (PS II), 101–103,
 102*f*, 103*f*
Phototropism, 460–461, 461*f*, 467, 467*f*
Phrenic nerve, 641*f*
pH scale, 31, 31*f*
Phycocyanin, 315
Phylogenetic cladistics, 294, 294*f*,
 295, 295*f*
Phylogenetic (evolutionary) trees, 291,
 291*f*, 295, 295*f*
Phylogeny
 definition of, 291
 Linnean classification and, 291
 phylogenetic trees, 291, 291*f*,
  295, 295*f*
 tracing of, 292–293, 292*f*, 293*f*
Phylum, 8, 8*t*, 290, 290*f*
Phytochrome, 471, 472*f*
Phytophthora infestans, 331
Phytoplankton, 332, 752, 772, 773
Phytoremediation, 447
PID. *See* Pelvic inflammatory disease
Pigs
 feral, environmental damage from,
  779, 786
 as organ source, 506
Pikaia, 275
Pileated woodpecker, 391*f*
Pillbugs, 651
Pilocene epoch, 285*t*
Pineal gland, 524*f*, 526, 668, 668*f*, 669*t*,
 675, 675*f*
Pine trees, life cycle of, 351, 351*f*
Pinkeye, viral, 305
Pinna, 544, 544*f*
Pinocytosis, 90
Pinworms, 378

Pioneer species, 750
Pisum sativum (garden pea)
 anatomy, 254, 254*f*
 assortative mating in, 254, 254*f*
 chromosome number, 137*t*
 Mendel's experiments with,
  160–162, 160*f*, 161*f*, 162*f*, 164,
  164*f*, 166, 166*f*, 240
Pith, in angiosperms, 428, 429*f*,
 432, 432*f*
Pituitary glands, 515, 515*f*, 524*f*, 669*t*
 anterior, 669*t*, 670*f*, 671, 707
 posterior, 669*t*, 670–671, 670*f*
PKU. *See* Phenylketonuria
Placenta, 392, 683, 698*f*, 699, 700, 700*f*,
 702, 702*f*
Placental mammals, 287, 291, 392–393,
 393*f*, 683
Plague, 318
Planarians, 376, 376*f*, 514, 514*f*
 chemoreceptors in, 538
 circulatory system, lack of,
  574, 574*f*
 digestive system, 614, 614*f*
 excretory system, 651, 651*f*
 reproduction, 682
 respiration in, 634
Plankton, 324
Plant(s). *See also* Angiosperms;
 Bryophytes; Ferns; Gymnosperms;
 Photosynthetic plants; Vascular
 plants
 adaptation to land environment,
  242–243, 242*f*, 243*f*
 adaptive radiation in, 440–441
 as amino acid source, 301
 artificial selection in, 246, 246*f*
 asexual reproduction in, 487–488,
  487*f*, 488*f*
 biotic environment, response to,
  472–473, 472*f*, 473*f*
 carnivorous, 340–341, 341*f*, 342*f*,
  454, 454*f*
 characteristics of, 343
 chromosome numbers, 137*t*
 classification of, 296
 cloning of, 209, 209*f*, 488, 488*f*
 color of, 94–95
 of coniferous forest, 765
 defense mechanisms, 71, 436,
  472–473, 472*f*, 473*f*
 of desert, 767
 dioecious, 480
 energy storage in, 42, 42*f*, 43, 43*f*
 environmental response by, 466–473
  circadian rhythms,
   468–469, 469*f*
  biological clock and, 469
  definition of, 466
  and plant stoma, 446
  sleep movements, 468–469, 469*f*
  tropisms, 436, 436*f*, 466–467,
   466*f*, 467*f*
  turgor movements, 468, 468*f*
 evolution of, 95, 284, 285*t*, 341–342,
  341*f*, 342*f*, 343, 343*f*, 348,
  350, 353
 flowering. *See* Angiosperms
 genetically modified, 229, 229*f*,
  230, 230*f*
 global warming and, 420–421
 growth
  primary, 430–431, 430*f*, 431*f*
  secondary, 432–433, 432*f*, 433*f*
 herbaceous, 428
 homeostasis in, 435–436, 435*f*, 436*f*,
  492–493
 hormones. *See* Plant hormones
 as kingdom, 9, 9*f*
 life cycle of, 150, 150*f*, 344, 344*f*
 of marine ecosystems, 772, 772*f*

 mineral requirements, determining,
  450–451
 monoecious, 480
 mycorrhizae and, 359, 359*f*, 454,
  454*f*, 459, 748
 nutrients
  beneficial nutrients, 450
  essential nutrients,
   450–451, 451*f*
  and mutualistic relationships,
   454, 454*f*
 parasitism in, 746
 perennial, 423
 photoperiodism in, 470–471,
  470*f*, 471*f*
 phototropism in, 460–461, 461*f*
 phytoremediation, 447
 plant-animal mutualism, 473, 473*f*
 polyploid, 152, 271, 271*f*
 of savanna, 767
 senescence in, 463
 similarities with animals, 36–37
 species, number of, 341, 780*f*
 sporophyte dominance in, 344–345
 sucrose transport in, 41
 of temperate deciduous forest, 766
 of temperate grassland, 766, 766*f*
 and terrestrial ecosystems, 764
 of tropical rain forest, 768, 768*f*
 in tundra biosystem, 765
 variegation in, 304, 304*f*
 viral diseases, 304, 304*f*
 water transport in, 27, 27*f*
 wound responses in, 472–473, 473*f*
Plant cells
 anatomy, 61, 61*f*
 cell cycle in, 136–140, 136*f*,
  138*f*–139*f*, 140*f*
 cell wall of, 61, 61*f*, 72, 72*f*
 central vacuole of, 61*f*, 66–67, 66*f*
 cytokinesis in, 140, 140*f*
 mitosis in, 138–139, 138*f*–139*f*
 turgor pressure in, 89
Plant hormones, 436, 436*f*
 abscisic acid (ABA), 446, 464, 464*f*
 action of, 460, 460*f*
 auxins, 436, 459, 460–461, 460*f*,
  461*f*, 463, 465, 467, 492
 in cloning of plants, 488
 cytokinins, 463
 definition of, 460
 ethylene, 464, 465, 465*f*, 467
 florigen, 471
 functions of, 459
 gibberellins, 462, 462*f*, 465
 and homeostasis, 492–493
 interplay of, 463
 overview of, 436, 436*f*
Plant tissue culture, 463, 463*f*
Plaque
 in blood vessels, 582, 582*f*, 583,
  583*f*, 627
 dental, 617
Plasma, 584, 584*f*
Plasma cells, 600, 600*f*, 600*t*, 604, 604*f*
Plasma membrane, 85–90
 bacterial, 311, 311*f*
 differential permeability of, 86
 in eukaryotic cells, 60, 60*f*, 61, 61*f*
 origin of, 310
 phospholipid bilayer, structure of,
  85, 85*f*
 in prokaryotic cells, 59, 59*f*, 135,
  135*f*
 proteins embedded in, 85–87,
  85*f*, 86*f*
 of sensory receptors, 537
 structure of, 85–86, 85*f*
 transport across, 88–90
Plasma proteins, 584, 584*f*, 587, 620
Plasmids, 226, 311, 313

Plasmin, 585
Plasmodesma(ta), 72, 72*f*
Plasmodial slime molds, 325*f*, 331, 331*f*
Plasmodium, 322, 323*f*, 330, 747, 747*f*
Plasmolysis, 89
Plastids, 68
Platelets, 499, 499*f*
Plate tectonics, 287
Play, in animals, 736
Pleated sheet, 47, 47*f*
Pleiotropy, 173, 173*f*
Pleistocene epoch, 285*t*
Pleistocene overkill, 412
Pneumocystis carinii, 330
Pneumocystis pneumonia, 593, 593*f*
PNS. *See* Peripheral nervous system
Poachers, 787
Point mutations, 200, 200*f*
Poison(s)
 cellular respiration inhibitors, 120,
  121, 121*f*
 cytochrome blockers, 120
 endotoxins, 317
 exotoxins, 317
 mushrooms, 362
 as plant defense, 71, 436, 472
 shellfish poisoning, 325
Poison-dart frog, 394, 394*f*
Poison ivy, 606
Polar bodies, 170, 170*f*, 687, 687*f*
Polar covalent bonds, 26
Polar ice ecosystem, 764*f*
Polar nuclei, in angiosperms, 354,
 354*f*–355*f*
Polio, 305, 601
Pollen grains, 345, 351, 351*f*, 354,
 354*f*–355*f*, 481, 481*f*, 482*f*
Pollen sac, 482*f*
Pollen tubes, 481, 482*f*, 483
Pollination, 351, 351*f*, 354*f*–355*f*, 355,
 478–479, 482–483, 482*f*, 483*f*
Pollinators, 478–479, 482–483
 adaptation in, 242–243, 242*f*, 243*f*
 coevolution of, 748
 decline in, 355
 evolution of, 353
 mechanical isolation in, 266
 value of, 783, 783*f*
Pollution
 biogeochemical cycles and, 784
 and extinction, 787, 787*f*
 generation of, in more- vs. less-
  developed countries, 720
 nitrogen fertilizers and, 12
 and susceptibility to disease, 788
Poly-A tail, 195, 195*f*, 214
Polychaetes, 381, 634
Polydactylism, in Amish, 255, 255*f*
Polygenic inheritance, 172, 172*f*
Polymer(s), 40, 40*f*
Polymerase chain reaction (PCR),
 227, 227*f*
Polyp(s), in colon, 621
Polyp body form, in cnidarians,
 375, 375*f*
Polypeptide(s)
 definition of, 45
 folding of, 197, 199
 synthesis. *See* Translation
Polyploidy, 152, 271, 271*f*
Polyribosomes, 60*f*, 63, 197
Polysaccharides, 42, 42*f*, 612, 612*f*
Polysomes, 199
Ponds, 771
Pons, 524*f*, 526
Population(s), 711–714
 age structure diagrams, 713, 713*f*
 annual growth rate, 712
 biotic potential of, 712
 carrying capacity and, 714, 714*f*
 definition of, 4, 10, 710

ecological study of, 710
equilibrium, 717, 717f, 718, 749, 749f
growth, patterns of, 714, 714f
human, 719–720
  demographic transition, 719
  doubling time, 719
  and food capacity, 413
  growth of, 719, 719f, 720, 720f
  in more- vs. less-developed countries, 719, 719f, 720, 720f
  overpopulation, types of, 720
as level of organization, 4, 5f
limiting factors, 711
metapopulations, 718
predator-prey dynamics and, 744, 744f
regulation of
  density-dependent factors, 715f, 716, 716f, 717
  density-independent factors, 715, 715f, 717
sink, 789
source, 789
Population density, 711
Population distribution, 711, 711f
Population genetics
  Hardy-Weinberg principle, 253
  microevolution, detection of, 253
Porcine endogenous retrovirus (PERV), 506
Porcupines, 724, 724f, 745
Poriferans, 288f–289f, 374
Porphyra, 333
Porphyria, 173
Portal system, 580
Portuguese man-of-war, 375f
Positive feedback, hormones and, 671
Positive gravitropism, 466–467, 466f
Positive phototropism, 467, 467f
Positive tropism, 466, 466f
Positron-emission tomography (PET), 22, 22f
Posterior compartment, of eye, 540, 540f
Posterior pituitary, 669t, 670–671, 670f
Postnatal tail, as chordate characteristic, 386
Postreproductive age, 713, 713f, 720, 720f
Postsynaptic membrane, 520, 520f
Posttranscriptional control, 214, 214f, 215, 215f
Posttranscriptional processing, 195, 195f, 214, 214f, 215, 215f
Posttranslational controls, 214, 215, 215f
Posttranslational processing, 199, 199f
Postzygotic isolating mechanisms, 267, 267f, 269
Potassium
  as biologically important ion, 24t
  dietary requirements, 624
Potassium gates, action potential and, 519, 519f
Potatoes, genetically modified, 229
Potato famine (Ireland, 1840s), 229, 331
Potential energy, 78, 78f
Powdery mildews, 360
Prairie
  short-grass, 766
  tall-grass, 766
Prairie lupine (Pupinus lepidus), 459
Prayer plant (Maranta leuconeura), 469, 469f
Prebiotic soup hypothesis, 308, 308f
Precambrian time, 284, 285t
Precipitation. See Rain
Predation
  coevolution in, 747
  and jaws, evolution of, 387, 388

and population control, 716, 716f
population dynamics in, 744, 744f
Pregnancy. See also Sexual reproduction, in humans
  ectopic, 686
  false labor, 702
  health habits and, 699
  sexually-transmitted diseases and, 689
  smoking in, 639
  symptoms of, 699, 700
  trimesters, 697
Pregnancy tests, 604
Prehensile hands and feet, in primates, 400, 401f
Premolar teeth (bicuspids), 616, 616f
Preparatory (prep) reaction, 115, 118, 122, 122f
Prereproductive age, 713, 713f, 720, 720f
Pressoreceptors, 537
Press/pulse theory of mass extinctions, 288, 289
Pressure-flow model, 448–449, 449f
Pressure receptors, in skin, 544
Presumptive notochord, 694, 694f
Presynaptic membrane, 520, 520f
Preven, 690
Prey, defenses of, 744–745, 745f
Prezygotic isolating mechanisms, 266–267, 266f, 267f, 269
Primary follicle, 687, 687f
Primary growth, 430–431, 430f, 431f
Primary motor area, 525, 525f
Primary mRNA, 195, 195f, 215f
Primary oocyte, 687, 687f
Primary phloem, 431, 431f
Primary somatosensory area, 525, 525f
Primary structure, of protein, 47, 47f
Primary succession, 750
Primary visual area, 525, 525f
Primary xylem, 431, 431f
Primates
  characteristics of, 400–401, 404
  evolution of, 400
    common ancestor, 402f, 403
    evolutionary tree, 402f, 403
    human-chimpanzee differences, 404–405, 404f
  group living in, 732
  as order, 393
  reproduction, 401
Prime mover, 564
Primitive streak, 693, 693f, 699
Principles of Geology (Lyell), 245
Pritchardia kaalae, 718f
PRL. See Prolactin
Probability, Mendel's laws and, 165
Proboscidea, 393
Procambium, 430, 430f, 431, 431f
Proconsul (fossil), 403, 403f
Producers, 752, 753, 753f
Progesterone, 669t, 687, 688, 688f, 690, 691f, 700
Proglottids, 377, 377f
Prokaryotes, 311–318. See also Archaea; Bacteria
  autotrophic, 314
  chemoheterotrophic, 314
  chromosomes, 59, 135, 135f
  definition of, 8
  environmental importance of, 317
  evolution of, 284, 285t, 286, 286f
  in extreme environments, 83
  gene regulation in, 208, 208f
  initiation in, 198
  medical importance, 317, 317t
  nutrition, 314, 314f
  RBA splicing in, 195
  reproduction, 135, 135f, 312, 312f

shapes, common, 311, 311f
structure of, 311, 311f
Prokaryotic cells
  cell division in, 135, 135f
  protein synthesis in, 311
  ribosomes in, 63
  structure of, 59, 59f
Prolactin (PRL), 669t, 670f, 671, 787
Proliferative phase, of uterine cycle, 688, 688f
Proline, 46f
Promoters, 194, 208, 208f
Proofreading, of DNA, 191, 201
Property, definition of, 20
Prophages, 303
Prophase
  in meiosis
    prophase I, 148, 148f–149f, 151f
    prophase II, 148f–149f, 149
  in mitosis, 136f, 138f–139f, 151f
Proprioceptors, 537
Prosimians, 400, 400f, 401, 403
Prostate cancer, 684
Prostate gland, 684, 684f
Protease, 306
Protective proteins, 596–597, 596f, 597f
Protein(s)
  as amino acid sequence, 45
  carrier, 86, 86f, 88, 88f
  channel, 86, 86f, 89
  chaperone, 197
  as class of organic molecule, 38
  denaturation of, 47
  in diet, 623
  digestion of, 612, 612f, 619, 619f
  DNA-binding, 208, 208f
  as energy source, 125, 126
  enzymatic, 86, 86f
  folding of, 197, 199
  functions of, 45
  gene expression and, 192
  junction, 86, 86f
  levels of organization in, 47, 47f
  in phospholipid bilayer, 85–87, 85f, 86f
  protective, 596–597, 596f, 597f
  proteomics, 236
  receptor, 86, 86f
  similarity in all organisms, 252
  subunits of, 40f
Proteinase inhibitors, 472–473
Protein synthesis, 62–65. See also Transcription; Translation
  in cell cycle, 136
  endomembrane system in, 67, 67f
  endoplasmic reticulum in, 64
  genetic information and, 62
  Golgi apparatus in, 65, 65f
  overview of, 192, 192f, 199, 199f
  in prokaryotic cells, 311
  pulse-labeling in, 65, 65f
  ribosomes in, 63
  RNA and, 62
Proteinuria, 657
Proteome, 236
Proteomics, 236
Prothrombin activator, 585
Protists, 322–336
  characteristics of, 4, 4f, 8–9, 9f, 325f
  classification of, 8–9, 296, 325f, 326, 326f
  and disease, 322–323, 324, 328, 330, 330f, 357
  diversity of, 324, 325f
  evolution of, 324, 326, 326f, 369
  heterotrophic, 325f, 326
  mold-like, 331, 331f
  parasitism in, 746
  photosynthetic, 325f, 326
  reproduction, 324
  species, number of, 780f

Protocell(s), evolution of, 309, 309f, 310, 310f
Protoderm, 430, 430f, 431, 431f
Protons, 21, 21f, 21t
Proto-oncogenes, 218, 218f, 219, 219f
Protostomes, 372, 372f, 379, 379f
Protozoans, 327–330
  amoeboids, 324, 325f, 328, 328f
  characteristics of, 325f
  ciliates, 324, 325f, 329, 329f
  classification of, 324
  cysts, 324, 327, 328, 330
  and disease, 323, 327
  mutualism in, 748
  sporozoans, 325f, 330, 330f
  types of, 327
Proviruses, HIV, 306
Proximal convoluted tubule, 654, 655f, 656f, 657, 658
Prozac, 521
Pseudocoelomates, 371, 371f, 378, 574, 574f
Pseudomyrmex ferruginea, 473, 473f, 748
Pseudoplasmodium, 331
Pseudopods, 70, 328, 328f
Pseudostratified ciliated columnar epithelium, 497, 497f
Pseudostratified epithelium, 497
PS I. See Photosystem I
PS II. See Photosystem II
P (peptide) site, 198, 198f
PTH. See Parathyroid hormone
P-Tr (Permian-Triassic) extinction, 289
Puberty
  regulation of onset, 526
  and secondary sex characteristics, 685
Pubic symphysis, 558, 558f
Puffballs, 361, 361f
Pulmonary arteries, 577, 577f, 580f
Pulmonary circuit, 576, 576f, 580, 580f
Pulmonary disorders
  lung cancer, 133f, 144, 201, 639, 644, 644f
  obstructive, 644
  restrictive, 644
Pulmonary semilunar valve, 577
Pulmonary trunk, 577, 577f
Pulmonary veins, 577, 577f
Pulse, 578
Pulse-labeling, of protein secretion pathway, 65, 65f
Pulvinus, 468, 468f
Punctuated equilibrium model of speciation, 273, 273f
Punnett square, 164f, 165, 165f, 253, 253f
Pupae, fire ant, 3
Pupil, 540, 540f
Purines, 186, 186f
Pyloric sphincter, 618
Pyramid-shaped age structure diagrams, 713, 713f
Pyrenoid, 327, 334, 334f
Pyrimidines, 186, 186f
Pyruvate, 115, 115f, 116, 117f, 118, 123, 123f

Q

Quadriceps femoris group, 563f
Quadriplegia, 524
Quaternary period, 285t
Quaternary structure, of protein, 47, 47f
Quinine, 357, 474, 488
Quinton, Wayne, 660

R

Rabbits
  digestion in, 611, 615, 615f
  evolution of, 252

Rabies, 305, 307
Radial nerve, 516f
Radial symmetry, 369, 371, 385
Radiant energy, as energy type, 78
Radiation
    and cancer, 144
    health effects of, 22
    as mutagen, 201
    uses of, 21–22, 22f
Radicle, 484, 484f
Radiolarians, 328, 328f
Radiotherapy, 22, 22f
Radius, 556f, 558, 558f
Radon, and cancer, 144
Radula, 380, 380f
Raggiana Bird of Paradise, 730,
    730f, 735
Rain
    acid, 32, 32f, 757, 787
    mountains and, 770, 770f
    pH of, 32
Rain forest
    as biodiversity frontier, 781
    temperate, 764f, 765
    tropical, 764f, 764t, 768, 768f, 770,
        770f, 785, 785f, 789
        and global warming, 108,
            108f, 421
        human impact on, 10
        medicines and, 472, 474, 474f
Rain shadow, 770, 770f
Raloxifene, 559
Rana pipiens (leopard frog), 137t
Random mating, and microevolution,
    253, 254
Random population distribution,
    711, 711f
Range, of species, 711
RAS. See Reticular activating system
Ras gene, 218, 218f
Rat(s), 734, 736, 786
Rat poison, 84
Ray(s), 388, 555
Ray-finned fishes, 388–389
Reaction(s)
    coupled, 81, 81f
    endergonic, 80
    exergonic, 80
Reaction center, of photosystem, 101,
    101f, 102, 102f
Receptacle, 353, 353f, 480, 480f
Reception, in signal transduction
    pathway, 460, 460f
Receptor-mediated endocytosis, 90
Receptor proteins, 86, 86f
Recessive alleles, 163, 254
Reciprocal altruism, 733
Recombinant DNA technology, 226,
    226f, 228
Recombinant gametes. See
    Crossing-over
Recruitment, 565
Rectum, 616f, 621, 621f
Rectus abdominis, 563f
Red alders, 459
Red algae, 94–95, 325f, 333, 333f, 773
Red-backed cleaning shrimp, 383f
Red blood cells (erythrocytes)
    hemolysis, 89
    malaria and, 322, 323f, 330, 330f
    respiration and, 643
    in sickle-cell disease, 173, 173f, 192,
        192f, 258
    structure and function of, 499, 499f,
        507, 584, 584f
Red bone marrow
    functions of, 560, 560f, 584
    and immune system, 594f, 595,
        600, 602
    stem cells in, 136, 586
Red bread molds, 360

Red deer, 731, 731f, 732
Red-green color blindness, 87
Red kangaroo, 711
Redox reaction
    cellular respiration as, 114
    definition of, 98
    photosynthesis as, 98
Red pulp, of spleen, 595, 595f
Red tides, 325
Reduced hemoglobin, 643
Reduction, definition of, 98
Redwood trees, 765
Reeve, Christopher, 502, 502f
Refinement, of grains, 622
Reflex actions, 524, 529, 529f, 530,
    530f, 537, 617
Refractory period
    of axon, 519
    following male orgasm, 684
Regeneration, in sponges, 374
Regulator gene, 208, 208f
Reindeer (Rangifer), 716
Reindeer moss, 346
Reinforcement, of isolating
    mechanisms, 269
Release stage, of viral life cycle, 303,
    305, 305f, 306, 306f
Remnin, 124
Remoras, 746
Renal artery, 580, 580f
Renal calculus (kidney stones), 654
Renal cortex, 654, 655f
Renal medulla, 654, 655f, 658, 658f
Renal pelvis, 654, 655f
Renal vein, 580, 580f
Renin, 658, 673
Renin-angiotensin-aldosteron system,
    658, 659f, 673
Rennet, 228
Replacement model
    of human ethnic diversity, 414
    of human evolution, 411, 411f
Replacement reproduction, 720
Replication. See DNA replication
Replication fork, 191, 191f
Repolarization, in action potential,
    519, 519f
Repressible operon, 208, 208f
Reproduction. See also Asexual
    reproduction; Sexual reproduction
    animals
        asexual, 682, 682f
        overview of, 680–681
        sexual, 682–683, 682f, 683f
    as characteristic of life, 7
    marsupials, 392
    parthenogenesis, 682
    plants, adaptations to land
        environment, 344–345
    prokaryotes, 135, 135f, 312, 312f
    water as necessity for, 19
Reproduction rate, in primates, 401
Reproductive age, 713, 713f, 720, 720f
Reproductive behavior, as adaptive,
    730–731, 730f, 731f
Reproductive cloning, 210, 210f, 211
Reproductive system
    in flatworms, 376, 376f
    human, 505, 505f
        female, 686–688, 686f, 687f
        male, 684–685, 684f, 685f
    in mammals, 392
Reproductive technologies, 691, 691f
Reptiles
    characteristics of, 373, 390, 390f
    circulatory system, 576, 576f
    enzymatic reactions in, 83, 83f
    evolution of, 284, 285t, 387f, 390
    extraembryonic membranes, 697
    homeostasis in, 494
    kidneys of, 654, 655

osmoregulation in, 653
phylogeny, 295f
reproduction, 680–681, 683
respiration in, 640–641
RER. See Rough ER
Reservoir, in chemical cycling, 756, 756f
Resource(s)
    competition for, and
        biodiversity, 443
    definition of, 711
    exploitation of
        and extinctions, 785f, 787–788
        and sustainable
            development, 790
    and population, 711, 719
    resource partitioning, 742–743,
        742f, 743f
Respiration
    energy used in, 634, 635f
    external
        gas exchange in, 642, 642f,
            643, 643f
        overview of, 634, 634f
        types of, 634–635, 635f
    in humans, water loss in, 635
    internal
        gas exchange in, 642, 642f,
            643, 643f
        overview of, 634, 634f
    steps of, 634, 634f
Respiratory center, 641, 641f, 659
Respiratory diseases, bacterial, 317t
Respiratory pigment, 572–573, 633
Respiratory system
    and acid-base balance, 659
    amphibians, 389
    arthropods, 382
    birds, 391
    echinoderms, 385
    fish, 388–389
    and homeostasis, 507
    humans, 638–644
        breathing rate, regulation of,
            641, 641f, 659
        disorders of, 644, 644f
        gas exchange, 642, 642f,
            643, 643f
        overview of, 505, 505f, 638, 638f
        smoking and, 639
        ventilation, 640, 640f
        water loss, 635
    vertebrates, 387
Response, in signal transduction
    pathway, 460, 460f
Restoration ecology, 790
Restriction enzymes, 226, 226f
Restrictive pulmonary disorders, 644
Reticular activating system (RAS),
    526, 537
Reticular formation, 526, 526f
Retina, 540, 540f, 541, 542–543, 543f
    disorders of, 541
Retinal, 542–543
Retinoblastoma protein (RB), 219
Retroviruses, 306, 306f
    reverse transcriptase, 306, 306f
Reverse transcription, and origin of
    DNA, 310
Reznick, David, 257
R groups of amino acids, 46, 46f
Rhesus monkeys, 403
Rheumatic fever, 577, 605
Rheumatoid arthritis, 605, 605f
Rhinoceros, 788
Rhizobium, 454
Rhizomes, 487, 487f
Rhizopus stolonifer (Black bread mold),
    360, 360f
Rhodophyta, 326f
Rhodopsin, 542–543
Rh system, 588

Rib cage, 557, 557f
Riboflavin
    deficiency in, 83
    dietary requirements, 625t
Ribonucleic acid. See RNA
Ribose, 41, 48, 48f, 48t, 186, 186f
Ribosomal RNA. See rRNA
Ribosomes
    in animal cells, 60f
    attachment to ER, 63, 63f, 64, 64f
    in plant cells, 61f
    in prokaryotic cells, 59, 199,
        199f, 311
    in protein synthesis, 63, 197, 197f,
        198, 198f
    structure of, 197, 197f, 198, 198f
Ribozymes, 195, 309, 310
Ribs, 556f
Rice (Oryza), 356, 356f, 425, 425f,
    782–783
Rice blast disease, 362
Rickets, 24, 625
Rieppel, Olivier C., 292
Rift Valley fever, 318
Right atrium, 577, 577f
Right hemisphere, 525
Right lymphatic duct, 594
Right ventricle, 577, 577f
Right whales, 736
Rigor mortis, 567
Ring-tailed lemur, 400f
Ringworm, 362, 362f
Rivers, 771, 771f
RNA (ribonucleic acid)
    digestion of, 619
    enzymatic function of, 195, 309, 310
    as first macromolecule, 309, 310
    function of, 48
    messenger (mRNA)
        function of, 62, 214, 241
        mature, 195, 195f, 215f
        posttranscriptional processing
            of, 195, 195f, 214, 214f,
            215, 215f
        primary, 195, 195f, 215f
        in transcription, 192, 192f, 194,
            194f, 199, 199f
        in translation, 62, 63, 63f,
            196–198, 196f, 197f, 199,
            199f
        and translational controls, 214
    production of, 62
    ribosomal (rRNA)
        function of, 197
        production of, 62, 197
    structure
        bases, 48, 48t, 186, 186f
        chemical structure, 48, 48f, 48t,
            186, 186f
        overview, 48, 48f
    transfer (tRNA)
        function of, 62, 196, 198, 198f
        initiator, 198, 198f
        structure of, 196, 196f
RNA polymerase, 191, 191f, 194, 194f
    DNA-binding proteins and, 208,
        208f, 213, 213f
RNA primer, 191, 191f
RNA world, 195
Robust, defined, 406
Rockweed (Fucus), 333, 333f, 336
Rocky Mountain spotted fever, 383
Rocky shores, 772, 772f
Rod cells, 401, 534, 535, 536,
    542–543, 543f
Rodents, as order, 393
Rodhocetus, 250, 264f
Rodriguez, Eloy, 474, 474f
Rohypnol (roofies), 522
Roofies (rohypnol), 522
Root cap, 430, 430f

Root hairs, 422*f*, 423, 423*f*, 426, 426*f*, 430–431, 430*f*, 452, 452*f*
Root nodules, 454, 454*f*
Root pressure, 444
Root system
   eudicots, 424, 424*f*, 429, 430–431, 430*f*
   growth, 430–431, 430*f*
   monocots, 424, 424*f*
   structure and function of, 422, 422*f*, 423, 423*f*
   tissues, 428–429, 428*f*, 429*f*
   water and mineral uptake, 452, 452*f*
   and water transport in plants, 445, 445*f*
Root tip, 422*f*, 423
Rose gardener's disease, 362
Rosenberg, Steven A., 233, 233*f*
Rosy periwinkle (Catharanthus roseus), 782, 782*f*
Rotational balance, 546–547, 547*f*
Rough ER (RER), 60*f*, 61*f*, 64, 64*f*, 67, 67*f*
Round window, 544, 544*f*
Roundworms, 372*f*
   anatomy, 378
   characteristics of, 373
   circulatory system, 574, 574*f*
   evolution of, 370*f*, 372*f*
   as parasite, 378, 378*f*
   as pseudocoelomate, 371, 371*f*
   reproduction, 682
rRNA (ribosomal RNA)
   function of, 197
   production of, 62, 197
RU-486, 690
Rubber, 357, 357*f*
Rubber trees (Hevea), 783*f*
Rubella, 305
RuBP, 104–105, 104*f*, 105*f*
RuBP carboxylase, 104–105, 104*f*
Rugae, 618, 618*f*
Rumen, 615, 615*f*
Ruminants, digestion in, 42, 611, 615, 615*f*
Runner's high, 521
Rwanda, biogas in, 741

S

Sac body plan, 375, 376
Saccharomyces cerevisiae, 137*t*, 234*t*
Saccule, 546–547
Sac fungi, 360–361, 360*f*
Sacral nerves, 516*f*
Sacral vertebrae, 557
Sacrum, 556*f*, 558
Saddle joints, 561
Safrole, 201
Sahelanthropus tchadensis, 405, 405*f*
St. Louis encephalitis, 307
St. Paul Island, 716
Salamanders, 268, 268*f*, 389, 389*f*, 635
Salicylic acid, 473, 473*f*
Salinization, of soil, 229
Saliva, 616
Salivary amylase, 616
Salivary glands, 616, 616*f*
Sally lightfoot crab, 383*f*
Salmon, 268–269, 269*f*
Salmonella, 317
Salt(s), 24
Saltatory conduction, 519, 519*f*
Salt-cured foods, and cancer, 144
Salt marshes, 771, 771*f*, 772
Salt-tolerant crops, 229
Salt-water balance, maintenance of, 658–659, 658*f*, 659*f*
San Andreas fault, 287, 287*f*
Sand, in soil, 453
Sand dollar, 385*f*
Sandworms, reproduction, 682

Sandy shores, 772
Sanger, Frederick, 45
SA (sinoatrial) node, 578, 578*f*
Saprolegnia, 331
Saprotrophs, 314, 741
Sap wood, 432, 433*f*
SAR. See Systemic acquired resistance
Sarcolemma, 566, 566*f*, 567*f*
Sarcomas, 143, 592*f*, 593
Sarcomeres, 566, 566*f*
Sarcoplasmic reticulum, 566, 566*f*, 567
Sargassum, 333
Sarin, 84, 84*f*
SARS. See Severe acute respiratory syndrome
Sartorius, 563*f*
Satin bowerbirds, 730, 730*f*
Saturated fatty acids, 43, 583, 623, 623*f*, 627
Savannas, 764*f*, 767, 767*f*
Scanning electron micrograph (SEM), 58
Scanning electron microscopes, 58, 58*f*
Scapula, 556*f*, 558, 558*f*
Scarlet kingsnake, 745
Scavengers, 740–741, 740*f*, 752, 753, 753*f*
Schistosoma, 377, 377*f*
Schistosoma japonicum, 474
Schistosomiasis, 377, 377*f*
Schleiden, Matthias, 54–55
Schwann, Theodor, 54–55
Schwann cells, 517, 517*f*
Sciatic nerve, 516*f*
SCID (severe combined immunodeficiency), treatment of, 232, 233
Science
   and ethics, 14
   models in, 21
Scientific method
   sample study, 12–13, 13*f*
   steps in, 11, 11*f*
Scientific names, 9
Scientific process, 11
Scientific theory, 11
Sclera, 540, 540*f*
Sclerenchyma cells, 426*f*, 427, 428, 434, 434*f*
Sclerids (stone cells), 427
Scolex, 377, 377*f*
Scoliosis, 557
Scorpions, 383, 651
Scribner, Belding, 660
Scrotum, 684, 684*f*
S-cyclin, 142, 142*f*
Sea anemone, 375*f*, 514, 746, 746*f*, 748
Sea breezes, origin of, 770
Sea cucumber, 385*f*
Seagrass beds, destruction of, 785
Seagulls, osmoregulation in, 653, 653*f*
Sea horses, 683
Sea lettuce (Ulva), 335, 335*f*, 336, 336*f*
Sea lily, 385*f*
Sea lions, and overfishing, 788
Seals
   elephant, 664
   northern elephant (Mirounga angustirostris), 632, 632*f*, 633
   and overfishing, 788
   Weddell (Leptonychotes weddelli), 632
Sea otters, 788
Seasickness, 548
Seasons, source of, 769, 769*f*
Sea squirts (tunicates), 386, 386*f*, 387*f*
Sea stars, 385, 385*f*, 683
Sea urchins, 385*f*, 512, 512*f*, 788
Seaweeds, 333, 334
Secondary follicle, 687, 687*f*
Secondary growth, 432–433, 432*f*, 433*f*

Secondary metabolites, plant, 472
Secondary oocyte, 687, 687*f*
Secondary sex characteristics, 685, 687
Secondary structure, of protein, 47, 47*f*
Secondary succession, 750, 750*f*, 751*f*
Second law of thermodynamics, 79, 753
Second messenger, 667, 667*f*
Second polar body, 687, 687*f*
Secretin, 621, 621*f*
Secretion (exocytosis), 65, 65*f*, 67, 67*f*, 90
Secretory phase, of uterine cycle, 688, 688*f*
Sedimentation, and fossils, 249
Seed(s)
   anatomy of, 483*f*, 486*f*
   dispersal of, 478–479
   embryo development in, 484, 484*f*
   evolution of, 343, 343*f*, 350
   formation of, 480–483, 482*f*
   germination, 486, 486*f*
   naked, 350, 351
Seed coat, 483, 483*f*
Seed ferns, 352, 352*f*
Seedless vascular plants, 348–349, 348*f*
   characteristics of, 342*f*, 343
   diversity of, 348, 348*f*
   economic value of, 348–349
   evolution of, 343, 343*f*
   life cycle, 349, 349*f*
Segmentation
   in development, 216
   evolution of, 370*f*, 371
   in lancelets, 386
segment-polarity genes, 216, 216*f*
Sehelanthropus tchadensis, 399*f*
Selam (fossil), 399, 406
Selection
   artificial, 246, 246*f*, 272, 272*f*
   natural. See Natural selection
Selenium, dietary requirements, 624
Self-antigens, 599
Self-pollination, 482
SEM. See Scanning electron micrograph
Semen (seminal fluid)
   characteristics of, 684
   propulsion through reproductive tract, 671
Semicircular canals, 544, 544*f*, 546–547, 547*f*
Semiconservative replication, 190, 190*f*
Semidesert, 764*f*
Semilunar valves, 577, 577*f*, 578
Seminal fluid. See Semen
Seminal vesicles, 684
Seminiferous tubules, 685, 685*f*
Senescence
   of cells, 142, 191
   in plants, 463
Sensation, as brain phenomenon, 501, 537
Sense organs, in flatworms, 376, 376*f*
Sensory adaptation, 537
Sensory (afferent) neurons, 517
Sensory receptors
   bilateral symmetry in, 515
   categories of, 536
   chemoreceptors, 536, 537, 538–539, 538*f*, 539*f*
   communication with CNS, 537, 537*f*
   electromagnetic receptors, 536, 536*f*
   mechanoreceptors, 536, 544, 545
   olfactory receptors, 539, 539*f*
   pain receptors (nociceptors), 536
   photoreceptors, 536, 540 (See also Cone cells; Rod cells)
   in skin, 503, 503*f*
   taste receptors, 536, 538–539, 538*f*
   thermoreceptors, 536, 536*f*
Sensory system, in humans, 504, 504*f*
Sepal, 353, 353*f*, 480, 480*f*

Septa, 381, 381*f*
Septate hypha, 358, 358*f*
September 11th terrorist attacks
   dust from, and lung disease, 644, 644*f*
   identification of remains from, 69, 227
Septum, heart, 577, 577*f*
Sequential hermaphroditism, 682
SER. See Smooth ER
Serine, 46*f*
Serotonin, 521, 522
Sertoli cells, 685, 685*f*
Serum, 585
Sessile leaves, 423
Setae, 381, 381*f*
Set point, 508
Severe acute respiratory syndrome (SARS), 300*f*, 307, 307*f*
Severe combined deficiency syndrome (SCID), 232, 233
Sex characteristics, secondary, 685, 687
Sex chromosomes, 145, 145*f*
   abnormalities in, 153, 153*t*
Sex hormones, 669*t*
   adrenal cortex and, 672
   and bone strength, 559
   and development, 700
   female, 687
   male, 685
   production of, 685
Sex pili, 59, 59*f*, 311, 311*f*, 313
Sex reversal, 682
Sexual development, regulation of, 675
Sexually-transmitted diseases, 317*t*, 327, 689, 689*t*
Sexual recombination, and microevolution, 254
Sexual reproduction. See also Meiosis
   advantages and disadvantages of, 147
   in algae, 334, 334*f*, 336, 336*f*
   in angiosperms, 345, 480–486
   in animals, 682–683, 682*f*, 683*f*
   cell division in, 134
   in ciliates, 329, 329*f*
   in diatoms, 332
   in dinoflagellates, 325
   in fungi, 358, 360, 360*f*, 361, 361*f*
   and genetic variation, 146–147, 147*f*, 150
   in humans, 684–691
      birth control, 690, 690*f*, 690*t*
      and disease. See Sexually-transmitted diseases
      female reproductive system, 505, 505*f*, 686–688, 686*f*, 687*f*
      male reproductive system, 505, 505*f*, 684–685, 684*f*, 685*f*
      reproductive technologies, 691, 691*f*
      water and, 19, 19*f*
   in insects, 384
   in protists, 324
   in reptiles, 390
   in sponges, 374
   in sporozoans, 330
Sexual satisfaction, hormones and, 671
Sexual selection, 730–731, 730*f*, 731*f*
Shape, of organism, genetic expression and, 276–277, 277*f*
Sharks, 388, 388*f*, 555, 650, 746
Shell, of mollusc, 380, 380*f*
Shellfish poisoning, 325
Shin, 558
Shingles, 305, 593
Shock, anaphylactic, 606
Shoot system, 422–423, 422*f*, 423*f*, 431, 431*f*
Shores
   rocky, 772, 772*f*
   sandy, 772

Short-day plants, 470, 471
Short-grass prairie, 766
Short-term memory, 527
Shrimps, 383, 383f, 651
Shrubland, 764f
Siberian tigers, 788
Sickle-cell disease
    causes and symptoms, 168
    as genetic disease, 49, 200
    hemoglobin in, 192, 192f, 258
    malaria and, 173, 258, 258f
    pleiotropy in, 173, 173f
Sickle-cell trait, 168
SIDS (sudden infant death
    syndrome), 639
Sierra Nevada mountains, 770
Sieve plate, 427, 427f, 442
Sieve tube(s), 427, 427f, 448
Sieve-tube members, 427, 427f,
    442, 448
Sieve-tube plates, 449
Sigmoid colon, 616f, 621
Signaling molecules, 142, 142f
Signal transduction pathways
    and cell cycle control, 142, 142f
    oncogenes and, 219, 219f
    plant hormones and, 460, 460f
    in plant wound responses,
        472–473, 473f
Silkworm, 664
Silt, in soil, 453
Silurian period, 285t
Silversword alliance, 440–441,
    440f–441f, 718f
Simple diffusion, 88, 88f
Simple goiter, 676, 676f
Simple muscle twitch, 564, 564f
Single nucleotide polymorphisms
    (SNPs), 234
Sink, in pressure-flow model, 449
Sink population, 789
Sinoris, 250
Sinuses
    cranial, 556
    in lymph nodes, 595
Sister chromatids, 137, 137f, 138f–139f,
    139, 145, 145f
Skates, 388
Skeletal muscle, 500, 500f, 563–568
    antagonistic pairs, 564, 564f
    contraction of, 81, 81f, 564–565,
        566–568, 566f, 567f
    fast- and slow-twitch fibers,
        568, 568f
    functions, 563, 563f
    motor units, 564–565, 564f
    muscle cells, 566, 566f
    neuromuscular junction, 567, 567f
Skeleton
    appendicular, 556, 558–559, 558f
    axial, 556–557, 556f, 557f
    endoskeleton, 555, 555f
        advantages of, 371
        jointed, as vertebrate
            characteristic, 387
        of sponge, 374
    exoskeleton, 382, 384, 544f, 554
    and homeostasis, 560
    human, 504, 504f
        appendicular, 556,
            558–559, 558f
        axial, 556–557, 556f, 557f
        functions of, 555
        gender differences, 553
    hydrostatic, 379, 554, 554f
    mammal, functions of, 555
    of vertebrates, 387, 555
Skill memory, 527
Skin
    anatomy of, 503, 503f
    artificial, 506, 506f

and homeostasis, 508
    immune system and, 596, 596f
S-kinase, 142, 142f
Skin cancer
    DNA repair and, 201
    treatment of, 233, 233f
    UV radiation and, 503
    xeroderma pigmentosum and, 201
Skin diseases, bacterial, 317t
Skin gills, 385, 385f
Skinner, B. F., 729, 736
Skull, human, 524f, 556–557, 556f, 557f
Sky burials, 740, 740f
Sleep movements, in plants,
    468–469, 469f
Slime layer, 59, 311
Slime molds, 324, 325f, 331, 331f
Slow-twitch muscle fibers, 568, 568f
Small intestine, human, 619, 619f
Smallpox, 305, 307, 318, 601
Smell receptors, 536, 539, 539f
Smoked foods, and cancer, 144
Smoking
    and cancer, 144, 201, 639
    and cardiovascular disease,
        582–583
    and chronic bronchitis, 497
    effects of nicotine, 522
    health effects, 639, 644
    quitting, 522
Smooth ER (SER), 60f, 61f, 64, 64f,
    67, 67f
Smooth muscle, 500, 500f
Snails, 248, 257, 257f, 548
Snakes, 390, 390f
    behavior, genetically-based,
        726–727, 727f
    brown tree snake, 786
    chemoreceptors in, 538
    coral snake, 745
    diamondback rattlesnake, 390f
    evolution of, 251
    fer-de-lance pit viper, 394
    garter (Thamnophis elegans), 683,
        726–727, 727f
    mimicry in, 745
    scarlet kingsnake, 745
    thermoreceptors in, 536, 536f
Snoring, 616
Snowshoe hare, 744, 744f
SNPs. See Single nucleotide
    polymorphisms
Snuff, 639
Social behavior, as adaptive, 732–733,
    732f, 733f
Society, 734
Sociobiology, 732
Sodium
    as biologically important ion, 24t
    dietary requirements, 624, 624t
Sodium chloride, formation of, 24, 24f
Sodium gates, and action potential,
    518–519, 519f
Sodium hydroxide, dissociation of, 30
Sodium-potassium pump, and neuron
    resting potential, 518, 518f
Soft palate, 616, 616f, 638
Soft spots, 561
Softwood, 350
Soil
    composition of, 453
    erosion, prevention of, 784
    formation of, 315
    salinization, 229
Soil profiles, 453, 453f
Solanum tuberosum (potato), 137t
Solar energy
    and climate, 769, 769f
    and energy flow in ecosystems,
        753, 753f
Solute, definition of, 28, 89

Solution(s)
    definition of, 28, 89
    hypertonic, 89, 89f
    hypotonic, 89, 89f
    isotonic, 89, 89f
Solvent, definition of, 89
Somaclonal variations, 488
Somatic cells
    cell cycle in, 136–140, 136f,
        138f–139f
    definition of, 134
Somatic embryogenesis, 488, 488f
Somatic mutations, 200
Somatic system, 529, 529f
Somites, 694, 694f
Song learning, in birds, 728
Sorus (sori), 349, 349f
Sound (auditory) communication, 734
Source, in pressure-flow model, 449
Source population, 789
South Africa, as biodiversity
    hotspot, 781
South America, Humboldt Current and,
    774, 774f
South Beach diet, 622
Southern adder's tongue fern
    (Ophioglossum vulgatum), 137t
Space-filling model, 25, 25f, 26f
Spanish moss, 346
Specialization. See also Differentiation
    of eukaryotic cells, 132–133,
        209, 209f
    segmentation and, 371
Special K (ketamine), 522
Speciation
    adaptive radiation, 270, 270f
    definition of, 264
    gradualist model, 273, 273f
    punctuated equilibrium model of,
        273, 273f
    reinforcement of, by natural
        selection, 269
    sympatric, 271–272
    undirected nature of, 278, 278f
Species
    biological species concept, 264–265
    definition of, 7, 264–265, 290, 290f
    evolutionary species concept, 264
    hybridized, 262–263, 262f, 263f
    keystone, 789
    life history of, 717
    postzygotic isolating mechanisms,
        267, 267f, 269
    prezygotic isolating mechanisms,
        266–267, 266f, 267f, 269
    reproductive barriers, 266–267,
        266f, 267f
    as taxon, 8, 8t
Species richness, 749
Specific immunity. See Immune system
Spemann, Hans, 695, 696
Sperm
    human, 684, 685, 685f
    plant, 481, 481f, 482f, 483
Spermatids, 685, 685f
Spermatogenesis, 150, 685, 685f
Spermicidal jelly, 690f, 690t
Sphagnum, 346
Sphenoid bones, 556, 556f, 557f
Sphincters, 617
Spices, 356
Spicules, 374, 374f
Spider(s), 383, 383f, 651
Spider mites, 473
Spider monkeys, 403
Spinal cord, 515f, 524f. See also Central
    nervous system
    anatomy of, 524
    development of, 515, 694, 699
    function of, 524
    injury to

effects of, 524
        nerve regeneration, 502, 502f
Spinal nerves, 516, 516f, 528, 528f
Spindle fibers, 138, 138f–139f
Spindle poles, 138f–139f
Spinks, Lorna, 522
Spiny anteater, 392
Spiracles, in insect tracheal system,
    637, 637f
Spirillum (spirilla), 311, 311f
Spirogyra, 334, 334f
Spleen, 594f, 595, 595f, 633
Spliceosomes, 195
Sponges, 374, 374f, 514
    characteristics of, 373
    evolution of, 370f, 372f
    reproduction, 682
Spongin, 374
Spongy bone, 560, 560f
Spongy mesophyll, 428, 434, 434f, 435f
Sporangium, 331, 331f, 349, 349f,
    360, 360f
Spores
    of fungus, 358, 358f
    of mosses, 347, 347f
    of pine, 351, 351f
    in plant life cycle, 344, 344f
    of protists, 324
    of slime mold, 331, 331f
Sporophytes
    in algae, 336, 336f
    in angiosperms, 354f–355f, 481,
        481f, 482f
    in bryophytes, 346, 347, 347f
    dominance, as land adaptation,
        344–345, 344f
    embryo development, 484, 484f
    in ferns, 344, 349, 349f
    of pine, 351, 351f
    in plant life cycle, 150, 150f, 344
Sporozoans, 325f, 330, 330f
Sporozoites, 330f
Springboks, 745, 747
Spring wood, 432
Spurges, phylogeny, 292, 292f
Squamous epithelium, 497, 497f
Squids, 380, 514–515, 514f, 534, 573,
    575, 636
Squirrel monkey, 393f
SRY (sex-determining region of Y
    chromosome), 231, 231f
S-shaped curve, characteristics of,
    714, 714f
S stage, 136, 136f
Stabilizing selection, 256, 256f, 258
Stable equilibrium phase, of S-shaped
    curve, 714, 714f
Stamen, 353, 353f, 354f–355f, 480,
    480f, 482f
Stanley, Stephen, 407
Stanley, Wendell, 304
Stapes (stirrup), 544, 544f, 545,
    545f, 546
Starch, 42, 42f
    dietary, sources of, 622
    digestion of, 616, 619, 619f
Starfish, 555, 555f
StarLink corn, 230, 230f
Start codons, 198
Statocysts, 548, 548f
Statoliths, 466–467, 548, 548f
Stem(s), angiosperm
    elongation hormones, 462, 462f
    eudicots, 424, 424f, 428, 429f
    monocots, 424f, 428, 429f
    structure and function, 422, 422f,
    tissues, 428, 428f, 429f
    and water transport in plants,
        445, 445f
Stem cells, 210
    adult, 132, 136, 210, 586

and animal cloning, 211
embryonic, 136, 210, 586
ethics of research on, 210, 211, 586
hematopoietic, 586, 586*f*
and nerve regeneration, 502, 502*f*
and tissue engineering, 605
umbilical cord, 586
Stentor, 329, 329*f*
Stents, 582, 582*f*
Stereocilia, 545, 545*f*, 546–547, 547*f*
Stereoscopic vision, 400–401, 401*f*, 535, 543
Sterility, of hybridized species, 263, 267, 267*f*
Sterilization, sexual, 690*t*
Sternum, 556*f*, 558
Steroid(s)
in phospholipid bilayer, 85
structure of, 44, 44*f*
Steroid hormones, action of, 667, 667*f*
Stickleback fish, 277
Sticky ends, of DNA, 226
Stigma, 353, 353*f*, 354*f*–355*f*, 480, 480*f*, 482*f*, 483
Stimuli
definition of, 536
response to, as characteristic of life, 6, 6*f*
Stinging, as defense mechanisms, 2, 3*f*
Stinkhorns, 361
Stirrup (stapes), 544, 544*f*, 545, 545*f*, 546
Stolons, 487, 487*f*
Stomach, human, 618, 618*f*, 621*f*
hormones produced by, 621
hydrochloric acid in, 30
ulcers, 618
Stomata (stoma)
functions of, 97, 102, 345, 345*f*, 428, 434
in leaf anatomy, 97, 97*f*, 429*f*, 434*f*
opening and closing of, 345, 426, 435–436, 435*f*, 446, 446*f*, 464, 464*f*
in plant anatomy, 442*f*, 445, 445*f*
structure of, 426, 426*f*
and water loss, 445
Stone cells (sclerids), 427
Stone fly, larva of, 771, 771*f*
Stoneworts, 326, 326*f*, 335, 335*f*, 341, 341*f*
Stop codons, 197, 198
Strata (stratum), 245, 249
Stratification
of habitat, 749
in temperate deciduous forest, 766
Stratified epithelium, 497
Strauss, Levi, 357
Strawberry plants, asexual reproduction in, 487, 487*f*
Streams, 771, 771*f*
Streptococcus pneumoniae, Griffith's research on, 184, 184*f*, 313
Streptomycin, 546, 782
Stress, adrenal glands and, 671*f*, 672–673
Stress hormone, in plants, 464, 464*f*
Striation, in muscle fiber, 500, 500*f*, 566
Stroke, 582
Stroma, 68, 68*f*, 97, 97*f*, 99, 104, 104*f*
Stromatolites, 315
Strong acids, 30
Strong bases, 30
Structural genes, 208, 208*f*
Structural polysaccharides, 42, 42*f*
Style, 353, 353*f*, 354*f*–355*f*, 480, 480*f*, 482*f*, 483
Subatomic particles, 21, 21*f*, 21*t*
Subclavian veins, 594*f*
Subcutaneous layer, 503, 503*f*

Subduction zone, 287
Suberin, 426, 428, 432
Suboxone, 523
Substance P, 521
Substrate feeders, 613, 613*f*
Substrate-level ATP synthesis, 116, 116*f*, 117*f*
Substrates, 82, 82*f*
Succession, ecological, 750–751, 750*f*, 751*f*
Succulents, photosynthesis in, 107
Suckers, 487
Sucrose, 41
Suctoria, 329
Sudden infant death syndrome (SIDS), 639
Sugar(s)
conversion to glucose, 622
in diet, 622, 622*t*
disaccharides, 41, 41*f*
monsaccharides, 41
polysaccharides, 42, 42*f*
Sugar beets, 356
Sugarcane, 356
Sulcus (sulci), 525, 525*f*
Sulfhydryl groups, 39*f*
Sulfur
carbon binding with, 38
and life, 20, 20*f*
Sulfur dioxide, and acid deposition, 757, 787
Summer wood, 432
Sundew plants, 341, 341*f*
Sunglasses, categories of, 541
Sunlight
absorption spectrums of photosynthetic pigments, 100, 100*f*
electromagnetic spectrum, 100
as source of energy on Earth, 7, 10, 10*f*, 76–77, 96
Sunscreen, 144
Superior vena cava, 577*f*, 580*f*
Superorganisms, fire ants as, 3
Surface-area-to-volume ratio
of cells, 57, 57*f*
of eukaryotic cells, 60
of human bodies, 414
Surface tension, of water, 27, 27*f*
Surgery, arthroscopic, 562, 562*f*
Surrogate mothers, 691
Survival of the fittest, 247
Survivorship, 712, 712*f*
Survivorship curve, 712–713, 713*f*
Suspension feeders, 374
Sustainable development, 790
Swallowing, 617
Swallows, 743
Swayback (lordosis), 557
Sweat glands, 503, 503*f*, 508
Swifts, 743
Swim bladder, 388
Swimmerets, of crayfish, 382*f*, 383
Swine fever, 318
Symbiosis, 746, 746*f*
and coevolution, 747
cyanobacteria, 315
in dinoflagellates, 325
in green algae, 334
lichens, 315
in protists, 324
Symmetry
bilateral, 369, 371, 376, 514, 515
radial, 369, 371, 385
Sympathetic division, 530, 530*f*
Sympatric speciation, 271–272
Synapses, neural, 520, 520*f*, 527
Synapsids, 250
Synapsis, of homologous chromosomes, 146, 146*f*
Synaptic cleft, 520, 520*f*, 567, 567*f*

Syndactyly, 217
Syndrome(s)
chromosome number abnormalities, 153
chromosome structure abnormalities, 154, 154*f*
definition of, 153
Synovial fluid, 561
Synovial joints, 561, 561*f*
Synura, 325*f*
Syphilis, 689, 782
Systematics. *See also* Classification of organisms; Phylogeny
definition of, 291
evolutionary, 294–295, 295*f*
phylogenetic cladistics, 294, 294*f*, 295, 295*f*
Systematics and the Origin of Species (Mayr), 268
Systemic acquired resistance (SAR), 473
Systemic circuit, 576, 576*f*, 580, 580*f*
Systemic diseases, bacterial, 317*t*
Systemic lupus erythematosus (lupus), 605, 605*f*
Systemin, 437*f*, 472–473
Systole, 578
Systolic pressure, 581

T

T2 bacteriophage, Hershey-Chase research on, 184–185, 185*f*
Tacrolimus, 605
Tactile communication, 735, 735*f*
Tadpoles, 368, 368*f*, 389
Taenia solium (pork tapeworm), 377, 377*f*
Taiga, 764*f*, 764*t*, 765, 765*f*
Tailbone (coccyx), 556*f*, 557
Tall-grass prairie, 766
Tamoxifen, 559
Tannins, 472
Tapeworms, 377, 377*f*, 682
Tarbela Dam, Pakistan, 784
Target cells, 666
Tarsal bones, 556*f*, 558*f*, 559
Tarsiers, 400*f*, 402*f*
Tasmanian devil, 718*f*
Tasmanian wolf/tiger, 392
Taste buds, 538–539, 538*f*
Taste receptors, 536, 538–539, 538*f*
Tatum Edward, 192
Taxol, 472
Taxon (taxa), 8, 8*t*, 290, 290*f*
Taxonomy, 8–9
binomial naming system, 290
definition of, 8, 290
taxa, 8, 8*t*, 290, 290*f*
Tay-Sachs disease, 66, 168, 171
Tbx5 gene, 276
T-cell leukemias, 305
T cell receptors (TCR), 600, 602, 602*f*
T cells, 585, 592, 592*f*, 595, 600, 600*f*, 600*t*, 602–603, 603*f*
TCR. *See* T cell receptors
Tea, 356
Tectonic plates, 287
Teeth
carnivores, 393, 610
decay in, 617
herbivores, 393, 611
human, 611, 616–617, 616*f*
omnivores, 610, 611
Telomerase, 191
Telomeres, 142, 191
Telophase
in meiosis
telophase I, 148, 148*f*–149*f*, 151*f*
telophase II, 148*f*–149*f*
in mitosis, 136*f*, 138*f*–139*f*, 151*f*
Telson, 382*f*, 383

TEM. *See* Transmission electron micrograph
Temperate deciduous forest, 764*f*, 764*t*, 766, 766*f*, 770, 770*f*
Temperate grassland, 764*f*, 766, 766*f*
Temperate rain forest, 764*f*, 765
Temperate zone, origin of, 769
Temperature
and enzymatic reactions, 83, 83*f*
global warming, 108, 108*f*
Template strand, 190, 190*f*, 194, 194*f*
Temporal bone, 556, 556*f*, 557*f*
Temporal isolation, as reproductive barrier, 266, 267*f*
Temporal lobe, 525, 525*f*
Tendons, 498*f*, 499, 555, 561, 564
Teosinte, 356
Terminal bud, 422, 422*f*, 431, 431*f*
Terminal bud scale scars, 431, 431*f*
Termination stage, in translation, 198, 199*f*
Termites, 42, 734, 748, 768, 786
Terrestrial ecosystems, 764–768, 764*t*
Terrestrial life, evolution of, 286, 286*f*
Territoriality, 731
Territory, of animal, 731
Terrorism, biological weapons and, 318
Tertiary period, 285*t*
Tertiary structure, of protein, 47, 47*f*
Test(s), of amoeboids, 328, 328*f*
Testcrosses, 166
one-trait, 166, 166*f*
two-trait, 166, 166*f*
Testes, 684, 684*f*
control of function, 685
development of, 700
in endocrine system, 668, 668*f*, 669*t*
functions of, 682
and hormones, 667
structure and function of, 685, 685*f*
Testing, in scientific method, 11, 11*f*
Testosterone
and bone strength, 559
functional groups and, 39, 39*f*
functions of, 44, 685
Tetanus (disease), 314, 564, 601
Tetanus (of muscle), 564, 564*f*
Tetracycline, 782
Tetrads, 146, 146*f*, 148, 148*f*–149*f*, 151, 151*f*
Tetraploids, 152, 271, 272
Tetrapods, 387
Thailand cobra, 394
Thalamus, 515, 515*f*, 524*f*, 526, 526*f*, 527*f*, 537
Thalidomide, 700
THC (tetrahydrocannabinol), 523
Therapeutic cloning, 210, 210*f*, 211
Therapsids, 392
Thermoacidophiles, 296, 316, 316*f*
Thermodynamics
first law of, 79, 753
second law of, 79, 753
Thermophiles, 315
Thermoreceptors, 536, 536*f*
Thiamine, dietary requirements, 625*t*
Thigmotropism, 467, 467*f*
Thisbe irenea, 473
Thompson, D'Arcy, 276
Thoracic cage, 556*f*
Thoracic cavity, 640
Thoracic duct, 594
Thoracic nerves, 516*f*
Thoracic vertebrae, 556*f*, 557
Thorax, of insect, 384
Threatened species, 780, 787
Three-domain system, 296, 296*f*
3PG, 104*f*, 105, 105*f*, 106, 116*f*
3′ end, 191, 191*f*, 195, 195*f*
Threonine, 46*f*

Threshold, of action potential, 518–519, 518f, 519f, 521, 521f
Thrombin, 585
Thromboembolism, 582
Thrombus, 582
Thylakoid(s)
  in photosynthesis, 97, 99, 101–103, 102f, 103f
  in photosynthetic cyanobacteria, 59
  in plant anatomy, 68, 68f, 97, 97f
Thylakoid membrane
  in photosynthesis, 97, 99, 101–103, 102f, 103f
  in plant anatomy, 68, 97, 97f
Thylakoid space, 68, 68f, 97, 97f, 102, 103, 103f
Thymic hormones, 595
Thymine (T), 48, 48t, 186, 186f, 187, 187f
Thymine dimers, 201, 201f
Thymosins, 595, 669t
Thymus gland, 594f, 595, 595f, 602, 668, 668f, 669t
Thyroid gland, 668, 668f, 669t, 676, 676f
Thyroid hormones, 667
Thyroid-stimulating hormone (TSH), 669t, 670f, 671, 676
Thyroid tumors, 676
Thyroxine (T₄), 669t, 676
Tibet, sky burials in, 740, 740f
Tibia, 556f, 558, 558f, 559
Ticks, 383
Tidal ventilation mechanism, 640–641, 640f
Tigers
  Bengal, 718, 718f, 788
  Siberian, 788
  Tasmanian wolf/tiger, 392
Tight junctions, in animal cells, 72, 72f, 501–502
Tigons, 263
Tilman, G. David, 443
Tineas, 362, 362f
Tinsel flagellum, 327
Tissue(s)
  of angiosperms, 426–429, 426f–429f
  animal, 496–501
    connective tissue, 498–499, 498f
    epithelial tissue, 496–497, 497f
    muscular tissue, 500, 500f
    nervous tissue, 501, 501f
  evolution of, 370f, 371
  as level of organization, 4, 5f, 496, 496f
Tissue culture, 488, 562
Tissue engineering, 506, 506f, 586, 605
Tissue fluid, 505, 507, 507f, 587, 587f
Tissue stage, of development, 692f, 693, 693f
TMV. See Tobacco mosaic virus
TNF. See Tumor necrosis factor
Toads, 389
Tobacco, chromosome number, 137t
Tobacco mosaic virus (TMV), 302, 304, 304f, 436
Toes, opposable, 400
Tolerance model of succession, 751
Tomatoes, 229, 485
Tongue, 616, 616f
Tonsilitis, 616
Tonsillectomy, 616
Tonsils, 594f, 595, 616, 616f
Tools, use by early humans, 409, 410, 412, 412f, 413
Tooth. See Teeth
Topography, and climate, 770, 770f
Torn cartilage, 561, 562
Torpedo stage, of angiosperm embryo development, 484, 484f
Tortoises, giant, on Galápagos, 786

Totipotency
  of eukaryotic cells, 209
  of plant cells, 488
Touch, sense of, 537
Toxicysts, 329
Toxoplasma gondii, 330
Toxoplasmic encephalitis, 593
Toxoplasmosis, 330
Tracers, 22, 185
Trachea
  in birds, 641f
  in humans, 497, 497f, 617, 617f, 638, 638f
Tracheal gills, 637
Tracheal system
  in arthropods, 382
  in insects, 575, 635, 635f, 637, 637f
Tracheids, 427, 427f, 432, 432f, 442, 444, 444f
Tracheoles, 637, 637f
Tracts, in white matter, 524
Trade winds, 769
Transcription, 192–194
  genetic code, 193, 193f
  overview of, 192, 192f, 199, 199f
  posttranscriptional processing of mRNA, 195, 195f
  process of, 194, 194f
Transcription activators, 213, 213f
Transcriptional control, 213, 213f, 215, 215f
Transcription factors, 213, 213f
Transduction
  in bacteria, 313
  in signal transduction pathway, 460, 460f
Trans fats, 583, 623, 627, 627f
Transfer RNA. See tRNA
Transformation, in bacteria, 313, 313f
Transform boundary, 287, 287f
Transgenic organisms, 228, 394. See also Genetically modified organisms (GMOs)
Transitional fossils, 250, 250f
Translation, 196–198
  elongation, 198, 198f, 199f
  initiation, 198, 198f, 199f
  overview of, 192, 192f, 199, 199f
  termination, 19f, 198
  translocation, 198
Translational controls, 214, 214f, 215, 215f
Translocation
  of chromosome segment, 154, 154f
  in translation, 198
Transmission electron micrograph (TEM), 58
Transmission electron microscopes, 58, 58f
Transpiration, 445, 445f
Transplantation
  bone marrow, 586
  and tissue rejection, 605
  xenotransplantation, 231, 394, 506, 605
Transport
  active, 90, 90f
  bulk, 90, 90f
  passive, 88, 88f
Transport systems, in humans, 505, 505f
Transport vesicles, 60–61, 64, 65, 65f, 67, 67f
Transposons, 200, 202, 202f
Transverse colon, 616f, 621
Transverse nerves, 376, 376f
Trapezius, 563f
Tree(s). See also Forest(s); Wood(s)
  annual rings, 432–433
  deciduous, 423
  evergreen, 359, 423

fungal diseases, 360–361
  trunk of, 426, 426f
Tree ferns, 352, 352f
Tree frog, 389f
Treponema pallidum, 689
Triassic period, 285t, 288, 289f, 350
Triceps brachii, 563f, 564, 564f
Trichinosis, 378, 378f
Trichocysts, 329, 329f
Trichomes, 472
Trichomonas vaginalis, 327, 689
Trichomoniasis, 689
Tricuspid valve, 577, 577f
Trigger hairs, 468, 468f
Triglycerides, 43
  in diet, 623, 623t
  digestion of, 612f
Triiodothyronine (T₃), 669t, 676
Triplet code, genetic code as, 193
Triploid(s), 152, 271
Triploid endosperm, in angiosperms, 354f–355f, 355
Trisomy, 152
Trisomy 21 (Down syndrome), 152, 153, 153f, 153t, 154
tRNA (transfer RNA)
  function of, 62, 196, 198, 198f
  initiator, 198, 198f
  structure of, 196, 196f
Trochophore larva, 372
Trophic level, 755
Trophoblast, 698, 698f, 699
Trophozoites, 330f
Tropical deciduous forest, 764f
Tropical rain forests, 764f, 764t, 768, 768f, 770, 770f, 785, 785f, 789
  and global warming, 108, 108f, 421
  human impact on, 10
  medicines and, 472, 474, 474f
Tropical seasonal forest, 764f
Tropic hormones, 671, 671f
Tropisms, 436, 436f, 466–467, 466f, 467f
Trout, 771, 771f
trp operon, 208
Truffles, 359, 360
Trumpet manucodes, 732
Trypanosoma, 323
Trypanosoma brucei, 323, 327
Trypanosoma cruzi, 323, 327
Trypsin, 619, 619f
Tsetse fly, 323
TSH. See Thyroid-stimulating hormone
T system, 566
T tubules, 566, 566f, 567
Tuataras, 390, 390f
Tubal ligation, 690t
Tuberculosis, 317, 606
Tubers, 487, 487f
Tube-within-a-tube body plan, 376
Tubular reabsorption, 656–657, 656f
Tubular secretion, 656f, 657
Tubulin, 70, 70f
Tularemia, 318
Tulips (Tulipa), 357f
Tumor(s)
  benign, 143
  formation of, 132–133, 143
  invasive, 143, 143f
  metastatic, 143, 143f
  in neurofibromatosis, 169
  thyroid, 676
Tumor necrosis factor (TNF), 603
Tumor suppressor genes, 218, 218f, 219, 219f
Tundra, 764f, 764t, 765, 765f, 770, 770f
Tunicates (sea squirts), 386, 386f, 387f
Turgor movements, in plant cells, 468, 468f
Turgor pressure, in plant cells, 89, 468, 468f

Turkey vulture, 391f
Turner syndrome, 153, 153t
Turtles, 292, 390, 390f
24-hour geological clock, 286, 286f
Two-trait testcrosses, 166, 166f
Tympanic canal, 545f
Tympanic membrane, 544, 544f, 545
Typhlosole, 614, 614f
Typhoid fever, heterozygote advantage and, 258
Tyrannosaurus rex, 390
Tyrosine, 46f, 200

U

Ulcers, stomach, 618
Ulna, 556f, 558, 558f
Ulnar nerve, 516f
Ultraviolet (UV) radiation
  and eye disease, 541
  and genetic mutation, 201
  insect vision and, 534, 536
  ozone layer depletion and, 787
  and skin cancer, 503
Ulva (sea lettuce), 335, 335f, 336, 336f
Umbilical cord, 698f, 699, 700
Umbilical cord stem cells, 586
U.N. Food and Agriculture Organization, 788
Unami, 539
Uncoating, of virus, 305, 305f, 306, 306f
Understory, 766
Uniformitarianism, 245
Uniform population distribution, 711, 711f
Unsaturated fatty acids, 43, 623, 627
Unwinding of DNA, 190, 190f, 194, 194f
Upright stance, evolution of, 405, 406–407, 406f
Upwelling, 774, 774f
Uracil (U), 48, 48f, 48t, 186, 186f, 193, 193f
Urea, 507, 620, 623, 650, 650f
Urea cycle, 650
Uremia, 660
Ureters, 504, 654, 654f
Urethra, 654, 654f, 684, 684f
Urey, Harold, 308, 308f
Uric acid, 650, 650f
Urinalysis, 657
Urinary bladder, 497, 654, 654f
Urinary system. See also Kidneys
  components of, 496, 496f
  functions of, 504
  human, 505, 505f, 654, 654f
Urine
  concentration of, 658–659, 658f, 659f
  formation of, 557f, 654, 654f, 656–657
  pH of, 659
Urn-shaped age structure diagrams, 713, 713f
Uropods, of crayfish, 382f, 383
Uterine cycle, 688, 688f
Uterine tubes. See Oviducts
Uterus, 392, 686, 686f
Utricle, 546–547
Uvula, 616, 616f

V

V1RL1 gene, 665
Vaccines, 599, 599f, 601
Vacuole, central, 89, 89f
Vacuoles
  central, of plant cell, 61f, 66–67, 66f
  functions of, 66–67
Vagina, 686, 686f
Vaginal orifice, 686f
Vaginal sponge, 690t
Vaginitis, 327
Vagus nerve, 528, 528f

Valence shell, 23
Valine, 46f, 192
Valium, 521
Valves
    heart, 577, 577f
    in invertebrate circulatory
        system, 575
    in lymphatic vessels, 594, 594f
    in veins, 579, 579f, 581, 581f
Vampire bats, 613, 613f, 733
Van Helmont, Jean-Baptiste, 450
van Niel, C. B., 98
Varicose veins, 581
Variegation, in plants, 304, 304f
Variola virus, 318
Vascular bundles, 428, 429f
Vascular cambium, 431, 431f, 432, 432f
Vascular plants, 348–357. See also
    Angiosperms; Gymnosperms
    evolution of, 348
    leaf structure, 345, 345f
    seedless, 348–349, 348f
        characteristics of, 342f, 343
        diversity of, 348, 348f
        economic value, 348–349
        evolution of, 343, 343f
        life cycle, 349, 349f
Vascular tissue, of angiosperm, 422,
    422f, 427, 427f
    monocot vs. eudicot, 424, 424f
Vas deferens, 684, 684f
Vasectomy, 690t
Vasopressin, 669t
Vector, definition of, 226
Vegetables
    artificial selection in, 246, 246f
    in diet, 625
Vegetal pole, of embryo, 692
Vegetarian diets, 623
Vegetative reproduction. See Asexual
    reproduction
VEGF (vascular endothelial growth
    factor), 232
Veins
    in human circulatory system, 577,
        579, 579f, 581, 581f
    in leaves, 97, 97f, 424, 428, 429f,
        434, 434f, 435f
Vena cava (venae cavae), 577f,
    580, 580f
Venter, Craig, 234
Ventilation (breathing), 634, 634f,
    640–641, 641f, 642f
    and pH levels, 31
    rate, regulation of, 641, 641f, 659
Ventricles
    brain, 524, 524f
    heart, 577, 577f
Ventricular fibrillation, 578
Venules, 579
Venus flytrap (Dionaea muscipula),
    340–341, 340f, 423, 423f, 454,
    454f, 468, 468f
Vermiform appendix, 595, 621, 621f
Vernix caseosa, 701
Veronia, 474
Vertebrae, 556f, 557
    caudal, 418
Vertebral column, 387, 557
Vertebrate(s), 386–394
    aquatic
        osmoregulation in, 652, 652f
        respiration in, 634
    blood of, 572–573
    brain, organization of, 513, 513f,
        515, 515f
    characteristics of, 387
    chemoreceptors in, 538, 538f
    chordate characteristics in,
        386, 386f
    circulatory system, 576, 576f

development in, 692–696
    cellular stage, 692, 692f
    organ stages, 694, 694f
    tissue stage, 692f, 693, 693f
evolution of
    ancestral chordate, 275
    embryological evidence,
        251, 251f
    evolutionary tree, 387, 387f
    eye of, 534–535, 540–543
    forelimbs of, as homologous
        structures, 251, 251f, 292
    genomes of, similarities in,
        236, 236f
    kidneys of, 654
    major phyla, 373
    medical importance, 394, 394f
    nervous system, 515, 515f
        central nervous system (CNS),
            515, 515f, 516, 516f, 521, 524
        peripheral nervous system
            (PNS), 516, 516f, 528–530,
            528f, 529f, 530f
        autonomic system, 530, 530f
        somatic system, 529, 529f
    organ systems of, 504–505,
        504f, 505f
    skeleton of, 387, 555
    terrestrial
        osmoregulation in, 653, 653f
        respiration in, 635, 635f,
            640–641, 640f, 641f
Vertigo, 546, 548
Vesicles, 60f
    lysosomes, 66, 66f
    peroxisomes, 66
    transport, 60–61, 64, 65, 65f, 67, 67f
Vesicular (Graafian) follicle, 687, 687f
Vessel elements, 427, 427f, 442,
    444, 444f
Vestibular canal, 545f
Vestibule, 544, 544f
Vestigial structures, as evidence for
    evolution, 251, 418
Viceroy butterfly, 745
Vicia faba (broad bean), 137t
Villus (villi), 619, 619f
Vinca alkaloids, 474
Virchow, Rudolph, 131
Virginia creeper (Parthenocissus
    quinquefolia), 290, 290f
Virus(es), 302–307
    as biological weapons, 318
    classification of, 302
    and disease, 300
        in animals, 305–307
        in plants, 304, 304f
    enveloped, 302, 320f
    genetic engineering and, 300
    genetic material of, 302, 304
    infection by, 302–303, 303f, 305,
        305f, 306, 306f
    life cycle, 303, 305, 305f, 306, 306f
    naked, 302, 320f
    parasitism in, 746
    structure of, 300, 302, 302f
    T2 bacteriophage, 184–185, 185f
Visceral mass, of mollusc, 380, 380f
Visceral muscle, 500, 500f
Vision. See also Eye(s)
    color, 400–401, 401f, 543 (See also
        Cone cells)
    image processing, 542–543
    panoramic, 535
    photoreceptors and, 536, 540 (See
        also Cone cells; Rod cells)
    in primates, 400–401, 401f
    stereoscopic (three-dimensional),
        400–401, 401f, 535, 543
Visual accommodation, 541, 541f
Visual communication, 734–735

Visual cortex, eye communication with,
    542–543
Vitamin(s)
    dietary requirements, 625, 625t
    and synthesis of coenzymes, 83
Vitamin A, 144
    dietary requirements, 625, 625t
    and vision, 542–543
Vitamin B_1, dietary requirements, 625t
Vitamin B_2, dietary requirements, 625t
Vitamin B_6, dietary requirements, 625t
Vitamin B_{12}
    dietary requirements, 625t
    sources of, 314
Vitamin C
    and cancer prevention, 144
    dietary requirements, 625, 625t
Vitamin D
    and absorption of calcium, 676
    dietary requirements, 625, 625t
    and sunlight, 414, 559
Vitamin-deficiency disorders, 83
Vitamin E, dietary requirements,
    625, 625t
Vitamin K
    dietary requirements, 625t
    sources of, 314, 621
Vitreous humor, 540, 540f
Viviparous animals, 683
VNO. See Vomeronasal organ
Vocal cords, 638
Voice box (larynx), 617, 638, 638f
Voluntary muscle, 500, 500f
Volvox, 335, 335f, 369
Vomeronasal organ (VNO), 538, 665
von Recklinghausen disease
    (neurofibromatosis), 169
von Sachs, Julius, 450–451
Vulcanization, 357
Vultures, 740, 740f
Vulva, 686, 686f

W

Wachtershaüser, Gunter, 309
Waggle dance, 735, 735f
Walcott, Charles Doolittle, 274
Walking sticks, 745
Wallace, Alfred Russel, 247
Warblers, niche specialization in,
    743, 743f
Warfarin, 84
Warning coloration, 745, 745f
Warren, Robin, 618
Warts, 305, 689
Wasp, 384f
Waste disposal, biogeochemical cycles
    and, 784
Waste products, types and
    characteristics, 650, 650f
Water
    and animal development, 697
    bodies of, and climate, 770
    in cellular respiration, 114, 115,
        120, 120f
    and chemical reactions, facilitation
        of, 28
    dissociation of, 30, 30f
    eutrophication of, 757
    frozen, density of, 29, 29f
    life and, 18–19, 19f, 27–29
    and photosynthesis, 98, 102, 445
    plants, water transport in, 27, 27f
    properties of, 27–29
    purification of, biogeochemical
        cycles and, 784
    root uptake of, 452, 452f
    surface tension of, 27, 27f
    temperature conservation properties
        of, 28
    water-salt balance, maintenance of,
        658–659, 658f, 659f

Water column, in xylem, 444, 445f
Water cycle, and waste disposal, 784
Water hyacinth, 786
Water intoxication, 660
Water molds, 324, 325f, 331
Water molecules
    cohesion of, 27
    models of, 26f
Water pollution, nitrogen fertilizers
    and, 12
Water striders, 27, 27f
Watson, James D., 188, 189f, 191, 241
Waxes, structure of, 44, 44f
Weaverbirds, 732
Weddell seal (Leptonychotes
    weddelli), 632
Wegener, Alfred, 286
Weight (human). See also Obesity
    excessive, and cardiovascular
        disease, 583
    management strategies for, 126,
        126f, 622, 622f
Weinberg, W., 253
Weisheit, Chris, 173
Welwitschia, 718f
Went, Frits, 460
Wernicke area, 525f
West, Ian, 288, 289
Western lowland gorilla, 401f, 402f
West Nile virus, 307
West wind drift, 774f
Wetlands, 771, 785
Whales
    ancestors of, 250, 250f, 251,
        264, 264f
    baleen, 613, 613f
    communication in, 734
    courtship in, 736
    evolution of, 277
    forelimbs, 393
    gray, 536f
    humpback, 734
    killer (orca; Orcinus orca), 264f,
        393f, 788
    osmoregulation in, 653
    right, 736
Wheat (Triticum), 356, 356f, 425, 425f
Whitebark pine, 748, 748f
White blood cells (leukocytes)
    genetic modification of, in cancer
        treatment, 233, 233f
    movement of, 71
    phagocytosis in, 90
    structure and function of, 66, 499,
        499f, 507, 585
    types of, 584f, 585
White Cliffs of Dover, 328, 328f
White-crowned sparrows, 728
White-faced monkey, 400f
White-handed gibbon, 400f
White matter, 524
White pulp, of spleen, 595, 595f
White-tailed deer, 393f, 708–709,
    708f, 709f
Wieschaus, Eric, 216
Wildlife, value to humans, 782–783,
    782f, 783f
Wilkins, Maurice H. F., 188
Williams syndrome, 154
Wilson, E. O., 749
Winds
    Earth rotation as source of,
        769, 769f
    ocean currents and, 774
Wine and winemaking, 124, 361
Wings
    of moths and butterflies, 206–207,
        206f, 207f
    vertebrate, as homologous structure,
        251, 251f, 292
Wisdom teeth, 616

Wobble hypothesis, 196
Woese, Carl, 296
Wolves, 736
Women. *See* Female(s), human
Women's Health Institute, 144
Wood(s)
  heartwood, 432–433, 433*f*
  lumber, 356–357
  sap, 432, 433*f*
  as secondary growth, 432–433, 432*f*
  soft and hard, 350
  uses of, 433, 433*f*
Woodpecker, pileated, 391*f*
Woody eudicots, secondary growth in,
  432–433, 432*f*, 433*f*
Woody plants, dormancy, 493
Woody twigs, anatomy of, 431, 431*f*
Wound responses, in plants,
  472–473, 473*f*
Wrangham, Richard, 474
Wrasses, 682
Wrinkles, in skin, 503
Würsig, B., 736

X

X chromosomes, 145, 145*f*
Xenotransplantation, 231, 394,
  506, 605
Xeroderma pigmentosum, 201, 201*f*
X-linked alleles, 174, 174*f*
X-linked genetic disorders, 175, 175*f*
X-ray crystallography, DNA diffraction
  pattern, 188, 188*f*
X-rays, and cancer, 144, 201
Xylem, 424, 424*f*, 427, 427*f*, 429, 429*f*,
  431, 431*f*, 432, 432*f*, 436, 442,
  442*f*, 444–445, 445*f*, 452, 452*f*
  in angiosperms, 430*f*
Xylem rays, 432, 432*f*
Xylem sap, 442

Y

Y chromosomes, 145, 145*f*
Yeast(s)
  budding, 361, 361*f*
  classification of, 361

fermentation and, 123, 123*f*,
  124, 361
  genome of, 234*t*
Yeast infections, 362
Yellow bone marrow, 560, 560*f*
Yellow fever, 305, 307
Yellow jacket wasp, 745, 745*f*
Yellow trumpet pitcher (*Sarracenia
  flava*), 341, 341*f*
Yersinia pestis, 318
Yoho National Park, 274–275
Yolk, 683, 692
Yolk plug, 693, 693*f*
Yolk sac, 390*f*, 697, 697*f*, 698–699

Z

Zea mays. See Corn
Zeatin, 463
Zebra mussels, 786
Zero population growth, 720, 720*f*
Zinc, dietary requirements, 624
Ziram, 201

Z lines, 566, 566*f*
Zone of cell division, 430, 430*f*
Zone of elongation, 430, 430*f*
Zone of maturation, 430–431, 430*f*
Zooflagellates, 325*f*, 327
Zoopharmacognosy, 474
Zooplankton, 328, 773
Zoospores, 334, 334*f*, 336, 336*f*
Zooxanthellae, 325, 773
Zygomatic bones, 557, 557*f*
Zygomycota, 360
Zygospore(s), 334, 334*f*
Zygospore fungi, 360
Zygote
  in algae life cycle, 336, 336*f*
  in angiosperm life cycle, 484, 484*f*
  development of, 692, 692*f*
  in human life cycle, 687
  in life cycle, generally, 150, 150*f*

Applications in Biology

HOW BIOLOGY IMPACTS OUR LIVES

1.9 DNA barcoding of life may become a reality 14
2.3 Radioactive isotopes have many medical uses 22
2.16 Acid deposition has many harmful effects 32
3.12 The Human Genome Project may lead to new disease treatments 49
4.17 Malfunctioning mitochondria can cause human diseases 69
5.9 Enzyme inhibitors can spell death 84
5.12 Malfunctioning plasma membrane proteins can cause human diseases 87
7.10 Fermentation helps produce numerous food products 124
7.12 Exercise burns fat 126
8.10 Protective behaviors and diet help prevent cancer 144
9.11 Genetic disorders may now be detected early on 170
10.19 Many agents can cause mutations 201
11.5 Animal cloning has benefits and drawbacks 211
12.6 Are genetically engineered foods safe? 230
12.11 New cures are on the horizon 235
13.17 Stabilizing selection helps maintain harmful alleles 258
16.6 Humans suffer from emerging viral diseases 307
16.16 Prokaryotes have environmental and medical importance 317
16.17 Disease-causing microbes can be biological weapons 318
18.8 Carboniferous forests became the coal we use today 352
18.11 Flowering plants provide many services 356
18.15 Fungi have economic and medical importance 362
19.26 Many vertebrates provide medical treatments for humans 394
20.5 Origins of the genus Homo 407
21.3 Monocots serve humans well 425
21.8 Wood has been a part of human history 433
22.5 Plants can clean up toxic messes 447
23.13 Eloy Rodriguez has discovered many medicinal plants 474
25.6 Will nerve regeneration reverse a spinal cord injury? 502

26.10 Drugs that interfere with neurotransmitter release or uptake may be abused 522
27.7 Protect your eyes from the sun 541
27.9 The inability to form a clear image can be corrected 542
27.13 Protect your ears from loud noises 546
27.15 Motion sickness can be disturbing 548
28.5 Avoidance of osteoporosis requires good nutrition and exercise 559
28.8 Joint disorders can be repaired 562
28.11 Exercise has many benefits 565
29.10 Blood vessel deterioration results in cardiovascular disease 582
29.11 Cardiovascular disease can often be prevented 582
30.4 A fever can be beneficial 597
30.11 Monoclonal antibodies have many uses 604
30.14 Allergic reactions can be debilitating and even fatal 606
31.7 Bacteria contribute to the cause of ulcers 618
31.17 Nutritional labels allow evaluation of a food's content 626
31.18 Certain disorders are associated with obesity 626
31.19 Eating disorders appear to have a psychological component 628
32.6 Questions about tobacco, smoking, and health 639
32.11 Respiratory disorders have resulted from breathing 9/11 dust 644
33.8 Urinalysis can detect drug use 657
33.12 Dehydration and water intoxication occur in humans 660
34.7 Glucocorticoid therapy can lead to Cushing syndrome 673
34.9 Diabetes is becoming a very common ailment 674
35.8 Sexual activity can transmit disease 689
35.9 Numerous birth control methods are available 690
35.10 Reproductive technologies are available to help the infertile 691

HOW SCIENCE PROGRESSES

4.3 Microscopes allow us to see cells 58
4.10 Pulse-labeling allows observation of the secretory pathway 65
6.4 Experiments showed that the O_2 released by photosynthesis comes from water 98
6.7 Fall temperatures cause leaves to change color 100
6.17 Destroying tropical rain forests contributes to global warming 108
9.20 Thomas Hunt Morgan is commonly called "the fruit fly guy" 178
10.20 Transposons are "jumping genes" 202
12.4 Making cheese—Genetic engineering comes to the rescue 228
12.9 Gene therapy trials have varying degrees of success 233
13.5 Wallace independently formulated a natural selection hypothesis 247
13.6 Natural selection can be witnessed 248
14.6 Artificial selection produced corn 272
14.8 The Burgess Shale hosts a diversity of life 274
15.2 The geologic clock can help put Earth's history in perspective 286

16.3 Viruses are responsible for a number of plant diseases 304
17.3 How can the protists be classified? 326
17.12 Life cycles among the algae have many variations 336
19.4 Molecular data suggest a new evolutionary tree for animals 372
20.7 Biocultural evolution began with Homo 409
20.10 Cro-Magnons made good use of tools 412
20.11 Agriculture made modern civilizations possible 412
22.2 Competition for resources is one aspect of biodiversity 443
25.9 Organs for transplant may come from various sources 506
29.14 Adult stem cells include blood stem cells 586
33.11 The artificial kidney machine makes up for faulty kidneys 660
37.7 Do animals have emotions? 736
38.5 Coevolution requires interaction between two species 747
38.7 The study of island biogeography pertains to biodiversity 749

Connecting the Concepts

Each chapter ends with a *CONNECTING THE CONCEPTS* reading that reviews the concepts of the chapter and tells how they relate to other concepts in the book.

CONNECTING THE CONCEPTS

All cells receive DNA from pre-existing cells through the process of cell division. Cell division ensures that DNA is passed on to the next generation of cells and to the next generation of organisms.

The end product of ordinary cell division (i.e., mitosis) is two new cells, each with the same number and kinds of chromosomes as the parent cell. Division of the cell cycle sequences i

chronization. Knowing how the cell cycle is regulated has contributed greatly to our knowledge of cancer and other disorders.

In contrast to mitosis, meiosis is part of the production of gametes, which have half the number of chromosomes as the parent cell. Through the mechanics of meiosis, sexually reproducing species have a greater likelihood of genetic variations

Genetic variations are essential to the process of evolution, which is discussed in Part III. In the meantime, Chapter 9 reviews the fundamental laws of genetics established by Gregor Mendel. Although Mendel had no knowledge of chromosome behavior, modern students have the advantage of being able to apply their knowledge of meiosis to their understanding of

CONNECTING THE CONCEPTS

Energy from the sun flows through all living things with the participation of chloroplasts and mitochondria. Through the process of photosynthesis, chloroplasts in plants and algae capture solar energy and use it to produce carbohydrates, which are broken down to carbon dioxide and water in the mitochondria of all organisms. The energy released when carbohydrates (and other organic molecules) are oxidized is used to produce ATP molecules. When the cell uses ATP to do cellular work, all the captured energy dissipates as heat.

During cellular respiration, oxidation by removal of hydrogen atoms (e^- + H^+) occurs during glycolysis, the prep reaction, and the citric acid cycle. The electrons are carried by NADH and FADH$_2$ to the electron transport chain (ETC) on the cristae of mitochondria. Oxygen serves as the final acceptor of electrons, and water is produced. Th
by the ETC
leads to AT

With this chapter, we have completed our study of the cell, and the Biological Viewpoint on pages 130–31 reviews for

you some of the major concepts regarding the cell theory.

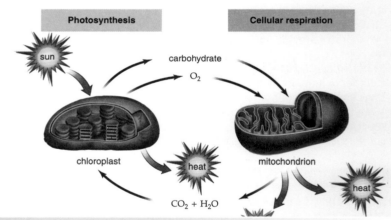

CONNECTING THE CONCEPTS

The protists we study today are not expected to include the direct ancestors to fungi, plants, and animals. Instead, they may be related to the other eukaryotic groups by way of common ancestors that have not been discovered in the fossil record. Today, nucleic acid sequencing alone tells us which protists are most closely related to the fungi, plants, and animals.

Perhaps today's protists represent an adaptive radiation experienced by the first eukaryotic cell to evolve. While certain

structures present in eukaryotic cells may have been unique to them alone, others appear to be endosymbionts, present only because they were engulfed by a much larger cell. Mutualism is a powerful force that shaped the eukaryotic cell and also shapes all sorts of relationships in the living world. For example, we have already mentioned that mutualism between flowers and their pollinators has contributed to the success of flowering plants.

All possible forms of reproduction and nutrition are present among the protists,

but each of the other eukaryotic groups specializes in a particular type of reproduction and a particular method of acquiring needed nutrients. Fungi, as we shall see, reproduce by means of windblown spores during both an asexual and sexual life cycle, and they are saprotrophic. Plants have the alternation of generations life cycle and are photosynthetic. Animals have the diploid life cycle and are heterotrophic. Chapter 18 pertains to the evolution of plants and fungi, while Chapter 19 discusses the evolution of animals.